Interest Boxes

Organic Chemistry

Organic Chemistry

Fifth Edition

Paula Yurkanis Bruice

University of California, Santa Barbara

PEARSON

Prentice Hall

Upper Saddle River, NJ 07458

Library of Congress Cataloging-in-Publication Data

Bruice, Paula Yurkanis
  Organic chemistry / Paula Yurkanis Bruice.—5th ed.
    p. cm.
  Includes index.
  ISBN 0-13-196316-3
  1. Chemistry, Organic. I. Title.

QD251.3.B78 2007
547—dc22   2006001413

Executive Editor: Nicole Folchetti
Editor in Chief, Development: Ray Mullaney
Editor in Chief, Science: Dan Kaveney
Development Editor: Moira Lerner-Nelson
Media Editor: Michael J. Richards
Project Manager: Kristen Kaiser
Art Director: John Christiana
Executive Managing Editor: Kathleen Schiaparelli
Assistant Managing Editor, Science Media: Nicole M. Jackson
Assistant Managing Editor, Science Supplements:
  Karen Bosch
National Sales Director for Key Markets: David Theisen
Editorial Assistant: Timothy Murphy
Production Editor: Donna King
Creative Director: Juan Lopez
Director, Creative Services: Paul Belfanti
Purchasing Manager: Alexis Heydt-Long
Manufacturing Buyer: Alan Fischer

Senior Managing Editor, AV Production &
  Management: Patricia Burns
Manager, Production Technologies: Mathew Haas
Managing Editor, Art Management: Abigail Bass
AV Art Editor: Connie Long
Art Studio: Artworks
Contributing Art Studio: Wavefunction
Spectra: Reproduced by permission of Aldrich Chemical Co.
Cover and Interior Designer: Jonathan Boylan
Director, Image Resource Center: Melinda Reo
Manager, Rights and Permissions: Zina Arabia
Interior Image Specialist: Beth Boyd-Brenzel
Cover Image Specialist: Karen Sanatar
Cover Photo: Getty Images–Photonica Amana America, Inc.
Image Permission Coordinator: Michelina Viscusi
Photo Researcher: Truitt & Marshall
Production Services/Composition: Progressive Publishing
  Alternatives/Progressive Information Technologies

© 2007, 2004, 2001, 1998, 1995 by Pearson Education, Inc.
Pearson Prentice Hall
Pearson Education, Inc.
Upper Saddle River, NJ 07458

Pearson Prentice Hall™ is a trademark of Pearson Education, Inc.

Spectra: © Sigma-Aldrich Co.

Printed in the United States of America
10 9 8 7 6 5 4 3

ISBN 0-13-196316-3

Pearson Education Ltd., *London*
Pearson Education Australia Pty. Ltd., *Sydney*
Pearson Education Singapore, Pte. Ltd.
Pearson Education North Asia Ltd., *Hong Kong*
Pearson Education Canada, Inc., *Toronto*
Pearson Educación de Mexico, S.A. de C.V.
Pearson Education—Japan, *Tokyo*
Pearson Education Malaysia, Pte. Ltd.

To Meghan, Kenton, and Alec
with love and immense respect
and to Tom, my best friend

Brief Contents

Contents

Part 1   An Introduction to the Study of Organic Chemistry   1

1 Electronic Structure and Bonding • Acids and Bases   2

5 Stereochemistry: The Arrangement of Atoms in Space; The Stereochemistry of Addition Reactions 200

6 The Reactions of Alkynes: An Introduction to Multistep Synthesis 258

10 Reactions of Alcohols, Amines, Ethers, Epoxides, and Sulfur-Containing Compounds • Organometallic Compounds   429

11 Radicals • Reactions of Alkanes   481

Part 5   Aromatic Compounds   639

14   Aromaticity • Reactions of Benzene   640

15   Reactions of Substituted Benzenes   677

18 Carbonyl Compounds III: Reactions at the α-Carbon   850

Part 7   More About Oxidation–Reduction Reactions and Amines   907

19 More About Oxidation–Reduction Reactions   908

20 More About Amines • Heterocyclic Compounds   943

Part 8   Bioorganic Compounds   977

21 Carbohydrates   978

Part 9 Special Topics in Organic Chemistry 1231

Preface

TO THE INSTRUCTOR

My guiding principle in writing this book has been to present organic chemistry as an exciting and vitally important science. Too many students look upon organic chemistry as a necessary evil, a course they have to take for reasons that to them are unknown. Some suspect it has been designed as a kind of test of their ability to stick things out. Others equate learning organic chemistry with learning the language of a foreign country that they will never visit. They come to organic chemistry having taken other science courses, but they probably have not had the opportunity to study a science that is not a collection of individual topics, but one that unfolds and grows, allowing them to use what they learn at the beginning of the course to explain and predict what follows. To counter the impression that the study of organic chemistry consists primarily of memorizing a diverse collection of molecules and reactions, this book is organized around shared features and unifying concepts, and emphasizes principles that can be applied again and again. I want students to learn how to apply what they have learned to a new setting, reasoning their way to a solution rather than memorizing a multitude of facts. I also want them to see that organic chemistry is integral to biology as well as to their daily lives.

The comments I have received from colleagues and students who used previous editions suggest the book is working in the way I had hoped. As much as I enjoy having faculty tell me that their students are scoring higher than ever before on tests, nothing is more rewarding than hearing positive comments from the students themselves. Many students have generously credited their success in organic chemistry to this book, not giving themselves nearly enough credit for how hard they studied to achieve that success. They always seem surprised that they have come to love "orgo" (on the East Coast) or "o-chem" (on the West Coast). I also hear from many premedical students who say that the book gave them the permanent understanding of organic chemistry that made the organic chemistry section of the MCAT the easiest part for them.

In striving to make this edition of *Organic Chemistry* even more useful to students, I have relied on constructive comments from many of you. For these I am extremely grateful. I also kept a journal of questions students had when they came to my office. These questions let me know what sections in the book needed clarifying and what answers in the *Study Guide and Solutions Manual* needed more detailed explanations. Most importantly, they showed me where I should create new problems to reduce the likelihood that students using the new edition will ask the same questions. Because I teach large classes, I have a vested interest in foreseeing and preventing potential confusion before it arises. In this edition, many sections have been rewritten to optimize readability and comprehension. There are new in-chapter and end-of-chapter problems to enhance students' mastery through problem solving. There are also new interest boxes to show students the relevance of organic chemistry and additional margin notes to remind them of important concepts and principles.

I hope you find this edition even more appealing to your students than earlier versions have been. As always, I am eager to hear your comments, bearing in mind that positive comments are the most fun, but critical comments are the most useful.

A Functional Group Approach with a Mechanistic Organization That Ties Together Synthesis and Reactivity

As mentioned earlier, this book is designed to discourage rote memorization. To this end, the presentation of functional groups is organized around mechanistic similarities: electrophilic additions, nucleophilic substitutions, eliminations, radical additions and substitutions, electrophilic aromatic substitutions, nucleophilic acyl substitutions, and nucleophilic additions. This organization allows a great deal of material to be understood in light of unifying principles of reactivity.

Many organic chemistry textbooks discuss the synthesis of a functional group and the reactivity of that group in the same place; but these two groups of reactions generally have little to do with one another. Instead, when I discuss a functional group's reactivity, I cover the synthesis of compounds that are formed as a result of that reactivity, often by having students design synthetic schemes. In Chapter 4, for example, students learn about the reactions of alkenes but do *not* learn at this point about the synthesis of alkenes. Instead, they learn about the synthesis of alkyl halides, alcohols, ethers, epoxides, and alkanes—the compounds formed when alkenes react. Because alkenes are synthesized from the reactions of alkyl halides and alcohols, the synthesis of alkenes is covered when the reactions of alkyl halides and alcohols are discussed. The strategy of tying together the reactivity of a functional group and the synthesis of compounds resulting from its reactivity prevents the student from having to memorize lists of unrelated reactions. It also results in a certain economy of presentation, allowing more material to be covered in less time.

Although memorizing different ways a particular functional group can be prepared can be counterproductive to mastering organic chemistry, it is useful to have such a compilation of reactions when designing multistep syntheses. For this reason, lists of reactions that yield particular functional groups are compiled in Appendix IV. In the course of learning how to design syntheses, students come to appreciate the importance of reactions that change the carbon skeleton of a molecule; these reactions are compiled in Appendix V.

A Modular Format

Because different instructors teach organic chemistry in different ways, I have tried to make the book as modular as possible. For example, the spectroscopy chapters (Chapters 12 and 13) have been composed so that they may be covered at any time in the course. For those who prefer to teach spectroscopy at the beginning—or in a separate laboratory course—I have put a table of the functional groups at the beginning of Chapter 12. For those who prefer to cover carbonyl chemistry earlier in the course, Part 6 (Carbonyl Compounds) can be covered before Part 5 (Aromatic Compounds). I anticipate that most instructors will cover the first 22 chapters during a year-long course and will then choose among the remaining chapters depending on personal preference and the interests of the students enrolled in their class. Those whose students are primarily interested in the biological sciences might be more inclined to cover Chapter 23 (Catalysis), Chapter 24 (The Organic Mechanisms of the Coenzymes), Chapter 25 (The Chemistry of Metabolism), Chapter 26 (Lipids), and Chapter 27 (Nucleosides, Nucleotides, and Nucleic Acids). Those teaching courses designed for chemistry or engineering majors may focus instead on Chapter 28 (Synthetic Polymers) and Chapter 29 (Pericyclic Reactions). The book ends with a chapter on drug discovery and design, a topic that in my experience interests students sufficiently that they will choose to read it on their own, even if it is not assigned by their instructor.

A Bioorganic Emphasis

Today, many students taking organic chemistry are interested in the biological sciences. I have therefore introduced bioorganic material throughout the text to show these students that organic chemistry and biochemistry are not separate entities but are closely related on a continuum of knowledge. Once students learn how, for example, electron delocalization, leaving-group tendency, electrophilicity, and nucleophilicity affect the reactions of simple organic compounds, they can appreciate how these same factors influence the reactions of more complicated organic molecules such as enzymes, nucleic acids, and vitamins. I have found that the economy of presentation achieved in the first 20 chapters of this text (explained above) makes it possible to devote time to bioorganic topics.

In the first two-thirds of the book, the bioorganic material is limited mostly to the last sections of the chapters. Thus, the material is available to the curious student without requiring the instructor to introduce bioorganic topics into the course. For example, after the stereochemistry of organic reactions is presented, the stereo-chemistry of enzymatic reactions is discussed; after alkyl halides are discussed, biological compounds used to methylate substrates are examined; after the methods chemists use to activate carboxylic acids are presented, the methods cells use to activate these acids are explained; after condensation reactions are discussed, examples of biological condensation reactions are shown.

In addition, seven chapters in the last part of the book (Chapters 21–27) focus on bio-organic chemistry. These chapters have the unique distinction of containing more chem-istry than is typically found in the corresponding parts of a biochemistry text. Chapter 23 on catalysis, for example, explains the various modes of catalysis seen in organic reactions and then shows that they are identical to the modes of catalysis found in enzymatic reac-tions. All of this is presented in a way that allows students to understand the lightning-fast rates of enzymatic reactions. Chapter 24 on coenzymes emphasizes the role of vitamin B_1 as an electron delocalizer, vitamin K as a strong base, vitamin B_{12} as a radical initiator, and biotin as a compound that can transfer a carboxyl group, and describes how the many different reactions of vitamin B_6 are controlled by the overlap of p orbitals. Chapter 25 on metabolism explains the chemical function of ATP. (Its role is not to provide a magic shot of energy that allows an endothermic reaction to take place—commonly called coupled reactions. Rather, its role is to provide a reaction pathway involving a good leaving group for a reaction that cannot occur because of a poor leaving group.) Chapter 26 on lipids presents the mechanisms for prostaglandin formation (allowing students to understand how aspirin works), fat breakdown, and terpene biosynthesis. In Chapter 27, students learn that DNA contains thymine instead of uracil because of imine hydrolysis, and they see how DNA strands are synthesized in the laboratory. Thus, these chapters do not replicate what will be covered in a biochemistry course, but they bridge the two disciplines, allowing students to see that a knowledge of organic chemistry is central to the understanding of biological processes.

With the conviction that learning should be fun, certain biologically oriented discussions presented in interest boxes serve as intriguing asides. Examples include: why Dalmatians are the only mammals that excrete uric acid; why life is based on carbon instead of silicon; how a microorganism has learned to use industrial waste as a source of carbon; the chemistry associated with SAMe, a product prominently displayed in health food stores; and trans fats.

An Early and Consistent Emphasis on Organic Synthesis

Students are introduced to synthetic chemistry and retrosynthetic analysis early in the book (Chapters 4 and 6, respectively), so they can use that technique throughout the course to design multistep syntheses. Nine special sections on synthesis design, each with a different focus, are introduced at appropriate intervals. One of them, for example, dis-cusses the proper choice of reagents and reaction conditions for maximizing the yield of the target molecule (Chapter 9); one introduces disconnections, synthons, and synthetic equivalents (Chapter 17); one discusses making new carbon–carbon bonds (Chapter 18); and one focuses on controlling stereochemistry (Chapter 19). The use of combinatorial methods in organic synthesis is described in Chapter 30, which examines drug discovery and design.

PEDAGOGICAL FEATURES

The pedagogical features that appealed to students in earlier editions have been retained and enhanced.

Margin Notes and Boxed Material to Engage the Student

Each chapter starts with a list of fundamental principles that have already been discussed in earlier chapters and that serve as a foundation for further learning in the present chapter (Building on Fundamentals). Margin notes and biographical sketches appear throughout the text. The margin notes encapsulate key points that students should remember, and the biographical sketches give students some appreciation of the history of chemistry and the people who contributed to that history. Interest boxes connect chemistry to real life (discussing, for example, Semisynthetic Drugs, Measuring Toxicity, Chimney Sweeps and Cancer, Ultraviolet Light and Sunscreens, and Penicillin and Drug Resistance) or provide additional instruction (as in Calculating Kinetic Parameters, A Few Words About Curved Arrows, and Incipient Primary Carbocations).

Summaries and Voice Boxes to Help the Student

Each chapter concludes with a Summary to help students synthesize the key points, as well as a list of cross-referenced Key Terms. Chapters that cover reactions conclude with a Summary of Reactions. Annotations (voice boxes) help students focus on points being discussed.

Problems, Solved Problems, and Problem-Solving Strategies

The book contains more than 1800 problems, many with multiple parts. The answers (and explanations, when needed) to all the problems are in the accompanying *Study Guide and Solutions Manual*, which I authored to ensure consistency with the text. The problems within each chapter are primarily drill problems that allow students to test themselves on material just covered before moving on to the next section. Selected problems are accompanied by worked-out solutions to provide insight into problem-solving techniques. Short answers provided at the end of the book for problems marked with a diamond give students immediate feedback concerning their mastery of a skill or concept. Most chapters also contain at least one Problem-Solving Strategy, a feature that teaches how to approach various kinds of problems. For example, the Problem-Solving Strategy in Chapter 8 teaches students how to determine whether a reaction will be more apt to take place by an S_N1 or an S_N2 pathway. Each Problem-Solving Strategy is followed by an exercise giving the student an opportunity to use the problem-solving skill just learned.

   The end-of-chapter problems vary in difficulty. They begin with drill problems that integrate material from the entire chapter, requiring the student to think in terms of all the material in the chapter rather than focusing on individual sections. The problems become more challenging as the student proceeds, often reinforcing concepts from prior chapters. The net result for the student is a progressive building of both problem-solving ability and confidence.

Companion Website

Part of the revamped and bolstered Companion Website (see the description on page xxix) is a newly assembled **Student Tutorials** quiz gallery. Each question contains a visual animation or interactive tutorial, then asks a question about it, and provides feedback that includes links back to the relevant section of the book. Each student tutorial has a WWW icon in the textbook to indicate where the tutorial relates to the content, as well as where the tutorial appears within the Activebook.

Art Program: Rich in Three-Dimensional, Computer-Generated Structures

This edition continues to present energy-minimized, three-dimensional structures throughout the text that give students an appreciation of the three-dimensional shapes of organic molecules. Color in the illustration program is used not simply for show but to highlight and organize the information; in fact, the use of highlights to draw attention to points of interest has been increased in this edition. I have attempted to employ specific colors in consistent ways (for example, mechanism arrows are always red), but there is no need for a student to memorize a color palette.

A CLOSER LOOK AT COVERAGE AND ORGANIZATION

This book is divided into nine parts, each starting with a brief overview so students can understand where they are "going." The first chapter in Part 1 provides a summary of material that students need to recall from general chemistry. The review of acids and bases emphasizes the relationship between acidity and the stability of the conjugate base, a theme recurrently raised throughout the text. (Acids and bases are covered even more extensively in the *Study Guide and Solutions Manual*.) In Chapter 2, students learn how to name five classes of organic compounds—those that will be the products of the reactions in the nine chapters that immediately follow. Chapter 2 also covers topics that must be mastered before the study of reactions can begin—structures, conformations, and physical properties of organic compounds.

The five chapters of Part 2 deal with electrophilic addition, stereochemistry, and electron delocalization. Chapter 3 sets the stage for the study of organic reactions by giving the students the knowledge of thermodynamics and kinetics they will need as they progress through the course. Rate equations are derived in an appendix for those who wish to take a more mathematical approach to kinetics, and the *Study Guide and Solutions Manual* contains a section on calculating kinetic parameters. Chapter 3 also introduces students to the concept of "curved arrows." (*The Study Guide and Solutions Manual* contains an extensive exercise on "electron pushing." I have found this exercise to be very successful in making my students comfortable with a topic that should be easy, but somehow perplexes even the best of them unless they get sufficient practice.)

I lead off the study of reactions with those of alkenes because of their simplicity. Thus, although Chapter 4 covers a wide variety of alkene reactions, they all have similar mechanisms—an electrophile adds to the least substituted sp^2 carbon and a nucleophile adds to the other sp^2 carbon—and differ only in the nature of the electrophile and the nucleophile. Because organic chemistry is all about the interactions of electrophiles and nucleophiles, it makes sense to begin the study of organic reactions by introducing students to a variety of electrophiles and nucleophiles. The reactions in Chapter 4 are discussed without regard to stereochemistry because I have found that students do well as long as only one new concept is introduced at a time. Understanding the mechanisms of the reactions is enough at this point.

Chapter 5 reviews the isomers introduced in Chapters 2 and 3 (constitutional isomers and cis–trans isomers) and then discusses isomers that result from an asymmetric center—one kind of stereogenic center. The *Study Guide and Solutions Manual* has an exercise that encourages students to open their box of molecular models and see how model building can help them learn. In addition, the Companion Website gives students the opportunity to manipulate many molecules in three dimensions. Now that the students are comfortable with both isomers and electrophilic addition reactions, the two topics are considered together at the end of Chapter 5, where the stereochemistry of the electrophilic addition reactions covered in Chapter 4 is presented. My students have become much more comfortable with stereochemistry now that I introduce it in the context of reactions with which they are familiar. Chapter 6 covers alkynes. This chapter builds the students' confidence because of the similarity of the material to that in Chapter 4.

Understanding electron delocalization is vitally important in organic chemistry, so this subject is covered in its own chapter (Chapter 7), which is a continuation of the introduction to this topic found in Chapter 1. Students see how electron delocalization affects stability and pK_a, and can affect the products of electrophilic addition reactions.

Part 3 covers substitution and elimination reactions at an sp^3 hybridized carbon. First, the substitution reactions of alkyl halides are discussed in Chapter 8. Chapter 9 presents elimination reactions of alkyl halides and then goes on to consider competition between substitution and elimination. Chapter 10 covers substitution and elimination reactions at an sp^3 hybridized carbon when a group other than a halogen is the leaving group—reactions of alcohols, ethers, epoxides, arene oxides, thiols, and sulfides. Organometallic compounds and transition metal-catalyzed coupling reactions are also introduced. Chapter 11 covers radical chemistry. Students learn that alkanes are largely unreactive due to the absence of a functional group, underscoring the importance of a functional group to chemical reactivity.

The two chapters in Part 4 discuss instrumental techniques. Mass spectrometry, IR spectroscopy, and UV/Vis spectroscopy are discussed in Chapter 12. Chapter 13 discusses NMR spectroscopy. Each spectral technique is written as a "stand-alone topic" so it can be covered independently at any time during the course. The first of these chapters opens with a table of functional groups for instructors who want to cover spectroscopy before students have been introduced to all the functional groups.

Part 5 is all about aromatic compounds. Chapter 14 covers aromaticity and the reactions of benzene. Chapter 15 discusses the reactions of substituted benzenes. Because not all organic courses cover the same amount of material in a semester, Chapters 12–15 have been strategically placed to come near the end of the first semester. Ending a semester before Chapter 12, Chapter 13, or Chapter 14, or after Chapter 15 will not leave students suspended halfway through a topic.

Part 6 is devoted primarily to the chemistry of carbonyl compounds. Instructors who had never covered carboxylic acid derivatives before aldehydes and ketones were initially skeptical about this order. However, when the opinions of those who had used previous editions of the book were sought, almost all preferred this order. Chapter 16 starts the treatment of carbonyl chemistry by discussing the reactions of carboxylic acids and their derivatives with oxygen and nitrogen nucleophiles. In this way, students are introduced to carbonyl chemistry by learning how tetrahedral intermediates partition. The first part of Chapter 17 discusses the reactions of carboxylic acid derivatives, aldehydes, and ketones with carbon and hydrogen nucleophiles. By studying all these carbonyl compounds together, students see how the reactions of aldehydes and ketones differ from those of carboxylic acid derivatives. Then, when they move on to study the formation and hydrolysis of imines, enamines, and acetals in the second part of Chapter 17, they can easily understand these mechanisms because they are well versed in how tetrahedral intermediates partition. Over the years I have experimented with my classes, and I believe this to be the most effective and easiest way to teach carbonyl chemistry. That being said, Chapters 16 and 17 can easily be switched as long as Sections 17.4, 17.5, and 17.6 are skipped and then covered with Chapter 16. Chapter 18 deals with reactions at the α-carbon of carbonyl compounds.

Chapter 19, the first chapter in Part 7, revisits oxidation and reduction reactions and discusses new ones. Students are much better able to understand oxidation reactions when they are presented together as a unit, rather than being introduced one at a time as each functional group is introduced. Chapter 20 reviews material on amines that was covered in previous chapters—structure and physical properties, acid–base properties, nomenclature, reactivity, and synthesis—and then explores those topics in greater detail. The chapter concludes with a discussion of heterocyclic compounds.

Part 8 covers bioorganic topics. Part 9 covers synthetic polymers, pericyclic reactions, and drug discovery and design.

CHANGES TO THIS EDITION

Responses from instructors and students have led to adjustments in the coverage and distribution of certain topics and encouraged expansion of the book's most successful pedagogical features.

Content and Organization

The discussions of nucleophilic substitution and elimination have been moved forward, so they are now Chapters 8 and 9, instead of Chapters 10 and 11. Radical addition reactions have been removed from the chapters on alkenes (Chapter 4) and alkynes (Chapter 6) and placed with radical substitution, so all the radical chemistry is now in a single chapter (Chapter 11). Epoxide formation has been added to the chapter on electrophilic addition of alkenes (Chapter 4); and the oxidation of alcohols has been added to the chapter that discusses the reactions of alcohols (Chapter 10). Both of these changes allow for a greater variety of synthetic problems earlier in the course. The section on how to draw compounds with delocalized electrons has been shortened because the *Study Guide and Solutions Manual* contains a new and somewhat extensive tutorial on this topic. A new chapter on metabolism (Chapter 25) emphasizes that the reactions and mechanisms encountered in metabolic processes are similar to the reactions and mechanisms encountered in an organic chemistry laboratory. There is also new material on PCR and genetic engineering. Lastly, much of the book has been rewritten to facilitate student understanding. In particular, the sections on electrophilic addition reactions, electron delocalization, and nucleophilic addition reactions have been reorganized and clarified.

Synthesis Sections

This edition has nine "Designing a Synthesis" sections, including a new one that helps students understand when retrosynthetic analysis should be used. There are also many new synthetic problems.

Pedagogical Elements

To develop students' understanding of organic chemistry as a discipline that builds on experience and knowledge, each chapter (after Chapter 1) starts with a list of fundamental principles students have seen previously that help explain the information presented in the coming pages. The mechanisms in this edition are presented in the form of bulleted steps to make the individual steps stand out more clearly. They are not, however, presented in boxes, as that might interfere with the student's recognition that mechanisms are central to understanding organic chemistry and are not extraneous or optional material. This edition also includes more voice boxes to aid student learning. In response to student comments, there are also more margin notes that succinctly repeat key points and facilitate review. Furthermore, there are 20 new interest boxes in this edition, including one on dissolving sutures, one on trans fats, and one on resistance to herbicides.

Problem Sets

This edition has more than 200 new problems, both in-chapter and end-of-chapter. They include new solved problems, new problem-solving strategies, and new problems incorporating information from more than one chapter.

FOR THE INSTRUCTOR

Instructor Resource Center on CD/DVD (0-13-196317-1). This lecture resource provides a fully searchable and integrated collection of resources to help you make efficient and effective use of your lecture preparation time, as well as to enhance your classroom presentations and assessment efforts. This resource features almost all the art from the text, including tables, in .JPG, .PDF and PowerPoint^{TM} formats; all interactive and dynamic media objects from the Companion Website; and three pre-built PowerPoint^{TM} presentations. The first contains a complete lecture outline. The second presentation contains the in-chapter **PROBLEM** boxes from the text, and the third contains questions you can use in-class along with your Classroom Response System (CRS). This CD/DVD also features a search-engine tool that enables you to find relevant resources via key terms, learning objectives, figure numbers, and resource types. This CD/DVD set also contains TestGen test-generation software and a TestGen version of the Test Item File that enables professors to create and tailor exams to their needs, or to create online quizzes for delivery in WebCT, Blackboard, or CourseCompass. Word^{TM} files of the Test Item File are also included.

Transparency Pack (0-13-222046-6) by Paula Yurkanis Bruice, University of California, Santa Barbara. This set features 285 full-color images from the text.

Test Item File (0-13-222047-4) by Gary Hollis, Roanoke College. Includes a selection of over 2,500 multiple choice, short answer, and essay test questions.

Course and Homework Management Tools

Blackboard, **WebCT**, and Pearson Education's own **CourseCompass** offer the best teaching and learning resources for your course. Conveniently organized by textbook chapters for *Organic Chemistry, 5th Edition*, these resources help you save time and help your students reinforce and apply what they have learned in class. Along with all the material from the Companion Website (*www.prenhall.com/bruice*), resources from the *Instructor Resource Center on CD/DVD* and ACE Organic are also included.

ACE – Achieving Chemistry Excellence Prentice Hall, in conjunction with Robert B. Grossman, Raphael A. Finkel, and a team of programmers at the University of Kentucky, has developed a homework system for organic chemistry that can finally support the types of problems assigned in organic chemistry. *Achieving Chemistry Excellence: Organic Chemistry*, or ACE, is a Web-based program that has been developed specifically for instructors and students of introductory organic chemistry. In the typical organic chemistry course, students work questions out of a textbook for which a solutions manual is available. Students struggle with a question for a few minutes, look up the answer, and then feel as if they understand how to answer the question. It isn't until exam time that they realize that learning an answer to a question and knowing how to answer a question are two entirely different things.

Computer programs that can tell students that their responses are incorrect without giving away the correct answer are an ideal solution to this problem. Many Web-based chemistry homework programs exist, but most require text-based, numeric, or multiple-choice responses. By contrast, ACE permits students to draw structural responses to posed questions with a graphical structure-drawing interface. They submit their structure to a server, and, if their response is incorrect, they are given feedback and asked to try again. Moreover, ACE is the only program that provides specific feedback for incorrect structural responses, effectively teaching students how to solve the problem without giving away correct answers. ACE is available for

Organic Chemistry, *Fifth Edition*, as part of one of the OneKey courses described above, and also available as a standalone study resource. Please contact your local Prentice Hall representative for more information.

FOR THE STUDENT

Study Guide and Solutions Manual (0-13-196328-7) by Paula Yurkanis Bruice. This *Study Guide and Solutions Manual* contains complete and detailed explanations of the solutions to the problems in the text and definitions of all the key terms used in each chapter. In addition, you'll find a section on advanced acid–base chemistry with an additional set of problems, two tutorials on "pushing electrons," a section on molecular orbital theory, a tutorial on drawing resonance contributors, exercises on building molecular models and calculating kinetic parameters, as well as 22 practice tests.

Virtual ChemLab Workbook and CD (0-13-238827-8) Prepared by Brian F. Woodfield of Brigham Young University. This virtual laboratory is a realistic, simulated laboratory environment where students can get a feel for what to expect in a real wet lab, or conduct experiments that are not included in the department's lab program. Students may experiment on their own in the full virtual lab environment, or choose pre-arranged labs that are referenced in the workbook and at the end of the chapter in the textbook. Virtual ChemLab CD can be run directly from the CD or installed on the student's computer.

Molecular Modeling Workbook & CD (0-13-222044-X) Features SpartanView™ and SpartanBuild™ software. This workbook includes a software tutorial and numerous challenging exercises students can tackle to solve problems involving structure building and analysis using the tools included in the two pieces of Spartan software. Contact your Prentice Hall representative for details.

Companion Website with GradeTracker (*http://www.prenhall.com/bruice*) Now with GradeTracker to provide flexible gradebook functionality for instructors who wish to use it. The text-specific Companion Website for *Organic Chemistry, Fifth Edition* has been thoroughly updated and enhanced from the previous edition, and provides students with a collection of learning and assessment tools for each chapter. The Companion Website for the Fifth Edition includes:

- The newly assembled **Student Tutorials quiz** gallery. Each question contains a visual animation or interactive tutorial, then asks a question about it, and provides feedback that includes links back to the relevant section of the book. By providing assessment tightly wrapped around the tutorial elements, including feedback on incorrect responses, and references to the book, the concept is reinforced. See the associated icons within the text, at point of use.

- A **Molecule Gallery**, with a vast selection of 3-D (Chime) renderings of most of the important molecules in the text. Students can rotate the molecule in three dimensions, change its representation, and explore its structure in detail.

- **Exercise Sets** and **Quizzes** with hints and instant feedback, including links to relevant sections from the book.

- The **MCAT Study Guide** consists of 200 questions, and can be a vital career resource.

- For students who are using ChemOffice Ltd., a workbook featuring **ChemOffice Exercises** is also available on the Website.

Collaborators for this edition's Website are Christine Hermann (Radford University), Brian Groh (Minnesota State University, Mankato), Bette A. Kreuz (The University of Michigan–Dearborn), Ron Wikholm (The University of Connecticut), and Christopher Hadad (The Ohio State University).

Activebook A full electronic version of *Organic Chemistry, Fifth Edition* is available as an Activebook. This online resource includes animations, tutorials, molecules, and glossary terms at point of reference in the textbook. A custom study plan allows the students to test their understanding and identify the topics they need to further study. Access the Activebook at *www.prenhall.com/bruice*.

ChemOffice Student CD This software includes ChemDraw LTD and Chem3D LTD. A free workbook is available on the Companion Website that includes a software tutorial and numerous exercises for each chapter of the text. The software is sold at the instructor's discretion and is available at a substantial discount if purchased with the textbook.

Prentice Hall Molecular Model Kit (0-205-508136-3) This best-selling model kit allows students to build space-filling and ball-and-stick models of common organic molecules. It allows accurate depiction of double and triple bonds, including heteroatomic molecules (which some model kits cannot handle well).

Prentice Hall Framework Molecular Model Kit (0-13-330076-5) This model kit allows students to build scale models that show the mutual relations of atoms in organic molecules, including precise interatomic distances and bond angles. This is the most accurate model kit available.

TO THE STUDENT

Welcome to organic chemistry. You are about to embark on an exciting journey. This book has been written with students like you in mind—those who are encountering the subject for the first time. The journey into organic chemistry can be an exciting one. The book's central goal is to make this journey both stimulating and enjoyable by helping you understand the central principles of the subject and asking you to apply them as you progress through the pages. You will be reminded about these principles at frequent intervals—for example, in references to sections you have already mastered and in the Building on Fundamentals list at the beginning of each chapter.

You should start by familiarizing yourself with the book. The inside front and back covers display information you may want to refer to often during the course. Chapter Summaries, Key Terms, and Reaction Summaries at each chapter's end provide helpful checklists of the terms and concepts you should understand after studying the chapter. The Glossary at the end of the book can also be a useful study aid, as can the Appendices, which were written to consolidate useful categories of information. The molecular models and electrostatic potential maps that you will find throughout the book are provided to give you an appreciation of what molecules look like in three dimensions and to show how charge is distributed within a molecule. Think of the margin notes as the author's opportunity to inject personal reminders of ideas and facts that are important to remember. Be sure to read them.

Work all the problems within each chapter. These are drill problems that allow you to check whether you have mastered the skills and concepts the chapter is teaching. Some of them (or parts of them) are solved for you in the text. Short answers to some of the others—those marked with a diamond—are provided at the end of the book. Do not overlook the "Problem-Solving Strategies" that are also sprinkled throughout the text; they provide practical suggestions on the best way to approach important types of problems.

In addition to the within-chapter problems, work as many end-of-chapter problems as you can. The more problems you work, the more comfortable you will be with the subject matter and the better prepared you will be for the material in subsequent chapters.

Do not let any problem frustrate you. If you cannot figure out the answer in a reasonable amount of time, turn to the *Study Guide and Solutions Manual* to learn how you should have approached the problem. Later on, go back and try to work the problem on your own again. Be sure to visit the Companion Website (*www.prenhall.com/bruice*) to try out some of the study tools like the Student Tutorials, Molecule Gallery, and Exercise Sets and Quizzes.

The most important advice to remember (and follow) in studying organic chemistry is DO NOT FALL BEHIND! The individual steps to learning organic chemistry are quite simple; each by itself is relatively easy to master. But they are numerous, and the subject can quickly become overwhelming if you do not keep up.

Before many of the theories and mechanisms were figured out, organic chemistry was a discipline that could be mastered only through memorization. Fortunately, that is no longer true. You will find many unifying ideas that allow you to use what you have learned in one situation to predict what will happen in other situations. So, as you read the book and study your notes, always try to understand *why* each chemical event or behavior happens. For example, when the reasons behind reactivity are understood, most reactions can be predicted. Approaching the course with the misconception that to succeed you must memorize hundreds of unrelated reactions could be your downfall. There is simply too much material to memorize. Understanding and reasoning, not memorization, provides the necessary foundation on which to lay subsequent learning. Nevertheless, from time to time some memorization will be required: some fundamental rules have to be memorized, and you will need to memorize the common names of a number of organic compounds. But the latter should not be a problem; after all, your friends have common names that you have been able to learn.

Students who study organic chemistry to gain entrance into medical school sometimes wonder why medical schools pay so much attention to this topic in particular. The importance of organic chemistry is not in the subject matter alone, however. Mastering organic chemistry requires a thorough understanding of certain fundamental principles and the ability to use those fundamentals to analyze, classify, and predict. The study of medicine makes similar demands: a physician uses an understanding of certain fundamental principles to analyze, classify, and diagnose.

Good luck in your study. I hope you will enjoy your course in organic chemistry and learn to appreciate the logic of this fascinating discipline. If you have any comments about the book or any suggestions for improving it, I would love to hear from you. Remember, positive comments are the most fun, but negative comments are the most useful.

Paula Yurkanis Bruice
pybruice@chem.ucsb.edu

ACKNOWLEDGMENTS

It gives me great pleasure to acknowledge the dedicated efforts of many good friends that made this book a reality. The many contributions of Ron Magid of the University of Tennessee, Ed Skibo of Arizona State University, Paul Papadopoulos of the University of New Mexico, Ron Starkey of the University of Wisconsin–Green Bay, and Jack Kirsch, University of California, Berkeley, persist in this edition. Particular thanks go to David Yerzley, M.D., for his assistance with the section on MRI; Warren Hehre of Wavefunction, Inc., and Alan Shusterman of Reed College for their advice on the electrostatic potential maps that appear in the book. I am also very grateful to my students, who pointed out sections that needed clarification, worked the problems, and searched for errors.

The following reviewers have played an enormously important role in the development of this textbook.

Fifth Edition Manuscript Reviewers

Jon C. Antilla, *University of Mississippi*
Arthur J. Ashe, III, *University of Michigan*
William F. Bailey, *University of Connecticut*
Chad Booth, *Texas State University*
Gary Breton, *Berry College*
Lale Aka Burk, *Smith College*
Dorian Canelas, *North Carolina State University*
Robert Coleman, *Ohio State University*
S. Todd Deal, *Georgia Southern University*
Malcolm D.E. Forbes, *University of North Carolina*
Annaliese Franz, *Harvard University*
Alison J. Frontier, *University of Rochester*
Albert Fry, *Wesleyan University*
Jose J. Gutierrez, *The University of Texas Pan American*
Christopher M. Hadad, *Ohio State University*
C. Frederick Jury, *Collin County Community College*
Bob Kane, *Baylor University*
Angela King, *Wake Forest University*
Irene Lee, *Case Western Reserve University*
Philip Lukeman, *New York University*
Neil Miranda, *University of Illinois at Chicago*
Robert P. O'Fee, *The College of New Jersey*
JaimeLee Iolani Rizzo, *Pace University*
Alexander J. Seed, *Kent State University*
William N. Setzer, *University of Alabama-Huntsville*
Thomas E. Sorensen, *University of Wisconsin-Milwaukee*
Jennifer A. Tripp, *University of Scranton*
Linda Waldman, *Cerritos College*
Ronald Wikholm, *University of Connecticut*
Emel L. Yakali, *Raymond Walters College*

Accuracy Reviewers

S. Todd Deal, *Georgia Southern University*
Malcolm Forbes, *University of North Carolina*
Steven Graham, *St. John's University*
Christopher Roy, *Duke University*
Susan Schelble, *University of Colorado at Denver*

Fourth Edition Manuscript Reviewers

Merritt Andrus, *Brigham Young University*
Daniel Appella, *Northwestern University*
George Bandik, *University of Pittsburgh*
Daniel Blanchard, *Kutztown University*
Ron Blankespoor, *Calvin College*
Paul Buonora, *California State University, Long Beach*
Robert Chesnut, *Eastern Illinois University*
Michael Chong, *University of Waterloo*
Robert Coleman, *Ohio State University*
David Collard, *Georgia Institute of Technology*
Debbie Crans, *Colorado State University*
Malcolm Forbes, *University of North Carolina, Chapel Hill*
Deepa Godambe, *Harper College*
Fathi Halaweish, *South Dakota State University*

Steve Hardinger, *University of California, Los Angeles*
Alvan Hengge, *Utah State University*
Steve Holmgren, *University of Montana*
Nichole Jackson, *Odessa College*
Carl Kemnitz, *California State University, Bakersfield*
Keith Krumpe, *University of North Carolina, Asheville*
Michael Kurz, *Illinois State University*
Li, Yuzhuo, *Clarkson University*
Janis Louie, *University of Utah*
Charles Lovelette, *Columbus State University*
Ray Lutgring, *University of Evansville*
Janet Maxwell, *Angelo State University*
Mark McMills, *Ohio University*
Andrew Morehead, *University of Maryland*
John Olson, *Augustana University*
Brian Pagenkopf, *University of Texas, Austin*
Joanna Petridou, *Spokane Falls Community College*
Michael Rathke, *Michigan State University*
Christopher Roy, *Duke University*
Tomikazu Sasaki, *University of Washington*
David Soriano, *University of Pittsburgh*
Jon Stewart, *University of Florida*
John Taylor, *Rutgers University*
Carl Wamser, *Portland State University*
Marshall Werner, *Lake Superior State University*
Catherine Woytowicz, *George Washington University*
Zhaohui Sunny Zhou, *Washington State University*

Critique Reviewers

Neil Allison, *University of Arkansas*
Joseph W. Bausch, *Villanova University*
Dana Chatellier, *University of Delaware*
Steven Fleming, *Brigham Young University*
Malcolm Forbes, *University of North Carolina, Chapel Hill*
Chuck Garner, *Baylor University*
Andrew Knight, *Loyola University*
Joe LeFevre, *State University of New York, Oswego*
Charles Liotta, *Georgia Institute of Technology*
Andrew Morehead, *University of Maryland*
Richard Pagni, *University of Tennessee*
Jimmy Rogers, *University of Texas, Arlington*
Richard Theis, *Oregon State University*
Peter J. Wagner, *Michigan State University*
John Williams, *Temple University*
Catherine Woytowicz, *George Washington University*

Accuracy Reviewers

Bruce Banks, *University of North Carolina, Greensboro*
Debra Bautista, *Eastern Kentucky University*
Vladimir Benin, *University of Dayton*
Linda Betz, *Widener University*
Anthony Bishop, *Amherst College*
Phil Brown, *Brigham Young University*
Sushama Dandekar, *University of North Texas*
S. Todd Deal, *Georgia Southern University*

Michael Detty, *University of Buffalo*
Matthew Dintzner, *DePaul University*
Nicholas Drapela, *Oregon State University*
Jeffrey Elbert, *University of Northern Iowa*
Mark Forman, *Saint Joseph's University*
Joe Fox, *University of Delaware*
Anne Gaquere, *State University of West Georgia*
Chuck Garner, *Baylor University*
Scott Goodman, *Buffalo State College*
Steven Graham, *St. John's University*
Christian Hamann, *Albright College*
Cliff Harris, *Albion College*
Alfred Hortmann, *Washington University*
Floyd Klavetter, *Indiana University of Pennsylvania*

Thomas Lectka, *Johns Hopkins University*
Len MacGillivray, *University of Iowa*
Jerry Manion, *University of Central Arkansas*
Alan P. Marshand, *University of North Texas*
Przemyslaw Maslak, *Pennsylvania State University*
Michael McKinney, *Marquette University*
Alex Nickon, *Johns Hopkins University*
Patrick O'Connor, *Rutgers University*
Kenneth Overly, *Providence College*
Cass Parker, *Clark Atlanta University*
Christopher Roy, *Duke University*
Susan Schelble, *University of Colorado, Denver*
Chris Spilling, *University of Missouri*
Janet Stepanek, *Colorado College*

I am deeply grateful to my editor, Nicole Folchetti, whose talents guided this book and caused it to be as good as it could be and whose gentle prodding made it happen. I also want to thank the other talented and dedicated people at Prentice Hall whose contributions made this book a reality. A huge thank you goes to Ray Mullaney who has been a part of all editions of this book, keeping me on track with a great deal of patience. I particularly want to thank David Theisen, National Sales Director for Key Markets, for his keen understanding of the book and his belief in it. I am enormously grateful to Moira Lerner Nelson, the developmental editor, both for her creativity and uncanny attention to detail. Her ability to make sense out of whatever I wrote has made a big difference to the quality of the book. And thank you to Michael J. Richards and Benjamin Paris, the creative brains behind the technology, and to Kristen Kaiser who built the student and instructor resources to accompany the book.

I particularly want to thank the many wonderful and talented students I have had over the years who taught me how to be a teacher. And I want to thank my children, from whom I may have learned the most.

To make this textbook as user friendly as possible, I would appreciate any comments that will help me achieve this goal in future editions. If you find sections that could be clarified or expanded, or examples that could be added, please let me know. Finally, this edition has been painstakingly combed for typographical errors. Any that remain are my responsibility; if you find any, please send me a quick email so they can be corrected in future printings.

Paula Yurkanis Bruice
University of California, Santa Barbara
pybruice@chem.ucsb.edu

Highlights of *Organic Chemistry,* Fifth Edition

Focus on Mechanistic Similarities

The unique organization of the text, functional groups assembled by mechanistic similarities, allows for the sequential presentation of reactions that occur by similar mechanisms.

Tying Together Synthesis and Reactivity

Tying together the reactivity of a functional group and the synthesis of compounds that result from that reactivity (instead of teaching the reactivity and synthesis of a particular functional group) decreases the need to memorize unrelated reactions and allows for a certain economy of presentation.

NEW Building on Fundamentals

Found at the beginning of the chapters, these lists show how key topics in the chapter are connected to material already learned. They reveal organic chemistry as a unified, unfolding story, rather than a series of disparate topics. ▼

BUILDING ON FUNDAMENTALS

SECTIONS 4.1–4.10 We saw the first example of an electrophilic addition reaction in Chapter 3 (Section 3.6). Now we will look at several more electrophilic addition reactions—the characteristic reaction of alkenes.

SECTION 4.2 Hyperconjugation—delocalization of electrons by the overlap of a σ bond orbital with an orbital on an adjacent carbon—explained why the staggered conformation of ethane is more stable than the eclipsed conformation. Now we will see that it also explains why alkyl groups stabilize carbocations (2.10).

SECTION 4.4 The product with the smallest free energy of activation for its formation is the one that is formed the fastest (3.7).

SECTION 4.5 A compound with an acidic proton loses that proton in solutions that have pH values greater than the compound's pK_a value (1.24).

SECTION 4.7 The octet rule helps explain why a cyclic bromonium ion rather than a carbocation is formed when an alkene reacts with a halogen (Br_2 or Cl_2) (1.4).

SECTION 4.10 Steric effects are partially responsible for the regioselectivity of hydroboration–oxidation (2.10).

SECTION 4.11 The steric strain that causes a gauche conformation to be less stable than an anti conformation, also causes a cis isomer to be less stable than a trans isomer (2.10).

NEW Stepped-out mechanisms clearly delineate the mechanisms of the reactions in a way that is integral to the text, but significantly highlighted. ▼

Mechanism for an electrophilic addition reaction

an electrophile

$$CH_3C{\equiv}CCH_3 \ + \ H{-}\ddot{C}l: \longrightarrow CH_3\overset{+}{C}{=}CHCH_3 \ + \ :\ddot{C}l:^- \longrightarrow CH_3\overset{\overset{\displaystyle Cl}{|}}{C}{=}CHCH_3$$

a nucleophile   an electrophile   a nucleophile

- The relatively weak π bond breaks because the π electrons are attracted to the electrophilic proton.
- The positively charged carbocation intermediate reacts rapidly with the negatively charged chloride ion.

Bioorganic Emphasis

Coverage of bioorganic chemistry connects organic chemistry to biochemistry to present them as a continuum of knowledge. This material is incorporated throughout the text in several ways: special interest boxes, specific chapter sections, and chapters that focus on bioorganic topics. ▶

PENICILLIN AND DRUG RESISTANCE

Penicillin contains an amide in a strained β-lactam ring. The strain in this four-membered ring increases the amide's reactivity. It is thought that the antibiotic activity of penicillin results from its ability to acylate (put an acyl group on) a CH_2OH group of an enzyme that has a role in the synthesis of bacterial cell walls. Acylation inactivates the enzyme, and actively growing bacteria die because they are unable to synthesize functional cell walls. Penicillin has no effect on mammalian cells because mammalian cells are not enclosed by cell walls. Penicillins are stored at cold temperatures to minimize hydrolysis of the β-lactam ring.

Bacteria that are resistant to penicillin secrete penicillinase, an enzyme that catalyzes the hydrolysis of penicillin's β-lactam ring. The ring-opened product has no antibacterial activity.

NEW Chapter on metabolism emphasizes that the reactions and mechanisms encountered in metabolic processes are similar to the reactions and mechanisms encountered in an organic chemistry laboratory.

Getting the Big Picture

NEW Concept Headings frame the context of the discussion to follow rather than merely title the section. ▼

8.4 The Reversibility of an S$_N$2 Reaction Depends on the Basicities of the Leaving Groups in the Forward and Reverse Directions

Many different kinds of nucleophiles can react with alkyl halides. Therefore, a wide range of organic compounds can be synthesized by means of S$_N$2 reactions.

Biographical Sketches give students an appreciation of the history of chemistry and the people who contributed to that history. ▶

Margin Notes emphasize core ideas and remind students of important principles. ▼

BIOGRAPHY

Francois Auguste Victor Grignard (1871–1935) *was born in France, the son of a sailmaker. He received a Ph.D. from the University of Lyons in 1901. His synthesis of the first Grignard reagent was announced in 1900. During the next five years, some 200 papers were published about Grignard reagents. Grignard was a professor of chemistry at the University of Nancy and later at the University of Lyons. He shared the Nobel Prize in chemistry in 1912 with Paul Sabatier (p. 188). During World War I, he was drafted into the French army, where he developed a method to detect war gases.*

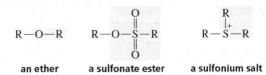

(X⁻); because they are stronger bases, they are poorer rder to displace. Consequently, alcohols and ethers are in substitution and elimination reactions. We will see basic leaving groups, alcohols and ethers have to be ndergo a substitution or an elimination reaction. In ulfonium salts have weakly basic leaving groups, so ons with ease.

The weaker the base, the more easily it can be displaced.

The stronger the acid, the weaker is its conjugate base.

an ether   a sulfonate ester   a sulfonium salt

SUMMARY OF REACTIONS

1. Electrophilic aromatic substitution reactions:
   a. Halogenation (Section 14.11)

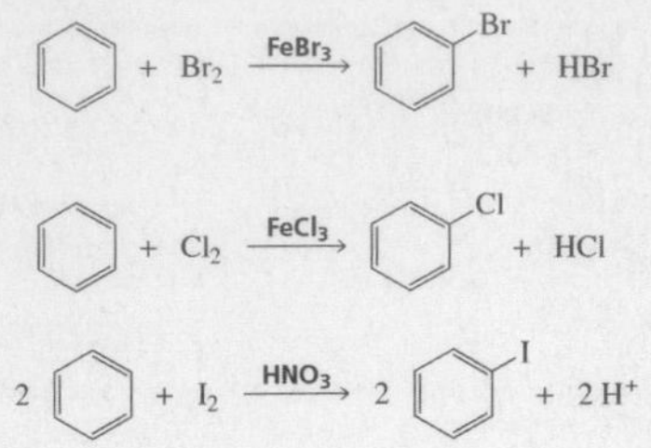

◀ Summary of Reactions at the end of each chapter list reactions that have been covered for student review. Cross references make it easy to locate the section covering a specific reaction.

Special Topics and Tutorials in the *Study Guide and Solutions Manual*:

- Acids/Bases: pH, pK_a, and Buffers
- Drawing Curved Arrows (now 2 tutorials)
- Kinetics
- Exercise in Model Building
- NEW Molecular Orbital Theory
- NEW Drawing Resonance Contributors

Problem Solving

Solved Problems throughout the text carefully walk students through the steps involved in solving a particular type of problem. ▼

PROBLEM 28 · SOLVED

Early chemists could envision several possible mechanisms for hydroxide-ion-promoted ester hydrolysis:

1. a nucleophilic acyl substitution reaction

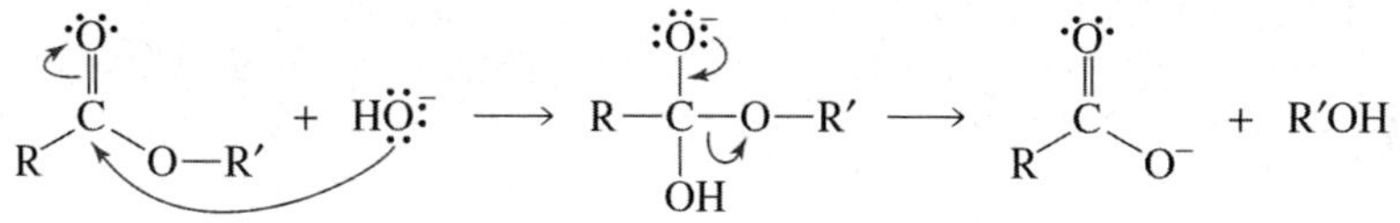

Problem-Solving Strategies in many chapters teach students how to approach a variety of problems, organize their thoughts, and improve their problems-solving abilities. Every strategy is followed by an exercise that allows students to practice the strategy just discussed. ▼

PROBLEM-SOLVING STRATEGY

Planning the Synthesis of an Alkyl Halide

a. What alkene should be used to synthesize 3-bromohexane?

$$? \ + \ HBr \ \longrightarrow \ CH_3CH_2CHCH_2CH_2CH_3$$

| Br

3-bromohexane

Over 200 new problems in the text offer even more opportunities for students to practice.

Designing a Synthesis sections appear throughout the text and help students learn to design multistep syntheses. Many synthesis problems include the synthesis of compounds that students recognize, such as Novocaine, Valium, Tagamet, and ibuprofen. ▶

18.21 Designing a Synthesis VII: Making New Carbon–Carbon Bonds

When you are planning the synthesis of a compound that requires the formation of a new carbon–carbon bond, first locate the new bond that must be made. For example, in the synthesis of the following β-diketone, the new bond is the one that makes the second five-membered ring:

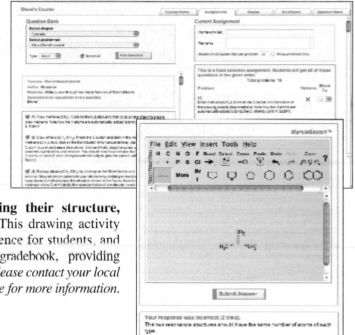

MEDIA RESOURCES

ACE Organic (*www.prenhall.com/aceorganic*) ▶

Prentice Hall, in conjunction with Robert B. Grossman and Raphael A. Finkel and a team of programmers at the University of Kentucky, has developed a homework system for organic chemistry that can finally support the types of problems assigned in organic chemistry. ACE (Achieving Chemistry Excellence) is a Web-based program developed specifically for instructors and students of introductory organic chemistry.

ACE Organic contains hundreds of organic chemistry structure-drawing problems. **ACE permits students to construct structural responses to questions with a graphical structure-drawing interface. After submitting their structure, students receive feedback specific to their response.** This drawing activity with immediate feedback results in a better learning experience for students, and automatically stores student grades and activity in a gradebook, providing instructors with a better way to assess student understanding. *Please contact your local Prentice Hall representative or see the description in the preface for more information.*

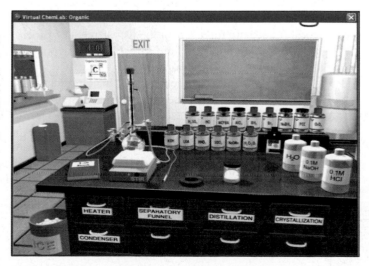

Virtual ChemLab Workbook and CD ▶

(0-13-238827-8) Created by Brian F. Woodfield of Brigham Young University. This virtual laboratory is a realistic, simulated laboratory environment where students can get a feel for what to expect in a real wet lab, or conduct short experiments that are designed to enhance the material students learn in lecture. Virtual ChemLab software may be run directly from the CD or installed on the student's computer. *Please contact your local Prentice Hall representative or see the description in the preface for more information.*

About the Author

Paula Bruice with Zeus and Abigail

Paula Yurkanis Bruice was raised primarily in Massachusetts. After graduating from the Girls' Latin School in Boston, she earned an A.B. from Mount Holyoke College and a Ph.D. in chemistry from the University of Virginia. She then received an NIH postdoctoral fellowship for study in the Department of Biochemistry at the University of Virginia Medical School and held a postdoctoral appointment in the Department of Pharmacology at Yale Medical School.

Paula has been a member of the faculty at the University of California, Santa Barbara since 1972, where she has received the Associated Students Teacher of the Year Award, the Academic Senate Distinguished Teaching Award, two Mortar Board Professor of the Year Awards, and the UCSB Alumni Association Teaching Award. Her research interests center on the mechanism and catalysis of organic reactions, particularly those of biological significance. Paula has a daughter and a son who are physicians and a son who is a lawyer. Her main hobbies are reading mystery and suspense novels and enjoying her pets (two dogs, two cats, and a parrot).

PART 1

An Introduction to the Study of Organic Chemistry

The first two chapters of the text cover a variety of topics that you need to be familiar with in order to get started.

CHAPTER 1
Electronic Structure and Bonding • Acids and Bases

Chapter 1 reviews the topics from general chemistry that will be important to your study of organic chemistry. The chapter starts with a description of the structure of atoms and then proceeds to a description of the structure of molecules. Molecular orbital theory is introduced. Acid–base chemistry, which is central to understanding many organic reactions, is reviewed. You will see how the structure of a molecule affects its acidity and how the acidity of a solution affects molecular structure.

CHAPTER 2
An Introduction to Organic Compounds: Nomenclature, Physical Properties, and Representation of Structure

To discuss organic compounds, you must be able to name them and to visualize their structures when you read or hear their names. In **Chapter 2**, you will learn how to name five different classes of organic compounds. This will give you a good understanding of the basic rules for naming compounds. Because the compounds examined in the chapter are reactants or products of many of the reactions presented in the next 10 chapters, you will have numerous opportunities to review the nomenclature of these compounds as you proceed through those chapters. Chapter 2 also compares and contrasts the structures and physical properties of these compounds, which makes learning about them a little easier than if each compound were presented separately. Because organic chemistry is a study of compounds that contain carbon, the last part of Chapter 2 discusses the spatial arrangement of the atoms in both chains and rings of carbon atoms.

CHAPTER 1

Electronic Structure and Bonding • Acids and Bases

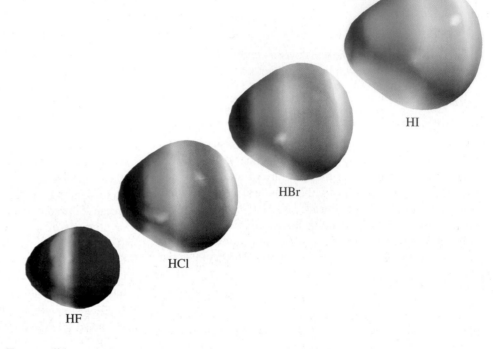

HI

HBr

HCl

HF

**Jöns Jakob Berzelius
(1779–1848),** *an important figure
in the development of modern
chemistry, was born in Sweden.
When Berzelius was two, his father
died. His mother remarried, but
died only two years after her
first husband. Unhappy in his
stepfather's home, Berzelius left
when he was 14 and supported
himself by working as a tutor and
laboring in the fields. Berzelius
not only coined the terms* organic
and inorganic, *but also invented
the system of chemical symbols
still used today. He published the
first list of accurate atomic
weights and proposed the idea that
atoms carry an electric charge. He
purified or discovered the elements
cerium, selenium, silicon, thorium,
titanium, and zirconium.*

To stay alive, early humans must have been able to tell the difference between the kinds of materials in their world. "You can live on roots and berries," they might have said, "but you can't eat dirt. You can stay warm by burning tree branches, but you can't burn rocks."

By the early eighteenth century, scientists thought they had grasped the nature of that difference and, in 1807, Jöns Jakob Berzelius gave names to the two kinds of materials. Compounds derived from living organisms were believed to contain an unmeasurable vital force—the essence of life. These he called "organic." Compounds derived from minerals—those lacking that vital force—were "inorganic."

Because chemists could not create life in the laboratory, they assumed they could not create compounds that had a vital force. Since this was their mind-set, you can imagine how surprised chemists were in 1828 when Friedrich Wöhler produced urea—a compound known to be excreted by mammals—by heating ammonium cyanate, an inorganic mineral.

$$\overset{+}{N}H_4 \ \overset{-}{O}CN \xrightarrow{\text{heat}} \begin{array}{c} O \\ \| \\ H_2N-C-NH_2 \end{array}$$

ammonium cyanate   **urea**

For the first time, an "organic" compound had been obtained from something other than a living organism and certainly without the aid of any kind of vital force. Clearly, chemists needed a new definition for "organic compounds." **Organic compounds** are now defined as *compounds that contain carbon.*

Why is an entire branch of chemistry devoted to the study of carbon-containing compounds? We study organic chemistry because just about all of the molecules that make life possible—proteins, enzymes, vitamins, lipids, carbohydrates, and nucleic acids—contain carbon; thus the chemical reactions that take place in living systems, including our own bodies, are reactions of organic compounds. Most of the compounds found in nature—those we rely on for food, medicine, clothing (cotton, wool, silk), and energy (natural gas, petroleum)—are organic as well.

Organic compounds are not, however, limited to those found in nature. Chemists have learned to synthesize millions of organic compounds never found in nature, including synthetic fabrics, plastics, synthetic rubber, medicines, and even things like photographic film and Super Glue. Many of these synthetic compounds prevent shortages of naturally occurring products. For example, it has been estimated that if synthetic materials were not available for clothing, all of the arable land in the United States would have to be used for the production of cotton and wool just to provide enough material to clothe us. Currently, there are about 16 million known organic compounds, and many more are possible.

What makes carbon so special? Why are there so many carbon-containing compounds? The answer lies in carbon's position in the periodic table. Carbon is in the center of the second row of elements. We will see that the atoms to the left of carbon have a tendency to give up electrons, whereas the atoms to the right have a tendency to accept electrons (Section 1.3).

the second row of the periodic table

Because carbon is in the middle, it neither readily gives up nor readily accepts electrons. Instead, it shares electrons. Carbon can share electrons with several different kinds of atoms, and it can also share electrons with other carbon atoms. Consequently, carbon is able to form millions of stable compounds with a wide range of chemical properties simply by sharing electrons.

When we study organic chemistry, we study how organic compounds react. When an organic compound reacts, some bonds break and some new bonds form. Bonds form when two atoms share electrons, and bonds break when two atoms no longer share electrons. How readily a bond forms and how easily it breaks depend on the particular electrons that are shared, which, in turn, depend on the atoms to which the electrons belong. So if we are going to start our study of organic chemistry at the beginning, we must start with an understanding of the structure of an atom—what electrons an atom has and where they are located.

German chemist **Friedrich Wöhler (1800–1882)** *began his professional life as a physician and later became a professor of chemistry at the University of Göttingen. Wöhler codiscovered the fact that two different chemicals could have the same molecular formula. He also developed methods of purifying aluminum—at the time, the most expensive metal on Earth—and beryllium.*

NATURAL VERSUS SYNTHETIC

It is a popular belief that natural substances—those made in nature—are superior to synthetic ones—those made in the laboratory. Yet when a chemist synthesizes a compound, such as penicillin or estradiol, it is exactly the same in all respects as the compound synthesized in nature. Sometimes chemists can improve on nature. For example, chemists have synthesized analogs of morphine—compounds with structures similar to but not identical to that of morphine—that have pain-killing effects like morphine but, unlike morphine, are not habit forming. Chemists have synthesized analogs of penicillin that do not produce the allergic responses that a significant fraction of the population experiences from naturally produced penicillin, or that do not have the bacterial resistance of the naturally produced antibiotic.

A field of poppies growing in Afghanistan. Commercial morphine is obtained from opium, the juice obtained from this species of poppy.

1.1 **The Structure of an Atom**

An atom consists of a tiny dense nucleus surrounded by electrons that are spread throughout a relatively large volume of space around the nucleus. The nucleus contains *positively charged protons* and *neutral neutrons*, so it is positively charged. The *electrons are negatively charged.* They are moving continuously. Like anything that moves, electrons have kinetic energy, and this energy is what counters the attractive force of the positively charged protons that would otherwise pull the negatively charged electrons into the nucleus.

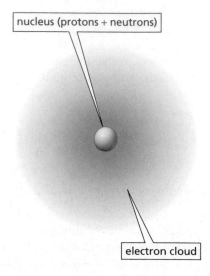

nucleus (protons + neutrons)

electron cloud

Because the amount of positive charge on a proton equals the amount of negative charge on an electron, a neutral atom has an equal number of protons and electrons. Atoms can gain electrons and thereby become negatively charged, or they can lose electrons and become positively charged. However, the number of protons in an atom does not change.

Protons and neutrons have approximately the same mass and are about 1800 times more massive than an electron. This means that most of the mass of an atom is in its nucleus. However, most of the *volume* of an atom is occupied by its electrons, and that is where our focus will be because it is the electrons that form chemical bonds.

The **atomic number** of an atom equals the *number of protons* in its nucleus. The atomic number is also the number of electrons that surround the nucleus of a neutral atom. For example, the atomic number of carbon is 6, which means that a neutral carbon atom has six protons and six electrons.

The **mass number** of an atom is the *sum of its protons and neutrons.* All carbon atoms have the same atomic number because they all have the same number of protons. They do not all have the same mass number because they do not all have the same number of neutrons. For example, 98.89% of naturally occurring carbon atoms have six neutrons—giving them a mass number of 12—and 1.11% have seven neutrons—giving them a mass number of 13. These two different kinds of carbon atoms (^{12}C and ^{13}C) are called isotopes. **Isotopes** have the same atomic number (that is, the same number of protons), but different mass numbers because they have different numbers of neutrons.

Naturally occurring carbon also contains a trace amount of ^{14}C, which has six protons and eight neutrons. This isotope of carbon is radioactive, decaying with a half-life of 5730 years. (The *half-life* is the time it takes for one-half of the nuclei to decay.) As long as a plant or animal is alive, it takes in as much ^{14}C as it excretes or exhales. When it dies, it no longer takes in ^{14}C, so the ^{14}C in the organism slowly decreases. Therefore, the age of an organic substance can be determined by its ^{14}C content.

The **atomic weight** of a naturally occurring element is the *average weighted mass of its atoms*. Because an *atomic mass unit (amu)* is defined as exactly 1/12 of the mass of ^{12}C, the atomic mass of ^{12}C is 12.0000 amu; the atomic mass of ^{13}C is 13.0034 amu. Therefore, the atomic weight of carbon is 12.011 amu $[(0.9889 \times 12.0000) + (0.0111 \times 13.0034)] = 12.011$. The **molecular weight** is the *sum of the atomic weights* of all the atoms in the molecule.

PROBLEM 1◆

Oxygen has three isotopes with mass numbers of 16, 17, and 18. The atomic number of oxygen is eight. How many protons and neutrons does each of the isotopes have?

1.2 How the Electrons in an Atom Are Distributed

For a long time, electrons were perceived to be particles—infinitesimal "planets" orbiting the nucleus of an atom. In 1924, however, a French physicist named Louis de Broglie showed that electrons also have wavelike properties. He did this by combining a formula developed by Albert Einstein that relates mass and energy with a formula developed by Max Planck that relates frequency and energy. The realization that electrons have wavelike properties spurred physicists to propose a mathematical concept known as quantum mechanics.

Quantum mechanics uses the same mathematical equations that describe the wave motion of a guitar string to characterize the motion of an electron around a nucleus. The version of quantum mechanics most useful to chemists was proposed by Erwin Schrödinger in 1926. According to Schrödinger, the behavior of each electron in an atom or a molecule can be described by a **wave equation**. The solutions to the Schrödinger equation are called **wave functions** or **orbitals**. They tell us the *energy* of the electron and the *volume of space* around the nucleus where an electron is most likely to be found.

According to quantum mechanics, the electrons in an atom can be thought of as occupying a set of concentric shells that surround the nucleus. The first shell is the one closest to the nucleus. The second shell lies farther from the nucleus, and even farther out lie the third and higher numbered shells. Each shell contains subshells known as **atomic orbitals**. Each atomic orbital has a characteristic shape and energy and occupies a characteristic volume of space, which is predicted by the Schrödinger equation. An important point to remember is that *the closer the atomic orbital is to the nucleus, the lower is its energy*.

The first shell consists of only an *s* atomic orbital; the second shell consists of *s* and *p* atomic orbitals; the third shell consists of *s*, *p*, and *d* atomic orbitals; and the fourth and higher shells consist of *s*, *p*, *d*, and *f* atomic orbitals (Table 1.1).

Table 1.1	Distribution of Electrons in the First Four Shells That Surround the Nucleus			
	First shell	**Second shell**	**Third shell**	**Fourth shell**
Atomic orbitals	*s*	*s, p*	*s, p, d*	*s, p, d, f*
Number of atomic orbitals	1	1, 3	1, 3, 5	1, 3, 5, 7
Maximum number of electrons	2	8	18	32

The closer the orbital is to the nucleus, the lower is its energy.

ALBERT EINSTEIN

Albert Einstein (1879–1955) was born in Germany. When he was in high school, his father's business failed and his family moved to Milan, Italy. Although Einstein wanted to join his family in Italy, he had to stay behind because German law required compulsory military service after high school. To help him, his high school mathematics teacher wrote a letter saying that Einstein could have a nervous breakdown without his family and also that there was nothing left to teach him. Eventually, Einstein was asked to leave the school because of his disruptive behavior. Popular folklore says he left because of poor grades in Latin and Greek, but his grades in those subjects were fine.

Einstein was visiting the United States when Hitler came to power, so he accepted a position at the Institute for Advanced Study in Princeton, N.J., becoming a U.S. citizen in 1940. Although a lifelong pacifist, he wrote a letter to President Roosevelt warning of ominous advances in German nuclear research. This led to the creation of the Manhattan Project, which developed the atomic bomb and tested it in New Mexico in 1945.

The bronze sculpture of Albert Einstein on the grounds of the National Academy of Sciences in Washington, D.C., measures 21 feet from the top of the head to the tip of the feet and weighs 7000 pounds. In his left hand, Einstein holds the mathematical equations that represent his three most important contributions to science: the photoelectric effect, the equivalency of energy and matter, and the theory of relativity. At his feet is a map of the sky.

Degenerate orbitals are orbitals that have the same energy.

Each shell contains one *s* orbital. The second and higher shells—in addition to their *s* orbital—each contain three *degenerate p* orbitals. **Degenerate orbitals** are orbitals that have the same energy. The third and higher shells—in addition to their *s* and *p* orbitals—also contain five degenerate *d* orbitals, and the fourth and higher shells also contain seven degenerate *f* orbitals. Because a maximum of two electrons can coexist in an atomic orbital (see the Pauli exclusion principle, below), the first shell, with only one atomic orbital, can contain no more than two electrons. The second shell, with four atomic orbitals—one *s* and three *p*—can have a total of eight electrons. Eighteen electrons can occupy the nine atomic orbitals—one *s*, three *p*, and five *d*—of the third shell, and 32 electrons can occupy the 16 atomic orbitals of the fourth shell. In studying organic chemistry, we will be concerned primarily with atoms that have electrons only in the first and second shells.

The **ground-state electronic configuration** of an atom describes the orbitals occupied by the atom's electrons when they are all in the available orbitals with the lowest energy. If energy is applied to an atom in the ground state, one or more electrons can jump into a higher energy orbital. The atom then would be in an **excited-state electronic configuration**. The ground-state electronic configurations of the 11 smallest atoms are shown in Table 1.2. (Each arrow—whether pointing up or down—represents one electron.) The following principles are used to determine which orbitals electrons occupy:

1. The **aufbau principle** (*aufbau* is German for "building up") tells us the first thing we need to know to be able to assign electrons to the various atomic orbitals. According to this principle, an electron always goes into the available orbital with the lowest energy. Because a 1*s* orbital is closer to the nucleus, it is lower in

MAX KARL ERNST LUDWIG PLANCK

Max Planck (1858–1947) was born in Germany, the son of a professor of civil law. He himself was a professor at the Universities of Kiel (1885–1889) and Berlin (1889–1926). Two of his daughters died in childbirth, and one of his sons was killed in action in World War I. In 1918, Planck received the Nobel Prize in physics for his development of quantum theory. He became president of the Kaiser Wilhelm Society of Berlin—later renamed the Max Planck Society—in 1930. Planck felt that it was his duty to remain in Germany during the Nazi era, but he never supported the Nazi regime. He unsuccessfully interceded with Hitler on behalf of his Jewish colleagues and, as a consequence, was forced to resign from the presidency of the Kaiser Wilhelm Society in 1937. A second son was accused of taking part in the plot to kill Hitler and was executed. Planck lost his home to Allied bombings. He was rescued by Allied forces during the final days of the war.

energy than a $2s$ orbital, which is lower in energy—and is closer to the nucleus—than a $3s$ orbital. Comparing atomic orbitals in the same shell, we see that an s orbital is lower in energy than a p orbital, and a p orbital is lower in energy than a d orbital.

Relative energies of atomic orbitals: $1s < 2s < 2p < 3s < 3p < 4s < 3d < 4p < 5s < 4d < 5p < 6s < 4f < 5d < 6p < 7s < 5f$

2. The **Pauli exclusion principle** states that (a) no more than two electrons can occupy each atomic orbital, and (b) the two electrons must be of opposite spin. It is called an exclusion principle because it states that only so many electrons can occupy any particular shell. Notice in Table 1.2 that spin in one direction is designated by an upward-pointing arrow, and spin in the opposite direction by a downward-pointing arrow.

Atom	Name of element	Atomic number	$1s$	$2s$	$2p_x$	$2p_y$	$2p_z$	$3s$
H	Hydrogen	1	↑					
He	Helium	2	↑↓					
Li	Lithium	3	↑↓	↑				
Be	Beryllium	4	↑↓	↑↓				
B	Boron	5	↑↓	↑↓	↑			
C	Carbon	6	↑↓	↑↓	↑	↑		
N	Nitrogen	7	↑↓	↑↓	↑	↑	↑	
O	Oxygen	8	↑↓	↑↓	↑↓	↑	↑	
F	Fluorine	9	↑↓	↑↓	↑↓	↑↓	↑	
Ne	Neon	10	↑↓	↑↓	↑↓	↑↓	↑↓	
Na	Sodium	11	↑↓	↑↓	↑↓	↑↓	↑↓	↑

Table 1.2   The Electronic Configurations of the Smallest Atoms

As a teenager, Austrian **Wolfgang Pauli (1900–1958)** *wrote articles on relativity that caught the attention of Albert Einstein. Pauli went on to teach physics at the University of Hamburg and at the Zurich Institute of Technology. When World War II broke out, he immigrated to the United States, where he joined the Institute for Advanced Study at Princeton.*

From these first two rules, we can assign electrons to atomic orbitals for atoms that contain one, two, three, four, or five electrons. The single electron of a hydrogen atom occupies a $1s$ orbital, the second electron of a helium atom fills the $1s$ orbital, the third electron of a lithium atom occupies a $2s$ orbital, the fourth electron of a beryllium atom fills the $2s$ orbital, and the fifth electron of a boron atom occupies one of the $2p$

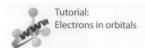

Tutorial:
Electrons in orbitals

orbitals. (The subscripts x, y, and z distinguish the three $2p$ orbitals.) Because the three p orbitals are degenerate, the electron can be put into any one of them. Before we can continue to atoms containing six or more electrons, we need Hund's rule:

3. **Hund's rule** states that when there are degenerate orbitals—two or more orbitals with the same energy—an electron will occupy an empty orbital before it will pair up with another electron. In this way, electron repulsion is minimized. The sixth electron of a carbon atom, therefore, goes into an empty $2p$ orbital, rather than pairing up with the electron already occupying a $2p$ orbital. (See Table 1.2.) There is one more empty $2p$ orbital, so that is where nitrogen's seventh electron goes. The eighth electron of an oxygen atom pairs up with an electron occupying a $2p$ orbital rather than going into a higher energy $3s$ orbital.

The locations of the electrons in the remaining elements can be assigned using these three rules.

Electrons in inner shells (those below the outermost shell) are called **core electrons**. Core electrons do not participate in chemical bonding. Electrons in the outermost shell are called **valence electrons**. Carbon, for example, has two core electrons and four valence electrons (Table 1.2). Lithium and sodium each have one valence electron. If you examine the periodic table inside the back cover of this book, you will see that lithium and sodium are in the same column. Elements in the same column of the periodic table have the same number of valence electrons. Because the number of valence electrons is the major factor determining an element's chemical properties, elements in the same column of the periodic table have similar chemical properties. Thus, the chemical behavior of an element depends on its electronic configuration.

B I O G R A P H Y

Friedrich Hermann Hund (1896–1997) *was born in Germany. He was a professor of physics at several German universities, the last being the University of Göttingen. He spent a year as a visiting professor at Harvard University. In February 1996, the University of Göttingen held a symposium to honor Hund on his 100th birthday.*

PROBLEM 2◆

How many valence electrons do the following atoms have?

a. boron **b.** nitrogen **c.** oxygen **d.** fluorine

PROBLEM 3◆

a. Find potassium (K) in the periodic table and predict how many valence electrons it has.

b. What orbital does the unpaired electron occupy?

PROBLEM 4◆

a. Write electronic configurations for chlorine (atomic number 17), bromine (atomic number 35), and iodine (atomic number 53).

b. How many valence electrons do chlorine, bromine, and iodine have?

PROBLEM 5

Compare the ground-state electronic configurations of the following atoms, and check the relative positions of the atoms in the periodic table.

a. carbon and silicon

b. oxygen and sulfur

c. fluorine and bromine

d. magnesium and calcium

1.3 Ionic and Covalent Bonds

In trying to explain why atoms form bonds, G. N. Lewis proposed that *an atom is most stable if its outer shell is either filled or contains eight electrons, and it has no electrons of higher energy.* According to Lewis's theory, an atom will give up, accept, or share electrons in order to achieve a filled outer shell or an outer shell that contains eight electrons. This theory has come to be called the **octet rule** (even though hydrogen has only two electrons in its filled outer shell).

Lithium (Li) has a single electron in its $2s$ orbital. If it loses this electron, the lithium atom ends up with a filled outer shell—a stable configuration. Lithium, therefore, loses an electron relatively easily. Sodium (Na) has a single electron in its $3s$ orbital, so it too loses an electron easily. Elements (such as lithium and sodium) that readily lose an electron and thereby become positively charged are said to be **electropositive**. The elements in the first column of the periodic table are all electropositive—each readily loses an electron because each has a single electron in its outermost shell.

When we draw the electrons around an atom, as in the following equations, core electrons are not shown; only valence electrons are shown because only valence electrons are used in bonding. Each valence electron is shown as a dot. Notice that when the single valence electron of lithium or sodium is removed, the species that is formed is called an ion because it carries a charge.

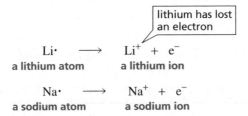

Fluorine has seven valence electrons (Table 1.2). Consequently, it readily acquires an electron in order to have an outer shell of eight electrons. Elements in the same column as fluorine (for example, chlorine, bromine, and iodine) also need only one electron to have an outer shell of eight, so they, too, readily acquire an electron. Elements that readily acquire an electron are said to be **electronegative**—they acquire an electron easily and thereby become negatively charged.

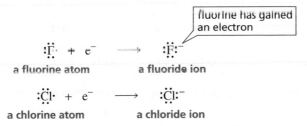

Ionic Bonds Are Formed by the Transfer of Electrons

We have just seen that sodium gives up an electron easily and chlorine readily acquires an electron. Therefore, when sodium metal and chlorine gas are mixed, each sodium atom transfers an electron to a chlorine atom, and crystalline sodium chloride (table salt) is formed as a result. The positively charged sodium ions and negatively charged chloride ions are independent species held together by the attraction of opposite charges (Figure 1.1). A **bond** is an attractive force between two atoms or between ions. Attractive forces between opposite charges are called **electrostatic attractions**. A bond that is the result of only electrostatic attractions is called an ionic bond. Thus, an **ionic bond** is formed when there is a *transfer of electrons*, causing one atom to become a positively charged ion and the other to become a negatively charged ion.

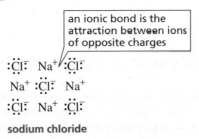

sodium chloride

Figure 1.1 ▶
(a) Crystalline sodium chloride.
(b) The electron-rich chloride ions are red and the electron-poor sodium ions are blue. Each chloride ion is surrounded by six sodium ions, and each sodium ion is surrounded by six chloride ions. Ignore the sticks holding the balls together; they are there only to keep the model from falling apart.

a.

b.

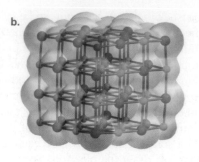

Sodium chloride is an example of an ionic compound. **Ionic compounds** are formed when an element on the left side of the periodic table (an electropositive element) transfers one or more electrons to an element on the right side of the periodic table (an electronegative element).

Covalent Bonds Are Formed by Sharing Electrons

Instead of giving up or acquiring electrons, an atom can achieve a filled outer shell by sharing electrons. For example, two fluorine atoms can each attain a filled second shell by sharing their unpaired valence electrons. A bond formed as a result of *sharing electrons* is called a **covalent bond**.

$$:\!\ddot{F}\!\cdot \ + \ \cdot\!\ddot{F}\!: \ \longrightarrow \ :\!\ddot{F}\!:\!\ddot{F}\!:$$

a covalent bond is formed by sharing electrons

Two hydrogen atoms can form a covalent bond by sharing electrons. As a result of covalent bonding, each hydrogen acquires a stable, filled first shell.

$$\mathrm{H}\!\cdot \ + \ \cdot\!\mathrm{H} \ \longrightarrow \ \mathrm{H}\!:\!\mathrm{H}$$

Similarly, hydrogen and chlorine can form a covalent bond by sharing electrons. In doing so, hydrogen fills its only shell and chlorine achieves an outer shell of eight electrons.

$$\mathrm{H}\!\cdot \ + \ \cdot\!\ddot{\mathrm{C}}\mathrm{l}\!: \ \longrightarrow \ \mathrm{H}\!:\!\ddot{\mathrm{C}}\mathrm{l}\!:$$

A hydrogen atom can achieve a completely empty shell by losing an electron. Loss of its sole electron results in a positively charged **hydrogen ion**. A positively charged hydrogen ion is called a **proton** because when a hydrogen atom loses its valence electron, only the hydrogen nucleus—which consists of a single proton—remains. A hydrogen atom can achieve a filled outer shell by gaining an electron, thereby forming a negatively charged hydrogen ion, called a **hydride ion**.

$$\underset{\text{a hydrogen atom}}{\mathrm{H}\!\cdot} \ \longrightarrow \ \underset{\text{a proton}}{\mathrm{H}^+} \ + \ \mathrm{e}^-$$

$$\underset{\text{a hydrogen atom}}{\mathrm{H}\!\cdot} \ + \ \mathrm{e}^- \ \longrightarrow \ \underset{\text{a hydride ion}}{\mathrm{H}\!:^-}$$

Because oxygen has six valence electrons, it needs to form two covalent bonds to achieve an outer shell of eight electrons. Nitrogen, with five valence electrons, must form three covalent bonds, and carbon, with four valence electrons, must form four covalent bonds to achieve a filled outer shell. Notice that all the atoms in water, ammonia, and methane have filled outer shells.

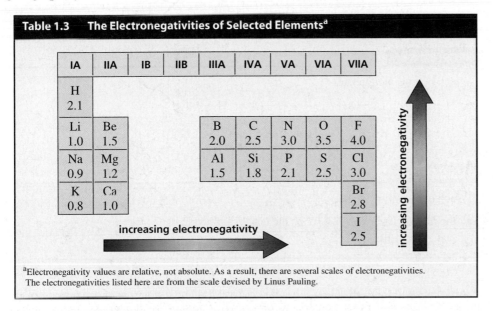

Polar Covalent Bonds

The atoms that share the bonding electrons in the F — F and H — H covalent bonds are identical. Therefore, they share the electrons equally; that is, each electron spends as much time in the vicinity of one atom as in the other. Such a bond is called a **nonpolar covalent bond**.

In contrast, the bonding electrons in hydrogen chloride, water, and ammonia are more attracted to one atom than to another because the atoms that share the electrons in these molecules are different and have different electronegativities. **Electronegativity** is a measure of the ability of an atom to pull the bonding electrons toward itself. The bonding electrons in hydrogen chloride, water, and ammonia are more attracted to the atom with the greater electronegativity. Therefore, the bonds in these compounds are polar covalent bonds. A **polar covalent bond** is a covalent bond between atoms of different electronegativities. The electronegativities of some of the elements are shown in Table 1.3. Notice that electronegativity increases from left to right across a row of the periodic table or going up any of the columns.

Student Tutorial:
Periodic trends in electronegativity

Table 1.3	The Electronegativities of Selected Elements[a]

IA	IIA	IB	IIB	IIIA	IVA	VA	VIA	VIIA
H 2.1								
Li 1.0	Be 1.5			B 2.0	C 2.5	N 3.0	O 3.5	F 4.0
Na 0.9	Mg 1.2			Al 1.5	Si 1.8	P 2.1	S 2.5	Cl 3.0
K 0.8	Ca 1.0							Br 2.8
								I 2.5

increasing electronegativity

increasing electronegativity

[a]Electronegativity values are relative, not absolute. As a result, there are several scales of electronegativities. The electronegativities listed here are from the scale devised by Linus Pauling.

A polar covalent bond has a slight positive charge on one end and a slight negative charge on the other. Polarity in a covalent bond is indicated by the symbols $\delta+$ and $\delta-$, which denote partial positive and partial negative charges, respectively. The negative end of the bond is the end that has the more electronegative atom. The greater the difference in electronegativity between the bonded atoms, the more polar the bond will be. (Notice that a pair of shared electrons can also be shown as a line between two atoms.)

$$\overset{\delta+}{H}\!-\!\overset{\delta-}{\ddot{\underset{..}{C}l}}\!:\qquad \overset{\delta+}{H}\!-\!\overset{\delta-}{\ddot{O}}\!:\qquad \overset{\delta+}{H}\!-\!\overset{\delta-}{N}\!-\!\overset{\delta+}{H}$$
$$\underset{\underset{\delta+}{H}}{|}\qquad\quad \underset{\underset{\delta+}{H}}{|}$$

The direction of bond polarity can be indicated with an arrow. By convention, chemists draw the arrow so that it points in the direction in which the electrons are pulled. Thus, the head of the arrow is at the negative end of the bond; a short perpendicular line near the tail of the arrow marks the positive end of the bond. (Physicists draw the arrow in the opposite direction.)

$$\overset{\longrightarrow}{H}\!-\!\ddot{\underset{..}{C}l}\!:\quad\boxed{\text{the negative end of the bond}}$$

You can think of ionic bonds and nonpolar covalent bonds as being at the opposite ends of a continuum of bond types. At one end is an ionic bond, a bond in which there is no sharing of electrons. At the other end is a nonpolar covalent bond, a bond in which the electrons are shared equally. Polar covalent bonds fall somewhere in between, and the greater the difference in electronegativity between the atoms forming the bond, the closer the bond is to the ionic end of the continuum. C—H bonds are relatively nonpolar, because carbon and hydrogen have similar electronegativities (electronegativity difference = 0.4; see Table 1.3); N—H bonds are more polar (electronegativity difference = 0.9), but not as polar as O—H bonds (electronegativity difference = 1.4). Even closer to the ionic end of the continuum is the bond between sodium and chloride ions (electronegativity difference = 2.1), but sodium chloride is not as ionic as potassium fluoride (electronegativity difference = 3.2).

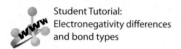

Student Tutorial:
Electronegativity differences and bond types

continuum of bond types

ionic bond	polar covalent bond	nonpolar covalent bond
K^+F^- Na^+Cl^-	O—H N—H	C—H C—C

PROBLEM 6♦

Which bond is more polar?

a. H—CH₃ or Cl—CH₃   **c.** H—Cl or H—F

b. H—OH or H—H   **d.** Cl—Cl or Cl—CH₃

PROBLEM 7♦

Which of the following has

a. the most polar bond?   **b.** the least polar bond?

    NaI   LiBr   Cl₂   KCl

A polar bond has a **dipole**—it has a negative end and a positive end. The size of the dipole is indicated by the dipole moment, symbolized by the Greek letter μ. The **dipole moment** of a bond is equal to the magnitude of the charge (e) on the atom (either the partial positive charge or the partial negative charge, because they have the same magnitude) times the distance between the two charges (d):

$$\text{dipole moment} = \mu = e \times d$$

A dipole moment is reported in a unit called a **debye (D)** (pronounced de-bye). Because the charge on an electron is 4.80×10^{-10} electrostatic units (esu) and the distance between charges in a polar bond is on the order of 10^{-8} cm, the product of charge and distance is on the order of 10^{-18} esu cm. A dipole moment of 1.5×10^{-18} esu cm can be more simply stated as 1.5 D. The dipole moments of some bonds commonly found in organic compounds are listed in Table 1.4.

Table 1.4 The Dipole Moments of Some Commonly Encountered Bonds			
Bond	**Dipole moment (D)**	**Bond**	**Dipole moment (D)**
H—C	0.4	C—C	0
H—N	1.3	C—N	0.2
H—O	1.5	C—O	0.7
H—F	1.7	C—F	1.6
H—Cl	1.1	C—Cl	1.5
H—Br	0.8	C—Br	1.4
H—I	0.4	C—I	1.2

When a molecule has only one covalent bond and the bond is polar, the molecule has a dipole moment that is identical to the dipole moment of the bond. For example, the dipole moment of hydrogen chloride (HCl) is 1.1 D because the dipole moment of the H—Cl bond is 1.1 D. The dipole moment of a molecule with more than one covalent bond depends on the dipole moments of all the bonds in the molecule and the geometry of the molecule. We will examine the dipole moments of molecules with more than one covalent bond in Section 1.15 after you learn about the geometry of molecules.

Peter Debye (1884–1966) *was born in the Netherlands. He taught at the Universities of Zürich (succeeding Einstein), Leipzig, and Berlin, but returned to his homeland in 1939 when the Nazis ordered him to become a German citizen. Upon visiting Cornell to give a lecture, he decided to stay in the United States, and he became a U.S. citizen in 1946. He received the Nobel Prize in chemistry in 1936 for his work on dipole moments and on the diffraction of X-rays and electrons in gases.*

PROBLEM 8   **SOLVED**

Determine the partial negative charge on the fluorine atom in a C—F bond. The bond length is 1.39 Å* and the bond dipole moment is 1.60 D.

Solution   If there were a full negative charge on the fluorine atom, the dipole moment would be

$$(4.80 \times 10^{-10} \text{ esu})(1.39 \times 10^{-8} \text{ cm}) = 6.97 \text{ esu cm} = 6.97 \text{ D}$$

Knowing that the dipole moment is 1.60 D, we calculate that the partial negative charge on the fluorine atom is about 0.23 of a full charge:

$$\frac{1.60}{6.97} = 0.23$$

PROBLEM 9◆

Use the symbols $\delta+$ and $\delta-$ to show the direction of polarity of the indicated bond in each of the following compounds (for example),

$$\overset{\delta+ \quad \delta-}{H_3C - OH}$$

a. HO—H

b. F—Br

c. H_3C—NH_2

d. H_3C—Cl

e. HO—Br

f. H_3C—MgBr

g. I—Cl

h. H_2N—OH

---

* The angstrom (Å) is not a Système International unit. Those who opt to adhere strictly to SI units can convert it into picometers: 1 picometer (pm) $= 10^{-12}$ m; 1 Å $= 10^{-10}$ m = 100 pm. Because the angstrom continues to be used by many organic chemists, we will use angstroms in this book.

Understanding bond polarity is critical to understanding how organic reactions occur, because a central rule governing the reactivity of organic compounds is that *electron-rich atoms or molecules are attracted to electron-deficient atoms or molecules* (Section 3.6). **Electrostatic potential maps** (often called simply potential maps) are models that show how charge is distributed in the molecule under the map. They, therefore, show the kind of electrostatic attraction an atom or molecule has for another atom or molecule, and as a result they can be used to predict chemical reactions. The potential maps for LiH, H_2, and HF are shown below.

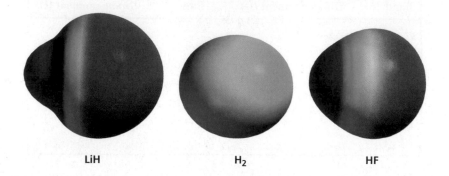

LiH H_2 HF

The colors on a potential map indicate the degree to which a molecule or an atom in a molecule attracts charged particles. Red, signifying the most negative electrostatic potential, is used for regions that attract positively charged molecules most strongly. Blue is used for areas with the most positive electrostatic potential, regions that attract negatively charged molecules most strongly. Other colors indicate intermediate levels of attraction.

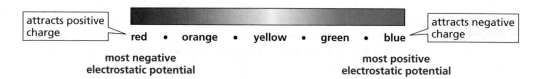

| attracts positive charge | | red • orange • yellow • green • blue | | attracts negative charge |

most negative electrostatic potential   most positive electrostatic potential

The colors on a potential map can also be used to estimate charge distribution. For example, the potential map for LiH indicates that the hydrogen atom has greater electron density than the lithium atom. By comparing the three maps, we can tell that the hydrogen in LiH has greater electron density than a hydrogen in H_2, and the hydrogen in HF has less electron density than a hydrogen in H_2.

A molecule's size and shape are determined by the number of electrons in the molecule and by the way these electrons move. Because a potential map roughly marks the "edge" of the molecule's electron cloud, the map tells us something about the relative size and shape of the molecule. Notice that a given kind of atom can have different sizes in different molecules. The negatively charged hydrogen in LiH is bigger than a neutral hydrogen in H_2, which, in turn, is bigger than the positively charged hydrogen in HF.

PROBLEM 10◆

After examining the potential maps for LiH, HF, and H_2, answer the following questions:

a. Which compounds are polar?

b. Why does LiH have the largest hydrogen?

c. Which compound has the hydrogen that would be most apt to attract a negatively charged molecule?

1.4 How the Structure of a Compound Is Represented

First we will see how compounds are drawn using Lewis structures. Then we will look at the kinds of structures that are used more commonly for organic compounds.

Lewis Structures

The chemical symbols we have been using, in which the valence electrons are represented as dots, are called **Lewis structures**. Lewis structures are useful because they show us which atoms are bonded together and tell us whether any atoms possess *lone-pair electrons* or have a *formal charge*, two concepts we describe below. The Lewis structures for H_2O, H_3O^+, HO^-, and H_2O_2 are

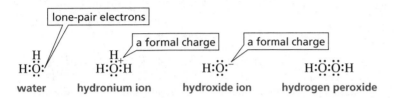

water   hydronium ion   hydroxide ion   hydrogen peroxide

Notice that the atoms in Lewis structures are always lined up linearly or at right angles, telling us nothing about the bond angles in the actual molecule.

When you draw a Lewis structure, make sure that hydrogen atoms are surrounded by no more than two electrons and that C, O, N, and halogen (F, Cl, Br, I) atoms are surrounded by no more than eight electrons, in accordance with the octet rule. Valence electrons not used in bonding are called **nonbonding electrons** or **lone-pair electrons**.

Once you have the atoms and the electrons in place, you must examine each atom to see whether a formal charge should be assigned to it. A **formal charge** is the *difference* between the number of valence electrons an atom has when it is not bonded to any other atoms and the number of electrons it "owns" when it is bonded. An atom "owns" all of its lone-pair electrons and half of its bonding (shared) electrons.

formal charge = number of valence electrons
− (number of lone-pair electrons + 1/2 number of bonding electrons)

For example, an oxygen atom has six valence electrons (Table 1.2). In water (H_2O), oxygen "owns" six electrons (four lone-pair electrons and half of the four bonding electrons). Because the number of electrons it "owns" is equal to the number of its valence electrons ($6 - 6 = 0$), the oxygen atom in water has no formal charge. The oxygen atom in the hydronium ion (H_3O^+) "owns" five electrons: two lone-pair electrons plus three (half of six) bonding electrons. Because the number of electrons it "owns" is one less than the number of its valence electrons ($6 - 5 = 1$), its formal charge is $+1$. The oxygen atom in the hydroxide ion (HO^-) "owns" seven electrons: six lone-pair electrons plus one (half of two) bonding electron. Because it "owns" one more electron than the number of its valence electrons ($6 - 7 = -1$), its formal charge is -1.

Student Tutorial:
Formal charges

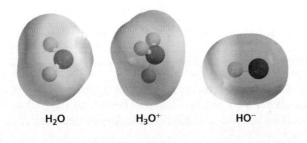

H_2O H_3O^+ HO^-

PROBLEM 11◆

A formal charge does not necessarily indicate that the atom has greater or less electron density than atoms in the molecule without formal charges. We can see this by examining the potential maps for H_2O, H_3O^+, and HO^-.

a. Which atom bears the formal negative charge in the hydroxide ion?
b. Which atom has the greater electron density in the hydroxide ion?
c. Which atom bears the formal positive charge in the hydronium ion?
d. Which atom has the least electron density in the hydronium ion?

Nitrogen has five valence electrons (Table 1.2). Prove to yourself that the appropriate formal charges have been assigned to the nitrogen atoms in the following Lewis structures:

H:N̈:H   H:N̈:H⁺   H:N̈:⁻   H:N̈:N̈:H

ammonia   **ammonium ion**   **amide anion**   **hydrazine**

Carbon has four valence electrons. Take a moment to make sure you understand why the carbon atoms in the following Lewis structures have the indicated formal charges:

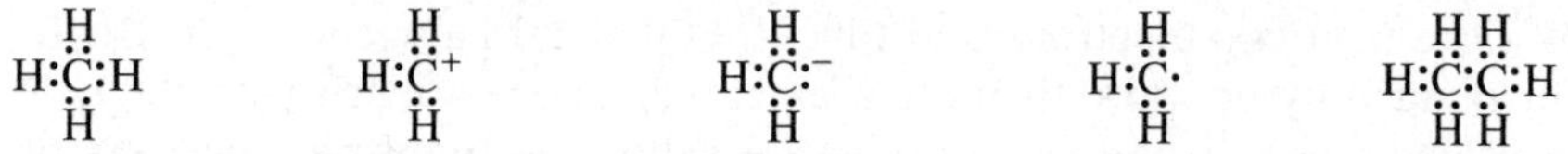

H:C̈:H   H:C̈⁺   H:C̈:⁻   H:C̈·   H:C̈:C̈:H

methane   **methyl cation**   **methyl anion**   **methyl radical**   **ethane**
      a carbocation   **a carbanion**

A species containing a positively charged carbon atom is called a **carbocation**, and a species containing a negatively charged carbon atom is called a **carbanion**. (Recall that a *cation* is a positively charged ion and an *anion* is a negatively charged ion.) A species containing an atom with a single unpaired electron is called a **radical** (often called a **free radical**).

Hydrogen has one valence electron, and each halogen (F, Cl, Br, I) has seven valence electrons, so the following species have the indicated formal charges:

H^+ H:⁻   H·   :B̈r:⁻   :B̈r·   :B̈r:B̈r: :C̈l:C̈l:

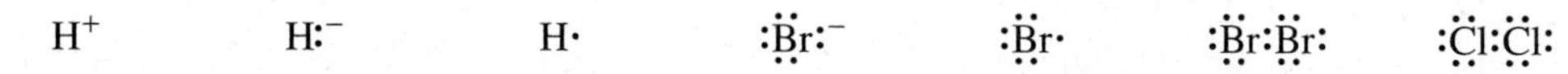

hydrogen ion   **hydride ion**   **hydrogen radical**   **bromide ion**   **bromine radical**   **bromine**   **chlorine**

PROBLEM 12◆

Give each atom the appropriate formal charge:

a. 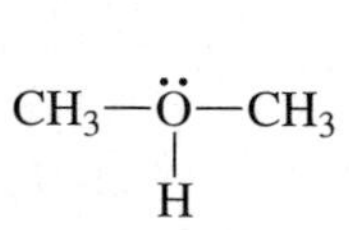 $CH_3—\ddot{O}—CH_3$
            |
            H

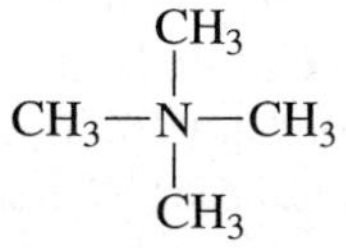

b. 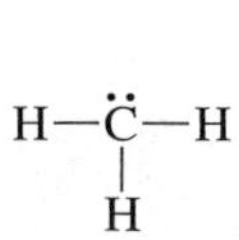 $H—\ddot{C}—H$
        |
        H

c.
           CH_3
           |
   $CH_3—N—CH_3$
           |
           CH_3

d. 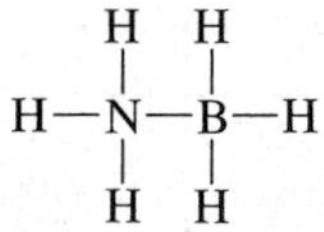
      H   H
      | |
   $H—N—B—H$
      | |
      H   H

In studying the molecules in this section, notice that when the atoms do not bear a formal charge or an unpaired electron, hydrogen and the halogens always have *one* covalent bond, oxygen always has *two* covalent bonds, nitrogen always has *three* covalent bonds, and carbon has *four* covalent bonds. Atoms that have more bonds or fewer bonds than the number required for a neutral atom will have either a formal charge or an unpaired electron. These numbers are very important to remember when you are first drawing structures of organic compounds because they provide a quick way to recognize when you have made a mistake.

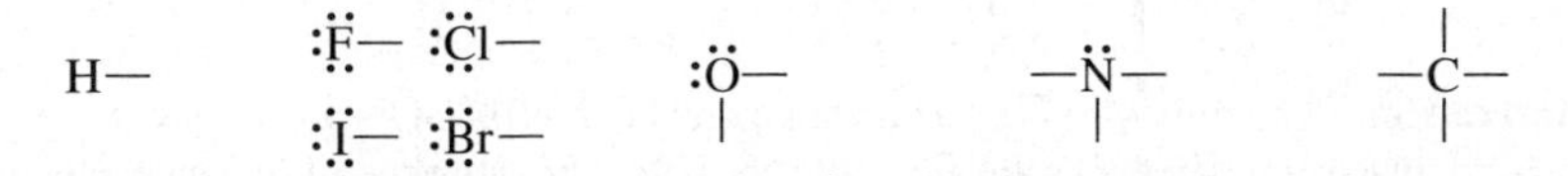

H—   :F̈— :C̈l—   :Ö—   —N̈—   —Ċ—
                                                    | | |
             :Ï— :B̈r— | | |

one bond   **one bond**   **two bonds**   **three bonds**   **four bonds**

In the following Lewis structures notice that each atom has a filled outer shell. Also notice that since none of the molecules has a formal charge or an unpaired electron, H and Br each form one bond, O forms two bonds (these can be two single bonds or one double bond), N forms three bonds (these can be three single bonds, one double bond and one single bond, or one triple bond), and C forms a total of four bonds. (In drawing the Lewis structure for a compound that has two or more oxygen atoms, avoid oxygen–oxygen single bonds. These are weak bonds, and few compounds have them.)

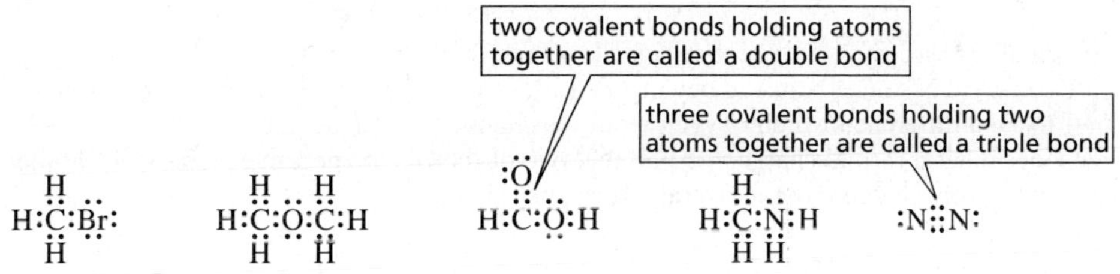

two covalent bonds holding atoms
together are called a double bond

three covalent bonds holding two
atoms together are called a triple bond

H   H H   :Ö: H
H:C:Br: H:C:O:C:H   H:C:Ö:H   H:C:N:H   :N:::N:
H   H H   H H

We have seen that a pair of shared electrons can also be shown as a line between two atoms (Section 1.3). Compare the preceding structures with the following ones:

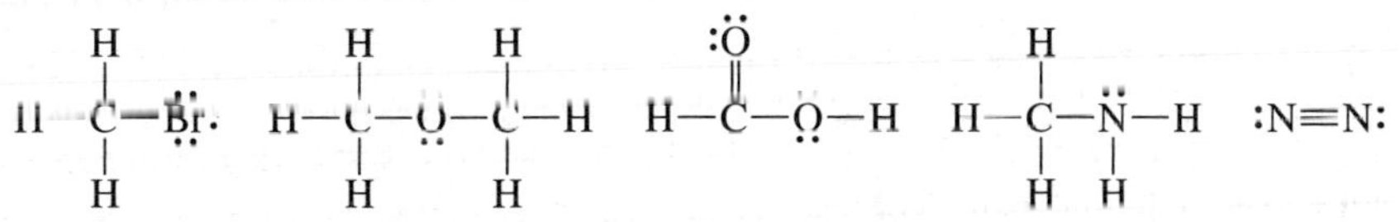

         H   H   H   :Ö: H
         | | | ‖ |
H — C — B̈r: H — C — O — C — H   H — C — Ö — H   H — C — N̈ — H   :N≡N:
         | | | | |
         H   H   H   H H

PROBLEM-SOLVING STRATEGY

Drawing Lewis Structures

Draw the Lewis structure for HNO_2.

1. Determine the total number of valence electrons (1 for H, 5 for N, and 6 for each O adds up to $1 + 5 + 12 = 18$.

2. Use the total number of valence electrons to form bonds and fill octets with lone-pair electrons.

3. If after all the electrons have been assigned, an atom (other than hydrogen) does not have a complete octet, use a lone pair to form a double bond to that atom.

4. Assign a formal charge to any atom whose number of valence electrons is not equal to the number of its lone-pair electrons plus one-half its bonding electrons. (None of the atoms in HNO_2 has a formal charge.)

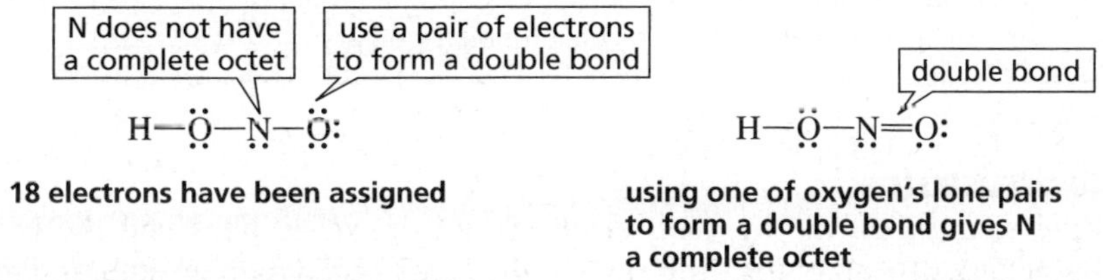

N does not have   use a pair of electrons
a complete octet   to form a double bond   double bond

         H—Ö—N̈—Ö: H—Ö—N=Ö:

18 electrons have been assigned   **using one of oxygen's lone pairs
to form a double bond gives N
a complete octet**

Now continue on to Problem 13.

Student Tutorial:
Lewis structure

PROBLEM 13 **SOLVED**

Draw the Lewis structure for each of the following:

a. NO_3^- c. $^-C_2H_5$ e. $CH_3\overset{+}{N}H_3$ g. HCO_3^-
b. NO_2^+ d. $^+C_2H_5$ f. NaOH h. H_2CO

Solution to 13a The only way we can arrange one N and three O's and avoid O—O single bonds is to place the three O's around the N. The total number of valence electrons is 23 (5 for N, and 6 for each of the three O's). Because the species has one negative charge, we must add 1 to the number of valence electrons, for a total of 24. We then use the 24 electrons to form bonds and fill octets with lone-pair electrons.

$$
\begin{array}{c}
:\ddot{O}: \\
| \\
:\ddot{O}-N-\ddot{O}: \\
\end{array}
$$

[incomplete octet]

When all 24 electrons have been assigned, we see that N does not have a complete octet. We complete N's octet by using one of oxygen's lone pairs to from a double bond. (It does not make a difference which oxygen atom we choose.) When we check each atom to see whether it has a formal charge, we find that two of the O's are negatively charged and the N is positively charged, for an overall charge of -1.

$$
\begin{array}{c}
:\ddot{O} \\
\| \\
^-:\ddot{O}-\underset{+}{N}-\ddot{O}:^- \\
\end{array}
$$

Solution to 13b The total number of valence electrons is 17 (5 for N and 6 for each of the two O's). Because the species has one positive charge, we must subtract 1 from the number of valence electrons, for a total of 16. The 16 electrons are used to form bonds and then fill octets with lone-pair electrons.

[incomplete octet]

$$:\ddot{O}-N-\ddot{O}:$$

Two double bonds are necessary to complete N's octet. The N has a formal charge of $+1$.

$$:\ddot{O}=\overset{+}{N}=\ddot{O}:$$

PROBLEM 14◆

a. Draw two Lewis structures for C_2H_6O.
b. Draw three Lewis structures for C_3H_8O.

(*Hint:* The two Lewis structures in part a are **constitutional isomers**, molecules that have the same atoms, but differ in the way the atoms are connected; see page 200. The three Lewis structures in part b are also constitutional isomers.)

Kekulé Structures

In **Kekulé structures**, the bonding electrons are drawn as lines and the lone-pair electrons are usually left out entirely, unless they are needed to draw attention to some chemical property of the molecule. (Although lone-pair electrons are not shown, you should remember that neutral nitrogen, oxygen, and halogen atoms always have them:

one pair in the case of nitrogen, two pairs in the case of oxygen, and three pairs in the case of a halogen.)

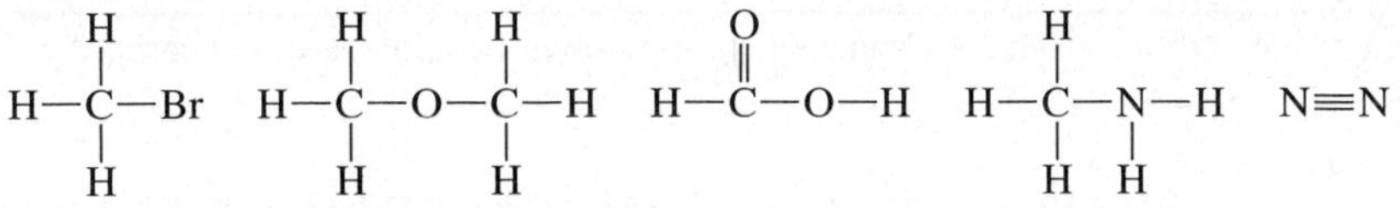

$$H-\overset{\displaystyle H}{\underset{\displaystyle H}{C}}-Br \quad H-\overset{\displaystyle H}{\underset{\displaystyle H}{C}}-O-\overset{\displaystyle H}{\underset{\displaystyle H}{C}}-H \quad H-\overset{\displaystyle O}{C}-O-H \quad H-\overset{\displaystyle H}{\underset{\displaystyle H\ H}{C}}-N-H \quad N\equiv N$$

Condensed Structures

Frequently, structures are simplified by omitting some (or all) of the covalent bonds and listing atoms bonded to a particular carbon (or nitrogen or oxygen) next to it with subscripts as necessary. These structures are called **condensed structures**. Compare the following examples with the Kekulé structures shown earlier:

$$CH_3Br \qquad CH_3OCH_3 \qquad HCO_2H \qquad CH_3NH_2 \qquad N_2$$

You can find more examples of condensed structures and the conventions commonly used to create them in Table 1.5. Notice that since none of the molecules in Table 1.5 has a formal charge or an unpaired electron, each C has four bonds, each N has three bonds, each O has two bonds, and each H or halogen has one bond.

PROBLEM 15◆

Draw the lone-pair electrons that are not shown in the following structures:

a. $CH_3CH_2NH_2$ c. CH_3CH_2OH e. CH_3CH_2Cl

b. CH_3NHCH_3 d. CH_3OCH_3 f. $HONH_2$

PROBLEM 16◆

Draw condensed structures for the compounds represented by the following models (black = C, white = H, red = O, blue = N, green = Cl):

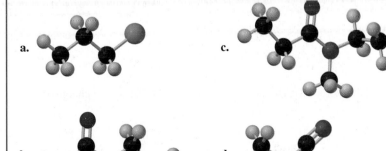

a.

b.

c.

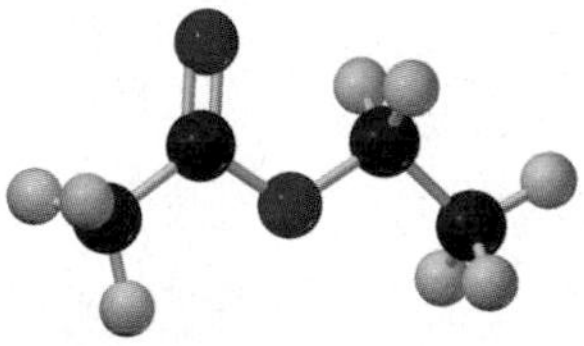

d.

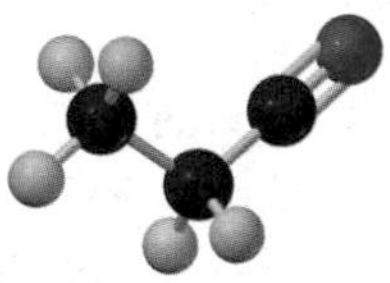

PROBLEM 17◆

Which of the atoms in the molecular models in Problem 16 have:

a. three lone pairs   b. two lone pairs   c. one lone pair   d. no lone pairs

---

PROBLEM 18

Expand the following condensed structures to show the covalent bonds and lone-pair electrons:

a. $CH_3NH(CH_2)_2CH_3$ c. $(CH_3)_3CBr$

b. $(CH_3)_2CHCl$ d. $(CH_3)_3C(CH_2)_3CH(CH_3)_2$

Table 1.5 **Kekulé and Condensed Structures**

Kekulé structure	Condensed structures

Atoms bonded to a carbon are shown to the right of the carbon. Atoms other than H can be shown hanging from the carbon.

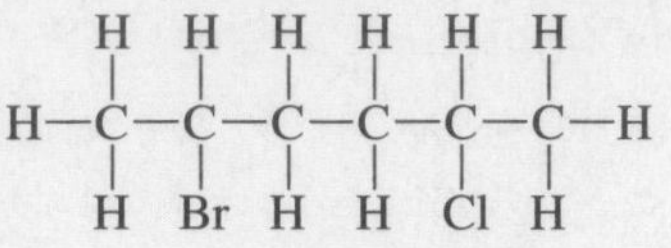

$CH_3CHBrCH_2CH_2CHClCH_3$ or $CH_3CHCH_2CH_2CHCH_3$ with Br and Cl hanging

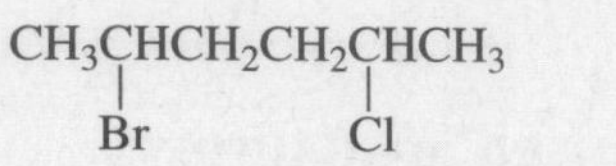

Repeating CH_2 groups can be shown in parentheses.

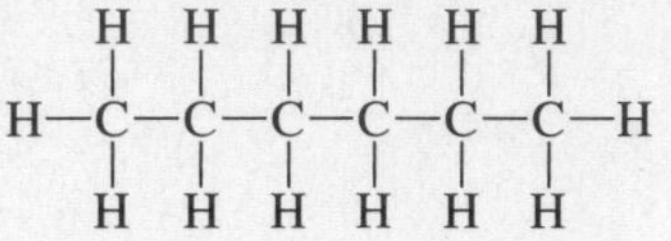

$CH_3CH_2CH_2CH_2CH_2CH_3$ or $CH_3(CH_2)_4CH_3$

Groups bonded to a carbon can be shown (in parentheses) to the right of the carbon, or hanging from the carbon.

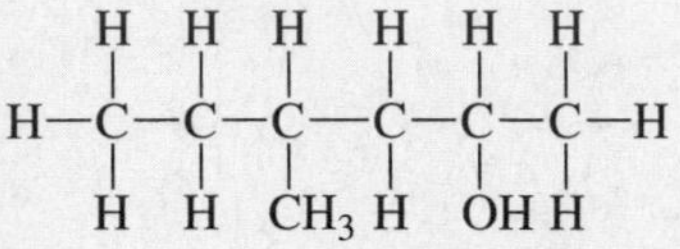

$CH_3CH_2CH(CH_3)CH_2CH(OH)CH_3$ or $CH_3CH_2CHCH_2CHCH_3$ with CH_3 and OH hanging

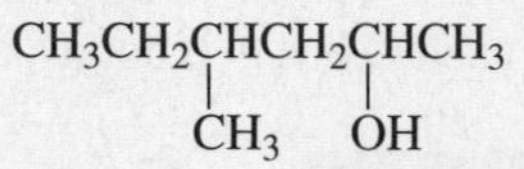

Groups bonded to the far-right carbon are not put in parentheses.

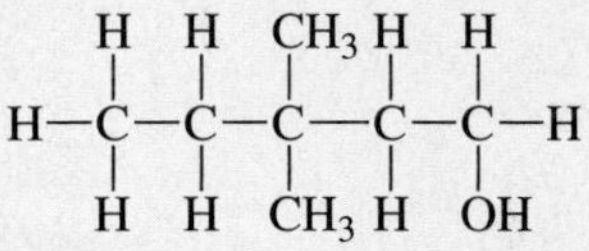

$CH_3CH_2C(CH_3)_2CH_2CH_2OH$ or $CH_3CH_2CCH_2CH_2OH$ with CH_3 above and CH_3 below

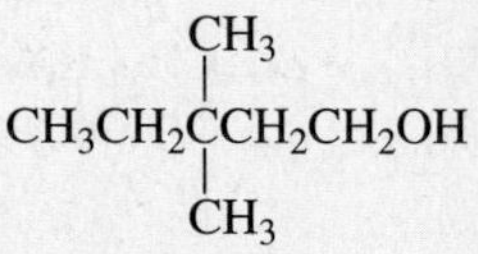

Two or more identical groups considered bonded to the "first" atom on the left can be shown (in parentheses) to the left of that atom, or hanging from the atom.

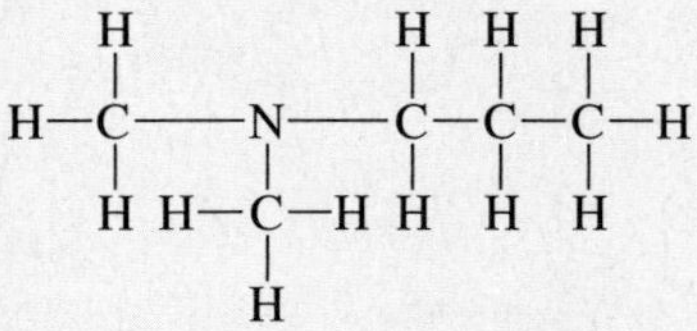

$(CH_3)_2NCH_2CH_2CH_3$ or $CH_3NCH_2CH_2CH_3$ with CH_3 hanging

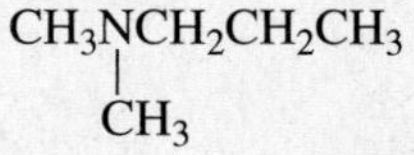

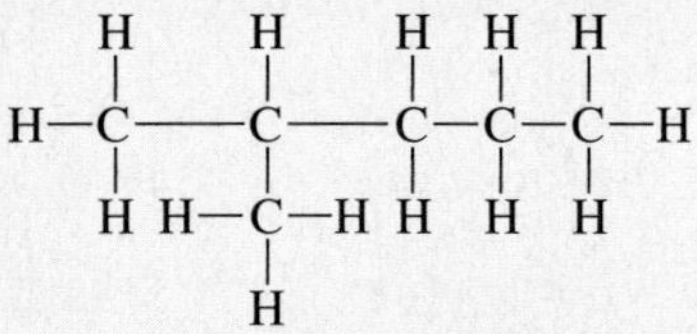

$(CH_3)_2CHCH_2CH_2CH_3$ or $CH_3CHCH_2CH_2CH_3$ with CH_3 hanging

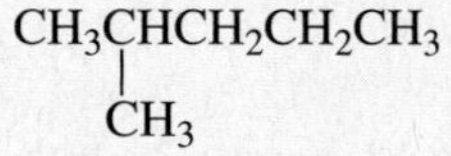

An oxygen doubly bonded to a carbon can be shown hanging off the carbon or to the right of the carbon.

$$\overset{\overset{\displaystyle O}{\|}}{CH_3CH_2CCH_3} \quad \text{or} \quad CH_3CH_2COCH_3 \quad \text{or} \quad CH_3CH_2C(=O)CH_3$$

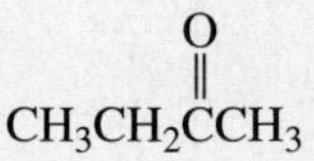

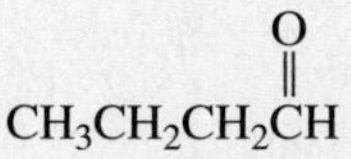

$$\overset{\overset{\displaystyle O}{\|}}{CH_3CH_2CH_2CH} \quad \text{or} \quad CH_3CH_2CH_2CHO \quad \text{or} \quad CH_3CH_2CH_2CH=O$$

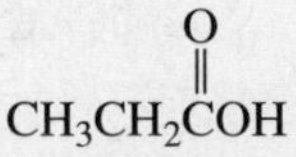

$$\overset{\overset{\displaystyle O}{\|}}{CH_3CH_2COH} \quad \text{or} \quad CH_3CH_2CO_2H \quad \text{or} \quad CH_3CH_2COOH$$

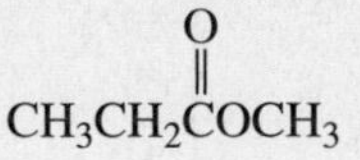

$$\overset{\overset{\displaystyle O}{\|}}{CH_3CH_2COCH_3} \quad \text{or} \quad CH_3CH_2CO_2CH_3 \quad \text{or} \quad CH_3CH_2COOCH_3$$

1.5 **Atomic Orbitals**

We have seen that electrons are distributed into different atomic orbitals (Table 1.2), three-dimensional regions around the nucleus where electrons are most likely to be found. The **Heisenberg uncertainty principle**, however, states that both the precise location and the momentum of an atomic particle cannot be simultaneously determined. This means that we can never say precisely where an electron is—we can only describe its probable location. Mathematical calculations indicate that an *s* atomic orbital is a sphere with the nucleus at its center, and experimental evidence supports this theory. Thus, when we say that an electron occupies a 1*s* orbital, we mean that there is a greater than 90% probability that the electron is in the space defined by the sphere.

An orbital tells us the volume of space around the nucleus where an electron is most likely to be found.

Because the second shell lies farther from the nucleus than the first shell (Section 1.2), the average distance from the nucleus is greater for an electron in a 2*s* orbital than it is for an electron in a 1*s* orbital. A 2*s* orbital, therefore, is represented by a larger sphere. Because of the greater size of a 2*s* orbital, the average electron density there is less than the average electron density in a 1*s* orbital.

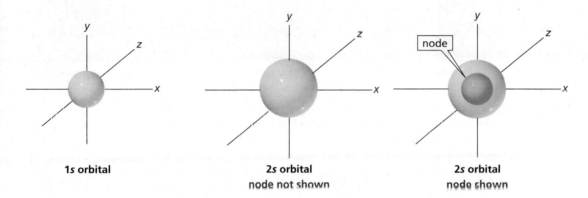

1*s* orbital

2*s* orbital
node not shown

2*s* orbital
node shown

An electron in a 1*s* orbital can be anywhere within the 1*s* sphere, but a 2*s* orbital has a region where the probability of finding an electron falls to zero. This is called a **node**, or, more precisely—since this absence of electron density lies at one set distance from the nucleus—a **radial node**. So a 2*s* electron can be found anywhere within the 2*s* sphere—including the region of space defined by the 1*s* sphere—except at the node.

To understand why nodes occur, you need to remember that electrons have both particlelike and wavelike properties. A node is a consequence of the wavelike properties of an electron. There are two types of waves: traveling waves and standing waves. Traveling waves move through space; light is an example of a traveling wave. A standing wave, in contrast, is confined to a limited space. A vibrating string of a guitar is an example of a standing wave—the string moves up and down, but does not travel through space. If you were to write a wave equation for the guitar string, the wave function would be $(+)$ in the region above where the guitar string is at rest and $(-)$ in the region below where the guitar string is at rest—the regions are of opposite phase. The regions where the guitar string has no transverse displacement are called *nodes*. A node is at the region where a standing wave has an amplitude of zero.

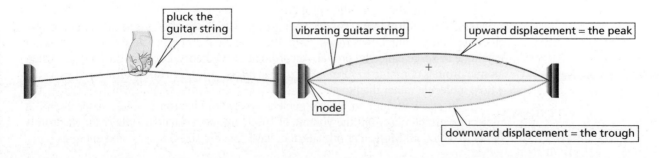

An electron behaves like a standing wave, but, unlike the wave created by a vibrating guitar string, it is three dimensional. This means that the node of a 2s orbital is actually a spherical surface within the 2s orbital. Because the electron wave has zero amplitude at the node, there is zero probability of finding an electron at the node.

Unlike s orbitals, which resemble spheres, p orbitals have two lobes. Generally, the lobes are depicted as teardrop shaped, but computer-generated representations reveal that they are shaped more like doorknobs. Like the vibrating guitar string, the lobes are of opposite phase, which can be designated by plus (+) and minus (−) or by two different colors. (In this context, + and − indicate the phase of the orbital; they do not indicate charge.) The node of the p orbital is a plane—called a **nodal plane**—that passes through the center of the nucleus, between its two lobes. There is zero probability of finding an electron in the nodal plane of the p orbital.

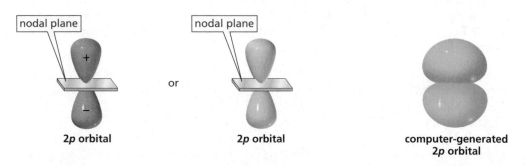

| 2p orbital | 2p orbital | computer-generated 2p orbital |

In Section 1.2, we saw that the second and higher numbered shells each contain three degenerate p orbitals. The p_x orbital is symmetrical about the x-axis, the p_y orbital is symmetrical about the y-axis, and the p_z orbital is symmetrical about the z-axis. This means that each p orbital is perpendicular to the other two p orbitals. The energy of a 2p orbital is slightly greater than that of a 2s orbital because the average location of an electron in a 2p orbital is farther away from the nucleus.

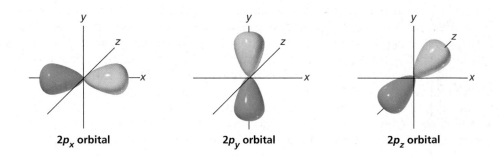

| $2p_x$ orbital | $2p_y$ orbital | $2p_z$ orbital |

1.6 An Introduction to Molecular Orbital Theory

How do atoms form covalent bonds in order to form molecules? The Lewis model, which shows atoms attaining a complete octet by sharing electrons, tells only part of the story. A drawback of the model is that it treats electrons like particles and does not take into account their wavelike properties.

Molecular orbital (MO) theory combines the tendency of atoms to fill their octets by sharing electrons (the Lewis model) with their wavelike properties, assigning electrons to a volume of space called an orbital. According to MO theory, covalent bonds result when atomic orbitals combine to form *molecular orbitals*. A **molecular orbital** belongs to the whole molecule rather than to a single atom. Like an atomic orbital, which describes the volume of space around an atom's nucleus where an electron is likely to be found, a molecular orbital describes the volume of space around a molecule where an electron is likely to be found. Molecular orbitals, too, have specific sizes, shapes, and energies.

Let's look first at the bonding in a hydrogen molecule (H_2). Imagine a meeting of two separate H atoms. As one atom with its $1s$ atomic orbital approaches the other with its $1s$ atomic orbital, the orbitals begin to overlap. The atoms continue to move closer, and the amount of overlap increases until the orbitals combine to form a molecular orbital. The covalent bond that is formed when the two s orbitals overlap is called a **sigma (σ) bond**. A σ bond is cylindrically symmetrical—the electrons in the bond are symmetrically distributed about an imaginary line connecting the centers of the two atoms joined by the bond.

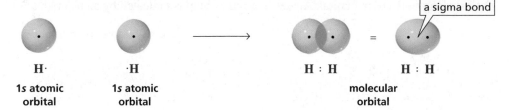

As the two orbitals start to overlap, energy is released (and stability increased) because the electron in each atom is attracted both to its own nucleus and to the positively charged nucleus of the other atom (Figure 1.2). The attraction of the negatively charged electrons for the two positively charged nuclei is what holds the atoms together. The more the orbitals overlap, the more the energy decreases until the atoms are so close that their positively charged nuclei start to repel each other. This repulsion causes a large increase in energy. Figure 1.2 shows that maximum stability (minimum energy) is achieved when the nuclei are a certain distance apart. This distance is the **bond length** of the new covalent bond. The bond length of the H—H bond is 0.74 Å.

Maximum stability corresponds to minimum energy.

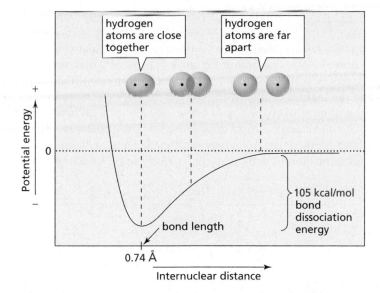

Student Tutorial.
H_2 bond formation

◀ **Figure 1.2**
The change in energy that occurs as two $1s$ atomic orbitals approach each other. The internuclear distance at minimum energy is the length of the H—H covalent bond.

As Figure 1.2 shows, energy is released when a covalent bond forms. When the H—H bond forms, 105 kcal/mol (or 439 kJ/mol)* of energy is released. Breaking the bond requires precisely the same amount of energy. Thus, the **bond strength**, also called the **bond dissociation energy**, is the energy required to break a bond, or the energy released when a bond is formed. Every covalent bond has a characteristic bond length and bond strength.

Orbitals are conserved. In other words, the number of molecular orbitals formed must equal the number of atomic orbitals combined. In describing the formation of an H—H

* Joules are the Système International (SI) units for energy, although many chemists use calories (1 kcal = 4.184 kJ). We will use both in this book.

bond we combined two atomic orbitals, but discussed only one molecular orbital. Where is the other molecular orbital? We will see that it is there, but it does not contain any electrons.

Atomic orbitals can combine in two different ways: constructively and destructively. They can combine in a constructive, additive manner, just as two light waves or sound waves may reinforce each other (Figure 1.3). This constructive combination is called a σ (sigma) **bonding molecular orbital**. Atomic orbitals can also combine in a destructive way, canceling each other. The cancellation is similar to the darkness that results when two light waves cancel each other or to the resulting silence when two sound waves cancel each other (Figure 1.3). This destructive combination is called a σ^* **antibonding molecular orbital**. An antibonding orbital is indicated by an asterisk (*).

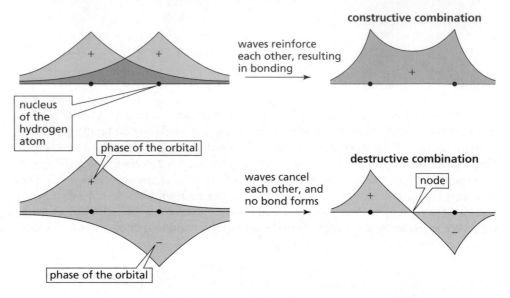

Figure 1.3 ▶
The wave functions of two hydrogen atoms can interact to reinforce each other (*top*) or can interact to cancel each other (*bottom*). Note that waves that interact constructively are in-phase, whereas waves that interact destructively are out-of-phase.

The σ bonding molecular orbital and σ^* antibonding molecular orbital are shown in the molecular orbital (MO) diagram in Figure 1.4. In an MO diagram, the energies of the orbitals are represented as horizontal lines; the bottom line is the lowest energy level, the top line the highest energy level. We see that any electrons in the bonding molecular orbital will most likely be found between the nuclei, attracting both nuclei simultaneously. This increased electron density between the nuclei is what binds the atoms together. Because there is a node between the nuclei in the antibonding molecular orbital, any electrons in that orbital are more likely to be found anywhere except

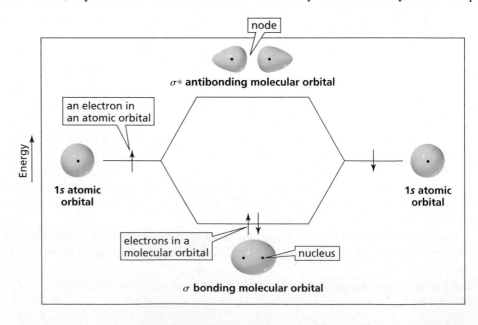

Figure 1.4 ▶
Atomic orbitals of H_2 and molecular orbitals of H_2. Before covalent bond formation, each electron is in an atomic orbital. After covalent bond formation, both electrons are in the bonding molecular orbital. The antibonding molecular orbital is empty.

between the nuclei, so the nuclei are more exposed to one another and will be forced apart by electrostatic repulsion. Thus, electrons that occupy this orbital detract from, rather than aid, the formation of a bond between the atoms.

The MO diagram shows that the bonding molecular orbital is lower in energy, and therefore more stable, than the individual atomic orbitals. This is because the more nuclei an electron "feels," the more stable it is. The antibonding molecular orbital, with less electron density between the nuclei, is less stable—is of higher energy—than the atomic orbitals.

After the sequence of horizontal lines in an MO diagram is constructed, representing orbitals, the electrons are assigned to the molecular orbitals. The aufbau principle and the Pauli exclusion principle, used to assign electrons to atomic orbitals, also are used to assign electrons to molecular orbitals: electrons always occupy available orbitals with the lowest energy, and no more than two electrons can occupy a molecular orbital. Thus, the two electrons of the H—H bond occupy the lower energy bonding molecular orbital (Figure 1.4), where they are attracted to both positively charged nuclei. It is this electrostatic attraction that gives a covalent bond its strength. Therefore, the greater the overlap of the atomic orbitals, the stronger is the covalent bond. *The strongest covalent bonds are formed by electrons that occupy the molecular orbitals with the lowest energy.*

The MO diagram in Figure 1.4 allows us to predict that H_2^+ would not be as stable as H_2 because H_2^+ has only one electron in the bonding orbital. We can also predict that He_2 does not exist: because each He atom would bring two electrons, He_2 would have four electrons—two filling the lower energy bonding molecular orbital and the remaining two filling the higher energy antibonding molecular orbital. The two electrons in the antibonding molecular orbital would cancel the advantage to bonding gained by the two electrons in the bonding molecular orbital.

> **When two atomic orbitals overlap, two molecular orbitals are formed—one lower in energy and one higher in energy than the atomic orbitals.**

PROBLEM 19◆

Predict whether or not He_2^+ exists.

Two p atomic orbitals can overlap either end-on or side-to-side. Let's first look at end-on overlap. End-on overlap forms a cylindrically symmetrical bond that is therefore a σ bond. If the overlapping lobes of the p orbitals are in-phase (as in Figure 1.5 where a blue lobe of one p orbital overlaps a blue lobe of the other p orbital), a σ bonding molecular orbital is formed. The electron density of the

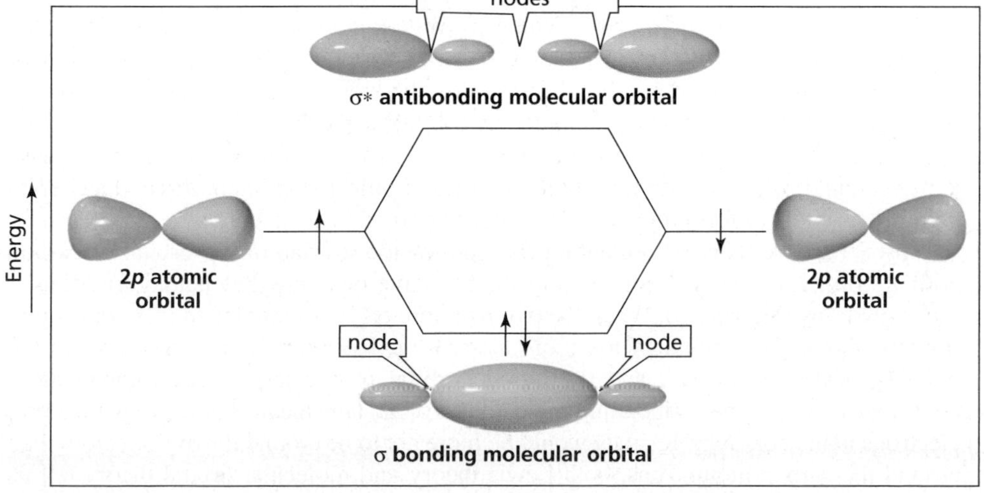

▲ **Figure 1.5**
End-on overlap of two p orbitals to form a σ bonding molecular orbital and a σ^* antibonding molecular orbital.

σ bonding molecular orbital is concentrated between the nuclei, which causes the back lobes (the nonoverlapping green lobes) of the molecular orbital to be quite small. The σ bonding molecular orbital has two nodes—a nodal plane passing through each of the nuclei.

If the overlapping lobes of the p orbitals are out-of-phase (a blue lobe of one p orbital overlaps a green lobe of the other p orbital), a σ^* antibonding molecular orbital is formed. The σ^* antibonding molecular orbital has *three* nodes. Notice in Figure 1.5 that the phase (color) of the molecular orbital is different on either side of each node.

Unlike the σ bond formed as a result of end-on overlap, side-to-side overlap of two p atomic orbitals forms a **pi (π) bond** (Figure 1.6). Side-to-side overlap of two in-phase p atomic orbitals forms a π bonding molecular orbital, whereas side-to-side overlap of two out-of-phase p orbitals forms a π^* antibonding molecular orbital. The π bonding molecular orbital has one node—a nodal plane that passes through both nuclei. The π^* antibonding molecular orbital has two nodal planes. Although σ bonds have cylindrically symmetrical molecular orbitals, π bonds do not.

> **In-phase overlap forms a bonding MO; out-of-phase overlap forms an antibonding MO.**

> **Side-to-side overlap of two p atomic orbitals forms a π bond. All other covalent bonds in organic molecules are σ bonds.**

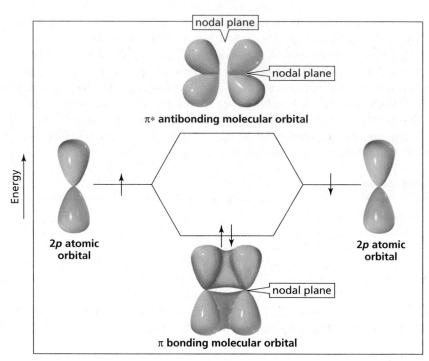

π∗ antibonding molecular orbital

2p atomic orbital

2p atomic orbital

π bonding molecular orbital

Figure 1.6 ▶
Side-to-side overlap of two parallel p orbitals to form a π bonding molecular orbital and a π^* antibonding molecular orbital.

> **A σ bond is stronger than a π bond.**

The extent of overlap is greater when p orbitals overlap end-on than when they overlap side-to-side. This means that a σ bond formed by the end-on overlap of p orbitals is stronger than a π bond formed by the side-to-side overlap of p orbitals. It also means that a σ bonding molecular orbital is more stable than a π bonding molecular orbital because the stronger the bond, the more stable it is. Figure 1.7 shows a molecular orbital diagram of two identical atoms using their three degenerate p atomic orbitals to form three bonds—one σ bond and two π bonds.

Now let's look at the MO diagram for side-to-side overlap of a p orbital of carbon with a p orbital of oxygen—the orbitals are the same type, but they belong to different kinds of atoms (Figure 1.8). When the two p atomic orbitals combine to form molecular orbitals, the atomic orbital of the more electronegative atom contributes more to the bonding molecular orbital, and the atomic orbital of the less electronegative atom contributes more to the antibonding molecular orbital. This means that if we were to put electrons in the bonding MO, they would be more apt to be around the oxygen atom than around the carbon atom. Thus, both Lewis theory and molecular orbital theory tell us that the electrons shared by carbon and oxygen are not shared equally—the oxygen atom of a carbon–oxygen bond has a partial negative charge and the carbon atom has a partial positive charge.

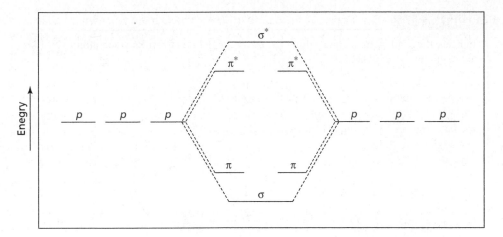

◀ **Figure 1.7**
p Orbitals can overlap end-on to form σ bonding and σ* antibonding molecular orbitals, or they can overlap side-to-side to form π bonding and π* antibonding molecular orbitals. The relative energies of the molecular orbitals are $\sigma < \pi < \pi^* < \sigma^*$.

Organic chemists find that the information obtained from MO theory, where valence electrons occupy bonding and antibonding molecular orbitals, does not always yield the needed information about the bonds in a molecule. The **valence-shell electron-pair repulsion (VSEPR) model** combines the Lewis concept of shared electron pairs and lone-pair electrons with the concept of atomic orbitals and adds a third principle: *the minimization of electron repulsion*. In this model, atoms share electrons by overlapping their atomic orbitals, and because electron pairs repel each other, the bonding electrons and lone-pair electrons around an atom are positioned as far apart as possible.

Because organic chemists generally think of chemical reactions in terms of the changes that occur in the bonds of the reacting molecules, the VSEPR model often provides the easiest way to visualize chemical change. However, the model is inadequate for some molecules because it does not allow for antibonding orbitals. We will use both the MO and the VSEPR models in this book. Our choice will depend on which model provides the best description of the molecule under discussion. We will use the VSEPR model in Sections 1.7–1.13.

A more extensive discussion of molecular orbital theory can be found in Special Topic VI in the *Study Guide and Solutions Manual*.

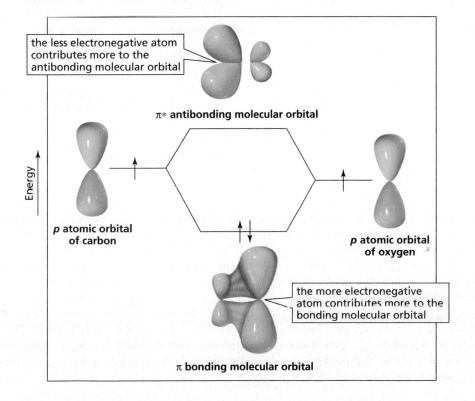

the less electronegative atom contributes more to the antibonding molecular orbital

π* antibonding molecular orbital

Energy

***p* atomic orbital of carbon**

***p* atomic orbital of oxygen**

the more electronegative atom contributes more to the bonding molecular orbital

π bonding molecular orbital

◀ **Figure 1.8**
Side-to-side overlap of a *p* orbital of carbon with a *p* orbital of oxygen to form a π bonding molecular orbital and a π* antibonding molecular orbital.

Indicate the kind of molecular orbital (σ, σ^*, π, or π^*) that results when atomic orbitals are combined as indicated:

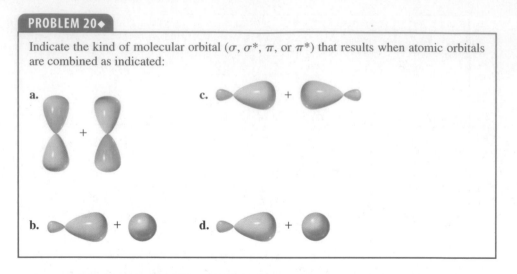

1.7 How Single Bonds Are Formed in Organic Compounds

We will begin the discussion of bonding in organic compounds by looking at the bonding in methane, a compound with only one carbon atom. Then we will examine the bonding in ethane, a compound with two carbons attached by a carbon–carbon single bond.

The Bonds in Methane

Methane (CH_4) has four covalent C—H bonds. Because all four bonds have the same length (1.10 Å) and all the bond angles are the same (109.5°), we can conclude that the four C—H bonds in methane are identical. Four different ways to represent a methane molecule are shown here.

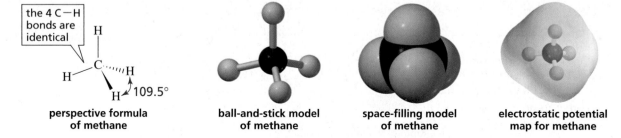

| perspective formula of methane | ball-and-stick model of methane | space-filling model of methane | electrostatic potential map for methane |

In a **perspective formula**, bonds in the plane of the paper are drawn as solid lines, bonds protruding out of the plane of the paper toward the viewer are drawn as solid wedges, and those projecting back from the plane of the paper away from the viewer are drawn as hatched wedges.

The potential map of methane shows that neither carbon nor hydrogen carries much of a charge: there are neither red areas, representing partially negatively charged atoms, nor blue areas, representing partially positively charged atoms. (Compare this map with the potential map for water on p. 37.) The absence of partially charged atoms can be explained by the similar electronegativities of carbon and hydrogen, which cause them to share their bonding electrons relatively equally. Methane is therefore a **nonpolar molecule**.

You may be surprised to learn that carbon forms four covalent bonds since you know that carbon has only two unpaired electrons in its ground-state electronic configuration (Table 1.2). But if carbon formed only two covalent bonds, it would not complete its octet. We therefore need to come up with an explanation that accounts for carbon's forming four covalent bonds.

If one of the electrons in carbon's $2s$ orbital were promoted into the empty $2p$ orbital, the new electronic configuration would have four unpaired electrons; thus, four covalent bonds could be formed.

If carbon used an s orbital and three p orbitals to form these four bonds, the bond formed with the s orbital would be different from the three bonds formed with p orbitals. What could account for the fact that the four C—H bonds in methane are identical if they are made using one s and three p orbitals? The answer is that carbon uses *hybrid orbitals*.

Hybrid orbitals are mixed orbitals that result from combining atomic orbitals. The concept of combining orbitals, called **hybridization**, was first proposed by Linus Pauling in 1931. If the one s and three p orbitals of the second shell are all combined and then apportioned into four equal orbitals, each of the four resulting orbitals will be one part s and three parts p. This type of mixed orbital is called an sp^3 (pronounced "s-p-three," not "s-p-cubed") orbital. (The superscript 3 means that three p orbitals were mixed with one s orbital to form the hybrid orbitals.) Each sp^3 orbital has 25% s character and 75% p character. The four sp^3 orbitals are degenerate—they each have the same energy.

Like a p orbital, an sp^3 orbital has two lobes. The lobes differ in size, however, because the s orbital adds to one lobe of the p orbital and subtracts from the other lobe of the p orbital (Figure 1.9). The stability of an sp^3 orbital reflects its composition; it is more stable than a p orbital, but not as stable as an s orbital (Figure 1.10). The larger lobe of the sp^3 orbital is used in covalent bond formation.

The four sp^3 orbitals adopt a spatial arrangement that keeps them as far away from each other as possible (Figure 1.11a). They do this because electrons repel each other, and moving as far from each other as possible minimizes the repulsion (Section 1.6). When four orbitals move as far from each other as possible, they point toward the corners of a regular tetrahedron (a pyramid with four faces, each an equilateral triangle). Each of the four C—H bonds in methane is formed from overlap of an sp^3 orbital of carbon with the s orbital of a hydrogen (Figure 1.11b). This explains why the four C—H bonds are identical.

BIOGRAPHY

Linus Carl Pauling (1901–1994) *was born in Portland, Oregon. A friend's home chemistry laboratory sparked Pauling's early interest in science. He received a Ph.D. from the California Institute of Technology and remained there for most of his academic career. He received the Nobel Prize in chemistry in 1954 for his work on molecular structure. Like Einstein, Pauling was a pacifist, winning the 1962 Nobel Peace Prize for his work on behalf of nuclear disarmament.*

Electron pairs stay as far from each other as possible.

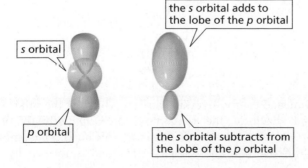

◀ **Figure 1.9**
The s orbital adds to one lobe of the p orbital and subtracts from the other lobe of the p orbital.

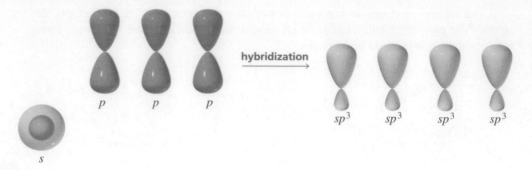

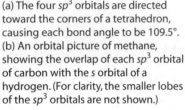

Figure 1.10 ▶
An *s* orbital and three *p* orbitals hybridize to form four *sp*³ orbitals. An *sp*³ orbital is more stable than a *p* orbital, but not as stable as an *s* orbital.

Figure 1.11 ▶
(a) The four *sp*³ orbitals are directed toward the corners of a tetrahedron, causing each bond angle to be 109.5°. (b) An orbital picture of methane, showing the overlap of each *sp*³ orbital of carbon with the *s* orbital of a hydrogen. (For clarity, the smaller lobes of the *sp*³ orbitals are not shown.)

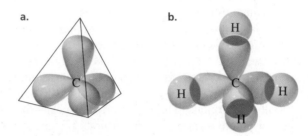

The angle between any two lines that point from the center to the corners of a tetrahedron is 109.5°. The bond angles in methane therefore are 109.5°. This is called a **tetrahedral bond angle**. A carbon, such as the one in methane, that forms covalent bonds using four equivalent *sp*³ orbitals is called a **tetrahedral carbon**.

Hybrid orbital theory may appear to have been contrived just to make things fit— and that is exactly the case. Nevertheless, it gives us a very good picture of the bonding in organic compounds.

Note to the student

It is important to understand what molecules look like in three dimensions. Therefore be sure to visit the textbook's Web site (http://www.prenhall.com/bruice) and look at the three-dimensional representations of molecules that can be found in the molecule gallery prepared for each chapter.

The Bonds in Ethane

The two carbon atoms in ethane (CH_3CH_3) are tetrahedral. Each carbon uses four *sp*³ orbitals to form four covalent bonds (Figure 1.12):

$$H-\overset{\overset{\displaystyle H}{|}}{\underset{\underset{\displaystyle H}{|}}{C}}-\overset{\overset{\displaystyle H}{|}}{\underset{\underset{\displaystyle H}{|}}{C}}-H$$

ethane

One *sp*³ orbital of one carbon overlaps an *sp*³ orbital of the other carbon to form the C—C bond. Each of the remaining three *sp*³ orbitals of each carbon overlaps the *s* orbital of a hydrogen to form a C—H bond. Thus, the C—C bond is formed by *sp*³–*sp*³ overlap, and each C—H bond is formed by *sp*³–*s* overlap.

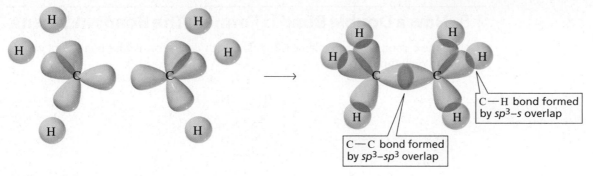

▲ **Figure 1.12**
An orbital picture of ethane. The C—C bond is formed by sp^3–sp^3 overlap, and each C—H bond is formed by sp^3–s overlap. (The smaller lobes of the sp^3 orbitals are not shown.)

Each of the bond angles in ethane is nearly the tetrahedral bond angle of 109.5°, and the length of the C—C bond is 1.54 Å. Ethane, like methane, is a nonpolar molecule.

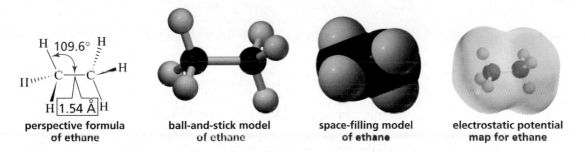

| perspective formula of ethane | ball-and-stick model of ethane | space-filling model of ethane | electrostatic potential map for ethane |

All the bonds in methane and ethane are sigma (σ) bonds because they are all formed by the end-on overlap of atomic orbitals. One bond connecting two atoms is called a **single bond**. *All single bonds found in organic compounds are sigma bonds.*

All single bonds found in organic compounds are sigma bonds.

PROBLEM 21◆

What orbitals are used to form the 10 covalent bonds in propane ($CH_3CH_2CH_3$)?

The MO diagram illustrating the overlap of an sp^3 orbital of one carbon with an sp^3 orbital of another carbon (Figure 1.13) is similar to the MO diagram for the end-on overlap of two p orbitals, which should not be surprising since sp^3 orbitals have 75% p character.

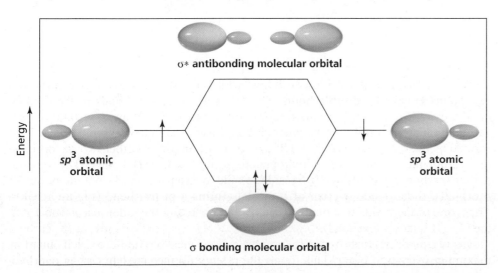

σ* antibonding molecular orbital

sp^3 atomic orbital

sp^3 atomic orbital

σ bonding molecular orbital

◀ **Figure 1.13**
End-on overlap of two sp^3 orbitals to form a σ bonding molecular orbital and a σ* antibonding molecular orbital.

1.8 How a Double Bond Is Formed: The Bonds in Ethene

Each of the carbon atoms in ethene (also called ethylene) forms four bonds, but each is bonded to only three atoms:

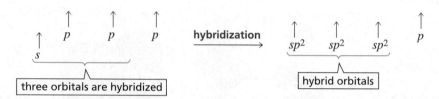

ethene
ethylene

To bond to three atoms, each carbon hybridizes three atomic orbitals: an *s* orbital and two of the *p* orbitals. Because three orbitals are hybridized, three hybrid orbitals are obtained. These are called sp^2 orbitals. After hybridization, each carbon atom has three degenerate sp^2 orbitals and one *p* orbital:

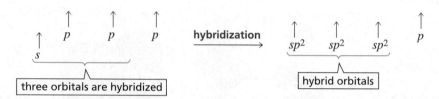

To minimize electron repulsion, the three sp^2 orbitals need to get as far from each other as possible. Therefore, the axes of the three orbitals lie in a plane, directed toward the corners of an equilateral triangle with the carbon nucleus at the center. This means that the bond angles are all close to 120°. Because the sp^2 hybridized carbon atom is bonded to three atoms that define a plane, it is called a **trigonal planar carbon**. The unhybridized *p* orbital is perpendicular to the plane defined by the axes of the sp^2 orbitals (Figure 1.14).

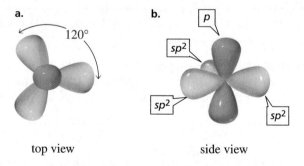

a. 120° top view

b. *p* sp^2 sp^2 sp^2 side view

Figure 1.14 ▶
(a) The three degenerate sp^2 orbitals lie in a plane. (b) The unhybridized *p* orbital is perpendicular to the plane. (The smaller lobes of the sp^2 orbitals are not shown.)

The carbons in ethene form two bonds with each other. Two bonds connecting two atoms is called a **double bond**. The two carbon–carbon bonds in the double bond are not identical. One of them results from the overlap of an sp^2 orbital of one carbon with an sp^2 orbital of the other carbon; this is a sigma (σ) bond because it is formed by end-on overlap (Figure 1.15a). Each carbon uses its other two sp^2 orbitals to overlap the *s* orbital of a hydrogen to form the C—H bonds. The second carbon–carbon bond results from side-to-side overlap of the two unhybridized *p* orbitals. Side-to-side overlap of *p* orbitals forms a pi (π) bond (Figure 1.15b). Thus, one of the bonds in a double bond is a σ bond and the other is a π bond. All the C—H bonds are σ bonds.

The two *p* orbitals that overlap to form the π bond must be parallel to each other for maximum overlap to occur. This forces the triangle formed by one carbon and two

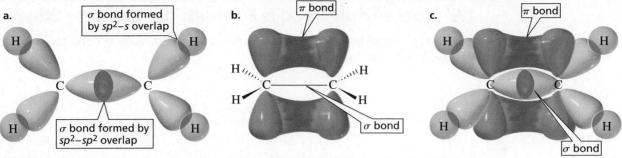

a.

σ bond formed by sp²–s overlap

σ bond formed by sp²–sp² overlap

b.

π bond

σ bond

c.

π bond

σ bond

▲ **Figure 1.15**

(a) One C—C bond in ethene is a σ bond formed by sp²–sp² overlap, and the C—H bonds are formed by sp²–s overlap. (b) The second C—C bond is a π bond formed by side-to-side overlap of a p orbital of one carbon with a p orbital of the other carbon. (c) There is an accumulation of electron density above and below the plane containing the two carbons and four hydrogens.

hydrogens to lie in the same plane as the triangle formed by the other carbon and two hydrogens. As a result, all six atoms of ethene lie in the same plane, and the electrons in the p orbitals occupy a volume of space above and below the plane (Figure 1.15c). The potential map for ethene shows that it is a nonpolar molecule with a slight accumulation of negative charge (the pale orange area) above the two carbons. (If you could turn the potential map over to show the hidden side, a similar accumulation of negative charge would be found there.)

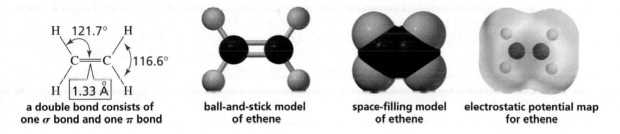

H 121.7° H
C=C 116.6°
H 1.33 Å H

a double bond consists of one σ bond and one π bond

ball-and-stick model of ethene

space-filling model of ethene

electrostatic potential map for ethene

Four electrons hold the carbons together in a carbon–carbon double bond; only two electrons hold the carbons together in a carbon–carbon single bond. This means that a carbon–carbon double bond is stronger (174 kcal/mol or 728 kJ/mol) and shorter (1.33 Å) than a carbon–carbon single bond (90 kcal/mol or 377 kJ/mol, and 1.54 Å).

DIAMOND, GRAPHITE, AND BUCKMINSTERFULLERENE: SUBSTANCES CONTAINING ONLY CARBON ATOMS

Diamond is the hardest of all substances. Graphite, in contrast, is a slippery, soft solid most familiar to us as the "lead" in pencils. Both materials, in spite of their very different physical properties, contain only carbon atoms. The two substances differ solely in the nature of the bonds holding the carbon atoms together. Diamond consists of a rigid three-dimensional network of carbon atoms, with each bonded to four others via sp³ orbitals. The carbon atoms in graphite, on the other hand, are sp² hybridized, so each bonds to only three other carbons. This trigonal planar arrangement causes the atoms in graphite to lie in flat, layered sheets that can shear off of neighboring sheets. When you write with a pencil, sheets of carbon atoms are shearing off to leave a thin trail of graphite. A compound called buckminsterfullerene is a third substance found in nature that contains only carbon atoms. Like graphite, buckminsterfullerene contains only sp² hybridized carbons, but instead of forming planar sheets, they join to form spherical structures. (Buckminsterfullerene is discussed in more detail in Section 14.2.)

1.9 How a Triple Bond Is Formed: The Bonds in Ethyne

The carbon atoms in ethyne (also called acetylene) are each bonded to only two atoms—a hydrogen and another carbon:

H—C≡C—H
ethyne
acetylene

Because each carbon forms covalent bonds with two atoms, only two of each carbon's orbitals are hybridized: an *s* and a *p*. Two degenerate *sp* orbitals result. Each carbon atom in ethyne, therefore, has two *sp* orbitals and two unhybridized *p* orbitals (Figure 1.16).

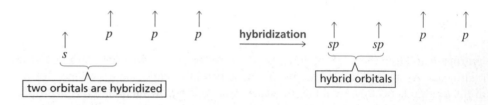

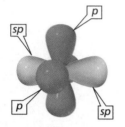

▲ **Figure 1.16**
The two *sp* orbitals are oriented 180° away from each other, perpendicular to the two unhybridized *p* orbitals. (The smaller lobes of the *sp* orbitals are not shown.)

To minimize electron repulsion, the two *sp* orbitals point in opposite directions.

The carbon atoms in ethyne are held together by three bonds. Three bonds connecting two atoms is called a **triple bond**. One of the *sp* orbitals of one carbon in ethyne overlaps an *sp* orbital of the other carbon to form a carbon–carbon σ bond. The other *sp* orbital of each carbon overlaps the *s* orbital of a hydrogen to form a C—H σ bond (Figure 1.17a). Because the two *sp* orbitals point in opposite directions, the bond angles are 180°. The two unhybridized *p* orbitals are perpendicular to each other, and both are perpendicular to the *sp* orbitals. Each of the unhybridized *p* orbitals engages in side-to-side overlap with a parallel *p* orbital on the other carbon, with the result that two π bonds are formed (Figure 1.17b).

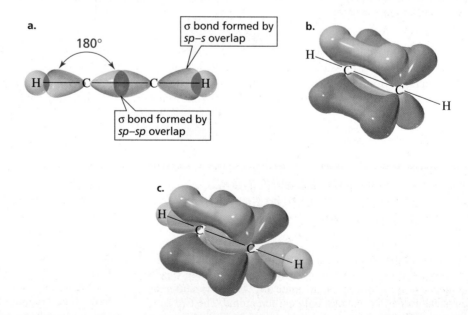

Figure 1.17 ▶
(a) The C—C σ bond in ethyne is formed by *sp–sp* overlap, and the C—H bonds are formed by *sp–s* overlap. The carbon atoms and the atoms bonded to them are in a straight line. (b) The two carbon–carbon π bonds are formed by side-to-side overlap of the *p* orbitals of one carbon with the *p* orbitals of the other carbon. (c) The triple bond has an electron-dense region above and below and in front of and in back of the internuclear axis of the molecule.

A triple bond therefore consists of one σ bond and two π bonds. The two unhybridized *p* orbitals on each carbon are perpendicular to each other, creating regions of high electron density above and below *and* in front of and back of the internuclear axis of the molecule (Figure 1.17c). The potential map for ethyne shows that negative charge accumulates in a cylinder that wraps around the egg-shaped molecule.

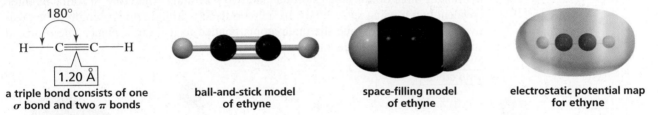

| a triple bond consists of one σ bond and two π bonds | ball-and-stick model of ethyne | space-filling model of ethyne | electrostatic potential map for ethyne |

Because the two carbon atoms in a triple bond are held together by six electrons, a triple bond is stronger (231 kcal/mol or 967 kJ/mol) and shorter (1.20 Å) than a double bond (174 kcal/mol or 728 kJ/mol, and 1.33 Å).

PROBLEM 22 **SOLVED**

For each of the given species:

a. Draw its Lewis structure.

b. Describe the orbitals used by each carbon atom in bonding and indicate the approximate bond angles.

   1. HCOH   **2.** CCl_4 **3.** CH_3COH **4.** HCN

Solution to 22a1. Because HCOH is neutral, we know that each H forms one bond, the oxygen forms two bonds, and the carbon forms four bonds. Our first attempt at a Lewis structure (drawing the atoms in the order given by the Kekulé structure) shows that carbon is the only atom that does not form the needed number of bonds.

$$H—C—O—H$$

If we place a double bond between carbon and oxygen and move an H, all the atoms end up with the correct number of bonds. Lone pair electrons are used to give each atom a filled outer shell. When we check to see if any atom needs to be assigned a formal charge, we find that none of them does.

$$\begin{array}{c} :O: \\ \parallel \\ H—C—H \end{array}$$

Solution to 22b1. Because the carbon atom forms a double bond, we know that carbon uses sp^2 orbitals (as it does in ethene) to bond to the two hydrogens and the oxygen. It uses its "left over" p orbital to form the second bond to oxygen. Because carbon is sp^2 hybridized, the bond angles are approximately 120°.

$$\begin{array}{c} \ddot{O} \\ 120° \nearrow \parallel \nwarrow 120° \\ C \\ H \quad H \\ 120° \end{array}$$

1.10 The Bonds in the Methyl Cation, the Methyl Radical, and the Methyl Anion

Not all carbon atoms form four bonds. A carbon with a positive charge, a negative charge, or an unpaired electron forms only three bonds. Now we will see what orbitals carbon uses when it forms three bonds.

The Methyl Cation ($^+CH_3$)

The positively charged carbon in the methyl cation is bonded to three atoms, so it hybridizes three orbitals—an s orbital and two p orbitals. Therefore, it forms its three covalent bonds using sp^2 orbitals. Its unhybridized p orbital remains empty. The positively charged carbon and the three atoms bonded to it lie in a plane. The p orbital stands perpendicular to the plane.

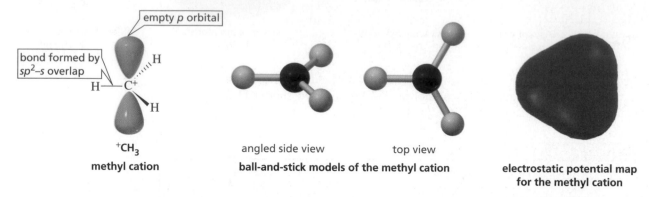

$^+CH_3$
methyl cation

ball-and-stick models of the methyl cation

angled side view   top view

**electrostatic potential map
for the methyl cation**

The Methyl Radical ($\cdot CH_3$)

The carbon atom in the methyl radical is also sp^2 hybridized. The methyl radical differs by one unpaired electron from the methyl cation. That electron is in the p orbital, with half of the electron density in each lobe. Notice the similarity in the ball-and-stick models of the methyl cation and the methyl radical. The potential maps, however, are quite different because of the additional electron in the methyl radical.

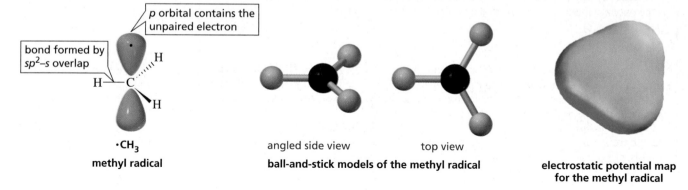

$\cdot CH_3$
methyl radical

ball-and-stick models of the methyl radical

angled side view   top view

**electrostatic potential map
for the methyl radical**

The Methyl Anion ($\bar{\vdots} CH_3$)

The negatively charged carbon in the methyl anion has three pairs of bonding electrons and one lone pair. The four pairs of electrons are farthest apart when the four orbitals containing the bonding and lone-pair electrons point toward the corners of a tetrahedron. In other words, a negatively charged carbon is sp^3 hybridized. In the methyl anion, three of carbon's sp^3 orbitals each overlap the s orbital of a hydrogen, and the fourth sp^3 orbital holds the lone pair.

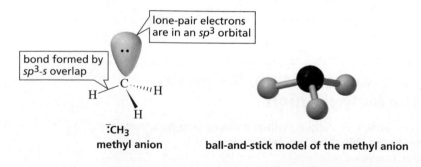

$\bar{\vdots}CH_3$
methyl anion

ball-and-stick model of the methyl anion

**electrostatic potential map
for the methyl anion**

Take a moment to compare the potential maps for the methyl cation, the methyl radical, and the methyl anion.

1.11 The Bonds in Water

The oxygen atom in water (H_2O) forms two covalent bonds. Because the electronic configuration of oxygen shows that it has two unpaired electrons (Table 1.2), oxygen does not need to promote an electron to form the number (two) of covalent bonds required to achieve an outer shell of eight electrons (that is, to complete its octet). If we assume that oxygen uses p orbitals to form the two O—H bonds, as predicted by oxygen's ground-state electronic configuration, we would expect a bond angle of about 90° because the two p orbitals are at right angles to each other. However, the experimentally observed bond angle is 104.5°.

To explain the observed bond angle, oxygen must use hybrid orbitals to form covalent bonds—just as carbon does. The s orbital and the three p orbitals must hybridize to produce four identical sp^3 orbitals.

The bond angles in a molecule indicate which orbitals are used in bond formation.

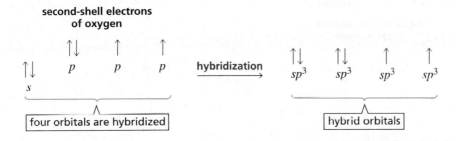

Each of the two O—H bonds is formed by the overlap of an sp^3 orbital of oxygen with the s orbital of a hydrogen. A lone pair occupies each of the two remaining sp^3 orbitals.

The bond angle in water (104.5°) is a little smaller than the bond angle in methane (109.5°) presumably because each of lone pairs are held by only one nucleus, which makes a lone pair more diffuse than a bonding pair that is shared by two nuclei and is therefore relatively confined between them. Consequently, lone pairs exert more electron repulsion, causing the O—H bonds to squeeze closer together, thereby decreasing the bond angle.

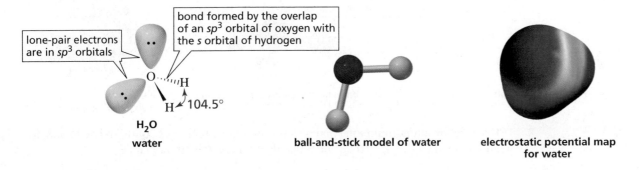

water   ball-and-stick model of water   electrostatic potential map for water

Compare the potential map for water with that for methane. Water is a polar molecule; methane is nonpolar.

PROBLEM 23◆

The bond angles in H_3O^+ are greater than _____ and less than _____.

WATER—A UNIQUE COMPOUND
Water is the most abundant compound found in living organisms. Its unique properties have allowed life to originate and evolve. Its high heat of fusion (the heat required to convert a solid to a liquid) protects organisms from freezing at low temperatures because a lot of heat must be removed from water to freeze it. Its high heat capacity (the heat required to raise the temperature of a substance by a given amount) minimizes temperature changes in organisms, and its high heat of vaporization (the heat required to convert a liquid to a gas) allows animals to cool themselves with a minimal loss of body fluid. Because liquid water is denser than ice, ice formed on the surface of water floats and insulates the water below. That is why oceans and lakes don't freeze from the bottom up. It is also why plants and aquatic animals can survive when the ocean or lake they live in freezes.

1.12 The Bonds in Ammonia and in the Ammonium Ion

The experimentally observed bond angles in NH_3 are 107.3°. The bond angles indicate that nitrogen also uses hybrid orbitals when it forms covalent bonds. Like carbon and oxygen, the one s and three p orbitals of the second shell of nitrogen hybridize to form four degenerate sp^3 orbitals:

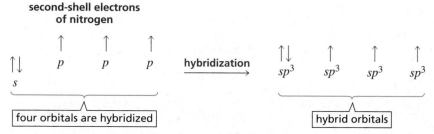

Each of the N—H bonds in NH_3 is formed from the overlap of an sp^3 orbital of nitrogen with the s orbital of a hydrogen. The single lone pair occupies an sp^3 orbital. The bond angle (107.3°) is smaller than the tetrahedral bond angle (109.5°) because of the relatively diffuse lone pair. Notice that the bond angles in NH_3 (107.3°) are larger than the bond angles in H_2O (104.5°) because nitrogen has only one lone pair, whereas oxygen has two lone pairs.

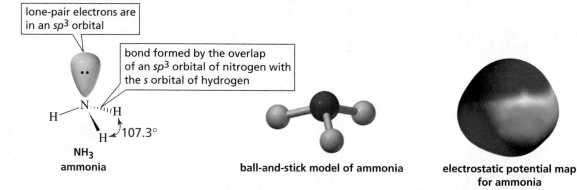

ball-and-stick model of ammonia

electrostatic potential map for ammonia

Because the ammonium ion ($^+NH_4$) has four identical N—H bonds and no lone pairs, all the bond angles are 109.5°, just like the bond angles in methane.

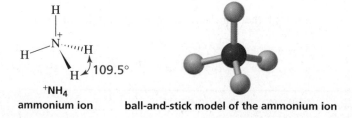

ball-and-stick model of the ammonium ion

electrostatic potential map for the ammonium ion

PROBLEM 24◆

According to the potential map for the ammonium ion, which atom(s) has (have) the least electron density?

PROBLEM 25◆

Compare the potential maps for methane, ammonia, and water. Which is the most polar molecule? Which is the least polar?

electrostatic potential map for methane

electrostatic potential map for ammonia

electrostatic potential map for water

PROBLEM 26◆

Predict the approximate bond angles in the methyl carbanion.

1.13 **The Bonds in the Hydrogen Halides**

Fluorine, chlorine, bromine, and iodine are known as the halogens, so HF, HCl, HBr, and HI are called hydrogen halides. Bond angles will not help us determine the orbitals that form a hydrogen halide bond, as they did with other molecules, because hydrogen halides have only one bond. We do know, however, that a halogen's three lone pairs are identical, and that lone-pair electrons position themselves to minimize electron repulsion (Section 1.6). Both of these observations suggest that the halogen's three lone pairs are in sp^3 orbitals. Therefore, we will assume that the hydrogen–halogen bond is formed by the overlap of an sp^3 orbital of the halogen with the s orbital of hydrogen.

hydrogen fluoride

hydrogen chloride

hydrogen bromide

hydrogen iodide

H—F̈:
hydrogen fluoride

ball-and-stick model of hydrogen fluoride

electrostatic potential map for hydrogen fluoride

In the case of fluorine, the sp^3 orbital used in bond formation belongs to the second shell of electrons. In chlorine, the sp^3 orbital belongs to the third shell of electrons. Because the average distance from the nucleus is greater for an electron in the third shell than it is for an electron in the second shell, the average electron density is less in a $3sp^3$ orbital than it is in a $2sp^3$ orbital. This means that the electron density in the region where the s orbital of hydrogen overlaps the sp^3 orbital of the halogen decreases as the size of the halogen increases (Figure 1.18). Therefore,

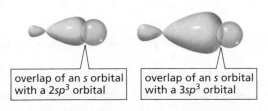

overlap of an *s* orbital with a $2sp^3$ orbital

overlap of an *s* orbital with a $3sp^3$ orbital

◀ **Figure 1.18**
There is greater electron density in the region of overlap of an *s* orbital with a $2sp^3$ orbital than in the region of overlap of an *s* orbital with a $3sp^3$ orbital.

the hydrogen–halogen bond becomes longer and weaker as the size (atomic weight) of the halogen increases (Table 1.6).

Table 1.6	Hydrogen–Halogen Bond Lengths and Bond Strengths		
Hydrogen halide	**Bond length (Å)**	**Bond strength**	
		kcal/mol	**kJ/mol**
H—F	0.917	136	571
H—Cl	1.2746	103	432
H—Br	1.4145	87	366
H—I	1.6090	71	298

PROBLEM 27◆

a. Predict the relative lengths and strengths of the bonds in Cl_2 and Br_2.
b. Predict the relative lengths and strengths of the carbon–halogen bonds in CH_3F, CH_3Cl, and CH_3Br.

PROBLEM 28◆

a. Which bond would be longer?
b. Which bond would be stronger?
    1. C—Cl or C—I   **2.** C—C or C—Cl   **3.** H—Cl or H—H

Student Tutorial:
Hybridization 1

The hybridization of a C, O, or N is $sp^{(3\text{—the number of } p \text{ orbitals})}$.

Student Tutorial:
Hybridization 2

1.14 Summary: Hybridization, Bond Lengths, Bond Strengths, and Bond Angles

All **single bonds** are σ bonds. All **double bonds** are composed of one σ bond and one π bond. All **triple bonds** are composed of one σ bond and two π bonds. The easiest way to determine the hybridization of a carbon, oxygen, or nitrogen atom is to look at the number of π bonds it forms: if it forms no π bonds, it is sp^3 hybridized; if it forms one π bond, it is sp^2 hybridized; if it forms two π bonds, it is sp hybridized. The exceptions are carbocations and carbon radicals, which are sp^2 hybridized—not because they form a π bond, but because they have an empty or a half-filled p orbital (Section 1.10).

$$CH_3—\ddot{N}H_2 \qquad \begin{matrix} CH_3 \\ \diagdown \\ C=\ddot{N}—\ddot{N}H_2 \\ \diagup \\ CH_3 \end{matrix} \qquad CH_3—C\equiv N\colon \qquad CH_3—\ddot{O}H \qquad \begin{matrix} \colon\!\ddot{O} \leftarrow sp^2 \\ \parallel \\ C \\ \diagup \;\; \diagdown \\ CH_3 \quad CH_3 \end{matrix} \qquad \colon\!\ddot{O}=C=\ddot{O}\colon$$

$$\begin{matrix} \uparrow & \uparrow \\ sp^3 & sp^3 \end{matrix} \qquad \begin{matrix} \uparrow & \uparrow & \uparrow & \uparrow \\ sp^3 & sp^2 & sp^2 & sp^3 \end{matrix} \qquad \begin{matrix} \uparrow & \uparrow & \uparrow \\ sp^3 & sp & sp \end{matrix} \qquad \begin{matrix} \uparrow & \uparrow \\ sp^3 & sp^3 \end{matrix} \qquad \begin{matrix} \uparrow & \uparrow & \uparrow \\ sp^3 & sp^2 & sp^3 \end{matrix} \qquad \begin{matrix} \uparrow & \uparrow & \uparrow \\ sp^2 & sp & sp^2 \end{matrix}$$

Table 1.7   Comparison of the Bond Angles and the Lengths and Strengths of the Carbon–Carbon and Carbon–Hydrogen Bonds in Ethane, Ethene, and Ethyne

Molecule	Hybridization of carbon	Bond angles	Length of C—C bond (Å)	Strength of C—C bond (kcal/mol)	(kJ/mol)	Length of C—H bond (Å)	Strength of C—H bond (kcal/mol)	(kJ/mol)
H H H—C—C—H H H **ethane**	sp^3	109.5°	1.54	90	377	1.10	101	423
H H C=C H H **ethene**	sp^2	120°	1.33	174	728	1.08	111	466
H—C≡C—H **ethyne**	sp	180°	1.20	231	967	1.06	131	548

In comparing the lengths and strengths of carbon–carbon single, double, and triple bonds, we see that the more bonds holding two carbon atoms together, the shorter and stronger is the carbon–carbon bond (Table 1.7): triple bonds are shorter and stronger than double bonds, which are shorter and stronger than single bonds.

The data in Table 1.7 indicate that a C—H σ bond is shorter and stronger than a C—C σ bond. This is because the s orbital of hydrogen is closer to the nucleus than is the sp^3 orbital of carbon. Consequently, the nuclei are closer together in a bond formed by sp^3–s overlap than they are in a bond formed by sp^3–sp^3 overlap. In addition to being shorter, a C—H σ bond is stronger than a C—C σ bond because there is greater electron density in the region of overlap of an sp^3 orbital with the s orbital than in the region of overlap of an sp^3 orbital with an sp^3 orbital.

The length and strength of a C—H bond depend on the hybridization of the carbon atom to which the hydrogen is attached. The more s character in the orbital used by carbon to form the bond, the shorter and stronger is the bond—again, because an s orbital is closer to the nucleus than is a p orbital. So a C—H bond formed by an sp hybridized carbon (50% s) is shorter and stronger than a C—H bond formed by an sp^2 hybridized carbon (33.3% s), which in turn is shorter and stronger than a C—H bond formed by an sp^3 hybridized carbon (25% s).

A double bond (a σ bond plus a π bond) is stronger than a single bond (a σ bond), but it is not twice as strong, so we can conclude that the π bond of a double bond is weaker than the σ bond. Table 1.7 indicates that the strength of a C—C σ bond formed by sp^3–sp^3 overlap is 90 kcal/mol. However, a C—C σ bond formed by sp^2–sp^2 overlap is expected to be stronger because of the greater s character in the overlapping orbitals; it has been estimated to be ~112 kcal/mol. We can conclude then, that the π bond strength of the π bond of ethene is about 174 kcal/mol – 112 kcal/mol = 62 kcal/mol (or 259 kJ/mol). We expected the π bond to be weaker than the σ bond because side-to-side overlap that forms a π bond is less effective than the end-on overlap that forms a σ bond (Section 1.6).

The bond angle, too, depends on the orbital used by carbon to form the bond. The greater the amount of s character in the orbital, the larger is the bond angle. For example, sp hybridized carbons have bond angles of 180°, sp^2 hybridized carbons have bond angles of 120°, and sp^3 hybridized carbons have bond angles of 109.5°.

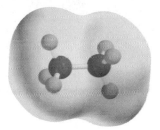

Ethane

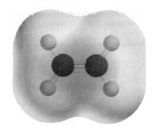

Ethene

Ethyne

The shorter the bond, the stronger it is.

The greater the electron density in the region of orbital overlap, the stronger is the bond.

The more s character, the shorter and stronger is the bond.

The more s character, the larger the bond angle.

You may wonder how an electron "knows" what orbital it should go into. In fact, electrons know nothing about orbitals. They simply occupy the space around atoms in the most stable arrangement possible. It is chemists who use the concept of orbitals to explain this arrangement.

PROBLEM 29◆

Which of the bonds in a carbon–oxygen double bond has more effective orbital–orbital overlap, the σ bond or the π bond?

PROBLEM 30◆

Would you expect a C—C σ bond formed by sp^2–sp^2 overlap to be stronger or weaker than a σ bond formed by sp^3–sp^3 overlap?

PROBLEM 31

a. What is the hybridization of each of the carbon atoms in the following compound?

$$CH_3CHCH{=}CHCH_2C{\equiv}CCH_3$$
$$\underset{\textstyle CH_3}{|}$$

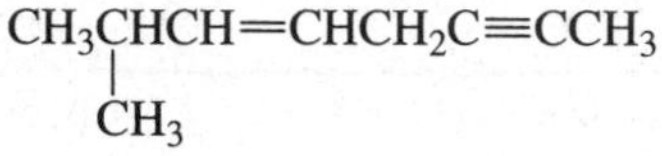

b. What is the hybridization of each of the carbon and oxygen atoms in the following compounds?

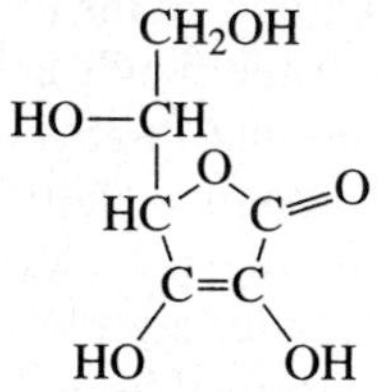

vitamin C

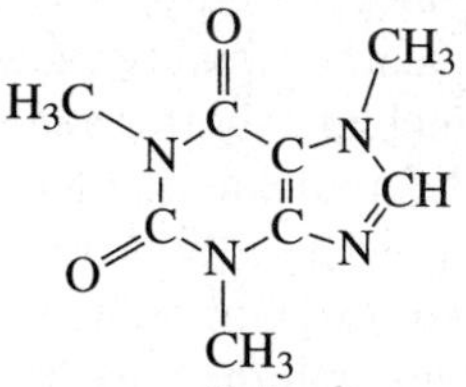

caffeine

PROBLEM-SOLVING STRATEGY

Predicting Bond Angles

Predict the approximate bond angle of the C—N—H bond in $(CH_3)_2NH$.

First we need to determine the hybridization of the central atom. Because the nitrogen atom forms only single bonds, we know it is sp^3 hybridized. Next we look to see if there are lone pairs that will affect the bond angle. A neutral nitrogen has one lone pair. Based on these observations, we can predict that the C—N—H bond angle will be about 107.3°, the same as the H—N—H bond angle in NH_3, another compound with a neutral sp^3 hybridized nitrogen.

Now continue on to Problem 32.

PROBLEM 32◆

Predict the approximate bond angles:

a. the C—N—C bond angle in $(CH_3)_2\overset{+}{N}H_2$
b. the C—C—N bond angle in $CH_3CH_2NH_2$
c. the H—C—N bond angle in $(CH_3)_2NH$
d. the H—C—O bond angle in CH_3OCH_3

PROBLEM 33

Describe the orbitals used in bonding and the bond angles in the following compounds. (*Hint:* See Table 1.7.)

a. BeH_2 **b.** BH_3 **c.** CCl_4 **d.** CO_2 **e.** HCOOH   **f.** N_2

1.15 **The Dipole Moments of Molecules**

In Section 1.3, we saw that in molecules with one covalent bond, the dipole moment of the bond is identical to the dipole moment of the molecule. When molecules have more than one covalent bond, the geometry of the molecule must be taken into account because both the *magnitude* and the *direction* of the individual bond dipole moments (the vector sum) determine the overall dipole moment of the molecule. Totally symmetrical molecules, therefore, have no dipole moment. For example, let's look at the dipole moment of carbon dioxide (CO_2). Because the carbon atom is bonded to two atoms, it uses *sp* orbitals to form the C—O σ bonds. The remaining two *p* orbitals on carbon form the two C—O π bonds. The *sp* orbitals form a bond angle of 180°, and so the individual carbon–oxygen bond dipole moments cancel each other. Therefore, carbon dioxide has a dipole moment of 0 D. Another symmetrical molecule is carbon tetrachloride (CCl_4). The four atoms bonded to the sp^3 hybridized carbon atom are identical and project symmetrically out from the carbon atom. Thus, as with CO_2, the symmetry of the molecule causes the bond dipole moments to cancel. Methane also has no dipole moment.

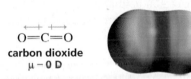

O=C=O
carbon dioxide
$\mu = 0$ D

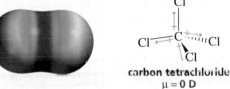

carbon tetrachloride
$\mu = 0$ D

The dipole moment of chloromethane (CH_3Cl) is greater (1.87 D) than the dipole moment of the C—Cl bond (1.5 D) because the C—H dipoles are oriented so that they reinforce the dipole of the C—Cl bond: all the electrons are pulled in the same relative direction. The dipole moment of water (1.85 D) is greater than the dipole moment of a single O—H bond (1.5 D) because the dipoles of the two O—H bonds reinforce each other. The lone-pair electrons also contribute to the dipole moment. Similarly, the dipole moment of ammonia (1.47 D) is greater than the dipole moment of a single N—H bond (1.3 D).

 Using Thin Layer Chromatography

chloromethane
$\mu = 1.87$ D

water
$\mu = 1.85$ D

ammonia
$\mu = 1.47$ D

PROBLEM 34

Account for the difference in the shape and color of the potential maps for ammonia and the ammonium ion in Section 1.12.

PROBLEM 35◆

Which of the following molecules would you expect to have a dipole moment of zero? To answer parts g and h, you may need to consult your answers to Problem 33a and b.

a. CH_3CH_3 **c.** CH_2Cl_2 **e.** $H_2C{=}CH_2$ **g.** $BeCl_2$
b. $H_2C{=}O$ **d.** NH_3 **f.** $H_2C{=}CHBr$ **h.** BF_3

1.16 An Introduction to Acids and Bases

BIOGRAPHY

Born in Denmark, **Johannes Nicolaus Brønsted (1879–1947)** *studied engineering before he switched to chemistry, soon becoming a professor of chemistry at the University of Copenhagen. During World War II, he became known for his anti-Nazi position, and in consequence was elected to the Danish parliament in 1947. He died before he could take his seat.*

BIOGRAPHY

Thomas M. Lowry (1874–1936) *was born in England, the son of an army chaplain. He earned a Ph.D. at Central Technical College, London (now Imperial College). He was head of chemistry at Westminster Training College and, later, at Guy's Hospital in London. In 1920, he became a professor of chemistry at Cambridge University.*

Early chemists called any compound that tasted sour an acid (from *acidus*, Latin for "sour"). Some familiar acids are citric acid (found in lemons and other citrus fruits), acetic acid (found in vinegar), and hydrochloric acid (found in stomach acid—the sour taste associated with vomiting). Compounds that neutralize acids, eliminating their acidic properties, were called bases, or alkaline compounds ("ash" in Arabic is *al kalai*). Glass cleaners and solutions designed to unclog drains are familiar alkaline solutions.

The terms *acid* and *base* can be defined by the Brønsted–Lowry definitions or the Lewis definitions (Section 1.26). In the Brønsted–Lowry definitions, an **acid** is a species that donates a proton, and a **base** is a species that accepts a proton. (Remember that positively charged hydrogen ions are called protons.) In the reaction shown below, hydrogen chloride (HCl) is an acid because it donates a proton to water, and water is a base because it accepts a proton from HCl. Water can accept a proton because it has two lone pairs, either of which can form a covalent bond with a proton. In the reverse reaction, H_3O^+ is an acid because it donates a proton to Cl^-, and Cl^- is a base because it accepts a proton from H_3O^+. The reaction of an acid with a base is called an **acid–base reaction** or a **proton-transfer reaction**. Both an acid and a base must be present in an acid–base reaction, because an acid cannot donate a proton unless a base is present to accept it.

$$\text{H}\ddot{\text{C}}\text{l}: + \text{H}_2\ddot{\text{O}}: \rightleftharpoons :\ddot{\text{C}}\text{l}:^- + \text{H}_3\ddot{\text{O}}^+$$
$$\text{an acid} \quad \text{a base} \quad\quad \text{a base} \quad \text{an acid}$$

Notice that according to the Brønsted–Lowry definitions, *any species that has a hydrogen can potentially act as an acid, and any compound possessing a lone pair can potentially act as a base.*

When a compound loses a proton, the resulting species is called its **conjugate base**. Thus, Cl^- is the conjugate base of HCl, and H_2O is the conjugate base of H_3O^+. When a compound accepts a proton, the resulting species is called its **conjugate acid**. Thus, HCl is the conjugate acid of Cl^-, and H_3O^+ is the conjugate acid of H_2O.

In a reaction between ammonia and water, ammonia (NH_3) is a base because it accepts a proton, and water is an acid because it donates a proton. Thus, HO^- is the conjugate base of H_2O, and ($^+NH_4$) is the conjugate acid of (NH_3). In the reverse reaction, ammonium ion ($^+NH_4$) is an acid because it donates a proton, and hydroxide ion (HO^-) is a base because it accepts a proton.

$$\ddot{\text{N}}\text{H}_3 + \text{H}_2\ddot{\text{O}}: \rightleftharpoons {}^+\text{NH}_4 + \text{H}\ddot{\text{O}}:^-$$
$$\text{a base} \quad \text{an acid} \quad\quad \text{an acid} \quad \text{a base}$$

Notice that water can behave as either an acid or a base. It can behave as an acid because it has a proton that it can donate, but it can also behave as a base because it has a lone pair that can accept a proton. In Section 1.19, we will see how we can predict that water acts as a base in the first reaction and acts as an acid in the second reaction.

Acidity is a measure of the tendency of a compound to give up a proton. **Basicity** is a measure of a compound's affinity for a proton. A strong acid is one that has a strong tendency to give up its proton. This means that its conjugate base must be weak because it has little affinity for the proton. A weak acid has little tendency to give up its proton, indicating that its conjugate base is strong because it has a high affinity for the proton. Thus, the following important relationship exists between an acid and its conjugate base: *the stronger the acid, the weaker is its conjugate base*. For example, since HBr is a stronger acid than HCl, we know that Br$^-$ is a weaker base than Cl$^-$.

The stronger the acid, the weaker is its conjugate base.

PROBLEM 36◆

a. What is the conjugate acid of each of the following:
 1. NH_3 **2.** Cl$^-$ **3.** HO$^-$ **4.** H_2O
b. What is the conjugate base of each of the following:
 1. NH_3 **2.** HBr   **3.** HNO_3 **4.** H_2O

1.17 pK_a and pH

When a strong acid such as hydrogen chloride is dissolved in water, almost all the molecules dissociate (break into ions), which means that the *products* are favored at equilibrium—the equilibrium lies to the right. When a much weaker acid, such as acetic acid, is dissolved in water, very few molecules dissociate, so the *reactants* are favored at equilibrium—the equilibrium lies to the left. Two half-headed arrows are used to designate equilibrium reactions. A longer arrow is drawn toward the species favored at equilibrium.

Student Tutorial:
Introduction to aqueous acids

$$H\ddot{C}l: \ + \ H_2\ddot{O}: \ \rightleftharpoons \ H_3\ddot{O}^+ \ + \ :\ddot{C}l\ddot{:}^-$$
hydrogen
chloride

acetic acid

The degree to which an acid (HA) dissociates is indicated by the **equilibrium constant** of the reaction, K_{eq}. Brackets are used to indicate the concentration in moles/liter, i.e., the molarity (M).

$$HA + H_2O \rightleftharpoons H_3O^+ + A^-$$

$$K_{eq} = \frac{[H_3O^+][A^-]}{[H_2O][HA]}$$

The degree to which an acid (HA) dissociates is normally determined in a dilute solution, so the concentration of water remains nearly constant. The equilibrium expression, therefore, can be rewritten using a new constant called the **acid dissociation constant, (K_a)**.

$$K_a = \frac{[H_3O^+][A^-]}{[HA]} = K_{eq}\,[H_2O]$$

The acid dissociation constant is the equilibrium constant multiplied by the molar concentration of water (55.5 M).

The larger the acid dissociation constant, the stronger is the acid—that is, the greater is its tendency to give up a proton. Hydrogen chloride, with an acid dissociation constant of 10^7, is a stronger acid than acetic acid, with an acid dissociation constant of only 1.74×10^{-5}. For convenience, the strength of an acid is generally indicated by its **pK_a** value rather than its K_a value, where

$$pK_a = -\log K_a$$

The pK_a of hydrogen chloride is −7 and the pK_a of acetic acid, a much weaker acid, is 4.76. Notice that the smaller the pK_a, the stronger the acid.

Student Tutorial:
Important pK_a values

very strong acids	pK_a < 1
moderately strong acids	pK_a = 1–3
weak acids	pK_a = 3–5
very weak acids	pK_a = 5–15
extremely weak acids	pK_a > 15

The stronger the acid, the smaller is its pK_a.

Unless otherwise stated, the pK_a values given in this text indicate the strength of the acid *in water*. Later (in Section 8.10), you will see how the pK_a of an acid is affected when the solvent is changed.

The concentration of positively charged hydrogen ions in a solution is indicated by **pH**. This concentration can be described as either [H$^+$] or, because a hydrogen ion in water is solvated, as [H$_3$O$^+$].

$$pH = -\log [H^+]$$

The lower the pH, the more acidic is the solution. Acidic solutions have pH values less than 7; basic solutions have pH values greater than 7. The pH values of some commonly encountered solutions are shown in the margin on page 47. The pH of a solution can be changed simply by adding acid or base to the solution. Do not confuse pH and pK_a: the pH scale is used to describe the acidity of a *solution*; the pK_a is characteristic of a particular *compound*, much like a melting point or a boiling point—it indicates the

ACID RAIN

Rain is mildly acidic (pH = 5.5) because when the CO$_2$ in the air reacts with water, a weak acid—carbonic acid (pK_a = 6.4)—is formed.

$$CO_2 + H_2O \rightleftharpoons H_2CO_3$$
carbonic acid

In some parts of the world, rain has been found to be much more acidic—with pH values as low as 4.3. Acid rain is formed where sulfur dioxide and nitrogen oxides are produced, because when these gases react with water, strong acids—sulfuric acid

(pK_a = −5.0) and nitric acid (pK_a = −1.3)—are formed. Burning fossil fuels for the generation of electric power is the factor most responsible for forming these acid-producing gases.

Acid rain has many deleterious effects. It can destroy aquatic life in lakes and streams; it can make soil so acidic that crops cannot grow; and it can cause the deterioration of paint and building materials, including monuments and statues that are part of our cultural heritage. Marble—a form of calcium carbonate—decays because acid reacts with CO$_3^{2-}$ to form carbonic acid, which decomposes to CO$_2$ and H$_2$O, the reverse of the reaction shown above.

$$CO_3^{2-} \overset{H^+}{\rightleftharpoons} HCO_3^- \overset{H^+}{\rightleftharpoons} H_2CO_3 \rightleftharpoons CO_2 + H_2O$$

photo taken in 1935

photo taken in 1994

Statue of George Washington in Washington Square Park, in Greenwich Village, New York.

tendency of the compound to give up its proton. The importance of organic acids and bases will become clear when we discuss how and why organic compounds react.

PROBLEM 37◆

a. Which is a stronger acid, one with a pK_a of 5.2 or one with a pK_a of 5.8?

b. Which is a stronger acid, one with an acid dissociation constant of 3.4×10^{-3} or one with an acid dissociation constant of 2.1×10^{-4}?

PROBLEM 38◆

An acid has a K_a of 4.53×10^{-6} in water. What is its K_{eq}? ($[H_2O] = 55.5$ M)

PROBLEM-SOLVING STRATEGY

Determining K_a from pK_a

Vitamin C has a pK_a value of 4.17. What is its K_a value?

You will need a calculator to answer this question. Remember that $pK_a = -\log K_a$.

1. Enter the pK_a value on your calculator.
2. Multiply it by -1.
3. Determine the inverse log by pressing the key labeled 10^x.

You should find that vitamin C has a K_a value of 6.76×10^{-5}.

Now continue on to Problem 39.

PROBLEM 39◆

Butyric acid, the compound responsible for the unpleasant odor and taste of sour milk, has a pK_a value of 4.82. What is its K_a value? Is it a stronger or a weaker acid than vitamin C?

PROBLEM 40

Antacids are compounds that neutralize stomach acid. Write the equations that show how Milk of Magnesia, Alka-Seltzer, and Tums remove excess acid.

a. Milk of Magnesia: $Mg(OH)_2$
b. Alka-Seltzer: $KHCO_3$ and $NaHCO_3$
c. Tums: $CaCO_3$

PROBLEM 41◆

Are the following body fluids acidic or basic?

a. bile (pH = 8.4)   **b.** urine (pH = 5.9)   **c.** spinal fluid (pH = 7.4)

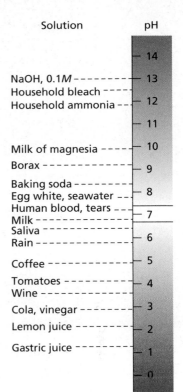

Solution	pH
	— 14
NaOH, 0.1M --------	— 13
Household bleach ----	
Household ammonia --	— 12
	— 11
Milk of magnesia -----	— 10
Borax ------------	— 9
Baking soda --------	
Egg white, seawater --	— 8
Human blood, tears ---	— 7
Milk --------------	
Saliva -------------	— 6
Rain --------------	
Coffee ------------	— 5
Tomatoes ---------	— 4
Wine -------------	
Cola, vinegar -------	— 3
Lemon juice --------	— 2
Gastric juice -------	— 1
	— 0

1.18 Organic Acids and Bases

The most common organic acids are carboxylic acids—compounds that have a COOH group. Acetic acid and formic acid are examples of carboxylic acids. Carboxylic acids have pK_a values ranging from about 3 to 5. (They are weak acids.) The pK_a values of a wide variety of organic compounds are given in Appendix II.

$$\underset{\substack{\text{acetic acid} \\ pK_a = \textbf{4.76}}}{H_3C-\overset{\displaystyle O}{\overset{\|}{C}}-OH} \qquad \underset{\substack{\text{formic acid} \\ pK_a = \textbf{3.75}}}{H-\overset{\displaystyle O}{\overset{\|}{C}}-OH}$$

Alcohols—compounds that have an OH group—are much weaker acids than carboxylic acids, with pK_a values close to 16. Methyl alcohol and ethyl alcohol are examples of alcohols.

$$CH_3OH \qquad CH_3CH_2OH$$

methyl alcohol   ethyl alcohol

pK_a = 15.5   pK_a = 15.9

We have seen that water can behave both as an acid and as a base. An alcohol behaves similarly: it can behave as an acid and donate a proton, or as a base and accept a proton.

a curved arrow indicates where the electrons start from and where they end up

$$CH_3\ddot{O}-H \;+\; H-\ddot{O}:^- \;\rightleftharpoons\; CH_3\ddot{O}:^- \;+\; H-\ddot{O}-H$$

an acid

$$CH_3\ddot{O}-H \;+\; H-\overset{+}{\ddot{O}}-H \;\rightleftharpoons\; CH_3\overset{+}{\ddot{O}}-H \;+\; H-\ddot{O}-H$$
$$\qquad\qquad\qquad\quad\; H \qquad\qquad\quad H$$

a base

Chemists frequently use curved arrows to indicate the bonds that are broken and formed as reactants are converted into products. The arrows show where the electrons start from and where they end up. In an acid–base reaction, they are drawn from a lone pair on the base (the tail of the arrow) to the proton of the acid (the point of the arrow). These are called "curved" arrows to distinguish them from the "straight" arrows used to link reactants with products in the equation for a chemical reaction.

A carboxylic acid also can behave as an acid and donate a proton, or as a base and accept a proton.

$$H_3C-\overset{\ddot{O}:}{\underset{\ddot{O}-H}{C}} \;+\; H-\ddot{O}:^- \;\rightleftharpoons\; H_3C-\overset{\ddot{O}:}{\underset{\ddot{O}:^-}{C}} \;+\; H-\ddot{O}-H$$

an acid

$$H_3C-\overset{\ddot{O}:}{\underset{\ddot{O}-H}{C}} \;+\; H-\overset{+}{\ddot{O}}-H \;\rightleftharpoons\; H_3C-\overset{\overset{+}{\ddot{O}}-H}{\underset{O-H}{C}} \;+\; H-\ddot{O}-H$$
$$\qquad\qquad\qquad\qquad\qquad H$$

a base

Notice that it is the sp^2 oxygen of the carboxylic acid that is protonated (acquires the proton). We will see why this is so in Section 16.11.

A *protonated* compound is a compound that has gained an additional proton. Protonated alcohols and protonated carboxylic acids are very strong acids. For example, protonated methyl alcohol has a pK_a of -2.5, protonated ethyl alcohol has a pK_a of -2.4, and protonated acetic acid has a pK_a of -6.1.

$$CH_3\overset{+}{\underset{H}{O}H} \qquad\qquad CH_3CH_2\overset{+}{\underset{H}{O}H} \qquad\qquad H_3C-\overset{\overset{+}{O}H}{\underset{OH}{C}}$$

protonated methanol   protonated ethanol   protonated acetic acid

pK_a = -2.5 pK_a = -2.4 pK_a = -6.1

A compound with an NH_2 group is an amine. An amine can behave as an acid and donate a proton, or it can behave as a base and accept a proton.

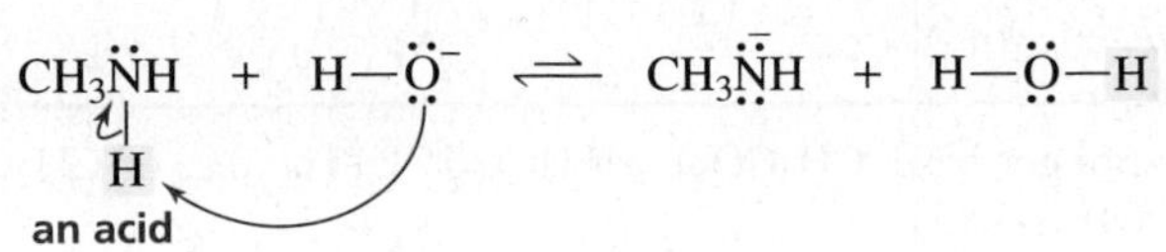

$$CH_3\overset{..}{N}H + H-\overset{..}{\underset{..}{O}}{}^- \rightleftharpoons CH_3\overset{..}{N}H + H-\overset{..}{\underset{..}{O}}-H$$

$$\underset{\overset{|}{H}}{}$$

an acid

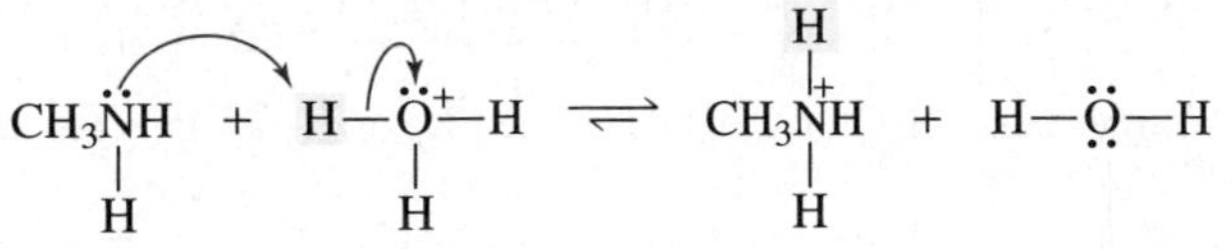

$$CH_3\overset{..}{N}H + H-\overset{+}{\underset{|}{O}}-H \rightleftharpoons CH_3\overset{\overset{\overset{H}{|}}{+}}{N}H + H-\overset{..}{\underset{..}{O}}-H$$

a base

Amines, however, have such high pK_a values that they rarely behave as acids. Ammonia also has a high pK_a value.

CH_3NH_2	NH_3
methylamine	ammonia
$pK_a = 40$	$pK_a = 36$

Amines are much more likely to act as bases. In fact, amines are the most common organic bases. Instead of talking about the strength of a base in terms of a pK_b value, it is easier to talk about the strength of its conjugate acid as indicated by its pK_a value, remembering that the stronger the acid, the weaker is its conjugate base. For example, from their pK_a values we see that protonated methylamine is a stronger acid than protonated ethylamine, which means that methylamine is a weaker base than ethylamine. Notice that the pK_a values of protonated amines are in the range of 10 to 11.

$CH_3\overset{+}{N}H_3$	$CH_3CH_2\overset{+}{N}H_3$
protonated methylamine	protonated ethylamine
$pK_a = 10.7$	$pK_a = 11.0$

It is important to know the approximate pK_a values of the various classes of compounds we have discussed. An easy way to remember them is in units of five, as shown in Table 1.8. (R is used when the particular carboxylic acid, alcohol, or amine is not specified.) Protonated alcohols, protonated carboxylic acids, and protonated water have pK_a values less than 0, carboxylic acids have pK_a values of about 5, protonated amines have pK_a values of about 10, and alcohols and water have pK_a values of about 15. These values are also listed inside the back cover of this book for easy reference.

Be sure to learn the approximate pK_a values given in Table 1.8.

Table 1.8 Approximate pK_a Values			
$pK_a < 0$	$pK_a \sim 5$	$pK_a \sim 10$	$pK_a \sim 15$
$R\overset{+}{O}H_2$ a protonated alcohol	$\underset{R}{\overset{O}{\overset{\|}{\underset{}{C}}}}{}^{OH}$ a carboxylic acid	$R\overset{+}{N}H_3$ a protonated amine	ROH an alcohol
$\underset{R}{\overset{+OH}{\overset{\|}{\underset{}{C}}}}{}^{OH}$ a protonated carboxylic acid			H_2O water
H_3O^+ protonated water			

PROBLEM 42◆

a. Which is a stronger base, CH_3COO^- or $HCOO^-$? (The pK_a of CH_3COOH is 4.8; the pK_a of $HCOOH$ is 3.8.)

b. Which is a stronger base, HO^- or $^-NH_2$? (The pK_a of H_2O is 15.7; the pK_a of NH_3 is 36.)

c. Which is a stronger base, H_2O or CH_3OH? (The pK_a of H_3O^+ is -1.7; the pK_a of $CH_3\overset{+}{O}H_2$ is -2.5.)

PROBLEM 43◆

Using the pK_a values in Section 1.18, rank the following species in order of decreasing base strength (that is, list the strongest base first):

$$CH_3NH_2 \quad CH_3NH^- \quad CH_3OH \quad CH_3O^- \quad CH_3\overset{\displaystyle O}{\overset{\|}{C}}O^-$$

1.19 How to Predict the Outcome of an Acid–Base Reaction

Now let's see how we can predict that water will act as a base in the first reaction in Section 1.16 and as an acid in the second reaction. To determine which of the two reactants of the first reaction will be the acid, we need to compare their pK_a values: the pK_a of hydrogen chloride is -7 and the pK_a of water is 15.7. Because hydrogen chloride is the stronger acid, it will donate a proton to water. Water, therefore, is a base in this reaction. When we compare the pK_a values of the two reactants of the second reaction, we see that the pK_a of ammonia is 36 and the pK_a of water is 15.7. In this case, water is the stronger acid, so it donates a proton to ammonia. Water, therefore, is an acid in this reaction.

To determine the position of equilibrium for an acid–base reaction (that is, whether reactants or products are favored at equilibrium), we need to compare the pK_a value of the acid on the left of the arrow with the pK_a value of the acid on the right of the arrow. The equilibrium favors *reaction* of the stronger acid and *formation* of the weaker acid. In other words, *strong reacts to form weak*. Thus, the equilibrium lies away from the stronger acid and toward the weaker acid. Notice that the stronger acid has the weaker conjugate base.

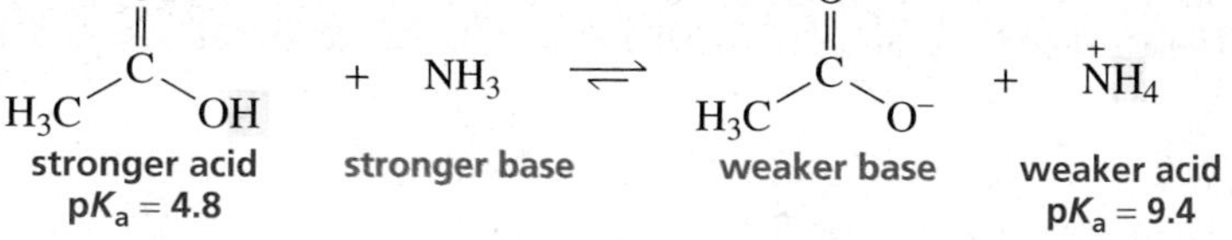

Strong reacts to form weak.

PROBLEM 44

a. Write an equation showing CH_3OH reacting as an acid with NH_3 and an equation showing it reacting as a base with HCl.

b. Write an equation showing NH_3 reacting as an acid with CH_3O^- and an equation showing it reacting as a base with HBr.

PROBLEM 45

a. For each of the acid–base reactions in Section 1.18, compare the pK_a values of the acids on either side of the equilibrium arrows to prove that the equilibrium lies in the direction indicated. (The pK_a values you need can be found in Section 1.18 or in Problem 42.)

b. Do the same for the equilibria in Section 1.16. (The pK_a of $^+NH_4$ is 9.4.)

PROBLEM 46

Ethyne has a pK_a value of 25, water has a pK_a value of 15.7, and ammonia (NH_3) has a pK_a value of 36. Draw the equation, showing equilibrium arrows that indicate whether reactants or products are favored, for the acid–base reaction of ethyne with:

a. HO^- **b.** $^-NH_2$

c. Which would be a better base to use if you wanted to remove a proton from ethyne, HO^- or $^-NH_2$?

The precise value of the equilibrium constant can be calculated by dividing the K_a of the reactant acid by the K_a of the product acid.

$$K_{eq} = \frac{K_a \text{ reactant acid}}{K_a \text{ product acid}}$$

Thus, the equilibrium constant for the reaction of acetic acid with ammonia is 4.0×10^4, and the equilibrium constant for the reaction of ethyl alcohol with methylamine is 6.3×10^{-6}. The calculations are

reaction of acetic acid with ammonia:

$$K_{eq} = \frac{10^{-4.8}}{10^{-9.4}} = 10^{4.6} = 4.0 \times 10^4$$

reaction of ethyl alcohol with methylamine:

$$K_{eq} = \frac{10^{-15.9}}{10^{-10.7}} = 10^{-5.2} = 6.3 \times 10^{-6}$$

Student Tutorial:
Acid base reaction

PROBLEM 47◆

Calculate the equilibrium constant for the acid–base reaction between the reactants in each of the following pairs:

a. $HCl + H_2O$ **c.** $CH_3NH_2 + H_2O$

b. $CH_3COOH + H_2O$ **d.** $CH_3\overset{+}{N}H_3 + H_2O$

1.20 How the Structure of an Acid Affects Its Acidity

The strength of an acid is determined by the stability of the conjugate base formed when the acid gives up its proton: the more stable the base, the stronger is its conjugate acid. (The reason for this will be explained in Section 3.7.) A stable base is a base that readily bears the electrons it formerly shared with a proton. In other words, stable bases are weak bases—they do not share their electrons well—allowing us to say, *the weaker the base, the stronger is its conjugate acid*, or *the more stable the base, the stronger is its conjugate acid*.

The weaker the base, the stronger is its conjugate acid.

Stable bases are weak bases.

The more stable the base, the stronger is its conjugate acid.

Two factors that affect the stability of a base are its *size* and its *electronegativity*. The elements in the second row of the periodic table are all similar in size, but they have very different electronegativities, which increase across the row from left to right. Of the atoms shown, carbon is the least electronegative and fluorine is the most electronegative.

relative electronegativities: $\quad C < N < O < F$

most electronegative

If we look at the acids formed by attaching hydrogens to these elements, we see that the most acidic compound is the one that has its hydrogen attached to the most electronegative atom. Thus, HF is the strongest acid and methane is the weakest acid.

relative acidities: $\quad CH_4 < NH_3 < H_2O < HF$

strongest acid

If we look at the stabilities of the conjugate bases of these acids, we find that they too increase from left to right because an atom that is more electronegative is better able to bear its negative charge. Thus, we see that the strongest acid has the most stable conjugate base.

relative stabilities: $\quad {}^-CH_3 < {}^-NH_2 < HO^- < F^-$

most stable

When atoms are similar in size, the strongest acid will have its hydrogen attached to the most electronegative atom.

We therefore can conclude that *when the atoms are similar in size, the strongest acid will have its hydrogen attached to the most electronegative atom.*

The effect that the electronegativity of the atom bonded to a hydrogen has on the acidity of that hydrogen can be appreciated when the pK_a values of alcohols and amines are compared. Because oxygen is more electronegative than nitrogen, an alcohol is more acidic than an amine.

$$CH_3OH \qquad\qquad CH_3NH_2$$
methyl alcohol $\qquad$ methylamine
$pK_a = 15.5$ $\qquad\qquad$ $pK_a = 40$

Similarly, a protonated alcohol is more acidic than a protonated amine.

$$CH_3\overset{+}{O}H_2 \qquad\qquad CH_3\overset{+}{N}H_3$$
protonated methyl alcohol $\quad$ protonated methylamine
$pK_a = -2.5$ $\qquad\qquad$ $pK_a = 10.7$

The hybridization of an atom affects the acidity of a hydrogen bonded to the atom because the electronegativity of an atom depends on its hybridization: an sp hybridized atom is more electronegative than the same atom that is sp^2 hybridized, which is more electronegative than the same atom that is sp^3 hybridized.

relative electronegativities of carbon atoms

most electronegative $\quad sp > sp^2 > sp^3$ $\quad$ least electronegative

An sp hybridized carbon is more electronegative than an sp^2 hybridized carbon, which is more electronegative than an sp^3 hybridized carbon.

Because the electronegativity of carbon atoms follows the order $sp > sp^2 > sp^3$, ethyne is a stronger acid than ethene, and ethene is a stronger acid than ethane—the most acidic compound is the one with the hydrogen attached to the most electronegative atom.

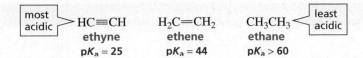

Why does the hybridization of the atom affect its electronegativity? Electronegativity is a measure of the ability of an atom to pull the bonding electrons toward itself. Thus, the most electronegative atom will be the one with its bonding electrons closest to the nucleus. The average distance of a $2s$ electron from the nucleus is less than the average distance of a $2p$ electron from the nucleus. Therefore, an sp hybridized atom with 50% s character is the most electronegative, an sp^2 hybridized atom with 33.3% s character is next, and an sp^3 hybridized atom with 25% s character is the least electronegative.

In comparing atoms that are very different in size, the *size* of the atom is more important than its *electronegativity* in determining how well it bears its negative charge. For example, as we proceed down a column in the periodic table, the elements get larger and their electronegativity *decreases*. But the stability of the bases increase down the column, so the strength of the conjugate acids also *increase*. Thus, HI is the strongest acid of the hydrogen halides, even though iodine is the least electronegative of the halogens. Therefore, *when atoms are very different in size, the strongest acid will have its hydrogen attached to the largest atom.*

Size overrides electronegativity.

When atoms are very different in size, the strongest acid will have its hydrogen attached to the largest atom.

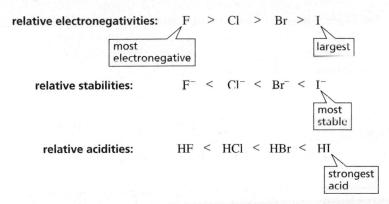

Why does the size of an atom have such a significant effect on the stability of the base that it more than overcomes any difference in electronegativity? The valence electrons of F^- are in a $2sp^3$ orbital, the valence electrons of Cl^- are in a $3sp^3$ orbital, those of Br^- are in a $4sp^3$ orbital, and those of I^- are in a $5sp^3$ orbital. The volume of space occupied by a $3sp^3$ orbital is significantly greater than the volume of space occupied by a $2sp^3$ orbital because a $3sp^3$ orbital extends out farther from the nucleus. Because its negative charge is spread over a larger volume of space, Cl^- is more stable than F^-.

Thus, as the halide ion increases in size, its stability increases because its negative charge is spread over a larger volume of space (its electron density decreases). Therefore, HI is the strongest acid of the hydrogen halides because I^- is the most stable halide ion, even though iodine is the least electronegative of the halogens (Table 1.9). The potential maps illustrate the large difference in size of the hydrogen halides:

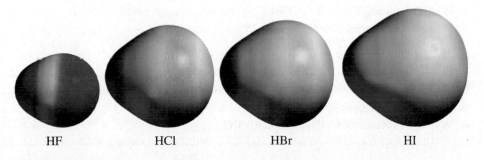

HF   HCl   HBr   HI

Table 1.9 The pK_a Values of Some Simple Acids			
CH_4 $pK_a = 60$	NH_3 $pK_a = 36$	H_2O $pK_a = 15.7$	HF $pK_a = 3.2$
		H_2S $pK_a = 7.0$	HCl $pK_a = -7$
			HBr $pK_a = -9$
			HI $pK_a = -10$

In summary, as we move across a row of the periodic table, the atoms' orbitals have approximately the same volume, so it is the electronegativity of the proton-accepting element that determines the stability of the base and, therefore, the acidity of a proton bonded to that base. As we move down a column of the periodic table, the volume of the orbitals increases. The increase in volume causes the electron density of the orbital to decrease. The electron density of the orbitals is more important than electronegativity in determining the stability of a base and, therefore, the acidity of its conjugate acid. That is, *the lower the electron density of an element, the more stable it is as a base and the stronger is its conjugate acid.*

PROBLEM 48◆

For each of the following pairs, indicate which is the stronger acid:

a. HCl or HBr

b. $CH_3CH_2CH_2\overset{+}{N}H_3$ or $CH_3CH_2CH_2\overset{+}{O}H_2$

c. 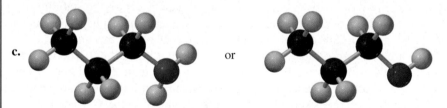 or

black = C, white = H, blue = N, red = O

d. 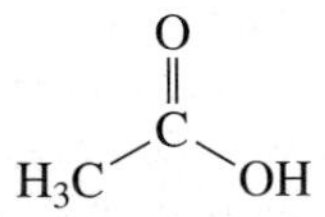 H₃C—C(=O)—OH or H₃C—C(=O)—SH

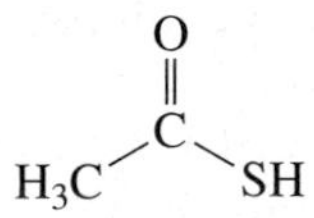

PROBLEM 49◆

a. Which of the halide ions (F^-, Cl^-, Br^-, I^-) is the strongest base?

b. Which is the weakest base?

PROBLEM 50◆

a. Which is more electronegative, oxygen or sulfur?

b. Which is a stronger acid, H_2O or H_2S?

c. Which is a stronger acid, CH_3OH or CH_3SH?

PROBLEM 51◆

For each of the following pairs, indicate which is the stronger base:

$$\overset{\displaystyle O}{\overset{\|}{}}$$

a. H_2O or HO^- **b.** H_2O or NH_3 **c.** $CH_3\overset{O}{\overset{\|}{C}}O^-$ or CH_3O^- **d.** CH_3O^- or CH_3S^-

---

1.21 How Substituents Affect the Strength of an Acid

Although the acidic proton of each of the following four carboxylic acids is attached to the same atom (an oxygen atom), the four compounds have different acidities:

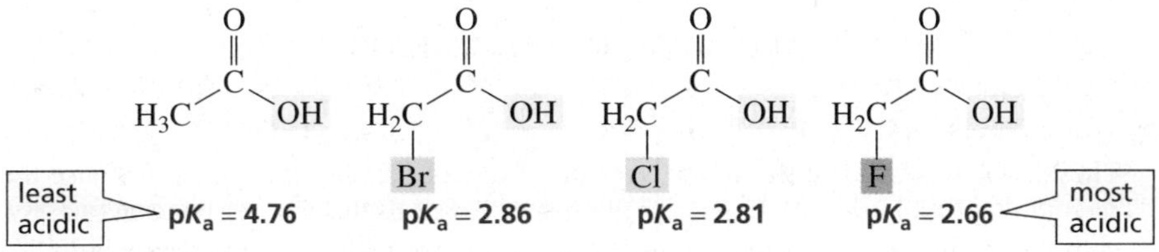

least acidic $pK_a = 4.76$ $pK_a = 2.86$ $pK_a = 2.81$ $pK_a = 2.66$ most acidic

This difference indicates that in addition to the nature of the atom to which the hydrogen is bonded, there must be another factor that affects acidity.

From the pK_a values of the four carboxylic acids, we see that replacing one of the hydrogen atoms of the CH_3 group with a halogen atom affects the acidity of the compound. (The term for replacing an atom in a compound is *substitution*, and the new atom is called a *substituent*.) The reason is that the halogens are more electronegative than hydrogen. An electronegative halogen atom pulls the bonding electrons toward itself. Pulling electrons through sigma (σ) bonds is called **inductive electron withdrawal**. If we look at the conjugate base of a carboxylic acid, we see that inductive electron withdrawal will stabilize it by *decreasing the electron density* about the oxygen atom. Stabilizing a base increases the acidity of its conjugate acid.

> **Inductive electron withdrawal increases the strength of an acid.**

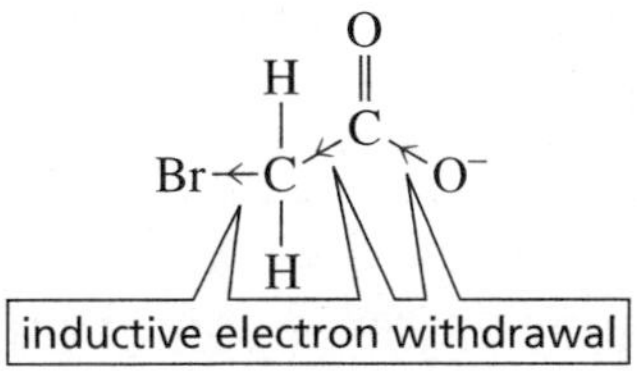

inductive electron withdrawal

As the pK_a values of the four carboxylic acids show, inductive electron withdrawal increases the acidity of a compound. The greater the electron-withdrawing ability (electronegativity) of the halogen substituent, the more the acidity is increased because the more its conjugate base is stabilized.

The effect of a substituent on the acidity of a compound decreases as the distance between the substituent and the acidic proton increases.

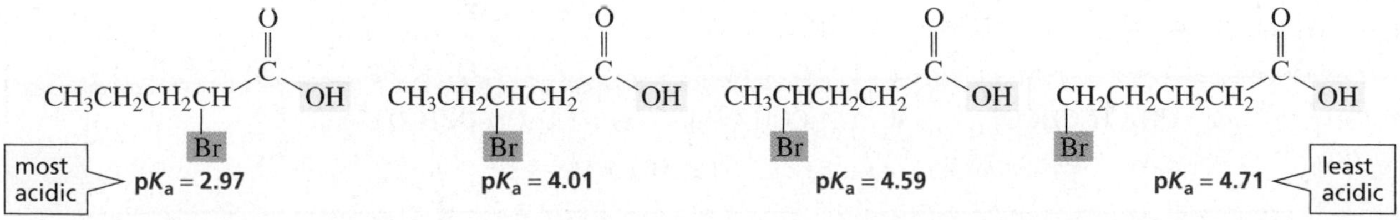

most acidic $pK_a = 2.97$ $pK_a = 4.01$ $pK_a = 4.59$ $pK_a = 4.71$ least acidic

---

PROBLEM-SOLVING STRATEGY

Determining Relative Acid Strength from Structure

a. Which is a stronger acid?

$$CH_3CHCH_2OH \quad or \quad CH_3CHCH_2OH$$
$$\quad\;\; | \qquad\qquad\qquad\quad | $$
$$\quad\;\; F \qquad\qquad\qquad\quad Br$$

When you are asked to compare two items, pay attention to how they differ; ignore where they are the same. These two compounds differ only in the halogen atom attached to the middle carbon of the molecule. Because fluorine is more electronegative than bromine, there is greater electron withdrawal from the oxygen atom in the fluorinated compound. The fluorinated compound, therefore, will have the more stable conjugate base, so it will be the stronger acid.

b. Which is a stronger acid?

$$\qquad\quad Cl \qquad\qquad\quad Cl$$
$$\qquad\quad | \qquad\qquad\qquad | $$
$$CH_3CCH_2OH \quad or \quad CH_2CHCH_2OH$$
$$\qquad\quad | \qquad\qquad\qquad | $$
$$\qquad\quad Cl \qquad\qquad\qquad Cl$$

These two compounds differ in the location of one of the chlorine atoms. Because the second chlorine in the compound on the left is closer to the O—H bond than is the chlorine in the compound on the right, the former is more effective at withdrawing electrons from the oxygen atom. Thus, the compound on the left will have the more stable conjugate base, so it will be the stronger acid.

Now continue on to Problem 52.

PROBLEM 52◆

For each of the following pairs, indicate which is the stronger acid:

a. $CH_3OCH_2CH_2OH$ or $CH_3CH_2CH_2CH_2OH$

b. $CH_3CH_2CH_2\overset{+}{N}H_3$ or $CH_3CH_2CH_2\overset{+}{O}H_2$

c. $CH_3OCH_2CH_2CH_2OH$ or $CH_3CH_2OCH_2CH_2OH$

d.
$$\qquad O \qquad\qquad\qquad\quad O$$
$$\qquad \| \qquad\qquad\qquad\quad \|$$
$$CH_3CCH_2OH \quad or \quad CH_3CH_2COH$$

PROBLEM 53◆

List the following compounds in order of decreasing acidity:

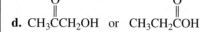

$$CH_3CHCH_2OH \qquad CH_3CH_2CH_2OH \qquad CH_2CH_2CH_2OH \qquad CH_3CHCH_2OH$$
$$\quad\;\; | \qquad\qquad\qquad\qquad\qquad\qquad\qquad\qquad\qquad | \qquad\qquad\qquad\qquad | $$
$$\quad\;\; F \qquad\qquad\qquad\qquad\qquad\qquad\qquad\qquad\qquad\; Cl \qquad\qquad\qquad\quad Cl$$

PROBLEM 54◆

For each of the following pairs, indicate which is the stronger base:

a.
$$\qquad\quad O \qquad\qquad\qquad O$$
$$\qquad\quad \| \qquad\qquad\qquad \|$$
$$CH_3CHCO^- \quad or \quad CH_3CHCO^-$$
$$\qquad | \qquad\qquad\qquad\quad | $$
$$\qquad Br \qquad\qquad\qquad\quad F$$

c.
$$\qquad\qquad O \qquad\qquad\qquad\quad O$$
$$\qquad\qquad \| \qquad\qquad\qquad\quad \|$$
$$BrCH_2CH_2CO^- \quad or \quad CH_3CH_2CO^-$$

b.
$$\qquad\qquad\quad O \qquad\qquad\qquad\quad O$$
$$\qquad\qquad\quad \| \qquad\qquad\qquad\quad \|$$
$$CH_3CHCH_2CO^- \quad or \quad CH_3CH_2CHCO^-$$
$$\qquad\quad | \qquad\qquad\qquad\qquad\quad | $$
$$\qquad\quad Cl \qquad\qquad\qquad\qquad\quad Cl$$

d.
$$\qquad\quad O \qquad\qquad\qquad\qquad O$$
$$\qquad\quad \| \qquad\qquad\qquad\qquad \|$$
$$CH_3CCH_2CH_2O^- \quad or \quad CH_3CH_2CCH_2O^-$$

PROBLEM 55 **SOLVED**

If HCl is a weaker acid than HBr, why is ClCH$_2$COOH a stronger acid than BrCH$_2$COOH?

Solution   To compare the acidities of HCl and HBr, we need to compare the stabilities of their conjugate bases, Cl$^-$ and Br$^-$. Because we know that size is more important than electronegativity in determining stability, we know that Br$^-$ is more stable than Cl$^-$. Therefore, HBr is a stronger acid than HCl. In comparing the acidities of the two carboxylic acids, we need to compare the stabilities of their conjugate bases, RCOO$^-$ and R$'$COO$^-$. (Notice that an O—H bond is broken in both compounds.) The only way the conjugate bases differ is in the electronegativity of the atom that is pulling electrons away from the negatively charged oxygen. Because Cl is more electronegative than Br, Cl exerts greater inductive electron withdrawal. Thus, it has a greater stabilizing effect on the base that is formed when the proton leaves, so the chloro-substituted compound is the stronger acid.

1.22 An Introduction to Delocalized Electrons

We have seen that a carboxylic acid has a pK_a of about 5, whereas the pK_a of an alcohol is about 15. Because a carboxylic acid is a much stronger acid than an alcohol, we know that a carboxylic acid has a considerably more stable conjugate base.

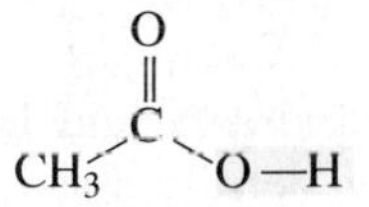

$$CH_3 \overset{\overset{\displaystyle O}{\|}}{C} O—H \qquad\qquad CH_3CH_2O—H$$

pK_a – 4.76   pK_a = 15.9

There are two factors causing the conjugate base of a carboxylic acid to be more stable than the conjugate base of an alcohol. First, the conjugate base of a carboxylic acid has a doubly bonded oxygen in place of two hydrogens in the conjugate base of an alcohol. Inductive electron withdrawal by this electronegative oxygen decreases the electron density of the negatively charged oxygen. Second, the electron density is further decreased by *electron delocalization.*

When an alcohol loses a proton, the negative charge resides on its single oxygen atom: electrons that belong to only one atom are called *localized* electrons. In contrast, when a carboxylic acid loses a proton, the negative charge is shared by both oxygen atoms because the electrons are *delocalized.* **Delocalized electrons** are shared by more than two atoms.

Delocalized electrons are shared by more than two atoms.

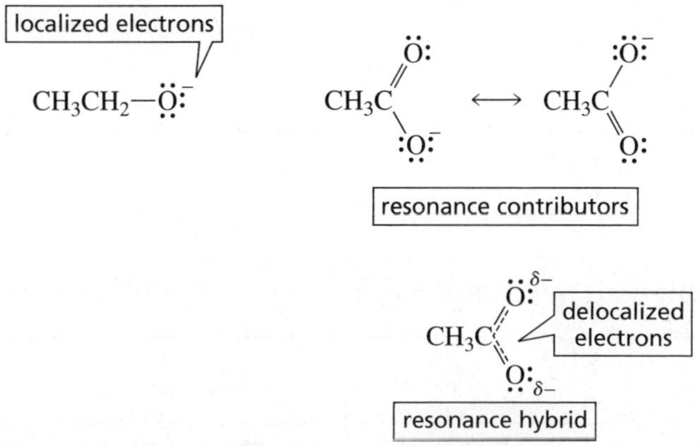

The two structures shown for the conjugate base of the carboxylic acid are called **resonance contributors**. Neither resonance contributor alone represents the actual structure of the conjugate base. Instead, the actual structure—called a **resonance hybrid**—is a composite of the two resonance contributors. The double-headed arrow between the two resonance contributors is used to indicate that the actual structure is a hybrid. Notice that the two resonance contributors differ only in the location of their π electrons and lone-pair electrons—all the atoms stay in the same place. In the resonance hybrid, an electron pair is spread over two oxygens and a carbon. The negative charge is shared equally by the two oxygens, and both carbon–oxygen bonds are the same length—they are not as long as a single bond, but they are longer than a double bond. A resonance hybrid can be drawn by using dotted lines to show the delocalized electrons.

The following potential maps show that there is less electron density on the oxygen atoms in the carboxylate ion (orange region) than on the oxygen atom of the alkoxide ion (red region):

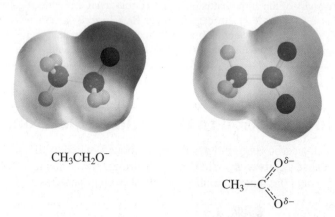

$CH_3CH_2O^-$

$$CH_3-C\underset{O^{\delta-}}{\overset{O^{\delta-}}{\lessgtr}}$$

Thus, the combination of inductive electron withdrawal and the ability of two atoms to share the negative charge decrease the electron density, making the conjugate base of the carboxylic acid more stable than the conjugate base of the alcohol.

We will discuss delocalized electrons in greater detail in Chapter 7. By that time, you will be thoroughly comfortable with compounds that have only localized electrons, and we can then further explore how delocalized electrons affect the stability and reactivity of organic compounds.

PROBLEM 56◆

Which compound would you expect to be a stronger acid? Why?

$$CH_3\overset{O}{\overset{\|}{C}}-O-H \quad or \quad CH_3\overset{O}{\underset{O}{\overset{\|}{S}}}-O-H$$

PROBLEM 57◆

Draw resonance contributors for the following compounds:

a.

$$\overset{\overset{..}{O}:}{\underset{:\overset{..}{O}:^- \quad \overset{..}{O}:^-}{\overset{\|}{C}}}$$

b.

$$\overset{\overset{..}{O}:}{\underset{:\overset{..}{O}: \quad \overset{..}{O}:^-}{\overset{\|}{\underset{+}{N}}}}$$

1.23 A Summary of the Factors That Determine Acid Strength

We have seen that the strength of an acid depends on five factors: the *size* of the atom to which the hydrogen is attached, the *electronegativity* of the atom to which the hydrogen is attached, the *hybridization* of the atom to which the hydrogen is attached, *inductive effects*, and *electron delocalization*. All five factors affect acidity by stabilizing the conjugate base.

1. **Size:** As the atom attached to the hydrogen increases in size (going down a column of the Periodic Table), the strength of the acid increases.

2. **Electronegativity:** As the atom attached to the hydrogen increases in electronegativity (going across a row of the Periodic Table), the strength of the acid increases.

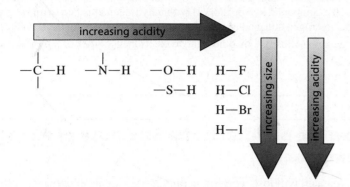

3. **Hybridization:** The relative electronegativities of an atom are: $sp > sp^2 > sp^3$. Because hybridization affects the electronegativity of an atom, a hydrogen attached to an sp carbon is the most acidic and a hydrogen attached to an sp^3 carbon is the least acidic.

most acidic	HC≡CH	>	$H_2C=CH_2$	>	CH_3CH_3	least acidic
	sp		sp^2		sp^3	

4. **Inductive effect:** An electron-withdrawing group increases acid strength: the more electronegative the electron-withdrawing group and the closer it is to the acidic hydrogen, the stronger is the acid.

most acidic ⟩ CH_3CHCH_2OH > CH_3CHCH_2OH > CH_3CHCH_2OH > $CH_3CH_2CH_2OH$ ⟨ least acidic
          F   Cl   Br

most acidic ⟩ CH_3CHCH_2OH > $CH_2CH_2CH_2OH$ > $CH_3CH_2CH_2OH$ ⟨ least acidic
          F   F

5. **Electron delocalization:** An acid whose conjugate base has delocalized electrons is more acidic than a similar acid with a base in which all the electrons are localized.

more acidic ⟩ $R-\overset{O}{\overset{\|}{C}}-OH$ RCH_2OH ⟨ less acidic

more stable ⟩ $R-\overset{O^{\delta-}}{\overset{\|}{C}}=O^{\delta-}$ RCH_2O^- ⟨ less stable

PROBLEM 58◆

Using the table of pK_a values given in Appendix II, answer the following:

a. Which is the most acidic organic compound in the table?

b. Which is the least acidic organic compound in the table?

c. Which is the most acidic carboxylic acid in the table?

d. Which is more electronegative, an sp^3 hybridized oxygen or an sp^2 hybridized oxygen? (*Hint:* Pick a compound in Appendix II with a hydrogen attached to an sp^2 oxygen and one with a hydrogen attached to an sp^3 oxygen, and compare their pK_a values.)

e. What compounds illustrate that the relative electronegativities of a hybridized nitrogen atom are $sp > sp^2 > sp^3$?

f. Which is more acidic, HNO_3 or HNO_2? Why?

1.24 How the pH Affects the Structure of an Organic Compound

Whether a given acid will lose a proton in an aqueous solution depends both on the pK_a of the acid and the pH of the solution. The relationship between the two is given by the **Henderson–Hasselbalch equation**. This is an extremely useful equation because it tells us whether a compound will exist in its acidic form (with its proton retained) or in its basic form (with its proton removed) at a particular pH.

the Henderson–Hasselbalch equation

$$pK_a = pH + \log \frac{[HA]}{[A^-]}$$

The Henderson–Hasselbalch equation tells us that when the pH of a solution equals the pK_a of the compound that undergoes dissociation, the concentration of the compound in its acidic form [HA] will equal the concentration of the compound in its basic form [A⁻] (because log 1 = 0). If the pH of the solution is less than the pK_a of the compound, the compound will exist primarily in its acidic form. If the pH of the solution is greater than the pK_a of the compound, the compound will exist primarily in its basic form. In other words, *compounds exist primarily in their acidic forms in solutions that are more acidic than their pK_a values and primarily in their basic forms in solutions that are more basic than their pK_a values.*

If we know the pH of the solution and the pK_a of the compound, the Henderson–Hasselbalch equation allows us to calculate precisely how much of the compound will be in its acidic form, and how much will be in its basic form. For example, when a compound

A compound will exist primarily in its acidic form if the pH of the solution is less than the compound's pK_a.

A compound will exist primarily in its basic form if the pH of the solution is greater than the compound's pK_a.

DERIVATION OF THE HENDERSON–HASSELBALCH EQUATION

The Henderson–Hasselbalch equation can be derived from the expression that defines the acid dissociation constant:

$$K_a = \frac{[H_3O^+][A^-]}{[HA]}$$

Taking the logarithms of both sides of the equation and, in the next step, multiplying both sides of the equation by -1, we obtain

$$\log K_a = \log [H_3O^+] + \log \frac{[A^-]}{[HA]}$$

and then

$$-\log K_a = -\log [H_3O^+] - \log \frac{[A^-]}{[HA]}$$

Substituting and remembering that when a fraction is inverted, the sign of its log changes, we get

$$pK_a = pH + \log \frac{[HA]}{[A^-]}$$

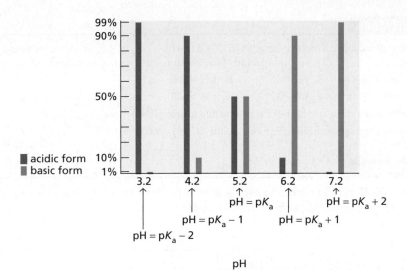

◀ **Figure 1.19**
The relative amounts of a compound with a pK_a of 5.2 in the acidic and basic forms at different pH values.

with a pK_a of 5.2 is in a solution of pH 5.2, half the compound will be in the acidic form and the other half will be in the basic form (Figure 1.19). If the pH is one unit less than the pK_a of the compound (pH = 4.2), there will be 10 times more compound present in the acidic form than in the basic form (because log 10 = 1). If the pH is two units less than the pK_a of the compound (pH = 3.2), there will be 100 times more compound present in the acidic form than in the basic form (because log 100 – 2). If the pH is 6.2, there will be 10 times more compound in the basic form than in the acidic form, and at pH = 7.2 there will be 100 times more compound present in the basic form than in the acidic form.

PROBLEM-SOLVING STRATEGY

Determining Structure at a Particular pH

Write the form in which the following compounds will predominate in a solution of pH 5.5:

a. CH_3CH_2OH (pK_a = 15.9)

b. $CH_3CH_2\overset{+}{O}H_2$ (pK_a = 2.5)

c. $CH_3\overset{+}{N}H_3$ (pK_a = 11.0)

To answer this kind of question, we need to compare the pH of the solution with the pK_a value of the compound's dissociable proton.

a. The pH of the solution is more acidic (5.5) than the pK_a value of the OH group (15.9). Therefore, the compound will exist primarily as CH_3CH_2OH (with its proton).

b. The pH of the solution is more basic (5.5) than the pK_a value of the $^+OH_2$ group (−2.5). Therefore, the compound will exist primarily as CH_3CH_2OH (without its proton).

c. The pH of the solution is more acidic (5.5) than the pK_a value of the $^+NH_3$ group (11.0). Therefore, the compound will exist primarily as $CH_3\overset{+}{N}H_3$ (with its proton).

Now continue on to Problem 59.

PROBLEM 59◆

For each of the following compounds, shown in their acidic forms, write the form that will predominate in a solution of pH = 5.5:

a. CH_3COOH (pK_a = 4.76)

b. $CH_3CH_2\overset{+}{N}H_3$ (pK_a = 11.0)

c. H_3O^+ (pK_a = −1.7)

d. HBr (pK_a = −9)

e. $^+NH_4$ (pK_a = 9.4)

f. $HC{\equiv}N$ (pK_a = 9.1)

g. HNO_2 (pK_a = 3.4)

h. HNO_3 (pK_a = −1.3)

PROBLEM 60◆

a. Indicate whether a carboxylic acid (RCOOH) with a pK_a of 4.5 will have more charged molecules or more neutral molecules in a solution with the following pH value:

1. pH = 1   3. pH = 5   5. pH = 10
2. pH = 3   4. pH = 7   6. pH = 13

b. Answer the same question for a protonated amine (R$\overset{+}{\text{N}}$H$_3$) with a pK_a of 9.

c. Answer the same question for an alcohol (ROH) with a pK_a of 15.

PROBLEM 61◆

A naturally occurring amino acid such as alanine has both a carboxylic acid group and an amine group. The pK_a values of the two groups are shown.

$$\underset{\substack{\displaystyle|\\[-2pt]\text{}^{+}\text{NH}_3}}{\text{CH}_3\text{CHCOH}} \quad \overset{\displaystyle\text{O}}{\overset{\|}{}}$$

pKₐ = 2.34

pKₐ = 9.69

alanine
an amino acid

a. Draw the structure of alanine in a solution at physiological pH (pH 7.3).

b. Is there a pH at which alanine will be neutral (neither group will have a charge)?

c. At what pH will alanine have no net charge (the amount of negative charge will be the same as the amount of positive charge)?

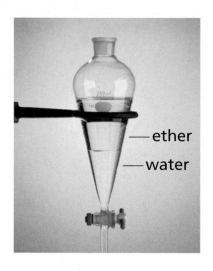

— ether

— water

The Henderson–Hasselbalch equation can be very useful in the laboratory for separating the compounds in a mixture. Water and diethyl ether are not miscible liquids and, therefore, will form two layers when combined; the ether layer will lie above the more dense water layer. Charged compounds are more soluble in water, whereas neutral compounds are more soluble in diethyl ether. Two compounds, such as a carboxylic acid (RCOOH) with a pK_a of 5.0 and a protonated amine (RNH$_3$$^+$) with a p$K_a$ of 10.0, dissolved in a mixture of water and diethyl ether, can be separated by adjusting the pH of the water layer. For example, if the pH of the water layer is 2, the carboxylic acid and the amine will both be in their acidic forms because the pH of the water is less than the pK_a values of both compounds. The acidic form of a carboxylic acid is neutral, whereas the acidic form of an amine is charged. Therefore, the carboxylic acid will be more soluble in the ether layer, and the protonated amine will be more soluble in the water layer.

$$\begin{array}{ccc} \text{acidic form} & & \text{basic form} \\ \text{RCOOH} & \rightleftharpoons & \text{RCOO}^- + \text{H}^+ \\ \text{R}\overset{+}{\text{N}}\text{H}_3 & \rightleftharpoons & \text{RNH}_2 + \text{H}^+ \end{array}$$

For the most effective separation, the pH of the water layer should be at least two units away from the pK_a values of the compounds being separated. Then the relative amounts of the compounds in their acidic and basic forms will be at least 100:1 (Figure 1.19).

You are what you're in: a compound will be mostly in the acidic form in an acidic solution (pH < pK_a) and mostly in the basic form in a basic solution (pH > pK_a).

PROBLEM 62◆ SOLVED

a. At what pH will the concentration of a compound with a pK_a of 8.4 be 100 times greater in its basic form than in its acidic form?

b. At what pH will the concentration of a compound with a pK_a of 3.7 be 10 times greater in its acidic form than in its basic form?

c. At what pH will the concentration of a compound with a pK_a of 8.4 be 100 times greater in its acidic form than in its basic form?

d. At what pH will 50% of a compound with a pK_a of 7.3 be in its basic form?

e. At what pH will the concentration of a compound with a pK_a of 4.6 be 10 times greater in its basic form than in its acidic form?

Solution to 62a   If the concentration in the basic form is 100 times greater than the concentration in the acidic form, the Henderson–Hasselbalch equation becomes

$$pK_a = pH + \log 1/100$$
$$8.4 = pH + \log .01$$
$$8.4 = pH - 2.0$$
$$pH = 10.4$$

There is a faster way to get the answer: if there is 100 times more compound present in the basic form than in the acidic form, the pH will be two units more basic than the pK_a. Thus, $pH = 8.4 + 2.0 = 10.4$.

Solution to 62b   If there is 10 times more compound present in the acidic form than in the basic form, the pH will be one unit more acidic than the pK_a. Thus, $pH = 3.7 - 1.0 = 2.7$.

PROBLEM 63◆

For each of the following compounds, indicate the pH at which

a. 50% of the compound will be in a form that possesses a charge.

b. more than 99% of the compound will be in a form that possesses a charge.
   1. CH_3CH_2COOH ($pK_a = 4.9$)
   2. $CH_3\overset{+}{N}H_3$ ($pK_a = 10.7$)

PROBLEM 64◆

As long as the pH is greater than _____, more than 50% of a protonated amine with a pK_a of 10.4 will be in its neutral, nonprotonated form.

1.25 **Buffer Solutions**

A solution of a weak acid (HA) and its conjugate base (A^-) is called a **buffer solution**. A buffer solution will maintain nearly constant pH when small amounts of acid or base are added to it, because the weak acid can donate a proton to any HO^- added to the solution, and its conjugate base can accept any H^+ that is added to the solution.

can donate an H^+
to HO^-

$$HA + HO^- \longrightarrow A^- + H_2O$$
$$A^- + H_3O^+ \longrightarrow HA + H_2O$$

can accept an H^+ from H_3O^+

Note to the student

Buffer solutions are discussed in detail in Special Topic I in the *Study Guide and Solutions Manual*. By working the problems you will find there, you will see the usefulness of the Henderson–Hasselbalch equation for dealing with buffer solutions.

BLOOD: A BUFFERED SOLUTION

Blood is the fluid that transports oxygen to all the cells of the human body. The normal pH of human blood is 7.3 to 7.4. Death will result if this pH decreases to a value less than ~6.8 or increases to a value greater than ~8.0 for even a few seconds.

Oxygen is carried to cells by a protein in the blood called hemoglobin (HbH^+). When hemoglobin binds O_2, hemoglobin loses a proton, which would make the blood more acidic if it did not contain a buffer to maintain its pH.

$$HbH^+ + O_2 \rightleftharpoons HbO_2 + H^+$$

A carbonic acid/bicarbonate (H_2CO_3/HCO_3^-) buffer controls the pH of blood. An important feature of this buffer is that carbonic acid decomposes to CO_2 and H_2O:

$$CO_2 + H_2O \rightleftharpoons \underset{\text{carbonic acid}}{H_2CO_3} \rightleftharpoons \underset{\text{bicarbonate}}{HCO_3^-} + H^+$$

During exercise our metabolism speeds up, producing large amounts of CO_2. The increased concentration of CO_2 shifts the equilibrium between carbonic acid and bicarbonate to the right, which increases the concentration of H^+. Significant amounts of lactic acid are also produced during exercise, and this further increases the concentration of H^+. Receptors in the brain respond to the increased concentration of H^+ by triggering a reflex that increases the rate of breathing. Hemoglobin then releases more oxygen to the cells and more CO_2 is eliminated by exhalation. Both processes decrease the concentration of H^+ in the blood by shifting the equilibrium to the left.

Thus, any disorder that decreases the rate and depth of ventilation, such as emphysema, will decrease the pH of the blood—a condition called acidosis. In contrast, any excessive increase in the rate and depth of ventilation, as with hyperventilation due to anxiety, will increase the pH of blood—a condition called alkalosis.

PROBLEM 65◆

Write the equation that shows how a buffer made by dissolving CH_3COOH and $CH_3COO^-Na^+$ in water prevents the pH of a solution from changing when:

a. a small amount of H^+ is added to the solution.
b. a small amount of HO^- is added to the solution.

1.26 The Second Definition of Acid and Base: Lewis Acids and Bases

In 1923, G. N. Lewis (page 9) offered new definitions for the terms *acid* and *base*. He defined an acid as a species that accepts a share in an electron pair and a base as a species that donates a share in an electron pair. All proton-donating acids fit the Lewis definition because all proton-donating acids lose a proton and the proton accepts a share in an electron pair.

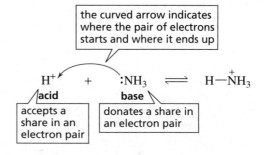

Lewis acids, however, are not limited to compounds that donate protons. According to the Lewis definition, compounds such as aluminum chloride ($AlCl_3$), boron trifluoride (BF_3), and borane (BH_3) are acids because they have unfilled valence orbitals and thus can accept a share in an electron pair. These compounds react with a compound that has a lone pair, just as a proton reacts with ammonia. Thus, the Lewis definition of an acid includes all proton-donating compounds and some additional compounds that do not have protons. Throughout this text, the term *acid* is used to mean a Brønsted proton-donating acid, and the term **Lewis acid** is used to refer to non-proton-donating acids such as $AlCl_3$ or BF_3.

Lewis acid: Need two from you.

All bases are **Lewis bases** because they all have a pair of electrons that they can share, either with a proton or with an atom such as aluminum or boron.

Lewis base: Have pair, will share.

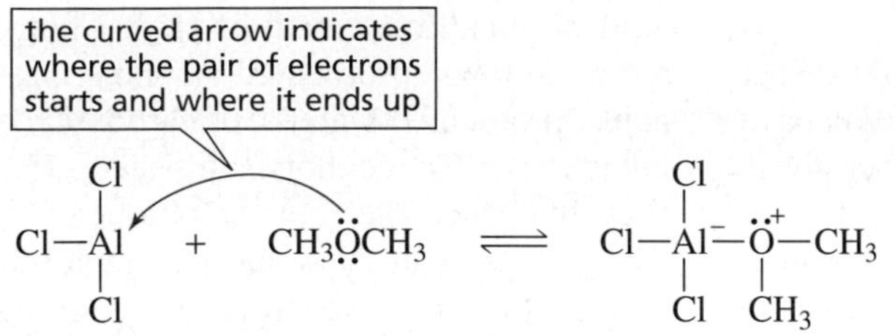

the curved arrow indicates where the pair of electrons starts and where it ends up

aluminum trichloride   dimethyl ether
a Lewis acid   a Lewis base

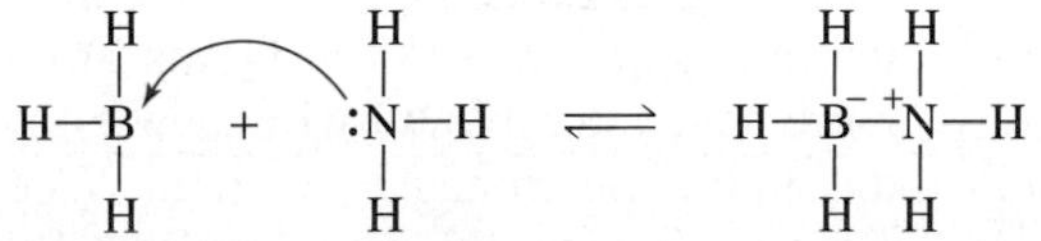

borane   ammonia
a Lewis acid   a Lewis base

PROBLEM 66

Give the products of the following reactions using arrows to show where the pair of electrons start and where it ends up.

a. $ZnCl_2 + CH_3\ddot{O}H \rightleftharpoons$

b. $FeBr_3 + :\ddot{Br}:^- \rightleftharpoons$

c. $AlCl_3 + :\ddot{Cl}:^- \rightleftharpoons$

PROBLEM 67

Show how each of the following compounds reacts with HO^-:

a. CH_3OH **c.** $CH_3\overset{+}{N}H_3$ **e.** $^+CH_3$ **g.** $AlCl_3$

b. $^+NH_4$ **d.** BF_3 **f.** $FeBr_3$ **h.** CH_3COOH

SUMMARY

Organic compounds are compounds that contain carbon. The **atomic number** of an atom equals the number of protons in its nucleus. The **mass number** of an atom is the sum of its protons and neutrons. **Isotopes** have the same atomic number, but different mass numbers.

An **atomic orbital** indicates where there is a high probability of finding an electron. The closer the atomic orbital is to the nucleus, the lower is its energy. **Degenerate orbitals** have the same energy. Electrons are assigned to orbitals following the **aufbau principle**, the **Pauli exclusion principle**, and **Hund's rule**.

The **octet rule** states that an atom will give up, accept, or share electrons in order to fill its outer shell or attain an outer shell with eight electrons. **Electropositive** elements readily lose electrons; **electronegative** elements readily acquire electrons. The **electronic configuration** of an atom describes the orbitals occupied by the atom's electrons. Electrons in inner shells are called **core electrons**; electrons in the outermost shell are called **valence electrons**. **Lone-pair electrons** are valence electrons that are not used in bonding. Attractive forces between opposite charges are called **electrostatic attractions**. An **ionic bond** is formed by a transfer of electrons; a **covalent bond** is formed by sharing electrons. A **polar covalent bond** is a covalent bond between atoms with different **electronegativites**. Therefore, a polar covalent bond has a **dipole**, measured by a **dipole moment**. The **dipole moment** of a molecule depends on the magnitudes and directions of the bond dipole moments.

Lewis structures indicate which atoms are bonded together and show **lone pairs** and **formal charges**. A **carbocation** has a positively charged carbon, a **carbanion** has a negatively charged carbon, and a **radical** has an unpaired electron.

According to **molecular orbital (MO) theory**, covalent bonds result when atomic orbitals combine to form **molecular orbitals**. Atomic orbitals combine to give a **bonding MO** and a higher energy **antibonding MO**. Cylindrically symmetrical bonds are called **sigma (σ) bonds**; **pi (π) bonds** form when p orbitals overlap side-to-side. Bond strength is measured by the **bond dissociation energy**. A σ bond is stronger than a π bond. All **single bonds** in organic compounds are σ bonds, a **double bond** consists of one σ bond and one π bond, and a **triple bond** consists of one σ bond and two π bonds. Triple bonds are shorter and stronger than double bonds, which are shorter and stronger than single bonds. To form four bonds, carbon promotes an electron from a $2s$ to a $2p$ orbital. C, N, and O form bonds using **hybrid orbitals**. The **hybridization** of C, N, or O depends on the number of π bonds the atom forms: no π bonds means that the atom is sp^3 **hybridized**, one π bond indicates that it is sp^2 **hybridized**, and two π bonds signifies that it is sp **hybridized**. Exceptions are carbocations and carbon radicals, which are sp^2 hybridized. The more s character in the orbital used to form a bond, the shorter and stronger the bond is and the larger the bond angle. Bonding and lone-pair electrons around an atom are positioned as far apart as possible.

An **acid** is a species that donates a proton, and a **base** is a species that accepts a proton. A **Lewis acid** is a species that accepts a share in an electron pair; a **Lewis base** is a species that donates a share in an electron pair.

Acidity is a measure of the tendency of a compound to give up a proton. **Basicity** is a measure of a compound's affinity for a proton. The stronger the acid, the weaker is its conjugate base. The strength of an acid is given by the **acid dissociation constant (K_a)**. Approximate pK_a values are as follows: protonated alcohols, protonated carboxylic acids, protonated water < 0; carboxylic acids ~5; protonated amines ~10; alcohols and water ~15. The **pH** of a solution indicates the concentration of positively charged hydrogen ions in the solution. In **acid–base reactions**, the equilibrium favors reaction of the strong and formation of the weak. Curved arrows indicate the bonds that are broken and formed as reactants are converted into products.

The strength of an acid is determined by the stability of its conjugate base: the more stable the base, the stronger is its conjugate acid. When atoms are similar in size, the more acidic compound will have its hydrogen attached to the more electronegative atom. When atoms are very different in size, the more acidic compound will have its hydrogen attached to the larger atom. Hybridization affects acidity because an sp hybridized atom is more electronegative than an sp^2 hybridized atom, which is more electronegative than an sp^3 hybridized atom. **Inductive electron withdrawal** increases acidity: the more electronegative the electron-withdrawing group and the closer it is to the acidic hydrogen, the stronger is the acid.

Delocalized electrons, which are electrons shared by more than two atoms, stabilize a compound. A **resonance hybrid** is a composite of the **resonance contributors**, structures that differ only in the location of their lone-pair and π electrons.

The **Henderson–Hasselbalch equation** gives the relationship between pK_a and pH: a compound exists primarily in its acidic form in solutions more acidic than its pK_a value and primarily in its basic form in solutions more basic than its pK_a value.

KEY TERMS

acid (p. 44)	carbanion (p. 16)	electrostatic attraction (p. 9)
acid–base reaction (p. 44)	carbocation (p. 16)	electrostatic potential map (p. 14)
acid dissociation constant (K_a) (p. 45)	condensed structure (p. 19)	equilibrium constant (p. 45)
acidity (p. 44)	conjugate acid (p. 44)	excited-state electronic configuration (p. 6)
antibonding molecular orbital (p. 24)	conjugate base (p. 44)	formal charge (p. 15)
atomic number (p. 4)	constitutional isomer (p. 18)	free radical (p. 16)
atomic orbital (p. 5)	core electrons (p. 8)	ground-state electronic configuration (p. 6)
atomic weight (p. 5)	covalent bond (p. 10)	
aufbau principle (p. 6)	degenerate orbitals (p. 6)	Heisenberg uncertainty principle (p. 21)
base (p. 44)	delocalized electrons (p. 57)	Henderson–Hasselbalch equation (p. 60)
basicity (p. 44)	debye (D) (p. 13)	Hund's rule (p. 8)
bond (p. 9)	dipole (p. 12)	hybrid orbital (p. 29)
bond dissociation energy (p. 23)	dipole moment (μ) (p. 12)	hybridization (p. 29)
bond length (p. 23)	double bond (p. 32)	hydride ion (p. 10)
bonding molecular orbital (p. 24)	electronegative (p. 9)	hydrogen ion (p. 10)
bond strength (p. 23)	electronegativity (p. 11)	inductive electron withdrawal (p. 55)
buffer solution (p. 63)	electropositive (p. 9)	ionic bond (p. 9)

ionic compound (p. 10)
isotopes (p. 4)
Kekulé structure (p. 18)
Lewis acid (p. 65)
Lewis base (p. 65)
Lewis structure (p. 15)
lone-pair electrons (p. 15)
mass number (p. 4)
molecular weight (p. 5)
molecular orbital (p. 22)
molecular orbital (MO) theory (p. 22)
node (p. 21)
nodal plane (p. 22)
nonbonding electrons (p. 15)
nonpolar covalent bond (p. 11)

nonpolar molecule (p. 28)
octet rule (p. 9)
orbital (p. 5)
orbital hybridization (p. 29)
organic compound (p. 2)
Pauli exclusion principle (p. 7)
pH (p. 46)
perspective formula (p. 28)
pi (π) bond (p. 26)
pK_a (p. 45)
polar covalent bond (p. 11)
proton (p. 10)
proton-transfer reaction (p. 44)
quantum mechanics (p. 5)
radial node (p. 21)

radical (p. 16)
resonance contributors (p. 57)
resonance hybrid (p. 57)
sigma (σ) bond (p. 23)
sigma bonding molecular orbital (p. 24)
single bond (p. 40)
tetrahedral bond angle (p. 30)
tetrahedral carbon (p. 30)
trigonal planar carbon (p. 32)
triple bond (p. 34)
valence electrons (p. 8)
valence-shell electron-pair repulsion
   (VSEPR) model (p. 27)
wave equation (p. 5)
wave function (p. 5)

PROBLEMS

68. Draw a Lewis structure for each of the following species:
   a. H_2CO_3
   b. CO_3^{2-}
   c. H_2CO
   d. N_2H_4
   e. CH_3NH_2
   f. $CH_3N_2^+$
   g. CO_2
   h. H_2NO^-

69. Give the hybridization of the central atom of each of the following species, and tell whether the bond arrangement around it is linear, trigonal planar, or tetrahedral:
   a. NH_3
   b. BH_3
   c. $^-CH_3$
   d. $\cdot CH_3$
   e. $^+NH_4$
   f. $^+CH_3$
   g. HCN
   h. $C(CH_3)_4$
   i. H_3O^+

70. Draw the condensed structure of a compound that contains only carbon and hydrogen atoms and that has
   a. three sp^3 hybridized carbons.
   b. one sp^3 hybridized carbon and two sp^2 hybridized carbons.
   c. two sp^3 hybridized carbons and two sp hybridized carbons.

71. Predict the approximate bond angles:
   a. the $C—N—H$ bond angle in $(CH_3)_2NH$
   b. the $C—N—C$ bond angle in $(CH_3)_2NH$
   c. the $C—N—C$ bond angle in $(CH_3)_2\overset{+}{N}H_2$
   d. the $C—O—C$ bond angle in CH_3OCH_3
   e. the $C—O—H$ bond angle in CH_3OH
   f. the $H—C—H$ bond angle in $H_2C—O$
   g. the $F—B—F$ bond angle in $^-BF_4$
   h. the $C—C—N$ bond angle in $CH_3C≡N$
   i. the $C—C—N$ bond angle in $CH_3CH_2NH_2$

72. Draw the ground-state electronic configuration for:
   a. Ca
   b. Ca^{2+}
   c. Ar
   d. Mg^{2+}

73. List the bonds in order of decreasing polarity (i.e., list the most polar bond first).
   a. $C—O, C—F, C—N$
   b. $C—Cl, C—I, C—Br$
   c. $H—O, H—N, H—C$
   d. $C—H, C—C, C—N$

74. Which is the stronger base?
   a. HS^- or HO^-
   b. CH_3O^- or $CH_3\overset{-}{N}H$
   c. CH_3OH or CH_3O^-
   d. Cl^- or Br^-

75. Write the Kekulé structure for each of the following compounds:
   a. CH_3CHO
   b. CH_3OCH_3
   c. CH_3COOH
   d. $(CH_3)_3COH$
   e. $CH_3CH(OH)CH_2CN$
   f. $(CH_3)_2CHCH(CH_3)CH_2C(CH_3)_3$

76. Show the direction of the dipole moment in each of the following bonds (use the electronegativities given in Table 1.3):
   a. $H_3C—Br$
   b. $H_3C—Li$
   c. $HO—NH_2$
   d. $I—Br$
   e. $H_3C—OH$
   f. $(CH_3)_2N—H$

77. What is the hybridization of the indicated atom in each of the following molecules?

a. $CH_3\overset{\downarrow}{CH}=CH_2$
c. $CH_3CH_2\overset{\downarrow}{O}H$
e. $CH_3CH=\overset{\downarrow}{N}CH_3$

b. $CH_3\overset{O\leftarrow}{\overset{\|}{C}}CH_3$
d. $CH_3C\overset{\downarrow}{\equiv}N$
f. $CH_3\overset{\downarrow}{O}CH_2CH_3$

78. Draw the missing lone-pair electrons and assign the missing formal charges.

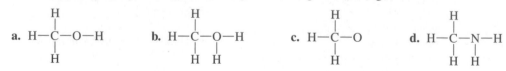

a. $H-\overset{H}{\underset{H}{\overset{|}{C}}}-O-H$
b. $H-\overset{H}{\underset{H}{\overset{|}{C}}}-\overset{}{\underset{H}{\overset{|}{O}}}-H$
c. $H-\overset{H}{\underset{H}{\overset{|}{C}}}-O$
d. $H-\overset{H}{\underset{H}{\overset{|}{C}}}-\overset{}{\underset{H}{\overset{|}{N}}}-H$

79. a. List the following carboxylic acids in order of decreasing acidity:

$$CH_3CH_2CH_2COOH$$
$$K_a = 1.52 \times 10^{-5}$$

$$CH_3CH_2\underset{Cl}{\overset{|}{CH}}COOH$$
$$K_a = 1.39 \times 10^{-3}$$

$$ClCH_2CH_2CH_2COOH$$
$$K_a = 2.96 \times 10^{-5}$$

$$CH_3\underset{Cl}{\overset{|}{CH}}CH_2COOH$$
$$K_a = 8.9 \times 10^{-5}$$

b. How does the presence of an electronegative substituent such as Cl affect the acidity of a carboxylic acid?
c. How does the location of the substituent affect the acidity of the carboxylic acid?

80. a. Which of the indicated bonds in each molecule is shorter?
b. Indicate the hybridization of the C, O, and N atoms in each of the molecules.

1. $CH_3\overset{\downarrow}{CH}=\overset{\downarrow}{CH}C\equiv CH$

4. $\overset{H}{\underset{H}{>}}C=CHC\equiv \overset{\downarrow}{C}-H$

2. $CH_3\overset{O}{\overset{\|}{\underset{\to}{C}}}CH_2\overset{\downarrow}{-}OH$

5. $\overset{H}{\underset{H}{>}}C=CHC\equiv C-\overset{CH_3}{\underset{CH_3}{\overset{|}{C}}}-H$

3. $CH_3\overset{\downarrow}{NH}-CH_2CH_2\overset{\downarrow}{N}=CHCH_3$

81. For each of the following compounds, draw the form in which it will predominate at pH = 3, pH = 6, pH = 10, and pH = 14:
a. CH_3COOH
   $pK_a = 4.8$
b. $CH_3CH_2\overset{+}{N}H_3$
   $pK_a = 11.0$
c. CF_3CH_2OH
   $pK_a = 12.4$

82. a. List the following alcohols in order of decreasing acidity:

$$CCl_3CH_2OH$$
$$K_a = 5.75 \times 10^{-13}$$

$$CH_2ClCH_2OH$$
$$K_a = 1.29 \times 10^{-13}$$

$$CHCl_2CH_2OH$$
$$K_a = 4.90 \times 10^{-13}$$

b. Explain the relative acidities.

83. Do the sp^2 carbons and the indicated sp^3 carbons lie in the same plane?

84. Give the products of the following acid–base reactions, and indicate whether reactants or products are favored at equilibrium (use the pK_a values that are given in Section 1.18):

a. $CH_3\overset{O}{\overset{\|}{C}}OH + CH_3O^- \rightleftharpoons$

b. $CH_3CH_2OH + {}^-NH_2 \rightleftharpoons$

c. $CH_3\overset{O}{\overset{\|}{C}}OH + CH_3NH_2 \rightleftharpoons$

d. $CH_3CH_2OH + HCl \rightleftharpoons$

85. Which of the following species have a tetrahedral atom?

$$H_2O \quad H_3O^+ \quad {}^+CH_3 \quad BF_3 \quad NH_3 \quad {}^+NH_4 \quad {}^-CH_3$$

86. For each of the following molecules, indicate the hybridization of each carbon atom and give the approximate values of all the bond angles:

a. $CH_3C\equiv CH$ **c.** $CH_3CH_2CH_3$
b. $CH_3CH\!=\!CH_2$ **d.** $CH_2\!=\!CH\!-\!CH\!=\!CH_2$

87. a. Estimate the pK_a value of each of the following acids without using a calculator (i.e., between 3 and 4, between 9 and 10, etc.):

     1. nitrous acid (HNO_2), $K_a = 4.0 \times 10^{-4}$ **4.** hydrogen cyanide (HCN), $K_a = 7.9 \times 10^{-10}$
     2. nitric acid (HNO_3), $K_a = 22$ **5.** formic acid (HCOOH), $K_a = 2.0 \times 10^{-4}$
     3. bicarbonate (HCO_3^-), $K_a = 6.3 \times 10^{-11}$

     b. Determine the pK_a values, using a calculator.
     c. Which is the strongest acid?

88. a. Locate the three nitrogen atoms in the electrostatic potential map of histamine, the compound that causes the symptoms associated with the common cold and with allergic responses. Which of the two nitrogen atoms in the ring is the most basic? Explain its greater basicity.

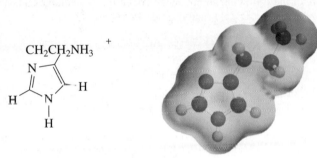

histamine

     b. If the potential map were not available, could you have predicted the most basic nitrogen from the structure of histamine?

89. a. List the following carboxylic acids in order of decreasing acidity:

$CH_3CH_2CHCOOH$	$CH_3CH_2CH_2COOH$	$CH_3CH_2CHCOOH$	$CH_3CH_2CHCOOH$
Br		OH	Cl
$K_a = 1.02 \times 10^{-3}$	$K_a = 1.51 \times 10^{-5}$	$K_a = 6.03 \times 10^{-5}$	$K_a = 1.45 \times 10^{-3}$

     b. Which is a more electronegative substituent, Cl or OH?

90. Draw a Lewis structure for each of the following species:

     a. $CH_3N_2^+$ **b.** CH_2N_2 **c.** N_3^- **d.** N_2O (arranged NNO)

91. a. For each of the following pairs of reactions, indicate which one has the more favorable equilibrium constant (that is, which one most favors products):

     1. $CH_3CH_2OH + NH_3 \rightleftharpoons CH_3CH_2O^- + \overset{+}{N}H_4$
     or
     $CH_3OH + NH_3 \rightleftharpoons CH_3O^- + \overset{+}{N}H_4$

     2. $CH_3CH_2OH + NH_3 \rightleftharpoons CH_3CH_2O^- + \overset{+}{N}H_4$
     or
     $CH_3CH_2OH + CH_3NH_2 \rightleftharpoons CH_3CH_2O^- + CH_3\overset{+}{N}H_3$

     b. Which of the four reactions has the most favorable equilibrium constant?

92. The following compound has two isomers:

$$ClCH\!=\!CHCl$$

One isomer has a dipole moment of 0 D, and the other has a dipole moment of 2.95 D. Propose structures for the two isomers that are consistent with these data.

93. Explain why the following compound is not stable:

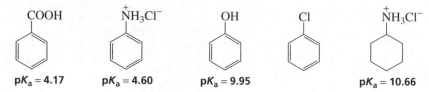

94. Given that pH + pOH = 14 and that the concentration of water in a solution of water is 55.5 M, show that the pK_a of water is 15.7. (*Hint:* pOH = −log [HO⁻].)

95. Water and diethyl ether are immiscible liquids. Charged compounds dissolve in water, and uncharged compounds dissolve in ether. Given that $C_6H_{11}COOH$ has a pK_a of 4.8 and $C_6H_{11}NH_3$ has a pK_a of 10.7.
 a. What pH would you make the water layer in order to cause both compounds to dissolve in it?
 b. What pH would you make the water layer in order to cause the acid to dissolve in the water layer and the amine to dissolve in the ether layer?
 c. What pH would you make the water layer in order to cause the acid to dissolve in the ether layer and the amine to dissolve in the water layer?

96. How could you separate a mixture of the following compounds? The reagents available to you are water, ether, 1.0 M HCl, and 1.0 M NaOH. (*Hint:* See Problem 95.)

COOH $\overset{+}{N}H_3Cl^-$ OH Cl $\overset{+}{N}H_3Cl^-$

pK_a = 4.17   pK_a = 4.60   pK_a = 9.95   pK_a = 10.66

97. Using molecular orbital theory, explain why shining light on Br_2 causes it to break apart into atoms, but shining light on H_2 does not break the molecule apart.

98. Show that $K_{eq} = \dfrac{K_a \text{ reactant acid}}{K_a \text{ product acid}} = \dfrac{[\text{products}]}{[\text{reactants}]}$

For help in answering Problems 99–101, see Special Topic I in the *Study Guide and Solutions Manual*.

99. Carbonic acid has a pK_a of 6.1 at physiological temperature. Is the carbonic acid/bicarbonate buffer system that maintains the pH of the blood at 7.3 better at neutralizing excess acid or excess base?

100. a. If an acid with a pK_a of 5.3 is in an aqueous solution of pH 5.7, what percentage of the acid is present in the acidic form?
 b. At what pH will 80% of the acid exist in the acidic form?

101. Calculate the pH values of the following solutions:
 a. a 1.0 M solution of acetic acid (pK_a = 4.76)
 b. a 0.1 M solution of protonated methylamine (pK_a = 10.7)
 c. a solution containing 0.3 M HCOOH and 0.1 M HCOO⁻ (pK_a of HCOOH = 3.76)

CHAPTER 2

An Introduction to Organic Compounds
Nomenclature, Physical Properties, and Representation of Structure

CH$_3$CH$_2$Cl   CH$_3$CH$_2$OH

CH$_3$OCH$_3$

CH$_3$CH$_2$NH$_2$ CH$_3$CH$_2$Br

BUILDING ON FUNDAMENTALS

SECTION 2.8 Alcohols and ethers are similar in structure to water (1.11); amines are similar in structure to ammonia	(1.12); and alkyl halides are similar in structure to alkanes (Section 1.7).

In this book, the presentation of organic chemistry is organized according to how organic compounds react. In studying how compounds react, you must not forget that whenever a compound undergoes a reaction, a new compound is synthesized. In other words, while you are learning how organic compounds react, you will simultaneously be learning how to synthesize organic compounds.

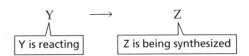

Y is reacting   Z is being synthesized

The main classes of compounds that are synthesized by the reactions we will look at in Chapters 3–10 are alkanes, alkyl halides, ethers, alcohols, and amines. As we learn how to synthesize compounds, we will need to be able to refer to them by name, so we will begin the study of organic chemistry by learning how to name these five classes of compounds.

First we will see how *alkanes* are named because their nomenclature forms the basis for the names of almost all organic compounds. **Alkanes** are composed of only carbon atoms and hydrogen atoms and contain only single bonds. Compounds that contain only carbon and hydrogen are called **hydrocarbons**, so an alkane is a hydrocarbon that has only single bonds. Alkanes in which the carbons form a continuous chain with no branches are called **straight-chain alkanes**. The names of several straight-chain alkanes are given in Table 2.1. It is important that you learn the names of at least the first 10.

Table 2.1	Nomenclature and Physical Properties of Straight-Chain Alkanes					
Number of carbons	Molecular formula	Name	Condensed structure	Boiling point (°C)	Melting point (°C)	Density[a] (g/mL)
1	CH_4	methane	CH_4	−167.7	−182.5	
2	C_2H_6	ethane	CH_3CH_3	−88.6	−183.3	
3	C_3H_8	propane	$CH_3CH_2CH_3$	−42.1	−187.7	
4	C_4H_{10}	butane	$CH_3CH_2CH_2CH_3$	−0.5	−138.3	
5	C_5H_{12}	pentane	$CH_3(CH_2)_3CH_3$	36.1	−129.8	0.5572
6	C_6H_{14}	hexane	$CH_3(CH_2)_4CH_3$	68.7	−95.3	0.6603
7	C_7H_{16}	heptane	$CH_3(CH_2)_5CH_3$	98.4	−90.6	0.6837
8	C_8H_{18}	octane	$CH_3(CH_2)_6CH_3$	125.7	−56.8	0.7026
9	C_9H_{20}	nonane	$CH_3(CH_2)_7CH_3$	150.8	−53.5	0.7177
10	$C_{10}H_{22}$	decane	$CH_3(CH_2)_8CH_3$	174.0	−29.7	0.7299
11	$C_{11}H_{24}$	undecane	$CH_3(CH_2)_9CH_3$	195.8	−25.6	0.7402
12	$C_{12}H_{26}$	dodecane	$CH_3(CH_2)_{10}CH_3$	216.3	−9.6	0.7487
13	$C_{13}H_{28}$	tridecane	$CH_3(CH_2)_{11}CH_3$	235.4	−5.5	0.7546
⋮	⋮	⋮	⋮	⋮	⋮	⋮
20	$C_{20}H_{42}$	eicosane	$CH_3(CH_2)_{18}CH_3$	343.0	36.8	0.7886
21	$C_{21}H_{44}$	heneicosane	$CH_3(CH_2)_{19}CH_3$	356.5	40.5	0.7917
⋮	⋮	⋮	⋮	⋮	⋮	⋮
30	$C_{30}H_{62}$	triacontane	$CH_3(CH_2)_{28}CH_3$	449.7	65.8	0.8097

[a]Density is temperature dependent. The densities given are those determined at 20°C ($d^{20°}$).

The family of alkanes shown in the table is an example of a homologous series. A **homologous series** (*homos* is Greek for "the same as") is a family of compounds in which each member differs from the one before it in the series by one **methylene (CH_2) group**. The members of a homologous series are called **homologs**. Propane ($CH_3CH_2CH_3$) and butane ($CH_3CH_2CH_2CH_3$) are homologs.

If you look at the relative numbers of carbon and hydrogen atoms in the alkanes listed in Table 2.1, you will see that the general molecular formula for an alkane is C_nH_{2n+2}, where n is any integer. So, if an alkane has one carbon atom, it must have four hydrogen atoms; if it has two carbon atoms, it must have six hydrogens atoms.

We have seen that carbon forms four covalent bonds and hydrogen forms only one covalent bond (Section 1.4). This means that there is only one possible structure for an alkane with molecular formula CH_4 (methane) and only one structure for an alkane with molecular formula C_2H_6 (ethane). We examined the structures of these compounds in Section 1.7. There is also only one possible structure for an alkane with molecular formula C_3H_8 (propane).

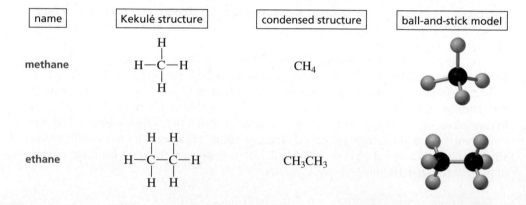

propane $\begin{array}{c} H\ H\ H \\ | \ | \ | \\ H-C-C-C-H \\ | \ | \ | \\ H\ H\ H \end{array}$ $CH_3CH_2CH_3$

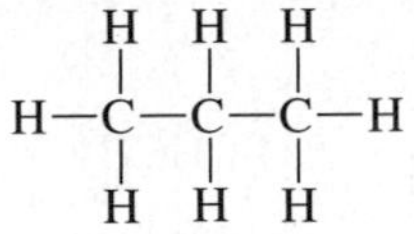

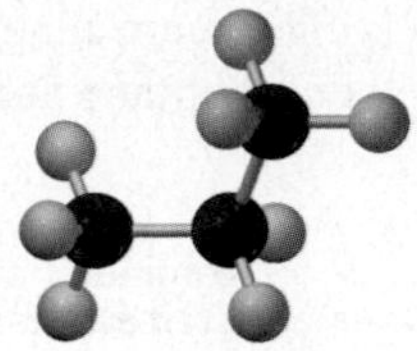

butane $\begin{array}{c} H\ H\ H\ H \\ | \ | \ | \ | \\ H-C-C-C-C-H \\ | \ | \ | \ | \\ H\ H\ H\ H \end{array}$ $CH_3CH_2CH_2CH_3$

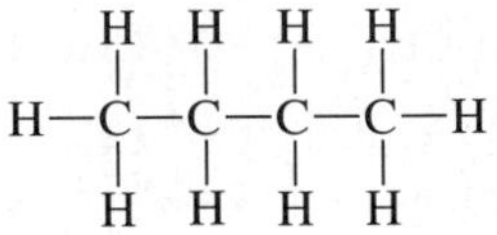

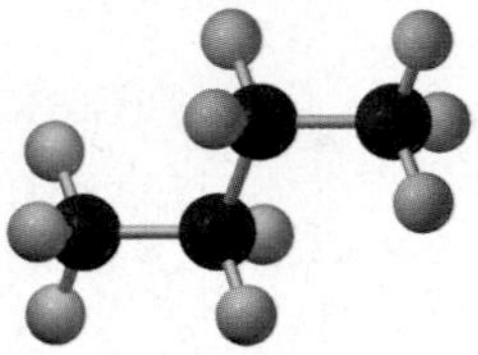

As the number of carbons in an alkane increases beyond three, the number of possible structures increases. There are two possible structures for an alkane with molecular formula C_4H_{10}. In addition to butane—a straight-chain alkane—there is a branched butane called isobutane. Both of these structures fulfill the requirement that each carbon forms four bonds and each hydrogen forms only one bond.

Compounds such as butane and isobutane that have the same molecular formula but differ in the order in which the atoms are connected are called **constitutional isomers**—their molecules have different constitutions. In fact, isobutane got its name because it is an "iso"mer (a word meaning "same parts") of butane. The structural unit consisting of a carbon bonded to a hydrogen and two CH_3 groups—which occurs in isobutane—has come to be called "iso." Thus, the name isobutane tells you that the compound is a four-carbon alkane with an iso structural unit.

$\underset{\text{butane}}{CH_3CH_2CH_2CH_3}$ $\underset{\underset{\underset{\text{isobutane}}{CH_3}}{|}}{CH_3CHCH_3}$ $\underset{\underset{\underset{\underset{\text{structural unit}}{\text{an "iso"}}}{CH_3}}{|}}{CH_3CH-}$

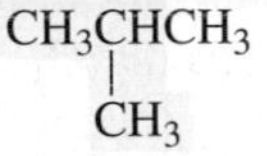

There are three alkanes with molecular formula C_5H_{12}. You have already learned how to name two of them. Pentane is the straight-chain alkane. Isopentane, as its name indicates, has an iso structural unit and five carbon atoms. We cannot name the other branched-chain alkane without defining a name for a new structural unit. (For now, ignore the names written in blue.)

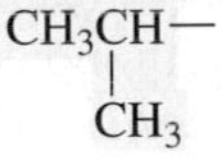

$\underset{\text{pentane}}{CH_3CH_2CH_2CH_2CH_3}$ $\underset{\underset{\underset{\text{isopentane}}{CH_3}}{|}}{CH_3CHCH_2CH_3}$ $\underset{\underset{\underset{\text{2,2-dimethylpropane}}{CH_3}}{|}}{\overset{\overset{CH_3}{|}}{CH_3CCH_3}}$

There are five constitutional isomers with molecular formula C_6H_{14}. Again, we are able to name only two of them, unless we define new structural units.

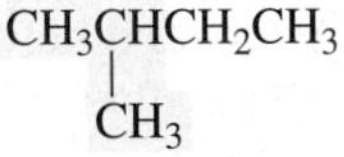

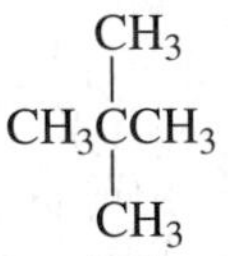

common name:
systematic name:

$\underset{\underset{\text{hexane}}{\text{hexane}}}{CH_3CH_2CH_2CH_2CH_2CH_3}$ $\underset{\underset{\underset{\underset{\text{2-methylpentane}}{\text{isohexane}}}{CH_3}}{|}}{CH_3CHCH_2CH_2CH_3}$ $\underset{\underset{\underset{\text{2,2-dimethylbutane}}{CH_3}}{|}}{\overset{\overset{CH_3}{|}}{CH_3CCH_2CH_3}}$

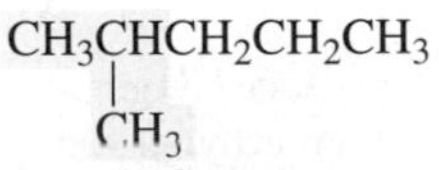

$\underset{\underset{\underset{\text{3-methylpentane}}{CH_3}}{|}}{CH_3CH_2CHCH_2CH_3}$ $\underset{\underset{\underset{\text{2,3-dimethylbutane}}{CH_3\ \ CH_3}}{|\ \ \ \ |}}{CH_3CH-CHCH_3}$

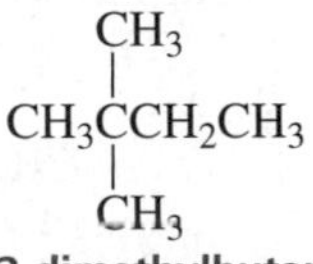

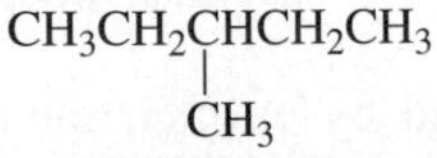

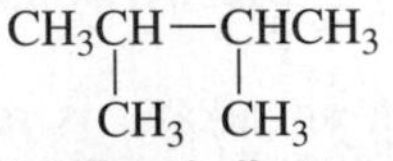

There are nine alkanes with molecular formula C_7H_{16}. We can name only two of them (heptane and isoheptane).

$CH_3CH_2CH_2CH_2CH_2CH_2CH_3$ $CH_3CHCH_2CH_2CH_2CH_3$

common name: heptane |
systematic name: heptane CH_3

                                                            isoheptane
                                                            2-methylhexane

$CH_3CH_2CHCH_2CH_2CH_3$ $CH_3CH—CHCH_2CH_3$ $CH_3CHCH_2CHCH_3$
             | | | | |
           CH_3 CH_3 CH_3 CH_3 CH_3
      3-methylhexane   **2,3-dimethylpentane**   **2,4-dimethylpentane**

            CH_3 CH_3
             | |
$CH_3CCH_2CH_2CH_3$ $CH_3CH_2CCH_2CH_3$ $CH_3CH_2CHCH_2CH_3$
             | | |
           CH_3 CH_3 CH_2CH_3
      2,2-dimethylpentane   **3,3-dimethylpentane**   **3-ethylpentane**

                        CH_3 CH_3
                          | |
                  $CH_3C—CHCH_3$
                          |
                        CH_3
                  2,2,3-trimethylbutane

The number of constitutional isomers increases rapidly as the number of carbons in an alkane increases. For example, there are 75 alkanes with molecular formula $C_{10}H_{22}$ and 4347 alkanes with molecular formula $C_{15}H_{32}$. To avoid having to memorize the names of thousands of structural units, chemists have devised rules for creating systematic names that describe the compound's structure. That way, only the rules have to be learned. Because the name describes the structure, these rules make it possible to deduce the structure of a compound from its name.

This method of nomenclature is called **systematic nomenclature**. It is also called **IUPAC nomenclature** because it was designed by a commission of the International Union of Pure and Applied Chemistry (abbreviated IUPAC and pronounced "eye-you-pack") at a meeting in Geneva, Switzerland, in 1892. The IUPAC rules have been continually revised by the commission since then. A name such as isobutane—a nonsystematic name—is called a **common name**. When both names are shown in this book, common names will be shown in red and systematic (IUPAC) names will be shown in blue. Before we can understand how a systematic name for an alkane is constructed, we must learn how to name alkyl substituents.

2.1 How Alkyl Substituents Are Named

Removing a hydrogen from an alkane results in an **alkyl substituent** (or an alkyl group). Alkyl substituents are named by replacing the "ane" ending of the alkane with "yl." The letter "R" is used to indicate any alkyl group.

$CH_3—$ $CH_3CH_2—$ $CH_3CH_2CH_2—$ $CH_3CH_2CH_2CH_2—$
a methyl group   **an ethyl group**   **a propyl group**   **a butyl group**

          $CH_3CH_2CH_2CH_2CH_2—$ $R—$
              a pentyl group   **any alkyl group**

If a hydrogen in an alkane is replaced by an OH, the compound becomes an **alcohol**; if it is replaced by an NH_2, the compound becomes an **amine**; if it is replaced by a halogen, the compound becomes an **alkyl halide**, and if it replaced by an OR, the compound becomes an **ether**.

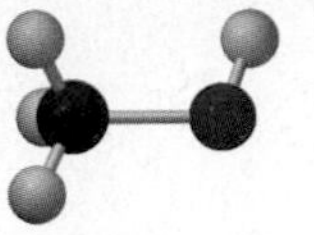

R—OH   R—NH₂   R—X | X = F, Cl, Br, or I | R—O—R
an alcohol   an amine   an alkyl halide   an ether

The alkyl group name followed by the name of the class of the compound (alcohol, amine, etc.) yields the common name of the compound. The two alkyl groups in ethers are cited in alphabetical order. The following examples show how alkyl group names are used to build common names:

CH₃OH	CH₃CH₂NH₂	CH₃CH₂CH₂Br	CH₃CH₂CH₂CH₂Cl
methyl alcohol	ethylamine	propyl bromide	butyl chloride
CH₃I	CH₃CH₂OH	CH₃CH₂CH₂NH₂	CH₃CH₂OCH₃
methyl iodide	ethyl alcohol	propylamine	ethyl methyl ether

Notice that there is a space between the name of the alkyl group and the name of the class of compound, except in the case of amines.

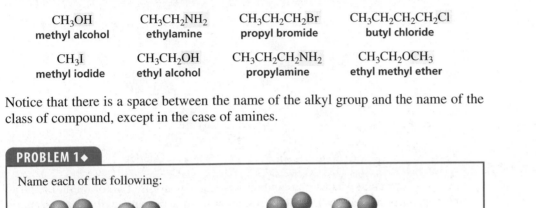

methyl alcohol

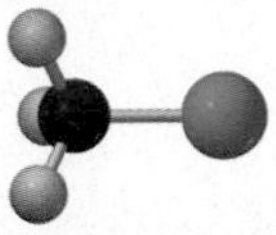

methyl chloride

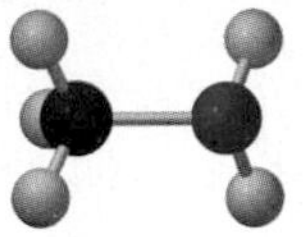

methylamine

PROBLEM 1◆

Name each of the following:

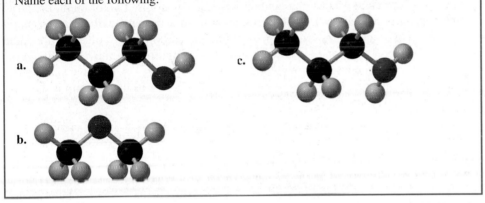

a.

b.

c.

Two alkyl groups—the propyl group and the isopropyl group—have three carbon atoms. A propyl group is obtained when a hydrogen is removed from a *primary carbon* of propane. A **primary carbon** is a carbon bonded to only one other carbon. An isopropyl group is obtained when a hydrogen is removed from the *secondary carbon* of propane. A **secondary carbon** is a carbon bonded to two other carbons. Notice that an isopropyl group, as its name indicates, has its three carbon atoms arranged as an iso structural unit.

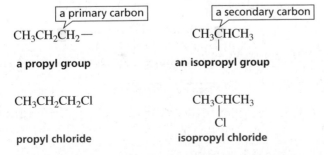

a primary carbon		a secondary carbon

CH₃CH₂CH₂—   CH₃CHCH₃

a propyl group   **an isopropyl group**

CH₃CH₂CH₂Cl   CH₃CHCH₃
                                      |
                                     Cl

propyl chloride   **isopropyl chloride**

Molecular structures can be drawn in different ways. Isopropyl chloride, for example, is drawn here in two different ways. Both representations depict the same compound, although, at first glance, the two-dimensional representations appear to be different: the methyl groups are placed at opposite ends in one structure and at right angles in the

Build models of the two representations of isopropyl chloride so you can see that they represent the same compound.

other. The structures are identical, however, because carbon is tetrahedral. The four groups bonded to the central carbon—a hydrogen, a chlorine, and two methyl groups—point to the corners of a tetrahedron. If you rotate the three-dimensional model on the right 90° in a clockwise direction, you should be able to see that the two models are the same. (If you visit the Molecule Gallery for Chapter 2 on the Web site (http://www.prenhall.com/bruice), you will be able to actually rotate isopropyl chloride to prove that the two molecules are identical.)

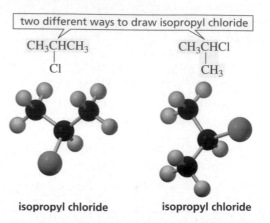

two different ways to draw isopropyl chloride

CH_3CHCH_3 $\quad$ CH_3CHCl
$\qquad$ Cl $\qquad\qquad\qquad$ CH_3

isopropyl chloride $\qquad$ **isopropyl chloride**

There are four alkyl groups that contain four carbon atoms. Two of them, the butyl and isobutyl groups, have a hydrogen removed from a primary carbon. A *sec*-butyl group has a hydrogen removed from a secondary carbon (*sec*-, often abbreviated *s*-, stands for secondary), and a *tert*-butyl group has a hydrogen removed from a tertiary carbon (*tert*-, sometimes abbreviated *t*-, stands for tertiary). A **tertiary carbon** is a carbon that is bonded to three other carbons. Notice that the isobutyl group is the only group with an iso structural unit.

> **A primary carbon is bonded to one carbon, a secondary carbon is bonded to two carbons, and a tertiary carbon is bonded to three carbons.**

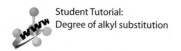

Student Tutorial:
Degree of alkyl substitution

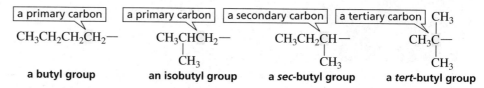

a primary carbon	a primary carbon	a secondary carbon	a tertiary carbon
$CH_3CH_2CH_2CH_2$—	CH_3CHCH_2—	CH_3CH_2CH—	CH_3C—
	CH_3	CH_3	CH_3
a butyl group	**an isobutyl group**	**a *sec*-butyl group**	**a *tert*-butyl group**

Names of straight-chain alkyl groups often have the prefix "*n*" (for "normal"), to emphasize that the carbon atoms are in an unbranched chain. If a name does not have a prefix such as "*n*" or "iso," we assume that the carbons are in an unbranched chain.

$CH_3CH_2CH_2CH_2Br$ $\qquad\qquad$ $CH_3CH_2CH_2CH_2CH_2F$

butyl bromide $\qquad\qquad\qquad$ **pentyl fluoride**

or $\qquad\qquad\qquad\qquad$ or

***n*-butyl bromide** $\qquad\qquad$ ***n*-pentyl fluoride**

Like the carbons, the hydrogens in a molecule are also referred to as primary, secondary, and tertiary. **Primary hydrogens** are attached to primary carbons, **secondary hydrogens** to secondary carbons, and **tertiary hydrogens** to tertiary carbons.

> **Primary hydrogens are attached to a primary carbon, secondary hydrogens to a secondary carbon, and tertiary hydrogens to a tertiary carbon.**

primary hydrogens	tertiary hydrogen	secondary hydrogens
$CH_3CH_2CH_2CH_2OH$	CH_3CHCH_2OH	CH_3CH_2CHOH
	CH_3	CH_3

A chemical name must specify one compound only. The prefix "*sec*," therefore, can only be used with *sec*-butyl compounds. The name "*sec*-pentyl," for example, cannot be used because pentane has two different secondary carbon atoms, which means that removing a hydrogen from a secondary carbon of pentane produces one of two different alkyl groups, depending on which hydrogen is removed. As a result, *sec*-pentyl chloride would specify two different alkyl chorides and, therefore, it is not a correct name.

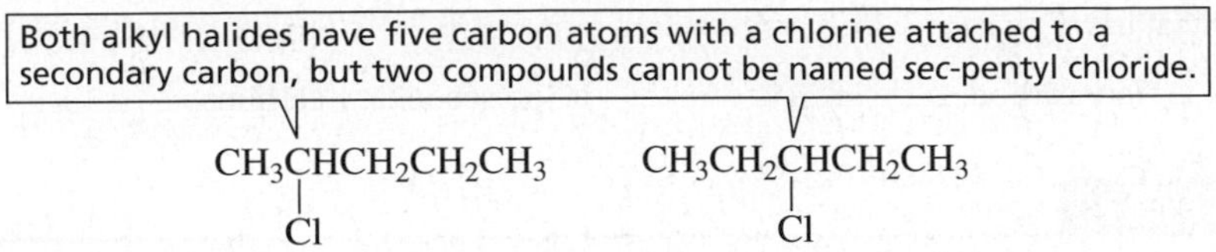

A name must specify one compound only.

> Both alkyl halides have five carbon atoms with a chlorine attached to a secondary carbon, but two compounds cannot be named *sec*-pentyl chloride.

$$CH_3CHCH_2CH_2CH_3 \qquad CH_3CH_2CHCH_2CH_3$$
$$\underset{Cl}{|} \qquad\qquad\qquad \underset{Cl}{|}$$

The prefix "*tert*" is used for *tert*-butyl and *tert*-pentyl compounds because each of these substituent names describes only one alkyl group. The name "*tert*-hexyl" cannot be used because it describes two different alkyl groups.

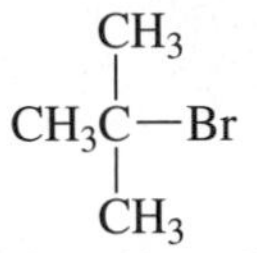

$$\overset{CH_3}{\underset{CH_3}{\overset{|}{CH_3C}}}-Br \qquad \overset{CH_3}{\underset{CH_2CH_3}{\overset{|}{CH_3C}}}-Br \qquad \overset{CH_2CH_3}{\underset{CH_3}{\overset{|}{CH_3CH_2C}}}-Br \qquad \overset{CH_3}{\underset{CH_3}{\overset{|}{CH_3CH_2CH_2C}}}-Br$$

tert-butyl bromide 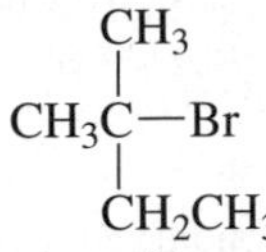   **tert-pentyl bromide**

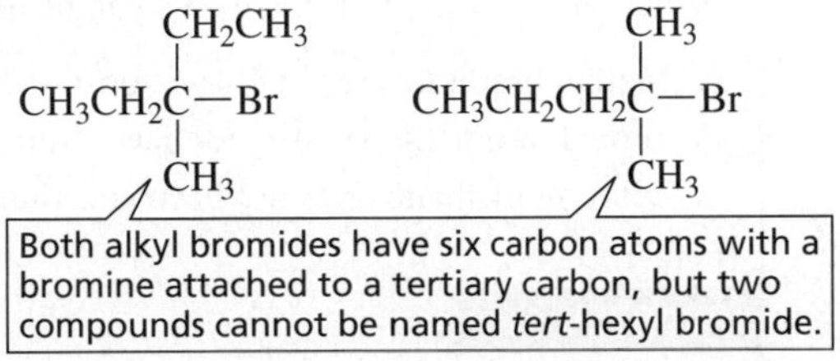

> Both alkyl bromides have six carbon atoms with a bromine attached to a tertiary carbon, but two compounds cannot be named *tert*-hexyl bromide.

If you examine the following structures, you will see that whenever the prefix "iso" is used, the iso structural unit will be at one end of the molecule and any group replacing a hydrogen will be at the other end:

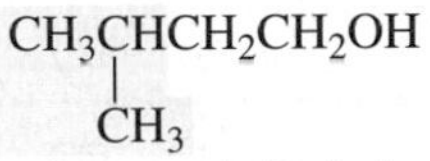

$$\underset{\underset{CH_3}{|}}{CH_3CHCH_2CH_2OH} \qquad \underset{\underset{CH_3}{|}}{CH_3CHCH_2CH_2CH_2Cl} \qquad \underset{\underset{CH_3}{|}}{CH_3CHCH_2NH_2}$$

isopentyl alcohol    **isohexyl chloride** 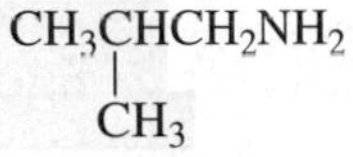   **isobutylamine**

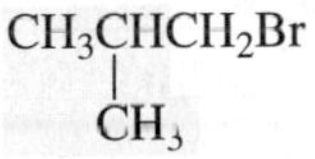

$$\underset{\underset{CH_3}{|}}{CH_3CHCH_2Br} \qquad \underset{\underset{CH_3}{|}}{CH_3CHCH_2CH_2OH} \qquad \underset{\underset{CH_3}{|}}{CH_3CHBr}$$

isobutyl bromide 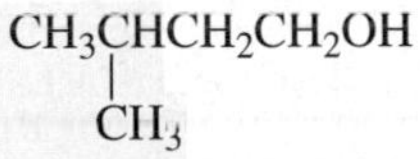   **isopentyl alcohol** 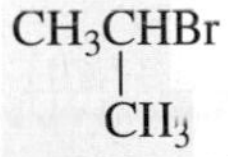   **isopropyl bromide**

Notice that an iso group has a methyl group on the next-to-the-last carbon in the chain. Notice also that all isoalkyl compounds have the substituent (OH, Cl, NH_2, etc.) on a primary carbon, except for isopropyl, which has the substituent on a secondary carbon. The isopropyl group could have been called a *sec*-propyl group. Either name would have been appropriate because the group has an iso structural unit and a hydrogen has been removed from a secondary carbon. Chemists decided to call it isopropyl, however, which means that "*sec*" is used only for *sec*-butyl.

Alkyl group names are used so frequently that you need to learn what they mean. Some of the most common alkyl group names are compiled in Table 2.2 for your convenience.

Table 2.2	Names of Some Common Alkyl Groups				
methyl	CH_3-	isobutyl	CH_3CHCH_2- $\quad\underset{CH_3}{|}$	pentyl	$CH_3CH_2CH_2CH_2CH_2-$
ethyl	CH_3CH_2-			isopentyl	$CH_3CHCH_2CH_2-$ $\quad\underset{CH_3}{|}$
propyl	$CH_3CH_2CH_2-$	*sec*-butyl	CH_3CH_2CH- $\qquad\underset{CH_3}{|}$		
isopropyl	CH_3CH- $\quad\underset{CH_3}{|}$			hexyl	$CH_3CH_2CH_2CH_2CH_2CH_2-$
			$\underset{CH_3}{|}$	isohexyl	$CH_3CHCH_2CH_2CH_2-$ $\quad\underset{CH_3}{|}$
butyl	$CH_3CH_2CH_2CH_2-$	*tert*-butyl	CH_3C- $\quad\underset{CH_3}{|}$		

PROBLEM 2◆

Draw the structure and give the systematic name of a compound with molecular formula C_5H_{12} that has

a. one tertiary carbon.

b. no secondary carbons.

PROBLEM 3◆

Draw the structures and name the four constitutional isomers with molecular formula C_4H_9Br.

PROBLEM 4◆

Which of the following statements can be used to prove that carbon is tetrahedral?

a. Methyl bromide does not have constitutional isomers.
b. Tetrachloromethane does not have a dipole moment.
c. Dibromomethane does not have constitutional isomers.

PROBLEM 5◆

Write a structure for each of the following:

a. isopropyl alcohol   **c.** *sec*-butyl iodide   **e.** *tert*-butylamine
b. isopentyl fluoride   **d.** *tert*-pentyl alcohol   **f.** *n*-octyl bromide

PROBLEM 6◆

Name the following compounds:

a. $CH_3OCH_2CH_3$

c. $CH_3CH_2CHNH_2$
         |
         CH_3

e. CH_3CHCH_2Br
         |
         CH_3

b. $CH_3OCH_2CH_2CH_3$

d. $CH_3CH_2CH_2CH_2OH$

f. CH_3CH_2CHCl
         |
         CH_3

2.2 The Nomenclature of Alkanes

The systematic name of an alkane is obtained using the following rules:

1. Determine the number of carbons in the longest continuous carbon chain. This chain is called the **parent hydrocarbon**. The name that indicates the number of carbons in the parent hydrocarbon becomes the alkane's "last name." For example, a parent hydrocarbon with eight carbons would be called *octane*. The longest continuous chain is not always a straight chain; sometimes you have to "turn a corner" to obtain the longest continuous chain.

First, determine the number of carbons in the longest continuous chain.

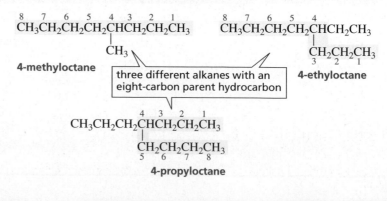

2. The name of any alkyl substituent hanging off the parent hydrocarbon is placed in front of the name of the parent hydrocarbon, together with a number to designate the carbon to which the alkyl substituent is attached. The carbons in the parent chain are numbered from the direction that gives the substituent as low a number as possible. The substituent's name and the name of the parent hydrocarbon are joined into one word, preceded by a hyphen connecting the number and the substituent's name.

Number the chain in the direction that gives the substituent the lower number.

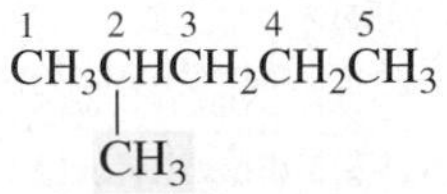

$$\overset{1}{C}H_3\overset{2}{C}H\overset{3}{C}H_2\overset{4}{C}H_2\overset{5}{C}H_3 \qquad \overset{6}{C}H_3\overset{5}{C}H_2\overset{4}{C}H_2\overset{3}{C}H\overset{2}{C}H_2\overset{1}{C}H_3$$
$$\underset{CH_3}{|} \qquad\qquad \underset{CH_2CH_3}{|}$$

2-methylpentane 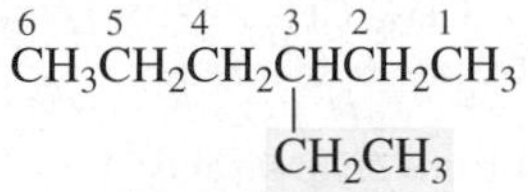   **3-ethylhexane**

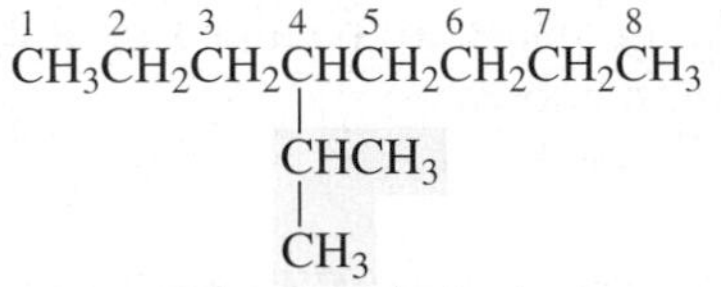

$$\overset{1}{C}H_3\overset{2}{C}H_2\overset{3}{C}H_2\overset{4}{C}H\overset{5}{C}H_2\overset{6}{C}H_2\overset{7}{C}H_2\overset{8}{C}H_3$$
$$\underset{CHCH_3}{|}$$
$$\underset{CH_3}{|}$$

4-isopropyloctane

Notice that only systematic names have numbers; common names never contain numbers.

Numbers are used only for systematic names, never for common names.

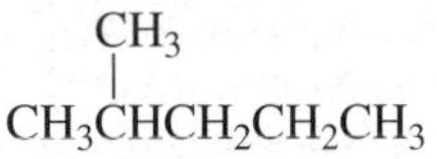

$$\underset{CH_3CHCH_2CH_2CH_3}{\overset{CH_3}{|}}$$

common name: isohexane
systematic name: 2-methylpentane

3. If more than one substituent is attached to the parent hydrocarbon, the chain is numbered in the direction that will produce a name containing the lowest of the possible numbers. The substituents are listed in alphabetical (not numerical) order, with each substituent preceded by the appropriate number. In the following example, the correct name (5-ethyl-3-methyloctane) contains a 3 as its lowest number, whereas the incorrect name (4-ethyl-6-methyloctane) contains a 4 as its lowest number:

Substituents are listed in alphabetical order.

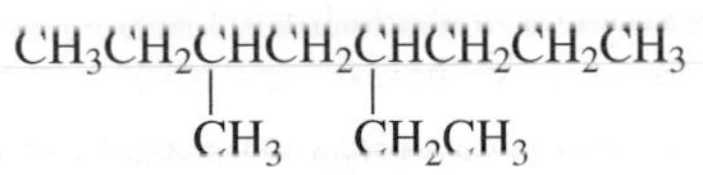

$$CH_3CH_2CHCH_2CHCH_2CH_2CH_3$$
$$\underset{CH_3}{|} \quad \underset{CH_2CH_3}{|}$$

5-ethyl-3-methyloctane
not
4-ethyl-6-methyloctane
because 3 < 4

If two or more substituents are the same, the prefixes "di," "tri," and "tetra" are used to indicate how many identical substituents the compound has. The numbers indicating the locations of the identical substituents are listed together, separated by commas. Notice that there must be as many numbers in a name as there are substituents. The prefixes "di," "tri," "tetra," "*sec*," and "*tert*" are ignored in alphabetizing substituent groups, but the prefixes "iso" and "cyclo" ("cyclo" is introduced below) are not ignored.

A number and a word are separated by a hyphen; numbers are separated by a comma.

di, tri, tetra, *sec*, and *tert* are ignored in alphabetizing.

iso and cyclo are not ignored in alphabetizing.

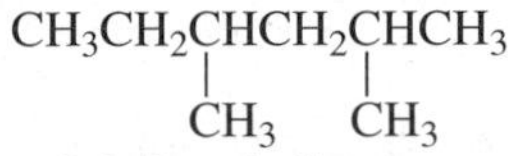

$$CH_3CH_2CHCH_2CHCH_3$$
$$\underset{CH_3}{|} \quad \underset{CH_3}{|}$$

2,4-dimethylhexane

$$CH_3CH_2CCH_2CH_2CHCH_3$$
$$\overset{CH_2CH_3}{|}$$
$$\underset{CH_3}{|} \quad \underset{CH_3}{|}$$

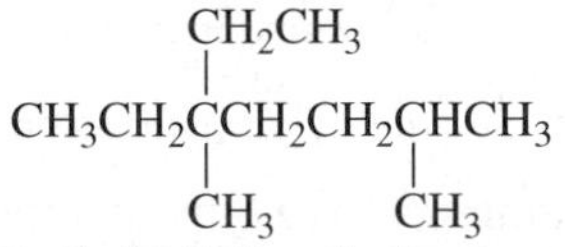

5-ethyl-2,5-dimethylheptane

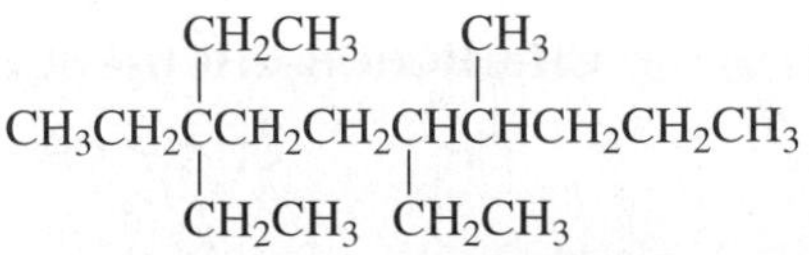

$$CH_3CH_2CCH_2CH_2CHCHCH_2CH_3$$
$$\overset{CH_2CH_3}{|} \quad \overset{CH_3}{|}$$
$$\underset{CH_2CH_3}{|} \quad \underset{CH_2CH_3}{|}$$

3,3,6-triethyl-7-methyldecane

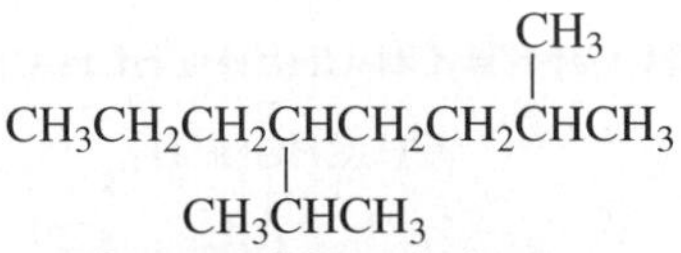

$$CH_3CH_2CH_2CHCH_2CH_2CHCH_3$$
$$\overset{CH_3}{|}$$
$$\underset{CH_3CHCH_3}{|}$$

5-isopropyl-2-methyloctane

4. When numbering in either direction leads to the same lowest number for one of the substituents, the chain is numbered in the direction that gives the lowest possible number to one of the remaining substituents.

$$CH_3CCH_2CHCH_3$$

2,2,4-trimethylpentane
not
2,4,4-trimethylpentane
because 2 < 4

$$CH_3CH_2CHCHCH_2CHCH_2CH_3$$

6-ethyl-3,4-dimethyloctane
not
3-ethyl-5,6-dimethyloctane
because 4 < 5

Only if the same set of numbers is obtained in both directions does the first group cited get the lower number.

5. If the same substituent numbers are obtained in both directions, the first group cited receives the lower number.

$$CH_3CHCHCH_3$$

2-bromo-3-chlorobutane
not
3-bromo-2-chlorobutane

$$CH_3CH_2CHCH_2CHCH_2CH_3$$

3-ethyl-5-methylheptane
not
5-ethyl-3-methylheptane

In the case of two hydrocarbon chains with the same number of carbons, choose the one with the most substituents.

6. If a compound has two or more chains of the same length, the parent hydrocarbon is the chain with the greatest number of substituents.

$$\overset{3}{CH_3}\overset{4}{CH_2}\overset{5}{CH}\overset{6}{CH_2CH_2CH_3}$$
$$\overset{2}{CHCH_3}$$
$$\overset{1}{CH_3}$$

3-ethyl-2-methylhexane (two substituents)

$$\overset{1}{CH_3}\overset{2}{CH_2}\overset{3}{CH}\overset{4}{CH_2}\overset{5}{CH_2}\overset{6}{CH_3}$$
$$CHCH_3$$
$$CH_3$$

not
3-isopropylhexane (one substituent)

7. Names such as "isopropyl," "*sec*-butyl," and "*tert*-butyl" are acceptable substituent names in the IUPAC system of nomenclature, but systematic substituent names are preferred. Systematic substituent names are obtained by numbering the alkyl substituent starting at the carbon attached to the parent hydrocarbon. This means that the carbon attached to the parent hydrocarbon is always the number-1 carbon of the substituent. In a compound such as 4-(1-methylethyl)octane, the substituent name is in parentheses; the number inside the parentheses indicates a position on the substituent, whereas the number outside the parentheses indicates a position on the parent hydrocarbon. (Note that if a prefix such as "di" is part of a branch name, it *is* included in the alphabetization.)

$$CH_3CH_2CH_2CHCH_2CH_2CH_3$$
$$\overset{1}{CHCH_3}$$
$$CH_3$$

4-isopropylheptane
or
4-(1-methylethyl)heptane

$$CH_3CH_2CH_2CH_2CHCH_2CH_2CH_2CH_3$$
$$\overset{1}{CH_2}\overset{2}{CH}\overset{3}{CH_3}$$
$$CH_3$$

5-isobutyldecane
or
5-(2-methylpropyl)decane

$$CH_3CH_2$$
$$CH_3CH_2CHCH_2CHCH_2CH_2CH_3$$
$$CH_3CCH_3$$
$$CH_3$$

5-*tert*-butyl-3-ethyloctane
or
5-(1,1-dimethylethyl)-3-ethyloctane

Some substituents have only a systematic name.

$$CH_3CH_2CH_2CH_2CHCH_2CHCH_2CH_3$$
$$CH_3\overset{2}{CH}\overset{3}{CH}CH_3$$
$$\overset{1}{CH_3}$$

6-(1,2-dimethylpropyl)-4-propyldecane

$$CH_3CHCHCH_2CHCH_2CHCH_2CH_3$$
$$CH_3 \qquad CH_2CH_2CH_2CH_2CH_3$$

2,3-dimethyl-5-(2-methylbutyl)decane

These rules will allow you to name thousands of alkanes, and eventually you will learn the additional rules necessary to name many other kinds of compounds. The rules are important for looking up a compound in the scientific literature, because it usually will be listed by its systematic name. Nevertheless, you must still learn common names because they are so entrenched in chemists' vocabularies that they are widely used in scientific conversation and are often found in the literature.

Look at the systematic names (the ones written in blue) for the isomeric hexanes and isomeric heptanes shown at the beginning of this chapter to make sure you understand how they are constructed.

Student Tutorial:
Basic nomenclature of alkanes

PROBLEM 7♦

Draw the structure for each of the following:

a. 2,3-dimethylhexane

b. 4-isopropyl-2,4,5-trimethylheptane

c. 4,4-diethyldecane

d. 2,2-dimethyl-4-propyloctane

e. 4-isobutyl-2,5-dimethyloctane

f. 4-(1,1-dimethylethyl)octane

PROBLEM 8   *SOLVED*

a. Draw the 18 isomeric octanes.

b. Give each isomer its systematic name.

c. How many isomers have common names?

d. Which isomers contain an isopropyl group?

e. Which isomers contain a *sec*-butyl group?

f. Which isomers contain a *tert*-butyl group?

Solution to 8a   Start with the isomer with an eight-carbon continuous chain. Then draw isomers with a seven-carbon continuous chain plus one methyl group. Next, draw isomers with a six-carbon continuous chain plus two methyl groups or one ethyl group. Then draw isomers with a five-carbon continuous chain plus three methyl groups or one methyl group and one ethyl group. Finally, draw a four-carbon continuous chain with four methyl groups. (You will be able to tell whether you have drawn duplicate structures by your answers to Problem 8b, because if two structures have the same systematic name, they represent the same compound.)

PROBLEM 9♦

Give the systematic name for each of the following:

a. 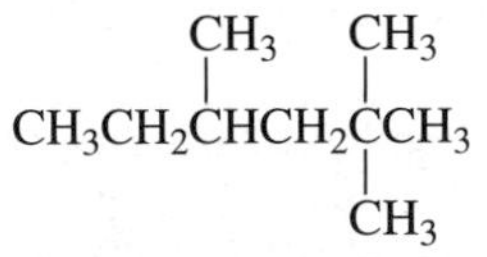 $CH_3CH_2\overset{\overset{\displaystyle CH_3}{|}}{C}HCH_2\underset{\underset{\displaystyle CH_3}{|}}{\overset{\overset{\displaystyle CH_3}{|}}{C}}CH_3$

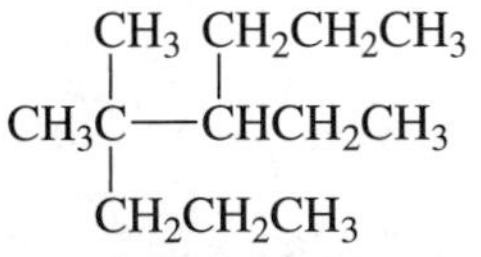

b. $CH_3CH_2C(CH_3)_3$

c. $CH_3CH_2CH_2\underset{\underset{\displaystyle CH_3CHCH_2CH_3}{|}}{C}HCH_2CH_2CH_3$

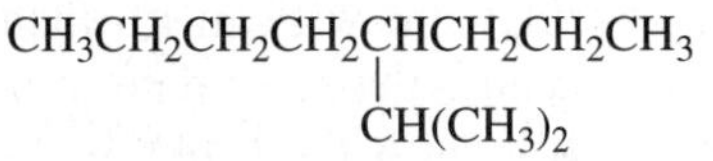

d. 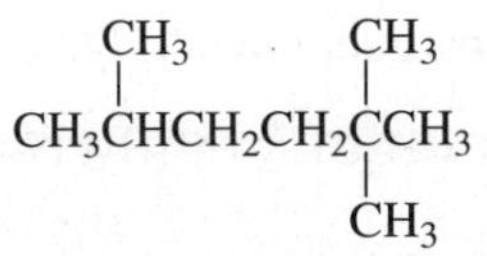 $CH_3\overset{\overset{\displaystyle CH_3}{|}}{C}HCH_2CH_2\underset{\underset{\displaystyle CH_3}{|}}{\overset{\overset{\displaystyle CH_3}{|}}{C}}CH_3$

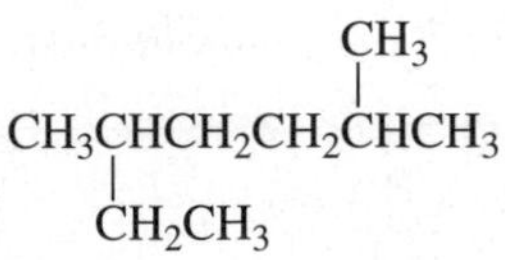

e. $CH_3CH_2C(CH_2CH_3)_2CH(CH_3)CH(CH_2CH_2CH_3)_2$

f. $CH_3\underset{\underset{\displaystyle CH_2CH_2CH_3}{|}}{\overset{\overset{\displaystyle CH_3}{|}}{C}}-\underset{}{\overset{\overset{\displaystyle CH_2CH_2CH_3}{|}}{C}}HCH_2CH_3$

g. $CH_3CH_2C(CH_2CH_3)_2CH_2CH_2CH_3$

h. $CH_3CH_2CH_2CH_2\underset{\underset{\displaystyle CH(CH_3)_2}{|}}{C}HCH_2CH_2CH_3$

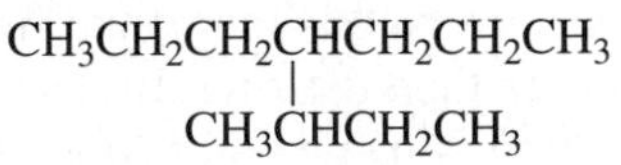

i. $CH_3\overset{\overset{\displaystyle CH_3}{|}}{C}HCH_2CH_2\underset{\underset{\displaystyle CH_2CH_3}{|}}{C}HCH_3$

PROBLEM 10♦

Draw the structure and give the systematic name of a compound with molecular formula C_5H_{12} that has

a. only primary and secondary hydrogens.   c. one tertiary hydrogen.
b. only primary hydrogens.   d. two secondary hydrogens.

2.3 Nomenclature of Cycloalkanes • Skeletal Structures

Cycloalkanes are alkanes with their carbon atoms arranged in a ring. Because of the ring, a cycloalkane has two fewer hydrogens than an acyclic (noncyclic) alkane with the same number of carbons. This means that the general molecular formula for a cycloalkane is C_nH_{2n}. Cycloalkanes are named by adding the prefix "cyclo" to the alkane name that signifies the number of carbon atoms in the ring.

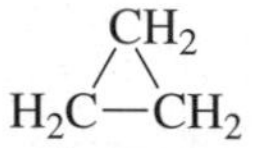

cyclopropane 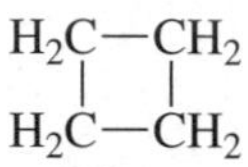   cyclobutane 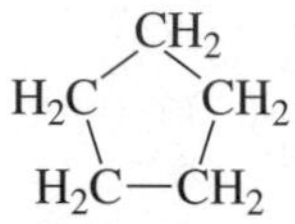   cyclopentane 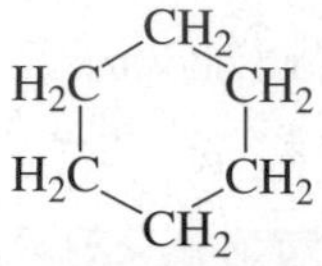   cyclohexane

Cycloalkanes are almost always written as **skeletal structures**. Skeletal structures show the carbon–carbon bonds as lines, but do not show the carbons or the hydrogens bonded to carbons. Atoms other than carbon are shown, and hydrogens bonded to atoms other than carbon are shown. Each vertex in a skeletal structure represents a carbon, and each carbon is understood to be bonded to the appropriate number of hydrogens to give the carbon four bonds.

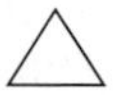

cyclopropane 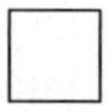   cyclobutane 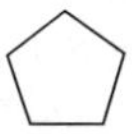   cyclopentane    cyclohexane

Acyclic molecules can also be represented by skeletal structures. In skeletal structures of acyclic molecules, the carbon chains are represented by zigzag lines. Again, each vertex represents a carbon, and carbons are assumed to be present where a line begins or ends.

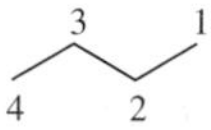

butane 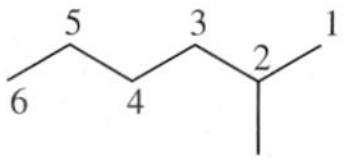   2-methylhexane 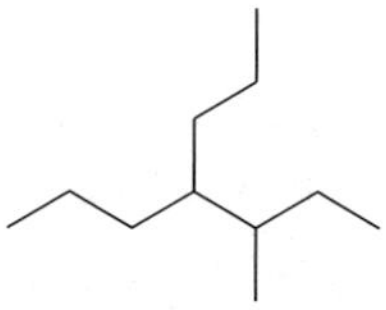   3-methyl-4-propylheptane 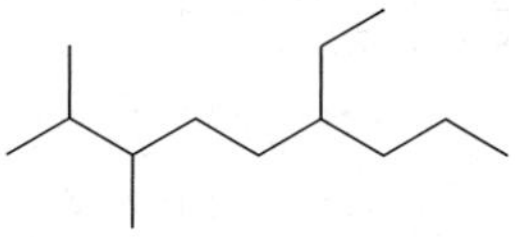   6-ethyl-2,3-dimethylnonane

The rules for naming cycloalkanes resemble the rules for naming acyclic alkanes:

1. In a cycloalkane with an attached alkyl substituent, the ring is the parent hydrocarbon unless the substituent has more carbon atoms than the ring. In that case, the substituent is the parent hydrocarbon and the ring is named as a substituent. There is no need to number the position of a single substituent on a ring.

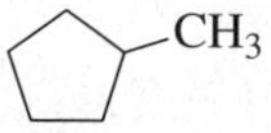

If there is only one substituent on a ring, do not give that substituent a number.

methylcyclopentane 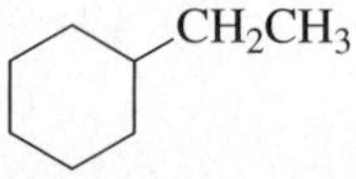   ethylcyclohexane 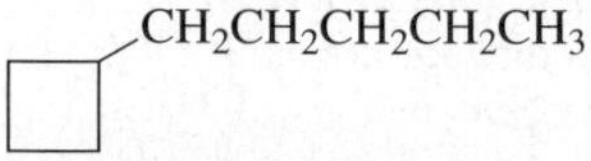   1-cyclobutylpentane

2. If the ring has two different substituents, they are cited in *alphabetical order* and the number-1 position is given to the substituent cited first.

Student Tutorial:
Alkyl group nomenclature

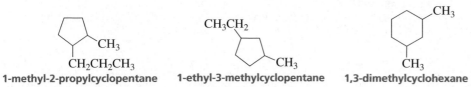

1-methyl-2-propylcyclopentane

1-ethyl-3-methylcyclopentane

1,3-dimethylcyclohexane

3. If there are more than two substituents on the ring, they are cited in alphabetical order. The substituent given the number-1 position is the one that results in a second substituent getting as low a number as possible. If two substituents have the same low number, the ring is numbered—either clockwise or counterclockwise—in the direction that gives the third substituent the lowest possible number. For example, the correct name of the following compound is 4-ethyl-2-methyl-1-propylcyclohexane, not 5-ethyl-1-methyl-2-propylcyclohexane:

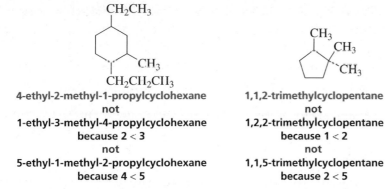

4-ethyl-2-methyl-1-propylcyclohexane
not
1-ethyl-3-methyl-4-propylcyclohexane
because 2 < 3
not
5-ethyl-1-methyl-2-propylcyclohexane
because 4 < 5

1,1,2-trimethylcyclopentane
not
1,2,2-trimethylcyclopentane
because 1 < 2
not
1,1,5-trimethylcyclopentane
because 2 < 5

Student Tutorial:
Advanced alkane nomenclature

PROBLEM-SOLVING STRATEGY

Interpreting a Skeletal Structure

Write the number of hydrogens attached to each of the indicated carbon atoms in the following compound:

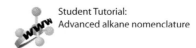

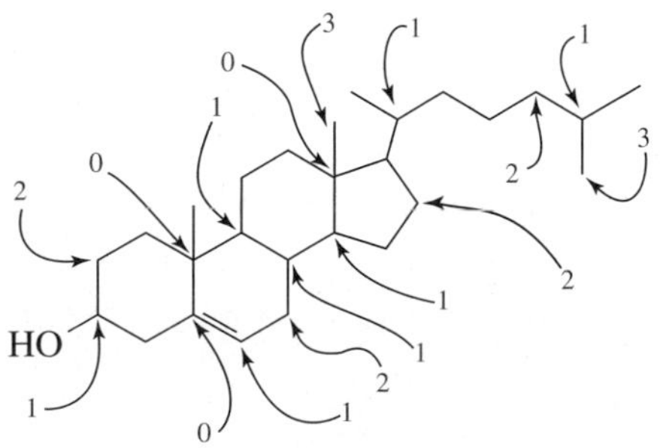

cholesterol

All the carbon atoms in the compound are neutral, so each needs to be bonded to four atoms. Thus if the carbon has only one bond showing, it must be attached to three hydrogens that are not shown; if the carbon has two bonds showing, it must be attached to two hydrogens that are not shown, etc.

Now continue on to Problem 11.

PROBLEM 11

Write the number of hydrogens attached to each of the indicated carbon atoms in the following compound.

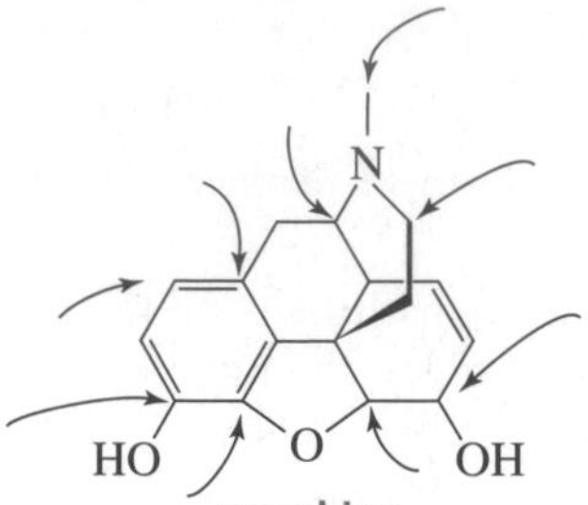

morphine

PROBLEM 12◆

Convert the following condensed structures into skeletal structures (remember that condensed structures show atoms but few, if any, bonds, whereas skeletal structures show bonds but few, if any, atoms):

a. $CH_3CH_2CH_2CH_2CH_2CH_2OH$

b. $CH_3CH_2CH_2CH_2CH_2CH_3$

c. $CH_3CH_2\overset{\displaystyle CH_3}{\underset{|}{C}H}CH_2\overset{\displaystyle CH_3}{\underset{|}{C}H}CH_2CH_3$

d. $CH_3CH_2CH_2CH_2OCH_3$

e. $CH_3CH_2NHCH_2CH_2CH_3$

f. $CH_3\overset{\displaystyle CH_3}{\underset{|}{C}H}CH_2CH_2\underset{\underset{\displaystyle Br}{|}}{C}HCH_3$

PROBLEM 13

Convert the structures in Problem 9 into skeletal structures.

PROBLEM 14◆

Give the systematic name for each of the following:

a. CH_2CH_3
        CH_3

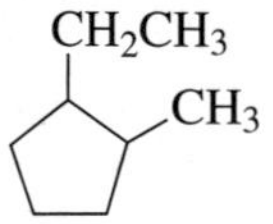

b. CH_2CH_3

c. H_3C
        H_3C CH_2CH_3

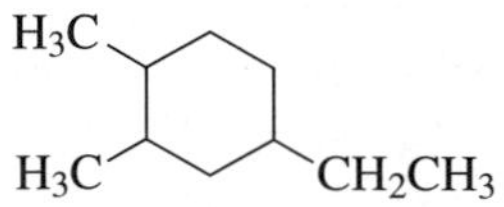

d.

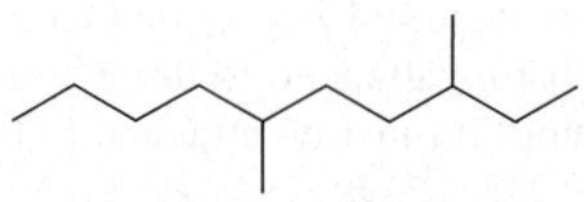

e. $CH_3CHCH_2CH_2CH_3$

f. CH_2CH_3
        CH_2CHCH_3
        CH_3

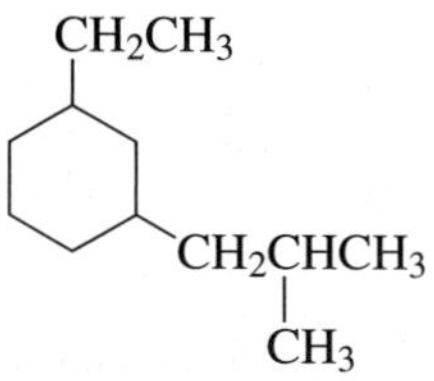

g. $CH_3CH_2CHCH_3$

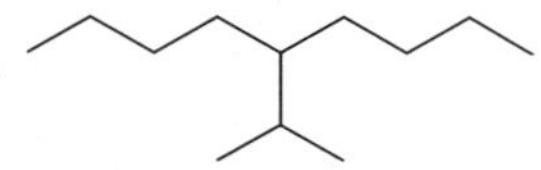

h. $CH_3CH_2CHCH_3$

        CH_3CHCH_3

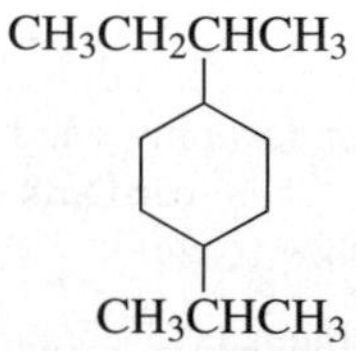

2.4 **Nomenclature of Alkyl Halides**

An **alkyl halide** is a compound in which a hydrogen of an alkane has been replaced by a halogen. Alkyl halides are classified as primary, secondary, or tertiary, depending on the carbon to which the halogen is attached. **Primary alkyl halides** have a halogen bonded to a primary carbon, **secondary alkyl halides** have a halogen bonded to a secondary carbon, and **tertiary alkyl halides** have a halogen bonded to a tertiary carbon (Section 2.1). The lone-pair electrons on the halogens are generally not shown unless they are needed to draw your attention to some chemical property of the atom.

The number of alkyl groups attached to the carbon to which the halogen is bonded determines whether an alkyl halide is primary, secondary, or tertiary.

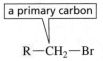

a primary carbon	a secondary carbon	a tertiary carbon
R—CH$_2$—Br	R—CH—R Br	R—C—R (with R above and Br below)
a primary alkyl halide	a secondary alkyl halide	a tertiary alkyl halide

The common names of alkyl halides consist of the name of the alkyl group, followed by the name of the halogen—with the "ine" ending of the halogen name replaced by "ide" (i.e., fluoride, chloride, bromide, iodide).

	CH$_3$Cl	CH$_3$CH$_2$F	CH$_3$CHI CH$_3$	CH$_3$CH$_2$CHBr CH$_3$
common name:	methyl chloride	ethyl fluoride	isopropyl iodide	*sec*-butyl bromide
systematic name:	chloromethane	fluoroethane	2-iodopropane	2-bromobutane

CH_3F
methyl fluoride

CH_3Cl
methyl chloride

CH_3Br
methyl bromide

CH_3I
methyl iodide

In the IUPAC system, alkyl halides are named as substituted alkanes. The prefixes for the halogens end with "o" (i.e., "fluoro," "chloro," "bromo," "iodo"). Therefore, alkyl halides are often called haloalkanes. Notice that although a name must specify only one compound, a compound can have more than one name.

CH$_3$
CH$_3$CH$_2$CHCH$_2$CH$_2$CHCH$_3$
Br
2-bromo-5-methylheptane

CH$_3$
CH$_3$CCH$_2$CH$_2$CH$_2$CH$_2$Cl
CH$_3$
1-chloro-5,5-dimethylhexane

1-ethyl-2-iodocyclopentane

4-bromo-2-chloro-1-methylcyclohexane

PROBLEM 15◆

Give two names for each of the following, and tell whether each alkyl halide is primary, secondary, or tertiary:

a. CH$_3$CH$_2$CHCH$_3$
           Cl

b. CH$_3$CHCH$_2$CH$_2$CH$_2$Cl
           CH$_3$

c. (cyclohexane with Br)

d. CH$_3$CHCH$_3$
           F

Student Tutorial:
Advanced alkyl nomenclature

A compound can have more than one name, but a name must specify only one compound.

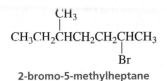

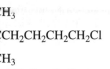

Draw the structures and provide systematic names for parts a, b, and c by substituting a chlorine for a hydrogen of methylcyclohexane:

a. a primary alkyl halide
b. a tertiary alkyl halide
c. three secondary alkyl halides

2.5 Nomenclature of Ethers

An **ether** is a compound in which an oxygen is bonded to two alkyl substituents. If the alkyl substituents are identical, the ether is a **symmetrical ether**. If the substituents are different, the ether is an **unsymmetrical ether**.

Chemists sometimes neglect the prefix "di" when they name symmetrical ethers. Try not to make this oversight a habit.

R—O—R   R—O—R′

a symmetrical ether   **an unsymmetrical ether**

The common name of an ether consists of the names of the two alkyl substituents (in alphabetical order), followed by the word "ether." The smallest ethers are almost always named by their common names.

dimethyl ether

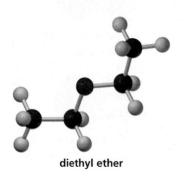

diethyl ether

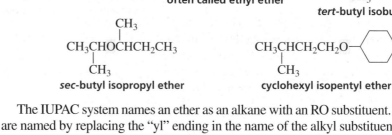

CH₃OCH₂CH₃   CH₃CH₂OCH₂CH₃
ethyl methyl ether   **diethyl ether**
often called ethyl ether

$$CH_3CHCH_2O\overset{\overset{\displaystyle CH_3}{|}}{C}CH_3$$
$$\underset{CH_3}{|}\quad\underset{CH_3}{|}$$
***tert*-butyl isobutyl ether**

$$CH_3CHOCHCH_2CH_3$$
with CH₃ groups
***sec*-butyl isopropyl ether**

$$CH_3CHCH_2CH_2O-\bigcirc$$
with CH₃
cyclohexyl isopentyl ether

The IUPAC system names an ether as an alkane with an RO substituent. The substituents are named by replacing the "yl" ending in the name of the alkyl substituent with "oxy."

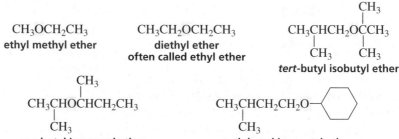

CH₃O—   CH₃CH₂O—   CH₃CHO—   CH₃CH₂CHO—   CH₃CO—
methoxy   **ethoxy**   with CH₃ **isopropoxy**   with CH₃ ***sec*-butoxy**   with CH₃ and CH₃ ***tert*-butoxy**

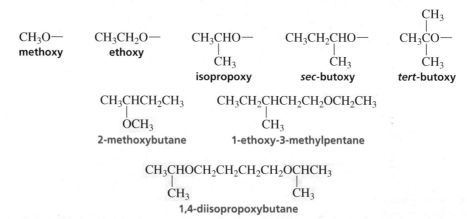

CH₃CHCH₂CH₃   CH₃CH₂CHCH₂CH₂OCH₂CH₃
with OCH₃   with CH₃
2-methoxybutane   **1-ethoxy-3-methylpentane**

CH₃CHOCH₂CH₂CH₂CH₂OCHCH₃
with CH₃ and CH₃
1,4-diisopropoxybutane

a. Give the systematic (IUPAC) name for each of the following ethers:

   1. CH₃OCH₂CH₃

   2. CH₃CH₂OCH₂CH₃

   3. CH₃CH₂CH₂CH₂CHCH₂CH₂CH₃
                   OCH₃

   4. CH₃CHOCH₂CH₂CHCH₃
           CH₃   CH₃

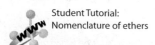

Student Tutorial:
Nomenclature of ethers

Student Tutorial:
Ether nomenclature

5. $CH_3CH_2CH_2OCH_2CH_2CH_2CH_3$ **6.** $CH_3\overset{\overset{\displaystyle CH_3}{|}}{C}HOCHCH_2CH_2CH_3$
$\qquad\qquad\qquad\qquad\qquad\qquad\qquad\qquad\underset{|}{}$
$\qquad\qquad\qquad\qquad\qquad\qquad\qquad\qquad CH_3$

b. Do all of these ethers have common names?

c. What are their common names?

2.6 Nomenclature of Alcohols

An **alcohol** is a compound in which a hydrogen of an alkane has been replaced by an OH group. Alcohols are classified as **primary alcohols**, **secondary alcohols**, or **tertiary alcohols**, depending on whether the OH group is bonded to a primary, secondary, or tertiary carbon—just like the way alkyl halides are classified.

> The number of alkyl groups attached to the carbon to which the OH group is attached determines whether an alcohol is primary, secondary, or tertiary.

$$R-CH_2-OH \qquad R-\overset{\overset{\displaystyle R}{|}}{C}H-OH \qquad R-\overset{\overset{\displaystyle R}{|}}{\underset{\underset{\displaystyle R}{|}}{C}}-OH$$

a primary alcohol   a secondary alcohol   a tertiary alcohol

The common name of an alcohol consists of the name of the alkyl group to which the OH group is attached, followed by the word "alcohol."

CH_3CH_2OH $\qquad$ $CH_3CH_2CH_2OH$ $\qquad$ $CH_3\underset{\underset{\displaystyle CH_3}{|}}{C}HOH$ $\qquad$ $CH_3\underset{\underset{\displaystyle CH_3}{|}}{C}HCH_2OH$
ethyl alcohol $\qquad$ propyl alcohol $\qquad\qquad$ isopropyl alcohol $\qquad$ isobutyl alcohol

The **functional group** is the center of reactivity in an organic molecule. In an alcohol, the OH is the functional group. The IUPAC system uses *suffixes* to denote certain functional groups. The systematic name of an alcohol, for example, is obtained by replacing the "e" at the end of the name of the parent hydrocarbon with the suffix "ol."

CH_3OH $\qquad$ CH_3CH_2OH
methanol $\qquad$ ethanol

When necessary, the position of the functional group is indicated by a number immediately preceding the name of the alcohol or immediately preceding the suffix. The most recently approved IUPAC names are those with the number immediately preceding the suffix. However, names with the number preceding the name of the alcohol have been in use for a long time, so those are the ones most likely to appear in the literature, on reagent bottles, and on standardized tests. They will also be the ones that appear most often in this book.

$CH_3CH_2\underset{\underset{\displaystyle OH}{|}}{C}HCH_2CH_3$

3-pentanol
or
pentan-3-ol

methyl alcohol

ethyl alcohol

propyl alcohol

The following rules are used to name a compound that has a functional group suffix:

1. The parent hydrocarbon is the longest continuous chain *containing the functional group*.

2. The parent hydrocarbon is numbered in the direction that gives the *functional group suffix the lowest possible number*.

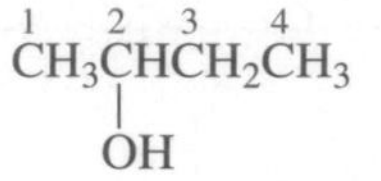

$$\overset{1}{C}H_3\overset{2}{C}H\overset{3}{C}H_2\overset{4}{C}H_3$$
|
OH

2-butanol
or
butan-2-ol

$$\overset{5}{C}H_3\overset{4}{C}H_2\overset{3}{C}H_2\overset{2}{C}H\overset{1}{C}H_2OH$$
|
CH_2CH_3

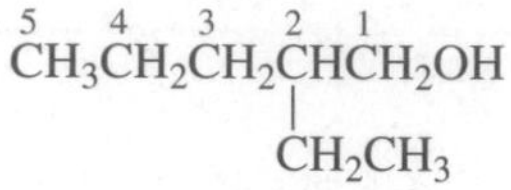

2-ethyl-1-pentanol
or
2-ethylpentan-1-ol

$$CH_3CH_2CH_2CH_2O\overset{3}{C}H_2\overset{2}{C}H_2\overset{1}{C}H_2OH$$

3-butoxy-1-propanol
or
3-butoxypropan-1-ol

The longest continuous chain has six carbons, but the longest continuous chain containing the OH functional group has five carbons so the compound is named as a pentanol.

The longest continuous chain has four carbons, but the longest continuous chain containing the OH functional group has three carbons, so the compound is named as a propanol.

When there is only a substituent, the substituent gets the lowest possible number.

When there is only a functional group suffix, the functional group suffix gets the lowest possible number.

When there is both a functional group suffix and a substituent, the functional group suffix gets the lowest possible number.

3. If there is a functional group suffix and a substituent, the functional group suffix gets the lowest possible number.

$$HO\overset{1}{C}H_2\overset{2}{C}H_2\overset{3}{C}H_2Br$$

3-bromo-1-propanol

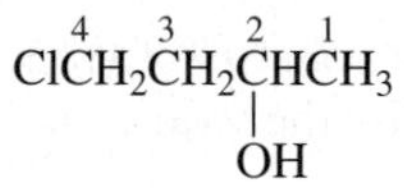

$$Cl\overset{4}{C}H_2\overset{3}{C}H_2\overset{2}{C}H\overset{1}{C}H_3$$
|
OH

4-chloro-2-butanol

$$\overset{5}{C}H_3\overset{4}{C}\overset{3}{C}H_2\overset{2}{C}H\overset{1}{C}H_3$$
| |
CH_3 OH

(with CH_3 above)

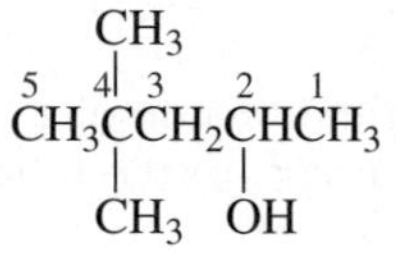

4,4-dimethyl-2-pentanol

4. If counting in either direction gives the same number for the functional group suffix, the chain is numbered in the direction that gives a substituent the lowest possible number. Notice that a number is not needed to designate the position of a functional group suffix in a cyclic compound, because it is assumed to be at the 1-position.

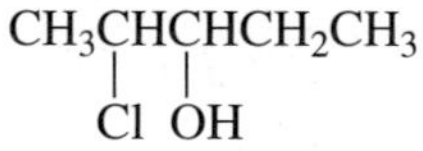

$$CH_3CHCHCH_2CH_3$$
| |
Cl OH

2-chloro-3-pentanol
not
4-chloro-3-pentanol

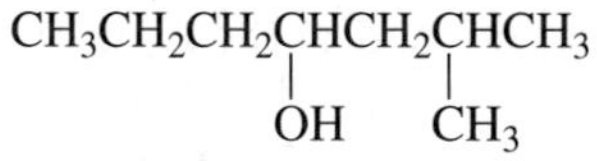

$$CH_3CH_2CH_2CHCH_2CHCH_3$$
| |
OH CH_3

2-methyl-4-heptanol
not
6-methyl-4-heptanol

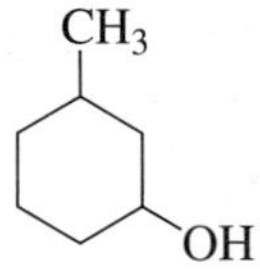

3-methylcyclohexanol
not
5-methylcyclohexanol

Student Tutorial:
Nomenclature of alcohols

5. If there is more than one substituent, the substituents are cited in alphabetical order.

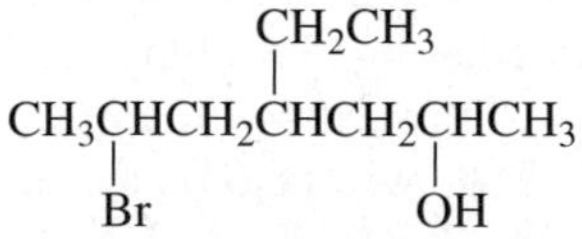

$$CH_3CHCH_2CHCH_2CHCH_3$$
| |
Br OH

(with CH_2CH_3 above)

6-bromo-4-ethyl-2-heptanol

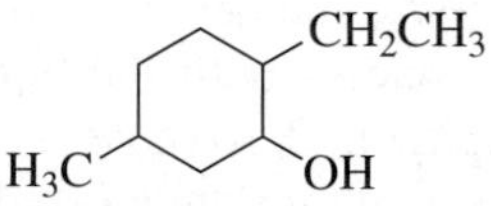

2-ethyl-5-methylcyclohexanol

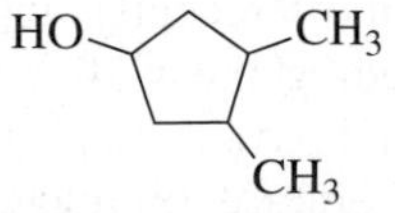

3,4-dimethylcyclopentanol

Remember that the name of a substituent is stated *before* the name of the parent hydrocarbon, and the functional group suffix is stated *after* the name of the parent hydrocarbon.

[substituent] [parent hydrocarbon] [functional group suffix]

PROBLEM 18

Draw the structures of a homologous series of alcohols that have from one to six carbons, and then give each of them a common name and a systematic name.

PROBLEM 19◆

Give each of the following a systematic name, and indicate whether each is a primary, secondary, or tertiary alcohol:

a. $CH_3CH_2CH_2CH_2CH_2OH$

d. $CH_3CHCH_2CHCH_2CH_3$
        | |
       CH_3 OH

b.

e.

c. $CH_3CCH_2CH_2CH_2Cl$ with CH_3 above and OH below

f. $CH_3CHCH_2CHCH_2CHCH_2CH_3$
       | | |
     CH_3 OH CH_3

PROBLEM 20◆

Write the structures of all the tertiary alcohols with molecular formula $C_6H_{14}O$, and give each a systematic name.

Student Tutorial:
Summary of systematic nomenclature

2.7 Nomenclature of Amines

An **amine** is a compound in which one or more hydrogens of ammonia have been replaced by alkyl groups. There are **primary amines**, **secondary amines**, and **tertiary amines**. The classification depends on how many alkyl groups are bonded to the nitrogen. Primary amines have one alkyl group bonded to the nitrogen, secondary amines have two, and tertiary amines have three.

NH_3 $R—NH_2$ $R—NH$ with R above $R—N—R$ with R above

ammonia   **a primary amine**   **a secondary amine**   **a tertiary amine**

Notice that the number of alkyl groups *attached to the nitrogen* determines whether an amine is primary, secondary, or tertiary. For an alkyl halide or an alcohol, on the other hand, the number of alkyl groups *attached to the carbon* to which the halogen or the OH is bonded determines the classification (Sections 2.4 and 2.6).

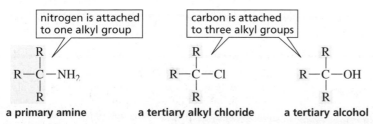

nitrogen is attached to one alkyl group

carbon is attached to three alkyl groups

$R—C—NH_2$ $R—C—Cl$ $R—C—OH$

a primary amine   **a tertiary alkyl chloride**   **a tertiary alcohol**

The number of alkyl groups attached to the nitrogen determines whether an amine is primary, secondary, or tertiary.

The common name of an amine consists of the names of the alkyl groups bonded to the nitrogen, in alphabetical order, followed by "amine." The entire name is written as one word (unlike the common names of alcohols, ethers, and alkyl halides, in which "alcohol," "ether," and "halide" are separate words).

$$CH_3NH_2$$
methylamine

$$CH_3NHCH_2CH_2CH_3$$
methylpropylamine

$$CH_3CH_2NHCH_2CH_3$$
diethylamine

$$\overset{\displaystyle CH_3}{\underset{\displaystyle |}{CH_3NCH_3}}$$
trimethylamine

$$\overset{\displaystyle CH_3}{\underset{\displaystyle |}{CH_3NCH_2CH_2CH_3}}$$
butyldimethylamine

$$\overset{\displaystyle CH_3}{\underset{\displaystyle |}{CH_3CH_2NCH_2CH_2CH_3}}$$
ethylmethylpropylamine

The IUPAC system uses the suffix "amine" to denote the amine functional group. The "e" at the end of the name of the parent hydrocarbon is replaced by "amine"—similar to the way in which alcohols are named. A number identifies the carbon to which the nitrogen is attached. The number can appear before the name of the parent hydrocarbon or before "amine." The name of any alkyl group bonded to nitrogen is preceded by an "*N*" (in italics) to indicate that the group is bonded to a nitrogen rather than to a carbon.

$$\overset{4\quad3\quad2\quad1}{CH_3CH_2CH_2CH_2NH_2}$$
1-butanamine
or
butan-1-amine

$$\overset{1\quad2\quad3\quad4\quad5\quad6}{CH_3CH_2CHCH_2CH_2CH_3}$$
$$|$$
$$NHCH_2CH_3$$
***N*-ethyl-3-hexanamine**
or
***N*-ethylhexan-3-amine**

$$\overset{3\quad2\quad1}{CH_3CH_2CH_2NCH_2CH_3}$$
$$|$$
$$CH_3$$
***N*-ethyl-*N*-methyl-1-propanamine**
or
***N*-ethyl-*N*-methylpropan-1-amine**

The substituents—regardless of whether they are attached to the nitrogen or to the parent hydrocarbon—are listed in alphabetical order, and then a number or an "*N*" is assigned to each one. The chain is numbered in the direction that gives the functional group suffix the lowest number.

$$\overset{4\quad3\quad2\quad1}{CH_3CHCH_2CH_2NHCH_3}$$
$$|$$
$$Cl$$
3-chloro-*N*-methyl-1-butanamine

$$\overset{1\quad2\quad3\quad4\quad5\quad6}{CH_3CH_2CHCH_2CHCH_3}$$
$$\overset{\displaystyle CH_3}{\underset{\displaystyle |}{}}$$
$$|$$
$$NHCH_2CH_3$$
***N*-ethyl-5-methyl-3-hexanamine**

$$\overset{5\quad4\quad3\quad2\quad1}{CH_3CHCH_2CHCH_3}$$
$$\overset{\displaystyle Br}{\underset{\displaystyle |}{}}$$
$$|$$
$$CH_3NCH_3$$
4-bromo-*N*,*N*-dimethyl-2-pentanamine

2-ethyl-*N*-propylcyclohexanamine

Nitrogen compounds with four alkyl groups bonded to the nitrogen—thereby giving the nitrogen a positive formal charge—are called **quaternary ammonium salts**. Their names consist of the names of the alkyl groups in alphabetical order, followed by "ammonium" (all in one word), and then the name of the counterion (the accompanying anion) as a separate word.

$$\overset{\displaystyle CH_3}{\underset{\displaystyle |}{CH_3-N^{\pm}-CH_3}}\ \ HO^-$$
$$|$$
$$CH_3$$
tetramethylammonium hydroxide

$$\overset{\displaystyle CH_3}{\underset{\displaystyle |}{CH_3CH_2CH_2-N^{\pm}-CH_3}}\ \ Cl^-$$
$$|$$
$$CH_2CH_3$$
ethyldimethylpropylammonium chloride

BAD-SMELLING COMPOUNDS

Amines are associated with some of nature's unpleasant odors. Amines with relatively small alkyl groups, for example, have a fishy smell. Thus, fermented shark, a traditional dish in Iceland, smells exactly like triethylamine. The amines putrescine and cadavarine are poisonous compounds formed when amino acids are degraded. Because the body excretes them in the quickest ways possible, their odors may be detected in the urine and breath. They are also responsible for the odor of decaying flesh.

putrescine

cadavarine

Table 2.3 summarizes the ways in which alkyl halides, ethers, alcohols, and amines are named.

Table 2.3 Summary of Nomenclature		
	Systematic name	**Common name**
Alkyl halide	substituted alkane CH_3Br bromomethane CH_3CH_2Cl chloroethane	alkyl group to which halogen is attached, plus *halide* CH_3Br methyl bromide CH_3CH_2Cl ethyl chloride
Ether	substituted alkane CH_3OCH_3 methoxymethane $CH_3CH_2OCH_3$ methoxyethane	alkyl groups attached to oxygen, plus *ether* CH_3OCH_3 dimethyl ether $CH_3CH_2OCH_3$ ethyl methyl ether
Alcohol	functional group suffix is *ol* CH_3OH methanol CH_3CH_2OH ethanol	alkyl group to which OH is attached, plus *alcohol* CH_3OH methyl alcohol CH_3CH_2OH ethyl alcohol
Amine	functional group suffix is *amine* $CH_3CH_2NH_2$ ethanamine $CH_3CH_2CH_2NHCH_3$ *N*-methyl-1-propanamine	alkyl groups attached to N, plus *amine* $CH_3CH_2NH_2$ ethylamine $CH_3CH_2CH_2NHCH_3$ methylpropylamine

PROBLEM 21◆

Tell whether the following are primary, secondary, or tertiary:

a. 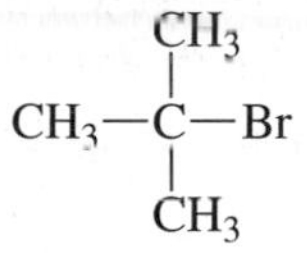
$$CH_3-\underset{\underset{CH_3}{|}}{\overset{\overset{CH_3}{|}}{C}}-Br$$

b.
$$CH_3-\underset{\underset{CH_3}{|}}{\overset{\overset{CH_3}{|}}{C}}-OH$$

c.
$$CH_3-\underset{\underset{CH_3}{|}}{\overset{\overset{CH_3}{|}}{C}}-NH_2$$

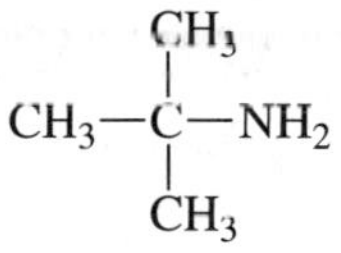

PROBLEM 22◆

Give a common (if it has one) and a systematic name for each of the following and tell whether each is a primary, secondary, or tertiary amine:

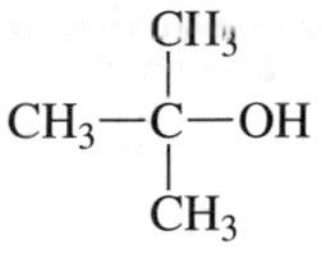

a. $CH_3CH_2CH_2CH_2CH_2CH_2NH_2$

b. 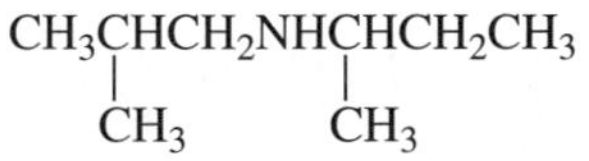 $CH_3\underset{\underset{CH_3}{|}}{CH}CH_2NH\underset{\underset{CH_3}{|}}{CH}CH_2CH_3$

c. 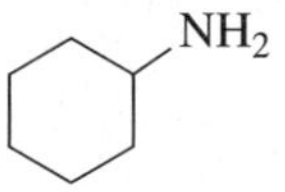
$$\text{(cyclohexyl)}-NH_2$$

d. $CH_3CH_2CH_2NHCH_2CH_2CH_2CH_3$

e. 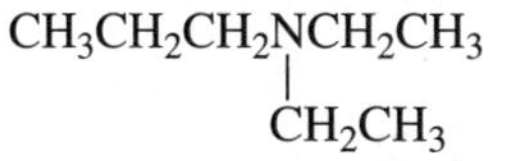 $CH_3CH_2CH_2N\underset{\underset{CH_2CH_3}{|}}{CH_2CH_3}$

f. 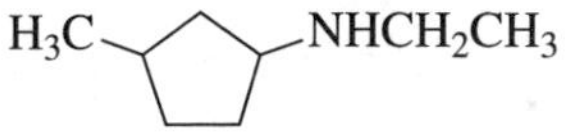 $H_3C-\text{(cyclopentyl)}-NHCH_2CH_3$

PROBLEM 23◆

Draw the structure of each of the following:

a. 2-methyl-*N*-propyl-1-propanamine

b. *N*-ethylethanamine

c. 5-methyl-1-hexanamine

d. methyldipropylamine

e. *N,N*-dimethyl-3-pentanamine

f. cyclohexylethylmethylamine

For each of the following, give the systematic name and the common name (if it has a common name), and indicate whether it is a primary, secondary, or tertiary amine:

a. $CH_3CHCH_2CH_2CH_2CH_2CH_2NH_2$
        $|$
       CH_3

c. $(CH_3CH_2)_2NCH_3$

b. $CH_3CH_2CH_2NHCH_2CH_2CHCH_3$
                            $|$
                         CH_3

d. (structure of a cyclohexane ring with CH_3, H_3C, and NH_2 substituents)

2.8 The Structures of Alkyl Halides, Alcohols, Ethers, and Amines

The classes of compounds we have been looking at in this chapter have structural resemblances to the simpler compounds introduced in Chapter 1. Let's begin by looking at alkyl halides and their resemblance to alkanes. Both classes of compounds have the same geometry; the only difference is a C—X bond (where X denotes a halogen) has replaced a C—H bond (Section 1.7). The C—X bond of an alkyl halide is formed from the overlap of an sp^3 orbital of the halogen with an sp^3 orbital of carbon. Fluorine uses a $2sp^3$ orbital to overlap with a $2sp^3$ orbital of carbon, chlorine uses a $3sp^3$ orbital, bromine a $4sp^3$ orbital, and iodine a $5sp^3$ orbital. Because the electron density of the orbital decreases with increasing volume, the C—X bond becomes longer and weaker as the size of the halogen increases (Table 2.4). Notice that this is the same trend shown by the H—X bond of hydrogen halides (Table 1.6, page 40).

Student Tutorial:
Functional groups

Table 2.4	Carbon–Halogen Bond Lengths and Bond Strengths			
	Orbital interactions	**Bond lengths**	**Bond strength**	
			kcal/mol	**kJ/mol**
H_3C-F		1.39 Å	108	451
H_3C-Cl		1.78 Å	84	350
H_3C-Br		1.93 Å	70	294
H_3C-I		2.14 Å	57	239

Now consider the geometry of the oxygen in an alcohol, which is the same as the geometry of the oxygen in water (Section 1.11). In fact, an alcohol molecule can be thought of structurally as a water molecule with an alkyl group in place of one of the hydrogens. The oxygen atom in an alcohol is sp^3 hybridized, as it is in water. One of the sp^3 orbitals of oxygen overlaps an sp^3 orbital of a carbon, one sp^3 orbital overlaps the s orbital of a hydrogen, and the other two sp^3 orbitals each contain a lone pair.

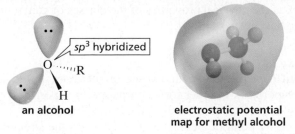

an alcohol

electrostatic potential
map for methyl alcohol

The oxygen of an ether also has the same geometry as the oxygen in water. An ether molecule can be thought of structurally as a water molecule with alkyl groups in place of both hydrogens.

an ether

electrostatic potential
map for dimethyl ether

The nitrogen of an amine has the same geometry as the nitrogen in ammonia (Section 1.12). It is sp^3 hybridized as in ammonia, with one, two, or three of the hydrogens replaced by alkyl groups. Remember that the number of hydrogens replaced by alkyl groups determines whether the amine is primary, secondary, or tertiary (Section 2.7).

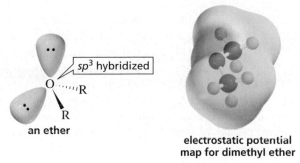

methylamine
a primary amine

dimethylamine
a secondary amine

trimethylamine
a tertiary amine

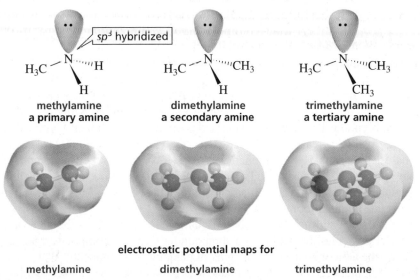

electrostatic potential maps for

methylamine   dimethylamine   trimethylamine

PROBLEM 25◆

Predict the approximate size of the following bond angles. (*Hint:* See Sections 1.11 and 1.12.)

a. the C—O—C bond angle in an ether
b. the C—N—C bond angle in a secondary amine
c. the C—O—H bond angle in an alcohol
d. the C—N—C bond angle in a quaternary ammonium salt

2.9 The Physical Properties of Alkanes, Alkyl Halides, Alcohols, Ethers, and Amines

Now we will look at the physical properties of the classes of compounds whose names and structures we have just examined.

Boiling Points

The **boiling point (bp)** of a compound is the temperature at which the liquid form becomes a gas (vaporizes). In order for a compound to vaporize, the forces that hold the individual molecules close to each other in the liquid must be overcome. This means that the boiling point of a compound depends on the strength of the attractive forces between the individual molecules. If the molecules are held together by strong forces, a lot of energy will be needed to pull the molecules away from each other and the compound will have a high boiling point. In contrast, if the molecules are held together by weak forces, only a small amount of energy will be needed to pull the molecules away from each other and the compound will have a low boiling point.

The attractive forces between alkane molecules are relatively weak. Alkanes contain only carbon and hydrogen atoms, and the electronegativities of carbon and hydrogen are similar. As a result, the bonds in alkanes are nonpolar—there are no significant partial charges on any of the atoms—and so alkanes are neutral (nonpolar) molecules.

However, it is only the average charge distribution over the alkane molecule that is neutral. The electrons are moving continuously, and at any instant the electron density on one side of the molecule can be slightly greater than on the other side, causing the molecule to have a temporary dipole. Recall that a molecule with a dipole has a negative end and a positive end (Section 1.1).

A temporary dipole in one molecule can induce a temporary dipole in a nearby molecule. As a result, the (temporarily) negative side of one molecule ends up adjacent to the (temporarily) positive side of another, as shown in Figure 2.1. Because the dipoles in the molecules are induced, the interactions between the molecules are called **induced-dipole–induced-dipole interactions**. The molecules of an alkane are held together by these induced-dipole–induced-dipole interactions, also known as **van der Waals forces**. Van der Waals forces are the weakest of all the intermolecular attractions.

The magnitude of the van der Waals forces that hold alkane molecules together depends on the area of contact between the molecules. The greater the area of contact, the stronger the van der Waals forces and the greater the amount of energy needed to overcome them. If you look at the homologous series of alkanes in Table 2.1, you will see that their boiling points increase as their size increases. This relationship holds because each additional methylene (CH_2) group increases the area of contact between the molecules. The four smallest alkanes have boiling points below room temperature (which is about 25 °C), so they exist as gases at room temperature. Pentane (bp = 36.1 °C) is the smallest alkane that is a liquid at room temperature.

Because the strength of the van der Waals forces depends on the area of contact between the molecules, branching in a compound lowers the compound's boiling point by reducing the area of contact. If you think of pentane, an unbranched alkane, as a cigar and think of branched 2,2-dimethylpropane as a tennis ball, you can see that branching decreases the area of contact between molecules: two cigars make contact over a greater area than do two tennis balls. Thus, if two alkanes have the same molecular weight, the more highly branched alkane will have a lower boiling point.

Johannes Diderik van der Waals (1837–1923) *was a Dutch physicist. He was born in Leiden, the son of a carpenter, and was largely self-taught when he entered the University of Leiden, where he earned a Ph.D. Van der Waals was a professor of physics at the University of Amsterdam from 1877 to 1907. He won the 1910 Nobel Prize in physics for his research on the gaseous and liquid states of matter.*

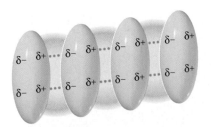

▲ **Figure 2.1**
Van der Waals forces are induced-dipole–induced–dipole interactions.

$$CH_3CH_2CH_2CH_2CH_3$$
pentane
bp = 36.1 °C

$$CH_3CHCH_2CH_3$$
$$|$$
$$CH_3$$
2-methylbutane
bp = 27.9 °C

$$\begin{array}{c} CH_3 \\ | \\ CH_3CCH_3 \\ | \\ CH_3 \end{array}$$
2,2-dimethylpropane
bp = 9.5 °C

The boiling points of the compounds in any homologous series increase as their molecular weights increase because of the increase in van der Waals forces. Thus the boiling points of the compounds in a homologous series of ethers, alkyl halides, alcohols, or amines increase with increasing molecular weight. (See Appendix I.) The boiling points of these compounds, however, are also affected by the polar character of the C—Z bond (where Z denotes N, O, F, Cl, or Br). The C—Z bond is polar because nitrogen, oxygen, and the halogens are more electronegative than the carbon to which they are attached.

> **The boiling point of a compound depends on the strength of the attractive forces between the individual molecules.**

$$R-\overset{|}{\underset{|}{C}}-\overset{\delta+\ \delta-}{Z} \qquad Z = N, O, F, Cl, \text{ or } Br$$

The magnitude of the charge differential between the two bonded atoms is indicated by the bond dipole moment (Section 1.3).

$$\begin{array}{ccc} H_3C-O-CH_3 & H_3C-OH & H_3C-NH_2 \\ \text{0.7 D} & \text{0.7 D} & \text{0.2 D} \end{array}$$

$$\begin{array}{cccc} H_3C-F & H_3C-Cl & H_3C-Br & H_3C-I \\ \text{1.6 D} & \text{1.5 D} & \text{1.4 D} & \text{1.2 D} \end{array}$$

Molecules with dipole moments are attracted to one another because they can align themselves in such a way that the positive end of one dipole is adjacent to the negative end of another dipole. These electrostatic attractive forces, called **dipole–dipole interactions**, are stronger than van der Waals forces, but not as strong as ionic or covalent bonds.

> **The dipole moment of a bond is equal to the magnitude of the charge on one of the bonded atoms times the distance between the bonded atoms.**

dipole–dipole interaction

Ethers generally have higher boiling points than alkanes of comparable molecular weight because both van der Waals forces and dipole–dipole interactions must be overcome for an ether to boil (Table 2.5).

cyclopentane
bp = 49.3 °C

tetrahydrofuran
bp = 65 °C

Table 2.5 Comparative Boiling Points (°C)			
Alkanes	**Ethers**	**Alcohols**	**Amines**
$CH_3CH_2CH_3$	CH_3OCH_3	CH_3CH_2OH	$CH_3CH_2NH_2$
−42.1	−23.7	78	16.6
$CH_3CH_2CH_2CH_3$	$CH_3OCH_2CH_3$	$CH_3CH_2CH_2OH$	$CH_3CH_2CH_2NH_2$
−0.5	10.8	97.4	47.8

As the table shows, alcohols have much higher boiling points than alkanes or ethers with similar molecular weights because, in addition to van der Waals forces and the dipole–dipole interactions of the C—O bond, alcohols can form **hydrogen bonds**. A hydrogen bond is a special kind of dipole–dipole interaction that occurs between a hydrogen that is bonded to an oxygen, nitrogen, or fluorine and the lone-pair electrons of an oxygen, nitrogen, or fluorine in another molecule.

The length of the covalent bond between oxygen and hydrogen is 0.96 Å. The hydrogen bond between an oxygen of one molecule and a hydrogen of another molecule is almost twice as long (1.69−1.79 Å), which means that a hydrogen bond is not as strong as an O—H covalent bond. A hydrogen bond, however, is

Hydrogen bonds are stronger than other dipole–dipole interactions, which are stronger than van der Waals forces.

stronger than other dipole–dipole interactions. The strongest hydrogen bonds are linear, meaning the two electronegative atoms and the hydrogen between them lie on a straight line.

$$H-\ddot{O}:\text{-----}H-\ddot{O}:\text{-----}H-\ddot{O}:\text{-----}H-\ddot{O}:$$

hydrogen bond
H   H   H   H

1.69 to 1.79 Å

$$H-\ddot{O}:\text{-----}H-\ddot{O}:\text{-----}H-\ddot{O}:\text{-----}H-\ddot{O}:$$

hydrogen bond
H   H   H   H

0.96 Å

H   H   H

hydrogen bonds

$$H-\ddot{N}-H\text{-----}:\ddot{N}-H\text{-----}:\ddot{N}-H\text{-----}:\ddot{N}-H \qquad H-\ddot{F}:\text{-----}H-\ddot{F}:\text{-----}H-\ddot{F}:$$

H   H   H   H

Although each individual hydrogen bond is weak, requiring about 5 kcal/mol (or 21 kJ/mol) to break, there are many such bonds holding alcohol molecules together. The extra energy required to break these hydrogen bonds is the reason alcohols have much higher boiling points than either alkanes or ethers with similar molecular weights.

The boiling point of water illustrates the dramatic effect that hydrogen bonding has on boiling points. Water has a molecular weight of 18 and a boiling point of 100 °C. The alkane nearest in size is methane, with a molecular weight of 16. Methane boils at −167.7 °C.

Primary and secondary amines also form hydrogen bonds, so these amines have higher boiling points than alkanes with similar molecular weights. Nitrogen is not as electronegative as oxygen, however, which means that the hydrogen bonds between amine molecules are weaker than the hydrogen bonds between alcohol molecules. An amine, therefore, has a lower boiling point than an alcohol with a similar molecular weight (Table 2.5).

Because primary amines have two N—H bonds, hydrogen bonding is more significant in primary amines than in secondary amines. Tertiary amines cannot form hydrogen bonds between their own molecules because they do not have a hydrogen attached to the nitrogen. Consequently, when we compare amines with the same molecular weight and similar structures, we find that a primary amine has a higher boiling point than a secondary amine and a secondary amine has a higher boiling point than a tertiary amine.

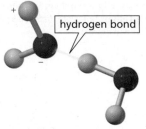

hydrogen bonding in water

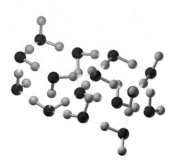

$$\overset{\displaystyle CH_3}{\underset{\displaystyle \vert}{}}$$
CH₃CH₂CHCH₂NH₂

a primary amine
bp = 97 °C

$$\overset{\displaystyle CH_3}{\underset{\displaystyle \vert}{}}$$
CH₃CH₂CHNHCH₃

a secondary amine
bp = 84 °C

$$\overset{\displaystyle CH_3}{\underset{\displaystyle \vert}{}}$$
CH₃CH₂NCH₂CH₃

a tertiary amine
bp = 65 °C

Hydrogen bonds play an important role in biology. We will see that proteins are shaped by hydrogen bonding (Section 22.14), and DNA relies on hydrogen bonding to copy all its hereditary information (Section 27.5).

Both van der Waals forces and dipole–dipole interactions must be overcome for an alkyl halide to boil. Moreover, as the halogen atom increases in size, these interactions become stronger. A larger electron cloud means that the van der Waals contact area is greater and the electron cloud's polarizability is also greater. **Polarizability** indicates how readily an electron cloud can be distorted. The larger the atom, the more loosely it holds the electrons in its outermost shell, and the more they can be distorted to create a strong induced dipole. Therefore, an alkyl fluoride has a lower boiling point than an alkyl chloride with the same alkyl group. Similarly, alkyl chlorides have lower boiling points than alkyl bromides, which have lower boiling points than alkyl iodides (Table 2.6).

More extensive tables of physical properties can be found in Appendix I.

Table 2.6 Comparative Boiling Points of Alkanes and Alkyl Halides (°C)

	H	F	Y Cl	Br	I
CH_3—Y	−161.7	−78.4	−24.2	3.6	42.4
CH_3CH_2—Y	−88.6	−37.7	12.3	38.4	72.3
$CH_3CH_2CH_2$—Y	−42.1	−2.5	46.6	71.0	102.5
$CH_3CH_2CH_2CH_2$—Y	−0.5	32.5	78.4	101.6	130.5
$CH_3CH_2CH_2CH_2CH_2$—Y	36.1	62.8	107.8	129.6	157.0

PROBLEM-SOLVING STRATEGY

Predicting Hydrogen Bonding

a. Which of the following compounds will form hydrogen bonds between its molecules?
   1. $CH_3CH_2CH_2OH$ 2. $CH_3CH_2CH_2F$ 3. $CH_3OCH_2CH_3$
b. Which of these compounds will form hydrogen bonds with a solvent such as ethanol?

To solve this type of question, start by defining the kind of compound that will do what is being asked.

a. A hydrogen bond forms when a hydrogen attached to an O, N, or F of one molecule interacts with a lone pair on an O, N, or F of another molecule. Therefore, a compound that will form hydrogen bonds with itself must have a hydrogen bonded to an O, N, or F. Only compound 1 will be able to form hydrogen bonds with itself.

b. Ethanol has an H bonded to an O, so it will be able to form hydrogen bonds with a compound that has a lone pair on an O, N, or F. All three compounds will be able to form hydrogen bonds with ethanol.

Now continue on to Problem 26.

PROBLEM 26♦

a. Which of the following will form hydrogen bonds between its molecules?
   1. $CH_3CH_2CH_2COOH$ 4. $CH_3CH_2CH_2NHCH_3$
   2. $CH_3CH_2N(CH_3)_2$ 5. $CH_3CH_2OCH_2CH_2OH$
   3. $CH_3CH_2CH_2CH_2Br$ 6. $CH_3CH_2CH_2CH_2F$
b. Which of the preceding compounds form hydrogen bonds with a solvent such as ethanol?

PROBLEM 27

Explain why

a. H_2O has a higher boiling point than CH_3OH (65 °C).
b. H_2O has a higher boiling point than NH_3 (−33 °C).
c. H_2O has a higher boiling point than HF (20 °C).

PROBLEM 28♦

List the following compounds in order of decreasing boiling point:

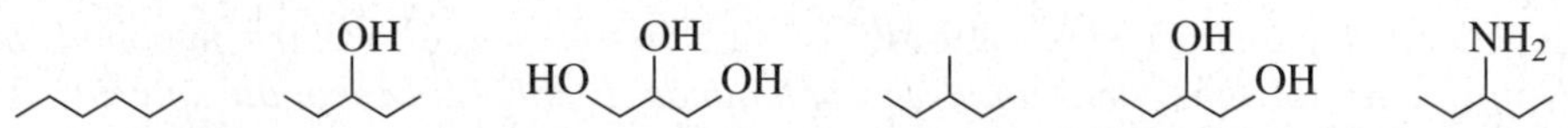

PROBLEM 29◆

List the compounds in each set in order of decreasing boiling point:

a. $CH_3CH_2CH_2CH_2CH_2CH_2Br$ $CH_3CH_2CH_2CH_2Br$ $CH_3CH_2CH_2CH_2CH_2Br$

b. $CH_3CHCH_2CH_2CH_2CH_2CH_3$
$\qquad$ |
$\qquad CH_3$

$CH_3CH_2CH_2CH_2CH_2CH_2CH_2CH_3$

$$CH_3C \overset{\overset{\displaystyle CH_3}{\vert}}{\underset{\underset{\displaystyle CH_3}{\vert}}{}} \!\!\! \overset{\overset{\displaystyle CH_3}{\vert}}{\underset{\underset{\displaystyle CH_3}{\vert}}{}} \!\!\! CCH_3$$

$CH_3CH_2CH_2CH_2CH_2CH_2CH_2CH_2CH_3$

c. $CH_3CH_2CH_2CH_2CH_3$ $CH_3CH_2CH_2CH_2OH$ $CH_3CH_2CH_2CH_2Cl$
$CH_3CH_2CH_2CH_2CH_2OH$

Melting Points

The **melting point (mp)** of a compound is the temperature at which its solid form is converted into a liquid. If you examine the melting points of the alkanes listed in Table 2.1, you will see that they increase (with a few exceptions) within a homologous series as the molecular weight increases. The increase in melting point is less regular than the increase in boiling point because, in addition to the intermolecular attractions we considered above, the melting point is influenced by the type of **packing**, that is, the arrangement, including the closeness and compactness, of the molecules in the crystal lattice. The tighter the fit, the more energy is required to break the lattice and melt the compound.

Figure 2.2 shows that the melting points of alkanes with even numbers of carbon atoms fall on a smooth curve (the red line). The melting points of alkanes with odd numbers of

Figure 2.2 ▶
Melting points of straight-chain alkanes. Alkanes with even numbers of carbon atoms fall on a melting-point curve that is higher than the melting-point curve for alkanes with odd numbers of carbon atoms.

carbon atoms also fall on a smooth curve (the green line). The two curves do not overlap, however, because alkanes with an odd number of carbon atoms pack less tightly than alkanes with an even number of carbon atoms. Alkanes with odd numbers of carbon atoms pack less tightly because the molecules, each a zigzag chain with its ends tilted the same way, can lie next to each other with a methyl group on the end of one facing and repelling the methyl group on the end of the other, thus increasing the average distance between the chains. Consequently, alkane molecules with odd numbers of carbon atoms have weaker intermolecular attractions and correspondingly lower melting points.

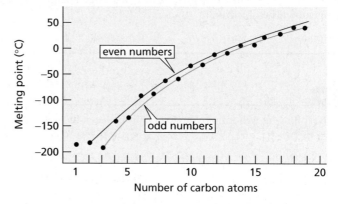

odd number of carbons   **even number of carbons**

Solubility

"Like" dissolves "like."

The general rule that governs **solubility** is "like dissolves like." *Polar compounds dissolve in polar solvents, and nonpolar compounds dissolve in nonpolar solvents.* The reason "polar dissolves polar" is that a **polar solvent**, such as water, has partial charges that can interact with the partial charges on a polar compound. The negative poles of the solvent molecules surround the positive pole of the polar *solute*, and the positive poles of the solvent molecules surround the negative pole of the polar *solute*. Clustering of the solvent molecules around the solute molecules separates solute molecules from each other, which is what makes them dissolve. The interaction between solvent molecules and solute molecules is called **solvation**.

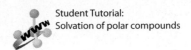
Student Tutorial:
Solvation of polar compounds

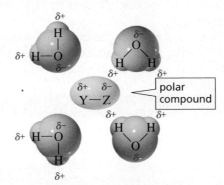

solvation of a polar compound by water

Because nonpolar compounds have no net charge, polar solvents are not attracted to them. In order for a nonpolar molecule to dissolve in a polar solvent such as water, the nonpolar molecule would have to push the water molecules apart, disrupting their hydrogen bonding. Hydrogen bonding is strong enough to exclude the nonpolar compound. In contrast, nonpolar solutes dissolve in nonpolar solvents because the van der Waals interactions between solvent and solute molecules are about the same as between solvent–solvent and solute–solute molecules.

Alkanes are nonpolar, which causes them to be soluble in **nonpolar solvents** and insoluble in polar solvents such as water. The densities of alkanes (Table 2.1) increase with increasing molecular weight, but even a 30-carbon alkane such as triacontane (density at 20 °C, or $d^{20°} = 0.8097$ g/mL) is less dense than water ($d^{20°} = 1.00$ g/mL). This means that a mixture of an alkane and water will separate into two distinct layers, with the less dense alkane floating on top. The Alaskan oil spill of 1989, the Persian Gulf spill of 1991, and the spill off the northwest coast of Spain in 2002 are large-scale examples of this phenomenon. (Crude oil is primarily a mixture of alkanes.)

Oil from a 70,000-ton oil spill in 1996 off the coast of Wales.

An alcohol has both a nonpolar alkyl group and a polar OH group. So is an alcohol molecule nonpolar or polar? Is it soluble in a nonpolar solvent, or is it soluble in water? The answer depends on the size of the alkyl group. As the alkyl group increases in size, becoming a more significant fraction of the alcohol molecule, the compound becomes less and less soluble in water. In other words, the molecule becomes more and more like an alkane. Groups consisting of four carbons tend to straddle the dividing line at room temperature: alcohols with fewer than four carbons are soluble in water, but alcohols with more than four carbons are insoluble in water. In other words, an OH group can drag about three or four carbons into solution in water.

The four-carbon dividing line is only an approximate guide because the solubility of an alcohol also depends on the structure of the alkyl group. Alcohols with branched alkyl groups are more soluble in water than alcohols with nonbranched alkyl groups having the same number of carbons, because branching minimizes the contact surface of the nonpolar portion of the molecule. So *tert*-butyl alcohol is more soluble than *n*-butyl alcohol in water.

Similarly, the oxygen atom of an ether can drag only about three carbons into solution in water (Table 2.7). We have already seen (photo on page 62) that diethyl ether—an ether with four carbons—is not soluble in water.

Table 2.7 Solubilities of Ethers in Water		
2 C's	CH_3OCH_3	soluble
3 C's	$CH_3OCH_2CH_3$	soluble
4 C's	$CH_3CH_2OCH_2CH_3$	slightly soluble (10 g/100 g H_2O)
5 C's	$CH_3CH_2OCH_2CH_2CH_3$	minimally soluble (1.0 g/100 g H_2O)
6 C's	$CH_3CH_2CH_2OCH_2CH_2CH_3$	insoluble (0.25 g/100 g H_2O)

Low-molecular-weight amines are soluble in water because amines can form hydrogen bonds with water. Comparing amines with the same number of carbons, we find that primary amines are more soluble than secondary amines because primary amines have two hydrogens that can engage in hydrogen bonding. Tertiary amines, like primary and secondary amines, have lone-pair electrons that can accept hydrogen bonds, but unlike primary and secondary amines, tertiary amines do not have hydrogens to donate for hydrogen bonds. Tertiary amines, therefore, are less soluble in water than are secondary amines with the same number of carbons.

Alkyl halides have some polar character, but only alkyl fluorides have an atom that can form a hydrogen bond with water. This means that alkyl fluorides are the most water soluble of the alkyl halides. The other alkyl halides are less soluble in water than ethers or alcohols with the same number of carbons (Table 2.8).

Table 2.8 Solubilities of Alkyl Halides in Water			
CH_3F very soluble	CH_3Cl soluble	CH_3Br slightly soluble	CH_3I slightly soluble
CH_3CH_2F soluble	CH_3CH_2Cl slightly soluble	CH_3CH_2Br slightly soluble	CH_3CH_2I slightly soluble
$CH_3CH_2CH_2F$ slightly soluble	$CH_3CH_2CH_2Cl$ slightly soluble	$CH_3CH_2CH_2Br$ slightly soluble	$CH_3CH_2CH_2I$ slightly soluble
$CH_3CH_2CH_2CH_2F$ insoluble	$CH_3CH_2CH_2CH_2Cl$ insoluble	$CH_3CH_2CH_2CH_2Br$ insoluble	$CH_3CH_2CH_2CH_2I$ insoluble

PROBLEM 30◆

Rank the following groups of compounds in order of decreasing solubility in water:

a. $CH_3CH_2CH_2OH$ $CH_3CH_2CH_2CH_2Cl$
$CH_3CH_2CH_2CH_2OH$ $HOCH_2CH_2CH_2OH$

b. CH_3 NH_2 OH

PROBLEM 31◆

In which solvent would cyclohexane have the lowest solubility, 1-pentanol, diethyl ether, ethanol, or hexane?

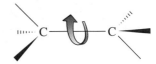

2.10 Rotation Occurs About Carbon–Carbon Bonds

We have seen that a carbon–carbon single bond (a σ bond) is formed when an sp^3 orbital of one carbon overlaps an sp^3 orbital of a second carbon (Section 1.7). Because σ bonds are cylindrically symmetrical (that is, symmetrical about an imaginary line connecting the centers of the two atoms joined by the σ bond), rotation about a carbon–carbon single bond can occur without any change in the amount of orbital overlap (Figure 2.3). The different spatial arrangements of the atoms that result from rotation about a single bond are called **conformations**.

The conformations produced by rotation about the carbon–carbon bond of ethane, represent a continuum between the two extremes shown below; a *staggered conformation* and an *eclipsed conformation*. An infinite number of conformations between these two extremes are also possible.

Our drawings of molecules are two-dimensional attempts to communicate three-dimensional structures. Chemists commonly use *perspective formulas*, introduced in Section 1.7, and *Newman projections* to represent the three-dimensional spatial arrangements resulting from rotation about a σ bond. A **Newman projection** assumes that the viewer is looking along the longitudinal axis of a particular C—C bond. The carbon in front is represented by a point (where three lines are seen to intersect), and the carbon at the back is represented by a circle. The three lines emanating from each of the carbons represent its other three bonds.

▲ **Figure 2.3**
A carbon–carbon bond is formed by the overlap of cylindrically symmetrical sp^3 orbitals. Therefore, rotation about the bond can occur without changing the amount of orbital overlap.

Melvin S. Newman (1908–1993) *was born in New York. He received a Ph.D. from Yale University in 1932 and was a professor of chemistry at Ohio State University from 1936 to 1973. He first suggested his technique for drawing organic molecules in 1952.*

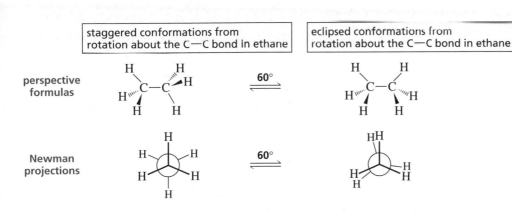

A **staggered conformation** is more stable, and therefore lower in energy, than an **eclipsed conformation**. Because of this energy difference, rotation about a carbon–carbon single bond is not completely free. The eclipsed conformation is higher in energy, so an energy barrier must be overcome when rotation about the C—C bond occurs (Figure 2.4). However, the barrier in ethane is small enough (2.9 kcal/mol or 12 kJ/mol) to allow continuous rotation. A molecule's conformation changes from staggered to eclipsed millions of times per second at room temperature. Because of this continuous interconversion, the conformations cannot be separated from each other, although we will see that some are likely to be more prevalent than others. The investigation of the various conformations of a compound and their relative stabilities is called **conformational analysis**.

A staggered conformation is more stable than an eclipsed conformation.

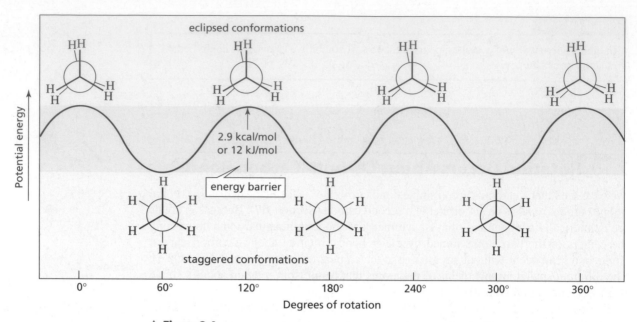

▲ **Figure 2.4**
Potential energy of ethane as a function of the angle of rotation about the carbon–carbon bond.

Figure 2.4 shows the potential energies of all the conformations of ethane obtained in one complete 360° rotation about the C—C bond. Notice that staggered conformations are at energy minima, whereas eclipsed conformations are at energy maxima. Conformations at energy minima are called **conformers**. Thus, ethane has three conformers.

Why is the staggered conformation more stable than the eclipsed conformation? The major contribution to the energy difference between them is a stabilizing interaction between the C—H σ bonding orbital on one carbon and the C—H σ^* antibonding orbital on the other carbon: the electrons in the filled bonding orbital move partially into the unoccupied antibonding orbital. This interaction is greatest in the staggered conformation because only in this conformation are the two orbitals parallel. Such a delocalization of electrons by the overlap of a σ orbital with an empty orbital is called **hyperconjugation**.

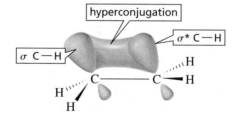

Butane has three carbon–carbon single bonds, and rotation can occur about each of them. The following Newman projections show staggered and eclipsed conformations from rotation about the C-1—C-2 bond:

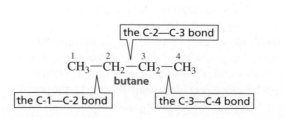

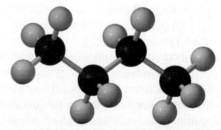

ball-and-stick model of butane

staggered conformation from rotation about the C-1—C-2 bond in butane	eclipsed conformation from rotation about the C-1—C-2 bond in butane

Note that the carbon with the lower number is placed in the foreground in a Newman projection.

Although the staggered conformations resulting from rotation about the C-1—C-2 bond in butane all have the same energy, the staggered conformations resulting from rotation about the C-2—C-3 bond do not. The staggered and eclipsed conformations from rotation about the C-2—C-3 bond in butane are:

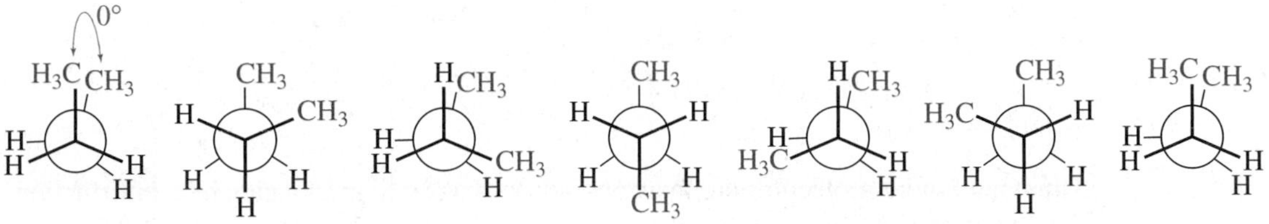

A   B   C   D   E   F   A

Of the staggered conformations, D, in which the two methyl groups are as far apart as possible, is more stable than the other two staggered conformations (B and F). The most stable of a bond's conformations (in this case, D) is called the **anti conformation**, and the other two staggered conformations (in this case, B and F) are called **gauche** ("goesh") **conformations**. (*Anti* is Greek for "opposite of"; *gauche* is French for "left.") In the anti conformation, the largest substituents are opposite each other; in a gauche conformation, they are adjacent. The two gauche conformations have the same energy, and it is higher than the energy of the anti conformation.

The anti and gauche conformations do not have the same energy because of steric strain. **Steric strain** is the strain (that is, the extra energy) experienced by a molecule when atoms or groups are close to one another, causing their electron clouds to repel each other. There is more steric strain in a gauche conformation than in the anti conformation because the two substituents (in butane, the two methyl groups) are closer together in a gauche conformation. This type of steric strain is called a **gauche interaction**. In general, steric strain in molecules increases as the size of the interacting atoms or groups increases.

The eclipsed conformations resulting from rotation about the C-2—C-3 bond in butane also have different energies. The eclipsed conformation in which the two methyl groups are closest to each other (A) is less stable than the eclipsed conformation in which they are farther apart (C and E).

The energies of the conformations obtained from rotation about the C-2—C-3 bond of butane are shown in Figure 2.5. The letters in Figure 2.5 correspond to the letters that identify the above structures. The degree of rotation of each conformation is identified by the dihedral angle, the angle formed in a Newman projection by the C—H bond on the nearer carbon and the C—H bond on the far carbon. The conformation in which one methyl group stands directly in front of the other—the least stable conformation—has a dihedral angle of 0°.

Because there is continuous rotation about all the C—C single bonds in a molecule, organic molecules with C—C single bonds are not static balls and sticks—they have many interconvertible conformations. The relative numbers of molecules in particular conformations at any one time depends on the stability of the conformations: the more stable the conformation, the greater the fraction of molecules that will be in that conformation. Most molecules, therefore, are in staggered conformations at any given

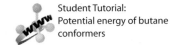

Student Tutorial:
Potential energy of butane conformers

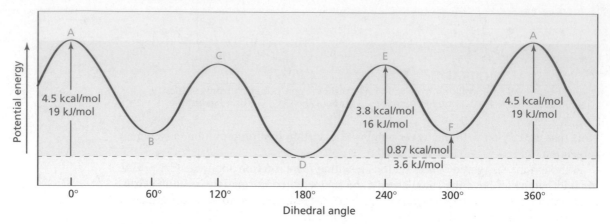

▲ **Figure 2.5**
Potential energy of butane as a function of the degree of rotation about the C-2 — C-3 bond. Green letters refer to the conformations (A–F) shown on page 103.

instant, and more molecules are in an anti conformation than in a gauche conformation. The lower energy of a staggered conformation gives carbon chains the tendency to adopt zigzag arrangements, as seen in the ball-and-stick model of decane.

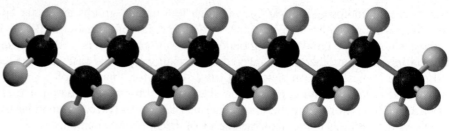

ball-and-stick model of decane

PROBLEM 32

a. Draw all the staggered and eclipsed conformations that result from rotation about the C-2 — C-3 bond of pentane.

b. Draw a potential-energy diagram for rotation of the C-2 — C-3 bond of pentane through 360°, starting with the least stable conformation.

PROBLEM 33◆

Using Newman projections, draw the most stable conformation for each of the following:

a. 3-methylpentane, viewed along the C-2 — C-3 bond

b. 3-methylhexane, viewed along the C-3 — C-4 bond

c. 3,3-dimethylhexane, viewed along the C-3 — C-4 bond

2.11 **Some Cycloalkanes Have Angle Strain**

Early chemists observed that cyclic compounds found in nature generally have five- or six-membered rings. Compounds with three- and four-membered rings are much less common. This observation suggests that compounds with five- and six-membered rings are more stable than compounds with three- or four-membered rings.

In 1885, the German chemist Adolf von Baeyer proposed that the instability of three- and four-membered rings is due to angle strain. We know that, ideally, an sp^3 carbon has bond angles of 109.5° (Section 1.7). Baeyer suggested that the stability of a cycloalkane could be predicted by assessing the difference between this ideal bond angle and the bond angle in the planar cycloalkane. The angles in an equilateral triangle, for example, are 60°, representing a 49.5° deviation from the tetrahedral 109.5°. According to Baeyer, this deviation causes **angle strain**, and thus instability, in cyclopropane.

The angle strain in a three-membered ring can be understood by looking at the overlap of the orbitals that form the σ bonds in cyclopropane (Figure 2.6). Normal σ bonds are formed by the overlap of two sp^3 orbitals that point directly at each other. In cyclopropane, the overlapping orbitals cannot point directly at each other, so the amount of overlap between them is less than in a normal C—C bond. The smaller degree of overlap causes the C—C bonds in cyclopropane to be weaker than normal C—C bonds. This weakness is what we have described as angle strain.

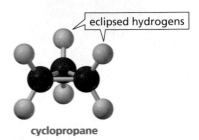

eclipsed hydrogens

cyclopropane

a.

good overlap
strong bond

b.

poor overlap
weak bond

◀ **Figure 2.6**
(a) Overlap of sp^3 orbitals in a normal σ bond. (b) Overlap sp^3 of orbitals in cyclopropane.

Because the C—C bonding orbitals in cyclopropane cannot point directly at each other, the bonds they form have shapes that resemble bananas and, consequently, are often called **banana bonds**. In addition to the angle strain of the C—C bonds, all the adjacent C—H bonds in cyclopropane are eclipsed rather than staggered, creating even more strain.

The bond angles in a hypothetical planar cyclobutane molecule would have to be compressed from 109.5° to 90°, the bond angle associated with a planar four-membered ring. Planar cyclobutane would therefore have less angle strain than cyclopropane because the bond angles in cyclobutane are only 19.5° away from the ideal bond angle. It would, however, have eight pairs of eclipsed hydrogens, compared with the six pairs in cyclopropane. Because of the eclipsed hydrogens, a cyclobutane molecule is not planar. (Its actual structure is explained and shown on page 106.)

HIGHLY STRAINED HYDROCARBONS

Organic chemists have been able to synthesize some highly strained cyclic hydrocarbons, such as bicyclo[1.1.0]butane, cubane, and prismane. Bicyclo[1.1.0] butane was synthesized by David Lemal, Fredric Menger, and George Clark at the University of Wisconsin. Cubane was synthesized by Philip Eaton and Thomas Cole Jr. at the University of Chicago. Prismane was synthesized by Thomas Katz and Nancy Acton at Columbia University.

bicyclo[1.1.0]butane   cubane   prismane

Philip Eaton, the first to synthesize cubane, also synthesized octanitrocubane, which is cubane with an NO_2 group bonded to each of the eight corners. It turned out to be less powerful an explosive than expected.

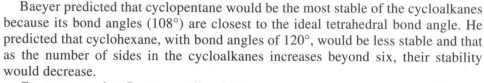

PROBLEM 34◆

The bond angles in a regular polygon with n sides are equal to

$$180° - \frac{360°}{n}$$

a. What are the bond angles in a regular octagon?

b. What are the bond angles in a regular nonagon?

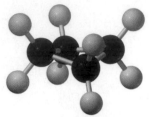

cyclobutane

Baeyer predicted that cyclopentane would be the most stable of the cycloalkanes because its bond angles (108°) are closest to the ideal tetrahedral bond angle. He predicted that cyclohexane, with bond angles of 120°, would be less stable and that as the number of sides in the cycloalkanes increases beyond six, their stability would decrease.

Contrary to what Baeyer predicted, however, cyclohexane is more stable than cyclopentane. Furthermore, cyclic compounds do not thereafter become less and less stable as the number of sides increases. The mistake Baeyer made was to assume that all cyclic molecules are planar. Because three points define a plane, the carbons of cyclopropane must lie in a plane. The other cycloalkanes, however, are not planar but twist and bend in order to attain a structure that maximizes their stability by minimizing both ring strain and the number of eclipsed hydrogens. Therefore, instead of being planar as in the hypothetical example described on page 105, the cyclobutane molecule is bent, with one of its methylene groups extending at an angle of about 25° from the plane defined by the other three carbon atoms.

cyclopentane

If the cyclopentane ring were planar, as Baeyer had predicted, it would have essentially no angle strain, but it would have 10 pairs of eclipsed hydrogens. So cyclopentane puckers, allowing some of the hydrogens to become nearly staggered; but, in the process, the molecule acquires some angle strain. The puckered form of cyclopentane is called the *envelope conformation*, because the shape of the ring resembles a squarish envelope with the flap up.

VON BAEYER AND BARBITURIC ACID

Johann Friedrich Wilhelm Adolf von Baeyer (1835–1917), a German chemist, was a professor of chemistry at the University of Strasbourg and later at the University of Munich. In 1864, he discovered barbituric acid—the first of a group of sedatives known as barbiturates—and named it after a woman named Barbara. Who Barbara was is not certain. Some say she was his girlfriend, but because Baeyer discovered barbituric acid in the same year that Prussia defeated Denmark, some believe he named it after Saint Barbara, the patron saint of artillerymen. Baeyer is also known as the first to synthesize indigo, the dye used in the manufacture of blue jeans. He received the Nobel Prize in chemistry in 1905 for his work in synthetic organic chemistry.

PROBLEM 35◆

The effectiveness of a barbiturate as a sedative is related to its ability to penetrate the nonpolar membrane of a cell. Which of the following barbiturates would you expect to be the more effective sedative?

hexethal

barbital

2.12 **Conformations of Cyclohexane**

The cyclic compounds most commonly found in nature contain six-membered rings because carbon rings of that size can exist in a conformation—called a *chair conformation*—that is almost completely free of strain. All the bond angles in a **chair conformer** are 111°, which is very close to the ideal tetrahedral bond angle of 109.5°, and all the adjacent bonds are staggered (Figure 2.7).

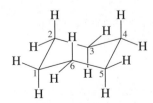

chair conformer of
cyclohexane

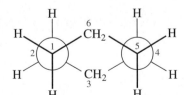

**Newman projection of
the chair conformer**

**ball-and-stick model of the
chair conformer of cyclohexane**

◀ **Figure 2.7**
The chair conformer of cyclohexane, a Newman projection of the chair conformer, and a ball-and-stick model showing that all the bonds are staggered.

The chair conformer is such an important one that you should learn how to draw it:

1. Draw two parallel lines of the same length, slanted upward and beginning at the same level.

2. Connect the tops of the lines with a V whose left-hand side is slightly longer than the right-hand side. Connect the bottoms of the lines with the same V inverted and reversed. (The bottom-left and top-right lines should be parallel; and the top-left and bottom-right lines should be parallel. This completes the framework of the six-membered ring.

3. Each carbon has an axial bond and an equatorial bond. The **axial bonds** (red lines) are vertical and alternate above and below the ring. The axial bond on one of the uppermost carbons is up, the next is down, the next is up, and so on.

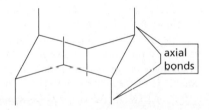

4. The **equatorial bonds** (red lines with blue balls) point outward from the ring. Because the bond angles are greater than 90°, the equatorial bonds are on a slant. If the axial bond points up, the equatorial bond on the same carbon is on

a downward slant. If the axial bond points down, the equatorial bond on the same carbon is on an upward slant.

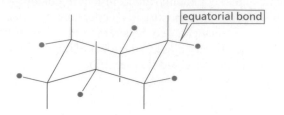

Notice that each equatorial bond is parallel to two ring bonds (two carbons over).

Remember that in this depiction cyclohexane is viewed edge-on. The lower bonds of the ring are in front and the upper bonds of the ring are in back.

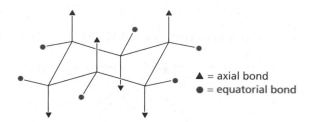

▲ = axial bond
● = equatorial bond

PROBLEM 36

Draw 1,2,3,4,5,6-hexamethylcyclohexane with

a. all the methyl groups in axial positions.
b. all the methyl groups in equatorial positions.

PROBLEM-SOLVING STRATEGY

Calculating the Strain Energy of a Cycloalkane

If we assume that cyclohexane is completely free of strain, we can use the **heat of formation**—the heat given off when a compound is formed from its elements under standard conditions—to calculate the total strain energy of the other cycloalkanes. Taking the heat of formation of cyclohexane (Table 2.9) and dividing by 6 for its six CH_2 groups gives us a value of −4.92 kcal/mol (or −20.6 kJ/mol) for a "strainless" CH_2 group (−29.5/6 = −4.92). With this value, we can calculate the heat of formation of any other "strainless" cycloalkane: we simply multiply the number of CH_2 groups in its ring by −4.92 kcal/mol. The total strain in the compound is the difference between its "strainless" heat of formation and its actual heat of formation (Table 2.9). For example, cyclopentane has a "strainless" heat of formation of (5)(−4.92) = −24.6 kcal/mol. Because its actual heat of formation is −18.4 kcal/mol, cyclopentane has a total strain energy of 6.2 kcal/mol, because [−18.4 − (−24.6) = 6.2]. (Multiplying by 4.184 converts kcal into kJ.)

Now continue on to Problem 37.

Table 2.9 Heats of Formation and Total Strain Energies of Cycloalkanes					
	Heat of formation		**"Strainless" Heat of formation**		**Total strain energy**
	(kcal/mol)	**(kJ/mol)**	**(kcal/mol)**	**(kJ/mol)**	**(kcal/mol)**	**(kJ/mol)**
Cylopropane	+12.7	53.1	−14.6	−61.1	27.3	114.2
Cyclobutane	+6.8	28.5	−19.7	−82.4	26.5	110.9
Cyclopentane	−18.4	−77.0	−24.6	−102.9	6.2	25.9
Cyclohexane	−29.5	−123.4	−29.5	−123.4	0	0
Cycloheptane	−28.2	−118.0	−34.4	−143.9	6.2	25.9
Cyclooctane	−29.7	−124.3	−39.4	−164.8	9.7	40.6
Cyclononane	−31.7	−132.6	−44.3	−185.4	12.6	52.7
Cyclodecane	−36.9	−154.4	−49.2	−205.9	12.3	51.5
Cycloundecane	−42.9	−179.5	−54.1	−226.4	11.2	46.9

PROBLEM 37◆

Calculate the total strain energy of cycloheptane.

Cyclohexane rapidly interconverts between two stable chair conformers because of the ease of rotation about its C—C bonds. This interconversion is called **ring flip** (Figure 2.8). When the two chair conformers interconvert, bonds that are equatorial in one chair conformer become axial in the other chair conformer and vice versa.

Bonds that are equatorial in one chair conformer are axial in the other chair conformer.

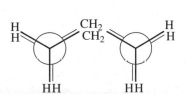

ring flip

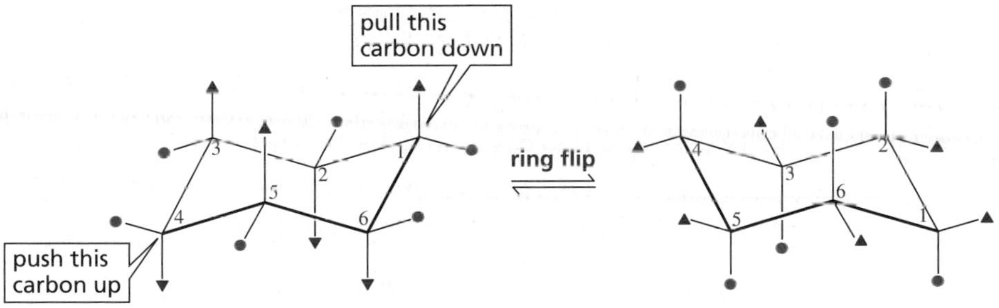

◀ **Figure 2.8**
The bonds that are axial in one chair conformer are equatorial in the other chair conformer. The bonds that are equatorial in one chair conformer are axial in the other chair conformer.

Cyclohexane can also exist in a **boat conformation**, shown in Figure 2.9. Like the chair conformation, the boat conformation is free of angle strain. However, the boat conformation is not as stable because some of the bonds are eclipsed. The boat conformation is further destabilized by the close proximity of the **flagpole hydrogens**—the hydrogens at the "bow" and "stern" of the boat—which causes steric strain.

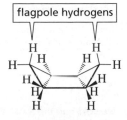

boat conformer of cyclohexane

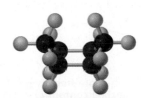

Newman projection of the boat conformer

ball-and-stick model of the boat conformer of cyclohexane

▲ **Figure 2.9**
The boat conformer of cyclohexane, a Newman projection of the boat conformer, and a ball-and-stick model showing that some of the bonds are eclipsed.

The conformations that cyclohexane can assume when interconverting from one chair conformer to the other are shown in Figure 2.10. To convert from the boat conformation to a chair conformation, one of the two topmost carbons of the boat conformation must be pulled down so that it becomes the bottommost carbon of the chair. When the carbon is pulled down just a little, the **twist-boat** (or **skew-boat**) **conformation** is obtained. The twist-boat conformation is more stable than the boat conformation because the flagpole hydrogens have moved away from each other, thus relieving some of the steric strain. When the carbon is pulled down to the point where it is in the same plane as the sides of the boat, the very unstable **half-chair conformation** is obtained. Pulling the carbon down farther produces the *chair conformation*. The graph in Figure 2.10 shows the energy of a cyclohexane molecule as it interconverts from one chair conformer to the other; the energy barrier for interconversion is 12.1 kcal/mol (50.6 kJ/mol). From this value, it can be calculated that cyclohexane undergoes 10^5 ring flips per second at room temperature. In other words, the two chair conformers are in rapid equilibrium.

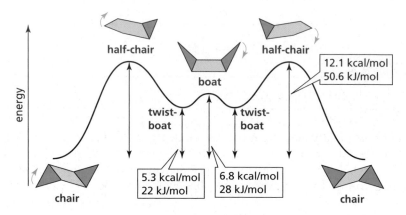

Figure 2.10 ▶
The conformers of cyclohexane—and their relative energies—as one chair conformer interconverts to the other chair conformer.

Go to the Web site to see three-dimensional representations of the conformers of cyclohexane.

The chair conformers are the most stable of cyclohexane's conformations, so at any instant more molecules of cyclohexane are in chair conformations than in any other. For every 10,000 molecules of cyclohexane in a chair conformation, there is no more than one molecule in the next most stable conformation, the twist-boat.

2.13 **Conformers of Monosubstituted Cyclohexanes**

Cyclohexane has two equivalent chair conformers, but the two chair conformers of a monosubstituted cyclohexane, such as methylcyclohexane, are not equivalent. The methyl substituent is in an equatorial position in one conformer and in an axial position in the other (Figure 2.11), because as we have just seen, hydrogens (or substituents) that are equatorial in one chair conformer are axial in the other (Figure 2.8).

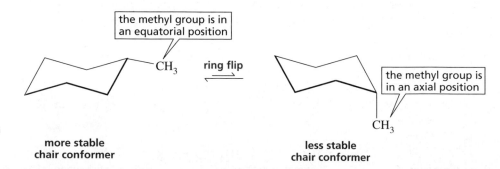

Figure 2.11 ▶
A substituent is in an equatorial position in one chair conformer and in an axial position in the other. The conformer with the substituent in the equatorial position is more stable.

The chair conformer with the methyl substituent in an equatorial position is the more stable one because a substituent has more room and, therefore, fewer steric interactions when it is in an equatorial position. This can be best understood from a drawing like Figure 2.12, which shows that when the methyl group is in an equatorial position, it is anti to the C-3 and C-5 carbons. Therefore, the substituent extends into space, away from the rest of the molecule.

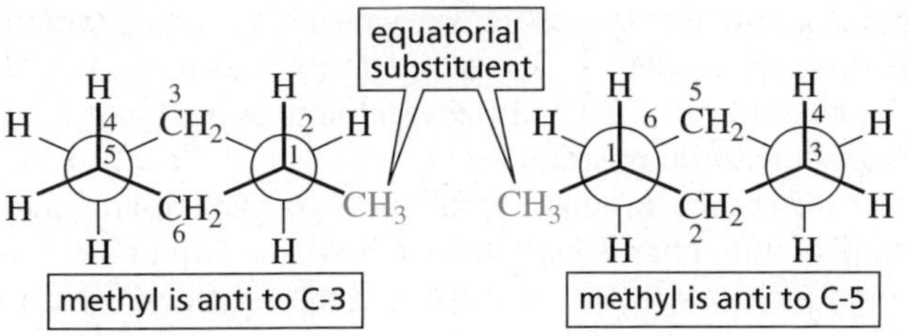

equatorial substituent

methyl is anti to C-3   methyl is anti to C-5

◀ **Figure 2.12**
An equatorial substituent on the C-1 carbon is anti to the C-3 and C-5 carbons.

In contrast, when the methyl group is in an axial position, it is gauche to the C-3 and C-5 carbons (Figure 2.13). As a result, there are unfavorable steric interactions between the axial methyl group and both the axial hydrogen on C-3 and the axial

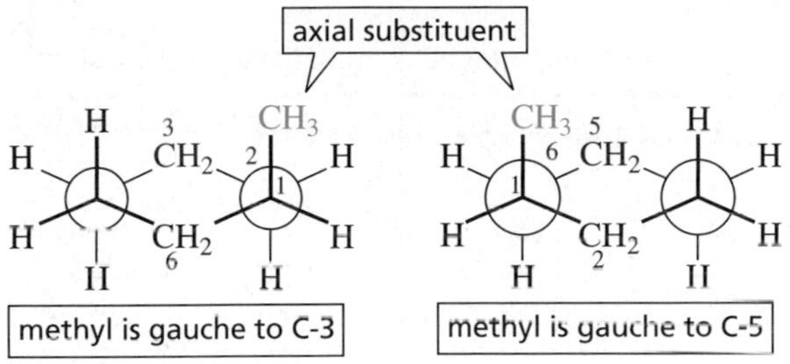

axial substituent

methyl is gauche to C-3   methyl is gauche to C-5

◀ **Figure 2.13**
An axial substituent on the C-1 carbon is gauche to the C-3 and C-5 carbons.

hydrogen on C-5. In other words, the three axial bonds on the same side of the ring are parallel to each other, so any axial hydrogen or substituent will be relatively close to the axial hydrogens or substituents on the other two carbons. Because the interacting hydrogens or substituents are in 1,3-positions relative to each other, these unfavorable steric interactions are called **1,3-diaxial interactions**. If you take a few minutes to build models, you will see that a hydrogen or a substituent has more room if it is in an equatorial position than if it is in an axial position.

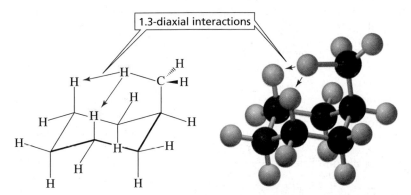

1.3-diaxial interactions

The gauche conformer of butane and the axially substituted conformer of methylcyclohexane are compared in Figure 2.14. Notice that the gauche interaction in butane

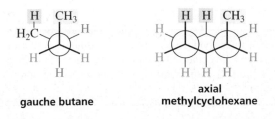

gauche butane   **axial methylcyclohexane**

◀ **Figure 2.14**
The steric strain of gauche butane is the same as the steric strain between an axial methyl group of methylcyclohexane and one of its axial hydrogens. Butane has one gauche interaction between a methyl group and a hydrogen; methylcyclohexane has two.

is the same as a 1,3-diaxial interaction in methylcyclohexane. Butane has one gauche interaction and methylcyclohexane has two 1,3-diaxial interactions.

In Section 2.10, we saw that the gauche interaction between the methyl groups of butane caused the gauche conformation to be 0.87 kcal/mol (3.6 kJ/mol) less stable than the anti conformation. Because there are two such gauche interactions in the chair conformer of methylcyclohexane when the methyl group is in an axial position, this conformer is 1.74 kcal/mol (7.2 kJ/mol) less stable than the chair conformer with the methyl group in the equatorial position.

Because of the difference in stability of the two chair conformers, a sample of methylcyclohexane or any other substituted cycloalkane will at any point in time contain more chair conformers with the substituent in the equatorial position than with the substituent in the axial position. The relative amounts of the two chair conformers depend on the substituent (Table 2.10).

Table 2.10 Equilibrium Constants for Several Monosubstituted Cyclohexanes at 25 °C

Substituent	Axial $\underset{\Longleftarrow}{\overset{K_{eq}}{\Longrightarrow}}$ Equatorial	Substituent	Axial $\underset{\Longleftarrow}{\overset{K_{eq}}{\Longrightarrow}}$ Equatorial
H	1	CN	1.4
CH_3	18	F	1.5
CH_3CH_2	21	Cl	2.4
$CH_3\overset{\overset{CH_3}{|}}{CH}$	35	Br	2.2
		I	2.2
$CH_3\overset{\overset{CH_3}{|}}{\underset{\underset{CH_3}{|}}{C}}$	4800	HO	5.4

The substituent with the greater bulk in the vicinity of the 1,3-diaxial hydrogens will have a greater preference for the equatorial position because it will have stronger 1,3-diaxial interactions. For example, the experimental equilibrium constant (K_{eq}) for the conformers of methylcyclohexane (Table 2.10) indicates that 95% of methylcyclohexane molecules have the methyl group in the equatorial position at 25 °C:

$$K_{eq} = \frac{[\text{equatorial conformer}]}{[\text{axial conformer}]} = \frac{18}{1}$$

$$\% \text{ of equatorial conformer} = \frac{[\text{equatorial conformer}]}{[\text{equatorial conformer}] + [\text{axial conformer}]} \times 100$$

$$\% \text{ of equatorial conformer} = \frac{18}{18 + 1} \times 100 = 95\%$$

The larger the substituent on a cyclohexane ring, the more the equatorial-substituted conformer will be favored.

In *tert*-butylcyclohexane, where the 1,3-diaxial interactions are even more destabilizing because a *tert*-butyl group is larger than a methyl group, more than 99.9% of the molecules have the *tert*-butyl group in the equatorial position.

PROBLEM 38◆

The chair conformer of fluorocyclohexane is 0.25 kcal/mol (1.0 kJ/mol) more stable when the fluoro substituent is in the equatorial position than when it is in the axial position. How much more stable is the anti conformation of 1-fluoropropane compared with a gauche conformation?

PROBLEM 39◆

From the data in Table 2.10, calculate the percentage of molecules of cyclohexanol that have the OH group in the equatorial position.

2.14 Conformers of Disubstituted Cyclohexanes

If a cyclohexane ring has two substituents, we must take both substituents into account when predicting which of the two chair conformers is more stable. Let's use 1,4-dimethylcyclohexane as an example. First of all, note that there are two different dimethylcyclohexanes. One has both methyl substituents on the *same side* of the cyclohexane ring (both point downward); it is called the **cis isomer** (*cis* is Latin for "on this side"). The other has the two methyl substituents on *opposite sides* of the ring (one points upward and one points downward); it is called the **trans isomer** (*trans* is Latin for "across").

The cis isomer of a disubstituted cyclic compound has its substituents on the same side of the ring.

The trans isomer of a disubstituted cyclic compound has its substituents on opposite sides of the ring.

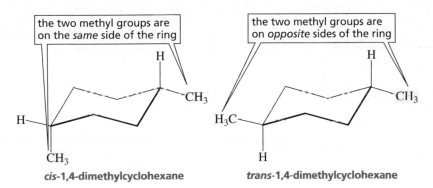

the two methyl groups are on the *same* side of the ring

the two methyl groups are on *opposite* sides of the ring

cis-**1,4-dimethylcyclohexane**

trans-**1,4-dimethylcyclohexane**

cis-1,4-Dimethylcyclohexane and *trans*-1,4-dimethylcyclohexane are examples of **cis–trans isomers** or **geometric isomers**; **cis–trans isomers** are compounds containing the same atoms, and the atoms are linked in the same order, but the atoms exhibit two different spatial arrangements. The cis and trans isomers are different compounds with different melting and boiling points. They can, therefore, be separated from one another.

PROBLEM-SOLVING STRATEGY

Differentiating Cis–Trans Isomers

Is the conformer of 1,2-dimethylcyclohexane with one methyl group in an equatorial position and the other in an axial position the cis isomer or the trans isomer?

Is this the cis isomer or the trans isomer?

To solve this kind of problem we need to determine whether the two substituents are on the same side of the ring (cis) or on opposite sides of the ring (trans). If the bonds bearing the substituents are both pointing upward or both pointing downward, the compound is the cis isomer; if one bond is pointing upward and the other downward, the compound is the trans isomer. Because the conformer in question has both methyl groups attached to downward-pointing bonds, it is the cis isomer.

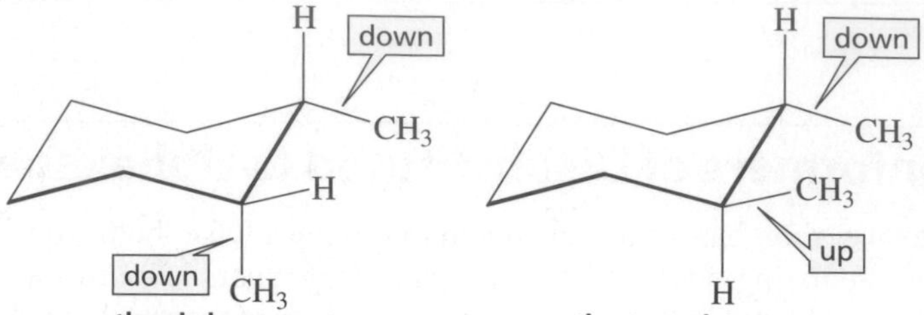

the cis isomer **the trans isomer**

The isomer that is the most misleading when drawn in two dimensions is a *trans*-1,2-disubstituted isomer. At first glance, the methyl groups of *trans*-1,2-dimethylcyclohexane (on the right, above) appear to be oriented in the same direction, so you might think the compound is the cis isomer. Closer inspection shows, however, that one bond points upward and the other downward, so we know that it is the trans isomer. (If you build a model of the compound, you can tell more easily that it is the trans isomer.)

Now continue on to Problem 40.

PROBLEM 40◆

Determine whether each of the following is a cis isomer or a trans isomer:

a. 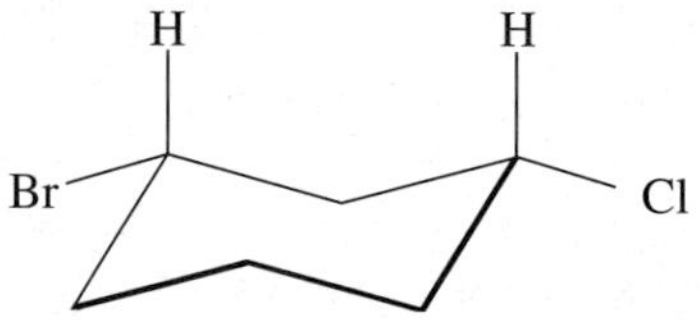

d. 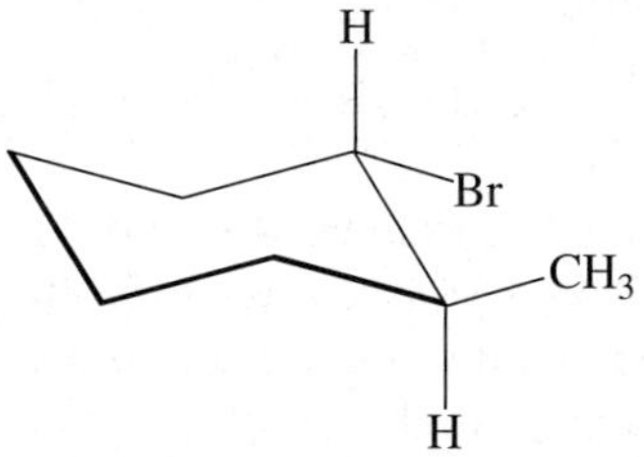

b. 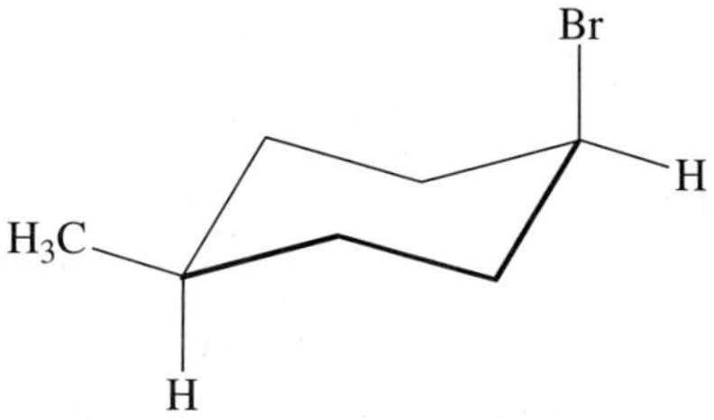

e. 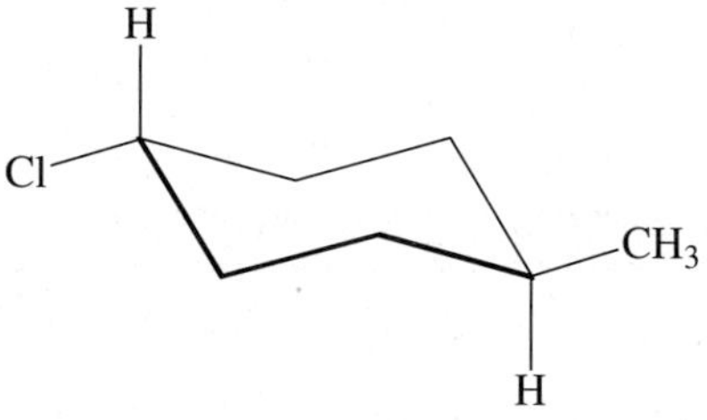

c. 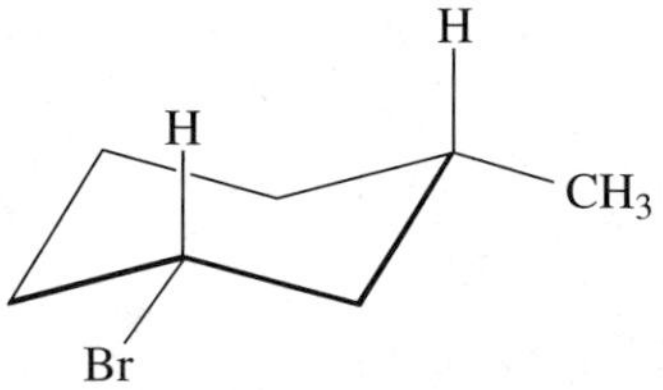

f. 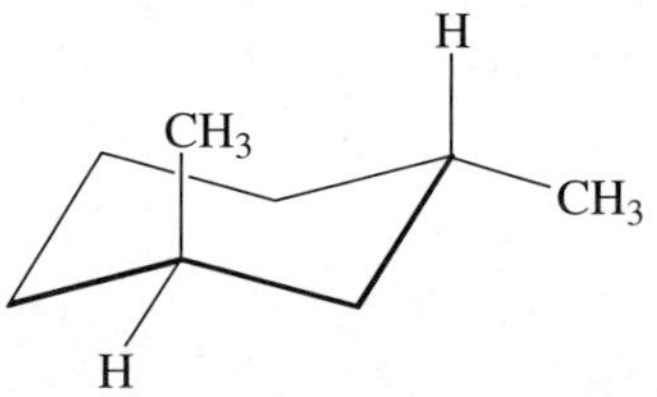

Every compound with a cyclohexane ring has two chair conformers; thus, the cis and trans isomers of disubstituted cyclohexanes each have two chair conformers. Let's compare the structures of the two chair conformers of *cis*-1,4-dimethylcyclohexane to see if we can predict any difference in their stabilities.

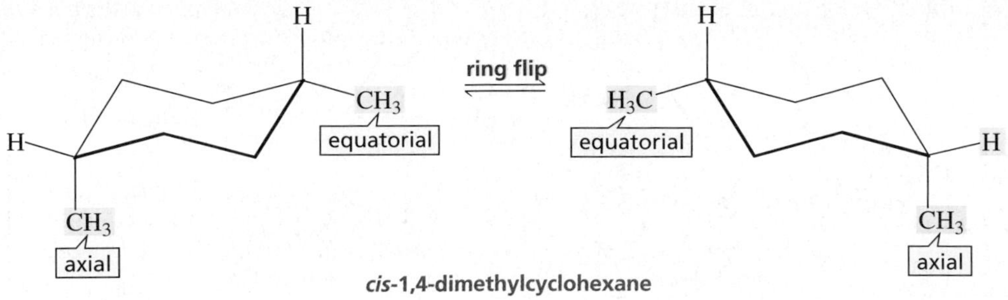

cis-**1,4-dimethylcyclohexane**

The conformer shown on the left has one methyl group in an equatorial position and one methyl group in an axial position. The conformer shown on the right also has one methyl group in an equatorial position and one methyl group in an axial position. Therefore, both chair conformers are equally stable.

In contrast, the two chair conformers of *trans*-1,4-dimethylcyclohexane have different stabilities because one has both methyl substituents in equatorial positions and the other has both methyl groups in axial positions. The conformer with both substituents in equatorial positions is more stable.

more stable   **less stable**

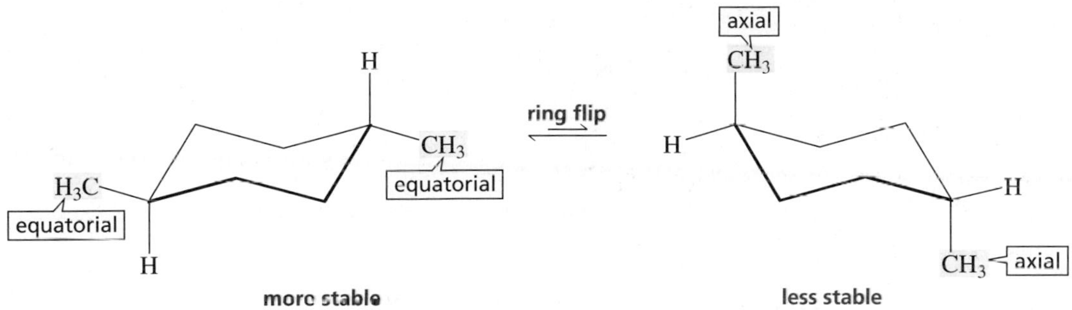

trans-**1,4-dimethylcyclohexane**

The chair conformer with both substituents in axial positions has four 1,3-diaxial interactions, causing it to be about 4×0.87 kcal/mol $= 3.5$ kcal/mol (or 14.6 kJ/mol) less stable than the chair conformer with both methyl groups in equatorial positions. We can, therefore, predict that *trans*-1,4-dimethylcyclohexane will exist almost entirely in the more stable diequatorial conformation.

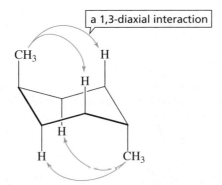

this chair conformer has four 1,3-diaxial interactions

Now let's look at the geometric isomers of 1-*tert*-butyl-3-methylcyclohexane. Both substituents of the cis isomer are in equatorial positions in one conformer and in axial positions in the other conformer. The conformer with both substituents in equatorial positions is more stable.

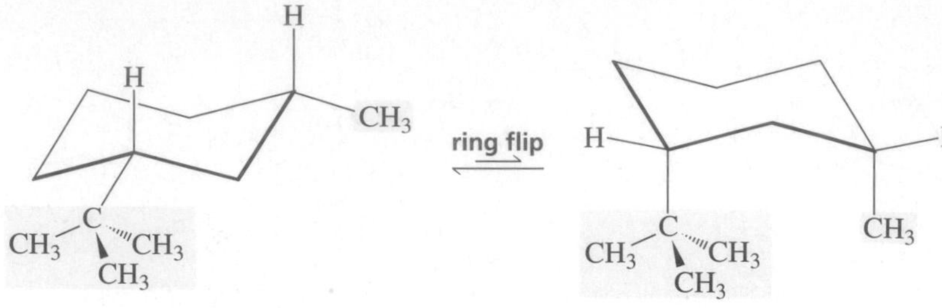

cis-1-*tert*-butyl-3-methylcyclohexane

Both conformers of the trans isomer have one substituent in an equatorial position and the other in an axial position. Because the *tert*-butyl group is larger than the methyl group, the 1,3-diaxial interactions will be stronger when the *tert*-butyl group is in the axial position. Therefore, the conformer with the *tert*-butyl group in the equatorial position is more stable.

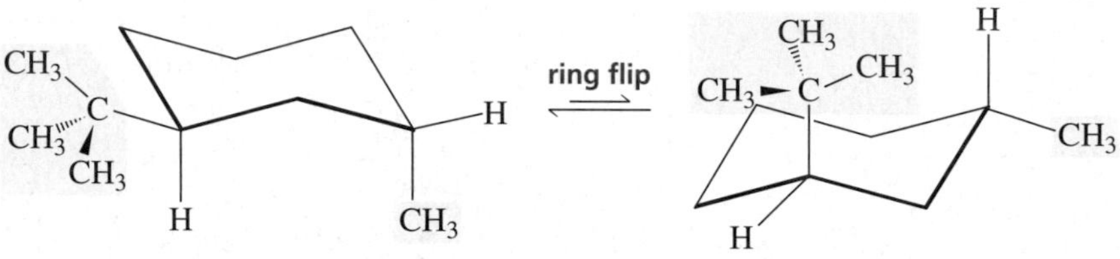

trans-1-*tert*-butyl-3-methylcyclohexane

PROBLEM 41◆

Which will have a higher percentage of the diequatorial-substituted conformer, compared with the diaxial-substituted conformer: *trans*-1,4-dimethylcyclohexane or *cis*-1-*tert*-butyl-3-methylcyclohexane?

PROBLEM 42 **SOLVED**

a. Draw the more stable chair conformer of *cis*-1-ethyl-2-methylcyclohexane.
b. Draw the more stable conformer of *trans*-1-ethyl-2-methylcyclohexane.
c. Which is more stable, *cis*-1-ethyl-2-methylcyclohexane or *trans*-1-ethyl-2-methylcyclohexane?

Solution to 42a If the two substituents of a 1,2-disubstituted cyclohexane are to be on the same side of the ring, one must be in an equatorial position and the other must be in an axial position. The more stable chair conformer is the one in which the larger of the two substituents (the ethyl group) is in the equatorial position.

PROBLEM 43◆

For each of the following disubstituted cyclohexanes, indicate whether the substituents in the two chair conformers would be both equatorial in one chair conformer and both axial in the other *or* one equatorial and one axial in each of the chair conformers:

a. *cis*-1,2-   c. *cis*-1,3-   e. *cis*-1,4-
b. *trans*-1,2-   d. *trans*-1,3-   f. *trans*-1,4-

PROBLEM 44◆

a. Calculate the energy difference between the two chair conformers of *trans*-1,4-dimethylcyclohexane.

b. What is the energy difference between the two chair conformers of *cis*-1,4-dimethylcyclohexane?

SUMMARY

Alkanes are **hydrocarbons** that contain only single bonds. Their general molecular formula is C_nH_{2n+2}. **Constitutional isomers** have the same molecular formula, but their atoms are linked differently. Alkanes are named by determining the number of carbons in their **parent hydrocarbon**—the longest continuous chain. Substituents are listed in alphabetical order, with a number to designate their position on the chain. When there is only a substituent, the substituent gets the lower of the possible numbers; when there is only a functional group suffix, the functional group suffix gets the lower of the possible numbers; when there is both a functional group suffix and a substituent, the functional group suffix gets the lower of the possible numbers. A **functional group** is a center of reactivity in a molecule.

Alkyl halides and **ethers** are named as substituted alkanes. **Alcohols** and **amines** are named using a functional group suffix. **Systematic names** can contain numbers; **common names** never do. A compound can have more than one name, but a name must specify only one compound. Whether alkyl halides or alcohols are **primary, secondary,** or **tertiary** depends on whether the X (halogen) or OH group is bonded to a primary, secondary, or tertiary carbon. A **primary carbon** is bonded to one carbon, a **secondary carbon** is bonded to two carbons, and a **tertiary carbon** is bonded to three carbons. Whether amines are **primary, secondary,** or **tertiary** depends on the number of alkyl groups bonded to the nitrogen. Compounds with four alkyl groups bonded to nitrogen are called **quaternary ammonium salts.**

The oxygen of an alcohol or an ether has the same geometry as the oxygen in water; the nitrogen of an amine has the same geometry as the nitrogen in ammonia. The greater the attractive forces between molecules—**van der Waals forces, dipole–dipole interactions, hydrogen bonds**—the higher is the **boiling point** of the compound. A **hydrogen bond** is an interaction between a hydrogen bonded to an O, N, or F and a lone pair of an O, N, or F in another molecule. In a series of homologs, the boiling point increases with increasing molecular weight. Branching lowers the boiling point. **Polarizability** indicates the ease with which an electron cloud can be distorted: larger atoms are more polarizable.

Polar compounds dissolve in polar solvents, and nonpolar compounds dissolve in nonpolar solvents. The interaction between a solvent and a molecule or an ion dissolved in that solvent is called **solvation.** The oxygen of an alcohol or an ether can usually drag three or four carbons into solution in water.

Rotation about a C—C bond results in two extreme conformations, staggered and eclipsed, that rapidly interconvert. A **staggered conformation** is more stable than an **eclipsed conformation** because of **hyperconjugation.** There can be two different staggered conformers: the **anti conformer** is more stable than the **gauche conformer** because of **steric strain**—repulsion between the electron clouds of atoms or groups. The steric strain in a gauche conformer is called a **gauche interaction.**

Five- and six-membered rings are more stable than three- and four membered rings because of the **angle strain** that results when bond angles deviate from the ideal bond angle of 109.5°. In a process called **ring flip,** cyclohexane rapidly interconverts between two stable chair conformations. **Bonds** that are **axial** in one chair conformer are **equatorial** in the other and vice versa. The chair conformer with a substituent in the equatorial position is more stable, because there is more room, and hence less steric strain, in an equatorial position. A substituent in an axial position experiences unfavorable **1,3-diaxial interactions.** In the case of disubstituted cyclohexanes, the more stable conformer will have its larger substituent in the equatorial position. Cis and trans isomers are called **geometric isomers** or **cis–trans isomers.** A **cis isomer** has its two substituents on the same side of the ring; a **trans isomer** has its substituents on opposite sides of the ring. Cis and trans isomers are different compounds. Conformers are different conformations of the same compound.

KEY TERMS

alcohol (p. 87)
alkane (p. 71)
alkyl halide (p. 85)
alkyl substituent (p. 74)
amine (p. 89)

angle strain (p. 105)
anti conformer (p. 103)
axial bond (p. 107)
banana bond (p. 105)
boat conformation (p. 109)

boiling point (bp) (p. 94)
chair conformer (p. 107)
cis isomer (p. 113)
cis–trans isomers (p. 113)
common name (p. 74)

conformation (p. 101)
conformational analysis (p. 101)
conformer (p. 102)
constitutional isomers (p. 73)
cycloalkane (p. 82)
1,3-diaxial interaction (p. 111)
dipole–dipole interaction (p. 95)
eclipsed conformation (p. 101)
equatorial bond (p. 107)
ether (p. 86)
flagpole hydrogen (p. 109)
functional group (p. 87)
gauche conformer (p. 103)
gauche interaction (p. 103)
geometric isomers (p. 113)
half-chair conformation (p. 110)
heat of formation (p. 108)
homolog (p. 72)
homologous series (p. 72)
hydrocarbon (p. 71)
hydrogen bond (p. 95)

hyperconjugation (p. 102)
induced-dipole–induced-dipole
  interaction (p. 94)
IUPAC nomenclature (p. 74)
melting point (mp) (p. 98)
methylene (CH_2) group (p. 72)
Newman projection (p. 101)
packing (p. 98)
parent hydrocarbon (p. 78)
perspective formula (p. 101)
polarizability (p. 96)
primary alcohol (p. 87)
primary alkyl halide (p. 85)
primary amine (p. 89)
primary carbon (p. 75)
primary hydrogen (p. 76)
quaternary ammonium salt (p. 90)
ring flip (p. 109)
secondary alcohol (p. 87)
secondary alkyl halide (p. 85)
secondary amine (p. 89)

secondary carbon (p. 75)
secondary hydrogen (p. 76)
skeletal structure (p. 82)
skew-boat conformation (p. 110)
solubility (p. 98)
solvation (p. 99)
staggered conformer (p. 101)
steric strain (p. 101)
straight-chain alkane (p. 71)
symmetrical ether (p. 86)
systematic nomenclature (p. 74)
tertiary alcohol (p. 87)
tertiary alkyl halide (p. 85)
tertiary amine (p. 89)
tertiary carbon (p. 76)
tertiary hydrogen (p. 76)
trans isomer (p. 113)
twist-boat conformation (p. 110)
unsymmetrical ether (p. 86)
van der Waals forces (p. 94)

PROBLEMS

45. Write a structural formula for each of the following:
- **a.** *sec*-butyl *tert*-butyl ether
- **b.** isoheptyl alcohol
- **c.** *sec*-butylamine
- **d.** isopentyl bromide
- **e.** 1,1-dimethylcyclohexane
- **f.** 4,5-diisopropylnonane
- **g.** triethylamine
- **h.** cyclopentylcyclohexane
- **i.** 4-*tert*-butylheptane
- **j.** 5,5-dibromo-2-methyloctane
- **k.** 1-methylcyclopentanol
- **l.** 3-ethoxy-2-methylhexane
- **m.** 5-(1,2-dimethylpropyl)nonane
- **n.** 3,4-dimethyloctane
- **o.** 4-(1-methylethyl)nonane

46. a. Give the systematic name for each of the following:

1. $(CH_3)_3CCH_2CH_2CH_2CH(CH_3)_2$

2. $CH_3CHCH_2CH_2CHCH_2CH_2CH_3$ with Br on the fourth carbon and CH_3 on the second carbon

3. $CH_3CHCH_2CHCH_2CH_3$ with CH_3 and OH substituents

4. $(CH_3CH_2)_4C$

5. $BrCH_2CH_2CH_2CH_2CH_2NHCH_2CH_3$

6. $CH_3CHCH_2CHCHCH_3$ with CH_3, CH_3, CH_3 substituents

7. $CH_3CH_2CHOCH_2CH_3$ with $CH_2CH_2CH_2CH_3$ substituent

8. $CH_3OCH_2CH_2CH_2OCH_3$

9. (cyclohexane with NCH_3 and CH_3 group)

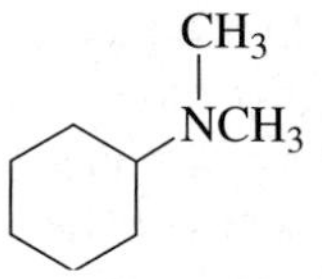

10. (cyclohexane with CH_2CH_3 and OH groups)

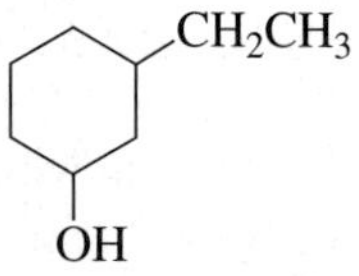

11. (cyclohexane with CH_3 and Br groups)

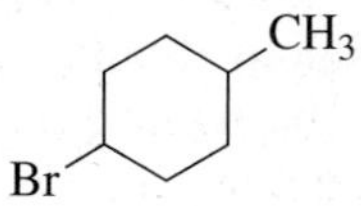

b. Draw skeletal structures for these compounds.

47. a. How many primary carbons does the following compound have?

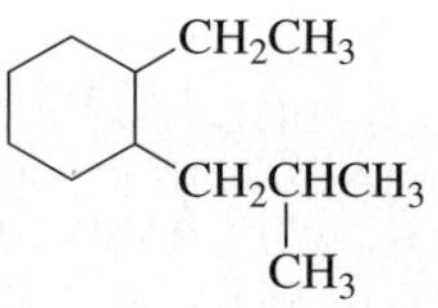

(cyclohexane with CH_2CH_3 and CH_2CHCH_3 with CH_3 substituents)

b. How many secondary carbons does the compound have?
c. How many tertiary carbons does it have?

48. Which of the following conformers of isobutyl chloride is the most stable?

49. Draw the structural formula of an alkane that has
   a. six carbons, all secondary
   b. eight carbons and only primary hydrogens
   c. seven carbons with two isopropyl groups

50. Give two names for each of the following:

a. $CH_3CH_2CHCH_3$
           |
          NH_2

b. $CH_3CH_2CHCH_3$
           |
          Cl

c. $CH_3CH_2CH(CH_3)NHCH_2CH_3$

d. $CH_3CH_2CH_2OCH_2CH_3$

e. $CH_3CHCH_2CH_2CH_3$
           |
          CH_3

f. CH_3CHNH_2
           |
          CH_3

g. CH_3
          |
        CH_3CBr
          |
        CH_2CH_3

h. $CH_3CHCH_2CH_2CH_2OH$
           |
          CH_3

i. 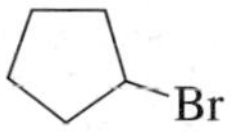 ⬠–Br

j. 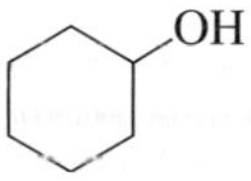 ⬡–OH

51. Which of the following pairs has
   a. the higher boiling point: 1-bromopentane or 1-bromohexane?
   b. the higher boiling point: pentyl chloride or isopentyl chloride?
   c. the greater solubility in water: 1-butanol or 1-pentanol?
   d. the higher boiling point: 1-hexanol or 1-methoxypentane?
   e. the higher melting point: hexane or isohexane?
   f. the higher boiling point: 1-chloropentane or 1-pentanol?
   g. the higher boiling point: 1-bromopentane or 1-chloropentane?
   h. the higher boiling point: diethyl ether or butyl alcohol?
   i. the greater density: heptane or octane?
   j. the higher boiling point: isopentyl alcohol or isopentylamine?
   k. the higher boiling point: hexylamine or dipropylamine?

52. Draw the nine isomeric heptanes and name each isomer.

53. Ansaid and Motrin belong to the group of drugs known as nonsteroidal anti-inflammatory drugs (NSAIDs). Both are only slightly soluble in water, but one is a little more soluble than the other. Which of the drugs has the greater solubility in water?

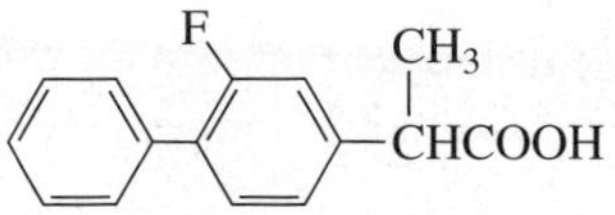

Ansaid®

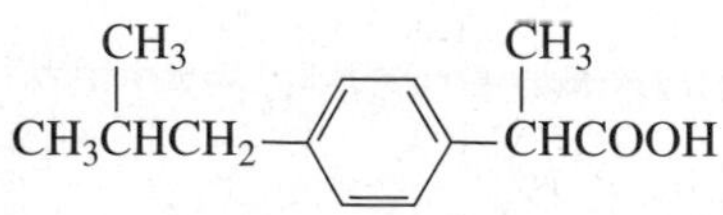

Motrin®

54. Al Kane was given the structural formulas of several compounds and was asked to give them systematic names. How many did Al name correctly? Correct those that are misnamed.
- **a.** 4-bromo-3-pentanol
- **b.** 2,2-dimethyl-4-ethylheptane
- **c.** 5-methylcyclohexanol
- **d.** 1,1-dimethyl-2-cyclohexanol
- **e.** 5-(2,2-dimethylethyl)nonane
- **f.** isopentyl bromide
- **g.** 3,3-dichlorooctane
- **h.** 5-ethyl-2-methylhexane
- **i.** 1-bromo-4-pentanol
- **j.** 3-isopropyloctane
- **k.** 2-methyl-2-isopropylheptane
- **l.** 2-methyl-N,N-dimethyl-4-hexanamine

55. Which of the following conformers has the highest energy?

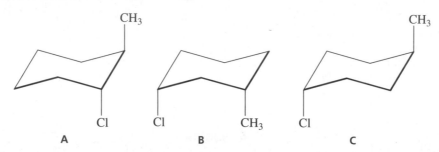

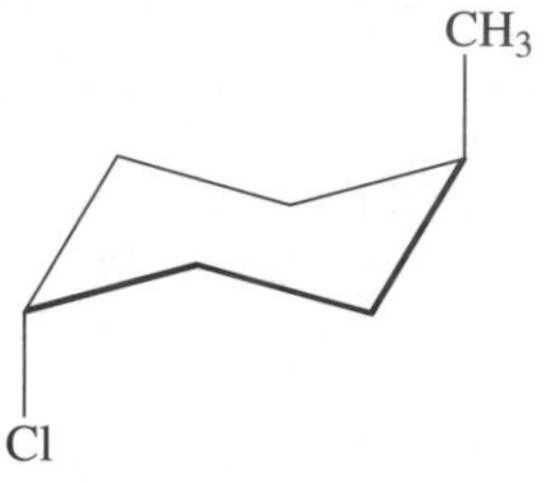

56. Give systematic names for all the alkanes with molecular formula C_7H_{16} that do not have any secondary hydrogens.

57. Draw skeletal structures for the following:
- **a.** 5-ethyl-2-methyloctane
- **b.** 1,3-dimethylcyclohexane
- **c.** 2,3,3,4-tetramethylheptane
- **d.** propylcyclopentane
- **e.** 2-methyl-4-(1-methylethyl)octane
- **f.** 2,6-dimethyl-4-(2-methylpropyl)decane

58. For rotation about the C-3 — C-4 bond of 2-methylhexane:
- **a.** Draw the Newman projection of the most stable conformation.
- **b.** Draw the Newman projection of the least stable conformation.
- **c.** About which other carbon–carbon bonds may rotation occur?
- **d.** How many of the carbon–carbon bonds in the compound have staggered conformers that are all equally stable?

59. Which of the following structures represents a cis isomer?

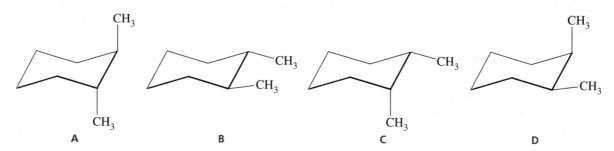

60. Draw all the isomers that have molecular formula $C_5H_{11}Br$. (*Hint:* There are eight such isomers.)
- **a.** Give the systematic name for each of the isomers.
- **b.** Give a common name for each isomer that has one.
- **c.** How many isomers do not have common names?
- **d.** How many of the isomers are primary alkyl halides?
- **e.** How many of the isomers are secondary alkyl halides?
- **f.** How many of the isomers are tertiary alkyl halides?

61. Give the systematic name for each of the following:

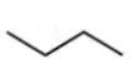

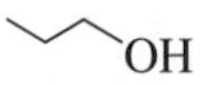

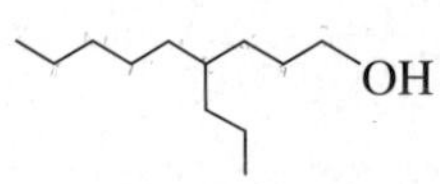

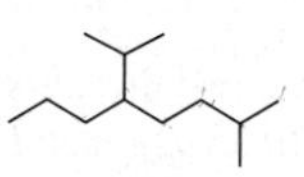

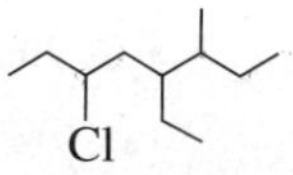

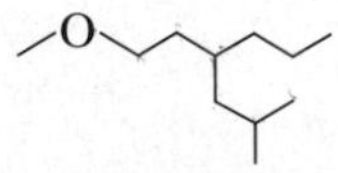

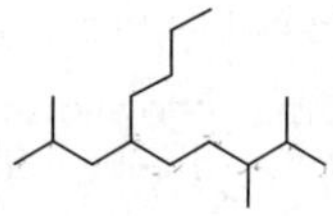

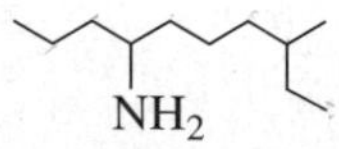

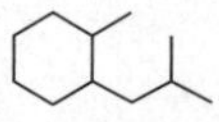

62. Draw the two chair conformers for each of the following, and indicate which conformer is more stable:
  a. *cis*-1-ethyl-3-methylcyclohexane
  b. *trans*-1-ethyl-2-isopropylcyclohexane
  c. *trans*-1-ethyl-2-methylcyclohexane
  d. *trans*-1-ethyl-3-methylcyclohexane
  e. *cis*-1-ethyl-3-isopropylcyclohexane
  f. *cis*-1-ethyl-4-isopropylcyclohexane

63. Why are alcohols of lower molecular weight more soluble in water than those of higher molecular weight?

64. How many ethers have molecular formula $C_5H_{12}O$? Draw their structures and give each a systematic name. What are their common names?

65. Draw the most stable conformer of the following molecule:

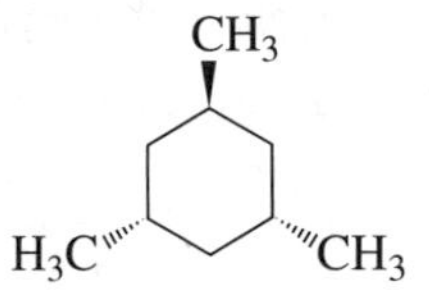

66. Give the systematic name for each of the following:

  a. $CH_3CH_2CHCH_2CH_2CHCH_3$
           $NHCH_3$ CH_3

  b. $CH_3CH_2CHCH_2CHCH_2CH_3$
            CH_3
            $CHCH_3$
            CH_3

  c. $CH_3CHCHCH_2Cl$
            Cl

  d. $CH_3CH_2CHCH_3$
            $CHCH_3$
            CH_3

  e. $CH_3CH_2CH_2CH_2CHCH_2CH_2CH_2CH_3$
            $CH_3CCH_2CH_3$
            CH_3

  f. $CH_3CH_2CH_2CH_2CH_2CHCH_2CHCH_2CH_3$
            CH_2CH_3
            CH_2
            CH_3CCH_3
            CH_2CH_3

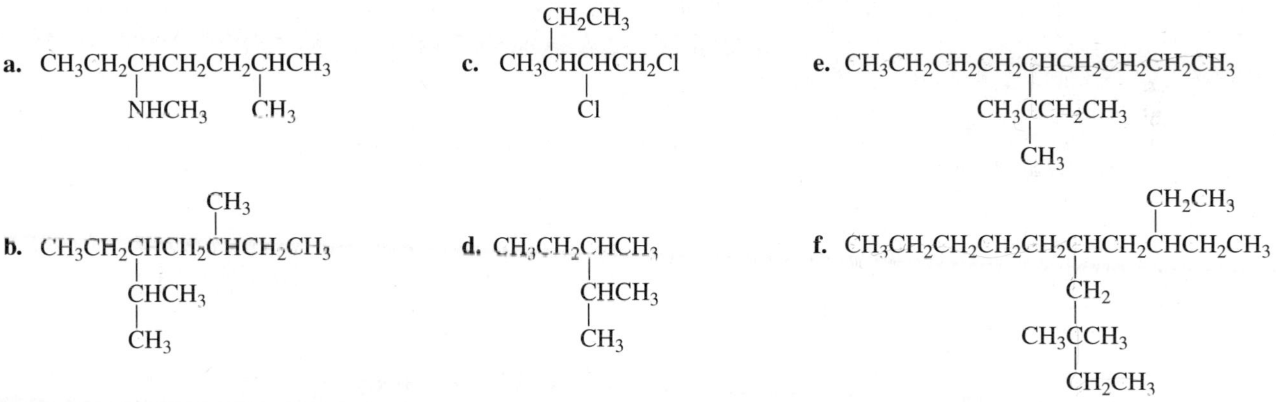

67. Calculate the energy difference between the two chair conformers of *trans*-1,2-dimethylcyclohexane.

68. The most stable form of glucose (blood sugar) is a six-membered ring in a chair conformation with its five substituents all in equatorial positions. Draw the most stable form of glucose by putting the OH groups on the appropriate bonds in the chair conformer.

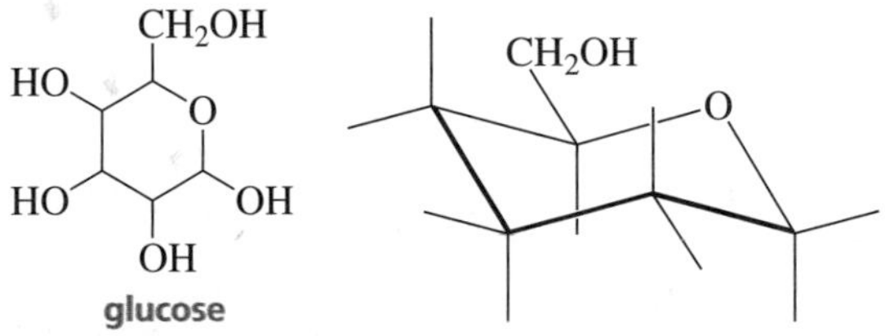

glucose

69. Give the systematic name for each of the following:

  a. $CH_3CHCH_2CHCH_2CH_3$
            CH_3 OH

  b.
            Br
            CH_3

  c. $CH_3CHCHCH_2CH_3$
            OH

  d. $CH_3CHCH_2CH_2CHCH_2CH_2CH_3$
            CH_3 Br

  e. $CH_3CH_2CH_2CHCHCH_2CH_2CH_2CH_3$
            CH_2CH_3
            $CH_2CH_2CH_2CH_3$

  f. $CH_3CHCH_2CH_2CH_2CH_2Br$
            OH

  g. HO
            CH_2CH_3
            CH_3

  h. Br
            CH_2CH_3
            CH_3

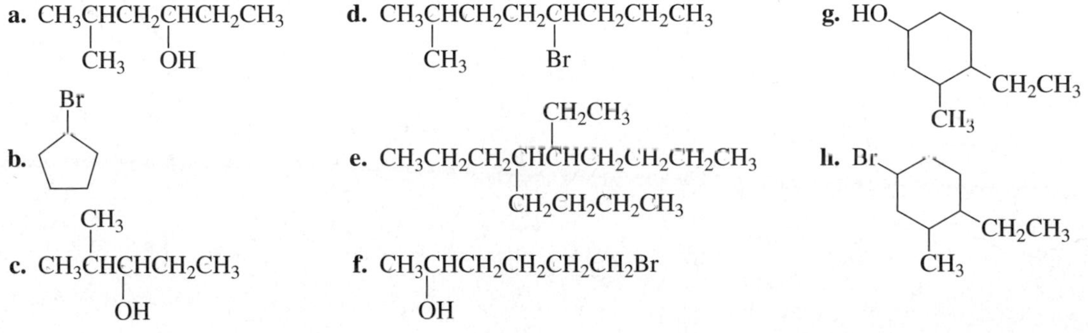

70. Explain the following facts:
   a. 1-Hexanol has a higher boiling point than 3-hexanol.
   b. Diethyl ether has very limited solubility in water, but tetrahydrofuran is essentially completely soluble.

tetrahydrofuran

71. One of the chair conformers of *cis*-1,3-dimethylcyclohexane has been found to be 5.4 kcal/mol (or 23 kJ/mol) less stable than the other. How much steric strain does a 1,3-diaxial interaction between two methyl groups introduce into the conformer?

72. Bromine is a larger atom than chlorine, but the equilibrium constants in Table 2.10 indicate that a chloro substituent has a greater preference for the equatorial position than a bromo substituent does. Suggest an explanation for this fact.

73. Calculate the amount of steric strain in each of the chair conformers of 1,1,3-trimethylcyclohexane. Which conformer would predominate at equilibrium?

Electrophilic Addition Reactions, Stereochemistry, and Electron Delocalization

The reactions of organic compounds can be divided into three main types: **addition reactions, substitution reactions**, and **elimination reactions**. The particular type of reaction a compound undergoes depends on the functional group in the compound. **Part 2** discusses the reactions of compounds that have carbon–carbon double bonds and carbon–carbon triple bonds. We will see that these compounds undergo addition reactions, or, more precisely, **electrophilic addition reactions**. **Part 2** also examines two topics that can be important when determining the outcome of a reaction. The first of these is stereochemistry and the second is electron delocalization.

CHAPTER 3
Alkenes: Structure, Nomenclature, and an Introduction to Reactivity • Thermodynamics and Kinetics

Chapter 3 begins with a look at the structure, nomenclature, and stability of alkenes—*compounds that contain carbon–carbon double bonds*—and then introduces some fundamental principles that govern the reactions of organic compounds. You will learn how to draw curved arrows to show how electrons move during the course of a reaction as new covalent bonds are formed and existing covalent bonds are broken. This chapter also discusses the principles of thermodynamics and kinetics, which are central to an understanding of how and why organic reactions take place.

CHAPTER 4
The Reactions of Alkenes

Organic compounds can be divided into families, and fortunately, all members of a family react in the same way. In **Chapter 4**, you will learn how the family of compounds known as alkenes reacts and what kinds of products are formed from the reactions. Although many different reactions are covered, you will see that they all take place by a similar pathway.

CHAPTER 5
Stereochemistry: The Arrangement of Atoms in Space; The Stereochemistry of Addition Reactions

Chapter 5 is all about stereochemistry. After learning about the different kinds of isomers that are possible for organic compounds, you will revisit the reactions that you learned in Chapter 4 to see whether the products can exist as isomers and, if so, which isomers are formed.

CHAPTER 6
The Reactions of Alkynes • An Introduction to Multistep Synthesis

Chapter 6 covers the reactions of alkynes—*compounds that contain carbon–carbon triple bonds*. Because alkenes and alkynes both have reactive carbon–carbon π bonds, you will discover that their reactions have many similarities. This chapter will also introduce you to some of the techniques chemists use to design syntheses of organic compounds, and you will then have your first opportunity to design a multistep synthesis.

CHAPTER 7
Delocalized Electrons and Their Effect on Stability, Reactivity, and pKa • More About Molecular Orbital Theory

In **Chapter 7**, you will learn more about delocalized electrons, which were introduced in Chapter 1. You will see how they influence some of the chemical properties with which you are already familiar—acidity, the stability of carbocations, and the reactions of alkenes. At this point we will turn to the reactions of dienes, compounds *that have two carbon–carbon double bonds*. You will see that if the two double bonds in a diene are sufficiently separated, the reactions of the diene are identical to the reactions of alkenes (Chapter 4); however, if the double bonds are separated by only one single bond, electron delocalization plays an important role in the reactions of the diene.

CHAPTER 3

Alkenes
Structure, Nomenclature, and an Introduction to Reactivity • Thermodynamics and Kinetics

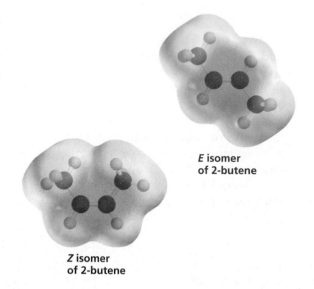

E isomer
of 2-butene

Z isomer
of 2-butene

BUILDING ON FUNDAMENTALS

SECTION 3.2	Alkenes are named using a functional group suffix, the same way alcohols (2.6) and amines (2.7) are named.
SECTION 3.3	The structures of alkenes are similar to the structure of ethene (1.8).
SECTION 3.4	We encountered cis–trans isomers when we looked at the structures of disubstituted cyclohexanes (2.14). Now we will see that alkenes can also have cis–trans isomers.
SECTION 3.5	Because a π bond is weaker than a σ bond (1.14), the π bond is the one that breaks when an alkene undergoes a reaction.

SECTION 3.6	The curved arrows that were used to show the bond-making and bond-breaking processes in acid–base reactions (1.18) will be used to show the bond-making and bond-breaking processes in a reaction of an alkene.
SECTION 3.7	We have seen that solvation causes a polar molecule to dissolve in a polar solvent (2.9). Now we will see that solvation affects both $\Delta H°$ and $\Delta S°$.

Ethene is the hormone that causes
tomatoes to ripen.

In Chapter 2, we saw that alkanes are hydrocarbons that contain only carbon–carbon *single* bonds. Hydrocarbons that contain a carbon–carbon *double* bond are called **alkenes**. Early chemists noted that an oily substance was formed when ethene ($H_2C\!=\!CH_2$), the smallest alkene, reacted with chlorine. On the basis of this observation, alkenes were originally called *olefins* ("oil forming").

Alkenes play many important roles in biology. Ethene, for example, is a plant hormone—a compound that controls growth and other changes in the plant's tissues. Ethene affects seed germination, flower maturation, and fruit ripening.

The chemicals that organisms release as a way of communicating with each other are called pheromones. Many of the sex, alarm, and trail pheromones are alkenes. Interfering with an insect's ability to send or receive chemical signals is an environmentally safe way to control insect populations. For example, traps scented with synthetic sex attractants have been used to capture such crop-destroying insects as the gypsy moth and the boll weevil.

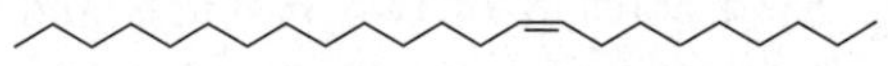

muscalure
sex attractant of the housefly

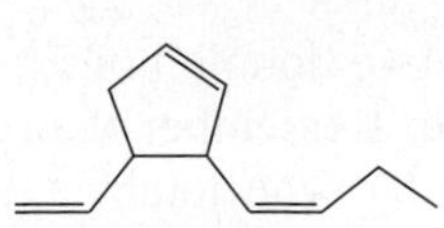

multifidene
**sex attractant of
brown algae**

Many of the flavors and fragrances produced by plants also belong to the alkene family.

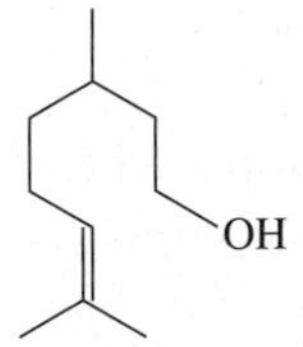

citronellol
**in rose and
geranium oils**

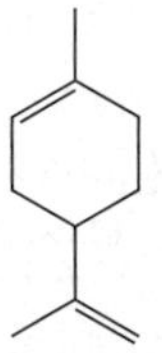

limonene
**in lemon and
orange oils**

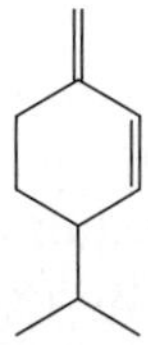

β-phellandrene
oil of eucalyptus

We will begin our study of alkenes by looking at their structures and how they are named. Then we will examine a reaction of an alkene, paying close attention to the steps by which it occurs and the energy changes that accompany them. You will see that some of the discussion in the chapter revolves around concepts with which you are already familiar, while some of the information is new and will broaden the foundation of knowledge that you will be building on in subsequent chapters.

3.1 Molecular Formulas and the Degree of Unsaturation

We have seen that the general molecular formula for a noncyclic alkane is C_nH_{2n+2} (Section 2.0). We have also seen that the general molecular formula for a cyclic alkane is C_nH_{2n} because the cyclic structure reduces the number of hydrogens by two (Section 2.2). Noncyclic compounds are also called **acyclic** compounds (the prefix "*a*" is Greek for "non" or "not").

The general molecular formula for an *acyclic alkene* is also C_nH_{2n} because, as a result of the double bond, an alkene has two fewer hydrogens than an alkane with the same number of carbons. Thus, the general molecular formula for a *cyclic alkene* must be C_nH_{2n-2}. We can, therefore, make the following statement: *the general molecular formula for a hydrocarbon is C_nH_{2n+2} minus two hydrogens for every π bond or ring in the molecule.*

> The general molecular formula for a hydrocarbon is C_nH_{2n+2} minus two hydrogens for every π bond or ring present in the molecule.

$CH_3CH_2CH_2CH_2CH_3$
an alkane
C_5H_{12}
C_nH_{2n+2}

$CH_3CH_2CH_2CH=CH_2$
an alkene
C_5H_{10}
C_nH_{2n}

a cyclic alkane
C_5H_{10}
C_nH_{2n}

a cyclic alkene
C_5H_8
C_nH_{2n-2}

Therefore, if we know the molecular formula of a hydrocarbon, we can determine how many rings and π bonds it has because for every *two* hydrogens missing from the general molecular formula (C_nH_{2n+2}), a hydrocarbon has either a π bond or a ring. For example, a compound with a molecular formula of C_8H_{14} needs four more hydrogens to become C_8H_{18} ($C_8H_{(2\times8)+2}$). Consequently, the compound has either (A) two double bonds, (B) a ring and a double bond, (C) two rings, or (D) a triple bond. Remember that a triple bond consists of two π bonds and a σ bond (Section 1.9). The total number of π bonds and rings in an alkene is called the alkene's **degree of unsaturation**. Thus, C_8H_{14} has two degrees of unsaturation.

several compounds with molecular formula C_8H_{14}:

$$CH_3CH=CH(CH_2)_3CH=CH_2$$

$$CH_3(CH_2)_5C\equiv CH$$

A   B   C   D

Because alkanes contain the maximum number of C—H bonds possible—that is, they are saturated with hydrogen—they are called **saturated hydrocarbons**. In contrast, alkenes are called **unsaturated hydrocarbons** because they have fewer than the maximum number of hydrogens.

$$CH_3CH_2CH_2CH_3$$
a saturated hydrocarbon

$$CH_3CH=CHCH_3$$
an unsaturated hydrocarbon

PROBLEM 1◆ *SOLVED*

Determine the molecular formula for each of the following:

a. a 5-carbon hydrocarbon with one π bond and one ring.
b. a 4-carbon hydrocarbon with two π bonds and no rings.
c. a 10-carbon hydrocarbon with one π bond and two rings.
d. an 8-carbon hydrocarbon with three π bonds and one ring.

Solution to 1a For a 5-carbon hydrocarbon with no π bonds and no rings, $C_nH_{2n+2} = C_5H_{12}$. A 5-carbon hydrocarbon with one π bond and one ring has four fewer hydrogens, because two hydrogens are subtracted for every π bond or ring in the molecule. The molecular formula, therefore, is C_5H_8.

PROBLEM 2◆ *SOLVED*

Determine the degree of unsaturation for hydrocarbons with the following molecular formulas:

a. $C_{10}H_{16}$ **b.** $C_{20}H_{34}$ **c.** C_8H_{16} **d.** $C_{12}H_{20}$ **e.** $C_{40}H_{56}$

Solution to 2a For a 10-carbon hydrocarbon with no π bonds and no rings, $C_nH_{2n+2} = C_{10}H_{22}$. A 10-carbon compound with molecular formula $C_{10}H_{16}$ has six fewer hydrogens. The degree of unsaturation, therefore, is $6/2 = 3$.

PROBLEM 3

Determine the degree of unsaturation, and then draw possible structures, for compounds with the following molecular formulas:

a. C_3H_6 **b.** C_3H_4 **c.** C_4H_6

3.2 The Nomenclature of Alkenes

We have seen that the IUPAC system uses a suffix to denote certain functional groups (Sections 2.6 and 2.7). The double bond is the functional group of an alkene, and its presence is denoted by the suffix "ene." Therefore, the systematic (IUPAC) name of an alkene is obtained by replacing the "ane" ending of the corresponding alkane with "ene." For example, a two-carbon alkene is called ethene and a three-carbon alkene is called propene. Ethene also is frequently called by its common name: ethylene.

$$H_2C=CH_2 \qquad CH_3CH=CH_2$$

systematic name: ethene   propene   cyclopentene   cyclohexene
common name: ethylene   propylene

Most alkene names need a number to indicate the position of the double bond. (The four names above do not, because there is no ambiguity.) The IUPAC rules you learned in Chapter 2 apply to alkenes as well:

1. The longest continuous chain containing the functional group (in this case, the carbon–carbon double bond) is numbered in a direction that gives the functional group suffix the lowest possible number. For example, 1-butene signifies that the double bond is between the first and second carbons of butene; 2-hexene signifies that the double bond is between the second and third carbons of hexene.

> **Number the longest continuous chain containing the functional group in the direction that gives the functional group suffix the lowest possible number.**

$$\overset{4}{C}H_3\overset{3}{C}H_2\overset{2}{C}H=\overset{1}{C}H_2 \qquad \overset{1}{C}H_3\overset{2}{C}H=\overset{3}{C}H\overset{4}{C}H_3 \qquad \overset{1}{C}H_3\overset{2}{C}H=\overset{3}{C}H\overset{4}{C}H_2\overset{5}{C}H_2\overset{6}{C}H_3$$

1-butene   2-butene   2-hexene

$$\overset{6}{C}H_3\overset{5}{C}H_2\overset{4}{C}H_2\overset{3}{C}H_2\overset{2}{C}CH_2CH_2CH_3$$
$$\underset{1\ CH_2}{\overset{\|}{\ }}$$

2-propyl-1-hexene

> the longest continuous chain has eight carbons but the longest continuous chain containing the functional group has six carbons, so the parent name of the compound is hexene

Notice that 1-butene does not have a common name. You might be tempted to call it "butylene," which is analogous to "propylene" for propene, but butylene is not an appropriate name. A name must be unambiguous, and "butylene" could signify either 1-butene or 2-butene.

2. For a compound with two double bonds, the suffix is "diene."

$$\overset{1}{C}H_2=\overset{2}{C}H-\overset{3}{C}H_2-\overset{4}{C}H-\overset{5}{C}H_2 \qquad \overset{1}{C}H_3\overset{2}{C}H-\overset{3}{C}H-\overset{4}{C}H=\overset{5}{C}H\overset{6}{C}H_2\overset{7}{C}H_3 \qquad \overset{5}{C}H_3\overset{4}{C}H=\overset{3}{C}H-\overset{2}{C}H=\overset{1}{C}H_2$$

1,4-pentadiene   2,4-heptadiene   1,3-pentadiene

3. The name of a substituent is cited before the name of the longest continuous chain that contains the functional group, together with a number to designate the carbon to which the substituent is attached. Notice that *if a compound contains both a functional group suffix and a substituent, the functional group suffix gets the lowest possible number*.

> **When there are both a functional group suffix and a substituent, the functional group suffix gets the lowest possible number.**

$$\overset{1}{C}H_3\overset{2}{C}H=\overset{3}{C}H\overset{4}{\underset{|}{C}H}\overset{5}{C}H_3$$
$$\underset{CH_3}{}$$

4-methyl-2-pentene

$$\overset{2}{C}H_2\overset{1}{C}H_3$$
$$\overset{3}{\underset{|}{C}}$$
$$CH_3\overset{}{C}=\overset{4}{C}H\overset{5}{C}H_2\overset{6}{C}H_2\overset{7}{C}H_3$$

3-methyl-3-heptene

$$CH_3CH_2CH_2CH_2CH_2OCH_2CH_2\overset{4}{C}H_2\overset{3}{C}H_2\overset{2}{C}H=\overset{1}{C}H_2$$

4-pentoxy-1-butene

$$\overset{CH_3}{\underset{|}{}}$$
$$CH_3\overset{4}{C}=\overset{3}{C}H\overset{2}{C}H=\overset{1}{C}H_2$$

4-methyl-1,3-pentadiene

4. If a chain has more than one substituent, the substituents are cited in alphabetical order, using the same rules for alphabetizing discussed in Section 2.2. Then the appropriate number is assigned to each substituent.

> **Substituents are cited in alphabetical order.**

3,6-dimethyl-3-octene

$$\underset{7}{C}H_3\underset{6}{C}H_2\underset{5}{\overset{Br}{\underset{|}{C}}}H\underset{4}{\overset{Cl}{\underset{|}{C}}}H\underset{3}{C}H_2\underset{2}{C}H=\underset{1}{C}H_2$$

5-bromo-4-chloro-1-heptene

5. If counting in either direction results in the same number for the alkene functional group suffix, the correct name is the one containing the lowest substituent number. For example, 2,5-dimethyl-4-octene is a 4-octene whether the longest continuous chain is numbered from left to right or from right to left. If you number from left to right, the substituents are at positions 4 and 7, but if you number from right to left, they are at positions 2 and 5. Of those four substituent numbers, 2 is the lowest, so the compound is named 2,5-dimethyl-4-octene, *not* 4,7-dimethyl-4-octene.

> **A substituent receives the lowest possible number only if there is no functional group suffix or if the same number for the functional group suffix is obtained in both directions.**

$$CH_3CH_2CH_2C=CHCH_2CHCH_3$$
$$\quad\quad\quad\quad\quad | \quad\quad\quad | $$
$$\quad\quad\quad\quad\quad CH_3 \quad\quad CH_3$$

2,5-dimethyl-4-octene
not
4,7-dimethyl-4-octene
because 2 < 4

$$CH_3CHCH=CCH_2CH_3$$
$$\quad | \quad\quad\quad | $$
$$\quad Br \quad\quad CH_3$$

2-bromo-4-methyl-3-hexene
not
5-bromo-3-methyl-3-hexene
because 2 < 3

Student Tutorial:
Alkene nomenclature

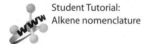

6. A number is not needed to denote the position of the double bond in a cyclic alkene because the ring is always numbered so that the double bond is between carbons 1 and 2. To assign numbers to any substituents, count around the ring in the direction (clockwise or counterclockwise) that puts the lowest number into the name.

3-ethylcyclopentene   **4,5-dimethylcyclohexene**   **4-ethyl-3-methylcyclohexene**

For example, 1,6-dichlorocyclohexene is *not* called 2,3-dichlorocyclohexene because 1,6-dichlorocyclohexene has the lowest substituent number (1), even though it does not have the lowest sum of substituent numbers (1 + 6 = 7 versus 2 + 3 = 5).

1,6-dichlorocyclohexene
not
2,3-dichlorocyclohexene
because 1 < 2

5-ethyl-1-methylcyclohexene
not
4-ethyl-2-methylcyclohexene
because 1 < 2

7. If counting in either direction leads to the same number for the alkene functional group suffix and the same lowest number or numbers for one or more of the substituents, then ignore those substituents and choose the direction that gives the lowest number to one of the remaining substituents.

$$\quad\quad\quad\quad\quad\quad Br$$
$$\quad\quad\quad\quad\quad\quad |$$
$$CH_3CHCH_2CH=CCH_2CHCH_3$$
$$\quad | \quad\quad\quad\quad\quad\quad | $$
$$\quad CH_3 \quad\quad\quad CH_2CH_3$$

2-bromo-4-ethyl-7-methyl-4-octene
not
7-bromo-5-ethyl-2-methyl-4-octene
because 4 < 5

6-bromo-3-chloro-4-methylcyclohexene
not
3-bromo-6-chloro-5-methylcyclohexene
because 4 < 5

The sp^2 carbons of an alkene are called **vinylic carbons**. An sp^3 carbon that is adjacent to a vinylic carbon is called an **allylic carbon**.

vinylic carbons
$$RCH_2-CH=CH-CH_2R$$
allylic carbons

Two groups containing a carbon–carbon double bond are used in common names—the **vinyl group** and the **allyl group**. The vinyl group is the smallest possible group containing a vinylic carbon; the allyl group is the smallest possible group containing an allylic carbon. When "vinyl" or "allyl" is used in a name, the substituent must be attached to the vinylic or allylic carbon, respectively.

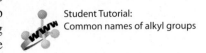

Student Tutorial:
Common names of alkyl groups

$$H_2C=CH—$$
the vinyl group

$$H_2C=CHCH_2—$$
the allyl group

$$H_2C=CHCl$$ $$H_2C=CHCH_2Br$$

systematic name: chloroethene   3-bromopropene
common name: vinyl chloride   allyl bromide

PROBLEM 4♦

Draw the structure for each of the following:

a. 3,3-dimethylcyclopentene   **c.** ethyl vinyl ether
b. 6-bromo-2,3-dimethyl-2-hexene   **d.** allyl alcohol

PROBLEM 5♦

Give the systematic name for each of the following:

a. 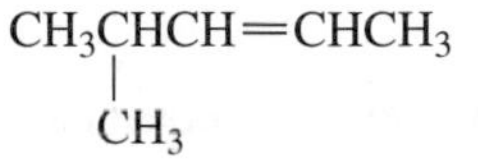 $CH_3CHCH=CHCH_3$
          |
          CH_3

b. 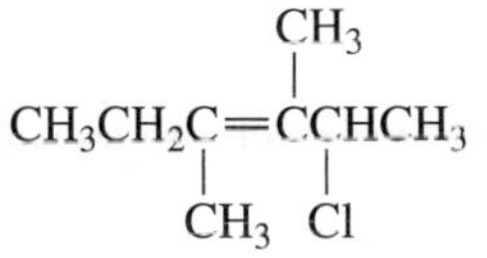 CH_3
               |
     $CH_3CH_2C=CCHCH_3$
                |
               CH_3 Cl

c. 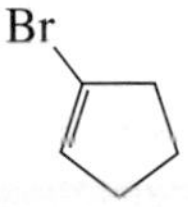 Br (cyclopentene with Br)

d. 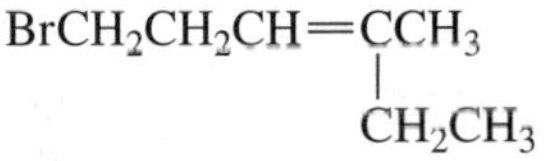 $BrCH_2CH_2CH=CCH_3$
                    |
                    CH_2CH_3

e. 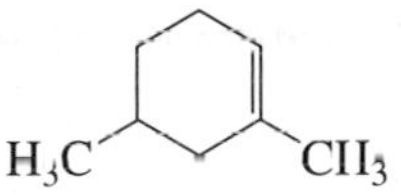 (cyclohexene ring) H_3C CH_3

f. $CH_3CH=CHOCH_2CH_2CH_2CH_3$

3.3 **The Structures of Alkenes**

The structure of the smallest alkene (ethene) was described in Section 1.8. Other alkenes have similar structures. Each double-bonded carbon of an alkene has three sp^2 orbitals that lie in a plane with angles of 120°. Each of these sp^2 orbitals overlaps an orbital of another atom to form a σ bond. Thus, one of the carbon–carbon bonds in a double bond is a σ bond, formed by the overlap of an sp^2 orbital of one carbon with an sp^2 orbital of the other carbon. The second carbon–carbon bond in the double bond (the π bond) is formed from side-to-side overlap of the remaining p orbital of one of the sp^2 carbons with the remaining p orbital of the other sp^2 carbon. Because three points determine a plane, each sp^2 carbon and the two atoms singly bonded to it lie in a plane. In order to achieve maximum orbital–orbital overlap, the two p orbitals must be parallel to each other. Therefore, all six atoms of the double-bond system are in the same plane.

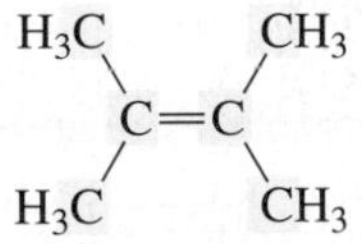

$$H_3C \qquad CH_3$$
$$\diagdown \qquad \diagup$$
$$C=C$$
$$\diagup \qquad \diagdown$$
$$H_3C \qquad CH_3$$

the six carbon atoms
are in the same plane

It is important to remember that the π bond consists of the cloud of electrons distributed above and below the plane defined by the two sp^2 carbons and the four atoms bonded to them.

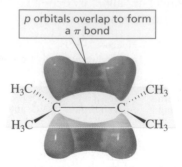

> *p* orbitals overlap to form a π bond

PROBLEM 6◆ *SOLVED*

For each of the following compounds, tell how many of its carbon atoms lie in the same plane:

a.   **b.**   **c.**   **d.**

Solution to 6a The two sp^2 carbons (blue dots) and the three carbons bonded to the sp^2 carbons (red dots) lie in the same plane. Therefore, five carbons lie in the same plane.

3.4 Alkenes Can Have Cis and Trans Isomers

We have just seen that the two *p* orbitals forming the π bond must be parallel to achieve maximum overlap. Therefore, rotation about a double bond does not readily occur. If rotation were to occur, the two *p* orbitals would cease to overlap, and the π bond would break (Figure 3.1). The energy barrier to rotation about a double bond is about 62 kcal/mol or 259 kJ/mol (Section 1.14). Compare this to the energy barrier of 2.9 kcal/mol or 12 kJ/mol to rotation about a carbon–carbon single bond (Section 2.10).

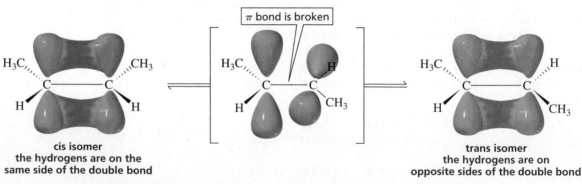

cis isomer
the hydrogens are on the
same side of the double bond

> π bond is broken

trans isomer
the hydrogens are on
opposite sides of the double bond

▲ **Figure 3.1**
Rotation about the carbon–carbon double bond would break the π bond.

Because of the high energy barrier to rotation about a carbon–carbon double bond, an alkene such as 2-butene can exist in two distinct forms: the hydrogens bonded to the sp^2 carbons can be on the same side of the double bond or on opposite sides of the double bond.

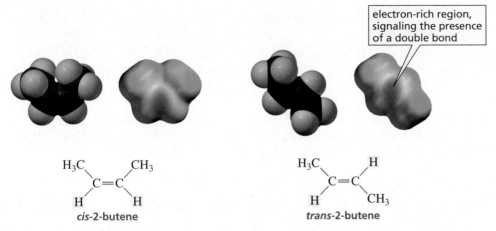

electron-rich region, signaling the presence of a double bond

cis-2-butene

trans-2-butene

The isomer with the hydrogens on the same side of the double bond is called the **cis isomer**, and the isomer with the hydrogens on opposite sides of the double bond is called the **trans isomer**. A pair of isomers such as *cis*-2-butene and *trans*-2-butene are called **cis–trans isomers** or **geometric isomers**. This should remind you of the cis–trans isomers of 1,2-disubstituted cyclohexanes you encountered in Section 2.14—the cis isomer had its substituents on the same side of the ring, and the trans isomer had its substituents on opposite sides of the ring. Notice that cis–trans isomers have the same molecular formula and the same bonds but differ in the way the atoms are oriented in space.

If one of the sp^2 carbons of the double bond is attached to two identical substituents, there is only one possible structure for the alkene. In other words, cis and trans isomers are not possible for an alkene that has identical substituents attached to one of the sp^2 carbons.

cis and trans isomers are not possible for these compounds because two substituents on an sp^2 carbon are the same

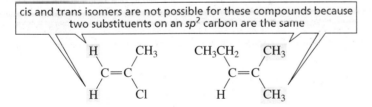

PROBLEM 7◆

a. Which of the following compounds can exist as cis–trans isomers?
b. For those compounds, draw and label the cis and trans isomers.
   1. $CH_3CH=CHCH_2CH_2CH_3$ **3.** $CH_3CH=CHCH_3$

   2. $CH_3CH_2C=CHCH_3$ **4.** $CH_3CH_2CH=CH_2$
          |
        CH_2CH_3

PROBLEM 8◆

Draw three alkenes with molecular formula C_5H_{10} that do not have cis–trans isomers.

Because of the energy barrier to rotation about a double bond, cis and trans isomers of alkenes cannot interconvert (except under conditions extreme enough to overcome the energy barrier and break the π bond). This means that they can be separated from each other. In other words, the two isomers are different compounds with different physical properties, such as different boiling points and different dipole moments. Notice that *trans*-2-butene and *trans*-1,2-dichloroethene, unlike their respective cis isomers, have dipole moments (μ) of zero because the bond dipole moments cancel (Section 1.15).

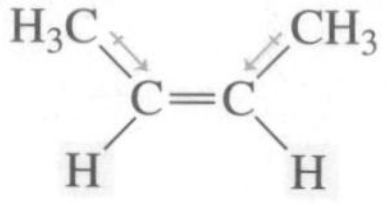

cis-2-butene
bp = 3.7 °C
μ = 0.33 D

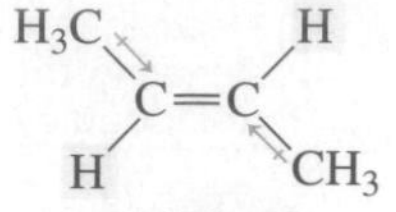

trans-2-butene
bp = 0.9 °C
μ = 0 D

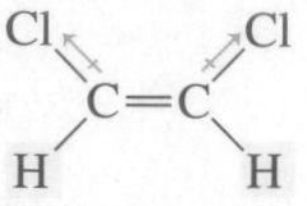

cis-1,2-dichloroethene
bp = 60.3 °C
μ = 2.95 D

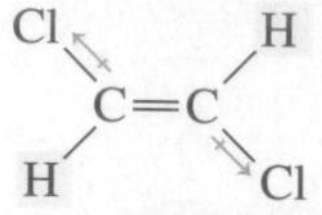

trans-1,2-dichloroethene
bp = 47.5 °C
μ = 0 D

Because cis and trans isomers can be interconverted (in the absence of any added reagents) only when the molecule absorbs sufficient heat or light energy to cause the π bond to break, cis–trans interconversion is not a practical laboratory process.

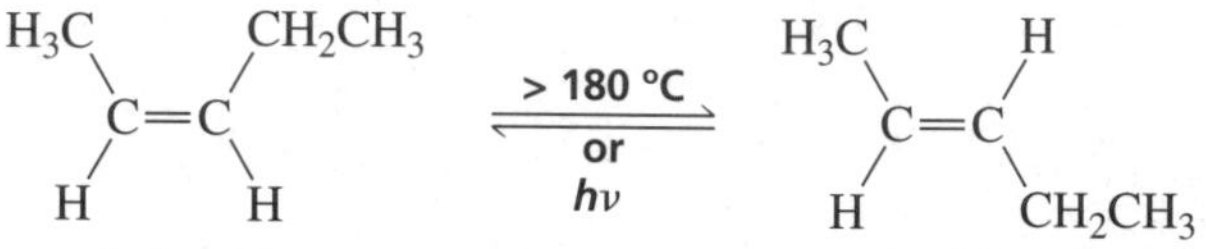

> 180 °C
or
$h\nu$

cis-2-pentene

trans-2-pentene

CIS–TRANS INTERCONVERSION IN VISION

The ability to see depends in part on a cis–trans interconversion taking place in our eyes. A protein called opsin binds to *cis*-retinal in the rod cells of the retina to form rhodopsin. When rhodopsin absorbs light, a double bond interconverts between the cis and trans configurations, triggering a nerve impulse that plays an important role in vision (Section 26.7). *trans*-Retinal is then released from opsin, isomerizing back to *cis*-retinal, and another cycle begins. To trigger the nerve impulse, a group of about 500 rod cells must register 5 to 7 rhodopsin isomerizations within a few tenths of a second.

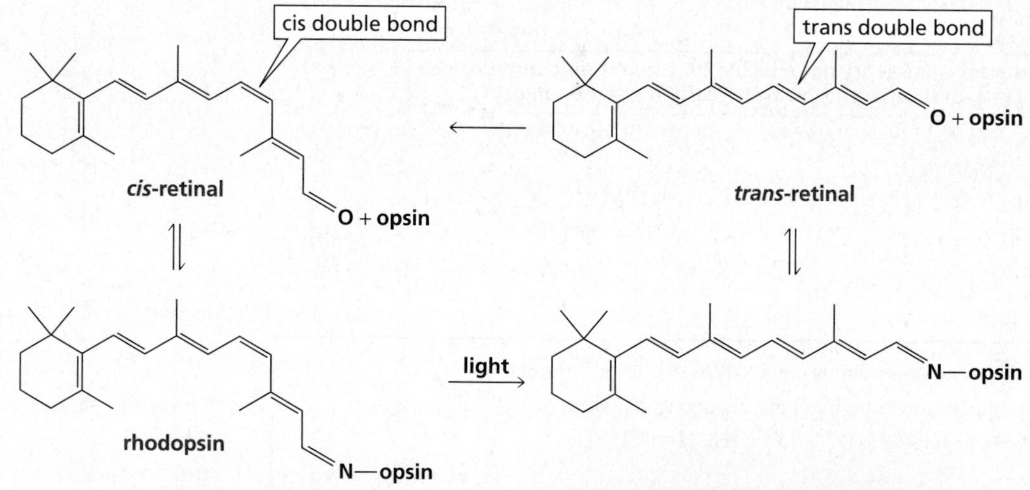

cis double bond

cis-retinal

O + opsin

trans double bond

trans-retinal

O + opsin

rhodopsin

N—opsin

light

N—opsin

—OPSIN

—OPSIN

cis form

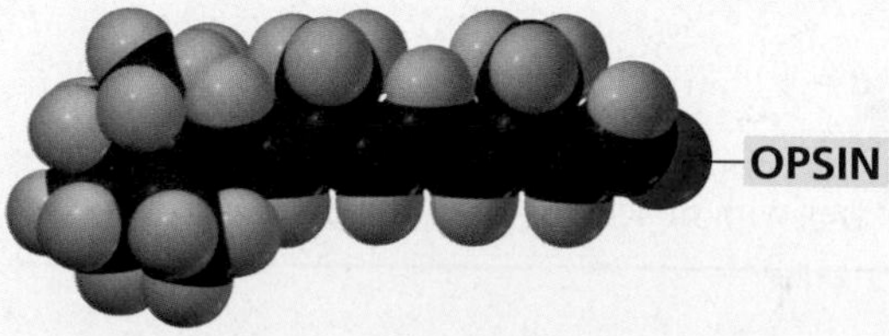

trans form

Which of the following compounds have a dipole moment of zero?

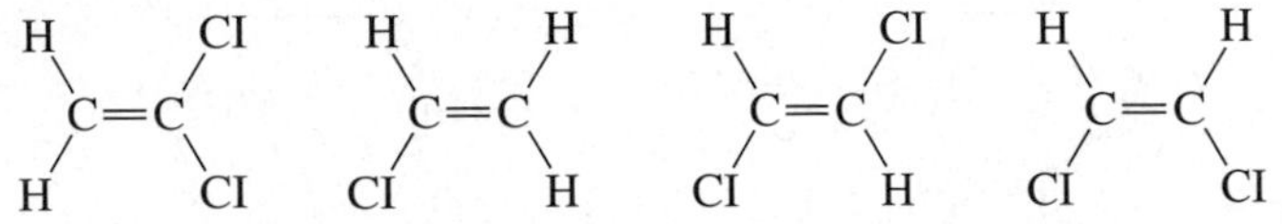

A   B   C   D

3.5 **Naming Alkenes Using the *E,Z* System**

As long as each of the sp^2 carbons of an alkene is bonded to only one substituent, we can use the terms "cis" and "trans" to designate the structure of the alkene: *if the hydrogens are on the same side of the double bond, it is the cis isomer; if they are on opposite sides of the double bond, it is the trans isomer.* But how do we designate the isomers of a compound such as 1-bromo-2-chloropropene?

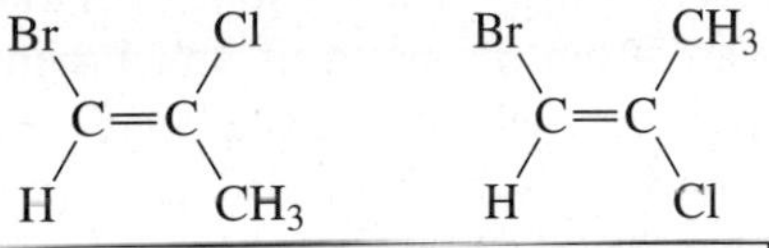

Which isomer is cis and which is trans?

For a compound such as 1-bromo-2-chloropropene, the cis–trans system of nomenclature cannot be used because there are four different groups on the two vinylic carbons. The *E,Z* system of nomenclature was devised for these kinds of compounds.*

To name an isomer by the *E,Z* system, we first determine the relative priorities of the two groups bonded to one of the sp^2 carbons and then the relative priorities of the two groups bonded to the other sp^2 carbon. (The rules for assigning relative priorities are explained below.) If the two higher-priority groups (one from each carbon) are on the same side of the double bond, the isomer has the *Z* configuration (*Z* is for *zusammen*, German for "together"). If the higher-priority groups are on opposite sides of the double bond, the isomer has the *E* configuration (*E* is for *entgegen*, German for "opposite").

The *Z* isomer has the high-priority groups on the same side.

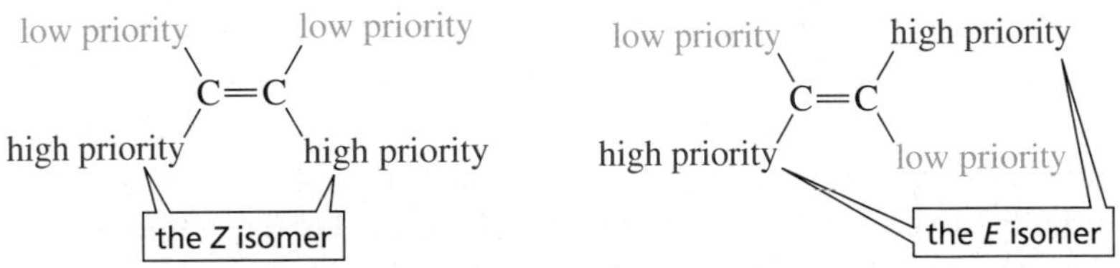

The relative priorities of the two groups bonded to an sp^2 carbon are determined using the following rules:

1. The relative priorities of the two groups depend on the atomic numbers of the atoms bonded directly to the sp^2 carbon. The greater the atomic number, the higher the priority.

The greater the atomic number of the atom bonded to the sp^2 carbon, the higher is the priority of the substituent.

---

* The IUPAC prefers the *E* and *Z* designations because they can be used for all alkene isomers. Many chemists, however, continue to use the "cis" and "trans" designations for simple molecules.

For example, in the following compounds, one of the sp^2 carbons is bonded to a Br and to an H:

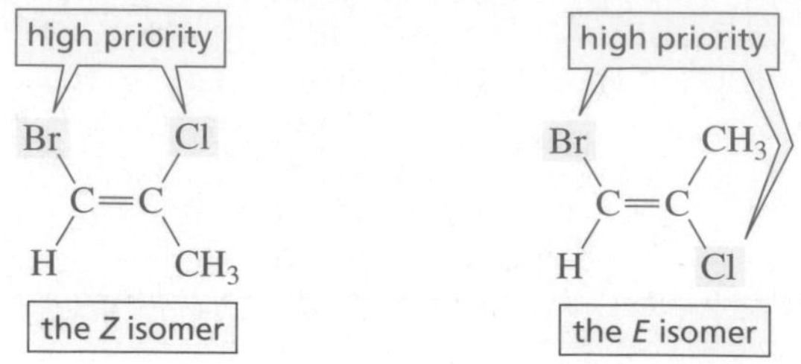

Bromine has a greater atomic number than hydrogen, so **Br** has a higher priority than **H**. The other sp^2 carbon is bonded to a Cl and to a C. Chlorine has the greater atomic number, so **Cl** has a higher priority than **C**. (Notice that you use the atomic number of C, not the mass of the CH_3 group, because the priorities are based on the atomic numbers of atoms, *not* on the masses of groups.) The isomer on the left has the high-priority groups (Br and Cl) on the same side of the double bond, so it is the **Z isomer**. (Zee groups are on Zee Zame Zide.) The isomer on the right has the high-priority groups on opposite sides of the double bond, so it is the **E isomer**.

If the atoms attached to an sp^2 carbon are the same, the atoms attached to the tied atoms are compared; the one with the greatest atomic number belongs to the group with the higher priority.

2. If the two groups bonded to an sp^2 carbon start with the same atom (there is a tie), you then move outward from the point of attachment and consider the atomic numbers of the atoms that are attached to the "tied" atoms.

In the following compounds, both atoms bonded to the sp^2 carbon on the left are C's (in a CH_2Cl group and a CH_2CH_2Cl group), so there is a tie.

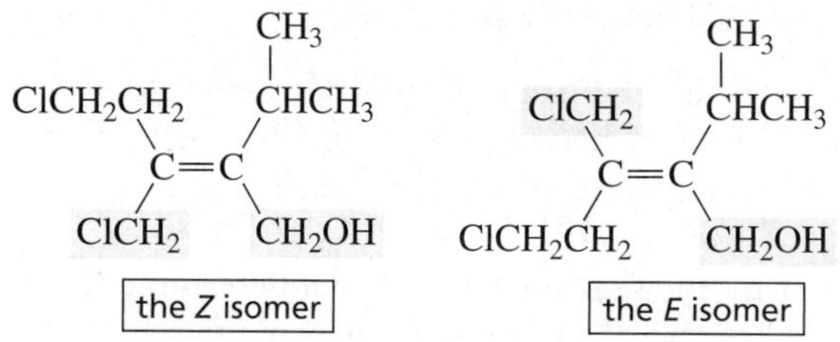

The C of the CH_2Cl group is bonded to **Cl, H, H**, and the C of the CH_2CH_2Cl group is bonded to **C, H, H**. Cl has a greater atomic number than C, so the CH_2Cl group has the higher priority. Both atoms bonded to the other sp^2 carbon are C's (in a CH_2OH group and a $CH(CH_3)_2$ group), so there is a tie on that side as well. The C of the CH_2OH group is bonded to **O, H**, and **H**, and the C of the $CH(CH_3)_2$ group is bonded to **C, C**, and **H**. Of these six atoms, O has the greatest atomic number, so **CH_2OH** has a higher priority than **$CH(CH_3)_2$**. (Note that you do not add the atomic numbers; you consider the single atom with the greatest atomic number.) The E and Z isomers are as shown above.

If an atom is doubly bonded to another atom, treat it as if it were singly bonded to two of those atoms.

If an atom is triply bonded to another atom, treat it as if it were singly bonded to three of those atoms.

3. If an atom is doubly bonded to another atom, the priority system treats it as if it were singly bonded to two of those atoms. If an atom is triply bonded to another atom, the priority system treats it as if it were singly bonded to three of those atoms.

For example, one of the sp^2 carbons in the following pair of isomers is bonded to a CH_2CH_2OH group and to a $CH_2C{\equiv}CH$ group:

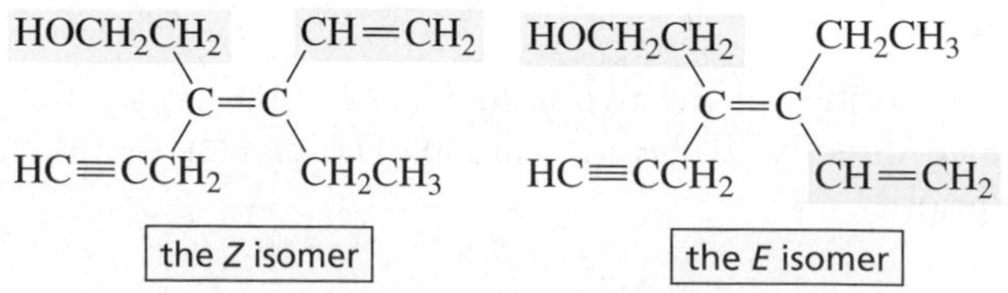

Because the carbons immediately bonded to the sp^2 carbon on the left are both bonded to C, H, and H, we ignore them and turn our attention to the groups attached to them. One of these is CH_2OH and the other is $C\equiv CH$. The triple-bonded C is considered to be bonded to **C, C,** and **C**; the other C is bonded to **O, H,** and **H**. Of the six atoms, O has the greatest atomic number, so CH_2OH has a higher priority than $C\equiv CH$. Both atoms that are bonded to the other sp^2 carbon are C's, so they are tied. The first carbon of the CH_2CH_3 group is bonded to **C, H,** and **H**. The first carbon of the $CH\!=\!CH_2$ group is bonded to an H and doubly bonded to a C, so it is considered to be bonded to **H, C,** and **C**. One C cancels in each of the two groups, leaving H and H in the CH_2CH_3 group and H and C in the $CH\!=\!CH_2$ group. C has a greater atomic number than H, so $CH\!=\!CH_2$ has a higher priority than CH_2CH_3.

> Cancel atoms that are identical in the two groups; use the remaining atoms to determine the group with the higher priority.

4. If two isotopes (atoms with the same atomic number, but different mass numbers) are being compared, the mass number is used to determine the relative priorities. In the following structures, for example, one of the sp^2 carbons is bonded to a deuterium (D) and a hydrogen (H):

> If atoms have the same atomic number, but different mass numbers, the one with the greater mass number has the higher priority.

Student Tutorial:
E and *Z* Nomenclature

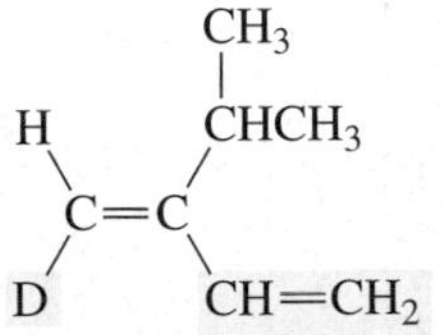

$$\underset{\textbf{the Z isomer}}{\overset{\displaystyle CH_3}{\underset{\displaystyle D}{\overset{\displaystyle \underset{H}{\overset{}{}}\quad CHCH_3}{C\!=\!C}}\quad CH\!=\!CH_2}}$$

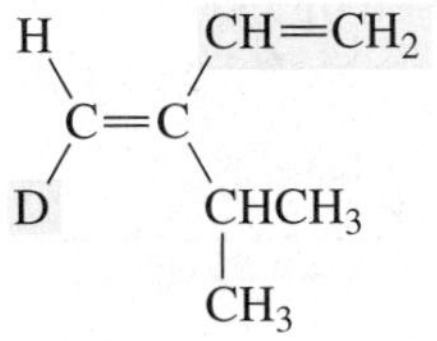

the Z isomer   the E isomer

Deuterium and hydrogen have the same atomic number, but D has a greater mass number, so **D** has a higher priority than **H**. The C's that are bonded to the other sp^2 carbon are *both* bonded to **C, C,** and **H**, so we must go to the next set of atoms to break the tie. The second carbon of the $CH(CH_3)_2$ group is bonded to **H, H,** and **H**, whereas the second carbon of the $CH\!=\!CH_2$ group is bonded to **H, H,** and **C**. Therefore, $CH\!=\!CH_2$ has a higher priority than $CH(CH_3)_2$.

PROBLEM 10◆

Assign relative priorities to each set of substituents:

a. $-Br, -I, -OH, -CH_3$
b. $-CH_2CH_2OH, -OH, -CH_2Cl, -CH\!=\!CH_2$

PROBLEM 11

Draw and label the *E* and *Z* isomers for each of the following:

a. $CH_3CH_2CH\!=\!CHCH_3$

c. $CH_3CH_2CH_2CH_2$
$\qquad\quad CH_3CH_2C\!=\!CCH_2Cl$
$\qquad\qquad\quad CH_3CHCH_3$

b. $CH_3CH_2C\!=\!CHCH_2CH_3$
$\qquad\quad\; Cl$

d. $HOCH_2CH_2C\!=\!CC\!\equiv\!CH$
$\qquad\qquad\; O\!=\!CH\;\; C(CH_3)_3$

PROBLEM 12◆

Name each of the following:

a. 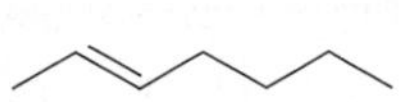   **b.** 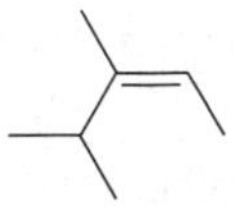   **c.**

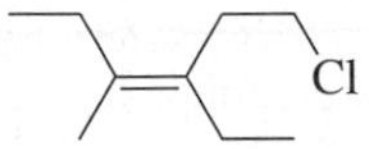

PROBLEM-SOLVING STRATEGY

Drawing E,Z Structures

Draw the structure of (E)-1-bromo-2-methyl-2-butene.

First draw the compound without specifying the isomer so you can see what substituents are bonded to the sp^2 carbons. Then determine the relative priorities of the two groups on each of the sp^2 carbons.

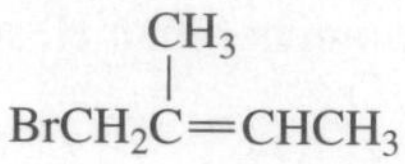

$$\overset{\displaystyle CH_3}{\underset{\displaystyle |}{BrCH_2C}}=CHCH_3$$

One sp^2 carbon is attached to a CH_3 and an H: CH_3 has the higher priority. The other sp^2 carbon is attached to a CH_3 and a CH_2Br: CH_2Br has the higher priority. To get the E isomer, draw the compound with the two high priority substituents on opposite sides of the double bond.

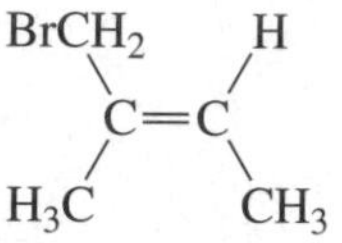

Now continue on to Problem 13.

PROBLEM 13

Draw the structure of (Z)-3-isopropyl-2-heptene.

PROBLEM-SOLVING STRATEGY

Drawing Isomers for Compounds with Two Double Bonds

How many isomers does the following compound have?

$$ClCH_2CH{=}CHCH{=}CHCH_2CH_3$$

It has four isomers because each of its double bonds can have either the E or the Z configuration. Thus, there are E-E, Z-Z, E-Z, and Z-E isomers.

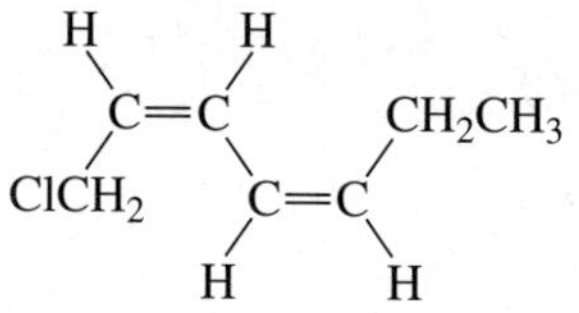

(2Z,4Z)-1-chloro-2,4-heptadiene   (2Z,4E)-1-chloro-2,4-heptadiene

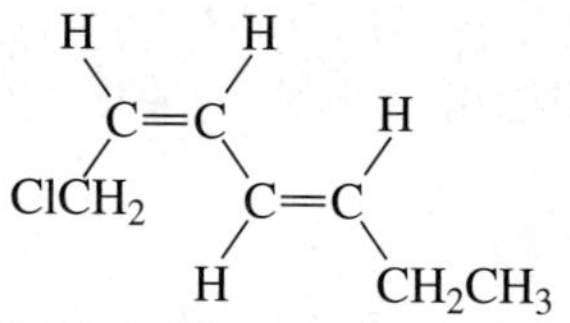

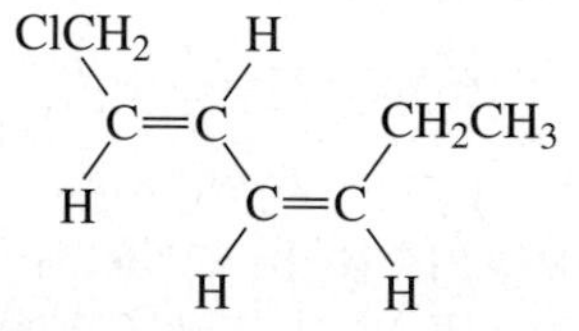

(2E,4Z)-1-chloro-2,4-heptadiene   (2E,4E)-1-chloro-2,4-heptadiene

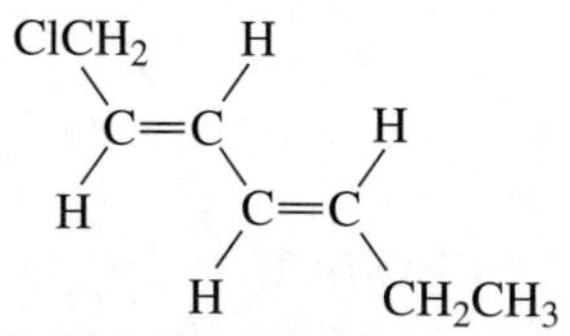

Now continue on to Problem 14.

PROBLEM 14

Draw the isomers for the following compounds, and name each one:

a. 2-methyl-2,4-hexadiene   **b.** 2,4-heptadiene   **c.** 1,3-pentadiene

3.6 How Alkenes React • Curved Arrows Show the Flow of Electrons

There are many millions of organic compounds. If you had to memorize how each of them reacts, studying organic chemistry would not be a very pleasant experience. Fortunately, organic compounds can be divided into families, and all the members of a family react in similar ways. What makes learning organic chemistry even easier is that there are only a few rules that govern the reactivity of each family.

The family that an organic compound belongs to is determined by its functional group. The **functional group** is the center of reactivity of a molecule (Section 2.6). You will find a table of common functional groups inside the back cover of this book. You are already familiar with the functional group of an alkene: the carbon–carbon double bond. All compounds with a carbon–carbon double bond react in similar ways, whether the compound is a small molecule like ethene or a large molecule like cholesterol.

First we need to understand *why* a functional group reacts the way it does. It is not sufficient to know that a compound with a carbon–carbon double bond reacts with HBr to form a product in which the H and Br atoms have taken the place of the π bond; we also need to understand *why* the compound reacts with HBr. In each chapter that discusses the reactivity of a particular functional group, we will see how the nature of the functional group allows us to predict the kind of reactions it will undergo. Then, when you are confronted with a reaction you have never seen before, knowledge of how the structure of the molecule affects its reactivity will help you predict the products of the reaction.

In essence, organic chemistry is all about the interaction between electron-rich atoms or molecules and electron-deficient atoms or molecules. These are the forces that make chemical reactions happen. From this observation follows a very important rule for predicting the reactivity of organic compounds: *electron-rich atoms or molecules are attracted to electron-deficient atoms or molecules.* Each time you study a new functional group, remember that the reactions it undergoes can be explained by this very simple rule.

Electron-rich atoms or molecules are attracted to electron-deficient atoms or molecules.

Therefore, to understand how a functional group reacts, you must first learn to recognize electron-deficient and electron-rich atoms and molecules. An electron-deficient atom or molecule is called an **electrophile**. Literally, "electrophile" means "electron loving" (*phile* is the Greek suffix for "loving"). An electrophile looks for a pair of electrons.

$$H^+ \qquad CH_3\overset{+}{C}H_2 \qquad BH_3$$

these are electrophiles because they can accept a pair of electrons

An electron-rich atom or molecule is called a **nucleophile**. A nucleophile has a pair of electrons it can share. Because a nucleophile has electrons to share and an electrophile is seeking electrons, it should not be surprising that they attract each other. Thus, the preceding rule can be restated as *a nucleophile reacts with an electrophile.*

A nucleophile reacts with an electrophile.

$$HO\overset{..}{\underset{..}{:}}^{-} \qquad \overset{..}{\underset{..}{:}}Cl\overset{..}{\underset{..}{:}}^{-} \qquad CH_3\overset{..}{N}H_2 \qquad H_2\overset{..}{\underset{..}{O}}:$$

these are nucleophiles because they have a pair of electrons to share

---

PROBLEM 15

Identify the nucleophile and the electrophile in the following acid–base reactions.

a. $AlCl_3 \; + \; NH_3 \; \rightleftharpoons \; Cl_3\overset{-}{Al} - \overset{+}{N}H_3$

b. $H-Br \; + \; HO^- \; \rightleftharpoons \; Br^- \; + \; H_2O$

We have seen that the π bond of an alkene consists of a cloud of electrons above and below the σ bond. As a result of this cloud of electrons, an alkene is an electron-rich molecule—it is a nucleophile. (Notice the relatively electron-rich pale orange area in the electrostatic potential maps for *cis*- and *trans*-2-butene in Section 3.4.) We have also seen that a π bond is weaker than a σ bond (Section 1.14). The π bond, therefore, is the bond that is most easily broken when an alkene undergoes a reaction. For these reasons, we can predict that an alkene will react with an electrophile and, in the process, the π bond will break. So if a reagent such as hydrogen bromide is added to an alkene, the alkene (a nucleophile) will react with the partially positively charged hydrogen (an electrophile) of hydrogen bromide and a carbocation will be formed. In the second step of the reaction, the positively charged carbocation (an electrophile) will react with the negatively charged bromide ion (a nucleophile) to form an alkyl halide.

$$CH_3CH = CHCH_3 \; + \; \overset{\delta+}{H} - \overset{\delta-}{Br} \; \longrightarrow \; CH_3\underset{+}{CH} - \underset{\underset{H}{|}}{CHCH_3} \; + \; Br^- \; \longrightarrow \; CH_3\underset{\underset{Br}{|}}{CH} - \underset{\underset{H}{|}}{CHCH_3}$$

a carbocation

2-bromobutane
an alkyl halide

The step-by-step description of the process by which reactants (e.g., alkene + HBr) are changed into products (e.g., alkyl halide) is called the **mechanism of the reaction**. To help us understand a mechanism, curved arrows are drawn to show how the electrons move as new covalent bonds are formed and existing covalent bonds are broken. Each arrow represents the simultaneous movement of two electrons (an electron pair) from an electron-rich center (at the tail of the arrow) and toward an electron-deficient center (at the point of the arrow). In this way, the arrows show which bonds are formed and which are broken.

Curved arrows show the flow of electrons; they are drawn from an electron-rich center to an electron-deficient center.

An arrowhead with two barbs signifies the movement of two electrons.

π bond has broken

$$CH_3CH = CHCH_3 \; + \; \overset{\delta+}{H} - \overset{\delta-}{\overset{..}{\underset{..}{Br}}}: \; \longrightarrow \; CH_3\underset{+}{CH} - \underset{\underset{H}{|}}{CHCH_3} \; + \; :\overset{..}{\underset{..}{Br}}:^-$$

new σ bond has formed

For the reaction of 2-butene with HBr, an arrow is drawn to show that the two electrons of the π bond of the alkene are attracted to the partially positively charged hydrogen of HBr. The hydrogen is not immediately free to accept this pair of electrons because it is already bonded to a bromine, and hydrogen can be bonded to only one atom at a time (Section 1.4); however, as the π electrons of the alkene move toward the hydrogen, the H—Br bond breaks, with bromine keeping the bonding electrons. Notice that the π electrons are pulled away from one carbon, but remain attached to the other. Thus, the two electrons that originally formed the π bond now form a σ bond between carbon and the hydrogen from HBr. The product of this first step in the reaction is a carbocation, because the sp^2 carbon that did not form the new bond with hydrogen has lost a share in an electron pair (the electrons of the π bond) and is, therefore, positively charged.

In the second step of the reaction, a lone pair on the negatively charged bromide ion forms a bond with the positively charged carbon of the carbocation. Notice that in both steps of the reaction *an electrophile reacts with a nucleophile*.

$$CH_3CH\overset{+}{-}CHCH_3 \;+\; :\overset{..}{\underset{..}{Br}}\overset{..}{:}^- \longrightarrow CH_3CH-CHCH_3$$

with "H" below the carbocation carbon, a curved arrow to "new σ bond", and product bearing $:\overset{..}{\underset{..}{Br}}:$ and H.

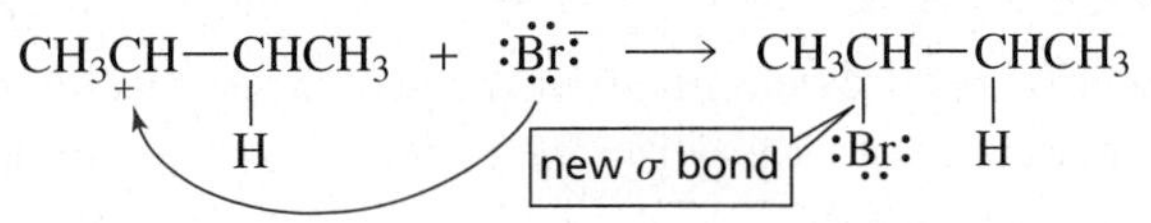

Solely from the knowledge that an electrophile reacts with a nucleophile and a π bond is the weakest bond in an alkene, we have been able to predict that the product of the reaction of 2-butene and HBr is 2-bromobutane. The overall reaction involves the addition of 1 mol of HBr to 1 mol of the alkene. The reaction, therefore, is called an **addition reaction**. Because the first step of the reaction is the addition of an electrophile (H^+) to the alkene, the reaction is more precisely called an **electrophilic addition reaction**. *Electrophilic addition reactions are the characteristic reactions of alkenes.*

At this point, you may think it would be easier just to memorize the fact that 2-bromobutane is the product of the reaction, without trying to understand the mechanism that explains why 2-bromobutane is formed. Keep in mind, however, that you will soon be encountering a great many reactions, and you will not be able to memorize them all. If you strive to understand the mechanism of each reaction, however, the unifying principles of organic chemistry will soon be clear to you, making mastery of the material much easier and a lot more fun.

PROBLEM 16◆

Which of the following are electrophiles, and which are nucleophiles?

$$H^- \qquad CH_3O^- \qquad CH_3C\equiv CH \qquad CH_3\overset{+}{C}HCH_3 \qquad NH_3$$

PROBLEM 17

Draw the consequence of following the incorrect arrows in item 1 of the box on page 140, "A Few Words About Curved Arrows." What is wrong with the structures that you obtain?

PROBLEM 18

Use curved arrows to show the movement of electrons in each of the following reaction steps. (*Hint:* Look at the starting material and look at the products, then draw the arrows.)

a.
$$CH_3\overset{O}{\overset{\|}{C}}-O-H \;+\; H\overset{..}{\underset{..}{O}}:^- \longrightarrow CH_3\overset{O}{\overset{\|}{C}}-O^- \;+\; H_2\overset{..}{\underset{..}{O}}:$$

b.
$$\text{(cyclohexene)} \;+\; \overset{..}{\underset{..}{Br}}:^+ \longrightarrow \text{(cyclohexyl cation with } \overset{..}{\underset{..}{Br}}: \text{ and } +)$$

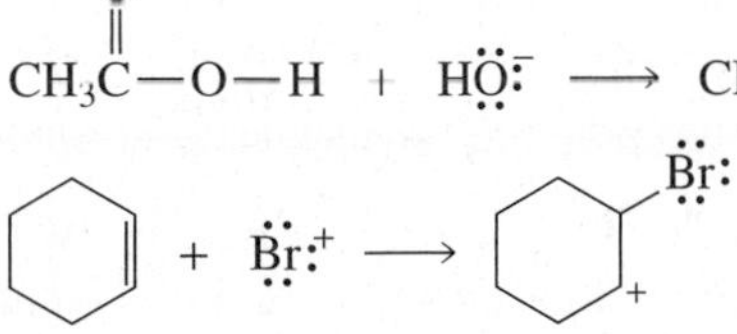

c. 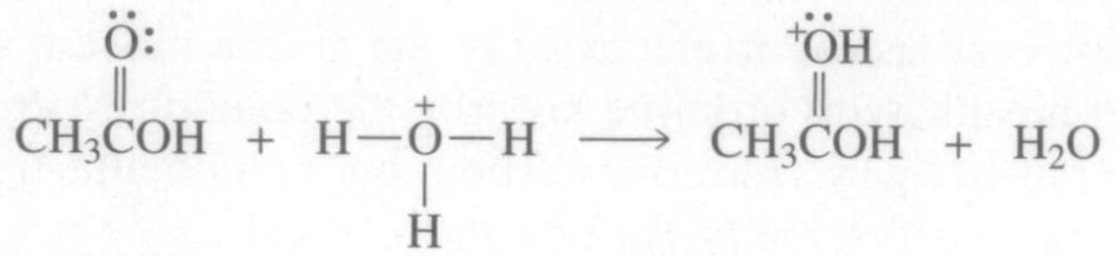
$$
\underset{\overset{\displaystyle \ddot{O}:}{\parallel}}{CH_3\overset{\displaystyle}{C}OH} \ + \ H\!-\!\overset{+}{\underset{H}{\ddot{O}}}\!-\!H \ \longrightarrow \ \underset{\overset{\displaystyle \overset{+}{\ddot{O}}H}{\parallel}}{CH_3\overset{\displaystyle}{C}OH} \ + \ H_2O
$$

d. 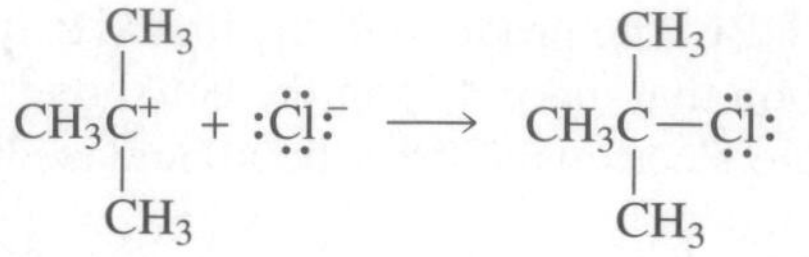
$$
\underset{\overset{\displaystyle CH_3}{|}}{\overset{\overset{\displaystyle CH_3}{|}}{CH_3\overset{+}{C}}} \ + \ :\!\overset{..}{\underset{..}{Cl}}\!:^- \ \longrightarrow \ \underset{\overset{\displaystyle CH_3}{|}}{\overset{\overset{\displaystyle CH_3}{|}}{CH_3\overset{}{C}\!-\!\overset{..}{\underset{..}{Cl}}\!:}}
$$

A FEW WORDS ABOUT CURVED ARROWS

1. Draw the arrows so that they point in the direction of the electron flow and never against the flow. This means that an arrow points away from a negative charge and toward a positive charge. An arrow is used to show both the bond that forms and the bond that breaks.

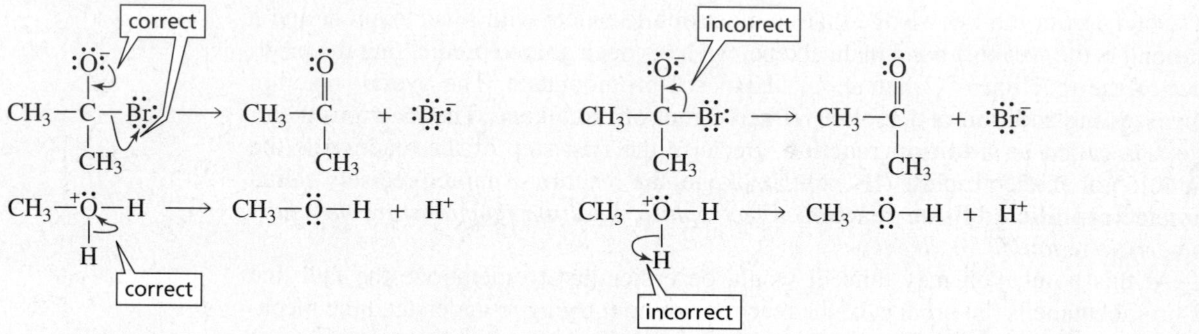

correct | incorrect

incorrect

$$CH_3\!-\!\overset{+}{\underset{H}{\ddot{O}}}\!-\!H \ \longrightarrow \ CH_3\!-\!\ddot{O}\!-\!H \ + \ H^+$$

correct

2. Curved arrows are meant to indicate the movement of electrons. Never use a curved arrow to indicate the movement of an atom. For example, do not use an arrow as a lasso to remove a proton, as shown here:

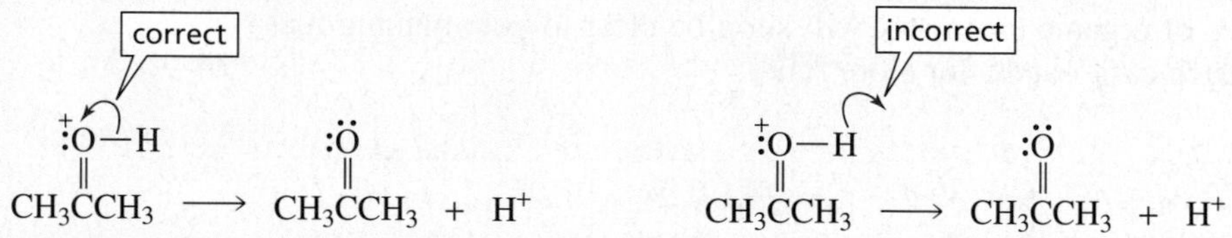

correct | incorrect

$$\underset{CH_3\overset{\displaystyle}{C}CH_3}{\overset{+}{\underset{\parallel}{\overset{..}{O}}}\!-\!H} \ \longrightarrow \ \underset{CH_3\overset{\displaystyle}{C}CH_3}{\overset{..}{\underset{\parallel}{\overset{..}{O}}}} \ + \ H^+$$

3. A head of a curved arrow always points at an atom or at a bond. Never draw the head pointing out into space.

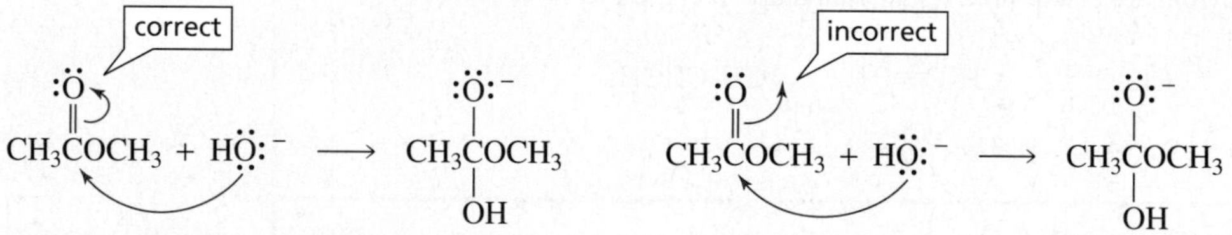

correct | incorrect

$$CH_3\overset{\displaystyle :\ddot{O}}{\underset{\parallel}{C}}OCH_3 \ + \ H\ddot{O}:^- \ \longrightarrow \ \underset{OH}{CH_3\overset{\displaystyle :\ddot{O}:^-}{\underset{|}{C}}OCH_3}$$

4. The arrow starts at the electron source; it does not start at an atom. In the following example, the arrow starts at the electron-rich π bond, not at a carbon atom:

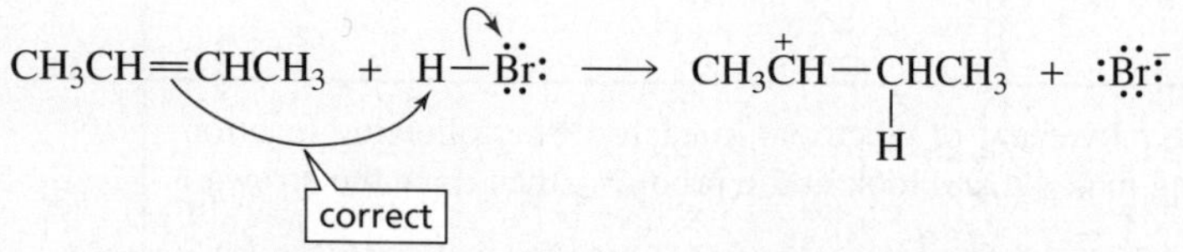

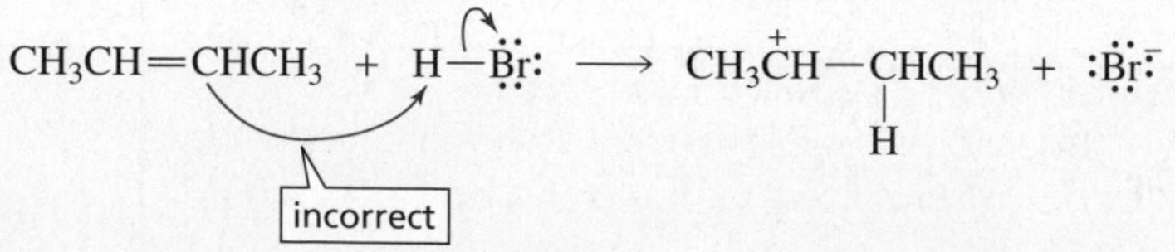

$$CH_3CH\!=\!CHCH_3 \ + \ H\!-\!\ddot{B}r: \ \longrightarrow \ \underset{H}{CH_3\overset{+}{C}H\!-\!\overset{}{C}HCH_3} \ + \ :\ddot{B}r:^-$$

correct

$$CH_3CH\!=\!CHCH_3 \ + \ H\!-\!\ddot{B}r: \ \longrightarrow \ \underset{H}{CH_3\overset{+}{C}H\!-\!\overset{}{C}HCH_3} \ + \ :\ddot{B}r:^-$$

incorrect

PROBLEM 19

For each of the reactions in Problem 18, indicate which reactant is the nucleophile and which is the electrophile.

Note to the student

It is critically important that you learn how to draw curved arrows. Be sure to do the exercise in Special Topic II in the *Study Guide and Solutions Manual*. It should take no more than 15 minutes, yet it can make an enormous difference to your success in the course.

3.7 **Thermodynamics and Kinetics**

To understand the energy changes that take place in a reaction such as the addition of HBr to an alkene, you need to understand some of the basic concepts of *thermodynamics*, which describes a reaction at equilibrium, and *kinetics*, which explains the rates of chemical reactions.

Consider a reaction in which Y is converted to Z: the *thermodynamics* of the reaction tells us the relative amounts of reactants (Y) and products (Z) present when the reaction has reached equilibrium, whereas the *kinetics* of the reaction tells us how fast reactants are converted into products.

$$Y \; \rightleftharpoons \; Z$$

thermodynamics: how much?

kinetics: how fast?

A Reaction Coordinate Diagram Describes the Reaction Pathway

The mechanism of a reaction, as we have shown, describes the steps known to occur as reactants are converted into products. A **reaction coordinate diagram** shows the energy changes that take place in each of these steps. In a reaction coordinate diagram, the total energy of all species is plotted against the progress of the reaction. A reaction progresses from left to right as written in a chemical equation, so the energy of the reactants is plotted on the left-hand side of the *x*-axis and the energy of the products is plotted on the right-hand side. A typical reaction coordinate diagram is shown in Figure 3.2. It describes the reaction of A—B with C to form A and B—C. Remember that *the more stable the species, the lower is its energy*.

The more stable the species, the lower is its energy.

$$A—B \; + \; C \; \rightleftharpoons \; A \; + \; B—C$$

reactants   products

As the reactants are converted into products, the reaction passes through a *maximum* energy state called a **transition state**. The height of the transition state (the difference between the energy of the reactants and the energy of the transition state) tells us how likely it is that the reaction will occur; if the height is too great, the reactants will not be able to be converted into products, so no reaction will take place. The structure of the transition state lies somewhere between the structure of the reactants and the structure of the products. Bonds that break and bonds that form, as reactants are converted to products, are partially broken and partially formed in the transition state. Dashed lines are used to show partially broken or partially formed bonds.

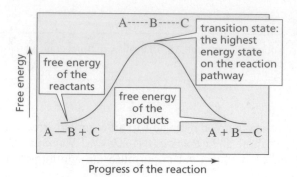

Figure 3.2 ▶
A reaction coordinate diagram. The dashed lines in the transition state indicate bonds that are partially formed or partially broken.

Thermodynamics: How Much Product Is Formed?

Thermodynamics is the field of chemistry that describes the properties of a system at equilibrium. The relative concentrations of reactants and products at equilibrium can be expressed numerically by an equilibrium constant, K_{eq} (Section 1.17). For example, in a reaction in which m moles of A react with n moles of B to form s moles of C and t moles of D, K_{eq} is equal to the relative concentrations of products and reactants at equilibrium.

$$m\,A \;+\; n\,B \;\rightleftharpoons\; s\,C \;+\; t\,D$$

$$K_{eq} = \frac{[\text{products}]}{[\text{reactants}]} = \frac{[C]^s\,[D]^t}{[A]^m\,[B]^n}$$

The more stable the compound, the greater is its concentration at equilibrium.

The relative concentrations of products and reactants at equilibrium depend on their relative stabilities: *the more stable the compound, the greater is its concentration at equilibrium*. Thus, if the products are more stable (have a lower free energy) than the reactants (Figure 3.3a), there will be a higher concentration of products than reactants at equilibrium, and K_{eq} will be greater than 1. On the other hand, if the reactants are more stable than the products (Figure 3.3b), there will be a higher concentration of reactants than products at equilibrium, and K_{eq} will be less than 1. Now you can understand why the strength of an acid is determined by the stability of its conjugate base (Section 1.20). In other words, as the base becomes more stable, the equilibrium constant (K_a) for its formation becomes larger.

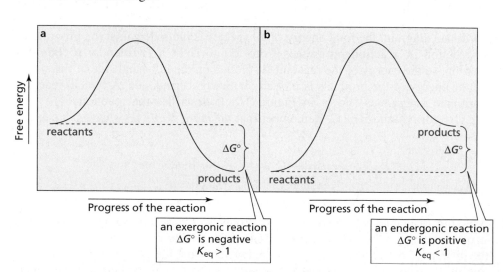

Figure 3.3 ▶
Reaction coordinate diagrams for (a) a reaction in which the products are more stable than the reactants (an exergonic reaction) and (b) a reaction in which the products are less stable than the reactants (an endergonic reaction).

The difference between the free energy of the products and the free energy of the reactants under standard conditions is called the **Gibbs free-energy change**, $\Delta G°$. The symbol ° indicates standard conditions: all species at a concentration of 1 M, a temperature of 25 °C, and a pressure of 1 atm.

$$\Delta G° = (\text{free energy of the products}) - (\text{free energy of the reactants})$$

From this equation, we can see that $\Delta G°$ will be negative if the products have a lower free energy (are more stable) than the reactants. In other words, the reaction will release more energy than it will consume, the definition of an **exergonic reaction** (Figure 3.3a). If the products have a higher free energy (are less stable) than the reactants, $\Delta G°$ will be positive, and the reaction will consume more energy than it will release, the definition of an **endergonic reaction** (Figure 3.3b). (Notice that the terms *exergonic* and *endergonic* refer to whether the reaction has a negative $\Delta G°$ or a positive $\Delta G°$, respectively. Do not confuse these terms with *exothermic* and *endothermic*, which are defined later.)

A successful reaction is one in which the products are favored at equilibrium. We have seen that whether reactants or products are favored at equilibrium can be indicated either by the equilibrium constant (K_{eq}) or by the change in free energy ($\Delta G°$). These two quantities are related by the equation

$$\Delta G° = -RT \ln K_{eq}$$

where R is the gas constant (1.986×10^{-3} kcal mol^{-1} K^{-1}, or 8.314×10^{-3} kJ mol^{-1} K^{-1}, because 1 kcal = 4.184 kJ) and T is the temperature in kelvins (K = °C + 273; therefore, 25 °C = 298 K).

A small difference in $\Delta G°$ gives rise to a large difference in K_{eq} and, therefore, a large difference in the relative concentrations of reactants and products (Section 3.1). For example, for a reaction carried out at 25 °C, the ratio of product to reactant changes by a factor of 10 when $\Delta G°$ becomes 1.36 kcal/mol (or 5.7 kJ/mol) more negative (Table 3.1). When you solve Problem 21, you will see that a small difference in $\Delta G°$ makes a big difference in the amount of product obtained at equilibrium.

When products are favored at equilibrium, $\Delta G°$ is negative and K_{eq} is greater than 1.

When reactants are favored at equilibrium, $\Delta G°$ is positive and K_{eq} is less than 1.

BIOGRAPHY

Josiah Willard Gibbs (1839–1903), *born in New Haven, Connecticut, was the son of a Yale professor. In 1863, he obtained the first Ph.D. awarded by Yale in engineering. After studying in France and Germany, he returned to Yale to become a professor of mathematical physics. His work on free energy received little attention for more than 20 years because few chemists could understand the mathematics and because Gibbs published it in* Transactions of the Connecticut Academy of Sciences, *a relatively obscure journal. In 1950, he was elected to the Hall of Fame for Great Americans.*

Table 3.1	The Relationship Between $\Delta G°$ and K_{eq} at 25 °C	
		B $\rightleftharpoons$ C
$\Delta G°$ (kcal/mol)	K_{eq}	% C at equilibrium
−0.1	1.2	54.5
−1	5.4	84.4
−2.36	5.4×10	98.1
−5	4.6×10^3	99.98
−5.36	4.6×10^4	99.9998

PROBLEM 20◆

a. Which of the monosubstituted cyclohexanes in Table 2.10 has a negative $\Delta G°$ for the conversion of an axial-substituted chair conformer to an equatorial-substituted chair conformer?

b. Which monosubstituted cyclohexane has the most negative value of $\Delta G°$?

c. Which monosubstituted cyclohexane has the greatest preference for the equatorial position?

d. Calculate $\Delta G°$ for conversion of "axial" methylcyclohexane to "equatorial" methylcyclohexane at 25 °C.

PROBLEM 21 **SOLVED**

a. The $\Delta G°$ for the conversion of "axial" fluorocyclohexane to "equatorial" fluorocyclohexane at 25 °C is −0.25 kcal/mol. Calculate the percentage of fluorocyclohexane molecules that have the fluoro substituent in the equatorial position at equilibrium.

b. Do the same calculation for isopropylcyclohexane (its $\Delta G°$ at 25 °C is −2.1 kcal/mol).

c. Why does isopropylcyclohexane have a greater percentage of the conformer with the substituent in the equatorial position?

The gas constant R is thought to be named after **Henri Victor Regnault (1810–1878)**, *who was commissioned by the French Minister of Public Works in 1842 to recalculate all the physical constants required for the design and operation of the steam engine. Regnault was known for his work on the thermal properties of gases. Later, while studying the thermodynamics of dilute solutions, van't Hoff (p. 213) found that R could be used for all chemical equilibria.*

If an equilibrium is disturbed, the system will adjust to offset the disturbance.

Solution to 21a

$$\text{fluorocyclohexane} \rightleftharpoons \text{fluorocyclohexane}$$
$$\text{axial} \qquad\qquad \text{equatorial}$$
$$\Delta G^\circ = -0.25 \text{ kcal/mol at 25 °C}$$
$$\Delta G^\circ = -RT \ln K_{eq}$$
$$-0.25 \frac{\text{kcal}}{\text{mol}} = -1.986 \times 10^{-3} \frac{\text{kcal}}{\text{mol K}} \times 298 \text{ K} \times \ln K_{eq}$$
$$\ln K_{eq} = 0.422$$
$$K_{eq} = 1.53 = \frac{[\text{fluorocyclohexane}]_{\text{equatorial}}}{[\text{fluorocyclohexane}]_{\text{axial}}} = \frac{1.53}{1}$$

Now we must determine the percentage of the total that is equatorial:

$$\frac{[\text{fluorocyclohexane}]_{\text{equatorial}}}{[\text{fluorocyclohexane}]_{\text{equatorial}} + [\text{fluorocyclohexane}]_{\text{axial}}} = \frac{1.53}{1.53 + 1} = \frac{1.53}{2.53} = 0.60 \text{ or } 60\%$$

Fortunately, there are ways to increase the amount of product formed in a reaction. **Le Châtelier's principle** states that *if an equilibrium is disturbed, the system will adjust to offset the disturbance.* In other words, if the concentration of C or D is decreased, A and B will react to form more C and D to maintain the value of the equilibrium constant. Thus if a product crystallizes out of solution as it is formed, or if it can be driven off as a gas, the reactants will continue to react to replace the departing product, so the value of the equilibrium constant can be maintained. More products can also be formed if the concentration of one of the reactants is increased.

$$A + B \rightleftharpoons C + D$$

$$K_{eq} = \frac{[C][D]}{[A][B]}$$

The Gibbs standard free-energy change (ΔG°) has an enthalpy (ΔH°) component and an entropy (ΔS°) component (T is the temperature in Kelvins):

$$\Delta G^\circ = \Delta H^\circ - T\Delta S^\circ$$

The **enthalpy** term (ΔH°) is the heat given off or the heat consumed during the course of a reaction. Atoms are held together by bonds. Heat is given off when bonds are formed, and heat is consumed when bonds are broken. Thus, ΔH° is a measure of the energy of the bond-making and bond-breaking processes that occur as reactants are converted into products.

$$\Delta H^\circ = \text{(energy of the bonds being broken)} - \text{(energy of the bonds being formed)}$$

If the bonds that are formed in a reaction are stronger than the bonds that are broken, more energy will be released in the bond-forming process than will be consumed in the bond-breaking process, and ΔH° will be negative. A reaction with a negative ΔH° is called an **exothermic reaction**. If the bonds that are formed are weaker than those that are broken, ΔH° will be positive. A reaction with a positive ΔH° is called an **endothermic reaction**.

Entropy is a measure of the freedom of motion of a system.

Entropy (ΔS°) is a measure of the freedom of movement in a system. Restricting the freedom of movement of a molecule decreases its entropy. For example, in a reaction in which two molecules come together to form a single molecule, the entropy in the product will be less than the entropy in the reactants because two separate molecules can move in ways that are not possible when they are bound together in a single molecule. In such a reaction, $\Delta S°$ will be negative. In a reaction in which a single molecule is cleaved into two separate molecules, the products will have greater freedom of motion than the reactant, and $\Delta S°$ will be positive.

$$\Delta S° = \text{(freedom of motion of the products)} - \text{(freedom of motion of the reactants)}$$

> **PROBLEM 22♦**
>
> **a.** For which reaction in each set below will $\Delta S°$ be more significant?
> **b.** For which reaction will $\Delta S°$ be positive?
>   1. $A \rightleftharpoons B$ or $A + B \rightleftharpoons C$
>   2. $A + B \rightleftharpoons C$ or $A + B \rightleftharpoons C + D$

A reaction with a negative $\Delta G°$ has a favorable equilibrium constant ($K_{eq} > 1$); that is, the reaction is favored as written from left to right because the products are more stable than the reactants. If you examine the expression for the Gibbs standard free-energy change, you will find that negative values of $\Delta H°$ and positive values of $\Delta S°$ contribute to make $\Delta G°$ negative. In other words, *the formation of products with stronger bonds and greater freedom of motion causes $\Delta G°$ to be negative.* Notice that the entropy term becomes more important as the temperature increases. Therefore, a reaction with a positive $\Delta S°$ may be endergonic at low temperatures, but exergonic at high temperatures.

> **PROBLEM 23♦**
>
> **a.** For a reaction with $\Delta H° = -12$ kcal/mol and $\Delta S° = 0.01$ kcal mol^{-1} K^{-1}, calculate the $\Delta G°$ and the equilibrium constant at (1.) 30 °C and (2.) 150 °C.
> **b.** How does $\Delta G°$ change as T increases?
> **c.** How does K_{eq} change as T increases?

BIOGRAPHY

Henri Louis Le Châtelier (1850–1936), *born in France, was the son of France's inspector-general of mines. He studied mining engineering and, understandably considering his father's position, was particularly interested in mine safety and learning how to prevent explosions. Le Châtelier's research into explosions led him to study heat and its measurement, which stimulated his interest in thermodynamics.*

The formation of products with stronger bonds and greater freedom of movement causes $\Delta G°$ to be negative.

Values of $\Delta H°$ are relatively easy to calculate, so organic chemists frequently evaluate reactions in terms of that quantity alone. However, you can ignore the entropy term only if the reaction involves just a small change in entropy, because then the $T\Delta S°$ term will be small and the value of $\Delta H°$ will be very close to the value of $\Delta G°$. Remember, though, that many organic reactions occur with a significant change in entropy or occur at high temperatures and so have significant $T\Delta S°$ terms; ignoring the entropy term in these cases can lead to wrong conclusions. It is permissible to use $\Delta H°$ values to *approximate* whether a reaction has a favorable equilibrium constant, but if a precise answer is needed, $\Delta G°$ values must be used. When $\Delta G°$ values are used to construct reaction coordinate diagrams, the *y*-axis represents free energy; when $\Delta H°$ values are used, the *y*-axis represents potential energy.

Values of $\Delta H°$ can be calculated from bond dissociation energies (Table 3.2), as shown in the following example. The bond dissociation energy of the π bond of ethene has been determined to be 62 kcal/mol. The bond dissociation energy is indicated by the special term *DH*. Recall from Section 3.4 that the barrier to rotation about the π bond of ethene was also described as 62 kcal/mol. In other words, rotation requires breaking the π bond. (Notice that this value cannot be obtained by subtracting the bond dissociation energy of the C—C σ bond of ethane from ethene's total bond energy, because a σ bond formed from sp^2—sp^2 overlap is stronger than σ bond formed from sp^3—sp^3 overlap.)

π bond of ethene $DH = 62$ kcal/mol $\qquad$ C—H $\quad DH = 101$ kcal/mol

H—Br $\quad \underline{DH = 87 \text{ kcal/mol}} \qquad$ C—Br $\quad \underline{DH = 72 \text{ kcal/mol}}$

$DH_{\text{total}} = 149$ kcal/mol $\qquad DH_{\text{total}} = 173$ kcal/mol

$\Delta H°$ for the reaction $= DH$ for bonds being broken $- DH$ for bonds being formed

$= 149$ kcal/mol $- 173$ kcal/mol

$= -24$ kcal/mol

The value of -24 kcal/mol for $\Delta H°$—calculated by subtracting the sum of the $\Delta H°$ values for the bonds being formed from the sum of the $\Delta H°$ values for the bonds being broken—indicates that the addition of HBr to ethene is an exothermic reaction. But does this mean that the $\Delta G°$ for the reaction is also negative? In other words, is the reaction exergonic as well as exothermic? Because $\Delta H°$ has a significant negative value (-24 kcal/mol), we can assume that $\Delta G°$ is also negative. If the value of $\Delta H°$ were close to zero, however, we could no longer assume that $\Delta H°$ has the same sign as $\Delta G°$.

Keep in mind that two conditions must be met to justify using $\Delta H°$ values to predict values of $\Delta G°$. The first condition is that the entropy change in the reaction must be small, causing $T\Delta S°$ to be close to zero and, therefore, the value of $\Delta H°$ to be very close to the value of $\Delta G°$; the second is that the reaction must take place in the gas phase.

Table 3.2 Bond Dissociation Energies for Y — Z → Y · + · Z

Bond	DH kcal/mol	DH kJ/mol	Bond	DH kcal/mol	DH kJ/mol
CH$_3$—H	105	439	H—H	104	435
CH$_3$CH$_2$—H	101	423	F—F	38	159
CH$_3$CH$_2$CH$_2$—H	101	423	Cl—Cl	58	242
(CH$_3$)$_2$CH—H	99	414	Br—Br	46	192
(CH$_3$)$_3$C—H	97	406	I—I	36	150
			H—F	136	571
CH$_3$—CH$_3$	90.1	377	H—Cl	103	432
CH$_3$CH$_2$—CH$_3$	89.0	372	H—Br	87	366
(CH$_3$)$_2$CH—CH$_3$	88.6	371	H—I	71	298
(CH$_3$)$_3$C—CH$_3$	87.5	366			
			CH$_3$—F	115	481
H$_2$C=CH$_2$	174	728	CH$_3$—Cl	84	350
HC≡CH	231	966	CH$_3$CH$_2$—Cl	85	356
			(CH$_3$)$_2$CH—Cl	85	356
HO—H	119	497	(CH$_3$)$_3$C—Cl	85	356
CH$_3$O—H	105	439	CH$_3$—Br	72	301
CH$_3$—OH	92	387	CH$_3$CH$_2$—Br	72	301
			(CH$_3$)$_2$CH—Br	74	310
			(CH$_3$)$_3$C—Br	73	305
			CH$_3$—I	58	243
			CH$_3$CH$_2$—I	57	238

S. J. Blanksby and G. B. Ellison, *Acc. Chem. Res.*, **2003**, *36*, 255.

When reactions occur in solution, which is the case for the vast majority of organic reactions, the solvent molecules can interact with the reagents and products. Polar solvent molecules cluster around a charge (either a full charge or a partial charge) on a reactant or a product, turning so that their negative poles surround the positive charge and their positive poles surround the negative charge. The interaction between a solvent and a species (a molecule or ion) in solution is called **solvation** (Section 8.10).

Solvation can have a large effect on both the $\Delta H°$ and the $\Delta S°$ of a reaction. For example, in a reaction of a solvated polar reagent, the $\Delta H°$ for breaking the dipole–dipole interactions between the solvent and the reagent has to be taken into account, and in a reaction having a polar product that becomes solvated, the $\Delta H°$ for forming the dipole–dipole interactions between the solvent and the product has to be taken into account. In addition, solvation of a polar reagent or a polar product by a polar solvent can greatly reduce the freedom of movement of the solvent molecules, thus affecting the value of $\Delta S°$.

PROBLEM 24◆

a. Using the bond dissociation energies in Table 3.2, calculate the $\Delta H°$ for the addition of HCl to ethene.

b. Calculate the $\Delta H°$ for the addition of H_2 to ethene.

c. Are the reactions exothermic or endothermic?

d. Do you expect the reactions to be exergonic or endergonic?

Kinetics: How Fast Is the Product Formed?

Knowing whether a given reaction is exergonic or endergonic will not tell us how fast the reaction occurs, because the $\Delta G°$ of a reaction describes only the difference between the stability of the reactants and the stability of the products; it does not indicate anything about the energy barrier of the reaction, which is the energy "hill" that must be climbed for the reactants to be converted into products. The higher the energy barrier, the slower is the reaction. **Kinetics** is the field of chemistry that studies the rates of chemical reactions and the factors that affect those rates.

> The higher the energy barrier, the slower is the reaction.

The energy barrier of a reaction, indicated in Figure 3.4 by $\Delta G^{\ddagger}$, is called the **free energy of activation**. It is the difference between the free energy of the transition state and the free energy of the reactants:

$$\Delta G^{\ddagger} = \text{(free energy of the transition state)} - \text{(free energy of the reactants)}$$

The smaller the $\Delta G^{\ddagger}$, the faster is the reaction. Thus, *anything that destabilizes the reactant or stabilizes the transition state will make the reaction go faster.*

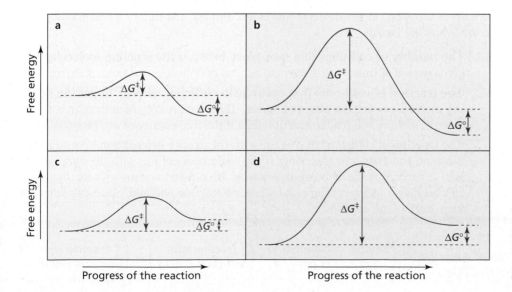

◀ Figure 3.4
Reaction coordinate diagrams for (a) a fast exergonic reaction, (b) a slow exergonic reaction, (c) a fast endergonic reaction, and (d) a slow endergonic reaction. (The four reaction coordinates are drawn on the same scale.)

Like $\Delta G°$, $\Delta G^{\ddagger}$ has both an enthalpy component and an entropy component. Notice that any quantity that refers to the transition state is represented by a double-dagger superscript (‡).

$$\Delta G^{\ddagger} = \Delta H^{\ddagger} - T\Delta S^{\ddagger}$$
$$\Delta H^{\ddagger} = \text{(enthalpy of the transition state)} - \text{(enthalpy of the reactants)}$$
$$\Delta S^{\ddagger} = \text{(entropy of the transition state)} - \text{(entropy of the reactants)}$$

Some exergonic reactions have small free energies of activation and therefore can take place at room temperature (Figure 3.4a). In contrast, some exergonic reactions have free energies of activation that are so large that the reaction cannot take place unless energy is supplied in addition to the energy provided by the existing thermal conditions (Figure 3.4b). Endergonic reactions can also have either small free energies of activation, as in Figure 3.4c, or large free energies of activation, as in Figure 3.4d.

Notice that $\Delta G°$ is related to the *equilibrium constant* of the reaction, whereas $\Delta G^{\ddagger}$ is related to the *rate* of the reaction. **Thermodynamic stability** is indicated by $\Delta G°$. If $\Delta G°$ is negative, for example, the product is *thermodynamically stable* compared with the reactant, and if $\Delta G°$ is positive, the product is *thermodynamically unstable* compared with the reactant. **Kinetic stability** is indicated by $\Delta G^{\ddagger}$. If $\Delta G^{\ddagger}$ for a reaction is large, the compound is *kinetically stable* because it does not undergo the reaction rapidly. If $\Delta G^{\ddagger}$ is small, the compound is *kinetically unstable*—it undergoes the reaction rapidly. Generally, when chemists use the term "stability," they are referring to thermodynamic stability.

PROBLEM 25◆

a. Which of the reactions in Figure 3.4 has a product that is thermodynamically stable compared with the reactant?

b. Which of the reactions in Figure 3.4 has the most kinetically stable product?

c. Which of the reactions in Figure 3.4 has the least kinetically stable product?

PROBLEM 26

Draw a reaction coordinate diagram for a reaction in which

a. the product is thermodynamically unstable and kinetically unstable.

b. the product is thermodynamically unstable and kinetically stable.

The rate of a chemical reaction is the speed at which the reacting substances are used up or the speed at which the products are formed. The rate of a reaction depends on the following factors:

1. **The number of collisions that take place between the reacting molecules in a given period of time.** The greater the number of collisions, the faster is the reaction.

2. **The fraction of collisions that occur with sufficient energy to get the reacting molecules over the energy barrier.** If the free energy of activation is small, more collisions will lead to reaction than if the free energy of activation is large.

3. **The fraction of collisions that occur with the proper orientation.** For example, 2-butene and HBr will react only if the molecules collide with the hydrogen of HBr approaching the π bond of 2-butene. If collision occurs with the hydrogen approaching a methyl group of 2-butene, no reaction will take place, regardless of the energy of the collision.

$$\text{rate of a reaction} = \left(\begin{array}{c}\text{number of collisions}\\ \text{per unit of time}\end{array}\right) \times \left(\begin{array}{c}\text{fraction with}\\ \text{sufficient energy}\end{array}\right) \times \left(\begin{array}{c}\text{fraction with}\\ \text{proper orientation}\end{array}\right)$$

Increasing the concentration of the reactants increases the rate of a reaction because it increases the number of collisions that occur in a given period of time. Increasing the temperature at which the reaction is carried out also increases the rate of a reaction because it increases the kinetic energy of the molecules, which increases both the frequency of collisions (molecules that are moving faster collide more frequently) and the number of collisions that have sufficient energy to get the reacting molecules over the energy barrier (Figure 3.5).

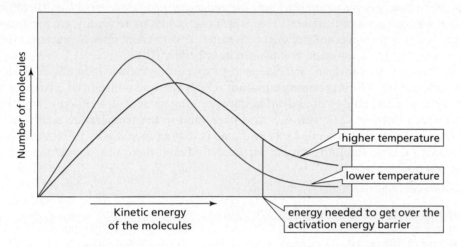

◀ **Figure 3.5**
Boltzmann distribution curves at two different temperatures. The curve shows the distribution of molecules with a particular kinetic energy. The energy of most molecules cluster about an average, but there are some with much lower and some with much higher energy. At a higher temperature, there will be more molecules with sufficient energy to get over the energy barrier.

For a reaction in which a single reactant molecule A is converted into a product molecule B, the rate of the reaction is proportional to the concentration of A. If the concentration of A is doubled, the rate of the reaction will double; if the concentration of A is tripled, the rate of the reaction will triple; and so on. Because the rate of this reaction is proportional to the concentration of only *one* reactant, it is called a **first-order reaction**.

$$\text{a first-order reaction:} \quad A \longrightarrow B$$

$$\text{rate} \propto [A]$$

We can replace the proportionality symbol ($\propto$) with an equals sign if we use a proportionality constant k, which is called a **rate constant**. The rate constant of a first-order reaction is called a **first-order rate constant**.

The smaller the rate constant, the slower is the reaction.

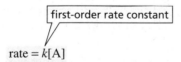

$$\text{rate} = k[A]$$

A reaction whose rate depends on the concentrations of *two* reactants, A and B, is called a **second-order reaction**. If the concentration of either A or B is doubled, the rate of the reaction will double; if the concentrations of both A and B are doubled, the rate of the reaction will quadruple; and so on. In this case, the rate constant k is a **second-order rate constant**.

$$\text{a second-order reaction:} \quad A + B \longrightarrow C + D$$

$$\text{rate} = k[A][B]$$

second-order rate constant

A reaction in which two molecules of A combine to form a molecule of B is also a second-order reaction: if the concentration of A is doubled, the rate of the reaction will quadruple.

$$\text{a second-order reaction:} \quad A + A \longrightarrow B$$

$$\text{rate} = k[A]^2$$

Swedish chemist **Svante August Arrhenius (1859–1927)** *received a Ph.D. from the University of Uppsala. Threatened with a low passing grade on his dissertation because his examiners did not understand his thesis on ionic dissociation, he sent the work to several influential scientists, who subsequently defended it. This dissertation eventually earned him the 1903 Nobel Prize in chemistry. Arrhenius was the first to describe the "greenhouse" effect, predicting that as concentrations of atmospheric carbon dioxide (CO_2) increase, so will Earth's surface temperature (Section 11.0).*

Do not confuse the *rate constant* of a reaction (k) with the *rate* of a reaction. The *rate constant* tells us how easy it is to reach the transition state (how easy it is to get over the energy barrier). Low energy barriers are associated with large rate constants (Figure 3.4a and c), whereas high energy barriers have small rate constants (Figure 3.4b and d). The reaction *rate* is a measure of the amount of product that is formed per unit of time. The preceding equations show that the *rate* is the product of the *rate constant and the reactant concentration or concentrations. Thus, reaction rates depend on concentration, whereas rate constants are independent of concentration.* Therefore, when we compare two reactions to see which one occurs more readily, we must compare their rate constants and not their concentration-dependent rates of reaction. (How rate constants are determined is explained in Appendix III.)

Although rate constants are independent of concentration, they are dependent on temperature. The **Arrhenius equation** relates the rate constant of a reaction to the experimental energy of activation (an approximate activation energy; see "The Difference Between $\Delta G^{\ddagger}$ and E_a," next page) and to the temperature at which the reaction is carried out. A good rule of thumb is that an increase of 10 °C in temperature will double the rate constant for a reaction and, therefore, double the rate of the reaction.

the Arrhenius equation

$$k = Ae^{-E_a/RT}$$

where k is the rate constant, E_a is the experimental energy of activation, R is the gas constant (1.986×10^{-3} kcal/mol·K, or 8.314×10^{-3} kJ/mol·K), T is the absolute temperature (K), and A is the frequency factor. The frequency factor represents the fraction of collisions that occurs with the proper orientation for reaction. The term $e^{-E_a/RT}$ is the fraction of collisions with the minimum energy (E_a) needed to react. Taking the logarithm of both sides of the Arrhenius equation, we obtain

$$\ln k = \ln A - \frac{E_a}{RT}$$

Problem 54 on page 158 shows how this equation is used to calculate values of E_a, $\Delta G^{\ddagger}$, $\Delta H^{\ddagger}$, and $\Delta S^{\ddagger}$ for a reaction.

PROBLEM 27 **SOLVED**

At 30 °C, the second-order rate constant for the reaction of methyl chloride and HO^- is 1.0×10^{-5} M^{-1} s^{-1}.

a. What is the rate of the reaction when $[CH_3Cl] = 0.10$ M and $[HO^-] = 0.10$ M?

b. If the concentration of methyl chloride is decreased to 0.01 M, what will be the effect on the *rate* of the reaction?

c. If the concentration of methyl chloride is decreased to 0.01 M, what will be the effect on the *rate constant* of the reaction?

Solution to 27a The rate of the reaction is given by

$$\text{rate} = k[\text{methyl chloride}]\,[HO^-]$$

Substituting the given rate constant and reactant concentrations yields

$$\text{rate} = 1.0 \times 10^{-5}\ M^{-1}s^{-1}\,[0.10\ M][0.10\ M]$$

$$= 1.0 \times 10^{-7}\ Ms^{-1}$$

PROBLEM 28◆

The rate constant for a reaction can be increased by _____ the stability of the reactant or by _____ the stability of the transition state.

THE DIFFERENCE BETWEEN $\Delta G^{\ddagger}$ AND E_a

Do not confuse the *free energy of activation*, $\Delta G^{\ddagger}$, with the **experimental energy of activation**, E_a, in the Arrhenius equation. The free energy of activation ($\Delta G^{\ddagger} = \Delta H^{\ddagger} - T\Delta S^{\ddagger}$) has both an enthalpy component and an entropy component, whereas the experimental energy of activation ($E_a = \Delta H^{\ddagger} + RT$) has only an enthalpy component since the entropy component is implicit in the A term in the Arrhenius equation. Therefore, the experimental energy of activation is an approximate energy barrier to a reaction. The true energy barrier to a reaction is given by $\Delta G^{\ddagger}$ because, although some reactions are driven by a change in enthalpy and some by a change in entropy, most reactions depend on changes in both of these properties.

PROBLEM 29◆

From the Arrhenius equation, predict how

a. increasing the experimental activation energy will affect the rate constant of a reaction.

b. increasing the temperature will affect the rate constant of a reaction.

The next question to consider is, how are the rate constants for a reaction related to the equilibrium constant? At equilibrium, the rate of the forward reaction must be equal to the rate of the reverse reaction because the amounts of reactants and products are not changing:

$$A \underset{k_{-1}}{\overset{k_1}{\rightleftharpoons}} B$$

forward rate = reverse rate

$$k_1[A] = k_{-1}[B]$$

Therefore,

$$K_{eq} = \frac{k_1}{k_{-1}} = \frac{[B]}{[A]}$$

From this equation, we can see that the equilibrium constant for a reaction can be determined from the relative concentrations of the products and reactants at equilibrium or from the relative rate constants for the forward and reverse reactions. The reaction shown in Figure 3.3a has a large equilibrium constant because the products are much more stable than the reactants. We could also say that it has a large equilibrium constant because the rate constant of the forward reaction is much greater than the rate constant of the reverse reaction.

PROBLEM 30◆

a. Which reaction has the greater equilibrium constant, one with a rate constant of $1 \times 10^{-3} \sec^{-1}$ for the forward reaction and a rate constant of $1 \times 10^{-5} \sec^{-1}$ for the reverse reaction or one with a rate constant of $1 \times 10^{-2} \sec^{-1}$ for the forward reaction and a rate constant of $1 \times 10^{-3} \sec^{-1}$ for the reverse reaction?

b. If both reactions start with a reactant concentration of 1.0 M, which reaction will form the most product?

Note to the student
Additional kinetics problems and their solutions can be found in Special Topic III in the *Study Guide and Solutions Manual*.

3.8 **Using a Reaction Coordinate Diagram to Describe a Reaction**

We have seen that the addition of HBr to 2-butene is a two-step process (Section 3.6). In each step, the reactants pass through a transition state as they are converted into products. The structure of the transition state for each of the steps is shown below in brackets. Notice that the bonds that break and the bonds that form during the course of the reaction are partially broken and partially formed in the transition state, as indicated by dashed lines. Similarly, atoms that either become charged or lose their charge during the course of the reaction are partially charged in the transition state. Transition states are always shown in brackets with a double-dagger superscript.

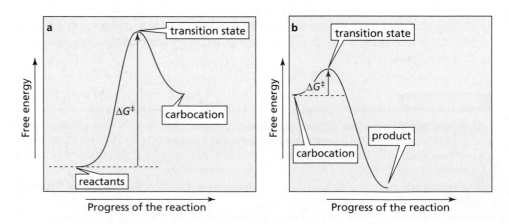

A reaction coordinate diagram can be drawn for each of the steps in the reaction (Figure 3.6). In the first step of the reaction, the alkene is converted into a carbocation that is less stable than the reactants. The first step, therefore, is endergonic ($\Delta G°$ is positive). In the second step of the reaction, the carbocation reacts with a nucleophile to form a product that is more stable than the carbocation reactant. This step, therefore, is exergonic ($\Delta G°$ is negative).

Figure 3.6 ▶
Reaction coordinate diagrams for the two steps in the addition of HBr to 2-butene: (a) the first step; (b) the second step.

Because the products of the first step are the reactants in the second step, we can hook the two reaction coordinate diagrams together to obtain the reaction coordinate diagram for the overall reaction (Figure 3.7). The $\Delta G°$ for the overall reaction is the difference between the free energy of the final products and the free energy of the initial reactants. The figure shows that $\Delta G°$ for the overall reaction is negative. Therefore, the overall reaction is exergonic.

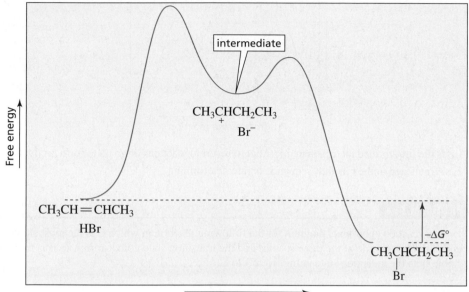

◀ **Figure 3.7**
Reaction coordinate diagram for the addition of HBr to 2-butene.

A chemical species that is a product of one step of a reaction and a reactant for the next step is called an **intermediate**. The carbocation intermediate in this reaction is too unstable to be isolated, but some reactions have more stable intermediates that can be isolated. **Transition states**, in contrast, represent the highest-energy structures that are involved in the reaction. They exist only fleetingly and can never be isolated. Do not confuse transition states with intermediates: *transition states have partially formed bonds, whereas intermediates have fully formed bonds.*

We can see from the reaction coordinate diagram (Figure 3.7) that the free energy of activation for the first step of the reaction is greater than the free energy of activation for the second step. In other words, the rate constant for the first step is smaller than the rate constant for the second step. This is what we would expect, considering that the molecules in the first step of this reaction must collide with sufficient energy to break covalent bonds, whereas no bonds are broken in the second step.

If a reaction has two or more steps, the step that has its transition state *at the highest point on the reaction coordinate* is called the **rate-determining step** or **rate-limiting step**. The rate-determining step controls the overall rate of the reaction because the overall rate of a reaction such as that shown in Figure 3.7 cannot exceed the rate of the rate-determining step. In Figure 3.7, the rate-determining step is the first step—the addition of the electrophile (the proton) to the alkene.

Reaction coordinate diagrams can also be used to explain why a given reaction forms a certain product and not others. We will see the first example of this use in Section 4.3.

Transition states have partially formed bonds. Intermediates have fully formed bonds.

PROBLEM 31

Draw a reaction coordinate diagram for a two-step reaction in which the first step is endergonic, the second step is exergonic, and the overall reaction is endergonic. Label the reactants, products, intermediates, and transition states.

PROBLEM 32◆

a. Which step in the reaction diagrammed below has the greatest free energy of activation?

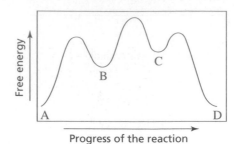

b. Is the first-formed intermediate more apt to revert to reactants or go on to form products?

c. Which step in the reaction sequence is rate determining?

PROBLEM 33◆

Draw a reaction coordinate diagram for the following reaction in which C is the most stable and B the least stable of the three species and the transition state going from A to B is more stable than the transition state going from B to C:

$$A \underset{k_{-1}}{\overset{k_1}{\rightleftharpoons}} B \underset{k_{-2}}{\overset{k_2}{\rightleftharpoons}} C$$

a. How many intermediates are there?

b. How many transition states are there?

c. Which step has the greater rate constant in the forward direction?

d. Which step has the greater rate constant in the reverse direction?

e. Of the four steps, which has the greatest rate constant?

f. Which is the rate-determining step in the forward direction?

g. Which is the rate-determining step in the reverse direction?

SUMMARY

Alkenes are hydrocarbons that contain a double bond. The double bond is the **functional group**, or center of reactivity of the alkene. The **functional group suffix** of an alkene is "ene." The general molecular formula for a hydrocarbon is C_nH_{2n+2}, minus two hydrogens for every π bond or ring in the molecule. The number of π bonds and rings is called the **degree of unsaturation**. Because alkenes contain fewer than the maximum number of hydrogens, they are called **unsaturated hydrocarbons**.

Rotation about the double bond is restricted, so an alkene can exist as **cis–trans isomers**. The **cis isomer** has its hydrogens on the same side of the double bond; the **trans isomer** has its hydrogens on opposite sides of the double bond. The **Z isomer** has the high-priority groups on the same side of the double bond; the **E isomer** has the high-priority groups on opposite sides of the double bond. The relative priorities depend on the atomic numbers of the atoms bonded directly to the sp^2 carbon.

All compounds with a particular **functional group** react similarly. Due to the cloud of electrons above and below its π bond, an alkene is an electron-rich species, or a **nucleophile**.

Nucleophiles are attracted to electron-deficient species, called **electrophiles**. Alkenes undergo **electrophilic addition reactions**. The description of the step-by-step process by which reactants are changed into products is called the **mechanism of the reaction**. **Curved arrows** show which bonds are formed and which are broken and the direction of the electron flow that accompanies these changes.

Thermodynamics describes a reaction at equilibrium; **kinetics** describes how fast the reaction occurs. A **reaction coordinate diagram** shows the energy changes that take place in a reaction. The more stable a species, the lower is its energy. As reactants are converted into products, a reaction passes through a maximum-energy **transition state**. An **intermediate** is a product of one step of a reaction and a reactant in the next step. Transition states have partially formed bonds; intermediates have fully formed bonds. The **rate-determining step** has its transition state at the highest point on the reaction coordinate.

The relative concentrations of reactants and products at equilibrium are given by the equilibrium constant K_{eq}. The more stable the product, the greater is its concentration at equilibrium. Le Châtelier's principle states that if an equilibrium is disturbed, the system will adjust to offset the disturbance. If products are more stable than reactants, K_{eq} is >1, $\Delta G°$ is negative, and the reaction is **exergonic**; if reactants are more stable than products, $K_{eq} < 1$, $\Delta G°$ is positive, and the reaction is **endergonic**. $\Delta G°$ is the **Gibbs free-energy change**, with $\Delta G° = \Delta H° - T\Delta S°$. $\Delta H°$ is the change in **enthalpy**, the heat given off or consumed as a result of bond making and bond breaking. An **exothermic reaction** has a negative $\Delta H°$; an **endothermic reaction** has a positive $\Delta H°$. $\Delta S°$ is the change in **entropy**, the change in the freedom of movement of the system. A reaction with a negative $\Delta G°$ has a **favorable equilibrium constant**: the formation of products with stronger bonds and greater freedom of motion causes $\Delta G°$ to be negative. $\Delta G°$ and K_{eq} are related by the formula $\Delta G° = -RT \ln K_{eq}$. The interac-

tion between a solvent and a species in solution is called **solvation**.

The **free energy of activation**, $\Delta G^{\ddagger}$, is the energy barrier of a reaction. It is the difference between the free energy of the reactants and the free energy of the transition state. The smaller the $\Delta G^{\ddagger}$, the faster is the reaction. Anything that destabilizes the reactant or stabilizes the transition state makes the reaction go faster. **Kinetic stability** is given by $\Delta G^{\ddagger}$, **thermodynamic stability** by $\Delta G°$. The **rate** of a reaction depends on the concentration of the reactants, the temperature, and the rate constant. The **rate constant**, which is independent of concentration, indicates how easy or difficult it is for reactants to reach the transition state. A **first-order reaction** depends on the concentration of one reactant, a **second-order reaction** on the concentration of two reactants.

KEY TERMS

acyclic (p. 125)
addition reaction (p. 139)
alkene (p. 124)
allyl group (p. 129)
allylic carbon (p. 128)
Arrhenius equation (p. 150)
cis isomer (p. 131)
cis–trans isomers (p. 131)
degree of unsaturation (p. 127)
E isomer (p. 134)
electrophile (p. 137)
electrophilic addition reaction (p. 139)
endergonic reaction (p. 143)
endothermic reaction (p. 144)
enthalpy (p. 144)
entropy (p. 144)

exergonic reaction (p. 143)
exothermic reaction (p. 144)
experimental energy of
   activation (p. 151)
first-order rate constant (p. 149)
first-order reaction (p. 149)
free energy of activation (p. 147)
functional group (p. 137)
geometric isomers (p. 131)
Gibbs free-energy change (p. 142)
intermediate (p. 153)
kinetics (p. 147)
kinetic stability (p. 148)
Le Châtelier's principle (p. 144)
mechanism of the reaction (p. 138)
nucleophile (p. 138)

rate constant (p. 149)
rate-determining step (p. 153)
rate-limiting step (p. 153)
reaction coordinate diagram (p. 141)
saturated hydrocarbon (p. 127)
second-order rate constant (p. 149)
second-order reaction (p. 149)
solvation (p. 147)
thermodynamics (p. 142)
thermodynamic stability (p. 148)
trans isomer (p. 131)
transition state (p. 141)
unsaturated hydrocarbon (p. 127)
vinyl group (p. 129)
vinylic carbon (p. 128)
Z isomer (p. 134)

PROBLEMS

34. Give the systematic name for each of the following compounds:

a. $CH_3CH_2CHCH=CHCH_2CH_2CHCH_3$ with Br, Br substituents

c. structure with CH$_3$ groups

e. structure with CH$_3$

b. structure with H$_3$C, CH$_2$CH$_3$, CH$_3$CH$_2$, CH$_2$CH$_2$CHCH$_3$, CH$_3$

d. structure with H$_3$C, CH$_2$CH$_3$, H$_3$C, CH$_2$CH$_2$CH$_2$CH$_3$

f. structure with CH$_2$CH$_3$, CH$_3$

35. Give the structure of a hydrocarbon that has six carbon atoms and
   a. three vinylic hydrogens and two allylic hydrogens.
   b. three vinylic hydrogens and one allylic hydrogen.
   c. three vinylic hydrogens and no allylic hydrogens.

36. Draw the structure for each of the following:
   a. (Z)-1,3,5-tribromo-2-pentene
   b. (Z)-3-methyl-2-heptene
   c. (E)-1,2-dibromo-3-isopropyl-2-hexene
   d. vinyl bromide
   e. 1,2-dimethylcyclopentene
   f. diallylamine

37. a. Give the structures and the systematic names for all alkenes with molecular formula C_6H_{12}, ignoring cis–trans isomers. (*Hint:* There are 13.)
   b. Which of the compounds have *E* and *Z* isomers?

38. Name the following compounds:

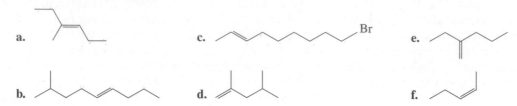

a.

b.

c.

d.

e.

f.

39. Draw curved arrows to show the flow of electrons responsible for the conversion of the reactants into the products:

$$H-\ddot{\underset{..}{O}}{:}^{-} + H-\underset{\underset{H}{|}}{\overset{\overset{H}{|}}{C}}-\underset{\underset{Br}{|}}{\overset{\overset{H}{|}}{C}}-H \longrightarrow H_2O + \underset{\underset{H}{}\,\,\,\,\,\underset{H}{}}{\overset{H\,\,\,\,\,\,H}{C=C}} + Br^-$$

40. Tamoxifen slows the growth of some breast tumors by binding to estrogen receptors. Is tamoxifen an *E* or a *Z* isomer?

OCH₂CH₂NCH₃

$OCH_2CH_2NCH_3$

CH_3

$C=C$

CH_3CH_2

tamoxifen

41. In a reaction in which reactant A is in equilibrium with product B at 25 °C, what are the relative amounts of A and B present at equilibrium if $\Delta G°$ at 25 °C is
   a. 2.72 kcal/mol?
   b. 0.65 kcal/mol?
   c. −2.72 kcal/mol?
   d. −0.65 kcal/mol?

42. Several studies have shown that β-carotene, a precursor of vitamin A, may play a role in preventing cancer. β-Carotene has a molecular formula of $C_{40}H_{56}$ and contains two rings and no triple bonds. How many double bonds does it have?

43. For each pair of bonds, which has the greater strength? Briefly explain why.
   a. $CH_3{-}Cl$ or $CH_3{-}Br$
   b. $I{-}Br$ or $Br{-}Br$

44. Tell whether each of the following compounds has the *E* or the *Z* configuration:
   a.
   $H_3C \quad\quad CH_2CH_3$
   $C=C$
   $CH_3CH_2 \quad\quad CH_2CH_2Cl$

   c.
   $H_3C \quad\quad CH_2Br$
   $C=C$
   $Br \quad\quad CH_2CH_2CH_2CH_3$

   b.
   $CH_3CH_2CH_2 \quad\quad CH(CH_3)_2$
   $C=C$
   $HC{\equiv}CCH_2 \quad\quad CH_2CH=CH_2$

   d.
   $\overset{\overset{\displaystyle O}{\|}}{CH_3C} \quad\quad CH_2Br$
   $C=C$
   $HOCH_2 \quad\quad CH_2CH_2Cl$

45. Squalene, a hydrocarbon with molecular formula $C_{30}H_{50}$, is obtained from shark liver. (*Squalus* is Latin for "shark.") If squalene is an acyclic compound, how many π bonds does it have?

46. Assign relative priorities to each set of substituents:
   a. $-CH_2CH_2CH_3, -CH(CH_3)_2, -CH=CH_2, -CH_3$
   b. $-CH_2NH_2, -NH_2, -OH, -CH_2OH$
   c. $-C(=O)CH_3, -CH=CH_2, -Cl, -C{\equiv}N$

47. Draw the configurational isomers for the following compounds, and name each one:
   a. 2-methyl-2,4-hexadiene   **b.** 1,5-heptadiene   **c.** 1,4-pentadiene

48. Molly Kule was a lab technician who was asked by her supervisor to add names to the labels on a collection of alkenes that showed only structures on the labels. How many did Molly get right? Correct the incorrect names.
   a. 3-pentene   **d.** 1-ethyl-1-pentene   **g.** 2-ethyl-2-butene
   b. 2-octene   **e.** 5-ethylcyclohexene   **h.** (E)-2-methyl-1-hexene
   c. 2-vinylpentane   **f.** 5-chloro-3-hexene   **i.** 2-methylcyclopentene

49. Draw structures for the following compounds:
   a. (2E,4E)-1-chloro-3-methyl-2,4-hexadiene   **c.** (3Z,5Z)-4,5-dimethyl-3,5-nonadiene
   b. (3Z,5E)-4-methyl-3,5-nonadiene   **d.** (3E,5E)-2,5-dibromo-3,5-octadiene

50. Given the following reaction coordinate diagram for the reaction of A to give D, answer the following questions:

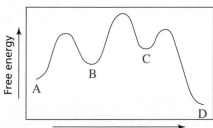

   a. How many intermediates are formed in the reaction?   **e.** What is the reactant of the rate-determining step?
   b. How many transition states are there?   **f.** Is the first step of the reaction exergonic or endergonic?
   c. What is the fastest step in the reaction?   **g.** Is the overall reaction exergonic or endergonic?
   d. Which is more stable, A or D?   **h.** Which is the more stable intermediate?

51. a. What is the equilibrium constant of a reaction that is carried out at 25 °C (298 K) with $\Delta H° = 20$ kcal/mol and $\Delta S° = 25$ kcal mol^{-1} K^{-1}?
   b. What is the equilibrium constant of the same reaction carried out at 125 °C?

52. a. For a reaction that is carried out at 25 °C, how much must $\Delta G°$ change in order to increase the equilibrium constant by a factor of 10?
   b. How much must $\Delta H°$ change if $\Delta S° = 0$ kcal mol^{-1} K^{-1}?
   c. How much must $\Delta S°$ change if $\Delta H° = 0$ kcal mol^{-1}?

53. Given that the twist-boat conformation of cyclohexane is 5.3 kcal/mole (or 15.9 kJ/mol) higher in free energy than the chair conformation, calculate the percentage of twist-boat conformations present in a sample of cyclohexane at 25 °C. Does your answer agree with the statement made in Section 2.12 about the relative number of molecules in these two conformations?

CALCULATING KINETIC PARAMETERS

After obtaining rate constants at several temperatures, you can calculate E_a, $\Delta H^{\ddagger}$, $\Delta G^{\ddagger}$, and $\Delta S^{\ddagger}$ for a reaction as follows:

- The Arrhenius equation allows E_a to be obtained from the slope of a plot of ln k versus $1/T$, because

$$\ln k_2 - \ln k_1 = -E_a/R \left(\frac{1}{T_2} - \frac{1}{T_1} \right)$$

- You can determine $\Delta H^{\ddagger}$ at a given temperature from E_a because $\Delta H^{\ddagger} = E_a - RT$.

- You can determine $\Delta G^{\ddagger}$, in kJ/mol, from the following equation, which relates $\Delta G^{\ddagger}$ to the rate constant at a given temperature:

$$-\Delta G^{\ddagger} = RT \ \ln \frac{kh}{Tk_B}$$

   In this equation, h is Planck's constant (6.62608×10^{-34} Js) and k_B is the Boltzmann constant (1.38066×10^{-23} JK^{-1})

- You can determine the entropy of activation from the other two kinetic parameters via the formula $\Delta S^{\ddagger} = (\Delta H^{\ddagger} - \Delta G^{\ddagger})/T$.

Use this information to answer Problem 54.

54. From the following rate constants, determined at five temperatures, calculate the experimental energy of activation and $\Delta G^{\ddagger}$, $\Delta H^{\ddagger}$, and $\Delta S^{\ddagger}$ for the reaction at 30 °C:

Temperature	Observed rate constant
31.0 °C	2.11×10^{-5} s^{-1}
40.0 °C	4.44×10^{-5} s^{-1}
51.5 °C	1.16×10^{-4} s^{-1}
59.8 °C	2.10×10^{-4} s^{-1}
69.2 °C	4.34×10^{-4} s^{-1}

The Reactions of Alkenes

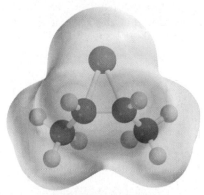

a cyclic bromonium ion

BUILDING ON FUNDAMENTALS

SECTIONS 4.1–4.10 We saw the first example of an electrophilic addition reaction in Chapter 3 (Section 3.6). Now we will look at several more electrophilic addition reactions—the characteristic reaction of alkenes.

SECTION 4.2 Hyperconjugation—delocalization of electrons by the overlap of a σ bond orbital with an orbital on an adjacent carbon—explained why the staggered conformation of ethane is more stable than the eclipsed conformation. Now we will see that it also explains why alkyl groups stabilize carbocations (2.10).

SECTION 4.4 The product with the smallest free energy of activation for its formation is the one that is formed fastest (3.7).

SECTION 4.5 A compound with an acidic proton loses that proton in solutions that have pH values greater than the compound's pK_a value (1.24).

SECTION 4.7 The octet rule helps explain why a cyclic bromonium ion rather than a carbocation is formed when an alkene reacts with a halogen (Br_2 or Cl_2) (1.4).

SECTION 4.10 Steric effects are partially responsible for the regioselectivity of hydroboration–oxidation (2.10).

SECTION 4.11 The steric strain that causes a gauche conformation to be less stable than an anti conformation, also causes a cis isomer to be less stable than a trans isomer (2.10).

We have seen that an **alkene** such as 2-butene undergoes an **electrophilic addition reaction** with HBr (Section 3.6). The first step of the reaction is a relatively slow addition of the electrophilic proton to the nucleophilic alkene to form a **carbocation intermediate**. In the second step, the positively charged carbocation intermediate (an electrophile) reacts rapidly with the negatively charged bromide ion (a nucleophile).

the electrophile adds to an sp^2 carbon of the alkene

a carbocation intermediate

the nucleophile adds to the carbocation

A curved arrow with a two-barbed arrowhead signifies the movement of two electrons.

In this chapter, we will look at a wide variety of alkene reactions. You will see that some of the reactions form carbocation intermediates like the one formed when

159

HBr reacts with an alkene, some form other kinds of intermediates, and some form no intermediates at all. At first, the reactions covered in this chapter might appear to be quite diverse, but you will see that they all occur by similar mechanisms. As you study each reaction, look for the feature that all alkene reactions have in common: *the relatively loosely held π electrons of the carbon–carbon double bond are attracted to an electrophile. Thus, each reaction starts with the addition of an electrophile to one of the sp^2 carbons of the alkene and concludes with the addition of a nucleophile to the other* sp^2 carbon. The end result is that the π bond breaks and the electrophile and nucleophile form new σ bonds with the sp^2 carbons. Notice that the sp^2 carbons in the reactant have become sp^3 carbons in the product.

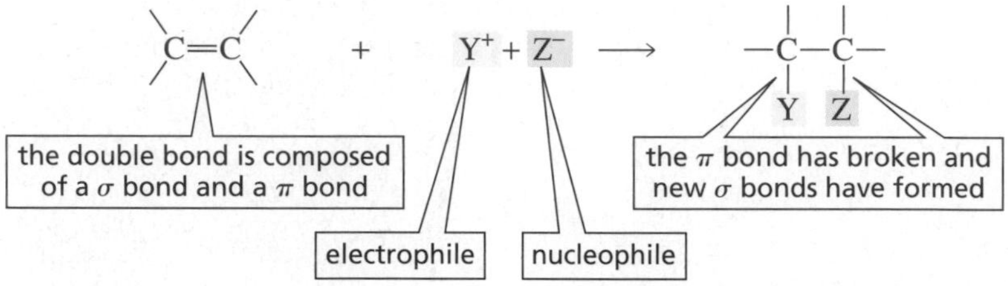

This reactivity makes alkenes an important class of organic compounds, because they can be used to synthesize a wide variety of other compounds. For example, we will see that alkyl halides, alcohols, ethers, epoxides, and alkanes all can be synthesized from alkenes by electrophilic addition reactions. The particular product obtained depends only on the *electrophile* and the *nucleophile* used in the addition reaction.

4.1 Addition of a Hydrogen Halide to an Alkene

If the electrophilic reagent that adds to an alkene is a hydrogen halide (HF, HCl, HBr, or HI), then the product of the reaction will be an alkyl halide:

$$CH_2{=}CH_2 + HCl \longrightarrow CH_3CH_2Cl$$

ethene   chloroethane

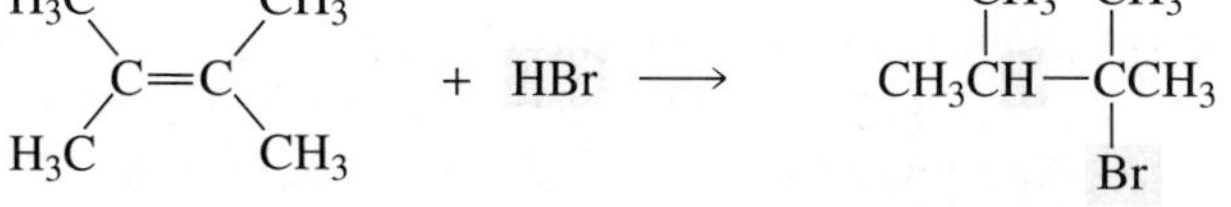

2,3-dimethyl-2-butene   2-bromo-2,3-dimethylbutane

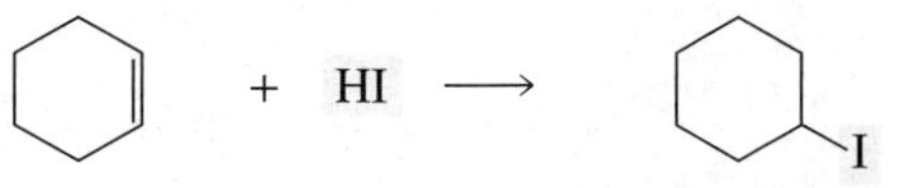

cyclohexene   iodocyclohexane

Because the alkenes in the preceding reactions have the same substituents on both sp^2 carbons, it is easy to predict the product of the reaction: the electrophile (H$^+$) adds to one of the sp^2 carbons, and the nucleophile (X$^-$) adds to the other sp^2 carbon. It doesn't matter which sp^2 carbon the electrophile attaches to, because the same product will be obtained in either case.

But what happens if the alkene does not have the same substituents on both sp^2 carbons? Which sp^2 carbon gets the hydrogen? For example, does the addition of HCl to 2-methylpropene produce *tert*-butyl chloride or isobutyl chloride?

$$
\underset{\textbf{2-methylpropene}}{\overset{\overset{\displaystyle CH_3}{|}}{CH_3C}=CH_2} + HCl \longrightarrow \underset{\textbf{\textit{tert}-butyl chloride}}{\overset{\overset{\displaystyle CH_3}{|}}{\underset{\underset{\displaystyle Cl}{|}}{CH_3CCH_3}}} \quad or \quad \underset{\textbf{isobutyl chloride}}{\overset{\overset{\displaystyle CH_3}{|}}{CH_3CHCH_2Cl}}
$$

To answer this question, we need to carry out the reaction, isolate the products, and identify them. When we do, we find that the only product of the reaction is *tert*-butyl chloride. Now we need to find out why that compound is the only product so we can use this knowledge to predict the products of other alkene reactions. To do this, we need to look again at the **mechanism of the reaction** (Section 3.6).

Recall that the first step of the reaction—the addition of H$^+$ to an sp^2 carbon to form either the *tert*-butyl cation or the isobutyl cation—is the rate-determining step (Section 3.7). If there is any difference in the rate of formation of these two carbocations, then the one that is formed faster will be the predominant product of the first step. Moreover, because carbocation formation is rate determining, the particular carbocation that is formed in the first step determines the final product of the reaction. That is, if the *tert*-butyl cation is formed, it will react rapidly with Cl$^-$ to form *tert*-butyl chloride. On the other hand, if the isobutyl cation is formed, it will react rapidly with Cl$^-$ to form isobutyl chloride. Since the only product of the reaction is *tert*-butyl chloride, we know that the *tert*-butyl cation is formed much faster than the isobutyl cation.

George Olah *was born in Hungary in 1927 and received a doctorate from the Technical University of Budapest in 1949. The Hungarian revolution caused him to emigrate to Canada in 1956, where he worked as a scientist at the Dow Chemical Company until joining the faculty of Case Western Reserve University in 1965. In 1977, he became a professor of chemistry at the University of Southern California. In 1994, he received the Nobel Prize in chemistry for his work on carbocations.*

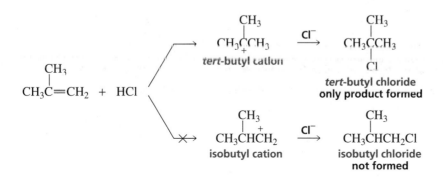

The question now is, why is the *tert*-butyl cation formed faster than the isobutyl cation? To answer this, we need to look at the factors that affect the stability of carbocations and thus the ease with which they are formed.

The sp^2 carbon that does *not* become attached to the proton is the carbon that is positively charged in the carbocation.

4.2 Carbocation Stability Depends on the Number of Alkyl Groups Attached to the Positively Charged Carbon

Carbocations are classified according to the number of alkyl substituents that are bonded to the positively charged carbon: a **primary carbocation** has one alkyl substituent, a **secondary carbocation** has two, and a **tertiary carbocation** has three. The stability of a carbocation increases as the number of alkyl substituents bonded to the positively charged carbon increases. Thus, tertiary carbocations are more stable than secondary carbocations, and secondary carbocations are more stable than primary carbocations. Keep in mind, however, that these stabilities are relative stabilities. Carbocations are rarely stable enough to isolate.

relative stabilities of carbocations

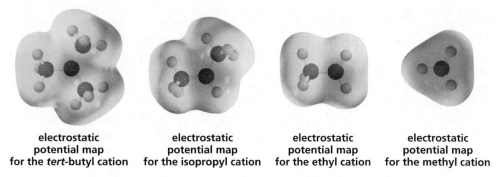

| a tertiary carbocation | a secondary carbocation | a primary carbocation | methyl cation |

The greater the number of alkyl substituents bonded to the positively charged carbon, the more stable the carbocation is.

The reason for this pattern of decreasing stability is that alkyl groups bonded to the positively charged carbon decrease the concentration of positive charge on the carbon, and decreasing the concentration of positive charge makes the carbocation more stable. Notice that the blue (representing positive charge) in these electrostatic potential maps is most intense for the least stable methyl cation and is least intense for the most stable *tert*-butyl cation.

electrostatic potential map for the *tert*-butyl cation | electrostatic potential map for the isopropyl cation | electrostatic potential map for the ethyl cation | electrostatic potential map for the methyl cation

Carbocation stability: 3° > 2° > 1°

How do alkyl groups decrease the concentration of positive charge on the carbon? Recall that the positive charge on a carbon signifies an empty *p* orbital (Section 1.10). Figure 4.1 shows that in the ethyl cation, the orbital of an adjacent C—H σ bond can overlap with the empty *p* orbital. No such overlap is possible in the methyl cation.

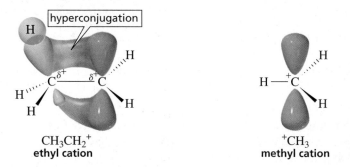

Figure 4.1 ▶
Stabilization of a carbocation by hyperconjugation: the electrons of an adjacent C—H σ bond orbital of the ethyl cation spread into the empty *p* orbital. Hyperconjugation cannot occur in a methyl cation.

Movement of electrons from the σ bond orbital toward the vacant *p* orbital of the ethyl cation decreases the charge on the sp^2 carbon and causes a partial positive charge to develop on the carbon bonded by the σ bond. Therefore, the positive charge is no longer concentrated solely on one atom, but is delocalized, spread out over a larger volume of space. This dispersion of positive charge stabilizes the carbocation because a charged species is more stable if its charge is spread out (delocalized) over more than one atom (Section 1.22). Delocalization of electrons by the overlap of a σ bond orbital with an orbital on an adjacent carbon is called **hyperconjugation** (Section 2.10). The simple molecular orbital diagram in Figure 4.2 is another way of depicting the stabilization achieved by the overlap of a filled C—H σ bond orbital with an empty *p* orbital.

Hyperconjugation occurs only if the σ bond orbital and the empty *p* orbital have the proper orientations. The proper orientations are easily achieved because there is free

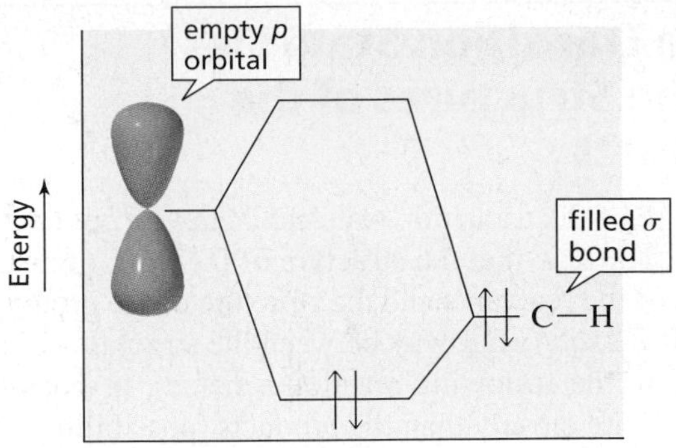

◀ **Figure 4.2**
A molecular orbital diagram showing the stabilization achieved by overlapping the electrons of a C—H bond with an empty *p* orbital.

rotation about the carbon–carbon σ bond (Section 2.10). Notice that the σ bonds capable of overlapping with the empty *p* orbital are those *attached to an atom that is attached to the positively charged carbon.* In the *tert*-butyl cation, nine σ bond orbitals can potentially overlap with the empty *p* orbital of the positively charged carbon. (The nine σ bonds are indicated below by red dots.) The isopropyl cation has six such orbitals, and the ethyl and propyl cations each have three. Therefore, there is greater stabilization through hyperconjugation in the tertiary *tert*-butyl cation than in the secondary isopropyl cation, and greater stabilization in the secondary isopropyl cation than in the primary ethyl or propyl cation. Notice that the ethyl and propyl cations each have about the same stability, because both C—H and C—C σ bond orbitals can overlap with the empty *p* orbital.

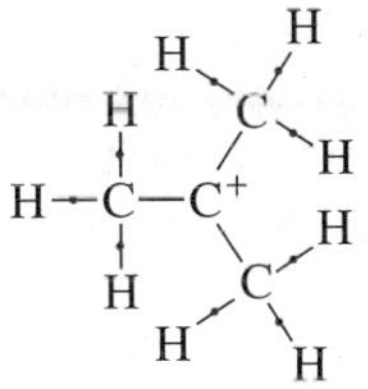

tert-butyl cation

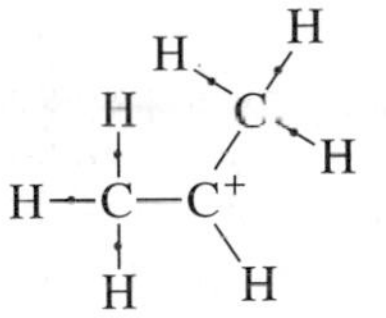

isopropyl cation

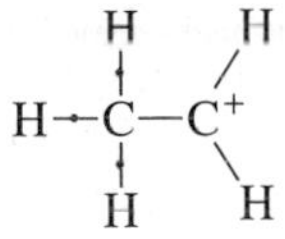

ethyl cation

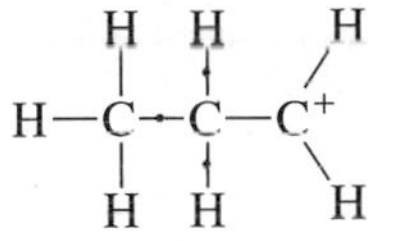

propyl cation

PROBLEM 1◆

a. How many σ bond orbitals are available for overlap with the vacant *p* orbital in the methyl cation?

b. Which is more stable, a methyl cation or an ethyl cation?

PROBLEM 2◆

a. How many σ bond orbitals are available for overlap with the vacant *p* orbital in **1.** the isobutyl cation **2.** the *n*-butyl cation **3.** the *sec*-butyl cation

b. Which is most stable?

Student Tutorial: First step in addition of hydrogen halides to alkenes

PROBLEM 3◆

List the following carbocations in order of decreasing stability:

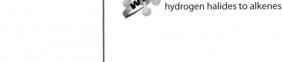

a. $CH_3CH_2\underset{+}{\overset{CH_3}{C}}CH_3$ $CH_3CH_2\underset{+}{C}HCH_3$ $CH_3CH_2CH_2\overset{+}{C}H_2$

b. $CH_3\underset{Cl}{C}HCH_2\overset{+}{C}H_2$ $CH_3\underset{CH_3}{C}HCH_2\overset{+}{C}H_2$ $CH_3\underset{F}{C}HCH_2\overset{+}{C}H_2$

Student Tutorial: Second step in addition of hydrogen halides to alkenes

4.3 The Structure of the Transition State Lies Partway Between the Structures of the Reactants and Products

Knowing something about the structure of the transition state can help you predict the products of a reaction. In Section 3.7, you saw that the structure of the transition state lies somewhere between the structure of the reactants and the structure of the products. But where exactly is "somewhere"? Is it *exactly* halfway between the structures of the reactants and products (possibility II in the following reaction scheme), or does the transition state resemble the reactants more closely than the products (possibility I), or the products more closely than the reactants (possibility III)?

$$
A{-}B\ +\ C\ \left[\begin{array}{l}\text{(I)}\quad A\text{----}B\text{----------}C\\[4pt]\text{(II)}\quad A\text{------}B\text{-------}C\\[4pt]\text{(III)}\quad A\text{----------}B\text{----}C\end{array}\right]^{\ddagger}\longrightarrow\ A\ +\ B{-}C
$$

‡ symbolizes the transition state

reactants

transition state

products

This question is answered by the **Hammond postulate**, which says that *the transition state is more similar in structure to the species to which it is more similar in energy.* In an exergonic reaction, the transition state is more similar in energy to the reactant than to the product (Figure 4.3, curve I). Therefore, the structure of the transition state will more closely resemble the structure of the reactant than that of the product, as in possibility I. In an endergonic reaction (Figure 4.3, curve III), the transition state is more similar in energy to the product, so the structure of the transition state will more closely resemble the structure of the product (possibility III). Only when the reactant and the product have identical energies (Figure 4.3, curve II) would we expect the structure of the transition state to be exactly halfway between the structures of the reactant and the product (possibility II).

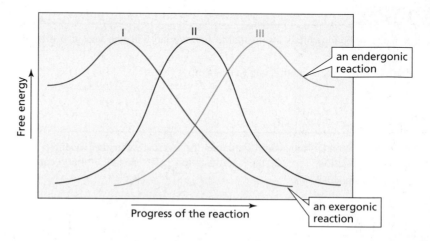

Figure 4.3 ▶
Reaction coordinate diagrams for reactions with an early transition state (I), a midway transition state (II), and a late transition state (III).

The transition state is more similar in structure to the species to which it is more similar in energy.

Now we can understand why the *tert*-butyl cation is formed faster than the isobutyl cation when 2-methylpropene reacts with HCl. Because the formation of a carbocation is an endergonic reaction (Figure 4.4), the structure of the transition state will resemble the structure of the carbocation product. This means that the transition state will have a significant amount of positive charge on a carbon. The same factors that stabilize the positively charged carbocation stabilize the partially positively charged transition state. We know that the *tert*-butyl cation (a tertiary carbocation) is more stable than the isobutyl cation (a primary carbocation). Therefore,

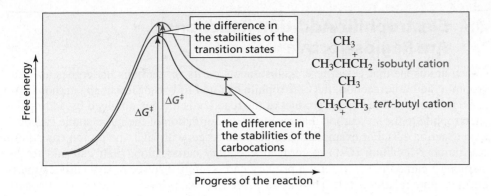

◀ **Figure 4.4**
Reaction coordinate diagram for the addition of H^+ to 2-methylpropene to form the primary isobutyl cation and the tertiary *tert*-butyl cation.

the transition state leading to the *tert*-butyl cation is more stable (lower in energy) than the transition state leading to the isobutyl cation. Note, however, that because the amount of positive charge in the transition state is not as great as the amount of positive charge in the carbocation product, the difference in the stabilities of the two transition states is not as great as the difference in the stabilities of the two carbocation products (Figure 4.4).

We have seen that the rate of a reaction is determined by the free energy of activation, which is the difference between the free energy of the transition state and the free energy of the reactant: the more stable the transition state, the smaller is the free energy of activation, and therefore, the faster is the reaction (Section 3.7). Thus, the *tert*-butyl cation will be formed faster than the isobutyl cation. *In an electrophilic addition reaction, the more stable carbocation will be the one that is formed more rapidly.*

Because the formation of the carbocation is the rate-limiting step of the reaction, the relative rates of formation of the two carbocations determine the relative amounts of the products that are formed. If the difference in the rates is small, both products will be formed, but the major product will be the one formed from reaction of the nucleophile with the more stable carbocation. If the difference in the rates is sufficiently large, the product formed from reaction of the nucleophile with the more stable carbocation will be the only product of the reaction. For example, when HCl adds to 2-methylpropene, the rates of formation of the two possible carbocation intermediates—one primary and the other tertiary—are sufficiently different to cause *tert*-butyl chloride to be the only product of the reaction.

PROBLEM 4◆

For each of the following reaction coordinate diagrams, tell whether the structure of the transition state will more closely resemble the structure of the reactants or the structure of the products:

a. 　**b.** 　**c.** 　**d.**

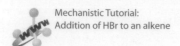
4.4 Electrophilic Addition Reactions Are Regioselective

When an alkene that has different substituents on its sp^2 carbons undergoes an electrophilic addition reaction, the electrophile can add to two different sp^2 carbons. We have just seen that the major product of the reaction is the one obtained by adding the electrophile to the sp^2 carbon that results in the formation of the more stable carbocation (Section 4.3). For example, when propene reacts with HCl, the proton can add to the number-1 carbon (C-1) to form a secondary carbocation, or it can add to the number-2 carbon (C-2) to form a primary carbocation. The secondary carbocation is formed more rapidly because it is more stable than the primary carbocation. (Primary carbocations are so unstable that they form only with great difficulty.) The product of the reaction, therefore, is 2-chloropropane.

The major product obtained from the addition of HI to 2-methyl-2-butene is 2-iodo-2-methylbutane; only a small amount of 2-iodo-3-methylbutane is obtained. The major product obtained from the addition of HBr to 1-methylcyclohexene is 1-bromo-1-methylcyclohexane. In both cases, the more stable tertiary carbocation is formed more rapidly than the less stable secondary carbocation, so the major product of each reaction is the one that results from forming the tertiary carbocation.

The two different products in each of these reactions are constitutional isomers. We saw in Chapter 2 that **constitutional isomers** have the same molecular formula, but differ in how their atoms are connected. A reaction (such as either of those just shown) in which two or more constitutional isomers could be obtained as products, but one of them predominates, is called a **regioselective reaction**.

There are degrees of **regioselectivity**: a reaction can be *moderately regioselective, highly regioselective,* or *completely regioselective.* In a completely regioselective reaction, one of the possible products is not formed at all. The addition of a hydrogen halide to 2-methylpropene (where the two possible carbocations are tertiary and primary) is more highly regioselective than the addition of a hydrogen halide to 2-methyl-2-butene (where the two possible carbocations are tertiary and secondary) because the two carbocations formed from 2-methyl-2-butene are closer in stability.

Regioselectivity is the preferential formation of one constitutional isomer over another.

The addition of HBr to 2-pentene is not regioselective. Because the addition of a proton to either of the sp^2 carbons produces a secondary carbocation, both carbocation intermediates have the same stability, so both form with equal ease. Thus, approximately equal amounts of the two alkyl halides are formed.

$$CH_3CH=CHCH_2CH_3 \ + \ HBr \ \longrightarrow \ CH_3\overset{\underset{\mid}{Br}}{C}HCH_2CH_2CH_3 \ + \ CH_3CH_2\overset{\underset{\mid}{Br}}{C}HCH_2CH_3$$

2-pentene   2-bromopentane   3-bromopentane
                      50%   50%

From the alkene reactions we have seen so far, we can devise a rule that applies to *all* alkene electrophilic addition reactions: *the electrophile adds to the* sp^2 *carbon that is bonded to the greater number of hydrogens.* Vladimir Markovnikov was the first to recognize that in the addition of a hydrogen halide to an alkene, the H^+ adds to the sp^2 carbon that is bonded to the greater number of hydrogens. As a result, this rule is often referred to as **Markovnikov's rule.**

The electrophile adds to the sp^2 carbon that is bonded to the greater number of hydrogens.

The rule is simply a quick way to compare the relative stabilities of the intermediates that could be formed in the rate-determining step. You will get the same answer, whether you identify the major product of an electrophilic addition reaction by using the rule or whether you identify it by determining relative carbocation stabilities. In the following reaction, for example, H^+ is the electrophile:

$$CH_3CH_2\overset{2}{C}H=\overset{1}{C}H_2 \ + \ HCl \ \longrightarrow \ CH_3CH_2\overset{\underset{\mid}{Cl}}{C}HCH_3$$

We can say that H^+ adds preferentially to C-1 because C-1 is bonded to two hydrogens, whereas C-2 is bonded to only one hydrogen. Or we can say that H^+ adds to C-1 because that results in the formation of a secondary carbocation, which is more stable than the primary carbocation that would have to be formed if H^+ added to C-2.

B I O G R A P H Y

Vladimir Vasilevich Markovnikov (1837–1904) *was born in Russia, the son of an army officer, and was a professor of chemistry at Kazan, Odessa, and Moscow Universities. By synthesizing rings containing four carbons and seven carbons, he disproved the notion that carbon could form only five- and six-membered rings.*

PROBLEM 5♦

What would be the major product obtained from the addition of HBr to each of the following compounds?

a. $CH_3CH_2CH=CH_2$

b. $CH_3CH=\overset{\underset{\mid}{CH_3}}{C}CH_3$

c. 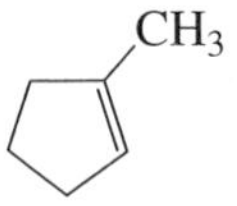 (cyclopentene with CH₃ substituent)

d. $CH_2=\overset{\underset{\mid}{CH_3}}{C}CH_2CH_2CH_3$

e. 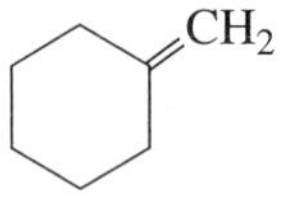 (cyclohexane with =CH₂)

f. $CH_3CH=CHCH_3$

PROBLEM-SOLVING STRATEGY

Planning the Synthesis of an Alkyl Halide

a. What alkene should be used to synthesize 3-bromohexane?

$$? \ + \ HBr \ \longrightarrow \ CH_3CH_2\overset{\underset{\mid}{}}{C}HCH_2CH_2CH_3$$
                                    Br

                                3-bromohexane

The best way to answer this kind of question is to begin by listing all the alkenes that could be used. Because you want to synthesize an alkyl halide that has a bromo substituent at the 3-position, the alkene should have an sp^2 carbon at that position. Two alkenes fit the description: 2-hexene and 3-hexene.

$$CH_3CH=CHCH_2CH_2CH_3 \qquad CH_3CH_2CH=CHCH_2CH_3$$
$$\text{2-hexene} \qquad\qquad\qquad \text{3-hexene}$$

Because there are two possibilities, we next need to decide whether there is any advantage to using one over the other. The addition of H^+ to 2-hexene can form two different carbocations, but they are both secondary carbocations. Because they have the same stability, approximately equal amounts of each will be formed. Therefore, half of the product will be 3-bromohexane and half will be 2-bromohexane.

$$CH_3CH=CHCH_2CH_2CH_3$$
$$\text{2-hexene}$$

HBr → $CH_3CH_2\overset{+}{C}HCH_2CH_2CH_3$ [secondary carbocation] $\xrightarrow{Br^-}$ $CH_3CH_2\underset{Br}{\overset{|}{C}}HCH_2CH_2CH_3$ 3-bromohexane

HBr → $CH_3\overset{+}{C}HCH_2CH_2CH_2CH_3$ [secondary carbocation] $\xrightarrow{Br^-}$ $CH_3\underset{Br}{\overset{|}{C}}HCH_2CH_2CH_2CH_3$ 2-bromohexane

The addition of H^+ to either of the sp^2 carbons of 3-hexene, on the other hand, forms the same carbocation because the alkene is symmetrical. Therefore, all of the product will be the desired 3-bromohexane.

$$CH_3CH_2CH=CHCH_2CH_3 \xrightarrow{HBr} CH_3CH_2\overset{+}{C}HCH_2CH_2CH_3 \xrightarrow{Br^-} CH_3CH_2\underset{Br}{\overset{|}{C}}HCH_2CH_2CH_3$$
$$\text{3-hexene} \qquad\qquad \text{[only one carbocation is formed]} \qquad \text{3-bromohexane}$$

Because all the alkyl halide formed from 3-hexene is 3-bromohexane, but only half the alkyl halide formed from 2-hexene is 3-bromohexane, 3-hexene is the best alkene to use to prepare 3-bromohexane.

b. What alkene should be used to synthesize 2-bromopentane?

$$? \quad + \quad HBr \quad\longrightarrow\quad CH_3\underset{Br}{\overset{|}{C}}HCH_2CH_2CH_3$$
$$\text{2-bromopentane}$$

Either 1-pentene or 2-pentene could be used because both have an sp^2 carbon at the 2-position.

$$CH_2=CHCH_2CH_2CH_3 \qquad CH_3CH=CHCH_2CH_3$$
$$\text{1-pentene} \qquad\qquad \text{2-pentene}$$

When H^+ adds to 1-pentene, one of the carbocations that could be formed is secondary and the other is primary. A secondary carbocation is more stable than a primary carbocation, which is so unstable that little, if any, will be formed. Thus, 2-bromopentane will be the only product of the reaction.

$$CH_2=CHCH_2CH_2CH_3$$
$$\text{1-pentene}$$

HBr → $CH_3\overset{+}{C}HCH_2CH_2CH_3$ $\xrightarrow{Br^-}$ $CH_3\underset{Br}{\overset{|}{C}}HCH_2CH_2CH_3$ 2-bromopentane

HBr ⤢ $\overset{+}{C}H_2CH_2CH_2CH_2CH_3$

When H^+ adds to 2-pentene, on the other hand, each of the two carbocations that can be formed is secondary. Both are equally stable, so they will be formed in approximately equal amounts. Thus, only about half of the product of the reaction will be 2-bromopentane. The other half will be 3-bromopentane.

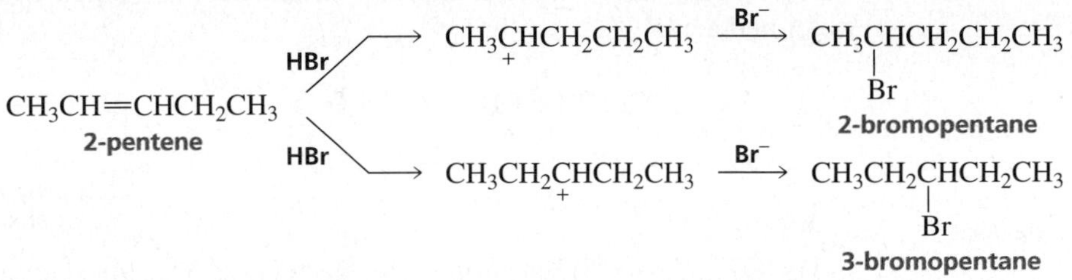

$$CH_3CH=CHCH_2CH_3 \xrightarrow{HBr} CH_3\overset{+}{C}HCH_2CH_2CH_3 \xrightarrow{Br^-} CH_3CHCH_2CH_2CH_3$$

CH₃CHCH₂CH₂CH₃ with Br
2-bromopentane

$$\xrightarrow{HBr} CH_3CH_2\overset{+}{C}HCH_2CH_3 \xrightarrow{Br^-} CH_3CH_2CHCH_2CH_3$$

with Br
3-bromopentane

2-pentene

Because all the alkyl halide formed from 1-pentene is 2-bromopentane, but only half the alkyl halide formed from 2-pentene is 2-bromopentane, 1-pentene is the best alkene to use to prepare 2-bromopentane.

Now continue on to Problem 6.

PROBLEM 6♦

What alkene should be used to synthesize each of the following alkyl bromides?

a. 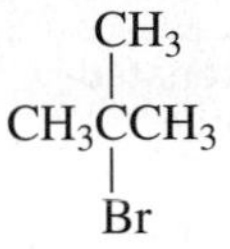 CH₃CCH₃ with CH₃ and Br

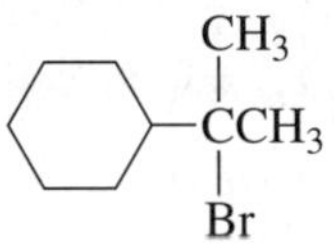

b. 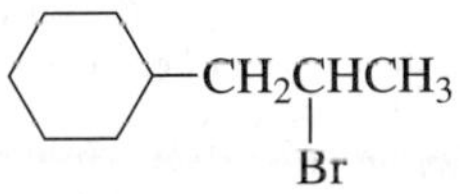 ⬡—CH₂CHCH₃ with Br

c. ⬡—CCH₃ with CH₃ and Br

d. 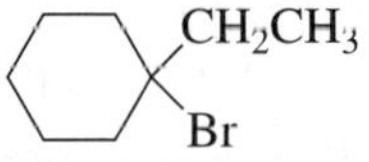 ⬡ with CH₂CH₃ and Br

PROBLEM 7♦

To which compound in each of the following pairs is the addition of HBr more highly regioselective?

a. 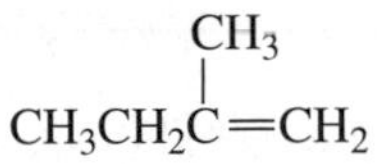 CH₃CH₂C=CH₂ with CH₃   or   CH₃C=CHCH₃ with CH₃

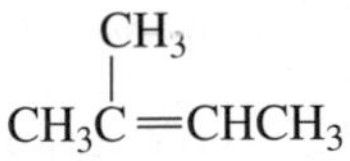

b. 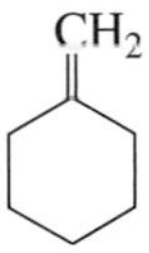 ⬡=CH₂   or   ⬡ with CH₃

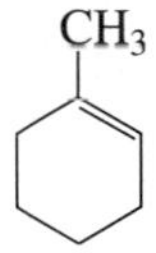

VCL **Alkene Halogenation-2**

4.5 Acid-Catalyzed Addition Reactions

Water and alcohols add to alkenes only if an acid is present.

Addition of Water to an Alkene

An alkene does not react with water, because there is no electrophile present to start a reaction by adding to the nucleophilic alkene. The O—H bonds of water are too strong—water is too weakly acidic—to allow the hydrogen to act as an electrophile for this reaction.

$$CH_3CH=CH_2 + H_2O \longrightarrow \text{no reaction}$$

If an acid (the acid used most often is H_2SO_4) is added to the solution, then the outcome is much different: a reaction will occur because the acid provides an electrophile (H^+). The product of the reaction is an alcohol. The addition of water to a molecule is

called **hydration**, so we can say that an alkene will be *hydrated* in the presence of water and acid.*

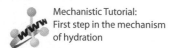

H₂SO₄ (pK_a = −5) is a strong acid, so it dissociates almost completely in an aqueous solution (Section 1.24). The acid that participates in the reaction, therefore, is most apt to be a hydrated proton—called a hydronium ion, which we will write as (H₃O⁺).

$$CH_3CH=CH_2 + H_2O \xrightarrow{H_2SO_4} CH_3CH-CH_2$$

$$H_2SO_4 + H_2O \rightleftharpoons H_3O^+ + HSO_4^-$$

hydronium ion

Mechanistic Tutorial:
First step in the mechanism of hydration

Notice that the first two steps of the *mechanism for the acid-catalyzed addition of water to an alkene* are essentially the same as the two steps of the *mechanism for the addition of a hydrogen halide to an alkene*:

Mechanism for the acid-catalyzed addition of water

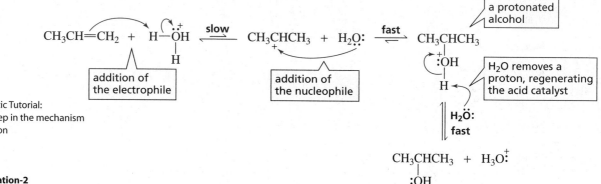

Mechanistic Tutorial:
Second step in the mechanism of hydration

Mechanistic Tutorial:
Third step in the mechanism of hydration

Mechanistic Tutorial:
First step in the mechanism of hydration showing orientation

Do not memorize the products of alkene addition reactions. Instead, for each reaction, ask yourself, "What is the electrophile?" and "What nucleophile is present in the greatest concentration?"

Mechanistic Tutorial:
Second step in the mechanism of hydration showing orientation

Mechanistic Tutorial:
Addition of water to an alkene

- The electrophile (H⁺) adds to the sp^2 carbon that is bonded to the greater number of hydrogens.
- The nucleophile (H₂O) adds to the carbocation, forming a protonated alcohol.
- The protonated alcohol loses a proton because the pH of the solution is greater than the pK_a of the protonated alcohol (Section 1.24). (We have seen that protonated alcohols are very strong acids; Section 1.18).

As we saw in Section 3.7, the addition of the electrophile to the alkene is relatively slow, whereas the subsequent addition of the nucleophile to the carbocation occurs rapidly. The reaction of the carbocation with a nucleophile is so fast, in fact, that the carbocation combines with whatever nucleophile it collides with: note that there are two nucleophiles in solution, water and the counterion of the acid (HSO₄⁻) that was used to start the reaction. (The reason HO⁻ is not a nucleophile in this reaction is that there is no appreciable concentration of HO⁻ in an acidic solution.)† Because the concentration of water is much greater than the concentration of the counterion, the carbocation is much more likely to collide with water. The final product of the addition reaction, therefore, is an alcohol.

A proton adds to the alkene in the first step, but a proton is returned to the reaction mixture in the final step. Overall, then, a proton is not consumed. A species that

---

* Because an alkene is not soluble in water, water is not the only solvent used in this reaction. A second solvent, such as DMSO (Section 8.3), is required that will not react with the reactants, the products, or any intermediates formed in the reaction but in which both the alkene and water can dissolve.
† At a pH of 4, for example, the concentration of HO⁻ is 1×10^{-10} M, whereas the concentration of water in a dilute aqueous solution is 55.5 M.

increases the rate of a reaction but is not consumed during the course of the reaction is called a **catalyst**. Catalysts increase the reaction rate by decreasing the free energy of activation (Section 3.7). They do *not* affect the equilibrium constant of the reaction. In other words, a catalyst increases the *rate* at which a product is formed, but does not affect the *amount* of product formed when the reaction has reached equilibrium. The catalyst employed in the hydration of an alkene is an acid, so the hydration reaction is called an **acid-catalyzed reaction**.

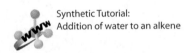

Synthetic Tutorial:
Addition of water to an alkene

PROBLEM 8◆

The pK_a of a protonated alcohol is about -2.5 and the pK_a of an alcohol is about 15. Therefore, as long as the pH of the solution is greater than _____ and less than _____, more than 50% of 2-propanol (the product of the previous reaction) will be in its neutral, nonprotonated form.

VCL **Alkene Hydration-3**

PROBLEM 9◆

Answer the following questions about the mechanism for the acid-catalyzed hydration of an alkene:

a. How many transition states are there?
b. How many intermediates are there?
c. Which step in the forward direction has the smallest rate constant?

PROBLEM 10◆

Give the major product obtained from the acid-catalyzed hydration of each of the following alkenes:

a. $CH_3CH_2CH_2CH\!=\!CH_2$

c. $CH_3CH_2CH_2CH\!=\!CHCH_3$

b.

d. 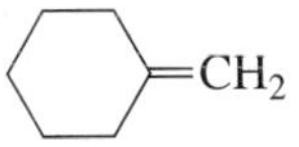 $=CH_2$

Addition of an Alcohol to an Alkene

Alcohols react with alkenes in the same way that water does, so this reaction, too, requires an acid catalyst. The product of the reaction is an ether.

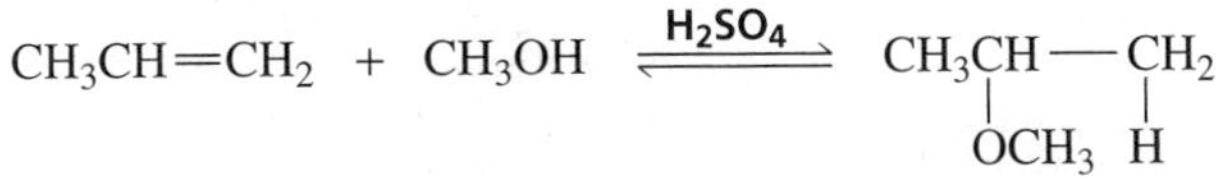

$$CH_3CH\!=\!CH_2 \;+\; CH_3OH \;\overset{H_2SO_4}{\rightleftharpoons}\; CH_3CH\!-\!CH_2$$

$$\underset{\substack{\text{2-methoxypropane}\\\text{an ether}}}{\overset{\displaystyle\;\;}{\underset{\displaystyle OCH_3\;\;H}{\big|\qquad\big|}}}$$

The *mechanism for the acid-catalyzed addition of an alcohol* is essentially the same as the *mechanism for the acid-catalyzed addition of water*. The only difference is the nucleophile is ROH instead of HOH.

Mechanism for the acid-catalyzed addition of an alcohol

$$CH_3CH\!=\!CH_2 \;+\; H\!-\!\overset{+}{\underset{H}{\overset{\displaystyle ..}{O}}}CH_3 \;\overset{slow}{\rightleftharpoons}\; CH_3\overset{+}{C}HCH_3 \;+\; CH_3\overset{..}{\overset{..}{O}}H \;\overset{fast}{\rightleftharpoons}\; CH_3CHCH_3$$

$$\underset{H}{\overset{+}{\underset{|}{:OCH_3}}}$$

$$\Bigg\Vert\,CH_3\overset{..}{\underset{..}{O}}H \quad \text{fast}$$

$$CH_3CHCH_3 \;+\; CH_3\overset{..}{\overset{+}{O}}H$$

$$\underset{:OCH_3}{|}\qquad\qquad\underset{H}{|}$$

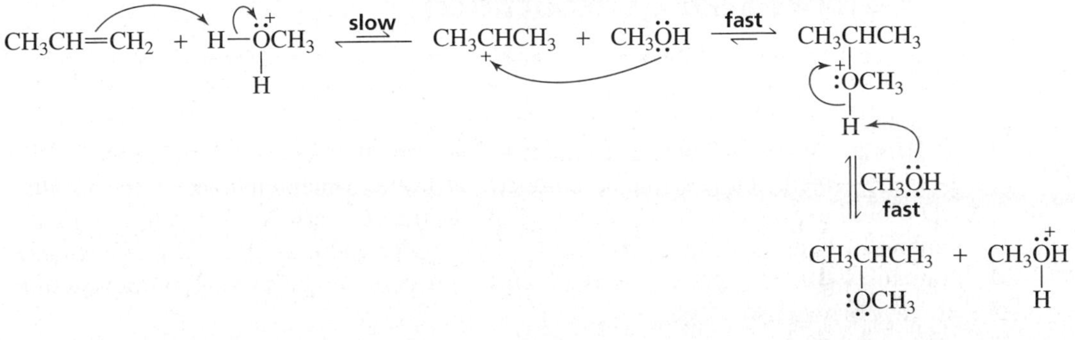

- The electrophile (H^+) adds to the sp^2 carbon that is bonded to the greater number of hydrogens.
- The nucleophile (CH_3OH) adds to the carbocation, forming a protonated ether.
- The protonated ether loses a proton, because the pH of the solution is greater than the pK_a of the protonated ether ($pK_a \sim -3.6$).

VCL **Etherification-1**

PROBLEM 11

a. Give the major product of each of the following reactions:

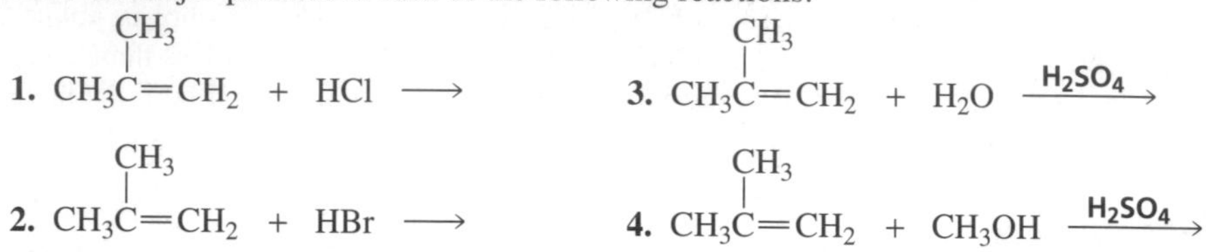

$$\overset{CH_3}{\underset{|}{1.\ CH_3C}}=CH_2 \ + \ HCl \ \longrightarrow \qquad \overset{CH_3}{\underset{|}{3.\ CH_3C}}=CH_2 \ + \ H_2O \ \xrightarrow{H_2SO_4}$$

$$\overset{CH_3}{\underset{|}{2.\ CH_3C}}=CH_2 \ + \ HBr \ \longrightarrow \qquad \overset{CH_3}{\underset{|}{4.\ CH_3C}}=CH_2 \ + \ CH_3OH \ \xrightarrow{H_2SO_4}$$

b. What do all the reactions have in common?

c. How do all the reactions differ?

PROBLEM 12

How could the following compounds be prepared, using an alkene as one of the starting materials?

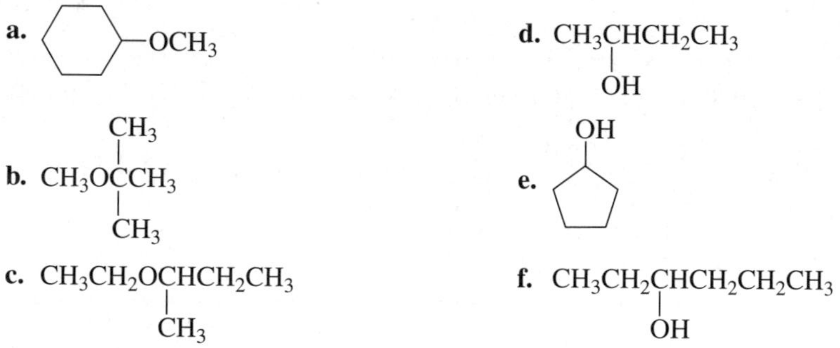

a. ⬡—OCH_3

$$\text{b. } CH_3O\overset{CH_3}{\underset{\underset{CH_3}{|}}{\overset{|}{C}}}CH_3$$

c. $CH_3CH_2O\overset{|}{\underset{CH_3}{C}}HCH_2CH_3$

d. $CH_3\overset{|}{\underset{OH}{C}}HCH_2CH_3$

e. ⬠—OH (cyclopentanol with OH)

f. $CH_3CH_2\overset{|}{\underset{OH}{C}}HCH_2CH_2CH_3$

Synthetic Tutorial:
Addition of alcohol to an alkene

PROBLEM 13

Propose a mechanism for the following reaction (remember to use curved arrows when showing a mechanism):

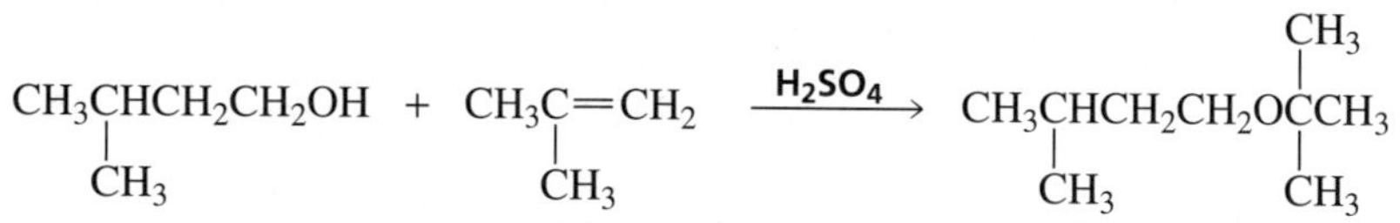

$$CH_3\overset{|}{\underset{CH_3}{C}}HCH_2CH_2OH \ + \ CH_3\overset{|}{\underset{CH_3}{C}}=CH_2 \ \xrightarrow{H_2SO_4} \ CH_3\overset{|}{\underset{CH_3}{C}}HCH_2CH_2O\overset{CH_3}{\underset{\underset{CH_3}{|}}{\overset{|}{C}}}CH_3$$

4.6 A Carbocation Will Rearrange If It Can Form a More Stable Carbocation

Some electrophilic addition reactions give products that clearly depart from the pattern of an electrophile attaching to the sp^2 carbon bonded to the greater number of hydrogens and a nucleophile attaching to the other sp^2 carbon. For example, the addition of HBr to 3-methyl-1-butene forms 2-bromo-3-methylbutane (minor product) and 2-bromo-2-methylbutane (major product). 2-Bromo-3-methylbutane is the product you would expect from the addition of H^+ to the sp^2 carbon bonded to the greater number of hydrogens and of Br^- to the other sp^2 carbon. 2-Bromo-2-methylbutane, however, is an "unexpected" product, and yet it is the major product of the reaction.

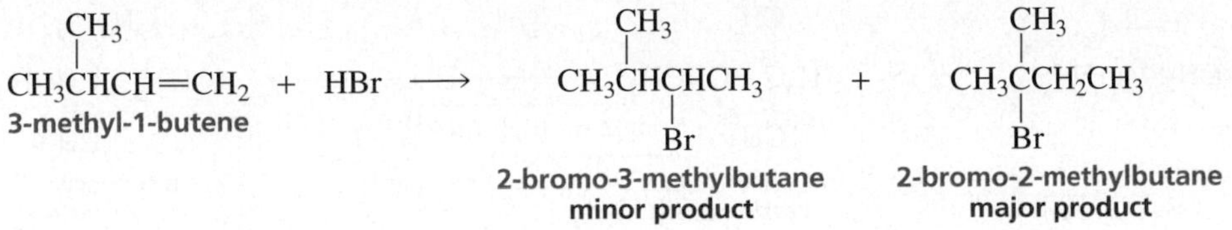

In another example, the addition of HCl to 3,3-dimethyl-1-butene forms both 3-chloro-2,2-dimethylbutane (an expected product) and 2-chloro-2,3-dimethylbutane (an unexpected product). Again, the unexpected product is obtained in greater yield.

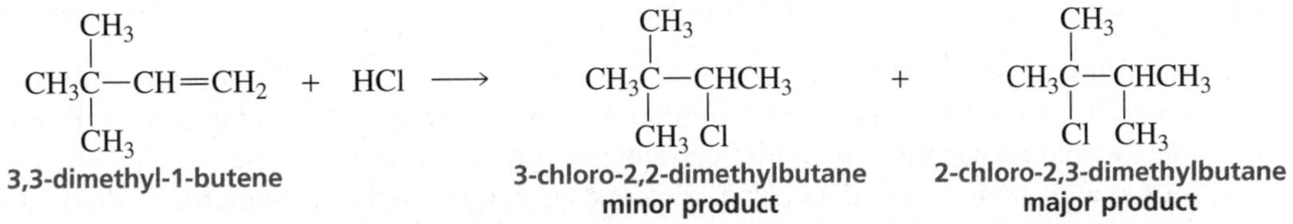

F. C. Whitmore was the first to suggest that the unexpected product results from a *rearrangement* of the carbocation intermediate. Not all carbocations rearrange. In fact, none of the carbocations that we have seen up to this point rearranges. Carbocations rearrange only if they become more stable as a result of the rearrangement. For example, when an electrophile adds to 3-methyl-1-butene, a *secondary* carbocation is formed at first. However, the secondary carbocation has a hydrogen that can shift with its pair of electrons to the adjacent positively charged carbon, creating a more stable *tertiary* carbocation.

> **Rearrangement results in a change in the way the atoms are connected.**

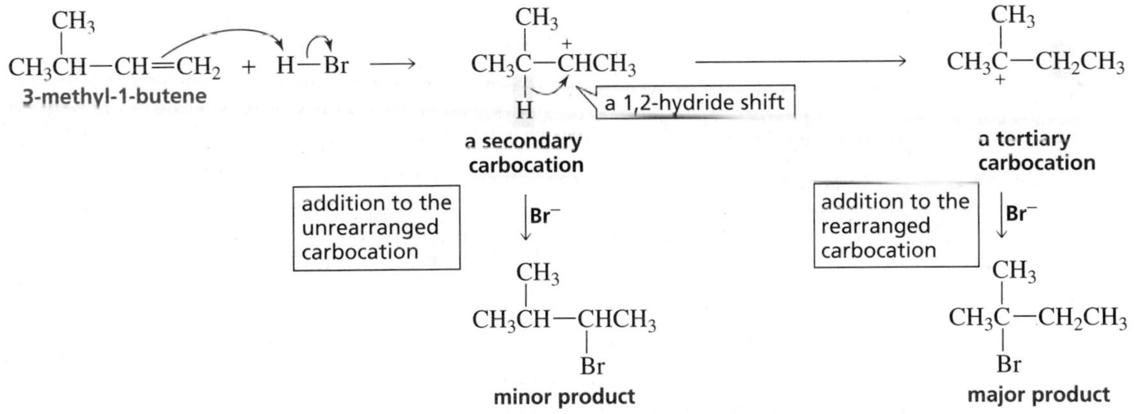

As a result of the **carbocation rearrangement**, two alkyl halides are formed, one from the addition of the nucleophile to the unrearranged carbocation and one from the addition of the nucleophile to the rearranged carbocation. The major product is the one containing the rearrangement. Because it entails the shifting of a hydrogen with its pair of electrons, the rearrangement is called a hydride shift. (Recall that H:⁻ is a hydride ion.) More specifically, it is called a **1,2-hydride shift** because the hydride ion moves from one carbon to an *adjacent* carbon. (Notice that this does not mean it moves from C-1 to C-2.)

After 3,3-dimethyl-1-butene acquires an electrophile to form a *secondary* carbocation, one of the methyl groups, with its pair of electrons, shifts to the adjacent positively charged carbon to form a more stable *tertiary* carbocation. This kind of shift is called a **1,2-methyl shift**. (It should have been called a 1,2-methide shift to match the term 1,2-hydride shift, but, for some reason, it was not.)

$$CH_3C-CH=CH_2 + H-Cl \longrightarrow CH_3C-\overset{+}{C}HCH_3 \longrightarrow CH_3\overset{CH_3}{\underset{CH_3}{\overset{|}{C}}}-CHCH_3$$

3,3-dimethyl-1-butene   a secondary   a 1,2-methyl shift   a tertiary
                               carbocation   carbocation

addition to the | Cl⁻   addition to the | Cl⁻
unrearranged   rearranged
carbocation   carbocation

$$CH_3\overset{CH_3}{\underset{H_3C \quad Cl}{\overset{|}{C}}}-CHCH_3 \qquad\qquad CH_3\overset{CH_3}{\underset{Cl \quad CH_3}{\overset{|}{C}}}-CHCH_3$$

minor product   major product

VCL Alkene Hydration-4

A shift is the movement of a species from one carbon to an adjacent carbon; 1,3-shifts normally do not occur. Furthermore, if a rearrangement does not lead to a more stable carbocation, then the rearrangement does not occur. For example, when a proton adds to 4-methyl-1-pentene, a secondary carbocation is formed. A 1,2-hydride shift would form a different secondary carbocation, but because both carbocations are equally stable, there is no energetic advantage to the shift. Consequently, the rearrangement does not occur, and only one alkyl halide is formed.

the carbocation does
not rearrange

$$CH_3\overset{CH_3}{\overset{|}{C}}HCH_2CH=CH_2 + HBr \longrightarrow CH_3\overset{CH_3}{\overset{|}{C}}HCH_2\overset{+}{C}HCH_3 \;\not\longrightarrow\; CH_3\overset{CH_3}{\overset{|}{C}}HCHCH_2CH_3$$

4-methyl-1-pentene

$$\downarrow Br^-$$

$$CH_3\overset{CH_3}{\underset{Br}{\overset{|}{C}}}HCH_2CHCH_3$$

Carbocation rearrangements also can occur by *ring expansion*, another type of 1,2-shift.

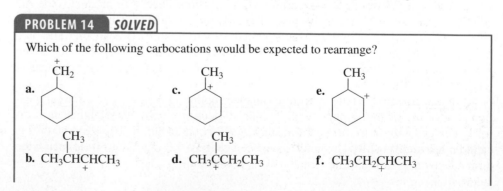

In this example, ring expansion produces a carbocation that is more stable because it is tertiary rather than secondary and because a five-membered ring has less angle strain than a four-membered ring (Section 2.11).

In subsequent chapters, you will study other reactions that form carbocation intermediates. Keep in mind that *whenever a reaction leads to the formation of a carbocation, you must check its structure for the possibility of rearrangement.*

Whenever a reaction forms a carbocation intermediate, check to see if the carbocation will rearrange.

PROBLEM 14 **SOLVED**

Which of the following carbocations would be expected to rearrange?

a. [cyclohexane with $\overset{+}{C}H_2$ substituent]

c. [cyclohexane with CH_3 and + on ring carbon]

e. [cyclohexane with CH_3 substituent and + on adjacent ring carbon]

b. $CH_3\overset{CH_3}{\overset{|}{C}}H\overset{+}{C}HCH_3$

d. $CH_3\overset{CH_3}{\overset{|}{\underset{+}{C}}}CH_2CH_3$

f. $CH_3CH_2\overset{+}{C}HCH_3$

Solution

a. This primary carbocation will rearrange because a 1,2-hydride shift converts it into a tertiary carbocation.

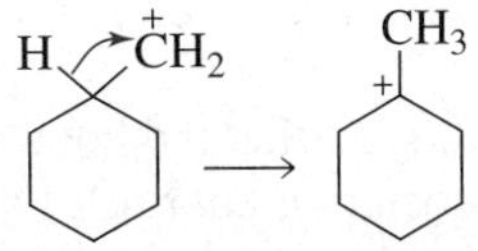

b. This secondary carbocation will rearrange because a 1,2-hydride shift converts it into a tertiary carbocation.

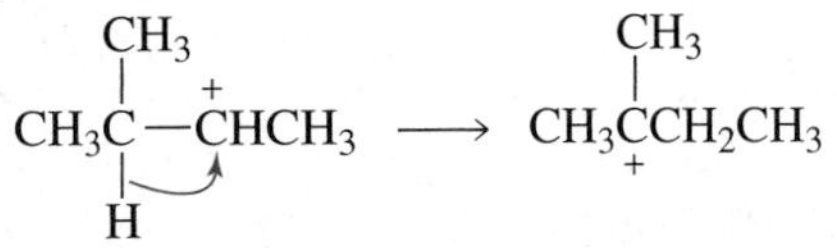

c. This carbocation will not rearrange because it is tertiary and its stability cannot be improved by a carbocation rearrangement.

d. This carbocation will not rearrange because it is tertiary and its stability cannot be improved by a carbocation rearrangement.

e. This secondary carbocation will rearrange because a 1,2-hydride shift converts it into a tertiary carbocation.

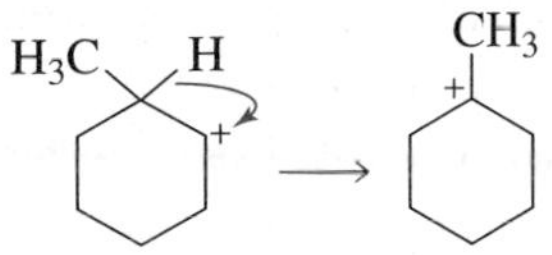

f. This carbocation will not rearrange because it is secondary and a carbocation rearrangement would only yield another secondary carbocation.

PROBLEM 15◆

Give the major product obtained from the reaction of HBr with each of the following:

a. $CH_3CHCH=CH_2$
       |
       CH_3

d. $CH_3CHCH_2CH=CH_2$
          |
          CH_3

b.

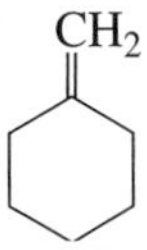

e. $CH_2=CHCCH_3$
         |
         CH_3
(with CH_3 above the central C)

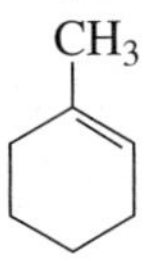

c.

f.

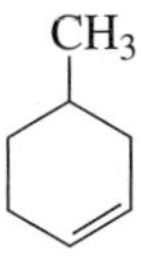

4.7 Addition of a Halogen to an Alkene

The halogens Br_2 and Cl_2 add to alkenes. This may surprise you because it is not immediately apparent that an electrophile—which is necessary to start an electrophilic addition reaction—is present.

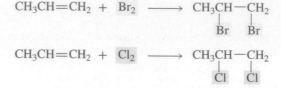

Alkene Halogenation-3

Alkene Halogenation-4

Mechanistic Tutorial:
Addition of halogens to alkenes

Mechanistic Tutorial:
Formation of bromonium ion

The reaction is made possible by the fact that the bond joining the two halogen atoms is relatively weak (see the bond dissociation energies listed in Table 3.2) and, therefore, is easily broken. The mechanism for the addition of bromine (the addition of chlorine is analogous) to an alkene is shown below.

Mechanism for the addition of bromine to an alkene

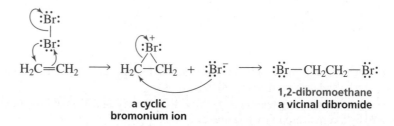

a cyclic
bromonium ion

1,2-dibromoethane
a vicinal dibromide

cyclic bromonium ion
of ethene

cyclic bromonium ion
of *cis*-2-butene

- As the π electrons of the alkene approach a molecule of Br_2, one of the bromine atoms accepts those electrons and releases the electrons of the Br—Br bond to the other bromine atom. Notice that a lone pair on bromine is the nucleophile that attaches to the other sp^2 carbon. Thus, the electrophile and the nucleophile add to the double bond in the same step.

- The intermediate, a cyclic bromonium ion, is unstable because there is still considerable positive charge on what was the sp^2 carbon. (See the potential maps.) Therefore, the cyclic bromonium ion reacts with a nucleophile, the bromide ion. The product is a *vicinal dibromide*. **Vicinal** indicates that the two bromines are on adjacent carbons (*vicinus* is the Latin word for "near").

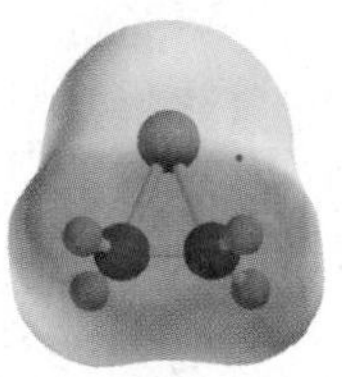

The product of the first step is a cyclic bromonium ion rather than a carbocation, because bromine's electron cloud is close enough to the other sp^2 carbon to engage in bond formation. The cyclic bromonium ion is not stable but is more stable than the carbocation would have been, since all the atoms (except hydrogen) in the bromonium ion have complete octets, whereas the positively charged carbon of the carbocation does not. (To review the octet rule, see Section 1.3.)

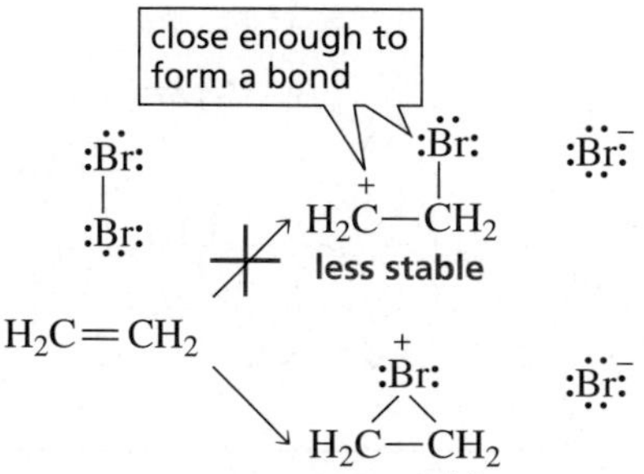

The electrostatic potential maps for the cyclic bromonium ions show that the electron-deficient region (the blue area) encompasses the carbons, even though the formal positive charge is on the bromine.

When Cl_2 adds to an alkene, a cyclic chloronium ion is formed. The final product of the reaction is a vicinal dichloride. Reactions of alkenes with Br_2 or Cl_2 are generally carried out by mixing the alkene and the halogen in an inert solvent, such as dichloromethane (CH_2Cl_2) that readily dissolves both reactants, but does not participate in the reaction.

$$CH_3C{=}CH_2 \ + \ Cl_2 \ \xrightarrow{CH_2Cl_2} \ CH_3CCH_2Cl$$

2-methylpropene

1,2-dichloro-2-methylpropane
a vicinal dichloride

Because a carbocation is not formed when Br_2 or Cl_2 adds to an alkene, carbocation rearrangements do not occur in these reactions.

$$CH_3CHCH{=}CH_2 \ + \ Br_2 \ \xrightarrow{CH_2Cl_2} \ CH_3CHCHCH_2Br$$

the carbon skeleton does not rearrange

3-methyl-1-butene

1,2-dibromo-3-methylbutane
a vicinal dibromide

PROBLEM 16◆

What would have been the product of the preceding reaction if HBr had been used as a reagent instead of Br_2?

PROBLEM 17

a. How does the first step in the reaction of propene with Br_2 differ from the first step in the reaction of propene with HBr?

b. To understand why Br^- attacks a carbon atom of the bromonium ion rather than the positively charged bromine atom, draw the product that would be obtained if Br^- *did* attack the bromine atom.

 Alkene Halogenation-5

The foregoing examples illustrate the way organic reactions are typically written. The reactants are placed to the left of the reaction arrow, and the products are placed to the right of the arrow. Any conditions that should be stipulated, such as the solvent, the temperature, or a necessary catalyst, are written above or below the arrow. Sometimes only the organic (carbon-containing) reagent is placed to the left of the arrow and any other reagents are written above or below the arrow.

$$CH_3CH{=}CHCH_3 \ \xrightarrow[CH_2Cl_2]{Cl_2} \ CH_3CHCHCH_3$$

$$\underset{Cl \ \ Cl}{}$$

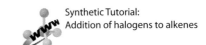

Synthetic Tutorial:
Addition of halogens to alkenes

Although F_2 and I_2 are halogens, they are not used as reagents in electrophilic addition reactions. Fluorine reacts explosively with alkenes, so the electrophilic addition of F_2 is not useful for synthesizing new compounds. The addition of I_2 to an alkene is thermodynamically unfavorable. The vicinal diiodides are unstable at room temperature, decomposing back to the alkene and I_2.

$$CH_3CH{=}CHCH_3 \ + \ I_2 \ \underset{CH_2Cl_2}{\rightleftharpoons} \ CH_3CHCHCH_3$$

$$\underset{I \ \ I}{}$$

unstable

If H_2O rather than CH_2Cl_2 is used as a solvent, the major product of the reaction will be a vicinal halohydrin (or, more specifically, a bromohydrin or a chlorohydrin). A **halohydrin** is an organic molecule that contains both a halogen and an OH group. In a vicinal halohydrin, the halogen and the OH group are bonded to adjacent carbons.

$$CH_3CH=CH_2 + Br_2 \xrightarrow{H_2O} \underset{\substack{OH \\ \text{a bromohydrin} \\ \text{major product}}}{CH_3CHCH_2Br} + \underset{\substack{Br \\ \text{minor product}}}{CH_3CHCH_2Br} + HBr$$

propene

$$\underset{\substack{| \\ CH_3CH=CCH_3 \\ \text{2-methyl-2-butene}}}{\overset{CH_3}{}} + Cl_2 \xrightarrow{H_2O} \underset{\substack{| \\ Cl \ OH \\ \text{a chlorohydrin} \\ \text{major product}}}{\overset{CH_3}{CH_3CHCCH_3}} + \underset{\substack{| \\ Cl \ Cl \\ \text{minor product}}}{\overset{CH_3}{CH_3CHCCH_3}} + HCl$$

The mechanism for halohydrin formation has three steps.

Mechanism for halohydrin formation

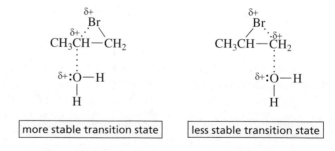

- A cyclic bromonium ion (or chloronium ion) is formed in the first step because Br^+ (or Cl^+) is the only electrophile in the reaction mixture.

- The relatively unstable cyclic bromonium ion rapidly reacts with any nucleophile it bumps into. Two nucleophiles—H_2O and Br^-—are present in the solution, but because H_2O is the solvent, its concentration far exceeds that of Br^-. Consequently, the bromonium ion is more likely to collide with water than with Br^-.

- The protonated halohydrin is a strong acid (Section 1.18), so it loses a proton.

Halohydrin Formation-1

How can we explain the regioselectivity of the preceding addition reaction? In other words, why does the electrophile (Br^+) end up on the sp^2 carbon bonded to the greater number of hydrogens? In both possible transition states for the second step of the reaction, the C—Br bond has broken to a greater extent than the C—O bond has formed. As a result, there is a partial positive charge on the carbon that is attacked by the nucleophile.

| more stable transition state | less stable transition state |

Therefore, the more stable transition state is the one achieved by adding the nucleophile to the more substituted sp^2 carbon—the carbon bonded to the fewer hydrogens—because in that case the partial positive charge is on a secondary carbon rather than on a primary carbon. Thus, this reaction, too, follows the general rule for electrophilic addition reactions: the electrophile (here, Br^+) adds to the sp^2 carbon that is bonded to the greater number of hydrogens.

When nucleophiles other than H_2O are added to the reaction mixture, they, too, change the product of the reaction, just as water changed the product of Br_2 addition from a vicinal dibromide to a vicinal bromohydrin. Because the concentration of the added nucleophile will be greater than the concentration of the halide ion generated from Br_2 or Cl_2, the added nucleophile will be the one most likely to participate in the second step of the reaction.

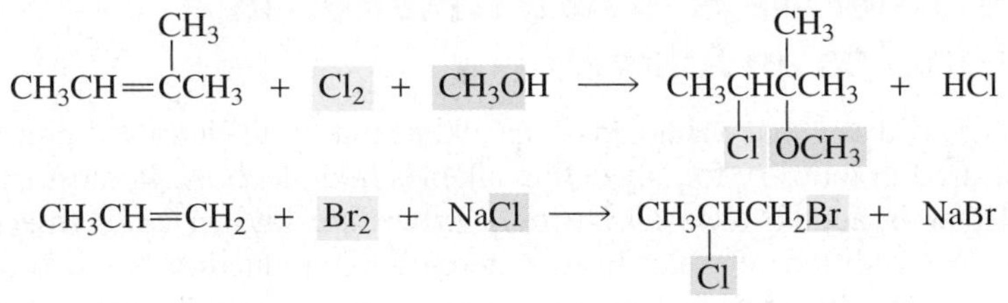

$$CH_3CH=\overset{\underset{|}{CH_3}}{C}CH_3 \ + \ Cl_2 \ + \ CH_3OH \ \longrightarrow \ CH_3CH\underset{\underset{Cl}{|}}{\overset{}{C}}\underset{\underset{OCH_3}{|}}{\overset{}{C}}CH_3 \ + \ HCl$$

$$CH_3CH=CH_2 \ + \ Br_2 \ + \ NaCl \ \longrightarrow \ CH_3\underset{\underset{Cl}{|}}{CH}CH_2Br \ + \ NaBr$$

Remember that ions such as Na^+ and K^+ cannot form covalent bonds, so they do not react with organic compounds. They serve only as counterions to negatively charged species, so their presence generally is ignored in writing chemical equations.

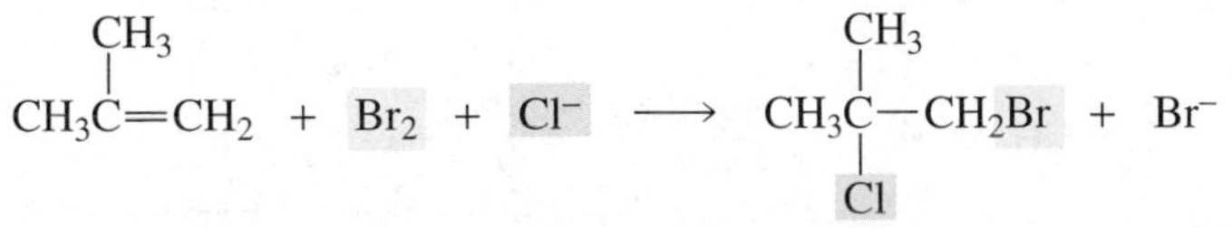

$$CH_3\overset{\underset{|}{CH_3}}{C}=CH_2 \ + \ Br_2 \ + \ Cl^- \ \longrightarrow \ CH_3\underset{\underset{Cl}{|}}{\overset{\underset{|}{CH_3}}{C}}-CH_2Br \ + \ Br^-$$

There are two nucleophiles in each of the following reactions:

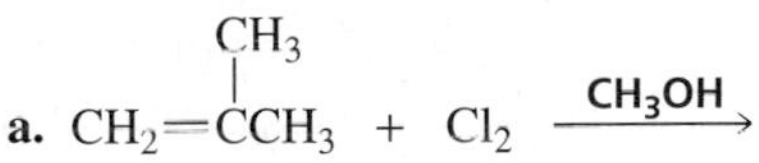

a. $CH_2=\overset{\underset{|}{CH_3}}{C}CH_3 \ + \ Cl_2 \ \xrightarrow{\ CH_3OH\ }$

b. $CH_2-CHCH_3 \ + \ 2\,NaI \ + \ HBr \ \longrightarrow$

c. $CH_3CH=CHCH_3 \ + \ HCl \ \xrightarrow{\ H_2O\ }$

d. $CH_3CH=CHCH_3 \ + \ HBr \ \xrightarrow{\ CH_3OH\ }$

For each reaction, explain why there is a greater concentration of one nucleophile than the other. What will be the major product of each reaction?

Why are Na^+ and K^+ unable to form covalent bonds?

What is the product of the addition of I—Cl to 1-butene? [*Hint*: Chlorine is more electronegative than iodine (Table 1.3).]

Synthetic Tutorial:
Halohydrin reaction

What would be the major product obtained from the reaction of Br_2 with 1-butene if the reaction were carried out in

a. dichloromethane?
b. water?
c. ethyl alcohol?
d. methyl alcohol?

4.8 Oxymercuration–Reduction and Alkoxymercuration–Reduction Are Other Ways to Add Water or an Alcohol to an Alkene

In Section 4.5, you saw that water adds to an alkene if an acid catalyst is present. This is the process used in industry for converting alkenes into alcohols. In most laboratories, however, water is added to an alkene by a procedure known as **oxymercuration–reduction**. The addition of water by oxymercuration–reduction has two advantages over acid-catalyzed addition: it does not require acidic conditions, which are harmful to many organic molecules; and because carbocation intermediates are not formed, carbocation rearrangements do not occur.

In oxymercuration, the alkene is treated with mercuric acetate in aqueous tetrahydrofuran (THF). After that reaction is finished, sodium borohydride is added to the reaction mixture.

$$R-CH=CH_2 \quad \xrightarrow[\text{2. NaBH}_4]{\text{1. Hg(OAc)}_2,\ \text{H}_2\text{O/THF}} \quad R-\underset{\underset{\text{OH}}{|}}{CH}-CH_3$$

(The numbers in front of the reagents above and below the reaction arrow indicate that two reactions are run in sequence; the second reagent is not added until reaction with the first reagent is completed.)

Mechanism for oxymercuration

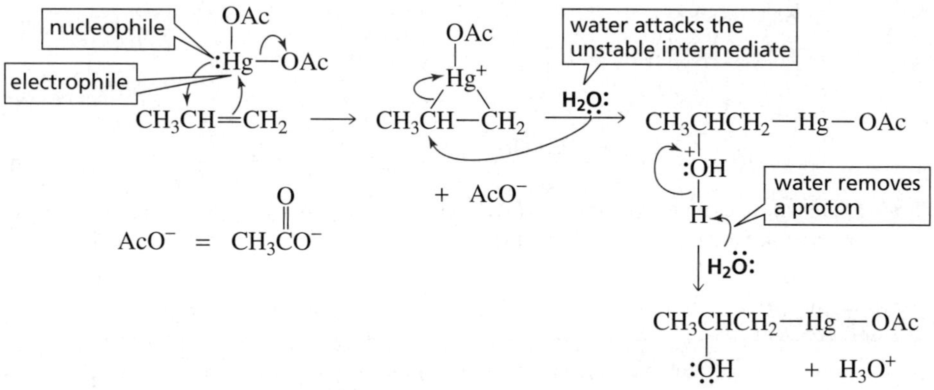

The steps of the oxymercuration mechanism are

- The electrophilic mercury of mercuric acetate adds to the double bond. (Two of mercury's $5d$ electrons are shown.) Because chemists have found that carbocation rearrangements do not occur, we can conclude that the product of the addition reaction is a cyclic mercurinium ion rather than a carbocation. As in the formation of a cyclic bromonium ion (Section 4.7), the electrophile has a lone pair that serves as the nucleophile.

- Water attacks the more substituted carbon of the unstable cyclic mercurinium ion—the carbon bonded to fewer hydrogens—for the same reason that water attacks the more substituted carbon of the cyclic bromonium ion in halohydrin formation (Section 4.7). That is, attacking at the more substituted carbon leads to the more stable transition state.

$$\underset{\text{more stable transition state}}{\overset{\begin{array}{c}\text{OAc}\\ {}^{\delta+}|\\ \text{Hg}\\ {}^{\delta+}\!\cdot\!\cdot\!\diagup\ \diagdown\\ \text{CH}_3\overset{}{\text{CH}}-\text{CH}_2\\ \vdots\\ {}^{\delta+}\!\ddot{\text{O}}-\text{H}\\ |\\ \text{H}\end{array}}{}} \qquad \underset{\text{less stable transition state}}{\overset{\begin{array}{c}\text{OAc}\\ {}^{\delta+}|\\ \text{Hg}\\ \diagup\ \diagdown\!\cdot\!\cdot^{+\delta}\\ \text{CH}_3\overset{}{\text{CH}}-\text{CH}_2\\ \vdots\\ {}^{\delta+}\!\ddot{\text{O}}-\text{H}\\ |\\ \text{H}\end{array}}{}}$$

- The protonated alcohol loses a proton because the pH of the solution is greater than the pK_a of the protonated alcohol group (Section 1.24).

In the second of the reactions, sodium borohydride ($NaBH_4$) converts the C—Hg bond into a C—H bond. A reaction that increases the number of C—H bonds or decreases the number of C—O, C—N, or C—X bonds in a compound (X denotes a halogen) is a **reduction reaction**. Consequently, the reaction with sodium borohydride is a reduction reaction. The mechanism of this reaction is not fully understood.

> **Reduction increases the number of C—H bonds or decreases the number of C—O, C—N, or C—X bonds.**

$$\underset{\underset{\text{OH}}{|}}{\text{CH}_3\text{CHCH}_2}-\text{Hg}-\text{OAc} \xrightarrow{\text{NaBH}_4} \underset{\underset{\text{OH}}{|}}{\text{CH}_3\text{CHCH}_3} + \text{Hg} + \text{AcO}^-$$

The overall reaction (oxymercuration–reduction) forms the same product that would be obtained from the acid-catalyzed addition of water: the hydrogen adds to the sp^2 carbon bonded to the greater number of hydrogens, and OH adds to the other sp^2 carbon.

We have seen that alkenes react with alcohols in the presence of an acid catalyst to form ethers (Section 4.5). Just as there are advantages to adding water in the presence of mercuric acetate rather than in the presence of a strong acid, there are also advantages to adding an alcohol in the presence of mercuric acetate. [Mercuric trifluoroacetate, $Hg(O_2CCF_3)_2$, works even better.] This reaction is called **alkoxymercuration–reduction**.

$$\underset{\textbf{1-methylcyclohexene}}{\overset{\text{CH}_3}{\bigcirc}} \xrightarrow[\textbf{2. NaBH}_4]{\textbf{1. Hg(O}_2\textbf{CCF}_3)_2\textbf{, CH}_3\textbf{OH}} \underset{\substack{\textbf{1-methoxy-1-methylcyclohexane}\\ \textbf{an ether}}}{\overset{\text{H}_3\text{C}\quad\text{OCH}_3}{\bigcirc}}$$

The mechanisms for oxymercuration and alkoxymercuration are essentially identical; the only difference is that water is the nucleophile in oxymercuration and an alcohol is the nucleophile in alkoxymercuration. Therefore, the product of oxymercuration–reduction is an alcohol, whereas the product of alkoxymercuration–reduction is an ether.

PROBLEM 22

How could the following compounds be synthesized from an alkene?

a. ![cyclopentane with OCH₂CH₃ substituent] —OCH$_2$CH$_3$

b. ![cyclohexane with CH₃ and OH substituents] CH$_3$ OH

c. $\underset{\underset{\text{OCH}_2\text{CH}_3}{|}}{\text{CH}_3\text{CHCH}_2\text{CH}_3}$

d. $\underset{\underset{\text{OCH}_3}{|}}{\overset{\overset{\text{CH}_3}{|}}{\text{CH}_3\text{CCH}_2\text{CH}_3}}$

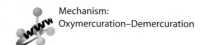

Mechanism:
Oxymercuration–Demercuration

PROBLEM 23

How could the following compounds be synthesized from 3-methyl-1-butene?

a. 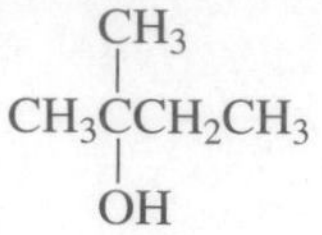 $CH_3CCH_2CH_3$ with CH_3 and OH groups

b. 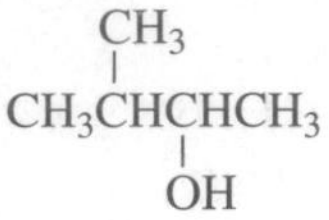 $CH_3CHCHCH_3$ with CH_3 and OH groups

4.9 Addition of a Peroxyacid to an Alkene

An alkene can be converted to an *epoxide* by a peroxyacid. An **epoxide** is an ether in which the oxygen atom is incorporated into a three-membered ring; a **peroxyacid** is a carboxylic acid with an extra oxygen atom. The overall reaction amounts to the transfer of an oxygen from the peroxyacid to the alkene; it is an oxidation reaction. An **oxidation reaction** increases the number of C—O, C—N, or C—X bonds in a compound (X denotes a halogen), or decreases the number of C—H bonds.

> **Oxidation decreases the number of C—H bonds or increases the number of C—O, C—N, or C—X bonds.**

$$RCH{=}CH_2 \ + \ \underset{\textbf{a peroxyacid}}{RCOOH} \ \longrightarrow \ \underset{\textbf{an epoxide}}{RCH{-}CH_2} \ + \ \underset{\textbf{a carboxylic acid}}{RCOH}$$

an alkene (two oxygens) (one oxygen)

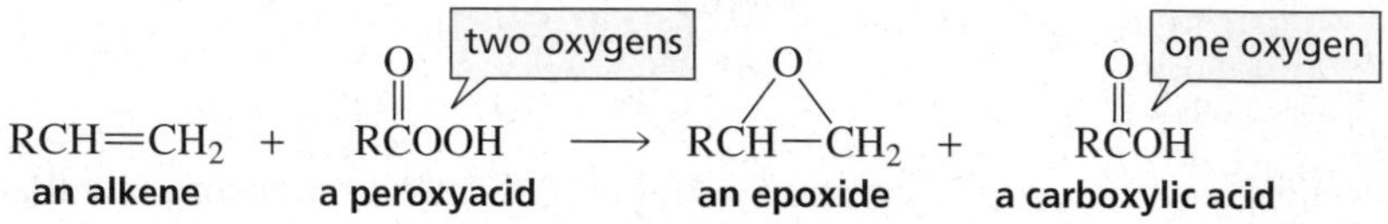

Recall that an O—O bond is weak and, therefore, easily broken (Section 1.4).

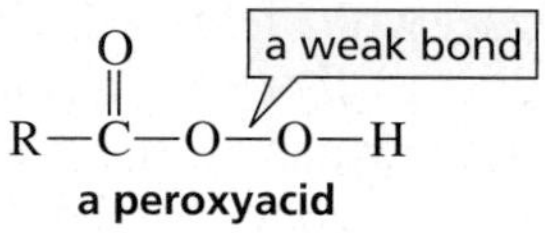

$$R{-}\overset{O}{\underset{}{C}}{-}O{-}O{-}H$$

a weak bond

a peroxyacid

Mechanism for epoxidation of an alkene

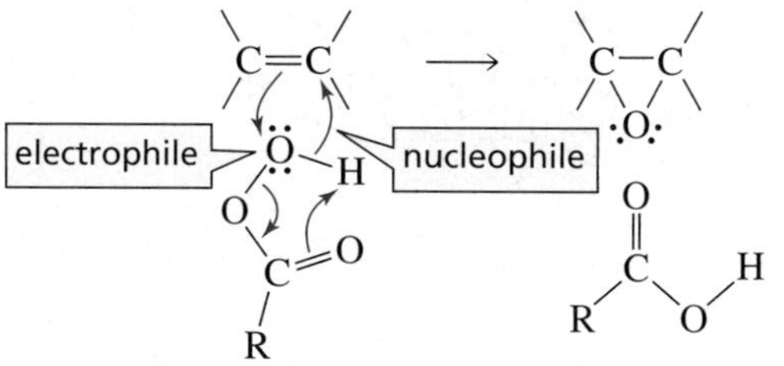

- The oxygen atom of the OH group of the peroxyacid is electron deficient and thus it is an electrophile. It accepts a pair of electrons from the π bond of the alkene, which causes the weak O—O bond to break. The electrons from the O—O bond are delocalized (Section 1.22). The electrons left behind as the O—H bond breaks add to the other sp^2 carbon of the alkene.

The mechanism for formation of the epoxide shows that it is a *concerted* reaction. A **concerted reaction** is a reaction in which all the bond-making and bond-breaking processes occur in a single step (all the events occur "in concert").

> **In a concerted reaction, all the bond-making and bond-breaking processes occur in the same step.**

The mechanism for the addition of oxygen to a double bond to form an epoxide is analogous to the mechanism we just saw for formation of a cyclic bromonium ion (Section 4.7) or a cyclic mercurinium ion (Section 4.8).

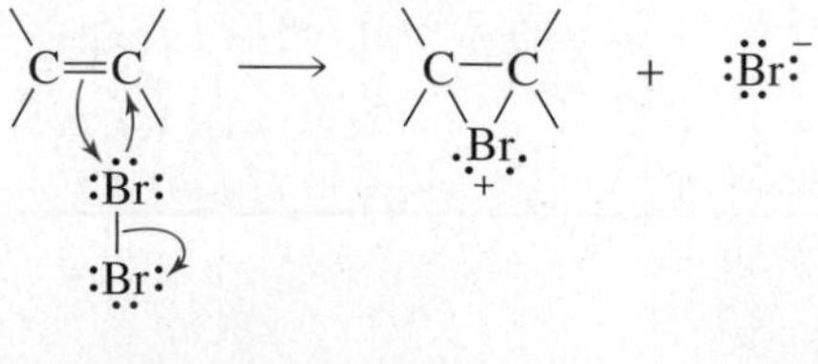

In all three reactions, a lone pair from the electrophile is the nucleophile that adds to the other sp^2 carbon. However, unlike the cyclic bromonium or mercurinium ions, the epoxide is stable enough to isolate because none of the ring atoms has a positive charge.

The common name of an epoxide is obtained by adding "oxide" to the common name of the alkene, assuming that the oxygen atom is where the π bond of an alkene would be. The simplest epoxide is ethylene oxide.

$$H_2C=CH_2$$
ethylene

$$H_2C-CH_2$$ (O)
ethylene oxide

$$H_2C=CHCH_3$$
propylene

$$H_2C-CHCH_3$$ (O)
propylene oxide

There are two systematic ways to name epoxides. One method calls the three-membered oxygen-containing ring "oxirane," and gives oxygen the 1-position in the ring. Thus, 2-ethyloxirane has an ethyl substituent at the 2-position of the oxirane ring. Alternatively, an epoxide can be named as an alkane, with an "epoxy" prefix that identifies the carbons to which the oxygen is attached.

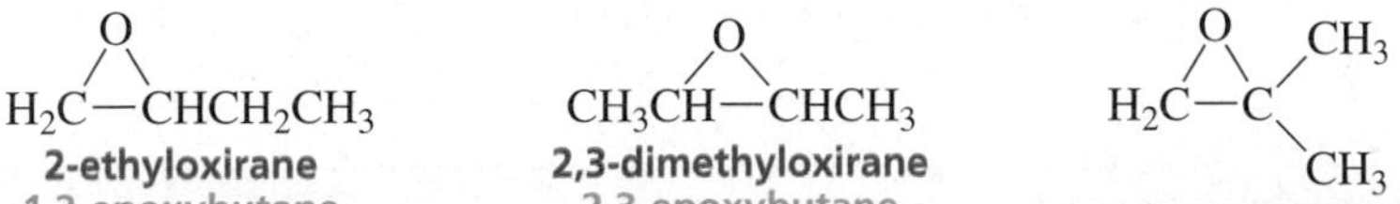

$$H_2C-CHCH_2CH_3$$ (O)
2-ethyloxirane
1,2-epoxybutane

$$CH_3CH-CHCH_3$$ (O)
2,3-dimethyloxirane
2,3-epoxybutane

$$H_2C-C\begin{smallmatrix}CH_3\\CH_3\end{smallmatrix}$$ (O)
2,2-dimethyloxirane
1,2-epoxy-2-methylpropane

PROBLEM 24◆

Draw the structure of the following compounds:

a. 2-propyloxirane

b. cyclohexene oxide

c. 2,2,3,3-tetramethyloxirane

d. 2,3-epoxy-2-methylpentane

PROBLEM 25◆

What alkene would you treat with a peroxyacid in order to obtain each of the following epoxides?

a.

b.  $H_2C-CHCH_2CH_3$ (O)

VCL **Epoxidation-1**

PROBLEM-SOLVING STRATEGY

Proposing a Mechanism

A **carbene** is an unusual carbon-containing species: it has a carbon with a lone pair of electrons and an empty orbital. The empty orbital makes the carbene highly reactive. The simplest carbene, methylene (:CH$_2$), is generated by heating diazomethane. Given this information, propose a mechanism for the following reaction:

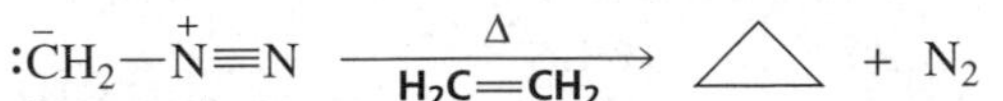

$$:\overset{-}{C}H_2-\overset{+}{N}\equiv N \xrightarrow[H_2C=CH_2]{\Delta} \triangle + N_2$$
diazomethane

The information provided is all you need. First, because you know the structure of methylene, you can see that it can be generated by breaking the C—N bond of diazomethane. Second, because methylene has an empty orbital, it is an electrophile and, therefore, will react with ethene (a nucleophile). Now, the question is, what nucleophile reacts with the other sp^2 carbon of the alkene? Because you know that cyclopropane (a three-carbon compound) is the product of the reaction, you also know that the nucleophile must be the lone-pair electrons of methylene.

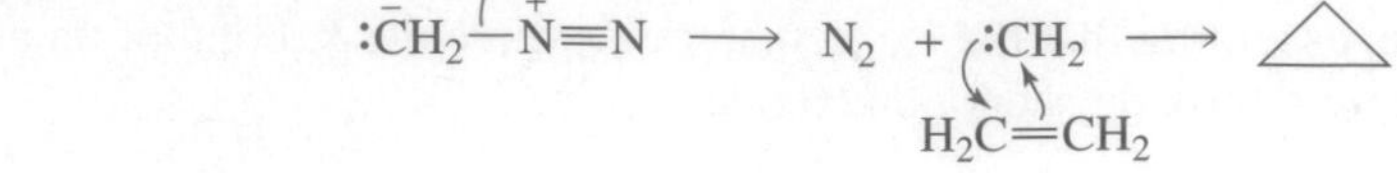

:CH2—N≡N → N2 + :CH2 →
H2C=CH2

(*Note:* Diazomethane is a gas that must be handled with great care because it is both explosive and toxic.)

Now continue on to Problem 26.

PROBLEM 26

Propose a mechanism for the following reaction:

CH3 CH3
CH2=CCH2CHCH2OH
HCl
H3C
H3C
O
CH3

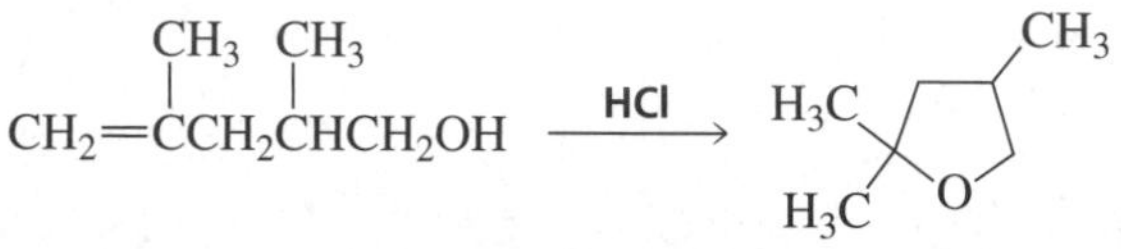

4.10 Addition of Borane to an Alkene: Hydroboration–Oxidation

An atom or a molecule does not have to be positively charged or even partially positively charged to be an electrophile. Borane (BH_3), a neutral molecule, is an electrophile because its boron atom has only six shared electrons in its valence shell rather than a complete octet. We have seen that boron forms bonds using sp^2 orbitals (Chapter 1, Problem 33b); thus, like a methyl cation, it has an empty *p* orbital that can accept electrons.

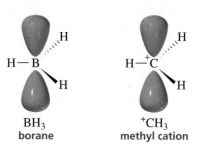

H—B
H
H
H—C
H
H

BH_3
borane

$^+CH_3$
methyl cation

BIOGRAPHY

Herbert Charles Brown (1921–2004) *was born in London, where his parents had emigrated from the Ukraine. When Brown was a toddler, his father, a cabinetmaker, moved the family to Chicago, where he opened a hardware store. When Brown was 14, his father died, and Brown dropped out of school for a short time to help run the family business. After receiving both a B.S. and a Ph.D. from the University of Chicago, he started his career as a professor at Wayne State University, and then was a professor of chemistry at Purdue University from 1947 until his retirement in 1978. For his studies on boron-containing organic compounds, he shared the 1979 Nobel Prize in chemistry with G. Wittig (page 822).*

Because boron readily accepts a pair of electrons, an alkene can undergo an electrophilic addition reaction, with borane serving as the electrophile. When the addition reaction is over, an aqueous solution of sodium hydroxide and hydrogen peroxide is added to the reaction mixture. The product of this reaction is an alcohol. The overall process—the addition of borane to an alkene, followed by reaction with hydroxide ion and hydrogen peroxide—is called **hydroboration–oxidation**. It was first reported by H. C. Brown in 1959.

hydroboration–oxidation

CH2=CH2
1. BH3/THF
2. HO−, H2O2, H2O
CH2—CH2
H OH

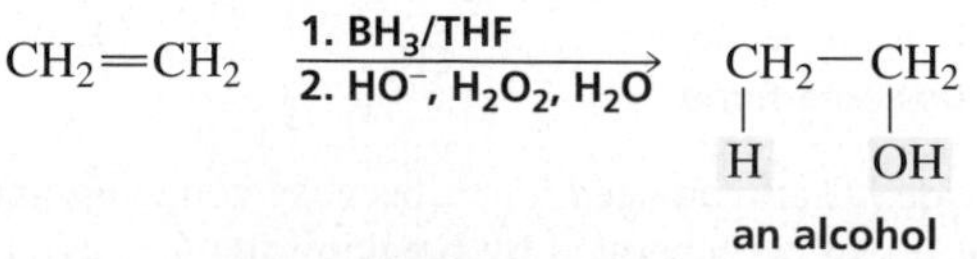

an alcohol

As the following reactions show, the alcohol formed from the hydroboration–oxidation of an alkene has the H and OH groups switched in comparison with the alcohol formed from the acid-catalyzed addition of water (Section 4.5). However, in both reactions *the electrophile adds to the* sp^2 *carbon that is bonded to the greater number of hydrogens.* In the acid-catalyzed addition of water, H^+ is the electrophile and H_2O is the nucleophile, whereas in hydroboration–oxidation, as we will see, BH_3 is the electrophile (with HO subsequently taking its place) and H^- is the nucleophile.

In the acid-catalyzed addition of water, H^+ is the electrophile; in hydroboration–oxidation, H^- is the nucleophile.

$$CH_3CH=CH_2 \xrightarrow[\text{H}_2\text{O}]{\text{H}_2\text{SO}_4} CH_3CHCH_3$$

propene $\underset{\text{OH}}{|}$

2-propanol

$$CH_3CH=CH_2 \xrightarrow[\text{2. HO}^-\text{, H}_2\text{O}_2\text{, H}_2\text{O}]{\text{1. BH}_3\text{/THF}} CH_3CH_2CH_2OH$$

propene   **1-propanol**

BORANE AND DIBORANE

Borane exists primarily as a colorless gas called diborane. Diborane is a **dimer**—a molecule formed by the joining of two identical molecules. Because boron is surrounded by only six electrons, it has a strong tendency to acquire an additional electron pair. In the dimer, therefore, two boron atoms share the two electrons in a hydrogen–boron bond by means of unusual half-bonds. These

hydrogen–boron bonds are shown as dotted lines to indicate that they consist of fewer than the normal two electrons.

borane   **diborane**

Because diborane (B_2H_6), the source of borane, is a flammable, toxic, explosive gas, a solution of borane—prepared by dissolving diborane in an ether such as THF—is a more convenient and less dangerous reagent. One of the ether oxygen's lone pairs satisfies boron's requirement for an additional two electrons, producing a borane–THF complex that is the actual source of BH_3 for hydroboration.

tetrahydrofuran (THF)

diborane

Student Tutorial:
Borane–THF complex

To understand why the hydroboration–oxidation of propene forms 1-propanol, we must look at the mechanism of the reaction. As the electrophilic boron accepts the π electrons from the nucleophilic alkene and forms a bond with one sp^2 carbon, it donates a hydride ion to the other sp^2 carbon. The addition of borane to an alkene is another example of a concerted reaction (Section 4.9).

$$CH_3CH=CH_2 \longrightarrow CH_3CH-CH_2$$

nucleophile $H-BH_2$ $\underset{H}{|}$ $\underset{BH_2}{|}$

electrophile   **an alkylborane**

Because both the electrophilic boron and the nucleophilic hydride ion are added to the alkene in one step, an intermediate is not formed.

Boron adds to the sp^2 carbon that is bonded to the greater number of hydrogens. The electrophiles in the other addition reactions we have looked at (such as H^+) also added to the sp^2 carbon bonded to the greater number of hydrogens. We discovered that they did so in order to form the most stable carbocation intermediate. But no intermediate is formed in a concerted reaction, so how can we explain the regioselectivity of the addition of boron? Why does boron add preferentially to the sp^2 carbon bonded to the greater number of hydrogens?

If we examine the two possible transition states for the addition of borane, we see that the C—B bond has formed to a greater extent than the C—H bond. Consequently, the sp^2 carbon that does not become attached to boron has a partial positive charge. The partial positive charge is on a secondary carbon if boron adds to the sp^2 carbon bonded to the greater number of hydrogens. The partial positive charge is on a primary carbon if boron adds to the other sp^2 carbon. So, even though a carbocation intermediate is not formed, a carbocation-like transition state is formed. Thus, the addition of borane and the addition of an electrophile such as H^+ take place at the same sp^2 carbon for the same reason: to form the more stable transition state.

addition of BH₃		addition of HBr	

$$CH_3-\overset{\overset{H}{\underset{\delta+}{|}}}{C}\!\!=\!\!=\!\!=\!\!\overset{\overset{H}{|}}{C}-H \qquad CH_3-\overset{\overset{H}{|}}{C}\!\!=\!\!=\!\!=\!\!\overset{\overset{H}{\underset{\delta+}{|}}}{C}-H \qquad\qquad CH_3-\overset{\overset{H}{\underset{\delta+}{|}}}{C}\!\!=\!\!=\!\!=\!\!\overset{\overset{H}{|}}{C}-H \qquad CH_3-\overset{\overset{H}{|}}{C}\!\!=\!\!=\!\!=\!\!\overset{\overset{H}{\underset{\delta+}{|}}}{C}-H$$

$$\underset{\delta-}{H\text{-----}BH_2} \qquad\qquad \underset{\delta-}{H_2B\text{-----}H} \qquad\qquad\qquad \underset{\delta-}{Br\text{------}H} \qquad\qquad\quad \underset{\delta-}{H\text{-----}Br}$$

more stable transition state	**less stable transition state**	**more stable transition state**	**less stable transition state**

The alkylborane that is formed in the first step of the reaction reacts with another molecule of alkene to form a dialkylborane, which then reacts with yet another molecule of alkene to form a trialkylborane. In each of these reactions, boron adds to the sp^2 carbon bonded to the greater number of hydrogens and the hydride ion adds to the other sp^2 carbon.

$$CH_3CH\!\!=\!\!CH_2 \longrightarrow CH_3\overset{\;}{CH}-CH_2-BHR$$
$$\underset{\textbf{an alkylborane}}{H\frown BHR} \qquad\qquad\qquad \underset{\textbf{a dialkylborane}}{\overset{|}{H}}$$

$$CH_3CH\!\!=\!\!CH_2 \longrightarrow CH_3\overset{\;}{CH}-CH_2-BR_2$$
$$\underset{\textbf{a dialkylborane}}{H\frown BR_2} \qquad\qquad\qquad \underset{\textbf{a trialkylborane}}{\overset{|}{H}}$$

The alkylborane (RBH_2) is a bulkier molecule than BH_3 because R is a larger substituent than H. The dialkylborane with two R groups (R_2BH) is even bulkier than the alkylborane. Thus, there are now two reasons for the alkylborane and the dialkylborane to add to the sp^2 carbon that is bonded to the greater number of hydrogens: first, to achieve the *most stable carbocation-like transition state*, and second, because there is *more room* at this carbon for the bulky group to attach itself. **Steric effects** are space-filling effects (recall steric strain in Section 2.10). **Steric hindrance** is a steric effect caused by bulky groups at the site of the reaction that make it difficult for the reactants to approach each other. Steric hindrance associated with the alkylborane—and particularly with the dialkylborane—causes the electrophile to add to the sp^2 carbon that is bonded to the greater number of hydrogens because that is the least sterically hindered of the two sp^2 carbons. Therefore, in each of the three successive additions to the alkene, boron adds to the sp^2 carbon that is bonded to the greater number of hydrogens and H^- adds to the other sp^2 carbon.

When the hydroboration reaction is over, aqueous sodium hydroxide and hydrogen peroxide (HOOH) are added to the reaction mixture to bring about replacement of boron by an OH group. Because replacing boron by an OH group is an *oxidation reaction* (the number of C—O bonds has increased), the overall reaction is called hydroboration–oxidation.

$$\underset{\underset{R}{|}}{\overset{\overset{R}{|}}{R-B}} \xrightarrow{\ \text{HO}^-,\ \text{H}_2\text{O}_2,\ \text{H}_2\text{O}\ } 3\ \boxed{R-OH} + {}^-\text{B(OH)}_4$$

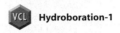 **Hydroboration-1**

Mechanism for the oxidation reaction

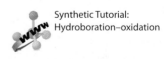

The mechanism for the oxidation reaction shows that:

- A hydrogen peroxide ion (a nucleophile) adds to R_3B (an electrophile).
- A 1,2-alkyl shift displaces a hydroxide ion. These first two steps are repeated two more times, so the three R groups all become OR groups.
- Hydroxide ion (a nucleophile) adds to $(RO)_3B$ (an electrophile).
- An alkoxide ion is eliminated.
- Protonation of the alkoxide ion forms the alcohol. The last three steps are repeated two more times, so all three alkoxide ions are expelled from boron and three molecules of alcohol are formed.

We have seen that, in the overall hydroboration–oxidation reaction, 1 mol of BH_3 reacts with 3 mol of alkene to form 3 mol of alcohol. The OH ends up on the sp^2 carbon that was bonded to the greater number of hydrogens because it replaces boron, which was the original electrophile in the reaction.

$$3\ CH_3CH{=}CH_2\ +\ BH_3 \xrightarrow{\ \textbf{THF}\ } (CH_3CH_2CH_2)_3B \xrightarrow[\textbf{H}_2\textbf{O}]{\textbf{HO}^-,\ \textbf{H}_2\textbf{O}_2} 3\ CH_3CH_2CH_2OH\ +\ {}^-B(OH)_4$$

Because carbocation intermediates are not formed in the hydroboration reaction, carbocation rearrangements do not occur.

$$\underset{\textbf{3-methyl-1-butene}}{\overset{\overset{CH_3}{|}}{CH_3CHCH{=}CH_2}} \xrightarrow[\textbf{2. HO}^-,\ \textbf{H}_2\textbf{O}_2,\ \textbf{H}_2\textbf{O}]{\textbf{1. BH}_3/\textbf{THF}} \underset{\textbf{3-methyl-1-butanol}}{\overset{\overset{CH_3}{|}}{CH_3CHCH_2CH_2OH}}$$

Synthetic Tutorial:
Hydroboration–oxidation

$$\underset{\underset{\textbf{3,3-dimethyl-1-butene}}{\underset{CH_3}{|}}}{\overset{\overset{CH_3}{|}}{CH_3CCH{=}CH_2}} \xrightarrow[\textbf{2. HO}^-,\ \textbf{H}_2\textbf{O}_2,\ \textbf{H}_2\textbf{O}]{\textbf{1. BH}_3/\textbf{THF}} \underset{\underset{\textbf{3,3-dimethyl-1-butanol}}{\underset{CH_3}{|}}}{\overset{\overset{CH_3}{|}}{CH_3CCH_2CH_2OH}}$$

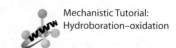 Mechanistic Tutorial:
Hydroboration–oxidation

PROBLEM 27◆

How many mol of BH_3 are needed to react with 2 mol of 1-pentene?

PROBLEM 28◆

What product would be obtained from hydroboration–oxidation of the following alkenes?

a. 2-methyl-2-butene　　　　**b.** 1-methylcyclohexene

4.11 Addition of Hydrogen to an Alkene • The Relative Stabilities of Alkenes

In the presence of a metal catalyst such as platinum, palladium, or nickel, hydrogen (H_2) adds to the double bond of an alkene to form an alkane. The reaction is a *reduction* reaction because there are more C—H bonds in the product than in the reactant (Section 4.8). Without the catalyst, the energy barrier to the reaction is enormous because the H—H bond is so strong (Table 3.2). The catalyst decreases the energy of activation by weakening the H—H bond. Platinum and palladium are used in a finely divided state adsorbed on charcoal (Pt/C, Pd/C). Often PtO_2, known as Adams catalyst, is used as the platinum catalyst.

$$CH_3CH{=}CHCH_3 \; + \; H_2 \; \xrightarrow{\text{Pt/C}} \; CH_3CH_2CH_2CH_3$$

$$\underset{\textbf{2-butene}}{} \qquad\qquad\qquad\qquad \underset{\textbf{butane}}{}$$

$$\overset{\displaystyle CH_3}{\underset{\textbf{2-methylpropene}}{CH_3C{=}CH_2}} \; + \; H_2 \; \xrightarrow{\text{Pd/C}} \; \overset{\displaystyle CH_3}{\underset{\textbf{2-methylpropane}}{CH_3CHCH_3}}$$

cyclohexene + H_2 $\xrightarrow{\text{Ni}}$ cyclohexane

The addition of hydrogen is called **hydrogenation**. Because hydrogenation reactions require a catalyst, they are **catalytic hydrogenations**. The metal catalysts described above are insoluble in the reaction mixture and therefore are classified as **heterogeneous catalysts**. A heterogeneous catalyst can easily be separated from the reaction mixture by filtration. It can then be recycled, which is fortunate, since metal catalysts tend to be expensive.

The details of the mechanism of catalytic hydrogenation are not completely understood. We know that hydrogen is adsorbed on the surface of the metal and that the alkene complexes with the metal by overlapping its own *p* orbitals with vacant orbitals of the metal. All the bond-breaking and bond-forming events occur on the surface of the metal. As the alkane product forms, it diffuses away from the metal surface (Figure 4.5).

Although it is not an accurate picture of what actually happens, we can think of the reaction as occurring in the following way: the H—H bond breaks and the π bond breaks, and the hydrogen radicals add to the carbon radicals.

$$CH_3CH{=}CHCH_3 \; \longrightarrow \; CH_3\overset{\centerdot}{C}H{-}\overset{\centerdot}{C}HCH_3 \; \longrightarrow \; \underset{\displaystyle H \qquad H}{CH_3CH{-}CHCH_3}$$

$$\underset{\displaystyle H{-}H}{} \qquad\qquad\qquad \underset{\displaystyle H\centerdot \quad \centerdot H}{}$$

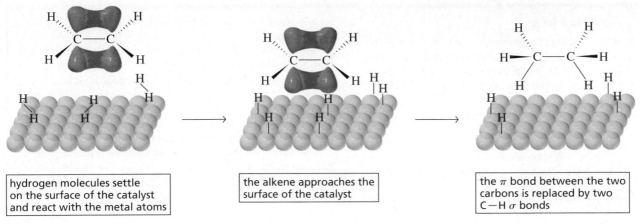

hydrogen molecules settle on the surface of the catalyst and react with the metal atoms

the alkene approaches the surface of the catalyst

the π bond between the two carbons is replaced by two C—H σ bonds

▲ **Figure 4.5**
Catalytic hydrogenation of an alkene.

The heat released in a hydrogenation reaction is called the **heat of hydrogenation**. It is customary to give it a positive value. Hydrogenation reactions, however, are exothermic (they have negative $\Delta H°$ values), so the heat of hydrogenation is the positive value of the $\Delta H°$ of the reaction.

The most stable alkene has the smallest heat of hydrogenation.

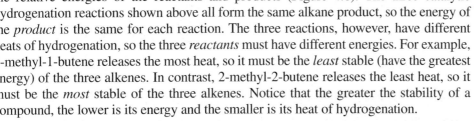

		$\Delta H°$	
	heat of hydrogenation	kcal/mol	kJ/mol
2-methyl-2-butene	26.9 kcal/mol	−26.9	−113
2-methyl-1-butene	28.5 kcal/mol	−28.5	−119
3-methyl-1-butene	30.3 kcal/mol	−30.3	−127

Because we do not know the precise mechanism of a hydrogenation reaction, we cannot draw a reaction coordinate diagram for it. We can, however, draw a diagram showing the relative energies of the reactants and products (Figure 4.6). The three catalytic hydrogenation reactions shown above all form the same alkane product, so the energy of the *product* is the same for each reaction. The three reactions, however, have different heats of hydrogenation, so the three *reactants* must have different energies. For example, 3-methyl-1-butene releases the most heat, so it must be the *least* stable (have the greatest energy) of the three alkenes. In contrast, 2-methyl-2-butene releases the least heat, so it must be the *most* stable of the three alkenes. Notice that the greater the stability of a compound, the lower is its energy and the smaller is its heat of hydrogenation.

Student Tutorial:
Catalytic hydrogenation of ethylene

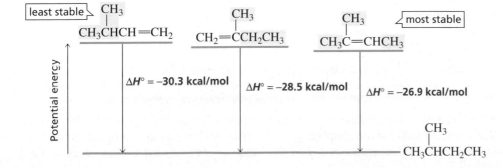

◄ **Figure 4.6**
The relative energies (stabilities) of three alkenes that can be catalytically hydrogenated to 2-methylbutane.

TRANS FATS

Fats are solids at room temperature, whereas oils are liquids at that temperature because they contain more carbon–carbon double bonds (Section 26.1). Oils are said to be polyunsaturated since they have many double bonds.

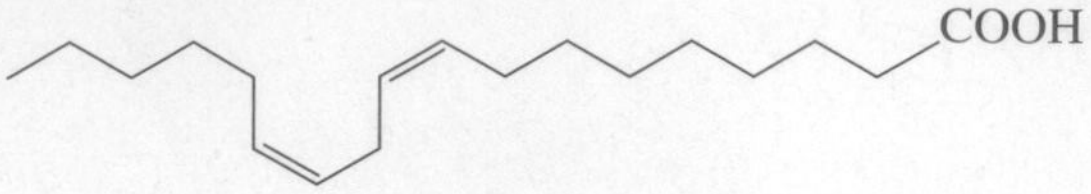

COOH

linoleic acid
an 18-carbon fatty acid with two cis double bonds

Some or all of the double bonds in oils can be reduced by catalytic hydrogenation. For example, margarine and shortening are prepared by hydrogenating vegetable oils, such as soybean

oil and safflower oil, until they have the desired creamy, solid consistency.

All the double bonds in naturally occurring fats and oils have the cis configuration. The heat used in the hydrogenation process breaks the π bond of these double bonds. Sometimes, instead of becoming hydrogenated, a double bond reforms; if the sigma bond rotated while the π bond was broken, the double bond may reform in the trans configuration (Section 3.4), forming what is known as a trans fat.

One reason trans fats are a health concern is that they do not have the same shape as natural cis fats but are able to take the place of cis fats in cell membranes, thereby affecting the ability of the membranes to control the flow of molecules into and out of our cells.

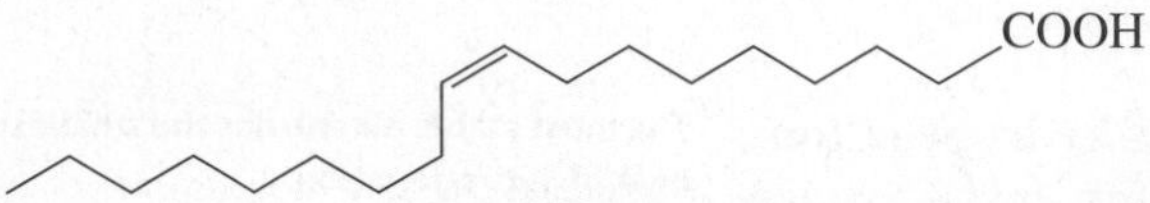

COOH

oleic acid
an 18-carbon fatty acid with one cis double bond
before being heated

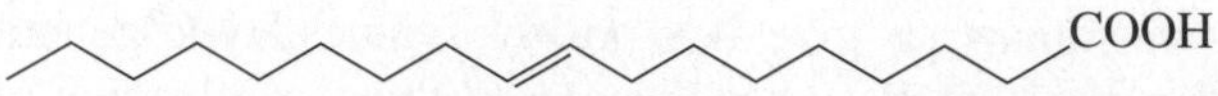

COOH

an 18-carbon fatty acid with one trans double bond
after being heated

PROBLEM-SOLVING STRATEGY

Choosing the Reactant for a Synthesis

What alkene would you start with if you wanted to synthesize methylcyclohexane?

You need to choose an alkene with the same number of carbons, attached in the same way, as those in the desired product. Several alkenes could be used for this synthesis, because the double bond can be located anywhere in the molecule.

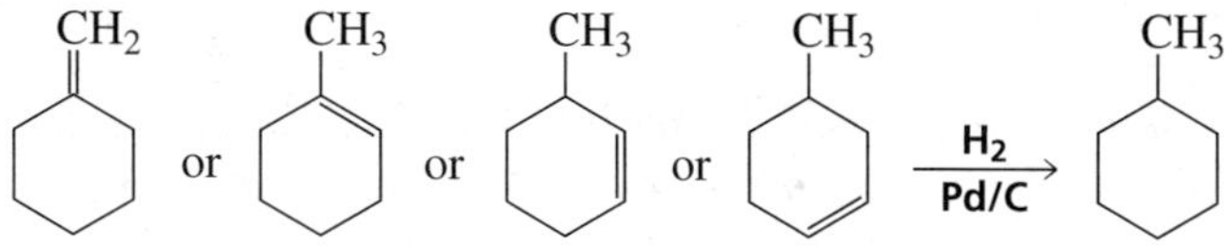

$$\text{CH}_2 \quad \text{or} \quad \text{CH}_3 \quad \text{or} \quad \text{CH}_3 \quad \text{or} \quad \text{CH}_3 \quad \xrightarrow[\text{Pd/C}]{\text{H}_2} \quad \text{CH}_3$$

Now continue on to Problem 29.

PROBLEM 29

What alkene would you start with if you wanted to synthesize:

a. pentane?　　**b.** methylcyclopentane?

PROBLEM 30

How many different alkenes can be hydrogenated to form:

a. butane?　　**b.** methylcyclohexane　　**c.** hexane?

If you look at the structures of the three alkene reactants in Figure 4.6, you will see that the most stable alkene has two alkyl substituents bonded to one of the sp^2 carbons and one alkyl substituent bonded to the other sp^2 carbon, for a total of three alkyl substituents (three methyl groups) bonded to its two sp^2 carbons. The alkene of intermediate stability has a total of two alkyl substituents (a methyl group and an ethyl group) bonded to its sp^2 carbons, and the least stable of the three alkenes has only one alkyl substituent (an isopropyl group) bonded to an sp^2 carbon. Thus, it is apparent that alkyl substituents bonded to the sp^2 carbons of an alkene have a stabilizing effect on the alkene. We can, therefore, make the following statement: *the more alkyl substituents bonded to the sp² carbons of an alkene, the greater is its stability.* (Some students find it easier to look at the number of hydrogens bonded to the sp² carbons. In terms of hydrogens, the statement is: *the fewer hydrogens bonded to the sp² carbons of an alkene, the greater is its stability.*)

The greater the number of alkyl groups bonded to the *sp²* carbons of an alkene, the more stable it is.

Alkyl substituents stabilize both alkenes *and* carbocations.

relative stabilities of alkyl-substituted alkenes

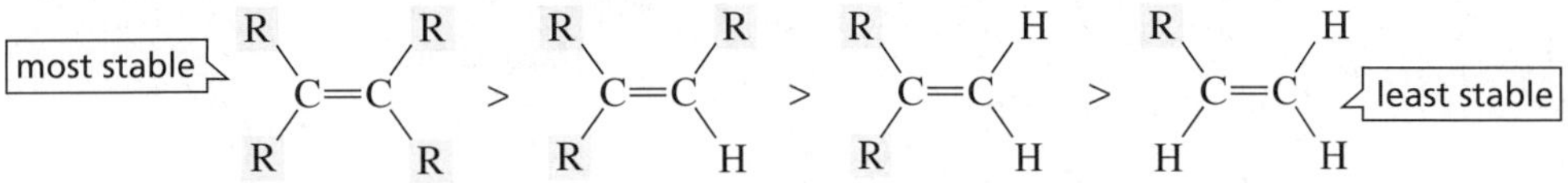

PROBLEM 31◆

The same alkane is obtained from the catalytic hydrogenation of both alkene A and alkene B. The heat of hydrogenation of alkene A is 29.8 kcal/mol (125 kJ/mol), and the heat of hydrogenation of alkene B is 31.4 kcal/mol (131 kJ/mol). Which alkene is more stable?

PROBLEM 32◆

a. Which of the following compounds is the most stable?

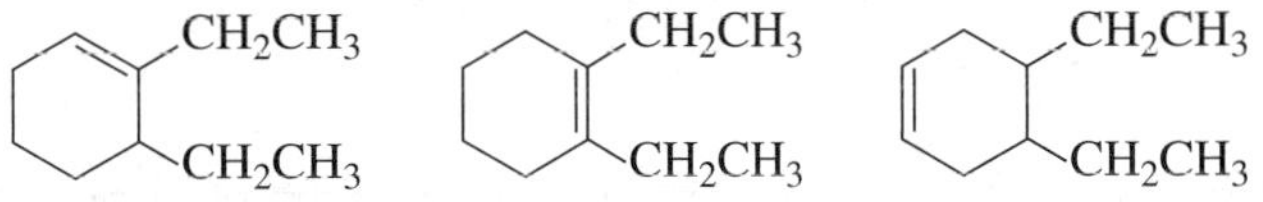

b. Which is the least stable?
c. Which has the smallest heat of hydrogenation?

Both *trans*-2-butene and *cis*-2-butene have two alkyl groups bonded to their sp^2 carbons, but *trans*-2-butene has a smaller heat of hydrogenation. This means that the trans isomer, in which the large substituents are farther apart, is more stable than the cis isomer, in which the large substituents are closer together.

	heat of hydrogenation	$\Delta H°$ kcal/mol	kJ/mol
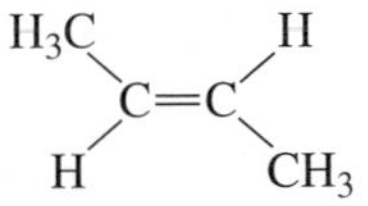 *trans*-2-butene	27.6	−27.6	−115
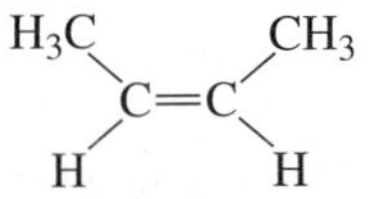 *cis*-2-butene	28.6	−28.6	−120

When the large substituents are on the same side of the molecule, their electron clouds can interfere with each other, causing steric strain in the molecule (Section 2.10) and making it less stable. When the large substituents are on opposite sides of the molecule, their electron clouds cannot interact, and the molecule has less steric strain and is therefore more stable.

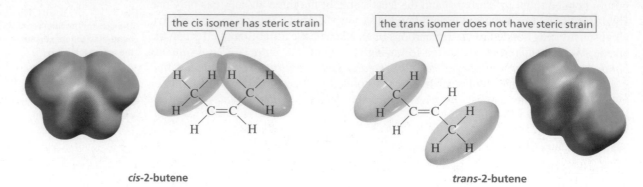

cis-2-butene

trans-2-butene

The heat of hydrogenation of *cis*-2-butene, in which the two alkyl substituents are on the *same side* of the double bond, is similar to that of 2-methylpropene, in which the two alkyl substituents are on the *same carbon*. The three dialkyl-substituted alkenes are all *less* stable than a trialkyl-substituted alkene and are all *more* stable than a monoalkyl-substituted alkene.

relative stabilities of dialkyl-substituted alkenes

| alkyl substituents are trans | alkyl substituents are cis | alkyl substituents are on the same sp^2 carbon |

$$> \qquad \sim$$

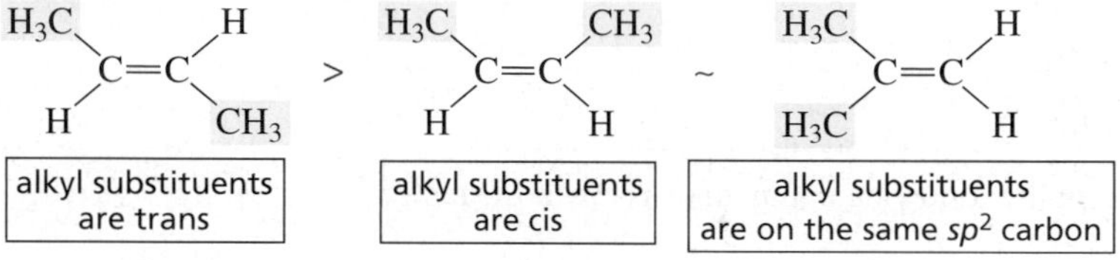

PROBLEM 33◆

Rank the following compounds in order of decreasing stability:
trans-3-hexene, *cis*-3-hexene, *cis*-2,5-dimethyl-3-hexene, *cis*-3,4-dimethyl-3-hexene

---

4.12 **Reactions and Synthesis**

This chapter has been concerned with the reactions of alkenes. You have seen why alkenes react, the kinds of reagents with which they react, the mechanisms by which the reactions occur, and the products that are formed. It is important to remember that when you are studying reactions, you are simultaneously studying synthesis. When you learn that compound A reacts with a certain reagent to form compound B, you are learning not only about the reactivity of A, but also about one way that compound B can be synthesized.

$$A \longrightarrow B$$
A reacts;
B is synthesized

For example, you have seen that many different reagents can add to alkenes and that compounds such as alkyl halides, vicinal dihalides, halohydrins, alcohols, ethers, epoxides, and alkanes are synthesized as a result.

Although you have seen how alkenes react and have learned about the kinds of compounds that are synthesized when alkenes undergo reactions, you have not yet seen how alkenes are synthesized. Reactions of alkenes involve the *addition* of atoms (or groups of atoms) to the two sp^2 carbons of the double bond. Reactions that lead to

PESTICIDES: NATURAL AND SYNTHETIC

Long before chemists learned to create compounds that would protect plants from predators, plants were doing the job themselves. Plants had every incentive to synthesize pesticides. When you cannot run, you need to find another way to protect yourself. But which pesticides are more harmful, those synthesized by chemists or those synthesized by plants? Unfortunately, we do not know the answer because although federal laws require all human-made pesticides to be tested for any cancer-causing effects, they do not require testing of plant-made pesticides. Besides, risk evaluations of chemicals are usually done on rats, and something that is carcinogenic in a rat may or may not be carcinogenic in a human. Furthermore, when rats are tested, they are exposed to much greater concentrations of the chemical than would be experienced by a human, and some chemicals are harmful only at high doses. For example, we all need sodium chloride for survival, but high concentrations are poisonous; and although we associate alfalfa sprouts with healthy eating, monkeys fed very large amounts of alfalfa sprouts develop an immune system disorder.

the synthesis of alkenes are exactly the opposite; they involve the *elimination* of atoms (or groups of atoms) from two adjacent sp^3 carbons.

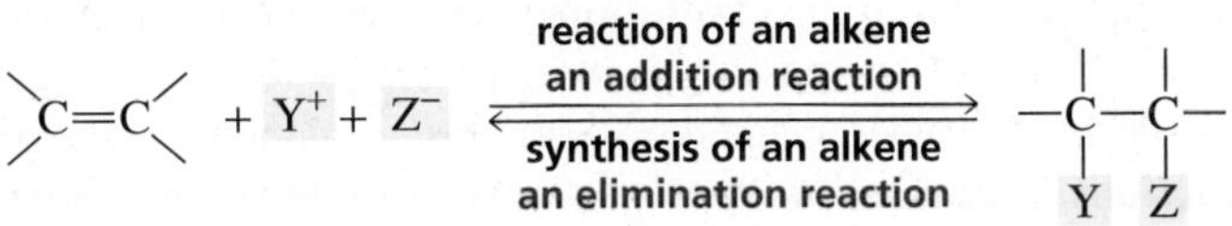

$$\text{C=C} + \text{Y}^+ + \text{Z}^- \underset{\substack{\text{synthesis of an alkene} \\ \text{an elimination reaction}}}{\overset{\substack{\text{reaction of an alkene} \\ \text{an addition reaction}}}{\rightleftharpoons}} \underset{\substack{\text{Y} \quad \text{Z}}}{-\text{C}-\text{C}-}$$

You will learn how alkenes are synthesized when you study compounds that undergo elimination reactions. The various reactions that result in the synthesis of alkenes are listed in Appendix IV.

PROBLEM 34 *SOLVED*

Indicate how each of the following compounds can be synthesized from an alkene:

a.

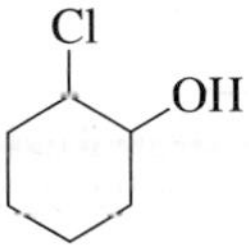

b.

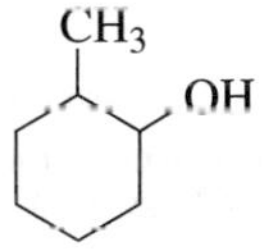

Solution

a. The only alkene that can be used for this synthesis is cyclohexene. To get the desired substituents on the ring, cyclohexene must react with Cl_2 in an aqueous solution so that water will be the nucleophile.

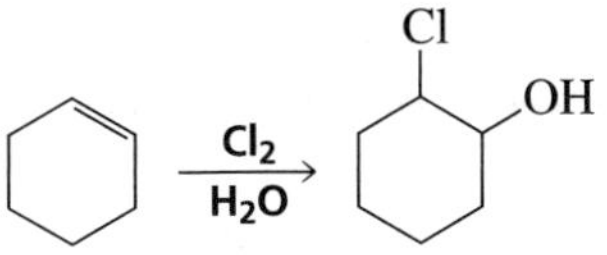

b. The alkene that should be used here is 1-methylcyclohexene. To get the substituents in the desired locations, the electrophile must be BH_3, with HO taking its place in the subsequent oxidation reaction.

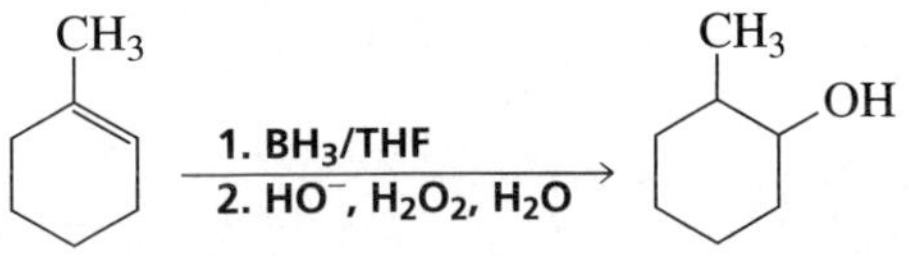

PROBLEM 35◆

Why should 3-methylcyclohexene not be used as the starting material in Problem 34b?

PROBLEM 36

Indicate how each of the following compounds can be synthesized from an alkene:

a. CH$_3$CHOCH$_3$
     |
     CH$_3$

c. (cyclohexane with CH$_2$OH substituent)

e. (cyclohexane with Br and CH$_3$ substituents)

b. CH$_3$CH$_2$CHCHCH$_3$
     | with Br substituents

d. (cyclohexane with CH$_3$O and CH$_3$ substituents)

f. (cyclohexane with OCH$_2$CH$_2$CH$_3$ substituent)

SUMMARY

Alkenes undergo **electrophilic addition reactions**. These start with the addition of an electrophile to one of the sp^2 carbons and concludes with the addition of a nucleophile to the other sp^2 carbon. In all electrophilic addition reactions, the *electrophile* adds to the sp^2 carbon bonded to the greater number of hydrogens.

The addition of hydrogen halides and the acid-catalyzed addition of water and alcohols form **carbocation intermediates**. **Hyperconjugation** causes **tertiary carbocations** to be more stable than **secondary carbocations**, which are more stable than **primary carbocations**. A carbocation will rearrange if it becomes more stable as a result of the rearrangement. **Carbocation rearrangements** occur by **1,2-hydride shifts**, **1,2-methyl shifts**, and **ring expansion**.

Oxymercuration, **alkoxymercuration**, and the addition of Br$_2$ or Cl$_2$ form cyclic intermediates. **Hydroboration** is a **concerted reaction** and does not form an **intermediate**. Because these reactions do not form carbocation intermediates, carbocation rearrangements do not occur.

The products of oxymercuration and alkoxymercuration are subjected to a reduction reaction. **Reduction** increases the number of C—H bonds or decreases the number of C—O, C—N, or C—X bonds (where X denotes a halogen). **Hydroboration** products are subjected to an oxidation

reaction. **Oxidation** decreases the number of C—H bonds or increases the number of C—O, C—N, or C—X bonds (where, again, X denotes a halogen). The reaction of an alkene with a peroxyacid to form an epoxide is another example of an oxidation reaction.

The **Hammond postulate** states that a transition state is more similar in structure to the species to which it is more similar in energy. Thus, the more stable product will have the more stable transition state and will lead to the major product of the reaction. **Regioselectivity** is the preferential formation of one **constitutional isomer** over another.

The addition of H$_2$ to an alkene is called **hydrogenation**. The **heat of hydrogenation** is the heat released in a hydrogenation reaction. The *greater* the *stability* of a compound, the *lower* is its *energy* and the *smaller* is its *heat of hydrogenation*. The more alkyl substituents bonded to the sp^2 carbons of an alkene, the greater is its stability. Hence, carbocations and alkenes are both stabilized by alkyl substituents. **Trans alkenes** are more stable than **cis alkenes** because of steric strain.

Electrophilic addition reactions of alkenes lead to the **synthesis** of **alkyl halides**, **vicinal dihalides**, **halohydrins**, **alcohols**, **ethers**, **epoxides**, and **alkanes**.

SUMMARY OF REACTIONS

As you review the reactions of alkenes, keep in mind the feature that is common to all of them: the first step of each reaction is the addition of an electrophile to the sp^2 carbon that is bonded to the greater number of hydrogens.

1. Electrophilic addition reactions

    a. Addition of hydrogen halides (H$^+$ is the electrophile; Section 4.1)

$$RCH{=}CH_2 + HX \longrightarrow RCHCH_3$$
$$\qquad\qquad\qquad\qquad\qquad |$$
$$\qquad\qquad\qquad\qquad\qquad X$$

HX = HF, HCl, HBr, HI

b. Acid-catalyzed addition of water and alcohols (H^+ is the electrophile; Section 4.5)

$$RCH{=}CH_2 \ + \ H_2O \ \underset{}{\overset{H_2SO_4}{\rightleftharpoons}} \ \underset{\underset{OH}{|}}{RCHCH_3}$$

$$RCH{=}CH_2 \ + \ CH_3OH \ \underset{}{\overset{H_2SO_4}{\rightleftharpoons}} \ \underset{\underset{OCH_3}{|}}{RCHCH_3}$$

c. Addition of halogen (Br^+ or CI^+ is the electrophile; Section 4.7)

$$RCH{=}CH_2 \ + \ Cl_2 \ \xrightarrow{CH_2Cl_2} \ \underset{\underset{Cl}{|}}{RCHCH_2Cl}$$

$$RCH{=}CH_2 \ + \ Br_2 \ \xrightarrow{CH_2Cl_2} \ \underset{\underset{Br}{|}}{RCHCH_2Br}$$

$$RCH{=}CH_2 \ + \ Br_2 \ \xrightarrow{H_2O} \ \underset{\underset{OH}{|}}{RCHCH_2Br}$$

d. Oxymercuration–reduction and alkoxymercuration–reduction (Hg is the electrophile and is subsequently replaced by II; Section 4.8)

$$RCH{=}CH_2 \ \xrightarrow[\text{2. NaBH}_4]{\text{1. Hg(OAc)}_2, \text{ H}_2\text{O,THF}} \ \underset{\underset{OH}{|}}{RCHCH_3}$$

$$RCH{=}CH_2 \ \xrightarrow[\text{2. NaBH}_4]{\text{1. Hg(O}_2\text{CCF}_3)_2, \text{ CH}_3\text{OH}} \ \underset{\underset{OCH_3}{|}}{RCHCH_3}$$

e. Addition of a peroxyacid (O is the electrophile; Section 4.9)

$$\underset{\text{an alkene}}{RCH{=}CH_2} \ + \ \underset{\text{a peroxyacid}}{\overset{\overset{\displaystyle O}{\|}}{RCOOH}} \ \longrightarrow \ \underset{\text{an epoxide}}{RCH{-}CH_2} \ + \ \underset{\text{a carboxylic acid}}{\overset{\overset{\displaystyle O}{\|}}{RCOH}}$$

f. Hydroboration–oxidation (B is the electrophile and is subsequently replaced by OH; Section 4.10)

$$RCH{=}CH_2 \ \xrightarrow[\text{2. HO}^-, \text{ H}_2\text{O}_2, \text{ H}_2\text{O}]{\text{1. BH}_3/\text{THF}} \ RCH_2CH_2OH$$

2. Addition of hydrogen (Section 4.11)

$$RCH{=}CH_2 \ + \ H_2 \ \xrightarrow{\text{Pd/C, Pt/C, or Ni}} \ RCH_2CH_3$$

KEY TERMS

acid-catalyzed reaction (p. 171)
alkene (p. 159)
alkoxymercuration–reduction (p. 181)
carbene (p. 183)
carbocation intermediate (p. 159)
carbocation rearrangement (p. 173)
catalyst (p. 171)
catalytic hydrogenation (p. 188)
concerted reaction (p. 182)
constitutional isomers (p. 166)
dimer (p. 185)
electrophilic addition
  reaction (p. 159)

epoxide (p. 182)
halohydrin (p. 177)
Hammond postulate (p. 164)
heat of hydrogenation (p. 189)
heterogeneous catalyst (p. 188)
hydration (p. 170)
hydroboration–oxidation (p. 184)
hydrogenation (p. 188)
hyperconjugation (p. 162)
Markovnikov's rule (p. 167)
mechanism of the
  reaction (p. 161)
1,2-hydride shift (p. 173)

1,2-methyl shift (p. 173)
oxidation reaction (p. 182)
oxymercuration–reduction (p. 180)
peroxyacid (p. 182)
primary carbocation (p. 161)
reduction reaction (p. 181)
regioselective reaction (p. 166)
regioselectivity (p. 166)
secondary carbocation (p. 161)
steric effects (p. 186)
steric hindrance (p. 186)
tertiary carbocation (p. 161)
vicinal (p. 176)

PROBLEMS

37. Give the major product of each of the following reactions:

a.

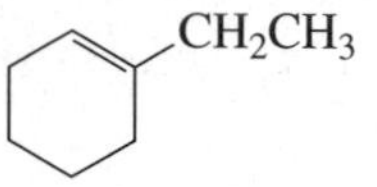

$$\text{cyclohexene with } CH_2CH_3 \text{ substituent} \quad + \quad HBr \quad \longrightarrow$$

b. $CH_2={\overset{\overset{\displaystyle CH_3}{|}}{C}}CH_2CH_3 \quad + \quad HBr \quad \longrightarrow$

c.

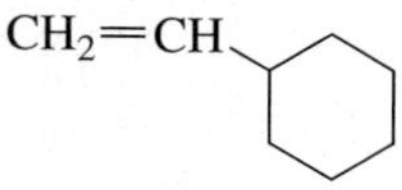

$$CH_2=CH-\text{cyclohexane} \quad + \quad HBr \quad \longrightarrow$$

d. $CH_3CH_2{\overset{\overset{\displaystyle CH_3}{|}}{\underset{\underset{\displaystyle CH_3}{|}}{C}}}CH=CH_2 \quad + \quad HBr \quad \longrightarrow$

38. Identify the electrophile and the nucleophile in each of the following reaction steps. Then draw curved arrows to illustrate the bond-making and bond-breaking processes.

a. $CH_3\overset{+}{C}HCH_3 \quad + \quad :\overset{..}{\underset{..}{Cl}}:^- \quad \longrightarrow \quad CH_3\overset{\overset{\displaystyle |}{}}{C}HCH_3$
$$\qquad\qquad\qquad\qquad\qquad\qquad\qquad\qquad :\overset{..}{\underset{..}{Cl}}:$$

b. $CH_3CH=CH_2 \quad + \quad H-Br \quad \longrightarrow \quad CH_3\overset{+}{C}H-CH_3 \quad + \quad Br^-$

c. $CH_3CH=CH_2 \quad + \quad BH_3 \quad \longrightarrow \quad CH_3CH_2-CH_2BH_2$

39. What will be the major product of the reaction of 2-methyl-2-butene with each of the following reagents?

a. HBr
b. a peroxyacid
c. HI
d. Cl_2/CH_2Cl_2
e. ICl

f. H_2/Pd
g. Br_2 + excess NaCl
h. $Hg(OAc)_2$, H_2O followed by $NaBH_4$
i. H_2O + trace H_2SO_4
j. Br_2/CH_2Cl_2

k. Br_2/H_2O
l. Br_2/CH_3OH
m. BH_3/THF, followed by $H_2O_2/HO^-/H_2O$
n. $Hg(O_2CCF_3)_2$ + CH_3OH, followed by $NaBH_4$

40. Give two names for each of the following compounds:

a.

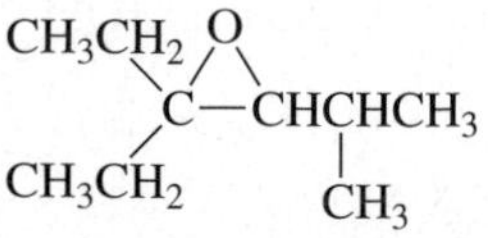

$$CH_3CH_2\overset{\displaystyle}{\underset{\displaystyle CH_3CH_2}{C}}\!\!-\!\!\overset{O}{\triangle}\!\!-\!\!CHCHCH_3 \text{ with } CH_3 \text{ substituent}$$

b.

$$\overset{\displaystyle H_3C}{\underset{\displaystyle H_3C}{C}}\!\!-\!\!\overset{O}{\triangle}\!\!-\!\!CHCH_2CH_3$$

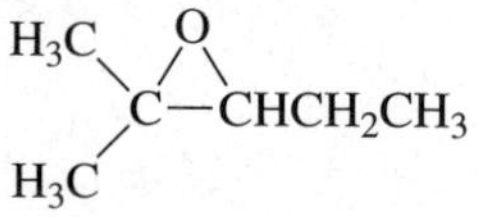

41. **a.** Which compound is the most stable: 3,4-dimethyl-2-hexene; 2,3-dimethyl-2-hexene; 4,5-dimethyl-2-hexene?
b. Which would you expect to have the largest heat of hydrogenation?
c. Which would you expect to have the lowest heat of hydrogenation?

42. What reagents are required to synthesize the following alcohols?

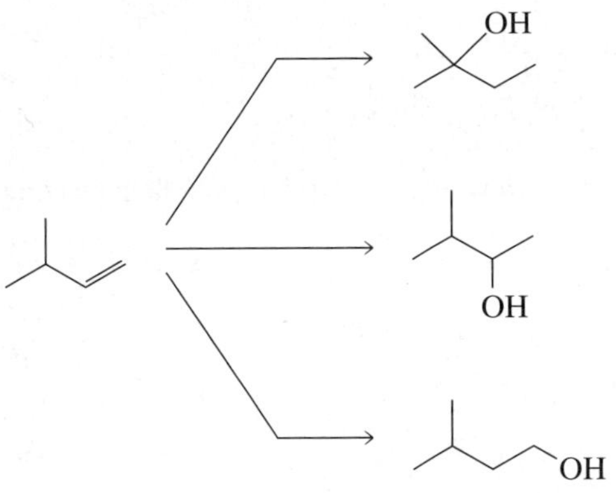

43. When 3-methyl-1-butene reacts with HBr, two alkyl halides are formed: 2-bromo-3-methylbutane and 2-bromo-2-methylbutane. Propose a mechanism that explains the formation of these products.

44. Problem 37 in Chapter 3 asked you to give the structures of all alkenes with molecular formula C_6H_{12}. Use those structures to answer the following questions:
a. Which of the compounds is the most stable?
b. Which of the compounds is the least stable?

45. Draw curved arrows to show the flow of electrons responsible for the conversion of reactants into products.

a. $CH_3-\underset{\underset{CH_3}{|}}{\overset{\overset{:\ddot{O}:^-}{|}}{C}}-OCH_3 \longrightarrow CH_3-\overset{\overset{:\ddot{O}}{\|}}{C}-CH_3 + CH_3O^-$

b. $CH_3C{\equiv}C-H + {}^{-}\!:\!\ddot{N}H_2 \longrightarrow CH_3C{\equiv}C^- + \ddot{N}H_3$

c. $CH_3CH_2-Br + CH_3\ddot{O}:^- \longrightarrow CH_3CH_2-\ddot{O}CH_3 + Br^-$

46. Give the reagents that would be required to carry out the following syntheses:

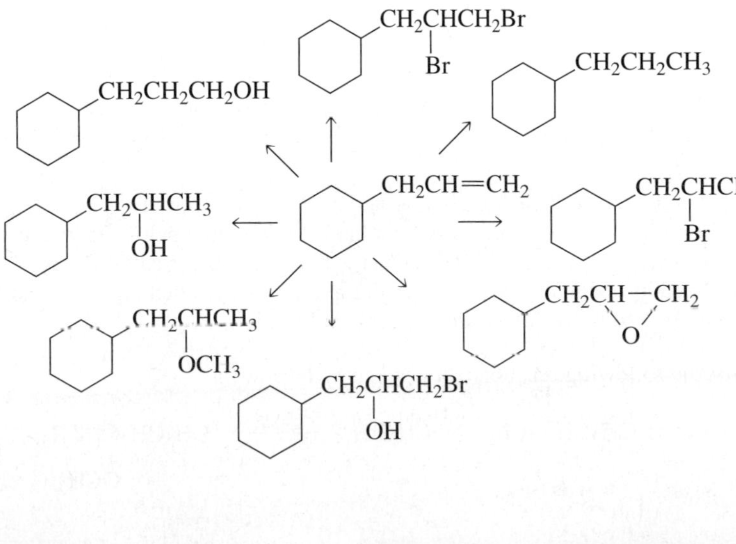

47. Give the major product of each of the following reactions:

a. [cyclohexene] $\xrightarrow{\text{HCl}}$ **c.** [methylenecyclohexane] $\xrightarrow{\text{RCOOH}}$ **e.** [1-methylcyclohexene] $\xrightarrow[\text{H}_2\text{O}]{\text{Cl}_2}$ **g.** [1-methylcyclohexene] $\xrightarrow[\text{CH}_3\text{OH}]{\text{H}_2\text{SO}_4}$

b. [methylenecyclohexane] $\xrightarrow[\text{CH}_3\text{OH}]{\text{Br}_2}$ **d.** [1-methylcyclohexene] $\xrightarrow[\text{H}_2\text{O}]{\text{H}_2\text{SO}_4}$ **f.** [1-methylcyclohexene] $\xrightarrow{\text{HBr}}$ **h.** [1-methylcyclohexene] $\xrightarrow[\text{CH}_2\text{Cl}_2]{\text{Cl}_2}$

48. Using any alkene and any other reagents, how would you prepare the following compounds?

a. [cyclohexane]

c. [cyclohexyl-CH$_2$OH]

e. [cyclohexyl-CH$_2$CHCH$_3$ with OH]

b. $CH_3CH_2CH_2CHCH_3$ with Cl

d. $CH_3CH_2CHCHCH_2CH_3$ with Br OH

f. $CH_3CH_2CHCHCH_2CH_3$ with Br Cl

49. Identify two alkenes that react with HBr to give 1-bromo-1-methylcyclohexane without undergoing a carbocation rearrangement.

50. For each of the following pairs, indicate which member is the more stable:

a. $CH_3\overset{+}{\underset{\underset{CH_3}{|}}{C}}CH_3$ or $CH_3\overset{+}{C}HCH_2CH_3$

c. $CH_3\overset{\underset{CH_3}{|}}{C}=CHCH_2CH_3$ or $CH_3CH=\overset{\underset{CH_3}{|}}{C}HCHCH_3$

b. $CH_3\overset{+}{C}HCH_3$ or $CH_3\overset{+}{C}HCH_2Cl$

d. [methylcyclohexene A] or [methylcyclohexene B]

51. The second-order rate constant (in units of $M^{-1}s^{-1}$) for acid-catalyzed hydration at 25 °C is given for each of the following alkenes:

[structures with values]
4.95×10^{-8} 8.32×10^{-8} 3.51×10^{-8} 2.15×10^{-4} 3.42×10^{-4}

a. Calculate the relative rates of hydration of the alkenes.
b. Why does (Z)-2-butene react faster than (E)-2-butene?
c. Why does 2-methyl-2-butene react faster than (Z)-2-butene?
d. Why does 2,3-dimethyl-2-butene react faster than 2-methyl-2-butene?

52. Which compound has the greater dipole moment?

a. [Cl,H / H,Cl alkene] or [H,H / H,Cl alkene]

b. [Cl,H / H,CH$_3$ alkene] or [Cl,CH$_3$ / H,H alkene]

53. Mark Onikoff was about to turn in the products he had obtained from the reaction of HI with 3,3,3-trifluoropropene when he realized that the labels had fallen off his flasks and he did not know which label belonged where. Another student reminded him of the rule that says the electrophile adds to the sp^2 carbon that is bonded to the greater number of hydrogens; thus, he should put the label 1,1,1-trifluoro-2-iodopropane on the flask containing the most product and label the flask with the least product 1,1,1-trifluoro-3-iodopropane. Should Mark follow this student's advice?

54. a. Propose a mechanism for the following reaction (show all curved arrows):

$$CH_3CH_2CH=CH_2 + CH_3OH \xrightarrow{\text{H}_2\text{SO}_4} CH_3CH_2CHCH_3$$
$$\underset{\underset{OCH_3}{|}}{}$$

b. Which step is the rate-determining step?
c. What is the electrophile in the first step?
d. What is the nucleophile in the first step?
e. What is the electrophile in the second step?
f. What is the nucleophile in the second step?

55. Give the major products for each of the following reactions:

a. $HOCH_2CH_2CH_2CH=CH_2$ + Br_2 $\xrightarrow{CH_2Cl_2}$

b. $HOCH_2CH_2CH_2CH_2CH=CH_2$ + Br_2 $\xrightarrow{CH_2Cl_2}$

56. **a.** What product is obtained from the reaction of HCl with 1-butene? With 2-butene?
    b. Which of the two reactions has the greater free energy of activation?
    c. Which of the two alkenes reacts more rapidly with HCl?
    d. Which compound reacts more rapidly with HCl, (Z)-2-butene or (E)-2-butene?

57. **a.** How many alkenes could you treat with H_2/Pt in order to prepare methylcyclopentane?
    b. Which of the alkenes is the most stable?
    c. Which of the alkenes has the smallest heat of hydrogenation?

58. **a.** Propose a mechanism for the following reaction:

+ HBr ⟶

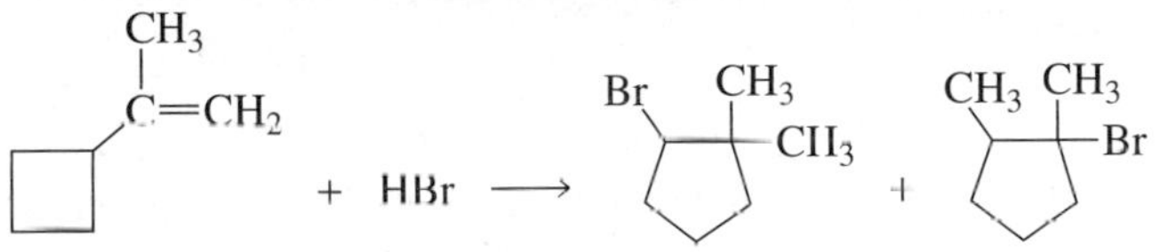

b. Is the initially formed carbocation primary, secondary, or tertiary?
c. Is the rearranged carbocation primary, secondary, or tertiary?
d. Why does the rearrangement occur?

59. When the following compound is hydrated in the presence of acid, the unreacted alkene is found to have retained the deuterium atoms:

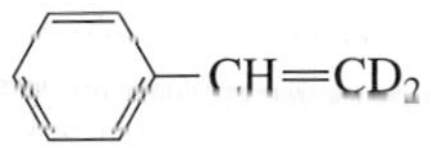

What does the preceding statement tell you about the mechanism of hydration?

60. Propose a mechanism for the following reaction:

+ H_2O $\xrightarrow{H_2SO_4}$

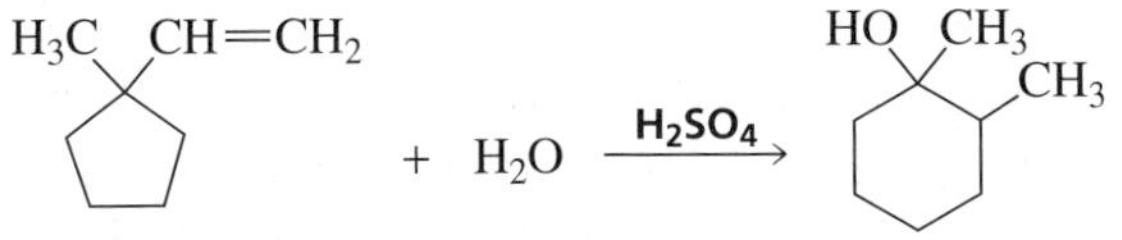

61. **a.** Dichlorocarbene can be generated by heating chloroform with HO^-. Propose a mechanism for the reaction.

$$CHCl_3 \ + \ HO^- \ \xrightarrow{\Delta} \ Cl_2C: \ + \ H_2O \ + \ Cl^-$$
chloroform   dichlorocarbene

b. Dichlorocarbene can also be generated by heating sodium trichloroacetate. Propose a mechanism for the reaction.

$$Cl_3C\overset{O}{\overset{\|}{C}}O^- \ Na^+ \ \xrightarrow{\Delta} \ Cl_2C: \ + \ CO_2 \ + \ Na^+ \ Cl^-$$
sodium trichloroacetate

CHAPTER 5

Stereochemistry

The Arrangement of Atoms in Space; The Stereochemistry of Addition Reactions

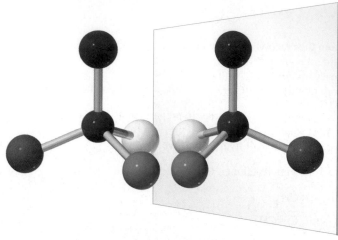

**nonsuperimposable
mirror images**

BUILDING ON FUNDAMENTALS

SECTION 5.1 Compounds with restricted rotation due to a cyclic structure or a double bond can have cis–trans isomers (2.14 and 3.4). We will now see that cis–trans isomers are one of two kinds of stereoisomers.

SECTION 5.7 Stereoisomers with asymmetric centers are named using the same system of priorities used for determining E and Z isomers (3.5).

SECTION 5.19 In Chapter 4, you learned how to determine the major product formed when an alkene undergoes an electrophilic addition reaction. Now you will learn how to determine which stereoisomers of that product are formed.

SECTION 5.19 A positively charged carbon is sp^2 hybridized, causing the three atoms bonded to it to lie in a plane (1.10). This geometry influences the stereoisomers formed in reactions that form a carbocation intermediate.

Compounds that have the same molecular formula but do not have identical structures are called **isomers**. Isomers fall into two main classes: *constitutional isomers* and *stereoisomers*. **Constitutional isomers** differ in the way their atoms are connected (Section 2.0). For example, ethanol and dimethyl ether are constitutional isomers with molecular formula C_2H_6O. The oxygen in ethanol is bonded to a carbon and to a hydrogen, whereas the oxygen in dimethyl ether is bonded to two carbons.

constitutional isomers

CH_3CH_2OH and CH_3OCH_3
ethanol   dimethyl ether

$CH_3CH_2CH_2CH_2Cl$ and $CH_3CH_2\overset{\underset{\displaystyle |}{Cl}}{C}HCH_3$
1-chlorobutane   2-chlorobutane

$CH_3CH_2CH_2CH_2CH_3$ and $CH_3\overset{\underset{\displaystyle |}{CH_3}}{C}HCH_2CH_3$
pentane   isopentane

$CH_3\overset{\displaystyle O}{\overset{\|}{C}}CH_3$ and $CH_3CH_2\overset{\displaystyle O}{\overset{\|}{C}}H$
acetone   propionaldehyde

Unlike constitutional isomers, the atoms in stereoisomers are connected in the same way. **Stereoisomers** (also called **configurational isomers**) differ in the way their atoms are arranged in space. Like constitutional isomers, stereoisomers can be separated because they are different compounds that do not readily interconvert. There are two kinds of stereoisomers: *cis–trans isomers* and isomers that contain *asymmetric centers*.

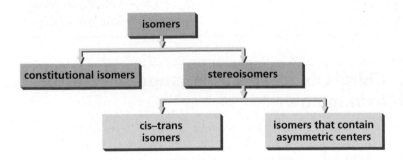

Student Tutorial:
Isomerism

After reviewing cis–trans isomers, we will look at isomers that contain asymmetric centers, the only stereoisomers that you have not seen previously. Then we will go back and look at the reactions you learned in Chapter 4 and, for those reactions with products that can have stereoisomers, see what specific stereo-isomers are formed.

PROBLEM 1◆

a. Draw three constitutional isomers with molecular formula C_3H_8O.

b. How many constitutional isomers can you draw for $C_4H_{10}O$?

5.1 Cis–Trans Isomers Result from Restricted Rotation

Cis–trans isomers (also called **geometric isomers**) result from restricted rotation. Restricted rotation can be caused either by a *double bond* or a *cyclic structure*. We have seen that owing to the restricted rotation about its carbon–carbon double bond, an alkene such as 2-pentene can exist as cis and trans isomers (Section 3.4). The **cis isomer** has the hydrogens on the *same side* of the double bond and the **trans isomer** has the hydrogens on *opposite sides* of the double bond. (Remember that Z and E are used instead of cis and trans for more complex molecules.)

As a result of restricted rotation about the bonds in a ring, cyclic compounds also have cis and trans isomers (Section 2.14). The cis isomer has the hydrogens on the same side of the ring, whereas the trans isomer has the hydrogens on opposite sides of the ring.

cis-2-pentene

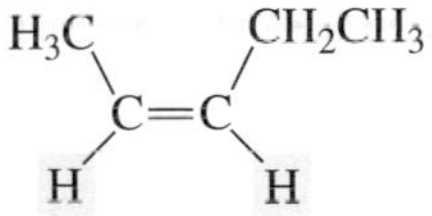

cis-2-pentene

trans-2-pentene

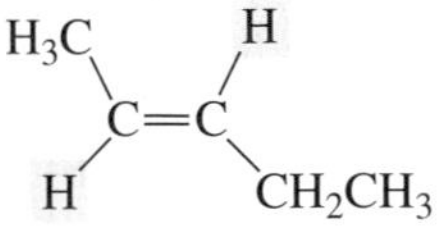

trans-2-pentene

cis-1-bromo-3-chlorocyclobutane

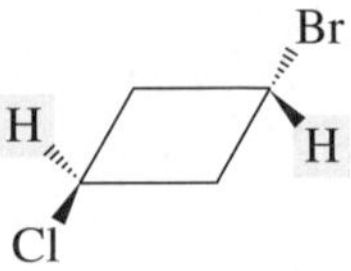

trans-1-bromo-3-chlorocyclobutane

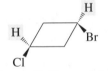

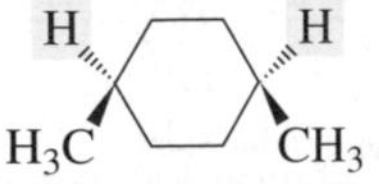

cis-1,4-dimethylcyclohexane

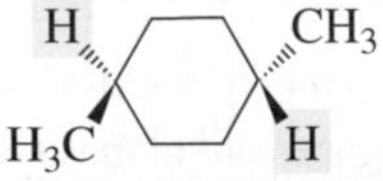

trans-1,4-dimethylcyclohexane

5.2 A Chiral Object Has a Nonsuperimposable Mirror Image

Why can't you put your right shoe on your left foot? Why can't you put your right glove on your left hand? It is because hands, feet, gloves, and shoes have right-handed and left-handed forms. An object with a right-handed and a left-handed form is said to be **chiral** (ky-ral), a word derived from the Greek word *cheir*, which means "hand."

A chiral object has a *nonsuperimposable mirror image*. In other words, its mirror image is not the same as an image of the object itself. A hand is chiral because when you look at your right hand in a mirror, you see not a right hand but a left hand (Figure 5.1). In contrast, a chair is not chiral; the reflection of the chair in the mirror looks the same as the chair itself. Objects that are not chiral are said to be **achiral**. An achiral object has a *superimposable mirror image*. Some other achiral objects are a table, a bowl, and a balloon (assuming they are simple and unadorned).

chiral objects

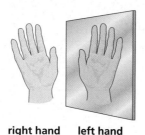

right hand   **left hand**

achiral objects

Figure 5.1 ▶
Using a mirror to test for chirality. A chiral object is not the same as its mirror image—they are nonsuperimposable. An achiral object is the same as its mirror image—they are superimposable.

5.3 An Asymmetric Center Is a Cause of Chirality in a Molecule

Objects are not the only things that can be chiral. Molecules can be chiral too. The usual *cause of chirality in a molecule is an asymmetric center.* (Other features that cause chirality are relatively uncommon and beyond the scope of this book, but you can see one example in Problem 96 at the end of this chapter.)

An **asymmetric center** (or chiral center) is a tetrahedral atom bonded to four different groups. Each of the compounds shown below has an asymmetric center indicated by a star. For example, the starred carbon in 4-octanol is an asymmetric center because it is bonded to four different groups (H, OH, $CH_2CH_2CH_3$, and $CH_2CH_2CH_2CH_3$). Notice that the atoms immediately bonded to the asymmetric center are not necessarily different from one another; the propyl and butyl groups are different even though the point at which they differ is several atoms away from the asymmetric center. The starred carbon in 2,4-dimethylhexane is an asymmetric center because it too is bonded to four different groups—(methyl, ethyl, isobutyl, and hydrogen).

A molecule with an asymmetric center is chiral.

Student Tutorial: Identification of asymmetric centers I

Student Tutorial: Identification of asymmetric centers II

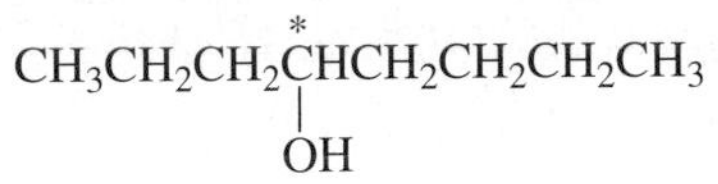

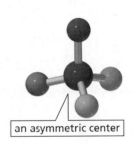

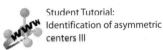
an asymmetric center

Student Tutorial: Identification of asymmetric centers III

$$\overset{*}{CH_3CH_2CH_2CHCH_2CH_2CH_2CH_3}$$
$$|$$
$$OH$$
4-octanol

$$CH_3\overset{*}{CHCH_2CH_3}$$
$$|$$
$$Br$$
2-bromobutane

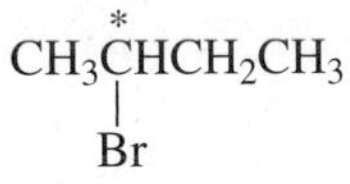

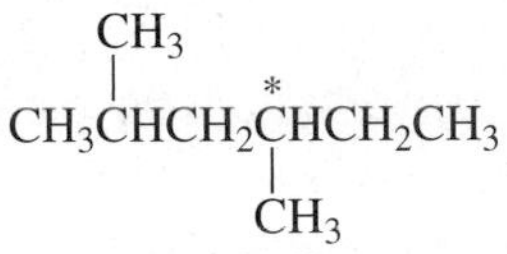

$$CH_3$$
$$|$$
$$CH_3\overset{*}{CHCH_2CHCH_2CH_3}$$
$$|$$
$$CH_3$$
2,4-dimethylhexane

PROBLEM 5◆

Which of the following compounds have an asymmetric center?

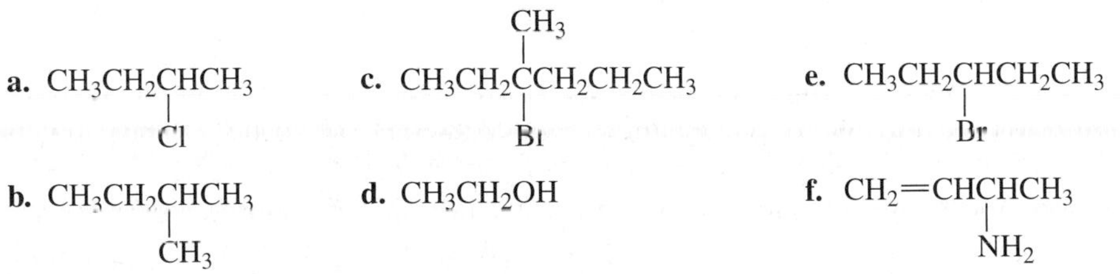

a. $CH_3CH_2\overset{|}{\underset{Cl}{C}}HCH_3$

b. $CH_3CH_2\overset{|}{\underset{CH_3}{C}}HCH_3$

c. $CH_3CH_2\overset{\overset{CH_3}{|}}{\underset{Br}{C}}CH_2CH_2CH_3$

d. CH_3CH_2OH

e. $CH_3CH_2\overset{|}{\underset{Br}{C}}HCH_2CH_3$

f. $CH_2{=}CHCH_3$
$$|$$
$$NH_2$$

PROBLEM 6   SOLVED

Tetracycline is called a broad-spectrum antibiotic because it is active against a wide variety of bacteria. How many asymmetric centers does tetracycline have?

Solution Only sp^3 carbons can be asymmetric centers, because an asymmetric center must have four different groups attached to it. Therefore, we start by locating all the sp^3 carbons in tetracycline. (They are numbered in red.) Tetracycline has nine sp^3 carbons. Four of them (numbers 1, 2, 5, and 8) are not asymmetric centers because they are not bonded to four different groups. Tetracycline, therefore, has five asymmetric centers.

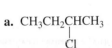
tetracycline

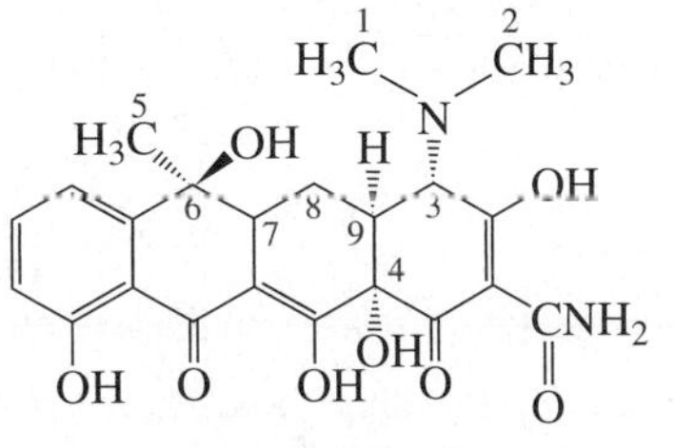

5.4 **Isomers with One Asymmetric Center**

A compound with one asymmetric center, such as 2-bromobutane, can exist as two stereoisomers. The two stereoisomers are analogous to a left and a right hand. If we imagine a mirror between the two stereoisomers, we can see they are mirror images of each other. Moreover, they are nonsuperimposable mirror images and thus different molecules.

$$CH_3\overset{*}{C}HCH_2CH_3$$
$$Br$$

2-bromobutane

Student Tutorial: Nonsuperimposable mirror image

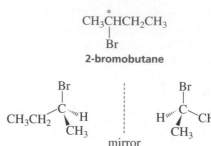

the two isomers of 2-bromobutane
enantiomers

Note to the student

Prove to yourself that the two 2-bromobutane isomers are not identical by building ball-and-stick models to represent them and trying to superimpose one on the other; use four different-colored balls to represent the four different groups bonded to the asymmetric center. Special Topic IV in the *Study Guide and Solutions Manual* tells you what other models you should build as you go through the chapter.

Molecules that are nonsuperimposable mirror-images of each other are called **enantiomers** (from the Greek *enantion*, which means "opposite"). Thus, the two stereoisomers of 2-bromobutane are enantiomers. A molecule that has a *nonsuperimposable* mirror image, like an object that has a *nonsuperimposable* mirror image, is *chiral*. Therefore, each member of a pair of enantiomers is chiral. Notice that chirality is a property of an entire object or an entire molecule.

A chiral molecule has a nonsuperimposable mirror image.

A molecule that has a *superimposable* mirror image, like an object that has a *superimposable* mirror image, is *achiral*. To see that the achiral molecule below is superimposable on its mirror image (that they are identical molecules), mentally rotate it clockwise.

An achiral molecule has a superimposable mirror image.

a chiral molecule	nonsuperimposable mirror image		an achiral molecule	superimposable mirror image
enantiomers			identical molecules	

PROBLEM 7◆

Which of the compounds in Problem 5 can exist as enantiomers?

5.5 Asymmetric Centers and Stereocenters

A **stereocenter** (or **stereogenic center**) is an atom at which the interchange of two groups produces a stereoisomer. Thus stereocenters include both *asymmetric centers*—where the interchange of two groups produces an enantiomer—and the sp^2 or sp^3 carbons—where the interchange of two groups converts a cis isomer to a trans isomer (or a Z isomer to an E isomer) or vice versa. This means that although *all asymmetric centers are stereocenters*, not all stereocenters are asymmetric centers.

5.6 How to Draw Enantiomers

Chemists draw enantiomers using either *perspective formulas* or *Fischer projections*. A **perspective formula** shows two of the bonds to the asymmetric center in the plane of the paper, one bond as a solid wedge protruding forward out of the paper, and the fourth bond as a hatched wedge extending behind the paper. The solid wedge and the hatched wedge must be adjacent. When you draw the first enantiomer, the four groups bonded to the asymmetric center can be placed around it in any order. You then draw the second enantiomer by drawing the mirror image of the first.

A solid wedge represents a bond that extends out of the plane of the paper toward the viewer.

A hatched wedge represents a bond that points back from the plane of the paper away from the viewer.

When you draw a perspective formula, make sure that the two bonds in the plane of the paper are adjacent to one another; neither the solid wedge nor the hatched wedge should be drawn between them.

perspective formulas of the enantiomers of 2-bromobutane

A **Fischer projection** is a shortcut for showing the three-dimensional arrangement of groups bonded to an asymmetric center. Devised in the late 1800s by Emil Fischer (Section 21.0), it represents an asymmetric center as the point of intersection of two perpendicular lines. Horizontal lines represent the bonds that project out of the plane of the paper toward the viewer, and vertical lines represent the bonds that extend back from the plane of the paper away from the viewer. The carbon chain is usually drawn vertically, with C-1 at the top.

In a Fischer projection, horizontal lines project out of the plane of the paper toward the viewer and vertical lines extend back from the plane of the paper away from the viewer.

Fischer projections of the enantiomers of 2-bromobutane

When you draw enantiomers using a Fischer projection, you can draw the first by placing the four atoms or groups bonded to the asymmetric center around that center in any order. Then draw the second enantiomer by interchanging two of the atoms or groups. It does not matter which two you interchange. (Make models to convince yourself that this is true.) It is best to interchange the groups on the two horizontal bonds, because then the enantiomers look like mirror images on your paper.

Whether you are drawing perspective formulas or Fischer projections, the interchanging of two atoms or groups will produce the other enantiomer. Interchanging two atoms or groups a second time, brings you back to the original molecule.

PROBLEM 8

Draw enantiomers for each of the following compounds using:

a. perspective formulas.

b. Fischer projections.

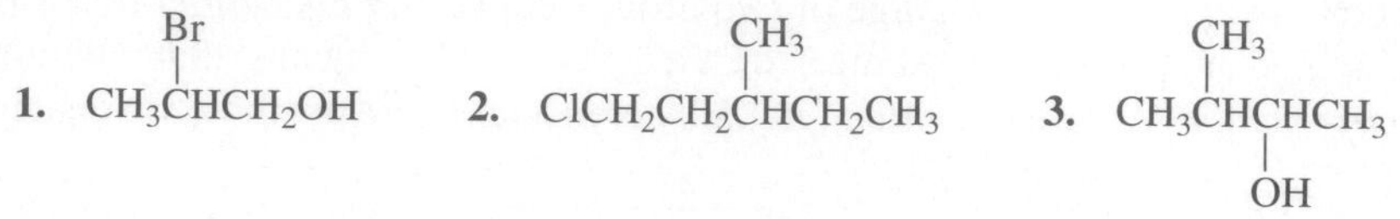

Br
1. CH_3CHCH_2OH
CH_3
2. $ClCH_2CH_2CHCH_2CH_3$
CH_3
3. $CH_3CHCHCH_3$
OH

5.7 Naming Enantiomers by the *R,S* System

BIOGRAPHY

Robert Sidney Cahn (1899–1981) *was born in England and received an M.A. from Cambridge University and a doctorate in natural philosophy in France. He edited the* Journal of the Chemical Society *(London).*

BIOGRAPHY

Sir Christopher Ingold (1893–1970) *was born in Ilford, England, and was knighted by Queen Elizabeth II. He was a professor of chemistry at Leeds University (1924–1930) and at University College, London (1930–1970).*

BIOGRAPHY

Vladimir Prelog (1906–1998) *was born in Sarajevo, Bosnia. In 1929 he received a Dr. Ing. degree from the Institute of Technology in Prague, Czechoslovakia. He taught at the University of Zagreb from 1935 until 1941, when he fled to Switzerland just ahead of the invading German army. He was a professor at the Swiss Federal Institute of Technology (ETH). For work that contributed to an understanding of how living organisms carry out chemical reactions, he shared the 1975 Nobel Prize in chemistry with John Cornforth (page 246).*

How do we name the different stereoisomers of a compound like 2-bromobutane so that we know which one we are talking about? We need a system of nomenclature that indicates the arrangement of the atoms or groups about the asymmetric center. Chemists use the letters *R* and *S* for this purpose. For any pair of enantiomers with one asymmetric center, one member will have the ***R* configuration** and the other will have the ***S* configuration**. The *R,S* system was devised by R. S. Cahn, C. Ingold, and V. Prelog.

Let's first look at how we can determine the configuration of a compound if we have a three-dimensional model.

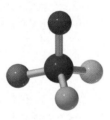

1. **Rank the groups (or atoms) bonded to the asymmetric center in order of priority.** The atomic numbers of the atoms directly attached to the asymmetric center determine the relative priorities. The higher the atomic number, the higher the priority. This should remind you of the way that relative priorities are determined for *E* and *Z* isomers (Section 3.5), because the system of priorities was originally devised for the *R,S* system of nomenclature and was later adopted for the *E,Z* system.

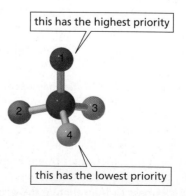

this has the highest priority
1
2
3
4
this has the lowest priority

2. **Orient the molecule so that the group (or atom) with the lowest priority (4) is directed away from you. Then draw an imaginary arrow from the group (or atom) with the highest priority (1) to the group (or atom) with the next highest priority (2).** If the arrow points clockwise, the asymmetric center

has the *R* configuration (*R* is for *rectus*, which is Latin for "right"). If the arrow points counterclockwise, the asymmetric center has the *S* configuration (*S* is for *sinister*, which is Latin for "left").

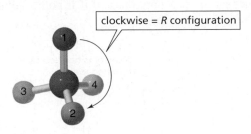

clockwise = *R* configuration

The molecule is oriented so the group with the lowest priority points away from the viewer. If an arrow drawn from the highest priority group to the next-highest priority group points clockwise, the molecule has the *R* configuration.

left turn

right turn

If you forget which direction corresponds to which configuration, imagine driving a car and turning the steering wheel clockwise to make a right turn or counterclockwise to make a left turn.

PROBLEM 9◆

Which of the following molecular models are identical?

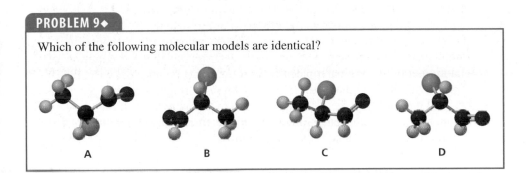

A   B   C   D

If you are able to visualize spatial relationships easily, the above two rules are all you need to determine whether the asymmetric center of a molecule written on a two-dimensional piece of paper has the *R* or the *S* configuration. Mentally rotate the molecule so that the group (or atom) with the lowest priority (4) is directed away from you, then draw an imaginary arrow from the group (or atom) with the highest priority to the group (or atom) with the next highest priority.

If you have trouble visualizing spatial relationships and you don't have access to a model, the following sets of instructions will allow you to determine the configuration about an asymmetric center without having to rotate the molecule mentally.

First, let's look at how you can determine the configuration of a compound drawn as a perspective formula. We will use the enantiomers of 2-bromobutane as an example.

$$ \underset{\text{the enantiomers of 2-bromobutane}}{CH_3CH_2\overset{\displaystyle \overset{Br}{|}}{\underset{CH_3}{C}}{}^{\text{....}}H \qquad H^{\text{....}}\overset{\displaystyle \overset{Br}{|}}{\underset{CH_3}{C}}CH_2CH_3} $$

1. Rank the groups (or atoms) that are bonded to the asymmetric center in order of priority. In our example, bromine has the highest priority (1), the ethyl group has the second highest priority (2), the methyl group is next (3), and hydrogen has the lowest priority (4). (Revisit Section 3.5 if you don't understand how these priorities are assigned.)

$$ CH_3\underset{2}{CH_2}\overset{\displaystyle \overset{1}{\overset{Br}{|}}}{\underset{\underset{3}{CH_3}}{C}}{}^{\text{....}}\overset{4}{H} \qquad \overset{4}{H}^{\text{....}}\overset{\displaystyle \overset{1}{\overset{Br}{|}}}{\underset{\underset{3}{CH_3}}{C}}\underset{2}{CH_2CH_3} $$

2. If the group (or atom) with the lowest priority is bonded by a hatched wedge, draw an arrow from the group (or atom) with the highest priority (1) to the group (or atom) with the second highest priority (2). If the arrow points clockwise, the compound has the *R* configuration, and if it points counterclockwise, the compound has the *S* configuration.

the group with the lowest priority is bonded by a hatched wedge

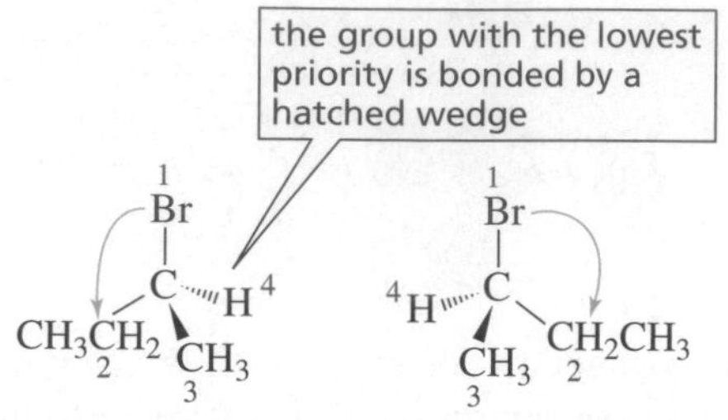

(*S*)-2-bromobutane   (*R*)-2-bromobutane

3. If the group with the lowest priority (4) is not bonded by a hatched wedge, interchange two groups, so group 4 is bonded by a hatched wedge. Then proceed as in step 2: Draw an arrow from the group (or atom) with the highest priority (1) to the group (or atom) with the second-highest priority (2). Because you have interchanged two groups, you are now determining the configuration of the enantiomer of the original molecule. So if the arrow points clockwise, the *enantiomer* (with the interchanged groups) has the *R* configuration, which means the original molecule has the *S* configuration. On the other hand, if the arrow points counterclockwise, the enantiomer (with the interchanged groups) has the *S* configuration, which means the original molecule has the *R* configuration.

Clockwise specifies *R* if the lowest priority substituent is on a hatched wedge.

Counterclockwise specifies *S* if the lowest priority substituent is on a hatched wedge.

what is its configuration?

switch
CH₃ and H

this molecule has the *R* configuration; therefore, the molecule had the *S* configuration before the groups were switched

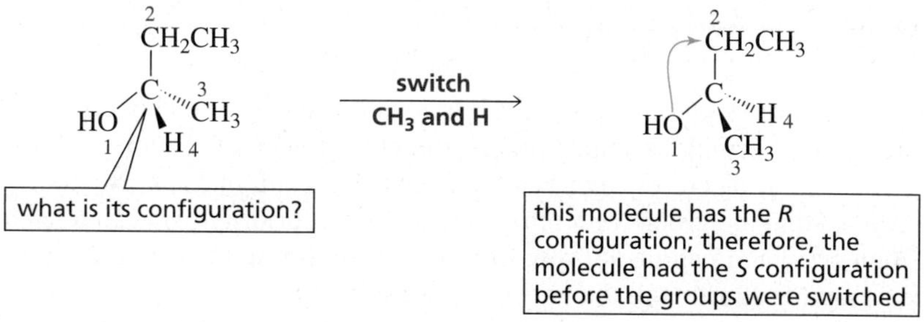

4. In drawing the arrow from group 1 to group 2, you can draw past the group with the lowest priority (4), but never draw past the group with the next-lowest priority (3).

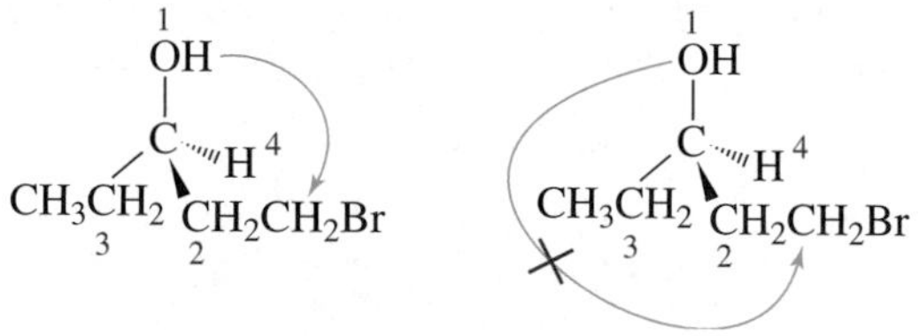

PROBLEM 10◆

Indicate the configuration of each of the following structures:

a.

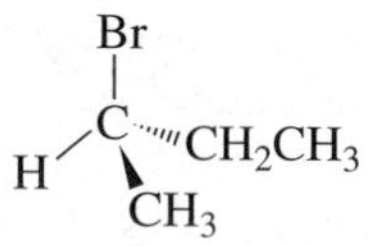

b.

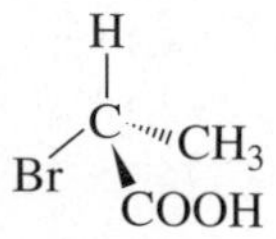

c.

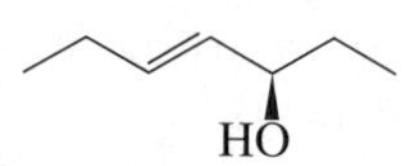

d.

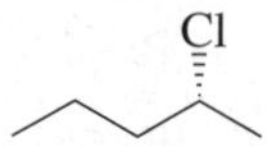

Now let's see how to determine the configuration of a compound drawn as a Fischer projection.

1. Rank the groups (or atoms) that are bonded to the asymmetric center in order of priority.

2. Draw an arrow from the group (or atom) with the highest priority (1) to the group (or atom) with the next highest priority (2). If the arrow points clockwise, the enantiomer has the *R* configuration; if it points counterclockwise, the enantiomer has the *S* configuration, *provided that the group with the lowest priority (4) is on a vertical bond.*

$\overset{1}{\text{Cl}}$

$\underset{3}{\text{CH}_3\text{CH}_2}\!\!-\!\!\underset{2}{\text{CH}_2\text{CH}_2\text{CH}_3}$
$\underset{4}{\text{H}}$

(R)-3-chlorohexane

$\overset{1}{\text{Cl}}$

$\underset{2}{\text{CH}_3\text{CH}_2\text{CH}_2}\!\!-\!\!\underset{3}{\text{CH}_2\text{CH}_3}$
$\underset{4}{\text{H}}$

(S)-3-chlorohexane

3. If the group (or atom) with the lowest priority is on a *horizontal* bond, the answer you get from the direction of the arrow will be the opposite of the correct answer. For example, if the arrow points clockwise, suggesting that the asymmetric center has the *R* configuration, it actually has the *S* configuration; if the arrow points counterclockwise, suggesting that the asymmetric center has the *S* configuration, it actually has the *R* configuration. In the following example, the group with the lowest priority is on a horizontal bond, so clockwise signifies the *S* configuration, not the *R* configuration.

Clockwise specifies *R* if the lowest priority substituent is on a vertical bond.

Clockwise specifies *S* if the lowest priority substituent is on a horizontal bond.

$\overset{3}{\text{CH}_3}$
$\underset{4}{\text{H}}\!-\!\!\overset{1}{\text{OH}}$
$\underset{2}{\text{CH}_2\text{CH}_3}$

(S)-2-butanol

$\overset{3}{\text{CH}_3}$
$\underset{1}{\text{HO}}\!-\!\!\underset{4}{\text{H}}$
$\underset{2}{\text{CH}_2\text{CH}_3}$

(R)-2-butanol

4. In drawing the arrow from group 1 to group 2, you can draw past the group (or atom) with the lowest priority (4), but never draw past the group (or atom) with the next-lowest priority (3).

$\overset{\text{O}}{\overset{\|}{\underset{2}{\text{C}}\text{OH}}}$
$\underset{4}{\text{H}}\!-\!\!\underset{3}{\text{CH}_3}$
$\underset{1}{\text{OH}}$

(S)-lactic acid

$\overset{\text{O}}{\overset{\|}{\underset{2}{\text{C}}\text{OH}}}$
$\underset{3}{\text{CH}_3}\!-\!\!\underset{4}{\text{H}}$
$\underset{1}{\text{OH}}$

(R)-lactic acid

Note to the student

When comparing two Fischer projections to see if they are the same or different, never rotate one 90° or flip it "front-to-back," because that is a quick way to get a wrong answer. A Fischer projection can be rotated 180° in the plane of the paper, but that is the only way you can move it without risking an incorrect answer.

PROBLEM 11◆

Indicate the configuration of each of the following structures:

a. 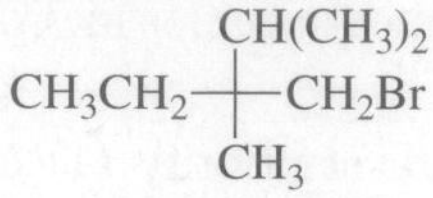 $\text{CH}_3\text{CH}_2 \underset{\overset{|}{\text{CH}_3}}{\overset{\overset{\text{CH(CH}_3)_2}{|}}{-\!\!\!-\!\!\!-}} \text{CH}_2\text{Br}$

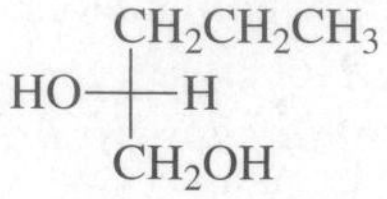

c. 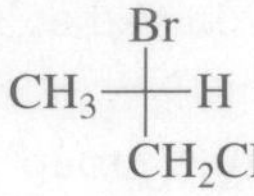 $\text{CH}_3 \underset{\overset{|}{\text{CH}_2\text{CH}_3}}{\overset{\overset{\text{Br}}{|}}{-\!\!\!-\!\!\!-}} \text{H}$

b. $\text{HO} \underset{\overset{|}{\text{CH}_2\text{OH}}}{\overset{\overset{\text{CH}_2\text{CH}_2\text{CH}_3}{|}}{-\!\!\!-\!\!\!-}} \text{H}$

d. 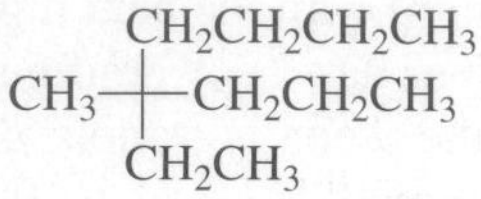 $\text{CH}_3 \underset{\overset{|}{\text{CH}_2\text{CH}_3}}{\overset{\overset{\text{CH}_2\text{CH}_2\text{CH}_2\text{CH}_3}{|}}{-\!\!\!-\!\!\!-}} \text{CH}_2\text{CH}_2\text{CH}_3$

PROBLEM-SOLVING STRATEGY

Recognizing Pairs of Enantiomers

Do each of the following pairs of structures represent identical molecules or a pair of enantiomers?

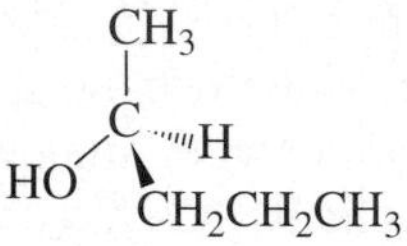

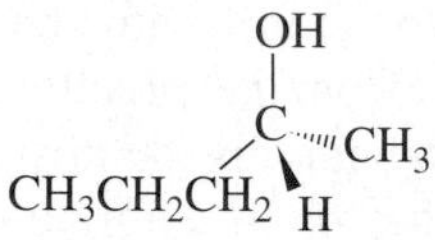

The easiest way to find out whether two molecules are enantiomers or identical molecules is to determine their configurations. If one has the *R* configuration and the other has the *S* configuration, they are enantiomers. If they both have the *R* configuration or both have the *S* configuration, they are identical molecules. Because the structure on the left has the *S* configuration and the structure on the right has the *R* configuration, we know that they represent a pair of enantiomers.

Now continue on to Problem 12.

PROBLEM 12◆ *SOLVED*

Do the following structures represent identical molecules or a pair of enantiomers?

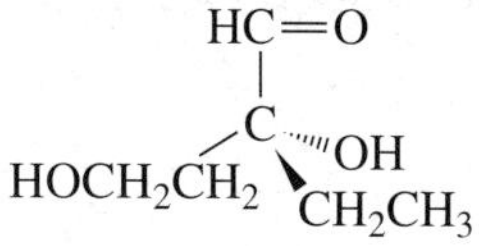

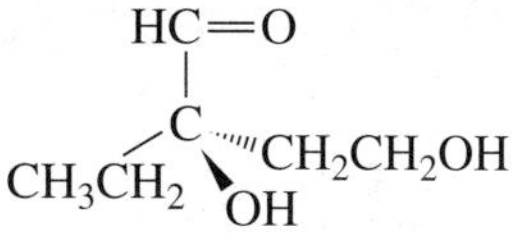

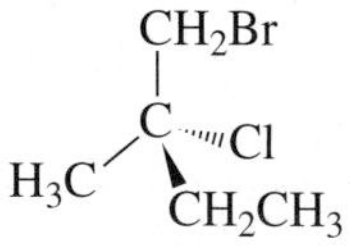

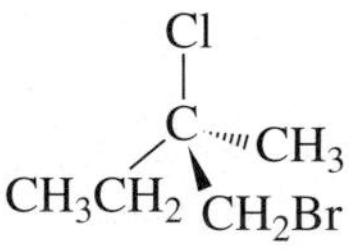

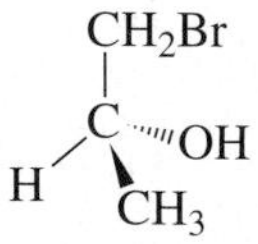

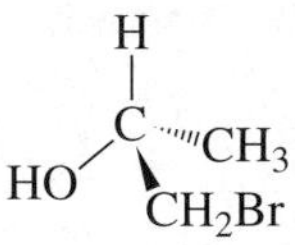

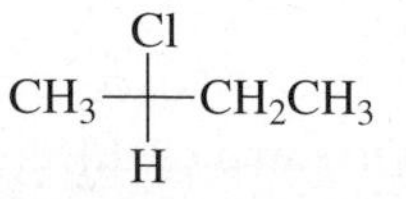

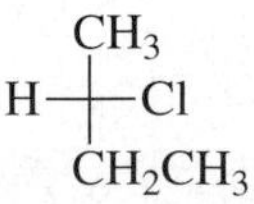

Solution to 12a The first structure shown in part a has the *R* configuration, and the second structure also has the *R* configuration. Because they both have the same configuration, the structures represent identical molecules.

PROBLEM-SOLVING STRATEGY

Drawing an Enantiomer with a Desired Configuration

(*S*)-Alanine is a naturally occurring amino acid. Draw its structure using a perspective formula.

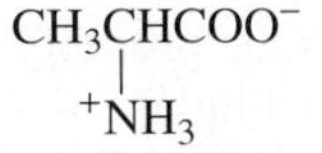

$$CH_3CHCOO^-$$
$$|$$
$$^+NH_3$$
alanine

First draw the bonds about the asymmetric center. Remember that the two bonds in the plane of the paper must be adjacent to one another.

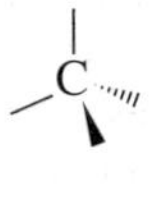

Put the group with the lowest priority on the hatched wedge. Put the group with the highest priority on any remaining bond.

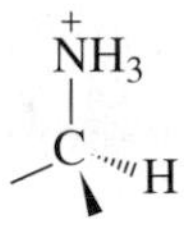

Because you have been asked to draw the *S* enantiomer, draw an arrow counterclockwise from the group with the highest priority to the next available bond and put the group with the next highest priority on that bond.

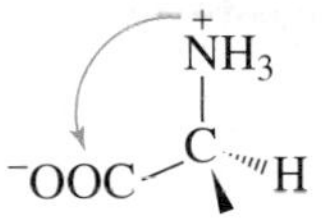

Put the remaining substituent on the last available bond.

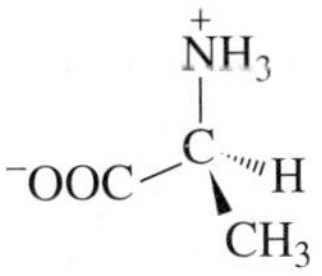

Now continue on to Problem 13.

PROBLEM 13

Draw perspective formulas for the following:

a. (*S*)-2-chlorobutane

b. (*R*)-1,2-dibromobutane

PROBLEM 14◆

Assign relative priorities to the groups or atoms within each of the following sets:

a. $-CH_2OH$ $-CH_3$ $-CH_2CH_2OH$ $-H$

b. $-CH{=}O$ $-OH$ $-CH_3$ $-CH_2OH$

c. $-CH(CH_3)_2$ $-CH_2CH_2Br$ $-Cl$ $-CH_2CH_2CH_2Br$

d. $-CH{=}CH_2$ $-CH_2CH_3$ $-CH_3$

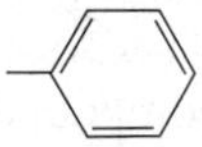

5.8 Chiral Compounds Are Optically Active

Enantiomers share many of the same properties; they have the same boiling points, the same melting points, and the same solubilities. In fact, all the physical properties of enantiomers are the same except those that stem from how groups bonded to the asymmetric center are arranged in space. One property that enantiomers do not share is the way they interact with plane-polarized light.

Normal light, such as that coming from a light bulb or the sun, consists of rays that oscillate in all directions. In contrast, all the rays in a beam of **plane-polarized light** (or simply polarized light) oscillate in a single plane. Plane-polarized light is produced by passing normal light through a polarizer, such as a polarized lens or a Nicol prism. Only light oscillating in a certain plane can pass through.

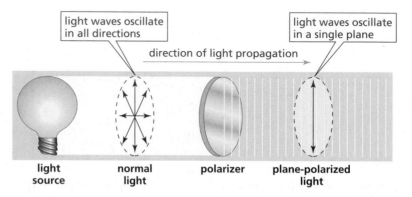

You can experience the effect of a polarized lens by wearing a pair of polarized sunglasses. Polarized sunglasses allow only light oscillating in a single plane to pass through, which is why they block reflections (glare) more effectively than nonpolarized sunglasses do.

In 1815, the physicist Jean-Baptiste Biot discovered that certain naturally occurring organic substances, such as camphor and oil of turpentine, are able to rotate the plane of polarization of polarized light. He noted that some compounds rotated the plane clockwise and others counterclockwise, while some did not rotate the plane of polarization at all. He predicted that the ability to rotate the plane of polarization was attributable to some asymmetry in the molecules. Van't Hoff and Le Bel later determined that the asymmetry was associated with compounds having one or more asymmetric centers.

When polarized light passes through a solution of achiral molecules, the light emerges from the solution with its plane of polarization unchanged. *An achiral compound does not rotate the plane of polarization. It is optically inactive.*

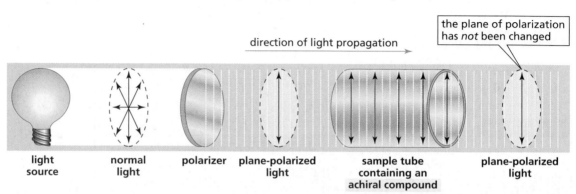

However, when polarized light passes through a solution of a chiral compound, the light emerges with its plane of polarization changed. Thus, *a chiral compound rotates the plane of polarization.* A chiral compound can rotate the plane of polarization clockwise or counterclockwise. If one enantiomer rotates the plane of polarization clockwise, its mirror image will rotate the plane of polarization exactly the same amount counterclockwise.

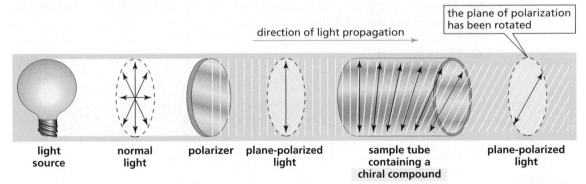

| light source | normal light | polarizer | plane-polarized light | sample tube containing a chiral compound | plane-polarized light |

A compound that rotates the plane of polarization is said to be **optically active**. In other words, chiral compounds are optically active and achiral compounds are **optically inactive**.

If an optically active compound rotates the plane of polarization clockwise, the compound is said to be **dextrorotatory**, indicated in the compound's name by the prefix (+). If it rotates the plane of polarization counterclockwise, it is said to be **levorotatory**, indicated by (−). *Dextro* and *levo* are Latin prefixes for "to the right" and "to the left," respectively. Sometimes lowercase *d* and *l* are used instead of (+) and (−).

Do not confuse (+) and (−) with *R* and *S*. The (+) and (−) symbols indicate the direction in which an optically active compound rotates the plane of polarization, whereas *R* and *S* indicate the arrangement of the groups about an asymmetric center. Some compounds with the *R* configuration are (+) and some are (−).

We can tell by looking at the structure of a compound whether it has the *R* or the *S* configuration, but the only way we can tell whether a compound is dextrorotatory (+) or levorotatory (−) is to put the compound in a polarimeter, an instrument that measures the direction and the amount the plane-polarized light is rotated. For example, (*S*)-lactic acid and (*S*)-sodium lactate both have an *S* configuration, but (*S*)-lactic acid has been found to be dextrorotatory whereas (*S*)-sodium lactate is levorotatory. When we know the direction an optically active compound rotates the plane of polarization, we can incorporate (+) or (−) into its name.

CH₃   CH₃
| |
C–H   C–H
HO   COOH   HO   COO⁻Na⁺

(*S*)-(+)-lactic acid   (*S*)-(−)-sodium lactate

PROBLEM 15◆

a. Is (*R*)-lactic acid dextrorotatory or levorotatory?

b. Is (*R*)-sodium lactate dextrorotatory or levorotatory?

5.9 How Specific Rotation Is Measured

Figure 5.2 provides a simplified description of how a **polarimeter** functions. The amount of rotation caused by an optically active compound will vary with the wavelength of the light being used, so the light source in a polarimeter must produce monochromatic (single wavelength) light. Most polarimeters use light from a sodium arc (called the sodium D-line; wavelength = 589 nm; see Section 12.16). In a polarimeter, monochromatic light passes through a polarizer and emerges as polarized light. The polarized light then passes through a sample tube. If the tube is empty, or filled with

Some molecules with the *R* configuration are (+), and some are (−). Likewise, some molecules with the *S* configuration are (+), and some are (−).

Student Tutorial:
Optical activity

an optically inactive solvent, the light emerges from it with the plane of polarization unchanged. The light then passes through an analyzer, which is a second polarizer mounted on an eyepiece with a dial marked in degrees. Before the start of an experiment, the user looks through the eyepiece and rotates the analyzer until he or she sees total darkness. At this point the analyzer is at a right angle to the first polarizer, so no light passes through. This analyzer setting corresponds to zero rotation.

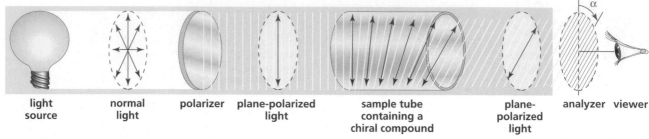

▲ **Figure 5.2**
A schematic drawing of a polarimeter.

When light is filtered through two polarized lenses at a 90° angle to one another, none of the light passes through.

The sample to be measured is then placed in the sample tube. If the sample is optically active, it will rotate the plane of polarization. The analyzer, therefore, will no longer block all the light, so some light reaches the user's eye. The user now rotates the analyzer again until no light passes through. The degree to which the analyzer is rotated can be read from the dial and represents the difference between an optically inactive sample and the optically active sample. This value, which is measured in degrees, is called the **observed rotation** (α). It is dependent on the number of optically active molecules that the light encounters in the sample, which depends in turn on the concentration of the sample and the length of the sample tube. The observed rotation also depends on the temperature and the wavelength of the light source.

Each optically active compound has a characteristic specific rotation. A compound's **specific rotation** is the rotation caused by a solution of 1.0 g of the compound per milliliter of solution in a sample tube 1.0 dm long at a specified temperature and wavelength.* The specific rotation can be calculated from the observed rotation using the following formula:

$$[\alpha]_\lambda^T = \frac{\alpha}{l \times c}$$

where $[\alpha]$ is the specific rotation; T is temperature in °C; λ is the wavelength of the incident light (when the sodium D-line is used, λ is indicated as D); α is the observed rotation; l is the length of the sample tube in decimeters; and c is the concentration of the sample in grams per milliliter of solution.

If one enantiomer has a specific rotation of +5.75, the specific rotation of the other enantiomer must be −5.75, because the mirror-image molecule rotates the plane of polarization the same amount but in the opposite direction. The specific rotations of some common compounds are shown in Table 5.1.

Table 5.1 Specific Rotation of Some Naturally Occurring Compounds	
Cholesterol	−31.5
Cocaine	−16
Codeine	−136
Morphine	−132
Penicillin V	+233
Progesterone (female sex hormone)	+172
Sucrose (table sugar)	+66.5
Testosterone (male sex hormone)	+109

(R)-2-methyl-1-butanol

$[\alpha]_D^{20\ °C} = +5.75$

(S)-2-methyl-1-butanol

$[\alpha]_D^{20\ °C} = -5.75$

---

* Unlike observed rotation, which is measured in degrees, specific rotation has units of 10^{-1} deg cm^2 g^{-1}. In this book, values of specific rotation will be given without units.

> **PROBLEM 16◆**
>
> The observed rotation of 2.0 g of a compound in 50 mL of solution in a polarimeter tube 20-cm long is +13.4°. What is the specific rotation of the compound?

A mixture of equal amounts of two enantiomers—such as (R)-(−)-lactic acid and (S)-(+)-lactic acid—is called a **racemic mixture** or a **racemate**. Racemic mixtures do not rotate the plane of polarized light. They are optically inactive because for every molecule in a racemic mixture that rotates the plane of polarization in one direction, there is a mirror-image molecule that rotates the plane in the opposite direction. As a result, the light emerges from a racemic mixture with its plane of polarization unchanged. The symbol (±) is used to specify a racemic mixture. Thus, (±)-2-bromobutane indicates a mixture of 50% (+)-2-bromobutane and 50% (−)-2-bromobutane.

> **PROBLEM 17◆**
>
> (S)-(+)-Monosodium glutamate (MSG) is a flavor enhancer used in many foods. Some people have an allergic reaction to MSG (headache, chest pain, and an overall feeling of weakness). "Fast food" often contains substantial amounts of MSG, which is widely used in Chinese food as well. (S)-(+)-MSG has a specific rotation of +24.
>
> $$COO^- Na^+$$
> $$|$$
> $$C\text{······}H$$
> $$^-OOCCH_2CH_2 \quad NH_3$$
> $$+$$
>
> **(S)-(+)-monosodium glutamate**
>
> **a.** What is the specific rotation of (R)-(−)-monosodium glutamate?
> **b.** What is the specific rotation of a racemic mixture of MSG?

5.10 Enantiomeric Excess

Whether a particular sample consists of a single enantiomer or a mixture of enantiomers can be determined by its **observed specific rotation**, the specific rotation measured for a particular sample. For example, if a sample of (S)-(+)-2-bromobutane is **enantiomerically pure**—meaning only one enantiomer is present, it will have an *observed specific rotation* of +23.1 because the *specific rotation* of (S)-(+)-2-bromobutane is +23.1. If, however, the sample of 2-bromobutane is a racemic mixture, it will have an observed specific rotation of 0. If the observed specific rotation is positive but less than +23.1, we will know that the sample is a mixture of enantiomers, and that the mixture contains more of the enantiomer with the *S* configuration than the enantiomer with the *R* configuration. The **enantiomeric excess (ee)** tells us how much of an excess of one enantiomer is in the mixture. It can be calculated from the observed specific rotation:

$$\text{enantiomeric excess} = \frac{\text{observed specific rotation}}{\text{specific rotation of the pure enantiomer}} \times 100\%$$

For example, if the sample of 2-bromobutane has an observed specific rotation of +9.2, the enantiomeric excess is 40%. In other words, the excess of one of the enantiomers comprises 40% of the mixture.

$$\text{enantiomeric excess} = \frac{+9.2}{+23.1} \times 100\% = 40\%$$

If the mixture has a 40% enantiomeric excess, 40% of the mixture is excess S enantiomer and 60% is a racemic mixture. Half of the racemic mixture plus the amount of excess S enantiomer equals the amount of the S enantiomer present in the mixture. Thus, 70% of the mixture is the S enantiomer $[(1/2 \times 60) + 40]$ and 30% is the R enantiomer.

PROBLEM 18◆

(+)-Mandelic acid has a specific rotation of +158. What would be the observed specific rotation of each of the following mixtures?

a. 25% (−)-mandelic acid and 75% (+)-mandelic acid
b. 50% (−)-mandelic acid and 50% (+)-mandelic acid
c. 75% (−)-mandelic acid and 25% (+)-mandelic acid

PROBLEM 19◆

Naproxen, a nonsteroidal anti-inflammatory drug that is the active ingredient in Aleve, has a specific rotation of +66. One commercial preparation results in a mixture that has a 97% enantiomeric excess.

a. Does naproxen have the R or the S configuration?
b. What percent of each enantiomer is obtained from the commercial preparation?

PROBLEM 20 **SOLVED**

A solution prepared by mixing 10 mL of a 0.10 M solution of the R enantiomer of a compound and 30 mL of a 0.10 M solution of the S enantiomer was found to have an observed specific rotation of +4.8. What is the specific rotation of each of the enantiomers? (*Hint:* mL × M = millimole, abbreviated mmol)

Solution One (10.0 mL × 0.10 M) mmol of the R enantiomer is mixed with 3 (30.0 mL × 0.10 M) mmol of the S enantiomer; 1 mmol of the R enantiomer plus 1 mmol of the S enantiomer will form 2 mmol of a racemic mixture, so there will be 2 mmol of S enantiomer left over. Because 2 mmol out of 4 mmol is excess S enantiomer (2/4 = 0.50), the solution has a 50% enantiomeric excess.

$$\text{enantiomer excess} = \frac{\text{observed specific rotation}}{\text{specific rotation of the pure enantiomer}} \times 100\%$$

$$50\% = \frac{+4.8}{x} \times 100\%$$

$$x = +4.8 \times 2$$

$$x = +9.6$$

The S enantiomer has a specific rotation of +9.6; the R enantiomer has a specific rotation of −9.6.

5.11 Isomers with More than One Asymmetric Center

Many organic compounds have more than one asymmetric center. The more asymmetric centers a compound has, the more stereoisomers it can have. If we know the number of asymmetric centers, we can calculate the maximum number of stereoisomers for that compound: *a compound can have a maximum of* 2^n *stereoisomers, where* n *equals the number of asymmetric centers* (provided it does not also have stereocenters that would cause it to have cis–trans isomers; see Problem 18). For example, 3-chloro-2-butanol has two asymmetric centers. Therefore, it can have a maximum of four $(2^2 = 4)$

stereoisomers. The four stereoisomers are shown both as perspective formulas and as Fischer projections.

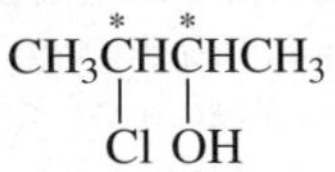

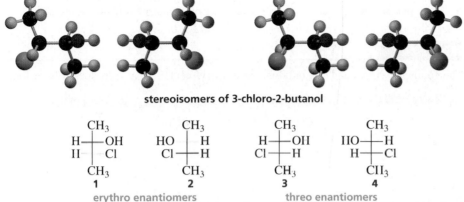

perspective formulas of the stereoisomers of 3-chloro-2-butanol (staggered)

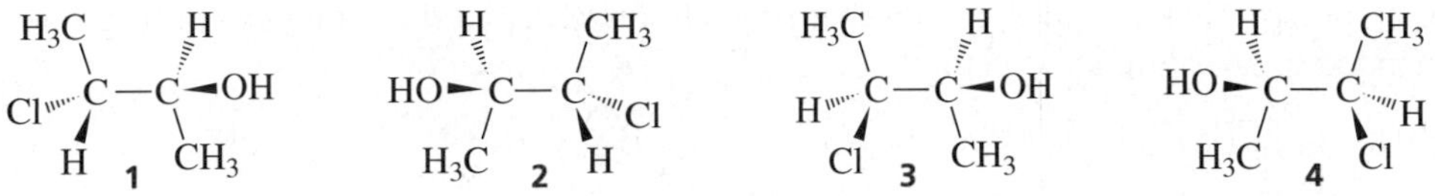

stereoisomers of 3-chloro-2-butanol

Fisher projections of the stereoisomers of 3-chloro-2-butanol

The four stereoisomers of 3-chloro-2-butanol consist of two pairs of enantiomers. Stereoisomers **1** and **2** are nonsuperimposable mirror images. They, therefore, are enantiomers. Stereoisomers **3** and **4** are also enantiomers. Stereoisomers **1** and **3** are not identical, and they are not mirror images. Such stereoisomers are called **diastereomers**. *Diastereomers are stereoisomers that are not enantiomers.* Stereoisomers **1** and **4**, **2** and **3**, and **2** and **4** are also pairs of diastereomers. Notice that the configuration of one of the asymmetric centers is the same in both of a pair of diastereomers but the configuration of the other asymmetric center is different. (Cis–trans isomers are also considered to be diastereomers, because they are stereoisomers that are not enantiomers.)

> **Diastereomers are stereoisomers that are not enantiomers.**

Enantiomers have *identical physical properties* (except for the way they interact with polarized light) and *identical chemical properties*, so they react at the same rate with a given achiral reagent. Diastereomers have *different physical properties*, meaning different melting points, boiling points, solubilities, specific rotations, and so on, and *different chemical properties*, so they react with a given achiral reagent at different rates.

When Fischer projections are drawn for stereoisomers with two adjacent asymmetric centers (such as those for 3-chloro-2-butanol), the enantiomers with similar groups on the same side of the carbon chain are called the **erythro enantiomers** (Section 21.3). Those with similar groups on opposite sides are called the **threo enantiomers**. Therefore, **1** and **2** are the erythro enantiomers of 3-chloro-2-butanol (the hydrogens are on the same side), whereas **3** and **4** are the threo enantiomers. In each of the Fischer projections shown here, the horizontal bonds project out of the paper toward the viewer and the vertical bonds extend behind the paper away from the viewer. Groups can rotate freely about the carbon–carbon single bonds, but Fischer projections show the stereoisomers in their eclipsed conformations.

Because a Fischer projection does not show the three-dimensional structure of the molecule, and because it represents the molecule in a relatively unstable eclipsed conformation, most chemists prefer to use perspective formulas. Perspective formulas show the molecule's three-dimensional structure in a stable, staggered conformation, so they provide a more accurate representation of structure. When perspective formulas are drawn to show the stereoisomers in their less stable eclipsed conformations, we can easily see that the erythro isomers have similar groups on the same side. We will use both perspective formulas and Fischer projections to depict the arrangement of groups bonded to an asymmetric center.

erythro enantiomers   threo enantiomers

perspective formulas of the stereoisomers of 3-chloro-2-butanol (eclipsed)

PROBLEM 21

The following compound has only one asymmetric center. Why then does it have four stereoisomers?

$$CH_3CH_2\overset{*}{C}HCH_2CH{=}CHCH_3$$
$$|$$
$$Br$$

PROBLEM 22◆

Is the following statement correct?

A compound can have a maximum of 2^n stereoisomers, where n equals the number of stereocenters?

PROBLEM 23◆

a. Stereoisomers with two asymmetric centers are called ——— if the configuration of both asymmetric centers in one stereoisomer is the opposite of the configuration of the asymmetric centers in the other stereoisomer.

b. Stereoisomers with two asymmetric centers are called ——— if the configuration of both asymmetric centers in one stereoisomer is the same as the configuration of the asymmetric centers in the other stereoisomer.

c. Stereoisomers with two asymmetric centers are called ——— if one of the asymmetric centers has the same configuration in both stereoisomers and the other asymmetric center has the opposite configuration in the two stereoisomers.

PROBLEM 24◆

The stereoisomer of cholesterol found in nature is

cholesterol

a. How many asymmetric centers does cholesterol have?
b. What is the maximum number of stereoisomers that cholesterol can have?

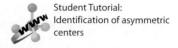

Student Tutorial:
Identification of asymmetric centers

PROBLEM 25

Draw the stereoisomers of the following amino acids. Indicate pairs of enantiomers and pairs of diastereomers.

$$CH_3CHCH_2—CHCOO^-$$
$$| \qquad | $$
$$CH_3 \qquad {}^+NH_3$$
leucine

$$CH_3CH_2CH—CHCOO^-$$
$$| \qquad |$$
$$CH_3 \; {}^+NH_3$$
isoleucine

1-Bromo-2-methylcyclopentane also has two asymmetric centers and four stereoisomers. Because the compound is cyclic, the substituents can be either cis or trans. The cis isomer exists as a pair of enantiomers, and the trans isomer exists as a pair of enantiomers.

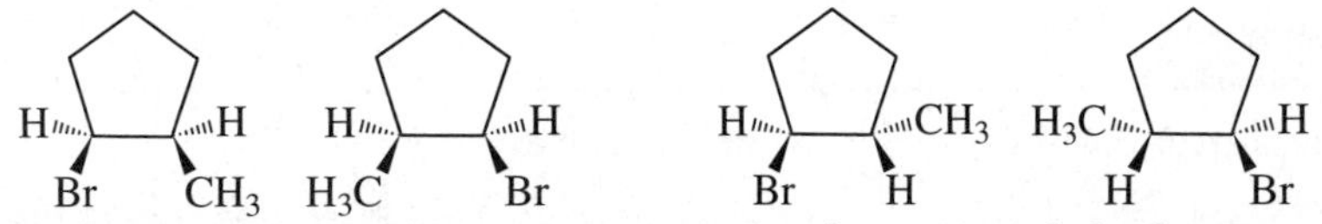

cis-1-bromo-2-methylcyclopentane   trans-1-bromo-2-methylcyclopentane

1-Bromo-3-methylcyclobutane does not have any asymmetric centers. The C-1 carbon has a bromine and a hydrogen attached to it, but its other two groups [—CH$_2$CH(CH$_3$)CH$_2$—] are identical; C-3 has a methyl group and a hydrogen attached to it, but its other two groups [—CH$_2$CH(Br)CH$_2$—] are identical. Because the compound does not have a carbon with four different groups attached to it, it has only two stereoisomers, the cis isomer and the trans isomer. The cis and trans isomers do not have enantiomers.

cis-1-bromo-3-methylcyclobutane   trans-1-bromo-3-methylcyclobutane

1-Bromo-3-methylcyclohexane has two asymmetric centers. The carbon that is bonded to a hydrogen and a bromine is also bonded to two different carbon-containing groups (—CH$_2$CH(CH$_3$)CH$_2$CH$_2$— and —CH$_2$CH$_2$CH$_2$CH(CH$_3$)CH$_2$—), so it is an asymmetric center. The carbon that is bonded to a hydrogen and a methyl group is also bonded to two different carbon-containing groups, so it too is an asymmetric center.

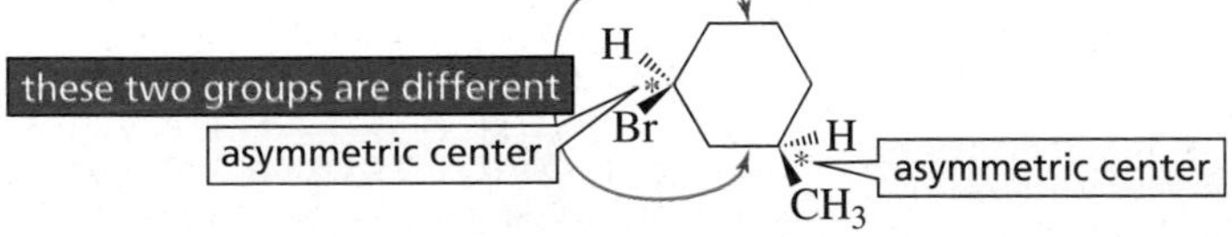

Because the compound has two asymmetric centers, it has four stereoisomers. Enantiomers can be drawn for the cis isomer, and enantiomers can be drawn for the trans isomer. Each of the stereoisomers is a chiral molecule.

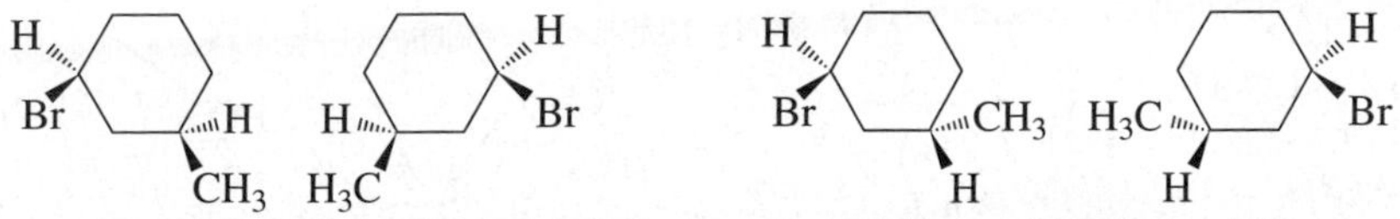

cis-1-bromo-3-methylcyclohexane   trans-1-bromo-3-methylcyclohexane

1-Bromo-4-methylcyclohexane has no asymmetric centers. Therefore, the compound has only one cis isomer and one trans isomer. Each of the stereoisomers is an achiral molecule.

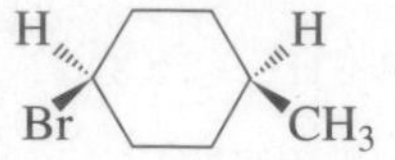

cis-**1-bromo-4-methylcyclohexane** 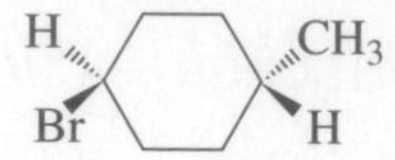   *trans*-**1-bromo-4-methylcyclohexane**

PROBLEM 26

Draw all possible stereoisomers for each of the following compounds:

a. 2-chloro-3-hexanol   **c.** 2,3-dichloropentane
b. 2-bromo-4-chlorohexane   **d.** 1,3-dibromopentane

PROBLEM 27

Draw the stereoisomers of 1-bromo-3-chlorocyclohexane.

PROBLEM 28◆

Of all the possible cyclooctanes that have one chloro substituent and one methyl substituent, which ones do not have any asymmetric centers?

PROBLEM-SOLVING STRATEGY

Drawing Enantiomers and Diastereomers

Draw an enantiomer and a diastereomer for the following compound:

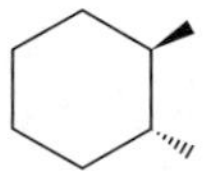

You can draw an enantiomer in one of two ways. You can change the configuration of all the asymmetric centers by changing all wedges to dashes and all dashes to wedges as in **A**. Or, you can draw a mirror image of the compound as in **B**. Notice that since **A** and **B** are each an enantiomer of the given compound, **A** and **B** are identical molecules.

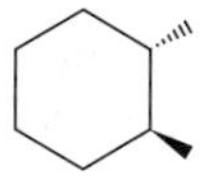

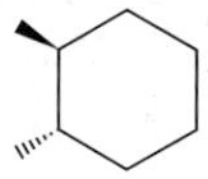

A   B

You can draw a diastereomer by changing the configuration of only one of the asymmetric centers as in **C**.

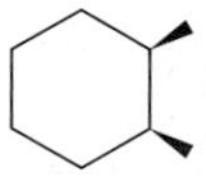

Now continue on to Problem 29.

PROBLEM 29

Draw a diastereomer for each of the following.

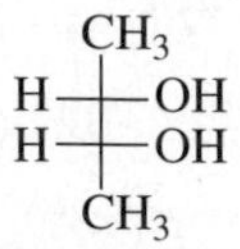

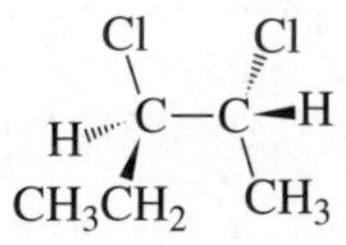

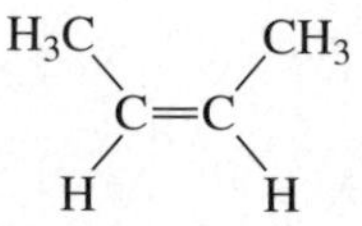

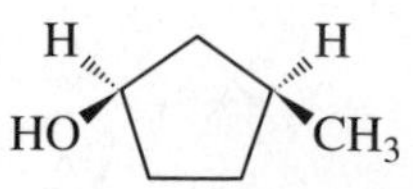

5.12 Meso Compounds Have Asymmetric Centers but Are Optically Inactive

In the examples we have just seen, the compounds with two asymmetric centers had four stereoisomers. However, some compounds with two asymmetric centers have only three stereoisomers. This is why we emphasized in Section 5.11 that the *maximum* number of stereoisomers a compound with *n* asymmetric centers can have is 2^n (unless it has other stereocenters), instead of stating that a compound with *n* asymmetric centers has 2^n stereoisomers.

An example of a compound with two asymmetric centers that has only three stereoisomers is 2,3-dibromobutane.

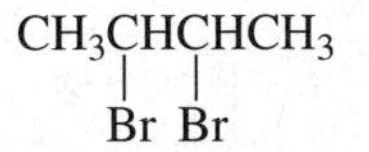

$$CH_3CHCHCH_3$$
$$||$$
$$Br\ Br$$

2,3-dibromobutane

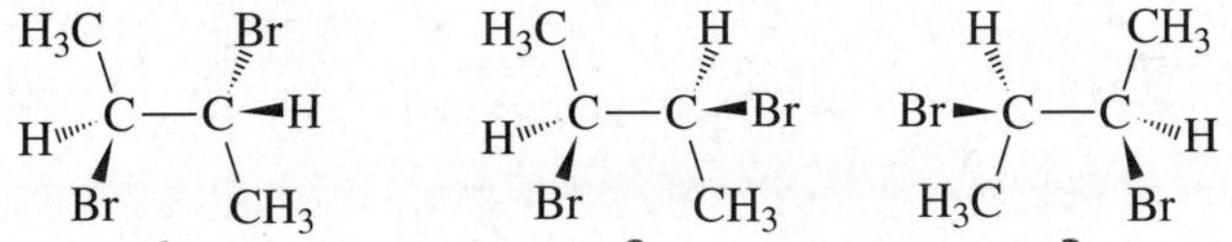

perspective formulas of the stereoisomers of 2,3-dibromobutane (staggered)

The "missing" isomer is the mirror image of **1**, because **1** and its mirror image are the same molecule. This can be seen more clearly when the perspective formulas are drawn in eclipsed conformations or when Fischer projections are used.

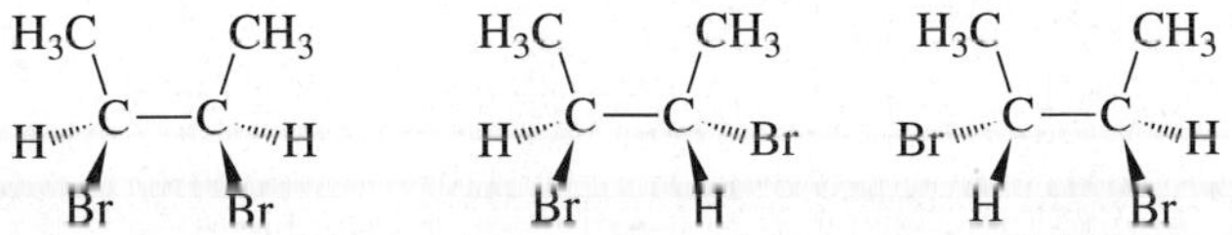

perspective formulas of the stereoisomers of 2,3-dibromobutane (eclipsed)

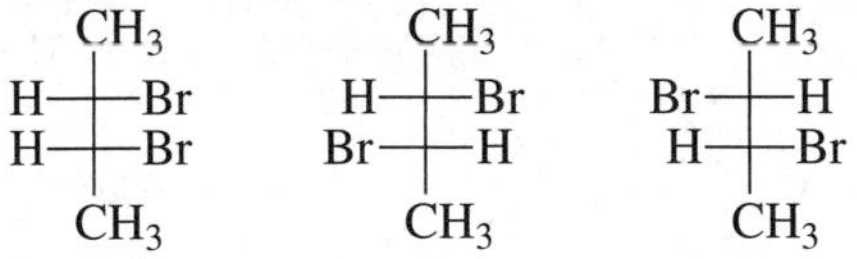

Fischer projections of the stereoisomers of 2,3-dibromobutane

Looking at the perspective formula for the eclipsed conformation, one can easily see that **1** and its mirror image are identical. To convince yourself that the Fischer projection of **1** and its mirror image are identical, rotate the mirror image 180°. (*Remember, you can move Fischer projections only by rotating them 180° in the plane of the paper.*)

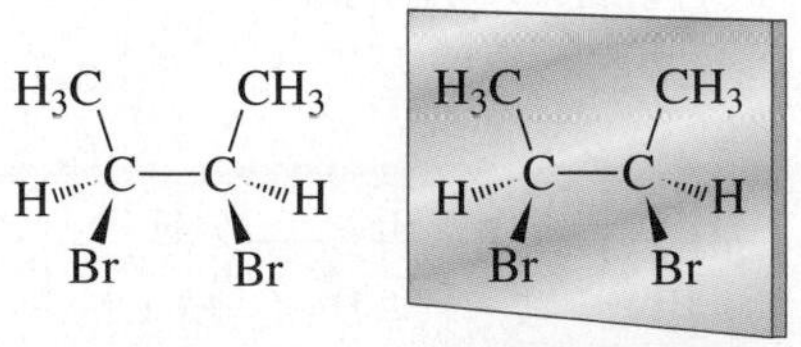

superimposable mirror image

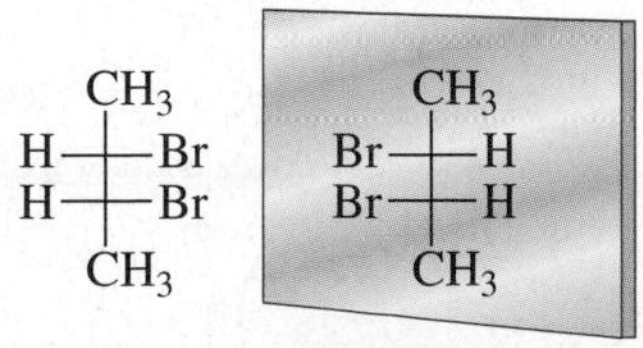

superimposable mirror image

A meso compound is achiral.

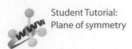

Student Tutorial:
Plane of symmetry

A meso compound has two or more asymmetric centers and a plane of symmetry.

A chiral compound cannot have a plane of symmetry.

If a compound has a plane of symmetry, it will not be optically active even though it has asymmetric centers.

If a compound with two asymmetric centers has the same four groups bonded to each of the asymmetric centers, one of its stereoisomers will be a meso compound.

Stereoisomer **1** is called a meso compound. Even though a **meso** (mee-zo) **compound** has asymmetric centers, it is achiral. A meso compound does not rotate plane-polarized light because it is superimposable on its mirror image. *Mesos* is the Greek word for "middle."

A meso compound can be recognized by the fact that it has two or more asymmetric centers and a plane of symmetry. A **plane of symmetry** cuts the molecule in half so that one half is the mirror image of the other. A molecule with a plane of symmetry does not have an enantiomer. Compare stereoisomer **1**, which has a plane of symmetry and thus no enantiomer, with stereoisomer **2**, which does not have a plane of symmetry and there-fore *does* have an enantiomer. *If a compound has a plane of symmetry, it will not be optically active and will not have an enantiomer, even though it has asymmetric centers.*

meso compounds

It is easy to recognize when a compound with two asymmetric centers has a stereoisomer that is a meso compound—the four atoms or groups bonded to one asymmetric center are identical to the four atoms or groups bonded to the other asymmetric center. *A compound with the same four atoms or groups bonded to two different asymmetric centers will have three stereoisomers: one will be a meso compound, and the other two will be enantiomers.*

a meso compound

enantiomers

a meso compound

enantiomers

In the case of cyclic compounds, the cis isomer will be the meso compound and the trans isomer will exist as enantiomers.

cis-1,3-dimethylcyclopentane
a meso compound

trans-1,3-dimethylcyclopentane
a pair of enantiomers

cis-1,2-dibromocyclohexane
a meso compound

trans-1,2-dibromocyclohexane
a pair of enantiomers

In the perspective formula shown above, *cis*-1,2-dibromocyclohexane appears to have a plane of symmetry. Remember, however, that cyclohexane is not a planar hexagon but exists preferentially in the chair conformation, and the chair conformation of *cis*-1,2-dibromocyclohexane does not have a plane of symmetry. Only the much less stable boat conformation of *cis*-1,2-dibromocyclohexane has a plane of symmetry. This issue raises the question of whether *cis*-1,2-dibromocyclohexane is a meso compound. The answer is yes. As long as one conformation of a compound has a plane of symmetry, the compound will be achiral, and an achiral compound with two asymmetric centers is a meso compound.

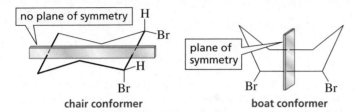

chair conformer   boat conformer

This rule holds for acyclic compounds as well. We have just seen that 2,3-dibromobutane is an achiral meso compound because it has a plane of symmetry. To find its plane of symmetry, however, we had to look at a relatively unstable eclipsed conformation. The more stable staggered conformation does not have a plane of symmetry. 2,3-Dibromobutane is still a meso compound, however, because it has a conformation that has a plane of symmetry.

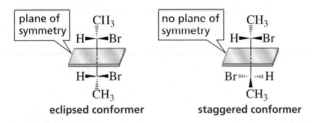

eclipsed conformer   staggered conformer

PROBLEM-SOLVING STRATEGY

Recognizing Whether a Compound Has a Stereoisomer That Is a Meso Compound

Which of the following compounds has a stereoisomer that is a meso compound?

A 2,3-dimethylbutane

B 3,4-dimethylhexane

C 2-bromo-3-methylpentane

D 1,3-dimethylcyclohexane

E 1,4-dimethylcyclohexane

F 1,2-dimethylcyclohexane

G 3,4-diethylhexane

H 1-bromo-2-methylcyclohexane

Check each compound to see if it has the necessary requirements for having a stereoisomer that is a meso compound. That is, does it have two or more asymmetric centers and if it has two asymmetric centers, do they each have the same four substituents attached to them?

Compounds **A**, **E**, and **G** do *not* have a stereoisomer that is a meso compound because they do not have any asymmetric centers.

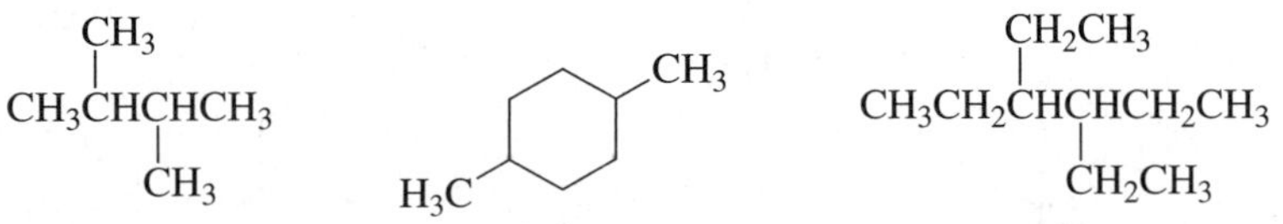

$$CH_3CHCHCH_3$$ with CH_3 above the second carbon and CH_3 below the third carbon

A

$$H_3C \quad \text{(cyclohexane)} \quad CH_3$$

E

$$CH_3CH_2CHCHCH_2CH_3$$ with CH_2CH_3 above the third carbon and CH_2CH_3 below the fourth carbon

G

Compounds **C** and **H** each have two asymmetric centers. They do *not* have a stereoisomer that is a meso compound, however, because each of the asymmetric centers is *not* bonded to the same four substituents.

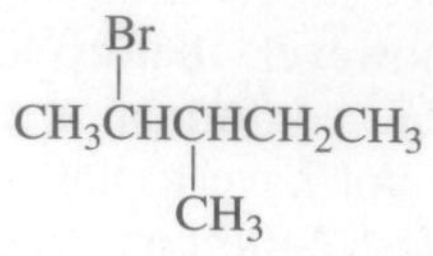

$$\underset{\underset{\overset{|}{CH_3}}{\overset{\overset{Br}{|}}{CH_3CHCHCH_2CH_3}}}{}$$

C

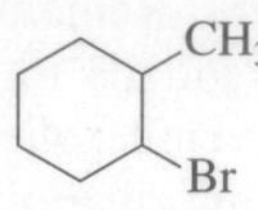

H

Compounds **B, D,** and **F** have two asymmetric centers and each asymmetric center is bonded to the same four atoms or groups. Therefore, these compounds have a stereoisomer that is a meso compound.

$$\underset{\underset{CH_3}{|}}{CH_3CH_2\overset{\overset{CH_3}{|}}{C}HCHCH_2CH_3}$$

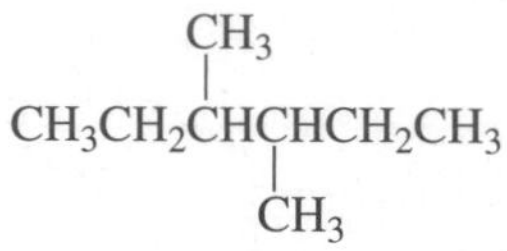

B

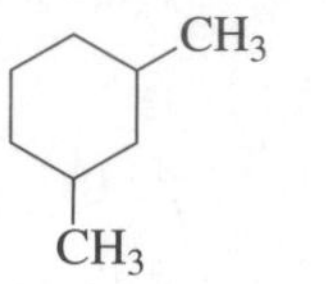

D

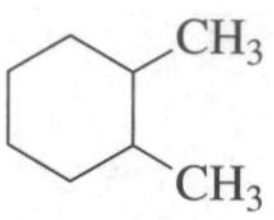

F

In the case of the acyclic compound, the isomer that is the meso compound is the one that has a plane of symmetry when the compound is drawn in its eclipsed conformation (**B**). For the cyclic compounds, the cis isomer is the meso compound (**D** and **F**).

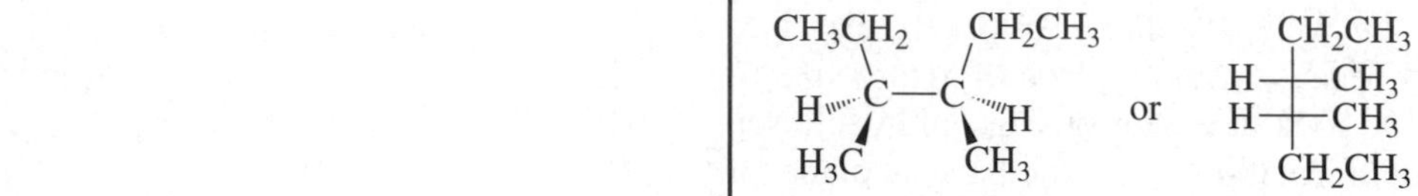

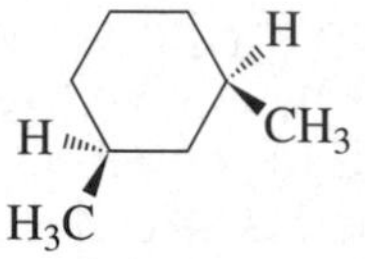

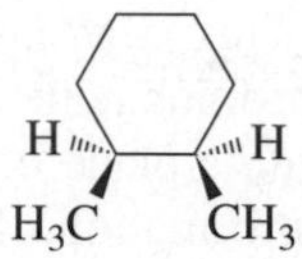

Now continue on to Problem 30.

PROBLEM 30◆

Which of the following compounds has a stereoisomer that is a meso compound?

a. 2,4-dibromohexane

b. 2,4-dibromopentane

c. 2,4-dimethylpentane

d. 1,3-dichlorocyclohexane

e. 1,4-dichlorocyclohexane

f. 1,2-dichlorocyclobutane

PROBLEM 31   *SOLVED*

Which of the following are chiral?

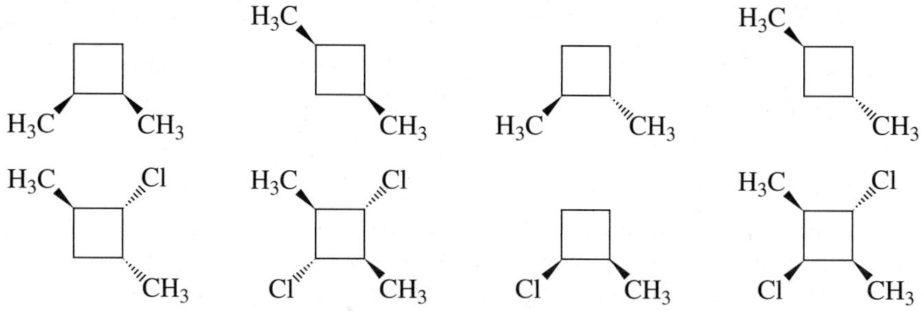

Solution   A chiral molecule does not have plane of symmetry. Therefore, only the following compounds are chiral.

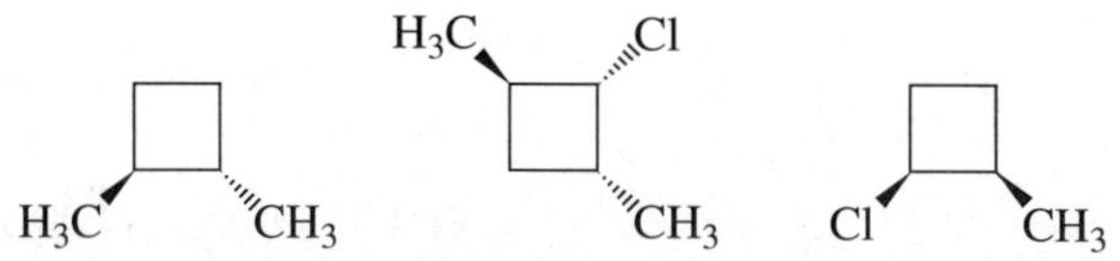

In the top row of compounds, only the third compound is chiral. The first, second, and fourth compounds each have a plane of symmetry. In the bottom row, the first and third compounds are chiral. The second and fourth compounds each have a plane of symmetry.

Draw all the stereoisomers for each of the following compounds:

a. 1-bromo-2-methylbutane

b. 1-chloro-3-methylpentane

c. 2-methyl-1-propanol

d. 2-bromo-1-butanol

e. 3-chloro-3-methylpentane

f. 3-bromo-2-butanol

g. 3,4-dichlorohexane

h. 2,4-dichloropentane

i. 2,4-dichloroheptane

j. 1,2-dichlorocyclobutane

k. 1,3-dichlorocyclohexane

l. 1,4-dichlorocyclohexane

m. 1-bromo-2-chlorocyclobutane

n. 1-bromo-3-chlorocyclobutane

5.13 How to Name Isomers with More than One Asymmetric Center

If a compound has more than one asymmetric center, the steps used to determine whether an asymmetric center has the R or the S configuration must be applied to each of the asymmetric centers individually. As an example, let's name one of the stereoisomers of 3-bromo-2-butanol.

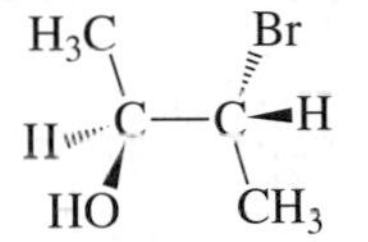

a stereoisomer of 3-bromo-2-butanol

First, we will determine the configuration at C-2. The OH group has the highest priority, the C-3 carbon (the C attached to Br, C, H) has the next-highest priority, CH_3 is next, and H has the lowest priority. Because the group with the lowest priority is bonded by a hatched wedge, we can immediately draw an arrow from the group with the highest priority to the group with the next highest priority. That arrow points counterclockwise, so the configuration at C-2 is S.

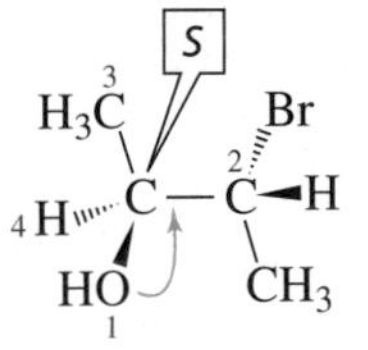

Now we need to determine the configuration at C-3. Because the group with the lowest priority (H) is not bonded by a hatched wedge, we must put it there by temporarily switching two groups.

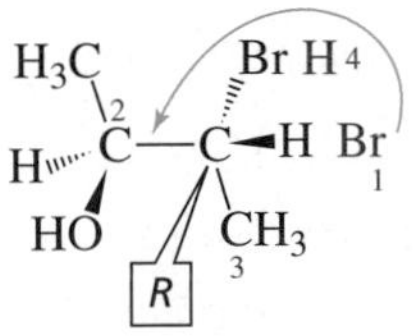

The arrow going from the highest priority group (Br) to the next-highest priority group (the C attached to O, C, H) points counterclockwise, suggesting it has the S configuration. However, because we switched two groups before we drew the arrow, C-3 has the opposite configuration—it has the R configuration. Thus, the isomer is named (2S,3R)-3-bromo-2-butanol.

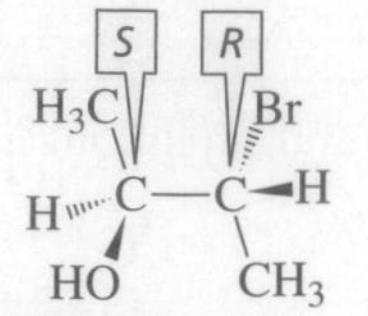

(2S,3R)-3-bromo-2-butanol

When Fischer projections are used, the procedure is similar. Just apply the steps to each asymmetric center that you learned for a Fischer projection with one asymmetric center. At C-2, the arrow from the group with the highest priority to the group with the next highest priority points clockwise, which would suggest an *R* configuration. However, the group with the lowest priority is on a horizontal bond, leading us to conclude that C-2 has the *S* configuration (Section 5.7).

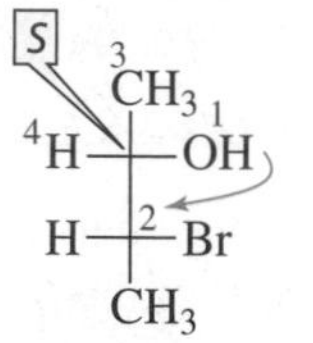

Repeating these steps for C-3 identifies that asymmetric center as having the *R* configuration. Thus, the isomer is named (2S,3R)-3-bromo-2-butanol.

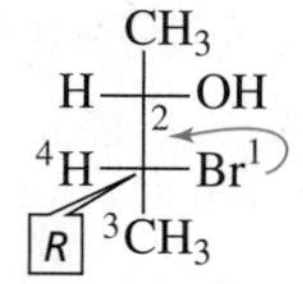

(2S,3R)-3-bromo-2-butanol

The four stereoisomers of 3-bromo-2-butanol are named as shown here. Take a few minutes to verify the names.

| (2S,3R)-3-bromo-2-butanol | (2R,3S)-3-bromo-2-butanol | (2S,3S)-3-bromo-2-butanol | (2R,3R)-3-bromo-2-butanol |

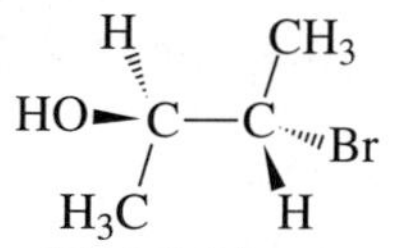

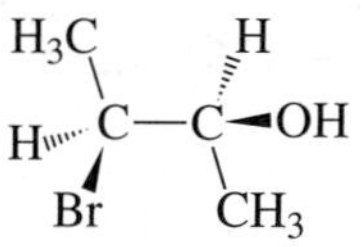

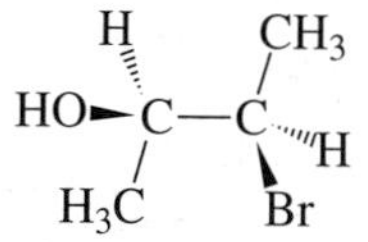

perspective formulas of the stereoisomers of 3-bromo-2-butanol

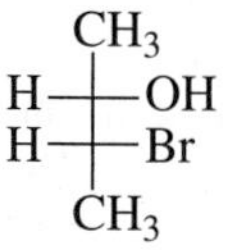

| (2S,3R)-3-bromo-2-butanol | (2R,3S)-3-bromo-2-butanol | (2S,3S)-3-bromo-2-butanol | (2R,3R)-3-bromo-2-butanol |

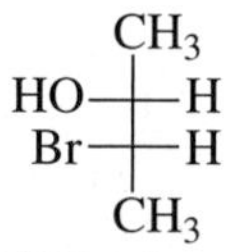

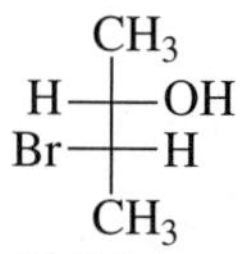

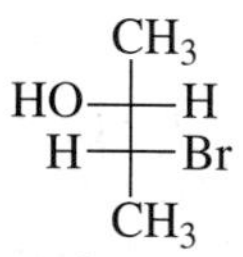

Fischer projections of the stereoisomers of 3-bromo-2-butanol

Notice that enantiomers have the opposite configuration at both asymmetric centers, whereas diastereomers have the same configuration at one asymmetric center and the opposite configuration at the other.

PROBLEM 33

Draw and name the four stereoisomers of 1,3-dichloro-2-butanol using:

a. perspective formulas   **b.** Fischer projections

Tartaric acid has three stereoisomers because each of its two asymmetric centers has the same set of four substituents. The meso compound and the pair of enantiomers are named as shown.

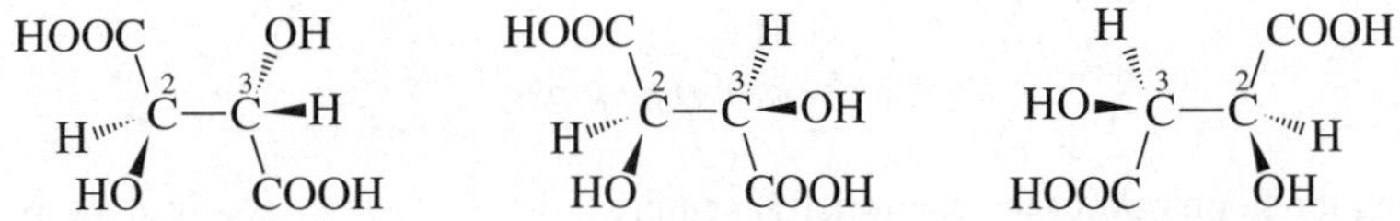

(2R,3S)-tartaric acid
a meso compound

(2R,3R)-tartaric acid
a pair of enantiomers

(2S,3S)-tartaric acid

perspective formulas of the stereoisomers of tartaric acid

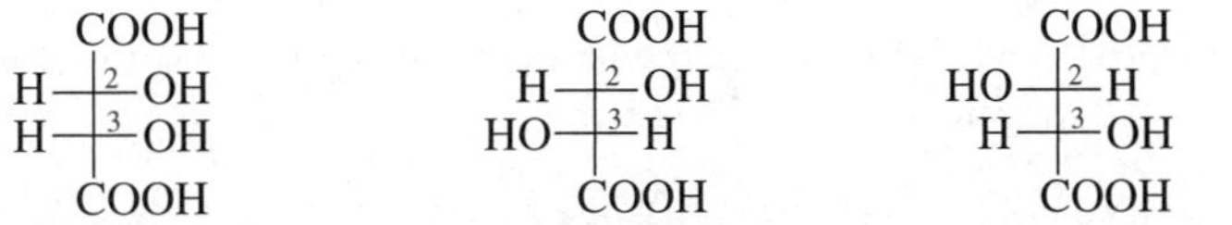

(2R,3S)-tartaric acid
a meso compound

(2R,3R)-tartaric acid
a pair of enantiomers

(2S,3S)-tartaric acid

Fischer projections of the stereoisomers of tartaric acid

The physical properties of the three stereoisomers of tartaric acid are listed in Table 5.2. The meso compound and either of the enantiomers are diastereomers. Notice that the physical properties of the enantiomers are the same, whereas the physical properties of the diastereomers are different. Also notice that the physical properties of the racemic mixture differ from the physical properties of the enantiomers.

Table 5.2   Physical Properties of the Stereoisomers of Tartaric Acid

	Melting point, °C	Specific rotation	Solubility, g/100 g H_2O at 15 °C
(2R,3R)-(+)-Tartaric acid	171	+11.98	139
(2S,3S)-(−)-Tartaric acid	171	−11.98	139
(2R,3S) Tartaric acid (meso)	146	0	125
(±)-Tartaric acid	206	0	

PROBLEM 34◆

Chloramphenicol is a broad-spectrum antibiotic that is particularly useful against typhoid fever. What is the configuration of each asymmetric center in chloramphenicol?

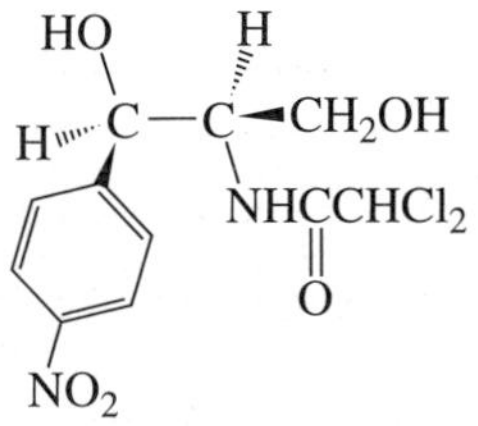

chloramphenicol

PROBLEM-SOLVING STRATEGY

Drawing a Perspective Formula for a Compound with Two Asymmetric Centers

Draw a perspective formula for (2S,3R)-3-chloro-2-pentanol.

First write a condensed structure for the compound, ignoring the configuration at the asymmetric centers.

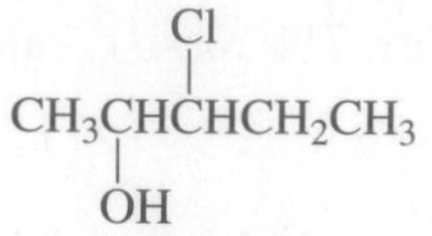

$$CH_3CHCHCH_2CH_3$$

with Cl above and OH below the chain

3-chloro-2-pentanol

Now draw the bonds about the asymmetric centers.

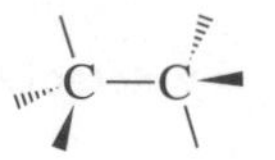

At each asymmetric center, put the group with the lowest priority on the hatched wedge.

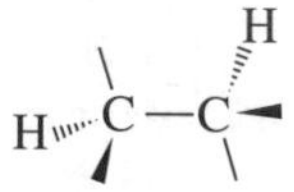

At each asymmetric center, put the group with the highest priority on a bond so that an arrow would point clockwise, if you want the *R* configuration, or counterclockwise, if you want the *S* configuration, to the group with the next highest priority.

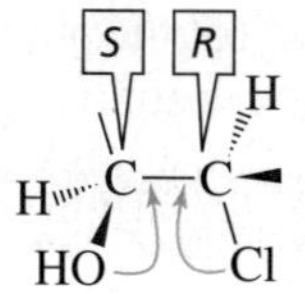

Put the remaining substituents on the last available bonds.

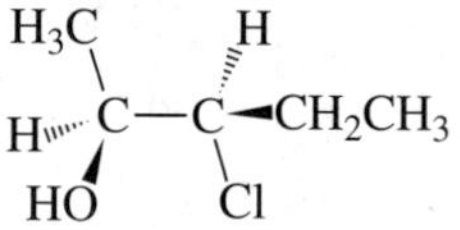

(2S,3R)-3-chloro-2-pentanol

Now continue on to Problem 35.

PROBLEM 35

Draw perspective formulas for the following compounds:

a. (*S*)-3-chloro-1-pentanol

b. (*2R,3R*)-2,3-dibromopentane

c. (*2S,3R*)-3-methyl-2-pentanol

d. (*R*)-1,2-dibromobutane

PROBLEM 36◆

Threonine, an amino acid, has four stereoisomers. The isomer found in nature is (*2S,3R*)-threonine. Which of the following structures represent the naturally occurring amino acid?

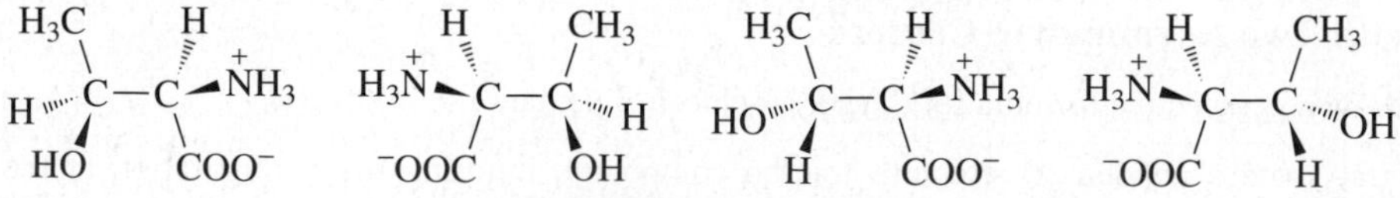

stereosiomers of threonine

PROBLEM 37◆

Name the following compounds:

a.

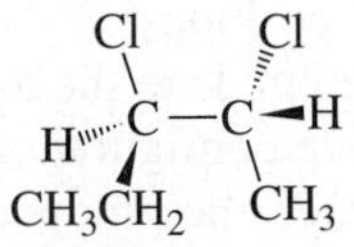

b.

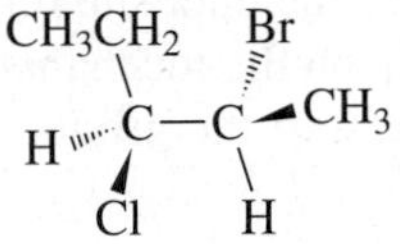

c.

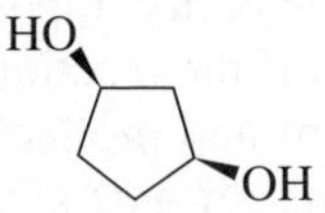

d.

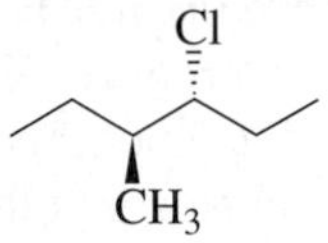

5.14 Reactions of Compounds That Contain an Asymmetric Center

When a compound containing an asymmetric center undergoes a reaction, the effect on the configuration of the asymmetric center depends on the reaction. If the reaction does not break any of the four bonds to the asymmetric center, then the relative positions of the groups bonded to the asymmetric center will not change. For example, when (*S*)-1-chloro-3-methylpentane reacts with hydroxide ion, OH substitutes for Cl. (We will see why OH substitutes for Cl in Section 8.2.) Because the reaction does not break any of the bonds to the asymmetric center, the reactant and product have the same **relative configuration**, meaning the groups maintain their relative positions, so that the CH_3CH_2 group is on the left, the CH_3 group is attached to a solid wedge, and the H is attached to a hatched wedge.

If a reaction does not break a bond to the asymmetric center, the reactant and the product will have the same relative configurations.

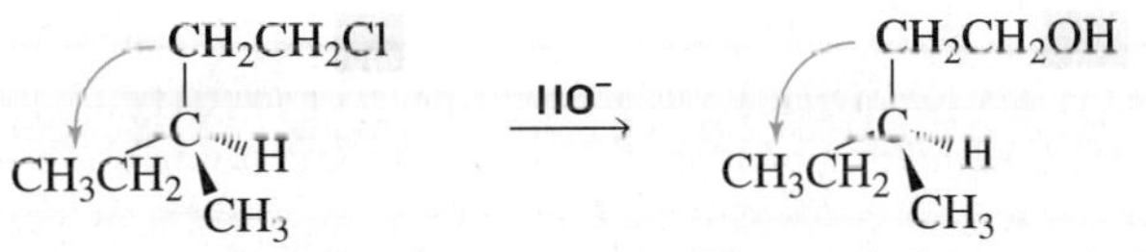

(*S*)-1-chloro-3-methylpentane → (*S*)-3-methyl-1-pentanol

A Word of Warning. Even if the four groups bonded to the asymmetric center maintain their relative positions, an *S* reactant does not always form an *S* product as occurred in the preceding reaction, and an *R* reactant does not always form an *R* product. In the following example, the groups maintain their relative positions during the reaction, giving the reactant and the product the same *relative configurations*. Nevertheless, the reactant has the *S* configuration but the product has the *R* configuration. Although, the groups have maintained their relative positions, their relative priorities—as defined by the Cahn–Ingold–Prelog rules—have changed (the vinyl group has the highest priority in the reactant, whereas the propyl group has the highest priority in the product; Section 3.5). The change in relative priorities—not any change in position of the groups or atoms—is what caused the *S* reactant to become an *R* product.

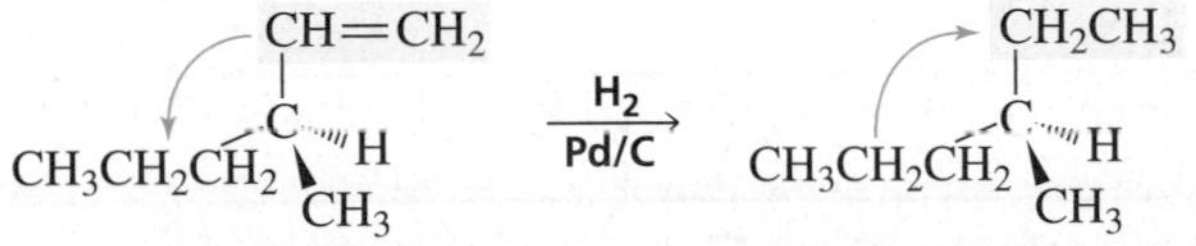

(*S*)-3-methylhexene → (*R*)-3-methylhexane

The reactant and product in this example have the same relative configuration, but because the reactant has the *S* configuration and the product has the *R* configuration, they have different absolute configurations. The **absolute configuration** is the compound's *actual* configuration. In other words, the configuration is known in an absolute sense rather than in a relative one. To know the *absolute configuration* of a compound is to know whether it has the *R* or the *S* configuration. In contrast, to know that two compounds have the same *relative configuration* is to know that the groups or atoms attached to the asymmetric center in both compounds have the same relative positions.

We have just seen that if the reaction does not break any of the bonds to the asymmetric center, the reactant and product will have the same relative configuration. In contrast, if the reaction *does break* a bond to the asymmetric center, the product can have the same relative configuration as the reactant or it can have the opposite relative configuration. Which of the products is actually formed depends on the mechanism of the reaction. Therefore, we cannot predict what the configuration of the product will be unless we know the mechanism of the reaction.

If a reaction does break a bond to the asymmetric center, you cannot predict the configuration of the product unless you know the mechanism of the reaction.

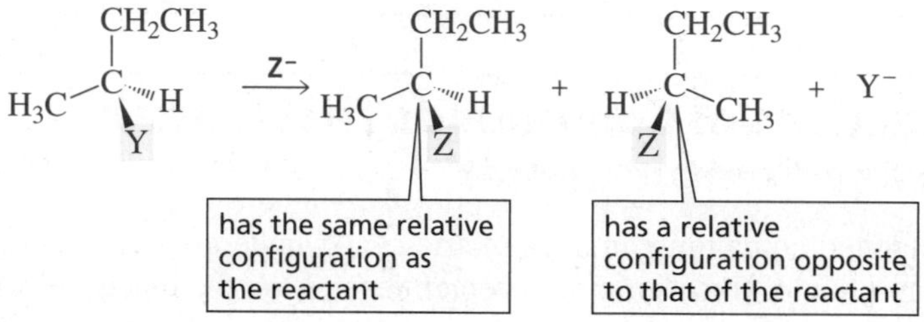

PROBLEM 38 ☐ SOLVED

(*S*)-(−)-2-Methyl-1-butanol can be converted to (+)-2-methylbutanoic acid without breaking any of the bonds to the asymmetric center. What is the configuration of (−)-2-methylbutanoic acid?

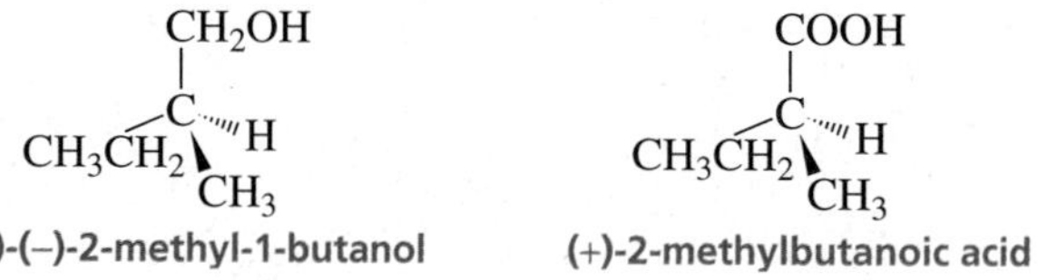

(S)-(−)-2-methyl-1-butanol   (+)-2-methylbutanoic acid

Solution We know that (+)-2-methylbutanoic acid has the relative configuration shown because it was formed from (*S*)-(−)-2-methyl-1-butanol without breaking any bonds to the asymmetric center. Therefore, we know that (+)-2-methylbutanoic acid has the *S* configuration. We can conclude then that (−)-2-methylbutanoic acid has the *R* configuration.

PROBLEM 39◆

The stereoisomer of 1-iodo-2-methylbutane with the *R* configuration rotates the plane of polarized light clockwise. The reaction shown below results in an alcohol that rotates the plane of polarized light counterclockwise. What is the configuration of (+)-2-methyl-1-butanol?

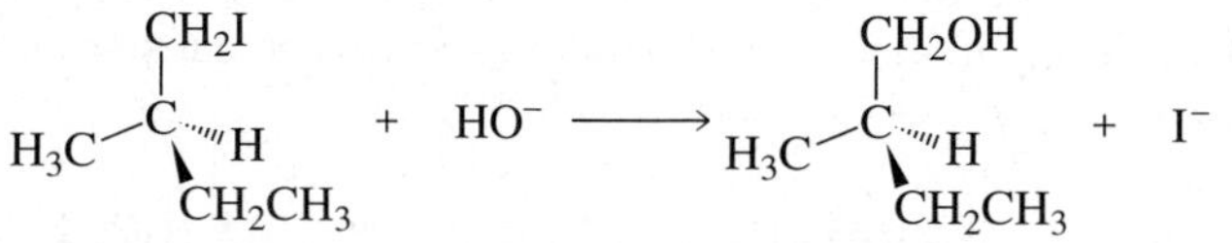

5.15 The Absolute Configuration of (+)-Glyceraldehyde

Glyceraldehyde has one asymmetric center and, therefore, has two stereoisomers. Their absolute configurations were not known until 1951. Thus, before that time, chemists did not know whether (+)-glyceraldehyde had the *R* or the *S* configuration, although they had arbitrarily assigned it the *R* configuration. They had a 50–50 chance of being correct.

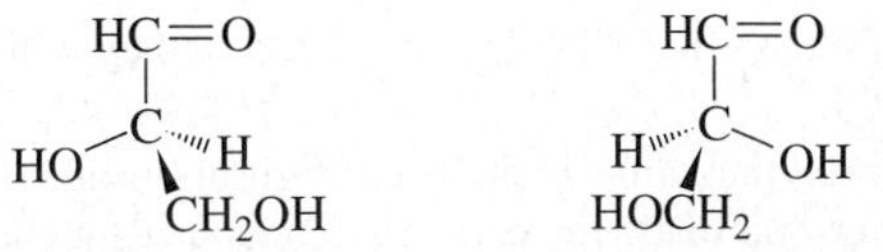

(*R*)-(+)-glyceraldehyde   (*S*)-(–)-glyceraldehyde

The configurations of many organic compounds were originally deduced based on whether they could be synthesized from (+)- or from (−)-glyceraldehyde, or converted to (+)- or to (−)-glyceraldehyde, in reactions that did not break any of the bonds to the asymmetric center. For example, since chemists assumed that the configuration of (+)-glyceraldehyde was the one shown above, they could assume that (−)-lactic acid had the configuration shown below, because a series of reactions that do not break any bonds to the asymmetric center show that (−)-lactic acid and (+)-glyceraldehyde have the same relative configuration. Yet, because chemists did not know for sure that (+)-glyceraldehyde had the *R* configuration, the configurations assigned to these molecules were relative configurations, not absolute configurations. They were relative to (+)-glyceraldehyde, and were based on the *assumption* that (+)-glyceraldehyde had the *R* configuration.

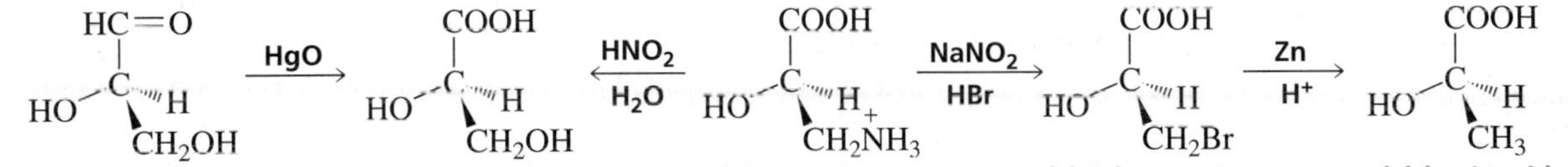

(+)-glyceraldehyde   (–)-glyceric acid   (+)-isoserine   (–)-3-bromo-2-hydroxypropanoic acid   (–)-lactic acid

In 1951, the Dutch chemists J. M. Bijvoet, A. F. Peerdeman, and A. J. van Bommel, using X-ray crystallography and a new technique known as anomalous dispersion, determined that the sodium rubidium salt of (+)-tartaric acid had the *R,R* configuration. Because (+)-tartaric acid could be synthesized from (−)-glyceraldehyde, (−)-glyceraldehyde had to be the *S* enantiomer. The assumption, therefore, that (+)-glyceraldehyde had the *R* configuration was correct!

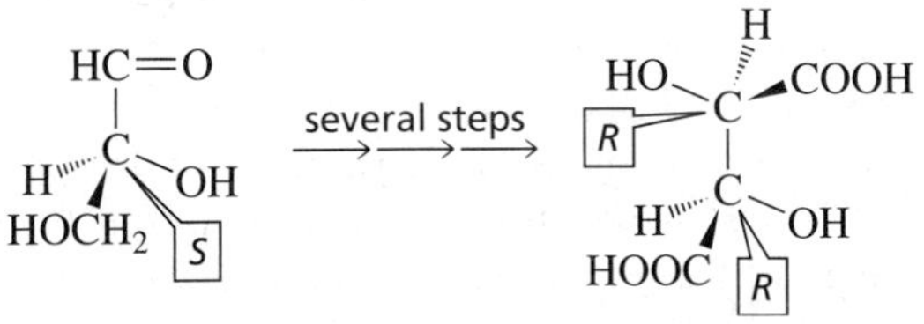

The work of these chemists immediately provided absolute configurations for all the compounds whose relative configurations had been determined by syntheses from or conversions to (+)-glyceraldehyde. Thus, (−)-lactic acid has the configuration shown above. If (+)-glyceraldehyde had been the *S* enantiomer (that is, if it had the configuration opposite to the one shown above), (−)-lactic acid would also have had the configuration opposite to the one shown.

PROBLEM 40◆

What is the absolute configuration of each of the following compounds?

a. (−)-glyceric acid   **c.** (−)-glyceraldehyde

b. (+)-isoserine   **d.** (+)-lactic acid

The French chemist and microbiologist **Louis Pasteur (1822–1895)** *was the first to demonstrate that microbes cause specific diseases. Asked by the French wine industry to find out why wine often went sour while aging, he showed that the microorganisms that cause grape juice to ferment, producing wine, also cause wine to become sour. Gently heating the wine after fermentation, a process called pasteurization, kills the organisms so they cannot sour the wine.*

Eilhardt Mitscherlich (1794–1863), *a German chemist, studied medicine so that he could travel to Asia in order to satisfy his interest in Oriental languages. He later became fascinated by chemistry. He was a professor of chemistry at the University of Berlin and wrote a successful chemistry textbook that was published in 1829.*

PROBLEM 41◆

Which of the following is true?

a. If two compounds have the same relative configuration, they will have the same absolute configuration.

b. If two compounds have the same relative configuration and you know the absolute configuration of either one of them, you can determine the absolute configuration of the other.

c. An *R* reactant always forms an *S* product.

5.16 How Enantiomers Can Be Separated

Enantiomers cannot be separated by the usual separation techniques such as fractional distillation or crystallization because their identical boiling points and solubilities cause them to distill or crystallize simultaneously. Louis Pasteur was the first to succeed in separating a pair of enantiomers. While working with crystals of sodium ammonium tartrate, he noted that the crystals were not identical—some were "right-handed" and some were "left-handed." After painstakingly separating the two kinds of crystals with a pair of tweezers, he found that a solution of the right-handed crystals rotated plane-polarized light clockwise, whereas a solution of the left-handed crystals rotated plane-polarized light counterclockwise.

$$Na^+{}^-OOC \overset{H}{\underset{HO}{\overset{|}{C}}} OH \qquad HO \overset{H}{\underset{H_4N^+{}^-OOC}{\overset{|}{C}}} COO^- Na^+$$

left-handed crystals   **right-handed crystals**

Pasteur, only 26 years old at the time and unknown in scientific circles, was concerned about the accuracy of his observations because a few years earlier, the well-known German organic chemist Eilhardt Mitscherlich had reported that crystals of sodium ammonium tartrate were all identical. Pasteur immediately reported his findings to Jean-Baptiste Biot (Section 5.8) and repeated the experiment with Biot present. Biot was convinced that Pasteur had successfully separated the enantiomers of sodium ammonium tartrate. Pasteur's experiment also gave rise to a new chemical term. Tartaric acid is obtained from grapes, so it was also called racemic acid (*racemus* is Latin for "a bunch of grapes"). This is why a mixture of equal amounts of enantiomers came to be known as a **racemic mixture** (Section 5.9). Separation of enantiomers is called the **resolution of a racemic mixture**.

Crystals of potassium hydrogen tartrate, a naturally occurring salt found in wines. Most fruits produce citric acid, but grapes produce large quantities of tartaric acid instead. Potassium hydrogen tartrate, also called cream of tartar, is used in place of vinegar or lemon juice in some recipes.

Later, chemists recognized how lucky Pasteur had been. Sodium ammonium tartrate forms asymmetric crystals only under the precise conditions that Pasteur happened to employ. Under other conditions, the symmetrical crystals that had fooled Mitscherlich are formed. But, to quote Pasteur, "Chance favors the prepared mind."

Separating enantiomers by hand, as Pasteur did, is not a universally useful method because few compounds form asymmetric crystals. Until relatively recently, separating enantiomers was a very tedious process. The enantiomers had to be chemically converted to diastereomers, which could be separated because they have different physical properties (Section 5.11). After separation, the individual diastereomers had to be chemically converted back to the original enantiomers.

Fortunately enantiomers can now be separated relatively easily by a technique called **chromatography**. In this method, the mixture to be separated is dissolved in a solvent and the solution is passed through a column packed with a chiral material that tends to adsorb organic compounds. The two enantiomers can be expected to move through the column at different rates because they will have different affinities for the chiral material—just as a right hand prefers a right-hand glove to a left-hand glove—so one enantiomer will emerge from the column before the other. Because it is now so much easier to separate enantiomers, many drugs are being sold as single enantiomers rather than as racemic mixtures (Section 5.10).

The chiral material used in chromatography is one example of a **chiral probe**, something capable of distinguishing between enantiomers. A polarimeter is another example of a chiral probe (Section 5.9). In Section 5.21, you will see two kinds of biological molecules—enzymes and receptors—that are chiral probes.

5.17 Nitrogen and Phosphorus Atoms Can Be Asymmetric Centers

Atoms other than carbon can be asymmetric centers. Any atom, such as nitrogen or phosphorus, that has four different groups or atoms attached to it is an asymmetric center. For example, the following pairs of compounds are enantiomers.

enantiomers   enantiomers

If one of the four "groups" attached to nitrogen is a lone pair, the enantiomers cannot be separated because they interconvert rapidly at room temperature. This rapid interconversion is called **amine inversion**. One way to picture amine inversion is to think of an umbrella that turns inside out in a windstorm.

amine inversion

Amine inversion takes place through a transition state in which the sp^3 nitrogen becomes an sp^2 nitrogen. The three groups bonded to the sp^2 nitrogen are coplanar in the transition state with bond angles of 120°, and the lone pair is in a p orbital. The "inverted" and "non-inverted" amine molecules are enantiomers, but they cannot be separated because amine inversion is so rapid. The energy required for amine inversion is approximately 6 kcal/mol (or 25 kJ/mol), about twice the amount of energy required for rotation about a carbon–carbon single bond, but still low enough to allow the enantiomers to interconvert rapidly at room temperature. The lone pair is necessary for inversion: quaternary ammonium ions—ions with four bonds to nitrogen and hence no lone pair—do not invert.

PROBLEM 42

Compound A has two stereoisomers, but compounds B and C exist as single compounds. Explain.

CH=CH$_2$	CH=CH$_2$	
H$_3$C—N$^+$—H Cl$^-$	H$_3$C—N$^+$—H Cl$^-$	H$_3$C—N̈—H
CH$_2$CH$_3$	CH$_3$	CH=CH$_2$
A	**B**	**C**

5.18 Stereochemistry of Reactions: Regioselective, Stereoselective, and Stereospecific Reactions

When we looked at the electrophilic addition reactions that alkenes undergo (Chapter 4), we examined the step-by-step process by which each reaction occurs (the mechanism of the reaction), and we determined what products are formed. However, we did not consider the stereochemistry of the reactions.

Stereochemistry is the field of chemistry that deals with the structures of molecules in three dimensions. When we study the stereochemistry of a reaction, we are concerned with the following questions:

1. If the *product* of a reaction can exist as two or more stereoisomers, does the reaction produce a single stereoisomer, a set of particular stereoisomers, or all possible stereoisomers?

2. If the *reactant* can exist as two or more stereoisomers, do all stereoisomers of the reactant form the same stereoisomers of the product, or does each stereoisomer of the reactant form a different stereoisomer or set of stereoisomers of the product?

Before we examine the stereochemistry of electrophilic addition reactions, we need to become familiar with some terms used in describing the stereochemistry of a reaction.

> **A regioselective reaction forms more of one constitutional isomer than of another.**

We have seen that a **regioselective** reaction is one in which two *constitutional isomers* can be obtained as products, but more of one is obtained than the other (Section 4.4). In other words, a regioselective reaction selects for a particular constitutional isomer. Recall that a reaction can be *moderately regioselective*, *highly regioselective*, or *completely regioselective* depending on the relative amounts of the constitutional isomers formed in the reaction.

a regioselective reaction

$$A \longrightarrow \underbrace{B \ + \ C}_{\text{constitutional isomers}}$$

more B is formed than C

Stereoselective is a similar term, but it refers to the preferential formation of a *stereoisomer* rather than a *constitutional isomer*. If a reaction that generates a carbon–carbon double bond or an asymmetric center in a product forms one stereoisomer preferentially over another, it is a stereoselective reaction. In other words, it selects for a particular stereoisomer. Depending on the degree of preference for a particular stereoisomer, a reaction can be described as being *moderately stereoselective*, *highly stereoselective*, or *completely stereoselective*.

a stereoselective reaction

$$A \longrightarrow \boxed{\text{stereoisomers}} \\ A \longrightarrow B + C$$

more B is formed than C

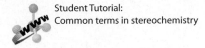

Student Tutorial:
Common terms in stereochemistry

A stereoselective reaction forms more of one stereoisomer than of another.

A reaction is **stereospecific** if the *reactant* can exist as stereoisomers and each stereoisomer of the reactant forms a different stereoisomer or different set of stereoisomers of the product.

In a stereospecific reaction each stereoisomer forms a different stereoisomeric product or a different set of stereoisomeric products.

stereospecific reactions

$$\boxed{\text{stereoisomers}} \Bigg\langle \begin{array}{ccc} A & \longrightarrow & B \\ C & \longrightarrow & D \end{array} \Bigg\rangle \boxed{\text{stereoisomers}}$$

In the preceding reaction, stereoisomer A forms stereoisomer B but does not form D, so the reaction is stereoselective in addition to being stereospecific. *All stereospecific reactions are also stereoselective. However, not all stereoselective reactions are stereospecific*, because there are stereoselective reactions in which the reactant does not have a carbon–carbon double bond or an asymmetric center, so it does not have stereoisomers.

A stereospecific reaction is also stereoselective. A stereoselective reaction is not necessarily stereospecific.

5.19 **The Stereochemistry of Electrophilic Addition Reactions of Alkenes**

Now that you are familiar with electrophilic addition reactions *and* with stereoisomers, we can combine the two topics and look at the stereochemistry of electrophilic addition reactions. In other words, we will look at the stereoisomers that are formed in the electrophilic addition reactions that were discussed in Chapter 4.

We have seen that when an alkene reacts with an electrophilic reagent such as HBr, the major product of the addition reaction is the one obtained by adding the electrophile (H^+) to the sp^2 carbon bonded to the greater number of hydrogens and adding the nucleophile (Br^-) to the other sp^2 carbon (Section 4.4). For example, the major product obtained from the reaction of propene with HBr is 2-bromopropane. This particular product does not have stereoisomers because it does not have an asymmetric center. Therefore, we do not have to be concerned with the stereochemistry of this reaction.

$$CH_3CH{=}CH_2 \xrightarrow{\ \textbf{HBr}\ } CH_3\overset{+}{C}HCH_3 \longrightarrow CH_3\underset{\underset{Br}{|}}{C}HCH_3$$
$$\text{propene} \qquad\qquad\qquad Br^- \qquad\qquad \textbf{2-bromopropane} \\ \textbf{major product}$$

If, however, the reaction creates a product with an asymmetric center, we need to know which stereoisomers are formed. For example, the reaction of HBr with 1-butene forms 2-bromobutane, a compound with an asymmetric center. But what is the configuration of the product? Do we get the *R* enantiomer, the *S* enantiomer, or both?

$$CH_3CH_2CH{=}CH_2 \xrightarrow{\ \textbf{HBr}\ } CH_3CH_2\overset{+}{C}HCH_3 \longrightarrow CH_3CH_2\underset{\underset{Br}{|}}{C}HCH_3$$
$$\text{1-butene} \qquad\qquad\qquad Br^- \qquad\qquad\qquad \boxed{\text{asymmetric center}} \\ \textbf{2-bromobutane}$$

We will begin our discussion of the stereochemistry of electrophilic addition reactions by looking at reactions that form a product with one asymmetric center. Then we will look at reactions that form a product with two asymmetric centers.

Addition Reactions that Form a Product with One Asymmetric Center

When a reactant that does not have an asymmetric center undergoes a reaction that forms a product with *one* asymmetric center, the product will always be a racemic mixture. For example, the reaction of 1-butene with HBr forms identical amounts of (*R*)-2-bromobutane and (*S*)-2-bromobutane. Thus, an electrophilic addition reaction that forms a compound with one asymmetric center from a reactant without any asymmetric centers is not stereoselective because it does not select for a particular stereoisomer. Why is this so?

We can see why a racemic mixture is obtained if we examine the structure of the carbocation formed in the first step of the reaction. The positively charged carbon is sp^2 hybridized, so the three atoms to which it is bonded lie in a plane (Section 1.10). When the bromide ion approaches the positively charged carbon from above the plane, one enantiomer is formed, but when it approaches from below the plane, the other enantiomer is formed. Because the bromide ion has equal access to both sides of the plane, identical amounts of the *R* and *S* enantiomers are obtained from the reaction.

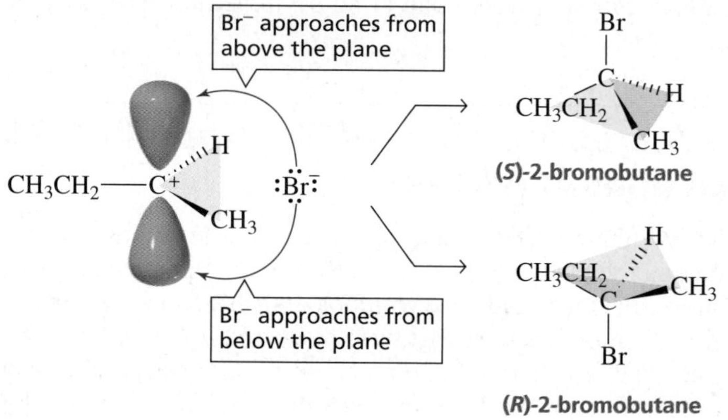

Br⁻ approaches from above the plane

CH_3CH_2—C⁺

:Br̈:⁻

(*S*)-2-bromobutane

Br⁻ approaches from below the plane

(*R*)-2-bromobutane

PROBLEM 43

a. Is the reaction of 2-butene with HBr regioselective?

b. Is it stereoselective?

c. Is it stereospecific?

d. Is the reaction of 1-butene with HBr regioselective?

e. Is it stereoselective?

f. Is it stereospecific?

PROBLEM 44

What stereoisomers are obtained from each of the following reactions?

a. $CH_3CH_2CH_2CH{=}CH_2$ $\xrightarrow{\text{HCl}}$

c. 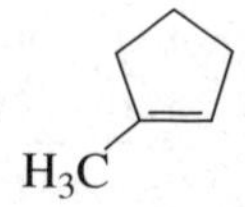 $\xrightarrow{\text{HBr}}$

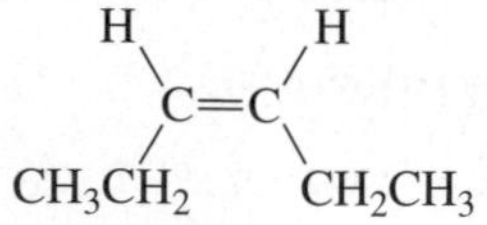

b. $\xrightarrow[\text{H}_2\text{O}]{\text{H}^+}$

d. 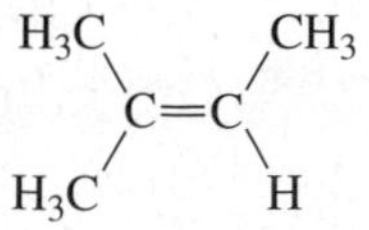 $\xrightarrow{\text{HBr}}$

If an addition reaction creates an asymmetric center in a compound that already has an asymmetric center (and the reactant is a single enantiomer of that compound), a pair of diastereomers will be formed. For example, let's look at the reaction shown below. Because none of the bonds to the asymmetric center in the reactant is broken during the addition of HBr, the configuration of this asymmetric center does not change. The bromide ion can approach the planar carbocation intermediate from either the top or the bottom in the process of creating the new asymmetric center, so two stereoisomers result. The stereoisomers are diastereomers because one of the asymmetric centers has the same configuration in both stereoisomers and the other has opposite configurations.

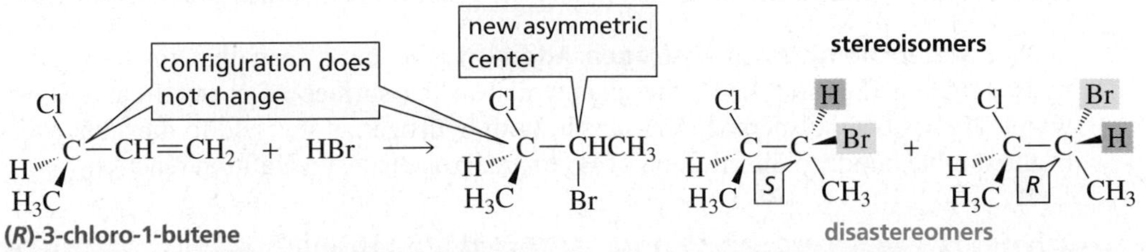

configuration does not change   **new asymmetric center**   **stereoisomers**

(R)-3-chloro-1-butene

disastereomers

Because the products of the preceding reaction are diastereomers, the transition states that lead to them are also diastereomeric. The two transition states, therefore, will not have the same stability, so different amounts of the two diastereomers will be formed. Thus, the reaction is stereoselective—it forms more of one stereoisomer than of the other. The reaction is also stereospecific; (R)-3-chloro-1-butene forms a different pair of diastereomers than (S)-3-chloro-1-butene forms.

Addition Reactions That Form Products with Two Asymmetric Centers

When a reactant that does not have an asymmetric center undergoes a reaction that forms a product with *two* asymmetric centers, the stereoisomers that are formed depend on the mechanism of the reaction.

Addition Reactions That Form a Carbocation Intermediate. If two asymmetric centers are created as the result of an addition reaction that forms a carbocation intermediate, four stereoisomers can be obtained as products.

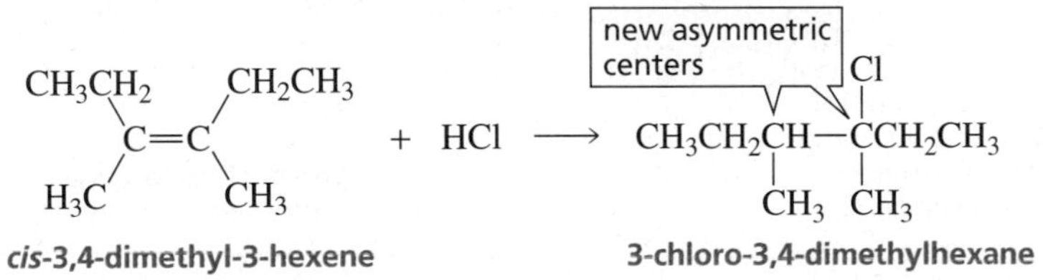

new asymmetric centers

cis-3,4-dimethyl-3-hexene   3-chloro-3,4-dimethylhexane

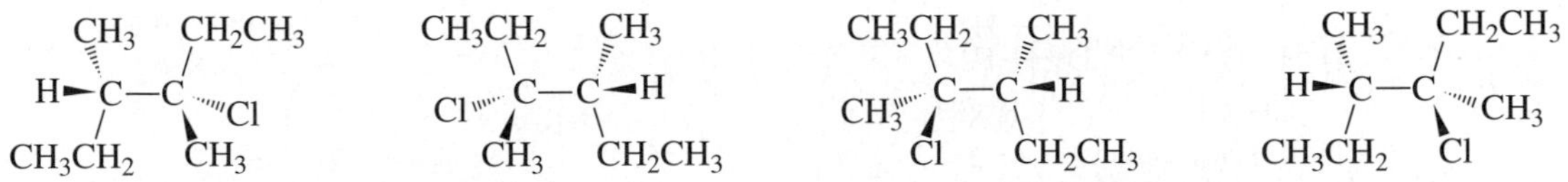

perspective formulas of the stereoisomers of the product

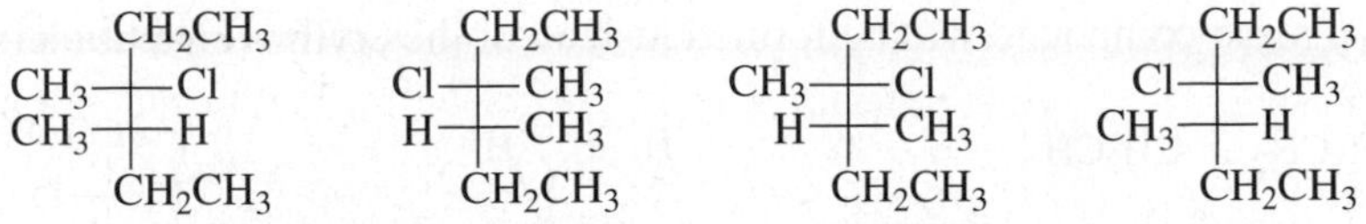

Fischer projections of the stereoisomers of the product

In the first step of the reaction, the proton can approach the plane containing the double-bonded carbons of the alkene from either above or below to form the carbocation. Once the carbocation is formed, the chloride ion can approach the positively charged carbon from above or below. As a result, four stereoisomers are obtained as products: the addition of the proton and the chloride ion can be described as above-above (both adding from above), above-below, below-above, or below-below. When two substituents add to the same side of a double bond, the addition is called **syn addition**. When two substituents add to opposite sides of a double bond, the addition is called **anti addition**. Both syn and anti additions occur in alkene addition reactions that form a carbocation intermediate. Equal amounts of the four stereoisomers are obtained so the reaction is not stereoselective and, because the four stereoisomers formed by the cis alkene are identical to the four stereoisomers formed by the trans alkene, the reaction is also not stereospecific.

The Stereochemistry of Hydrogen Addition. We have seen that in a catalytic hydrogenation (Section 4.11), the alkene sits on the surface of a metal catalyst on which H_2 has been dispersed. As a result, both hydrogen atoms add to the same side of the double bond, so the addition of H_2 to an alkene is a syn addition reaction.

addition of H_2 is a syn addition

If addition of hydrogen to an alkene forms a product with two asymmetric centers, then only two of the four possible stereoisomers are obtained because only syn addition can occur. (The other two stereoisomers would have to come from anti addition.) One of the two stereoisomers results from addition of both hydrogens from above the plane of the double bond, and the other results from addition of both hydrogens from below the plane, because the alkene can sit on the metal catalyst with either side facing up. The particular pair of stereoisomers that is formed depends on whether the reactant is a *cis*-alkene or a *trans*-alkene. Syn addition of H_2 to a *cis*-alkene forms only the erythro enantiomers. (In Section 5.11, we saw that the erythro enantiomers are the ones with identical groups on the same side of the carbon chain in the eclipsed conformers.)

cis-2,3-dideuterio-2-pentene

erythro enantiomers
**perspective formulas
(eclipsed conformers)**

erythro enantiomers
**perspective formulas
(staggered conformers)**

erythro enantiomers
Fischer projections

If each of the two asymmetric centers in the product is bonded to the same four substituents, a meso compound will be formed instead of the erythro enantiomers.

cis-2,3-dideutero-2-pentene

meso compound

In contrast, syn addition of H_2 to a *trans*-alkene forms only the threo enantiomers. Thus, the addition of hydrogen is a stereospecific reaction—the product obtained from addition to the cis isomer is different from the product obtained from addition to the trans isomer. It is also a stereoselective reaction because all possible stereoisomers are not formed; for example, only the threo enantiomers are formed in the following reaction.

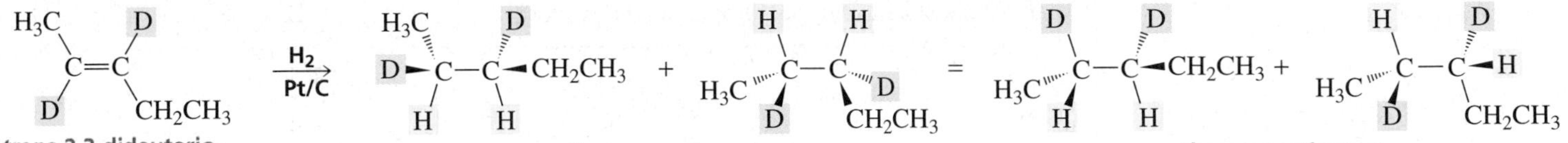

trans-2,3-dideuterio-2-pentene

threo enantiomers
perspective formulas
(eclipsed conformers)

threo enantiomers
perspective formulas
(staggered conformers)

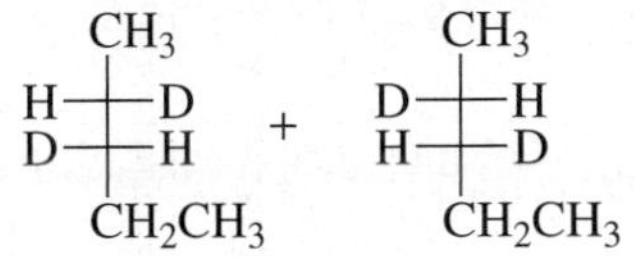

threo enantiomers
Fischer projections

If the product is a cyclic compound, the addition of H_2 will form the cis enantiomers since the two hydrogens add to the same side of the double bond.

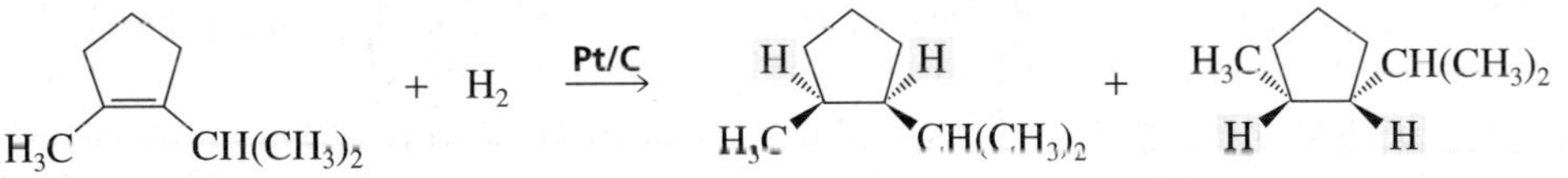

1-isopropyl-2-methyl-cyclopentene

Each of the two asymmetric centers in the product of the following reaction is bonded to the same four substituents. Therefore, syn addition forms a meso compound.

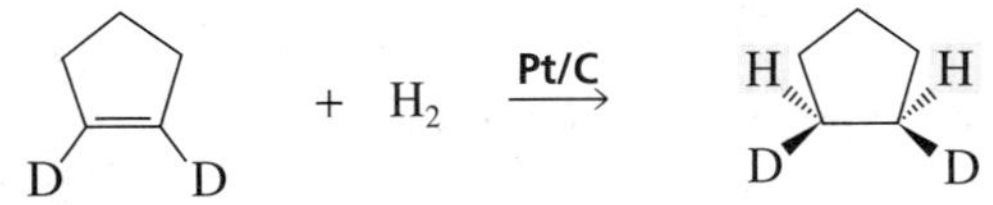

1,2-dideuterio-cyclopentene

CYCLIC ALKENES

Cyclic alkenes with fewer than eight carbons in the ring, such as cyclopentene and cyclohexene, can exist only in the cis configuration because they do not have the number of carbons they would need for a trans double bond. Therefore, it is not necessary to use the cis designation with their names. Both cis and trans isomers are possible for cyclic alkenes containing eight or more carbons, however, so the configuration of the compound must be specified in its name.

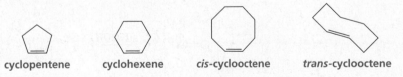

cyclopentene   cyclohexene   *cis*-cyclooctene   *trans*-cyclooctene

PROBLEM 45

a. What stereoisomers are formed in the following reaction?

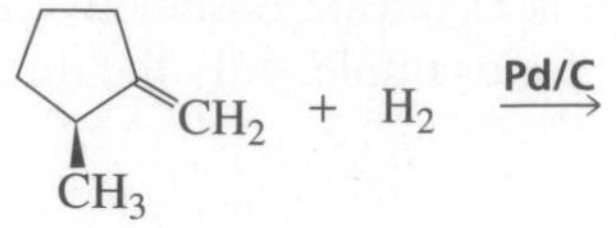

b. What stereoisomer is formed in greater yield?

The Stereochemistry of Peroxyacid Addition. The addition of a peroxyacid to an alkene to from an epoxide (Section 4.9) is a concerted reaction: the oxygen atom adds to the two sp^2 carbons at the same time. Therefore, it must be a syn addition.

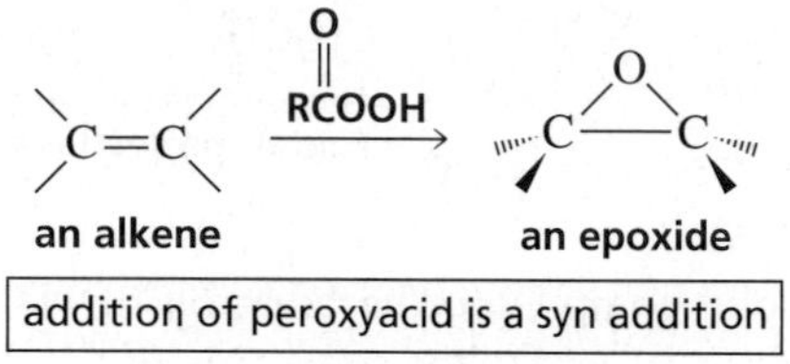

an alkene   an epoxide

addition of peroxyacid is a syn addition

The oxygen can add from above or from below the plane containing the double bond. Therefore, addition of a peroxyacid to an alkene forms two stereoisomers. Syn addition to a cis alkene forms the cis enantiomers. Because only syn addition occurs, the reaction is stereoselective.

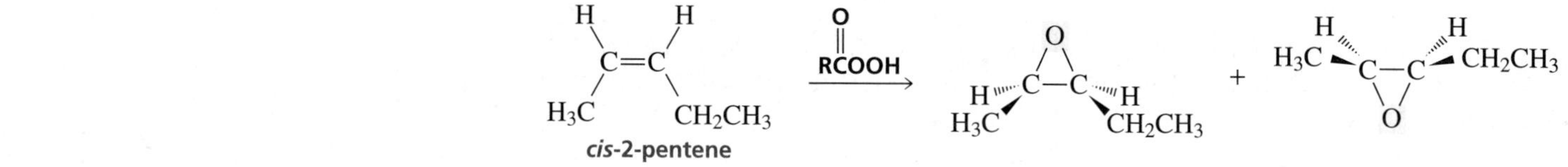

Syn addition to a trans alkene forms the trans enantiomers.

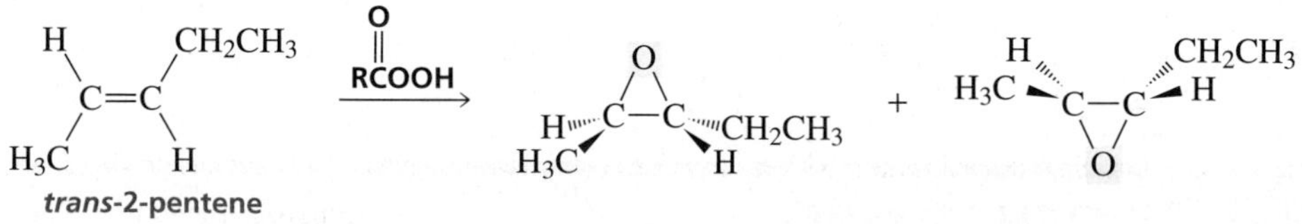

Addition of a peroxyacid to *cis*-2-butene forms a meso compound—each of the two asymmetric centers is attached to the same four groups (Section 5.12).

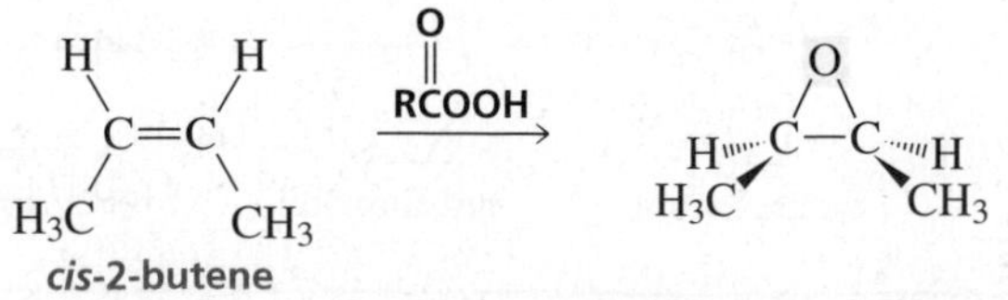

PROBLEM 46◆

a. What alkene would be required to synthesize the following compounds?

1. 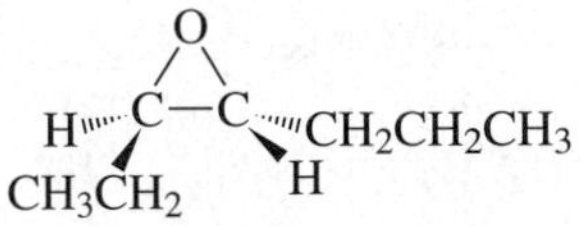 $\underset{CH_3CH_2}{\overset{O}{\underset{}{H\cdots C - C\cdots CH_2CH_2CH_3}}}$ 2. $\underset{CH_3CH_2}{\overset{O}{\underset{CH_2CH_2CH_3}{H\cdots C - C\cdots H}}}$

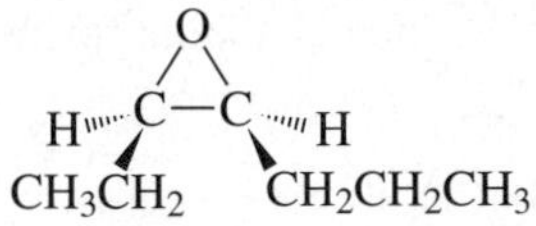

b. What other epoxide would be found?

The Stereochemistry of Hydroboration–Oxidation. The addition of borane to an alkene (Section 4.10) is also a concerted reaction. The boron and the hydride ion add to the two sp^2 carbons of the double bond at the same time. Because the two species add simultaneously, they must add to the same side of the double bond, so the addition of borane to an alkene, like the addition of hydrogen and the addition of a peroxyacid, is a syn addition.

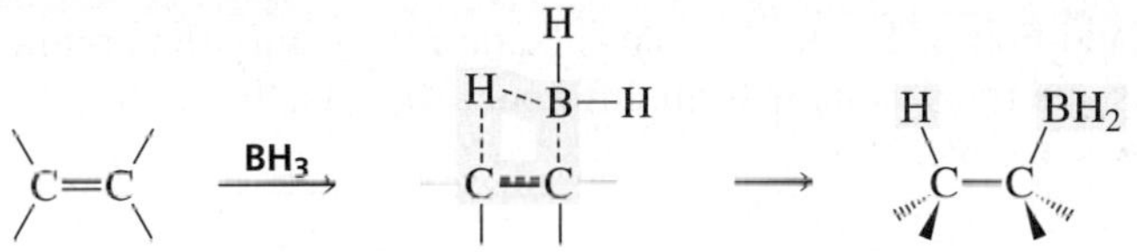

$$\overset{}{\underset{}{C=C}} \xrightarrow{BH_3} \overset{}{\underset{}{C\cdots C}} \longrightarrow \overset{}{\underset{}{C-C}}$$

addition of borane is a syn addition

VCL   **Hydroboration-2**

When the resulting alkylborane is oxidized by reaction with hydrogen peroxide and hydroxide ion, the OH group ends up in the same position as the boron group it replaces. Consequently, the overall reaction, called hydroboration–oxidation, amounts to a syn addition of water to a carbon–carbon double bond.

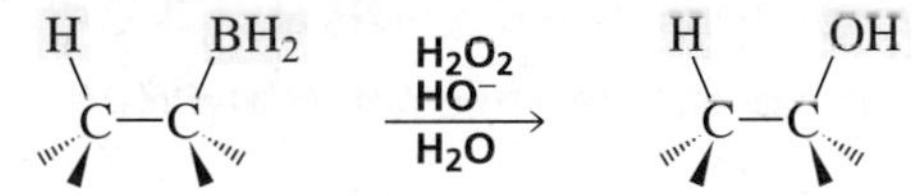

$$\overset{H \qquad BH_2}{\underset{}{C-C}} \xrightarrow[\substack{H_2O_2 \\ HO^- \\ H_2O}]{} \overset{H \qquad OH}{\underset{}{C-C}}$$

an alkyl borane   **an alcohol**

hydroboration–oxidation is a syn addition of water

Because only syn addition occurs, hydroboration–oxidation is stereoselective— only two of the four possible stereoisomers are formed. As we saw when we looked at the addition of H_2 or a peroxyacid, if the product is cyclic, syn addition results in the formation of only the pair of enantiomers that has the added groups on the same side of the ring.

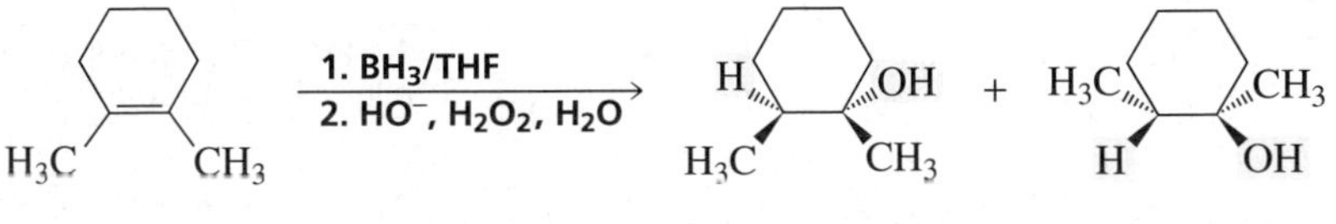

$$\underset{H_3C \qquad CH_3}{\bigcirc} \xrightarrow[\text{2. HO}^-, H_2O_2, H_2O]{\text{1. BH}_3/\text{THF}} \underset{H_3C \qquad CH_3}{\overset{H \qquad OH}{\bigcirc}} + \underset{H \qquad OH}{\overset{H_3C \qquad CH_3}{\bigcirc}}$$

PROBLEM 47◆

What stereoisomers would be obtained from hydroboration–oxidation of the following compounds?

a. cyclohexene

b. 1-ethylcyclohexene

c. 1,2-dimethylcyclopentene

d. *cis*-2-butene

Addition Reactions that Form a Cyclic Bromonium Ion Intermediate. If two asymmetric centers are created as the result of an addition reaction that forms a bromonium ion intermediate, only one pair of enantiomers will be formed. The addition of Br_2 to the cis alkene forms only the threo enantiomers.

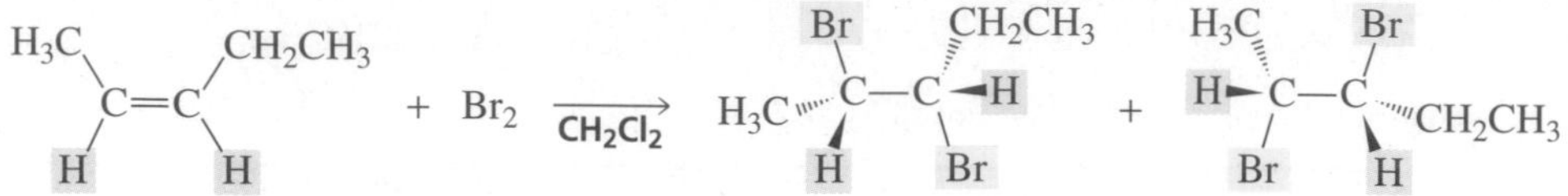

threo enantiomers
perspective formulas

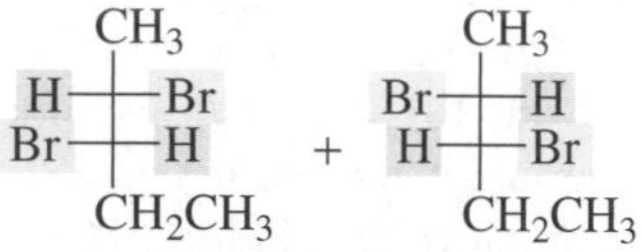

threo enantiomers
Fischer projections

If you have trouble determining the configuration of a product, make a model.

Similarly, the addition of Br_2 to the trans alkene forms only the erythro enantiomers. Because the cis and trans isomers form different products, the reaction is stereospecific as well as stereoselective.

Alkene Bromination-1

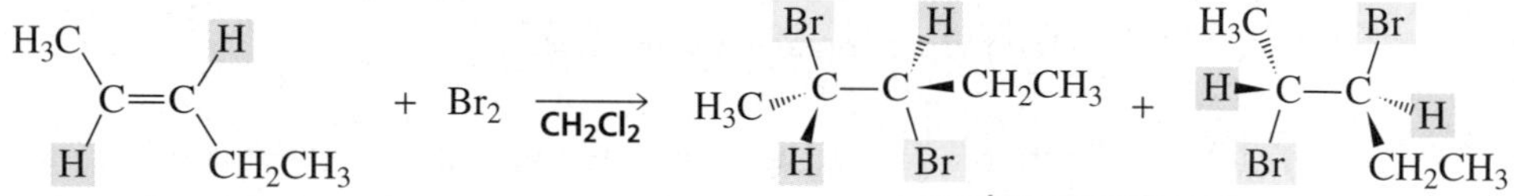

erythro enantiomers
perspective formulas

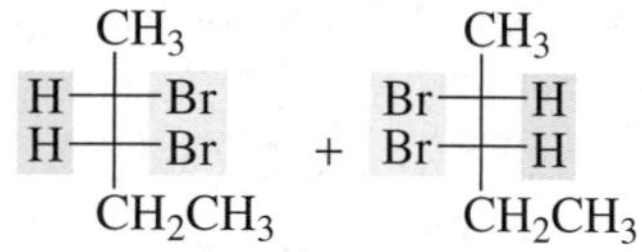

erythro enantiomers
Fischer projections

Because the addition of Br_2 to the *cis alkene* forms the threo enantiomers, we know that anti addition must have occurred, because we have just seen that syn addition would have formed the erythro enantiomers. The addition of Br_2 is anti because the reaction intermediate is a cyclic bromonium ion (Section 4.7). Once the bromonium ion is formed, the bridged bromine atom blocks that side of the ion. As a result, the negatively charged bromide ion must approach from the opposite side (following either the green arrows *or* the red arrows as shown below). Thus, the two bromine atoms add to opposite sides of the double bond. Because only anti addition of Br_2 can occur, only two of the four possible stereoisomers are obtained.

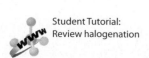

the cyclic bromonium ion formed from the reaction of Br_2 with *cis*-2-butene

Student Tutorial: Review halogenation

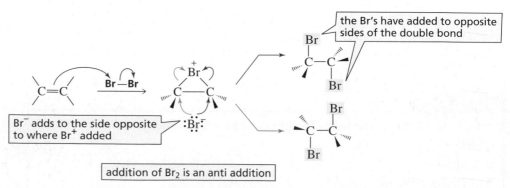

If the two asymmetric centers in the product each have the same four substituents, the erythro isomers are identical and constitute a meso compound. Therefore, addition of Br_2 to *trans*-2-butene forms a meso compound.

***trans*-2-butene** — a meso compound, **perspective formula**

a meso compound
Fischer projection

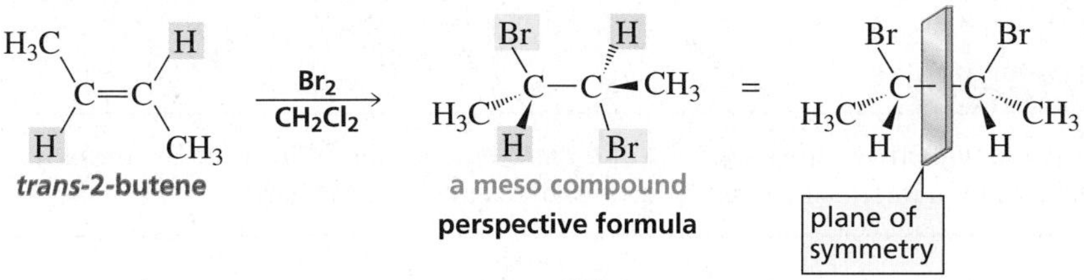

VCL **Alkene Bromination-2**

Because only anti addition occurs, addition of Br_2 to a cyclohexene forms only the enantiomers that have the added bromine atoms on opposite sides of the ring.

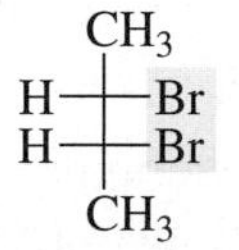

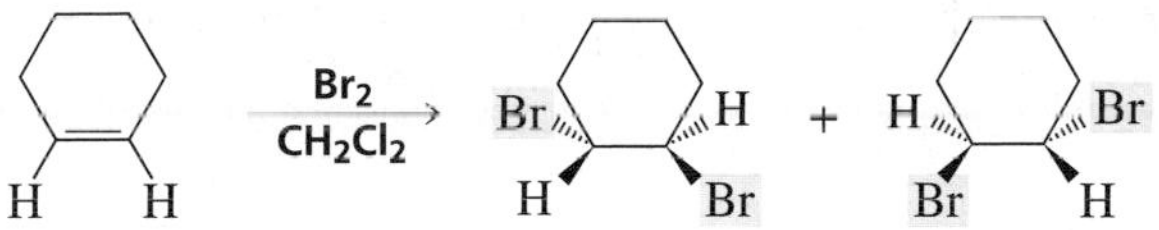

A summary of the stereochemistry of the products obtained from addition reactions to alkenes is given in Table 5.3.

Table 5.3 Stereochemistry of Alkene Addition Reactions

Reaction	Type of addition	Stereoisomers formed
Addition reactions that create one asymmetric center in the product		1. If the reactant does not have an asymmetric center, a pair of enantiomers will be obtained (equal amounts of *R* and *S*). 2. If the reactant has an asymmetric center, unequal amounts of a pair of diastereomers will be obtained.
Addition reactions that create two asymmetric centers in the product		
Addition of reagents that form a carbocation intermediate	syn and anti	Four stereoisomers can be obtained* (the cis and trans isomers form the same products)
Addition of H_2 Addition of borane Addition of a peroxyacid	syn	cis ⟶ erythro or cis enantiomers* trans ⟶ threo or trans enantiomers
Addition of Br_2, $Br_2 + H_2O$, $Br_2 + ROH$ (any reaction that forms a cyclic bromonium ion intermediate)	anti	cis ⟶ threo or trans enantiomers trans ⟶ erythro or cis enantiomers*

* If the two asymmetric centers have the same substituents, a meso compound will be obtained instead of the pair of erythro enantiomers.

PROBLEM 48

Reaction of 2-ethyl-1-pentene with Br_2, with H_2 + Pt/C, or with BH_3 followed by HO^- + H_2O_2 leads to a racemic mixture. Explain why a racemic mixture is obtained in each case.

PROBLEM 49

How could you prove, using a sample of *trans*-2-butene, that addition of Br_2 forms a cyclic bromonium ion intermediate rather than a carbocation intermediate?

 **Halohydrin Formation-2**

One way that can be used to determine what stereoisomers are obtained from a reaction that creates a product with two asymmetric centers is the mnemonic **CIS-SYN-ERYTHRO** or **-CIS**. (The third term is "erythro" if the product is acyclic, and "cis" if it is cyclic.) The three terms are easy to remember because they all mean "on the same side." You can change any two of the terms but you can't change just one. For example, **TRANS-ANTI-ERYTHRO**, **TRANS-SYN-THREO**, and **CIS-ANTI-THREO** are allowed, but **TRANS-SYN-ERYTHRO** is not allowed. Thus, if you have a cis reactant that undergoes addition of Br_2 (which is anti), the threo products are obtained if the products are acyclic and the trans products are obtained if the product is cyclic. This mnemonic will work for all reactions that have products with structures that can be described by erythro and threo or cis and trans.

PROBLEM-SOLVING STRATEGY

Predicting the Stereoisomers Obtained from Addition Reactions of Alkenes

What stereoisomers are obtained from the following reactions:

a. 1-butene + HCl

b. cyclohexene + HBr

c. *cis*-3-heptene + Br_2

d. *trans*-3-hexene + Br_2

Start by drawing the product without regard to its configuration to check whether the reaction has created any asymmetric centers. Then determine the configurations of the products, paying attention to the configuration (if any) of the reactant, how many asymmetric centers are formed, and the mechanism of the reaction. Let's start with part a.

a. $CH_3CH_2\overset{\underset{|}{Cl}}{C}HCH_3$

The product has one asymmetric center, so equal amounts of the *R* and *S* enantiomers will be formed.

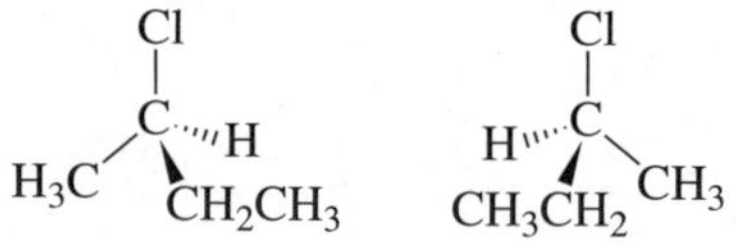

b.

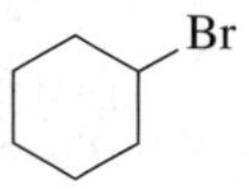

The product does not have an asymmetric center, so it has no stereoisomers.

c. 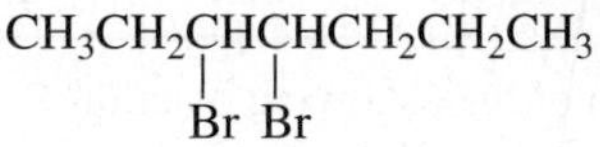 CH₃CH₂CHCHCH₂CH₂CH₃
              | |
             Br Br

Two asymmetric centers have been created in the product. Because the reactant is cis and addition of Br₂ is anti, the threo enantiomers are formed.

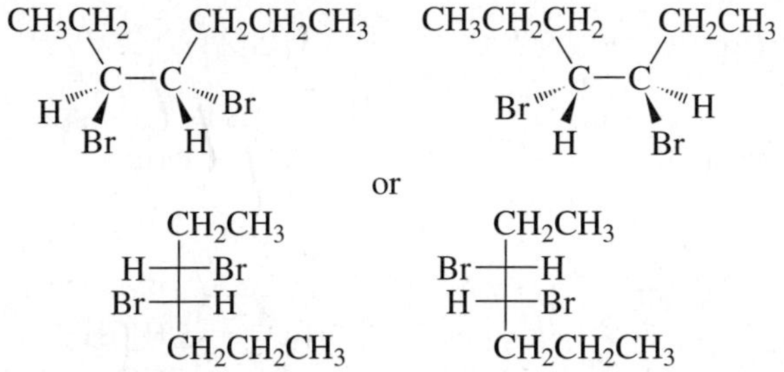

or

d.  CH₃CH₂CHCHCH₂CH₃
              | |
             Br Br

Two asymmetric centers have been created in the product. Because the reactant is trans and addition of Br₂ is anti, one would expect the erythro enantiomers. However, the two asymmetric centers are bonded to the same four groups, so the erythro product is a meso compound. Thus, only one stereoisomer is formed.

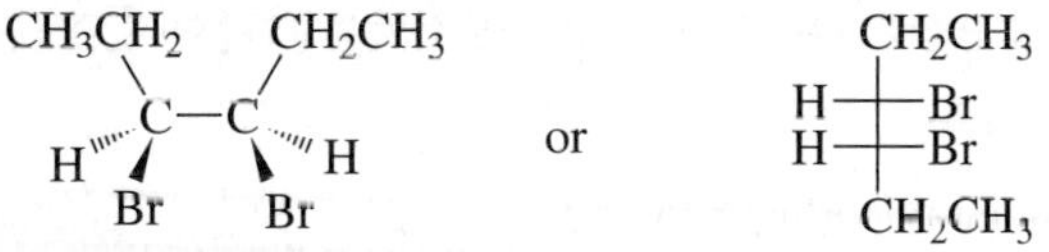

or

Now continue on to Problem 50.

PROBLEM 50

What stereoisomers are obtained from the following reactions:

a. *trans*-2-butene + HBr

b. (Z)-3-methyl-2-pentene + HBr

c. (E)-3-methyl-2-pentene + HBr

d. *cis*-3-hexene + HBr

e. *cis*-2-pentene + Br₂

f. 1-hexene + Br₂

PROBLEM 51

When Br₂ adds to an alkene that has different substituents on each of the two sp^2 carbons, such as *cis*-2-heptene, identical amounts of the two threo enantiomers are obtained even though Br is more likely to attack the less sterically hindered carbon atom of the bromonium ion. Explain why identical amounts of the stereoisomers are obtained.

PROBLEM 52

a. What products would be obtained from the addition of Br₂ to cyclohexene if the solvent were H₂O instead of CH₂Cl₂?

b. Propose a mechanism for the reaction.

PROBLEM 53

What stereoisomers would you expect to obtain from each of the following reactions?

a. 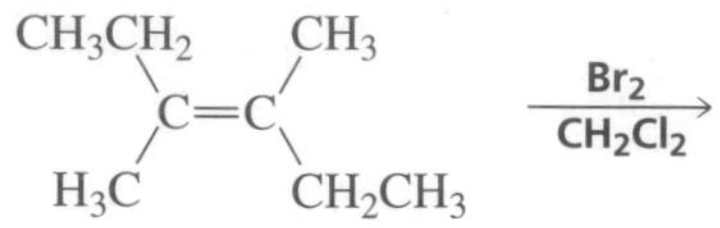
$$CH_3CH_2 \quad CH_3$$
$$C=C$$
$$H_3C \quad CH_2CH_3$$
$$\xrightarrow[CH_2Cl_2]{Br_2}$$

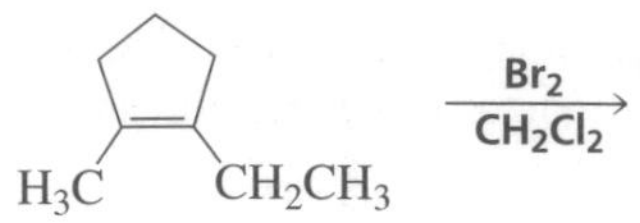

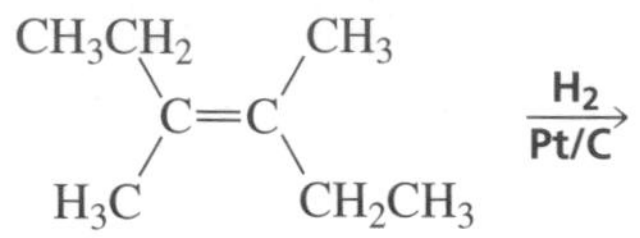

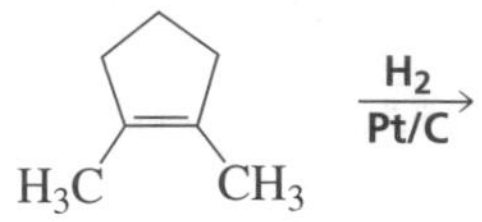

d.
$$\xrightarrow[CH_2Cl_2]{Br_2}$$
$$H_3C \quad CH_2CH_3$$

b.
$$CH_3CH_2 \quad CH_3$$
$$C=C$$
$$H_3C \quad CH_2CH_3$$
$$\xrightarrow[Pt/C]{H_2}$$

e.
$$\xrightarrow[Pt/C]{H_2}$$
$$H_3C \quad CH_3$$

c.
$$\xrightarrow[CH_2Cl_2]{Br_2}$$
$$H_3C \quad CH_3$$

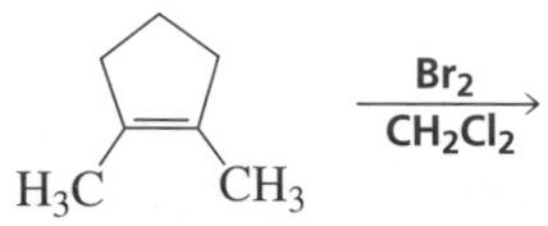

f. 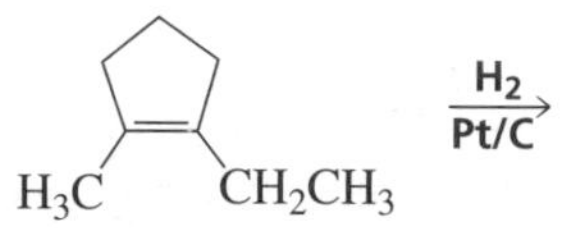
$$\xrightarrow[Pt/C]{H_2}$$
$$H_3C \quad CH_2CH_3$$

PROBLEM 54◆

a. What is the major product obtained from the reaction of propene and Br_2 plus excess Cl^-?

b. Indicate the relative amounts of the stereoisomers obtained.

5.20 The Stereochemistry of Enzyme-Catalyzed Reactions

BIOGRAPHY

For studies on the stereochemistry of enzyme-catalyzed reactions, **Sir John Cornforth** *received the Nobel Prize in chemistry in 1975 (sharing it with Vladimir Prelog, page 206). Born in Australia in 1917, Conforth studied at the University of Sydney and received a Ph.D. from Oxford. His major research was carried out in laboratories at Britain's Medical Research Council and at Shell Research Ltd. He was knighted in 1977.*

The chemistry associated with living organisms is called **biochemistry**. When you study biochemistry, you study the structures and functions of the molecules found in the biological world and the reactions involved in the synthesis and degradation of these molecules. Because the compounds in living organisms are organic compounds, it is not surprising that many of the reactions encountered in organic chemistry also occur in biological systems. Living cells do not contain molecules such as Cl_2, HBr, or BH_3, so you would not expect to find the addition of such reagents to alkenes in biological systems. However, living cells do contain water and acid catalysts, so some alkenes found in biological systems undergo the acid-catalyzed addition of water (Section 4.5).

Reactions that occur in biological systems are catalyzed by proteins called **enzymes**. When an enzyme catalyzes a reaction that forms a product with an asymmetric center, only one stereoisomer is formed, because enzyme-catalyzed reactions are completely stereoselective. For example, the enzyme fumarase catalyzes the addition of water to fumarate to form malate, a compound with an asymmetric center.

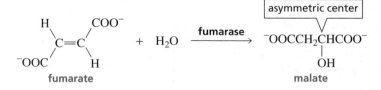

However, the reaction forms only (*S*)-malate; the *R* enantiomer is not formed.

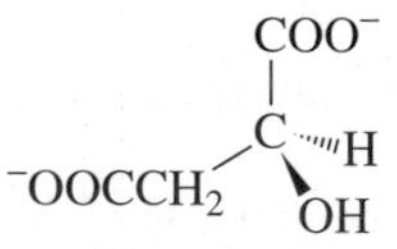

$$COO^-$$
$$|$$
$$C\text{''''}H$$
$$^-OOCCH_2 \quad OH$$
(*S*)-malate

An enzyme-catalyzed reaction forms only one stereoisomer because an enzyme's binding site restricts delivery of the reagents to only one side of the reactant's functional group.

Enzyme-catalyzed reactions are also stereospecific; an enzyme typically catalyzes the reaction of only one stereoisomer. For example, fumarase catalyzes the addition of water to fumarate (the trans isomer) but not to maleate (the cis isomer).

$$^-OOC\underset{H}{\overset{}{C}}{=}\underset{H}{\overset{COO^-}{C}} \quad + \quad H_2O \quad \xrightarrow{\text{fumarase}} \quad \text{no reaction}$$

maleate

An enzyme is able to differentiate between the two stereoisomers because only one of them has the structure that allows it to fit into the enzyme's binding site.

BIOGRAPHY

Frank H. Westheimer *did fundamental work on the stereochemistry of enzyme-catalyzed reactions. Born in Baltimore in 1912, he received his graduate training at Harvard University, served on the faculty of the University of Chicago, and subsequently returned to Harvard as a professor of chemistry.*

PROBLEM 55◆

a. What would be the product of the reaction of fumarate and H₂O if H⁺ were used as a catalyst instead of fumarase?

b. What would be the product of the reaction of maleate and H₂O if H⁺ were used as a catalyst instead of fumarase?

5.21 Enantiomers Can Be Distinguished by Biological Molecules

Enzymes and receptors can tell the difference between enantiomers because enzymes and receptors are proteins, and proteins are chiral molecules.

Enzymes

When enantiomers react with *achiral* reagents, both enantiomers react at the same rate because they have the same chemical properties. Thus, hydroxide ion (an achiral reagent) reacts with (R)-2-bromobutane at the same rate that it reacts with (S)-2-bromobutane.

Because an enzyme is a *chiral reagent*, not only can it distinguish between cis–trans isomers, such as maleate and fumarate (Section 5.20), it can also distinguish between enantiomers and catalyze the reaction of only one of them. Chemists use an enzyme's specificity for a particular enantiomer to separate enantiomers. For example, the enzyme D-amino acid oxidase catalyzes only the reaction of the R enantiomer and leaves the S enantiomer unchanged. The product of the enzyme-catalyzed reaction can be easily separated from the unreacted enantiomer.

$$\underset{\substack{R \text{ enantiomer}}}{\overset{\substack{COO^-}}{R{\overset{|}{\underset{H}{C}}}NH_2}} \quad + \quad \underset{\substack{S \text{ enantiomer}}}{\overset{\substack{COO^-}}{H_2N{\overset{|}{\underset{H}{C}}}R}} \quad \xrightarrow[\text{oxidase}]{\text{D-amino acid}} \quad \underset{\substack{\text{oxidized} \\ R \text{ enantiomer}}}{\overset{\substack{^-OOC}}{\underset{R}{C}{=}NH}} \quad + \quad \underset{\substack{\text{unreacted} \\ S \text{ enantiomer}}}{\overset{\substack{COO^-}}{H_2N{\overset{|}{\underset{H}{C}}}R}}$$

An enzyme is able to differentiate between enantiomers and between cis and trans isomers as we saw above, because its binding site is chiral. The enzyme will therefore bind only the stereoisomer whose substituents are in the correct positions to interact with substituents in the chiral binding site. In Figure 5.3 the enzyme binds the *R* enantiomer but not the *S* enantiomer. The latter does not have its substituents in the proper positions, so they cannot bind efficiently to the enzyme. Like a right-handed glove, which fits only the right hand, an enzyme forms only one stereoisomer and reacts with only one stereoisomer (Section 5.20).

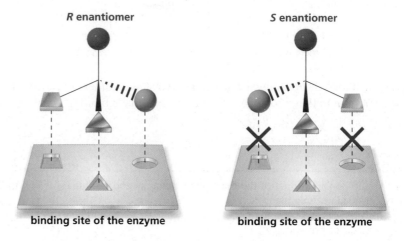

Figure 5.3 ▶
Schematic diagram showing why only one enantiomer is bound by a enzyme. One enantiomer fits into the binding site and one does not.

An achiral reagent reacts identically with both enantiomers. A sock, which is achiral, fits on either foot.

A chiral reagent reacts differently with each enantiomer. A shoe, which is chiral, fits on only one foot.

The problem of having to separate enantiomers can be avoided if a synthesis is carried out that forms one of the enantiomers preferentially. **Chiral catalysts** are being developed that will synthesize one enantiomer in great excess over the other. For example, the catalytic hydrogenation of 2-ethyl-1-pentene forms equal amounts of two enantiomers because H_2 can be delivered equally easily to both faces of the double bond (Section 5.19).

If, however, the metal catalyst is complexed to a chiral organic molecule, H_2 will be delivered to only one face of the double bond. One such chiral catalyst—using Ru(II) as the metal and BINAP (2,2′-bis(diphenylphosphino)-1,1′-binaphthyl) as the chiral molecule—has been used to synthesize (*S*)-naproxen, the active ingredient in Aleve and in several other over-the-counter nonsteroidal anti-inflammatory drugs, in greater than 98% enantiomeric excess.

PROBLEM 56◆

What percent of naproxen is obtained as the *S* enantiomer in the above synthesis?

Receptors

A **receptor** is a protein that binds a particular molecule. Because a receptor is chiral, it will bind one enantiomer better than the other, just like an enzyme binds one enantiomer better than another.

The fact that a receptor typically recognizes only one enantiomer causes enantiomers to have different physiological properties. For example, receptors located on the exteriors of nerve cells in the nose are able to perceive and differentiate the estimated 10,000 smells to which the cells are exposed. The reason that (*R*)-(−)-carvone (found in spearmint oil) and (*S*)-(+)-carvone (the main constituent of caraway seed oil) have such different odors is that each enantiomer fits into a different receptor.

$$\text{(}R\text{)-(}-\text{)-carvone}$$
spearmint oil

$$[\alpha]_D^{20\ °C} = -62.5$$

(*S*)-(+)-carvone
caraway seed oil

$$[\alpha]_D^{20\ °C} = +62.5$$

Many drugs exert their physiological activity by binding to cell-surface receptors. If the drug has an asymmetric center, the receptor can bind one of the enantiomers preferentially. Thus, enantiomers of a drug can have the same physiological activities, different degrees of the same activity, or very different activities, depending on the drug.

Dr. Frances O. Kelsey receives a medal from President John F. Kennedy for preventing the sale of thalidomide.

THE ENANTIOMERS OF THALIDOMIDE

Thalidomide was developed in West Germany and was first marketed in 1957 for insomnia and morning sickness. At that time it was available in more than 40 countries but had not been approved for use in the United States because Frances O. Kelsey, a physician for the Food and Drug Administration (FDA), had insisted upon additional tests (See also Section 30.4).

The dextrorotatory isomer has stronger sedative properties but the commercial drug was a racemic mixture. No one knew that the levorotatory isomer was highly teratogenic—causes horrible birth defects—until women who had been given the drug during the first three months of pregnancy gave birth to babies with a wide variety of defects, deformed limbs being the most common. About 10,000 children were damaged by the drug. It was eventually determined that the dextrorotatory

isomer also has mild teratogenic activity and that each of the enantiomers can racemize (interconvert) in the body. Thus, it is not clear whether the birth defects would have been less severe if the women had been given the dextrorotatory isomer only. Thalidomide recently has been approved—with restrictions—to treat leprosy and melanomas.

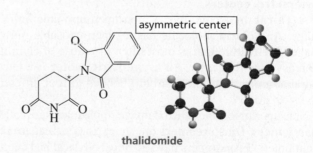

asymmetric center

thalidomide

CHIRAL DRUGS

Until relatively recently, most drugs have been marketed as racemic mixtures because of the difficulty of synthesizing single enantiomers and the high cost of separating enantiomers. In 1992, however, the Food and Drug Administration issued a policy statement encouraging drug companies to use recent advances in synthesis and separation techniques to develop single-enantiomer drugs. Now one-third of all drugs sold are single enantiomers. Drug companies have been able to extend their patents by developing a drug as a single enantiomer that was marketed previously as a racemate (Section 30.13).

If a drug is sold as a racemate, the FDA requires that both enantiomers be tested, because the enantiomers of a drug can have similar or very different properties. Examples are numerous. The *S* isomer of Prozac, an antidepressant, is better at blocking serotonin, but is used up faster than the *R* isomer. Testing has shown that (*S*)-(+)-ketamine is four times more

potent an anesthetic than (*R*)-(−)-ketamine and the disturbing side effects are apparently associated only with the (*R*)-(−)-enantiomer. Only the *S* isomer of the beta-blocker propanolol shows activity; the *R* isomer is inactive. The activity of ibuprofen, the popular analgesic marketed as Advil, Nuprin, and Motrin, resides primarily in the (*S*)-(+)-enantiomer. Heroin addicts can be maintained with (−)-α-acetylmethadol for a 72-hour period compared to 24 hours with racemic methadone. This means less frequent visits to an outpatient clinic; a single dose can keep an addict stable through an entire weekend.

Prescribing a single enantiomer prevents the patient from having to metabolize the less potent enantiomer and decreases the chance of unwanted drug interactions. Drugs that could not be given as racemates because of the toxicity of one of the enantiomers can now be used. For example, (*S*)-penicillamine can be used to treat Wilson's disease even though (*R*)-penicillamine causes blindness.

PROBLEM 57◆

Limonene exists as two different stereoisomers. The *R* enantiomer is found in oranges and the *S* enantiomer is found in lemons. Which of the following molecules is found in oranges?

(+)-limonene   (−)-limonene

SUMMARY

Stereochemistry is the field of chemistry that deals with the structures of molecules in three dimensions. Compounds that have the same molecular formula but are not identical are called **isomers**; they fall into two classes: constitutional isomers and stereoisomers. **Constitutional isomers** differ in the way their atoms are connected. **Stereoisomers** differ in the way their atoms are arranged in space. There are two kinds of stereoisomers: **cis–trans isomers** and isomers that contain **asymmetric centers**.

A **chiral** molecule has a nonsuperimposable mirror image. An **achiral** molecule has a superimposable mirror image. The feature that is most often the cause of chirality is an asymmetric center. An **asymmetric center** is a tetrahedral atom (most often a carbon) bonded to four different atoms or groups.

Nonsuperimposable mirror-image molecules are called **enantiomers**. **Diastereomers** are stereoisomers that are not enantiomers. Enantiomers have identical physical and chemical properties; diastereomers have different physical and chemical properties. An achiral reagent reacts identically

with both enantiomers; a chiral reagent reacts differently with each enantiomer. A mixture of equal amounts of two enantiomers is called a **racemic mixture**.

The letters ***R*** and ***S*** indicate the **configuration** about an asymmetric center. If one molecule has the *R* and the other has the *S* configuration, they are enantiomers; if they both have the *R* or both have the *S* configuration, they are identical molecules.

Chiral compounds are **optically active**, meaning that they rotate the plane of polarized light; achiral compounds are **optically inactive**. If one enantiomer rotates the plane of polarization clockwise (+), its mirror image will rotate the plane of polarization the same amount counterclockwise (−). Each optically active compound has a characteristic **specific rotation**. A **racemic mixture** is optically inactive. A **meso compound** has two or more asymmetric centers and a plane of symmetry; it is an achiral molecule. A compound with the same four groups bonded to two different asymmetric centers will have three stereoisomers, a meso compound and a pair of enantiomers.

If a reaction does not break any bonds to the asymmetric center, the reactant and product will have the same **relative configuration**—their substituents will have the same relative positions. The **absolute configuration** is the actual configuration. If a reaction does break a bond to the asymmetric center, the configuration of the product will depend on the mechanism of the reaction.

A **regioselective** reaction selects for a particular constitutional isomer; a **stereoselective** reaction selects for a particular stereoisomer. A reaction is **stereospecific** if the reactant can exist as stereoisomers and each stereoisomer of the reactant forms a different or different set of stereoisomers. When a reactant that does not have an asymmetric center forms a product with one asymmetric center, the product will be a racemic mixture.

In **syn addition** the substituents add to the same side of a double bond; in **anti addition** they add to opposite sides of the double bond. Both syn and anti addition occur in electrophilic addition reactions that form a carbocation intermediate. Addition of H_2 to an alkene is a syn addition reaction; addition of a peroxyacid to an alkene is a syn addition of an oxygen atom; hydroboration–oxidation is overall a syn addition of water. Addition of Br_2 is an anti addition reaction. An enzyme-catalyzed reaction forms only one stereoisomer; an enzyme typically catalyzes the reaction of only one stereoisomer.

KEY TERMS

absolute configuration (p. 229)
achiral (p. 202)
amine inversion (p. 233)
anti addition (p. 238)
asymmetric center (p. 203)
biochemistry (p. 246)
chiral (p. 202)
chiral catalyst (p. 248)
chiral probe (p. 233)
cis isomer (p. 201)
cis–trans isomers (p. 201)
chromatography (p. 233)
configuration (p. 206)
configurational isomers (p. 200)
constitutional isomers (p. 200)
dextrorotatory (p. 213)
diastereomer (p. 217)
enantiomer (p. 204)

enantiomerically pure (p. 215)
enantiomeric excess (ee) (p. 215)
enzyme (p. 246)
erythro enantiomers (p. 217)
Fischer projection (p. 205)
geometric isomers (p. 201)
isomers (p. 200)
levorotatory (p. 213)
meso compound (p. 222)
observed rotation (p. 214)
observed specific rotation (p. 215)
optically active (p. 213)
optically inactive (p. 213)
perspective formula (p. 205)
plane of symmetry (p. 222)
plane-polarized light (p. 212)
polarimeter (p. 213)
racemate (p. 215)

racemic mixture (p. 215)
receptor (p. 249)
R configuration (p. 206)
regioselective (p. 234)
relative configuration (p. 229)
resolution of a racemic
    mixture (p. 232)
S configuration (p. 206)
specific rotation (p. 214)
stereocenter (p. 205)
stereochemistry (p. 234)
stereogenic center (p. 205)
stereoisomers (p. 200)
stereoselective (p. 234)
stereospecific (p. 235)
syn addition (p. 238)
threo enantiomers (p. 217)
trans isomer (p. 201)

PROBLEMS

58. Disregarding stereoisomers, give the structures of all compounds with molecular formula C_5H_{10}. Which ones can exist as stereoisomers?

59. Draw all possible stereoisomers for each of the following compounds. Indicate if no stereoisomers are possible.
  a. 1-bromo-2-chlorocyclohexane
  b. 2-bromo-4-methylpentane
  c. 1,2-dichlorocyclohexane
  d. 2-bromo-4-chloropentane
  e. 3-heptene
  f. 1-bromo-4-chlorocyclohexane
  g. 1,2-dimethylcyclopropane
  h. 4-bromo-2-pentene
  i. 3,3-dimethylpentane
  j. 3-chloro-1-butene
  k. 1-bromo-2-chlorocyclobutane
  l. 1-bromo-3-chlorocyclobutane

60. Which of the following are optically active?

$$CHBr_2Cl \qquad BHFCl \qquad CH_3CHCl_2 \qquad CHFBrCl \qquad BeHCl$$

61. Name each of the following compounds using R,S and E,Z (Section 3.5) designations where necessary:

a.
$$\begin{array}{c} HC{=}CH_2 \\ | \\ CH_3CH_2 \overset{}{\underset{H}{\overset{C}{\diagdown}}} CH_3 \end{array}$$

b.
$$\underset{Cl}{\overset{F}{\diagup}}C{=}C\underset{Br}{\overset{I}{\diagdown}}$$

c.
$$\begin{array}{c} CH_3 \\ HO{-}\!\!-\!\!|{-}\!\!-\!\!H \\ H{-}\!\!-\!\!|{-}\!\!-\!\!Cl \\ CH_2CH_3 \end{array}$$

d.

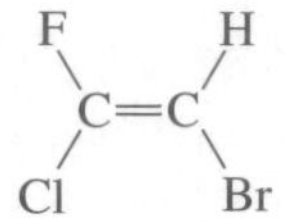

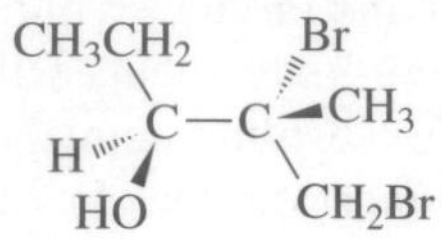

e.

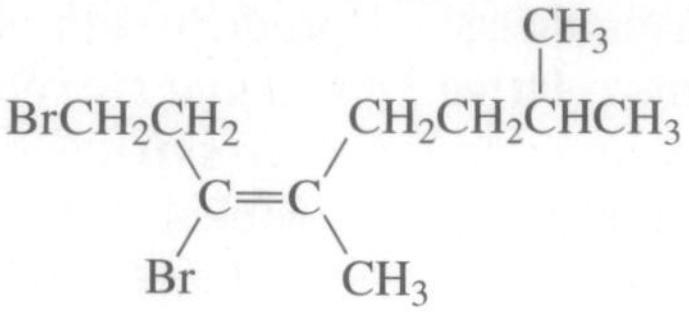

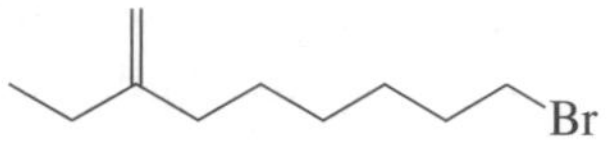

f.

g.

h.

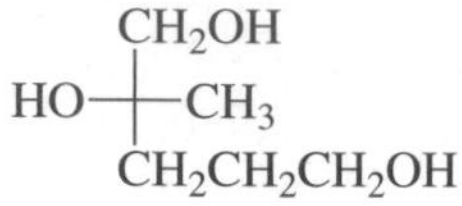

i.

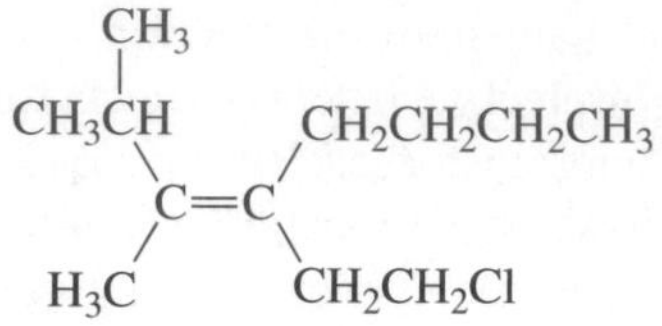

62. Mevacor, is used clinically to lower serum cholesterol levels. How many asymmetric centers does Mevacor have?

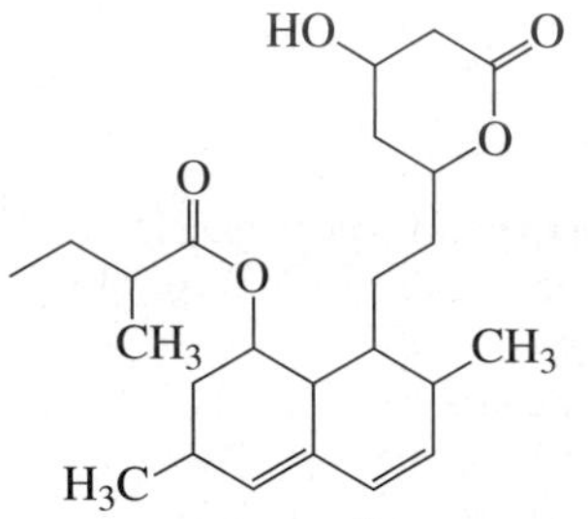

Mevacor®

63. Indicate whether each of the following pairs of compounds are identical or are enantiomers, diastereomers, or constitutional isomers:

a.

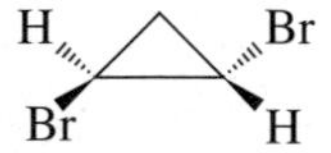

and

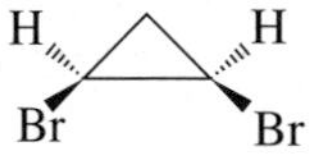

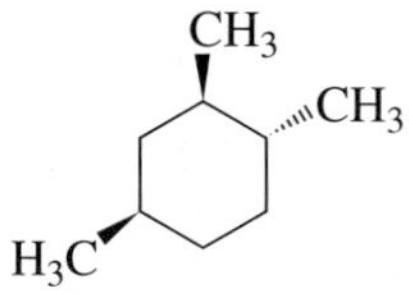

e.

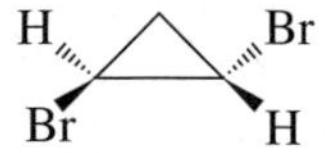

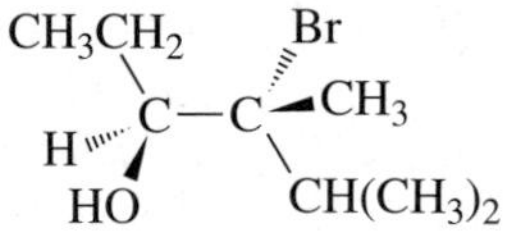

and

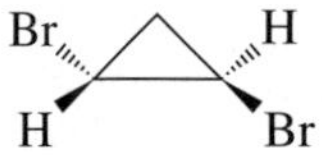

b.

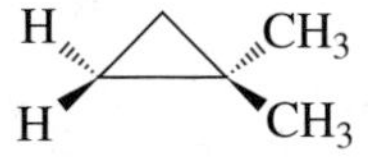

and

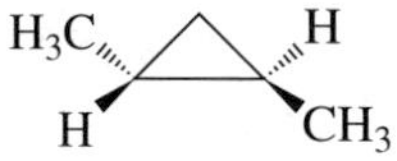

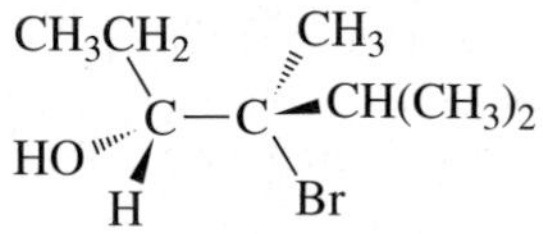

f.

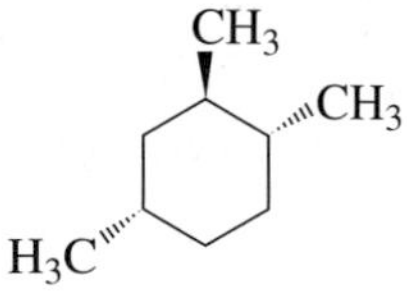

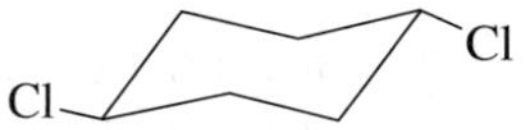

and

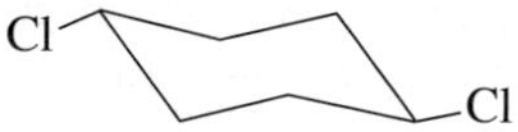

c.

and

g.

and

d.

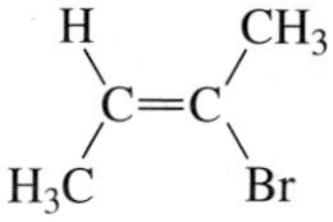

and

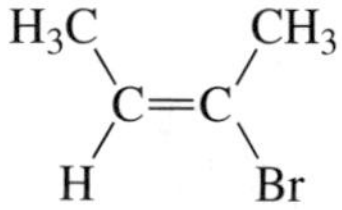

h.

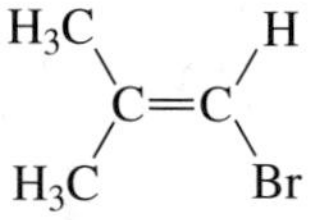

and

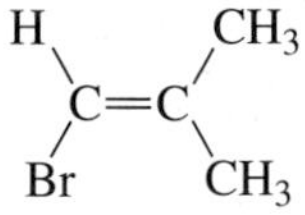

64. a. Give the product or products that would be obtained from the reaction of *cis*-2-butene and *trans*-2-butene with each of the following reagents. If a product can exist as stereoisomers, show which stereoisomers are obtained.
   1. HCl
   2. BH_3/THF followed by HO^-, H_2O_2, H_2O
   3. a peroxyacid
   4. Br_2 in CH_2Cl_2
   5. $Br_2 + H_2O$
   6. $H_2 + Pt/C$
   7. $HCl + H_2O$
   8. $HCl + CH_3OH$
   b. With which reagents do the two alkenes react to give different products?

65. For many centuries, the Chinese have used extracts from a group of herbs known as ephedra to treat asthma. A compound named ephedrine that has been isolated from these herbs was found to be a potent dilator of air passages in the lungs.

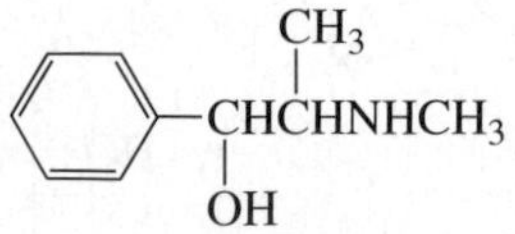

ephedrine

a. How many stereoisomers are possible for ephedrine?

b. The stereoisomer shown below is the one that is pharmacologically active. What is the configuration of each of the asymmetric centers?

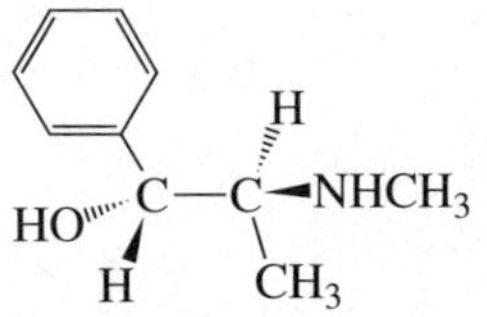

66. Which of the following compounds have an achiral stereoisomer?

a. 2,3-dichlorobutane
b. 2,3-dichloropentane
c. 2,3-dichloro-2,3-dimethylbutane
d. 1,3-dichlorocyclopentane
e. 1,3-dibromocyclobutane

f. 2,4-dibromopentane
g. 2,3-dibromopentane
h. 1,4-dimethylcyclohexane
i. 1,2-dimethylcyclopentane
j. 1,2-dimethylcyclobutane

67. Indicate whether each of the following pairs of compounds are identical or are enantiomers, diastereomers, or constitutional isomers:

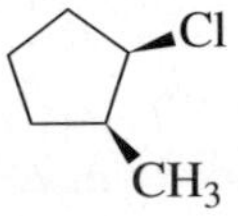

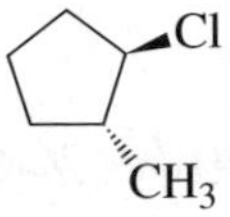

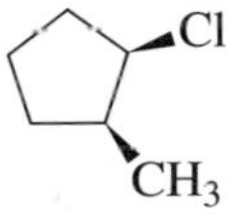

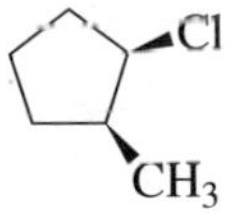

68. Give the products, including their configurations, obtained from the reaction of 1-ethylcyclohexene with the following reagents:

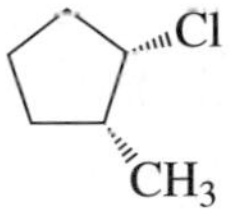

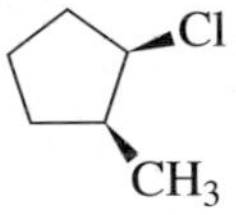

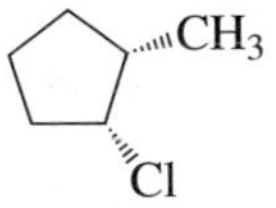

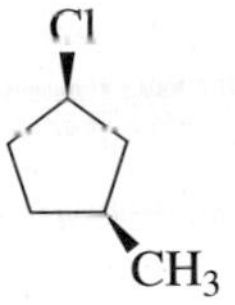

a. HBr b. H$_2$, Pt/C c. BH$_3$/THF followed by HO$^-$, H$_2$O$_2$, H$_2$O d. Br$_2$/CH$_2$Cl$_2$

69. Citrate synthase, one of the enzymes in the series of enzyme-catalyzed reactions known as the citric acid cycle (Section 25.10), catalyzes the synthesis of citric acid from oxaloacetic acid and acetyl-CoA. If the synthesis is carried out with acetyl-CoA that contains radioactive carbon (^{14}C) in the indicated position (Section 1.1), the isomer shown here is obtained.

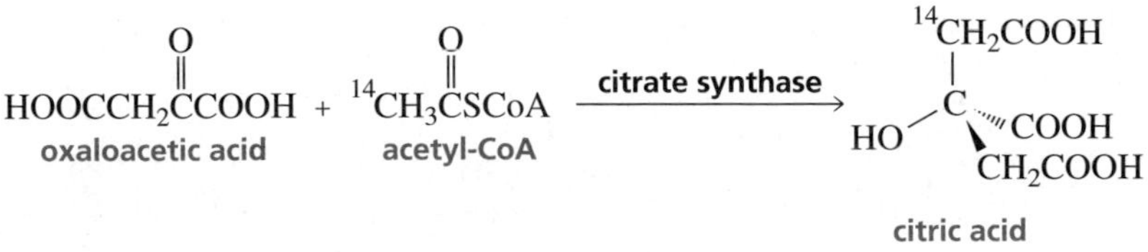

a. Which stereoisomer of citric acid is synthesized, R or S?
b. Why is the other stereoisomer not obtained?
c. If the acetyl-CoA used in the synthesis does not contain ^{14}C (Section 1.1), will the product of the reaction be chiral or achiral?

70. Give the products of the following reactions. If the products can exist as stereoisomers, show which stereoisomers are obtained.

a. cis-2-pentene + HCl
b. trans-2-pentene + HCl
c. 1-ethylcyclohexene + H$_3$O$^+$, H$_2$O
d. 2,3-dimethyl-3-hexene + H$_2$, Pt/C
e. 1,2-dimethylcyclohexene + HCl

f. 1,2-dideuteriocyclohexene + H$_2$, Pt/C
g. 3,3-dimethyl-1-pentene + Br$_2$/CH$_2$Cl$_2$
h. (E)-3,4-dimethyl-3-heptene + H$_2$, Pt/C
i. (Z)-3,4-dimethyl-3-heptene + H$_2$, Pt/C
j. 1-chloro-2-ethylcyclohexene + H$_2$, Pt/C

71. Indicate whether each of the following pairs of compounds are identical or are enantiomers, diastereomers, or constitutional isomers:

a. 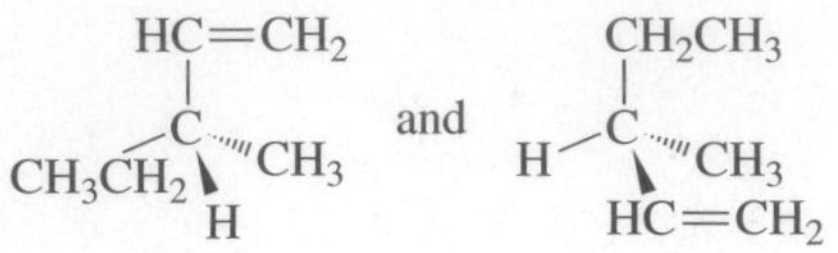
$$
\begin{array}{c}
HC{=}CH_2\\
|\\
CH_3CH_2\overset{C}{\underset{H}{\diagdown}}{}^{\cdots}CH_3
\end{array}
\quad\text{and}\quad
\begin{array}{c}
CH_2CH_3\\
|\\
H{-}\overset{C}{\underset{HC{=}CH_2}{-}}{}^{\cdots}CH_3
\end{array}
$$

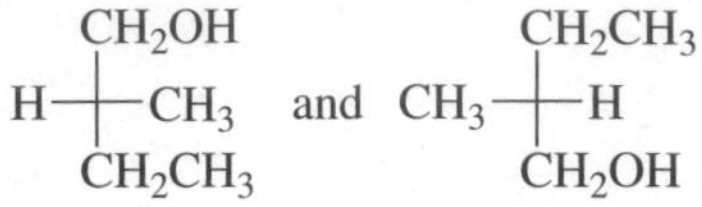

b.
$$
\begin{array}{c}
CH_2OH\\
H{-}\!\!\!-\!\!\!{-}CH_3\\
CH_2CH_3
\end{array}
\quad\text{and}\quad
\begin{array}{c}
CH_2CH_3\\
CH_3{-}\!\!\!-\!\!\!{-}H\\
CH_2OH
\end{array}
$$

c.
$$
\begin{array}{c}
CH_3\\
HO{-}\!\!\!-\!\!\!{-}H\\
H{-}\!\!\!-\!\!\!{-}Cl\\
CH_3
\end{array}
\quad\text{and}\quad
\begin{array}{c}
CH_3\\
H{-}\!\!\!-\!\!\!{-}OH\\
Cl{-}\!\!\!-\!\!\!{-}H\\
CH_3
\end{array}
$$

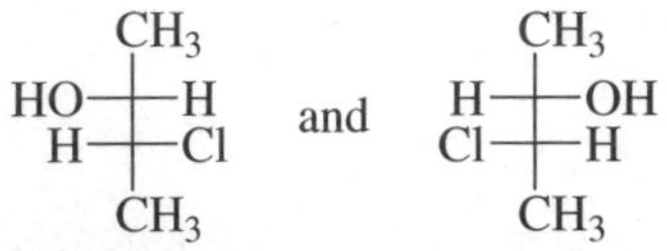

d. 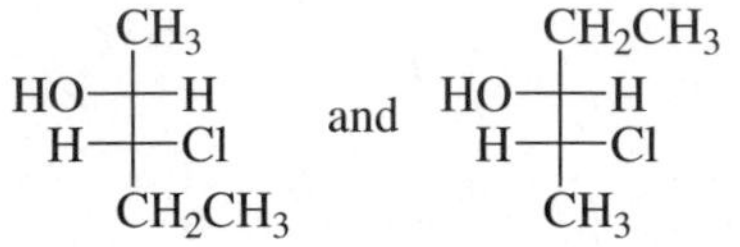
$$
\begin{array}{c}
CH_3\\
HO{-}\!\!\!-\!\!\!{-}H\\
H{-}\!\!\!-\!\!\!{-}Cl\\
CH_2CH_3
\end{array}
\quad\text{and}\quad
\begin{array}{c}
CH_2CH_3\\
HO{-}\!\!\!-\!\!\!{-}H\\
H{-}\!\!\!-\!\!\!{-}Cl\\
CH_3
\end{array}
$$

e.

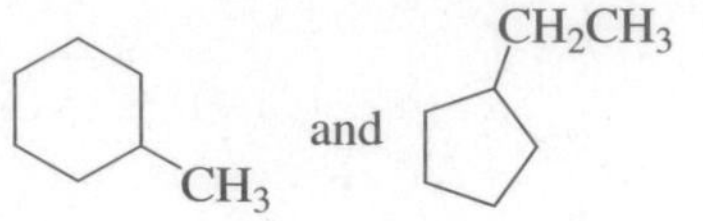

f.

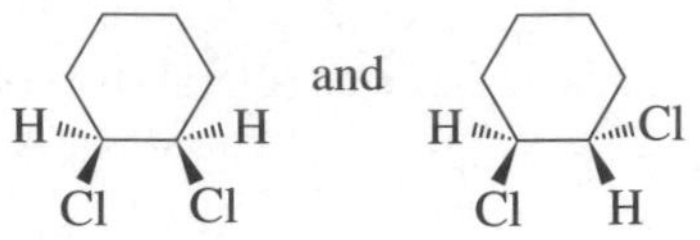

g.

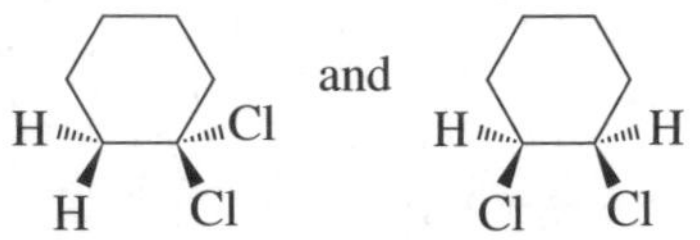

h.

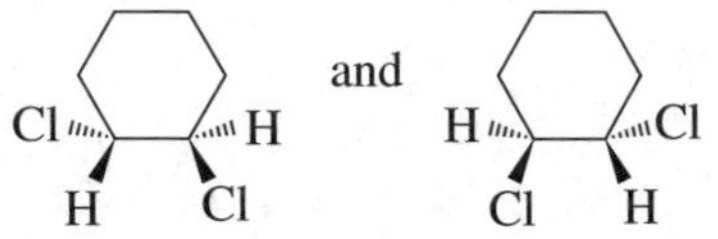

72. The specific rotation of (R)-(+)-glyceraldehyde is +8.7. If the observed specific rotation of a mixture of (R)-glyceraldehyde and (S)-glyceraldehyde is +1.4, what percent of glyceraldehyde is present as the R enantiomer?

73. Indicate whether each of the following structures is (R)-2-chlorobutane or (S)-2-chlorobutane. (Use models, if necessary.)

a. 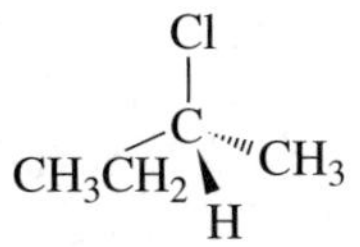
$$
\begin{array}{c}
Cl\\
|\\
CH_3CH_2\overset{C}{\underset{H}{\diagdown}}{}^{\cdots}CH_3
\end{array}
$$

b. 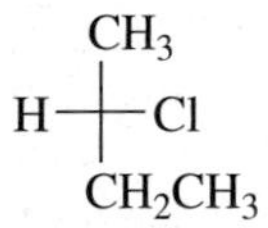
$$
\begin{array}{c}
CH_3\\
H{-}\!\!\!-\!\!\!{-}Cl\\
CH_2CH_3
\end{array}
$$

c.

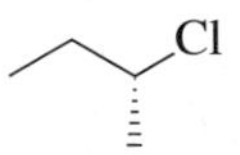

d.

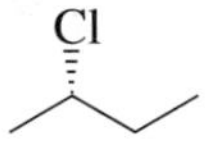

e.

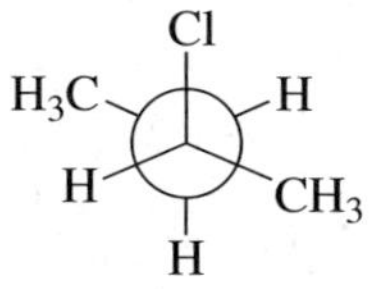

f.

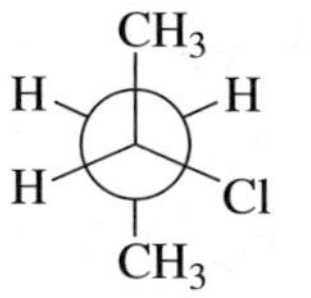

74. A solution of an unknown compound (3.0 g of the compound in 20 mL of solution), when placed in a polarimeter tube 2.0 dm long, was found to rotate the plane of polarized light 1.8° in a counterclockwise direction. What is the specific rotation of the compound?

75. Butaclamol is a potent antipsychotic that has been used clinically in the treatment of schizophrenia. How many asymmetric centers does it have?

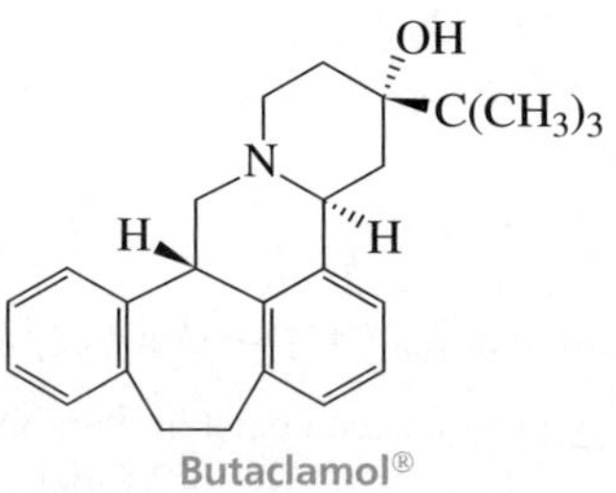

Butaclamol®

76. Explain how R and S are related to (+) and (−).

77. Give the products of the following reactions. If the products can exist as stereoisomers, show which stereoisomers are obtained.
a. *cis*-2-pentene + Br_2/CH_2Cl_2
b. *trans*-2-pentene + Br_2/CH_2Cl_2
c. 1-butene + HCl
d. methylcyclohexene + HBr
e. *trans*-3-hexene + Br_2/CH_2Cl_2
f. *cis*-3-hexene + Br_2/CH_2Cl_2

g. 3,3-dimethyl-1-pentene + HBr
h. *cis*-2-butene + HBr
i. (*Z*)-2,3-dichloro-2-butene + H_2, Pt/C
j. (*E*)-2,3-dichloro-2-butene + H_2, Pt/C
k. (*Z*)-3,4-dimethyl-3-hexene + H_2, Pt/C
l. (*E*)-3,4-dimethyl-3-hexene + H_2, Pt/C

78. a. Draw all possible stereoisomers for the following compound.

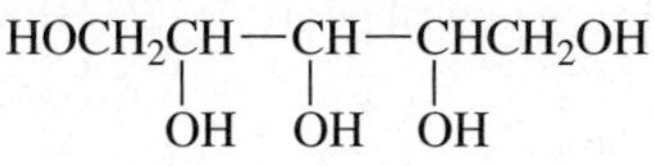

$$HOCH_2CH-CH-CHCH_2OH$$
$$\quad\quad\quad | \quad\ | \quad\ |$$
$$\quad\quad\ OH\ \ OH\ \ OH$$

b. Which isomers are optically inactive (will not rotate plane-polarized light)?

79. Indicate the configuration of the asymmetric centers in the following molecules:

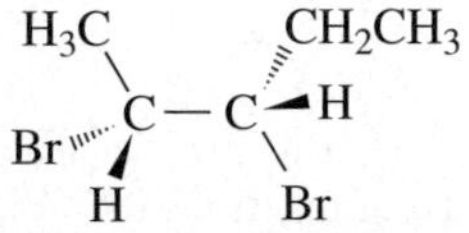

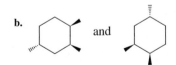

80. a. Draw all the isomers with molecular formula C_6H_{12} that contain a cyclobutane ring. (*Hint:* There are seven.)
b. Name the compounds without specifying the configuration of any asymmetric centers.
c. Identify:

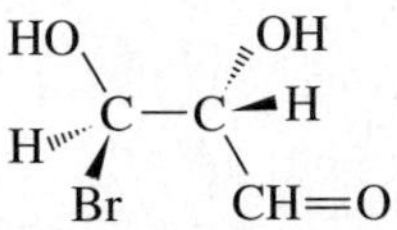

1. constitutional isomers
2. stereoisomers
3. cis–trans isomers
4. chiral compounds

5. achiral compounds
6. meso compounds
7. enantiomers
8. diastereomers

81. A compound has a specific rotation of −39.0. A solution of the compound (0.187 g/mL) has an observed rotation of −6.52 when placed in a polarimeter tube 10 cm long. What is the percent of each enantiomer in the solution?

82. Indicate whether each of the following pairs of compounds are identical or are enantiomers, diastereomers, or constitutional isomers:

a. 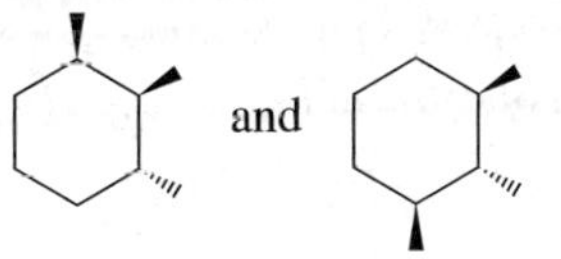 [structures] and [structures]   c. 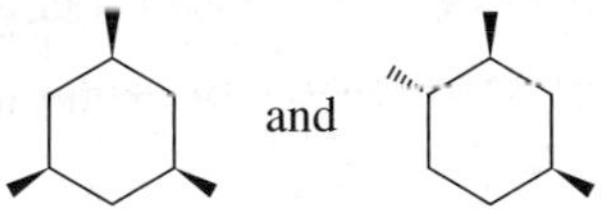 [structures] and [structures]

b. [structures] and [structures]   d. 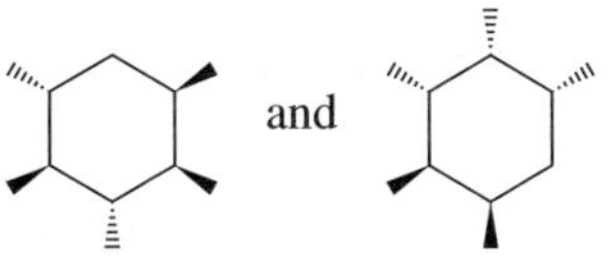 [structures] and [structures]

83. Draw structures for each of the following molecules:
a. (*S*)-1-bromo-1-chlorobutane
b. (2*R*,3*R*)-2,3-dichloropentane
c. an achiral isomer of 1,2-dimethylcyclohexane
d. a chiral isomer of 1,2-dibromocyclobutane
e. two achiral isomers of 3,4,5-trimethylheptane

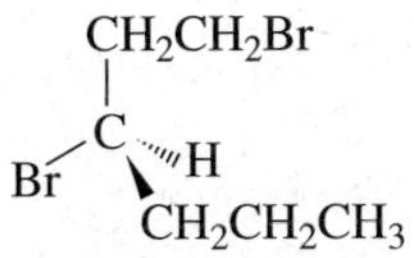

84. Explain why the enantiomers of 1,2-dimethylaziridine can be separated, even though one of the "groups" attached to nitrogen is a lone pair.

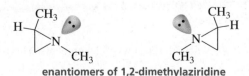

enantiomers of 1,2-dimethylaziridine

85. Of the possible products shown for the following reaction, are there any that would not be formed?

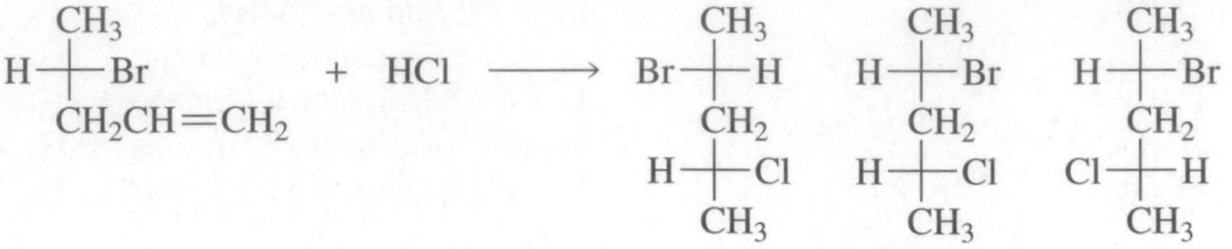

$$
\begin{array}{c}
\text{CH}_3 \\
\text{H} \!-\!\!\!+\!\!\!-\! \text{Br} \\
\text{CH}_2\text{CH}\!=\!\text{CH}_2
\end{array}
\quad + \text{ HCl} \longrightarrow
\begin{array}{c}
\text{CH}_3 \\
\text{Br}\!-\!\!\!+\!\!\!-\!\text{H} \\
\text{CH}_2 \\
\text{H}\!-\!\!\!+\!\!\!-\!\text{Cl} \\
\text{CH}_3
\end{array}
\quad
\begin{array}{c}
\text{CH}_3 \\
\text{H}\!-\!\!\!+\!\!\!-\!\text{Br} \\
\text{CH}_2 \\
\text{H}\!-\!\!\!+\!\!\!-\!\text{Cl} \\
\text{CH}_3
\end{array}
\quad
\begin{array}{c}
\text{CH}_3 \\
\text{H}\!-\!\!\!+\!\!\!-\!\text{Br} \\
\text{CH}_2 \\
\text{Cl}\!-\!\!\!+\!\!\!-\!\text{H} \\
\text{CH}_3
\end{array}
$$

86. A sample of (S)-$(+)$-lactic acid was found to have an optical purity of 72%. How much R isomer is present in the sample?

87. Is the following compound optically active?

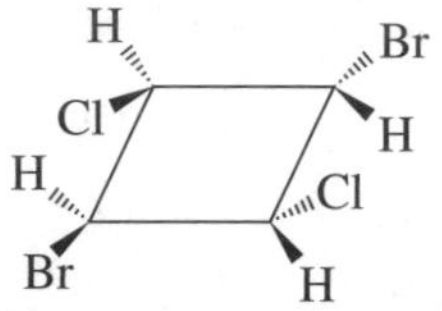

88. Give the products of the following reactions including their configurations:

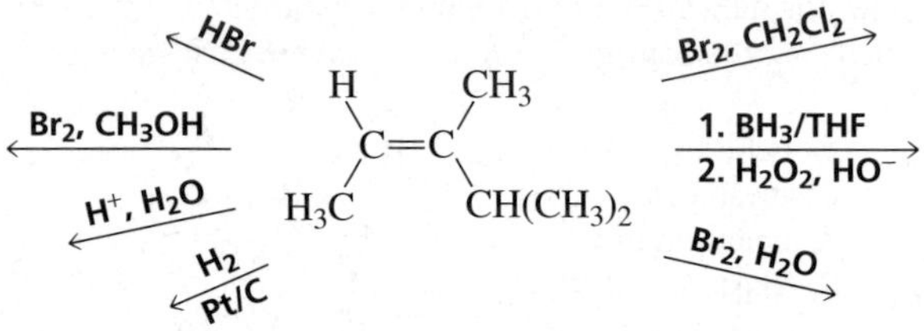

89. a. Using the wedge-and-dash notation, draw the nine stereoisomers of 1,2,3,4,5,6-hexachlorocyclohexane.
b. From the nine stereoisomers, identify one pair of enantiomers.
c. Draw the most stable conformation of the most stable stereoisomer.

90. Sherry O. Eismer decided that the configuration of the asymmetric centers in a sugar such as D-glucose could be determined rapidly by simply assigning the R configuration to an asymmetric center with an OH group on the right and the S configuration to an asymmetric center with an OH group on the left. Is she correct? (We will see in Chapter 21 that the "D" in D-glucose means the OH group on the bottommost asymmetric center is on the right.)

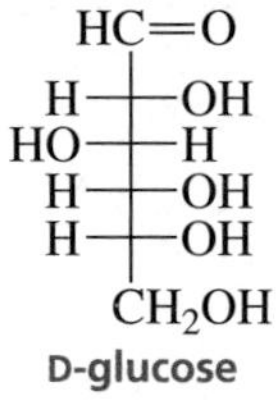

$$
\begin{array}{c}
\text{HC}\!=\!\text{O} \\
\text{H}\!-\!\!\!+\!\!\!-\!\text{OH} \\
\text{HO}\!-\!\!\!+\!\!\!-\!\text{H} \\
\text{H}\!-\!\!\!+\!\!\!-\!\text{OH} \\
\text{H}\!-\!\!\!+\!\!\!-\!\text{OH} \\
\text{CH}_2\text{OH}
\end{array}
$$

D-glucose

91. Cyclohexene exists only in the cis form, whereas cyclodecene exists in both the cis and trans forms. Explain. (*Hint:* Molecular models are helpful for this problem.)

92. When fumarate reacts with D_2O in the presence of the enzyme fumarase, only one isomer of the product is formed, as shown below. Is the enzyme catalyzing a syn or an anti addition of D_2O?

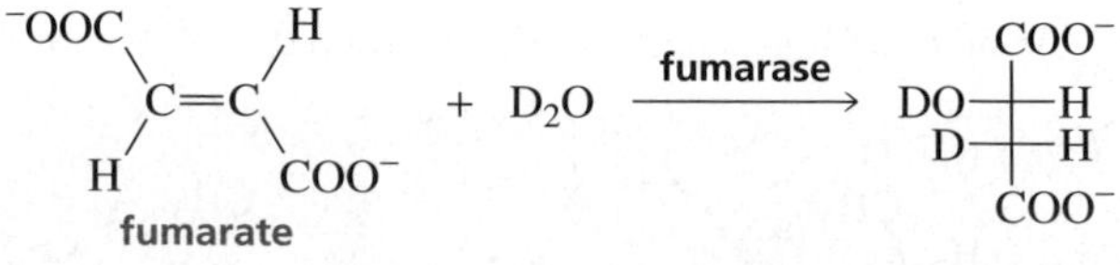

fumarate

93. When (S)-$(+)$-1-chloro-2-methylbutane reacts with chlorine, one of the products is $(-)$-1,4-dichloro-2-methylbutane. Does this product have the R or the S configuration?

94. Indicate the configuration of the asymmetric centers in the following molecules:

a. 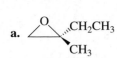   b. 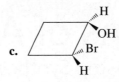   c.

95. a. Draw the two chair conformers for each of the stereoisomers of *trans*-1-*tert*-butyl-3-methylcyclohexane.
   b. For each pair, indicate which conformer is more stable.

96. a. Do the following compounds have any asymmetric centers?
   1. CH_2=C=CH_2
   2. CH_3CH=C=$CHCH_3$
   b. Are the compounds chiral? (*Hint:* Make models.)

CHAPTER 6

The Reactions of Alkynes
An Introduction to Multistep Synthesis

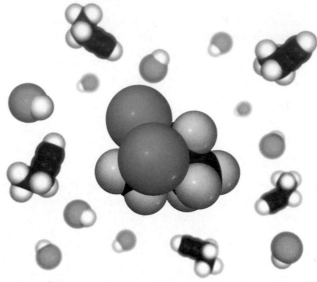

1-butyne   +   2 HCl   ⟶   2,2-dichlorobutane

BUILDING ON FUNDAMENTALS

SECTION 6.1   Just like alcohols (2.6), amines (2.7), and alkenes (3.2), alkynes are named using a functional group suffix.

SECTION 6.4   The structures of alkynes are similar to the structure of ethyne, the first alkyne we looked at (1.9).

SECTION 6.3   The physical properties of alkenes and alkynes are similar to the physical properties of alkanes (2.9). In other words, all hydrocarbons have similar physical properties.

SECTION 6.4   We will see that alkyl groups stabilize alkynes, just like they stabilize carbocations (4.2) and alkenes (4.11).

SECTION 6.5   Alkynes, like alkenes, have a relatively weak π bond that causes them to undergo electrophilic addition reactions, just like alkenes do (Chapter 4). However,

because alkynes have two π bonds, they can undergo some of these reactions twice.

SECTION 6.5   Since we know that reactivity depends on the $\Delta G^{\ddagger}$ for the rate-limiting step of the reaction (3.7) we can understand why alkynes are less reactive than alkenes.

SECTION 6.6   Alkenes and alkynes have the same regioselectivity in electrophilic addition reactions: the electrophile adds to whichever sp^2 carbon (in alkenes) or sp carbon (in alkynes) is bonded to the greater number of hydrogens.

SECTION 6.10   Because an sp hybridized carbon is more electronegative than other carbon atoms (1.20), terminal alkynes have acidic properties that are useful for synthetic schemes in which the number of carbons in a reactant needs to be increased.

An **alkyne** is a hydrocarbon that contains a carbon–carbon triple bond. Because of its triple bond, an alkyne has four fewer hydrogens than an alkane with the same number of carbons. Therefore, while the general molecular formula for an alkane is C_nH_{2n+2}, the general molecular formula for an acyclic (noncyclic) alkyne is C_nH_{2n-2} and that for a cyclic alkyne is C_nH_{2n-4}.

There are only a few naturally occurring alkynes. Examples include capillin, which has fungicidal activity, and ichthyothereol, a convulsant used by the indigenous people of the Amazon for poisoned arrowheads. A class of naturally occurring compounds called enediynes has been found to have powerful antibiotic and anticancer properties. These compounds all have a nine- or ten-membered ring that contains two triple bonds separated by a double bond. Some enediynes are currently being tested in clinical trials (Section 30.13).

$$CH_3C\equiv C-C\equiv C-\overset{O}{\overset{\|}{C}}-\text{(phenyl)}$$
capillin

$$CH_3C\equiv C-C\equiv C-C\equiv C-\overset{H}{\underset{H}{\overset{|}{C}}}$$
ichthyothereol

an enediyne

The few drugs that contain alkyne functional groups are not naturally occurring compounds; they exist only because chemists have been able to synthesize them. Their trade names are shown below, in green. Trade names are always capitalized; only the company that holds the patent for a product can use the product's trade name for commercial purposes (Section 30.1).

Parsal®
Sinovial®
parsalmide
an analgesic

Eudatin®
Supirdyl®
pargyline
an antihypertensive

Norquen®
Ovastol®
mestranol
a component in oral contraceptives

Acetylene (HC≡CH) the common name for the smallest alkyne, may be familiar to you because of the oxyacetylene torch used in welding. Acetylene is supplied to the torch from a high-pressure gas tank, and oxygen is supplied from another. Burning acetylene produces a high-temperature flame capable of melting and vaporizing iron and steel.

ETHYNE CHEMISTRY OR THE FORWARD PASS?

Father Julius Arthur Nieuwland (1878–1936) did much of the early work that led to the synthesis of a polymer called neoprene, a synthetic rubber. He formed the required starting material by reacting vinylacetylene with HCl. Because of the vow of poverty he took as a priest, he refused to accept any royalties for his discovery.

$$HC\equiv CCH=CH_2 \xrightarrow{\textbf{HCl}} CH_2=\overset{Cl}{\overset{|}{C}}CH=CH_2 \longrightarrow$$
vinylacetylene   **2-chloro-1,3-butadiene**
chloroprene

neoprene

Father Nieuwland was born in Belgium but settled with his parents in South Bend, Indiana, two years later. After becoming a priest, he was a professor of botany and chemistry at the University of Notre Dame, where Knute Rockne—the inventor of the forward pass—worked for him as a research assistant. Rockne also taught chemistry at Notre Dame, but when he received an offer to coach the football team, he switched fields, in spite of Father Nieuwland's attempts to convince him to continue his work as a scientist.

Rockne became one of the most famous coaches in college football; his Notre Dame teams won six national championships. During half-time at the game against Army in 1928, Rockne inspired his team to victory with his "Win one for the Gipper" story.

Knute Rockne in his uniform during the year he was captain of the Notre Dame football team.

6.1 The Nomenclature of Alkynes

The systematic name of an alkyne is obtained by replacing the "ane" ending of the alkane name with "yne." Analogous to the way compounds with other functional groups are named (Sections 2.6, 2.7, 3.2), the longest continuous chain containing the functional group, in this case the carbon–carbon triple bond, is numbered in the direction that gives the alkyne functional group suffix as low a number as possible. If the triple bond is at the end of the chain, the alkyne is classified as a **terminal alkyne**. Alkynes with triple bonds located elsewhere along the chain are called **internal alkynes**. For example, 1-butyne is a terminal alkyne, whereas 2-pentyne is an internal alkyne.

**1-hexyne
a terminal alkyne**

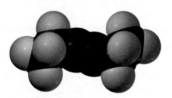

**3-hexyne
an internal alkyne**

HC≡CH

Systematic: ethyne
Common: acetylene

$\overset{4}{C}H_3\overset{3}{C}H_2\overset{2}{C}≡\overset{1}{C}H$

**1-butyne
ethylacetylene
a terminal alkyne**

$\overset{1}{C}H_3\overset{2}{C}≡\overset{3}{C}\overset{4}{C}H_2\overset{5}{C}H_3$

**2-pentyne
ethylmethylacetylene
an internal alkyne**

$\overset{5}{C}H_2\overset{6}{C}H_3$
$\overset{4}{C}H_3\overset{3}{C}H\overset{}{C}≡\overset{2}{C}\overset{1}{C}H_3$

**4-methyl-2-hexyne
sec-butylmethyl-
acetylene**

In common nomenclature, alkynes are named as *substituted acetylenes*. The common name is obtained by citing the names of the alkyl groups, in alphabetical order, that have replaced the hydrogens of acetylene. Acetylene is an unfortunate common name for the smallest alkyne because its "ene" ending is characteristic of a double bond rather than a triple bond.

If counting from either direction leads to the same number for the functional group suffix, the correct systematic name is the one that contains the lowest substituent number. If the compound contains more than one substituent, the substituents are listed in alphabetical order.

Cl Br
| |
$\overset{1}{C}H_3\overset{2}{C}H\overset{3}{C}H\overset{4}{C}≡\overset{5}{C}\overset{6}{C}H_2\overset{7}{C}H_2\overset{8}{C}H_3$

3-bromo-2-chloro-4-octyne
not **6-bromo-7-chloro-4-octyne**
because 2 < 6

CH₃
|
$\overset{6}{C}H_3\overset{5}{C}H\overset{4}{C}≡\overset{3}{C}\overset{2}{C}H_2\overset{1}{C}H_2Br$

1-bromo-5-methyl-3-hexyne
not **6-bromo-2-methyl-3-hexyne**
because 1 < 2

> **A substituent receives the lowest possible number only if there is no functional group suffix or if counting from either direction leads to the same number for the functional group suffix.**

The triple-bond-containing propargyl group is used in common nomenclature. It is analogous to the double-bond-containing allyl group that you saw in Section 3.2.

HC≡CCH₂—   H₂C=CHCH₂—
propargyl group   **allyl group**

HC≡CCH₂Br   H₂C=CHCH₂OH
propargyl bromide   **allyl alcohol**

PROBLEM 3◆

Name the following compounds:

a. [structure with Br]   b. [structure]   c. [structure]

PROBLEM 4◆

Draw the structures and give the common and systematic names for the seven alkynes with molecular formula C_6H_{10}.

PROBLEM 5◆

Give the systematic name for each of the following compounds:

a. $BrCH_2CH_2C{\equiv}CCH_3$

b. $CH_3CH_2\underset{\underset{\textstyle Br}{|}}{C}HC{\equiv}CCH_2\underset{\underset{\textstyle Cl}{|}}{C}HCH_3$

c. $CH_3OCH_2C{\equiv}CCH_2CH_3$

d. $CH_3CH_2\underset{\underset{\textstyle CH_2CH_2CH_3}{|}}{C}HC{\equiv}CH$

6.2 How to Name a Compound That Has More than One Functional Group

To create the systematic name of a compound with two double bonds, we identify the longest continuous chain that contains both double bonds by its alkane name and replace the "ne" ending with "diene." We number the chain in the direction that gives the lowest possible numbers to the carbons participating in the double bonds. Then we place the numbers indicating the locations of the double bonds either before the name of the parent compound or before the suffix. Substituents are cited in alphabetical order. Similar rules are followed for compounds with two triple bonds, using the ending "diyne."

$CH_2{=}C{=}CH_2$

systematic: propadiene
common: allene

$\overset{1}{C}H_2{=}\overset{2}{\underset{\underset{\textstyle CH_3}{|}}{C}}{-}\overset{3}{C}H{=}\overset{4}{C}H_2$

2-methyl-1,3-butadiene
or
2-methylbuta-1,3-diene
isoprene

[structure of 5-bromo-1,3-cyclohexadiene]

5-bromo-1,3-cyclohexadiene
or
5-bromocyclohexa-1,3-diene

$\overset{6}{C}H_3\overset{5}{C}H{=}\overset{4}{C}H\overset{3}{C}H_2\overset{2}{\underset{\underset{\textstyle CH_3}{|}}{C}}{=}\overset{1}{C}H_2$

2-methyl-1,4-hexadiene
or
2-methylhexa-1,4-diene

$CH_3\underset{\underset{\textstyle CH_3}{|}}{C}HC{\equiv}CCH_2C{\equiv}CH$

6-methyl-1,4-heptadiyne
or
6-methylhepta-1,4-diyne

To name an alkene in which the second functional group is not another double bond but is also identified by a functional group suffix, find the longest continuous chain containing both functional groups, and put both suffixes at the end of the name. The "ene" ending should be first, with the terminal "e" omitted to avoid two adjacent vowels. The number indicating the location of the first cited functional group is usually placed before the name of the parent chain. The number indicating the location of the second-cited functional group is placed immediately before the suffix for that functional group.

If the two functional groups are a *double bond* and a *triple bond*, the chain is numbered in the direction that produces a name containing the lower number. Thus, in the

When the functional groups are a double bond and a triple bond, the chain is numbered in the direction that produces the name containing the lowest possible number, regardless of which functional group gets the lower number.

following examples, the lower number is given to the alkyne suffix in the compound on the left and to the alkene suffix in the compound on the right.

$$\overset{7}{C}H_3\overset{6}{C}H=\overset{5}{C}H\overset{4}{C}H_2\overset{3}{C}H_2\overset{2}{C}\equiv\overset{1}{C}H \qquad \overset{1}{C}H_2=\overset{2}{C}H\overset{3}{C}H_2\overset{4}{C}H_2\overset{5}{C}\equiv\overset{6}{C}\overset{7}{C}H_3$$

5-hepten-1-yne
not **2-hepten-6-yne**
because 1 < 2

1-hepten-5-yne
not **6-hepten-2-yne**
because 1 < 2

$$\overset{}{C}H_2CH_2CH_2CH_3$$
$$|$$
$$\overset{1}{C}H_2=\overset{2}{C}H\overset{3}{C}H\overset{4}{C}\equiv\overset{5\ 6}{C}CH_3$$

3-butyl-1-hexen-4-yne

> the longest continuous chain has eight carbons, but the 8-carbon chain does not contain both functional groups; therefore, the compound is named as a hexenyne because the longest continuous chain containing both functional groups has six carbons

If there is a tie between a double bond and a triple bond, the double bond gets the lower number.

If the same low number is obtained in both directions, the chain is numbered in the direction that gives the double bond the lower number.

$$\overset{1}{C}H_3\overset{2}{C}H=\overset{3}{C}H\overset{4}{C}\equiv\overset{5\ 6}{C}CH_3 \qquad H\overset{6}{C}\equiv\overset{5}{C}\overset{4}{C}H_2\overset{3}{C}H_2\overset{2}{C}H=\overset{1}{C}H_2$$

2-hexen-4-yne
not **4-hexen-2-yne**

1-hexen-5-yne
not **5-hexen-1-yne**

A chain is numbered to give the lowest possible number to the functional group with the higher priority.

The relative priorities of the functional group suffixes are shown in Table 6.1. If the second functional group suffix has a higher priority than the alkene suffix, the chain is numbered in the direction that assigns the lower number to the functional group with the higher-priority suffix.

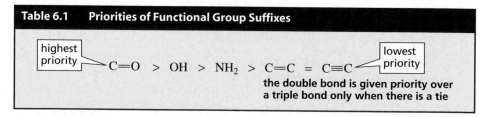

$CH_2=CHCH_2OH$
2-propen-1-ol
not **1-propen-3-ol**

$CH_3\overset{\overset{\displaystyle CH_3}{|}}{C}=CHCH_2CH_2OH$
4-methyl-3-penten-1-ol

$CH_2=CHCH_2CH_2CH_2\overset{\overset{\displaystyle NH_2}{|}}{C}HCH_3$
6-hepten-2-amine

$CH_3CH_2CH=CH\overset{\overset{\displaystyle}{|}}{\underset{OH}{C}}HCH_3$
3-hexen-2-ol

6-methyl-2-cyclohexenol

3-cyclohexenamine

Table 6.1   Priorities of Functional Group Suffixes

highest priority ─── $C=O$ > OH > NH_2 > $C=C$ = $C\equiv C$ ─── lowest priority

the double bond is given priority over a triple bond only when there is a tie

---

PROBLEM 6♦

Give the systematic name for each of the following compounds:

a. $CH_2=CHCH_2C\equiv CCH_2CH_3$

b. $CH_3CH=\overset{\overset{\displaystyle CH_3}{|}}{C}CH_2CH=CH_2$

c. $CH_3CH_2CH=\overset{\overset{\displaystyle CH=CH_2}{|}}{C}CH_2CH_2C\equiv CH$

d. $HOCH_2CH_2C\equiv CH$

e. $CH_3CH=CHCH=CHCH=CH_2$

f. $CH_3CH=\overset{\overset{\displaystyle CH_3\ \ \ CH_3}{|\ \ \ \ \ \ |}}{C}CH_2CHCH_2OH$

Many species of mushrooms synthesize 1-octen-3-ol, which serves as a repellent that drives off predatory slugs. Such mushrooms can be recognized by small bite marks on their caps, where the slug started to nibble before the volatile compound was released. Humans are not put off by the release of this compound because to them 1-octen-3-ol smells like a mushroom. 1-Octen-3-ol also has antibacterial properties that may protect the mushroom from organisms that would otherwise invade the wound made by the slug. Not surprisingly, the species of mushroom that banana slugs commonly eat cannot synthesize 1-octen-3-ol.

$$H_2C=CHCHCH_2CH_2CH_2CH_2CH_3$$
$$|$$
$$OH$$
1-octen-3-ol

6.3 The Physical Properties of Unsaturated Hydrocarbons

All hydrocarbons have similar physical properties. In other words, alkenes and alkynes have physical properties similar to those of alkanes (Section 2.9). All are insoluble in water and soluble in nonpolar solvents such as benzene and diethyl ether. They are less dense than water and, like all other homologous series, have boiling points that increase with increasing molecular weight (Table 6.2). Alkynes are more linear than alkenes, and a triple bond is more polarizable than a double bond (Section 2.9). These two features cause alkynes to have stronger van der Waals interactions. As a result, an alkyne has a higher boiling point than an alkene containing the same number of carbons.

Table 6.2   Boiling Points of the Smallest Hydrocarbons

	bp (°C)		bp (°C)		bp (°C)
CH_3CH_3 ethane	−88.6	$H_2C=CH_2$ ethene	−104	$HC\equiv CH$ ethyne	−84
$CH_3CH_2CH_3$ propane	−42.1	$CH_3CH=CH_2$ propene	−47	$CH_3C\equiv CH$ propyne	−23
$CH_3CH_2CH_2CH_3$ butane	−0.5	$CH_3CH_2CH=CH_2$ 1-butene	−6.5	$CH_3CH_2C\equiv CH$ 1-butyne	8
$CH_3(CH_2)_3CH_3$ pentane	36.1	$CH_3CH_2CH_2CH=CH_2$ 1-pentene	30	$CH_3CH_2CH_2C\equiv CH$ 1-pentyne	39
$CH_3(CH_2)_4CH_3$ hexane	68.7	$CH_3CH_2CH_2CH_2CH=CH_2$ 1-hexene	63.5	$CH_3CH_2CH_2CH_2C\equiv CH$ 1-hexyne	71
		$CH_3CH=CHCH_3$ cis-2-butene	3.7	$CH_3C\equiv CCH_3$ 2-butyne	27
		$CH_3CH=CHCH_3$ trans-2-butene	0.9	$CH_3CH_2C\equiv CCH_3$ 2-pentyne	55

Internal alkenes have higher boiling points than terminal alkenes. Similarly, internal alkynes have higher boiling points than terminal alkynes. Notice that the boiling point of cis-2-butene is slightly higher than that of trans-2-butene because the cis isomer has a small dipole moment, whereas the dipole moment of the trans isomer is zero (Section 3.4).

PROBLEM 7◆

What is the smallest alkane, the smallest alkene, and the smallest alkyne in Table 6.2 that are liquids at room temperature?

6.4 The Structure of Alkynes

The structure of ethyne was discussed in Section 1.9. We saw that each carbon is *sp* hybridized, so each has two *sp* orbitals and two *p* orbitals. One *sp* orbital overlaps the *s* orbital of a hydrogen, and the other overlaps an *sp* orbital of the other carbon. Because the *sp* orbitals are oriented as far from each other as possible to minimize electron repulsion, ethyne is a linear molecule with bond angles of 180°.

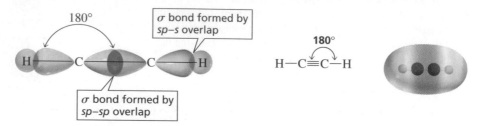

The two remaining *p* orbitals on each carbon are oriented at right angles to one another and to the *sp* orbitals (Figure 6.1). Each of the two *p* orbitals on one carbon overlaps the parallel *p* orbital on the other carbon to form two π bonds. One pair of overlapping *p* orbitals results in a cloud of electrons above and below the σ bond, and the other pair results in a cloud of electrons in front of and behind the σ bond. The electrostatic potential map of ethyne (displayed above) shows that the end result can be thought of as a cylinder of electrons wrapped around the σ bond.

A triple bond is composed of a σ bond and two π bonds.

Figure 6.1 ▶
(a) Each of the two π bonds of a triple bond is formed by side-to-side overlap of a *p* orbital of one carbon with a parallel *p* orbital of the adjacent carbon. (b) A triple bond consists of a σ bond formed by *sp*–*sp* overlap (yellow) and two π bonds formed by *p*–*p* overlap (blue and purple).

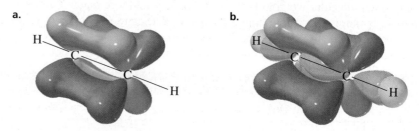

We have seen that a carbon–carbon triple bond is shorter and stronger than a carbon–carbon double bond, which in turn, is shorter and stronger than a carbon–carbon single bond. We have also seen that a π bond is weaker than a σ bond (Section 1.14). The relatively weak π bonds allow alkynes to react easily. Alkyl groups stabilize alkynes, like they stabilize alkenes, by hyperconjugation. Internal alkynes, therefore, are more stable than terminal alkynes. We now have seen that *alkyl groups stabilize alkenes, alkynes,* and *carbocations.*

PROBLEM 8◆

What orbitals are used to form the carbon–carbon σ bond between the highlighted carbons?

a. $CH_3CH=CHCH_3$

b. $CH_3CH=CHCH_3$

c. $CH_3CH=C=CH_2$

d. $CH_3C\equiv CCH_3$

e. $CH_3C\equiv CCH_3$

f. $CH_2=CHCH=CH_2$

g. $CH_3CH=CHCH_2CH_3$

h. $CH_3C\equiv CCH_2CH_3$

i. $CH_2=CHC\equiv CH$

6.5 How Alkynes React

With a cloud of electrons completely surrounding the σ bond, an alkyne is an electron-rich molecule. In other words, it is a nucleophile and consequently it will react with an electrophile.

Mechanism for an electrophilic addition reaction

$$CH_3C\equiv CCH_3 + H-\overset{..}{\underset{..}{Cl}}: \longrightarrow CH_3\overset{+}{C}=CHCH_3 + :\overset{..}{\underset{..}{Cl}}:^- \longrightarrow CH_3\overset{\overset{\textstyle Cl}{|}}{C}=CHCH_3$$

a nucleophile | an electrophile — an electrophile — a nucleophile

- The relatively weak π bond breaks because the π electrons are attracted to the electrophilic proton.
- The positively charged carbocation intermediate reacts rapidly with the negatively charged chloride ion.

Thus alkynes, like alkenes, undergo electrophilic addition reactions. We will see that the same electrophilic reagents that add to alkenes also add to alkynes and that—again like alkenes—electrophilic addition to a *terminal* alkyne is regioselective: when an electrophile adds to a terminal alkyne, it adds to the *sp* carbon that is bonded to the hydrogen. The addition reactions of alkynes, however, have a feature that alkenes do not have: because the product of the addition of an electrophilic reagent to an alkyne is an alkene, a second electrophilic addition reaction can occur.

The electrophile adds to the *sp* carbon of a terminal alkyne that is bonded to the hydrogen.

$$CH_3C\equiv CCH_3 \xrightarrow{\ \textbf{HCl}\ } CH_3\overset{\overset{\textstyle Cl}{|}}{C}=CHCH_3 \xrightarrow{\ \textbf{HCl}\ } CH_3\overset{\overset{\textstyle Cl}{|}}{\underset{\underset{\textstyle Cl}{|}}{C}}CH_2CH_3$$

a second electrophilic addition reaction occurs

An alkyne is *less* reactive than an alkene in electrophilic addition reactions. This might at first seem surprising because an alkyne is less stable than an alkene (Figure 6.2). However, reactivity depends on $\Delta G^{\ddagger}$, which in turn depends on the stability of the reactant *and* the stability of the transition state (Section 3.7). For an alkyne to be both less stable and less reactive than an alkene, two conditions must hold: the transition state for the rate-limiting step (the first step) of an electrophilic addition reaction for an alkyne must be less stable than the transition state for the first step of an electrophilic addition reaction for an alkene, *and* the difference in the stabilities of the transition states must be greater than the difference in the stabilities of the reactants so that $\Delta G^{\ddagger}_{\text{alkyne}} > \Delta G^{\ddagger}_{\text{alkene}}$ (Figure 6.2).

Alkynes are less reactive than alkenes in electrophilic addition reactions.

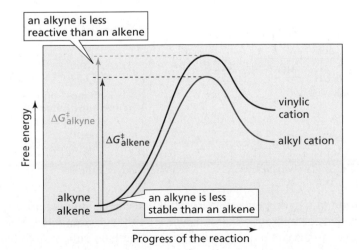

◀ **Figure 6.2**
Comparison of the free energies of activation for the addition of an electrophile to an alkyne and to an alkene. Because an alkyne is less reactive than an alkene toward electrophilic addition, we know that $\Delta G^{\ddagger}$ for the reaction of an alkyne is greater than the $\Delta G^{\ddagger}$ for the reaction of an alkene.

Why is the transition state for the first step of an electrophilic addition reaction for an alkyne less stable than that for an alkene? The Hammond postulate predicts that the structure of the transition state will resemble the structure of the intermediate (Section 4.3). The intermediate that is formed when a proton adds to an alkyne is a vinylic cation, whereas the intermediate that is formed when a proton adds to an alkene is an alkyl cation. A **vinylic cation** has a positive charge on a vinylic carbon. A vinylic cation is less stable than a similarly substituted alkyl cation. In other words, a primary vinylic cation is less stable than a primary alkyl cation, and a secondary vinylic cation is less stable than a secondary alkyl cation.

relative stabilities of carbocations

$$\boxed{\text{most stable}} \quad R\!-\!\overset{R}{\underset{R}{\overset{|}{\underset{|}{\overset{+}{C}}}}} \;>\; R\!-\!\overset{R}{\underset{H}{\overset{|}{\underset{|}{\overset{+}{C}}}}} \;>\; RCH\!=\!\overset{+}{C}\!-\!R \;\approx\; R\!-\!\overset{H}{\underset{H}{\overset{|}{\underset{|}{\overset{+}{C}}}}} \;>\; RCH\!=\!\overset{+}{C}\!-\!H \;\approx\; H\!-\!\overset{H}{\underset{H}{\overset{|}{\underset{|}{\overset{+}{C}}}}} \quad \boxed{\text{least stable}}$$

|a tertiary carbocation | a secondary carbocation | a secondary vinylic cation | a primary carbocation | a primary vinylic cation | the methyl cation |

A vinylic cation is less stable because the positive charge is on an sp carbon, which we have seen is more electronegative than the sp^2 carbon of an alkyl cation (Section 1.20). Therefore, a vinylic cation is less able to bear a positive charge. In addition, hyperconjugation is less effective in stabilizing the charge on a vinylic cation than on an alkyl cation (Section 4.2).

PROBLEM 9◆

Under what circumstances can you assume that the less stable of two compounds will be the more reactive compound?

6.6 Addition of Hydrogen Halides and Addition of Halogens to an Alkyne

Student Tutorial:
Addition of HCl to an alkyne

We have just seen that an alkyne is a nucleophile and that in the first step of the reaction of an alkyne with a hydrogen halide, the electrophilic H^+ adds to the alkyne. If the alkyne is a *terminal* alkyne, the H^+ will add to the sp carbon bonded to a hydrogen, because the secondary vinylic cation that results is more stable than the primary vinylic cation that would be formed if the H^+ added to the other sp carbon. (Recall that alkyl groups stabilize positively charged carbon atoms; Section 4.2.)

$$\boxed{\text{more reactive}} \quad CH_3CH_2C\equiv CH \xrightarrow{\ \textbf{HBr}\ } CH_3CH_2\overset{+}{C}=CH \longrightarrow \underset{\substack{\text{2-bromo-1-butene}\\ \text{a halo-substituted alkene}}}{CH_3CH_2\overset{Br}{\underset{}{C}}=\overset{H}{\underset{}{CH}}} \quad \boxed{\text{less reactive}}$$

the electrophile adds here

Br^-

$$\boxed{\text{more stable}} \quad \underset{\text{a secondary vinylic cation}}{CH_3CH_2\overset{+}{C}=CH_2} \qquad \underset{\text{a primary vinylic cation}}{CH_3CH_2CH=\overset{+}{CH}} \quad \boxed{\text{less stable}}$$

The addition of a hydrogen halide to an alkyne can be stopped after the addition of one equivalent of hydrogen halide because, although an alkyne is less reactive than an alkene, an alkyne is more reactive than the halo-substituted alkene that is the product of the first addition reaction. The halo-substituted alkene is less reactive because a halogen substituent withdraws electrons inductively (through the σ bond), thereby decreasing the nucleophilic character of the double bond.

Although the addition of a hydrogen halide to an alkyne can generally be stopped after the addition of one equivalent of hydrogen halide, a second addition reaction will take place if excess hydrogen halide is present. The product of the second addition reaction is a **geminal dihalide**, a molecule with two halogens on the same carbon. "Geminal" comes from *geminus*, which is Latin for "twin."

electrophile adds here

Br
|
$CH_3CH_2C{=}CH_2$ $\xrightarrow{\text{HBr}}$ $CH_3CH_2CCH_3$
                                        |
                                        Br

2-bromo-1-butene   **2,2-dibromobutane**
                            a geminal dihalide

When the second equivalent of hydrogen halide adds to the double bond, the electrophile (H^+) adds to the sp^2 carbon bonded to the greater number of hydrogens—as predicted by the rule that governs electrophilic addition reactions (Section 4.4). The carbocation that results is more stable than the carbocation that would have been formed if H^+ had added to the other sp^2 carbon because bromine can share the positive charge with carbon by overlapping one of its orbitals that contains a lone pair with the empty $2p$ orbital of the positively charged carbon.

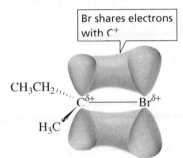

In describing the mechanism for addition of a hydrogen halide, we have shown the intermediate as a vinylic cation. This mechanism may not be completely correct. A secondary vinylic cation is about as stable as a primary carbocation, and generally primary carbocations are too unstable to be formed. Some chemists, therefore, think that a **π-complex** rather than a vinylic cation is formed as an intermediate.

$\delta{-}$Cl
|
$\delta{+}$H
|
$HC{\equiv}CH$

a π-complex

Support for the intermediate's being a π-complex comes from the observation that many (but not all) alkyne addition reactions are stereoselective. For example, the addition of HCl to 2-butyne forms only (Z)-2-chloro-2-butene, which means that only anti addition of H and Cl occurs. Clearly, the nature of the intermediate in alkyne addition reactions is not completely understood.

anti addition

$CH_3C{\equiv}CCH_3$ $\xrightarrow{\text{HCl}}$

H   CH$_3$
 \   /
  C{=}C
 /   \
H$_3$C   Cl

2-butyne   **(Z)-2-chloro-2-butene**

Addition of a hydrogen halide to an *internal* alkyne forms two geminal dihalides because the initial addition of the proton can occur with equal ease to either of the *sp* carbons.

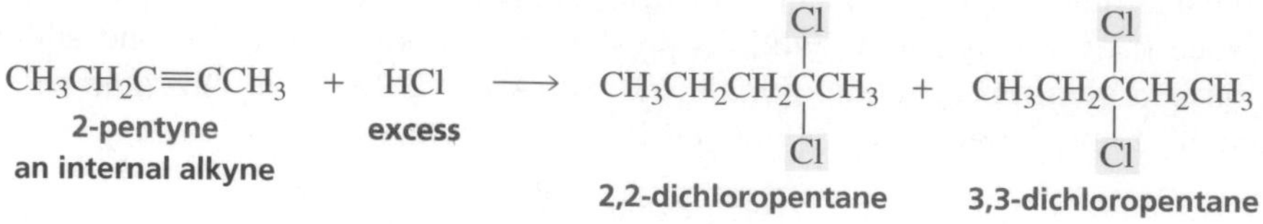

$$CH_3CH_2C{\equiv}CCH_3 \; + \; HCl \longrightarrow CH_3CH_2CH_2\overset{\overset{\textstyle Cl}{|}}{\underset{\underset{\textstyle Cl}{|}}{C}}CH_3 \; + \; CH_3CH_2\overset{\overset{\textstyle Cl}{|}}{\underset{\underset{\textstyle Cl}{|}}{C}}CH_2CH_3$$

2-pentyne   excess   2,2-dichloropentane   3,3-dichloropentane
an internal alkyne

Note, however, that if the same group is attached to each of the *sp* carbons of the internal alkyne, only one geminal dihalide will be obtained.

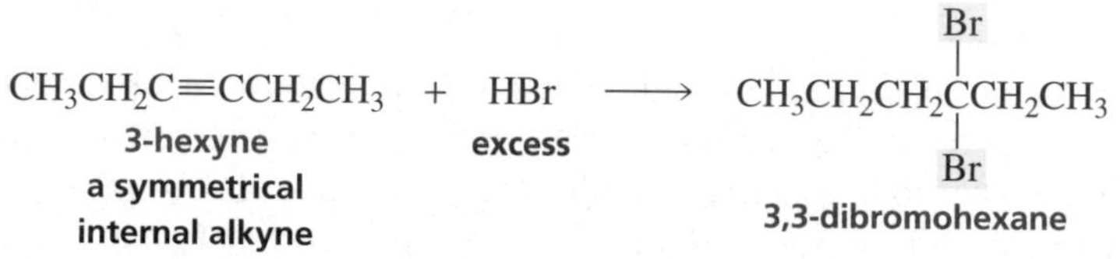

$$CH_3CH_2C{\equiv}CCH_2CH_3 \; + \; HBr \longrightarrow CH_3CH_2CH_2\overset{\overset{\textstyle Br}{|}}{\underset{\underset{\textstyle Br}{|}}{C}}CH_2CH_3$$

3-hexyne   excess
a symmetrical   3,3-dibromohexane
internal alkyne

The halogens Cl_2 and Br_2 also add to alkynes. In the presence of excess halogen, a second addition reaction occurs. Typically, the solvent is CH_2Cl_2.

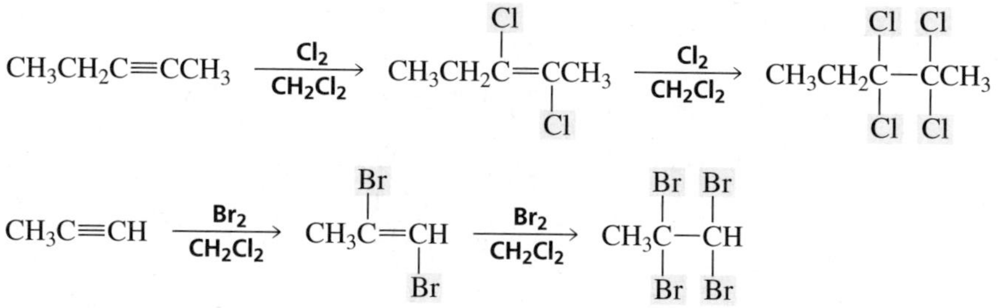

$$CH_3CH_2C{\equiv}CCH_3 \xrightarrow[CH_2Cl_2]{Cl_2} CH_3CH_2\overset{\overset{\textstyle Cl}{|}}{C}{=}\underset{\underset{\textstyle Cl}{|}}{C}CH_3 \xrightarrow[CH_2Cl_2]{Cl_2} CH_3CH_2\overset{\overset{\textstyle Cl\;\;Cl}{|\;\;\;|}}{\underset{\underset{\textstyle Cl\;\;Cl}{|\;\;\;|}}{C-C}}CH_3$$

$$CH_3C{\equiv}CH \xrightarrow[CH_2Cl_2]{Br_2} CH_3\overset{\overset{\textstyle Br}{|}}{C}{=}\underset{\underset{\textstyle Br}{|}}{C}H \xrightarrow[CH_2Cl_2]{Br_2} CH_3\overset{\overset{\textstyle Br\;\;Br}{|\;\;\;|}}{\underset{\underset{\textstyle Br\;\;Br}{|\;\;\;|}}{C-C}}H$$

PROBLEM 10◆

Give the major product of each of the following reactions:

a. $HC{\equiv}CCH_3 \xrightarrow{HBr}$

b. $HC{\equiv}CCH_3 \xrightarrow{\text{excess} \atop HBr}$

c. $CH_3C{\equiv}CCH_3 \xrightarrow[CH_2Cl_2]{Br_2}$

d. $HC{\equiv}CCH_3 \xrightarrow[CH_2Cl_2]{\text{excess } Br_2}$

e. $CH_3C{\equiv}CCH_3 \xrightarrow{\text{excess} \atop HBr}$

f. $CH_3C{\equiv}CCH_2CH_3 \xrightarrow{\text{excess} \atop HBr}$

PROBLEM 11◆

From what you know about the stereochemistry of alkene addition reactions, predict the configuration of the product that would be obtained from the reaction of 2-butyne with one equivalent of Br_2 in CH_2Cl_2.

6.7 Addition of Water to an Alkyne

In Section 4.5, we saw that alkenes undergo the acid-catalyzed addition of water. The product of the reaction is an alcohol.

$$CH_3CH_2CH{=}CH_2 \;+\; H_2O \xrightarrow{\;H_2SO_4\;} CH_3CH_2CH{-}CH_2$$
$$\underset{\textbf{1-butene}}{} \qquad\qquad \underset{\textbf{2-butanol}}{\overset{\displaystyle ||}{OH\;\;H}}$$

Alkynes also undergo the acid-catalyzed addition of water. The initial product of the reaction is an enol. An **enol** is a compound with a carbon–carbon double bond and an OH group bonded to one of the sp^2 carbons. (The syllable "ene" signifies the double bond, and "ol" the OH group. When the two syllables are joined, the second e of "ene" is dropped to avoid two consecutive vowels, but the word is pronounced as if the second e were still there: "ene-ol.")

$$CH_3C{\equiv}CCH_3 \;+\; H_2O \xrightarrow{\;H_2SO_4\;} \underset{\textbf{an enol}}{CH_3\overset{\displaystyle OH}{\overset{\displaystyle |}{C}}{=}CHCH_3} \;\rightleftharpoons\; \underset{\textbf{a ketone}}{CH_3\overset{\displaystyle O}{\overset{\displaystyle \|}{C}}{-}CH_2CH_3}$$

The enol immediately rearranges to a *ketone*, a compound with the general structure shown below. A carbon doubly bonded to an oxygen is called a **carbonyl** ("car-bo-nil") **group**; and a **ketone** is a compound that has two alkyl groups bonded to a carbonyl group. An **aldehyde** is a compound that has at least one hydrogen bonded to a carbonyl group.

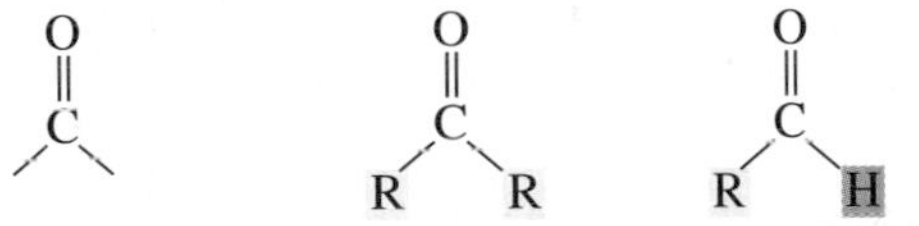

a carbonyl group   a ketone   an aldehyde

A ketone and an enol differ only in the location of a double bond and a hydrogen. A ketone and its corresponding enol are called **keto–enol tautomers**. **Tautomers** ("taw-toe-mers") are isomers that are in rapid equilibrium. Interconversion of the tautomers is called **tautomerization**. We will examine the mechanism of keto-enol interconversion in Chapter 18. For now, the important thing to remember is that the keto and enol tautomers come to equilibrium in solution, and the keto tautomer, because it is usually much more stable than the enol tautomer, predominates.

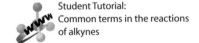

Student Tutorial:
Common terms in the reactions
of alkynes

$$\underset{\textbf{keto tautomer}}{RCH_2{-}\overset{\displaystyle O}{\overset{\displaystyle \|}{C}}{-}R} \;\rightleftharpoons\; \underset{\textbf{enol tautomer}}{RCH{=}\overset{\displaystyle OH}{\overset{\displaystyle |}{C}}{-}R}$$
$$\textbf{tautomerization}$$

The addition of water to an internal alkyne that has the same group attached to each of the *sp* carbons forms a single ketone as a product. But if the two groups are not identical, two ketones are formed because the initial addition of the proton can occur to either of the *sp* carbons.

$$\underset{\substack{\textbf{a symmetrical}\\\textbf{internal alkyne}}}{CH_3CH_2C{\equiv}CCH_2CH_3} \;+\; H_2O \xrightarrow{\;H_2SO_4\;} CH_3CH_2\overset{\displaystyle O}{\overset{\displaystyle \|}{C}}CH_2CH_2CH_3$$

$$\underset{\substack{\textbf{an unsymmetrical}\\\textbf{internal alkyne}}}{CH_3C{\equiv}CCH_2CH_3} \;+\; H_2O \xrightarrow{\;H_2SO_4\;} CH_3\overset{\displaystyle O}{\overset{\displaystyle \|}{C}}CH_2CH_2CH_3 \;+\; CH_3CH_2\overset{\displaystyle O}{\overset{\displaystyle \|}{C}}CH_2CH_3$$

Terminal alkynes are less reactive than internal alkynes toward the addition of water. Terminal alkynes will add water if mercuric ion (Hg^{2+}) is added to the acidic mixture. The mercuric ion is a catalyst—it increases the rate of the addition reaction.

$$CH_3CH_2C\equiv CH + H_2O \xrightarrow[\text{HgSO}_4]{\text{H}_2\text{SO}_4} \underset{\text{an enol}}{CH_3CH_2\overset{\overset{\displaystyle OH}{|}}{C}=CH_2} \rightleftharpoons \underset{\text{a ketone}}{CH_3CH_2\overset{\overset{\displaystyle O}{\|}}{C}-CH_3}$$

The first step in the *mechanism for the mercuric-ion-catalyzed hydration* of an alkyne should remind you of the cyclic bromonium and mercurinium ions formed as intermediates in electrophilic addition reactions of alkenes (Sections 4.7 and 4.8).

Mechanism for the mercuric-ion-catalyzed hydration of an alkyne

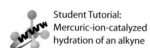

water attacks the more substituted carbon

a mercuric enol

a mercuric ketone

- Reaction of the alkyne with mercuric ion (Hg^{2+}) forms a cyclic mercurinium ion. (Two of the electrons in mercury's filled $5d$ atomic orbital are shown.)
- Water attacks the more substituted carbon of the cyclic intermediate (Section 4.8).
- The protonated OH group, which is a very strong acid, loses a proton to form a mercuric enol, which immediately rearranges to a mercuric ketone.
- Loss of the mercuric ion forms an enol, which rearranges to a ketone.

Notice that the overall addition of water follows the general rule for electrophilic addition reactions: the electrophile (H^+) adds to the *sp* carbon bonded to the greater number of hydrogens.

Student Tutorial:
Mercuric-ion-catalyzed
hydration of an alkyne

PROBLEM 12◆

What ketones would be formed from the acid-catalyzed hydration of 3-heptyne?

PROBLEM 13◆

Which alkyne would be the best reagent to use for the synthesis of each of the following ketones?

a. $CH_3\overset{\overset{\displaystyle O}{\|}}{C}CH_3$

b. $CH_3CH_2\overset{\overset{\displaystyle O}{\|}}{C}CH_2CH_2CH_3$

c. $CH_3\overset{\overset{\displaystyle O}{\|}}{C}-$⬡

PROBLEM 14◆

Draw all the enol tautomers for each of the ketones in Problem 13.

6.8 Addition of Borane to an Alkyne: Hydroboration–Oxidation

Borane adds to alkynes in the same way it adds to alkenes. That is, boron is the electrophile and H^- is the nucleophile, and one mole of BH_3 reacts with three moles of alkyne to form one mole of boron-substituted alkene (Section 4.10). When the addition reaction is over, aqueous sodium hydroxide and hydrogen peroxide are added to the reaction mixture. The end result, as in the case of alkenes, is replacement of the boron by an OH group. The enol product immediately rearranges to a ketone.

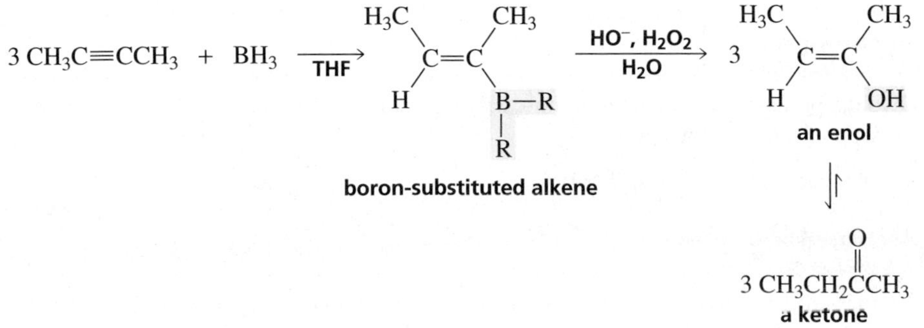

In order to obtain the enol as the product of the addition reaction, only one equivalent of borane can be allowed to add to the alkyne. In other words, the reaction must stop at the alkene stage. In the case of internal alkynes, the substituents on the boron-substituted alkene prevent the second addition from occurring. Terminal alkynes, however, have less steric hindrance, so it is harder to stop the addition reaction at the alkene stage. A special reagent called disiamylborane has been developed for use with terminal alkynes ("siamyl" stands for secondary **isoamyl**; amyl is a common name for a five-carbon fragment). The bulky alkyl groups of disiamylborane prevent a second addition to the boron-substituted alkene. Thus, borane can be used to hydrate internal alkynes, but disiamylborane is preferred for the hydration of terminal alkynes.

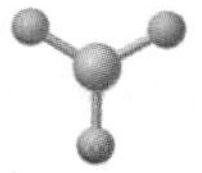

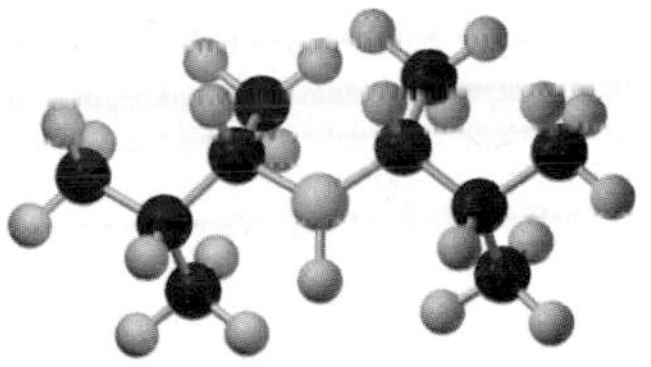

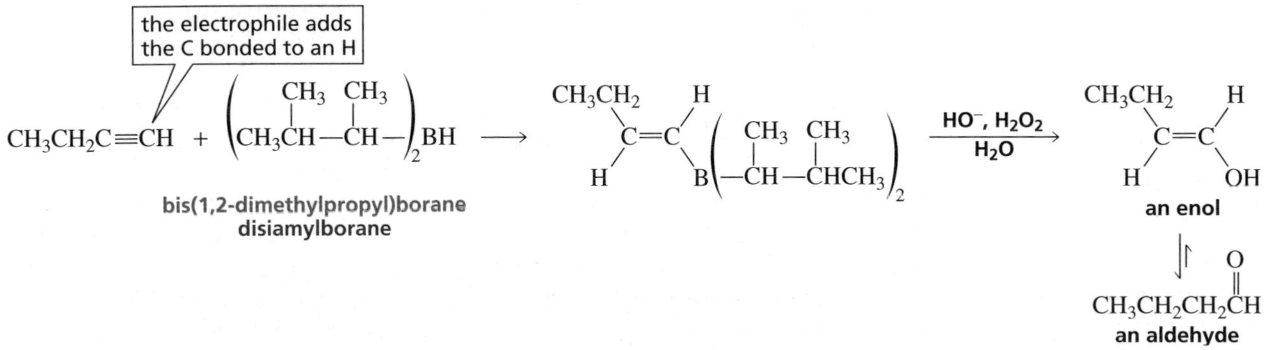

The addition of borane (or disiamylborane) to a terminal alkyne exhibits the same regioselectivity seen in borane addition to an alkene. Boron, with its electron-seeking empty orbital, adds preferentially to the sp carbon bonded to the hydrogen, and the boron-containing group is subsequently replaced by an OH group. Thus, the reaction follows the general rule for electrophilic addition reactions: the electrophile (BH_3) adds to the sp carbon bonded to the greater number of hydrogens. Consequently, mercuric-ion-catalyzed addition of water to a terminal alkyne produces a *ketone* (the carbonyl group is *not* on the terminal carbon), whereas

Hydroboration–oxidation of a terminal alkyne forms an aldehyde.

hydroboration–oxidation of a terminal alkyne produces an *aldehyde* (the carbonyl group *is* on the terminal carbon).

Addition of water to a terminal alkyne forms a ketone.

$$CH_3C{\equiv}CH$$

$$\xrightarrow[\text{HgSO}_4]{\text{H}_2\text{O, H}_2\text{SO}_4} \quad \underset{}{CH_3\overset{\overset{\text{OH}}{|}}{C}{=}CH_2} \;\rightleftharpoons\; \underset{\textbf{a ketone}}{CH_3\overset{\overset{\text{O}}{\|}}{C}CH_3}$$

$$\xrightarrow[\text{2. HO}^-,\ \text{H}_2\text{O}_2,\ \text{H}_2\text{O}]{\text{1. disiamylborane}} \quad \underset{}{CH_3CH{=}\overset{\overset{\text{OH}}{|}}{C}H} \;\rightleftharpoons\; \underset{\textbf{an aldehyde}}{CH_3CH_2\overset{\overset{\text{O}}{\|}}{C}H}$$

PROBLEM 15◆

Give the products of (1) mercuric-ion-catalyzed addition of water and (2) hydroboration–oxidation for the following:

a. 1-butyne **b.** 2-butyne **c.** 2-pentyne

PROBLEM 16◆

There is only one alkyne that forms an aldehyde when it undergoes either acid-catalyzed or mercuric-ion-catalyzed addition of water. Identify the alkyne.

6.9 Addition of Hydrogen to an Alkyne

$$(CH_3COO^-)_2Pb^{2+}$$
lead(II) acetate

quinoline

Hydrogen adds to an alkyne in the presence of a metal catalyst such as palladium, platinum, or nickel in the same manner that it adds to an alkene (Section 4.11). The initial product is an alkene, but it is difficult to stop the reaction at that stage because of hydrogen's strong tendency to add to alkenes in the presence of these efficient metal catalysts. The ultimate product of the hydrogenation reaction, therefore, is an alkane.

> an alkyne is converted to an alkane

$$\underset{\textbf{alkyne}}{CH_3CH_2C{\equiv}CH} \xrightarrow{\overset{\text{H}_2}{\text{Pt/C}}} \underset{\textbf{alkene}}{CH_3CH_2CH{=}CH_2} \xrightarrow{\overset{\text{H}_2}{\text{Pt/C}}} \underset{\textbf{alkane}}{CH_3CH_2CH_2CH_3}$$

BIOGRAPHY

Herbert H. M. Lindlar *was born in Switzerland in 1909 and received a Ph.D. from the University of Bern. He worked at Hoffmann–La Roche and Co. in Basel, Switzerland, and he authored many patents. His last patent was a procedure for isolating the carbohydrate xylose from the waste produced in paper mills.*

The reaction can be stopped at the alkene stage if a "poisoned" (partially deactivated) metal catalyst is used. The most commonly used partially deactivated metal catalyst is called Lindlar catalyst, which is prepared by precipitating palladium on calcium carbonate and treating it with lead(II) acetate and quinoline. This treatment modifies the surface of the palladium, making it much more effective at catalyzing the addition of hydrogen to a triple bond than to a double bond.

Because the alkyne sits on the surface of the metal catalyst and the hydrogens are delivered to the triple bond from the surface of the catalyst, both hydrogens are delivered to the same side of the double bond. In other words, only syn addition of hydrogen occurs (Section 5.19). Syn addition of hydrogen to an internal alkyne forms a *cis alkene*.

Student Tutorial:
Hydrogenation/Lindlar catalyst

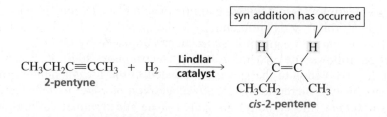

$$\underset{\textbf{2-pentyne}}{CH_3CH_2C{\equiv}CCH_3} + H_2 \xrightarrow[\textbf{catalyst}]{\textbf{Lindlar}}$$

> syn addition has occurred

cis-2-pentene

Internal alkynes can be converted into *trans alkenes* using sodium (or lithium) in liquid ammonia. The reaction stops at the alkene stage because sodium (or lithium) reacts more rapidly with triple bonds than with double bonds. Ammonia is a gas at room temperature (bp = −33 °C), so it is kept in the liquid state by use of a dry ice/acetone mixture (bp = −78 °C).

$$CH_3C\equiv CCH_3 \xrightarrow[\substack{\text{NH}_3 \text{ (liq)} \\ -78 \text{ °C}}]{\text{Na or Li}}$$

2-butyne

trans-2-butene

Mechanism for conversion of an alkyne to a trans alkene

sodium gives up a 1s electron · a strong base · sodium gives up a 1s electron · a strong base

$$CH_3-C\equiv C-CH_3 + Na\cdot \longrightarrow CH_3-\dot{C}=\ddot{C}-CH_3 \xrightarrow{H-NH_2} CH_3-\dot{C}=C\cdot \xrightarrow{Na\cdot} CH_3-\ddot{C}=C \xrightarrow{H-NH_2}$$

a radical anion **a vinylic radical** **a vinylic anion** **a trans alkene**

+ Na⁺   + ⁻NH₂   + Na⁺   + ⁻NH₂

The steps in the mechanism for the conversion of an internal alkyne to a trans alkene are:

- The *s* orbital electron from sodium (or lithium) is transferred to an *sp* carbon to form a **radical anion**—a species with a negative charge and an unpaired electron. Notice that the transfer of a single electron is represented by an arrow head with a single barb. (Recall that sodium and lithium have a strong tendency to lose the single electron in their outer-shell *s* orbital; Section 1.3.)
- The radical anion is such a strong base that it can remove a proton from ammonia. This results in the formation of a **vinylic radical**—the unpaired electron is on a vinylic carbon.
- Another single-electron transfer from sodium (or lithium) to the vinylic radical forms a vinylic anion.
- The vinylic anion is also a strong base; it removes a proton from another molecule of ammonia. The product is a trans alkene.

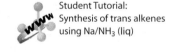

Student Tutorial:
Synthesis of trans alkenes using Na/NH₃ (liq)

The vinylic anion can have either the cis or the trans configuration. The cis and trans anions are in equilibrium, but the equilibrium favors the more stable trans isomer because in that configuration the relatively bulky alkyl groups are as far from each other as possible.

trans vinylic anion
more stable

cis vinylic anion
less stable

Describe the alkyne you would start with and the reagents you would use if you wanted to synthesize

a. pentane.   **b.** *cis*-2-butene.   **c.** *trans*-2-pentene.   **d.** 1-hexene.

6.10 A Hydrogen Bonded to an *sp* Carbon Is "Acidic"

We have seen that the hybridization of an atom affects the acidity of a hydrogen bonded to it because the electronegativity of an atom depends on its hybridization. For example, an *sp* hybridized carbon is more electronegative than an sp^2 hybridized carbon, which is more electronegative than an sp^3 hybridized carbon (Section 1.20).

> **An *sp* carbon is more electronegative than an sp^2 carbon, which is more electronegative than an sp^3 carbon.**

relative electronegativities of carbon atoms

$$\boxed{\text{most electronegative}} \quad sp \; > \; sp^2 \; > \; sp^3 \quad \boxed{\text{least electronegative}}$$

Because the most acidic compound is the one with the hydrogen attached to the most electronegative atom (when the atoms are the same size), ethyne is a stronger acid than ethene, and ethene is a stronger acid than ethane.

HC≡CH	H₂C=CH₂	CH₃CH₃
ethyne	ethene	ethane
pK_a = 25	**pK_a = 44**	**pK_a > 60**

> **The stronger the acid, the weaker its conjugate base.**

In order to remove a proton from an acid (in a reaction that strongly favors products), the base that removes the proton must be stronger than the base that is generated as a result of removing the proton (Section 1.19). In other words, you must start with a stronger base than the base that will be formed. Because NH_3 is a weaker acid (pK_a = 36) than a terminal alkyne (pK_a = 25), an amide ion ($^-NH_2$) is a stronger base than the carbanion—called an acetylide ion—that is formed when a hydrogen is removed from the *sp* carbon of a terminal alkyne. Therefore, an amide ion can be used to remove a proton from a terminal alkyne to prepare an **acetylide ion**.

> **To remove a proton from an acid in a reaction that favors products, the base that removes the proton must be stronger than the base that is formed.**

$$RC\equiv CH \; + \; ^-NH_2 \; \rightleftharpoons \; RC\equiv C^- \; + \; NH_3$$

	amide ion		acetylide ion	
stronger acid	stronger base		weaker base	weaker acid

If hydroxide ion were used as the base, the reaction would strongly favor the reactants because hydroxide ion is a much weaker base than the acetylide ion that would be formed.

$$RC\equiv CH \; + \; HO^- \; \rightleftharpoons \; RC\equiv C^- \; + \; H_2O$$

	hydroxide anion		acetylide anion	
weaker acid	weaker base		stronger base	stronger acid

An amide ion ($^-NH_2$) cannot remove a hydrogen bonded to an sp^2 or an sp^3 carbon. Only a hydrogen bonded to an *sp* carbon is sufficiently acidic to be removed by an amide ion. Consequently, a hydrogen bonded to an *sp* carbon sometimes is referred to as an "acidic" hydrogen. The "acidic" property of terminal alkynes is one way their reactivity differs from that of alkenes. Be careful not to misinterpret what is meant when we say that a hydrogen bonded to an *sp* carbon is "acidic." It is more acidic than

---

SODIUM AMIDE AND SODIUM

Take care not to confuse the compound sodium amide ($Na^+ \ ^-NH_2$) with a mixture of sodium (Na) in liquid ammonia. Sodium amide is the strong base used to remove a proton from a terminal alkyne. Sodium in ammonia is used as a source of electrons and protons, respectively, to convert an internal alkyne to a trans alkene (Section 6.9).

---

most other carbon-bound hydrogens but it is much less acidic than a hydrogen of a water molecule, and water is only a very weakly acidic compound ($pK_a = 15.7$).

relative acid strengths

strongest acid							weakest acid
HF	>	H_2O	>	$HC\equiv CH$	>	NH_3	>
$pK_a = 3.2$		$pK_a = 15.7$		$pK_a = 25$		$pK_a = 36$	

strongest acid > HF > H_2O > $HC\equiv CH$ > NH_3 > $H_2C=CH_2$ > CH_3CH_3 < weakest acid

$pK_a = 3.2$ $pK_a = 15.7$ $pK_a = 25$ $pK_a = 36$ $pK_a = 44$ $pK_a > 60$

PROBLEM 18◆

Any base whose conjugate acid has a pK_a greater than _____ can remove a proton from a terminal alkyne to form an acetylide ion (in a reaction that favors products).

PROBLEM 19◆

Which carbocation in each of the following pairs is more stable?

a. $\overset{+}{C}H_3CH_2$ or $H_2C=\overset{+}{C}H$

b. $H_2\overset{+}{C}=CH$ or $HC\equiv \overset{+}{C}$

PROBLEM 20◆

Explain why sodium amide cannot be used to form a carbanion from an alkane in a reaction that favors products.

PROBLEM-SOLVING STRATEGY

Comparing the Acidities of Compounds

a. List the following compounds in order of decreasing acidity:

$$CH_3CH_2\overset{+}{N}H_3 \qquad CH_3CH=\overset{+}{N}H_2 \qquad CH_3C\equiv\overset{+}{N}H$$

To compare the acidities of a group of compounds, first look at how the compounds differ. These three compounds differ in the hybridization of the nitrogen to which the acidic hydrogen is attached. Now, recall what you know about hybridization and acidity. You know that hybridization of an atom affects its electronegativity (*sp* is more electronegative than *sp²*, and *sp²* is more electronegative than *sp³*), and you know that the more electronegative the atom to which a hydrogen is attached, the more acidic the hydrogen. Now you can answer the question.

relative acidities $CH_3C\equiv\overset{+}{N}H \;>\; CH_3CH=\overset{+}{N}H_2 \;>\; CH_3CH_2\overset{+}{N}H_3$

b. Draw the conjugate bases of the above compounds and list them in order of decreasing basicity.

First remove a proton from each acid to obtain the structures of the conjugate bases. The stronger the acid, the weaker its conjugate base, so using the relative acid strengths obtained in part a, we find the order of decreasing basicity is

relative basicities $CH_3CH_2NH_2 > CH_3CH=NH > CH_3C\equiv N$

Now continue on to Problem 21.

PROBLEM 21◆

List the following species in order of decreasing basicity:

a. $CH_3CH_2CH=\bar{C}H$ $CH_3CH_2C\equiv C^-$ $CH_3CH_2CH_2\bar{C}H_2$

b. $CH_3CH_2O^-$ F^- $CH_3C\equiv C^-$ $^-NH_2$

6.11 Synthesis Using Acetylide Ions

Reactions that form carbon–carbon bonds are important in the synthesis of organic compounds because, without such reactions, we could not convert molecules with small carbon skeletons into molecules with larger carbon skeletons. Instead, the product of a reaction would always have the same number of carbons as the starting material.

One reaction that forms a carbon–carbon bond is the reaction of an acetylide ion with an alkyl halide. Only primary alkyl halides or methyl halides should be used in this reaction.

$$CH_3CH_2C\equiv C^- + CH_3CH_2CH_2Br \longrightarrow CH_3CH_2C\equiv CCH_2CH_2CH_3 + Br^-$$
3-heptyne

The mechanism of this reaction is well understood. Bromine is more electronegative than carbon, and as a result, the electrons in the C—Br bond are not shared equally by the two atoms. There is a partial positive charge on carbon and a partial negative charge on bromine.

$$CH_3CH_2C\equiv\ddot{C}^- + CH_3CH_2CH_2\overset{\delta+}{—}\overset{\delta-}{Br} \longrightarrow CH_3CH_2C\equiv CCH_2CH_2CH_3 + Br^-$$

The negatively charged acetylide ion (a nucleophile) is attracted to the partially positively charged carbon (an electrophile) of the alkyl halide. As the electrons of the acetylide ion approach the carbon to form the new C—C bond, they push out the bromine and its bonding electrons because carbon can bond to no more than four atoms at a time.

This is an example of an *alkylation reaction*. An **alkylation reaction** attaches an alkyl group to the starting material. The mechanism of this and similar reactions will be discussed in greater detail in Chapter 8. At that time we will also see why the reaction works best with primary alkyl halides and methyl halides.

We can convert terminal alkynes into internal alkynes of any desired chain length, simply by choosing an alkyl halide of the appropriate structure. Just count the number of carbons in the terminal alkyne and the number of carbons in the product to see how many carbons are needed in the alkyl halide.

$$CH_3CH_2CH_2C\equiv CH \xrightarrow{\text{NaNH}_2} CH_3CH_2CH_2C\equiv C^- \xrightarrow{\text{CH}_3\text{CH}_2\text{Br}} CH_3CH_2CH_2C\equiv CCH_2CH_3$$
1-pentyne   **3-heptyne**

PROBLEM 22 **SOLVED**

A chemist wants to synthesize 3-heptyne but cannot find any 1-pentyne, the starting material used in the synthesis shown above. How else can 3-heptyne be synthesized?

Solution The *sp* carbons of 3-heptyne are bonded to a propyl group and to an ethyl group. Therefore to produce 3-heptyne, the acetylide ion of 1-pentyne can react with an ethyl halide or the acetylide ion of 1-butyne can react with a propyl halide. Since 1-pentyne is not available, the chemist should use 1-butyne and a propyl halide.

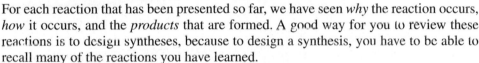

$$CH_3CH_2C\equiv CH \xrightarrow[\text{2. CH}_3\text{CH}_2\text{CH}_2\text{Cl}]{\text{1. NaNH}_2} CH_3CH_2CH_2C\equiv CCH_2CH_3$$

1-butyne   3-heptyne

(Remember that the numbers 1 and 2 in front of the reagents above and below the reaction arrow indicate two sequential reactions; the second reagent is not added until reaction with the first reagent is completely over.)

6.12 Designing a Synthesis I: An Introduction to Multistep Synthesis

For each reaction that has been presented so far, we have seen *why* the reaction occurs, *how* it occurs, and the *products* that are formed. A good way for you to review these reactions is to design syntheses, because to design a synthesis, you have to be able to recall many of the reactions you have learned.

Synthetic chemists consider time, cost, and yield in designing syntheses. In the interest of time, a well-designed synthesis will consist of as few steps (sequential reactions) as possible, and each of those steps will be a reaction that is easy to carry out. If two chemists in a pharmaceutical company were each asked to prepare a new drug, and one synthesized the drug in three simple steps while the other used 20 difficult steps, which chemist would not get a raise? The costs of the starting materials are also taken into consideration in designing a synthesis. Moreover, each step in the synthesis should provide the greatest possible yield of the desired product. The more reactant needed to synthesize 1 gram of product, the more expensive the product is to produce. Sometimes, a synthesis involving several steps is preferred because the starting materials are inexpensive, the reactions are easy to carry out, and the yield of each step is high. Such a synthesis is better than one with fewer steps if those steps require expensive starting materials and consist of reactions that are more difficult to run or give lower yields. At this point in your chemical education, however, you are not yet familiar with the costs of different chemicals or the difficulty of carrying out specific reactions. So, for the time being, when you design a synthesis, just focus on finding the route with the fewest steps.

The following examples will give you an idea of the type of thinking required for designing a successful synthesis. Problems of this kind will appear repeatedly throughout the book, because working them is a good way to learn organic chemistry.

Example 1. Starting with 1-butyne, how could you make the ketone shown below? You can use any organic and inorganic reagents.

$$CH_3CH_2C\equiv CH \xrightarrow{?} CH_3CH_2\overset{\overset{\displaystyle O}{\|}}{C}CH_2CH_2CH_3$$

1-butyne

Many chemists find that the easiest way to design a synthesis is to work backward. Instead of looking at the starting material and deciding how to do the first step of the synthesis, look at the product and decide how to do the last step. The product is a ketone. At this point, the only reactions that you know that form a ketone are the acid-catalyzed addition of water to an alkyne and the hydroboration–oxidation of an alkyne. Either of these methods can be used to get the desired ketone. (As the number of reactions that you know increases, you may find it helpful to consult Appendix IV when designing syntheses; it lists the various methods used to synthesize particular functional groups.) If the alkyne used in the reaction has identical substituents on both *sp* carbons, only one ketone will be obtained. Thus, 3-hexyne is the alkyne that should be used for the synthesis of the desired ketone.

$$CH_3CH_2C{\equiv}CCH_2CH_3 \xrightarrow[\text{H}_2\text{SO}_4]{\text{H}_2\text{O}} \underset{\text{OH}}{CH_3CH_2\overset{|}{C}{=}CHCH_2CH_3} \rightleftharpoons \underset{\text{O}}{CH_3CH_2\overset{\|}{C}CH_2CH_2CH_3}$$
3-hexyne

3-Hexyne can be obtained from the starting material by removing the proton from its *sp* carbon, followed by alkylation. To produce the desired product, a two-carbon alkyl halide must be used in the alkylation reaction.

$$CH_3CH_2C{\equiv}CH \xrightarrow[\text{2. CH}_3\text{CH}_2\text{Br}]{\text{1. NaNH}_2} CH_3CH_2C{\equiv}CCH_2CH_3$$
1-butyne   **3-hexyne**

Designing a synthesis by working backward from product to reactant is not simply a technique taught to organic chemistry students. It is used so frequently by experienced synthetic chemists that it has been given a name—**retrosynthetic analysis**. Chemists use open arrows when they write retrosynthetic analyses, to indicate they are working backward. Typically, the reagents needed to carry out each step are not specified until the reaction is written in the forward direction. For example, the ketone synthesis discussed above can be arrived at by the following retrosynthetic analysis.

retrosynthetic analysis

$$\underset{\text{O}}{CH_3CH_2\overset{\|}{C}CH_2CH_2CH_3} \Longrightarrow CH_3CH_2C{\equiv}CCH_2CH_3 \Longrightarrow CH_3CH_2C{\equiv}CH$$

Once the complete sequence of reactions has been worked out by retrosynthetic analysis, the synthetic scheme can be written by reversing the steps and including the reagents required for each step.

synthesis

$$CH_3CH_2C{\equiv}CH \xrightarrow[\text{2. CH}_3\text{CH}_2\text{Br}]{\text{1. NaNH}_2} CH_3CH_2C{\equiv}CCH_2CH_3 \xrightarrow[\text{H}_2\text{SO}_4]{\text{H}_2\text{O}} \underset{\text{O}}{CH_3CH_2\overset{\|}{C}CH_2CH_2CH_3}$$

Example 2. Starting with ethyne, how could you make 2-bromopentane?

$$HC{\equiv}CH \xrightarrow{?} \underset{\underset{\text{Br}}{|}}{CH_3CH_2CH_2CHCH_3}$$
ethyne   **2-bromopentane**

The desired product can be prepared from 1-pentene, which can be prepared from 1-pentyne. 1-Pentyne can be prepared from ethyne and an alkyl halide with three carbons.

retrosynthetic analysis

$$\underset{\underset{\text{Br}}{|}}{CH_3CH_2CH_2CHCH_3} \Longrightarrow CH_3CH_2CH_2CH{=}CH_2 \Longrightarrow CH_3CH_2CH_2C{\equiv}CH \Longrightarrow HC{\equiv}CH$$

Now we can write the synthetic scheme:

synthesis

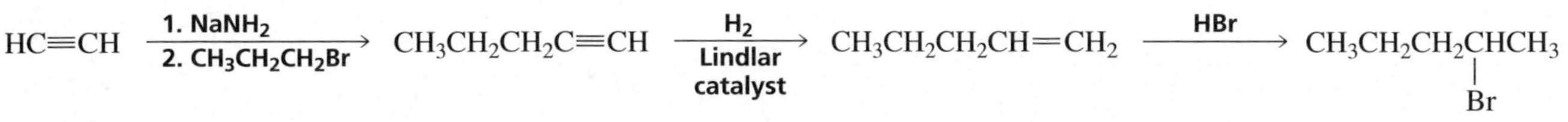

$$HC\equiv CH \xrightarrow[\text{2. CH}_3\text{CH}_2\text{CH}_2\text{Br}]{\text{1. NaNH}_2} CH_3CH_2CH_2C\equiv CH \xrightarrow[\substack{\text{Lindlar} \\ \text{catalyst}}]{\text{H}_2} CH_3CH_2CH_2CH=CH_2 \xrightarrow{\text{HBr}} CH_3CH_2CH_2\underset{\underset{Br}{|}}{C}HCH_3$$

Example 3. How could 2,6-dimethylheptane be prepared from an alkyne and an alkyl halide? (The prime in R′ signifies that R and R′ can be different alkyl groups.)

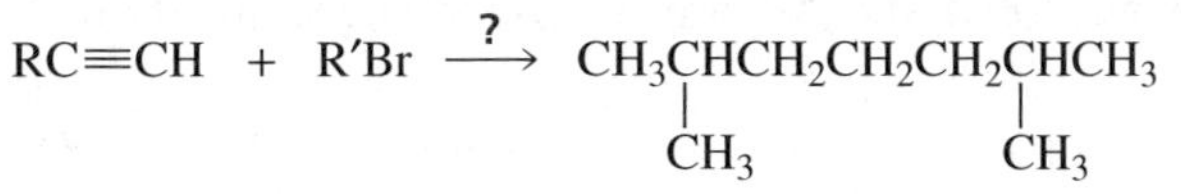

$$RC\equiv CH \ + \ R'Br \xrightarrow{?} CH_3\underset{\underset{CH_3}{|}}{C}HCH_2CH_2CH_2\underset{\underset{CH_3}{|}}{C}HCH_3$$

2,6-dimethylheptane

2,6-Dimethyl-3-heptyne is the only alkyne that will form the desired alkane upon hydrogenation. This alkyne can be dissected in two different ways: the alkyne could be prepared from the reaction of an acetylide ion with a primary alkyl halide (isobutyl bromide) or from the reaction of an acetylide ion with a secondary alkyl halide (isopropyl bromide).

retrosynthetic analysis

$$CH_3\underset{\underset{CH_3}{|}}{C}HCH_2CH_2CH_2\underset{\underset{CH_3}{|}}{C}HCH_3 \implies CH_3\underset{\underset{CH_3}{|}}{C}HCH_2C\equiv C\underset{\underset{CH_3}{|}}{C}HCH_3$$

$$CH_3\underset{\underset{CH_3}{|}}{C}HCH_2Br \ + \ HC\equiv C\underset{\underset{CH_3}{|}}{C}HCH_3 \quad \text{or} \quad CH_3\underset{\underset{CH_3}{|}}{C}HBr \ + \ HC\equiv CCH_2\underset{\underset{CH_3}{|}}{C}HCH_3$$

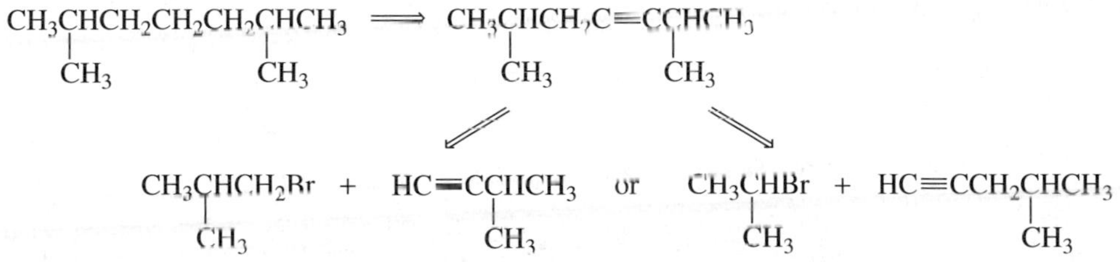

Because we know that the reaction of an acetylide ion with an alkyl halide works best with primary alkyl halides and methyl halides, we know how to proceed:

synthesis

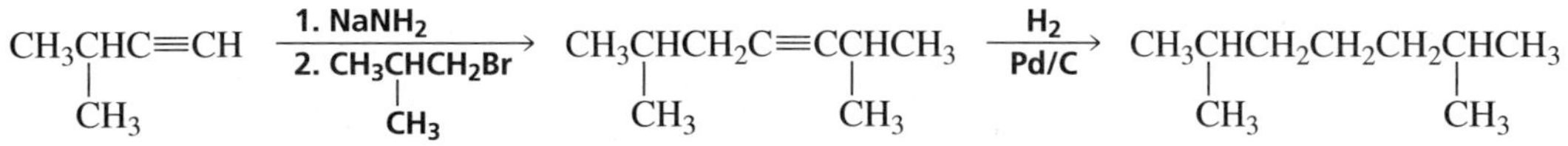

$$CH_3\underset{\underset{CH_3}{|}}{C}HC\equiv CH \xrightarrow[\text{2. CH}_3\text{CHCH}_2\text{Br}]{\text{1. NaNH}_2} CH_3\underset{\underset{CH_3}{|}}{C}HCH_2C\equiv C\underset{\underset{CH_3}{|}}{C}HCH_3 \xrightarrow[\text{Pd/C}]{\text{H}_2} CH_3\underset{\underset{CH_3}{|}}{C}HCH_2CH_2CH_2\underset{\underset{CH_3}{|}}{C}HCH_3$$

Example 4. How could you carry out the following synthesis using the given starting material?

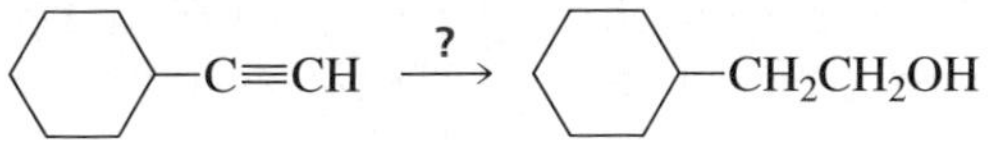

$$\bigcirc\!\!-C\equiv CH \xrightarrow{?} \bigcirc\!\!-CH_2CH_2OH$$

An alcohol can be prepared from an alkene, and an alkene can be prepared from an alkyne.

retrosynthetic analysis

$$\bigcirc\!\!-CH_2CH_2OH \implies \bigcirc\!\!-CH=CH_2 \implies \bigcirc\!\!-C\equiv CH$$

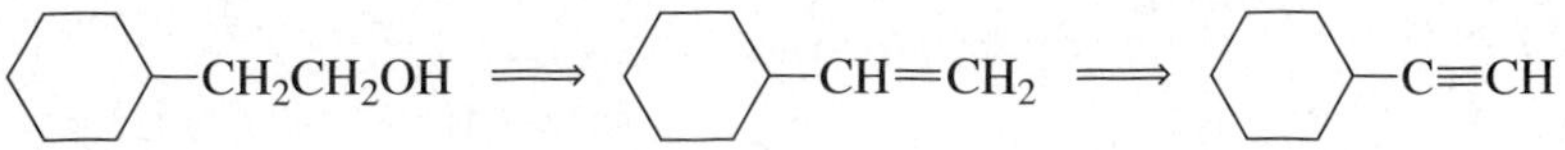

You can use either of the two methods you know for converting an alkyne into an alkene, because the desired alkene does not have cis–trans isomers. Hydroboration–oxidation must be used to convert the alkene into the desired alcohol, because acid-catalyzed addition of water would not form the desired alcohol.

synthesis

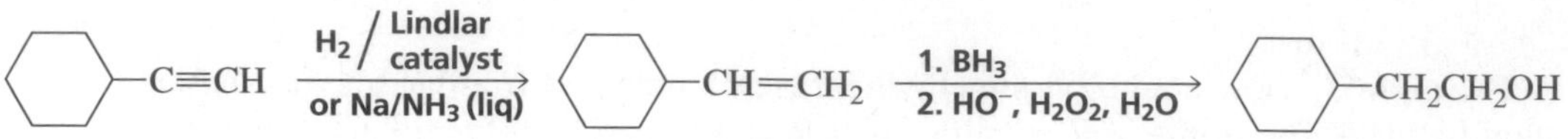

Example 5. How could you prepare (*E*)-2-pentene from ethyne?

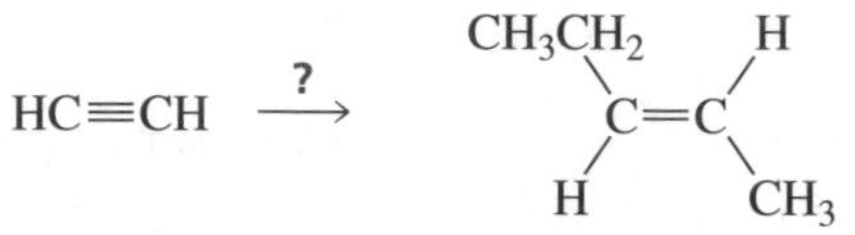

(*E*)-2-pentene

A trans alkene can be prepared from the reaction of an internal alkyne with sodium and ammonia. The alkyne needed to synthesize the desired alkene can be prepared from 1-butyne and a methyl halide. 1-Butyne can be prepared from ethyne and an ethyl halide.

retrosynthetic analysis

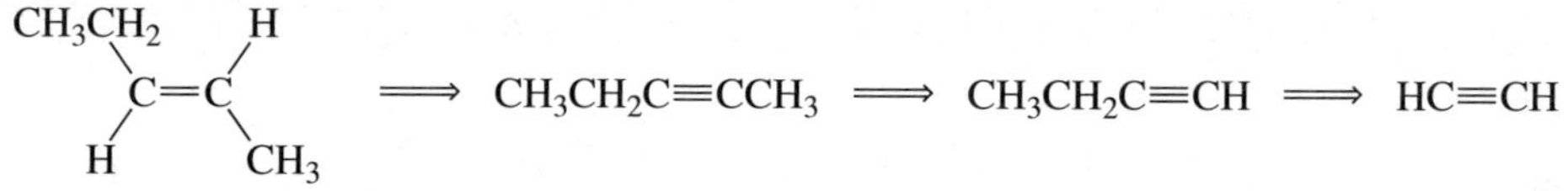

synthesis

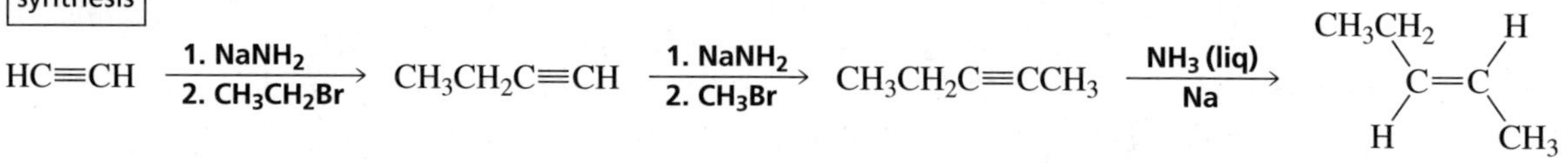

Example 6. How could you prepare *cis*-2,3-diethyloxirane from ethyne?

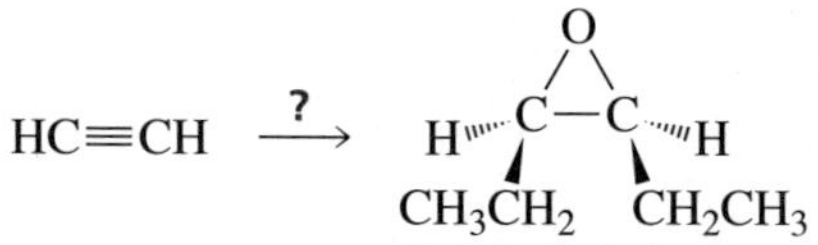

cis-2,3-diethyloxirane

A cis epoxide can be prepared from a cis alkene and a peroxyacid. The cis alkene that should be used is *cis*-3-hexene, which can be obtained from 3-hexyne. 3-Hexyne can be prepared from 1-butyne, and 1-butyne can be prepared from ethyne.

retrosynthetic analysis

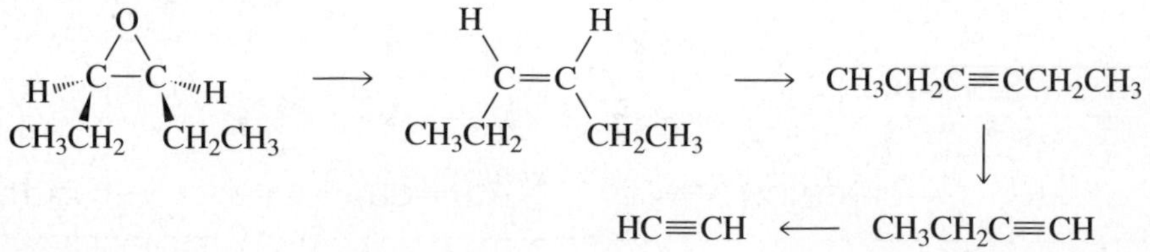

The alkyne must be converted to the alkene using H_2 and Lindlar catalyst, so the desired cis alkene is formed.

synthesis

$$HC{\equiv}CH \xrightarrow[\text{2. CH}_3\text{CH}_2\text{Br}]{\text{1. NaNH}_2} CH_3CH_2C{\equiv}CH \xrightarrow[\text{2. CH}_3\text{CH}_2\text{Br}]{\text{1. NaNH}_2} CH_3CH_2C{\equiv}CCH_2CH_3$$

$$\downarrow H_2 \mid \text{Lindlar catalyst}$$

$$\underset{CH_3CH_2 \quad CH_2CH_3}{\overset{O}{\underset{H}{\overset{\diagup\,\diagdown}{C-C}}}} \xleftarrow{CH_3COOH} \underset{CH_3CH_2 \qquad CH_2CH_3}{\overset{H \qquad\quad H}{C=C}}$$

Example 7. How could you prepare 3,3-dibromohexane from reagents that contain no more than two carbon atoms?

$$\text{reagents with no more than 2 carbon atoms} \xrightarrow{\;?\;} \underset{\underset{\text{Br}}{|}}{\overset{\overset{\text{Br}}{|}}{CH_3CH_2CCH_2CH_2CH_3}}$$

3,3-dibromohexane

A geminal dibromide can be prepared from an alkyne. 3-Hexyne is the alkyne of choice because it will form one geminal bromide, whereas 2-hexyne would form two different geminal dibromides. 3-Hexyne can be prepared from 1-butyne and ethyl bromide, and 1-butyne can be prepared from ethyne and ethyl bromide.

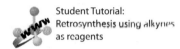
Student Tutorial:
Retrosynthesis using alkynes
as reagents

retrosynthetic analysis

$$\underset{\underset{\text{Br}}{|}}{\overset{\overset{\text{Br}}{|}}{CH_3CH_2CCH_2CH_2CH_3}} \Longrightarrow CH_3CH_2C{\equiv}CCH_2CH_3 \Longrightarrow CH_3CH_2C{\equiv}CH \Longrightarrow HC{\equiv}CH$$

synthesis

$$HC{\equiv}CH \xrightarrow[\text{2. CH}_3\text{CH}_2\text{Br}]{\text{1. NaNH}_2} CH_3CH_2C{\equiv}CH \xrightarrow[\text{2. CH}_3\text{CH}_2\text{Br}]{\text{1. NaNH}_2} CH_3CH_2C{\equiv}CCH_2CH_3 \xrightarrow{\text{excess HBr}} \underset{\underset{\text{Br}}{|}}{\overset{\overset{\text{Br}}{|}}{CH_3CH_2CCH_2CH_2CH_3}}$$

IDENTIFYING COMPOUNDS

After completing a synthesis, chemists must identify the final product of the reaction to make sure that what they think they synthesized has actually been produced. There are several quick chemical tests that can be used to help identify organic compounds. One is to treat the product with a small amount of Br_2.

$$CH_3CH_2CH{=}CH_2 + Br_2 \longrightarrow \underset{\underset{\text{Br}}{|}}{CH_3CH_2CHCH_2Br}$$

an alkene   reddish-brown   colorless

$$CH_3CH_2C{\equiv}CH + Br_2 \longrightarrow \underset{\underset{\text{Br}}{|}}{CH_3CH_2C{=}CHBr}$$

an alkyne   reddish-brown   colorless

Br_2 has a reddish-brown color. If the reddish colors disappears, we can conclude that the product is a compound that reacts with Br_2; if the color persists, we can conclude that the product is not a compound that can react with Br_2.

PROBLEM 23

Starting with acetylene, how could the following compounds be synthesized?

a. $CH_3CH_2CH_2C≡CH$

b. $CH_3CH=CH_2$

c.

$$\underset{H}{\overset{H_3C}{}}C=C\underset{H}{\overset{CH_3}{}}$$

d.
$$CH_3CH_2CH_2CH_2\overset{\displaystyle O}{\overset{\|}{C}}H$$

e.
$$CH_3\underset{Br}{\overset{}{C}}HCH_3$$

f.
$$CH_3\underset{Cl}{\overset{Cl}{C}}CH_3$$

SUMMARY

An **alkyne** is a hydrocarbon that contains a carbon–carbon triple bond. A triple bond can be thought of as a cylinder of electrons wrapped around the σ bond. The functional group suffix of an alkyne is "yne." A **terminal alkyne** has the triple bond at the end of the chain; an **internal alkyne** has the triple bond located elsewhere along the chain. Internal alkynes, with two alkyl substituents bonded to the sp carbons, are more stable than terminal alkynes. We now have seen that alkyl groups stabilize carbocations, alkenes, and alkynes.

Alkynes, like alkenes, undergo electrophilic addition reactions. The same reagents that add to alkenes add to alkynes. Although an alkyne is less stable than an alkene, it is less reactive because a **vinylic cation** is less stable than a similarly substituted alkyl cation. Electrophilic addition to a terminal alkyne is regioselective; the electrophile adds to the sp carbon that is bonded to the hydrogen because the intermediate formed—a secondary vinylic cation—is more stable than a primary vinylic cation. If excess reagent is available, alkynes undergo a second addition reaction with hydrogen halides and halogens because the product of the first reaction is an alkene.

When an alkyne undergoes the acid-catalyzed addition of water, the product of the reaction is an enol. The enol immediately rearranges to a ketone. A **ketone** is a compound that has two alkyl groups bonded to a **carbonyl** (C=O) **group**. An **aldehyde** is a compound that has at least one hydrogen bonded to a carbonyl group. The ketone and enol are called **keto–enol tautomers**; they differ in the location of a double

bond and a hydrogen. Interconversion of the tautomers is called **tautomerization**. The keto tautomer predominates at equilibrium. Terminal alkynes add water if mercuric ion is added to the acidic mixture. In hydroboration–oxidation, H^+ is not the electrophile, $H:^-$ is the nucleophile. Consequently, mercuric-ion-catalyzed addition of water to a terminal alkyne produces a *ketone*, whereas hydroboration–oxidation of a terminal alkyne produces an *aldehyde*.

Hydrogen adds to an alkyne in the presence of a metal catalyst (Pd, Pt, or Ni) to form an alkane. Addition of hydrogen to an alkyne in the presence of Lindlar catalyst forms *an alkene*; if it is an internal alkyne it forms a *cis alkene*. Sodium in liquid ammonia converts an internal alkyne to a *trans alkene*.

The electronegativities of carbon atoms decrease in the order: $sp > sp^2 > sp^3$. Ethyne is, therefore, a stronger acid than ethene, and ethene is a stronger acid than ethane. An amide ion can remove a hydrogen bonded to an sp carbon of a terminal alkyne because it is a stronger base than the **acetylide ion** that is formed. An acetylide ion can undergo an alkylation reaction with a methyl halide or a primary alkyl halide to form an internal alkyne. An **alkylation reaction** attaches an alkyl group to the starting material.

Designing a synthesis by working backward is called **retrosynthetic analysis**. Chemists use open arrows to indicate they are working backward. The reagents needed to carry out each step are not included until the reaction is written in the forward direction.

SUMMARY OF REACTIONS

1. Electrophilic addition reactions

a. Addition of hydrogen halides (H^+ is the electrophile; Section 6.6)

$$RC≡CH \xrightarrow{\text{HX}} \underset{X}{\overset{}{R}}C=CH_2 \xrightarrow{\text{excess HX}} \underset{X}{\overset{X}{R}}C-CH_3$$

$$HX = HF, HCl, HBr, HI$$

b. Addition of halogens (Section 6.6)

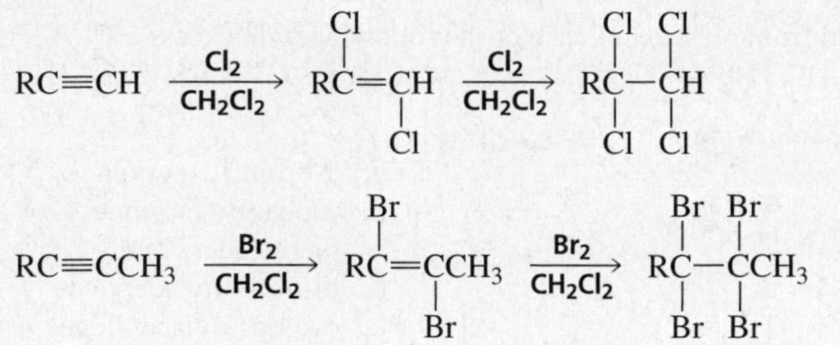

$$RC \equiv CH \xrightarrow[CH_2Cl_2]{Cl_2} \underset{Cl}{\overset{Cl}{RC = CH}} \xrightarrow[CH_2Cl_2]{Cl_2} \underset{Cl \ Cl}{\overset{Cl \ Cl}{RC - CH}}$$

$$RC \equiv CCH_3 \xrightarrow[CH_2Cl_2]{Br_2} \underset{Br}{\overset{Br}{RC = CCH_3}} \xrightarrow[CH_2Cl_2]{Br_2} \underset{Br \ Br}{\overset{Br \ Br}{RC - CCH_3}}$$

c. Acid-catalyzed addition of water/hydroboration–oxidation (Sections 6.7 and 6.8)

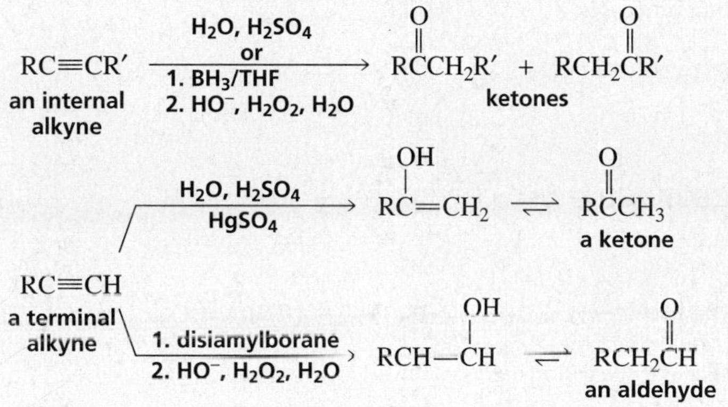

$$RC \equiv CR' \xrightarrow[\begin{array}{c}\text{1. BH}_3/\text{THF}\\\text{2. HO}^-, \text{H}_2\text{O}_2, \text{H}_2\text{O}\end{array}]{\begin{array}{c}\text{H}_2\text{O, H}_2\text{SO}_4\\\text{or}\end{array}} \underset{\text{ketones}}{\overset{O}{RCCH_2R'} + \overset{O}{RCH_2CR'}}$$

an internal alkyne

$$RC \equiv CH \begin{array}{c} \xrightarrow[\text{HgSO}_4]{\text{H}_2\text{O, H}_2\text{SO}_4} \underset{}{\overset{OH}{RC = CH_2}} \rightleftharpoons \underset{\text{a ketone}}{\overset{O}{RCCH_3}} \\ \\ \xrightarrow[\text{2. HO}^-, \text{H}_2\text{O}_2, \text{H}_2\text{O}]{\text{1. disiamylborane}} \underset{}{\overset{OH}{RCH - CH}} \rightleftharpoons \underset{\text{an aldehyde}}{\overset{O}{RCH_2CH}} \end{array}$$

a terminal alkyne

2. Addition of hydrogen (Section 6.9)

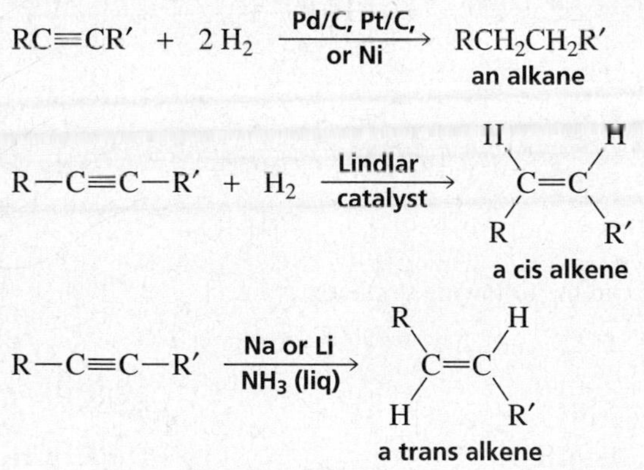

$$RC \equiv CR' + 2 H_2 \xrightarrow[\text{or Ni}]{\text{Pd/C, Pt/C,}} \underset{\text{an alkane}}{RCH_2CH_2R'}$$

$$R - C \equiv C - R' + H_2 \xrightarrow[\text{catalyst}]{\text{Lindlar}} \underset{R \quad\quad R'}{\overset{H \quad\quad H}{C = C}}$$

a cis alkene

$$R - C \equiv C - R' \xrightarrow[\text{NH}_3 \text{ (liq)}]{\text{Na or Li}} \underset{H \quad\quad R'}{\overset{R \quad\quad H}{C = C}}$$

a trans alkene

3. Removal of a proton from a terminal alkyne, followed by alkylation (Sections 6.10 and 6.11)

$$RC \equiv CH \xrightarrow{\text{NaNH}_2} RC \equiv C^- \xrightarrow{R'CH_2Br} RC \equiv CCH_2R'$$

KEY TERMS

acetylide ion (p. 274)
aldehyde (p. 269)
alkylation reaction (p. 276)
alkyne (p. 258)
carbonyl group (p. 269)
enol (p. 269)

geminal dihalide (p. 267)
internal alkyne (p. 260)
keto–enol tautomers (p. 269)
ketone (p. 269)
π-complex (p. 267)
radical anion (p. 273)

retrosynthetic analysis (p 278)
tautomerization (p. 269)
tautomers (p. 269)
terminal alkyne (p. 260)
vinylic cation (p. 266)
vinylic radical (p. 273)

PROBLEMS

24. Give the major product obtained from reaction of each of the following with excess HCl:
    a. $CH_3CH_2C\equiv CH$ **b.** $CH_3CH_2C\equiv CCH_2CH_3$ **c.** $CH_3CH_2C\equiv CCH_2CH_2CH_3$

25. Draw a structure for each of the following:
    a. 2-hexyne
    b. 5-ethyl-3-octyne
    c. methylacetylene
    d. vinylacetylene
    e. methoxyethyne
    f. *sec*-butyl-*tert*-butylacetylene
    g. 1-bromo-1-pentyne
    h. propargyl bromide
    i. diethylacetylene
    j. di-*tert*-butylacetylene
    k. cyclopentylacetylene
    l. 5,6-dimethyl-2-heptyne

26. Identify the electrophile and the nucleophile in each of the following reaction steps. Then draw curved arrows to illustrate the bond-making and bond-breaking processes.

$$CH_3CH_2\overset{+}{C}=CH_2 \quad + \quad :\overset{..}{\underset{..}{C}}l:^- \longrightarrow CH_3CH_2C=CH_2$$
$$\underset{\displaystyle :\overset{..}{\underset{..}{C}}l:}{|}$$

$$CH_3C\equiv CH \quad + \quad H-Br \longrightarrow CH_3\overset{+}{C}=CH_2 \quad + \quad Br^-$$

$$CH_3C\equiv C-H \quad + \quad :\overset{..}{N}H_2^- \longrightarrow CH_3C\equiv C:^- \quad + \quad \overset{..}{N}H_3$$

27. Give the systematic name for each of the following:

    a. $CH_3C\equiv CCH_2\underset{\underset{\displaystyle Br}{|}}{C}HCH_3$

    b. $CH_3C\equiv CCH_2\underset{\underset{\displaystyle CH_2CH_2CH_3}{|}}{C}HCH_3$

    c. $CH_3C\equiv CCH_2\overset{\overset{\displaystyle CH_3}{|}}{\underset{\underset{\displaystyle CH_3}{|}}{C}}CH_3$

    d. $CH_3\underset{\underset{\displaystyle Cl}{|}}{C}HCH_2C\equiv C\underset{\underset{\displaystyle CH_3}{|}}{C}HCH_3$

    e. (cyclooctadiene structure)

    f. (cyclohexadiene with two CH₃ groups)

28. What reagents could be used to carry out the following syntheses?

(central $RC\equiv CH$ with arrows to the following products)

RCH_2CH_3

$RCH=CH_2$

$R\underset{\underset{\displaystyle Br}{|}}{C}H=CH_2$ → $R\underset{\underset{\displaystyle Br}{|}}{C}CH_3$

$R\underset{\underset{\displaystyle Br}{|}}{C}=CH_2$

$RCH\overset{O}{-}CH_2$

$R\underset{\underset{\displaystyle Br}{|}}{C}HCH_3$

$R\overset{\overset{\displaystyle O}{\|}}{C}CH_3$

$RCH_2\overset{\overset{\displaystyle O}{\|}}{C}H$

29. Al Kyne was given the structural formulas of several compounds and was asked to give them systematic names. How many did Al name correctly? Correct those that are misnamed.
    a. 4-ethyl-2-pentyne   **b.** 1-bromo-4-heptyne   **c.** 2-methyl-3-hexyne   **d.** 3-pentyne

30. Draw the structures and give the common and systematic names for alkynes with molecular formula C_7H_{12}.

31. Explain why the following reaction results in the product that is shown:

$$CH_3CH_2CH_2C\equiv CH \xrightarrow[\text{H}_2\text{O}]{\text{Br}_2} CH_3CH_2CH_2\overset{\overset{\displaystyle O}{\|}}{C}CH_2Br$$

32. How could the following compounds be synthesized, starting with a hydrocarbon that has the same number of carbon atoms as the desired product?

a. $CH_3CH_2CH_2CH_2\overset{\displaystyle O}{\overset{\|}{C}}H$
b. $CH_3CH_2CH_2CH_2OH$
c. $CH_3CH_2CH_2\overset{\displaystyle O}{\overset{\|}{C}}CH_2CH_2CH_2CH_3$

33. What reagents would you use for the following syntheses?
  a. (*Z*)-3-hexene from 3-hexyne
  b. (*E*)-3-hexene from 3-hexyne
  c. hexane from 3-hexyne

34. Give the systematic name for each of the following:

  a. $CH_3C\equiv CCH_2CH_2CH_2CH=CH_2$

  d. [structure: cyclohexadiene with Cl]

  b. $\begin{array}{c} HOCH_2CH_2 \quad\quad CH_2CH_3 \\ \diagdown \quad\quad\quad \diagup \\ C=C \\ \diagup \quad\quad\quad \diagdown \\ H \quad\quad\quad\quad H \end{array}$

  e. [structure: cycloheptatriene with CH₃]

  c. $CH_3CH_2C\equiv CCH_2CH_2C\equiv CH$

35. What is the molecular formula of a hydrocarbon that has 1 triple bond, 2 double bonds, 1 ring, and 32 carbons?

36. What will be the major product of the reaction of 1 mol of propyne with each of the following reagents?
  a. HBr (1 mol)
  b. HBr (2 mol)
  c. Br_2 (1 mol)/CH_2Cl_2
  d. Br_2 (2 mol)/CH_2Cl_2
  e. aqueous H_2SO_4, $HgSO_4$
  f. disiamylborane followed by H_2O_2/HO^-.
  g. excess H_2, Pt/C
  h. H_2/Lindlar catalyst
  i. sodium in liquid ammonia
  j. sodium amide
  k. product of part j followed by 1-chloropentane

37. Answer Problem 36 using 2-butyne as the starting material instead of propyne.

38. a. Starting with 3-methyl-1-butyne, how could you prepare the following alcohols?
    1. 2-methyl-2-butanol   **2.** 3-methyl-1-butanol
  b. In each case, a second alcohol would also be obtained. What alcohol would it be?

39. How many of the following names are correct? Correct the incorrect names.
  a. 4-heptyne
  b. 2-ethyl-3-hexyne
  c. 4-chloro-2-pentyne
  d. 2,3-dimethyl-5-octyne
  e. 4,4-dimethyl-2-pentyne
  f. 2,5-dimethyl-3-hexyne

40. Which of the following pairs are keto–enol tautomers?

  a. $CH_3CH_2CH=CHCH_2OH$ and $CH_3CH_2CH_2CH_2\overset{\displaystyle O}{\overset{\|}{C}}H$

  b. $CH_3\overset{\displaystyle OH}{\overset{|}{C}}HCH_3$ and $CH_3\overset{\displaystyle O}{\overset{\|}{C}}CH_3$

  c. $CH_3CH_2CH=CHOH$ and $CH_3CH_2CH_2\overset{\displaystyle O}{\overset{\|}{C}}H$

  d. $CH_3CH_2CH_2CH=CHOH$ and $CH_3CH_2CH_2\overset{\displaystyle O}{\overset{\|}{C}}CH_3$

  e. $CH_3CH_2CH_2\overset{\displaystyle OH}{\overset{|}{C}}=CH_2$ and $CH_3CH_2CH_2\overset{\displaystyle O}{\overset{\|}{C}}CH_3$

41. Using ethyne as the starting material, how can the following compounds be prepared?

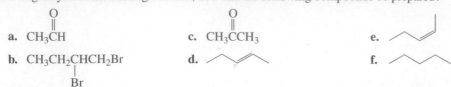

a. CH₃CH=O

b. CH₃CH₂CHCH₂Br
       |
       Br

c. CH₃CCH₃ (ketone)

d. (cis-2-butene structure)

e. (cis-pentene structure)

f. (trans-pentene structure)

42. Give the stereoisomers obtained from the reaction of 2-butyne with the following reagents:
   a. 1. H_2/Lindlar catalyst 2. Br_2/CH_2Cl_2
   b. 1. Na/NH_3(liq) 2. Br_2/CH_2Cl_2
   c. 1. Cl_2/CH_2Cl_2 2. Br_2/CH_2Cl_2

43. Draw the keto tautomer for each of the following:

a. CH₃CH=CCH₃ (with OH on the C)

b. CH₃CH₂CH₂C=CH₂ (with OH)

c. (cyclohexene)—OH

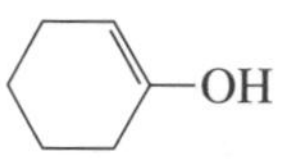

d. 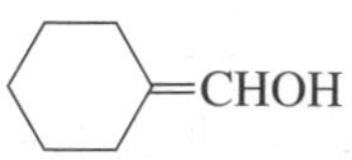 (cyclohexane)=CHOH

44. Show how each of the following compounds could be prepared using the given starting material, any necessary inorganic reagents, and any necessary organic compound that has no more than four carbon atoms:

a. HC≡CH ⟶ CH₃CH₂CH₂CH₂CCH₃ (ketone at end)

b. HC≡CH ⟶ CH₃CH₂CHCH₃
                         |
                         Br

c. HC≡CH ⟶ CH₃CH₂CH₂CHCH₃
                           |
                           OH

d. 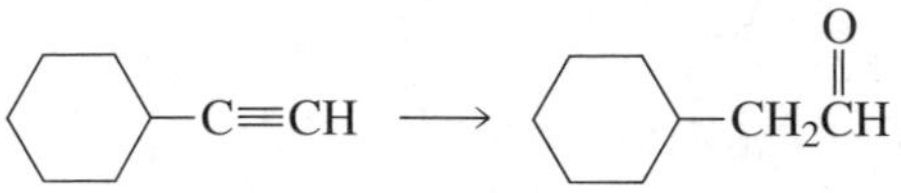 (cyclohexyl)—C≡CH ⟶ (cyclohexyl)—CH₂CH=O

e. 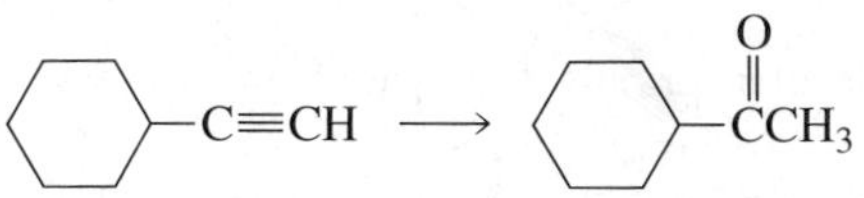 (cyclohexyl)—C≡CH ⟶ (cyclohexyl)—CCH₃ (ketone)

f. 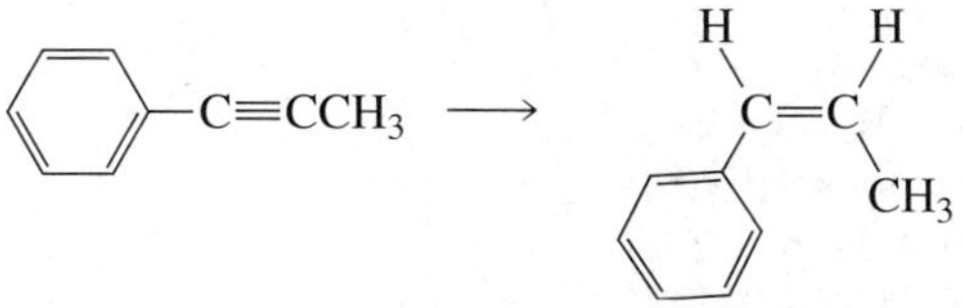 (phenyl)—C≡CCH₃ ⟶ (phenyl)(H)C=C(H)(CH₃)

45. Dr. Polly Meher was planning to synthesize 3-octyne by adding 1-bromobutane to the product obtained from the reaction of 1-butyne with sodium amide. Unfortunately, however, she had forgotten to order 1-butyne. How else can she prepare 3-octyne?

46. a. Explain why a single pure product is obtained from hydroboration–oxidation of 2-butyne, whereas two products are obtained from hydroboration–oxidation of 2-pentyne.
   b. Name two other internal alkynes that will yield only one product on hydroboration–oxidation.

47. Give the configurations of the products obtained from the following reactions:

   a. CH₃CH₂C≡CCH₂CH₃ $\xrightarrow{\text{1. Na, NH}_3\text{(liq)} \quad \text{2. D}_2\text{, Pd/C}}$

   b. CH₃CH₂C≡CCH₂CH₃ $\xrightarrow{\text{1. H}_2\text{/Lindlar catalyst} \quad \text{2. D}_2\text{, Pd/C}}$

48. In Section 6.5 it was stated that hyperconjugation is less effective in stabilizing the charge on a vinylic cation than the charge on an alkyl cation. Why do you think this is so?

49. α-Farnesene is a dodecatetraene found in the waxy coating of apple skins. What is its systematic name? Include E and Z where necessary to indicate the configuration of the double bonds.

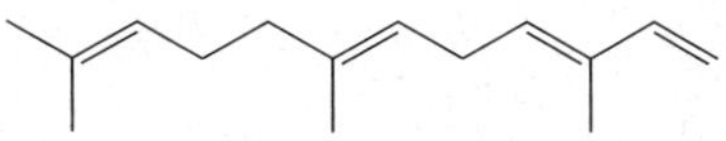

α-farnesene

50. Show how the following compounds could be synthesized starting with acetylene:
   a. *cis*-2-octene
   b. *trans*-3-heptene
   c. 4-bromo-3-hexanol

Delocalized Electrons and Their Effect on Stability, Reactivity, and pK_a • More About Molecular Orbital Theory

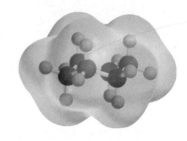

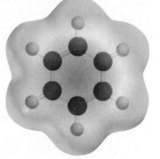

benzene   cyclohexane

BUILDING ON FUNDAMENTALS

SECTIONS 7.1–7.10 We continue the discussion of delocalized electrons (1.22).	**SECTION 7.10** The relative stabilities of dienes, like the relative stabilities of alkenes (4.11), can be determined by the $-\Delta H°$ values for hydrogenation.
SECTION 7.8 Isolated dienes react just like alkenes (4.4).	
SECTION 7.8 The closer the electrons are to the nucleus, the shorter and stronger is the bond (1.14).	**SECTION 7.12** In looking at the stereochemistry of the Diels–Alder reactions, we will see that if the reaction forms a product with one asymmetric center, a racemic mixture will be formed; if it forms a product with two asymmetric centers, the stereochemistry of the products will depend on the mechanism of the reaction (5.19).
SECTION 7.8 We continue the discussion of molecular orbital theory (1.6).	
SECTION 7.9 Two more classes of compounds are added to the list of compounds whose approximate pK_a values you should know (Table 1.8 in Section 1.18).	

Electrons that are restricted to a particular region are called **localized electrons**. Localized electrons either belong to a single atom or are confined to a bond between two atoms.

$$CH_3\!-\!\ddot{N}H_2 \qquad CH_3\!-\!CH\!=\!CH_2$$

localized electrons   localized electrons

Not all electrons are confined to a single atom or bond. Many organic compounds contain *delocalized* electrons. **Delocalized electrons** neither belong to a single atom nor are confined to a bond between two atoms, but are shared by three or more atoms. You were first introduced to delocalized electrons in Section 1.22, where you saw that the two electrons represented by the π bond of the COO^- group are shared by three atoms—the carbon and both oxygen atoms. The dashed lines in the chemical structures indicate that the two electrons are delocalized over three atoms.

$$CH_3C\underset{\overset{\|}{\underset{\ddot{O}:}{}}^{\delta-}}{\overset{\overset{\ddot{O}:^{\delta-}}{/\!/}}{}}$$

delocalized electrons

In this chapter, you will learn to recognize compounds that contain delocalized electrons and to draw structures that represent the electron distribution in molecules with delocalized electrons. You will also be introduced to some of the special characteristics of compounds that have delocalized electrons. You will then be able to understand the wide-ranging effects that delocalized electrons have on the reactivity of organic compounds. We begin by looking at benzene, a compound whose properties chemists could not explain until they recognized that electrons in organic molecules could be delocalized.

7.1 Benzene Has Delocalized Electrons

Because they didn't know about delocalized electrons, benzene's structure puzzled early organic chemists. They knew that benzene had a molecular formula of C_6H_6, that it was an unusually stable compound, and that it did not undergo the addition reactions characteristic of alkenes (Section 3.6). They also knew that when a different atom is substituted for any one of the hydrogens of benzene, only one product is obtained and when the substituted product undergoes a second substitution, three products are obtained.

$$C_6H_6 \xrightarrow[\text{with an X}]{\text{replace a hydrogen}} C_6H_5X \xrightarrow[\text{with an X}]{\text{replace a hydrogen}} C_6H_4X_2 + C_6H_4X_2 + C_6H_4X_2$$

one monosubstituted compound   three disubstituted compounds

For every two hydrogens that are missing from the general molecular formula, C_nH_{2n+2}, a hydrocarbon has either a π bond or a ring.

What kind of structure might we predict for benzene if we knew only what the early chemists knew? The molecular formula (C_6H_6) tells us that benzene has eight fewer hydrogens than an acyclic alkane with six carbons ($C_nH_{2n+2} = C_6H_{14}$). Benzene, therefore, has a degree of unsaturation of four (Section 3.1). This means that benzene is either an acyclic compound with four π bonds, a cyclic compound with three π bonds, a bicyclic compound with two π bonds, a tricyclic compound with one π bond, or a tetracyclic compound.

Because only one product is obtained regardless of which of the six hydrogens is replaced with another atom, we know that all the hydrogens must be identical. Two structures that fit these requirements are shown here:

$$CH_3C \equiv C - C \equiv CCH_3$$

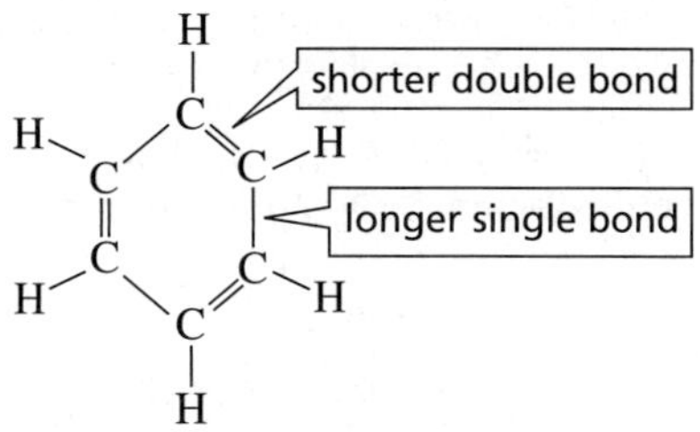

shorter double bond

longer single bond

Neither of these structures is consistent with the observation that three compounds are obtained if a second hydrogen is replaced with another atom. The acyclic structure yields two disubstituted products.

$$CH_3C \equiv C - C \equiv CCH_3 \xrightarrow[\text{with Br's}]{\text{replace 2 H's}} CH_3C \equiv C - C \equiv CCHBr \quad \text{and} \quad BrCH_2C \equiv C - C \equiv CCH_2Br$$
$$| \\ Br$$

The cyclic structure, with alternating single and slightly shorter double bonds, yields four disubstituted products—a 1,3-disubstituted product, a 1,4-disubstituted product, and two 1,2-disubstituted products—because the two substituents can be placed either on two adjacent carbons joined by a single bond or on two adjacent carbons joined by a double bond.

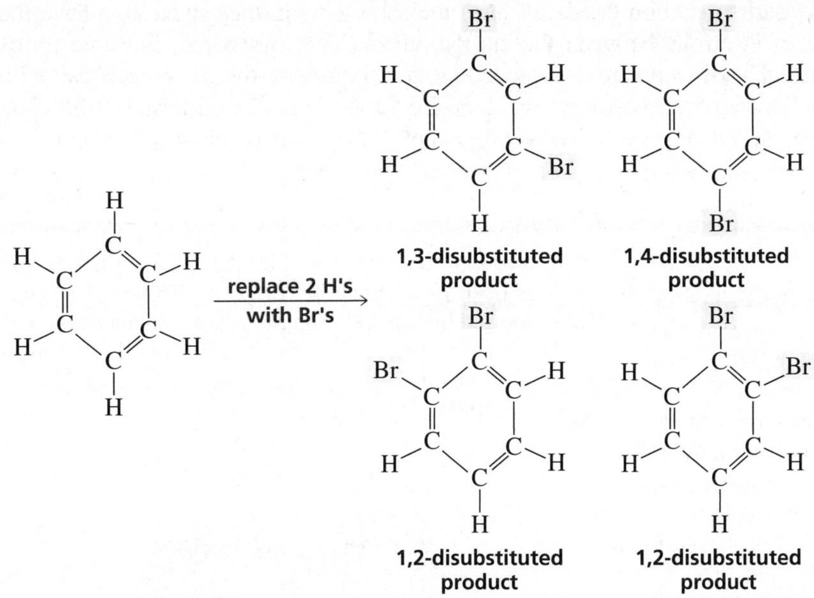

| 1,3-disubstituted product | 1,4-disubstituted product |
| 1,2-disubstituted product | 1,2-disubstituted product |

In 1865, the German chemist Friedrich Kekulé suggested a way of resolving this dilemma. He proposed that benzene was not a single compound, but a mixture of two compounds in rapid equilibrium.

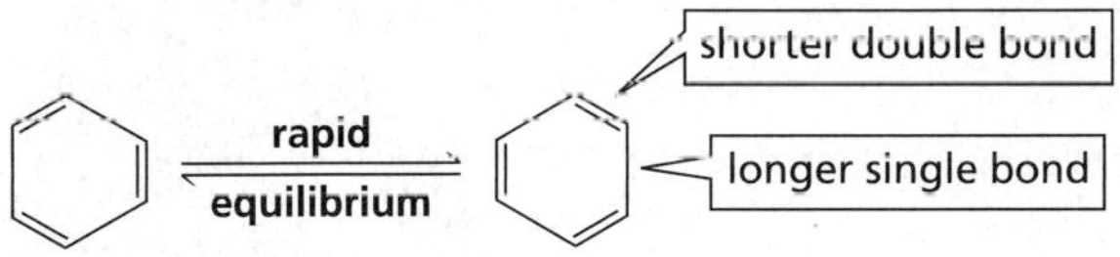

Kekulé structures of benzene

Kekulé's proposal explained why only three disubstituted products are obtained when a monosubstituted benzene undergoes a second substitution. According to Kekulé, there actually *are* four disubstituted products, but the two 1,2-disubstituted products interconvert too rapidly to be distinguished and separated from each other.

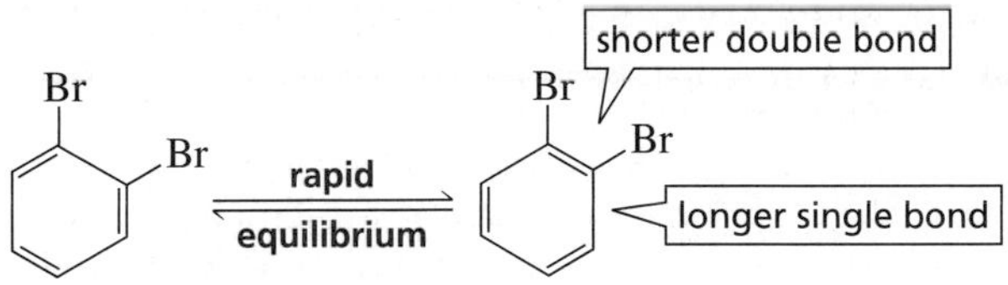

The Kekulé structures of benzene account for the molecular formula of benzene and for the number of isomers obtained as a result of substitution. However, they fail to account for the unusual stability of benzene that prevents the double bonds of benzene from undergoing the addition reactions characteristic of alkenes. That benzene has a six-membered ring was confirmed in 1901, when Paul Sabatier (Section 4.11) found that the hydrogenation of benzene produced cyclohexane. This, however, still did not solve the puzzle of benzene's structure.

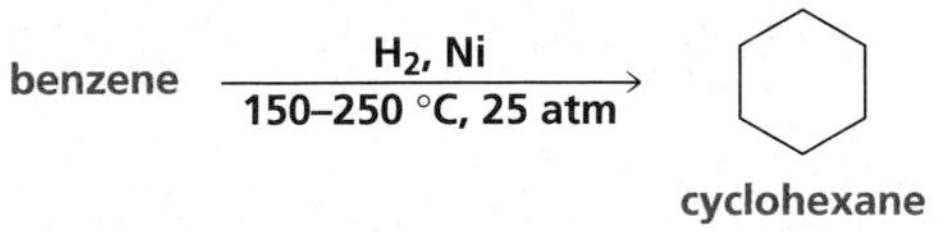

benzene $\xrightarrow[\text{150–250 °C, 25 atm}]{\text{H}_2,\ \text{Ni}}$

cyclohexane

Controversy over the structure of benzene continued until the 1930s, when the new techniques of X-ray and electron diffraction produced a surprising result: they showed that *benzene is a planar molecule and that the six carbon–carbon bonds have the same length.* The length of each carbon–carbon bond is 1.39 Å, which is shorter than a carbon–carbon single bond (1.54 Å) but longer than a carbon–carbon double bond (1.33 Å; Section 1.14). In other words, benzene does not have alternating single and double bonds.

If the carbon–carbon bonds all have the same length, they must also have the same number of electrons between the carbon atoms. This can be so, however, only if the π electrons of benzene are delocalized around the ring, rather than each pair of π electrons being localized between two carbon atoms. To better understand the concept of delocalized electrons, we will now take a close look at the bonding in benzene.

KEKULÉ'S DREAM

Friedrich August Kekulé von Stradonitz (1829–1896) was born in Germany. He entered the University of Giessen to study architecture, but switched to chemistry after taking a course in the subject. He was a professor of chemistry at the University of Heidelberg, at the University of Ghent in Belgium, and then at the University of Bonn. In 1890, he gave an extemporaneous speech at the twenty-fifth-anniversary celebration of his first paper on the cyclic structure of benzene. In this speech, he claimed that he had arrived at the Kekulé structures as a result of dozing off in front of a fire while working on a textbook. He dreamed of chains of carbon atoms twisting and turning in a snakelike motion, when suddenly the head of one snake seized hold of its own tail and formed a spinning ring. Recently, the veracity of his snake story has been questioned by those who point out that there is no written record of the dream from the time he experienced it in 1861 until the time he related it in 1890. Others counter that dreams are not the kind of evidence one publishes in scientific papers, although it is not uncommon for scientists to report creative ideas emerging from their subconscious at moments when they were not thinking about science. Also, Kekulé warned against publishing dreams when he said, "Let us learn to dream, and perhaps then we shall learn the truth. But let us also beware not to publish our dreams until they have been examined by the wakened mind." In 1895, he was made a nobleman by Emperor William II of Germany. This allowed him to add "von Stradonitz" to his name. Kekulé's students received three of the first five Nobel Prizes in chemistry: van't Hoff in 1901 (page 213), Fischer in 1902 (page 991), and Baeyer in 1905 (page 920).

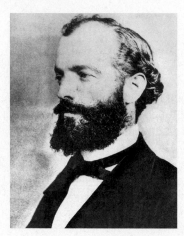

Friedrich August Kekulé von Stradonitz

BIOGRAPHY

Sir James Dewar (1842–1923) *was born in Scotland, the son of an innkeeper. After studying under Kekulé, he became a professor at Cambridge University and then at the Royal Institution in London. Dewar's most important work was in the field of low-temperature chemistry. He used double-walled flasks with evacuated space between the walls in order to reduce heat transmission. These flasks are now called Dewar flasks—better known to nonchemists as thermos bottles.*

BIOGRAPHY

Albert Ladenburg (1842–1911) *was born in Germany. He was a professor of chemistry at the University of Kiel.*

PROBLEM 1◆

a. How many monosubstituted products would each of the following compounds have? (Notice that each compound has the same molecular formula as benzene.)

1. $HC{\equiv}CC{\equiv}CCH_2CH_3$ **2.** $CH_2{=}CHC{\equiv}CCH{=}CH_2$

b. How many disubstituted products would each of the preceding compounds have? (Do not include stereoisomers.)

c. How many disubstituted products would each of the compounds have if stereoisomers are included?

PROBLEM 2

Between 1865 and 1890, other possible structures were proposed for benzene, two of which are shown here:

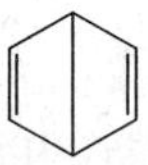

Dewar benzene

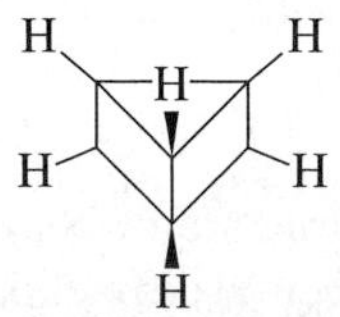

Ladenburg benzene

Considering what nineteenth-century chemists knew about benzene, which is a better proposal for benzene's structure, Dewar benzene or Ladenburg benzene? Why?

7.2 **The Bonding in Benzene**

Each of benzene's six carbons is sp^2 hybridized. An sp^2 carbon has bond angles of 120°—identical to the size of the angles of a planar hexagon. Thus, benzene is a planar molecule. Each of the carbons in benzene uses two sp^2 orbitals to bond to two other carbons; the third sp^2 orbital of each carbon overlaps the s orbital of a hydrogen (Figure 7.1a). Each carbon also has a p orbital at right angles to the sp^2 orbitals. Because benzene is planar, the six p orbitals are parallel (Figure 7.1b). The p orbitals are close enough for side-to-side overlap, so each p orbital overlaps the p orbitals on *both* adjacent carbons. As a result, the overlapping p orbitals form a continuous doughnut-shaped cloud of electrons above, and another doughnut-shaped cloud of electrons below, the plane of the benzene ring (Figure 7.1c). The electrostatic potential map (Figure 7.1d) shows that all the carbon–carbon bonds have the same electron density.

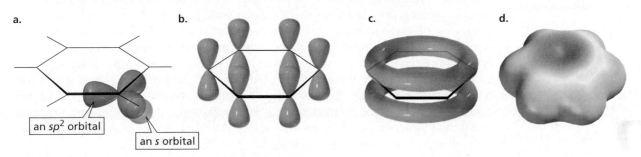

a. an sp^2 orbital   an s orbital   b.   c.   d.

▲ **Figure 7.1**
(a) The carbon–carbon and carbon–hydrogen σ bonds in benzene.
(b) The p orbital on each carbon of benzene can overlap two adjacent p orbitals.
(c) The clouds of electrons above and below the plane of the benzene ring.
(d) The electrostatic potential map for benzene.

Each of the six π electrons, therefore, is localized neither on a single carbon nor in a bond between two carbons (as in an alkene). Instead, each π electron is shared by all six carbons. In other words, the six π electrons are delocalized—they roam freely within the doughnut-shaped clouds that lie over and under the ring of carbon atoms. Benzene is often drawn as a hexagon containing either dashed lines or a circle to symbolize the six delocalized π electrons.

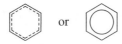

or

This type of representation makes it clear that there are no double bonds in benzene. We see now that Kekulé's structure was very nearly correct. The actual structure of benzene is a Kekulé structure with delocalized electrons.

7.3 **Resonance Contributors and the Resonance Hybrid**

A disadvantage of using dashed lines to represent delocalized electrons is that those lines do not tell us how many π electrons they represent. The dashed lines inside the hexagon indicate that the π electrons are shared equally by all six carbons of benzene and that all the carbon–carbon bonds have the same length, but they do not show how many π electrons are in the ring. Consequently, chemists prefer to use structures that portray the electrons as localized (and show the numbers of electrons), even though the electrons in the compound's actual structure are delocalized. The *approximate* structure with localized electrons is called a **resonance contributor**, a **resonance structure**, or a **contributing resonance structure**. The *actual* structure with delocalized electrons is called a **resonance hybrid**. Notice that it is easy to see that there are six π electrons in the ring of each resonance contributor.

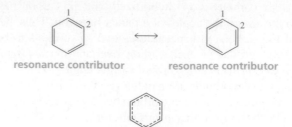

resonance contributor   resonance contributor

resonance hybrid

Electron delocalization is shown by double-headed arrows (⟷). Equilibrium is shown by two arrows pointing in opposite directions (⇌).

Resonance contributors are shown with a double-headed arrow between them. The double-headed arrow does *not* mean that the structures are in equilibrium with one another. Rather, it indicates that the actual structure lies somewhere between the structures of the resonance contributors. Resonance contributors are merely a convenient way to show the π electrons; they do not depict any real electron distribution. For example, the bond between C-1 and C-2 in benzene is not a double bond, although the resonance contributor on the left implies that it is. Nor is it a single bond, as represented by the resonance contributor on the right. Neither of the contributing resonance structures accurately represents the structure of benzene. The actual structure of benzene—the resonance hybrid—is obtained by mentally averaging the two resonance contributors.

The following analogy illustrates the difference between resonance contributors and the resonance hybrid. Imagine that you are trying to describe to a friend what a rhinoceros looks like. You might tell your friend that a rhinoceros looks like a cross between a unicorn and a dragon. Like resonance contributors, the unicorn and the dragon do not really exist. Further, like resonance contributors, they are not in equilibrium: a rhinoceros does not change back and forth between the two forms, looking like a unicorn one instant and a dragon the next. The unicorn and dragon are simply ways to describe what the actual structure—the rhinoceros—looks like. *Resonance contributors, like unicorns and dragons, are imaginary, not real. Only the resonance hybrid, like the rhinoceros, is real.*

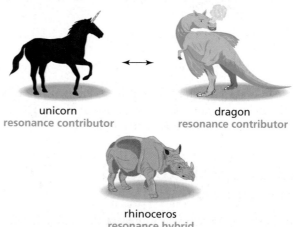

unicorn
resonance contributor

dragon
resonance contributor

rhinoceros
resonance hybrid

Electron delocalization is most effective if all the atoms sharing the delocalized electrons lie in or close to the same plane, so that their p orbitals can maximally overlap. For example, cyclooctatetraene is not planar; its sp^2 carbons have bond angles of 120°, whereas a planar eight-membered ring would have bond angles of 135° (another reason causing it to be non-planar will be discussed in Section 14.6). Because the ring is not planar, a p orbital can overlap with one adjacent p orbital, but it can have little overlap with the other adjacent p orbital. Therefore, the eight electrons are not delocalized over the entire cyclooctatetraene ring, and its carbon–carbon bonds do not all have the same length.

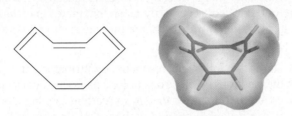

7.4 **How to Draw Resonance Contributors**

We have seen that an organic compound with delocalized electrons is generally represented as a structure with localized electrons to let us know how many π electrons are in the molecule. For example, nitroethane is represented as having a nitrogen–oxygen double bond and a nitrogen–oxygen single bond.

$$CH_3CH_2 \!-\! \overset{+}{N} \!\!\begin{array}{c} \ddots\!\ddot{O} \\ \diagdown \\ :\ddot{O}:^- \end{array}$$

nitroethane

However, the two nitrogen–oxygen bonds in nitroethane are actually identical; they each have the same bond length. A more accurate description of the molecule's structure is obtained by drawing the two resonance contributors. Both resonance contributors show the compound with a nitrogen–oxygen double bond and a nitrogen–oxygen single bond, but they indicate that the electrons are delocalized by depicting the double bond in one contributor as a single bond in the other.

$$CH_3CH_2 \!-\! \overset{+}{N} \!\!\begin{array}{c} \cdot\ddot{O} \\ \diagup\!\!\diagup \\ :\ddot{O}:^- \end{array} \qquad \longleftrightarrow \qquad CH_3CH_2 \!-\! \overset{+}{N} \!\!\begin{array}{c} :\ddot{O}:^- \\ \diagup\!\!\diagup \\ \ddot{O}: \end{array}$$

resonance contributor   **resonance contributor**

The resonance hybrid, in contrast, shows that the p orbital of nitrogen overlaps the p orbital of each oxygen. In other words, it shows that the two π electrons are shared by three atoms. The resonance hybrid also shows that the two nitrogen–oxygen bonds are identical and that the negative charge is shared equally by both oxygen atoms. Thus, we need to visualize and mentally average both resonance contributors to appreciate what the actual molecule—the resonance hybrid—looks like.

> **Delocalized electrons result from a p orbital overlapping the p orbitals of more than one adjacent atom.**

$$CH_3CH_2 \!-\! \overset{+}{N} \!\!\begin{array}{c} O\!:^{\delta-} \\ \diagup\!\!\diagup \\ O\!:^{\delta-} \end{array}$$

resonance hybrid

Rules for Drawing Resonance Contributors

To draw a set of resonance contributors, we draw a Lewis structure for the molecule—it becomes our first resonance contributor—and then we move the electrons, following the rules listed below, to generate the next resonance contributor.

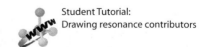

Student Tutorial:
Drawing resonance contributors

1. Only electrons move. Atoms never move.

2. Only π electrons (electrons in π bonds) and lone-pair electrons can move; σ electrons never move.

3. The total number of electrons in the molecule does not change. Therefore, each of the resonance contributors for a particular compound must have the same net charge. If one has a net charge of 0, all the others must also have net charges of 0. (A net charge of 0 does not necessarily mean that there is no charge on any of the atoms: a molecule with a positive charge on one atom and a negative charge on another atom has a net charge of 0.)

> To draw resonance contributors, move only π electrons or lone pairs toward an sp^2 (or sp) carbon.

Notice, as you study the following resonance contributors and practice drawing them yourself, that the electrons (π electrons or lone pairs) are always moved toward an sp^2 (or sp) hybridized atom. Remember that an sp^2 carbon is either a positively charged carbon (Section 1.10) or a double-bonded carbon, and that an sp carbon has two π bonds and is, therefore, generally a triple-bonded carbon. Electrons cannot be moved toward an sp^3 carbon, because an sp^3 carbon has a complete octet so it cannot accommodate any more electrons.

The following carbocation has delocalized electrons. To draw its resonance contributor, we *move the π electrons toward an sp^2 carbon*. A curved arrow can help you decide how to draw the next contributor. Remember the tail of the curved arrow shows where the electrons start from, and the head shows where the electrons are going. The resonance hybrid shows that the positive charge is shared by two carbons.

an sp^2 carbon

$$CH_3CH{=}CH{-}\overset{+}{C}HCH_3 \longleftrightarrow CH_3\overset{+}{C}H{-}CH{=}CHCH_3$$

resonance contributors

$$CH_3\overset{\delta+}{C}H{=\!=\!=}CH{=\!=\!=}\overset{\delta+}{C}HCH_3$$

resonance hybrid

Let's compare this carbocation with a similar compound in which all the electrons are localized. The π electrons in the compound shown below cannot move, because the carbon they would move to is an sp^3 carbon, and sp^3 carbons cannot accept electrons.

an sp^3 carbon
cannot accept electrons

$$CH_2{=}CH{-}CH_2\overset{+}{C}HCH_3$$

localized electrons

In the next example, *π electrons again move toward an sp^2 carbon*. The resonance hybrid shows that the positive charge is shared by three carbons.

an sp^2 carbon

$$CH_3CH{=}CH{-}CH{=}CH{-}\overset{+}{C}H_2 \longleftrightarrow CH_3CH{=}CH{-}\overset{+}{C}H{-}CH{=}CH_2 \longleftrightarrow CH_3\overset{+}{C}H{-}CH{=}CH{-}CH{=}CH_2$$

resonance contributors

$$CH_3\overset{\delta+}{C}H{=\!=\!=}CH{=\!=\!=}\overset{\delta+}{C}H{=\!=\!=}CH{=\!=\!=}\overset{\delta+}{C}H_2$$

resonance hybrid

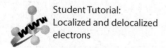

Student Tutorial:
Localized and delocalized
electrons

The resonance contributor for the next compound is obtained by *moving lone-pair electrons toward an sp^2 carbon*. The sp^2 carbon can accommodate the new electrons by breaking a π bond.

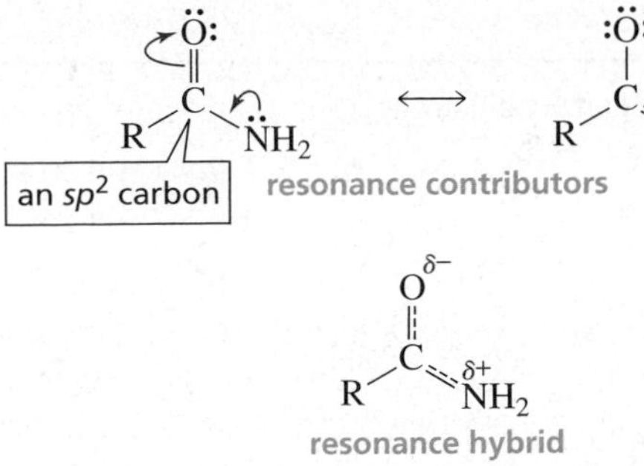

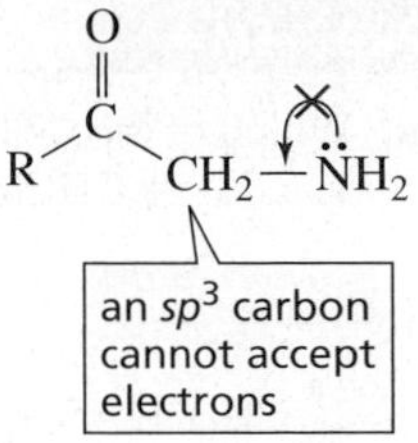

In the next example, *lone pair electrons move toward an sp carbon.*

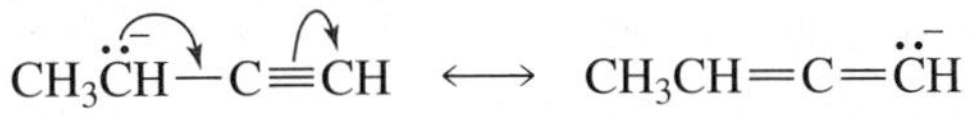

$$CH_3\ddot{C}H-C\equiv CH \longleftrightarrow CH_3CH=C=\ddot{C}H$$

The resonance contributor for the next compound is obtained by *moving π electrons toward an sp carbon.*

$$CH_2=CH-C\equiv N \longleftrightarrow \overset{+}{C}H_2-CH=C=\ddot{N}$$

PROBLEM 3♦

a. Predict the relative bond lengths of the three carbon–oxygen bonds in the carbonate ion $\left(CO_3{}^{2-}\right)$.

b. What would you expect the charge to be on each oxygen atom?

PEPTIDE BONDS

Every third bond in a protein is a peptide bond. A resonance contributor can be drawn for a peptide bond by moving the lone pair on nitrogen toward the sp^2 carbon.

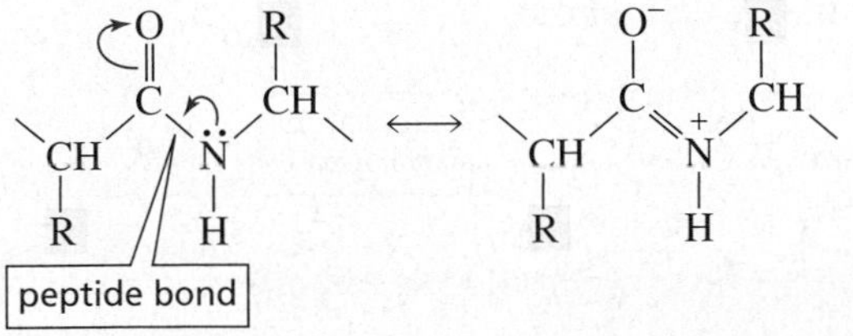

Because of the partial double bond character of the peptide bond, the carbon and nitrogen atoms and the two atoms bonded to each are held rigidly in a plane, as represented below by the blue and green boxes. This planarity affects the way proteins can fold, so it has important implications for the three-dimensional shape of these biological molecules.

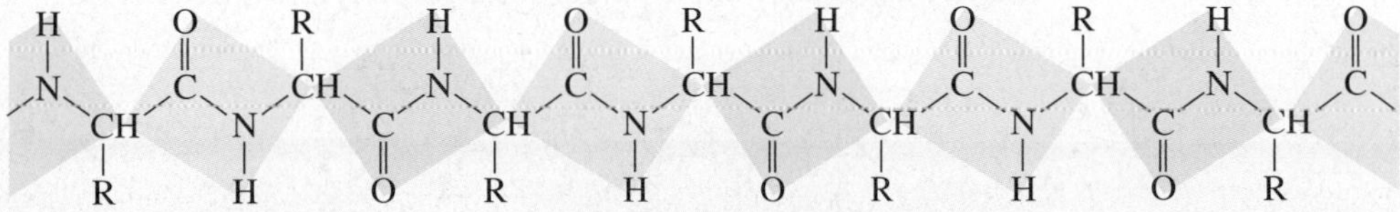

peptide bond

a segment of a protein

PROBLEM 4

a. Which of the following compounds have delocalized electrons?

1. $CH_2=CHCH_2CH=CH_2$

2. $CH_3CH=CHCH=\overset{+}{C}HCH_2$

3. $CH_3CH_2\overset{..}{N}HCH_2CH=CH_2$

4. [pyrrole ring structure with N–H]

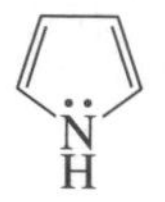

5. [pyran ring structure with O]

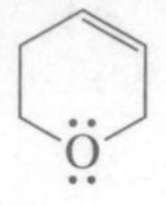

6. [cyclohexene ring with $-CH_2\overset{..}{N}H_2$]

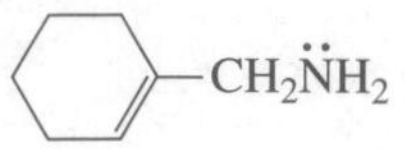

7. [cyclohexene ring with $-\overset{..}{N}H_2$]

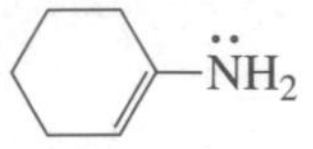

b. Draw the contributing resonance structures for those compounds.

7.5 The Predicted Stabilities of Resonance Contributors

All resonance contributors do not necessarily contribute equally to the resonance hybrid. The degree to which each resonance contributor contributes depends on its predicted stability. Because resonance contributors are not real, their stabilities cannot be measured. Therefore, the stabilities of resonance contributors have to be predicted based on molecular features found in real molecules. *The greater the predicted stability of the resonance contributor, the more it contributes to the structure of the resonance hybrid; and the more it contributes to the structure of the resonance hybrid, the more similar the contributor is to the real molecule.* The examples that follow illustrate these points.

The two resonance contributors for a carboxylic acid are shown below, labeled **A** and **B**. Structure **B** has two features that make it less stable than structure **A**: one of the oxygen atoms has a positive charge—not a comfortable situation for an electronegative atom—and the structure has separated charges. A molecule with **separated charges** has a positive charge and a negative charge that can be neutralized by the movement of electrons. Resonance contributors with separated charges are relatively unstable (relatively high in energy) because energy is required to keep opposite charges separated. Structure **A**, therefore, is predicted to be more stable than structure **B**. Consequently, **A** makes a greater contribution to the resonance hybrid, so the resonance hybrid looks more like **A** than like **B**.

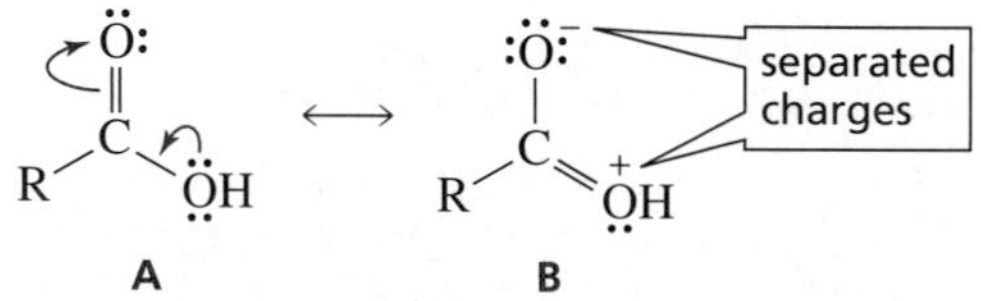

> The greater the predicted stability of the resonance contributor, the more it contributes to the structure of the resonance hybrid.

[resonance structures A and B of a carboxylic acid, with "separated charges" labeled]

A   **B**

a carboxylic acid

The two resonance contributors for a carboxylate ion are shown next.

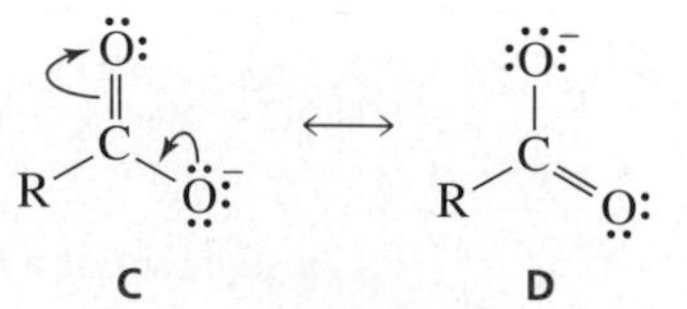

[resonance structures C and D of a carboxylate ion]

C   **D**

a carboxylate ion

Structures **C** and **D** are predicted to be equally stable and therefore are expected to contribute equally to the resonance hybrid.

When electrons can be moved in more than one direction, the most stable resonance contributor is obtained by moving them toward the more electronegative atom. For instance, structure **G** in the next example results from moving the π electrons toward oxygen—the most electronegative atom in the molecule. In contrast, structure **E** results from moving the π electrons away from oxygen.

an incomplete octet

an incomplete octet

E

F

G

resonance contributor obtained by moving π electrons away from the most electronegative atom

resonance contributor obtained by moving π electrons toward the most electronegative atom

an insignificant resonance contributor

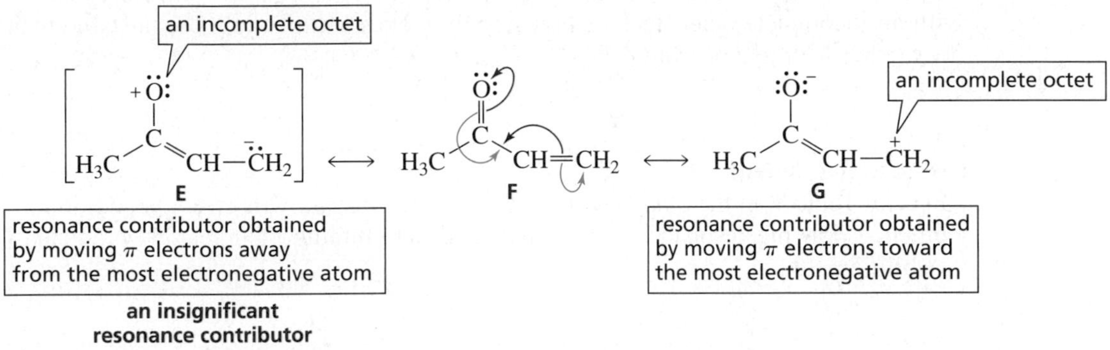

We can predict that structure **G** will make only a small contribution to the resonance hybrid because it has separated charges, as well as an atom with an incomplete octet. Structure **E** also has separated charges and an atom with an incomplete octet, but its predicted stability is even less than that of structure **G** because it has a positive charge on the electronegative oxygen. Its contribution to the resonance hybrid is so insignificant that we do not need to include it as one of the resonance contributors. The resonance hybrid, therefore, looks very much like structure **F**.

The only time you need to show a resonance contributor obtained by moving electrons away from the most electronegative atom is when that is the only way the electrons can be moved. In other words, movement of electrons away from the most electronegative atom is better than no movement at all, because electron delocalization makes a molecule more stable (as we will see in Section 7.6). For example, the only resonance contributor that can be drawn for the following molecule requires movement of the electrons away from oxygen:

$$CH_2{=}CH{-}\ddot{O}CH_3 \quad \longleftrightarrow \quad \bar{C}H_2{-}CH{=}\overset{+}{O}CH_3$$

H   **I**

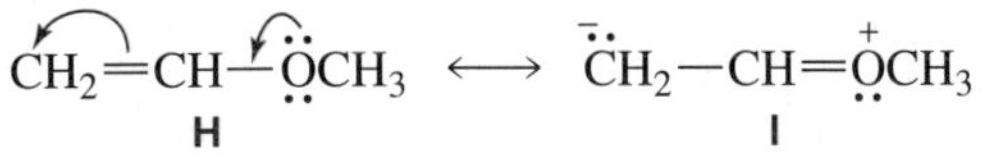

Structure **I** is predicted to be relatively unstable because it has separated charges and its most electronegative atom is the atom with the positive charge. Therefore, the structure of the resonance hybrid is similar to structure **H**, with only a small contribution from structure **I**.

Let's now see which of the resonance contributors shown below has a greater predicted stability. Structure **J** has a negative charge on carbon, and structure **K** has a negative charge on oxygen. Oxygen is more electronegative than carbon, so oxygen can accommodate the negative charge better. Consequently, structure **K** is predicted to be more stable than structure **J**. The resonance hybrid, therefore, more closely resembles structure **K**; that is, the resonance hybrid has a greater concentration of negative charge on the oxygen atom than on the carbon atom.

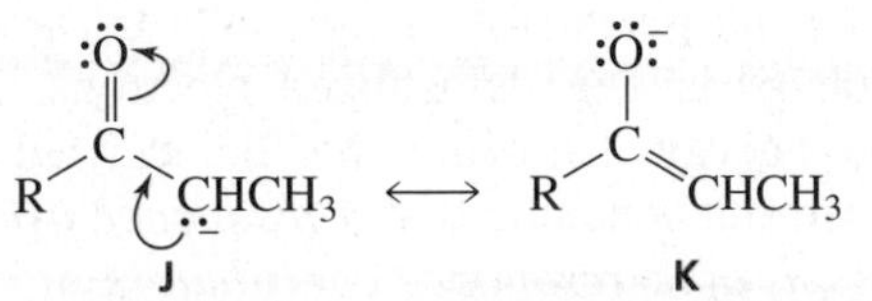

J   **K**

We can summarize the features that decrease the predicted stability of a contributing resonance structure as follows:

1. an atom with an incomplete octet
2. a negative charge that is not on the most electronegative atom or a positive charge that is not on the least electronegative (most electropositive) atom
3. charge separation

When we compare the relative stabilities of contributing resonance structures, an atom with an incomplete octet (feature 1) generally makes a structure more unstable than does either feature 2 or feature 3.

Note to the student

Special Topic V in the *Study Guide and Solutions Manual* will give you additional practice drawing resonance contributors and determining their relative predicted stabilities.

PROBLEM 5 SOLVED

Draw contributing resonance structures for each of the following species, and rank the structures in order of decreasing contribution to the hybrid:

a. $CH_3\overset{+}{C}-CH=CHCH_3$
  |
  CH_3

b. $CH_3\overset{O}{\overset{\|}{C}}OCH_3$

c. (cyclohexenone ring with $\overset{\cdot\cdot}{\overset{-}{O}}$)

d. (cyclohexadienone ring with $=O$)

e. $CH_3-\overset{\overset{+OH}{\|}}{C}-NHCH_3$

f. $CH_3\overset{+}{C}H-CH=CHCH_3$

Solution to 5a   Structure **A** is more stable than structure **B** because the positive charge is on a tertiary carbon in **A** and on a secondary carbon in **B** (Section 4.2).

$$CH_3\overset{+}{C}\overset{\frown}{-}CH=CHCH_3 \longleftrightarrow CH_3C=CH-\overset{+}{C}HCH_3$$
$$\quad\quad | \quad\quad\quad\quad\quad\quad\quad\quad\quad |$$
$$\quad\quad CH_3 \quad\quad\quad\quad\quad\quad\quad\quad CH_3$$
$$\quad\quad\quad A \quad\quad\quad\quad\quad\quad\quad\quad\quad\quad B$$

PROBLEM 6

Draw the resonance hybrid for each of the species in Problem 5.

7.6 Delocalization Energy Is the Additional Stability Delocalized Electrons Give to a Compound

The delocalization energy is a measure of how much more stable a compound with delocalized electrons is than it would be if its electrons were localized.

A resonance hybrid is more stable than any of its resonance contributors is predicted to be.

Delocalized electrons stabilize a compound. The extra stability a compound gains from having delocalized electrons is called the **delocalization energy**. **Electron delocalization** is also called **resonance**, so delocalization energy is also called **resonance energy**. Since the resonance energy tells us how much more stable a compound is as a result of having delocalized electrons, it is sometimes called **resonance stabilization energy**. Knowing that delocalized electrons increase the stability of a molecule, we can conclude that *a resonance hybrid is more stable than the predicted stability of any of its resonance contributors.*

The delocalization energy associated with a compound that has delocalized electrons depends on the number *and* predicted stability of the resonance contributors: *the greater the number of relatively stable resonance contributors, the greater is the delocalization energy.* For example, the delocalization energy of a carboxylate ion with two relatively stable resonance contributors is significantly greater than the delocalization energy of a carboxylic acid with only one relatively stable resonance contributor.

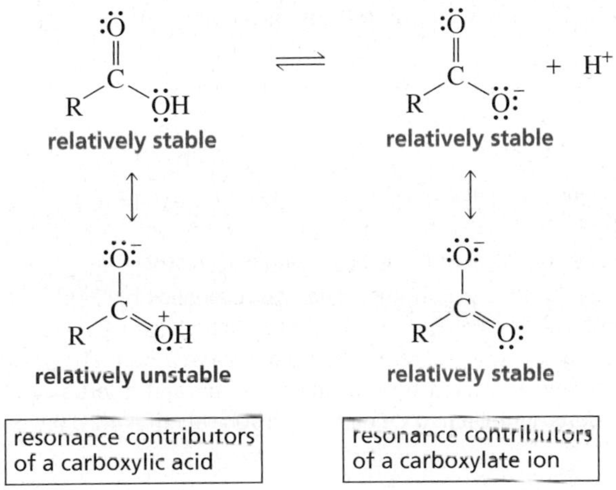

Notice that it is the number of *relatively stable* resonance contributors—not the total number of resonance contributors—that is important in determining the delocalization energy. For example, the delocalization energy of a carboxylate ion with two relatively stable resonance contributors is greater than the delocalization energy of the compound in the following example because even though the latter has three resonance contributors, only one of them is relatively stable:

The greater the number of relatively stable resonance contributors, the greater is the delocalization energy.

$$\overset{-}{C}H_2-CH=CH-\overset{+}{C}H_2 \longleftrightarrow CH_2=CH-CH=CH_2 \longleftrightarrow \overset{+}{C}H_2-CH=CH-\overset{-}{C}H_2$$

relatively unstable relatively stable relatively unstable

The more nearly equivalent the resonance contributors are in structure, the greater is the delocalization energy. The carbonate dianion is particularly stable because it has three equivalent resonance contributors.

The more nearly equivalent the resonance contributors are in structure, the greater is the delocalization energy.

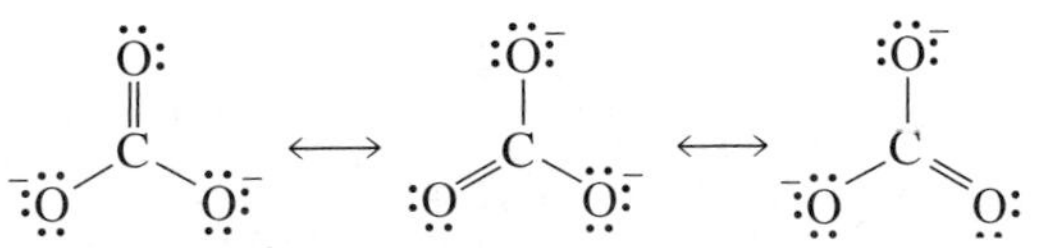

We can now summarize what we know about contributing resonance structures:*

1. The greater the predicted stability of a resonance contributor, the more it contributes to the resonance hybrid.
2. The greater the number of relatively stable resonance contributors, the greater is the delocalization energy.
3. The more nearly equivalent the resonance contributors, the greater is the delocalization energy.

---

* In cyclic systems, there is another factor that profoundly influences the predicted stability of resonance contributors. See Sections 14.1–14.7 for details.

PROBLEM-SOLVING STRATEGY

Determining Relative Stabilities

Which carbocation is more stable?

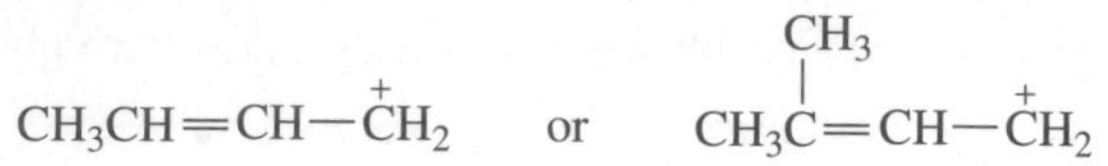

$$CH_3CH=CH-\overset{+}{C}H_2 \quad \text{or} \quad CH_3\overset{\overset{\displaystyle CH_3}{|}}{C}=CH-\overset{+}{C}H_2$$

Start by drawing the resonance contributor for each carbocation.

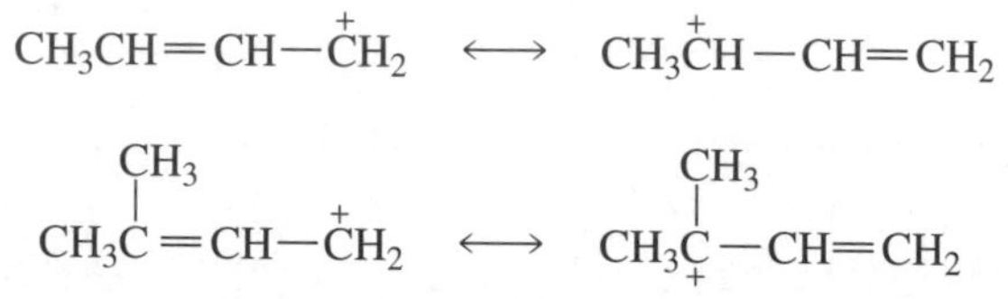

$$CH_3CH=CH-\overset{+}{C}H_2 \quad \longleftrightarrow \quad CH_3\overset{+}{C}H-CH=CH_2$$

$$CH_3\overset{\overset{\displaystyle CH_3}{|}}{C}=CH-\overset{+}{C}H_2 \quad \longleftrightarrow \quad CH_3\overset{\overset{\displaystyle CH_3}{|}}{\underset{+}{C}}-CH=CH_2$$

Now look at ways the two sets of resonance contributors differ and think about how those differences affect the relative stabilities of the two resonance hybrids.

Each carbocation has two resonance contributors. The positive charge of the carbocation on the left is shared by a primary carbon and a secondary carbon. The positive charge of the carbocation on the right is shared by a primary carbon and a tertiary carbon. Because a tertiary carbon is more stable than a secondary carbon (Section 4.2), the carbocation on the right is more stable.

Now continue on to Problem 7.

PROBLEM 7◆

Which species is more stable?

a. 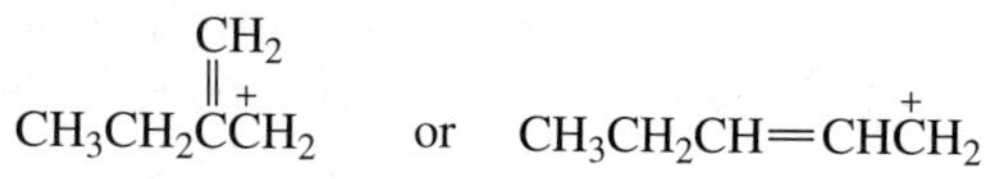 $$CH_3CH_2\overset{\overset{\displaystyle CH_2}{\|}}{\underset{+}{C}}CH_2 \quad \text{or} \quad CH_3CH_2CH=\overset{+}{C}HCH_2$$

b. $$CH_3\overset{\overset{\displaystyle O}{\|}}{C}CH=CH_2 \quad \text{or} \quad CH_3\overset{\overset{\displaystyle O}{\|}}{C}CH=CHCH_3$$

c. $$CH_3\overset{\overset{\displaystyle O^-}{|}}{C}HCH=CH_2 \quad \text{or} \quad CH_3\overset{\overset{\displaystyle O^-}{|}}{C}=CHCH_3$$

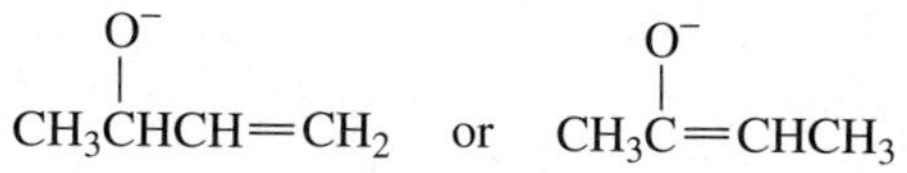

d. 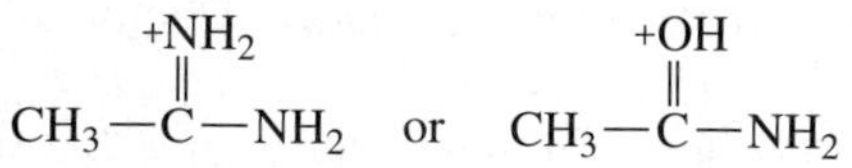 $$CH_3-\overset{\overset{\displaystyle +NH_2}{\|}}{C}-NH_2 \quad \text{or} \quad CH_3-\overset{\overset{\displaystyle +OH}{\|}}{C}-NH_2$$

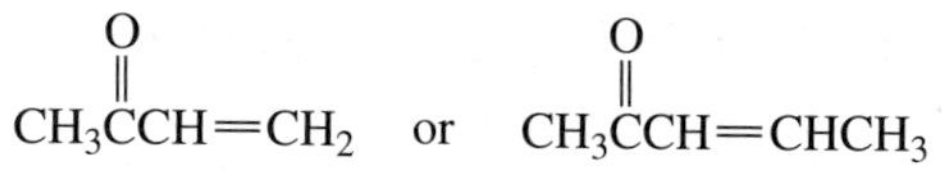

PROBLEM 8◆

Which species has the greatest delocalization energy?

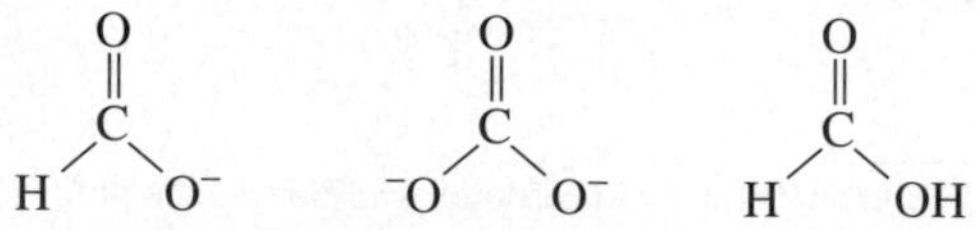

$$\overset{\overset{\displaystyle O}{\|}}{\underset{H \quad O^-}{C}} \qquad \overset{\overset{\displaystyle O}{\|}}{\underset{-O \quad O^-}{C}} \qquad \overset{\overset{\displaystyle O}{\|}}{\underset{H \quad OH}{C}}$$

7.7 Examples Illustrating the Effect of Delocalized Electrons on Stability

We will now look at two examples that illustrate the extra stability a molecule acquires as a result of having delocalized electrons.

Stability of Dienes

Dienes are hydrocarbons with two double bonds. **Isolated dienes** have isolated double bonds: **isolated double bonds** are separated by more than one single bond. **Conjugated dienes** have conjugated double bonds: **conjugated double bonds** are separated by one single bond.

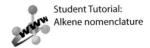

Student Tutorial:
Alkene nomenclature

double bonds are separated by more than one single bond

double bonds are separated by one single bond

$CH_2=CH-CH_2-CH=CH_2$
an isolated diene

$CH_3CH=CH-CH=CHCH_3$
a conjugated diene

We saw in Section 4.11 that the relative stabilities of alkenes can be determined by their $-\Delta H°$ values for catalytic hydrogenation. Remember that the most stable alkene has the smallest $-\Delta H°$ value; it gives off the least heat when it is hydrogenated, because it has less energy to begin with. The $-\Delta H°$ value for hydrogenation of 1,3-pentadiene (a conjugated diene) is smaller than that of 1,4-pentadiene (an isolated diene). We can conclude, therefore, that conjugated dienes are more stable than isolated dienes.

The most stable alkene has the smallest $-\Delta H°$ value.

$$CH_2=CH-CH_2-CH=CH_2 \ + \ 2\,H_2 \ \xrightarrow{\text{Pt/C}} \ CH_3CH_2CH_2CH_2CH_3 \qquad \Delta H° = \text{−60.2 kcal/mol (−252 kJ/mol)}$$

1,4-pentadiene
an isolated diene

$$CH_2=CH-CH=CHCH_3 \ + \ 2\,H_2 \ \xrightarrow{\text{Pt/C}} \ CH_3CH_2CH_2CH_2CH_3 \qquad \Delta H° = \text{−54.1 kcal/mol (−226 kJ/mol)}$$

1,3-pentadiene
a conjugated diene

Why is a conjugated diene more stable than an isolated diene? Two factors contribute to the difference. The first is *electron delocalization*. The π electrons in each of the double bonds of an isolated diene are *localized* between two carbons. In contrast, the π electrons in a conjugated diene are *delocalized*. As we discovered in Section 7.6, electron delocalization stabilizes a molecule. The resonance hybrid shows that the single bond in 1,3-butadiene is not a pure single bond, but has partial double-bond character as a result of electron delocalization. (Notice that because the compound does not have an electronegative atom that would determine the direction in which the electrons move, they can move both to the left and to the right.)

An increase in delocalization energy means an increase in stability.

$$\overset{-}{C}H_2-CH=CH-\overset{+}{C}H_2 \ \longleftrightarrow \ CH_2=CH-CH=CH_2 \ \longleftrightarrow \ \overset{+}{C}H_2-CH=CH-\overset{-}{C}H_2$$

resonance contributors

delocalized electrons

$$CH_2{=\!=\!=}CH{=\!=\!=}CH{=\!=\!=}CH_2$$
1,3-butadiene
resonance hybrid

The *hybridization of the orbitals* forming the carbon–carbon single bonds also causes a conjugated diene to be more stable than an isolated diene. The carbon–carbon single bond in 1,3-butadiene is formed from the overlap of an sp^2 orbital with another sp^2 orbital, whereas the carbon–carbon single bonds in 1,4-pentadiene are formed from the overlap of an sp^3 orbital with an sp^2 orbital.

> single bond formed by
> sp^2–sp^2 overlap

> single bonds formed by
> sp^3–sp^2 overlap

$$CH_2=CH-CH=CH_2 \qquad CH_2=CH-CH_2-CH=CH_2$$

1,3-butadiene **1,4-pentadiene**

In Section 1.14, we saw that the length and strength of a bond depend on how close the electrons in the bonding orbital are to the nucleus: *the closer the electrons are to the nucleus, the shorter and stronger is the bond.* Because a 2s electron is closer, on average, to the nucleus than is a 2p electron, a bond formed by sp^2–sp^2 overlap is shorter and stronger than one formed by sp^3–sp^2 overlap (Table 7.1). (An sp^2 orbital has 33.3% *s* character, whereas an sp^3 orbital has only 25% *s* character.) Thus, a conjugated diene has one stronger single bond than an isolated diene, and stronger bonds cause a compound to be more stable.

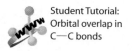

Student Tutorial:
Orbital overlap in
C—C bonds

Table 7.1	Dependence of the Length of a Carbon–Carbon Single Bond on the Hybridization of the Orbitals Used in Its Formation	
Compound	**Hybridization**	**Bond length (Å)**
H_3C-CH_3	sp^3–sp^3	1.54
$H_3C-\overset{\overset{H}{\mid}}{C}=CH_2$	sp^3–sp^2	1.50
$H_2C=\overset{\overset{H}{\mid}}{C}-\overset{\overset{H}{\mid}}{C}=CH_2$	sp^2–sp^2	1.47
$H_3C-C\equiv CH$	sp^3–sp	1.46
$H_2C=\overset{\overset{H}{\mid}}{C}-C\equiv CH$	sp^2–sp	1.43
$HC\equiv C-C\equiv CH$	sp–sp	1.37

Allenes are compounds that have **cumulated double bonds**. These are double bonds that are adjacent to one another. The cumulated double bonds give allenes an unusual geometry because of the *sp* hybridization of the central carbon. One of the *p* orbitals of the central carbon of the cumulated double bond overlaps a *p* orbital of an adjacent sp^2 carbon. The second *p* orbital of the central carbon overlaps a *p* orbital of the other sp^2 carbon (Figure 7.2a).

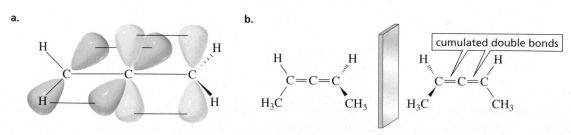

▲ **Figure 7.2**
(a) Double bonds are formed by *p* orbital–*p* orbital overlap. The two *p* orbitals on the central carbon are perpendicular, causing allene to be a nonplanar molecule. (b) 2,3-Pentadiene has a nonsuperimposable mirror image. It is, therefore, a chiral molecule, even though it does not have an asymmetric center.

The two *p* orbitals of the central carbon are perpendicular. Therefore, the plane containing one H—C—H group is perpendicular to the plane containing the other H—C—H group. Thus, a substituted allene such as 2,3-pentadiene has a nonsuperimposable mirror image (Figure 7.2b), so it is a chiral molecule, even though it does not have any asymmetric centers.

PROBLEM 9◆

The $-\Delta H°$ value for hydrogenation of 2,3-pentadiene, a cumulated diene is 70.5 kcal/mol. What are the relative stabilities of cumulated, conjugated, and isolated dienes?

PROBLEM 10◆

Name the following dienes and rank them in order of increasing stability. (Alkyl groups stabilize dienes in the same way that they stabilize alkenes; see Section 4.11.)

$CH_3CH{=}CHCH{=}CHCH_3$ $\quad CH_2{=}CHCH_2CH{=}CH_2$

$CH_3C(CH_3){=}CHCH{=}C(CH_3)CH_3$ $\quad CH_3CH{=}CHCH{=}CH_2$

Stability of Allylic and Benzylic Cations

We now look at two classes of carbocations that have delocalized electrons and are therefore more stable than similar carbocations with localized electrons. An **allylic cation** is a carbocation with the positive charge on an allylic carbon; an **allylic carbon** is a carbon adjacent to an sp^2 carbon of an alkene (Section 3.2). A **benzylic cation** is a carbocation with the positive charge on a benzylic carbon; a **benzylic carbon** is a carbon adjacent to an sp^2 carbon of a benzene ring.

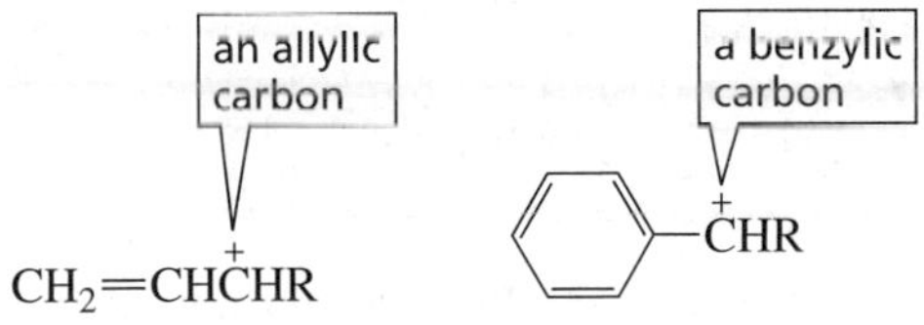

an allylic cation a benzylic cation

The *allyl cation* is an unsubstituted allylic cation, and the *benzyl cation* is an unsubstituted benzylic cation.

the allyl cation the benzyl cation

An allylic cation has two resonance contributors. The positive charge is not localized on a single carbon, but is shared by two carbons.

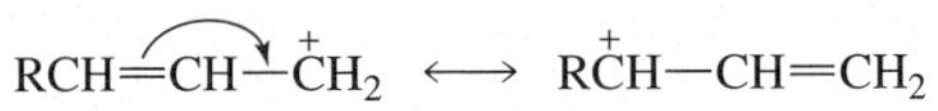

an allylic cation

A benzylic cation has five resonance contributors. Notice that the positive charge is shared by four carbons.

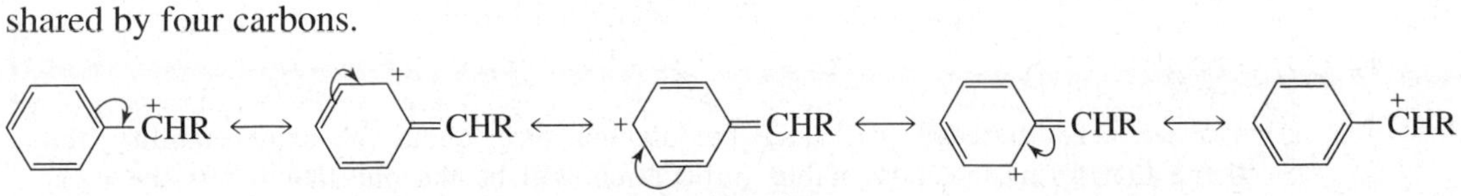

a benzylic cation

Because the allyl and benzyl cations have delocalized electrons, they are more stable than other primary carbocations. (Indeed, they have about the same stability as secondary alkyl carbocations.) We can add the benzyl and allyl cations to the list of carbocations whose relative stabilities were shown in Sections 4.2 and 6.5.

relative stabilities of carbocations

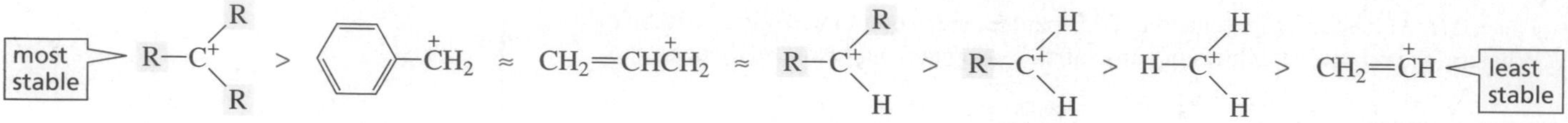

most stable → a tertiary carbocation > benzyl cation ≈ allyl cation ≈ a secondary carbocation > a primary carbocation > methyl cation > vinyl cation ← least stable

Student Tutorial:
Common terms

Not all allylic and benzylic cations have the same stability. Just as a tertiary alkyl carbocation is more stable than a secondary alkyl carbocation, a tertiary allylic cation is more stable than a secondary allylic cation, which in turn is more stable than the (primary) allyl cation. Similarly, a tertiary benzylic cation is more stable than a secondary benzylic cation, which is more stable than the (primary) benzyl cation.

relative stabilities

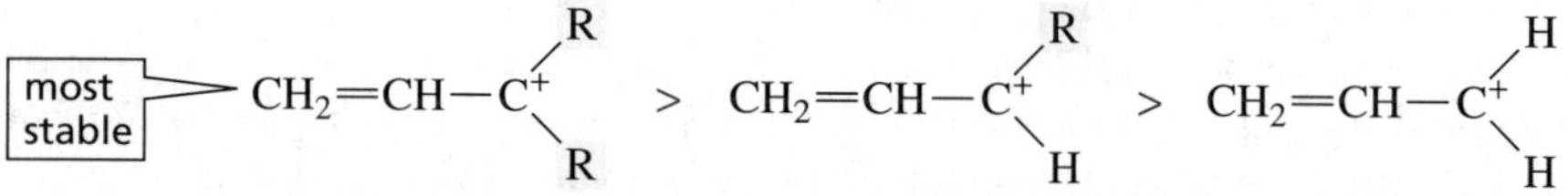

most stable → tertiary allylic cation > secondary allylic cation > allyl cation

most stable → tertiary benzylic cation > secondary benzylic cation > benzyl cation

Notice that it is the *primary* benzyl and the *primary* allyl cations that have about the same stability as *secondary* alkyl carbocations. Secondary benzylic and allylic cations, as well as tertiary benzylic and allylic cations, are even more stable than primary benzyl and allyl cations.

PROBLEM-SOLVING STRATEGY

Determining Relative Reactivities

What of the following alkenes reacts the fastest with HBr?

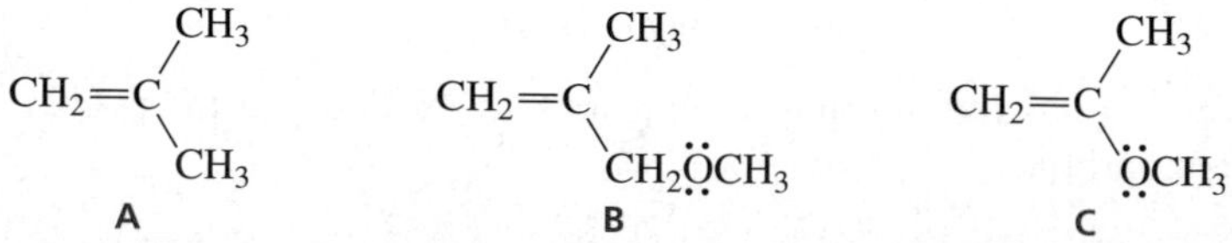

A: $CH_2{=}C$ with CH_3 and CH_3

B: $CH_2{=}C$ with CH_3 and $CH_2\ddot{O}CH_3$

C: $CH_2{=}C$ with CH_3 and $\ddot{O}CH_3$

First we need to recall that since the alkenes have about the same stability, the alkene that forms the most stable carbocation will be the one that reacts the most rapidly with HBr (Section 4.4). We therefore start by drawing the carbocation that would be formed by each of the three alkenes and then we will analyze them for their relative stabilities.

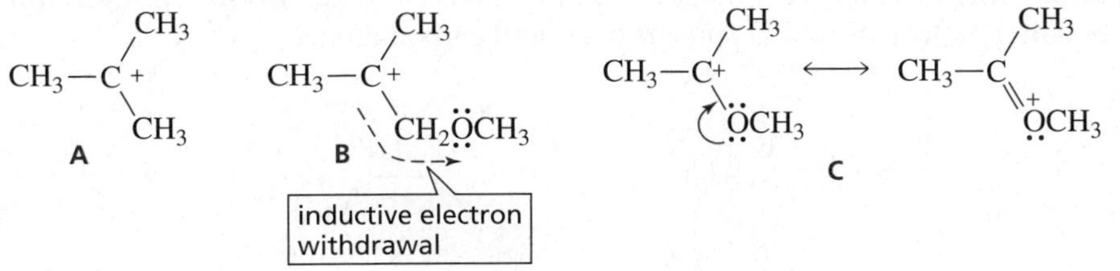

inductive electron withdrawal

Structure A is more stable than B, because oxygen withdraws electron inductively, increasing the size of the positive charge, which destabilizes the carbocation. Structure C is more stable than A because the positive charge on carbon is reduced since it is shared with oxygen. Therefore the alkene that forms C is the most reactive.

Now continue on to Problem 11

PROBLEM 11♦

Which carbocation in each of the following pairs is more stable?

a.

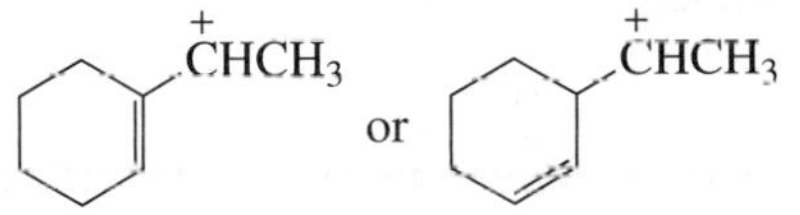

b. $CH_3\overset{+}{O}CH_2$ or $CH_3\overset{+}{N}HCH_2$

c.

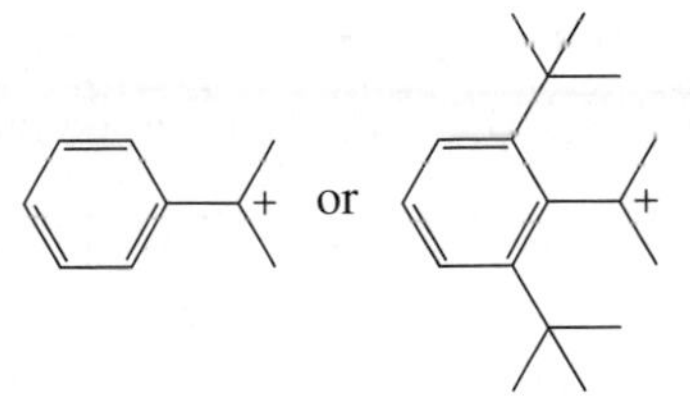

7.8 A Molecular Orbital Description of Stability

So far in our discussion we have used contributing resonance structures to show why compounds are stabilized by electron delocalization. The stabilization of compounds by electron delocalization can also be explained by molecular orbital (MO) theory.

We saw in Section 1.5 that the two lobes of a *p* orbital have opposite phases. We also saw that when two in-phase *p* orbitals overlap, a covalent bond is formed, and when two out-of-phase *p* orbitals overlap, they cancel each other and produce a node between the two nuclei (Section 1.6). Recall that a *node* is a region where there is zero probability of finding an electron.

Let's review how the π molecular orbitals of ethene are constructed. An MO description of ethene is shown in Figure 7.3. The two *p* orbitals can be either in-phase or out-of-phase. (The different phases are indicated by different colors.) Notice that the number of orbitals is conserved—the number of molecular orbitals equals the

Take a few minutes to review Section 1.6. See Special Topic VI in the *Study Guide and Solutions Manual* for additional information on molecular orbital theory.

number of atomic orbitals that produced them. Thus, the two atomic p orbitals of ethene overlap to produce two molecular orbitals. Side-to-side overlap of in-phase p orbitals (lobes of the same color) produces a **bonding molecular orbital** designated ψ_1 (the Greek letter psi). The bonding molecular orbital is lower in energy than the p atomic orbitals, and it encompasses both carbons. In other words, each electron in the bonding molecular orbital spreads over both carbon atoms.

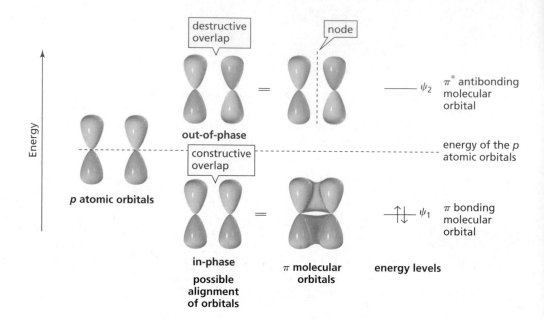

Figure 7.3 ▶
The distribution of electrons in ethene. Overlapping of in-phase p orbitals produces a bonding molecular orbital that is lower in energy than the p atomic orbitals. Overlapping of out-of-phase p orbitals produces an antibonding molecular orbital that is higher in energy than the p atomic orbitals.

Side-to-side overlap of out-of-phase p orbitals produces an **antibonding molecular orbital**, ψ_2, which is higher in energy than the p atomic orbitals. The antibonding molecular orbital has a node between the lobes of opposite phases. Recall that the overlap of in-phase orbitals holds atoms together; it is a bonding interaction. In contrast, the overlap of out-of-phase orbitals pulls atoms apart; it is an antibonding interaction.

The π electrons are placed in molecular orbitals according to the same rules that govern the placement of electrons in atomic orbitals (Section 1.2): the aufbau principle (orbitals are filled in order of increasing energy), the Pauli exclusion principle (each orbital can hold two electrons of opposite spin), and Hund's rule (an electron will occupy an empty degenerate orbital before it will pair up with an electron that is already present in an orbital).

1,3-Butadiene and 1,4-Pentadiene

The π electrons in 1,3-butadiene are delocalized over four sp^2 carbons (Section 7.7). In other words, there are four carbons in the π system.

$$\overset{\cdot\cdot}{\text{C}}\text{H}_2-\text{CH}=\text{CH}-\overset{+}{\text{C}}\text{H}_2 \quad \longleftrightarrow \quad \text{CH}_2=\text{CH}-\text{CH}=\text{CH}_2 \quad \longleftrightarrow \quad \overset{+}{\text{C}}\text{H}_2-\text{CH}=\text{CH}-\overset{\cdot\cdot}{\text{C}}\text{H}_2$$

1,3-butadiene
resonance contributors

$$\text{CH}_2\text{---}\text{CH}\text{---}\text{CH}\text{---}\text{CH}_2$$

resonance hybrid

A molecular orbital description of 1,3-butadiene is shown in Figure 7.4. Each of the four carbons contributes one p atomic orbital, and the four p atomic orbitals combine to produce four π molecular orbitals: ψ_1, ψ_2, ψ_3, and ψ_4, in order of increasing

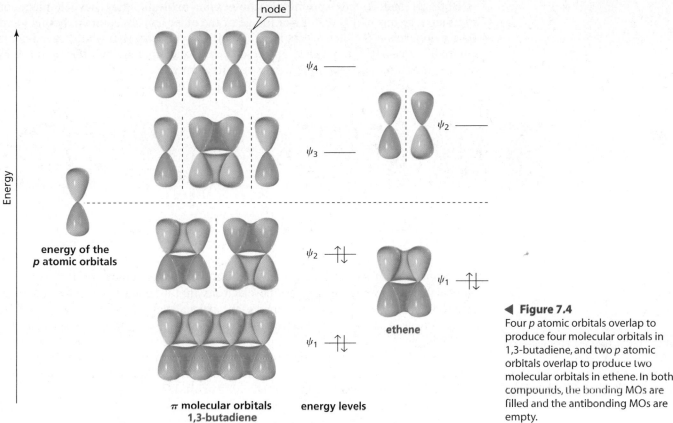

node

energy of the
p atomic orbitals

ψ_4 ___

ψ_3 ___

ψ_2 ___

ψ_1 ___

ψ_2 ⇅

ψ_1 ⇅

ethene

π **molecular orbitals**
1,3-butadiene

energy levels

◀ **Figure 7.4**
Four *p* atomic orbitals overlap to
produce four molecular orbitals in
1,3-butadiene, and two *p* atomic
orbitals overlap to produce two
molecular orbitals in ethene. In both
compounds, the bonding MOs are
filled and the antibonding MOs are
empty.

Student Tutorial:
Molecular orbitals

energy. Thus, we see that a molecular orbital results from the **linear combination of atomic orbitals (LCAO)**. Half of the MOs are π bonding MOs (ψ_1 and ψ_2), and the other half are π* antibonding MOs (ψ_3 and ψ_4). The energies of the bonding and antibonding MOs are symmetrically distributed above and below the energy of the *p* atomic orbitals.

Notice that as the MOs increase in energy, the number of nodes within them increases and the number of bonding interactions decreases. The lowest-energy MO (ψ_1) has only the node that bisects the *p* orbitals (it has no nodes between the nuclei because all the blue lobes overlap on one face of the molecule and all the green lobes overlap on the other face) and three bonding interactions; ψ_2 has one node between the nuclei and two bonding interactions (for a net result of one bonding interaction); ψ_3 has two nodes between the nuclei and one bonding interaction (for a net result of one antibonding interaction); and ψ_4 has three nodes between the nuclei—three antibonding interactions—and no bonding interactions. The four π electrons of 1,3-butadiene reside in ψ_1 and ψ_2.

1,3-Butadiene's lowest-energy MO (ψ_1) is particularly stable because it has three bonding interactions and its two electrons are delocalized over all four nuclei, encompassing all the carbons in the π system. The MO next in energy (ψ_2) is also a bonding MO because it has one more bonding interaction than antibonding interaction; it is not as strongly bonding or as low in energy as ψ_1. These two bonding MOs show that the greatest π electron density in a compound with two double bonds joined by one single bond is between C-1 and C-2 and between C-3 and C-4, but there is some π electron density between C-2 and C-3—just as the resonance contributors show. They also show why 1,3-butadiene is most stable in a planar conformation: if 1,3-butadiene weren't planar, there would be little or no overlap between C-2 and C-3. Overall, ψ_3 is an antibonding MO: it has one more antibonding interaction than bonding interaction, but it is not as strongly antibonding as ψ_4, which has no bonding interactions and three antibonding interactions.

Both ψ_1 and ψ_3 are **symmetric molecular orbitals**; they have a plane of symmetry, so one half is the mirror image of the other half. In contrast, both ψ_2 and ψ_4 are *antisymmetric*; they do not have a plane of symmetry (but would have one, if one half of the MO were turned upside down). Notice that as the MOs increase in energy, they alternate from symmetric to antisymmetric.

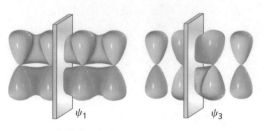

symmetric molecular orbitals

antisymmetric molecular orbitals

The energies of the MOs of 1,3-butadiene and ethene are compared in Figure 7.4. Notice that the average energy of the electrons in 1,3-butadiene is lower than the electrons in ethene. This lower energy is the delocalization energy. In other words, 1,3-butadiene is stabilized by electron delocalization.

The highest-energy molecular orbital of 1,3-butadiene that contains electrons is ψ_2. Therefore, ψ_2 is called the **highest occupied molecular orbital (HOMO)**. The lowest-energy molecular orbital of 1,3-butadiene that does not contain electrons is ψ_3; ψ_3 is called the **lowest unoccupied molecular orbital (LUMO)**.

HOMO = the highest occupied molecular orbital.

LUMO = the lowest unoccupied molecular orbital.

The molecular orbital description of 1,3-butadiene shown in Figure 7.4 represents the electronic configuration of the molecule in its *ground state*. If the molecule absorbs light of an appropriate wavelength, the light will promote an electron from its HOMO to its LUMO (from ψ_2 to ψ_3). The molecule then is in an *excited* state (Section 1.2). We will see that the excitation of an electron from the HOMO to the LUMO is the basis of ultraviolet and visible spectroscopy (Section 12.16).

PROBLEM 12◆

What is the total number of nodes in the ψ_3 and ψ_4 molecular orbitals of 1,3-butadiene?

PROBLEM 13◆

Answer the following questions for the π molecular orbitals of 1,3-butadiene:

a. Which are the bonding MOs and which are the antibonding MOs?
b. Which MOs are symmetric and which are antisymmetric?
c. Which MO is the HOMO and which is the LUMO in the ground state?
d. Which MO is the HOMO and which is the LUMO in the excited state?
e. What is the relationship between the HOMO and the LUMO and symmetric and antisymmetric orbitals?

Now let's look at the π molecular orbitals of 1,4-pentadiene.

$$CH_2{=}CHCH_2CH{=}CH_2$$
1,4-pentadiene

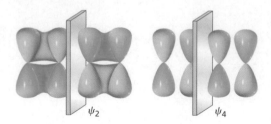

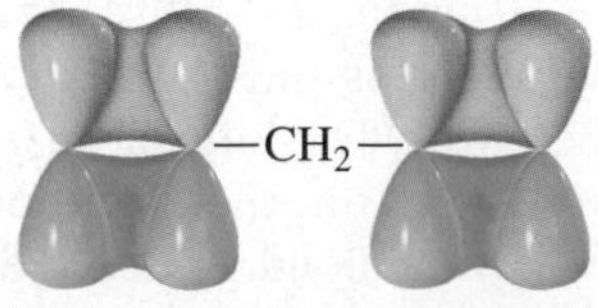

molecular orbitals of 1,4-pentadiene

1,4-Pentadiene, like 1,3-butadiene, has four π electrons. However, unlike the delocalized π electrons in 1,3-butadiene, the π electrons in 1,4-pentadiene are completely separate from one another. In other words, the electrons are localized. The molecular orbitals of 1,4-pentadiene have the same energy as those of ethene, a compound with one pair of localized π electrons. Thus, molecular orbital theory and contributing resonance structures are two different ways to show that the π electrons in 1,3-butadiene are delocalized and that electron delocalization stabilizes a molecule.

1,3,5-Hexatriene and Benzene

1,3,5-Hexatriene, with six carbon atoms, has six p atomic orbitals.

$$CH_2\!=\!CH\!-\!CH\!=\!CH\!-\!CH\!=\!CH_2$$
1,3,5-hexatriene

$$CH_2\!=\!\!=\!CH\!=\!\!=\!CH\!=\!\!=\!CH\!=\!\!=\!CH\!=\!\!=\!CH_2$$
resonance hybrid of
1,3,5-hexatriene

The six p atomic orbitals combine to produce six π molecular orbitals: ψ_1, ψ_2, ψ_3, ψ_4, ψ_5, and ψ_6 (Figure 7.5). Half of the MOs (ψ_1, ψ_2, and ψ_3) are bonding and the other half (ψ_4, ψ_5, and ψ_6) are antibonding. 1,3,5-Hexatriene's six π electrons occupy the three bonding MOs (ψ_1, ψ_2, and ψ_3), and two of the electrons (those in ψ_1) are delocalized over all six carbons. Thus, molecular orbital theory and resonance contributors are two

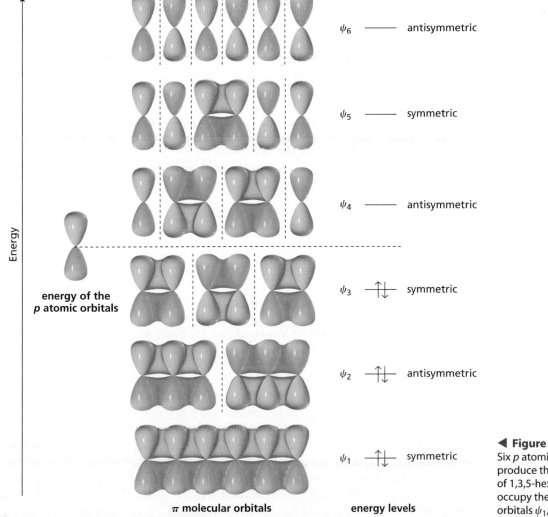

ψ_6 —— antisymmetric

ψ_5 —— symmetric

ψ_4 —— antisymmetric

energy of the
p atomic orbitals

ψ_3 symmetric

ψ_2 antisymmetric

ψ_1 symmetric

Energy

π **molecular orbitals**   **energy levels**

◀ **Figure 7.5**
Six p atomic orbitals overlap to produce the six π molecular orbitals of 1,3,5-hexatriene. The six electrons occupy the three bonding molecular orbitals ψ_1, ψ_2, and ψ_3.

different ways of showing that the π electrons in 1,3,5-hexatriene are delocalized. Notice in Figure 7.5 that as the MOs increase in energy, the number of nodes increases, the number of bonding interactions decreases, and the MOs alternate from symmetric to antisymmetric.

PROBLEM 14◆

Answer the following questions for the π molecular orbitals of 1,3,5-hexatriene:

a. Which are the bonding MOs and which are the antibonding MOs?
b. Which MOs are symmetric and which are antisymmetric?
c. Which MO is the HOMO and which is the LUMO in the ground state?
d. Which MO is the HOMO and which is the LUMO in the excited state?
e. What is the relationship between the HOMO and the LUMO and symmetric and antisymmetric orbitals?

Like 1,3,5-hexatriene, benzene has a six-carbon π system. The six-carbon π system in benzene, however, is cyclic. The six p atomic orbitals combine to produce six π molecular orbitals (Figure 7.6). Three of the MOs are bonding (ψ_1, ψ_2, and ψ_3) and three are antibonding (ψ_4, ψ_5, and ψ_6). Benzene's six π electrons occupy the three lowest-energy MOs (the bonding MOs). The method used to determine the relative energies of the MOs of compounds with cyclic π systems is described in Section 14.7.

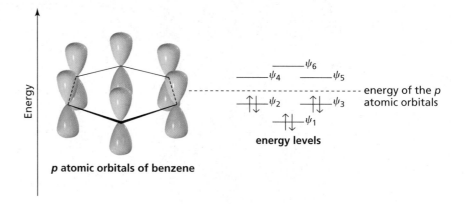

Figure 7.6 ▶
Benzene has six π molecular orbitals, three bonding (ψ_1, ψ_2, ψ_3) and three antibonding (ψ_4, ψ_5, ψ_6). The six π electrons occupy the three bonding molecular orbitals.

Figure 7.7 shows that there are *six* bonding interactions in the lowest-energy MO (ψ_1) of benzene—one more than in the lowest-energy MO of 1,3,5-hexatriene (Figure 7.5). In other words, when the three double bonds of benzene are arranged in a ring, the molecule's stability increases. The other two bonding MOs of benzene (ψ_2 and ψ_3) are degenerate: ψ_2 has four bonding interactions and two antibonding interactions, for a net result of two bonding interactions, and ψ_3 also has two bonding interactions. Thus, ψ_2 and ψ_3 are bonding MOs, but they are not as strongly bonding as ψ_1.

The energy levels of the MOs of ethene, 1,3-butadiene, 1,3,5-hexatriene, and benzene are compared in Figure 7.8. You can see that benzene is a particularly stable molecule—more stable than 1,3,5-hexatriene and much more stable than a molecule with one or more isolated double bonds. Compounds such as benzene that are unusually stable because of large delocalization energies are called **aromatic compounds**. Benzene's large delocalization energy is what prevents it from undergoing the addition reactions characteristic of alkenes, one of the observations that puzzled nineteenth-century chemists (Section 7.1). The structural features that make a compound aromatic will be discussed in Section 14.2.

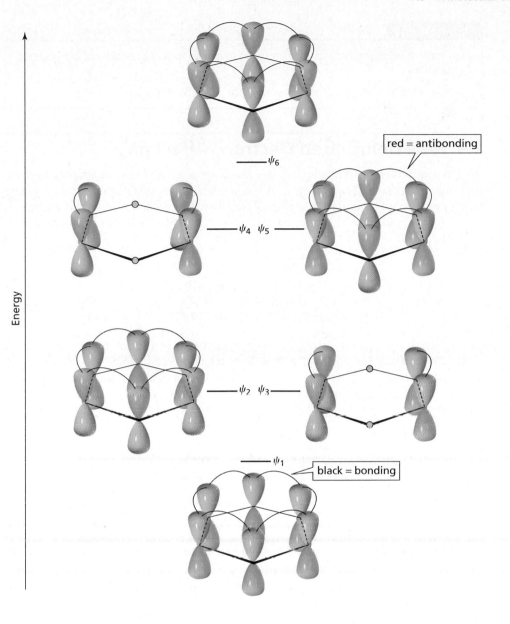

red = antibonding

ψ_6

$\psi_4 \quad \psi_5$

$\psi_2 \quad \psi_3$

ψ_1

black = bonding

◀ Figure 7.7
As the energy of the π molecular orbitals increases, the net number of bonding interactions decreases.

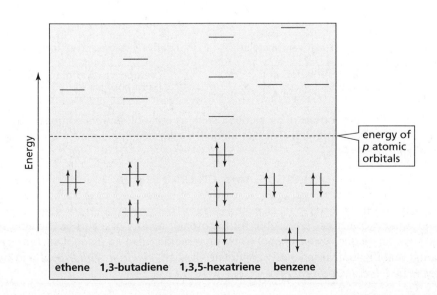

energy of p atomic orbitals

ethene   1,3-butadiene   1,3,5-hexatriene   benzene

◀ Figure 7.8
A comparison of the energy levels of the π molecular orbitals of ethene, 1,3-butadiene, 1,3,5-hexatriene, and benzene.

PROBLEM 15◆

How many bonding interactions are there in the ψ_1 and ψ_2 molecular orbitals of the following compounds?

a. 1,3-butadiene   **b.** 1,3,5,7-octatetraene

7.9 How Delocalized Electrons Affect pK_a

We have seen that a carboxylic acid is a much stronger acid than an alcohol because the conjugate base of a carboxylic acid is considerably more stable than the conjugate base of an alcohol (Section 1.22). (Recall that the more stable the base, the stronger is its conjugate acid.) For example, the pK_a of acetic acid is 4.76, whereas the pK_a of ethanol is 15.9.

$$
\begin{array}{cc}
\overset{\displaystyle O}{\underset{\displaystyle \parallel}{CH_3\overset{}{C}OH}} & CH_3CH_2OH \\
\textbf{acetic acid} & \textbf{ethanol} \\
\textbf{p}K_a = \textbf{4.76} & \textbf{p}K_a = \textbf{15.9}
\end{array}
$$

The difference in stability of the two conjugate bases is attributable to two factors. First, the carboxylate ion has a double-bonded oxygen atom that replaces the two hydrogens of the alkoxide ion. Electron withdrawal by this electronegative oxygen stabilizes the ion by decreasing the electron density of the negatively charged oxygen.

Electron withdrawal increases the stability of an anion.

$$
\begin{array}{cc}
\overset{\displaystyle O}{\underset{\displaystyle \parallel}{CH_3\overset{}{C}O^-}} & CH_3CH_2O^- \\
\textbf{a carboxylate ion} & \textbf{an alkoxide ion}
\end{array}
$$

The other factor responsible for the increased stability of the carboxylate ion is its *greater delocalization energy* relative to that of its conjugate acid. The carboxylate ion has greater delocalization energy because the ion has two equivalent resonance contributors that are predicted to be relatively stable, whereas the carboxylic acid has only one (Section 7.6). Therefore, loss of a proton from a carboxylic acid is accompanied by increased delocalization energy—in other words, an increase in stability.

relatively stable	**relatively unstable**	**relatively stable**	**relatively stable**
resonance contributors of a carboxylic acid		resonance contributors of a carboxylate ion	

In contrast, all the electrons in an alcohol, such as ethanol, and its conjugate base are localized, so loss of a proton from an alcohol is not accompanied by an increase in delocalization energy.

$$CH_3CH_2OH \;\rightleftharpoons\; CH_3CH_2O^- \;+\; H^+$$
ethanol

The same two factors responsible for the greater acidity of a carboxylic acid compared with an alcohol cause phenol to be more acidic than an alcohol such as cyclohexanol: stabilization of phenol's conjugate base by *electron withdrawal* and by an increase in *delocalization energy*.

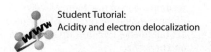
Student Tutorial:
Acidity and electron delocalization

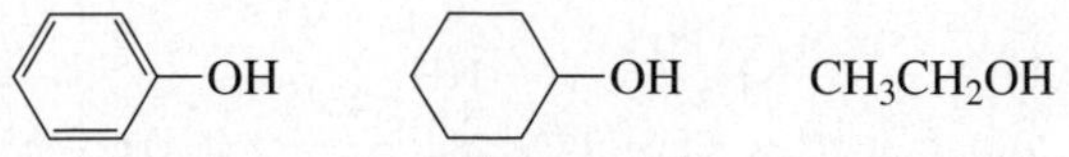

phenol	cyclohexanol	ethanol
pK_a = 10	pK_a = 16	pK_a = 16

The OH group of phenol is attached to an sp^2 carbon, which is more electronegative than the sp^3 carbon to which the OH group of cyclohexanol is attached (Section 1.20). The greater *electron withdrawal* by the sp^2 carbon stabilizes the conjugate base by decreasing the electron density of its negatively charged oxygen. While both phenol and the phenolate ion have delocalized electrons, the delocalization energy of the phenolate ion is greater than that of phenol because three of phenol's resonance contributors have separated charges. The loss of a proton from phenol, therefore, is accompanied by an increase in delocalization energy. Donation of electrons through π bonds is called **donation of electrons by resonance** or **resonance electron donation**.

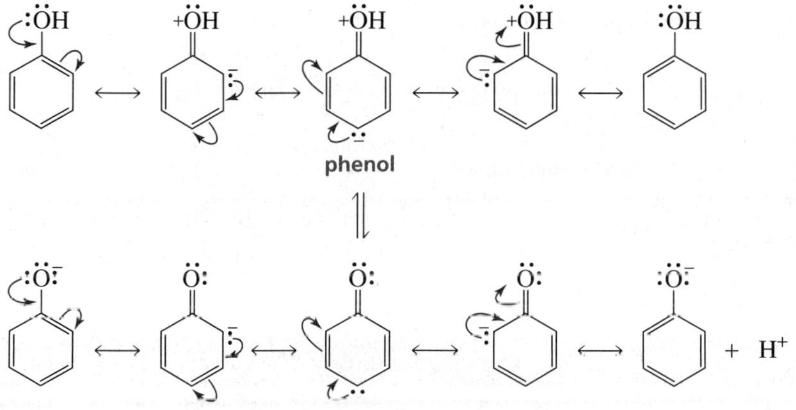

phenol

phenolate ion

In contrast, neither cyclohexanol nor its conjugate base has delocalized electrons to stabilize it.

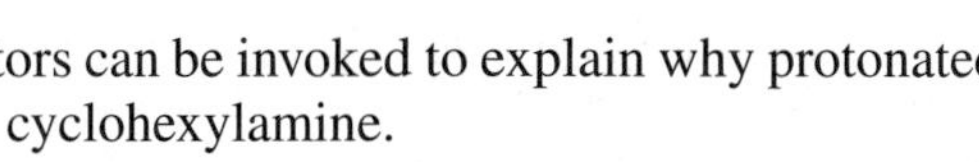

cyclohexanol

Phenol is a weaker acid than a carboxylic acid because electron withdrawal from the oxygen in the phenolate ion is not as great as in the carboxylate ion. In addition, the increased delocalization energy resulting from loss of a proton is not as great in a phenolate ion as in a carboxylate ion, where the negative charge is shared equally by two oxygens.

The same two factors can be invoked to explain why protonated aniline is a stronger acid than protonated cyclohexylamine.

protonated aniline	protonated cyclohexylamine
pK_a = 4.60	pK_a = 11.2

First, the nitrogen atom of aniline is attached to an sp^2 carbon, whereas the nitrogen atom of cyclohexylamine is attached to a less electronegative sp^3 carbon. Second, the nitrogen atom of protonated aniline lacks a lone pair that can be delocalized. However, when the nitrogen loses a proton, the lone pair that formerly held the proton can be delocalized. Loss of a proton, therefore, is accompanied by an increase in delocalization energy.

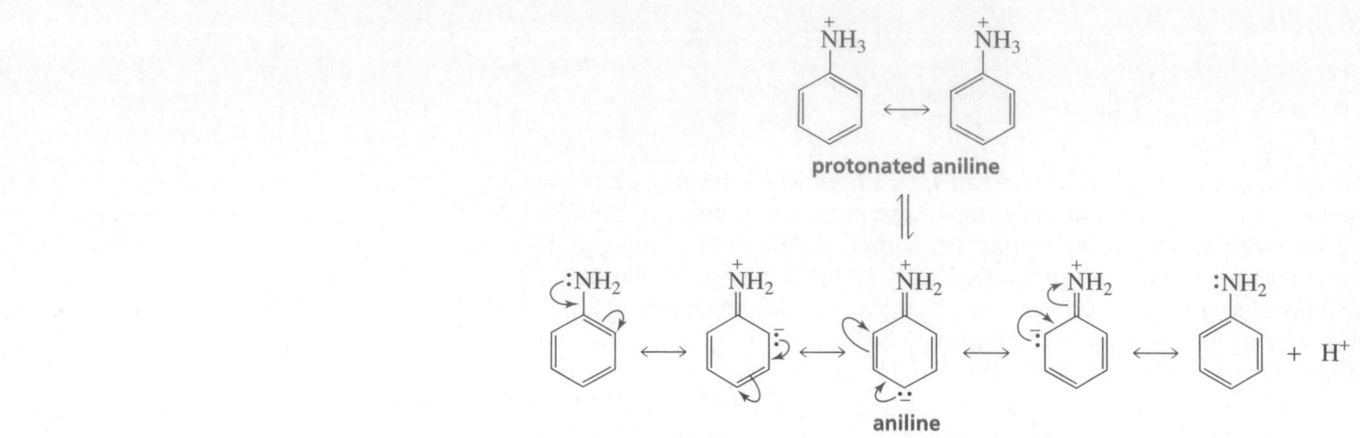

protonated aniline

aniline

An amine such as cyclohexylamine has no delocalized electrons either in the protonated form or in the unprotonated form to stabilize it.

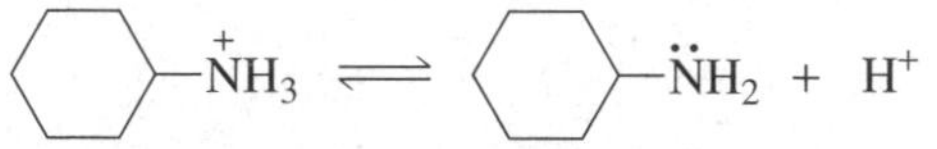

protonated cyclohexylamine   **cyclohexylamine**

We can now add phenol and protonated aniline to the list of organic compounds whose approximate pK_a values you should know (Table 7.2). They are also listed inside the back cover for easy reference.

Table 7.2	Approximate pK_a Values		
pK_a < 0	**pK_a ≈ 5**	**pK_a ≈ 10**	**pK_a ≈ 15**
$R\overset{+}{\underset{H}{O}}H$	$\overset{O}{\overset{\|}{R C O H}}$	$R\overset{+}{N}H_3$	ROH
$\overset{+OH}{\underset{R C O H}{\|}}$	⬡$-\overset{+}{N}H_3$	⬡$-OH$	H_2O
H_3O^+			

PROBLEM 16◆

Which is a stronger acid?

a. $CH_3CH_2CH_2OH$ or $CH_3CH{=}CHOH$

b. $\overset{O}{\overset{\|}{H C}}CH_2OH$ or $\overset{O}{\overset{\|}{CH_3 C}}OH$

c. $CH_3CH{=}CHCH_2OH$ or $CH_3CH{=}CHOH$

d. $CH_3CH_2CH_2\overset{+}{N}H_3$ or $CH_3CH{=}CH\overset{+}{N}H_3$

PROBLEM 17◆

Which is a stronger base?

a. ethylamine or aniline

b. ethylamine or ethoxide ion ($CH_3CH_2O^-$)

c. phenolate ion or ethoxide ion

PROBLEM 18◆

Rank the following compounds in order of decreasing acid strength:

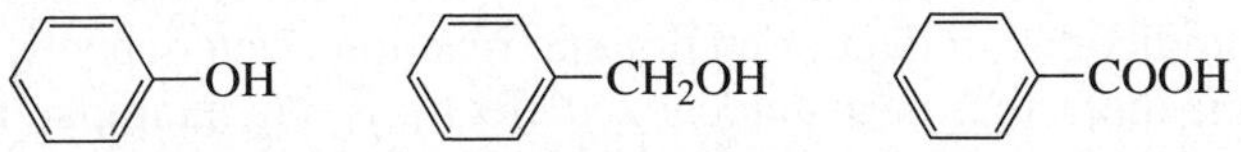

⬡—OH ⬡—CH$_2$OH ⬡—COOH

PROBLEM 19 *SOLVED*

Which of the following would you predict to be the stronger acid?

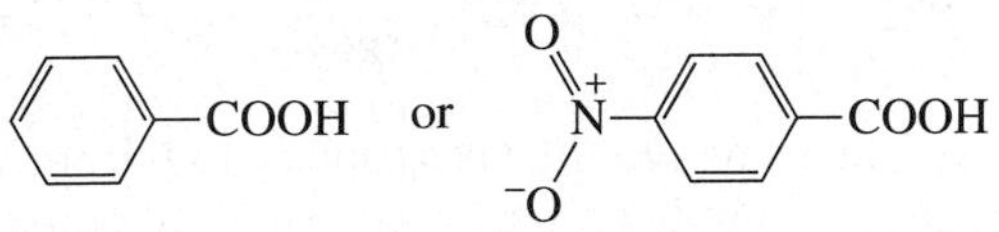

⬡—COOH or (O$_2$N)⬡—COOH

Solution The nitro-substituted compound is the stronger acid because the nitro substituent withdraws electrons inductively (through the σ bonds) and also **withdraws electrons by resonance** (through the π bonds). We have seen that electron-withdrawing substituents increase the acidity of a compound by stabilizing its conjugate base.

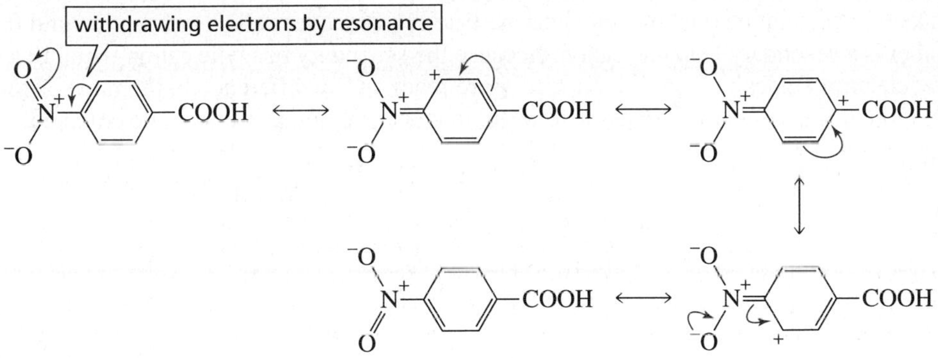

withdrawing electrons by resonance

PROBLEM 20◆

A methoxy substituent (CH$_3$O) attached to a benzene ring withdraws electrons inductively (through the σ bonds) because oxygen is more electronegative than carbon. The group also **donates electrons by resonance** (through the π bonds).

donating electrons by resonance

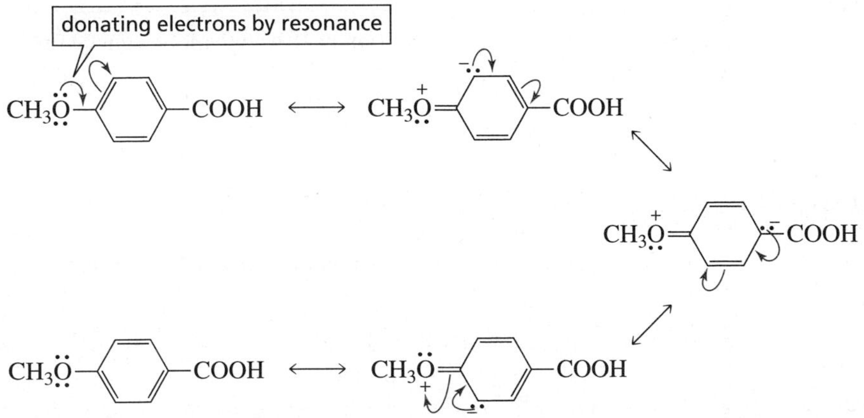

From the pK_a values of the unsubstituted and methoxy-substituted carboxylic acids, predict which is a more important effect, inductive electron withdrawal or resonance electron donation.

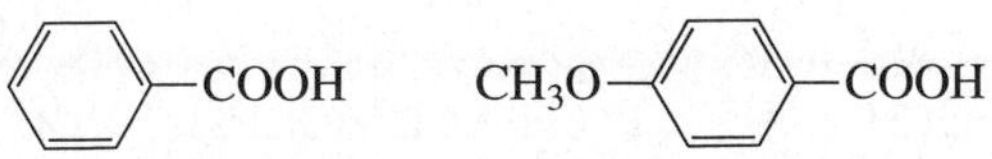

⬡—COOH   CH$_3$O—⬡—COOH

pK_a = 4.20   pK_a = 4.47

7.10 Delocalized Electrons Can Affect the Product of a Reaction

Our ability to predict the product of an organic reaction often depends upon recognizing when organic molecules have delocalized electrons. For example, in the following reaction, both sp^2 carbons of the alkene are bonded to the same number of hydrogens:

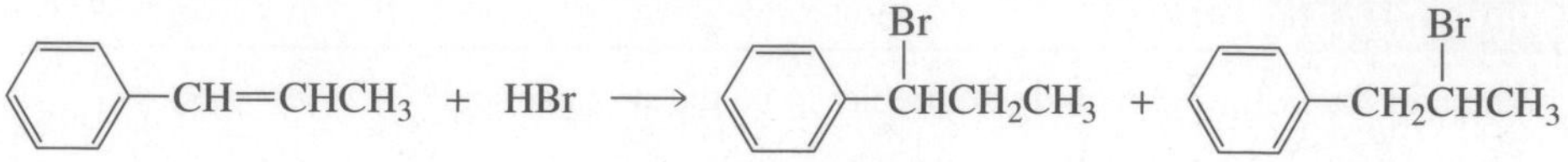

$$\text{Ph}-CH=CHCH_3 + HBr \longrightarrow \text{Ph}-\overset{\overset{\displaystyle Br}{|}}{C}HCH_2CH_3 + \text{Ph}-CH_2\overset{\overset{\displaystyle Br}{|}}{C}HCH_3$$

100%   0%

Therefore, the rule that tells us to add the electrophile to the sp^2 carbon bonded to the greater number of hydrogens predicts that approximately equal amounts of the two products will be formed. When the reaction is carried out, however, only one of the products is obtained.

The rule leads us to an incorrect prediction of the reaction product because it does not take electron delocalization into consideration. It presumes that both carbocation intermediates are equally stable since they are both secondary carbocations. The rule does not take into account the fact that one intermediate is a secondary alkyl carbocation and the other is a secondary benzylic cation. Because the secondary benzylic cation is stabilized by electron delocalization, it is formed more readily. The difference in the rates of formation of the two carbocations is sufficient to cause only one product to be obtained.

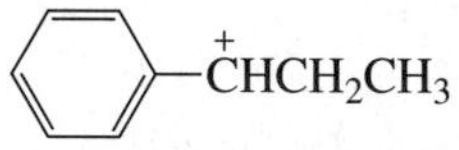

$$\text{Ph}-\overset{+}{C}HCH_2CH_3 \qquad\qquad \text{Ph}-CH_2\overset{+}{C}HCH_3$$

a secondary benzylic cation 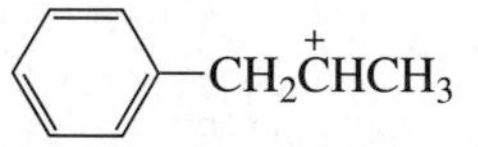   **a secondary carbocation**

This example serves as a warning. The rule describing the sp^2 carbon to which the electrophile becomes attached cannot be used for reactions involving carbocations that can be stabilized by electron delocalization. In such cases, you must look at the relative stabilities of the individual carbocations to predict the major product of the reaction.

PROBLEM 21◆

What is the major product obtained from the addition of HBr to the following compound?

$$\text{Ph}-CH_2CH=CH_2$$

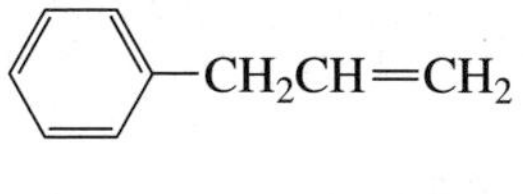

PROBLEM 22 | **SOLVED**

Predict the sites where protonation can occur on each of the following compounds.

a. $CH_3CH=CHOCH_3 + H^+$ **b.** 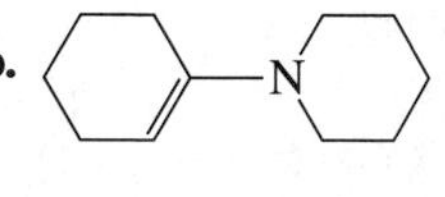 $\text{Ph}-N\text{(ring)} + H^+$

Solution to 22a The contributing resonance structures reveal that there are two sites that can be protonated: the lone pair on oxygen and the lone pair on carbon.

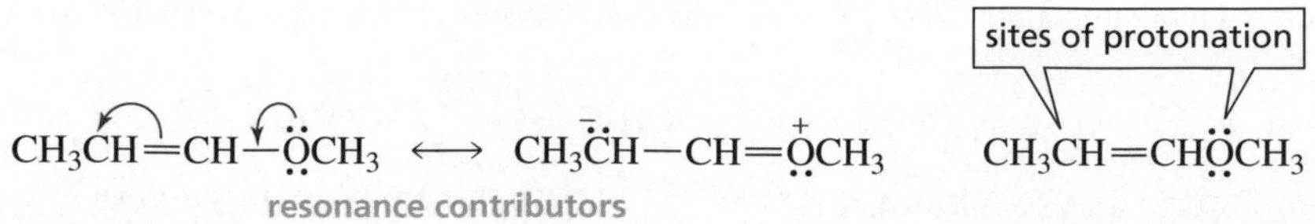

sites of protonation

$$CH_3\overset{\frown}{C}H=CH\overset{\frown}{-}\overset{..}{\underset{..}{O}}CH_3 \longleftrightarrow CH_3\overset{-}{C}H-CH=\overset{+}{\underset{..}{O}}CH_3 \qquad CH_3CH=CH\overset{..}{\underset{..}{O}}CH_3$$

resonance contributors

Let's now compare the products formed when *isolated dienes* (dienes that have only localized electrons) undergo electrophilic addition reactions to the products formed when *conjugated dienes* (dienes that have delocalized electrons) undergo the same reactions.

$$CH_2\!=\!CHCH_2CH_2CH\!=\!CH_2 \qquad CH_3CH_2\!=\!CH\!-\!CH\!=\!CHCH_3$$

an isolated diene   **a conjugated diene**

Reactions of Isolated Dienes

The reactions of *dienes with isolated double bonds* are like the reactions of alkenes. If an excess of the electrophilic reagent is present, two independent addition reactions will occur. In each reaction, the electrophile adds to the sp^2 carbon that is bonded to the greater number of hydrogens.

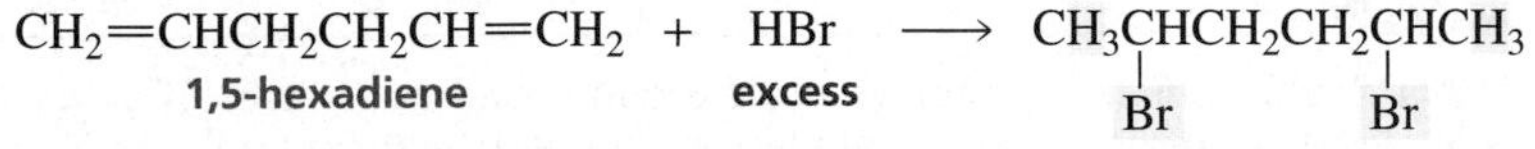

$$CH_2\!=\!CHCH_2CH_2CH\!=\!CH_2 \; + \; HBr \; \longrightarrow \; CH_3CHCH_2CH_2CHCH_3$$

1,5-hexadiene   **excess**   Br   Br

The reaction proceeds exactly as we would predict from our knowledge of the mechanism for the reaction of alkenes with electrophilic reagents.

Mechanism for the reaction of an isolated diene with excess HBr

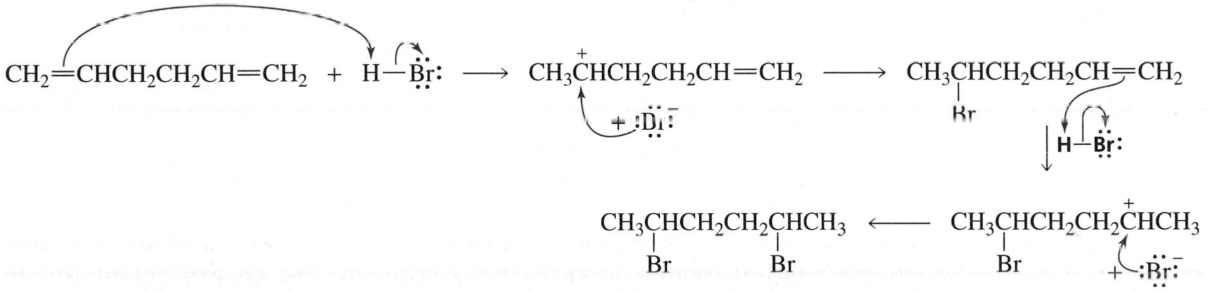

- The electrophile (H^+) adds to the electron-rich double bond in a manner that produces the more stable carbocation (Section 4.4).
- The bromide ion then adds to the carbocation.
- Because there is an excess of the electrophilic reagent, there is enough reagent to add to the other double bond.

If there is only enough electrophilic reagent to add to one of the double bonds, it will add preferentially to the more reactive double bond. For example, in the reaction of 2-methyl-1,5-hexadiene with HCl, addition of HCl to the double bond on the left forms a secondary carbocation, whereas addition of HCl to the double bond on the right forms a tertiary carbocation. Because the transition state leading to formation of a tertiary carbocation is more stable than that leading to a secondary carbocation, the tertiary carbocation is formed faster (Section 4.4). Therefore, in the presence of a limited amount of HCl, the major product of the reaction will be 5-chloro-5-methyl-1-hexene.

$$CH_2\!=\!CHCH_2CH_2\overset{\overset{\displaystyle CH_3}{|}}{C}\!=\!CH_2 \; + \; HCl \; \longrightarrow \; CH_2\!=\!CHCH_2CH_2\overset{\overset{\displaystyle CH_3}{|}}{\underset{\underset{\displaystyle Cl}{|}}{C}}CH_3$$

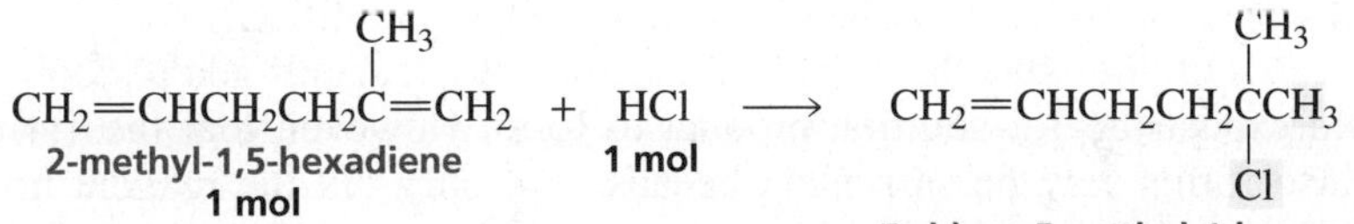

2-methyl-1,5-hexadiene   **1 mol**

1 mol

                                        5-chloro-5-methyl-1-hexene

                                             major product

PROBLEM 23

Give the major product of each of the following reactions, assuming that one equivalent of each reagent is used in every reaction.

a. CH_2=$CHCH_2CH_2CH$=$\overset{\overset{\displaystyle CH_3}{|}}{C}CH_3$ $\xrightarrow{\text{HBr}}$

b. HC≡CCH_2CH_2CH=CH_2 $\xrightarrow{\text{Cl}_2}$

c. 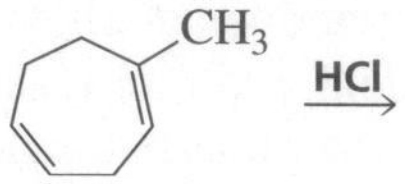 [structure: methylcycloheptadiene with CH$_3$] $\xrightarrow{\text{HCl}}$

PROBLEM 24◆

Which of the double bonds in zingiberene, the compound responsible for the odor of ginger, is the most reactive in an electrophilic addition reaction?

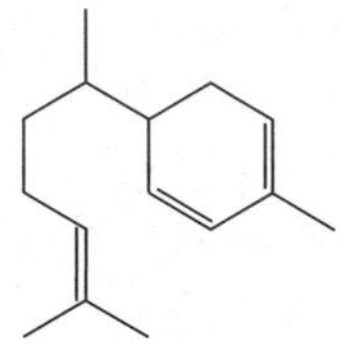

[structure of zingiberene]

zingiberene

Reactions of Conjugated Dienes

When a diene with *conjugated double bonds*, such as 1,3-butadiene, reacts with a limited amount of electrophilic reagent, so that addition can occur at only one of the double bonds, two addition products are formed. One is a **1,2-addition product**, which is a result of addition at the 1- and 2-positions. The other is a **1,4-addition product**, the result of addition at the 1- and 4-positions.

$$\overset{1}{CH_2}=\overset{2}{CH}-\overset{3}{CH}=\overset{4}{CH_2} + Cl_2 \longrightarrow CH_2-CH-CH=CH_2 + CH_2-CH=CH-CH_2$$

$$\underset{\text{1,3-butadiene}}{} \quad \underset{\text{1 mol}}{} \qquad \underset{Cl \quad Cl}{} \qquad \qquad \underset{Cl \qquad\qquad Cl}{}$$

1,3-butadiene   1 mol
1 mol

3,4-dichloro-1-butene   **1,4-dichloro-2-butene**
1,2-addition product   **1,4-addition product**

$$CH_2=CH-CH=CH_2 + HBr \longrightarrow CH_3CH-CH=CH_2 + CH_3-CH=CH-CH_2$$

$$\underset{\text{1,3-butadiene}}{} \quad \underset{\text{1 mol}}{} \qquad\qquad \underset{Br}{} \qquad\qquad\qquad \underset{Br}{}$$

1,3-butadiene   1 mol
1 mol

3-bromo-1-butene   **1-bromo-2-butene**
1,2-addition product   **1,4-addition product**

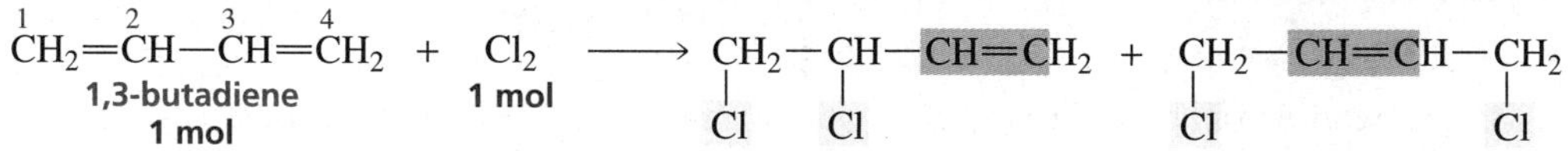

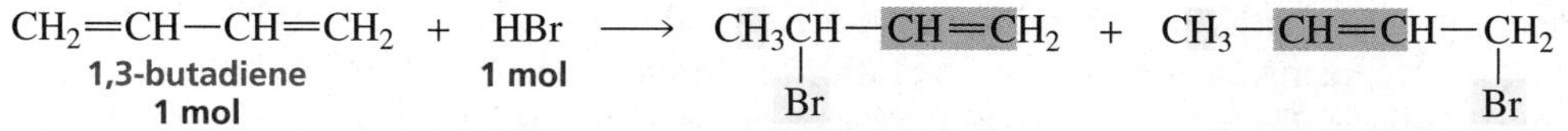

An isolated diene undergoes only 1,2-addition.

A conjugated diene undergoes both 1,2- and 1,4-addition.

Addition at the 1- and 2-positions is called **1,2-addition** or **direct addition**. Addition at the 1- and 4-positions is called **1,4-addition** or **conjugate addition**.

On the basis of your knowledge of how electrophilic reagents add to double bonds, you would expect the 1,2-addition product to form. However, that the 1,4-addition product also forms may be surprising because not only did the reagent not add to adjacent carbons, but a double bond has changed its position. The double bond in the

1,4-product is between the 2- and 3-positions, whereas the reactant had a single bond in this position.

When we talk about addition at the 1- and 2-positions or at the 1- and 4-positions, the numbers refer to the four carbons of the conjugated system. Thus, the carbon in the 1-position is one of the sp^2 carbons at the end of the conjugated system—it is not necessarily the first carbon in the molecule.

$$\overset{1}{R-CH}=\overset{2}{CH}-\overset{3}{CH}=\overset{4}{CH}-R$$

the conjugated system

$$CH_3CH=CH-CH=CHCH_3 \xrightarrow{\textbf{Br}_2} CH_3CH-CH-CH=CHCH_3 + CH_3CH-CH=CH-CHCH_3$$

| | |
| Br Br | Br Br |

2,4-hexadiene   **4,5-dibromo-2-hexene**   **2,5-dibromo-3-hexene**
1,2-addition product   **1,4-addition product**

To understand why an electrophilic addition reaction to a conjugated diene forms both 1,2-addition and 1,4-addition products, we need to look at the mechanism of the reaction.

Mechanism for the reaction of a conjugated diene with HBr

$$CH_2=CH-CH=CH_2 + H-\ddot{B}r: \longrightarrow CH_3-\overset{+}{CH}-CH-CH_2 \longleftrightarrow CH_3-CH=CH \overset{+}{CH_2}$$

1,3-butadiene $+ :\ddot{Br}:^-$ an allylic cation $+ :\ddot{Br}:^-$

$$\overset{+}{CH_2}-CH_2-CH=CH_2$$

a primary carbocation

$$CH_3-CH-CH=CH_2 + CH_3-CH=CH-CH_2$$

| | |
| Br | Br |

3-bromo-1-butene   **1-bromo-2-butene**
1,2-addition product   **1,4-addition product**

- The proton adds to C-1, forming an allylic cation. The π electrons of the allylic cation are delocalized, so that the positive charge is shared by two carbons. (Notice that because 1,3-butadiene is symmetrical, adding to C-1 is the same as adding to C-4.) The proton does not add to C-2 or C-3, because doing so would form a primary carbocation. The π electrons of a primary carbocation would be localized; thus, it is not as stable as the delocalized allylic cation.

- The contributing resonance structures of the allylic cation show that the positive charge on the carbocation is not localized on C-2 but is shared by C-2 and C-4. Consequently, the halide ion can attack either C-2 or C-4 to form the 1,2-addition product or the 1,4-addition product, respectively.

$$CH_3\overset{\delta+}{-CH}=\!\!=\!\!CH\overset{\delta+}{-CH_2}$$

As we look at more examples, notice that the first step in all electrophilic additions to conjugated dienes is addition of the electrophile to one of the sp^2 carbons at the end of the conjugated system. This is the only way to form a carbocation that is stabilized by electron delocalization. If the electrophile were to add to one of the internal sp^2 carbons, the resulting carbocation would have only localized electrons.

Diene Halogenation-1

PROBLEM 25◆

Give the products of the following reactions, assuming that one equivalent of each reagent is used in each reaction:

a. $CH_3CH{=}CH{-}CH{=}CHCH_3 \xrightarrow{Cl_2}$

b. 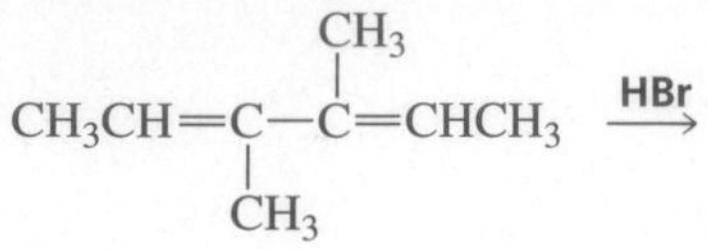 $CH_3CH{=}\overset{\overset{\displaystyle CH_3}{|}}{C}{-}\underset{\underset{\displaystyle CH_3}{|}}{C}{=}CHCH_3 \xrightarrow{HBr}$

c. [cyclopentadiene structure] $\xrightarrow{Br_2}$

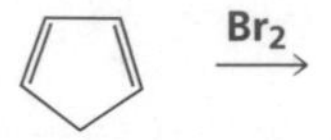

PROBLEM 26

What stereoisomers are obtained from the reactions shown on pages 316 and 317. (*Hint:* Review Section 5.19.)

Let's review the information in this section by comparing the result of electrophilic addition to an isolated diene with the result of electrophilic addition to a conjugated diene. The carbocation formed by addition of an electrophile to an isolated diene is not stabilized by resonance. The positive charge is localized on a single carbon, so only direct (1,2-) addition occurs.

addition to an isolated diene

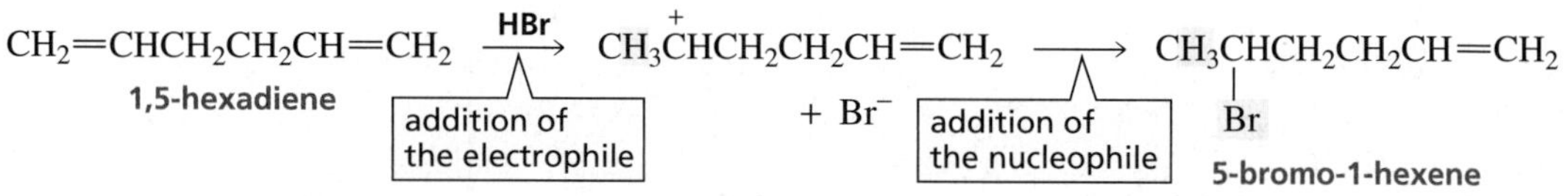

$$CH_2{=}CHCH_2CH_2CH{=}CH_2 \xrightarrow{HBr} \overset{+}{CH_3}CHCH_2CH_2CH{=}CH_2 \longrightarrow CH_3\underset{\underset{\displaystyle Br}{|}}{CH}CH_2CH_2CH{=}CH_2$$

1,5-hexadiene

addition of the electrophile $+ Br^-$ addition of the nucleophile   **5-bromo-1-hexene**

The carbocation formed by addition of an electrophile to a conjugated diene, in contrast, is stabilized by electron delocalization. The positive charge is shared by two carbons, and as a result, both direct (1,2-) and conjugate (1,4-) addition occur.

addition to a conjugated diene

the carbocation is stabilized by electron delocalization

$$CH_3CH{=}CH{-}CH{=}CHCH_3 \xrightarrow{HBr} CH_3CH_2{-}\overset{+}{CH}{-}CH{=}CHCH_3 \longleftrightarrow CH_3CH_2{-}CH{=}CH{-}\overset{+}{CH}CH_3$$

2,4-hexadiene

addition of the electrophile $+ Br^-$ $+ Br^-$ addition of the nucleophile

$$CH_3CH_2{-}\underset{\underset{\displaystyle Br}{|}}{CH}{-}CH{=}CHCH_3 \qquad CH_3CH_2{-}CH{=}CH{-}\underset{\underset{\displaystyle Br}{|}}{CH}CH_3$$

4-bromo-2-hexene   **2-bromo-3-hexene**
1,2-addition product   **1,4-addition product**

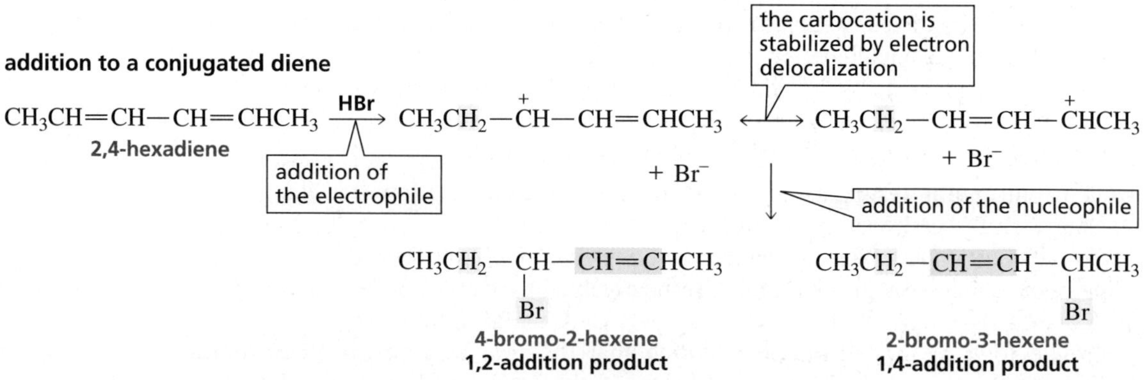

If the conjugated diene is not symmetrical, the major products of the reaction are those obtained by adding the electrophile to whichever terminal sp^2 carbon results in formation of the more stable carbocation. For example, in the reaction of 2-methyl-1,3-butadiene with HBr, the proton adds preferentially to C-1 because the positive charge on the resulting carbocation is shared by a tertiary allylic and a primary allylic carbon. Adding the proton to C-4 would form a carbocation in which the positive charge is shared by a secondary allylic and a primary allylic carbon. Because addition to C-1 forms the more stable carbocation, 3-bromo-3-methyl-1-butene and 1-bromo-3-methyl-2-butene are the major products of the reaction.

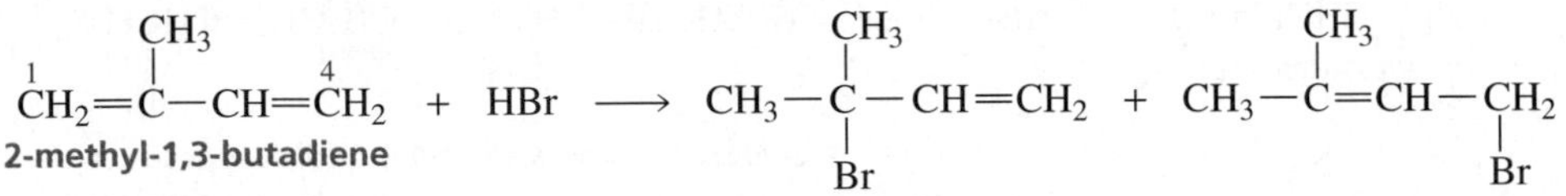

$$CH_2{=}\overset{\overset{\displaystyle CH_3}{|}}{\underset{\displaystyle 4}{C}}{-}CH{=}\overset{4}{C}H_2 \ + \ HBr \ \longrightarrow \ CH_3{-}\overset{\overset{\displaystyle CH_3}{|}}{\underset{\underset{\displaystyle Br}{|}}{C}}{-}CH{=}CH_2 \ + \ CH_3{-}\overset{\overset{\displaystyle CH_3}{|}}{C}{=}CH{-}\underset{\underset{\displaystyle Br}{|}}{C}H_2$$

2-methyl-1,3-butadiene   **3-bromo-3-methyl-1-butene**   **1-bromo-3-methyl-2-butene**

Etherification-2

$$CH_3\overset{\overset{\displaystyle CH_3}{|}}{\underset{+}{C}}{-}CH{=}CH_2 \ \longleftrightarrow \ CH_3\overset{\overset{\displaystyle CH_3}{|}}{C}{=}CH{-}\underset{+}{C}H_2$$

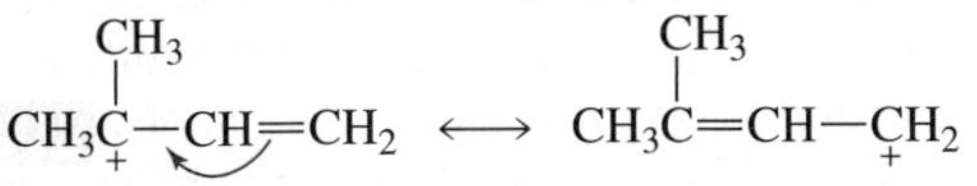

carbocation formed by adding H⁺ to C-1

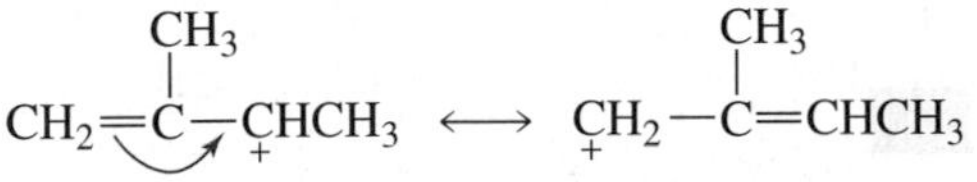

$$CH_2{=}\overset{\overset{\displaystyle CH_3}{|}}{\underset{+}{C}}{-}CHCH_3 \ \longleftrightarrow \ \underset{+}{C}H_2{-}\overset{\overset{\displaystyle CH_3}{|}}{C}{=}CHCH_3$$

carbocation formed by adding H⁺ to C-4

PROBLEM 27

What products would be obtained from the reaction of 1,3,5- hexatriene with one equivalent of HBr? Disregard stereoisomers.

PROBLEM 28

Give the products of the following reactions, disregarding stereoisomers. (One equivalent of each reagent is used in each reaction.)

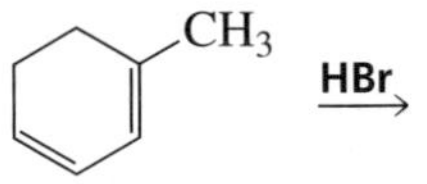

a. 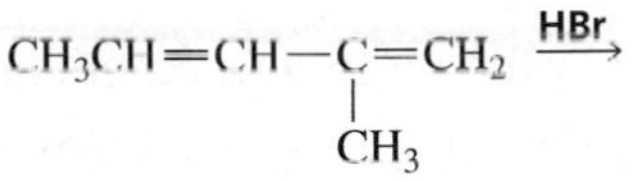 $CH_3CH{=}CH{-}\overset{\overset{\displaystyle CH_3}{|}}{C}{=}CH_2 \ \xrightarrow{\ HBr\ }$

b. [cyclohexadiene with CH₃ substituent] $\xrightarrow{\ HBr\ }$

c. $CH_3CH{=}CH{-}\overset{\overset{\displaystyle CH_3}{|}}{C}{=}CHCH_3 \ \xrightarrow{\ HBr\ }$

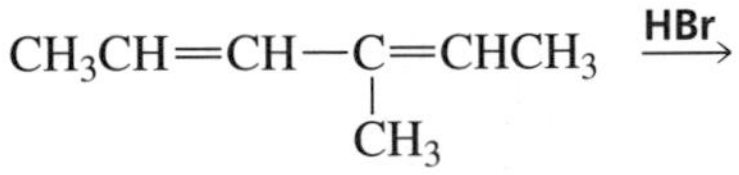

7.11 Thermodynamic Versus Kinetic Control of Reactions

When a conjugated diene undergoes an electrophilic addition reaction, two factors—the temperature at which the reaction is carried out and the structure of the reactant—determine whether the 1,2-addition product or the 1,4-addition product will be the major product of the reaction.

When a reaction produces more than one product, *the most rapidly formed product* is called the **kinetic product**, and *the most stable product* is called the **thermodynamic product**. Reactions that produce the kinetic product as the major product are said to be *kinetically controlled*. Reactions that produce the thermodynamic product as the major product are said to be *thermodynamically controlled*.

The kinetic product is the product that is formed most rapidly.

The thermodynamic product is the most stable product.

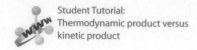

Student Tutorial:
Thermodynamic product versus
kinetic product

In many organic reactions, the most stable product is also the one that is formed most rapidly. In other words, the kinetic product and the thermodynamic product are one and the same. Electrophilic addition to 1,3-butadiene is an example of a reaction in which the kinetic product and the thermodynamic product are *not* the same: the 1,2-addition product is the kinetic product, and the 1,4-addition product is the thermodynamic product.

$$CH_2=CHCH=CH_2 \ + \ HBr \ \longrightarrow \ CH_3CHCH=CH_2 \ + \ CH_3CH=CHCH_2$$

1,3-butadiene
$\quad\quad\quad\quad\quad\quad\quad\quad$ Br $\quad\quad\quad\quad\quad\quad\quad$ Br

$\quad\quad\quad\quad\quad\quad\quad\quad$ **1,2-addition product** $\quad\quad$ **1,4-addition product**
$\quad\quad\quad\quad\quad\quad\quad\quad\quad$ **kinetic product** $\quad\quad\quad$ **thermodynamic product**

The kinetic product predominates when the reaction is irreversible.

For a reaction in which the kinetic and thermodynamic products are not the same, the product that predominates depends on the conditions under which the reaction is carried out. If the reaction is carried out under conditions that are sufficiently mild (low in temperature) to cause the reaction to be *irreversible*, the major product will be the *kinetic product*. For example, when addition of HBr to 1,3-butadiene is carried out at −80 °C, the major product is the *1,2-addition product*, the product that is formed more rapidly.

$$CH_2=CHCH=CH_2 \ + \ HBr \ \xrightarrow{-80\ °C} \ CH_3CHCH=CH_2 \ + \ CH_3CH=CHCH_2$$

$\quad\quad\quad\quad\quad\quad\quad\quad\quad\quad\quad\quad\quad\quad\quad$ Br $\quad\quad\quad\quad\quad\quad\quad$ Br

$\quad\quad\quad\quad\quad\quad\quad\quad\quad\quad$ **1,2-addition product** $\quad\quad$ **1,4-addition product**
$\quad\quad\quad\quad\quad\quad\quad\quad\quad\quad\quad\quad$ **80%** $\quad\quad\quad\quad\quad\quad\quad$ **20%**

The thermodynamic product predominates when the reaction is reversible.

If, on the other hand, the reaction is carried out under conditions that are sufficiently vigorous (high in temperature) to cause the reaction to be *reversible*, the major product will be the *thermodynamic product*. When the same reaction is carried out at 45 °C, the major product is the *1,4-addition product*—the product that is more stable.

$$CH_2=CHCH=CH_2 \ + \ HBr \ \underset{}{\overset{45\ °C}{\rightleftharpoons}} \ CH_3CHCH=CH_2 \ + \ CH_3CH=CHCH_2$$

$\quad\quad\quad\quad\quad\quad\quad\quad\quad\quad\quad\quad\quad\quad\quad$ Br $\quad\quad\quad\quad\quad\quad\quad$ Br

$\quad\quad\quad\quad\quad\quad\quad\quad\quad\quad$ **1,2-addition product** $\quad\quad$ **1,4-addition product**
$\quad\quad\quad\quad\quad\quad\quad\quad\quad\quad\quad\quad$ **15%** $\quad\quad\quad\quad\quad\quad\quad$ **85%**

A reaction coordinate diagram helps explain why different products predominate under different reaction conditions (Figure 7.9). The first step of the addition reaction

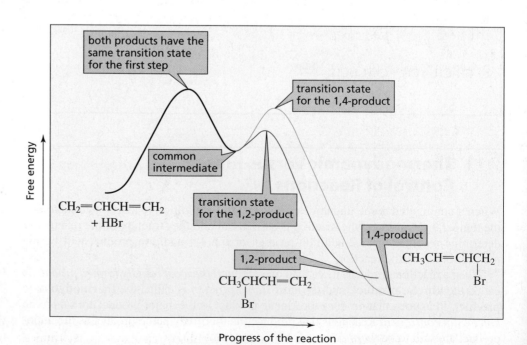

Figure 7.9 ▶
A reaction coordinate diagram for the addition of HBr to 1,3-butadiene.

is the same whether the 1,2-addition product or the 1,4-addition product is ultimately formed: a proton adds to C-1. The second step of the reaction is the one that determines whether the nucleophile (Br⁻) attacks C-2 or C-4. Because the 1,2-addition product forms more rapidly, we know that the transition state for its formation is more stable than the transition state for formation of the 1,4-addition product. This is the first time we have seen a reaction in which the less stable product has the more stable transition state!

At low temperatures (−80 °C), there is enough energy for the reactants to overcome the energy barrier for the first step of the reaction, and there is enough energy for the intermediate formed in the first step to form the two addition products. However, there is not enough energy for the reverse reaction to occur: the products cannot overcome the large energy barriers separating them from the intermediate. Consequently, the relative amounts of the two products obtained at −80 °C reflect the relative energy barriers for the second step of the reaction. The energy barrier to formation of the 1,2-addition product is lower than the energy barrier to formation of the 1,4-addition product, so the major product is the 1,2-addition product.

In contrast, at 45 °C, there is enough energy for one or more of the products to go back to the intermediate. The intermediate is called a **common intermediate** because it is an intermediate that both products have in common. The ability to return to a common intermediate allows the products to interconvert. When two products can interconvert, their relative amounts at equilibrium depend on their relative stabilities.

Thus, a reaction that is *irreversible* under the conditions employed in the experiment, is said to be under *kinetic control*. When a reaction is under **kinetic control**, the relative amounts of the products *depend on the rates* at which they are formed.

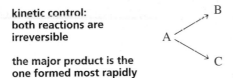

kinetic control:
both reactions are
irreversible

the major product is the
one formed most rapidly

A reaction is said to be under *thermodynamic control* when sufficient energy is available to make the reaction *reversible*. When a reaction is under **thermodynamic control**, the relative amounts of the products *depend on their stabilities*. Because a reaction must be reversible to be under thermodynamic control, thermodynamic control is also called **equilibrium control**.

thermodynamic control:
one or both reactions
are reversible

the major product is the
one that is the most stable

For each reaction that is irreversible under mild conditions and reversible under more vigorous conditions, there is a temperature at which the change from irreversible to reversible occurs. The temperature at which a reaction changes from being kinetically controlled to being thermodynamically controlled depends on the reactants. For example, the reaction of 1,3-butadiene with HCl remains under kinetic control at 45 °C, even though addition of HBr to 1,3-butadiene is under thermodynamic control at that temperature. Because a C—Cl bond is stronger than a C—Br bond (Table 3.2), a higher temperature is required for the products containing a C—Cl bond to undergo the reverse reaction. (Remember, thermodynamic control is achieved only when there is sufficient energy to reverse one or both of the reactions.)

For the reaction of 1,3-butadiene with HBr, why is the 1,4-addition product more stable? We saw in Section 4.11 that the relative stability of an alkene is determined by the number of alkyl groups bonded to its sp^2 carbons: the greater the number of alkyl groups, the more stable is the alkene. The two products formed from the reaction of 1,3-butadiene with one equivalent of HBr have different stabilities since the 1,2-addition product has one alkyl group bonded to its sp^2 carbons, whereas the 1,4-product has two alkyl groups bonded to its sp^2 carbons. The 1,4-addition product, therefore, is more stable. Thus, it is the thermodynamic product.

CH$_3$CHCH=CH$_2$
|
Br
1,2-addition product
kinetic product

CH$_3$CH=CHCH$_2$
|
Br
1,4-addition product
thermodynamic product

Diene Halogenation-2

The next question we need to answer is, why is the 1,2-addition product formed faster? In other words, why is the transition state for formation of the 1,2-addition product more stable than the transition state for formation of the 1,4-addition product? For many years, chemists thought it was because the transition state for formation of the 1,2-addition product resembles the contributing resonance structure in which the positive charge is on a secondary allylic carbon. In contrast, the transition state for formation of the 1,4-addition product resembles the contributing resonance structure in which the positive charge is on a less stable primary allylic carbon.

| secondary allylic carbon | | primary allylic carbon |

CH$_2$=CHCH=CH$_2$ →(HBr) CH$_3$\overset{+}{C}HCH=CH$_2$ ⟷ CH$_3$CH=CH\overset{+}{C}H$_2$

+ Br$^-$ + Br$^-$

$$\left[\begin{array}{c} \overset{\delta+}{CH_3CHCH=CH_2} \\ \overset{\delta-}{:Br:} \end{array} \right]^{\ddagger}$$

transition state for formation of the 1,2-addition product

$$\left[\begin{array}{c} \overset{\delta+}{CH_3CH=CHCH_2} \\ \overset{\delta-}{:Br:} \end{array} \right]^{\ddagger}$$

transition state for formation of the 1,4-addition product

However, when the reaction of 1,3-pentadiene with DCl is carried out under kinetic control, essentially the same relative amounts of 1,2- and 1,4-addition products are obtained as are obtained from the kinetically controlled reaction of 1,3-butadiene with HBr. If the original hypothesis were correct, the amounts of the two products obtained from the reaction of 1,3-pentadiene would be the same, because the transition states for formation of the 1,2- and 1,4-addition products from 1,3-pentadiene would both have the same stability (because both resemble a contributing resonance structure in which the positive charge is on a secondary allylic carbon). Why, then, is the 1,2-addition product formed faster in this reaction as well?

CH$_2$=CHCH=CHCH$_3$ + DCl →(−78 °C) CH$_2$CHCH=CHCH$_3$ + CH$_2$CH=CHCHCH$_3$
1,3 pentadiene | | | |
                                            D Cl   D   Cl
                                   1,2-addition product   **1,4-addition product**
                                            78%   **22%**

When the π electrons of the diene abstract D$^+$ from a molecule of undissociated DCl, the chloride ion can more easily stabilize a positive charge at C-2 than at

C-4 simply because when the chloride ion is first produced, it is closer to C-2 than to C-4. So it is a *proximity effect* that causes the 1,2-addition product to be formed faster. A **proximity effect** is an effect caused by one species being close to another.

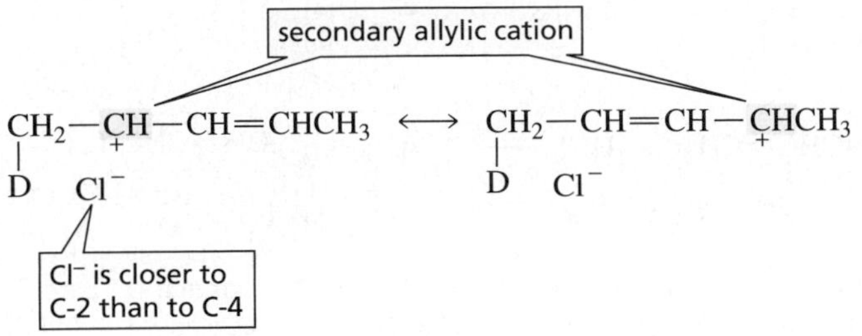

Because the greater proximity of the nucleophile to C-2 contributes to the faster rate of formation of the 1,2-addition product, the 1,2-addition product is the kinetic product for essentially all conjugated dienes. Do *not* assume, however, that the 1,4-addition product is *always* the thermodynamic product. The structure of the conjugated diene is what ultimately determines the thermodynamic product. For example, the 1,2-addition product is both the kinetic product and the thermodynamic product in the reaction of 4-methyl-1,3-pentadiene with HBr, because not only is the 1,2-product formed faster, it is more stable than the 1,4-product.

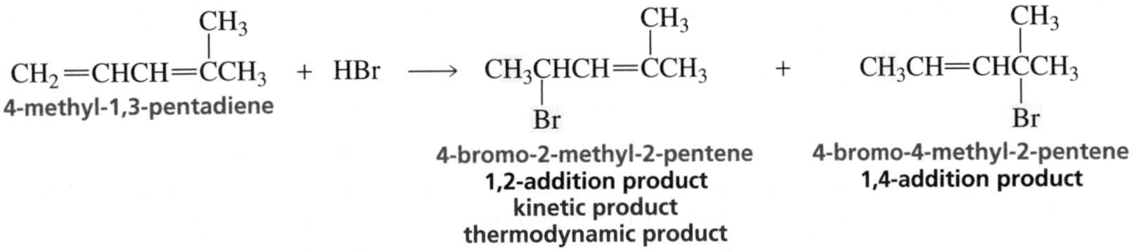

PROBLEM 29◆

a. Why does deuterium add to C-1 rather than to C-4 in the preceding reaction?

b. Why was DCl rather than HCl used in the reaction?

PROBLEM 30◆

a. When HBr adds to a conjugated diene, what is the rate-determining step?

b. When HBr adds to a conjugated diene, what is the product-determining step?

The 1,2- and 1,4-addition products obtained from the reaction of 2,4-hexadiene with HCl have the same stability, because both have the same number of alkyl groups bonded to their sp^2 carbons. Thus, neither product is thermodynamically controlled.

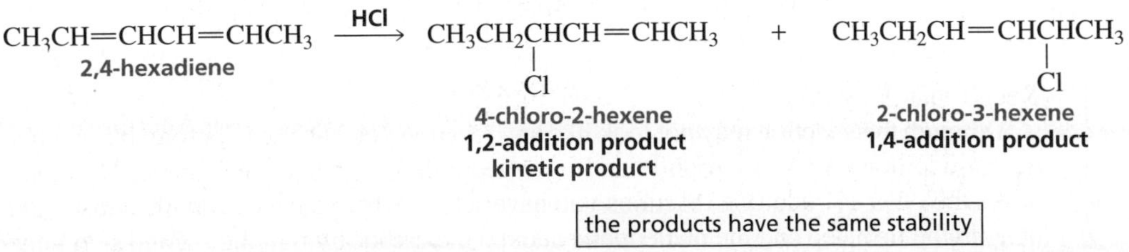

PROBLEM 31 **SOLVED**

For each of the following reactions, (1) give the major 1,2- and 1,4-addition products and (2) indicate which is the kinetic product and which is the thermodynamic product:

a.

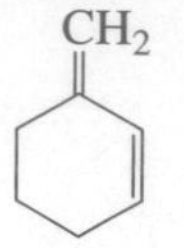

CH$_2$ ring + HCl ⟶

c.

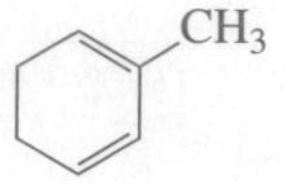

CH$_3$ benzene ring + HCl ⟶

b. CH$_3$CH=CHC=CH$_2$ + HCl ⟶
     |
     CH$_3$

d.

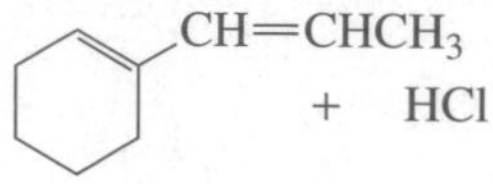

ring with CH=CHCH$_3$ + HCl ⟶

Solution to 31a First we need to determine which of the terminal sp^2 carbons of the conjugated system is to be the C-1 carbon. The proton will be more apt to add to the sp^2 carbon indicated below because the carbocation that then is formed shares its positive charge with a tertiary allylic and a secondary allylic carbon. If the proton were to add to the sp^2 carbon at the other end of the conjugated system, the carbocation that would be formed would be less stable because its positive charge is shared by a primary allylic and a secondary allylic carbon. Therefore, 3-chloro-3-methylcyclohexene is the 1,2-addition product and 3-chloro-1-methylcyclohexene is the 1,4-addition product. 3-Chloro-3-methylcyclohexene is the kinetic product because of the chloride ion's proximity to C-2, and 3-chloro-1-methyl-cyclohexene is the thermodynamic product because its more highly substituted double bond makes it more stable.

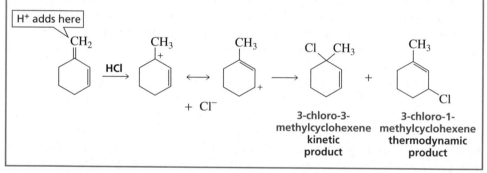

3-chloro-3-
methylcyclohexene
**kinetic
product**

3-chloro-1-
methylcyclohexene
**thermodynamic
product**

7.12 The Diels–Alder Reaction Is a 1,4-Addition Reaction

Reactions that create new carbon–carbon bonds are very important to synthetic organic chemists because it is only through such reactions that small carbon skeletons can be converted into larger ones (Section 6.11). The Diels–Alder reaction is particularly important because it creates *two* new carbon–carbon bonds and in the process forms a cyclic molecule. In recognition of the importance of this reaction to synthetic organic chemistry, Otto Diels and Kurt Alder shared the Nobel Prize in chemistry in 1950.

In a **Diels–Alder reaction**, a conjugated diene reacts with a compound containing a carbon–carbon double bond. The latter compound is called a **dienophile** because it "loves a diene."

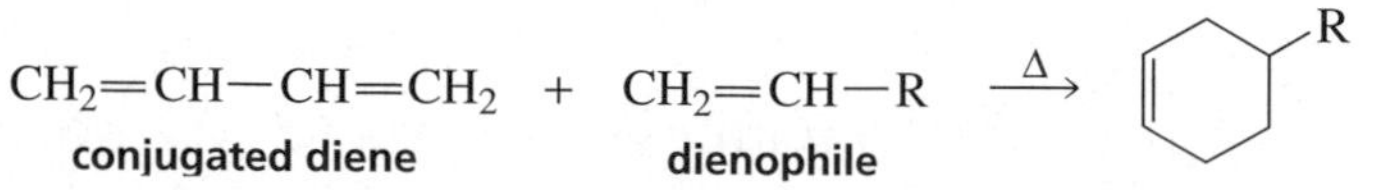

CH$_2$=CH—CH=CH$_2$ + CH$_2$=CH—R $\xrightarrow{\Delta}$ [cyclohexene ring with R]
conjugated diene   **dienophile**

(Recall that Δ signifies heat.)

Although this reaction may not look like any reaction you have seen before, it is simply the 1,4-addition of an electrophile and a nucleophile to a conjugated diene. However, unlike the other 1,4-addition reactions you have seen—where the electrophile adds to the diene in the first step and the nucleophile adds to the carbocation in the second step—the

Diels–Alder reaction is a **concerted reaction**: the addition of the electrophile and the nucleophile occurs in a single step. The reaction looks odd at first glance because the electrophile and the nucleophile that add to the conjugated diene are the adjacent sp^2 carbons of a double bond. As with other 1,4-addition reactions, the double bond in the product is between C-2 and C-3 of what was the conjugated diene.

> The Diels–Alder reaction is a 1,4-addition of a dienophile to a conjugated diene.

nucleophile	electrophile

new σ bond

diene
four π electrons

dienophile
two π electrons

transition state
six π electrons

new **double bond**   new σ bond

The Diels–Alder reaction is a pericyclic reaction. A **pericyclic reaction** is a reaction that takes place in one step by a cyclic shift of electrons. It is also a cycloaddition reaction. A **cycloaddition reaction** is a reaction in which two reactants form a cyclic product. More precisely, the Diels–Alder reaction is a **[4 + 2] cycloaddition reaction** because, of the six π electrons participating in the cyclic transition state, *four* come from the conjugated diene and *two* come from the dienophile. The reaction, in essence, converts two π bonds into two σ bonds.

The reactivity of the dienophile is increased if one or more electron-withdrawing groups are attached to its sp^2 carbons.

nucleophile	electron-withdrawing group

30 °C

a 1,4-addition reaction to 1,3-butadiene

An electron-withdrawing group, such as a carbonyl group (C—O) or a cyano (C≡N) group, withdraws electrons from the double bond. This puts a partial positive charge on one of its sp^2 carbons, making the Diels–Alder reaction easier to initiate (Figure 7.10).

$$CH_2{=}CH{-}CCH_3 \longleftrightarrow {}^{+}CH_2{-}CH{=}CCH_3$$

resonance contributors of the dienophile

$$\overset{\delta^+}{CH_2}{=\!=}CH{=\!=}\overset{\overset{\delta^-}{O}}{\underset{\|}{C}}CH_3$$

resonance hybrid

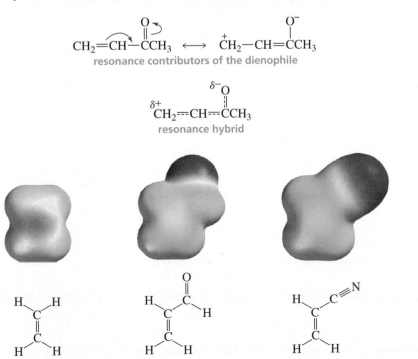

B I O G R A P H Y

Kurt Alder (1902–1958) *was born in a part of Germany that is now Poland. After World War I, he and his family moved to Germany, having been expelled from their home region when it was ceded to Poland. After receiving his Ph.D. under Diels in 1926, Alder continued working with him, and in 1928 they discovered the Diels–Alder reaction. Alder was a professor of chemistry at the University of Kiel and at the University of Cologne. He received the 1950 Nobel Prize in chemistry, sharing it with his mentor, Otto Diels. (Alder is seen accepting his Nobel Prize on the right in the photo.)*

◀ **Figure 7.10**
By comparing these electrostatic potential maps you can see that an electron-withdrawing substituent decreases the electron density of the carbon–carbon double bond.

The partially positively charged sp^2 carbon of the dienophile can be likened to the electrophile that is attacked by π electrons from C-1 of the conjugated diene. The other sp^2 carbon of the dienophile is the nucleophile that adds to C-4 of the diene.

A Molecular Orbital Description of the Diels–Alder Reaction

The two new σ bonds formed in a Diels–Alder reaction result from a transfer of electron density between the reactants. Molecular orbital theory provides insight into this process. In a cycloaddition reaction, the orbitals of one reactant must overlap the orbitals of the second reactant. Because the new σ bonds in the product are formed by donation of electron density from one reactant to the other, we must consider the HOMO of one reactant and the LUMO of the other, because only an empty orbital can accept electrons (Section 7.11). It does not matter whether we envision the overlap as between the HOMO of the dienophile and the LUMO of the diene or vice versa. We just need to consider the HOMO of one and the LUMO of the other.

To construct the HOMO and LUMO needed to illustrate the transfer of electrons in a Diels–Alder reaction, we need to look back at Figures 7.3 and 7.4. They show that the HOMO of the diene and the LUMO of the dienophile are antisymmetric (Figure 7.11a) and that the LUMO of the diene and the HOMO of the dienophile are symmetric (Figure 7.11b).

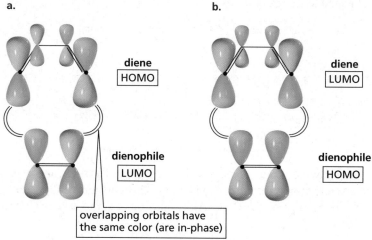

Figure 7.11 ▶
The new σ bonds formed in a Diels–Alder reaction result from overlap of in-phase orbitals. (a) Overlap of the HOMO of the diene and the LUMO of the dienophile. (b) Overlap of the HOMO of the dienophile and the LUMO of the diene.

   Diels Alder-1

Pericyclic reactions, such as the Diels–Alder reaction, can be described by a theory called the *conservation of orbital symmetry*. This simple theory says that pericyclic reactions occur as a result of the overlap of in-phase orbitals. The phase of the orbitals in Figure 7.11 is indicated by their color. Thus, each new σ bond formed in a Diels–Alder reaction must be created by the overlap of orbitals of the same color. Because two new σ bonds are formed, we need to have four orbitals in the correct place and with the correct symmetry (color). The figure shows that, regardless of which pair of HOMO and LUMO we choose, the overlapping orbitals have the same color. In other words, a Diels–Alder reaction occurs with relative ease. The Diels–Alder reaction and other cycloaddition reactions will be discussed in greater detail in Section 29.4.

A wide variety of cyclic compounds can be obtained by varying the structures of the conjugated diene and the dienophile. Notice that compounds containing carbon–carbon triple bonds can also be used as dienophiles in Diels–Alder reactions to prepare compounds with two isolated double bonds.

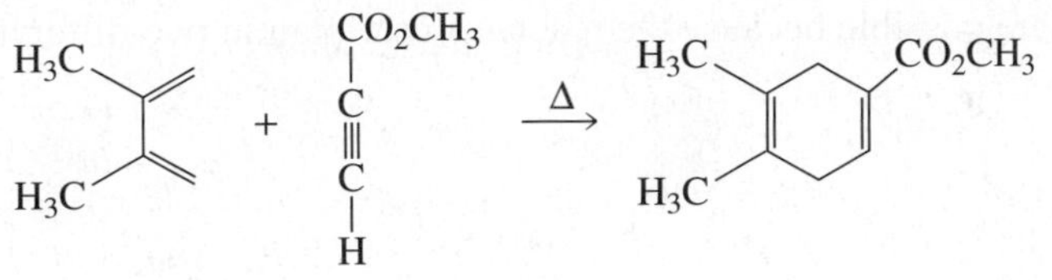

If the dienophile has two carbon–carbon double bonds, two successive Diels–Alder reactions can occur if excess diene is available.

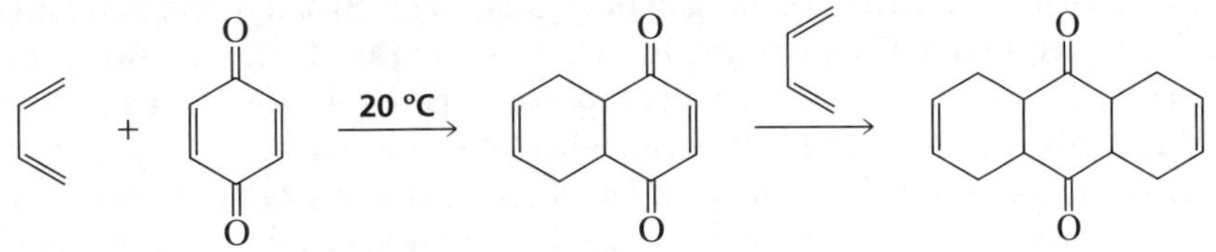

PROBLEM 32

Explain why a [2 + 2] cycloaddition reaction will not occur without the absorption of light. Recall that unless molecules absorb light, they have a ground-state electronic configuration; that is, the electrons are in the available orbitals with the lowest energy.

PROBLEM 33♦

Give the products of each of the following reactions:

a. $CH_2=CH-CH=CH_2$ + $CH_3\overset{O}{\overset{\|}{C}}-C\equiv C-\overset{O}{\overset{\|}{C}}CH_3$ $\xrightarrow{\Delta}$

b. $CH_2=CH-CH=CH_2$ + $HC\equiv C-C\equiv N$ $\xrightarrow{\Delta}$

c. 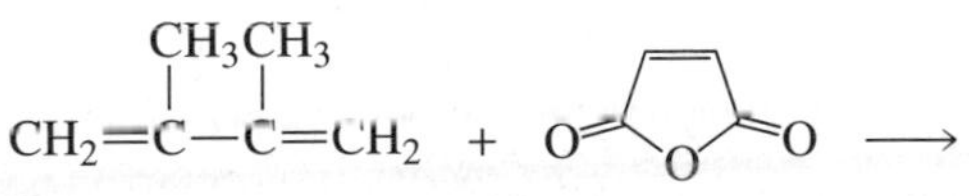 $CH_2=\overset{CH_3}{\underset{|}{C}}-\overset{CH_3}{\underset{|}{C}}=CH_2$ + $\longrightarrow$

d. 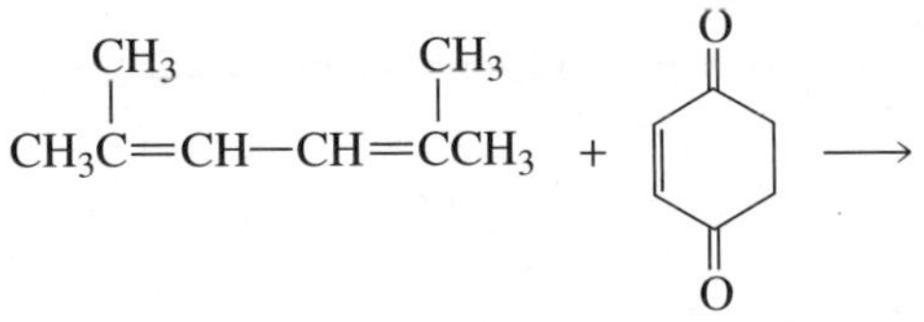 $CH_3\overset{CH_3}{\underset{|}{C}}=CH-CH=\overset{CH_3}{\underset{|}{C}}CH_3$ +

VCL   **Diels Alder-2**

Predicting the Product When Both Reagents Are Unsymmetrically Substituted

In each of the preceding Diels–Alder reactions, only one product is formed (disregarding stereoisomers) because at least one of the reacting molecules is symmetrically substituted. If both the diene and the dienophile are unsymmetrically substituted, however, two products are possible. The products are constitutional isomers.

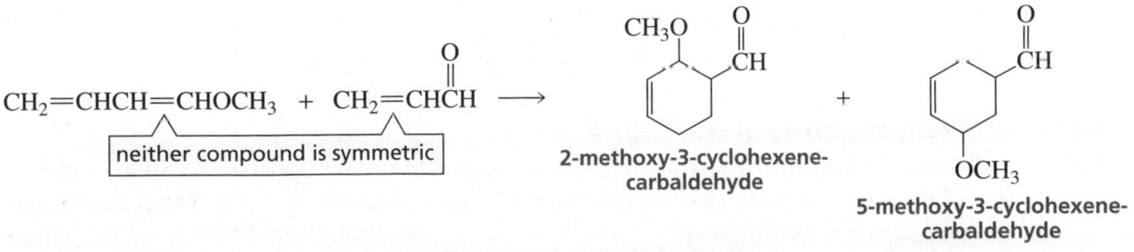

$CH_2=CHCH=CHOCH_3$ + $CH_2=CH\overset{O}{\overset{\|}{C}}H$ $\longrightarrow$

neither compound is symmetric

2-methoxy-3-cyclohexene-carbaldehyde

+

5-methoxy-3-cyclohexene-carbaldehyde

Diels Alder-3

Two products are possible because the reactants can align in two different ways.

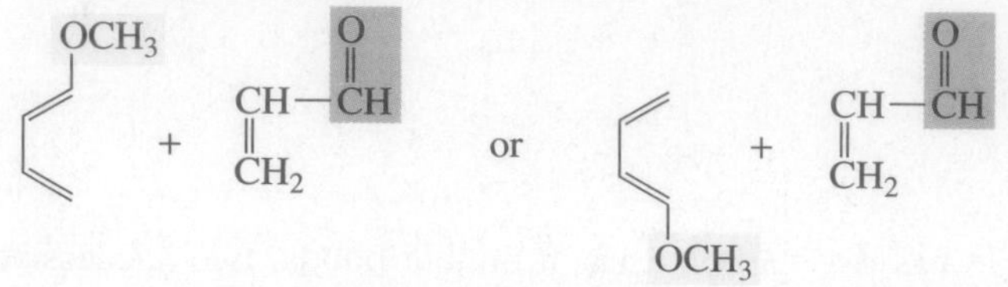

Which of the two products will be formed (or will be formed in greater yield) depends on the charge distribution in each of the reactants. To determine the charge distribution, we need to draw contributing resonance structures. In the reaction presented above, the methoxy group of the diene is capable of *donating electrons by resonance* (see Problem 20). As a result, the terminal carbon atom bears a partial negative charge. The aldehyde group of the dienophile, on the other hand, *withdraws electrons by resonance* (see Problem 19), so its terminal carbon has a partial positive charge.

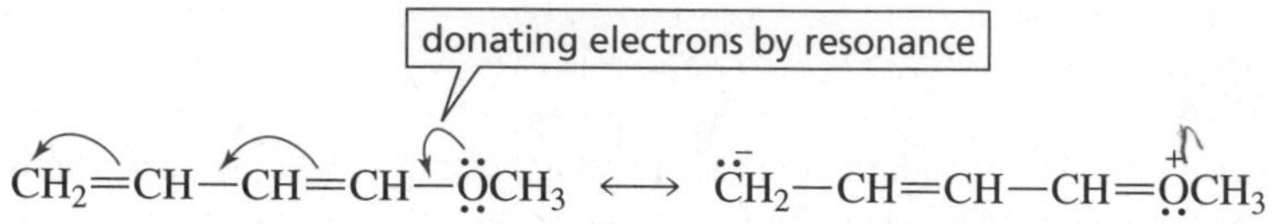

donating electrons by resonance

$CH_2=CH-CH=CH-\ddot{O}CH_3 \longleftrightarrow \ddot{C}H_2-CH=CH-CH=\overset{+}{O}CH_3$

resonance contributors of the diene

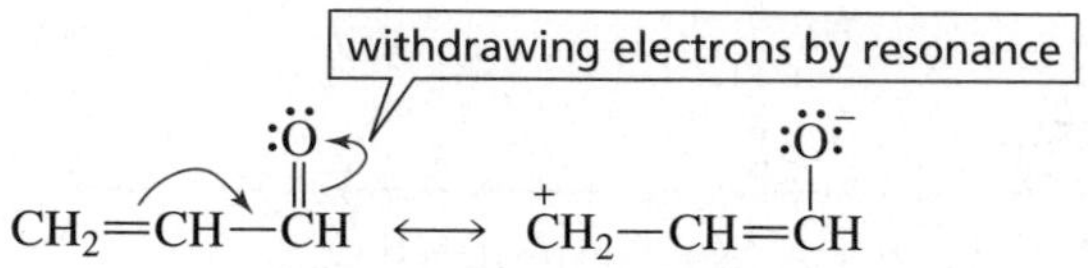

withdrawing electrons by resonance

$CH_2=CH-CH \longleftrightarrow \overset{+}{C}H_2-CH=CH$

resonance contributors of the dienophile

The partially positively charged carbon atom of the dienophile will bond preferentially to the partially negatively charged carbon of the diene. Therefore, 2-methoxy-3-cyclohexenecarbaldehyde will be the major product.

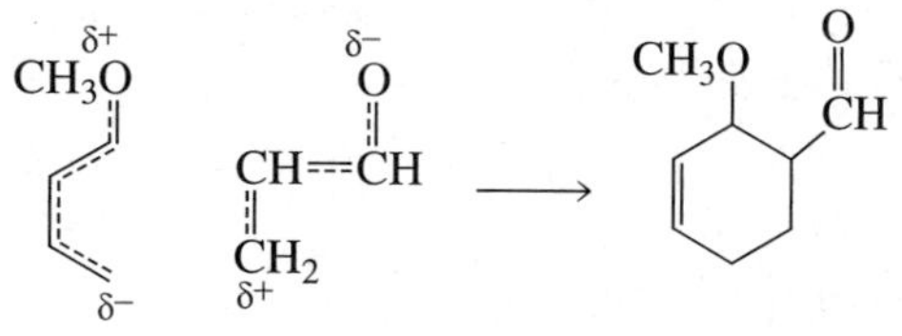

Diels Alder-4

PROBLEM 34◆

What would be the major product if the methoxy substituent in the preceding reaction were bonded to C-2 of the diene rather than to C-1?

PROBLEM 35◆

Give the products of each of the following reactions:

a. $CH_2=CH-CH=CH-CH_3 + HC\equiv C-C\equiv N \xrightarrow{\Delta}$

b. $CH_2=CH-\underset{\underset{CH_3}{|}}{C}=CH_2 + HC\equiv C-C\equiv N \xrightarrow{\Delta}$

Conformations of the Diene

We saw in Section 7.8 that a conjugated diene such as 1,3-butadiene is most stable in a planar conformation. A conjugated diene can exist in two different planar conformations: an *s*-cis conformation and an *s*-trans conformation. (Recall that a conformation results from rotation about single bonds; Section 2.10.) In the **s-cis conformation**, the double bonds are cis about the single bond (s = single), whereas they are trans about the single bond in the **s-trans conformation**. The s-trans conformation is little more stable (by 2.3 kcal/mol or 9.6 kJ/mol) than the s-cis conformation because the close proximity of the hydrogens causes some steric strain (Section 2.10). The rotational barrier between the s-cis and s-trans conformations is low enough (4.9 kcal/mol or 20.5 kJ/mole) to allow them to interconvert rapidly at room temperature.

s-trans conformation   **s-cis conformation**

In order to participate in a Diels–Alder reaction, the conjugated diene must be in an s-cis conformation because in an s-trans conformation, the number-1 and number-4 carbons are too far apart to react with the dienophile. A conjugated diene that is locked in an s-trans conformation cannot undergo a Diels–Alder reaction.

locked in an
s-trans conformation

$$+ \quad \overset{CH_2}{\underset{CHCO_2CH_3}{\|}} \quad \longrightarrow \quad \text{no reaction}$$

A conjugated diene that is locked in an s-cis conformation, such as 1,3-cyclopentadiene, is highly reactive in a Diels–Alder reaction. When the diene is a cyclic compound, the product of a Diels–Alder reaction is a **bridged bicyclic compound**—a compound that contains two rings that share two nonadjacent carbons.

locked in an
s-cis conformation

both rings share these carbons

1,3-cyclopentadiene

$$+ \quad \overset{CH_2}{\underset{CHCO_2CH_3}{\|}} \quad \longrightarrow$$

81%   **19%**

bridged bicyclic compounds

There are two possible configurations for bridged bicyclic compounds, because the substituent (R) can point away from the double bond (the **exo** configuration) or toward the double bond (the **endo** configuration).

endo   exo

points away from
the double bond

points toward
the double bond

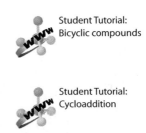

Student Tutorial:
Bicyclic compounds

Student Tutorial:
Cycloaddition

The endo product is formed faster when the dienophile has a substituent with π electrons. Recent studies suggest that the increased rate of endo product formation is due to steric and electrostatic effects.

When the dienophile has π electrons (other than the π electrons of its carbon–carbon double bond), more of the endo product is formed.

PROBLEM 36◆

Which of the following conjugated dienes would not react with a dienophile in a Diels–Alder reaction?

a. 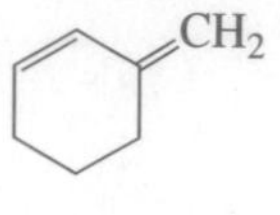 [structure with CH₂]   c. 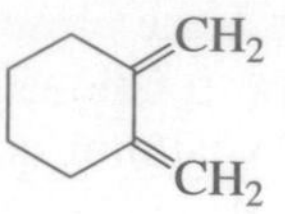 [structure with CH₂ and CH₂]   e. 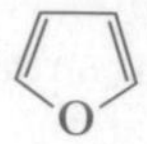 [furan structure with O]

b. 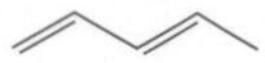 [diene structure]   d. 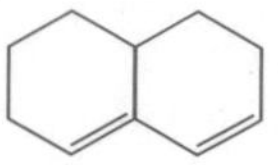 [fused ring structure]   f. 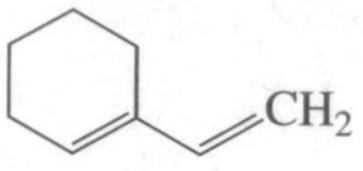 [ring structure with CH₂]

PROBLEM 37 **SOLVED**

List the following dienes in order of decreasing reactivity in a Diels–Alder reaction:

[structures: H₃C diene with H; cyclopentane with =CH₂; ring with CH₃ and H; cyclopentadiene]

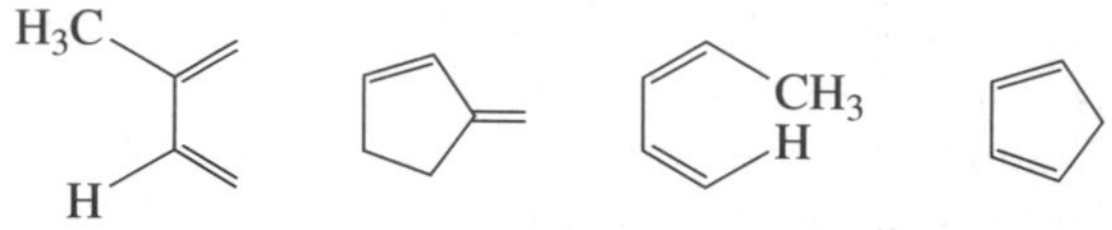

Solution The most reactive diene has the double bonds locked in an *s*-cis conformation, whereas the least reactive diene cannot achieve the required *s*-cis conformation because it is locked in an *s*-trans conformation. 2-Methyl-1,3-butadiene and 1,3-pentadiene are of intermediate reactivity because they can exist in both *s*-cis and *s*-trans conformations. 1,3-Pentadiene is less apt to be in the required *s*-cis conformation because of steric interference between the hydrogen and the methyl group. Consequently, 1,3-pentadiene is less reactive than 2-methyl-1,3-butadiene.

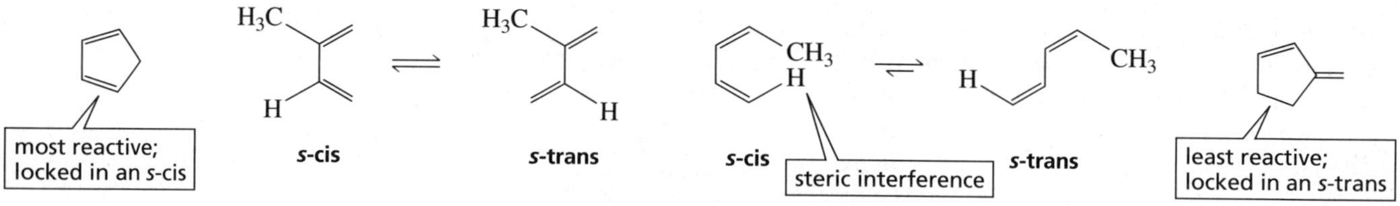

[structures diagram]

| most reactive; locked in an *s*-cis | *s*-cis ⇌ *s*-trans | *s*-cis (steric interference) ⇌ *s*-trans | least reactive; locked in an *s*-trans |

Thus, the four dienes have the following order of decreasing reactivity:

[cyclopentadiene] > [H₃C diene with H] > [ring with CH₃ and H] > [cyclopentane with =CH₂]

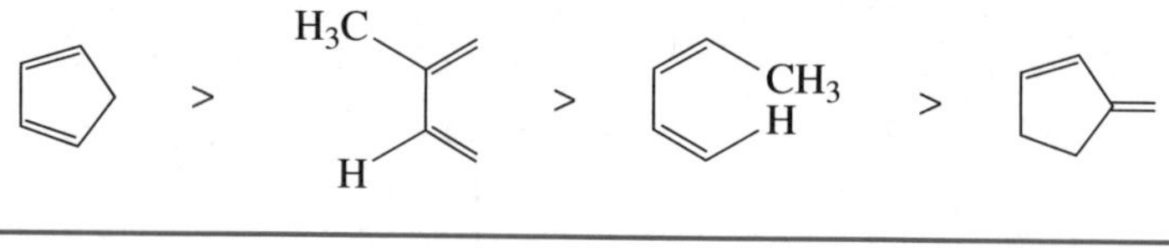

The Stereochemistry of the Diels–Alder Reaction

If a Diels–Alder reaction creates a product with an asymmetric center, identical amounts of the *R* and *S* enantiomers will be formed. In other words, the product will be a racemic mixture (Section 5.19).

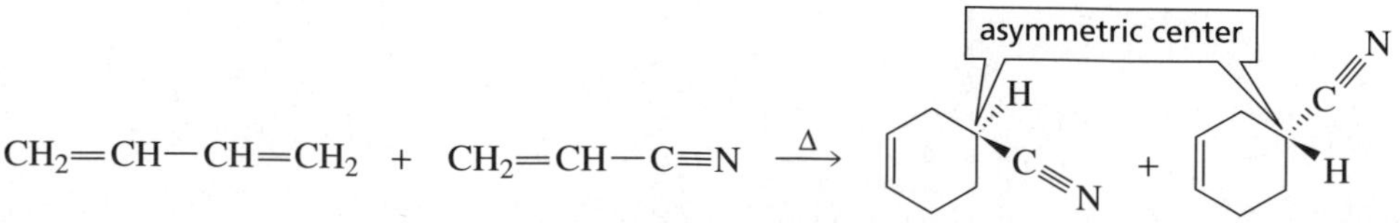

$$CH_2=CH-CH=CH_2 + CH_2=CH-C\equiv N \xrightarrow{\Delta}$$ [product structures with asymmetric center]

The Diels–Alder reaction is a syn addition reaction with respect to both the diene and the dienophile: one face of the diene adds to one face of the dienophile.

Therefore, if the substituents in the *dienophile* are cis, they will be cis in the product; if the substituents in the *dienophile* are trans, they will be trans in the product.

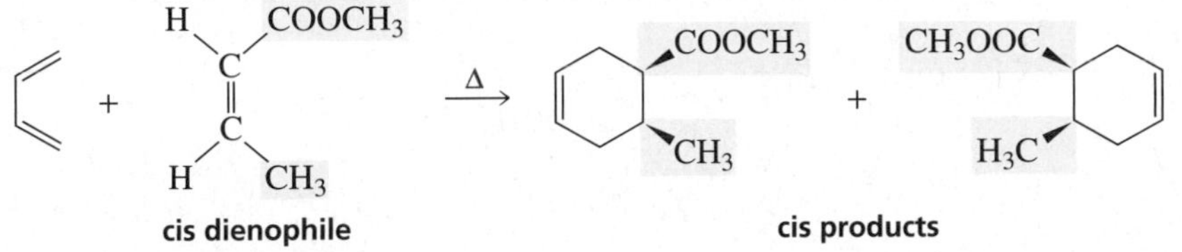

cis dienophile   cis products

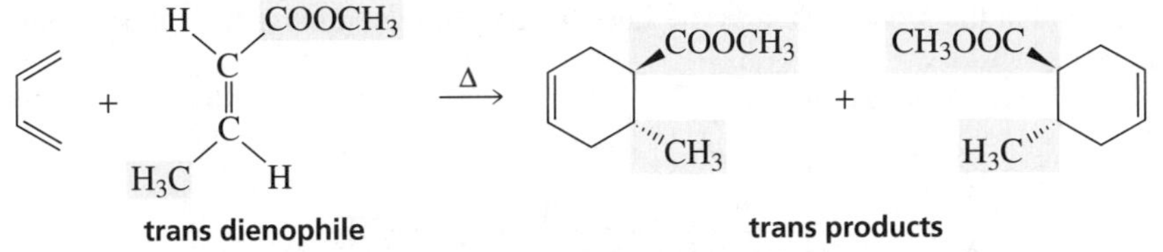

trans dienophile   trans products

The substituents in the *diene* will also maintain their relative configurations in the products. For example in the following reactions, if the hydrogens are on the same side of the conjugated system in the reactant, they will be on the same side in the product; if the hydrogens are on opposite sides of the conjugated system in the reactant, they will be on opposite sides in the product.

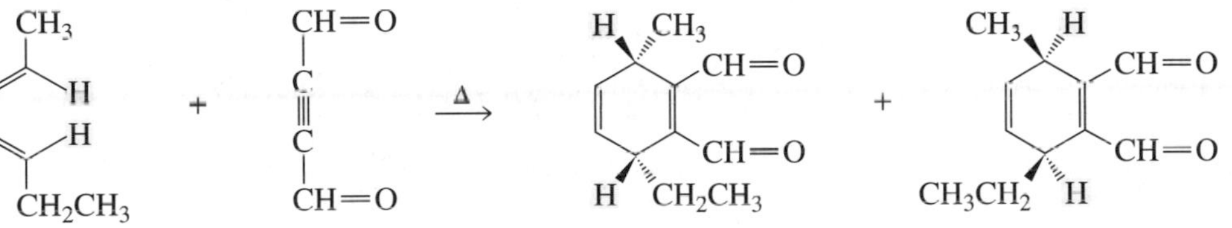

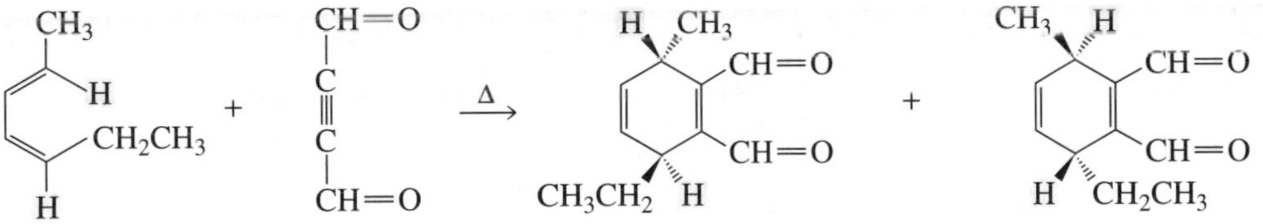

Each of the four preceding reactions forms a product with two new asymmetric centers. Thus, each product has four possible stereoisomers. However, because only syn addition occurs, each reaction forms only two of the stereoisomers (Section 5.19). Thus, the Diels–Alder reaction is stereospecific: each stereoisomeric reactant forms a different set of stereoisomeric products, because the configuration of the reactants is maintained during the course of the reaction. The stereochemistry of the reaction will be discussed in more detail in Section 29.4.

PROBLEM 38◆

Explain why the following products are not optically active:

a. the product obtained from the reaction of 1,3-butadiene with *cis*-1,2-dichloroethene.

b. the product obtained from the reaction of 1,3-butadiene with *trans*-1,2-dichloroethene.

VCL **Diels Alder-5**

PROBLEM-SOLVING STRATEGY

Analyzing a Diels–Alder Product

What diene and what dienophile were used to synthesize the following compound?

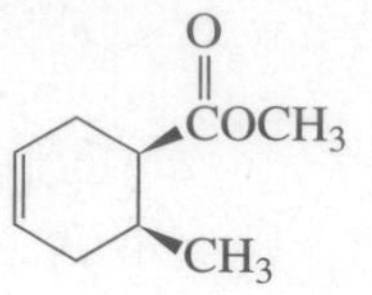

The diene that was used to form the cyclic product had double bonds on either side of the double bond in the product, so draw in those bonds and then remove the π bond between them.

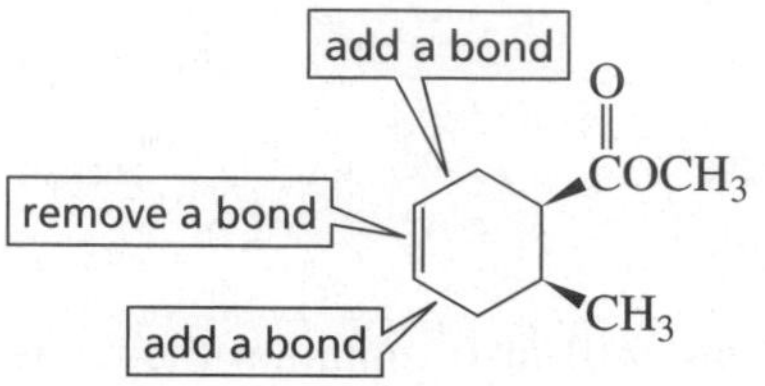

The new σ bonds are now on either side of the double bonds.

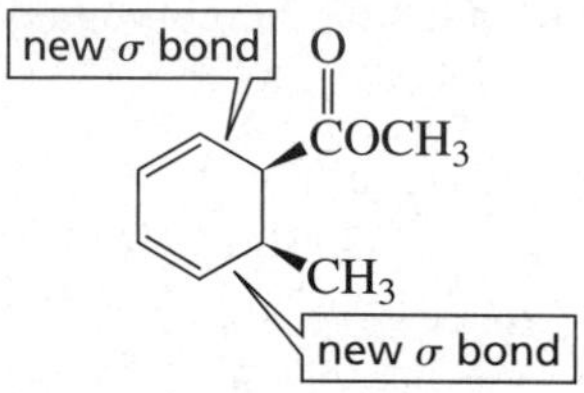

Erasing these σ bonds and putting a π bond between the two carbons that were attached by the σ bonds gives the diene and the dienophile.

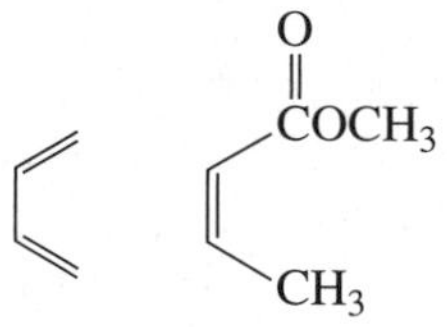

Now continue on to Problem 39.

PROBLEM 39◆

What diene and what dienophile should be used to synthesize the following compounds?

a.

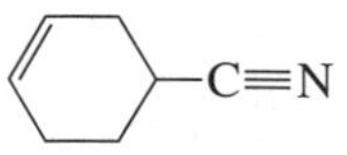

c.

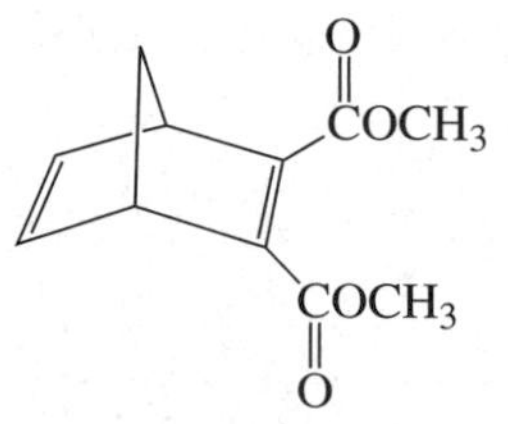

e.

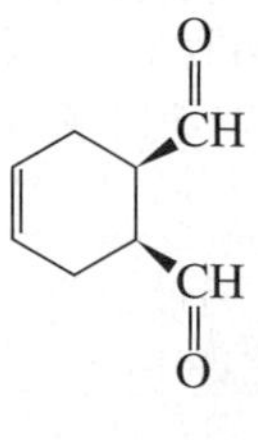

b.

d.

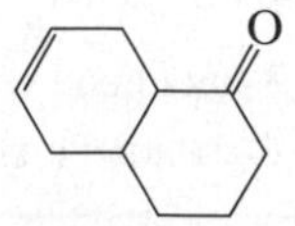

f.

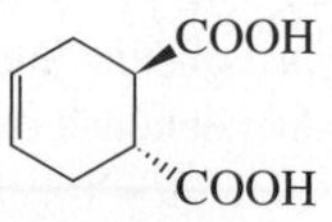

SUMMARY

Localized electrons belong to a single atom or are confined to a bond between two atoms. **Delocalized electrons** are shared by more than two atoms; they result when a p orbital overlaps the p orbitals of more than one adjacent atom. Electron delocalization occurs only if all the atoms sharing the delocalized electrons lie in or close to the same plane.

Each of benzene's six carbons is sp^2 hybridized, with bond angles of $120°$. A p orbital of each carbon overlaps the p orbitals of both adjacent carbons. The six π electrons are shared by all six carbons. Thus, benzene is a planar molecule with six delocalized π electrons.

Chemists use **resonance contributors**—structures with localized electrons—to approximate the actual structure of a compound that has delocalized electrons: the **resonance hybrid**. To draw resonance contributors, move only π electrons or unpaired electrons toward an sp^2 or sp hybridized atom. The total number of electrons and the numbers of paired and unpaired electrons do not change.

The greater the predicted stability of the resonance contributor, the more it contributes to the structure of the hybrid and the more similar its structure is to the real molecule. The predicted stability is decreased by (1) an atom with an incomplete octet, (2) a negative (positive) charge not on the most electronegative (electropositive) atom, or (3) charge separation. A resonance hybrid is more stable than the predicted stability of any of its resonance contributors.

The extra stability that a compound gains from having delocalized electrons is called **delocalization energy**, **resonance energy**, or **resonance stabilization**. It tells us how much more stable a compound with delocalized electrons is than it would be if its electrons were localized. The greater the number of relatively stable resonance contributors and the more nearly equivalent they are, the greater is the resonance energy of the compound. Allylic and benzylic cations have delocalized electrons, so they are more stable than similarly substituted carbocations with localized electrons. Compounds such as benzene that are unusually stable because of large delocalization energies are called **aromatic compounds**.

Conjugated double bonds are separated by one single bond. **Isolated double bonds** are separated by more than one single bond. Because dienes with conjugated double bonds have delocalized electrons, they are more stable than dienes with isolated double bonds. The least stable alkene has the greatest $-\Delta H°$ value for hydrogenation.

A molecular orbital results from the **linear combination of atomic orbitals (LCAO)**. The number of orbitals is conserved; in other words, the number of molecular orbitals equals the number of atomic orbitals that produced them. Side-to-side overlap of in-phase p orbitals produces a **bonding molecular orbital**, which is more stable than the atomic orbitals. Side-to-side overlap of out-of-phase p orbitals produces an **antibonding molecular orbital**, which is less stable than the atomic orbitals. The **highest occupied molecular orbital (HOMO)** is the highest-energy MO that contains electrons. The **lowest unoccupied molecular orbital (LUMO)** is the lowest-energy MO that does not contain electrons.

As the MOs increase in energy, the number of nodes increases, the number of bonding interactions decreases, and the orbitals alternate from **symmetric** to antisymmetric. Molecular orbital theory and contributing resonance structures both show that electrons are delocalized and that electron delocalization makes a molecule more stable.

Electron delocalization can affect the pK_a of a compound. A carboxylic acid and a phenol are more acidic than an alcohol such as ethanol, and a protonated aniline is more acidic than a protonated amine because electron withdrawal stabilizes their conjugate bases and the loss of a proton is accompanied by an increase in delocalization energy. Donation of electrons through π bonds is called **resonance electron donation**; withdrawal of electrons through π bonds is called **resonance electron withdrawal**.

Electron delocalization can affect the nature of the product formed in a reaction. An isolated diene, like an alkene, undergoes only 1,2-addition. If there is only enough electrophilic reagent to add to one of the double bonds, it will add preferentially to the one that forms the more stable carbocation. A conjugated diene reacts with a limited amount of electrophilic reagent to form a **1,2-addition product** and a **1,4-addition product**. The first step is addition of the electrophile to one of the sp^2 carbons at the end of the conjugated system.

When a reaction produces more than one product, the product formed most rapidly is the **kinetic product**; the most stable product is the **thermodynamic product**. If the reaction is carried out under mild conditions so that it is irreversible, the major product will be the kinetic product; if the reaction is carried out under more vigorous conditions so that it is reversible, the major product will be the thermodynamic product. When a reaction is under **kinetic control**, the relative amounts of the products depend on the rates at which they are formed; when a reaction is under **thermodynamic control**, the relative amounts of the products depend on their stabilities. A **common intermediate** is an intermediate that both products have in common. In electrophilic addition to a conjugated diene, the 1,2-product is always the kinetic product; either the 1,2- or the 1,4-product can be the thermodynamic product, depending on the structures of the products.

In a **Diels–Alder reaction**, a **conjugated diene** reacts with a dienophile to form a cyclic compound; in this concerted **[4 + 2] cycloaddition reaction**, two new σ bonds are formed at the expense of two π bonds. The conjugated diene must be in an **s-cis conformation**. The reactivity of the **dienophile** is increased by electron-withdrawing groups attached to the sp^2 carbons. In the molecular orbital description of the reaction, the **HOMO** of one reactant and the **LUMO** of the other are used to show the transfer of electrons between the molecules. According to the conservation of orbital symmetry, **pericyclic reactions** occur as a result of the overlap of in-phase orbitals. If both the diene and the dienophile are unsymmetrically substituted, two products are possible because the reactants can be aligned in two different ways. The Diels–Alder reaction is stereospecific; it is a syn addition reaction with respect to both the diene and the dienophile. In **bridged bicyclic compounds**, a substituent can be **endo** or **exo**; endo is favored if the dienophile's substituent has π electrons.

SUMMARY OF REACTIONS

1. In the presence of excess electrophilic reagent, both double bonds of an *isolated diene* will undergo electrophilic addition.

$$CH_2{=}CHCH_2CH_2\overset{\overset{\displaystyle CH_3}{|}}{C}{=}CH_2 \;+\; \underset{\textbf{excess}}{HBr} \longrightarrow CH_3\underset{\underset{\displaystyle Br}{|}}{C}HCH_2CH_2\underset{\underset{\displaystyle Br}{|}}{\overset{\overset{\displaystyle CH_3}{|}}{C}}CH_3$$

In the presence of only one equivalent of electrophilic reagent, only the most reactive double bond of an *isolated diene* will undergo electrophilic addition (Section 7.10).

$$CH_2{=}CHCH_2CH_2\overset{\overset{\displaystyle CH_3}{|}}{C}{=}CH_2 \;+\; HBr \longrightarrow CH_2{=}CHCH_2CH_2\underset{\underset{\displaystyle Br}{|}}{\overset{\overset{\displaystyle CH_3}{|}}{C}}CH_3$$

2. *Conjugated dienes* undergo 1,2- and 1,4-addition in the presence of one equivalent of an electrophilic reagent (Section 7.10).

$$RCH{=}CHCH{=}CHR \;+\; HBr \longrightarrow \underset{\underset{\displaystyle Br}{|}}{RCH_2CH}CH{=}CHR \;+\; RCH_2CH{=}\underset{\underset{\displaystyle Br}{|}}{CHCHR}$$

$$\qquad\qquad\qquad\qquad\qquad\qquad\;\textbf{1,2-addition product} \qquad\qquad \textbf{1,4-addition product}$$

3. *Conjugated dienes* undergo 1,4-addition with a dienophile (a Diels–Alder reaction) (Section 7.12).

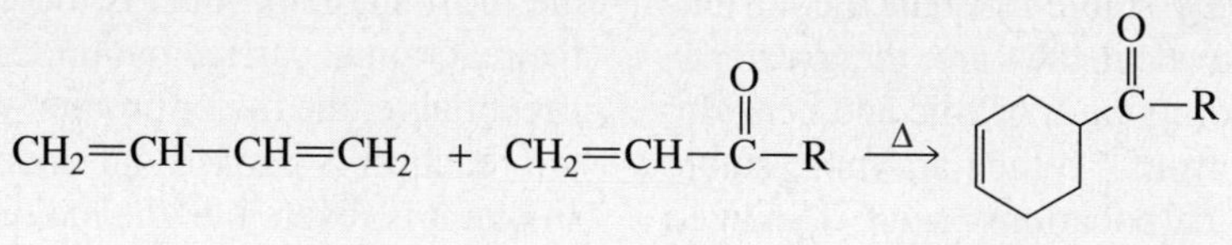

$$CH_2{=}CH{-}CH{=}CH_2 \;+\; CH_2{=}CH{-}\overset{\overset{\displaystyle O}{\|}}{C}{-}R \overset{\Delta}{\longrightarrow}$$

KEY TERMS

lowest unoccupied molecular orbital (LUMO) (p. 308)
pericyclic reaction (p. 327)
proximity effect (p. 325)
resonance (p. 298)
resonance contributor (p. 291)
resonance electron donation (p. 313)

resonance electron withdrawal (p. 315)
resonance energy (p. 298)
resonance hybrid (p. 292)
resonance structure (p. 291)
resonance stabilization energy (p. 298)
s-cis conformation (p. 331)
separated charges (p. 296)

s-trans conformation (p. 331)
symmetric molecular orbital (p. 308)
thermodynamic control (p. 323)
thermodynamic product (p. 321)
withdrawal of electrons by resonance (p. 315)

PROBLEMS

40. Which of the following compounds have delocalized electrons?

a.
$$CH_2=CHCCH_3$$ (with O double bonded)

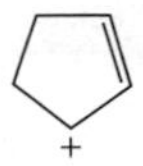

f. $CH_3\overset{CH_3}{\underset{+}{C}}CH_2CH=CH_2$

k. (cyclohexene ring structure)

b. $CH_3CH=CHOCH_2CH_3$

g. $CH_2=CHCH_2CH=CH_2$

l. $CH_3CH_2\overset{+}{C}HCH=CH_2$

c. $CH_3\overset{+}{C}HCH_2CH=CH_2$

h. $CH_3CH_2NHCH_2CH=CHCH_3$

m. $CH_3CH_2\overset{+}{N}HCH=CHCH_3$

d. (bicyclic ring structure)

i. (cyclohexadiene ring)

n. 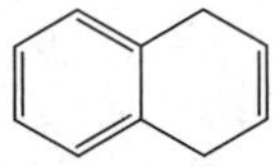 (bicyclic ring structure)

e. (cyclopentyl cation)

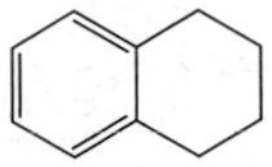

j. (cyclopentadienyl cation)

41. a. Draw resonance contributors for the following species, showing all the lone pairs:
    1. CH_2N_2 **2.** N_2O **3.** NO_2^-
   b. For each species, indicate the most stable resonance contributor.

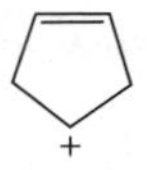

42. Give the major product of each of the following reactions, assuming the presence of one equivalent of each reagent:

a. 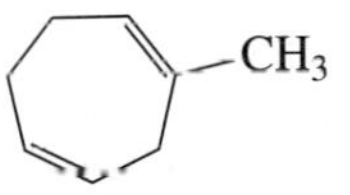 (cyclooctatriene ring with —CH_3) $+$ HBr $\longrightarrow$

b. 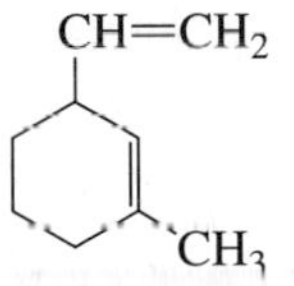 (cyclohexene ring with $CH=CH_2$ and CH_3 substituents) $+$ HBr $\longrightarrow$

43. Draw resonance contributors for the following ions:

a. 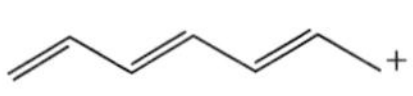 (hexadienyl cation) $+$

b. 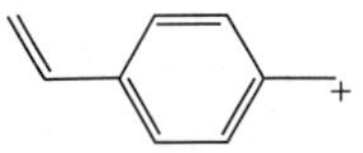 (vinyl benzyl cation) $+$

c. 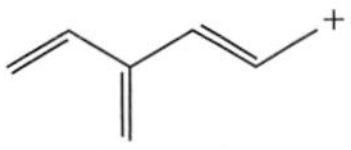 (branched pentadienyl cation) $+$

d. 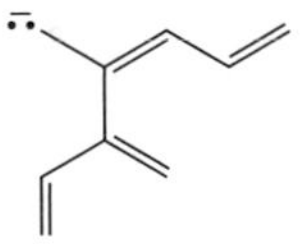 (divinyl cation with lone pair)

44. Give all the products of the following reaction.

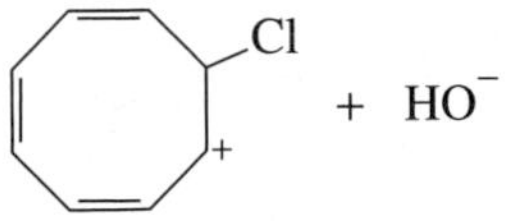

(cyclooctadiene ring with Cl) $+$ HO^-

45. Are the following pairs of structures resonance contributors or different compounds?

a. $CH_3\overset{O}{\overset{||}{C}}CH_2CH_3$ and $CH_3\overset{OH}{\overset{|}{C}}=CHCH_3$

b. $CH_3\overset{+}{C}HCH=CHCH_3$ and $CH_3CH=CHCH_2\overset{+}{C}H_2$

c. $CH_3CH=CH\overset{+}{C}HCH=CH_2$ and $CH_3\overset{+}{C}HCH=CHCH=CH_2$

d. 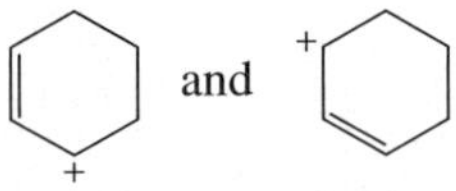 (cyclohexene ring with $+$) and (cyclohexene ring with $+$)

e. 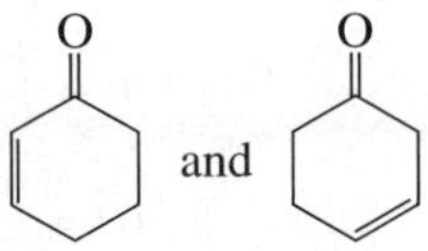 (cyclohexenone) and (cyclohexenone)

46. a. How many linear dienes have molecular formula C_6H_{10}? (Disregard cis–trans isomers.)
 b. How many of the linear dienes in part a are conjugated dienes?
 c. How many are isolated dienes?

47. a. Draw resonance contributors for the following species. Do not include structures that are so unstable that their contributions to the resonance hybrid would be negligible. Indicate which are major contributors and which are minor contributors to the resonance hybrid.

1. $CH_3CH=CHOCH_3$

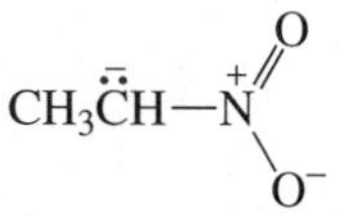

6. $CH_3CH=CH\overset{+}{C}H_2$

11. $CH_3\overset{..}{\overset{-}{C}}HC\equiv N$

2. [benzene ring with $CH_2\overset{..}{N}H_2$ substituent]

7. 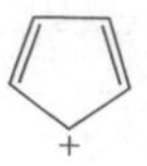 [cyclopentadienyl cation with +]

12. [benzene ring with $\overset{..}{\underset{..}{O}}CH_3$ substituent]

3. [benzene ring with C(=O)CH₃ substituent]

8. $CH_3CH_2\overset{O}{\overset{||}{C}}OCH_2CH_3$

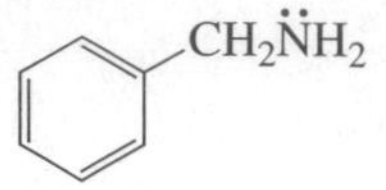

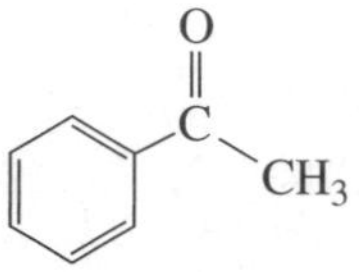

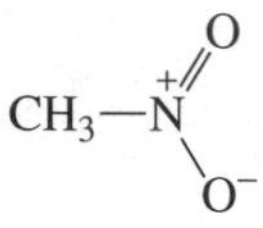

13. $H\overset{O}{\overset{||}{C}}NHCH_3$

4. $CH_3-\overset{+}{N}\begin{smallmatrix}O\\O^-\end{smallmatrix}$

9. $CH_3CH=CHCH=CH\overset{+}{C}H_2$

14. $H\overset{O}{\overset{||}{C}}CH=CH\overset{..}{\overset{-}{C}}H_2$

5. $CH_3\overset{..}{\overset{-}{C}}H-\overset{+}{N}\begin{smallmatrix}O\\O^-\end{smallmatrix}$

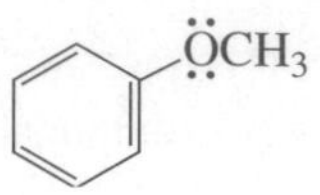

10. $\overset{..}{\overset{-}{C}}H_2\overset{O}{\overset{||}{C}}CH_2CH_3$

15. $CH_3\overset{O}{\overset{||}{C}}\overset{..}{\overset{-}{C}}H\overset{O}{\overset{||}{C}}CH_3$

 b. Do any of the species have resonance contributors that all contribute equally to the resonance hybrid?

48. Which compound would you expect to have the greater $-\Delta H°$ value for hydrogenation, 1,2-pentadiene or 1,4-pentadiene?

49. Which resonance contributor makes the greater contribution to the resonance hybrid?

 a. $CH_3\overset{+}{C}HCH=CH_2$ or $CH_3CH=CH\overset{+}{C}H_2$

 b. 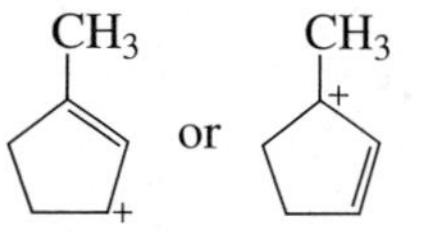 [cyclopentene ring with CH₃ and + substituents] or [cyclopentene ring with CH₃ and + substituents]

 c. 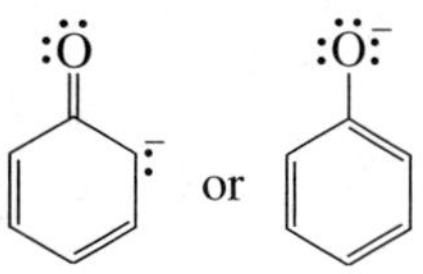 [cyclohexadienyl with :Ö and :⁻] or [cyclohexadienyl with :Ö:⁻]

50. a. Which oxygen atom has the greater electron density?

$$CH_3\overset{O}{\overset{||}{C}}OCH_3$$

 b. Which compound has the greater electron density on its nitrogen atom?

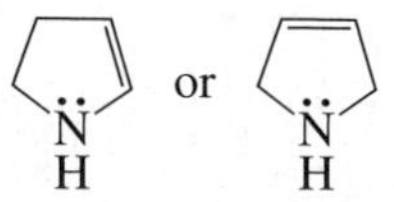

[pyrrole ring with N–H] or [pyrrolidine ring with N–H]

 c. Which compound has the greater electron density on its oxygen atom?

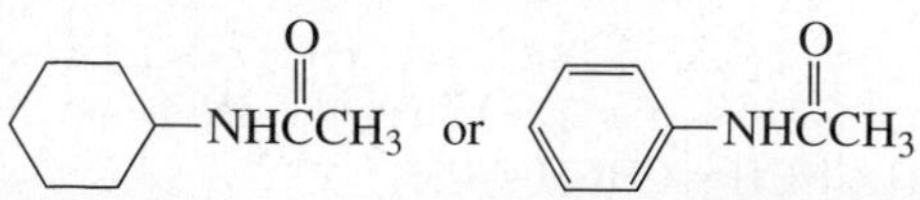

[cyclohexyl–NHC(=O)CH₃] or [phenyl–NHC(=O)CH₃]

51. Which can lose a proton more readily, a methyl group bonded to cyclohexane or a methyl group bonded to benzene?

52. The triphenylmethyl cation is so stable that a salt such as triphenylmethyl chloride can be isolated and stored. Why is this carbocation so stable?

triphenylmethyl chloride

53. Draw the contributing resonance structures for the following anion, and rank them in order of decreasing stability:

$$CH_3CH_2\ddot{\underset{\cdot\cdot}{O}}-\overset{\overset{\ddot{O}:}{\|}}{C}-\underset{\cdot}{C}H-C\equiv N:$$

54. Rank the following compounds in order of decreasing acidity:

55. Which species in each pair is more stable?

a. $H\overset{O}{\overset{\|}{C}}CH_2O^-$ or $CH_3\overset{O}{\overset{\|}{C}}O^-$

c. $CH_3\overset{\cdot\cdot}{C}HCH_2\overset{O}{\overset{\|}{C}}CH_3$ or $CH_3CH_2\overset{\cdot\cdot}{C}H\overset{O}{\overset{\|}{C}}CH_3$

b. $CH_3\overset{O}{\overset{\|}{C}}\overset{\cdot\cdot}{C}HCH_2\overset{O}{\overset{\|}{C}}H$ or $CH_3\overset{O}{\overset{\|}{C}}\overset{\cdot\cdot}{C}H\overset{O}{\overset{\|}{C}}CH_3$

d.

56. Which species in each of the pairs in Problem 55 is the stronger base?

57. Why is the resonance energy of pyrrole (21 kcal/mol) greater than the resonance energy of furan (16 kcal/mol)?

pyrrole   **furan**

58. Rank the following compounds in order of decreasing acidity of the indicated hydrogen:

$$CH_3\overset{O}{\overset{\|}{C}}CH_2CH_2\overset{O}{\overset{\|}{C}}CH_3 \qquad CH_3\overset{O}{\overset{\|}{C}}CH_2CH_2CH_2\overset{O}{\overset{\|}{C}}CH_3 \qquad CH_3\overset{O}{\overset{\|}{C}}CH_2\overset{O}{\overset{\|}{C}}CH_3$$

59. Answer the following questions for the π molecular orbitals of 1,3,5,7-octatetraene:
a. How many π MOs does the compound have?
b. Which are the bonding MOs and which are the antibonding MOs?
c. Which MOs are symmetric and which are antisymmetric?
d. Which MO is the HOMO and which is the LUMO in the ground state?
e. Which MO is the HOMO and which is the LUMO in the excited state?
f. What is the relationship between HOMO and LUMO and symmetric and antisymmetric orbitals?
g. How many nodes does the highest-energy π molecular orbital of 1,3,5,7-octatetraene have between the nuclei?

60. Diane O'File treated 1,3-cyclohexadiene with Br_2 and obtained two products (disregarding stereoisomers). Her lab partner treated 1,3-cyclohexadiene with HBr and was surprised to find that he obtained only one product (disregarding stereoisomers). Account for these results.

61. How could the following compounds be synthesized using a Diels–Alder reaction?

a.

c.

b. .CO$_2$CH$_3$

d.

62. a. How could each of the following compounds be prepared from a hydrocarbon in a single step?

1. OH

2. Br CH$_3$

3. CH$_3$

b. What other organic compound would be obtained from each synthesis?

63. a. Give the products obtained from the reaction of 1 mol HBr with 1 mol 1,3,5-hexatriene.
b. Which product(s) will predominate if the reaction is under kinetic control?
c. Which product(s) will predominate if the reaction is under thermodynamic control?

64. How would the following substituents affect the rate of a Diels–Alder reaction?
a. an electron-donating substituent in the diene
b. an electron-donating substituent in the dienophile
c. an electron-withdrawing substituent in the diene

65. Give the major products obtained from the reaction of one equivalent of HCl with the following compounds. For each reaction, indicate the kinetic and thermodynamic products.
a. 2,3-dimethyl-1,3-pentadiene
b. 2,4-dimethyl-1,3-pentadiene

66. The acid dissociation constant (K_a) for loss of a proton from cyclohexanol is 1×10^{-16}.
a. Draw an energy diagram for loss of a proton from cyclohexanol.

$$\text{⬡—OH} \quad \underset{}{\overset{K_a = 1 \times 10^{-16}}{\rightleftharpoons}} \quad \text{⬡—O}^- + \text{H}^+$$

b. Draw the contributing resonance structures for phenol.
c. Draw the contributing resonance structures for the phenolate ion.
d. On the same plot with the energy diagram for loss of a proton from cyclohexanol, draw an energy diagram for loss of a proton from phenol.

$$\text{⬡—OH} \rightleftharpoons \text{⬡—O}^- + \text{H}^+$$

e. Which has a greater K_a, cyclohexanol or phenol?
f. Which is a stronger acid, cyclohexanol or phenol?

67. Protonated cyclohexylamine has a $K_a = 1 \times 10^{-11}$. Using the same sequence of steps as in Problem 66, determine which is a stronger base, cyclohexylamine or aniline.

$$\text{⬡—}\overset{+}{\text{N}}\text{H}_3 \rightleftharpoons \text{⬡—NH}_2 + \text{H}^+$$

$$\text{⬡—}\overset{+}{\text{N}}\text{H}_3 \rightleftharpoons \text{⬡—NH}_2 + \text{H}^+$$

68. Give the product or products that would be obtained from each of the following reactions:

a. 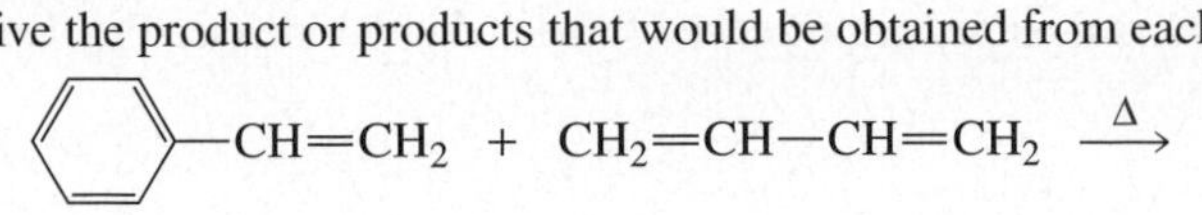 [benzene ring]$-CH=CH_2$ + $CH_2=CH-CH=CH_2$ $\xrightarrow{\Delta}$

b. 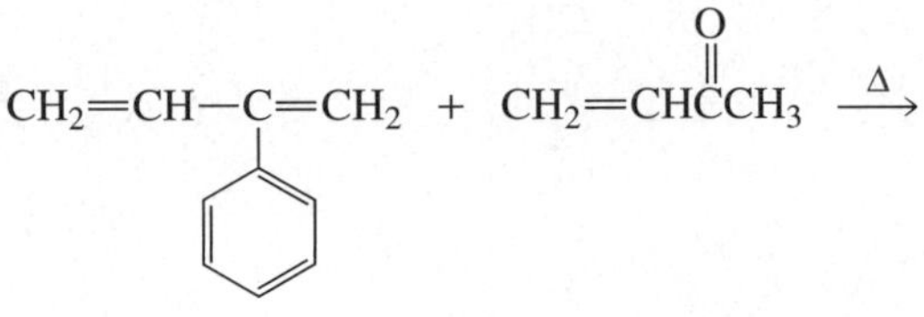 $CH_2=CH-C=CH_2$ + $CH_2=CHCCH_3$ (with O double bond) $\xrightarrow{\Delta}$
[benzene ring attached below]

c. [benzene ring]$-CH=CH_2$ + $CH_2=CH-\overset{\overset{CH_3}{|}}{C}=CH_2$ $\xrightarrow{\Delta}$

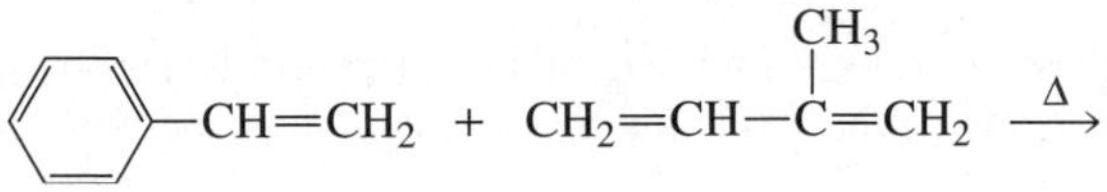

69. What two sets of a conjugated diene and a dienophile could be used to prepare the following compound?

[structure: cyclohexadiene with $\overset{O}{\overset{||}{C}}CH_3$ group]

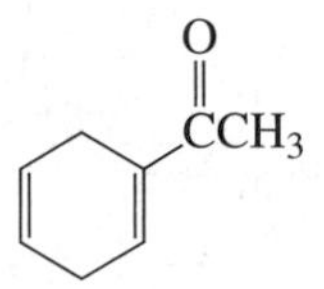

70. a. Which dienophile in each pair is more reactive in a Diels–Alder reaction?

1. CH_2-CHCH (with O double bond) or CH_2-CHCH_2CH (with O double bond)　　**2.** $CH_2=CHCH$ or $CH_2=CHCH_3$

b. Which diene is more reactive in a Diels Alder reaction?

$CH_2=CHCH=CHOCH_3$ or $CH_2=CHCH=CHCH_2OCH_3$

71. Cyclopentadiene can react with itself in a Diels–Alder reaction. Draw the endo and exo products.

72. Which diene and which dienophile could be used to prepare each of the following compounds?

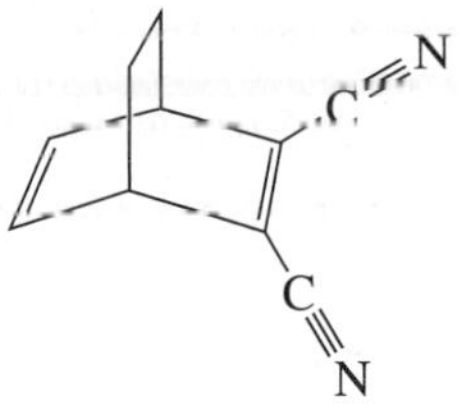

a. 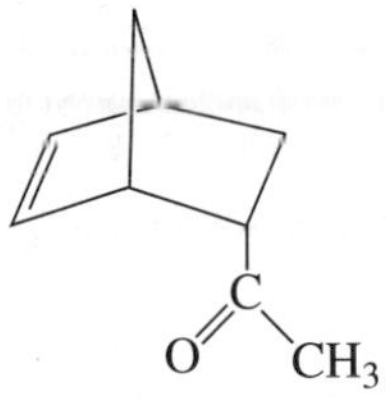 [bicyclic structure with $\overset{C}{\underset{O}{}}CH_3$ group]　**b.** [bicyclic structure with two Cl groups]　**c.** [bicyclic structure with two C≡N groups]　**d.** 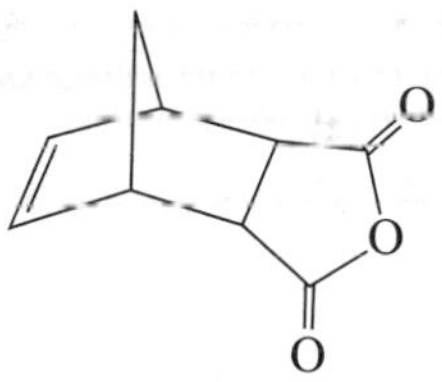 [bicyclic structure with anhydride, two O double bonds and ring O]

73. a. Give the products of the following reaction:

[cyclohexene with vinyl group] + Br_2 $\longrightarrow$

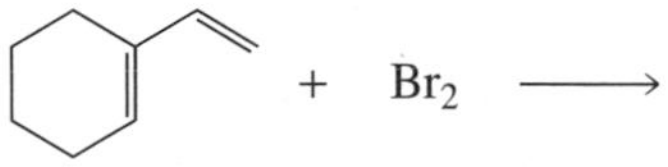

b. How many stereoisomers of each product could be obtained?

74. As many as 18 different Diels–Alder products could be obtained by heating a mixture of 1,3-butadiene and 2-methyl-1,3-butadiene. Identify the products.

75. On a single graph, draw the reaction coordinate for the addition of one equivalent of HBr to 2-methyl-1,3-pentadiene and one equivalent of HBr to 2-methyl-1,4-pentadiene. Which reaction is faster?

76. While attempting to recrystallize maleic anhydride, Professor Nots O. Kareful dissolved it in freshly distilled cyclopentadiene rather than in freshly distilled cyclopentane. Was his recrystallization successful?

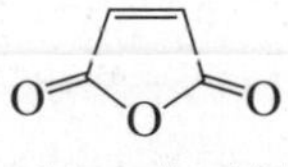

[structure: maleic anhydride]

maleic anhydride

77. Reverse Diels–Alder reactions can occur at high temperatures. Why are high temperatures required?

78. The following equilibrium is driven to the right if the reaction is carried out in the presence of maleic anhydride (see Problem 76):

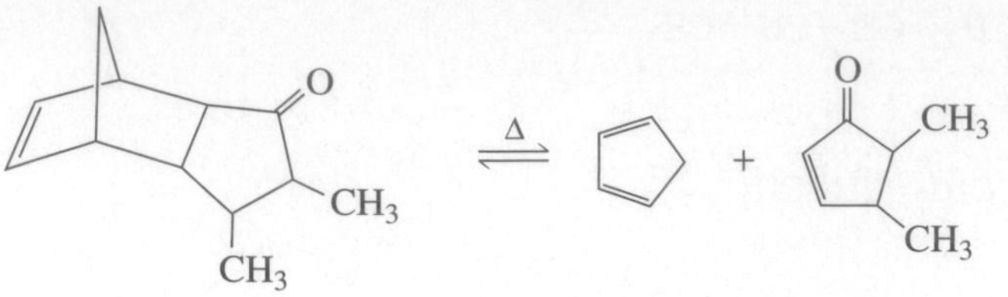

What is the function of maleic anhydride?

79. In 1935, J. Bredt, a German chemist, proposed that a bicycloalkene could not have a double bond at a bridgehead carbon unless one of the rings contains at least eight carbon atoms. This is known as Bredt's rule. Explain why there cannot be a double bond at this position.

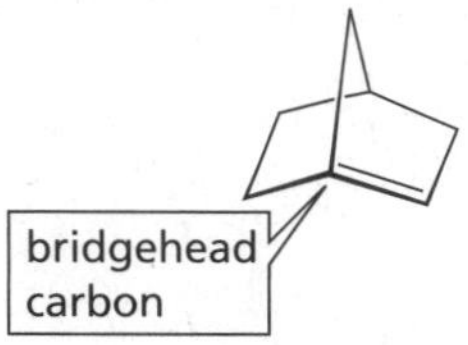

bridgehead
carbon

80. The experiment shown below and discussed in Section 7.11, showed that the proximity of the chloride ion to C-2 in the transition state caused the 1,2-addition product to be formed faster than the 1,4-addition product:

$$CH_2{=}CHCH{=}CHCH_3 \quad + \quad DCl \quad \xrightarrow{-78\,°C} \quad CH_2CHCH{=}CHCH_3 \quad + \quad CH_2CH{=}CHCHCH_3$$
$$\qquad\qquad\qquad\qquad\qquad\qquad\qquad\qquad\qquad\qquad\quad\; |\;\;\; | \qquad\qquad\qquad\qquad |\qquad\qquad\quad |$$
$$\qquad\qquad\qquad\qquad\qquad\qquad\qquad\qquad\qquad\qquad\quad D\;\; Cl \qquad\qquad\qquad\qquad D\qquad\qquad\quad Cl$$

  a. Why was it important for the investigators to know that the reaction was being carried out under kinetic control?

  b. How could the investigators determine that the reaction was being carried out under kinetic control?

81. A student wanted to know whether the greater proximity of the nucleophile to the C-2 carbon in the transition state is what causes the 1,2-addition product to be formed faster when 1,3-butadiene reacts with HCl. Therefore, he decided to investigate the reaction of 2-methyl-1,3-cyclohexadiene with HCl. His friend, Noel Noall, told him that he should use 1-methyl-1,3-cyclohexadiene instead. Should he follow his friend's advice?

Substitution and Elimination Reactions

The first three chapters in Part 3 discuss the reactions of compounds that have an electron-withdrawing atom or group—a potential leaving group—bonded to an sp^3 hybridized carbon. These compounds can undergo substitution or elimination reactions or both substitution and elimination reactions. The fourth and last chapter in Part 3 discusses the reactions of alkanes, compounds that do not have a leaving group but that can undergo a radical substitution reaction under extreme conditions.

CHAPTER 8
Substitution Reactions of Alkyl Halides

Chapter 8 discusses the substitution reactions of alkyl halides. Of the different compounds that undergo substitution and elimination reactions, alkyl halides are examined first because they have relatively good leaving groups. The chapter also describes the kinds of compounds that biological organisms use in place of alkyl halides, since alkyl halides are not readily available in nature.

CHAPTER 9
Elimination Reactions of Alkyl Halides • Competition Between Substitution and Elimination

Chapter 9 covers the elimination reactions of alkyl halides. Because many alkyl halides can undergo both substitution and elimination reactions, this chapter also discusses the factors that determine whether a given alkyl halide will undergo a substitution reaction, an elimination reaction, or both substitution and elimination reactions.

CHAPTER 10
Reactions of Alcohols, Amines, Ethers, Epoxides, and Sulfur-Containing Compounds • Organometallic Compounds

Chapter 10 discusses compounds other than alkyl halides that undergo substitution and elimination reactions. There you will see that because alcohols and ethers have relatively poor leaving groups compared with the leaving groups of alkyl halides, alcohols and ethers must be activated before they can be substituted or eliminated. Several methods commonly used to activate leaving groups will be examined. Amines have such poor leaving groups that they cannot undergo substitution and elimination reactions, although they do undergo other important reactions. The chapter also compares the reactions of thiols and sulfides with those of alcohols and ethers; looks at the reactions of epoxides, which show how ring strain affects leaving ability; and discusses how the carcinogenicity of arene oxides is related to carbocation stability. Finally, this chapter introduces you to organometallic compounds, a class of compounds that is very important to synthetic organic chemists.

CHAPTER 11
Radicals • Reactions of Alkanes

Chapter 11 discusses the substitution reactions of alkanes, hydrocarbons that contain only single bonds. In previous chapters, we have seen that when a compound reacts, the weakest bond in the molecule breaks first. Alkanes, however, have only strong bonds. Therefore, conditions vigorous enough to generate radicals are required for alkanes to react. The chapter also looks at radical reactions of alkenes and alkynes. It concludes with a discussion of some radical reactions that occur in the biological world.

CHAPTER 8

Substitution Reactions of Alkyl Halides

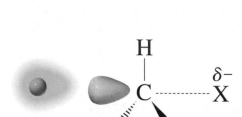

BUILDING ON FUNDAMENTALS

SECTION 8.1 When we looked at the addition reactions of alkene, alkynes, and dienes, we saw that they were called *electrophilic* addition reactions because those compounds react with electrophiles. In this chapter we will see that the substitution reactions of alkyl halides are called *nucleophilic* substitution reactions because alkyl halides react with nucleophiles.

SECTION 8.3 When you study S_N2 reactions, they may seem familiar because you have seen an S_N2 reaction previously, although it was not identified as such (6.11).

SECTION 8.3 We saw that steric effects cause a gauche conformer to be less stable than an anti conformer (2.10). Now we will see that steric effects can influence the rates of certain reactions.

SECTIONS 8.3 AND 8.6 Iodide ion is the weakest base of the halide ions (1.20). Now we will see that because the leaving ability of a group bonded to an sp^3 carbon depends on the group's basicity, iodide ion is the best leaving group of the halide ions.

SECTION 8.4 Le Châtelier's principle (3.7) can be used to drive a reaction toward the desired products.

SECTION 8.5 A carbocation intermediate is formed in an S_N1 reaction. As with other reactions that have carbocation intermediates (4.6), the carbocation will rearrange if a more stable carbocation can be formed.

SECTION 8.6 We saw that an atom's polarizability can affect a compound's boiling point (2.9). Here we will see that it also affects the atom's nucleophilicity.

SECTION 8.8 Electron delocalization in a carbocation intermediate formed during the addition of an electrophile to a conjugated diene leads to two addition products (7.10). Now we will see that electron delocalization in an allylic carbocation formed during an S_N1 reaction can also lead to two substitution products.

SECTION 8.10 Section 1.17 stated that most of the pK_a values in this book were determined in water. Now we will see why pK_a values are different in different solvents.

SECTION 8.10 In determining how a change in solvent affects the rate of a reaction, we will recall that the rate of a reaction depends on the relative stabilities of the reactants and transition state of the rate-limiting step. In determining how a change in solvent affects a pK_a value, we will recall that the equilibrium constant depends on the relative stabilities of the reactants and products (3.7).

Organic compounds that have an electronegative atom or an electron-withdrawing group bonded to an sp^3 hybridized carbon undergo substitution reactions and/or elimination reactions.

In a **substitution reaction**, the electronegative atom or electron-withdrawing group is replaced by another atom or group. In an **elimination reaction**, the electronegative atom or electron-withdrawing group is eliminated, along with a hydrogen from an adjacent carbon. The atom or group that is *substituted* or *eliminated* in these reactions is called a **leaving group**. The substitution reaction is more precisely called a **nucleophilic substitution reaction** because the atom or group that replaces the leaving group is a nucleophile.

$$RCH_2CH_2X \ + \ Y^-$$

a substitution reaction $\longrightarrow$ $RCH_2CH_2Y \ + \ X^-$

an elimination reaction $\longrightarrow$ $RCH{=}CH_2 \ + \ HY \ + \ X^-$

the leaving group

This chapter focuses on the substitution reactions of alkyl halides—compounds in which the leaving group is a halide ion (F^-, Cl^-, Br^-, or I^-). The nomenclature of alkyl halides was discussed in Section 2.4.

alkyl halides

R—F	R—Cl	R—Br	R—I
an alkyl fluoride	**an alkyl chloride**	**an alkyl bromide**	**an alkyl iodide**

In Chapter 9, we will discuss the elimination reactions of alkyl halides and the factors that determine whether substitution or elimination will prevail when an alkyl halide undergoes a reaction.

Alkyl halides are a good family of compounds with which to start the study of substitution and elimination reactions because they have relatively good leaving groups; that is, the halide ions are easily displaced. After studying the reactions of alkyl halides, you will be prepared to look, in Chapter 10, at the substitution and elimination reactions of compounds with poor leaving groups—those that are more difficult to displace.

Substitution reactions are important in organic chemistry because they make it possible to convert readily available alkyl halides into a wide variety of other compounds. Substitution reactions are also important in the cells of plants and animals. We will see, however, that because alkyl halides are insoluble in water, and cells exist in predominantly aqueous environments, biological systems use compounds in which the group that is replaced is more polar than a halogen and, therefore, more soluble in water.

SURVIVAL COMPOUNDS

Several marine organisms, including sponges, corals, and algae, synthesize organohalides (halogen-containing organic compounds) that they use to deter predators. For example, red algae synthesize a toxic, foul-tasting organohalide that keeps predators from eating them. One predator that is not deterred, however, is a mollusk called a sea hare. After consuming red algae, a sea hare converts the algae's organohalide into a structurally similar compound it uses for its own defense. Unlike other mollusks, a sea hare does not have a shell. Its method of defense is to surround itself with a slimy substance that contains the organohalide, thereby protecting itself from carnivorous fish.

synthesized by red algae　　**synthesized by the sea hare**

a sea hare

8.1 How Alkyl Halides React

Fluorine, chlorine, and bromine are all more electronegative than carbon. Consequently, when carbon is bonded to any of these elements, the two atoms do not share their bonding electrons equally. Because the more electronegative halogen has a larger share of the electrons, it has a partial negative charge (δ^-) and the carbon to which it is bonded has a partial positive charge (δ^+).

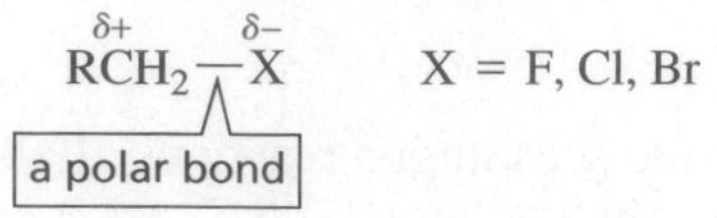

$$\overset{\delta^+}{RCH_2}\!-\!\overset{\delta^-}{X} \qquad X = F, Cl, Br$$

a polar bond

It is this polar carbon–halogen bond that causes alkyl halides to undergo substitution and elimination reactions. There are two important mechanisms for nucleophilic substitution reactions:

1. A nucleophile is attracted to the partially positively charged carbon (an electrophile). As the nucleophile approaches the carbon and forms a new bond, the carbon–halogen bond breaks heterolytically (the halogen takes both of the bonding electrons).

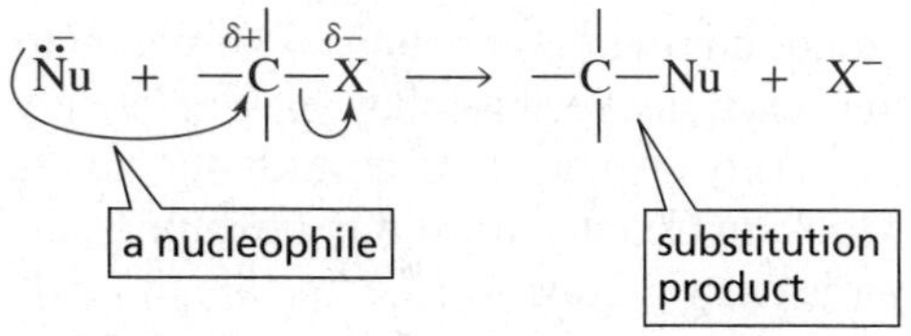

$$\overset{..}{\underset{..}{Nu}}{}^- + \ \overset{\delta^+}{-\underset{|}{\overset{|}{C}}}\!-\!\overset{\delta^-}{X} \ \longrightarrow \ -\underset{|}{\overset{|}{C}}\!-\!Nu \ + \ X^-$$

a nucleophile   substitution product

2. The carbon–halogen bond breaks heterolytically without any assistance from the nucleophile, forming a carbocation. The carbocation—an electrophile—then reacts with the nucleophile to form the substitution product.

$$\overset{\delta^+}{-\underset{|}{\overset{|}{C}}}\!-\!\overset{\delta^-}{X} \ \longrightarrow \ -\underset{|}{\overset{|}{C}}{}^+ \ + \ X^-$$

$$-\underset{|}{\overset{|}{C}}{}^+ \ + \ \overset{..}{\underset{..}{Nu}} \ \longrightarrow \ -\underset{|}{\overset{|}{C}}\!-\!Nu$$

substitution product

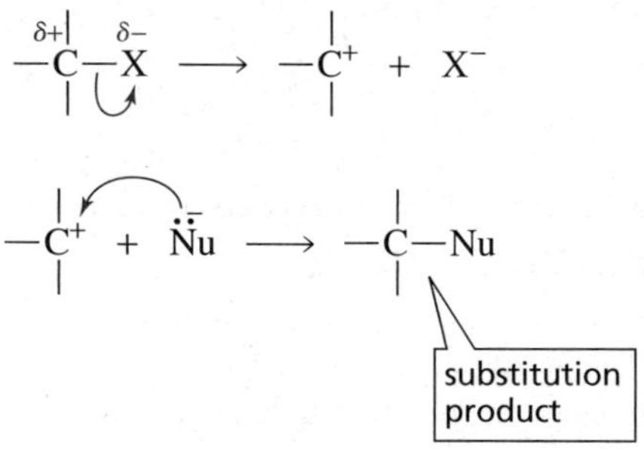

We will see that the mechanism *that predominates* depends on the following factors:

- the structure of the alkyl halide
- the reactivity of the nucleophile
- the concentration of the nucleophile
- the solvent in which the reaction is carried out

8.2 The Mechanism of an S_N2 Reaction

Perhaps you have been wondering how the mechanism of a reaction is determined. We can learn a great deal about a reaction's mechanism by studying its **kinetics**—the factors that affect the rate of the reaction.

The rate of a nucleophilic substitution reaction, such as the reaction of bromomethane with hydroxide ion, depends on the concentrations of both reagents. If the concentration of bromomethane in the reaction mixture is doubled, the rate of the reaction doubles. Likewise, if the concentration of the nucleophile (hydroxide ion) is doubled, the rate of the reaction doubles. If the concentrations of both reactants are doubled, the rate of the reaction quadruples.

$$CH_3Br \;+\; HO^- \;\longrightarrow\; CH_3OH \;+\; Br^-$$
$$\textbf{bromomethane} \qquad\qquad \textbf{methanol}$$

When you know the relationship between the rate of a reaction and the concentration of the reactants, you can write a **rate law** for the reaction. Because the rate of the reaction of bromomethane with hydroxide ion is dependent on the concentration of both reactants, the rate law for the reaction is

rate ∝ [alkyl halide][nucleophile]

As we saw in Section 3.7, a proportionality sign ($\propto$) can be replaced by an equals sign and a proportionality constant. The proportionality constant (k) is called the **rate constant**. The magnitude of the rate constant for a particular reaction indicates how difficult it is for the reactants to overcome the energy barrier of the reaction—how hard it is to reach the transition state. The larger the rate constant, the lower is the energy barrier and, therefore, the easier it is for the reactants to reach the transition state (see Figure 8.3, page 350).

$$\text{rate} = k\,[\text{alkyl halide}][\text{nucleophile}]$$

the rate constant

Because the rate of this reaction depends on the concentration of two reactants, the reaction is a **second-order reaction**.

The rate law tells us which molecules are involved in the transition state of the rate-determining step of the reaction. From the rate law for the reaction of bromomethane with hydroxide ion, for example, we know that *both* bromomethane and hydroxide ion are involved in the rate-determining transition state.

PROBLEM 1◆

How is the rate of the reaction affected if the concentration of bromomethane is changed from 1.00 M to 0.05 M?

The reaction of bromomethane with hydroxide ion is an example of an **S$_N$2 reaction**, where "S" stands for substitution, "N" for nucleophilic, and "2" for bimolecular. **Bimolecular** means that two molecules are involved in the transition state of the rate-determining step. In 1937, Edward Hughes and Christopher Ingold proposed a mechanism for an S$_N$2 reaction. Remember that a mechanism describes the step-by-step process by which reactants are converted into products. It is a theory that fits the accumulated experimental evidence pertaining to the reaction. Hughes and Ingold based their mechanism for an S$_N$2 reaction on the following three pieces of experimental evidence:

1. The rate of the reaction depends on the concentration of the alkyl halide *and* on the concentration of the nucleophile. This means that both reactants are involved in the transition state of the rate-determining step.

2. As the hydrogens of bromomethane are successively replaced with methyl groups, the rate of the reaction with a given nucleophile becomes progressively slower (Table 8.1).

3. The reaction of an alkyl halide in which the halogen is bonded to an asymmetric center leads to the formation of only one stereoisomer, and the configuration of the asymmetric center in the product is inverted relative to its configuration in the reacting alkyl halide.

Table 8.1	Relative Rates of S$_N$2 Reactions for Several Alkyl Halides	
$R-Br + Cl^- \xrightarrow{S_N2} R-Cl + Br^-$		
Alkyl halide	**Class of alkyl halide**	**Relative rate**
CH_3-Br	methyl	1200
CH_3CH_2-Br	primary	40
$CH_3CH_2CH_2-Br$	primary	16
CH_3CH-Br $\quad\vert$ $\quad CH_3$	secondary	1
$\quad CH_3$ $\quad\vert$ CH_3C-Br $\quad\vert$ $\quad CH_3$	tertiary	too slow to measure

Hughes and Ingold proposed that an S$_N$2 reaction is a *concerted* reaction (thus taking place in a single step), so no intermediates are formed. The nucleophile attacks the carbon bearing the leaving group and displaces the leaving group.

Mechanism for the S$_N$2 reaction of an alkyl halide

$$HO\!:^- + CH_3-Br\!: \longrightarrow CH_3-OH + :Br\!:^-$$

leaving group

An S$_N$2 reaction is a one-step reaction.

A productive collision is one that leads to the formation of the product. A productive collision in an S$_N$2 reaction requires the nucleophile to hit the carbon on the side opposite the side bonded to the leaving group. Therefore, the carbon is said to undergo **back-side attack**. Why must the nucleophile attack from the back side? The simplest explanation is that the leaving group blocks the approach of the nucleophile to the front side of the molecule.

Molecular orbital theory also explains back-side attack. Recall from Section 7.12 that to form a bond, the LUMO (lowest unoccupied molecular orbital) of one species must interact with the HOMO (highest occupied molecular orbital) of the other. When the nucleophile approaches the alkyl halide, the filled nonbonding molecular orbital (the HOMO) of the nucleophile must interact with the empty $\sigma*$ antibonding molecular orbital (the LUMO) associated with the C—Br bond. Figure 8.1a shows that in a back-side attack, a bonding interaction occurs between the nucleophile and the larger lobe of $\sigma*$. Compare this with what happens when the nucleophile approaches the front side of the carbon (Figure 8.1b): Both a bonding and an antibonding interaction occur, and the two cancel each other. Consequently, the best overlap of the interacting orbitals is achieved through back-side attack. In fact, a nucleophile always approaches an sp^3 hybridized carbon from the back side. (We saw back-side attack previously in the reaction of a bromide ion with a cyclic bromonium ion; Section 5.19.)

A nucleophile attacks the back side of the carbon that is bonded to the leaving group.

How does Hughes and Ingold's mechanism account for the three pieces of experimental evidence? The mechanism shows the alkyl halide and the nucleophile coming together in the transition state of the one-step reaction. Therefore, increasing the concentration of either of them makes their collision more probable. Thus, the reaction will follow second-order kinetics, exactly as observed.

$$HO^- + \underset{}{{}^{\prime\prime\prime\prime}C-Br} \longrightarrow \left[\overset{\delta-}{HO}\text{---}\underset{|}{C}\text{---}\overset{\delta-}{Br} \right]^{\ddagger} \longrightarrow HO-C^{\prime\prime\prime\prime} + Br^-$$

transition state

a. Back-side attack

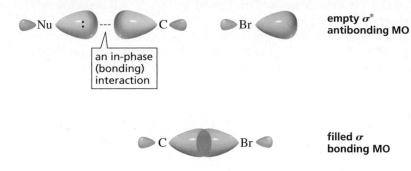

an in-phase (bonding) interaction

empty σ^* antibonding MO

filled σ bonding MO

b. Front-side attack

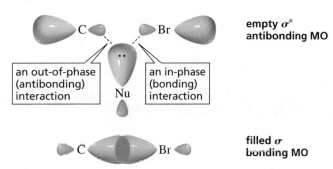

empty σ^* antibonding MO

an out-of-phase (antibonding) interaction

an in-phase (bonding) interaction

Nu

filled σ bonding MO

◀ **Figure 8.1**
(a) Back-side attack results in a bonding interaction between the HOMO (the filled nonbonding orbital) of the nucleophile and the LUMO (the empty σ^* antibonding orbital) of C—Br.
(b) Front-side attack would result in both a bonding and an antibonding interaction that would cancel out.

Because the nucleophile attacks the back side of the carbon bonded to the halogen, bulky substituents attached to this carbon will decrease the nucleophile's access to the back side of the carbon and will therefore decrease the rate of the reaction (Figure 8.2). This explains why substituting methyl groups for the hydrogens in bromomethane progressively slows the rate of the substitution reaction (Table 8.1).

▲ **Figure 8.2**
The approach of HO⁻ to a methyl halide, a primary alkyl halide, a secondary alkyl halide, and a tertiary alkyl halide. Increasing the bulk of the substituents bonded to the carbon that is undergoing nucleophilic attack decreases access to the back side of the carbon, thereby decreasing the rate of the S_N2 reaction.

Steric effects are effects caused by the fact that groups occupy a certain volume of space (Section 2.10). A steric effect that decreases reactivity is called **steric hindrance**. Steric hindrance occurs when groups are in the way at a reaction site. Steric hindrance causes alkyl halides to have the following relative reactivities in an S_N2 reaction because, *generally*, primary alkyl halides are less sterically hindered than secondary alkyl halides, which, in turn, are less hindered than tertiary alkyl halides:

Steric hindrance causes methyl halides and primary alkyl halides to be the most reactive alkyl halides in S_N2 reactions.

relative reactivities of alkyl halides in an S_N2 reaction

| most reactive | > methyl halide > 1° alkyl halide > 2° alkyl halide > 3° alkyl halide | too unreactive to undergo an S_N2 reaction |

Tertiary alkyl halides cannot undergo S$_N$2 reactions.

The three alkyl groups of a tertiary alkyl halide make it impossible for the nucleophile to come within bonding distance of the tertiary carbon, so tertiary alkyl halides are unable to undergo S$_N$2 reactions.

The reaction coordinate diagrams for the S$_N$2 reaction of *unhindered* bromomethane (Figure 8.3a) and for that of a *sterically hindered* secondary alkyl bromide (Figure 8.3b) show that steric hindrance raises the energy of the transition state, slowing the reaction.

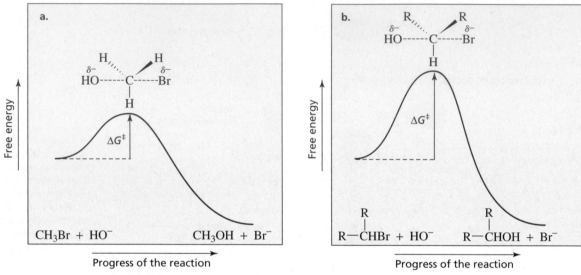

▲ **Figure 8.3**
Reaction coordinate diagrams for
(a) the S$_N$2 reaction of bromomethane with hydroxide ion;
(b) an S$_N$2 reaction of a sterically hindered secondary alkyl bromide with hydroxide ion.

The rate of an S$_N$2 reaction depends not only on the *number* of alkyl groups attached to the carbon that is undergoing nucleophilic attack, but also on their size. For example, while bromoethane and 1-bromopropane are both primary alkyl halides, bromoethane is more than twice as reactive in an S$_N$2 reaction (Table 8.1) because the bulkier group on the carbon undergoing nucleophilic attack in 1-bromopropane provides more steric hindrance to back-side attack. Also, although 1-bromo-2,2-dimethylpropane is a primary alkyl halide, it undergoes S$_N$2 reactions very slowly because its single alkyl group is unusually bulky.

$$\begin{array}{c} CH_3 \\ | \\ CH_3CCH_2Br \\ | \\ CH_3 \end{array}$$

1-bromo-2,2-dimethylpropane

As for the third piece of evidence explained by Hughes and Ingold's mechanism, Figure 8.4 shows that as the nucleophile approaches the back side of the carbon of bromomethane, the C—H bonds begin to move away from the nucleophile and its attacking electrons. By the time the transition state is reached, the C—H bonds are all in the same plane and the carbon is pentacoordinate (fully bonded to three atoms and partially bonded to two) rather than tetrahedral. As the nucleophile gets closer to the carbon and the bromine moves farther away from it, the C—H bonds continue to move in the same direction. Eventually, the bond between the carbon and the nucleophile is fully formed, and the bond between the carbon and bromine is completely broken, so the carbon is once again tetrahedral.

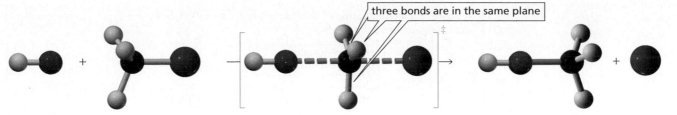

three bonds are in the same plane

▲ **Figure 8.4**
An S$_N$2 reaction between hydroxide ion and bromomethane.

The carbon at which substitution occurs has inverted its configuration during the course of the reaction, just like an umbrella tends to invert in a windstorm. This **inversion of configuration** is called a *Walden inversion*, after Paul Walden, who first discovered that the configuration of a compound becomes inverted during the course of an S$_N$2 reaction.

   Because an S$_N$2 reaction takes place with inversion of configuration, only one substitution product is formed when an alkyl halide whose halogen atom is bonded to an asymmetric center undergoes an S$_N$2 reaction. The configuration of that product is inverted relative to the configuration of the alkyl halide. For example, the substitution product obtained from the reaction of hydroxide ion with (*R*)-2-bromopentane is (*S*)-2-pentanol. Thus, the proposed mechanism also accounts for the observed configuration of the product.

Student Tutorial:
S$_N$2

Student Tutorial:
S$_N$2 a concerted reaction

To draw the inverted product of an S$_N$2 reaction, draw the mirror image of the reactant and replace the halogen with the nucleophile.

the configuration of the product is inverted relative to the configuration of the reactant

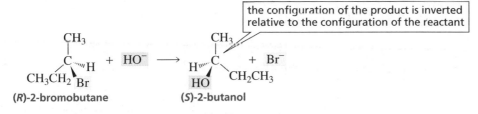

(*R*)-2-bromobutane   (*S*)-2-butanol

PROBLEM 2◆

Does increasing the energy barrier to an S$_N$2 reaction increase or decrease the magnitude of the rate constant for the reaction?

PROBLEM 3◆

Arrange the following alkyl bromides in order of decreasing reactivity in an S$_N$2 reaction: 1-bromo-2-methylbutane, 1-bromo-3-methylbutane, 2-bromo-2-methylbutane, and 1-bromopentane.

PROBLEM 4◆   *SOLVED*

Determine the product that would be formed from the S$_N$2 reaction of

a. 2-bromobutane and hydroxide ion.

b. (*R*)-2-bromobutane and hydroxide ion.

c. (*S*)-3-chlorohexane and hydroxide ion.

d. 3-iodopentane and hydroxide ion.

Solution to 4a   The product is 2-butanol. Because the reaction is an S$_N$2 reaction, we know that the configuration of the product is inverted relative to the configuration of the

Paul Walden (1863–1957) *was born in Cesis, Latvia, the son of a farmer. His parents died when he was a child and he supported himself at Riga University and St. Petersburg University by working as a tutor. Walden received a Ph.D. from the University of Leipzig and returned to Latvia to become a professor of chemistry at Riga University. Following the Russian Revolution, he went back to Germany to be a professor at the University of Rostock and, later, at the University of Tübingen.*

reactant. The configuration of the reactant is not specified, however, so we cannot specify the configuration of the product.

the configuration
is not specified

$$CH_3CHCH_2CH_3 \ + \ HO^- \ \longrightarrow \ CH_3CHCH_2CH_3 \ + \ Br^-$$
$$\qquad\ |$$
$$\qquad Br \qquad\qquad\qquad\qquad\qquad\qquad\ OH$$

8.3 Factors That Affect S$_N$2 Reactions

We will now look at how the nature of the leaving group and the nucleophile affect the S$_N$2 reaction.

The Leaving Group in an S$_N$2 Reaction

If an alkyl iodide, an alkyl bromide, an alkyl chloride, and an alkyl fluoride all having the same alkyl group were allowed to react with the same nucleophile under the same conditions, we would find that the alkyl iodide is the most reactive and the alkyl fluoride is the least reactive.

		relative rates of reaction
HO^- + RCH_2I $\longrightarrow$ RCH_2OH + I^-		30,000
HO^- + RCH_2Br $\longrightarrow$ RCH_2OH + Br^-		10,000
HO^- + RCH_2Cl $\longrightarrow$ RCH_2OH + Cl^-		200
HO^- + RCH_2F $\longrightarrow$ RCH_2OH + F^-		1

The only difference between these four reactions is the nature of the leaving group. From the relative reaction rates, we can see that the iodide ion is the best leaving group and the fluoride ion is the worst. This brings us to an important rule in organic chemistry—one that you will encounter frequently: *the weaker the basicity of a group, the better is its leaving ability.* The reason leaving ability depends on basicity is because *weak bases are stable bases;* they readily bear the electrons they formerly shared with a proton. Because weak bases do not share their electrons well, a weak base is not bonded as strongly to the carbon as a strong base would be, and a weaker bond is more easily broken (Section 1.20).

The weaker the base, the better it is as a leaving group.

Stable bases are weak bases.

We have seen that the iodide ion is the weakest base of the halide ions and the fluoride ion is the strongest (Section 1.20).

relative basicities of the halide ions

$$I^- \ < \ Br^- \ < \ Cl^- \ < \ F^-$$

weakest base,
most stable base

strongest base,
least stable base

Student Tutorial:
Steric hindrance in the S$_N$2
tertiary cannot undergo S$_N$2

Therefore, alkyl iodides are the most reactive of the alkyl halides and alkyl fluorides are the least reactive. In fact, the fluoride ion is such a strong base that alkyl fluorides essentially do not undergo S$_N$2 reactions.

relative reactivities of alkyl halides in an S$_N$2 reaction

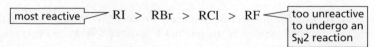

most reactive $\rightarrow$ RI > RBr > RCl > RF $\leftarrow$ too unreactive to undergo an S$_N$2 reaction

In Section 8.1, we saw that it is the polar carbon–halogen bond that causes alkyl halides to undergo substitution reactions. Carbon and iodine, however, have the same electronegativity. (See Table 1.3 on page 11.) Why, then, does an alkyl iodide undergo a substitution reaction? We know that larger atoms are more polarizable than smaller atoms. (Recall from Section 2.9 that polarizability is a measure of how easily an atom's electron cloud can be distorted.) The high polarizability of the large iodine atom causes it to react as if it were polar even though, on the basis of the electronegativity of the carbon and iodine atoms, the bond is nonpolar.

The Nucleophile in an S$_N$2 Reaction

When we talk about atoms or molecules that have lone-pair electrons, sometimes we call them bases and sometimes we call them nucleophiles. What is the difference between a base and a nucleophile?

Basicity is a measure of how well a compound (a **base**) shares its lone pair. The stronger the base, the better it shares its electrons. Basicity is measured by an *equilibrium constant* (the acid dissociation constant, K_a) that indicates the tendency of the conjugate acid of the base to lose a proton (Section 1.17).

Nucleophilicity is a measure of how readily a compound (a **nucleophile**) is able to attack an electron-deficient atom. Nucleophilicity is measured by a *rate constant* (k). In the case of an S$_N$2 reaction, nucleophilicity is a measure of how readily the nucleophile attacks an sp^3 carbon bonded to a leaving group.

When we compare molecules that have *the same attacking atom*, we find a direct relationship between basicity and nucleophilicity: *stronger bases are better nucleophiles*. For example, a species with a negative charge is a stronger base *and* a better nucleophile than a species that has the same attacking atom but that is neutral. Thus, HO$^-$ is a stronger base and a better nucleophile than H$_2$O. Notice that bases are described as being strong or weak; nucleophiles are described as being good or poor.

stronger base,   weaker base,
better nucleophile   poorer nucleophile

HO$^-$ > H$_2$O
CH$_3$O$^-$ > CH$_3$OH
$^-$NH$_2$ > NH$_3$
CH$_3$CH$_2$NH$^-$ > CH$_3$CH$_2$NH$_2$

When we compare molecules whose *attacking atoms are approximately the same in size*, we discover that the *stronger bases are again the better nucleophiles*. The atoms in the second row of the periodic table have approximately the same size. If hydrogens are attached to the second-row elements, the resulting compounds have the following relative acidities (Section 1.20):

relative acid strengths

$$\boxed{\text{weakest acid}} \rightarrow \quad NH_3 \ < \ H_2O \ < \ HF$$

Consequently, the conjugate bases have the following relative base strengths and relative nucleophilicities:

relative base strengths and relative nucleophilicities

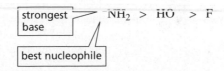

$$\boxed{\text{strongest base}} \rightarrow \quad ^-NH_2 \ > \ HO^- \ > \ F^-$$

$$\boxed{\text{best nucleophile}}$$

Note that the amide anion is the strongest base, as well as the best nucleophile.

In comparing molecules whose *attacking atoms are very different in size,* another factor comes into play: the polarizability of the atom. Because the electrons are farther away in the larger atom, they are not held as tightly and can, therefore, move more freely toward a positive charge. As a result, the electrons are able to overlap the orbital of carbon from farther away, as shown in Figure 8.5. This results in a greater degree of bonding in the transition state, making the transition state more stable.

Figure 8.5 ▶
An iodide ion is larger and more polarizable than a fluoride ion. Therefore, when an iodide ion attacks a carbon, the relatively loosely held electrons of the ion can overlap the orbital of carbon from farther away. The tightly bound electrons of the fluoride ion cannot start to overlap the orbital of carbon until the reactants are closer together.

Now the question becomes, does the greater polarizability that helps the larger atoms to be better nucleophiles make up for the decreased basicity that causes them to be poorer nucleophiles? We will see that the answer depends on the conditions under which the reaction is carried out.

If the reaction is carried out in an **aprotic polar solvent**—meaning the solvent molecules *do not have* a hydrogen bonded to an oxygen or to a nitrogen—the direct relationship between basicity and nucleophilicity is maintained: the stronger bases are still the best nucleophiles. In other words, the greater polarizability of the larger atoms does not make up for their decreased basicity. *Therefore iodide ion is the poorest nucleophile of the halide ions in an aprotic polar solvent.* If, however, the reaction is carried out in a **protic solvent**—meaning the solvent molecules *have* a hydrogen bonded to an oxygen or to a nitrogen—the relationship between basicity and nucleophilicity becomes inverted. The largest atom is the best nucleophile even though it is the weakest base. *Therefore, iodide ion is the best nucleophile of the halide ions in a protic solvent.*

An aprotic solvent does not contain a hydrogen bonded to either an oxygen or a nitrogen; it is not a hydrogen bond donor.

A protic solvent contains a hydrogen bonded to an oxygen or a nitrogen; it is a hydrogen bond donor.

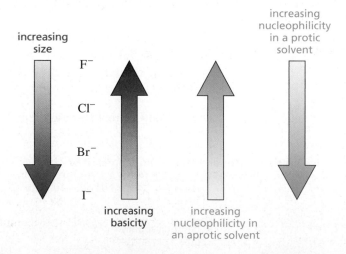

a. Which is a stronger base, RO$^-$ or RS$^-$?

b. Which is a better nucleophile in an aqueous solution?

Nucleophilicity Is Affected by the Solvent

Why, in a protic solvent, is the smallest atom the poorest nucleophile even though it is the strongest base? *How does a protic solvent make strong bases less nucleophilic*? When a negatively charged species is placed in a protic solvent, the ion becomes solvated (Section 2.9). Protic solvents are hydrogen bond donors, so the solvent molecules arrange themselves with their partially positively charged hydrogens pointing toward the negatively charged species. The interaction between the ion and the dipole of the protic solvent is called an **ion–dipole interaction**.

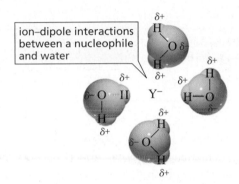

ion–dipole interactions between a nucleophile and water

Because the solvent shields the nucleophile, at least one of the ion–dipole interactions must be broken before the nucleophile can participate in an S$_N$2 reaction. Weak bases interact weakly with protic solvents, whereas strong bases interact more strongly because they are better at sharing their electrons. It is easier, therefore, to break the ion–dipole interactions between an iodide ion (a weak base) and the solvent than between a fluoride ion (a stronger base) and the solvent. As a result, in a protic solvent, an iodide ion is a better nucleophile than a fluoride ion (Table 8.2).

Table 8.2	Relative Nucleophilicity Toward CH$_3$I in Methanol

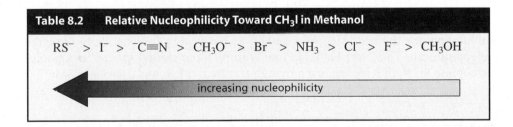

RS$^-$ > I$^-$ > $^-$C≡N > CH$_3$O$^-$ > Br$^-$ > NH$_3$ > Cl$^-$ > F$^-$ > CH$_3$OH

increasing nucleophilicity

Fluoride ion would be a better nucleophile in a *nonpolar solvent* than in a polar solvent because there would be no ion–dipole interactions between the ion and the nonpolar solvent; but ionic compounds are insoluble in most nonpolar solvents. They can, however, dissolve in aprotic polar solvents, such as dimethylformamide (DMF) or dimethylsulfoxide (DMSO). An aprotic polar solvent is not a hydrogen bond donor because it does not have a hydrogen attached to an oxygen or to a nitrogen, so there are no positively charged hydrogens to form ion–dipole interactions. The molecules of an aprotic polar solvent have a partial negative charge on their surface that can solvate cations, but the partial positive charge is on the *inside* of the molecule, and therefore less accessible. Fluoride ion, therefore, is a good nucleophile in DMSO and a poor nucleophile in water.

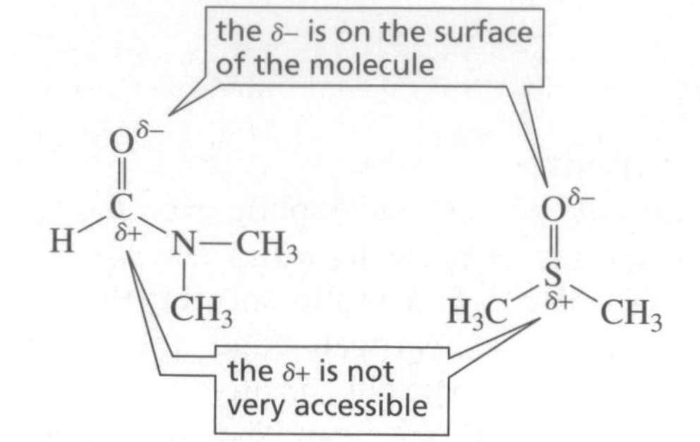

the δ– is on the surface of the molecule

the δ+ is not very accessible

N,N-dimethylformamide
DMF

dimethyl sulfoxide
DMSO

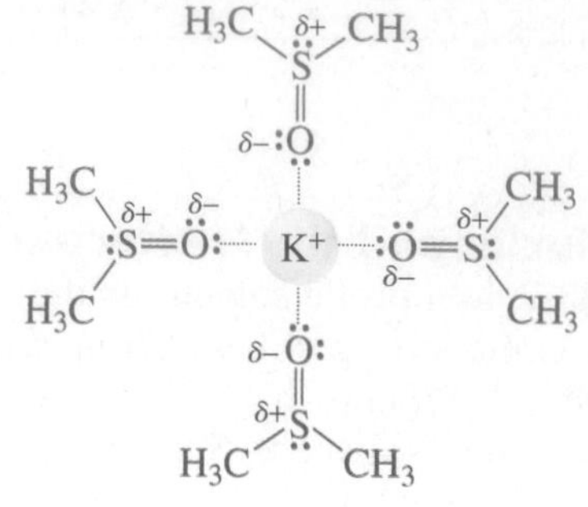

DMSO can solvate a cation better than it can solvate an anion

---

PROBLEM 6◆

Indicate whether each of the following solvents is protic or aprotic:

a. chloroform ($CHCl_3$)
b. diethyl ether ($CH_3CH_2OCH_2CH_3$)

c. acetic acid (CH_3COOH)
d. hexane [$CH_3(CH_2)_4CH_3$]

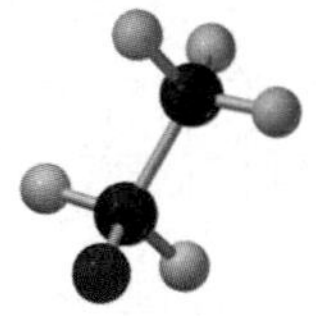

---

ethoxide ion

tert-butoxide ion

Nucleophilicity Is Affected by Steric Effects

Base strength is relatively unaffected by steric effects because a base removes a relatively unhindered proton. The strength of a base depends only on how well the base shares its electrons with a proton. Thus, *tert*-butoxide ion, in spite of its bulkier substituents, is a stronger base than ethoxide ion since *tert*-butanol ($pK_a = 18$) is a weaker acid than ethanol ($pK_a = 15.9$).

$$CH_3CH_2O^-$$

ethoxide ion
better nucleophile

$$CH_3\overset{\displaystyle CH_3}{\underset{\displaystyle CH_3}{C}}O^-$$

tert-butoxide ion
stronger base

Steric effects, on the other hand, do affect nucleophilicity. A bulky nucleophile cannot approach the back side of a carbon as easily as a less sterically hindered nucleophile can. Thus, *tert*-butoxide ion, with its three methyl groups, is a poorer nucleophile than ethoxide ion, even though *tert*-butoxide ion is a stronger base.

PROBLEM 7 **SOLVED**

List the following species in order of *decreasing* nucleophilicity in an aqueous solution:

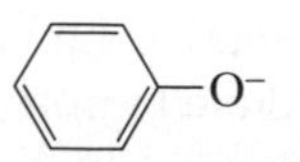

$$\text{⟨benzene⟩—O}^- \quad CH_3OH \quad HO^- \quad CH_3\overset{\displaystyle O}{\overset{\|}{C}}O^- \quad CH_3S^-$$

Solution Let's first divide the nucleophiles into groups. There is one nucleophile with a negatively charged sulfur, three with negatively charged oxygens, and one with a neutral oxygen. We know that in the polar aqueous solvent, the compound with the negatively charged sulfur is the most nucleophilic because sulfur is larger than oxygen. We also know that the poorest nucleophile is the one with the neutral oxygen atom. To complete the problem, we need to rank the three nucleophiles with negatively charged oxygens, which we can do by looking at the pK_a values of their conjugate acids. A carboxylic acid is a stronger acid than phenol, which is a stronger acid than water (Section 7.9). Because water is the weakest acid, its conjugate base is the strongest base and the best nucleophile. Thus, the relative nucleophilicities are:

$$CH_3S^- \quad > \quad HO^- \quad > \quad \underset{\text{[benzene ring]}}{}-O^- \quad > \quad CH_3\overset{\overset{O}{\|}}{C}O^- \quad > \quad CH_3OH$$

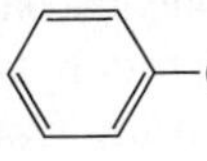

PROBLEM 8◆

For each of the following pairs of S$_N$2 reactions, indicate which reaction occurs faster:

a. $CH_3CH_2Br + H_2O$ or $CH_3CH_2Br + HO^-$

b. $CH_3\underset{\overset{|}{CH_3}}{C}HCH_2Br + HO^-$ or $CH_3CH_2\underset{\overset{|}{CH_3}}{C}HBr + HO^-$

c. $CH_3CH_2Cl + CH_3O^-$ or $CH_3CH_2Cl + CH_3S^-$
(in ethanol)

d. $CH_3CH_2Cl + I^-$ or $CH_3CH_2Br + I^-$

8.4   The Reversibility of an S$_N$2 Reaction Depends on the Basicities of the Leaving Groups in the Forward and Reverse Directions

Many different kinds of nucleophiles can react with alkyl halides. Therefore, a wide range of organic compounds can be synthesized by means of S$_N$2 reactions.

$$CH_3CH_2Cl + HO^- \longrightarrow \underset{\text{an alcohol}}{CH_3CH_2OH} + Cl^-$$

$$CH_3CH_2Br + HS^- \longrightarrow \underset{\text{a thiol}}{CH_3CH_2SH} + Br^-$$

$$CH_3CH_2I + RO^- \longrightarrow \underset{\text{an ether}}{CH_3CH_2OR} + I^-$$

$$CH_3CH_2Br + RS^- \longrightarrow \underset{\text{a thioether}}{CH_3CH_2SR} + Br^-$$

$$CH_3CH_2Cl + {^-}NH_2 \longrightarrow \underset{\text{a primary amine}}{CH_3CH_2NH_2} + Cl^-$$

$$CH_3CH_2Br + {^-}C{\equiv}CR \longrightarrow \underset{\text{an alkyne}}{CH_3CH_2C{\equiv}CR} + Br^-$$

$$CH_3CH_2I + {^-}C{\equiv}N \longrightarrow \underset{\text{a nitrile}}{CH_3CH_2C{\equiv}N} + I^-$$

Notice that the second-to-last reaction is the reaction of an alkyl halide with an acetylide ion. This is the reaction we looked at in Section 6.11 that is used to create longer carbon chains. Now we know that it is an S$_N$2 reaction.

It may seem that the reverse of each of these reactions can also occur via nucleophilic substitution. In the first reaction, for example, ethyl chloride reacts with hydroxide ion to form ethyl alcohol and a chloride ion. The reverse reaction would also appear to satisfy the requirements for a nucleophilic substitution reaction, given that chloride ion is a nucleophile and ethyl alcohol has an HO^- leaving group. But ethyl alcohol and chloride ion do *not* react.

Why does a nucleophilic substitution reaction take place in one direction, but not in the other? We can answer this question by comparing the leaving tendency of Cl^- in the forward direction and the leaving tendency of HO^- in the reverse direction. To compare leaving tendencies means comparing basicities. Because HCl is a much stronger acid than H_2O (Table 8.3), Cl^- is a much weaker base than HO^-; and because it is a weaker base, Cl^- is a better leaving group. Consequently, HO^- can displace Cl^- in the forward reaction, but Cl^- cannot displace HO^- in the reverse reaction. This is shown graphically in Figure 8.6a: the free energy of activation in the forward direction is much smaller than the free energy of activation in the reverse direction. Thus, an S_N2 reaction proceeds in the direction that allows the stronger base to displace the weaker base (the better leaving group).

An S_N2 reaction proceeds in the direction that allows the stronger base to displace the weaker base.

Table 8.3	The Acidities of the Conjugate Acids of Some Leaving Groups	
Acid	**pK_a**	**Conjugate base (leaving group)**
HI	−10.0	I^-
HBr	−9.0	Br^-
HCl	−7.0	Cl^-
C$_6$H$_5$—SO$_3$H	−6.5	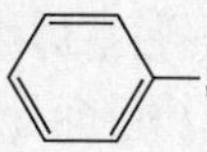 C$_6$H$_5$—SO$_3^-$
H_2SO_4	−5.0	$^-OSO_3H$
$CH_3\overset{+}{O}H_2$	−2.5	CH_3OH
H_3O^+	−1.7	H_2O
HF	3.2	F^-
$CH_3\overset{O}{\overset{\|}{C}}OH$	4.8	$CH_3\overset{O}{\overset{\|}{C}}O^-$
H_2S	7.0	HS^-
$HC{\equiv}N$	9.1	$^-C{\equiv}N$
$\overset{+}{N}H_4$	9.4	NH_3
CH_3CH_2SH	10.5	$CH_3CH_2S^-$
$(CH_3)_3\overset{+}{N}H$	10.8	$(CH_3)_3N$
CH_3OH	15.5	CH_3O^-
H_2O	15.7	HO^-
$HC{\equiv}CH$	25	$HC{\equiv}C^-$
NH_3	36	$^-NH_2$
H_2	~40	H^-

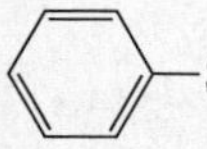

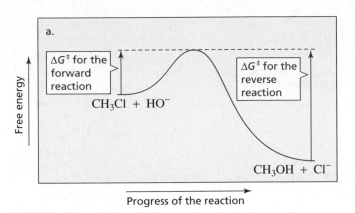

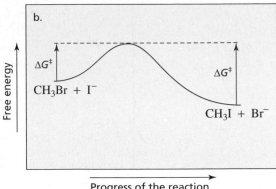

▲ **Figure 8.6**
(a) A reaction coordinate diagram for an irreversible S$_N$2 reaction.
(b) A reaction coordinate diagram for an reversible S$_N$2 reaction.

If the difference between the basicities of the nucleophile and the leaving group is not very large, the reaction will be reversible (Figure 8.6b). For example, in the reaction of ethyl bromide with iodide ion, Br$^-$ is the leaving group in one direction and I$^-$ is the leaving group in the other direction. The reaction is reversible because the pK_a values of the conjugate acids of the two leaving groups are similar (pK_a of HBr $= -9$; pK_a of HI $= -10$; see Table 8.3).

$$CH_3CH_2Br \; + \; I^- \; \rightleftharpoons \; CH_3CH_2I \; + \; Br^-$$

> an S$_N$2 reaction is reversible when the basicities of the leaving groups are similar

We have seen that a reversible reaction can be driven toward the desired products by removing one of the products as it is formed (Le Châtelier's principle; Section 3.7). Thus, if the concentration of product C is decreased, A and B will react to form more C and D to maintain the value of the equilibrium constant.

$$A \; + \; B \; \rightleftharpoons \; C \; + \; D$$

$$K_{eq} = \frac{[C][D]}{[A][B]}$$

For example, the reaction of chloroethane with methanol is reversible because the difference between the basicities of the nucleophile and the leaving group is not very large. However, if the reaction is carried out in a neutral solution, the protonated product will lose a proton (Section 1.24), disturbing the equilibrium and driving the reaction toward the products.

If an equilibrium is disturbed, the system will adjust to offset the disturbance.

$$CH_3CH_2Cl \; + \; CH_3OH \; \rightleftharpoons \; CH_3CH_2\overset{+}{O}CH_3 \; \xrightarrow{\text{fast}} \; CH_3CH_2OCH_3 \; + \; H^+$$
$$\overset{|}{H} \; + \; Cl^-$$

PROBLEM 9   SOLVED

The reaction of methyl chloride with hydroxide ion at 30 °C has a $\Delta G°$ value of -21.7 kcal/mol. What is the equilibrium constant for the reaction?

Solution The equation needed to calculate K_{eq} from the change in free energy is given in Section 3.7.

$$\ln K_{eq} = \frac{-\Delta G°}{RT}$$

$$\ln K_{eq} = \frac{-(-21.7 \text{ kcal mol}^{-1})}{0.001986 \text{ kcal mol}^{-1} \text{ K}^{-1} \times 303 \text{ K}} = \frac{21.7}{0.60}$$

$$\ln K_{eq} = 36.1$$

$$K_{eq} = 4.8 \times 10^{15}$$

As expected, this highly exergonic reaction (Figure 8.6a) has a very large equilibrium constant.

PROBLEM 10

Check to see whether the pK_a values listed in Table 8.3 predict correctly the direction of each of the reactions on page 357.

PROBLEM 11◆

What is the product of the reaction of ethyl bromide with each of the following nucleophiles?

a. $CH_3CH_2CH_2O^-$ **b.** $CH_3C{\equiv}C^-$ **c.** $(CH_3)_3N$ **d.** $CH_3CH_2S^-$

PROBLEM 12 *SOLVED*

What product is obtained when ethylamine reacts with excess methyl iodide in a basic solution of potassium carbonate?

$$CH_3CH_2\ddot{N}H_2 \ + \ CH_3{-}I \ \xrightarrow{\textbf{K}_2\textbf{CO}_3} \ ?$$
$$\textbf{excess}$$

Solution Methyl iodide and ethylamine undergo an S_N2 reaction. The product of the reaction is a secondary amine that is predominantly in its basic (neutral) form since the reaction is carried out in a basic solution. The secondary amine can undergo an S_N2 reaction with another equivalent of methyl iodide, forming a tertiary amine. The tertiary amine can react with methyl iodide in yet another S_N2 reaction. The final product of the reaction is a quaternary ammonium iodide.

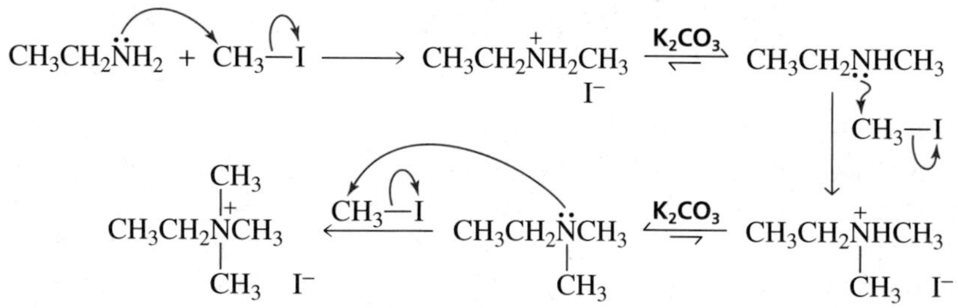

PROBLEM 13

a. Explain why the reaction of an alkyl halide with ammonia gives a low yield of primary amine.

b. Explain why a much better yield of primary amine is obtained from the reaction of an alkyl halide with azide ion ($^-N_3$), followed by catalytic hydrogenation. (*Hint:* An alkyl azide is not nucleophilic.)

$$CH_3CH_2CH_2Br \xrightarrow{^-N_3} CH_3CH_2CH_2N{=}\overset{+}{N}{=}N^- \xrightarrow[\text{Pt}]{H_2} CH_3CH_2CH_2NH_2 \ + \ N_2$$
$$\textbf{an alkyl azide}$$

8.5 The Mechanism of an S$_N$1 Reaction

Given our understanding of S$_N$2 reactions, we would expect the rate of the reaction of 2-bromo-2-methylpropane with water to be very slow because water is a poor nucleophile and 2-bromo-2-methylpropane is sterically hindered to attack by a nucleophile. It turns out, however, that the reaction is surprisingly fast. In fact, it is over 1 million times faster than the reaction of bromomethane (a compound with no steric hindrance) with water (Table 8.4). Clearly, the reaction must be taking place by a mechanism different from that of an S$_N$2 reaction.

$$
\begin{array}{ccc}
\underset{\underset{\text{2-bromo-2-methylpropane}}{}}{\overset{\text{CH}_3}{\underset{\text{CH}_3}{\text{CH}_3\text{C}-\text{Br}}}} + \text{H}_2\text{O} & \longrightarrow & \underset{\underset{\text{2-methyl-2-propanol}}{}}{\overset{\text{CH}_3}{\underset{\text{CH}_3}{\text{CH}_3\text{C}-\text{OH}}}} + \text{HBr}
\end{array}
$$

Table 8.4	Relative Rates of S$_N$1 Reactions for Several Alkyl Bromides (solvent is H_2O, nucleophile is H_2O)	
Alkyl bromide	**Class of alkyl bromide**	**Relative rate**
$\overset{\text{CH}_3}{\underset{\text{CH}_3}{\text{CH}_3\text{C}-\text{Br}}}$	tertiary	1,200,000
$\underset{\text{CH}_3}{\text{CH}_3\text{CH}-\text{Br}}$	secondary	11.6
$\text{CH}_3\text{CH}_2-\text{Br}$	primary	1.00*
CH_3-Br	methyl	1.05*

*Although the rate of the S$_N$1 reaction of this compound with water is 0, a small rate is observed as a result of an S$_N$2 reaction.

We have seen that in order to determine the mechanism of a reaction, we need to find out what factors affect the rate of the reaction, and we need to know the configuration of the products of the reaction. Finding that doubling the concentration of the alkyl halide doubles the rate of the reaction but changing the concentration of the nucleophile has no effect on the rate of the reaction, allows us to write a rate law for the reaction:

$$\text{rate } = k[\text{alkyl halide}]$$

The rate of the reaction depends on the concentration of only one reactant. Thus, the reaction is a **first-order reaction** (Section 3.7).

Because the rate law for the reaction of 2-bromo-2-methylpropane with water differs from the rate law for the reaction of bromomethane with hydroxide ion (Section 8.2), the two reactions must have different mechanisms. We have seen that the reaction between bromomethane and hydroxide ion is an S_N2 reaction. The reaction between 2-bromo-2-methylpropane and water is an **S_N1 reaction**, where "S" stands for substitution, "N" stands for nucleophilic, and "1" stands for unimolecular. **Unimolecular** means that only one molecule is involved in the transition state of the rate-determining step. The mechanism of an S_N1 reaction is based on the following experimental evidence:

1. The rate law shows that the rate of the reaction depends only on the concentration of the alkyl halide. This means that the transition state of the rate-determining step involves only the alkyl halide.

2. When the methyl groups of 2-bromo-2-methylpropane are successively replaced by hydrogens, the rate of the S_N1 reaction decreases progressively (Table 8.4). This is the opposite of the pattern of reactivity exhibited by alkyl halides in S_N2 reactions (Table 8.1).

3. The substitution reaction of an alkyl halide in which the halogen is bonded to an asymmetric center forms two stereoisomers: one with the same relative configuration at the asymmetric center as the reacting alkyl halide, and the other with the inverted configuration.

Unlike an S_N2 reaction, where the leaving group departs and the nucleophile approaches *at the same time*, the leaving group in an S_N1 reaction departs *before* the nucleophile approaches.

Mechanism for the S_N1 reaction of an alkyl halide

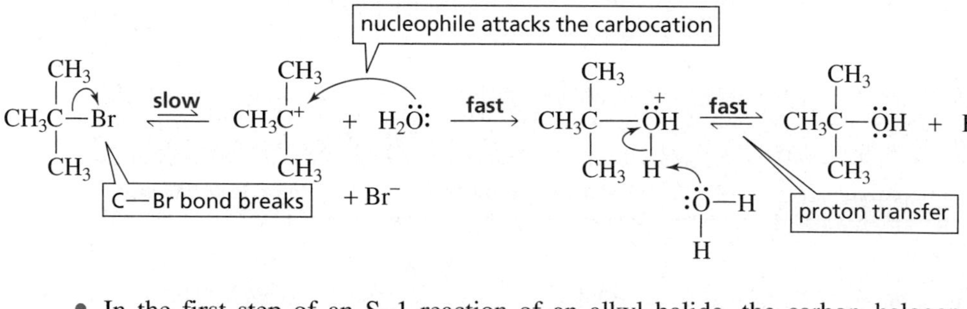

An S_N1 reaction is a two-step reaction.

*   In the first step of an S_N1 reaction of an alkyl halide, the carbon–halogen bond breaks and the previously shared pair of electrons stays with the halogen. As a result, a carbocation intermediate is formed.

*   In the second step, the nucleophile reacts rapidly with the carbocation to form a protonated alcohol.

*   Whether the alcohol product will exist in its protonated (acidic) form or neutral (basic) form depends on the pH of the solution. At pH = 7, the alcohol will exist predominantly in its neutral form (Section 1.24).

Because the rate of an S$_N$1 reaction depends only on the concentration of the alkyl halide, the first step must be the slow (rate-determining) step. The nucleophile is not involved in the rate-determining step, so its concentration has no effect on the rate of the reaction. If you look at the reaction coordinate diagram in Figure 8.7, you will see why increasing the rate of the second step will not make an S$_N$1 reaction go any faster.

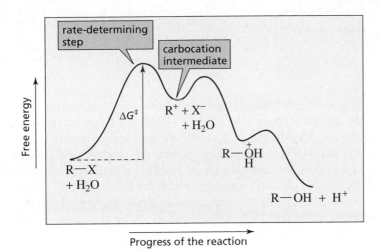

◀ **Figure 8.7**
A reaction coordinate diagram for an S$_N$1 reaction.

How does the mechanism for an S$_N$1 reaction account for the three pieces of experimental evidence? First, because the alkyl halide is the only species that participates in the rate-determining step, the mechanism agrees with the observation that the rate of the reaction depends only on the concentration of the alkyl halide; it does not depend on the concentration of the nucleophile.

Second, the mechanism shows that a carbocation is formed in the rate-determining step. We know that a tertiary carbocation is more stable and is therefore formed more easily than a secondary carbocation, which in turn is more stable and formed more easily than a primary carbocation (Section 4.2). Tertiary alkyl halides, therefore, are more reactive than secondary alkyl halides, which are more reactive than primary alkyl halides. This relative order of reactivity agrees with the observation that the rate of an S$_N$1 reaction decreases as the methyl groups of 2-bromo-2-methylpropane are successively replaced by hydrogens (Table 8.4).

Carbocation stability: 3° > 2° > 1°.

Primary alkyl halides and methyl halides cannot undergo S$_N$1 reactions.

relative reactivities of alkyl halides in an S$_N$1 reaction

most reactive ▷ 3° alkyl halide > 2° alkyl halide > 1° alkyl halide ◁ too unreactive to undergo an S$_N$1 reaction

In reality, primary carbocations and methyl cations are so unstable that primary alkyl halides and methyl halides do not undergo S$_N$1 reactions. (The very slow reactions reported for ethyl bromide and bromomethane in Table 8.4 are S$_N$2 reactions.)

Third, the positively charged carbon of the carbocation intermediate is sp^2 hybridized, which means the three bonds connected to it are in the same plane. In the second step of the S$_N$1 reaction, the nucleophile can approach the carbocation from either side of the plane.

Student Tutorial: S$_N$1

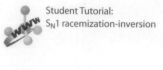

Student Tutorial:
S_N1 racemization-inversion

Student Tutorial:
S_N1 racemization-retention

If the nucleophile attacks the side of the carbon from which the leaving group departed (labeled b in the depiction above), the product will have the same relative configuration as that of the reacting alkyl halide. If, however, the nucleophile attacks the opposite side of the carbon (labeled a in the depiction above), the product will have the inverted configuration relative to the configuration of the alkyl halide. We can now understand why an S_N1 reaction of an alkyl halide in which the leaving group is attached to an asymmetric center forms two stereoisomers: attack of the nucleophile on one side of the planar carbocation forms one stereoisomer and attack on the other side produces the other stereoisomer.

if the leaving group in an S_N1 reaction is attached to an asymmetric center, a pair of enantiomers will be formed as products

---

PROBLEM 14◆

Arrange the following alkyl bromides in order of decreasing reactivity in an S_N1 reaction: 2-bromopropane, 1-bromopropane, 2-bromo-2-methylpropane, bromomethane.

PROBLEM 15

Explain why the rates given in Table 8.4 for the S_N2 reactions of bromoethane and bromomethane are so slow.

8.6 Factors That Affect S_N1 Reactions

We will now look at how the nature of the leaving group and the nucleophile affect S_N1 reactions, and we will see that these reactions can undergo carbocation rearrangements.

The Leaving Group in S_N1 Reactions

Because the rate-determining step of an S_N1 reaction is the dissociation of the alkyl halide to form a carbocation, two factors affect the rate of the reaction: (1) the ease with which the leaving group dissociates from the carbon and (2) the stability of the carbocation that is formed. In the preceding section, we saw that tertiary alkyl halides are more reactive than secondary alkyl halides because tertiary carbocations are more stable than secondary carbocations and, therefore, more easily formed. We also saw that primary carbocations are so unstable that primary alkyl halides do not undergo S_N1 reactions.

But how do we rank the relative reactivity of a series of alkyl halides with different leaving groups that dissociate to form the same carbocation? As in the case of an S$_N$2 reaction, there is a direct relationship between basicity and leaving ability in an S$_N$1 reaction: the weaker the base, the less tightly it is bonded to the carbon and the more easily the carbon–halogen bond can be broken. As a result, an alkyl iodide is the most reactive and an alkyl fluoride is the least reactive of the alkyl halides in both S$_N$1 and S$_N$2 reactions.

relative reactivities of alkyl halides in an S$_N$1 reaction

most reactive > RI > RBr > RCl > RF < least reactive

The Nucleophile in S$_N$1 Reactions

We have seen that the rate-determining step of an S$_N$1 reaction is formation of the carbocation. Because the nucleophile does not participate until *after* the rate-determining step, the reactivity of the nucleophile has no effect on the rate of an S$_N$1 reaction (Figure 8.7).

In most S$_N$1 reactions, the solvent is the nucleophile. For example, the relative rates given in Table 8.4 are for the reactions of alkyl halides with water in water. Water serves as both the nucleophile and the solvent. Reaction with a solvent is called **solvolysis**. Thus, the relative rates in Table 8.4 are for the solvolysis of the indicated alkyl bromides in water.

Carbocation Rearrangements

In Section 4.6, we saw that a carbocation will rearrange if it becomes more stable in the process. If the carbocation formed in an S$_N$1 reaction rearranges, S$_N$1 and S$_N$2 reactions of the same alkyl halide can produce different constitutional isomers as products, since a carbocation is not formed in an S$_N$2 reaction and therefore the carbon skeleton cannot rearrange. For example, the product obtained when OH is substituted for the Br of 2-bromo-3-methylbutane by an S$_N$1 reaction is different from the product obtained by an S$_N$2 reaction. When the reaction is carried out under conditions that favor an S$_N$1 reaction, the secondary carbocation formed initially undergoes a 1,2-hydride shift to form a more stable tertiary carbocation. In Sections 8.9 and 8.10, you will see that we can exercise some control over whether an S$_N$1 or an S$_N$2 reaction takes place by selecting appropriate reaction conditions.

> When a reaction forms a carbocation intermediate, always check for the possibility of a carbocation rearrangement.

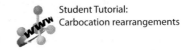

Student Tutorial:
Carbocation rearrangements

Arrange the following alkyl halides in order of decreasing reactivity in an S$_N$1 reaction: 2-bromopentane, 2-chloropentane, 1-chloropentane, 3-bromo-3-methylpentane.

PROBLEM 17◆

Which of the following alkyl halides form a substitution product in an S_N1 reaction that is different from the substitution product formed in an S_N2 reaction?

a. $CH_3\overset{CH_3}{\underset{CH_3}{\underset{|}{\overset{|}{C}H}}}\overset{Br}{\underset{|}{CH}}CHCH_3$

c. $CH_3CH_2\overset{CH_3}{\underset{CH_3\ Br}{\overset{|}{C}}}{-}\overset{}{\underset{}{C}H}CH_3$

e. cyclohexane with Br—CHCH₃

b. cyclohexane with CH₃ and Cl

d. $CH_3\overset{}{\underset{Cl}{\overset{|}{C}H}}CH_2\overset{CH_3}{\underset{CH_3}{\overset{|}{C}}}CH_3$

f. cyclohexane with CH₃ and Br

PROBLEM 18◆

Two substitution products result from the reaction of 3-chloro-3-methyl-1-butene with sodium acetate ($CH_3COO^-\ Na^+$) in acetic acid under conditions that favor an S_N1 reaction. Identify the products.

8.7 More About the Stereochemistry of S$_N$2 and S$_N$1 Reactions

We have seen that when an alkyl halide with a leaving group attached to an asymmetric center undergoes an S_N2 reaction, the product has an inverted configuration, but when it undergoes an S_N1 reaction, the product is a pair of enantiomers. Let's now look at some examples of the stereoisomers formed in S_N2 and S_N1 reactions.

The Stereochemistry of S$_N$2 Reactions

The reaction of 2-bromopropane with hydroxide ion forms a substitution product without any asymmetric centers. The product, therefore, has no stereoisomers.

$$CH_3\underset{Br}{\overset{|}{C}H}CH_3 + HO^- \longrightarrow CH_3\underset{OH}{\overset{|}{C}H}CH_3 + Br^-$$

2-bromopropane   **2-propanol**

The reaction of 2-bromobutane with hydroxide ion forms a substitution product with an asymmetric center. The product, therefore, has stereoisomers.

$$CH_3\underset{Br}{\overset{|}{C}H}CH_2CH_3 + HO^- \longrightarrow CH_3\underset{OH}{\overset{|}{C}H}CH_2CH_3 + Br^-$$

asymmetric center   asymmetric center

2-bromobutane   **2-butanol**

However we cannot predict which stereoisomers of the product will be formed unless we know the configuration of the reactant *and* whether the reaction is an S_N2 or an S_N1 reaction.

For example, when (S)-2-bromobutane undergoes an S_N2 reaction, we know that the product will be (R)-2-butanol, because in an S_N2 reaction, the incoming nucleophile attacks the back side of the carbon that is attached to the halogen (Section 8.2). Therefore, the configuration of the product will be inverted relative to that of the reactant. (Recall that an S_N2 reaction takes place with *inversion of configuration*.)

An S$_N$2 reaction takes place with inversion of configuration.

the configuration is inverted
relative to that of the reactant

CH$_2$CH$_3$ | C—H | H$_3$C | Br
(S)-2-bromobutane

+ HO⁻ — S$_N$2 conditions →

CH$_2$CH$_3$ | H—C | HO CH$_3$
(R)-2-butanol

+ Br⁻

The Stereochemistry of S$_N$1 Reactions

In contrast to the S$_N$2 reaction, the S$_N$1 reaction of (S)-2-bromobutane forms two substitution products: one with the same relative configuration as the reactant and the other with the inverted configuration because, in an S$_N$1 reaction, the nucleophile can attack either side of the planar carbocation intermediate (Section 8.5).

An S$_N$1 reaction takes place with racemization.

CH$_2$CH$_3$ | C—H | H$_3$C Br
(S)-2-bromobutane

+ H$_2$O — S$_N$1 conditions →

product with inverted configuration

CH$_2$CH$_3$ | H—C | HO CH$_3$
(R)-2-butanol

+

product with retained configuration

CH$_2$CH$_3$ | C—H | H$_3$C OH
(S)-2-butanol

+ HBr

Although you might expect that equal amounts of both products should be formed in an S$_N$1 reaction, a greater amount of the product with the inverted configuration is obtained in most cases. Typically, 50–70% of the product of an S$_N$1 reaction is the inverted product. If the reaction leads to equal amounts of the two stereoisomers, the reaction is said to take place with **complete racemization**. When more of one of the products is formed, the reaction is said to take place with **partial racemization**.

Saul Winstein was the first to explain why a greater amount of inverted product is generally formed in an S$_N$1 reaction. He postulated that dissociation of the alkyl halide initially results in the formation of an **intimate ion pair**. In an intimate ion pair, the bond between the carbon and the leaving group has broken, but the cation and anion remain next to each other. When they move slightly father apart, they become a *solvent-separated ion pair*, meaning an ion pair with one or more solvent molecules between the cation and the anion. As the ions separate further, they become dissociated ions.

If the leaving group is attached to an asymmetric center, an S$_N$2 reaction forms the stereoisomer with the inverted configuration.

If the leaving group is attached to an asymmetric center, an S$_N$1 reaction forms a pair of enantiomers.

R—X ⟶ R⁺ X⁻ ⟶ R⁺ [solvent] X⁻ ⟶ R⁺ [solvent] X⁻

undissociated molecule | intimate ion pair | solvent-separated ion pair | dissociated ions

The nucleophile can attack any of these four species. If the nucleophile attacks only the completely dissociated carbocation, the product will be completely racemized. If the nucleophile attacks the carbocation of either the intimate ion pair or the solvent-separated ion pair, the leaving group will be in position to partially block the approach of the nucleophile to that side of the carbocation and more of the product with the inverted configuration will be formed.

H$_2$O ⤻ C⁺ ⤸ H$_2$O
H$_3$C H

Br⁻ has diffused away, giving H$_2$O equal access to both sides of the carbocation

H$_2$O ⤻ C⁺ Br⁻ ⤫ H$_2$O
H$_3$C H

Br⁻ has not diffused away, so it blocks the approach of H$_2$O to one side of the carbocation

B I O G R A P H Y

Saul Winstein (1912–1969) *was born in Montreal, Canada. He received a Ph.D. from the California Institute of Technology and was a professor of chemistry at the University of California, Los Angeles, from 1942 until his death.*

(Notice that if the nucleophile attacks the undissociated molecule, the reaction will be an S_N2 reaction and all of the product will have the inverted configuration.)

The difference between the products obtained from an S_N1 reaction and from an S_N2 reaction is a little easier to visualize in the case of cyclic compounds. For example, when *cis*-1-bromo-4-methylcyclohexane undergoes an S_N2 reaction, only the trans product is obtained because the carbon bonded to the leaving group is attacked by the nucleophile only on its back side.

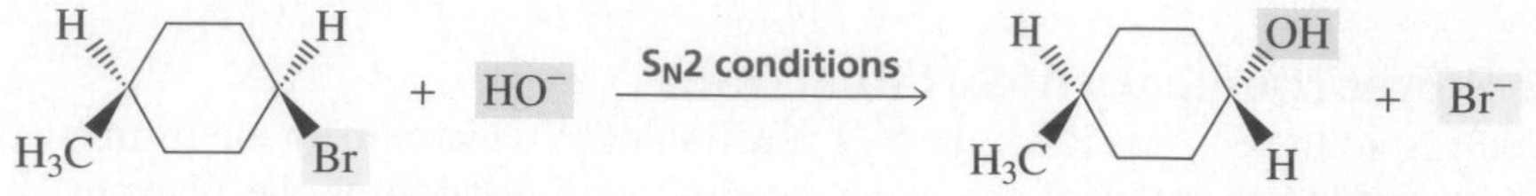

cis-1-bromo-4-methylcyclohexane   *trans*-4-methylcyclohexanol

However, when *cis*-1-bromo-4-methylcyclohexane undergoes an S_N1 reaction, both the cis and the trans products are formed because the nucleophile can approach the carbocation intermediate from either side.

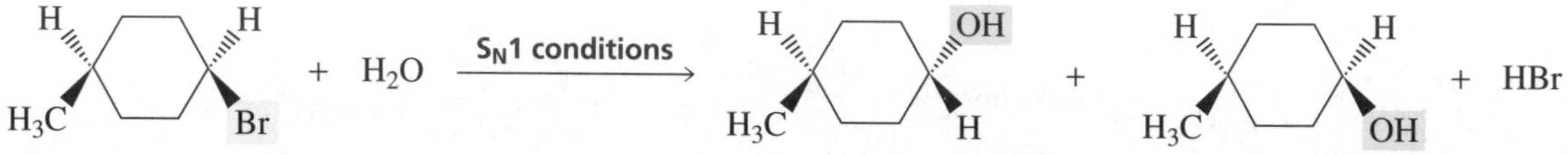

cis-1-bromo-4-methylcyclohexane   *trans*-4-methylcyclohexanol   *cis*-4-methylcyclohexanol

---

PROBLEM 19

Give the substitution products that will be formed from the following reactions if:

a. the reaction is carried out under conditions that favor an S_N2 reaction.

b. the reaction is carried out under conditions that favor an S_N1 reaction.
   1. *trans*-1-iodo-4-ethylcyclohexane + sodium methoxide/methanol
   2. *cis*-1-chloro-3-methylcyclobutane + sodium hydroxide/water

---

8.8 Benzylic Halides, Allylic Halides, Vinylic Halides, and Aryl Halides

Up to this point, our discussion of substitution reactions has been limited to methyl halides and primary, secondary, and tertiary alkyl halides. But what about benzylic, allylic, vinylic, and aryl halides?

Let's first consider benzylic and allylic halides. Unless they are tertiary, these halides readily undergo S_N2 reactions. Tertiary benzylic and tertiary allylic halides, like other tertiary halides, are unreactive in S_N2 reactions because of steric hindrance.

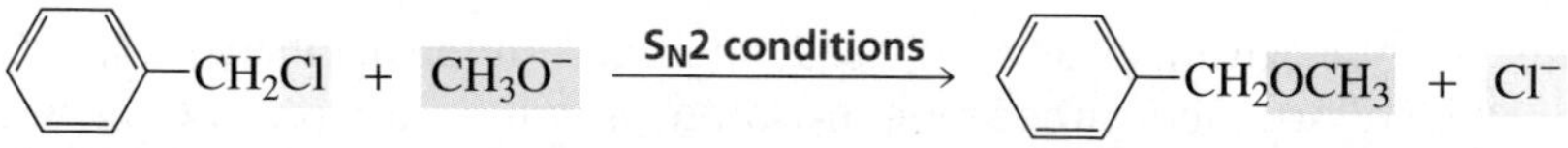

benzyl chloride   benzyl methyl ether

$$CH_3CH{=}CHCH_2Br \ + \ HO^- \xrightarrow{\text{S}_N\text{2 conditions}} CH_3CH{=}CHCH_2OH \ + \ Br^-$$
1-bromo-2-butene   **2-buten-1-ol**
an allylic halide

Benzylic and allylic halides readily undergo S_N1 reactions, as well, because they form relatively stable carbocations. While primary alkyl halides (such as CH_3CH_2Br and $CH_3CH_2CH_2Br$) cannot undergo S_N1 reactions because their carbocations are too unstable, primary benzylic and primary allylic halides readily undergo S_N1 reactions because their carbocations are stabilized by electron delocalization (Section 7.7).

Benzylic and allylic halides undergo S_N1 and S_N2 reactions.

$$\text{C}_6\text{H}_5-\text{CH}_2\text{Cl} \xrightleftharpoons{\text{S}_\text{N}1} \text{C}_6\text{H}_5-\overset{+}{\text{CH}}_2 + \text{Cl}^- \xrightarrow{\text{CH}_3\text{OH}} \text{C}_6\text{H}_5-\text{CH}_2\text{OCH}_3 + \text{H}^+$$

$$\text{CH}_2=\text{CHCH}_2\text{Br} \xrightleftharpoons{\text{S}_\text{N}1} \text{CH}_2=\text{CH}\overset{+}{\text{C}}\text{H}_2 + \text{Br}^- \longleftrightarrow \overset{+}{\text{C}}\text{H}_2\text{CH}=\text{CH}_2 \xrightarrow{\text{H}_2\text{O}} \text{CH}_2=\text{CHCH}_2\text{OH} + \text{H}^+$$

If the resonance contributors of the allylic carbocation intermediate have different groups bonded to their sp^2 carbons, two substitution products will be formed. This is another example of how electron delocalization can affect the nature of the products formed in a reaction (Section 7.10).

$$\text{CH}_3\text{CH}=\text{CHCH}_2\text{Br} \xrightleftharpoons{\text{S}_\text{N}1} \text{CH}_3\text{CH}=\text{CH}\overset{+}{\text{C}}\text{H}_2 \longleftrightarrow \text{CH}_3\overset{+}{\text{C}}\text{HCH}=\text{CH}_2 + \text{Br}^-$$

$$\downarrow \text{H}_2\text{O} \qquad\qquad\qquad \downarrow \text{H}_2\text{O}$$

$$\text{CH}_3\text{CH}=\text{CHCH}_2\text{OH} \qquad\qquad \underset{\underset{\text{OH}}{|}}{\text{CH}_3\text{CHCH}=\text{CH}_2}$$
$$+ \text{ H}^+ \qquad\qquad\qquad + \text{ H}^+$$

Vinylic halides and aryl halides (compounds in which the halogen is attached to an aromatic ring such as benzene) do not undergo S_N2 or S_N1 reactions. They do not undergo S_N2 reactions because, as the nucleophile approaches the back side of the sp^2 carbon, it is repelled by the π electron cloud of the double bond or the aromatic ring.

Vinylic and aryl halides undergo neither S_N1 nor S_N2 reactions.

a nucleophile is repelled by the π electron cloud

a vinylic halide   an aryl halide

There are two reasons vinylic halides and aryl halides do not undergo S_N1 reactions. First, vinylic and aryl cations are even more unstable than primary carbocations (Section 6.5) because the positive charge is on an sp carbon. Since sp carbons are more electronegative than the sp^2 carbons that carry the positive charge of alkyl carbocations, sp carbons are more resistant to becoming positively charged. Second, we have seen that sp^2 carbons form stronger bonds than do sp^3 carbons (Section 1.14). As a result, when a halogen is bonded to an sp^2 carbon, it is harder to break the carbon–halogen bond.

$$\text{RCH}=\text{CH}-\text{Cl} \overset{sp^2 \text{ hybridized}}{\nrightarrow} \text{RCH}=\overset{+}{\text{CH}} \overset{sp \text{ hybridized}}{} + \text{Cl}^-$$

vinylic cation
too unstable to be formed

$$\text{C}_6\text{H}_5-\text{Br} \overset{sp^2 \text{ hybridized}}{\nrightarrow} \text{C}_6\text{H}_5^+ \overset{sp \text{ hybridized}}{} + \text{Br}^-$$

aryl cation
too unstable to be formed

PROBLEM-SOLVING STRATEGY

Predicting Relative Reactivities

Which alkyl halide would you expect to be more reactive in an S_N1 solvolysis reaction?

$$CH_3\overset{\cdot\cdot}{\underset{\cdot\cdot}{O}}-CH=CH-CH_2Br \quad \text{or} \quad CH_3\overset{\cdot\cdot}{\underset{\cdot\cdot}{O}}CH_2-CH=CH-CH_2Br$$

When asked to determine the relative reactivities of two compounds, we need to compare the $\Delta G^{\ddagger}$ values of their rate-determining steps. The faster-reacting compound will have the *smaller* $\Delta G^{\ddagger}$ value, that is, the smaller difference between its free energy and the free energy of its rate-determining transition state. Both alkyl halides have approximately the same stability, so the difference in their reaction rates will be due to the difference in the stabilities of the transition states for their rate-determining steps. The rate-determining step is carbocation formation, so the compound that forms the more stable carbocation will be the one that has the faster rate of solvolysis. The compound on the left forms the more stable carbocation because it has three resonance contributors, whereas the other carbocation has only two significant resonance contributors.

$$CH_3\overset{\cdot\cdot}{\underset{\cdot\cdot}{O}}-CH=CH-\overset{+}{C}H_2 \longleftrightarrow CH_3\overset{\cdot\cdot}{\underset{\cdot\cdot}{O}}-\overset{+}{C}H-CH=CH_2 \longleftrightarrow CH_3\overset{\cdot\cdot}{\underset{\cdot\cdot}{O}}{}^{+}=CH-CH=CH_2$$

$$CH_3\overset{\cdot\cdot}{\underset{\cdot\cdot}{O}}CH_2-CH=CH-\overset{+}{C}H_2 \longleftrightarrow CH_3\overset{\cdot\cdot}{\underset{\cdot\cdot}{O}}CH_2-\overset{+}{C}H-CH=CH_2$$

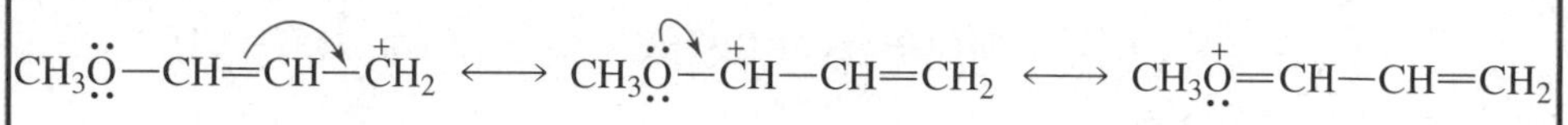

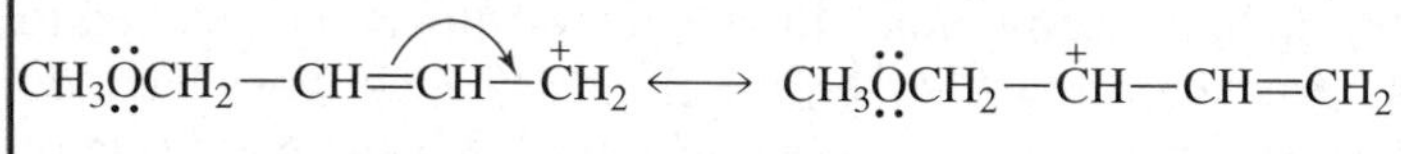

Now continue on to Problems 20–22.

PROBLEM 20◆

Which alkyl halide would you expect to be more reactive in an S_N1 solvolysis reaction?

or

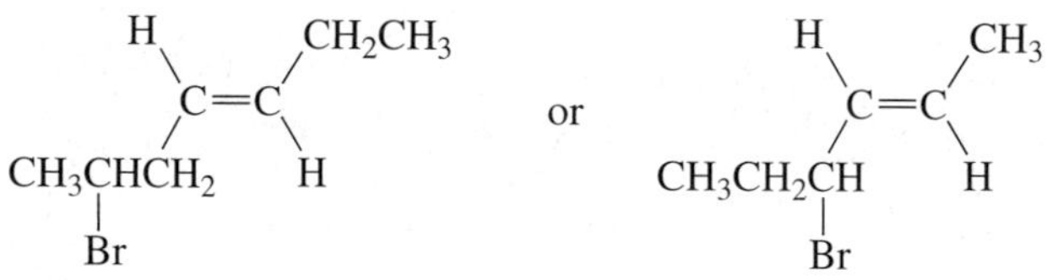

PROBLEM 21◆

Which alkyl halide would you expect to be more reactive in an S_N2 reaction with a given nucleophile? In each case, you can assume that both alkyl halides have the same stability.

a. 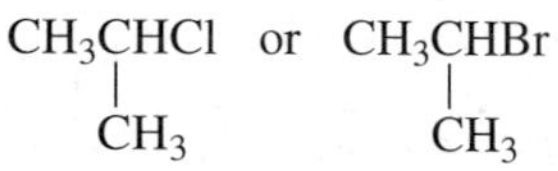 $CH_3\underset{\underset{CH_3}{|}}{C}HCl$ or $CH_3\underset{\underset{CH_3}{|}}{C}HBr$

b. $CH_3CH_2\underset{\overset{|}{CH_3}}{C}HBr$ or $CH_3CH_2\underset{\overset{|}{CH_2CH_3}}{C}HBr$

c. $CH_3CH_2CH_2\underset{\overset{|}{CH_3}}{C}HBr$ or $CH_3CH_2\underset{\overset{|}{CH_3}}{C}HCH_2Br$

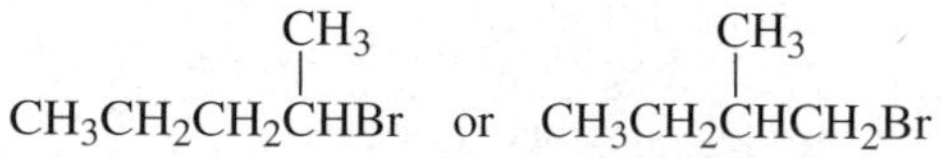

d. 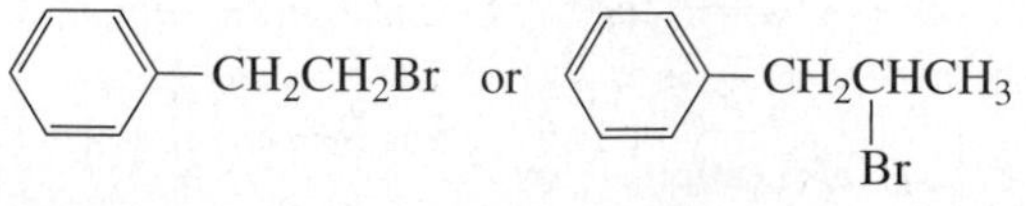 $-CH_2CH_2Br$ or $-CH_2\underset{\underset{Br}{|}}{C}HCH_3$

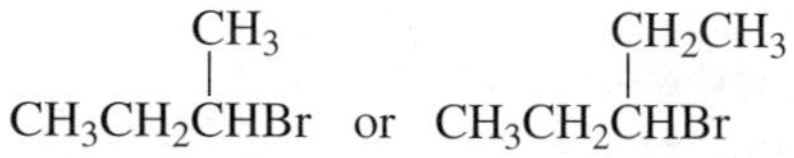

e. 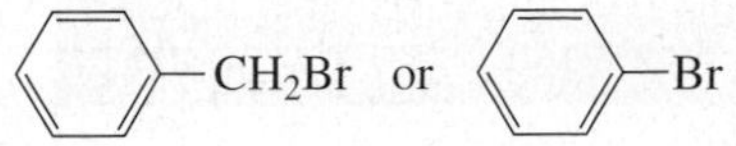 [benzene ring]—CH$_2$Br or [benzene ring]—Br

f. CH$_3$CH=CCH$_3$ or CH$_3$CH=CHCHCH$_3$
         | |
         Br   Br

g. CH$_3$CH$_2$CH$_2$Cl or CH$_3$OCH$_2$Cl

PROBLEM 22◆

For pairs a–f in Problem 21, which compound would be more reactive in an S_N1 reaction?

PROBLEM 23

Give the products obtained from the following reaction and show their configurations:

a. under conditions that favor an S_N2 reaction
b. under conditions that favor an S_N1 reaction

$$CH_3CH=CHCH_2Br \ + \ CH_3O^- \ \xrightarrow{CH_3OH}$$

8.9 Competition Between S_N2 and S_N1 Reactions

The characteristics of S_N2 and S_N1 reactions are compared in Table 8.5. Remember that the "2" in "S_N2" and the "1" in "S_N1" refer to the molecularity of the reaction (the number of molecules involved in the transition state of the rate-determining step). Thus, the rate-determining step of an S_N2 reaction is bimolecular, whereas the rate-determining step of an S_N1 reaction is unimolecular. These numbers do not refer to the number of steps in the mechanism. In fact, the opposite is true: an S_N2 reaction proceeds by a *one*-step concerted mechanism, and an S_N1 reaction proceeds by a *two*-step mechanism with a carbocation intermediate.

Table 8.5 Comparison of S_N2 and S_N1 Reactions	
S_N2	**S_N1**
A one-step mechanism	A stepwise mechanism that forms a carbocation intermediate
A bimolecular rate-determining step	A unimolecular rate-determining step
No carbocation rearrangements	Carbocation rearrangements
Product has inverted configuration relative to the reactant	Products have both retained and inverted configurations relative to the reactant
Reactivity order: methyl > 1° > 2° > 3°	Reactivity order: 3° > 2° > 1° > methyl

We have seen that *methyl halides and primary alkyl halides undergo only S_N2 reactions* because methyl cations and primary carbocations, which would have to be formed in an S_N1 reaction, are too unstable to be formed. *Tertiary alkyl halides undergo only S_N1 reactions* because steric hindrance makes them unreactive in S_N2 reactions. *Secondary alkyl halides as well as benzylic and allylic halides* (unless they are tertiary) *can undergo both S_N1 and S_N2 reactions* because they form relatively stable carbocations and the steric hindrance associated with these alkyl halides is generally not very great. *Vinylic and aryl halides do not undergo either S_N1 or S_N2 reactions*. These results are summarized in Table 8.6.

Table 8.6 Summary of the Reactivity of Alkyl Halides in Nucleophilic Substitution Reactions

Methyl and 1° alkyl halides	S_N2 only
Vinylic and aryl halides	Neither S_N1 nor S_N2
2° alkyl halides	S_N1 and S_N2
1° and 2° benzylic and 1° and 2° allylic halides	S_N1 and S_N2
3° alkyl halides	S_N1 only
3° benzylic and 3° allylic halides	S_N1 only

VCL **Alkyl Halide Solvolysis**

When an alkyl halide can undergo both an S_N1 reaction *and* an S_N2 reaction, both reactions take place simultaneously. The conditions under which the reaction is carried out determine which of the reactions predominates. Therefore, we have some control over which reaction takes place.

What conditions favor an S_N2 reaction? These are important questions to synthetic chemists because an S_N2 reaction forms a single substitution product, whereas an S_N1 reaction can form two substitution products if the leaving group is bonded to an asymmetric center. An S_N1 reaction is further complicated by carbocation rearrangements. In other words, an S_N2 reaction is a synthetic chemist's friend, but an S_N1 reaction can be a synthetic chemist's nightmare.

When the *structure* of the alkyl halide allows it to undergo both S_N2 and S_N1 reactions, three conditions determine which reaction will predominate: (1) the *concentration* of the nucleophile, (2) the *reactivity* of the nucleophile, and (3) the *solvent* in which the reaction is carried out. To understand how the concentration and the reactivity of the nucleophile affect whether an S_N2 or an S_N1 reaction predominates, we must examine the rate laws for the two reactions (Sections 8.2 and 8.5). The rate constants have been given subscripts that indicate the reaction order.

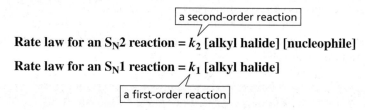

a second-order reaction

Rate law for an S_N2 reaction = k_2 [alkyl halide] [nucleophile]

Rate law for an S_N1 reaction = k_1 [alkyl halide]

a first-order reaction

The rate law for the reaction of an alkyl halide that can undergo both S_N2 and S_N1 reactions simultaneously is the sum of the individual rate laws.

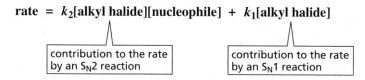

rate = k_2[alkyl halide][nucleophile] + k_1[alkyl halide]

contribution to the rate by an S_N2 reaction

contribution to the rate by an S_N1 reaction

From the rate law, you can see that increasing the *concentration* of the nucleophile increases the rate of an S_N2 reaction but has no effect on the rate of an S_N1 reaction. Therefore, when both reactions occur simultaneously, increasing the concentration of the nucleophile increases the fraction of the reaction that takes place by an S_N2 pathway. In contrast, decreasing the concentration of the nucleophile decreases the fraction of the reaction that takes place by an S_N2 pathway.

The slow (and only) step of an S_N2 reaction is attack of the nucleophile on the alkyl halide. Increasing the *reactivity* of the nucleophile increases the rate of an S_N2 reaction by increasing the value of the rate constant (k_2), because a more reactive nucleophile is better able to displace the leaving group. The slow step of an S_N1 reaction is the dissociation of the alkyl halide. The carbocation formed in the slow step reacts rapidly in a second step with any nucleophile present in the reaction mixture. Increasing the rate of the fast step does not affect the rate of the prior slow, carbocation-forming step. This means that increasing the reactivity of the nucleophile has no effect on the rate of an S_N1 reaction. A good nucleophile, therefore, favors an S_N2 reaction over an S_N1 reaction. A poor nucleophile favors an S_N1 reaction, not by increasing the rate of the S_N1 reaction itself, but by decreasing the rate of the competing S_N2 reaction. In summary:

Nucleophilic Substitution-1

- An S_N2 reaction is favored by a high concentration of a good nucleophile.
- An S_N1 reaction is favored by a poor nucleophile.

Look back at the S_N1 reactions in previous sections and notice that every one of them has a poor nucleophile (H_2O, CH_3OH), whereas all the S_N2 reactions have good nucleophiles (HO^-, CH_3O^-). In other words, a poor nucleophile encourages an S_N1 reaction, and a good nucleophile encourages an S_N2 reaction. In Section 8.10, we will look at the third factor that influences whether an S_N2 or an S_N1 reaction will predominate—the solvent in which the reaction is carried out.

PROBLEM-SOLVING STRATEGY

Predicting Whether a Substitution Reaction Is an S_N1 or an S_N2 Reaction

This problem will give you practice determining whether a substitution reaction will take place by an S_N1 or an S_N2 pathway. (Keep in mind that good nucleophiles encourage S_N2 reactions, whereas poor nucleophiles encourage S_N1 reactions.)

Give the configuration(s) of the substitution product(s) that will be formed from the reactions of the following secondary alkyl halides with the indicated nucleophile:

a.

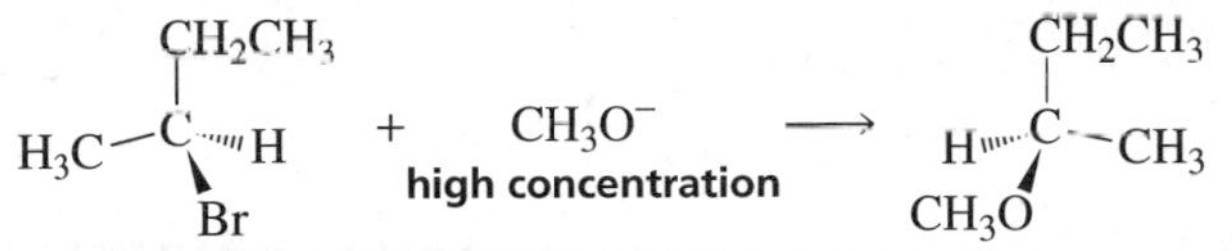

$$H_3C-\overset{\overset{\displaystyle CH_2CH_3}{|}}{\underset{\underset{\displaystyle Br}{|}}{C}}\cdots H \quad + \quad CH_3O^- \quad \xrightarrow{\text{high concentration}} \quad H\cdots\overset{\overset{\displaystyle CH_2CH_3}{|}}{\underset{\underset{\displaystyle CH_3O}{}}{C}}-CH_3$$

Because a high concentration of a good nucleophile is used, we can predict that the reaction is an S_N2 reaction. Therefore, the product will have the inverted configuration relative to the configuration of the reactant. (An easy way to draw the inverted product is to draw the mirror image of the reacting alkyl halide and then put the nucleophile in the same location as the leaving group.)

b.

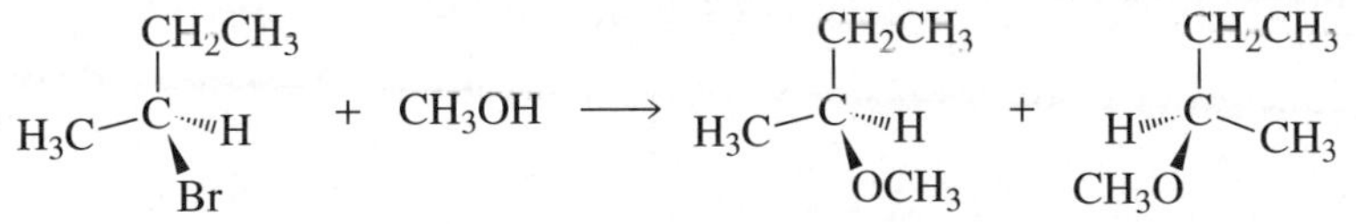

$$H_3C-\overset{\overset{\displaystyle CH_2CH_3}{|}}{\underset{\underset{\displaystyle Br}{|}}{C}}\cdots H \quad + \quad CH_3OH \quad \longrightarrow \quad H_3C-\overset{\overset{\displaystyle CH_2CH_3}{|}}{\underset{\underset{\displaystyle OCH_3}{}}{C}}\cdots H \quad + \quad H\cdots\overset{\overset{\displaystyle CH_2CH_3}{|}}{\underset{\underset{\displaystyle CH_3O}{}}{C}}-CH_3$$

Because a poor nucleophile is used, we can predict that an S_N1 reaction predominates. Therefore, we will obtain two substitution products, one with the retained configuration and one with the inverted configuration, relative to the configuration of the reactant.

c.
$$CH_3CH_2\overset{|}{\underset{|}{C}}HCH_2CH_3 \quad + \quad CH_3OH \quad \longrightarrow \quad CH_3CH_2\overset{|}{\underset{\underset{\displaystyle OCH_3}{}}{C}}HCH_2CH_3$$
with I below the carbon on the left structure

The poor nucleophile allows us to predict that the reaction is an S_N1 reaction. However, the product does not have an asymmetric center, so it does not have stereoisomers. (The same substitution product would have been obtained if the reaction had been an S_N2 reaction.)

d.
$$CH_3CH_2\overset{|}{\underset{\underset{\displaystyle Cl}{|}}{C}}HCH_3 \quad + \quad HO^- \quad \xrightarrow{\text{high concentration}} \quad CH_3CH_2\overset{|}{\underset{\underset{\displaystyle OH}{|}}{C}}HCH_3$$

Because a high concentration of a good nucleophile is employed, we can predict that the reaction is an S_N2 reaction. Therefore, the product will have the inverted configuration relative to the configuration of the reactant. But since the configuration of the reactant is not indicated, we do not know the configuration of the product.

e. 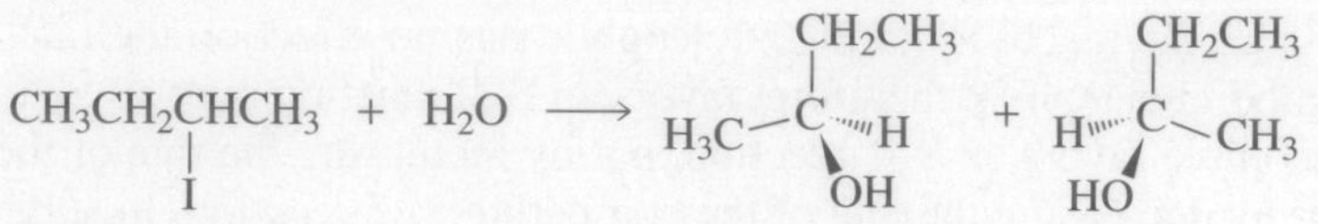 $CH_3CH_2CHCH_3$ + H_2O $\longrightarrow$

Because the nucleophile is poor, we can predict that the reaction is an S_N1 reaction. Therefore, both stereoisomers will be formed, regardless of the configuration of the reactant.

Now continue on to Problem 24.

PROBLEM 24

VCL   Williamson Ether Synthesis-1

Give the configuration(s) of the substitution product(s) that will be formed from the reactions of the following secondary alkyl halides with the indicated nucleophile:

a. 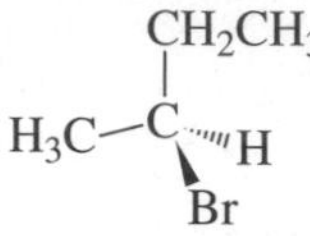 + $CH_3CH_2CH_2O^-$ **high concentration** $\longrightarrow$

b. 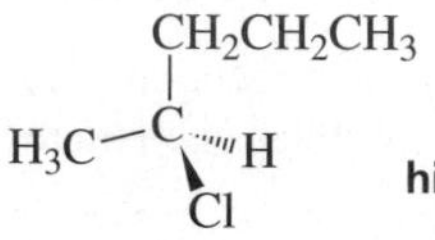 + NH_3 **high concentration** $\longrightarrow$

c. 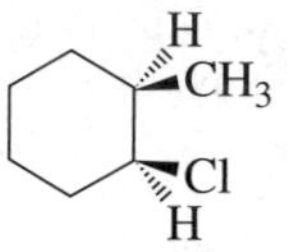 + CH_3O^- **high concentration** $\longrightarrow$

d. 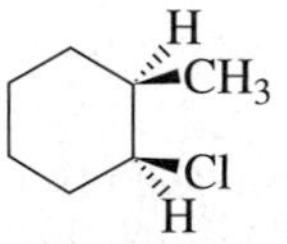 + CH_3OH $\longrightarrow$

e. 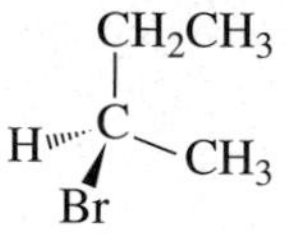 + CH_3O^- **high concentration** $\longrightarrow$

f. 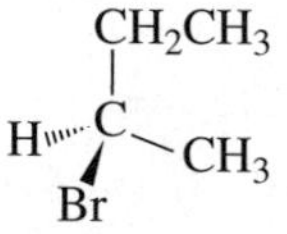 + CH_3OH $\longrightarrow$

PROBLEM 25◆

Which of the following reactions will go faster if the concentration of the nucleophile is increased?

a. 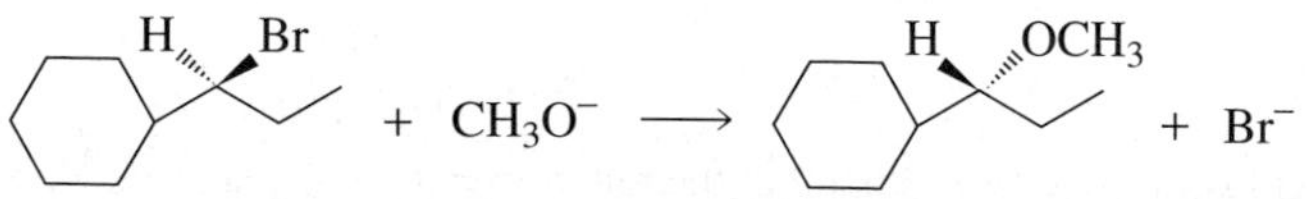 + CH_3O^- $\longrightarrow$ + Br^-

b.  + CH_3S^- $\longrightarrow$ + Br^-

c. 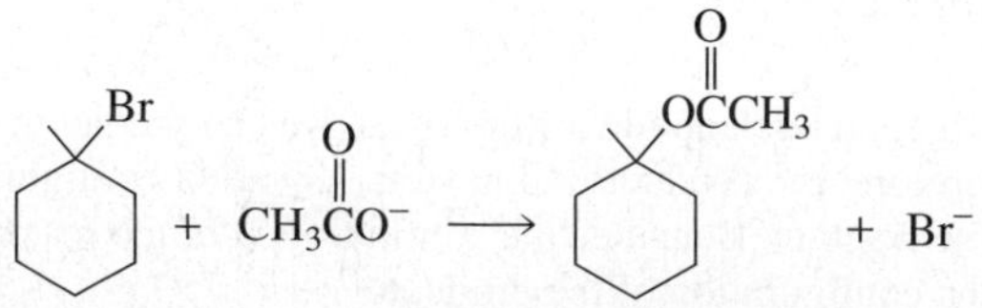 + $CH_3\overset{O}{\overset{\|}{C}}O^-$ $\longrightarrow$ + Br^-

PROBLEM 26 **SOLVED**

The rate law for the substitution reaction of 2-bromobutane and HO^- in 75% ethanol and 25% water at 30 °C is

$$\text{rate} = 3.20 \times 10^{-5}[\text{2-bromobutane}][HO^-] + 1.5 \times 10^{-6}[\text{2-bromobutane}]$$

What percentage of the reaction takes place by an S_N2 pathway when the following conditions are met?

a. $[HO^-] = 1.00$ M   **b.** $[HO^-] = 0.001$ M

Solution to 26a

$$\text{percentage by } S_N2 = \frac{\text{amount by } S_N2}{\text{amount by } S_N2 + \text{amount by } S_N1} \times 100$$

$$= \frac{3.20 \times 10^{-5}[\text{2-bromobutane}](1.00) \times 100}{3.20 \times 10^{-5}[\text{2-bromobutane}](1.00) + 1.5 \times 10^{-6}[\text{2-bromobutane}]}$$

$$= \frac{3.20 \times 10^{-5}}{3.20 \times 10^{-5} + 0.15 \times 10^{-5}} \times 100 = \frac{3.20 \times 10^{-5}}{3.35 \times 10^{-5}} \times 100$$

$$= 96\%$$

8.10 **The Role of the Solvent in S_N2 and S_N1 Reactions**

The solvent in which a nucleophilic substitution reaction is carried out also influences whether an S_N2 or an S_N1 reaction will predominate. Before we can understand how a particular solvent favors one reaction over another, however, we must understand how solvents stabilize organic molecules.

The **dielectric constant** of a solvent is a measure of how well the solvent can insulate opposite charges from one another. Solvent molecules insulate a charge by clustering around it, so that the positive poles of solvent molecules surround negative charges while the negative poles of solvent molecules surround positive charges. Recall that the interaction between a solvent and an ion or a molecule dissolved in that solvent is called *solvation* (Section 2.9). When an ion interacts with a polar solvent, the charge is no longer localized solely on the ion, but is spread out to the surrounding solvent molecules. Spreading out the charge stabilizes the charged species.

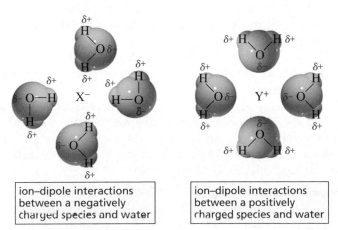

ion–dipole interactions between a negatively charged species and water

ion–dipole interactions between a positively charged species and water

Polar solvents have high dielectric constants and thus are very good at insulating (solvating) charges. Nonpolar solvents have low dielectric constants and are poor insulators. The dielectric constants of some common solvents are listed in Table 8.7 where they are divided into two groups: protic solvents and aprotic solvents. Recall that

Table 8.7 The Dielectric Constants of Some Common Solvents

Solvent	Structure	Abbreviation	Dielectric constant (ε, at 25 °C)	Boiling point (°C)
Protic solvents				
Water	H_2O	—	79	100
Formic acid	HCOOH	—	59	100.6
Methanol	CH_3OH	MeOH	33	64.7
Ethanol	CH_3CH_2OH	EtOH	25	78.3
tert-Butyl alcohol	$(CH_3)_3COH$	*tert*-BuOH	11	82.3
Acetic acid	CH_3COOH	HOAc	6	117.9
Aprotic solvents				
Dimethyl sulfoxide	$(CH_3)_2SO$	DMSO	47	189
Acetonitrile	CH_3CN	MeCN	38	81.6
Dimethylformamide	$(CH_3)_2NCHO$	DMF	37	153
Hexamethylphosphoric acid triamide	$[(CH_3)_2N]_3PO$	HMPA	30	233
Acetone	$(CH_3)_2CO$	Me_2CO	21	56.3
Dichloromethane	CH_2Cl_2	—	9.1	40
Tetrahydrofuran	[structure]	THF	7.6	66
Ethyl acetate	$CH_3COOCH_2CH_3$	EtOAc	6	77.1
Diethyl ether	$CH_3CH_2OCH_2CH_3$	Et_2O	4.3	34.6
Benzene	[structure]	—	2.3	80.1
Hexane	$CH_3(CH_2)_4CH_3$	—	1.9	68.7

protic solvents contain a hydrogen bonded to an oxygen or a nitrogen, so protic solvents are hydrogen bond donors. **Aprotic solvents**, on the other hand, do not have a hydrogen bonded to an oxygen or a nitrogen, so they are *not* hydrogen bond donors.

Stabilization of charges by solvent interaction plays an important role in organic reactions. For example, when an alkyl halide undergoes an S_N1 reaction, the first step is dissociation of the carbon–halogen bond to form a carbocation and a halide ion. Energy is required to break the bond, but with no bonds being formed, where does the energy come from? If the reaction is carried out in a polar solvent, the ions that are produced are solvated. The energy associated with a single ion–dipole interaction is small, but the additive effect of all the ion–dipole interactions that take place when a solvent stabilizes a charged species represents a great deal of energy. These ion–dipole interactions provide much of the energy necessary for dissociation of the carbon–halogen bond. So, in an S_N1 reaction, the alkyl halide does not fall apart spontaneously, but instead, polar solvent molecules pull it apart. An S_N1 reaction, therefore, cannot take place in a nonpolar solvent. It also cannot take place in the gas phase, where there are no solvent molecules and, consequently, no solvation effects.

SOLVATION EFFECTS

The tremendous amount of energy provided by solvation can be appreciated by considering the energy required to break the crystal lattice of sodium chloride (table salt) (Figure 1.1). In the absence of a solvent, sodium chloride must be heated to more than 800 °C to overcome the forces that hold the oppositely charged ions together. However, sodium chloride readily dissolves in water at room temperature because solvation of the Na^+ and Cl^- ions by water provides the energy necessary to separate the ions.

How a Solvent Affects Reaction Rates in General

One simple rule describes how a change in solvent will affect the rate of most chemical reactions: *increasing the polarity of the solvent will decrease the rate of the reaction if one or more reactants in the rate-determining step are charged and will increase the rate of the reaction if none of the reactants in the rate-determining step is charged.*

Now let's see why this rule is true. The rate of a reaction depends on the difference between the free energy of the reactants and the free energy of the transition state for the rate-determining step of the reaction. We can, therefore, predict how changing the polarity of the solvent will affect the rate of a reaction simply by looking at the charge on the reactant or reactants of the rate-determining step and the charge on the transition state of the rate-determining step, to see which of these species will be more stabilized by a polar solvent.

The greater the charge on a solvated molecule, the stronger will be its interactions with a polar solvent and more the charge will be stabilized. Therefore, if the charge on the reactants is greater than the charge on the transition state, a polar solvent will stabilize the reactants more than it will stabilize the transition state, increasing the difference in energy ($\Delta G^{\ddagger}$) between them. Consequently, *increasing the polarity of the solvent will decrease the rate of the reaction*, as shown in Figure 8.8.

> **Increasing the polarity of the solvent will decrease the rate of the reaction if one or more reactants in the rate-determining step are charged.**

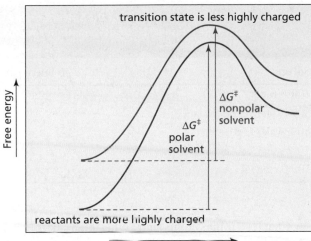

◀ **Figure 8.8**
A reaction coordinate diagram for a reaction in which the charge on the reactants is greater than the charge on the transition state.

On the other hand, if the charge on the transition state is greater than the charge on the reactants, a polar solvent will stabilize the transition state more than it will stabilize the reactants. Therefore, *increasing the polarity of the solvent* will decrease the difference in energy ($\Delta G^{\ddagger}$) between transition state and reactants, which *will increase the rate of the reaction*, as shown in Figure 8.9.

> **Increasing the polarity of the solvent will increase the rate of the reaction if none of the reactants in the rate-determining step is charged.**

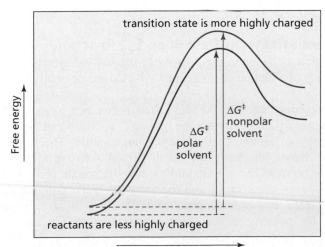

◀ **Figure 8.9**
A reaction coordinate diagram for a reaction in which the charge on the transition state is greater than the charge on the reactants.

How a Solvent Affects the Rate of an S$_N$1 Reaction

Now let's look at specific kinds of reactions, beginning with an S$_N$1 reaction of an alkyl halide. The alkyl halide, which is the only reactant in the rate-determining step of an S$_N$1 reaction, is a neutral molecule with a small dipole moment. The rate-determining transition state has a greater charge because as the carbon–halogen bond breaks, the carbon becomes more positive and the halogen becomes more negative. Since the charge on the transition state is greater than the charge on the reactant, increasing the polarity of the solvent will stabilize the transition state more than the reactant and will increase the rate of the S$_N$1 reaction (Figure 8.9 and Table 8.8).

rate-determining step of an S$_N$1 reaction

reactant   transition state   products

Table 8.8	The Effect of the Polarity of the Solvent on the Rate of Reaction of 2-Bromo-2-methylpropane in an S$_N$1 Reaction
Solvent	**Relative rate**
100% water	1200
80% water / 20% ethanol	400
50% water / 50% ethanol	60
20% water / 80% ethanol	10
100% ethanol	1

We will see in Chapter 10 that compounds other than alkyl halides also undergo S$_N$1 reactions. As long as the compound undergoing an S$_N$1 reaction is neutral, increasing the polarity of the solvent will *increase* the rate of the S$_N$1 reaction because the polar solvent will stabilize the dispersed charges on the transition state more than it will stabilize the relatively neutral reactant (Figure 8.9). If, however, the compound undergoing an S$_N$1 reaction is charged, increasing the polarity of the solvent will *decrease* the rate of the reaction because the more polar solvent will stabilize the full charge on the reactant to a greater extent than it will stabilize the dispersed charge on the transition state (Figure 8.8).

How a Solvent Affects the Rate of an S$_N$2 Reaction

The way in which a change in the polarity of the solvent affects the rate of an S$_N$2 reaction depends on whether the reactants are charged or neutral, just as in an S$_N$1 reaction.

Most S$_N$2 reactions of alkyl halides occur between a neutral alkyl halide and a charged nucleophile. Increasing the polarity of a solvent will have a strong stabilizing effect on the negatively charged nucleophile. The transition state also has a negative charge, but that charge is dispersed over two atoms. Consequently, the interactions between the solvent and the transition state are not as strong as the interactions between the solvent and the fully charged nucleophile. Therefore, a polar solvent stabilizes the nucleophile more than it stabilizes the transition state, so increasing the polarity of the solvent will decrease the rate of the reaction (Figure 8.8).

If, however, the S_N2 reaction occurs between an alkyl halide and a neutral nucleophile, the charge on the transition state will be larger than the charge on the neutral reactants, so increasing the polarity of the solvent will increase the rate of the substitution reaction (Figure 8.9).

In summary, the way a change in solvent affects the rate of a substitution reaction does not depend on the mechanism of the reaction. It depends *only* on whether a reactant in the rate-determining step is charged. *If a reactant in the rate-determining step is charged, increasing the polarity of the solvent will decrease the rate of the reaction. If none of the reactants in the rate-determining step is charged, increasing the polarity of the solvent will increase the rate of the reaction.*

When we discussed the solvation of charged species by a polar solvent, the polar solvents we considered were hydrogen bond donors (protic polar solvents) such as water and alcohols. Some polar solvents—for example, *N,N*-dimethylformamide (DMF), dimethyl sulfoxide (DMSO), and hexamethylphosphoric acid triamide (HMPA)—are not hydrogen bond donors; they are aprotic polar solvents (Table 8.7).

Because a polar solvent decreases the rate of an S_N2 reaction when the nucleophile is negatively charged, we would like to carry out such a reaction in a nonpolar solvent. However, negatively charged nucleophiles generally will not dissolve in nonpolar solvents. Instead, an aprotic polar solvent is used. Because aprotic polar solvents are not hydrogen bond donors, they are less effective than polar protic solvents at solvating negative charges; indeed, we have seen that they solvate negative charges very poorly (Section 8.3). Thus, the rate of an S_N2 reaction with a negatively charged nucleophile will be greater in an aprotic polar solvent than in a protic polar solvent. Consequently, an aprotic polar solvent is the solvent of choice for an S_N2 reaction in which the nucleophile is negatively charged, whereas a protic polar solvent is used if the nucleophile is a neutral molecule.

We have now seen that when an alkyl halide can undergo both S_N2 and S_N1 reactions, the S_N2 reaction will be favored by a high concentration of a good (negatively charged) nucleophile in an aprotic polar solvent, whereas the S_N1 reaction will be favored by a poor (neutral) nucleophile in a protic polar solvent.

An S_N2 reaction of an alkyl halide is favored by a high concentration of a good nucleophile in an aprotic polar solvent.

An S_N1 reaction of an alkyl halide is favored by a poor nucleophile in a protic polar solvent.

ENVIRONMENTAL ADAPTATION

The microorganism *Xanthobacter* has learned to use alkyl halides that reach the ground as industrial pollutants as a source of carbon. The microorganism synthesizes an enzyme that uses the alkyl halide as a starting material to produce other carbon-containing compounds that it

needs. This enzyme has several nonpolar groups at its active site (the pocket in the enzyme where the reaction it catalyzes takes place). The first step of the enzyme-catalyzed reaction is an S_N2 reaction with a charged nucleophile. The nonpolar groups on the surface of the enzyme provide the nonpolar environment needed to increase the rate of the reaction.

PROBLEM 27◆

If each of the following S_N2 reactions is run in a protic polar solvent, how will the rate change if the polarity of the solvent is increased?

Student Tutorial:
S_N2 Promoting factors

a. $CH_3CH_2CH_2CH_2Br \ + \ HO^- \longrightarrow CH_3CH_2CH_2CH_2OH \ + \ Br^-$

b. $CH_3\overset{+}{\underset{\underset{CH_3}{|}}{S}}CH_3 \ + \ CH_3O^- \longrightarrow CH_3OCH_3 \ + \ CH_3SCH_3$

c. $CH_3CH_2I \ + \ NH_3 \longrightarrow CH_3CH_2\overset{+}{N}H_3 \ I^-$

Student Tutorial:
Common terms

PROBLEM 28◆

Which reaction in each of the following pairs will take place more rapidly?

a. $CH_3Br + HO^- \longrightarrow CH_3OH + Br^-$

$CH_3Br + H_2O \longrightarrow CH_3OH + HBr$

b. $CH_3I + HO^- \longrightarrow CH_3OH + I^-$

$CH_3Cl + HO^- \longrightarrow CH_3OH + Cl^-$

c. $CH_3Br + NH_3 \longrightarrow CH_3\overset{+}{N}H_3 + Br^-$

$CH_3Br + H_2O \longrightarrow CH_3OH + Br^-$

d. $CH_3Br + HO^- \xrightarrow{\textbf{DMSO}} CH_3OH + Br^-$

$CH_3Br + HO^- \xrightarrow{\textbf{EtOH}} CH_3OH + Br^-$

e. $CH_3Br + NH_3 \xrightarrow{\textbf{Et}_2\textbf{O}} CH_3\overset{+}{N}H_3 + Br^-$

$CH_3Br + NH_3 \xrightarrow{\textbf{EtOH}} CH_3\overset{+}{N}H_3 + Br^-$

PROBLEM 29   *SOLVED*

Most of the pK_a values given throughout this text are values determined in water. How would the pK_a values of the following classes of compounds change if they were determined in a solvent less polar than water such as 50% water/50% dioxane: carboxylic acids, alcohols, phenols, ammonium ions (RNH_3^+), and anilinium ions ($C_6H_5NH_3^+$)?

Solution   A pK_a is the negative logarithm of an equilibrium constant, K_a (Section 1.17). Because we are determining how changing the polarity of a solvent affects an equilibrium constant, we must look at how changing the polarity of the solvent affects the stability of the reactants and products (Section 3.7).

$$K_a = \frac{[\textbf{B}^-]\,[\textbf{H}^+]}{[\textbf{HB}]} \qquad K_a = \frac{[\textbf{B}]\,[\textbf{H}^+]}{[\textbf{HB}^+]}$$

| a neutral acid | a positively charged acid |

Carboxylic acids, alcohols, and phenols are neutral in their acidic forms (HB) and charged in their basic forms (B^-). A polar protic solvent will stabilize B^- and H^+ more than it will stabilize HB, thereby increasing K_a. Therefore, K_a will be larger in water than in a less polar solvent, so the pK_a values of carboxylic acids, alcohols, and phenols will be higher (they will be weaker acids) in a less polar solvent.

Ammonium ions and anilinium ions are charged in their acidic forms (HB^+) and neutral in their basic forms (B). A polar solvent will stabilize HB^+ and H^+ more than it will stabilize B. Because HB^+ is stabilized slightly more than H^+, K_a will be smaller in water than in a less polar solvent, so the pK_a values of ammonium ions and anilinium ions will be lower (they will be stronger acids) in a less polar solvent.

PROBLEM 30◆

Would you expect acetate ion ($CH_3CO_2^-$) to be a more reactive nucleophile in an S_N2 reaction carried out in methanol or in dimethyl sulfoxide?

PROBLEM 31◆

Under which of the following reaction conditions would (*R*)-2-chlorobutane form the most (*R*)-2-butanol: HO⁻ in 50% water and 50% ethanol or HO⁻ in 100% ethanol?

8.11 Intermolecular Versus Intramolecular Reactions

A molecule with two functional groups is called a **bifunctional molecule**. If the two functional groups are able to react with each other, two kinds of reactions can occur. Let's say the two functional groups are those that can react in an S_N2 reaction: a strong nucleophile and an alkyl halide. The nucleophile of one molecule of the compound can displace the bromide ion of a second molecule. Such a reaction is called an inter-molecular reaction. *Inter* is Latin for "between": an **intermolecular reaction** takes place between two molecules. If the product of this reaction subsequently reacts with a third bifunctional molecule (and then a fourth, and so on), a polymer will be formed. A polymer is a large molecule formed by linking together repeating units of small molecules (Chapter 28).

an intermolecular reaction

$$BrCH_2(CH_2)_nCH_2\ddot{\underset{..}{O}}\colon^- \quad Br^-CH_2(CH_2)_nCH_2\ddot{\underset{..}{O}}\colon^- \longrightarrow BrCH_2(CH_2)_nCH_2\ddot{\underset{..}{O}}CH_2(CH_2)_nCH_2\ddot{\underset{..}{O}}\colon^- \ + \ Br^-$$

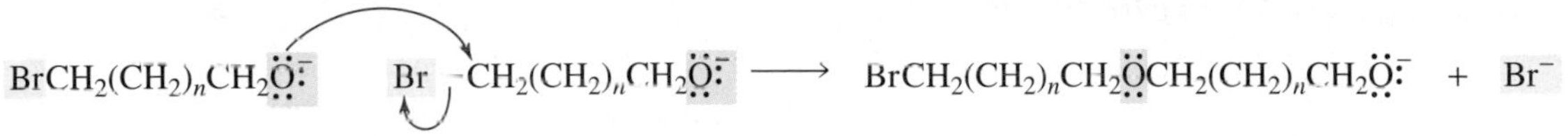

Alternatively, the nucleophile of a molecule can displace the bromide ion of the same molecule, thereby forming a cyclic compound. Such a reaction is called an intramolecular reaction. *Intra* is Latin for "within": an **intramolecular reaction** takes place within a single molecule.

an intramolecular reaction

$$Br-CH_2(CH_2)_nCH_2\ddot{\underset{..}{O}}\colon^- \longrightarrow \ \begin{array}{c} (CH_2)_n \\ H_2C \quad\quad CH_2 \\ \ddot{\underset{..}{O}} \end{array} \ + \ Br^-$$

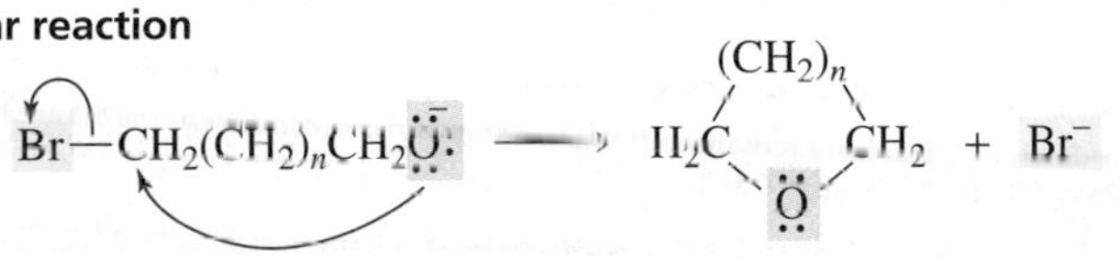

Which reaction is more likely to occur, the intermolecular reaction or the intramolecular reaction? The answer depends on the *concentration* of the bifunctional molecule and the *size of the ring* that would be formed in the intramolecular reaction.

The intramolecular reaction has an advantage: the reacting groups are tethered together, so they don't have to diffuse through the solvent to find a group with which to react. Therefore, a low concentration of reactant favors an intramolecular reaction because the two functional groups have a better chance of finding each other if they are in the same molecule. A high concentration of reactant helps compensate for the advantage gained by tethering, increasing the likelihood of intermolecular reactions.

How much of an advantage an intramolecular reaction has over an intermolecular reaction depends on the size of the ring that is formed—that is, on the length of the tether. If the intramolecular reaction forms a five- or six-membered ring, it will be highly favored over the intermolecular reaction because five- and six-membered rings are stable and, therefore, easily formed.

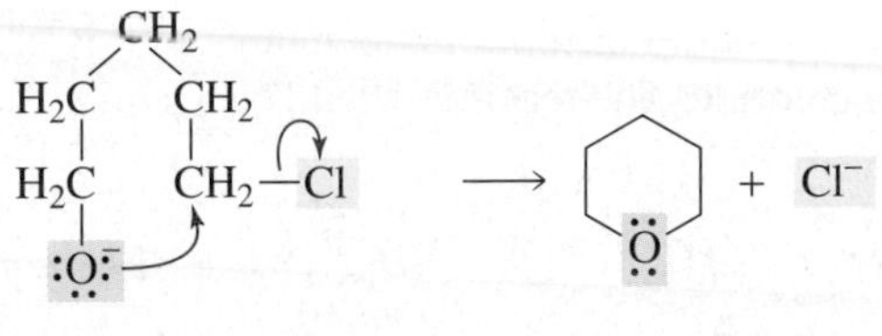

$$\begin{array}{c} CH_2 \\ H_2C \quad CH_2 \\ H_2C \quad CH_2-Cl \\ \vdots\ddot{O}\colon^- \end{array} \longrightarrow \ \begin{array}{c} \\ \ddot{\underset{..}{O}} \end{array} \ + \ Cl^-$$

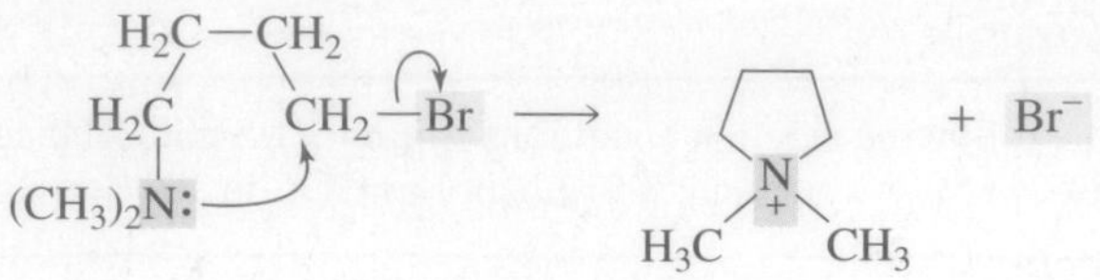

$$\text{(diagram of pyrrolidine ring formation)} \quad + \quad Br^-$$

Three- and four-membered rings are strained, so they are less stable than five- and six-membered rings and, therefore, are less easily formed. The higher activation energy for three- and four-membered ring formation cancels some of the advantage gained by tethering.

Three-membered ring compounds are generally formed more easily than four-membered ring compounds. For a cyclic ether of any size to form, the nucleophilic oxygen atom must be oriented so that it can attack the back side of the carbon bonded to the halogen. Rotation about a C—C bond can produce conformations in which the groups point away from one another and are not able to react. The molecule that forms a three-membered ring ether has only one C—C bond that can rotate, whereas the molecule that forms a four-membered ring has two C—C bonds that can rotate. Therefore, the molecule that forms the three-membered rings is more apt to have its reacting groups in the conformation required for reaction.

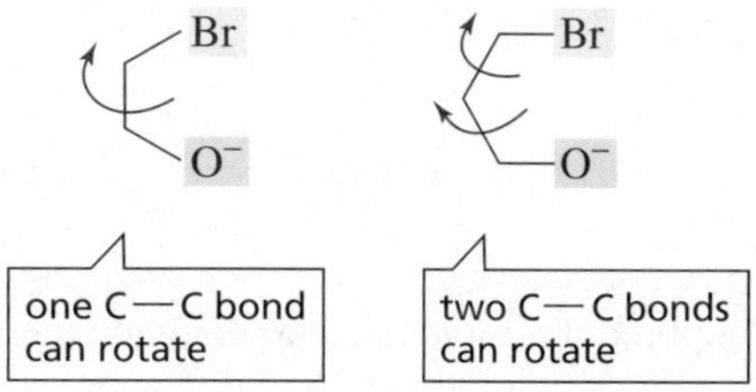

| one C—C bond can rotate | two C—C bonds can rotate |

The likelihood of the reacting groups finding each other decreases sharply when the groups are in compounds that would form seven-membered and larger rings. Therefore, the intramolecular reaction becomes less favored as the ring size increases beyond six members.

PROBLEM 32◆

Which compound in each pair, after removing a proton from the OH group, would form a cyclic ether more rapidly?

a. 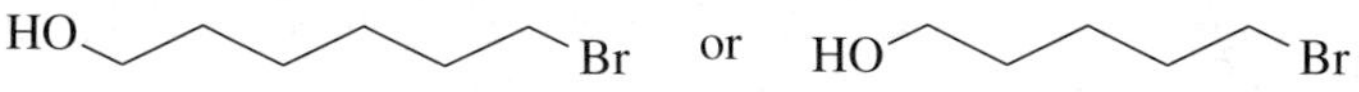 $HO \diagdown\diagup\diagdown\diagup\diagdown\diagup Br$ or $HO \diagup\diagdown\diagup\diagdown\diagup Br$

b. $HO \diagup\diagdown\diagup Br$ or $HO \diagdown\diagup\diagdown\diagup Br$

c. 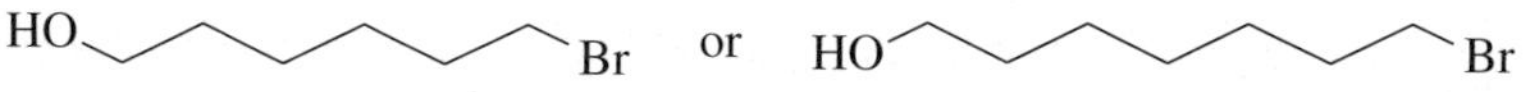 $HO \diagdown\diagup\diagdown\diagup\diagdown Br$ or $HO \diagup\diagdown\diagup\diagdown\diagup\diagdown Br$

8.12 Biological Methylating Reagents Have Good Leaving Groups

If an organic chemist wanted to put a methyl group on a nucleophile, methyl iodide would most likely be used as the methylating agent. Of the methyl halides, methyl iodide has the most easily displaced leaving group because I^- is the weakest base of the halide ions. In addition, methyl iodide is a liquid at room temperature, so it is easier to handle than methyl bromide or methyl chloride. The reaction would be a simple S_N2 reaction.

$$\overset{-}{Nu} + CH_3{-}I \longrightarrow CH_3{-}Nu + I^-$$

In a living cell, however, methyl iodide is not available. It is only slightly soluble in water, so it is not found in the predominantly aqueous environments in biological systems. Instead, biological systems use S-adenosylmethionine (SAM) and N^5-methyltetrahydrofolate as methylating agents; both of these compounds are soluble in water. Although they look much more complicated than methyl iodide, they perform the same function: they transfer a methyl group to a nucleophile. Notice that the methyl group in each of these methylating agents is attached to a positively charged atom. The positively charged atom readily accepts the electrons when the leaving group is displaced. In other words, the methyl groups are attached to very good leaving groups, allowing biological methylation to take place at a reasonable rate.

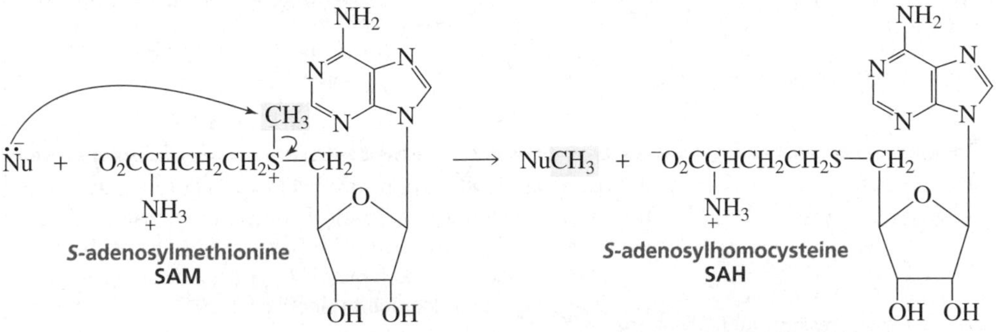

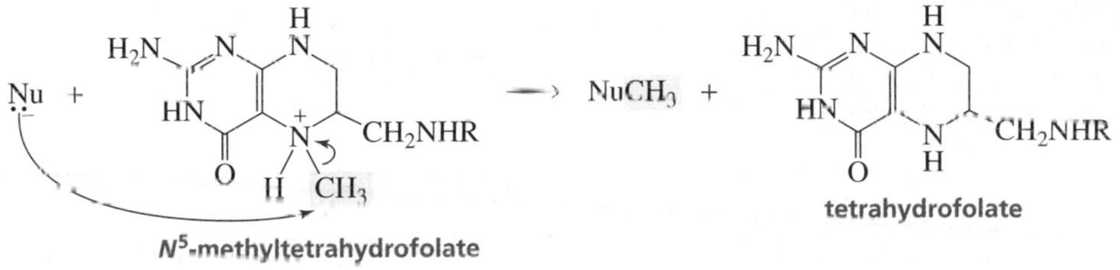

A specific example of a methylation reaction that takes place in biological systems is the conversion of noradrenaline (norepinephrine) to adrenaline (epinephrine). The reaction uses SAM to provide the methyl group. Noradrenaline and adrenaline are hormones that control glycogen metabolism; they are released into the bloodstream in response to stress. Adrenaline is more potent than noradrenaline.

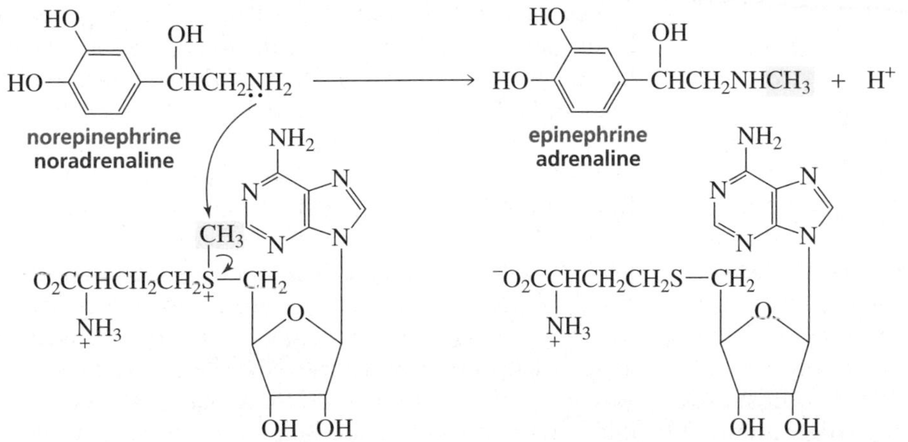

ERADICATING TERMITES

Alkyl halides can be very toxic to biological organisms. For example, bromomethane is used to kill termites and other pests. Bromomethane works by methylating the NH_2 and SH groups of enzymes, thereby destroying the enzymes' ability to catalyze necessary biological reactions. Unfortunately, bromomethane has been found to deplete the ozone layer (Section 11.11), so its production has recently been banned in developed countries, and developing countries will have until 2015 to phase out its use.

The conversion of phosphatidylethanolamine, a component of cell membranes, into phosphatidylcholine, another component of cell membranes, requires three methylations by three equivalents of SAM. (Biological cell membranes will be discussed in Section 26.4; the use of N^5-methyltetrahydrofolate as a biological methylating agent will be discussed in more detail in Section 24.8.)

phosphatidylethanolamine   phosphatidylcholine

S-ADENOSYLMETHIONINE: A NATURAL ANTIDEPRESSANT

Marketed under the name SAMe (pronounced Sammy), S-adenosylmethionine is sold in many health food and drug stores as a treatment for depression and arthritis. Although SAMe has been used clinically in Europe for more than two decades, it has not been rigorously evaluated in the United States and thus has not been approved by the FDA. It can be sold, however, because the FDA does not prohibit the sale of most naturally occurring substances as long as the marketer does not make therapeutic claims. SAMe has also been found to be effective in the treatment of liver diseases, such as diseases caused by alcohol and the hepatitis C virus. The attenuation of liver injuries is accompanied by increased levels of glutathione in the liver. Glutathione is an important biological antioxidant (Section 22.9). SAM is required for the biosynthesis of cysteine, an amino acid that, in turn, is required for the synthesis of glutathione.

SUMMARY

Alkyl halides undergo two kinds of **nucleophilic substitution reactions**: S_N2 and S_N1. In both reactions, a nucleophile substitutes for a halogen. An S_N2 reaction is bimolecular: two molecules are involved in the transition state of the rate-limiting step. An S_N1 reaction is unimolecular; one molecule is involved in the transition state of the rate-limiting step.

The rate of an **S_N2 reaction** depends on the concentration of both the alkyl halide and the nucleophile. An S_N2 reaction has a one-step mechanism: the nucleophile attacks the back side of the carbon that is attached to the halogen. The reaction proceeds in the direction that allows the stronger base to displace the weaker base; it is reversible only if the difference between the basicities of the nucleophile and the leaving group is small. The rate of an S_N2 reaction is affected by steric hindrance: the bulkier the groups at the back side of the carbon undergoing attack, the slower is the reaction. Tertiary carbocations, therefore, cannot undergo S_N2 reactions. An S_N2 reaction takes place with **inversion of configuration**.

The rate of an **S_N1 reaction** depends only on the concentration of the alkyl halide. The halogen departs in the first step, forming a carbocation that is attacked by a nucleophile in the second step. Consequently, carbocation rearrangements can occur. The rate of an S_N1 reaction depends on the ease of carbocation formation. Tertiary alkyl halides, therefore, are more reactive than secondary alkyl halides because tertiary carbocations are more stable than secondary carbocations.

Primary carbocations are so unstable that primary alkyl halides cannot undergo S_N1 reactions. An S_N1 reaction is accompanied by racemization. Most S_N1 reactions are **solvolysis** reactions: the solvent is the nucleophile.

The rates of both S_N2 and S_N1 reactions are influenced by the nature of the leaving group. Weak bases are the best leaving groups because weak bases form the weakest bonds. Thus, the weaker the basicity of the leaving group, the faster the reaction will occur. Therefore, the relative reactivities of alkyl halides that differ only in the halogen atom are $RI > RBr > RCl > RF$ in both S_N2 and S_N1 reactions.

Basicity is a measure of how well a compound shares its lone pair with a proton. **Nucleophilicity** is a measure of how readily a species is able to attack an electron-deficient atom. In a comparison of two molecules with the same attacking atom or with attacking atoms of the same size, the stronger base is a better nucleophile. If the attacking atoms are very different in size, the relationship between basicity and nucleophilicity depends on the solvent. In protic solvents, stronger bases are poorer nucleophiles because of **ion–dipole interactions** between the ion and the solvent.

Methyl halides and primary alkyl halides undergo only S_N2 reactions; tertiary alkyl halides undergo only S_N1 reactions; vinylic and aryl halides undergo neither S_N2 nor S_N1 reactions; and secondary alkyl halides and benzylic and allylic halides (unless they are tertiary) undergo both S_N1 and

S_N2 reactions. When the structure of the alkyl halide allows it to undergo both S_N2 and S_N1 reactions, the S_N2 reaction is favored by a high concentration of a good nucleophile in an aprotic polar solvent, whereas the S_N1 reaction is favored by a poor nucleophile in a protic polar solvent.

Protic solvents (H_2O, ROH) are hydrogen bond donors; **aprotic solvents** (DMF, DMSO) are not hydrogen bond donors. The **dielectric constant** of a solvent tells how well the solvent insulates opposite charges from one another. Increasing the polarity of the solvent will decrease the rate of the reaction if one or more reactants in the rate-determining step are charged and will increase the rate of the reaction if none of the reactants in the rate-determining step are charged.

If the two functional groups of a **bifunctional molecule** can react with each other, both **intermolecular** and **intramolecular reactions** can occur. The reaction that is more likely to occur depends on the concentration of the bifunctional molecule and the size of the ring that would be formed in the intramolecular reaction.

SUMMARY OF REACTIONS

1. S_N2 reaction: a one-step mechanism

$$\overset{-}{\ddot{N}u} + \ -\overset{|}{\underset{|}{C}}-X \ \longrightarrow \ -\overset{|}{\underset{|}{C}}-Nu \ + \ X^-$$

Relative reactivities of alkyl halides: $CH_3X > 1° > 2° > 3°$.

Only the inverted product is formed.

2. S_N1 reaction: a two-step mechanism with a carbocation intermediate

$$-\overset{|}{\underset{|}{C}}-X \ \longrightarrow \ -\overset{|}{\underset{|}{C^+}} \ \overset{\overset{-}{\ddot{N}u}}{\longrightarrow} \ -\overset{|}{\underset{|}{C}}-Nu$$

$$+ \ X^-$$

Relative reactivities of alkyl halides: $3° > 2° > 1° > CH_3X$.

Both the inverted and noninverted products are formed.

KEY TERMS

aprotic solvent (p. 376)
aprotic polar solvent (p. 354)
base (p. 353)
basicity (p. 353)
back-side attack (p. 348)
bifunctional molecule (p. 381)
bimolecular (p. 347)
complete racemization (p. 367)
dielectric constant (p. 375)
elimination reaction (p. 344)
first-order reaction (p. 362)

intermolecular reaction (p. 381)
intimate ion pair (p. 367)
intramolecular reaction (p. 381)
inversion of configuration (p. 351)
ion–dipole interaction (p. 355)
kinetics (p. 346)
leaving group (p. 345)
nucleophile (p. 353)
nucleophilicity (p. 353)
nucleophilic substitution reaction (p. 345)
partial racemization (p. 367)

protic solvent (p. 354)
rate constant (p. 347)
rate law (p. 347)
second-order reaction (p. 347)
S_N1 reaction (p. 362)
S_N2 reaction (p. 347)
solvolysis (p. 365)
steric effects (p. 349)
steric hindrance (p. 349)
substitution reaction (p. 344)
unimolecular (p. 362)

PROBLEMS

33. Give the product of the reaction of bromomethane with each of the following nucleophiles:
    a. HO^- b. $^-NH_2$ c. H_2S d. HS^- e. CH_3O^- f. CH_3NH_2

34. Indicate how each of the factors listed below affects
    a. an S_N1 reaction.
    b. an S_N2 reaction.
        1. the structure of the alkyl halide   3. the concentration of the nucleophile
        2. the reactivity of the nucleophile   4. the solvent

35. Which is a better nucleophile in methanol?
   a. H_2O or HO^-
   b. NH_3 or H_2O
   c. H_2O or H_2S
   d. HO^- or HS^-
   e. I^- or Br^-
   f. Cl^- or Br^-

36. Indicate which member of each pair in Problem 35 is a better leaving group.

37. What nucleophiles could be used to react with butyl iodide to prepare the following compounds?
   a. $CH_3CH_2CH_2CH_2OH$
   b. $CH_3CH_2CH_2CH_2OCH_3$
   c. $CH_3CH_2CH_2CH_2SH$
   d. $CH_3CH_2CH_2CH_2SCH_2CH_3$
   e. $CH_3CH_2CH_2CH_2NHCH_3$
   f. $CH_3CH_2CH_2CH_2C\equiv N$
   g. $CH_3CH_2CH_2CH_2O\overset{\displaystyle O}{\overset{\displaystyle \|}{C}}CH_3$
   h. $CH_3CH_2CH_2CH_2C\equiv CCH_3$

38. Starting with cyclohexene, how could the following compounds be prepared?
   a. methoxycyclohexane
   b. dicyclohexyl ether
   c. cyclohexylmethylamine

39. Rank the following compounds in order of *decreasing* nucleophilicity:

   a. $CH_3\overset{\displaystyle O}{\overset{\displaystyle \|}{C}}O^-$, $CH_3CH_2S^-$, $CH_3CH_2O^-$ in methanol

   c. H_2O and NH_3 in methanol

   b. ⬡—O^- and ⬡—O^- in DMSO

   d. Br^-, Cl^-, I^- in methanol

40. The pK_a of acetic acid in water is 4.76 (Section 1.18). What effect would a decrease in the polarity of the solvent have on the pK_a? Why?

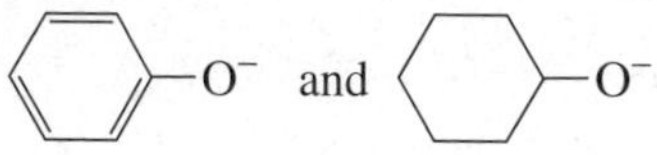

41. a. Identify the products that form when 2-bromo-2-methylpropane is dissolved in a mixture of 80% ethanol and 20% water.
   b. Explain why the same products are obtained when 2-chloro-2-methylpropane is dissolved in a mixture of 80% ethanol and 20% water.

42. For each of the following reactions, give the substitution products; if the products can exist as stereoisomers, show what stereoisomers are obtained:
   a. (R)-2-bromopentane + high concentration of CH_3O^-
   b. (R)-2-bromopentane + CH_3OH
   c. *trans*-1-chloro-2-methylcyclohexane + high concentration of CH_3O^-
   d. *trans*-1-chloro-2-methylcyclohexane + CH_3OH
   e. 3-bromo-2-methylpentane + CH_3OH
   f. 3-bromo-3-methylpentane + CH_3OH

43. Give the products obtained from the solvolysis of each of the following compounds in ethanol:

   a. 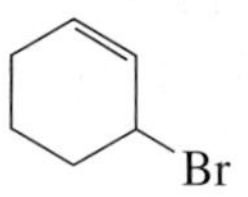 [cyclohexene with Br]   b. 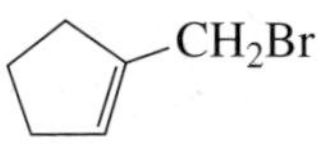 [cyclopentene with CH₂Br]   c. 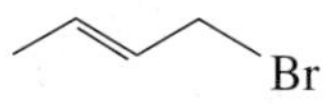 [butene with Br]

44. Would you expect methoxide ion to be a better nucleophile if it were dissolved in CH_3OH or if it were dissolved in dimethyl sulfoxide (DMSO)? Why?

45. Which reaction in each of the following pairs will take place more rapidly?

   a. 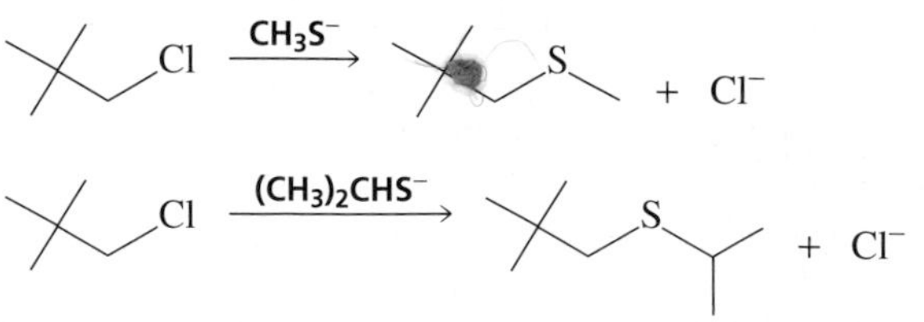 [structure] $\xrightarrow{CH_3S^-}$ [structure] + Cl^-

   [structure] $\xrightarrow{(CH_3)_2CHS^-}$ [structure] + Cl^-

   b. 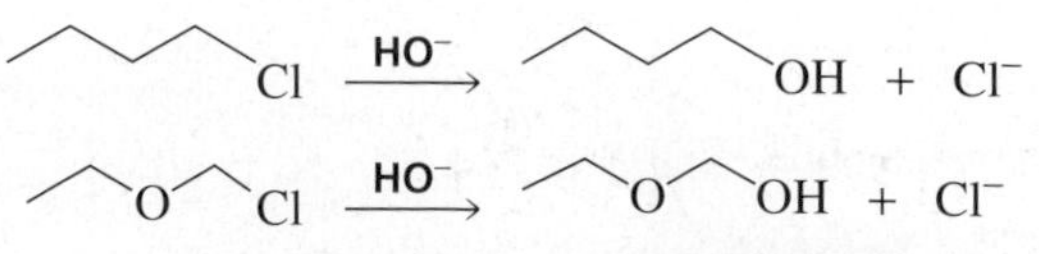 [structure]—Cl $\xrightarrow{HO^-}$ [structure]—OH + Cl^-

   [structure]—Cl $\xrightarrow{HO^-}$ [structure]—OH + Cl^-

   c. 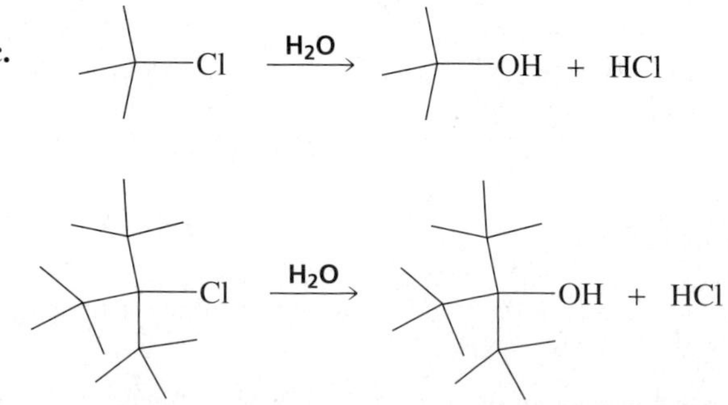 [structure]—Cl $\xrightarrow{H_2O}$ [structure]—OH + HCl

   [structure]—Cl $\xrightarrow{H_2O}$ [structure]—OH + HCl

   d. $(CH_3)_3CBr$ $\xrightarrow{H_2O}$ $(CH_3)_3COH$ + HBr

   $(CH_3)_3CBr$ $\xrightarrow{CH_3CH_2OH}$ $(CH_3)_3COCH_2CH_3$ + HBr

46. The reaction of an alkyl chloride with potassium iodide is generally carried out in acetone to maximize the amount of alkyl iodide that is formed. Why does the solvent increase the yield of alkyl iodide? (*Hint:* Potassium iodide is soluble in acetone, but potassium chloride is not.)

47. In Section 8.12, we saw that *S*-adenosylmethionine (SAM) methylates the nitrogen atom of noradrenaline to form adrenaline, a more potent hormone. If SAM methylates an OH group on the benzene ring instead, it completely destroys noradrenaline's activity. Give the mechanism for the methylation of the OH group by SAM.

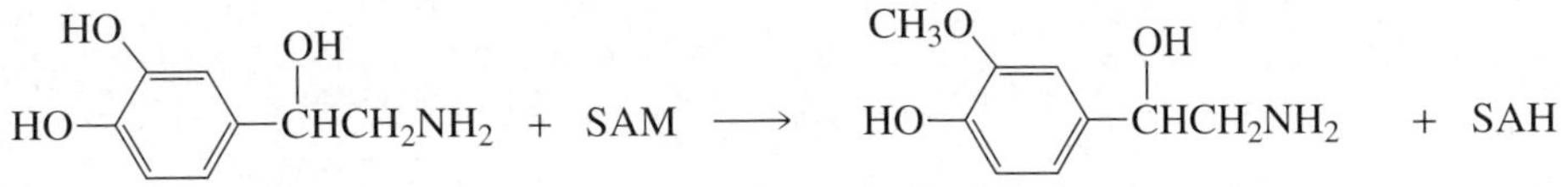

noradrenaline   a biologically inactive compound

48. *tert*-Butyl chloride undergoes solvolysis in both acetic acid and formic acid.

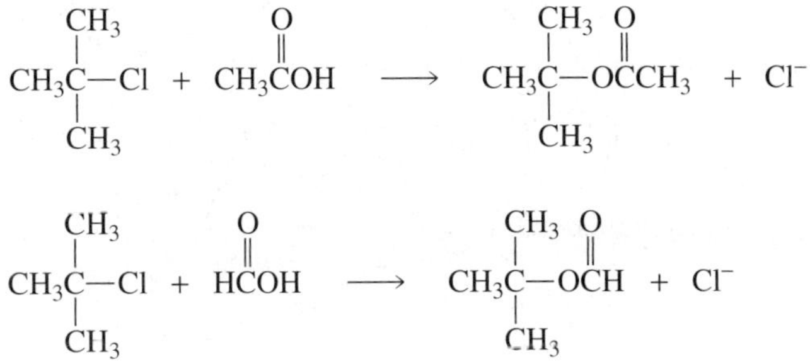

Solvolysis occurs 5,000 time faster in one of the acids than in the other. In which solvent is solvolysis fater? Explain your answer. (*Hint:* See Table 8.3.)

49. For each of the following reactions, give the substitution products; if the products can exist as stereoisomers, show what stereoisomers are obtained:
   a. (2*S*,3*S*)-2-chloro-3-methylpentane + high concentration of CH_3O^-
   b. (2*S*,3*R*) 2-chloro-3-methylpentane + high concentration of CH_3O^-
   c. (2*R*,3*S*)-2-chloro-3-methylpentane + high concentration of CH_3O^-
   d. (2*R*,3*R*)-2-chloro-3-methylpentane + high concentration of CH_3O^-
   e. 3-chloro-2,2-dimethylpentane + CH_3CH_2OH
   f. benzyl bromide + CH_3CH_2OH

50. Give the substitution products obtained when each of the following compounds is added to a solution of sodium acetate in acetic acid.
   a. 2-chloro-2-methyl-3-hexene   **b.** 3-bromo-1-methylcyclohexene

51. In which solvent would the equilibrium for the following S_N2 reaction lie farther to the right, ethanol or diethyl ether?

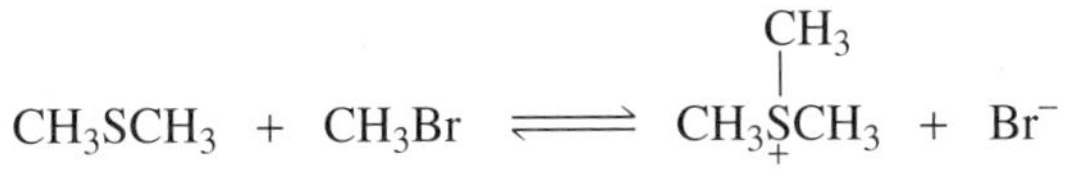

52. The rate of reaction of methyl iodide with quinuclidine was measured in nitrobenzene, and then the rate of reaction of methyl iodide with triethylamine was measured in the same solvent.
   a. Which reaction had the larger rate constant?
   b. The same experiment was done using isopropyl iodide instead of methyl iodide. Which reaction had the larger rate constant?
   c. Which alkyl halide has the larger $k_{triethylamine}/k_{quinuclidine}$ ratio?

quinuclidine 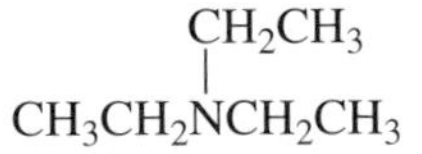   triethylamine

53. Only one bromoether (disregarding stereoisomers) is obtained from the reaction of the following alkyl dihalide with methanol:

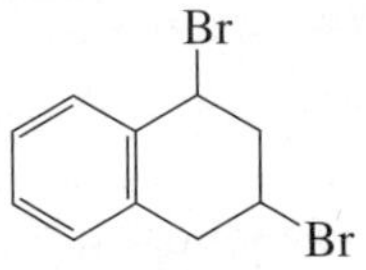

Give the structure of the ether.

54. When equivalent amounts of methyl bromide and sodium iodide are dissolved in methanol, the concentration of iodide ion quickly decreases and then slowly returns to its original concentration. Account for this observation.

55. The rate constant of an intramolecular reaction depends on the size of the ring (n) that is formed. Explain the relative rates of formation of the cyclic secondary ammonium ions.

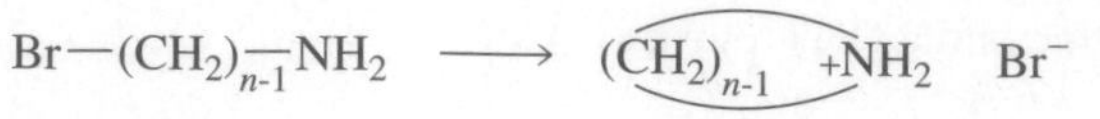

$$\text{Br}-(\text{CH}_2)_{\overline{n-1}}\text{NH}_2 \longrightarrow (\text{CH}_2)_{n-1}\ ^+\text{NH}_2 \quad \text{Br}^-$$

$n =$	3	4	5	6	7	10
relative rate:	1×10^{-1}	2×10^{-3}	100	1.7	3×10^{-3}	1×10^{-8}

56. For each of the following reactions, give the substitution products, assuming that all the reactions are carried out under S_N2 conditions; if the products can exist as stereoisomers, show what stereoisomers are formed:
  a. (3S,4S)-3-bromo-4-methylhexane + CH_3O^-
  b. (3S,4R)-3-bromo-4-methylhexane + CH_3O^-
  c. (3R,4R)-3-bromo-4-methylhexane + CH_3O^-
  d. (3R,4S)-3-bromo-4-methylhexane + CH_3O^-

57. Explain why tetrahydrofuran can solvate a positively charged species better than diethyl ether can.

tetrahydrofuran $CH_3CH_2OCH_2CH_3$
                        diethyl ether

58. Propose a mechanism for each of the following reactions:

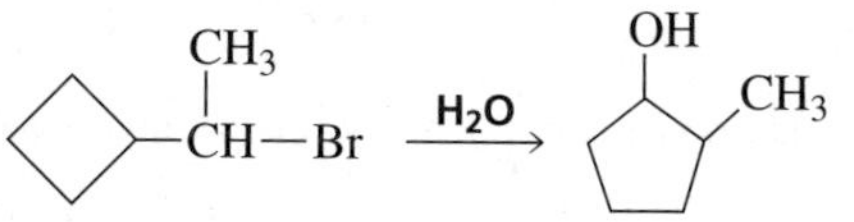

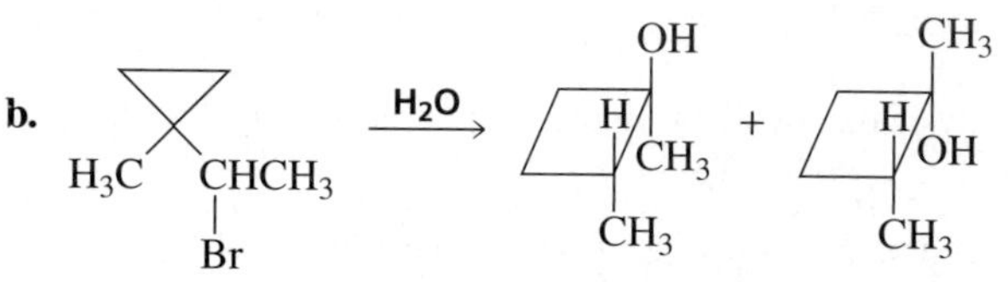

59. Which of the following will react faster in an S_N1 reaction, *cis*-1-bromo-4-*tert*-butylcyclohexane or *trans*-1-bromo-4-*tert*-butylcyclohexane?

60. Alkyl halides have been used as insecticides since the discovery of DDT in 1939. DDT was the first compound to be found that had a high toxicity to insects and a relatively low toxicity to mammals. In 1972, DDT was banned in the United States because it is a long-lasting compound and its widespread use was resulting in the accumulation of substantial concentrations in wildlife. Chlordane is an alkyl halide insecticide used to protect wooden buildings from termites. Chlordane can be synthesized from two reactants in a one-step reaction. One of the reactants is hexachlorocyclopentadiene. What is the other reactant? (*Hint*: See Section 7.12.)

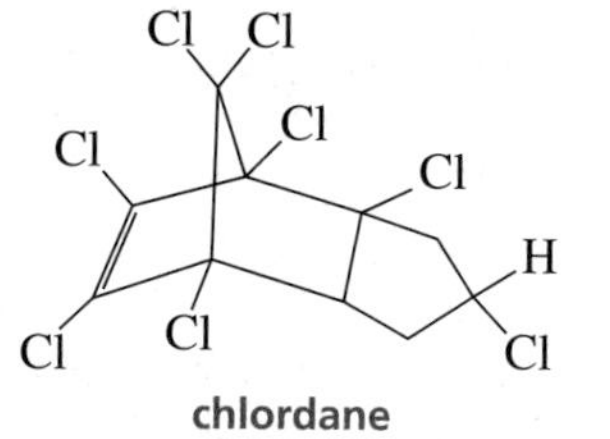

chlordane

61. Explain why the following alkyl halide does not undergo a substitution reaction, regardless of the conditions under which the reaction is run.

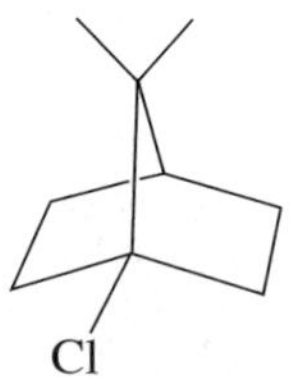

Elimination Reactions of Alkyl Halides • Competition Between Substitution and Elimination

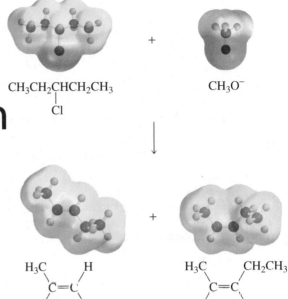

$$CH_3CH_2CHCH_2CH_3 \quad + \quad CH_3O^-$$
$$\underset{Cl}{|}$$

BUILDING ON FUNDAMENTALS

SECTIONS 9.1 AND 9.3   The rate law helps determine the mechanism of a reaction (8.2 and 8.5).

SECTION 9.1   The elimination of HX from an alkyl halide to form an alkene is compared with the reverse reaction, the addition of HX to an alkene to form an alkyl halide (4.1).

SECTION 9.2   Carbocation stability follows the order $3° > 2° > 1°$ because electron-donating alkyl groups stabilize the positive charge (4.2). Now we will see that carbanion stability follows the opposite order because alkyl group destabilize the negative charge.

SECTION 9.2   Dienes are stabilized by alkyl groups bonded to the sp^2 carbons (4.11); dienes are also stabilized by conjugation (7.7).

SECTION 9.3   Hyperconjugation explains why the staggered conformation of ethane is more stable than the eclipsed conformation (2.10), and why alkyl groups stabilize carbocations (4.2). Now we will see that it also helps explain why a proton is so easy to remove from the β-carbon of a carbocation.

SECTION 9.4   The reaction conditions that favor S_N2 reactions also favor E2 reactions, and those that favor S_N1 reactions also favor E1 reactions (8.9).

SECTION 9.5   Anti elimination requires the molecule to be in the more stable staggered conformation (2.10) and is therefore favored in an E2 reaction.

SECTION 9.6   When substituted cyclohexanes undergo ring-flip, equatorial bonds become axial bonds and vice versa (2.13). This is important when we determine the products formed when substituted cyclohexanes undergo E2 reactions.

In addition to undergoing the nucleophilic substitution reactions described in Chapter 8, alkyl halides also undergo elimination reactions. In an **elimination reaction**, atoms or groups are removed from a reactant. For example, when an alkyl halide undergoes an elimination reaction, the halogen (X) is removed from one carbon and a hydrogen is removed from an adjacent carbon. A double bond is formed between the two carbons from which the atoms are eliminated. Therefore, when alkyl halides undergo elimination reactions, *the product is an alkene.*

The product of an elimination reaction is an alkene.

$$\text{substitution} \longrightarrow \text{CH}_3\text{CH}_2\text{CH}_2\text{Y} + \text{X}^-$$

$$\text{CH}_3\text{CH}_2\text{CH}_2\text{X} + \text{Y}^-$$

$$\text{elimination} \longrightarrow \text{CH}_3\text{CH}{=}\text{CH}_2 + \text{HY} + \text{X}^-$$

new double bond

We will start this chapter by looking at the elimination reactions of alkyl halides. Then we will examine the factors that determine whether an alkyl halide will undergo a substitution reaction, an elimination reaction, or both substitution and elimination reactions.

INVESTIGATING NATURALLY OCCURRING ORGANOHALIDES

Like many other natural products—that is, compounds produced in nature—certain organohalides found in marine organisms have interesting and potent biological activity. Cyclocinamide A, one such organohalide, comes from a creature called an orange-encrusting sponge. This compound, as well as a host of analogs, has impressive antitumor properties that are currently being exploited in the development of new anticancer drugs.

Jasplankinolide, another organohalide found in a sponge, modulates the formation and depolymerization of actin microtubules. Microtubules are found in all cells and are used for motile events, such as transportation of vesicles, migration, and cell division. Jasplankinolide is therefore being used to further our understanding of these processes. Notice that cyclocinamide A has four asymmetric centers and jasplankinolide has six. In view of the great diversity of marine life, the ocean probably contains many compounds with useful medicinal properties waiting to be discovered by scientists.

cyclocinamide A

jasplankinolide

9.1 The E2 Reaction

Just as there are two important nucleophilic substitution reactions, S_N1 and S_N2, there are two important elimination reactions, E1 and E2. The reaction of 2-bromo-2-methylpropane with hydroxide ion is an example of an **E2 reaction**; "E" stands for *elimination* and "2" stands for *bimolecular* (Section 8.2).

$$\begin{array}{c} \text{CH}_3 \\ | \\ \text{CH}_3{-}\text{C}{-}\text{CH}_3 \\ | \\ \text{Br} \end{array} + \text{HO}^- \longrightarrow \begin{array}{c} \text{CH}_3 \\ | \\ \text{CH}_2{=}\text{C}{-}\text{CH}_3 \end{array} + \text{H}_2\text{O} + \text{Br}^-$$

2-bromo-2-methylpropane   **2-methylpropene**

The rate of an E2 reaction depends on the concentrations of both the alkyl halide and hydroxide ion. It is, therefore, a second-order reaction.

$$\textbf{rate} = k[\textbf{alkyl halide}][\textbf{base}]$$

The rate law tells us that both the alkyl halide and hydroxide ion are involved in the transition state of the rate-determining step. The following mechanism—portraying an E2 reaction as a concerted one-step reaction—agrees with the observed second-order kinetics:

Mechanism of the E2 reaction

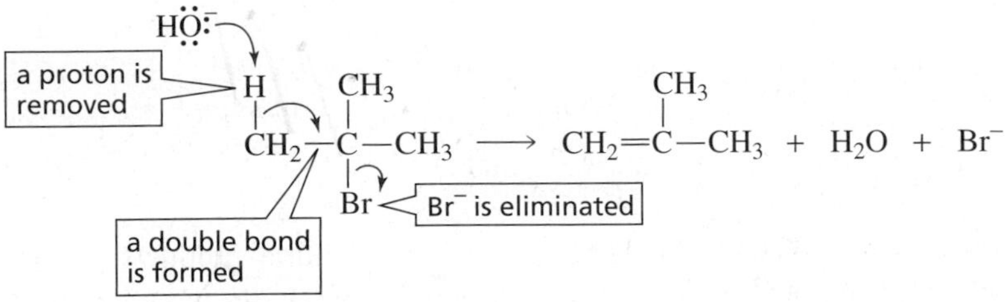

- The base removes a proton from a carbon that is adjacent to the carbon bonded to the halogen. As the proton is removed, the electrons that it shared with carbon move toward the adjacent carbon that is bonded to the halogen. As these electrons move toward the carbon, the halogen leaves, taking its bonding electrons with it.

When the reaction is over, the electrons that were originally bonded to the hydrogen in the reactant have formed a π bond in the product. Removal of a proton and a halide ion is called **dehydrohalogenation**.

The carbon to which the halogen is attached is called the α-carbon. A carbon adjacent to the α-carbon is called a β carbon. Because the elimination reaction is initiated by removing a proton from a β-carbon, an E2 reaction is sometimes called a β-**elimination reaction**. It is also called a **1,2-elimination reaction** because the atoms being removed are on adjacent carbons.

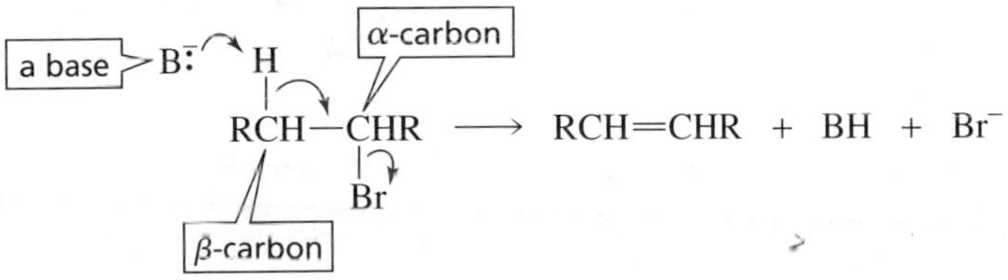

Student Tutorial:
E2 dehydrohalogenation

In a series of alkyl halides that have the same alkyl group, alkyl iodides are the most reactive and alkyl fluorides the least reactive in E2 reactions because weaker bases are better leaving groups (Section 8.3).

The weaker the base, the better a leaving group it is.

relative reactivities of alkyl halides in an E2 reaction

$$\text{most reactive} \quad RI \;>\; RBr \;>\; RCl \;>\; RF \quad \text{least reactive}$$

9.2 An E2 Reaction Is Regioselective

An alkyl halide such as 2-bromopropane has two β-carbons from which a proton can be removed in an E2 reaction. Because the two β-carbons are identical, the proton can be removed with equal ease from either one. The product of this elimination reaction is propene.

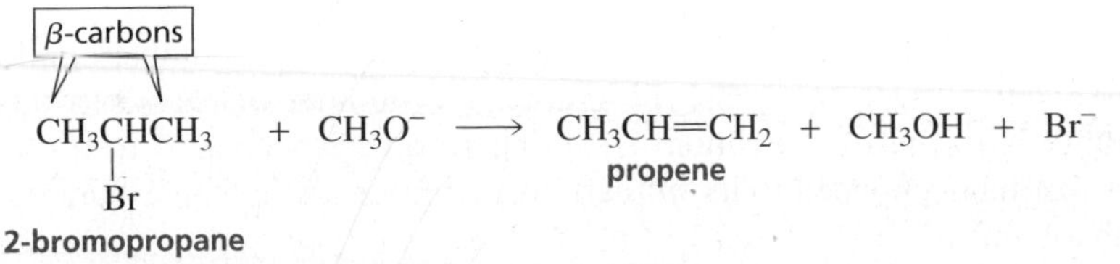

In contrast, 2-bromobutane has two structurally different β-carbons from which a proton can be removed. Therefore when 2-bromobutane reacts with a base, two elimination products are formed: 2-butene (80%) and 1-butene (20%). This E2 reaction is *regioselective* because more of one constitutional isomer is formed than the other.

β-carbons

$$CH_3CHCH_2CH_3 \;+\; CH_3O^- \xrightarrow[CH_3OH]{} \; CH_3CH\!=\!CHCH_3 \;+\; CH_2\!=\!CHCH_2CH_3 \;+\; CH_3OH \;+\; Br^-$$

$\underset{Br}{\mid}$	

2-bromobutane
2-butene 80% (mixture of *E* and *Z*)
1-butene 20%

What factors dictate which of the two elimination products will be formed in greater yield? In other words, what causes regioselectivity in an E2 reaction? We can answer this question by looking at the reaction coordinate diagram (Figure 9.1) for the E2 reaction of 2-bromobutane.

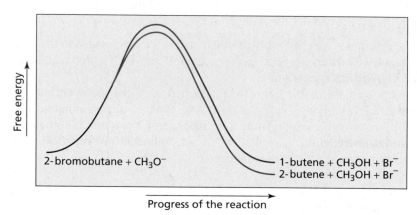

▲ **Figure 9.1**
A reaction coordinate diagram for the E2 reaction of 2-bromobutane and methoxide ion.

In the transition state leading to an alkene, the C—H and C—Br bonds are partially broken and the double bond is partially formed (the partially broken and partially formed bonds are indicated below by dashed lines), giving the transition state an alkene-like structure. Because the transition state has an alkene-like structure, any factors that stabilize the alkene will also stabilize the transition state leading to its formation, allowing the alkene to be formed faster. The difference in the rate of formation of the two alkenes is not very great. Consequently, both products are formed, but the *more stable* of the two alkenes will generally be the major product of the reaction.

transition state leading to 2-butene	transition state leading to 1-butene
more stable	less stable

The major product of an E2 reaction is the most stable alkene.

We have seen that the stability of an alkene depends on the number of alkyl substituents bonded to its sp^2 carbons: the greater the number of alkyl substituents, the more stable is the alkene (Section 4.11). Therefore, 2-butene, with a total of two methyl substituents bonded to its sp^2 carbons, is more stable than 1-butene, with one ethyl substituent.

The reaction of 2-bromo-2-methylbutane with hydroxide ion produces two elimination products, 2-methyl-2-butene and 2-methyl-1-butene. Because 2-methyl-2-butene is the more substituted alkene (it has a greater number of alkyl substituents bonded to its sp^2 carbons), it is the more stable of the two alkenes and, therefore, is the major product of the elimination reaction.

$$\underset{\substack{\text{2-bromo-2-methylbutane}}}{\overset{\substack{CH_3 \\ |}}{CH_3\underset{|}{C}CH_2CH_3}} + HO^- \xrightarrow{H_2O} \underset{\substack{\text{2-methyl-2-butene} \\ 70\%}}{\overset{\substack{CH_3 \\ |}}{CH_3C=CHCH_3}} + \underset{\substack{\text{2-methyl-1-butene} \\ 30\%}}{\overset{\substack{CH_3 \\ |}}{CH_2=CCH_2CH_3}} + H_2O + Br^-$$

Alexander M. Zaitsev, a nineteenth-century Russian chemist, devised a shortcut to predict the more substituted alkene product. He pointed out *that the more substituted alkene is obtained when a proton is removed from the β-carbon that is bonded to the fewest hydrogens.* This is called Zaitsev's rule. In 2-chloropentane, for example, one β-carbon is bonded to three hydrogens and the other β-carbon is bonded to two hydrogens. According to **Zaitsev's rule**, the more substituted alkene will be the one formed by removing a proton from the β-carbon bonded to two hydrogens. Therefore, 2-pentene (a disubstituted alkene) is the major product, and 1-pentene (a monosubstituted alkene) is the minor product.

Student Tutorial:
E2 stereochemistry

$$\boxed{\text{2 β-hydrogens}} \overset{Cl}{\underset{}{\quad}} \boxed{\text{3 β-hydrogens}}$$
$$\underset{\substack{\text{2-chloropentane}}}{CH_3CH_2CH_2\underset{|}{C}HCH_3} + HO^- \longrightarrow \underset{\substack{\text{2-pentene} \\ 67\%}}{\overset{\boxed{\text{disubstituted}}}{CH_3CH_2CH=CHCH_3}} + \underset{\substack{\text{1-pentene} \\ 33\%}}{\overset{\boxed{\text{monosubstituted}}}{CH_3CH_2CH_2CH=CH_2}}$$
$$\text{(mixture of } E \text{ and } Z\text{)}$$

Because elimination from a tertiary alkyl halide typically leads to a more substituted alkene than does elimination from a secondary alkyl halide, and elimination from a secondary alkyl halide generally leads to a more substituted alkene than does elimination from a primary alkyl halide, the relative reactivities of alkyl halides in an E2 reaction are as follows:

Student Tutorial:
E2 elimination regiochemistry

relative reactivities of alkyl halides in an E2 reaction

tertiary alkyl halide > **secondary alkyl halide** > **primary alkyl halide**

$$\underset{\substack{\downarrow}}{\overset{\substack{R \\ |}}{RCH_2\underset{|}{C}R}} \qquad \underset{\substack{\downarrow}}{RCH_2\underset{|}{C}HR} \qquad \underset{\substack{\downarrow}}{RCH_2CH_2Br}$$
$$\underset{\substack{Br}}{} \qquad \underset{\substack{Br}}{}$$

$$\underset{\boxed{\text{three alkyl substituents}}}{\overset{\substack{R \\ |}}{RCH=CR}} \qquad \underset{\boxed{\text{two alkyl substituents}}}{RCH=CHR} \qquad \underset{\boxed{\text{one alkyl substituent}}}{RCH=CH_2}$$

PROBLEM 1♦

What would be the major elimination product obtained from the reaction of each of the following alkyl halides with hyroxide ion?

a. $\underset{\substack{| \\ I}}{\overset{\substack{CH_3 \\ |}}{CH_3\underset{}{C}CH_2CH_3}}$ **b.** $\underset{\substack{| \\ Br}}{\overset{\substack{CH_3 \\ |}}{CH_3CHCHCH_3}}$ **c.** $\underset{\substack{| \quad | \\ CH_3 \ Br}}{\overset{\substack{CH_3 \\ |}}{CH_3CH_2C-CHCH_3}}$

BIOGRAPHY

Alexander M. Zaitsev (1841–1910) *was born in Kazan, Russia. (The German transliteration, Saytzeff, is sometimes used for his family name). He received a Ph.D. from the University of Leipzig in 1866. Zaitsev was a professor of chemistry first at the University of Kazan and later at the University of Kiev.*

Keep in mind that the major product of an E2 reaction is the *more stable alkene,* and Zaitsev's rule is just a shortcut to determine which of the possible alkene products is the *more substituted alkene.* The more substituted alkene is not, however, always the more stable alkene. In the following reactions, the conjugated alkene is the more stable alkene, even though it is not the most substituted alkene. The major product of each reaction is, therefore, the conjugated alkene because, being more stable, it is more easily formed.

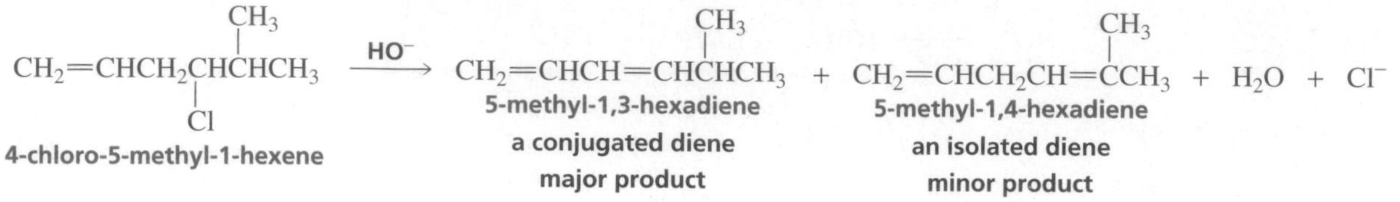

$$CH_2=CHCH_2CHCHCH_3 \quad \xrightarrow{HO^-} \quad CH_2=CHCH=CHCHCH_3 \quad + \quad CH_2=CHCH_2CH=CCH_3 \quad + \quad H_2O \quad + \quad Cl^-$$

4-chloro-5-methyl-1-hexene	**5-methyl-1,3-hexadiene** **a conjugated diene** **major product**	**5-methyl-1,4-hexadiene** **an isolated diene** **minor product**

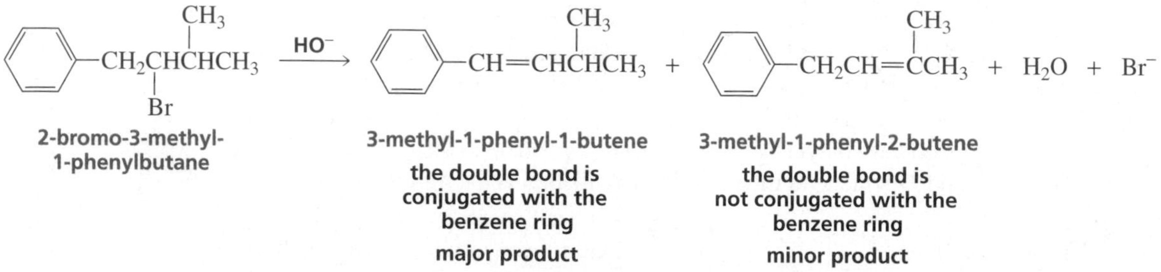

$$\text{(benzene ring)}-CH_2CHCHCH_3 \quad \xrightarrow{HO^-} \quad \text{(benzene ring)}-CH=CHCHCH_3 \quad + \quad \text{(benzene ring)}-CH_2CH=CCH_3 \quad + \quad H_2O \quad + \quad Br^-$$

2-bromo-3-methyl- **1-phenylbutane**	**3-methyl-1-phenyl-1-butene** **the double bond is** **conjugated with the** **benzene ring** **major product**	**3-methyl-1-phenyl-2-butene** **the double bond is** **not conjugated with the** **benzene ring** **minor product**

Zaitsev's rule leads to the most substituted alkene.

Zaitsev's rule cannot be used to predict the major products of the foregoing reactions because it does not take account of the fact that conjugated double bonds are more stable than isolated double bonds (Section 7.7). Therefore, if the alkyl halide has a double bond or a benzene ring, do not use Zaitsev's rule to predict the most stable (major) product of an elimination reaction.

The most stable alkene is generally (but not always) the most substituted alkene.

In some elimination reactions, the more stable alkene is not the major product. For example, if the base in an E2 reaction is bulky and the approach to the alkyl halide is sterically hindered, the base will preferentially remove the most accessible hydrogen. In the following reaction, it is easier for the bulky *tert*-butoxide ion to remove one of the more exposed terminal hydrogens, which leads to formation of the less substituted alkene. Because the less substituted alkene is more easily formed, it is the major product of the reaction.

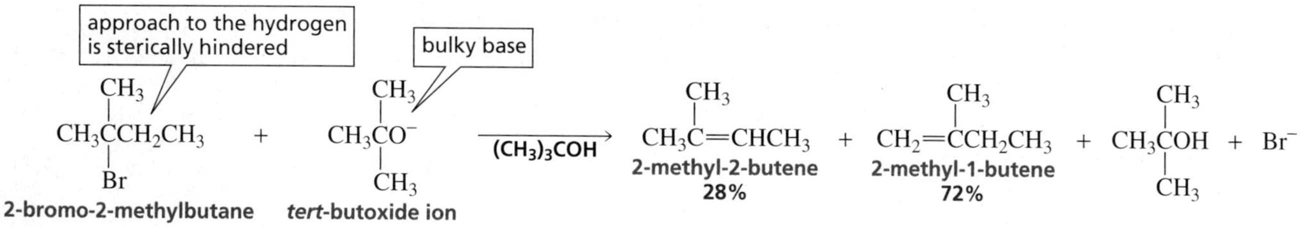

approach to the hydrogen is sterically hindered   bulky base

$$CH_3CCH_2CH_3 \quad + \quad CH_3CO^- \quad \xrightarrow{(CH_3)_3COH} \quad CH_3C=CHCH_3 \quad + \quad CH_2=CCH_2CH_3 \quad + \quad CH_3COH \quad + \quad Br^-$$

2-bromo-2-methylbutane	*tert*-**butoxide ion**	**2-methyl-2-butene** **28%**	**2-methyl-1-butene** **72%**	

The data in Table 9.1, describing the results of E2 reactions between the same sterically hindered alkyl halide and various alkoxide ions of different sizes, show that the percentage of the less substituted alkene increases as the size of the base increases.

If the alkyl halide is not sterically hindered and the base is only moderately hindered, the major product will be the more stable product, in accordance with the general rule. For example, the major product obtained from the reaction of 2-iodobutane and *tert*-butoxide ion is 2-butene. In other words, it takes a lot of hindrance for the less stable product to be the major product.

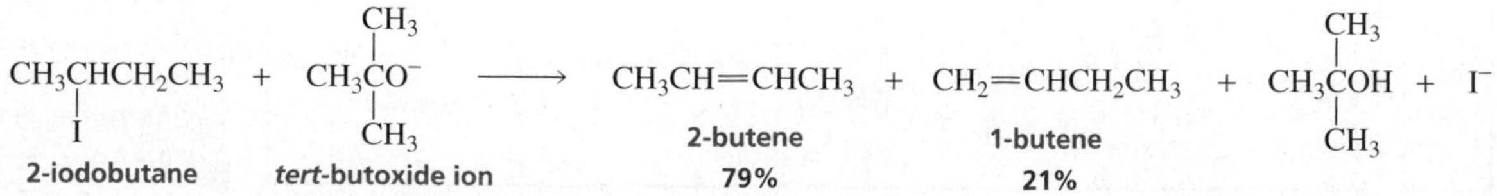

$$CH_3CHCH_2CH_3 \quad + \quad CH_3CO^- \quad \longrightarrow \quad CH_3CH=CHCH_3 \quad + \quad CH_2=CHCH_2CH_3 \quad + \quad CH_3COH \quad + \quad I^-$$

2-iodobutane	*tert*-**butoxide ion**	**2-butene** **79%**	**1-butene** **21%**	

Table 9.1 Effect of the Steric Properties of the Base on the Distribution of Products in an E2 Reaction

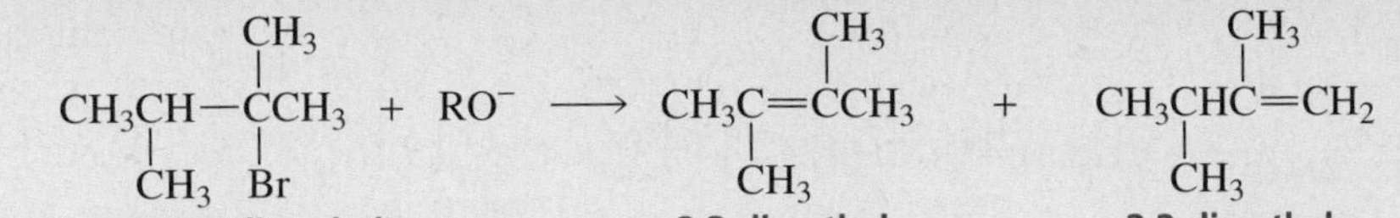

$$CH_3CH-CCH_3 \ (with \ CH_3 \ above, \ CH_3 \ and \ Br \ below) \ + \ RO^- \longrightarrow CH_3C=CCH_3 \ (with \ CH_3 \ above, \ CH_3 \ below) \ + \ CH_3CHC=CH_2 \ (with \ CH_3 \ above, \ CH_3 \ below)$$

2-bromo-2,3-dimethyl-butane   2,3-dimethyl-2-butene   2,3-dimethyl-1-butene

Base	More substituted alkene	Less substituted alkene
$CH_3CH_2O^-$	79%	21%
CH_3CO^- (with CH_3 above and CH_3 below)	27%	73%
CH_3CO^- (with CH_3 above and CH_2CH_3 below)	19%	81%
$CH_3CH_2CO^-$ (with CH_2CH_3 above and CH_2CH_3 below)	8%	92%

PROBLEM 2◆

Which of the alkyl halides in each pair is more reactive in an E2 reaction?

a. $CH_3CH_2CH_2CH_2Br$ or $CH_3CH_2CHCH_3$ (with Br below)

b. 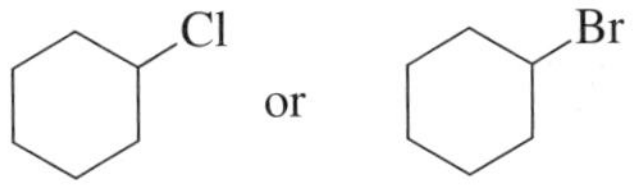 (cyclohexane with Cl)   or   (cyclohexane with Br)

c. $CH_3CHCH_2CHCH_3$ (with CH_3 above, Br below)   or $CH_3CH_2CH_2CCH_3$ (with CH_3 above, Br below)

d. $CH_3CH_2CCH_2Cl$ (with CH_3 above, CH_3 below)   or $CH_3CCH_2CH_2Cl$ (with CH_3 above, CH_3 below)

PROBLEM 3

Draw a reaction coordinate diagram for the E2 reaction of 2-bromo-2,3-dimethylbutane with sodium *tert*-butoxide.

Although the major product of an E2 dehydrohalogenation of alkyl chlorides, alkyl bromides, and alkyl iodides is normally the more substituted alkene, the major product of the E2 dehydrohalogenation of alkyl fluorides is the less substituted alkene.

$$\underset{\substack{\text{2-fluoropentane}}}{\overset{\overset{\displaystyle F}{|}}{CH_3CHCH_2CH_2CH_3}} + \underset{\substack{\text{methoxide}\\\text{ion}}}{CH_3O^-} \xrightarrow{CH_3OH} \underset{\substack{\text{2-pentene}\\30\%\\\text{(mixture of } E \text{ and } Z)}}{CH_3CH{=}CHCH_2CH_3} + \underset{\substack{\text{1-pentene}\\70\%}}{CH_2{=}CHCH_2CH_2CH_3} + CH_3OH + F^-$$

When a hydrogen and a chlorine, bromine, or iodine are eliminated from an alkyl halide, the halogen starts to leave as soon as the base begins to remove the proton. Consequently, a negative charge does not build up on the carbon that is losing the proton. Thus, the transition state resembles an alkene rather than a carbanion (Section 9.1). The fluoride ion, however, is the strongest base of the halide ions and therefore the poorest leaving group. So when a base begins to remove a proton from an alkyl fluoride, the fluoride ion does not have as strong a tendency to leave as another halide ion would have. As a result, negative charge develops on the carbon that is losing the proton, causing the transition state to resemble a carbanion rather than an alkene. To determine which of the carbanion-like transition states is more stable, we look to see which carbanion would be more stable.

transition state leading to
1-pentene
more stable

transition state leading to
2-pentene
less stable

Carbocation stability: 3° is more stable than 1°.

We have seen that carbocations, because they are positively charged, are stabilized by electron-donating alkyl groups. Thus, tertiary carbocations are the most stable and methyl cations are the least stable (Section 4.2).

relative stabilities of carbocations

Carbanion stability: 1° is more stable than 3°.

Carbanions, on the other hand, are negatively charged, so they are destabilized by electron-donating alkyl groups. Therefore, methyl anions are the most stable and tertiary carbanions are the least stable.

relative stabilities of carbanions

The developing negative charge in the transition state leading to 1-pentene is on a primary carbon, which is more stable than the transition state leading to 2-pentene, in which the developing negative charge is on a secondary carbon. Because the transition state leading to 1-pentene is more stable, 1-pentene is formed more rapidly and is the major product of the E2 reaction of 2-fluoropentane.

The data in Table 9.2 show that as the halide ion increases in basicity (decreases in leaving ability), the yield of the more substituted alkene product decreases. However, the more substituted alkene remains the major elimination product in all cases, except when the halogen is fluorine.

Table 9.2   Products Obtained from the E2 Reaction of CH_3O^- and 2-Halohexanes

	More substituted product	Less substituted product
$\overset{X}{\underset{\mid}{CH_3CHCH_2CH_2CH_2CH_3}}$ + CH_3O^- $\longrightarrow$	$CH_3CH{=}CHCH_2CH_2CH_3$ 2-hexene (mixture of *E* and *Z*)	+ $CH_2{=}CHCH_2CH_2CH_2CH_3$ 1-hexene

Leaving group	Conjugate acid	pK_a		
X = I	HI	−10	81%	19%
X = Br	HBr	−9	72%	28%
X = Cl	HCl	−7	67%	33%
X = F	HF	3.2	30%	70%

We can summarize by saying that *the major product of an E2 elimination reaction is the more stable alkene* except *if the reactants are sterically hindered or the leaving group is poor* (for example, a fluoride ion), in which case the major product will be the less stable alkene. In Section 9.6, you will see that in the case of certain cyclic compounds, the more stable alkene is not always the major product.

PROBLEM 4◆

Give the major elimination product obtained from an E2 reaction of each of the following alkyl halides with hydroxide ion:

a. $\overset{}{\underset{\underset{Cl}{\mid}}{CH_3CHCH_2CH_3}}$

b. $\overset{CH_3}{\underset{\underset{Cl}{\mid}}{CH_3CHCHCH_2CH_3}}$

c. $\overset{}{\underset{\underset{Cl}{\mid}}{CH_3CHCH_2CH{=}CH_2}}$

d. $\overset{}{\underset{\underset{F}{\mid}}{CH_3CHCH_2CH_3}}$

e. 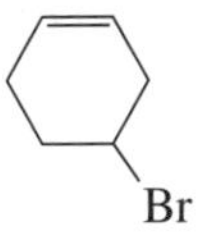 (cyclohexene with Br substituent)

f. $\overset{CH_3}{\underset{\underset{F}{\mid}}{CH_3CHCHCH_2CH_3}}$

PROBLEM 5◆

Which alkyl halide in each pair would you expect to be more reactive in an E2 reaction?

a. $\overset{CH_3}{\underset{\underset{Br}{\mid}}{CH_3CHCHCH_2CH_3}}$ or $\overset{CH_3}{\underset{\underset{Br}{\mid}}{CH_3CHCH_2CHCH_3}}$

b.

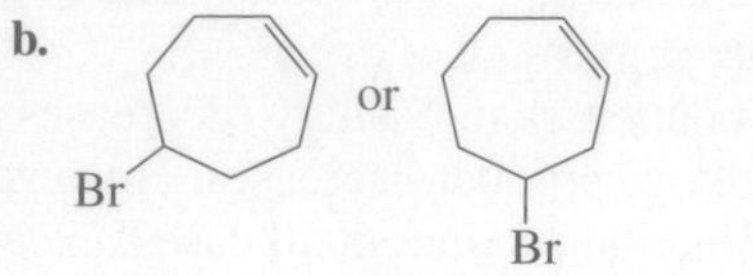

c. CH₃CH₂CH₂CHCH₃   or   CH₃CH₂CHCH₂CH₃
               | |
               Br   Br

d.

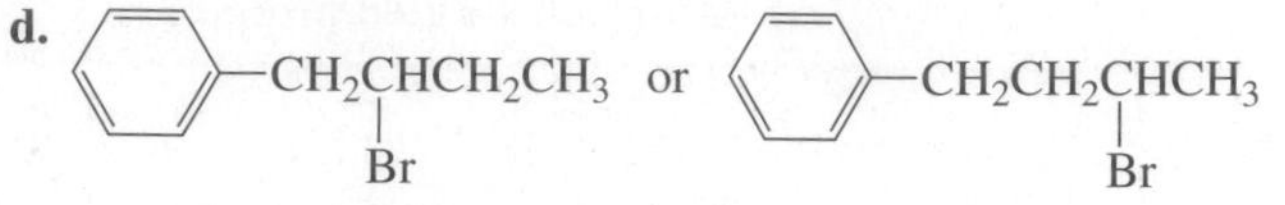

9.3 The E1 Reaction

The second kind of elimination reaction that alkyl halides can undergo is an E1 reaction. The reaction of 2-bromo-2-methylpropane with water to form 2-methylpropene is an example of an **E1 reaction**; "E" stands for *elimination* and "1" stands for *unimolecular*.

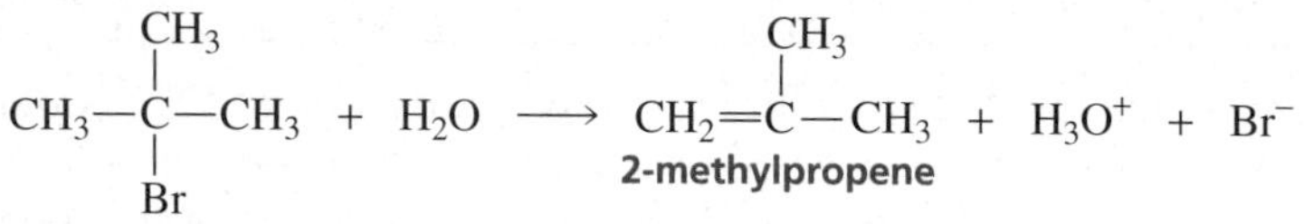

$$\begin{array}{ccc}
\quad\quad CH_3 & & CH_3 \\
\quad\quad | & & | \\
CH_3{-}C{-}CH_3 + H_2O \longrightarrow & CH_2{=}C{-}CH_3 + H_3O^+ + Br^- \\
\quad\quad | & & \text{2-methylpropene} \\
\quad\quad Br & & \\
\text{2-bromo-2-methylpropane} & &
\end{array}$$

An E1 reaction is a first-order elimination reaction because the rate of the reaction depends only on the concentration of the alkyl halide.

rate = k[alkyl halide]

Student Tutorial:
E1 mechanism

We therefore know that only the alkyl halide takes part in the rate-determining step of the reaction. Therefore, an E1 reaction must have at least two steps. The following mechanism agrees with the observed first-order kinetics. Because the first step is the rate-determining step, an increase in the concentration of the base—which participates in the second step of the reaction only—has no effect on the rate of the reaction.

Mechanism of the E1 reaction

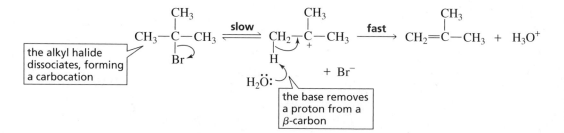

- The alkyl halide dissociates, forming a carbocation.
- The base forms the elimination product by removing a proton from a carbon that is adjacent to the positively charged carbon (i.e., from the β-carbon).

We have seen that the pK_a of a compound such as ethane that has hydrogens attached only to sp^3 carbons is > 60 (Section 6.10). How, then, can a weak base like water remove a proton from an sp^3 carbon in the second step of the reaction? First of all, the

pK_a is greatly reduced by the positively charged carbon that can accept the electrons left behind when the proton is removed from an adjacent carbon. Second, the carbon adjacent to the positively charged carbon shares the positive charge as a result of hyperconjugation. Hyperconjugation thus drains electron density from the C—H bond, thereby weakening it. Recall that hyperconjugation—where the σ electrons of a bond attached to a carbon adjacent to the positively charged carbon spread into the empty p orbital—is responsible for the greater stability of a tertiary carbocation compared with a secondary carbocation (Section 4.2).

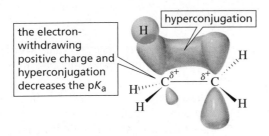

When more than one alkene can be formed, the E1 reaction, like the E2 reaction is regioselective. The major product is generally the more substituted alkene.

$$
\begin{array}{c}
\overset{\displaystyle CH_3}{\underset{\displaystyle Cl}{CH_3CH_2\overset{|}{\underset{|}{C}}CH_3}} + H_2O \longrightarrow \underset{\substack{\text{2-methyl-2-butene}\\\text{major product}}}{CH_3CH=\overset{CH_3}{\overset{|}{C}}CH_3} + \underset{\substack{\text{2-methyl-1-butene}\\\text{minor product}}}{CH_3CH_2\overset{CH_3}{\overset{|}{C}}=CH_2} + H_3O^+ + Cl^-
\end{array}
$$

2-chloro-2-methylbutane

The more substituted alkene is the more stable of the two alkenes formed in the preceding reaction, and therefore, it has the more stable transition state leading to its formation (Figure 9.2). As a result, the more substituted alkene is formed more rapidly, so it is the major product. The more substituted alkene is formed by removing the hydrogen from the β-carbon bonded to the fewest hydrogens, in accordance with Zaitsev's rule.

Because the first step is the rate-determining step, the rate of an E1 reaction depends both on the ease with which the carbocation is formed *and* on how readily the leaving group leaves. The more stable the carbocation, the more easily it is formed, because more stable carbocations have more stable transition states leading to their formation. Therefore, the relative reactivities of a series of alkyl halides that all have

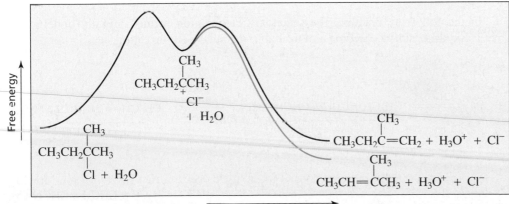

◀ **Figure 9.2**
A reaction coordinate diagram for the E1 reaction of 2-chloro-2-methylbutane. The major product is the more substituted alkene because its greater stability causes the transition state leading to its formation to be more stable.

the same leaving group parallel the relative stabilities of the carbocations. A tertiary benzylic halide is the most reactive alkyl halide because a tertiary benzylic cation—the most stable carbocation—is the easiest to form (Sections 7.7).

relative reactivities of alkyl halides in an E1 reaction = relative stabilities of carbocations

3° benzylic ≈ 3° allylic > 2° benzylic ≈ 2° allylic ≈ 3° > 1° benzylic ≈ 1° allylic ≈ 2° > 1° > vinyl

    ⌐ most stable ¬   ⌐ least stable ¬

Notice that tertiary alkyl halides are more reactive than secondary alkyl halides, which, in turn, are more reactive than primary alkyl halides, in both E1 and E2 reactions (Section 9.2).

We have seen that the weakest bases are the best leaving groups (Section 9.3). Therefore, for series of alkyl halides with the same alkyl group, alkyl iodides are the most reactive and alkyl fluorides the least reactive in E1 reactions.

relative reactivities of alkyl halides in an E1 reaction

most reactive ⟩ RI > RBr > RCl > RF ⟨ least reactive

Thus for *both* E2 and E1 reactions: the major product is the most stable alkene, tertiary alkyl halides are the most reactive and primary alkyl halides are the least reactive, and an alkyl iodide is more reactive than an alkyl chloride with the same alkyl group.

Because the E1 reaction forms a carbocation intermediate, the carbon skeleton can rearrange before the proton is lost, if rearrangement leads to a more stable carbocation. For example, the secondary carbocation that is formed when a chloride ion dissociates from 3-chloro-2-methyl-2-phenylbutane undergoes a 1,2-methyl shift to become a more stable tertiary benzylic cation, which then undergoes deprotonation to form the alkene.

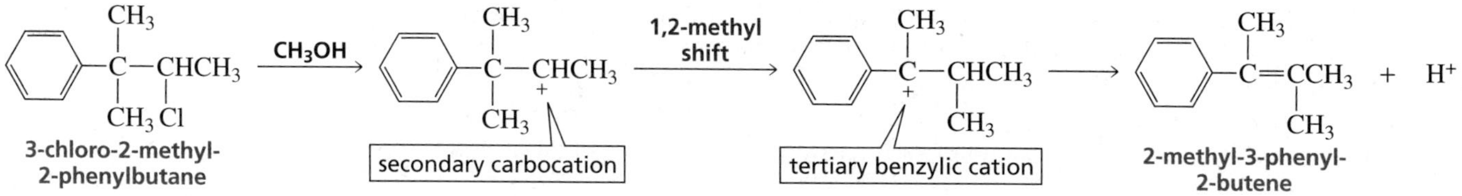

In the following reaction, the secondary carbocation formed initially undergoes a 1,2-hydride shift to become a more stable secondary allylic cation:

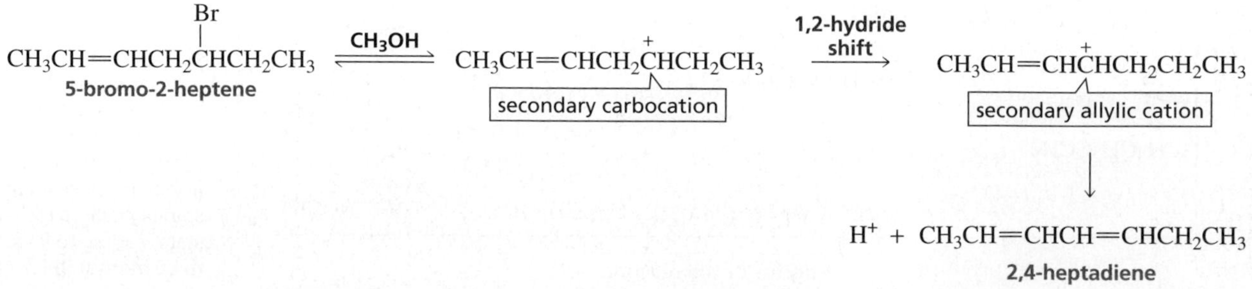

Let's take a minute to compare elimination from an alkyl halide to form an alkene with the reverse reaction, which you studied in Section 4.4: addition to an alkene to form an alkyl halide. The elimination reaction requires a base to remove a proton from the carbocation to form the alkene. The addition reaction requires an acid to react with the nucleophilic alkene to form the carbocation.

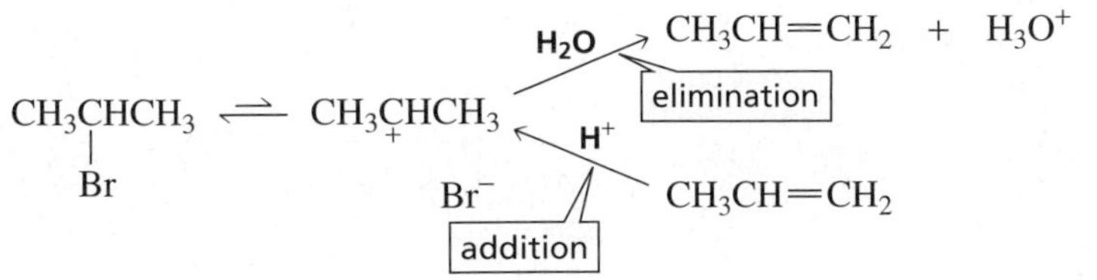

$$CH_3CHCH_3 \; \rightleftharpoons \; CH_3\overset{+}{C}HCH_3 \quad \begin{array}{c} \xrightarrow{\text{H}_2\text{O}} \; CH_3CH{=}CH_2 \; + \; H_3O^+ \quad \boxed{\text{elimination}} \\[2mm] \xleftarrow{\text{H}^+} \\[2mm] \underset{Br^-}{} \quad CH_3CH{=}CH_2 \end{array}$$

Br \qquad\qquad \boxed{\text{addition}}

PROBLEM 6◆

Four alkenes are formed from the E1 reaction of 3-bromo-2,3-dimethylpentane. Give the structures of the alkenes, and rank them according to the amounts that would be formed.

PROBLEM 7◆

If 2-fluoropentane were to undergo an E1 reaction, would you expect the major product to be the one predicted by Zaitsev's rule? Explain.

PROBLEM 8◆

Which of the following compounds would react faster in an

a. E1 reaction? **b.** E2 reaction? **c.** S$_N$1 reaction? **d.** S$_N$2 reaction?

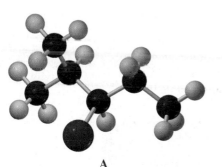

A B

PROBLEM-SOLVING STRATEGY

Proposing a Mechanism

Propose a mechanism for the following reaction:

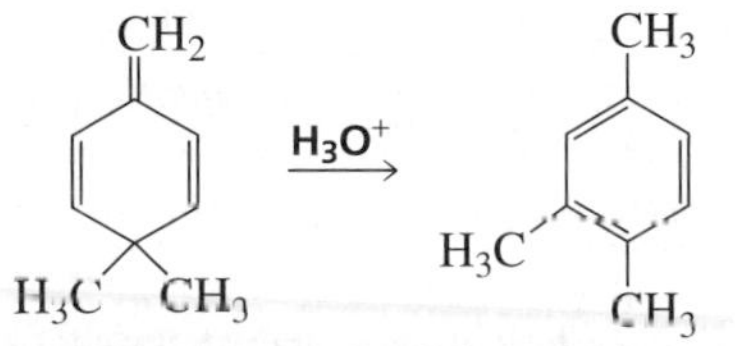

$$\underset{\substack{H_3C \quad CH_3}}{\overset{\overset{\displaystyle CH_2}{\bigcirc}}{}} \quad \xrightarrow{\text{H}_3\text{O}^+} \quad \underset{\substack{H_3C \\ CH_3}}{\overset{CH_3}{\bigcirc}}$$

Because the reagent used in the reaction is an acid, we start by protonating the molecule at the position that allows the most stable carbocation to be formed. Protonating the CH_2

group forms a tertiary allylic carbocation in which the positive charge is delocalized over two other carbons. Now we move the π electrons so that the 1,2-methyl shift required to form the product can take place. Loss of a proton gives the final product.

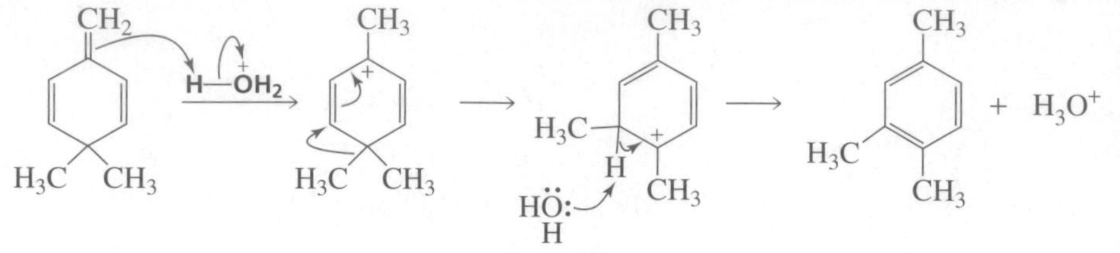

Now continue on to Problem 9.

PROBLEM 9

Propose a mechanism for the following reaction:

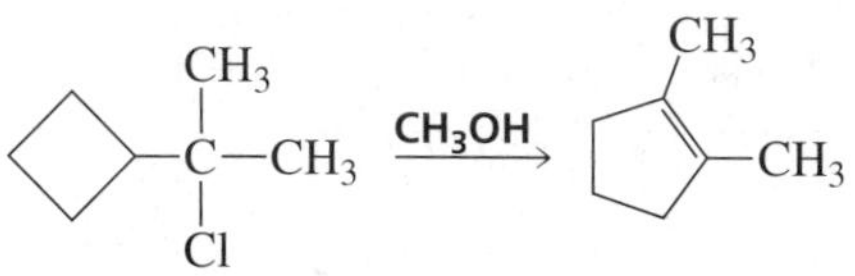

9.4 Competition Between E2 and E1 Reactions

Primary alkyl halides undergo only E2 reactions; they cannot undergo E1 reactions because primary carbocations are too unstable to be formed. *Secondary* and *tertiary* alkyl halides undergo both E2 and E1 reactions (Table 9.3).

Table 9.3	Summary of the Reactivity of Alkyl Halides in Elimination Reactions
Primary alkyl halide	E2 only
Secondary alkyl halide	E1 and E2
Tertiary alkyl halide	E1 and E2

Student Tutorial: Common terms in E1 and E2 reactions

An S_N2 reaction of an alkyl halide is favored by a high concentration of a good nucleophile in an aprotic polar solvent.

An S_N1 reaction of an alkyl halide is favored by a poor nucleophile in a protic polar solvent.

For those alkyl halides that can undergo both E2 and E1 reactions, the E2 reaction is favored by the same factors that favor an S_N2 reaction and the E1 reaction is favored by the same factors that favor an S_N1 reaction. Thus, *an E2 reaction is favored by a high concentration of a strong base and an aprotic polar solvent (such as DMSO or DMF), whereas an E1 reaction is favored by a weak base and a protic polar solvent (such as H_2O or ROH).* How the solvent favors one reaction over another was discussed in Section 8.10.

Notice that a tertiary alkyl halide and a strong base were chosen to illustrate the E2 reaction in Section 9.1, whereas a tertiary alkyl halide and a weak base were used to illustrate the E1 reaction in Section 9.3.

PROBLEM 10◆

For each of the following reactions, (1) decide which kind of elimination will occur, E2 or E1, and (2) give the major elimination product of each reaction, disregarding stereoisomers:

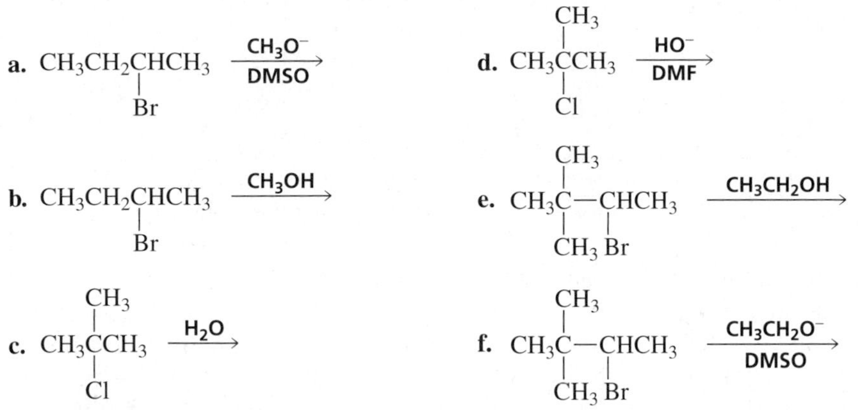

a. $CH_3CH_2CHCH_3$ $\xrightarrow[\text{DMSO}]{CH_3O^-}$
    |
    Br

b. $CH_3CH_2CHCH_3$ $\xrightarrow{CH_3OH}$
    |
    Br

c. $CH_3\overset{\displaystyle CH_3}{\underset{\displaystyle Cl}{\overset{|}{\underset{|}{C}}}}CH_3$ $\xrightarrow{H_2O}$

d. $CH_3\overset{\displaystyle CH_3}{\underset{\displaystyle Cl}{\overset{|}{\underset{|}{C}}}}CH_3$ $\xrightarrow[\text{DMF}]{HO^-}$

e. $CH_3\overset{\displaystyle CH_3}{\overset{|}{C}}{-}\overset{}{\underset{\displaystyle Br}{\overset{}{\underset{|}{CHCH_3}}}}$ $\xrightarrow{CH_3CH_2OH}$
       |
       CH_3

f. $CH_3\overset{\displaystyle CH_3}{\overset{|}{C}}{-}\overset{}{\underset{\displaystyle Br}{\overset{}{\underset{|}{CHCH_3}}}}$ $\xrightarrow[\text{DMSO}]{CH_3CH_2O^-}$
       |
       CH_3

PROBLEM 11◆

The rate law for the reaction of HO$^-$ with *tert*-butyl bromide to form an elimination product in 75% ethanol/25% water at 30 °C is the sum of the rate laws for the E2 and E1 reactions:

$$\text{rate} = 7.1 \times 10^{-5}\,[\textit{tert}\text{-butyl bromide}]\,[HO^-] + 1.5 \times 10^{-5}\,[\textit{tert}\text{-butyl bromide}]$$

What percentage of the reaction takes place by the E2 pathway when the following conditions exist?

a. $[HO^-] = 5.0$ M

b. $[HO^-] = 0.0025$ M

9.5 E2 and E1 Reactions Are Stereoselective

We have seen that E2 and E1 reactions are regioselective: more of one constitutional isomer is formed than the other. For example, the major product formed from the E2 or E1 reaction of 2-bromopentane is 2-pentene (Sections 9.1 and 9.3). Now we will see that E2 and E1 reactions also are stereoselective.

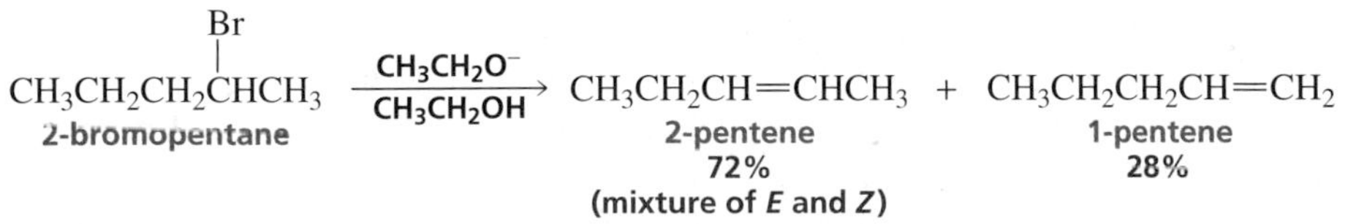

$$CH_3CH_2CH_2\overset{\displaystyle Br}{\underset{}{\overset{|}{CH}}}CH_3 \xrightarrow[CH_3CH_2OH]{CH_3CH_2O^-} CH_3CH_2CH{=}CHCH_3 \;+\; CH_3CH_2CH_2CH{=}CH_2$$

2-bromopentane   2-pentene   1-pentene
                    72%   28%
          (mixture of *E* and *Z*)

The Stereoisomers Formed in an E2 Reaction

An E2 reaction removes two groups from adjacent carbons. It is a concerted reaction, because the two groups are eliminated in the same step. The bonds to the groups to be eliminated (identified as H and X in the example below) must be in the same plane because the sp^3 orbital of the carbon bonded to H and the sp^3 orbital of the carbon bonded to X become overlapping p orbitals in the alkene product. Therefore, the orbitals must overlap in the transition state. For the overlap to be optimal, the orbitals must be parallel (that is, in the same plane).

There are two ways in which the C—H and C—X bonds can be in the same plane. They can either be parallel to one another on the same side of the molecule—an arrangement called **syn-periplanar**—or they can be parallel to one another on opposite sides of the molecule—an arrangement called **anti-periplanar**.

substituents are syn-periplanar	substituents are anti-periplanar
an eclipsed conformation	**a staggered conformation**

If an elimination reaction removes two substituents from the same side of the C—C bond, it is called a **syn elimination**. If the substituents are removed from opposite sides of the C—C bond, the reaction is called an **anti elimination**. Both types of elimination can occur, but syn elimination is much slower, so anti elimination is highly favored in an E2 reaction. One reason anti elimination is favored is that syn elimination requires the molecule to be in an eclipsed conformation, whereas anti elimination requires it to be in a more stable, staggered conformation.

Sawhorse-projections, which show the the C—C bond from an oblique angle, reveal another reason that anti elimination is favored. In syn elimination, the electrons of the departing hydrogen move to the *front* side of the carbon bonded to X, whereas in anti elimination, the electrons move to the *back* side of the carbon bonded to X.

syn elimination	anti elimination
front-side attack	**back-side attack**

We have seen that displacement reactions occur through back-side attack because in that way they achieve the best overlap of the interacting orbitals (Section 8.2). Finally, in anti elimination, the electron-rich base is spared the repulsion it experiences when it is on the same side of the molecule as the electron-rich departing halide ion.

Anti elimination predominates in an E2 reaction.

Because of the factors that favor anti elimination, an E2 reaction is *stereoselective;* which means that more of one stereoisomer is formed than the other. For example, the 2-pentene obtained as the major product from the elimination reaction of 2-bromopentane can exist as a pair of stereoisomers, and more (*E*)-2-pentene is formed than (*Z*)-2-pentene.

(*E*)-2-pentene	(*Z*)-2-pentene
major product	**minor product**

We can make the following general statement about the stereoselectivity of E2 reactions: if the reactant has two hydrogens bonded to the carbon from which a hydrogen is to be removed, both the E and Z products will be formed, because the reactant has two conformers in which the groups to be eliminated are anti.

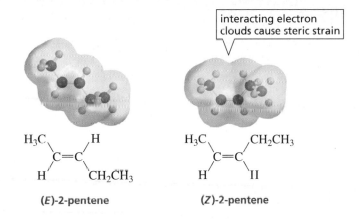

(E)-2-pentene
more stable

(Z)-2-pentene
less stable

The alkene with the *bulkiest groups on opposite sides of the double bond* will be formed in greater yield because it is the more stable alkene. Remember that the alkene with the bulkiest groups on the *same* side of the double bond is less stable because the electron clouds of the large substituents can interfere with each other, causing steric strain (Section 4.11).

When two hydrogens are bonded to the β-carbon, the major product of an E2 reaction is the alkene with the bulkiest substituents on opposite sides of the double bond.

interacting electron clouds cause steric strain

(E)-2-pentene

(Z)-2-pentene

The more stable alkene has the more stable transition state and therefore is formed more rapidly (Figure 9.3).

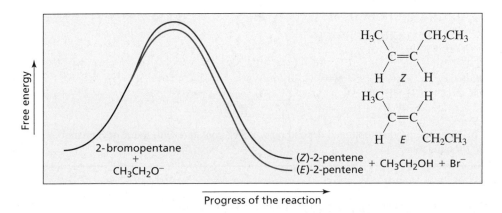

◀ **Figure 9.3**
A reaction coordinate diagram for the E2 reaction of 2-bromopentane and ethoxide ion.

Thus, elimination of HBr from 3-bromo-2,2,3-trimethylpentane leads predominantly to the *E* isomer because this stereoisomer has the methyl group, the bulkier group on one sp^2 carbon, opposite the *tert*-butyl group, the bulkier group on the other sp^2 carbon.

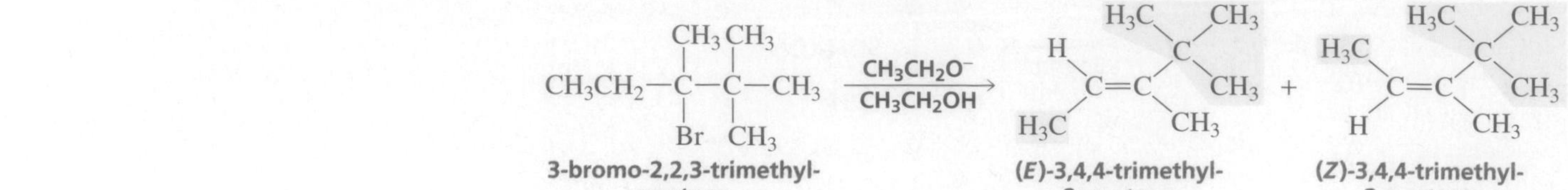

3-bromo-2,2,3-trimethyl-
pentane

(E)-3,4,4-trimethyl-
2-pentene
major product

(Z)-3,4,4-trimethyl-
2-pentene
minor product

On the other hand, if the β-carbon from which a hydrogen is to be removed is bonded to only one hydrogen, there is only one conformer in which the groups to be eliminated are anti. Therefore, only one alkene product can be formed. The particular isomer formed depends on the configuration of the reactant. For example, anti elimination of HBr from (2S,3S)-2-bromo-3-phenylbutane forms the E isomer, whereas anti elimination of HBr from (2S,3R)-2-bromo-3-phenylbutane forms the Z isomer. Notice that the groups that are not eliminated retain their relative positions. (If you have trouble understanding how to decide which product is formed, see the answer to Problem 42d in the *Study Guide and Solutions Manual*.)

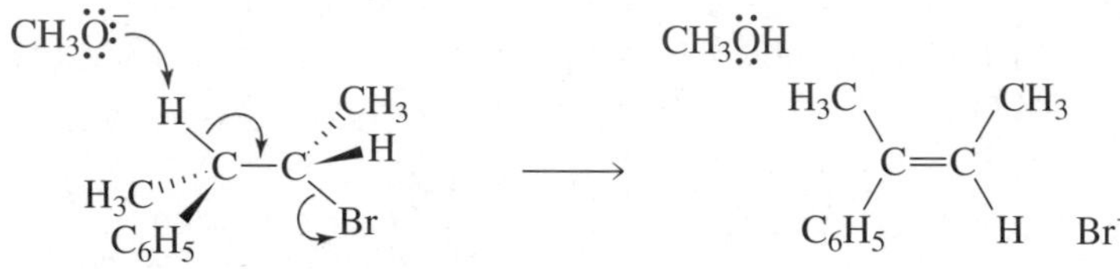

When only one hydrogen is bonded to the β-carbon, the major product of an E2 reaction depends on the structure of the alkene.

Molecular models can be helpful whenever complex stereochemistry is involved.

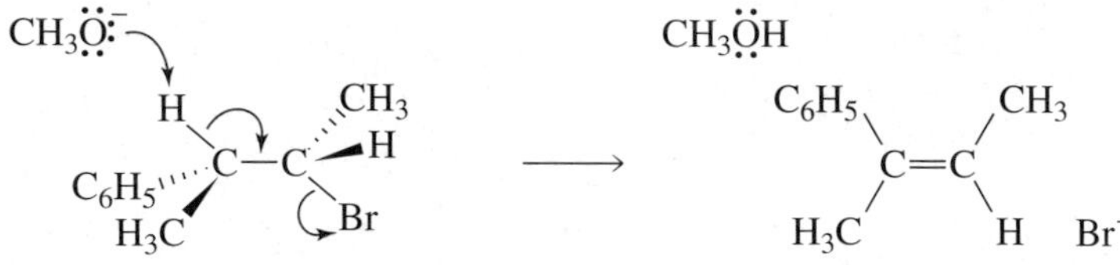

(2S,3S)-2-bromo-3-phenylbutane

(E)-2-phenyl-2-butene

(2S,3R)-2-bromo-3-phenylbutane

(Z)-2-phenyl-2-butene

PROBLEM 12◆

a. For each of the following alkyl halides, give the structure of the major product that would be obtained from an E2 reaction, indicating its configuration:

1. 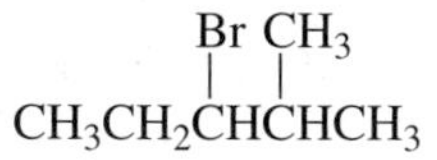 CH₃CH₂CHCHCH₃ (with Br CH₃ substituents)

2. 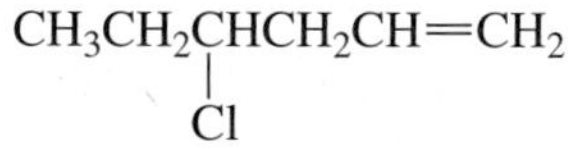 CH₃CH₂CHCH₂CH=CH₂ (with Cl substituent)

3. 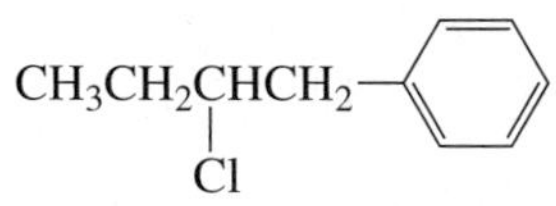 CH₃CH₂CHCH₂—C₆H₅ (with Cl substituent)

b. Does the product obtained depend on whether you start with the R or S enantiomer of the reactant?

The Stereoisomers Formed in an E1 Reaction

The major stereoisomer obtained in an E1 reaction is the alkene in which the bulkiest substituents are on opposite sides of the double bond.

An E1 reaction, like an E2 reaction, is stereoselective; both the E and Z products will be formed, but the major product will be the one with the *bulkiest groups on opposite sides of the double bond.* Let's see why this is so.

We have seen that an E1 reaction takes place in two steps. The leaving group leaves in the first step, and a proton is lost from an adjacent carbon in the second step, following Zaitsev's rule for forming the more stable alkene. The carbocation created in the first step is planar, so the electrons from a departing proton can move toward the positively charged carbon from *either side*. Therefore, both syn and anti elimination can occur.

Because both syn and anti elimination can occur in an E1 reaction, both the *E* and *Z* products are formed, regardless of whether the β-carbon from which the proton is removed is bonded to one or two hydrogens. The major product is the one with the bulkiest groups on opposite sides of the double bond, because that is the more stable alkene.

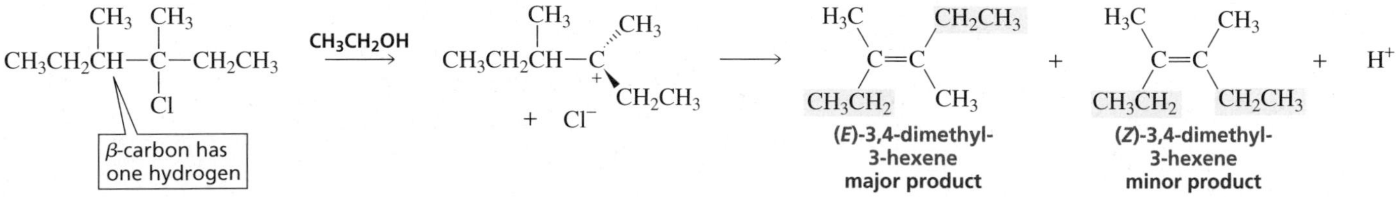

In contrast, we just saw that an E2 reaction forms both the *E* and *Z* products only if the β-carbon from which the proton is removed is bonded to two hydrogens. If the β-carbon is bonded to only one hydrogen, an E2 reaction forms only one product because anti elimination is favored.

PROBLEM 13◆ **SOLVED**

For each of the following alkyl halides, determine the major product formed when that alkyl halide undergoes an E1 reaction:

a. 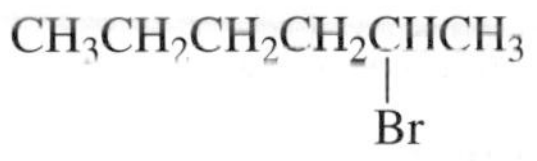 $CH_3CH_2CH_2CH_2CHICH_3$
      |
      Br

          CH₃
          |
c. 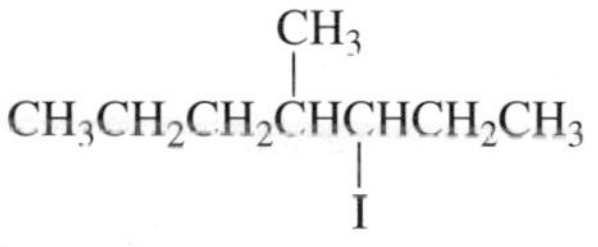 $CH_3CH_2CH_2CHCHCH_2CH_3$
                    |
                    I

          CH₃
          |
b. 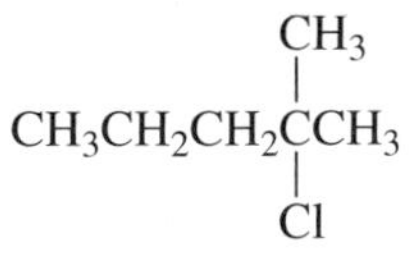 $CH_3CH_2CH_2CCH_3$
          |
          Cl

d.

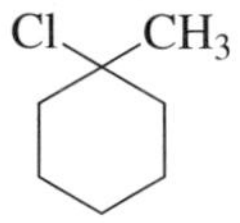

Solution to 13a   First, we need to consider the regiochemistry of the reaction: more 2-hexene will be formed than 1-hexene because 2-hexene is more stable.

$CH_3CH_2CH_2CH_2CHCH_3$ →(E1) $CH_3CH_2CH_2CH\!=\!CHCH_3$ + $CH_3CH_2CH_2CH_2CH\!=\!CH_2$
          |
          Br   **2-hexene**   **1-hexene**

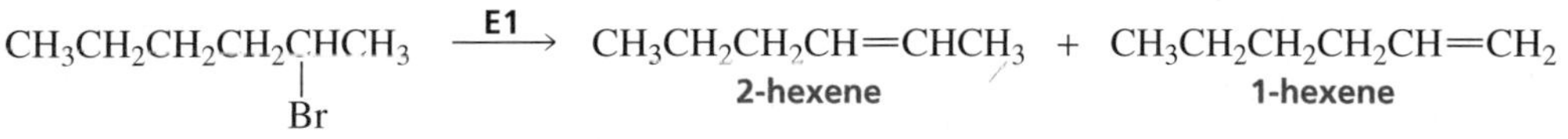

Next, we need to consider the stereochemistry of the reaction: of the 2-hexene that is formed, more will be (*E*)-2-hexene than (*Z*)-2-hexene, because (*E*)-2-hexene is more stable. Thus, (*E*)-2-hexene is the major product of the reaction.

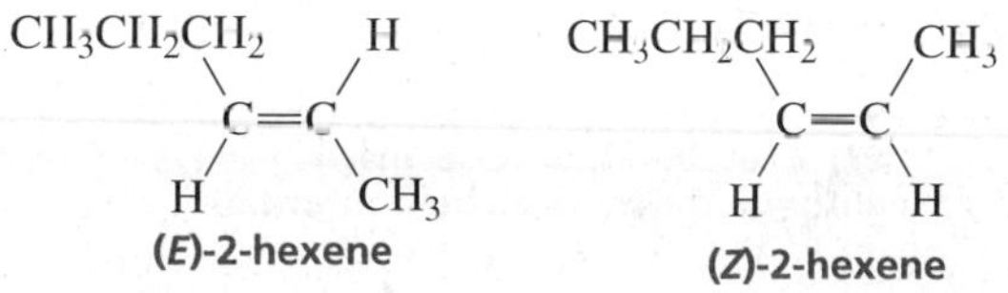

9.6 Elimination from Substituted Cyclohexanes

Elimination from substituted cyclohexanes follows the same stereochemical rules as elimination from open-chain compounds.

E2 Reactions of Substituted Cyclohexanes

In an E2 reaction of a substituted cyclohexane, the groups being eliminated must both be in axial positions.

We just saw that to achieve the anti-periplanar geometry favored in an E2 reaction, the two groups being eliminated must be parallel (Section 9.5). For two groups on a cyclohexane ring to be parallel, both must be *in axial positions*.

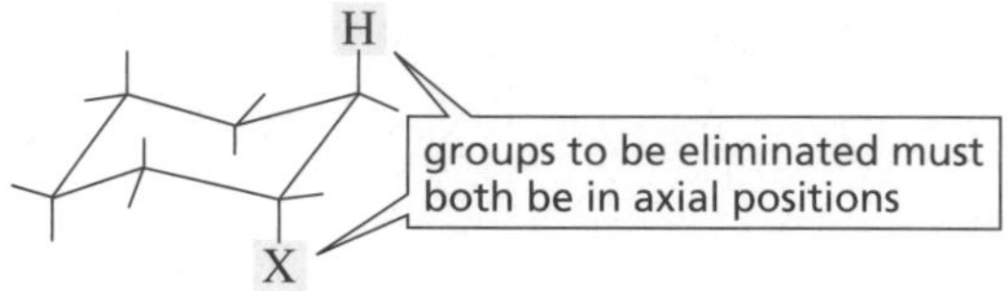

H

groups to be eliminated must both be in axial positions

X

The more stable conformer of chlorocyclohexane does not undergo an E2 reaction, because the chloro substituent is in an equatorial position. (Recall from Section 2.13 that the more stable conformer of a monosubstituted cyclohexane is the one in which the substituent is in an equatorial position because there is more room for a substituent in that position.) The less stable conformer, with the chloro substituent in the axial position, readily undergoes an E2 reaction.

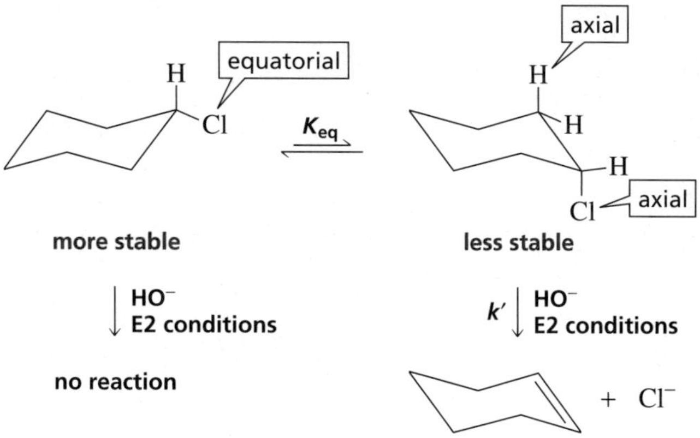

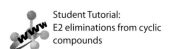

Student Tutorial:
E2 eliminations from cyclic
compounds

H | equatorial
Cl

K_{eq}

H | axial
H
H
Cl | axial

more stable   **less stable**

HO⁻
E2 conditions

k' | HO⁻
E2 conditions

no reaction

+ Cl⁻

Because one of the two conformers does not undergo an E2 reaction, the rate constant for the elimination reaction is given by $k'K_{eq}$. Therefore, the reaction is faster if K_{eq} is large.

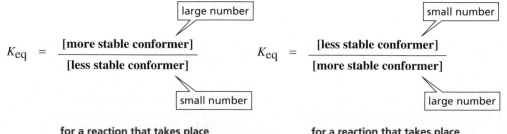

$$K_{eq} = \frac{[\text{more stable conformer}]}{[\text{less stable conformer}]}$$

large number

small number

$$K_{eq} = \frac{[\text{less stable conformer}]}{[\text{more stable conformer}]}$$

small number

large number

for a reaction that takes place through the more stable conformer

for a reaction that takes place through the less stable conformer

Most molecules are in the more stable conformer at any given time. Therefore, K_{eq} will be large if elimination takes place by way of the more stable conformer, and will be small if elimination takes place by way of the less stable conformer. For example, neomenthyl chloride undergoes an E2 reaction with ethoxide ion about 200 times faster than menthyl chloride does. The conformer of neomenthyl chloride that undergoes elimination is the *more* stable conformer because when the Cl and H are in the required axial positions, the methyl and isopropyl groups are in the equatorial positions.

BIOGRAPHY

Sir Derek H. R. Barton (1918–1998) *was the first to point out that the chemical reactivity of substituted cyclohexanes is controlled by their conformation. Barton was born in Gravesend, Kent, England. He received a Ph.D. and a D.Sc. from Imperial College, London, in 1942 and became a faculty member there three years later. Subsequently, he was a professor at the University of London, the University of Glasgow, the Institut de Chimie des Substances Naturelles, and Texas A&M University. He received the 1969 Nobel Prize in chemistry for his work on the relationship of the three-dimensional structures of organic compounds to their chemical reactivity. Barton was knighted by Queen Elizabeth II in 1972.*

In contrast, the conformer of menthyl chloride that undergoes elimination is the *less* stable conformer because when the Cl and H are in the required axial positions, the methyl and isopropyl groups are also in axial positions.

Notice that when menthyl chloride (above) or *trans*-1-chloro-2-methylcyclohexane (below) undergoes an E2 reaction, the hydrogen that is eliminated is not removed from the β-carbon bonded to the fewest hydrogens. This may seem like a violation of Zaitsev's rule, but the rule states that when there is *more than one* β-carbon from which a hydrogen can be removed, the hydrogen is removed from the β-carbon bonded to the fewest hydrogens. The hydrogen that is removed has to be in an axial position, and menthyl chloride and trans-1-chloro-2-methylcyclohexane have only one β-carbon with a hydrogen in an axial position. Therefore, that hydrogen is the one that is removed, even though it is not bonded to the β-carbon with the fewest hydrogens.

a.

b.

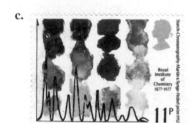

c.

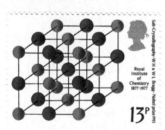

d.

Stamps issued in honor of English Nobel laureates: (a) Sir Derek Barton for conformational analysis, 1969; (b) Sir Walter Haworth for the synthesis of vitamin C, 1937; (c) A. J. P. Martin and Richard L. M. Synge for chromatography, 1952; (d) William H. Bragg and William L. Bragg for crystallography, 1915 (the only father and son to receive a Nobel Prize).

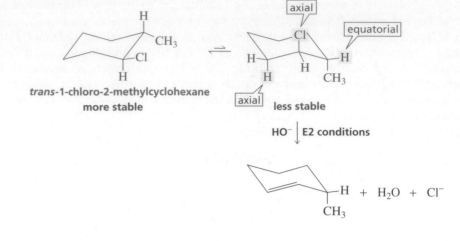

trans-1-chloro-2-methylcyclohexane
more stable   **less stable**

HO^- | E2 conditions

$+ H_2O + Cl^-$

PROBLEM 14◆

Why do *cis*-1-bromo-2-ethylcyclohexane and *trans*-1-bromo-2-ethylcyclohexane form different major products when they undergo an E2 reaction?

PROBLEM 15◆

Which isomer reacts more rapidly in an E2 reaction, *cis*-1-bromo-4-*tert*-butylcyclohexane or *trans*-1-bromo-4-*tert*-butylcyclohexane? Explain your answer.

E1 Reactions of Substituted Cyclohexanes

When a substituted cyclohexane undergoes an E1 reaction, the two groups that are eliminated do not both have to be in axial positions, because the elimination reaction is not concerted. In the following reaction, a carbocation is formed in the first step. It then loses a proton from the adjacent carbon that is bonded to the fewest hydrogens—in other words, Zaitsev's rule is followed.

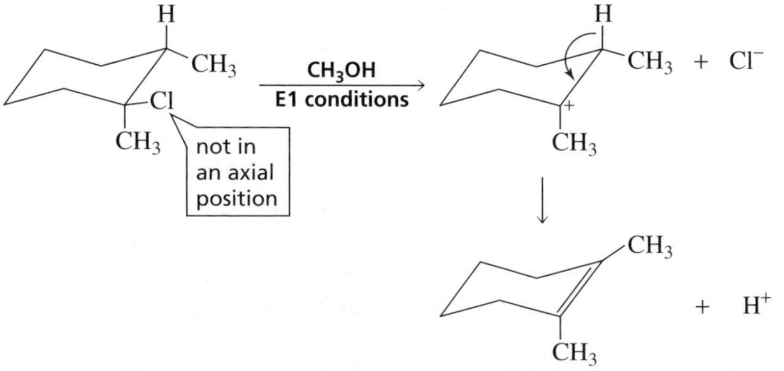

An E1 reaction includes both syn and anti elimination.

Because a carbocation is formed in an E1 reaction, you must check for the possibility of a carbocation rearrangement before you use Zaitsev's rule to determine the elimination product. In the following reaction, the secondary carbocation undergoes a 1,2-hydride shift, thus forming a more stable tertiary carbocation.

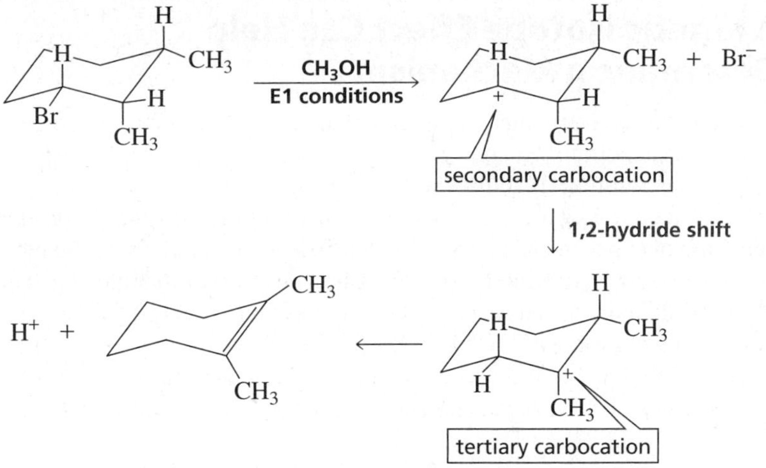

Table 9.4 summarizes the stereochemical outcomes of substitution and elimination reactions.

Table 9.4   Stereochemistry of Substitution and Elimination Reactions

Reaction	Products
S_N1	Both stereoisomers (R and S) are formed (more inverted than retained).
E1	Both E and Z stereoisomers are formed (more of the stereoisomer with the bulkiest groups on opposite sides of the double bond).
S_N2	Only the inverted product is formed.
E2	Both E and Z stereoisomers are formed (more of the stereoisomer with the bulkiest groups on opposite sides of the double bond) unless the β-carbon of the reactant is bonded to only one hydrogen, in which case only one stereoisomer is formed; its configuration depends on the configuration of the reactant.

PROBLEM 16

Give the substitution and elimination products for the following reactions, showing the configuration of each product:

a. (S)-2-chlorohexane $\xrightarrow[\text{S}_N\text{2/E2 conditions}]{\text{CH}_3\text{O}^-}$

b. (S)-2-chlorohexane $\xrightarrow[\text{S}_N\text{1/E1 conditions}]{\text{CH}_3\text{OH}}$

c. *trans*-1-chloro-2-methylcyclohexane $\xrightarrow[\text{S}_N\text{2/E2 conditions}]{\text{CH}_3\text{O}^-}$

d. *trans*-1-chloro-2-methylcyclohexane $\xrightarrow[\text{S}_N\text{1/E1 conditions}]{\text{CH}_3\text{OH}}$

e. 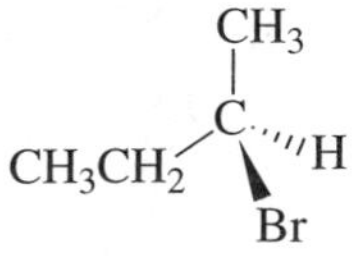 $\text{CH}_3\text{CH}_2\text{—}\overset{\overset{\displaystyle\text{CH}_3}{|}}{\underset{\underset{\displaystyle\text{Br}}{}}{\text{C}}}\text{···}^{,,,}\text{H}$ $\xrightarrow[\text{S}_N\text{2/E2 conditions}]{\text{CH}_3\text{O}^-}$

f. 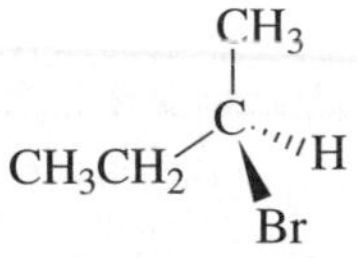 $\text{CH}_3\text{CH}_2\text{—}\overset{\overset{\displaystyle\text{CH}_3}{|}}{\underset{\underset{\displaystyle\text{Br}}{}}{\text{C}}}\text{···}^{,,,}\text{H}$ $\xrightarrow[\text{S}_N\text{1/E1 conditions}]{\text{CH}_3\text{OH}}$

9.7 A Kinetic Isotope Effect Can Help Determine a Mechanism

We have seen that a mechanism is a model that accounts for all the experimental evidence that has accumulated concerning a reaction. For example, the mechanisms of the S_N1, S_N2, E1, and E2 reactions are based on the rate law of each reaction, the relative reactivities of specific reactants, and the structures of specific products.

Another kind of experimental evidence that is helpful in determining the mechanism of a reaction is the **deuterium kinetic isotope effect**, the ratio of the rate constant observed for a compound containing hydrogen to the rate constant observed for an identical compound in which one or more of the hydrogens has been replaced by deuterium, an isotope of hydrogen. Recall that the nucleus of a deuterium atom has one proton and one neutron, whereas the nucleus of a hydrogen atom contains only a proton (Section 1.1).

$$\text{deuterium kinetic isotope effect} = \frac{k_H}{k_D} = \frac{\text{rate constant for H-containing reactant}}{\text{rate constant for D-containing reactant}}$$

The chemical properties of deuterium and hydrogen are similar; however, a $C—D$ bond is about 1.2 kcal/mol (5 kJ/mol) stronger than a $C—H$ bond. Therefore, a $C—D$ bond is more difficult to break than a corresponding $C—H$ bond.

When the rate constant (k_H) for elimination of HBr from 1-bromo-2-phenylethane is compared with the rate constant (k_D) for elimination of DBr from 2-bromo-1, 1-dideuterio-1-phenylethane (determined under identical conditions), k_H is found to be 7.1 times greater than k_D. The deuterium kinetic isotope effect for this reaction, therefore, is 7.1. The difference in the reaction rates is due to the difference in energy required to break a $C—H$ bond compared with a $C—D$ bond.

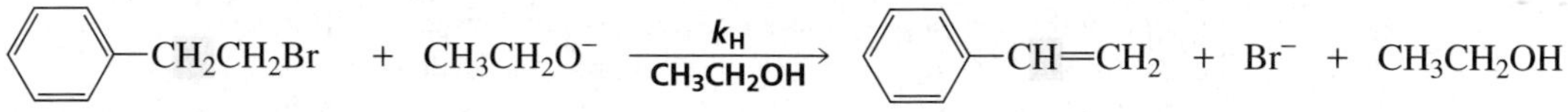

1-bromo-2-phenylethane

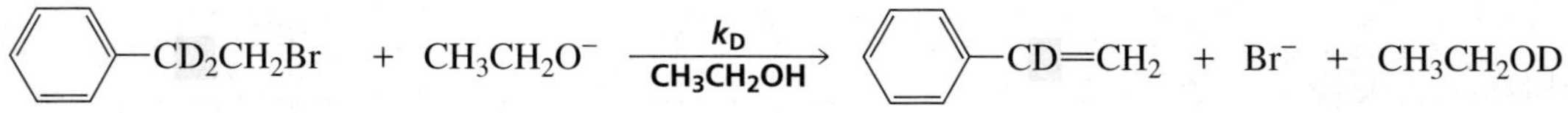

2-bromo-1,1-dideuterio-1-phenylethane

Because the deuterium kinetic isotope effect is greater than unity for this reaction, indicating that the change from hydrogen to deuterium affects the rate of the reaction, we know that the $C—H$ (or $C—D$) bond must be broken in the rate-determining step, a fact consistent with the mechanism proposed for an E2 reaction.

PROBLEM 17◆

List the following compounds in order of decreasing reactivity in an E2 reaction:

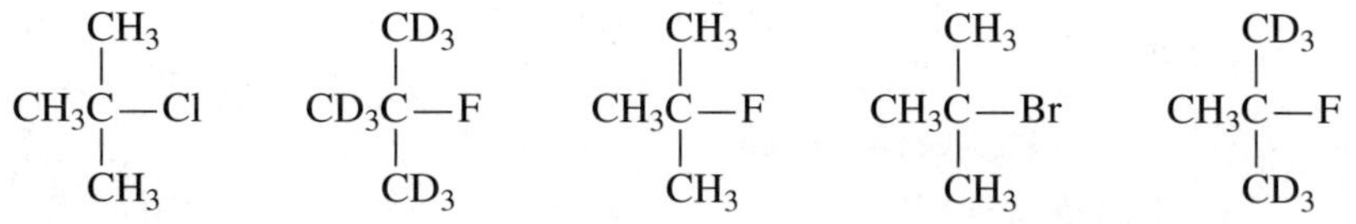

PROBLEM 18◆

If the two reactions described in this section were E1 elimination reactions, what value would you expect to obtain for the deuterium kinetic isotope effect?

9.8 **Competition Between Substitution and Elimination**

We have seen that alkyl halides can undergo four types of reactions: S_N2, S_N1, E2, and E1. As a result, you may feel a bit overwhelmed when you are asked to predict the products of the reaction of a given alkyl halide and a nucleophile/base. Let's therefore pause to organize what we know about the reactions of alkyl halides to make it a little easier for you to predict their products. Notice, in the following discussion, that HO^- is called a *nucleophile* in a substitution reaction (because it attacks a carbon) and a *base* in an elimination reaction (because it removes a proton).

To predict the products of an alkyl halide reaction, the first thing you must decide is whether the reaction conditions favor S_N2/E2 or S_N1/E1 reactions. (Recall that the conditions that favor an S_N2 reaction also favor an E2 reaction, and the conditions that favor an S_N1 reaction also favor an E1 reaction.)

If the reactant is a *primary* alkyl halide, it undergoes only S_N2/E2 reactions. (Remember that primary carbocations are too unstable to be formed, so primary alkyl halides cannot undergo S_N1/E1 reactions.) If the reactant is a *secondary* or a *tertiary* alkyl halide, it may undergo either S_N2/E2 or S_N1/E1 reactions: S_N2/E2 reactions are favored by a high concentration of a good nucleophile/strong base, whereas S_N1/E1 reactions are favored by a poor nucleophile/weak base (Sections 8.9 and 9.4). In addition, the solvent in which the reaction is carried out can influence the mechanism (Section 8.10).

Look back at the S_N1/E1 reactions in previous sections and notice that they all have poor nucleophiles/weak bases (H_2O, CH_3OH), whereas the S_N2/E2 reactions have good nucleophiles/strong bases (HO^-, CH_3O^-, NH_3). In other words, a good nucleophile/strong base encourages S_N2/E2 reactions, and a poor nucleophile/weak base encourages S_N1/E1 reactions by discouraging S_N2/E2 reactions.

Having decided whether the reaction conditions favor S_N2/E2 reactions or S_N1/E1 reactions, you must next decide whether the reagent will form the substitution product, the elimination product, or both substitution and elimination products. The relative amounts of substitution and elimination products depend on whether the alkyl halide is primary, secondary, or tertiary, and on the nature of the nucleophile/base. This question is discussed next and is summarized in Table 9.6 on page 417.

An S_N2/E2 reaction of an alkyl halide is favored by a high concentration of a good nucleophile/strong base.

An S_N1/E1 reaction of an alkyl halide is favored by a poor nucleophile/weak base.

S_N2/E2 Conditions

Let's first consider conditions that lead to S_N2/E2 reactions: a high concentration of a good nucleophile/strong base. A negatively charged species can act as a nucleophile and attack the back side of the α-carbon to form the substitution product, or it can act as a base and remove a β-hydrogen, leading to the elimination product. Thus, the two reactions compete with each other. In fact, they both occur for the same reason: the electron-withdrawing halogen gives the carbon to which it is bonded a partial positive charge.

The relative reactivities of alkyl halides in S_N2 and E2 reactions are shown in Table 9.5. Because a *primary* alkyl halide is the most reactive in an S_N2 reaction

Primary alkyl halides undergo primarily substitution under S$_N$2/E2 conditions.

(because the back side of the α-carbon is unhindered) and the least reactive in an E2 reaction, a primary alkyl halide forms principally the substitution product in a reaction carried out under conditions that favor S$_N$2/E2 reactions. In other words, substitution wins the competition.

a primary alkyl halide

$$CH_3CH_2CH_2Br \ + \ CH_3O^- \ \xrightarrow{\ CH_3OH\ } \ CH_3CH_2CH_2OCH_3 \ + \ CH_3CH{=}CH_2 \ + \ CH_3OH \ + \ Br^-$$

propyl bromide methyl propyl ether 90% propene 10%

VCL Nucleophilic Substitution - 2

Table 9.5	Relative Reactivities of Alkyl Halides		
In an S$_N$2 reaction:	1° > 2° > 3°	In an S$_N$1 reaction:	3° > 2° > 1°
In an E2 reaction:	3° > 2° > 1°	In an E1 reaction:	3° > 2° > 1°

However, if either the primary alkyl halide or the nucleophile/base is sterically hindered, the nucleophile will have difficulty getting to the back side of the α-carbon. As a result, elimination will win the competition, so the elimination product will predominate.

the primary alkyl halide is sterically hindered

$$\underset{\substack{\text{1-bromo-2-methyl-}\\ \text{propane}}}{CH_3\overset{\overset{\displaystyle CH_3}{|}}{CH}CH_2Br} \ + \ CH_3O^- \ \xrightarrow{\ CH_3OH\ } \ \underset{\substack{\text{isobutyl methyl ether}\\ \text{40\%}}}{CH_3\overset{\overset{\displaystyle CH_3}{|}}{CH}CH_2OCH_3} \ + \ \underset{\substack{\text{2-methylpropene}\\ \text{60\%}}}{CH_3\overset{\overset{\displaystyle CH_3}{|}}{C}{=}CH_2} \ + \ CH_3OH \ + \ Br^-$$

the nucleophile is sterically hindered

$$\underset{\text{1-bromopentane}}{CH_3CH_2CH_2CH_2CH_2Br} \ + \ CH_3\overset{\overset{\displaystyle CH_3}{|}}{\underset{\underset{\displaystyle CH_3}{|}}{C}}O^- \ \xrightarrow{\ (CH_3)_3COH\ } \ \underset{\substack{\text{\textit{tert}-butyl pentyl ether}\\ \text{15\%}}}{CH_3CH_2CH_2CH_2CH_2O\overset{\overset{\displaystyle CH_3}{|}}{\underset{\underset{\displaystyle CH_3}{|}}{C}}CH_3} \ + \ \underset{\substack{\text{1-pentene}\\ \text{85\%}}}{CH_3CH_2CH_2CH{=}CH_2} \ + \ CH_3\overset{\overset{\displaystyle CH_3}{|}}{\underset{\underset{\displaystyle CH_3}{|}}{C}}OH \ + \ Br^-$$

A strong base encourages elimination over substitution.

A bulky base encourages elimination over substitution.

A *secondary* alkyl halide can form both substitution and elimination products under S$_N$2/E2 conditions. The relative amounts of the two products depend on the strength and bulk of the nucleophile/base. *The stronger and bulkier the base, the greater the percentage of elimination product.* For example, acetic acid is a stronger acid (pK_a = 4.76) than ethanol (pK_a = 15.9), which means that acetate ion is a weaker base than ethoxide ion. The elimination product is the main product formed from the reaction of 2-chloropropane with the strongly basic ethoxide ion, whereas no elimination product is formed with the weakly basic acetate ion. The percentage of elimination product produced would be increased further if the bulky *tert*-butoxide ion were used instead of ethoxide ion (Section 8.3).

a secondary alkyl halide

a strong base favors the elimination product

$$\underset{\substack{\text{2-chloropropane}}}{\overset{\overset{\displaystyle Cl}{|}}{CH_3CHCH_3}} + \underset{\substack{\text{ethoxide ion}}}{CH_3CH_2O^-} \xrightarrow[CH_3CH_2OH]{} \underset{\substack{\text{2-ethoxypropane}\\25\%}}{\overset{\overset{\displaystyle OCH_2CH_3}{|}}{CH_3CHCH_3}} + \underset{\substack{\text{propene}\\75\%}}{CH_3CH=CH_2} + CH_3CH_2OH + Cl^-$$

$$\underset{\substack{\text{2-chloropropane}}}{\overset{\overset{\displaystyle Cl}{|}}{CH_3CHCH_3}} + \underset{\substack{\text{acetate ion}}}{\overset{\overset{\displaystyle O}{\parallel}}{CH_3CO^-}} \xrightarrow[\text{acetic acid}]{} \underset{\substack{\text{isopropyl acetate}\\100\%}}{\overset{\overset{\displaystyle O}{\overset{\parallel}{\underset{|}{OCCH_3}}}}{CH_3CHCH_3}} + Cl^-$$

a weak base favors the substitution product

The relative amounts of substitution and elimination products are also affected by temperature; higher temperatures favor elimination because of the greater $\Delta S°$ value for the elimination reaction since an elimination reaction forms more product molecules than does a substitution reaction (Section 3.7).

A high temperature favors elimination over substitution.

$$\underset{\substack{}}{\overset{\overset{\displaystyle Br}{|}}{CH_3CHCH_3}} + HO^- \xrightarrow[CH_3CH_2OH/H_2O]{45°C} \underset{\substack{\text{2-propanol}\\47\%}}{\overset{\overset{\displaystyle OH}{|}}{CH_3CHCH_3}} + \underset{\substack{\text{propene}\\53\%}}{CH_3CH=CH_2} + H_2O + Br^-$$

a higher temperature favors elimination

$$\underset{\substack{}}{\overset{\overset{\displaystyle Br}{|}}{CH_3CHCH_3}} + HO^- \xrightarrow[CH_3CH_2OH/H_2O]{100°C} \underset{\substack{\text{2-propanol}\\29\%}}{\overset{\overset{\displaystyle OH}{|}}{CH_3CHCH_3}} + \underset{\substack{\text{propene}\\71\%}}{CH_3CH=CH_2} + H_2O + Br^-$$

A *tertiary* alkyl halide is the least reactive of the alkyl halides in an S_N2 reaction and the most reactive in an E2 reaction (Table 9.5). Consequently, only the elimination product is formed when a tertiary alkyl halide reacts with a nucleophile/base under S_N2/E2 conditions.

Tertiary alkyl halides undergo only elimination under S_N2/E2 conditions.

a tertiary alkyl halide

$$\underset{\substack{\text{2-bromo-2-methyl-}\\\text{propane}}}{\overset{\overset{\displaystyle CH_3}{|}}{\underset{\underset{\displaystyle CH_3}{|}}{CH_3CBr}}} + CH_3CH_2O^- \xrightarrow[CH_3CH_2OH]{} \underset{\substack{\text{2-methylpropene}\\97\%}}{\overset{\overset{\displaystyle CH_3}{|}}{CH_3C=CH_2}} + CH_3CH_2OH + Br^-$$

PROBLEM 19◆

In what way would you expect the ratio of substitution product to elimination product formed from the reaction of propyl bromide with CH_3O^- in methanol to change if the nucleophile were changed to CH_3S^-?

PROBLEM 20◆

Explain why only a substitution product and no elimination product is obtained when the following compound is treated with sodium methoxide:

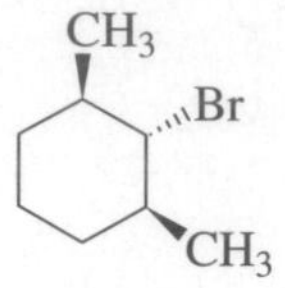

S$_N$1/E1 Conditions

Now let's look at what happens when conditions favor S$_N$1/E1 reactions, conditions characterized by a poor nucleophile/weak base. In S$_N$1/E1 reactions, the alkyl halide dissociates to form a carbocation, which can then either combine with the nucleophile to form the substitution product or lose a proton to form the elimination product.

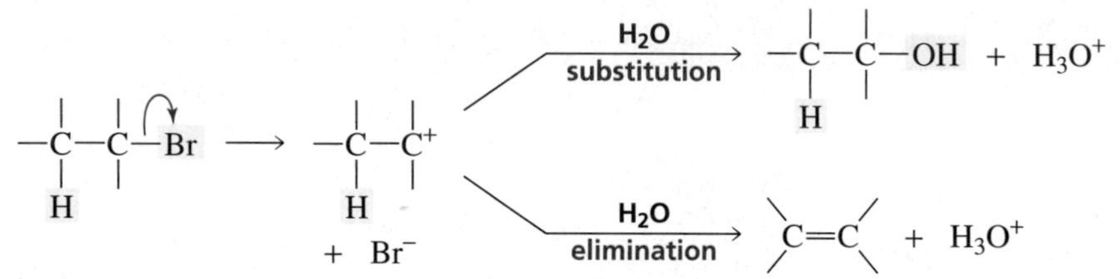

Alkyl halides have the same order of reactivity in S$_N$1 reactions as they do in E1 reactions, because both reactions have the same rate-determining step—dissociation of the alkyl halide to form a carbocation (Table 9.5). This means that all alkyl halides that react under S$_N$1/E1 conditions will give both substitution and elimination products. Remember that primary alkyl halides do not undergo S$_N$1/E1 reactions because primary carbocations are too unstable to be formed. Fortunately, S$_N$1/E1 reactions of tertiary alkyl halides favor the substitution product, because the elimination product is the only product obtained under S$_N$2/E2 conditions.

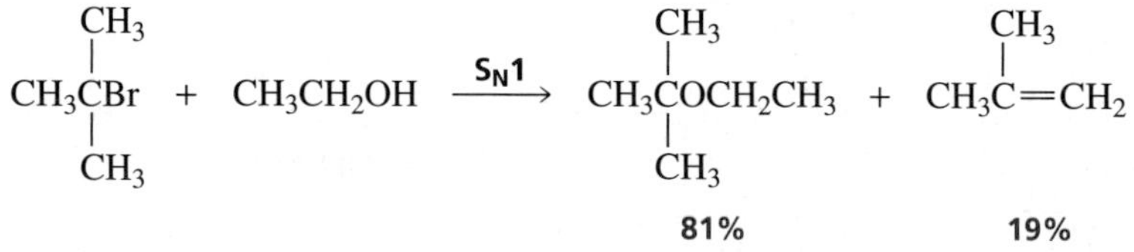

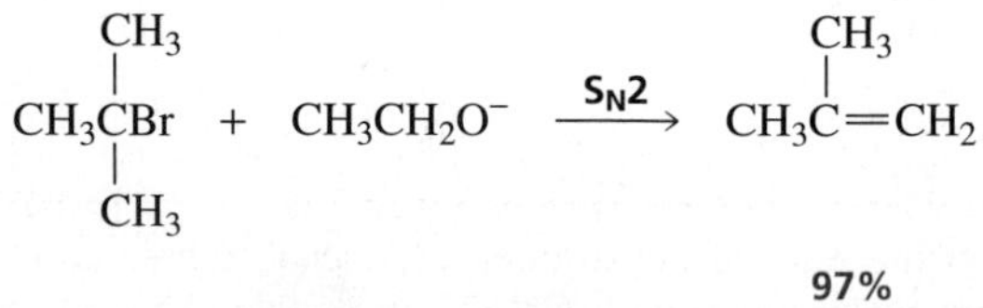

Primary alkyl halides do not form carbocations; therefore, they cannot undergo S$_N$1 and E1 reactions.

Table 9.6 summarizes the products obtained when alkyl halides react with nucleophiles/bases under S$_N$2/E2 and under S$_N$1/E1 conditions.

Table 9.6 Summary of the Products Expected in Substitution and Elimination Reactions

Class of alkyl halide	S_N2 versus E2	S_N1 versus E1
Primary alkyl halide	Primarily substitution, unless there is steric hindrance in the alkyl halide or nucleophile, in which case elimination is favored	Cannot undergo S_N1/E1 reactions
Secondary alkyl halide	Both substitution and elimination; the stronger and bulkier the base and the higher the temperature, the greater the percentage of elimination	Both substitution and elimination
Tertiary alkyl halide	Only elimination	Both substitution and elimination

PROBLEM 21◆

a. Which reacts faster in an S_N2 reaction?

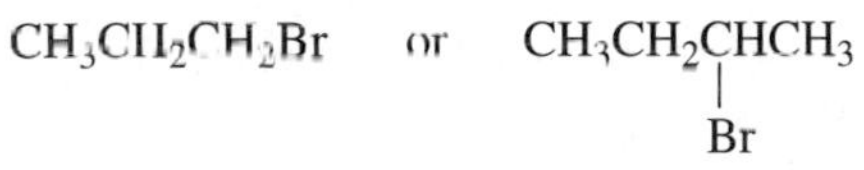

$$CH_3CH_2CH_2Br \quad \text{or} \quad CH_3CH_2CHCH_3$$
$$| $$
$$Br$$

b. Which reacts faster in an E1 reaction?

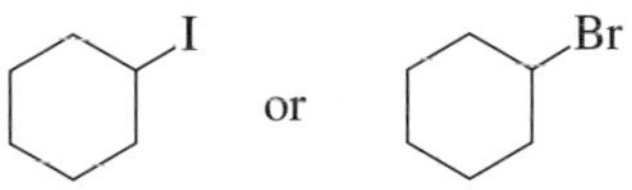

c. Which reacts faster in an S_N1 reaction?

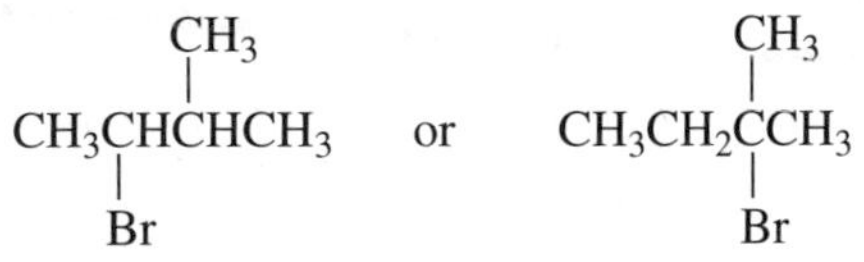

$$\begin{array}{cc} CH_3 & CH_3 \\ | & | \\ CH_3CHCHCH_3 & \text{or} \quad CH_3CH_2CCH_3 \\ | & | \\ Br & Br \end{array}$$

d. Which reacts faster in an E2 reaction?

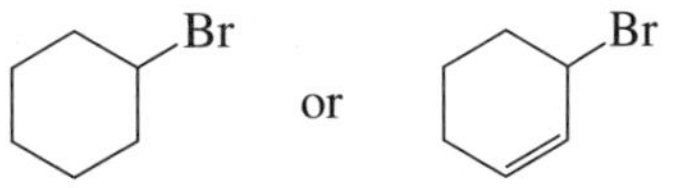

PROBLEM 22◆

Indicate whether the alkyl halides listed will give primarily substitution products, only elimination products, both substitution and elimination products, or no products when they are treated with the following:

a. methanol under S_N1/E1 conditions
b. sodium methoxide under S_N2/E2 conditions
   1. 1-bromobutane
   2. 1-bromo-2-methylpropane
   3. 2-bromobutane
   4. 2-bromo-2-methylpropane

PROBLEM 23◆

1-Bromo-2,2-dimethylpropane has difficulty undergoing either S_N2 or S_N1 reactions.

a. Explain why.
b. Can it undergo E2 and E1 reactions?

9.9 Substitution and Elimination Reactions in Synthesis

When substitution or elimination reactions are used in synthesis, care must be taken to choose reactants and reaction conditions that will maximize the yield of the desired product.

Using Substitution Reactions to Synthesize Compounds

In Section 8.4, you saw that nucleophilic substitution reactions of alkyl halides can lead to a wide variety of organic compounds. For example, ethers are synthesized by the reaction of an alkyl halide with an alkoxide ion. This reaction, discovered by Alexander Williamson in 1850, is still considered one of the best ways to synthesize an ether.

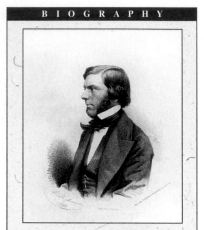
Williamson ether synthesis

$$R-Br \ + \ R-O^- \ \longrightarrow \ R-O-R \ + \ Br^-$$
alkyl halide   alkoxide ion   ether

The alkoxide ion (RO^-) for the **Williamson ether synthesis** is prepared by using sodium metal or sodium hydride (NaH) to remove a proton from an alcohol.

$$ROH \ + \ Na \ \longrightarrow \ RO^- \ + \ Na^+ \ + \ \tfrac{1}{2}H_2$$

$$ROH \ + \ NaH \ \longrightarrow \ RO^- \ + \ Na^+ \ + \ H_2$$

The Williamson ether synthesis is a nucleophilic substitution reaction. It requires a high concentration of a good nucleophile, which indicates that it is an S_N2 reaction. If you want to synthesize an ether such as butyl propyl ether, you have a choice of starting materials: you can use either a propyl halide and butoxide ion or a butyl halide and propoxide ion.

$$CH_3CH_2CH_2Br \ + \ CH_3CH_2CH_2CH_2O^- \ \longrightarrow \ CH_3CH_2CH_2OCH_2CH_2CH_2CH_3 \ + \ Br^-$$
propyl bromide   butoxide ion   butyl propyl ether

$$CH_3CH_2CH_2CH_2Br \ + \ CH_3CH_2CH_2O^- \ \longrightarrow \ CH_3CH_2CH_2OCH_2CH_2CH_3 \ + \ Br^-$$
butyl bromide   propoxide ion   butyl propyl ether

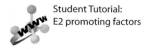

Student Tutorial:
E2 promoting factors

However, if you want to synthesize *tert*-butyl ethyl ether, the starting materials must be an ethyl halide and *tert*-butoxide ion. If you tried to use a *tert*-butyl halide and ethoxide ion as reactants, you would not obtain any ether, because the reaction of a tertiary alkyl halide under S_N2/E2 conditions forms only the elimination product. Consequently, a Williamson ether synthesis should be designed in such a way that the less hindered alkyl group is provided by the alkyl halide and the more hindered alkyl group comes from the alkoxide ion.

CH₃CH₂Br + CH₃CO⁻ with CH₃ groups (tert-butoxide ion) ⟶ CH₃CH₂OCCH₃ with CH₃ groups (tert-butyl ethyl ether) + CH₂=CH₂ (ethene) + CH₃COH with CH₃ groups + Br⁻

ethyl bromide

In ether synthesis, the less hindered group should be provided by the alkyl halide.

CH₃CH₂O⁻ (ethoxide ion) + CH₃CBr with CH₃ groups (tert-butyl bromide) ⟶ CH₂=CCH₃ with CH₃ (2-methylpropene) + CH₃CH₂OH + Br⁻

We saw in Section 6.11 that alkynes can be synthesized by the reaction of an acetylide anion with an alkyl halide.

CH₃CH₂C≡C⁻ + CH₃CH₂CH₂Br ⟶ CH₃CH₂C≡CCH₂CH₂CH₃ + Br⁻

Now that you know that this is an S$_N$2 reaction (the alkyl halide reacts with a high concentration of a good nucleophile), you can understand why you were told that it is best to use primary alkyl halides and methyl halides in the reaction. These alkyl halides are the only ones that form primarily the desired substitution product. A tertiary alkyl halide would form only the elimination product, and a secondary alkyl halide would form mainly the elimination product because the acetylide ion is a very strong base.

PROBLEM 24◆

What other organic product will be formed when the alkyl halide used in the synthesis of butyl propyl ether is

a. propyl bromide?   **b.** butyl bromide?

PROBLEM 25

How could the following ethers be prepared using an alkyl halide and an alcohol?

a. CH₃CH₂CHOCH₂CH₂CH₃ with CH₃ group

c. CH₃CH₂OCH₂CHCH₂CH₂CH₃ with CH₃ group

b. ⬡—OCH₃

d. ⬡—CH₂O—⬡

Using Elimination Reactions to Synthesize Compounds

If you want to synthesize an alkene, you should choose the most hindered alkyl halide possible in order to maximize the elimination product and minimize the substitution product. For example, 2-bromopropane would be a better starting material than 1-bromopropane for the synthesis of propene because the secondary alkyl halide would give a higher yield of the desired elimination product and a lower yield of the competing substitution product. The percentage of alkene could be further increased by using a sterically hindered base such as *tert*-butoxide ion instead of hydroxide ion.

Student Tutorial:
Synthesis of alkenes

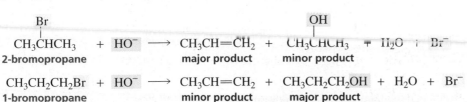

CH₃CHCH₃ with Br + HO⁻ ⟶ CH₃CH=CH₂ + CH₃CHCH₃ with OH + H₂O + Br⁻

2-bromopropane   major product   minor product

CH₃CH₂CH₂Br + HO⁻ ⟶ CH₃CH=CH₂ + CH₃CH₂CH₂OH + H₂O + Br⁻

1-bromopropane   minor product   major product

To synthesize 2-methyl-2-butene from 2-bromo-2-methylbutane, you would use $S_N2/E2$ conditions (a high concentration of HO^- in an aprotic polar solvent) because a tertiary alkyl halide gives *only* the elimination product under those conditions. If $S_N1/E1$ conditions were used (a low concentration of HO^- in water), both elimination and substitution products would be obtained.

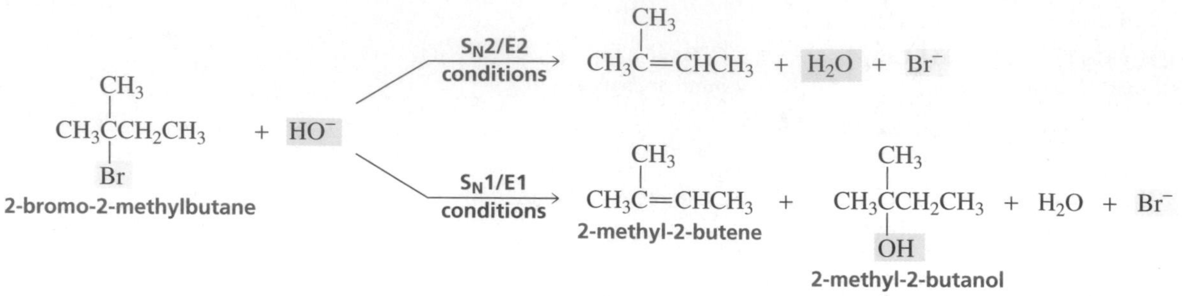

PROBLEM 26◆

Identify the three products formed when 2-bromo-2-methylpropane is dissolved in a mixture of 80% ethanol and 20% water.

PROBLEM 27

a. What products (including stereoisomers, if applicable) would be formed from the reaction of 3-bromo-2-methylpentane with HO^- under $S_N2/E2$ conditions and under $S_N1/E1$ conditions?

b. Answer the same question for 3-bromo-3-methylpentane.

9.10 Consecutive E2 Elimination Reactions

Alkyl dihalides can undergo two consecutive dehydrohalogenations, giving products that contain two double bonds. In the following example, Zaitsev's rule predicts the most stable product of the first dehydrohalogenation, but not the most stable product of the second. The reason Zaitsev's rule fails in the second reaction is that a conjugated diene is more stable than an isolated diene (Section 7.7).

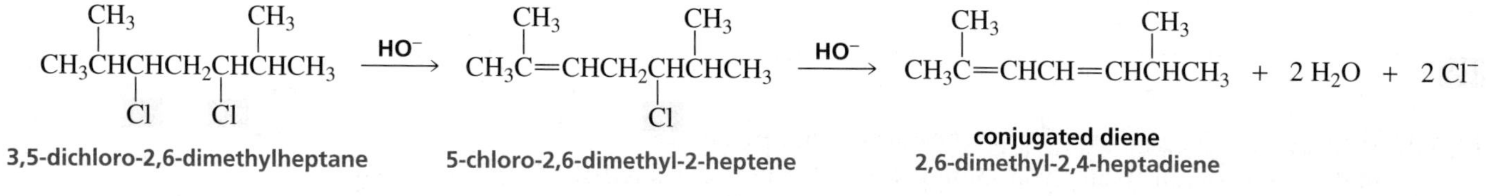

If the two halogens are on the same carbon (geminal dihalides) or on adjacent carbons (vicinal dihalides), the two consecutive E2 dehydrohalogenations can result in the formation of a triple bond. This is how alkynes are commonly synthesized.

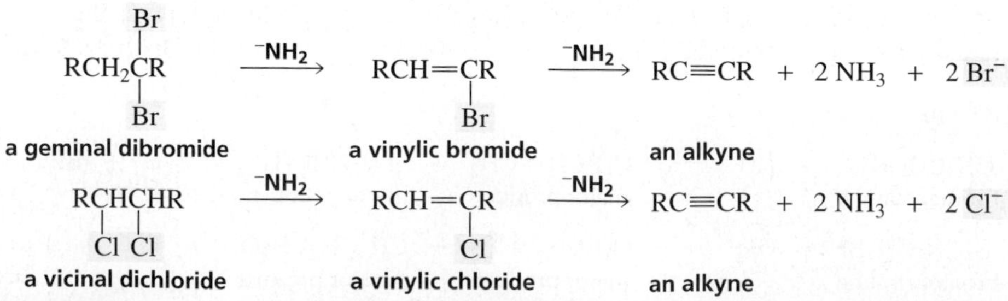

The vinylic halide intermediates in the preceding reactions are relatively unreactive. Consequently, a very strong base, such as $^-NH_2$, is needed for the second elimination. If a weaker base, such as HO^-, is used at room temperature, the reaction will stop at the vinylic halide and no alkyne will be formed.

Because a vicinal dihalide is formed from the reaction of an alkene with Br_2 or Cl_2, you have just learned how to convert a double bond into a triple bond.

$$CH_3CH{=}CHCH_3 \xrightarrow[\text{CH}_2\text{Cl}_2]{\text{Br}_2} \underset{\underset{\text{Br Br}}{|\ \ |}}{CH_3CHCHCH_3} \xrightarrow{^-NH_2} \underset{\underset{\text{Br}}{|}}{CH_3CH{=}CCH_3} \xrightarrow{^-NH_2} CH_3C{\equiv}CCH_3 + 2\ NH_3 + 2\ Br^-$$

2-butene   2-butyne

PROBLEM 28

Why isn't a cumulated diene formed in the preceding reaction?

9.11 Designing a Synthesis II: Approaching the Problem

When you are asked to design a synthesis, one way to approach the task is to think about the starting material you have been given and ask yourself if there is an obvious series of reactions beginning with the starting material that can get you on the road to the **target molecule** (the desired product). Sometimes this is the best way to approach a *simple* synthesis. The following examples will give you practice employing this strategy.

Example 1. Using the given starting material, how could you prepare the target molecule?

$$\bigcirc \xrightarrow{?} \bigcirc{-}C{\equiv}N$$

Adding HBr to the alkene would form a compound with a leaving group that can be replaced by a nucleophile. Because $^-C{\equiv}N$ is a relatively weak base (the pK_a of $HC{\equiv}N$ is 9), the desired substitution will be favored over the competing elimination reaction.

synthesis

$$\bigcirc \xrightarrow{\text{HBr}} \bigcirc{-}Br \xrightarrow{^-C{\equiv}N} \bigcirc{-}C{\equiv}N$$

Example 2. Starting with 1-bromo-1-methylcyclohexane, how could you prepare *trans*-2-methylcyclohexanol?

$$\bigcirc\!\!\overset{CH_3}{\underset{Br}{\big\langle}} \xrightarrow{?} \bigcirc\!\!\overset{CH_3}{\underset{\text{'''OH}}{\big\langle}}$$

The elimination reaction is carried out under E2 conditions, because under these conditions tertiary alkyl halides undergo only elimination, so there will be no competing substitution product. Because, overall, hydroboration–oxidation results in the syn addition of water, the target molecule (as well as its enantiomer) is obtained.

synthesis

$$\bigcirc\!\!\overset{CH_3}{\underset{Br}{\big\langle}} \xrightarrow[\text{CH}_3\text{CH}_2\text{OH}]{\substack{\text{high} \\ \text{concentration} \\ \text{CH}_3\text{CH}_2\text{O}^-}} \bigcirc\!\!{-}CH_3 \xrightarrow[\text{2. H}_2\text{O}_2,\ \text{HO}^-,\ \text{H}_2\text{O}]{\text{1. BH}_3/\text{THF}} \bigcirc\!\!\overset{CH_3}{\underset{\text{'''OH}}{\big\langle}} + \bigcirc\!\!\overset{\text{'''}CH_3}{\underset{\text{OH}}{\big\langle}}$$

As you saw in Section 6.12, working backward can be a useful way to design a synthesis, particularly when the starting material does not clearly indicate how to proceed. Look at the target molecule and ask yourself how it could be prepared. Once you have an answer, look at the precursor you have identified for the target molecule and ask yourself how that precursor could be prepared. Keep working backward one step at a time, until you get to a readily available starting material. This technique is called *retrosynthetic analysis*.

Example 3. How could you prepare ethyl methyl ketone from 1-bromobutane?

$$CH_3CH_2CH_2CH_2Br \xrightarrow{\ ?\ } CH_3CH_2\overset{\overset{\displaystyle O}{\|}}{C}CH_3$$

At this point in your study of organic chemistry, you know only two ways to synthesize a ketone: (1) the addition of water to an alkyne (Section 6.7) and (2) hydroboration–oxidation of an alkyne (Section 6.8). The alkyne can be prepared from two successive E2 reactions of a vicinal dihalide, which in turn can be synthesized from an alkene. The desired alkene can be prepared from the given starting material by an elimination reaction.

retrosynthetic analysis

$$CH_3CH_2\overset{\overset{\displaystyle O}{\|}}{C}CH_3 \implies CH_3CH_2C\equiv CH \implies \underset{\underset{\displaystyle Br}{|}}{CH_3CH_2CHCH_2Br} \implies CH_3CH_2CH=CH_2 \implies CH_3CH_2CH_2CH_2Br$$

target molecule

open arrow indicates
you are working backwards

Now you can write the reaction sequence in the forward direction, along with the reagents needed to carry out each reaction. Notice that a bulky base is used in the elimination reaction in order to maximize the amount of elimination product.

synthesis

$$CH_3CH_2CH_2CH_2Br \xrightarrow[\textbf{\textit{tert}-BuOH}]{\overset{\textbf{high}}{\underset{}{\overset{\textbf{concentration}}{\textbf{\textit{tert}-BuO}^-}}}} CH_3CH_2CH=CH_2 \xrightarrow[\textbf{CH}_2\textbf{Cl}_2]{\textbf{Br}_2} \underset{\underset{\displaystyle Br}{|}}{CH_3CH_2CHCH_2Br} \xrightarrow{\overline{\textbf{NH}}_2} CH_3CH_2C\equiv CH$$

$$\xrightarrow[\textbf{HgSO}_4]{\overset{\textbf{H}_2\textbf{SO}_4}{}\ \big|\ \textbf{H}_2\textbf{O}}$$

$$CH_3CH_2\overset{\overset{\displaystyle O}{\|}}{C}CH_3$$

target molecule

Example 4. How could the following cyclic ether be prepared from the given starting material?

$$BrCH_2CH_2CH_2CH=CH_2 \xrightarrow{\ ?\ } \quad \text{(cyclic ether)}{-}CH_3$$

To prepare a cyclic ether, the alkyl halide and alcohol required for ether synthesis must both be part of the same molecule. Since the target ether has a five-membered ring, the carbons bearing the two reacting groups must be separated by two additional carbons. Addition of water to the given starting material will create the required bifunctional compound, which will then form the cyclic ether through an intramolecular reaction.

retrosynthetic analysis

CH3
O
BrCH2CH2CH2CHCH3
OH
BrCH2CH2CH2CH=CH2

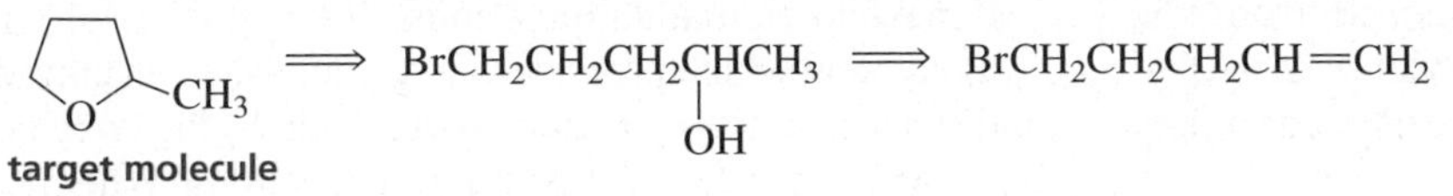

target molecule

synthesis

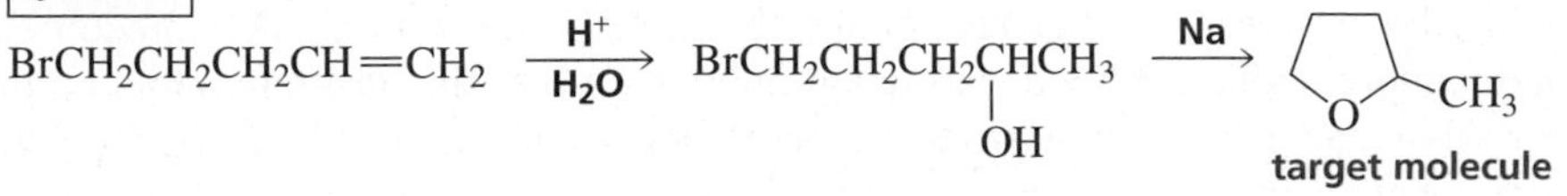

BrCH2CH2CH2CH=CH2
H+
H2O
BrCH2CH2CH2CHCH3
OH
Na
O
CH3

target molecule

PROBLEM 29

How could you have prepared the target molecule in the preceding synthesis using 4-penten-1-ol as the starting material? Which synthesis would give you a higher yield of the target molecule?

PROBLEM 30

For each of the following target compounds, design a multistep synthesis to show how it could be prepared from the given starting material:

a.
Br
O

b.
CH=CH2
O
CH2CH

c.
Br
Br
Br

d.
CH=CH2
CH2CH2CH2CH3

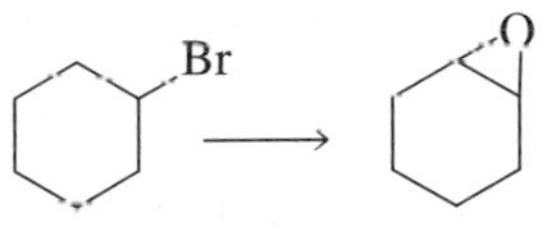

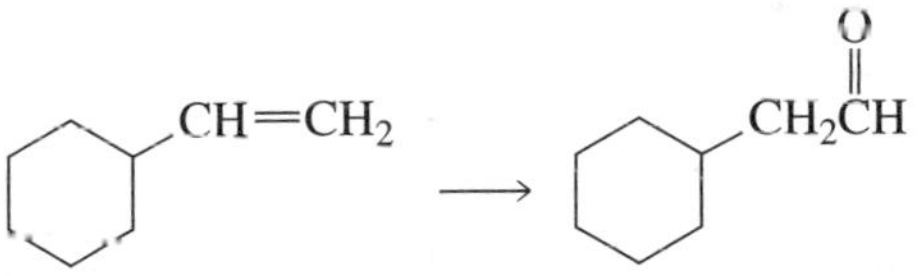

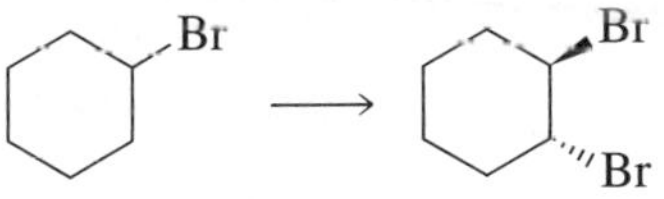

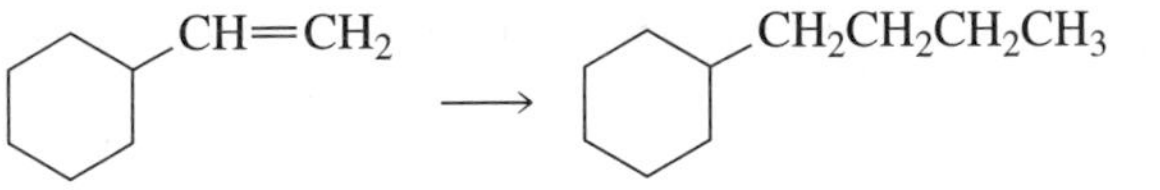

SUMMARY

In addition to undergoing nucleophilic substitution reactions, alkyl halides undergo β-elimination reactions, in which the halogen is removed from one carbon, a hydrogen is removed from an adjacent carbon, and a double bond is formed between the two carbons from which the atoms were eliminated. The product of an **elimination reaction** is therefore an alkene. Removal of a proton and a halide ion is called **dehydrohalogenation**. There are two important **β-elimination reactions**, E1 and E2.

An **E2 reaction** is a concerted, one-step reaction; the proton and halide ion are removed in the same step, so no intermediate is formed. In an **E1 reaction**, the alkyl halide dissociates, forming a carbocation intermediate. In a second step, a base removes a proton from a carbon that is adjacent to the positively charged carbon. Because the E1 reaction forms a carbocation intermediate, the carbon skeleton can rearrange before the proton is lost.

Primary alkyl halides undergo only E2 elimination reactions. Secondary and tertiary alkyl halides undergo both E2 and E1 reactions. For alkyl halides that can undergo both E2 and E1 reactions, the E2 reaction is favored by the same factors that favor an S_N2 reaction—a high concentration of a strong base and an aprotic polar solvent—and the E1 reaction is favored by the same factors that favor an S_N1 reaction—a weak base and a protic polar solvent.

An E2 reaction is regioselective; the major product is the more stable alkene, unless the reactants are sterically hindered or the leaving group is poor. The more stable alkene is generally (but not always) the more substituted alkene. The more substituted alkene is predicted by Zaitsev's rule: it is the alkene formed when a proton is removed from the β-carbon that is bonded to the fewest hydrogens. Alkyl substitution increases the stability of a carbocation and decreases the stability of a carbanion.

An E2 reaction is also stereoselective: **anti elimination** is favored. If the β-carbon has two hydrogens, both E and Z products will be formed, but the one with the bulkiest groups on opposite sides of the double bond is more stable and will be formed in greater yield. If the β-carbon has only one hydrogen, only one alkene is formed, since there is only one conformer in which the groups to be eliminated are anti. The two groups eliminated from a six-membered ring must both be in axial positions; elimination is more rapid when H and X are diaxial in the more stable conformer.

An E1 reaction is regioselective: the major product is the most stable alkene, which is generally the most substituted alkene. An E1 reaction is stereoselective: the major product is the alkene with the bulkiest groups on opposite sides of the double bond. The carbocation formed in the first step

can undergo both syn and anti elimination; therefore, the two groups eliminated from a six-membered ring do not both have to be in axial positions.

Predicting which products are formed when an alkyl halide undergoes a reaction begins with determining whether the conditions favor $S_N2/E2$ or $S_N1/E1$ reactions. When $S_N2/E2$ reactions are favored, primary alkyl halides form primarily substitution products unless the nucleophile/base is sterically hindered, in which case elimination products predominate. Secondary alkyl halides form both substitution and elimination products; the stronger and bulkier the base, the greater is the percentage of the elimination product. Tertiary alkyl halides form only elimination products. When $S_N1/E1$ conditions are favored, secondary and tertiary alkyl halides form both substitution and elimination products; primary alkyl halides do not undergo $S_N1/E1$ reactions.

The **Williamson ether synthesis** involves the reaction of an alkyl halide with an alkoxide ion. If two halogens are on the same or adjacent carbons, two consecutive E2 dehydrohalogenations can result in the formation of a triple bond.

SUMMARY OF REACTIONS

1. E2 reaction: a one-step mechanism

Relative reactivities of alkyl halides: $3° > 2° > 1°$

Anti elimination: both E and Z stereoisomers are formed. The isomer with the bulkiest groups on opposite sides of the double bond will be formed in greater yield. If the β-carbon from which the hydrogen is removed is bonded to only one hydrogen, only one elimination product is formed. Its configuration depends on the configuration of the reactant.

2. E1 reaction: a two-step mechanism with a carbocation intermediate

Relative reactivities of alkyl halides: $3° > 2° > 1°$

Anti and syn elimination: both E and Z stereoisomers are formed. The isomer with the bulkiest groups on opposite sides of the double bond will be formed in greater yield.

Competing S_N2 and E2 Reactions

Primary alkyl halides: primarily substitution
Secondary alkyl halides: substitution and elimination
Tertiary alkyl halides: only elimination

Competing S_N1 and E1 Reactions

Primary alkyl halides: cannot undergo S_N1 or E1 reactions
Secondary alkyl halides: substitution and elimination
Tertiary alkyl halides: substitution and elimination

KEY TERMS

anti elimination (p. 404)
anti-periplanar (p. 404)
dehydrohalogenation (p. 391)
deuterium kinetic isotope
    effect (p. 412)
elimination reaction (p. 389)

β-elimination reaction (p. 391)
E1 reaction (p. 398)
E2 reaction (p. 390)
1,2-elimination reaction (p. 391)
sawhorse projection (p. 404)
syn elimination (p. 404)

syn-periplanar (p. 404)
target molecule (p. 421)
Williamson ether
    synthesis (p. 418)
Zaitsev's rule (p. 393)

PROBLEMS

31. Give the major product obtained when each of the following alkyl halides undergoes an E2 reaction:

a. $CH_3CHCH_2CH_3$
         |
         Br

c. $CH_3CHCH_2CH_3$
         |
         Cl

e. $CH_3CHCH_2CH_2CH_3$
         |
         Cl

b.

d.

f.

32. Give the major product obtained when each alkyl halide in Problem 31 undergoes an E1 reaction.

33. a. Indicate how each of the following factors affects an E1 reaction:
    1. the structure of the alkyl halide   **3.** the concentration of the base
    2. the strength of the base   **4.** the solvent
    b. Indicate how each of the same factors affects an E2 reaction.

34. Which species in each pair is more stable?

a. $CH_3\overset{-}{C}HCH_2CH_3$ or $CH_3CH_2CH_2\overset{-}{C}H_2$

b. $CH_3\overset{+}{C}HCH_2CH_3$ or $CH_3CH_2CH_2\overset{+}{C}H_2$

c. $\overset{+}{C}H_2CH_2CH=CH_2$ or $CH_3\overset{+}{C}HCH=CH_2$

d. $\overset{-}{C}H_2CH_2CH=CH_2$ or $CH_3\overset{-}{C}HCH=CH_2$

e. $CH_3CHCH=CH_2$ or $CH_3CH_2C=CH_2$
       | |
      CH_3 CH_3

f. $CH_3CH\overset{-}{C}HCH_3$ or $CH_3\overset{-}{C}CH_2CH_3$
      | |
     CH_3 CH_3

g. $CH_3CH\overset{+}{C}HCH_3$ or $CH_3\overset{+}{C}CH_2CH_3$
      | |
     CH_3 CH_3

35. Dr. Don T. Doit wanted to synthesize the anesthetic 2-ethoxy-2-methylpropane. He used ethoxide ion and 2-chloro-2-methylpropane for his synthesis and ended up with very little ether. What was the predominant product of his synthesis? What reagents should he have used?

36. Which reactant in each of the following pairs will undergo an elimination reaction more rapidly? Explain your choice.

a. $(CH_3)_3CCl$ $\xrightarrow[\text{H}_2\text{O}]{\text{HO}^-}$ or $(CH_3)_3CBr$ $\xrightarrow[\text{H}_2\text{O}]{\text{HO}^-}$

b. $\xrightarrow[\text{CH}_3\text{OH}]{\text{CH}_3\text{O}^-}$ or $\xrightarrow[\text{CH}_3\text{OH}]{\text{CH}_3\text{O}^-}$

37. For each of the following reactions, give the major elimination product; if the product can exist as stereoisomers, indicate which stereoisomer is obtained in greater yield.
   a. (*R*)-2-bromohexane + high concentration of HO⁻
   b. (*R*)-2-bromohexane + H_2O
   c. *trans*-1-chloro-2-methylcyclohexane + high concentration of CH_3O^-
   d. *trans*-1-chloro-2-methylcyclohexane + CH_3OH
   e. 3-bromo-3-methylpentane + high concentration of HO⁻
   f. 3-bromo-3-methylpentane + H_2O

38. a. Which reacts faster in an E2 reaction, 3-bromocyclohexene or bromocyclohexane?
   b. Which reacts faster in an E1 reaction?

39. Starting with an alkyl halide, how could the following compounds be prepared?
   a. 2-methoxybutane   **b.** 1-methoxybutane   **c.** butylmethylamine

40. Indicate which of the compounds in each pair will give a higher substitution-product to elimination-product ratio when it reacts with isopropyl bromide:
   a. ethoxide ion or *tert*-butoxide ion   **c.** Cl⁻ or Br⁻
   b. ⁻OCN or ⁻SCN   **d.** CH_3S^- or CH_3O^-

41. Rank the following compounds in order of decreasing reactivity in an E2 reaction:

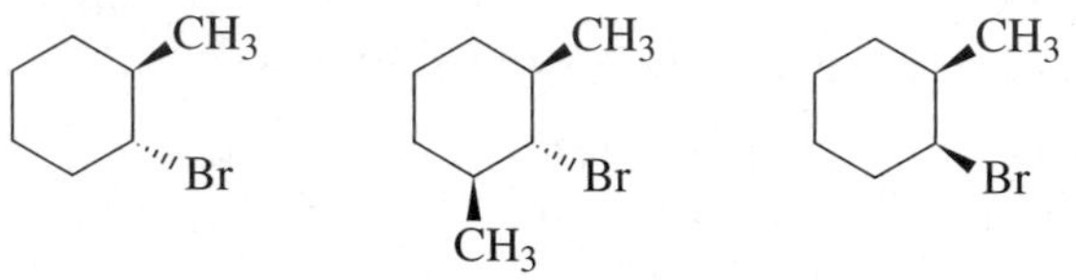

42. For each of the following alkyl halides, indicate what stereoisomer would be obtained in greatest yield from the reaction of the alkyl halide with a high concentration of hydroxide ion.
   a. 3-bromo-2,2,3-trimethylpentane   **c.** 3-bromo-2,3-dimethylpentane
   b. 4-bromo-2,2,3,3-tetramethylpentane   **d.** 3-bromo-3,4-dimethylhexane

43. When 2-bromo-2,3-dimethylbutane reacts with a base under E2 conditions, two alkenes (2,3-dimethyl-1-butene and 2,3-dimethyl-2-butene) are formed.
   a. Which of the bases shown below would give the highest percentage of the 1-alkene?
   b. Which would give the highest percentage of the 2-alkene?

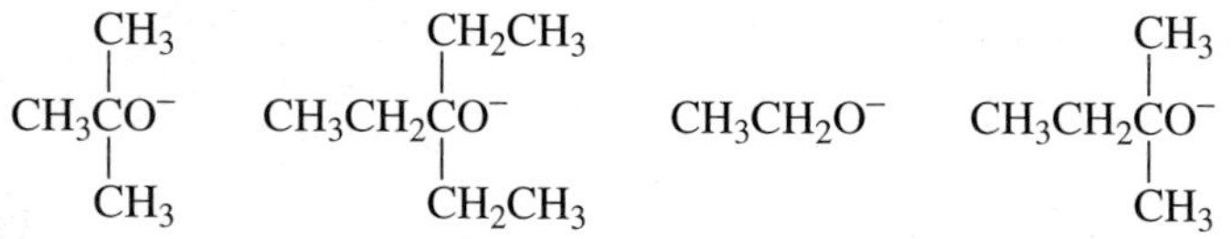

44. a. Give the structures of the products obtained from the reaction of each enantiomer of *cis*-1-chloro-2-isopropylcyclopentane with a high concentration of sodium methoxide in methanol.
   b. Are all the products optically active?
   c. How would the products differ if the starting material were the trans isomer? Are all these products optically active?
   d. Will the cis enantiomers or the trans enantiomers form substitution products more rapidly?
   e. Will the cis enantiomers or the trans enantiomers form elimination products more rapidly?

45. When the following compound undergoes solvolysis in ethanol, three products are obtained. Propose a mechanism to account for the formation of these products.

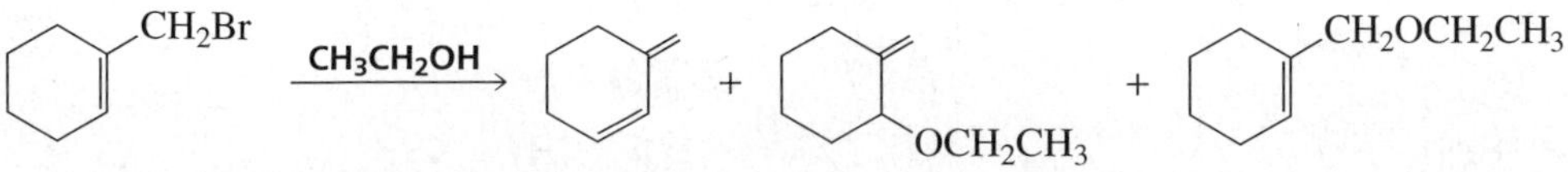

46. *cis*-1-Bromo-4-*tert*-butylcyclohexane and *trans*-1-bromo-4-*tert*-butylcyclohexane both react with sodium ethoxide in ethanol to give 4-*tert*-butylcyclohexene. Explain why the cis isomer reacts much more rapidly than the trans isomer.

47. For each of the following reactions, give the elimination products; if the products can exist as stereoisomers, indicate which stereoisomers are obtained.
    a. (2S,3S)-2-chloro-3-methylpentane + high concentration of CH_3O^-
    b. (2S,3R)-2-chloro-3-methylpentane + high concentration of CH_3O^-
    c. (2R,3S)-2-chloro-3-methylpentane + high concentration of CH_3O^-
    d. (2R,3R)-2-chloro-3-methylpentane + high concentration of CH_3O^-
    e. 3-chloro-3-ethyl-2,2-dimethylpentane + high concentration of $CH_3CH_2O^-$

48. Give the major elimination product that would be obtained from each of the following reactions:

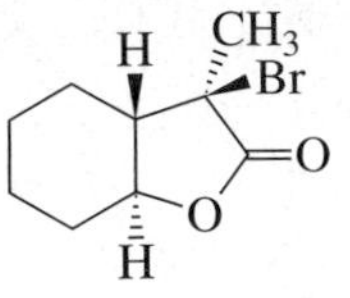

49. Which of the following hexachlorocyclohexanes is the least reactive in an E2 reaction?

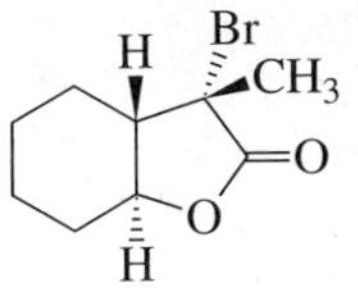

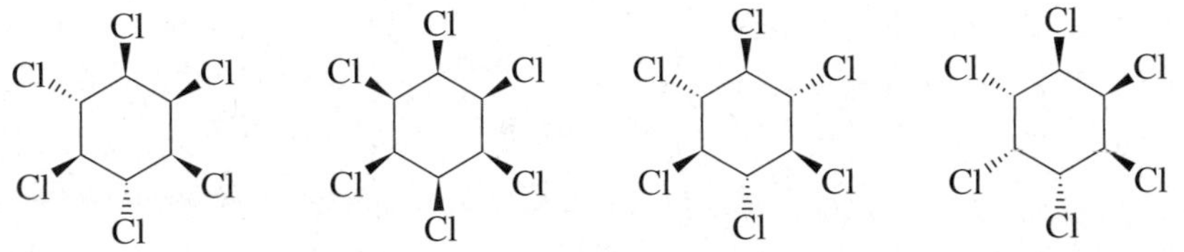

50. Explain why the rate of the reaction of 1-bromo-2-butene with ethanol is increased if silver nitrate is added to the reaction mixture.

51. Give the products for each of the following reactions carried out under S_N2/E2 conditions. If the products can exist as stereoisomers, show which stereoisomers are formed.
    a. (3S,4S)-3-bromo-4-methylhexane + CH_3O^- **c.** (3S,4R)-3-bromo-4-methylhexane + CH_3O^-
    b. (3R,4R)-3-bromo-4-methylhexane + CH_3O^- **d.** (3R,4S)-3-bromo-4-methylhexane + CH_3O^-

52. Two elimination products are obtained from the following E2 reaction:

$$CH_3CH_2CHDCH_2Br \xrightarrow{\text{HO}^-}$$

    a. What are the elimination products?
    b. Which is formed in greater yield? Explain.

53. How could you prepare the following compounds from the given starting materials?

    a. $CH_3CH_2CH_2CH_2Br \longrightarrow CH_3CH_2CCH_2CH_2CH_3$ **b.** $BrCH_2CH_2CH_2CH_2Br \longrightarrow$

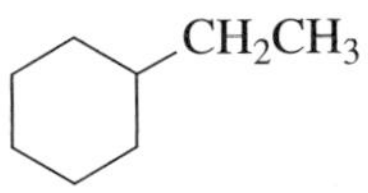

54. *cis*-4-Bromocyclohexanol and *trans*-4-bromocyclohexanol form the same elimination product but a different substitution product when they react with HO^-.

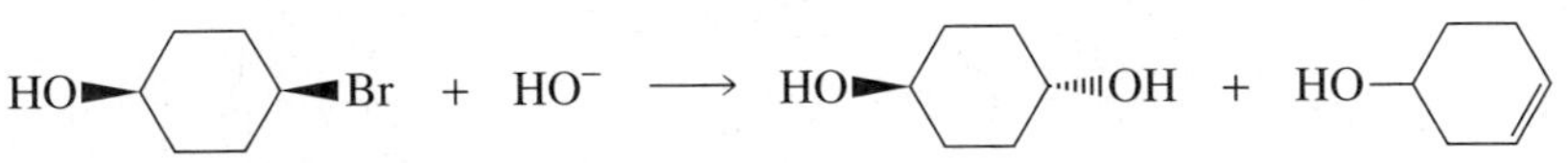

cis-**4-bromocyclohexanol**

trans-**4-bromocyclohexanol**

    a. Explain, by showing the mechanisms, why different substitution products are obtained.
    b. How many stereoisomers of the product does each reaction form?

55. Three substitution products and three elimination products are obtained from the following reaction:

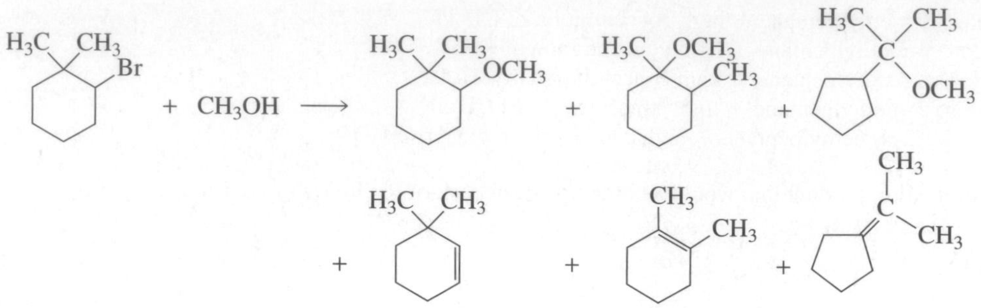

Account for the formation of these products.

56. Only one product is formed when the following stereoisomer of 2-chloro-1,3-dimethylcyclohexane reacts with methoxide ion in a solvent that encourages $S_N2/E2$ reactions:

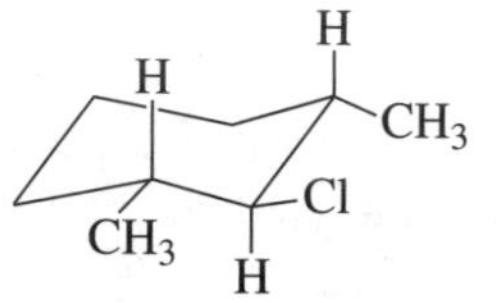

When the same compound reacts with methoxide ion in a solvent that favors $S_N1/E1$ reactions, twelve products are formed. Identify the products that are formed under the two sets of conditions.

57. For each of the following compounds, give the product that will be formed in an E2 reaction and indicate the configuration of the product:

a. (1S,2S)-1-bromo-1,2-diphenylpropane

b. (1S,2R)-1-bromo-1,2-diphenylpropane

Reactions of Alcohols, Amines, Ethers, Epoxides, and Sulfur-Containing Compounds • Organometallic Compounds

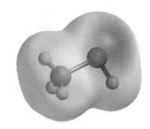

BUILDING ON FUNDAMENTALS

SECTIONS 10.1 AND 10.7 Alcohols and ethers undergo nucleophilic substitution reactions (S_N1 and S_N2) like alkyl halides do (8.2 and 8.5); however, alcohols and ethers must be activated before they react, because they have much poorer leaving groups. There are several ways to activate an alcohol, but only one way to activate an ether.

SECTION 10.4 When activated, alcohols also undergo elimination reactions (E1 and E2) like alkyl halides do (9.1 and 9.3).

SECTION 10.4 The dehydration of alcohols is the reverse of the acid-catalyzed addition of water to an alkene (4.5).

SECTION 10.5 We have seen two oxidation reactions (4.8 and 4.9); now we will see some others.

SECTION 10.8 We have seen that an alkene can be converted into an epoxide (4.9). Now you will learn how epoxides are named and how they react.

SECTION 10.11 The physical properties of thiols are compared with the physical properties of alcohols (2.9).

We have seen that alkyl halides undergo substitution and elimination reactions because of their electron-withdrawing halogen atoms (Chapters 8 and 9). Compounds with other electron-withdrawing groups also undergo substitution and elimination reactions. The relative reactivity of these compounds depends on the electron-withdrawing group.

Alcohols and ethers have electron-withdrawing groups (HO^-, RO^-) that are stronger bases than halide ions (X^-); because they are stronger bases, they are poorer leaving groups and therefore harder to displace. Consequently, alcohols and ethers are less reactive than alkyl halides in substitution and elimination reactions. We will see that because of their strongly basic leaving groups, alcohols and ethers have to be "activated" before they can undergo a substitution or an elimination reaction. In contrast, sulfonate esters and sulfonium salts have weakly basic leaving groups, so they undergo substitution reactions with ease.

The weaker the base, the more easily it can be displaced.

The stronger the acid, the weaker is its conjugate base.

R—X	R—O—H	R—O—R	R—O—$\overset{\overset{\displaystyle O}{\|}}{\underset{\underset{\displaystyle O}{\|}}{S}}$—R	R—$\overset{\displaystyle R}{\underset{\displaystyle \|^+}{S}}$—R
an alkyl halide X = F, Cl, Br, I	an alcohol	an ether	a sulfonate ester	a sulfonium salt

10.1 Nucleophilic Substitution Reactions of Alcohols: Forming Alkyl Halides

An **alcohol** has a strongly basic leaving group (HO^-) that cannot be displaced by a nucleophile. Therefore, an alcohol cannot undergo a nucleophilic substitution reaction.

a strongly basic leaving group

$$CH_3-\ddot{O}H \ + \ Br^- \ \not\rightarrow \ CH_3-Br \ + \ HO^-$$
strong base

However, if the alcohol's OH group is converted into a group that is a weaker base (and therefore a better leaving group), a nucleophilic substitution reaction can occur. One way to convert an OH group into a weaker base is to protonate it by adding acid to the solution. Protonation changes the leaving group from HO^- to H_2O, which is a weak enough base to be displaced by a nucleophile. The substitution reaction is slow and requires heat (except in the case of tertiary alcohols) if it is to take place in a reasonable period of time.

a weakly basic leaving group

$$CH_3-\ddot{O}H \ + \ HBr \ \rightleftharpoons \ CH_3-\overset{H}{\underset{Br^-}{\overset{|}{\ddot{O}H}}} \ \overset{\Delta}{\rightarrow} \ CH_3-Br \ + \ H_2O$$
weak base

poor leaving group

good leaving group

Because the OH group of the alcohol has to be protonated before it can be displaced by a nucleophile, only weakly basic nucleophiles (I^-, Br^-, Cl^-) can be used in the substitution reaction. Moderately and strongly basic nucleophiles (NH_3, RNH_2, CH_3O^-) cannot be used because they too would be protonated in the acidic solution and, once protonated would no longer be nucleophiles ($^+NH_4$, RNH_3^+) or would be poor nucleophiles (CH_3OH).

PROBLEM 1◆

Why are NH_3 and CH_3NH_2 no longer nucleophiles when they are protonated?

Primary, secondary, and tertiary alcohols all undergo nucleophilic substitution reactions with HI, HBr, and HCl to form alkyl halides.

$$CH_3CH_2CH_2OH \ + \ HI \ \overset{\Delta}{\rightarrow} \ CH_3CH_2CH_2I \ + \ H_2O$$
1-propanol
a primary alcohol

1-iodopropane

cyclohexanol
a secondary alcohol

bromocyclohexane

$$\underset{\underset{CH_3}{|}}{\overset{\overset{CH_3}{|}}{CH_3CH_2COH}} \ + \ HBr \ \longrightarrow \ \underset{\underset{CH_3}{|}}{\overset{\overset{CH_3}{|}}{CH_3CH_2CBr}} \ + \ H_2O$$
2-methyl-2-butanol
a tertiary alcohol

2-bromo-2-methylbutane

The mechanism of the substitution reaction depends on the structure of the alcohol. Secondary and tertiary alcohols undergo S_N1 reactions.

Secondary and tertiary alcohols undergo S_N1 reactions with hydrogen halides.

Mechanism for the S_N1 reaction of an alcohol

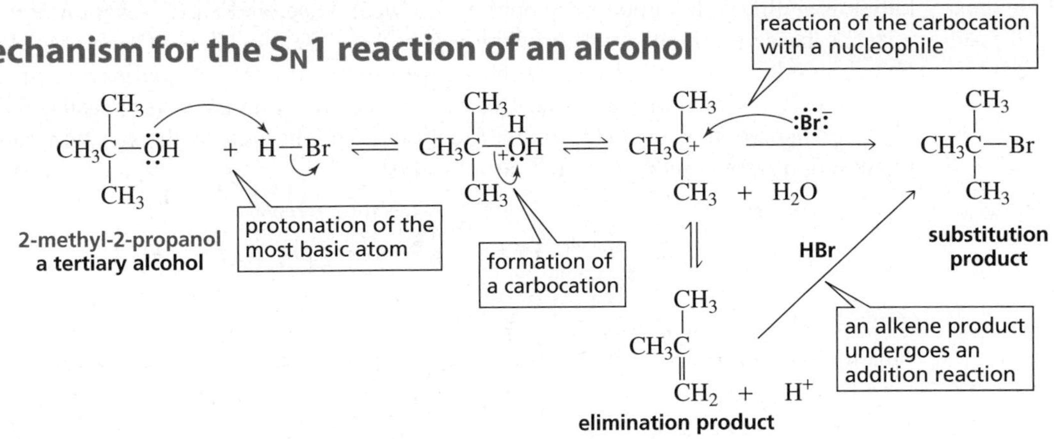

reaction of the carbocation with a nucleophile

2-methyl-2-propanol
a tertiary alcohol

protonation of the most basic atom

formation of a carbocation

substitution product

HBr

an alkene product undergoes an addition reaction

elimination product

- An acid always reacts with an organic molecule in the same way: it protonates the most basic atom in the molecule.
- Weakly basic water is the leaving group that is expelled, forming a carbocation.
- The carbocation has two possible fates: it can combine with a nucleophile and form a substitution product, or it can lose a proton and form an elimination product.

Although the reaction can form both a substitution product and an elimination product, only the substitution product is actually obtained because any alkene formed in an elimination reaction will undergo a subsequent addition reaction with HBr to form more of the substitution product.

Tertiary alcohols undergo substitution reactions with hydrogen halides faster than secondary alcohols do, because tertiary carbocations are easier to form than secondary carbocations. (Recall that alkyl groups stabilize carbocations by hyperconjugation; Section 4.2) Thus, the reaction of a tertiary alcohol with a hydrogen halide proceeds readily at room temperature, whereas the reaction of a secondary alcohol with a hydrogen halide has to be heated to have the reaction occur at the same rate.

An acid protonates the most basic atom in a molecule.

Carbocation stability: 3° > 2° > 1°

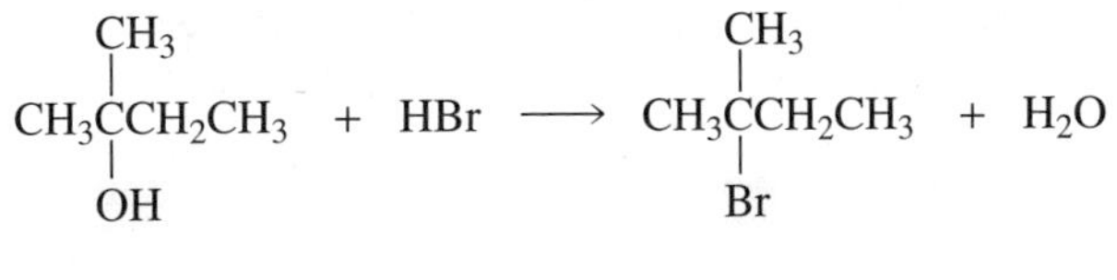

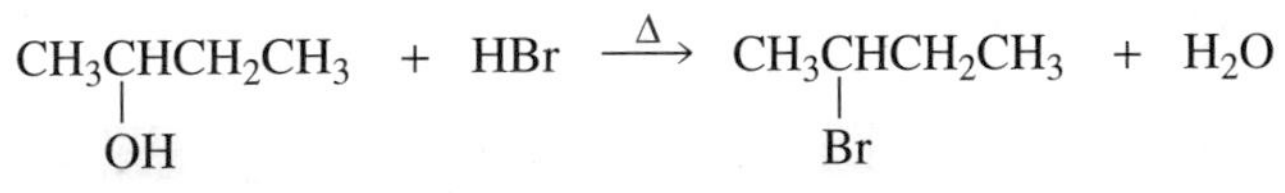

Primary alcohols cannot undergo S_N1 reactions because primary carbocations are too unstable to be formed (Section 8.5). Therefore, when a primary alcohol reacts with a hydrogen halide, it must do so by an S_N2 reaction.

Mechanism for the S_N2 reaction of an alcohol

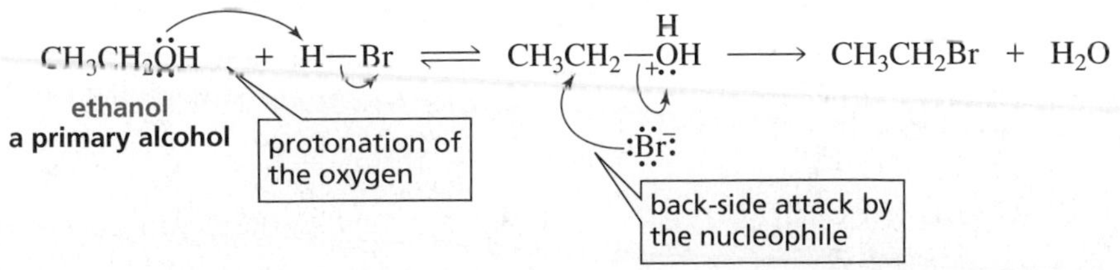

ethanol
a primary alcohol

protonation of the oxygen

back-side attack by the nucleophile

Primary alcohols undergo S$_N$2 reactions with hydrogen halides.

- The acid protonates the most basic atom in the reactant.
- The nucleophile hits the back side of the carbon and displaces the leaving group.

Only a substitution product is obtained. No elimination product is formed because the halide ion, although a good nucleophile, is a weak base, and an E2 reaction requires a strong base to remove a proton from a β-carbon (Section 9.4). (Remember that a β-carbon is the carbon adjacent to the carbon that is attached to the leaving group.)

When HCl is used instead of HBr or HI, the S$_N$2 reaction is slower because Cl$^-$ is a poorer nucleophile than Br$^-$ or I$^-$ (Section 8.3). The rate of the reaction can be increased, however, by using ZnCl$_2$ as a catalyst.

$$CH_3CH_2CH_2OH \ + \ HCl \ \xrightarrow[\Delta]{ZnCl_2} \ CH_3CH_2CH_2Cl \ + \ H_2O$$

ZnCl$_2$ is a Lewis acid that complexes strongly with the lone-pair electrons on oxygen. This interaction weakens the C—O bond and creates a better leaving group.

$$CH_3CH_2CH_2\overset{..}{\underset{..}{O}}H \ + \ \underset{\overset{|}{Cl}}{ZnCl} \longrightarrow CH_3CH_2CH_2-\overset{\overset{ZnCl}{\overset{|}{+}}}{\underset{..}{O}}H \longrightarrow CH_3CH_2CH_2Cl \ + \ HOZnCl$$
$$:\overset{..}{\underset{..}{Cl}}:^-$$

THE LUCAS TEST

The Lucas test determines whether an alcohol is primary, secondary, or tertiary by taking advantage of the relative rates at which the three classes of alcohols react with HCl/ZnCl$_2$. The alcohol is added to a mixture of HCl and ZnCl$_2$, known as the Lucas reagent. Low-molecular-weight alcohols are soluble in the Lucas reagent, but the alkyl halide products are not, so they cause the solution to turn cloudy. If the alcohol is tertiary, the solution turns cloudy immediately. If the alcohol is secondary, the solution will turn cloudy in about five minutes. A primary alcohol will turn cloudy only if the solution is heated. Because the test relies on the complete solubility of the alcohol in the Lucas reagent, it is limited to alcohols with fewer than six carbons.

BIOGRAPHY

Howard J. Lucas (1885–1963) *was born in Ohio and earned B.S. and M.S. degrees from Ohio State University. He published a description of the Lucas test in 1930. He was a professor of chemistry at the California Institute of Technology.*

PROBLEM 2 *SOLVED*

Using the pK_a values of the conjugate acids of the leaving groups (the pK_a of HBr is -9; the pK_a of H$_2$O is 15.7; the pK_a, of H$_3$O$^+$ is -1.7), explain the difference in reactivity in substitution reactions between:

a. CH$_3$Br and CH$_3$OH

b. CH$_3$OH$_2$$^+$ and CH$_3$OH

Solution to 2a The conjugate acid of the leaving group of CH$_3$Br is HBr; the conjugate acid of the leaving group of CH$_3$OH is H$_2$O. Because HBr is a much stronger acid (p$K_a = -9$) than H$_2$O (p$K_a = 15.7$), Br$^-$ is a much weaker base than HO$^-$. (Recall that the stronger the acid, the weaker its conjugate base.) Therefore, Br$^-$ is a much better leaving group than HO$^-$, causing CH$_3$Br to be much more reactive than CH$_3$OH.

PROBLEM 3 *SOLVED*

Show how 1-butanol can be converted into the following compounds:

a. CH$_3$CH$_2$CH$_2$CH$_2$OCH$_3$

c. CH$_3$CH$_2$CH$_2$CH$_2$NHCH$_2$CH$_3$

b. CH$_3$CH$_2$CH$_2$CH$_2$O$\overset{\overset{\textstyle O}{\|}}{C}CH_2CH_3$

d. CH$_3$CH$_2$CH$_2$CH$_2$C$\equiv$N

Solution to 3a Because the OH group of 1-butanol is too basic to be substituted, the alcohol must first be converted into an alkyl halide. The alkyl halide has a leaving group that can be substituted by CH_3O^-, the nucleophile required to obtain the desired product.

$$CH_3CH_2CH_2CH_2OH \xrightarrow[\Delta]{\textbf{HBr}} CH_3CH_2CH_2CH_2Br \xrightarrow[\Delta]{\textbf{CH}_3\textbf{O}^-} CH_3CH_2CH_2CH_2OCH_3$$

PROBLEM 4◆

The observed relative reactivities of primary, secondary, and tertiary alcohols with a hydrogen halide are $3° > 2° > 1°$. If secondary alcohols underwent an S_N2 reaction rather than an S_N1 reaction with a hydrogen halide, what would be the relative reactivities of the three classes of alcohols?

Because the reaction of a secondary or a tertiary alcohol with a hydrogen halide is an S_N1 reaction, a carbocation is formed as an intermediate. Therefore, we must check for the possibility of a carbocation rearrangement when predicting the product of the substitution reaction. Remember that a carbocation rearrangement will occur if it leads to formation of a more stable carbocation (Section 4.6). For example, the major product of the reaction of 3-methyl-2-butanol with HBr is 2-bromo-2-methylbutane, because a 1,2-hydride shift converts the secondary carbocation formed initially into a more stable tertiary carbocation.

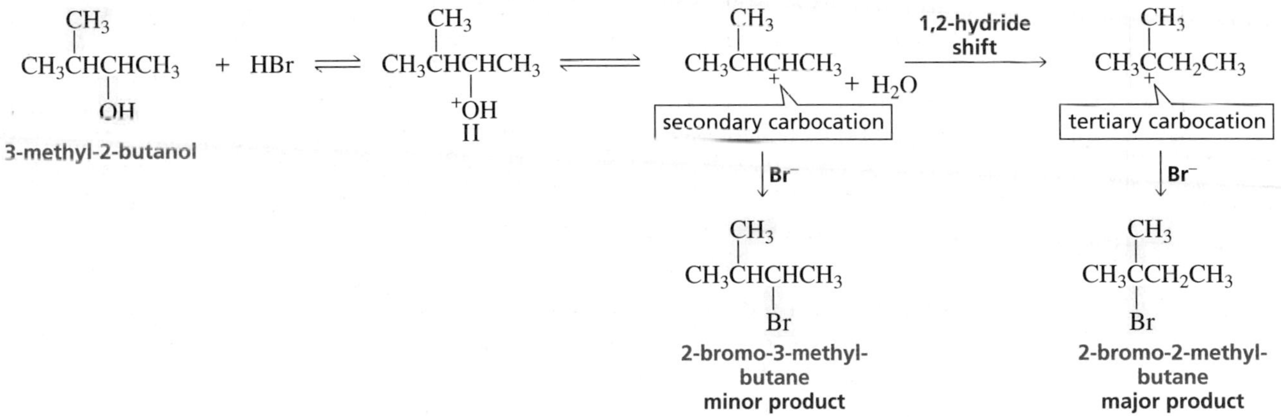

PROBLEM 5◆

Give the major product of each of the following reactions:

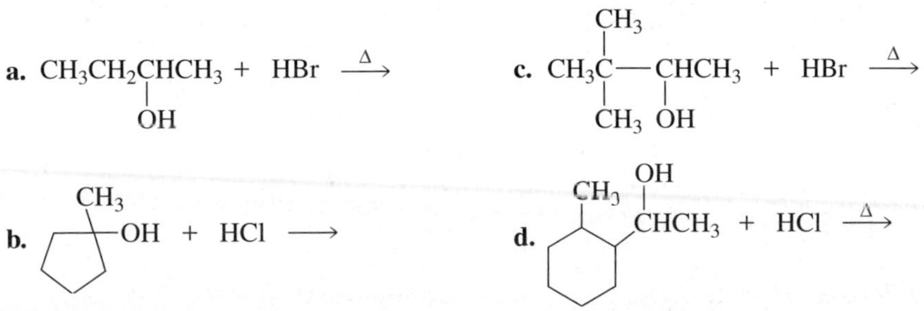

a. $CH_3CH_2\underset{\underset{OH}{|}}{C}HCH_3$ + HBr $\xrightarrow{\Delta}$

c. $CH_3\underset{\underset{CH_3}{|}}{\overset{\overset{CH_3}{|}}{C}}-\underset{\underset{OH}{|}}{C}HCH_3$ + HBr $\xrightarrow{\Delta}$

b. [cyclopentane ring with CH₃ and OH substituents] + HCl $\longrightarrow$

d. [cyclohexane ring with CH₃, OH, CHCH₃ substituents] + HCl $\xrightarrow{\Delta}$

GRAIN ALCOHOL AND WOOD ALCOHOL

When ethanol is ingested, it acts on the central nervous system. Moderate amounts affect judgment and lower inhibitions. Higher amounts interfere with motor coordination and cause slurred speech and amnesia. Still higher amounts cause nausea and loss of consciousness. Ingesting very large amounts of ethanol interferes with spontaneous respiration and can be fatal.

The ethanol in alcoholic beverages is produced by the fermentation of glucose, generally obtained from grapes or from grains such as corn, rye, and wheat (which is why ethanol is also known as grain alcohol). Grains are cooked in the presence of malt (sprouted barley) to convert much of their starch into glucose. Yeast is added to convert the glucose into ethanol and carbon dioxide (Section 25.8).

$$C_6H_{12}O_6 \xrightarrow{\text{yeast enzymes}} 2\ CH_3CH_2OH\ +\ 2\ CO_2$$

glucose   ethanol

The kind of beverage produced (white or red wine, beer, scotch, bourbon, champagne) depends on the plant species providing the glucose, whether the CO_2 formed in the fermentation is allowed to escape, whether other substances are added, and how the beverage is purified (by sedimentation, for wines; by distillation, for scotch and bourbon).

The tax imposed on liquor would make ethanol a prohibitively expensive laboratory reagent. Laboratory alcohol, therefore, is not taxed, because ethanol is needed in a wide variety of commercial processes. Although this alcohol is not taxed, it is carefully regulated by the federal government to make certain that it is not used for the preparation of alcoholic beverages. Denatured alcohol— ethanol that has been made undrinkable by the addition of a denaturant such as benzene or methanol—is not taxed, but the added impurities make it unfit for many laboratory uses.

Methanol, also known as wood alcohol because at one time it was obtained by heating wood in the absence of oxygen, is highly toxic. Ingesting even very small amounts can cause blindness, and ingesting as little as an ounce has been fatal. The antidote to methanol poisoning will be discussed in Section 24.2.

10.2 Other Methods for Converting Alcohols into Alkyl Halides

Alcohols are inexpensive and readily available compounds. As we have just seen, they do not undergo nucleophilic substitution because the ⁻OH group is too basic to be displaced by a nucleophile. Chemists, therefore, need ways to convert readily available but unreactive alcohols into reactive alkyl halides that can be used as starting materials for the preparation of many organic compounds (Section 8.4).

$$R{-}OH \xrightarrow[\Delta]{\textbf{HX}} R{-}X \xrightarrow{\textbf{Nu}} R{-}Nu$$

alcohol   alkyl halide
              X = Cl, Br, I

We have just seen that an alcohol can be converted into an alkyl halide by treating it with a hydrogen halide. However, better yields are obtained and carbocation rearrangements can be avoided if a phosphorus trihalide (PCl_3, PBr_3, or PI_3)* or thionyl chloride ($SOCl_2$) is used instead. These reagents all act in the same way: they convert the alcohol into an intermediate possessing a leaving group that is readily displaced by a halide ion. For example, phosphorus tribromide converts the OH group of an alcohol into a bromophosphite group that can be easily displaced by a bromide ion.

* Because of its instability, PI_3 is generated in situ (in the reaction mixture) from the reaction of phosphorus with iodine.

Thionyl chloride converts an OH group into a chlorosulfite group that can be displaced by Cl^-.

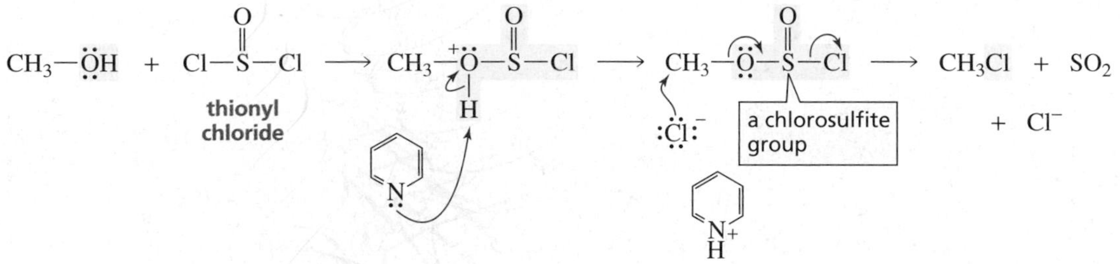

Pyridine is generally used as the solvent in these reactions because it prevents the buildup of HBr or HCl and is a relatively poor nucleophile.

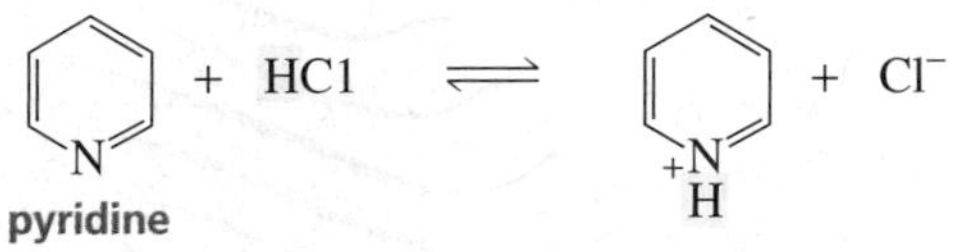
pyridine

VCL Alcohol Halogenation-1

The foregoing reactions work well for primary and secondary alcohols, but tertiary alcohols give poor yields because the intermediate formed by a tertiary alcohol is sterically hindered to attack by the halide ion.

Table 10.1 summarizes some of the commonly used methods for converting alcohols into alkyl halides.

VCL Alcohol Halogenation-2

Table 10.1 Commonly Used Methods for Converting Alcohols into Alkyl Halides

ROH + HBr	$\xrightarrow{\Delta}$	RBr
ROH + HI	$\xrightarrow{\Delta}$	RI
ROH + HCl	$\xrightarrow[\Delta]{ZnCl_2}$	RCl
ROH + PBr_3	$\xrightarrow[\text{pyridine}]{}$	RBr
ROH + PCl_3	$\xrightarrow[\text{pyridine}]{}$	RCl
ROH + $SOCl_2$	$\xrightarrow[\text{pyridine}]{}$	RCl

10.3 Converting Alcohols into Sulfonate Esters

Besides being converted into an alkyl halide, another way an alcohol can be activated for subsequent reaction with a nucleophile is to convert it into a sulfonate ester. A **sulfonate ester** is formed when an alcohol reacts with a sulfonyl chloride.

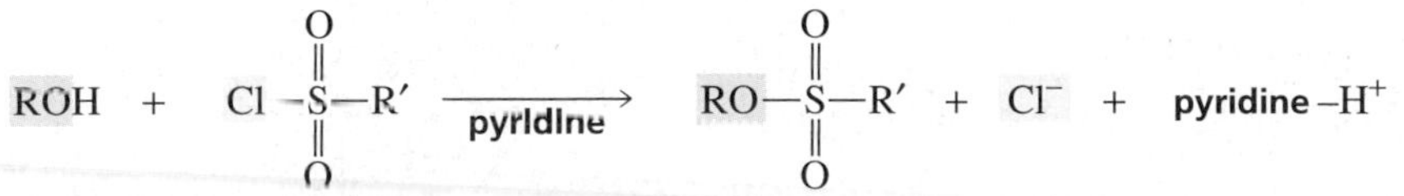
a sulfonyl chloride a sulfonate ester

The reaction is a nucleophilic substitution reaction. The alcohol displaces the chloride ion. Pyridine is used both as the solvent and to prevent HCl from building up.

$$Cl-\overset{O}{\underset{O}{\overset{\|}{S}}}-R' \ + \ R\overset{..}{\overset{}{O}}H \ \longrightarrow \ R\overset{+}{\overset{}{O}}-\overset{O}{\underset{O}{\overset{\|}{S}}}-R' \ \longrightarrow \ RO-\overset{O}{\underset{O}{\overset{\|}{S}}}-R' \ + \ Cl^-$$

A sulfonic acid is a very strong acid ($pK_a = -6.5$) because its conjugate base is particularly stable due to delocalization of its negative charge over three oxygen atoms. (Recall from Section 7.6 that electron delocalization stabilizes a charged species.) Since a sulfonic acid is a strong acid, its conjugate base is weak, giving the sulfonate ester an excellent leaving group. (Notice that sulfur has an expanded valence shell; it is surrounded by 12 electrons.)

$$R-\overset{O}{\underset{O}{\overset{\|}{S}}}-OH \ \longrightarrow \ \left[R-\overset{O}{\underset{O}{\overset{\|}{S}}}-O^- \ \longleftrightarrow \ R-\overset{O}{\underset{O^-}{\overset{\|}{S}}}=O \ \longleftrightarrow \ R-\overset{O^-}{\underset{O}{\overset{\|}{S}}}=O \right] \ + \ H^+$$

resonance contributors

VCL **Alcohol Halogenation-3**

Several sulfonyl chlorides are available to activate OH groups. The most common one is *para*-toluenesulfonyl chloride.

$$Cl-\overset{O}{\underset{O}{\overset{\|}{S}}}-\text{⬡}-CH_3 \qquad Cl-\overset{O}{\underset{O}{\overset{\|}{S}}}-CH_3 \qquad Cl-\overset{O}{\underset{O}{\overset{\|}{S}}}-CF_3$$

***para*-toluenesulfonyl chloride**   **methanesulfonyl chloride**   **trifluoromethanesulfonyl chloride**

Once the alcohol has been activated by being converted into a sulfonate ester, the appropriate nucleophile is added, generally under conditions that favor S_N2 reactions. The reactions take place readily at room temperature because the leaving group is so good. For example, a *para*-toluenesulfonate ion is about 100 times better than a chloride ion as a leaving group. Sulfonate esters react with a wide variety of nucleophiles, so they can be used to synthesize a wide variety of compounds.

$$CH_3\overset{..}{\underset{..}{S}}{}^- \ + \ CH_3CH_2CH_2CH_2-O-\overset{O}{\underset{O}{\overset{\|}{S}}}-\text{⬡}-CH_3 \ \longrightarrow \ CH_3CH_2CH_2CH_2SCH_3 \ + \ {}^-O-\overset{O}{\underset{O}{\overset{\|}{S}}}-\text{⬡}-CH_3$$

ROTs   **$^-$OTs**

$$\overset{..}{C}{\equiv}N \ + \ CH_3CH_2CH_2-O-\overset{O}{\underset{O}{\overset{\|}{S}}}-\text{⬡}-CH_3 \ \longrightarrow \ CH_3CH_2CH_2C{\equiv}N \ + \ {}^-O-\overset{O}{\underset{O}{\overset{\|}{S}}}-\text{⬡}-CH_3$$

ROTs   **$^-$OTs**

para-Toluenesulfonyl chloride is called tosyl chloride and abbreviated TsCl; the product of the reaction of *para*-toluenesulfonyl chloride and an alcohol is called an **alkyl tosylate** and is abbreviated ROTs. The leaving group, therefore, is ⁻OTs. The product of the reaction of trifluoromethanesulfonyl chloride and an alcohol is called an **alkyl triflate** and is abbreviated ROTf.

Student Tutorial:
Leaving groups

CH₃CH₂CH₂OTs + ⁻C≡N ⟶ CH₃CH₂CH₂C≡N + ⁻OTs
an alkyl tosylate

CH₃CH₂CH₂OTf + CH₃NH₂ ⟶ CH₃CH₂CH₂N⁺H₂CH₃ + ⁻OTf
an alkyl triflate

PROBLEM 6 **SOLVED**

Explain why the ether obtained by treating an optically active alcohol with PBr₃ followed by sodium methoxide has the same configuration as the alcohol, whereas the ether obtained by treating the alcohol with tosyl chloride followed by sodium methoxide has a configuration opposite that of the alcohol.

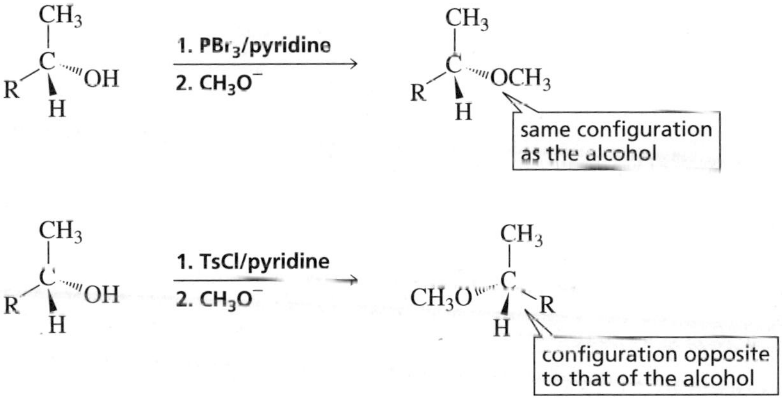

CH₃
|
R⎯C⸝⸝⸝OH
   |
   H

1. PBr₃/pyridine
2. CH₃O⁻

CH₃
|
R⎯C⸝⸝⸝OCH₃
   |
   H

same configuration as the alcohol

CH₃
|
R⎯C⸝⸝⸝OH
   |
   H

1. TsCl/pyridine
2. CH₃O⁻

CH₃
|
CH₃O⸝⸝⸝C⎯R
      |
      H

configuration opposite to that of the alcohol

Solution Conversion of the alcohol to the ether by way of the alkyl halide requires two successive S_N2 reactions: (1) attack of Br⁻ on the bromophosphite and (2) attack of CH₃O⁻ on the alkyl halide. Each S_N2 reaction takes place with inversion of configuration, so the final product has the same configuration as the starting material. In contrast, conversion of the alcohol to the ether by way of the alkyl tosylate requires only one S_N2 reaction: attack of CH₃O⁻ on the alkyl tosylate. Therefore, the final product and the starting material have opposite configurations.

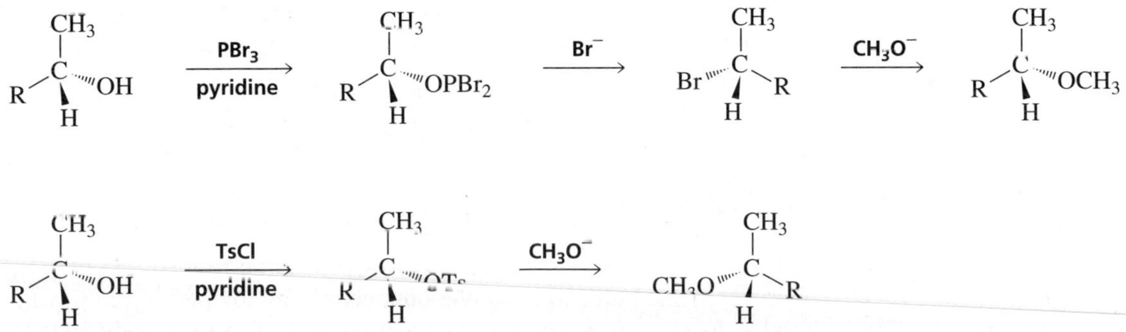

CH₃
|
R⎯C⸝⸝⸝OH
   |
   H

PBr₃
pyridine

CH₃
|
R⎯C⸝⸝⸝OPBr₂
   |
   H

Br⁻

CH₃
|
Br⸝⸝⸝C⎯R
    |
    H

CH₃O⁻

CH₃
|
R⎯C⸝⸝⸝OCH₃
   |
   H

CH₃
|
R⎯C⸝⸝⸝OH
   |
   H

TsCl
pyridine

CH₃
|
R⎯C⸝⸝⸝OTs
   |
   H

CH₃O⁻

CH₃
|
CH₃O⸝⸝⸝C⎯R
      |
      H

PROBLEM 7

Show how 1-propanol can be converted into the following compounds by means of a sulfonate ester:

a. $CH_3CH_2CH_2SCH_2CH_3$

b. $CH_3CH_2CH_2OCH_2CHCH_3$
$\quad\quad\quad\quad\quad\quad\quad\quad\quad |$
$\quad\quad\quad\quad\quad\quad\quad\quad\quad CH_3$

10.4 Elimination Reactions of Alcohols: Dehydration

An alcohol can undergo an elimination reaction by losing an OH from one carbon and an H from an adjacent carbon. The product of the reaction is an alkene. Overall, this amounts to the elimination of a molecule of water. Loss of water from a molecule is called **dehydration**. Dehydration of an alcohol requires an acid catalyst and heat. Sulfuric acid (H_2SO_4) and phosphoric acid (H_3PO_4) are the most commonly used acid catalysts. Recall that a catalyst increases the rate of a reaction but is not consumed during the course of a reaction (Section 4.5). Thus, dehydration of an alcohol is an acid catalyzed reaction; the acid is not consumed. In contrast, the reaction of an alcohol with HBr to form an alkyl halide is not an acid catalyzed reaction (Section 10.1); the acid is consumed.

$$CH_3CH_2CHCH_3 \underset{\Delta}{\overset{H_2SO_4}{\rightleftharpoons}} CH_3CH=CHCH_3 + H_2O$$
$$\quad\quad\quad | $$
$$\quad\quad\quad OH$$

Dehydration of secondary and tertiary alcohols are E1 reactions.

 Student Tutorial:
Alkene synthesis by dehydration I

Mechanism for the E1 dehydration of an alcohol

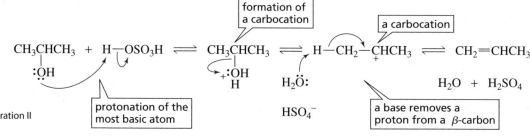

 Student Tutorial:
Alkene synthesis by dehydration II

- The acid protonates the most basic atom in the reactant. As we saw earlier, protonation converts the very poor leaving group (HO^-) into a good leaving group (H_2O).
- Water departs, leaving behind a carbocation.
- A base in the reaction mixture (water is the base that is present in the highest concentration) removes a proton from a β-carbon (a carbon adjacent to the positively charged carbon), forming an alkene and regenerating the acid catalyst. Notice that the dehydration reaction is an E1 reaction of a protonated alcohol.

Secondary and tertiary alcohols undergo dehydration by an E1 pathway.

When more than one elimination product can be formed, the major product is the more substituted alkene—the one obtained by removing a proton from the β-carbon bonded to the fewest hydrogens (Section 9.2). The more substituted alkene is the major product because it is the more stable alkene, so it has the more stable transition state leading to its formation (Figure 10.1).

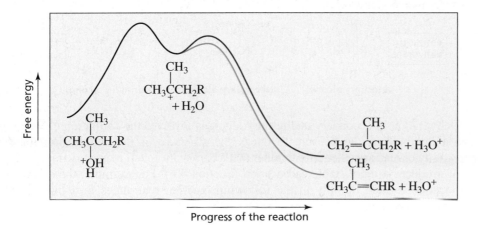

$$CH_3\overset{\underset{\displaystyle OH}{|}}{\underset{\displaystyle CH_3}{\overset{|}{C}}}CH_2CH_3 \;\rightleftharpoons\; \overset{CH_3}{CH_3C}=CHCH_3 \;+\; CH_2=\overset{CH_3}{\overset{|}{C}}CH_2CH_3 \;+\; H_2O$$

84%   16%

H₂SO₄ / Δ

93%   7%   + H₂O

We have seen that an alkene is hydrated (undergoes the addition of a water molecule) in the presence of an acid catalyst, thereby forming an alcohol (Section 4.5). The hydration of an alkene is the reverse of the acid-catalyzed dehydration of an alcohol.

$$RCH_2\overset{\underset{\displaystyle OH}{|}}{C}HR \;+\; H^+ \;\underset{\text{hydration}}{\overset{\text{dehydration}}{\rightleftharpoons}}\; RCH=CHR \;+\; H_2O \;+\; H^+$$

To prevent the alkene formed in the dehydration reaction from adding water and reforming the alcohol, the alkene can be removed by distillation as it is formed, because it has a much lower boiling point than the alcohol (Section 2.9). Removing a product displaces the reaction to the right. (See Le Châtelier's principle, Section 3.7.)

PROBLEM 8◆

Which of the following alcohols would dehydrate the most rapidly when heated with acid?

A   B   C   D

PROBLEM 9

Explain why the acid-catalyzed dehydration of an alcohol is a reversible reaction, whereas the base-promoted dehydrohalogenation of an alkyl halide is an irreversible reaction.

◀ **Figure 10.1**
The reaction coordinate diagram for the dehydration of a protonated alcohol. The major product is the more substituted alkene because the transition state leading to its formation is more stable, allowing that alkene to be formed more rapidly.

Because the rate-determining step in the dehydration of a secondary or a tertiary alcohol is formation of a carbocation intermediate, the rate of dehydration reflects the ease with which the carbocation is formed: tertiary alcohols are the easiest to dehydrate because tertiary carbocations are more stable and are therefore more easily formed than secondary and primary carbocations (Section 4.2). In order to undergo dehydration, tertiary alcohols must be heated to about 50 °C in 5% H_2SO_4, secondary alcohols must be heated to about 100 °C in 75% H_2SO_4, and primary alcohols can be dehydrated only under extreme conditions (170 °C in 95% H_2SO_4) and by a different mechanism because primary carbocations are too unstable to be formed (Section 8.5).

relative ease of dehydration

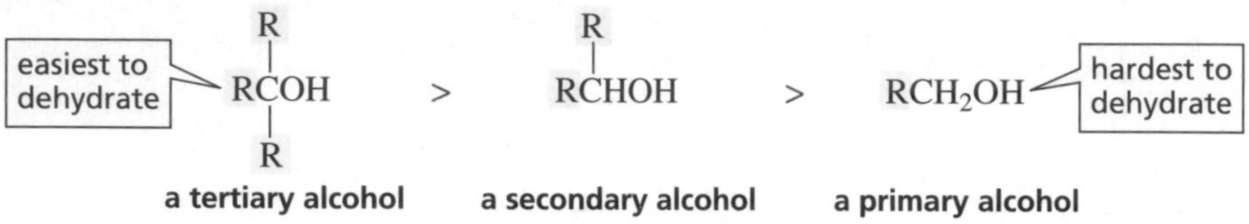

easiest to dehydrate — $\overset{R}{\underset{R}{RCOH}}$ > $\overset{R}{RCHOH}$ > RCH_2OH — hardest to dehydrate

a tertiary alcohol   **a secondary alcohol**   **a primary alcohol**

Dehydration of secondary and tertiary alcohols involves the formation of a carbocation intermediate, so be sure to check the structure of the carbocation for the possibility of rearrangement. Remember that a carbocation will rearrange if rearrangement produces a more stable carbocation (Section 4.6). For example, the secondary carbocation formed initially in the following reaction rearranges to a more stable tertiary carbocation:

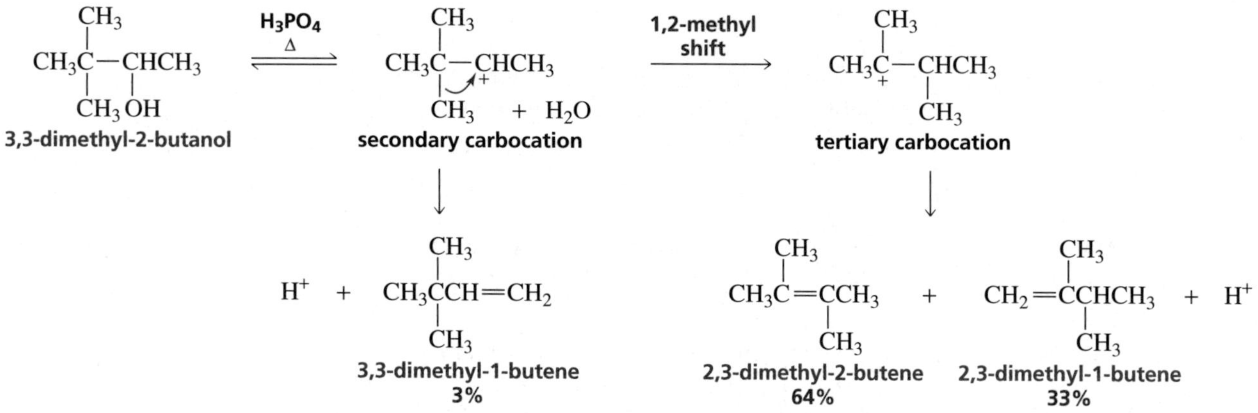

The following reaction provides an example of a **ring-expansion rearrangement**. Both the carbocation formed initially and the carbocation to which it rearranges are secondary carbocations, but the carbocation formed initially is less stable because of the strain in its four-membered ring (Section 2.11). Rearrangement into a larger ring relieves this strain. The rearranged secondary carbocation can rearrange by a 1,2-hydride shift to an even more stable tertiary carbocation.

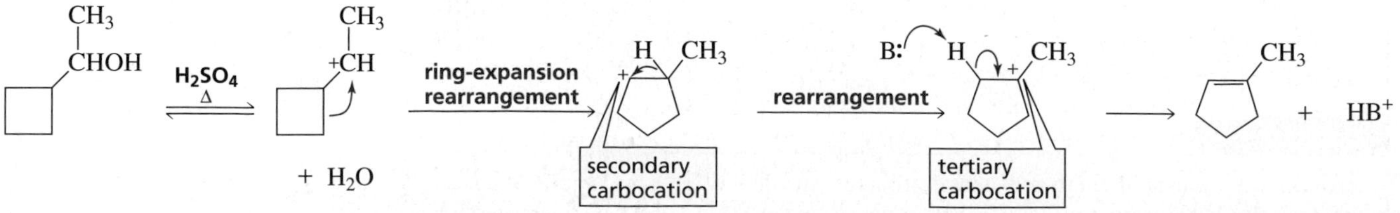

where B: is any base present in the solution.

PROBLEM 10♦

What product would be formed if the four-membered ring alcohol shown above were heated with an equivalent amount of HBr rather than with a catalytic amount of H_2SO_4?

PROBLEM 11♦

List the following alcohols in order of decreasing rate of dehydration in the presence of acid:

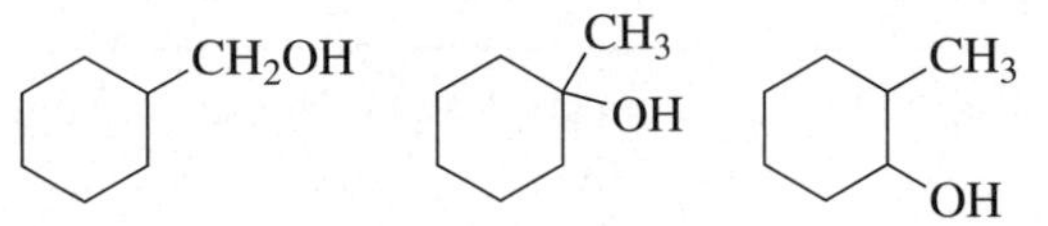

While the dehydration of a tertiary or a secondary alcohol is an E1 reaction, the dehydration of a primary alcohol is an E2 reaction, because primary carbocations are too unstable to be formed. Any base (B:) in the reaction mixture (ROH, ROR, H_2O, HSO_4^-) can remove the proton in the elimination reaction. An ether is also obtained; it is the product of a competing S_N2 reaction, since primary alcohols are the ones most likely to form substitution products in $S_N2/E2$ reactions (Section 9.8).

Primary alcohols undergo dehydration by an E2 pathway.

Mechanism for the E2 dehydration of an alcohol and competing substitution (S_N2)

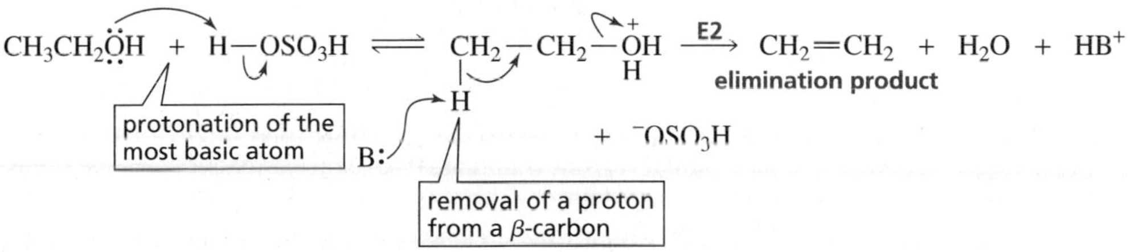

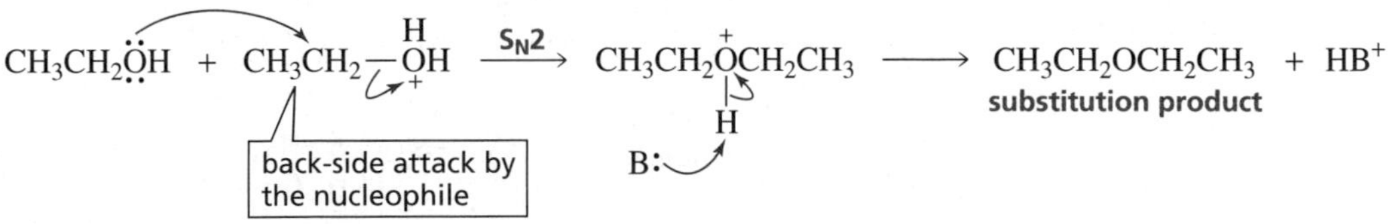

Although the dehydration of a primary alcohol is an E2 reaction and therefore does not form a carbocation intermediate, the product obtained in most cases is identical to the product that would be obtained if a carbocation had been formed in an E1 reaction and then had rearranged. For example, we would expect 1-butene to be the product of the E2 dehydration of 1-butanol. However, we find that the product is actually 2-butene, which would have been the product if an E1 reaction had occurred and the initially formed primary carbocation intermediate had rearranged to a more stable secondary carbocation. 2-Butene is the product of the reaction, not because an E1 reaction occurs, but because, after the E2 product (1-butene) forms, a proton from the acidic solution adds to the double bond, adding to the sp^2 carbon bonded to the greater number of hydrogens (in accordance with the rule that governs electrophilic addition reactions; Section 4.4) to form a carbocation. Loss of a proton from the carbocation—from the β-carbon bonded to the fewest hydrogens (in accordance with Zaitsev's rule)—gives 2-butene, the final product of the reaction.

$$CH_3CH_2CH_2CH_2OH \underset{\Delta}{\overset{H_2SO_4}{\rightleftharpoons}} CH_3CH_2CH=CH_2 \overset{H^+}{\rightleftharpoons} CH_3CH_2\overset{+}{C}HCH_3 \rightleftharpoons CH_3CH=CHCH_3 + H^+$$

1-butanol　　　　**1-butene**　　　　　　　　　　　　**2-butene**

$+\ H_2O$

PROBLEM 12

Heating an alcohol with sulfuric acid is a good way to prepare a symmetrical ether such as diethyl ether.

a. Explain why it is not a good way to prepare an unsymmetrical ether such as ethyl propyl ether.

b. How would you synthesize ethyl propyl ether?

The stereochemical outcome of the E1 dehydration of an alcohol is identical to the stereochemical outcome of the E1 dehydrohalogenation of an alkyl halide. That is, both the E and Z isomers are obtained as products. The reaction produces more of the stereoisomer in which the bulkier group on each of the sp^2 carbons are on opposite sides of the double bond; that stereoisomer, being more stable, is formed more rapidly, since the transition state leading to its formation is more stable (Section 9.5).

$$\underset{\underset{\text{2-butanol}}{\underset{|}{OH}}}{CH_3CH_2CHCH_3} \underset{\Delta}{\overset{H_2SO_4}{\rightleftharpoons}} \underset{\underset{+\ H_2O}{}}{CH_3CH_2\overset{+}{C}HCH_3} \longrightarrow$$

$$\underset{\underset{74\%}{\textit{trans}\text{-2-butene}}}{\overset{H_3C}{\underset{H}{}}C=C\overset{H}{\underset{CH_3}{}}} + \underset{\underset{23\%}{\textit{cis}\text{-2-butene}}}{\overset{H_3C}{\underset{H}{}}C=C\overset{CH_3}{\underset{H}{}}} + \underset{\underset{3\%}{\text{1-butene}}}{CH_3CH_2CH=CH_2} + H^+$$

Alcohols and ethers undergo S$_N$1/E1 reactions unless they would have to form a primary carbocation, in which case they undergo S$_N$2/E2 reactions.

We can summarize what we have learned about the mechanisms by which alcohols undergo substitution and elimination reactions: they react by S$_N$1 and E1 pathways, unless they cannot. In other words, 3° and 2° alcohols undergo S$_N$1 and E1 reactions; 1° alcohols, because they cannot form primary carbocations, have to undergo S$_N$2 and E2 reactions.

The relatively harsh conditions (acid and heat) required for alcohol dehydration, and the structural changes resulting from carbocation rearrangements, may result in low yields of the desired alkene. Dehydration, however, can be carried out under milder conditions by using phosphorus oxychloride (POCl$_3$) and pyridine.

$$\underset{\underset{OH}{\underset{|}{}}}{CH_3CH_2CHCH_3} \xrightarrow[\text{pyridine, 0 °C}]{POCl_3} CH_3CH=CHCH_3$$

Reaction with POCl$_3$ converts the OH group of the alcohol into OPOCl$_2$, a good leaving group. The basic reaction conditions favor an E2 reaction, so a carbocation is not formed and carbocation rearrangements do not occur. Pyridine serves as a base to remove the proton in the elimination reaction and to prevent the buildup of HCl, which would add to the alkene.

BIOLOGICAL DEHYDRATIONS

Dehydration reactions occur in many important biological processes. Instead of being catalyzed by strong acids, which would not be available to a cell, they are catalyzed by enzymes. Fumarase, for example, is the enzyme that catalyzes the dehydration of malate in the citric acid cycle (Section 25.10). The citric acid cycle is a series of reactions that oxidize compounds derived from carbohydrates, fatty acids, and amino acids.

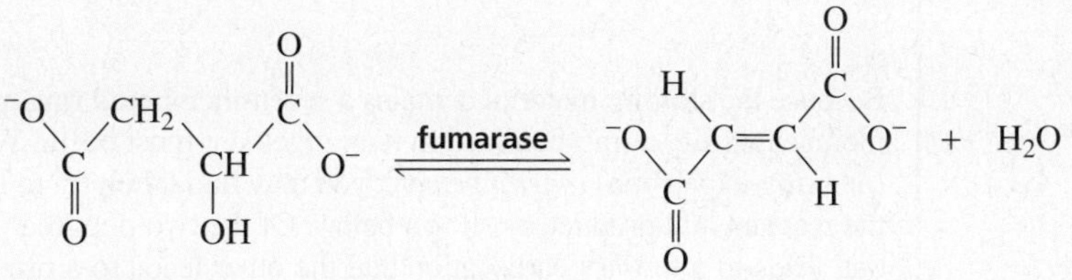

malate   fumarate

Enolase, another enzyme, catalyzes the dehydration of α-phosphoglycerate in glycolysis (Section 25.7). Glycolysis is a series of reactions that prepare glucose for entry into the citric acid cycle.

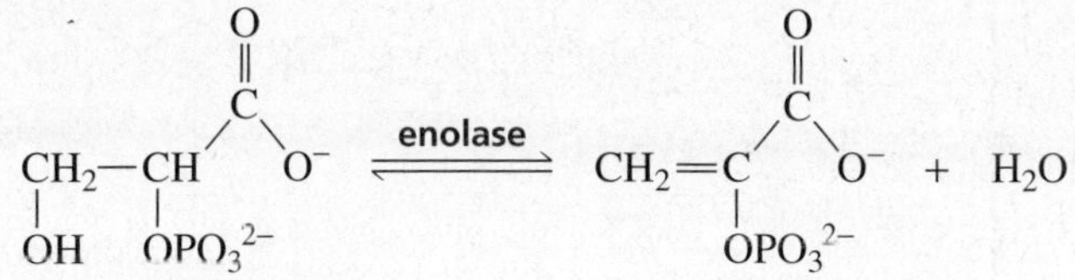

α-phosphoglycerate   phosphoenolpyruvate

PROBLEM 13◆

What alcohol would you treat with phosphorus oxychloride and pyridine to form each of the following alkenes?

a. $CH_3CH_2\overset{\underset{\displaystyle CH_3}{|}}{C}=CH_2$

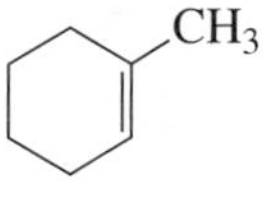

c. $CH_3CH=CHCH_2CH_3$

b.

d.

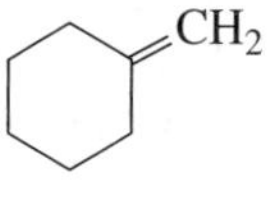

PROBLEM-SOLVING STRATEGY

Proposing a Mechanism

Propose a mechanism for the following reaction:

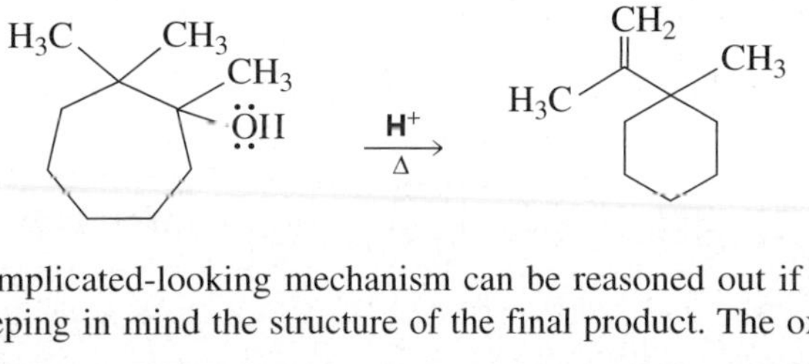

Even the most complicated-looking mechanism can be reasoned out if you proceed one step at a time, keeping in mind the structure of the final product. The oxygen is the only basic atom in the starting material, so that is where protonation occurs. Loss of water forms a tertiary carbocation.

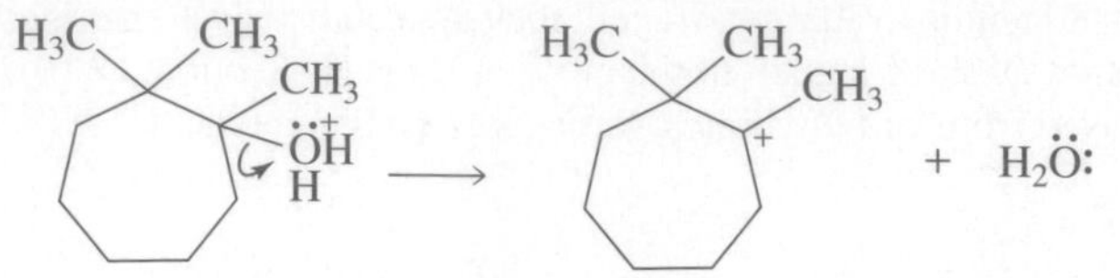

Because the starting material contains a seven-membered ring and the final product has a six-membered ring, a ring-contraction rearrangement must occur. When doing a *ring-contraction* (or a ring-expansion) *rearrangement*, you may find it helpful to label the equivalent carbons in the reactant and product, as shown below. Of the two possible pathways for ring contraction, one leads to a tertiary carbocation and the other leads to a primary carbocation. The correct pathway must be the one that leads to the tertiary carbocation, since that carbocation has the same arrangement of atoms as the product, and since the primary carbocation would be too unstable to form.

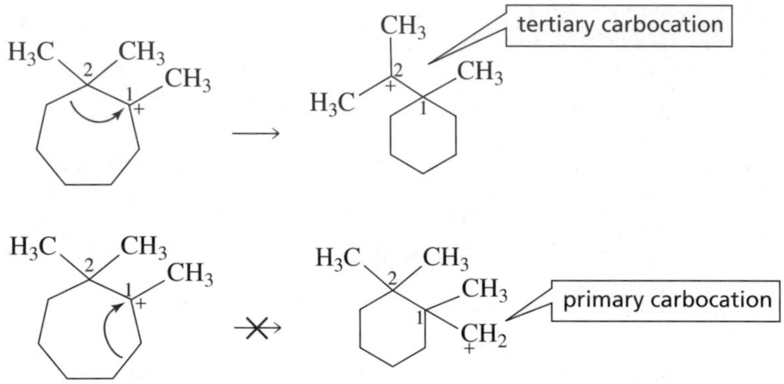

The final product can now be obtained by removing a proton from the rearranged carbocation.

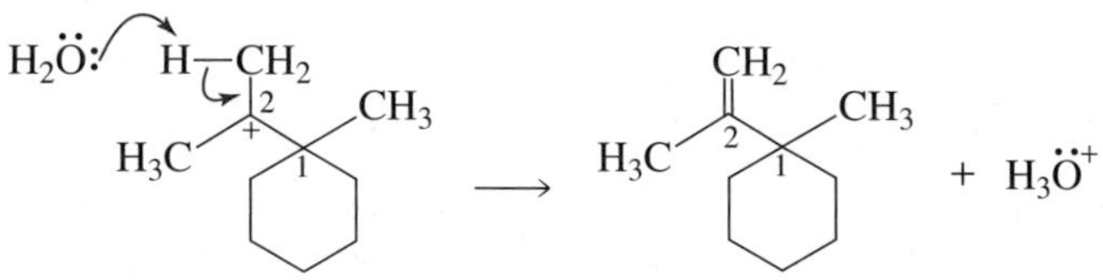

Now continue on to Problem 14.

PROBLEM 14

Propose a mechanism for each of the following reactions:

a.

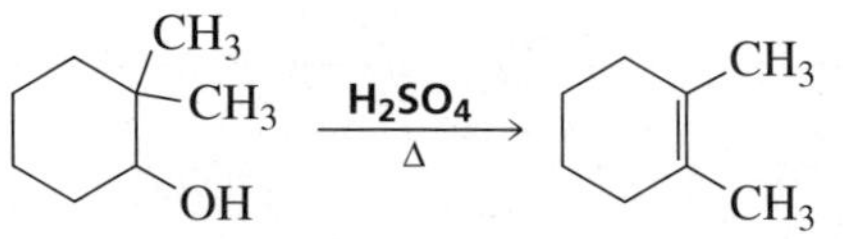

b.

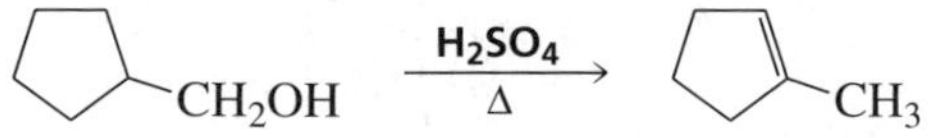

c.

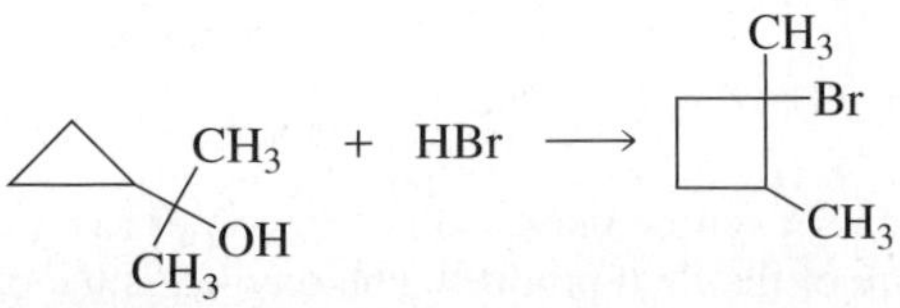

PROBLEM 15◆

Give the major product formed when each of the following alcohols is heated in the presence of H_2SO_4:

a. 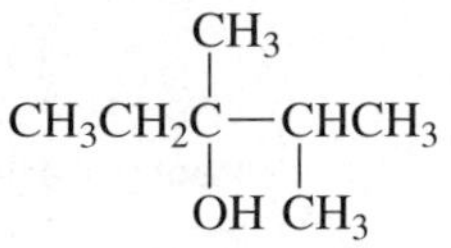 CH₃CH₂C—CHCH₃
with CH₃ above and OH CH₃ below

d. 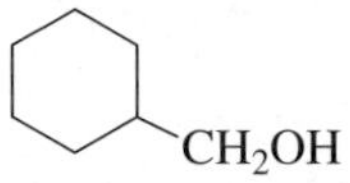
cyclohexane–CH₂OH

b. 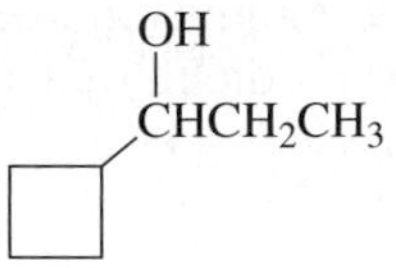 OH
cyclobutyl–CHCH₂CH₃

e. 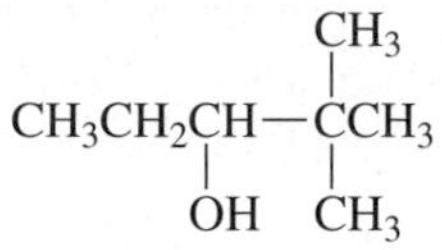 CH₃CH₂CH—CCH₃
with CH₃ above and OH CH₃ below

c. 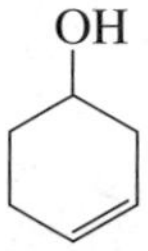 OH
cyclohexene with OH

f. CH₃CH₂CH₂CH₂CH₂OH

PROBLEM 16

When the following compound is heated in the presence of H_2SO_4:

a. What constitutional isomer is produced in greatest yield?
b. What stereoisomer in the answer for part a is produced in greater yield?

10.5 Oxidation of Alcohols

A reagent often used to oxidize alcohols is chromic acid (H_2CrO_4), which is formed when chromium trioxide (CrO_3) or sodium dichromate ($Na_2Cr_2O_7$) is dissolved in aqueous acid. *Secondary alcohols* are oxidized to *ketones*. These reactions are easily recognized as oxidations because the number of C—H bonds in the reactant decreases and the number of C—O bonds increases (Section 4.9).

Student Tutorial:
Changes in oxidation state

secondary alcohols   **ketones**

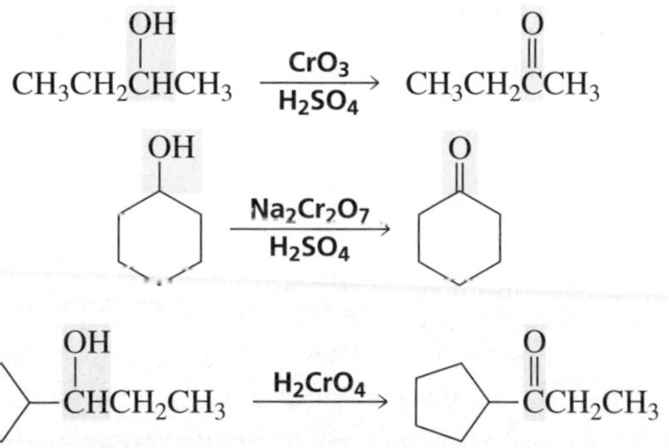

OH
CH₃CH₂CHCH₃ $\xrightarrow[\text{H}_2\text{SO}_4]{\text{CrO}_3}$ CH₃CH₂CCH₃ (with O above)

OH
cyclohexane with OH $\xrightarrow[\text{H}_2\text{SO}_4]{\text{Na}_2\text{Cr}_2\text{O}_7}$ cyclohexanone (with O)

Secondary alcohols are oxidized to ketones.

OH
cyclopentyl–CHCH₂CH₃ $\xrightarrow{\text{H}_2\text{CrO}_4}$ cyclopentyl–CCH₂CH₃ (with O above)

Primary alcohols are initially oxidized to aldehydes by these reagents. The reaction, however, does not stop at the aldehyde. Instead, the aldehyde is further oxidized (the number of C—O bonds increases) to a carboxylic acid.

$$CH_3CH_2CH_2CH_2OH \xrightarrow{H_2CrO_4} \left[\underset{\text{an aldehyde}}{CH_3CH_2CH_2\overset{\displaystyle O}{\overset{\|}{C}H}} \right] \xrightarrow[\text{oxidation}]{\text{further}} \underset{\text{a carboxylic acid}}{CH_3CH_2CH_2\overset{\displaystyle O}{\overset{\|}{C}OH}}$$

a primary alcohol

Primary alcohols are oxidized to carboxylic acids and aldehydes.

The oxidation of a primary alcohol will stop at the aldehyde if pyridinium chlorochromate (PCC) is used as the oxidizing agent in a solvent such as dichloromethane (CH_2Cl_2). We will see why this reaction stops at the aldehyde in Section 19.2.

pyridinium
chlorochromate
PCC

$$CH_3CH_2CH_2CH_2OH \xrightarrow[CH_2Cl_2]{PCC} \underset{\text{an aldehyde}}{CH_3CH_2CH_2\overset{\displaystyle O}{\overset{\|}{C}H}}$$

a primary alcohol

Notice that in the oxidation of both a primary and a secondary alcohol, a hydrogen is removed from the carbon to which the OH is attached. The carbon bearing the OH group in a tertiary alcohol is not bonded to a hydrogen, so its OH group cannot be oxidized to a carbonyl group.

$$CH_3-\underset{\underset{\displaystyle CH_3}{|}}{\overset{\overset{\displaystyle CH_3}{|}}{C}}-OH$$

cannot be oxidized to a carbonyl group

a tertiary alcohol

Both an S_N2 and an E2 reaction are involved in the overall oxidation reaction.

Mechanism for alcohol oxidation by chromic acid

$$HO-\overset{\displaystyle O}{\underset{\displaystyle O}{\overset{\|}{Cr}}}-\ddot{O}H \rightleftharpoons HO-\overset{\displaystyle O}{\overset{\|}{Cr}}-\overset{+}{\overset{H}{\ddot{O}}}H \longrightarrow RCH_2-\overset{+}{\ddot{O}}-\overset{\displaystyle O}{\overset{\|}{Cr}}-OH \rightleftharpoons RCH-O-\overset{\displaystyle O}{\overset{\|}{Cr}}-OH \longrightarrow RCH=O + H_2CrO_3$$

an S_N2 reaction   an E2 reaction

a chromate ester

$$HB^+ + H_2O$$

- An oxygen of chromic acid is protonated in the acidic solution.
- The alcohol molecule displaces a molecule of water in an S_N2 reaction on chromium.
- A base present in the reaction mixture (H_2O, ROH) removes a proton from the strongly acidic species.
- A base removes a proton from the chromate ester in an E2 reaction, forming the carbonyl compound and regenerating the acid used in the first step.

PROBLEM 17◆

Give the product formed from the reaction of each of the following compounds with an acidic solution of sodium dichromate:

a. 3-pentanol   **c.** 2-methyl-2-pentanol   **e.** cyclohexanol

b. 1-pentanol   **d.** 2,4-hexanediol   **f.** 1,4-butanediol

BLOOD ALCOHOL CONTENT

As blood passes through the arteries in our lungs, an equilibrium is established between alcohol in the blood and alcohol in the breath. Therefore, if the concentration of one is known, the concentration of the other can be estimated. The test that law enforcement agencies use to approximate a person's blood alcohol level is based on the oxidation of breath ethanol. An oxidizing agent impregnated onto an inert material is enclosed within a sealed glass tube. When the test is to be administered, the ends of the tube are broken off and replaced with a mouthpiece at one end and a balloon-type bag at the other. The person being tested blows into the mouthpiece until the bag is filled with air.

Any ethanol in the breath is oxidized as it passes through the column. When ethanol is oxidized, the red-orange dichromate ion is reduced to green chromic ion (Cr^{3+}). The greater the concentration of alcohol in the breath, the farther the green color spreads through the tube.

$$CH_3CH_2OH \ + \ \underset{\text{red orange}}{Cr_2O_7{}^{2-}} \ \xrightarrow{\ H^+\ } \ \underset{}{CH_3\overset{\displaystyle O}{\overset{\|}{C}}OH} \ + \ \underset{\text{green}}{Cr^{3+}}$$

If the person fails this test—determined by the extent to which the green color spreads through the tube—a more accurate Breathalyzer™ test is administered. The Breathalyzer test also depends on the oxidation of breath ethanol, but it provides more accurate results because it is quantitative. In the test, a known volume of breath is bubbled through an acidic solution of sodium dichromate, and the concentration of the green chromic ion is measured precisely with a spectrophotometer (Section 12.17).

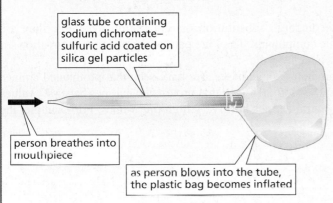

glass tube containing sodium dichromate–sulfuric acid coated on silica gel particles

person breathes into mouthpiece

as person blows into the tube, the plastic bag becomes inflated

---

PROBLEM 18

Propose a mechanism for the chromic acid oxidation of 1-propanol to propanal.

10.6 Amines Do Not Undergo Substitution or Elimination Reactions but Are the Most Common Organic Bases

We have just seen that alcohols are much less reactive than alkyl halides in substitution and elimination reactions. Amines are even *less reactive* than alcohols. The relative reactivities of an alkyl fluoride (the least reactive of the alkyl halides because it has the poorest leaving group), an alcohol, and an amine can be appreciated by comparing the pK_a values of the conjugate acids of their leaving groups, recalling that the weaker the acid, the stronger its conjugate base and the poorer the base is as a leaving group. The leaving group of an amine ($^-NH_2$) is such a strong base that amines cannot undergo substitution or elimination reactions.

The stronger the base, the poorer it is as a leaving group.

relative reactivities

| most reactive | RCH_2F | > | RCH_2OH | > | RCH_2NH_2 | least reactive |

| | HF | | H_2O | | NH_3 | |
| | $pK_a = 3.2$ | | $pK_a = 15.7$ | | $pK_a = 36$ | |

Protonation of the amino group makes it a better leaving group, but not nearly as good as a protonated alcohol, which is ~13 pK_a units more acidic than a protonated amine.

$$CH_3CH_2\overset{+}{O}H_2 \quad > \quad CH_3CH_2\overset{+}{N}H_3$$
$$pK_a = -2.4 \qquad\qquad pK_a = 11.2$$

Therefore, unlike the leaving group of a protonated alcohol, the leaving group of a protonated amine cannot dissociate to form a carbocation or be replaced by a halide ion. Protonated amino groups also cannot be displaced by strongly basic nucleophiles such as HO^- because the nucleophile would react immediately with the acidic hydrogen of the $^+NH_3$ group and thereby be converted to water, a poor nucleophile.

$$CH_3CH_2\overset{+}{N}H_3 \;+\; HO^- \;\rightleftharpoons\; CH_3CH_2NH_2 \;+\; H_2O$$

Although amines cannot undergo substitution or elimination reactions, they are extremely important organic compounds. The lone pair on the nitrogen atom allows it to act as both a base and as a nucleophile.

Amines are the most common organic bases. We have seen that protonated amines have pK_a values of about 11 (Section 1.18) and that protonated anilines have pK_a values of about 5 (Section 7.9). Neutral amines have very high pK_a values. For example, the pK_a of methylamine is 40.

$CH_3CH_2CH_2\overset{+}{N}H_3$ $\qquad$ $CH_3\overset{+}{N}H_2$ | CH_3 $\qquad$ $CH_3CH_2\overset{+}{N}H$ with CH_2CH_3 / CH_2CH_3 $\qquad$ phenyl-$\overset{+}{N}H_3$ $\qquad$ CH_3-phenyl-$\overset{+}{N}H_3$ $\qquad$ CH_3NH_2

$pK_a = 10.8$ $\qquad$ $pK_a = 10.9$ $\qquad$ $pK_a = 11.1$ $\qquad$ $pK_a = 4.58$ $\qquad$ $pK_a = 5.07$ $\qquad$ $pK_a = 40$

Amines react as nucleophiles in a wide variety of reactions. For example, they react as nucleophiles with alkyl halides in S_N2 reactions.

an S_N2 reaction

$$CH_3CH_2Br \;+\; CH_3NH_2 \;\longrightarrow\; CH_3CH_2\overset{+}{N}H_2CH_3 \;+\; Br^-$$

We will see that they also react as nucleophiles with a wide variety of carbonyl compounds (Sections 16.8–16.10, 17.8).

ALKALOIDS

Alkaloids are amines found in the leaves, bark, roots, or seeds of many plants. Examples include caffeine (found in tea leaves, coffee beans, and cola nuts), nicotine (found in tobacco leaves), and cocaine (obtained from the coca bush in the rainforest areas of Columbia, Peru, and Bolivia). Ephedrine, a bronchodilator, is an alkaloid obtained from *Ephedra sinica*, a plant found in China. Morphine is an alkaloid obtained from opium, the juice derived from a species of poppy (Section 30.3).

caffeine $\qquad$ nicotine $\qquad$ ephedrine $\qquad$ morphine

PROBLEM 19

Why can a halide ion such as Br$^-$ undergo an S_N2 reaction with a protonated primary alcohol, but not with a protonated primary amine?

10.7 Nucleophilic Substitution Reactions of Ethers

The OR group of an **ether** and the OH group of an alcohol have nearly the same basicity, because the conjugate acids of these two groups have similar pK_a values. (The pK_a of CH_3OH is 15.5 and the pK_a of H_2O is 15.7.) Both groups are strong bases, so both are very poor leaving groups. Consequently, ethers, like alcohols, need to be activated before they can undergo nucleophilic substitution.

$$R—\ddot{O}—H \qquad R—\ddot{O}—R$$

an alcohol   **an ether**

Ethers, like alcohols, can be activated by protonation. Ethers, therefore, can undergo nucleophilic substitution reactions with HBr or HI. As with alcohols, the reaction of ethers with hydrogen halides is slow. The reaction mixture must be heated to cause the reaction to occur at a reasonable rate.

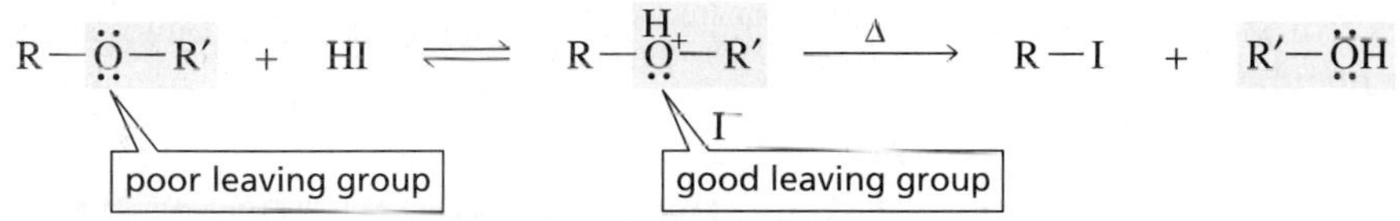

$$R—\ddot{O}—R' \; + \; HI \; \rightleftharpoons \; R—\overset{H}{\underset{I^-}{\overset{+}{\ddot{O}}}}—R' \; \overset{\Delta}{\longrightarrow} \; R—I \; + \; R'—\ddot{O}H$$

poor leaving group   good leaving group

- Again we see that the first step in a mechanism, in which an acid is one of the reagents, is protonation of the most basic atom, in this case oxygen. Protonation converts the very basic RO$^-$ leaving group into the less basic ROH leaving group.

- What happens next depends on the structure of the ether. If departure of ROH creates a relatively stable carbocation (such as a tertiary carbocation), an S_N1 reaction occurs: the leaving group departs, and the halide ion combines with the carbocation.

Ether cleavage: an S_N1 reaction

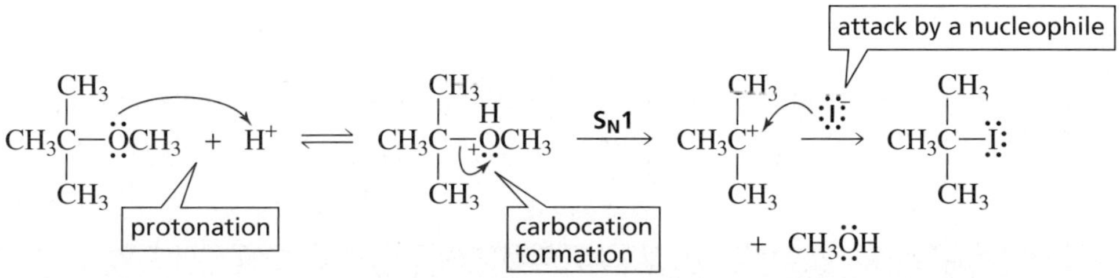

attack by a nucleophile

protonation   carbocation formation $+ \; CH_3\ddot{O}H$

- However, if departure of the leaving group would create an unstable carbocation (such as a methyl, vinyl, aryl, or primary carbocation), the leaving group cannot depart. It has to be displaced by the halide ion. In other words, an S_N2 reaction occurs. In the S_N2 reaction, the halide ion preferentially attacks the less sterically hindered of the two alkyl groups.

Ether cleavage: an S$_N$2 reaction

$$CH_3\ddot{O}CH_2CH_2CH_3 \ + \ H^+ \ \rightleftharpoons \ CH_3 \overset{H}{\underset{+}{-}} \ddot{O}CH_2CH_2CH_3 \ \xrightarrow{S_N2} \ CH_3\ddot{I}\colon \ + \ CH_3CH_2CH_2OH$$

protonation

nucleophile attacks the less sterically hindered carbon

Ethers are cleaved by an S$_N$1 pathway unless the instability of the carbocation requires an S$_N$2 pathway.

In summary, ethers are cleaved by an S$_N$1 pathway unless the instability of the carbocation causes the reaction to follow an S$_N$2 pathway; S$_N$1 cleavage is faster than S$_N$2 cleavage. S$_N$2 cleavage with HI is more rapid than S$_N$2 cleavage with HBr because I$^-$ is a better nucleophile than Br$^-$. Harsher conditions must be used if HCl is employed, because Cl$^-$ is even a poorer nucleophile.

The cleavage reaction forms only a substitution product because the bases in the reaction mixture (halide ions and H$_2$O) are too weak to remove a proton in an E2 reaction, and any alkene formed in an E1 reaction would undergo electrophilic addition with HBr or HI to form the same alkyl halide as would be obtained from the substitution reaction.

Many of the reagents used to activate alcohols so they can undergo nucleophilic substitution (such as SOCl$_2$, PCl$_3$) cannot be used to activate ethers. When an alcohol reacts with an activating agent such as a sulfonyl chloride, a proton dissociates from the intermediate in the second step of the reaction and a stable sulfonate ester results.

$$\underset{\textbf{an alcohol}}{ROH} \ + \ R'-\overset{O}{\underset{O}{\overset{\|}{\underset{\|}{S}}}}-Cl \ \longrightarrow \ R'-\overset{O}{\underset{O}{\overset{\|}{\underset{\|}{S}}}}-\overset{+}{O}R \ \underset{H \ + \ Cl^-}{|} \ \rightleftharpoons \ \underset{\textbf{a sulfonate ester}}{R'-\overset{O}{\underset{O}{\overset{\|}{\underset{\|}{S}}}}-OR} \ + \ H^+$$

However, when an ether reacts with a sulfonyl chloride, the oxygen atom does not have a proton that can dissociate. The alkyl group (R) cannot dissociate, so a stable sulfonate ester cannot be formed. Instead, the more stable starting materials are reformed.

$$\underset{\textbf{an ether}}{ROR} \ + \ R'-\overset{O}{\underset{O}{\overset{\|}{\underset{\|}{S}}}}-Cl \ \rightleftharpoons \ R'-\overset{O}{\underset{\underset{R}{|}}{\overset{\|}{\underset{O}{\|}}}}\overset{+}{O}R \ + \ Cl^-$$

Because the only reagents that react with ethers are hydrogen halides, ethers are frequently used as solvents. Some common ether solvents are shown in Table 10.2.

Table 10.2 Some Ethers Are Used as Solvents					
$CH_3CH_2OCH_2CH_3$	(structure)	(structure)	(structure)	$CH_3OCH_2CH_2OCH_3$	$CH_3OC(CH_3)_3$
diethyl ether "ether"	tetrahydrofuran THF	tetrahydropyran	1,4-dioxane	1,2-dimethoxyethane DME	*tert*-butyl methyl ether MTBE

ANESTHETICS

Because diethyl ether (commonly known as ether) is a short-lived muscle relaxant, it was at one time widely used as an inhalation anesthetic. However, it takes effect slowly and has a slow and unpleasant recovery period, so other compounds, such as enflurane, isoflurane, and halothane, have replaced it as an anesthetic. Even so, diethyl ether is still used where trained anesthesiologists are scarce, because it is the safest anesthetic for an untrained person to administer. Anesthetics interact with the nonpolar molecules of cell membranes, causing the membranes to swell, which interferes with their permeability.

"ether"   **isoflurane**   **enflurane**   **halothane**

Sodium pentothal (also called thiopental sodium) is commonly used as an intravenous anesthetic. The onset of anesthesia and the loss of consciousness occur within seconds of its administration. Care must be taken when administering sodium pentothal because the dose for effective anesthesia is 75% of the lethal dose. Because of this high level of toxicity, it cannot be used as the sole anesthetic but instead is generally used to induce anesthesia before an inhalation anesthetic is administered. Propofol, in contrast, has all the properties of the "perfect anesthetic": it can be administered as the sole anesthetic by intravenous drip, and it has a rapid and pleasant induction period and a wide margin of safety. Furthermore, recovery from the drug is also rapid and pleasant.

sodium pentothal   **propofol**

Amputation of a leg without anesthetic in 1528.

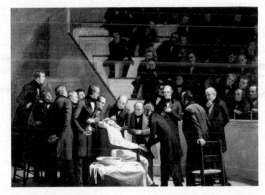

A painting showing the first use of anesthesia in 1846 at Massachusetts General Hospital by surgeon John Collins Warren.

PROBLEM 20 *SOLVED*

Explain why methyl propyl ether forms both methyl iodide and propyl iodide when it is heated with excess HI.

Solution   We have just seen that the S_N2 reaction of methyl propyl ether with an equivalent amount of HI forms methyl iodide and propyl alcohol because the methyl group is less sterically hindered to attack by the iodide ion. When there is excess HI, the alcohol product of this first reaction can react with HI in another S_N2 reaction (Section 10.1). Thus, the products are methyl iodide and propyl iodide.

$$CH_3CH_2CH_2OCH_3 \xrightarrow{HI} CH_3CH_2CH_2OH \xrightarrow{HI} CH_3CH_2CH_2I + H_2O$$
$$+ CH_3I$$

PROBLEM 21◆

Can HF be used to cleave ethers? Explain.

PROBLEM 22 **SOLVED**

Give the major products obtained from heating each of the following ethers with one equivalent of HI:

a. $CH_3CH{=}CHOCH_2CH_3$

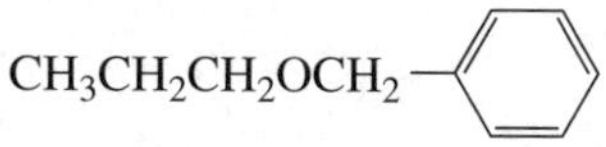

d.

b.

e.

c. $CH_3CH_2CH_2OCH_2{-}$

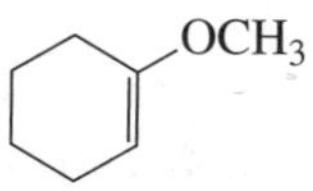

f.

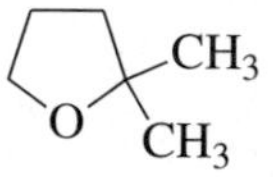

Solution to 22a The reaction takes place by an S_N2 pathway because neither alkyl group will form a relatively stable carbocation. Iodide ion attacks the carbon of the ethyl group because otherwise it would have to attack a vinyl carbon, and vinyl carbons are generally not attacked by nucleophiles (Section 8.8). Thus, the major products are ethyl iodide and an enol that immediately rearranges to an aldehyde (Section 6.7).

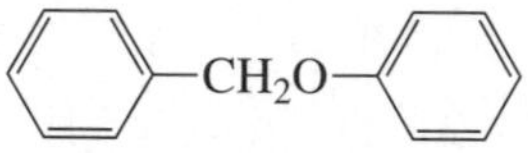

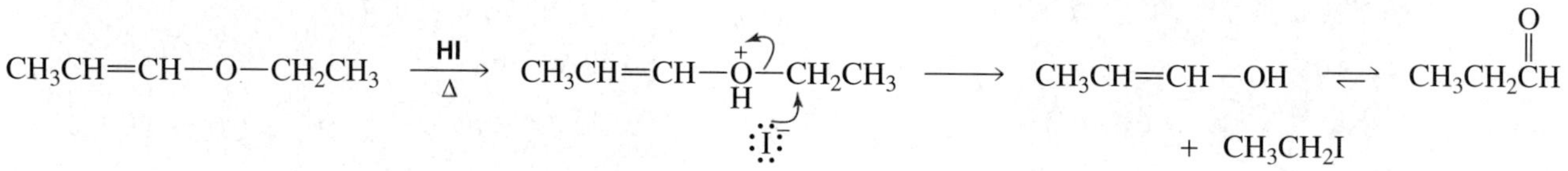

10.8 Nucleophilic Substitution Reactions of Epoxides

We have seen that an alkene can be converted to an **epoxide** by a peroxyacid (Section 4.9).

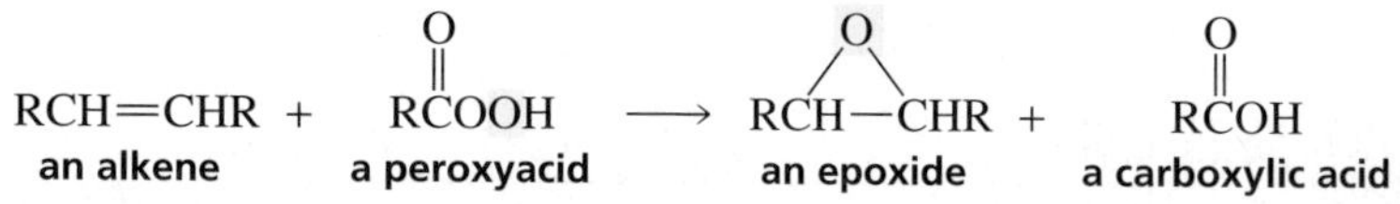

$$RCH{=}CHR + \underset{\text{a peroxyacid}}{\overset{O}{\underset{\|}{RCOOH}}} \longrightarrow \underset{\text{an epoxide}}{RCH{-}CHR} + \underset{\text{a carboxylic acid}}{\overset{O}{\underset{\|}{RCOH}}}$$

an alkene

Although an epoxide and an ether have the same leaving group, epoxides are much more reactive than ethers in nucleophilic substitution reactions because the strain in their three-membered ring is relieved when the ring opens (Figure 10.2). Epoxides, therefore, undergo nucleophilic substitution reactions with a wide variety of nucleophiles.

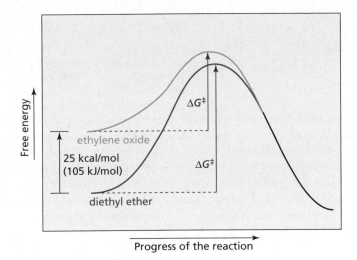

◀ **Figure 10.2**
The reaction coordinate diagrams for nucleophilic attack of hydroxide ion on ethylene oxide and on diethyl ether. The greater reactivity of the epoxide is a result of the ring strain in the three-membered ring, which increases the epoxide's free energy.

Epoxides, like other ethers, react with hydrogen halides.

- The acid protonates the oxygen atom of the epoxide.
- The protonated epoxide undergoes back-side attack by the halide ion.

Because epoxides are so much more reactive than ethers, the reaction takes place readily at room temperature, unlike the reaction of an ether with a hydrogen halide, which requires heat.

Protonated epoxides are so reactive that they can be opened by poor nucleophiles, such as H_2O and alcohols,

where HB^+ is any acid in the solution and B: is any base.

If different substituents are attached to the two carbons of the protonated epoxide, and the nucleophile is something other than H_2O, the product obtained from nucleophilic attack on the 2-position of the oxirane ring will be different from that obtained from nucleophilic attack on the 3-position. The major product is the one resulting from nucleophilic attack on the *more substituted* carbon.

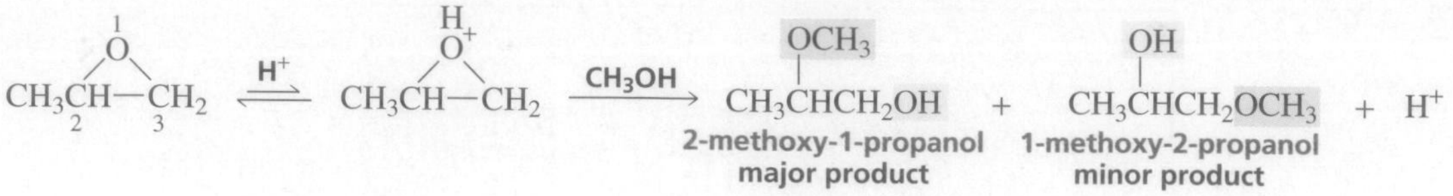

$$CH_3CH\overset{O}{-}CH_2 \xrightleftharpoons[]{H^+} CH_3CH\overset{\overset{+}{O}H}{-}CH_2 \xrightarrow{CH_3OH} \underset{\substack{\text{2-methoxy-1-propanol}\\\text{major product}}}{CH_3CHCH_2OH} + \underset{\substack{\text{1-methoxy-2-propanol}\\\text{minor product}}}{CH_3CHCH_2OCH_3} + H^+$$

The more substituted carbon is more likely to be attacked because, after the epoxide is protonated, it is so reactive that one of the C—O bonds begins to break even before the nucleophile has an opportunity to attack. As the C—O bond starts to break, a partial positive charge develops on the carbon that is losing its share of oxygen's electrons. The protonated epoxide breaks preferentially in the direction that puts the partial positive charge on the more substituted carbon, because a more substituted carbocation is more stable. (Recall that tertiary carbocations are more stable than secondary carbocations, which are more stable than primary carbocations.)

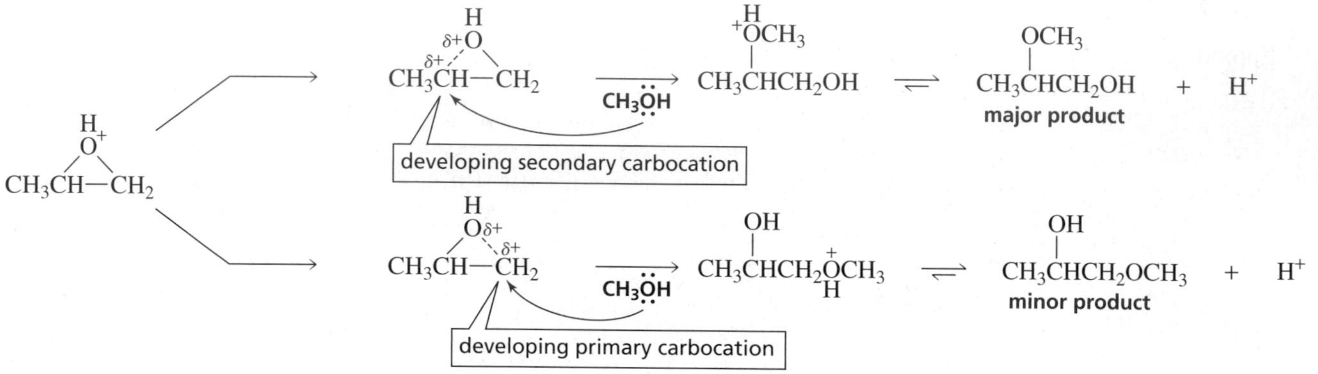

The best way to describe the reaction is to say that it occurs by a pathway that is partially S_N1 and partially S_N2. It is not a pure S_N1 reaction because a carbocation intermediate is not fully formed; it is not a pure S_N2 reaction because the leaving group begins to depart before the compound is attacked by the nucleophile.

Although an ether must be protonated before it can undergo a nucleophilic substitution reaction (Section 10.7), epoxides can undergo nucleophilic substitution reactions without first being protonated because of the strain in the three-membered ring (Figure 10.2). When a nucleophile attacks an unprotonated epoxide, the reaction is a pure S_N2 reaction.

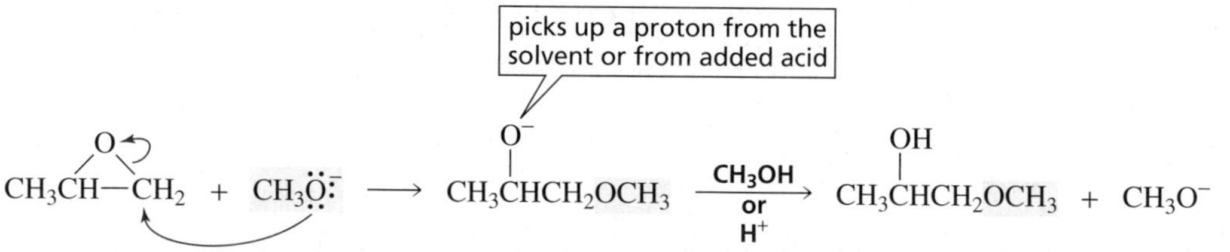

Under acidic conditions, the nucleophile attacks the most substituted ring carbon.

Under basic conditions, the nucleophile attacks the less sterically hindered ring carbon.

- The C—O bond does not begin to break until the carbon is attacked by the nucleophile. The nucleophile is more likely to attack the *less substituted* carbon because the less substituted carbon is more accessible to attack, because it is less sterically hindered.

- The alkoxide ion picks up a proton from the solvent or from an acid added after the reaction is over.

Thus, the site of nucleophilic attack on an unsymmetrical epoxide under neutral or basic conditions (when the epoxide *is not* protonated) is different from the site of nucleophilic attack under acidic conditions (when the epoxide *is* protonated).

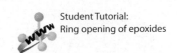

Student Tutorial:
Ring opening of epoxides

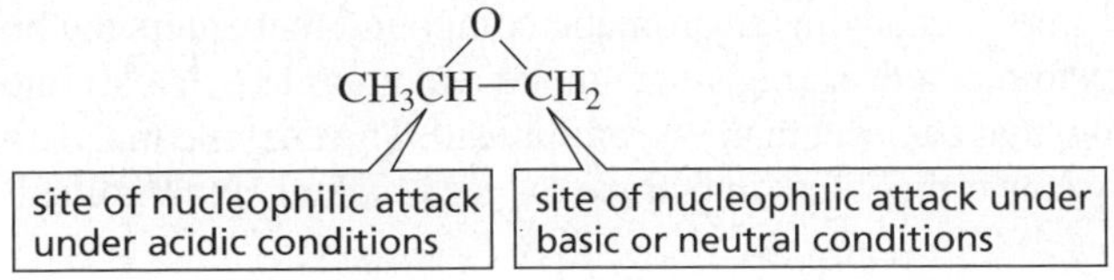

$$CH_3CH-CH_2$$

| site of nucleophilic attack under acidic conditions | site of nucleophilic attack under basic or neutral conditions |

Epoxides are useful reagents because they can react with a wide variety of nucleophiles, leading to the formation of a wide variety of products.

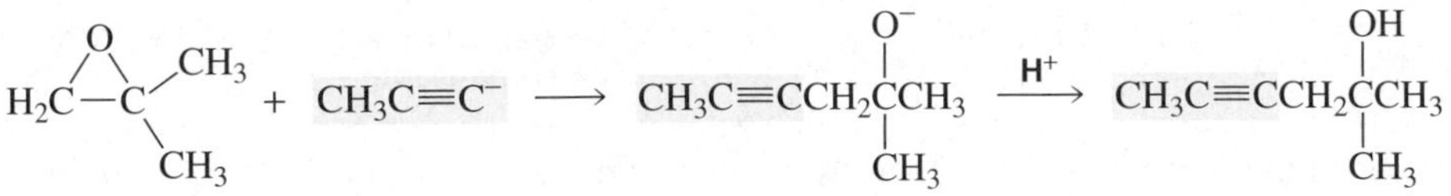

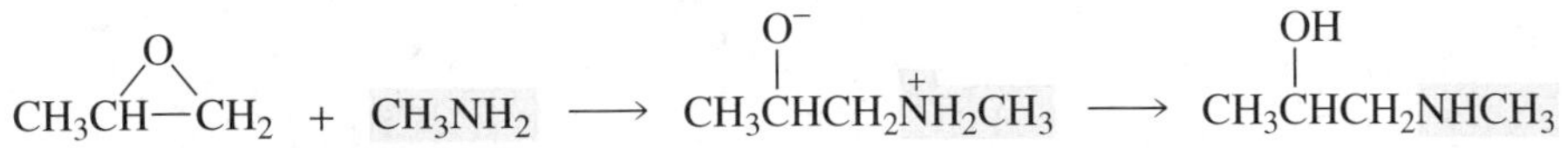

$$H_2C-C\overset{CH_3}{\underset{CH_3}{}} + CH_3C\equiv C^- \longrightarrow CH_3C\equiv CCH_2\overset{O^-}{\underset{CH_3}{C}}CH_3 \xrightarrow{H^+} CH_3C\equiv CCH_2\overset{OH}{\underset{CH_3}{C}}CH_3$$

$$CH_3CH-CH_2 + CH_3NH_2 \longrightarrow CH_3\overset{O^-}{C}HCH_2\overset{+}{N}H_2CH_3 \longrightarrow CH_3\overset{OH}{C}HCH_2NHCH_3$$

Epoxides also are important in biological processes, because they are reactive enough to be attacked by nucleophiles under the conditions found in living systems (Section 10.9).

Notice that the reaction of cyclohexene oxide with a nucleophile leads to trans products because the S_N2 reaction involves back-side nucleophilic attack.

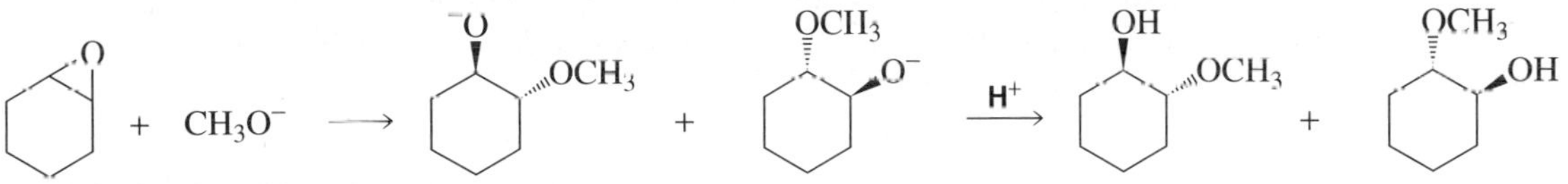

The reaction forms two new asymmetric centers, so back-side attack forms two stereoisomers.

Give the major product of each of the following reactions:

a. 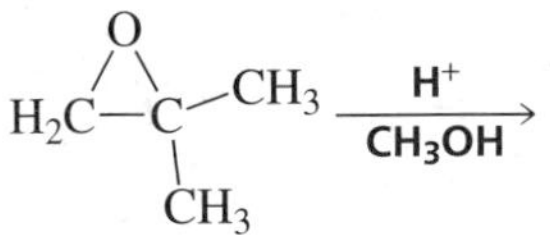 $H_2C-C\overset{O}{\underset{CH_3}{}}-CH_3 \xrightarrow[CH_3OH]{H^+}$

c. $H-C\overset{O}{\underset{H_3C}{}}C\overset{}{\underset{CH_3}{}}-CH_3 \xrightarrow[CH_3OH]{H^+}$

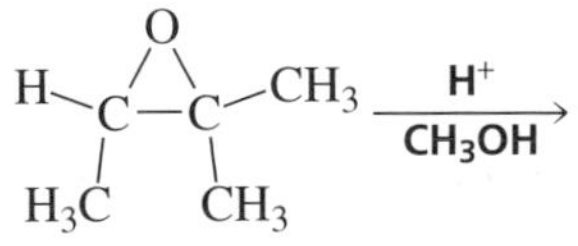

b. 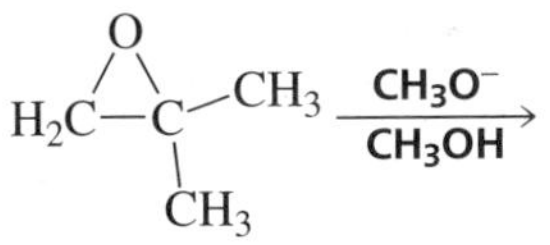 $H_2C-C\overset{O}{\underset{CH_3}{}}-CH_3 \xrightarrow[CH_3OH]{CH_3O^-}$

d. 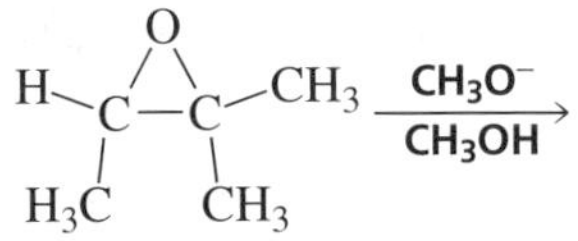 $H-C\overset{O}{\underset{H_3C}{}}C\overset{}{\underset{CH_3}{}}-CH_3 \xrightarrow[CH_3OH]{CH_3O^-}$

Would you expect the reactivity of a five-membered ring ether such as tetrahydrofuran (Table 10.2) to be more similar to an epoxide or to a noncyclic ether?

10.9 Arene Oxides

After an aromatic hydrocarbon such as benzene is ingested or inhaled, it is converted into an *arene oxide* by an enzyme. An **arene oxide** is a compound in which one of the "double bonds" of the aromatic ring has been converted into an epoxide. Formation of an arene

benzene

benzene oxide

oxide is the first step in changing an aromatic compound that enters the body as a foreign substance (for example, a drug, cigarette smoke, automobile exhaust) into a more water-soluble compound that can eventually be eliminated. The enzyme that detoxifies aromatic hydrocarbons by converting them into arene oxides is called cytochrome P_{450}.

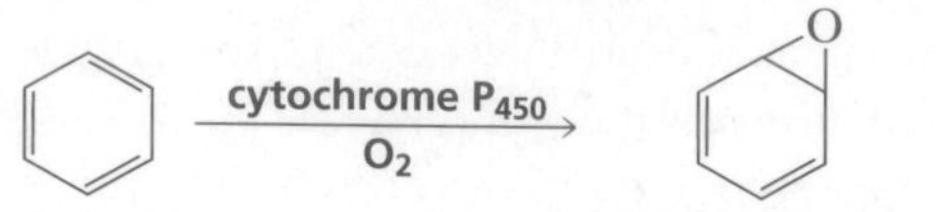

benzene $\xrightarrow[\text{O}_2]{\text{cytochrome P}_{450}}$ **benzene oxide**
an arene oxide

Certain arene oxides are important intermediates in the biosynthesis of biochemically important phenols such as tyrosine (an amino acid) and serotonin (a vasoconstrictor).

tyrosine
an amino acid

serotonin
a vasoconstrictor

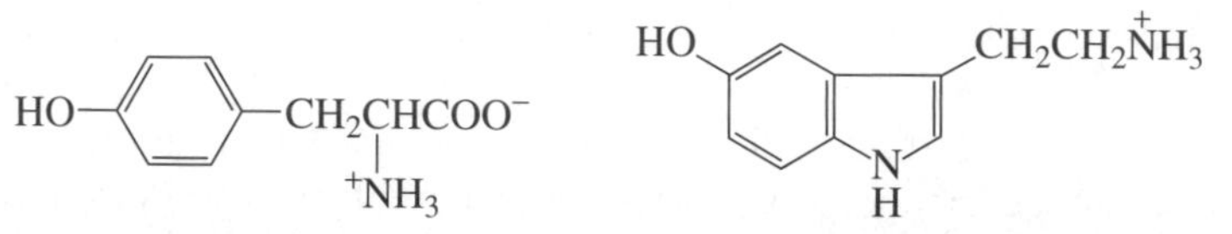

An arene oxide can react in two ways. It can react as a typical epoxide, undergoing attack by a nucleophile to form addition products (Section 10.8). Alternatively, it can rearrange to form a phenol, which epoxides cannot do.

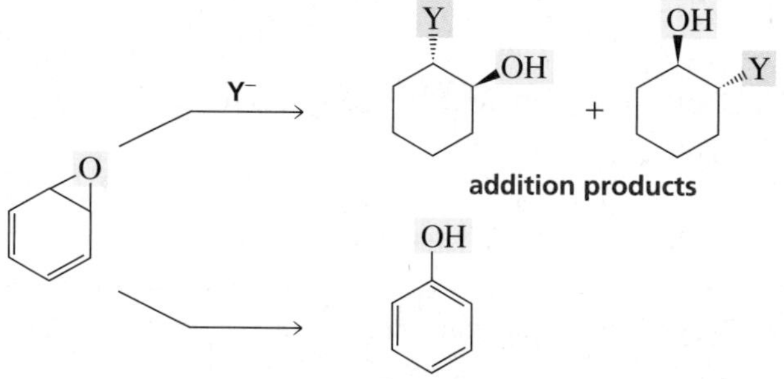

Y^-

addition products

rearranged product

The mechanism for the rearrangement is shown below.

Mechanism for arene oxide rearrangement

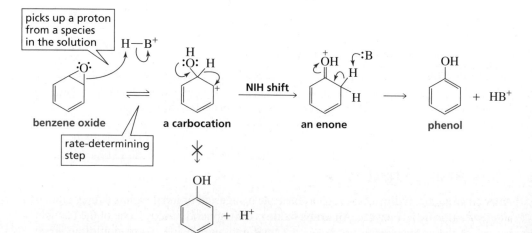

picks up a proton from a species in the solution

benzene oxide

rate-determining step

a carbocation

NIH shift

an enone

phenol $+ HB^+$

- The three-membered ring opens, picking up a proton from a species (such as water) in the solution.
- Instead of immediately losing a proton to form phenol, the carbocation forms an *enone* as a result of a 1,2-hydride shift. This is called an *NIH shift* because it was first observed in a laboratory at the National Institutes of Health.
- Removal of a proton from the enone forms phenol.

Because formation of the carbocation is the rate-determining step, the rate of phenol formation depends on the stability of the carbocation. The more stable the carbocation, the more easily the ring opens to form phenol.

Only one arene oxide can be formed from naphthalene because the "double bond" shared by the two rings cannot be epoxidized. Remember that benzene rings are particularly stable, so the molecule is much more likely to be epoxidized at a position that leaves one of the benzene rings intact (Section 7.8).

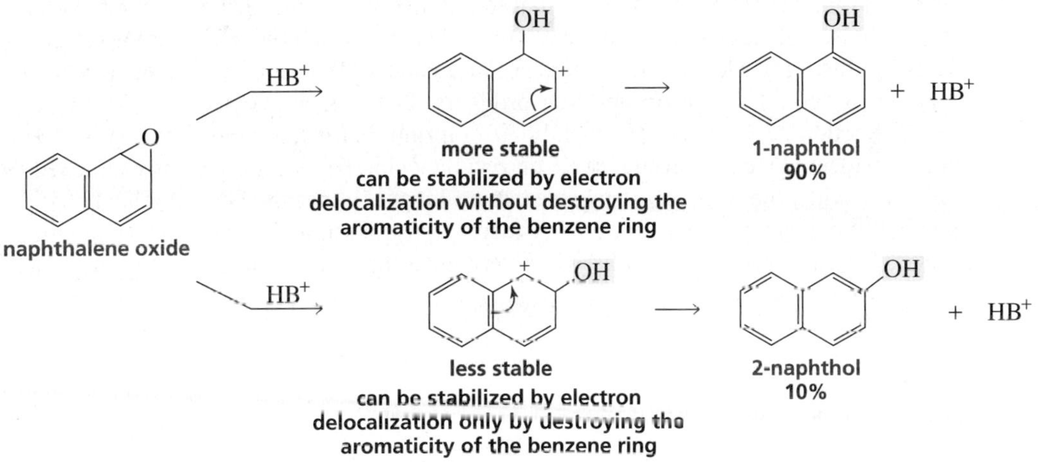

naphthalene oxide

more stable
can be stabilized by electron delocalization without destroying the aromaticity of the benzene ring

1-naphthol
90%

less stable
can be stabilized by electron delocalization only by destroying the aromaticity of the benzene ring

2-naphthol
10%

Naphthalene oxide can rearrange to form either 1-naphthol or 2-naphthol. The carbocation leading to 1-naphthol is more stable because its positive charge can be stabilized by electron delocalization without destroying the aromaticity of the benzene ring on the left of the structure. In contrast, the positive charge on the carbocation leading to 2-naphthol can be stabilized by electron delocalization only if the aromaticity of the benzene ring is destroyed. Consequently, rearrangement leads predominantly to 1-naphthol.

PROBLEM 25

Draw all possible resonance contributors for the two carbocations in the preceding reaction. Use the resonance contributors to explain why 1-naphthol is the major product of the reaction.

PROBLEM 26◆

The existence of the NIH shift was established by determining the major product obtained from rearrangement of the following arene oxide, in which a hydrogen has been replaced by a deuterium:

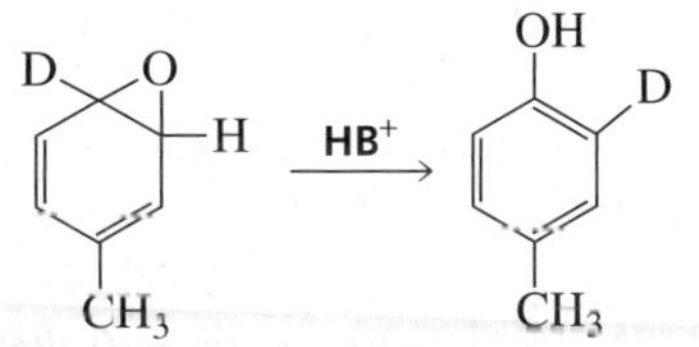

What would be the major product if the NIH shift did not occur? (*Hint:* Recall from Section 9.7 that a C—H bond is easier to break than a C—D bond.)

PROBLEM 27◆

How would the major products obtained from rearrangement of the following arene oxides differ?

A segment of DNA

Some aromatic hydrocarbons are carcinogens—compounds that cause cancer. Investigation has revealed, however, that the hydrocarbons themselves are not carcinogenic; the actual carcinogens are the arene oxides into which the hydrocarbons are converted in the body. How do arene oxides cause cancer? We have seen that nucleophiles react with epoxides to form addition products. 2′-Deoxyguanosine, a component of DNA (Section 27.1), has a nucleophilic NH_2 group that is known to react with certain arene oxides. Once a molecule of 2′-deoxyguanosine becomes covalently attached to an arene oxide, the 2′-deoxyguanosine can no longer fit into the DNA double helix. As a result, the genetic code will not be properly transcribed, which can lead to mutations that cause cancer. Cancer results when cells lose their ability to control their growth and reproduction.

an arene oxide

2′-deoxyguanosine

covalently attached to the arene oxide

Not all arene oxides are carcinogenic. Whether a particular arene oxide is carcinogenic depends on the relative rates of its two reaction pathways: rearrangement and reaction with a nucleophile. Arene oxide rearrangement leads to phenols that are not carcinogenic, whereas formation of addition products from nucleophilic attack by DNA can lead to cancer-causing products. Thus, if the rate of arene oxide rearrangement is faster than the rate of nucleophilic attack by DNA, the arene oxide will be harmless. However, if the rate of nucleophilic attack is faster than the rate of rearrangement, the arene oxide will likely be a carcinogen.

Because the rate of arene oxide rearrangement depends on the stability of the carbocation formed in the first step of the rearrangement, *an arene oxide's cancer-causing potential depends on the stability of this carbocation.* If the carbocation is relatively stable, rearrangement will be fast and the arene oxide will most likely not be carcinogenic. On the other hand, if the carbocation is relatively unstable, rearrangement will be slow and the arene oxide will more likely exist long enough to be attacked by nucleophiles and, therefore, be carcinogenic. This means that the more reactive the arene oxide (the more easily it opens to form a carbocation), the less likely it is to be carcinogenic.

The more stable the carbocation formed when the epoxide ring of an arene oxide opens, the less likely it is that the arene oxide is carcinogenic.

BENZO[*a*]PYRENE AND CANCER

Benzo[*a*]pyrene is one of the most carcinogenic of the aromatic hydrocarbons. It is formed whenever an organic compound is not completely burned. For example, benzo[*a*]pyrene is found in cigarette smoke, automobile exhaust, and charcoal-broiled meat. Several arene oxides can be formed from benzo[*a*]pyrene. The two most harmful are the 4,5-oxide and the 7,8-oxide. It has been suggested that people who develop lung cancer as a result of smoking may have a higher than normal concentration of cytochrome P_{450} in their lung tissue.

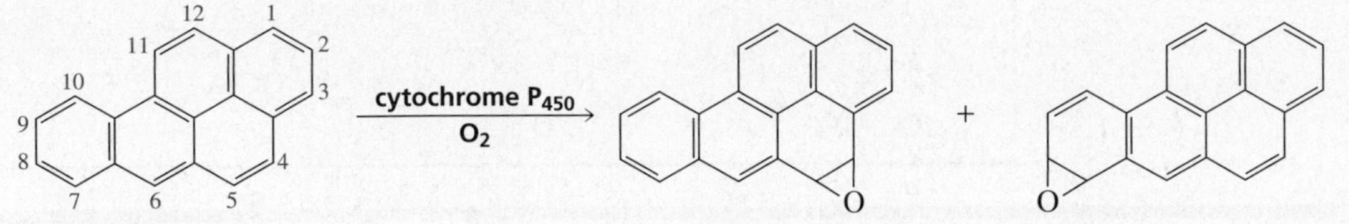

benzo[*a*]pyrene   4,5-benzo[*a*]pyrene oxide   7,8-benzo[*a*]pyrene oxide

The 4,5-oxide is harmful because it forms a carbocation that cannot be stabilized by electron delocalization without destroying the aromaticity of an adjacent benzene ring. Thus, the carbocation is relatively unstable, so the epoxide tends not to open until it is attacked by a nucleophile (the carcinogenic pathway). The 7,8-oxide is harmful because it reacts with water to form a diol, which then forms a diol epoxide. The diol epoxide does not readily undergo rearrangement (the harmless pathway), because it opens to a carbocation that is destabilized by the electron-withdrawing OH groups. Therefore, the diol epoxide can exist long enough to be attacked by nucleophiles (the carcinogenic pathway).

a diol epoxide

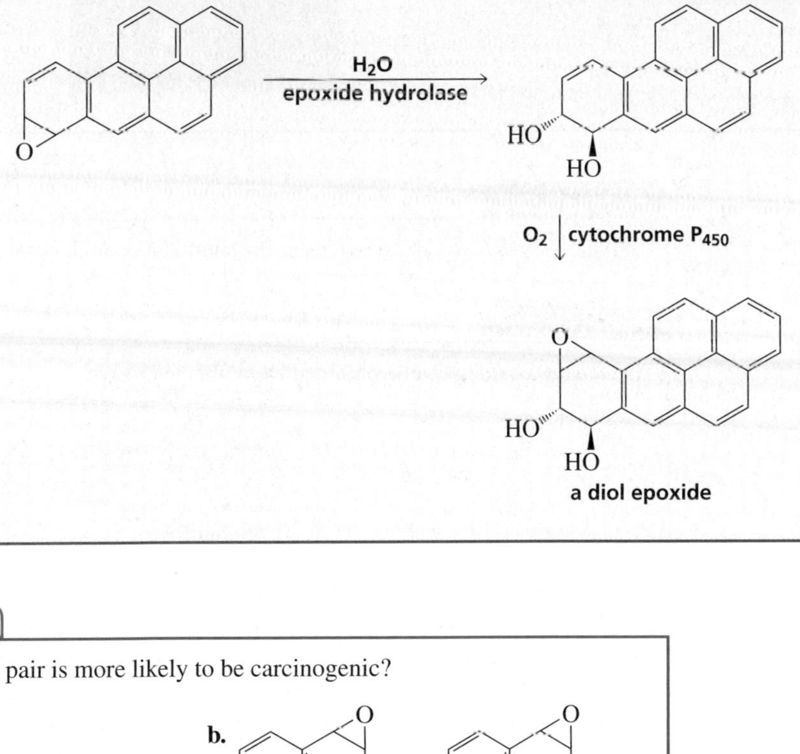

PROBLEM 28   *SOLVED*

Which compound in each pair is more likely to be carcinogenic?

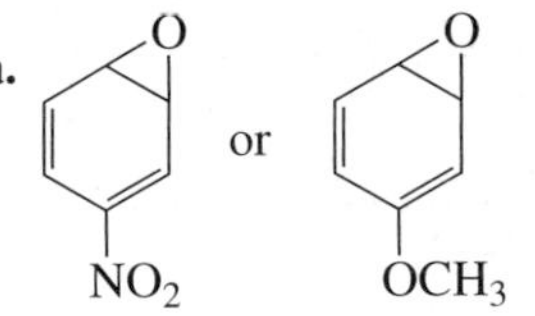

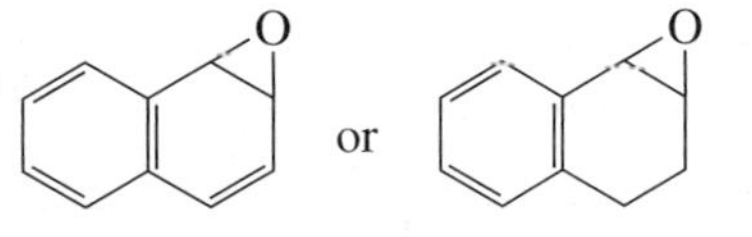

Solution to 28a   The nitro-substituted compound is more likely to be carcinogenic. The nitro group destabilizes the carbocation formed when the ring opens by withdrawing electrons from the ring by resonance (Chapter 7, Problem 19). In contrast, the methoxy group stabilizes the carbocation by donating electrons into the ring by resonance (Chapter 7, Problem 20). Carbocation formation leads to the harmless product, so the nitro-substituted compound with a less stable (less easily formed) carbocation will be less likely to undergo rearrangement to a harmless product. In addition, the electron-withdrawing nitro group increases the arene oxide's susceptibility to nucleophilic attack, which is the cancer-causing pathway.

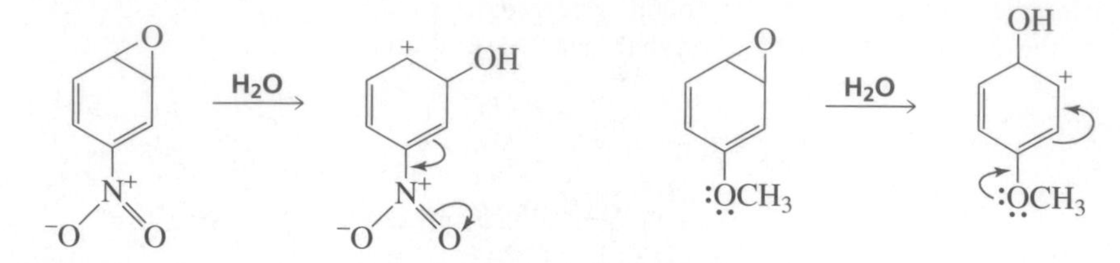

CHIMNEY SWEEPS AND CANCER

In 1775, British physician Percival Potts became the first to recognize that environmental factors can cause cancer when he observed that chimney sweeps had a higher incidence of scrotum cancer than the male population as a whole. He theorized that something in the chimney soot was causing cancer. We now know that it was benzo[*a*]pyrene.

Titch Cox, the chimney sweep responsible for cleaning the 800 chimneys at Buckingham Palace.

PROBLEM 29

Explain why the two arene oxides in Problem 28a open in opposite directions.

PROBLEM 30

Three arene oxides can be obtained from phenanthrene.

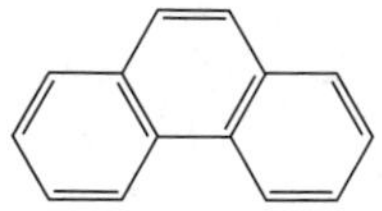

phenanthrene

a. Give the structures of the three phenanthrene oxides.
b. Give the structures of the phenols that can be obtained from each phenanthrene oxide.
c. If a phenanthrene oxide can lead to the formation of more than one phenol, which phenol will be obtained in greater yield?
d. Which of the three phenanthrene oxides is most likely to be carcinogenic?

10.10 Crown Ethers

Crown ethers are cyclic compounds containing several ether linkages around a central cavity. A crown ether specifically binds certain metal ions or organic molecules, depending on the cavity's size. The crown ether is called the "host" and the species it binds is called the "guest." Because the ether linkages are chemically inert, the crown ether can bind the guest without reacting with it. The *crown–guest complex* is called an **inclusion compound**. Crown ethers are named [X]-crown-Y, where X is the total number of atoms in the ring and Y is the number of oxygen atoms in the ring.

[15]-Crown-5 selectively binds Na^+ because that crown ether has a cavity diameter of 1.7 to 2.2 Å and Na^+ has an ionic diameter of 1.80 Å. Binding occurs through interaction of the positively charged ion with the lone-pair electrons of the oxygen atoms that point into the cavity.

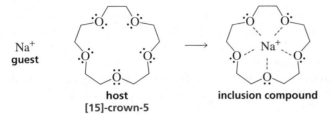

Na^+
guest

host
[15]-crown-5

inclusion compound

With an ionic diameter of 1.20 Å, Li^+ is too small to be bound by [15]-crown-5, but it binds neatly in [12]-crown-4. On the other hand, K^+, with an ionic diameter of 2.66 Å, is too large to fit into [15]-crown-5, but is bound specifically by [18]-crown-6.

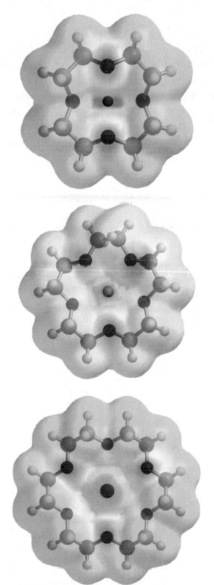

The ability of a host to bind only certain guests is an example of **molecular recognition**, the recognition of one molecule for another as a result of specific interactions. Molecular recognition explains how enzymes recognize their substrates, how antibodies recognize antigens, how drugs recognize receptors, and many other biochemical processes. Only recently have chemists been able to design and synthesize organic molecules that exhibit molecular recognition, although the specificity of these synthetic compounds for their guests is generally less highly developed than the specificity exhibited by biological molecules.

A remarkable property of crown ethers is that they enable inorganic salts to dissolve in nonpolar organic solvents, thus permitting many reactions to be carried out in nonpolar solvents that otherwise would not be able to take place. For example, the S_N2 reaction of 1-bromohexane with acetate ion poses a problem because the alkyl halide is soluble only in a nonpolar solvent, whereas potassium acetate, an ionic compound, is soluble only in water. In addition, acetate ion is a poor nucleophile.

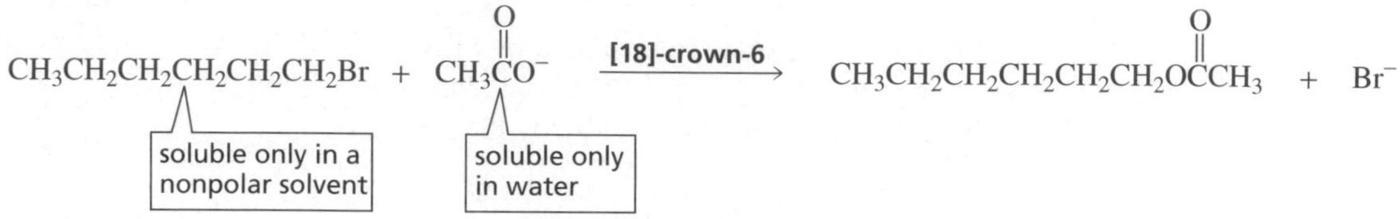

$$CH_3CH_2CH_2CH_2CH_2CH_2Br \ + \ CH_3\overset{O}{\overset{||}{C}}O^- \ \xrightarrow{\text{[18]-crown-6}} \ CH_3CH_2CH_2CH_2CH_2CH_2O\overset{O}{\overset{||}{C}}CH_3 \ + \ Br^-$$

soluble only in a nonpolar solvent

soluble only in water

However, potassium acetate will dissolve in a nonpolar solvent if [18]-crown-6 is added to the solution. The crown ether binds the positively charged potassium ion in its cavity, and the entire nonpolar crown ether–potassium complex dissolves in the nonpolar solvent, accompanied by the acetate ion to maintain electrical neutrality. Because acetate is not solvated in the nonpolar solvent, it is a much better nucleophile than it would be in a protic polar solvent (Section 8.3). The crown ether is a **phase-transfer catalyst**—it brings a reactant into the phase where it is needed. More examples of phase transfer catalysts will be discussed in Section 20.5.

AN IONOPHOROUS ANTIBIOTIC

An antibiotic is a compound that interferes with the growth of microorganisms. Nonactin is a naturally occurring antibiotic that owes its biological activity to its ability to disrupt the carefully maintained levels of electrolytes on the inside and outside of a cell. Normal cell function requires gradients between the concentrations of potassium and sodium ions inside the cell and those outside the cell. To achieve these gradients, potassium ions are pumped in and sodium ions are pumped out.

Nonactin disrupts one of these gradients by acting like a crown ether. Nonactin's diameter is such that it specifically binds potassium ions. The eight oxygen atoms that point into the cavity and interact with K^+ are highlighted in the structure shown below. The outside of nonactin is nonpolar, so it can easily transport K^+ ions out of the cell through the nonpolar cell membrane. The consequent decrease in the concentration of K^+ within the cell causes the bacterium to die. Nonactin is an example of an *ionophorous* antibiotic. An *ionophore* is a compound that transports metal ions by binding them tightly.

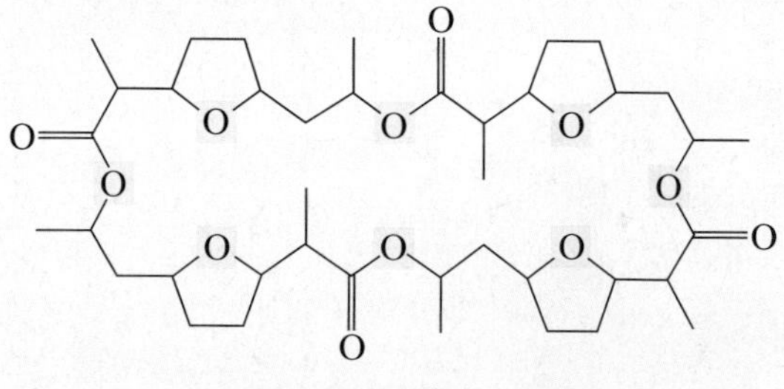

nonactin

10.11 Thiols, Sulfides, and Sulfonium Salts

Thiols are sulfur analogs of alcohols. They used to be called mercaptans because they form strong complexes with heavy metal cations such as arsenic and mercury (they capture mercury).

$$2 \text{ CH}_3\text{CH}_2\ddot{\text{S}}\text{H} + \text{Hg}^{2+} \longrightarrow \text{CH}_3\text{CH}_2\ddot{\text{S}}\text{—Hg—}\ddot{\text{S}}\text{CH}_2\text{CH}_3 + 2 \text{ H}^+$$
a thiol   **mercuric ion**

Thiols are named by adding the suffix "thiol" to the name of the parent hydrocarbon. If there is a second functional group in the molecule that is identified with a suffix, the SH group can be indicated by its substituent name, "mercapto." Like other substituent names, it is placed before the name of the parent hydrocarbon.

		$\underset{\text{CH}_3}{\overset{\text{CH}_3}{|}}$	
CH$_3$CH$_2$SH	CH$_3$CH$_2$CH$_2$SH	CH$_3$CHCH$_2$CH$_2$SH	HSCH$_2$CH$_2$OH
ethanethiol	**1-propanethiol**	**3-methyl-1-butanethiol**	**2-mercaptoethanol**

Low-molecular-weight thiols are noted for their strong and pungent odors, such as the odors associated with onions, garlic, and skunks. Natural gas is completely odorless and can cause deadly explosions if a leak goes undetected. Therefore, a small amount of a thiol is added to natural gas to give it an odor so that gas leaks can be detected.

Sulfur atoms are larger than oxygen atoms, so the negative charge of the thiolate ion is spread over a larger volume of space than the charge of an alkoxide ion, causing the thiolate ion to be more stable. (Recall that the more stable the base, the stronger is its conjugate acid; Section 1.18.) Thiols, therefore, are stronger acids (pK_a = 10) than alcohols, and thiolate ions are weaker bases than alkoxide ions. The larger thiolate ions are less well solvated than alkoxide ions, so in protic solvents thiolate ions are better nucleophiles than alkoxide ions (Section 8.3).

$$\text{CH}_3\ddot{\text{S}}\text{:}^- + \text{CH}_3\text{CH}_2\text{—Br} \xrightarrow{\text{CH}_3\text{OH}} \text{CH}_3\ddot{\text{S}}\text{CH}_2\text{CH}_3 + \text{Br}^-$$

Because sulfur is not as electronegative as oxygen, thiols are not good at hydrogen bonding. Consequently, they have weaker intermolecular attractions and, therefore, considerably lower boiling points than alcohols (Section 2.9). For example, the boiling point of CH$_3$CH$_2$OH is 78 °C, whereas the boiling point of CH$_3$CH$_2$SH is 37 °C.

The sulfur analogs of ethers are called **sulfides** or **thioethers**. Sulfur is an excellent nucleophile because its electron cloud is polarizable (Section 8.3). As a result, sulfides react readily with alkyl halides to form **sulfonium salts**, whereas ethers do not react with alkyl halides because oxygen is not as nucleophilic as sulfur and cannot accommodate a charge as well as sulfur can.

$$\text{CH}_3\ddot{\text{S}}\text{CH}_3 + \text{CH}_3\text{—I} \longrightarrow \underset{\text{CH}_3}{\overset{\text{CH}_3}{\text{CH}_3\overset{+}{\underset{\text{...}}{\text{S}}}\text{CH}_3}} \text{ I}^-$$
dimethyl   **trimethylsulfonium iodide**
sulfide   **a sulfonium salt**

Because it has a weakly basic leaving group, a sulfonium ion readily undergoes nucleophilic substitution reactions. Like other S$_N$2 reactions, the reaction works best if the group undergoing nucleophilic attack is a methyl group or a primary alkyl group. In Section 8.12, we saw that SAM, a biological methylating agent, is a sulfonium salt.

$$\text{HÖ:}^- + \text{CH}_3\overset{+}{\underset{\text{..}}{\text{S}}}\text{CH}_3 \longrightarrow \text{CH}_3\ddot{\text{O}}\text{H} + \text{CH}_3\ddot{\text{S}}\text{CH}_3$$

PROBLEM 31

Using an alkyl halide and a thiol as starting materials, how would you prepare the following compounds?

a. $CH_3CH_2SCH_2CH_3$

b. $CH_3\overset{\overset{\displaystyle CH_3}{|}}{\underset{\underset{\displaystyle CH_3}{|}}{S}}CCH_3$

c. $CH_2{=}CHCHSCH_2CH_3$ (with CH_3 on the CH)

d. 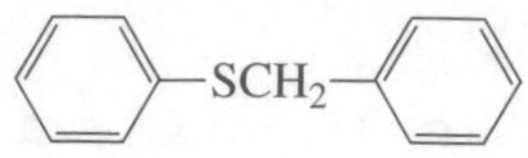 ⟨benzene⟩—SCH_2—⟨benzene⟩

MUSTARD—A CHEMICAL WARFARE AGENT

Chemical warfare occurred for the first time in 1915, when Germany released chlorine gas against French and British forces in the battle of Ypres. For the remainder of World War I, both sides used a variety of chemical agents as weapons. One of the more common was mustard, a reagent that produces blisters over the surface of the body. Mustard is a very reactive compound because its highly nucleophilic sulfur atom easily displaces a chloride ion by an intramolecular S_N2 reaction, forming a cyclic sulfonium salt that reacts rapidly with a nucleophile. The sulfonium salt is particularly reactive because of its strained three-membered ring and the excellent (positively charged) leaving group.

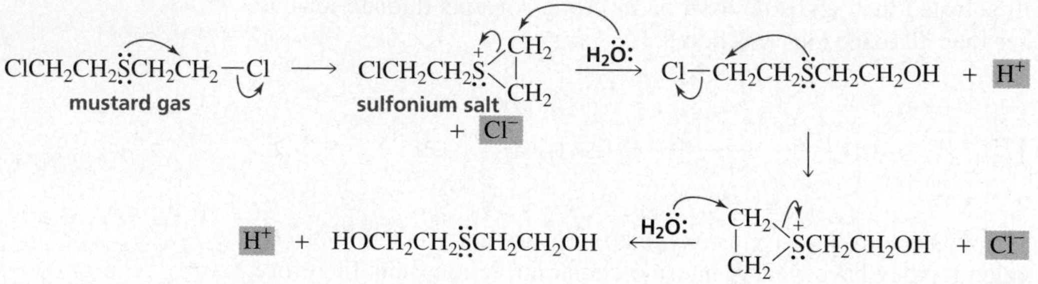

The blistering caused by mustard results from the high local concentrations of HCl produced when mustard comes into contact with water—or any other nucleophile—on skin or in lung tissue. Autopsies of soldiers killed by mustard in World War I—estimated to be about 400,000 in number—revealed that they had extremely low white blood cell counts, indicating that mustard interfered with bone marrow development. An international treaty in the 1980s banned the use of mustard and required that all stockpiles be destroyed.

ANTIDOTE TO A CHEMICAL WARFARE AGENT

Lewisite is a chemical warfare agent developed in 1917 by W. Lee Lewis, an American scientist. It rapidly penetrates clothing and skin and is poisonous because it contains arsenic, which inactivates enzymes by combining with their thiol groups (Section 22.8). During World War II, the Allies were concerned that the Germans would use lewisite, so British scientists developed an antidote that the Allies called "British anti-lewisite" (BAL). BAL contains two thiol groups that react with lewisite, thereby preventing it from reacting with the thiol groups of enzymes.

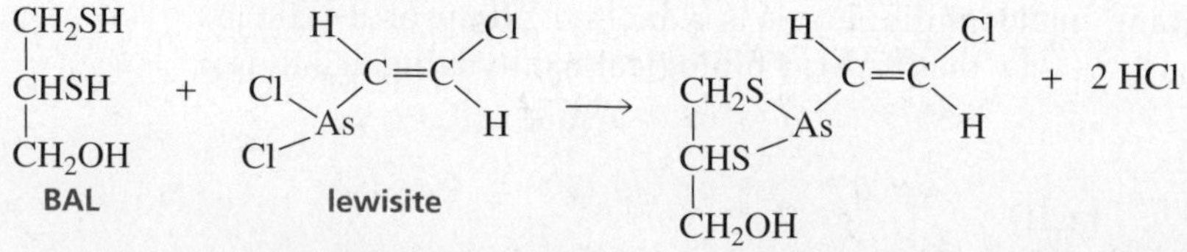

PROBLEM 32◆

The discovery that mustard interfered with bone marrow development caused chemists to look for less reactive mustards that might be used clinically. The following three compounds were studied:

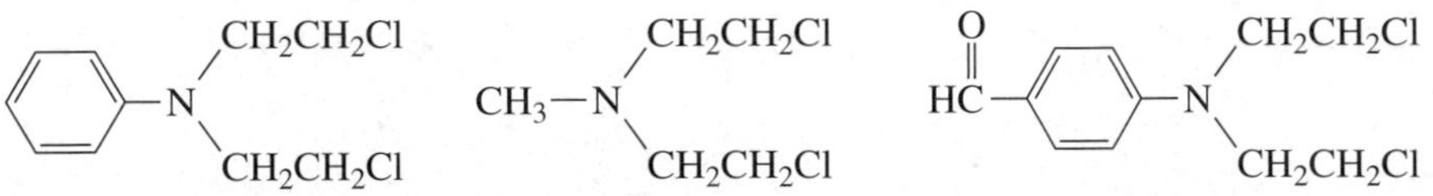

The structures show: a phenyl group attached to N with two CH₂CH₂Cl groups; CH₃—N with two CH₂CH₂Cl groups; and HC(=O)— attached to a benzene ring attached to N with two CH₂CH₂Cl groups.

One was found to be too reactive, one was found to be too unreactive, and one was found to be too insoluble in water to be injected intravenously. Which is which? (*Hint:* Draw resonance contributors.)

10.12 Organometallic Compounds

Alcohols, ethers, and alkyl halides all contain a carbon atom that is bonded to a *more* electronegative atom. The carbon atom, therefore, is *electrophilic* and reacts with a nucleophile.

> A carbon is an electrophile if it is attached to a more electronegative atom.

$$CH_3CH_2 \overset{\delta+}{\underset{}{}}\!\!-\!\!\overset{\delta-}{Z} + Y \longrightarrow CH_3CH_2\!-\!Y + Z^-$$

Z is labeled "more electronegative than carbon"; CH_3CH_2—Z is the **electrophile**, Y is the **nucleophile**.

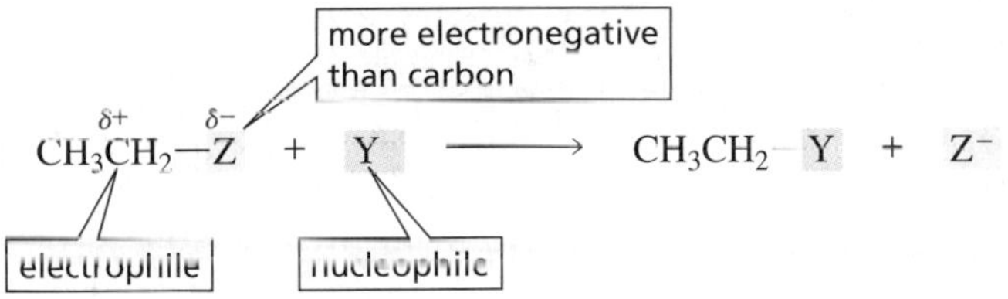

But what if you wanted a carbon atom to react with an electrophile? For that, you would need a compound with a nucleophilic carbon atom. To be *nucleophilic*, carbon would have to be bonded to a *less* electronegative atom.

> A carbon is a nucleophile if it is attached to a less electronegative atom.

$$CH_3CH_2 \overset{\delta-}{\underset{}{}}\!\!-\!\!\overset{\delta+}{M} + E^+ \longrightarrow CH_3CH_2\!-\!E + M^+$$

M is labeled "less electronegative than carbon"; CH_3CH_2—M is the **nucleophile**, E^+ is the **electrophile**.

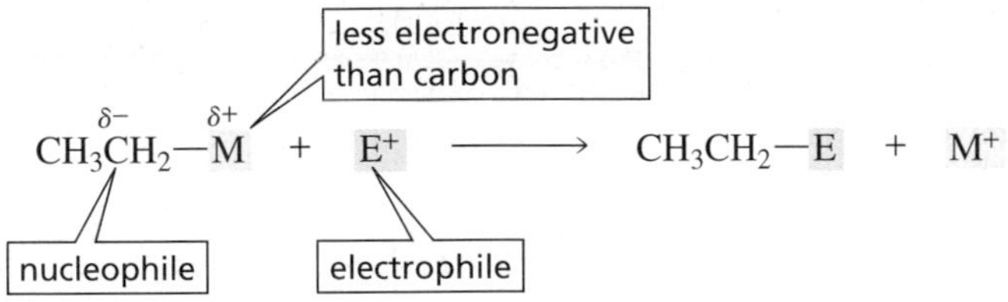

A carbon bonded to a metal is nucleophilic because most metals are less electronegative than carbon (Table 10.3). An **organometallic compound** is a compound that contains a carbon–metal bond. *Organolithium compounds* and *organomagnesium compounds* are two of the most common organometallic compounds. The electrostatic potential maps show that the carbon attached to the halogen in the alkyl halide is an electrophile (it is blue-green), whereas the carbon attached to the metal ion in the organometallic compound is a nucleophile (it is red).

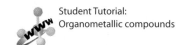

Student Tutorial:
Organometallic compounds

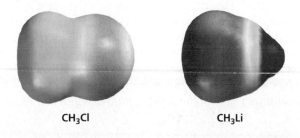

CH₃Cl   **CH₃Li**

Table 10.3		The Electronegativities of Some of the Elements[a]						
IA	IIA	IB	IIB	IIIA	IVA	VA	VIA	VIIA
H 2.1								
Li 1.0	Be 1.5			B 2.0	C 2.5	N 3.0	O 3.5	F 4.0
Na 0.9	Mg 1.2			Al 1.5	Si 1.8	P 2.1	S 2.5	Cl 3.0
K 0.8	Ca 1.0	Cu 1.8	Zn 1.7	Ga 1.8	Ge 2.0			Br 2.8
		Ag 1.4	Cd 1.5		Sn 1.7			I 2.5
			Hg 1.5		Pb 1.6			

[a]From the scale devised by Linus Pauling

BIOGRAPHY

Francois Auguste Victor Grignard (1871–1935) *was born in France, the son of a sailmaker. He received a Ph.D. from the University of Lyons in 1901. His synthesis of the first Grignard reagent was announced in 1900. During the next five years, some 200 papers were published about Grignard reagents. Grignard was a professor of chemistry at the University of Nancy and later at the University of Lyons. He shared the Nobel Prize in chemistry in 1912 with Paul Sabatier (p. 188). During World War I, he was drafted into the French army, where he developed a method to detect war gases.*

Organolithium compounds are prepared by adding lithium to an alkyl halide in a nonpolar solvent such as hexane.

$$CH_3CH_2CH_2CH_2Br + 2\ Li \xrightarrow{\text{hexane}} CH_3CH_2CH_2CH_2Li + LiBr$$
1-bromobutane   **butyllithium**

Cl + 2 Li $\xrightarrow{\text{hexane}}$ Li + LiCl
chlorobenzene   **phenyllithium**

Organomagnesium compounds, commonly called **Grignard reagents** after their discoverer, are prepared by adding an alkyl halide to magnesium shavings being stirred in diethyl ether or THF under anhydrous conditions. The reaction inserts magnesium between the carbon and the halogen.

—Br + Mg $\xrightarrow{\text{diethyl ether}}$ —MgBr
cyclohexyl bromide   **cyclohexylmagnesium bromide**

$$CH_2{=}CHBr + Mg \xrightarrow{\text{THF}} CH_2{=}CHMgBr$$
vinyl bromide   **vinylmagnesium bromide**

The solvent (usually diethyl ether or tetrahydrofuran) plays a crucial role in the formation of a Grignard reagent. The magnesium atom in a Grignard reagent is surrounded by only four electrons, so it needs two more pairs of electrons to form an octet. Solvent molecules provide these electrons by coordinating to the metal. Coordination allows the Grignard reagent to dissolve in the solvent and prevents it from coating the magnesium shavings, which would make them unreactive.

Because carbon is more electronegative than the metal atom (Li or Mg) to which it is attached, organolithium and organomagnesium compounds react as if they were carbanions.

$$CH_3CH_2MgBr \qquad \text{reacts as if it were} \qquad CH_3\overset{..}{\underset{-}{C}}H_2 \ \overset{+}{MgBr}$$
ethylmagnesium bromide

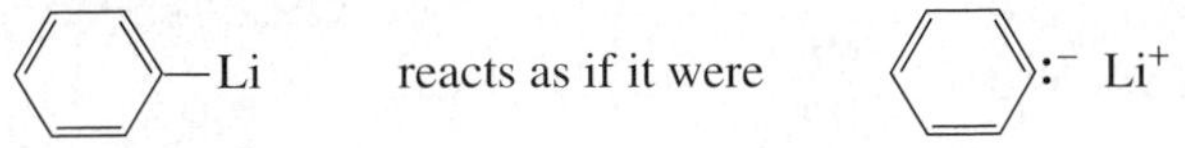

—Li   reacts as if it were   :$^-$ Li$^+$
phenyllithium

Alkyl halides, vinyl halides, and aryl halides can all be used to form organolithium and organomagnesium compounds. Alkyl bromides are the alkyl halides most often used to form organometallic compounds because they react more readily than alkyl chlorides and are less expensive than alkyl iodides.

The nucleophilic carbon of an organometallic compound reacts with an electrophile. For example, in the following reactions, the organolithium reagent and the Grignard reagent each react with an epoxide.

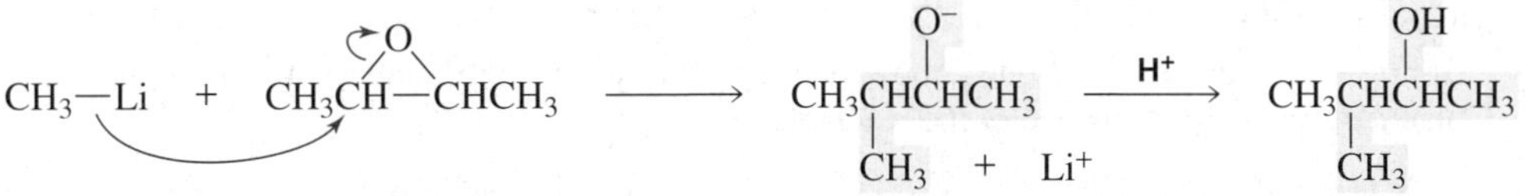

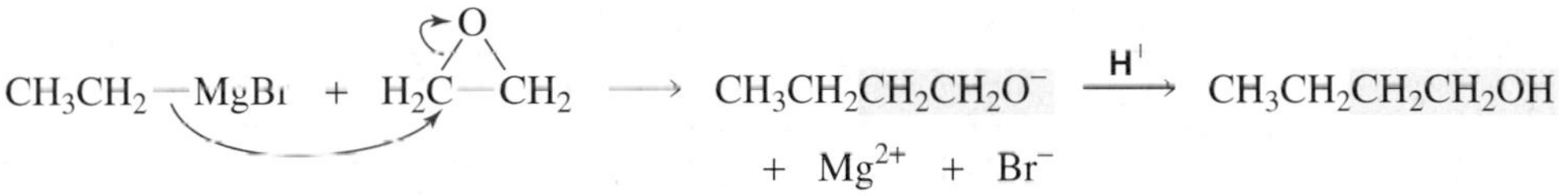

Notice that when an organometallic compound reacts with ethylene oxide, a primary alcohol containing two more carbons than the organometallic compound is formed.

PROBLEM 33◆

What alcohols would be formed from the reaction of ethylene oxide with the following Grignard reagents?

a. $CH_3CH_2CH_2MgBr$

b. 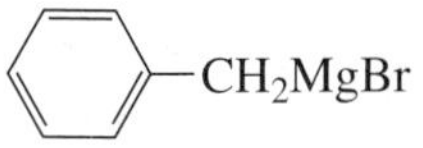 —CH_2MgBr

c. 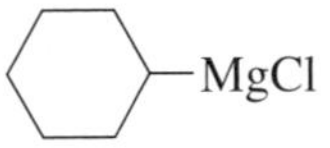 —$MgCl$

PROBLEM 34

How could the following compounds be prepared, using cyclohexene as a starting material?

a. 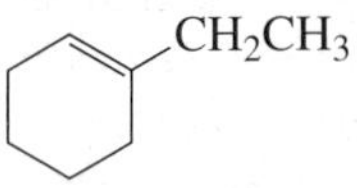 **b.** 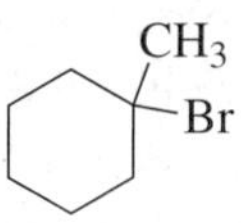 **c.**

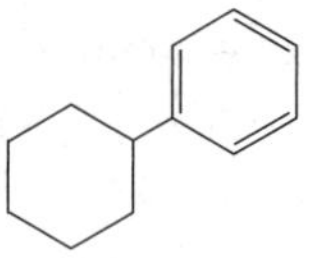

Organomagnesium and organolithium compounds are such strong bases that they will react immediately with any acid present in the reaction mixture, even with very weak acids such as water and alcohols. When this happens, the organometallic compound is converted into an alkane. If D_2O is used instead of H_2O, a deuterated compound will be obtained.

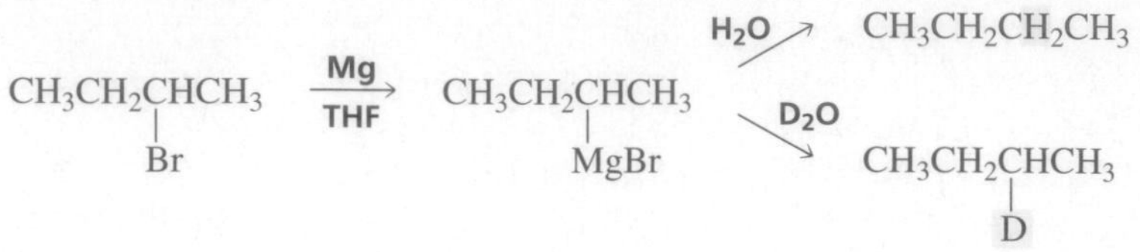

$$CH_3CH_2CHCH_3 \underset{\text{Br}}{|} \xrightarrow[\text{THF}]{\text{Mg}} CH_3CH_2CHCH_3 \underset{\text{MgBr}}{|} \xrightarrow[\text{D}_2\text{O}]{\text{H}_2\text{O}} \begin{matrix} CH_3CH_2CH_2CH_3 \\ \\ CH_3CH_2CHCH_3 \underset{\text{D}}{|} \end{matrix}$$

This means that Grignard reagents and organolithium compounds cannot be prepared from compounds that contain acidic groups ($-OH$, $-NH_2$, $-NHR$, $-SH$, $-C\equiv CH$, or $-COOH$ groups). Because even trace amounts of moisture can destroy an organometallic compound, it is important that all reagents are dry when organometallic compounds are being synthesized and when they react with other reagents.

PROBLEM 35 **SOLVED**

Using any necessary reagents, show how the following compounds could be prepared using ethylene oxide as one of the reactants:

a. $CH_3CH_2CH_2CH_2OH$

c. $CH_3CH_2CH_2CH_2D$

b. $CH_3CH_2CH_2CH_2Br$

d. $CH_3CH_2CH_2CH_2CH_2CH_2OH$

Solution

a. $CH_3CH_2Br \xrightarrow[\text{Et}_2\text{O}]{\text{Mg}} CH_3CH_2MgBr \xrightarrow[\text{2. H}^+]{1. \overset{\text{O}}{\triangle}} CH_3CH_2CH_2CH_2OH$

b. product of a $\xrightarrow{\text{PBr}_3} CH_3CH_2CH_2CH_2Br$

c. product of b $\xrightarrow[\text{Et}_2\text{O}]{\text{Mg}} CH_3CH_2CH_2CH_2MgBr \xrightarrow{\text{D}_2\text{O}} CH_3CH_2CH_2CH_2D$

d. the same reaction sequence as in part a, but with butyl bromide in the first step.

PROBLEM 36◆

Which of the following reactions will occur? (For the pK_a values necessary to do this problem, see Appendix II.)

$CH_3MgBr + H_2O \longrightarrow CH_4 + HOMgBr$

$CH_3MgBr + CH_3OH \longrightarrow CH_4 + CH_3OMgBr$

$CH_3MgBr + NH_3 \longrightarrow CH_4 + H_2NMgBr$

$CH_3MgBr + CH_3NH_2 \longrightarrow CH_4 + CH_3NMgBr$

$CH_3MgBr + HC\equiv CH \longrightarrow CH_4 + HC\equiv CMgBr$

There are many organometallic compounds. As long as the carbon is more electronegative than the metal, the carbon bonded to the metal will be nucleophilic.

$$\overset{\delta-}{C}\,\overset{\delta+}{-Mg} \qquad \overset{\delta-}{C}\,\overset{\delta+}{-Li} \qquad \overset{\delta-}{C}\,\overset{\delta+}{-Cu} \qquad \overset{\delta-}{C}\,\overset{\delta+}{-Cd} \qquad \overset{\delta-}{C}\,\overset{\delta+}{-Si}$$

$$\overset{\delta-}{C}\,\overset{\delta+}{-Zn} \qquad \overset{\delta-}{C}\,\overset{\delta+}{-Al} \qquad \overset{\delta-}{C}\,\overset{\delta+}{-Pb} \qquad \overset{\delta-}{C}\,\overset{\delta+}{-Hg} \qquad \overset{\delta-}{C}\,\overset{\delta+}{-Sn}$$

The reactivity of an organometallic compound toward an electrophile depends on the polarity of the carbon–metal bond: the greater the polarity of the bond, the more reactive the compound is as a nucleophile. The polarity of the bond depends on the difference in electronegativity between the metal and carbon (Table 10.3). For example, magnesium has an electronegativity of 1.2, compared with 2.5 for carbon. This large difference in electronegativity makes the carbon–magnesium bond highly polar (it is about 52% ionic). Lithium (1.0) is even less electronegative than magnesium (1.2). Thus, the carbon–lithium bond is more polar than the carbon–magnesium bond, so an organolithium reagent is a more reactive nucleophilic reagent than a Grignard reagent.

The name of an organometallic compound usually begins with the name of the alkyl group, followed by the name of the metal.

CH_3CH_2MgBr $CH_3CH_2CH_2CH_2Li$ $(CH_3CH_2CH_2)_2Cd$ $(CH_3CH_2)_4Pb$
ethylmagnesium   **butyllithium**   **dipropylcadmium**   **tetraethyllead**
bromide

A Grignard reagent will undergo **transmetallation** (metal exchange) if it is added to a metal halide whose metal is more electronegative than magnesium. In other words, metal exchange will occur if it results in a less polar carbon–metal bond. For example, cadmium (1.5) is more electronegative than magnesium (1.2). Consequently, a carbon–cadmium bond is less polar than a carbon–magnesium bond, so metal exchange occurs.

$$2\ CH_3CH_2MgCl\ +\ CdCl_2\ \longrightarrow\ (CH_3CH_2)_2Cd\ +\ 2\ MgCl_2$$
ethylmagnesium   **diethylcadmium**
chloride

PROBLEM 37◆

What organometallic compound will be formed from the reaction of methylmagnesium chloride and $SiCl_4$? (*Hint:* See Table 10.3.)

10.13 **Coupling Reactions**

New carbon–carbon bonds can be made using an organometallic reagent in which the metal ion is a transition metal. Transition metals are indicated by purple in the periodic table at the back of this book. The reactions are called **coupling reactions** because two groups (any two alkyl, aryl, or vinyl groups) are joined (coupled) together.

The first organometallic compounds used in coupling reactions were **Gilman reagents**, also called **organocuprates**. They are prepared by the reaction of an organolithium compound with cuprous iodide in diethyl ether or THF.

Coupling reactions join two CH-containing groups.

$$2\ CH_3Li\ +\ CuI\ \xrightarrow{\text{THF}}\ (CH_3)_2CuLi\ +\ LiI$$
organolithium   **Gilman reagent**
reagent

Gilman reagents are very useful to synthetic chemists. When a Gilman reagent reacts with an alkyl halide (with the exception of alkyl fluorides, which do not undergo this reaction), one of the alkyl groups of the Gilman reagent replaces the halogen. This means that an alkane can be formed from two alkyl halides—one alkyl halide is used to form the Gilman reagent, which then reacts with the second alkyl halide. The precise mechanism of the reaction is unknown, but is thought to involve radicals.

> **BIOGRAPHY**
>
> **Henry Gilman (1893–1986)**, *born in Boston, received his B.A. and Ph.D. degrees from Harvard University. He joined the faculty at Iowa State University in 1919 and remained there for his entire career. He published over 1000 research papers, more than half of them after he lost almost all of his sight as the result of a detached retina and glaucoma in 1947. His wife, Ruth, acted as his eyes for 40 years.*

$$CH_3CH_2CH_2CH_2Br\ +\ (CH_3CH_2CH_2)_2CuLi\ \xrightarrow{\text{THF}}\ CH_3CH_2CH_2CH_2CH_2CH_2CH_3\ +\ CH_3CH_2CH_2Cu$$
                                                  heptane $+\ LiBr$

Gilman reagents can be used to prepare compounds that cannot be prepared by using nucleophilic substitution reactions. For example, S_N2 reactions cannot be used to prepare the following compounds because vinyl and aryl halides cannot undergo nucleophilic attack (Section 8.8):

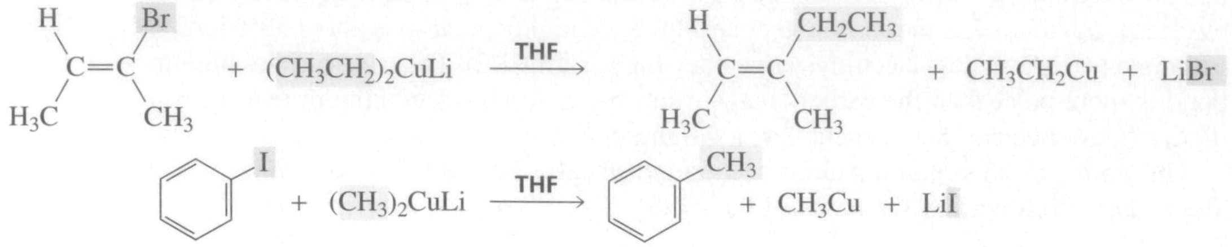

Gilman reagents can even replace halogens in compounds that contain other functional groups.

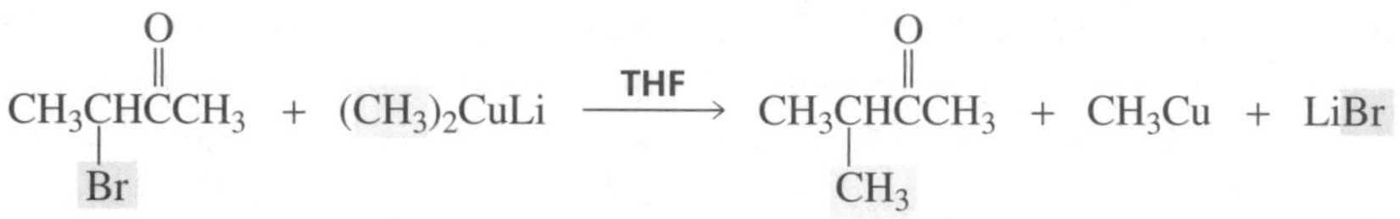

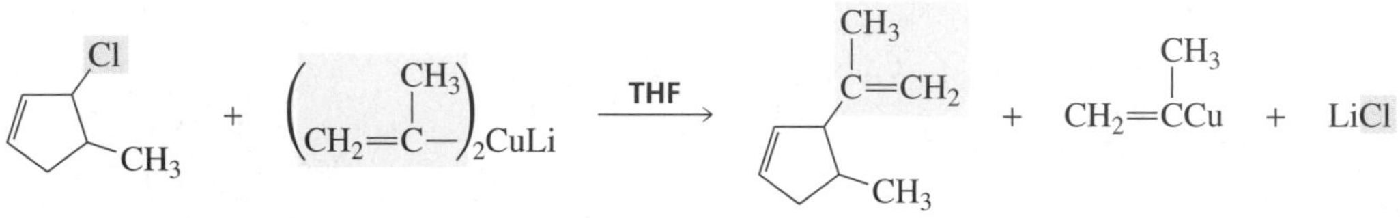

PROBLEM 38

Explain why tertiary alkyl halides cannot be used in coupling reactions with Gilman reagents.

PROBLEM 39

Muscalure is the sex attractant of the common housefly. Flies are lured to traps filled with fly bait containing muscalure and an insecticide. Eating the bait is fatal.

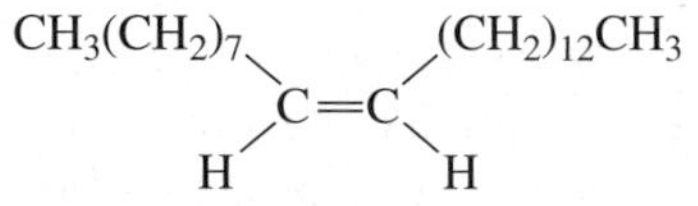

muscalure

How could you synthesize muscalure using the following reagents?

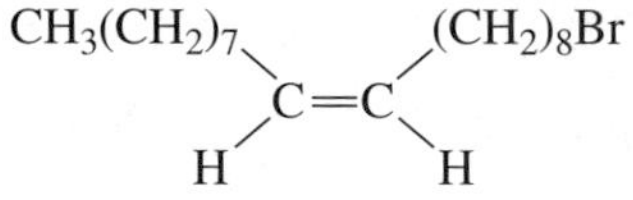

and $CH_3(CH_2)_4Br$

The Heck, Stille, and Suzuki reactions are also coupling reactions that use an organometallic reagent in which the metal ion is a transition metal (palladium). In all three reactions, the reagent is palladium ligated to four molecules of triphenyl phosphine. These reactions give high yields of products (80–98%).

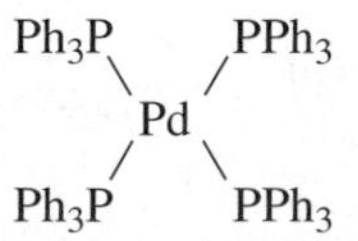

transition-metal catalyst

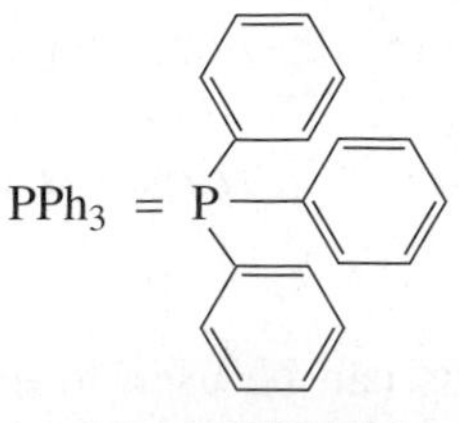

triphenyl phosphine

BIOGRAPHY

John K. Stille (1930–1989) *received a Ph.D. from the University of Arizona. He was a professor of chemistry at Colorado State University until his untimely death in a commercial air crash on his way to a scientific meeting in Switzerland.*

The **Heck reaction** couples an aryl, benzylic, or vinylic halide or triflate (Section 10.3) with an alkene in a basic solution.

Heck reaction

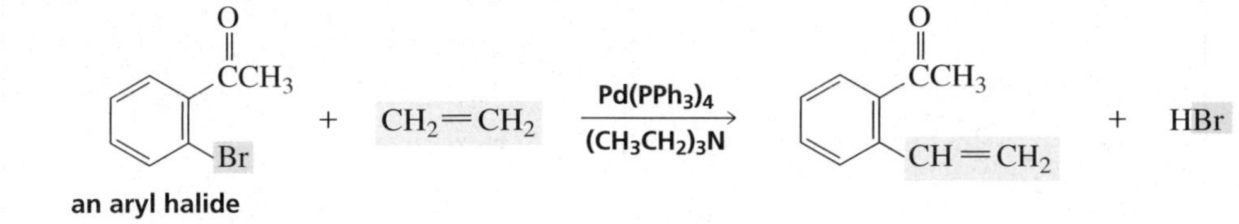

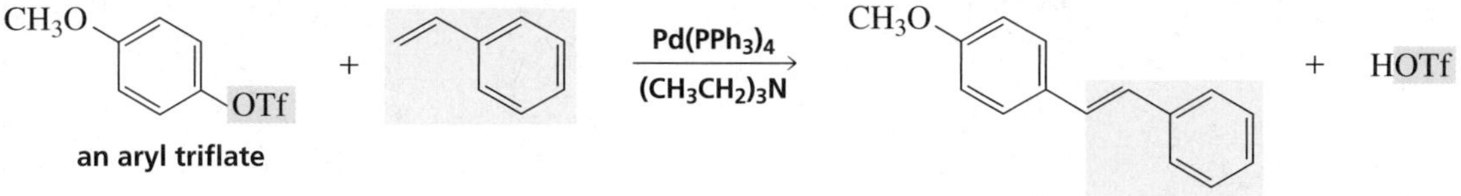

The **Stille reaction** couples an aryl, benzylic, or vinylic halide or triflate with a stannane.

Stille reaction

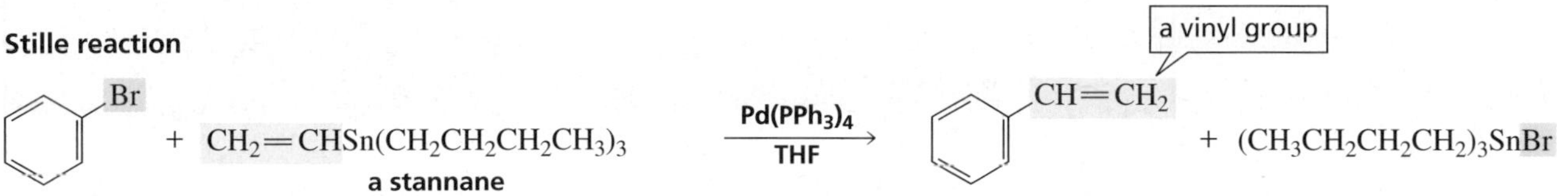

Aryl and vinylic groups are coupled preferentially, but an alkyl group can be coupled if a tetraalkylstannane is used.

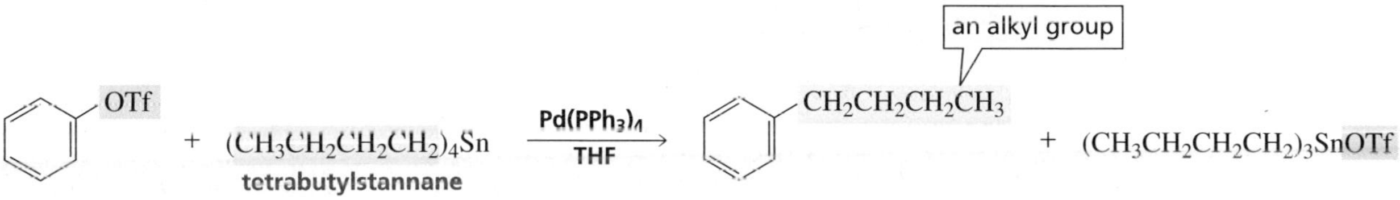

The **Suzuki reaction** couples an aryl, benzylic, or vinylic halide with an organoborane in a basic solution.

Suzuki reaction

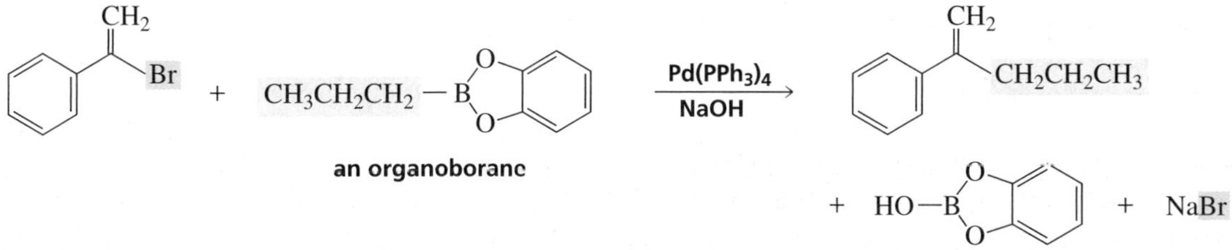

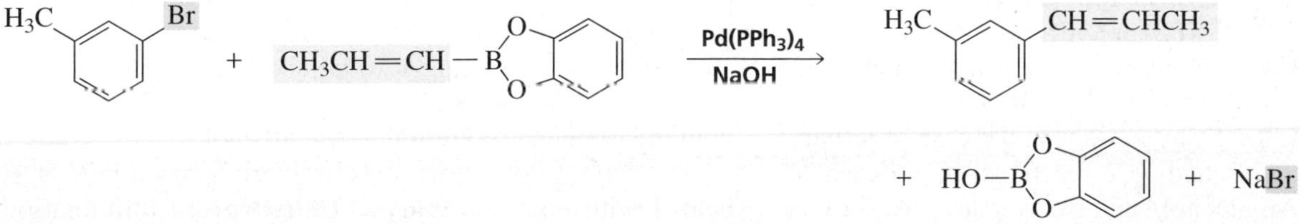

PROBLEM 40◆

What alkyl halides would you utilize to synthesize the following compounds, using the organoborane shown?

a.

c.

b.

PROBLEM 41◆

The organoborane used in a Suzuki reaction is prepared by the reaction of catecholborane with an alkene or an alkyne.

$$RCH{=}CH_2 \ + \ H{-}B \qquad \longrightarrow \qquad RCH_2CH_2{-}B$$

catecholborane

What hydrocarbon would you use to prepare the organoborane of Problem 40?

PROBLEM 42◆

Identify two sets of an alkyl bromide and an alkene that could be used in a Heck reaction to prepare the following compound:

$$CH_3\overset{\overset{\displaystyle O}{\|}}{C}{-}\!\!\!\!\!\!\!\!\!\bigcirc\!\!\!\!\!\!\!\!\!{-}CH{=}CH{-}\!\!\!\!\!\!\!\!\!\bigcirc\!\!\!\!\!\!\!\!\!{-}OCH_3$$

Student Tutorial:
Common terms

SUMMARY

The leaving groups of **alcohols** and **ethers** are stronger bases than halide ions are, so alcohols and ethers are less reactive than alkyl halides and have to be "activated" before they can undergo a substitution or an elimination reaction. **Epoxides** do not have to be activated, because ring strain increases their reactivity. **Sulfonate esters** and **sulfonium salts** have weakly basic leaving groups, so they undergo substitution reactions with ease. Amines cannot undergo substitution or elimination reactions because their leaving groups ($^-NH_2$, ^-NHR, $^-NR_2$) are very strong bases.

Primary, secondary, and tertiary alcohols undergo nucleophilic substitution reactions with HI, HBr, and HCl to form alkyl halides. These are S_N1 reactions in the case of tertiary and secondary alcohols and S_N2 reactions in the case of primary alcohols. An alcohol can also be converted into an

alkyl halide by phosphorus trihalides or thionyl chloride. These reagents convert the alcohol into an intermediate that has a leaving group easily displaced by a halide ion.

Converting an alcohol to a **sulfonate ester** is another way to activate an alcohol for subsequent reaction with a nucleophile. Because a sulfonic acid is a strong acid, its conjugate base is weak. Activating an alcohol by converting it to a sulfonate ester followed by reaction with a nucleophile forms a substitution product with a configuration opposite to that of the alcohol, whereas activating an alcohol by converting it to an alkyl halide followed by reaction with a nucleophile forms a substitution product with the same configuration as the alcohol.

An alcohol undergoes an elimination reaction if it is heated with an acid catalyst. **Dehydration** (elimination of a water molecule) is an E1 reaction in the case of secondary and tertiary alcohols and an E2 reaction in the case of primary alcohols. Tertiary alcohols are the easiest to dehydrate and primary alcohols are the hardest. The major product is the more substituted alkene. If the alkene has stereoisomers, the stereoisomer in which the bulkiest groups are on opposite sides of the double bond predominates. E1 reactions form carbocation intermediates, so carbocation and **ring-expansion (or ring-contraction) rearrangements** can occur.

Chromic acid oxidizes secondary alcohols to ketones and primary alcohols to carboxylic acids. PCC oxidizes primary alcohols to aldehydes.

Ethers can undergo nucleophilic substitution reactions with HBr or HI; if departure of the leaving group creates a relatively stable carbocation, an S_N1 reaction occurs; otherwise, an S_N2 reaction occurs.

Epoxides undergo nucleophilic substitution reactions. Under basic conditions, the least sterically hindered carbon is attacked; under acidic conditions, the most substituted carbon is attacked. **Arene oxides** undergo rearrangement to form phenols or nucleophilic attack to form addition products. An arene oxide's cancer-causing potential depends on the stability of the carbocation formed during rearrangement.

A **crown ether** specifically binds certain metal ions or organic molecules, depending on the size of its cavity,

forming an **inclusion compound**. The ability of a host to bind only certain guests is an example of **molecular recognition**. The crown ether can act as a **phase-transfer catalyst**.

Thiols are sulfur analogs of alcohols. They are stronger acids and have lower boiling points than alcohols. Thiolate ions are weaker bases and better nucleophiles in protic solvents than alkoxide ions. Sulfur analogs of ethers are called **sulfides** or **thioethers**. Sulfides react with alkyl halides to form **sulfonium salts**.

Organomagnesium compounds (Grignard reagents) and **organolithium compounds** are the most common **organometallic compounds**—compounds that contain a carbon–metal bond. They cannot be prepared from compounds that contain acidic groups. The carbon atom attached to the halogen in the alkyl halide is an electrophile, whereas the carbon atom attached to the metal ion in the organometallic compound is a nucleophile. The greater the polarity of the carbon–metal bond, the more reactive the organometallic compound is as a nucleophile.

New carbon–carbon bonds can be made using transition metal organometallic reagents. The reactions are called **coupling reactions** because two carbon-containing groups are joined together. **Gilman reagents** were the first organometallic compounds used in coupling reactions. The **Heck, Stille,** and **Suzuki reactions** are coupling reactions.

SUMMARY OF REACTIONS

1. Converting an *alcohol* to an *alkyl halide* (Sections 10.1 and 10.2).

$$ROH + HBr \xrightarrow{\Delta} RBr$$

$$ROH + HI \xrightarrow{\Delta} RI$$

$$ROH + HCl \xrightarrow{\Delta} RCl$$

relative rate: tertiary > secondary > primary

$$ROH + PBr_3 \xrightarrow{\text{pyridine}} RBr$$

$$ROH + PCl_3 \xrightarrow{\text{pyridine}} RCl$$

$$ROH + SOCl_2 \xrightarrow{\text{pyridine}} RCl$$

only for primary and secondary alcohols

2. Converting an alcohol to a *sulfonate ester* (Section 10.3).

$$ROH + R'-\overset{O}{\underset{O}{\overset{\|}{\underset{\|}{S}}}}-Cl \xrightarrow{\text{pyridine}} RO-\overset{O}{\underset{O}{\overset{\|}{\underset{\|}{S}}}}-R' + HCl$$

3. Converting an *activated alcohol* (an alkyl halide or a sulfonate ester) to a *compound with a new group bonded to the sp³ carbon* (Section 10.3).

$$RBr \ + \ Y^- \ \longrightarrow \ RY \ + \ Br^-$$

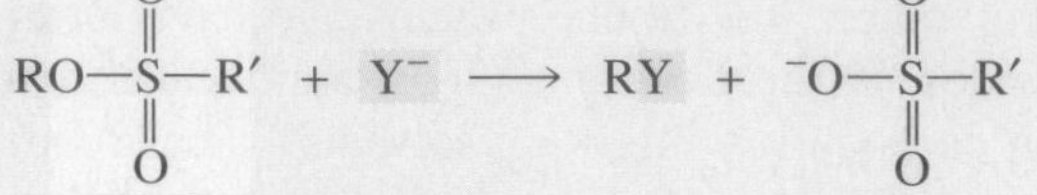

4. Elimination reactions of alcohols: dehydration (Section 10.4).

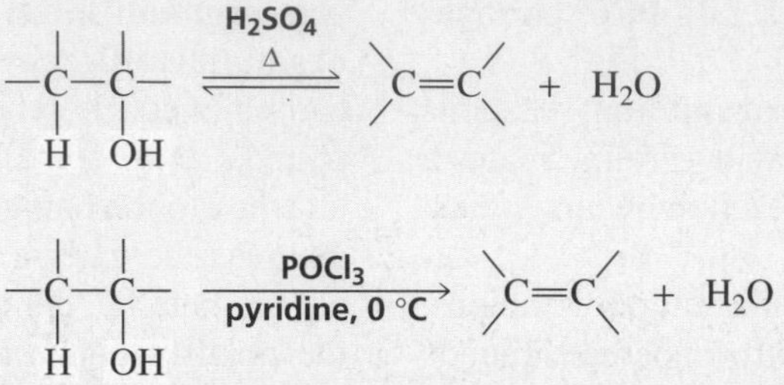

relative rate: **tertiary > secondary > primary**

5. Oxidation of alcohols (Section 10.5).

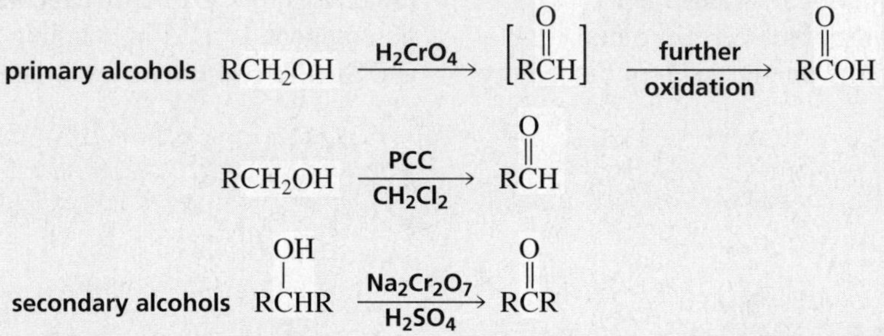

6. Nucleophilic substitution reactions of ethers (Section 10.7).

$$ROR' \ + \ HX \ \overset{\Delta}{\longrightarrow} \ ROH \ + \ R'X$$

HX = HBr or HI

7. Nucleophilic substitution reactions of *epoxides* (Section 10.8)

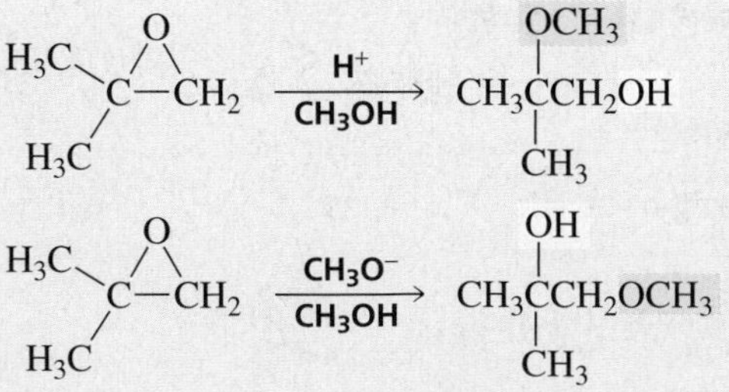

under acidic conditions, the nucleophile attacks the more substituted ring carbon

under neutral or basic conditions, the nucleophile attacks the less sterically hindered ring carbon

8. Reactions of *arene oxides*: ring opening and rearrangement (Section 10.9).

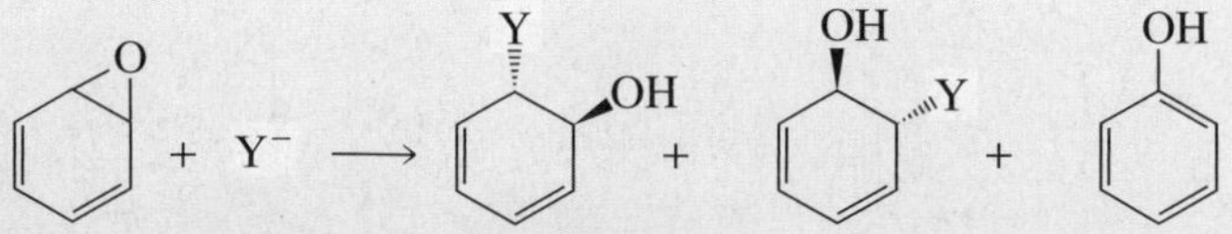

9. Reactions of thiols, sulfides, and sulfonium salts (Section 10.11).

$$2\,RSH \;+\; Hg^{2+} \;\longrightarrow\; RS{-}Hg{-}SR \;+\; 2\,H^+$$

$$RS^- \;+\; R'{-}Br \;\longrightarrow\; RSR' \;+\; Br^-$$

$$RSR \;+\; R'I \;\longrightarrow\; R\overset{R'}{\underset{+}{S}}R \;\; I^-$$

$$R\overset{R}{\underset{+}{S}}R \;+\; Y^- \;\longrightarrow\; RY \;+\; RSR$$

10. Formation of an *organolithium* or *organomagnesium* compound and its reaction with an *epoxide* (Section 10.12).

$$RBr \;\xrightarrow[\textbf{diethyl ether}]{\textbf{Mg}}\; RMgBr$$

a Grignard reagent

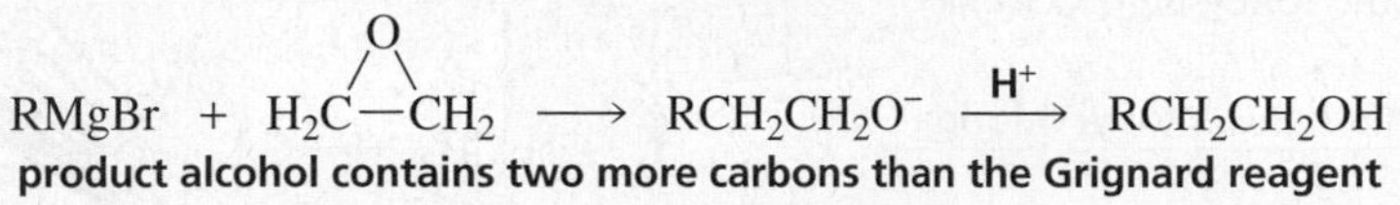

$$RMgBr \;+\; H_2C\overset{O}{-}CH_2 \;\longrightarrow\; RCH_2CH_2O^- \;\xrightarrow{H^+}\; RCH_2CH_2OH$$

product alcohol contains two more carbons than the Grignard reagent

11. Formation of a *Gilman reagent* and its reaction with an *alkyl halide* (Section 10.13).

$$2\,RLi \;+\; CuI \;\xrightarrow{\textbf{THF}}\; R_2CuLi \;+\; LiI$$

Gilman reagent

$$CH_3CH_2CH_2X \;+\; R_2CuLi \;\xrightarrow{\textbf{THF}}\; CH_3CH_2CH_2R \;+\; RCu \;+\; LiX$$

X = Cl, Br, or I

12. Reaction of an *aryl*, *benzyl*, or *vinyl halide* or *triflate* with an *alkene*: the Heck reaction (Section 10.13).

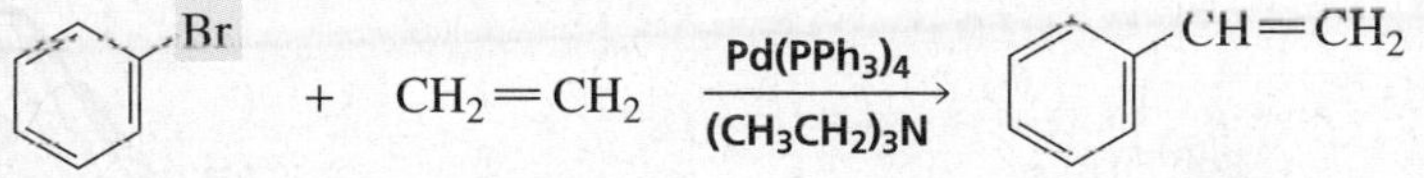

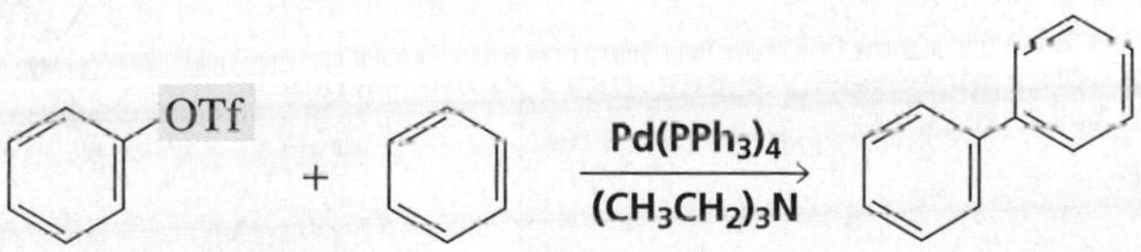

13. Reaction of an *aryl*, *benzyl*, or *vinyl halide* or *triflate* with a *stannane*: the Stille reaction (Section 10.13).

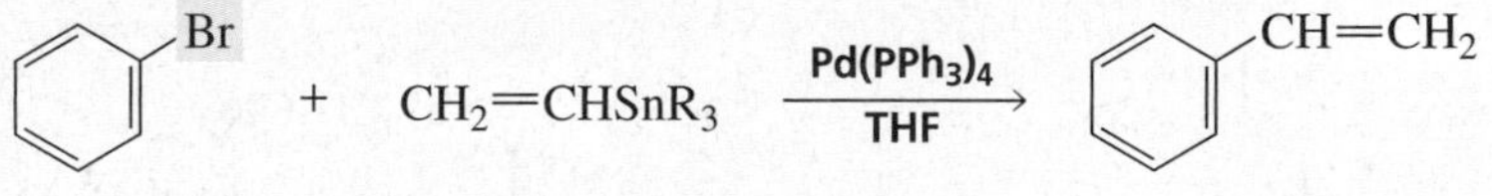

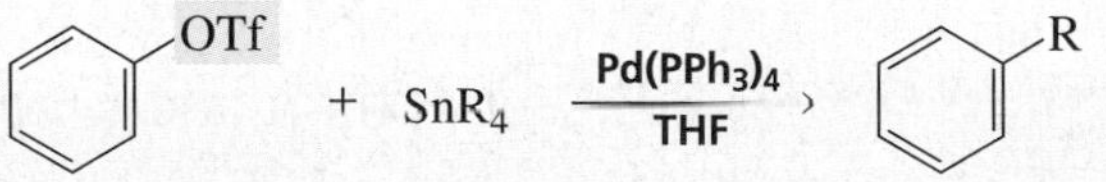

14. Reaction of an *aryl*, *benzyl*, or *vinyl halide* with an *organoborane*: the Suzuki reaction (Section 10.13).

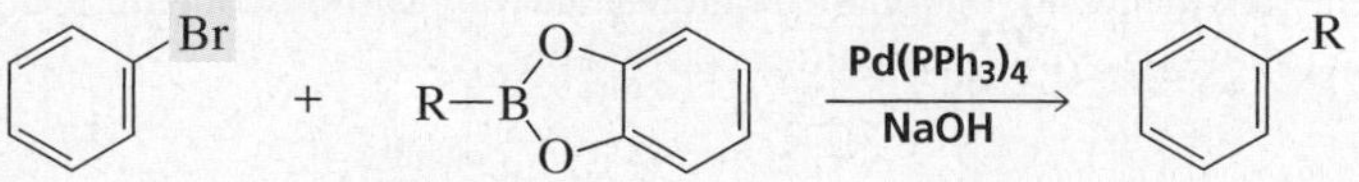

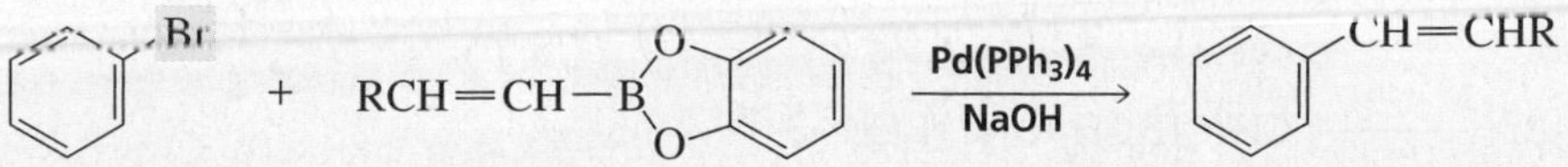

KEY TERMS

alcohol (p. 430)
alkaloid (p. 448)
alkyl tosylate (p. 436)
alkyl triflate (p. 437)
arene oxide (p. 455)
coupling reaction (p. 469)
crown ether (p. 460)
dehydration (p. 438)
epoxide (p. 452)
ether (p. 449)

Gilman reagent (p. 469)
Grignard reagent (p. 466)
Heck reaction (p. 471)
inclusion compound (p. 460)
molecular recognition (p. 461)
organocuprate (p. 469)
organolithium compound (p. 466)
organomagnesium compound (p. 466)
organometallic compound (p. 465)
phase-transfer catalyst (p. 462)

ring-expansion
  rearrangement (p. 440)
Stille reaction (p. 471)
sulfide (p. 463)
sulfonate ester (p. 435)
sulfonium salt (p. 463)
Suzuki reaction (p. 471)
thioether (p. 463)
thiol (p. 462)
transmetallation (p. 469)

PROBLEMS

43. Give the product of each of the following reactions:

a. $CH_3CH_2CH_2OH$ $\xrightarrow{\text{1. methanesulfonyl chloride}}_{\text{2. } CH_3CO^-}$

b. $CH_3CH_2CH_2CH_2OH + PBr_3$ $\xrightarrow{\text{pyridine}}$

c. 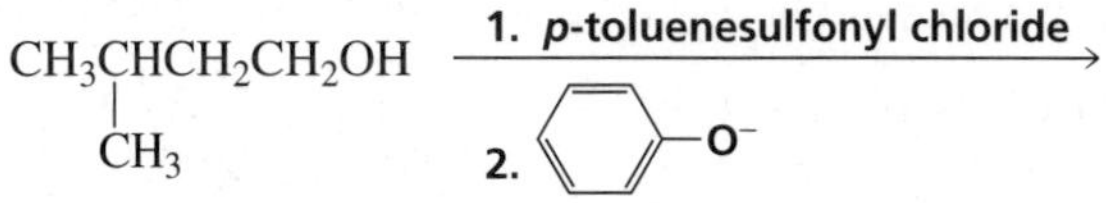 $CH_3CHCH_2CH_2OH$ | CH_3 $\xrightarrow{\text{1. } p\text{-toluenesulfonyl chloride}}_{\text{2. } \bigcirc\!-O^-}$

d. 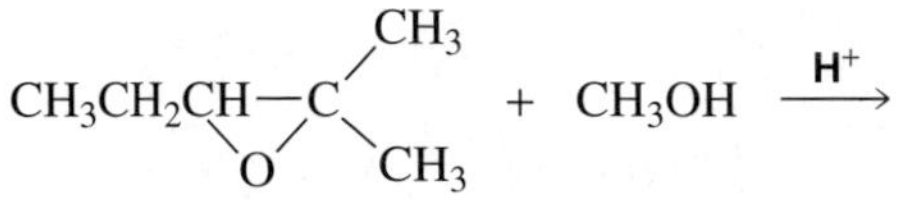 $CH_3CH_2CH{-}C(CH_3)_2$ (epoxide with CH_3) $+ CH_3OH$ $\xrightarrow{H^+}$

e. 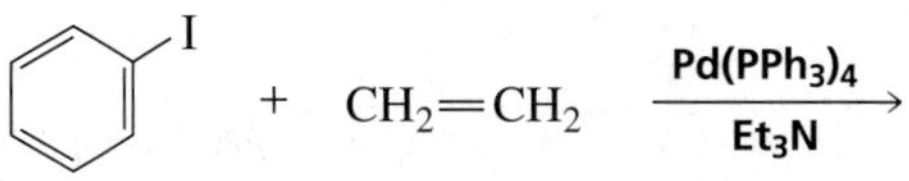 $\bigcirc\!-I + CH_2{=}CH_2$ $\xrightarrow[\text{Et}_3N]{Pd(PPh_3)_4}$

f. 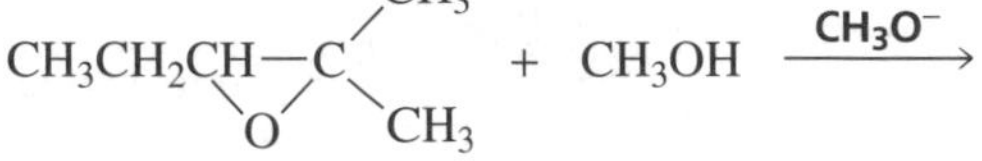 $CH_3CH_2CH{-}C(CH_3)_2{-}O{-}CH_3$ $+ CH_3OH$ $\xrightarrow{CH_3O^-}$

g. $CH_3CH_2CH_2CH_2OH$ $\xrightarrow[\Delta]{H_2SO_4}$

h. $CH_3CHCH_2CH_2OH$ | CH_3 $\xrightarrow[\text{pyridine}]{SOCl_2}$

i. 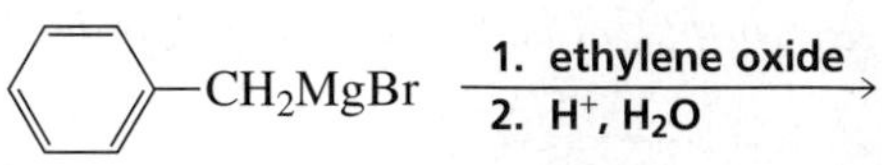 $\bigcirc\!-CH_2MgBr$ $\xrightarrow{\text{1. ethylene oxide}}_{\text{2. H}^+, H_2O}$

j. 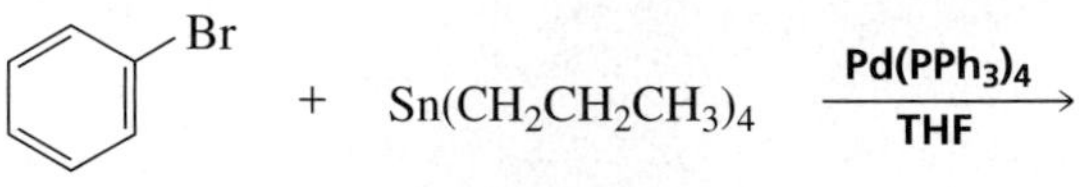 $\bigcirc\!-Br + Sn(CH_2CH_2CH_3)_4$ $\xrightarrow[\text{THF}]{Pd(PPh_3)_4}$

44. Indicate which alcohol in each pair will undergo dehydration more rapidly when heated with H_2SO_4.

a. 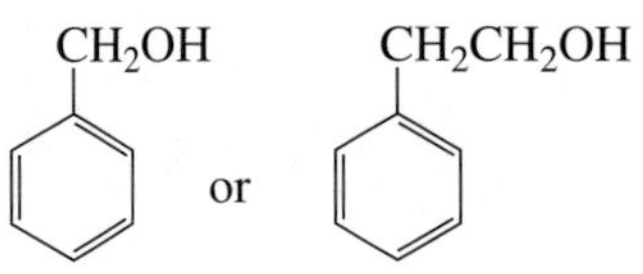 $\bigcirc\!-CH_2OH$ or $\bigcirc\!-CH_2CH_2OH$

b. (cyclohexanol) or (cyclohexenol)

c. H_3C OH (cyclohexane) or CH_3 OH (cyclohexane)

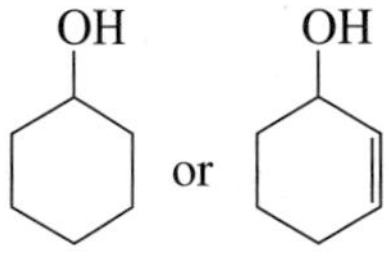

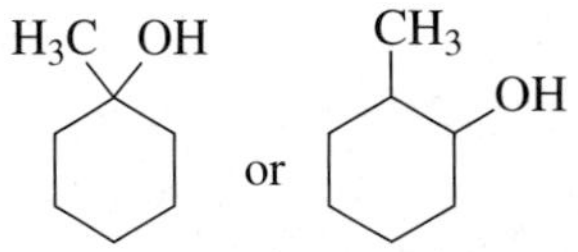

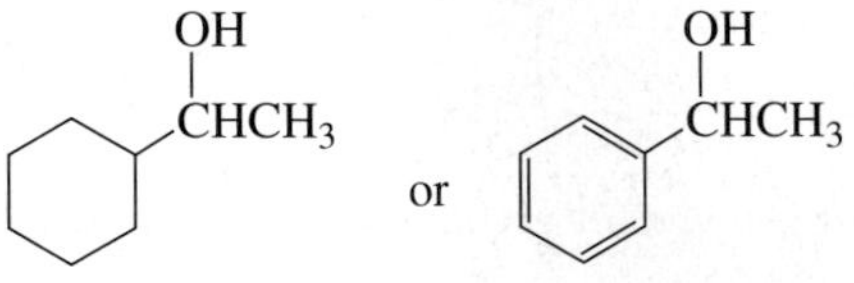

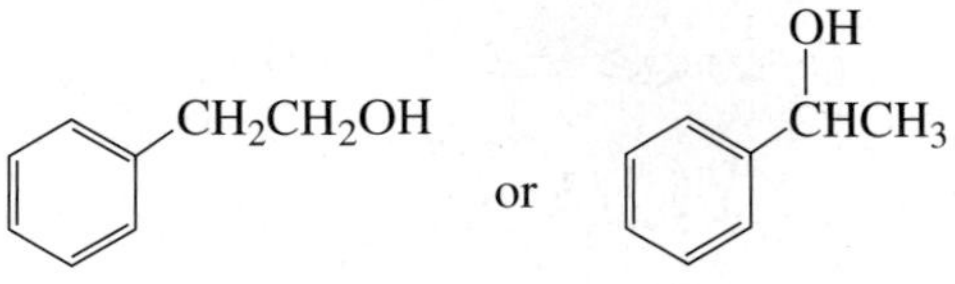

d. (cyclohexyl)$-CHCH_3$ with OH or $\bigcirc\!-CHCH_3$ with OH

e. $\bigcirc\!-CH_2CH_2OH$ or $\bigcirc\!-CHCH_3$ with OH

f. $CH_3CH_2CHCH_3$ | OH or $CH_3CCH_2CH_3$ | CH_3 | OH

45. Using the given starting material, any necessary inorganic reagents, and any carbon-containing compounds with no more than two carbon atoms, indicate how the following syntheses could be carried out:

a. $CH_3CH_2CH_2CH_2Br \longrightarrow CH_3CH_2CH_2\overset{O}{\overset{\|}{C}}OH$

b. $CH_3CH_2CH_2CH_2Br \longrightarrow CH_3CH_2CH_2CH_2CH_2\overset{O}{\overset{\|}{C}}H$

c. $CH_3CH = CHCH_3 \longrightarrow CH_3CH = CCH_2CH_3$
with CH_3 substituent

d. $CH_3CH = CHCH_3 \longrightarrow CH_3CHCCH_3$
with O (double bond) and CH_3 substituent

46. Which of the following alkyl halides could be successfully used to form a Grignard reagent?

a. $HOCH_2CH_2CH_2CH_2Br$ b. $BrCH_2CH_2CH_2\overset{O}{\overset{\|}{C}}OH$ c. $CH_3\underset{\underset{CH_3}{|}}{N}CH_2CH_2CH_2Br$ d. $H_2NCH_2CH_2CH_2Br$

47. Starting with (R)-1-deuterio-1-propanol, how could you prepare:
a. (S)-1-deuterio-1-propanol?   b. (S)-1-deuterio-1-methoxypropane?   c. (R)-1-deuterio-1-methoxypropane?

48. What alkenes would you expect to be obtained from the acid-catalyzed dehydration of 1-hexanol?

49. Give the product of each of the following reactions:

a. $CH_3\underset{\underset{CH_3}{|}}{\overset{\overset{CH_3}{|}}{C}}OCH_2CH_3 + HBr \overset{\Delta}{\longrightarrow}$

b. $CH_3\underset{\underset{CH_3}{|}}{C}HCH_2OCH_3 + HI \overset{\Delta}{\longrightarrow}$

c. $CH_3CH_2\underset{\underset{OHCH_3}{|}}{C}H\overset{\overset{CH_3}{|}}{C}CH_3 \overset{H_2SO_4}{\underset{\Delta}{\longrightarrow}}$

d. cyclohexane-$CH_2CH_2OH \overset{H_2CrO_4}{\longrightarrow}$

e. spiro epoxide $\overset{H^+}{\underset{CH_3OH}{\longrightarrow}}$

f. spiro epoxide $\overset{CH_3O^-}{\underset{CH_3OH}{\longrightarrow}}$

g. $\underset{H_3C}{\overset{H}{}}$cyclohexane$\overset{H}{\underset{OH}{}}$ $\overset{1.\ TsCl/pyridine}{\underset{2.\ NaC\equiv N}{\longrightarrow}}$

h. cyclohexene-Cl $\overset{(CH_3CH_2CH_2)_2CuLi}{\longrightarrow}$

i. cyclohexane-$\underset{\underset{}{}}{\overset{\overset{OH}{|}}{C}HCH_3} \overset{H_2CrO_4}{\longrightarrow}$

50. When heated with H_2SO_4, both 3,3-dimethyl-2-butanol and 2,3-dimethyl-2-butanol are dehydrated to form 2,3-dimethyl-2-butene. Which alcohol dehydrates more rapidly?

51. Show how bromocyclohexane can be converted into the following compounds:

a. cyclohexanone
b. cyclohexane with OH, OH (cis)
c. cyclohexane with OH, CH_3
d. cyclohexene with CH_3

52. Propose a mechanism for the following reaction:

$$CH_3\underset{\underset{Cl}{|}}{CH}\overset{O}{\overset{}{CH}}-CH_2 + CH_3O^- \overset{}{\underset{CH_3OH}{\longrightarrow}} CH_3\overset{O}{\overset{}{CH}}-CHCH_2OCH_3 + Cl^-$$

53. When deuterated phenanthrene oxide undergoes a rearrangement in water to form a phenol, 81% of the deuterium is retained in the product.

a. What percentage of the deuterium will be retained if an NIH shift occurs?
b. What percentage of the deuterium will be retained if an NIH shift does not occur?

54. Explain why (S)-2-butanol forms a racemic mixture when it is heated in sulfuric acid.

55. Using the given starting material, any necessary inorganic reagents, and any carbon-containing compounds with no more than two carbon atoms, indicate how the following syntheses could be carried out:

a.

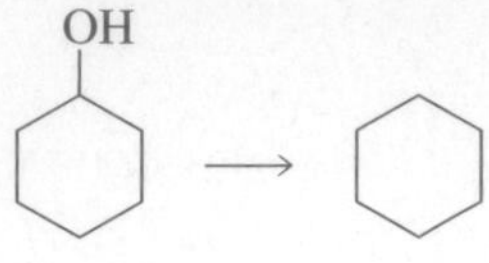

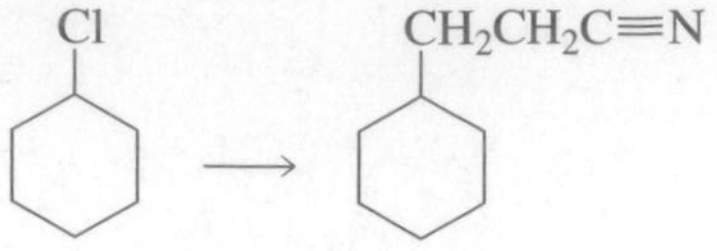

b.

c.

d.

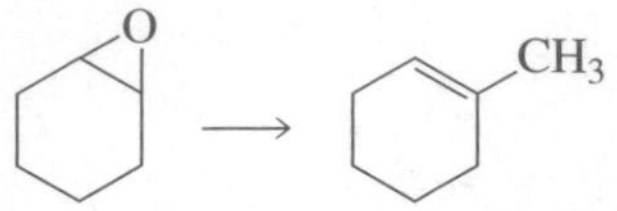

e. $CH_3CH_2C\equiv CH \longrightarrow CH_3CH_2C\equiv CCH_2CH_2OH$

f. $CH_3CHCH_2OH \longrightarrow CH_3CHCH_2CH_2CH_2OH$
   with CH_3 substituents

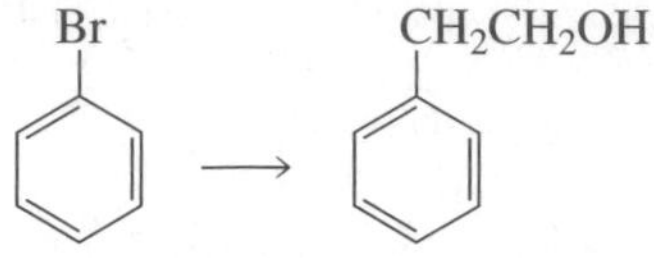

56. When 3-methyl-2-butanol is heated with concentrated HBr, a rearranged product is obtained. When 2-methyl-1-propanol reacts under the same conditions, a rearranged product is not obtained. Explain.

57. When the following seven-membered ring alcohol is dehydrated, three alkenes are formed:

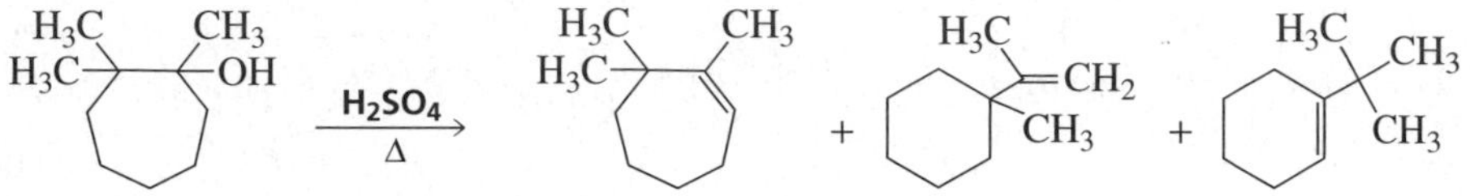

Propose a mechanism for their formation.

58. How could you synthesize isopropyl propyl ether, using isopropyl alcohol as the only carbon-containing reagent?

59. Ethylene oxide reacts readily with HO^- because of the strain in the three-membered ring. Explain why cyclopropane, a compound with approximately the same amount of strain, does not react with HO^-.

60. Which of the following ethers would be obtained in greatest yield directly from alcohols?

$$CH_3OCH_2CH_2CH_3 \qquad CH_3CH_2OCH_2CH_2CH_3 \qquad CH_3CH_2OCH_2CH_3 \qquad CH_3OCCH_3$$

(the fourth structure has two CH_3 groups on the central carbon)

61. Propose a mechanism for each of the following reactions:

a. $HOCH_2CH_2CH_2CH_2OH \xrightarrow[\Delta]{H^+}$ $+ H_2O$

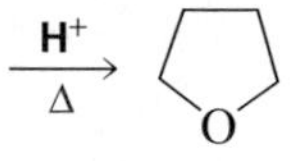

b. 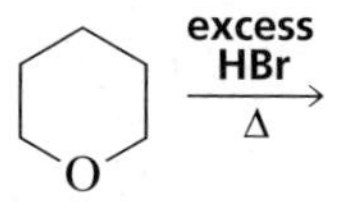 $\xrightarrow[\Delta]{\textbf{excess } \textbf{HBr}} BrCH_2CH_2CH_2CH_2CH_2Br + H_2O$

62. Indicate how each of the following compounds could be prepared, using the given starting material:

a.

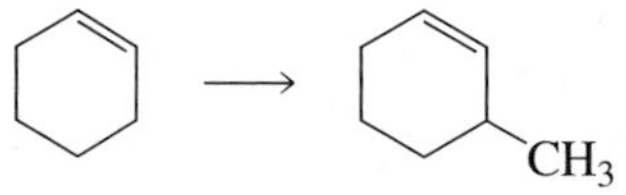

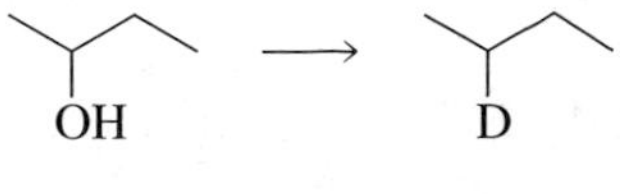

b.

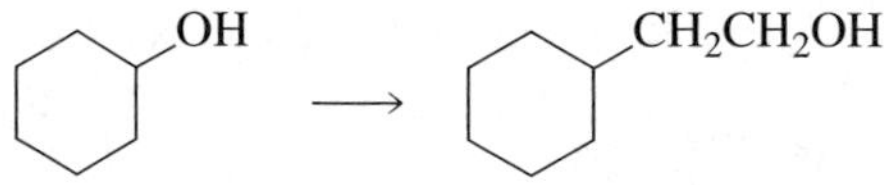

c. $CH_3CH_2CH=CH_2 \longrightarrow CH_3CH_2CH_2CH_2CH_2CH_2CH_2CH_3$

d.

e. $CH_3CCH_2CH_2CH_3 \longrightarrow CH_3CHCHCH_2CH_3$
   (with CH_3 and OH on left; CH_3 and Br on right)

63. Triethylene glycol is one of the products obtained from the reaction of ethylene oxide and hydroxide ion. Propose a mechanism for its formation.

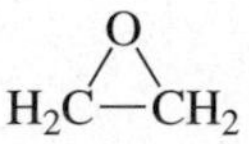

$$H_2C-CH_2 + HO^- \longrightarrow HOCH_2CH_2OCH_2CH_2OCH_2CH_2OH$$
triethylene glycol

64. Give the major product expected from the reaction of 2-ethyloxirane with each of the following reagents:
a. 0.1 M HCl
b. CH_3OH/H^+
c. ethyl magnesium bromide in ether, followed by 0.1 M HCl
d. 0.1 M NaOH
e. CH_3OH/CH_3O^-

65. When ethyl ether is heated with excess HI for several hours, the only organic product obtained is ethyl iodide. Explain why ethyl alcohol is not obtained as a product.

66. a. Propose a mechanism for the following reaction:

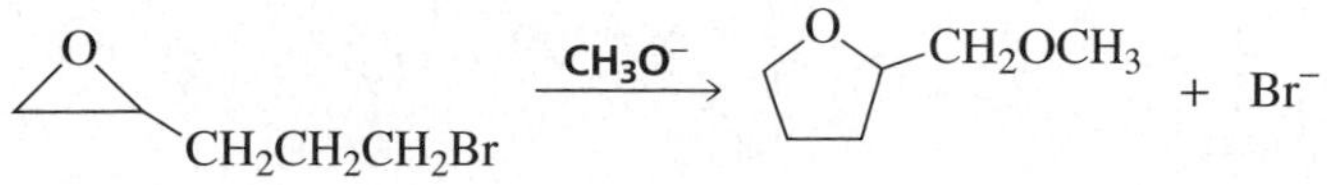

b. A small amount of a product containing a six-membered ring is also formed. Give the structure of that product.
c. Why is so little six-membered ring product formed?

67. Identify A through H.

$$CH_3Br \xrightarrow[B]{A} C \xrightarrow[2.\ E]{1.\ D} CH_3CH_2CH_2OH \xrightarrow[2.\ G \quad 3.\ H]{1.\ F} CH_3CH_2CH_2OCH_2CH_2OH$$

68. Greg Nard added an equivalent of 3,4-epoxy-4-methylcyclohexanol to an ether solution of methyl magnesium bromide and then added dilute hydrochloric acid. He expected that the product would be 1,2-dimethyl-1,4-cyclohexanediol. He did not get any of the expected product. What product did he get?

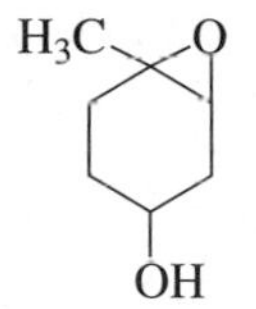

3,4-epoxy-4-methyl-cyclohexanol

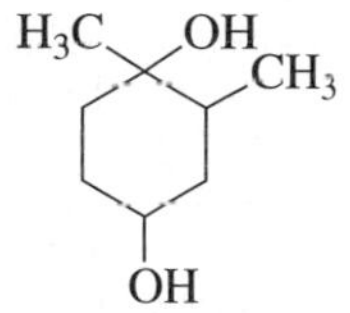

1,2-dimethyl-1,4-cyclohexanediol

69. An ion with a positively charged nitrogen atom in a three-membered ring is called an aziridinium ion. The following aziridinium ion reacts with sodium methoxide to form compounds A and B:

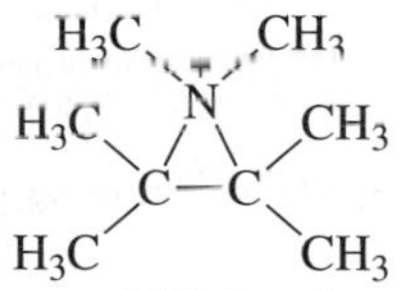

an aziridinium ion

If a small amount of aqueous Br_2 is added to A, the reddish color of Br_2 persists, but the color disappears when Br_2 is added to B. When the aziridinium ion reacts with methanol, only A is formed. Identify A and B.

70. Dimerization is a side reaction that occurs during the preparation of a Grignard reagent. Propose a mechanism that accounts for the formation of the dimer.

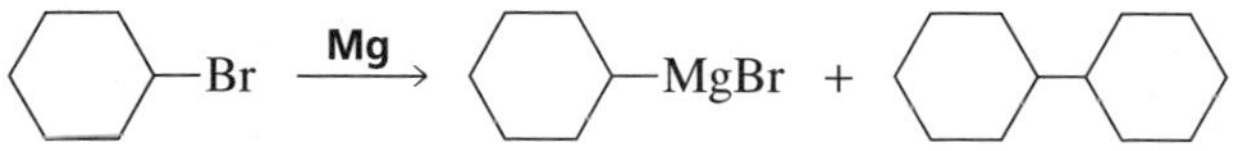

a dimer

71. Propose a mechanism for each of the following reactions:

a.

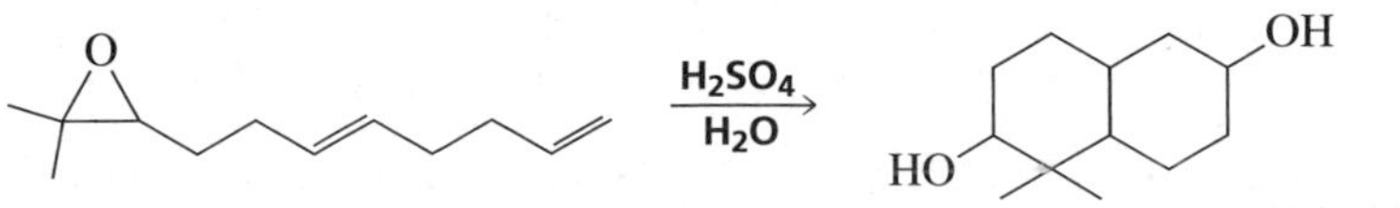

c.

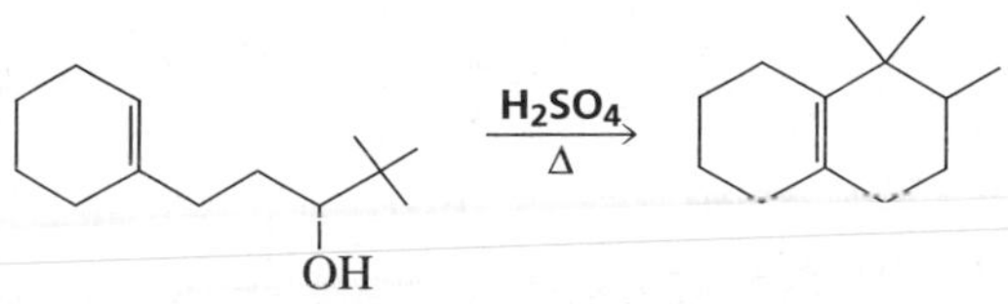

b.

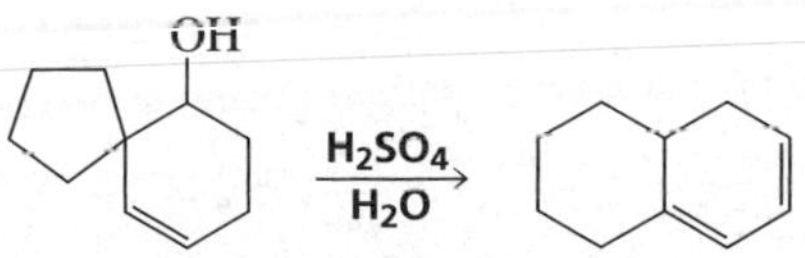

72. One method used to synthesize an epoxide is to treat an alkene with an aqueous solution of Br_2, followed by an aqueous solution of sodium hydroxide.
    a. Propose a mechanism for the conversion of cyclohexene into cyclohexene oxide by this method.
    b. How many products are formed when cyclohexene oxide reacts with methoxide ion in methanol? Draw their structures.

73. Which of the following reactions occurs most rapidly? Why?

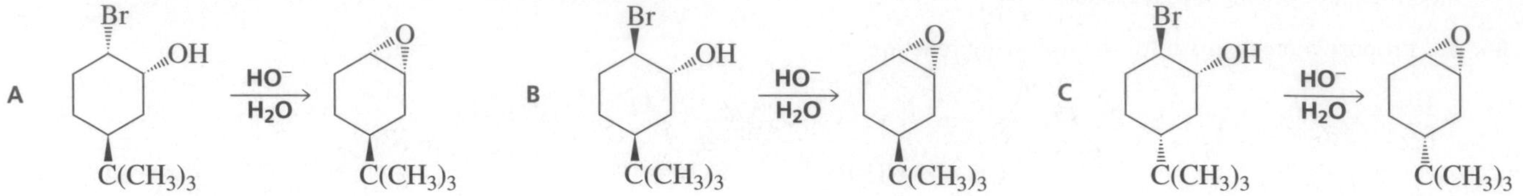

74. When bromobenzene reacts with propene in a Heck reaction, two constitutional isomers are obtained as products. Give the structures of the products and explain why two products are obtained.

75. The following reaction takes place several times faster than the rate of reaction of 2-chlorobutane with HO^-.

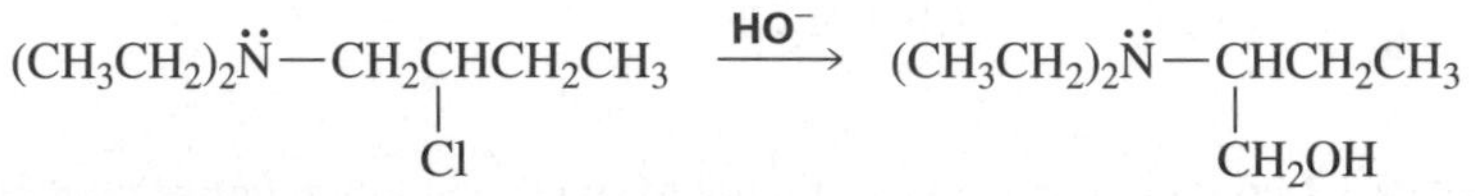

$$(CH_3CH_2)_2\ddot{N}-CH_2CHCH_2CH_3 \xrightarrow{HO^-} (CH_3CH_2)_2\ddot{N}-CHCH_2CH_3$$
$$\underset{Cl}{|} \qquad\qquad \underset{CH_2OH}{|}$$

    a. Explain the enhanced reaction rate.
    b. Explain why the OH group in the product is not bonded to the carbon that was bonded to the Cl group in the reactant.

76. A vicinal diol has OH groups on adjacent carbons. The dehydration of a vicinal diol is accompanied by a rearrangement called the pinacol rearrangement. Propose a mechanism for this reaction.

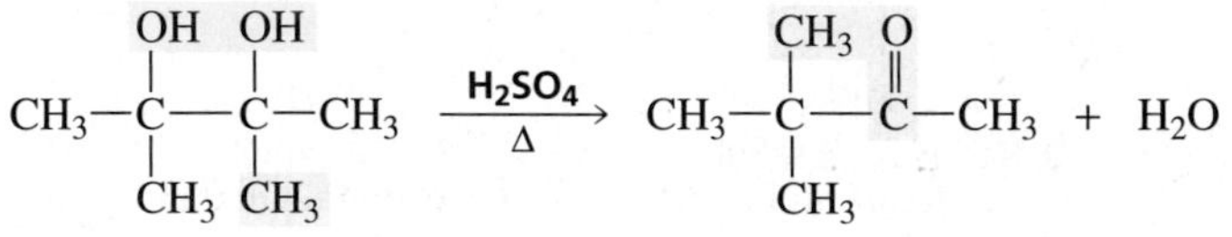

77. Although 2-methyl-1,2-propanediol is an unsymmetrical vicinal diol, only one product is obtained when it is dehydrated in the presence of acid.
    a. What is this product?
    b. Why is only one product formed?

78. What product is obtained when the following vicinal diol is heated in an acidic solution?

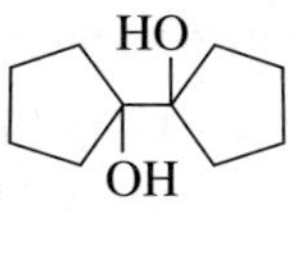

79. Two stereoisomers are obtained from the reaction of cyclopentene oxide and dimethylamine. The R,R-isomer is used in the manufacture of eclanamine, an antidepressant. What other isomer is obtained?

80. Propose a mechanism for each of the following reactions:

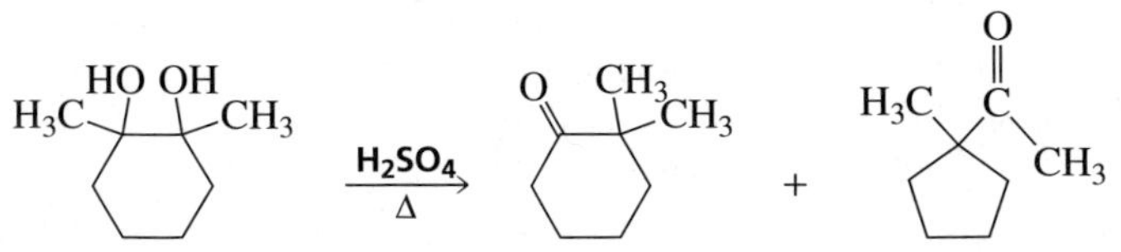

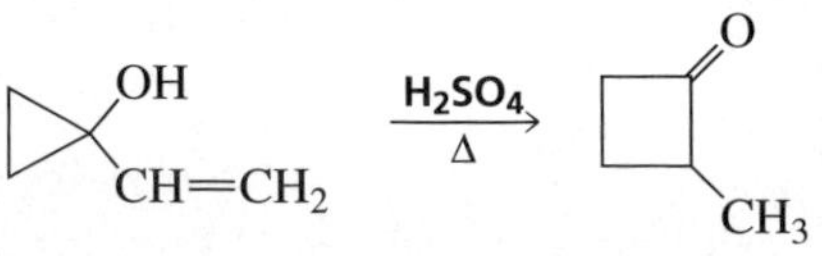

CHAPTER 11

Radicals • Reactions of Alkanes

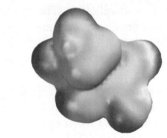

vitamin C

vitamin E

BUILDING ON FUNDAMENTALS

SECTION 11.3 Alkyl groups stabilize radicals in the same way that they stabilize carbocations (4.2).

SECTION 11.5 The Hammond postulate (1.3) and $\Delta H°$ values (3.7) are used to explain why a bromine radical is more selective than a chlorine radical.

SECTION 11.6 A peroxide reverses the "order of addition" of HBr to an alkene (4.4) because it causes a bromine radical rather than H^+ to be the electrophile.

SECTION 11.6 In a radical addition reaction to an alkene, the bromine radical (like other electrophiles) adds to the sp^2 carbon bonded to the greater number of hydrogens (4.4).

SECTION 11.7 When a reactant that does not have an asymmetric center undergoes a reaction that forms a product with one asymmetric center, the product will be a racemic mixture (5.19).

We have seen that there are three classes of hydrocarbons: *alkanes*, which contain only single bonds; *alkenes*, which contain double bonds; and *alkynes*, which contain triple bonds. Because **alkanes** do not contain any double or triple bonds, they are called **saturated hydrocarbons**, meaning they are saturated with hydrogen. A few examples of alkanes are

$CH_3CH_2CH_2CH_3$
butane

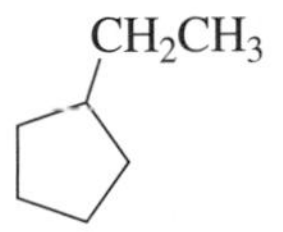

ethylcyclopentane

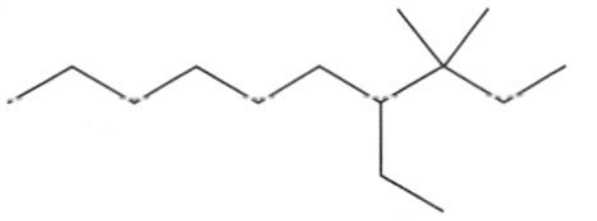

4-ethyl-3,3-dimethyldecane

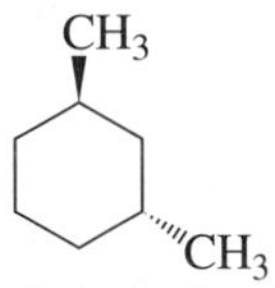

trans-1,3-dimethyl-cyclohexane

Alkanes are widespread both on Earth and on other planets. The atmospheres of Jupiter, Saturn, Uranus, and Neptune contain large quantities of methane (CH_4), the smallest alkane, an odorless and flammable gas. In fact, the blue colors of Uranus and Neptune are the result of methane in their atmospheres. Alkanes on Earth are found in natural gas and petroleum, which are formed by the decomposition of plant and animal material that have been buried for long periods of time in the Earth's crust, where oxygen is scarce. Natural gas and petroleum, therefore, are known as *fossil fuels*.

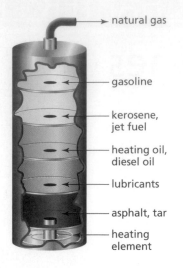

natural gas

gasoline

kerosene,
jet fuel

heating oil,
diesel oil

lubricants

asphalt, tar

heating
element

Natural gas is approximately 75% methane. The remaining 25% is composed of other small alkanes such as ethane, propane, and butane. In the 1950s, natural gas replaced coal as the main energy source for domestic and industrial heating in many parts of the United States.

Petroleum is a complex mixture of alkanes and cycloalkanes that can be separated into fractions by distillation. The fraction that boils off at the lowest temperature (hydrocarbons containing three and four carbons) is a gas that can be liquefied under pressure. This gas is used as a fuel for cigarette lighters, camp stoves, and barbecues. The fraction that boils at somewhat higher temperatures (hydrocarbons containing 5 to 11 carbons) is gasoline; the next fraction (9 to 16 carbons) includes kerosene and jet fuel. The fraction with 15 to 25 carbons is used for heating oil and diesel oil, and the highest-boiling fraction is used for lubricants and greases. The nonpolar nature of these compounds gives them their oily feel. After distillation, a nonvolatile residue called asphalt or tar is left behind.

The 5- to 11-carbon fraction that is used for gasoline is, in fact, a poor fuel for internal combustion engines and requires a process known as catalytic cracking to become a high-performance gasoline. Catalytic cracking converts straight-chain

OCTANE NUMBER

When poor-quality fuels are used in an engine, combustion can begin before the spark plug fires. A pinging or knocking may then be heard in the running engine. As the quality of the fuel improves, the engine is less likely to knock. The quality of a fuel is indicated by its octane number. Straight-chain hydrocarbons have low octane numbers and make poor fuels. Heptane, for example, with an arbitrarily assigned octane number of 0, causes engines to knock badly. Branched-chain alkanes have more hydrogens on primary carbons. These are the C—H bonds that require the most energy to break and, therefore, make combustion more difficult to initiate, thereby reducing knocking. Consequently, branched-chain alkanes have high octane numbers. 2,2,4-Trimethylpentane, for example, does not cause

knocking and has arbitrarily been assigned an octane number of 100.

$$CH_3CH_2CH_2CH_2CH_2CH_2CH_3$$
heptane
octane number = 0

$$CH_3CCH_2CHCH_3$$ with CH_3, CH_3, CH_3 substituents
2,2,4-trimethylpentane
octane number = 100

The octane number of a gasoline is determined by comparing its knocking with the knocking of various mixtures of heptane and 2,2,4-trimethylpentane. The octane number given to the gasoline corresponds to the percent of 2,2,4-trimethylpentane in the matching mixture. The term "octane number" originated from the fact that 2,2,4-trimethylpentane contains eight carbons.

FOSSIL FUELS: A PROBLEMATIC ENERGY SOURCE

Modern society faces three major problems as a consequence of our dependence on fossil fuels for energy. First, these fuels are a nonrenewable resource and the world's supply is continually decreasing. Second, a group of Middle Eastern and South American countries controls a large portion of the world's supply of petroleum. These countries have formed a cartel called the *Organization of Petroleum Exporting Countries* (*OPEC*) that controls both the supply and the price of crude oil. Political instability in any OPEC country can seriously affect the world oil supply. Third, burning fossil fuels increases the concentration of CO_2 in the atmosphere; burning coal increases the concentration of both CO_2 and SO_2.

Scientists have established experimentally that atmospheric SO_2 causes "acid rain," a threat to the Earth's plants and, therefore, to our food and oxygen supplies (Section 1.17).

Since 1958, the concentration of atmospheric CO_2 at Mauna Loa, Hawaii, has been periodically measured. The concentration has increased 20% since the first measurements were taken, causing scientists to predict an increase in the Earth's temperature as a result of the absorption of infrared radiation by CO_2 (the *greenhouse effect*). A steady increase in the temperature of the Earth would have devastating consequences, including the formation of new deserts, massive crop failure, and the melting of glaciers with a concomitant rise in sea level. Clearly, what we need is a renewable, nonpolitical, nonpolluting, and economically affordable source of energy.

hydrocarbons that are poor fuels into branched-chain compounds that are high-performance fuels. Originally, cracking (also called *pyrolysis*) required heating the gasoline to very high temperatures in order to obtain hydrocarbons with three to five carbons. Modern cracking methods use catalysts to accomplish the same thing at much lower temperatures.

11.1 Alkanes Are Unreactive Compounds

The double and triple bonds of alkenes and alkynes are composed of strong σ bonds and weaker π bonds. We have seen that the reactivity of *alkenes* and *alkynes* is the result of an electrophile being attracted to the cloud of electrons that constitutes the π bond (Sections 3.6 and 6.5).

Alkanes have only strong σ bonds. In addition, the electrons in the C—H and C—C σ bonds are shared equally by the bonding atoms, so none of the atoms in an alkane has any significant charge. This means that neither nucleophiles nor electrophiles are attracted to them. Alkanes, therefore, are relatively unreactive compounds. The failure of alkanes to undergo reactions prompted early organic chemists to call them *paraffins*, from the Latin *parum affinis*, which means "little affinity" (for other compounds).

11.2 Chlorination and Bromination of Alkanes

Alkanes do react with chlorine (Cl_2) or bromine (Br_2) to form alkyl chlorides or alkyl bromides. These **halogenation reactions** take place only at high temperatures or in the presence of light. (Irradiation with light is symbolized by *hv*.) They are the only reactions that alkanes undergo, with the exception of **combustion** (burning), a reaction with oxygen that takes place at high temperatures and converts alkanes to carbon dioxide and water.

$$CH_4 + Cl_2 \xrightarrow[\substack{\text{or} \\ h\nu}]{\Delta} \underset{\text{chloromethane}}{CH_3Cl} + HCl$$

$$CH_3CH_3 + Br_2 \xrightarrow[\substack{\text{or} \\ h\nu}]{\Delta} \underset{\text{bromethane}}{CH_3CH_2Br} + HBr$$

When a bond breaks so that both of its electrons stay with one of the atoms, the process is called **heterolytic bond cleavage** or **heterolysis**. When a bond breaks so that each of the atoms retains one of the bonding electrons, the process is called **homolytic bond cleavage**, or **homolysis**. Notice that an arrowhead with one barb—sometimes called a fishhook—signifies the movement of one electron, whereas an arrowhead with two barbs signifies the movement of two electrons (Section 3.6).

| heterolytic bond cleavage | | arrowhead has 2 barbs |

$$A{-}B \longrightarrow A^+ + B^-$$

| homolytic bond cleavage | | arrowhead has one barb |

$$A{-}B \longrightarrow A\cdot + \cdot B$$

The mechanism for the halogenation of an alkane is well understood. As an example, let's look at the mechanism for the monochlorination of methane.

Mechanism for the monochlorination of methane

$$:\!\overset{..}{\underset{..}{Cl}}\!\!-\!\!\overset{..}{\underset{..}{Cl}}\!: \quad \xrightarrow[\text{or} \atop h\nu]{\Delta} \quad 2\ :\!\overset{..}{\underset{..}{Cl}}\!\cdot \quad \boxed{\text{initiation step}}$$

homolytic cleavage

$$:\!\overset{..}{\underset{..}{Cl}}\!\cdot \ + \ H\!-\!CH_3 \ \longrightarrow \ H\overset{..}{\underset{..}{Cl}}\!: \ + \ \cdot CH_3$$

a methyl radical

$$\cdot CH_3 \ + \ :\!\overset{..}{\underset{..}{Cl}}\!\!-\!\!\overset{..}{\underset{..}{Cl}}\!: \ \longrightarrow \ CH_3Cl \ + \ :\!\overset{..}{\underset{..}{Cl}}\!\cdot$$

propagation steps

$$:\!\overset{..}{\underset{..}{Cl}}\!\cdot \ + \ :\!\overset{..}{\underset{..}{Cl}}\!\cdot \ \longrightarrow \ Cl_2$$

$$\cdot CH_3 \ + \ \cdot CH_3 \ \longrightarrow \ CH_3CH_3$$

$$:\!\overset{..}{\underset{..}{Cl}}\!\cdot \ + \ \cdot CH_3 \ \longrightarrow \ CH_3Cl$$

termination steps

- Heat or light supplies the energy required to break the Cl—Cl (or Br—Br) bond homolytically. This is the **initiation step** of the reaction because it creates radicals. A **radical** (often called a **free radical**) is a species containing an atom with an unpaired electron. A radical is highly reactive because acquiring an electron will complete its octet.

- The chlorine radical formed in the initiation step removes a hydrogen atom from the alkane (in this case methane), forming HCl and a methyl radical.

- The methyl radical removes a chlorine atom from Cl_2, forming chloromethane and another chlorine radical, which can remove a hydrogen atom from another molecule of methane. Steps 2 and 3 are **propagation steps** because the radical created in the first propagation step reacts in the second propagation step to produce the radical that can repeat the first propagation step. A propagation step is one that propagates the chain. Thus, the two propagation steps are repeated over and over. The first propagation step is the rate-determining step of the overall reaction.

- Any two radicals in the reaction mixture can combine to form a molecule in which all the electrons are paired. The combination of two radicals is called a **termination step** because it helps bring the reaction to an end by decreasing the number of radicals available to propagate the reaction. Any two radicals present in the reaction mixture can combine in a termination step, so radical reactions produce a mixture of products.

Student Tutorial:
Radical chain reaction

Radical reactions have initiation, propagation, and termination steps.

The radical chlorination of alkanes other than methane has the same mechanism. Because the reaction has radical intermediates and repeating propagation steps, it is called a **radical chain reaction**. This particular radical chain reaction—the reaction of an alkane with chlorine (or bromine) to form an alkyl halide—is called a **radical substitution reaction** because it substitutes a halogen for one of the hydrogens of the alkane.

In order to maximize the amount of monohalogenated product obtained, a radical substitution reaction should be carried out in the presence of excess alkane. Excess alkane in the reaction mixture increases the probability that the halogen radical will collide with a molecule of alkane rather than with a molecule of alkyl halide—even toward the end of the reaction, by which time a considerable amount of alkyl halide will have been formed. If the halogen radical removes a hydrogen from a molecule of alkyl halide rather than from a molecule of alkane, a dihalogenated product will be obtained.

$$Cl\cdot \ + \ CH_3Cl \ \longrightarrow \ \cdot CH_2Cl \ + \ HCl$$

$$\cdot CH_2Cl \ + \ Cl_2 \ \longrightarrow \ CH_2Cl_2 \ + \ Cl\cdot$$

a dihalogenated compound

Bromination of alkanes has the same mechanism as chlorination.

Mechanism for the monobromination of ethane

$$Br \overbrace{} Br \xrightarrow[\text{or}]{\Delta}_{h\nu} 2\ Br\cdot \qquad \boxed{\text{initiation step}}$$

$$Br\cdot\ +\ H\overbrace{}CH_2CH_3 \longrightarrow CH_3\dot{C}H_2\ +\ HBr$$

$$CH_3\dot{C}H_2\ +\ Br\overbrace{}Br \longrightarrow CH_3CH_2Br\ +\ Br\cdot$$

$$\left. \right\} \boxed{\text{propagation steps}}$$

$$Br\cdot\ +\ Br\cdot \longrightarrow Br_2$$

$$CH_3\dot{C}H_2\ +\ CH_3\dot{C}H_2 \longrightarrow CH_3CH_2CH_2CH_3$$

$$CH_3\dot{C}H_2\ +\ Br\cdot \longrightarrow CH_3CH_2Br$$

$$\left. \right\} \boxed{\text{termination steps}}$$

PROBLEM 1

Show the initiation, propagation, and termination steps for the monochlorination of cyclohexane.

PROBLEM 2

Write the mechanism for the formation of tetrachloromethane, CCl_4, from the reaction of methane with $Cl_2 + h\nu$.

11.3 Radical Stability Depends on the Number of Alkyl Groups Attached to the Carbon with the Unpaired Electron

Like carbocations, radicals are stabilized by electron-donating alkyl groups. Therefore, the relative stabilities of **primary, secondary,** and **tertiary alkyl radicals** follow the same order as the relative stabilities of primary, secondary, and tertiary carbocations.

Alkyl radical stability: $3° > 2° > 1°$

relative stabilities of alkyl radicals

tertiary radical   secondary radical   primary radical   methyl radical

The differences in the relative stabilities of the radicals are much smaller than the differences in the relative stabilities of the carbocations, however, because alkyl groups do not stabilize radicals as well as they stabilize carbocations. Alkyl groups stabilize both carbocations and radicals by hyperconjugation (Section 4.2). Stabilization of a carbocation results from overlap between a filled $C-H$ or $C-C$ bond and an empty p orbital: a two-electron system (Figure 11.1a). In contrast, stabilization of a radical results from overlap between a filled $C-H$ or $C-C$ bond and a p orbital that contains an electron: a three-electron system (Figure 11.1b). In the two-electron system, both electrons are in a bonding molecular orbital, but in the three-electron system, one of the electrons has to go into an antibonding molecular orbital. The three-electron system is still stabilizing overall because there are more electrons in the bonding MO than in the antibonding MO, but it is not as stabilizing as the two-electron system, which does not have an electron in the antibonding MO. Consequently, alkyl groups stabilize carbocations about 5–10 times better than they stabilize radicals.

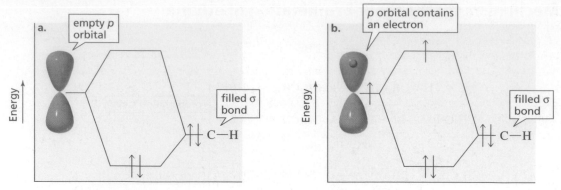

▲ **Figure 11.1**
Molecular orbital diagrams showing the stabilization achieved when the electrons of a C—H bond overlap with (a) an empty *p* orbital, (b) a *p* orbital that contains one electron.

PROBLEM 3◆

a. Which of the hydrogens in the following structure is the easiest for a chlorine radical to remove?

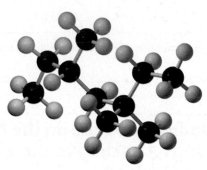

b. How many secondary hydrogens does the structure have?

11.4 The Distribution of Products Depends on Probability and Reactivity

Two different alkyl halides are obtained from the monochlorination of butane. Substitution of a hydrogen bonded to one of the terminal carbons produces 1-chlorobutane, whereas substitution of a hydrogen bonded to one of the internal carbons forms 2-chlorobutane.

$$CH_3CH_2CH_2CH_3 + Cl_2 \xrightarrow{h\nu} CH_3CH_2CH_2CH_2Cl + CH_3CH_2\overset{\displaystyle Cl}{\overset{|}{C}}HCH_3 + HCl$$

butane	1-chlorobutane	2-chlorobutane	
	expected = 60%	expected = 40%	
	experimental = 29%	experimental = 71%	

The expected (statistical) distribution of products is 60% 1-chlorobutane and 40% 2-chlorobutane because six of butane's 10 hydrogens can be substituted to form 1-chlorobutane, whereas only four can be substituted to form 2-chlorobutane. This assumes, however, that all of the C—H bonds in butane are equally easy to break. If that were the case, the relative amounts of the two products would depend only on the probability of a chlorine radical colliding with a primary hydrogen, compared with the probability of its colliding with a secondary hydrogen. When we carry out the reaction in the laboratory and analyze the product, however, we find 29% 1-chlorobutane and 71% 2-chlorobutane. Therefore, probability alone does not explain the product distribution.

When a chlorine radical reacts with butane, it can remove a hydrogen atom from an internal carbon, forming a secondary alkyl radical that leads to the formation of 2-chlorobutane, or it can remove a hydrogen atom from a terminal carbon, forming a primary alkyl radical that leads to the formation of 1-chlorobutane. Because more 2-chlorobutane is obtained than expected and *the rate-determining step of the overall reaction is the removal of the hydrogen atom*, we can conclude that a hydrogen atom is easier to remove from a secondary carbon than from a primary carbon, causing 2-chlorobutane to be formed faster.

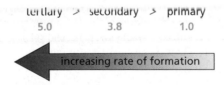

We should not be surprised that it is easier to remove a hydrogen atom from a secondary carbon to form a secondary radical than to remove it from a primary carbon to form a primary radical, because a secondary radical is more stable than a primary radical. The more stable the radical, the more easily it is formed, because the stability of the radical is reflected in the stability of the transition state leading to its formation (Section 4.3).

After experimentally determining the amount of each chlorination product obtained from various hydrocarbons, chemists were able to conclude that *at room temperature* it is 5.0 times easier for a chlorine radical to remove a hydrogen atom from a tertiary carbon than from a primary carbon, and it is 3.8 times easier to remove a hydrogen atom from a secondary carbon than from a primary carbon (see Problem 33). The precise ratios differ at different temperatures.

relative rates of alkyl radical formation by a chlorine radical at room temperature

tertiary	>	secondary	>	primary
5.0		3.8		1.0

increasing rate of formation

To determine the relative amounts of different products obtained from radical chlorination of an alkane, both *probability* (the number of hydrogens that can be removed that will lead to the formation of the particular product) and *reactivity* (the relative rate at which a particular hydrogen is removed) must be taken into account. When both factors are considered, the calculated amounts of 1-chlorobutane and 2-chlorobutane agree with the amounts obtained experimentally.

Both probability and reactivity must be taken into account when calculating the relative amounts of products.

relative amount of 1-chlorobutane

number of hydrogens × reactivity
$$6 \times 1.0 = 6.0$$

$$\text{percent yield} = \frac{6.0}{21} = 29\%$$

relative amount of 2-chlorobutane

number of hydrogens × reactivity
$$4 \times 3.8 = 15$$

$$\text{percent yield} = \frac{15}{21} = 71\%$$

The percent yield of each alkyl chloride is calculated by dividing the relative amount of the particular product by the sum of the relative amounts of all the alkyl chloride products (6 + 15 = 21).

Radical monochlorination of 2,2,5-trimethylhexane results in the formation of five monochlorination products. Because the relative amounts of the five alkyl chlorides total 35 (9.0 + 7.6 + 7.6 + 5.0 + 6.0 = 35), the percent yield of each product can be calculated as follows:

$$
\underset{\substack{\text{2,2,5-trimethylhexane}}}{\text{CH}_3\text{CCH}_2\text{CH}_2\text{CHCH}_3} + \text{Cl}_2 \xrightarrow{\Delta} \underset{\substack{9 \times 1.0 = 9.0 \\ \frac{9.0}{35} = 26\%}}{\text{CH}_2\text{Cl} \quad \text{CH}_3 \\ \text{CH}_3\text{CCH}_2\text{CH}_2\text{CHCH}_3 \\ \text{CH}_3} + \underset{\substack{2 \times 3.8 = 7.6 \\ \frac{7.6}{35} = 22\%}}{\text{CH}_3 \quad \text{CH}_3 \\ \text{CH}_3\text{C}\!-\!\text{CHCH}_2\text{CHCH}_3 \\ \text{CH}_3 \;\; \text{Cl}} +
$$

$$
\underset{\substack{2 \times 3.8 = 7.6 \\ \frac{7.6}{35} = 22\%}}{\text{CH}_3 \quad \text{CH}_3 \\ \text{CH}_3\text{CCH}_2\text{CHCHCH}_3 \\ \text{CH}_3 \;\; \text{Cl}} + \underset{\substack{1 \times 5.0 = 5.0 \\ \frac{5.0}{35} = 14\%}}{\text{CH}_3 \quad \text{CH}_3 \\ \text{CH}_3\text{CCH}_2\text{CH}_2\text{CCH}_3 \\ \text{CH}_3 \quad \text{Cl}} + \underset{\substack{6 \times 1.0 = 6.0 \\ \frac{6.0}{35} = 17\%}}{\text{CH}_3 \quad \text{CH}_3 \\ \text{CH}_3\text{CCH}_2\text{CH}_2\text{CHCH}_2\text{Cl} \\ \text{CH}_3} + \text{HCl}
$$

Because radical chlorination of an alkane can yield several different monosubstitution products as well as products that contain more than one chlorine atom, it is not the best method to use to synthesize an alkyl halide. Addition of a hydrogen halide to an alkene (Section 4.1) or conversion of an alcohol to an alkyl halide (Sections 10.1 and 10.2) is a much better way to make an alkyl halide. Radical halogenation of an alkane is nevertheless still a useful reaction because it is the only way to convert an inert alkane into a reactive compound; once the halogen is introduced into the alkane, it can be replaced by a variety of other substituents (Section 10.2).

PROBLEM 4

When 2-methylpropane is monochlorinated in the presence of light at room temperature, 36% of the product is 2-chloro-2-methylpropane and 64% is 1-chloro-2-methylpropane. From these data, calculate how much easier it is to remove a hydrogen atom from a tertiary carbon than from a primary carbon under these conditions.

PROBLEM 5♦

How many alkyl chlorides can be obtained from monochlorination of the following alkanes? Disregard stereoisomers.

a. $\text{CH}_3\text{CH}_2\text{CH}_2\text{CH}_2\text{CH}_3$

b. $\underset{\substack{\text{CH}_3 \qquad \text{CH}_3}}{\text{CH}_3\text{CHCH}_2\text{CH}_2\text{CHCH}_3}$

c. $\underset{\substack{\text{CH}_3}}{\text{CH}_3\text{CHCH}_2\text{CH}_2\text{CH}_3}$

d. (cyclohexane ring)

e. (methylcyclohexane ring with CH_3)

f. (1,2-dimethylcyclohexane ring with CH_3 and CH_3)

g. $\underset{\substack{\text{CH}_3 \quad \text{CH}_3}}{\overset{\substack{\text{CH}_3 \quad \text{CH}_3}}{\text{CH}_3\text{CCH}_2\text{CCH}_3}}$

h. $\underset{\substack{\text{CH}_3 \;\; \text{CH}_3}}{\overset{\substack{\text{CH}_3 \;\; \text{CH}_3}}{\text{CH}_3\text{C}\!-\!\text{CCH}_3}}$

i. $\underset{\substack{\text{CH}_3}}{\overset{\substack{\text{CH}_3 \;\; \text{CH}_3}}{\text{CH}_3\text{CCH}_2\text{CHCH}_3}}$

PROBLEM 6◆

Calculate the percent yield of each product obtained in Problem 5a, b, and c if chlorination is carried out in the presence of light at room temperature.

PROBLEM 7   *SOLVED*

If cyclopentane reacts with more than one equivalent of Cl_2 at a high temperature, how many dichlorocyclopentanes would you expect to obtain as products?

Solution   Seven dichlorocyclopentanes would be obtained as products. Only one isomer is possible for the 1,1-dichloro compound. The 1,2- and 1,3-dichloro compounds have two asymmetric centers. Each has three stereoisomers because the cis isomer is a meso compound and the trans isomer is a pair of enantiomers.

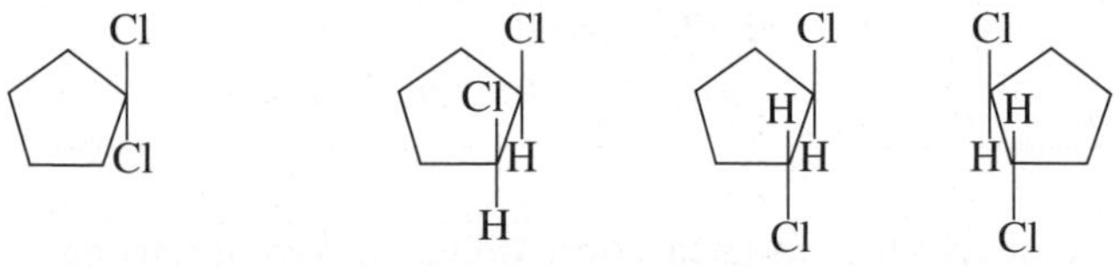

meso compound   enantiomers

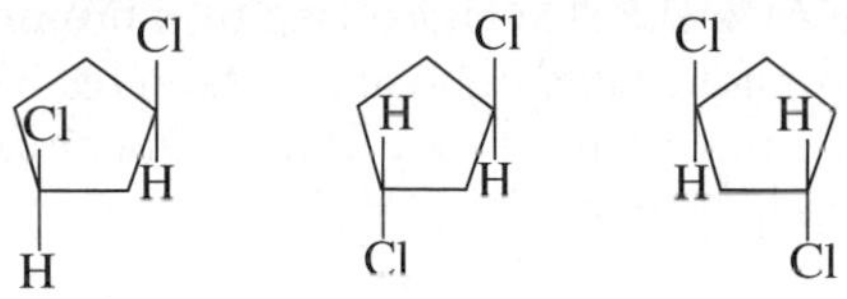

meso compound   enantiomers

11.5 The Reactivity–Selectivity Principle

The relative rates of radical formation by a bromine radical are different from the relative rates of radical formation by a chlorine radical. For example, at 125 °C, a bromine radical removes a hydrogen atom from a tertiary carbon 1600 times faster than from a primary carbon and removes a hydrogen atom from a secondary carbon 82 times faster than from a primary carbon.

relative rates of radical formation by a bromine radical at 125 °C

tertiary > secondary > primary
  1600   82   1

increasing rate of formation

When a bromine radical removes a hydrogen atom, the differences in reactivity are so great that the reactivity factor is vastly more important than the probability factor. For example, radical bromination of butane gives a 98% yield of 2-bromobutane, compared with the 71% yield of 2-chlorobutane obtained when butane is chlorinated (Section 11.4). In other words, bromination is more highly selective than chlorination.

$$CH_3CH_2CH_2CH_3 + Br_2 \xrightarrow{h\nu} CH_3CH_2CH_2CH_2Br + CH_3CH_2\overset{\overset{\displaystyle Br}{|}}{C}HCH_3 + HBr$$

1-bromobutane   **2-bromobutane**
2%   **98%**

Similarly, bromination of 2,2,5-trimethylhexane gives an 82% yield of the product in which bromine replaces the tertiary hydrogen. Chlorination of the same alkane results in only a 14% yield of the tertiary alkyl chloride (Section 11.4).

$$CH_3CCH_2CH_2CHCH_3 + Br_2 \xrightarrow{h\nu} CH_3CCH_2CH_2CCH_3 + HBr$$

2,2,5-trimethylhexane   2-bromo-2,5,5-trimethylhexane
                                                82%

PROBLEM 8◆

Carry out the calculations that predict

a. 2-bromobutane will be obtained in 98% yield.

b. 2-bromo-2,5,5-trimethylhexane will be obtained in 82% yield.

Why are the relative rates of radical formation so different when a bromine radical rather than a chlorine radical is used as the hydrogen-removing reagent? To answer this question, we must compare the $\Delta H°$ values for forming primary, secondary, and tertiary radicals by a chlorine radical and by a bromine radical. These $\Delta H°$ values can be calculated using the bond dissociation energies in Table 3.2 on page 146. (Remember that $\Delta H°$ is equal to the energy of the bond being broken minus the energy of the bond being formed; Section 3.7.)

			$\Delta H°$ (kcal/mol)	$\Delta H°$ (kJ/mol)
Cl· + $CH_3CH_2CH_3$	⟶	$CH_3CH_2\dot{C}H_2$ + HCl	$101 - 103 = -2$	−8
Cl· + $CH_3CH_2CH_3$	⟶	$CH_3\dot{C}HCH_3$ + HCl	$99 - 103 = -4$	−17
Cl· + CH_3CHCH_3	⟶	$CH_3\dot{C}CH_3$ + HCl	$97 - 103 = -6$	−25

			$\Delta H°$ (kcal/mol)	$\Delta H°$ (kJ/mol)
Br· + $CH_3CH_2CH_3$	⟶	$CH_3CH_2\dot{C}H_2$ + HBr	$101 - 87 = 14$	59
Br· + $CH_3CH_2CH_3$	⟶	$CH_3\dot{C}HCH_3$ + HBr	$99 - 87 = 12$	50
Br· + CH_3CHCH_3	⟶	$CH_3\dot{C}CH_3$ + HBr	$97 - 87 = 10$	42

We must also be aware that bromination is a much slower reaction than chlorination. The activation energy for removing a hydrogen atom by a bromine radical has been found experimentally to be about 4.5 times greater than that for removing a hydrogen atom by a chlorine radical. Using the calculated $\Delta H°$ values and the experimental activation energies, we can draw reaction coordinate diagrams for the formation of primary, secondary, and tertiary radicals by a chlorine radical (Figure 11.2a) and by a bromine radical (Figure 11.2b).

Because the reaction of a chlorine radical with an alkane to form a primary, secondary, or tertiary radical is exothermic, the transition states resemble the reactants more than they resemble the products (see the Hammond postulate, Section 4.3). The reactants all have approximately the same energy, so there is only a small difference in the activation energies for removal of a hydrogen atom from a primary, secondary, or tertiary carbon. In contrast, the reaction of a bromine radical with an alkane is endothermic, so the transition states resemble the products more than they resemble the reactants. Because there are significant differences between the energies of the product radicals—depending on whether they are primary, secondary, or tertiary—there are significant differences

A bromine radical is less reactive and more selective than a chlorine radical.

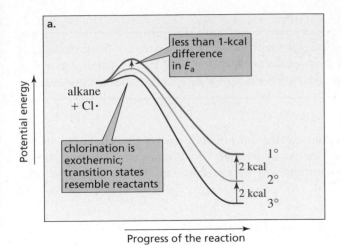

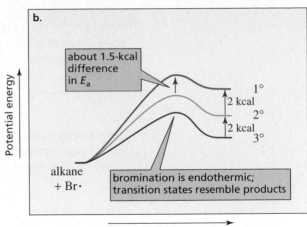

▲ **Figure 11.2**
(a) Reaction coordinate diagrams for the formation of primary, secondary, and tertiary alkyl radicals as a result of removal of a hydrogen atom by a chlorine radical. The transition states have relatively little radical character because they resemble the reactants. (b) Reaction coordinate diagrams for the formation of primary, secondary, and tertiary alkyl radicals as a result of removal of a hydrogen atom by a bromine radical. The transition states have a relatively high degree of radical character because they resemble the products.

between the activation energies. Therefore, a chlorine radical makes primary, secondary, and tertiary radicals with almost equal ease, whereas a bromine radical has a clear preference for forming the easiest-to-form tertiary radical (Figure 11.2). In other words, because a bromine radical is relatively unreactive, it is highly selective about which hydrogen atom it removes. In contrast, the much more reactive chlorine radical is considerably less selective. These observations illustrate the **reactivity–selectivity principle**, which states that *the greater the reactivity of a species, the less selective it will be*.

> **The more reactive a species is, the less selective it will be.**

Because chlorination is relatively nonselective, it is a useful reaction only when there is only one kind of hydrogen in the alkane.

$$\text{(cyclohexane)} + \text{Cl}_2 \xrightarrow{\ h\nu\ } \text{(chlorocyclohexane, Cl)} + \text{HCl}$$

PROBLEM-SOLVING STRATEGY

Planning the Synthesis of an Alkyl Halide

Would chlorination or bromination of methylcyclohexane produce a greater yield of 1-halo-1-methylcyclohexane?

To solve this kind of problem, first draw the structures of the compounds being discussed.

$$\overset{\text{CH}_3}{\text{(methylcyclohexane)}} \xrightarrow[h\nu]{\text{X}_2} \overset{\text{X}\quad\text{CH}_3}{\text{(1-halo-1-methylcyclohexane)}}$$

1-Halo-1-methylcyclohexane is a tertiary alkyl halide, so the question becomes, "Will bromination or chlorination produce a greater yield of a tertiary alkyl halide?" Because bromination is more selective, it will produce a greater yield of the desired compound. Chlorination will form some of the tertiary alkyl halide, but it will also form significant amounts of primary and secondary alkyl halides.

Now continue on to Problem 9.

> ### PROBLEM 9◆
>
> **a.** Would chlorination or bromination produce a greater yield of 1-halo-2,3-dimethylbutane?
> **b.** Would chlorination or bromination produce a greater yield of 2-halo-2,3-dimethylbutane?
> **c.** Would chlorination or bromination be a better way to make 1-halo-2,2-dimethylpropane?

By comparing the $\Delta H°$ values for the sum of the two propagating steps in the monohalogenation of methane, we can understand why alkanes undergo chlorination and bromination but not iodination and why fluorination is too violent a reaction to be useful.

F_2

Cl_2

Br_2

I_2
Halogens

$$F\cdot + CH_4 \longrightarrow \cdot CH_3 + HF \qquad 105 - 136 = -31$$
$$\cdot CH_3 + F_2 \longrightarrow CH_3F + F\cdot \qquad \underline{38 - 115 = -77}$$
$$\Delta H° = -108 \text{ kcal/mol} \quad (\text{or} -452 \text{ kJ/mol})$$

$$Cl\cdot + CH_4 \longrightarrow \cdot CH_3 + HCl \qquad 105 - 103 = \quad 2$$
$$\cdot CH_3 + Cl_2 \longrightarrow CH_3Cl + Cl\cdot \qquad \underline{58 - 84 = -26}$$
$$\Delta H° = \ -24 \text{ kcal/mol} \quad (\text{or} -100 \text{ kJ/mol})$$

$$Br\cdot + CH_4 \longrightarrow \cdot CH_3 + HBr \qquad 105 - 87 = \quad 18$$
$$\cdot CH_3 + Br_2 \longrightarrow CH_3Br + Br\cdot \qquad \underline{46 - 72 = -26}$$
$$\Delta H° = \ -8 \text{ kcal/mol} \quad (\text{or} -23 \text{ kJ/mol})$$

$$I\cdot + CH_4 \longrightarrow \cdot CH_3 + HI \qquad 105 - 71 = \quad 34$$
$$\cdot CH_3 + I_2 \longrightarrow CH_3I + I\cdot \qquad \underline{36 - 58 = -22}$$
$$\Delta H° = \ 12 \text{ kcal/mol} \quad (\text{or} 50 \text{ kJ/mol})$$

The fluorine radical is the most reactive of the halogen radicals, and it reacts violently with alkanes ($\Delta H° = -31$ kcal/mol). In contrast, the iodine radical is the least reactive of the halogen radicals. In fact, it is so unreactive ($\Delta H° = 34$ kcal/mol) that it is unable to remove a hydrogen atom from an alkane. Consequently, it reacts with another iodine radical and reforms I_2.

PROBLEM 10 **SOLVED**

How could butanone be prepared from butane?

$$CH_3CH_2CH_2CH_3 \xrightarrow{?} \underset{\text{butanone}}{CH_3\overset{\overset{\displaystyle O}{\|}}{C}CH_2CH_3}$$
$$\underset{\text{butane}}{}$$

Solution   We know that the first reaction has to be a radical halogenation because that is the only reaction that an alkane undergoes. Bromination will lead to a greater yield of the desired 2-halo-substituted compound than will chlorination because a bromine radical is more selective than a chlorine radical. A nucleophilic substitution reaction forms the alcohol, which forms the target compound when it is oxidized.

$$CH_3CH_2CH_2CH_3 \xrightarrow[h\nu]{Br_2} CH_3\overset{\overset{\displaystyle Br}{|}}{C}HCH_2CH_3 \xrightarrow{HO^-} CH_3\overset{\overset{\displaystyle OH}{|}}{C}HCH_2CH_3 \xrightarrow{H_2CrO_4} CH_3\overset{\overset{\displaystyle O}{\|}}{C}CH_2CH_3$$

PROBLEM 11◆

If 2-methylpropane is brominated at 125 °C in the presence of light, what percent of the product will be 2-bromo-2-methylpropane? Compare your answer with the percent given in Problem 4 for chlorination of 2-methylpropane.

PROBLEM 12◆

Take the same alkanes whose percentages of monochlorination products you calculated in Problem 6, and calculate what the percentages of monobromination products would be if bromination were carried out at 125 °C.

Student Tutorial:
Reactivity–selectivity

11.6 Addition of Radicals to an Alkene

The addition of HBr to 1-butene forms 2-bromobutane, because the electrophile (H^+), following the rule that applies to all electrophilic addition reactions (Section 4.4), adds to the sp^2 carbon bonded to the greater number of hydrogens. If, however, you want to synthesize 1-bromobutane, you need to find a way to make bromine an electrophile so it, instead of H^+, will add to the sp^2 carbon bonded to the greater number of hydrogens.

$$CH_3CH_2CH{=}CH_2 \ + \ HBr \longrightarrow \underset{\textbf{2-bromobutane}}{CH_3CH_2\overset{\overset{\textstyle Br}{|}}{C}HCH_3}$$

$$\underset{\textbf{1-butene}}{CH_3CH_2CH{=}CH_2} \ + \ HBr \xrightarrow{\textbf{peroxide}} \underset{\textbf{1-bromobutane}}{CH_3CH_2CH_2CH_2Br}$$

If an alkyl peroxide (ROOR) is added to the reaction mixture, the product of the addition reaction will be the desired 1-bromobutane. Thus, the peroxide changes the mechanism of the reaction in a way that causes Br· to be the electrophile. The following mechanism for the addition of HBr to an alkene in the presence of a peroxide shows that it is a radical chain reaction with characteristic initiation, propagation, and termination steps:

Mechanism for the addition of HBr to an alkene in the presence of a peroxide

$$\underset{\textbf{an alkyl peroxide}}{R\ddot{O}{-}\ddot{O}R} \xrightarrow[\Delta]{\textbf{light} \atop \textbf{or}} \underset{\textbf{alkoxyl radicals}}{2\ R\ddot{O}\cdot}$$

$$R{-}\ddot{O}\cdot \ + \ H{-}\ddot{B}r\text{:} \longrightarrow R{-}\ddot{O}{-}H \ + \ \underset{\substack{\textbf{a bromine}\\\textbf{radical}}}{\cdot\ddot{B}r\text{:}}$$

initiation steps

$$\text{:}\ddot{B}r\cdot \ + \ CH_2{=}CHCH_2CH_3 \longrightarrow \underset{\underset{\textbf{an alkyl radical}}{\overset{|}{\text{:}\ddot{B}r\text{:}}}}{CH_2\dot{C}HCH_2CH_3}$$

$$\underset{\overset{|}{Br}}{CH_2\dot{C}HCH_2CH_3} \ + \ H{-}\ddot{B}r\text{:} \longrightarrow \underset{\overset{|}{Br}\ \ \ \overset{|}{H}}{CH_2{-}CHCH_2CH_3} \ + \ \cdot\ddot{B}r\text{:}$$

propagation steps

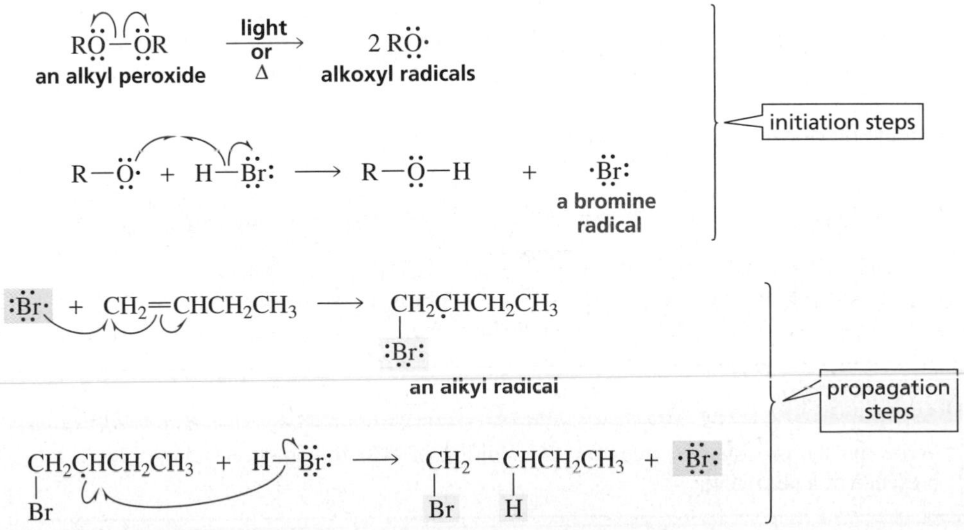

$$:\ddot{Br}\cdot + \cdot\ddot{Br}: \longrightarrow :\ddot{Br}-\ddot{Br}:$$

$$BrCH_2\dot{C}HCH_2CH_3 + :\ddot{Br}\cdot \longrightarrow \underset{\underset{Br}{|}}{BrCH_2CHCH_2CH_3}$$

$$2\ BrCH_2\dot{C}HCH_2CH_3 \longrightarrow \underset{\underset{BrCH_2}{|}\ \underset{CH_2Br}{|}}{CH_3CH_2CH-CHCH_2CH_3}$$

> termination steps

- The alkyl peroxide contains a weak oxygen–oxygen single bond that readily breaks homolytically in the presence of light or heat to form alkoxy radicals. This is an initiation step because it creates radicals.

- The alkoxy radical completes its octet by removing a hydrogen atom from a molecule of HBr, forming a bromine radical. This too is an initiation step because it forms the chain-propagating radical (Br·).

- The bromine radical now seeks an electron to complete its octet. Because the double bond of an alkene is electron rich, the bromine radical completes its octet by combining with one of the electrons of the π bond of the alkene to form a C—Br bond. The second electron of the π bond is the unpaired electron in the resulting alkyl radical. If the bromine radical adds to the sp^2 carbon of 1-butene that is bonded to the greater number of hydrogens, a secondary alkyl radical is formed. If the bromine radical adds to the other sp^2 carbon, a primary alkyl radical is formed. The bromine radical, therefore, adds to the sp^2 carbon that is bonded to the greater number of hydrogens in order to form the more stable intermediate.

- The alkyl radical removes a hydrogen atom from another molecule of HBr to produce a molecule of the alkyl halide product and another bromine radical. This step and the preceding step are propagation steps; in the first propagation step, a radical (Br·) reacts to produce another radical; in the second propagation step, the radical produced in the first propagation step reacts to form the radical (Br·) that was the reactant in the first propagation step.

- The last three steps are termination steps.

Because the first species that adds to the alkene is a radical (Br·), the addition of HBr in the presence of a peroxide is called a **radical addition reaction**.

When HBr reacts with an alkene in the absence of a peroxide, the electrophile—the first species to add to the alkene—is H^+. In the presence of a peroxide, the electrophile is Br·. In both cases, the electrophile adds to the sp^2 carbon that is bonded to the greater number of hydrogens, so both reactions follow the general rule for electrophilic addition reactions: *the electrophile adds to the sp^2 carbon that is bonded to the greater number of hydrogens.*

Because the addition of HBr in the presence of a peroxide forms a radical intermediate rather than a carbocation intermediate, the intermediate does not rearrange. Radicals do not rearrange as readily as carbocations.

The electrophile adds to the sp^2 carbon that is bonded to the greater number of hydrogens.

$$\underset{\textbf{3-methyl-1-butene}}{\underset{\underset{CH_3}{|}}{CH_3CHCH{=}CH_2}} + HBr \xrightarrow{\textbf{peroxide}} \underset{\textbf{1-bromo-3-methylbutane}}{\underset{\underset{CH_3}{|}}{CH_3CHCH_2CH_2Br}}$$

> carbon skeleton is not rearranged

PROBLEM 13

Write out the propagation steps for the addition of HBr to 1-methylcyclohexene in the presence of a peroxide.

An alkyl peroxide is a **radical initiator** because it creates radicals. Without a peroxide, the radical reaction we have just described would not occur. Any reaction that occurs in the presence of a radical initiator, but does not occur in its absence, must take place by a mechanism that involves radicals as intermediates. Any compound that can readily undergo homolysis (dissociate to form radicals) can act as a radical initiator. We will encounter radical initiators again when we discuss polymers in Chapter 28; examples of radical initiators are shown in Table 28.3.

While radical initiators cause radical reactions to occur, **radical inhibitors** have the opposite effect; they trap radicals as they are formed, preventing reactions that depend on the presence of radicals. How radical inhibitors trap radicals is discussed in Section 11.10.

A peroxide has no effect on the addition of HCl or HI to an alkene; the addition that occurs when a peroxide is present is the same as the addition that occurs when a peroxide is absent.

$$CH_3CH{=}CH_2 \ + \ HCl \ \xrightarrow{\text{peroxide}} \ CH_3\overset{|}{\underset{Cl}{C}}HCH_3$$

$$CH_3\overset{\overset{\displaystyle CH_3}{|}}{C}{=}CH_2 \ + \ HI \ \xrightarrow{\text{peroxide}} \ CH_3\overset{\overset{\displaystyle CH_3}{|}}{\underset{\underset{\displaystyle I}{|}}{C}}CH_3$$

Why is the **peroxide effect** observed for the addition of HBr, but not for the addition of HCl or HI? This question can be answered by calculating the $\Delta H°$ for the two propagation steps in the radical chain reaction (using the bond dissociation energies in Table 3.2).

$Cl{\cdot} \ + \ CH_2{=}CH_2 \ \longrightarrow \ ClCH_2\dot{C}H_2$ $\Delta H° = 63 - 85 = -22$ kcal/mol (or –91 kJ/mol) ⟨ exothermic ⟩

$ClCH_2\dot{C}H_2 \ + \ HCl \ \longrightarrow \ ClCH_2CH_3 \ + \ Cl{\cdot}$ $\Delta H° = 103 - 101 = +2$ kcal/mol (or +8 kJ/mol) ⟨ endothermic ⟩

$Br{\cdot} \ + \ CH_2{=}CH_2 \ \longrightarrow \ BrCH_2\dot{C}H_2$ $\Delta H° = 63 - 72 = -9$ kcal/mol (or –38 kJ/mol)

$BrCH_2\dot{C}H_2 \ + \ HBr \ \longrightarrow \ BrCH_2CH_3 \ + \ Br{\cdot}$ $\Delta H° = 87 - 101 = -14$ kcal/mol (or –59 kJ/mol) ⟩ exothermic

$I{\cdot} \ + \ CH_2{=}CH_2 \ \longrightarrow \ ICH_2\dot{C}H_2$ $\Delta H° = 63 - 57 = +6$ kcal/mol (or +25 kJ/mol) ⟨ endothermic ⟩

$ICH_2\dot{C}H_2 \ + \ HI \ \longrightarrow \ ICH_2CH_3 \ + \ I{\cdot}$ $\Delta H° = 71 - 101 = -30$ kcal/mol (or –126 kJ/mol) ⟨ exothermic ⟩

For the radical addition of HCl, the first propagation step is exothermic and the second is endothermic. For the radical addition of HI, the first propagation step is endothermic and the second is exothermic. Only for the radical addition of HBr are both propagation steps exothermic. In a radical reaction, the steps that propagate the chain reaction compete with the steps that terminate it. Termination steps are always exothermic, because only bond making (and no bond breaking) occurs. Therefore, only when both propagation steps are exothermic can propagation compete successfully with termination. When HCl or HI adds to an alkene in the presence of a peroxide, any chain reaction that is initiated is then terminated rather than propagated because propagation cannot compete successfully with termination. Consequently, the radical chain reaction does not take place, and the only reaction that occurs is ionic addition (H^+ followed by Cl^- or I^-).

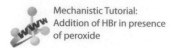

Mechanistic Tutorial:
Addition of HBr in presence
of peroxide

PROBLEM 14◆

What will be the major product of the reaction of 2-methyl-2-butene with each of the following reagents:

a. HBr

b. HCl

c. HBr + peroxide

d. HCl + peroxide

11.7 Stereochemistry of Radical Substitution and Addition Reactions

We have seen that when a reactant that does not have an asymmetric center undergoes a reaction that forms a product with one asymmetric center, the product will be a racemic mixture (Section 5.19). Thus, the following radical substitution reaction forms equal amounts of each enantiomer.

> **When a reactant that does not have an asymmetric center undergoes a reaction that forms a product with one asymmetric center, the product will be a racemic mixture.**

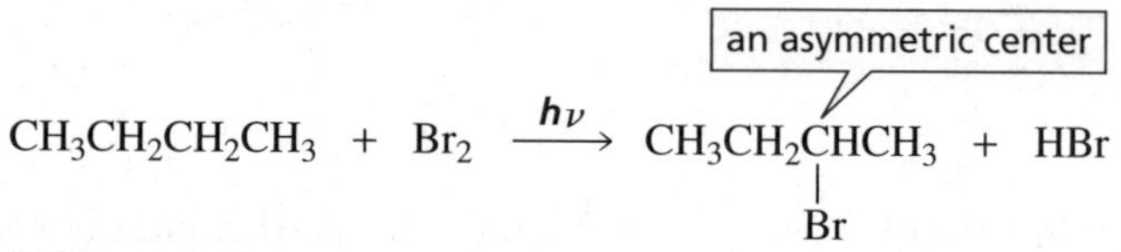

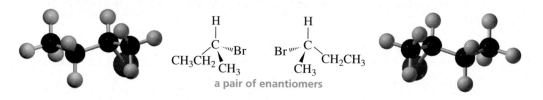

$$CH_3CH_2CH_2CH_3 + Br_2 \xrightarrow{h\nu} CH_3CH_2\underset{|}{\overset{}{C}}HCH_3 + HBr$$

an asymmetric center

Br

configuration of the products

a pair of enantiomers

Similarly, the product of the following radical addition reaction is a racemic mixture:

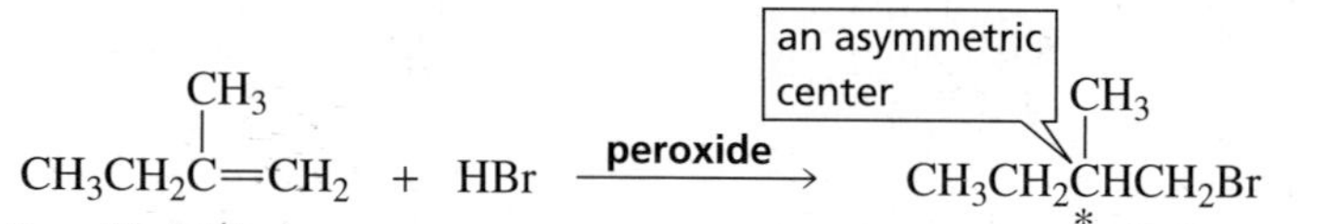

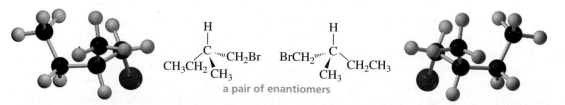

2-methyl-1-butene

peroxide

1-bromo-2-methylbutane

an asymmetric center

configuration of the products

a pair of enantiomers

Both the radical substitution and radical addition reactions form a racemic mixture because both reactions form a radical intermediate. The radical intermediate in the substitution reaction is formed when the bromine radical removes a hydrogen atom from the reactant; the radical intermediate in the addition reaction is formed when the bromine radical adds to one of the sp^2 carbons of the double bond.

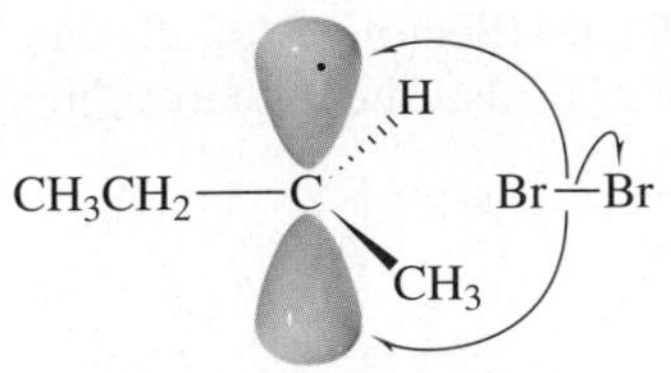

radical intermediate
in the substitution reaction

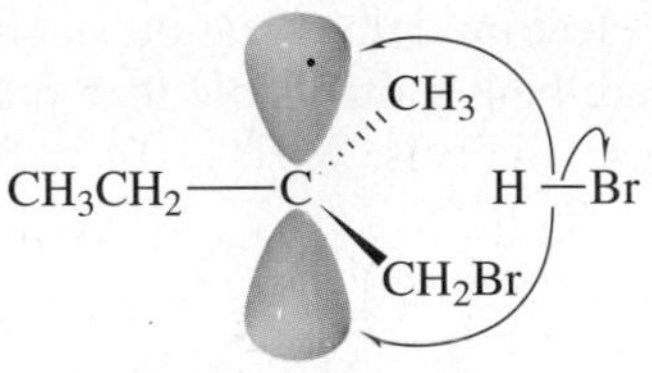

radical intermediate
in the addition reaction

The carbon that bears the unpaired electron in the radical intermediate is sp^2 hybridized; therefore, the three atoms to which it is bonded lie in a plane (Section 1.10). The incoming atom has equal access to both sides of the plane. Consequently, identical amounts of the R enantiomer and the S enantiomer are obtained in both the substitution and addition reactions.

To summarize, the stereochemical outcome of a *radical substitution reaction* is identical to the stereochemical outcome of a *radical addition reaction* because both reactions form a radical intermediate, and it is the reaction of the intermediate that determines the configuration of the products.

Identical amounts of the R and S enantiomers are also obtained if a hydrogen bonded to an asymmetric center is substituted by a halogen. Breaking the bond to the asymmetric center destroys the configuration at the asymmetric center and forms a planar radical intermediate. The incoming halogen has equal access to both sides of the plane, so identical amounts of the two enantiomers are obtained.

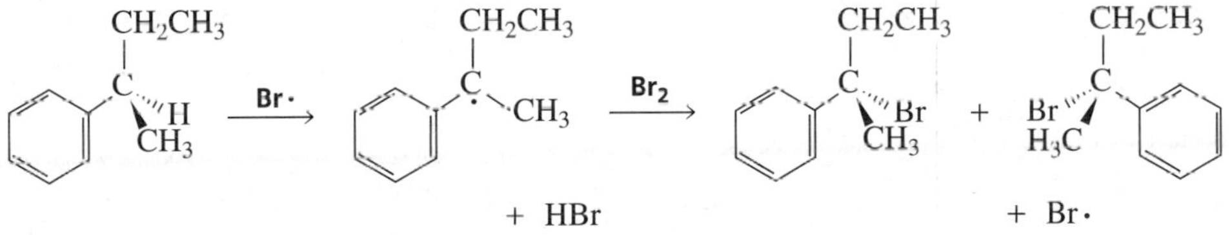

+ HBr

+ Br·

PROBLEM 15◆

a. What hydrocarbon with molecular formula C_4H_{10} forms only two monochlorinated products? Both products are achiral.

b. What hydrocarbon with the same molecular formula as in part a forms only three monochlorinated products? One is achiral and two are chiral.

11.8 Radical Substitution of Benzylic and Allylic Hydrogens

An **allylic radical** has an unpaired electron on an allylic carbon and, like an allylic cation, has two contributing resonance structures.

an allylic radical

A **benzylic radical** has an unpaired electron on a benzylic carbon and, like a benzylic cation, has five contributing resonance structures.

See Special Topic VII in the *Study Guide and Solutions Manual* for additional information on drawing curved arrows in radical systems.

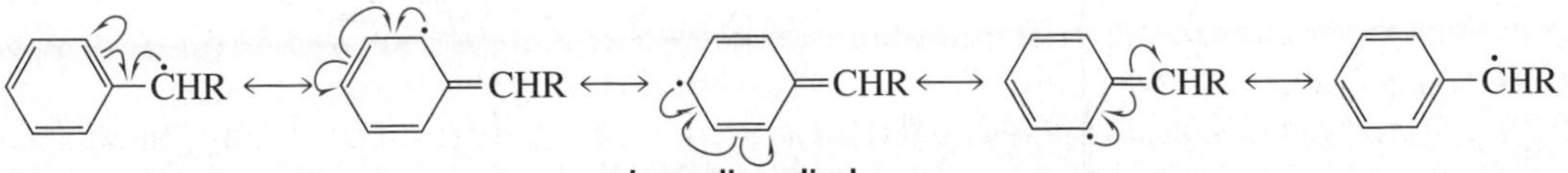

a benzylic radical

Electron delocalization increases the stability of a molecule.

Because electron delocalization stabilizes a molecule (Section 7.6), allyl and benzyl radicals are both more stable than other primary radicals. They are even more stable than tertiary radicals.

relative stabilities of radicals

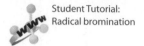

We know that the more stable the radical, the faster it can be formed. This means that a hydrogen bonded to either a benzylic carbon or an allylic carbon will be preferentially substituted in a halogenation reaction. Because bromination is more highly regioselective than chlorination, the percent of substitution at the benzylic or allylic carbon is greater for bromination.

We know that the more stable the radical, the faster it can be formed. This means that a hydrogen bonded to either a benzylic carbon or an allylic carbon will be preferentially substituted in a halogenation reaction. Because bromination is more highly regioselective than chlorination, the percent of substitution at the benzylic or allylic carbon is greater for bromination.

Student Tutorial:
Radical bromination

benzylic substituted product

allylic substituted product

N-Bromosuccinimide (NBS) is frequently used to brominate allylic carbons because it allows a radical substitution reaction to be carried out in the presence of a low concentration of Br_2 and a low concentration of HBr. If a high concentration of Br_2 is present, electrophilic addition of Br_2 to the double bond will compete with allylic substitution. If a high concentration of HBr is present, radical addition of HBr to the double bond will compete with allylic substitution.

NBS is used to brominate allylic carbons.

N-bromosuccinimide
NBS

succinimide

The reaction begins with homolytic cleavage of the N—Br bond in NBS. This generates the bromine radical needed to initiate the radical reaction. Light or heat and a radical initiator such as a peroxide are used to promote the reaction. The bromine radical removes an allylic hydrogen to form HBr and a resonance stabilized allylic radical. The allylic radical reacts with Br_2, forming the allyl bromide and a chain-propagating bromine radical.

The Br$_2$ used in the second step of the preceding reaction sequence is produced in low concentration from a fast ionic reaction of NBS with HBr.

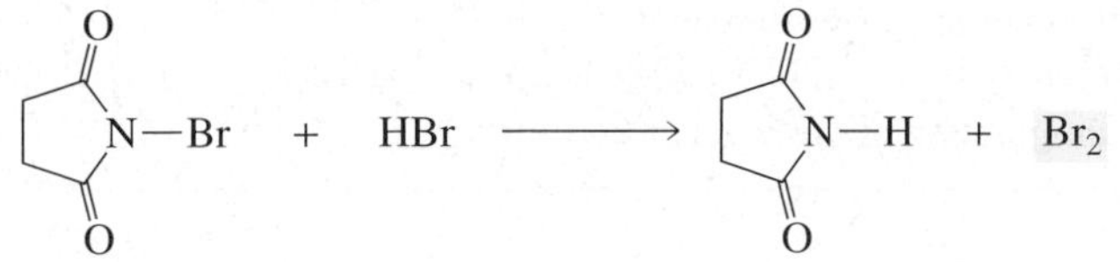

$$\text{N—Br} \quad + \quad \text{HBr} \quad \longrightarrow \quad \text{N—H} \quad + \quad \text{Br}_2$$

When a radical removes a hydrogen atom from an allylic carbon, the unpaired electron of the allylic radical is shared by two carbons. In other words, the allylic radical has two resonance contributors. In the following reaction, only one substitution product is formed, because the groups attached to the sp^2 carbons are the same in both resonance contributors:

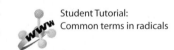

Student Tutorial:
Common terms in radicals

$$\text{Br} \cdot \ + \ \text{CH}_3\text{CH}=\text{CH}_2 \ \longrightarrow \ \dot{\text{C}}\text{H}_2\text{CH}=\text{CH}_2 \ \longleftrightarrow \ \text{CH}_2=\text{CH}\dot{\text{C}}\text{H}_2 \ + \ \text{HBr}$$

$$\bigg\downarrow \text{Br}_2$$

$$\text{BrCH}_2\text{CH}=\text{CH}_2 \ + \ \text{Br} \cdot$$
3-bromopropene

However, if the groups attached to the two sp^2 carbons of the allylic radical are *not* the same in both resonance contributors, two substitution products are formed:

$$\text{Br} \cdot \ + \ \text{CH}_3\text{CH}_2\text{CH}=\text{CH}_2 \ \longrightarrow \ \text{CH}_3\dot{\text{C}}\text{HCH}=\text{CH}_2 \ \longleftrightarrow \ \text{CH}_3\text{CH}=\text{CH}\dot{\text{C}}\text{H}_2 \ + \ \text{HBr}$$

$$\bigg\downarrow \text{Br}_2$$

$$\overset{\text{Br}}{\underset{|}{\text{CH}_3\text{CHCH}=\text{CH}_2}} \ + \ \text{CH}_3\text{CH}=\text{CHCH}_2\text{Br} \ + \ \text{Br} \cdot$$
3-bromo-1-butene   **1-bromo-2-butene**

PROBLEM 16

Two products are formed when methylenecyclohexane reacts with NBS. Explain how each is formed.

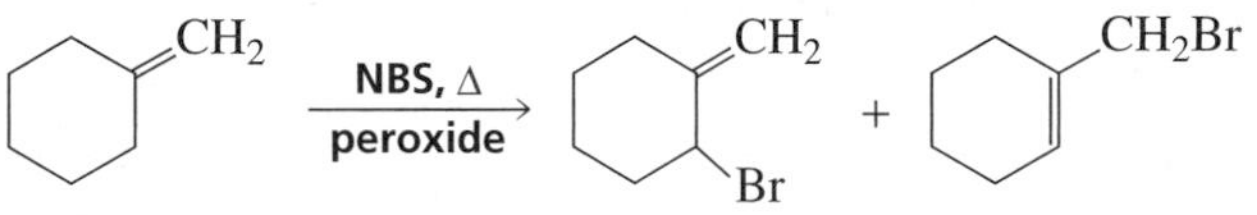

$$\text{[methylenecyclohexane]} \xrightarrow[\text{peroxide}]{\text{NBS, } \Delta} \text{[product with =CH}_2\text{ and Br]} \ + \ \text{[product with CH}_2\text{Br]}$$

PROBLEM 17   *SOLVED*

How many allylic substituted bromoalkenes are formed from the reaction of 2-pentene with NBS? Disregard stereoisomers.

Solution The bromine radical will more readily remove a secondary allylic hydrogen from C-4 of 2-pentene than a primary allylic hydrogen from C-1. The resonance contributors of the resulting radical intermediate have the same groups attached to the sp^2 carbons, so only one bromoalkene is formed. Because of the high selectivity of the bromine radical, an insignificant amount of radical will be formed by removal of a hydrogen from the less reactive primary allylic position.

$$\text{CH}_3\text{CH}=\text{CHCH}_2\text{CH}_3 \xrightarrow[\text{peroxide}]{\text{NBS, } \Delta} \text{CH}_3\text{CH}=\text{CH}\dot{\text{C}}\text{HCH}_3 \ \longleftrightarrow \ \text{CH}_3\dot{\text{C}}\text{HCH}=\text{CHCH}_3 \ + \ \text{HBr}$$

$$\bigg\downarrow$$

$$\underset{\underset{\text{Br}}{|}}{\text{CH}_3\text{CH}=\text{CHCHCH}_3}$$

PROBLEM 18

a. Give the major product(s) of the reaction of 1-methylcyclohexene with the following reagents, disregarding stereoisomers:
  1. NBS/Δ/peroxide   **2.** Br₂/CH₂Cl₂   **3.** HBr   **4.** HBr/peroxide
b. Give the configuration of the products.

Why does the bromine radical from NBS remove an allylic hydrogen whereas the bromine radical from HBr + peroxide adds to the double bond? The bromine radical can in fact do both. However, when NBS is used, there is little HBr present to complete the addition reaction after the bromine radical has added to the double bond. Therefore, the reactant is reformed, since the addition of a bromine radical to a double bond is reversible, and allylic substitution is the major product.

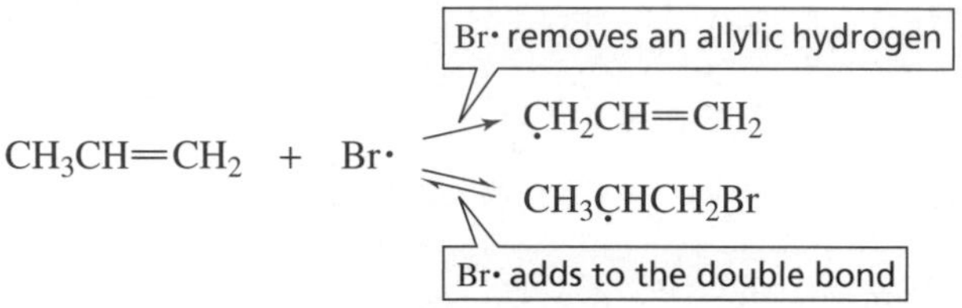

$$CH_3CH{=}CH_2 \;+\; Br\cdot$$

Br· removes an allylic hydrogen
$$\dot{C}H_2CH{=}CH_2$$
$$CH_3\dot{C}HCH_2Br$$
Br· adds to the double bond

CYCLOPROPANE

Although it is an alkane, cyclopropane undergoes electrophilic addition reactions as if it were an alkene.

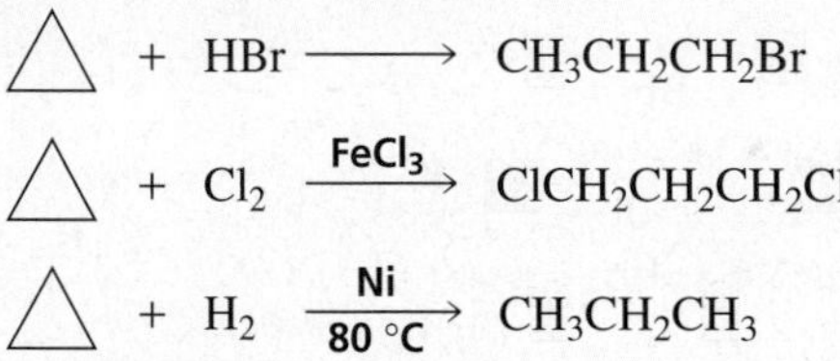

$$\triangle + HBr \longrightarrow CH_3CH_2CH_2Br$$

$$\triangle + Cl_2 \xrightarrow{FeCl_3} ClCH_2CH_2CH_2Cl$$

$$\triangle + H_2 \xrightarrow[80\,°C]{Ni} CH_3CH_2CH_3$$

Cyclopropane is more reactive than propene toward addition of acids such as HBr and HCl, but is less reactive toward addition of Cl₂ and Br₂, so a Lewis acid (FeCl₃ or FeBr₃) is needed to catalyze halogen addition (Section 1.26).

It is the strain in the small ring that makes it possible for cyclopropane to undergo electrophilic addition reactions (Section 2.11). Because of the 60° bond angles in its three-membered ring, the compound's sp^3 orbitals cannot overlap head-on; this decreases the effectiveness of the orbital overlap. Thus, the C—C "banana bonds" in cyclopropane are considerably weaker than normal C—C σ bonds (Figure 2.6 on page 105). Consequently, three-membered rings undergo ring-opening reactions with electrophilic reagents.

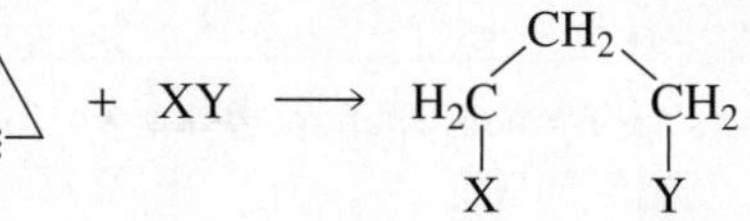

$$\triangle + XY \longrightarrow H_2C \underset{X}{\overset{CH_2}{|}} \underset{Y}{\overset{CH_2}{|}}$$

11.9 Designing a Synthesis III: More Practice with Multistep Synthesis

Now that the number of reactions that you are familiar with has increased, you can design the synthesis of a wide variety of compounds.

Example 1. Starting with the ether shown below, how could you prepare the carboxylic acid?

Heating the ether with one equivalent of HI forms an alcohol that, when oxidized, forms the desired carboxylic acid.

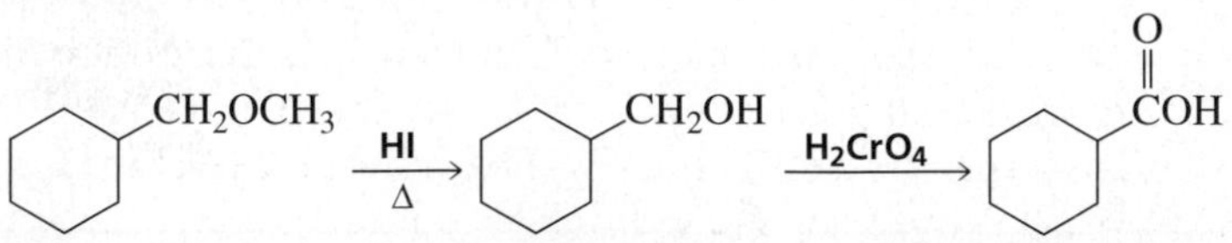

Example 2. Suggest a way to prepare 1,3-cyclohexadiene from cyclohexane.

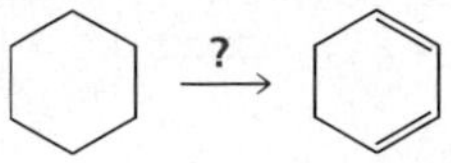

Deciding what the first reaction should be is easy, because the only reaction that an alkane can undergo is halogenation. Next, an E2 reaction, using a high concentration of a strong and bulky base and carried out at a relatively high temperature to encourage elimination over substitution, will form cyclohexene. *tert*-Butoxide ion would therefore be used as the base and *tert*-butyl alcohol as the solvent. Bromination of cyclohexene forms an allylic bromide, which will form the desired target molecule by undergoing another E2 reaction.

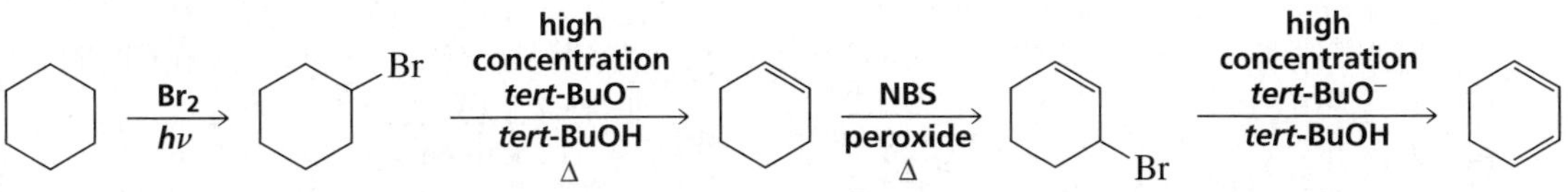

Example 3. Starting with methylcyclohexane, how could the following vicinal *trans*-dihalide be prepared?

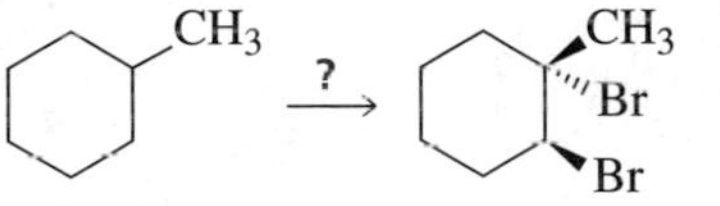

Again, since the starting material is an alkane, the first reaction must be a radical substitution: bromination leads to selective substitution of the tertiary hydrogen. Under E2 conditions, tertiary alkyl halides undergo only elimination, so there will be no competing substitution product formed in the next reaction. However, a relatively unhindered base should be used to favor removal of a proton from the secondary carbon over removal of a proton from the methyl group. The final step is addition of Br₂; only anti addition occurs in this reaction, so the target molecule (along with its enantiomer) is obtained.

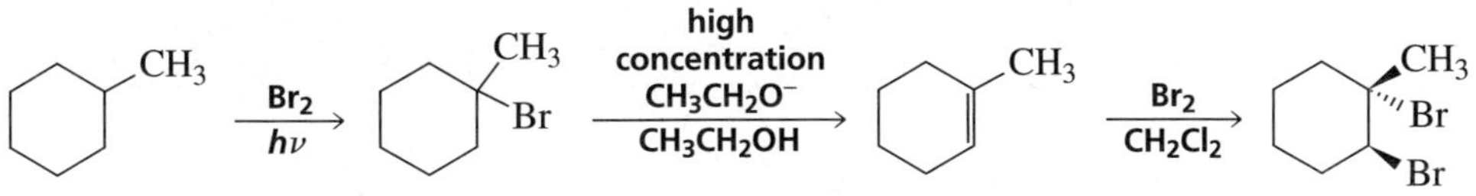

Example 4. Design a synthesis for the target molecule from the indicated starting material.

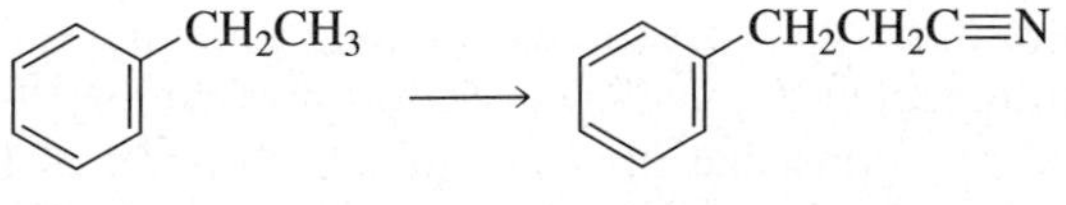

It is not immediately obvious how to carry out this synthesis, so let's use retrosynthetic analysis to find a way. The only method you know for introducing a C≡N group into a molecule is nucleophilic substitution. The alkyl halide for that substitution reaction can be obtained from the addition of HBr in the presence of a peroxide. The alkene for that addition reaction can be obtained from an elimination reaction using an alkyl halide obtained by benzylic substitution.

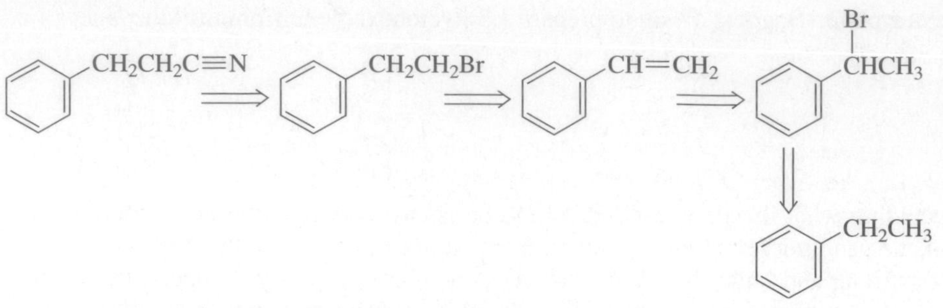

The reaction sequence can now be written in the forward direction. Notice that a bulky base is used to encourage elimination over substitution.

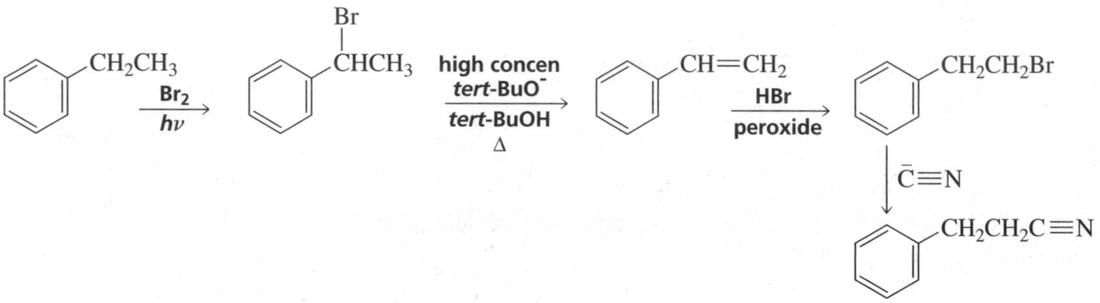

PROBLEM 19

Design a multistep synthesis to show how each of the following compounds could be prepared from the given starting material:

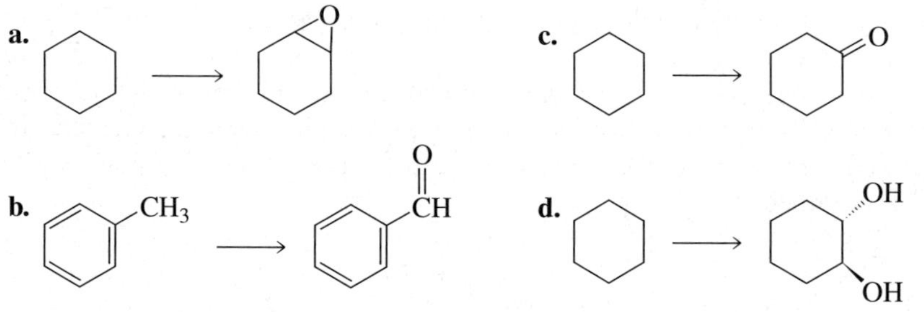

11.10 Radical Reactions Occur in Biological Systems

Because of the large amount of energy—heat or light—required to initiate a radical reaction, and because once a radical chain reaction is initiated the propagation steps are difficult to control, scientists assumed for a long time that radical reactions were not important in biological systems. However, it is now widely recognized that there are many biological reactions that involve radicals. Instead of being generated by heat or light, the radicals in these reactions are formed by the interaction of organic molecules with metal ions. The radical reactions take place at the active sites of enzymes (Section 23.9). Containing the reaction at a specific site allows the reaction to be controlled.

Water-soluble (polar) compounds are readily eliminated by the body. In contrast, water-insoluble (nonpolar) compounds are not readily eliminated but instead accumulate in the nonpolar components of cells. For cells to avoid becoming "toxic dumps," nonpolar compounds that are ingested (drugs, foods, environmental pollutants) must be converted into polar compounds that can be excreted.

A biological reaction carried out in the liver converts toxic nonpolar hydrocarbons into less toxic polar alcohols by substituting an H in the hydrocarbon with an OH. The reaction involves radicals and is catalyzed by an iron-containing enzyme called cytochrome P_{450}, the same enzyme that detoxifies aromatic hydrocarbons by converting them into arene oxides (Section 10.9). An alkyl radical intermediate is created when $Fe^{V}=O$ removes a hydrogen atom from an alkane. Then $Fe^{IV}OH$ dissociates homolytically into Fe^{III} and $HO^{\cdot}$, and the $HO^{\cdot}$ immediately combines with the radical intermediate to form the alcohol.

an alkane   a radical   an alcohol
                   intermediate

This reaction can also have the opposite toxicological effect. That is, substituting an OH for an H in some compounds causes a nontoxic compound to become toxic. For example, studies done on animals showed that substituting an OH for an H caused dichloromethane (CH_2Cl_2) to become a carcinogen when inhaled. This shows that compounds that are nontoxic *in vitro* (in a test tube) are not necessarily nontoxic *in vivo* (in an organism).

DECAFFEINATED COFFEE AND THE CANCER SCARE

The animal studies revealing that dichloromethane becomes a carcinogen when inhaled caused some concern, because dichloromethane was the solvent used to extract caffeine from coffee beans in the manufacture of decaffeinated coffee. However, when dichloromethane was added to the drinking water given to laboratory rats and mice, researchers found no toxic effects, even in rats that had consumed an amount of dichloromethane equivalent to the amount that would be ingested by drinking 120,000 cups of decaffeinated coffee per day and in mice that had consumed an amount equivalent to drinking 4.4 million cups of decaffeinated coffee per day. In addition, no increased risk of cancer was found in a study of thousands of workers exposed daily to inhaled dichloromethane. (This shows that the results of studies done on humans do not always agree with the results of those done on laboratory animals.) Because of the initial concern, however, researchers sought alternative methods for extracting caffeine from coffee beans. Extraction by liquid CO_2 at supercritical temperatures and pressures was found to be a better method because it extracts caffeine without simultaneously extracting some of the flavor compounds, as dichloromethane does. This was one of the first green (environmentally benign) commercial chemical processes to be developed; CO_2 is a relatively harmless waste product, whereas dichloromethane is not a substance we would want to release into the environment.

Fats and oils react with radicals to form compounds that have strong odors (Section 26.3). These compounds are responsible for the unpleasant taste and smell associated with sour milk and rancid butter. The molecules that form cell membranes can undergo this same radical reaction (Section 26.4). Furthermore, radical reactions in biological systems have been implicated in the aging process.

Clearly, unwanted radicals in biological systems must be destroyed before they do damage to cells. Unwanted radical reactions are prevented by **radical inhibitors**, compounds that destroy reactive radicals by converting them either into unreactive radicals or into compounds with only paired electrons. Hydroquinone is an example of a radical inhibitor. When hydroquinone traps a radical, it forms semiquinone, also a radical but it is stabilized by electron delocalization and is, therefore, less reactive than other radicals. Furthermore, semiquinone can trap another radical and form quinone, a compound whose electrons are all paired.

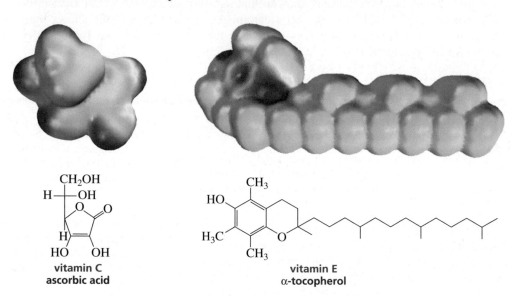

Two other examples of radical inhibitors in biological systems are vitamins C and E. Like hydroquinone, they form relatively stable (unreactive) radicals. Vitamin C (also called ascorbic acid) is a water-soluble compound that traps radicals formed in the cell and in blood plasma (both have aqueous environments). Vitamin E (also called α-tocopherol) is a water-insoluble (hence fat-soluble) compound that traps radicals formed in nonpolar membranes. Why one vitamin functions in aqueous environments and the other in nonaqueous environments should be apparent from their structures and electrostatic potential maps, which show that vitamin C is a relatively polar compound, whereas vitamin E is nonpolar.

vitamin C
ascorbic acid

vitamin E
α-tocopherol

FOOD PRESERVATIVES

Radical inhibitors that occur in or are added to foods are known as *preservatives* or *antioxidants*. They preserve food by preventing undesirable radical reactions. Vitamin E is a naturally occurring preservative found in vegetable oil. BHA and BHT are synthetic preservatives added to many packaged foods.

butylated hydroxyanisole
BHA

butylated hydroxytoluene
BHT

food preservatives

11.11 Radicals and Stratospheric Ozone

Ozone (O_3), a major constituent of smog, is a health hazard at ground level. In the stratosphere, however, a layer of ozone shields the Earth from harmful solar radiation. The ozone layer is thinnest at the equator and densest toward the poles, with the greatest concentrations lying between 12 and 15 miles above the Earth's surface. Ozone is formed in the atmosphere from the interaction of molecular oxygen with very short wavelength ultraviolet light.

$$O_2 \xrightarrow{h\nu} O + O$$

$$O + O_2 \longrightarrow O_3$$
ozone

In the stratosphere, ozone acts as a filter for biologically harmful ultraviolet radiation that otherwise would reach the surface of the Earth. Among other effects, high-energy short-wavelength ultraviolet light can damage DNA in skin cells, causing mutations that trigger skin cancer (Section 29.6). We owe our very existence to this protective ozone layer. According to current theories of evolution, life could not have developed on land without it. Instead, most if not all living things would have had to remain in the ocean, where water screens out the harmful ultraviolet radiation.

Since about 1985, scientists have noted a precipitous drop in stratospheric ozone over Antarctica. This area of ozone depletion, dubbed the "ozone hole," is unprecedented in the history of ozone observations. Scientists subsequently noted a similar decrease in ozone over Arctic regions; then, in 1988, they detected a depletion of ozone over the United States for the first time. Three years later, scientists determined that the rate of ozone depletion was two to three times faster than originally anticipated. Many in the scientific community blame recently observed increases in cataracts and skin cancer as well as diminished plant growth on the ultraviolet radiation that has penetrated the reduced ozone layer. Some predict that erosion of the protective ozone layer will cause an additional 200,000 deaths from skin cancer over the next 50 years.

Strong circumstantial evidence implicates synthetic chlorofluorocarbons (CFCs)—alkanes such as $CFCl_3$ and CF_2Cl_2, in which all the hydrogens have been replaced by fluorine and chlorine—as a major cause of ozone depletion. These gases, known commercially as Freon, have been used extensively as cooling fluids in refrigerators and air conditioners. They were also once widely used as propellants in aerosol spray cans (deodorant, hair spray, and so on) because of their odorless, nontoxic, and nonflammable properties and because, being chemically inert, they do not react with the contents of the can. Such use now, however, has been banned.

Chlorofluorocarbons remain very stable in the atmosphere until they reach the stratosphere. There they encounter wavelengths of ultraviolet light that cause the C—Cl bond to break homolytically, generating chlorine radicals.

$$\underset{\underset{F}{|}}{\overset{\overset{Cl}{|}}{F-C-Cl}} \xrightarrow{h\nu} \underset{\underset{F}{|}}{\overset{\overset{Cl}{|}}{F-C\cdot}} + Cl\cdot$$

These chlorine radicals are the ozone-removing agents. They react with ozone to form chlorine monoxide radicals and molecular oxygen. The chlorine monoxide radical then reacts with more ozone to form chlorine dioxide, which dissociates to regenerate a chlorine radical. These three steps—two of which each destroy an ozone

Student Tutorial:
Chlorofluorocarbons and ozone

Polar stratospheric clouds increase the rate of ozone destruction. These clouds form over Antarctica during the cold winter months. Ozone depletion in the Arctic is less severe because the temperature generally does not get low enough for stratospheric clouds to form there.

molecule—are repeated over and over. It has been calculated that a single chlorine atom destroys 100,000 ozone molecules!

$$Cl\cdot + O_3 \longrightarrow ClO\cdot + O_2$$

$$ClO\cdot + O_3 \longrightarrow \cdot ClO_2 + O_2$$

$$\cdot ClO_2 \longrightarrow Cl\cdot + O_2$$

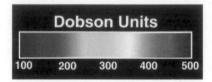

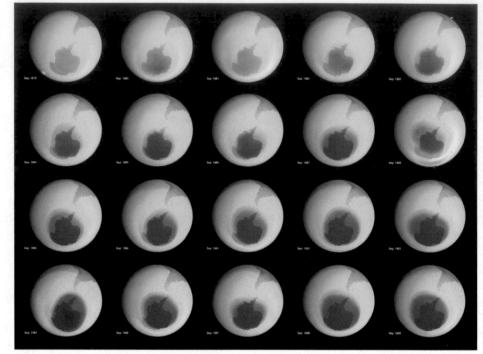

Growth of the Antarctic ozone hole, located mostly over the continent of Antarctica, since 1979. The images were made from data supplied by total ozone-mapping spectrometers (TOMSs). The color scale depicts the total ozone values in Dobson units*, with the lowest ozone densities represented by dark blue.

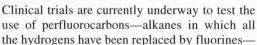

ARTIFICIAL BLOOD

Clinical trials are currently underway to test the use of perfluorocarbons—alkanes in which all the hydrogens have been replaced by fluorines—as a substitute for blood. One compound under study has been found to be more effective than hemoglobin in carrying oxygen to cells and in transporting carbon dioxide to the lungs. Artificial blood has several advantages: it is safe from disease, it can be administered to any blood type, its availability is not dependent on blood donors, and it can be stored longer than whole blood, which is good for only about 40 days.

SUMMARY

Alkanes are called **saturated hydrocarbons** because they do not contain any double or triple bonds. Since they have only strong σ bonds and atoms with no partial charges, alkanes are very unreactive. Alkanes undergo **radical substitution reactions** with chlorine (Cl_2) or bromine (Br_2) at high temperatures or in the presence of light to form alkyl chlorides or alkyl bromides. The substitution reaction is a **radical chain reaction** with **initiation**, **propagation**, and **termination steps**.

In **heterolytic bond cleavage**, a bond breaks so that both electrons stay with one of the atoms; in **homolytic bond cleavage**, a bond breaks so that each of the atoms retains one of the bonding electrons.

The rate-determining step of the radical substitution reaction is the formation of an alkyl **radical** by removal of a hydrogen atom. Radicals are stabilized by electron-donating alkyl groups. Thus, a **tertiary alkyl radical** is more stable than a **secondary alkyl radical**, which is more stable than a

---

* The ozone over a given area is compressed to 0 °C and 1 atm pressure and the thickness of the slab is measured. 1 Dobson unit = 0.01 mm thickness.

primary alkyl radical. The relative rates of radical formation are benzylic ~ allyl > 3° > 2° > 1° > vinyl ~ methyl. Calculation of the relative amounts of products obtained from the radical halogenation of an alkane must take into account both probability and the relative rate at which a particular hydrogen is removed. The **reactivity–selectivity principle** states that the more reactive a species is, the less selective it will be. A bromine radical is *less reactive* than a chlorine radical, so a bromine radical is *more selective* about which hydrogen atom it removes. *N*-Bromosuccinimide (NBS) is used to brominate allylic carbons. Unwanted radical reactions are prevented by **radical inhibitors**, compounds that destroy reactive radicals by creating radicals that are relatively stable or compounds that have only paired electrons.

An alkyl peroxide is a **radical initiator** because it creates radicals. **Radical addition reactions** are also chain reactions with **initiation**, **propagation**, and **termination**

steps. A peroxide reverses the order of addition of H and Br because it causes Br·, instead of H⁺, to be the electrophile. The **peroxide effect** is observed only for the addition of HBr.

If a reactant that does not have an asymmetric center undergoes a radical substitution or radical addition reaction that forms a product with an asymmetric center, a racemic mixture will be obtained. A racemic mixture is also obtained if a hydrogen bonded to an asymmetric center is substituted by a halogen.

Some biological reactions involve radicals formed by the interaction of organic molecules with metal ions. The reactions take place in the active site of an enzyme.

Strong circumstantial evidence implicates synthetic chlorofluorocarbons as being responsible for depletion of the ozone layer. The interaction of these compounds with UV light generates chlorine radicals, which are the ozone-removing agents.

SUMMARY OF REACTIONS

1. *Alkanes* undergo radical substitution reactions with Cl_2 or Br_2 in the presence of heat or light (Sections 11.2–11.5).

$$CH_3CH_3 + Cl_2 \xrightarrow{\Delta \text{ or } h\nu} CH_3CH_2Cl + HCl$$
excess

$$CH_3CH_3 + Br_2 \xrightarrow{\Delta \text{ or } h\nu} CH_3CH_2Br + HBr$$
excess

bromination is more selective than chlorination

2. Radical addition of hydrogen bromide to an *alkene* in the presence of a peroxide (Br· is the electrophile; Section 11.6)

$$RCH{=}CH_2 + HBr \xrightarrow{\text{peroxide}} RCH_2CH_2Br$$

3. *Alkyl-substituted benzenes* undergo radical substitution at the benzylic position (Section 11.8).

4. *Alkenes* undergo radical substitution at allylic carbons. NBS is used for bromination at allylic carbons (Section 11.8).

KEY TERMS

alkane (p. 481)
allylic radical (p. 497)
benzylic radical (p. 497)
combustion (p. 483)
free radical (p. 484)
halogenation reaction (p. 483)
heterolysis (p. 483)
heterolytic bond cleavage (p. 483)
homolysis (p. 483)

homolytic bond cleavage (p. 483)
initiation step (p. 484)
peroxide effect (p. 495)
primary alkyl radical (p. 485)
propagation step (p. 484)
radical (p. 484)
radical addition reaction (p. 494)
radical chain reaction (p. 484)
radical inhibitor (p. 495)

radical initiator (p. 495)
radical substitution
    reaction (p. 484)
reactivity–selectivity
    principle (p. 491)
saturated hydrocarbons (p. 481)
secondary alkyl radical (p. 485)
termination step (p. 484)
tertiary alkyl radical (p. 485)

PROBLEMS

20. Give the product(s) of each of the following reactions, disregarding stereoisomers:

a. $CH_2=CHCH_2CH_2CH_3$ + NBS $\xrightarrow[\text{peroxide}]{\Delta}$

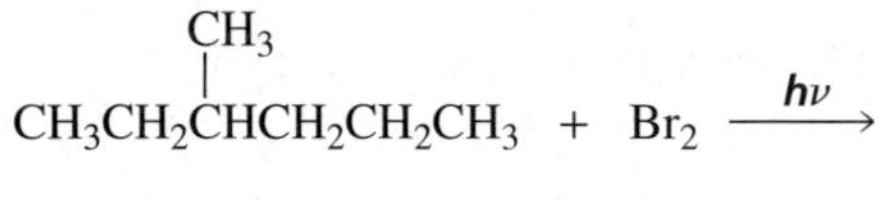

d. 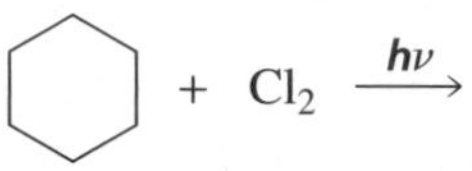 ⬡ + Cl_2 $\xrightarrow{h\nu}$

b. $CH_3\overset{\overset{\displaystyle CH_3}{|}}{C}=CHCH_3$ + NBS $\xrightarrow[\text{peroxide}]{\Delta}$

e. 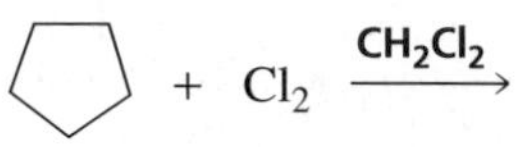 ⬠ + Cl_2 $\xrightarrow{CH_2Cl_2}$

c. $CH_3CH_2\overset{\overset{\displaystyle CH_3}{|}}{C}HCH_2CH_2CH_3$ + Br_2 $\xrightarrow{h\nu}$

f. 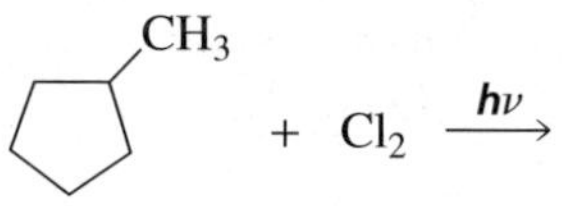 (cyclopentane with CH_3) + Cl_2 $\xrightarrow{h\nu}$

21. a. Identify an alkane with molecular formula C_5H_{12} that forms only one monochlorinated product when heated with Cl_2.
 b. Identify an alkane with molecular formula C_7H_{16} that forms seven monochlorinated products (disregarding stereoisomers) when heated with Cl_2.

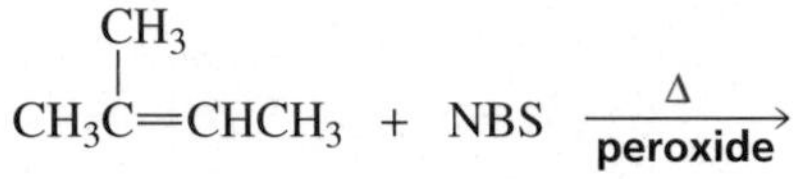

22. For each of the following compounds, give the major product that would be obtained from treating an excess of the compound with Cl_2 in the presence of light at room temperature. Disregard stereoisomers.

a. 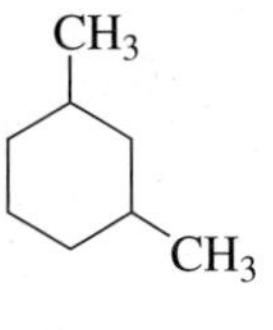 (cyclohexane with CH_3 and CH_3)

b. 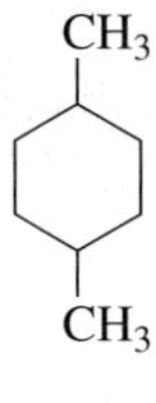 (cyclohexane with CH_3 and CH_3)

c. 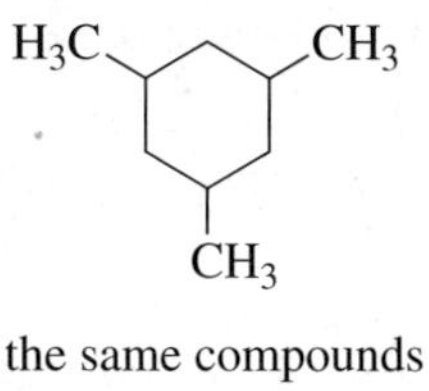 H_3C (cyclohexane with CH_3, CH_3)

23. What would the answers be to Problem 22 if the same compounds were treated with Br_2 at 125 °C?

24. Give the major product of each of the following reactions, disregarding stereoisomers:

a.  ⬠ + NBS $\xrightarrow[\text{peroxide}]{\Delta}$

d. 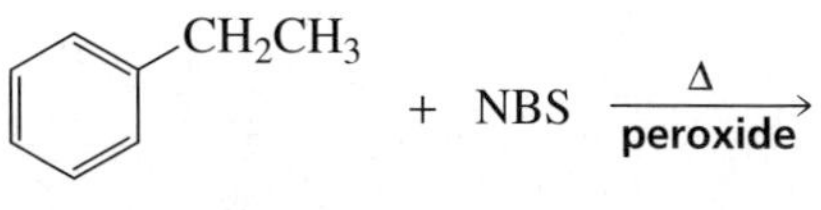 (benzene with CH_2CH_3) + NBS $\xrightarrow[\text{peroxide}]{\Delta}$

b. (cyclopentadiene) + NBS $\xrightarrow[\text{peroxide}]{\Delta}$

e. 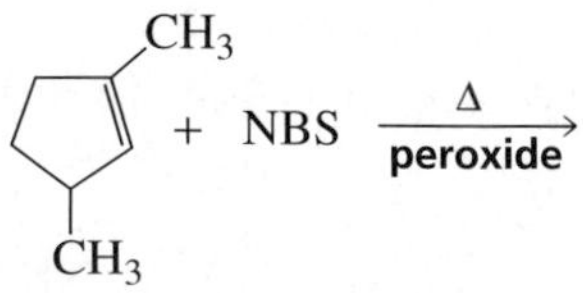 (cyclopentene with CH_3 and CH_3) + NBS $\xrightarrow[\text{peroxide}]{\Delta}$

c. (cyclohexene with CH_3) + NBS $\xrightarrow[\text{peroxide}]{\Delta}$

f. 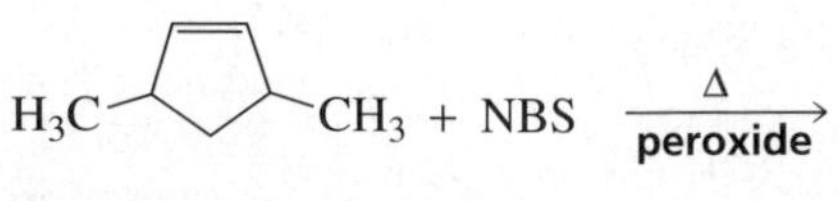 H_3C (cyclopentene) CH_3 + NBS $\xrightarrow[\text{peroxide}]{\Delta}$

25. Iodine (I_2) does not react with ethane, even though I_2 is more easily cleaved homolytically than the other halogens. Explain.

26. Propose a mechanism to account for the products formed in the following reaction:

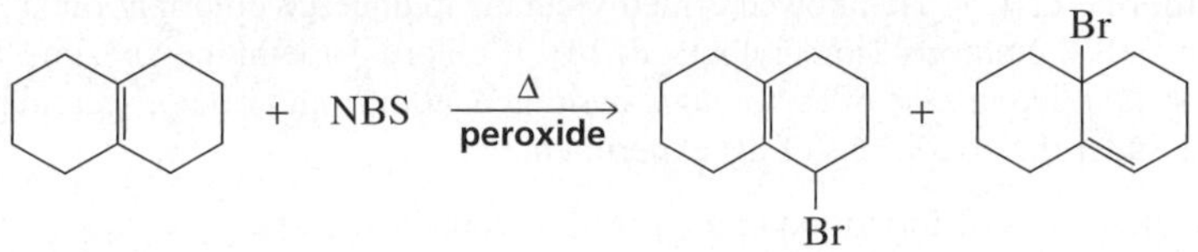

27. The deuterium kinetic isotope effect for chlorination of an alkane is defined in the following equation:

$$\textbf{deuterium kinetic isotope effect} = \frac{\textbf{rate of homolytic cleavage of a C—H bond by Cl}\cdot}{\textbf{rate of homolytic cleavage of a C—D bond by Cl}\cdot}$$

Predict whether chlorination or bromination would have a greater deuterium kinetic isotope effect.

28. **a.** How many monobromination products would be obtained from the radical bromination of methylcyclohexane? Disregard stereoisomers.
b. Which product would be obtained in greatest yield? Explain.
c. How many monobromination products would be obtained if all stereoisomers are included?

29. Give the major product of each of the following reactions, disregarding stereoisomers:

a. $CH_3CH(CH_3)CH_3 + Cl_2 \xrightarrow{h\nu}$

b. $CH_3CH(CH_3)CH_3 + Br_2 \xrightarrow{h\nu}$

c. 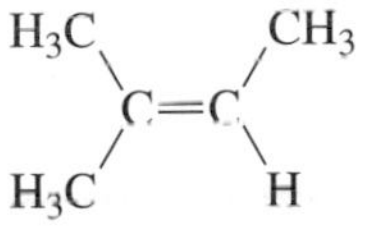 $\xrightarrow[\text{peroxide}]{\text{HBr}}$

d. $CH_2{=}CHCH_2CH_2C(CH_3){=}CH_2 \xrightarrow[\text{peroxide}]{\text{HBr}}$

e. 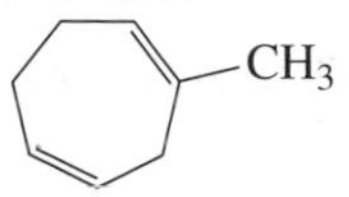 + HBr $\xrightarrow{\text{peroxide}}$

f. 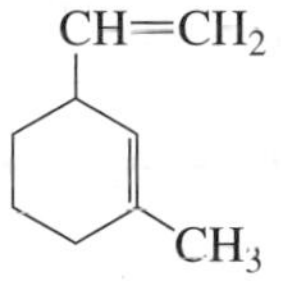 + HBr $\xrightarrow{\text{peroxide}}$

30. **a.** Propose a mechanism for the following reaction:

$$CH_3CH_3 + CH_3{-}C(CH_3)_2{-}OCl \xrightarrow{\Delta} CH_3CH_2Cl + CH_3{-}C(CH_3)_2{-}OH$$

b. Given that the $\Delta H°$ value for the reaction is -42 kcal/mol and the bond dissociation energies for the C—H, C—Cl, and O—H bonds are 101, 82, and 102 kcal/mol, respectively, calculate the bond dissociation energy of the O—Cl bond.
c. Which set of propagation steps is more likely to occur?

31. What stereoisomers would be obtained from the following reaction?

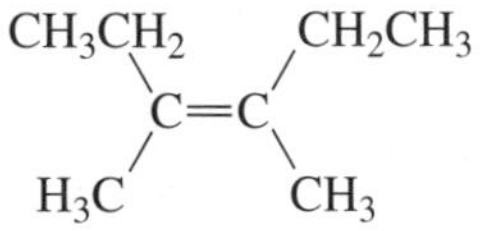 + HBr $\xrightarrow{H_2O_2}$

32. Using the given starting material and any necessary organic or inorganic reagents, indicate how the desired compounds could be synthesized:

a. $HC{\equiv}CH \longrightarrow CH_3CH_2CH_2CH_2CH_2Br$

b. $HOCH_2CH_2CH{=}CH_2 \longrightarrow$ 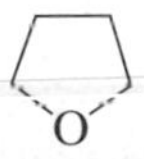

c. $CH_3CH_2CH{=}CH_2 \longrightarrow CH_2{=}CHCH{=}CH_2$

d. 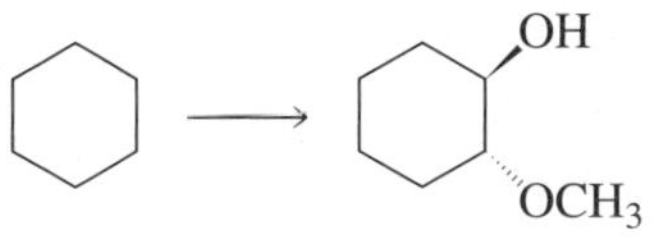

e. 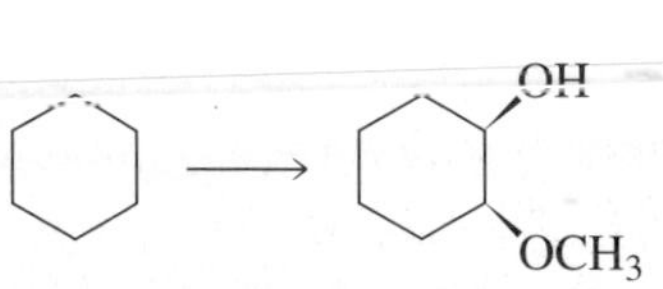

33. Dr. Al Cahall wanted to determine experimentally the relative ease of removal of a hydrogen atom from a tertiary, a secondary, and a primary carbon by a chlorine radical. He allowed 2-methylbutane to undergo chlorination at 300 °C and obtained as products 36% 1-chloro-2-methylbutane, 18% 2-chloro-2-methylbutane, 28% 2-chloro-3-methylbutane, and 18% 1-chloro-3-methylbutane. What values did he obtain for the relative ease of removing a hydrogen atom from tertiary, secondary, and primary hydrogen carbons by a chlorine radical under the conditions of his experiment?

34. At 600 °C, the ratio of the relative rates of formation of a tertiary, a secondary, and a primary radical by a chlorine radical is 2.6 : 2.1 : 1. Explain the change in the degree of regioselectivity compared with what Dr. Al Cahall found in Problem 33.

35. **a.** What five-carbon alkene will form the same product whether it reacts with HBr in the *presence* of a peroxide or with HBr in the *absence* of a peroxide?
    b. Give three 6-carbon alkenes that form the same product, whether they react with HBr in the *presence* of a peroxide or with HBr in the *absence* of a peroxide.

36. **a.** Calculate the $\Delta H°$ value for the following reaction:

$$CH_4 \ + \ Cl_2 \ \xrightarrow{h\nu} \ CH_3Cl \ + \ HCl$$

   b. Calculate the sum of the $\Delta H°$ values for the following two propagation steps:

$$CH_3{-}H \ + \ {}^{\cdot}Cl \ \longrightarrow \ {}^{\cdot}CH_3 \ + \ H{-}Cl$$
$$^{\cdot}CH_3 \ + \ Cl{-}Cl \ \longrightarrow \ CH_3{-}Cl \ + \ {}^{\cdot}Cl$$

   c. Why do both calculations give you the same value of $\Delta H°$?

37. A possible alternative mechanism to that shown in Problem 36 for the monochlorination of methane would involve the following propagation steps:

$$CH_3{-}H \ + \ {}^{\cdot}Cl \ \longrightarrow \ CH_3{-}Cl \ + \ {\cdot}H$$
$${\cdot}H \ + \ Cl{-}Cl \ \longrightarrow \ H{-}Cl \ + \ {}^{\cdot}Cl$$

   How do you know that the reaction does not take place by this mechanism?

38. Propose a mechanism for the following reaction:

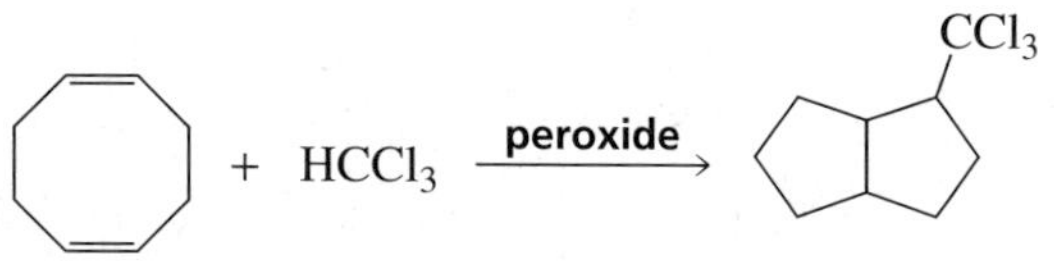

39. Using the given starting material and any necessary organic or inorganic reagents, indicate how the desired compounds could be synthesized:

   a.

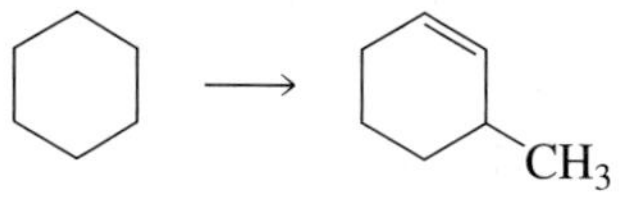

   c.

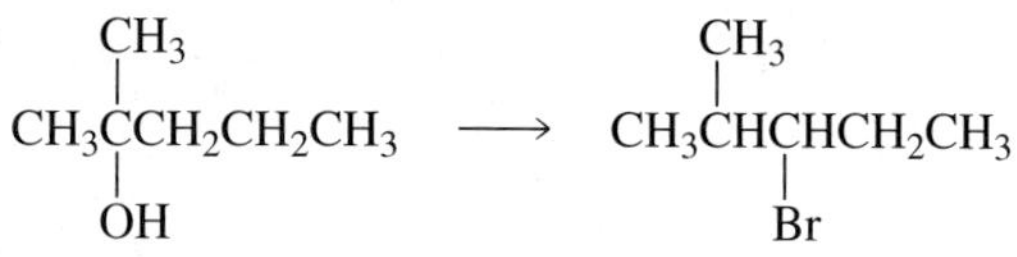

   b. $CH_3CH_2CH{=}CH_2 \ \longrightarrow \ CH_3CH_2CH_2CH_2CH_2CH_2CH_2CH_3$

40. Explain why the rate of bromination of methane decreases if HBr is added to the reaction mixture.

PART 4

Identification of Organic Compounds

By now you have worked through many problems that asked you to design the synthesis of an organic compound. But if you were actually to go into the laboratory to carry out a synthesis you designed, how would you know that the compound you obtained was the one you had set out to prepare? When a scientist discovers a new compound with physiological activity, its structure must be ascertained. Only after its structure is known, can methods to synthesize the compound be designed and studies to provide insights into its biological behavior be carried out. Clearly, chemists need to be able to determine the structures of compounds.

CHAPTER 12
Mass Spectrometry, Infrared Spectroscopy, and Ultraviolet/Visible Spectroscopy

In **Chapter 12** you will learn about three instrumental techniques that chemists use to analyze compounds. **Mass spectrometry** is used to find the molecular mass and the molecular formula of an organic compound and to identify certain structural features of the compound. **Infrared (IR) spectroscopy** allows us to identify the kinds of functional groups in an organic compound. **Ultraviolet/visible (UV/Vis)** spectroscopy provides information about organic compounds with conjugated double bonds.

CHAPTER 13
NMR Spectroscopy

Chapter 13 discusses **nuclear magnetic resonance (NMR) spectroscopy**, which provides information about the carbon–hydrogen framework of an organic compound.

CHAPTER 12

Mass Spectrometry, Infrared Spectroscopy, and Ultraviolet/Visible Spectroscopy

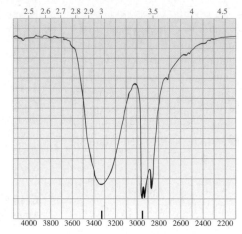

BUILDING ON FUNDAMENTALS

SECTION 12.3 Isotopes (1.1) are useful in analyzing mass spectra.

SECTION 12.9 We look again at dipole moments of bonds (1.3), since the greater the change in dipole moment, the greater is the intensity of an IR absorption band.

SECTION 12.11 The greater the *s* character of the carbon, the stronger is the bond that it forms (1.23). This relationship becomes important in IR spectroscopy, since stronger bonds absorb IR light of higher wavenumbers.

SECTION 12.11 We see more examples of electron donation by resonance (7.9) and of inductive electron withdrawal (1.21).

SECTION 12.11 Hydrogen bonding (2.9) affects the position of an IR absorption band.

SECTION 12.18 Conjugated double bonds (7.7) and molecular orbital concepts (7.12) are important in UV/Vis spectroscopy.

Determining the structures of organic compounds is an important part of organic chemistry. Whenever a chemist synthesizes a compound, its structure must be confirmed. For example, you were told that a ketone is formed when an alkyne undergoes the acid-catalyzed addition of water (Section 6.7). But how was it determined that the product of that reaction is actually a ketone?

Scientists search the world for new compounds with physiological activity. If a promising compound is found, its structure needs to be determined. Without knowing its structure, chemists cannot design ways to synthesize the compound, nor can they undertake studies to provide insights into its biological behavior.

Before the structure of a compound can be determined, the compound must be isolated. For example, if the product of a reaction carried out in the laboratory is to be identified, it must first be isolated from the solvent and from any unreacted starting materials, as well as from any side products that might have formed. A compound found in nature must be isolated from the organism that manufactures it.

Isolating products and figuring out their structures used to be daunting tasks. The only tools chemists had for isolating products were distillation (for liquids) and sublimation or fractional recrystallization (for solids). Today, a variety of chromatographic techniques allow compounds to be isolated relatively easily. You will learn about these techniques in your laboratory course.

At one time, determining the structure of an organic compound required finding out its molecular formula by elemental analysis, determining the compound's physical properties (its melting point, boiling point, and so on), and conducting simple chemical tests that indicate the presence (or absence) of certain functional groups. For example, when an aldehyde is added to a test tube containing a solution of silver oxide in ammonia, a silver mirror is formed on the inside of the test tube. Only aldehydes have this ability, so if a mirror forms, you can conclude that the unknown compound is an aldehyde; if a mirror does not form, the compound is not an aldehyde. Another example of a simple test is the Lucas test, which distinguishes primary, secondary, and tertiary alcohols by how rapidly the test solution turns cloudy after the addition of the Lucas reagent (Section 10.1). Unfortunately, these simple procedures were inadequate for characterizing molecules with complex structures, and because a relatively large sample of the compound was needed in order to perform all the tests, they were impractical for the analysis of compounds that were difficult to obtain.

Today, a number of different instrumental techniques are used to identify organic compounds. These techniques can be performed quickly on small amounts of a compound and can provide much more information about the compound's structure than simple chemical tests can give. In this chapter, we will look at three instrumental techniques:

- **Mass spectrometry** allows us to determine the *molecular mass* and the *molecular formula* of a compound, as well as some of its *structural features*.

- **Infrared spectroscopy** tells us the *kinds of functional groups* a compound has.

- **Ultraviolet/visible (UV/Vis) spectroscopy** provides information about organic compounds with conjugated double bonds.

Of these instrumental techniques, mass spectrometry is the only one that does not employ electromagnetic radiation. Thus, it is called *spectrometry*, whereas the others are called *spectroscopy*.

We will be referring to different classes of organic compounds as we discuss various instrumental techniques; they are listed in Table 12.1. (They are also listed inside the back cover of the book for easy reference.)

12.1 Mass Spectrometry

At one time, the molecular weight of a compound was determined by its vapor density or freezing-point depression, and molecular formulas were determined by elemental analysis, a technique for measuring the relative proportions of the elements in the compound. These were long and tedious procedures that required a relatively large amount of a very pure sample of the compound. Today, molecular weights and molecular formulas can be rapidly determined by mass spectrometry from a very small sample.

In mass spectrometry, a small amount of a compound is introduced into an instrument called a mass spectrometer, where it is vaporized and then ionized (an electron is removed from each molecule). Ionization can be accomplished in several ways. A common method bombards the vaporized molecules with a beam of high-energy electrons. The energy of the beam can be varied, but is typically about 70 electron volts (1614 kcal/mol). When the electron beam hits a molecule, it knocks out an electron, producing a **molecular ion**. A molecular ion is a **radical cation**, a species with an unpaired electron and a positive charge.

$$M \xrightarrow{\overset{\text{electron}}{\underset{}{\text{beam}}}} M^{+\cdot} \quad + \quad e^{-}$$

molecule   molecular ion   electron
              a radical cation

Electron bombardment injects so much kinetic energy into the molecular ions that most of them break apart into smaller cations, radicals, neutral molecules, and other

Table 12.1	Classes of Organic Compounds			
Alkane	$-\overset{\mid}{\underset{\mid}{C}}-$	contains only C—C and C—H bonds	Aldehyde	$\overset{O}{\overset{\|}{RCH}}$
Alkene	$\underset{\diagdown}{\overset{\diagup}{C}}=\underset{\diagup}{\overset{\diagdown}{C}}$		Ketone	$\overset{O}{\overset{\|}{RCR}}$
Alkyne	$-C\equiv C-$		Carboxylic acid	$\overset{O}{\overset{\|}{RCOH}}$
Nitrile	$-C\equiv N$		Ester	$\overset{O}{\overset{\|}{RCOR}}$
Alkyl halide	RX	X = F, Cl, Br, or I		$\overset{O}{\overset{\|}{RCNH_2}}$
Ether	ROR		Amides	$\overset{O}{\overset{\|}{RCNHR}}$
Alcohol	ROH			$\overset{O}{\overset{\|}{RCNR_2}}$
Phenol	ArOH	Ar = ⬡—	Amine (primary)	RNH_2
			Amine (secondary)	R_2NH
Aniline	$ArNH_2$		Amine (tertiary)	R_3N

radical cations. Not surprisingly, the bonds most likely to break are the weakest ones and those that result in the formation of the most stable products. All the *positively charged fragments* of the molecule are drawn between two negatively charged plates, which accelerate the fragments into an analyzer tube (Figure 12.1). Neutral fragments

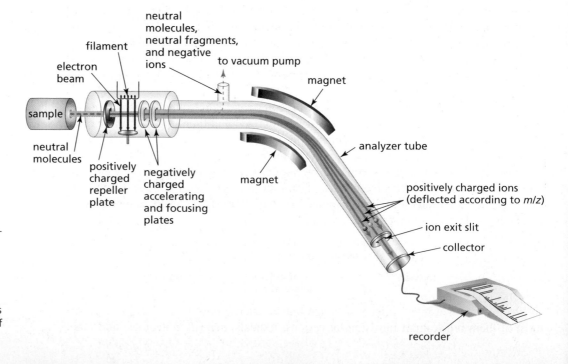

Figure 12.1 ▶
Schematic diagram of a mass spectrometer. A beam of high-energy electrons causes molecules to ionize and fragment. Positively charged fragments pass through the analyzer tube. Changing the magnetic field strength allows the separation of fragments of varying mass-to-charge ratio.

are not attracted to the negatively charged plates and therefore are not accelerated. They are eventually pumped out of the spectrometer.

The analyzer tube is surrounded by a magnet whose magnetic field deflects the positively charged fragments in a curved path. At a given magnetic field strength, the degree to which the path is curved depends on the mass-to-charge ratio (m/z) of the fragment: the path of a fragment with a smaller m/z value will bend more than that of a heavier fragment. In this way, the particles with the same m/z values can be separated from all the others. If a fragment's path matches the curvature of the analyzer tube, the fragment will pass through the tube and out the ion exit slit. A collector records the relative number of fragments with a particular m/z passing through the slit. The more stable the fragment, the more likely it is to arrive at the collector without breaking down further. The strength of the magnetic field is gradually increased, so fragments with progressively larger m/z values are guided through the tube and out the exit slit.

The mass spectrometer records a **mass spectrum**—a graph of the relative abundance of each fragment plotted against its m/z value. Because the charge (z) on essentially all the fragments that reach the collector plate is +1, m/z is the molecular mass (m) of the fragment. *Remember that only positively charged species reach the collector.*

> **A mass spectrum records only positively charged fragments.**

PROBLEM 1◆

Which of the following fragments produced in a mass spectrometer would be accelerated through the analyzer tube?

$$CH_3\overset{\cdot}{C}H_2 \qquad CH_3CH_2\overset{+}{C}H_2 \qquad [CH_3CH_2CH_3]^{\cdot +} \qquad \overset{\cdot}{C}H_2CH\!=\!CH_2 \qquad \overset{+}{C}H_2CH\!=\!CH_2$$

12.2 The Mass Spectrum • Fragmentation

The molecular ion produced in the spectrometer and the pattern of fragment ion peaks that the spectrometer records are unique for each compound. A mass spectrum, therefore, is like a fingerprint of the compound. A positive identification of a compound can be made by comparing its mass spectrum with the mass spectrum of the actual compound.

The mass spectrum of pentane is shown in Figure 12.2. Each m/z value in the spectrum is the **nominal molecular mass** of one of the fragments—the molecular mass to the nearest whole number. ^{12}C is defined as having a mass of 12.000 atomic mass units

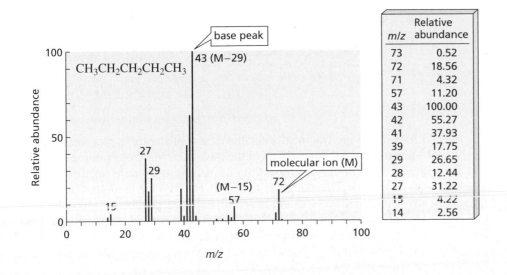

m/z	Relative abundance
73	0.52
72	18.56
71	4.32
57	11.20
43	100.00
42	55.27
41	37.93
39	17.75
29	26.65
28	12.44
27	31.22
15	4.22
14	2.56

◄ **Figure 12.2**
The mass spectrum of pentane, shown as a bar graph and in tabular form. The base peak represents the fragment that appears in greatest abundance. The m/z value of the molecular ion gives the molecular mass of the compound.

(amu), and the masses of other atoms are based on this standard. For example, a proton has a mass of 1.007825 amu. Pentane, therefore, has a *molecular mass* of 72.0939 and a *nominal molecular mass* of 72.

The peak with the highest m/z value in the spectrum—in this case, at $m/z = 72$—corresponds to the fragment that results when an electron is knocked out of a molecule of the injected sample—in this case, a pentane molecule. In other words, the peak with the highest m/z value represents the molecular ion (M) of pentane. (The extremely tiny peak at $m/z = 73$ will be explained later.) Since it is not known what bond loses the electron, the molecular ion is written in brackets and the positive charge and unpaired electron are assigned to the entire structure. *The m/z value of the molecular ion gives the molecular mass of the compound.* Peaks with smaller m/z values—called **fragment ion peaks**—represent positively charged fragments of the molecule.

> **The m/z value of the molecular ion gives the molecular mass of the compound.**

$$CH_3CH_2CH_2CH_2CH_3 \xrightarrow{\text{electron beam}} [CH_3CH_2CH_2CH_2CH_3]^{\ddagger} + e^-$$

molecular ion
$m/z = 72$

The **base peak** is the tallest peak, due to its having the greatest relative abundance. The base peak is assigned a relative abundance of 100%, and the relative abundance of each of the other peaks is reported as a percentage of the base peak. Mass spectra can be shown either as bar graphs or in tabular form.

A mass spectrum gives us structural information about the compound because *the m/z values and relative abundances of the fragments depend on the strength of the molecular ion's bonds and the stability of the fragments.* Weak bonds break in preference to strong bonds, and bonds that break to form more stable fragments break in preference to those that form less stable fragments.

For example, all the C—C bonds in the molecular ion formed from pentane have about the same strength. However, the C-2—C-3 bond is more likely to break than the C-1—C-2 bond because C-2—C-3 fragmentation leads to a *primary* carbocation and a *primary* radical, which together are more stable than the *primary* carbocation and *methyl* radical (or *primary* radical and *methyl* cation) obtained from C-1—C-2 fragmentation. Ions formed by C-2—C-3 fragmentation have $m/z = 43$ or 29, whereas ions formed by C-1—C-2 fragmentation have $m/z = 57$ or 15. The base peak of 43 in the mass spectrum of pentane indicates the greater likelihood of C-2—C-3 fragmentation. (See Sections 7.7 and 11.3 to review the relative stabilities of carbocations and radicals, respectively.)

$$[\overset{1}{C}H_3\overset{2}{C}H_2\overset{3}{C}H_2\overset{4}{C}H_2\overset{5}{C}H_3]^{\ddagger}$$
molecular ion
$m/z = 72$

$$\longrightarrow CH_3\dot{C}H_2 + \overset{+}{C}H_2CH_2CH_3$$
$m/z = 43$

$$\longrightarrow CH_3\overset{+}{C}H_2 + \dot{C}H_2CH_2CH_3$$
$m/z = 29$

$$\longrightarrow \dot{C}H_3 + \overset{+}{C}H_2CH_2CH_2CH_3$$
$m/z = 57$

$$\longrightarrow \overset{+}{C}H_3 + \dot{C}H_2CH_2CH_2CH_3$$
$m/z = 15$

A method for identifying fragment ions makes use of the difference between the m/z value of a given fragment ion and that of the molecular ion. For example, the fragment ion with $m/z = 43$ in the mass spectrum of pentane is 29 units smaller than the molecular ion ($72 - 43 = 29$). An ethyl radical (CH_3CH_2) has a molecular mass of 29 (because the mass numbers of C and H are 12 and 1, respectively). Thus, the peak at 43 can be attributed to the molecular ion minus an ethyl radical. Similarly, the fragment ion with $m/z = 57$ can be attributed to the molecular ion minus a methyl radical ($72 - 57 = 15$). Peaks at $m/z = 15$ and $m/z = 29$ are readily recognizable as being due to methyl and ethyl cations, respectively. Appendix VI contains a table of common fragment ions and a table of common fragments lost.

Peaks are commonly observed at m/z values two units below the m/z values of the carbocations, because the carbocations can lose two hydrogen atoms.

$$CH_3CH_2\overset{+}{C}H_2 \longrightarrow \overset{+}{C}H_2CH=CH_2 + H\cdot$$
$$m/z = 43 \qquad\qquad m/z = 41$$

2-Methylbutane has the same molecular formula as pentane, so it, too, has a molecular ion with $m/z = 72$ (Figure 12.3). Its mass spectrum is similar to that of pentane, with one notable exception: the peak at $m/z = 57$, indicating loss of a methyl radical, is much more intense.

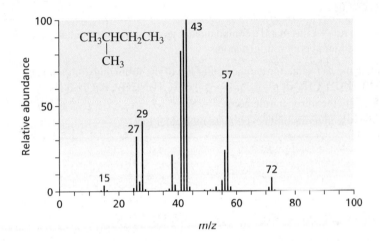

◀ **Figure 12.3**
The mass spectrum of 2-methylbutane.

2-Methylbutane is more likely than pentane to lose a methyl radical, because, when it does, a *secondary* carbocation is formed. In contrast, when pentane loses a methyl radical, a less stable *primary* carbocation is formed.

$$\begin{matrix} CH_3 \\ | \\ [CH_3CHCH_2CH_3]^{+\cdot} \end{matrix} \longrightarrow CH_3\overset{+}{C}HCH_2CH_3 + \overset{\cdot}{C}H_3$$
molecular ion $m/z = 57$
$m/z = 72$

PROBLEM 2

What would distinguish the mass spectrum of 2,2-dimethylpropane from those of pentane and 2-methylbutane?

PROBLEM 3◆

What is the likeliest m/z value for the base peak in the mass spectrum of 3-methylpentane?

PROBLEM 4   **SOLVED**

The mass spectra of two very stable cycloalkanes both show a molecular ion peak at $m/z = 98$. One spectrum shows a base peak at $m/z = 69$; the other shows a base peak at $m/z = 83$. Identify the cycloalkanes.

Solution   The molecular formula for a cycloalkane is C_nH_{2n}. Because the molecular mass of both cycloalkanes is 98, their molecular formulas must be C_7H_{14} ($7 \times 12 = 84$; $84 + 14 = 98$).

A base peak of 69 means the loss of an ethyl substituent ($98 - 69 = 29$), whereas a base peak of 83 means the loss of a methyl substituent ($98 - 83 = 15$). Because the two cyclo-alkanes are known to be very stable, we assume they do not have three- or four-membered rings. A seven-carbon cycloalkane with a base peak signifying the loss of an ethyl substituent must be ethylcyclopentane. A seven-carbon cycloalkane with a base peak signifying the loss of a methyl substituent must be methylcyclohexane.

PROBLEM 5

The "nitrogen rule" states that if a compound has an odd-mass molecular ion, the compound contains an odd number of nitrogen atoms.

a. Calculate the m/z value for the molecular ion of the following compounds:
   1. $CH_3CH_2CH_2CH_2NH_2$ 2. $H_2NCH_2CH_2CH_2NH_2$
b. Explain why the nitrogen rule holds.
c. State the rule in terms of an even-mass molecular ion.

12.3 Isotopes in Mass Spectrometry

Although the molecular ions of pentane and 2-methylbutane both have m/z values of 72, each spectrum shows a very small peak at $m/z = 73$ (Figures 12.2 and 12.3). This peak is called an M + 1 peak because the ion responsible for it is one unit heavier than the molecular ion. The M + 1 peak owes its presence to the fact that there are two naturally occurring isotopes of carbon: 98.89% of natural carbon is ^{12}C and 1.11% is ^{13}C (Section 1.1). Thus, 1.11% of the carbons in the molecular ions contain a ^{13}C instead of a ^{12}C and therefore appear at M + 1.

Peaks that are attributable to isotopes can help identify the compound responsible for a mass spectrum. For example, if a compound contains five carbon atoms, the relative abundance of the M + 1 ion is estimated to be $5(1.1\%) = 5(.011)$, multiplied by the relative abundance of the molecular ion. This means that the number of carbon atoms in a compound can be calculated if the relative intensities of both the M and M + 1 peaks are known.

$$\text{number of carbon atoms} = \frac{\text{relative intensity of M + 1 peak}}{.011 \times (\text{relative intensity of M peak})}$$

The isotopic distributions of several elements commonly found in organic compounds are shown in Table 12.2. From the isotopic distributions, we see that the M + 1 peak can be used to determine the number of carbon atoms in compounds containing H, O, or halogens because the contributions to the M + 1 peak by isotopes of H, O, and the halogens are very small or nonexistent. This formula does not work as well in predicting the number of carbon atoms in a nitrogen-containing compound because the natural abundance of ^{15}N is relatively high.

Mass spectra can show M + 2 peaks as a result of a contribution from ^{18}O or from having two heavy isotopes in the same molecule (say, ^{13}C and ^2H, or two ^{13}C's). Most

Table 12.2 The Natural Abundance of Isotopes Commonly Found in Organic Compounds

Element		Natural abundance		
Carbon	^{12}C 98.89%	^{13}C 1.11%		
Hydrogen	^{1}H 99.99%	^{2}H 0.01%		
Nitrogen	^{14}N 99.64%	^{15}N 0.36%		
Oxygen	^{16}O 99.76%	^{17}O 0.04%	^{18}O 0.20%	
Sulfur	^{32}S 95.0%	^{33}S 0.76%	^{34}S 4.22%	^{36}S 0.02%
Fluorine	^{19}F 100%			
Chlorine	^{35}Cl 75.77%		^{37}Cl 24.23%	
Bromine	^{79}Br 50.69%		^{81}Br 49.31%	
Iodine	^{127}I 100%			

of the time, the M + 2 peak is very small. The presence of a large M + 2 peak is evidence of a compound containing either chlorine or bromine, because each of these elements has a high percentage of a naturally occurring isotope that is two units heavier than the most abundant isotope. From the natural abundance of the isotopes of chlorine and bromine in Table 12.2, one can conclude that if the M + 2 peak is one-third the height of the molecular ion peak, then the compound contains one chlorine atom because the natural abundance of ^{37}Cl is one-third that of ^{35}Cl. If the M and M + 2 peaks are about the same height, then the compound contains one bromine atom because the natural abundances of ^{79}Br and ^{81}Br are about the same.

In calculating the molecular masses of molecular ions and fragments, the *atomic mass* of a single isotope of the atom must be used (Cl = 35 or 37, for example) because mass spectrometry measures the *m/z* value of an *individual* fragment. The *atomic weights* in the periodic table (Cl = 35.453) cannot be used, because they are the *weighted averages* of all the naturally occurring isotopes for that element.

PROBLEM 6◆

The mass spectrum of an unknown compound has a molecular ion peak with a relative abundance of 43.27% and an M + 1 peak with a relative abundance of 3.81%. How many carbon atoms are in the compound?

PROBLEM 7◆

Predict the relative intensities of the molecular ion peak, the M + 2 peak, and the M + 4 peak for CH_2Br_2.

12.4 High-Resolution Mass Spectrometry Can Reveal Molecular Formulas

All the mass spectra shown in this text were produced with a low-resolution mass spectrometer. Such spectrometers give the *nominal molecular mass* of a fragment—the mass to the nearest whole number. High-resolution mass spectrometers can determine the *exact molecular mass* of a fragment to a precision of 0.0001 amu, making it possible to distinguish between compounds with the same nominal mass. For example, the following listing shows six compounds having a nominal molecular mass of 122 amu, but each of them has a different exact molecular mass.

Some Compounds with a Nominal Molecular Mass of 122 amu and Their Exact Molecular Masses and Molecular Formulas

Exact molecular mass (amu)	122.1096	122.0845	122.0732	122.0368	122.0579	122.0225
Molecular formula	C_9H_{14}	$C_7H_{10}N_2$	$C_8H_{10}O$	$C_7H_6O_2$	$C_4H_{10}O_4$	$C_4H_{10}S_2$

The exact masses of some common isotopes are listed in Table 12.3. Some computer programs can determine the molecular formula of a compound from the compound's exact molecular mass.

Table 12.3	The Exact Masses of Some Common Isotopes		
Isotope	**Mass**	**Isotope**	**Mass**
^1H	1.007825 amu	^{32}S	31.9721 amu
^{12}C	12.00000 amu	^{35}Cl	34.9689 amu
^{14}N	14.0031 amu	^{79}Br	78.9183 amu
^{16}O	15.9949 amu		

PROBLEM 8◆

Which molecular formula has an exact molecular mass of 86.1096 amu: C_6H_{14}, $C_4H_{10}N_2$, or $C_4H_6O_2$?

12.5 Fragmentation Patterns of Functional Groups

Each functional group has characteristic fragmentation patterns that can help identify a compound. The patterns began to be recognized after the mass spectra of many compounds containing a particular functional group had been studied. We will look at the fragmentation patterns of alkyl halides, ethers, alcohols, and ketones as examples.

Alkyl Halides

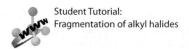
Student Tutorial:
Fragmentation of alkyl halides

Let's look first at the mass spectrum of 1-bromopropane, shown in Figure 12.4. The relative heights of the M and M + 2 peaks are about equal, which we have seen is characteristic of a compound containing a bromine atom. Electron bombardment is most likely to dislodge a lone-pair electron if the molecule has any, because a molecule does not hold onto its lone-pair electrons as tightly as it holds onto its bonding electrons. Thus, the molecular ion is created when electron bombardment dislodges one of bromine's lone-pair electrons.

$$CH_3CH_2CH_2 \overset{79}{\text{:}}\ddot{\text{Br}}\text{:} \ + \ CH_3CH_2CH_2 \overset{81}{\text{:}}\ddot{\text{Br}}\text{:} \ \xrightarrow{-e^-} \ CH_3CH_2CH_2 \overset{79}{\overset{+}{\text{Br}}}\text{:} \ + \ CH_3CH_2CH_2 \overset{81}{\overset{+}{\text{Br}}}\text{:} \ \longrightarrow \ CH_3CH_2\overset{+}{CH}_2 \ + \ ^{79}\text{:}\ddot{\text{Br}}\text{:} \ + \ ^{81}\text{:}\ddot{\text{Br}}\text{:}$$

1-bromopropane $m/z = 122$ $m/z = 124$ $m/z = 43$

The weakest bond in this molecular ion is the C—Br bond (the C—Br bond dissociation energy is 69 kcal/mol; the C—C bond dissociation energy is 85 kcal/mol; see Table 3.2 on page 146). When the C—Br bond breaks, it breaks heterolytically, with both electrons going to the more electronegative of the atoms that were joined by the bond, forming a propyl cation and a bromine atom. As a result, the base peak in the mass spectrum of 1-bromopropane is at $m/z = 43$ [M − 79, or (M + 2) − 81].

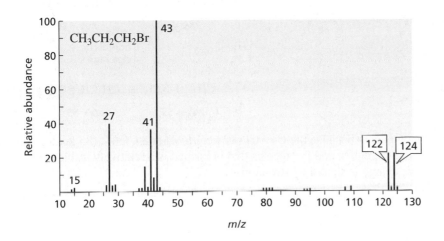

◀ **Figure 12.4**
The mass spectrum of 1-bromopropane.

The mass spectrum of 2-chloropropane is shown in Figure 12.5. We know that the compound contains a chlorine atom, because the M + 2 peak is one-third the height of the molecular ion peak (Section 12.3).

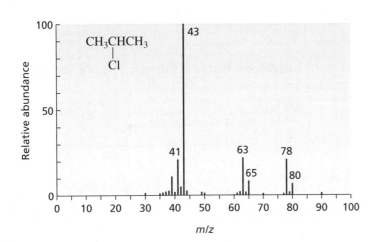

◀ **Figure 12.5**
The mass spectrum of 2-chloropropane.

The base peak at $m/z = 43$ results from *heterolytic cleavage* of the C—Cl bond. The peaks at $m/z = 63$ and $m/z = 65$ have a 3:1 ratio, indicating that these fragments contain a chlorine atom. They result from *homolytic cleavage* of a C—C bond at the α-carbon (the carbon bonded to the chlorine). This cleavage is known as **α-cleavage**.

The way a molecular ion fragments depends on the strength of its bonds and the stability of the fragments.

Recall that an arrowhead with one barb represents the movement of one electron.

α-Cleavage occurs because the C—Cl (82 kcal/mol) and C—C (85 kcal/mol) bonds have similar strengths, and the species that is formed is a relatively stable cation, since its positive charge is shared by two atoms:

$$CH_3CH=\overset{+}{\underset{..}{Cl}}: \longleftrightarrow CH_3\overset{+}{C}H-\overset{..}{\underset{..}{Cl}}$$

α-Cleavage is less likely to occur in alkyl bromides because the C—C bond is much stronger than the C—Br bond.

PROBLEM 9

Sketch the mass spectrum of 1-chloropropane.

Ethers

Student Tutorial:
Fragmentation of ethers

The mass spectrum of 2-isopropoxybutane is shown in Figure 12.6. The fragmentation pattern of an ether is similar to that of an alkyl halide.

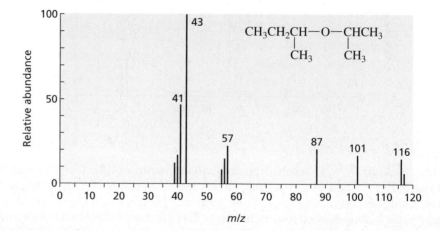

Figure 12.6 ▶
The mass spectrum of
2-isopropoxybutane.

1. Electron bombardment dislodges one of the lone-pair electrons from oxygen.
2. Fragmentation of the resulting molecular ion occurs mainly in two ways:
   a. A C—O bond is cleaved *heterolytically*, with the electrons going to the more electronegative oxygen atom.

$$CH_3CH_2CH(CH_3)-\ddot{O}-CHCH_3(CH_3) \xrightarrow{-e^-} CH_3CH_2CH(CH_3)-\overset{+}{\ddot{O}}-CHCH_3(CH_3)$$

2-isopropoxybutane   *m/z* = 116

$$CH_3CH_2\overset{+}{C}H(CH_3) + :\dot{O}-CHCH_3(CH_3)$$
m/z = 57

$$CH_3CH_2CH(CH_3)-\dot{O}: + \overset{+}{C}HCH_3(CH_3)$$
m/z = 43

   b. A C—C bond is cleaved *homolytically* at an α-position because it leads to a relatively stable cation in which the positive charge is shared by two atoms (a carbon and an oxygen). The alkyl group that breaks off to form the most stable radical is the alkyl group most easily cleaved. Thus, the peak at *m/z* = 87 is more abundant than the one at *m/z* = 101 because a primary radical is more stable than a methyl radical, even though the compound has three methyl groups bonded to α-carbons that can be cleaved to produce the peak at *m/z* = 101.

$$CH_3CH_2-CH(CH_3)-\overset{+}{\ddot{O}}-CHCH_3(CH_3) \xrightarrow{\alpha\text{-cleavage}} CH(CH_3)=\overset{+}{O}-CHCH_3(CH_3) + CH_3\dot{C}H_2$$
m/z = 87

$$CH_3CH_2CH(CH_3)-\overset{+}{\ddot{O}}-CHCH_3(CH_3) \xrightarrow{\alpha\text{-cleavage}} CH_3CH_2CH=\overset{+}{O}-CHCH_3(CH_3) + \dot{C}H_3$$
m/z = 101

$$CH_3CH_2CH(CH_3)-\overset{+}{\ddot{O}}-CHCH_3(CH_3) \xrightarrow{\alpha\text{-cleavage}} CH_3CH_2CH(CH_3)-\overset{+}{O}=CHCH_3 + \dot{C}H_3$$
m/z = 101

α-carbon

PROBLEM 10◆

The mass spectra of 1-methoxybutane, 2-methoxybutane, and 2-methoxy-2-methylpropane are shown in Figure 12.7. Match each compound with its spectrum.

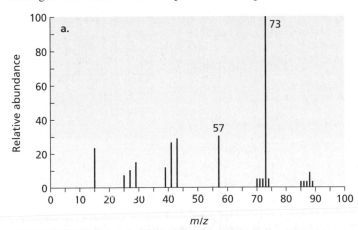

▲ **Figure 12.7**
The mass spectra referred to in Problem 10.

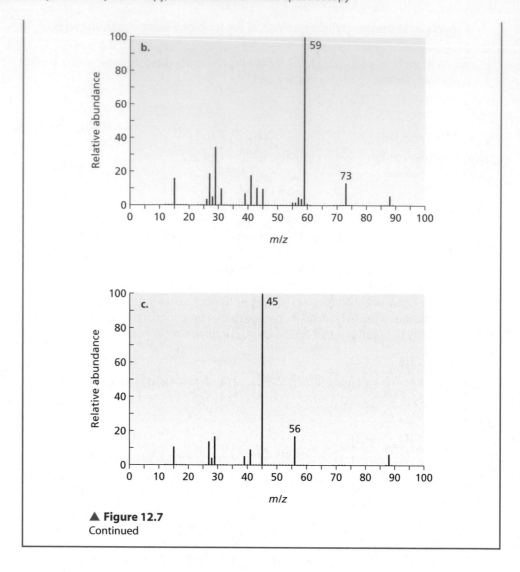

▲ **Figure 12.7**
Continued

Alcohols

The molecular ions obtained from alcohols fragment so readily that few of them survive to reach the collector. As a result, the mass spectra of alcohols show small molecular ion peaks. Notice the small molecular ion peak at $m/z = 102$ in the mass spectrum of 2-hexanol (Figure 12.8).

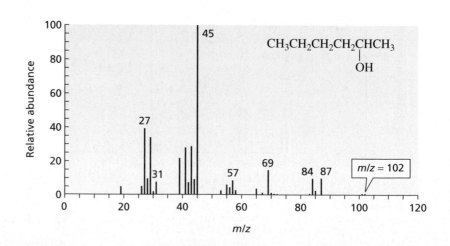

Figure 12.8 ▶
The mass spectrum of 2-hexanol.

Student Tutorial:
Fragmentation of alcohols

Like alkyl halides and ethers, alcohols undergo α-cleavage. Consequently, the mass spectrum of 2-hexanol shows a base peak at $m/z = 45$ (α-cleavage leading to a more stable butyl radical) and a smaller peak at $m/z = 87$ (α-cleavage leading to a less stable methyl radical).

$$CH_3CH_2CH_2CH_2\overset{:\ddot{O}H}{\underset{|}{CH}}CH_3 \xrightarrow{-e^-} CH_3CH_2CH_2CH_2\overset{^+\ddot{O}H}{\underset{|}{CH}}CH_3$$

2-hexanol $m/z = 102$

α carbon

α-cleavage → $CH_3CH_2CH_2\dot{C}H_2$ + $CH_3CH{=}\overset{+}{\ddot{O}}H$ $m/z = 45$

α-cleavage → $CH_3CH_2CH_2CH_2CH{=}\overset{+}{\ddot{O}}H$ + $\dot{C}H_3$ $m/z = 87$

In all the fragmentations we have seen so far, only one bond is broken. With alcohols, however, an important fragmentation occurs in which two bonds are broken. Two bonds break because the fragmentation forms a stable water molecule. The water is composed of the OH group and a γ-hydrogen. Loss of water results in a fragmentation peak at $m/z = M - 18$.

a γ-hydrogen is bonded to a γ-carbon

$$CH_3CH_2\overset{H}{\underset{\gamma}{CH}}CH_2\overset{^+\ddot{O}H}{\underset{\alpha}{CH}}CH_3 \longrightarrow CH_3CH_2\dot{C}HCH_2\overset{+}{CH}CH_3 + H_2O$$

$m/z = (102 - 18) = 84$

Notice that alkyl halides, ethers, and alcohols have the following fragmentation behavior in common:

1. A bond between carbon and a *more electronegative* atom (a halogen or an oxygen) breaks heterolytically.

2. A bond between carbon and an atom of *similar electronegativity* (a carbon or a hydrogen) breaks homolytically.

3. The bonds most likely to break are the weakest bonds and those that lead to formation of the most stable cation. (Look for fragmentation that results in a cation with a positive charge shared by two atoms.)

PROBLEM 11♦

Primary alcohols have a strong peak at $m/z = 31$. What fragment is responsible for this peak?

Ketones

The mass spectrum of a ketone generally has an intense molecular ion peak. A ketone fragments homolytically at the C—C bond adjacent to the C=O bond because in this way a cation with a positive charge shared by two atoms is formed. The alkyl group leading to the more stable radical is the one that is more easily cleaved.

Student Tutorial:
Fragmentation of ketones

$$CH_3CH_2CH_2\overset{\ddot{O}:}{\overset{||}{C}}CH_3 \xrightarrow{-e^-} CH_3CH_2CH_2\overset{\dot{\ddot{O}}^+}{\overset{||}{C}}CH_3$$

2-pentanone $m/z = 86$

→ $CH_3CH_2\dot{C}H_2$ + $CH_3C{\equiv}\overset{+}{O}:$ $m/z = 43$

→ $CH_3CH_2CH_2C{\equiv}\overset{+}{O}:$ + $\dot{C}H_3$ $m/z = 71$

If one of the alkyl groups attached to a carbonyl carbon has a γ-hydrogen, a cleavage, which goes through a favorable six-membered-ring transition state, may occur. This cleavage is known as a **McLafferty rearrangement**. In this rearrangement, the bond between the α-carbon and the β-carbon breaks homolytically and a hydrogen atom from the γ-carbon migrates to the oxygen atom. Again, fragmentation has occurred in a way that produces a cation with a positive charge shared by two atoms.

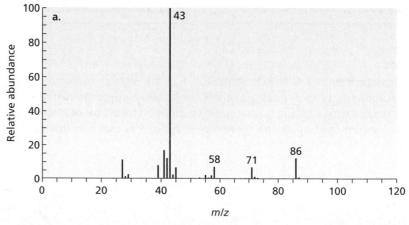

MASS SPECTROMETRY IN FORENSICS

Forensic science is the application of science for the purpose of justice. Mass spectrometry is an important tool of the forensic scientist. It is used to analyze body fluids for the presence and levels of drugs and other toxic substances. It can also identify the presence of drugs in hair, which increases the window of detection from hours and days (after which body fluids are no longer useful) to months and even years. It was employed for the first time in 1955 to detect drugs in athletes at a cycling competition in France. (Twenty percent of those tests were positive.) Mass spectrometry is also used to identify residues of arson fires and explosives from post-explosion residues, and to analyze paints, adhesives, and fibers.

PROBLEM 12

How could their mass spectra distinguish the following compounds?

$$\underset{\text{CH}_3\text{CH}_2\overset{\displaystyle O}{\overset{\|}{\text{C}}}\text{CH}_2\text{CH}_3}{}\qquad \underset{\text{CH}_3\overset{\displaystyle O}{\overset{\|}{\text{C}}}\text{CH}_2\text{CH}_2\text{CH}_3}{}\qquad \underset{\text{CH}_3\overset{\displaystyle O}{\overset{\|}{\text{C}}}\underset{\underset{\displaystyle \text{CH}_3}{|}}{\text{CHCH}_3}}{}$$

PROBLEM 13◆

Identify the ketones responsible for the mass spectra in Figure 12.9.

▲ **Figure 12.9**
The mass spectra referred to in Problem 13.

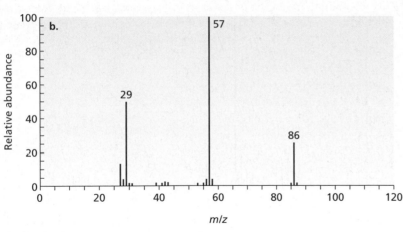

▲ **Figure 12.9**
Continued

PROBLEM 14

Using curved arrows, show the principal fragments that would be observed in the mass spectrum of each of the following compounds:

a. $CH_3CH_2CH_2CH_2CH_2OH$

b. $CH_3CH_2CHCH_2CH_2CH_2CH_3$
   |
   OH

c. $CH_3CH_2OCCH_2CH_2CH_3$
   | |
   CH_2CH_3 (above) CH_3 (below)

d. $CH_3\overset{O}{\overset{\|}{C}}CH_2CH_2CH_2CH_3$

e. CH_3CH_2CHCl
   |
   CH_3

f. $CH_3-\overset{CH_3}{\underset{CH_3}{\overset{|}{\underset{|}{C}}}}-Br$

PROBLEM 15◆

Two products are obtained from the reaction of (Z)-2-pentene with water and a trace of H_2SO_4. The mass spectra of these products are shown in Figure 12.10. Identify the compounds responsible for the spectra.

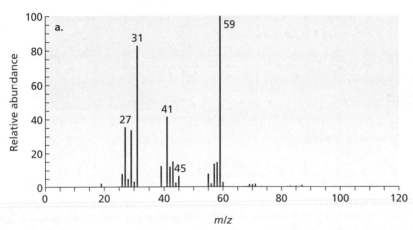

▲ **Figure 12.10**
The mass spectra referred to in Problem 15.

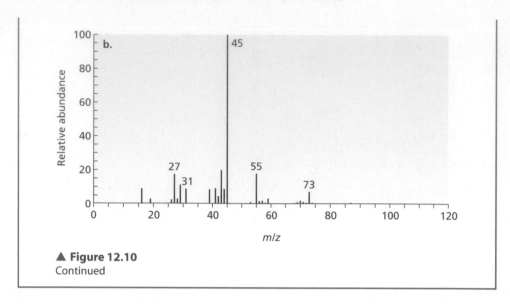

▲ **Figure 12.10**
Continued

12.6 **Spectroscopy and the Electromagnetic Spectrum**

Spectroscopy is the study of the interaction of matter and electromagnetic radiation. A continuum of different types of **electromagnetic radiation**—each associated with a particular energy range—makes up the electromagnetic spectrum (Figure 12.11). Visible light is the electromagnetic radiation we are most familiar with, but it represents only a fraction of the full electromagnetic spectrum. X-rays and radio waves are other familiar types electromagnetic radiation.

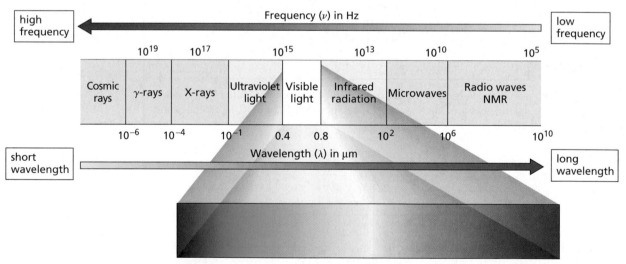

▲ **Figure 12.11**
The electromagnetic spectrum.

The various kinds of electromagnetic radiation can be characterized briefly as follows:

- *Cosmic rays* are discharged by the sun; they have the highest energy, the highest frequencies, and the shortest wavelengths.

- *γ-Rays* (gamma rays) are emitted by the nuclei of certain radioactive elements and, because of their high energy, can severely damage biological organisms.

- *X-Rays*, somewhat lower in energy than γ-rays, are less harmful, except in high doses. Low-dose X-rays are used to examine the internal structure of organisms. The denser the tissue, the more it blocks X-rays.

- *Ultraviolet (UV) light*, a component of sunlight, is responsible for sunburns, and repeated exposure to it can cause skin cancer by damaging DNA molecules in skin cells (Section 29.6).

- *Visible light* is the electromagnetic radiation we see.

- We feel *infrared radiation* as heat.

- We cook with *microwaves* and use them in radar.

- *Radio waves* have the lowest energy (lowest frequency) of the different kinds of electromagnetic radiation. We use them for radio and television communication, digital imaging, remote control devices, and wireless linkages for laptop computers. Radio waves are also used in NMR spectroscopy and in magnetic resonance imaging (MRI).

Each spectroscopic technique discussed in this book employs a different type of electromagnetic radiation. In this chapter, we will look at infrared (IR) spectroscopy and ultraviolet/visible (UV/Vis) spectroscopy, and in the next chapter we will see how compounds can be identified using nuclear magnetic resonance (NMR) spectroscopy.

Because electromagnetic radiation has wave-like properties, it can be characterized, as a wave can, by either its frequency (ν) or its wavelength (λ). **Frequency** is defined as the number of wave crests that pass by a given point in one second. Frequency has units of hertz (Hz). **Wavelength** is the distance from any point on one wave to the corresponding point on the next wave. Wavelength is generally measured in micrometers or nanometers. One micrometer (μm) is 10^{-6} meter; one nanometer (nm) is 10^{-9} meter.

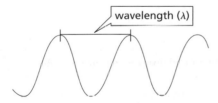

wavelength (λ)

The relationship between the energy (E) and the frequency (ν) or wavelength (λ) of the electromagnetic radiation is described by the equation

$$E = h\nu = \frac{hc}{\lambda}$$

where h is the proportionality constant called *Planck's constant*, named after the German physicist who discovered the relationship (Section 3.7), and c is the speed of light ($c = 3 \times 10^{10}$ cm/s). The equation shows that short wavelengths have high frequencies, and long wavelengths have low frequencies.

$$\nu = \frac{c}{\lambda}$$

Another way to describe the *frequency* of electromagnetic radiation—and the one most often used in infrared spectroscopy—is **wavenumber** ($\tilde{\nu}$), the number of waves in 1 cm. Therefore, it has units of reciprocal centimeters (cm^{-1}). The relationship between wavenumber (in cm^{-1}) and wavelength (in μm) is given by the equation

$$\widetilde{\nu}(\text{cm}^{-1}) = \frac{10^4}{\lambda(\mu\text{m})} \quad \text{(because 1 } \mu\text{m} = 10^{-4} \text{ cm)}$$

High frequencies, large wavenumbers, and short wavelengths are associated with high energy.

So *high frequencies, large wavenumbers*, and *short wavelengths* are associated with *high energy*.

PROBLEM 16◆

One of the following depicts the waves associated with infrared radiation, and one depicts the waves associated with ultraviolet light. Which is which?

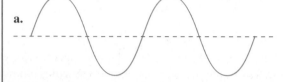

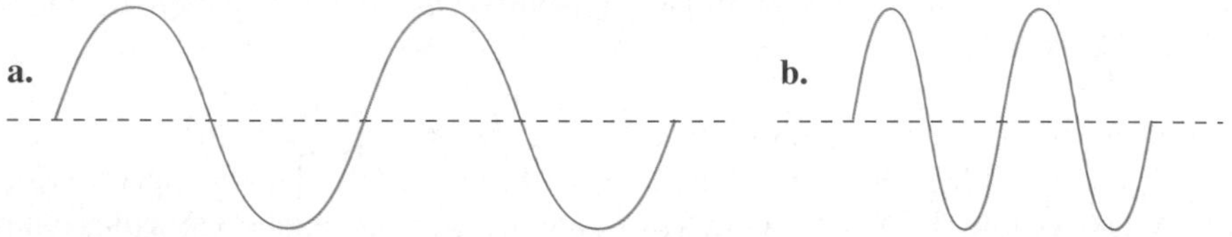

a.

b.

PROBLEM 17◆

a. Which is higher in energy, electromagnetic radiation with wavenumber 100 cm^{-1} or with wavenumber 2000 cm^{-1}?

b. Which is higher in energy, electromagnetic radiation with wavelength 9 μm or with wavelength 8 μm?

c. Which is higher in energy, electromagnetic radiation with wavenumber 3000 cm^{-1} or with wavelength 2 μm?

PROBLEM 18◆

a. What is the wavenumber of radiation that has a wavelength of 4 μm?

b. What is the wavelength of radiation that has a wavenumber of 200 cm^{-1}?

12.7 Infrared Spectroscopy

The length reported for a bond between two atoms is only an average length, because in reality a bond behaves as if it were a vibrating spring. A bond vibrates with both stretching and bending motions. A *stretch* is a vibration occurring along the line of the bond that changes the bond length. A *bend* is a vibration that does *not* occur along the line of the bond; bending vibrations change the bond angles. A diatomic molecule such as H—Cl can undergo only a **stretching vibration** because it has no bond angles.

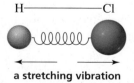

a stretching vibration

The vibrations of a molecule containing three or more atoms are more complex (Figure 12.12). Such molecules can experience symmetric and asymmetric stretches and bends, and their bending vibrations can be either in-plane or out-of-plane. **Bending vibrations** are often referred to by the descriptive terms *rock, scissor, wag,* and *twist*.

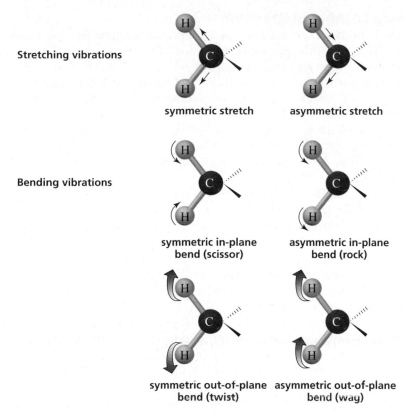

Student Tutorial:
IR stretching and bending

◄ **Figure 12.12**
Stretching and bending vibrations of bonds in organic molecules.

Each stretching and bending vibration of a given bond occurs with a characteristic frequency. **Infrared radiation** has just the right range of frequencies to correspond to the frequencies of the stretching and bending vibrations of organic molecules. Infrared radiation ranges in wavenumbers from 4000 to 600 cm^{-1}. It is just below the "red region" of visible light. (*Infra* is Latin for "below.").

When a molecule is bombarded with radiation of a frequency that exactly matches the frequency of one of its vibrations, the molecule absorbs energy. This allows the bonds to stretch and bend a bit more. By experimentally determining the wavenumbers of the energy absorbed by a particular compound, we can ascertain what kinds of bonds it has. For example, the stretching vibration of a C=O bond absorbs energy with wavenumber ~1700 cm^{-1}, whereas the stretching vibration of an O—H bond absorbs energy with wavenumber ~3450 cm^{-1} (Figure 12.13).

$$C=O \qquad O-H$$
$$\tilde{\nu} = \sim 1700 \text{ cm}^{-1} \qquad \tilde{\nu} = \sim 3400 \text{ cm}^{-1}$$

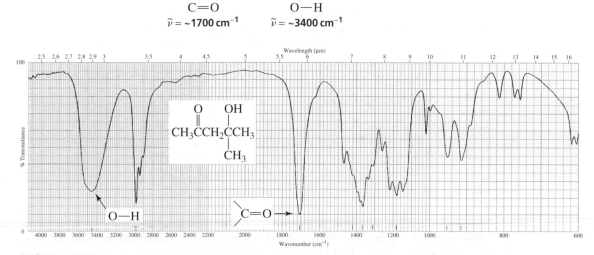

▲ **Figure 12.13**
The infrared spectrum of 4-hydroxy-4-methyl-2-pentanone. The spectrum shows the percent transmission of radiation versus the wavenumber or wavelength of the radiation.

Obtaining an Infrared Spectrum

An **infrared spectrum**, obtained by passing infrared radiation through a sample of a compound, is a plot of the percent transmission of radiation versus the wavenumber (or wavelength) of the radiation transmitted (Figure 12.13). The instrument used to obtain an infrared spectrum is called an IR spectrometer. At 100% transmission on an infrared spectrum, all the energy of the radiation (of a particular wavelength or wavenumber) passes through the molecule. Lower values of percent transmission mean that some of the energy is being absorbed by the compound. Each downward spike in the IR spectrum represents absorption of energy. The spikes are called **absorption bands**. Most chemists report the location of absorption bands using wavenumbers.

A newer type of IR spectrophotometer, called a Fourier transform IR (FT-IR) spectrophotometer, has several advantages over conventional spectrophotometers. Its sensitivity is better because, instead of scanning through the frequencies sequentially, it measures all frequencies simultaneously. The information is then digitized and Fourier transformed by a computer to produce the FT-IR spectrum. A conventional IR spectrophotometer can take 2 to 10 minutes to scan through all the frequencies. In contrast, FT-IR spectra can be taken in 1 to 2 seconds. The spectra shown in this text are FT-IR spectra.

An IR spectrum can be taken of a gas, a solid, or a liquid sample. Gases are expanded into an evacuated cell (a small container). Solids can be compressed with anhydrous KBr into a disc that is placed in the path of the light beam. A solid can also be examined as a mull. A mull is prepared by grinding a few milligrams of the solid in a mortar, adding a drop or two of mineral oil, and then grinding further. Liquids can be examined neat (undiluted); a few drops are put between two optically polished plates of NaCl that are placed in the light beam. Alternatively, a cell with optically polished NaCl or AgCl windows is used to hold samples dissolved in solvents. The discs, plates, and cells used to hold samples are constructed of ionic substances because, not having covalent bonds, they do not absorb IR radiation, whereas glass, quartz, and plastics have covalent bonds and so do absorb IR radiation.

When compounds are analyzed in solutions, they must be in solvents that have few absorption bands in the region of interest. Commonly used solvents are CH_2Cl_2 and $CHCl_3$. In a double-beam spectrophotometer, the IR radiation is split into two beams—one that passes through the sample cell and another that passes through a cell containing only the solvent. Any absorptions of the solvent are canceled out, so the absorption spectrum is that of the solute alone.

The Functional Group and Fingerprint Regions

An infrared spectrum can be divided into two areas. The left-hand two-thirds of the spectrum ($4000 - 1400 \text{ cm}^{-1}$) is where most of the functional groups show absorption bands. This is called the **functional group region**. The right-hand third ($1400 - 600 \text{ cm}^{-1}$) of the spectrum is called the **fingerprint region** because it is characteristic of the compound as a whole, just as a fingerprint is characteristic of an individual person. Even if two different molecules have the same functional groups, their IR spectra will not be identical, since the functional groups are not in exactly the same environment in each compound; this difference is reflected in the pattern of absorption bands in the fingerprint region, where every compound shows a unique pattern. For example, 2-pentanol and 3-pentanol have the same functional groups, so they show similar absorption bands in the functional group region. Their fingerprint regions are different, however, because the compounds are different (Figure 12.14). Thus, a compound can be positively identified by comparing its fingerprint region with the fingerprint region of the spectrum of a known sample of the compound.

a.

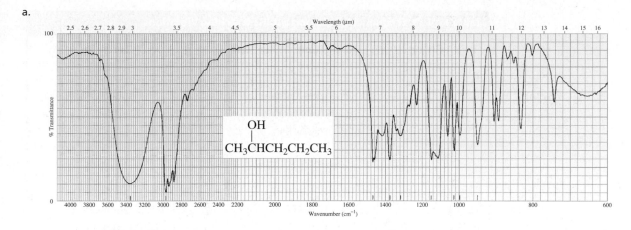

b.

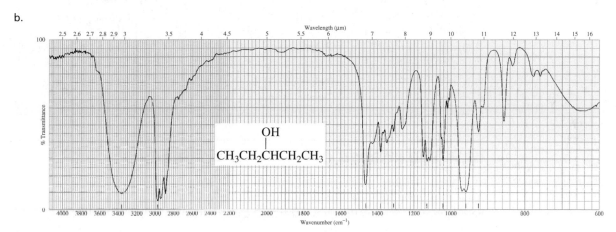

▲ **Figure 12.14**
The IR spectra of (a) 2-pentanol and (b) 3-pentanol. The functional group regions are very similar because the two compounds have the same functional group, but the fingerprint regions are unique for each compound.

12.8 **Characteristic Infrared Absorption Bands**

IR spectra can be quite complex because the stretching and bending vibrations of each bond in a molecule can produce an absorption band. Obviously, there is a lot more to infrared spectroscopy than we have room to cover here. However, organic chemists generally do not try to identify all the absorption bands in an IR spectrum. When identifying an unknown compound, they often use IR spectroscopy in conjunction with information obtained from other spectroscopic techniques. In this chapter, we will look at several characteristic absorption bands so you will be able to tell something about the structure of a compound that gives a particular IR spectrum. You can find an extensive table of characteristic functional group frequencies in Appendix VI. Some of the problems in this chapter and many of those in Chapter 13 provide practice in using information from two or more instrumental methods to identify compounds.

Because more energy is required to stretch a bond than to bend it, absorption bands for stretching vibrations are found in the functional group region ($4000 - 1400$ cm^{-1}), whereas absorption bands for bending vibrations are typically found in the fingerprint region ($1400 - 600$ cm^{-1}). Therefore, stretching vibrations are the ones most often used to determine what kinds of bonds a molecule has. The *frequencies of the stretching vibrations* associated with different types of bonds are shown in Table 12.4 and will be discussed in Sections 12.10 and 12.11.

It takes more energy to stretch a bond than to bend it.

Table 12.4	Frequencies of Important IR Stretching Vibrations	
Type of bond	**Wavenumber (cm^{-1})**	**Intensity**
C≡N	2260–2220	medium
C≡C	2260–2100	medium to weak
C=C	1680–1600	medium
C=N	1650–1550	medium
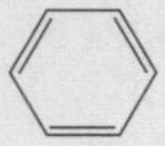 ⬡	~1600 and ~1500–1430	strong to weak
C=O	1780–1650	strong
C—O	1250–1050	strong
C—N	1230–1020	medium
O—H (alcohol)	3650–3200	strong, broad
O—H (carboxylic acid)	3300–2500	strong, very broad
N—H	3500–3300	medium, broad
C—H	3300–2700	medium

12.9 The Intensity of Absorption Bands

The greater the change in dipole moment, the more intense the absorption.

As a bond vibrates, its dipole moment changes. The intensity of the absorption band associated with the vibration depends on the size of this change in dipole moment: the greater the change, the more intense the absorption. When the bond stretches, the increasing distance between the atoms increases the dipole moment. (Recall that the dipole moment of a bond is equal to the magnitude of the charge on one of the bonded atoms, multiplied by the distance between the two charges; Section 1.3.) The stretching vibration of an O—H bond is associated with a greater change in dipole moment than that of an N—H bond because the O—H bond is more polar. Consequently, an O—H bond will show more intense absorption than an N—H bond. Likewise, an N—H bond will show more intense absorption than a C—H bond because the N—H bond is more polar.

The more polar the bond, the more intense the absorption.

relative bond polarities
relative intensities of IR absorption

The intensity of an absorption band also depends on the number of bonds responsible for the absorption. For example, the absorption band for the C—H stretch will be more intense for a compound such as octyl iodide, which has 17 C—H bonds, than for methyl iodide, which has only three C—H bonds. The concentration of the sample used to obtain an IR spectrum also affects the intensity of the absorption bands. Concentrated samples have greater numbers of absorbing molecules and, therefore, more intense absorption bands. In the chemical literature, you will find intensities referred to as strong (s), medium (m), weak (w), broad, and sharp.

PROBLEM 19◆

Which would be expected to be more intense, the stretching vibration of a C=O bond or the stretching vibration of a C=C bond?

12.10 The Position of Absorption Bands

The amount of energy required to stretch a bond depends on the *strength* of the bond and the *masses* of the bonded atoms. The stronger the bond, the greater the energy required to stretch it; think of a stronger bond as resembling a tighter spring. The frequency of the vibration (wavenumber of the absorption band) is inversely related to the mass of the atoms attached to the spring, so heavier atoms vibrate at lower frequencies.

Hooke's Law

The approximate wavenumber of an absorption band can be calculated from the following equation derived from **Hooke's law**, which describes the motion of a vibrating spring:

$$\widetilde{\nu} = \frac{1}{2\pi c}\left[\frac{f(m_1 + m_2)}{m_1 m_2}\right]^{1/2}$$

where c is the speed of light, $\widetilde{\nu}$ is wavenumber of an absorption band, f is the force constant of the bond (a measure of the strength of the bond), and m_1 and m_2 are the masses of the atoms (in grams) joined by the bond. This equation shows that *stronger bonds* and *lighter atoms* give rise to higher frequencies.

Lighter atoms show absorption bands at larger wavenumbers.

C—H
~3000 cm^{-1}
C—D
~2200 cm^{-1}
C—O
~1100 cm^{-1}
C—Cl
~700 cm^{-1}

THE ORIGINATOR OF HOOKE'S LAW

Robert Hooke (1635–1703) was born on the Isle of Wight off the southern coast of England. A brilliant scientist, he contributed to almost every scientific field. He was the first to suggest that light had wave-like properties. He discovered that Gamma Arietis is a double star, and he discovered Jupiter's Great Red Spot. In a lecture published posthumously, he suggested that earthquakes are caused by the cooling and contracting of the Earth. He examined cork under a microscope and coined the term "cell" to describe what he saw. He wrote about evolutionary development based on his studies of microscopic fossils, and he produced some highly regarded studies of insects. Hooke also invented the balance spring for watches and the universal joint currently used in cars.

Robert Hooke's drawing of a "blue fly" appeared in *Micrographia*, the first book on microscopy, published by Hooke in 1665.

The Effect of Bond Order

Bond order affects bond strength, and therefore it affects the position of absorption bands as predicted by Hooke's law. A C≡C bond is stronger than a C=C bond, so a C≡C bond stretches at a higher frequency (~2100 cm^{-1}) than does a C=C bond (~1650 cm^{-1}); C—C bonds show stretching vibrations in the region from 1200 to 800 cm^{-1}, but these vibrations are weak and very common, so they are of little value in identifying compounds. Similarly, a C=O bond stretches at a higher frequency (~1700 cm^{-1}) than does a C—O bond (~1100 cm^{-1}), and a C≡N bond stretches at a higher frequency (~2200 cm^{-1}) than does a C=N bond (~1600 cm^{-1}), which in turn stretches at a higher frequency than does a C—N bond (~1100 cm^{-1}) (Table 12.4).

Stronger bonds show absorption bands at larger wavenumbers.

C≡N
~2200 cm^{-1}
C=N
~1600 cm^{-1}
C—N
~1100 cm^{-1}

PROBLEM 20◆

a. Which will occur at a larger wavenumber
   1. a C≡C stretch or a C=C stretch?   3. a C—N stretch or a C=N stretch?
   2. a C—H stretch or a C—H bend?   4. a C=O stretch or a C—O stretch?
b. Assuming that the force constants are the same, which will occur at a larger wavenumber
   1. a C—O stretch or a C—Cl stretch?   2. a C—O stretch or a C—C stretch?

12.11 The Position of an Absorption Band Is Affected by Electron Delocalization, Electron Donation and Withdrawal, and Hydrogen Bonding

Table 12.4 shows a range of wavenumbers for the frequency of the stretching vibration for each functional group because the exact position of a group's absorption band in the spectrum of a given compound depends on other structural features of the molecule, such as electron delocalization, the electronic effect of neighboring substituents, and hydrogen bonding. In fact, important details about the structure of a compound can be revealed by the exact positions of its absorption bands.

For example, the IR spectrum in Figure 12.15 shows that the carbonyl group (C=O) of 2-pentanone absorbs at 1720 cm^{-1}, whereas the IR spectrum in Figure 12.16

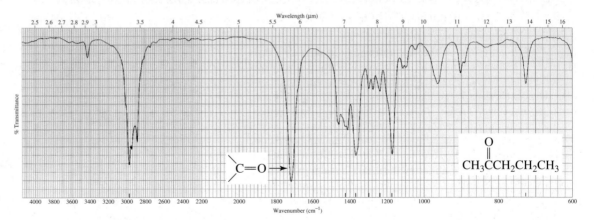

▲ **Figure 12.15**
The IR spectrum of 2-pentanone. The intense absorption band at ~1720 indicates a C=O bond.

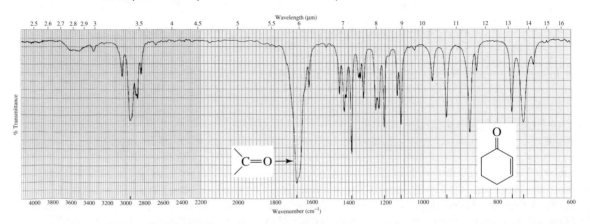

▲ **Figure 12.16**
The IR spectrum of 2-cyclohexenone. Electron delocalization gives its carbonyl group less double-bond character, so it absorbs at a lower frequency (~1680 cm^{-1}) than does a carbonyl group with localized electrons (~1720 cm^{-1}).

shows that the carbonyl group of 2-cyclohexenone absorbs at a lower frequency (1680 cm^{-1}). 2-Cyclohexenone's carbonyl group absorbs at a lower frequency because it has more single-bond character due to electron delocalization. A single bond is weaker than a double bond, so a carbonyl group with significant single-bond character will stretch at a lower frequency than will one with little or no single-bond character.

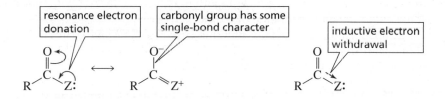

Putting an atom other than carbon next to the carbonyl group also causes the position of the carbonyl absorption band to shift. Whether it shifts to a lower or to a higher frequency depends on whether the predominant effect of the atom is to donate electrons by resonance or to withdraw electrons inductively.

The predominant effect of the nitrogen of an amide is electron donation by resonance (Section 7.9). In contrast, since oxygen is less able than nitrogen to accomodate a positive charge because of oxygen's greater electronegativity, the predominant effect of the oxygen of an ester is inductive electron withdrawal (Section 1.21); therefore, the resonance contributor with the C—O single bond contributes less to the hybrid. (See also Section 16.2.) As a result, the carbonyl group of an ester has less single-bond character, so it requires more energy to stretch (1740 cm^{-1} in Figure 12.17) than the carbonyl group of an amide (1660 cm^{-1} in Figure 12.18).

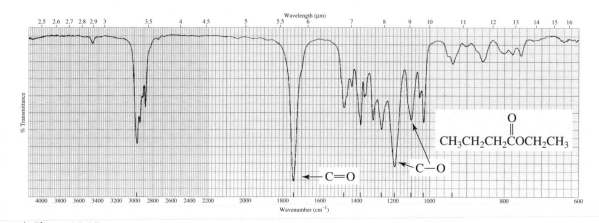

▲ **Figure 12.17**
The IR spectrum of ethyl butanoate. The electron-withdrawing oxygen atom makes the carbonyl group of an ester harder to stretch (~1740 cm^{-1}) than the carbonyl group of a ketone (~1720 cm^{-1}).

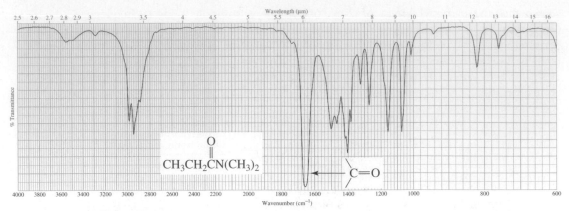

▲ **Figure 12.18**
The IR spectrum of *N,N*-dimethylpropanamide. The carbonyl group of an amide has less double-bond character than does the carbonyl group of a ketone, so the former stretches more easily (~1660 cm^{-1}).

When we compare the frequency for the stretching vibration of the carbonyl group of an ester (1740 cm^{-1} in Figure 12.17) with that of the carbonyl group of a ketone (1720 cm^{-1} in Figure 12.15), we can see how important inductive electron withdrawal is to the position of the stretching vibration of the carbonyl group.

A C—O bond shows a stretching vibration between 1250 and 1050 cm^{-1}. If the C—O bond is in an alcohol (Figure 12.19) or an ether, the stretch will lie toward the lower end of the range. If, however, the C—O bond is in a carboxylic acid (Figure 12.20), the stretch will lie at the higher end of the range. The position of the C—O absorption varies because the C—O bond in an alcohol is a pure single bond, whereas the C—O bond in a carboxylic acid has partial double-bond character due to electron donation by resonance. Esters show C—O stretches at both ends of the range (Figure 12.18) because esters have two C—O single bonds: one that is a pure single bond and one that has partial double-bond character.

CH$_3$CH$_2$—OH   CH$_3$CH$_2$—O—CH$_2$CH$_3$
~1050 cm^{-1} ~1050 cm^{-1}

~1250 cm^{-1} ~1250 cm^{-1} and ~1050 cm^{-1}

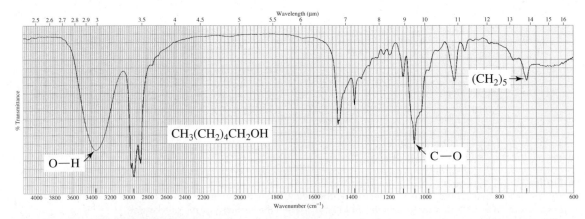

▲ **Figure 12.19**
The IR spectrum of 1-hexanol.

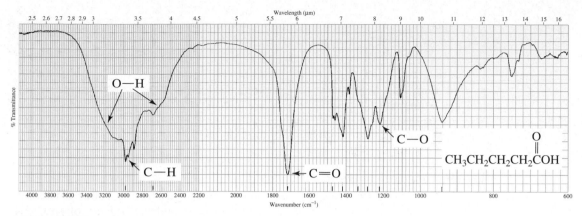

▲ **Figure 12.20**
The IR spectrum of pentanoic acid.

PROBLEM-SOLVING STRATEGY

Differences in IR Spectra

Which will occur at a larger wavenumber, the C—N stretch of an amine or the C—N stretch of an amide?

To answer this kind of question, we need to see if electron delocalization causes one of the bonds to be other than a pure single bond. When we do that we see that electron delocalization causes the C—N bond of the amide to have partial double-bond character, whereas there is no electron delocalization in the C—N bond of the amine. The C—N stretch of an amide, therefore, will occur at a larger wavenumber than that of an amine.

R—NH$_2$

no electron delocalization

electron delocalization causes the C—N bond to have partial double-bond character

Now continue on to Problem 21.

PROBLEM 21◆

Which will occur at a larger wavenumber

a. the C—O stretch of phenol or the C—O stretch of cyclohexanol?
b. the C=O stretch of a ketone or the C=O stretch of an amide?
c. the stretch or the bend of the C—O bond in ethanol?

PROBLEM 22◆

Which would show an absorption band at a larger wavenumber: a carbonyl group bonded to an sp^3 carbon or a carbonyl group bonded to an sp^2 carbon?

PROBLEM 23◆

List the following compounds in order of decreasing wavenumber of the C=O absorption band:

a. H$_3$C—C(=O)—CH$_3$ H—C(=O)—H   H$_3$C—C(=O)—H   **b.**

c.

(*Hint:* Recall that hyperconjugation causes an alkyl group to be more electron donating than a hydrogen; Section 4.2.)

O—H Absorption Bands

Because O—H bonds are polar, they show intense absorption bands that can be quite broad (Figures 12.19 and 12.20). The position and the breadth of an O—H absorption band depend on the concentration of the solution. The more concentrated the solution, the more likely it is for the OH-containing molecules to form intermolecular hydrogen bonds. It is easier for an O—H bond to stretch if it is hydrogen bonded, because the hydrogen is attracted to the oxygen of a neighboring molecule. Therefore, the O—H stretch in a concentrated (hydrogen-bonded) solution of an alcohol occurs at 3550 to 3200 cm^{-1}, whereas the O—H stretch in a dilute solution (with little or no hydrogen bonding) occurs at 3650 to 3590 cm^{-1}. Hydrogen-bonded OH groups also have broader absorption bands because the hydrogen bonds vary in strength (Section 2.9). The absorption bands of non-hydrogen-bonded OH groups are sharper.

```
        hydrogen bond
                  |          H
                   \         |
    R—O—H------O—R                    R—O—H
  concentrated solution              dilute solution
   3550–3200 cm⁻¹                  3650–3590 cm⁻¹
```

PROBLEM 24◆

Which will show an O—H stretch at a larger wavenumber, ethanol dissolved in carbon disulfide or an undiluted sample of ethanol?

PROBLEM 25◆

Why is the C—O absorption band of 1-hexanol at a smaller wavenumber (1060 cm^{-1}) than the C—O absorption band of pentanoic acid (1220 cm^{-1})?

Student Tutorial:
IR spectra

C—H Absorption Bands

Important information about the identity of a compound is provided by the stretching and bending vibrations of the C—H bonds.

Stretching Vibrations. The strength of a C—H bond depends on the hybridization of the carbon: the greater the *s* character of the carbon, the stronger is the bond that it forms (Section 1.14). Therefore, a C—H bond is stronger when the carbon is *sp* hybridized than when it is sp^2 hybridized, which in turn is stronger than when the carbon is sp^3 hybridized (See Table 1.7 on page 41). More energy is needed to stretch a stronger bond, and this is reflected in the absorption bands for a C—H stretch, which occur at ~3300 cm^{-1} for an *sp* carbon, at ~3100 cm^{-1} for an sp^2 carbon, and at ~2900 cm^{-1} for an sp^3 carbon (Table 12.5).

Table 12.5	IR Absorptions of Carbon–Hydrogen Bonds
Carbon–Hydrogen Stretching Vibrations	**Wavenumber (cm⁻¹)**
C≡C—H	~3300
C=C—H	3100–3020
C—C—H	2960–2850
R—C(=O)—H	~2820 and ~2720
Carbon–Hydrogen Bending Vibrations	**Wavenumber (cm⁻¹)**
CH_3— —CH_2— —CH—	1450–1420
CH_3—	1385–1365
C=C trans	980–960
C=C cis	730–675
C=C trisubstituted	840–800
C=C terminal alkene	890
C=C terminal alkene	990 and 910

A useful step in the analysis of a spectrum is to look at the absorption bands in the vicinity of 3000 cm⁻¹. Figures 12.21, 12.22, and 12.23 show the IR spectra for methylcyclohexane, cyclohexene, and ethylbenzene, respectively. The only absorption band in the vicinity of 3000 cm⁻¹ in Figure 12.21 is slightly to the right of that value. This tells us that the compound has hydrogens bonded to sp^3 carbons, but none bonded to sp^2 or to sp carbons. Both Figures 12.22 and 12.23 show absorption

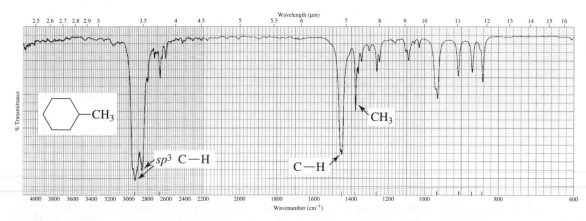

▲ **Figure 12.21**
The IR spectrum of methylcyclohexane.

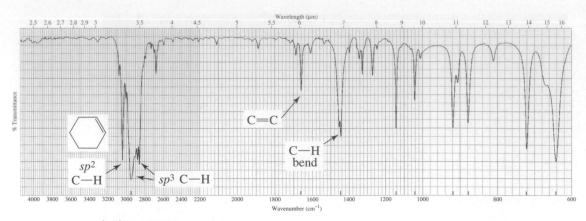

▲ **Figure 12.22**
The IR spectrum of cyclohexene.

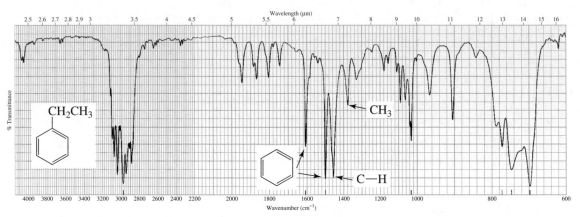

▲ **Figure 12.23**
The IR spectrum of ethylbenzene.

bands slightly to the left and slightly to the right of 3000 cm^{-1}, indicating that the compounds that produced those spectra contain hydrogens bonded to sp^2 and sp^3 carbons.

Once we know that a compound has hydrogens bonded to sp^2 carbons, we need to find out whether those carbons are the sp^2 carbons of an alkene or of a benzene ring. A benzene ring is indicated by sharp absorption bands at ~1600 cm^{-1} and 1500–1430 cm^{-1}, whereas an alkene is indicated by a band at ~1600 cm^{-1} only (Table 12.4). The compound whose spectrum is shown in Figure 12.22 is, therefore, an alkene, while the one whose spectrum is shown in Figure 12.23 has a benzene ring. Be aware that N—H bending vibrations also occur at 1600 cm^{-1}, so absorption at that wavelength does not always indicate a C=C bond. However, absorption bands resulting from N—H bends tend to be broader (due to hydrogen bonding) and more intense (due to being more polar) than those resulting from C=C stretches (see Figure 12.25), and they will be accompanied by N—H stretches at 3500–3300 cm^{-1} (Table 12.4).

The stretch of the C—H bond of an aldehyde group shows two absorption bands— one at ~2820 cm^{-1} and the other at ~2720 cm^{-1} (Figure 12.24). This makes aldehydes relatively easy to identify because essentially no other absorption occurs at these wavenumbers.

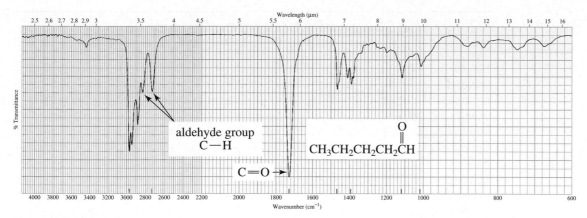

▲ **Figure 12.24**
The IR spectrum of pentanal. The absorptions at ~2820 and ~2720 cm⁻¹ readily identify an aldehyde group. Note also the intense absorption band at ~1730 cm⁻¹ indicating a C=O bond.

Bending Vibrations If a compound has sp^3 carbons, a look at the 1400 cm⁻¹ position on its IR spectrum will tell you whether the compound has a methyl group. All hydrogens bonded to sp^3 carbons show a C—H bending vibration slightly to the *left* of 1400 cm⁻¹. Only methyl groups show a C—H bending vibration slightly to the *right* of 1400 cm⁻¹. So, if a compound has a methyl group, absorption bands will appear *both* to the left and to the right of 1400 cm⁻¹; otherwise, only the band to the left of 1400 cm⁻¹ will be present. You can see evidence of a methyl group in Figure 12.21 (methylcyclohexane) and in Figure 12.23 (ethylbenzene), but not in Figure 12.22 (cyclohexene). Two methyl groups attached to the same carbon can sometimes be detected by a split in the methyl peak at ~1380 cm⁻¹ (Figure 12.25).

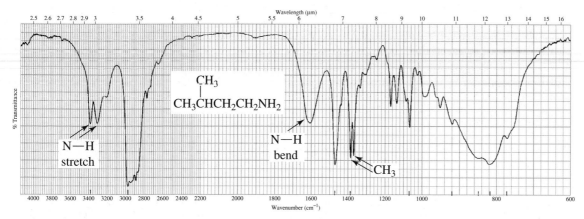

▲ **Figure 12.25**
The IR spectrum of isopentylamine. The double peak at ~1380 cm⁻¹ indicates the presence of an isopropyl group. Two N—H bonds are indicated as well at ~3350 cm⁻¹.

The C—H bending vibrations for hydrogens bonded to sp^2 carbons give rise to absorption bands in the 1000–600 cm⁻¹ region. As Table 12.5 shows, the frequency of the C—H bending vibration of an alkene depends on the number of alkyl groups attached to the double bond and on the configuration of the alkene. It is important to realize that these absorption bands can be shifted out of the characteristic regions if strongly electron-withdrawing or electron-donating substituents are close to the double bond (Section 12.10). Acyclic compounds with more than four adjacent methylene (CH₂) groups show a characteristic absorption band at 720 cm⁻¹ that results from in-phase rocking of the methylene groups (Figure 12.19).

12.12 The Shape of Absorption Bands

The position, intensity, and shape of an absorption band are helpful in identifying functional groups.

The shape of an absorption band can be helpful in identifying the compound responsible for an IR spectrum. For example, both O—H and N—H bonds stretch at wavenumbers above 3100 cm^{-1}, but the shapes of their stretching absorption bands are distinctive. Notice the difference in the shape of these absorption bands in the IR spectra of 1-hexanol (Figure 12.19), pentanoic acid (Figure 12.20), and isopentyl-amine (Figure 12.25). An N—H absorption band (~3300 cm^{-1}) is narrower and less intense than an O—H absorption band (~3300 cm^{-1}), and the O—H absorption band of a carboxylic acid (~3300–2500 cm^{-1}) is broader than the O—H absorption band of an alcohol (Sections 12.9 and 12.10). Notice that two absorption bands are detectable in Figure 12.25 for the N—H stretch because there are two N—H bonds in the compound.

> **PROBLEM 26◆**
>
> **a.** Why is an O—H stretch more intense than an N—H stretch?
> **b.** Why is the O—H stretch of a carboxylic acid broader than the O—H stretch of an alcohol?

12.13 The Absence of Absorption Bands

The absence of an absorption band can be as useful as the presence of a band in identifying a compound by IR spectroscopy. For example, the spectrum in Figure 12.26 shows a strong absorption at ~1100 cm^{-1}, indicating the presence of a C—O bond. Clearly, the compound is not an alcohol because there is no absorption above 3100 cm^{-1}. Nor is it an ester or any other kind of carbonyl compound because there is no absorption at ~1700 cm^{-1}. The compound has no C≡C, C=C, C≡N, C=N, or C—N bonds. We may deduce, then, that the compound is an ether. Its C—H absorption bands show that it has hydrogens only on sp^3 carbons and that it has a methyl group. We also know that the compound has fewer than four adjacent methylene groups, because there is no absorption at ~720 cm^{-1}. The compound is in fact diethyl ether.

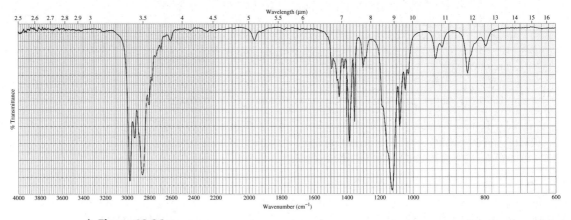

▲ **Figure 12.26**
The IR spectrum of diethyl ether.

> **PROBLEM 27◆**
>
> How do we know that the absorption band at ~1100 cm^{-1} in Figure 12.26 is due to a C—O bond and not to a C—N bond?

PROBLEM 28◆

a. An oxygen-containing compound shows an absorption band at ~ 1700 cm^{-1} and no absorption bands at ~ 3300 cm^{-1}, ~ 2700 cm^{-1}, or ~ 1100 cm^{-1}. What class of compound is it?

b. A nitrogen-containing compound shows no absorption band at ~ 3400 cm^{-1} and no absorption bands between 1700 cm^{-1} and 1600 cm^{-1}. What class of compound is it?

PROBLEM 29

How could IR spectroscopy distinguish between the following compounds?

a. a ketone and an aldehyde

b. a cyclic ketone and an open-chain ketone

c. benzene and cyclohexene

d. *cis*-2-hexene and *trans*-2-hexene

e. cyclohexene and cyclohexane

f. a primary amine and a tertiary amine

PROBLEM 30

For each of the following pairs of compounds, give one absorption band that could be used to distinguish between them:

a. $CH_3CH_2CH_2CH_3$ and $CH_3CH_2OCH_3$

d. $CH_3CH_2C\equiv CCH_3$ and $CH_3CH_2C\equiv CH$

b. $CH_3CH_2\overset{\overset{\displaystyle O}{\|}}{C}OCH_3$ and $CH_3CH_2\overset{\overset{\displaystyle O}{\|}}{C}OH$

e. $CH_3CH_2\overset{\overset{\displaystyle O}{\|}}{C}OH$ and $CH_3CH_2CH_2OH$

c. and

f. and

12.14 Some Vibrations Are Infrared Inactive

Not all vibrations give rise to absorption bands. In order for a bond's vibration to absorb IR radiation, the dipole moment of the bond must change when the bond vibrates. For example, the $C-C$ bond in 1-butene has a dipole moment because the molecule is not symmetrical about this bond. When the $C=C$ bond stretches, the increasing distance between the atoms increases the dipole moment. (Recall that the dipole moment is equal to the magnitude of the charge on the atoms times the distance between them; Section 1.3.) Because the dipole moment changes when the bond stretches, an absorption band is observed for the $C=C$ stretching vibration.

asymmetrical about the $C=C$ bond; has a dipole moment

symmetrical about the $C=C$ bond; has no dipole moment

1-butene **2,3-dimethyl-2-butene** **2,3-dimethyl-2-heptene**

2,3-Dimethyl-2-butene, in contrast, is a symmetrical molecule, so its $C=C$ bond has no dipole moment. When the bond stretches, it still has no dipole moment. Since stretching is not accompanied by a change in dipole moment, no absorption band is observed for that bond. The vibration is *infrared inactive*. 2,3-Dimethyl-2-heptene experiences a very small change in dipole moment when its $C=C$ bond stretches, so only an extremely weak absorption band (if any) will be detected for the stretching vibration of the bond.

PROBLEM 31◆

Which of the following compounds has a vibration that is infrared inactive: acetone, 1-butyne, 2-butyne, H_2, H_2O, Cl_2, ethene?

PROBLEM 32◆

The mass spectrum and infrared spectrum of an unknown compound are shown in Figures 12.27 and 12.28, respectively. Identify the compound.

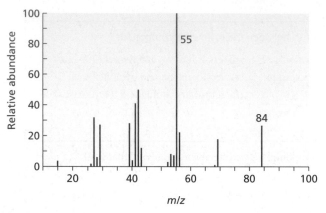

▲ **Figure 12.27**
The mass spectrum referred to in Problem 32.

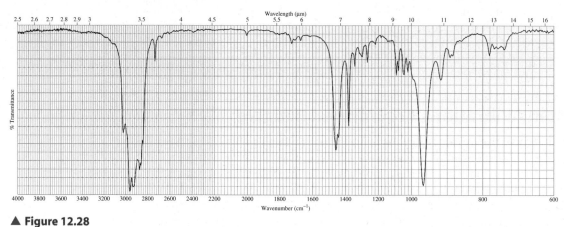

▲ **Figure 12.28**
The IR spectrum referred to in Problem 32.

12.15 A Lesson in Interpreting Infrared Spectra

Interpreting IR Spectra

We will now look at some IR spectra and see what we can deduce about the structures of the compounds that give rise to the spectra. We might not be able to identify the compound precisely, but when we are told what it is, its structure should fit our observations.

Compound 1. The absorptions in the 3000 cm^{-1} region in Figure 12.29 indicate that hydrogens are attached both to sp^2 carbons (3075 cm^{-1}) and to sp^3 carbons (2950 cm^{-1}). Now let's see if the sp^2 carbons belong to an alkene or to a benzene ring. The absorption at 1650 cm^{-1}, the absence of a band at 1500–1430 cm^{-1}, and the absorption at ~890 cm^{-1} (Table 12.5) suggest that the compound is a terminal alkene with two alkyl substituents at the 2-position. The absence of absorption at ~720 cm^{-1} indicates that the compound has fewer than four adjacent methylene groups. We are not surprised to find that the compound is 2-methyl-1-pentene.

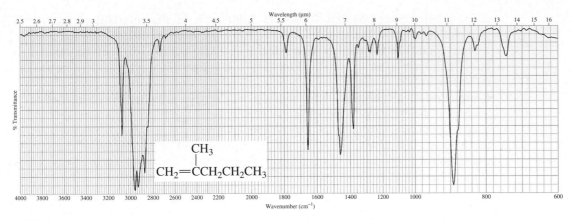

▲ **Figure 12.29**
The IR spectrum of Compound 1.

Compound 2. The absorption in the 3000 cm^{-1} region in Figure 12.30 indicates that hydrogens are attached to sp^2 carbons (3050 cm^{-1}) but not to sp^3 carbons. The absorptions at 1600 cm^{-1} and 1460 cm^{-1} indicate that the compound has a benzene ring. The absorptions at 2810 cm^{-1} and 2730 cm^{-1} show that the compound is an aldehyde. The absorption band for the carbonyl group (C=O) is lower (1700 cm^{-1}) than normal (1720 cm^{-1}) so the carbonyl group has partial single-bond character. Thus, it must be attached directly to the benzene ring. The compound is benzaldehyde.

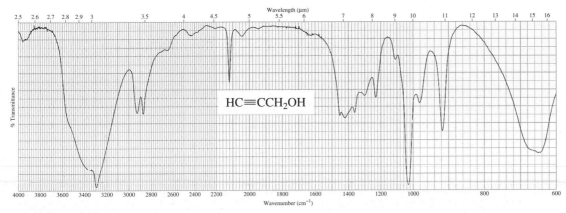

▲ **Figure 12.30**
The IR spectrum of Compound 2.

Compound 3. The absorptions in the 3000 cm^{-1} region in Figure 12.31 indicate that hydrogens are attached to sp^3 carbons (2950 cm^{-1}) but not to sp^2 carbons. The

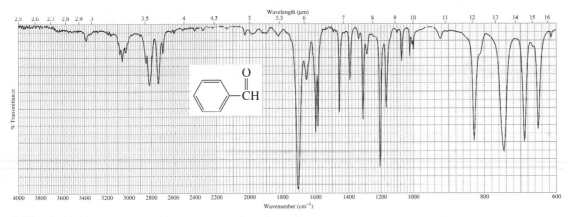

▲ **Figure 12.31**
The IR spectrum of Compound 3.

shape of the strong absorption band at 3300 cm^{-1} is characteristic of an O—H group of an alcohol. The absorption at ~2100 cm^{-1} indicates that the compound has a triple bond. The sharp absorption band at 3300 cm^{-1} indicates that the compound has a hydrogen on an *sp* carbon, so we know it is a terminal alkyne. The compound is 2-propyn-1-ol.

Compound 4. The absorption in the 3000 cm^{-1} region in Figure 12.32 indicates that hydrogens are attached to *sp*^{3} carbons (2950 cm^{-1}). The relatively strong absorption band at 3300 cm^{-1} suggests that the compound has one N—H bond. The presence of the N—H bond is confirmed by the absorption band at 1560 cm^{-1}. The C=O absorption at 1660 cm^{-1} indicates that the compound is an amide. The compound is *N*-methylacetamide.

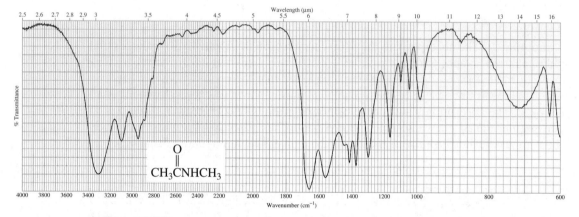

▲ **Figure 12.32**
The IR spectrum of Compound 4.

Compound 5. The absorptions in the 3000 cm^{-1} region in Figure 12.33 indicate that the compound has hydrogens attached to *sp*^{2} carbons (> 3000 cm^{-1}) and to *sp*^{3} carbons (< 3000 cm^{-1}). The absorptions at 1605 cm^{-1} and 1500 cm^{-1} indicate that the compound contains a benzene ring. The absorption at 1720 cm^{-1} for the carbonyl group indicates that the compound is a ketone and that the carbonyl group is not directly attached to the benzene ring. The absorption at ~1380 cm^{-1} indicates that the compound contains a methyl group. The compound is 1-phenyl-2-butanone.

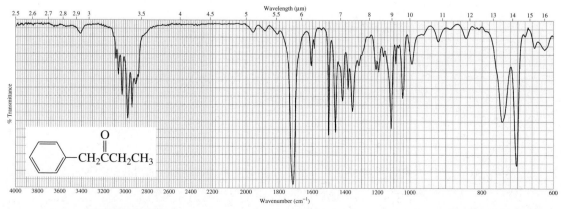

▲ **Figure 12.33**
The IR spectrum of Compound 5.

PROBLEM 33◆

A compound with molecular formula C_4H_6O gives the infrared spectrum shown in Figure 12.34. Identify the compound.

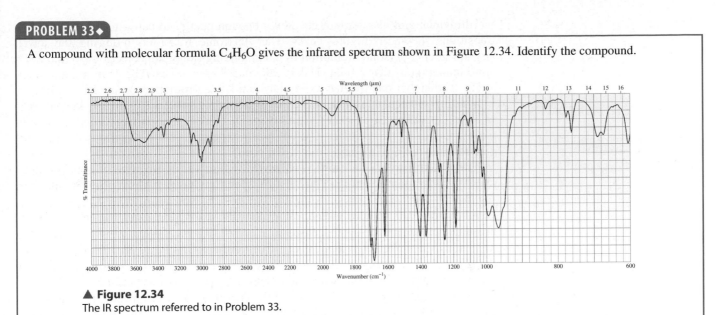

▲ **Figure 12.34**
The IR spectrum referred to in Problem 33.

12.16 **Ultraviolet and Visible Spectroscopy**

Ultraviolet and visible (UV/Vis) spectroscopy provides information about compounds that have conjugated double bonds. Ultraviolet light and visible light have just the right energy to cause an electronic transition in a molecule, that is, to promote an electron from one molecular orbital to another of higher energy. Depending on the energy needed for the electronic transition, a molecule will absorb either ultraviolet or visible light. If it absorbs **ultraviolet light**, a UV spectrum is obtained; if it absorbs **visible light**, a visible spectrum is obtained. Ultraviolet light is electromagnetic radiation with wavelengths ranging from 180 to 400 nm (nanometers); visible light has wavelengths ranging from 400 to 780 nm. Recall from Section 12.6 that **wavelength** (λ) is inversely related to the energy of the radiation: the shorter the wavelength, the greater the energy. Ultraviolet light, therefore, has greater energy than visible light.

> The shorter the wavelength, the greater is the energy of the radiation.

$$E = \frac{hc}{\lambda}$$

In the ground-state electronic configuration of a molecule, all the electrons are in the lowest-energy molecular orbitals. When a molecule absorbs light with the energy required to promote an electron to a higher-energy molecular orbital—when it undergoes an **electronic transition**—the molecule is then in an excited state. The relative energies of the bonding, nonbonding, and antibonding molecular orbitals are shown in Figure 12.35.

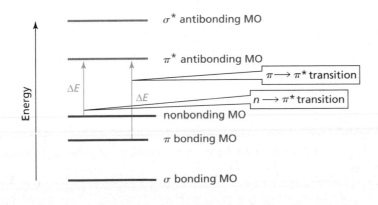

◀ **Figure 12.35**
Relative energies of the bonding, nonbonding, and antibonding orbitals.

Ultraviolet and visible light have only enough energy to cause the two electronic transitions shown in Figure 12.35. The electronic transition requiring the lower amount of energy is the promotion of a nonbonding (lone-pair) electron (n) into a π^* antibonding molecular orbital. This is called an $n \rightarrow \pi^*$ (stated as "n to π star") transition. The higher-energy electronic transition is the promotion of an electron from a π bonding molecular orbital into a π^* antibonding molecular orbital, known as a $\pi \rightarrow \pi^*$ (stated as "π to π star") transition. This means that *only organic compounds with π electrons can produce UV/Vis spectra* (because both transitions promote an electron into a π^* antibonding molecular orbital).

The UV spectrum of acetone is shown in Figure 12.36. Acetone has both π electrons and lone-pair electrons. Thus, its UV spectrum has two **absorption bands**: one for the $\pi \rightarrow \pi^*$ transition and one for the $n \rightarrow \pi^*$ transition. The λ_{max} (stated as "lambda max") is the wavelength at which the absorption band has its maximum absorbance. For the $\pi \rightarrow \pi^*$ transition, $\lambda_{max} = 195$ nm; for the $n \rightarrow \pi^*$ transition, $\lambda_{max} = 274$ nm. We know that the $\pi \rightarrow \pi^*$ transition in Figure 12.35 corresponds to the λ_{max} at the shorter wavelength because that transition requires more energy than the $n \rightarrow \pi^*$ transition. The absorption bands are broad because each electronic state has vibrational sublevels (Figure 12.37). Therefore, the electronic transitions are from and to different vibrational sublevels, so the electronic transitions span a range of wavelengths.

> **Only compounds with π electrons can produce UV/Vis spectra.**

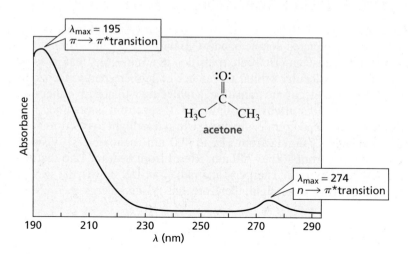

Figure 12.36 ▶
The UV spectrum of acetone.

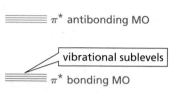

▲ Figure 12.37
UV/Vis absorption bands are broad because each electronic state has vibrational sublevels.

A **chromophore** is the part of a molecule that absorbs UV or visible light. The carbonyl group is the chromophore of acetone. The following four compounds all have the same chromophore, so they all have approximately the same λ_{max}.

PROBLEM 34◆

Explain why diethyl ether does not have a UV spectrum, even though it has lone-pair electrons.

ULTRAVIOLET LIGHT AND SUNSCREENS

Exposure to ultraviolet (UV) light stimulates specialized cells in the skin to produce a black pigment known as melanin, which causes the skin to look tan. Melanin absorbs UV light, so it protects our bodies from the harmful effects of the sun. If more UV light reaches the skin than the melanin can absorb, the light will burn the skin and cause photochemical reactions that can result in skin cancer (Section 29.6). UV-A is the lowest-energy UV light (315 to 400 nm) and does the least biological damage. Fortunately, most of the more dangerous, higher-energy UV light, UV-B (290 to 315 nm) and UV-C (180 to 290 nm), is filtered out by the ozone layer in the stratosphere. That is why we need to worry about the apparent thinning of the ozone layer (Section 11.11).

Applying a sunscreen can protect skin against UV light. The amount of protection provided by a particular sunscreen is indicated by its SPF (sun protection factor). The higher the SPF, the greater the protection. Some sunscreens contain an inorganic component, such as zinc oxide, that reflects the light as it reaches the skin. Others contain a compound that absorbs UV light. PABA was the first commercially available UV-absorbing sunscreen. PABA absorbs UV-B light, but is not very soluble in oily skin lotions. Less polar compounds, such as Padimate O, are now commonly used. Recent research has shown that sunscreens that absorb only UV-B light do not give adequate protection against skin cancer; both UV-A and UV-B protection are needed. Giv Tan F absorbs both UV-B and UV-A light, so it gives better protection.

para-aminobenzoic acid
PABA

2-ethylhexyl 4-(dimethylamino)benzoate
Padimate O

2-ethylhexyl (*E*)-3-(4-methoxyphenyl)-2-propenoate
Giv Tan F

12.17 The Beer–Lambert Law

Wilhelm Beer and Johann Lambert independently proposed that the absorbance of a sample at a given wavelength depends on the amount of absorbing species that the light encounters as it passes through a solution of the sample. In other words, absorbance depends on both the concentration of the sample and the length of the light path through the sample. The relationship between absorbance, concentration, and length of the light path, known as the **Beer–Lambert law**, is given by

$$A = \varepsilon cl$$

where

A = absorbance of the sample = $\log \dfrac{I_0}{I}$

I_0 = intensity of the radiation entering the sample

I = intensity of the radiation emerging from the sample

c = concentration of the sample, in moles/liter

l = length of the light path through the sample, in centimeters

ε = molar absorptivity (liter mol^{-1} cm^{-1})

The **molar absorptivity** (ε) of a compound is a constant that is characteristic of the compound at a particular wavelength. It is the absorbance that would be observed for a 1.00 M solution in a cell with a 1.00-cm path length. The molar absorptivity of acetone dissolved in hexane, for example, is 9000 M^{-1}cm^{-1} at 195 nm and 15 M^{-1}cm^{-1} at 274 nm. The solvent in which the sample is dissolved is reported because molar absorptivity is not

BIOGRAPHY

Wilhelm Beer (1797–1850), *born in Germany, was a banker whose hobby was astronomy. He was the first to make a map of the darker and lighter areas of Mars.*

BIOGRAPHY

Johann Heinrich Lambert (1728–1777), *a German-born mathematician, was the first to make accurate measurements of light intensities and to introduce hyperbolic functions into trigonometry. Although the equation that relates absorbance, concentration, and light path bears the names of Beer and Lambert, it is believed that* **Pierre Bouguer (1698–1758)**, *a French mathematician, first formulated the relationship in 1729.*

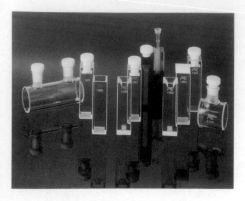

Cells used in UV/Vis spectroscopy.

exactly the same in all solvents. So the UV spectrum of acetone in hexane would be reported as: λ_{max} 195 nm ($\varepsilon_{max} = 9000$, hexane); λ_{max} 274 nm ($\varepsilon_{max} = 13.6$, hexane). Because absorbance is proportional to concentration, the concentration of a solution can be determined if the absorbance and molar absorptivity at a particular wavelength are known.

The two absorption bands of acetone in Figure 12.36 are very different in size because of the difference in molar absorptivity at the two wavelengths. Small molar absorptivities are characteristic of $n \rightarrow \pi^*$ transitions, so these transitions can be difficult to detect. Consequently, $\pi \rightarrow \pi^*$ transitions are more useful for analyzing spectra.

The solution being used to produce a UV or visible spectrum is put into a cell. Most cells have 1-cm path lengths. Either glass or quartz cells can be used for visible spectra, but quartz cells (made of high-purity fused silica) must be used for UV spectra because glass absorbs UV light.

PROBLEM 35♦

A solution of 4-methyl-3-penten-2-one in ethanol shows an absorbance of 0.52 at 236 nm in a cell with a 1-cm light path. Its molar absorptivity in ethanol at that wavelength is $12{,}600 \ M^{-1}cm^{-1}$. What is the concentration of the compound?

PROBLEM 36♦

A 4.0×10^{-5} M solution of nitrobenzene in hexane shows an absorbance of 0.40 at 252 nm in a cell with a 1-cm light path. What is the molar absorptivity of nitrobenzene in hexane at $\lambda_{max} = 252$ nm?

12.18 The Effect of Conjugation on λ_{max}

The more conjugated double bonds that a compound has, the longer is the wavelength at which the $n \rightarrow \pi^*$ and $\pi \rightarrow \pi^*$ transitions occur. For example, the $n \rightarrow \pi^*$ transition for methyl vinyl ketone is at 324 nm, and the $\pi \rightarrow \pi^*$ transition is at 219 nm. Both λ_{max} values are at longer wavelengths than the corresponding λ_{max} values of acetone because methyl vinyl ketone has two conjugated double bonds, whereas acetone has only one double bond.

The λ_{max} increases as the number of conjugated double bonds increases.

acetone

$n \longrightarrow \pi^*$ $\lambda_{max} = $ **274 nm** ($\varepsilon_{max} = $ **13.6**)
$\pi \longrightarrow \pi^*$ $\lambda_{max} = $ **195 nm** ($\varepsilon_{max} = $ **9000**)

methyl vinyl ketone

$\lambda_{max} = $ **324 nm** ($\varepsilon_{max} = 25$)
$\lambda_{max} = $ **219 nm** ($\varepsilon_{max} = 9600$)

The λ_{max} values of the $\pi \rightarrow \pi^*$ transition for several conjugated dienes are shown in Table 12.6. Notice that both the λ_{max} and the molar absorptivity increase as the

Table 12.6	Values of λ_{max} and ε for Ethylene and Conjugated Dienes	
Compound	λ_{max} **(nm)**	$\varepsilon(M^{-1} cm^{-1})$
$H_2C{=}CH_2$	165	15,000
	217	21,000
	256	50,000
	290	85,000
	334	125,000
	364	138,000

number of conjugated double bonds increases. Thus, the λ_{max} of a compound can be used to estimate the number of conjugated double bonds in a compound.

Conjugation raises the energy of the **HOMO (highest occupied molecular orbital)** and lowers the energy of the **LUMO (lowest unoccupied molecular orbital**; Section 7.12), so less energy is required for an electronic transition in a conjugated system than in a nonconjugated system (Figure 12.38). The more conjugated double bonds there are in a compound, the less energy is required for the electronic transition, and therefore the longer is the wavelength at which the electronic transition occurs.

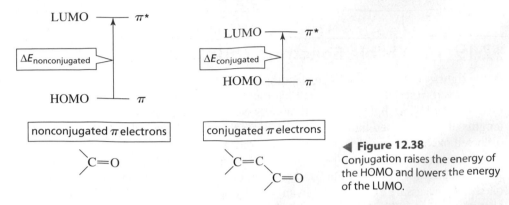

◀ Figure 12.38
Conjugation raises the energy of the HOMO and lowers the energy of the LUMO.

If a compound has enough conjugated double bonds, it will absorb visible light ($\lambda_{max} > 400$ nm), and the compound will be colored. Thus, β-carotene, a precursor of vitamin A—found in carrots, apricots, and sweet potatoes—is an orange substance. Lycopene—found in tomatoes, watermelon, and pink grapefruit—is red.

β-carotene
$\lambda_{max} = $ **455 nm**

lycopene
$\lambda_{max} = $ **474 nm**

An **auxochrome** is a substituent that when attached to a chromophore, alters the λ_{max} and the intensity of the absorption, usually increasing both. For example, OH and NH$_2$ groups are auxochromes. In the OH- and NH$_2$-containing compounds shown below, the lone-pair electrons on oxygen and nitrogen are available to interact with the π electron cloud of the benzene ring; such an interaction increases λ_{max}. Because the anilinium ion does not have an auxochrome, its λ_{max} is similar to that of benzene.

benzene
$\lambda_{max} = $ **255 nm**

phenol
270 nm

aniline
280 nm

anilinium ion
254 nm

Removing a proton from phenol, thereby forming a phenolate ion increases the λ_{max} because the resulting ion has an additional lone pair. Protonating aniline (and thereby forming the anilinium ion) decreases the λ_{max}, because the lone pair is no longer available to interact with the π cloud of the benzene ring.

PROBLEM 37◆

Rank the compounds in order of decreasing λ_{max}:

a.

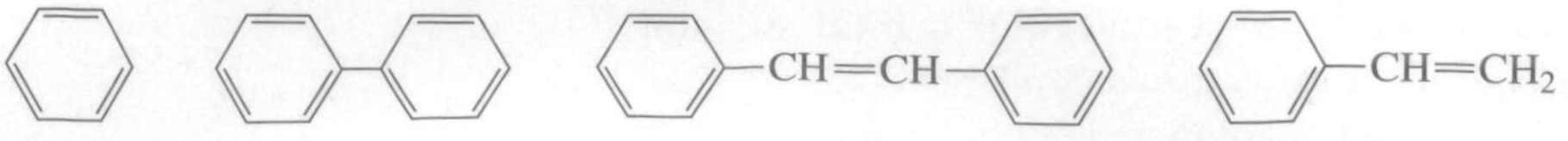

b.

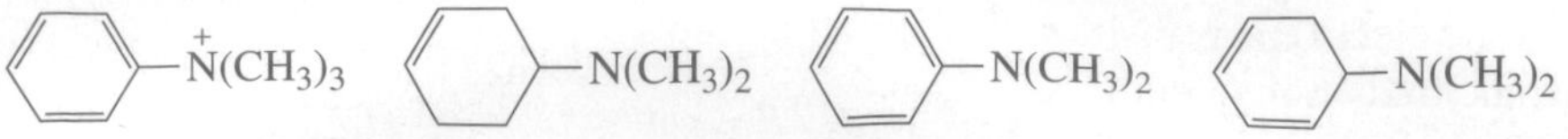

12.19 The Visible Spectrum and Color

White light is a mixture of all wavelengths of visible light. If any of these wavelengths are removed from white light, the remaining light is colored. Therefore, a compound that absorbs visible light is colored, and its color depends on the color of the wavelengths of the absorbed light. The wavelengths that the compound does not absorb are reflected back at the viewer, producing the color the viewer sees.

The relationship between the wavelengths of the light that a substance absorbs and the substance's absorbed color is shown in Table 12.7. Notice that two absorption bands are necessary to produce green. Most colored compounds have fairly broad absorption bands, although vivid colors have narrow absorption bands. The human eye is able to distinguish more than a million different shades of color!

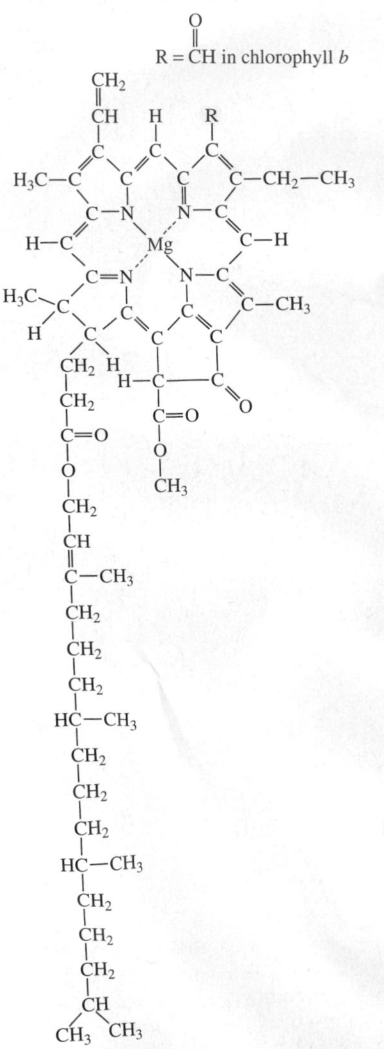

R = CH₃ in chlorophyll *a*

R = CH in chlorophyll *b*

Chlorophyll *a* and *b* are the pigments that make plants look green. These highly conjugated compounds absorb nongreen light. Therefore, the light reflected from the surface tissues of plants is green.

Table 12.7	Dependence of the Color Observed on the Wavelength of Light Absorbed	
Wavelengths absorbed (nm)	**Color absorbed**	**Color observed**
380–460	blue-violet	yellow
380–500	blue	orange
440–560	blue-green	red
480–610	green	purple
540–650	orange	blue
380–420 and 610–700	purple	green

Azobenzenes (benzene rings connected by an N=N bond) have an extended conjugated system that causes them to absorb light from the visible region of the spectrum. Some substituted azobenzenes, such as the two shown below, are used commercially as dyes. Variations in the extent of conjugation and the substituents attached to the conjugated system create a large number of different colors. Notice that the only difference between butter yellow and methyl orange is an $SO_3^- Na^+$ group. When margarine was first produced, butter. (White manufacturers colored it with butter yellow to make it look more like it was found to be care would not be very appetizing.) This dye was abandoned after Methyl orange is a commenic. β-Carotene is now used to color margarine (page 553). used acid–base indicator (see Problem 70a).

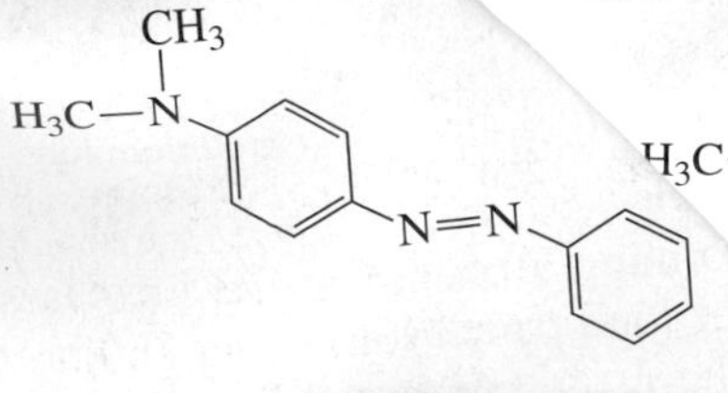

butter yellow
an azobenzene

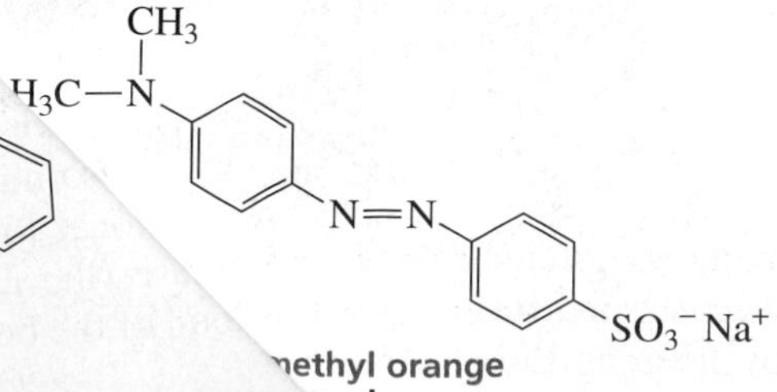

methyl orange
azobenzene

ANTHOCYANINS: A COLORFUL CLASS OF COMPOUNDS

A class of highly conjugated compounds called *anthocyanins* is responsible for the red, purple, and blue colors of many flowers (poppies, peonies, cornflowers), fruits (cranberries, rhubarb, strawberries, blueberries, red skin of apples, purple skin of grapes), and vegetables (beets, radishes, red cabbage). In a neutral or basic solution, the monocyclic fragment (on the right-hand side of the anthocyanin) is not conjugated with the rest of the molecule, so the anthocyanin does not absorb visible light and is therefore a colorless compound. In an acidic environment, however, the OH group becomes protonated and water is eliminated. (Recall that water, being a weak base, is a good leaving group.) Loss of water results in the third ring's becoming conjugated with the rest of the molecule. As a result of the extended conjugation, the anthocyanin absorbs visible light with wavelengths between 480 and 550 nm. The exact wavelength of light absorbed depends on the substituents (R and R') on the anthocyanin. Thus, the flower, fruit, or vegetable appears red, purple, or blue, depending on what the R groups are. You can see this color change if you alter the pH of cranberry juice so that it is no longer acidic.

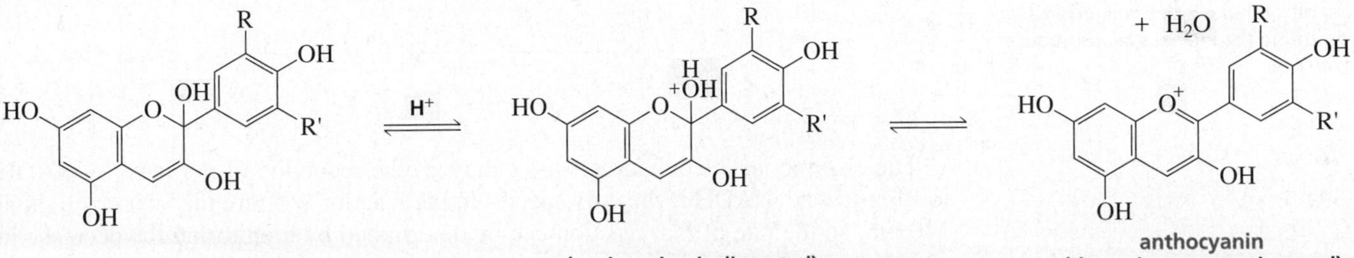

(conjugation is disrupted)
colorless

(conjugation is disrupted)
colorless

anthocyanin
(three rings are conjugated)
red, blue, or purple

R – H, OH, or OCH_3
R' = H, OH, or OCH_3

PROBLEM 38◆

a. At pH = 7, one of the ions shown below is purple and the other is blue. Which is which?

b. What would be the difference in the colors of the compounds at pH = 3?

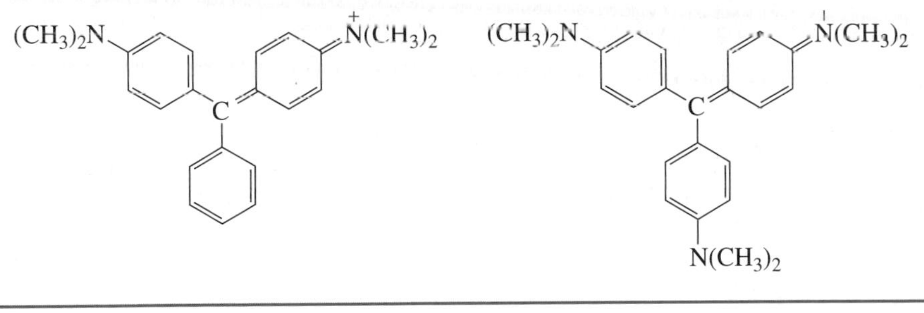

12.20 Some Uses of UV/Vis Spectroscopy

UV/Vis spectroscopy is often used to measure reaction rates. The rate of any reaction can be measured, as long as one of the reactants or one of the products absorbs UV or visible light at a wavelength at which the other reactants and products have little or no absorbance. For example, the anion of nitroethane has a λ_{max} at 240 nm, but neither the other product (H_2O) nor the reactants show any significant absorbance at that wavelength. In order to measure the rate at which hydroxide ion removes a proton from nitroethane (that is, the rate at which the nitroethane anion is formed), the UV spectrophotometer is adjusted to measure absorbance at 240 nm as a function of time instead of absorbance as a function of wavelength. Nitroethane is added to a quartz cell containing a basic solution, and the rate of the reaction is determined by monitoring the increase in absorbance at 240 nm (Figure 12.39).

Lycopene, β-carotene, and anthocyanins are found in the leaves of trees, but their characteristic colors are usually obscured by the green color of chlorophyll. In the fall, when chlorophyll degrades, the other colors become apparent.

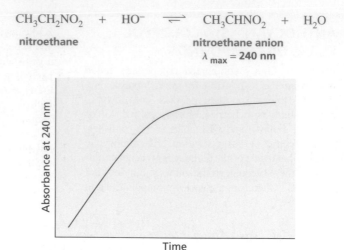

$$\text{CH}_3\text{CH}_2\text{NO}_2 \quad + \quad \text{HO}^- \quad \rightleftharpoons \quad \text{CH}_3\bar{\text{C}}\text{HNO}_2 \quad + \quad \text{H}_2\text{O}$$

nitroethane

nitroethane anion
$\lambda_{max} = \textbf{240 nm}$

Figure 12.39 ▶
The rate at which a proton is removed from nitroethane is determined by monitoring the increase in absorbance at 240 nm.

The enzyme lactate dehydrogenase catalyzes the reduction of pyruvate by NADH to form lactate. NADH is the only species in the reaction mixture that absorbs light at 340 nm, so the rate of that reaction can be determined by monitoring the decrease in absorbance at 340 nm (Figure 12.40).

$$\overset{\overset{\text{O}}{\|}}{\text{CH}_3\text{CCOO}^-} \; + \; \text{NADH} \; + \; \text{H}^+ \; \xrightarrow[\text{dehydrogenase}]{\text{lactate}} \; \overset{\overset{\text{OH}}{|}}{\text{CH}_3\text{CHCOO}^-} \; + \; \text{NAD}^+$$

pyruvate $\lambda_{max} = \textbf{340 nm}$

lactate

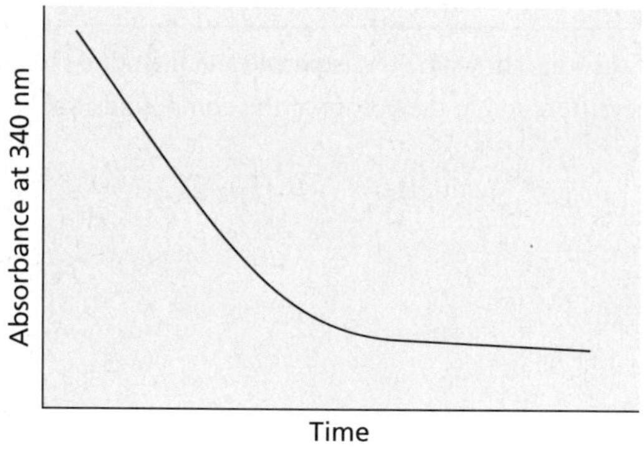

Figure 12.40 ▶
The rate of reduction of pyruvate by NADH is measured by monitoring the decrease in absorbance at 340 nm.

PROBLEM 39◆

Describe a way to determine the rate of the alcohol dehydrogenase catalyzed oxidation of ethanol by NAD$^+$ (Section 24.2).

The pK_a of a compound can be determined by UV/Vis spectroscopy if either the acidic form or the basic form of the compound absorbs UV or visible light. For example, the phenolate ion has a λ_{max} at 287 nm. If the absorbance at 287 nm is determined as a function of pH, the pK_a of phenol can be ascertained by determining the pH at which exactly one-half the increase in absorbance has occurred (Figure 12.41). At this pH, half of the phenol has been converted into phenolate ion. Recall that the Henderson–Hasselbalch equation (Section 1.24) states that the pK_a of a compound is the pH at which half the compound exists in its acidic form and half exists in its basic form ([HA] = [A$^-$]).

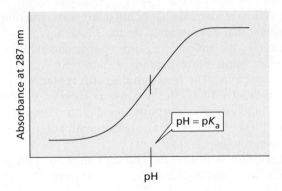

◀ **Figure 12.41**
The absorbance of an aqueous solution of phenol as a function of pH.

UV spectroscopy can also be used to estimate the nucleotide composition of DNA. The two strands of DNA are held together by hydrogen bonds between the bases in one strand and the bases in the other (see Figure 27.4 on page 1204): each guanine forms hydrogen bonds with a cytosine (a G–C pair); each adenine forms hydrogen bonds with a thymine (an A–T pair). When DNA is heated, the strands break apart and the absorbance increases because single-stranded DNA has a greater molar absorptivity at 260 nm than does double-stranded DNA. The melting temperature (T_m) of DNA is the midpoint of an absorbance-versus-temperature curve (Figure 12.42). For double-stranded DNA, T_m increases with increasing numbers of G–C pairs, because they are held together by three hydrogen bonds, whereas A–T pairs are held together by only two hydrogen bonds. Therefore, T_m can be used to estimate the number of G–C pairs. These are just a few examples of the many uses of UV/Vis spectroscopy.

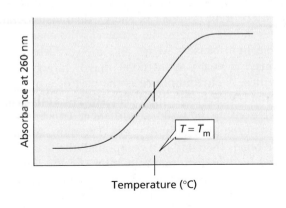

◀ **Figure 12.42**
The absorbance of a solution of DNA as a function of temperature.

PROBLEM 40◆

The absorbance of a solution of a weak acid was measured under the same conditions at a series of pH values. Its conjugate base is the only species in the solution that absorbs UV light at the wavelength used. Estimate the pK_a of the compound from the data obtained.

pH	1.0	2.0	3.0	4.0	5.0	6.0	7.0	8.0	9.0	10.0
Absorbance	0	0	0.10	0.50	0.80	1.10	1.50	1.60	1.60	1.60

SUMMARY

Mass spectrometry allows us to determine the *molecular mass* and the *molecular formula* of a compound, as well as some structural features of the compound. In this analytical method, a small sample of the compound is vaporized and then ionized as a result of an electron's being removed from each molecule, producing a **molecular ion**—a radical cation. Many of the molecular ions break apart into cations, radicals, neutral molecules, and other radical cations. The bonds most likely to break are the weakest ones and those that result in the formation of the most stable products.

The mass spectrometer records a **mass spectrum**, which is a graph of the relative abundance of each positively charged fragment plotted against its *m/z* value. The molecular ion peak (M) represents the molecular ion, and its *m/z* value gives the molecular mass of the compound. Peaks with smaller *m/z* values—**fragment ion peaks**—represent positively charged fragments of the molecule. The **base peak** is the peak with the greatest intensity. High-resolution mass spectrometers determine the exact molecular mass, which allows a compound's molecular formula to be determined. The "nitrogen rule" states that if a compound has an odd-mass molecular ion, the compound contains an odd number of nitrogen atoms.

The M + 1 peak occurs because of the naturally occurring ^{13}C isotope of carbon. The number of carbon atoms in a compound can be calculated from the relative intensities of the M and M + 1 peaks. A large M + 2 peak is evidence of a compound containing either chlorine or bromine; if it is one-third the height of the M peak, the compound contains one chlorine atom; if the M and M + 2 peaks are about the same height, the compound contains one bromine atom.

Characteristic fragmentation patterns are associated with specific functional groups. Electron bombardment is most likely to dislodge a lone-pair electron. A bond between carbon and a more electronegative atom breaks *heterolytically*, with the electrons going to the more electronegative atom, whereas a bond between carbon and an atom of similar electronegativity breaks *homolytically*. α-Cleavage occurs because the species it forms has a positive charge that is shared by two atoms.

Spectroscopy is the study of the interaction of matter and **electromagnetic radiation**. A continuum of different types of electromagnetic radiation constitutes the electromagnetic spectrum. High-energy radiation is associated with *high frequencies, large wavenumbers*, and *short wavelengths*.

Infrared spectroscopy identifies the kinds of functional groups in a compound. Bonds vibrate with stretching and bending motions. The stretching vibrations of a given bond have a characteristic frequency, as do the bending vibrations. It takes more energy to stretch a bond than to bend it. When a compound is bombarded with radiation of a frequency that exactly matches the frequency of one of its vibrations, the molecule absorbs energy and exhibits an **absorption band**, corresponding to that frequency in the infrared spectrum. The **functional group region** of an IR spectrum ($4000-1400$ cm^{-1}) is where most of the functional groups show absorption bands; the **fingerprint region** ($1400-600$ cm^{-1}) is characteristic of the compound as a whole.

The position, intensity, and shape of an absorption band help identify functional groups. The amount of energy required to stretch a bond depends on the *strength* of the bond: stronger bonds show absorption bands at larger wavenumbers. Therefore, the frequency of the absorption band depends on bond order, hybridization, electron donation and withdrawal, and electron delocalization. The frequency is inversely related to the *mass* of the atoms, so heavier atoms vibrate at lower frequencies. The intensity of an absorption band depends on the size of the change in dipole moment associated with the vibration and on the number of bonds vibrating at that frequency (the number of bonds of that type in the compound). In order for a vibration to absorb IR radiation, the dipole moment of the bond must change when the vibration occurs.

Ultraviolet and **visible (UV/Vis) spectroscopy** provide information about compounds with conjugated double bonds. UV light is higher in energy than visible light; the shorter the wavelength, the greater is the energy. UV and visible light cause $n \rightarrow \pi^*$ and $\pi \rightarrow \pi^*$ **electronic transitions**; $\pi \rightarrow \pi^*$ transitions have larger molar absorptivities. A **chromophore** is the part of a molecule that absorbs UV or visible light. The **Beer–Lambert law** is the relationship between absorbance, concentration, and length of the light path: $A = \varepsilon c l$. The more conjugated double bonds there are in a compound, the less energy is required for the electronic transition and the longer is the λ_{max} at which it occurs. Reaction rates and pK_a values are commonly measured using UV/Vis spectroscopy.

KEY TERMS

absorption band (p. 532)
auxochrome (p. 553)
base peak (p. 516)
Beer–Lambert law (p. 551)
bending vibration (p. 530)
chromophore (p. 550)
α-cleavage (p. 521)
electromagnetic radiation (p. 528)
electronic transition (p. 549)
fingerprint region (p. 532)
fragment ion peak (p. 516)
frequency (p. 529)

functional group region (p. 532)
highest occupied molecular orbital (HOMO) (p. 553)
Hooke's law (p. 535)
infrared radiation (p. 531)
infrared spectroscopy (p. 513)
infrared spectrum (p. 532)
lowest unoccupied molecular orbital (LUMO) (p. 553)
mass spectrometry (p. 513)
mass spectrum (p. 515)
McLafferty rearrangement (p. 526)

molar absorptivity (p. 551)
molecular ion (p. 513)
nominal molecular mass (p. 515)
radical cation (p. 513)
spectroscopy (p. 528)
stretching vibration (p. 530)
ultraviolet light (p. 549)
UV/Vis spectroscopy (p. 513)
visible light (p. 549)
wavelength (p. 529)
wavenumber (p. 529)

PROBLEMS

41. In the mass spectrum of the following compounds, which would be more intense—the peak at $m/z = 57$ or the peak at $m/z = 71$?
   a. 3-methylpentane   **b.** 2-methylpentane

42. List three factors that influence the intensity of an IR absorption band.

43. For each of the following pairs of compounds, identify one IR absorption band that could be used to distinguish between them:

a. $CH_3CH_2\overset{O}{\overset{\|}{C}}OCH_3$ and $CH_3CH_2\overset{O}{\overset{\|}{C}}CH_3$

g. $CH_3\overset{O}{\overset{\|}{C}}OCH_2CH_3$ and $CH_3\overset{O}{\overset{\|}{C}}CH_2OCH_3$

b. 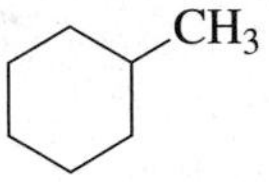 (cyclohexane with CH_3) and $CH_3CH_2CH_2CH_2CH_2CH_2CH_3$

h. (cyclohexenone structures) and

c. $CH_3CH_2CH_2OH$ and $CH_3CH_2OCH_3$

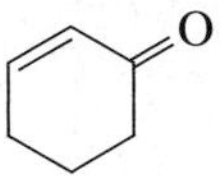

i. $CH_3CH_2CH=CHCH_3$ and $CH_3CH_2C\equiv CCH_3$

d. $CH_3CH_2\overset{O}{\overset{\|}{C}}NH_2$ and $CH_3CH_2\overset{O}{\overset{\|}{C}}OCH_3$

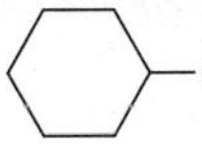

j. $CH_3CH_2\overset{O}{\overset{\|}{C}}H$ and $CH_3CH_2\overset{O}{\overset{\|}{C}}CH_3$

e. (cyclohexyl)$-CH_2CH_2OH$ and (cyclohexyl)$-\underset{OH}{CHCH_3}$

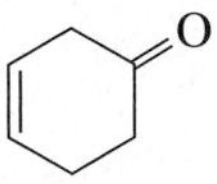

k. (cyclohexyl)$\overset{O}{\overset{\|}{C}}H$ and (phenyl)$\overset{O}{\overset{\|}{C}}H$

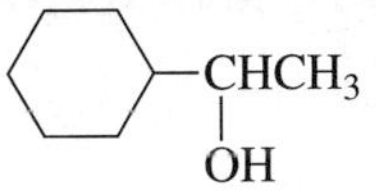

f. *cis*-2-butene   and   *trans*-2-butene

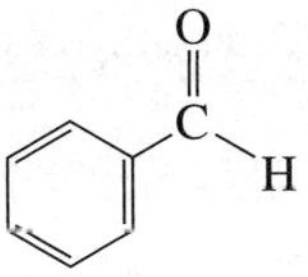

l. $CH_3CH_2CH=CH_2$ and $CH_3CH_2CH=\underset{CH_3}{\overset{|}{C}}CH_3$

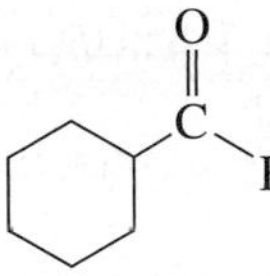

44. 4-Methyl-3-penten-2-one has two absorption bands in its UV spectrum, one at 236 nm and one at 314 nm.

$$CH_3\overset{O}{\overset{\|}{C}}CH=\underset{CH_3}{\overset{|}{C}}CH_3$$
4-methyl-3-penten-2-one

   a. Why are there two absorption bands?
   b. Which band shows the greater absorbance?

45. a. How could you use IR spectroscopy to determine whether the following reaction had occurred?

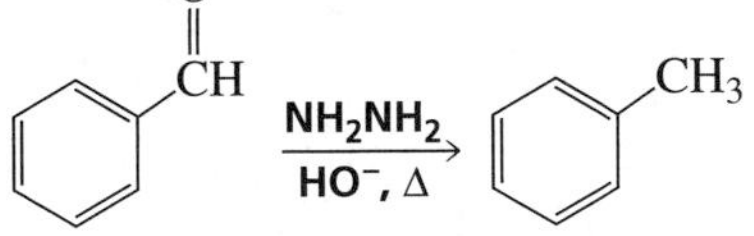

(benzaldehyde) $\overset{O}{\overset{\|}{C}}H$ $\xrightarrow[\text{HO}^-,\,\Delta]{\text{NH}_2\text{NH}_2}$ (toluene) CH_3

   b. After purifying the product, how could you determine whether the NH_2NH_2 had been removed?

46. How could you use UV spectroscopy to distinguish between the compounds in each of the following pairs?

a. 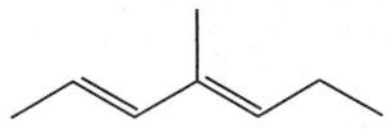 (diene structure) and (diene structure)

c. (phenylacetone) and (propiophenone)

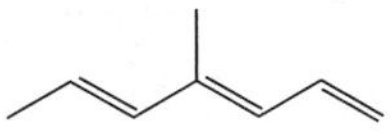

b. $CH_2=CHCH=CHCH=CH_2$ and $CH_2=CHCH=CH\overset{O}{\overset{\|}{C}}CH_3$

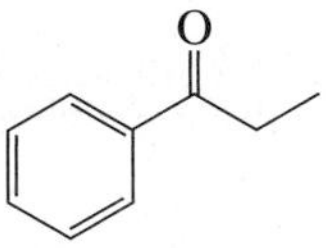

d. 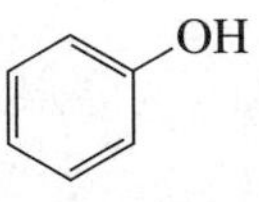 (phenol) OH and (anisole) OCH_3

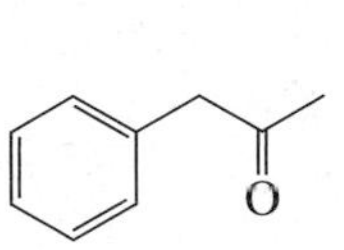

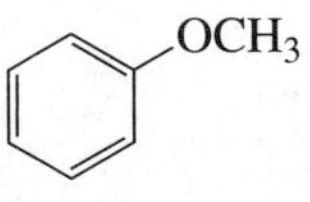

47. What identifying characteristics would be present in the mass spectrum of a compound containing two bromine atoms?

48. Assuming that the force constant is approximately the same for $C-C$, $C-N$, and $C-O$ bonds, predict the relative positions of their stretching vibrations.

49. In the following boxes, list the types of bonds and the approximate wavenumber at which each type of bond is expected to show an IR absorption:

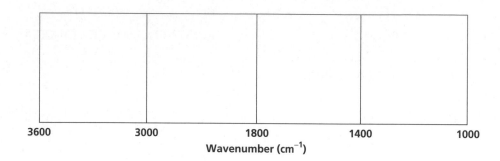

50. A mass spectrum shows significant peaks at m/z = 87, 115, 140, and 143. Which of the following compounds is responsible for that mass spectrum: 4,7-dimethyl-1-octanol, 2,6-dimethyl-4-octanol, or 2,2,4-trimethyl-4-heptanol?

51. How could IR spectroscopy distinguish between 1,5-hexadiene and 2,4-hexadiene?

52. A compound gives a mass spectrum with essentially only three peaks at m/z = 77 (40%), 112 (100%), 114 (33%). Identify the compound.

53. What hydrocarbons will have a molecular ion peak at m/z = 112?

54. Each of the IR spectra presented in Figures 12.43, 12.44, and 12.45 is accompanied by a set of four compounds. In each case, indicate which of the four compounds is responsible for the spectrum.

a. $CH_3CH_2CH_2C\equiv CCH_3$ $CH_3CH_2CH_2CH_2OH$ $CH_3CH_2CH_2CH_2C\equiv CH$ $CH_3CH_2CH_2\overset{\displaystyle O}{\overset{\displaystyle \|}{C}}OH$

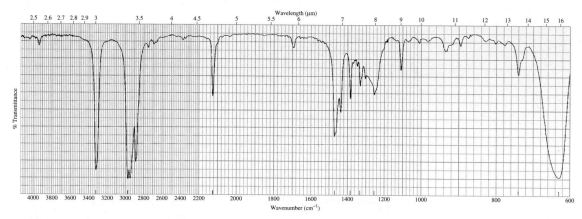

▲ **Figure 12.43**
The IR spectrum for Problem 54a.

b. $CH_3CH_2\overset{\displaystyle O}{\overset{\displaystyle \|}{C}}OH$ $CH_3CH_2\overset{\displaystyle O}{\overset{\displaystyle \|}{C}}OCH_2CH_3$ $CH_3CH_2\overset{\displaystyle O}{\overset{\displaystyle \|}{C}}H$ $CH_3CH_2\overset{\displaystyle O}{\overset{\displaystyle \|}{C}}CH_3$

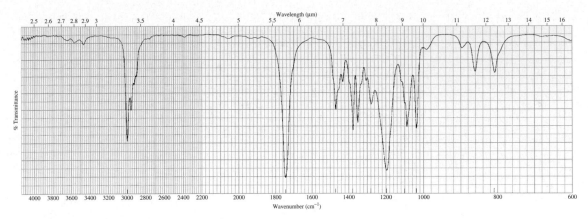

▲ Figure 12.44
The IR spectrum for Problem 54b.

c.

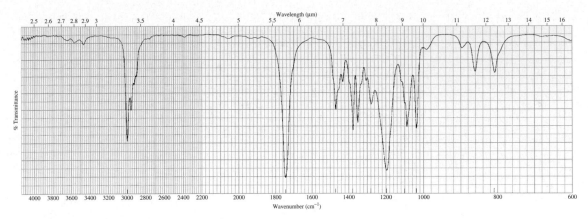

(Structures shown: C(CH₃)₃ on benzene, CH₂CH₂Br on benzene, CH=CH₂ on benzene, CH₂OH on benzene)

▲ Figure 12.45
The IR spectrum for Problem 54c.

55. What peaks in their mass spectra could be used to distinguish between 4-methyl-2-pentanone and 2-methyl-3-pentanone?

56. A compound is known to be one of those shown here. What absorption bands in its IR spectrum would allow you to identify this compound?

A   B   C

57. How could IR spectroscopy distinguish between 1-hexyne, 2-hexyne, and 3-hexyne?

58. Five compounds are shown for each of the IR spectra in Figures 12.46, 12.47, and 12.48. Indicate which of the five compounds is responsible for each spectrum.

a. $CH_3CH_2CH=CH_2$ $CH_3CH_2CH_2CH_2OH$ $CH_2=CHCH_2CH_2OH$ $CH_3CH_2CH_2OCH_3$ $CH_3CH_2CH_2\overset{\displaystyle O}{\overset{\displaystyle \|}{C}}OH$

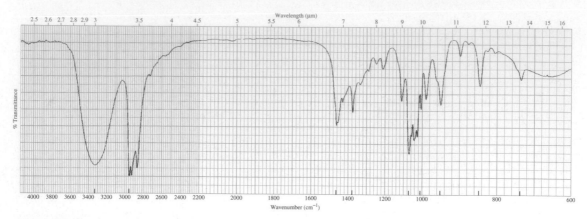

▲ Figure 12.46
The IR spectrum for Problem 58a.

b.

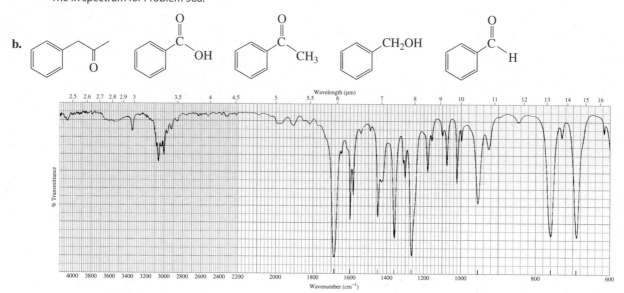

▲ Figure 12.47
The IR spectrum for Problem 58b.

c.

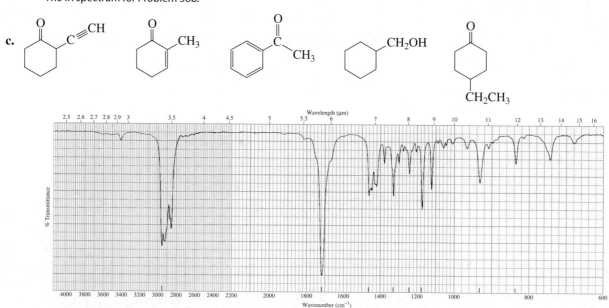

▲ Figure 12.48
The IR spectrum for Problem 58c.

59. A solution of ethanol has been contaminated with benzene—a technique employed to make ethanol unfit to drink. Benzene has a molar absorptivity of 230 at 260 nm in ethanol, and ethanol shows no absorbance at 260 nm. How could the concentration of benzene in the solution be determined?

60. Each of the IR spectra shown in Figure 12.49 is the spectrum of one of the following compounds. Identify the compound that produced each spectrum.

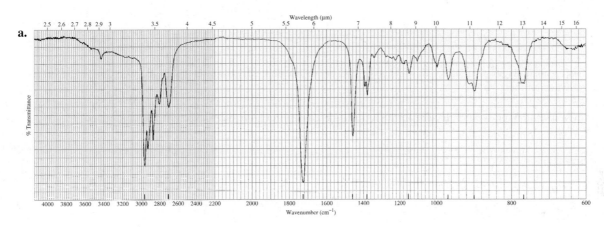

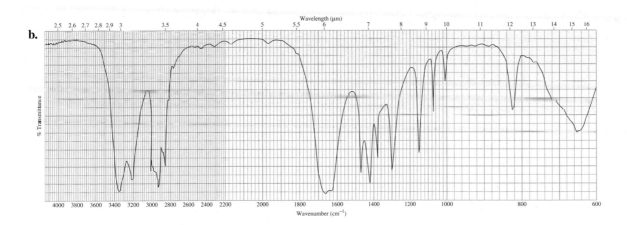

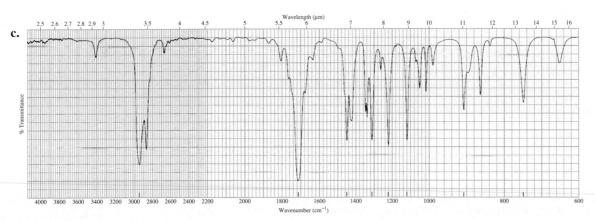

▲ **Figure 12.49**
The IR spectra for Problem 60.

61. Predict the major characteristic IR absorption bands that would be given by each of the following compounds:

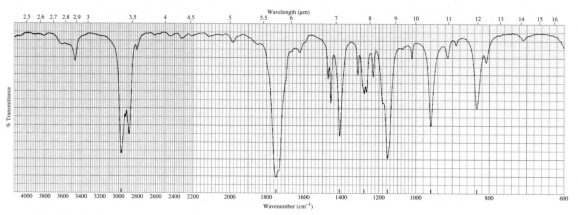

a. $CH_2{=}CHCH_2$ $\overset{\overset{\textstyle O}{\|}}{C}$ H

b. (benzene ring)$\overset{\overset{\textstyle O}{\|}}{C}$ OCH_2CH_3

c. (cyclohexane ring)CH_2CH_2OH

d. CH_3CH_2 $\overset{\overset{\textstyle O}{\|}}{C}$ $CH_2CH_2NH_2$

e. (cyclohexane ring)$CH_2C{\equiv}CH$

f. (cyclohexane ring)$CH_2\overset{\overset{\textstyle O}{\|}}{C}OH$

62. Given that the force constants are similar for C—H and C—C bonds, explain why the stretching vibration of a C—H bond occurs at a larger wavenumber.

63. The IR spectrum of a compound with molecular formula C_5H_8O is shown in Figure 12.50. Identify the compound.

▲ **Figure 12.50**
The IR spectrum for Problem 63.

64. Some credit-card sales slips have a top sheet of "carbonless paper" that transfers an imprint of your signature to a sheet lying underneath (the customer receipt). The paper contains tiny capsules filled with the following colorless compound:

When you press on the paper, the capsules burst, and the colorless compound comes into contact with the acid-treated bottom sheet, forming a highly colored compound. What is the structure of the colored compound?

65. Which one of the following compounds produced the IR spectrum shown in Figure 12.51?

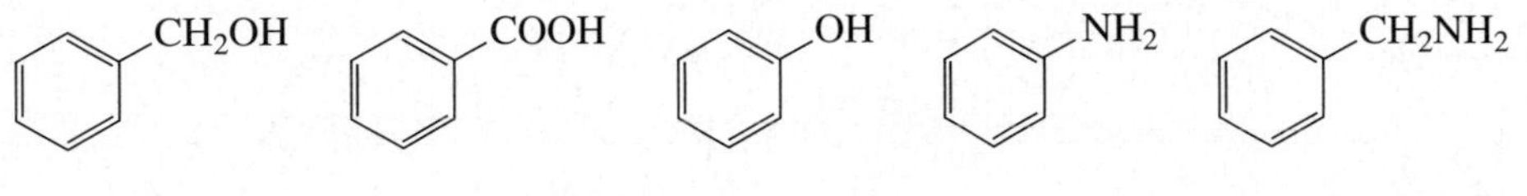

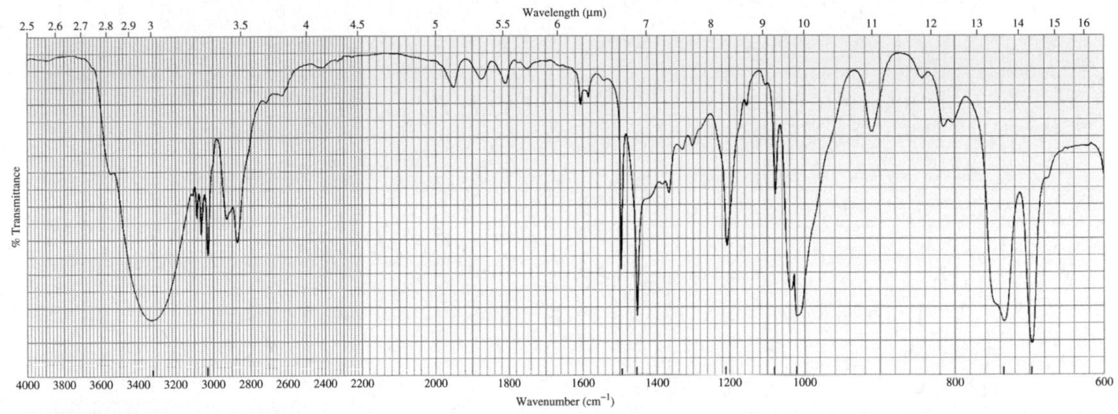

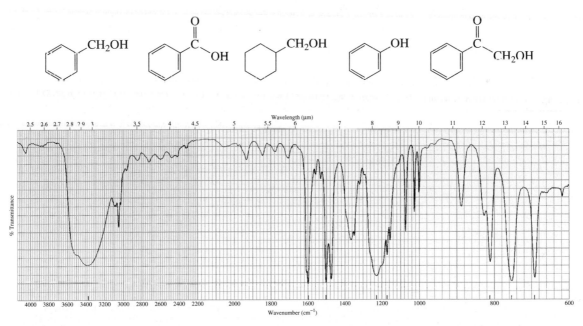

▲ Figure 12.51
The IR spectrum for Problem 65.

66. Which one of the following compounds produced the IR spectrum shown in Figure 12.52?

▲ Figure 12.52
The IR spectrum for Problem 66.

67. What is the molecular formula of a saturated acyclic hydrocarbon that has an M peak at $m/z = 100$ with a relative abundance of 27.32%, and an M + 1 peak with a relative abundance of 2.10%?

68. Calculate the approximate wavenumber at which a $C\!=\!C$ stretch will occur, given that the force constant for the $C\!=\!C$ bond is 10×10^5 gs^{-2}.

69. The IR and mass spectra for three different compounds are shown in Figures 12.53–12.55. Identify each compound.

a.

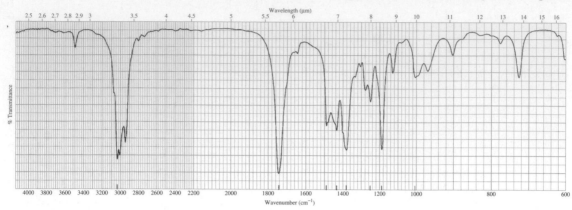

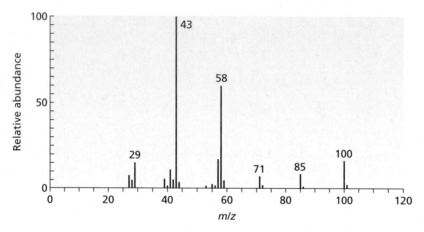

▲ **Figure 12.53**
The IR and mass spectra for Problem 69a.

b.

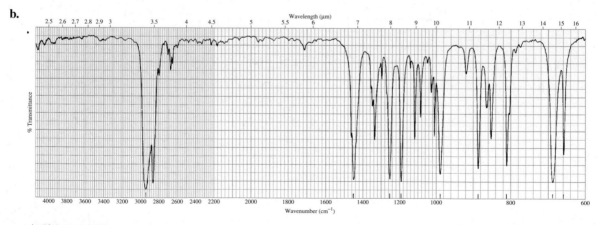

▲ **Figure 12.54**
The IR and mass spectra for Problem 69b.

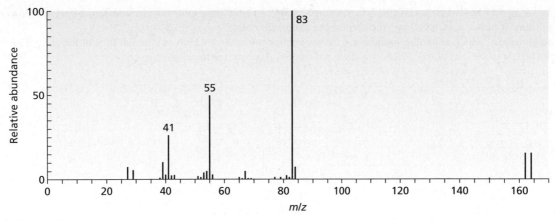

▲ **Figure 12.54**
Continued

c.

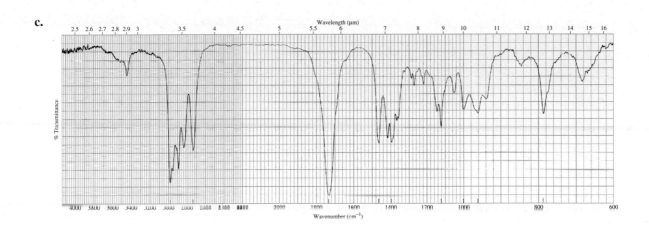

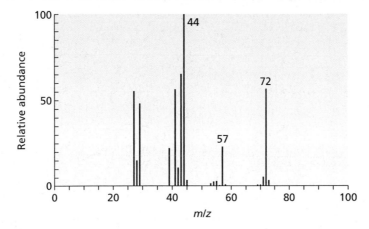

▲ **Figure 12.55**
The IR and mass spectra for Problem 69c.

70. a. Methyl orange (whose structure is given in Section 12.19) is an acid–base indicator. In solutions of pH < 4, it is red; in solutions of pH > 4, it is yellow. Account for the change in color.
   b. Phenolphthalein, which is another indicator, exhibits a much more dramatic color change. In solutions of pH < 8.5, it is colorless; in solutions of pH > 8.5, it is deep red-purple. Account for the change in color.

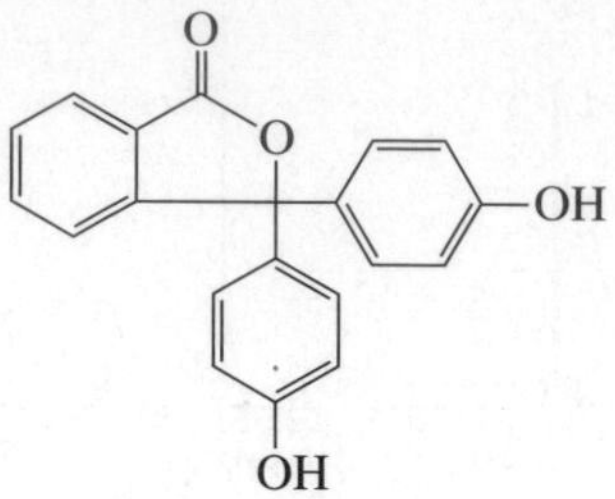

phenolphthalein

CHAPTER 13

NMR Spectroscopy

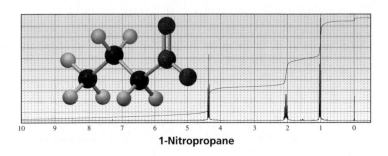

1-Nitropropane

BUILDING ON FUNDAMENTALS

SECTION 13.6 Inductive electron withdrawal (1.21) determines the relative positions of signals in an ^1H NMR spectrum.

SECTION 13.7 The chemical shifts show that carbon is more electronegative than hydrogen. We have previously seen that when there is an adjacent *p* orbital, alkyl groups are electron donating compared to hydrogens (4.2, 4.11, and 11.3). Evidently, hyperconjugation more than makes up for the greater electronegativity of carbon.

SECTION 13.18 ^{13}C NMR records signals from the isotope of carbon (^{13}C) that constitutes only 1.11% of carbon atoms (1.1).

SECTION 13.18 Inductive electron withdrawal (1.21) determines the relative positions of signals in a ^{13}C NMR spectrum.

Chapter 12 introduced three instrumental techniques used to determine the structures of organic compounds: mass spectrometry, IR spectroscopy, and UV/Vis spectroscopy. Now we will look at a fourth technique, *nuclear magnetic resonance (NMR) spectroscopy*. **NMR spectroscopy** helps to identify the carbon–hydrogen framework of an organic compound.

The advantage of NMR spectroscopy over the other instrumental techniques we have studied is that it not only makes it possible to identify the functionality at a specific carbon, but it also enables us to connect neighboring carbons. In many cases, NMR spectroscopy can be used to determine a molecule's entire structure.

13.1 An Introduction to NMR Spectroscopy

NMR spectroscopy was developed by physical chemists in the late 1940s to study the properties of atomic nuclei. In 1951, chemists realized that NMR spectroscopy could also be used to study the structures of organic compounds. Nuclei with an odd number of protons or an odd number of neutrons (or both) have magnetic properties, as indicated by a non-zero value for their spin quantum number. Such nuclei (^1H, ^{13}C, ^{15}N, ^{19}F, and ^{31}P) can be studied by NMR. In contrast, ^{12}C and ^{16}O do not have magnetic properties (each has a zero value for its spin quantum number) and therefore cannot be studied by NMR. Because hydrogen nuclei (protons) were the first nuclei studied by nuclear magnetic resonance, the acronym "NMR" is generally assumed to mean **^{1}H NMR (proton magnetic resonance)**.

A spinning nucleus with magnetic properties generates a magnetic field similar to the magnetic field generated by a small bar magnet. In the absence of an applied magnetic field, the magnetic moments associated with the nuclear spins are randomly oriented. However, when placed between the poles of a strong magnet (Figure 13.1), the nuclear magnetic moments align either *with* or *against* the applied magnetic field. Those that align with the field are in the lower-energy **α-spin state**; those that align against the field are in the higher-energy **β-spin state**, because more energy is needed to align against the field than with it. More nuclei are in the α-spin state than in the β-spin state. The difference in the populations is very small (about 20 out of 1 million protons), but it is sufficient to form the basis of NMR spectroscopy.

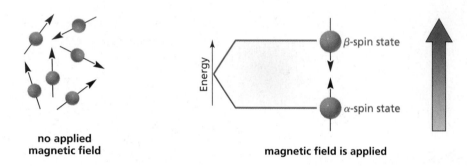

Figure 13.1 ▶
In the absence of an applied magnetic field, the magnetic moments of the nuclei are randomly oriented. In the presence of an applied magnetic field, the magnetic moments of the nuclei line up with or against the field.

The energy difference (ΔE) between the α- and β-spin states depends on the strength of the **applied magnetic field** (B_0); the greater the strength of the magnetic field, the greater is the difference in energy between the α- and β-spin states (Figure 13.2).

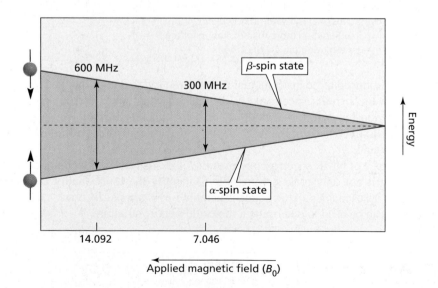

Figure 13.2 ▶
The greater the strength of the applied magnetic field, the greater is the difference in energy between the α- and β-spin states.

When a sample is subjected to a pulse of radiation whose energy corresponds to the difference in energy (ΔE) between the α- and β-spin states, nuclei in the α-spin state are promoted to the β-spin state. This transition is called "flipping" the spin. With currently available magnets, the energy difference between the α- and β-spin states is small, so only a small amount of energy is needed to flip the spin. The radiation used to supply this energy is in the radio frequency (rf) region of the electromagnetic spectrum and is called **rf radiation**. When the nuclei absorb rf radiation, they flip their spin, causing them to generate signals whose frequency depends on the difference in energy (ΔE) between the α- and β-spin states. The NMR spectrometer detects these signals and displays them as a plot of signal frequency versus intensity; this plot is the NMR spectrum. The nuclei are said to be *in resonance* with the rf radiation, hence the term "nuclear magnetic resonance." In this context, "resonance" refers to the flipping back and forth of nuclei between the α- and β-spin states in response to the rf radiation; it has nothing to do with the "resonance" associated with electron delocalization.

The equation displayed below shows that the energy difference between the spin states (ΔE) depends on the operating frequency of the spectrometer (v), which in turn depends on the strength of the magnetic field (B_0), measured in tesla (T),* and the *gyromagnetic ratio* (γ); h is Planck's constant (Section 12.6).

$$\Delta E = hv = h\frac{\gamma}{2\pi}B_0$$

The **gyromagnetic ratio**, also called the magnetogyric ratio, is a constant that depends on the magnetic moment of the particular kind of nucleus. In the case of a proton, the value of γ is 2.675×10^8 T^{-1} s^{-1}; in the case of a ^{13}C nucleus, it is 6.688×10^7 T^{-1} s^{-1}. Canceling Planck's constant on both sides of the equation gives

Earth's magnetic field is 5×10^{-5} T, measured at the equator. Its maximum surface magnetic field is 7×10^{-5} T, measured at the south magnetic pole.

$$v = \frac{\gamma}{2\pi}B_0$$

The following calculation shows that if an ^{1}H NMR spectrometer is equipped with a magnet that generates a magnetic field of 7.046 T, the spectrometer will require an operating frequency of 300 MHz (megahertz):

$$v = \frac{\gamma}{2\pi}B_0$$

$$= \frac{2.675 \times 10^8}{2(3.1416)} \text{ T}^{-1}\text{ s}^{-1} \times 7.046 \text{ T}$$

$$= 300 \times 10^6 \text{ Hz} = 300 \text{ MHz}$$

The equation shows that the strength of the *magnetic field (B_0) is proportional to the operating frequency* (MHz). Therefore, if the spectrometer has a more powerful magnet, it must have a higher **operating frequency**. For example, a magnetic field of 14.092 T requires an operating frequency of 600 MHz.

The magnetic field is proportional to the operating frequency.

Today's **NMR spectrometers** operate at frequencies between 60 and 950 MHz. The greater the operating frequency of the instrument—and the stronger the magnet— the better is the resolution of the NMR spectrum (we will discuss resolution in Section 13.17).

Because each kind of nucleus has its own gyromagnetic ratio, different energies are required to bring different kinds of nuclei into resonance. For example, an NMR spectrometer with a magnet requiring a frequency of 300 MHz to flip the spin of an ^{1}H nucleus requires a frequency of 75 MHz to flip the spin of a ^{13}C nucleus. NMR spectrometers are equipped with electromagnetic radiation sources that can be tuned to different frequencies so that they can be used to obtain NMR spectra of different kinds of nuclei (^{1}H, ^{13}C, ^{15}N, ^{19}F, ^{31}P).

---

* Until recently, the gauss (G) was the unit in which magnetic field strength was commonly measured (1 T = 10^4 G).

NIKOLA TESLA (1856–1943)

Nikola Tesla was born in Croatia, the son of a clergyman. He emigrated to the United States in 1884 and became a citizen in 1891. He was a proponent of the use of alternating current to distribute electricity and bitterly fought Thomas Edison, who promoted direct current. Although Tesla did not win his dispute with Guglielmo Marconi over which of them invented the radio, Tesla is given credit for developing neon and fluorescent lighting, the electron microscope, the refrigerator motor, and the Tesla coil, a type of transformer (for changing the voltage of alternating current).

Nikola Tesla in his laboratory

PROBLEM 1◆

What frequency (in MHz) is required to cause a proton to flip its spin when it is exposed to a magnetic field of 1 T ?

PROBLEM 2◆

a. Calculate the magnetic field (in tesla) required to flip an ^1H nucleus in an NMR spectrometer that operates at 360 MHz.

b. A magnetic field of what strength is required when a 500-MHz instrument is used for ^1H NMR?

13.2 Fourier Transform NMR

To obtain an **NMR spectrum**, a small amount of a compound is dissolved in about 0.5 mL of solvent. This solution is put into a long, thin glass tube, which is then placed within a powerful magnetic field (Figure 13.3). (The solvents used in NMR are discussed in Section 13.16.) Spinning the sample tube about its long axis averages

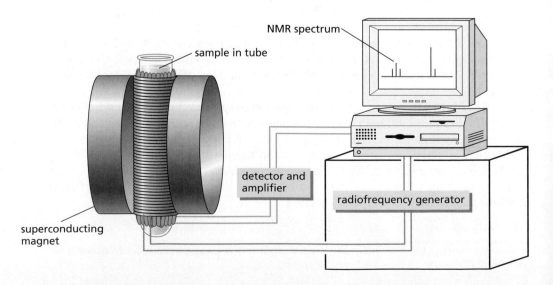

Figure 13.3 ▶
Schematic diagram of an NMR spectrometer.

the position of the molecules in the magnetic field and thus greatly increases the resolution of the spectrum.

In modern instruments called *pulsed Fourier transform (FT) spectrometers*, the magnetic field is held constant and an rf pulse of short duration excites all the protons simultaneously. Because the short rf pulse covers a range of frequencies, the individual protons absorb the frequency each requires to come into resonance (flip their spin) and produce a complex signal—called a free induction decay (FID)—at a frequency corresponding to ΔE. The intensity of the signal decays as the nuclei lose the energy they gained from the rf pulse. A computer measures the change in intensity over time and converts it into intensity-versus-frequency data, in a mathematical operation known as a *Fourier transform*, to produce a spectrum called a **Fourier transform NMR (FT–NMR) spectrum** (Section 12.7). An FT–NMR spectrum can be recorded in about 2 seconds—and large numbers of FIDs can be averaged in a few minutes—using less than 5 mg of compound. The NMR spectra in this book are FT–NMR spectra that were taken on a spectrometer with an operating frequency of 300 MHz. The discussions throughout the rest of this chapter pertain to FT–NMR, rather than to the older continuous wave (CW) NMR, because FT–NMR is both more modern and easier to understand.

BIOGRAPHY

The 1991 Nobel Prize in chemistry was awarded to **Richard R. Ernst** *for two important contributions: FT–NMR spectroscopy and an NMR tomography method that forms the basis of magnetic resonance imaging (MRI). Ernst was born in 1933, received a Ph.D. from the Swiss Federal Institute of Technology [Eidgenössische Technische Hochschule (ETH)] in Zurich, and became a research scientist at Varian Associates in Palo Alto, California. In 1968, he returned to the ETH to be professor of chemistry. Currently he spends his time at the Scripps Research Institute in La Jolla, California, and at the Swiss Academy of Science.*

13.3 Shielding Causes Different Hydrogens to Show Signals at Different Frequencies

The frequency of an NMR signal depends on the strength of the magnetic field (Figure 13.2) experienced by the nucleus. Thus, if all the hydrogens in an organic compound were to experience the applied magnetic field to the same degree, they would all give signals of the same frequency. If this were the case, all NMR spectra would consist of one signal, which would tell us nothing about the structure of the compound, except that it contains hydrogens.

A nucleus, however, is embedded in a cloud of electrons that partly *shields* it from the applied magnetic field. Fortunately for chemists, the **shielding** varies for different hydrogens within a molecule. In other words, all the hydrogens do not experience the same applied magnetic field.

What causes shielding? In a magnetic field, the electrons circulate about the nuclei and induce a local magnetic field that acts in opposition to the applied magnetic field and subtracts from it. The **effective magnetic field**—the amount of magnetic field that the nuclei actually "sense" through the surrounding electronic environment—is therefore, somewhat smaller than the applied field:

$$B_{\text{effective}} = B_{\text{applied}} - B_{\text{local}}$$

This means that the greater the electron density of the environment in which the proton* is located, the greater B_{local} is and the more the proton is shielded from the applied magnetic field. This type of shielding is called **diamagnetic shielding**. Thus, protons in electron-dense environments sense a *smaller effective magnetic field*. They, therefore, will require a *lower frequency* to come into resonance—that is, flip their spin—because ΔE is smaller (Figure 13.2). Protons in electron-poor environments sense a *larger effective magnetic field* and, therefore, will require a *higher frequency* to come into resonance, because ΔE is larger.

An NMR spectrum exhibits a signal for each proton in a different environment. Protons in electron-rich environments are more shielded and appear at lower frequencies (on the right-hand side of the spectrum) (Figure 13.4). Protons in electron-poor

The larger the magnetic field sensed by the proton, the higher is the frequency of the signal.

---

* The terms "proton" and "hydrogen" are both used to describe covalently bonded hydrogen in discussions of NMR spectroscopy.

deshielded = less shielded

environments are less shielded and appear at higher frequencies (on the left-hand side of the spectrum). For example, the signal for the methyl protons of CH_3F occurs at a higher frequency than the signal for the methyl protons of CH_3Br because fluorine is more electronegative and, therefore, withdraws electrons inductively more strongly than bromine does (Section 1.21); thus, the protons of CH_3F are in a less electron-rich environment. (Notice that high frequency in an NMR spectrum is on the left-hand side, just as it is in IR and UV/Vis spectra.)

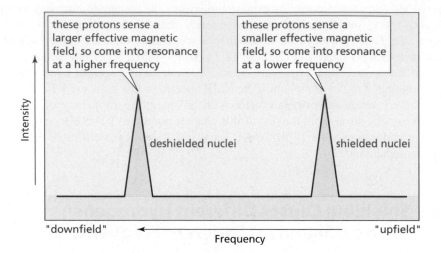

Figure 13.4 ▶
Shielded nuclei come into resonance at lower frequencies than deshielded nuclei.

The terms "upfield" and "downfield," which came into use when continuous wave (CW) spectrometers were used (before the advent of Fourier transform spectrometers), are so entrenched in the vocabulary of NMR that you need to know what they mean. **Upfield** means farther to the right-hand side of the spectrum, and **downfield** means farther to the left-hand side of the spectrum. In contrast to FT–NMR techniques, which keep the magnetic field strength constant and vary the frequency, continuous-wave techniques keep the frequency constant and vary the magnetic field. The magnetic field increases from left to right across a spectrum because higher magnetic fields are required to bring shielded protons into resonance at a given frequency (Figure 13.4). Therefore, *upfield* is toward the right and *downfield* is toward the left.

13.4 The Number of Signals in an ^1H NMR Spectrum

Protons in the same environment are called **chemically equivalent protons**. For example, 1-bromopropane has three different sets of chemically equivalent protons: (1) the three methyl protons are chemically equivalent because of rotation about the C—C bond; (2) the two methylene (CH_2) protons on the middle carbon are chemically equivalent; and (3) the two methylene protons on the carbon bonded to the bromine atom make up the third set of chemically equivalent protons.

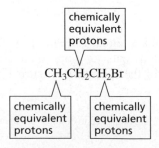

Each set of chemically equivalent protons in a compound gives rise to a signal in the ^{1}H NMR spectrum of that compound. (Sometimes the signals are not sufficiently separated and overlap each other. When this happens, one sees fewer signals than anticipated.) Because 1-bromopropane has three sets of chemically equivalent protons, it has three signals in its ^{1}H NMR spectrum.

2-Bromopropane has two sets of chemically equivalent protons and, therefore, it has two signals in its ^{1}H NMR spectrum; the six methyl protons in 2-bromopropane are equivalent so they give rise to only one signal; the hydrogen bonded to the middle carbon gives the second signal. Ethyl methyl ether has three sets of chemically equivalent protons: the methyl protons on the carbon adjacent to the oxygen, the methylene protons on the carbon adjacent to the oxygen, and the methyl protons on the carbon that is one carbon removed from the oxygen. The chemically equivalent protons in the following compounds are designated by the same letter:

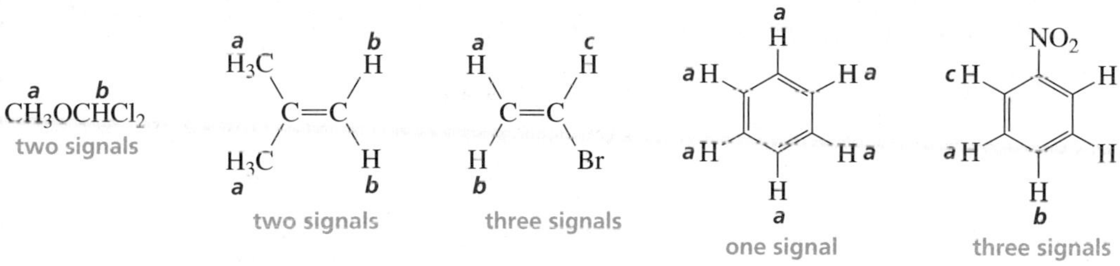

> Each set of chemically equivalent protons gives rise to a signal.

You can tell how many sets of chemically equivalent protons a compound has from the number of signals in its ^{1}H NMR spectrum.

Sometimes, two protons on the same carbon are not equivalent. For example, the ^{1}H NMR spectrum of chlorocyclobutane has five signals. Even though they are bonded to the same carbon, the H$_a$ and H$_b$ protons are not equivalent because they are not in the same environment: H$_a$ is trans to Cl and H$_b$ is cis to Cl. Similarly, the H$_c$ and H$_d$ protons are not equivalent.

its ^{1}H NMR spectrum has five signals

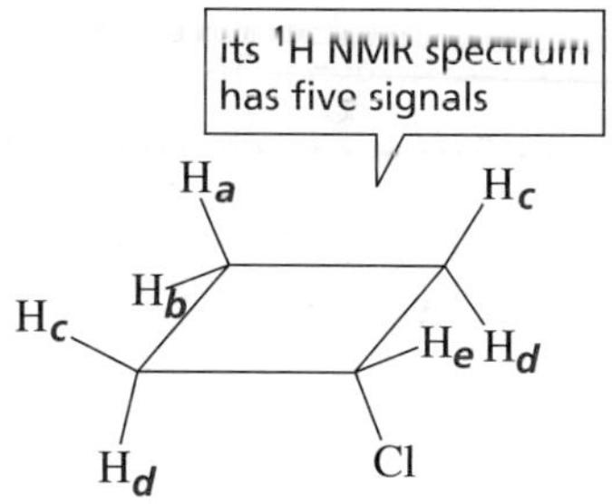

chlorocyclobutane
H$_a$ and H$_b$ are not equivalent
H$_c$ and H$_d$ are not equivalent

PROBLEM-SOLVING STRATEGY

Determining the Number of Signals in an ^{1}H NMR Spectrum

How many signals would you expect to see in the ^{1}H NMR spectrum of ethylbenzene?

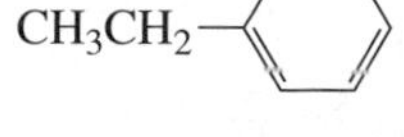

To determine the number of signals you would expect to see in the spectrum, replace each hydrogen in turn by another atom (here we use Br) and name the resulting compound. The number of different names corresponds to the number of signals in the ^{1}H NMR spectrum. We get five different names for the bromosubstituted cyclohexenes, so we expect to see five signals in the ^{1}H NMR spectrum of ethylbenzene.

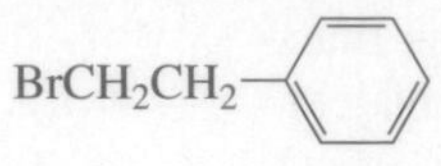

BrCH$_2$CH$_2$— (phenyl)
1-bromo-2-phenylethane

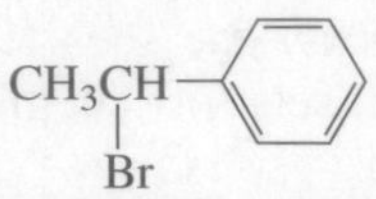

CH$_3$CH— (phenyl)
|
Br
1-bromo-1-phenylethane

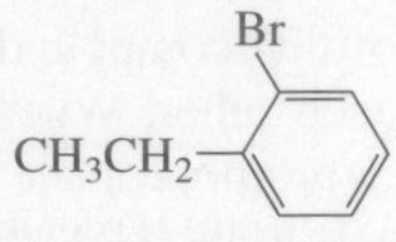

CH$_3$CH$_2$— (phenyl with Br)
1-bromo-2-ethylbenzene

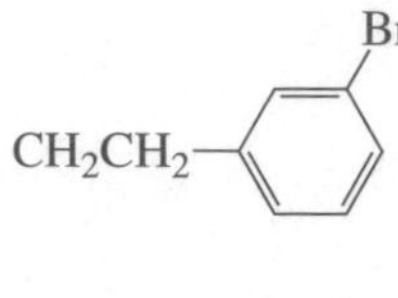

CH$_2$CH$_2$— (phenyl with Br)
1-bromo-3-ethylbenzene

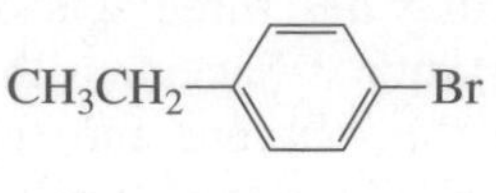

CH$_3$CH$_2$— (phenyl) —Br
1-bromo-4-ethylbenzene

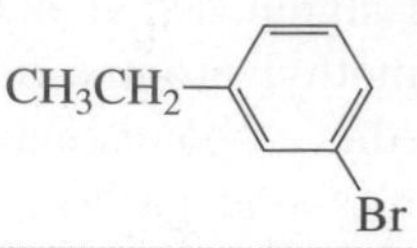

CH$_3$CH$_2$— (phenyl with Br)
1-bromo-3-ethylbenzene

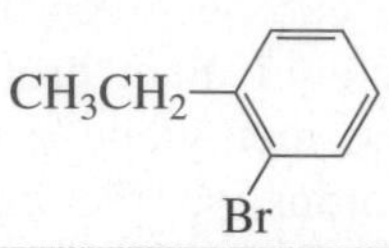

CH$_3$CH$_2$— (phenyl with Br)
1-bromo-2-ethylbenzene

Now continue on to Problem 3.

PROBLEM 3◆

How many signals would you expect to see in the ^{1}H NMR spectrum of each of the following compounds?

a. CH$_3$CH$_2$CH$_2$CH$_3$

b. BrCH$_2$CH$_2$Br

c. CH$_2$=CHCl

d.  (cyclohexadiene structure)

e.

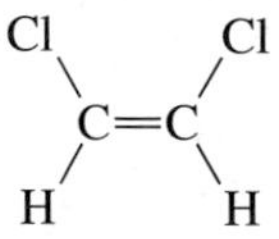

Cl   Cl
\   /
C=C
/   \
H   H

f.
$$\underset{\ }{\overset{O}{\parallel}}$$
CH$_3$CH$_2$CH$_2$CCH$_3$

g. CH$_3$CH$_2$CHCH$_2$CH$_3$
|
Cl

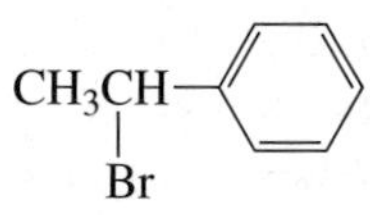

h. CH$_3$CHCH$_2$CHCH$_3$
| |
CH$_3$ CH$_3$

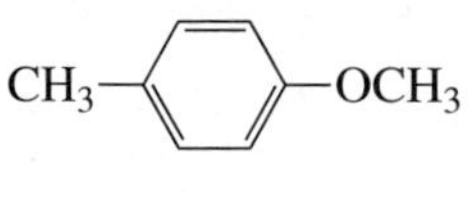

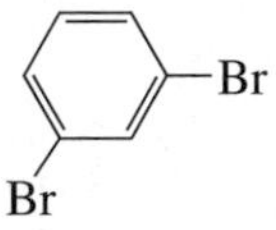

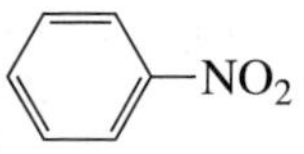

i. CH$_3$CH— (phenyl)
|
Br

j. CH$_3$— (phenyl) —OCH$_3$

k. CH$_3$— (phenyl) —CH$_3$

l. (phenyl with Br, Br)

m. (phenyl) —NO$_2$

n.
$$\overset{O}{\overset{\parallel}{\ }}$$
CH$_2$=CHCH

o.

Cl   CH$_3$
\   /
C=C
/   \
H   H

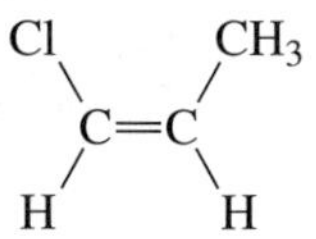

PROBLEM 4

How could you distinguish the ^{1}H NMR spectra of the following compounds?

a. CH$_3$OCH$_2$OCH$_3$

b. CH$_3$OCH$_3$

c. CH$_3$OCH$_2$CCH$_2$OCH$_3$
with CH$_3$ above and CH$_3$ below

PROBLEM 5◆

There are three isomeric dichlorocyclopropanes. Their ^{1}H NMR spectra show one signal for isomer 1, two signals for isomer 2, and three signals for isomer 3. Draw the structures of isomers 1, 2, and 3.

13.5 The Chemical Shift Tells How Far the Signal Is from the Reference Signal

A small amount of an inert **reference compound** is added to the sample tube containing the compound whose NMR spectrum is to be taken. The positions of the signals in an NMR spectrum are defined according to how far they are from the signal of the reference compound. The most commonly used reference compound is tetramethylsilane (TMS). Because TMS is a highly volatile compound (bp = 26.5 °C), it can easily be removed from the sample by evaporation after the NMR spectrum is taken.

The methyl protons of TMS are in a more electron-dense environment than are most protons in organic molecules because silicon is less electronegative than carbon (their electronegativities are 1.8 and 2.5, respectively). Consequently, the signal for the methyl protons of TMS is at a lower frequency than most other signals (that is, it appears to the right of the other signals).

The position at which a signal occurs in an NMR spectrum is called the *chemical shift*. The **chemical shift** is a measure of how far the signal is from the reference TMS signal. The most common scale for chemical shifts is the δ (delta) scale. The TMS signal is used to define the zero position on this scale. The chemical shift is determined by measuring the distance from the TMS peak in hertz and dividing by the operating frequency of the instrument in megahertz. Because the units are Hz/MHz, a chemical shift has units of parts per million (ppm) of the operating frequency:

$$\delta = \text{chemical shift (ppm)} = \frac{\text{distance downfield from TMS (Hz)}}{\text{operating frequency of the spectrometer (MHz)}}$$

Most proton chemical shifts are between 0 and 12 ppm.

The ^{1}H NMR spectrum for 1-bromo-2,2-dimethylpropane in Figure 13.5 shows that the chemical shift of the methyl protons is at 1.05 ppm and the chemical shift of the methylene protons is at 3.28 ppm. *Notice that low-frequency (upfield, shielded) signals have small δ (ppm) values, whereas high-frequency (downfield, deshielded) signals have large δ values.*

> **The greater the value of the chemical shift (δ), the higher the frequency.**

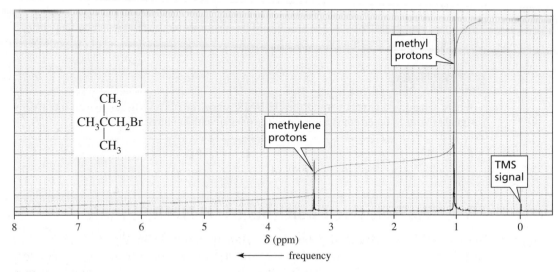

▲ **Figure 13.5**
The ^{1}H NMR spectrum of 1-bromo-2,2-dimethylpropane. The TMS signal is a reference signal from which chemical shifts are measured; it defines the zero position on the scale.

The advantage of the δ scale is that the chemical shift of a given nucleus is *independent of the operating frequency of the NMR spectrometer.* Thus, the chemical shift of the methyl protons of 1-bromo-2,2-dimethylpropane is at 1.05 ppm in both a 60-MHz and a 360-MHz instrument. If the chemical shift were instead reported in hertz, it would be at 63 Hz in a 60-MHz instrument and at 378 Hz in a 360-MHz

> **The chemical shift (δ) is independent of the operating frequency of the spectrometer.**

instrument (63/60 = 1.05; 378/360 = 1.05). The following diagram will help you keep track of the terms associated with NMR spectroscopy:

protons in electron-poor environments	protons in electron-dense environments
deshielded protons	shielded protons
downfield	upfield
high frequency	low frequency
large δ values	small δ values

$\longleftarrow$ δ ppm
$\longleftarrow$ frequency

PROBLEM 6◆

How many hertz downfield from the TMS signal would be the signal occurring at 1.0 ppm

a. in a 300-MHz spectrometer? **b.** in a 500-MHz spectrometer?

PROBLEM 7◆

A signal is seen at 600 Hz downfield from the TMS signal in an NMR spectrometer with a 300-MHz operating frequency.

a. What is the chemical shift of the signal?
b. What would its chemical shift be in an instrument operating at 100 MHz?
c. How many hertz downfield from TMS would the signal be in a 100-MHz spectrometer?

PROBLEM 8◆

a. If two signals differ by 1.5 ppm in a 300-MHz spectrometer, by how much do they differ in a 100-MHz spectrometer?
b. If two signals differ by 90 Hz in a 300-MHz spectrometer, by how much do they differ in a 100-MHz spectrometer?

PROBLEM 9◆

Where would you expect to find the ^{1}H NMR signal of $(CH_3)_2Mg$ relative to the TMS signal? (*Hint:* Magnesium is even less electronegative than silicon.)

13.6 The Relative Positions of ^{1}H NMR Signals

The ^{1}H NMR spectrum of 1-bromo-2,2-dimethylpropane in Figure 13.5 has two signals because the compound has two different kinds of protons. The methylene protons are in a less electron-dense environment than are the methyl protons because the methylene protons are closer to the electron-withdrawing bromine. Because the methylene protons are in a less electron-dense environment, they are less shielded from the applied magnetic field. The signal for these protons therefore occurs at a higher frequency than the signal for the more shielded methyl protons. *Remember that the right-hand side of an NMR spectrum is the low-frequency side, where protons in electron-dense environments (more shielded) show a signal. The left-hand side is the high-frequency side, where protons in electron-poor environments (less shielded) show a signal* (Figure 13.4).

Protons in electron-poor environments show signals at high frequencies.

We would expect the ^{1}H NMR spectrum of 1-nitropropane to have three signals because the compound has three different kinds of protons. The closer the protons are to the nitro group, which withdraws electrons inductively, the less they are shielded from the applied magnetic field, so the higher the frequency at which their signal will appear (the farther it will be downfield). Thus, the protons closest to the nitro group show a signal at the highest frequency (4.37 ppm), and the ones farthest from the nitro group show a signal at the lowest frequency (1.04 ppm).

Electron withdrawal causes NMR signals to appear at higher frequencies (at larger δ values).

$$\boxed{1.04\ ppm}\quad \boxed{2.07\ ppm}$$
$$\boxed{4.37\ ppm}$$
$$CH_3CH_2CH_2NO_2$$

Compare the chemical shifts of the methylene protons immediately adjacent to the halogen in each of the following alkyl halides. The position of the signal depends on the electronegativity of the halogen—the more electronegative the halogen, the higher is the frequency of the signal. Thus, the signal for the methylene protons adjacent to fluorine (the most electronegative of the halogens) occurs at the highest frequency, whereas the signal for the methylene protons adjacent to iodine (the least electronegative of the halogens) occurs at the lowest frequency.

$CH_3CH_2CH_2CH_2CH_2F$ $CH_3CH_2CH_2CH_2CH_2Cl$ $CH_3CH_2CH_2CH_2CH_2Br$ $CH_3CH_2CH_2CH_2CH_2I$

$\boxed{4.50\ ppm}$ $\boxed{3.50\ ppm}$ $\boxed{3.40\ ppm}$ $\boxed{3.20\ ppm}$

PROBLEM 10◆

a. Which set of protons in each of the following compounds is the least shielded?

1. $CH_3CH_2CH_2Cl$ **2.** $CH_3CH_2\overset{O}{\overset{||}{C}}OCH_3$ **3.** $CH_3CHCHBr$
                                                Br Br

b. Which set of protons in each compound is the most shielded?

13.7 Characteristic Values of Chemical Shifts

Approximate values of chemical shifts for different kinds of protons are shown in Table 13.1. (A more extensive compilation is given in Appendix VI.) An ^{1}H NMR spectrum can be divided into seven regions, one of which is empty. If you can remember the kinds of protons that appear in each region, you will be able to tell what kinds of protons a molecule has from a quick look at its NMR spectrum.

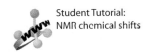

Student Tutorial:
NMR chemical shifts

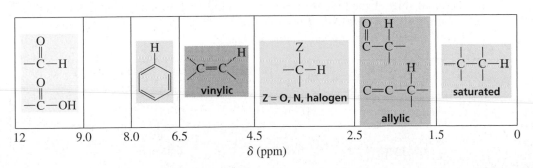

Table 13.1	Approximate Values of Chemical Shifts for ^{1}H NMR^a		
Type of proton	Approximate chemical shift (ppm)	Type of proton	Approximate chemical shift (ppm)
$-CH_3$	0.85	$I-\overset{\vert}{\underset{\vert}{C}}-H$	2.5–4
$-CH_2-$	1.20	$Br-\overset{\vert}{\underset{\vert}{C}}-H$	2.5–4
$-\overset{\vert}{\underset{\vert}{CH}}-$	1.55	$Cl-\overset{\vert}{\underset{\vert}{C}}-H$	3–4
$-\overset{\vert}{C}=\overset{\vert}{C}-CH_3$	1.7	$F-\overset{\vert}{\underset{\vert}{C}}-H$	4–4.5
$-\overset{O}{\overset{\|}{C}}-CH_3$	2.1	RNH_2	Variable, 1.5–4
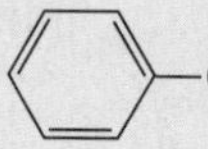		ROH	Variable, 2–5
⬡$-CH_3$	2.3	$ArOH$	Variable, 4–7
$-C\equiv C-H$	2.4	⬡$-H$	6.5–8
$R-O-CH_3$	3.3	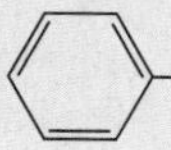	
$R-\underset{\underset{R}{\vert}}{C}=CH_2$	4.7	$-\overset{O}{\overset{\|}{C}}-H$	9.0–10
$R-\underset{\underset{R}{\vert}}{C}=\underset{\underset{R}{\vert}}{C}-H$	5.3	$-\overset{O}{\overset{\|}{C}}-OH$	Variable, 10–12
		$-\overset{O}{\overset{\|}{C}}-NH_2$	Variable, 5–8

^{a}The values are approximate because they are affected by neighboring substituents.

$-\overset{\vert}{\underset{\vert}{CH}}-$
methine

$-CH_2-$
methylene

$-CH_3$
methyl

In a similar environment, the signal for a methine proton occurs at a higher frequency than the signal for methylene protons, which in turn occurs at a higher frequency than the signal for methyl protons.

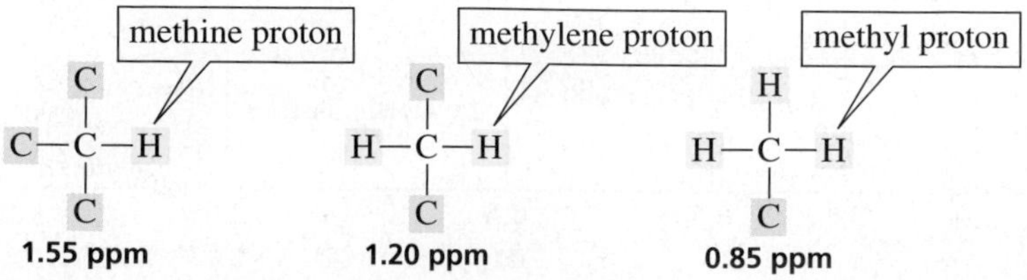

Carbon is more electronegative than hydrogen (Table 1.3 on page 11). Therefore, the chemical shift of a **methine proton** (a hydrogen attached to an sp^3 carbon that is attached to three carbons) is more deshielded and, therefore, shows a chemical shift at a higher frequency than the chemical shift of **methylene protons** (hydrogens attached to an sp^3 carbon that is attached to two carbons) in a similar environment. Similarly, the chemical shift of methylene protons is at a higher frequency than is the chemical shift of **methyl protons** (hydrogens attached to an sp^3 carbon that is attached to one carbon) in a similar environment (Table 13.1).

methine proton

methylene proton

methyl proton

$\underset{\underset{C}{|}}{\overset{\overset{C}{|}}{C}}-C-H$ $\underset{\underset{C}{|}}{\overset{\overset{C}{|}}{H}}-C-H$ $\underset{\underset{C}{|}}{\overset{\overset{H}{|}}{H}}-C-H$

1.55 ppm   **1.20 ppm**   **0.85 ppm**

For example, the ^{1}H NMR spectrum of butanone shows three signals. The signal at the lowest frequency is the signal for the *a* protons of butanone; these protons are farthest from the electron-withdrawing carbonyl group. (In correlating an NMR spectrum with a structure, the set of protons responsible for the signal at the lowest frequency will be labeled *a*, the next set will be labeled *b*, the next set *c*, and so on.) The *b* and *c* protons are the same distance from the carbonyl group, but the signal for the *b* protons is at a lower frequency because methyl protons appear at a lower frequency than do methylene protons in a similar environment.

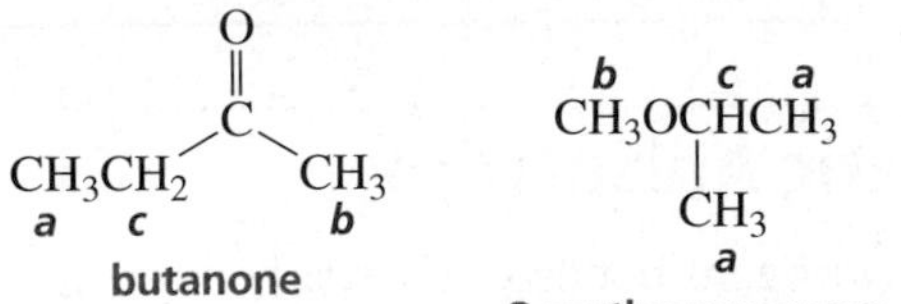

The signal for the *a* protons of 2-methoxypropane is the signal at the lowest frequency in the ^{1}H NMR spectrum of this compound because these protons are farthest from the electron-withdrawing oxygen. The *b* and *c* protons are the same distance from the oxygen, but the signal for the *b* protons appears at a lower frequency because, in a similar environment, methyl protons appear at a lower frequency than does a methine proton.

PROBLEM 11◆

In each of the following compounds, which of the underlined protons (or sets of protons) has the greater chemical shift (that is, the higher frequency signal)?

a. CH$_3$CHCHBr
     | |
    Br Br

c. CH$_3$CH$_2$CHCH$_3$
        |
       Cl

e. CH$_3$CH$_2$CH=CH$_2$

b. CH$_3$CHOCH$_3$
    |
   CH$_3$

d. CH$_3$CHCCH$_2$CH$_3$
      | ‖
     CH$_3$ O

f. CH$_3$OCH$_2$CH$_2$CH$_3$

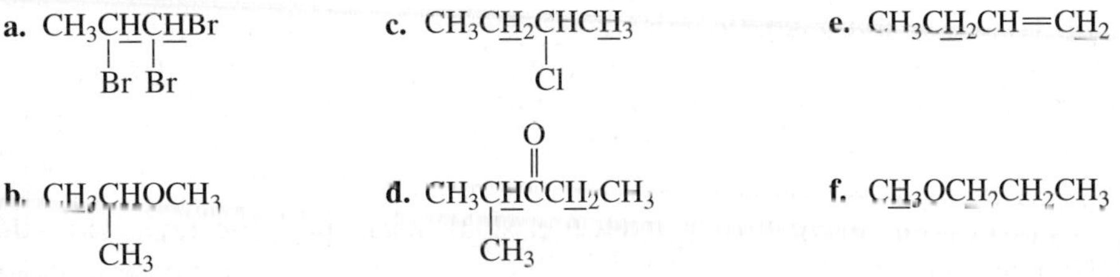

PROBLEM 12◆

In each of the following pairs of compounds, which of the underlined protons (or sets of protons) has the greater chemical shift (that is, the higher frequency signal)?

a. CH$_3$CH$_2$CH$_2$Cl or CH$_3$CH$_2$CH$_2$Br

b. CH$_3$CH$_2$CH$_2$Cl or CH$_3$CH$_2$CHCH$_3$
                             |
                             Cl

c. CH$_3$CH$_2$CH or CH$_3$CH$_2$COCH$_3$
        ‖   ‖
        O   O

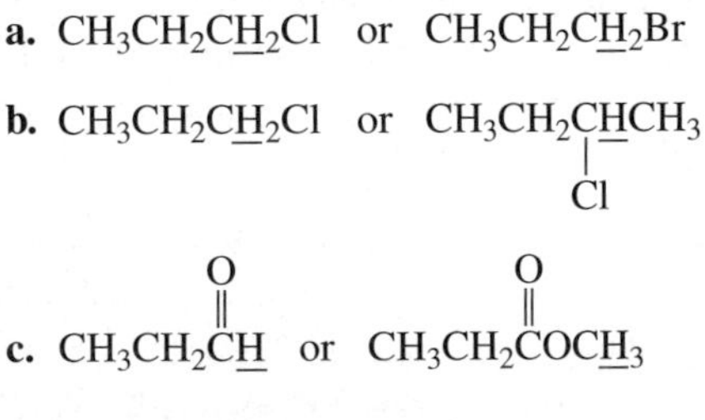

PROBLEM 13

For each compound, without referring to Table 13.1, label the proton or set of protons that gives the signal at the lowest frequency *a*, the next lowest *b*, and so on.

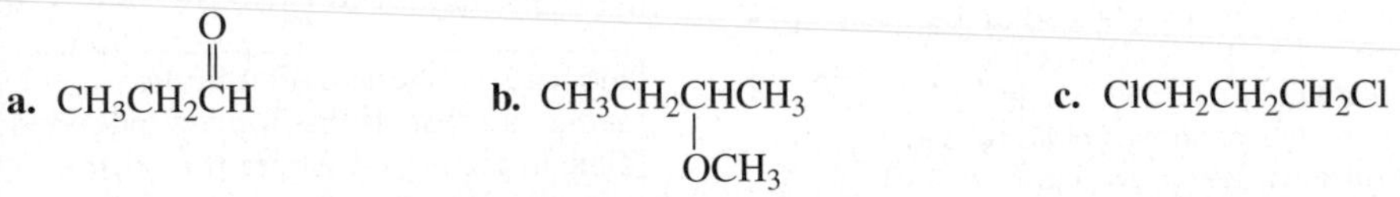

a. CH$_3$CH$_2$CH
         ‖
        O

b. CH$_3$CH$_2$CHCH$_3$
        |
     OCH$_3$

c. ClCH$_2$CH$_2$CH$_2$Cl

d. $CH_3CH_2CH_2\overset{\overset{\displaystyle O}{\|}}{C}OCH_3$ f. $CH_3CH_2CH_2OCHCH_3$ h. $CH_3CHCH_2OCH_3$
                              | |
                            CH_3 CH_3

                                  $\overset{\displaystyle O}{\|}$ CH_3

e. $CH_3CH_2CHCH_2CH_3$ g. $CH_3CH_2CH_2CCH_3$ i. $CH_3CHCHCH_3$
         | |
       OCH_3 Cl

13.8 Diamagnetic Anisotropy

The chemical shifts of hydrogens bonded to sp^2 carbons are at higher frequencies than one would predict from the electronegativities of the sp^2 carbons. For example, a hydrogen bonded to a benzene ring appears at 6.5 to 8.0 ppm, a hydrogen bonded to a terminal sp^2 carbon of an alkene appears at 4.7 to 5.3 ppm, and a hydrogen bonded to a carbonyl carbon appears at 9.0 to 10.0 ppm (Table 13.1).

5.3 ppm

$CH_3CH_2CH{=}CH_2$

$\underset{\boxed{7.3\ ppm}}{\text{benzene}}{-}H$ $\boxed{4.7\ ppm}$ $CH_3CH_2\overset{\overset{\displaystyle O}{\|}}{C}H$ $\boxed{9.0\ ppm}$

The unusual chemical shifts associated with hydrogens bonded to carbons that form π bonds are due to **diamagnetic anisotropy**. This term describes an environment in which different magnetic fields are found at different points in space. (*Anisotropic* is Greek for "different in different directions.") Because π electrons are less tightly held by nuclei than are σ electrons, π electrons are freer to move in response to a magnetic field. When a magnetic field is applied to a compound with π electrons, the π electrons move in a circular path that induces a small local magnetic field. How this induced magnetic field affects the chemical shift of a proton depends on the direction of the induced field in the region where the proton is located, relative to the direction of the applied magnetic field.

The magnetic field induced by the π electrons of a benzene ring in the region where benzene's protons are located is oriented in the same direction as the applied magnetic field (Figure 13.6). As a result, the protons sense a larger effective magnetic field—the sum of the strengths of the applied field and the induced field. Because frequency is proportional to the strength of the magnetic field experienced by the protons (Figure 13.2), the protons show signals at *higher frequencies* than they would if the π electrons did not induce a magnetic field.

Figure 13.6 ▶
The magnetic field induced by the π electrons of a benzene ring in the vicinity of the protons attached to the sp^2 carbons has the same direction as the applied magnetic field. Since a larger effective magnetic field is sensed by the protons, they show signals at higher frequencies.

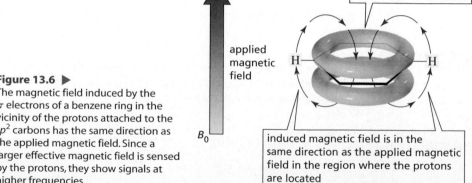

circulation of electrons

applied magnetic field

B_0

induced magnetic field is in the same direction as the applied magnetic field in the region where the protons are located

The magnetic field induced by the π electrons of an alkene or an aldehyde (in the region where the protons bonded to the sp^2 carbons of the alkene or to the sp^2 carbon of the aldehyde are located) is also oriented in the *same direction* as the applied magnetic field (Figure 13.7). These protons, too, show signals at higher than expected frequencies.

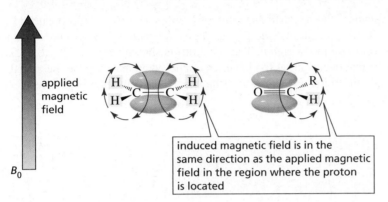

applied magnetic field

B_0

induced magnetic field is in the same direction as the applied magnetic field in the region where the proton is located

◀ **Figure 13.7**
The magnetic fields induced by the π electrons of an alkene and by the π electrons of a carbonyl group in the vicinity of the vinylic and aldehydic protons have the same direction as the applied magnetic field. Since a larger effective magnetic field is sensed by the protons, they show signals at higher frequencies.

In contrast, the chemical shift of a hydrogen bonded to an sp carbon is at a lower frequency than one would predict from the electronegativity of the sp carbon. In the region where the proton is located, the direction of the magnetic field induced by the alkyne's cylinder of π electrons is *opposite* to the direction of the applied magnetic field (Figure 13.8). Thus, the proton shows a signal at a *lower frequency* than it would if the π electrons did not induce a magnetic field.

induced magnetic field is in the direction opposite to that of the applied magnetic field in the region where the proton is located

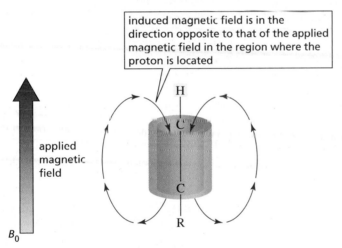

applied magnetic field

B_0

◀ **Figure 13.8**
The magnetic field induced by the π electrons of an alkyne in the vicinity of the proton bonded to the sp carbon is in a direction opposite to that of the applied magnetic field. Since a smaller effective magnetic field is sensed by the proton, it shows a signal at a lower frequency.

PROBLEM 14◆

[18]-Annulene shows two signals in its ^{1}H NMR spectrum: one at 9.25 ppm and the other very far upfield (to the right of the TMS signal) at -2.88 ppm. What hydrogens are responsible for each of the signals? (*Hint:* Look at the direction of the induced magnetic field outside and inside the benzene ring in Figure 13.6.)

13.9 The Integration of NMR Signals Reveals the Relative Number of Protons Causing the Signal

The two signals in the ^{1}H NMR spectrum of 1-bromo-2,2-dimethylpropane in Figure 13.5 are not the same size, because *the area under each signal is proportional to the number of protons giving rise to the signal*. The spectrum is shown again in Figure 13.9. The area under the signal occurring at the lower frequency is larger because the signal is caused by *nine* methyl protons, whereas the smaller, higher-frequency signal results from *two* methylene protons.

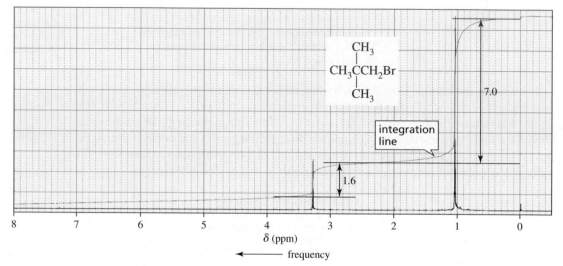

▲ **Figure 13.9**
Analysis of the integration line in the ^{1}H NMR spectrum of 1-bromo-2,2-dimethylpropane.

You probably remember from a calculus course that the area under a curve can be determined by an integral. An ^{1}H NMR spectrometer is equipped with a computer that calculates the integrals electronically and then displays them as an integral trace superimposed on the original spectrum (Figure 13.9). The height of each step in the integral trace is proportional to the area under the corresponding signal, which, in turn, is proportional to the number of protons giving rise to the signal. The heights of the integration steps in Figure 13.9, for example, tell us that the ratio of the integrals is approximately 1.6 : 7.0 = 1 : 4.4. We multiply the ratio by a number that will cause all the numbers making up the ratio to be close to whole numbers—in this case, we multiply by 2. That means that the ratio of protons in the compound is 2 : 8.8, which is rounded to 2 : 9, as there can be only whole numbers of protons. In this way we find that the ratio of protons in the compound is 2 : 9. (The measured integrals are approximate because of experimental error.) Modern spectrometers print the integrals as numbers on the spectrum; see Figure 13.11 on page 586.

The **integration** tells us the *relative* number of protons that give rise to each signal, not the *absolute* number. In other words, integration could not distinguish between 1,1-dichloroethane and 1,2-dichloro-2-methylpropane because both compounds would show an integral ratio of 1 : 3.

$$CH_3—CH—Cl$$
$$|$$
$$Cl$$

1,1-dichloroethane
ratio of protons = 1:3

$$\overset{\displaystyle CH_3}{\underset{\displaystyle Cl}{CH_3—\overset{|}{\underset{|}{C}}—CH_2Cl}}$$

1,2-dichloro-2-methylpropane
ratio of protons 2:6 = 1:3

PROBLEM 15◆

How would integration distinguish the ^{1}H NMR spectra of the following compounds?

$$CH_3-\underset{\underset{\displaystyle CH_3}{|}}{\overset{\overset{\displaystyle CH_3}{|}}{C}}-CH_2Br \qquad CH_3-\underset{\underset{\displaystyle Br}{|}}{\overset{\overset{\displaystyle CH_3}{|}}{C}}-CH_2Br \qquad CH_3-\underset{\underset{\displaystyle CH_2Br}{|}}{\overset{\overset{\displaystyle CH_2Br}{|}}{C}}-CH_2Br$$

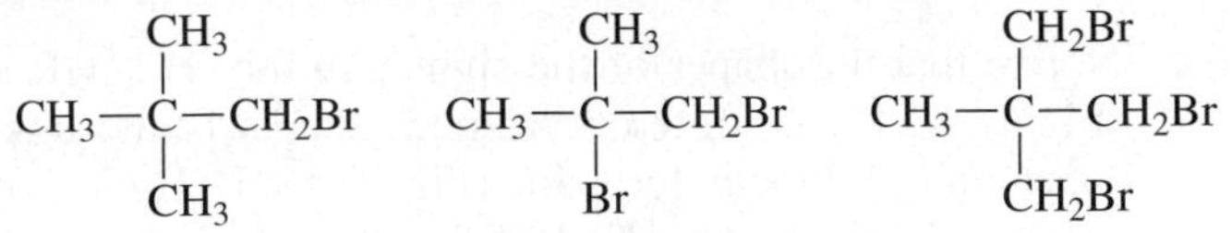

PROBLEM 16 **SOLVED**

a. Calculate the ratios of the different kinds of protons in a compound with an integral ratio of 6 : 4 : 18.4 (going from left to right across the spectrum).

b. Determine the structure of a compound that would give these relative integrals in the observed order.

Solution

a. Divide each number in the ratio by the smallest number:

$$\frac{6}{4} = 1.5 \qquad \frac{4}{4} = 1 \qquad \frac{18.4}{4} = 4.6$$

Multiply by a number that will cause all the resulting numbers to be close to whole numbers:

$$1.5 \times 2 = 3 \qquad 1 \times 2 = 2 \qquad 4.6 \times 2 = 9$$

The ratio 3 : 2 : 9 gives the relative numbers of the different kinds of protons. The actual ratio could be 6 : 4 : 18, or even some higher multiple, but let's not go there if we don't have to.

b. The "3" suggests the presence of a methyl group, the "2" a methylene group, and the "9" a *tert*-butyl group. The methyl group is closest to the group in the molecule that causes deshielding, and the *tert*-butyl group is farthest away from the group that causes deshielding. The following compound meets these requirements:

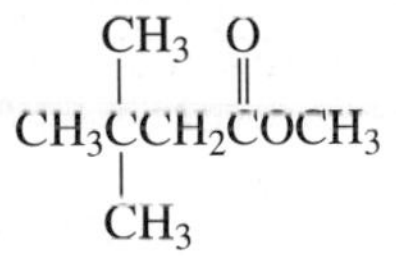

$$CH_3\underset{\underset{\displaystyle CH_3}{|}}{\overset{\overset{\displaystyle CH_3}{|}}{C}}CH_2\overset{\overset{\displaystyle O}{\|}}{C}OCH_3$$

PROBLEM 17◆

The ^{1}H NMR spectrum shown in Figure 13.10 corresponds to one of the following compounds. Which compound is responsible for this spectrum?

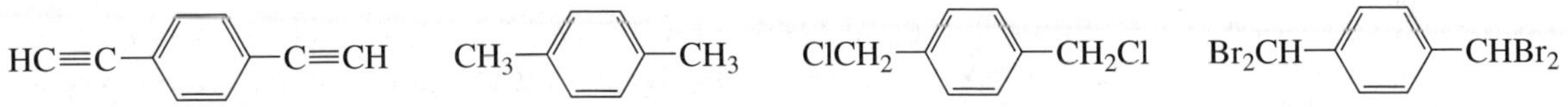

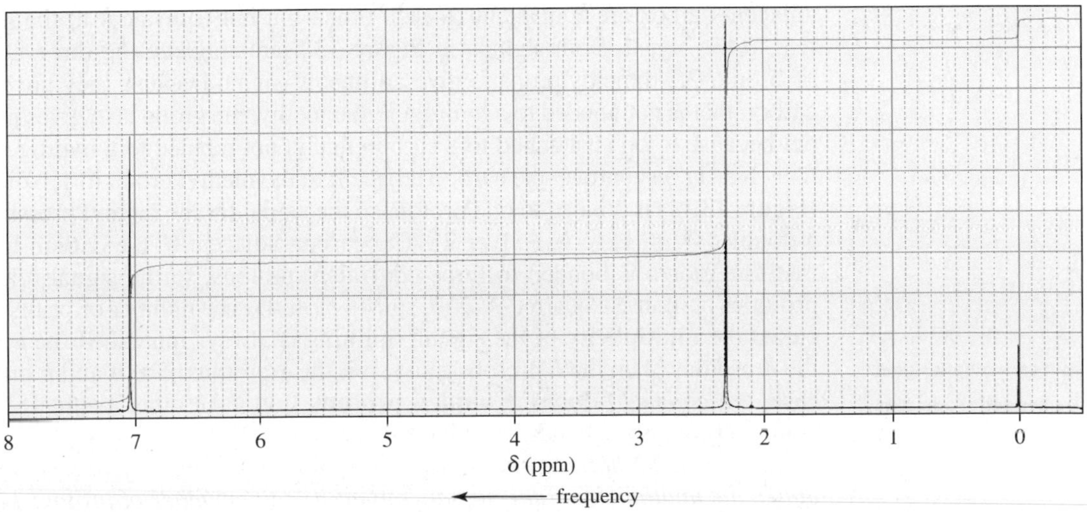

▲ Figure 13.10
The ^{1}H NMR spectrum for Problem 17.

13.10 Splitting of the Signals Is Described by the *N* + 1 Rule

Notice that the shapes of the signals in the ^{1}H NMR spectrum of 1,1-dichloroethane (Figure 13.11) are different from the shapes of the signals in the ^{1}H NMR spectrum of 1-bromo-2,2-dimethylpropane (Figure 13.5). Both signals in Figure 13.5 are **singlets**, meaning each is composed of a single peak. In contrast, the signal for the methyl protons of 1,1-dichloroethane (the lower-frequency signal) is split into two peaks (a **doublet**), and the signal for the methine proton is split into four peaks (a **quartet**). (Magnifications of the frequency axis for the doublet and quartet are shown as insets in Figure 13.11; integration numbers are given in green.)

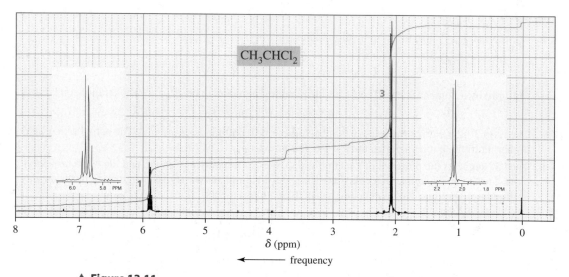

▲ **Figure 13.11**
The ^{1}H NMR spectrum of 1,1-dichloroethane. The higher-frequency signal is an example of a quartet; the lower-frequency signal is a doublet.

A ^{1}H NMR signal is split into *N* + 1 peaks, where *N* is the number of equivalent protons bonded to adjacent carbons.

Splitting is caused by protons bonded to adjacent carbons. The splitting of a signal is described by the ***N* + 1 rule**, where *N* is the number of *equivalent* protons bonded to *adjacent* carbons. By "equivalent protons," we mean that the protons bonded to an adjacent carbon are equivalent to each other, but not equivalent to the proton giving rise to the signal. Both signals in Figure 13.5 are singlets; the three methyl groups of bromo-2,2-dimethylpropane give an unsplit signal because they are attached to a carbon that is not bonded to a hydrogen; the methylene group also gives an unsplit signal because it too is attached to a carbon that is not bonded to a hydrogen ($N = 0$, so $N + 1 = 1$). In contrast, the carbon adjacent to the methyl group in 1,1-dichloroethane (Figure 13.11) is bonded to one proton, so the signal for the methyl protons is split into a doublet ($N = 1$, so $N + 1 = 2$). The carbon adjacent to the carbon bonded to the methine proton is bonded to three equivalent protons, so the signal for the methine proton is split into a quartet ($N = 3$, so $N + 1 = 4$). The number of peaks in a signal is called the **multiplicity** of the signal. Splitting is always mutual: if the ***a*** protons split the ***b*** protons, then the ***b*** protons must split the ***a*** protons. The ***a*** and ***b*** protons, in this case, are *coupled protons*. **Coupled protons** split each other's signal. Note that the coupled protons are bonded to adjacent carbons.

Keep in mind that it is not the number of protons giving rise to a signal that determines the multiplicity of the signal; rather, it is the number of protons bonded to the immediately adjacent carbons that determines the multiplicity. For example, the signal for the ***a*** protons in the following compound will be split into three peaks (a **triplet**) because the adjacent carbon is bonded to two hydrogens. The signal for the ***b*** protons

Coupled protons split each other's signal.

Coupled protons are bonded to adjacent carbons.

will appear as a quartet because the adjacent carbon is bonded to three hydrogens, and the signal for the **c** protons will be a singlet.

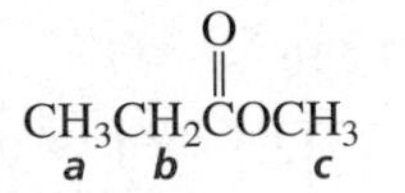

$$\underset{a \quad b \qquad \ c}{CH_3CH_2\overset{\overset{\displaystyle O}{\|}}{C}OCH_3}$$

A signal for a proton is never split by *equivalent* protons. For example, the ^{1}H NMR spectrum of bromomethane shows one singlet. The three methyl protons are chemically equivalent, and chemically equivalent protons do not split each other's signal. The four protons in 1,2-dichloroethane are also chemically equivalent, so its ^{1}H NMR spectrum also shows one singlet.

Equivalent protons do not split each other's signal.

$$CH_3Br \qquad\qquad ClCH_2CH_2Cl$$
bromomethane   **1,2-dichloroethane**

> each compound has an NMR spectrum that shows one singlet
> because equivalent protons do not split each other's signals

The splitting of signals occurs when different kinds of protons are close enough together for their magnetic fields to influence one another, a situation called **spin–spin coupling**. For example, the frequency at which the methyl protons of 1,1-dichloroethane show a signal is influenced by the magnetic field of the methine proton. If the magnetic field of the methine proton aligns *with* that of the applied magnetic field, it will add to the applied magnetic field, causing the methyl protons to show a signal at a slightly higher frequency. On the other hand, if the magnetic field of the methine proton aligns *against* the applied magnetic field, it will subtract from the applied magnetic field and the methyl protons will show a signal at a lower frequency (Figure 13.12). Therefore, the signal for the methyl protons is split into two peaks, one corresponding to the higher frequency and one corresponding to the lower frequency. Because each spin state has almost the same population, about half the methine protons are lined up with the applied magnetic field and about half are lined up against it. Therefore, the two peaks of the *doublet* have approximately the same height and area.

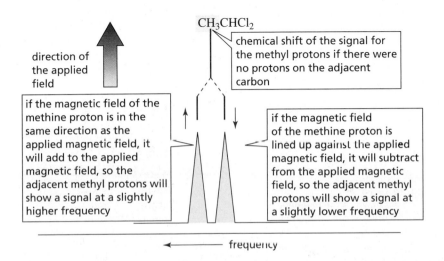

direction of the applied field

$$CH_3CHCl_2$$

> chemical shift of the signal for the methyl protons if there were no protons on the adjacent carbon

> if the magnetic field of the methine proton is in the same direction as the applied magnetic field, it will add to the applied magnetic field, so the adjacent methyl protons will show a signal at a slightly higher frequency

> if the magnetic field of the methine proton is lined up against the applied magnetic field, it will subtract from the applied magnetic field, so the adjacent methyl protons will show a signal at a slightly lower frequency

← frequency

◄ **Figure 13.12**
The signal for the methyl protons of 1,1-dichloroethane is split into a doublet by the methine proton.

Similarly, the frequency at which the methine proton shows a signal is influenced by the magnetic fields of the three protons bonded to the adjacent carbon. The magnetic fields of all three methyl protons can align with the applied magnetic field, two can align with the field and one against it, one can align with it and two against it, or all three can align against it. Because the magnetic field that the methine proton senses is affected in four different ways, its signal is a *quartet* (Figure 13.13).

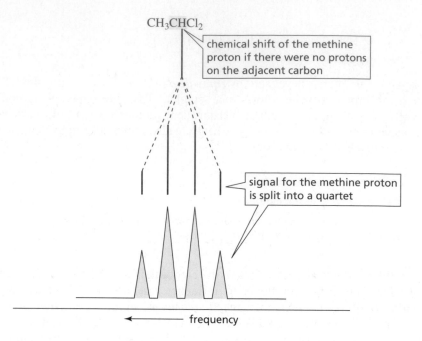

CH₃CHCl₂

chemical shift of the methine proton if there were no protons on the adjacent carbon

signal for the methine proton is split into a quartet

← frequency

Figure 13.13 ▶
The signal for the methine proton of 1,1-dichloroethane is split into a quartet by the methyl protons.

The relative intensities of the peaks in a signal reflect the number of ways the neighboring protons can be aligned relative to the applied magnetic field. For example, a quartet has relative peak intensities of 1 : 3 : 3 : 1. There is only one way to align the magnetic fields of three protons so that they are all oriented with the applied magnetic field and only one way to align them so that they are all oriented against the applied field. However, there are three ways to align the magnetic fields of three protons so that two are lined up with the applied field and one is lined up against it (Figure 13.14), and there are three ways to align them so that one is lined up with the applied field and two are lined up against it. These different possibilities explain the different peak intensities.

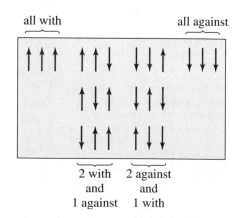

all with   all against

2 with   2 against
and   and
1 against   1 with

Figure 13.14 ▶
The ways in which the magnetic fields of three protons can be aligned.

The relative intensities obey the mathematical mnemonic known as *Pascal's triangle*. (You may remember this mnemonic device from one of your math classes.) According to Pascal, each number at the bottom of a triangle in the rightmost column of Table 13.2 is the sum of the two numbers to its immediate left and right in the row above it.

Table 13.2 Multiplicity of the Signal and Relative Intensities of the Peaks in the Signal

Number of equivalent protons causing splitting	Multiplicity of the signal	Relative peak intensities
0	singlet	1
1	doublet	1 : 1
2	triplet	1 : 2 : 1
3	quartet	1 : 3 : 3 : 1
4	quintet	1 : 4 : 6 : 4 : 1
5	sextet	1 : 5 : 10 : 10 : 5 : 1
6	septet	1 : 6 : 15 : 20 : 15 : 6 : 1

BIOGRAPHY

Blaise Pascal (1623–1662) *was born in France. At age 16, he published a book on geometry and, at 19, he invented a calculating machine. He propounded the modern theory of probability, developed the principle underlying the hydraulic press, and showed that atmospheric pressure decreases as altitude increases. In 1644, he narrowly escaped death when the horses leading a carriage in which he was riding bolted. That scare caused him to devote the rest of his life to meditation and religious writings.*

Normally, *nonequivalent* protons split each other's signal only if they are on *adjacent* carbons. Splitting is a "through-bond" effect, not a "through-space" effect; it is rarely observed if the protons are separated by more than three σ bonds. However, if they are separated by more than three bonds and one of the bonds is a double or triple bond, a small splitting is sometimes observed. This phenomenon is called **long-range coupling**.

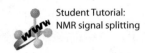

Student Tutorial: NMR signal splitting

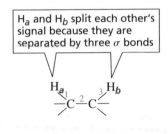

H_a and H_b split each other's signal because they are separated by three σ bonds

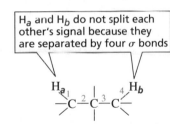

H_a and H_b do not split each other's signal because they are separated by four σ bonds

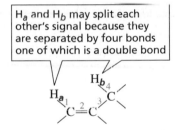

H_a and H_b may split each other's signal because they are separated by four bonds one of which is a double bond

PROBLEM 18◆

One of the spectra in Figure 13.15 is produced by 1-chloropropane, and the other by 1-iodopropane. Which is which?

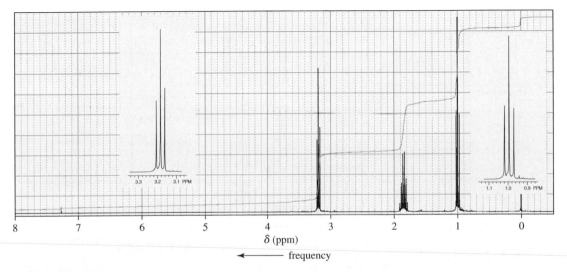

▲ **Figure 13.15**
The ¹H NMR spectrum for Problem 18.

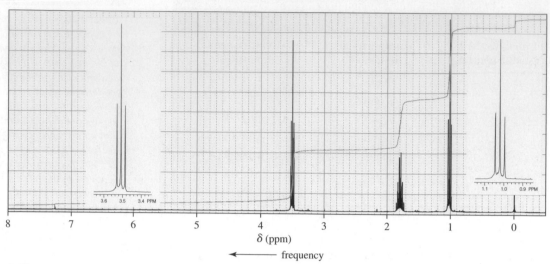

▲ **Figure 13.15**
Continued

Draw a diagram like the one in Figure 13.14 to predict

a. the relative intensities of the peaks in a triplet.
b. the relative intensities of the peaks in a quintet.

The ^{1}H NMR spectra of two carboxylic acids with molecular formula $C_3H_5O_2Cl$ are shown in Figure 13.16. Identify the carboxylic acids. (The "offset" notation means that the farthest-left signal has been moved to the right by the indicated amount in order to fit on the spectrum; thus, the signal at 9.8 ppm offset by 2.4 ppm has an actual chemical shift of 12.2 ppm.)

a.

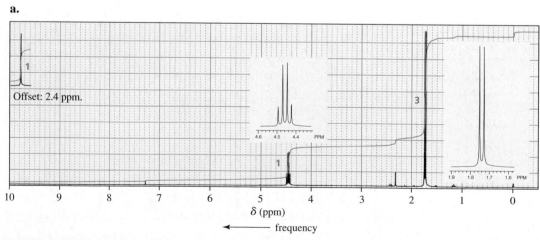

▲ **Figure 13.16a**
The ^{1}H NMR spectra for Problem 20.

b.

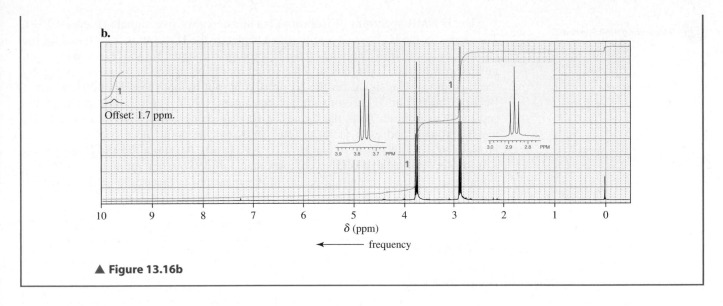

Offset: 1.7 ppm.

δ (ppm)

← frequency

▲ Figure 13.16b

13.11 More Examples of ^{1}H NMR Spectra

VCL **Interpreting NMR Spectra-1**

We will now look at a few more spectra to give you additional practice in analyzing ^{1}H NMR spectra.

There are two signals in the ^{1}H NMR spectrum of 1,3-dibromopropane (Figure 13.17). The signal for the H$_b$ protons is split into a triplet by the two hydrogens on the adjacent carbon. The carbons adjacent to the one bonded to the H$_a$ protons are both bonded to protons. The protons on one of these are equivalent to the protons on the other. Because the two sets of protons are equivalent, the $N + 1$ rule is applied to both sets at the same time. In other words, N is equal to the sum of the equivalent protons on both carbons. So the signal for the H$_a$ protons is split into a quintet ($4 + 1 = 5$). Integration confirms that two methylene groups contribute to the H$_b$ signal because it shows that twice as many protons give rise to that signal as to the H$_a$ signal.

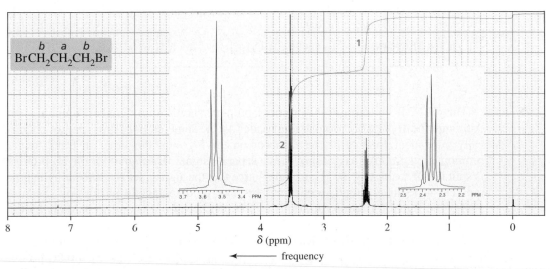

δ (ppm)

← frequency

▲ Figure 13.17
The ^{1}H NMR spectrum of 1,3-dibromopropane.

The ^{1}H NMR spectrum of isopropyl butanoate shows five signals (Figure 13.18). The signal for the H$_a$ protons is split into a triplet by the H$_c$ protons. The signal for the H$_b$ protons is split into a doublet by the H$_e$ proton. The signal for the H$_d$ protons is split into a triplet by the H$_c$ protons, and the signal for the H$_e$ proton is split into a septet by the H$_b$ protons. The signal for the H$_c$ protons is split by both the H$_a$ and H$_d$ protons. Because the H$_a$ and H$_d$ protons are not equivalent, the $N + 1$ rule has to be applied separately to each set. Thus, the signal for the H$_c$ protons will be split into a quartet by the H$_a$ protons, and each of these four peaks will be split into a triplet by the H$_d$ protons: $(N_a + 1)(N_d + 1) = (4)(3) = 12$. As a result, the signal for the H$_c$ protons is a **multiplet** (a signal that is more complex than a triplet, quartet, quintet, or such). The reason that 12 peaks are not apparent in the spectrum is that some of them overlap (Section 13.13).

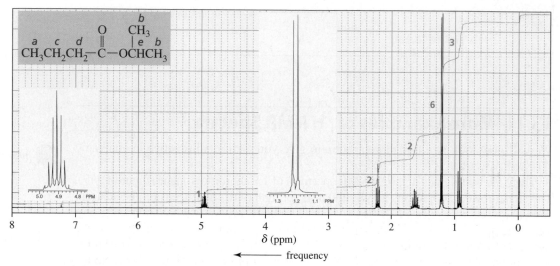

▲ **Figure 13.18**
The ^{1}H NMR spectrum of isopropyl butanoate.

PROBLEM 21

Indicate the number of signals and the multiplicity of each signal in the ^{1}H NMR spectrum of each of the following compounds:

a. $ICH_2CH_2CH_2Br$ **b.** $ClCH_2CH_2CH_2Cl$ **c.** $ICH_2CH_2CHBr_2$

The ^{1}H NMR spectrum of 3-bromo-1-propene shows four signals (Figure 13.19). Although the H$_b$ and H$_c$ protons are bonded to the same carbon, they are not chemically equivalent (one is cis to the bromomethyl group, the other is trans to the bromomethyl group), so each produces a separate signal. The signal for the H$_a$ protons is split into a doublet by the H$_d$ proton. Notice that the signals for the three vinylic protons are at relatively high frequencies because of diamagnetic anisotropy (Section 13.7). The signal for the H$_d$ proton is a multiplet because it is split separately by the H$_a$, H$_b$, and H$_c$ protons.

Because the H$_b$ and H$_c$ protons are not equivalent, they split one another's signal. This means that the signal for the H$_b$ proton is split into a doublet by the H$_d$ proton and that each of the peaks in the doublet is split into a doublet by the H$_c$ proton. The signal for the H$_b$ proton should therefore be what is called a **doublet of doublets**, and so should the signal for the H$_c$ proton. However, the mutual splitting of the

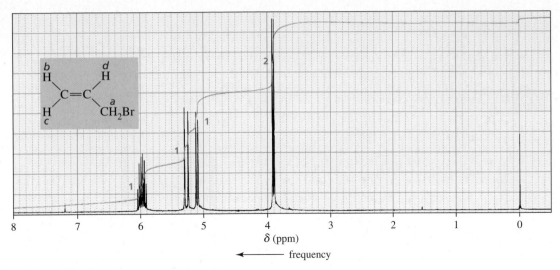

▲ **Figure 13.19**
The ¹H NMR spectrum of 3-bromo-1-propene.

signals of two nonidentical protons bonded to the same sp^2 carbon, called **geminal coupling**, is often too small to be observed (see Table 13.3). Therefore, the signals for the H_b and H_c protons in Figure 13.19 appear as doublets rather than as doublets of doublets. (If the signals were expanded along the frequency axis, the doublet of doublets would be observed.)

There is a clear difference between a quartet and a doublet of doublets, although both have four peaks. A quartet results from splitting by *three equivalent* adjacent protons; it therefore has relative peak intensities of 1 : 3 : 3 : 1, and the individual peaks are equally spaced. A doublet of doublets, on the other hand, results from splitting by *two nonequivalent* adjacent protons; it has relative peak intensities of 1 : 1 : 1 : 1, and the individual peaks are not necessarily equally spaced (see Figure 13.26 on page 601).

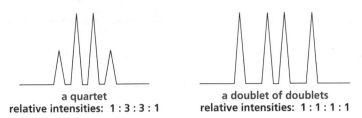

a quartet
relative intensities: 1 : 3 : 3 : 1

a doublet of doublets
relative intensities: 1 : 1 : 1 : 1

Ethylbenzene has five sets of chemically equivalent protons (Figure 13.20). We see the expected triplet for the H_a protons and the quartet for the H_b protons. (This is a characteristic pattern for an ethyl group.) We expect the signal for the H_c protons to be a doublet and the signal for the H_e proton to be a triplet. Because the H_c and H_e protons are not equivalent, they must be considered separately in determining the splitting of the signal for the H_d protons. Therefore, we expect the signal for the H_d protons to be split into a doublet by the H_c protons and each peak of the doublet to be split into another doublet by the H_e proton, forming a doublet of doublets. However, we do not see three distinct signals for the H_c, H_d, and H_e protons in Figure 13.20. Instead, we see overlapping signals. Apparently, the electronic effect (that is, the electron-donating or electron-withdrawing ability) of an ethyl substituent is not sufficiently different from that of a hydrogen to cause a difference in the environments of the H_c, H_d, and H_e protons that is large enough to allow them to appear as separate signals.

 **Interpreting NMR Spectra-4**

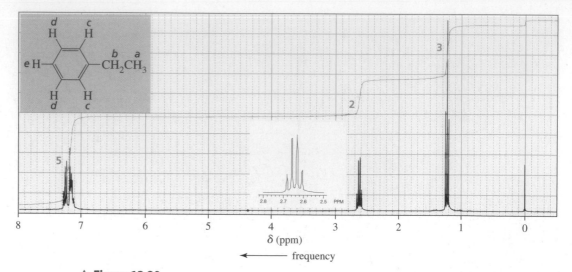

▲ **Figure 13.20**
The ^{1}H NMR spectrum of ethylbenzene. The signals for the H_c, H_d, and H_e protons overlap.

In contrast to the benzene protons of ethylbenzene (H_c, H_d, and H_e), the benzene protons of nitrobenzene (H_a, H_b, and H_c) show three distinct signals (Figure 13.21), and the multiplicity of each signal is what we would have predicted for the signals for the benzene ring protons in ethylbenzene (H_c is a doublet, H_b is a triplet, and H_a is a doublet of doublets). The nitro group is sufficiently electron withdrawing to cause the H_a, H_b, and H_c protons to be in sufficiently different environments for their signals not to overlap.

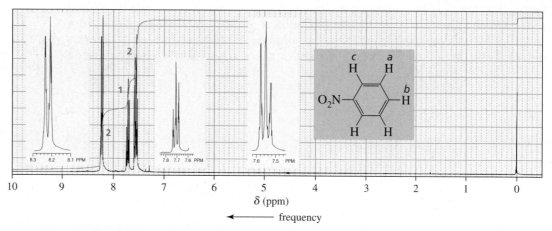

▲ **Figure 13.21**
The ^{1}H NMR spectrum of nitrobenzene. The signals for the H_a, H_b, and H_c protons do not overlap.

Student Tutorial:
NMR spectrum assignment

Notice that the signals for the benzene ring protons in Figures 13.20 and 13.21 occur in the 7.0 to 8.5 ppm region. Other kinds of protons usually do not resonate in this region, so signals in this region of an ^{1}H NMR spectrum indicate that the compound probably contains an aromatic ring.

PROBLEM 22

Explain why the signal for the protons identified as H_a in Figure 13.21 appears at the lowest frequency and the signal for the protons identified as H_c appears at the highest frequency. (*Hint:* Draw the contributing resonance structures.)

PROBLEM 23

How could their ^{1}H NMR spectra distinguish the following compounds?

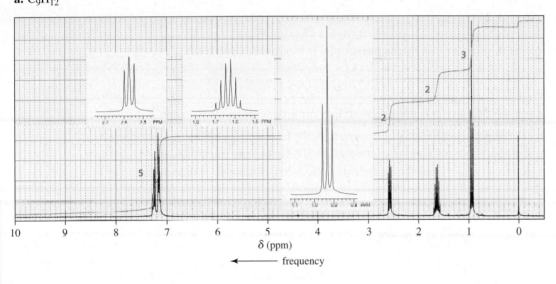

PROBLEM 24

How would the ^{1}H NMR spectra for the four compounds with molecular formula $C_3H_6Br_2$ differ?

PROBLEM 25◆

Identify each compound from its molecular formula and its ^{1}H NMR spectrum:

a. C_9H_{12}

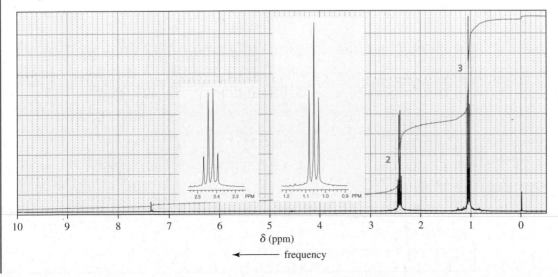

δ (ppm)

frequency

b. $C_5H_{10}O$

c. $C_9H_{10}O_2$

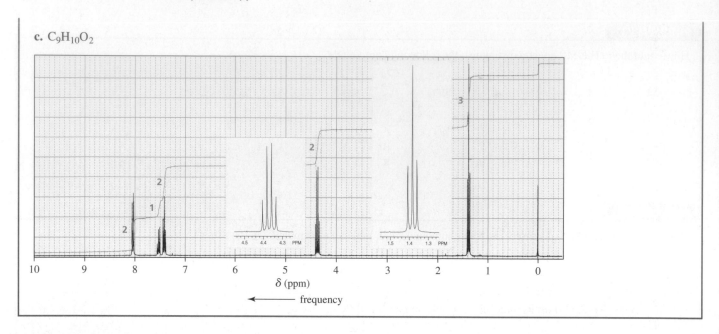

PROBLEM 26

Predict the splitting patterns for the signals given by compounds a–m in Problem 3.

PROBLEM 27◆

Identify the following compounds. (Relative integrals are given from left to right across the spectrum.)

a. The ^{1}H NMR spectrum of a compound with molecular formula $C_4H_{10}O_2$ has two singlets with an area ratio of 2 : 3.

b. The ^{1}H NMR spectrum of a compound with molecular formula $C_6H_{10}O_2$ has two singlets with an area ratio of 2 : 3.

c. The ^{1}H NMR spectrum of a compound with molecular formula $C_8H_6O_2$ has two singlets with an area ratio of 1 : 2.

Student Tutorial:
NMR spectrum interpretation

PROBLEM 28

Describe the ^{1}H NMR spectrum you would expect for each of the following compounds, using relative chemical shifts rather than absolute chemical shifts:

a. $BrCH_2CH_2CH_2CH_2Br$

f.
$$\begin{array}{c} H \\ \diagdown \\ C=C \\ \diagup \quad \diagdown \\ H \qquad Cl \end{array}$$ (with H on top left)

k. CH_3CHCH_2CH (with CH_3 branch and C=O)

b. $CH_3OCH_2CH_2CH_2Br$

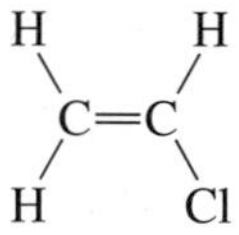

g. $CH_3CH_2OCH_2CH_3$

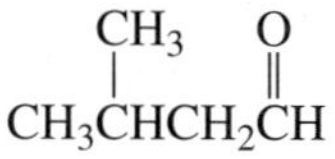

l. $CH_3OCH_2CH_2CH_2OCH_3$

c. $O=\hexagon=O$

h. $CH_3CH_2OCH_2Cl$

m.
$$\begin{array}{c} H \qquad Cl \\ \diagdown \quad \diagup \\ C=C \\ \diagup \quad \diagdown \\ H \qquad Cl \end{array}$$

d. $CH_3CCH_2CH_3$ (with CH_3 and Br branches)

i. $CH_3CHCHCl_2$ (with Cl branch)

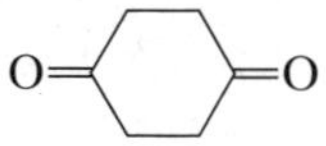

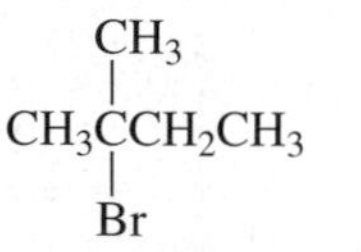

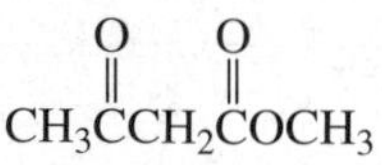

n.
$$\begin{array}{c} Cl \qquad H \\ \diagdown \quad \diagup \\ C=C \\ \diagup \quad \diagdown \\ H \qquad Cl \end{array}$$

e. $CH_3CCH_2COCH_3$ (with two C=O groups)

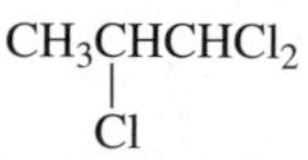

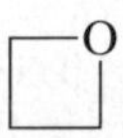

j. $\square{-}O$

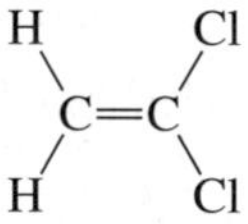

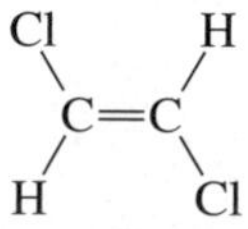

o. $\pentagon$

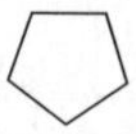

13.12 **Coupling Constants Identify Coupled Protons**

The distance, in hertz, between two adjacent peaks of a split NMR signal is called the **coupling constant** (denoted by J). The coupling constant for H_a being split by H_b is denoted by J_{ab}. The signals of coupled protons (protons that split each other's signal) have the same coupling constant; in other words, $J_{ab} = J_{ba}$ (Figure 13.22). Coupling constants are useful in analyzing complex NMR spectra because protons on adjacent carbons can be identified by their identical coupling constants.

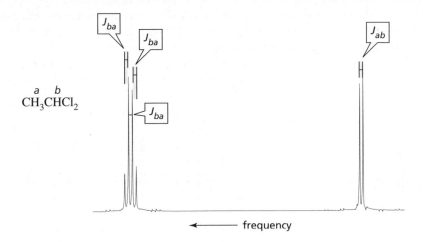

◀ **Figure 13.22**
The H_a and H_b protons of 1,1-dichloroethane are coupled protons, so their signals have the same coupling constant, $J_{ab} = J_{ba}$.

The magnitude of a coupling constant is independent of the operating frequency of the spectrometer: the same coupling constant is obtained from a 300-MHz instrument as from a 600-MHz instrument. The magnitude of a coupling constant is a measure of how strongly the nuclear spins of the coupled protons influence each other. It, therefore, depends on the number and type of bonds that connect the coupled protons, as well as the geometric relationship of the protons. Characteristic coupling constants are shown in Table 13.3; they range from 0 to 15 Hz.

Table 13.3	**Approximate Values of Coupling Constants**	
Approximate value of J_{ab} (Hz)	**Approximate value of J_{ab} (Hz)**	
$-\overset{\displaystyle H_a}{\underset{\displaystyle \vert}{\overset{\displaystyle \vert}{C}}}-H_b$ 12	$\overset{H_a}{\diagdown}C=C\diagup \diagdown H_b$	15 (trans)
$-\overset{H_a}{\underset{\vert}{\overset{\vert}{C}}}-\overset{H_b}{\underset{\vert}{\overset{\vert}{C}}}-$ 7	$\overset{H_a}{\diagdown}C=C\overset{H_b}{\diagup}$	10 (cis)
$-\overset{H_a}{\underset{\vert}{\overset{\vert}{C}}}-\overset{}{\underset{\vert}{\overset{\vert}{C}}}-\overset{H_b}{\underset{\vert}{\overset{\vert}{C}}}-$ 0	$\overset{H_a}{\diagdown}C=C\diagdown C\diagdown H_b$	1 (long-range coupling)
$\diagup C=C\overset{\diagup H_a}{\diagdown H_b}$ 2 (geminal coupling)		

The coupling constant for trans alkene protons is greater than the coupling constant for cis alkene protons.

The coupling constant for two nonequivalent hydrogens on the *same sp³* carbon is large. (Two hydrogens on the same *sp³* carbon will be nonequivalent if the *sp³* carbon is attached to an asymmetric center; Section 13.14.) In contrast, the coupling constant for two nonequivalent hydrogens on the *same sp²* carbon is often too small to see (Figure 13.19), but it is large if the nonequivalent hydrogens are bonded to *adjacent sp²* carbons. Apparently, the interaction between the hydrogens is strongly affected by the intervening π electrons. We have seen that π electrons also allow long-range coupling, that is, coupling through four or more bonds (Section 13.10).

Coupling constants can be used to distinguish between the ^{1}H NMR spectra of cis and trans alkenes. The coupling constant of trans vinylic protons is significantly greater than the coupling constant of cis vinylic protons (Figure 13.23), because the coupling constant depends on the dihedral angle between the two C—H bonds in the H—C=C—H unit (Section 2.10). The coupling constant is greatest when the angle between the two C—H bonds is 180° (trans) and smaller when the angle is 0° (cis). Notice the difference between J_{bd} and J_{cd} in the spectrum of 3-bromo-1-propene (Figure 13.19 on page 593).

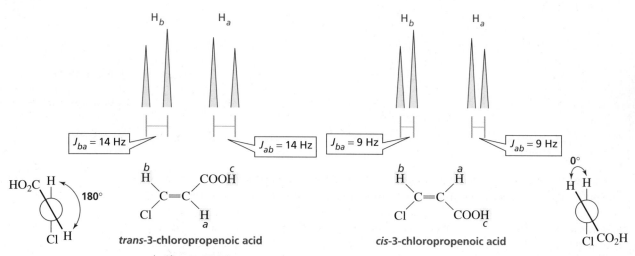

▲ **Figure 13.23**
The doublets observed for the H_a and H_b protons in the ^{1}H NMR spectra of *trans*-3-chloropropenoic acid and *cis*-3-chloropropenoic acid. The coupling constant for trans protons (14 Hz) is greater than the coupling constant for cis protons (9 Hz).

PROBLEM 29

Why is there no coupling between H_a and H_c or between H_b and H_c in *cis*- or *trans*-3-chloropropenoic acid?

Qualitative Analysis Unknowns

Let's now summarize the kind of information that can be obtained from an ^{1}H NMR spectrum:

1. The number of signals indicates the number of different kinds of protons in the compound.

2. The position of a signal indicates the kind of proton(s) responsible for the signal (methyl, methylene, methine, allylic, vinylic, aromatic, and so on) and the kinds of neighboring substituents.

3. The integration of the signal tells the relative number of protons responsible for the signal.

4. The multiplicity of the signal ($N + 1$) tells the number of protons (N) bonded to adjacent carbons.

5. The coupling constants identify coupled protons.

PROBLEM-SOLVING STRATEGY

Using IR and ^{1}H NMR Spectra to Deduce a Chemical Structure

Identify the compound with molecular formula $C_9H_{10}O$ that gives the IR and ^{1}H NMR spectra in Figure 13.24.

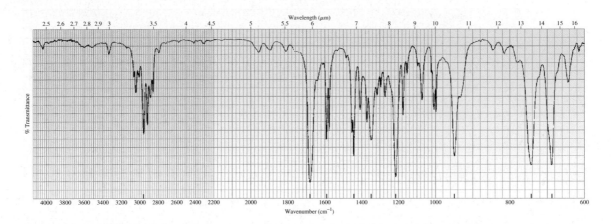

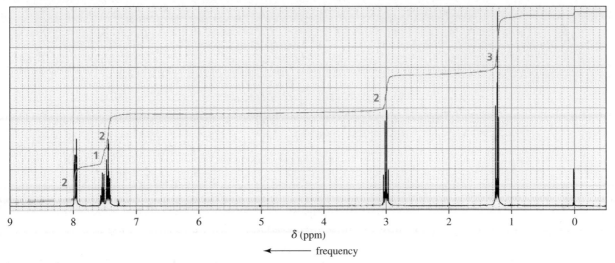

▲ **Figure 13.24**
The IR and ^{1}H NMR spectra for this problem-solving strategy.

The best way to approach this kind of problem is to identify whatever structural features you can from the molecular formula and IR spectrum and then use the information from the ^{1}H NMR spectrum to expand on that knowledge. From the molecular formula and IR spectrum, we learn that the compound is a ketone: it has a carbonyl group at ~1680 cm^{-1}, only one oxygen, and no absorption bands at ~2820 and ~2720 cm^{-1} that would indicate an aldehyde. The carbonyl group absorption band is at a lower frequency than is typical, which suggests that it has partial single-bond character as a result of electron delocalization, indicating that it is attached to an sp^2 carbon. The compound contains a benzene ring (absorption bands >3000 cm^{-1}, ~1600 cm^{-1}, and ~1440 cm^{-1}), and it has hydrogens bonded to sp^3 carbons (absorption in the region <3000 cm^{-1}). In the NMR spectrum, the triplet at ~1.2 ppm and the quartet at ~3.0 ppm indicate the presence of an ethyl group that is attached to an electron-withdrawing group. The signals in the 7.4 to 8.0 ppm region confirm the presence of a benzene ring. From this information, we can conclude that the compound is the ketone shown below. The integration ratio (5 : 2 : 3) confirms this answer.

Now continue on to Problem 30.

Identify the compound with molecular formula $C_8H_{10}O$ that gives the IR and ^1H NMR spectra shown in Figure 13.25.

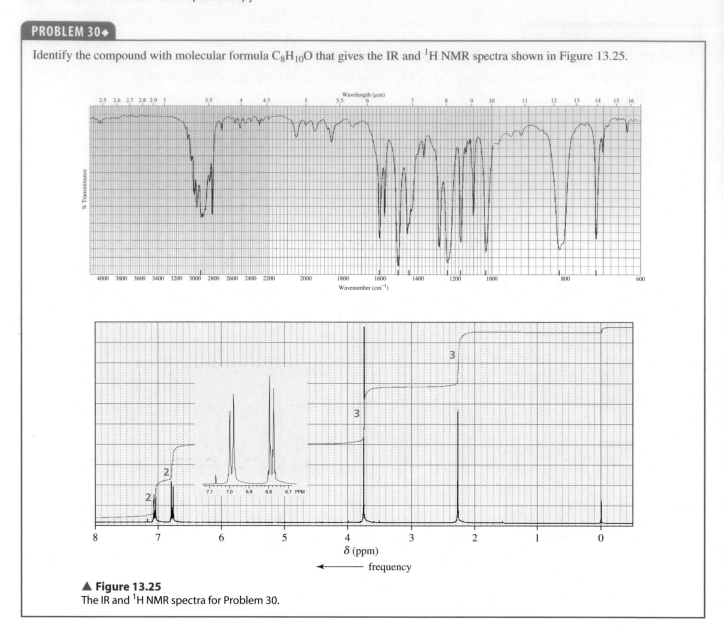

▲ **Figure 13.25**
The IR and ^1H NMR spectra for Problem 30.

13.13 Splitting Diagrams Explain the Multiplicity of a Signal

The splitting pattern obtained when a signal is split by more than one set of protons can best be understood by using a splitting diagram. In a **splitting diagram** (also called a **splitting tree**), the NMR peaks are shown as vertical lines, and the effect of each of the splittings is shown one at a time. The splitting diagram shown in Figure 13.26, for example, depicts the splitting of the signal for the H_c proton of 1,1,2-trichloro-3-methylbutane into a doublet of doublets by the H_b and H_d protons.

The signal for the H_b protons of 1-bromopropane is split into a quartet by the H_a protons (Figure 13.27), and each of the resulting four peaks is split into a triplet by

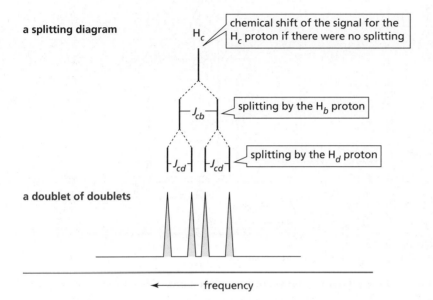

1,1,2-trichloro-3-methylbutane

a splitting diagram

a doublet of doublets

◀ Figure 13.26
A splitting diagram for a doublet
of doublets.

the H_c protons. How many of the 12 peaks are actually seen in the spectrum depends on the relative magnitudes of the two coupling constants, J_{ba} and J_{bc}. For example, Figure 13.27 shows that there are 12 peaks when J_{ba} is much greater than J_{bc}, 9 peaks when $J_{ba} = 2J_{bc}$, and only 6 peaks when $J_{ba} = J_{bc}$. The number of peaks observed depends on how many overlap with one another. When peaks overlap, their intensities add together.

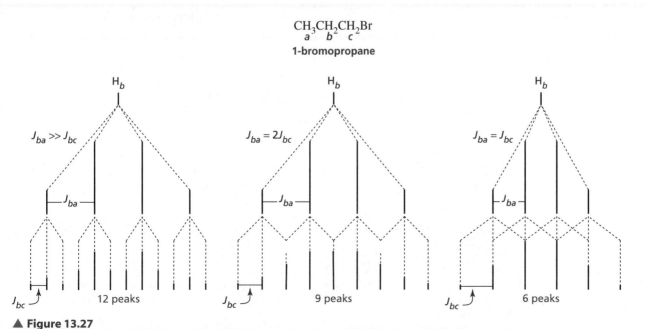

$$CH_3CH_2CH_2Br$$
$$a \quad b \quad c$$

1-bromopropane

▲ Figure 13.27
A splitting diagram for a quartet of triplets. The number of peaks actually observed when a signal
is split by two sets of protons depends on the relative magnitudes of the two coupling constants.

We expect the signal for the H$_a$ protons of 1-chloro-3-iodopropane to be a triplet of triplets (split into nine peaks) because the signal would be split into a triplet by the H$_b$ protons and each of the resulting peaks would be split into a triplet by the H$_c$ protons. The signal, however, is a quintet (Figure 13.28).

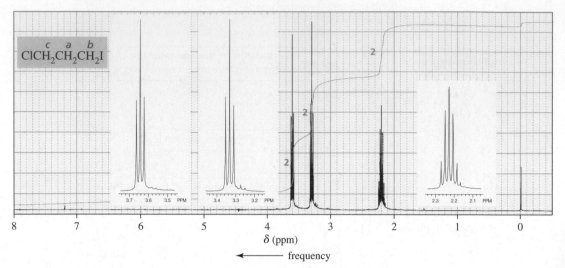

▲ **Figure 13.28**
The ^{1}H NMR spectrum of 1-chloro-3-iodopropane.

Finding that the signal for the H$_a$ protons of 1-chloro-3-iodopropane is a quintet indicates that J_{ab} and J_{ac} have about the same value. The splitting diagram shows that a quintet results if $J_{ab} = J_{ac}$.

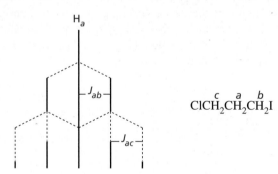

We can conclude that *when two different sets of protons split a signal, the multiplicity of the signal can be determined by using the N + 1 rule separately for each set of hydrogens, as long as the coupling constants for the two sets are different. When the coupling constants are similar, the multiplicity of a signal can be determined by treating both sets of adjacent hydrogens as if they were equivalent.* In other words, when the coupling constants are similar, the $N + 1$ rule can be applied to both sets of protons simultaneously.

PROBLEM 31

Draw a splitting diagram for H$_b$, where

a. $J_{ba} = 12$ Hz and $J_{bc} = 6$ Hz.   **b.** $J_{ba} = 12$ Hz and $J_{bc} = 12$ Hz.

$$-\overset{|}{\underset{|}{C}}-\overset{|}{\underset{|}{C}}-\overset{|}{\underset{|}{C}}-$$
$$\quad H_a \quad H_b \quad H_c$$

13.14 **Diastereotopic Hydrogens Are Not Chemically Equivalent**

If a carbon is bonded to two hydrogens and to two different groups, the two hydrogens are called **enantiotopic hydrogens**. For example, the two hydrogens (H_a and H_b) in the CH_2 group of ethanol are enantiotopic hydrogens because the other two groups bonded to the carbon (CH_3 and OH) are not identical. Replacing an enantiotopic hydrogen by a deuterium (or any other atom or group other than CH_3 or OH) forms a chiral molecule.

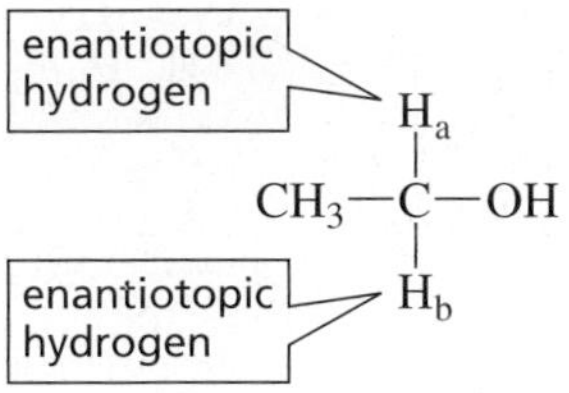

The carbon to which the enantiotopic hydrogens are attached is called a **prochiral carbon** because it will become a chirality center (an asymmetric center) if one of the hydrogens is replaced by a deuterium (or any group other than CH_3 or OH). If the H_a hydrogen is replaced by a deuterium, the asymmetric center will have the R configuration. Thus, the H_a hydrogen is called the **pro-R-hydrogen**. The H_b hydrogen is called the **pro-S-hydrogen** because if it is replaced by a deuterium, the asymmetric center will have the S configuration. The pro-R- and pro-S-hydrogens are chemically equivalent, so they give only one NMR signal.

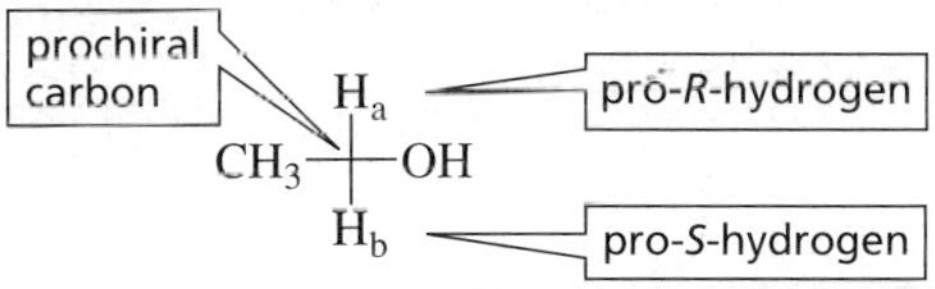

If a carbon is bonded to two hydrogens in a compound with an asymmetric center, the two hydrogens are called **diastereotopic hydrogens**, because replacing each of them in turn with deuterium (or another group) creates a pair of diastereomers.

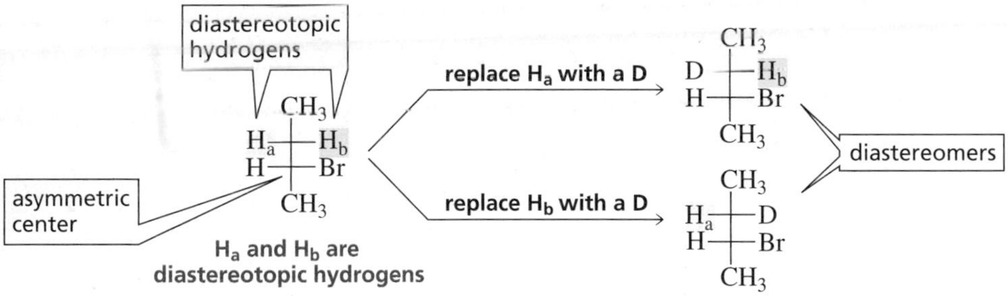

Diastereotopic hydrogens are not chemically equivalent. Therefore, they do not have the same reactivity with achiral reagents. For example, in Section 9.5 we saw that because *trans*-2-butene is more stable than *cis*-2-butene, removal of H_b and Br to form *trans*-2-butene occurs more rapidly than removal of H_a and Br to form *cis*-2-butene.

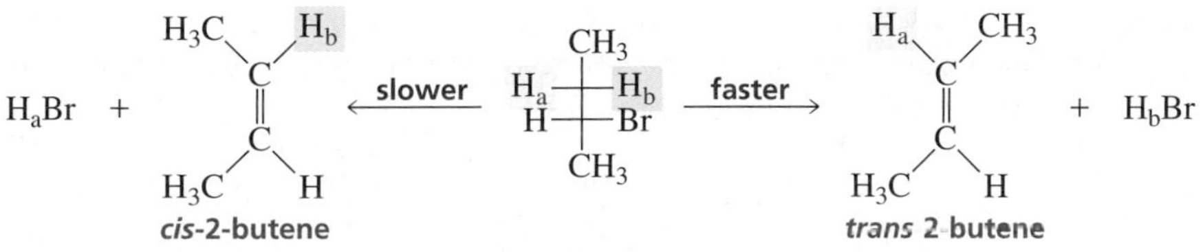

Because diastereotopic hydrogens are not chemically equivalent, they have different chemical shifts. Typically the chemical shifts are similar and, like other nonequivalent hydrogens, can even be the same by chance. The further the diastereotopic hydrogens are from the asymmetric center, the more similar their chemical shifts are expected to be. Because diastereotopic hydrogens are not equivalent, the $N + 1$ rule has to be applied to them separately.

Student Tutorial:
Stereochemical nonequivalence

PROBLEM 32◆

a. Which of the following compounds have enantiotopic hydrogens?

b. Which have diastereotopic hydrogens?

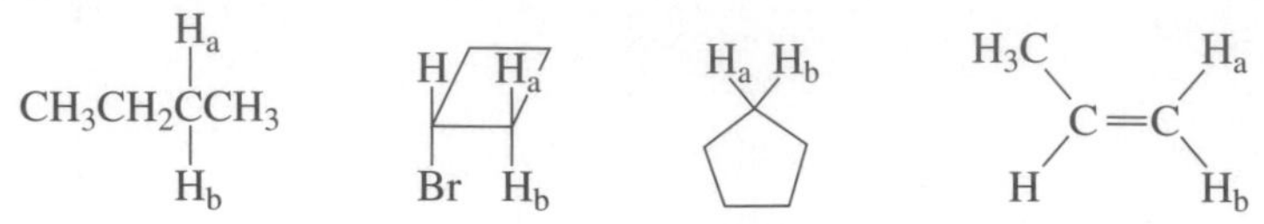

A   B   C   D

PROBLEM 33 **SOLVED**

Applying the $N + 1$ rule separately to the two diastereotopic hydrogens of 2-bromobutane tells us that the signal for the adjacent methyl hydrogens should be a doublet of doublets. The signal, however, is a triplet. Use a splitting diagram to explain why it is a triplet rather than a doublet of doublets.

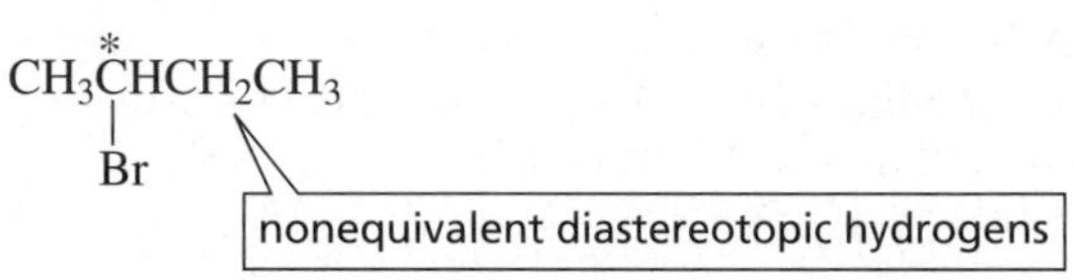

$$CH_3\overset{*}{C}HCH_2CH_3$$

Br

nonequivalent diastereotopic hydrogens

Solution   The observation of a triplet means that the $N + 1$ rule did not have to be applied to the diastereotopic hydrogens separately, in this case, but could have been applied to the two protons as a set ($N = 2$, so $N + 1 = 3$). This indicates that the coupling constant for splitting of the methyl signal by one of the diastereotopic hydrogens is similar to the coupling constant for splitting by the other diastereotopic hydrogen.

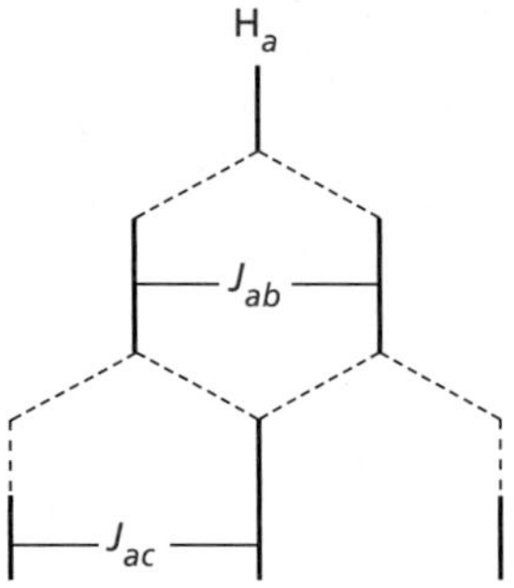

13.15 **The Time Dependence of NMR Spectroscopy**

We have seen that the three methyl hydrogens of ethyl bromide give rise to just one signal in the ^{1}H NMR spectrum because they are chemically equivalent due to rotation about the C—C bond. At any given instant, however, the three hydrogens can be in quite different environments: one can be anti to the bromine, one can be gauche to the bromine, one can be eclipsed with the bromine.

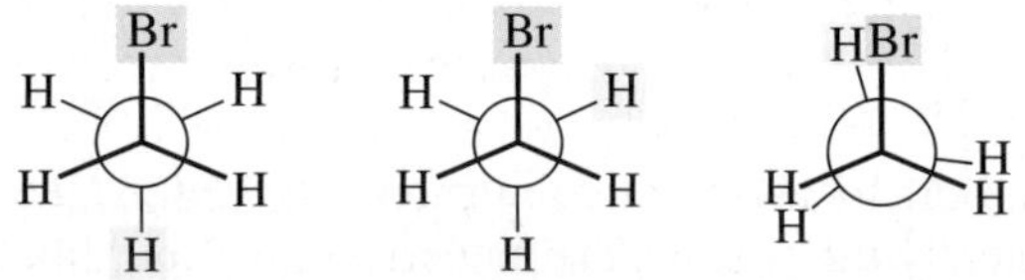

anti   gauche   eclipsed

An NMR spectrometer is very much like a camera with a slow shutter speed: it is too slow to be able to detect these different environments, so what it sees is an average of all of them. Because each of the three methyl hydrogens has the same average environment, we see one signal for the methyl group in the ^{1}H NMR spectrum.

Similarly, the ^{1}H NMR spectrum of cyclohexane shows only one signal, even though cyclohexane has both axial and equatorial hydrogens. There is only one signal because the chair conformers of cyclohexane intervert too rapidly at room temperature for the NMR spectrometer to detect them individually. Because axial hydrogens in one chair conformer are equatorial hydrogens in the other chair conformer (Section 2.13), all the hydrogens in cyclohexane have the same average environment on the NMR time scale, so the NMR spectrum shows one signal.

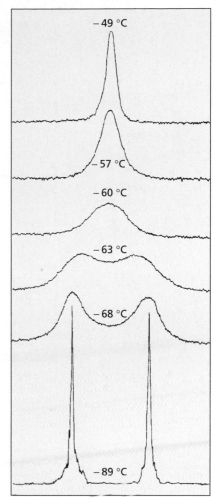

▲ **Figure 13.29**
A series of ^{1}H NMR spectra of cyclohexane-d_{11} obtained at various temperatures.

The rate of chair–chair interconversion is temperature-dependent: the lower the temperature, the slower is the rate. Cyclohexane-d_{11} has 11 deuterium atoms, which means that it has only one hydrogen. Several ^{1}H NMR spectra of cyclohexane-d_{11} taken at various temperatures are shown in Figure 13.29. Cyclohexane with only one hydrogen was used for this experiment in order to prevent splitting of the signal, which would have complicated the spectrum. Deuterium signals are not detectable in ^{1}H NMR, and splitting by a deuterium on the same or on an adjacent carbon is often too small to be detectable at the operating frequency of an ^{1}H NMR spectrometer.

At room temperature, the ^{1}H NMR spectrum of cyclohexane-d_{11} shows one sharp signal, which is an average for the axial hydrogen of one chair and the equatorial hydrogen of the other chair. As the temperature decreases, the signal becomes broader and eventually separates into two signals, which are equidistant from the original signal. At $-89\ °C$, two sharp singlets are observed because at that temperature, the rate of chair–chair interconversion has decreased sufficiently to allow the two kinds of hydrogens (axial and equatorial) to be individually detected on the NMR time scale.

13.16 Protons Bonded to Oxygen and Nitrogen

The chemical shift of a proton bonded to an oxygen or to a nitrogen depends on the degree of hydrogen bonding the proton experiences; the greater the extent of hydrogen bonding, the greater is the chemical shift, because the extent of hydrogen bonding affects the electron density around the proton. For example, the chemical shift of the OH proton of an alcohol ranges from 2 to 5 ppm; the chemical shift of the OH proton of a carboxylic acid, from 10 to 12 ppm; the chemical shift of the NH proton of an amine, from 1.5 to 4 ppm; and the chemical shift of the NH proton of an amide, from 5 to 8 ppm.

The ^{1}H NMR spectrum of pure dry ethanol is shown in Figure 13.30a, and the ^{1}H NMR spectrum of ethanol with a trace amount of acid is shown in Figure 13.30b. The spectrum shown in Figure 13.30a is what you would predict from what you have learned so far. The signal for the proton bonded to oxygen is farthest downfield and is split into a triplet by the neighboring methylene protons; the signal for the methylene protons is split into a multiplet by the combined effects of the methyl protons and the OH proton.

The spectrum shown in Figure 13.30(b) is the type of spectrum most often obtained for alcohols. The signal for the proton bonded to oxygen is not split, and this proton does not split the signal of the adjacent protons. Therefore, the signal for the OH proton is a singlet, and the signal for the methylene protons is a quartet because it is split only by the methyl protons.

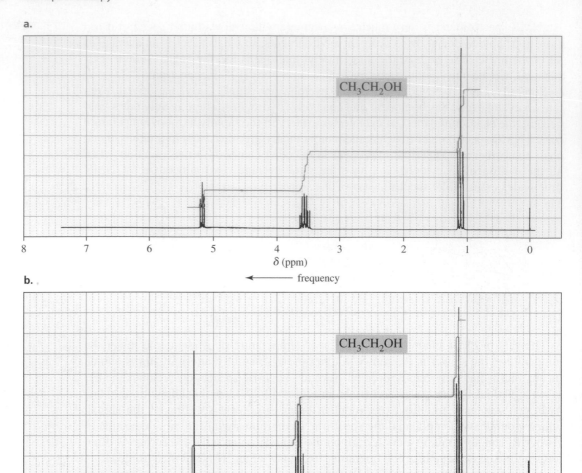

a.

δ (ppm)

← frequency

b.

CH₃CH₂OH

δ (ppm)

← frequency

▲ **Figure 13.30**
(a) The ¹H NMR spectrum of pure ethanol.
(b) The ¹H NMR spectrum of ethanol containing a trace amount of acid.

The two spectra differ because protons bonded to oxygen undergo **proton exchange**, which means that they are transferred from one molecule to another. Whether the OH proton and the methylene protons split each other's signals depends on how long a particular proton stays on the OH group.

In a sample of pure alcohol, the rate of proton exchange is very slow. This causes the spectrum to look no different from one that would be obtained if proton exchange did not occur. Acids and bases catalyze proton exchange, so if the alcohol is contaminated with just a trace of acid or base, proton exchange becomes rapid. When proton exchange is rapid, the spectrum records only an average of all possible environments. Therefore, a rapidly exchanging proton is recorded as a singlet. The effect of a rapidly exchanging proton on adjacent protons is also averaged. Thus, not only is its signal not split by adjacent protons, the rapidly exchanging proton does not cause splitting.

Unless the sample is pure, the hydrogen of an OH group is not split by its neighbors and does not split its neighbors.

Mechanism for acid-catalyzed proton exchange

RÖ—H + HOH ⇌ RO—H + HÖH ⇌ RO: + HOH

The signal for an OH proton is often easy to spot in an ^{1}H NMR spectrum because it is frequently somewhat broader than other signals (see the signal at δ 4.9 in Figure 13.32b on page 608). The broadening occurs because the rate of proton exchange is not slow enough to result in a cleanly split signal, as in Figure 13.28a, or fast enough for a cleanly averaged signal, as in Figure 13.30b. NH protons also show broad signals, not because of chemical exchange, which is generally quite slow for NH, but for reasons that are beyond the scope of this book.

PROBLEM 34

Explain why the chemical shift of the OH proton of a carboxylic acid is at a higher frequency than the chemical shift of an OH proton of an alcohol.

PROBLEM 35◆

Which would show a greater chemical shift for the OH proton, the ^{1}H NMR spectrum of pure ethanol or the ^{1}H NMR spectrum of ethanol dissolved in CH_2Cl_2?

PROBLEM 36

Propose a mechanism for base-catalyzed proton exchange.

PROBLEM 37◆

Identify the compound with molecular formula C_3H_7NO responsible for the ^{1}H NMR spectrum in Figure 13.31.

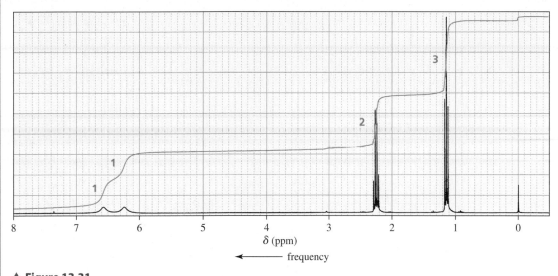

▲ **Figure 13.31**
The ^{1}H NMR spectrum for Problem 37.

13.17 **The Use of Deuterium in ^{1}H NMR Spectroscopy**

Because deuterium signals are not seen in an ^{1}H NMR spectrum, substituting a deuterium for a hydrogen is a technique used to identify signals and to simplify ^{1}H NMR spectra (Section 13.14).

To identify the OH signal in an alcohol's ^{1}H NMR spectrum, for example, a spectrum of the alcohol is taken, and then another spectrum taken after a few drops of D_2O are added to the sample. The OH signal will be the one that becomes less intense (or disappears) in the second spectrum because of the proton exchange process just discussed. This technique can be used with any proton that undergoes exchange.

$$R{-}O{-}H \ + \ D{-}O{-}D \ \longrightarrow \ R{-}O{-}D \ + \ D{-}O{-}H$$

seen in 'H NMR   not seen in 'H NMR

If the ^{1}H NMR spectrum of $CH_3CH_2OCH_3$ were compared with the ^{1}H NMR spectrum of $CH_3CD_2OCH_3$, the signal at the highest frequency in the first spectrum would be absent in the second spectrum, indicating that this signal corresponds to the methylene group.

The sample used to obtain an ^{1}H NMR spectrum is made by dissolving the compound in an appropriate solvent. Solvents with protons cannot be used since the signals for solvent protons would be very intense because there is more solvent than compound in a solution. Instead, deuterated solvents such as $CDCl_3$ (rather than $CHCl_3$) and D_2O (rather than H_2O) are commonly used in NMR spectroscopy.

13.18 The Resolution of ^{1}H NMR Spectra

The ^{1}H NMR spectrum of 2-*sec*-butylphenol taken on a 60-MHz NMR spectrometer is shown in Figure 13.32a; the ^{1}H NMR spectrum of the same compound taken on a

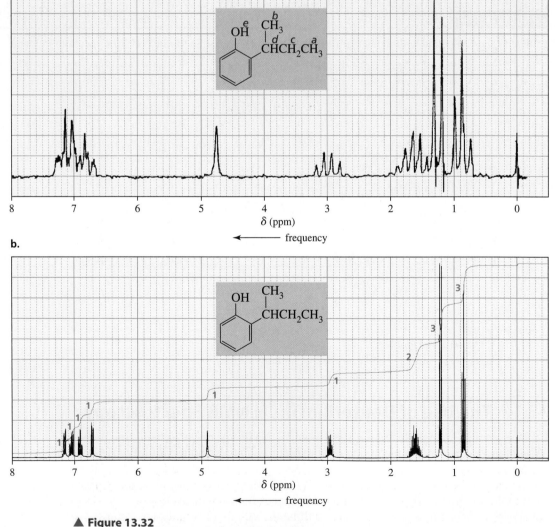

a.

b.

▲ **Figure 13.32**

(a) A 60-MHz ^{1}H NMR spectrum of 2-*sec*-butylphenol.
(b) A 300-MHz ^{1}H NMR spectrum of 2-*sec*-butylphenol.

300-MHz instrument is shown in Figure 13.32b. Why is the resolution of the second spectrum so much better?

To produce separate signals with "clean" splitting patterns, the difference in the chemical shifts (Δv in Hz) of two adjacent coupled protons must be at least 10 times the value of the coupling constant (J). The signals in Figure 13.33, produced by the H$_a$ and H$_b$ protons in an ethyl group, show that as $\Delta v/J$ decreases, the two signals appear closer to each other and the outer peaks of the signals become less intense while the inner peaks become more intense. The quartet and triplet of the ethyl group are clearly observed only when $\Delta v/J$ is greater than 10.

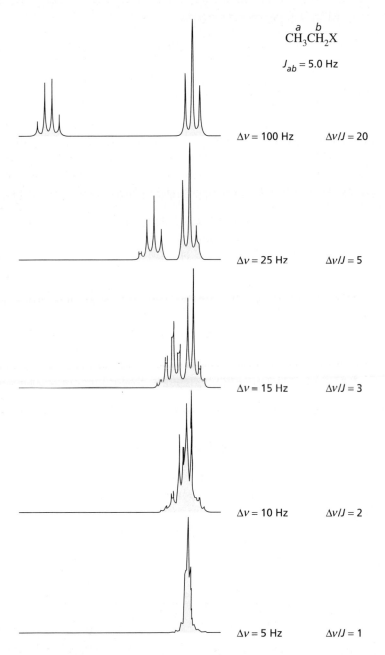

$$\overset{a}{\text{CH}_3}\overset{b}{\text{CH}_2}\text{X}$$

$$J_{ab} = 5.0 \text{ Hz}$$

$\Delta v = 100$ Hz $\Delta v/J = 20$

$\Delta v = 25$ Hz $\Delta v/J = 5$

$\Delta v = 15$ Hz $\Delta v/J = 3$

$\Delta v = 10$ Hz $\Delta v/J = 2$

$\Delta v = 5$ Hz $\Delta v/J = 1$

◀ **Figure 13.33**
The splitting pattern of an ethyl group as a function of the $\Delta v/J$ ratio.

The difference in the chemical shifts of the H$_a$ and H$_c$ protons of 2-*sec*-butylphenol is 0.8 ppm, which corresponds to 48 Hz in a 60-MHz spectrometer and 240 Hz in a 300-MHz spectrometer. (Recall that Δv values are dependent on the operating frequency of the spectrometer; Section 13.5.) In contrast, J values are independent of the operating frequency, so J_{ac} is 7 Hz, whether the spectrum is taken on a 60-MHz or a 300-MHz instrument. Only in the case of the 300-MHz spectrometer is the difference in chemical shift more than 10 times the value of the coupling constant, so only in the 300-MHz spectrum do the signals show clean splitting patterns.

in a 300-MHz spectrometer

$$\frac{\Delta v}{J} = \frac{240}{7} = 34$$

in a 60-MHz spectrometer

$$\frac{\Delta v}{J} = \frac{48}{7} = 6.9$$

13.19 ^{13}C NMR Spectroscopy

The number of signals in a ^{13}C NMR spectrum tells how many different kinds of carbons a compound has—just as the number of signals in an ^{1}H NMR spectrum tells how many different kinds of hydrogens a compound has. The principles behind ^{1}H NMR and ^{13}C NMR spectroscopy are essentially the same. There are, however, some differences that make ^{13}C NMR easier to interpret.

The use of ^{13}C NMR spectroscopy as a routine analytical procedure was not possible until computers were available that could carry out a Fourier transform (Section 13.2). ^{13}C NMR requires Fourier transform techniques because the signals obtained from a single scan are too weak to be distinguished from background electronic noise. However, FT–^{13}C NMR scans can be repeated rapidly, so a large number of scans can be recorded and added together. When hundreds of scans are combined, ^{13}C signals stand out because electronic noise is random, so its sum is close to zero. Without Fourier transform, it could take days to record the number of scans required for a ^{13}C NMR spectrum using a continuous wave (CW) spectrometer.

The individual ^{13}C signals are weak because the isotope of carbon (^{13}C) that gives rise to ^{13}C NMR signals constitutes only 1.11% of the carbon in nature (Section 13.3). (The most abundant isotope of carbon, ^{12}C, has no nuclear spin and therefore cannot produce an NMR signal.) The low abundance of ^{13}C means that the intensities of the signals in ^{13}C NMR are weaker than those in ^{1}H NMR by a factor of approximately 100. In addition, the gyromagnetic ratio (γ) of ^{13}C is about one-fourth that of ^{1}H, and the intensity of a signal is proportional to γ^3. Therefore, the overall intensity of a ^{13}C signal is about 6400 times ($100 \times 4 \times 4 \times 4$) weaker than the intensity of an ^{1}H signal.

One advantage to ^{13}C NMR spectroscopy is that the chemical shifts of carbon atoms range over about 220 ppm (Table 13.4), compared with about 12 ppm for protons (Table 13.1). This means that their signals are less likely to overlap. For example, the data in Table 13.4 show that aldehyde (190 to 200 ppm) and ketone (205 to 220 ppm) carbonyl groups can be easily distinguished from other carbonyl groups. The reference compound used in ^{13}C NMR is TMS, the same reference compound used in ^{1}H NMR. You will find it helpful when analyzing ^{13}C NMR spectra to divide them into five regions and remember the kind of carbons that show signals in each.

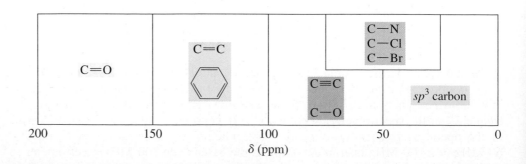

Table 13.4	Approximate Values of Chemical Shifts for ^{13}C NMR			
Type of carbon	**Approximate chemical shift (ppm)**		**Type of carbon**	**Approximate chemical shift (ppm)**
$(CH_3)_4Si$	0		C—I	0–40
R—CH$_3$	8–35		C—Br	25–65
R—CH$_2$—R	15–50		C—Cl / C—N / C—O	35–80 / 40–60 / 50–80
R—CH—R (R)	20–60		R₂N—C=O	165–175
R—C—R (R,R)	30–40		RO—C=O	165–175
C≡C	65–85		HO—C=O	175–185
C=C	100–150		H—C=O	190–200
(benzene) C	110–170		R—C=O (R)	205–220

A disadvantage of ^{13}C NMR spectroscopy is that, unless special techniques are used, the area under a ^{13}C NMR signal is *not* proportional to the number of carbons giving rise to the signal. Thus, the number of carbons giving rise to a ^{13}C NMR signal cannot routinely be determined by integration.

The ^{13}C NMR spectrum of 2-butanol is shown in Figure 13.34. 2-Butanol has carbons in four different environments, so there are four signals in the spectrum.

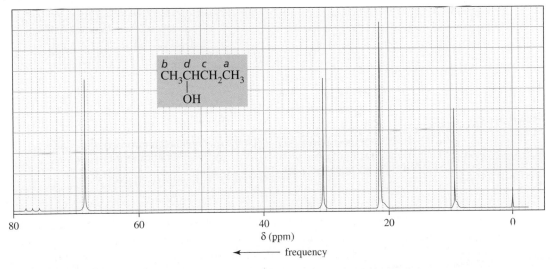

▲ **Figure 13.34**
A ^{13}C NMR spectrum of 2-butanol.

The relative positions of the signals depend on the same factors that determine the relative positions of the proton signals in an ^1H NMR spectrum. Carbons in electron-dense environments produce low-frequency signals, and carbons close to electron-withdrawing groups produce high-frequency signals. This means that the signals for the carbons of 2-butanol are in the same relative order as the signals of the protons bonded to those carbons in the ^1H NMR spectrum. Thus, the carbon of the methyl group farthest away from the electron-withdrawing OH group gives the lowest-frequency signal. The other methyl carbon comes next in order of increasing frequency, followed by the methylene carbon; the carbon attached to the OH group gives the highest-frequency signal.

The signals in ^{13}C NMR are not normally split by neighboring carbons because there is little likelihood of an adjacent carbon being a ^{13}C. ^{12}C, not having a magnetic moment, cannot split the signal of an adjacent ^{13}C, and the probability of two ^{13}C carbons being next to each other is $1.11\% \times 1.11\%$ (about 1 in 10,000). The signals in a ^{13}C NMR spectrum can be split by nearby hydrogens, but such splitting is not usually observed because the spectra are recorded using spin-decoupling, which obliterates the carbon–proton interactions. Thus, in an ordinary ^{13}C NMR spectrum, all the signals are singlets (Figure 13.34).

If the spectrometer is run in a *proton-coupled* mode, however, each signal will be split by the *hydrogens* bonded to the carbon that produces the signal. The multiplicity of the signal is determined by the $N + 1$ rule. The **proton-coupled ^{13}C NMR spectrum** of 2-butanol is shown in Figure 13.35. (The triplet at 78 ppm is produced by the solvent, $CDCl_3$.) The signals for the methyl carbons are each split into a quartet because each methyl carbon is bonded to three hydrogens ($3 + 1 = 4$). The signal for the methylene carbon is split into a triplet ($2 + 1 = 3$), and the signal for the carbon bonded to the OH group is split into a doublet ($1 + 1 = 2$).

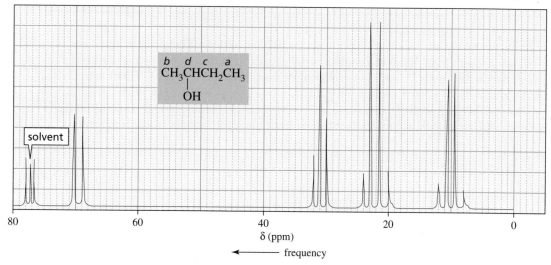

▲ **Figure 13.35**
A proton-coupled ^{13}C NMR spectrum of 2-butanol. If the spectrometer is run in a proton-coupled mode, splitting by the directly attached protons is observed in a ^{13}C NMR spectrum.

The ^{13}C NMR spectrum of 2,2-dimethylbutane is shown in Figure 13.36. The three methyl groups at one end of the molecule are equivalent, so they give one signal. Because the intensity of a signal is somewhat related to the number of carbons giving rise to it, the signal for these three methyl groups is the most intense

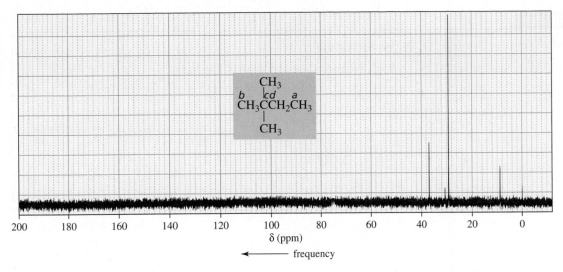

▲ Figure 13.36
A ^{13}C NMR spectrum of 2,2-dimethylbutane.

signal in the spectrum. The tiny signal at ≤31 ppm is for the quaternary carbon; carbons that are not attached to hydrogens give very small signals.

PROBLEM 38

Answer the following questions for each of the compounds:

a. How many signals are in its ^{13}C NMR spectrum?
b. Which signal is at the lowest frequency?

1. $CH_3CH_2CH_2Br$

4. $CH_3CH_2\overset{\displaystyle O}{\overset{\|}{C}}OCH_3$

7. $CH_3\overset{\displaystyle CH_3}{\underset{\displaystyle CH_3}{\overset{|}{\underset{|}{C}}}}OCH_3$

2. $(CH_3)_2C\!=\!CH_2$

5. $CH_3\overset{\displaystyle O}{\overset{\|}{C}}H\overset{\|}{C}H$
$\underset{\displaystyle CH_3}{|}$

8. $CH_3\overset{\displaystyle O}{\overset{\|}{C}}CH_2CH_2\overset{\displaystyle O}{\overset{\|}{C}}CH_3$

3. $CH_3\underset{\displaystyle Br}{\overset{\displaystyle |}{C}}HCH_3$

6.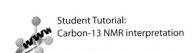

9. $CH_2\!=\!CHBr$

Student Tutorial:
Carbon-13 NMR interpretation

PROBLEM 39

Describe the proton-coupled ^{13}C NMR spectrum for compounds 1, 3, and 5 in Problem 38, showing relative values (not absolute values) of chemical shifts.

PROBLEM 40

How can 1,2-, 1,3-, and 1,4-dinitrobenzene be distinguished by

a. ^{1}H NMR spectroscopy?   **b.** ^{13}C NMR spectroscopy?

PROBLEM-SOLVING STRATEGY

Deducing a Chemical Structure from a ^{13}C NMR Spectrum

Identify the compound with molecular formula $C_9H_{10}O_2$ that gives the following ^{13}C NMR spectrum:

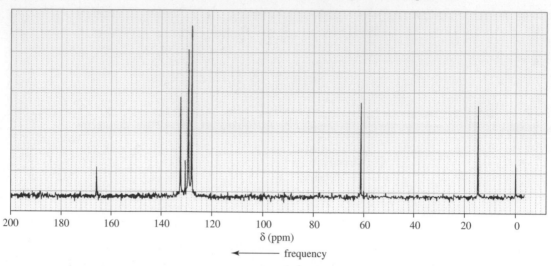

First, pick out the signals that can be identified with certainty. For example, the two oxygen atoms in the molecular formula and the carbonyl carbon signal at 166 ppm indicate that the compound is an ester. The four signals at about 130 ppm suggest that the compound has a benzene ring with a single substituent. (One signal is for the carbon to which the substituent is attached, one signal is for the two adjacent carbons, and so on.) Subtracting those fragments (C_6H_5 and CO_2) from the molecular formula of the compound leaves C_2H_5, the molecular formula of an ethyl substituent. Therefore, we know that the compound is either ethyl benzoate or phenyl propanoate.

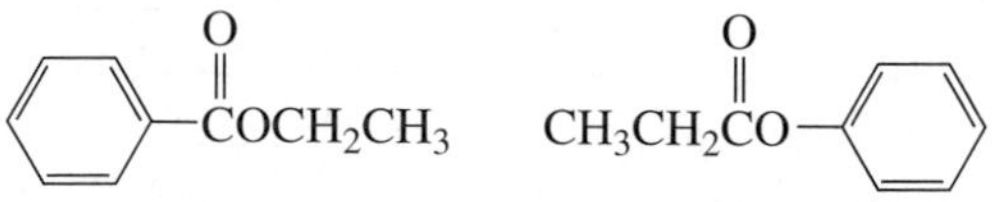

ethyl benzoate   **phenyl propanoate**

Since the signal for the methylene group is at about 60 ppm, we can conclude that it is adjacent to an oxygen. Thus, the compound must be ethyl benzoate.

Now continue on to Problem 41.

PROBLEM 41◆

Identify each compound in Figure 13.37 from its molecular formula and its ^{13}C NMR spectrum.

  a. $C_{11}H_{22}O$

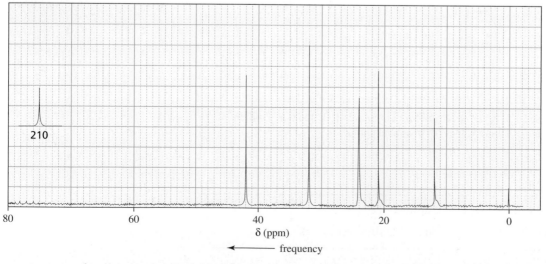

b. C_8H_9Br

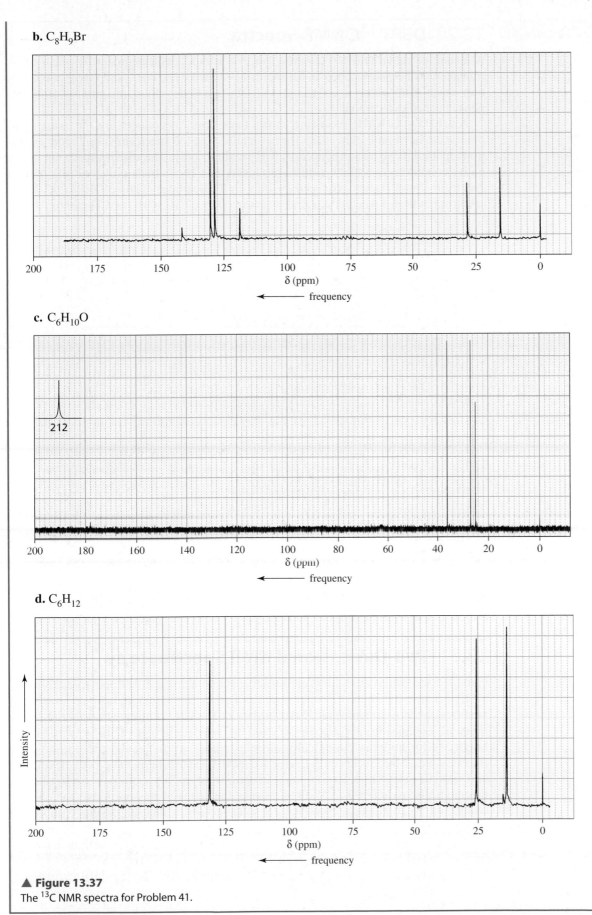

c. $C_6H_{10}O$

d. C_6H_{12}

▲ **Figure 13.37**
The ^{13}C NMR spectra for Problem 41.

13.20 DEPT ^{13}C NMR Spectra

A technique called DEPT ^{13}C NMR (for *distortionless enhancement by polarization transfer*) has been developed to distinguish between CH_3, CH_2, and CH groups. It is now much more widely used than proton coupling to determine the number of hydrogens attached to a carbon.

A **DEPT ^{13}C NMR** recording shows three spectra produced by the same compound, as shown in Figure 13.38 for citronellal. The top spectrum is run under conditions that allow signals produced by CH_3 carbons only. The middle spectrum is run under conditions that allow signals produced by CH_2 carbons only, and the bottom spectrum shows signals produced by only by CH carbons. Thus a DEPT ^{13}C NMR spectrum indicates whether a signal is produced by a CH_3, CH_2, or CH carbon.

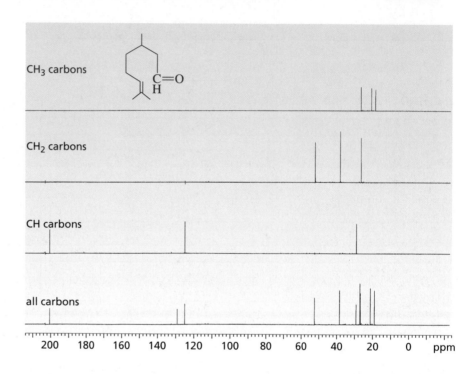

Figure 13.38 ▶
The DEPT ^{13}C NMR spectrum of citronellal.

A fourth spectrum can be run that shows signals for all carbons, thus allowing carbons not bonded to hydrogens to be detected.

13.21 Two-Dimensional NMR Spectroscopy

Student Tutorial:
2-D NMR

Complex molecules such as proteins and nucleic acids are difficult to analyze by NMR because the signals in their spectra overlap. Such compounds are now being analyzed by two-dimensional (2-D) NMR. **2-D NMR** techniques allow the structures of complex molecules in solution to be determined. This is particularly important for studying biological molecules whose properties depend on how they fold in water. More recently, 3-D and 4-D NMR spectroscopy have been developed and can be used to determine the structures of highly complex molecules. A thorough discussion of 2-D NMR is beyond the scope of this book, but the chapter would not be complete without a brief introduction to this increasingly important spectroscopic technique.

The ^{1}H NMR and ^{13}C NMR spectra discussed in the preceding sections have one frequency axis and one intensity axis; 2-D NMR spectra have two frequency axes and one intensity axis. The most common 2-D spectra involve ^{1}H–^{1}H shift correlations, which identify protons that are coupled (that split each other's signal). This is called ^1H–^1H shift-correlated spectroscopy, known by the acronym COSY.

A portion of the **COSY spectrum** of ethyl vinyl ether is shown in Figure 13.39a; it looks like a mountain range viewed from the air because intensity is the third axis. These "mountain-like" spectra, called *stack plots*, are not the spectra actually used to identify a compound. Instead, the compound is identified using a contour plot (Figure 13.37b) that represents each mountain in Figure 13.39a by a large dot (as if its top had been cut off). The two mountains shown in Figure 13.39a correspond to the dots labeled Y and Z in Figure 13.39b.

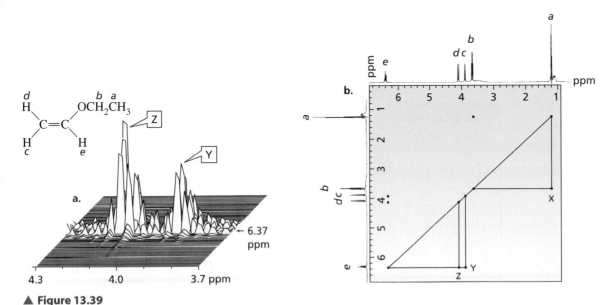

▲ **Figure 13.39**
(a) COSY spectrum of ethyl vinyl ether (stack plot).
(b) COSY spectrum of ethyl vinyl ether (contour plot). In a COSY spectrum, an ^1H NMR spectrum is plotted on both the *x*- and *y*-axes. Cross peaks Y and Z represent the mountains in (a).

In the contour plot (Figure 13.39b), the usual one-dimensional ^1H NMR spectrum is plotted on both the *x*- and *y*-axes. To analyze the spectrum, a diagonal line is drawn through the dots that bisect the spectrum. Dots that are *not* on the diagonal (X, Y, Z) are called *cross peaks*; they reveal pairs of protons that are coupled. For example, if we start at the cross peak labeled X and draw a straight line from X to the diagonal and parallel to the *y*-axis, we hit the dot on the diagonal at ~1.1 ppm produced by the H_a protons. If we then go back to X and draw a straight line from X to the diagonal and parallel to the *x*-axis, we hit the dot on the diagonal at ~3.8 ppm produced by the H_b protons. This means that the H_a and H_b protons are coupled. If we go next to the cross peak labeled Y and draw another two perpendicular lines back to the diagonal, we see that the H_c and H_e protons are coupled. Similarly, the cross peak labeled Z shows that the H_d and H_e protons are coupled. Although we have been using cross peaks below the diagonal as examples, the cross peaks above the diagonal give the same information. The absence of a cross peak due to coupling of H_c and H_d is consistent with expectations—two protons bonded to an sp^2 carbon are not coupled in the one-dimensional ^1H NMR spectrum shown in Figure 13.17. The power of a COSY experiment is that it reveals coupled protons without having to do coupling constant analysis.

The COSY spectrum of 1-nitropropane is shown in Figure 13.40. Cross peak X shows that the H_a and H_b protons are coupled, and cross peak Y shows that the H_b and H_c protons are coupled. Notice that the two triangles in the figure have a common vertex, since the H_b protons are coupled to both the H_a and H_c protons.

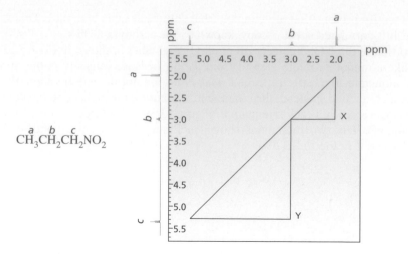

Figure 13.40 ▶
The COSY spectrum of 1-nitropropane.

$$\overset{a}{C}H_3\overset{b}{C}H_2\overset{c}{C}H_2NO_2$$

PROBLEM 42

Identify pairs of coupled protons in 2-methyl-3-pentanone, using the COSY spectrum in Figure 13.41.

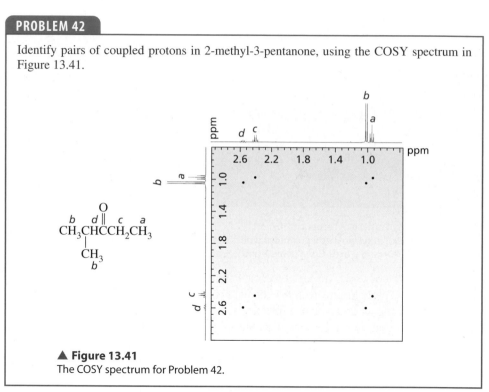

$$\overset{b}{C}H_3\overset{d}{C}H\overset{O}{\underset{\underset{\underset{b}{CH_3}}{|}}{\overset{\|}{C}}}\overset{c}{C}H_2\overset{a}{C}H_3$$

▲ **Figure 13.41**
The COSY spectrum for Problem 42.

HETCOR spectra (from *het*eronuclear *cor*relation) are 2-D NMR spectra that show $^{13}C-^{1}H$ shift correlations and thus reveal coupling between protons and the carbon to which they are attached.

In a HETCOR spectrum, a compound's ^{13}C NMR spectrum is shown on the *x*-axis, and its ^{1}H NMR spectrum is shown on the *y*-axis. The HETCOR spectrum of 2-methyl-3-pentanone is shown in Figure 13.42. The cross peaks identify which hydrogens are attached to which carbons. For example, in the HETCOR spectrum of 2-methyl-3-pentanone (Figure 13.42), cross peak X indicates that the hydrogens that produce a signal at ~0.9 ppm in the ^{1}H NMR spectrum are bonded to the carbon that produces a signal at ~6 ppm in the ^{13}C NMR spectrum. Cross peak Z shows that the hydrogens that produce a signal at ~2.5 ppm are bonded to the carbon that produces a signal at ~34 ppm.

Clearly, 2-D NMR techniques are not necessary for interpreting the NMR spectra of simple compounds such as 2-methyl-3-pentanone. However, in the case of many complicated molecules, signals cannot be assigned without the aid of 2-D NMR. These techniques include the use of 2-D ^{13}C INADEQUATE spectra, which show $^{13}C-^{13}C$

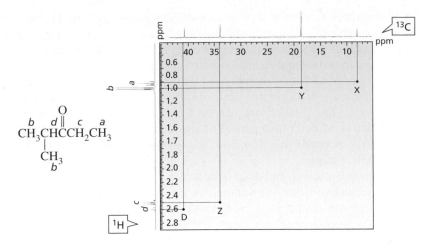

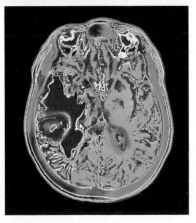

◀ Figure 13.42
HETCOR spectrum of 2-methyl-3-pentanone. A HETCOR spectrum indicates coupling between protons and the carbons to which they are attached.

shift correlations and thus identify directly bonded carbons; another technique plots chemical shifts on one frequency axis and coupling constants on the other; and techniques that use the nuclear Overhauser effect (NOESY for very large molecules, ROESY for mid-size molecules)* to locate protons that are close together in space.

13.22 NMR Used in Medicine Is Called Magnetic Resonance Imaging

NMR has become an important tool in medical diagnosis because it allows physicians to examine internal organs and structures without resorting to surgery or the harmful ionizing radiation of X-rays. When NMR was first introduced into clinical practice in 1981, the selection of an appropriate name was a matter of some debate. Because many non-scientists associate the word "nuclear" with harmful radiation (radioactivity), the "N" was dropped from the name for the medical application of NMR, which became known as **magnetic resonance imaging (MRI)**. The spectrometer is called an **MRI scanner**.

An MRI scanner consists of a magnet large enough to surround a person entirely, along with an apparatus for exciting the nuclei, modifying the magnetic field, and receiving signals. (By comparison, the NMR spectrometer used by chemists is only large enough to accommodate a 5-mm glass tube.) Different tissues yield different signals, which are separated into components by Fourier transform analysis. Each component can be attributed to a specific site of origin, allowing a cross-sectional image of the person's body to be constructed.

MRI can produce an image showing any cross-section of the body, irrespective of the person's position within the machine, allowing optimal visualization of the anatomical feature of interest. By contrast, the cross-section provided by a computed tomography (CT) scan, which uses X-rays, is defined by the person's position within the machine and is usually perpendicular to the long axis of the body. CT scans in other planes may be obtained only if the person is a skilled contortionist.

Most of the signals in an MRI scan originate from the hydrogens of water molecules because tissues contain far more of these hydrogens than they do hydrogens of organic compounds. The difference in the way water is bound in different tissues is what produces much of the variation in signal between different organs, as well as the variation between healthy and diseased tissue (Figure 13.43). MRI scans, therefore, can sometimes provide much more information than images obtained by other means. For example, MRI can provide detailed images of blood vessels. Flowing fluids, such as blood, respond differently to excitation in an MRI scanner than do stationary tissues, and normally do not produce a signal. However, the data may be processed to eliminate signals from motionless structures, thereby showing signals only from the fluids. This

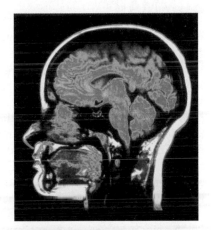

▲ Figure 13.43
(a) An MRI of a normal brain. The pituitary is highlighted (pink).
(b) An MRI of an axial section through a brain showing a tumor (purple) surrounded by damaged, fluid-filled tissue (red).

---

* NOESY and ROESY are the acronyms for nuclear Overhauser effect spectroscopy and rotation-frame Overhauser effect spectroscopy, respectively.

technique is sometimes used instead of more invasive methods to examine the vascular tree. It is now possible to completely suppress the signal from certain types of tissue (usually fat). It is also possible to differentiate intracellular and extracellular edema, which is important in assessing patients suspected of having suffered strokes.

The versatility of MRI has been enhanced by the use of gadolinium as a contrast reagent. Gadolinium, a paramagnetic metal, modifies the magnetic field in its immediate vicinity, altering the signal from nearby hydrogen nuclei. The distribution of gadolinium, which is infused into a patient's veins, may be affected by certain disease processes, such as cancer and inflammation. Any abnormal patterns of distribution are revealed in the MRI images.

Another promising research and diagnostic tool is ^{31}P NMR spectroscopy. The clinical use of this technique is not yet routine, but it is widely used in clinical research. Because the phosphorus-containing molecules ATP and ADP (Sections 16.22 and 25.2) are involved in most metabolic processes, ^{31}P NMR provides a way to investigate cellular metabolism.

SUMMARY

NMR spectroscopy is used to identify the carbon–hydrogen framework of an organic compound. When a sample is placed in a magnetic field, protons aligning with the field are in the lower-energy α-spin state, those aligning against the field are in the higher-energy β-spin state. The energy difference between the spin states depends on the strength of the **applied magnetic field**. When subjected to radiation with energy corresponding to the energy difference between the spin states, nuclei in the α-spin state are promoted to the β-spin state and emit signals whose frequency depends on the difference in energy between the spin states. An **NMR spectrometer** detects and displays these signals as a plot of their frequency versus their intensity—an **NMR spectrum**.

Each set of chemically equivalent protons gives rise to a signal, so the number of signals in an ^1H NMR spectrum indicates the number of different kinds of protons in a compound. The **chemical shift** is a measure of how far the signal is from the reference TMS signal. The chemical shift (δ) is independent of the operating frequency of the spectrometer.

The larger the magnetic field sensed by the proton, the higher is the frequency of the signal. The electron density of the environment in which the proton is located **shields** the proton from the applied magnetic field. Therefore, a proton in an electron-dense environment shows a signal at a lower frequency than a proton near electron-withdrawing groups. Low-frequency (upfield) signals have small δ (ppm) values; high-frequency (downfield) signals have large δ values. Thus, the position of a signal indicates the kind of proton(s) responsible for the signal and the kinds of neighboring substituents. In a similar environment, the chemical shift of methyl protons is at a lower frequency than that of methylene protons, which in turn is at a lower frequency than that of a methine proton. **Diamagnetic anisotropy** causes unusual chemical shifts for hydrogens

bonded to carbons that form π bonds. **Integration** tells us the relative number of protons that give rise to each signal.

The **multiplicity** of a signal (the number of peaks in the signal) indicates the number of protons bonded to adjacent carbons. Multiplicity is described by the **$N + 1$ rule**, where N is the number of equivalent protons bonded to adjacent carbons. A **splitting diagram** can help us understand the splitting pattern obtained when a signal is split by more than one set of protons. Deuterium substitution can be a helpful technique in the analysis of complicated ^1H NMR spectra.

The **coupling constant** (J) is the distance between two adjacent peaks of a split NMR signal. Coupling constants are independent of the operating frequency of the spectrometer. Coupled protons have the same coupling constant. The coupling constant for trans alkene protons is greater than that for cis alkene protons. When two different sets of protons split a signal, the multiplicity of the signal is determined by using the $N + 1$ rule separately for each set of hydrogens when the coupling constants for the two sets are different. When the coupling constants are similar, the $N + 1$ rule can be applied to both sets simultaneously.

The chemical shift of a proton bonded to an O or to an N depends on the degree to which the proton is hydrogen bonded. In the presence of trace amounts of acid or base, protons bonded to oxygen undergo **proton exchange**. In that case, the signal for a proton bonded to an O is not split and does not split the signal of adjacent protons.

The number of signals in a ^{13}C NMR spectrum tells how many kinds of carbons a compound has. Carbons in electron-dense environments produce low-frequency signals; carbons close to electron-withdrawing groups produce high-frequency signals. Chemical shifts for ^{13}C NMR range over about 220 ppm, compared with about 12 ppm for ^1H NMR. ^{13}C NMR signals are not split by attached protons unless the spectrometer is run in a proton-coupled mode.

KEY TERMS

applied magnetic field (p. 570)
chemically equivalent
    protons (p. 574)
chemical shift (p. 577)
^{13}C NMR (p. 610)
COSY spectrum (p. 616)
coupled protons (p. 586)
coupling constant (J) (p. 597)
DEPT ^{13}C NMR spectrum (p. 616)
diamagnetic anisotropy (p. 582)
diamagnetic shielding (p. 573)
diastereotopic hydrogens (p. 603)
2-D NMR (p. 616)
doublet (p. 586)
doublet of doublets (p. 592)
downfield (p. 574)
effective magnetic field (p. 573)
enantiotopic hydrogens (p. 603)

Fourier transform NMR (FT–NMR)
    spectrum (p. 573)
geminal coupling (p. 593)
gyromagnetic ratio (p. 571)
HETCOR spectrum (p. 618)
^{1}H NMR (proton magnetic
    resonance) (p. 570)
integration (p. 584)
long-range coupling (p. 589)
magnetic resonance imaging (MRI) (p. 619)
MRI scanner (p. 619)
multiplet (p. 592)
multiplicity (p. 586)
$N + 1$ rule (p. 586)
NMR spectrometer (p. 571)
NMR spectroscopy (p. 569)
NMR spectrum (p. 572)
operating frequency (p. 571)

prochiral carbon (p. 603)
pro-R-hydrogen (p. 603)
pro-S-hydrogen (p. 603)
proton-coupled ^{13}C NMR
    spectrum (p. 612)
proton exchange (p. 606)
quartet (p. 586)
reference compound (p. 577)
rf radiation (p. 571)
shielding (p. 573)
singlet (p. 586)
α-spin state (p. 570)
β-spin state (p. 570)
spin–spin coupling (p. 587)
splitting diagram (p. 600)
splitting tree (p. 600)
triplet (p. 586)
upfield (p. 574)

PROBLEMS

43. How many signals are produced by each of the following compounds in its
   a. ^{1}H NMR spectrum?   **b.** ^{13}C NMR spectrum?

1. CH_3—⬡—$OCHCH_3$
                 CH_3

2. ⬡—C(=O)—OCH_2CH_3

3. (cyclohexanone lactone, with =O and —O)

4. (oxetane with O)

5. (chlorocyclopropane, Cl)

6. (2,5-dimethyl-2,4-hexadiene structure)

44. Draw a splitting diagram for the H_b proton and give its multiplicity if
   a. $J_{ba} = J_{bc}$ **b.** $J_{ba} = 2J_{bc}$

$$H_a-\overset{\overset{\displaystyle H_a}{|}}{\underset{\underset{\displaystyle H_a}{|}}{C}}-\overset{\overset{\displaystyle X}{|}}{\underset{\underset{\displaystyle H_b}{|}}{C}}-\overset{\overset{\displaystyle X}{|}}{\underset{\underset{\displaystyle H_c}{|}}{C}}-X$$

45. Label each set of chemically equivalent protons, using a for the set that will be at the lowest frequency (farthest upfield) in the ^{1}H NMR spectrum, b for the next lowest, and so on. Give the multiplicity of each signal.

  a. CH_3CHNO_2
           $|$
          CH_3

  b. $CH_3CH_2CH_2OCH_3$

  c. CH_3CH—C(=O)—$CH_2CH_2CH_3$
           $|$
        CH_3

  d. $CH_3CH_2CH_2$—C(=O)—CH_2Cl

                 CH_3
  e. $ClCH_2\overset{|}{\underset{|}{C}}CHCl_2$
                 CH_3

  f. $ClCH_2CH_2CH_2CH_2CH_2Cl$

46. Match each of the ^{1}H NMR spectra with one of the following compounds:

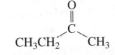
CH_3CH_2—C(=O)—CH_3

CH_3CNO_2 (with CH_3 above and CH_3 below)

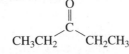
CH_3CH_2—C(=O)—CH_2CH_3

$CH_3CH_2CH_2NO_2$ $CH_3CH_2NO_2$ CH_3CHBr (with CH_3 above)

a.

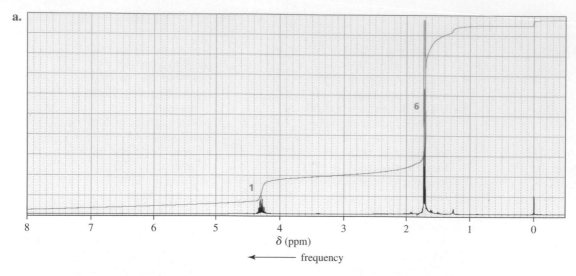

b.

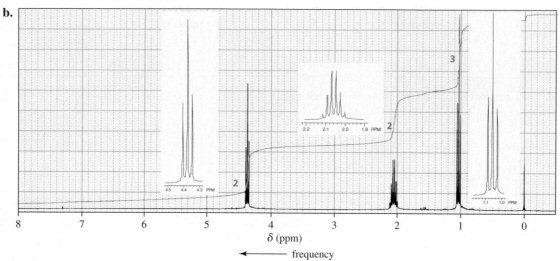

c.

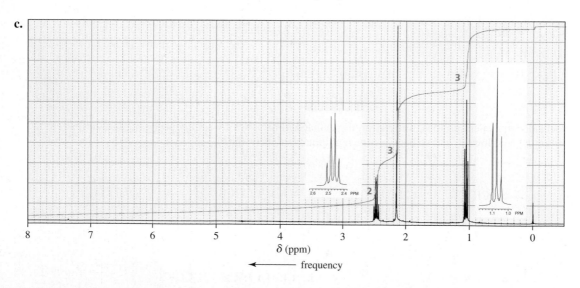

47. Determine the ratios of the chemically nonequivalent protons in a compound if the steps of the integration curves measure 40.5, 27, 13, and 118 mm, from left to right across the spectrum. Give the structure of a compound whose ^{1}H NMR spectrum would show these integrals in the observed order.

48. How could ^{1}H NMR distinguish between the compounds in each of the following pairs?

    a. $CH_3CH_2CH_2OCH_3$ and $CH_3CH_2OCH_2CH_3$ **b.** $BrCH_2CH_2CH_2Br$ and $BrCH_2CH_2CH_2NO_2$

c. CH₃CH—CHCH₃ and CH₃CCH₂CH₃
with CH₃ CH₃ substituents and CH₃

d. CH₃—C—C—OCH₃ and CH₃—C—CH₃
with CH₃O and CH₃ substituents, OCH₃ and OCH₃

e. CH₃O—⬡—CH₂CH₃ and CH₃—⬡—OCH₂CH₃

f. ⬡ and ⬡

g. CH₃CHCl and CH₃CDCl
with CH₃ and CH₃

h. H—CH₃ / D—H with Cl and Cl and D—CH₃ / H—H with Cl and Cl

i. triangle with H H / Cl Cl and triangle with H Cl / Cl H

j. CH₃—⬡—CCH₃ with CH₃ / CH₃ and ⬡—CH₂CCH₃ with CH₃ / CH₃

49. Answer the following questions:
 a. What is the relationship between chemical shift in ppm and operating frequency?
 b. What is the relationship between chemical shift in hertz and operating frequency?
 c. What is the relationship between coupling constant and operating frequency?
 d. How does the operating frequency in NMR spectroscopy compare with the operating frequency in IR and UV/Vis spectroscopy?

50. The ¹H NMR spectra of three isomers with molecular formula C₄H₉Br are shown here. Which isomer produces which spectrum?

a.

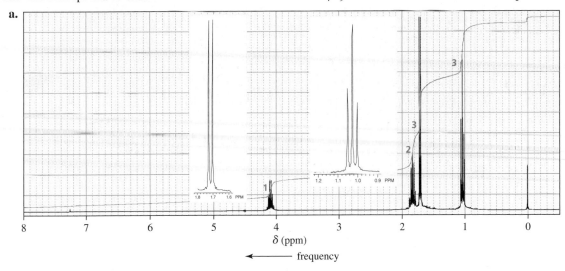

δ (ppm)

⟵ frequency

b.

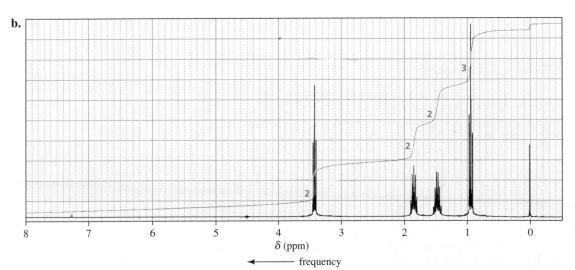

δ (ppm)

⟵ frequency

c.

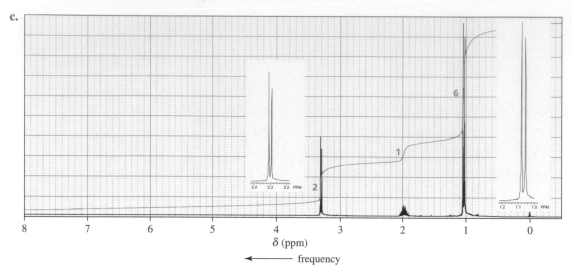

δ (ppm)

← frequency

51. Identify each of the following compounds from the ^{1}H NMR data and molecular formula. The number of hydrogens responsible for each signal is shown in parentheses.

   a. $C_4H_8Br_2$ 1.97 ppm (6) singlet   **b.** C_8H_9Br 2.01 ppm (3) doublet   **c.** $C_5H_{10}O_2$ 1.15 ppm (3) triplet
                3.89 ppm (2) singlet   5.14 ppm (1) quartet   1.25 ppm (3) triplet
                                        7.35 ppm (5) broad singlet   2.33 ppm (2) quartet
                                                                                     4.13 ppm (2) quartet

52. Identify the compound with molecular formula $C_7H_{14}O$ that gives the following proton-coupled ^{13}C NMR spectrum:

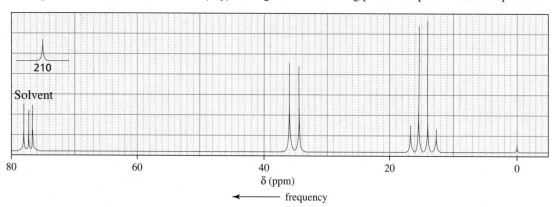

δ (ppm)

← frequency

53. Compound A, with molecular formula C_4H_9Cl, shows two signals in its ^{13}C NMR spectrum. Compound B, an isomer of compound A, shows four signals, and in the proton-coupled mode, the signal farthest downfield is a doublet. Identify compounds A and B.

54. The ^{1}H NMR spectra of three isomers with molecular formula $C_7H_{14}O$ are shown here. Which isomer produces which spectrum?

a.

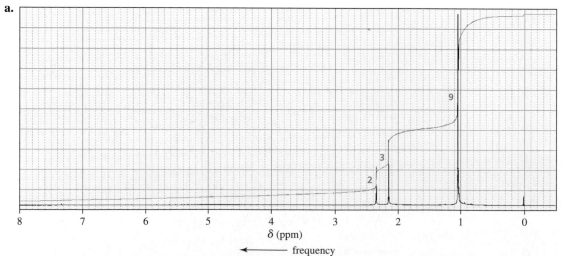

δ (ppm)

← frequency

b.

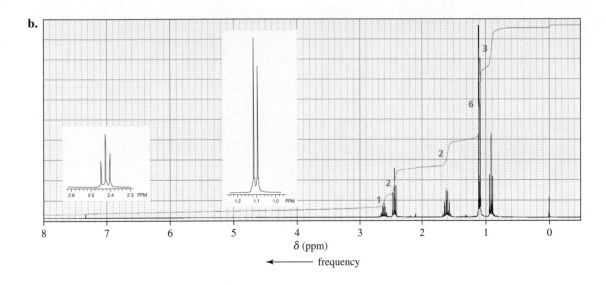

c.

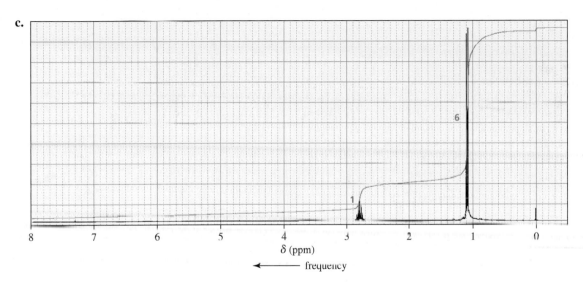

55. Would it be better to use ^{1}H NMR or ^{13}C NMR to distinguish between 1-butene, *cis*-2-butene, and 2-methylpropene? Explain your answer.

56. Determine the structure of each of the following unknown compounds based on its molecular formula and its IR and ^{1}H NMR spectra.

a. $C_5H_{12}O$

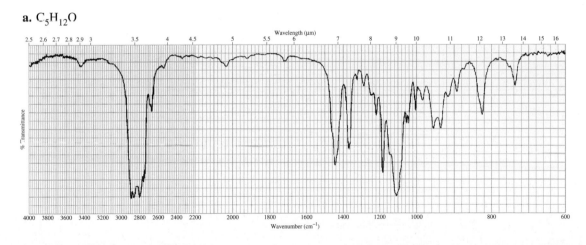

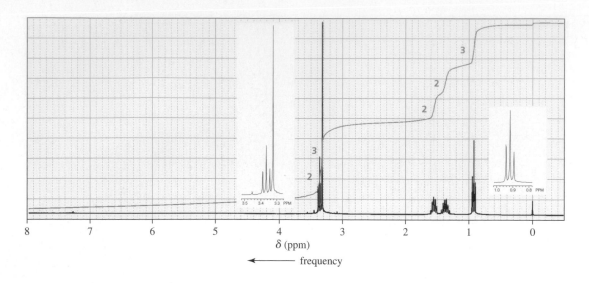

b. $C_6H_{12}O_2$

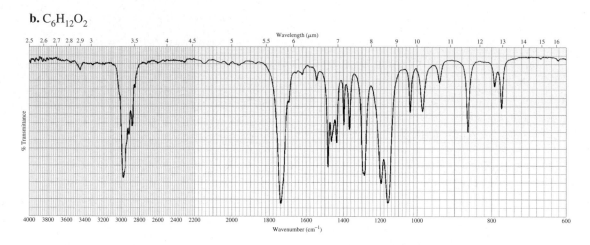

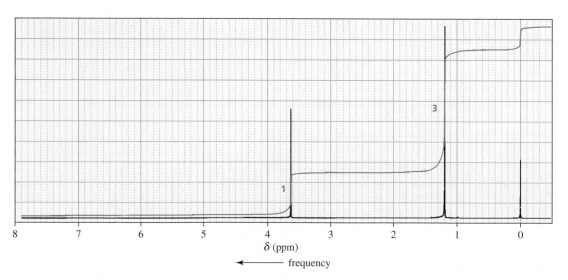

c. $C_4H_7ClO_2$

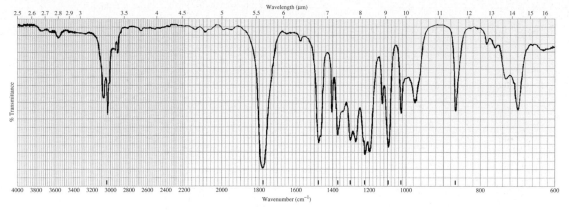

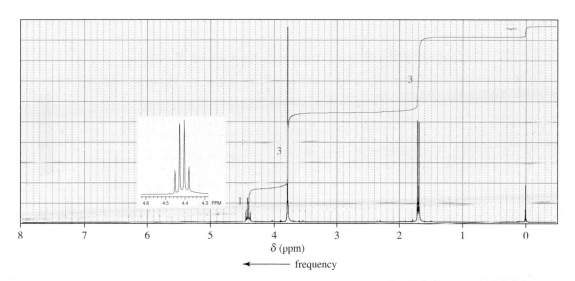

d. $C_4H_8O_2$

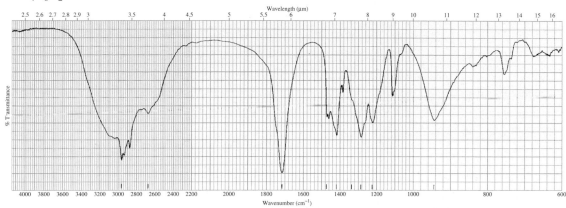

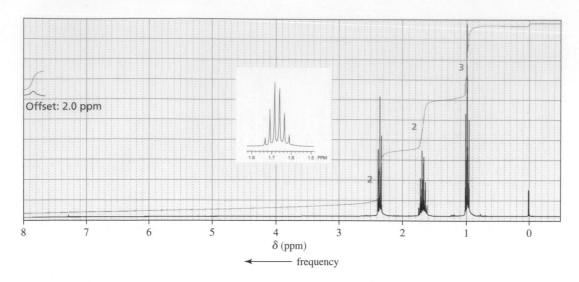

57. There are four esters with molecular formula $C_4H_8O_2$. How could they be distinguished by ^1H NMR?

58. An alkyl halide reacts with an alkoxide ion to form a compound whose ^1H NMR spectrum is shown here. Identify the alkyl halide and the alkoxide ion. (*Hint:* See Section 9.9.)

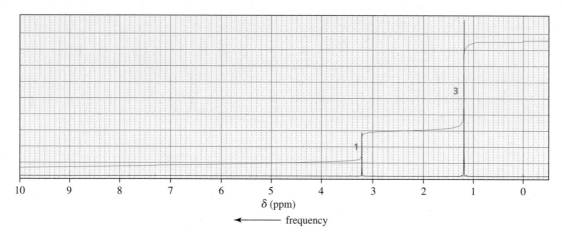

59. Determine the structure of each of the following compounds based on its molecular formula and its ^{13}C NMR spectrum:

   a. $C_4H_{10}O$

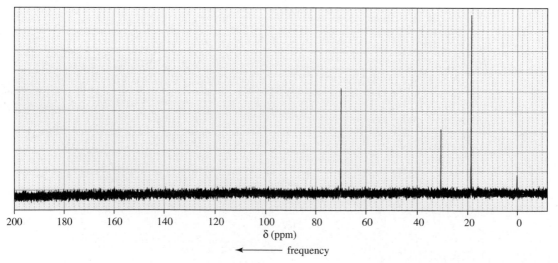

b. $C_6H_{12}O$

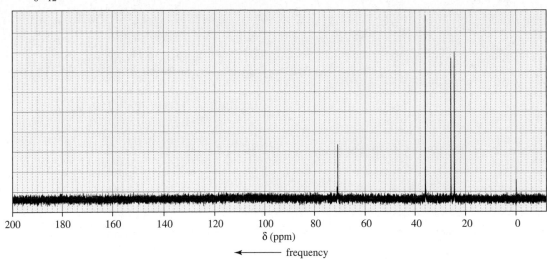

60. The ^1H NMR spectrum of 2-propen-1-ol is shown here. Indicate the protons in the molecule that give rise to each of the signals in the spectrum.

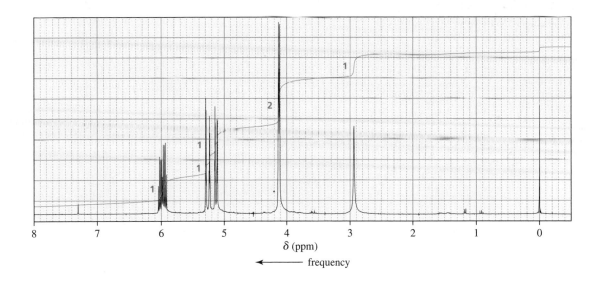

61. How could the signals in the 6.5 to 8.1-ppm region of their ^1H NMR spectra distinguish between the following compounds?

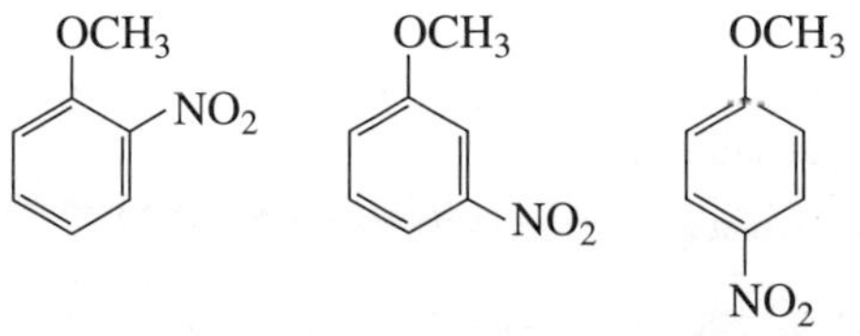

OCH_3
NO_2

OCH_3
NO_2

OCH_3
NO_2

62. The ^{1}H NMR spectra of two compounds, each with molecular formula $C_{11}H_{16}$, are shown here. Identify the compounds.

a.

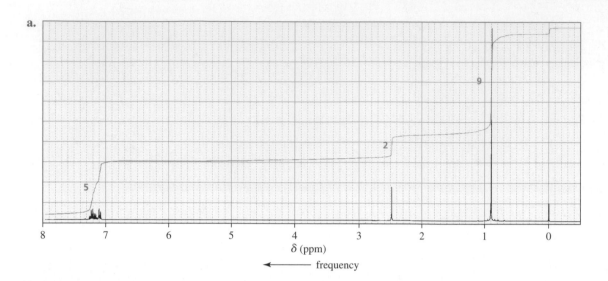

δ (ppm)

← frequency

b.

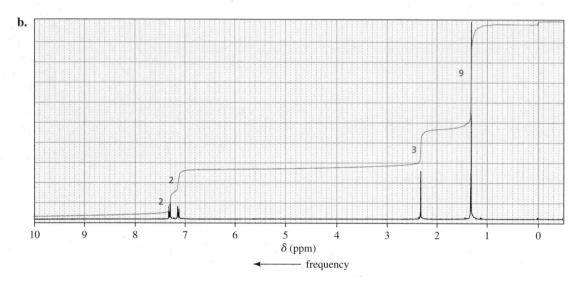

δ (ppm)

← frequency

63. Draw a splitting diagram for the H_b proton if $J_{bc} = 10$ and $J_{ba} = 5$.

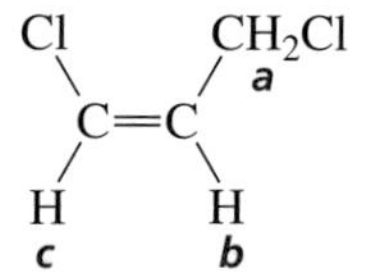

$$\underset{c}{H}\underset{}{\overset{Cl}{\diagdown}}C=C\overset{CH_2Cl}{\overset{a}{\diagup}}\underset{b}{H}$$

64. Sketch the following spectra that would be obtained for 2-chloroethanol:
   a. the ^{1}H NMR spectrum for a dry sample of the alcohol.
   b. the ^{1}H NMR spectrum for a sample of the alcohol that contains a trace amount of acid.
   c. the ^{13}C NMR spectrum.
   d. the proton-coupled ^{13}C NMR spectrum.
   e. the four parts of a DEPT ^{13}C NMR spectrum.

65. How could ^{1}H NMR be used to prove that the addition of HBr to propene follows the rule that says that the electrophile adds to the sp^2 carbon bonded to the greater number of hydrogens?

66. Identify each of the following compounds from its molecular formula and its ^1H NMR spectrum.

a. C_8H_8

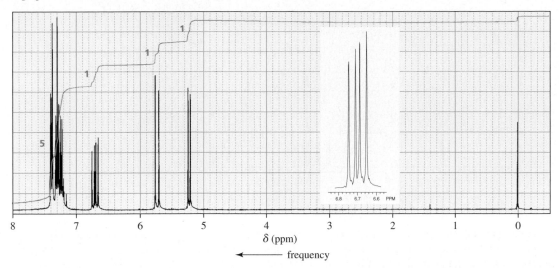

b. $C_6H_{12}O$

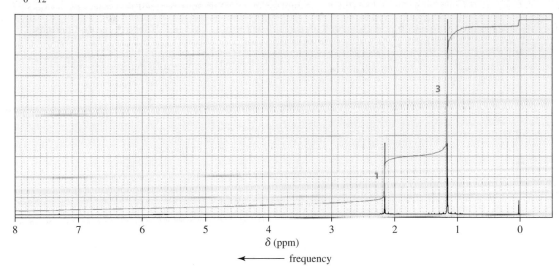

c. $C_9H_{18}O$

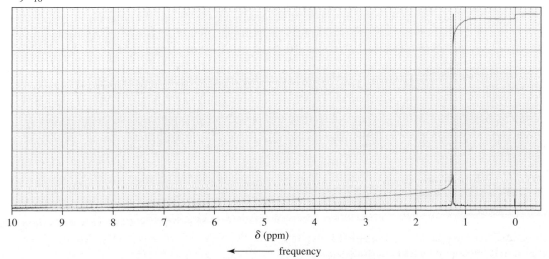

d. C_4H_8O

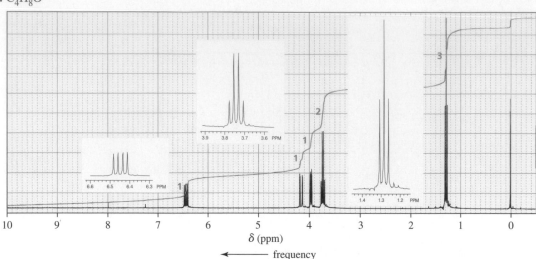

67. Dr. N. M. Arr was called in to help analyze the ^1H NMR spectrum of a mixture of compounds known to contain only C, H, and Br. The mixture showed two singlets—one at 1.8 ppm and the other at 2.7 ppm—with relative integrals of 1 : 6, respectively. Dr. Arr determined that the spectrum was that of a mixture of bromomethane and 2-bromo-2-methylpropane. What was the ratio of bromomethane to 2-bromo-2-methylpropane in the mixture?

68. Calculate the amount of energy (in calories) required to flip an ^1H nucleus in an NMR spectrometer that operates at 60 MHz.

69. The following ^1H NMR spectra are for four compounds each with molecular formula $C_6H_{12}O_2$. Identify the compounds.

a.

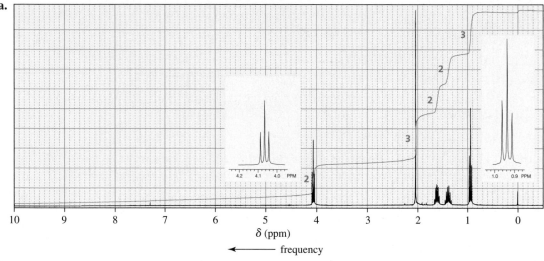

b.

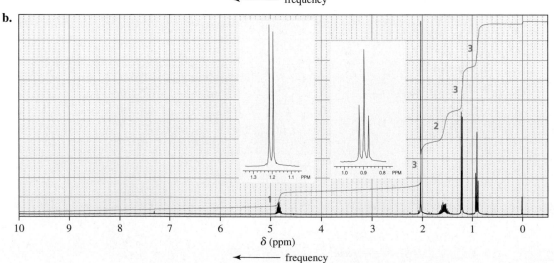

c.

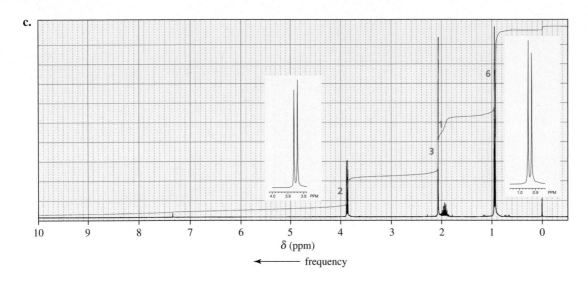

d.

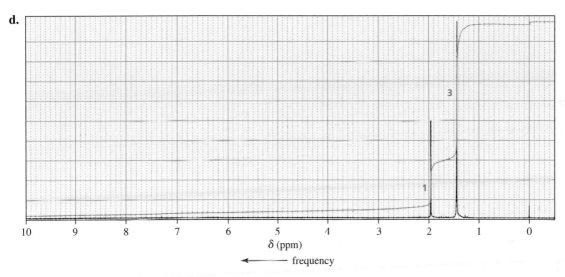

70. When compound A ($C_5H_{12}O$) is treated with HBr, it forms compound B ($C_5H_{11}Br$). The ^1H NMR spectrum of compound A has one singlet (1), two doublets (3, 6), and two multiplets (both 1). (The relative areas of the signals are indicated in parentheses.) The ^1H NMR spectrum of compound B has a singlet (6), a triplet (3), and a quartet (2). Identify compounds A and B.

71. Determine the structure of each of the following compounds, based on its molecular formula and its IR and ^1H NMR spectra.

a. $C_6H_{12}O$

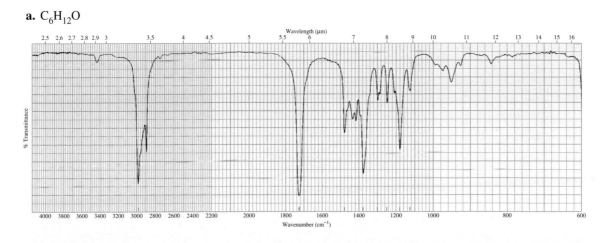

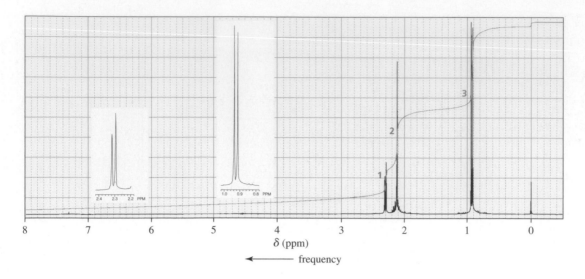

b. $C_6H_{14}O$

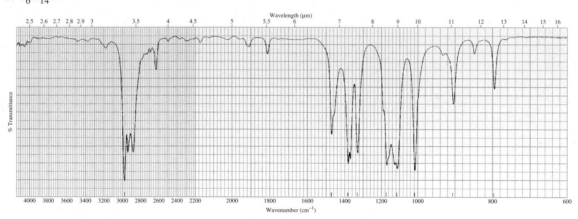

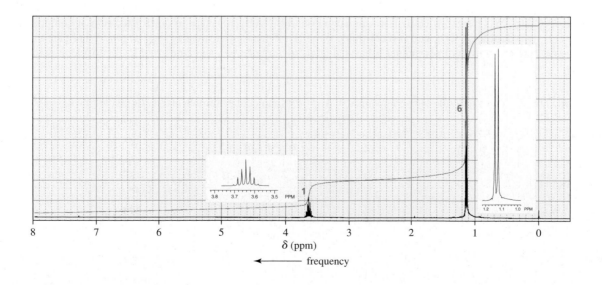

c. $C_{10}H_{13}NO_3$

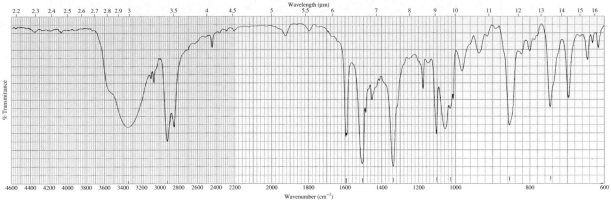

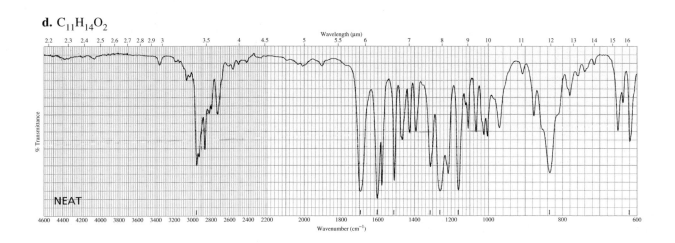

d. $C_{11}H_{14}O_2$

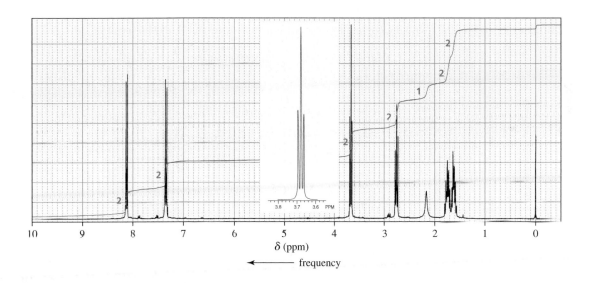

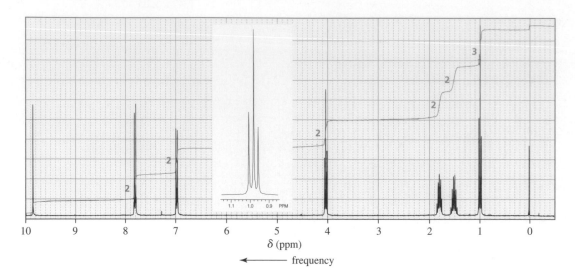

δ (ppm)

← frequency

72. Determine the structure of each of the following compounds, based on its mass, IR, and ^{1}H NMR spectra.

a.

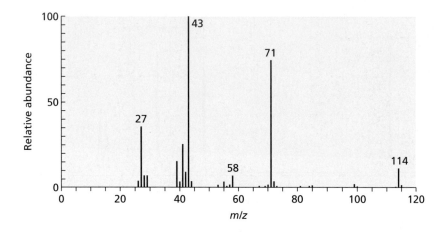

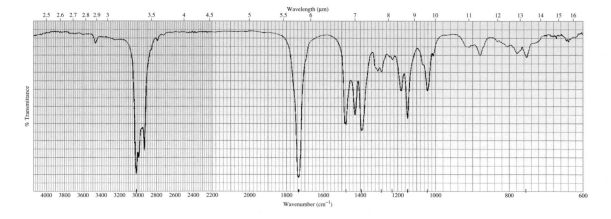

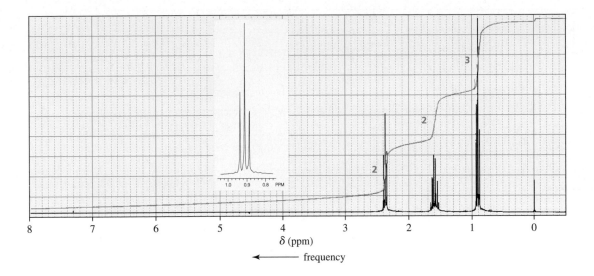

b.

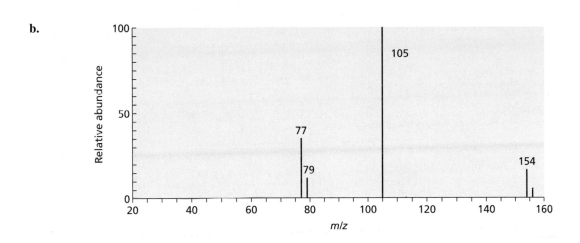

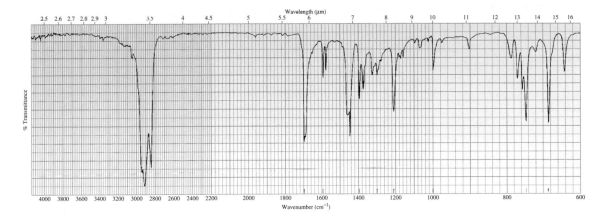

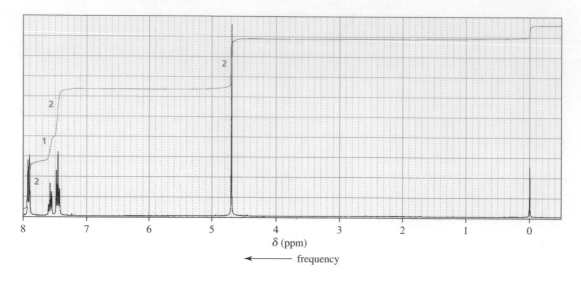

73. Identify the compound with molecular formula $C_6H_{10}O$ that is responsible for the following DEPT ^{13}C NMR spectrum:

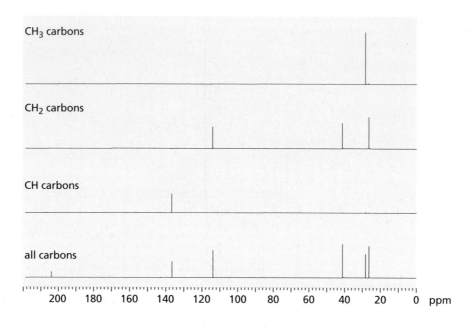

74. Identify the compound with molecular formula C_6H_{14} that is responsible for the following ^1H NMR spectrum:

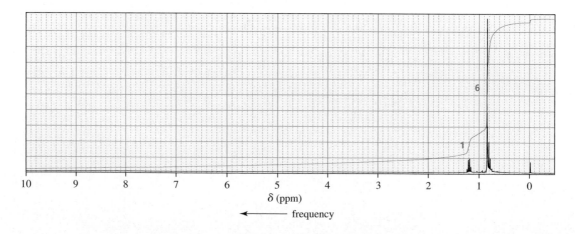

Aromatic Compounds

The two chapters in Part 5 deal with aromaticity and the reactions of aromatic compounds. Benzene is the most common aromatic compound. We looked at the structure of benzene in Chapter 7 and saw that it has a six-membered ring with three pairs of delocalized π electrons. Now we will look at the criteria that are used to classify a compound as aromatic, and then we will examine the kinds of reactions that aromatic compounds undergo. In Chapter 20, we will look at the reactions of aromatic compounds in which one of the ring atoms is an atom other than a carbon.

CHAPTER 14
Aromaticity • Reactions of Benzene

Chapter 14 examines the structural features that cause a compound to be aromatic, as well as the features that cause a compound to be antiaromatic. Then it looks at the reactions that benzene undergoes. You will see that although benzene, alkenes, and dienes are all nucleophiles (because they have carbon–carbon π bonds), benzene's aromaticity causes it to react in ways that are quite different from the reactions of alkenes and dienes. Once you have learned how to put a substituent on a benzene ring, you will look at some reactions that change the substituent.

CHAPTER 15
Reactions of Substituted Benzenes

Chapter 15 looks at the reactions of substituted benzenes. You will see how a substituent affects both the reactivity of a benzene ring and the placement of an incoming substituent. It also describes three types of reactions that can be used to synthesize substituted benzenes other than the reactions discussed in Chapter 14: reactions of arene diazonium salts, nucleophilic aromatic substitution reactions, and reactions that involve benzyne intermediates. At the chapter's end, you have the opportunity to design syntheses of compounds that contain benzene rings.

CHAPTER 14

Aromaticity • Reactions of Benzene

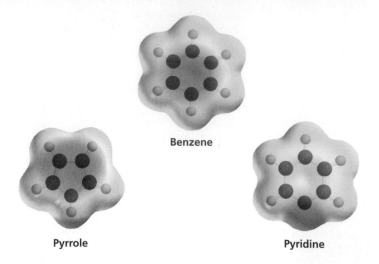

Benzene

Pyrrole

Pyridine

BUILDING ON FUNDAMENTALS

SECTION 14.1 The structure of benzene (7.2) is reviewed in preparation for a discussion of its stability and the kinds of reactions it undergoes.

SECTION 14.1 Because aromatic compounds have large delocalization energies (7.6), they are particularly stable compounds.

SECTION 14.9 A reaction coordinate diagram (3.7) shows why benzene undergoes electrophilic substitution reactions rather than electrophilic addition reactions (4.1).

SECTION 14.15 Because the alkylation of benzene forms a carbocation intermediate, we have to check for the possibility of a carbocation rearrangement (4.6) when determining the product of the reaction.

SECTION 14.17 Coupling reactions (10.13) are a useful way to synthesize alkyl-substituted benzenes.

BIOGRAPHY

Michael Faraday (1791–1867)
was born in England, a son of a blacksmith. At the age of 14, he was apprenticed to a bookbinder and educated himself by reading the books that he bound. He became an assistant to Sir Humphry Davy at the Royal Institution of Great Britain in 1812, where he taught himself chemistry. In 1825, he became director of the laboratory there, and, in 1833, a professor of chemistry. He is best known for his work on electricity and magnetism.

The compound we know as benzene was first isolated in 1825 by Michael Faraday. He extracted it from the liquid residue obtained after heating whale oil under pressure to produce the gas then used to illuminate buildings in London.

In 1834, Eilhardt Mitscherlich correctly determined the molecular formula (C_6H_6) of Faraday's compound and named it benzin because of its relationship to benzoic acid, a known substituted form of the compound. Later the name was changed to benzene.

Compounds like benzene that have relatively few hydrogens in relation to the number of carbons, are typically found in oils produced by trees and other plants. Early chemists called such compounds **aromatic compounds** because of their pleasing fragrances. In this way, they were distinguished from **aliphatic compounds**, which have higher hydrogen-to-carbon ratios. Today, chemists use the word "aromatic" to signify certain kinds of chemical structures. We will look at the features that cause a compound to be classified as aromatic in Section 14.2.

14.1 Aromatic Compounds Are Unusually Stable

In Chapter 7, we saw that benzene is a planar, cyclic compound with a cyclic cloud of delocalized electrons above and below the plane of the ring (Figure 14.1). Because its π electrons are delocalized, all the C—C bonds in benzene have

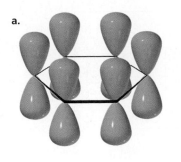
a.

b.

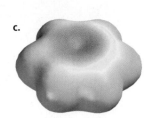
c.

◀ **Figure 14.1**
(a) Each carbon of benzene has a
p orbital. (b) The overlap of the
p orbitals forms a cloud of π electrons
above and below the plane of the
benzene ring. (c) The electrostatic
potential map for benzene shows that
all the carbon–carbon bonds have the
same electron density.

the same length, shorter than a typical single bond but longer than a typical double bond.

Benzene is a particularly stable compound because it has an unusually large delocalization energy. Most compounds with delocalized electrons have much smaller delocalization energies. Recall that delocalization energy (also called resonance energy) tells us how much more stable a compound with delocalized electrons is than it would be if all its electrons were localized (Section 7.6). Therefore, we can determine benzene's delocalization energy by comparing the stability of benzene—a compound with three pairs of delocalized π electrons—with the stability of "cyclohexatriene," an unknown, hypothetical compound with three localized π bonds.

The $\Delta H°$ for the hydrogenation of cyclohexene, a compound with one localized double bond, has been determined experimentally to be −28.6 kcal/mol. We would therefore expect the $\Delta H°$ for the hydrogenation of "cyclohexatriene," with three localized double bonds, to be three times that much, or $3 \times (-28.6) = -85.8$ kcal/mol (Section 4.11).

BIOGRAPHY

**Eilhardt Mitscherlich
(1794–1863)** *was born in
Germany. He studied oriental
languages at the University of
Heidelberg and the Sorbonne,
where he concentrated on Farsi,
hoping to be included in a
delegation Napoleon intended to
send to Persia. That ambition died
with Napoleon's defeat. After
returning to Germany to study
science, and simultaneously
receiving a doctorate in Persian
studies, Mitscherlich became a
professor of chemistry at the
University of Berlin.*

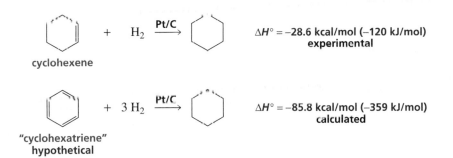

cyclohexene $\quad + \quad H_2 \quad \xrightarrow{\text{Pt/C}} \quad$ $\Delta H° = -28.6$ kcal/mol (−120 kJ/mol)
experimental

"cyclohexatriene"
hypothetical $\quad + \quad 3\,H_2 \quad \xrightarrow{\text{Pt/C}} \quad$ $\Delta H° = -85.8$ kcal/mol (−359 kJ/mol)
calculated

When the $\Delta H°$ for the hydrogenation of benzene was determined experimenally, it was found to be −49.8 kcal/mol, much less than that calculated for hypothetical "cyclohexatriene."

benzene $\quad + \quad 3\,H_2 \quad \xrightarrow{\text{Pt/C}} \quad$ $\Delta H° = -49.8$ kcal/mol (−208 kJ/mol)
experimental

Because the hydrogenation of "cyclohexatriene" and the hydrogenation of benzene both form cyclohexane, the difference in the $\Delta H°$ values can be accounted for only by a difference in the energies of "cyclohexatriene" and benzene. Figure 14.2 shows that benzene must be 36 kcal/mol (or 151 kJ/mole) more stable than "cyclohexatriene" because the experimental $\Delta H°$ for the hydrogenation of benzene is 36 kcal/mol less than that calculated for "cyclohexatriene."

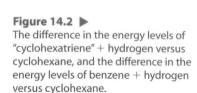

Figure 14.2 ▶
The difference in the energy levels of "cyclohexatriene" + hydrogen versus cyclohexane, and the difference in the energy levels of benzene + hydrogen versus cyclohexane.

Aromatic compounds are particularly stable.

We conclude from their different energies that benzene and "cyclohexatriene" are two different compounds, although the only difference in their structures is that benzene has six delocalized π electrons and "cyclohexatriene" has six localized π electrons. The difference in their energies is 36 kcal/mol. This is the delocalization energy of benzene—the extra stability a compound has as a result of having delocalized electrons. This is an unusually large delocalization energy.

Compounds with unusually large resonance energies, like benzene, are called **aromatic compounds**. Now we can understand why nineteenth-century chemists, who didn't know about delocalized electrons, were puzzled by benzene's unusual stability (Section 7.1).

14.2 The Two Criteria for Aromaticity

How can we tell whether a compound is aromatic by looking at its structure? In other words, what structural features do aromatic compounds have in common?

To be classified as aromatic, a compound must meet both of the following criteria:

For a compound to be aromatic, it must be cyclic and planar and have an uninterrupted cloud of π electrons. The π cloud must contain an odd number of pairs of π electrons.

1. *It must have an uninterrupted cyclic cloud of π electrons* (called a π cloud) *above and below the plane of the molecule.* Let's look a little more closely at what this means:

   For the π cloud to be cyclic, *the molecule must be cyclic.*
   For the π cloud to be uninterrupted, *every atom in the ring must have a p orbital.*
   For the π cloud to form, each *p* orbital must overlap with the *p* orbitals on either side of it. Therefore, *the molecule must be planar.*

2. *The π cloud must contain an odd number of pairs of π electrons.*

Therefore, benzene is an aromatic compound because it is cyclic and planar, every carbon in the ring has a *p* orbital, and the π cloud contains *three* pairs of π electrons.

The German physicist Erich Hückel was the first to recognize that an aromatic compound must have an odd number of pairs of π electrons. In 1931, he described this requirement in what has come to be known as **Hückel's rule**, or the **$4n + 2$ rule**. The rule states that for a planar, cyclic compound to be aromatic, its uninterrupted π cloud must contain ($4n + 2$) π electrons, where *n* is any whole number. According to Hückel's rule, then, an aromatic compound must have 2 ($n = 0$), 6 ($n = 1$), 10 ($n = 2$), 14 ($n = 3$), 18 ($n = 4$), and so on π electrons. Because there are two electrons in a pair, Hückel's rule requires that an aromatic compound have 1, 3, 5, 7, 9 and so on pairs of π electrons. Thus, Hückel's rule is just a mathematical way of saying that an aromatic compound must have an *odd* number of pairs of π electrons.

PROBLEM 1♦

a. What is the value of *n* in Hückel's rule when a compound has nine pairs of π electrons?

b. Is such a compound aromatic?

14.3 Applying the Criteria for Aromaticity

Monocyclic hydrocarbons with alternating single and double bonds are called **annulenes**. A prefix in brackets denotes the number of carbons in the ring. Cyclobutadiene, benzene, and cyclooctatetraene are examples of annulenes.

cyclobutadiene **benzene** **cyclooctatetraene**
[4]-annulene **[6]-annulene** **[8]-annulene**

Cyclobutadiene has two pairs of π electrons, and cyclooctatetraene has four pairs of π electrons. Unlike benzene, these compounds are *not* aromatic because they have an *even* number of pairs of π electrons. There is an additional reason why cyclooctatetraene is not aromatic—it is not planar, it is tub-shaped. (See page 293.) Earlier, we saw that, for an eight-membered ring to be planar, it must have bond angles of 135° (Chapter 2, Problem 34), and we know that sp^2 carbons have 120° bond angles. Therefore, if cyclooctatetraene were planar, it would have considerable angle strain. Because cyclobutadiene and cyclooctatetraene are not aromatic, they do not have the unusual stability of aromatic compounds.

Now let's look at some other compounds and determine whether they are aromatic. Cyclopentadiene is not aromatic because it does not have an uninterrupted ring of *p* orbital-bearing atoms. One of its ring atoms is sp^3 hybridized, and only sp^2 and sp hybridized carbons have *p* orbitals. Therefore, cyclopentadiene does not fulfill the first criterion for aromaticity.

cyclopentadiene **cyclopentadienyl** **cyclopentadienyl**
cation **anion**

The cyclopentadienyl cation also is not aromatic because, although it has an uninterrupted ring of *p* orbital-bearing atoms, its π cloud has *two* (an even number) pairs of π electrons. The cyclopentadienyl anion is aromatic: it has an uninterrupted ring of *p* orbital-bearing atoms and the π cloud contains *three* (an odd number) pairs of delocalized π electrons.

Notice that the negatively charged carbon in the cyclopentadienyl anion is sp^2 hybridized, because if it were sp^3 hybridized, the ion would not be aromatic. The resonance hybrid shows that all the carbons in the cyclopentadienyl anion are equivalent. Each carbon has exactly one-fifth of the negative charge associated with the anion.

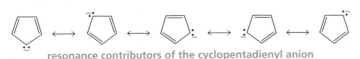

resonance contributors of the cyclopentadienyl anion

resonance hybrid

The criteria that determine whether a monocyclic hydrocarbon is aromatic can also be used to determine whether a polycyclic hydrocarbon is aromatic. Naphthalene (five pairs of π electrons), phenanthrene (seven pairs of π electrons), and chrysene (nine pairs of π electrons) are aromatic.

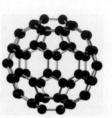

naphthalene   phenanthrene   chrysene

BUCKYBALLS AND AIDS

Diamond and graphite (Section 1.8) are two familiar forms of pure carbon. A third form was discovered unexpectedly in 1985, while scientists were conducting experiments designed to understand how long-chain molecules are formed in outer space. R. E. Smalley, R. F. Curl, Jr., and H. W. Kroto shared the 1996 Nobel Prize in chemistry for discovering this new form of carbon. They named the substance buckminsterfullerene (often shortened to fullerene) because its structure reminded them of the geodesic domes popularized by R. Buckminster Fuller, an American architect and philosopher. Its nickname is "buckyball."

Consisting of a hollow cluster of 60 carbons, fullerene is the most symmetrical large molecule known. Like graphite, fullerene has only sp^2 hybridized carbons, but instead of being arranged in layers, the carbons are arranged in rings that fit together like the seams of a soccer ball. Each molecule has 32 interlocking rings (20 hexagons and 12 pentagons). At first glance, fullerene would appear to be aromatic because of its benzene-like rings. However, the curvature of the ball prevents the molecule from fulfilling the first criterion for aromaticity—that it be planar. Therefore, fullerene is not aromatic.

Buckyballs have extraordinary chemical and physical properties. For example, they are exceedingly rugged, as shown by their ability to survive the extreme temperatures of outer space. Because they are essentially hollow cages, they can be manipulated to make materials never before known. For example, when a buckyball is "doped" by inserting potassium or cesium into its cavity, it becomes an excellent organic superconductor. These molecules are now being studied for use in many other applications, including the development of new polymers, catalysts, and drug delivery systems. The discovery of buckyballs is a strong reminder of the technological advances that can be achieved as a result of basic research.

Scientists have even turned to buckyballs in the quest for a cure for AIDS. An enzyme that the AIDS virus needs in order to reproduce has a nonpolar pocket in its three-dimensional structure. If this pocket is blocked, the production of the virus ceases. Because buckyballs are nonpolar and have approximately the same diameter as the pocket of the enzyme, they are being considered as possible blockers. The first step in

C_{60}
buckminsterfullerene
"buckyball"

A geodesic dome

exploring this possibility was to equip the buckyball with polar side chains to make it water soluble so that it could flow through the bloodstream. The next step was to modify the side chains so they would bind to the enzyme. This project is still a long way from providing a cure for AIDS, but it represents one example of the many and varied approaches that scientists are taking to find such a cure.

Richard E. Smalley (1943–2005) *was born in Akron, Ohio. He received a B.S. from the University of Michigan and a Ph.D. from Princeton University. He was a professor of chemistry and physics at Rice University.*

Robert F. Curl, Jr. *was born in Texas in 1933. He received a B.A. from Rice University and a Ph.D. from the University of California, Berkeley. He is a professor of chemistry at Rice University.*

Sir Harold W. Kroto *was born in 1939 in England and is a professor of chemistry at the University of Sussex.*

PROBLEM 2

a. Draw arrows to show the movement of electrons in going from one resonance contributor to the next in the cyclopentadienyl anion.

b. How many ring atoms share the negative charge?

When drawing resonance contributors, remember that only electrons move; atoms never move.

PROBLEM 3◆

Which compound in each set is aromatic? Explain your choice.

a.

cyclopropene 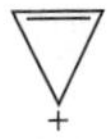   cyclopropenyl cation    cyclopropenyl anion

b.

cycloheptatriene    cycloheptatrienyl cation    cycloheptatrienyl anion

PROBLEM 4◆

Which of the following are aromatic?

a.

b.

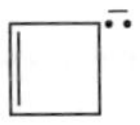

c.

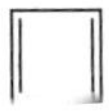

d.

e.

f. CH_2=CHCH=CHCH=CH_2

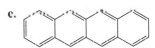
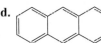

PROBLEM 5◆ **SOLVED**

a. How many monobromonaphthalenes are there?

b. How many monobromophenanthrenes are there?

Solution to 5a There are two monobromonaphthalenes. Substitution cannot occur at either of the carbons shared by both rings, because those carbons are not bonded to a hydrogen. Naphthalene is a flat molecule, so substitution for a hydrogen at any other carbon will result in one of the compounds shown.

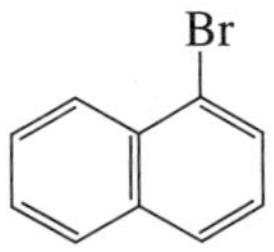

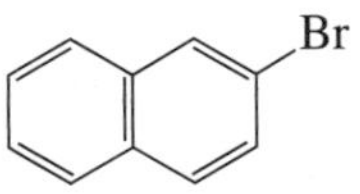

PROBLEM 6

The [10]- and [12]-annulenes have been synthesized, and neither has been found to be aromatic. Explain.

benzene

pyridine

pyrrole

14.4 Aromatic Heterocyclic Compounds

A compound does not have to be a hydrocarbon to be aromatic. Many *heterocyclic compounds* are aromatic. A **heterocyclic compound** is a cyclic compound in which one or more of the ring atoms is an atom other than carbon. The atom that is not carbon is called a **heteroatom**. The name comes from the Greek word *heteros,* which means "different." The most common heteroatoms are N, O, and S.

heterocyclic compounds

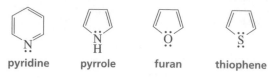

pyridine   pyrrole   furan   thiophene

Pyridine is an aromatic heterocyclic compound. Each of the six ring atoms of pyridine is sp^2 hybridized, which means that each has a p orbital; and the molecule contains three pairs of π electrons. (Do not be confused by the lone-pair electrons on the nitrogen; they are not π electrons.) Because nitrogen is sp^2 hybridized, it has three sp^2 orbitals and a p orbital. The p orbital is used to form the π bond. Two of nitrogen's sp^2 orbitals overlap the sp^2 orbitals of adjacent carbons, and nitrogen's third sp^2 orbital contains the lone pair.

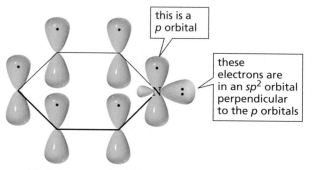

this is a
p orbital

these electrons are in an sp^2 orbital perpendicular to the p orbitals

orbital structure of pyridine

It is not immediately apparent that the electrons represented as lone-pair electrons on the nitrogen atom of pyrrole are π electrons. The resonance contributors, however, show that the nitrogen atom is sp^2 hybridized and uses its three sp^2 orbitals to bond to two carbons and one hydrogen. The lone-pair electrons, therefore, must be in a p orbital that overlaps the p orbitals of adjacent carbons, forming a π bond—thus, they are π electrons. Pyrrole, therefore, has three pairs of π electrons and is aromatic.

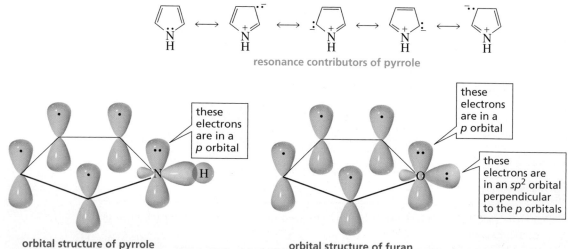

resonance contributors of pyrrole

these electrons are in a p orbital

orbital structure of pyrrole

these electrons are in a p orbital

these electrons are in an sp^2 orbital perpendicular to the p orbitals

orbital structure of furan

Similarly, furan and thiophene are stable aromatic compounds. Both the oxygen in the former and the sulfur in the latter are sp^2 hybridized and have one lone pair in an sp^2 orbital. The second lone pair is in a p orbital that overlaps the p orbitals of adjacent carbons, forming a π bond. Thus, they are π electrons.

resonance contributors of furan

Quinoline, indole, imidazole, purine, and pyrimidine are other examples of heterocyclic aromatic compounds. The heterocyclic compounds discussed in this section will be examined in greater detail in Chapter 20.

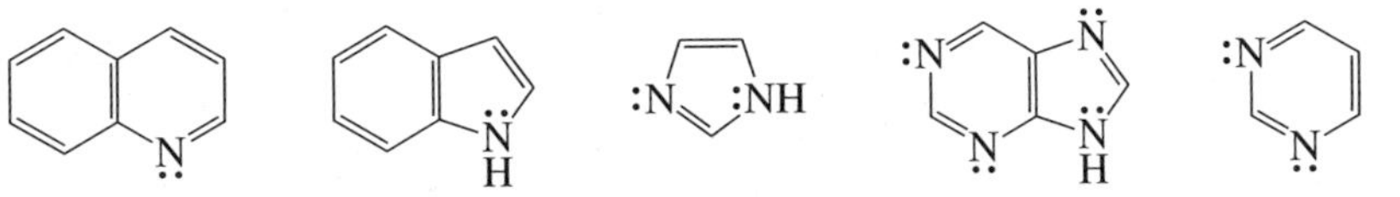

quinoline   indole   imidazole   purine   pyrimidine

PROBLEM 7

a. Draw arrows to show the movement of electrons in going from one resonance contributor to the next in pyrrole.

b. How many ring atoms share the negative charge?

PROBLEM 8◆

What orbitals contain the electrons represented as lone pairs in the structures of quinoline, indole, imidazole, purine, and pyrimidine?

PROBLEM 9

Refer to the electrostatic potential maps on page 646 to answer the following questions:

a. Why is the bottom part of the electrostatic potential map of pyrrole blue?

b. Why is the bottom part of the electrostatic potential map of pyridine red?

c. Why is the center of the electrostatic potential map of benzene more red than the center of the electrostatic potential map of pyridine?

14.5 **Some Chemical Consequences of Aromaticity**

The pK_a of cyclopentadiene is 15, which is extraordinarily acidic for a hydrogen that is bonded to an sp^3 hybridized carbon. Ethane, for example, has a p$K_a > 60$.

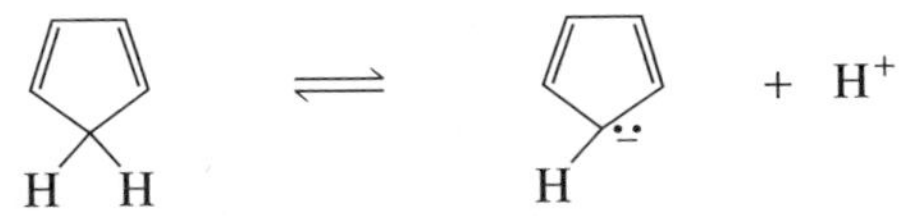

cyclopentadiene   cyclopentadienyl
pK_a = 15   anion

$$CH_3CH_3 \rightleftharpoons CH_3\ddot{C}H_2 + H^+$$

ethane   ethyl anion
p$K_a > 60$

Why is the pK_a of cyclopentadiene so much lower than that of ethane? To answer this question, we must look at the stabilities of the anions that are formed when the compounds lose a proton. (Recall that the strength of an acid is determined by the stability of its conjugate base: the more stable its conjugate base, the stronger is the acid; see Section 1.18.) All the electrons in the ethyl anion are localized. In contrast, the anion formed when cyclopentadiene loses a proton is aromatic (Section 14.2). As a result of its aromaticity, the cyclopentadienyl anion is an unusually stable carbanion. Thus, its conjugate acid has an unusually low pK_a.

Another example of the influence of aromaticity on chemical reactivity is the unusual chemical behavior exhibited by cycloheptatrienyl bromide. Recall from Section 2.9 that alkyl halides tend to be relatively nonpolar covalent compounds—they are soluble in nonpolar solvents and *insoluble in water*. Cycloheptatrienyl bromide, however, is an alkyl halide that behaves like an ionic compound—it is insoluble in nonpolar solvents, but readily *soluble in water.*

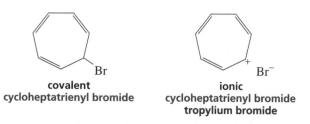

covalent	ionic
cycloheptatrienyl bromide	cycloheptatrienyl bromide
	tropylium bromide

Cycloheptatrienyl bromide is an ionic compound because its cation is aromatic. The alkyl halide is *not* aromatic in the covalent form because it has an sp^3 hybridized carbon and therefore does *not* have an uninterrupted ring of *p* orbital-bearing atoms. However, the cycloheptatrienyl cation (also known as the tropylium cation) *is* aromatic: it is a planar cyclic ion, all the ring atoms are sp^2 hybridized (which means that each ring atom has a *p* orbital), and it has three pairs of delocalized π electrons. The stability associated with the aromatic cation causes the alkyl halide to exist in the ionic form.

PROBLEM 10♦

Predict the relative pK_a values of cyclopentadiene and cycloheptatriene.

PROBLEM 11 *SOLVED*

a. Draw arrows to show the movement of electrons in going from one resonance contributor to the next in the cycloheptatrienyl cation.
b. How many ring atoms share the positive charge?

Solution to 11a

Solution to 11b
All seven ring atoms share the positive charge.

PROBLEM-SOLVING STRATEGY

Analyzing Electron Distribution in Compounds

Which of the following compounds has the greater dipole moment?

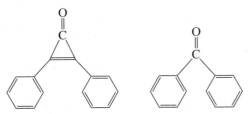

Before attempting to answer this kind of question, let's make sure we know exactly what the question is asking. We know that the dipole moment of these compounds results from the unequal sharing of electrons by carbon and oxygen. Therefore, the more unequal the sharing, the greater is the dipole moment. So now the question becomes, Which compound has a greater negative charge on its oxygen atom? To find out, let's draw the structures with separated charges and determine their relative stabilities. The three-membered ring in the compound on the left becomes aromatic when the charges are separated, but the structure on the right does not. Because being aromatic makes a species more stable, the compound on the left has the greater charge separation and, therefore, the greater dipole moment.

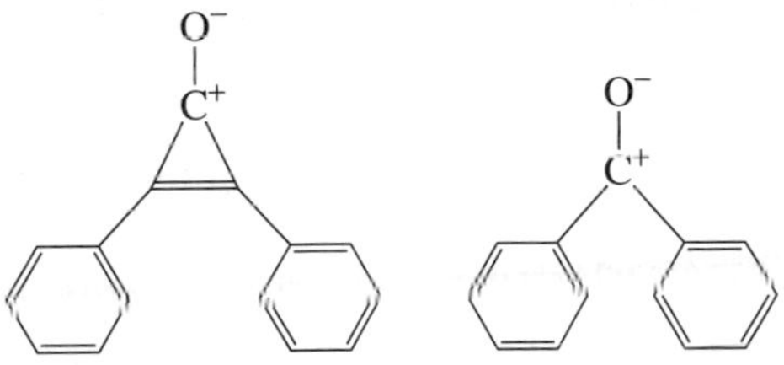

Now continue on to Problem 12.

PROBLEM 12

a. In what direction is the dipole moment in fulvene? Explain.
b. In what direction is the dipole moment in calicene? Explain.

fulvene   calicene

14.6 Antiaromaticity

An aromatic compound is *more stable* than an analogous cyclic compound with localized electrons. In contrast, an **antiaromatic compound** is *less stable* than an analogous cyclic compound with localized electrons. *Aromaticity is characterized by stability, whereas antiaromaticity is characterized by instability.*

Antiaromatic compounds are highly unstable.

relative stabilities

aromatic compound > cyclic compound with localized electrons > antiaromatic compound

increasing stability

A compound is classified as being antiaromatic if it fulfills the first criterion for aromaticity but does not fulfill the second. In other words, it must be a planar, cyclic compound with an uninterrupted ring of *p* orbital-bearing atoms, and the π cloud must contain an *even* number of pairs of π electrons. Hückel would state that the π cloud must contain $4n$ π electrons, where n is any whole number—a mathematical way of saying that the cloud must contain an *even* number of pairs of π electrons.

Cyclobutadiene is a planar, cyclic molecule with two pairs of π electrons. Hence, it is expected to be antiaromatic and highly unstable. In fact, it is too unstable to be isolated, although it has been trapped at very cold temperatures. The cyclopentadienyl cation also has two pairs of π electrons, so we can conclude that it too is antiaromatic and unstable.

cyclobutadiene   cyclopentadienyl
cation

PROBLEM 13◆

a. Predict the relative pK_a values of cyclopropene and cyclopropane.

b. Which is more soluble in water, 3-bromocyclopropene or bromocyclopropane?

PROBLEM 14◆

Which of the compounds in Problem 4 are antiaromatic?

14.7 A Molecular Orbital Description of Aromaticity and Antiaromaticity

Why are planar molecules with uninterrupted cyclic π electron clouds highly stable (aromatic) if they have an odd number of pairs of π electrons and highly unstable (antiaromatic) if they have an even number of pairs of π electrons? To answer this question, we must turn to molecular orbital theory.

The relative energies of the π molecular orbitals of a planar molecule with an uninterrupted cyclic π electron cloud can be determined—without having to use any math—by first drawing the cyclic compound with one of its vertices pointed down. The relative energies of the π molecular orbitals can be thought of as corresponding to the relative levels of the vertices (Figure 14.3). Molecular orbitals below the midpoint of

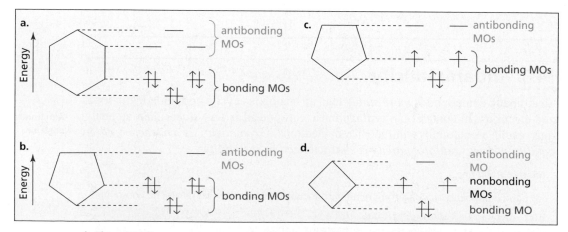

▲ **Figure 14.3**
The distribution of electrons in the π molecular orbitals of (a) benzene, (b) the cyclopentadienyl anion, (c) the cyclopentadienyl cation, and (d) cyclobutadiene. The relative energies of the π molecular orbitals in a cyclic compound correspond to the relative levels of the vertices. Molecular orbitals below the midpoint of the cyclic structure are bonding, those above the midpoint are antibonding, and those at the midpoint are nonbonding.

the cyclic structure are bonding molecular orbitals, those above the midpoint are anti-bonding molecular orbitals, and any at the midpoint are nonbonding molecular orbitals. This simple scheme is sometimes called a Frost device (or a Frost circle) in honor of Arthur A. Frost, the scientist who devised it. Notice that the number of π molecular orbitals is the same as the number of atoms in the ring because each ring atom contributes a p orbital. (Recall that orbitals are conserved; Section 1.6.)

The six π electrons of benzene occupy its three bonding π molecular orbitals, and the six π electrons of the cyclopentadienyl anion occupy *its* three bonding π molecular orbitals. Notice that there is always an odd number of bonding orbitals because one corresponds to the lowest vertex and the others come in degenerate pairs. Consequently, aromatic compounds—such as benzene and the cyclopentadienyl anion, with their odd number of pairs of π electrons, have completely filled bonding orbitals and no electrons in either nonbonding or antibonding orbitals. This is what gives aromatic molecules their stability. (A more detailed description of the molecular orbitals in benzene was given in Section 7.8.)

Antiaromatic compounds have an even number of pairs of π electrons. Therefore, either they are unable to fill their bonding orbitals (cylopentadienyl cation) or they have a pair of π electrons left over after the bonding orbitals are filled (cyclobutadiene). Hund's rule requires that these two electrons go into two degenerate orbitals (Section 1.2).

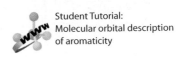

Student Tutorial:
Molecular orbital description of aromaticity

Aromatic compounds are stable because they have filled bonding π molecular orbitals.

PROBLEM 15◆

How many bonding, nonbonding, and antibonding π molecular orbitals does cyclobutadiene have? In which molecular orbitals are the π electrons?

PROBLEM 16◆

Can a radical be aromatic?

PROBLEM 17

Following the instructions for drawing the π molecular orbital energy levels of the compounds shown in Figure 14.3, draw the π molecular orbital energy levels for the cycloheptatrienyl cation, the cycloheptatrienyl anion, and the cyclopropenyl cation. For each compound, show the distribution of the π electrons. Which of the compounds are aromatic? Which are antiaromatic?

14.8 Nomenclature of Monosubstituted Benzenes

Some monosubstituted benzenes are named simply by attaching "benzene" after the name of the substituent

bromobenzene	chlorobenzene	nitrobenzene used as a solvent in shoe polish	ethylbenzene

Some monosubstituted benzenes have names that incorporate the substituent. Unfortunately, such names have to be memorized.

toluene	phenol	aniline	benzenesulfonic acid

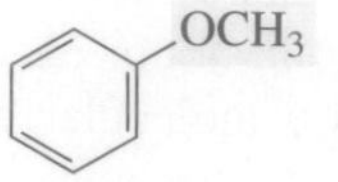

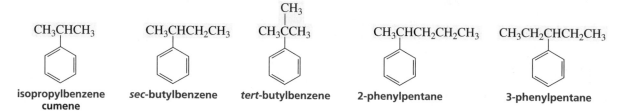

anisole 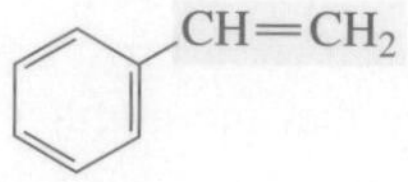   styrene 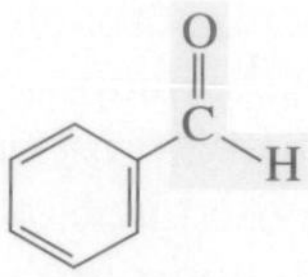   benzaldehyde 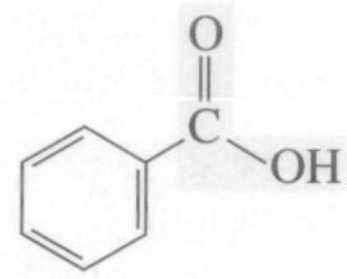   benzoic acid 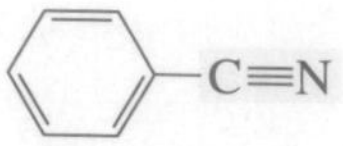   benzonitrile

When a benzene ring is a substituent, it is called a **phenyl group**. A benzene ring with a methylene group is called a **benzyl group**.

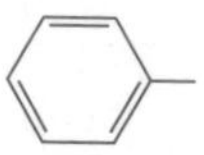

a phenyl group 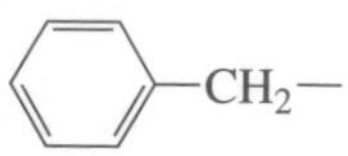   a benzyl group

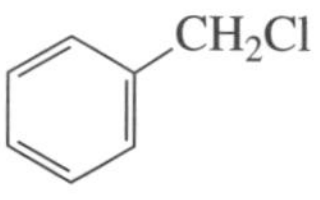

chloromethylbenzene
benzyl chloride   diphenyl ether    dibenzyl ether

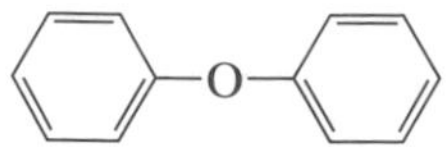

With the exception of toluene, benzene rings with an alkyl substituent are named as alkyl-substituted benzenes or as phenyl-substituted alkanes.

isopropylbenzene
cumene   *sec*-butylbenzene   *tert*-butylbenzene   2-phenylpentane   3-phenylpentane

Aryl group (Ar) is the general term for either a phenyl group or a substituted phenyl group, just as alkyl group (R) is the general term for a group derived from an alkane. In other words, ArOH could be used to designate any of the following phenols:

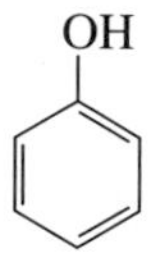

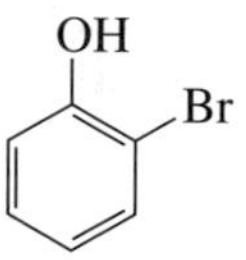

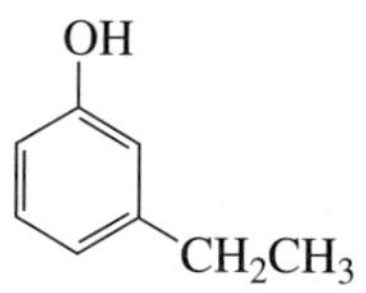

THE TOXICITY OF BENZENE

Benzene, which has been widely used in chemical synthesis and has been frequently used as a solvent, is a toxic substance. The major adverse effects of chronic exposure are seen in the central nervous system and bone marrow; it causes leukemia and aplastic anemia. A higher than average incidence of leukemia, for example, has been found in industrial workers with long-term exposure to as little as 1 ppm benzene in the atmosphere. Toluene has replaced benzene as a solvent because, although it too is a central nervous system depressant, it does not cause leukemia or aplastic anemia. "Glue sniffing," a highly dangerous activity, produces narcotic central nervous system effects because glue contains toluene.

PROBLEM 18◆

Draw the structure of each of the following:

a. 2-phenylhexane

b. benzyl alcohol

c. 3-benzylpentane

d. bromomethylbenzene

14.9 **How Benzene Reacts**

Aromatic compounds such as benzene undergo **electrophilic aromatic substitution reactions**: an electrophile substitutes for one of the hydrogens attached to the benzene ring.

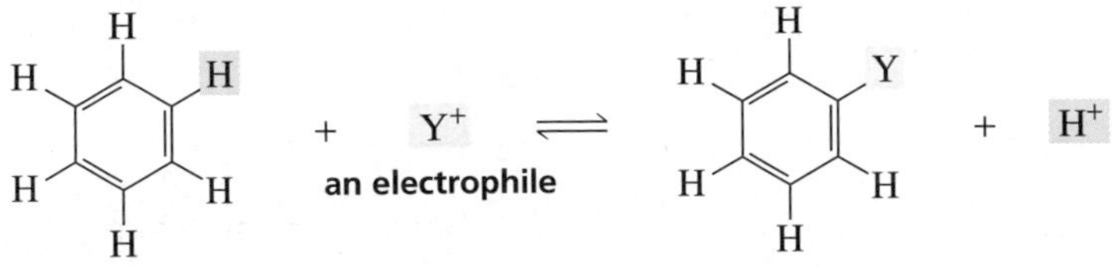

Now let's look at why this substitution reaction occurs. The cloud of π electrons above and below the plane of the ring causes benzene to be a nucleophile. It will, therefore, react with an electrophile (Y^+). When an electrophile attaches itself to a benzene ring, a carbocation intermediate is formed.

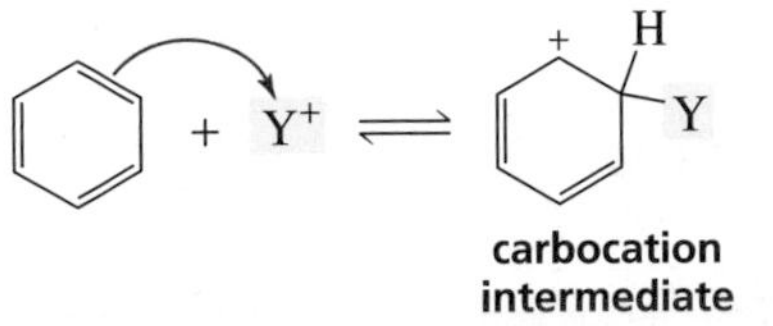

carbocation
intermediate

This description should remind you of the first step in an electrophilic addition reaction of an alkene: the nucleophilic alkene reacts with an electrophile and forms a carbocation intermediate (Section 3.6). In the second step of an electrophilic addition reaction of an alkene, the carbocation reacts with a nucleophile (Z^-) to form an addition product.

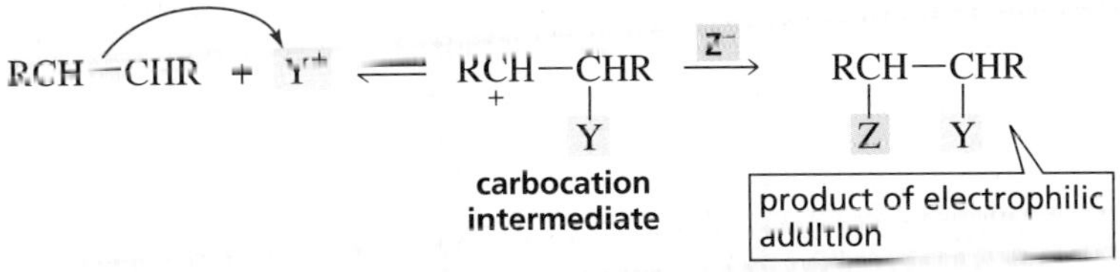

carbocation
intermediate

product of electrophilic
addition

If the carbocation intermediate that is formed from the reaction of benzene with an electrophile were to react similarly with a nucleophile (depicted as path *b* in Figure 14.4), the addition product would not be aromatic. But if the carbocation instead were to lose a proton from the site of electrophilic attack (depicted as path *a* in Figure 14.4), the aromaticity of the benzene ring would be restored.

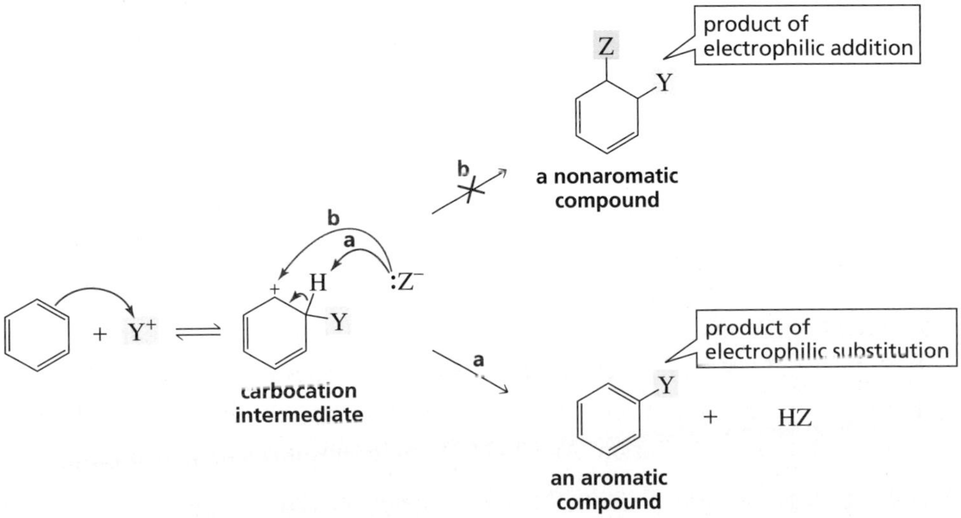

product of
electrophilic addition

a nonaromatic
compound

product of
electrophilic substitution

an aromatic
compound

◀ **Figure 14.4**
Reaction of benzene with an electrophile. Because the aromatic product is more stable, the reaction proceeds as (a) an electrophilic substitution reaction rather than (b) an electrophilic addition reaction.

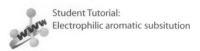

Because the aromatic substitution product is much more stable than the nonaromatic addition product (Figure 14.5), benzene undergoes *electrophilic substitution reactions* that preserve aromaticity, rather than *electrophilic addition reactions*—the reactions characteristic of alkenes—that would destroy aromaticity. The substitution reaction is more accurately called an **electrophilic aromatic substitution reaction**, since the electrophile substitutes for a hydrogen of an aromatic compound.

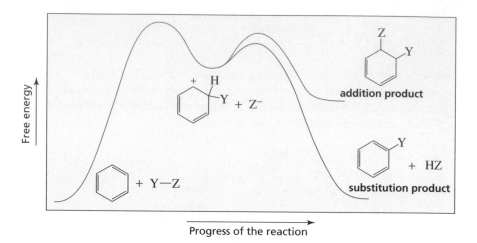

Figure 14.5 ▶
Reaction coordinate diagrams for electrophilic substitution of benzene and electrophilic addition to benzene.

PROBLEM 19

If electrophilic addition to benzene is an endergonic reaction overall, how can electrophilic addition to an alkene be an exergonic reaction overall?

14.10 The General Mechanism for Electrophilic Aromatic Substitution Reactions

In an electrophilic aromatic substitution reaction, an electrophile (Y⁺) becomes attached to a ring carbon and an H⁺ comes off the same ring carbon.

In an electrophilic aromatic substitution reaction, an electrophile becomes attached to a ring carbon and an H^+ comes off the same ring carbon.

an electrophilic aromatic substitution reaction

The following are the five most common electrophilic aromatic substitution reactions:

1. **Halogenation:** A bromine (Br), a chlorine (Cl), or an iodine (I) substitutes for a hydrogen.
2. **Nitration:** A nitro (NO_2) group substitutes for a hydrogen.
3. **Sulfonation:** A sulfonic acid (SO_3H) group substitutes for a hydrogen.
4. **Friedel–Crafts acylation:** An acyl ($RC{=}O$) group substitutes for a hydrogen.
5. **Friedel–Crafts alkylation:** An alkyl (R) group substitutes for a hydrogen.

All of these reactions take place by the same two-step mechanism.

General mechanism for electrophilic aromatic substitution

the proton is removed from the carbon that has formed the new bond with the electrophile

a base in the reaction mixture

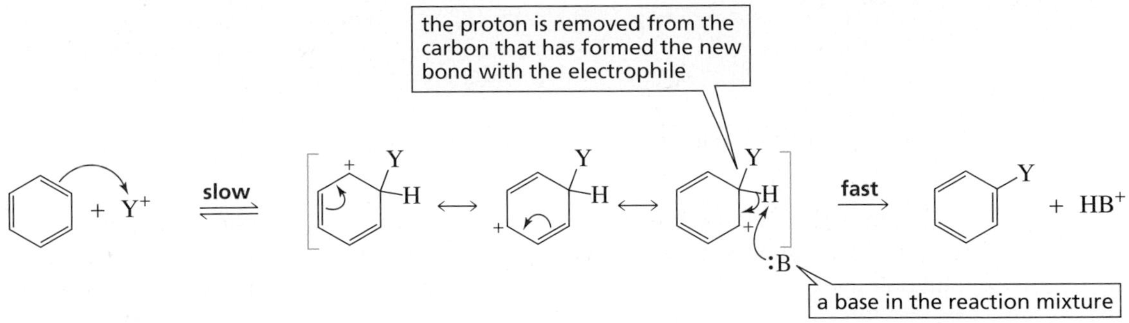

- Benzene reacts with an electrophile (Y^+) forming a carbocation intermediate. The structure of the carbocation intermediate can be approximated by three resonance contributors.
- A base in the reaction mixture (:B) pulls off a proton from the carbocation intermediate, and the electrons that held the proton move into the ring to reestablish its aromaticity. Notice that *the proton is always removed from the carbon that has formed the new bond with the electrophile.*

The first step is relatively slow and endergonic because an aromatic compound is being converted into a much less stable nonaromatic intermediate (Figure 14.5). The second step is fast and strongly exergonic because this step restores the stability-enhancing aromaticity.

We will look at each of these five electrophilic aromatic substitution reactions individually. As you study them, notice that they differ only in how the electrophile (Y^+) needed to start the reaction is generated. Once the electrophile is formed, all five reactions follow the same two-step mechanism for electrophilic aromatic substitution.

14.11 Halogenation of Benzene

The bromination or chlorination of benzene requires a Lewis acid catalyst such as ferric bromide or ferric chloride. Recall that a *Lewis acid* is a compound that accepts a share in an electron pair (Section 1.26).

bromination

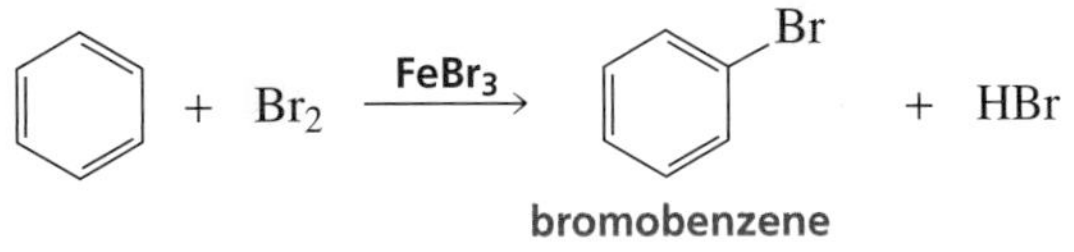

bromobenzene

chlorination

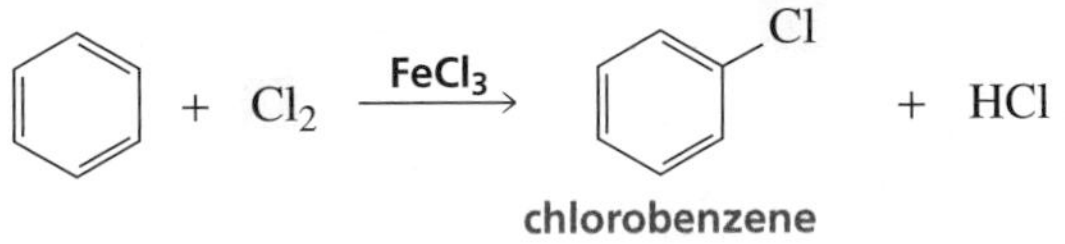

chlorobenzene

Why does the reaction of benzene with Br_2 or Cl_2 require a catalyst when the reaction of an alkene with these reagents does not require a catalyst? Benzene's aromaticity makes it much more stable and therefore much less reactive than an alkene. Thus, benzene requires a better electrophile. Donating a lone pair to the

Lewis acid weakens the Br—Br (or Cl—Cl) bond and thus makes Br_2 (or Cl_2) a better electrophile.

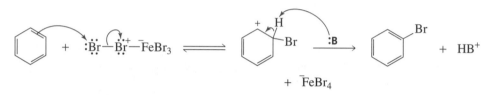

For the sake of clarity, only one of the three resonance contributors of the carbocation intermediate is shown in this and subsequent mechanisms for electrophilic aromatic substitution reactions. Bear in mind, however, that each carbocation intermediate actually has the three resonance contributors shown in Section 14.10.

Mechanism for bromination

Student Tutorial:
The bromination of benzene 1

Student Tutorial:
The bromination of benzene 2

Student Tutorial:
The bromination of benzene 3

- The electrophile attaches to the benzene ring.
- A base (:B) from the reaction mixture (such as Br^- or a solvent) removes a proton from the carbocation intermediate.

The following equation shows that the catalyst is regenerated:

$$\overline{F}eBr_4 \quad + \quad HB^+ \quad \longrightarrow \quad HBr \quad + \quad FeBr_3 \quad + \quad :B$$

Chlorination of benzene occurs by the same mechanism as bromination.

Mechanism for chlorination

Ferric bromide and ferric chloride react readily with moisture in the air during handling, which inactivates them as catalysts. Therefore, the actual salts are not used. Instead, ferric bromide or ferric chloride is generated in situ (in the reaction mixture) by adding iron filings and bromine or chlorine to the reaction mixture. Therefore, the halogen in the Lewis acid is the same as the reagent halogen.

$$2\ Fe \quad + \quad 3\ Br_2 \quad \longrightarrow \quad 2\ FeBr_3$$

$$2\ Fe \quad + \quad 3\ Cl_2 \quad \longrightarrow \quad 2\ FeCl_3$$

PROBLEM 20

Why does hydration inactivate $FeBr_3$?

Electrophilic iodine (I^+) is obtained by treating (I_2) with an oxidizing agent such as nitric acid.

iodination

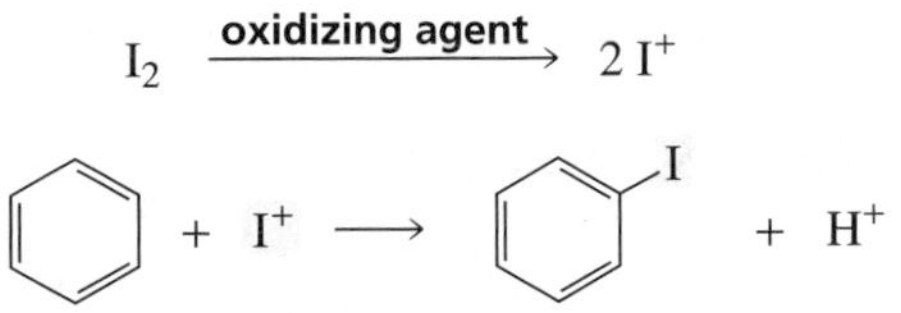

$$I_2 \xrightarrow{\text{oxidizing agent}} 2\,I^+$$

iodobenzene

Once the electrophile is formed, iodination of benzene occurs by the same mechanism as bromination and chlorination.

Mechanism for iodination

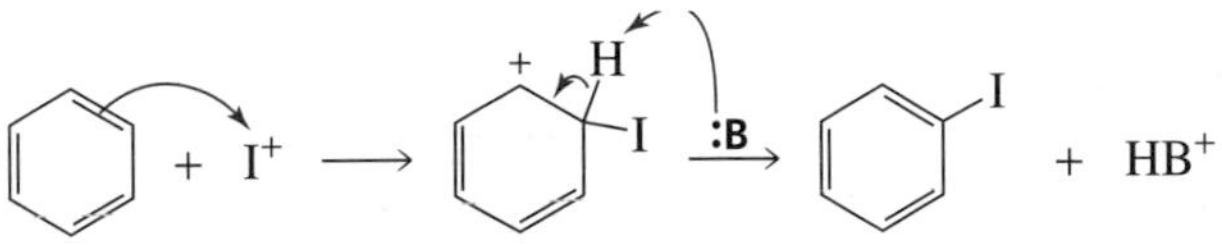

Once the electrophile is formed, iodination of benzene occurs by the same mechanism

---

THYROXINE

Thyroxine, which is a hormone produced by the thyroid gland, regulates the metabolic rate in the body, causing an increase in the rate at which fats, carbohydrates, and proteins are metabolized. Without thyroxine, development in the young comes to a halt. Humans obtain thyroxine from tyrosine (an amino acid) and iodine. The thyroid gland is the only part of the body that uses iodine, which we acquire primarily from the iodized salt in our diet. An enzyme called iodoperoxidase converts the I^- we ingest to I^+, the electrophile needed to place an iodo substituent on a benzene ring.

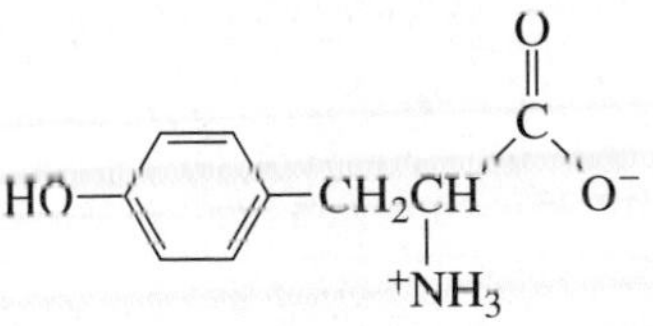

tyrosine

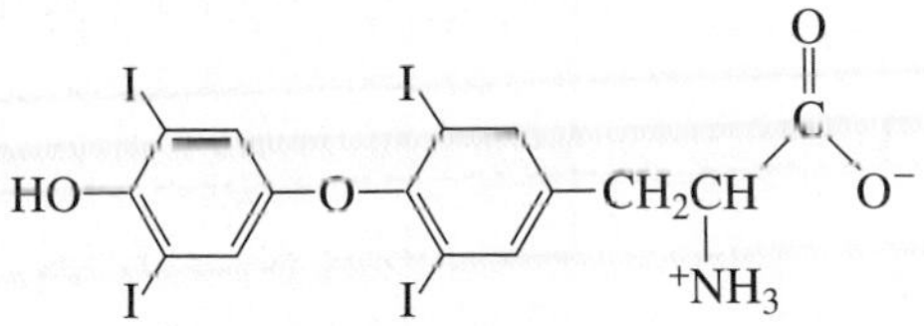

thyroxine

Chronically low levels of thyroxine cause enlargement of the thyroid gland, a condition known as goiter. Low thyroxine levels can be corrected by taking thyroxine orally. Synthroid, the most popular brand of thyroxine, is the fifth most-prescribed drug in the United States (see Table 30.1 on page 1294).

14.12 **Nitration of Benzene**

Nitration of benzene with nitric acid requires sulfuric acid as a catalyst.

nitration

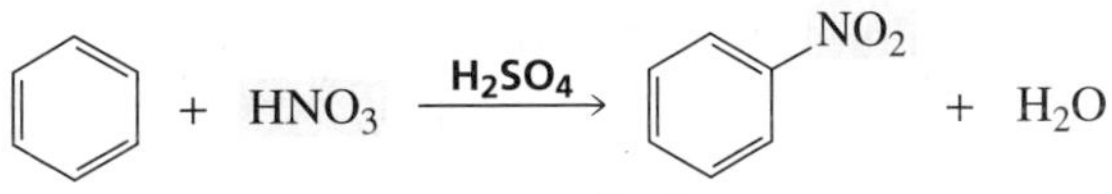

$$\text{benzene} + HNO_3 \xrightarrow{H_2SO_4} \text{nitrobenzene} + H_2O$$

nitrobenzene

To generate the necessary electrophile, sulfuric acid protonates nitric acid, which loses water to form a nitronium ion, the electrophile required for nitration.

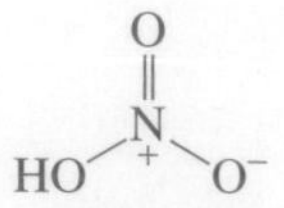

nitric acid

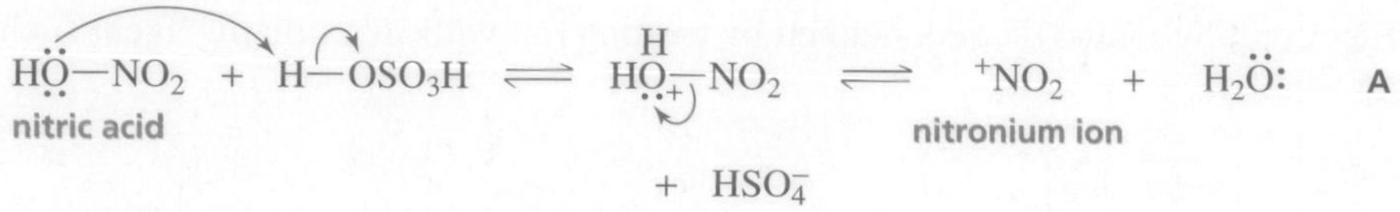

$O=\overset{+}{N}=O$

nitronium ion

$$\text{HO–NO}_2 + \text{H–OSO}_3\text{H} \rightleftharpoons \text{HO}\overset{\text{H}}{\underset{(+)}{-}}\text{NO}_2 \rightleftharpoons {}^+\text{NO}_2 + \text{H}_2\overset{..}{\text{O}}: \quad \textbf{A}$$

nitric acid — **nitronium ion**

$$+ \text{ HSO}_4^-$$

The mechanism for the electrophilic aromatic substitution reaction is the same as the mechanisms in Section 14.11.

Mechanism for nitration

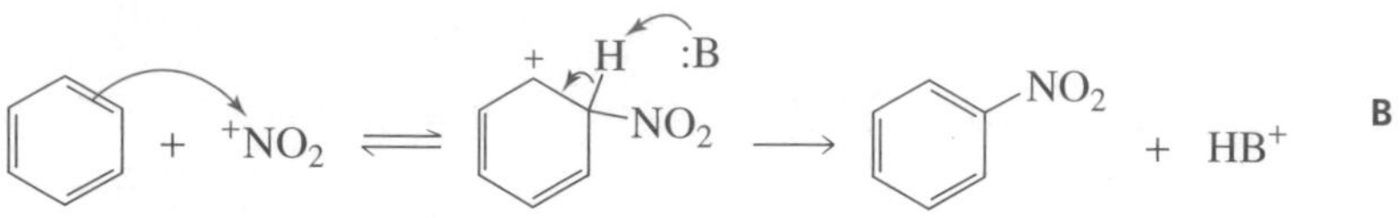

$$\text{⬡} + {}^+\text{NO}_2 \rightleftharpoons \text{⬡}\overset{+}{\underset{}{}}\text{(}\overset{\text{H}}{\text{NO}_2}\text{)} :\text{B} \longrightarrow \text{⬡–NO}_2 + \text{HB}^+ \quad \textbf{B}$$

- The electrophile attaches to the ring.
- Any base (:B) present in the reaction mixture (for example, H_2O, HSO_4^-, solvent) can remove the proton in the second step of the aromatic substitution reaction.

PROBLEM 21 **SOLVED**

Propose a mechanism for the following reaction:

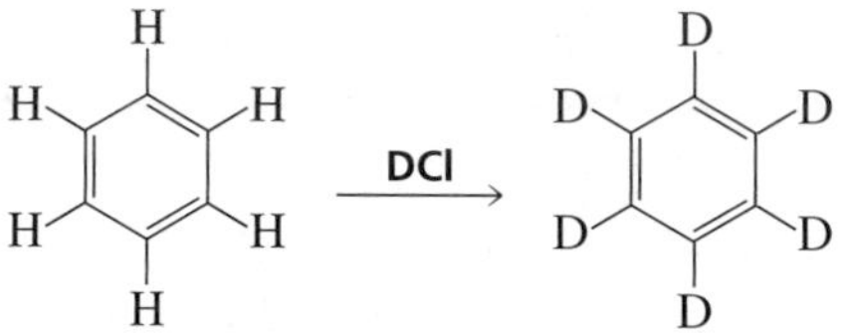

$$\text{⬡(H)} \xrightarrow{\text{DCl}} \text{⬡(D)}$$

Solution The only electrophile available is D^+. Therefore, D^+ attaches to a ring carbon and H^+ comes off the same ring carbon. This reaction can be repeated at each of the other five ring carbons.

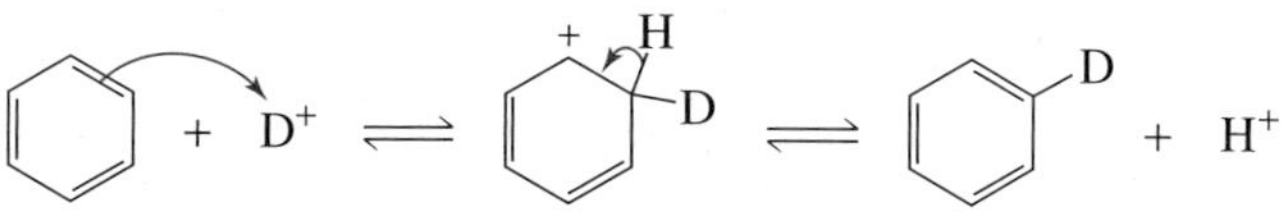

$$\text{⬡} + D^+ \rightleftharpoons \text{⬡}\overset{+}{\underset{}{}}\text{(}\overset{\text{H}}{\text{D}}\text{)} \rightleftharpoons \text{⬡–D} + H^+$$

14.13 Sulfonation of Benzene

Fuming sulfuric acid (a solution of SO_3 in sulfuric acid) or concentrated sulfuric acid is used to sulfonate aromatic rings.

sulfonation

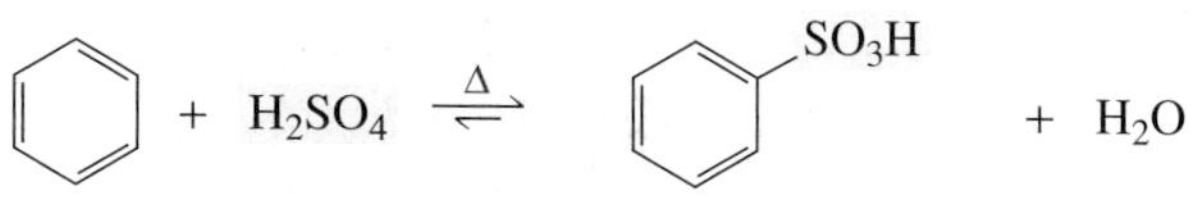

$$\text{⬡} + \text{H}_2\text{SO}_4 \overset{\Delta}{\rightleftharpoons} \text{⬡–SO}_3\text{H} + \text{H}_2\text{O}$$

benzenesulfonic acid

Take a minute to note the similarities in the mechanisms for forming the $^+SO_3H$ electrophile for sulfonation and the $^+NO_2$ electrophile for nitration. A substantial amount of electrophilic sulfur trioxide (SO_3) is generated when concentrated sulfuric acid is heated, as a result of the $^+SO_3H$ electrophile losing a proton.

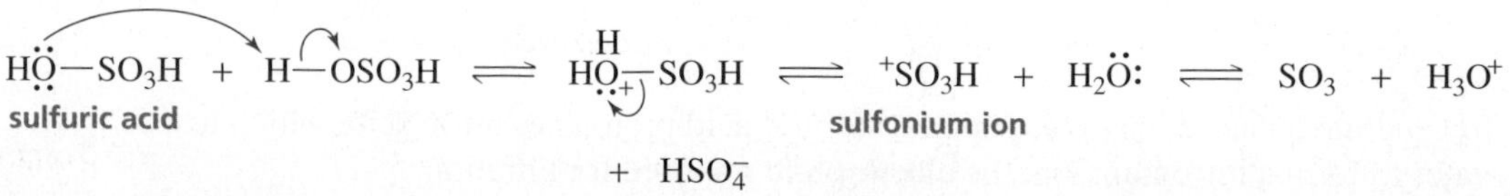

$$\text{HO–SO}_3\text{H} + \text{H–OSO}_3\text{H} \rightleftharpoons \text{HO}\overset{\text{H}}{\underset{(+)}{-}}\text{SO}_3\text{H} \rightleftharpoons {}^+\text{SO}_3\text{H} + \text{H}_2\overset{..}{\text{O}}: \rightleftharpoons \text{SO}_3 + \text{H}_3\text{O}^+$$

sulfuric acid — **sulfonium ion**

$$+ \text{ HSO}_4^-$$

Mechanism for sulfonation

A sulfonic acid is a strong acid because of the three electron-withdrawing oxygen atoms and the stability of its conjugate base—the electrons left behind when a proton is lost are shared by three oxygen atoms (Section 7.6).

benzenesulfonic acid   benzenesulfonate ion

Sulfonation is the only aromatic electrophilic substitution reaction that is reversible. If benzenesulfonic acid is heated in dilute acid, an H^+ adds to the ring and the sulfonic acid group comes off the ring.

Mechanism for desulfonation

The **principle of microscopic reversibility** applies to all reactions. It states that the mechanism of a reaction in the reverse direction must retrace each step of the mechanism in the forward direction in microscopic detail. This means that the forward and reverse reactions must have the same intermediates, and that the "hill" with the highest point on the reaction coordinate represents the rate-determining step in both the forward and reverse directions. For example, sulfonation is described by the reaction coordinate diagram in Figure 14.6, going from left to right. Therefore, desulfonation is described by the same reaction coordinate diagram going from right to left. In sulfonation, the rate-determining step is nucleophilic attack of benzene on the $^+SO_3H$ ion.

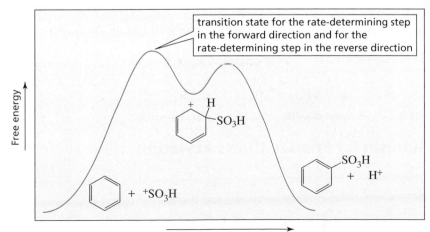

transition state for the rate-determining step in the forward direction and for the rate-determining step in the reverse direction

◀ **Figure 14.6**
Reaction coordinate diagram for the sulfonation of benzene (left to right) and the desulfonation of benzenesulfonic acid (right to left).

In desulfonation, the rate-limiting step is loss of the $^+SO_3H$ ion from the carbocation intermediate. An example of the usefulness of desulfonation to synthetic chemists is given in Chapter 15, Problem 19.

> **PROBLEM 22**
>
> The reaction coordinate diagram in Figure 14.6 shows that the rate-determining step for sulfonation is the slower of the two steps in the mechanism, whereas the rate-determining step for desulfonation is the faster of the two steps. Explain how the faster step can be the rate-determining step.

14.14 Friedel–Crafts Acylation of Benzene

Two electrophilic substitution reactions bear the names of chemists Charles Friedel and James Crafts. *Friedel–Crafts acylation* places an acyl group on a benzene ring, and *Friedel–Crafts alkylation* places an alkyl group on a benzene ring. These reactions are useful to synthetic chemists because they increase the number of carbons in the starting material (Section 6.11).

an acyl group   an alkyl group

Either an acyl chloride or an acid anhydride can be used for Friedel–Crafts acylation. An acyl chloride has a chlorine in place of the OH group of a carboxylic acid.

Friedel–Crafts acylation

The electrophile (an acylium ion) required for a Friedel–Crafts acylation reaction is formed by the reaction of an acyl chloride or an acid anhydride with $AlCl_3$, a Lewis acid. Oxygen and carbon share the positive charge in the acylium ion, which stabilizes it.

an acyl chloride   an acylium ion

Mechanism for Friedel–Crafts acylation

Because the product of a Friedel–Crafts acylation reaction contains a carbonyl group that can complex with AlCl$_3$, Friedel–Crafts acylation reactions must be carried out with more than one equivalent of AlCl$_3$. When the reaction is over, water is added to the reaction mixture to liberate the product from the complex.

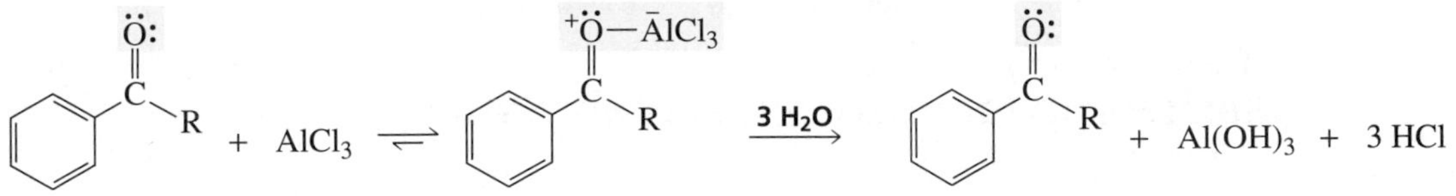

PROBLEM 23

Show the mechanism for the generation of the acylium ion if an acid anhydride is used instead of an acyl chloride in a Friedel–Crafts acylation reaction.

PROBLEM 24

Propose a mechanism for the following reaction:

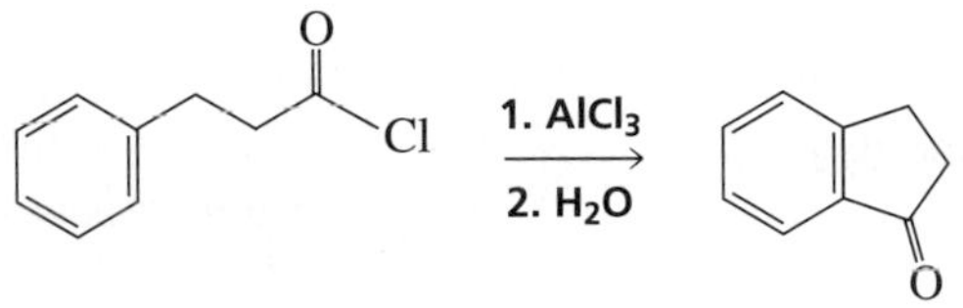

The synthesis of benzaldehyde from benzene poses a problem because formyl chloride, the acyl halide required for the reaction, is unstable and cannot be purchased. Formyl chloride can be prepared, however, by means of the **Gatterman–Koch** formylation **reaction**. This reaction uses a high-pressure mixture of carbon monoxide and HCl to generate formyl chloride; it also uses an aluminum chloride–cuprous chloride catalyst for the acylation reaction itself.

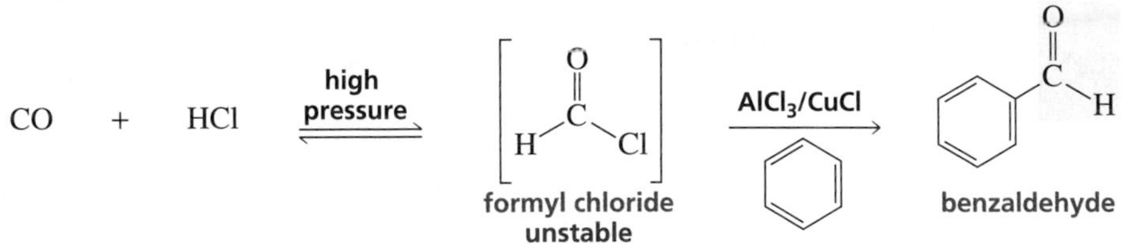

14.15 Friedel–Crafts Alkylation of Benzene

Friedel–Crafts alkylation substitutes an alkyl group for a hydrogen.

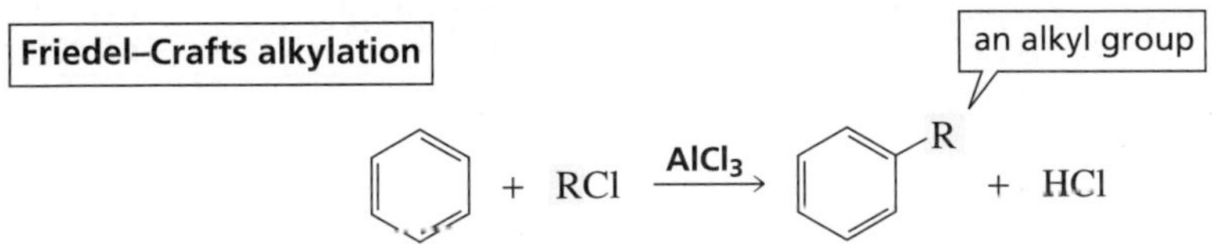

In the first step of the reaction, a carbocation is formed from the reaction of an alkyl halide with AlCl$_3$. Alkyl fluorides, alkyl chlorides, alkyl bromides, and alkyl iodides can all be used. Vinyl halides and aryl halides cannot be used because their carbocations are too unstable to be formed (Section 8.8).

$$R{-}\ddot{\underset{..}{Cl}}{:} + AlCl_3 \longrightarrow R^+ + {}^-AlCl_4$$

an alkyl halide   a carbocation

Mechanism for Friedel–Crafts alkylation

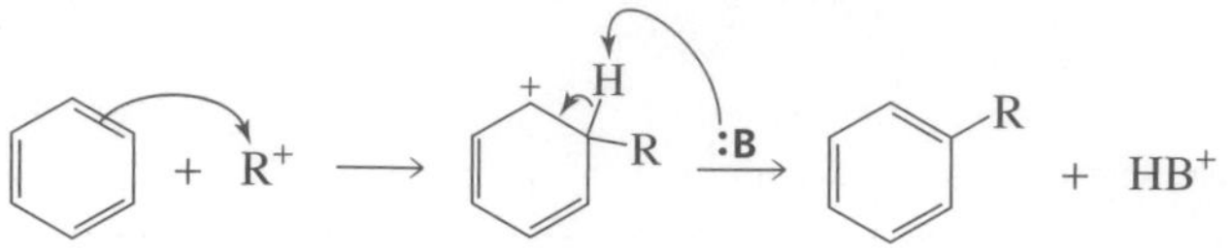

In Section 15.2, we will see that an alkyl-substituted benzene is more reactive than benzene. Therefore, to prevent further alkylation of the alkyl-substituted benzene, a large excess of benzene is used in Friedel–Crafts alkylation reactions. When benzene is in excess, the electrophile is more likely to encounter a molecule of benzene than a molecule of alkyl-substituted benzene.

Recall that a carbocation will rearrange if rearrangement leads to a more stable carbocation (Section 4.6). When the carbocation formed in a Friedel–Crafts alkylation reaction rearranges, the major product will be the product with the rearranged alkyl group on the benzene ring. For example, when benzene reacts with 1-chlorobutane, 60%–80% of the product (the actual percentage depends on the reaction conditions) is the one with the rearranged alkyl substituent.

1-chlorobutane   1-phenylbutane 35%   2-phenylbutane 65%

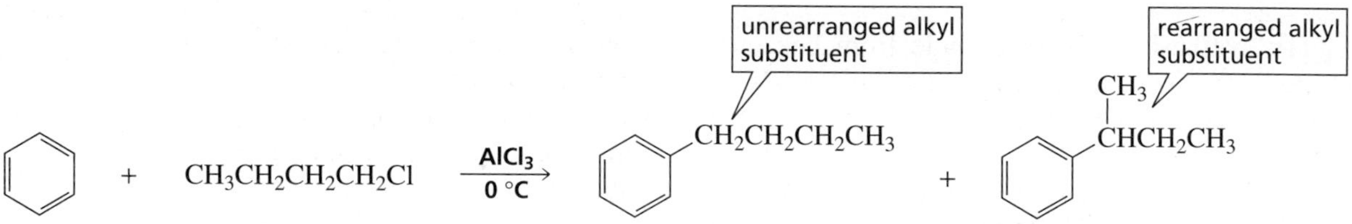

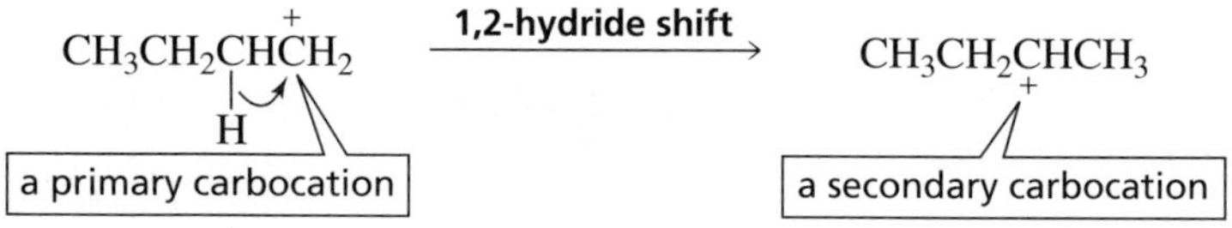

When benzene reacts with 1-chloro-2,2-dimethylpropane, 100% of the product (under all reaction conditions) has the rearranged alkyl substituent. Steric hindrance makes it difficult for benzene to react with the complexed unrearranged carbocation (see the box "Incipient Primary Carbocations"), so no unrearragned product is formed.

1-chloro-2,2-dimethylpropane   2,2-dimethyl-1-phenylpropane 0%   2-methyl-2-phenylbutane 100%

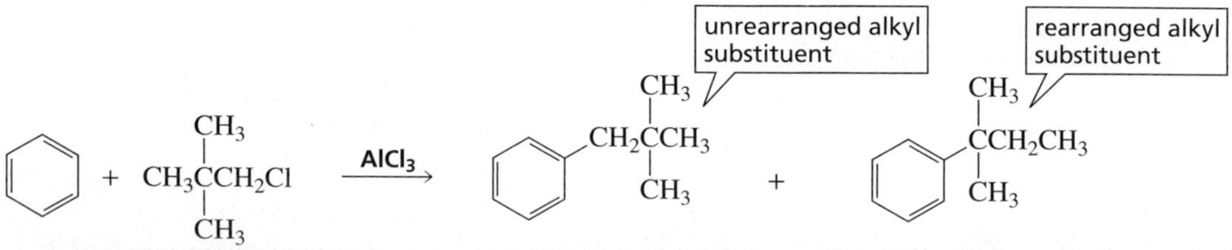

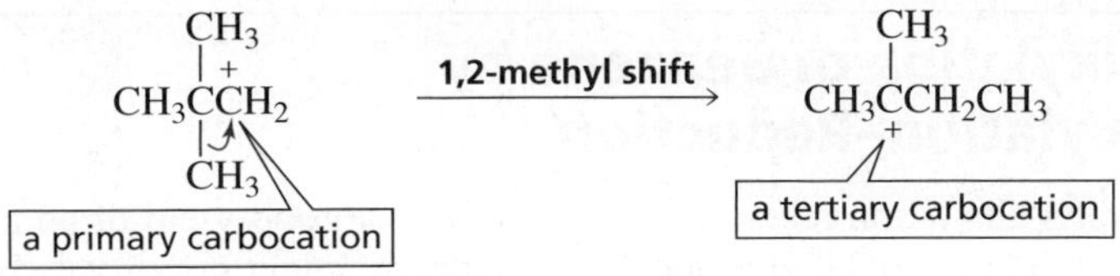

a primary carbocation → **1,2-methyl shift** → a tertiary carbocation

In addition to reacting with carbocations generated from alkyl halides, benzene can react with carbocations generated from the reaction of an alkene (Section 4.1) or an alcohol (Section 10.1) with an acid.

alkylation of benzene by an alkene

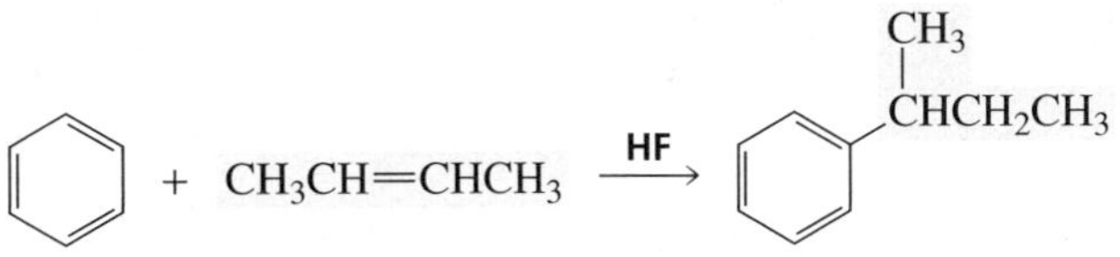

$$+ \quad CH_3CH{=}CHCH_3 \xrightarrow{\text{HF}}$$

sec-butylbenzene

alkylation of benzene by an alcohol

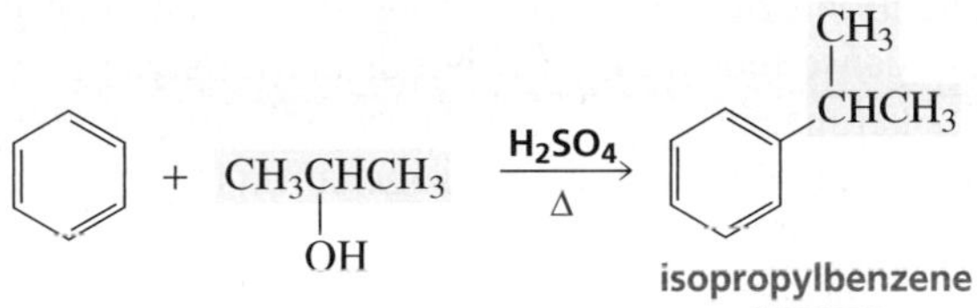

$$+ \quad CH_3CHCH_3 \xrightarrow[\Delta]{\text{H}_2\text{SO}_4}$$
         |
         OH

isopropylbenzene
cumene

INCIPIENT PRIMARY CARBOCATIONS

For simplicity, the two reactions on pages 662 and 663 that involve carbocation rearrangements were written showing the formation of a primary carbocation. However, as we saw in Section 8.5, primary carbocations are too unstable to be formed. The fact is, a true primary carbocation is never formed in a Friedel–Crafts alkylation reaction. Instead, the carbocation remains complexed with the Lewis acid; it is called an "incipient" carbocation. A carbocation rearrangement occurs because the incipient carbocation has sufficient carbocation character to permit the rearrangement.

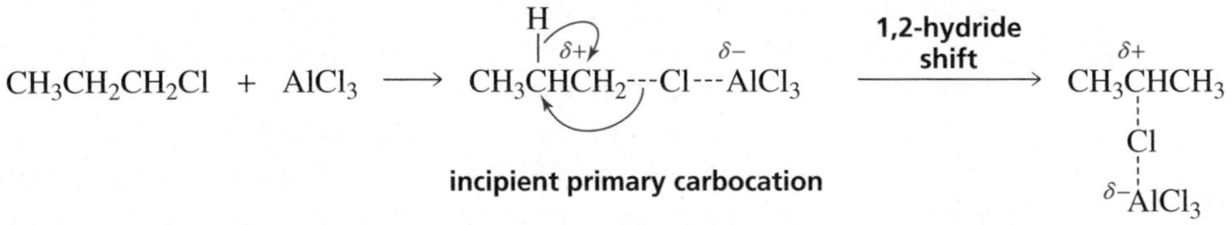

$$CH_3CH_2CH_2Cl + AlCl_3 \longrightarrow CH_3CHCH_2{\cdots}Cl{\cdots}AlCl_3 \xrightarrow[\text{shift}]{\text{1,2-hydride}} CH_3CHCH_3$$

incipient primary carbocation

PROBLEM 25

Show the mechanism for the alkylation of benzene by 2-butene + HF.

PROBLEM 26◆

What would be the major product of a Friedel–Crafts alkylation reaction using the following alkyl chlorides?

a. CH_3CH_2Cl **c.** $CH_3CH_2CH(Cl)CH_3$ **e.** $(CH_3)_2CHCH_2Cl$

b. $CH_3CH_2CH_2Cl$ **d.** $(CH_3)_3CCl$ **f.** $CH_2{=}CHCH_2Cl$

14.16 Alkylation of Benzene by Acylation–Reduction

A Friedel–Crafts alkylation reaction cannot produce a good yield of an alkylbenzene containing a straight-chain alkyl group, because the incipient primary carbocation will rearrange to a more stable carbocation.

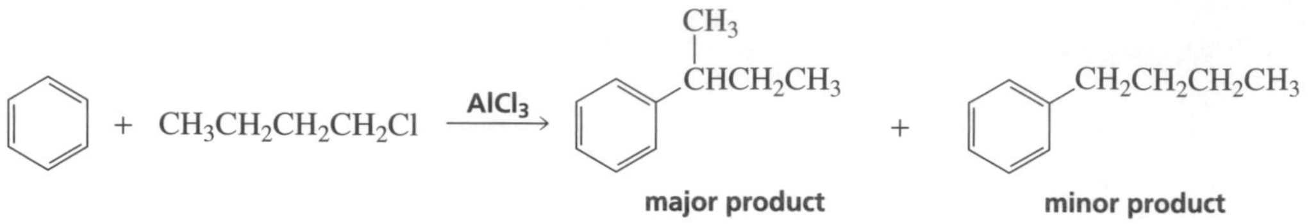

Acylium ions, however, do not rearrange. Consequently, a straight-chain alkyl group can be placed on a benzene ring by means of a Friedel–Crafts acylation reaction, followed by reduction of the carbonyl group to a methylene group. It is called a reduction reaction because the two C—O bonds are replaced by two C—H bonds (Section 4.11). Only a ketone carbonyl group that is adjacent to a benzene ring can be reduced to a methylene group by catalytic hydrogenation (H_2/Pd).

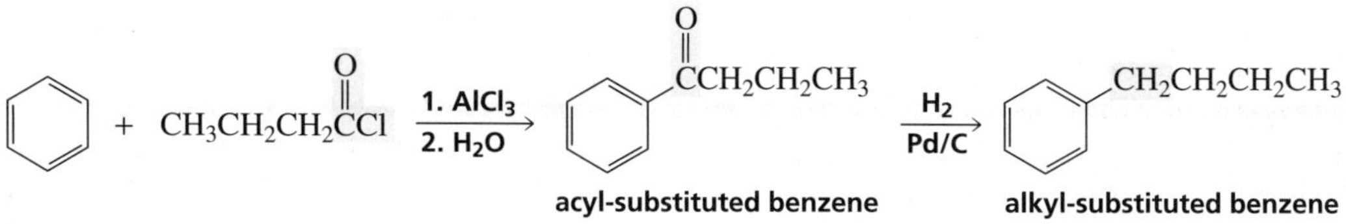

BIOGRAPHY

E. C. Clemmensen (1876–1941) *was born in Denmark and received a Ph.D. from the University of Copenhagen. He was a scientist at Clemmensen Corp. in Newark, New York.*

Ludwig Wolff (1857–1919) *was born in Germany. He received a Ph.D. from the University of Strasbourg. He was a professor at the University of Jena in Germany.*

N. M. Kishner (1867–1935) *was born in Moscow. He received a Ph.D. from the University of Moscow under the direction of Vladimir Markovnikov (page 167). He was a professor at the University of Tomsk and later at the University of Moscow.*

Besides avoiding carbocation rearrangements, another advantage to preparing alkyl-substituted benzenes by acylation–reduction rather than by direct alkylation is that a large excess of benzene does not have to be used (Section 14.15). Unlike alkyl-substituted benzenes, which are more reactive than benzene (Section 15.2), acyl-substituted benzenes are less reactive than benzene, so they will not undergo additional Friedel–Crafts reactions.

There are several other methods available for reducing a ketone carbonyl group to a methylene group. These methods reduce all ketone carbonyl groups, not just those adjacent to benzene rings. Two of the most effective are the Clemmensen reduction and the Wolff–Kishner reduction. The **Clemmensen reduction** uses an acidic solution of zinc dissolved in mercury as the reducing reagent. The **Wolff–Kishner reduction** employs hydrazine (H_2NNH_2) under basic conditions. The mechanism of the Wolff–Kishner reduction is shown in Section 17.8.

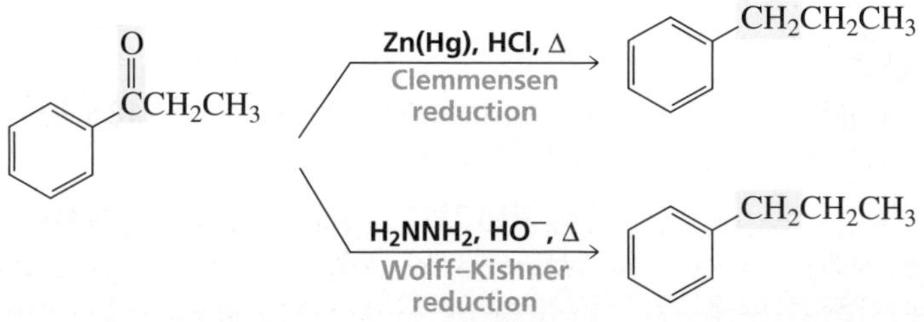

14.17 Using Coupling Reactions to Alkylate Benzene

Alkylbenzenes with straight-chain alkyl groups can also be prepared by means of the coupling reactions described in Section 10.13. One of the alkyl groups of a Gilman reagent can replace the halogen of an aryl halide.

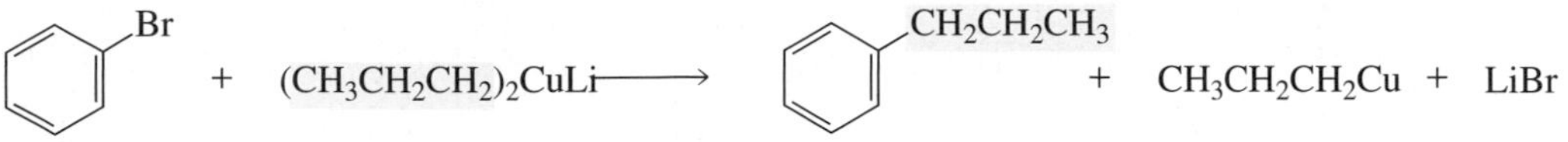

The **Stille reaction** couples an aryl halide with a stannane.

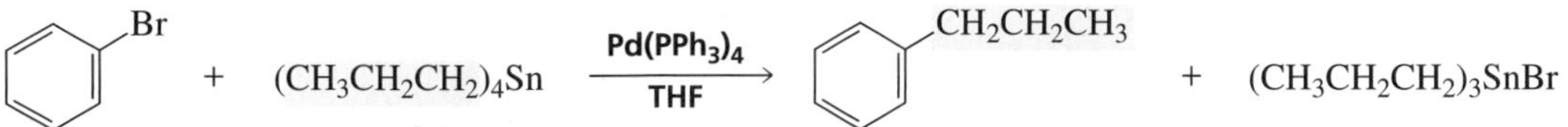

tetrapropylstannane

The **Suzuki reaction** couples an aryl halide with an organoborane.

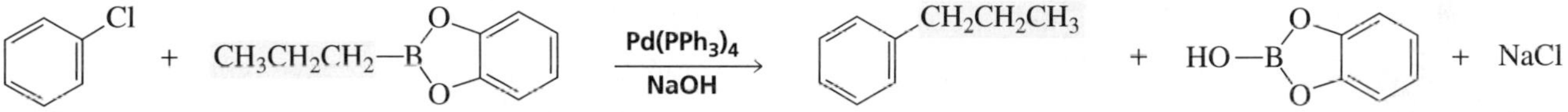

an organoborane   propylbenzene

The required organoborane is obtained from the reaction of an alkene with catecholborane. Because alkenes are readily available, this method can be used to prepare a wide variety of alkyl benzenes.

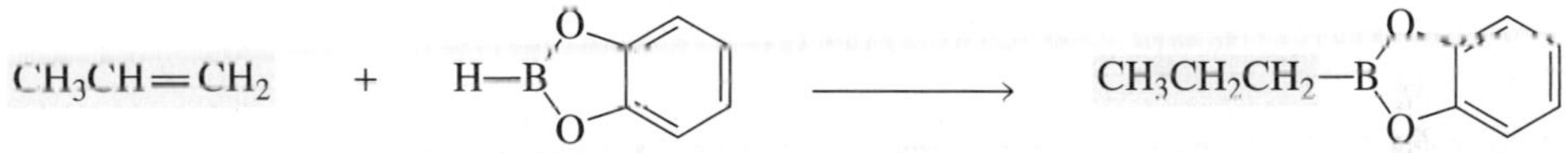

catecholborane

PROBLEM 27

Describe how the following compounds could be prepared from benzene:

a.

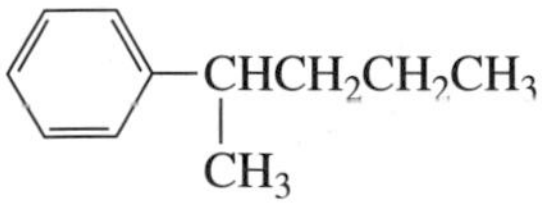

b.

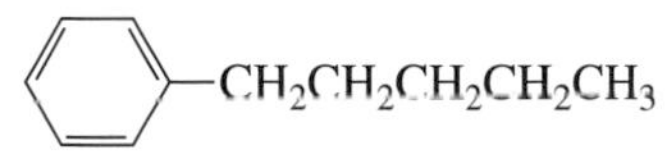

14.18 It Is Important to Have More Than One Way to Carry Out a Reaction

At this point, you may wonder why it is necessary to have more than one way to carry out the same reaction. Alternative methods are useful when there is another functional group in the molecule that could react with the reagents you are using to carry out the desired reaction. For example, heating the following compound with HCl (as required by the Clemmensen reduction) would cause the alcohol to undergo substitution (Section 10.1). Under the basic conditions of the Wolff–Kishner reduction, however, the alcohol group would remain unchanged.

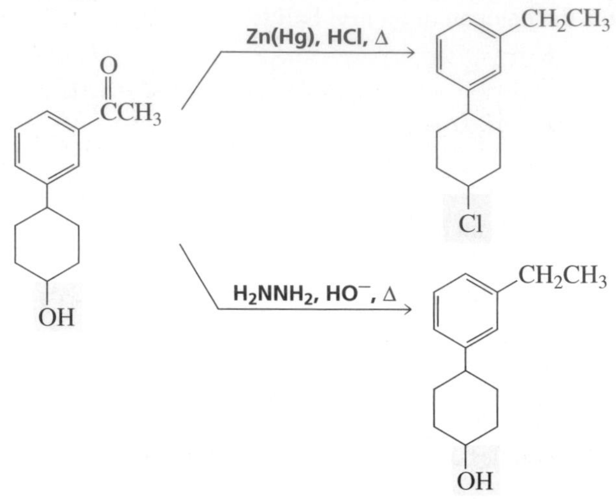

A compound could have a group that would require that both strong acids and strong bases be avoided. In such a case, rather than adding an alkyl substituent to a benzene ring by alkylation–reduction, one of the coupling reactions could be used.

14.19 How Some Substituents on a Benzene Ring Can Be Chemically Changed

Benzene rings with substituents other than the ones listed in Section 14.10 can be prepared by first synthesizing one of these substituted benzenes and then chemically changing the substituent. Several of these reactions should be familiar.

Reactions of Alkyl Substituents

We have seen that a bromine will selectively substitute for a benzylic hydrogen in a radical substitution reaction.

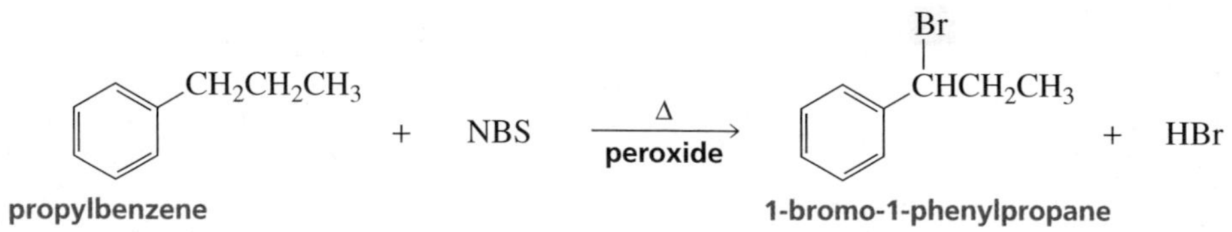

propylbenzene + NBS $\xrightarrow[\text{peroxide}]{\Delta}$ 1-bromo-1-phenylpropane + HBr

(NBS stands for *N*-bromosuccinimide; Section 11.8.)

Once a halogen has been placed in the benzylic position, it can be replaced by a nucleophile by means of an S_N2 or S_N1 reaction (Sections 8.3 and 8.6). A wide variety of substituted benzenes can be prepared this way.

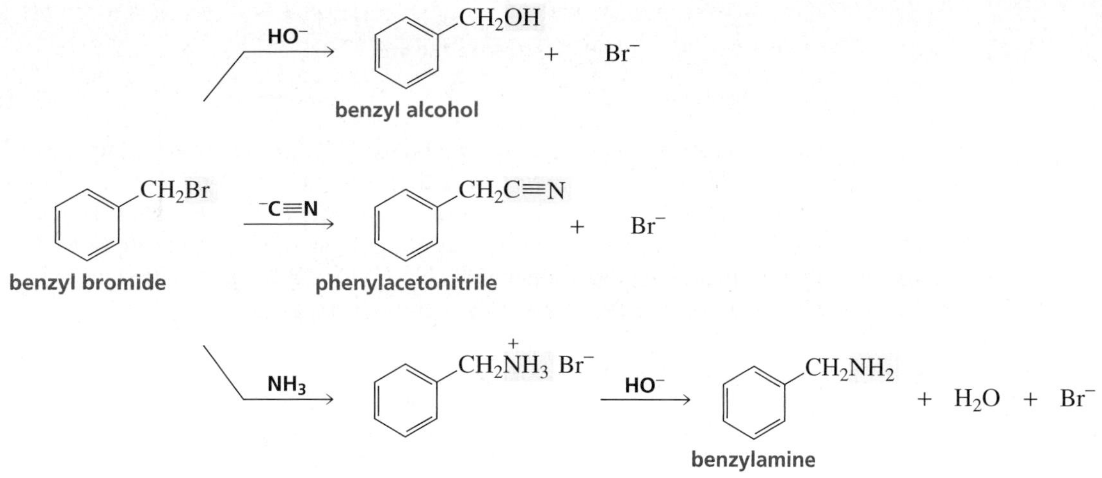
CH2OH
HO⁻
benzyl alcohol
CH2Br
benzyl bromide
C≡N
CH2C≡N
phenylacetonitrile
NH3
CH2NH3 Br⁻
HO⁻
CH2NH2
benzylamine
+ H2O + Br⁻

Remember that halo-substituted alkyl groups can also undergo E2 and E1 reactions (Section 9.8). Notice that a bulky base (*tert*-BuO$^-$) is used to encourage elimination over substitution.

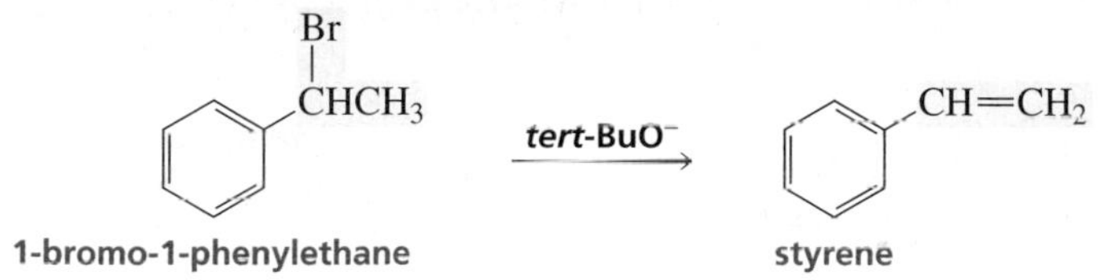
Br
CHCH3
tert-BuO⁻
CH=CH2
1-bromo-1-phenylethane
styrene

Substituents with double and triple bonds can undergo catalytic hydrogenation (Sections 4.11 and 6.9). Recall that addition of hydrogen to a double or triple bond is an example of a reduction reaction, a reaction that increases the number of C—H bonds or decreases the number of C—O, C—N, or C—X bonds in a compound (X denotes a halogen).

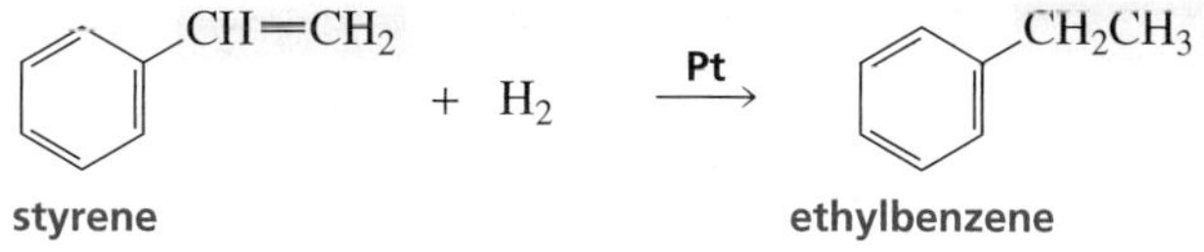
CH=CH2
+ H2
Pt
CH2CH3
styrene
ethylbenzene

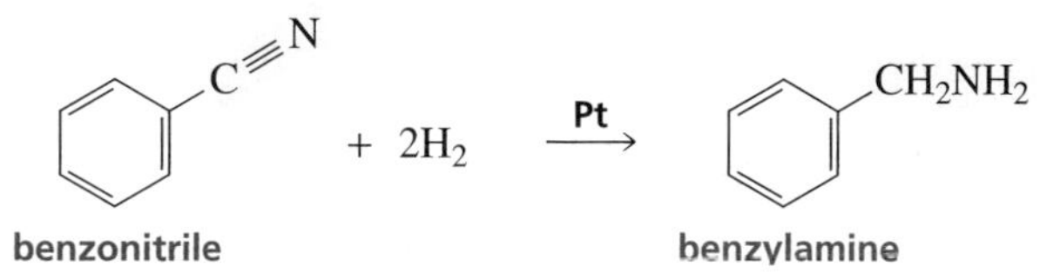
C≡N
+ 2H2
Pt
CH2NH2
benzonitrile
benzylamine

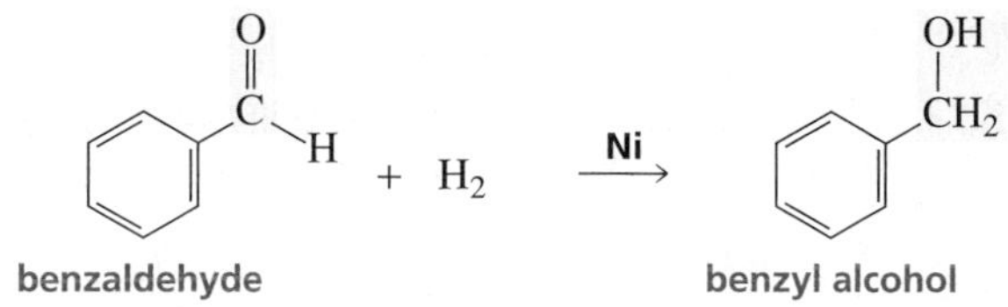
O
C
H
+ H2
Ni
OH
CH2
benzaldehyde
benzyl alcohol

Because benzene is an unusually stable compound (Section 14.1), it can be reduced only at high temperature and pressure.

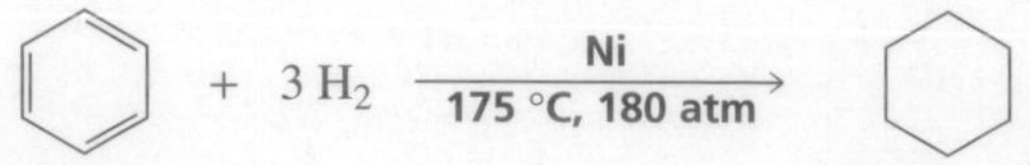

benzene + 3 H₂ $\xrightarrow[\text{175 °C, 180 atm}]{\text{Ni}}$ cyclohexane

An alkyl group bonded to a benzene ring can be oxidized to a carboxyl group. Recall that when an organic compound is *oxidized,* either the number of C—O, C—N, or C—X (X denotes a halogen) bonds increases or the number of C—H bonds decreases (Section 4.9). Commonly used oxidizing agents are potassium permanganate ($KMnO_4$) and acidic solutions of sodium dichromate (H^+, $Na_2Cr_2O_7$). Because the benzene ring is so stable, it will not be oxidized—only the alkyl group is oxidized.

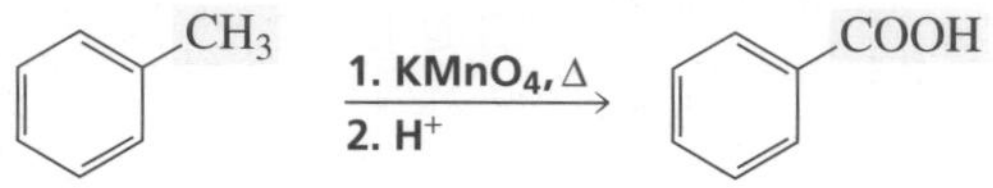

toluene $\xrightarrow[\text{2. H}^+]{\text{1. KMnO}_4, \Delta}$ benzoic acid

Regardless of the length of the alkyl substituent, it will be oxidized to a COOH group, provided that a hydrogen is bonded to the benzylic carbon.

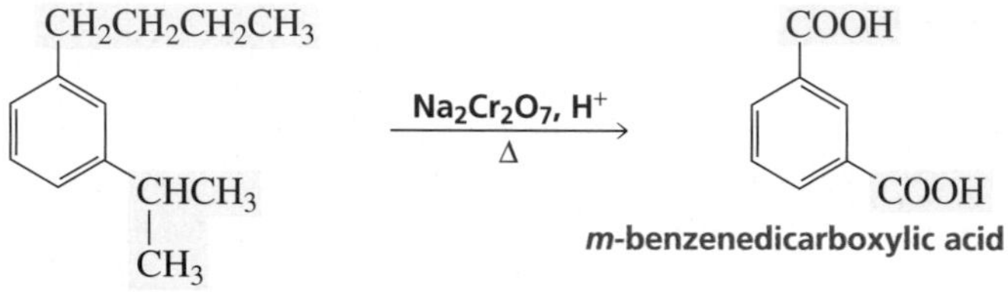

m-butylisopropylbenzene $\xrightarrow[\Delta]{\text{Na}_2\text{Cr}_2\text{O}_7, \text{H}^+}$ *m*-benzenedicarboxylic acid

If the alkyl group lacks a benzylic hydrogen, the oxidation reaction will not occur because the first step in the oxidation reaction is removal of a hydrogen from the benzylic carbon.

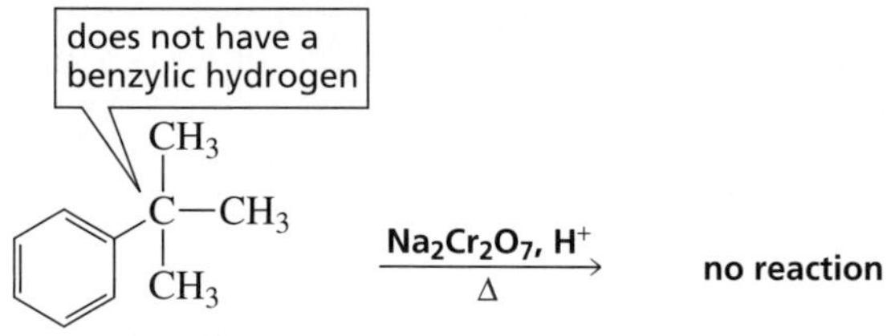

does not have a benzylic hydrogen

tert-butylbenzene $\xrightarrow[\Delta]{\text{Na}_2\text{Cr}_2\text{O}_7, \text{H}^+}$ no reaction

The same reagents that oxidize alkyl substituents will oxidize benzylic alcohols to benzoic acid.

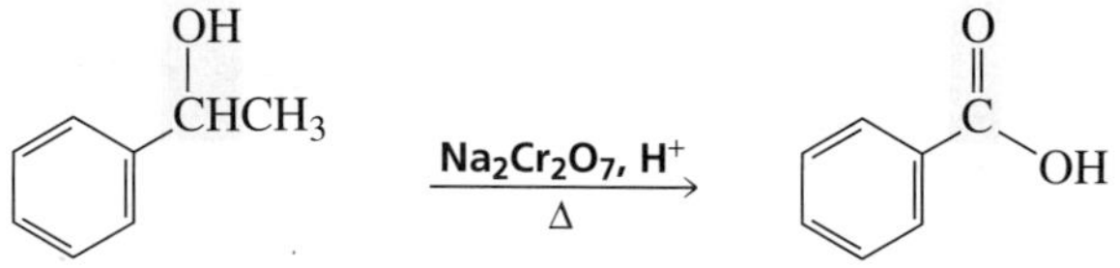

1-phenylethanol $\xrightarrow[\Delta]{\text{Na}_2\text{Cr}_2\text{O}_7, \text{H}^+}$ benzoic acid

If, however, a mild oxidizing agent such as MnO_2 is used, benzylic alcohols are oxidized to aldehydes or ketones.

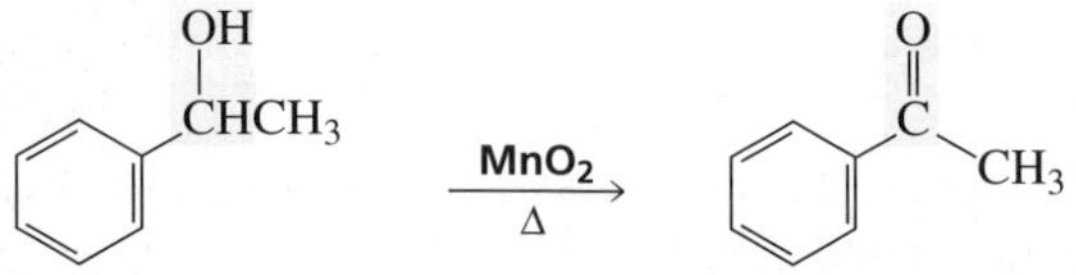

1-phenylethanol $\xrightarrow[\Delta]{\text{MnO}_2}$ acetophenone

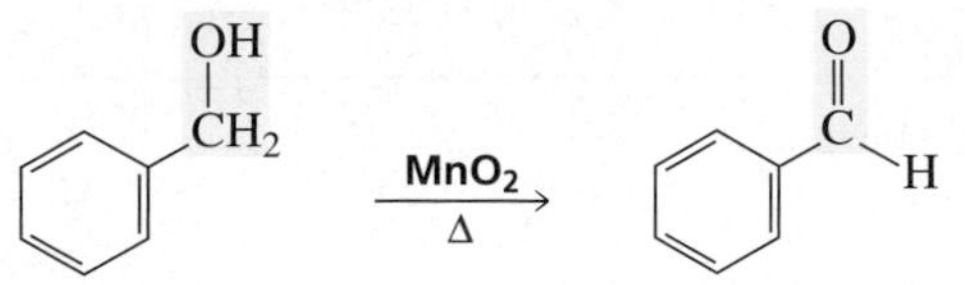

phenylmethanol
benzyl alcohol

benzaldehyde

Reducing a Nitro Substituent

A nitro substituent can be reduced to an amino substituent. Either a metal (tin, iron, or zinc) plus an acid (HCl) or catalytic hydrogenation can be used to carry out this reaction. Recall from Section 1.24 that if acidic conditions are employed, the product will be in its acidic form (anilinium ion). When the reaction is over, base can be added to convert the product into its basic form (aniline).

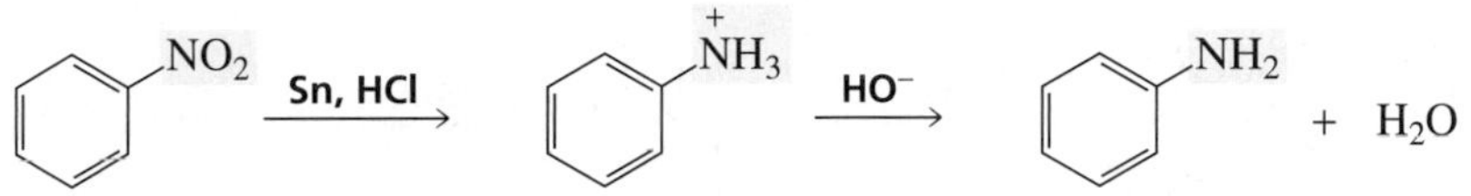

nitrobenzene

protonated
aniline
anilinium ion

aniline $+$ H_2O

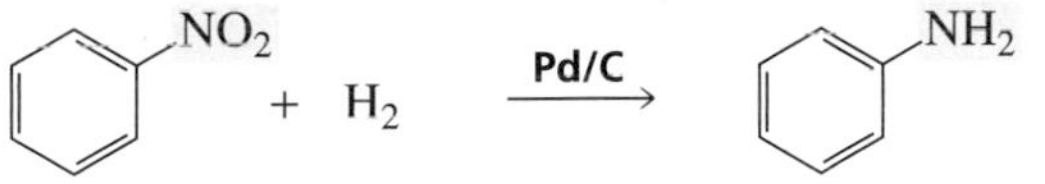

It is possible to selectively reduce just one of two nitro groups.

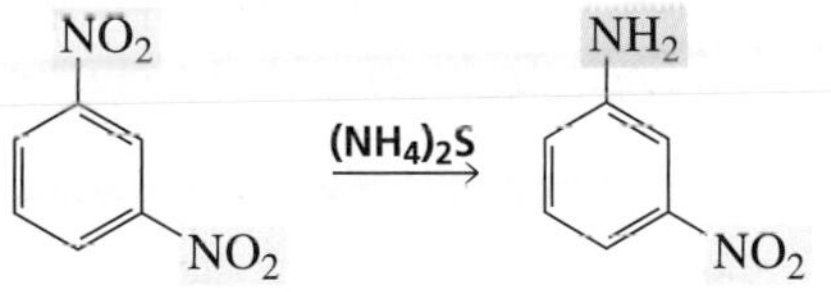

1,3-dinitrobenzene

3-nitroaniline

PROBLEM 28◆

Give the product of each of the following reactions:

a.
$Na_2Cr_2O_7, H^+$
Δ

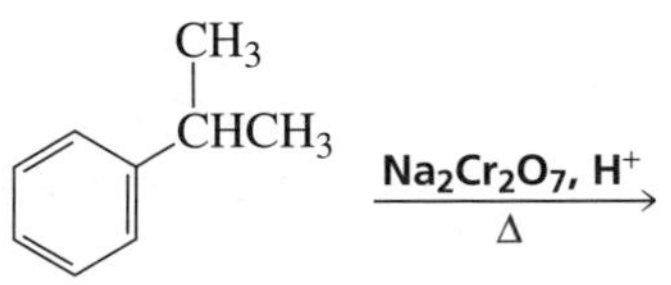

c.
1. NBS/Δ/peroxide
2. CH_3O^-

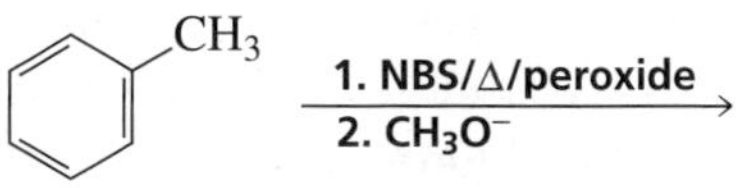

b.
$Na_2Cr_2O_7, H^+$
Δ

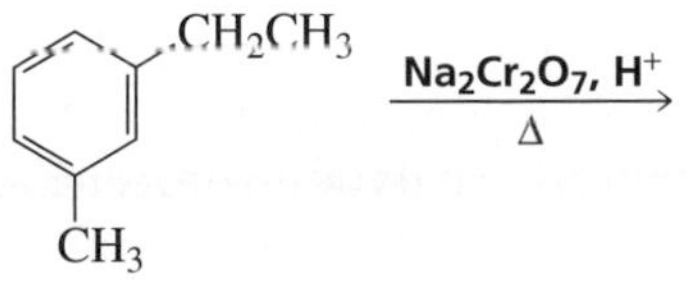

d.
1. NBS/Δ/peroxide
2. $^-C{\equiv}N$
3. H_2/Ni

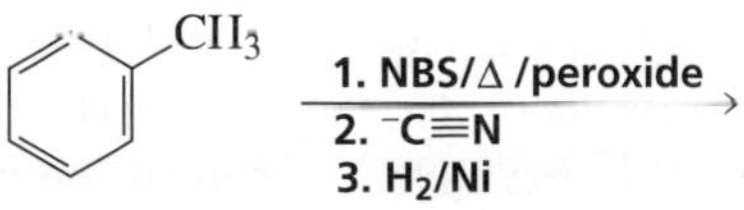

PROBLEM 29 **SOLVED**

Show how the following compounds could be prepared from benzene:

a. benzaldehyde   **c.** 1-bromo-2-phenylethane   **e.** aniline
b. styrene   **d.** 2-phenyl-1-ethanol   **f.** benzoic acid

Solution to 29a   Benzaldehyde cannot be prepared by a Friedel–Crafts acylation because formyl chloride, the required acyl halide, is unstable.

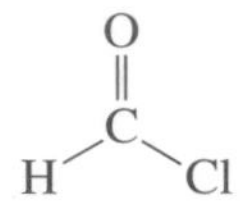

formyl chloride
unstable

Benzaldehyde can be prepared by the Gatterman–Koch reaction (page 661) or by the following sequence of reactions:

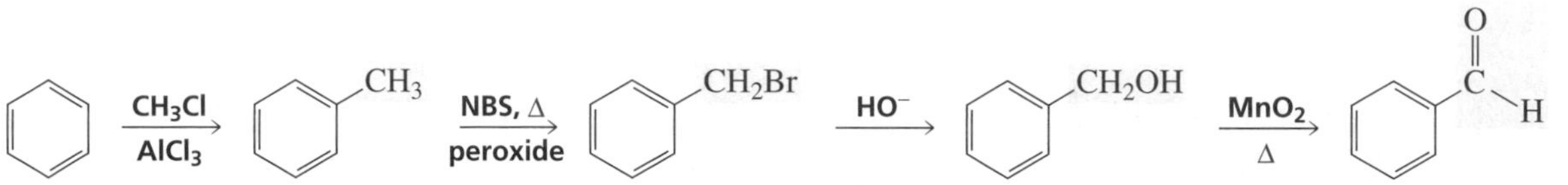

SUMMARY

To be classified as **aromatic**, a compound must have an uninterrupted cyclic cloud of π electrons that contains an *odd number of pairs* of π electrons. An **antiaromatic compound** has an uninterrupted cyclic cloud of π electrons with an *even number of pairs* of π electrons. Molecular orbital theory shows that aromatic compounds are stable because their bonding orbitals are completely filled, with no electrons in either nonbonding or antibonding orbitals; in contrast, antiaromatic compounds are unstable because they either are unable to fill their bonding orbitals or they have a pair of π electrons left over after the bonding orbitals are filled. As a result of their aromaticity, the cyclopentadienyl anion and the cycloheptatrienyl cation are unusually stable.

An **annulene** is a monocyclic hydrocarbon with alternating single and double bonds. A **heterocyclic compound** is a cyclic compound in which one or more of the ring atoms is a **heteroatom**, an atom other than carbon. Pyridine, pyrrole, furan, and thiophene are aromatic heterocyclic compounds.

Benzene's aromaticity causes it to undergo **electrophilic aromatic substitution reactions**. Electrophilic addition reactions that are characteristic of alkenes and dienes would lead to much less stable nonaromatic addition products. The most common electrophilic aromatic substitution reactions are halogenation, nitration, sulfonation, and Friedel–Crafts acylation and alkylation. Once the electrophile is generated, all electrophilic aromatic substitution reactions take place by

the same two-step mechanism: (1) the aromatic compound reacts with an electrophile, forming a carbocation intermediate; and (2) a base pulls off a proton from the carbon that formed the bond with the electrophile. The first step is relatively slow and endergonic because an aromatic compound is being converted into a much less stable nonaromatic intermediate; the second step is fast and strongly exergonic because the stability-enhancing aromaticity is being restored.

Some monosubstituted benzenes are named as substituted benzenes (for example, bromobenzene, nitrobenzene); some have names that incorporate the substituent (for example, toluene, phenol, aniline, anisole). Bromination or chlorination requires a Lewis acid catalyst; iodination requires an oxidizing agent. **Nitration** with nitric acid requires sulfuric acid as a catalyst. Either an acyl halide or an acid anhydride can be used for **Friedel–Crafts acylation**, a reaction that places an acyl group on a benzene ring. If the carbocation formed from the alkyl halide used in a **Friedel–Crafts alkylation** reaction can rearrange, the major product will be the product with the rearranged alkyl group. A straight-chain alkyl group can be placed on a benzene ring via a Friedel–Crafts acylation reaction, followed by reduction of the carbonyl group by catalytic hydrogenation, a **Clemmensen reduction**, or a **Wolff–Kishner reduction**. Alkylbenzenes with straight-chain alkyl groups can also be prepared by means of coupling reactions.

A benzene ring can be sulfonated with fuming or concentrated sulfuric acid. **Sulfonation** is a reversible reaction; heating benzenesulfonic acid in dilute acid removes the sulfonic acid group. The **principle of microscopic reversibility** states that the mechanism of a reaction in the reverse direction must retrace each step of the mechanism in the forward direction in microscopic detail.

Benzene rings with substituents other than halo, nitro, sulfonic acid, alkyl, and acyl can be prepared by synthesizing one of these substituted benzenes and then chemically changing the substituent.

SUMMARY OF REACTIONS

1. Electrophilic aromatic substitution reactions:
   a. Halogenation (Section 14.11)

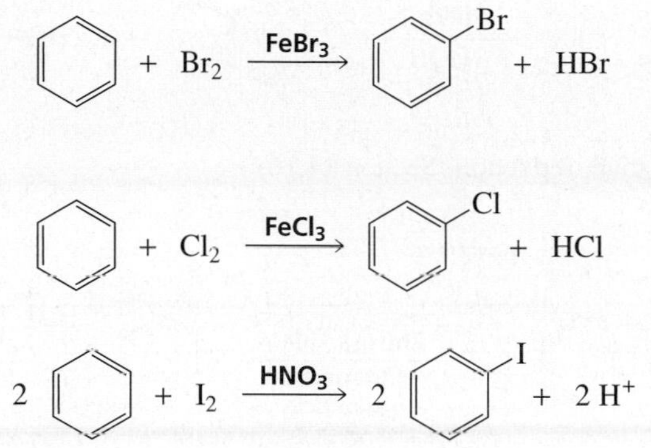

$$\text{benzene} + Br_2 \xrightarrow{FeBr_3} \text{bromobenzene} + HBr$$

$$\text{benzene} + Cl_2 \xrightarrow{FeCl_3} \text{chlorobenzene} + HCl$$

$$2 \text{ benzene} + I_2 \xrightarrow{HNO_3} 2 \text{ iodobenzene} + 2 H^+$$

   b. Nitration, sulfonation, and desulfonation (Sections 14.12 and 14.13)

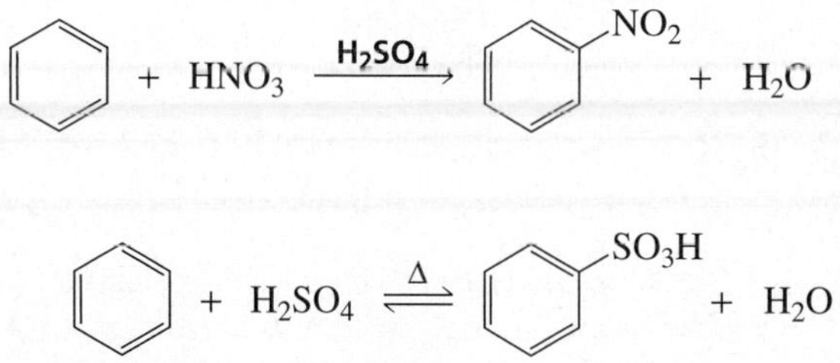

$$\text{benzene} + HNO_3 \xrightarrow{H_2SO_4} \text{nitrobenzene} + H_2O$$

$$\text{benzene} + H_2SO_4 \underset{\Delta}{\rightleftharpoons} \text{benzenesulfonic acid} + H_2O$$

   c. Friedel–Crafts acylation and alkylation (Sections 14.14 and 14.15)

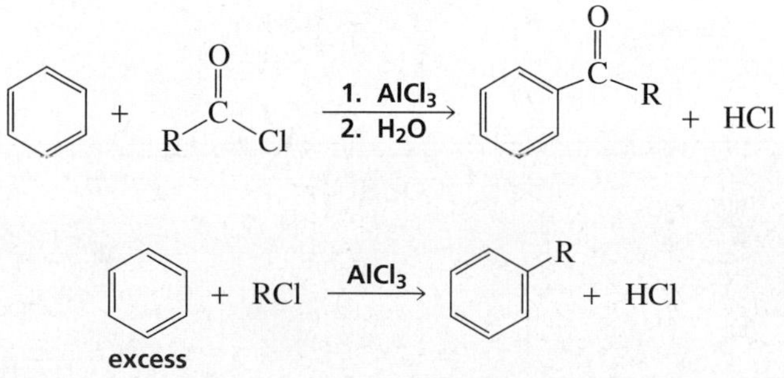

$$\text{benzene} + R\overset{O}{\underset{Cl}{C}} \xrightarrow[\text{2. } H_2O]{\text{1. } AlCl_3} \text{acylbenzene} + HCl$$

$$\text{benzene (excess)} + RCl \xrightarrow{AlCl_3} \text{alkylbenzene} + HCl$$

   d. Formation of benzaldehyde by a Gatterman–Koch reaction (Section 14.14)

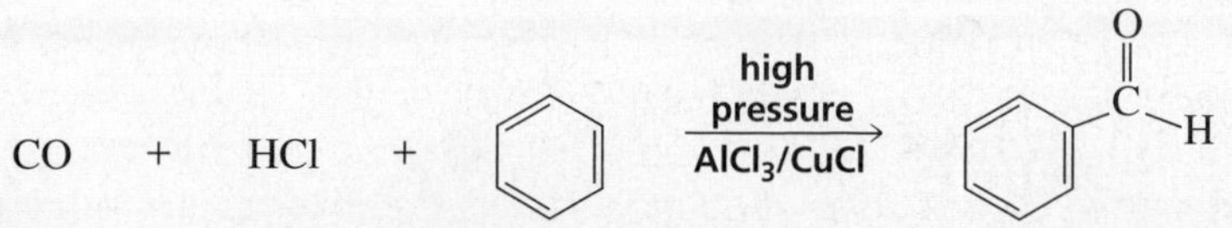

$$CO + HCl + \text{benzene} \xrightarrow[\text{AlCl}_3/\text{CuCl}]{\text{high pressure}} \text{benzaldehyde}$$

e. Alkylation with a Gilman reagent (14.17)

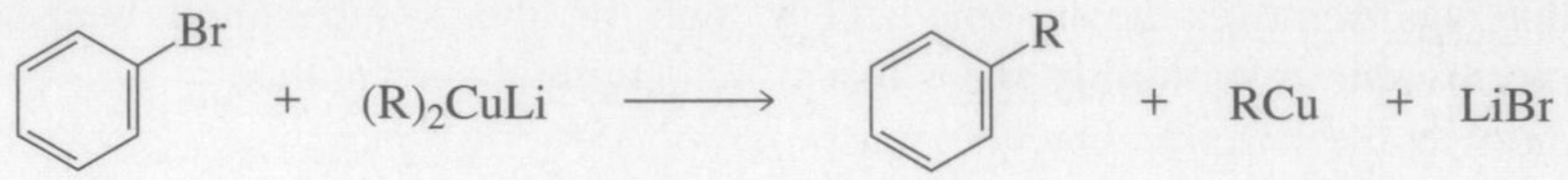

f. Alkylation by a Stille reaction (Section 14.17)

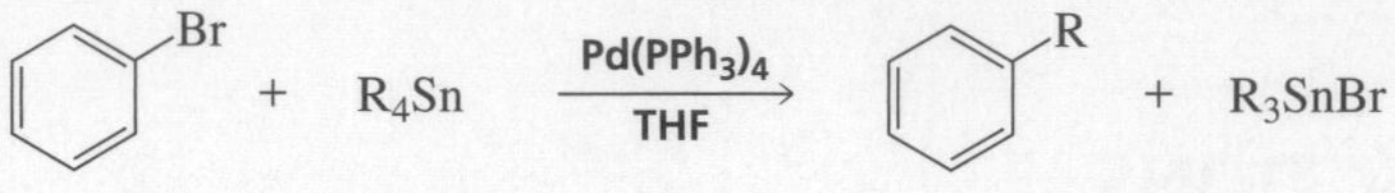

g. Alkylation by a Suzuki reaction (Section 14.17)

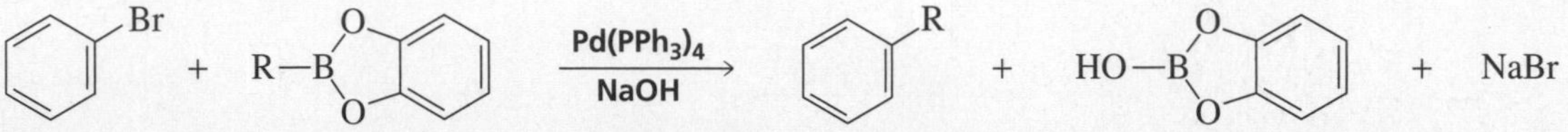

2. Clemmensen reduction and Wolff–Kishner reduction (Section 14.16)

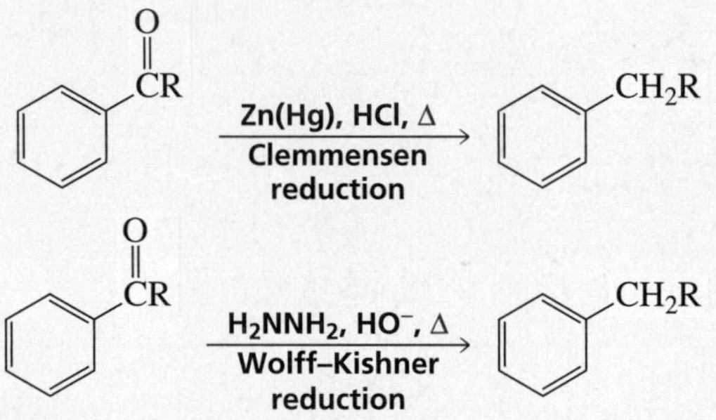

3. Reactions of substituents on a benzene ring (Section 14.19)

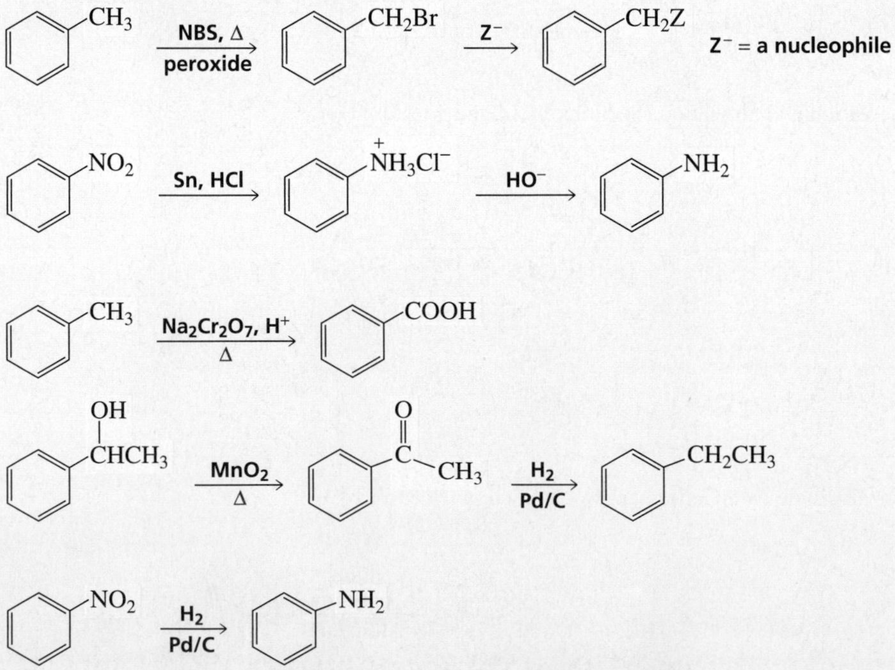

KEY TERMS

aliphatic compound (p. 640)
annulene (p. 643)
antiaromatic compound (p. 649)
aromatic compound (p. 642)
benzyl group (p. 652)
Clemmensen reduction (p. 664)
electrophilic aromatic substitution
  reaction (p. 653)

Friedel–Crafts acylation (p. 654)
Friedel–Crafts alkylation (p. 654)
Gatterman–Koch reaction (p. 661)
halogenation (p. 654)
heteroatom (p. 646)
heterocyclic compound (p. 646)
Hückel's rule, or the $4n + 2$
  rule (p. 642)

nitration (p. 654)
phenyl group (p. 652)
principle of microscopic
  reversibility (p. 659)
Stille reaction (p. 665)
sulfonation (p. 654)
Suzuki reaction (p. 665)
Wolff–Kishner reduction (p. 664)

PROBLEMS

30. Draw the structure for each of the following:
  a. phenol
  b. benzyl phenyl ether
  c. benzonitrile
  d. benzaldehyde
  e. anisole
  f. styrene
  g. toluene
  h. *tert*-butylbenzene
  i. benzyl chloride

31. Classify each of the following compounds as aromatic, nonaromatic, antiaromatic: (*Hint:* If possible, a ring will be nonplanar to avoid being antiaromatic.)

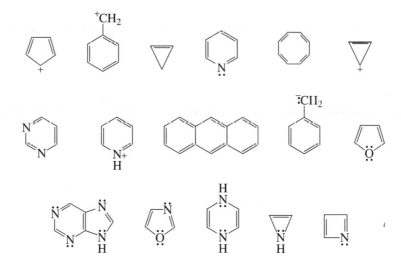

32. Give the product of the reaction of excess benzene with each of the following reagents:
  a. isobutyl chloride + AlCl$_3$
  b. propene + HF
  c. 1-chloro-2,2-dimethylpropane + AlCl$_3$
  d. dichloromethane + AlCl$_3$

33. Which ion in each of the following pairs is more stable and why?

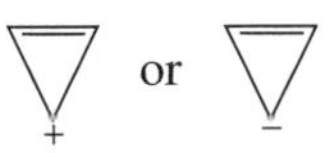

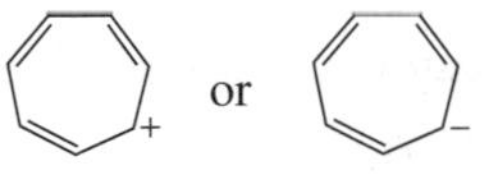

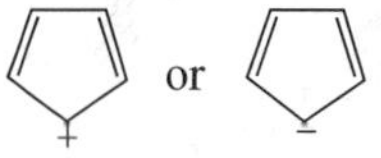

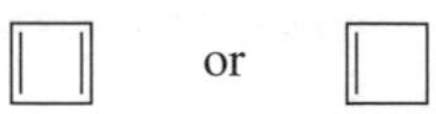

34. Which can lose a proton more readily, a methyl group bonded to cyclohexane or a methyl group bonded to benzene?

35. How could you prepare the following compounds with benzene as one of the starting materials?

a.

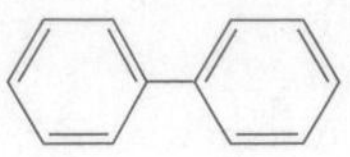

b.

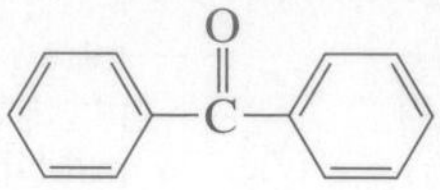

36. Benzene underwent a Friedel–Crafts acylation reaction followed by a Clemmensen reduction. The product gave the following ¹H NMR spectrum. What acyl chloride was used in the Friedel–Crafts acylation reaction?

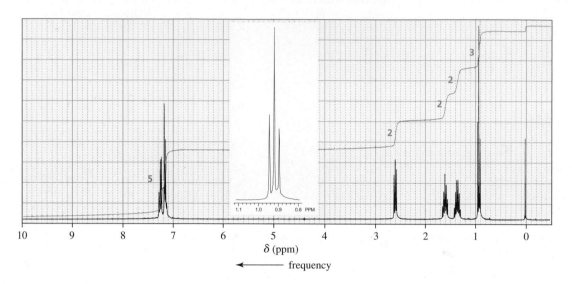

37. Give the products of the following reactions:

a.
CH_2CH_2CCl

$\xrightarrow{\text{1. AlCl}_3 \\ \text{2. H}_2\text{O}}$

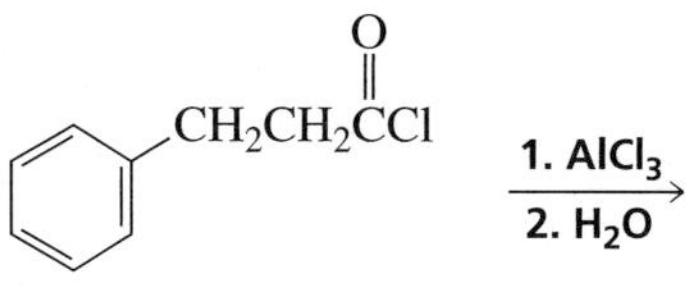

b.
$CH_2CH_2CH_2CCl$

$\xrightarrow{\text{1. AlCl}_3 \\ \text{2. H}_2\text{O}}$

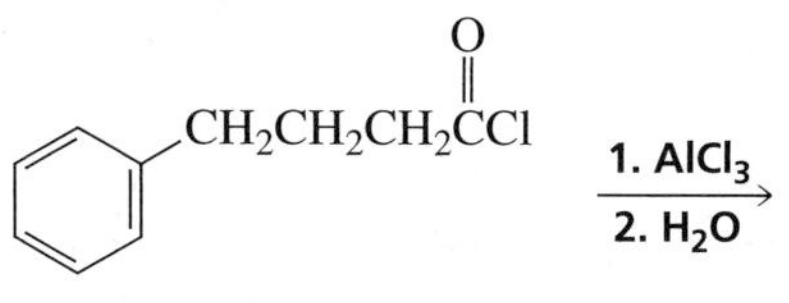

38. Which compound is a stronger base? Why?

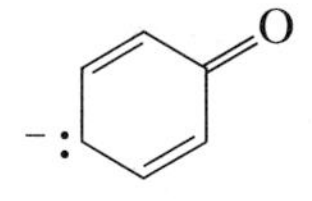

or

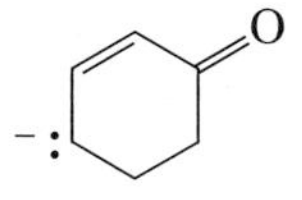

39. Which compound undergoes an S_N1 reaction more rapidly?

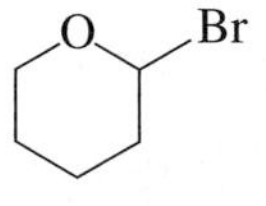

or

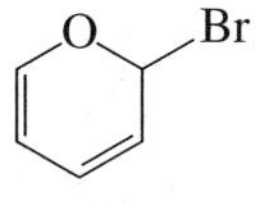

40. Purine is a heterocyclic compound with four nitrogen atoms.
  a. Which nitrogen is most apt to be protonated?   **b.** Which nitrogen is least apt to be protonated?

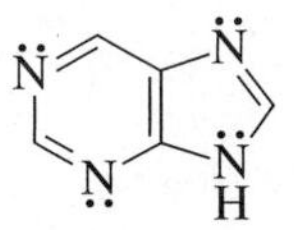

purine

41. Professor Orbie Tal isolated an aromatic compound with molecular formula $C_6H_4Br_2$. He treated this compound with nitric acid and sulfuric acid (conditions that replace an H with an NO_2 group) and isolated three different isomers with molecular formula $C_6H_3Br_2NO_2$. What was the structure of the original compound?

42. Propose a mechanism for each of the following reactions:

a.

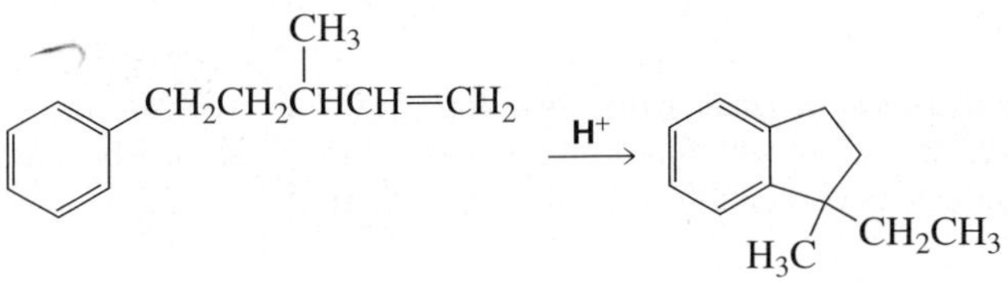

b.

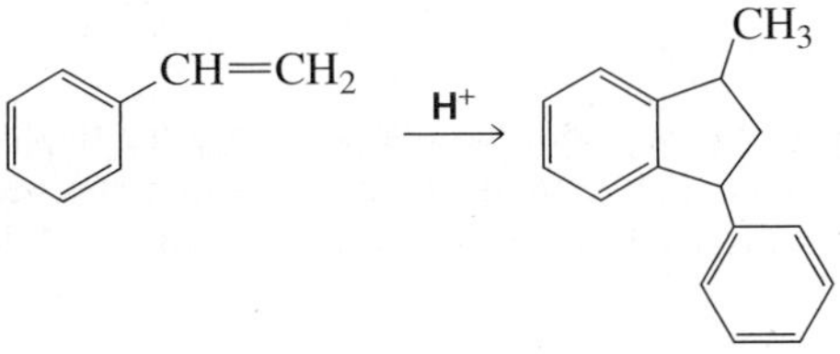

43. Give the product of each of the following reactions:

a.

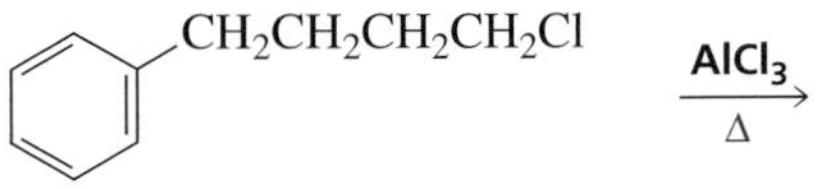

b.

44. Which of the following compounds is the strongest acid?

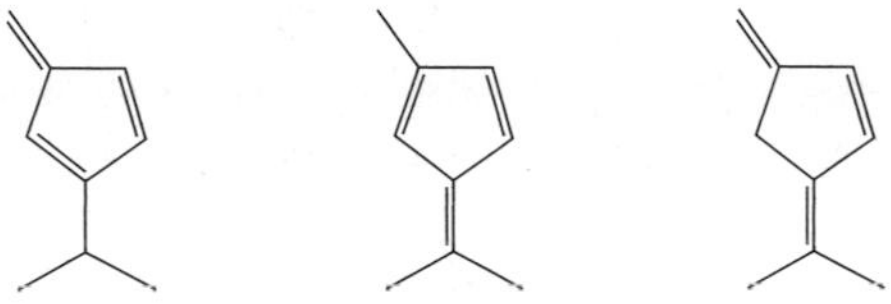

45. Show two ways that the following compound could be synthesized:

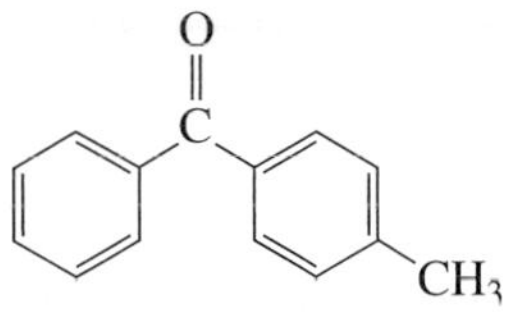

46. a. Propose a mechanism for the following reaction:

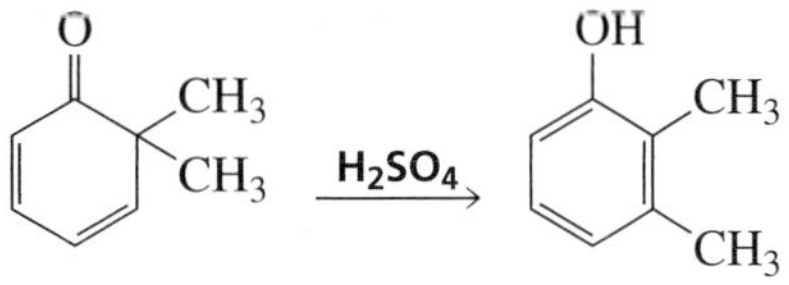

b. Give the product of the following reaction:

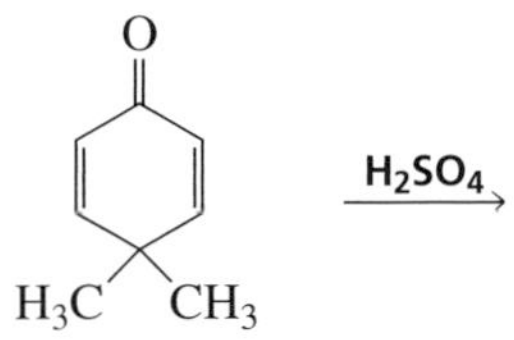

47. In a reaction called the Birch reduction, benzene can be partially reduced to 1,4-cyclohexadiene by an alkali metal (Na, Li, or K) in liquid ammonia and a low-molecular-weight alcohol. Propose a mechanism for this reaction. (*Hint:* See Section 6.9.)

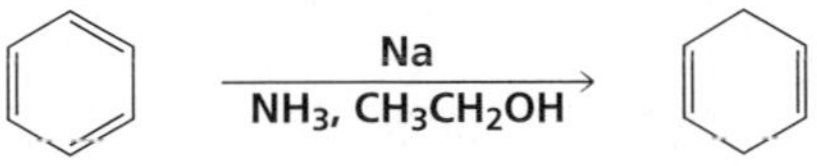

1,4-cyclohexadiene

48. Draw the resonance contributors of the cyclooctatrienyl dianion.
   a. Which of the resonance contributors is the least stable?
   b. Which of the resonance contributors makes the smallest contribution to the hybrid?

49. Investigation has shown that cyclobutadiene is actually a rectangular molecule rather than a square molecule. In addition, it has been established that there are two different 1,2-dideuterio-1,3-cyclobutadienes. Explain the reason for these unexpected observations.

cyclobutadiene

50. The *principle of least motion,* which states that the reaction that involves the least change in atomic positions or electronic configuration (all else being equal) is favored, has been suggested to explain why the Birch reduction forms only 1,4-hexadiene. How does this account for the observation that no 1,3-cyclohexadiene is obtained from a Birch reduction?

CHAPTER 15

Reactions of Substituted Benzenes

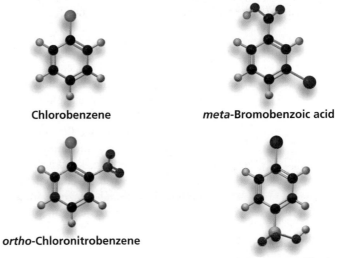

Chlorobenzene

meta-**Bromobenzoic acid**

ortho-**Chloronitrobenzene**

para-**Iodobenzenesulfonic acid**

BUILDING ON FUNDAMENTALS

SECTION 15.2 Inductive electron withdrawal (1.21) and electron donation by hyperconjugation (4.2) are two factors that help explain the reactivity of substituted benzenes.

SECTION 15.2 Resonance electron withdrawal and resonance electron donation (7.9) are also factors that explain the reactivity of substituted benzenes.

SECTION 15.3 Having a complete octet stabilizes a species (1.3, 7.5).

SECTION 15.4 Substituents affect the pK_a values of compounds (1.21, 7.9).

Many substituted benzenes are found in nature. A few that have physiological activity are shown here.

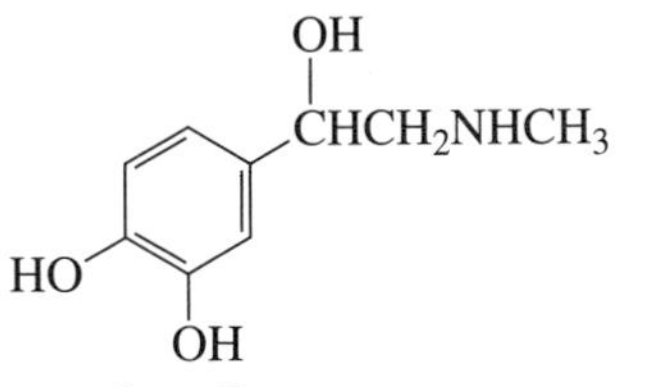

adrenaline
epinephrine
a hormone released
by the body in
response to stress

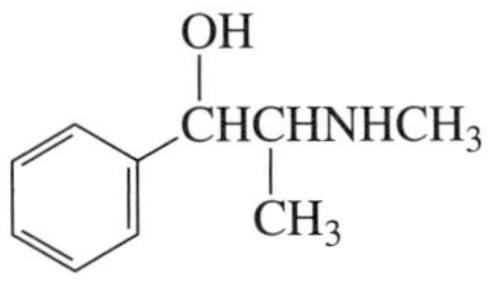

ephedrine
a bronchodilator

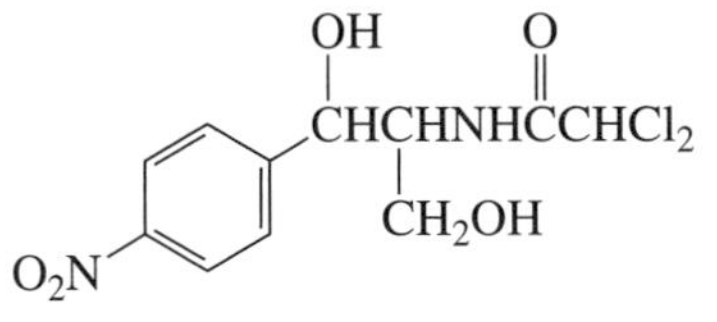

chloramphenicol
an antibiotic that is particularly
effective against typhoid fever

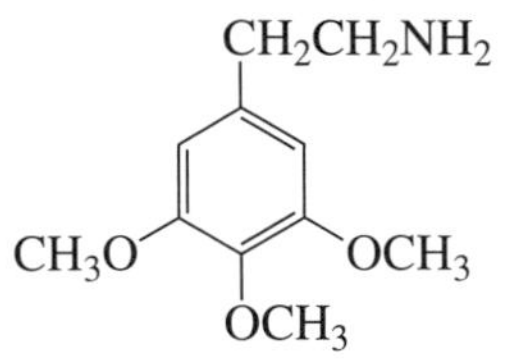

mescaline
active agent of
the peyote cactus

Many other physiologically active substituted benzenes are not found in nature, but exist because chemists have synthesized them. The now-banned diet drug "fen-phen" is a mixture of two synthetic substituted benzenes: fenfluramine and phentermine.

Two other synthetic substituted benzenes, BHA and BHT, are preservatives (Section 11.10) found in a wide variety of packaged foods.

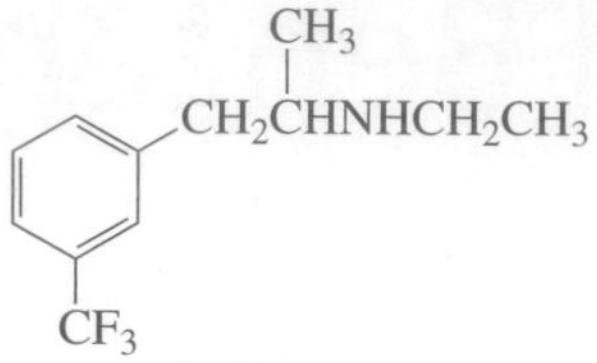

fenfluramine

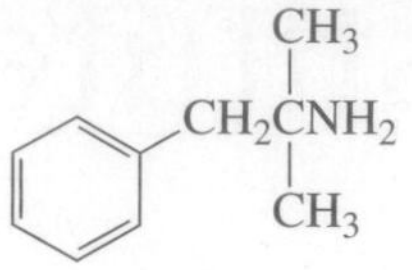

phentermine

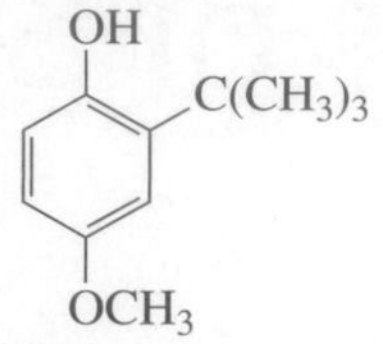

butylated
hydroxyanisole
BHA
a food antioxidant

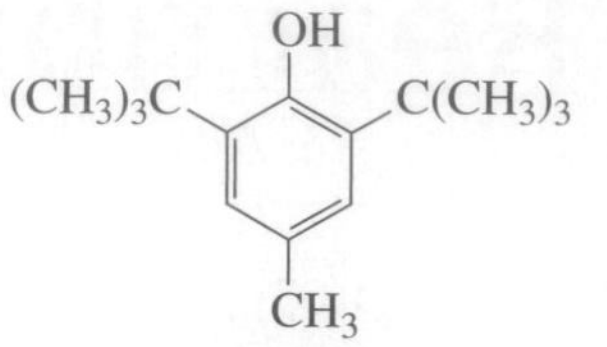

butylated
hydroxytoluene
BHT
a food antioxidant

When naturally occurring compounds are found to have desirable physiological activities, chemists will attempt to synthesize structurally similar compounds for their development into useful products. For example, chemists have synthesized compounds with structures similar to that of adrenaline, producing amphetamine, a central nervous system stimulant, and the closely related methamphetamine (methylated amphetamine). Amphetamine and methamphetamine are both used clinically as appetite suppressants. Methamphetamine, known as "speed," is also made and sold illegally because of its rapid and intense psychological effects. These compounds represent just a few of the many substituted benzenes that have been synthesized for commercial use by the chemical and pharmaceutical industries. The physical properties of several substituted benzenes are given in Appendix I.

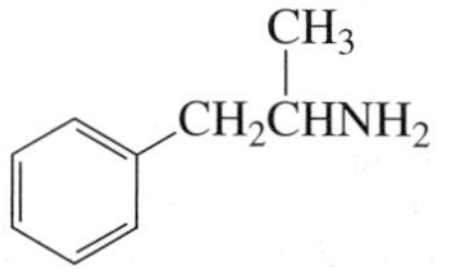

amphetamine
an appetite
suppressant

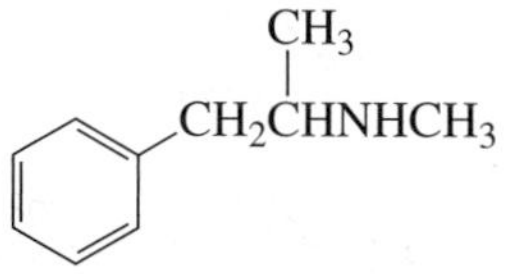

methamphetamine
"speed"

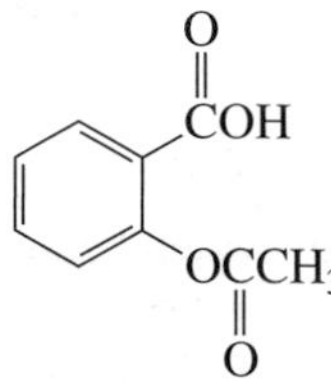

acetylsalicylic acid
aspirin

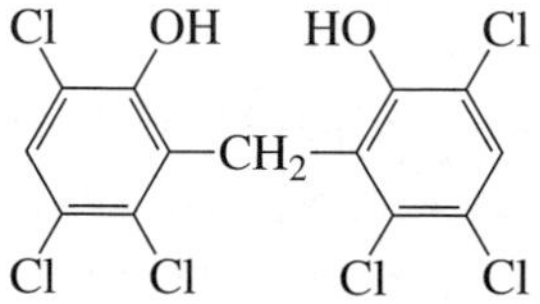

hexachlorophene
a disinfectant

and

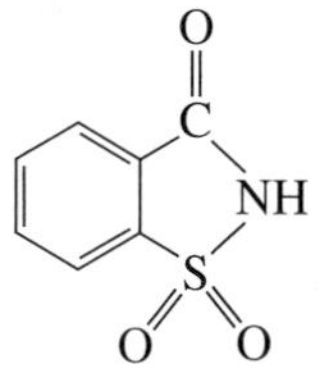

saccharin
an artificial
sweetener

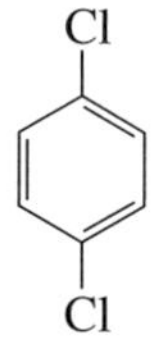

p-dichlorobenzene
in mothballs and
air fresheners

15.1 Nomenclature of Disubstituted and Polysubstituted Benzenes

Section 14.8 described how monosubstituted benzenes are named. Now we will look at how benzene rings that have more than one substituent are named.

Naming Disubstituted Benzenes

The relative positions of two substituents on a benzene ring can be indicated either by numbers or by the prefixes *ortho, meta,* and *para.* Adjacent substituents are called *ortho,* substituents separated by one carbon are called *meta,* and substituents located opposite one another are designated *para.* Often, the abbreviations for these prefixes (*o, m, p*) are used in compounds' names.

MEASURING TOXICITY

Agent Orange, a defoliant widely used in the Vietnam War, is a mixture of two synthetic substituted benzenes: 2,4-D and 2,4,5-T. Dioxin (TCDD), a contaminant formed during the manufacture of Agent Orange, has been implicated as the causative agent of the various symptoms suffered by those exposed to Agent Orange during the war.

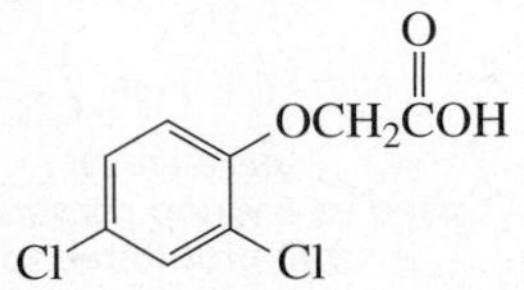

2,4-dichlorophenoxyacetic acid
2,4-D

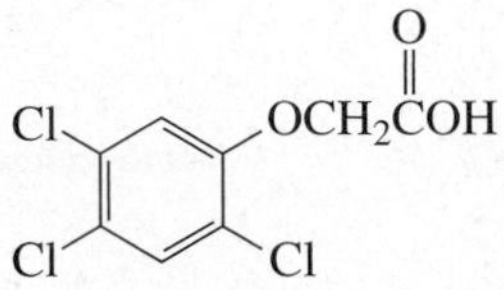

2,4,5-trichlorophenoxyacetic acid
2,4,5-T

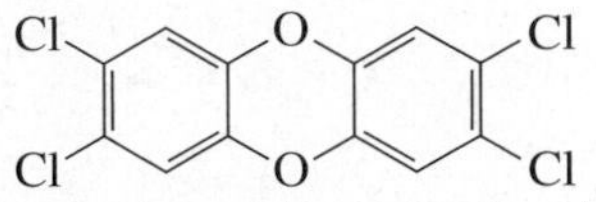

2,3,7,8-tetrachlorodibenzo[*b,e*][1,4]dioxin
TCDD

The toxicity of a compound is indicated by its LD_{50} value—the dosage found to kill 50% of the test animals exposed to it. Dioxin, with an LD_{50} value of 0.0006 mg/kg for guinea pigs, is an extremely toxic compound. Compare this with the LD_{50} values of some well-known but far less toxic poisons: 0.96 mg/kg for strychnine and 15 mg/kg for both arsenic trioxide and sodium cyanide. One of the most toxic agents known is botulism, with an LD_{50} value of about 1×10^{-8} mg/kg.

1,2-dibromobenzene
ortho-dibromobenzene
o-dibromobenzene

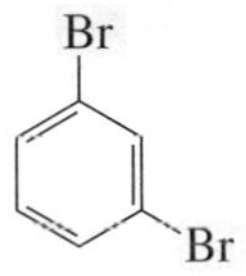

1,3-dibromobenzene
meta-dibromobenzene
m-dibromobenzene

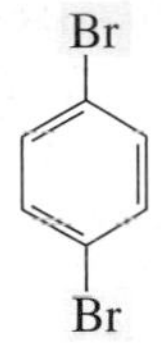

1,4-dibromobenzene
para-dibromobenzene
p-dibromobenzene

If the two substituents are different, they are listed in alphabetical order. The first mentioned substituent is given the 1-position, and the ring is numbered in the direction that gives the second substituent the lowest possible number.

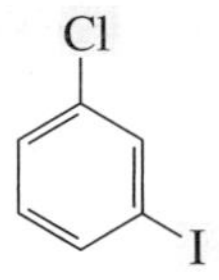

1-chloro-3-iodobenzene
meta-chloroiodobenzene
not
1-iodo-3-chlorobenzene or
meta-iodochlorobenzene

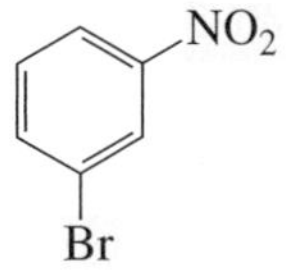

1-bromo-3-nitrobenzene
meta-bromonitrobenzene

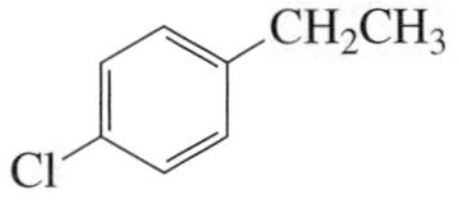

1-chloro-4-ethylbenzene
para-chloroethylbenzene

If one of the substituents can be incorporated into a name (Section 14.8), that name is used and the incorporated substituent is given the 1-position.

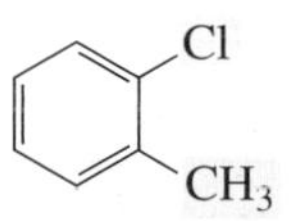

2-chlorotoluene
ortho-chlorotoluene
not
ortho-chloromethylbenzene

4-nitroaniline
para-nitroaniline
not
para-aminonitrobenzene

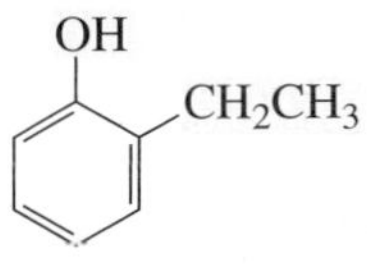

2-ethylphenol
ortho-ethylphenol
not
ortho-ethylhydroxybenzene

A few disubstituted benzenes have names that incorporate both substituents.

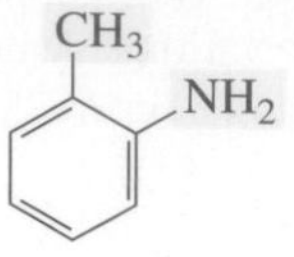

ortho-toluidine

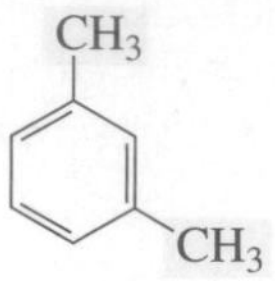

meta-xylene

para-cresol
**used as a wood preservative
until prohibited for
environmental reasons**

PROBLEM 1◆

Name the following:

a. 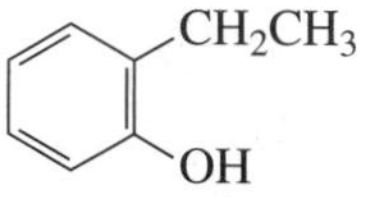 **b.** 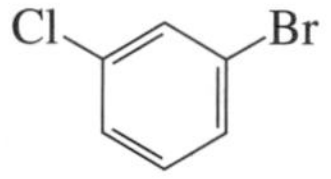 **c.** 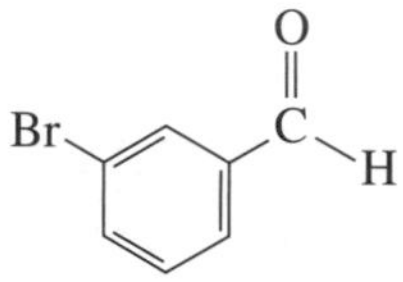 **d.**

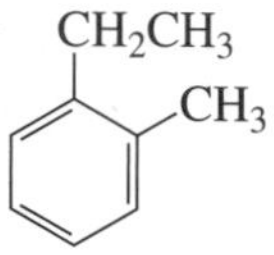

PROBLEM 2◆

Draw structures of the following:

a. *para*-toluidine

b. *meta*-cresol

c. *para*-xylene

d. *ortho*-chlorobenzenesulfonic acid

Naming Polysubstituted Benzenes

If the benzene ring has more than two substituents, the substituents are numbered in the direction that results in the lowest possible numbers. The substituents are listed in alphabetical order, each preceded by its assigned number.

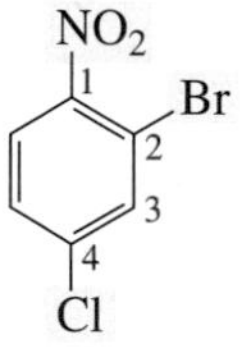

2-bromo-4-chloro-1-nitrobenzene 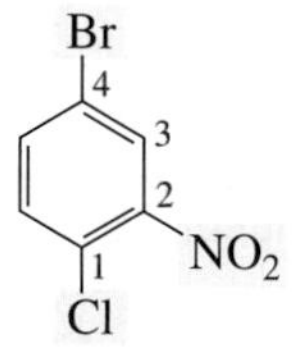 **4-bromo-1-chloro-2-nitrobenzene** 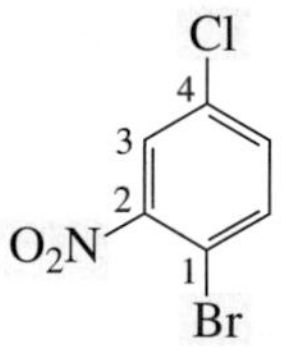 **1-bromo-4-chloro-2-nitrobenzene**

As with disubstituted benzenes, if one of the substituents can be incorporated into a name, that name is used and the incorporated substituent is given the 1-position. The ring is then numbered in the direction that results in the lowest possible numbers.

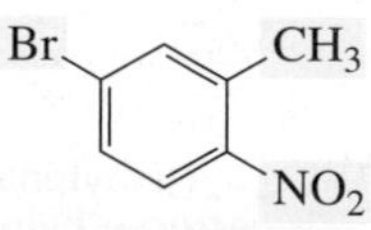

5-bromo-2-nitrotoluene 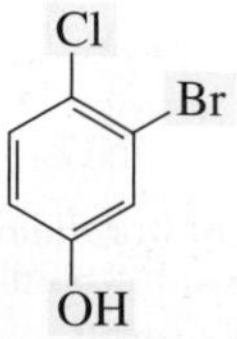 **3-bromo-4-chlorophenol** 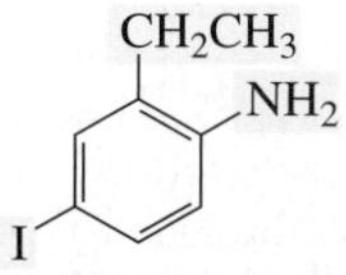 **2-ethyl-4-iodoaniline**

PROBLEM 3◆

Draw the structure of each of the following:

a. *m*-chlorotoluene
b. *p*-bromophenol
c. *o*-nitroaniline
d. *m*-chlorobenzonitrile

e. 2-bromo-4-iodophenol
f. *m*-dichlorobenzene
g. 2,5-dinitrobenzaldehyde
h. 4-bromo-3-chloroaniline

PROBLEM 4◆

Correct the following incorrect names:

a. 2,4,6-tribromobenzene
b. 3-hydroxynitrobenzene

c. *para*-methylbromobenzene
d. 1,6-dichlorobenzene

15.2 Some Substituents Increase the Reactivity of a Benzene Ring and Some Decrease Its Reactivity

Like benzene, substituted benzenes undergo the five electrophilic aromatic substitution reactions discussed in Chapter 14:

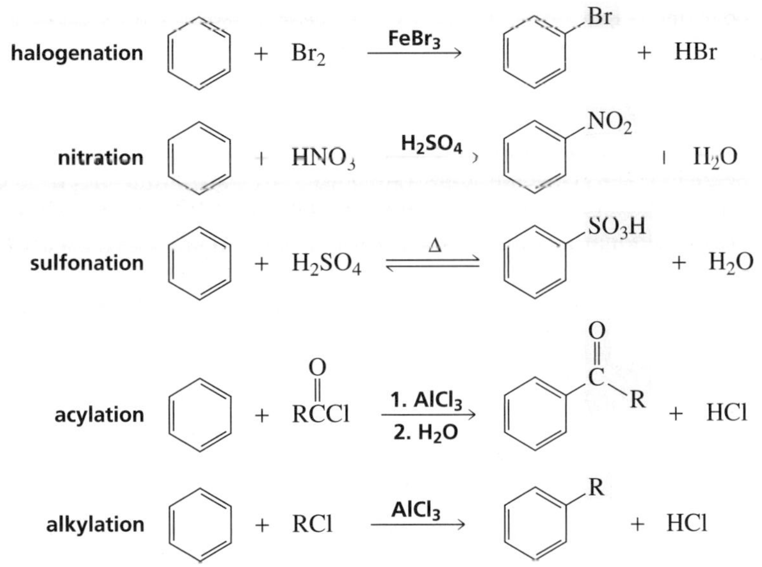

Now we need to find out whether a substituted benzene is more reactive or less reactive than benzene itself. The answer depends on the substituent. Some substituents make the ring more reactive toward electrophilic aromatic substitution than benzene and some make it less reactive than benzene.

The slow step of an electrophilic aromatic substitution reaction is the addition of an electrophile to the nucleophilic aromatic ring to form a carbocation intermediate (Section 14.10). Substituents that donate electrons into the benzene ring stabilize both the carbocation intermediate and the transition state leading to its formation (Section 4.3), thereby increasing the rate of electrophilic aromatic substitution; these are called **activating substituents**. In contrast, substituents that withdraw electrons from the benzene ring destabilize the carbocation intermediate and the transition state leading to its formation, thereby decreasing the rate of electrophilic aromatic substitution; these are called **deactivating substituents**. Before we see how the carbocation intermediate is stabilized by electron donation and destabilized by electron withdrawal, we will review the ways a substituent can donate or withdraw electrons.

relative rates of electrophilic aromatic substitution

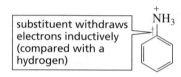

Inductive Electron Withdrawal

If a substituent that is bonded to a benzene ring is *more electron withdrawing than a hydrogen,* it will draw the σ electrons away from the benzene ring more strongly than a hydrogen will. Withdrawal of electrons through a σ bond is called **inductive electron withdrawal** (Section 1.21). The $^+NH_3$ group is an example of a substituent that withdraws electrons inductively because it is more electronegative than a hydrogen.

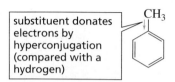

Electron Donation by Hyperconjugation

We have seen that alkyl substituents (such as CH_3) are more electron donating than a hydrogen because of hyperconjugation (Section 4.2). Notice that we compare the electron-donating ability of an alkyl group—not just a carbon—with the electron-donating ability of a hydrogen. Carbon is actually slightly more electron withdrawing than hydrogen (because C is more electronegative than H; see Table 1.3 and Section 13.6), but hyperconjugation more than makes up for this inductive effect.

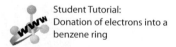

Resonance Electron Donation and Withdrawal

If a substituent has a lone pair on the atom directly attached to the benzene ring, the lone pair can be delocalized into the ring; these substituents are said to **donate electrons by resonance** (Section 7.9). Substituents such as NH_2, OH, OR, and Cl donate electrons by resonance. These substituents also withdraw electrons inductively because the atom attached to the benzene ring is more electronegative than a hydrogen.

donation of electrons into a benzene ring by resonance

anisole

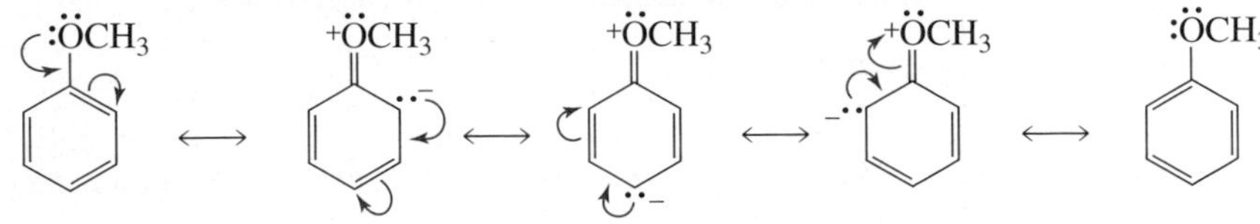

If a substituent is attached to the benzene ring by an atom that is doubly or triply bonded to a more electronegative atom, the π electrons of the ring can be delocalized onto the substituent; these substituents are said to **withdraw electrons by resonance**. Substituents such as $C=O$, $C\equiv N$, SO_3H, and NO_2 withdraw electrons by resonance. These substituents also withdraw electrons inductively because the atom attached to the benzene ring has a full or partial positive charge and, therefore, is more electronegative than a hydrogen.

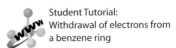

Student Tutorial:
Withdrawal of electrons from a benzene ring

withdrawal of electrons from a benzene ring by resonance

nitrobenzene

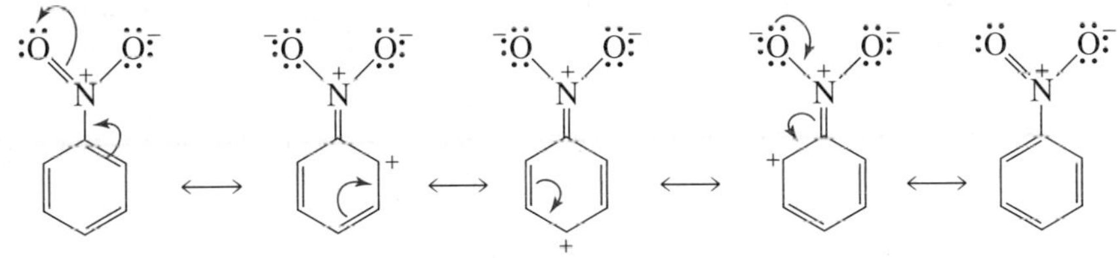

PROBLEM 5◆

For each of the following substituents, indicate whether it withdraws electrons inductively, donates electrons by hyperconjugation, withdraws electrons by resonance, or donates electrons by resonance. (Effects should be compared with that of a hydrogen; remember that many substituents can be characterized in more than one way.)

a. Br   b. CH_2CH_3 c. $\overset{O}{\overset{\|}{C}}CH_3$ d. $NHCH_3$ e. OCH_3 f. $\overset{+}{N}(CH_3)_3$

Relative Reactivity of Substituted Benzenes

The substituents shown in Table 15.1 are listed according to how they affect the reactivity of the benzene ring toward electrophilic aromatic substitution compared with benzene—in which the substituent is a hydrogen. *The activating substituents make the benzene ring more reactive toward electrophilic aromatic substitution; the deactivating substituents make the benzene ring less reactive.* Remember that activating substituents donate electrons into the ring and deactivating substituents withdraw electrons from the ring.

Electron-donating substituents increase the reactivity of the benzene ring toward electrophilic aromatic substitution.

Electron-withdrawing substituents decrease the reactivity of the benzene ring toward electrophilic aromatic substitution.

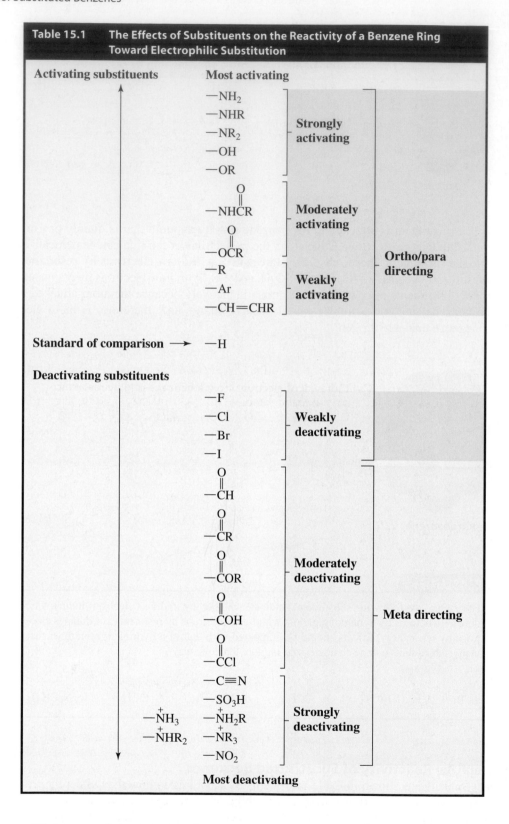

Table 15.1   The Effects of Substituents on the Reactivity of a Benzene Ring Toward Electrophilic Substitution

All the *strongly activating substituents* both donate electrons into the ring by resonance and withdraw electrons from the ring inductively. The fact that these substituents have been found experimentally to make the benzene ring more reactive indicates that their electron donation into the ring by resonance is more significant than their inductive electron withdrawal from the ring.

strongly activating substituents

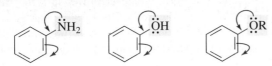

The *moderately activating substituents* also both donate electrons into the ring by resonance and withdraw electrons from the ring inductively. Because they are only moderately activating, we know that they donate electrons into the ring by resonance less effectively than do the strongly activating substituents.

moderately activating substituents

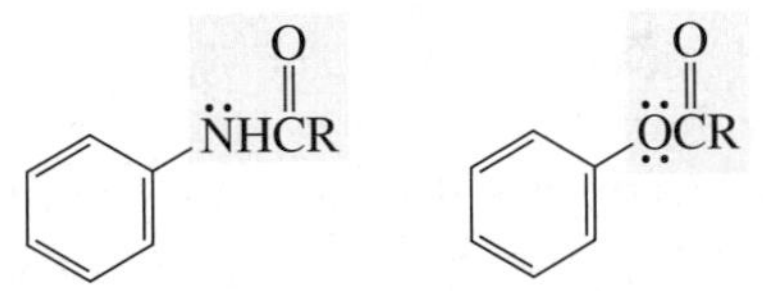

These substituents are less effective at donating electrons into the ring by resonance because, unlike the strongly activating substituents that donate electrons by resonance only *into* the ring, the moderately activating substituents can donate electrons by resonance in two competing directions: *into* the ring and *away from* the ring. The fact that these substituents increase the reactivity of the benzene ring indicates that, despite their diminished resonance electron donation into the ring, overall they donate electrons by resonance more strongly than they withdraw electrons inductively.

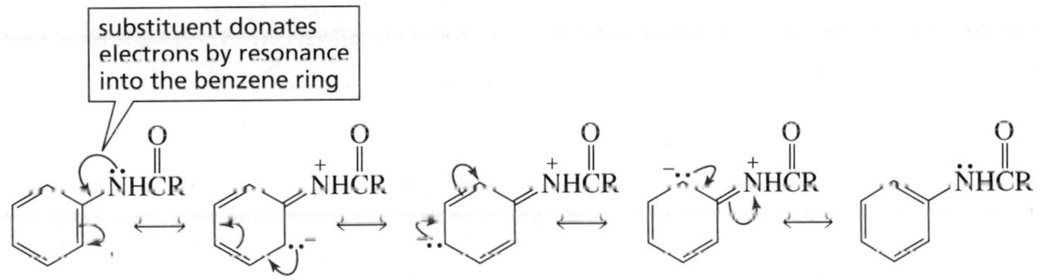

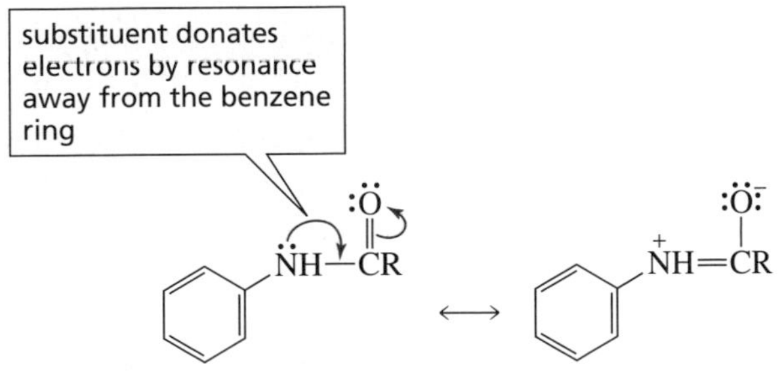

substituent donates
electrons by resonance
into the benzene ring

substituent donates
electrons by resonance
away from the benzene
ring

Alkyl, aryl, and CH=CHR groups are *weakly activating substituents*. We have seen that an alkyl substituent, compared with a hydrogen, is electron donating. Aryl and CH=CHR groups can donate electrons into the ring by resonance and can also withdraw electrons from the ring by resonance. The fact that they are weak activators indicates that they are slightly more electron donating than they are electron withdrawing.

weakly activating substituents

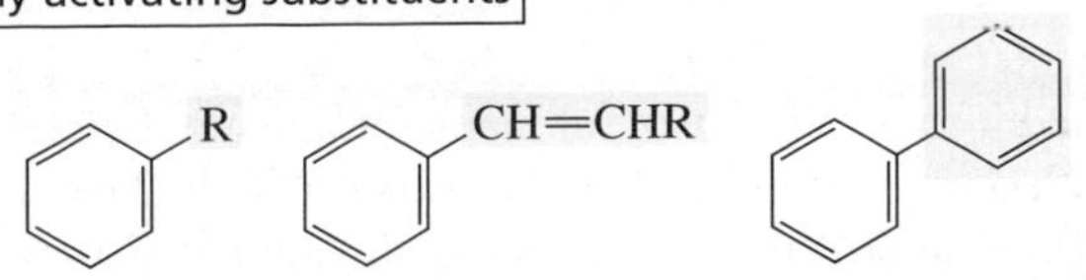

The halogens are *weakly deactivating substituents*. Like the strongly and moderately activating substituents, the halogens also both donate electrons into the ring by resonance and withdraw electrons from the ring inductively. Because the halogens have been found experimentally to make the benzene ring less reactive, we can conclude that they withdraw electrons inductively more strongly than they donate electrons by resonance.

weakly deactivating substituents

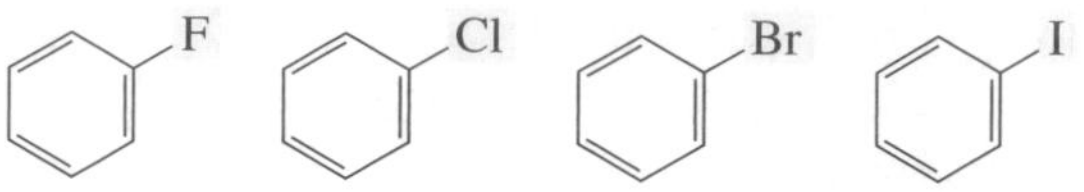

Let's look at why this is so. The electronegativities of chlorine and oxygen are similar, so they have similar inductive electron-withdrawing abilities. However, chlorine does not donate electrons by resonance as well as oxygen does because chlorine uses a $3p$ orbital to overlap with the $2p$ orbital of carbon. A $3p$–$2p$ orbital overlap is much less effective than the $2p$–$2p$ orbital overlap that occurs between oxygen and carbon. Fluorine, which uses a $2p$ orbital, donates electrons by resonance better than chlorine, but this is outweighed by fluorine's greater electronegativity, which causes it to strongly withdraw electrons inductively. Bromine and iodine are less effective than chlorine at withdrawing electrons inductively, but they are also less effective at donating electrons by resonance because they use $4p$ and $5p$ orbitals, respectively. Thus, all the halogens withdraw electrons inductively more strongly than they donate electrons by resonance.

The *moderately deactivating substituents* all have a carbonyl group directly attached to the benzene ring. A carbonyl group withdraws electrons from a benzene ring both inductively and by resonance.

moderately deactivating substituents

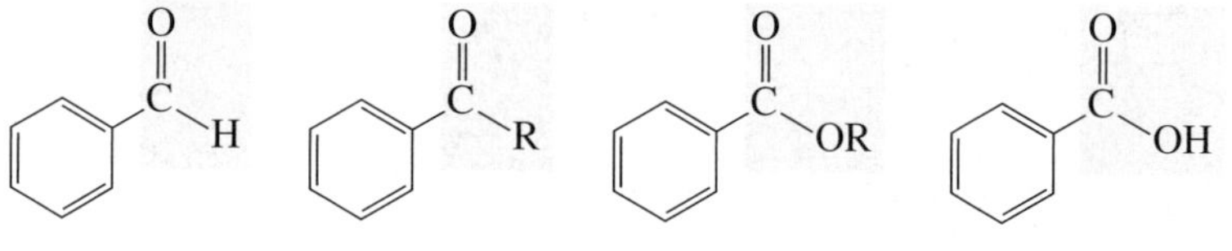

The *strongly deactivating substituents* are powerful electron withdrawers. Except for the ammonium ions ($^+NH_3$, $^+NH_2R$, $^+NHR_2$, and $^+NR_3$), these substituents withdraw electrons both inductively and by resonance. The ammonium ions have no resonance effect, but the positive charge on the nitrogen atom causes them to withdraw electrons strongly through the σ bond (inductively).

strongly deactivating substituents

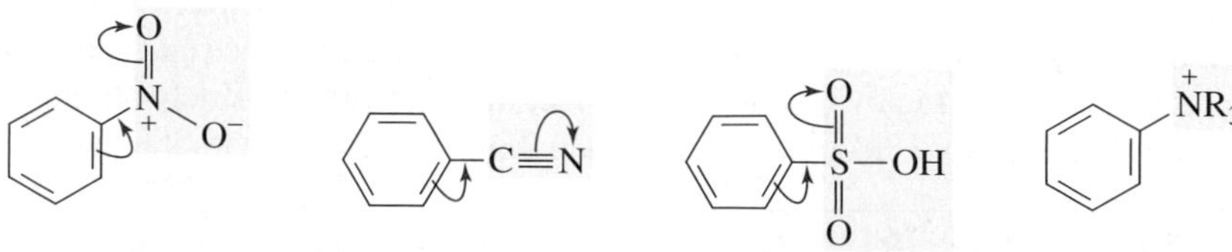

Take a minute to compare the electrostatic potential maps for anisole, benzene, and nitrobenzene. Notice that an electron-donating substituent (OCH_3) makes the ring more red (more negative), whereas an electron-withdrawing substituent (NO_2) makes the ring less red (less negative).

| anisole | benzene | nitrobenzene |

PROBLEM 6♦

List the compounds in each set in order of decreasing reactivity toward electrophilic aromatic substitution:

a. benzene, phenol, toluene, nitrobenzene, bromobenzene

b. dichloromethylbenzene, difluoromethylbenzene, toluene, chloromethylbenzene

PROBLEM 7   *SOLVED*

Explain why the halo-substituted benzenes have the relative reactivities shown in Table 15.1.

Solution   Table 15.1 shows that fluorine is the least deactivating of the halogen substituents and iodine is the most deactivating. We know that fluorine is the most electronegative of the halogens, which means that it is best at withdrawing electrons inductively. Fluorine is also best at donating electrons by resonance because its $2p$ orbital—compared with the $3p$ orbital of chlorine, the $4p$ orbital of bromine, or the $5p$ orbital of iodine—can better overlap with the $2p$ orbital of carbon. So the fluorine substituent is best both at donating electrons by resonance and at withdrawing electrons inductively. Because Table 15.1 shows that fluorine is the weakest deactivator of the halogens, we can conclude that electron donation by resonance is the more important factor in determining the relative reactivities of halo-substituted benzenes.

15.3 **The Effect of Substituents on Orientation**

When a substituted benzene undergoes an electrophilic substitution reaction, where does the new substituent attach itself? Is the product of the reaction the ortho isomer, the meta isomer, or the para isomer?

ortho isomer	**meta isomer**	**para isomer**

The substituent already attached to the benzene ring determines the location of the new substituent. The attached substituent will have one of two effects: it will direct an incoming substituent either to the ortho *and* para positions, or it will direct an incoming substituent to the meta position. All activating substituents and the weakly deactivating halogens are **ortho–para directors**, and all substituents that are more deactivating than the halogens are **meta directors**. Thus, the substituents can be divided into three groups:

All activating substituents are ortho–para directors.

1. All *activating substituents* direct an incoming electrophile to the ortho and para positions.

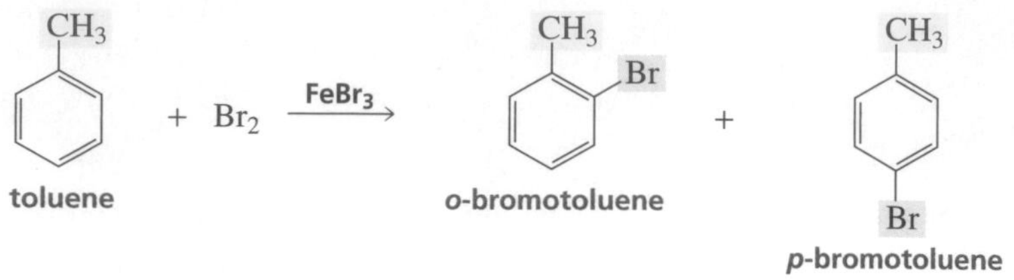
CH3
+ Br2
FeBr3
CH3
Br
+
CH3
Br
toluene
o-bromotoluene
p-bromotoluene

The weakly deactivating halogens are ortho–para directors.

2. The *weakly deactivating* halogens also direct an incoming electrophile to the ortho and para positions.

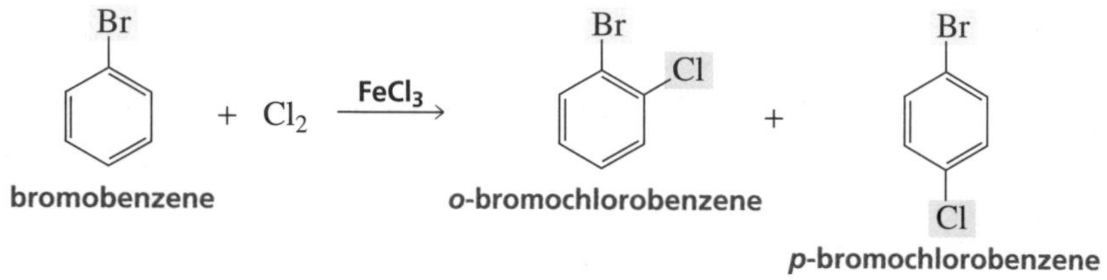
Br
+ Cl2
FeCl3
Br
Cl
+
Br
Cl
bromobenzene
o-bromochlorobenzene
p-bromochlorobenzene

All deactivating substituents (except the halogens) are meta directors.

3. All *moderately and strongly deactivating* substituents direct an incoming electrophile to the meta position.

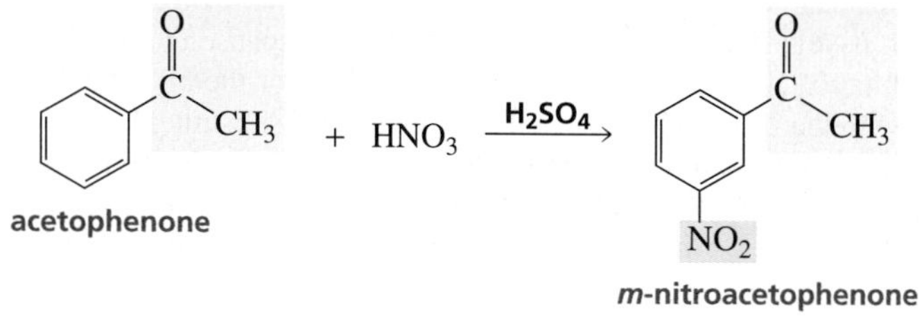
O
C
CH3
+ HNO3
H2SO4
O
C
CH3
NO2
acetophenone
m-nitroacetophenone

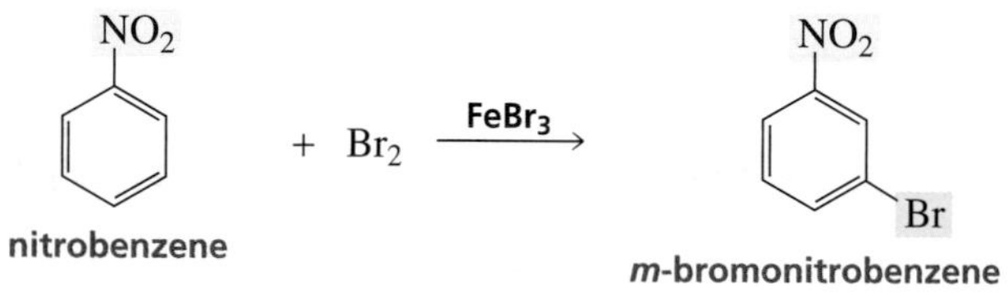
NO2
+ Br2
FeBr3
NO2
Br
nitrobenzene
m-bromonitrobenzene

To understand why a substituent directs an incoming electrophile to a particular position, we must look at the stability of the carbocation intermediate, because as Figure 14.5 on page 654 shows, formation of the carbocation is the rate-determining step. When a substituted benzene undergoes an electrophilic substitution reaction, three different carbocation intermediates can be formed: an *ortho*-substituted carbocation, a *meta*-substituted carbocation, and a *para*-substituted carbocation (Figure 15.1). The relative stabilities of the three carbocations enable us to determine the preferred pathway of the reaction because the more stable the carbocation, the more stable the transition state for its formation, and the more rapidly it will be formed (Section 4.3).

When the substituent is one that can donate electrons by *resonance,* the carbocations formed by putting the incoming electrophile on the ortho and para positions have a fourth resonance contributor (Figure 15.1). This is an especially stable resonance contributor because it is the only one whose atoms (except for hydrogen) all have complete octets; it is obtained only by directing an incoming substituent to the ortho and para positions. Therefore, *all substituents that donate electrons by resonance are ortho–para directors.*

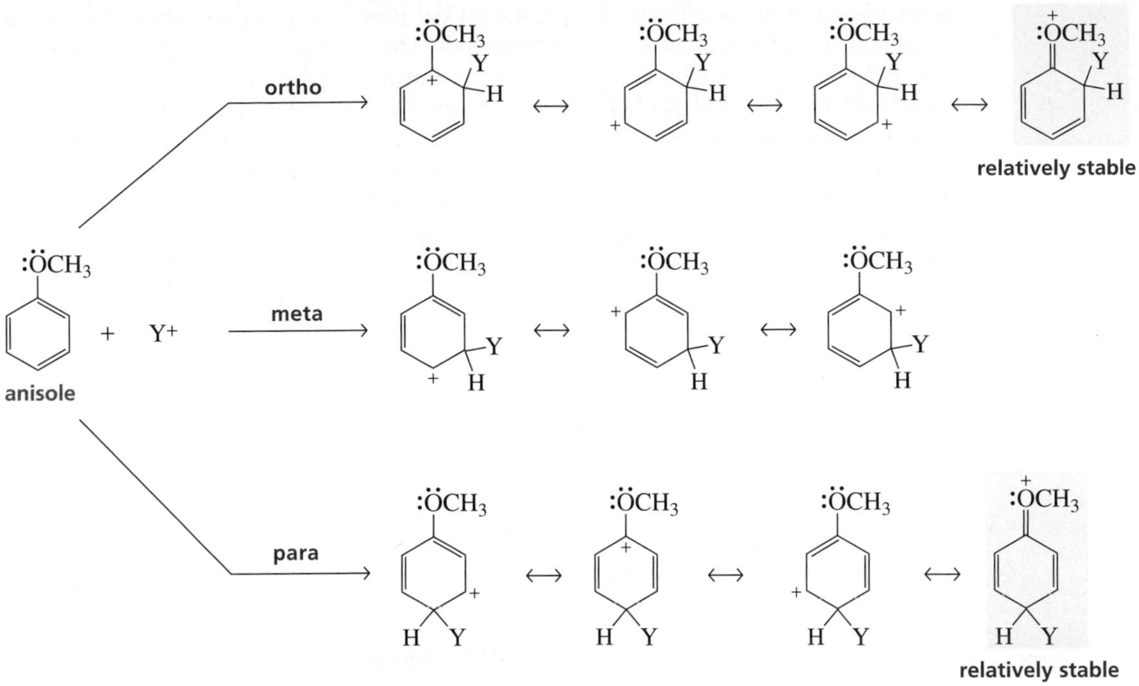

ortho →

relatively stable

:ÖCH₃
anisole   + Y⁺ ⟶

meta →

para →

relatively stable

▲ **Figure 15.1**
The structures of the carbocation intermediates formed from the reaction of an electrophile with anisole at the ortho, meta, and para positions.

When the substituent is an alkyl group, the resonance contributors that are highlighted in Figure 15.2 are the most stable. In those contributors, the alkyl group is attached directly to the positively charged carbon and can stabilize it by hyperconjugation. A relatively stable resonance contributor is obtained only when the incoming group is directed to an ortho or para position. Therefore, the most stable carbocations are obtained by directing the incoming group to the ortho and para positions. In short, *alkyl substituents are ortho–para directors because they donate electrons by hyperconjugation.*

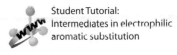
Student Tutorial:
Intermediates in electrophilic aromatic substitution

ortho →

most stable

Student Tutorial:
Nitration of anisole 1

CH₃
toluene   + Y⁺   meta ⟶

Student Tutorial:
Nitration of anisole 2

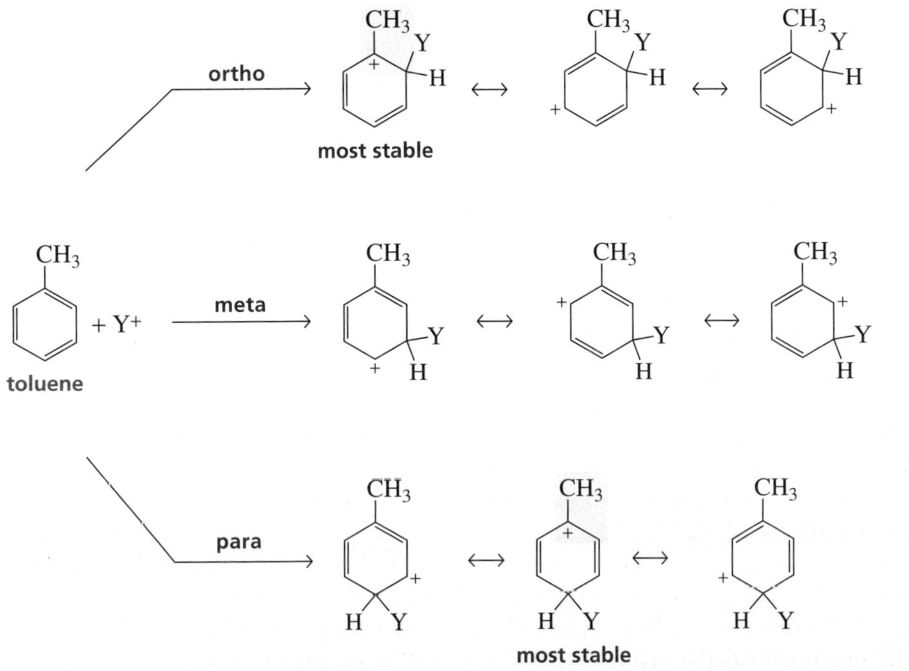

para →

most stable

◀ **Figure 15.2**
The structures of the carbocation intermediates formed from the reaction of an electrophile with toluene at the ortho, meta, and para positions.

Substituents with a positive charge or a partial positive charge on the atom attached to the benzene ring will withdraw electrons inductively from the benzene ring and most will withdraw electrons by resonance as well. For all such substituents, the resonance contributors highlighted in Figure 15.3 are the least stable because they have a positive charge on each of two adjacent atoms, so the most stable carbocation is formed when the incoming electrophile is directed to the meta position. Thus, *all substituents that withdraw electrons (except for the halogens, which are ortho–para directors because they donate electrons by resonance) are meta directors.*

VCL Benzene Nitration-1

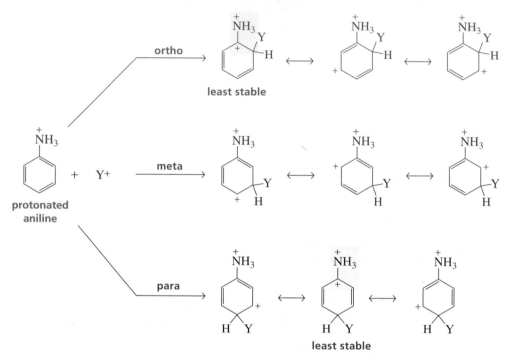

▲ **Figure 15.3**
The structures of the carbocation intermediates formed from the reaction of an electrophile with protonated aniline at the ortho, meta, and para positions.

Notice that the three possible carbocation intermediates in Figures 15.2 and 15.3 are the same, except for the substituent. The nature of that substituent determines whether the resonance contributors with the substituent directly attached to the positively charged carbon are the most stable (as with electron-donating substituents) or the least stable (as with electron-withdrawing substituents).

The only deactivating substituents that are ortho–para directors are the halogens, which are the weakest of the deactivators. We have seen that they are deactivators because they inductively withdraw electrons from the ring more strongly than they donate electrons by resonance. The halogens nevertheless are ortho–para directors because of their ability to donate electrons by resonance; they can stabilize the transition states leading to reaction at the ortho and para positions by resonance electron donation as the CH_3O substituent does in Figure 15.1.

In summary, as shown in Table 15.1, all the activating substituents and the weakly deactivating halogens are ortho–para directors. All substituents more deactivating than the halogens are meta directors. In other words, all substituents that donate electrons either by resonance or by hyperconjugation are ortho–para directors; all substituents that cannot donate electrons are meta directors.

You don't need to resort to memorization to be able to identify a substituent as an ortho–para director or a meta director. It is easy to tell them apart: all ortho–para directors, except for alkyl, aryl, and $CH=CHR$ groups, have at least one lone pair on the atom directly attached to the ring; all meta directors have a positive charge or a partial positive charge on the atom attached to the ring. Take a few minutes to examine the substituents listed in Table 15.1 to see that this is true.

All substituents that donate electrons either by resonance or by hyperconjugation are ortho–para directors.

All substituents that cannot donate electrons are meta directors.

PROBLEM 8

a. Draw the resonance contributors for nitrobenzene.

b. Draw the resonance contributors for chlorobenzene.

PROBLEM 9♦

What product(s) would result from nitration of each of the following?

a. propylbenzene   c. benzaldehyde   e. benzenesulfonic acid

b. bromobenzene   d. benzonitrile   f. cyclohexylbenzene

PROBLEM 10♦

Are the following substituents ortho–para directors or meta directors?

a. CH=CHC≡N   b. NO$_2$ c. CH$_2$OH   d. COOH   e. CF$_3$ f. N=O

PROBLEM 11   SOLVED

Give the product(s) obtained from the reaction of each of the following with one equivalent of Br$_2$ and employing the catalyst, FeBr$_3$:

a.

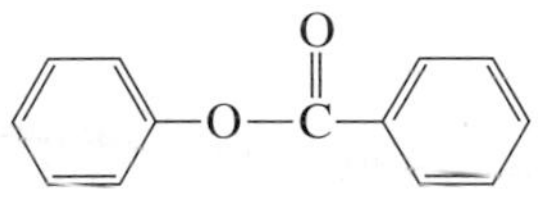

c.

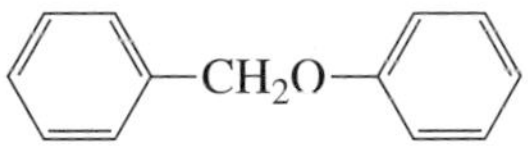

b. 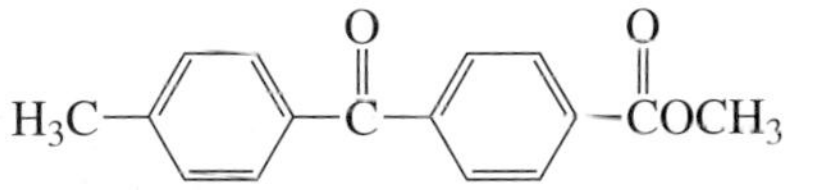 H$_3$C— —COCH$_3$

d. 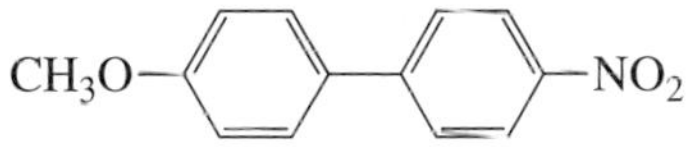 CH$_3$O— —NO$_2$

Solution to 11a   The left-hand ring is attached to a substituent that activates that ring by donating electrons into it by resonance. In contrast, the right-hand ring is attached to a substituent that deactivates that ring by withdrawing electrons from it by resonance.

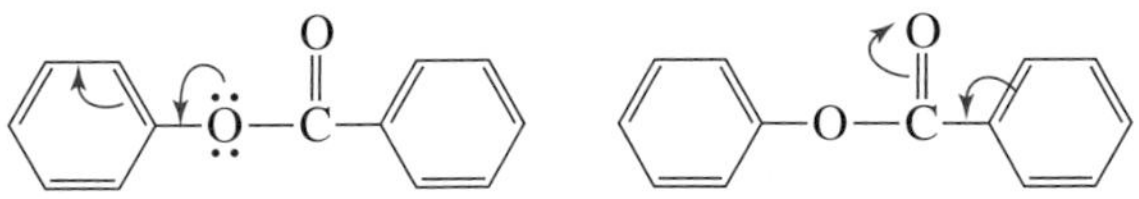

Thus, the left-hand ring is more reactive toward electrophilic aromatic substitution. The activating substituent will direct the bromine to the positions that are ortho and para to it.

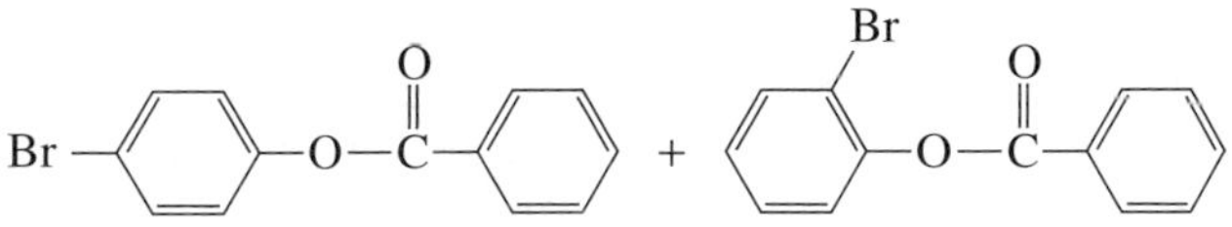

15.4 **The Effect of Substituents on pK_a**

When a substituent either withdraws electrons from or donates electrons into a benzene ring, the pK_a values of substituted phenols, benzoic acids, and protonated anilines will reflect this withdrawal or donation.

Electron-withdrawing groups stabilize a base and, therefore, increase the strength of its conjugate acid. Electron-donating groups destabilize a base and this decreases the strength of its conjugate acid (Section 1.21). (Remember the stronger the acid, the more stable its conjugate base.)

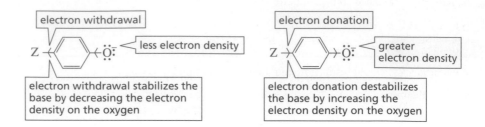

For example, the pK_a of phenol in H_2O at 25 °C is 9.95. The pK_a of *para*-nitrophenol is lower (7.14) because the nitro substituent withdraws electrons from the ring, whereas the pK_a of *para*-methyl phenol is higher (10.19) because the methyl substituent donates electrons into the ring.

OH OCH₃	OH CH₃	OH	OH Cl	OH HC=O	OH NO₂
$pK_a = 10.20$	$pK_a = 10.19$	$pK_a = 9.95$ phenol	$pK_a = 9.38$	$pK_a = 7.66$	$pK_a = 7.14$

Take a minute to compare the effect a substituent has on the reactivity of a benzene ring toward electrophilic aromatic substitution with the effect the substituent has on the pK_a of phenol. Notice that the more strongly deactivating the substituent, the lower the pK_a of the phenol; and the more strongly activating the substituent, the higher the pK_a of the phenol. In other words, *electron withdrawal decreases reactivity toward electrophilic substitution and increases acidity, whereas electron donation increases reactivity toward electrophilic substitution and decreases acidity.*

A similar substituent effect on pK_a is observed for substituted benzoic acids and substituted protonated anilines: electron withdrawing substituents increase acidity; electron-donating substituents decrease acidity.

The more deactivating (electron withdrawing) a substituent, the more it increases the acidity of a COOH, an OH, or an $^+NH_3$ group attached to a benzene ring.

COOH OCH₃	COOH CH₃	COOH	COOH Br	COOH CH₃C=O	COOH NO₂
$pK_a = 4.47$	$pK_a = 4.34$	$pK_a = 4.20$	$pK_a = 4.00$	$pK_a = 3.70$	$pK_a = 3.44$

The more activating (electron donating) the substituent, the more it decreases the acidity of a COOH, an OH, or an $^+NH_3$ group attached to a benzene ring.

$^+NH_3$ OCH₃	$^+NH_3$ CH₃	$^+NH_3$	$^+NH_3$ Br	$^+NH_3$ HC=O	$^+NH_3$ NO₂
$pK_a = 5.29$	$pK_a = 5.07$	$pK_a = 4.58$	$pK_a = 3.91$	$pK_a = 1.76$	$pK_a = 0.98$

PROBLEM 12♦

Which of the compounds in each of the following pairs is more acidic?

a. CH$_3$COH or ClCH$_2$COH

b. O$_2$NCH$_2$COH or O$_2$NCH$_2$CH$_2$COH

c. CH$_3$CH$_2$COH or H$_3$NCH$_2$COH

d. [structure: benzene ring with COOH and OCH$_3$] or [structure: benzene ring with COOH]

e. HOCCH$_2$COH or ⁻OCCH$_2$COH

f. HCOH or CH$_3$COH

g. FCH$_2$COH or ClCH$_2$COH

h. [structure: benzene ring with COOH and F] or [structure: benzene ring with COOH and Cl]

PROBLEM-SOLVING STRATEGY

Explaining the Effect of Substituents on pK_a

The *para*-nitroanilinium ion is 3.60 pK_a units more acidic than the anilinium ion (pK_a = 0.98 versus 4.58), but *para*-nitrobenzoic acid is only 0.76 pK_a unit more acidic than benzoic acid (pK_a = 3.44 versus 4.20). Explain why the nitro substituent causes a large change in pK_a in one case and a small change in pK_a in the other.

Do not expect to be able to solve this kind of problem simply by reading it. First, you need to remember that the acidity of a compound depends on the stability of its conjugate base (Sections 1.18 and 7.9). Next, draw structures of the conjugate bases in question in order to compare their stabilities.

[structures of conjugate bases]

When a proton is lost from the *para*-nitroanilinium ion, the electrons it leaves behind are shared by five atoms. (Draw resonance contributors if you want to see which atoms share the electrons.) In contrast, when a proton is lost from *para*-nitrobenzoic acid, the electrons it leaves behind are shared by two atoms. In other words, loss of a proton leads to greater electron delocalization in one base than in the other. Electron delocalization stabilizes a compound, so the difference in electron delocalization explains why the addition of a nitro substituent has a greater effect on the acidity of an anilinium ion than on benzoic acid.

Now continue on to Problem 13.

PROBLEM 13

Explain why *p*-nitrophenol has a pK_a of 7.14, whereas the pK_a of *m*-nitrophenol is 8.39.

15.5 The Ortho–Para Ratio

When a benzene ring with an ortho–para-directing substituent undergoes an electrophilic aromatic substitution reaction, what percentage of the product is the ortho isomer and what percentage is the para isomer? Solely on the basis of probability, one would expect more of the ortho product because there are two ortho positions available to the incoming electrophile and only one para position. The ortho position, however, is sterically hindered, whereas the para position is not. Consequently, the para isomer will be formed preferentially if either the substituent on the ring or the incoming electrophile is large. The following nitration reactions illustrate the decrease in the ortho–para ratio with an increase in the size of the alkyl substituent:

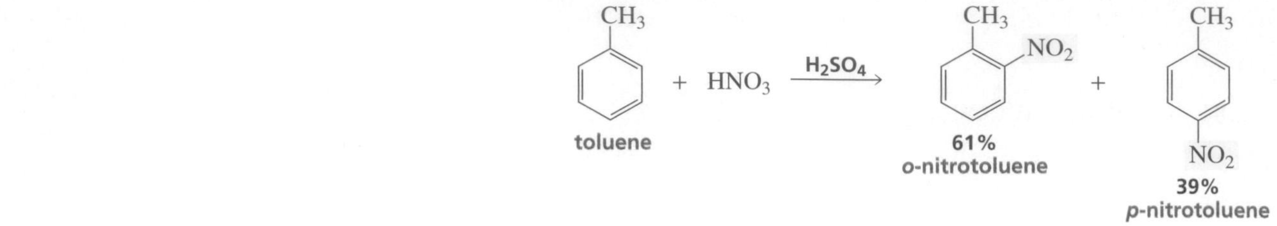

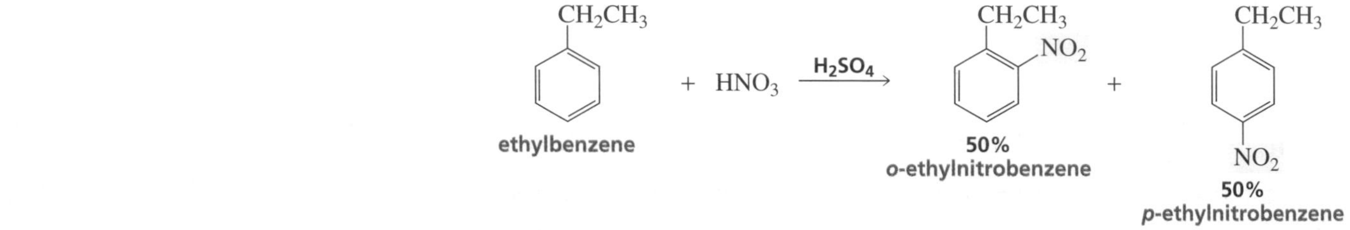

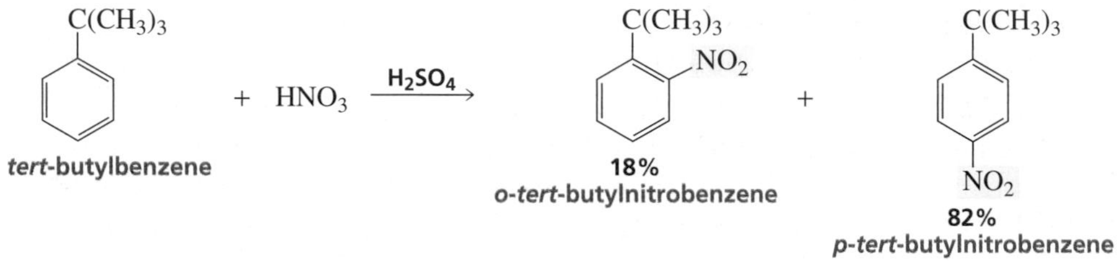

Fortunately, the difference in the physical properties of the ortho- and para-substituted isomers is sufficient to allow them to be easily separated. Consequently, electrophilic aromatic substitution reactions that lead to both ortho and para isomers are useful in syntheses because the desired product can be easily separated from the reaction mixture.

15.6 Additional Considerations Regarding Substituent Effects

It is important to know whether a substituent is activating or deactivating in order to determine the conditions needed to carry out a reaction. For example, methoxy and hydroxy substituents are so strongly activating that halogenation is carried out without the Lewis acid ($FeBr_3$ or $FeCl_3$) catalyst.

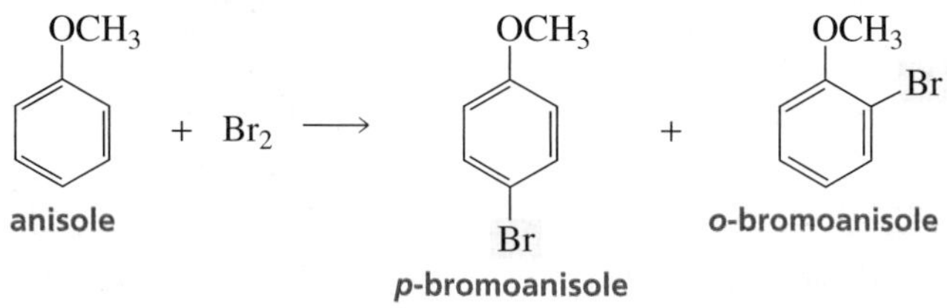

If the Lewis acid catalyst and excess bromine are used, the tribromide is obtained.

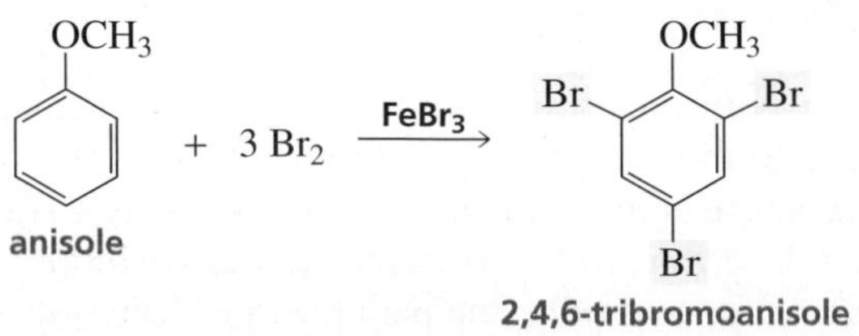

Friedel–Crafts reactions are the slowest of the electrophilic aromatic substitution reactions. Therefore, if a benzene ring has been moderately or strongly deactivated, (remember that all meta directors are moderate or strong deactivators), it will be too unreactive to undergo either Friedel–Crafts acylation or Friedel–Crafts alkylation. In fact, nitrobenzene is so unreactive that it is often used as a solvent for Friedel–Crafts reactions.

A benzene ring with a meta director cannot undergo a Friedel–Crafts reaction.

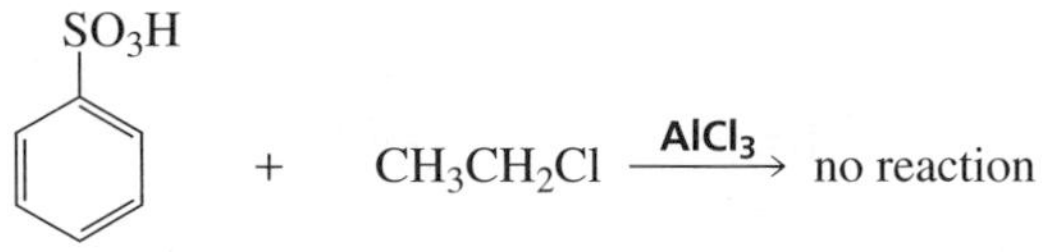

SO_3H

benzenesulfonic
acid

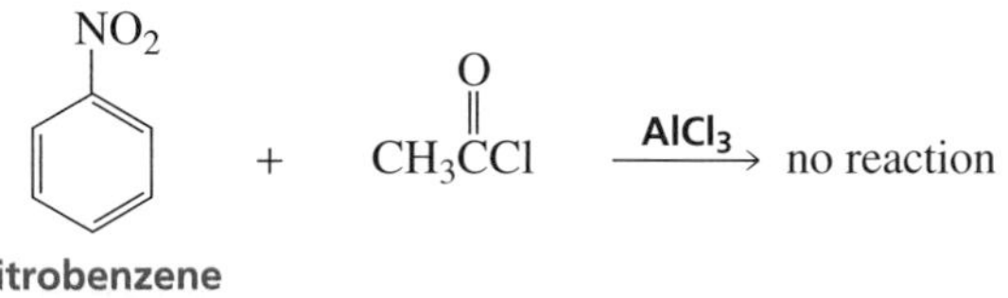

$+$ CH_3CH_2Cl $\xrightarrow{\text{AlCl}_3}$ no reaction

NO_2

nitrobenzene

$+$ $CH_3\overset{\overset{\displaystyle O}{\|}}{C}Cl$ $\xrightarrow{\text{AlCl}_3}$ no reaction

Aniline and *N*-substituted anilines also do not undergo Friedel–Crafts reactions. The lone pair on the amino group will complex with the Lewis acid catalyst ($AlCl_3$) needed to carry out the reaction, converting the NH_2 substituent into a deactivating meta director. As we have just seen, Friedel–Crafts reactions do not occur with benzene rings deactivated by meta-directing substituents.

Benzene Nitration-2

a meta director

$H_2N:$ $H_2\overset{+}{N}-\overset{-}{A}lCl_3$

$\xrightarrow{\text{AlCl}_3}$

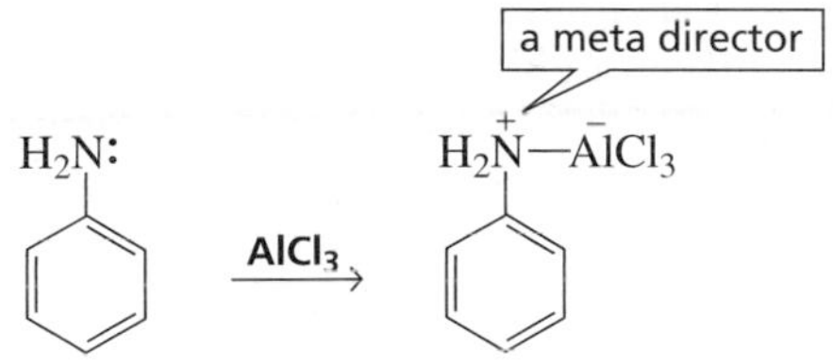

aniline

Phenol and anisole undergo Friedel–Crafts reactions—at the ortho and para positions—because oxygen, being a weaker base than nitrogen, does not complex with the Lewis acid.

Benzene Nitration-3

Aniline also cannot be nitrated, because nitric acid is an oxidizing agent and primary amines are easily oxidized. (Nitric acid and aniline can be an explosive combination.)

PROBLEM 14

Show how the following compounds could be synthesized from benzene:

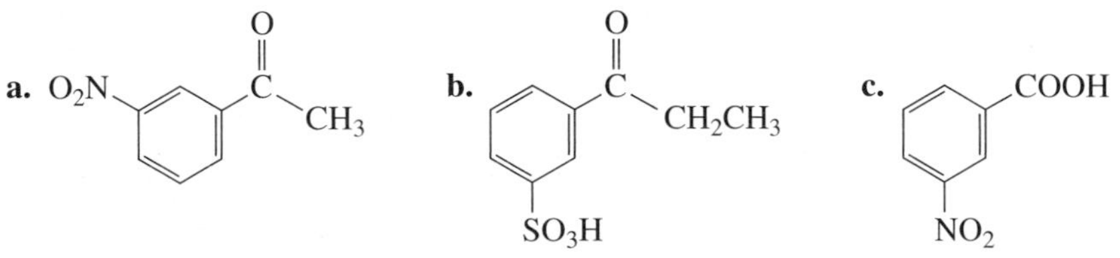

a. O_2N—⬡—$\overset{\overset{\displaystyle O}{\|}}{C}$—$CH_3$ **b.** ⬡—$\overset{\overset{\displaystyle O}{\|}}{C}$—$CH_2CH_3$ (with SO_3H)   **c.** ⬡—$COOH$ (with NO_2)

PROBLEM 15◆

Give the products, if any, of each of the following reactions:

a. benzonitrile + methyl chloride + $AlCl_3$ **c.** benzoic acid + CH_3CH_2Cl + $AlCl_3$

b. aniline + 3 Br_2 **d.** benzene + 2 CH_3Cl + $AlCl_3$

15.7 Designing a Synthesis IV: Synthesis of Monosubstituted and Disubstituted Benzenes

As the number of reactions with which we are familiar increases, the more possibilities we are able to choose from when we design a synthesis. For example, we can now design two very different routes for the synthesis of 2-phenylethanol from benzene.

Which route we choose depends on such factors as convenience, expense, and expected yield of the target molecule (the desired product). For example, the first route shown for the synthesis of 2-phenylethanol is the better procedure. The second route has more steps, requires excess benzene to prevent polyalkylation, and it uses a radical reaction that can produce unwanted side products. Moreover, the yield of the elimination reaction is not high (because some substitution product is formed as well), and hydroboration–oxidation is not an easy reaction to carry out.

Designing the synthesis of a disubstituted benzene requires careful consideration of the order in which the substituents are to be placed on the ring. For example, if you want to synthesize *meta*-bromobenzenesulfonic acid, the sulfonic acid group has to be placed on the ring first, because that group will direct the bromo substituent to the desired meta position.

However, if the desired product is *para*-bromobenzenesulfonic acid, the order of the two reactions must be reversed because only the bromo substituent is an ortho–para director.

Both substituents of *meta*-nitroacetophenone are meta directors. However, the Friedel–Crafts acylation reaction must be carried out first because the benzene ring of nitrobenzene is too deactivated to undergo a Friedel-Crafts reaction (Section 15.6).

Friedel-Crafts-1

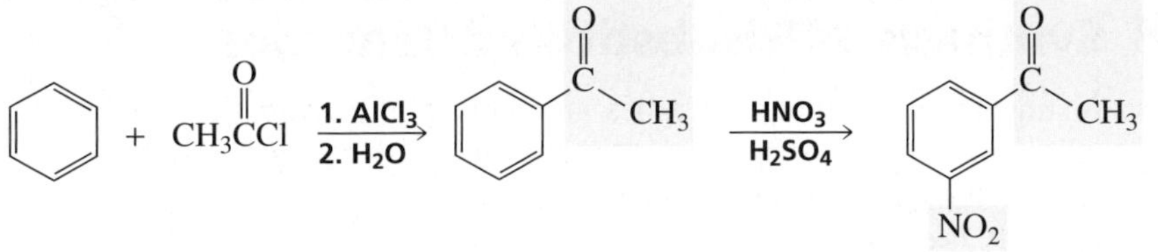

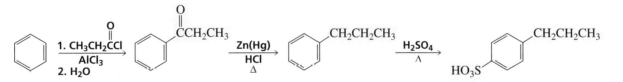

m-nitroacetophenone

Another question to consider is, at what point in a reaction sequence should a substituent be chemically modified? In the synthesis of *para*-chlorobenzoic acid from toluene, the methyl group is oxidized after it directs the chloro substituent to the para position. (*ortho*-Chlorobenzoic acid is also formed in this reaction.)

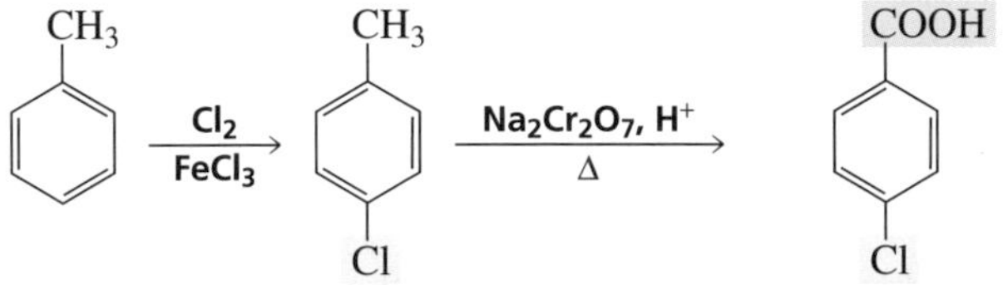

para-chlorobenzoic acid

In the synthesis of *meta*-chlorobenzoic acid, the methyl group is oxidized before chlorination because a meta director is needed to obtain the desired product.

meta-chlorobenzoic acid

VCL Friedel-Crafts-2

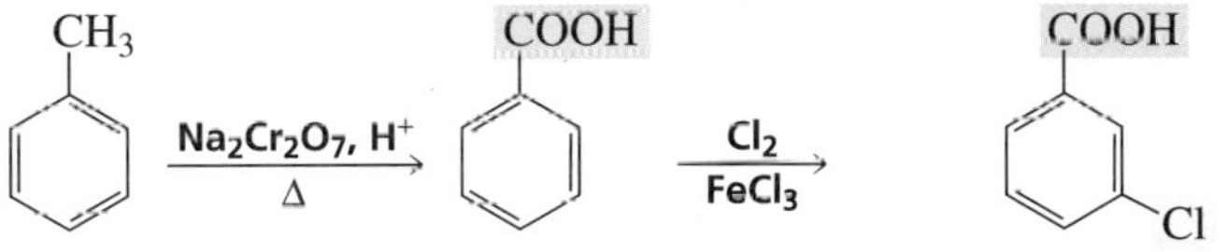

In the following synthesis of *para*-propylbenzenesulfonic acid, the type of reaction employed, the order in which the substituents are put on the benzene ring, and the point at which a substituent is chemically modified must all be considered. The straight-chain propyl substituent must be put on the ring by a Friedel–Crafts acylation rather than alkylation reaction, to avoid the carbocation rearrangement that would occur with the alkylation reaction. The Friedel-Crafts acylation must be carried out before sulfonation because acylation could not be carried out on a ring with a strongly deactivating sulfonic acid substituent, and also because the sulfonic acid group is a meta director. Finally, the sulfonic acid group must be put on the ring after the carbonyl group is reduced to a methylene group, so that the sulfonic acid group will be directed primarily to the para position by the alkyl group.

para-propylbenzenesulfonic acid

PROBLEM 16

Show how each of the following compounds can be synthesized from benzene:

a. *p*-chloroaniline
b. *m*-chloroaniline
c. *p*-nitrobenzoic acid
d. *m*-nitrobenzoic acid

e. *m*-bromopropylbenzene
f. *o*-bromopropylbenzene
g. 1-phenyl-2-propanol
h. 2-phenylpropene

Student Tutorial:
Multistep synthesis of disubstituted benzenes

Student Tutorial:
Synthesis of benzene derivatives

15.8 **Synthesis of Trisubstituted Benzenes**

When a disubstituted benzene undergoes an electrophilic aromatic substitution reaction, the directing effects of both substituents have to be considered. If both substituents direct the incoming substituent to the same position, the product of the reaction is easily predicted.

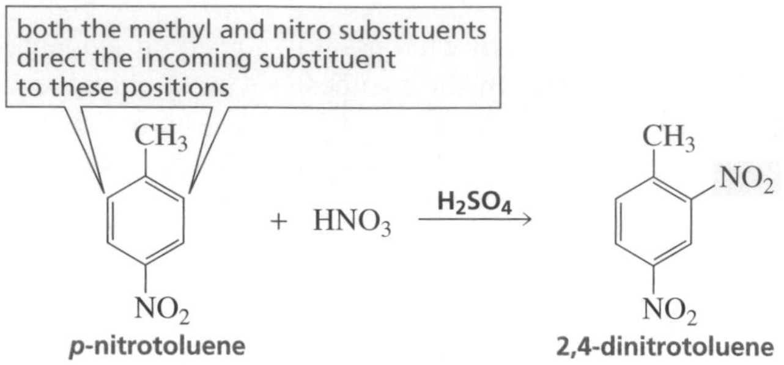

both the methyl and nitro substituents direct the incoming substituent to these positions

CH₃   +   HNO₃ $\xrightarrow{\text{H}_2\text{SO}_4}$

NO₂

p-nitrotoluene

CH₃   NO₂

NO₂

2,4-dinitrotoluene

Notice that three positions are activated in the following reaction, but the new substituent ends up on only two of the three. Steric hindrance makes the position between the substituents less accessible.

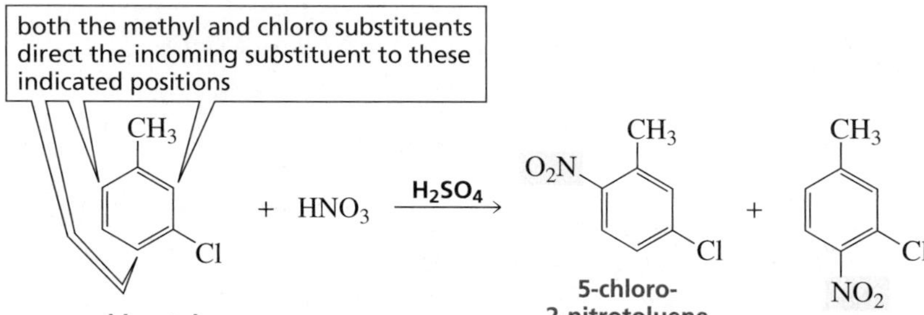

both the methyl and chloro substituents direct the incoming substituent to these indicated positions

CH₃   +   HNO₃ $\xrightarrow{\text{H}_2\text{SO}_4}$

Cl

m-chlorotoluene

O₂N   CH₃

Cl

5-chloro-2-nitrotoluene

+   CH₃

Cl

NO₂

3-chloro-4-nitrotoluene

 **Friedel-Crafts-3**

If the two substituents direct the new substituent to different positions, a strongly activating substituent will win out over a weakly activating substituent or a deactivating substituent.

 **Friedel-Crafts-4**

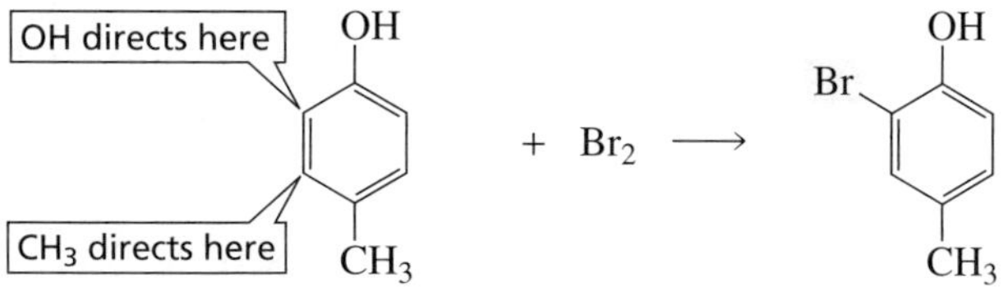

OH directs here   OH

CH₃ directs here   CH₃

p-methylphenol

+   Br₂   ⟶

OH

Br

CH₃

2-bromo-4-methylphenol major product

If the two substituents have similar activating properties, neither will dominate and a mixture of products will be obtained.

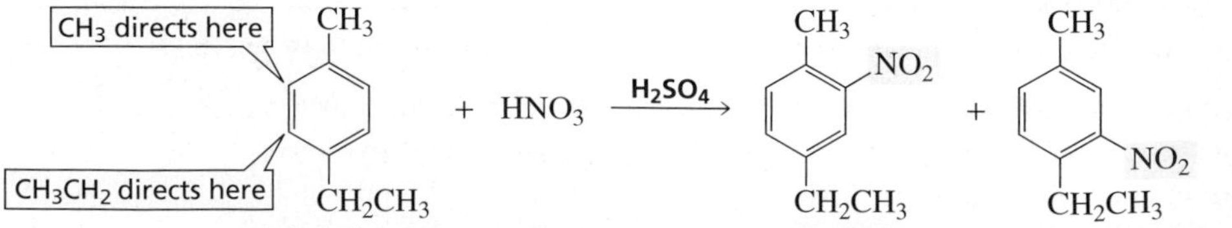

CH₃ directs here   CH₃

CH₃CH₂ directs here   CH₂CH₃

p-ethyltoluene

+   HNO₃ $\xrightarrow{\text{H}_2\text{SO}_4}$

CH₃   NO₂

CH₂CH₃

4-ethyl-2-nitrotoluene

+   CH₃

NO₂

CH₂CH₃

4-ethyl-3-nitrotoluene

PROBLEM 17◆

Give the major product(s) of each of the following reactions:

a. bromination of *p*-methylbenzoic acid

b. chlorination of *o*-benzenedicarboxylic acid

c. bromination of *p*-chlorobenzoic acid

d. nitration of *p*-fluoroanisole

e. nitration of *p*-methoxybenzaldehyde

f. nitration of *p*-*tert*-butyltoluene

PROBLEM 18◆

How many products are obtained from the chlorination of:

a. *o*-xylene? **b.** *p*-xylene? **c.** *m*-xylene?

PROBLEM 19 **SOLVED**

When phenol is treated with Br_2, a mixture of monobromo-, dibromo-, and tribromophenols is obtained. Design a synthesis that would convert phenol primarily to *ortho*-bromophenol.

Solution In the synthesis shown below, the bulky sulfonic acid group will add preferentially to the para position. Both the OH and SO_3H groups will direct bromine to the position ortho to the OH group. Heating in dilute acid removes the sulfonic acid group (Section 14.13).

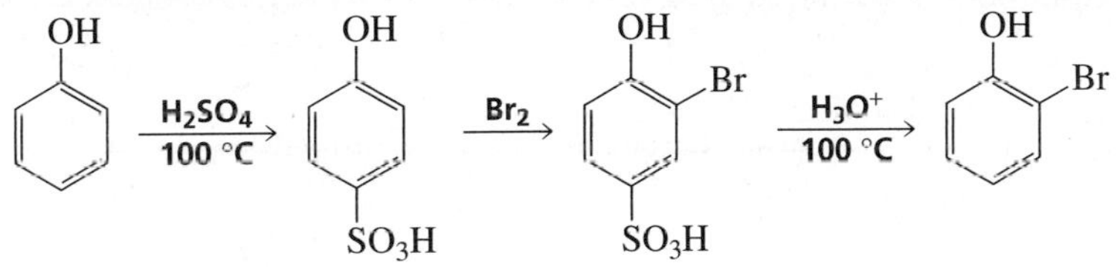

The use of a sulfonic acid group to block the para position is a common strategy for synthesizing high yields of ortho-substituted compounds.

15.9 Synthesis of Substituted Benzenes Using Arenediazonium Salts

So far, we have learned how to place a limited number of different substituents on a benzene ring—the substituents listed in Section 15.2 and those that can be obtained from these substituents by chemical conversion (Section 14.19). However, the list of substituents that can be placed on a benzene ring can be greatly expanded using **arenediazonium salts**.

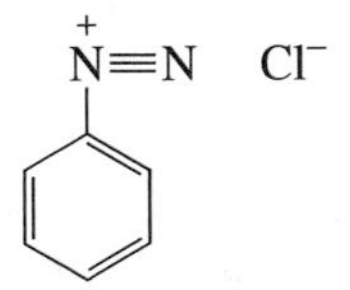

an arenediazonium salt

Displacing the leaving group of a diazonium ion by any of a wide variety of nucleophiles occurs readily because it results in the formation of a molecule of stable nitrogen gas. The mechanism by which the nucleophile displaces the diazonium group depends on the particular nucleophile: some displacements involve phenyl cations, whereas others involve radicals.

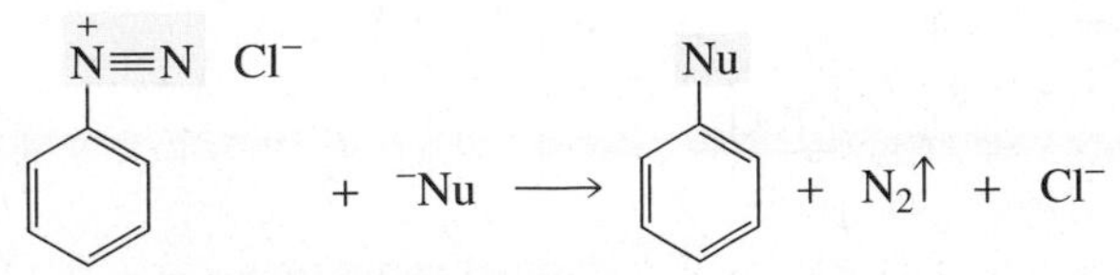

benzenediazonium chloride

Aniline can be converted into an arenediazonium salt by treatment with nitrous acid (HNO_2). Because nitrous acid is unstable, it must be formed in situ, using an aqueous solution of sodium nitrite and HCl or HBr; indeed, N_2 is such a good leaving group that the diazonium salt is synthesized at 0 °C and used immediately without isolation. [The mechanism for conversion of a primary amino group (NH_2) to a diazonium group ($^+N\equiv N$) is shown in Section 15.11.]

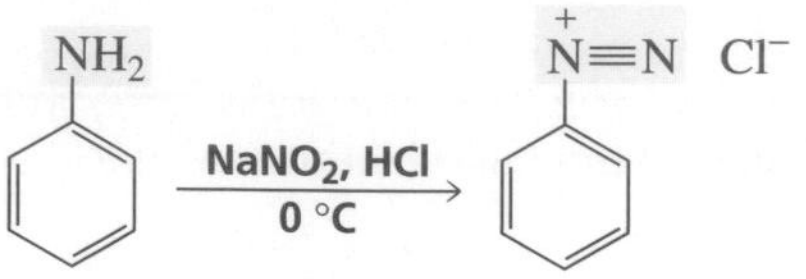

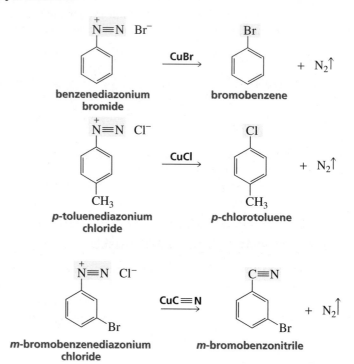

The nucleophiles $^-C\equiv N$, Cl^-, and Br^- will replace the diazonium group if the appropriate copper(I) salt is added to the solution containing the arenediazonium salt. The reaction of an arenediazonium salt with a copper(I) salt is known as a **Sandmeyer reaction**.

Sandmeyer reactions

Student Tutorial:
The Sandmeyer reaction

KCl and KBr cannot be used in place of CuCl and CuBr in Sandmeyer reactions; the cuprous salts are required. This indicates that the copper(I) ion has a role in the reaction. Although the precise mechanism is not known, it is thought that the copper(I) ion donates an electron to the diazonium salt, forming an aryl radical and nitrogen gas.

Although chloro and bromo substituents can be placed directly on a benzene ring by halogenation, the Sandmeyer reaction can be a useful alternative. For example, if you wanted to make *para*-chloroethylbenzene, the chlorination of ethylbenzene leads to a mixture of ortho and para isomers.

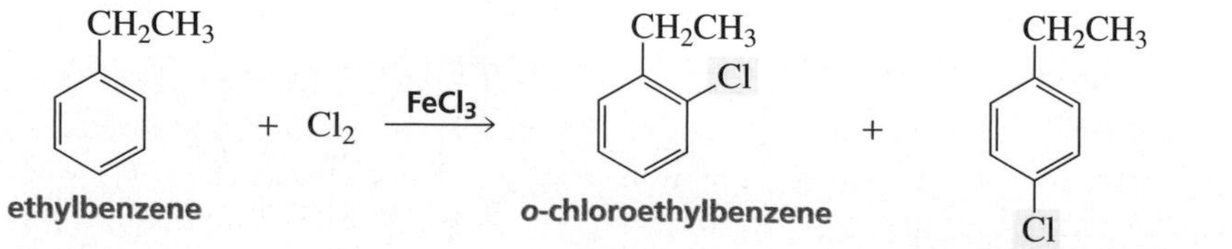

CH_2CH_3

ethylbenzene + Cl_2 $\xrightarrow{FeCl_3}$ CH_2CH_3

o-chloroethylbenzene + CH_2CH_3

p-chloroethylbenzene

If, however, you started with *para*-ethylaniline and used a Sandmeyer reaction for chlorination, only the desired para product would be formed.

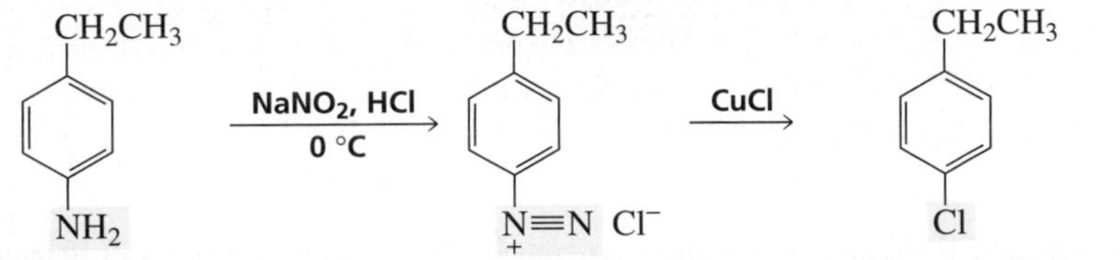

p-ethylaniline   **p-chloroethylbenzene**

PROBLEM 20

Explain why a diazonium group on a benzene ring cannot be used to direct an incoming substituent to the meta position.

PROBLEM 21

Explain why HBr should be used to generate the benzenediazonium salt if bromobenzene is the desired product of the Sandmeyer reaction, whereas HCl should be used if chlorobenzene is the desired product.

An iodo substituent will replace the diazonium group if potassium iodide is added to the solution containing the diazonium ion.

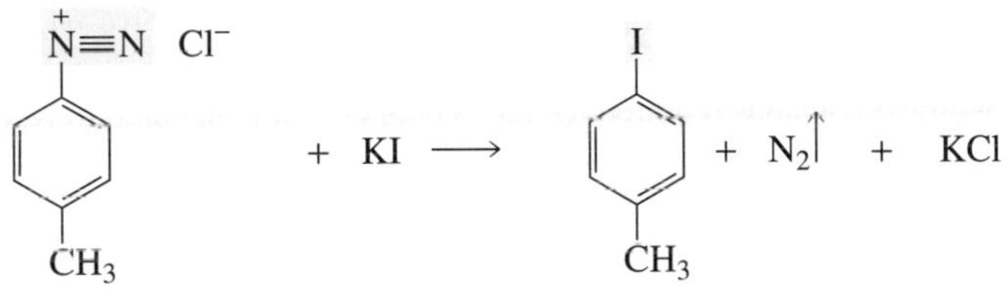

p-toluenediazonium   **p-iodotoluene**
chloride

Fluoro substitution occurs if the arenediazonium salt is heated with fluoroboric acid (HBF$_4$). This is known as the **Schiemann reaction**.

B I O G R A P H Y

Günther Schiemann
(1899–1969) *was born in Germany. He was a professor of chemistry at the Technologische Hochschule in Hannover, Germany.*

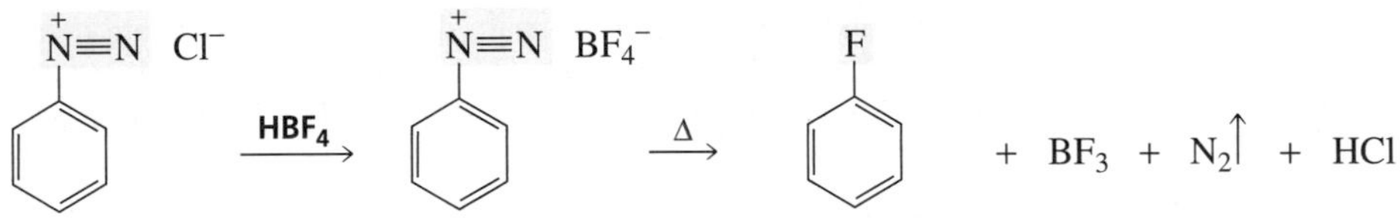

Schiemann reaction

fluorobenzene

If the aqueous solution in which the diazonium salt has been synthesized is acidified and heated, an OH group will replace the diazonium group. (H$_2$O is the nucleophile.)

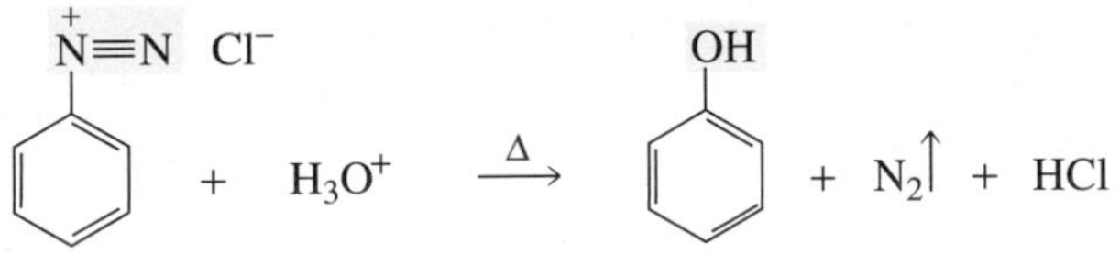

phenol

Adding copper (I) oxide and aqueous copper(II) nitrate to the solution is a faster way to form a phenol, and because the reaction occurs at room temperature, there is less competition from other nucleophiles in the solutions (such as Cl$^-$) that could replace the diazonium group.

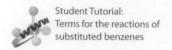

Student Tutorial:
Terms for the reactions of
substituted benzenes

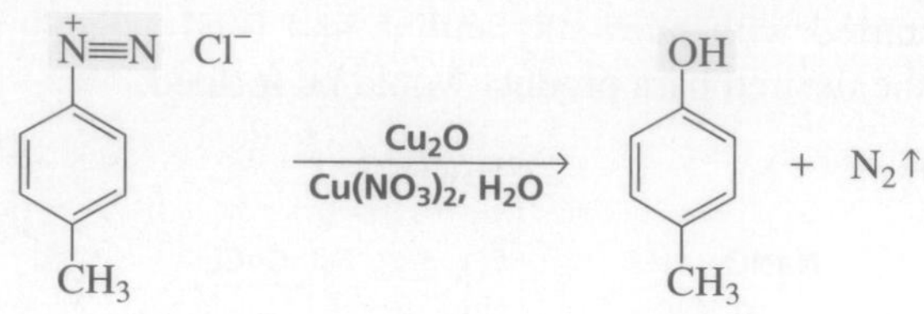

A hydrogen will replace a diazonium group if the arenediazonium salt is treated with hypophosphorous acid (H_3PO_2). This is a useful reaction if an amino group or a nitro group is needed for directing purposes and subsequently must be removed. It is difficult to envision how 1,3,5-tribromobenzene could be synthesized without such a reaction.

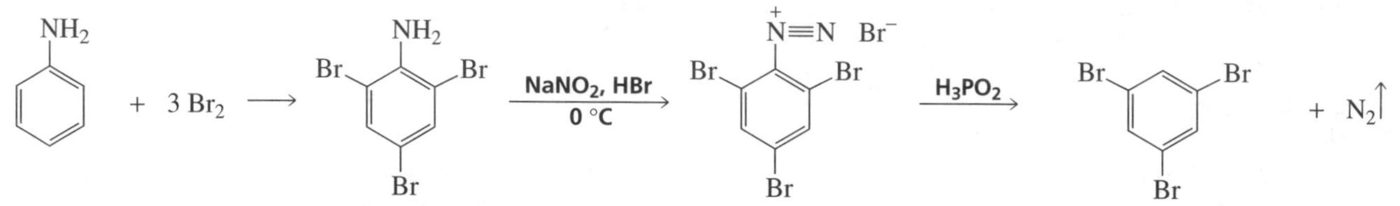

PROBLEM 22◆

Why is $FeBr_3$ not used as a catalyst in the above reaction?

PROBLEM 23

Write the sequence of steps required for the conversion of benzene into benzenediazonium chloride.

PROBLEM 24 *SOLVED*

Show how the following compounds could be synthesized from benzene:

a. *m*-dibromobenzene   **d.** *m*-nitrotoluene

b. *m*-bromophenol   **e.** *p*-methylbenzonitrile

c. *o*-chlorophenol   **f.** *m*-chlorobenzaldehyde

Solution to 24a A bromo substituent is an ortho–para director, so halogenation cannot be used to introduce both bromo substituents of *m*-dibromobenzene. Knowing that a bromo substituent can be placed on a benzene ring with a Sandmeyer reaction and the bromo substituent in a Sandmeyer reaction replaces what originally was a meta-directing nitro substituent, we have a route to the synthesis of the target compound.

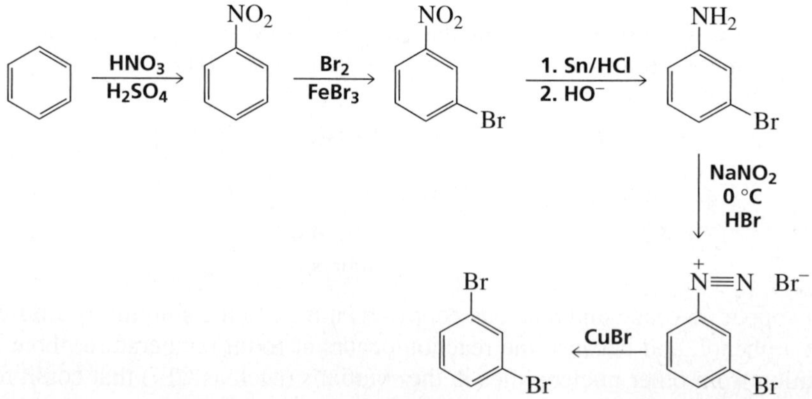

15.10 **The Arenediazonium Ion as an Electrophile**

In addition to being used to synthesize substituted benzenes, arenediazonium ions can be used as electrophiles in electrophilic aromatic substitution reactions. Because an arenediazonium ion is unstable at room temperature, it can be used as an electrophile only in reactions that can be carried out well below room temperature. In other words, only highly activated benzene rings (phenols, anilines, and *N*-alkylanilines) can undergo electrophilic aromatic substitution reactions with arenediazonium ion electrophiles. The product of the reaction is an *azo compound*. The N=N linkage is called an **azo linkage**. Because the electrophile is so large, substitution takes place preferentially at the less sterically hindered para position.

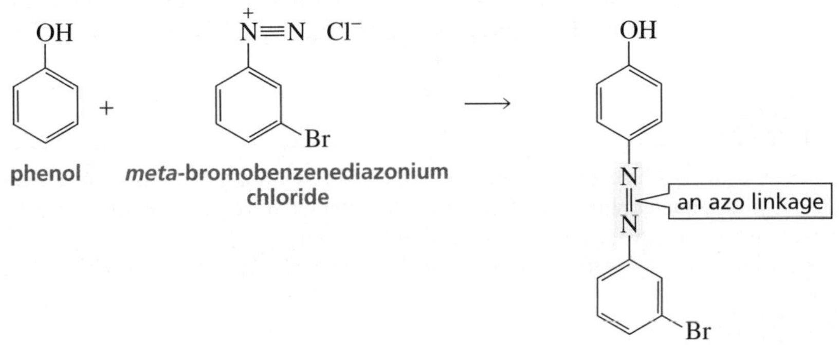

phenol   *meta*-bromobenzenediazonium chloride

an azo linkage

3-bromo-4'-hydroxyazobenzene
an azo compound

However, if the para position is blocked, substitution will occur at an ortho position.

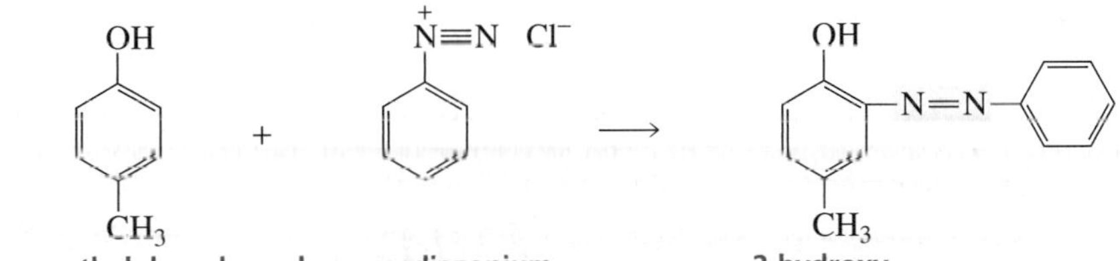

p-methylphenol   benzenediazonium chloride   2-hydroxy-5-methylazobenzene

The mechanism for electrophilic aromatic substitution with an arenediazonium ion electrophile is the same as the mechanism for electrophilic aromatic substitution with any other electrophile.

Mechanism for electrophilic aromatic substitution using an arenediazonium ion electrophile

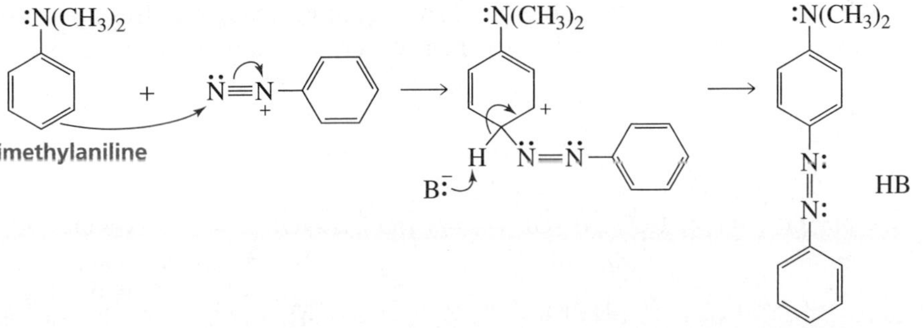

N,N-dimethylaniline

p-N,N-dimethylaminoazobenzene

Azo compounds, like alkenes, can exist in cis and trans forms. The trans isomer is considerably more stable than the cis isomer because the cis isomer has steric strain (Section 4.11).

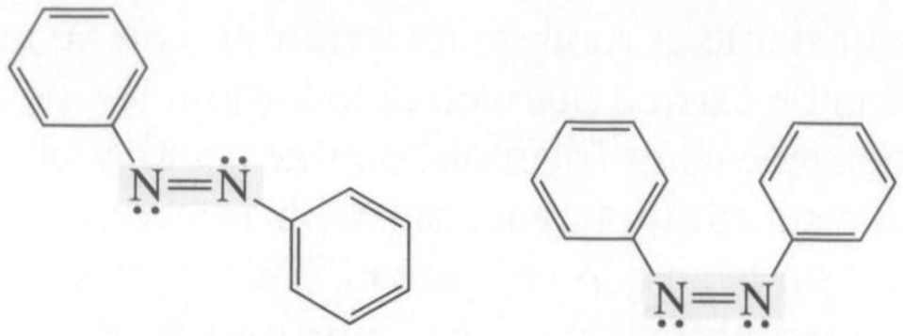

trans-azobenzene   **cis-azobenzene**

We have seen that azobenzenes are colored compounds because of their extended conjugation (Section 12.19). They are used commercially as dyes.

PROBLEM 25

In the mechanism for electrophilic aromatic substitution with a diazonium ion as the electrophile, why does nucleophilic attack occur on the terminal nitrogen atom of the diazonium ion rather than on the nitrogen atom bonded to the benzene ring?

PROBLEM 26

Give the structure of the activated benzene ring and the diazonium ion used in the synthesis of the following compounds:

a. butter yellow   **b.** methyl orange

(The structures of these compounds can be found in Section 12.19.)

15.11 The Mechanism for the Reaction of Amines with Nitrous Acid

We have seen that the reaction of an aniline with nitrous acid forms an arenediazonium salt. Both primary aryl amines and primary alkyl amines can form diazonium salts, and both do so by the same mechanism.

$$\text{aryl}-NH_2 \xrightarrow[\text{0 °C}]{\text{NaNO}_2,\ \text{HCl}} \text{aryl}-\overset{+}{N}\equiv N \quad Cl^-$$

$$\text{alkyl}-NH_2 \xrightarrow[\text{0 °C}]{\text{NaNO}_2,\ \text{HCl}} \text{alkyl}-\overset{+}{N}\equiv N \quad Cl^-$$

Conversion of a *primary* amino group to a diazonium group requires a nitrosonium ion. A nitrosonium ion is formed when water is eliminated from protonated nitrous acid.

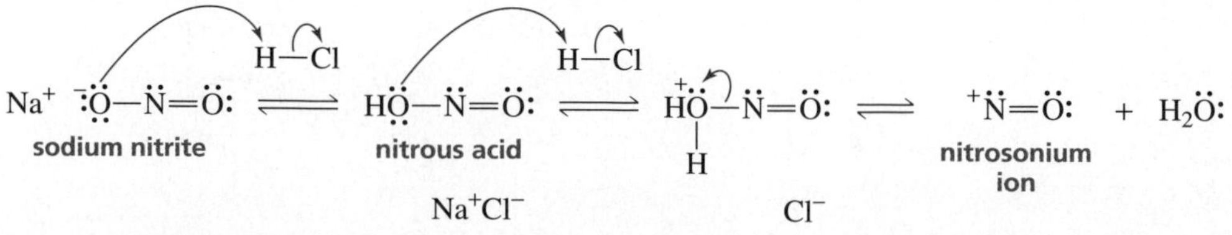

The mechanism for formation of a diazonium is shown below.

Mechanism for formation of a diazonium ion from aniline

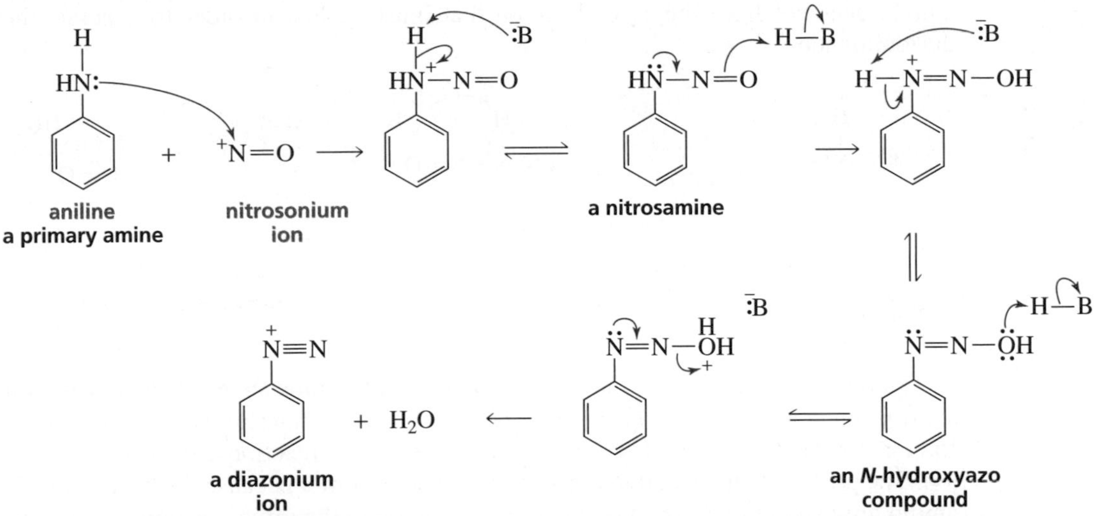

aniline
a primary amine

nitrosonium
ion

a nitrosamine

a diazonium
ion

an *N*-hydroxyazo
compound

- In the first step of the mechanism, aniline shares an electron pair with the nitrosonium ion.
- Loss of a proton from nitrogen forms a **nitrosamine** (also called an **N-nitroso compound** because a nitroso substituent is bonded to a nitrogen).
- Delocalization of nitrogen's lone pair and protonation of oxygen form a protonated *N*-hydroxyazo compound.
- The protonated *N*-hydroxyazo compound is in equilibrium with its nonprotonated form.
- The *N*-hydroxyazo compound can be reprotonated on nitrogen (reverse reaction) or protonated on oxygen (forward reaction).
- Elimination of water forms the diazonium ion.

Remember that reactions involving arenediazonium ions must be carried out at 0 °C because they are unstable at higher temperatures. Alkanediazonium ions are even less stable. They lose molecular N_2—even at 0 °C—as they are formed, reacting with whatever nucleophiles are present in the reaction mixture by both $S_N1/E1$ and $S_N2/E2$ mechanisms. Because of the mixture of products obtained, alkanediazonium ions are of limited synthetic use.

PROBLEM 27◆

What products would be formed from the reaction of isopropylamine with sodium nitrite and aqueous HCl?

PROBLEM 28

Diazomethane can be used to convert a carboxylic acid into a methyl ester. Propose a mechanism for this reaction.

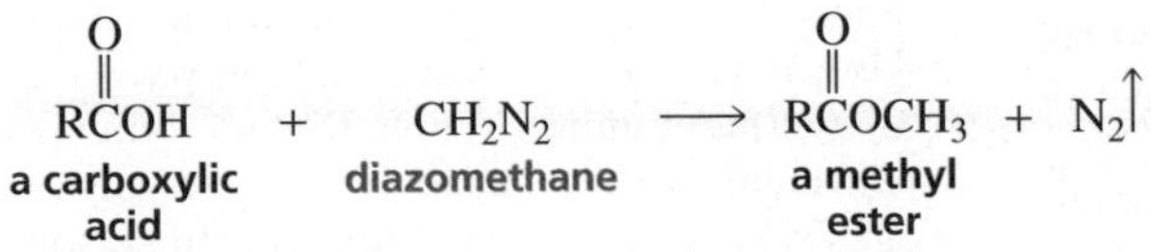

RCOH   + CH_2N_2 → $RCOCH_3$ + N_2↑

a carboxylic
acid

diazomethane

a methyl
ester

Secondary aryl and alkyl amines react with a nitrosonium ion to form nitrosamines rather than diazonium ions. The mechanism of the reaction is similar to that for the reaction of a primary amine with a nitrosonium ion, except that the reaction stops at the nitrosamine stage. The reaction stops because a secondary amine, unlike a primary amine, does not have the second proton that must be lost in order to generate the diazonium ion.

N-methylaniline
a secondary amine

N-methyl-N-nitrosoaniline
a nitrosamine

The product formed when the nitrogen of a *tertiary* amine shares its lone pair with a nitrosonium ion cannot be stabilized by loss of a proton. A tertiary aryl amine, therefore, can undergo an electrophilic aromatic substitution reaction with a nitrosonium ion. The product of the reaction is primarily the para isomer because the bulky dialkyl-amino group blocks approach of the nitrosonium ion to the ortho position.

N,N-dimethylaniline
a tertiary amine

para-nitroso-N,N-dimethylaniline
85%

NITROSAMINES AND CANCER

A 1962 outbreak of food poisoning in sheep in Norway was traced to their ingestion of nitrite-treated fish meal. This incident immediately raised concerns about human consumption of nitrite-treated foods, because sodium nitrite, a commonly used food preservative, can react with naturally occurring secondary amines present in food, to produce nitrosamines, which are known to be carcinogenic. Smoked fish, cured meats, and beer all contain nitrosamines. Nitrosamines are also found in cheese because some cheeses are preserved with nitrite and cheese is rich in secondary amines. When consumer groups in the United States asked the Food and Drug Administration to ban the use of sodium nitrite as a preservative, the request was vigorously opposed by the meat-packing industry. Despite extensive investigations, it has not yet been determined whether the small amounts of nitrosamines present in our food pose a hazard to our health. Until this question is answered, it will be hard to avoid sodium nitrite in our diet. Meanwhile, it is worrisome to note that Japan has both one of the highest gastric cancer rates and the highest average ingestion of sodium nitrite. Some good news, however, is that the concentration of nitrosamines present in bacon has been considerably reduced in recent years by the addition of ascorbic acid—a nitrosamine inhibitor—to the curing mixture. Also, improvements in the malting process have reduced the level of nitrosamines in beer. Dietary sodium nitrite does have a redeeming feature: there is some evidence that it protects against botulism, a type of severe food poisoning.

15.12 Nucleophilic Aromatic Substitution: An Addition–Elimination Mechanism

We have seen that aryl halides do not react with nucleophiles under standard reaction conditions because the π electron cloud repels the approach of a nucleophile (Section 8.8).

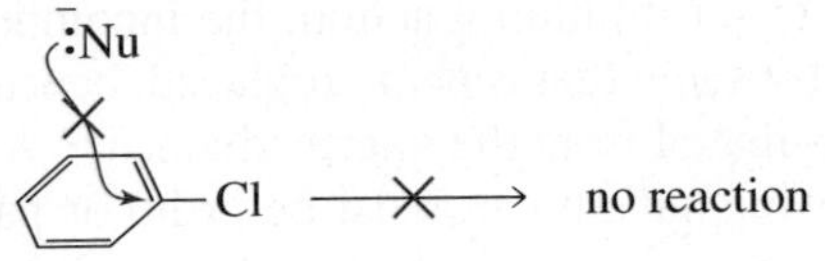

If, however, the aryl halide has one or more substituents that strongly withdraw electrons from the ring by resonance, **nucleophilic aromatic substitution** reactions can occur without using extreme conditions. The electron-withdrawing groups must be positioned ortho or para to the halogen. The greater the number of electron-withdrawing substituents, the more easily the nucleophilic aromatic substitution reaction occurs. Notice the different conditions under which the following reactions occur:

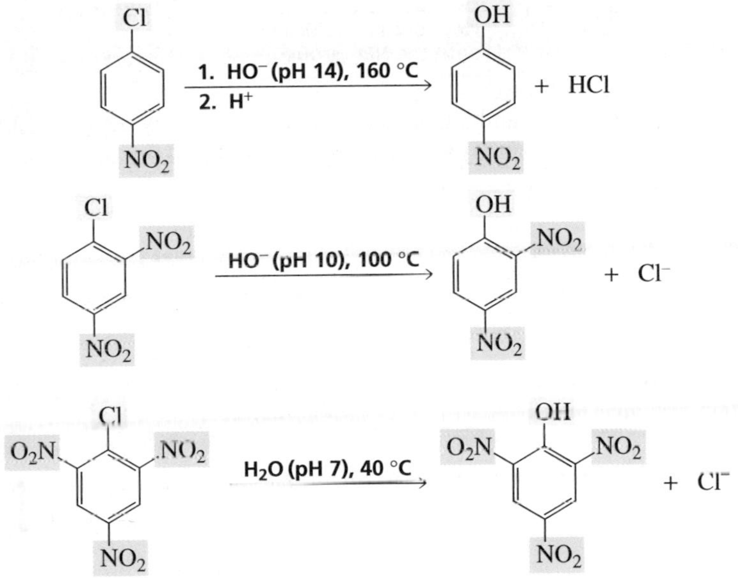

Notice also that the strongly electron-withdrawing substituents that *activate* the benzene ring toward *nucleophilic aromatic substitution* reactions are the same substituents that *deactivate* the ring toward *electrophilic aromatic substitution*. In other words, making the ring less electron rich makes it more reactive toward a nucleophile but less reactive toward an electrophile. Thus, any substituent that deactivates the benzene ring toward electrophilic substitution activates it toward nucleophilic substitution and vice versa.

Nucleophilic aromatic substitution takes place by a two-step mechanism. The reaction is called an **S_NAr reaction** (substitution nucleophilic aromatic).

> **Electron-withdrawing substituents increase the reactivity of the benzene ring toward nucleophilic substitution and decrease the reactivity of the benzene ring toward electrophilic substitution.**

> **A substituent that deactivates a ring toward electrophilic aromatic substitution, activates a ring toward nucleophilic aromatic substitution.**

General mechanism for nucleophilic aromatic substitution

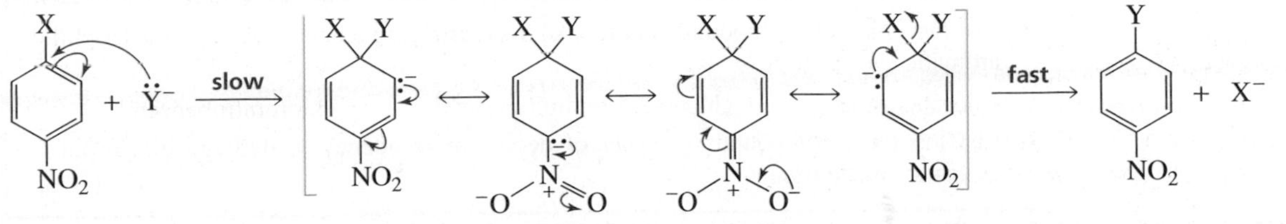

- The nucleophile attacks the carbon bearing the leaving group from a trajectory that is nearly perpendicular to the aromatic ring. (Recall from Section 8.8 that groups cannot be displaced from sp^2 carbon atoms by back-side attack.) Nucleophilic attack forms a resonance-stabilized carbanion intermediate called a *Meisenheimer complex*, after Jakob Meisenheimer (1876–1934).
- The leaving group departs, reestablishing the aromaticity of the ring.

In a nucleophilic aromatic substitution reaction, the incoming nucleophile must be a stronger base than the substituent that is being replaced, because the weaker of the two bases will be the one eliminated from the intermediate.

The electron-withdrawing substituent must be ortho or para to the site of nucleophilic attack because the electrons of the attacking nucleophile can be delocalized onto the substituent only if the substituent is in one of those positions.

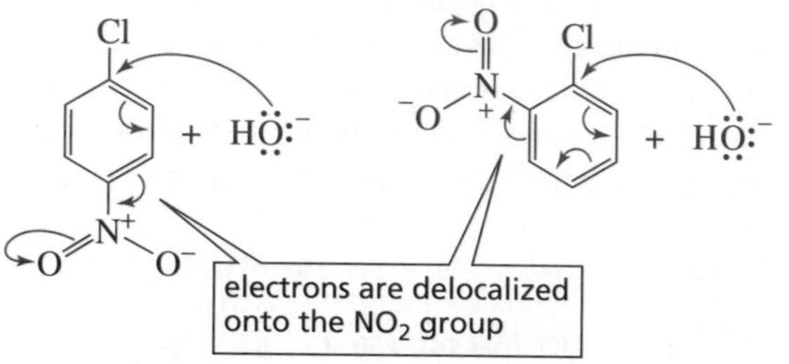

electrons are delocalized onto the NO_2 group

A variety of substituents can be placed on a benzene ring by means of nucleophilic aromatic substitution reactions. The only requirement is that the incoming group be a stronger base than the group that is being replaced.

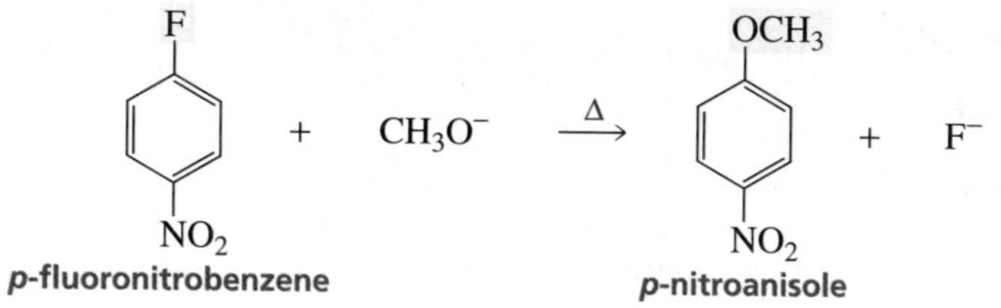

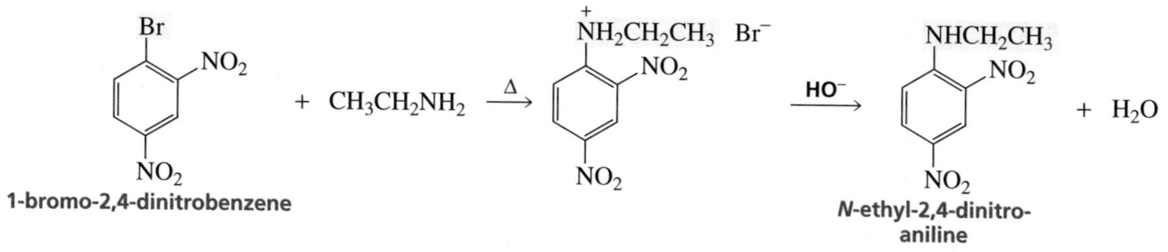

p-fluoronitrobenzene   *p*-nitroanisole

1-bromo-2,4-dinitrobenzene   *N*-ethyl-2,4-dinitro-aniline

PROBLEM 29

Draw resonance contributors for the carbanion that would be formed if *meta*-chloronitrobenzene were to react with hydroxide ion. Why does it not react?

PROBLEM 30◆

a. List the following compounds in order of decreasing tendency to undergo nucleophilic aromatic substitution:

chlorobenzene   1-chloro-2,4-dinitrobenzene   *p*-chloronitrobenzene

b. List the same compounds in order of decreasing tendency to undergo electrophilic aromatic substitution.

PROBLEM 31

Show how each of the following compounds could be synthesized from benzene:

a. *o*-nitrophenol   **b.** *p*-nitroaniline   **c.** *p*-bromoanisole   **d.** anisole

15.13 Nucleophilic Aromatic Substitution: An Elimination–Addition Mechanism That Forms a Benzyne Intermediate

An aryl halide such as chlorobenzene can undergo a nucleophilic substitution reaction in the presence of a very strong base such as $^-NH_2$. There are two surprising aspects to this reaction: the aryl halide does not have to contain an electron-withdrawing group, and the incoming substituent does not always end up on the carbon vacated by the leaving group. For example, when chlorobenzene—with the carbon to which the chlorine is attached isotopically labeled with ^{14}C—is treated with amide ion in liquid ammonia, aniline is obtained as the product. Half of the product has the amino group attached to the isotopically labeled carbon (denoted by the asterisk) as expected, but the other half has the amino group attached to the carbon adjacent to the labeled carbon.

BIOGRAPHY

Martin D. Kamen (1913–2002), born in Toronto, was the first person to isolate ^{14}C, which immediately became the most useful of all the isotopes in chemical and biochemical research. Kamen received a B.S. and a Ph.D. from the University of Chicago and became a U.S. citizen in 1938. He was a professor at the University of California, Berkeley; at Washington University, St. Louis; and at Brandeis University. He was also one of the founding professors of the University of California, San Diego. In later years, he became a member of the faculty at the University of Southern California. He received the Fermi medal in 1996.

chlorobenzene $+$ $^-NH_2$ $\xrightarrow{NH_3 \text{ (liq)}}$

approximately equal amounts of the two products are obtained

These are the only products formed. Anilines with the amino group two or three carbons removed from the labeled carbon are not formed.

The fact that the two products are formed in approximately equal amounts indicates that the reaction takes place by a mechanism that forms an intermediate in which the two carbons to which the amino group is attached in the product are equivalent. The mechanism that accounts for the experimental observations involves the formation of a **benzyne intermediate**. Benzyne has a triple bond between two adjacent carbon atoms of benzene.

- In the first step of the mechanism, the strong base ($^-NH_2$) removes a proton from the position ortho to the halogen.
- The resulting anion expels the halide ion, thereby forming benzyne.

$+$ $^-:NH_2$ $\longrightarrow$ $+$ NH_3 $\longrightarrow$ benzyne $+$ Cl^-

- The incoming nucleophile can attack either of the carbons of the "triple bond" of benzyne.
- Protonation of the resulting anion forms the substitution product.

benzyne $:NH_2$

The overall reaction is an elimination–addition reaction: benzyne is formed in an elimination reaction and immediately undergoes an addition reaction.

Substitution at the carbon that was attached to the leaving group is called **direct substitution**. Substitution at the adjacent carbon is called **cine substitution** (cine comes from *kinesis*, which is Greek for "movement"). In the following reaction, *o*-toluidine is the direct-substitution product; *m*-toluidine is the cine-substitution product.

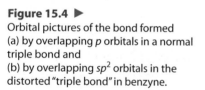

o-bromotoluene + NaNH₂ → NH₃ (liq)

o-toluidine
direct-substitution product

+ *m*-toluidine
cine-substitution product

Benzyne is an extremely reactive species. In Section 6.4, we saw that in a molecule with a triple bond, the two *sp* hybridized carbons and the atoms attached to these carbons (C—C≡C—C) are linear because the bond angles are 180°. Four linear atoms cannot be incorporated into a six-membered ring, so the C—C≡C—C system in benzyne is distorted: the original π bond is unchanged, but the *sp²* orbitals that form the new π bond are not parallel to one another (Figure 15.4). Therefore, they cannot overlap as well as the *p* orbitals that form a normal π bond, resulting in a much weaker and much more reactive bond.

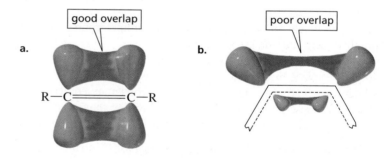

Figure 15.4 ▶
Orbital pictures of the bond formed
(a) by overlapping *p* orbitals in a normal triple bond and
(b) by overlapping *sp²* orbitals in the distorted "triple bond" in benzyne.

Although benzyne is too unstable to be isolated, evidence that it is formed can be obtained by a trapping experiment. When furan is added to a reaction that forms a benzyne intermediate, furan traps the benzyne intermediate by reacting with it in a Diels–Alder reaction (Section 7.12). The product of the Diels–Alder reaction can be isolated.

benzyne
a dienophile

furan
a diene

the product of the
Diels–Alder reaction

A synthesis employing a benzyne intermediate is a good way to prepare *meta*-amino-substituted aryl ethers (see Problem 66). The amide ion preferentially attacks the meta position because there is less steric hindrance at this position and the resulting negative charge can be stabilized by the adjacent electron-withdrawing oxygen.

PROBLEM 32

Propose a mechanism to account for the fact that when a basic solution of *p*-chlorotoluene is heated with sodium amide in liquid ammonia, both *p*-cresol and *m*-cresol are formed.

PROBLEM 33

Give the products that would be obtained from the reaction of the following compounds with sodium amide in liquid ammonia:

a. 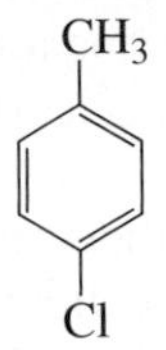 [structure: toluene with Cl at para position]
b.  [structure: ethylbenzene with Br at meta position]
c. 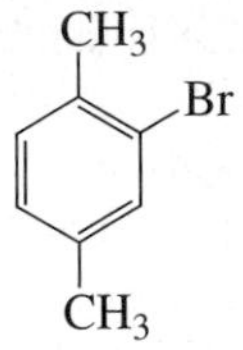 [structure: 1,4-dimethylbenzene with Br]
d. 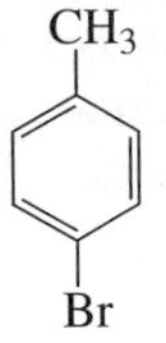 [structure: toluene with Br at para position]

15.14 Polycyclic Benzenoid Hydrocarbons

Polycyclic benzenoid hydrocarbons are compounds that contain two or more fused benzene rings. **Fused rings** share two adjacent carbons. Naphthalene has two fused rings, anthracene and phenanthrene have three fused rings, and tetracene, triphenylene, pyrene, and chrysene have four fused rings. There are many polycyclic benzenoid hydrocarbons with more than four fused rings.

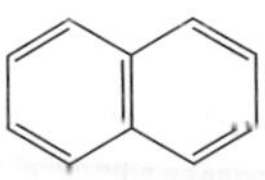

naphthalene 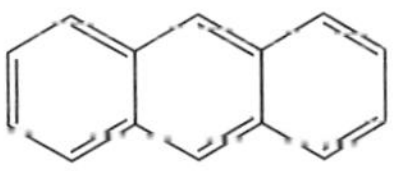   anthracene 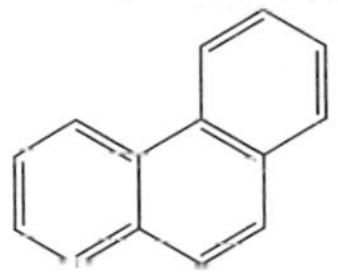   phenanthrene 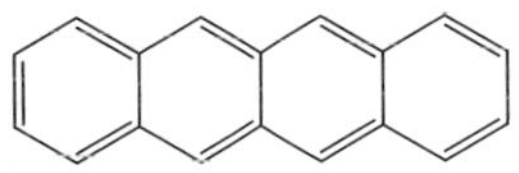   tetracene

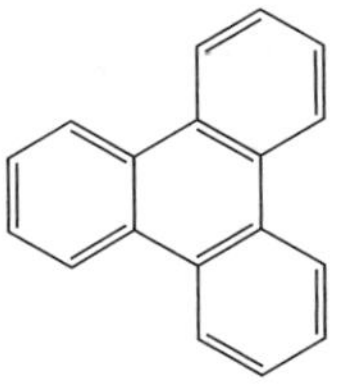

triphenylene 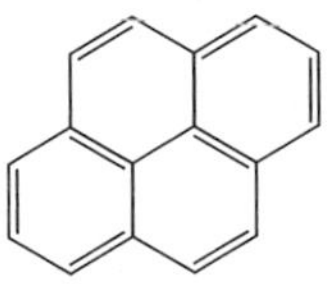   pyrene 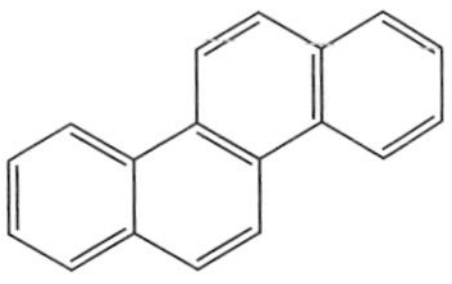   chrysene

Like benzene, all the larger polycyclic benzenoid hydrocarbons undergo electrophilic aromatic substitution reactions. Some of these compounds are well-known carcinogens. The chemical reactions responsible for causing cancer and how you can predict which compounds are carcinogenic were discussed in Section 10.9.

SUMMARY

The relative positions of two substituents on a benzene ring are indicated in the compound's name either by numbers or by the prefixes *ortho, meta,* and *para*. The nature of the substituent affects both the reactivity of the benzene ring and the placement of an incoming substituent: the rate of electrophilic aromatic substitution is increased by electron-donating substituents and decreased by electron-withdrawing

substituents. Substituents can donate or withdraw electrons **inductively** or by **resonance**, and can donate electrons by hyperconjugation.

The stability of the carbocation intermediate determines the position to which a substituent directs an incoming electrophile. All activating substituents and the weakly deactivating halogens are **ortho–para directors**; all substituents more deactivating than the halogens are **meta directors**. Ortho–para directors—with the exception of alkyl, aryl, and CH=CHR—have a lone pair on the atom attached to the ring; meta directors have a positive or partial positive charge on the atom attached to the ring. Ortho–para-directing substituents form the para isomer preferentially if either the substituent or the incoming electrophile is large.

In planning the synthesis of disubstituted benzenes, the order in which the substituents are placed on the ring and the point in a reaction sequence at which a substituent is chemically modified are important considerations. When a disubstituted benzene undergoes an electrophilic aromatic substitution reaction, the directing effect of both sub-stituents has to be considered.

RO- and HO-substituted benzenes are halogenated without the Lewis acid. Benzene rings with meta-directing substituents cannot undergo Friedel–Crafts reactions. Aniline and N-alkyl substituted anilines also cannot undergo Friedel–Crafts reactions.

The kinds of substituents that can be placed on benzene rings are greatly expanded by reactions of arene diazonium salts, nucleophilic aromatic substitution reactions, and reactions involving a benzyne intermediate. Aniline and substituted anilines react with nitrous acid to form **arenediazonium salts**; a diazonium group can be displaced by a nucleophile. Arenediazonium ions can be used as electrophiles with highly activated benzene rings to form **azo compounds** that can exist in cis and trans forms. Secondary amines react with nitrous acid to form **nitrosamines**.

An aryl halide with one or more substituents ortho or para to the leaving group that strongly withdraw electrons by resonance undergoes a **nucleophilic aromatic substitution** (S_NAr) reaction: the nucleophile forms a resonance-stabilized carbanion intermediate, and then the leaving group departs, reestablishing the aromaticity of the ring. The incoming nucleophile must be a stronger base than the substituent being replaced. A substituent that deactivates a benzene ring toward electrophilic substitution activates it toward nucleophilic substitution.

In the presence of a strong base, an aryl halide under-goes a nucleophilic substitution reaction via a **benzyne intermediate**. After a hydrogen halide is eliminated, the nucleophile can attack either of the carbons of the distort-ed "triple bond" in benzyne. **Direct substitution** is substi-tution at the carbon that was attached to the leaving group; **cine substitution** is substitution at the adjacent carbon.

The ability of a substituent to withdraw electrons from or donate electrons into a benzene ring is reflected in the pK_a values of substituted phenols, benzoic acids, and protonated anilines: electron withdrawal increases acidity; electron donation decreases acidity.

Polycyclic benzenoid hydrocarbons contain two or more fused benzene rings; **fused rings** share two adjacent carbons. Like benzene, polycyclic benzenoid hydrocarbons undergo electrophilic aromatic substitution reactions.

SUMMARY OF REACTIONS

1. Reactions of amines with nitrous acid (Section 15.11)

2. Replacement of a diazonium group (Section 15.9)

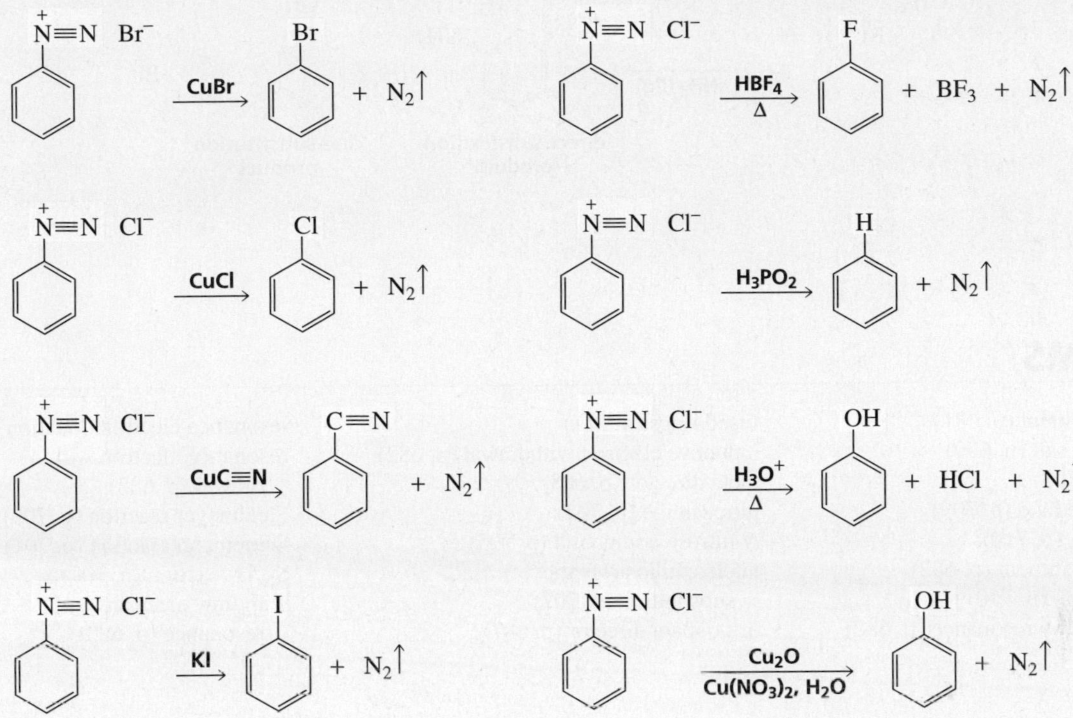

3. Formation of an azo compound (Section 15.10)

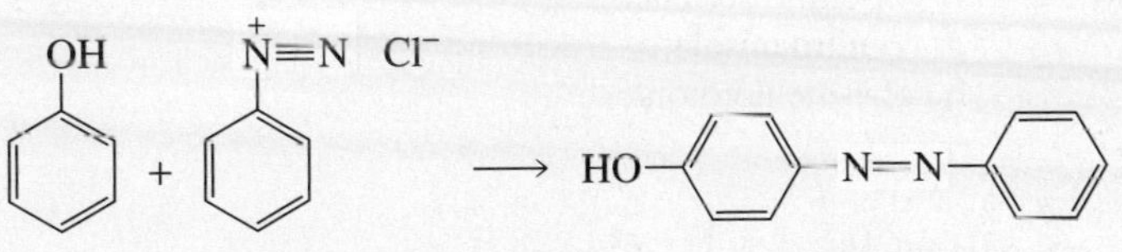

4. Nucleophilic aromatic substitution reactions (Section 15.12)

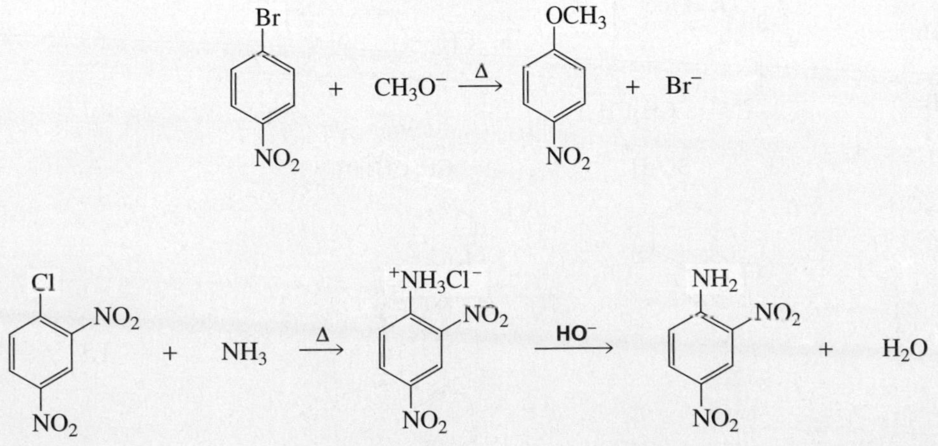

5. Nucleophilic substitution via a benzyne intermediate (Section 15.13)

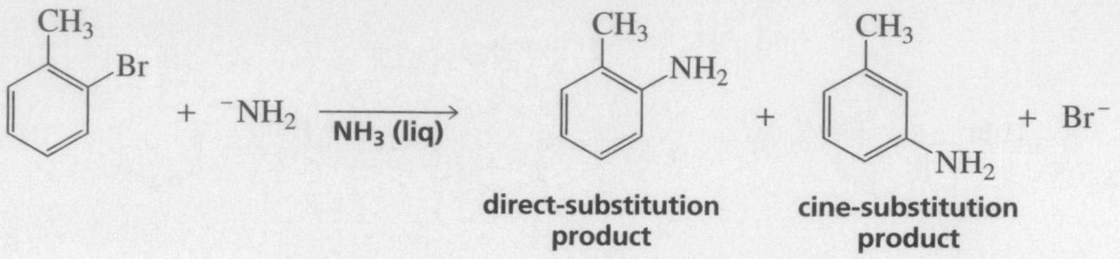

direct-substitution
product

cine-substitution
product

KEY TERMS

activating substituent (p. 681)
arenediazonium salt (p. 699)
azo linkage (p. 703)
benzyne intermediate (p. 709)
cine substitution (p. 710)
deactivating substituent (p. 682)
direct substitution (p. 710)
donate electrons by resonance (p. 682)

fused rings (p. 711)
inductive electron withdrawal (p. 682)
meta director (p. 688)
nitrosamine (p. 705)
N-nitroso compound (p. 705)
nucleophilic aromatic
　substitution (p. 707)
ortho–para director (p. 687)

resonance electron donation (p. 682)
resonance electron with-
　drawal (p. 682)
Sandmeyer reaction (p. 700)
Schiemann reaction (p. 701)
S_NAr reaction (p. 707)
withdraw electrons by
　resonance (p. 683)

PROBLEMS

34. Draw the structure of each of the following:
    a. *m*-ethylphenol
    b. *p*-nitrobenzenesulfonic acid
    c. (*E*)-2-phenyl-2-pentene
    d. *o*-bromoaniline

    e. 2-chloroanthracene
    f. *m*-chlorostyrene
    g. *o*-nitroanisole
    h. 2,4-dichlorotoluene

35. Name the following:

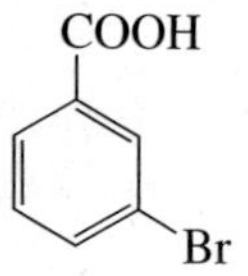

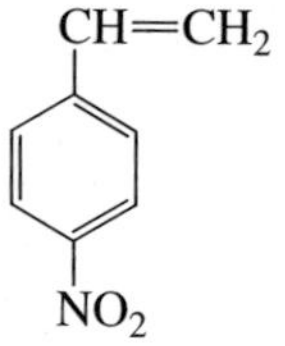

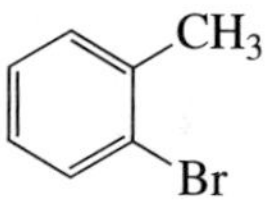

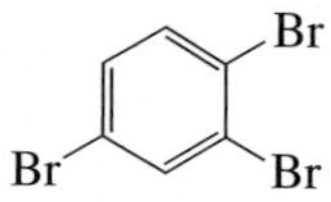

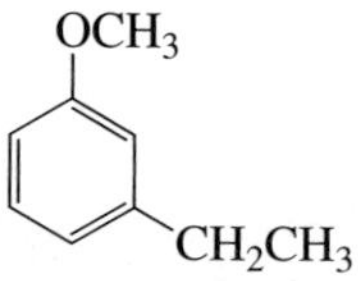

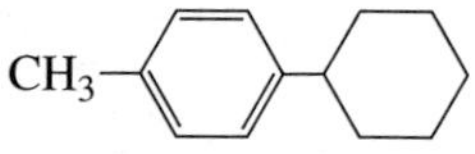

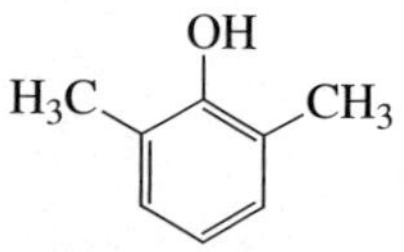

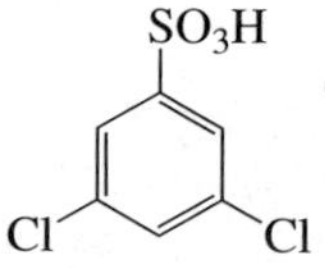

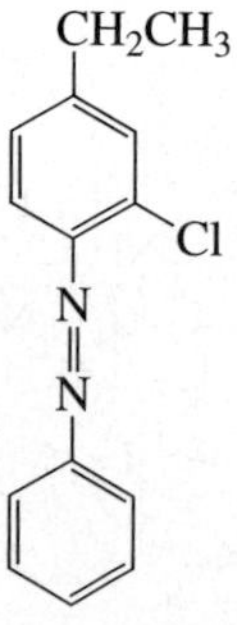

36. Provide the necessary reagents next to the arrows.

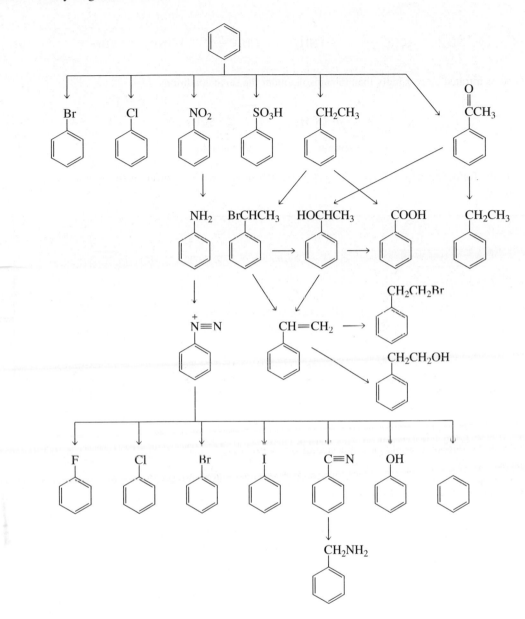

37. Answer the following questions, referring to the four structures shown below:

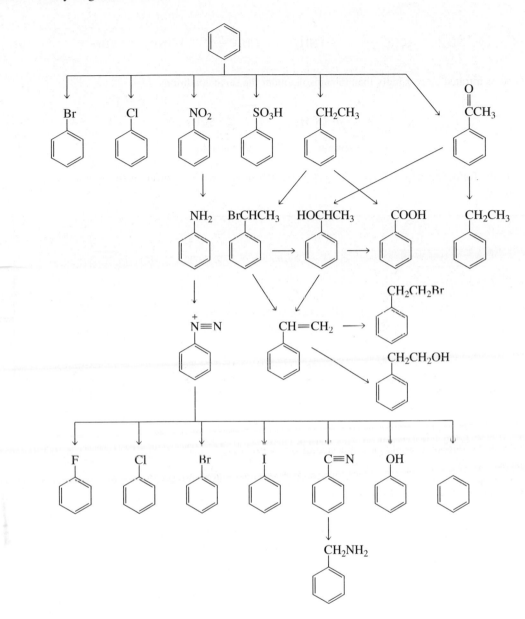

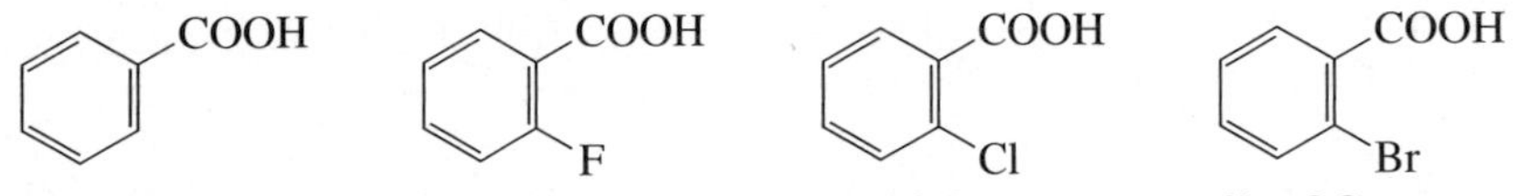

pKₐ = 4.2   pKₐ = 3.3   pKₐ = 2.9   pKₐ = 2.8

a. Why are the *ortho*-halo-substituted benzoic acids stronger acids than benzoic acid?
b. Why is *o*-fluorobenzoic acid the weakest of the *ortho*-halo-substituted benzoic acids?
c. Why do *o*-chlorobenzoic acid and *o*-bromobenzoic have similar pKₐ values?

38. Give the product(s) of each of the following reactions:
a. benzoic acid + HNO₃/H₂SO₄
b. isopropylbenzene + cyclohexene + HF
c. *p*-xylene + acetyl chloride + AlCl₃ followed by H₂O
d. *o*-methylaniline + benzenediazonium chloride
e. cyclohexyl phenyl ether + Br₂
f. phenol + H₂SO₄ + Δ
g. ethylbenzene + Br₂/FeBr₃
h. *m*-xylene + Na₂Cr₂O₇ + H⁺ + Δ

39. Rank the following substituted anilines in order of decreasing basicity:

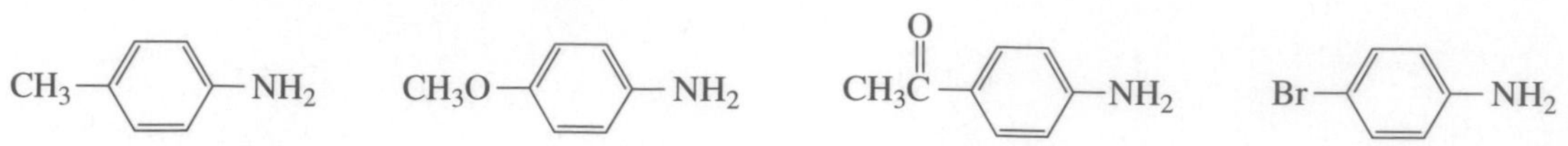

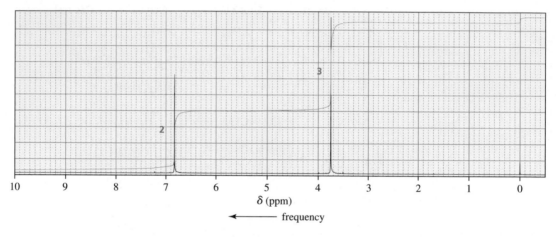

40. Explain why anisole is nitrated more rapidly than thioanisole under the same conditions.

anisole   thioanisole

41. The compound with the following ^{1}H NMR spectrum is known to be highly reactive toward electrophilic aromatic substitution. Identify the compound.

42. Show how the following compounds could be synthesized from benzene:
  a. *m*-chlorobenzenesulfonic acid   **d.** *m*-bromobenzonitrile   **g.** *m*-bromobenzoic acid
  b. *m*-chloroethylbenzene   **e.** 1-phenylpentane   **h.** *p*-cresol
  c. benzyl alcohol   **f.** *m*-hydroxybenzoic acid

43. For each horizontal row of substituted benzenes shown below, indicate
  a. the one that would be the most reactive in an electrophilic aromatic substitution reaction.
  b. the one that would be the least reactive in an electrophilic aromatic substitution reaction.
  c. the one that would yield the highest percentage of meta product.

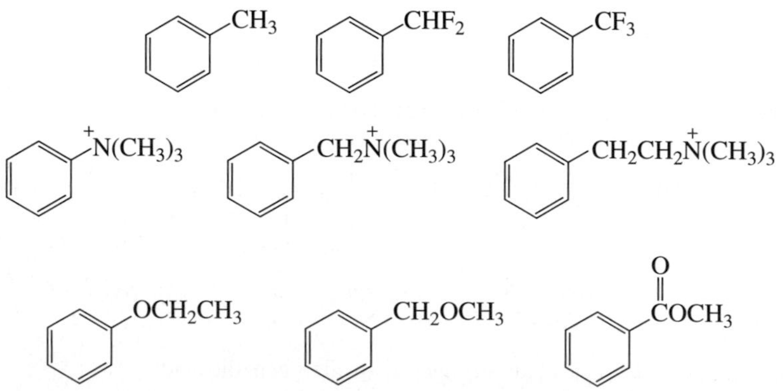

44. Arrange the following groups of compounds in order of decreasing reactivity toward electrophilic aromatic substitution:
  a. benzene, ethylbenzene, chlorobenzene, nitrobenzene, anisole
  b. 1-chloro-2,4-dinitrobenzene, 2,4-dinitrophenol, 2,4-dinitrotoluene
  c. toluene, *p*-cresol, benzene, *p*-xylene
  d. benzene, benzoic acid, phenol, propylbenzene
  e. *p*-nitrotoluene, 2-chloro-4-nitrotoluene, 2,4-dinitrotoluene, *p*-chlorotoluene
  f. bromobenzene, chlorobenzene, fluorobenzene, iodobenzene

45. Give the products of the following reactions:

a. 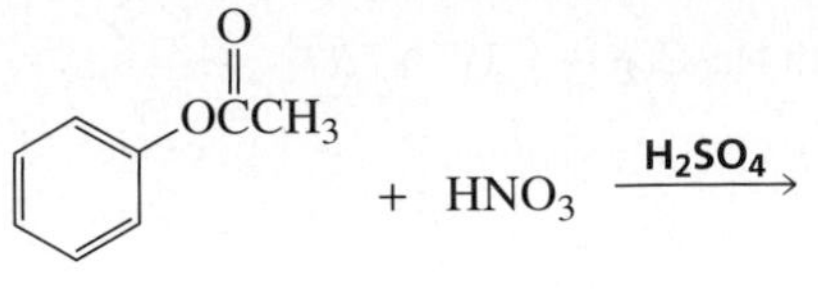 [benzene with OCCH₃ group] + HNO₃ →(H₂SO₄)

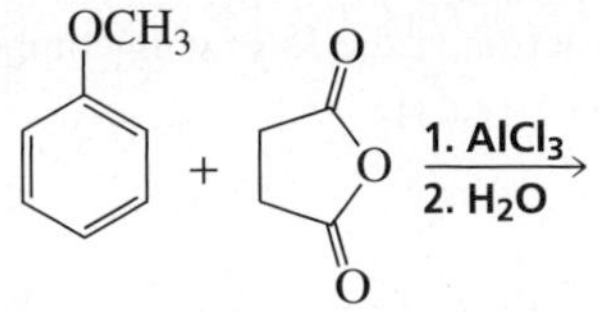

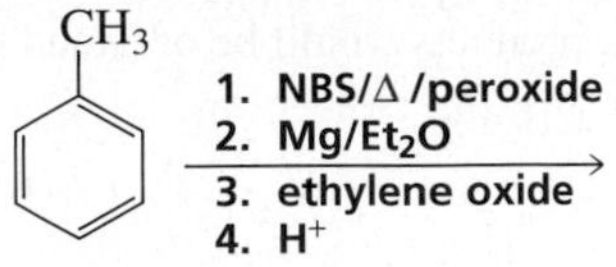

b. 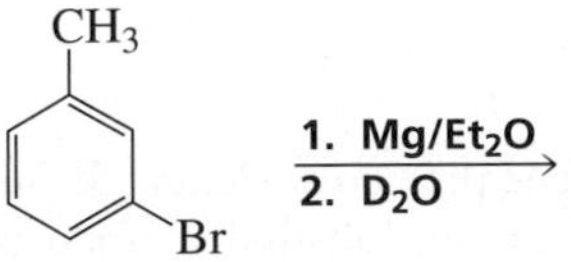 [CH₃ benzene with Br] →(1. Mg/Et₂O, 2. D₂O)

c. [OCH₃ benzene] + [anhydride] →(1. AlCl₃, 2. H₂O)

d. 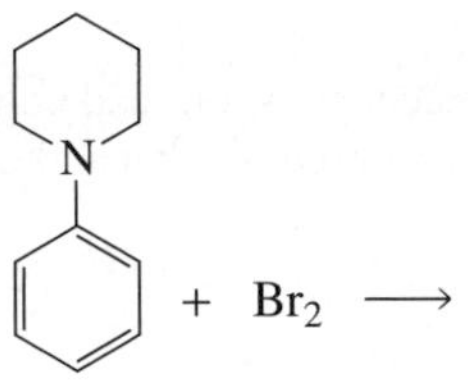 [N-phenyl piperidine] + Br₂ →

e. [CH₃ benzene] →(1. NBS/Δ /peroxide, 2. Mg/Et₂O, 3. ethylene oxide, 4. H⁺)

f. 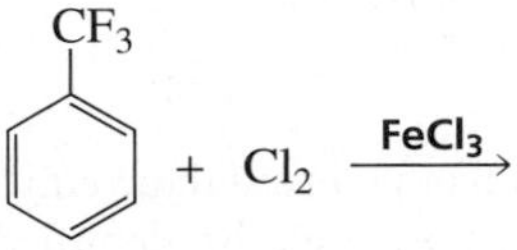 [CF₃ benzene] + Cl₂ →(FeCl₃)

46. For each of the statements in Column I, choose a substituent from Column II that fits the description for the following compound:

[Z-substituted benzene]

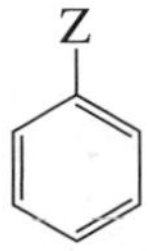

Column I	Column II
a. Z donates electrons by hyperconjugation and does not donate or withdraw electrons by resonance.	OH
	Br
b. Z withdraws electrons inductively and withdraws electrons by resonance.	⁺NH₃
	CH₂CH₃
c. Z deactivates the ring and directs ortho–para.	NO₂
d. Z withdraws electrons inductively, donates electrons by resonance, and activates the ring.	
e. Z withdraws electrons inductively and does not donate or withdraw electrons by resonance.	

47. Describe two ways to prepare anisole from benzene.

48. For each of the following compounds, indicate the ring carbon that would be nitrated if the compound is treated with HNO₃/H₂SO₄:

a. 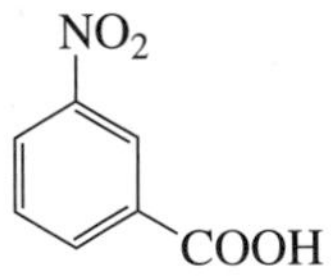 [benzene with NO₂ and COOH]

b. [benzene with OCCH₃ and CH₃OC=O groups]

c. [OCH₃ benzene with Cl]

d. 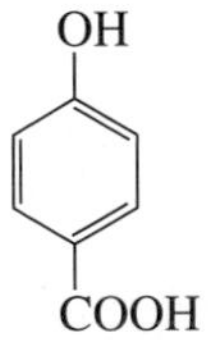 [OH benzene with COOH]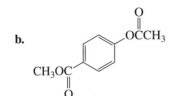

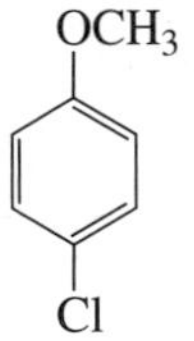

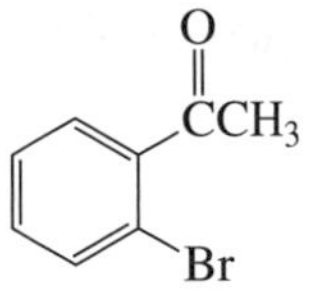

e. [benzene with CCH₃ (ketone) and Br]

f. 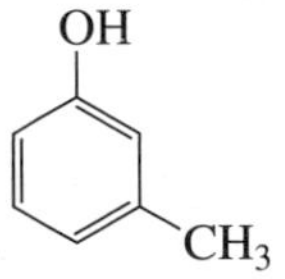 [OH benzene with CH₃]

g. [CH₃ benzene with NO₂]

h. 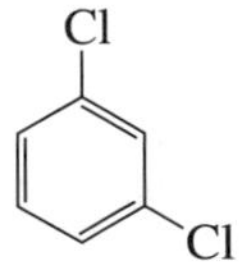 [benzene with two Cl groups]

49. Show how the following compounds could be synthesized from benzene:
a. *N,N,N*-trimethylanilinium iodide
b. benzyl methyl ether
c. *p*-benzylchlorobenzene
d. 2-methyl-4-nitrophenol
e. *p*-nitroaniline
f. *m*-bromoiodobenzene
g. *p*-dideuteriobenzene
h. *p*-nitro-*N*-methylaniline

50. Which of the following compounds will react with HBr more rapidly?

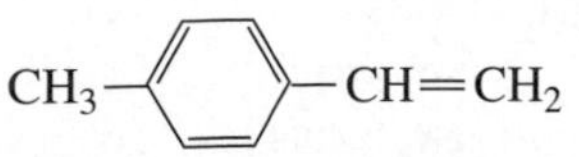

CH₃—[benzene]—CH=CH₂   or   CH₃O—[benzene]—CH=CH₂

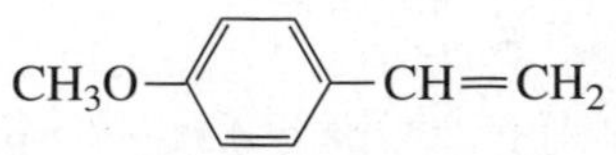

51. Would *m*-xylene or *p*-xylene react more rapidly with Cl_2 + $FeCl_3$? Explain.

52. What products would be obtained from the reaction of the following compounds with $Na_2Cr_2O_7$ + H^+ + Δ?

a.

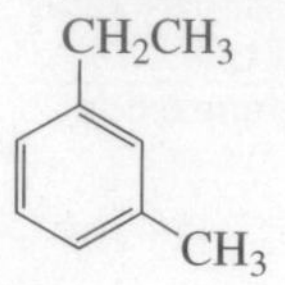

CH₂CH₃
CH₃

b.

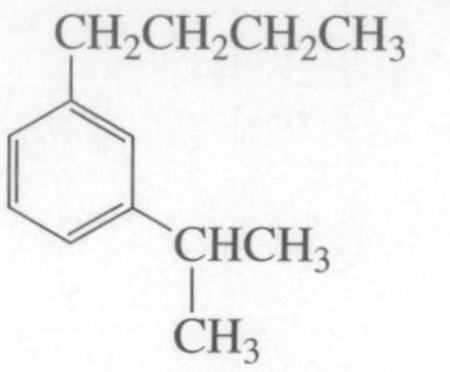

CH₂CH₂CH₂CH₃
CHCH₃
CH₃

c.

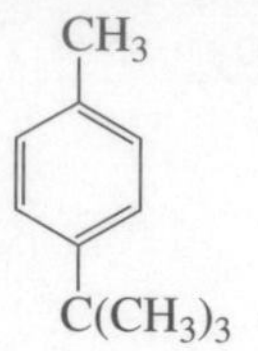

CH₃
C(CH₃)₃

53. A student had prepared three ethyl-substituted benzaldehydes, but had neglected to label them. The premed student at the next bench said they could be identified by brominating a sample of each and determining how many bromo-substituted products were formed. Is the premed student's advice sound?

54. Identify A–L.

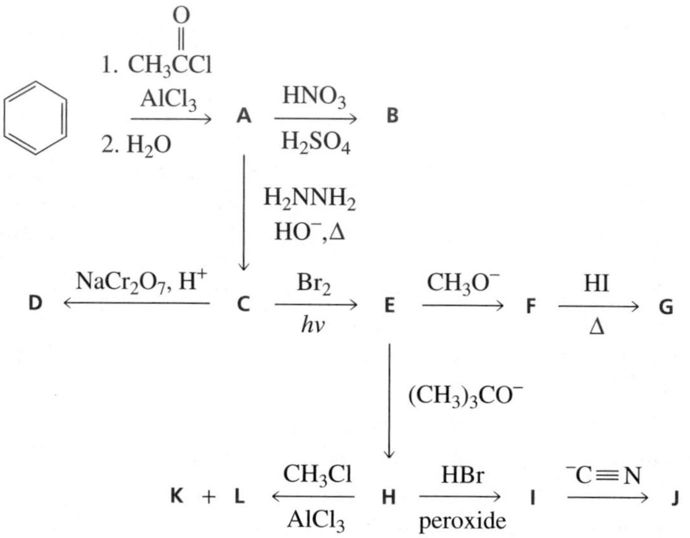

55. Explain, using resonance contributors for the intermediate carbocation, why a phenyl group is an ortho–para director.

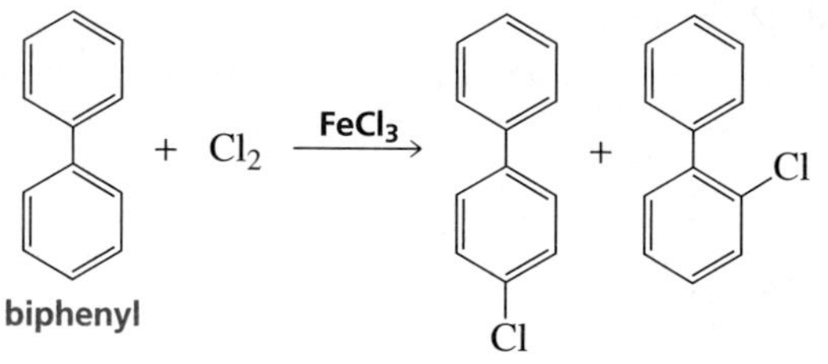

56. The pK_a values of a few ortho-, meta-, and para-substituted benzoic acids are shown here:

pK_a = 2.94   pK_a = 3.83   pK_a = 3.99   pK_a = 2.17   pK_a = 3.49   pK_a = 3.44

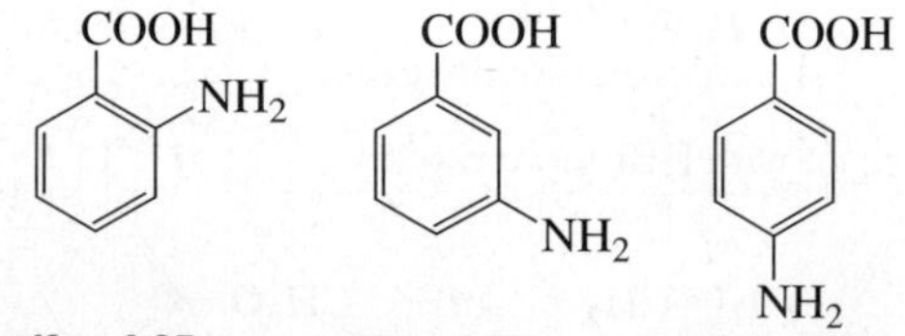

pK_a = 4.95   pK_a = 4.73   pK_a = 4.89

The relative pK_a values depend on the substituent. For chloro-substituted benzoic acids, the ortho isomer is the most acidic and the para isomer is the least acidic; for nitro-substituted benzoic acids, the ortho isomer is the most acidic and the meta isomer is the least acidic; and for amino-substituted benzoic acids, the meta isomer is the most acidic and the ortho isomer is the least acidic. Explain these relative acidities.

> Cl: ortho > meta > para   NO$_2$: ortho > para > meta   NH$_2$: meta > para > ortho

57. When heated with an acidic solution of sodium dichromate, compound A forms benzoic acid. Identify compound A from its ^{1}H NMR spectrum.

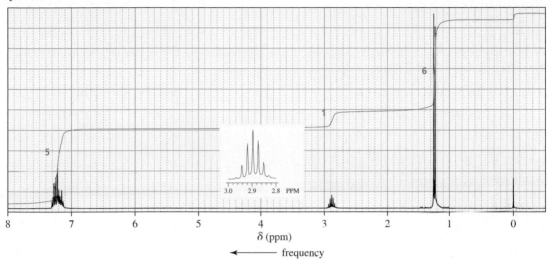

58. Describe two synthetic routes for the preparation of *p*-methoxyaniline with benzene as the starting material.

59. Which is a more stable intermediate in each pair?

a.

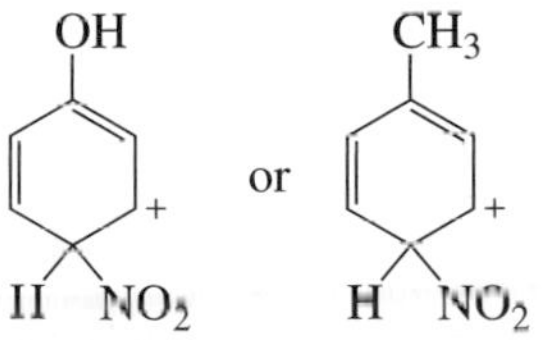

OH   CH$_3$

       or

H   NO$_2$ H   NO$_2$

b.

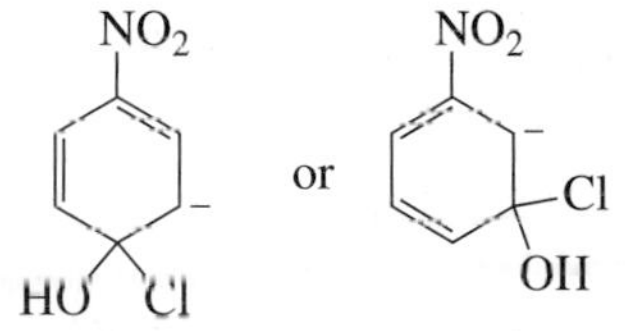

NO$_2$ NO$_2$

       or

HO   Cl   Cl   OH

60. If phenol is allowed to sit in D$_2$O that contains a small amount of D$_2$SO$_4$, what products will be formed?

61. Show how the following compounds could be prepared from benzene:

a.

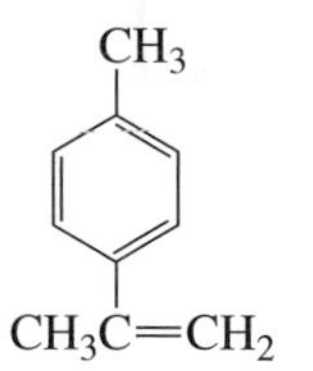

CH$_3$

CH$_3$C=CH$_2$

b.

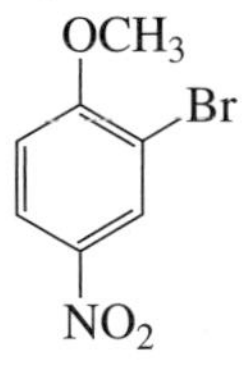

OCH$_3$
Br

NO$_2$

c.   CH$_3$C

O

SO$_3$H

CH$_2$CH(CH$_3$)$_2$

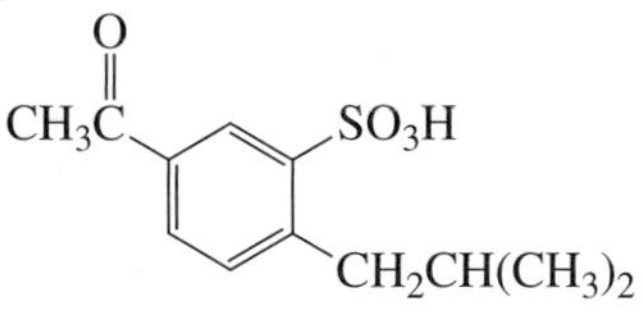

62. An unknown compound reacts with ethyl chloride and aluminum trichloride to form a compound that has the following ^{1}H NMR spectrum. Give the structure of the compound.

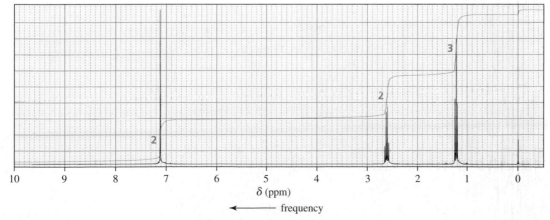

63. How could you distinguish among the following compounds, using
  a. their infrared spectra?
  b. their ^{1}H NMR spectra?

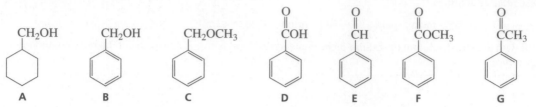

64. The following tertiary alkyl bromides undergo an S_N1 reaction in aqueous acetone to form the corresponding tertiary alcohols. List the alkyl bromides in order of decreasing reactivity.

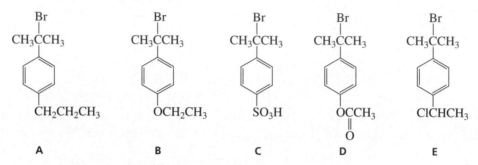

65. *p*-Fluoronitrobenzene is more reactive than *p*-chloronitrobenzene toward hydroxide ion. What does this tell you about the rate-determining step for nucleophilic aromatic substitution?

66. Starting with benzene, show how *meta*-aminoanisole could be prepared by a synthesis that
  a. involves a benzyne intermediate.
  b. does not involve a benzyne intermediate.

67. a. Explain why the following reaction leads to the products shown:

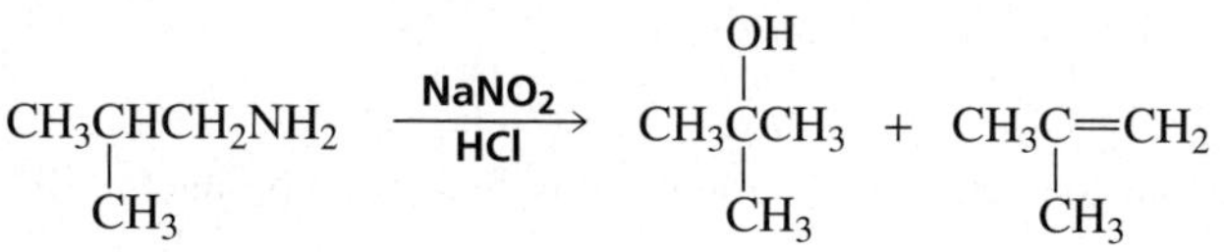

$$CH_3CHCH_2NH_2 \xrightarrow[\text{HCl}]{\text{NaNO}_2} CH_3CCH_3 + CH_3C{=}CH_2$$

with CH₃ substituent on the left reactant, OH and CH₃ on first product, CH₃ on second product.

  b. What product would be obtained from the following reaction?

$$CH_3-\underset{\underset{CH_3}{|}}{\overset{\overset{OH}{|}}{C}}-\underset{\underset{CH_3}{|}}{\overset{\overset{NH_2}{|}}{C}}-CH_3 \xrightarrow[\text{HCl}]{\text{NaNO}_2}$$

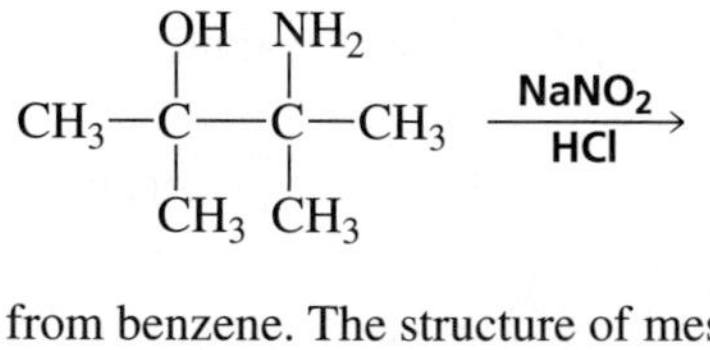

68. Describe how mescaline could be synthesized from benzene. The structure of mescaline is given on page 677.

69. Propose a mechanism for the following reaction that explains why the configuration of the asymmetric center in the reactant is retained in the product:

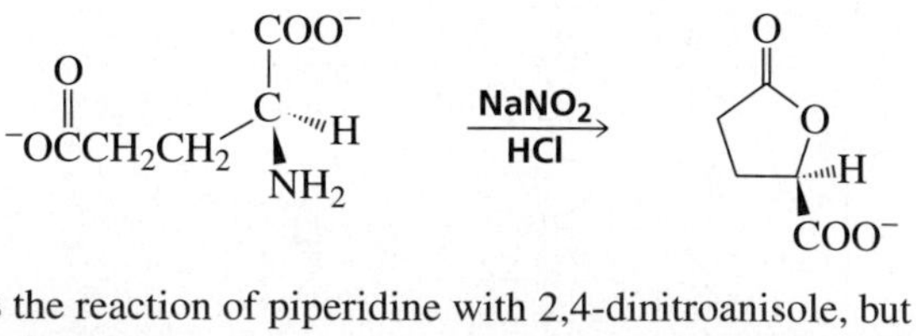

(reaction with $^-OCCH_2CH_2$ and COO^-, NH_2, H groups $\xrightarrow[\text{HCl}]{\text{NaNO}_2}$ lactone product with H, COO^-)

70. Explain why hydroxide ion catalyzes the reaction of piperidine with 2,4-dinitroanisole, but has no effect on the reaction of piperidine with 1-chloro-2,4-dinitrobenzene.

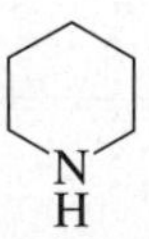

piperidine

Carbonyl Compounds

The three chapters in Part 6 focus on the reactions of compounds that contain a carbonyl group. Carbonyl compounds can be placed in one of two classes: those that contain a group that can be replaced by another group (class I) and those that contain a group that cannot be replaced by another group (class II).

CHAPTER 16
Carbonyl Compounds I: Nucleophilic Acyl Substitution

Class I carbonyl compounds undergo *nucleophilic acyl substitution* reactions. These reactions are discussed in **Chapter 16**, where you will see that all class I carbonyl compounds react with nucleophiles in the same way: they form an unstable tetrahedral intermediate that collapses by eliminating the weakest base. As a result, all you need to know to determine the product of one of these reactions—or even whether a reaction will occur—is the relative basicity of the groups in the tetrahedral intermediate.

CHAPTER 17
Carbonyl Compounds II: Reactions of Aldehydes and Ketones • More Reactions of Carboxylic Acid Derivatives • Reactions of α,β-Unsaturated Carbonyl Compounds

Chapter 17 compares class I and class II carbonyl compounds by discussing their reactions with good nucleophiles—carbon nucleophiles and hydride ion. It then looks at the reactions of class II carbonyl compounds with poor nucleophiles—nitrogen and oxygen nucleophiles. You will see that while class I carbonyl compounds undergo *nucleophilic acyl substitution* reactions with all nucleophiles, class II carbonyl compounds undergo *nucleophilic acyl addition* reactions with good nucleophiles and *nucleophilic addition elimination* reactions with poor nucleophiles (oxygen and nitrogen nucleophiles form unstable tetrahedral intermediates that preferentially eliminate the weakest base). Thus, what you learned about the partitioning of tetrahedral intermediates formed by class I carbonyl compounds is revisited with class II carbonyl compounds. The reactions of α,β-unsaturated class I and class II carbonyl compounds are also discussed.

CHAPTER 18
Carbonyl Compounds III: Reactions at the α-Carbon

Many carbonyl compounds have two sites of reactivity: the carbonyl group *and* the α-carbon. Chapter 16 and Chapter 17 discuss the reactions of carbonyl compounds that take place at the carbonyl group. **Chapter 18** examines the reactions of carbonyl compounds that take place at the α-carbon.

CHAPTER 16

Carbonyl Compounds I
Nucleophilic Acyl Substitution

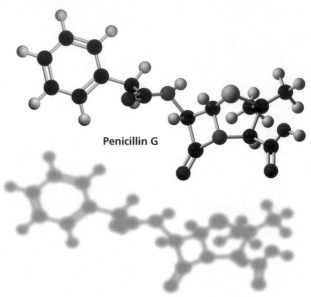

Penicillin G

BUILDING ON FUNDAMENTALS

SECTION 16.3	The greater the predicted stability of a resonance contributor, the more it contributes to the resonance hybrid (7.5).
SECTION 16.3	Hydrogen bonding increases the boiling point of a compound (2.9).
SECTION 16.4	The more electronegative the base, the more strongly it withdraws electrons inductively (1.21).
SECTION 16.5	The stronger the bond, the greater the wavenumber at which it absorbs IR radiation (12.10).
SECTION 16.5	The weaker the base, the better it is as a leaving group (8.3).

SECTION 16.11	When an acid is added to a reaction, the first thing that happens is the acid protonates the atom in the reactant that has the greatest electron density (10.4).
SECTION 16.11	A catalyst increases the rate of a reaction but is not consumed during the course of the reaction (4.5).
SECTION 16.11	An ester with a tertiary alkyl group undergoes an S_N1 reaction with water (8.5).
SECTION 16.21	A reagent such as $SOCl_2$ replaces the OH group of both carboxylic acids and alcohols with a Cl (10.2).

The **carbonyl group**—a carbon double bonded to an oxygen—is probably the most important functional group. Compounds containing carbonyl groups—called **carbonyl compounds**—are abundant in nature. Many play important roles in biological processes. Hormones, vitamins, amino acids, proteins, drugs, and flavorings are just a few of the carbonyl compounds that affect us daily.

An **acyl group** consists of a carbonyl group attached to an alkyl group (R) or to an aryl group (Ar).

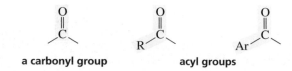

a carbonyl group   acyl groups

The group (or atom) attached to the acyl group strongly affects the reactivity of the carbonyl compound. In fact, carbonyl compounds can be divided into two classes determined by that group. Class I carbonyl compounds are those in which the acyl group is attached to a group (or atom) that *can* be replaced by another group. Carboxylic acids, acyl halides, acid anhydrides, esters, and amides belong to this class. All of these

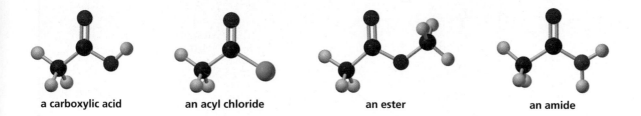

a carboxylic acid   an acyl chloride   an ester   an amide

compounds contain a group (OH, Cl, OR, NH_2, NHR, NR_2) that can be replaced by a nucleophile. Acyl halides, acid anhydrides, esters, and amides are all called **carboxylic acid derivatives** because they differ from a carboxylic acid only in the nature of the group or atom that has replaced the OH group of the carboxylic acid.

compounds with groups that can be replaced by a nucleophile

a carboxylic acid   an ester   an acid anhydride

an acyl chloride   an acyl bromide   amides
acyl halides

Class II carbonyl compounds are those in which the acyl group is attached to a group that *cannot* be readily replaced by another group. Aldehydes and ketones belong to this class. The H bonded to the acyl group of an aldehyde and the R group bonded to the acyl group of a ketone cannot be readily replaced by a nucleophile.

cannot be replaced by a nucleophile

an aldehyde   a ketone

We have seen that weak bases are good leaving groups and strong bases are poor leaving groups because weak bases do not share their electrons as well as strong bases do (Section 8.3). The pK_a values of the conjugate acids of the leaving groups of various carbonyl compounds are listed in Table 16.1. Notice that the acyl groups of Class I carbonyl compounds are attached to weaker bases than are the acyl groups of Class II carbonyl compounds. (Remember that the lower the pK_a, the stronger the acid and the weaker its conjugate base.) The hydrogen of an aldehyde and the alkyl group of a ketone are too basic to be replaced by another group.

Table 16.1 The pK_a Values of the Conjugate Acids of the Leaving Groups of Carbonyl Compounds

Carbonyl compound	Leaving group	Conjugate acid of the leaving group	pK$_a$
Class I			
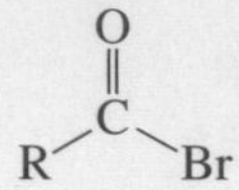 R–C(=O)–Br	Br$^-$	HBr	−9
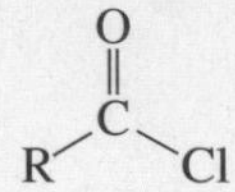 R–C(=O)–Cl	Cl$^-$	HCl	−7
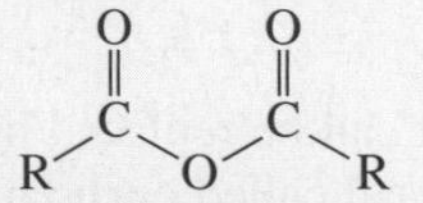 R–C(=O)–O–C(=O)–R	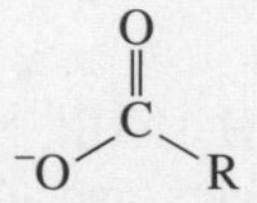 $^-$O–C(=O)–R	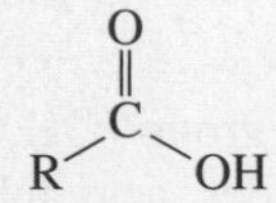 R–C(=O)–OH	~3–5
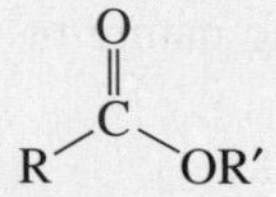 R–C(=O)–OR′	$^-$OR′	R′OH	~15–16
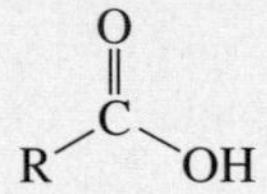 R–C(=O)–OH	$^-$OH	H$_2$O	15.7
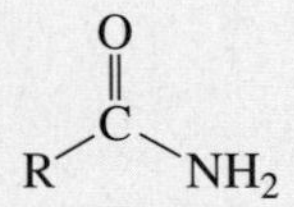 R–C(=O)–NH$_2$	$^-$NH$_2$	NH$_3$	36
Class II			
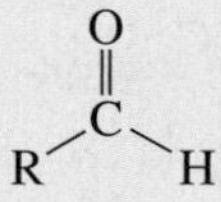 R–C(=O)–H	H$^-$	H$_2$	~40
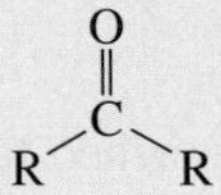 R–C(=O)–R	R$^-$	RH	>60

This chapter discusses the reactions of Class I carbonyl compounds. We will see that these compounds undergo substitution reactions, because they have an acyl group attached to a group that can be replaced by a nucleophile. The reactions of Class II carbonyl compounds—aldehydes and ketones—will be considered in Chapter 17, where we will see that these compounds *do not* undergo substitution reactions, because their acyl group is attached to a group that *cannot* be replaced by a nucleophile.

16.1 Nomenclature of Carboxylic Acids and Carboxylic Acid Derivatives

We will look at the names of carboxylic acids first because their names form the basis of the names of the other carbonyl compounds.

Naming Carboxylic Acids

In systematic (IUPAC) nomenclature, a **carboxylic acid** is named by replacing the terminal "e" of the alkane name with "oic acid." For example, the one-carbon alkane is methan*e*, so the one-carbon carboxylic acid is methan*oic acid*.

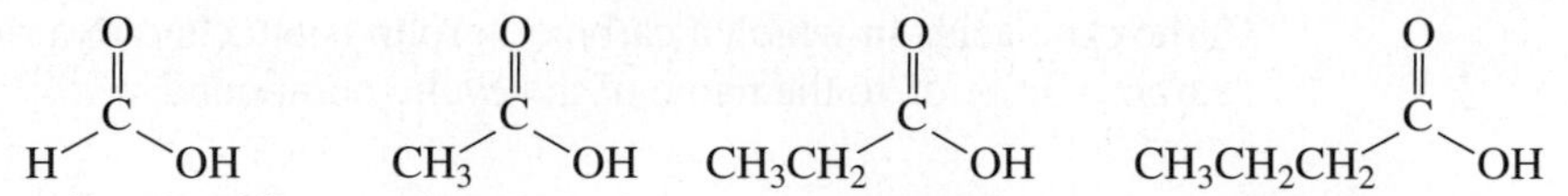

systematic name: methanoic acid	ethanoic acid	propanoic acid	butanoic acid
common name: formic acid	acetic acid	propionic acid	butyric acid

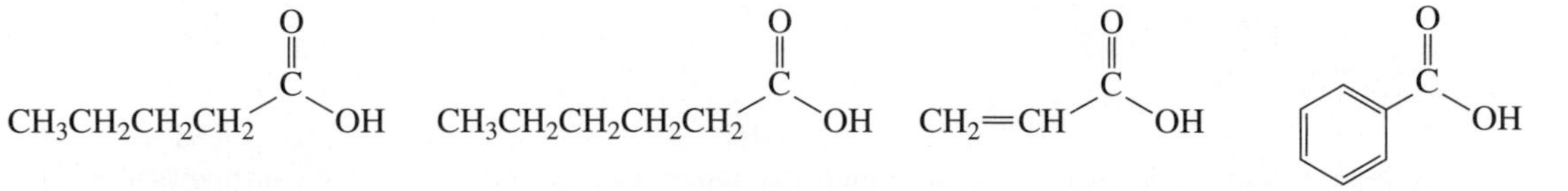

pentanoic acid	hexanoic acid	propenoic acid	benzenecarboxylic acid
valeric acid	caproic acid	acrylic acid	benzoic acid

Carboxylic acids containing six or fewer carbons are frequently called by their common names. These names were chosen by early chemists to describe some feature of the compound, usually its origin. For example, formic acid is found in ants, bees, and other stinging insects; its name comes from *formica*, which is Latin for "ant." Acetic acid—contained in vinegar—got its name from *acetum*, the Latin word for "vinegar." Propionic acid is the smallest acid that shows some of the characteristics of the larger fatty acids (Section 26.1); its name comes from the Greek words *pro* ("the first") and *pion* ("fat"). Butyric acid is found in rancid butter; the Latin word for "butter" is *butyrum*. Caproic acid is found in goat's milk; if you have the occasion to smell both a goat and caproic acid, you will find they have similar odors. *Caper* is the Latin word for "goat."

In systematic nomenclature, the position of a substituent is designated by a number. The carbonyl carbon is always the C-1 carbon. In common nomenclature, the position of a substituent is designated by a lowercase Greek letter, and the carbonyl carbon is not given a designation. The carbon adjacent to the carbonyl carbon is the **α-carbon**, the carbon adjacent to the α-carbon is the β-carbon, and so on.

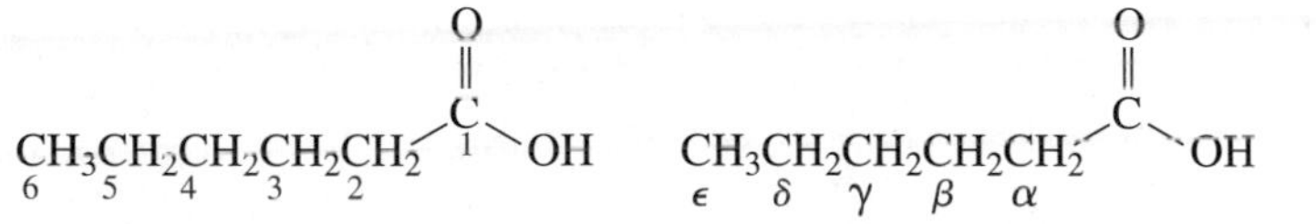

$CH_3CH_2CH_2CH_2CH_2$ 6 5 4 3 2 1 C OH	$CH_3CH_2CH_2CH_2CH_2$ ε δ γ β α C OH	α = alpha β = beta γ = gamma δ = delta ε = epsilon
systematic nomenclature	common nomenclature	

Take a careful look at the following examples to make sure that you understand the difference between systematic (IUPAC) and common nomenclature:

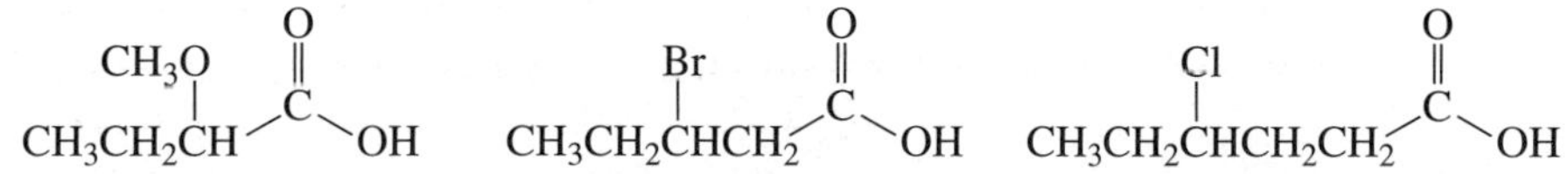

systematic name: 2-methoxybutanoic acid	3-bromopentanoic acid	4-chlorohexanoic acid
common name: α-methoxybutyric acid	β-bromovaleric acid	γ-chlorocaproic acid

The functional group of a carboxylic acid is called a **carboxyl group**.

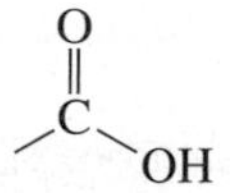

a carboxyl group   —COOH   —CO₂H

carboxyl groups are frequently shown in abbreviated forms

Carboxylic acids in which a carboxyl group is attached to a ring are named by adding "carboxylic acid" to the name of the cyclic compound.

cyclohexanecarboxylic acid	trans-3-methylcyclopentanecarboxylic acid	1,2,4-benzenetricarboxylic acid

Naming Acyl Halides

Acyl halides have a Cl or Br in place of the OH group of a carboxylic acid. The most common acyl halides are acyl chlorides. Acyl halides are named by using the acid name and replacing "ic acid" with "yl chloride" (or "yl bromide"). For acids ending with "carboxylic acid," "carboxylic acid" is replaced with "carbonyl chloride" (or "bromide").

systematic name: ethanoyl chloride	3-methylpentanoyl bromide	cyclopentanecarbonyl chloride
common name: acetyl chloride	β-methylvaleryl bromide	

Acid Anhydrides

Loss of water from two molecules of a carboxylic acid results in an **acid anhydride**. "Anhydride" means "without water."

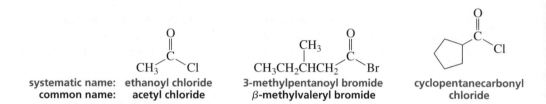

an acid anhydride

If the two carboxylic acid molecules forming the acid anhydride are the same, the anhydride is a **symmetrical anhydride**. If they are different, the anhydride is a **mixed anhydride**. Symmetrical anhydrides are named by taking the acid name and replacing "acid" with "anhydride." Mixed anhydrides are named by stating the names of both acids in alphabetical order, followed by "anhydride."

systematic name: ethanoic anhydride	ethanoic methanoic anhydride
common name: acetic anhydride	acetic formic anhydride
a symmetrical anhydride	a mixed anhydride

Naming Esters

An **ester** has an OR′ group in place of the OH group of a carboxylic acid. In naming an ester, the name of the group (R′) attached to the **carboxyl oxygen** is stated first, followed by the name of the acid, with "ic acid" replaced by "ate."

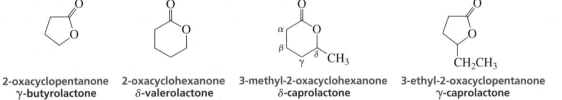

systematic name:	ethyl ethanoate	phenyl propanoate	methyl 3-bromobutanoate	ethyl cyclohexanecarboxylate
common name:	ethyl acetate	phenyl propionate	methyl β-bromobutyrate	

Salts of carboxylic acids are named in the same way. The cation is named first, followed by the name of the acid, again with "ic acid" replaced by "ate."

systematic name:	sodium methanoate	potassium ethanoate	sodium benzenecarboxylate
common name:	sodium formate	potassium acetate	sodium benzoate

Cyclic esters are called **lactones**. In systematic nomenclature, they are named as "2-oxacycloalkanones." Their common names are derived from the common name of the carboxylic acid, which designates the length of the carbon chain, and a Greek letter to indicate the carbon to which the carboxyl oxygen is attached. Thus, four-membered ring lactones are β-lactones (the carboxyl oxygen is on the β-carbon), five-membered ring lactones are γ-lactones, and six-membered ring lactones are δ-lactones.

2-oxacyclopentanone	2-oxacyclohexanone	3-methyl-2-oxacyclohexanone	3-ethyl-2-oxacyclopentanone
γ-butyrolactone	δ-valerolactone	δ-caprolactone	γ-caprolactone

PROBLEM 1

The word "lactone" has its origin in lactic acid, a three-carbon carboxylic acid with an OH group on the α-carbon. Ironically, lactic acid cannot form a lactone. Why not?

Naming Amides

An **amide** has an NH_2, NHR, or NR_2 group in place of the OH group of a carboxylic acid. Amides are named by taking the acid name and replacing "oic acid," "ic acid," or "ylic acid" with "amide."

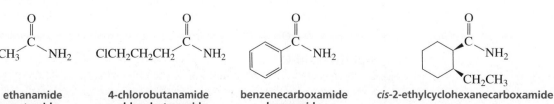

systematic name:	ethanamide	4-chlorobutanamide	benzenecarboxamide	cis-2-ethylcyclohexanecarboxamide
common name:	acetamide	γ-chlorobutyramide	benzamide	

If a substituent is bonded to the nitrogen, the name of the substituent is stated first (if there is more than one substituent bonded to the nitrogen, they are stated alphabetically), followed by the name of the amide. The name of each substituent is preceded by a capital *N* to indicate that the substituent is bonded to a nitrogen.

N-cyclohexylpropanamide   **N-ethyl-N-methylpentanamide**   **N,N-diethylbutanamide**

Cyclic amides are called **lactams**. Their nomenclature is similar to that of lactones. In systematic nomenclature, they are named as "2-azacycloalkanones" ("aza" is used to designate the nitrogen atom). For their common names, the length of the carbon chain is indicated by the common name of the carboxylic acid, and a Greek letter specifies the carbon to which the nitrogen is attached.

2-azacyclohexanone   **2-azacyclopentanone**   **2-azacyclobutanone**
δ-valerolactam   **γ-butyrolactam**   **β-propiolactam**

Naming Nitriles

Nitriles are compounds that contain a C≡N functional group, called a cyano group. They are considered carboxylic acid derivatives because, like all class I carbonyl compounds, they react with water to form carboxylic acids (Section 16.19). In systematic nomenclature, nitriles are named by adding "nitrile" to the parent alkane name. Notice that the triple-bonded carbon of the nitrile group is counted in the number of carbons in the longest continuous chain. In common nomenclature, nitriles are named by replacing "ic acid" of the carboxylic acid name with "onitrile." They can also be named as alkyl cyanides—using the name of the alkyl group that is attached to the C≡N group.

	CH₃C≡N	⟨benzene⟩–C≡N	CH₃CHCH₂CH₂CH₂C≡N (with CH₃)	CH₂=CHC≡N
systematic name:	ethanenitrile	benzenecarbonitrile	5-methylhexanenitrile	propenenitrile
common name:	acetonitrile	benzonitrile	δ-methylcapronitrile	acrylonitrile
	methyl cyanide	phenyl cyanide	isohexyl cyanide	

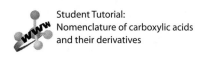
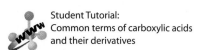

PROBLEM 2◆

Name the following:

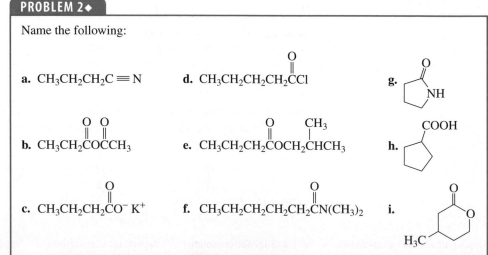

a. CH₃CH₂CH₂C≡N

b. CH₃CH₂C(O)C(O)CH₃

c. CH₃CH₂CH₂C(O)O⁻ K⁺

d. CH₃CH₂CH₂CH₂C(O)Cl

e. CH₃CH₂CH₂C(O)CH₂CHCH₃ (with CH₃)

f. CH₃CH₂CH₂CH₂CH₂C(O)N(CH₃)₂

g. (lactam structure with O and NH)

h. (cyclopentane with COOH)

i. (lactone structure with O, O, H₃C)

PROBLEM 3

Write the structure of each of the following:

a. phenyl acetate   d. *N*-benzylethanamide   g. β-bromobutyramide
b. γ-caprolactam   e. γ-methylcaproic acid   h. propanoic anhydride
c. butanenitrile   f. ethyl 2-chloropentanoate   i. cyclohexanecarbonyl chloride

16.2   Structures of Carboxylic Acids and Carboxylic Acid Derivatives

The **carbonyl carbon** in carboxylic acids and carboxylic acid derivatives is sp^2 hybridized. It uses its three sp^2 orbitals to form σ bonds to the carbonyl oxygen, the α-carbon, and a substituent (Y). The three atoms attached to the carbonyl carbon are in the same plane, and the bond angles are each approximately 120°.

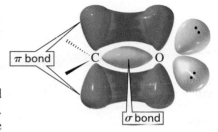

▲ **Figure 16.1**
Bonding in a carbonyl group. The π bond is formed by side-to-side overlap of a *p* orbital of carbon with a *p* orbital of oxygen.

The **carbonyl oxygen** is also sp^2 hybridized. One of its sp^2 orbitals forms a σ bond with the carbonyl carbon, and each of the other two sp^2 orbitals contains a lone pair. The remaining *p* orbital of the carbonyl oxygen overlaps the remaining *p* orbital of the carbonyl carbon to form a π bond (Figure 16.1).

Esters, carboxylic acids, and amides each have two major resonance contributors. The resonance contributor on the right is much less important for acyl chlorides and acid anhydrides, so those contributors are not shown here.

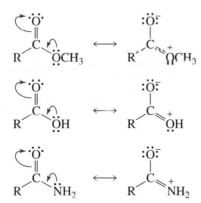

The resonance contributor on the right makes a greater contribution to the hybrid in the amide than in the ester, because the less electronegative nitrogen atom can better accommodate a positive charge.

PROBLEM 4

Which is longer, the carbon–oxygen single bond in a carboxylic acid or the carbon–oxygen bond in an alcohol? Why?

PROBLEM 5◆

There are three carbon–oxygen bonds in methyl acetate.

a. What are their relative lengths?
b. What are the relative infrared (IR) stretching frequencies of these bonds?

PROBLEM 6◆

Match the compound to the appropriate carbonyl IR absorption band:

acyl chloride	~ 1800 and 1750 cm^{-1}
acid anhydride	~ 1640 cm^{-1}
ester	~ 1730 cm^{-1}
amide	~1800 cm^{-1}

16.3 Physical Properties of Carbonyl Compounds

The acid properties of carboxylic acids were discussed in Sections 1.22 and 7.9. Recall that carboxylic acids have pK_a values of approximately 3–5 (Appendix II; see also "Special Topics I" in the *Study Guide and Solutions Manual).* The acid properties of dicarboxylic acids are discussed in Section 16.23. The boiling points and other physical properties of carbonyl compounds are listed in Appendix I. Carbonyl compounds have the following relative boiling points:

relative boiling points

amide > carboxylic acid > nitrile ≫ ester ~ acyl chloride ~ aldehyde ~ ketone

The boiling points of an ester, acyl chloride, aldehyde, and ketone are similar and are lower than the boiling point of the alcohol with a comparable molecular weight because only the alcohol molecules can form hydrogen bonds with each other. The boiling points of these four carbonyl compounds are higher than the boiling point of the same-size ether because of the polar carbonyl group.

$CH_3CH_2CH_2OH$
bp = 97.4 °C

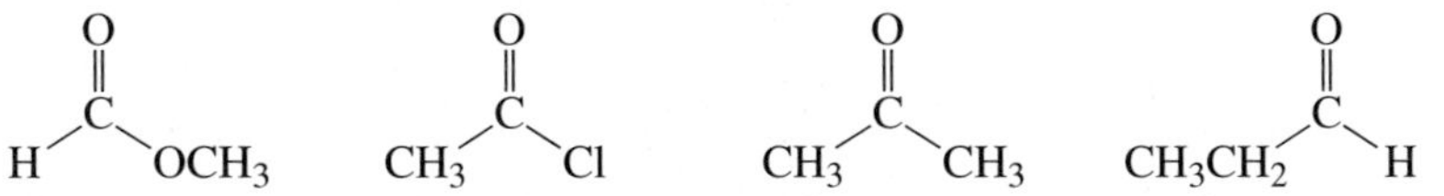

bp = 32 °C   bp = 51 °C   bp = 56 °C   bp = 49 °C

$CH_3CH_2OCH_3$
bp = 10.8 °C

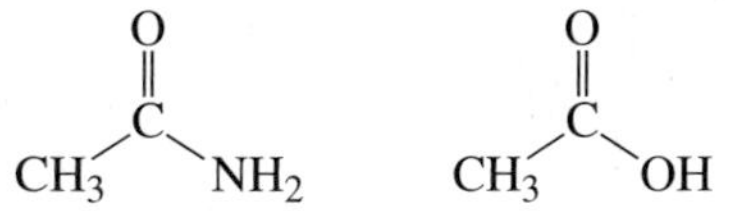

bp = 221 °C   bp = 118 °C

$CH_3CH_2C{\equiv}N$
bp = 97 °C

Carboxylic acids have relatively high boiling points because they form intermolecular hydrogen bonds, giving them larger effective molecular weights.

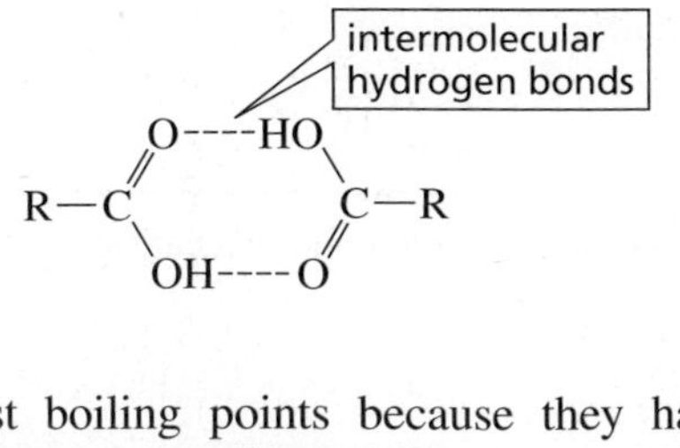

Amides have the highest boiling points because they have strong dipole–dipole interactions, since the resonance contributor with separated charges contributes significantly to the overall structure of the compound (Section 16.2). Additionally, if the nitrogen of an amide is bonded to a hydrogen, hydrogen bonds will form between the molecules. The strong dipole–dipole interactions of a nitrile give it a boiling point similar to that of an alcohol (Section 2.9).

Carboxylic acid derivatives are soluble in solvents such as ethers, chloroalkanes, and aromatic hydrocarbons. Like alcohols and ethers, carbonyl compounds with fewer than four carbons are soluble in water.

Esters, *N,N*-disubstituted amides, and nitriles are often used as solvents because they are polar but do not have reactive OH or NH$_2$ groups. We have seen that dimethylformamide (DMF) is a common aprotic polar solvent (Section 8.3).

16.4 Naturally Occurring Carboxylic Acids and Carboxylic Acid Derivatives

Acyl halides and acid anhydrides are much more reactive than carboxylic acids and esters, which, in turn, are more reactive than amides. We will see the reason for these differences in reactivity in Section 16.5.

Because of their high reactivity, acyl halides and acid anhydrides are not found in nature. Carboxylic acids, on the other hand, being less reactive *are* found widely in nature. For example, glucose is metabolized to pyruvic acid (Section 25.7). (*S*)-(+)-Lactic acid is the compound responsible for the burning sensation felt in muscles during anaerobic exercise, and it is also found in sour milk. Spinach and other leafy green vegetables are rich in oxalic acid. Succinic acid and citric acid are important intermediates in the citric acid cycle, a series of reactions in biological systems that oxidize the compounds formed from the metabolism of fatty acids, carbohydrates, and amino acids to CO$_2$ (Section 25.10). Citrus fruits are rich in citric acid; the concentration is greatest in lemons, less in grapefruit, and still less in oranges.

pyruvic acid   (*S*)-(+)-lactic acid   oxalic acid   succinic acid   citric acid

(*S*)-(−)-Malic acid is responsible for the sharp taste of unripe apples and pears. As these fruit ripen, the amount of malic acid decreases and the amount of sugar increases. The inverse relationship between the levels of malic acid and sugar is important for the propagation of the plant: malic acid prevents animals from eating the fruit until it becomes ripe, at which time its seeds are mature enough to germinate when they are scattered about. Prostaglandins are locally acting hormones that have several different physiological functions (Sections 16.10 and 26.5), such as stimulating inflammation, raising blood pressure, and producing pain and swelling.

(*S*)-(−)-malic acid   prostaglandin A$_2$ prostaglandin F$_{2\alpha}$

Esters are also commonly found in nature. The aromas of many flowers and fruits are due to esters. (See Problems 30 and 54.)

benzyl acetate
jasmine

isopentyl acetate
banana

methyl butyrate
apple

Carboxylic acids with an amino group on the α-carbon are commonly called **amino acids**. Amino acids are linked together by amide bonds to form peptides and proteins (Section 22.8).

an amino acid

general structure for a peptide or a protein

Caffeine, another naturally occurring amide, is found in cocoa and in coffee beans, and piperine is the major component of black pepper. Penicillin G, an antibacterial compound with two amide bonds (one of which is in a β-lactam ring), was first isolated from a mold in 1928 by Sir Alexander Fleming.

caffeine

piperine

penicillin G

THE DISCOVERY OF PENICILLIN

Sir Alexander Fleming was a professor of bacteriology at the University of London. The story is told that one day Fleming was about to throw away a culture of staphylococcal bacteria that had been contaminated by a rare strain of the mold *Penicillium notatum*. He noticed that the bacteria had disappeared wherever there was a particle of mold. This suggested to him that the mold must have produced an antibacterial substance. Ten years later, Howard Florey and Ernest Chain isolated the active substance—penicillin G (Section 16.17)—but the delay allowed the sulfa drugs to be the first antibiotics (Section 30.4). After penicillin G was found to cure bacterial infections in mice, it was used successfully in 1941 on nine cases of human bacterial infections. By 1943, it was being produced for the military and was first used for war casualties in Sicily and Tunisia. The drug became available to the civilian population in 1944. The pressure of the war made the determination of penicillin G's structure a priority because once its structure was determined, large quantities of the drug could conceivably be synthesized.

Fleming, Florey, and Chain shared the 1945 Nobel Prize in physiology or medicine. Chain also discovered penicillinase, the enzyme that destroys penicillin (Section 16.17). Although Fleming is generally given credit for the discovery of penicillin, there is clear evidence that the germicidal activity of the mold was recognized in the nineteenth century by Lord Joseph Lister (1827–1912), the English physician renowned for the introduction of aseptic surgery.

Sir Alexander Fleming (1881–1955) *was born in Scotland, the seventh of eight children of a farmer. In 1902, he received a legacy from an uncle that, together with a scholarship, allowed him to study medicine at the University of London. He subsequently became a professor there in 1928. He was knighted in 1944.*

Sir Howard W. Florey (1898–1968) *was born in Australia and received a medical degree from the University of Adelaide. He went to England as a Rhodes Scholar and studied at both Oxford and Cambridge Universities. He became a professor of pathology at the University of Sheffield in 1931 and then at Oxford in 1935. Knighted in 1944, he was given a peerage in 1965 that made him Baron Florey of Adelaide.*

Ernest B. Chain (1906–1979) *was born in Germany and received a Ph.D. from Friedrich-Wilhelm University in Berlin. In 1933, he left Germany for England because Hitler had come to power. He studied at Cambridge, and in 1935 Florey invited him to Oxford. In 1948, he became the director of an institute in Rome, but he returned to England in 1961 to become a professor at the University of London.*

DALMATIANS: THE HIGH COST OF BLACK SPOTS

When amino acids are metabolized, the excess nitrogen is concentrated into uric acid, a compound with five amide bonds. A series of enzyme-catalyzed reactions degrades uric acid all the way to ammonium ion. The extent to which uric acid is degraded depends on the species. Birds, reptiles, and insects excrete excess nitrogen as uric acid. Mammals excrete excess nitrogen as allantoin. Excess nitrogen in aquatic animals is excreted as allantoic acid, urea, or ammonium salts.

uric acid
excreted by:
birds, reptiles, insects

allantoin
mammals

allantoic acid
marine vertebrates

urea
cartilaginous fish, amphibia

$^{+}NH_4X^{-}$
**ammonium salt
marine invertebrates**

Dalmatians, unlike other mammals, excrete high levels of uric acid. This is because breeders of Dalmatians have selected dogs that have no white hairs in their black spots, and the gene that causes the white hairs is linked to the gene that causes uric acid to be converted to allantoin. Dalmatians are, therefore, susceptible to gout, a painful buildup of uric acid in joints.

16.5 How Class I Carbonyl Compounds React

The reactivity of carbonyl compounds is due to the polarity of the carbonyl group that results from oxygen being more electronegative than carbon. The carbonyl carbon is, therefore, electron deficient (an electrophile), so we can safely predict that it will be attacked by nucleophiles.

When a nucleophile attacks the carbonyl carbon of a carboxylic acid derivative, the weakest bond in the molecule—the carbon–oxygen π bond—breaks, and an intermediate is formed. The intermediate is called a **tetrahedral intermediate** because the sp^2 carbon in the reactant has become an sp^3 carbon in the intermediate. Generally, *a compound that has an sp^3 carbon bonded to an oxygen atom will be unstable if the sp^3 carbon is bonded to another electronegative atom.* The tetrahedral intermediate, therefore, is unstable because Y and Z are both electronegative atoms. A lone pair on the oxygen reforms the π bond, and either Y^- (k_2) or Z^- (k_{-1}) is expelled with its bonding electrons.

A compound that has an sp^3 carbon bonded to an oxygen atom generally will be unstable if the sp^3 carbon is bonded to another electronegative atom.

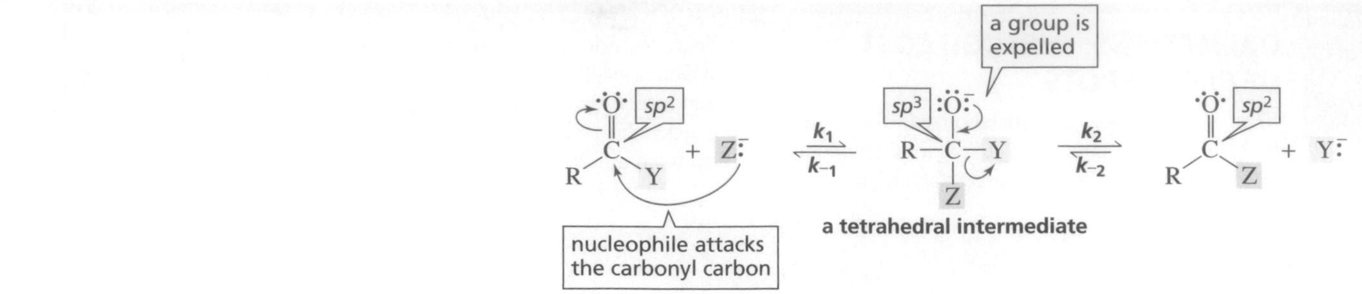

Let's compare this two-step reaction with an S_N2 reaction. When a nucleophile attacks an alkyl halide, the weakest bond in the molecule is the carbon–halogen bond (Section 8.2), so the incoming group can displace the leaving group in one step.

$$CH_3CH_2-Y \;+\; Z\!:^{-} \;\longrightarrow\; CH_3CH_2-Z \;+\; Y\!:^{-}$$

an S_N2 reaction

Whether Y^- or Z^- is expelled from the tetrahedral intermediate depends on their relative basicities. The weaker base is expelled preferentially, making this another example of the principle we first saw in Section 8.3: *the weaker the base, the better it is as a leaving group.* Because a weak base does not share its electrons as well as a strong base does, a weaker base forms a weaker bond—one that is easier to break. If Z^- is a much weaker base than Y^-, Z^- will be expelled. In such a case, $k_{-1} \gg k_2$, and the reaction can be written as follows:

The weaker the base, the better it is as a leaving group.

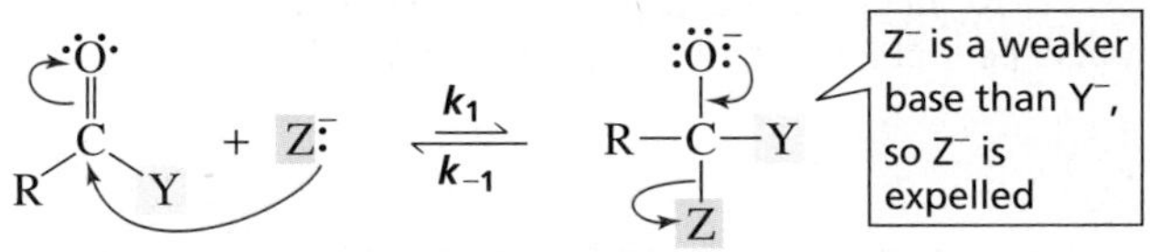

In this case, no new product is formed. The nucleophile attacks the carbonyl carbon, but the tetrahedral intermediate expels the attacking nucleophile and reforms the reactants.

On the other hand, if Y^- is a much weaker base than Z^-, Y^- will be expelled and a new product will be formed. In this case, $k_2 \gg k_{-1}$, and the reaction can be written as follows:

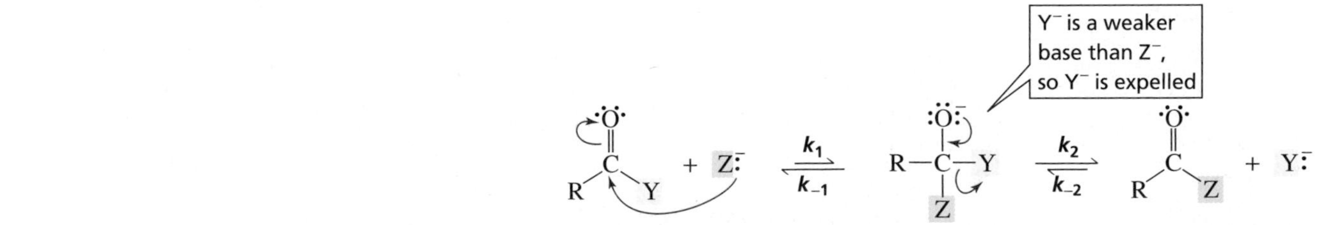

This reaction is called a **nucleophilic acyl substitution reaction** because a nucleophile (Z^-) has replaced the substituent (Y^-) that was attached to the acyl group in the reactant. It is also called an **acyl transfer reaction** because an acyl group has been transferred from one group (Y^-) to another (Z^-).

If the basicities of Y^- and Z^- are similar, the values of k_{-1} and k_2 will be similar. Therefore, some molecules of the tetrahedral intermediate will expel Y^- and others will expel Z^-. When the reaction is over, both reactant and product will be present. The relative amounts of each depend on the relative basicities of Y^- and Z^- (that is, the relative values of k_2 and k_{-1}) as well as the relative nucleophilicities of Y^- and Z^- (that is, the relative values of k_1 and k_{-2}).

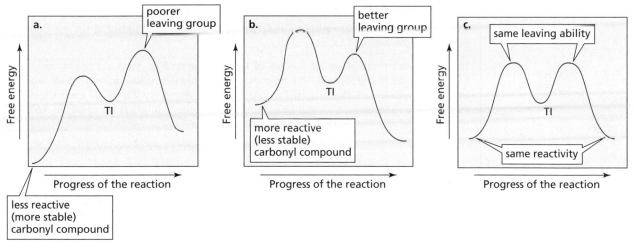

These three cases are illustrated by the reaction coordinate diagrams shown in Figure 16.2.

1. If the new group in the tetrahedral intermediate is a weaker base than the group that was attached to the acyl group in the reactant, the easier pathway—the lower energy hill—is for the tetrahedral intermediate (TI) to expel the newly added group and reform the reactants, so no reaction takes place (Figure 16.2a).

2. If the new group in the tetrahedral intermediate is a stronger base than the group that was attached to the acyl group in the reactant, the easier pathway is for the tetrahedral intermediate to expel the group that was attached to the acyl group in the reactant and form a substitution product (Figure 16.2b).

3. If both groups in the tetrahedral intermediate have similar basicities, the tetrahedral intermediate can expel either group with similar ease. A mixture of reactant and substitution product will result (Figure 16.2c).

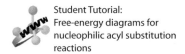

▲ Figure 16.2
Reaction coordinate diagrams for nucleophilic acyl substitution reactions in which (a) the nucleophile is a weaker base than the group attached to the acyl group in the reactant, (b) the nucleophile is a stronger base than the group attached to the acyl group in the reactant, and (c) the nucleophile and the group attached to the acyl group in the reactant have similar basicities. TI is the tetrahedral intermediate.

Student Tutorial:
Free-energy diagrams for
nucleophilic acyl substitution
reactions

We can therefore make the following general statement about the reactions of carboxylic acid derivatives: *a carboxylic acid derivative will undergo a nucleophilic acyl substitution reaction, provided that the newly added group in the tetrahedral intermediate is as strong a base or stronger than the group that was attached to the acyl group in the reactant.*

A carboxylic acid derivative will undergo a nucleophilic acyl substitution reaction, provided that the newly added group in the tetrahedral intermediate is as strong a base or stronger than the group that is attached to the acyl group in the reactant.

Let's now look at a molecular orbital description of how carbonyl compounds react. In Section 1.6, which first introduced you to molecular orbital theory, you saw that because oxygen is more electronegative than carbon, the $2p$ orbital of oxygen contributes more to the π bonding molecular orbital (it is closer to it in energy) and the $2p$ orbital of carbon contributes more to the π^* antibonding molecular orbital (see Figure 1.8). Therefore, the π^* antibonding orbital is largest at the carbon atom, so that is where the nucleophile's nonbonding orbital, in which the lone pair resides, overlaps. This allows the greatest amount of orbital overlap, and greater overlap means greater stability. When a filled orbital and an empty orbital overlap, the result is a molecular orbital—in this case, a σ molecular orbital—that is more stable than either of the overlapping orbitals (Figure 16.3).

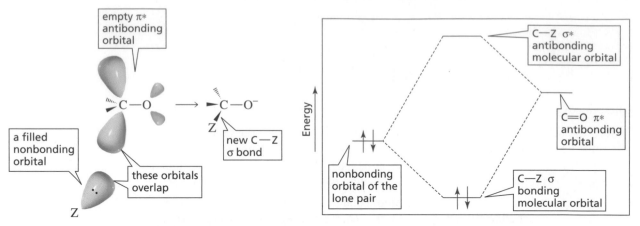

▲ **Figure 16.3**
The filled nonbonding orbital containing the nucleophile's lone pair overlaps the empty π* antibonding orbital of the carbonyl group, forming the new σ bond in the tetrahedral intermediate.

PROBLEM SOLVING STRATEGY

Using Basicity to Predict the Outcome of a Nucleophilic Acyl Substitution Reaction

The pK_a of HCl is -7; the pK_a of CH_3OH is 15.5. What is the product of the reaction of acetyl chloride with CH_3O^-?

In order to determine what the product of the reaction will be, we need to compare the basicities of the two groups that will be in the tetrahedral intermediate in order to see which one will be eliminated. Because HCl is a stronger acid than CH_3OH, Cl^- is a weaker base than CH_3O^-. Therefore, Cl^- will be eliminated from the tetrahedral intermediate, so the product of the reaction will be methyl acetate.

$$\underset{\text{acetyl chloride}}{CH_3\overset{\overset{\textstyle O}{\|}}{C}Cl} \;+\; CH_3O^- \;\longrightarrow\; CH_3-\underset{\underset{\textstyle OCH_3}{|}}{\overset{\overset{\textstyle O^-}{|}}{C}}-Cl \;\longrightarrow\; \underset{\text{methyl acetate}}{CH_3\overset{\overset{\textstyle O}{\|}}{C}OCH_3} \;+\; Cl^-$$

Now continue on to Problem 7.

PROBLEM 7◆

a. The pK_a of HCl is -7; the pK_a of H_2O is 15.7. What is the product of the reaction of acetyl chloride with HO^-?

b. The pK_a of NH_3 is 36; the pK_a of H_2O is 15.7. What is the product of the reaction of acetamide with HO^-?

PROBLEM 8◆

Is the following statement true or false?

If the newly added group in the tetrahedral intermediate is a stronger base than the group attached to the acyl group in the reactant, formation of the tetrahedral intermediate is the rate-limiting step of a nucleophilic acyl substitution reaction.

16.6 Relative Reactivities of Carboxylic Acids and Carboxylic Acid Derivatives

We have just seen that there are two steps in a nucleophilic acyl substitution reaction: formation of a tetrahedral intermediate and collapse of the tetrahedral intermediate. The weaker the base attached to the acyl group, the easier it is for *both steps* of the reaction to take place. In other words, the reactivity of a carboxylic acid derivative depends on the basicity of the substituent attached to the acyl group: the less basic the substituent, the more reactive the carboxylic acid derivative.

relative basicities of the leaving groups

relative reactivities of carboxylic acid derivatives

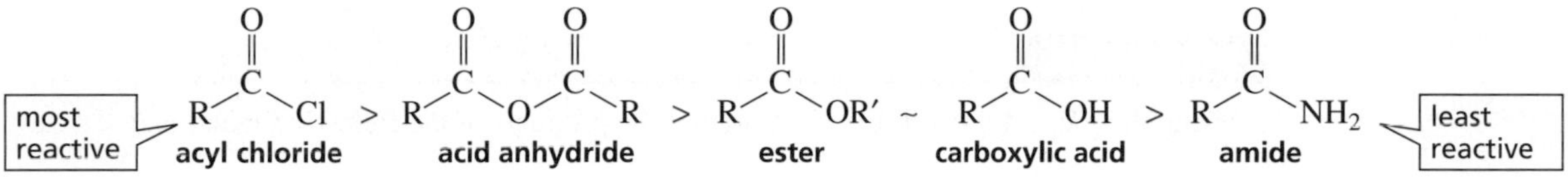

How does having a weak base attached to the acyl group make the *first* step of the nucleophilic acyl substitution reaction easier? First of all, a weaker base is a more electronegative base; that is, it is better able to accommodate a negative charge (Section 1.18). Thus, weaker bases are better at withdrawing electrons inductively from the carbonyl carbon (Section 1.21), and electron withdrawal increases the carbonyl carbon's susceptibility to nucleophilic attack.

relative reactivity: acyl chloride > anhydride > ester ~ carboxylic acid > amide

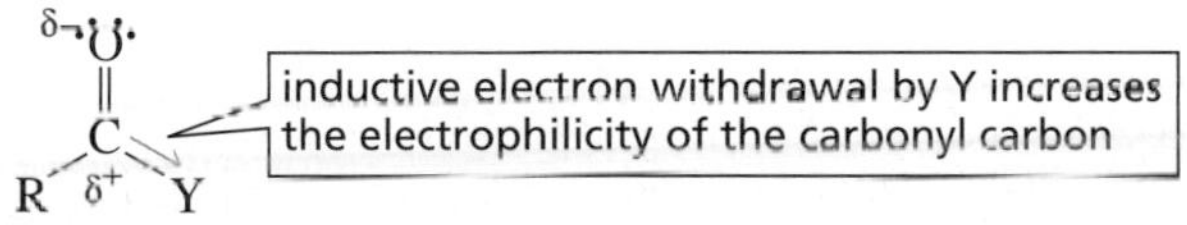

inductive electron withdrawal by Y increases the electrophilicity of the carbonyl carbon

Second, while weak bases are better able to accommodate a negative charge, they are less able to accommodate a positive charge. Therefore, the weaker the basicity of Y, the smaller is the contribution from the resonance contributor with a positive charge on Y (Section 16.2), and the less the carboxylic acid derivative is stabilized by electron delocalization, the more reactive it will be.

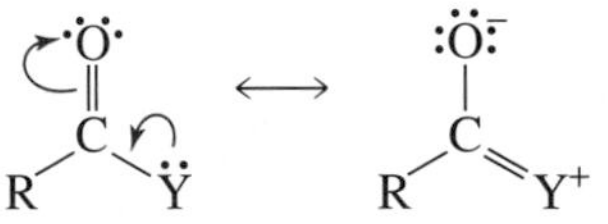

resonance contributors of a carboxylic acid or carboxylic acid derivative

A weak base attached to the acyl group also makes the *second* step of the nucleophilic acyl substitution reaction easier, because weak bases—which form weak bonds—are easier to expel when the tetrahedral intermediate collapses.

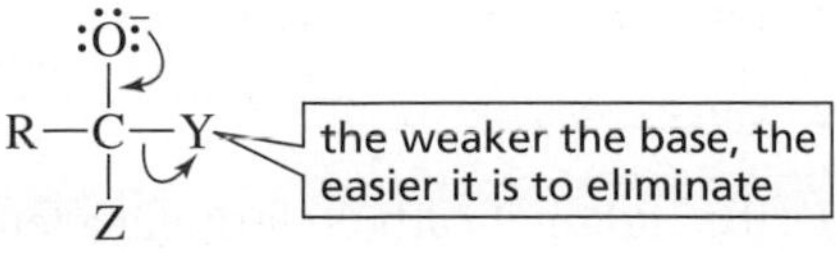

the weaker the base, the easier it is to eliminate

In Section 16.4 we saw that in a nucleophilic acyl substitution reaction, the nucleophile that forms the tetrahedral intermediate must be a stronger base than the base that is already there. This means that *a carboxylic acid derivative can be converted into a less reactive carboxylic acid derivative, but not into one that is more reactive.* For example, an acyl chloride can be converted into an ester because an alkoxide ion, such as methoxide ion, is a stronger base than a chloride ion.

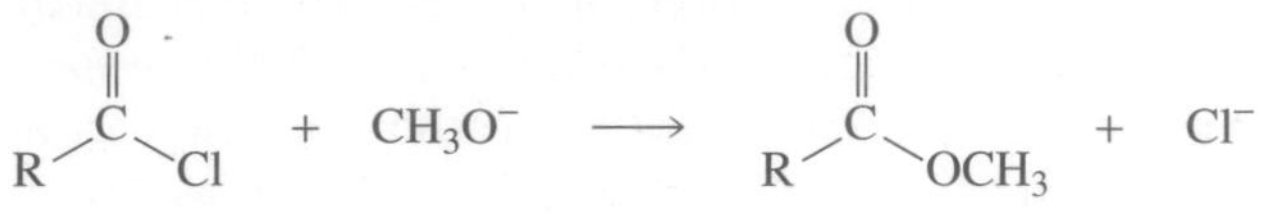

An ester, however, cannot be converted into an acyl chloride because a chloride ion is a weaker base than an alkoxide ion.

PROBLEM 9◆

Using the pK_a values in Table 16.1, predict the products of the following reactions:

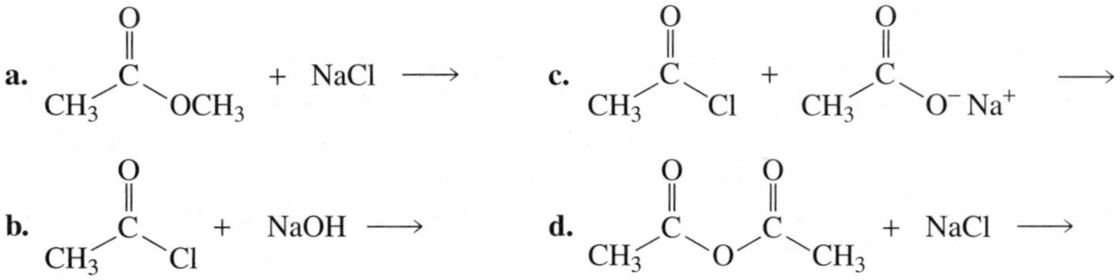

16.7 General Mechanism for Nucleophilic Acyl Substitution Reactions

All carboxylic acid derivatives react by the same mechanism.

All carboxylic acid derivatives undergo nucleophilic acyl substitution reactions by the same mechanism. If the nucleophile is negatively charged, the mechanism described in Section 16.5 is followed:

Student Tutorial:
A nucleophilic acyl substitution reaction

Mechanism for a nucleophilic acyl substitution reaction with a negatively charged nucleophile

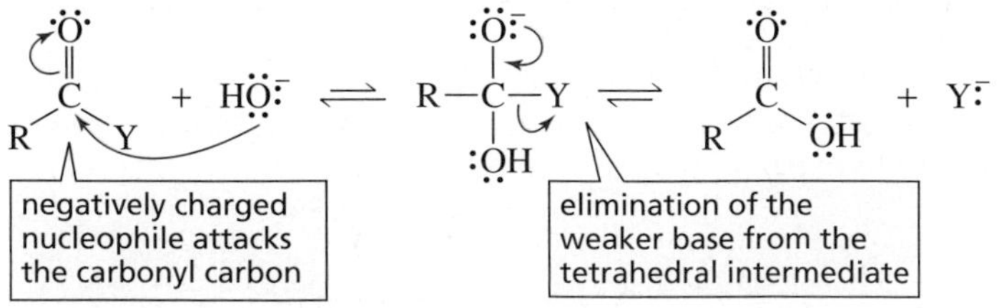

- The nucleophile attacks the carbonyl carbon, forming a tetrahedral intermediate.
- The tetrahedral intermediate collapses, eliminating the weaker base.

If the nucleophile is neutral, the mechanism has an additional step.

Mechanism for a nucleophilic acyl substitution reaction with a neutral nucleophile

$$\underset{\substack{R \quad Y \\ \text{neutral nucleophile} \\ \text{attacks the carbonyl} \\ \text{carbon}}}{\overset{\overset{\ddot{O}}{\|}}{C}} + H_2O: \ \rightleftharpoons \ \underset{\substack{+ \\ :OH \\ H \quad :B \\ \text{removal of} \\ \text{a proton from} \\ \text{the tetrahedral} \\ \text{intermediate}}}{R-\overset{\ddot{O}\bar{\cdot}}{\underset{|}{C}}-Y} \ \rightleftharpoons \ \underset{\substack{:OH \\ HB^+}}{R-\overset{:\ddot{O}\bar{\cdot}}{\underset{|}{C}}-Y} \ \longrightarrow \ \underset{\substack{R \quad OH \\ \text{elimination of the} \\ \text{weaker base from the} \\ \text{tetrahedral intermediate}}}{\overset{\overset{\ddot{O}}{\|}}{C}} + Y\bar{\cdot}$$

where :B represents any species in the solution that is capable of removing a proton; and HB$^+$ represents any species in the solution that is capable of donating a proton.

- The nucleophile attacks the carbonyl carbon, forming a tetrahedral intermediate.
- A proton is lost from the tetrahedral intermediate, resulting in a tetrahedral intermediate equivalent to the one formed by negatively charged nucleophile.
- This tetrahedral intermediate expels the weaker of the two bases—either the newly added group after it has lost a proton or the group that was attached to the acyl group in the reactant.

The tetrahedral intermediate eliminates the weakest base.

The remaining sections of this chapter show specific examples of these general principles. Keep in mind that *all the reactions follow the same mechanism.* Therefore, you can always determine the outcome of the reactions of carboxylic acids and carboxylic acid derivatives presented in this chapter by examining the tetrahedral intermediate and remembering that the weaker base is preferentially eliminated (Section 16.5).

PROBLEM 10◆

What will be the product of a nucleophilic acyl substitution reaction—a new carboxylic acid derivative, a mixture of two carboxylic acid derivatives, or no reaction—if the new group in the tetrahedral intermediate is the following?

a. a stronger base than the group that was already there
b. a weaker base than the group that was already there
c. similar in basicity to the group that was already there

16.8 Reactions of Acyl Halides

Acyl halides react with carboxylate ions to form anhydrides, with alcohols to form esters, with water to form carboxylic acids, and with amines to form amides because in each case the incoming nucleophile is a stronger base than the departing halide ion (Table 16.1). Notice that both alcohols and phenols can be used to prepare esters.

acetyl chloride

$$\underset{\substack{CH_3 \quad Cl \\ \textbf{acetyl chloride}}}{\overset{\overset{O}{\|}}{C}} + \underset{CH_3 \quad O^-}{\overset{\overset{O}{\|}}{C}} \ \longrightarrow \ \underset{\substack{CH_3 \quad O \quad CH_3 \\ \textbf{acetic anhydride}}}{\overset{\overset{O \quad\quad O}{\| \quad\quad \|}}{C \quad C}} + \ Cl^-$$

$$\underset{\textbf{benzoyl chloride}}{\overset{\overset{O}{\|}}{\underset{Cl}{C}}} + \ CH_3OH \ \longrightarrow \ \underset{\textbf{methyl benzoate}}{\overset{\overset{O}{\|}}{\underset{OCH_3}{C}}} + \ H^+ + \ Cl^-$$

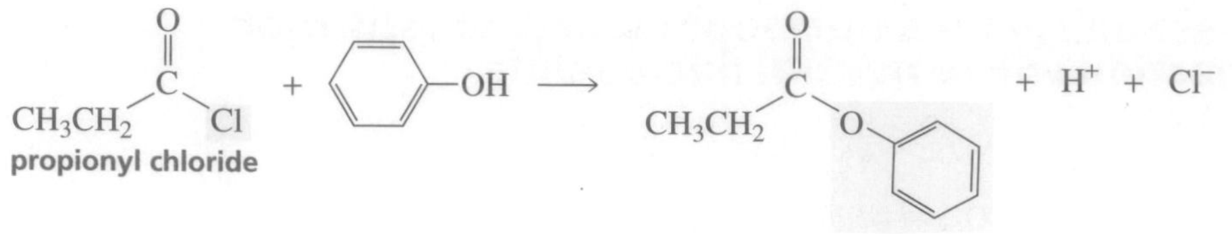

propionyl chloride + phenol → phenyl propionate + H⁺ + Cl⁻

butyryl chloride + H_2O → butyric acid + H⁺ + Cl⁻

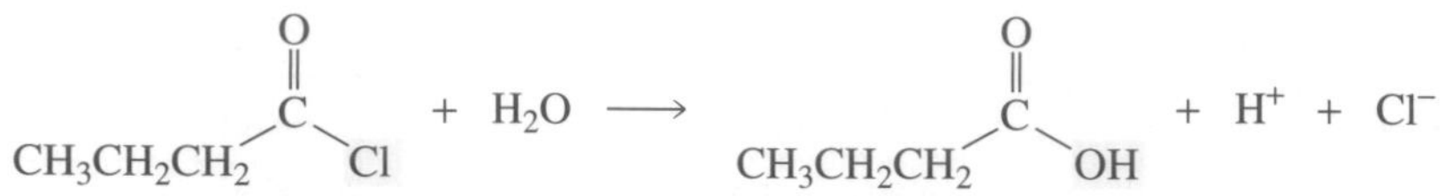

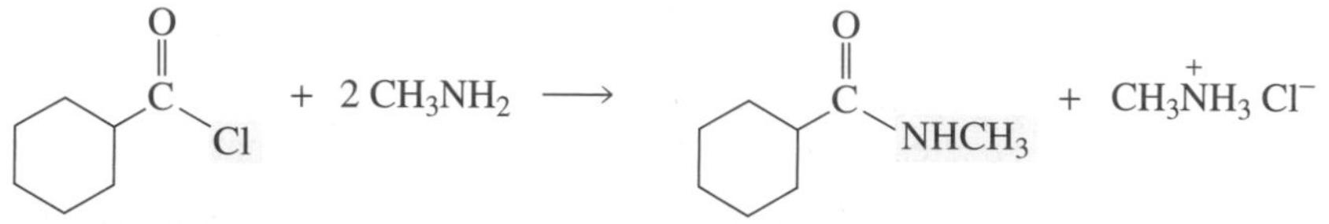

cyclohexanecarbonyl chloride + 2 CH_3NH_2 → N-methylcyclohexanecarboxamide + $CH_3\overset{+}{N}H_3\ Cl^-$

All the reactions follow the general mechanism described in Section 16.7.

Mechanism for the conversion of an acyl chloride into an acid anhydride

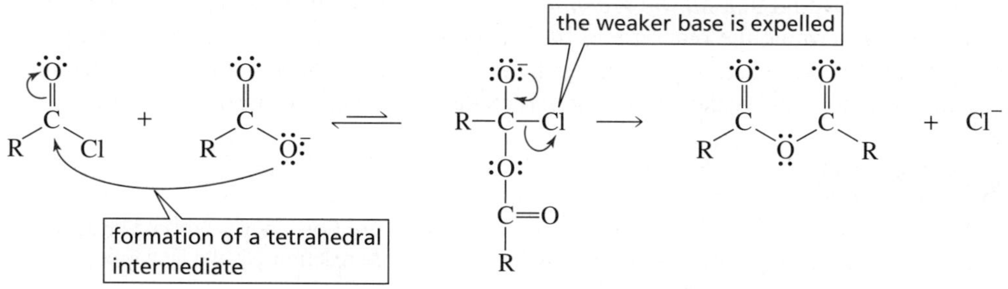

the weaker base is expelled

formation of a tetrahedral intermediate

- The nucleophilic carboxylate ion attacks the carbonyl carbon of the acyl chloride, forming a tetrahedral intermediate.
- The tetrahedral intermediate expels the chloride ion because it is a weaker base than the carboxylate ion. The final product is an anhydride.

The following mechanism has one more step than the one above because, since the nucleophile is neutral rather than negatively charged, a proton must be removed from the tetrahedral intermediate before the intermediate eliminates the leaving group.

Mechanism for the conversion of an acyl chloride into an ester

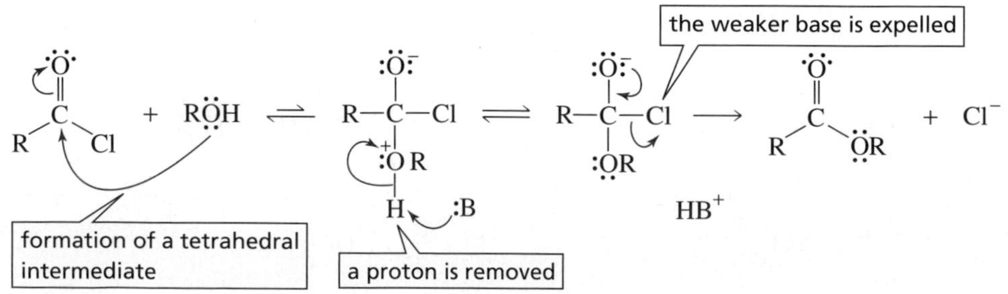

the weaker base is expelled

formation of a tetrahedral intermediate

a proton is removed

- The nucleophilic alcohol attacks the carbonyl carbon of the acyl chloride, forming a tetrahedral intermediate.
- Because the protonated ether group is a strong acid (Section 1.18), the tetrahedral intermediate loses a proton.
- Chloride ion is expelled from the deprotonated tetrahedral intermediate because chloride ion is a weaker base than the alkoxide ion.

The reaction of an acyl chloride with ammonia or with a primary or secondary amine forms an amide and HCl. The acid generated in the reaction will protonate unreacted ammonia or unreacted amine, and since a protonated amine is not a nucleophile, it cannot then react with the acyl chloride. The reaction, therefore, must be carried out with twice as much ammonia or amine as acyl chloride so that there will be enough unprotonated amine to react with all the acyl halide.

Amide Formation

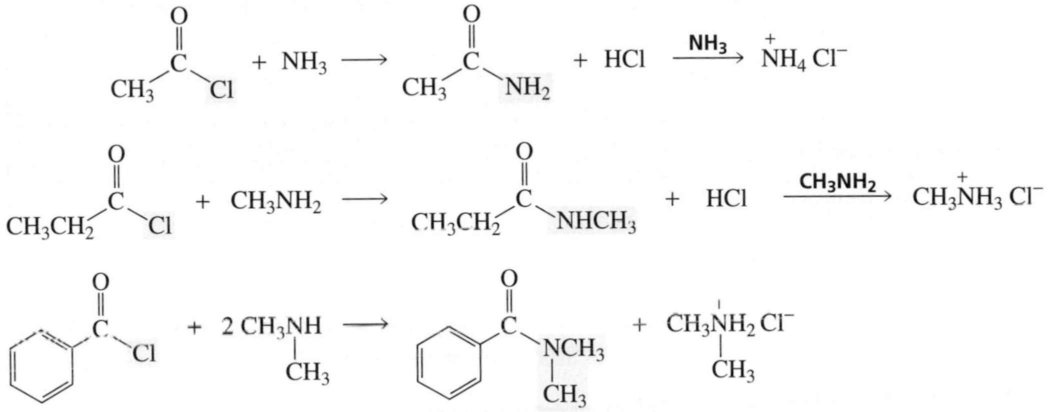

Because tertiary amines cannot form amides, an equivalent of a tertiary amine such as triethylamine can be used instead of excess amine.

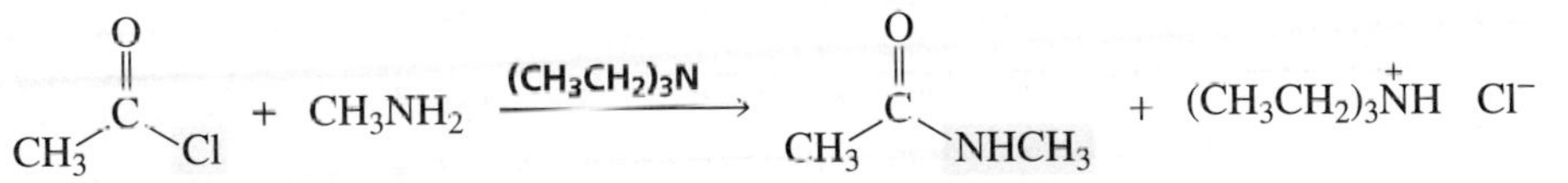

PROBLEM 11 *SOLVED*

a. Two amides are obtained from the reaction of acetyl chloride with a mixture of ethylamine and propylamine. Identify the amides.

b. Why is only one amide obtained from the reaction of acetyl chloride with a mixture of ethylamine and triethylamine?

Solution to 11a Either of the amines can react with acetyl chloride, so both *N*-ethylacetamide and *N*-propylacetamide are formed.

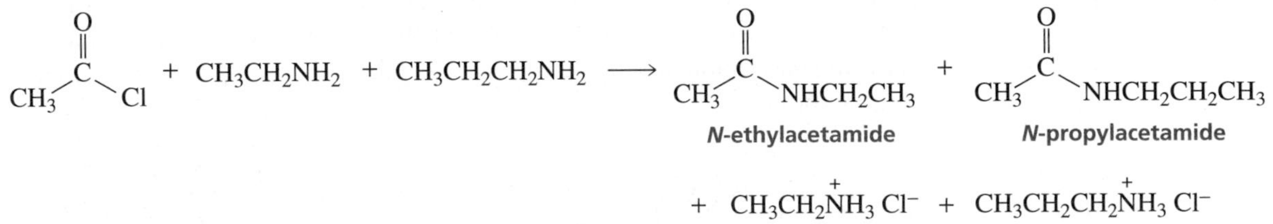

Solution to 11b Initially two amides are formed. However, the amide formed by triethylamine is very reactive because it has a positively charged nitrogen, which makes it an excellent leaving group. Therefore, it will react immediately with unreacted ethylamine, causing *N*-ethylacetamide to be the only amide formed in the reaction.

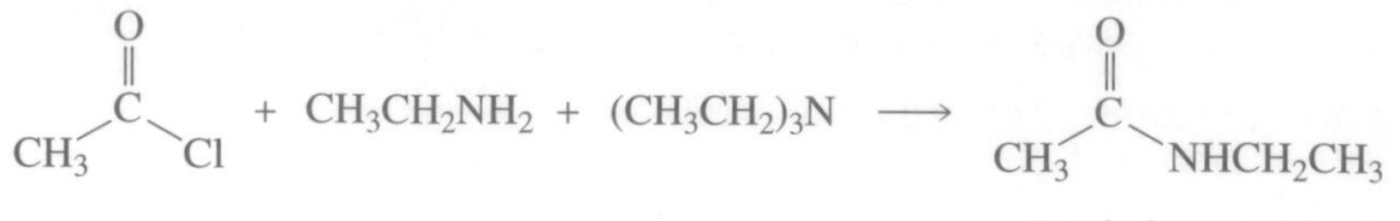

$$\underset{\substack{CH_3 \quad Cl}}{\overset{\displaystyle O}{\parallel}} + CH_3CH_2NH_2 + (CH_3CH_2)_3N \longrightarrow \underset{\substack{CH_3 \quad NHCH_2CH_3}}{\overset{\displaystyle O}{\parallel}}$$

N-ethylacetamide

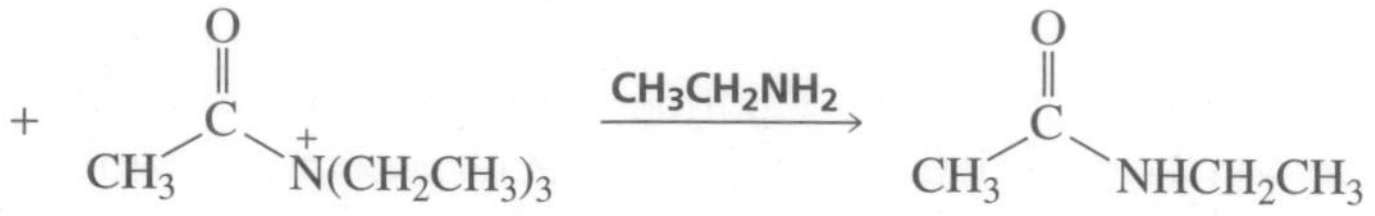

$$+ \underset{\substack{CH_3 \quad \overset{+}{N}(CH_2CH_3)_3}}{\overset{\displaystyle O}{\parallel}} \xrightarrow{\text{CH}_3\text{CH}_2\text{NH}_2} \underset{\substack{CH_3 \quad NHCH_2CH_3}}{\overset{\displaystyle O}{\parallel}}$$

N-ethylacetamide

PROBLEM 12

We have seen that it is necessary to use excess amine in the reaction of an acyl chloride with an amine. Explain why it is not necessary to use excess alcohol in the reaction of an acyl chloride with an alcohol.

PROBLEM 13

Write the mechanism for each of the following reactions:

a. the reaction of acetyl chloride with water to form acetic acid.
b. the reaction of acetyl bromide with excess methylamine to form *N*-methylacetamide.

PROBLEM 14◆

If you were starting with acetyl chloride, what nucleophile would you use to make each of the following compounds?

a. $CH_3\overset{\displaystyle O}{\overset{\parallel}{C}}OCH_2CH_2CH_3$

c. $CH_3\overset{\displaystyle O}{\overset{\parallel}{C}}N(CH_3)_2$

e. $CH_3\overset{\displaystyle O}{\overset{\parallel}{C}}O\overset{\displaystyle O}{\overset{\parallel}{C}}CH_3$

b. $CH_3\overset{\displaystyle O}{\overset{\parallel}{C}}NHCH_2CH_3$

d. $CH_3\overset{\displaystyle O}{\overset{\parallel}{C}}OH$

f. $CH_3\overset{\displaystyle O}{\overset{\parallel}{C}}O-\!\!\!\bigcirc\!\!\!-NO_2$

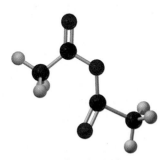

acetic anhydride

16.9 Reactions of Acid Anhydrides

Acid anhydrides do not react with sodium chloride (or with sodium bromide) because the incoming halide ion is a weaker base than the departing carboxylate ion (Table 16.1).

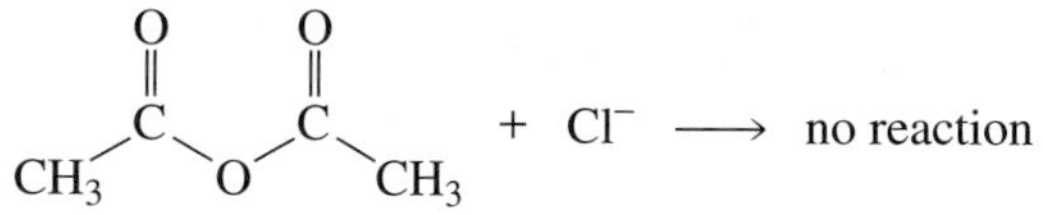

$$\underset{\substack{CH_3 \quad O \quad CH_3}}{\overset{\displaystyle O \qquad O}{\parallel \qquad \parallel}} + Cl^- \longrightarrow \text{no reaction}$$

Because the incoming halide ion is the weaker base, it will be the substituent expelled from the tetrahedral intermediate. (Recall that a carboxylic acid derivative cannot be converted into a more reactive carboxylic acid derivative, only to one that is less reactive; Section 16.6.)

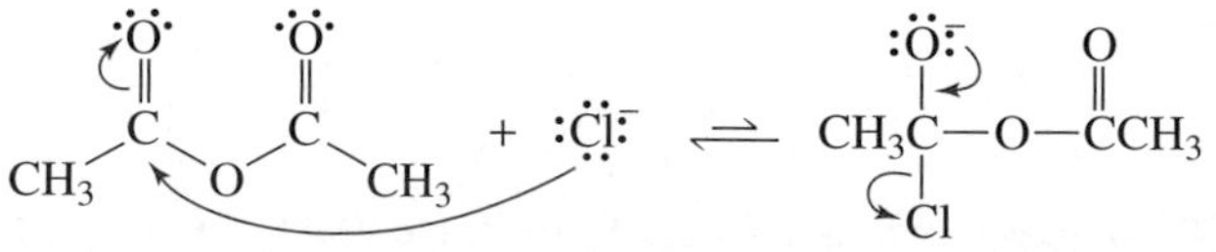

$$\underset{\substack{CH_3 \quad O \quad CH_3}}{\overset{\displaystyle \ddot{O} \qquad \ddot{O}}{\parallel \qquad \parallel}} + :\ddot{C}l:^- \rightleftharpoons CH_3\overset{:\ddot{O}:^-}{\underset{\overset{\displaystyle |}{Cl}}{\overset{\displaystyle |}{C}}}-O-\overset{\displaystyle O}{\overset{\parallel}{C}}CH_3$$

An acid anhydride reacts with an alcohol to form an ester and a carboxylic acid, with water to form two equivalents of a carboxylic acid, and with an amine to form an amide and a carboxylate ion. In each case, the incoming nucleophile—after it loses a proton—is a stronger base than the departing carboxylate ion. In the reaction of an amine with an anhydride, two equivalents of the amine or one equivalent of the amine plus one equivalent of a tertiary amine must be used so that sufficient amine will be present to react with both the carbonyl compound and the proton produced in the reaction.

$$\underset{\text{acetic anhydride}}{\text{CH}_3\text{—C(=O)—O—C(=O)—CH}_3} + \text{CH}_3\text{CH}_2\text{OH} \longrightarrow \underset{\text{ethyl acetate}}{\text{CH}_3\text{—C(=O)—OCH}_2\text{CH}_3} + \underset{\text{acetic acid}}{\text{CH}_3\text{—C(=O)—OH}}$$

$$\underset{\text{benzoic anhydride}}{(\text{C}_6\text{H}_5\text{—C(=O))}_2\text{O}} + \text{H}_2\text{O} \longrightarrow 2\ \underset{\text{benzoic acid}}{\text{C}_6\text{H}_5\text{—C(=O)—OH}}$$

$$\underset{\text{propionic anhydride}}{\text{CH}_3\text{CH}_2\text{—C(=O)—O—C(=O)—CH}_2\text{CH}_3} + 2\ \text{CH}_3\text{NH}_2 \longrightarrow \underset{\textit{N}\text{-methylpropionamide}}{\text{CH}_3\text{CH}_2\text{—C(=O)—NHCH}_3} + \underset{}{\text{CH}_3\text{CH}_2\text{—C(=O)—O}^-\ \text{H}_3\overset{+}{\text{N}}\text{CH}_3}$$

All these reactions follow the general mechanism described in Section 16.7. For example, compare the following mechanism for conversion of an acid anhydride into an ester with the mechanism for conversion of an acyl chloride into an ester presented on page 740.

Mechanism for the conversion of an acid anhydride into an ester (and a carboxylic acid)

$$\underset{}{\text{R—C(=O)—O—C(=O)—R}} + \text{R}\ddot{\text{O}}\text{H} \rightleftharpoons \text{R—C(O}^-)\text{(}^+\text{OR—H}:\text{B)}\text{—O—C(=O)—R} \rightleftharpoons \text{R—C(O}^-)\text{(OR)—O—C(=O)—R} + \text{HB}^+ \longrightarrow \underset{}{\text{R—C(=O)—OR}} + \underset{}{:\ddot{\text{O}}\text{—C(=O)—R}}$$

PROBLEM 15

a. Propose a mechanism for the reaction of acetic anhydride with water.

b. How does this mechanism differ from the mechanism for the reaction of acetic anhydride with an alcohol?

PROBLEM 16

We have seen that acid anhydrides react with alcohols, water, and amines. In which one of these three reactions does the tetrahedral intermediate not have to lose a proton before it eliminates the carboxylate ion? Explain.

16.10 Reactions of Esters

Esters do not react with halide ions or with carboxylate ions because these nucleophiles are much weaker bases than the RO$^-$ leaving group of the ester and therefore would be expelled from the tetrahedral intermediate (Table 16.1).

methyl acetate

An ester reacts with water to form a carboxylic acid and an alcohol. This is an example of a hydrolysis reaction. A **hydrolysis reaction** is a reaction with water that converts one compound into two compounds (*lysis* is Greek for "breaking down").

a hydrolysis reaction

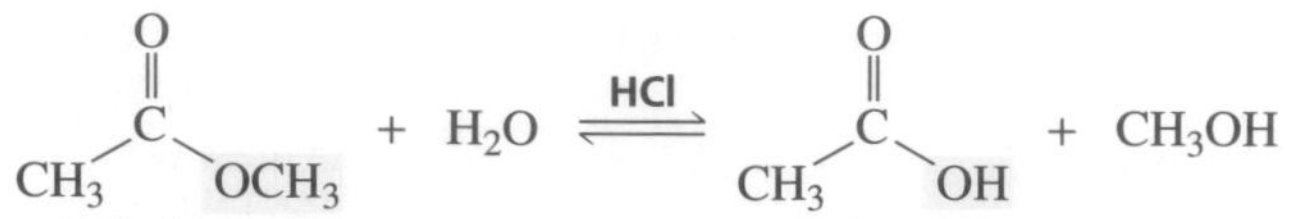

methyl acetate acetic acid

An ester reacts with an alcohol to form a new ester and a new alcohol. This is an example of an **alcoholysis reaction**—a reaction with an alcohol that converts one compound into two compounds. This particular alcoholysis reaction is also called a **transesterification reaction** because one ester is converted to another ester.

a transesterification reaction

methyl benzoate ethyl benzoate

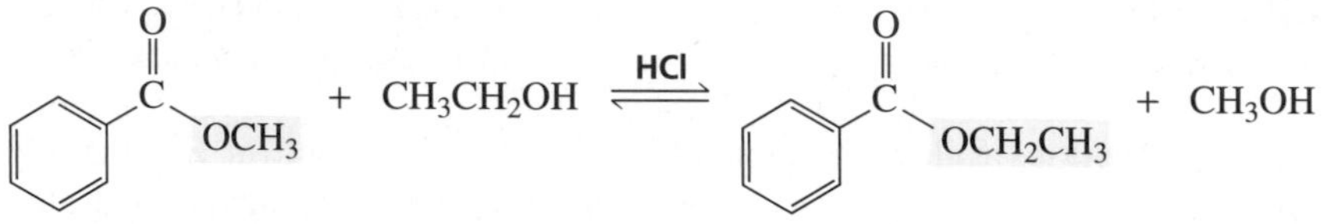

Both the hydrolysis and the alcoholysis of an ester are very slow reactions because water and alcohols are poor nucleophiles and esters have basic leaving groups. These reactions are therefore always catalyzed when carried out in the laboratory. Both hydrolysis and alcoholysis of an ester can be catalyzed by acids (Section 16.11). The rate of hydrolysis can also be increased by hydroxide ion (Section 16.12), and the rate of alcoholysis can be increased by the conjugate base (RO^-) of the reactant alcohol.

Esters also react with amines to form amides. A reaction with an amine that converts one compound into two compounds is called **aminolysis**. Notice that the aminolysis of an ester requires only one equivalent of amine, unlike the aminolysis of an acyl halide or an acid anhydride, which requires two equivalents (Sections 16.8 and 16.9). This is because the leaving group of an ester (RO^-) is more basic than the amine, so the alkoxide ion—rather than unreacted amine—picks up the proton generated in the reaction.

an aminolysis reaction

ethyl propionate *N*-methylpropionamide

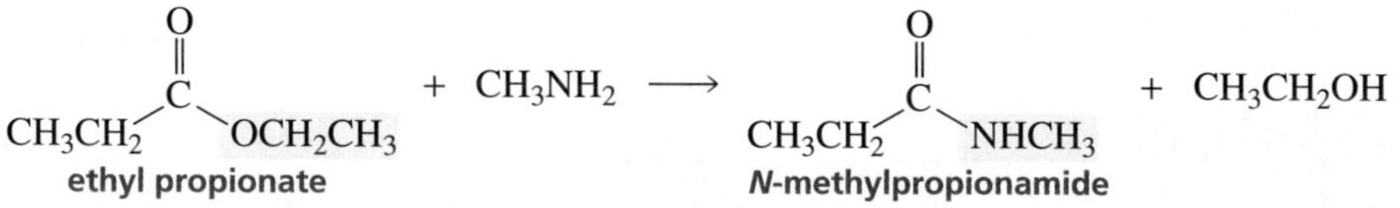

The reaction of an ester with an amine is not as slow as the reaction of an ester with water or an alcohol, because an amine is a better nucleophile. This is fortunate, because the reaction of an ester with an amine cannot be catalyzed by an acid. The acid would protonate the amine, and a protonated amine is not a nucleophile (Section 10.6). The reaction also cannot be catalyzed by HO^- or RO^-, because they are better nucleophiles than the amine.

In Section 7.9, we saw that phenols are stronger acids than alcohols. Therefore, phenolate ions (ArO^-) are weaker bases than alkoxide ions (RO^-), which means that phenyl esters are more reactive than alkyl esters.

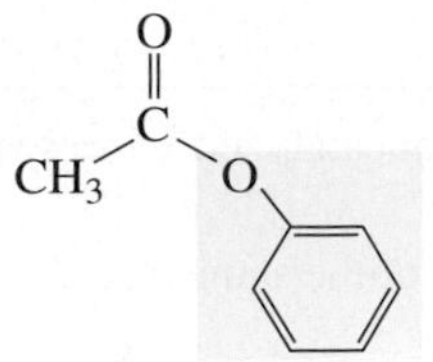

phenyl acetate **is more reactive than** methyl acetate

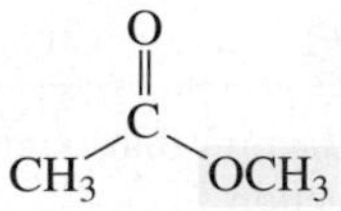

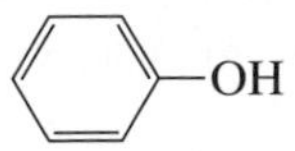

$pK_a = 10.0$ $pK_a = 15.5$

ASPIRIN

Aspirin, found naturally in willow bark and myrtle leaves, is one of the oldest and most commonly used drugs. As early as the fifth century BC, Hippocrates wrote about the curative powers of willow bark. Even so, aspirin's mode of action was not discovered until 1971, when John Vane found that the anti-inflammatory and fever-reducing activity of aspirin and related compounds called NSAIDs (nonsteroidal anti-inflammatory agents) were due to a transesterification reaction that blocks prostaglandin synthesis. Prostaglandins have several different biological functions, one being to stimulate inflammation and another to induce fever. The enzyme prostaglandin synthase catalyzes the conversion of arachidonic acid into PGH$_2$, a precursor of prostaglandins and the related thromboxanes (Section 26.5).

$$\text{arachidonic acid} \xrightarrow{\text{prostaglandin synthase}} \text{PGH}_2 \nearrow \text{prostaglandins} \searrow \text{thromboxanes}$$

Prostaglandin synthase is composed of two enzymes. One of them—cyclooxygenase—has a CH$_2$OH group (called a serine hydroxyl group because it is part of the amino acid called serine; See Table 22.1 on page 1019) that is necessary for enzymatic activity. The CH$_2$OH group reacts with aspirin (acetylsalicylic acid) in a transesterification reaction that inactivates the enzyme. When the enzyme is inactivated, prostaglandins cannot be synthesized, and inflammation is suppressed.

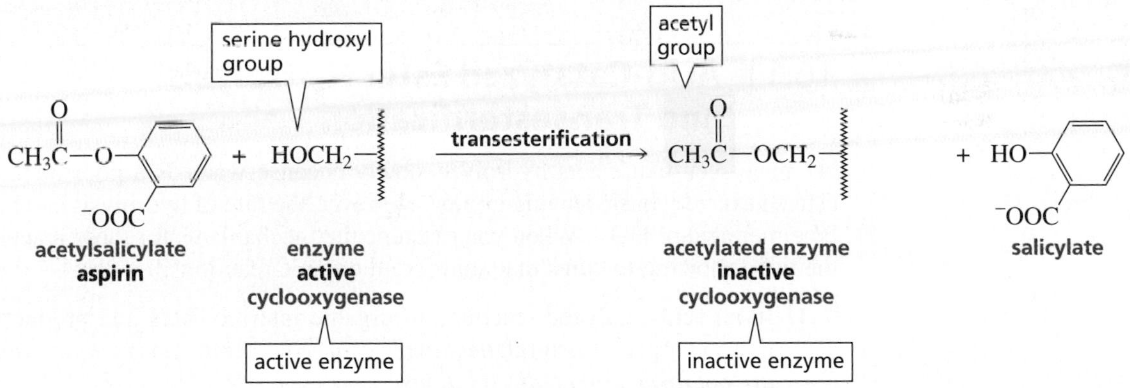

Thromboxanes stimulate platelet aggregation. Because aspirin inhibits the formation of PGH$_2$, it inhibits thromboxane production and, therefore, platelet aggregation. Presumably, this is why low levels of aspirin have been reported to reduce the incidence of strokes and heart attacks that result from blood clot formation. Aspirin's activity as an anticoagulant is why doctors caution patients not to take aspirin for several days before surgery.

PROBLEM 17

Write a mechanism for the following reactions:

a. the noncatalyzed hydrolysis of methyl propionate.

b. the aminolysis of phenyl formate, using methylamine.

PROBLEM 18◆

a. State three factors that contribute to making the noncatalyzed hydrolysis of an ester is a slow reaction.

b. Which is faster, hydrolysis of an ester or aminolysis of the same ester?

PROBLEM 19 **SOLVED**

List the following esters in order of decreasing reactivity toward hydrolysis:

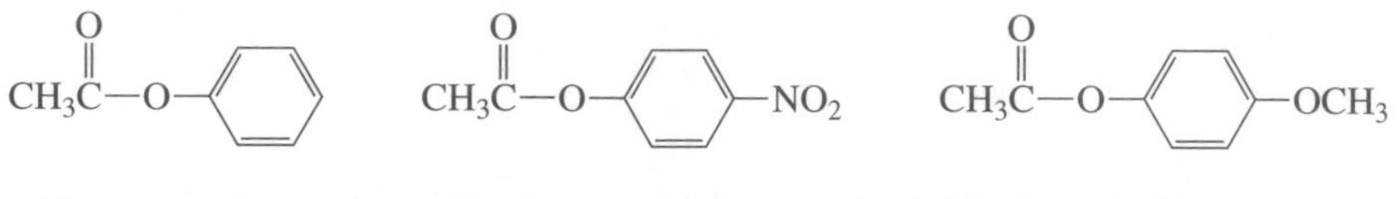

Solution   Both formation of the tetrahedral intermediate and collapse of the tetrahedral intermediate will be fastest for the ester with the electron-withdrawing nitro substituent and slowest for the ester with the electron-donating methoxy substituent. *Formation of the tetrahedral intermediate:* An electron-withdrawing substituent increases the ester's susceptibility to nucleophilic attack, and an electron-donating substituent decreases its susceptibility. *Collapse of the tetrahedral intermediate:* Electron withdrawal increases acidity and electron donation decreases acidity. Therefore, *para*-nitrophenol with its strong electron-withdrawing group is a stronger acid than phenol; and phenol is a stronger acid than *para*-methoxyphenol, which has a strong electron-donating group. Therefore, the *para*-nitrophenolate ion is the weakest base and the best leaving group of the three, whereas the *para*-methoxyphenolate ion is the strongest base and the worst leaving group. Thus,

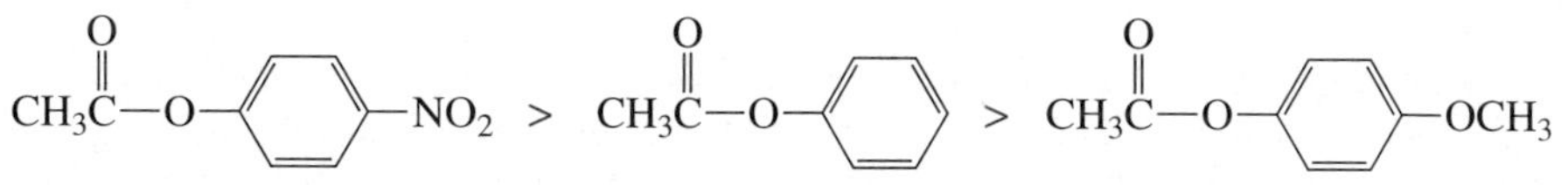

16.11 Acid-Catalyzed Ester Hydrolysis and Transesterification

We have seen that esters hydrolyze slowly because water is a poor nucleophile and esters have very basic leaving groups. However, the rate of hydrolysis can be increased by either acid or HO^-. When you examine the mechanisms for these reactions, notice the two following features that hold for all organic reactions:

1. In an acid-catalyzed reaction, all organic intermediates and products are positively charged or neutral; *negatively charged organic intermediates and products are not formed in acidic solutions.*

2. In a reaction in which HO^- is used to increase the rate of the reaction, all organic intermediates and products are negatively charged or neutral; *positively charged organic intermediates and products are not formed in basic solutions.*

Hydrolysis of Esters with Primary or Secondary Alkyl Groups

We have seen that when an acid is added to a reaction, the first thing that happens is the acid protonates the atom in the reactant that has the greatest electron density. Therefore, the acid protonates the carbonyl oxygen.

HB^+ **represents any species in the solution that is capable of donating a proton and :B represents any species in the solution that is capable of removing a proton.**

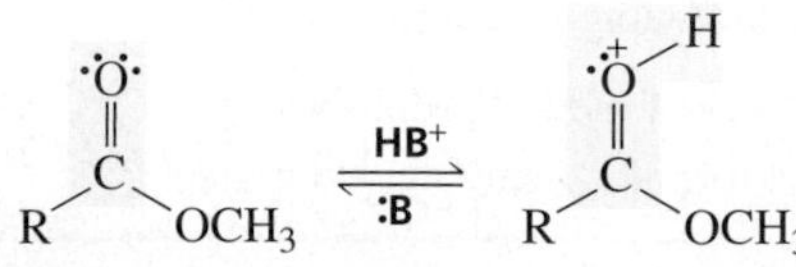

The resonance contributors of the ester show why the carbonyl oxygen is the atom with the greatest electron density.

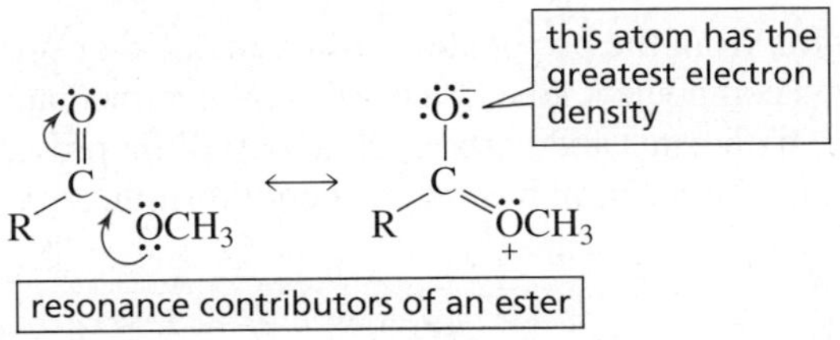

resonance contributors of an ester

The mechanism for the acid-catalyzed hydrolysis of an ester is shown below.

Mechanism for acid-catalyzed ester hydrolysis

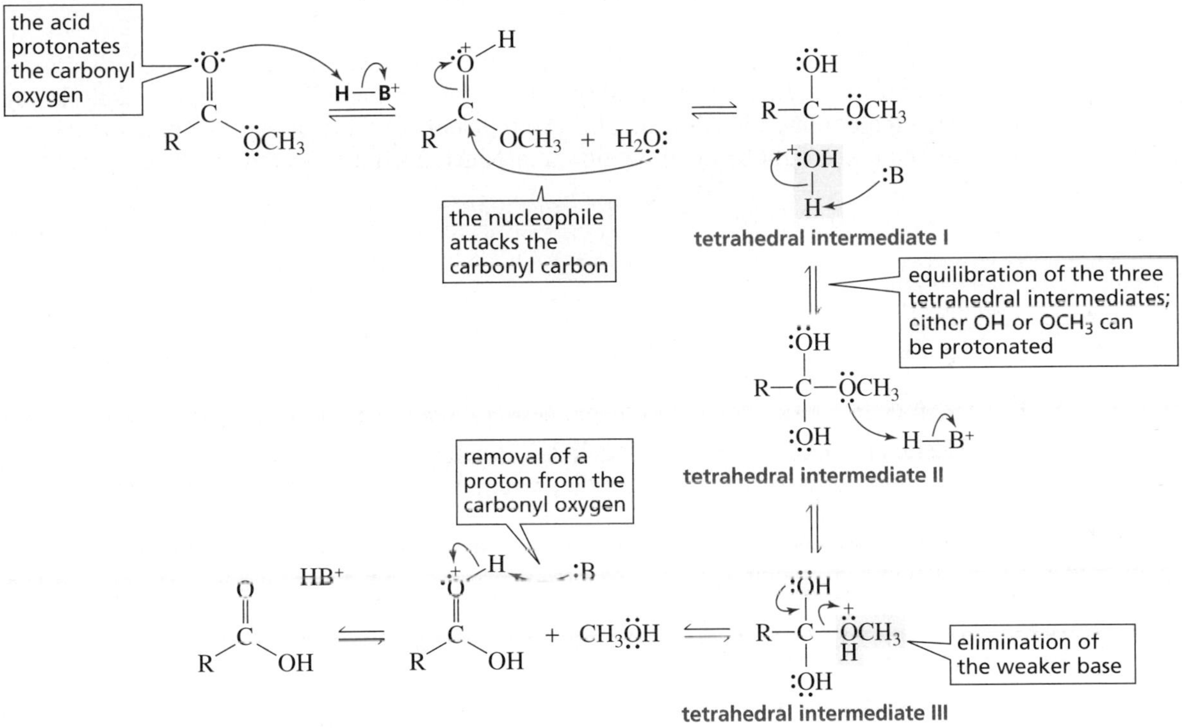

the acid protonates the carbonyl oxygen

the nucleophile attacks the carbonyl carbon

tetrahedral intermediate I

equilibration of the three tetrahedral intermediates; either OH or OCH_3 can be protonated

tetrahedral intermediate II

removal of a proton from the carbonyl oxygen

elimination of the weaker base

tetrahedral intermediate III

 **Ester Hydrolysis**

- The acid protonates the carbonyl oxygen.
- The nucleophile (H_2O) attacks the carbonyl carbon of the protonated carbonyl group, forming a protonated tetrahedral intermediate.
- The protonated tetrahedral intermediate (tetrahedral intermediate I) is in equilibrium with its nonprotonated form (tetrahedral intermediate II).
- Once the nonprotonated tetrahedral intermediate has been formed, either the OH or the OR group of this intermediate (in this case, OR is OCH_3) can be protonated, in the one case reforming tetrahedral intermediate I (OH is protonated), and in the other case forming tetrahedral intermediate III (OR is protonated). (From Section 1.24, we know that the relative amounts of the three tetrahedral intermediates depend on the pH of the solution and the pK_a values of the protonated intermediates.)
- When tetrahedral intermediate I collapses, it expels H_2O in preference to CH_3O^- (because H_2O is a weaker base), and reforms the ester. When tetrahedral intermediate III collapses, it expels CH_3OH rather than HO^- (because CH_3OH is a weaker base) and forms the carboxylic acid. Because H_2O and CH_3OH have approximately the same basicity, it will be as likely for tetrahedral intermediate I to collapse to reform

the ester as it will for tetrahedral intermediate III to collapse to form the carboxylic acid. (Tetrahedral intermediate II is much less likely to collapse, because both HO⁻ and CH₃O⁻ are strong bases.)

- Removal of a proton from the protonated carboxylic acid or protonated ester forms the neutral carbonyl compounds. Because tetrahedral intermediate I and III are equally likely to collapse, both ester and carboxylic acid will be present in approximately equal amounts when the reaction has reached equilibrium.

$$R-C(=O)-OCH_3 \ + \ H_2O \ \underset{HCl}{\rightleftharpoons} \ R-C(=O)-OH \ + \ CH_3OH$$

both ester and carboxylic acid will be present in approximately equal amounts when the reaction has reached equilibrium

Student Tutorial:
Manipulating the equilibrium

Excess water can be used to force the equilibrium to the right (Le Châtelier's principle; Section 3.7). Or, if the boiling point of the product alcohol is significantly lower than the boiling points of the other components of the reaction, the reaction can be driven to the right by distilling off the product alcohol as it is formed.

$$R-C(=O)-OCH_3 \ + \ H_2O_{\ \textbf{excess}} \ \underset{HCl}{\rightleftharpoons} \ R-C(=O)-OH \ + \ CH_3OH$$

The mechanism for the acid-catalyzed reaction of a carboxylic acid and an alcohol to form an ester and water is the exact reverse of the mechanism for the acid-catalyzed hydrolysis of an ester to form a carboxylic acid and an alcohol. If the ester is the desired product, the reaction should be carried out under conditions that will drive the equilibrium to the left (for example, using excess alcohol or removing water as it is formed (Section 16.14).

$$R-C(=O)-OCH_3 \ + \ H_2O \ \underset{HCl}{\rightleftharpoons} \ R-C(=O)-OH \ + \ CH_3OH_{\ \textbf{excess}}$$

PROBLEM 20◆

What products would be formed from the acid-catalyzed hydrolysis of the following esters?

a. (phenyl ketone with OCH₂CH₃)

b. $CH_3CH_2CH_2-C(=O)-OCH_3$

c. (cyclic lactone structure)

PROBLEM 21

Referring to the mechanism for the acid-catalyzed hydrolysis of an ester, write the mechanism—showing all the curved arrows—for the acid-catalyzed reaction of acetic acid and methanol to form methyl acetate. Use HB⁺ and :B to represent proton-donating and proton-removing species, respectively.

Now let's see how the acid increases the rate of ester hydrolysis. The acid is a catalyst. Recall that **catalyst** is a substance that increases the rate of a reaction without being consumed or changed in the overall reaction (Section 4.5). For a catalyst to increase the rate of a reaction, it must increase the rate of the slow step of the reaction, because changing the rate of a fast step will not affect the rate of the overall reaction. Four of the six steps in the mechanism for acid-catalyzed ester hydrolysis are proton transfer steps. Proton transfer to or from an electronegative atom such as oxygen or nitrogen is always a fast step. The other two steps in the mechanism—formation of the tetrahedral intermediate and collapse of the tetrahedral intermediate—are relatively slow. The acid increases the rates of both these steps.

The acid increases *the rate of formation of the tetrahedral intermediate* by protonating the carbonyl oxygen. Protonated carbonyl groups are more susceptible than nonprotonated carbonyl groups to nucleophilic attack, because a positively charged oxygen is more electron withdrawing than a neutral oxygen. Increased electron withdrawal by the oxygen makes the carbonyl carbon more electron deficient, which increases its attractiveness to nucleophiles.

protonation of the carbonyl oxygen increases the susceptibility of the carbonyl carbon to nucleophilic attack

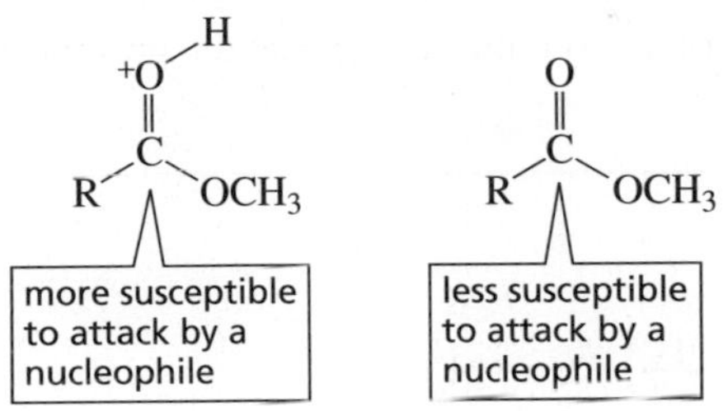

The acid increases *the rate of collapse of the tetrahedral intermediate* by decreasing the basicity of the leaving group, so the group is more easily eliminated. In the acid-catalyzed hydrolysis of an ester, the leaving group is ROH (in this case CH_3OH), a weaker base than the leaving group (RO^-; in this case CH_3O^-) in the uncatalyzed reaction.

An acid catalyst increases the reactivity of a carbonyl group.

An acid catalyst can make a group a better leaving group.

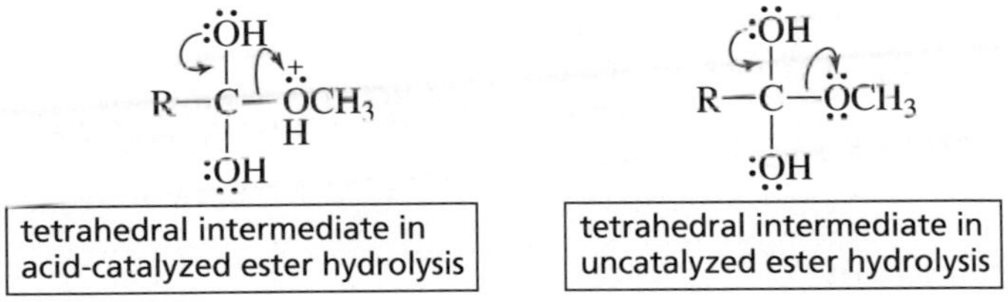

PROBLEM 22◆

In the mechanism for the acid-catalyzed hydrolysis of an ester:

a. what species could be represented by HB^+?
b. what species could be represented by $:B$?
c. what species is HB^+ most likely to be in a hydrolysis reaction?
d. what species is HB^+ most likely to be in the reverse reaction?

Hydrolysis of Esters with Tertiary Alkyl Groups

Esters with tertiary alkyl groups undergo acid-catalyzed hydrolysis much more rapidly than do other esters because they hydrolyze by a completely different mechanism, one that does not involve the formation of a tetrahedral intermediate. The hydrolysis of an ester with a tertiary alkyl group is an S_N1 reaction because when the carboxylic acid leaves, it leaves behind a relatively stable tertiary carbocation.

Mechanism for hydrolysis of an ester with a tertiary alkyl group

departure of the
leaving group to form
a tertiary carbocation

reaction of
the carbocation
with a nucleophile

- An acid protonates the carbonyl oxygen.
- The leaving group departs, forming a tertiary carbocation.
- A nucleophile reacts with the carbocation.
- A base removes a proton from the strongly acidic protonated alcohol.

PROBLEM 23 SOLVED

How could butanone be prepared from butane?

$$CH_3CH_2CH_2CH_3 \xrightarrow{\text{?}} CH_3\overset{O}{\overset{\|}{C}}CH_2CH_3$$

butane   butanone

Solved   We know that the first reaction has to be a radical halogenation because that is the only reaction that an alkane undergoes. Bromination will lead to a greater yield of the desired 2-halo-substituted compound than will chlorination because the bromine radical is more selective than a chlorine radical. To maximize the yield of substitution product (Section 9.8), the alkyl bromide is treated with acetate ion (a weak base) and the ester is hydrolyzed to the alcohol, which forms the target compound when it is oxidized.

$$CH_3CH_2CH_2CH_3 \xrightarrow[h\nu]{Br_2} CH_3\overset{Br}{\overset{|}{C}}HCH_2CH_3 \xrightarrow{CH_3COO^-} CH_3\overset{O-\overset{O}{\overset{\|}{C}}CH_3}{\overset{|}{C}}HCH_2CH_3 \quad CH_3\overset{OH}{\overset{|}{C}}HCH_2CH_3 \xrightarrow[H_2SO_4]{HCl,\ H_2O \quad Na_2Cr_2O_7} CH_3\overset{O}{\overset{\|}{C}}CH_2CH_3$$

 **Transesterification**

Transesterification

Transesterification—the reaction of an ester with an alcohol—is also catalyzed by acid. The mechanism for acid-catalyzed transesterification is identical to the mechanism for acid-catalyzed ester hydrolysis (for esters with primary or secondary alkyl groups), except that the nucleophile is ROH rather than H_2O. As in the hydrolysis of an ester, the leaving groups in the tetrahedral intermediate formed in transesterification have approximately the same basicity. Consequently, an excess of the reactant alcohol must be used to produce more of the desired product.

$$\overset{O}{\underset{CH_3}{\overset{\|}{C}}}_{OCH_3} + CH_3CH_2CH_2OH \underset{}{\overset{HCl}{\rightleftharpoons}} \overset{O}{\underset{CH_3}{\overset{\|}{C}}}_{OCH_2CH_2CH_3} + CH_3OH$$

methyl acetate   propyl alcohol   propyl acetate   methyl alcohol
                        excess

PROBLEM 24

Write the mechanism for the acid-catalyzed transesterification reaction of methyl acetate with ethanol.

16.12 **Hydroxide-Ion-Promoted Ester Hydrolysis**

The rate of hydrolysis of an ester can be increased by carrying out the reaction in a basic solution. Like an acid catalyst, hydroxide ion increases the rates of the two slow steps of the reaction—formation of the tetrahedral intermediate and collapse of the tetrahedral intermediate.

Mechanism for hydroxide-ion-promoted hydrolysis of an ester

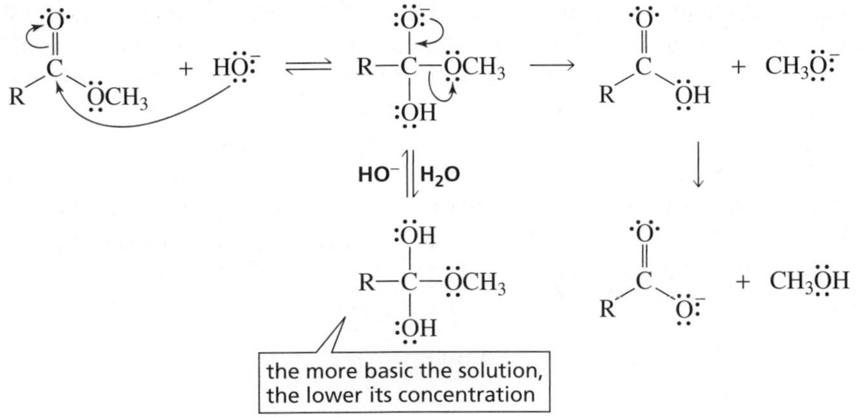

Hydroxide ion increases the rate of formation of the tetrahedral intermediate because HO^- is a better nucleophile than H_2O. Hydroxide ion increases the rate of collapse of the tetrahedral intermediate because a smaller fraction of the negatively charged tetrahedral intermediate becomes protonated in a basic solution; a negatively charged oxygen can more readily expel the very basic leaving group (RO^-) because the oxygen does not develop a partial positive charge in the transition state. (The relative amounts of the neutral and negatively charged tetrahedral intermediates depend on the pH of the solution and the pK_a value of the neutral tetrahedral intermediate; see Section 1.24.)

Hydroxide ion is a better nucleophile than water.

Notice that when CH_3O^- is expelled, the final products are not the carboxylic acid and methoxide ion because if only one species is protonated, it will be the more basic one. The final products are the carboxylate ion and methanol because CH_3O^- is more basic than CH_3COO^-. Since carboxylate ions are negatively charged, they are not attacked by nucleophiles.

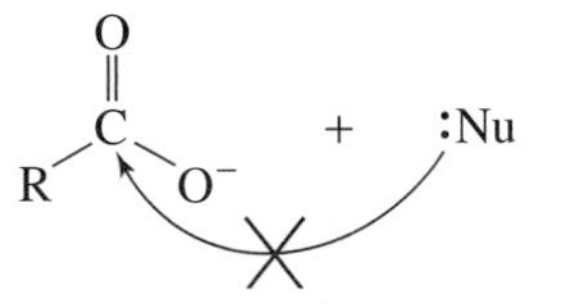

Therefore, the hydroxide-ion-promoted hydrolysis of an ester, unlike the acid-catalyzed hydrolysis of an ester, is *not* a reversible reaction.

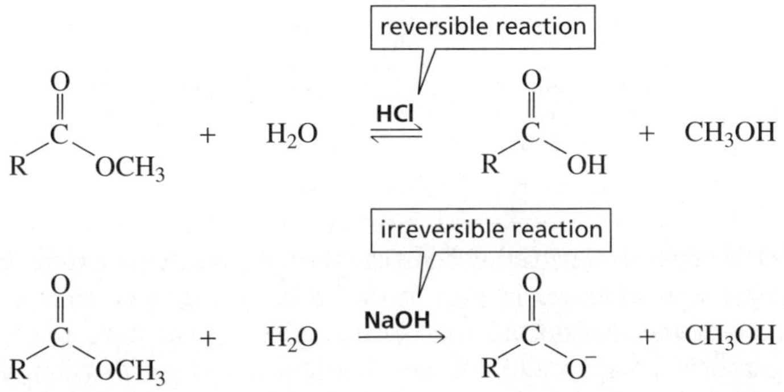

This reaction is called a *hydroxide-ion-promoted reaction* rather than a base-catalyzed reaction because hydroxide ion increases the rate of the first step of the reaction by being a better nucleophile than water—not by being a stronger base than water—and because hydroxide ion is consumed in the overall reaction. To be a catalyst, a species must not be changed by or consumed in the reaction (Section 16.11). Therefore, hydroxide ion is actually a reagent rather than a catalyst, so it is more accurate to call the reaction a hydroxide-ion-*promoted* reaction than a hydroxide-ion-*catalyzed* reaction.

Hydroxide ion promotes only hydrolysis reactions; it does not promote transesterification reactions or aminolysis reactions. Hydroxide ion cannot promote reactions of carboxylic acid derivatives with alcohols or with amines because one function of hydroxide ion is to provide a strong nucleophile for the first step of the reaction. When the nucleophile is supposed to be an alcohol or an amine, nucleophilic attack by hydroxide ion would form a different product from the one that would be formed by the alcohol or amine. Hydroxide can be used to promote a hydrolysis reaction because the same product is formed, whether the attacking nucleophile is H_2O or HO^-.

Reactions in which the nucleophile is an alcohol can be promoted by the conjugate base of the alcohol. The function of the alkoxide ion is to provide a strong nucleophile for the reaction, so only reactions in which the nucleophile is an alcohol can be promoted by the conjugate base of the alcohol.

$$\text{C}_6\text{H}_5\text{-C(=O)-OCH}_3 + \underset{\text{excess}}{\text{CH}_3\text{CH}_2\text{OH}} \underset{}{\overset{\text{CH}_3\text{CH}_2\text{O}^-}{\rightleftharpoons}} \text{C}_6\text{H}_5\text{-C(=O)-OCH}_2\text{CH}_3 + \text{CH}_3\text{OH}$$

PROBLEM 25

a. What species other than an acid can be used to increase the rate of a transesterification reaction that converts methyl acetate to propyl acetate?

b. Explain why the rate of aminolysis of an ester cannot be increased by H^+, HO^-, or RO^-.

16.13 How the Mechanism for Nucleophilic Acyl Substitution Reactions Was Confirmed

You have seen that nucleophilic acyl substitution reactions take place by a mechanism in which a tetrahedral intermediate is formed and subsequently collapses. The tetrahedral intermediate, however, is too unstable to be isolated. How, then, do we know that it is formed? How do we know that the reaction doesn't take place by a one-step direct-displacement mechanism (similar to the mechanism of an S_N2 reaction) in which the incoming nucleophile attacks the carbonyl carbon and displaces the leaving group—a mechanism that does not break the π bond so does not form a tetrahedral intermediate?

$$\underset{\underset{\text{R}}{|}}{\overset{\overset{\text{O}}{\|}}{\text{HO}\overset{\delta-}{\cdots}\text{C}\overset{\delta-}{\cdots}\text{OR}}}$$

transition state for a hypothetical one-step direct-displacement mechanism

Myron L. Bender (1924–1988)
was born in St. Louis. He was a professor of chemistry at the Illinois Institute of Technology and at Northwestern University.

To answer this question, Myron Bender investigated the hydroxide-ion-promoted hydrolysis of ethyl benzoate after first labeling the carbonyl oxygen of ethyl benzoate with ^{18}O. When he isolated ethyl benzoate from the reaction mixture, he found that some of the ester was no longer labeled. If the reaction had taken place by a one-step direct-displacement mechanism, all the isolated ester would have remained labeled because the carbonyl group would not have participated in the reaction. On the other hand, if the mechanism involved a tetrahedral intermediate, some of the isolated ester would no longer be labeled because some of the label would have been transferred to the hydroxide ion. By this experiment, Bender provided evidence for the reversible formation of a tetrahedral intermediate.

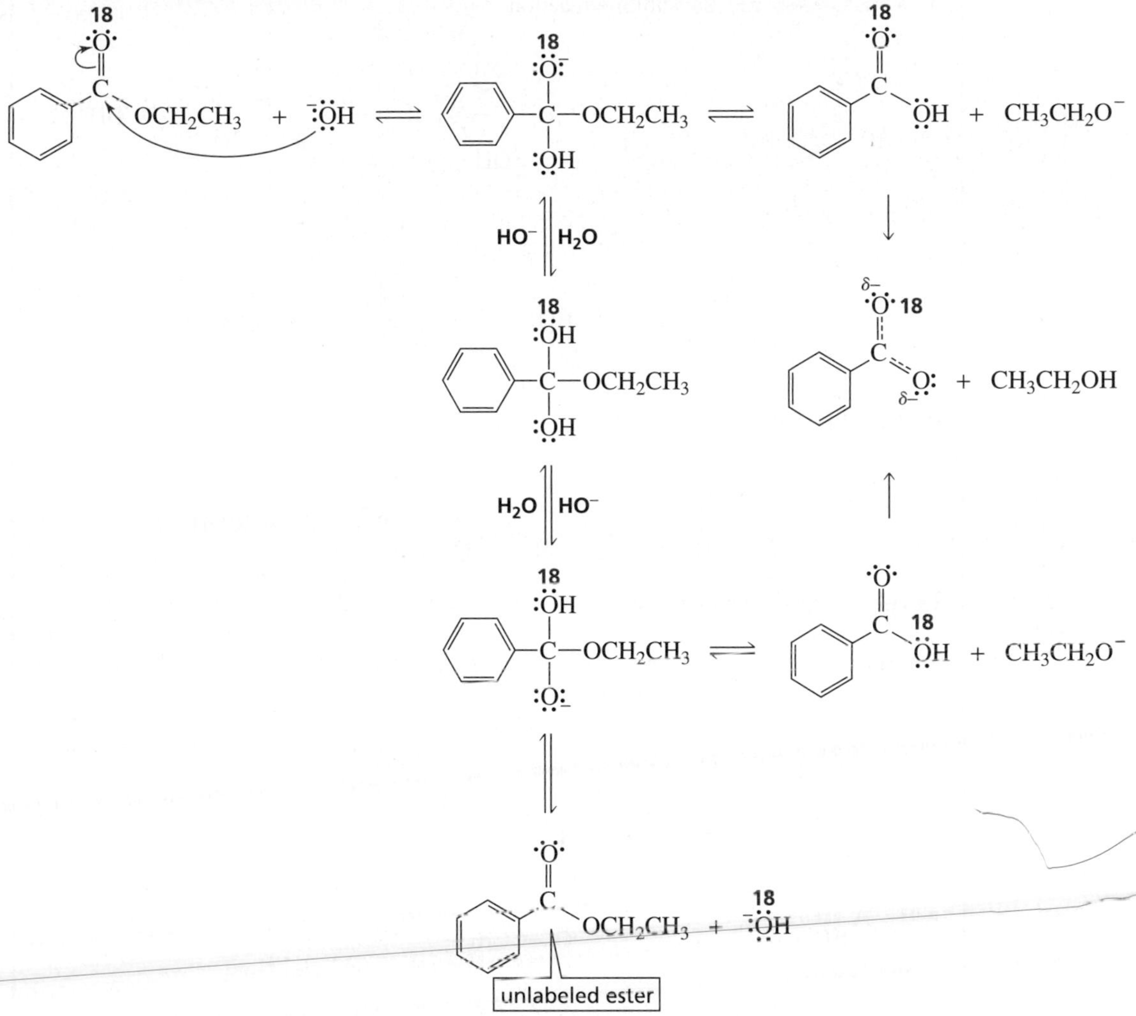

PROBLEM 26◆

If butanoic acid and ^{18}O labeled methanol are allowed to react under acidic conditions, what compounds will be labeled when the reaction has reached equilibrium?

PROBLEM 27◆

D. N. Kursanov, a Russian chemist, proved that the bond that is broken in the hydroxide-ion-promoted hydrolysis of an ester is the acyl C—O bond, rather than the alkyl C—O bond, by studying the reaction of the following ester with HO^-/H_2O:

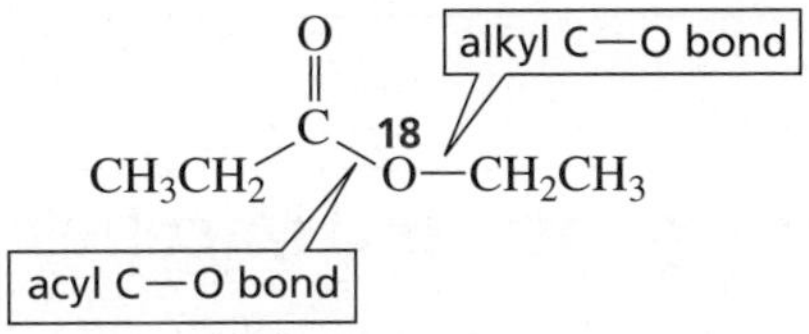

a. Which of the products contained the ^{18}O label?

b. What product would have contained the ^{18}O label if the alkyl C—O bond had broken?

PROBLEM 28 *SOLVED*

Early chemists could envision several possible mechanisms for hydroxide-ion-promoted ester hydrolysis:

1. a nucleophilic acyl substitution reaction

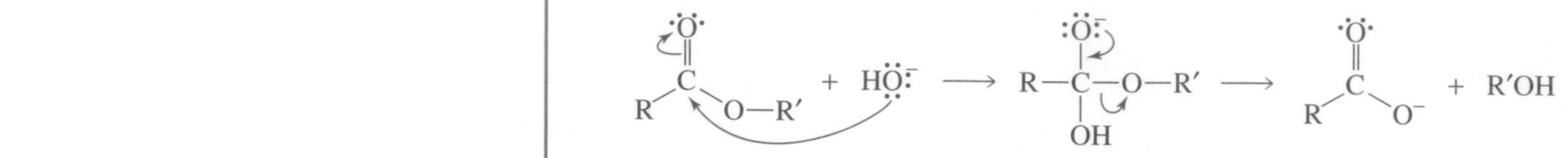

2. an S_N2 reaction

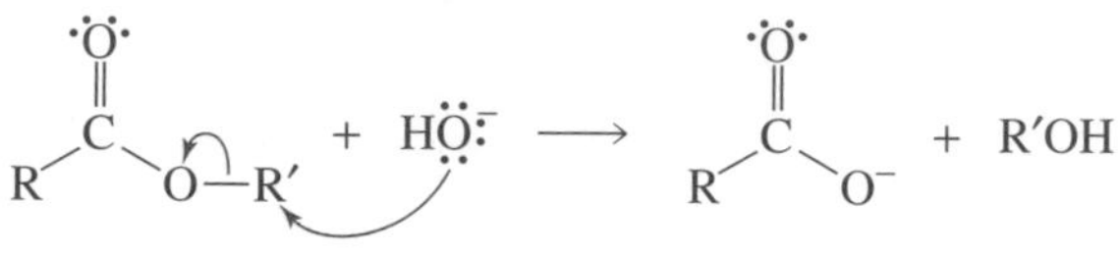

3. an S_N1 reaction

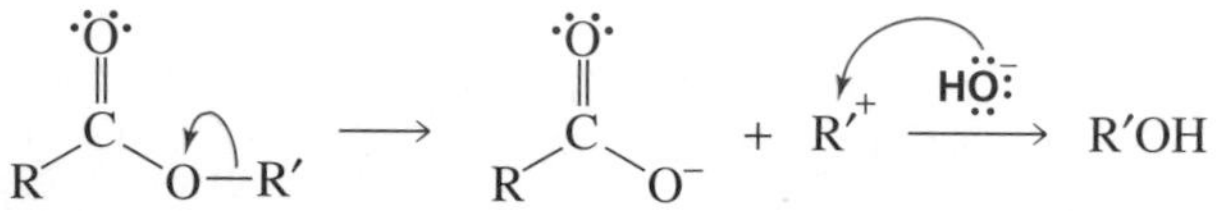

Devise an experiment that would show which was the actual mechanism.

Solution Start with a single stereoisomer of an alcohol whose OH group is bonded to an asymmetric center, and determine the specific rotation of the alcohol. Then convert the alcohol into an ester, using a method that does not break any bonds to the asymmetric center. Next, hydrolyze the ester, isolate the alcohol obtained from hydrolysis, and determine its specific rotation.

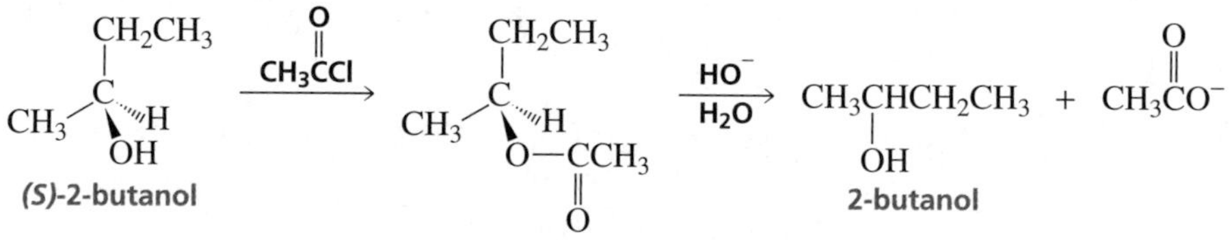

If the reaction is a nucleophilic acyl substitution reaction, the product alcohol will have the same specific rotation as the reactant alcohol because no bonds to the asymmetric center are broken (Section 5.14).

If the reaction is an S_N2 reaction, the product alcohol and the reactant alcohol will have opposite specific rotations because the mechanism requires back-side attack on the asymmetric center (Section 8.2).

If the reaction is an S_N1 reaction, the product alcohol will have a small (or zero) specific rotation because the mechanism requires carbocation formation, which leads to racemization of the alcohol (Section 8.7).

16.14 Soaps, Detergents, and Micelles

Fats and **oils** are triesters of glycerol. Glycerol contains three alcohol groups and therefore can form three ester groups. When the ester groups of a fat or an oil are hydrolyzed in a basic solution, glycerol and carboxylate ions are formed. The carboxylic acids that are bonded to glycerol in fats and oils have long, unbranched R groups. Because they are obtained from fats, long-chain unbranched carboxylic acids are called **fatty acids**. In Section 26.1, we will see that the difference between a fat and an oil resides in the structure of the fatty acids.

$$
\begin{array}{c}
\underset{\substack{| \\ \text{CH}_2\text{O}-\overset{\displaystyle \text{O}}{\overset{\|}{\text{C}}}-\text{R}^1 \\ | \\ \text{CHO}-\overset{\displaystyle \text{O}}{\overset{\|}{\text{C}}}-\text{R}^2 \\ | \\ \text{CH}_2\text{O}-\overset{\displaystyle \text{O}}{\overset{\|}{\text{C}}}-\text{R}^3}}{\underset{\textbf{a fat or an oil}}{}}
\; + \; \text{H}_2\text{O}
\; \xrightarrow{\textbf{NaOH}} \;
\underset{\substack{| \\ \text{CH}_2\text{OH} \\ | \\ \text{CHOH} \\ | \\ \text{CH}_2\text{OH}}}{\underset{\textbf{glycerol}}{}}
\; + \;
\underset{\substack{\text{R}^1-\overset{\displaystyle \text{O}}{\overset{\|}{\text{C}}}-\text{O}^-\,\text{Na}^+ \\ \text{R}^2-\overset{\displaystyle \text{O}}{\overset{\|}{\text{C}}}-\text{O}^-\,\text{Na}^+ \\ \text{R}^3-\overset{\displaystyle \text{O}}{\overset{\|}{\text{C}}}-\text{O}^-\,\text{Na}^+}}{\underset{\substack{\textbf{sodium salts of fatty acids}\\\textbf{soap}}}{}}
\end{array}
$$

   Soaps are sodium or potassium salts of fatty acids. Thus, soaps are obtained when fats or oils are hydrolyzed under basic conditions. The hydrolysis of an ester in a basic solution is called **saponification**—the Latin word for "soap" is *sapo*. Three of the most common soaps are:

$$\underset{\textbf{sodium stearate}}{\text{CH}_3(\text{CH}_2)_{16}\overset{\displaystyle \text{O}}{\overset{\|}{\text{C}}}\text{O}^-\,\text{Na}^+} \qquad \underset{\textbf{sodium oleate}}{\text{CH}_3(\text{CH}_2)_7\text{CH}=\text{CH}(\text{CH}_2)_7\overset{\displaystyle \text{O}}{\overset{\|}{\text{C}}}\text{O}^-\,\text{Na}^+}$$

$$\underset{\textbf{sodium linoleate}}{\text{CH}_3(\text{CH}_2)_4\text{CH}=\text{CHCH}_2\text{CH}=\text{CH}(\text{CH}_2)_7\overset{\displaystyle \text{O}}{\overset{\|}{\text{C}}}\text{O}^-\,\text{Na}^+}$$

PROBLEM 29 **SOLVED**

An oil obtained from coconuts is unusual in that all three fatty acid components are identical. The molecular formula of the oil is $C_{45}H_{86}O_6$. What is the molecular formula of the carboxylate ion obtained when the oil is saponified?

Solution When the oil is saponified, it forms glycerol and three equivalents of carboxylate ion. In losing glycerol, the fat loses three carbons and five hydrogens. Thus, the three equivalents of carboxylate ion have a combined molecular formula of $C_{42}H_{81}O_6$. Dividing by three gives a molecular formula of $C_{14}H_{27}O_2$ for the carboxylate ion.

   Long-chain carboxylate ions do not exist as individual ions in aqueous solution. Instead, they arrange themselves in spherical clusters called **micelles**, as shown in Figure 16.4. Each micelle contains 50 to 100 long-chain carboxylate ions and resembles a large ball. The polar heads of the carboxylate ions, each accompanied by a counterion, are on the outside of the ball because of their attraction for water, whereas the nonpolar tails are buried in the interior of the ball to minimize their contact with water. The attractive forces between hydrocarbon chains in water (such as those experienced by the tails of the carboxylate ions) are called **hydrophobic interactions** (Section 22.15). Soap has cleansing ability because nonpolar oil molecules that carry dirt dissolve in the nonpolar interior of the micelle and are washed away with the micelle during rinsing.

stearate

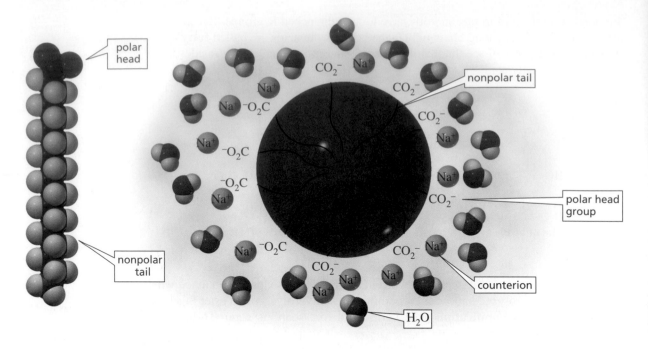

▲ **Figure 16.4**
In aqueous solution, soap forms micelles. The polar heads (carboxylate groups) of the soap molecules form the micelle's outer surface; the nonpolar tails (fatty acid R groups) extend into the micelle's interior.

Because the surface of the micelle is negatively charged, the individual micelles usually repel each other instead of clustering to form larger aggregates. However, in "hard" water—water containing high concentrations of calcium and magnesium ions (having leached them out of rocks and soils)—the micelles do form aggregates. In hard water, therefore, soaps form a precipitate that we know as "bathtub ring" or "soap scum."

MAKING SOAP

For thousands of years, soap was prepared by heating animal fat with wood ashes. Because the ashes contain potassium carbonate, the solution that results is basic. In the modern commercial method of making soap, fats or oils are boiled in aqueous sodium hydroxide. Sodium chloride is then added to precipitate the soap, which is dried and pressed into bars. Perfume can be added for scented soaps, dyes can be added for colored soaps, sand can be added for scouring soaps, and air can be blown into the soap to make it float in water.

Making soap

The formation of soap scum in hard water led to a search for synthetic materials that would have the cleansing properties of soap but would not form scum when they encountered calcium and magnesium ions. The synthetic "soaps" that were developed, known as **detergents** (from the Latin *detergere*, which means "to wipe off"), are salts of benzene sulfonic acids. Calcium and magnesium sulfonate salts do not form aggregates. After the initial introduction of detergents into the marketplace, it was discovered that straight-chain alkyl groups are biodegradable, whereas branched-chain alkyl groups are not. To prevent detergents from polluting rivers and lakes, they should be made only with straight-chain alkyl groups.

acetic acid

16.15 **Reactions of Carboxylic Acids**

Carboxylic acids can undergo nucleophilic acyl substitution reactions only when they are in their acidic forms. The basic form of a carboxylic acid cannot undergo nucleophilic acyl substitution reactions because the negatively charged carboxylate ion is resistant to nucleophilic attack. Thus, carboxylate ions are even less reactive in nucleophilic acyl substitution reactions than are amides.

relative reactivities toward nucleophilic acyl substitution

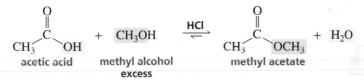

Carboxylic acids have approximately the same reactivity as esters, because the HO⁻ leaving group of a carboxylic acid has about the same basicity as the RO⁻ leaving group of an ester. Therefore, like esters, carboxylic acids do not react with halide ions or with carboxylate ions.

Carboxylic acids react with alcohols to form esters. The reaction must be carried out in an acidic solution, not only to catalyze the reaction but also to keep the carboxylic acid in its acidic form so that it will react with the nucleophile. Because the tetrahedral intermediate formed in this reaction has two potential leaving groups of approximately the same basicity, the reaction must be carried out with excess alcohol to drive it toward products. Emil Fischer (Section 5.6) was the first to discover that an ester could be prepared by treating a carboxylic acid with excess alcohol in the presence of an acid catalyst, so the reaction is called a **Fischer esterification**. Its mechanism is the exact reverse of the mechanism for the acid-catalyzed hydrolysis of an ester shown in Section 16.11. See also Problem 21.

Mechanism:
Fischer esterification

$$CH_3\!-\!\underset{\underset{OH}{}}{\overset{\overset{O}{\|}}{C}} \;+\; CH_3OH \;\underset{}{\overset{HCl}{\rightleftharpoons}}\; CH_3\!-\!\underset{\underset{OCH_3}{}}{\overset{\overset{O}{\|}}{C}} \;+\; H_2O$$

acetic acid   methyl alcohol   methyl acetate
                        excess

Carboxylic acids do not undergo nucleophilic acyl substitution reactions with amines. Because a carboxylic acid is an acid and an amine is a base, the carboxylic acid immediately donates a proton to the amine when the two compounds are mixed. The resulting ammonium carboxylate salt is the final product of the reaction; the carboxylate ion is unreactive and the protonated amine is not a nucleophile.

$$CH_3\!-\!\underset{\underset{OH}{}}{\overset{\overset{O}{\|}}{C}} \;+\; CH_3CH_2NH_2 \;\longrightarrow\; CH_3\!-\!\underset{\underset{O^-\; \overset{+}{H_3N}CH_2CH_3}{}}{\overset{\overset{O}{\|}}{C}}$$

**an ammonium
carboxylate salt**

$$CH_3CH_2\!-\!\underset{\underset{OH}{}}{\overset{\overset{O}{\|}}{C}} \;+\; NH_3 \;\longrightarrow\; CH_3CH_2\!-\!\underset{\underset{O^-\;\overset{+}{N}H_4}{}}{\overset{\overset{O}{\|}}{C}}$$

PROBLEM 30◆

Show how each of the following esters could be prepared using a carboxylic acid as one of the starting materials:

a. methyl butyrate (odor of apples)

b. octyl acetate (odor of oranges)

Proposing a Mechanism

Propose a mechanism for the following reaction:

$$
\text{[cyclohexene with CH}_2\text{COOH substituent]} \xrightarrow[\text{CH}_2\text{Cl}_2]{\text{Br}_2} \text{[bicyclic lactone with Br]} + \text{HBr}
$$

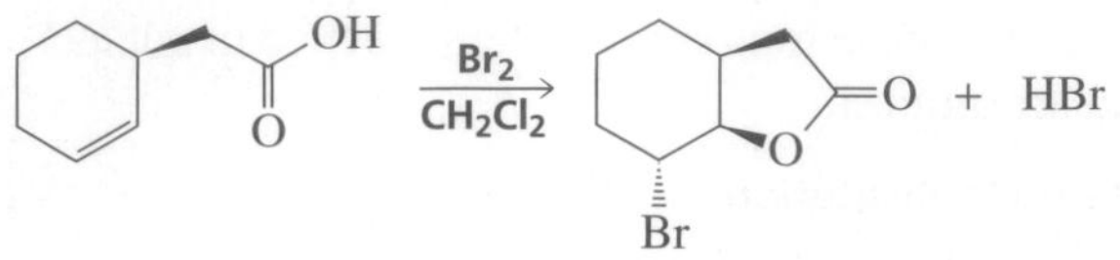

When you are asked to propose a mechanism, look carefully at the reagents to determine the first step of the mechanism. One of the reagents has two functional groups: a carboxyl group and a carbon–carbon double bond. The other reagent, Br_2, does not react with carboxylic acids, but does react with alkenes (Section 4.7). Approach to one side of the double bond is sterically hindered by the carboxyl group. Therefore, Br_2 will add to the other side of the double bond, forming a bromonium ion. We know that in the second step of this addition reaction, a nucleophile attacks the bromonium ion. Of the two nucleophiles present, the carbonyl oxygen is the one positioned to attack the back side of the bromonium ion, resulting in a compound with the observed configuration. Loss of a proton gives the final product of the reaction.

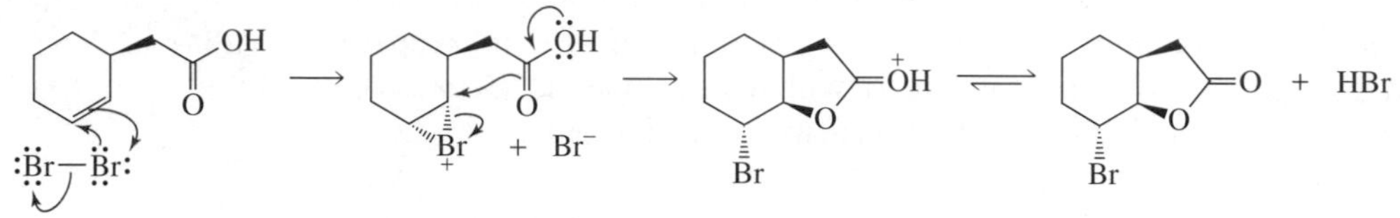

$$
\text{[mechanism sequence of structures]}
$$

Now continue on to Problem 31.

Propose a mechanism for the following reaction. (*Hint:* Number the carbons to help you see where they end up in the product.)

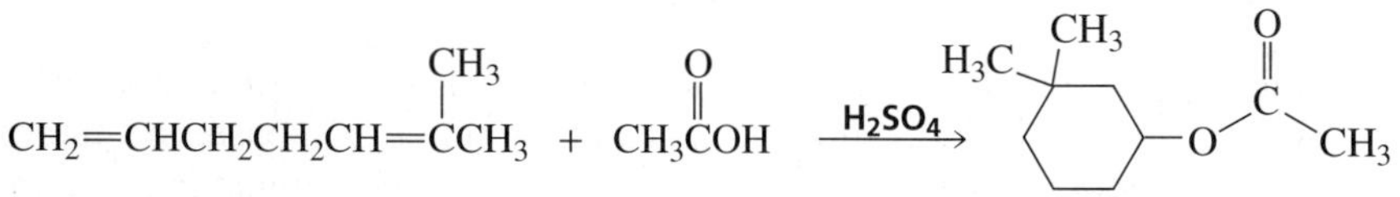

$$
CH_2{=}CHCH_2CH_2CH{=}\overset{\underset{\displaystyle CH_3}{|}}{C}CH_3 \;+\; CH_3\overset{\displaystyle O}{\overset{\|}{C}}OH \xrightarrow{H_2SO_4} \text{[cyclohexyl acetate product]}
$$

16.16 **Reactions of Amides**

Amides are very unreactive compounds, which is comforting, since proteins are composed of amino acids linked together by amide bonds (Sections 16.4 and 22.7). Amides do not react with halide ions, carboxylate ions, alcohols, or water because, in each case, the incoming nucleophile is a weaker base than the leaving group of the amide (Table 16.1).

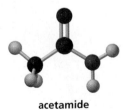

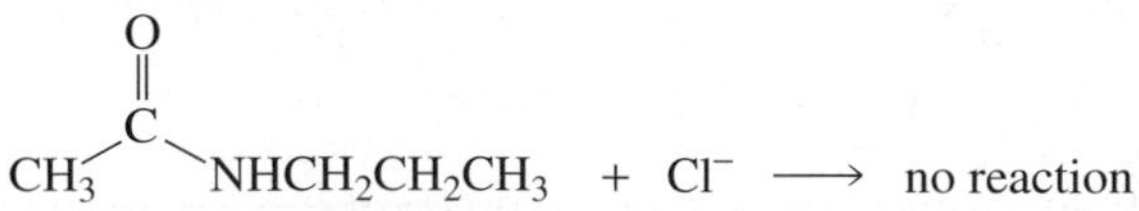

acetamide

$$
CH_3\overset{\displaystyle O}{\overset{\|}{C}}NHCH_2CH_2CH_3 \;+\; Cl^- \longrightarrow \text{no reaction}
$$
***N*-propylacetamide**

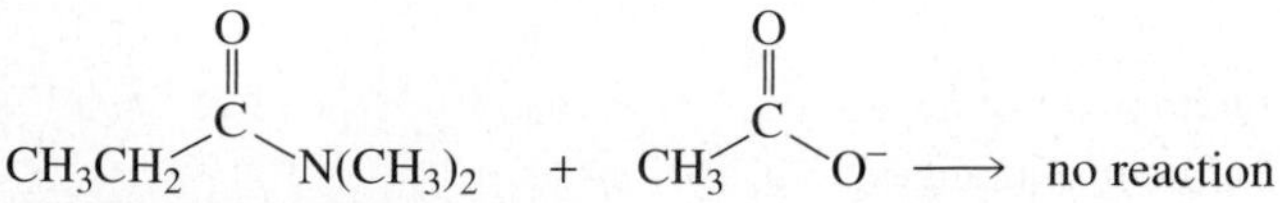

$$
CH_3CH_2\overset{\displaystyle O}{\overset{\|}{C}}N(CH_3)_2 \;+\; CH_3\overset{\displaystyle O}{\overset{\|}{C}}O^- \longrightarrow \text{no reaction}
$$
***N,N*-dimethylpropionamide**

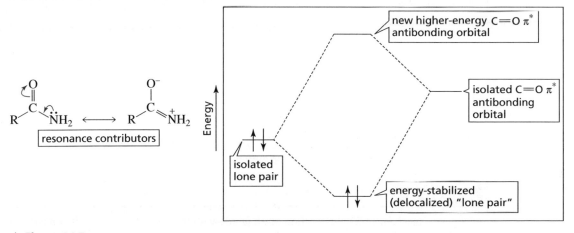

N-methylbenzamide   + CH₃OH ⟶ no reaction

N-ethylpropanamide   + H₂O ⟶ no reaction

Amides do, however, react with water and alcohols if the reaction mixture is heated in the presence of an acid. The reason for this will be explained in Section 16.17.

N-ethylacetamide   + H₂O $\xrightarrow{\text{HCl}}{\Delta}$ + CH₃CH₂⁺NH₃

N-methylbenzamide   + CH₃CH₂OH $\xrightarrow{\text{HCl}}{\Delta}$ + CH₃⁺NH₃

Molecular orbital theory can explain why amides are unreactive. In Section 16.2, we saw that amides have an important resonance contributor in which nitrogen shares its lone pair with the carbonyl carbon; the orbital that contains the lone pair overlaps the empty π* antibonding orbital of the carbonyl group (Figure 16.3). This overlap lowers the energy of the lone pair—so that it is neither basic nor nucleophilic—and raises the energy of the π* antibonding orbital of the carbonyl group, making it less reactive to nucleophiles (Figure 16.5).

resonance contributors

new higher-energy C=O π* antibonding orbital

isolated C=O π* antibonding orbital

Energy

isolated lone pair

energy-stabilized (delocalized) "lone pair"

▲ **Figure 16.5**
The filled nonbonding orbital containing nitrogen's lone pair overlaps the empty π* antibonding molecular orbital of the carbonyl group. This stabilizes the lone pair, making it less reactive, and raises the energy of the π* orbital of the carbonyl group, making it less able to react with nucleophiles.

An amide with an NH₂ group can be dehydrated to a nitrile. Dehydrating reagents commonly employed for this purpose are P₂O₅, POCl₃, and SOCl₂.

$\xrightarrow[\text{80 °C}]{\text{P}_2\text{O}_5}$ CH₃CH₂C≡N

NATURE'S SLEEPING PILL

Melatonin, a naturally occurring amide, is a hormone synthesized by the pineal gland from the amino acid tryptophan. Melatonin regulates the dark–light clock in our brains that governs such things as the sleep–wake cycle, body temperature, and hormone production.

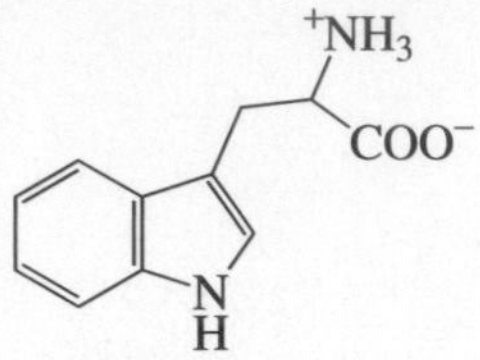

Melatonin levels increase from evening to night and then decrease as morning approaches. People with high levels of melatonin sleep longer and more soundly than those with low levels. The concentration of the hormone in our bodies varies with age—6-year-olds have more than five times the concentration that 80-year-olds have—which is one of the reasons why young people have less trouble sleeping than older people. Melatonin supplements are used to treat insomnia, jet lag, and seasonal affective disorder.

tryptophan
an amino acid

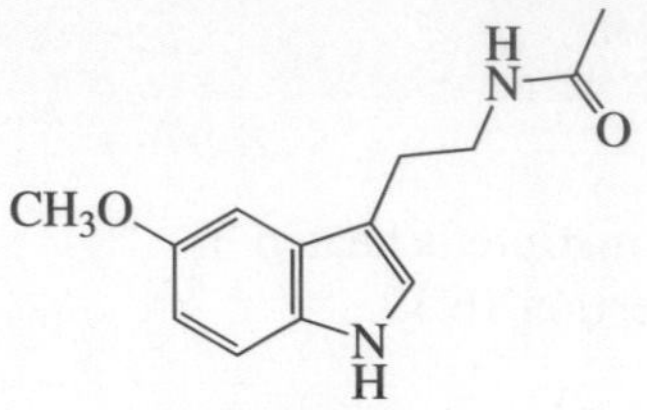

melatonin

PROBLEM 32

What acyl chloride and what amine would be required to synthesize the following amides?

a. *N*-ethylbutanamide

b. *N,N*-dimethylbenzamide

PROBLEM 33

a. Which of the following reactions would lead to the formation of an amide?

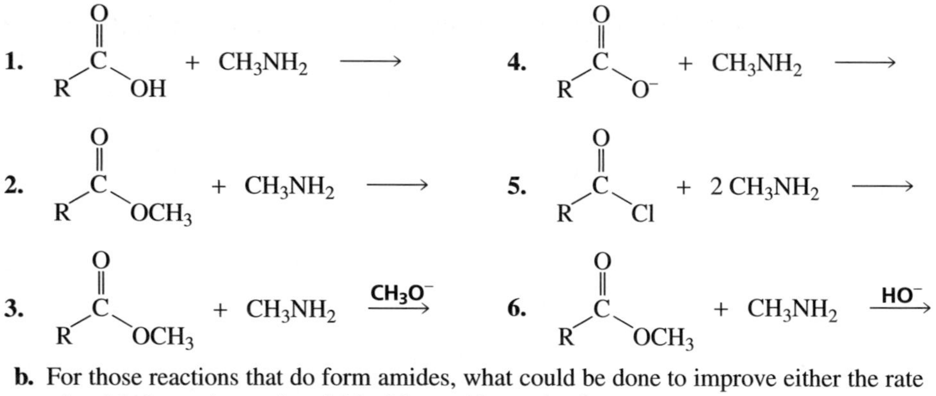

1. $R-C(=O)-OH$ + CH_3NH_2 ⟶

2. $R-C(=O)-OCH_3$ + CH_3NH_2 ⟶

3. $R-C(=O)-OCH_3$ + CH_3NH_2 $\xrightarrow{CH_3O^-}$

4. $R-C(=O)-O^-$ + CH_3NH_2 ⟶

5. $R-C(=O)-Cl$ + $2\,CH_3NH_2$ ⟶

6. $R-C(=O)-OCH_3$ + CH_3NH_2 $\xrightarrow{HO^-}$

b. For those reactions that do form amides, what could be done to improve either the rate of amide formation or the yield of the amide product?

PROBLEM 34

Propose a mechanism for the reaction of an amide with thionyl chloride to form a nitrile. (*Hint:* In the first step of the reaction, the amide is the nucleophile and thionyl chloride is the electrophile.)

16.17 The Hydrolysis of Amides Is Catalyzed by Acids

The mechanism for the acid-catalyzed hydrolysis of an amide is similar to the mechanism for the acid-catalyzed hydrolysis of an ester (Section 16.11).

Mechanism for acid-catalyzed hydrolysis of an amide

the acid protonates
the carbonyl
oxygen

the nucleophile
attacks the
carbonyl carbon

Mechanism:
Basic hydrolysis of an amide

tetrahedral intermediate I

either NH$_2$ or OH
can be protonated

tetrahedral intermediate II

the weaker base
is expelled

tetrahedral intermediate III

- The acid protonates the carbonyl oxygen, which increases the susceptibility of the carbonyl carbon to nucleophilic attack.
- Nucleophilic attack by water on the carbonyl carbon leads to tetrahedral intermediate I, which is in equilibrium with its nonprotonated form, tetrahedral intermediate II.
- Reprotonation can occur either on oxygen to reform tetrahedral intermediate I or on nitrogen to form tetrahedral intermediate III. Protonation on nitrogen is favored because the NH$_2$ group is a stronger base than the OH group.
- Of the two possible leaving groups in tetrahedral intermediate III (HO$^-$ and NH$_3$), NH$_3$ is the weaker base, so it is expelled, forming the carboxylic acid as the final product.
- Since the reaction is carried out in an acidic solution, NH$_3$ will be protonated after it is expelled from the tetrahedral intermediate. This prevents the reverse reaction from occurring, because $^+$NH$_4$ is not a nucleophile.

Let's take a minute to see why an amide cannot be hydrolyzed without a catalyst. In an uncatalyzed reaction, the amide would not be protonated. Therefore, water, a very poor nucleophile, would have to attack a neutral amide that is much less susceptible to nucleophilic attack than a protonated amide would be. In addition, the NH$_2$ group of the tetrahedral intermediate would not be protonated in the uncatalyzed reaction. Therefore, HO$^-$ is the group that would be expelled from the tetrahedral intermediate, because HO$^-$ is a weaker base than $^-$NH$_2$. This would reform the amide.

tetrahedral intermediate in
acid-catalyzed amide hydrolysis

tetrahedral intermediate in
uncatalyzed amide hydrolysis

An amide reacts with an alcohol in the presence of acid for the same reason that it reacts with water in the presence of acid.

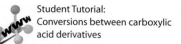

Student Tutorial:
Conversions between carboxylic
acid derivatives

PENICILLIN AND DRUG RESISTANCE

Penicillin contains an amide in a strained β-lactam ring. The strain in this four-membered ring increases the amide's reactivity. It is thought that the antibiotic activity of penicillin results from its ability to acylate (put an acyl group on) a CH_2OH group of an enzyme that has a role in the synthesis of bacterial cell walls. Acylation inactivates the enzyme, and actively growing bacteria die because they are unable to synthesize functional cell walls. Penicillin has no effect on mammalian cells because mammalian cells are not enclosed by cell walls. Penicillins are stored at cold temperatures to minimize hydrolysis of the β-lactam ring.

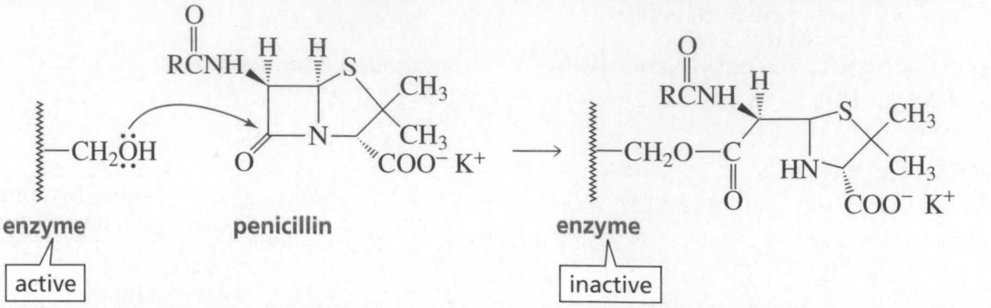

Bacteria that are resistant to penicillin secrete penicillinase, an enzyme that catalyzes the hydrolysis of penicillin's β-lactam ring. The ring-opened product has no antibacterial activity.

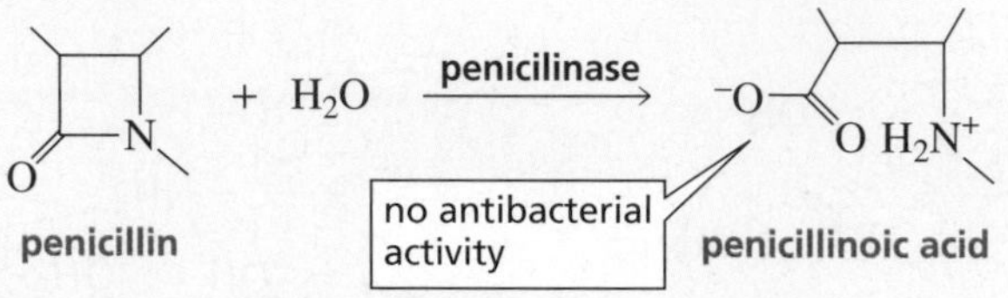

PENICILLINS IN CLINICAL USE

More than 10 different penicillins are currently in clinical use. They differ only in the group (R) attached to the carbonyl group. Some of these penicillins are shown here. In addition to their structural differences, the penicillins differ in the organisms against which they are most effective. They also differ in their susceptibility to penicillinase. For example, methicillin, a synthetic penicillin, is clinically effective against bacteria that are resistant to penicillin G, a naturally occurring penicillin. Almost 19% of humans are allergic to penicillin G.

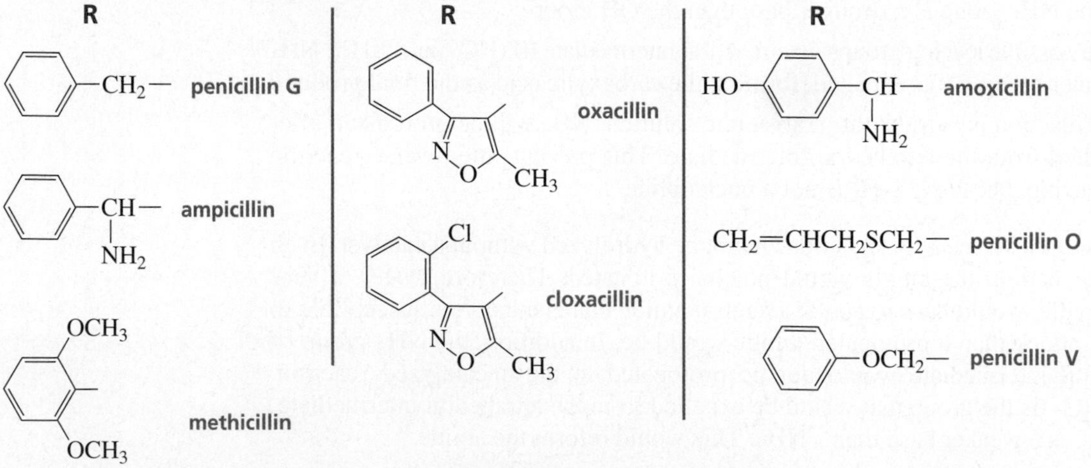

Penicillin V is a semisynthetic penicillin in clinical use. It is not a naturally occurring penicillin, but it is also not a true synthetic penicillin because chemists don't synthesize it. The *Penicillium* mold synthesizes it after being fed 2-phenoxyethanol, the compound needed for the side chain.

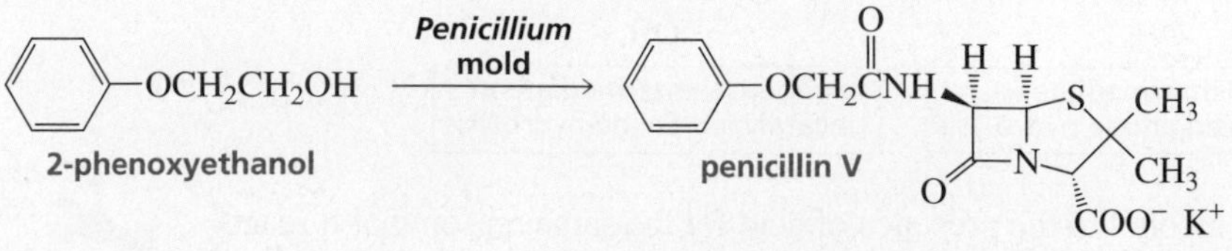

PROBLEM 35

Suggest reasons to explain why methicillin is effective in treating patients infected with bacteria resistant to penicillin G.

PROBLEM 36◆

List the following amides in order of decreasing reactivity toward acid-catalyzed hydrolysis:

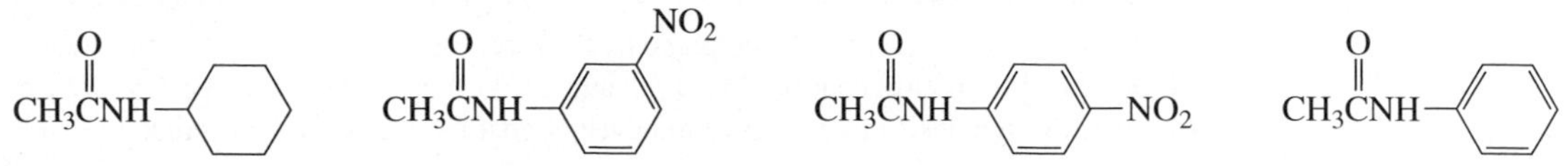

16.18 The Hydrolysis of an Imide: A Way to Synthesize Primary Amines

An **imide** is a compound with two acyl groups bonded to a nitrogen. The **Gabriel synthesis**, which converts alkyl halides into primary amines, involves the hydrolysis of an imide.

$$\text{RCH}_2\text{Br} \xrightarrow{\text{Gabriel synthesis}} \text{RCH}_2\text{NH}_2$$

alkyl halide — primary amine

The steps involved in the synthesis are shown below.

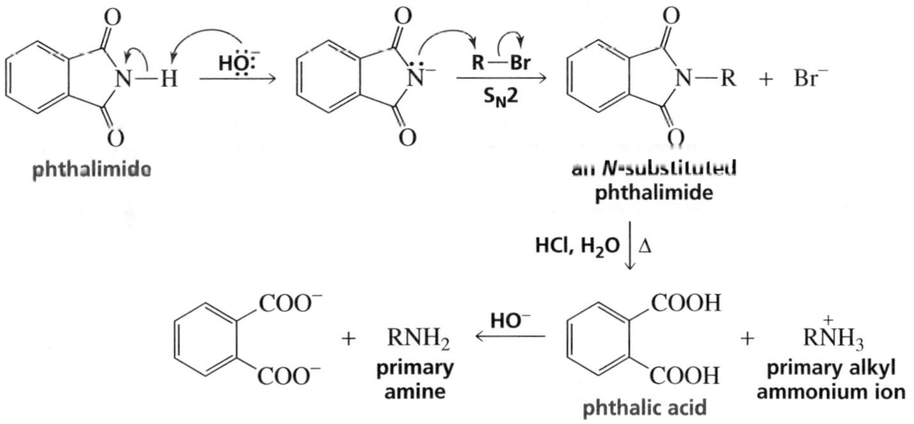

- A base removes a proton from the nitrogen of phthalimide.
- The resulting nucleophile reacts with an alkyl halide. Because this is an S_N2 reaction, it works best with primary alkyl halides (Section 8.2).
- Hydrolysis of the *N*-substituted phthalimide is catalyzed by acid. Because the solution is acidic, the final products are a primary alkyl ammonium ion and phthalic acid.
- Neutralization of the ammonium ion with base forms the primary amine.

Notice that the alkyl group of the primary amine is identical to the alkyl group of the alkyl halide.

Only one alkyl group can be placed on the nitrogen because there is only one hydrogen bonded to the nitrogen of phthalimide. This means that the Gabriel synthesis can be used only for the preparation of primary amines.

What alkyl bromide would you use in a Gabriel synthesis to prepare each of the following amines?

a. pentylamine   **b.** isohexylamine   **c.** benzylamine   **d.** cyclohexylamine

PROBLEM 38

Primary amines can also be prepared by the reaction of an alkyl halide with azide ion, followed by catalytic hydrogenation. What advantage do this method and the Gabriel synthesis have over the synthesis of a primary amine using an alkyl halide and ammonia?

$$CH_3CH_2CH_2Br \xrightarrow{^-N_3} CH_3CH_2CH_2N{=}\overset{+}{N}{=}\overset{-}{N} \xrightarrow[\text{Pt}]{H_2} CH_3CH_2CH_2NH_2 \ + \ N_2$$

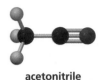

acetonitrile

16.19 **The Hydrolysis of Nitriles**

Nitriles are even harder to hydrolyze than amides, but they are slowly hydrolyzed to carboxylic acids when heated with water and an acid.

$$CH_3CH_2C{\equiv}N \xrightarrow[\Delta]{HCl,\ H_2O} \underset{CH_3CH_2}{\overset{\overset{\textstyle O}{\parallel}}{C}}{-}OH$$

The mechanism for the acid-catalyzed hydrolysis of a nitrile follows:

Mechanism for acid-catalyzed hydrolysis of a nitrile

nitrogen gains a proton

oxygen loses a proton

R—C—OH
‖
O:
a carboxylic acid

resonance contributors for a protonated amide

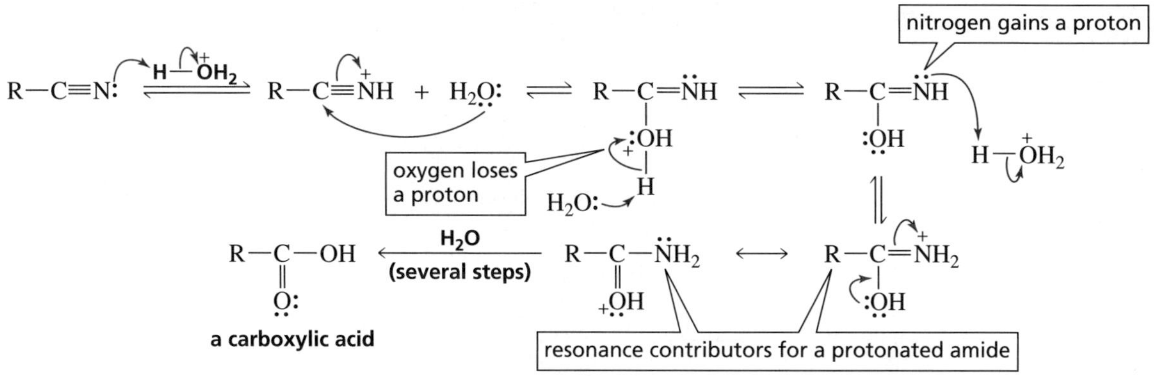

- The acid protonates the nitrogen of the cyano group (C≡N), making it easier for water to attack the carbon of the cyano group in the next step.
- Nucleophilic attack by water on the protonated cyano group is analogous to nucleophilic attack by water on a protonated carbonyl group.
- Because nitrogen is a stronger base than oxygen, oxygen loses a proton and nitrogen gains a proton, resulting in a protonated amide (whose two resonance contributors are shown).
- The amide is immediately hydrolyzed to a carboxylic acid—because an amide is easier to hydrolyze than a nitrile—by means of the acid-catalyzed mechanism shown in Section 16.17.

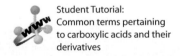

Student Tutorial:
Common terms pertaining
to carboxylic acids and their
derivatives

Nitriles can be prepared from an S$_N$2 reaction of alkyl halide with cyanide ion. Because a nitrile can be hydrolyzed to carboxylic acid, you now know how to convert an alkyl halide into a carboxylic acid. Notice that the carboxylic acid has one more carbon than the alkyl halide.

$$CH_3CH_2Br \xrightarrow[\text{DMF}]{\overset{\boxed{\text{an } S_N2 \text{ reaction}}}{^-C\equiv N}} CH_3CH_2C\equiv N \xrightarrow[\Delta]{\text{HCl, H}_2\text{O}} CH_3CH_2\overset{\overset{\displaystyle O}{\|}}{C}OH$$

A nitrile can be reduced to a primary amine by the same reagent that reduces an alkyne to an alkane (Section 6.9).

$$CH_3CH_2CH_2CH_2C\equiv N \xrightarrow[\text{Pt/C}]{\text{H}_2} CH_3CH_2CH_2CH_2CH_2NH_2$$

pentanenitrile   **pentylamine**

PROBLEM 39◆

Which alkyl halides form the carboxylic acids listed below after reaction with sodium cyanide followed by heating the product in an acidic aqueous solution?

a. butyric acid   **b.** isovaleric acid   **c.** cyclohexanecarboxylic acid

---

16.20 Designing a Synthesis V: The Synthesis of Cyclic Compounds

Most of the reactions we have been studying are intermolecular reactions: the two reacting groups are in different molecules. *Cyclic compounds are formed from intramolecular reactions*: the two reacting groups are in the same molecule. We have seen that intramolecular reactions are particularly favored if the reaction forms a compound with a five- or a six-membered ring (Section 8.11).

In designing the synthesis of a cyclic compound, we must examine the target molecule to determine what kinds of reactive groups are necessary for a successful synthesis. For example, we know that an ester is formed from the acid-catalyzed reaction of a carboxylic acid with an alcohol. Therefore, a cyclic ester (a lactone) can be prepared from a reactant that has a carboxylic acid group and an alcohol group in the same molecule. The size of the lactone ring will be determined by the number of carbon atoms between the carboxylic acid group and the alcohol group.

$$HOCH_2CH_2CH_2CH_2\overset{\overset{\displaystyle O}{\|}}{C}OH \xrightarrow{\text{HCl}}$$

$\underbrace{}$
$\boxed{\text{four intervening carbon atoms}}$

$$HOCH_2CH_2CH_2\overset{\overset{\displaystyle O}{\|}}{C}OH \xrightarrow{\text{HCl}}$$

$\underbrace{}$
$\boxed{\text{three intervening carbon atoms}}$

A compound with a ketone group attached to a benzene ring can be prepared using a Friedel–Crafts acylation reaction (Section 14.14). Therefore, a cyclic ketone will result if a Lewis acid (AlCl₃) is added to a compound that contains both a benzene ring and an acyl chloride group separated by the appropriate number of carbon atoms.

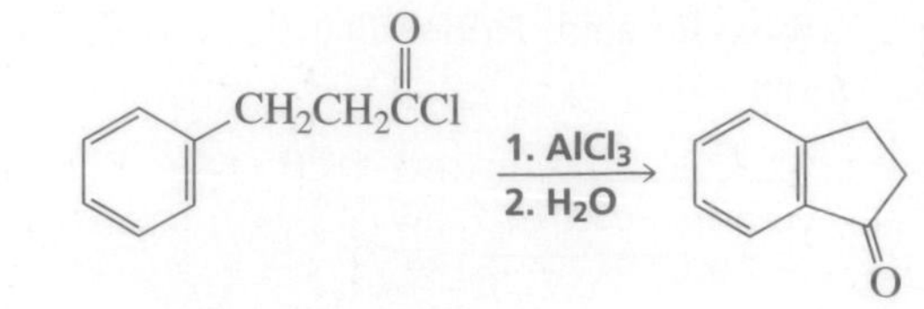

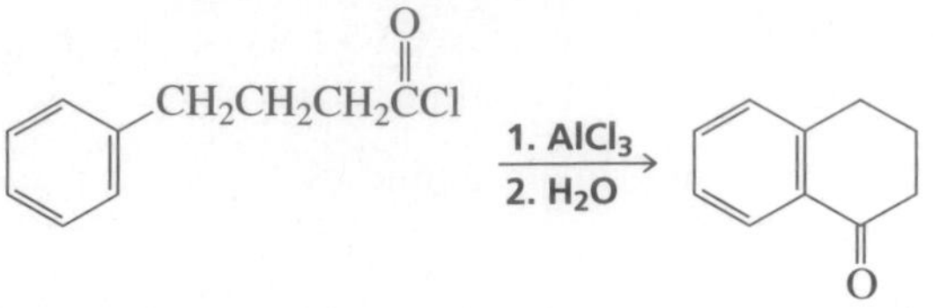

A cyclic ether can be prepared by an intramolecular Williamson ether synthesis (Section 9.9).

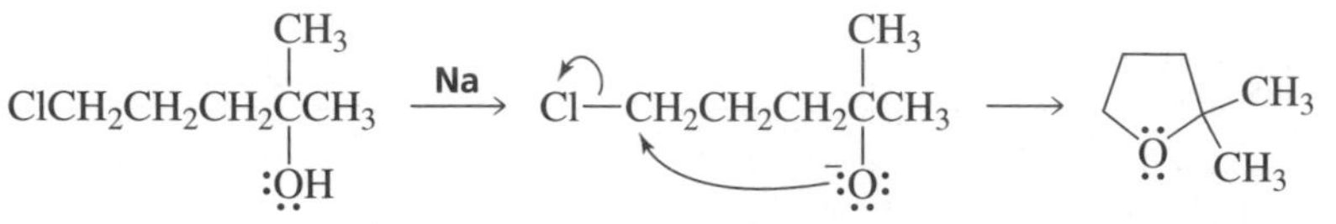

A cyclic ether can also be prepared by an intramolecular electrophilic addition reaction.

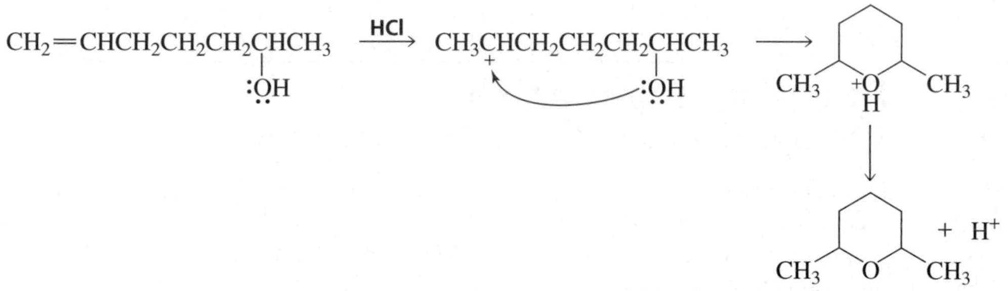

The product obtained from an intramolecular reaction can undergo further reactions allowing for the synthesis of many different compounds. For example, the alkyl bromide formed in the following reaction could undergo an elimination reaction, or could undergo substitution with a wide variety of nucleophiles, or could be converted into a Grignard reagent that could react with many different electrophiles (Section 10.12).

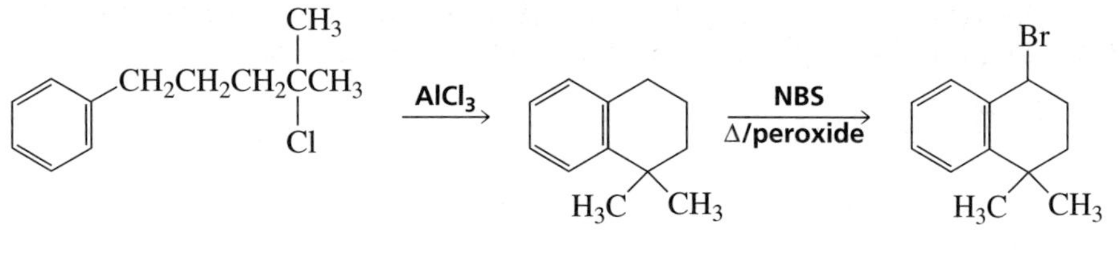

PROBLEM 40

Design a synthesis for each of the following compounds, using an intramolecular reaction:

a.

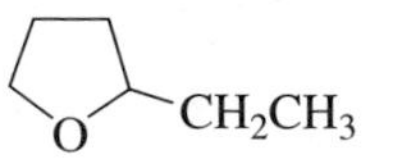

c.

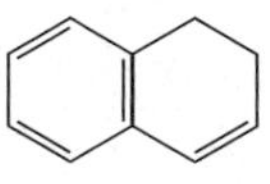

e.

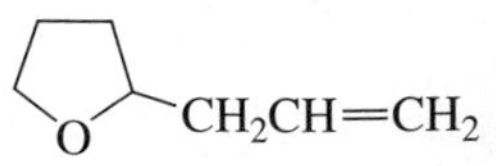

b.

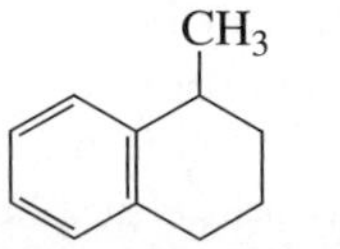

d.

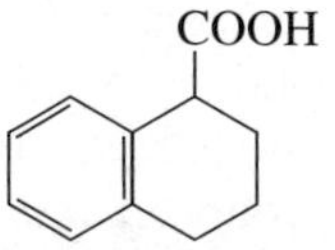

f.

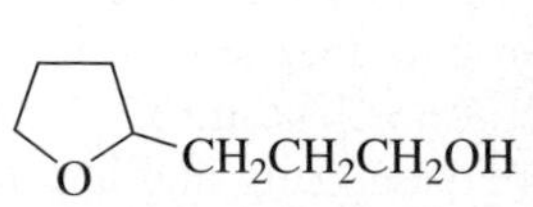

16.21 **How Chemists Activate Carboxylic Acids**

Of the various classes of carbonyl compounds discussed in this chapter—acyl halides, acid anhydrides, esters, carboxylic acids, and amides—carboxylic acids are the most commonly available both in the laboratory and in biological systems. This means that carboxylic acids are the reagents most likely to be available when a chemist or a cell needs to synthesize a carboxylic acid derivative. However, we have seen that carboxylic acids are relatively unreactive toward nucleophilic acyl substitution reactions because the OH group of a carboxylic acid is a strong base and therefore a poor leaving group. In neutral solutions (physiological pH = 7.3), a carboxylic acid is even more resistant to nucleophilic acyl substitution reactions because it exists predominantly in its unreactive, negatively charged basic form (Sections 1.24 and 16.15). Therefore, both organic chemists and cells need a way to activate carboxylic acids so that they can readily undergo nucleophilic acyl substitution reactions. First we will look at how chemists activate carboxylic acids, and then in Section 16.22 we will see how cells do it.

Because acyl halides are the most reactive of the carboxylic acid derivatives, the easiest way to synthesize any other carboxylic acid derivative is to add the appropriate nucleophile to an acyl halide. Consequently, organic chemists activate carboxylic acids by converting them into acyl halides.

A carboxylic acid can be converted into an acyl chloride by being heated either with thionyl chloride (SOCl$_2$) or with phosphorus trichloride (PCl$_3$). Acyl bromides can be synthesized by using phosphorus tribromide (PBr$_3$).

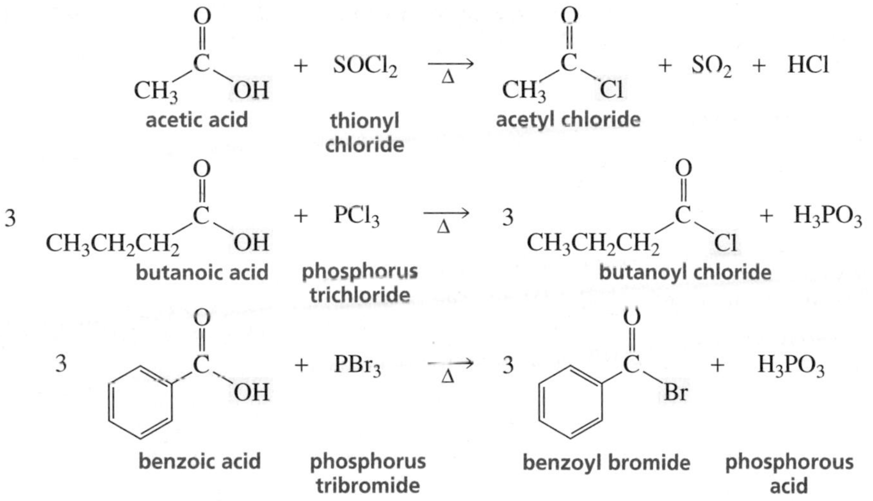

All these reagents convert the OH group of a carboxylic acid into a better leaving group than the halide ion.

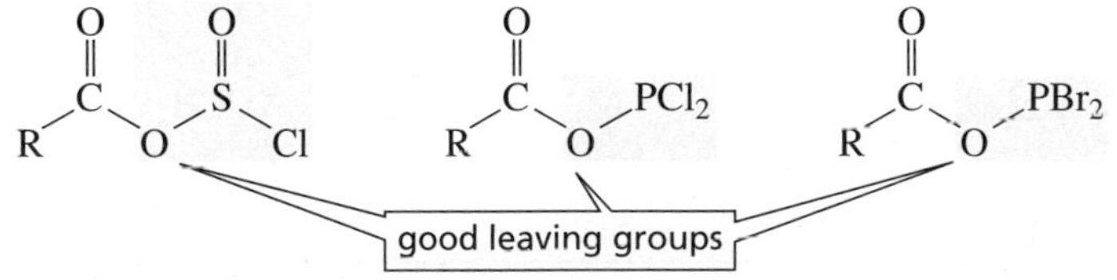

As a result, when the halide ion subsequently attacks the carbonyl carbon and forms a tetrahedral intermediate, the halide ion is *not* the group that is eliminated.

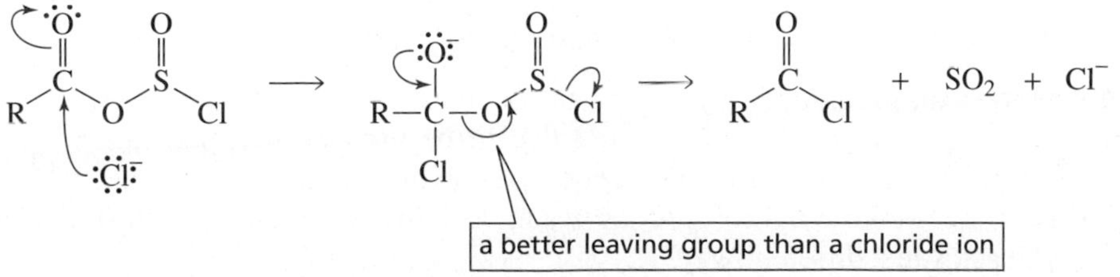

Notice that the reagents that cause the OH group of a carboxylic acid to be replaced by a halogen are the same reagents that cause the OH group of an alcohol to be replaced by a halogen (Section 10.2).

Once the acyl halide has been prepared, a wide variety of carboxylic acid derivatives can be synthesized by adding the appropriate nucleophile (Section 16.8).

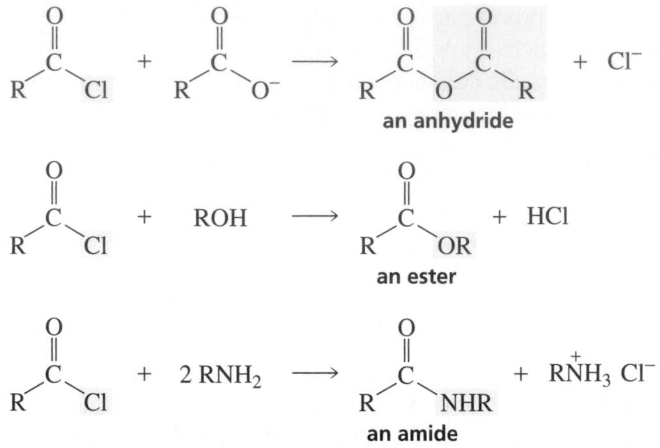

Carboxylic acids can also be activated for nucleophilic acyl substitution reactions by being converted into anhydrides. Treating a carboxylic acid with a strong dehydrating agent such as P_2O_5 yields an anhydride.

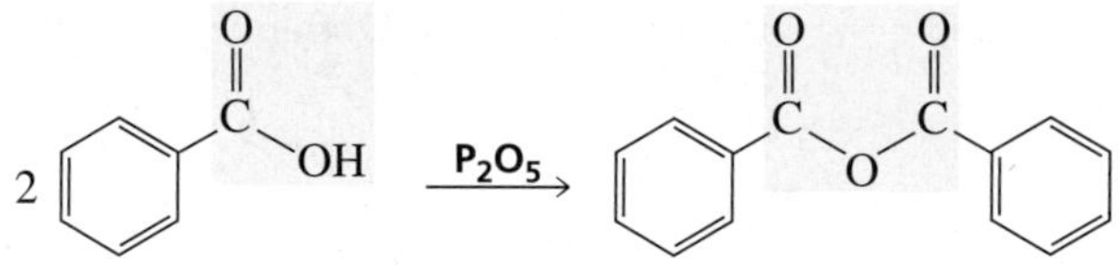

Carboxylic acids and carboxylic acid derivatives can also be prepared by methods other than nucleophilic acyl substitution reactions. A summary of the methods used to synthesize these compounds is provided in Appendix IV.

PROBLEM 41◆

How would you synthesize the following compounds starting with a carboxylic acid?

a. 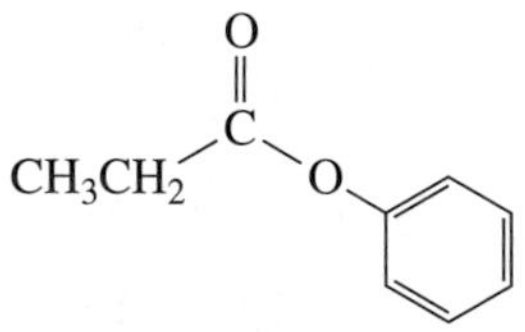 CH_3CH_2 ... O ...

b. 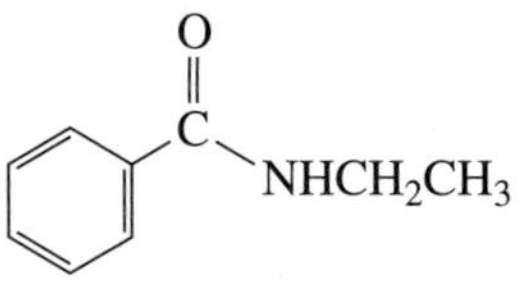 ... NHCH_2CH_3

16.22 How Cells Activate Carboxylic Acids

The synthesis of compounds by biological organisms is called **biosynthesis**. Acyl halides and acid anhydrides are too reactive to be used as reagents in biological systems. Cells live in a predominantly aqueous environment, and acyl halides and acid anhydrides are rapidly hydrolyzed in water. So living organisms must activate carboxylic acids in a different way.

One way living organisms activate carboxylic acids is to convert them into acyl phosphates, acyl pyrophosphates, and acyl adenylates.

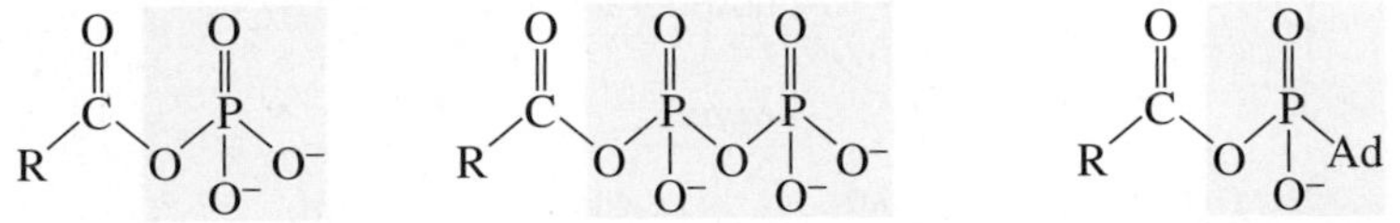

an acyl phosphate   **an acyl pyrophosphate**   **an acyl adenylate**

An **acyl phosphate** is a mixed anhydride of a carboxylic acid and phosphoric acid; and an **acyl pyrophosphate** is a mixed anhydride of a carboxylic acid and pyrophosphoric acid.

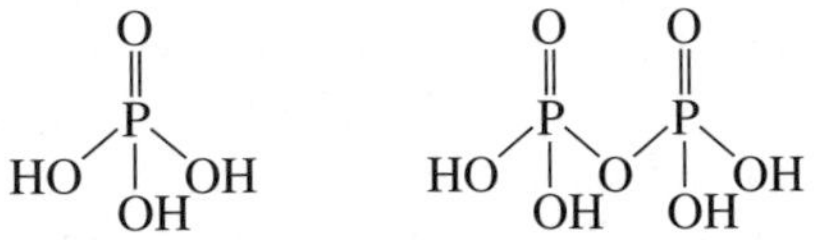

phosphoric acid   **pyrophosphoric acid**

An **acyl adenylate** is a mixed anhydride of a carboxylic acid and adenosine monophosphate (AMP). The structure of adenosine triphosphate (ATP) is shown below, both in its entirety and with "Ad" in place of the adenosyl group; adenosine monophosphate has two fewer phosphate groups than ATP.

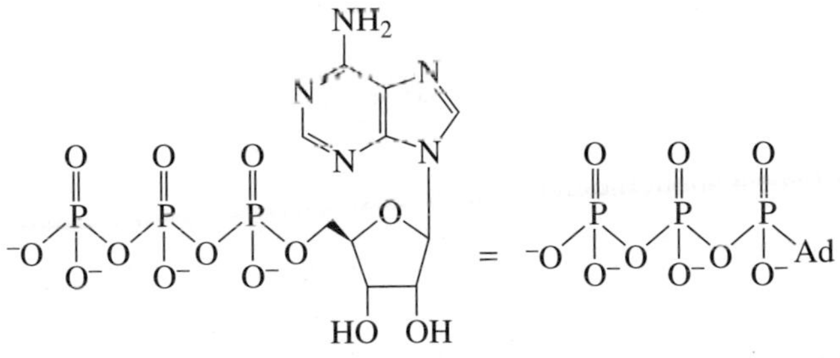

adenosine triphosphate
ATP

Acyl phosphates are formed by nucleophilic attack of a carboxylate ion on the γ-phosphorus (the terminal phosphorus) of ATP. Attack of a nucleophile on the P=O group breaks a **phosphoanhydride bond** (rather than the π bond), so an intermediate is not formed. Essentially, it is an S_N2 reaction with an adenosine pyrophosphate leaving group. This reaction and the ones that follow will be discussed in greater detail in Sections 25.2 and 25.3.

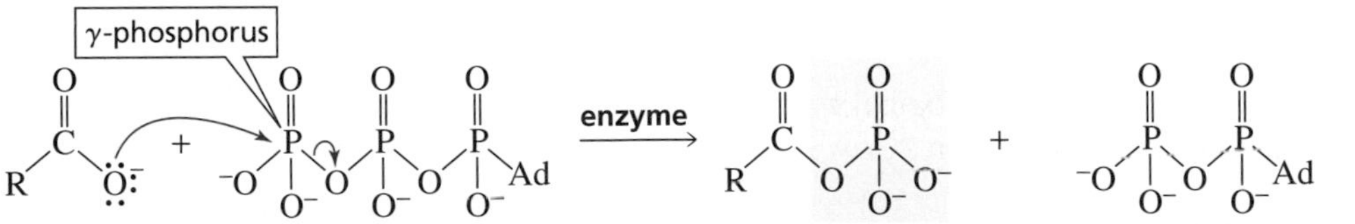

adenosine triphosphate   **an acyl phosphate**   **adenosine diphosphate**
ATP   **ADP**

Acyl pyrophosphates are formed by nucleophilic attack of a carboxylate ion on the β-phosphorus of ATP.

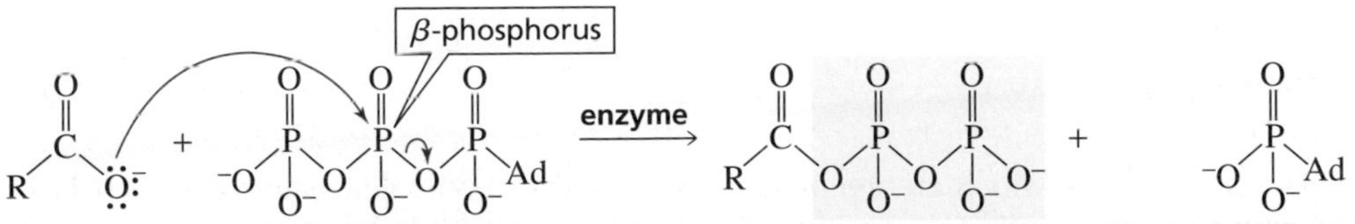

adenosine triphosphate   **an acyl pyrophosphate**   **adenosine monophosphate**
ATP   **AMP**

Acyl adenylates are formed by nucleophilic attack of a carboxylate ion on the α-phosphorus of ATP.

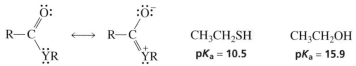

Which phosphorus atom the nucleophile attacks depends on the enzyme that catalyzes the reaction.

Because these mixed anhydrides are negatively charged, they are not readily approached by nucleophiles. Thus, they are used only in enzyme-catalyzed reactions. One of the functions of enzymes that catalyze biological nucleophilic acyl substitution reactions is to neutralize the negative charges of the mixed anhydride (Section 25.5). Another function of the enzyme is to exclude water from the site where the reaction takes place. Otherwise hydrolysis of the mixed anhydride would compete with the desired nucleophilic acyl substitution reaction.

A **thioester** is an ester with a sulfur atom in place of the oxygen atom between the acyl and alkyl groups.

$$\underset{\textbf{a thioester}}{\overset{\displaystyle \overset{O}{\underset{\|}{}}}{R-\overset{}{C}-SR'}}$$

Thioesters are the most common forms of activated carboxylic acids in a cell. Although thioesters hydrolyze at about the same rate as oxygen esters, they are much more reactive than oxygen esters toward attack by nitrogen and carbon nucleophiles. This allows a thioester to survive in the aqueous environment of the cell—without being hydrolyzed—while waiting to be used as a substrate in a nucleophilic acyl substitution reaction.

The carbonyl carbon of a thioester is more susceptible to nucleophilic attack than is the carbonyl carbon of an oxygen ester because the electron delocalization onto the carbonyl oxygen is weaker when Y is S than when Y is O. The electron delocalization is weaker because less overlap occurs between the $3p$ orbital of sulfur and the $2p$ orbital of carbon, compared with the amount of overlap between the $2p$ orbital of oxygen and the $2p$ orbital of carbon (Section 16.2). In addition, a thiolate ion is a weaker base and therefore a better leaving group than an alkoxide ion.

$$R-\overset{\overset{\displaystyle \ddot{O}:}{|}}{\underset{\overset{|}{\ddot{Y}R}}{C}} \longleftrightarrow R-\overset{\overset{\displaystyle :\ddot{O}:^{-}}{|}}{\underset{\overset{|}{\overset{+}{\ddot{Y}R}}}{C}} \qquad \underset{pK_a = 10.5}{CH_3CH_2SH} \qquad \underset{pK_a = 15.9}{CH_3CH_2OH}$$

The thiol used in biological systems for the formation of thioesters is coenzyme A. The compound is written "CoASH" to emphasize that the thiol group is the reactive part of the molecule.

coenzyme A
CoASH

The first step by which an organism converts a carboxylic acid into a thioester is to convert the carboxylic acid into an acyl adenylate. The acyl adenylate then reacts with CoASH to form the thioester. The most common thioester in cells is acetyl-CoA.

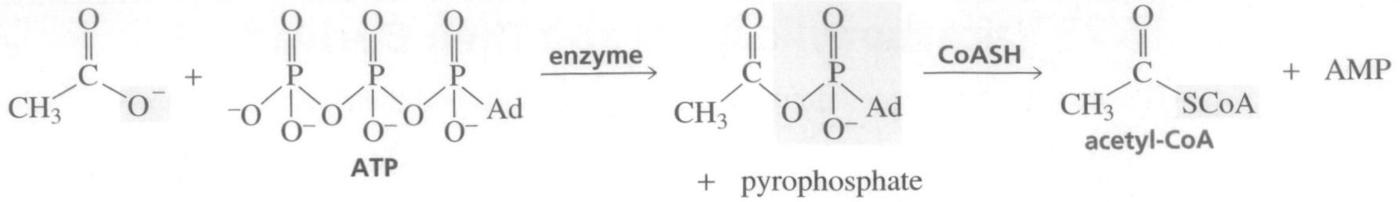

Acetylcholine, an ester, is one example of a carboxylic acid derivative that cells synthesize using acetyl-CoA. Acetylcholine is a *neurotransmitter*—a compound that transmits nerve impulses across the synapses (spaces) between nerve cells.

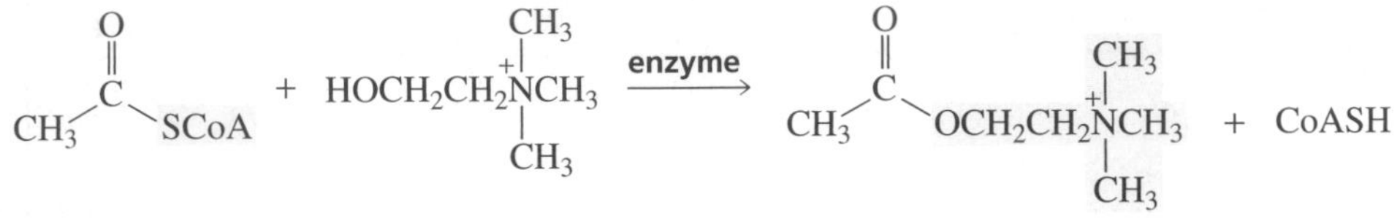

It is thought that one way genes are activated is by formation of an amide between acetyl-CoA and a lysine residue (an amine) of a DNA-bound protein.

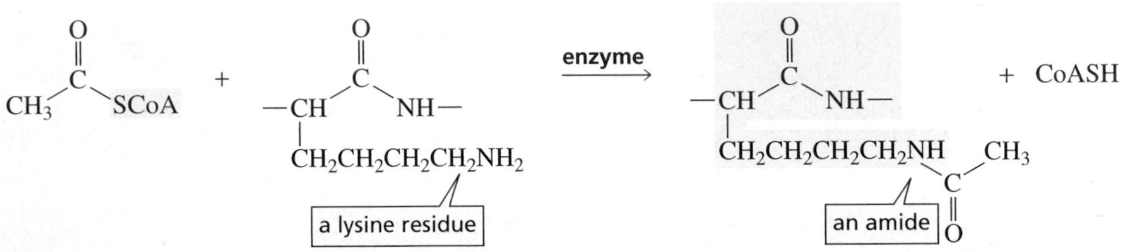

a lysine residue

an amide

NERVE IMPULSES, PARALYSIS, AND INSECTICIDES

After an impulse is transmitted between two nerve cells, acetylcholine must be hydrolyzed immediately to enable the recipient cell to receive another impulse.

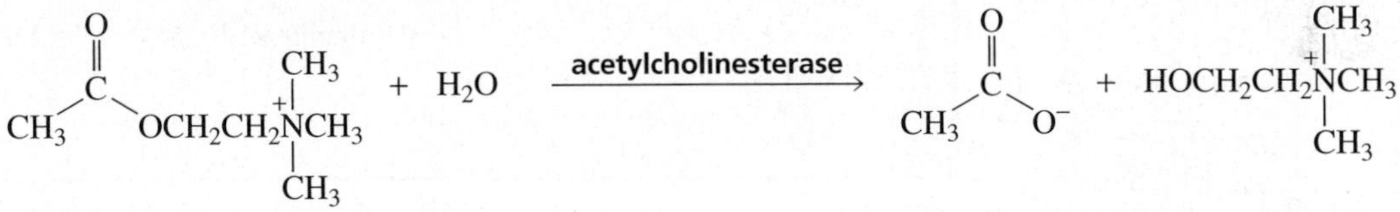

Acetylcholinesterase, the enzyme that catalyzes this hydrolysis, has a CH_2OH group that is necessary for its catalytic activity. Diisopropyl fluorophosphate (DFP), a military nerve gas used during World War II, irreversibly inhibits acetylcholinesterase by reacting with the CH_2OH group. When the enzyme is inhibited, the nerve impulses cannot be transmitted properly, and paralysis occurs. DFP is extremely toxic. Its LD_{50} (the lethal dose for 50% of the test animals) is only 0.5 mg/kg of body weight.

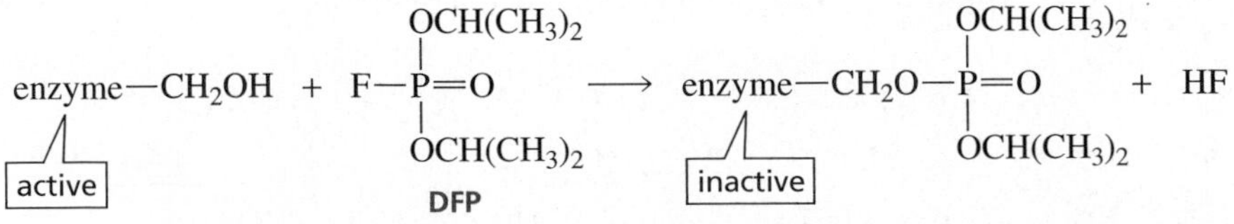

enzyme—CH_2OH + F—P=O → enzyme—CH_2O—P=O + HF

active

DFP

inactive

Malathion and parathion, widely used as insecticides, are compounds related to DFP. The LD_{50} of malathion is 2800 mg/kg. Parathion is more toxic, with an LD_{50} of 2 mg/kg.

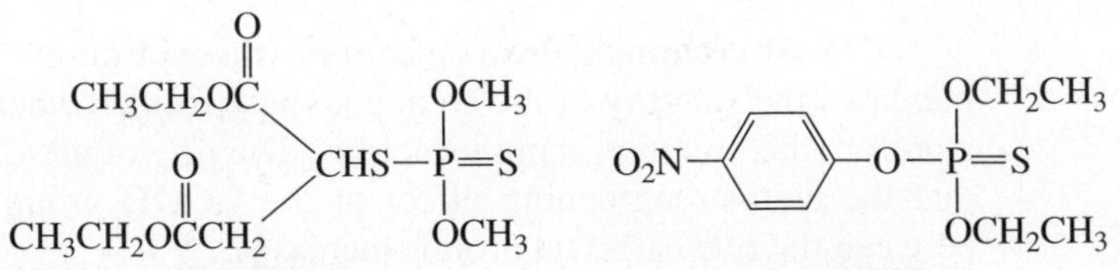

malathion

parathion

16.23 **Dicarboxylic Acids and Their Derivatives**

The structures of some common dicarboxylic acids and their pK_a values are shown in Table 16.2.

Table 16.2	Structures, Names, and pK_a Values of Some Simple Dicarboxylic Acids		
Dicarboxylic acid	**Common name**	**pK_{a1}**	**pK_{a2}**
$\underset{\displaystyle HOCOH}{\overset{\displaystyle O}{\|\|}}$	Carbonic acid	6.37	10.25
$\overset{\displaystyle O \quad O}{HOC-COH}$	Oxalic acid	1.27	4.27
$\overset{\displaystyle O \quad O}{HOCCH_2COH}$	Malonic acid	2.86	5.70
$\overset{\displaystyle O \qquad O}{HOCCH_2CH_2COH}$	Succinic acid	4.21	5.64
$\overset{\displaystyle O \qquad O}{HOCCH_2CH_2CH_2COH}$	Glutaric acid	4.34	5.27
$\overset{\displaystyle O \qquad O}{HOCCH_2CH_2CH_2CH_2COH}$	Adipic acid	4.41	5.28
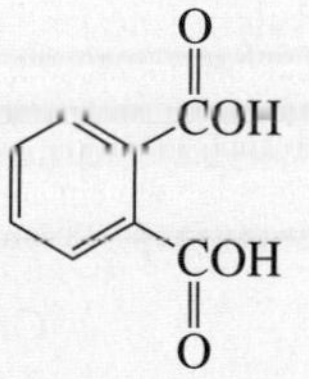 (benzene ring with two COOH groups ortho)	Phthalic acid	2.95	5.41

Although the two carboxyl groups of a dicarboxylic acid are identical, the two pK_a values are different because the protons are lost one at a time and therefore leave from different species. The first proton is lost from a neutral molecule, whereas the second proton is lost from a negatively charged ion.

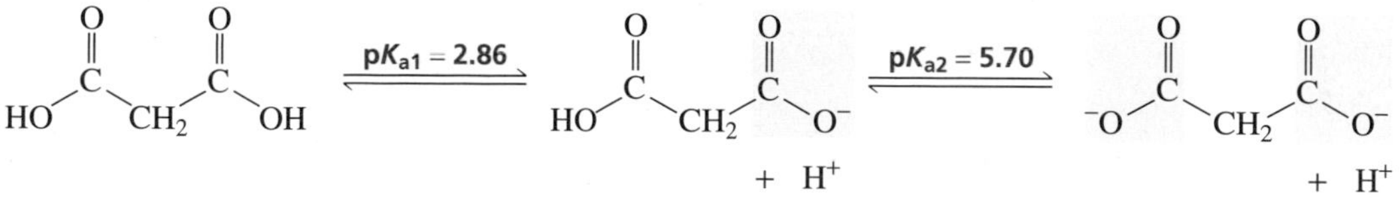

A COOH group withdraws electrons (more strongly than does an H) and therefore increases the stability of the conjugate base formed when the first COOH group loses a proton, thereby increasing its acidity. The pK_a values of the dicarboxylic acids show that the acid-strengthening effect of the COOH group decreases as the separation between the two carboxyl groups increases.

Dicarboxylic acids readily lose water when heated if they can form a cyclic anhydride with a five- or a six-membered ring.

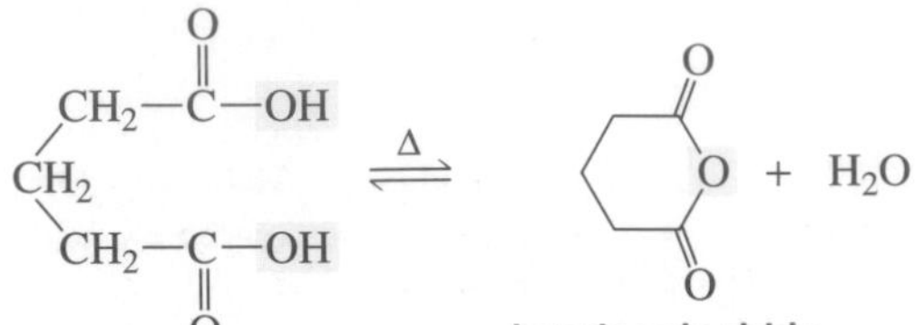

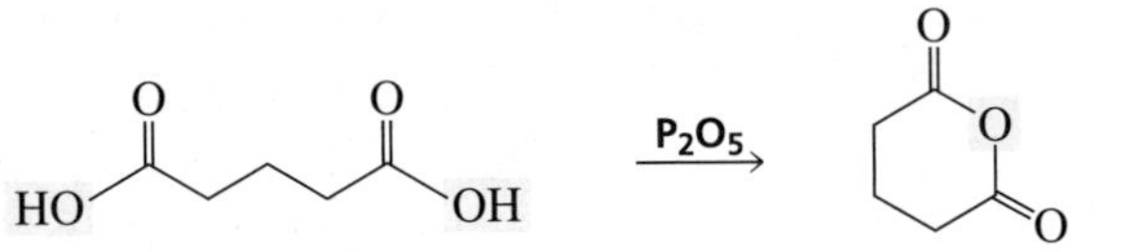

glutaric acid $\rightleftharpoons$ glutaric anhydride $+ H_2O$

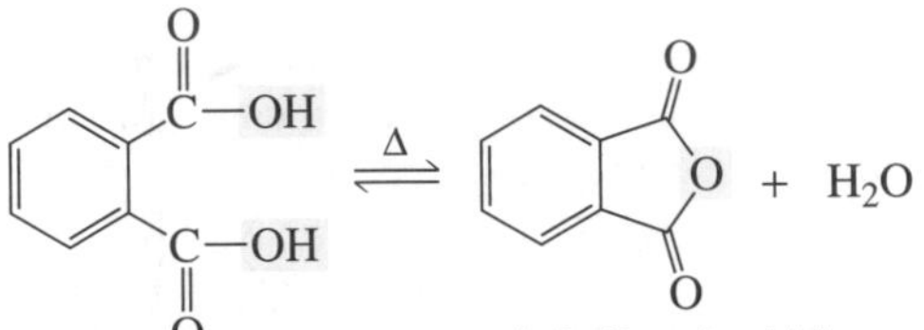

phthalic acid $\rightleftharpoons$ phthalic anhydride $+ H_2O$

However, cyclic anhydrides are more easily prepared if the dicarboxylic acid is heated in the presence of acetyl chloride or acetic anhydride or if it is treated with a strong dehydrating agent such as P_2O_5.

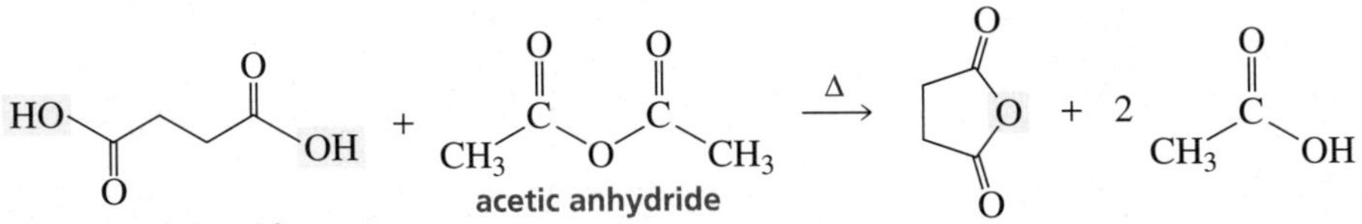

succinic acid + acetic anhydride $\xrightarrow{\Delta}$ succinic anhydride $+ 2\ CH_3$—C—OH

glutaric acid $\xrightarrow{P_2O_5}$ glutaric anhydride

PROBLEM 42

a. Propose a mechanism for the formation of succinic anhydride in the presence of acetic anhydride.

b. How does acetic anhydride help in the formation of succinic anhydride?

Carbonic acid—a compound with two OH groups bonded to the carbonyl carbon—is unstable, readily breaking down to CO_2 and H_2O. The reaction is reversible, so carbonic acid is formed when CO_2 is bubbled into water (Section 1.17).

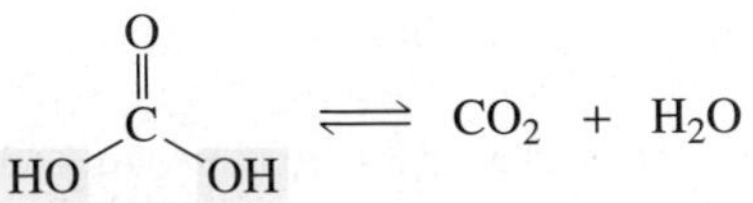

carbonic acid $\rightleftharpoons CO_2 + H_2O$

We have seen that the OH group of a carboxylic acid can be substituted to give a variety of carboxylic acid derivatives. Similarly, the OH groups of carbonic acid can be substituted by other groups.

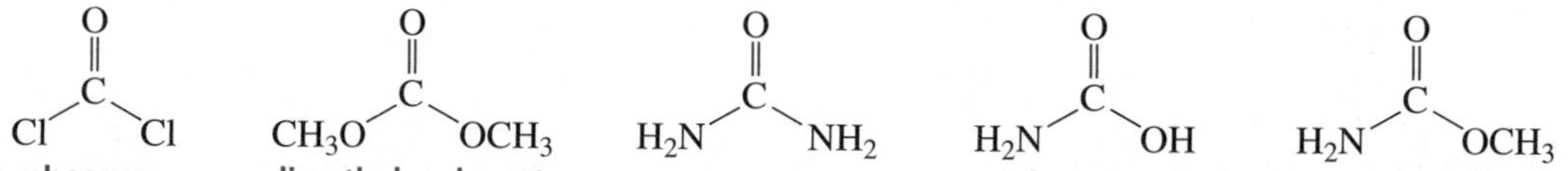

phosgene dimethyl carbonate urea carbamic acid methyl carbamate

SYNTHETIC POLYMERS

Synthetic polymers play important roles in our daily lives. Polymers are compounds that are made by linking together many small molecules called monomers. In many synthetic polymers, the monomers are held together by ester and amide bonds. For example, Dacron is a polyester and nylon is a polyamide.

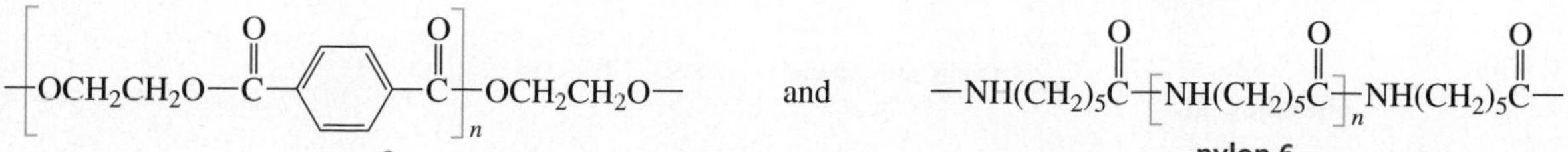

Dacron®   and   **nylon 6**

Synthetic polymers have taken the place of metals, fabrics, glass, wood, and paper, allowing us to have a greater variety and larger quantities of materials than nature could have provided. New polymers are continually being designed to fit human needs. For example, Kevlar has a tensile strength greater than steel. It is used for high-performance skis and bulletproof vests. Lexan is a strong and transparent polymer used for such things as traffic light lenses and compact disks.

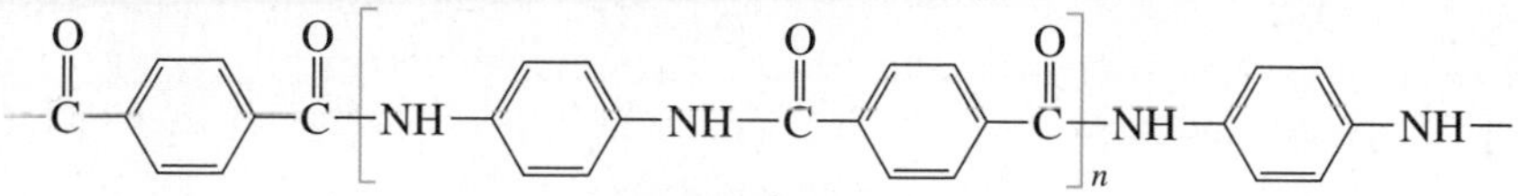

Kevlar®

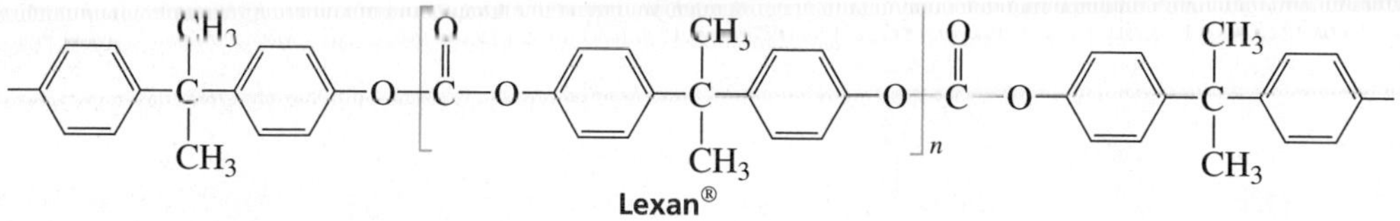

Lexan®

These and other synthetic polymers are discussed in detail in Chapter 28.

DISSOLVING SUTURES

Dissolving sutures, such as dexon and poly(dioxanone) (PDS), are synthetic polymers that are now routinely used in surgery. The many ester groups they contain are slowly hydrolyzed to small molecules that are then metabolized to compounds easily excreted by the body. Patients no longer have to undergo the second medical procedure that was required to remove the sutures when traditional suture materials were used.

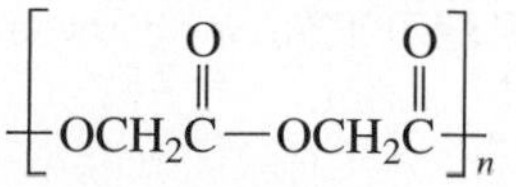

Dexon 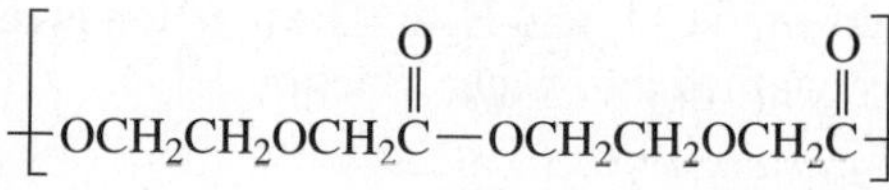   **PDS**

Depending on their structures, these synthetic sutures lose 50% of their strength after two to three weeks and are completely absorbed within 3 to 6 months.

PROBLEM 43◆

One of the two polymers for dissolving sutures shown in the box on page 774 loses 50% of its strength in two weeks, and the other loses that much strength in three weeks. Which is the suture material that lasts longer?

PROBLEM 44

What products would you expect to obtain from the following reactions?

a. phosgene + excess diethylamine

b. malonic acid + 2 acetyl chloride

c. methyl carbamate + methylamine

d. urea + water

e. urea + water + H$^+$

f. β-ethylglutaric acid + acetyl chloride + Δ

SUMMARY

A **carbonyl group** is a carbon double bonded to an oxygen; an **acyl group** is a carbonyl group attached to an alkyl or aryl group. **Acyl halides**, **acid anhydrides**, **esters**, and **amides** are called **carboxylic acid derivatives** because they differ from a carboxylic acid only in the nature of the group that has replaced the OH group of the carboxylic acid. Cyclic esters are called **lactones**; cyclic amides are **lactams**. There are **symmetrical anhydrides** and **mixed anhydrides**.

Carbonyl compounds can be placed in one of two classes. Class I carbonyl compounds contain a group that can be replaced by another group; carboxylic acids and carboxylic acid derivatives belong to this class. Class II carbonyl compounds do not contain a group that can be replaced by another group; aldehydes and ketones belong to this class.

The reactivity of carbonyl compounds resides in the polarity of the carbonyl group; the carbonyl carbon has a partial positive charge that is attractive to nucleophiles. Class I carbonyl compounds undergo **nucleophilic acyl substitution reactions**, in which a nucleophile replaces the substituent that was attached to the acyl group in the reactant. All class I carbonyl compounds react with nucleophiles in the same way: the nucleophile attacks the carbonyl carbon, forming an unstable **tetrahedral intermediate**, which reforms the carbonyl compound by eliminating the weakest base. (Generally, a compound with an sp^3 carbon bonded to an oxygen is unstable if the sp^3 carbon is bonded to another electronegative atom.)

A carboxylic acid derivative will undergo a nucleophilic acyl substitution reaction provided that the newly added group in the tetrahedral intermediate is not a much weaker base than the group that was attached to the acyl group in the reactant. The weaker the base attached to the acyl group, the more easily both steps of the nucleophilic acyl substitution reaction can take place. The relative reactivities toward nucleophilic acyl substitution are acyl halides > acid anhydrides > esters and carboxylic acids > amides > carboxylate ions.

Hydrolysis, **alcoholysis**, and **aminolysis** are reactions in which water, alcohols, and amines, respectively, convert one compound into two compounds. A **transesterification reaction** converts one ester to another ester. Treating a carboxylic acid with excess alcohol and an acid catalyst is called a **Fischer esterification**. An ester with a tertiary alkyl group hydrolyzes by an S_N1 reaction.

The rate of hydrolysis can be increased by either acid or HO$^-$; the rate of alcoholysis can be increased by either acid or RO$^-$. An acid increases the rate of formation of the tetrahedral intermediate by protonating the carbonyl oxygen, which increases the electrophilicity of the carbonyl carbon, and by decreasing the basicity of the leaving group, which makes it easier to eliminate. Hydroxide (or alkoxide) ion increases the rate of formation of the tetrahedral intermediate—it is a better nucleophile than water (or an alcohol)—and increases the rate of collapse of the tetrahedral intermediate. Hydroxide ion promotes only hydrolysis reactions; alkoxide ion promotes only alcoholysis reactions. In an acid-catalyzed reaction, all organic reactants, intermediates, and products are positively charged or neutral; in hydroxide-ion- or alkoxide-ion-promoted reactions, all organic reactants, intermediates, and products are negatively charged or neutral.

Fats and **oils** are triesters of glycerol. Hydrolyzing the ester groups in a basic solution forms glycerol and fatty acid salts (soaps). In water, long-chain carboxylate ions arrange themselves in spherical clusters called **micelles**. The attractive forces of hydrocarbon chains for each other in water are called **hydrophobic interactions**.

Amides are unreactive compounds but do react with water and alcohols if the reaction mixture is heated in the presence of an acid. Nitriles are harder to hydrolyze than amides. The **Gabriel synthesis**, which converts an alkyl halide into a primary amine, involves the hydrolysis of an **imide**.

Organic chemists activate carboxylic acids by converting them into acyl halides or acid anhydrides. Cells activate carboxylic acids by converting them into **acyl phosphates**, **acyl pyrophosphates**, **acyl adenylates**, and **thioesters**.

SUMMARY OF REACTIONS

1. Reactions of acyl halides (Section 16.8)

$$
\underset{\substack{\text{R}}}{\text{C}}\overset{\text{O}}{\text{Cl}} + \underset{\substack{\text{CH}_3}}{\text{C}}\overset{\text{O}}{\text{O}^-} \longrightarrow \underset{\substack{\text{R}}}{\text{C}}\overset{\text{O}}{\text{O}}\underset{\substack{\text{CH}_3}}{\text{C}}\overset{\text{O}}{} + \text{Cl}^-
$$

$$
\underset{\text{R}}{\text{C}}\overset{\text{O}}{\text{Cl}} + \text{CH}_3\text{OH} \longrightarrow \underset{\text{R}}{\text{C}}\overset{\text{O}}{\text{OCH}_3} + \text{HCl}
$$

$$
\underset{\text{R}}{\text{C}}\overset{\text{O}}{\text{Cl}} + \text{H}_2\text{O} \longrightarrow \underset{\text{R}}{\text{C}}\overset{\text{O}}{\text{OH}} + \text{HCl}
$$

$$
\underset{\text{R}}{\text{C}}\overset{\text{O}}{\text{Cl}} + 2\,\text{CH}_3\text{NH}_2 \longrightarrow \underset{\text{R}}{\text{C}}\overset{\text{O}}{\text{NHCH}_3} + \text{CH}_3\overset{+}{\text{NH}_3}\,\text{Cl}^-
$$

2. Reactions of acid anhydrides (Section 16.9)

$$
\underset{\text{R}}{\text{C}}\overset{\text{O}}{}\text{O}\underset{\text{R}}{\text{C}}\overset{\text{O}}{} + \text{CH}_3\text{OH} \longrightarrow \underset{\text{R}}{\text{C}}\overset{\text{O}}{\text{OCH}_3} + \text{RCOH}\overset{\text{O}}{}
$$

$$
\underset{\text{R}}{\text{C}}\overset{\text{O}}{}\text{O}\underset{\text{R}}{\text{C}}\overset{\text{O}}{} + \text{H}_2\text{O} \longrightarrow 2\,\underset{\text{R}}{\text{C}}\overset{\text{O}}{\text{OH}}
$$

$$
\underset{\text{R}}{\text{C}}\overset{\text{O}}{}\text{O}\underset{\text{R}}{\text{C}}\overset{\text{O}}{} + 2\,\text{CH}_3\text{NH}_2 \longrightarrow \underset{\text{R}}{\text{C}}\overset{\text{O}}{\text{NHCH}_3} + \text{RCO}^-\,\text{H}_3\overset{+}{\text{NCH}_3}
$$

3. Reactions of esters (Sections 16.10–16.13)

$$
\underset{\text{R}}{\text{C}}\overset{\text{O}}{\text{OR}} + \text{CH}_3\text{OH} \underset{\text{HCl}}{\rightleftharpoons} \underset{\text{R}}{\text{C}}\overset{\text{O}}{\text{OCH}_3} + \text{ROH}
$$

$$
\underset{\text{R}}{\text{C}}\overset{\text{O}}{\text{OR}} + \text{H}_2\text{O} \underset{\text{HCl}}{\rightleftharpoons} \underset{\text{R}}{\text{C}}\overset{\text{O}}{\text{OH}} + \text{ROH}
$$

$$
\underset{\text{R}}{\text{C}}\overset{\text{O}}{\text{OR}} + \text{H}_2\text{O} \overset{\text{HO}^-}{\longrightarrow} \underset{\text{R}}{\text{C}}\overset{\text{O}}{\text{O}^-} + \text{ROH}
$$

$$
\underset{\text{R}}{\text{C}}\overset{\text{O}}{\text{OR}} + \text{CH}_3\text{NH}_2 \longrightarrow \underset{\text{R}}{\text{C}}\overset{\text{O}}{\text{NHCH}_3} + \text{ROH}
$$

4. Reactions of carboxylic acids (Section 16.15)

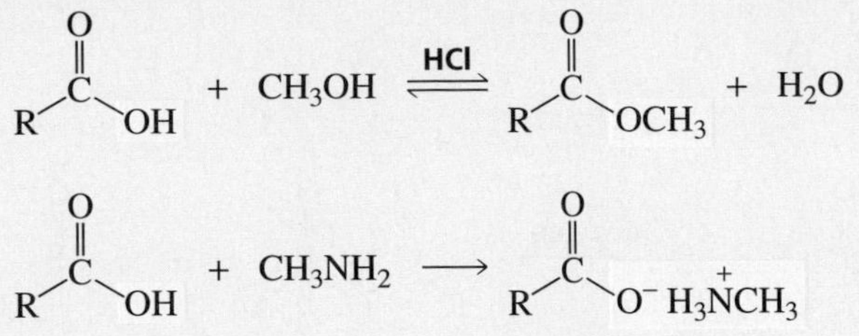

$$\text{RCOOH} + \text{CH}_3\text{OH} \underset{}{\overset{\text{HCl}}{\rightleftharpoons}} \text{RCOOCH}_3 + \text{H}_2\text{O}$$

$$\text{RCOOH} + \text{CH}_3\text{NH}_2 \longrightarrow \text{RCOO}^- \; \text{H}_3\overset{+}{\text{N}}\text{CH}_3$$

5. Reactions of amides (Sections 16.16 and 16.17)

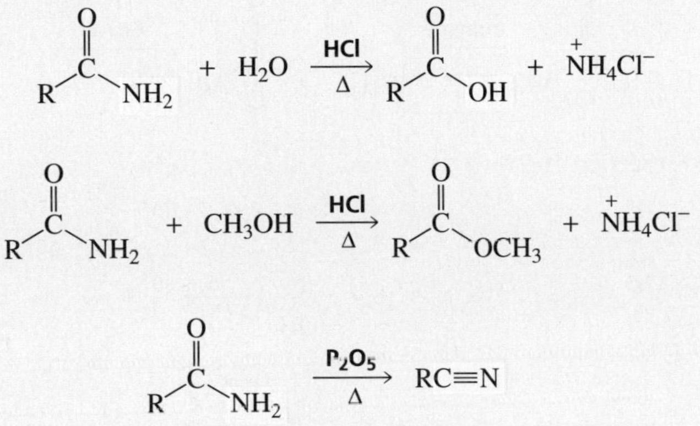

$$\text{RCONH}_2 + \text{H}_2\text{O} \xrightarrow[\Delta]{\text{HCl}} \text{RCOOH} + \overset{+}{\text{N}}\text{H}_4\text{Cl}^-$$

$$\text{RCONH}_2 + \text{CH}_3\text{OH} \xrightarrow[\Delta]{\text{HCl}} \text{RCOOCH}_3 + \overset{+}{\text{N}}\text{H}_4\text{Cl}^-$$

$$\text{RCONH}_2 \xrightarrow[\Delta]{\text{P}_2\text{O}_5} \text{RC}\!\equiv\!\text{N}$$

6. Gabriel synthesis of primary amines (Section 16.18)

$$\text{phthalimide} \xrightarrow[\begin{array}{l}\textbf{1. HO}^-\\ \textbf{2. RCH}_2\textbf{Br}\\ \textbf{3. HCl, H}_2\textbf{O, }\Delta\\ \textbf{4. HO}^-\end{array}]{} \text{RCH}_2\text{NH}_2$$

7. Hydrolysis of nitriles (Section 16.19)

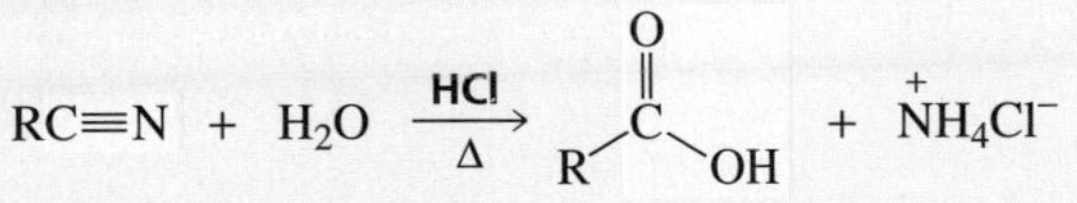

$$\text{RC}\!\equiv\!\text{N} + \text{H}_2\text{O} \xrightarrow[\Delta]{\text{HCl}} \text{RCOOH} + \overset{+}{\text{N}}\text{H}_4\text{Cl}^-$$

8. Activation of carboxylic acids by chemists (Section 16.21)

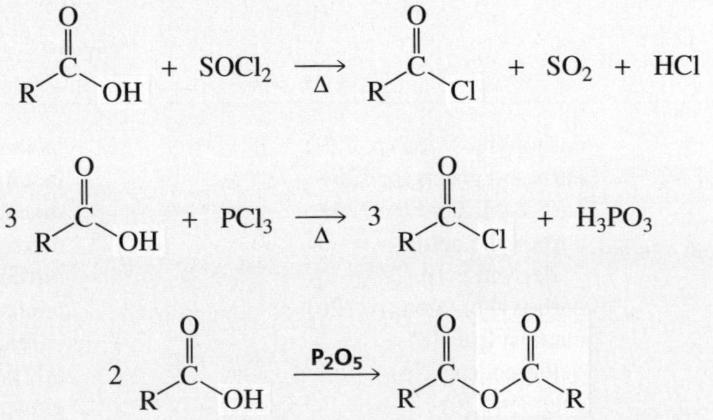

$$\text{RCOOH} + \text{SOCl}_2 \xrightarrow{\Delta} \text{RCOCl} + \text{SO}_2 + \text{HCl}$$

$$3\,\text{RCOOH} + \text{PCl}_3 \xrightarrow{\Delta} 3\,\text{RCOCl} + \text{H}_3\text{PO}_3$$

$$2\,\text{RCOOH} \xrightarrow{\text{P}_2\text{O}_5} \text{RCO-O-COR}$$

9. Activation of carboxylic acids by cells (Section 16.22)

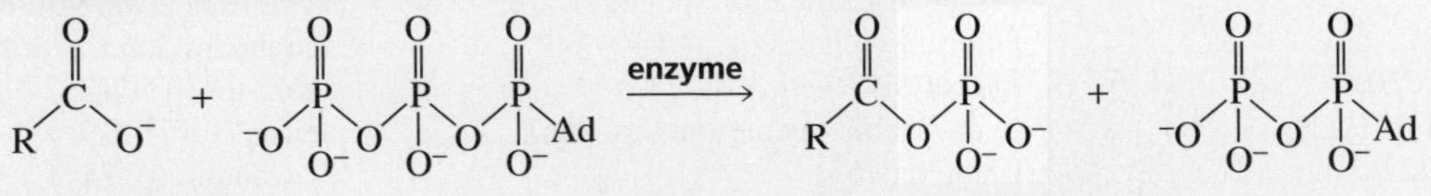

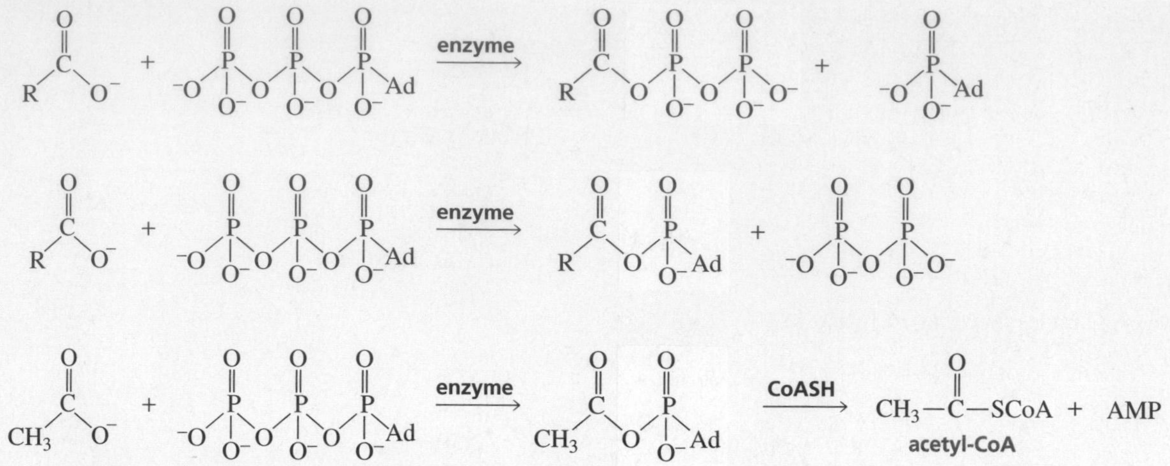

10. Dehydration of dicarboxylic acids (Section 16.23)

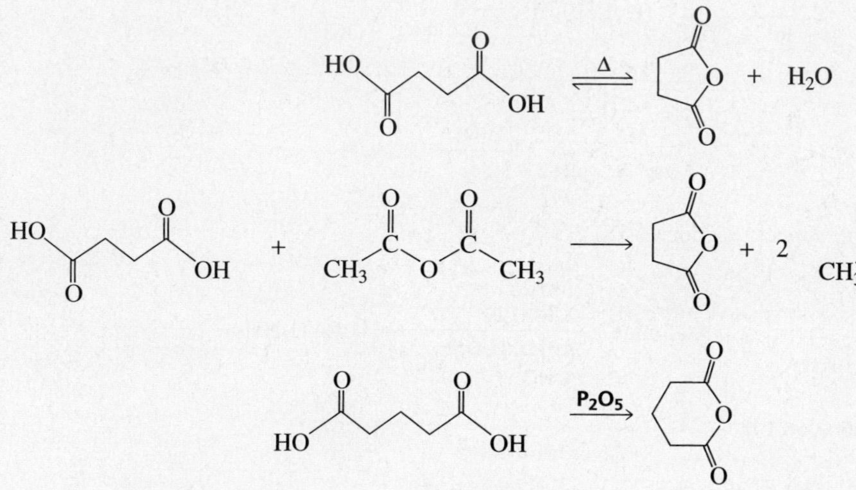

KEY TERMS

acid anhydride (p. 726)
acyl adenylate (p. 769)
acyl group (p. 722)
acyl halide (p. 726)
acyl phosphate (p. 769)
acyl pyrophosphate (p. 769)
acyl transfer reaction (p. 734)
alcoholysis (p. 744)
α-carbon (p. 725)
amide (p. 727)
amino acid (p. 732)
aminolysis (p. 744)
biosynthesis (p. 768)
carbonyl carbon (p. 729)
carbonyl compound (p. 722)
carbonyl group (p. 722)

carbonyl oxygen (p. 729)
carboxyl group (p. 725)
carboxylic acid (p. 724)
carboxylic acid
    derivative (p. 724)
carboxyl oxygen (p. 726)
catalyst (p. 748)
detergent (p. 756)
ester (p. 726)
fatty acid (p. 755)
fat (p. 754)
Fischer esterification (p. 757)
Gabriel synthesis (p. 763)
hydrolysis reaction (p. 743)
hydrophobic interactions (p. 755)
imide (p. 763)

lactam (p. 728)
lactone (p. 727)
micelle (p. 755)
mixed anhydride (p. 726)
nitrile (p. 728)
nucleophilic acyl substitution
    reaction (p. 734)
oil (p. 754)
phosphoanhydride bond (p. 769)
saponification (p. 755)
soap (p. 755)
symmetrical anhydride (p. 726)
tetrahedral intermediate (p. 733)
thioester (p. 770)
transesterification
    reaction (p. 744)

PROBLEMS

45. Write a structure for each of the following compounds:
- **a.** *N,N*-dimethylhexanamide
- **b.** 3,3-dimethylhexanamide
- **c.** cyclohexanecarbonyl chloride
- **d.** propanenitrile
- **e.** propionyl bromide
- **f.** sodium acetate
- **g.** benzoic anhydride
- **h.** β-valerolactone
- **i.** 3-methylbutanenitrile
- **j.** cycloheptanecarboxylic acid

46. Name the following compounds:

a.
$$CH_3CH_2\overset{\overset{\displaystyle CH_2CH_3}{|}}{CH}CH_2CH_2CH_2\overset{\overset{\displaystyle O}{||}}{C}OH$$

b.
$$CH_3CH_2\overset{\overset{\displaystyle O}{||}}{C}OCH_2CH_2CH_3$$

c.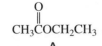

d.
$$CH_3CH_2CH_2\overset{\overset{\displaystyle O}{||}}{C}N(CH_3)_2$$

e.
$$CH_3CH_2CH_2CH_2\overset{\overset{\displaystyle O}{||}}{C}Cl$$

f.

g.
$$CH_2{=}CHCH_2\overset{\overset{\displaystyle O}{||}}{C}NHCH_3$$

h.
$$CH_3CH_2\overset{\overset{\displaystyle O}{||}}{C}O\overset{\overset{\displaystyle O}{||}}{C}CH_2CH_3$$

i.

47. What products would be formed from the reaction of benzoyl chloride with the following reagents?
- **a.** sodium acetate
- **b.** water
- **c.** excess dimethylamine
- **d.** aqueous HCl
- **e.** aqueous NaOH
- **f.** cyclohexanol
- **g.** excess benzylamine
- **h.** 4-chlorophenol
- **i.** isopropyl alcohol
- **j.** excess aniline

48. a. List the following esters in order of decreasing reactivity in the first slow step of a nucleophilic acyl substitution reaction (formation of the tetrahedral intermediate):

b. List the same esters in order of decreasing reactivity in the second slow step of a nucleophilic acyl substitution reaction (collapse of the tetrahedral intermediate).

49. Because bromocyclohexane is a secondary alkyl halide, both cyclohexanol and cyclohexene are formed when the alkyl halide reacts with hydroxide ion. Suggest a method to synthesize cyclohexanol from bromocyclohexane that would form little or no cyclohexene.

50. a. Which compound would you expect to have a higher dipole moment, methyl acetate or butanone?

$$\underset{\textbf{methyl acetate}}{CH_3\overset{\overset{\displaystyle O}{||}}{C}OCH_3} \qquad \underset{\textbf{butanone}}{CH_3\overset{\overset{\displaystyle O}{||}}{C}CH_2CH_3}$$

b. Which would you expect to have a higher boiling point?

51. How could you use ^{1}H NMR spectroscopy to distinguish between the following esters?

$$\underset{\textbf{A}}{CH_3\overset{\overset{\displaystyle O}{||}}{C}OCH_2CH_3} \qquad \underset{\textbf{B}}{H\overset{\overset{\displaystyle O}{||}}{C}OCH_2CH_2CH_3} \qquad \underset{\textbf{C}}{CH_3CH_2\overset{\overset{\displaystyle O}{||}}{C}OCH_3} \qquad \underset{\textbf{D}}{H\overset{\overset{\displaystyle O}{||}}{C}O\overset{\overset{\displaystyle CH_3}{|}}{C}HCH_3}$$

52. If propionyl chloride is added to one equivalent of methylamine, only a 50% yield of *N*-methylpropanamide is obtained. If, however, the acyl chloride is added to two equivalents of methylamine, the yield of *N*-methylpropanamide is almost 100%. Explain these observations.

53. a. When a carboxylic acid is dissolved in isotopically labeled water (H_2O^{18}), and an acid catalyst is added, the label is incorporated into both oxygens of the acid. Propose a mechanism to account for this.

$$\underset{CH_3}{\overset{\overset{\displaystyle O}{||}}{C}}{\diagdown}OH \;+\; \overset{18}{H_2O} \;\underset{}{\overset{H^+}{\rightleftharpoons}}\; \underset{CH_3}{\overset{\overset{\displaystyle O18}{||}}{C}}{\diagdown}18\text{--}OH \;+\; H_2O$$

b. If a carboxylic acid is dissolved in isotopically labeled methanol ($CH_3{}^{18}OH$) and an acid catalyst is added, where will the label reside in the product?

c. If an ester is dissolved in isotopically labeled water (H_2O^{18}) and an acid catalyst is added, where will the label reside in the product?

54. Using an alcohol for one method and an alkyl halide for the other, show two ways to make each of the following esters:

a. propyl acetate (odor of pears)

b. isopentyl acetate (odor of bananas)

c. ethyl butyrate (odor of pineapple)

d. methyl phenylethanoate (odor of honey)

55. What reagents would you use to convert methyl propanoate into the following compounds?

a. isopropyl propanoate

b. sodium propanoate

c. *N*-ethylpropanamide

d. propanoic acid

56. A compound with molecular formula $C_5H_{10}O_2$ gives the following IR spectrum:

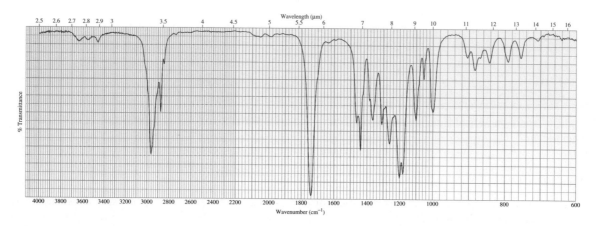

When it undergoes acid-catalyzed hydrolysis, the compound with the following ${}^1H$ NMR spectrum is formed. Identify the compounds.

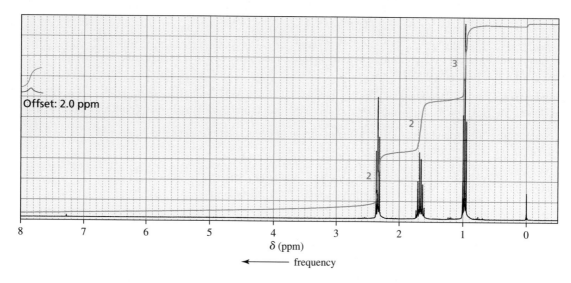

57. Aspartame, the sweetener used in the commercial products NutraSweet and Equal, is 160 times sweeter than sucrose. What products would be obtained if aspartame were hydrolyzed completely in an aqueous solution of HCl?

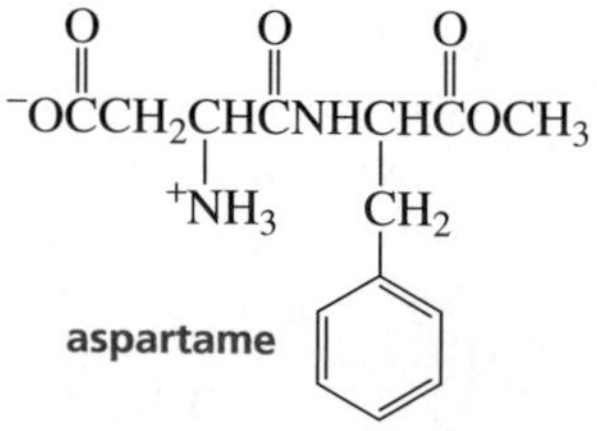

aspartame

58. a. Which of the following reactions will not give the carbonyl product shown?

1. $CH_3\overset{O}{\overset{\|}{C}}OH + CH_3\overset{O}{\overset{\|}{C}}O^- \longrightarrow CH_3\overset{O}{\overset{\|}{C}}O\overset{O}{\overset{\|}{C}}CH_3$

6. $CH_3\overset{O}{\overset{\|}{C}}OCH_3 + Cl^- \longrightarrow CH_3\overset{O}{\overset{\|}{C}}Cl$

2. $CH_3\overset{O}{\overset{\|}{C}}Cl + CH_3\overset{O}{\overset{\|}{C}}O^- \longrightarrow CH_3\overset{O}{\overset{\|}{C}}O\overset{O}{\overset{\|}{C}}CH_3$

7. $CH_3\overset{O}{\overset{\|}{C}}NHCH_3 + CH_3\overset{O}{\overset{\|}{C}}O^- \longrightarrow CH_3\overset{O}{\overset{\|}{C}}O\overset{O}{\overset{\|}{C}}CH_3$

3. $CH_3\overset{O}{\overset{\|}{C}}NH_2 + Cl^- \longrightarrow CH_3\overset{O}{\overset{\|}{C}}Cl$

8. $CH_3\overset{O}{\overset{\|}{C}}Cl + H_2O \longrightarrow CH_3\overset{O}{\overset{\|}{C}}OH$

4. $CH_3\overset{O}{\overset{\|}{C}}OH + CH_3NH_2 \longrightarrow CH_3\overset{O}{\overset{\|}{C}}NHCH_3$

9. $CH_3\overset{O}{\overset{\|}{C}}NHCH_3 + H_2O \longrightarrow CH_3\overset{O}{\overset{\|}{C}}OH$

5. $CH_3\overset{O}{\overset{\|}{C}}OCH_3 + CH_3NH_2 \longrightarrow CH_3\overset{O}{\overset{\|}{C}}NHCH_3$

10. $CH_3\overset{O}{\overset{\|}{C}}O\overset{O}{\overset{\|}{C}}CH_3 + CH_3OH \longrightarrow CH_3\overset{O}{\overset{\|}{C}}OCH_3$

b. Which of the reactions that do not occur can be made to occur if an acid catalyst is added to the reaction mixture?

59. 1,4-Diazabicyclo[2.2.2]octane (abbreviated DABCO) is a tertiary amine that catalyzes transesterification reactions. Propose a mechanism to show how it does this.

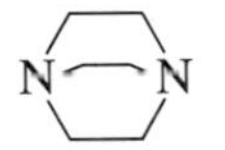

1,4-diazabicyclo[2.2.2]octane
DABCO

60. Two products, A and B, are obtained from the reaction of 1-bromobutane with NH_3. Compound A reacts with acetyl chloride to form C, and B reacts with acetyl chloride to form D. The IR spectra of C and D are shown. Identify A, B, C, and D.

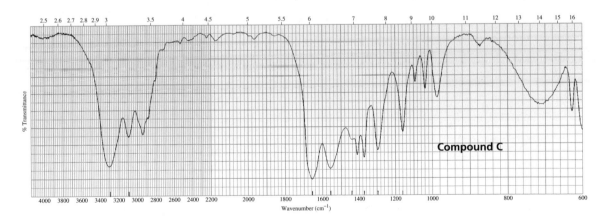

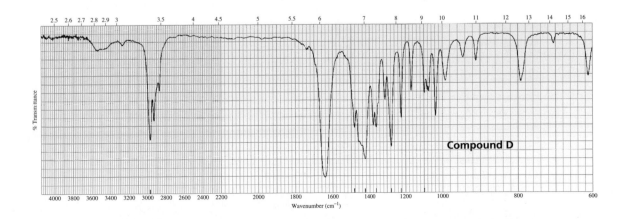

61. Phosgene (COCl$_2$) was used as a poison gas in World War I. Give the product that would be formed from the reaction of phosgene with each of the following reagents:

1. one equivalent of methanol

2. excess methanol

3. excess propylamine

4. one equivalent of ethanol followed by one equivalent of methylamine

62. What reagent should be used to carry out the following reaction?

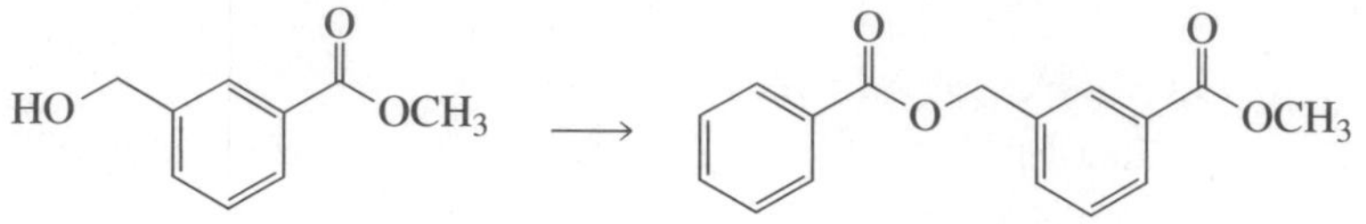

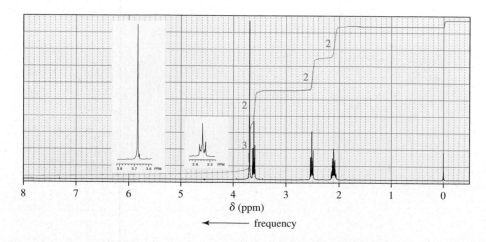

63. When Ethyl Ester treated butanedioic acid with thionyl chloride, she was surprised to find that the product she obtained was an anhydride rather than an acyl chloride. Propose a mechanism to explain why she obtained an anhydride.

64. Give the products of the following reactions:

a. 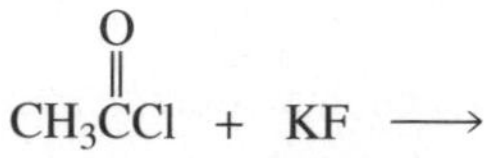 CH$_3$CCl + KF $\longrightarrow$

b. 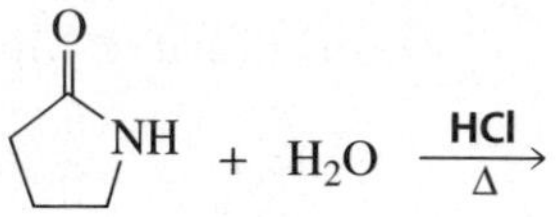 [pyrrolidinone] + H$_2$O $\xrightarrow{\text{HCl}}{\Delta}$

c. 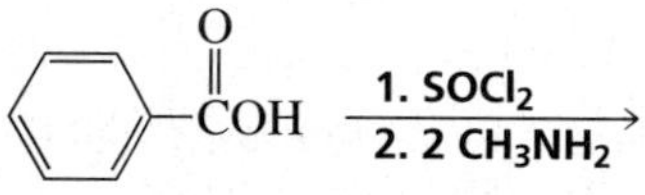 [benzoic acid] $\xrightarrow[\text{2. 2 CH}_3\text{NH}_2]{\text{1. SOCl}_2}$

d. 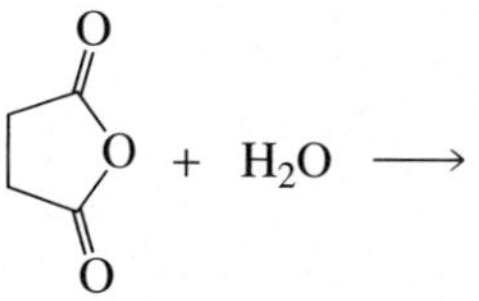 [succinic anhydride] + H$_2$O $\longrightarrow$

e. 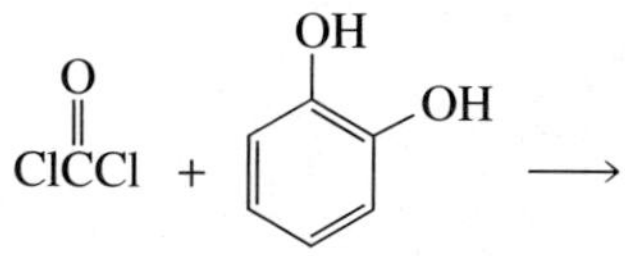 ClCCl + [catechol] $\longrightarrow$

f. 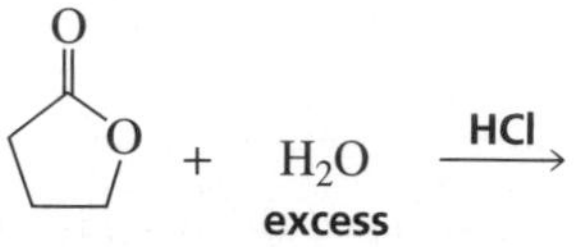 [γ-butyrolactone] + H$_2$O $\xrightarrow{\text{HCl}}$ excess

g. CH$_3$CCH$_2$OCCH$_3$ + CH$_3$OH $\xrightarrow{\text{CH}_3\text{O}^-}$ excess

h. [o-(carboxymethyl)benzoic acid] $\xrightarrow[\Delta]{(CH_3C)_2O}$

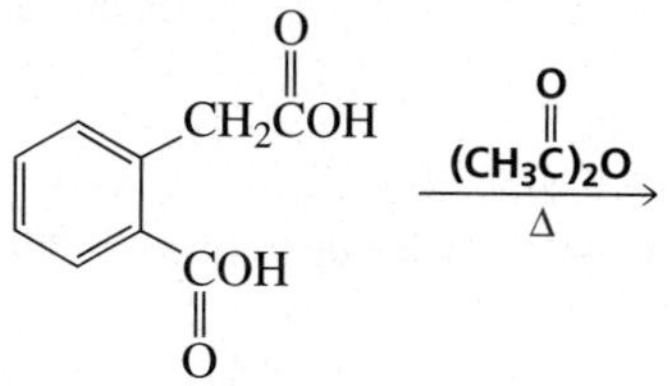

i. 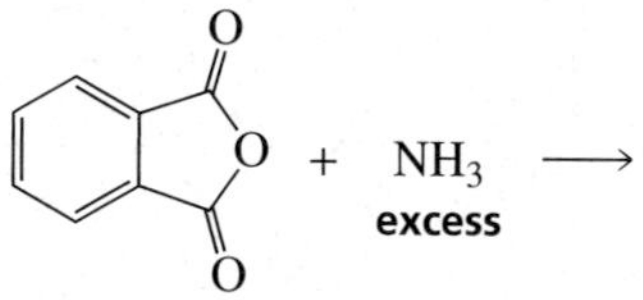 [phthalic anhydride] + NH$_3$ $\longrightarrow$ excess

j. 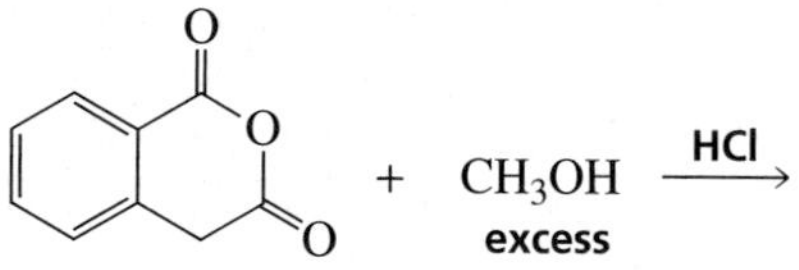 [isochromandione] + CH$_3$OH $\xrightarrow{\text{HCl}}$ excess

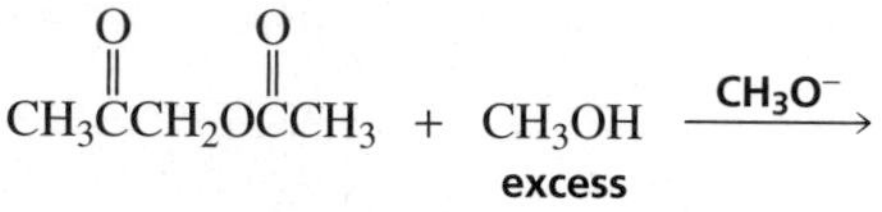

65. When treated with an equivalent of methanol, compound A, with molecular formula C$_4$H$_6$Cl$_2$O, forms the compound whose ^{1}H NMR spectrum is shown below. Identify compound A.

[NMR spectrum with δ (ppm) axis from 8 to 0, labeled "frequency"]

66. a. Identify the two products obtained from the following reaction:

$$\underset{\text{excess}}{CH_3COCCH_3} + CH_3CHCH_2CH_2OH \longrightarrow$$

(with O O above the first compound, and NH₂ above the second)

b. Eddie Amine carried out the preceding reaction, but stopped it before it was half over, whereupon he isolated the major product. He was surprised to find that the product he isolated was neither of the products obtained when the reaction was allowed to go to completion. What product did he isolate?

67. An aqueous solution of a primary or secondary amine reacts with an acyl chloride to form an amide as the major product. However, if the amine is tertiary, an amide is not formed. What product *is* formed? Explain.

68. Identify the major and minor products of the following reaction:

(structure of piperidine with CH₃, CHOH, CH₂CH₃ substituents) + CH₃CCl ⟶

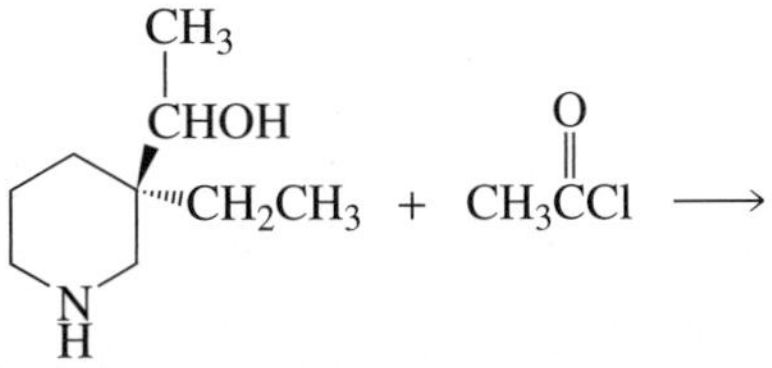

69. a. Ann Hydride did not obtain any ester when she added 2,4,6-trimethylbenzoic acid to an acidic solution of methanol. Why? (*Hint:* Build models.)
b. Would Ann have encountered the same problem if she had tried to synthesize the methyl ester of *p*-methylbenzoic acid in the same way?
c. How could she prepare the methyl ester of 2,4,6-trimethylbenzoic acid? (*Hint:* See Section 15.11.)

70. When a compound with molecular formula $C_{11}H_{14}O_2$ undergoes acid-catalyzed hydrolysis, one of the products that is isolated gives the following ^1H NMR spectrum. Identify the compound.

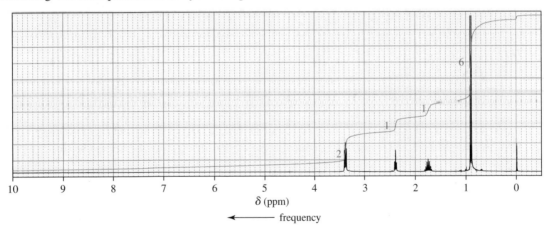

71. List the following compounds in order of decreasing frequency of the carbon–oxygen double-bond stretch:

$$CH_3COCH_3 \qquad CH_3CCl \qquad CH_3CH \qquad CH_3CNH_2$$

(each with O above the carbonyl carbon)

72. Cardura, a drug used to treat hypertension, is synthesized as shown below.

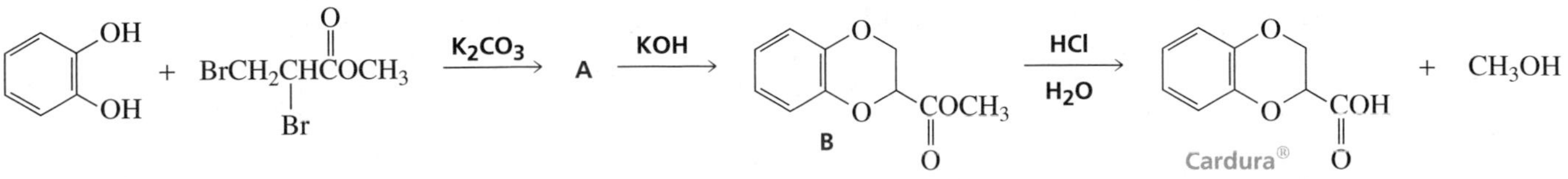

(reaction scheme: catechol + BrCH₂CHCOCH₃ with Br substituent, K₂CO₃ → **A** → KOH → benzodioxane intermediate **B** (COCH₃) → HCl, H₂O → Cardura® + CH₃OH)

a. Identify the intermediate (A) and show the mechanism for its formation.
b. Show the mechanism for conversion of A to B. Which would be formed most rapidly, A or B?

73. a. If the equilibrium constant for the reaction of acetic acid and ethanol to form ethyl acetate is 4.02, what will be the concentration of ethyl acetate at equilibrium if the reaction is carried out with equal amounts of acetic acid and ethanol?

b. What will be the concentration of ethyl acetate at equilibrium if the reaction is carried out with 10 times more ethanol than acetic acid? *Hint:* Recall the quadratic equation: for $ax^2 + bx + c = 0$,

$$x = \frac{-b \pm (b^2 - 4ac)^{1/2}}{2a}$$

c. What will be the concentration of ethyl acetate at equilibrium if the reaction is carried out with 100 times more ethanol than acetic acid?

74. The ^{1}H NMR spectra for two esters with molecular formula $C_8H_8O_2$ are shown below. If each of the esters is added to an aqueous solution with a pH of 10, which of the esters will be hydrolyzed more completely when the hydrolysis reactions have reached equilibrium?

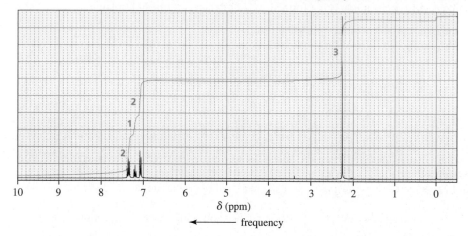

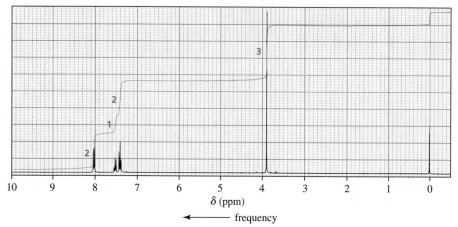

75. Show how the following compounds could be prepared from the given starting materials. You can use any necessary organic or inorganic reagents.

a. $CH_3CH_2\overset{O}{\overset{\|}{C}}NH_2 \longrightarrow CH_3CH_2\overset{O}{\overset{\|}{C}}Cl$

b. $CH_3CH_2CH_2CH_2OH \longrightarrow CH_3CH_2CH_2CH_2\overset{O}{\overset{\|}{C}}OH$

c. 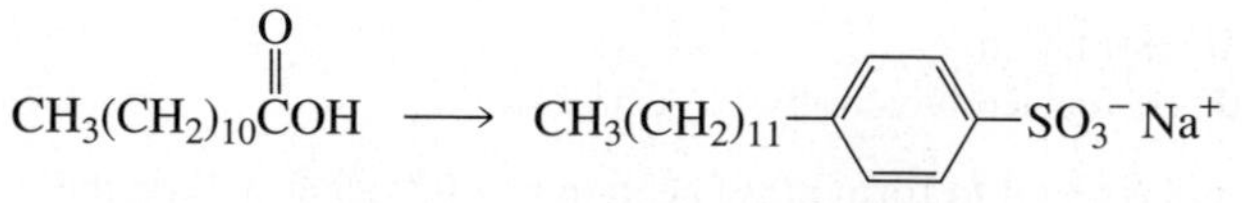 $CH_3(CH_2)_{10}\overset{O}{\overset{\|}{C}}OH \longrightarrow CH_3(CH_2)_{11}$—⟨benzene ring⟩—$SO_3^-$ Na^+

a detergent

d. 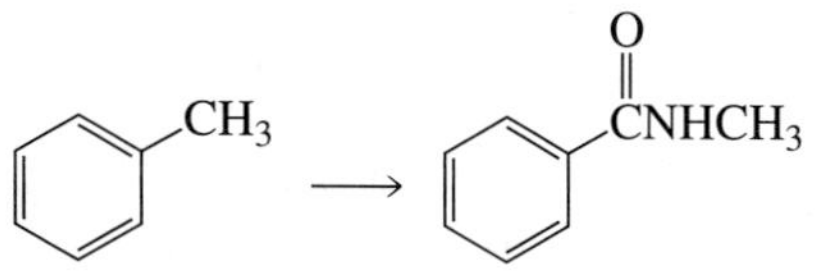 ⟨toluene⟩ $CH_3 \longrightarrow$ ⟨benzene⟩$\overset{O}{\overset{\|}{C}}NHCH_3$

e. 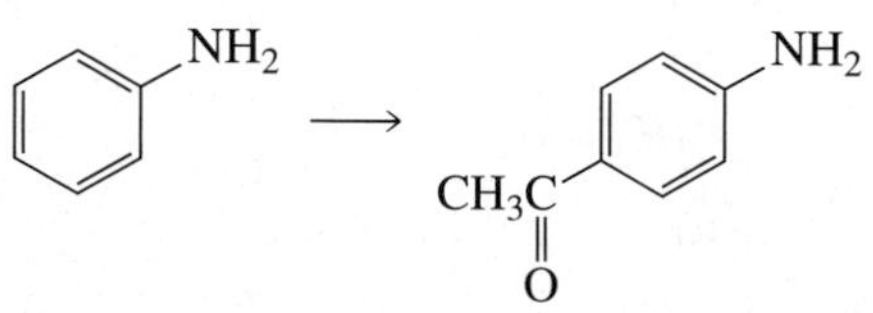 ⟨aniline⟩$NH_2 \longrightarrow$ $CH_3\overset{}{\underset{O}{C}}$—⟨benzene⟩$NH_2$

f. 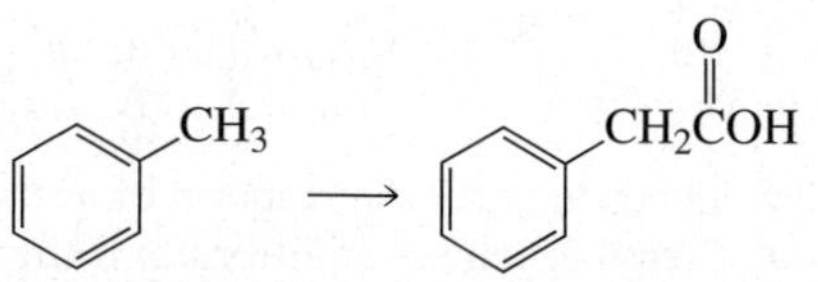 ⟨toluene⟩$CH_3 \longrightarrow$ ⟨benzene⟩$CH_2\overset{O}{\overset{\|}{C}}OH$

76. Is the acid-catalyzed hydrolysis of acetamide a reversible or an irreversible reaction? Explain.

77. What product would you expect to obtain from each of the following reactions?

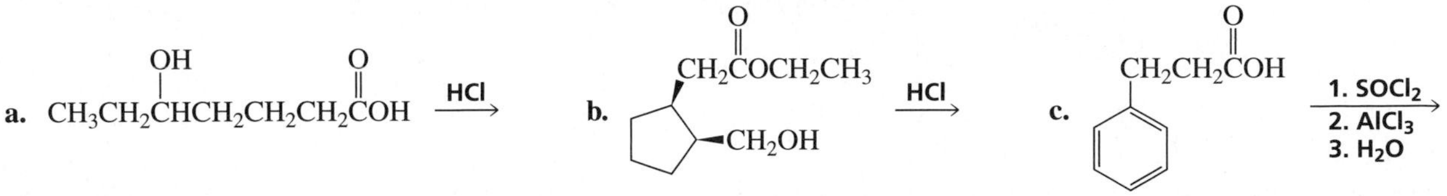

a. $\text{CH}_3\text{CH}_2\overset{\text{OH}}{\underset{|}{\text{CH}}}\text{CH}_2\text{CH}_2\text{CH}_2\overset{\text{O}}{\overset{\|}{\text{C}}}\text{OH}$ $\xrightarrow{\text{HCl}}$

b. (cyclopentane ring with $\text{CH}_2\overset{\text{O}}{\overset{\|}{\text{C}}}\text{OCH}_2\text{CH}_3$ and CH_2OH substituents) $\xrightarrow{\text{HCl}}$

c. (benzene ring with $\text{CH}_2\text{CH}_2\overset{\text{O}}{\overset{\|}{\text{C}}}\text{OH}$) $\xrightarrow[\begin{array}{l}\text{1. SOCl}_2\\ \text{2. AlCl}_3\\ \text{3. H}_2\text{O}\end{array}]{}$

78. Sulfonamides, the first antibiotics, were introduced clinically in 1934 (Sections 24.8 and 30.4). Show how a sulfonamide can be prepared from benzene.

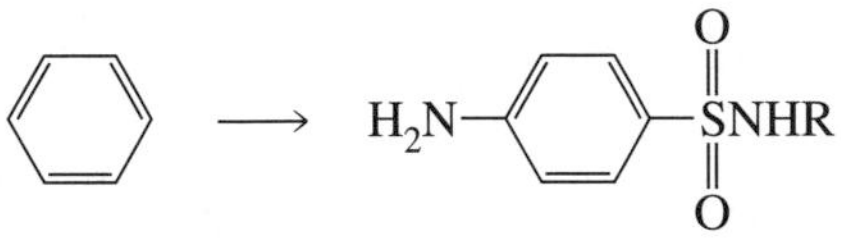

(benzene) $\longrightarrow$ $\text{H}_2\text{N}-$(benzene ring)$-\overset{\text{O}}{\underset{\text{O}}{\overset{\|}{\underset{\|}{\text{S}}}}}\text{NHR}$

a sulfonamide

79. The reaction of a nitrile with an alcohol in the presence of a strong acid forms a secondary amide. This reaction, known as the *Ritter reaction*, does not work with primary alcohols.

$$\text{RC}\equiv\text{N} + \text{R}'\text{OH} \xrightarrow{\text{H}^+} \text{R}\overset{\text{O}}{\overset{\|}{\text{C}}}\text{NHR}'$$

the Ritter reaction

a. Propose a mechanism for the Ritter reaction.
b. Why does the Ritter reaction not work with primary alcohols?
c. How does the Ritter reaction differ from the acid-catalyzed hydrolysis of a nitrile to form a primary amide?

80. The intermediate shown here is formed during the hydroxide-ion-promoted hydrolysis of the ester group. Propose a mechanism for the reaction.

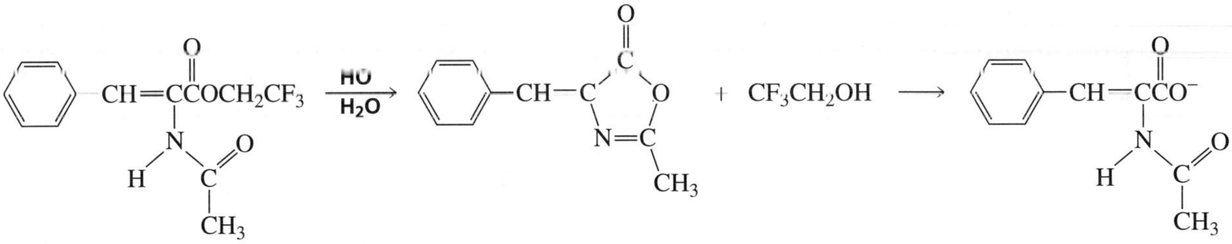

(benzene)$-\text{CH}=\overset{\text{O}}{\overset{\|}{\text{C}}}\text{OCH}_2\text{CF}_3$ with $\underset{\underset{\text{CH}_3}{\overset{\|}{\text{C}}}}{\overset{|}{\text{N}}}{-}\text{H}$ $\xrightarrow[\text{H}_2\text{O}]{\text{HO}^-}$ (benzene)$-\text{CH}-\text{C}$ ... $+ \text{CF}_3\text{CH}_2\text{OH} \longrightarrow$ (benzene)$-\text{CH}-\overset{\text{O}}{\overset{\|}{\text{C}}}\text{O}^-$

81. a. How could aspirin be synthesized, starting with benzene?
b. Ibuprofen is the active ingredient in pain relievers such as Advil, Mortin, and Nuprin. How could ibuprofen be synthesized, starting with benzene?
c. Acetaminophen is the active ingredient in Tylenol. How could acetominophen be synthesized, starting with benzene?

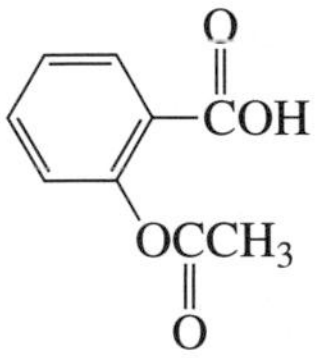

(benzene ring with $\overset{\text{O}}{\overset{\|}{\text{C}}}\text{OH}$ and OCCH_3 with O) — **aspirin**

$\text{CH}_3\text{CHCH}_2-$(benzene ring)$-\overset{\text{O}}{\overset{\|}{\text{CHC}}}\text{OH}$ with CH_3 groups — **ibuprofen**

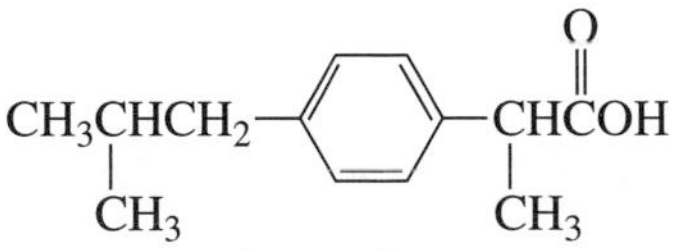

$\text{CH}_3\overset{\text{O}}{\overset{\|}{\text{C}}}\text{NH}-$(benzene ring)$-\text{OH}$ — **acetaminophen**

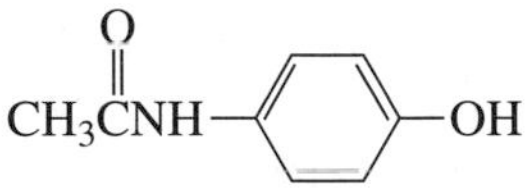

82. The following compound has been found to be an inhibitor of penicillinase. The enzyme can be reactivated by hydroxylamine (NH_2OH). Propose a mechanism to account for the inhibition and for the reactivation.

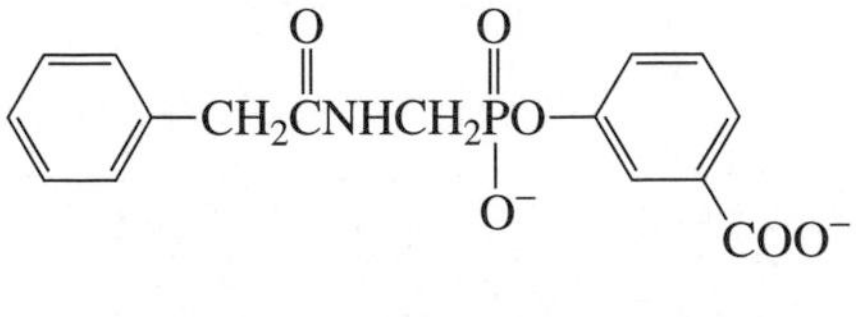

(benzene)$-\text{CH}_2\overset{\text{O}}{\overset{\|}{\text{C}}}\text{NHCH}_2\overset{\text{O}}{\overset{\|}{\text{P}}}\text{O}-$(benzene ring)$-\text{COO}^-$ with O^-

83. For each of the following reactions, propose a mechanism that will account for the formation of the product:

a.

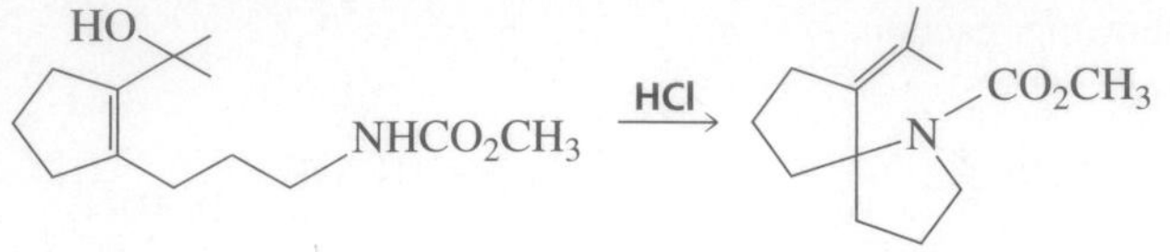

b.

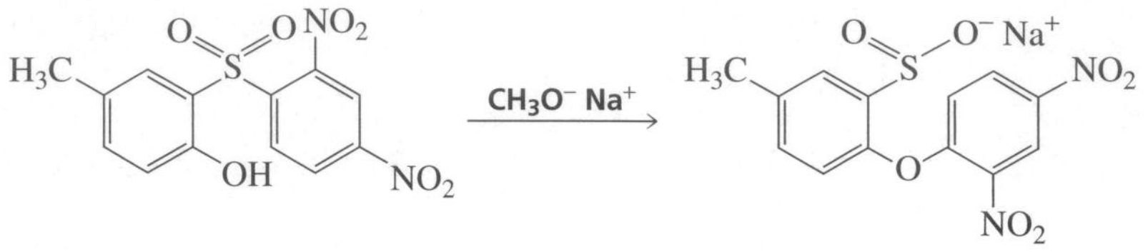

84. Show how Novocain, a painkiller used frequently by dentists (Section 30.3), can be prepared from benzene.

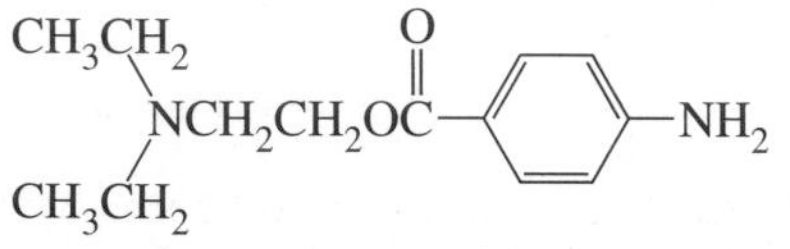

Novocain®

85. Catalytic antibodies catalyze a reaction by forcing the conformation of the substrate in the direction of the transition state. The synthesis of the antibody is carried out in the presence of a transition state analog—a stable molecule that structurally resembles the transition state. This causes an antibody to be generated that will recognize and bind to the transition state, thereby stabilizing it. For example, the following transition state analog has been used to generate a catalytic antibody that catalyzes the hydrolysis of the structurally similar ester:

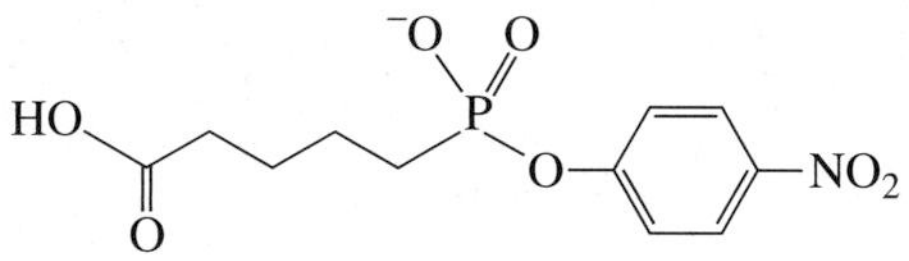

transition state analog

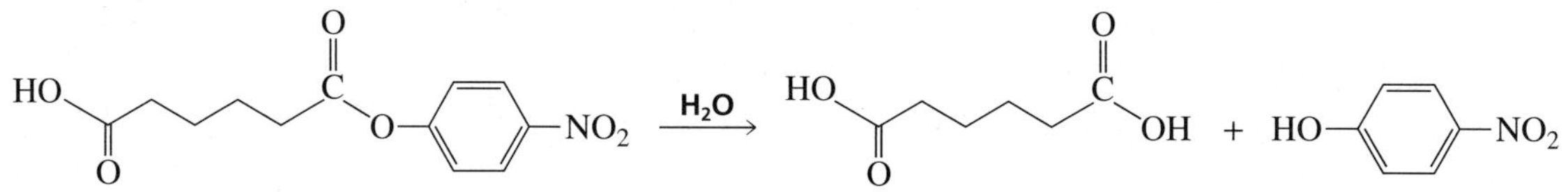

a. Draw the transition state for the hydrolysis reaction.

b. The following transition state analog is used to generate a catalytic antibody for the catalysis of ester hydrolysis. Give the structure of an ester whose rate of hydrolysis would be increased by this catalytic antibody.

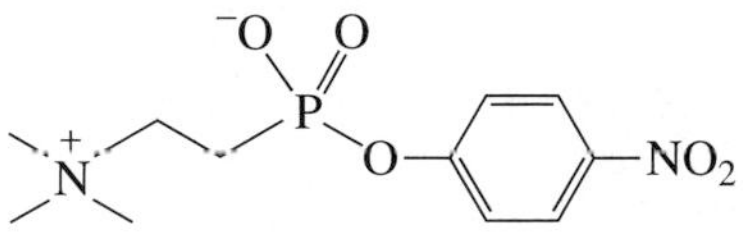

c. Design a transition state analog that would catalyze amide hydrolysis at the amide group indicated.

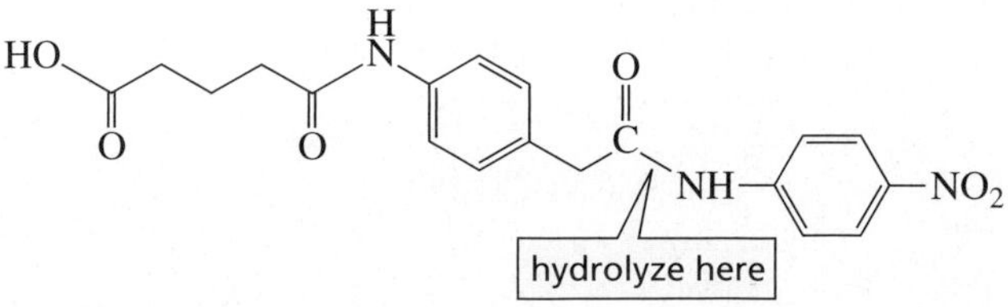

hydrolyze here

86. Show how lidocaine, one of the most widely used injectable anesthetics (Section 30.3), can be prepared from benzene and compounds containing no more than four carbons.

87. Saccharin, an artificial sweetener, is about 300 times sweeter than sucrose. Describe how saccharin could be prepared, using benzene as the starting material.

saccharin

88. Information about the mechanism of a reaction undergone by a series of substituted benzenes can be obtained by plotting the logarithm of the observed rate constant determined at a particular pH against the Hammett substituent constant (σ) for the particular substituent. The σ value for hydrogen is 0. Electron-donating substituents have negative σ values; electron-withdrawing substituents have positive σ values. The more strongly electron donating the substituent, the more negative its σ value will be; the more strongly electron withdrawing the substituent, the more positive its σ value will be. The slope of a plot of the logarithm of the rate constant versus σ is called the ρ (rho) value. The ρ value for the hydroxide-ion-promoted hydrolysis of a series of meta- and para-substituted ethyl benzoates is +2.46; the ρ value for amide formation for the reaction of a series of meta- and para-substituted anilines with benzoyl chloride is −2.78.

a. Why does one set of experiments give a positive ρ value while the other set of experiments gives a negative ρ value?

b. Why were ortho-substituted compounds not included in the experiment?

c. What would you predict the sign of the ρ value to be for the ionization of a series of meta- and para-substituted benzoic acids?

CHAPTER 17

Carbonyl Compounds II

Reactions of Aldehydes and
Ketones • More Reactions of
Carboxylic Acid Derivatives •
Reactions of α,β-Unsaturated
Carbonyl Compounds

formaldehyde

acetaldehyde

acetone

BUILDING ON FUNDAMENTALS

SECTION 17.7 The weaker the base, the better it is as a leaving group (8.3).

SECTION 17.1 A Grignard reagent reacts as if it were a carbanion (10.12).

SECTION 17.2 Alkyl groups stabilize carbocations (4.2), alkenes (4.11), alkynes (6.4), and radicals (11.3). We will now see that they stabilize carbonyl groups as well.

SECTION 17.3 Tetrahedral compounds are unstable if they have a group that is a weak enough base to be expelled (16.5).

SECTIONS 17.7 AND 17.8 Catalytic hydrogenation reduces carbon–carbon double and triple bonds (4.11 and 6.9). We will see that it also reduces carbon–nitrogen double (and triple) bonds.

SECTION 17.9 When an acid is added to a reaction, the first thing that happens is the acid protonates the atom in the reactant that has the greatest electron density (10.4).

SECTION 17.9 A catalyst increases the rate of a reaction but is not consumed during the course of the reaction (4.5).

SECTION 17.9 The relative stabilities of the reactants and products determine the equilibrium constant of the reaction (3.7).

SECTION 17.14 If a reaction forms a product with an asymmetric center from a reactant that does not have an asymmetric center, the product will be a racemic mixture (5.19).

SECTION 17.17 We will see another reaction that can be carried out under kinetic control (the major product is the more rapidly formed product) or under thermodynamic control (the major product is the more stable product) (7.11).

At the beginning of Chapter 16, we saw that carbonyl compounds—compounds that possess a carbonyl group (C=O)—can be divided into two classes: class I carbonyl compounds, which have a group that can be replaced by a nucleophile, and class II carbonyl compounds, which do not have a group that can be replaced by a nucleophile. Class II carbonyl compounds comprise aldehydes and ketones.

The carbonyl carbon of the simplest aldehyde, formaldehyde, is bonded to two hydrogens. The carbonyl carbon in all other **aldehydes** is bonded to a hydrogen and to an alkyl (or an aryl) group. The carbonyl carbon of a **ketone** is bonded to two alkyl (or aryl) groups. Aldehydes and ketones *do not have* a group that can be replaced by another group, because hydride ions (H⁻) and carbanions (R⁻) are too basic to be displaced by nucleophiles under normal conditions.

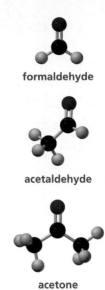

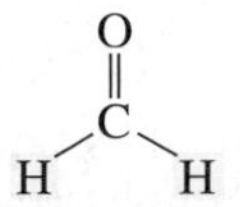

formaldehyde

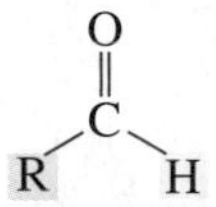

an aldehyde

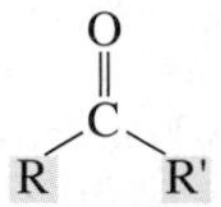

a ketone

formaldehyde

acetaldehyde

acetone

The physical properties of aldehydes and ketones are discussed in Section 16.3 (see also Appendix I), and the methods used to prepare aldehydes and ketones are summarized in Appendix IV.

Many compounds found in nature have aldehyde or ketone functional groups. Aldehydes have pungent odors, whereas ketones tend to smell sweet. Vanillin and cinnamaldehyde are examples of naturally occurring aldehydes. A whiff of vanilla extract will allow you to appreciate the pungent odor of vanilla. The ketones carvone and camphor are responsible for the characteristic sweet odors of spearmint leaves, caraway seeds, and the leaves of the camphor tree.

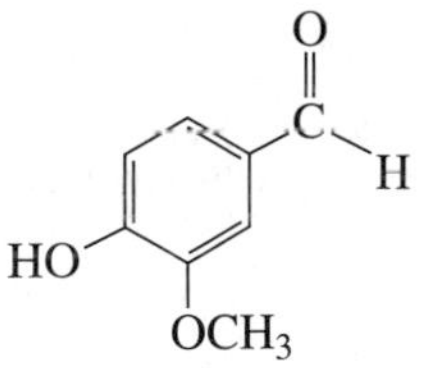

vanillin
vanilla flavoring

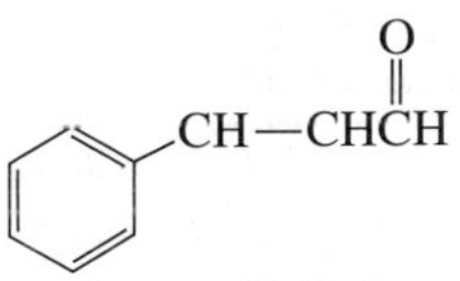

cinnamaldehyde
cinnamon flavoring

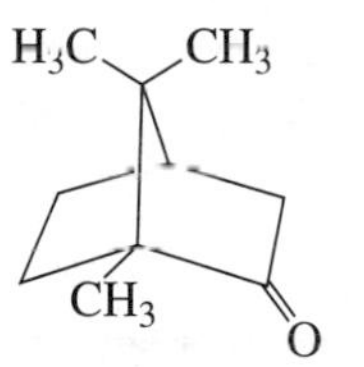

camphor

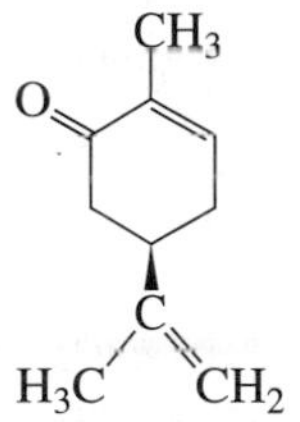

(R)-(−)-carvone
spearmint oil

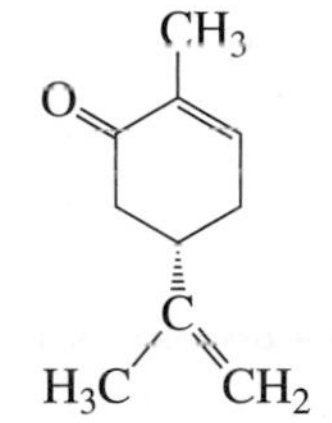

(S)-(+)-carvone
caraway seed oil

In ketosis, a pathological condition that can occur in people with diabetes, the body produces more acetoacetate than can be metabolized. The excess acetoacetate breaks down to acetone (a ketone) and CO_2. Ketosis can be recognized by the smell of acetone on a person's breath.

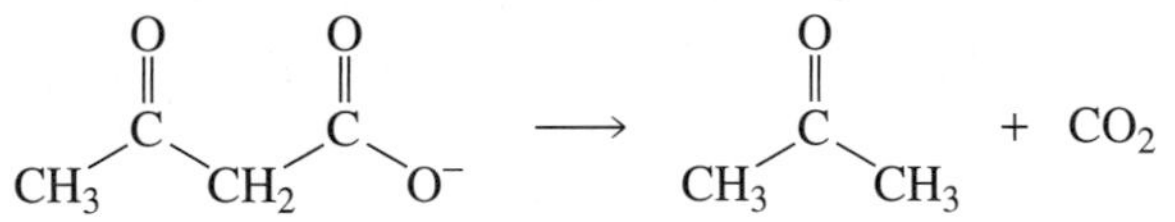

acetoacetate

acetone   + CO_2

Two ketones that are of biological importance illustrate how a small difference in structure can be responsible for a large difference in biological activity: progesterone is a sex hormone synthesized primarily in the ovaries, whereas testosterone is a sex hormone synthesized primarily in the testes.

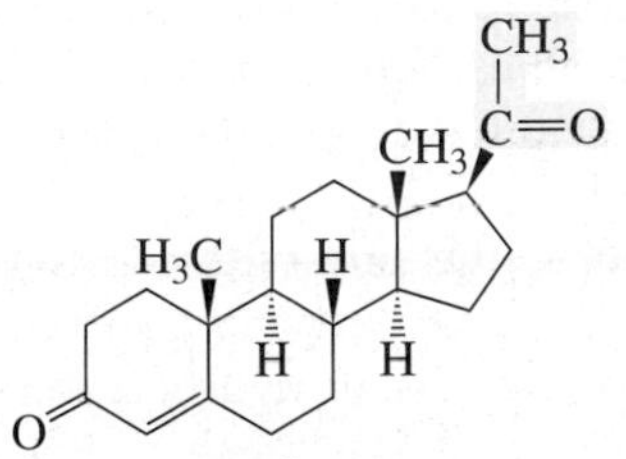

progesterone

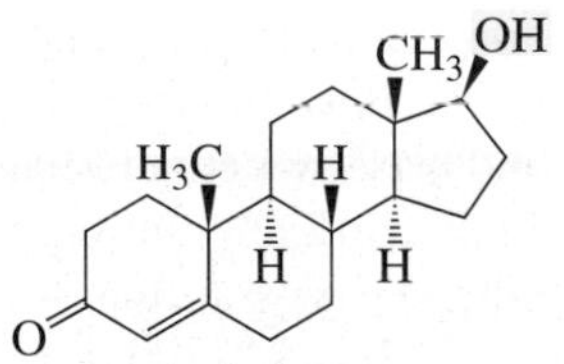

testosterone

17.1 Nomenclature of Aldehydes and Ketones

Both aldehydes and ketones are named using a functional group suffix.

Naming Aldehydes

The systematic (IUPAC) name of an aldehyde is obtained by replacing the final "e" on the name of the parent hydrocarbon with "al." For example, a one-carbon aldehyde is methanal; a two-carbon aldehyde is ethanal. The position of the carbonyl carbon does not have to be designated, because it is always at the end of the parent hydrocarbon and therefore always has the 1-position.

The common name of an aldehyde is the same as the common name of the corresponding carboxylic acid (Section 16.1), except that "aldehyde" is substituted for "ic acid" (or "oic acid"). When common names are used, the position of a substituent is designated by a lowercase Greek letter. The carbonyl carbon is not given a letter; the carbon adjacent to the carbonyl carbon is the α-carbon.

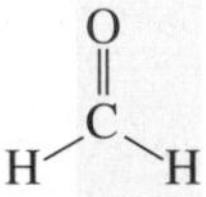

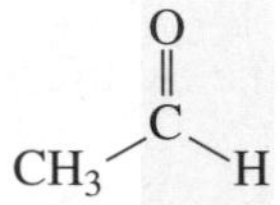

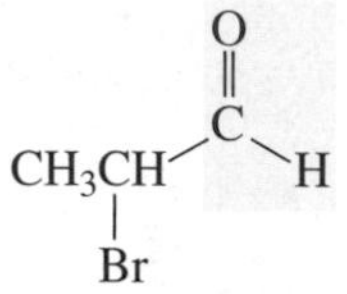

systematic name:	methanal	ethanal
common name:	**formaldehyde**	**acetaldehyde**

2-bromopropanal
α-bromopropionaldehyde

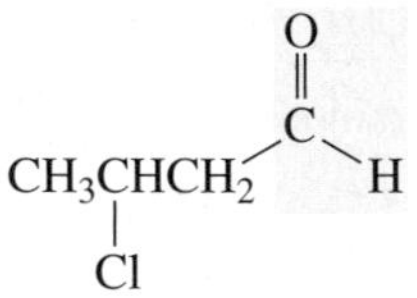

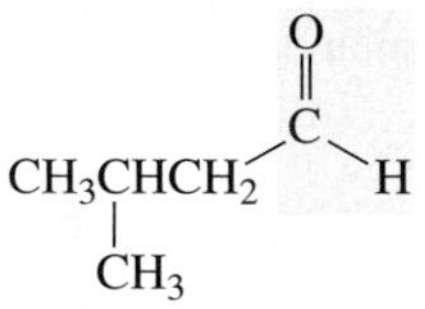

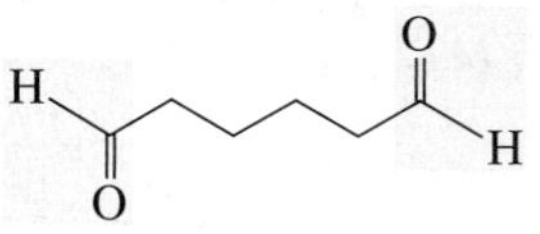

systematic name:	3-chlorobutanal	3-methylbutanal
common name:	β-chlorobutyraldehyde	isovaleraldehyde

hexanedial

Notice that the terminal "e" is not removed in hexanedial; the "e" is removed only to avoid two successive vowels.

If the aldehyde group is attached to a ring, the aldehyde is named by adding "carbaldehyde" to the name of the cyclic compound.

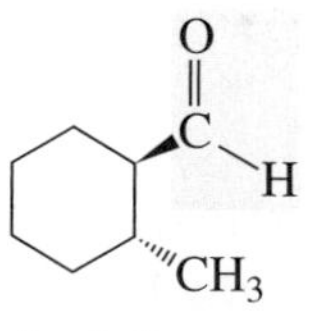

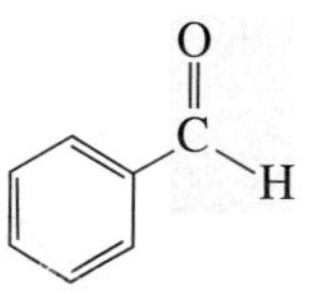

systematic name:	trans-2-methylcyclohexanecarbaldehyde	benzenecarbaldehyde
common name:		**benzaldehyde**

In Section 6.2, we saw that a carbonyl group has a higher nomenclature priority than an alcohol or an amine group. However, all carbonyl compounds do not have the same priority. Nomenclature priorities of the various functional groups, including the carbonyl groups, are shown in Table 17.1.

If a compound has two functional groups, the one with the lower priority is indicated by a prefix and the one with the higher priority by a suffix. Note that the prefix of an aldehyde oxygen that is part of the parent hydrocarbon is "oxo," and

Table 17.1	Summary of Functional Group Nomenclature		
	Class	**Suffix name**	**Prefix name**
	Carboxylic acid	-oic acid	Carboxy
	Ester	-oate	Alkoxycarbonyl
	Amide	-amide	Amido
	Nitrile	-nitrile	Cyano
	Aldehyde	-al	Oxo (=O)
	Aldehyde	-al	Formyl (CH=O)
	Ketone	-one	Oxo (=O)
	Alcohol	-ol	Hydroxy
	Amine	-amine	Amino
	Alkene	-ene	Alkenyl
	Alkyne	-yne	Alkynyl
increasing priority	Alkane	-ane	Alkyl
	Ether	—	Alkoxy
	Alkyl halide	—	Halo

prefix of a one-carbon aldehyde group that is not part of the parent hydrocarbon is "formyl."

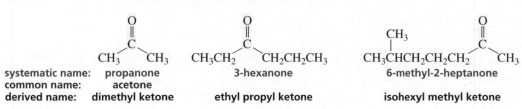

3-hydroxybutanal   methyl 5-oxopentanoate   ethyl 4-formylhexanoate

If the compound has both an alkene and an aldehyde functional group, the alkene functional group is cited first, with the "e" ending omitted to avoid two successive vowels (Section 6.2).

$$CH_3CH=CHCH_2 \overset{\overset{\displaystyle O}{\|}}{C} H$$

3-pentenal

Naming Ketones

The systematic name of a ketone is obtained by replacing the final "e" on the end of the name of the parent hydrocarbon with "one." The chain is numbered in the direction that gives the carbonyl carbon the smaller number. In the case of cyclic ketones, however, a number is not necessary because the carbonyl carbon is assumed to be at the 1-position. Derived names are often used for ketones: the substituents attached to the carbonyl group are cited in alphabetical order, followed by "ketone."

systematic name:	propanone	3-hexanone	6-methyl-2-heptanone
common name:	acetone		
derived name:	dimethyl ketone	ethyl propyl ketone	isohexyl methyl ketone

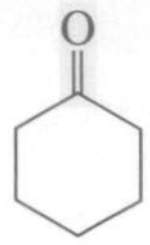

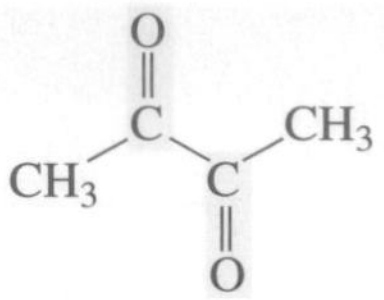

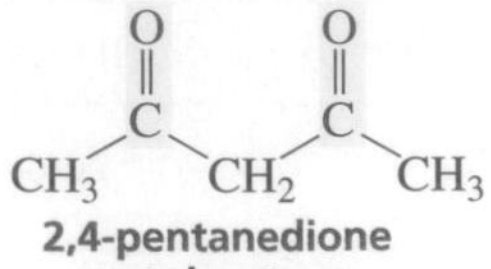

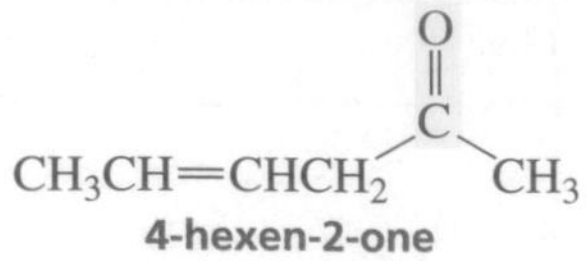

systematic name: cyclohexanone   butanedione   **2,4-pentanedione**   **4-hexen-2-one**
common name: **acetylacetone**

Only a few ketones have common names. The smallest ketone, propanone, is usually referred to by its common name, acetone. Acetone is a widely used laboratory solvent. Common names are also used for some phenyl-substituted ketones; the number of carbons (other than those of the phenyl group) is indicated by the common name of the corresponding carboxylic acid, substituting "ophenone" for "ic acid."

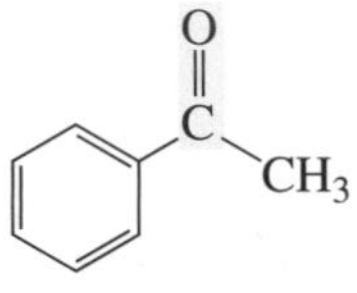

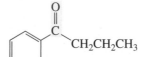

common name: **acetophenone**   **butyrophenone**   **benzophenone**
derived name: **methyl phenyl ketone**   **phenyl propyl ketone**   **diphenyl ketone**

Student Tutorial:
Nomenclature of aldehydes
and ketones

If the ketone has a second functional group of higher naming priority, the ketone oxygen is indicated by the prefix "oxo."

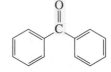

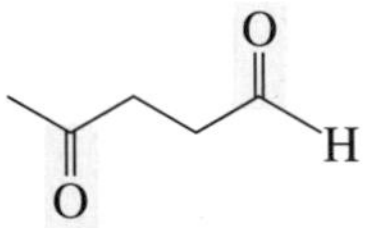

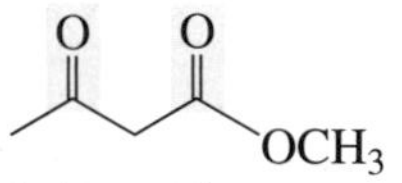

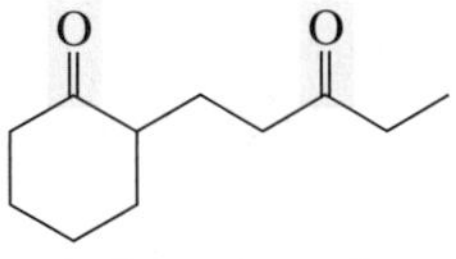

systematic name: **4-oxopentanal**   **methyl 3-oxobutanoate**   **2-(3-oxopentyl)-**
                                                            cyclohexanone

PROBLEM 1◆

Why are numbers not used to designate the positions of the functional groups in propanone and butanedione?

BUTANEDIONE: AN UNPLEASANT COMPOUND

Fresh perspiration is odorless. The smells we associate with perspiration result from a chain of events initiated by bacteria that are always present on our skin. These bacteria produce lactic acid, which creates an acidic environment that allows other bacteria to break down the components of perspiration, producing compounds with the unappealing odors we associate with armpits and sweaty feet. One such compound is butanedione.

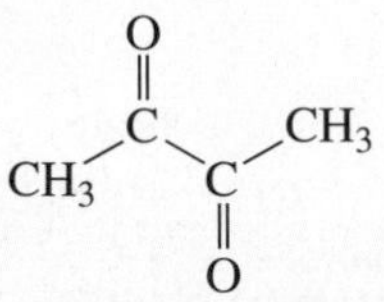

butanedione

PROBLEM 2◆

Give two names for each of the following:

a. CH₃CH₂CHCH₂CH
$\quad\quad\quad$ |$\quad$ ‖
$\quad\quad\quad$ CH₃$\quad$ O

d. ⬡—CH₂CH₂CH₂CH
$\quad\quad\quad\quad\quad\quad\quad\quad$ ‖
$\quad\quad\quad\quad\quad\quad\quad\quad$ O

b. CH₃CH₂CH₂CCH₂CH₂CH₃
$\quad\quad\quad\quad\quad\quad$ ‖
$\quad\quad\quad\quad\quad\quad$ O

e. CH₃CH₂CHCH₂CH₂CH
$\quad\quad\quad\quad$ |$\quad\quad\quad$ ‖
$\quad\quad\quad\quad$ CH₂CH₃$\quad$ O

c. CH₃CHCH₂CCH₂CH₂CH₃
$\quad\quad$ |$\quad\quad$ ‖
$\quad\quad$ CH₃$\quad$ O

f. CH₂=CHCCH₂CH₂CH₂CH₃
$\quad\quad\quad\quad$ ‖
$\quad\quad\quad\quad$ O

PROBLEM 3◆

Name the following:

a. CH₃CHCH₂CH₂CCH₂CH₃
$\quad\quad$ |$\quad\quad\quad$ ‖
$\quad\quad$ OH$\quad\quad$ O

b. [cyclohexanone with C≡N substituent]

c. CH₃CH₂CHCH₂CNH₂
$\quad\quad\quad\quad$ |$\quad\quad$ ‖
$\quad\quad\quad\quad$ HC=O$\quad$ O

17.2 **Relative Reactivities of Carbonyl Compounds**

We have seen that the carbonyl group is polar because oxygen, being more electronegative than carbon, has a greater share of the double bond's electrons (Section 16.5). The partial positive charge on the carbonyl carbon causes that carbon to be attacked by nucleophiles. The electron deficiency of the carbonyl carbon is indicated by the blue regions in the electrostatic potential maps.

formaldehyde

$$\overset{\delta-}{\underset{R \quad R(H)}{\overset{\ddot{O}}{\underset{\delta+}{C}}}} \quad \bar{:}Nu$$

An aldehyde has a greater partial positive charge on its carbonyl carbon than does a ketone because a hydrogen is electron withdrawing compared with an alkyl group (Section 4.2). An aldehyde, therefore, is more reactive than a ketone toward nucleophilic attack. We have now seen that alkyl groups stabilize carbonyl groups, alkenes, alkynes, carbocations, and radicals.

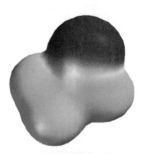

acetaldehyde

relative reactivities

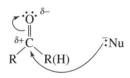

most reactive →
$$\underset{H \quad H}{\overset{O}{\overset{‖}{C}}} > \underset{R \quad H}{\overset{O}{\overset{‖}{C}}} > \underset{R \quad R'}{\overset{O}{\overset{‖}{C}}}$$ ← least reactive

$\quad$ **formaldehyde** $\quad$ **an aldehyde** $\quad$ **a ketone**

Steric factors also contribute to the greater reactivity of an aldehyde. The carbonyl carbon of an aldehyde is more accessible to a nucleophile than is the carbonyl carbon of a ketone because the hydrogen attached to the carbonyl carbon of an aldehyde is smaller than the second alkyl group attached to the carbonyl carbon of a ketone. Steric

acetone

factors also become important in the tetrahedral transition state because the bond angles are 109.5°, bringing the alkyl groups closer to one another than they are in the carbonyl compound, where the bond angles are 120°.

Aldehydes are more reactive than ketones.

Because ketones have greater steric crowding in their transition states, they have less stable transition states than aldehydes have. In summary, alkyl groups stabilize the reactant and destabilize the transition state, causing ketones to be less reactive than aldehydes.

Steric crowding also causes ketones with large alkyl groups bonded to the carbonyl carbon to be less reactive than ketones with small alkyl groups.

relative reactivities

PROBLEM 4◆

Which ketone in each pair is more reactive?

a. 2-heptanone or 4-heptanone   **b.** *p*-nitroacetophenone or *p*-methoxyacetophenone

---

Aldehydes and ketones are less reactive than acyl chlorides and acid anhydrides but more reactive than esters, carboxylic acids, and amides.

How does the reactivity of an aldehyde or a ketone toward nucleophiles compare with the reactivity of the carbonyl compounds whose reactions we looked at in Chapter 16? Aldehydes and ketones are right in the middle; they are less reactive than acyl halides and acid anhydrides, but are more reactive than esters, carboxylic acids, and amides.

relative reactivities of carbonyl compounds

acyl halide > acid anhydride > aldehyde > ketone > ester ~ carboxylic acid > amide > carboxylate ion

$\boxed{\text{most reactive}}$ $\boxed{\text{least reactive}}$

The carbonyl compounds discussed in Chapter 16 have a lone pair on an atom attached to the carbonyl group that can be shared with the carbonyl carbon by resonance electron donation; this makes the carbonyl carbon less electron deficient. We have seen that the reactivity of these carbonyl compounds is related to the basicity of Y^- (Section 16.5). The weaker the basicity of Y^-, the more reactive is the carbonyl group, because weak bases are less able to share their lone pair with the carbonyl carbon and are better able to withdraw electrons inductively from the carbonyl carbon (Section 16.6).

Consequently, aldehydes and ketones are not as reactive as carbonyl compounds in which Y^- is a very weak base (acyl halides and acid anhydrides), but are more reactive than carbonyl compounds in which Y^- is a relatively strong base (carboxylic acids, esters, and amides). A molecular orbital explanation of why resonance electron donation decreases the reactivity of the carbonyl group is presented in Section 16.16.

PROBLEM 8 *SOLVED*

a. Which of the following tertiary alcohols cannot be prepared from the reaction of an ester with excess Grignard reagent?

Student Tutorial:
Grignard reagents in synthesis

       OH
       |
1. CH₃CCH₃
       |
      CH₃

           OH
           |
3. CH₃CH₂CCH₂CH₂CH₃
           |
          CH₃

         OH
         |
5. CH₃CCH₂CH₂CH₂CH₃
        |
       CH₂CH₃

        OH
        |
2. CH₃CCH₂CH₃
       |
      CH₃

          OH
         |
4. CH₃CH₂CCH₂CH₃
         |
        CH₃

6. (two phenyl groups attached to a central C bearing OH and CH₃)

b. For those alcohols that can be prepared by the reaction of an ester with excess Grignard reagent, what ester and what Grignard reagent should be used?

Solution to 8a A tertiary alcohol is obtained from the reaction of an ester with two equivalents of a Grignard reagent. Therefore, tertiary alcohols prepared in this way must have two identical substituents on the carbon to which the OH is bonded, because two substituents come from the Grignard reagent. Alcohols 3 and 5 do not have two identical substituents, so they cannot be prepared in this way.

Solution to 8b (2) Methyl propanoate and excess methylmagnesium bromide.

PROBLEM 9♦

Which of the following secondary alcohols can be prepared from the reaction of methyl formate with excess Grignard reagent?

CH₃CH₂CHCH₃
      |
     OH

   A

CH₃CHCH₃
    |
   OH

  B

CH₃CHCH₂CH₂CH₃
    |
   OH

   C

CH₃CH₂CHCH₂CH₃
      |
     OH

    D

PROBLEM SOLVING STRATEGY

Predicting Reactions with Grignard Reagents

Why does a Grignard reagent not add to the carbonyl carbon of a carboxylic acid?

We know that Grignard reagents add to carbonyl carbons, so if we find that a Grignard reagent does not add to the carbonyl carbon of a carboxylic acid, we can conclude that it must react more rapidly with another part of the molecule. A carboxylic acid has an acidic proton that reacts rapidly with the Grignard reagent, converting it to an alkane.

$$
\underset{R}{\overset{O}{\underset{\|}{C}}}\!\!-\!O\!-\!H \;+\; CH_3CH_2\!-\!MgBr \;\longrightarrow\; \underset{R}{\overset{O}{\underset{\|}{C}}}\!\!-\!O^-\; {}^+MgBr \;+\; CH_3CH_3
$$

Now continue on to Problem 10.

PROBLEM 10♦

Which of the following compounds will not undergo a nucleophilic addition reaction with a Grignard reagent?

         O
         ‖
         C
CH₃CH₂   NHCH₃

     A

        O
        ‖
        C
CH₃CH₂   OCH₃

    B

        O
        ‖
        C
HOCH₂CH₂   OCH₃

    C

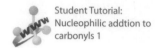

17.5 Reactions of Carbonyl Compounds with Acetylide Ions

We have seen that a terminal alkyne can be converted into an acetylide ion by a strong base (Section 6.10).

$$CH_3C{\equiv}CH \xrightarrow[\text{NH}_3]{\text{NaNH}_2} CH_3C{\equiv}C{:}^-$$

An acetylide ion is another example of a strongly basic nucleophile that reacts with a carbonyl compound to form a nucleophilic addition product. Because it is a reaction that forms a new C—C bond, it is an important reaction to synthetic chemists. When the reaction is over, a weak acid (one that will not react with the triple bond, such as pyridinium ion), is added to the reaction mixture to protonate the alkoxide ion.

$$CH_3CH_2{-}\underset{H}{\overset{O}{\underset{|}{C}}}\ +\ CH_3C{\equiv}C{:}^- \longrightarrow CH_3CH_2\underset{}{\overset{:\ddot{O}{:}^-}{CHC}}{\equiv}CCH_3 \xrightarrow{} CH_3CH_2\underset{}{\overset{:\ddot{O}H}{CHC}}{\equiv}CCH_3$$

PROBLEM 11

Show how the following compounds could be prepared, using ethyne as one of the starting materials. Explain why ethyne should be alkylated before, rather than after, nucleophilic addition.

a. 1-pentyn-3-ol   **b.** 1-phenyl-2-butyn-1-ol   **c.** 2-methyl-3-hexyn-2-ol

PROBLEM 12◆

What is the product of the reaction of an ester with excess acetylide ion followed by the addition of pyridinium ion?

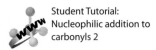

17.6 Reactions of Carbonyl Compounds with Hydride Ion

A hydride ion is another strongly basic nucleophile (Table 16.1 on page 724) that reacts with aldehydes and ketones to form nucleophilic addition products.

Mechanism for the reaction of an aldehyde or a ketone with hydride ion

$$\underset{R}{\overset{O}{\|}}{\underset{R'}{C}}\ +\ {:}H^- \longrightarrow R{-}\underset{H}{\overset{O^-}{\underset{|}{C}}}{-}R' \underset{:B}{\overset{HB^+}{\rightleftharpoons}} R{-}\underset{H}{\overset{OH}{\underset{|}{C}}}{-}R' \quad \boxed{\text{product of nucleophilic addition}}$$

- Addition of a hydride ion to an aldehyde or a ketone forms an alkoxide ion.
- Subsequent protonation by an acid produces an alcohol. The overall reaction adds H_2 to the carbonyl group.

Recall that the addition of hydrogen to a compound is a **reduction reaction** (Section 4.11). Aldehydes and ketones are generally reduced using sodium borohydride ($NaBH_4$) as the source of hydride ion. Aldehydes are reduced to primary alcohols, and ketones are reduced to secondary alcohols. Notice that the acid is not added to the reaction mixture until after the hydride ion has reacted with the carbonyl compound.

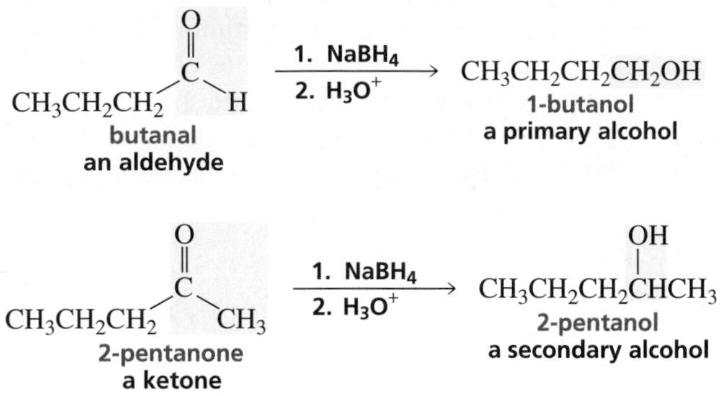

$$CH_3CH_2CH_2 \overset{\overset{\displaystyle O}{\|}}{\underset{\text{butanal}}{C}} H \xrightarrow[\text{2. H}_3\text{O}^+]{\text{1. NaBH}_4} CH_3CH_2CH_2CH_2OH$$

butanal
an aldehyde

1-butanol
a primary alcohol

VCL   **Carbonyl Reduction**

$$CH_3CH_2CH_2 \overset{\overset{\displaystyle O}{\|}}{\underset{}{C}} CH_3 \xrightarrow[\text{2. H}_3\text{O}^+]{\text{1. NaBH}_4} CH_3CH_2CH_2\overset{\overset{\displaystyle OH}{|}}{\underset{}{C}}HCH_3$$

2-pentanone
a ketone

2-pentanol
a secondary alcohol

PROBLEM 13◆

What alcohols are obtained from the reduction of the following compounds with sodium borohydride?

a. 2-methylpropanal **b.** cyclohexanone **c.** benzaldehyde **d.** acetophenone

---

Because class I carbonyl compounds have a group that can be replaced by another group, they undergo two successive reactions with hydride ion, just like they undergo two successive reactions with a Grignard reagent (Section 17.4.). For this reason, the reaction of an acyl chloride with sodium borohydride forms an alcohol.

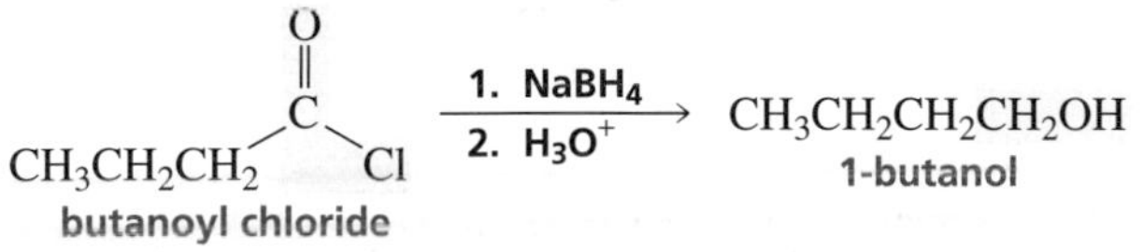

$$CH_3CH_2CH_2 \overset{\overset{\displaystyle O}{\|}}{\underset{}{C}} Cl \xrightarrow[\text{2. H}_3\text{O}^+]{\text{1. NaBH}_4} CH_3CH_2CH_2CH_2OH$$

butanoyl chloride

1-butanol

The mechanism of the reaction is shown below.

Mechanism for the reaction of an acyl chloride with hydride ion

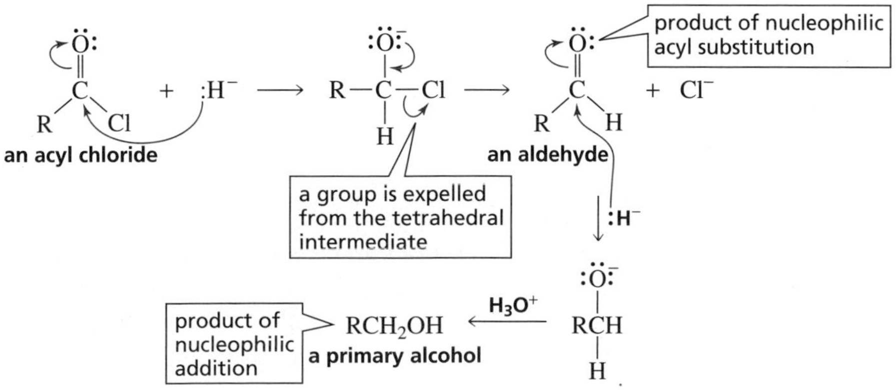

- The acyl chloride undergoes a nucleophilic acyl substitution reaction because it has a group (Cl^-) that can be replaced by hydride ion. The product of this reaction is an aldehyde.

- The aldehyde then undergoes a nucleophilic addition reaction with a second equivalent of hydride ion, forming an alkoxide ion, which, when protonated, gives a primary alcohol.

Sodium borohydride ($NaBH_4$) is not a sufficiently strong hydride donor to react with carbonyl compounds that are less reactive than aldehydes and ketones. Therefore, esters, carboxylic acids, and amides must be reduced with lithium aluminum hydride ($LiAlH_4$), a more reactive hydride donor. Because lithium aluminum hydride is more reactive than sodium borohydride, it is not as safe or as easy to use. It reacts violently with protic solvents, so it must be used in a dry, aprotic solvent.

The reaction of an ester with $LiAlH_4$ produces two alcohols, one corresponding to the acyl portion of the ester and one corresponding to the alkyl portion.

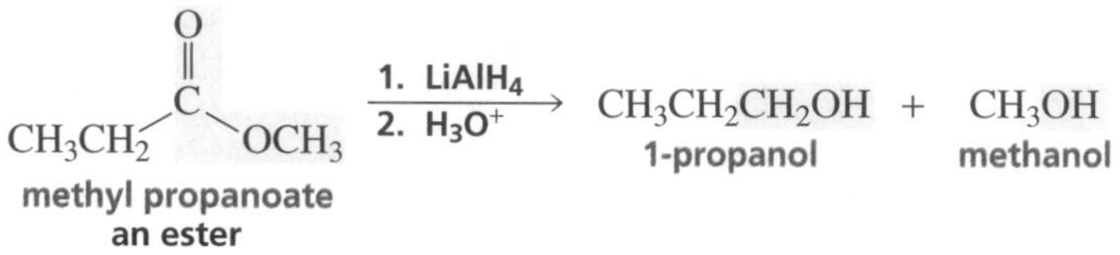

The mechanism for the reaction is shown below.

Mechanism for the reaction of an ester with hydride ion

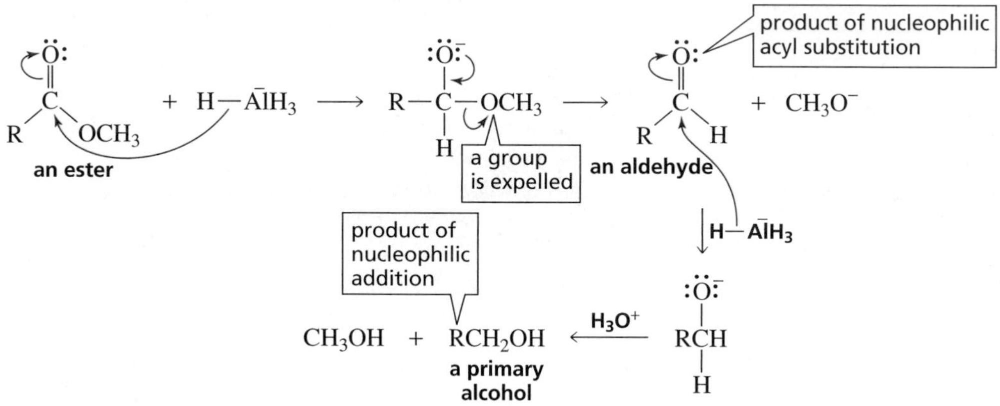

Esters and acyl chlorides undergo two successive reactions with hydride ion and with Grignard reagents.

- The ester undergoes a nucleophilic acyl substitution reaction because an ester has a group (CH_3O^-) that can be replaced by hydride ion. The product of this reaction is an aldehyde.

- The aldehyde then undergoes a nucleophilic addition reaction with a second equivalent of hydride ion, forming an alkoxide ion, which when protonated gives a primary alcohol. The reaction cannot be stopped at the aldehyde stage because an aldehyde is more reactive than an ester toward nucleophilic attack (Section 17.2).

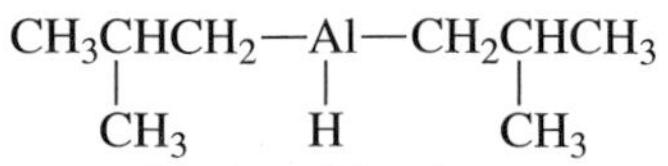

Chemists have found that if diisobutylaluminum hydride (DIBALH) is used as the hydride donor at a low temperature, the reaction can be stopped after the addition of one equivalent of hydride ion. This reagent, therefore, makes it possible to convert esters into aldehydes, which is initially surprising, since aldehydes are more reactive than esters toward hydride ion.

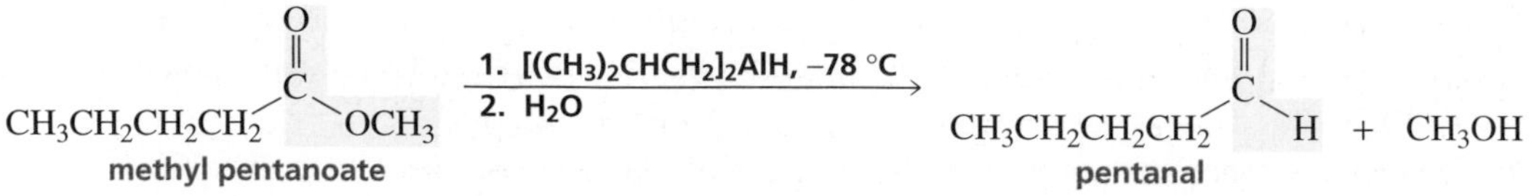

diisobutylaluminum
hydride
DIBALH

In addition to reacting with aldehydes and ketones (class II carbonyl compounds), Grignard reagents react with class I carbonyl compounds (carbonyl compounds that have a group that can be replaced by another group).

Class I carbonyl compounds undergo two successive reactions with the Grignard reagent. For example, when an ester reacts with a Grignard reagent, the first reaction is a *nucleophilic acyl substitution reaction* because an ester, unlike an aldehyde or a ketone, has a group that can be replaced by the Grignard reagent; the second reaction is a *nucleophilic addition reaction*. The mechanism for the overall reaction is shown below.

Mechanism for the reaction of an ester with a Grignard reagent

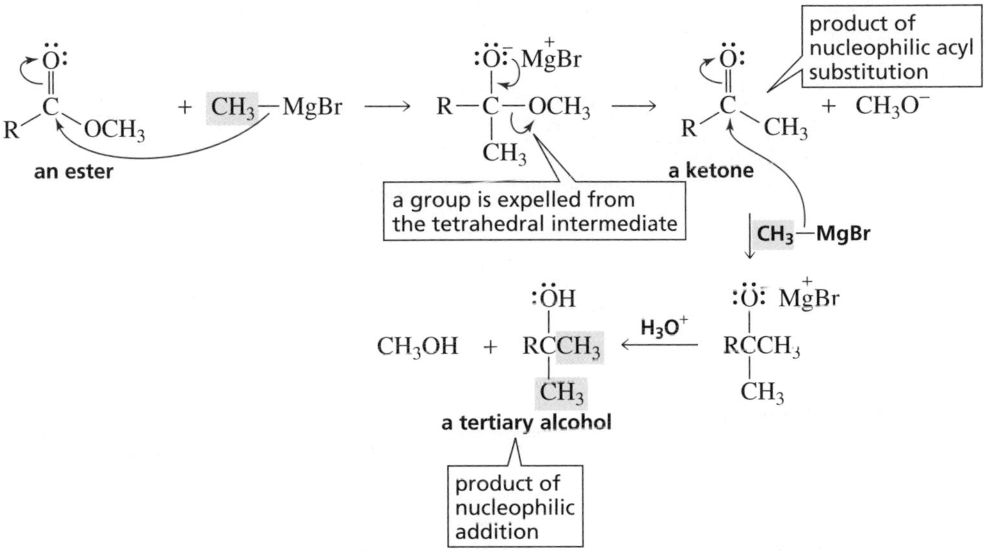

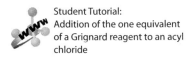

Student Tutorial:
Addition of the first equivalent of a Grignard reagent to an ester

- Nucleophilic attack by the Grignard reagent forms a tetrahedral intermediate that is unstable because it has a group that can be expelled.

- The tetrahedral intermediate expels methoxide ion, forming a ketone. The reaction does not stop at the ketone stage, however, because ketones are more reactive than esters toward nucleophilic attack (Section 17.2).

- Reaction of the ketone with a second molecule of the Grignard reagent, followed by protonation of the alkoxide ion, forms a tertiary alcohol.

Because the tertiary alcohol is formed as a result of two successive reactions with a Grignard reagent, the alcohol has two identical alkyl groups bonded to the tertiary carbon.

Student Tutorial:
Addition of the one equivalent of a Grignard reagent to an acyl chloride

Tertiary alcohols are also formed from the reaction of two equivalents of a Grignard reagent with an acyl chloride. The first equivalent replaces the Cl in a nucleophilic acyl substitution reaction; the second equivalent reacts in a nucleophilic addition reaction.

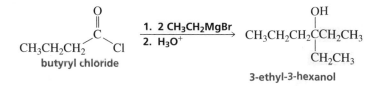

In theory, we should be able to stop this reaction at the ketone stage because a ketone is less reactive than an acyl halide. However, the Grignard reagent is so reactive that it can be prevented from reacting with the ketone only under very carefully controlled conditions. There are better ways to synthesize ketones (Appendix IV).

When a Grignard reagent reacts with a ketone, the addition product is a tertiary alcohol.

 **Grignard Addition-1**

2-pentanone + ethylmagnesium bromide $\longrightarrow$ $\xrightarrow{H_3O^+}$ 3-methyl-3-hexanol a tertiary alcohol

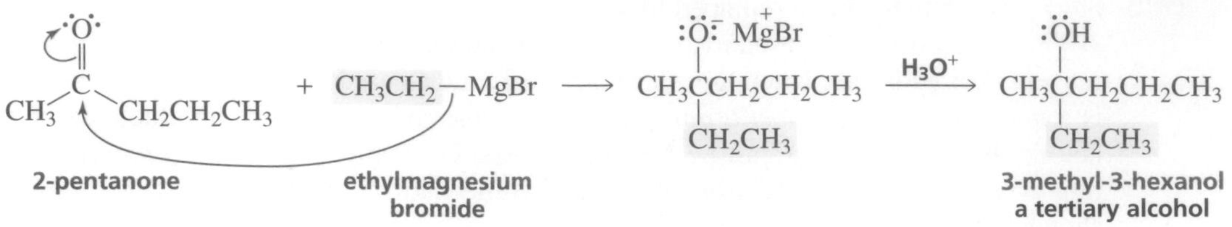

In the following reactions, the reagents are numbered in order of use, indicating that the acid is not added until after the Grignard reagent has reacted with the carbonyl compound.

3-pentanone $\xrightarrow[\text{2. } H_3O^+]{\text{1. } CH_3MgBr}$ 3-methyl-3-pentanol

Student Tutorial:
Addition of a Grignard reagent to a ketone

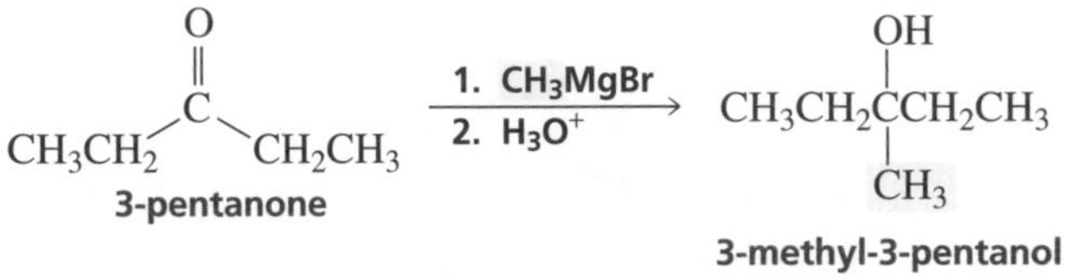

butanal $\xrightarrow[\text{2. } H_3O^+]{\text{1. } \text{—MgBr}}$ 1-phenyl-1-butanol

Grignard Addition-2

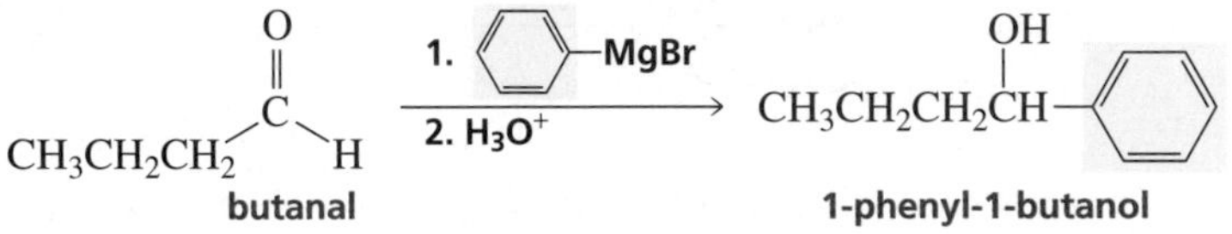

A Grignard reagent can also react with carbon dioxide. The product of the reaction is a carboxylic acid that has one more carbon atom than the Grignard reagent.

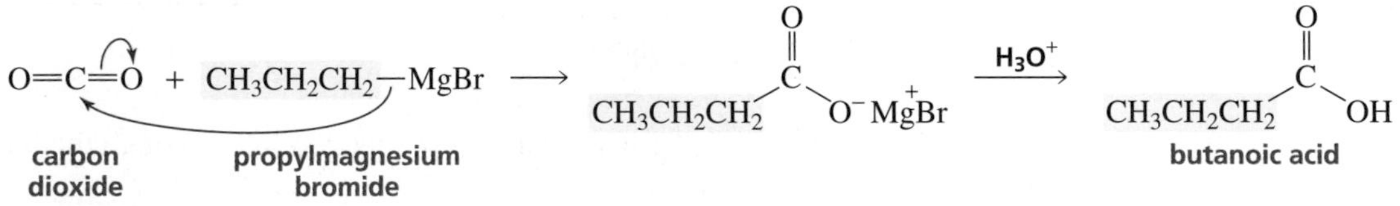

$O=C=O$ carbon dioxide + $CH_3CH_2CH_2$—MgBr propylmagnesium bromide $\longrightarrow$ $\xrightarrow{H_3O^+}$ butanoic acid

PROBLEM 5◆

What products would be formed when the following compounds react with CH_3MgBr, followed by the addition of acid? Disregard stereoisomers.

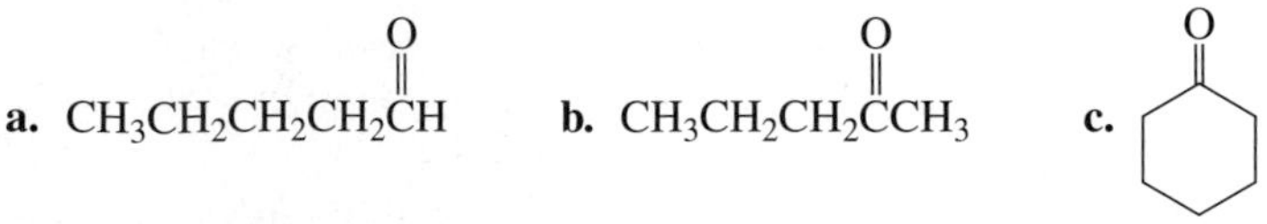

a. $CH_3CH_2CH_2CH_2CH$ (with O)   **b.** $CH_3CH_2CH_2CCH_3$ (with O)   **c.** (cyclohexanone structure)

PROBLEM 6◆

We saw that 3-methyl-3-hexanol can be synthesized from the reaction of 2-pentanone with ethylmagnesium bromide. What two other combinations of ketone and Grignard reagent could be used to prepare the same tertiary alcohol?

PROBLEM 7◆

a. How many isomers are obtained from the reaction of 2-pentanone with ethylmagnesium bromide followed by treatment with aqueous acid?

b. How many isomers are obtained from the reaction of 2-pentanone with methylmagnesium bromide followed by treatment with aqueous acid?

attached to the tetrahedral intermediate, either one of which can be protonated under the acidic conditions employed in the reaction and, therefore, expelled.

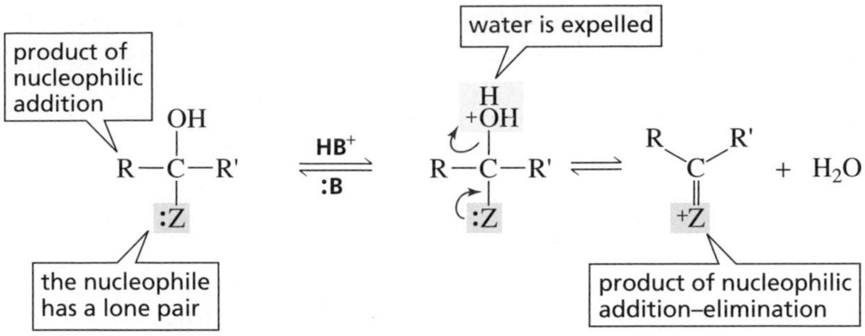

17.4 Reactions of Carbonyl Compounds with Grignard Reagents

Addition of a Grignard reagent to a carbonyl compound is a versatile reaction that leads to the formation of a new C—C bond. This reaction can produce compounds with a variety of structures because both the structure of the carbonyl compound and the structure of the Grignard reagent can be varied. We have seen that a Grignard reagent can be prepared by adding an alkyl halide to magnesium shavings in diethyl ether under anhydrous conditions (Section 10.12). We also saw that a Grignard reagent reacts as if it were a carbanion; therefore, this nucleophile is a strong base. Consequently, aldehydes and ketones undergo nucleophilic addition reactions with Grignard reagents (Section 17.3).

$$CH_3CH_2Br \xrightarrow[\text{Et}_2\text{O}]{\text{Mg}} CH_3CH_2MgBr$$

$$CH_3CH_2MgBr \quad \textbf{reacts as if it were} \quad CH_3\overset{\cdot\cdot}{C}H_2 \quad \overset{+}{M}gBr$$

Attack of a Grignard reagent on a carbonyl carbon forms an alkoxide ion that is complexed with magnesium ion. Addition of water or dilute acid breaks up the complex. When a Grignard reagent reacts with formaldehyde, the addition product is a primary alcohol.

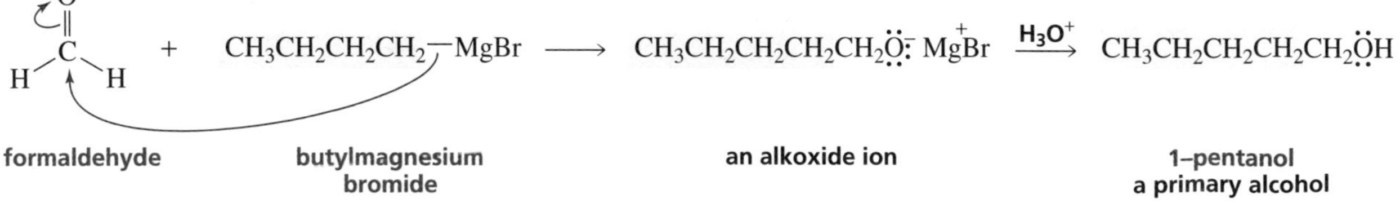

When a Grignard reagent reacts with an aldehyde other than formaldehyde, the addition product is a secondary alcohol.

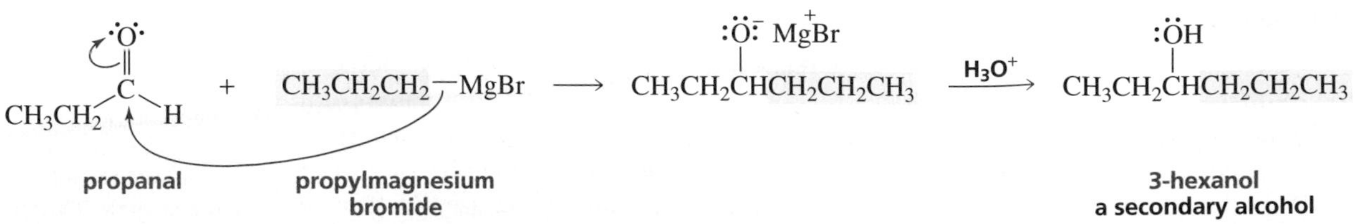

17.3 How Aldehydes and Ketones React

In Section 16.5, we saw that the carbonyl group of a carboxylic acid or a carboxylic acid derivative is attached to a group that can be replaced by another group. Therefore, these compounds undergo **nucleophilic acyl substitution reactions**.

Carboxylic acid derivatives undergo nucleophilic acyl substitution reactions with nucleophiles.

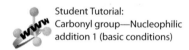

a group that can be replaced by another group

product of nucleophilic substitution

In contrast, the carbonyl group of an aldehyde or a ketone is attached to a group that is too strong a base (H^- or R^-) to be eliminated under normal conditions, so it cannot be replaced by another group. Consequently, aldehydes and ketones do not undergo acyl substitution reactions.

When a nucleophile adds to the carbonyl group of an aldehyde or a ketone, a tetrahedral compound is formed. If the nucleophile is a strong base, the tetrahedral compound does not have a group that can be expelled. The tetrahedral compound, therefore, is the final product of the reaction. The reaction is an *irreversible* **nucleophilic addition reaction**.

Aldehydes and ketones undergo nucleophilic addition reactions with nucleophiles that do not have a lone pair on the attacking atom.

a nucleophile that is a strong base

the reaction is irreversible because Z is too basic to be expelled

product of nucleophilic addition

HB^+ represents any species in the solution that is capable of donating a proton, and :B represents any species in the solution that is capable of removing a proton.

Student Tutorial:
Carbonyl group—Nucleophilic addition 1 (basic conditions)

If the nucleophile is a relatively weak base, the product of the reaction will again be the tetrahedral compound. However, the reaction will be a *reversible* **nucleophilic addition reaction** because the tetrahedral compound can expel the weak base and revert to the starting materials.

Student Tutorial:
Carbonyl group—Nucleophilic addition 2 (reverse conditions)

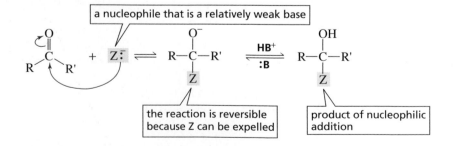

a nucleophile that is a relatively weak base

the reaction is reversible because Z can be expelled

product of nucleophilic addition

If the attacking atom of the nucleophile has a lone pair and there is sufficient acid to protonate the OH group of the tetrahedral compound, water can be eliminated from the addition product. This is called a **nucleophilic addition–elimination reaction**. We will see that the fate of the dehydrated product depends on the identity of Z. The nucleophilic addition–elimination reaction is reversible because there are two groups

Aldehydes and ketones undergo nucleophilic addition–elimination reactions with nucleophiles that have a lone pair on the attacking atom.

reaction. If H_3O^+ were used, the product would be an ammonium ion rather than an amine.)

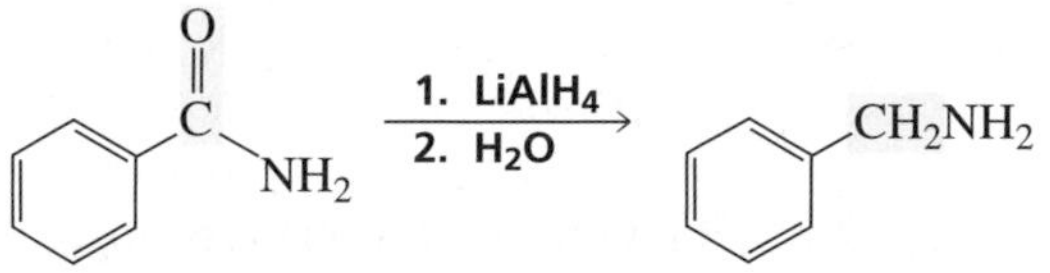

benzamide → benzylamine **a primary amine** (1. LiAlH₄, 2. H₂O)

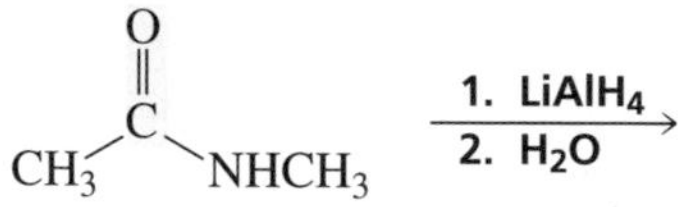

N-methylacetamide → $CH_3CH_2NHCH_3$ ethylmethylamine **a secondary amine** (1. LiAlH₄, 2. H₂O)

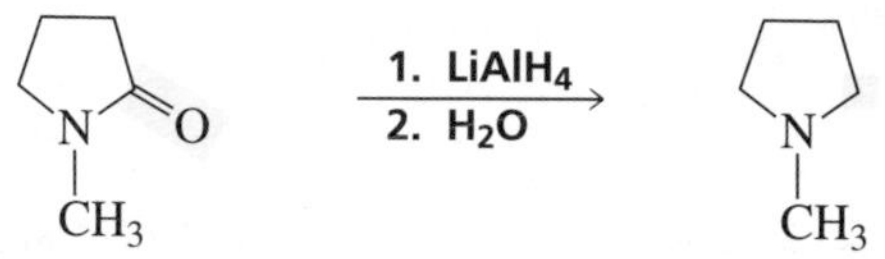

N-methyl-γ-butyrolactam → N-methylpyrrolidine **a tertiary amine** (1. LiAlH₄, 2. H₂O)

The mechanism of the reaction shows why the product of the reaction is an amine. Take a minute to note the similarities between this mechanism for the reaction of hydride ion with an *N*-substituted amide and the mechanism for the reaction of hydride ion with a carboxylic acid.

Mechanism for the reaction of an *N*-substituted amide with hydride ion

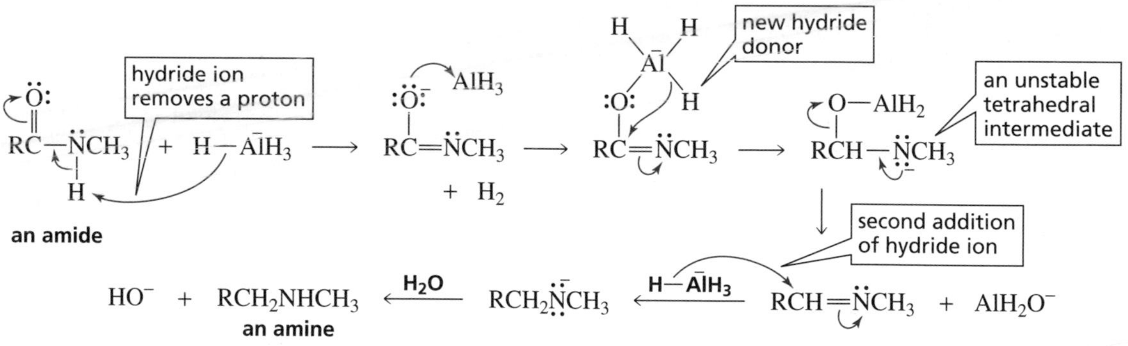

The mechanisms for the reaction of LiAlH₄ with unsubstituted and *N,N*-disubstituted amines are somewhat different, but have the same result: the conversion of a carbonyl group into a methylene group.

PROBLEM 14◆

What amides would you treat with LiAlH₄ in order to prepare the following amines?

a. benzylmethylamine   **b.** ethylamine   **c.** diethylamine   **d.** triethylamine

---

PROBLEM 15

Starting with *N*-benzylbenzamide, how would you make the following compounds?

a. dibenzylamine   **b.** benzoic acid   **c.** benzyl alcohol

The reaction is carried out at $-78\,°C$ (the temperature of a dry ice–acetone bath). At this cold temperature, the initially formed tetrahedral intermediate is stable, so it does not eliminate the alkoxide ion. If all of the unreacted hydride donor is removed before the solution warms up, there will be no reducing agent available to react with the aldehyde that is formed when the tetrahedral intermediate eliminates the alkoxide ion.

The reaction of a carboxylic acid with $LiAlH_4$ forms a single primary alcohol.

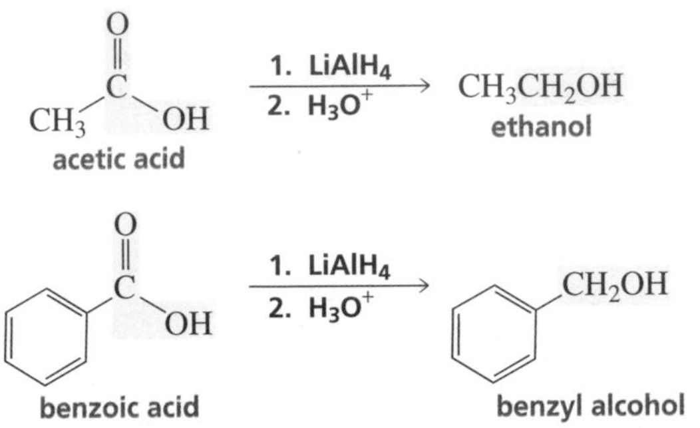

The mechanism for the reaction is shown below.

Mechanism for the reaction of a carboxylic acid with hydride ion

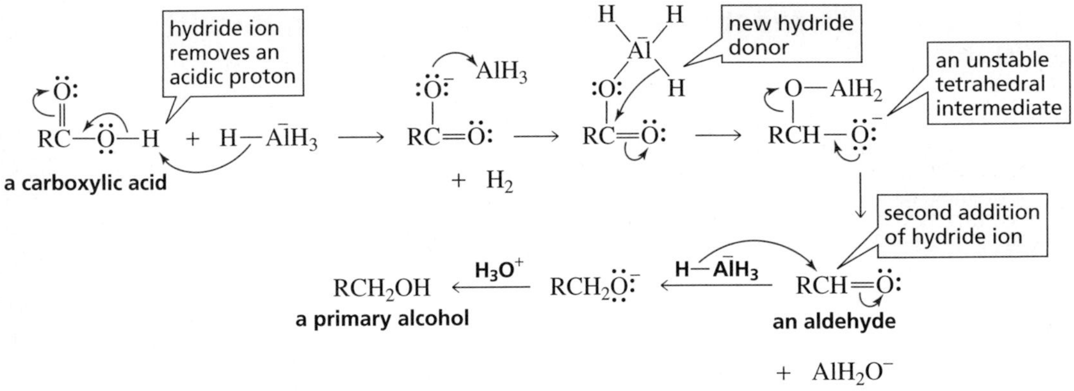

- In the first step, a hydride ion reacts with the acidic hydrogen of the carboxylic acid, forming H_2 and a carboxylate ion.
- We have seen that nucleophiles do not react with a carboxylate ion because of its negative charge. However, in this case, an electrophile (AlH_3) is present that accepts a pair of electrons from the carboxylate ion and forms a new hydride donor.
- Then, analogous to the reduction of an ester by $LiAlH_4$, two successive additions of hydride ion take place, with an aldehyde being formed as an intermediate on the way to the primary alcohol.

Amides also undergo two successive additions of hydride ion when they react with $LiAlH_4$. Overall, the reaction converts a carbonyl group into a methylene (CH_2) group. The product of the reaction is an amine. Primary, secondary, or tertiary amines can be formed, depending on the number of substituents bonded to the nitrogen of the amide. (Notice that H_2O rather than H_3O^+ is used in the second step of the

17.7 Reactions of Aldehydes and Ketones with Hydrogen Cyanide

Hydrogen cyanide adds to aldehydes and ketones to form **cyanohydrins**.

Mechanism for the reaction of an aldehyde or ketone with cyanide ion

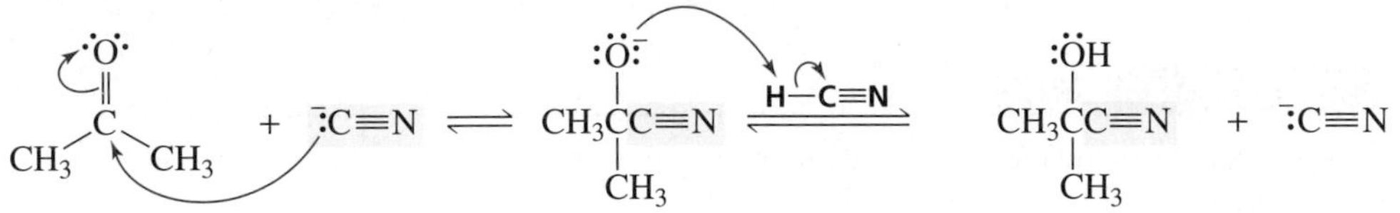

acetone   acetone cyanohydrin

- In the first step, the cyanide ion attacks the carbonyl carbon.
- The alkoxide ion accepts a proton from an undissociated molecule of hydrogen cyanide.

Because hydrogen cyanide is a toxic gas, the best way to carry out this reaction is to generate hydrogen cyanide during the reaction by adding HCl to a mixture of the aldehyde or ketone and excess sodium cyanide. Excess sodium cyanide is used in order to ensure that some cyanide ion is available to act as a nucleophile.

Compared with Grignard reagents and hydride ion, cyanide ion is a relatively weak base (the pK_a of $HC\equiv N$ is 9.14, the pK_a of $HC\equiv CH$ is 25, the pK_a of CH_3CH_3 is >60), which means that the cyano group can be eliminated from the addition product. Cyanohydrins, however, are stable: the OH group will not eliminate the cyano group because the transition state for that elimination reaction would be relatively unstable since the oxygen atom would bear a partial positive charge. If the OH group loses its proton, however, the cyano group will be eliminated, because the oxygen atom would then have a partial negative charge instead of a partial positive charge in the transition state. Therefore, in basic solutions, a cyanohydrin is converted back to the carbonyl compound.

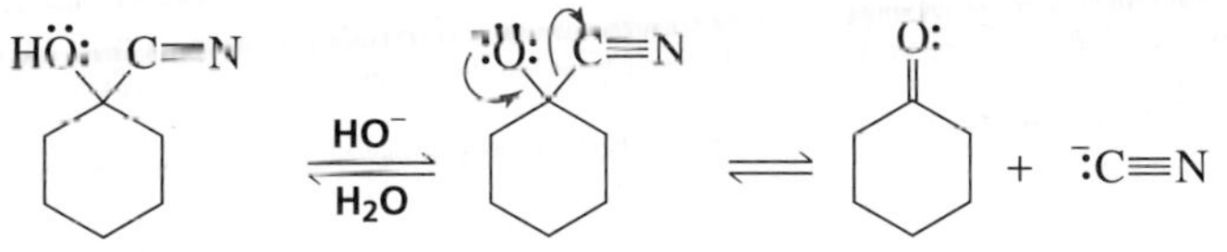

cyclohexanone
cyanohydrin

Cyanide ion does not react with esters because the cyanide ion is a weaker base than an alkoxide ion, so the cyanide ion would be eliminated from the tetrahedral intermediate.

The addition of hydrogen cyanide to aldehydes and ketones is a synthetically useful reaction because of the subsequent reactions that can be carried out on the cyanohydrin. For example, the acid-catalyzed hydrolysis of a cyanohydrin forms an α-hydroxycarboxylic acid (Section 16.19).

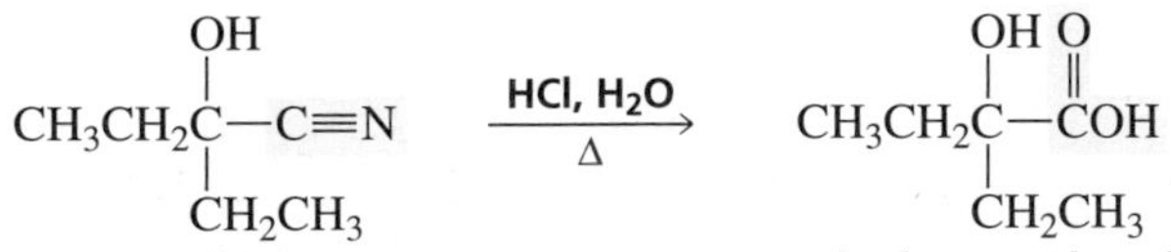

a cyanohydrin   an α-hydroxy carboxylic acid

The catalytic addition of hydrogen to the triple bond of a cyanohydrin produces a primary amine with an OH group on the β-carbon.

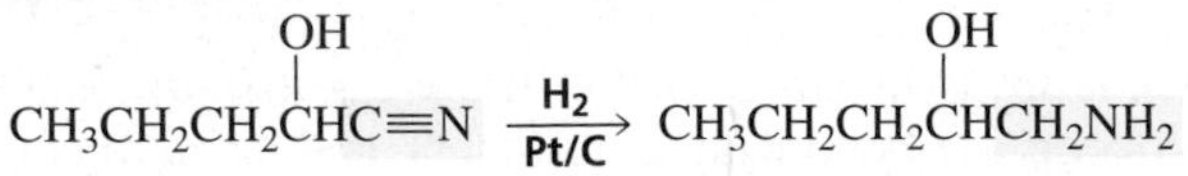

PROBLEM 16◆

Can a cyanohydrin be prepared by treating a ketone with sodium cyanide?

PROBLEM 17◆

Explain why aldehydes and ketones react with a weak acid such as hydrogen cyanide in the presence of $^-C \equiv N$, but do not react with strong acids such as HCl or H_2SO_4 in the presence of Cl^- or HSO_4^-.

PROBLEM 18 *SOLVED*

How can the following compounds be prepared, starting with a carbonyl compound that has one fewer carbon atoms than the desired product?

a. $HOCH_2CH_2NH_2$

b. 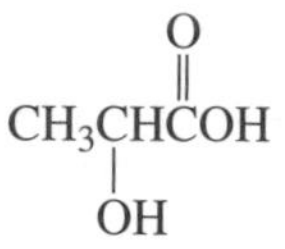 CH_3CHCOH (with $\overset{O}{\underset{\parallel}{}}$ and OH)

Solution to 18a   The starting material for the synthesis of this two-carbon compound must be formaldehyde. Addition of hydrogen cyanide followed by addition of H_2 to the triple bond of the cyanohydrin forms the target compound.

$$HCH \xrightarrow[\text{HCl}]{\text{NaC} \equiv \text{N}} HOCH_2C \equiv N \xrightarrow[\text{Pt/C}]{H_2} HOCH_2CH_2NH_2$$

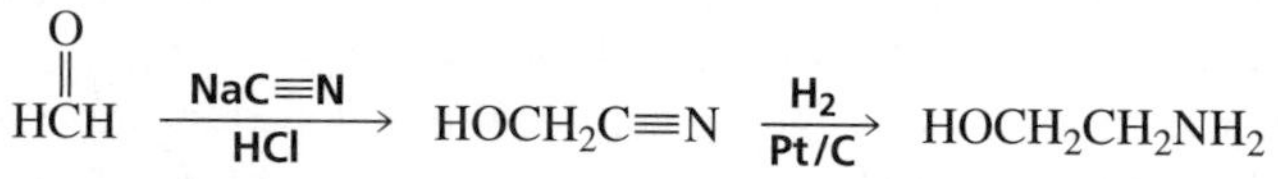

Solution to 18b   The addition of cyanide ion adds one carbon to the reactant, so the starting material for the synthesis of this three-carbon α-hydroxycarboxylic acid must be ethanal. Addition of hydrogen cyanide, followed by hydrolysis of the cyanohydrin, forms the target compound.

$$CH_3CH \xrightarrow[\text{HCl}]{\text{NaC} \equiv \text{N}} CH_3CHC \equiv N \xrightarrow[\Delta]{\text{HCl, } H_2O} CH_3CHCOH$$

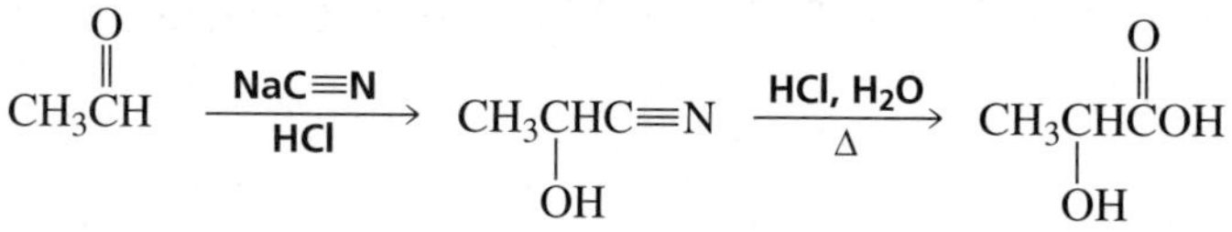

17.8 Reactions of Aldehydes and Ketones with Amines and Derivatives of Amines

An aldehyde or a ketone reacts with a *primary* amine to form an imine. An **imine** is a compound with a carbon–nitrogen double bond. The imine obtained from the reaction of a carbonyl compound and a primary amine is often called a **Schiff base**.

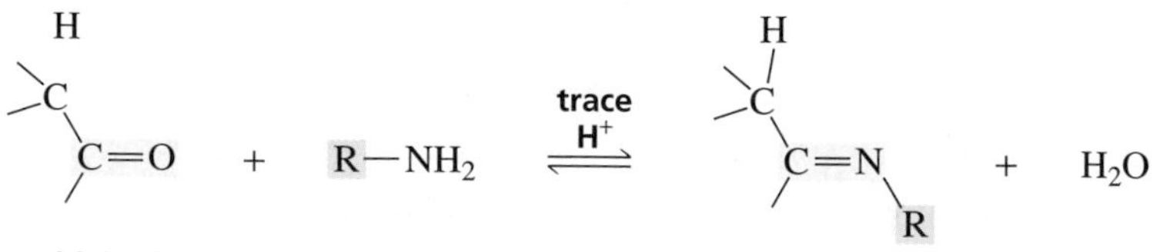

$$\overset{H}{\underset{}{}}C=O \quad + \quad R-NH_2 \quad \underset{}{\overset{\text{trace}}{\underset{H^+}{\rightleftarrows}}} \quad \overset{H}{\underset{}{}}C=N_R \quad + \quad H_2O$$

an aldehyde or   a primary amine   an imine
a ketone   a Schiff base

A $C=N$ group (Figure 17.1) is similar to a $C=O$ group (Figure 16.1 on page 729). The imine nitrogen is sp^2 hybridized. One of its sp^2 orbitals forms a σ bond with the imine carbon, one forms a σ bond with a substituent, and the third contains a lone pair. The p orbital of nitrogen and the p orbital of carbon overlap to form a π bond.

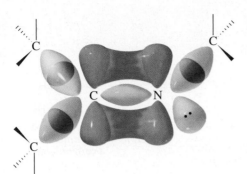

◀ **Figure 17.1**
Bonding in an imine

An aldehyde or a ketone reacts with a *secondary amine* to form an enamine (pronounced "ENE-amine"). An **enamine** is an α,β-unsaturated tertiary amine—a tertiary amine with a double bond in the α,β-position relative to the nitrogen atom. Notice that the double bond is in the part of the molecule that comes from the aldehyde or ketone, not from the part that is provided by the secondary amine. The name comes from "ene" + "amine," with the "e" omitted in order to avoid two successive vowels.

$$
\underset{\substack{\text{an aldehyde or} \\ \text{a ketone}}}{\text{H}\overset{|}{\underset{|}{\text{C}}}\text{C}=\text{O}} \quad + \quad \underset{\substack{\text{a secondary amine}}}{\overset{R}{\underset{R}{\text{NH}}}} \quad \underset{\substack{\text{trace} \\ \text{H}^+}}{\rightleftharpoons} \quad \underset{\substack{\text{an enamine}}}{-\text{C}\overset{\fbox{α, -position}}{\underset{|}{\text{C}-\text{N}\overset{R}{\underset{R}{}}}}} \quad + \quad \text{H}_2\text{O}
$$

When you first look at the products of imine and enamine formation, they appear to be quite different. However, when you look at the mechanisms for the reactions, you will see that they are exactly the same except for the site from which a proton is lost in the last step.

Primary Amines Form Imines

Aldehydes and ketones react with primary amines to form imines.

$$
\underset{\substack{\text{benzaldehyde} \\ \text{an aldehyde}}}{\text{C}=\text{O}} \quad + \quad \underset{\substack{\text{ethylamine} \\ \text{a primary amine}}}{\text{CH}_3\text{CH}_2\text{NH}_2} \quad \underset{\substack{\text{trace} \\ \text{H}^+}}{\rightleftharpoons} \quad \underset{\substack{\text{an imine}}}{\text{C}=\text{NCH}_2\text{CH}_3} \quad + \quad \text{H}_2\text{O}
$$

$$
\underset{\substack{\text{3-pentanone} \\ \text{a ketone}}}{\overset{\text{CH}_3\text{CH}_2}{\underset{\text{CH}_3\text{CH}_2}{\text{C}=\text{O}}}} \quad + \quad \underset{\substack{\text{benzylamine} \\ \text{a primary amine}}}{\text{H}_2\text{NCH}_2-} \quad \underset{\substack{\text{trace} \\ \text{H}^+}}{\rightleftharpoons} \quad \underset{\substack{\text{an imine}}}{\overset{\text{CH}_3\text{CH}_2}{\underset{\text{CH}_3\text{CH}_2}{\text{C}=\text{NCH}_2-}}} \quad + \quad \text{H}_2\text{O}
$$

The reaction requires a trace amount of acid; we will see that the pH of the reaction mixture must be carefully controlled. Compared to a Grignard reagent or a hydride ion, an amine is a relatively weak base and, because it has a lone pair on the attacking atom, water is eliminated from the tetrahedral intermediate.

Aldehydes and ketones react with primary amines to form imines.

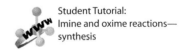

Student Tutorial:
Imine and oxime reactions—synthesis

Mechanism for imine formation

HB⁺ represents any species in the solution that is capable of donating a proton, and :B represents any species in the solution that is capable of removing a proton.

nucleophile attacks the carbonyl carbon

N–protonated carbinolamine

neutral tetrahedral intermediate
a carbinolamine

O–protonated carbinolamine

elimination of water

an imine   a protonated imine

removal of a proton

- The amine attacks the carbonyl carbon.
- Gain of a proton by the alkoxide ion and loss of a proton by the ammonium ion forms a neutral tetrahedral intermediate.
- The neutral tetrahedral intermediate, called a *carbinolamine*, is in equilibrium with two protonated forms because either the oxygen or the nitrogen can be protonated.
- Elimination of water from the oxygen-protonated intermediate forms a protonated imine that loses a proton to yield the imine.

Imine formation is reversible because there are two groups that can be expelled from the tetrahedral intermediate. The equilibrium favors the nitrogen-protonated tetrahedral intermediate because nitrogen is more basic than oxygen. However, the equilibrium can be forced toward the imine by removing water as it is formed or by precipitation of the imine.

Overall, the addition of an amine to an aldehyde or a ketone is a *nucleophilic addition–elimination reaction*: nucleophilic addition of an amine to form an unstable tetrahedral intermediate, followed by elimination of water. The tetrahedral intermediate is unstable because it contains groups that can be protonated and thereby become weak enough bases to be easily expelled.

The pH at which imine formation is carried out must be carefully controlled. There must be sufficient acid present to protonate the tetrahedral intermediate so that H_2O rather than the much more basic HO^- is the leaving group. However, if too much acid is present, it will protonate all of the reactant amine. Protonated amines are not nucleophiles, so they cannot react with carbonyl groups. Therefore, unlike the acid-catalyzed reactions we have seen previously (Sections 16.11 and 16.17), there is not sufficient acid present to protonate the carbonyl group in the first step of the reaction (see Problem 20).

A plot of the observed rate constant for the reaction of acetone with hydroxylamine as a function of the pH of the reaction mixture is shown in Figure 17.2. This type of plot is called a **pH–rate profile**. Here it is a bell-shaped curve with the maximum rate occurring at about pH 4.5, 1.5 pH units below the pK_a of protonated hydroxylamine ($pK_a = 6.0$). As the acidity increases below pH 4.5, the rate of the reaction decreases because more and more of the amine becomes protonated. As a result, less and less of the amine is present in the nucleophilic nonprotonated form. As the acidity decreases above pH 4.5, the rate decreases because less and less of the tetrahedral intermediate is present in the reactive protonated form.

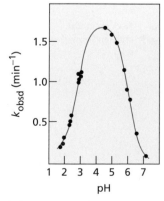

▲ **Figure 17.2**
A pH-rate profile for the reaction of acetone with hydroxylamine. It shows the dependence of the reaction rate on the pH of the reaction mixture.

Imine formation is reversible: in an acidic aqueous solution, an imine is hydrolyzed back to the carbonyl compound and amine.

An imine undergoes acid-catalyzed hydrolysis to form a carbonyl compound and a primary amine.

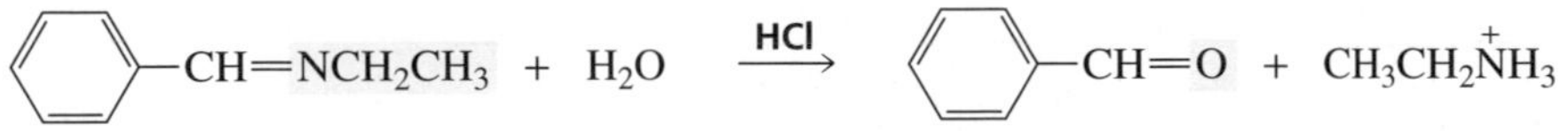

In an acidic solution, the amine is protonated and, therefore, is unable to react with the carbonyl compound to reform the imine.

Imine formation and hydrolysis are important reactions in biological systems (Sections 18.22, 23.9, and 24.6). For example, we will see that imine hydrolysis is why DNA contains A, G, C, and T nucleotides, whereas RNA contains A, G, C, and U nucleotides (Section 27.9).

PROBLEM 19◆

At what pH should imine formation be carried out if the amine's protonated form has a pK_a value of 10.0?

PROBLEM 20◆

The pK_a of protonated acetone is about -7.5 and the pK_a of protonated hydroxylamine is 6.0.

a. In a reaction with hydroxylamine at pH = 4.5 (Figure 17.2), what fraction of acetone will be present in its acidic, protonated form? (*Hint:* See Section 1.24.)

b. In a reaction with hydroxylamine at pH = 1.5, what fraction of acetone will be present in its acidic, protonated form?

c. In a reaction with acetone at pH = 1.5 (Figure 17.2), what fraction of hydroxylamine will be present in its reactive basic form?

PROBLEM 21

A ketone can be prepared from the reaction of a nitrile with a Grignard reagent. Describe the intermediate formed in this reaction, and explain how it can be converted to a ketone.

Secondary Amines Form Enamines

Aldehydes and ketones react with secondary amines to form enamines. Like imine formation, the reaction requires a trace amount of an acid catalyst.

Aldehydes and ketones react with secondary amines to form enamines.

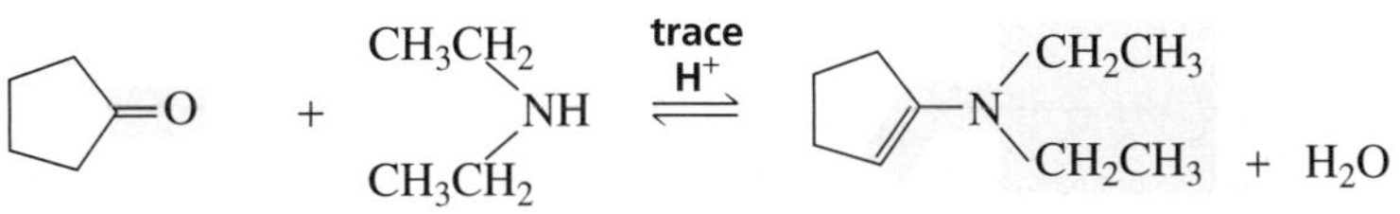

cyclopentanone — diethylamine, a secondary amine — an enamine

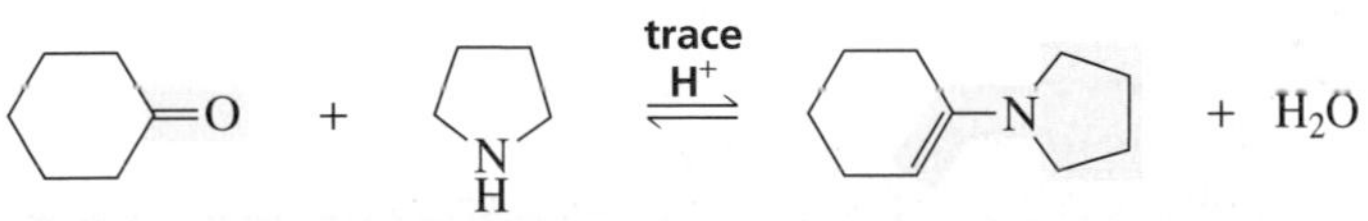

cyclohexanone — pyrrolidine, a secondary amine — an enamine

Notice that the mechanism for enamine formation is exactly the same as that for imine formation, except in the last step.

Mechanism for enamine formation

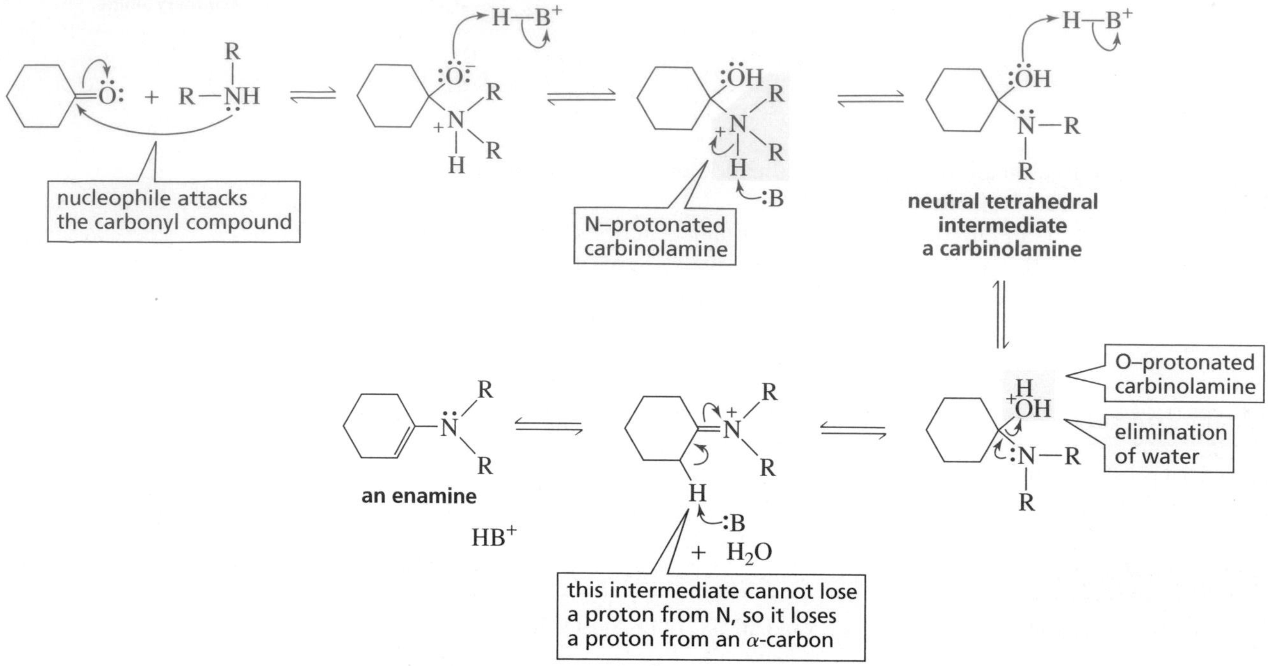

nucleophile attacks the carbonyl compound

N–protonated carbinolamine

neutral tetrahedral intermediate a carbinolamine

O–protonated carbinolamine

elimination of water

an enamine

this intermediate cannot lose a proton from N, so it loses a proton from an α-carbon

- The amine attacks the carbonyl carbon.
- Gain of a proton by the alkoxide ion and loss of a proton by the ammonium ion forms a neutral tetrahedral intermediate.
- The neutral tetrahedral intermediate is in equilibrium with two protonated forms because either the oxygen or the nitrogen can be protonated.
- When a primary amine reacts with an aldehyde or a ketone, the protonated imine loses a proton from nitrogen in the last step of the mechanism, forming a neutral imine. However, when the amine is secondary, the positively charged nitrogen is not bonded to a hydrogen. In this case, a stable neutral molecule is obtained by loss of a proton from the α-carbon of the compound derived from the carbonyl compound. An enamine is the result.

An enamine undergoes acid-catalyzed hydrolysis to form a carbonyl compound and a secondary amine.

In an aqueous acidic solution, an enamine is hydrolyzed back to the carbonyl compound and secondary amine, a reaction similar to the acid-catalyzed hydrolysis of an imine back to the carbonyl compound and primary amine.

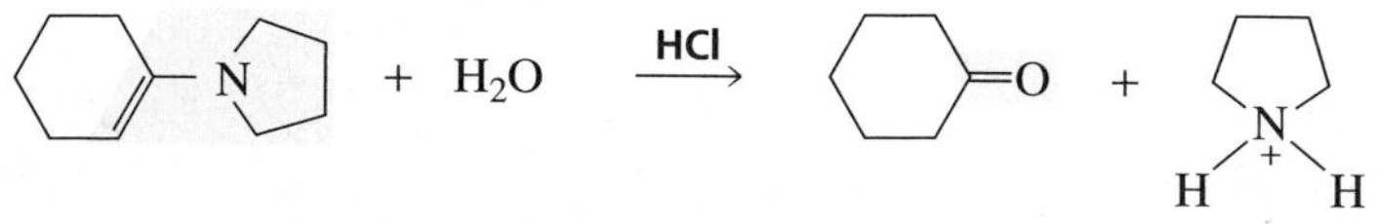

PROBLEM 22

a. Write the mechanism for the following reactions:
  1. the acid-catalyzed hydrolysis of an imine to a carbonyl compound and a primary amine
  2. the acid-catalyzed hydrolysis of an enamine to a carbonyl compound and a secondary amine
b. How do these mechanisms differ?

PROBLEM 23◆

Give the products of the following reactions. (A trace amount of acid is present in each case.)

a. cyclopentanone + ethylamine

b. cyclopentanone + diethylamine

c. acetophenone + hexylamine

d. acetophenone + cyclohexylamine

Formation of Imine Derivatives

Compounds such as hydroxylamine, phenylhydrazine, 2,4-dinitrophenylhydrazine, and semicarbazide (shown below) are similar to primary amines in that they all have an NH_2 group. Thus, like primary amines, they react with aldehydes and ketones to form imines—often called *imine derivatives* because the substituent attached to the imine nitrogen is not an R group. The imine obtained from the reaction with hydroxylamine is called an **oxime**, the imine obtained from the reaction with hydrazine is called a **hydrazone**, and the imine obtained from the reaction with semicarbazide is called a **semicarbazone**.

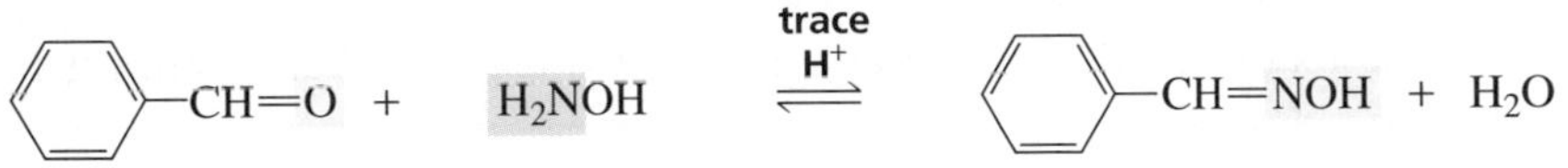

$$\text{Ph—CH=O} + \text{H}_2\text{NOH} \underset{\text{H}^+}{\overset{\text{trace}}{\rightleftharpoons}} \text{Ph—CH=NOH} + \text{H}_2\text{O}$$

hydroxylamine   an oxime

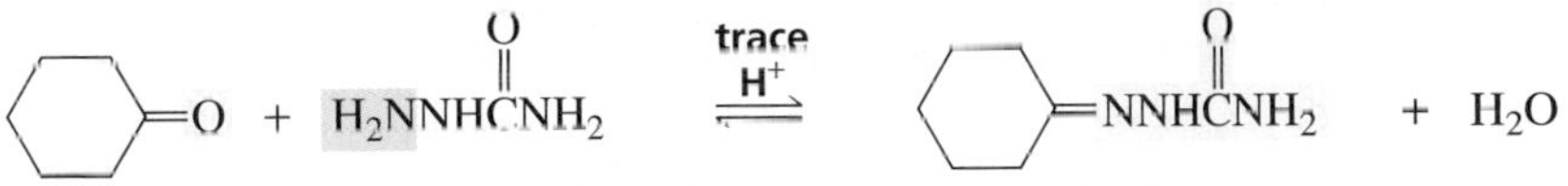

$$\text{cyclohexanone} + \text{H}_2\text{NNHCNH}_2 \underset{\text{H}^+}{\overset{\text{trace}}{\rightleftharpoons}} =\text{NNHCNH}_2 + \text{H}_2\text{O}$$

semicarbazide   a semicarbazone

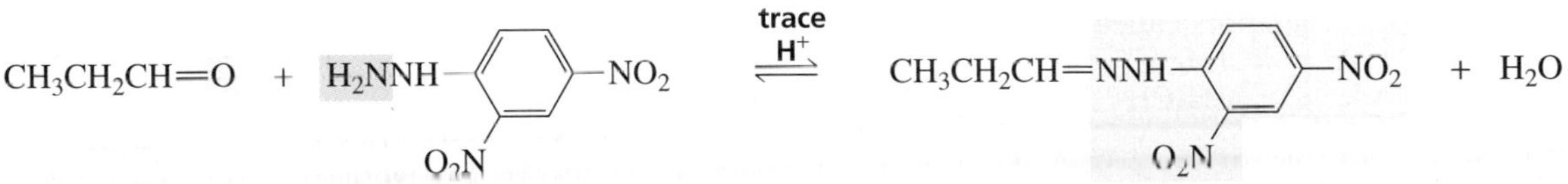

$$\text{CH}_3\text{CH}_2\text{CH=O} + \text{H}_2\text{NNH—}\underset{\text{O}_2\text{N}}{}—\text{NO}_2 \underset{\text{H}^+}{\overset{\text{trace}}{\rightleftharpoons}} \text{CH}_3\text{CH}_2\text{CH=NNH—}\underset{\text{O}_2\text{N}}{}—\text{NO}_2 + \text{H}_2\text{O}$$

2,4-dinitrophenylhydrazine   a 2,4-dinitrophenylhydrazone

IDENTIFYING OF ALDEHYDES AND KETONES WITHOUT USING SPECTROSCOPY

Before spectrophotometric techniques were available for analyzing compounds, unknown aldehydes and ketones were identified by preparing imine derivatives. For example,

suppose you have an unknown ketone whose boiling point you have determined to be 140 °C. This information allows you to narrow the possibilities to the five ketones (A to E) listed in the following table. (Ketones boiling at 139 °C and 141 °C cannot be excluded, unless your thermometer is calibrated perfectly and your laboratory technique is sensational.)

Ketone	bp (°C)	2,4-Dinitrophenylhydrazone mp (°C)	Oxime mp (°C)	Semicarbazone mp (°C)
A	140	94	57	98
B	140	102	68	123
C	139	121	79	121
D	140	101	69	112
E	141	90	61	101

Adding 2,4-dinitrophenylhydrazine to a sample of the unknown ketone produces crystals of a 2,4-dinitrophenylhydrazone that melt at 102 °C. You can now narrow the choice to two ketones: B and D. Preparing the oxime of the unknown ketone will not distinguish between B and D because the oximes of B and D

have similar melting points, but preparing the semicarbazone will allow you to identify the ketone. Finding that the semicarbazone of the unknown ketone has a melting point of 112 °C establishes that the unknown ketone is D.

PROBLEM 24

Imines can exist as stereoisomers. The isomers are named by the *E,Z* system of nomenclature. (The lone pair has the lowest priority.)

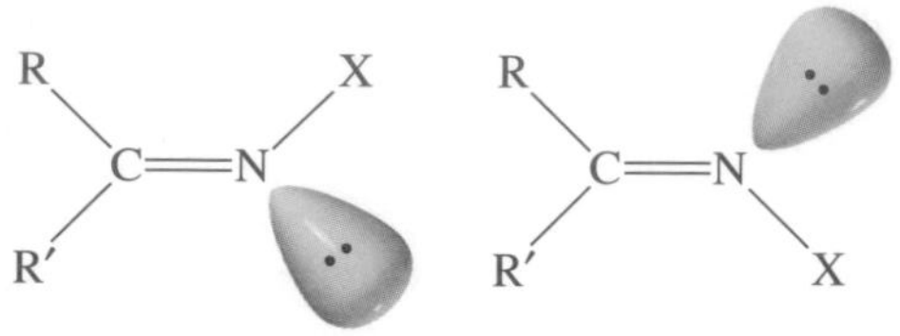

Draw the structure of each of the following compounds:

a. (*E*)-benzaldehyde semicarbazone
b. (*Z*)-propiophenone oxime
c. cyclohexanone 2,4-dinitrophenylhydrazone

PROBLEM 25

Semicarbazide has two NH_2 groups. Explain why only one of them forms an imine.

Reductive Amination

The imine formed from the reaction of an aldehyde or a ketone with ammonia is relatively unstable because it does not have a substituent other than a hydrogen bonded to the nitrogen. Nevertheless, such an imine is a useful intermediate. For example, if the reaction with ammonia is carried out in the presence of H_2 and an appropriate metal catalyst, H_2 will add to the $C=N$ bond as it is formed, forming a primary amine. The reaction of an aldehyde or a ketone with excess ammonia in the presence of a reducing agent is called **reductive amination**.

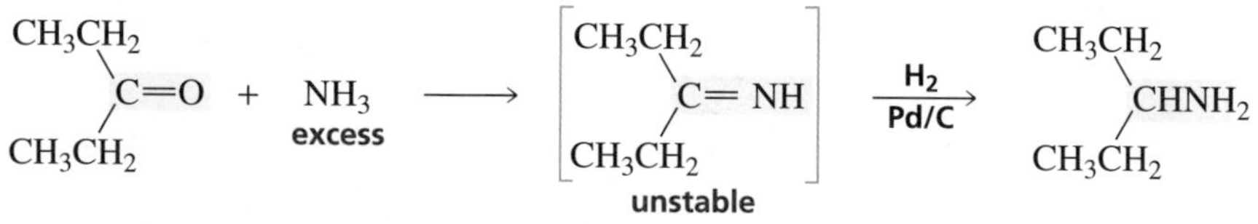

A primary amine can also be formed by reducing the oxime formed from the reaction of an aldehyde or a ketone with hydroxylamine.

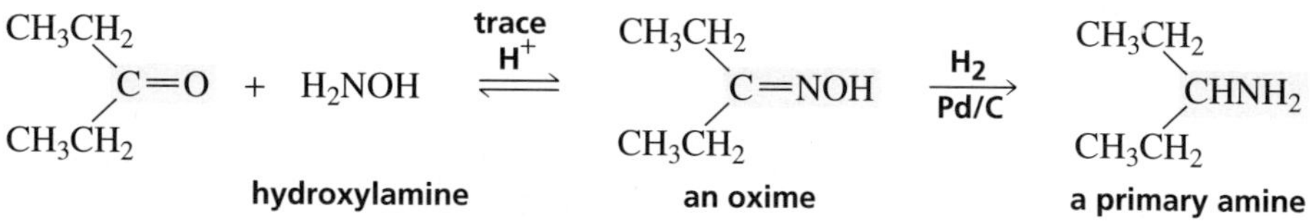

Secondary and tertiary amines can be prepared from imines and enamines by reducing the imine or enamine. Sodium triacetoxyborohydride is a commonly used reducing agent for this reaction.

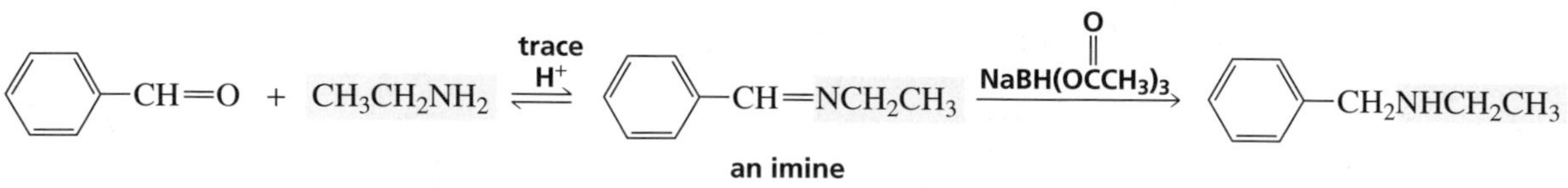

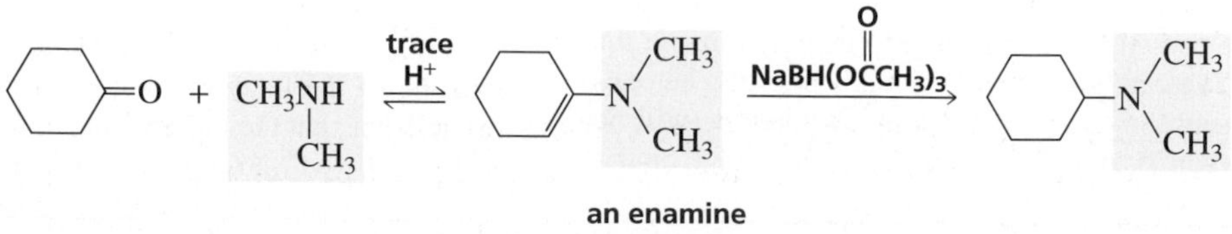

PROBLEM 26◆

Excess ammonia must be used when a primary amine is synthesized by reductive amination. What product will be obtained if the reaction is carried out with an excess of the carbonyl compound?

PROBLEM 27

The compounds commonly known as "amino acids" are actually α-aminocarboxylic acids (Section 22.0). What carbonyl compounds should be used to synthesize the following amino acids?

a. 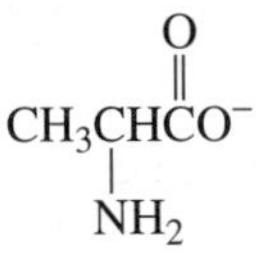 CH₃CHCO⁻ with O double bond and NH₂ group

b. 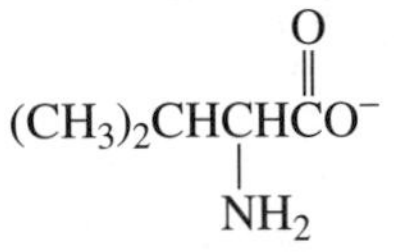 (CH₃)₂CHCHCO⁻ with O double bond and NH₂ group

The Wolff–Kishner Reduction

In Section 14.16, we saw that when a ketone or an aldehyde is heated in a basic solution of hydrazine, the carbonyl group is converted into a methylene group. This process is called **deoxygenation** because an oxygen is removed from the reactant. The reaction is known as the *Wolff Kishner reduction.*

$$\text{Ph–CCH}_3 \xrightarrow[\text{HO}^-,\ \Delta]{\text{NH}_2\text{NH}_2} \text{Ph–CH}_2\text{CH}_3$$

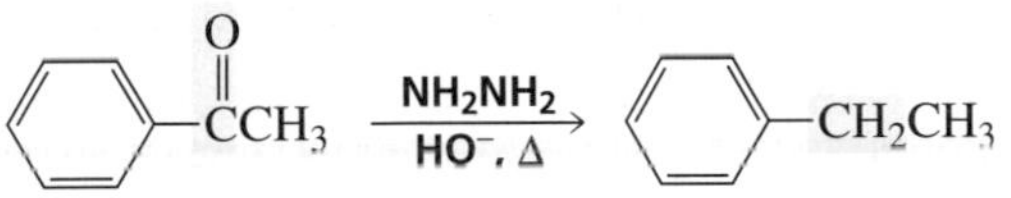

Hydroxide ion and heat differentiate the Wolff–Kishner reduction from ordinary hydrazone formation.

Mechanism for the Wolff–Kishner reduction

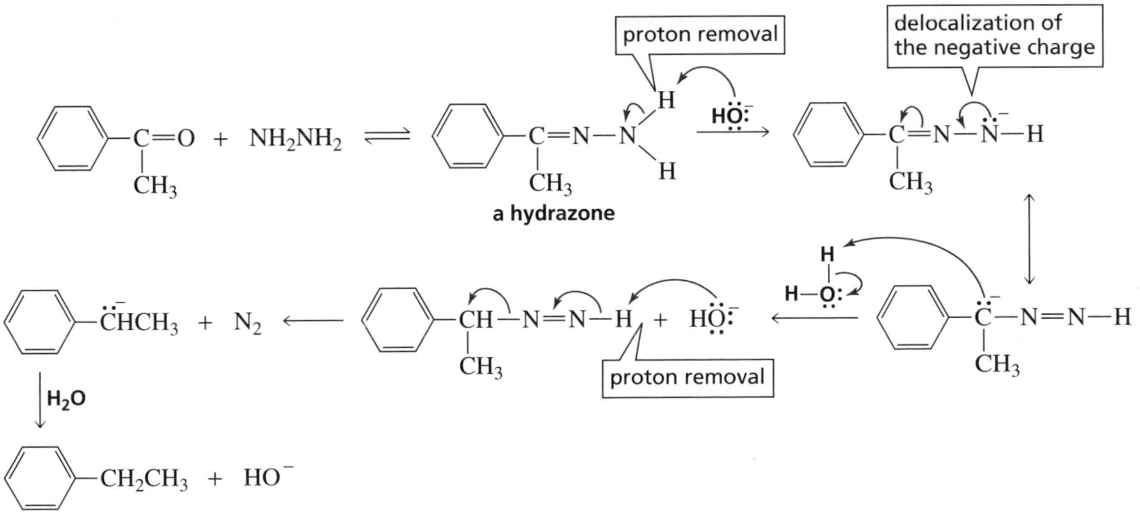

- Initially, the ketone reacts with hydrazine to form a hydrazone.
- Hydroxide ion removes a proton from the NH₂ group. The reaction requires heat because this proton is not easily removed.
- The negative charge can be delocaiized onto carbon, which abstracts a proton from water. The last two steps are repeated to form the deoxygenated product and nitrogen gas.

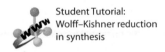

Student Tutorial:
Wolff–Kishner reduction
in synthesis

17.9 Reactions of Aldehydes and Ketones with Water

The addition of water to an aldehyde or a ketone forms a *hydrate*. A **hydrate** is a molecule with two OH groups on the same carbon. Hydrates are also called **gem-diols** (*gem* comes from *geminus*, Latin for "twin"). Hydrates of aldehydes or ketones are generally too unstable to be isolated.

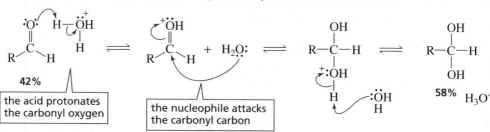

an aldehyde or a ketone   a *gem*-diol a hydrate

Water is a poor nucleophile and therefore adds relatively slowly to a carbonyl group. The rate of the reaction can be increased by an acid catalyst (Figure 17.3). Keep in mind that a catalyst has no effect on the position of the equilibrium. A catalyst affects the *rate* at which the equilibrium is achieved. In other words, the catalyst affects the rate at which an aldehyde or a ketone is converted to a hydrate; it has no effect on the *amount* of aldehyde or ketone converted to hydrate (Section 23.0).

Mechanism for acid-catalyzed hydrate formation

42%

the acid protonates the carbonyl oxygen

the nucleophile attacks the carbonyl carbon

58% H_3O^+

PROBLEM 28

Hydration of an aldehyde can also be catalyzed by hydroxide ion. Propose a mechanism for hydroxide-ion-catalyzed hydration.

The extent to which an aldehyde or a ketone is hydrated in an aqueous solution depends on the substituents attached to the carbonyl group. For example, only 0.2% of acetone is hydrated at equilibrium, but 99.9% of formaldehyde is hydrated. Why is there such a difference?

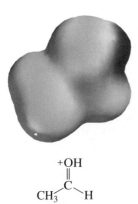

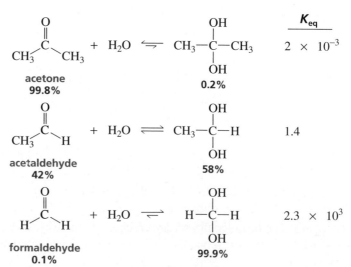

acetone
99.8%   0.2% 2×10^{-3}

acetaldehyde
42%   58%   1.4

formaldehyde
0.1%   99.9% 2.3×10^{3}

▲ **Figure 17.3**
The electrostatic potential maps show that the carbonyl carbon of the protonated aldehyde is more electrophilic (the blue is more intense) than the carbonyl carbon of the unprotonated aldehyde.

The equilibrium constant for a reaction depends on the relative stabilities of the reactants and products (Section 3.7). The equilibrium constant for hydrate formation, therefore, depends on the relative stabilities of the carbonyl compound and the hydrate. We have seen that electron-donating alkyl groups make a carbonyl compound *more stable* (less reactive) (Section 17.2).

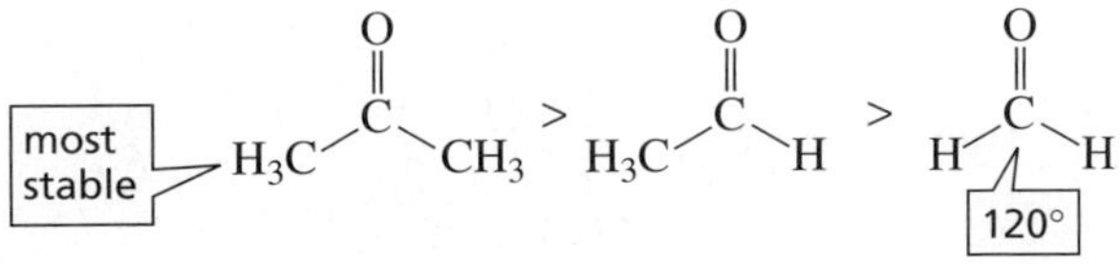

In contrast, alkyl groups make the hydrate *less stable* because of steric interactions between the alkyl groups when the bond angle changes from 120° to 109.5° (Section 17.2).

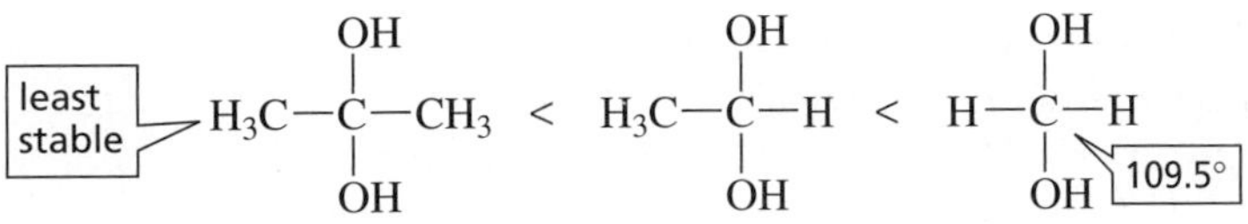

Alkyl groups, therefore, shift the equilibrium to the left (toward reactants) because they stabilize the carbonyl compound and destabilize the hydrate. As a result, less acetone than formaldehyde is hydrated at equilibrium.

In summary, the percentage of hydrate present in solution at equilibrium depends on both electronic and steric effects. Electron-donating substituents and bulky substituents (such as the methyl groups of acetate) *decrease* the percentage of hydrate present at equilibrium, whereas electron-withdrawing substituents and small substituents (the hydrogens of formaldehyde) *increase* it.

PRESERVING BIOLOGICAL SPECIMENS

A 37% solution of formaldehyde in water, known as *formalin*, was commonly used in the past to preserve biological specimens. Formaldehyde is an eye and skin irritant, however, so formalin has been replaced in most biology laboratories by other preservatives. One frequently used preparation is a solution of 2%–5% phenol in ethanol with added antimicrobial agents.

If the amount of hydrate formed from the reaction of water with a ketone is too small to detect, how do we know that the reaction has even occurred? We can prove that it occurs by adding the ketone to ^{18}O-labeled water and isolating the ketone after equilibrium has been established. Finding that the label has been incorporated into the ketone indicates that hydration has occurred.

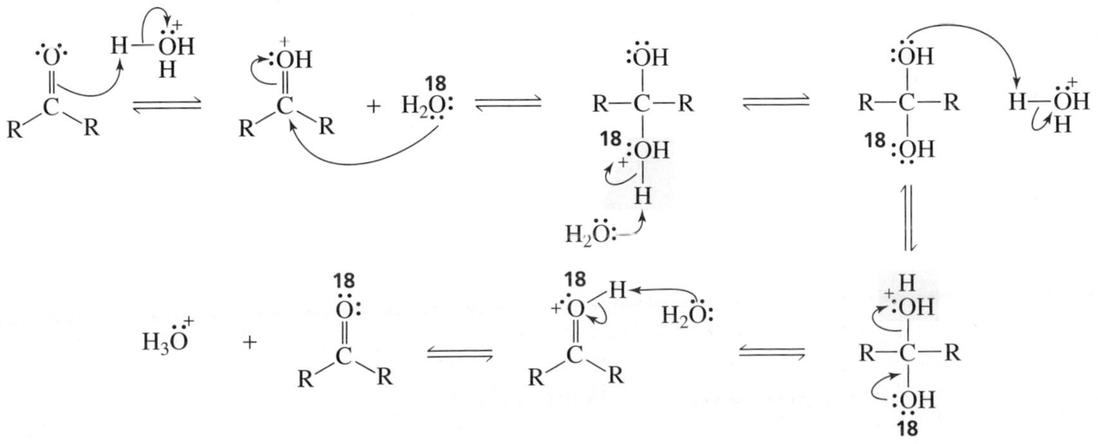

PROBLEM 29

When trichloroacetaldehyde is dissolved in water, almost all of it is converted to the hydrate. Chloral hydrate, the product of the reaction, is a sedative that can be lethal. A cocktail laced with it is known—in detective novels, at least—as a "Mickey Finn." Explain why an aqueous solution of trichloroacetaldehyde is almost all hydrate.

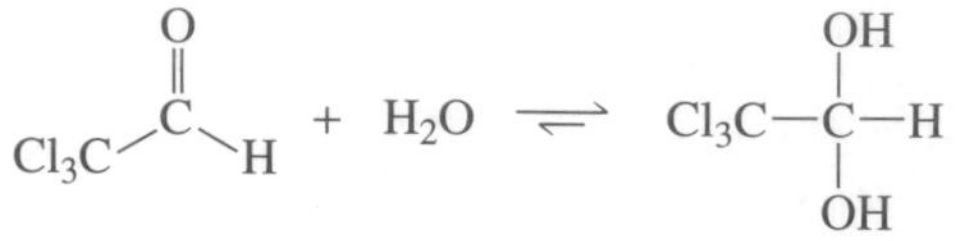

$$Cl_3C-\overset{\overset{\displaystyle O}{\|}}{C}-H \; + \; H_2O \; \rightleftharpoons \; Cl_3C-\overset{\overset{\displaystyle OH}{|}}{\underset{\underset{\displaystyle OH}{|}}{C}}-H$$

trichloroacetaldehyde   chloral hydrate

PROBLEM 30◆

Which of the following ketones forms the most hydrate in an aqueous solution?

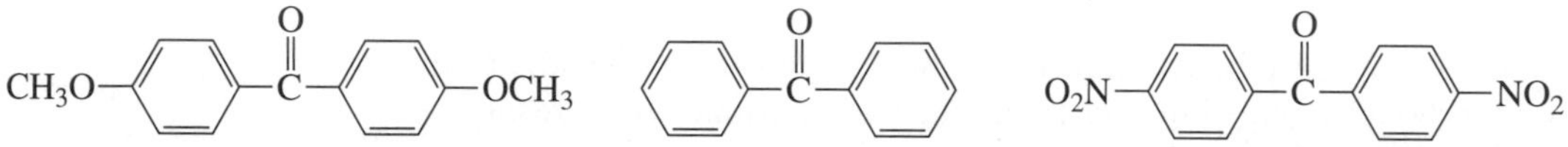

$$CH_3O-\bigcirc-\overset{O}{\overset{\|}{C}}-\bigcirc-OCH_3 \qquad \bigcirc-\overset{O}{\overset{\|}{C}}-\bigcirc \qquad O_2N-\bigcirc-\overset{O}{\overset{\|}{C}}-\bigcirc-NO_2$$

17.10 Reactions of Aldehydes and Ketones with Alcohols

The product formed when one equivalent of an alcohol adds to an *aldehyde* is called a **hemiacetal**. The product formed when a second equivalent of alcohol is added is called an **acetal** (ass-ett-AL). Like water, an alcohol is a poor nucleophile, so an acid catalyst is required for the reaction to take place at a reasonable rate.

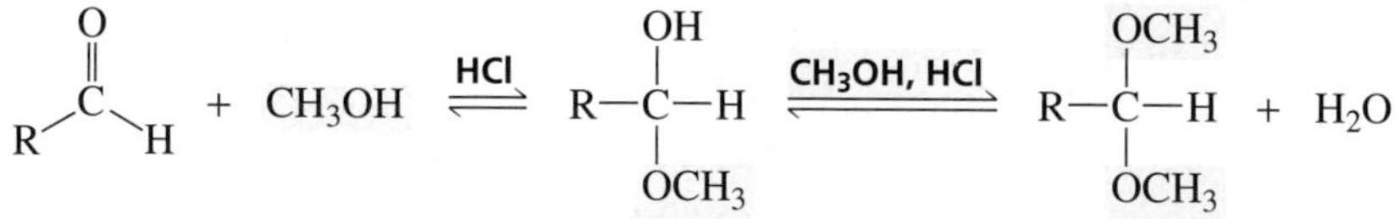

$$\underset{\text{an aldehyde}}{R-\overset{\overset{\displaystyle O}{\|}}{C}-H} + CH_3OH \overset{HCl}{\rightleftharpoons} \underset{\text{a hemiacetal}}{R-\overset{\overset{\displaystyle OH}{|}}{\underset{\underset{\displaystyle OCH_3}{|}}{C}}-H} \overset{CH_3OH,\,HCl}{\rightleftharpoons} \underset{\text{an acetal}}{R-\overset{\overset{\displaystyle OCH_3}{|}}{\underset{\underset{\displaystyle OCH_3}{|}}{C}}-H} + H_2O$$

When the carbonyl compound is a *ketone* instead of an aldehyde, the addition products are called a **hemiketal** and a **ketal**, respectively.

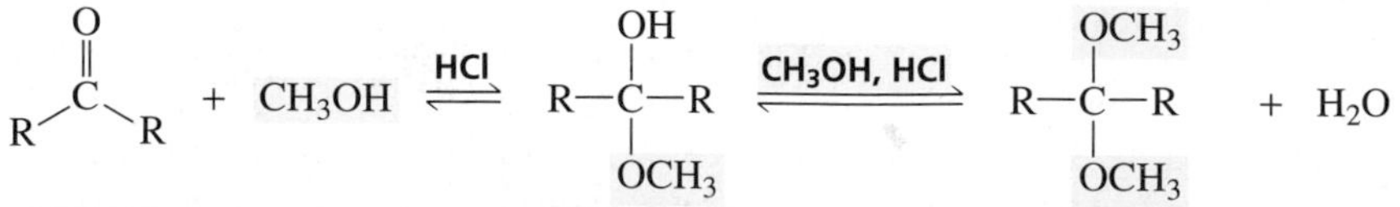

$$\underset{\text{a ketone}}{R-\overset{\overset{\displaystyle O}{\|}}{C}-R} + CH_3OH \overset{HCl}{\rightleftharpoons} \underset{\text{a hemiketal}}{R-\overset{\overset{\displaystyle OH}{|}}{\underset{\underset{\displaystyle OCH_3}{|}}{C}}-R} \overset{CH_3OH,\,HCl}{\rightleftharpoons} \underset{\text{a ketal}}{R-\overset{\overset{\displaystyle OCH_3}{|}}{\underset{\underset{\displaystyle OCH_3}{|}}{C}}-R} + H_2O$$

Hemi is the Greek word for "half." When one equivalent of alcohol has added to an aldehyde or a ketone, the compound is halfway to the final acetal or ketal, which contains groups from two equivalents of alcohol.

The mechanism for ketal (or acetal) formation is shown below.

Mechanism for acid-catalyzed acetal or ketal formation

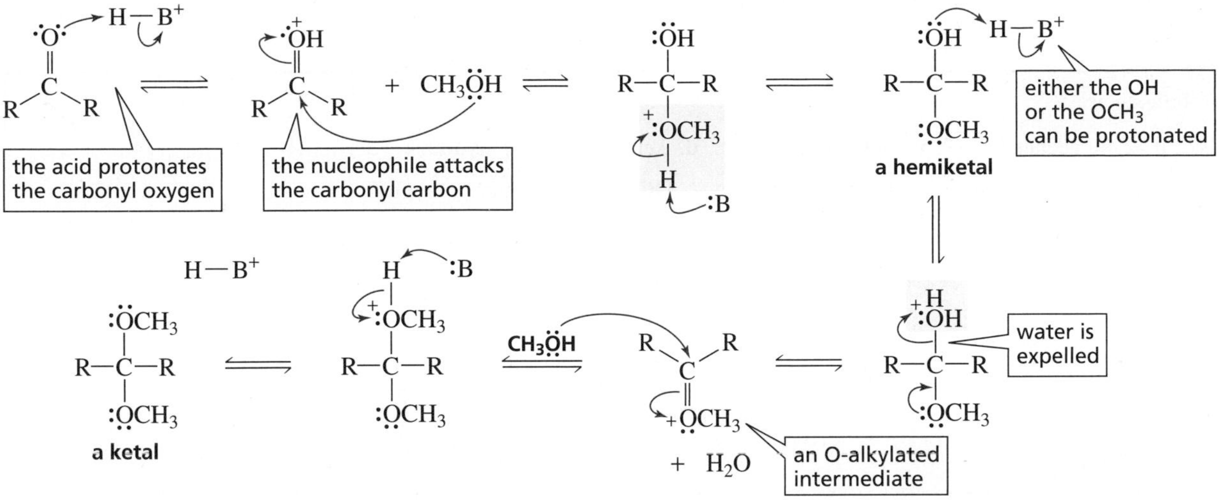

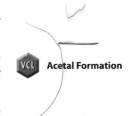

- The acid protonates the carbonyl oxygen, making the carbonyl carbon more susceptible to nucleophilic attack (Figure 17.9).
- Loss of a proton from the protonated tetrahedral intermediate gives the hemiketal (or hemiacetal).
- Because the reaction is carried out in an acidic solution, the hemiacetal (or hemiketal) is in equilibrium with its protonated form. The two oxygen atoms of the hemiketal (or hemiacetal) are equally basic, so either one can be protonated.
- Loss of water from the tetrahedral intermediate with a protonated OH group forms an O-alkylated intermediate that is very reactive because of its positively charged oxygen. Nucleophilic attack on this intermediate by a second molecule of alcohol, followed by loss of a proton, forms the acetal (or ketal).

VCL Acetal Formation

The acetal or ketal can be isolated if the water that is eliminated from the hemiacetal (or hemiketal) is removed from the reaction mixture. This is because, if water is not available, the only compound the acetal or ketal can form is the O-alkylated species, which is less stable than the acetal or ketal.

The acetal or ketal can be hydrolyzed back to the aldehyde or ketone in an acidic aqueous solution.

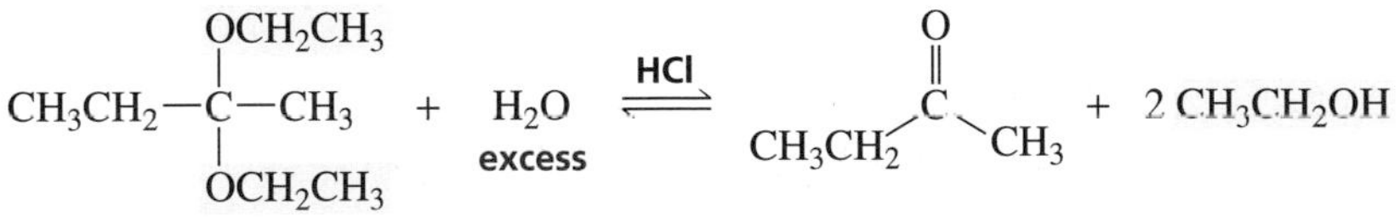

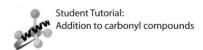

Notice that the mechanisms for imine, enamine, hydrate, and acetal (or ketal) formation are similar; they are all reactions with a nucleophile that has a lone pair on an attacking atom. After the nucleophile (a primary amine in the case of imine formation, a secondary amine in the case of enamine formation, water in the case of hydrate formation, and an alcohol in the case of acetal or ketal formation) has added to the carbonyl group, water is eliminated from the protonated tetrahedral intermediate, forming a positively charged intermediate. In imine and hydrate formation, a neutral product is achieved by loss of a proton from a nitrogen and an oxygen, respectively. In enamine formation, a neutral product is achieved by loss of a proton from an α-carbon.

Student Tutorial:
Addition to carbonyl compounds

Student Tutorial:
Common terms for the addition of nucleophiles to carbonyl compounds

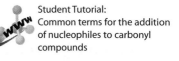

In acetal formation, a neutral compound is achieved by addition of a second equivalent of alcohol. Also notice that, because the nucleophile in hydrate formation is water, elimination of water gives back the original aldehyde or ketone.

PROBLEM 31◆

Which of the following are

a. hemiacetals?   **b.** acetals?   **c.** hemiketals?   **d.** ketals?   **e.** hydrates?

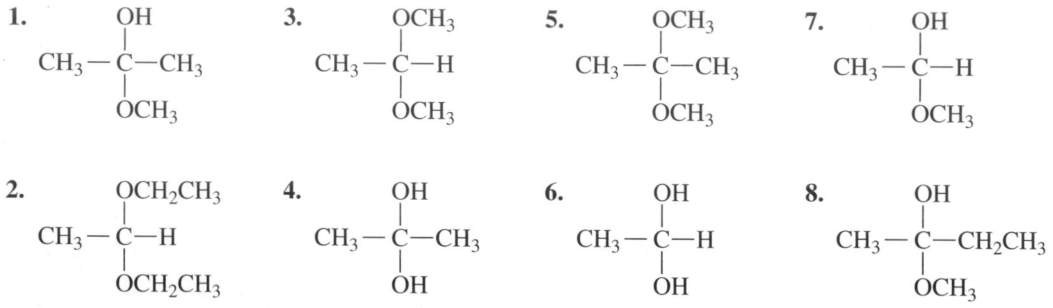

1.
$$CH_3-\overset{\overset{\displaystyle OH}{|}}{\underset{\underset{\displaystyle OCH_3}{|}}{C}}-CH_3$$

2.
$$CH_3-\overset{\overset{\displaystyle OCH_2CH_3}{|}}{\underset{\underset{\displaystyle OCH_2CH_3}{|}}{C}}-H$$

3.
$$CH_3-\overset{\overset{\displaystyle OCH_3}{|}}{\underset{\underset{\displaystyle OCH_3}{|}}{C}}-H$$

4.
$$CH_3-\overset{\overset{\displaystyle OH}{|}}{\underset{\underset{\displaystyle OH}{|}}{C}}-CH_3$$

5.
$$CH_3-\overset{\overset{\displaystyle OCH_3}{|}}{\underset{\underset{\displaystyle OCH_3}{|}}{C}}-CH_3$$

6.
$$CH_3-\overset{\overset{\displaystyle OH}{|}}{\underset{\underset{\displaystyle OH}{|}}{C}}-H$$

7.
$$CH_3-\overset{\overset{\displaystyle OH}{|}}{\underset{\underset{\displaystyle OCH_3}{|}}{C}}-H$$

8.
$$CH_3-\overset{\overset{\displaystyle OH}{|}}{\underset{\underset{\displaystyle OCH_3}{|}}{C}}-CH_2CH_3$$

PROBLEM-SOLVING STRATEGY

Analyzing the Behavior of Acetals and Ketals

Explain why acetals and ketals are hydrolyzed back to the aldehyde or ketone in acidic aqueous solutions, but are stable in basic aqueous solutions.

The best way to approach this kind of question is to write out the structures and the mechanism that describe the situation to which the question is referring. When the mechanism is written, the answer may become apparent. In an acidic solution, the acid protonates an oxygen of the acetal. This creates a weak base (CH_3OH) that can be expelled by the other CH_3O group. When the group is expelled, water can attack the reactive intermediate, and you are then on your way back to the ketone (or aldehyde).

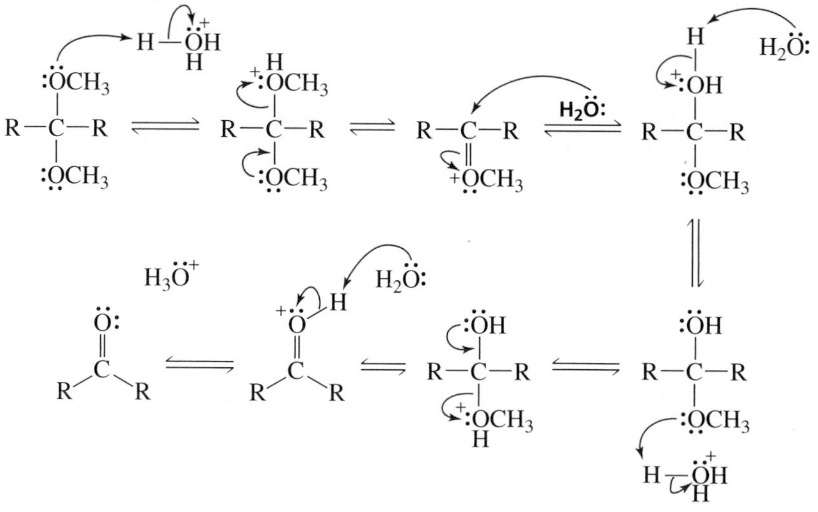

In a basic solution, the CH_3O group cannot be protonated. Therefore, the group that would have to be eliminated to reform the ketone (or aldehyde) would be a very basic CH_3O^- group. However, a CH_3O^- group is too basic to be eliminated by the other CH_3O group, which has little driving force because of the positive charge that would be placed on its oxygen atom if elimination were to occur.

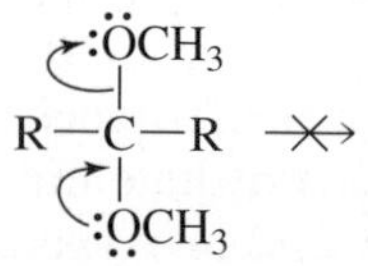

Now continue on to Problem 32.

a. Would you expect hemiacetals to be stable in basic solutions? Explain your answer.

b. Acetal formation must be catalyzed by an acid. Explain why it cannot be catalyzed by CH_3O^-.

c. Can the rate of hydrate formation be increased by hydroxide ion as well as by acid? Explain.

Explain why an acetal or ketal can be isolated but most hydrates cannot be isolated.

17.11 **Protecting Groups**

Ketones (or aldehydes) react with 1,2-diols to form five-membered ring ketals (or acetals) and with 1,3-diols to form six-membered ring ketals (or acetals). Recall that five- and six-membered rings are formed relatively easily (Section 8.11). The mechanism is the same as that shown in Section 17.10 for acetal formation, except that instead of reacting with two separate molecules of alcohol, the carbonyl compound reacts with the two alcohol groups of a single molecule of the diol.

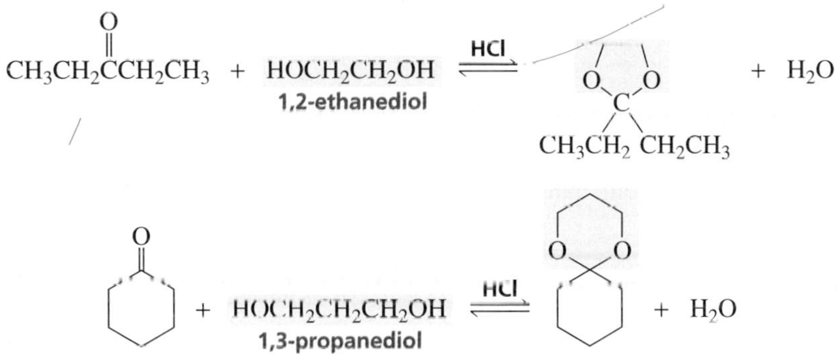

$$CH_3CH_2\overset{O}{\overset{\|}{C}}CH_2CH_3 \ + \ HOCH_2CH_2OH \ \underset{}{\overset{HCl}{\rightleftharpoons}} \ \ + \ H_2O$$

1,2-ethanediol

$CH_3CH_2 \quad CH_2CH_3$

$$ \ + \ HOCH_2CH_2CH_2OH \ \underset{}{\overset{HCl}{\rightleftharpoons}} \ \ + \ H_2O$$

1,3-propanediol

   If a compound has two functional groups that will react with a given reagent and you want only one of them to react, it is necessary to protect the other functional group from the reagent. A group that protects a functional group from a synthetic operation that it would not otherwise survive is called a **protecting group**.

   If you have ever painted a room with a spray gun, you may have taped over the things you did not want to paint, such as baseboards and window frames. In a similar way, 1,2-diols and 1,3-diols are used to protect ("tape over") the carbonyl group of aldehydes and ketones. For example, suppose you wanted to synthesize the following hydroxyketone from the keto ester. Both functional groups of the keto ester will be reduced by $LiAlH_4$, and the one that you don't want to react—the keto group—is the more reactive of the two.

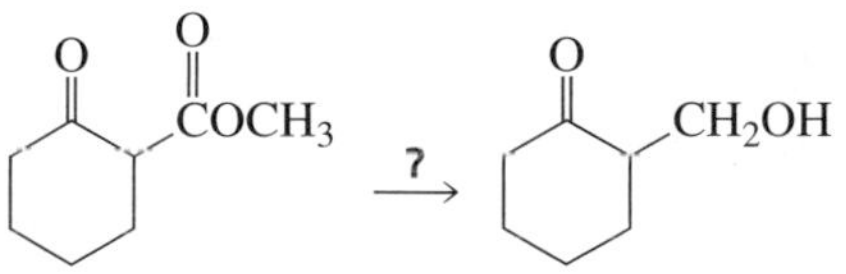

$$ \overset{?}{\longrightarrow} $$

However, if the keto group is first converted to a ketal, only the ester group will react with $LiAlH_4$. The protecting group can be removed by acid-catalyzed hydrolysis after the ester has been reduced. It is critical that the conditions used to remove a protecting group do not affect other groups in the molecule. Acetals and ketals are good protecting groups because, being ethers, they do not react with bases, reducing agents, or oxidizing agents.

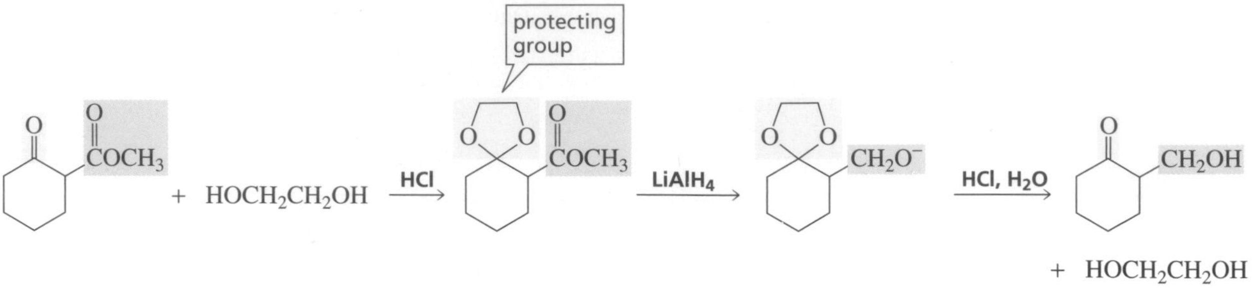

PROBLEM 34◆

a. What would have been the product of the preceding reaction with LiAlH₄ if the keto group had not been protected?

b. What reagent could you use to reduce only the keto group?

PROBLEM 35

Why don't acetals react with nucleophiles?

In the following reaction, the aldehyde reacts with the diol because aldehydes are more reactive than ketones. The Grignard reagent will then react only with the keto group. The protecting group can be removed by acid-catalyzed hydrolysis.

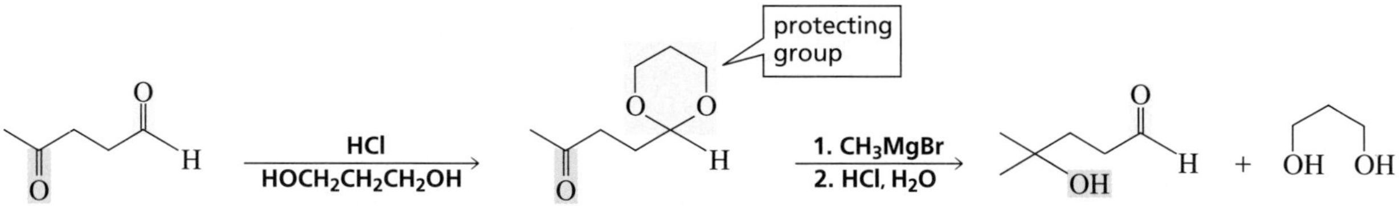

One of the best ways to protect an OH group of an alcohol is to convert it to a trimethylsilyl (TMS) ether by treating the alcohol with chlorotrimethylsilane and a tertiary amine. The ether is formed by an S_N2 reaction. Although a tertiary alkyl halide does not undergo an S_N2 reaction, the tertiary silyl compound does because Si—C bonds are longer than C—C bonds, reducing steric hindrance at the site of nucleophilic attack. The amine prevents the solution from becoming acidic by reacting with the HCl generated in the reaction. The TMS ether, which is stable in neutral and basic solutions, can have its protecting group removed with aqueous acid under mild conditions.

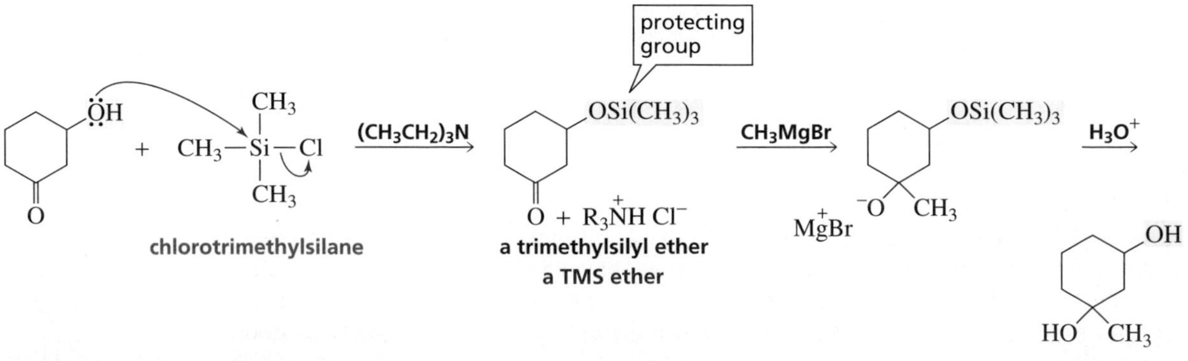

The OH group of a carboxylic acid group can be protected by converting the carboxylic acid into an ester.

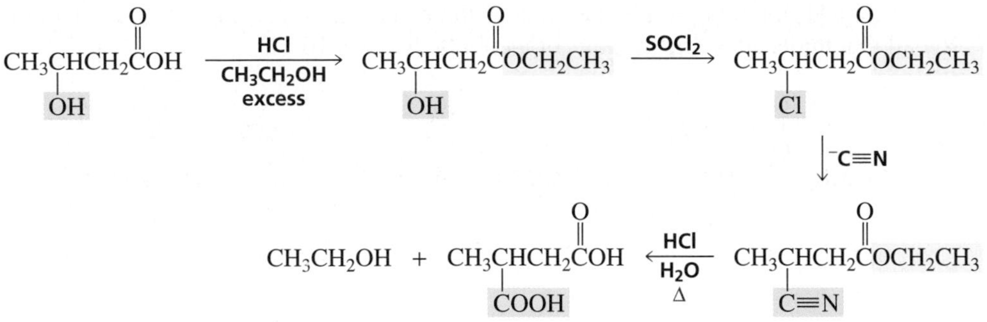

An amino group can be protected by being converted into an amide (Section 16.8). The acetyl group can subsequently be removed by acid-catalyzed hydrolysis (Section 16.17).

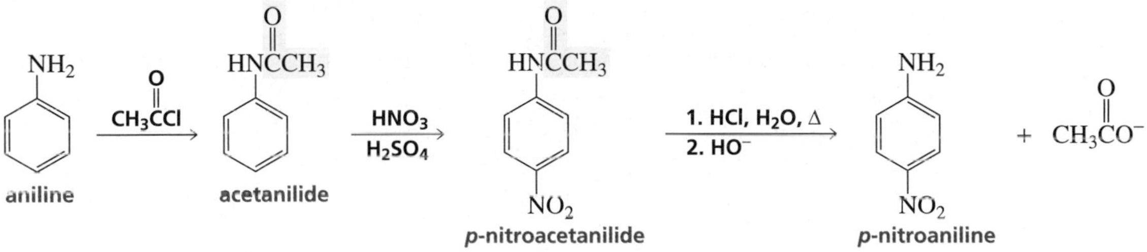

Protecting groups should be used only when absolutely necessary, because attaching and then removing them adds two steps to the synthesis, which decreases the overall yield of the target compound (the desired product).

PROBLEM 36

What products would be formed from the preceding reaction if aniline's amino group were not protected?

PROBLEM 37◆

a. In a six-step synthesis, what is the yield of the target compound if each of the reactions employed gives an 80% yield? (An 80% yield is a relatively high laboratory yield.)
b. What would the yield be if two more steps were added to the synthesis?

PROBLEM 38

Show how each of the following compounds can be prepared from the given starting material. In each case, you will need to use a protecting group.

a. $HOCH_2CH_2CH_2Br \longrightarrow HOCH_2CH_2CH_2CHCH_3$ with an OH on the CHCH_3 carbon

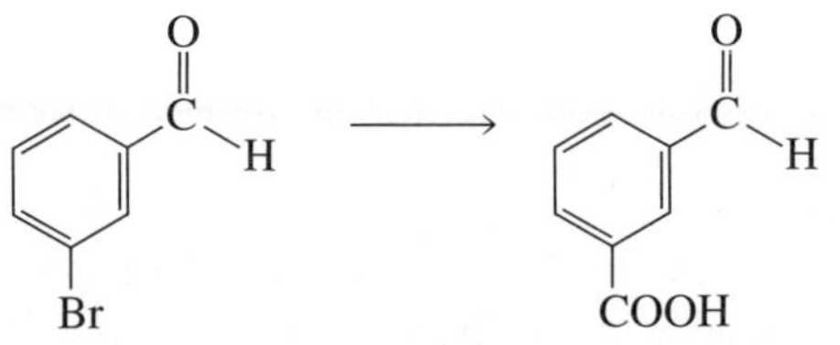

17.12 Addition of Sulfur Nucleophiles

Aldehydes and ketones react with thiols to form thioacetals and thioketals. The mechanism for addition of a thiol is the same as the mechanism for addition of an alcohol. Recall that thiols are sulfur analogs of alcohols (Section 10.11).

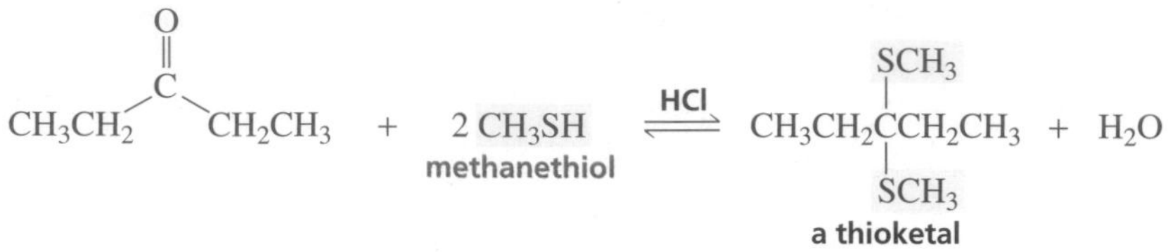

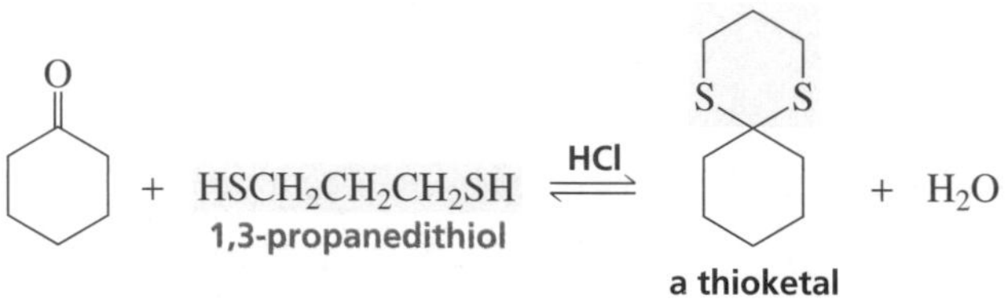

$$CH_3CH_2\overset{\overset{\displaystyle O}{\|}}{C}CH_2CH_3 \;+\; 2\;CH_3SH \;\underset{\text{methanethiol}}{\overset{\text{HCl}}{\rightleftharpoons}}\; CH_3CH_2\underset{\underset{\displaystyle SCH_3}{|}}{\overset{\overset{\displaystyle SCH_3}{|}}{C}}CH_2CH_3 \;+\; H_2O$$

a thioketal

$$\text{cyclohexanone} \;+\; HSCH_2CH_2CH_2SH \;\underset{\text{1,3-propanedithiol}}{\overset{\text{HCl}}{\rightleftharpoons}}\; \text{(thioketal ring)} \;+\; H_2O$$

a thioketal

Thioacetal (or thioketal) formation is useful in organic synthesis because a thioacetal (or thioketal) is desulfurized when it reacts with H_2 and Raney nickel. Desulfurization replaces the C—S bonds with C—H bonds.

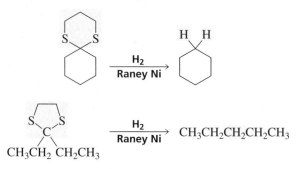

Thioketal formation followed by desulfurization provides us with a third method we can use to convert the carbonyl group of a ketone into a methylene group. The other two we have seen are the Clemmensen reduction and the Wolff–Kishner reduction (Section 14.16 and page 813.).

17.13 The Wittig Reaction Forms an Alkene

An aldehyde or a ketone reacts with a phosphonium ylide (pronounced "ILL-id") to form an alkene. This is called a **Wittig reaction**. Overall, it amounts to interchanging the double-bonded oxygen of the carbonyl compound and the double-bonded carbon group of the phosphonium ylide.

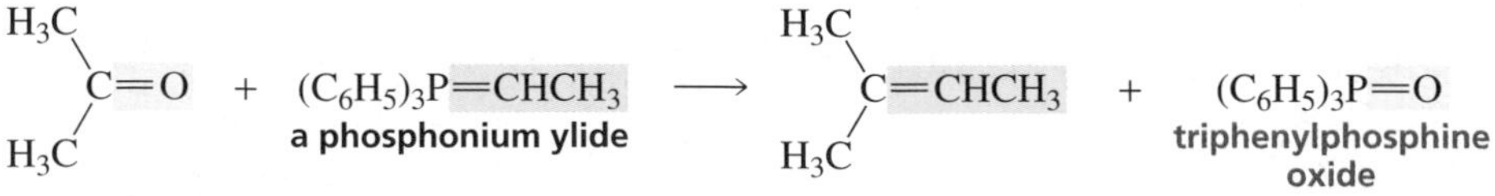

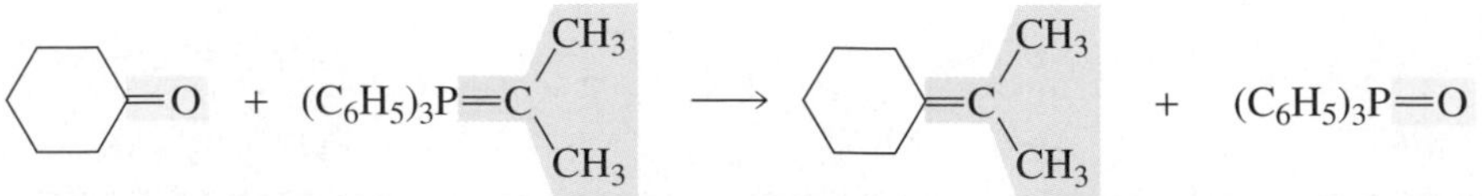

$$\underset{H_3C}{\overset{H_3C}{>}}C=O \;+\; \underset{\text{a phosphonium ylide}}{(C_6H_5)_3P=CHCH_3} \;\longrightarrow\; \underset{H_3C}{\overset{H_3C}{>}}C=CHCH_3 \;+\; \underset{\substack{\text{triphenylphosphine}\\\text{oxide}}}{(C_6H_5)_3P=O}$$

$$\text{cyclohexanone} =O \;+\; (C_6H_5)_3P=\underset{CH_3}{\overset{CH_3}{<}} \;\longrightarrow\; \text{(cyclohexylidene)}=C\underset{CH_3}{\overset{CH_3}{<}} \;+\; (C_6H_5)_3P=O$$

An **ylide** is a compound that has opposite charges on adjacent covalently bonded atoms that have complete octets. The ylide can also be written in the double-bonded form because phosphorus can have more than eight valence electrons.

$$(C_6H_5)_3\overset{+}{P}-\overset{..}{\overset{-}{C}}H_2 \quad \longleftrightarrow \quad (C_6H_5)_3P=CH_2$$
a phosphonium ylide

Evidence has accumulated that the Wittig reaction is a concerted [2 + 2] cycloaddition reaction. (Cycloaddition reactions were introduced in Section 7.12.) It is called a [2 + 2] cycloaddition reaction because, of the four π electrons involved in the cyclic transition state, two come from the carbonyl group and two come from the ylide.

Mechanism for the Wittig reaction

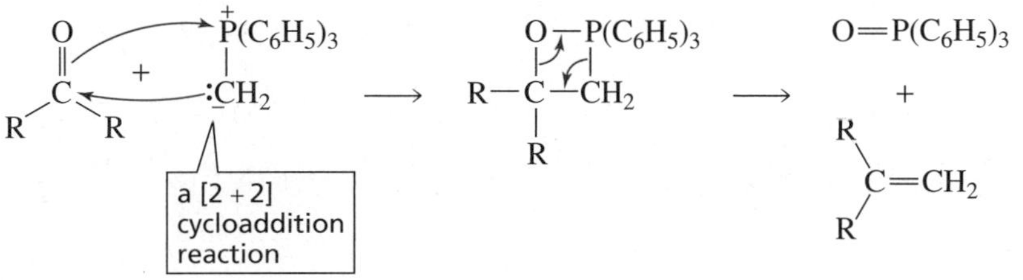

a [2 + 2] cycloaddition reaction

- The nucleophilic carbon of the ylide attacks the carbonyl carbon while the carbonyl oxygen attacks the electrophilic phosphorus.
- Elimination of triphenylphosphine oxide forms the alkene product.

The phosphonium ylide needed for a particular synthesis is obtained by an S_N2 reaction between triphenylphosphine and an alkyl halide with the appropriate number of carbon atoms. A proton on the carbon adjacent to the positively charged phosphorus atom is sufficiently acidic ($pK_a = 35$) to be removed by a strong base such as butyllithium (Section 10.12).

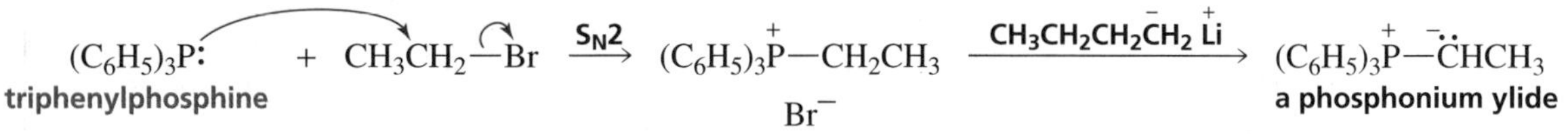

$$(C_6H_5)_3\overset{..}{P}: \quad + \quad CH_3CH_2-Br \quad \xrightarrow{S_N2} \quad (C_6H_5)_3\overset{+}{P}-CH_2CH_3 \quad \xrightarrow{CH_3CH_2CH_2CH_2\overset{-}{}\overset{+}{Li}} \quad (C_6H_5)_3\overset{+}{P}-\overset{..}{\overset{-}{C}}HCH_3$$
triphenylphosphine $\qquad\qquad Br^-$ $\qquad\qquad\qquad\qquad\qquad\qquad\qquad$ **a phosphonium ylide**

If two sets of reagents are available for the synthesis of an alkene, the better choice is one that requires the less sterically hindered alkyl halide for synthesis of the ylide. (Recall that the more sterically hindered the alkyl halide, the less reactive it is in an S_N2 reaction; Section 8.2). For example, it is better to use the three-carbon alkyl halide and the five-carbon carbonyl compound than the five-carbon alkyl halide and the three-carbon carbonyl compound for the synthesis of 3-ethyl-3-hexene because it would be easier to form an ylide from 1-bromopropane than from 3-bromopentane.

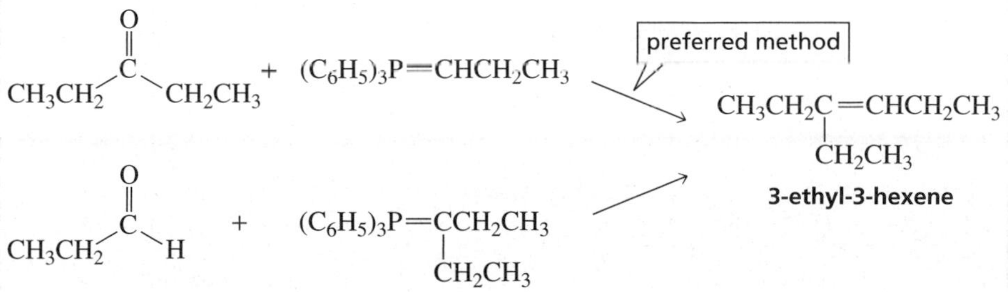

preferred method

$$CH_3CH_2C=CHCH_2CH_3$$
$$|$$
$$CH_2CH_3$$
3-ethyl-3-hexene

The Wittig reaction is a very powerful way to make an alkene because the reaction is completely regioselective—the double bond will be in only one place.

$$\text{\Large⬡}=O \ + \ (C_6H_5)_3P=CH_2 \ \longrightarrow \ \text{\Large⬡}=CH_2 \ + \ (C_6H_5)_3P=O$$

methylenecyclohexane

Student Tutorial:
Wittig reaction–synthesis

The Wittig reaction also is the best way to make a terminal alkene, such as the one shown above, because other methods would form a terminal alkene only as a minor product, if at all.

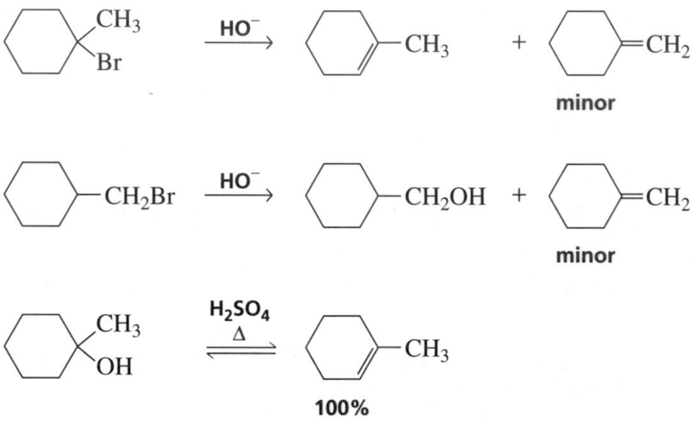

Stabilized ylides form primarily *E* isomers, and unstabilized ylides form primarily *Z* isomers.

The stereoselectivity of the Wittig reaction depends on the structure of the ylide. Ylides can be divided into two types: *stabilized ylides* have a group, such as a carbonyl group, that can share carbon's negative charge; *unstabilized ylides* do not have such a group.

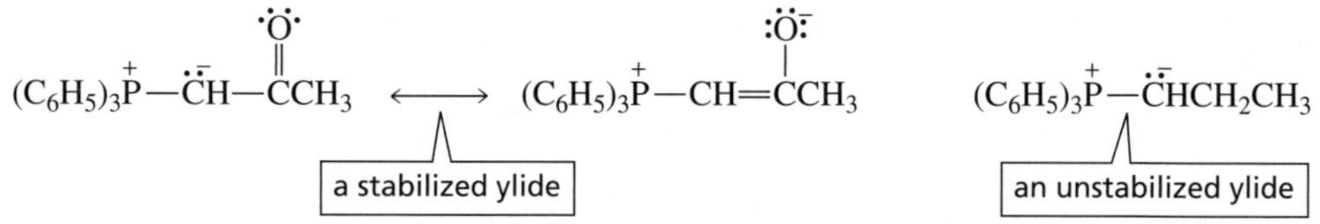

$$(C_6H_5)_3\overset{+}{P}-\overset{..}{\overset{-}{C}}H-\overset{\overset{\displaystyle ..\overset{..}{O}..}{\|}}{C}CH_3 \longleftrightarrow (C_6H_5)_3\overset{+}{P}-CH=\overset{\overset{\displaystyle :\overset{..}{O}:^-}{|}}{C}CH_3 \qquad (C_6H_5)_3\overset{+}{P}-\overset{..}{\overset{-}{C}}HCH_2CH_3$$

a stabilized ylide   an unstabilized ylide

Stabilized ylides form primarily *E* isomers, and unstabilized ylides form primarily *Z* isomers.

E alkene

Z alkene

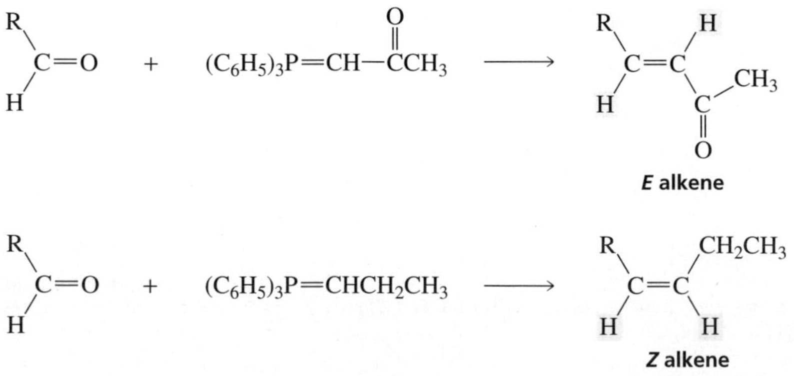

β-CAROTENE

β-Carotene is found in orange and yellow-orange fruits and vegetables such as apricots, mangoes, carrots, and sweet potatoes. The synthesis of β-carotene from vitamin A for use in foods is an important example of the use of the Wittig reaction in industry. Notice that the ylide is a stabilized one, so the product has the *E* configuration at the reaction site.

β-Carotene is used in the food industry to color margarine. Many people take β-carotene as a dietary supplement because there is some evidence associating high levels of β-carotene with a low incidence of cancer. More recent evidence, however, suggests that β-carotene taken in pill form does not have the cancer-preventing effects of β-carotene obtained from vegetables.

vitamin A aldehyde

β-carotene

PROBLEM 39 **SOLVED**

a. What carbonyl compound and what phosphonium ylide are required for the synthesis of the following alkenes?

1. $CH_3CH_2CH_2CH=CCH_3$
        $\quad\quad\quad\quad\quad\quad\quad\; |$
        $\quad\quad\quad\quad\quad\quad\quad CH_3$

3. $(C_6H_5)_2C=CHCH_3$

2. ⬡=CHCH$_2$CH$_3$

4. ⬡—CH=CH$_2$

b. What alkyl halide is required to prepare each of the phosphonium ylides in part a?

Solution to 39a (1) The atoms on either side of the double bond can come from the carbonyl compound, so there are two pairs of compounds that can be used.

$$CH_3\overset{O}{\overset{\|}{C}}CH_3 + (C_6H_5)_3P=CHCH_2CH_2CH_3 \quad \textbf{or} \quad CH_3CH_2CH_2\overset{O}{\overset{\|}{C}}H + (C_6H_5)_3P=CCH_3$$
$$\quad\quad\quad\quad\quad\quad\quad\quad\quad\quad\quad\quad\quad\quad\quad\quad\quad\quad\quad\quad\quad\quad\quad\quad\quad\quad\quad\quad\quad\quad\quad\quad\quad\quad\quad |$$
$$\quad\quad\quad\quad\quad\quad\quad\quad\quad\quad\quad\quad\quad\quad\quad\quad\quad\quad\quad\quad\quad\quad\quad\quad\quad\quad\quad\quad\quad\quad\quad\quad\quad CH_3$$

Solution to 39b (1) The alkyl halide required depends on which phosphonium ylide is used; it would be either 1-bromobutane or 2-bromopropane.

$$CH_3CH_2CH_2CH_2Br \quad \textbf{or} \quad CH_3CHCH_3$$
$$\quad\quad\quad\quad\quad\quad\quad\quad\quad\quad\quad\quad\quad\quad\quad\quad\quad |$$
$$\quad\quad\quad\quad\quad\quad\quad\quad\quad\quad\quad\quad\quad\quad\quad\quad\quad Br$$

The primary alkyl halide would be more reactive in the S_N2 reaction required to make the ylide, so the best method would be to use acetone and the ylide obtained from 1-bromobutane.

17.14 **Stereochemistry of Nucleophilic Addition Reactions: *Re* and *Si* Faces**

A carbonyl carbon bonded to two different substituents is a **prochiral carbonyl carbon** because it will become an asymmetric center if it acquires an additional group unlike either of the groups already bonded to it. Because the reaction forms an asymmetric center in a reactant that does not have an asymmetric center, the addition product will be a racemic mixture (Section 5.19).

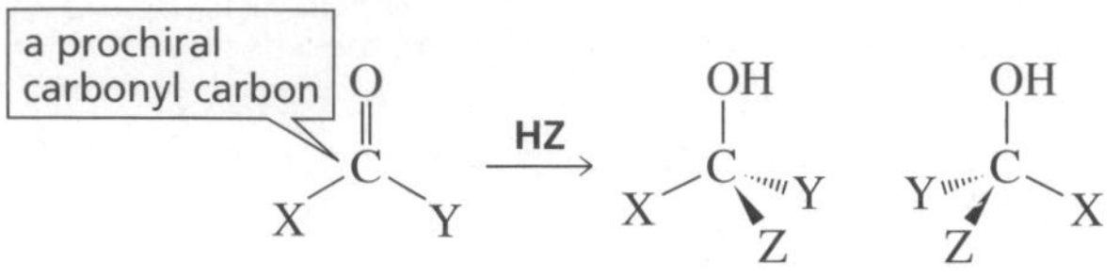

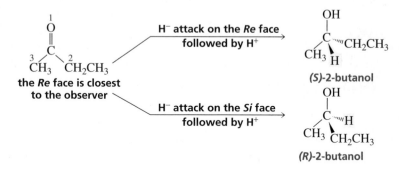

a pair of enantiomers

the *Si* face

the *Re* face

The carbonyl carbon and the three atoms attached to it define a plane. In a nucleophilic addition reaction, the nucleophile can approach either side of the plane. One side of the carbonyl compound is called the *Re* (pronounced "ree") face, and the other side is called the *Si* (pronounced "sigh") face; *Re* is for *rectus* and *Si* is for *sinister*—similar to *R* and *S*. To distinguish between the ***Re* and *Si* faces**, the three groups attached to the carbonyl carbon are assigned priorities using the Cahn–Ingold–Prelog system from *E,Z* and *R,S* nomenclature (Sections 3.5 and 5.7, respectively). The *Re* face is the face closest to the observer when priorities decrease (1 > 2 > 3) in a clockwise direction, and the *Si* face is the opposite face, the one closest to the observer when the priorities decrease in a counterclockwise direction.

Attack by a nucleophile on the *Re* face forms one enantiomer, whereas attack on the *Si* face forms the other enantiomer. For example, attack by hydride ion on the *Re* face of butanone forms (*S*)-2-butanol, and attack on the *Si* face forms (*R*)-2-butanol.

Whether attack on the *Re* face forms the *R* or *S* enantiomer depends on the priority of the attacking nucleophile relative to the priorities of the groups attached to the carbonyl carbon. For example, although attack by hydride ion on the *Re* face of butanone forms (*S*)-2-butanol (above), attack by a methyl Grignard reagent on the *Re* face of propanal forms (*R*)-2-butanol (below).

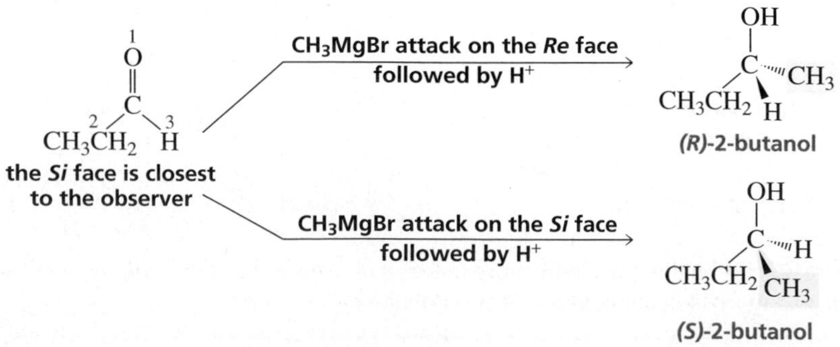

Because the carbonyl carbon and the three atoms attached to it define a plane, the *Re* and *Si* faces have an equal probability of being attacked. Consequently, an addition reaction forms equal amounts of the two enantiomers.

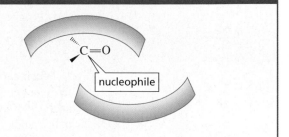

ENZYME-CATALYZED CARBONYL ADDITIONS

In an enzyme-catalyzed addition to a carbonyl compound, only one of the enantiomers is formed. The enzyme can block one face of the carbonyl compound so that it cannot be attacked, or it can position the nucleophile so it is able to attack the carbonyl group from only one side.

PROBLEM 40◆

Which enantiomer is formed when a methyl Grignard reagent attacks the *Re* face of each of the following carbonyl compounds?

a. propiophenone   **b.** benzaldehyde   **c.** 2-pentanone   **d.** 3-hexanone

---

17.15 Designing a Synthesis VI: Disconnections, Synthons, and Synthetic Equivalents

The route to the synthesis of a complicated molecule from simple starting materials is not always obvious. We have seen that it is often easier to work backward from the desired product—a process called *retrosynthetic analysis* (Section 6.12). In a retrosynthetic analysis, the chemist dissects a molecule into smaller and smaller pieces to arrive at readily available starting materials.

retrosynthetic analysis

target molecule $\Longrightarrow$ Y $\Longrightarrow$ X $\Longrightarrow$ W $\Longrightarrow$ starting materials

A useful step in a retrosynthetic analysis is a **disconnection**—breaking a bond to produce two fragments. Typically, one fragment is positively charged and one is negatively charged. The fragments of a disconnection are called **synthons**. Synthons are often not real compounds. For example, if we consider the retrosynthetic analysis of cyclohexanol, a disconnection gives two synthons—an α-hydroxycarbocation and a hydride ion.

retrosynthetic analysis

a disconnection

synthons

+ H⁻

open arrow represents a
retrosynthetic operation

A **synthetic equivalent** is the reagent that is actually used as the source of a synthon. In the synthesis of cyclohexanol, cyclohexanone is the synthetic equivalent for the α-hydroxycarbocation and sodium borohydride is the synthetic equivalent for hydride ion. Thus, cyclohexanol, the target molecule, can be prepared by treating cyclohexanone with sodium borohydride.

synthesis

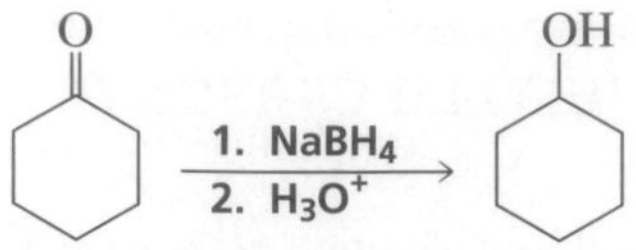

When carrying out a disconnection, you must decide, after breaking the bond, which fragment gets the positive charge and which gets the negative charge. In the retrosynthetic analysis of cyclohexanol, we could have given the positive charge to the hydrogen, and many acids (HCl, HBr, and so on) could have been used for the synthetic equivalent for H$^+$. However, we would have been at a loss to find a synthetic equivalent for an α-hydroxycarbanion. Therefore, when we carried out the disconnection, we assigned the positive charge to the carbon and the negative charge to the hydrogen.

Cyclohexanol can also be disconnected by breaking the C—O bond instead of the C—H bond, forming a carbocation and hydroxide ion.

retrosynthetic analysis

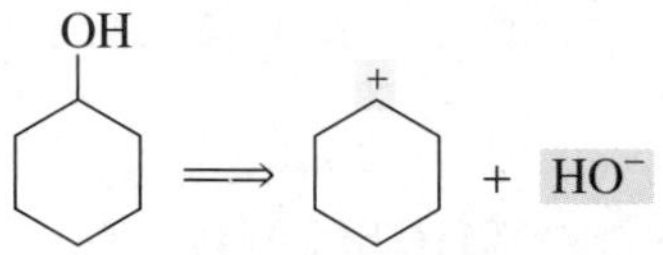

The problem then becomes choosing a synthetic equivalent for the carbocation. A synthetic equivalent for a positively charged synthon needs an electron-withdrawing group at just the right place. Cyclohexyl bromide, with an electron-withdrawing bromine, is a synthetic equivalent for the cyclohexyl carbocation. Cyclohexanol, therefore, can be prepared by treating cyclohexyl bromide with hydroxide ion. This method, however, is not as satisfactory as the first synthesis we proposed—reduction of cyclohexanone—because some of the alkyl halide is converted into an alkene, so the overall yield of the target compound is lower.

synthesis

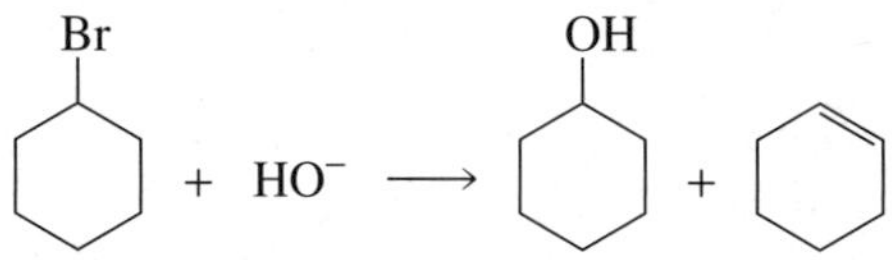

Retrosynthetic analysis shows that 1-methylcyclohexanol can be formed from the reaction of cyclohexanone, the synthetic equivalent for the α-hydroxycarbocation and methylmagnesium bromide, the synthetic equivalent for the methyl anion.

retrosynthetic analysis

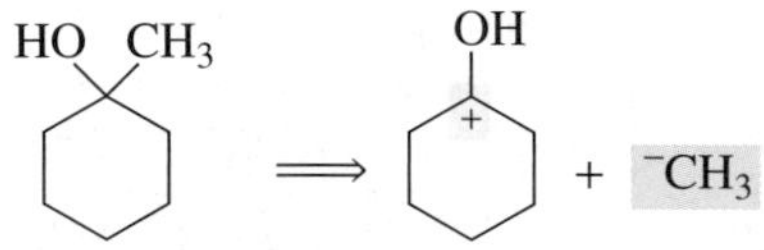

synthesis

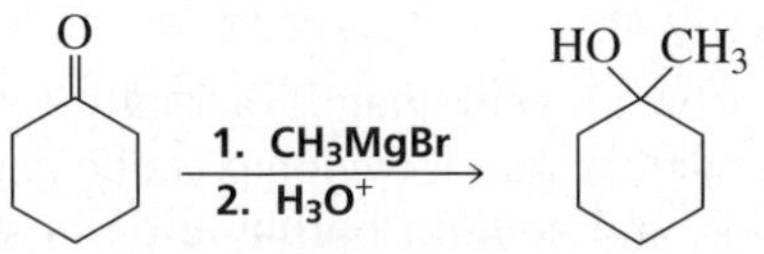

Other disconnections of 1-methylcyclohexanol are possible because any bond to carbon can serve as a disconnection site. For example, one of the ring C—C bonds could be broken. However, these are not useful disconnections, because the synthetic equivalents of the synthons they produce are not easily prepared. A retrosynthetic step must lead to readily obtainable starting materials.

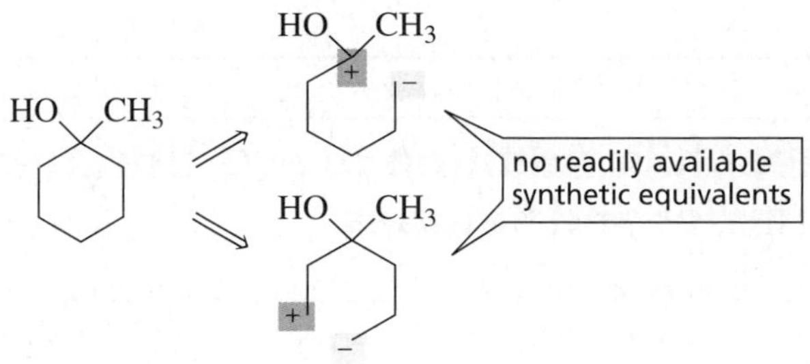

no readily available synthetic equivalents

PROBLEM 41

Using bromocyclohexane as a starting material, how could you synthesize the following compounds?

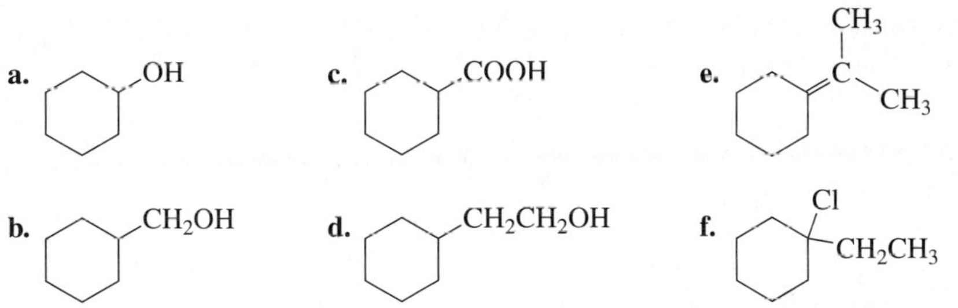

a. OH

b. CH₂OH

c. COOH

d. CH₂CH₂OH

e.

f.

SYNTHESIZING ORGANIC COMPOUNDS

Organic chemists synthesize compounds for many reasons: to study their properties, to answer a variety of chemical questions, or because they have useful properties. One reason chemists synthesize natural products is to provide us with larger supplies of those compounds than nature can produce. For example, Taxol—a compound that has been successful in treating ovarian cancer, breast cancer, and certain forms of lung cancer—is extracted from the bark of *Taxus*, the yew tree found in the Pacific Northwest. The supply of natural Taxol is limited because yew trees are uncommon, grow very slowly, and stripping the bark kills the tree. Moreover, the bark of a 40-foot tree, which may have taken 200 years to grow, provides only 0.5 g of the drug. In addition, *Taxus* forests serve as habitats for the spotted owl, an endangered species, so harvesting the trees would accelerate the owl's demise. Once chemists were successful in determining the structure of Taxol, efforts were undertaken to synthesize it in order to make it more widely available as an anticancer drug. Several syntheses have been successful.

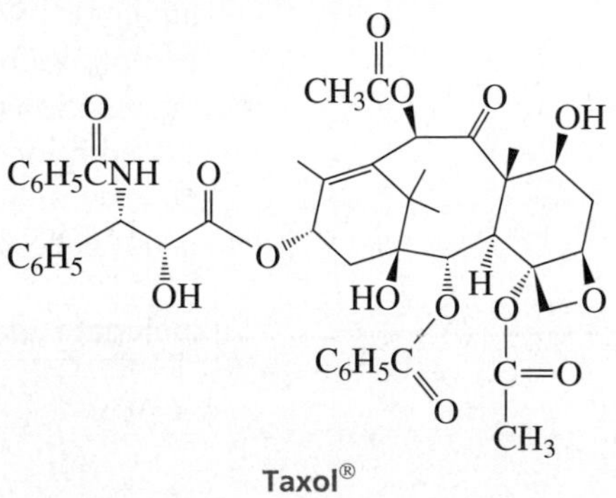

Taxol®

Once a compound has been synthesized, chemists can study its properties to learn how it works; then they can design and synthesize safer or more potent analogs. For example, chemists have found that the anticancer activity of Taxol is substantially reduced if its four ester groups are hydrolyzed. This gives us one small clue to how the molecule functions.

17.16 Nucleophilic Addition to α,β-Unsaturated Aldehydes and Ketones

The resonance contributors for an α,β-unsaturated carbonyl compound show that the molecule has two electrophilic sites: the carbonyl carbon and the β-carbon.

This means that if an aldehyde or a ketone has a double bond in the α,β-position a nucleophile can add either to the carbonyl carbon or to the β-carbon.

Nucleophilic addition to the carbonyl carbon is called **direct addition** or 1,2- addition.

Nucleophilic addition to the β-carbon is called **conjugate addition** or 1,4-addition, because it occurs at the 1- and 4-positions (that is, across the conjugated system). After 1,4-addition has occurred, the product—an enol—tautomerizes to a ketone (or to an aldehyde, Section 6.7), so the overall reaction amounts to addition to the carbon–carbon double bond, with the nucleophile adding to the β-carbon and a proton from the reaction mixture adding to the α-carbon. Compare these reactions with the 1,2- and 1,4-addition reactions you studied in Section 7.10. In direct and conjugate addition, as we saw in Section 7.11, the product that is formed more rapidly (the kinetic product) is not necessarily the more stable product (the thermodynamic product).

Whether the product obtained from nucleophilic addition to an α,β-unsaturated aldehyde or ketone is the direct addition product or the conjugate addition product depends on the nature of the nucleophile, the structure of the carbonyl compound, and the conditions under which the reaction is carried out.

If the nucleophile is a strong base, such as Grignard reagent or hydride ion, direct addition is irreversible. We have seen that if both competing reactions are irreversible, the reaction will be under kinetic control (Section 7.11). Thus, because conjugate addition is always irreversible, the kinetic product predominates. We will see that either the direct addition product or the conjugate addition product can be the kinetic product.

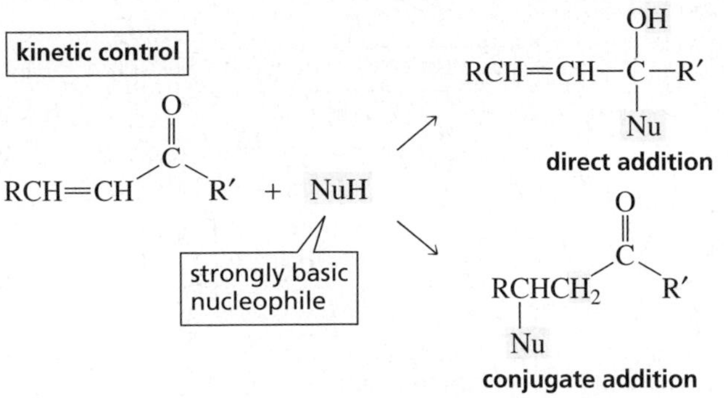

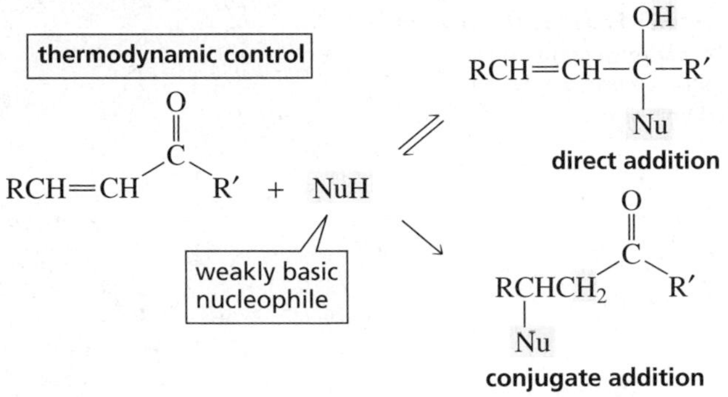

If the nucleophile is a weak base, such as a halide ion, cyanide ion, thiol, alcohol, or amine, direct addition is reversible. We have seen that if one of the competing reactions is reversible, the reaction will be under thermodynamic control. The conjugate addition product is always the thermodynamic product; it is the more stable product because it retains the very stable carbonyl group.

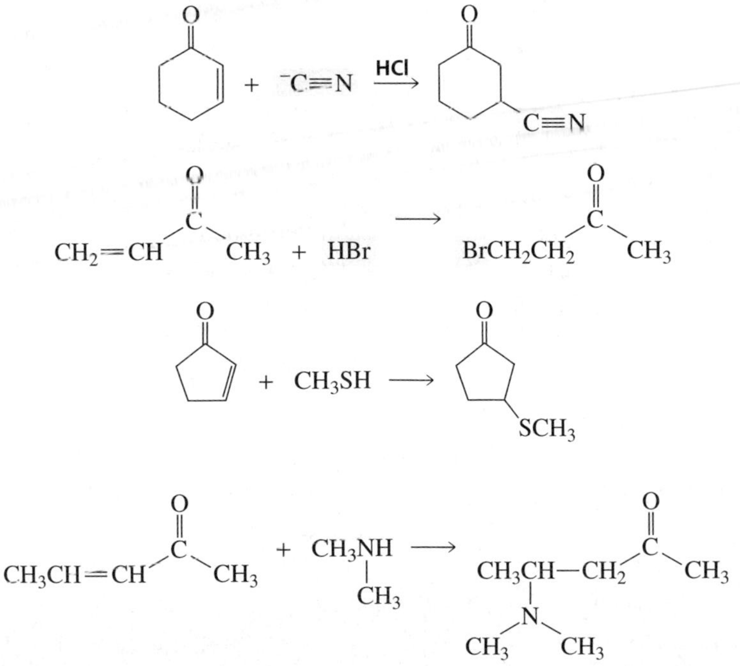

Weak bases form conjugate addition products.

The reaction that prevails when the reaction is under kinetic control is the one that is faster, so the product that is formed will depend on the reactivity of the carbonyl group. Compounds with reactive carbonyl groups form primarily direct addition products because for those compounds, direct addition is faster, whereas compounds with less reactive carbonyl groups can form conjugate addition products because for *those* compounds, conjugate addition is faster. For example, aldehydes have more reactive carbonyl groups than do ketones, so sodium borohydride forms primarily direct addition products with aldehydes. Compared with aldehydes, ketones form less of the direct addition product and more of the conjugate addition product. Ethanol (EtOH) is used to protonate the alkoxide ion formed in the first reaction.

Strong bases form direct addition products with reactive carbonyl groups and conjugate addition products with less reactive carbonyl groups.

product of direct addition
97%

product of direct addition
51%

+

product of conjugate addition
49%

Notice that a saturated alcohol is the final product of conjugate addition in the preceding reaction because the carbonyl group of the ketone will react with a second equivalent of hydride ion.

If direct addition is the desired outcome of hydride addition, it can be achieved by carrying out the reaction in the presence of cerium chloride, a Lewis acid that activates the carbonyl group toward nucleophilic attack by complexing with the carbonyl oxygen.

99%

Like hydride ions, Grignard reagents add irreversibly to carbonyl groups. Therefore, Grignard reagents react with α,β-unsaturated aldehydes and unhindered (reactive) α,β-unsaturated ketones to form direct addition products.

$$CH_2{=}CH{-}\overset{O}{\underset{}{C}}{-}H \quad \xrightarrow[\text{2. EtOH}]{\text{1. CH}_3\text{MgBr}} \quad CH_2{=}CH{-}\overset{OH}{\underset{}{C}}HCH_3$$

If, however, the rate of direct addition is slowed down by steric hindrance, a Grignard reagent will form a conjugate addition product because conjugate addition then becomes the faster reaction.

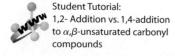
Student Tutorial:
1,2- Addition vs. 1,4-addition to α,β-unsaturated carbonyl compounds

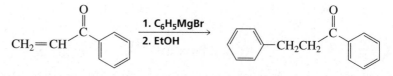

Only conjugate addition occurs when Gilman reagents (lithium dialkylcuprates, Section 10.13) react with α,β-unsaturated aldehydes and ketones. Therefore, Grignard reagents should be used when you want to add an alkyl group to the carbonyl carbon, whereas Gilman reagents should be used when you want to add an alkyl group to the β-carbon.

17.18 **Enzyme-Catalyzed Additions to α,β-Unsaturated Carbonyl Compounds**

Several reactions in biological systems involve addition to α,β-unsaturated carbonyl compounds. The following are examples of biological conjugate addition reactions. Notice that the carbonyl groups are either unreactive (such as COO^-) or have low reactivity (such as the CoA ester) toward the nucleophile, so conjugate addition occurs in each case. The last reaction is an important step in the metabolism of fatty acids (Section 25.6).

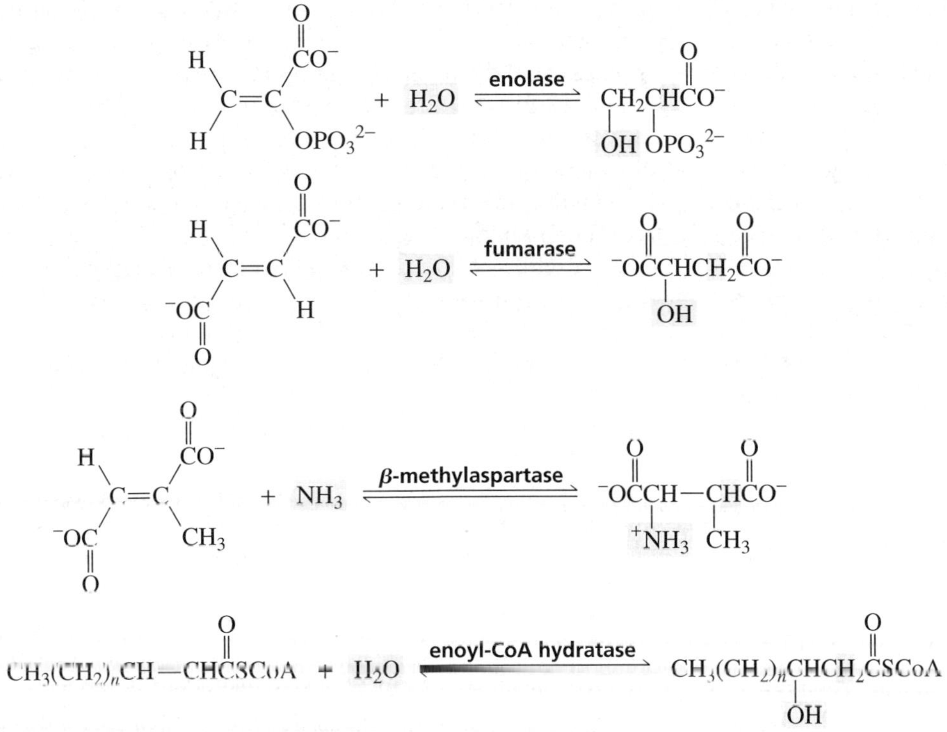

ENZYME-CATALYZED CIS–TRANS INTERCONVERSION

Enzymes that catalyze the interconversion of cis and trans isomers are called cis–trans isomerases. These isomerases are all known to contain thiol (SH) groups. Thiols are weak bases and therefore add to the β-carbon of an α,β-unsaturated carbonyl compound (conjugate addition), forming a carbon–carbon single bond that rotates before the enol is able to tautomerize to the ketone. When tautomerization occurs, the thiol is eliminated, leaving the compound as it was originally except for the configuration about the double bond.

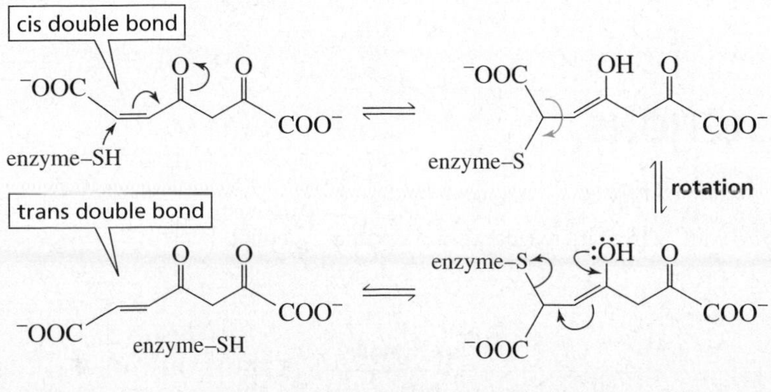

SUMMARY

Aldehydes and **ketones** are Class II carbonyl compounds; they have an acyl group attached to a group (H, R, or Ar) that cannot be readily replaced by another group. Aldehydes and ketones undergo **nucleophilic addition reactions** with strongly basic nucleophiles and **nucleophilic addition–elimination reactions** with less basic nucleophiles that have a lone pair on the attacking atom. With the exception of amides, a carboxylic acid derivative (a Class I carbonyl compound) undergoes a **nucleophilic acyl substitution** reaction with strongly basic nucleophiles to form a Class II carbonyl compound, which then undergoes a **nucleophilic addition** reaction with a second equivalent of the nucleophile.

Electronic and steric factors cause an aldehyde to be more reactive than a ketone toward nucleophilic attack. Aldehydes and ketones are less reactive than acyl halides and acid anhydrides and are more reactive than esters, carboxylic acids, and amides.

Grignard reagents react with aldehydes to form secondary alcohols, with ketones, esters, and acyl halides to form tertiary alcohols, and with carbon dioxide to form carboxylic acids. Aldehydes, acyl chlorides, and carboxylic acids are reduced by hydride ion to primary alcohols, ketones to secondary alcohols, and amides to amines.

Aldehydes and ketones react with primary amines to form **imines** and with secondary amines to form **enamines**. The mechanisms are the same, except for the site from which a proton is lost in the last step of the reaction. Imine and enamine formation are reversible; imines and enamines are hydrolyzed under acidic conditions back to the carbonyl compound and amine. A **pH–rate profile** is a plot of the observed rate constant as a function of the pH of the reaction mixture. Hydroxide ion and heat differentiate the Wolff–Kishner reduction from ordinary **hydrazone** formation.

Aldehydes and ketones undergo acid-catalyzed addition of water to form hydrates. Electron donating and bulky substituents decrease the percentage of hydrate present at equilibrium. Most hydrates are too unstable to be isolated. Acid-catalyzed addition of an alcohol to an aldehyde forms **hemiacetals** and **acetals**, and to a ketone forms **hemiketals** and **ketals**. Acetal and ketal formation are reversible. Cyclic acetals and ketals serve as **protecting groups** for aldehyde and ketone functional groups. Aldehydes and ketones react with thiols to form thioacetals and thioketals; desulfurization replaces the C—S bonds with C—H bonds.

An aldehyde or a ketone reacts with a phosphonium ylide in a **Wittig reaction** to form an alkene. A Wittig reaction is a concerted $[2 + 2]$ cycloaddition reaction; it is completely regioselective. Stabilized ylides form primarily E isomers; unstabilized ylides form primarily Z isomers.

A **prochiral carbonyl carbon** is a carbonyl carbon that is bonded to two different substituents. The Re face is the one closest to the observer when priorities decrease in a clockwise direction; the Si face is the opposite face. Attack by a nucleophile on either the Re face or the Si face forms a racemic mixture.

A useful step in a retrosynthetic analysis is a **disconnection**—breaking a bond to produce two fragments. **Synthons** are fragments produced by a disconnection. A **synthetic equivalent** is the reagent used as the source of a synthon.

Nucleophilic addition to the carbonyl carbon of an α,β-unsaturated Class II carbonyl compound is called **direct addition**; addition to the β-carbon is called **conjugate addition**. Whether direct or conjugate addition occurs depends on the nature of the nucleophile, the structure of the carbonyl compound, and the reaction conditions. Nucleophiles that form unstable direct addition products—halide ions, cyanide ion, thiols, alcohols, and amines—form conjugate addition products. Nucleophiles that form stable addition products—hydride ion and carbanions—form direct addition products with reactive carbonyl groups and conjugate addition products with less reactive carbonyl groups. A Grignard reagent with a highly polarized C—Mg bond reacts with the harder C=O bond; a Gilman reagent with a less polarized C—Cu bond reacts with the softer C=C bond.

Nucleophiles form nucleophilic acyl substitution products with α,β-unsaturated Class I carbonyl compounds that have reactive carbonyl groups and conjugate addition products with compounds with less reactive carbonyl groups.

SUMMARY OF REACTIONS

1. Reactions of *carbonyl compounds* with Grignard reagents (Section 17.4)

   a. Reaction of *formaldehyde* with a Grignard reagent forms a primary alcohol.

$$ \underset{H}{\overset{O}{\underset{\big|}{\overset{\|}{C}}}}{}_{H} \quad \xrightarrow[\text{2. H}_3\text{O}^+]{\text{1. CH}_3\text{MgBr}} \quad CH_3CH_2OH $$

b. Reaction of an *aldehyde* (other than formaldehyde) with a Grignard reagent forms a secondary alcohol.

$$
\underset{R}{\overset{O}{\underset{}{\overset{\parallel}{C}}}}\!-\!H \xrightarrow[\text{2. H}_3\text{O}^+]{\text{1. CH}_3\text{MgBr}} R\!-\!\underset{CH_3}{\overset{OH}{\underset{|}{\overset{|}{C}}}}\!-\!H
$$

c. Reaction of a *ketone* with a Grignard reagent forms a tertiary alcohol.

$$
\underset{R}{\overset{O}{\overset{\parallel}{C}}}\!-\!R' \xrightarrow[\text{2. H}_3\text{O}^+]{\text{1. CH}_3\text{MgBr}} R\!-\!\underset{CH_3}{\overset{OH}{\underset{|}{\overset{|}{C}}}}\!-\!R'
$$

d. Reaction of an *ester* with a Grignard reagent forms a tertiary alcohol with two identical subunits.

$$
\underset{R}{\overset{O}{\overset{\parallel}{C}}}\!-\!OR' \xrightarrow[\text{2. H}_3\text{O}^+]{\text{1. 2 CH}_3\text{MgBr}} R\!-\!\underset{CH_3}{\overset{OH}{\underset{|}{\overset{|}{C}}}}\!-\!CH_3
$$

e. Reaction of an *acyl chloride* with a Grignard reagent forms a tertiary alcohol with two identical substituents.

$$
\underset{R}{\overset{O}{\overset{\parallel}{C}}}\!-\!Cl \xrightarrow[\text{2. H}_3\text{O}^+]{\text{1. 2 CH}_3\text{MgBr}} R\!-\!\underset{CH_3}{\overset{OH}{\underset{|}{\overset{|}{C}}}}\!-\!CH_3
$$

f. Reaction of CO_2 with a Grignard reagent forms a carboxylic acid.

$$
O\!=\!C\!=\!O \xrightarrow[\text{2. H}_3\text{O}^+]{\text{1. CH}_3\text{MgBr}} CH_3\!-\!\overset{O}{\overset{\parallel}{C}}\!-\!OH
$$

2. Reaction of *carbonyl compounds* with acetylide ions (Section 17.5)

$$
\underset{R}{\overset{O}{\overset{\parallel}{C}}}\!-\!R \xrightarrow[\text{2. H}_3\text{O}^+]{\text{1. RC}\equiv\text{C}^-} R\!-\!\underset{R}{\overset{OH}{\underset{|}{\overset{|}{C}}}}\!-\!C\!\equiv\!CR
$$

3. Reactions of *carbonyl compounds* with hydride ion donors (Section 17.6)

a. Reaction of an *aldehyde* with sodium borohydride forms a primary alcohol.

$$
\underset{R}{\overset{O}{\overset{\parallel}{C}}}\!-\!H \xrightarrow[\text{2. H}_3\text{O}^+]{\text{1. NaBH}_4} RCH_2OH
$$

b. Reaction of a *ketone* with sodium borohydride forms a secondary alcohol.

$$
\underset{R}{\overset{O}{\overset{\parallel}{C}}}\!-\!R \xrightarrow[\text{2. H}_3\text{O}^+]{\text{1. NaBH}_4} R\!-\!\underset{}{\overset{OH}{\underset{|}{\overset{|}{CH}}}}\!-\!R
$$

c. Reaction of an *ester* with lithium aluminum hydride forms two alcohols.

$$
\underset{R}{\overset{O}{\overset{\parallel}{C}}}\!-\!OR' \xrightarrow[\text{2. H}_3\text{O}^+]{\text{1. LiAlH}_4} RCH_2OH \ + \ R'OH
$$

d. Reaction of an *ester* with diisobutylaluminum hydride forms an aldehyde.

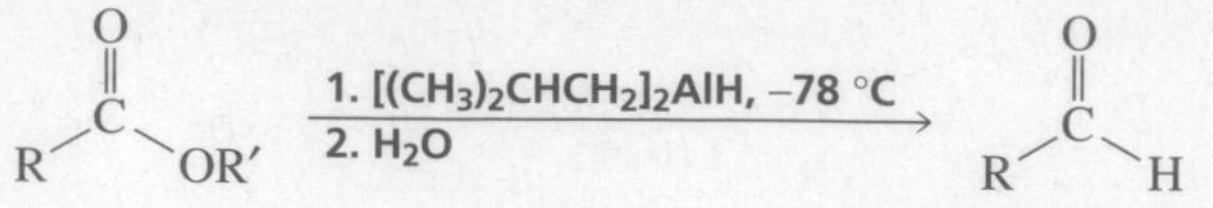

$$\underset{\substack{R \quad OR'}}{C} \xrightarrow[\textbf{2. H}_2\textbf{O}]{\textbf{1. [(CH}_3\textbf{)}_2\textbf{CHCH}_2\textbf{]}_2\textbf{AlH, } -78\ °\textbf{C}} \underset{\substack{R \quad H}}{C}$$

e. Reaction of a *carboxylic acid* with lithium aluminum hydride forms a primary alcohol.

$$\underset{\substack{R \quad OH}}{C} \xrightarrow[\textbf{2. H}_3\textbf{O}^+]{\textbf{1. LiAlH}_4} R-CH_2-OH$$

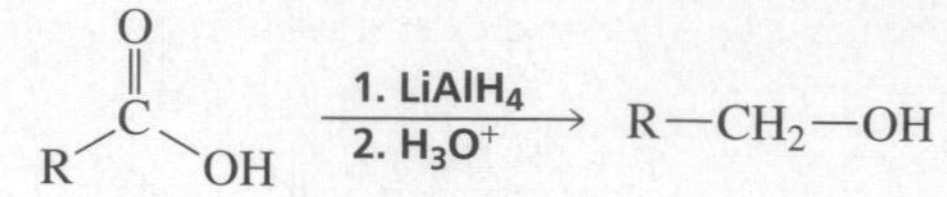

f. Reaction of an *acyl chloride* with sodium borohydride forms a primary alcohol.

$$\underset{\substack{R \quad Cl}}{C} \xrightarrow[\textbf{2. H}_3\textbf{O}^+]{\textbf{1. NaBH}_4} R-CH_2-OH$$

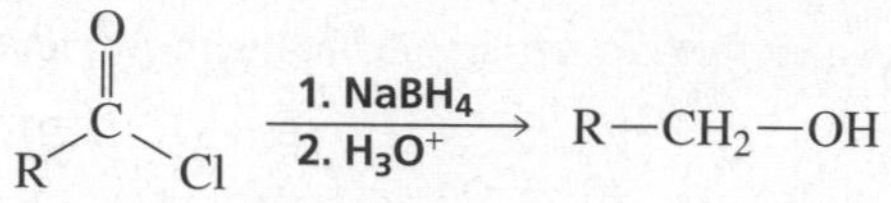

g. Reaction of an *amide* with lithium aluminum hydride forms an amine.

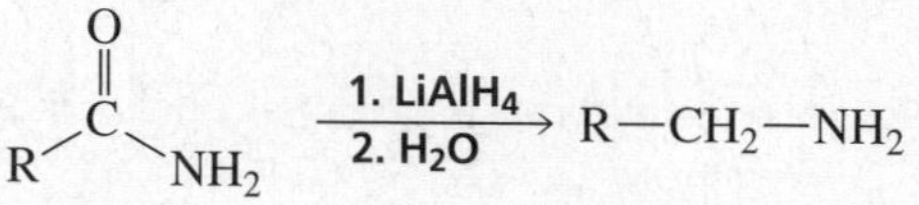

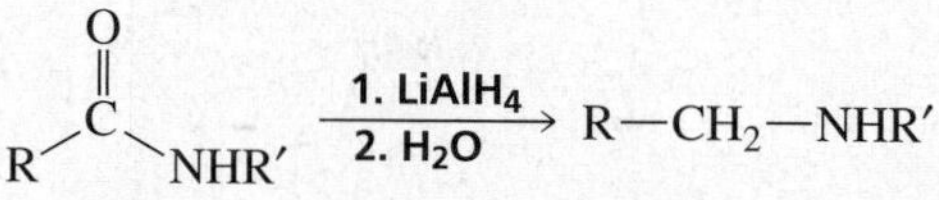

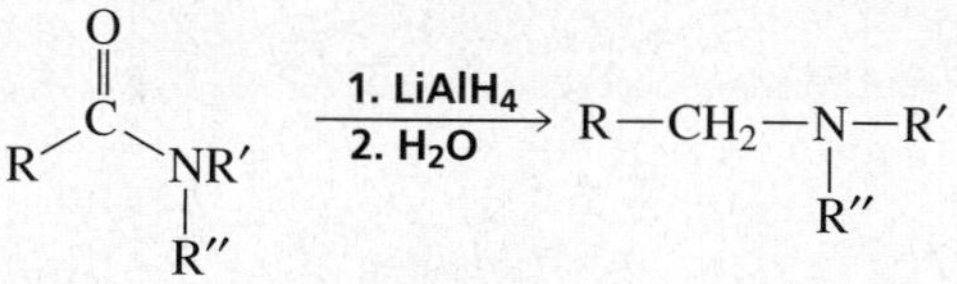

$$\underset{\substack{R \quad NH_2}}{C} \xrightarrow[\textbf{2. H}_2\textbf{O}]{\textbf{1. LiAlH}_4} R-CH_2-NH_2$$

$$\underset{\substack{R \quad NHR'}}{C} \xrightarrow[\textbf{2. H}_2\textbf{O}]{\textbf{1. LiAlH}_4} R-CH_2-NHR'$$

$$\underset{\substack{R \quad NR' \\ \quad R''}}{C} \xrightarrow[\textbf{2. H}_2\textbf{O}]{\textbf{1. LiAlH}_4} R-CH_2-\underset{\substack{R''}}{N}-R'$$

4. Reactions of *aldehydes* and *ketones* with cyanide ion (Section 17.7)

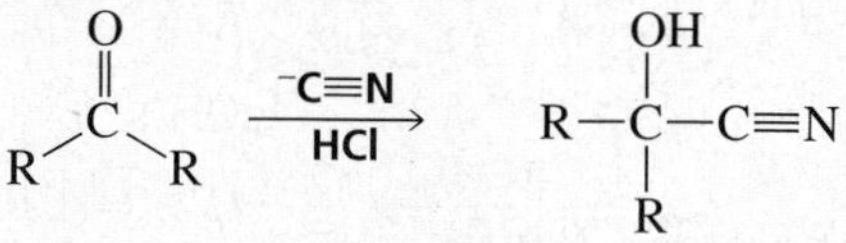

$$\underset{\substack{R \quad R}}{C} \xrightarrow[\textbf{HCl}]{^-C\equiv N} R-\underset{\substack{R}}{\overset{OH}{C}}-C\equiv N$$

5. Reactions of *aldehydes* and *ketones* with amines and amine derivatives (Section 17.8)

a. Reaction with a *primary amine* forms an imine.

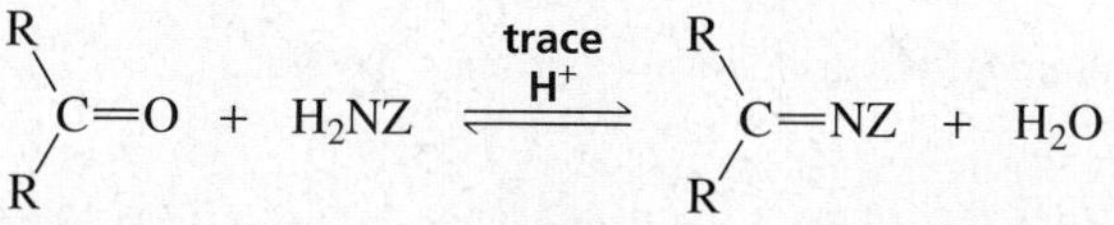

$$\underset{\substack{R}}{\overset{R}{C}}=O \ + \ H_2NZ \ \underset{}{\overset{\textbf{trace} \atop \textbf{H}^+}{\rightleftharpoons}} \ \underset{\substack{R}}{\overset{R}{C}}=NZ \ + \ H_2O$$

When Z=R, the product is a Schiff base; Z can also be OH, NH_2, NHC_6H_5, $NHC_6H_3(NO_2)_2$, or $NHCONH_2$.

b. Reaction with a *secondary amine* forms an enamine.

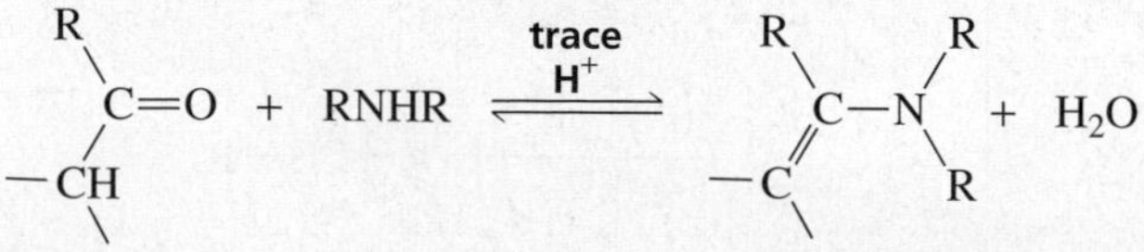

$$\underset{\substack{-CH}}{\overset{R}{C}}=O \ + \ RNHR \ \underset{}{\overset{\textbf{trace} \atop \textbf{H}^+}{\rightleftharpoons}} \ \underset{\substack{-C}}{\overset{R}{C}}-\underset{\substack{R}}{\overset{R}{N}} \ + \ H_2O$$

c. The Wolff–Kishner reduction converts a carbonyl group to a methylene group.

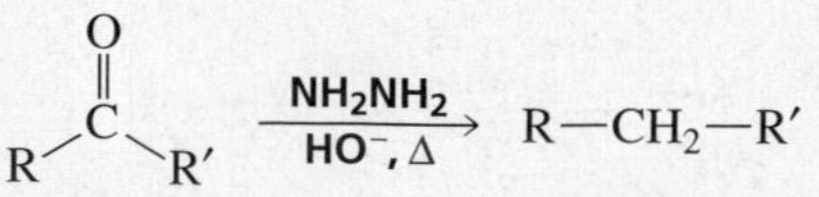

$$\underset{\substack{R \quad R'}}{C} \xrightarrow[\textbf{HO}^-, \Delta]{\textbf{NH}_2\textbf{NH}_2} R-CH_2-R'$$

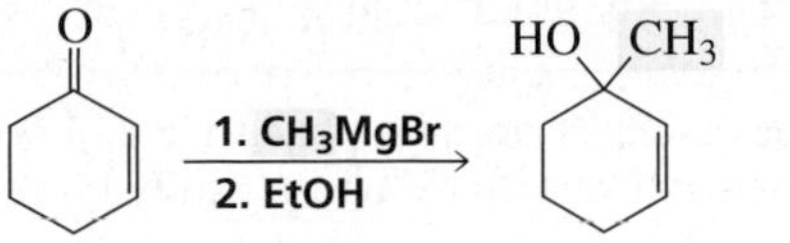

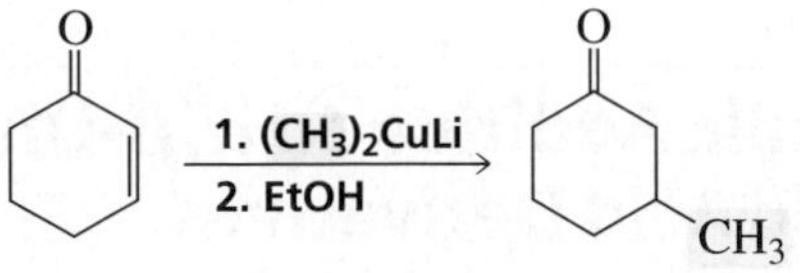

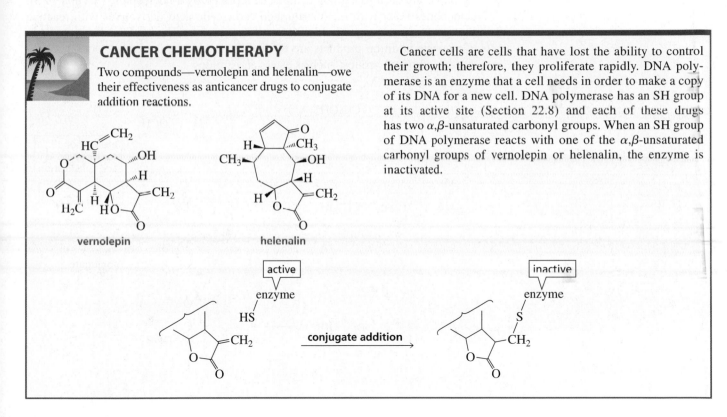

Electrophiles and nucleophiles can be classified as either *hard* or *soft*. Hard electrophiles and nucleophiles are more polarized than soft ones. Hard nucleophiles prefer to react with hard electrophiles, and soft nucleophiles prefer to react with soft electrophiles. Therefore, a Grignard reagent with a highly polarized C—Mg bond prefers to react with the harder C=O bond, whereas a Gilman reagent with a much less polarized C—Cu bond prefers to react with the softer C=C bond.

CANCER CHEMOTHERAPY

Two compounds—vernolepin and helenalin—owe their effectiveness as anticancer drugs to conjugate addition reactions.

Cancer cells are cells that have lost the ability to control their growth; therefore, they proliferate rapidly. DNA polymerase is an enzyme that a cell needs in order to make a copy of its DNA for a new cell. DNA polymerase has an SH group at its active site (Section 22.8) and each of these drugs has two α,β-unsaturated carbonyl groups. When an SH group of DNA polymerase reacts with one of the α,β-unsaturated carbonyl groups of vernolepin or helenalin, the enzyme is inactivated.

vernolepin

helenalin

active enzyme → conjugate addition → inactive enzyme

PROBLEM 42

Give the major product of each of the following reactions:

a. NaC≡N / HCl

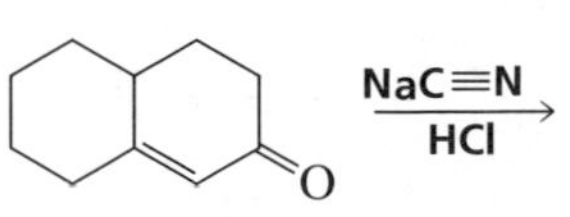

b. 1. NaBH₄ / 2. EtOH

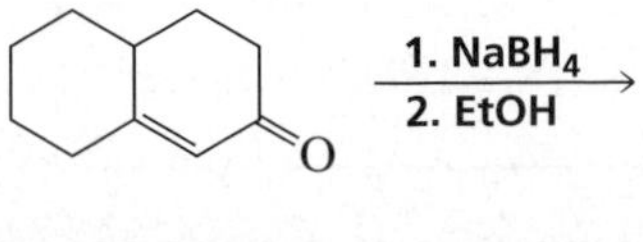

c. 1. CH₃MgBr / 2. EtOH

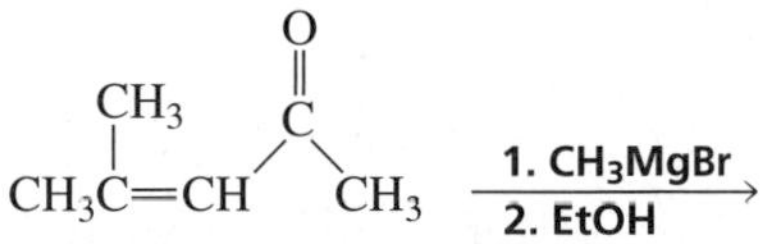

d. 1. NaBH₄ / 2. EtOH

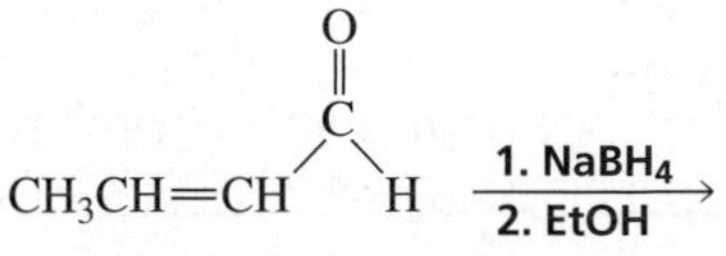

PROBLEM 43◆

Which would give a higher yield of an unsaturated alcohol when treated with sodium borohydride, a sterically hindered ketone or a nonsterically hindered ketone?

17.17 Nucleophilic Addition to α,β-Unsaturated Carboxylic Acid Derivatives

α,β-Unsaturated carboxylic acid derivatives, like α,β-unsaturated aldehydes and ketones, have two electrophilic sites for nucleophilic attack: they can undergo *conjugate addition* or *nucleophilic acyl substitution*. Notice that they undergo *nucleophilic acyl substitution* rather than *direct addition* because the α,β-unsaturated carbonyl compound had a group that can be replaced by a nucleophile. In other words, as with nonconjugated carbonyl compounds, nucleophilic acyl addition becomes nucleophilic acyl substitution if the carbonyl group is attached to a group that can be replaced by a nucleophile (Section 17.3).

Nucleophiles react with α,β-unsaturated carboxylic acid derivatives with reactive carbonyl groups, such as acyl chlorides, to form nucleophilic acyl substitution products. Conjugate addition products are formed from the reaction of nucleophiles with less reactive carbonyl groups, such as esters and amides.

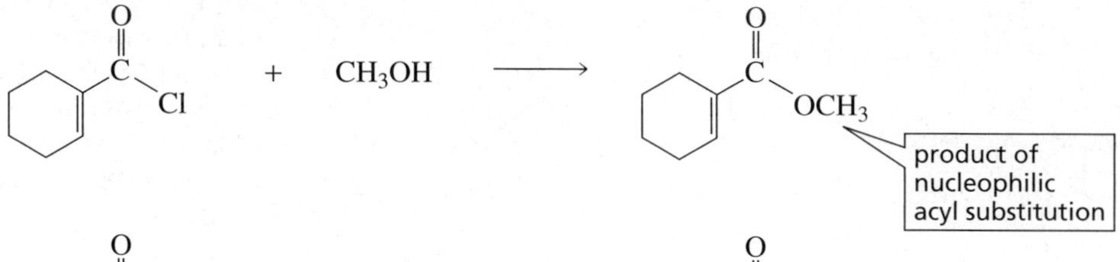

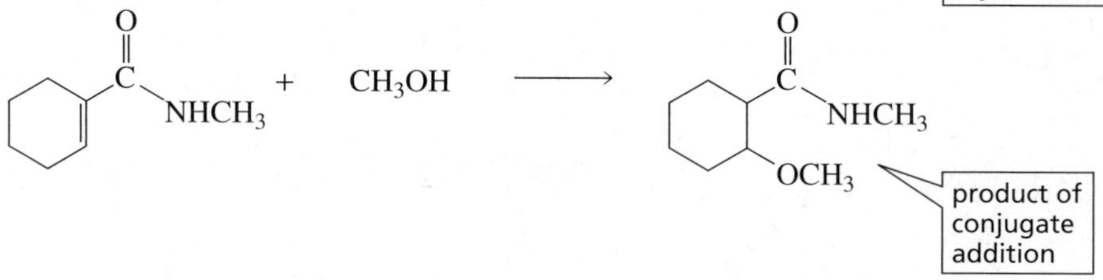

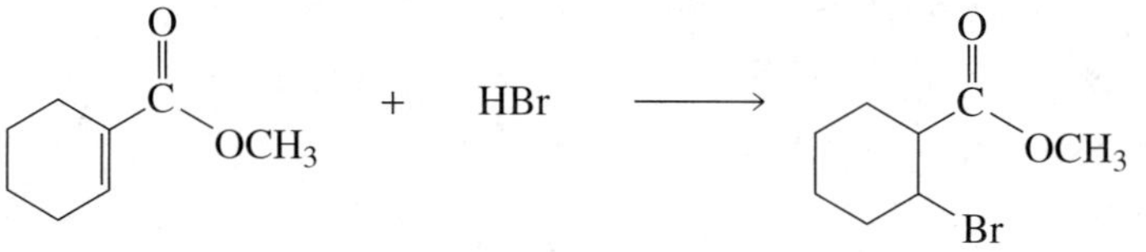

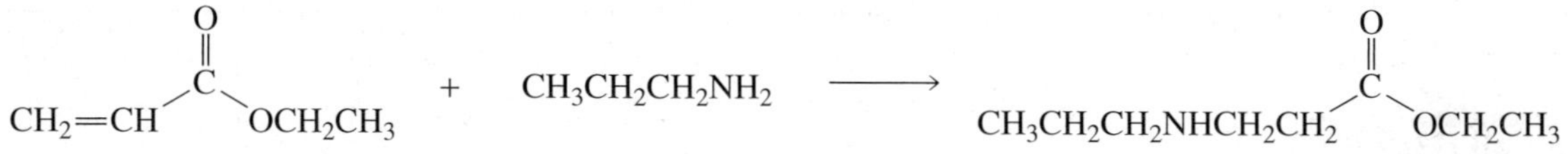

PROBLEM 44◆

Give the major product of each of the following reactions:

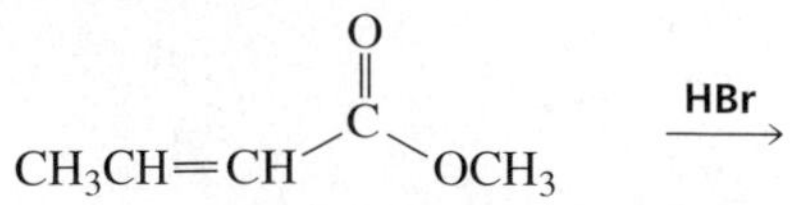

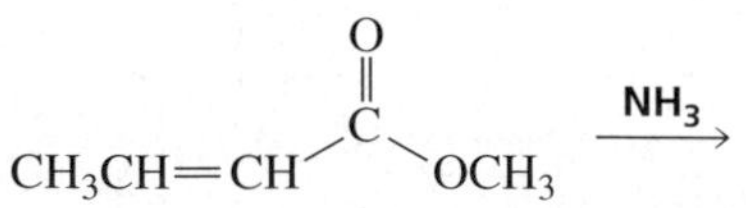

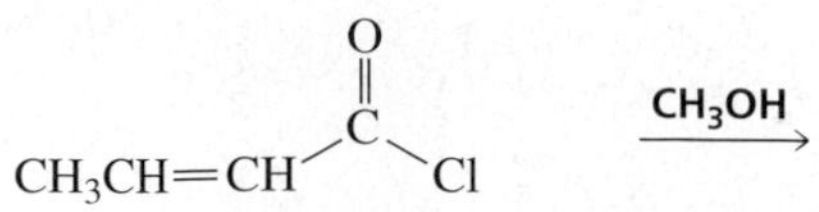

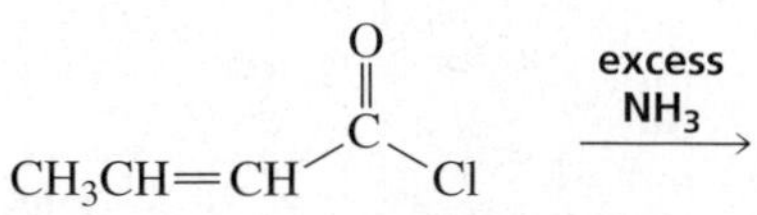

6. Reaction of an *aldehyde* or a *ketone* with water forms a hydrate (Section 17.9)

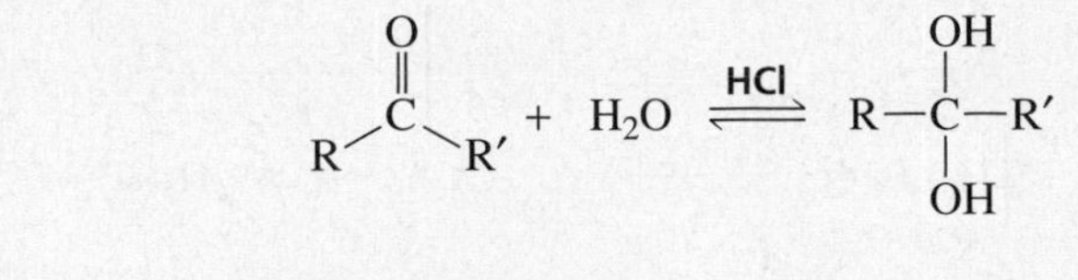

7. Reaction of an *aldehyde* or a *ketone* with excess alcohol forms an acetal or a ketal (Section 17.10).

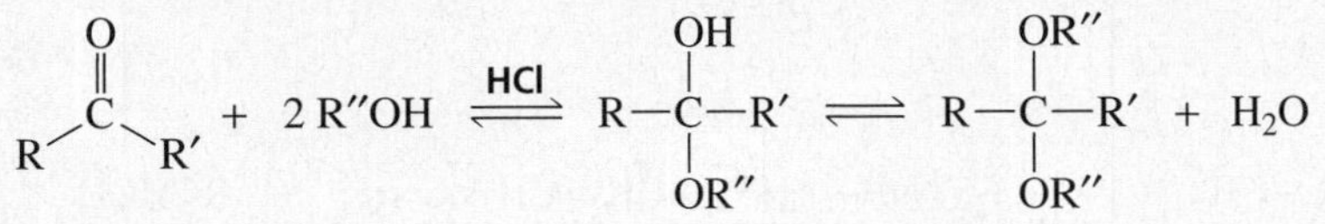

8. Protecting groups (Section 17.11)

    a. *Aldehydes* and *ketones* can be protected by being converted to acetals.

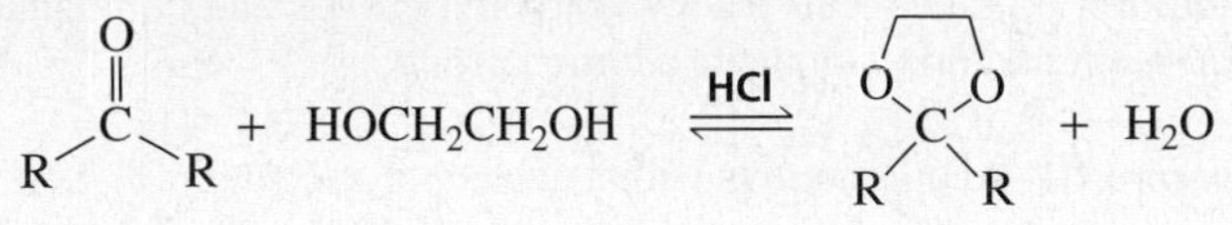

    b. The OH group of an *alcohol* can be protected by being converted to a TMS ether.

$$R-OH \quad + \quad (CH_3)_3SiCl \xrightarrow{(CH_3CH_2)_3N} R-OSi(CH_3)_3$$

    c. The OH group of a *carboxylic acid* can be protected by being converted to an ester.

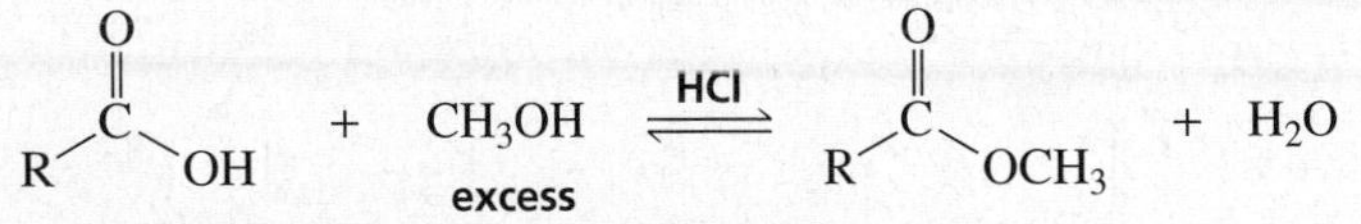

    d. An *amino group* can be protected by being converted to an amide.

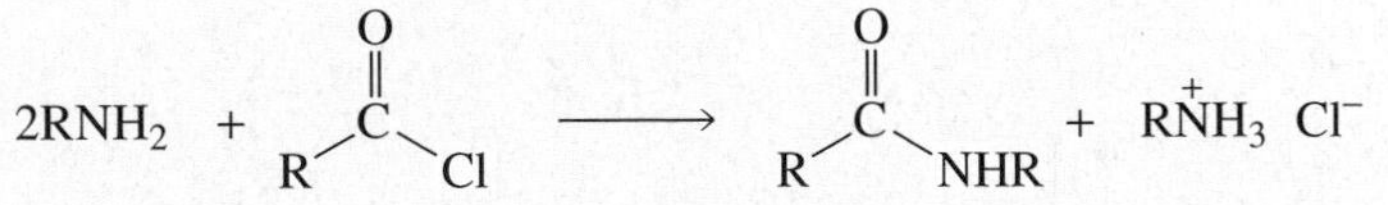

9. Reaction of an *aldehyde* or a *ketone* with a thiol forms a thioacetal or a thioketal (Section 17.12).

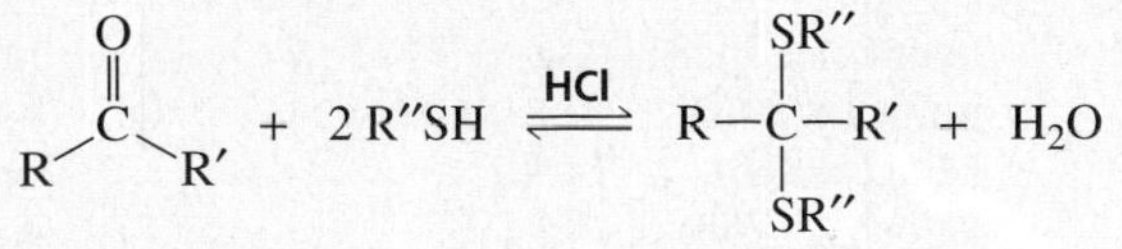

10. Desulfurization of a *thioacetal* or a *thioketal* forms an alkane (Section 17.12).

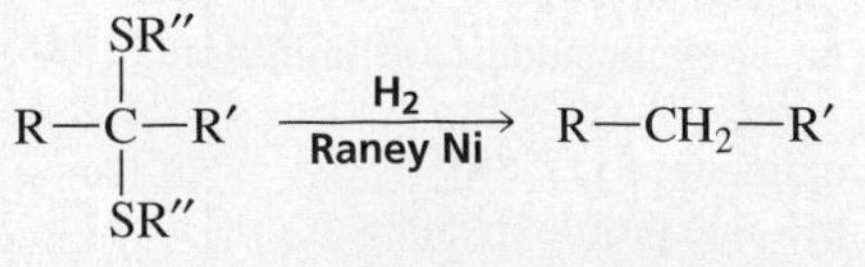

11. Reaction of an *aldehyde* or a *ketone* with a phosphonium ylide (a Wittig reaction) forms an alkene (Section 17.13).

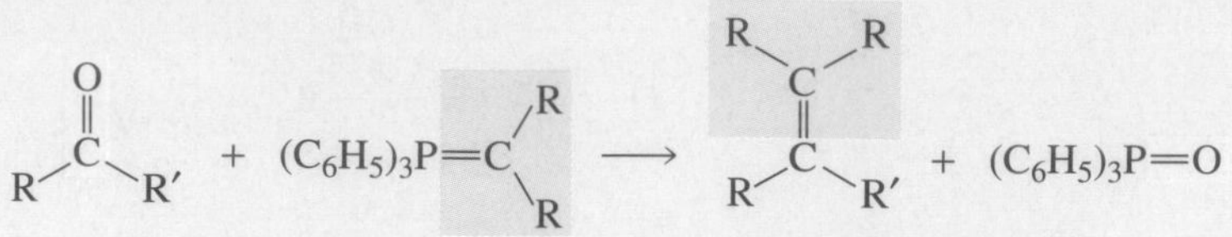

$$\underset{R}{\overset{O}{\underset{\parallel}{C}}}\!\!-R' \quad + \quad (C_6H_5)_3P{=}C\underset{R}{\overset{R}{\diagdown}} \quad \longrightarrow \quad \underset{R}{\overset{R\;\diagup\;C\;\diagdown\;R}{\underset{\parallel}{C}}}\!\!-R' \quad + \quad (C_6H_5)_3P{=}O$$

12. Reaction of an α,β-*unsaturated aldehyde* or *ketone* with a nucleophile (Section 17.16).

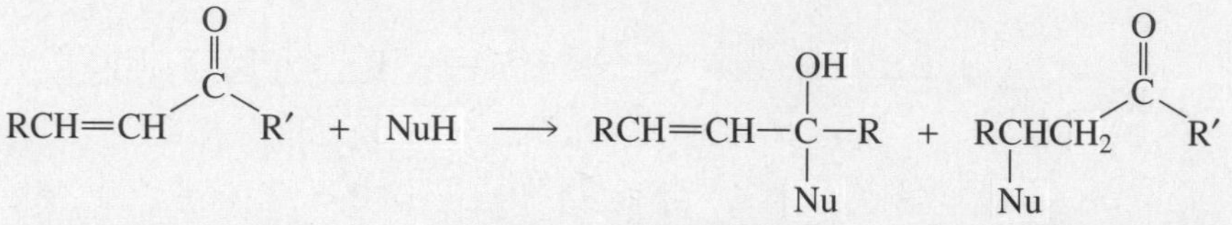

$$\underset{\text{}}{RCH{=}CH}\overset{O}{\underset{\parallel}{\overset{\diagup}{C}}}R' \;+\; NuH \;\longrightarrow\; RCH{=}CH\underset{\underset{Nu}{|}}{\overset{\overset{OH}{|}}{C}}\!-R \;+\; RCHCH_2\overset{O}{\underset{\parallel}{\overset{\diagup}{C}}}R'$$

direct addition   conjugate addition

Nucleophiles that are weak bases ($^-$CN, RSH, RNH$_2$, Br$^-$) form conjugate addition products. Nucleophiles that are strong bases (RLi, RMgBr, H$^-$) form direct addition products with reactive carbonyl groups and conjugate addition products with less reactive carbonyl groups. Gilman (R$_2$CuLi) reagents form conjugate addition products.

13. Reaction of an α,β-*unsaturated carboxylic acid derivative* with a nucleophile (Section 17.17).

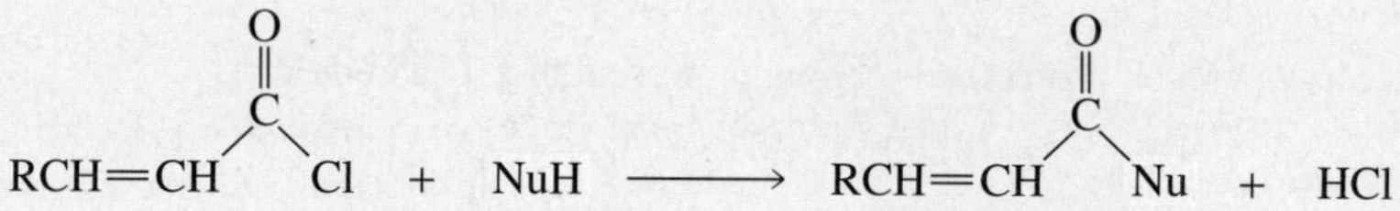

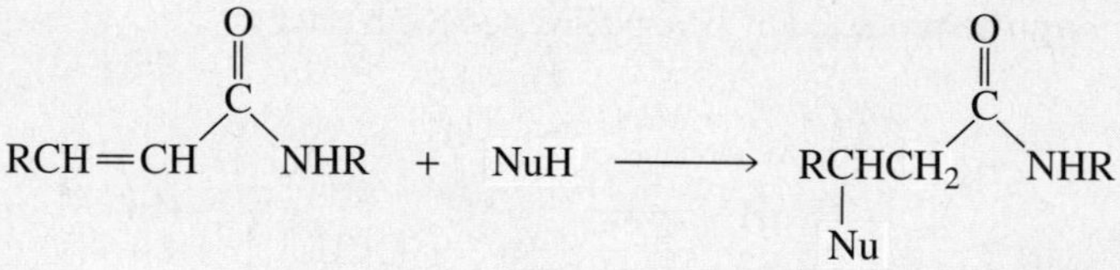

$$RCH{=}CH\overset{O}{\underset{\parallel}{\overset{\diagup}{C}}}Cl \;+\; NuH \;\longrightarrow\; RCH{=}CH\overset{O}{\underset{\parallel}{\overset{\diagup}{C}}}Nu \;+\; HCl$$

nucleophilic acyl
substitution

$$RCH{=}CH\overset{O}{\underset{\parallel}{\overset{\diagup}{C}}}NHR \;+\; NuH \;\longrightarrow\; RCHCH_2\overset{O}{\underset{\parallel}{\overset{\diagup}{C}}}NHR$$

conjugate addition

Nucleophiles form nucleophilic acyl substitution products with reactive carbonyl groups and conjugate addition products with less reactive carbonyl groups.

KEY TERMS

PROBLEMS

45. Draw the structure for each of the following compounds:
  a. isobutyraldehyde
  b. 4-hexenal
  c. diisopentyl ketone
  d. 3-methylcyclohexanone
  e. 2,4-pentanedione

  f. 4-bromo-3-heptanone
  g. γ-bromocaproaldehyde
  h. 2-ethylcyclopentanecarbaldehyde
  i. 4-methyl-5-oxohexanal
  j. benzene-1,3-dicarbaldehyde

46. Give the products of each of the following reactions:

a. $\underset{\text{O}}{\overset{\text{O}}{\text{CH}_3\text{CH}_2\overset{||}{\text{CH}}}}$ + CH$_3$CH$_2$OH $\xrightarrow{\text{HCl}}$
  excess

b. $\text{C}_6\text{H}_5\overset{\text{O}}{\overset{||}{\text{C}}}\text{CH}_2\text{CH}_3$ + NH$_2$NH$_2$ $\xrightarrow[\ ]{\overset{\text{trace}}{\text{H}^+}}$

c. $\text{C}_6\text{H}_5\overset{\text{O}}{\overset{||}{\text{C}}}\text{CH}_2\text{CH}_3$ + NH$_2$NH$_2$ $\xrightarrow[\Delta]{\text{HO}^-}$

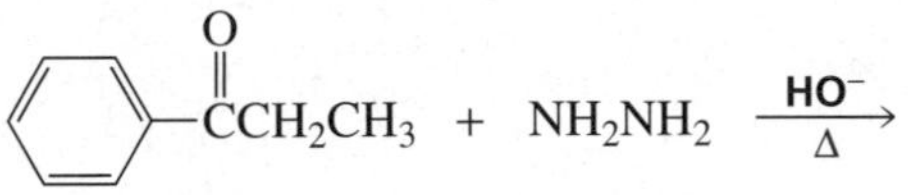

d. CH$_3$CH$_2\overset{\text{O}}{\overset{||}{\text{C}}}CH_3$ $\xrightarrow[\text{2. H}_3\text{O}^+]{\text{1. NaBH}_4}$

e. CH$_3$CH$_2\overset{\text{O}}{\overset{||}{\text{C}}}CH_2CH_3$ + NaC≡N $\xrightarrow{\text{HCl}}$
  excess

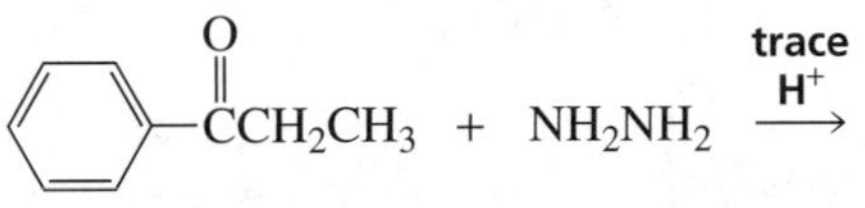

f. CH$_3$CH$_2$CH$_2\overset{\text{O}}{\overset{||}{\text{C}}}OCH_2CH_3$ $\xrightarrow[\text{2. H}_3\text{O}^+]{\text{1. LiAlH}_4}$

g. CH$_3$CH$_2$CH$_2\overset{\text{O}}{\overset{||}{\text{C}}}CH_3$ + HOCH$_2$CH$_2$OH $\xrightarrow{\text{HCl}}$

h. 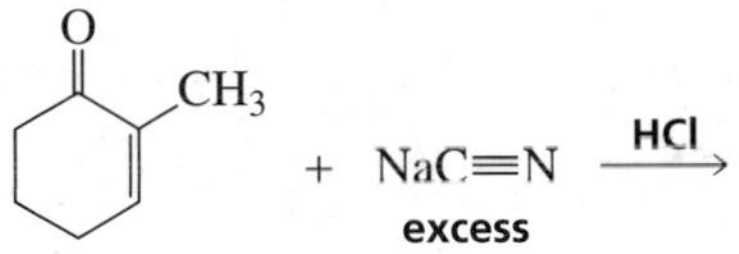 [2-methylcyclohex-2-enone] + NaC≡N $\xrightarrow{\text{HCl}}$
  excess

47. List the following compounds in order of decreasing reactivity toward nucleophilic attack:

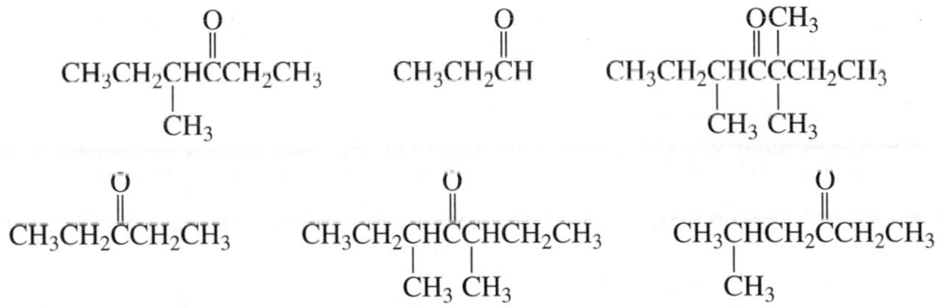

$\underset{\underset{\text{CH}_3}{|}}{\text{CH}_3\text{CH}_2\text{CH}}\overset{\text{O}}{\overset{||}{\text{C}}}\text{CH}_2\text{CH}_3$ $\text{CH}_3\text{CH}_2\overset{\text{O}}{\overset{||}{\text{CH}}}$ $\underset{\underset{\text{CH}_3\ \text{CH}_3}{|\ \ \ \ |}}{\text{CH}_3\text{CH}_2\text{CHC}}\overset{\text{OCH}_3}{\overset{||}{\text{C}}}\text{CCH}_2\text{CH}_3$

$\text{CH}_3\text{CH}_2\overset{\text{O}}{\overset{||}{\text{C}}}\text{CH}_2\text{CH}_3$ $\underset{\underset{\text{CH}_3\ \text{CH}_3}{|\ \ \ \ |}}{\text{CH}_3\text{CH}_2\text{CHC}}\overset{\text{O}}{\overset{||}{\text{C}}}\text{CHCH}_2\text{CH}_3$ $\underset{\underset{\text{CH}_3}{|}}{\text{CH}_3\text{CHCH}_2}\overset{\text{O}}{\overset{||}{\text{C}}}\text{CH}_2\text{CH}_3$

48. a. Show the reagents required to form the primary alcohol.

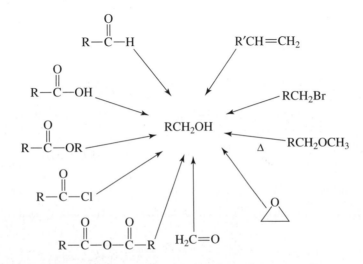

b. Which of the reactions cannot be used for the synthesis of isobutyl alcohol?
c. Which of the reactions changes the carbon skeleton of the starting material?

49. Using cyclohexanone as the starting material, describe how each of the following compounds could be synthesized:

a. 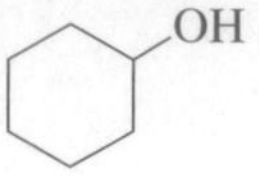 [cyclohexane with OH]

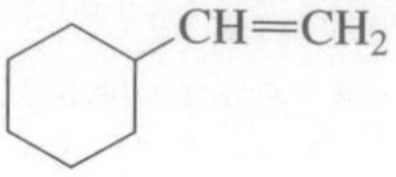

b. 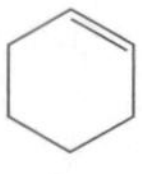 [cyclohexene]

c. 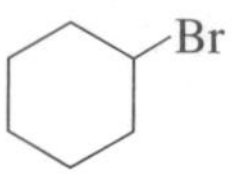 [cyclohexane with Br]

d. [cyclohexane with NH₂]

e. 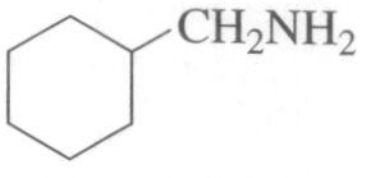 [cyclohexane with CH₂NH₂]

f. [cyclohexane with N(CH₃)₂]

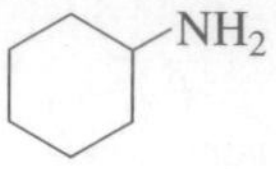

g. [cyclohexane with CH=CH₂]

h. 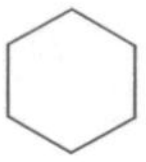 [cyclohexane] (show two methods)

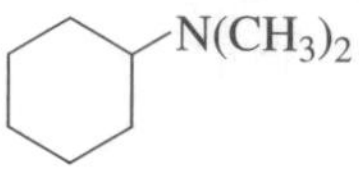

i. 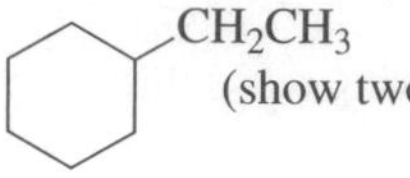 [cyclohexane with CH₂CH₃] (show two methods)

50. Propose a mechanism for the following reaction:

$$HOCH_2CH_2CH_2CH_2 \overset{O}{\underset{}{\overset{\|}{C}}} H \xrightarrow[CH_3OH]{HCl} \text{[tetrahydropyran with OCH}_3\text{]}$$

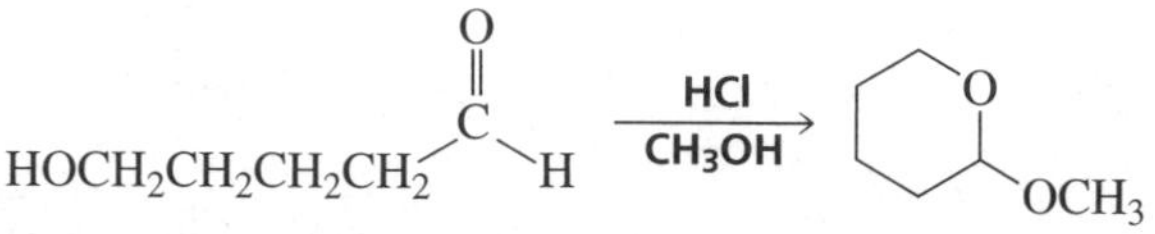

51. List the following compounds in order of decreasing K_{eq} for hydrate formation:

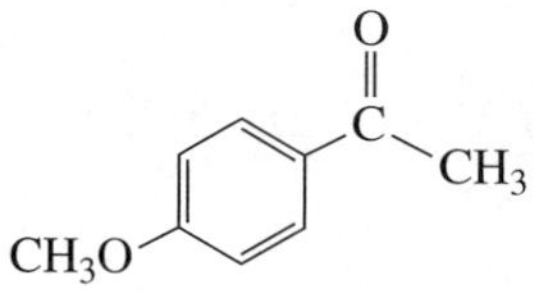

[four acetophenone derivatives: unsubstituted; Cl-substituted; O₂N-substituted; CH₃O-substituted]

52. Fill in the boxes:

a. $CH_3OH \xrightarrow{\square} CH_3Br \xrightarrow[\square]{\square} \square \xrightarrow[2.\square]{1.\square} CH_3CH_2OH$

b. $CH_4 \xrightarrow{\square} CH_3Br \xrightarrow[\square]{\square} \square \xrightarrow[2.\square]{1.\square} CH_3CH_2CH_2OH$

53. Give the products of the following reactions:

a. 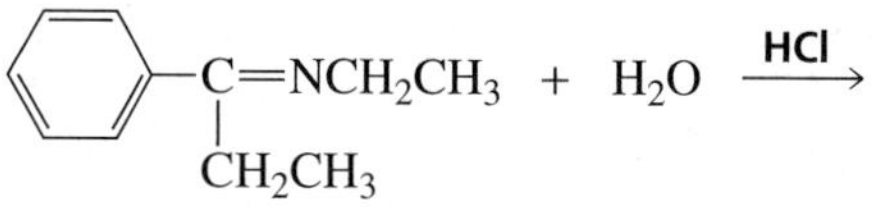 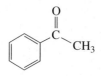 $\overset{\text{CH}_2\text{CH}_3}{\underset{}{\text{—C=NCH}_2\text{CH}_3}} + H_2O \xrightarrow{HCl}$

b. $CH_3CH_2\overset{O}{\overset{\|}{C}}CH_3 \xrightarrow[2.\ H_3O^+]{1.\ CH_3CH_2MgBr}$

c. 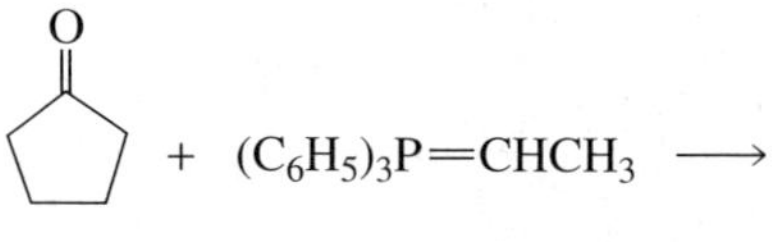 [cyclopentanone] $+ (C_6H_5)_3P=CHCH_3 \longrightarrow$

d. $CH_3CH_2\overset{O}{\overset{\|}{C}}OCH_3 \xrightarrow[2.\ H_3O^+]{\substack{1.\ CH_3CH_2MgBr \\ excess}}$

e. 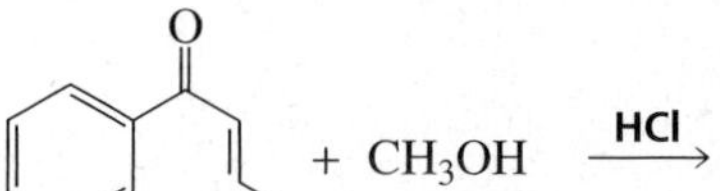 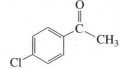 $+ CH_3OH \xrightarrow{HCl}$

f. 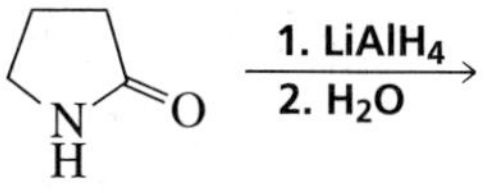 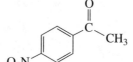 [2-pyrrolidinone] $\xrightarrow[2.\ H_2O]{1.\ LiAlH_4}$

g. 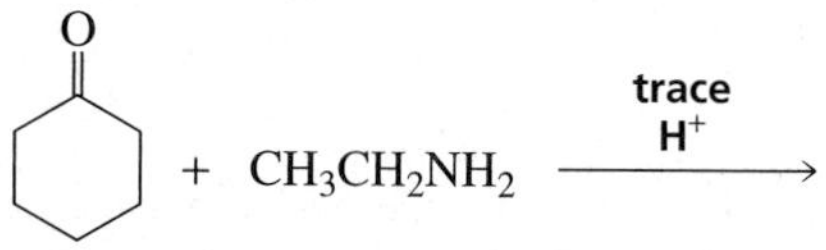 [cyclohexanone] $+ CH_3CH_2NH_2 \xrightarrow[H^+]{trace}$

h. [cyclohexanone] $+ (CH_3CH_2)_2NH \xrightarrow[H^+]{trace}$

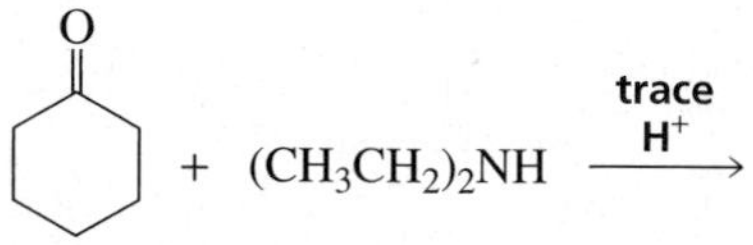

i. $CH_3\overset{CH_3}{\overset{|}{C}}=CH\overset{O}{\overset{\|}{C}}CH_3 + HBr \longrightarrow$

j. $2\ CH_2=CH-\overset{O}{\overset{\|}{C}}OCH_3 + CH_3NH_2 \longrightarrow$

54. Thiols can be prepared from the reaction of thiourea with an alkyl halide, followed by hydroxide-ion-promoted hydrolysis.

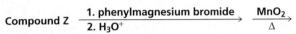

a. Propose a mechanism for the reaction.
b. What thiol would be formed if the alkyl halide employed were pentyl bromide?

55. The only organic compound obtained when compound Z undergoes the following sequence of reactions gives the ^{1}H NMR spectrum shown. Identify compound Z.

$$\text{Compound Z} \xrightarrow[\text{2. H}_3\text{O}^+]{\text{1. phenylmagnesium bromide}} \xrightarrow[\Delta]{\text{MnO}_2}$$

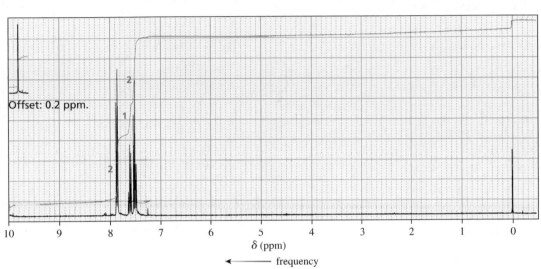

Offset: 0.2 ppm.

δ (ppm)

← frequency

56. Propose a mechanism for each of the following reactions:

a.

b.

c.

57. How many signals would the product of the following reaction show in the following spectra?
a. its ^{1}HNMR spectrum
b. its ^{13}C NMR spectrum

$$CH_3CCH_2CH_2COCH_3 \xrightarrow[\text{2. H}_3\text{O}^+]{\text{1. excess CH}_3\text{MgBr}}$$

58. Fill in the boxes:

59. How could you convert *N*-methylbenzamide into the following compounds?
a. *N*-methylbenzylamine
b. benzoic acid
c. methyl benzoate
d. benzyl alcohol

60. Give the products of the following rections. Show all stereoisomers that are formed.

a.

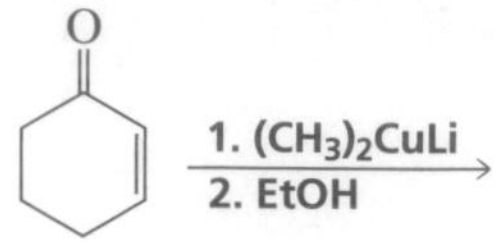

$\dfrac{\text{1. (CH}_3)_2\text{CuLi}}{\text{2. EtOH}}$

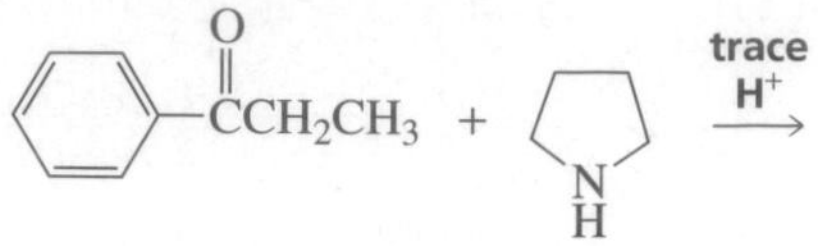

b.

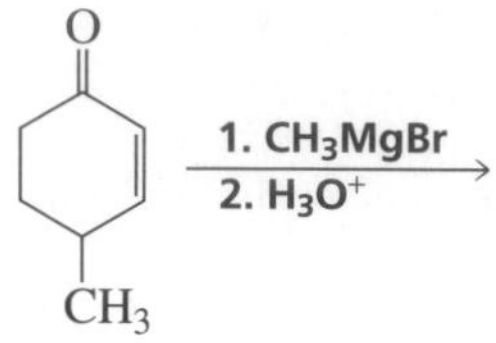

$\dfrac{\text{1. CH}_3\text{MgBr}}{\text{2. H}_3\text{O}^+}$

c.

+

$\dfrac{\text{trace}}{\text{H}^+}$

d. $CH_3CH_2CCH_2CH_2CH_2CH_3$ $\dfrac{\text{1. NaBH}_4}{\text{2. H}_3\text{O}^+}$

61. List three different sets of reagents (each set consisting of a carbonyl compound and a Grignard reagent) that could be used to prepare each of the following tertiary alcohols:

a. 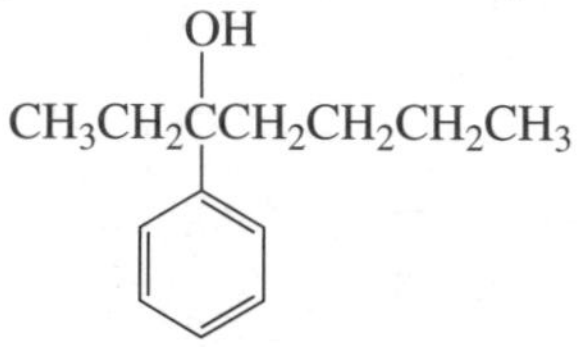 $CH_3CH_2CCH_2CH_2CH_3$ (with OH and phenyl substituent)

b. $CH_3CH_2CCH_2CH_2CH_3$ (with OH and CH_2CH_3)

62. Give the product of the reaction of 3-methyl-2-cyclohexenone with each of the following reagents:
   a. CH_3MgBr followed by H_3O^+
   b. excess NaCN, HCl
   c. H_2, Pd/C
   d. HBr
   e. $(CH_3CH_2)_2CuLi$ followed by H_3O^+
   f. CH_3CH_2SH

63. Norlutin and Enovid are ketones that suppress ovulation. Consequently, they have been used clinically as contraceptives. For which of these compounds would you expect the infrared carbonyl absorption (C=O stretch) to be at a higher frequency? Explain.

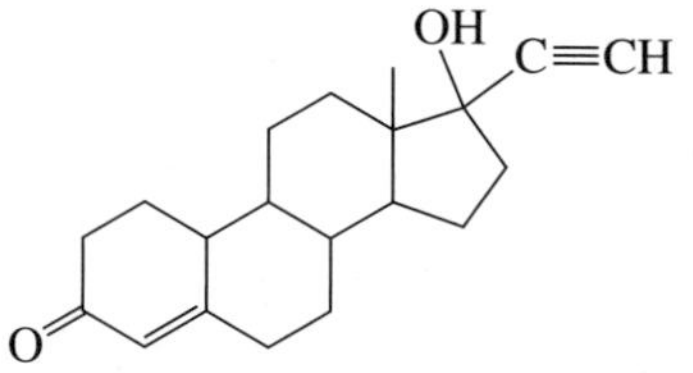

Norlutin® 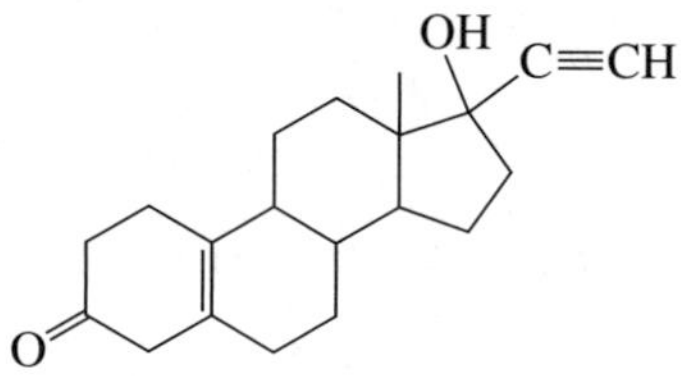   Enovid®

64. Give the product of each of the following reactions:

a.

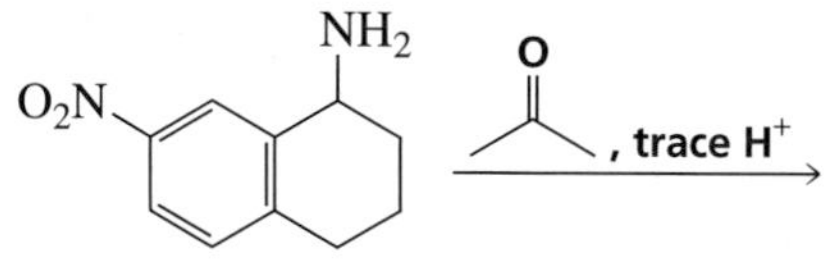

, trace H⁺

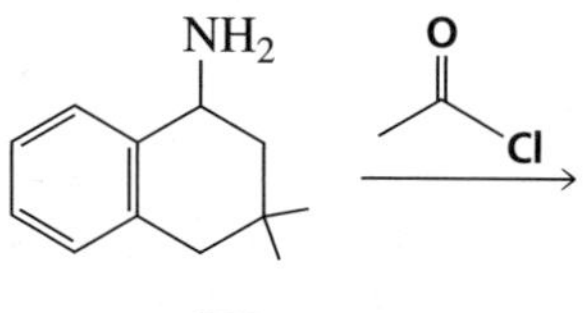

b.

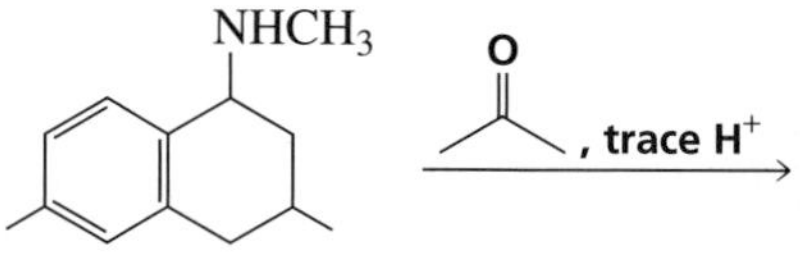

, trace H⁺

c.

, Cl

d.

, HCl

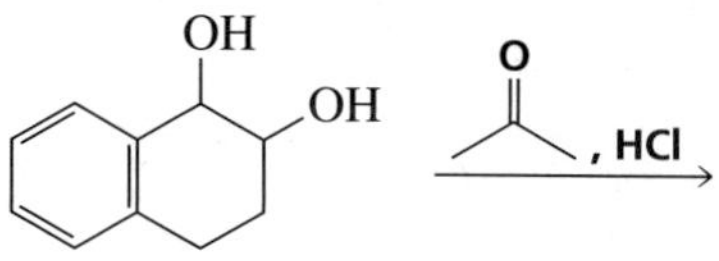

65. Propose a reasonable mechanism for each of the following reactions:

a. 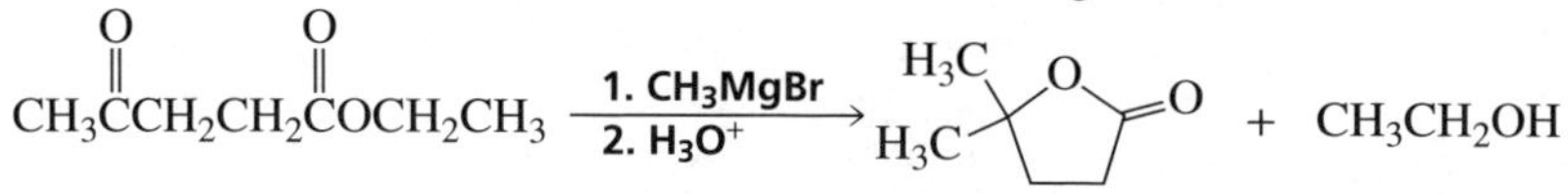 $CH_3CCH_2CH_2COCH_2CH_3$ $\dfrac{\text{1. CH}_3\text{MgBr}}{\text{2. H}_3\text{O}^+}$ + CH_3CH_2OH

b.

$\dfrac{\text{HCl}}{\text{CH}_3\text{OH}}$

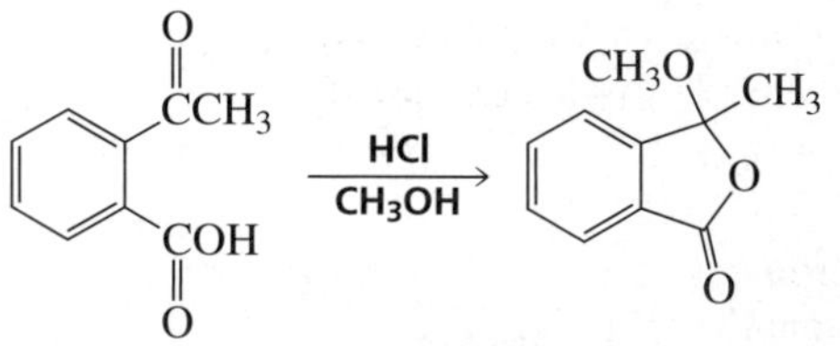

66. A compound gives the following IR spectrum. Upon reaction with sodium borohydride followed by acidification, it forms the product with the following ^{1}HNMR spectrum. Identify the compounds.

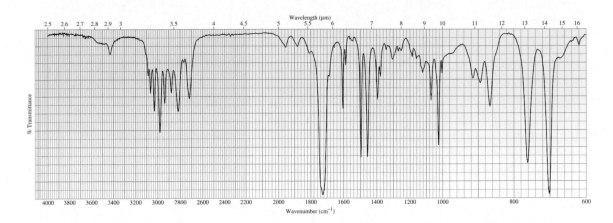

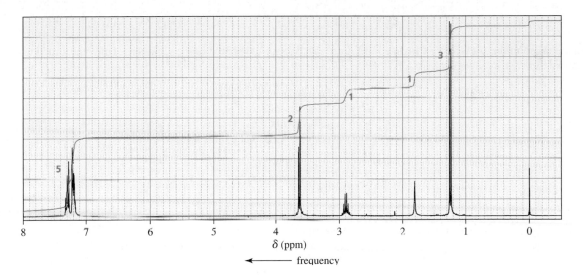

67. a. Propose a mechanism for the following reaction:

$$CH_2=CHCHC\equiv N \xrightarrow[H_2O]{HO^-} N\equiv CCH_2CH_2CH$$

with OH above the second carbon and O (double bond) above the CH, for the product.

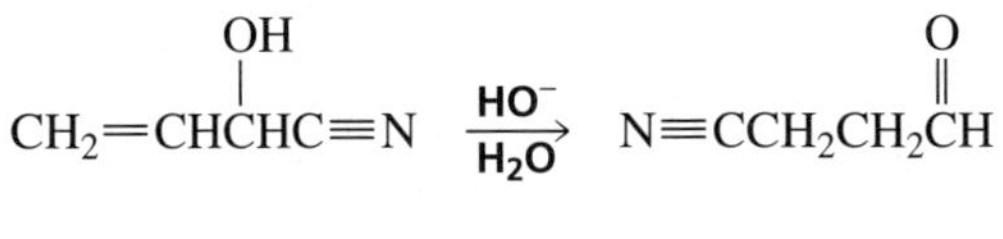

b. What is the product of the following reaction?

$$CH_3CCH=CH_2 \xrightarrow[H_2O]{HO^-}$$

with OH above the carbon and C≡N below it.

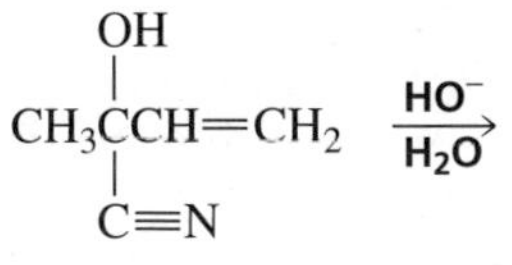

68. Unlike a phosphonium ylide, which reacts with an aldehyde or ketone to form an alkene, a sulfonium ylide reacts with an aldehyde or ketone to form an epoxide. Explain why one ylide forms an alkene, whereas the other forms an epoxide.

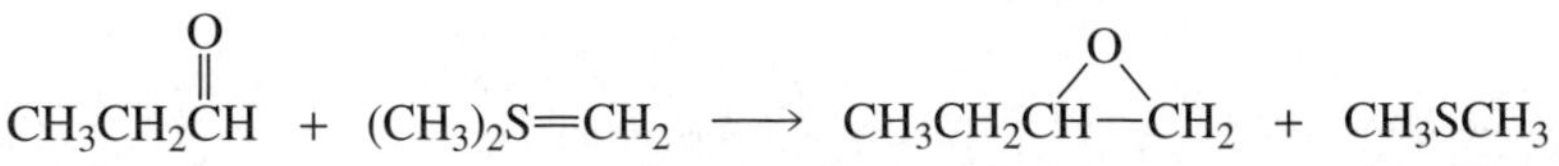

$$CH_3CH_2CH + (CH_3)_2S=CH_2 \longrightarrow CH_3CH_2CH-CH_2 + CH_3SCH_3$$

with O (double bond) on the first aldehyde and an epoxide O on the product.

69. Indicate how the following compounds could be prepared from the given starting materials:

a. 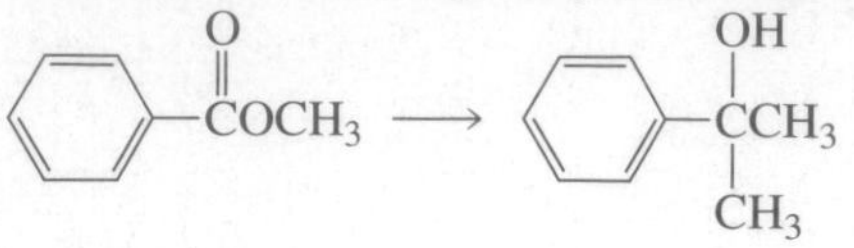 (phenyl)-COCH$_3$ ⟶ (phenyl)-C(OH)(CH$_3$)CH$_3$

b. (phenyl)-COCH$_3$ ⟶ (phenyl)-CH=O

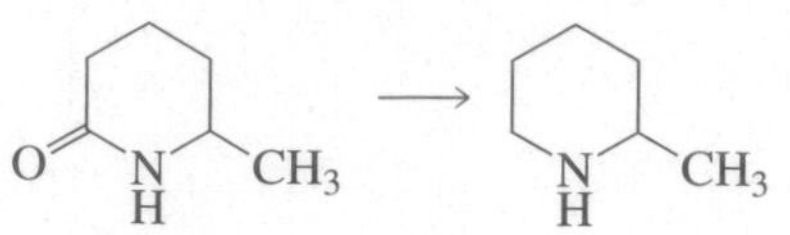

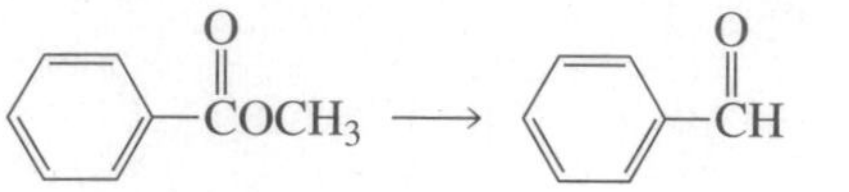

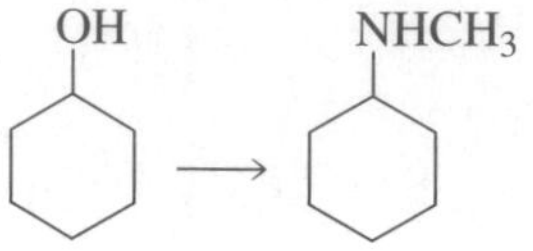

c. CH$_3$CH$_2$CH$_2$CH$_2$Br ⟶ CH$_3$CH$_2$CH$_2$CH$_2$COH

d. (piperidinone with CH$_3$) ⟶ (piperidine with CH$_3$)

e. OH-cyclohexane ⟶ NHCH$_3$-cyclohexane

70. Propose a reasonable mechanism for each of the following reactions:

a. 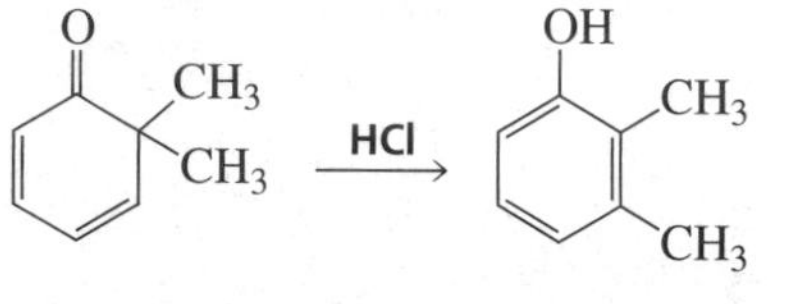 (cyclohexenone with two CH$_3$) + HCl ⟶ (phenol with two CH$_3$)

b. 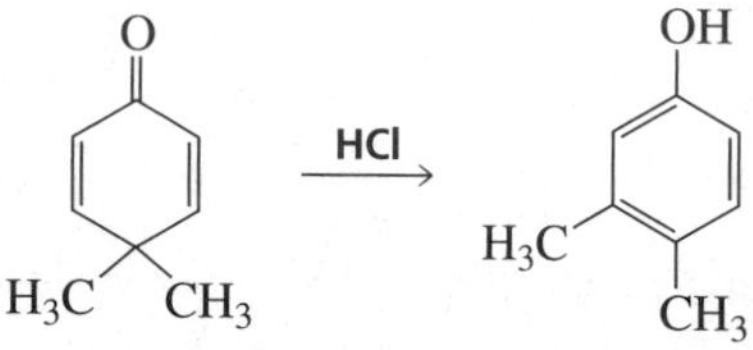 (cyclohexadienone with H$_3$C, CH$_3$) + HCl ⟶ (phenol with H$_3$C, CH$_3$)

71. a. In aqueous solution, D-glucose exists in equilibrium with two six-membered ring compounds. Draw the structures of these compounds.

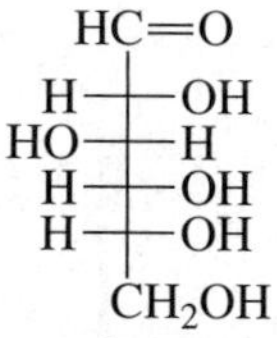

HC=O
H——OH
HO——H
H——OH
H——OH
CH$_2$OH
D-glucose

b. Which of the six-membered ring compounds will be present in greater amount?

72. The ^{1}H NMR spectrum of the alkyl bromide used to make the ylide to form a compound with molecular formula C$_{11}$H$_{14}$ is shown below. What product is obtained from the Wittig reaction?

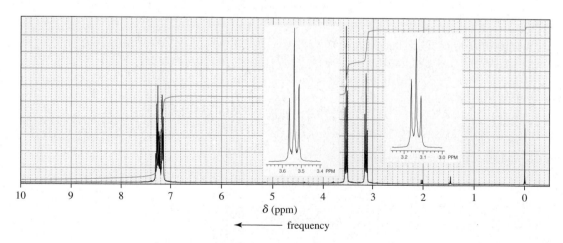

73. In the presence of an acid catalyst, acetaldehyde forms a trimer known as paraldehyde. Because it induces sleep when it is administered to animals in large doses, paraldehyde is used as a sedative or hypnotic. Propose a mechanism for the formation of paraldehyde.

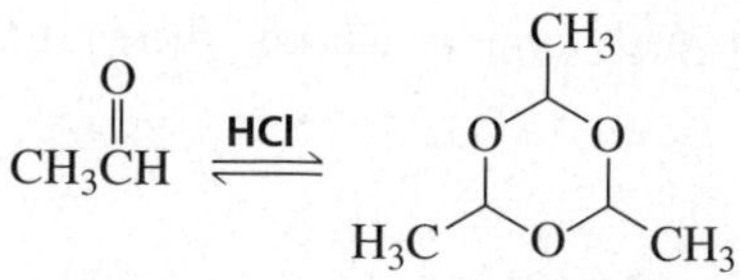

CH$_3$CH=O ⇌ (HCl) paraldehyde
paraldehyde

74. The addition of hydrogen cyanide to benzaldehyde forms a compound called mandelonitrile. (*R*)-Mandelonitrile is formed from the hydrolysis of amygdalin, a compound found in the pits of peaches and apricots. Amygdalin is the principal constituent of laetrile, a compound that was once highly touted as a treatment for cancer. The drug was subsequently found to be ineffective. Is (*R*)-mandelonitrile formed by attack of cyanide ion on the *Re* face or the *Si* face of benzaldehyde?

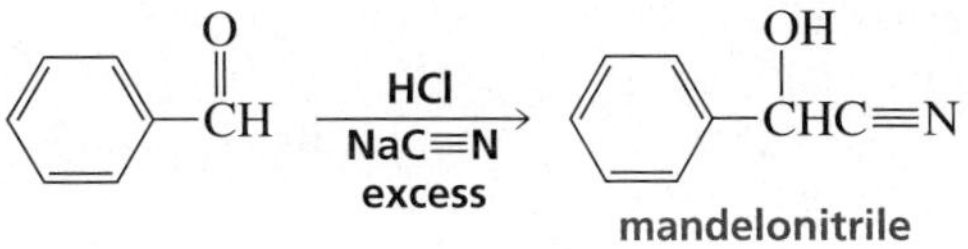

75. What carbonyl compound and what phosphonium ylide are needed to synthesize the following compounds?

a. 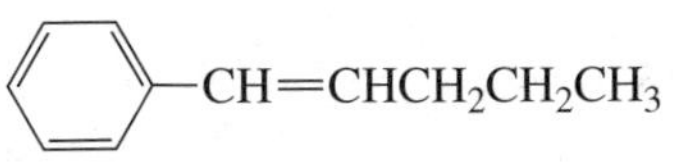 $\underset{}{\bigcirc}-CH=CHCH_2CH_2CH_3$

b.

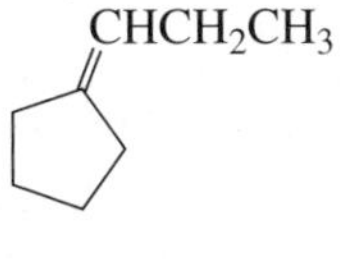

c. 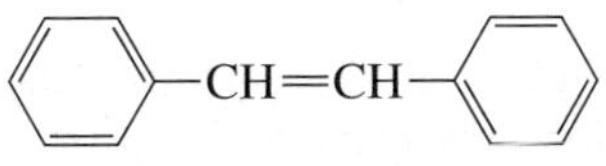 $\bigcirc-CH=CH-\bigcirc$

d. 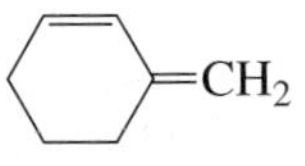 $\bigcirc=CH_2$

76. Identify compounds A and B:

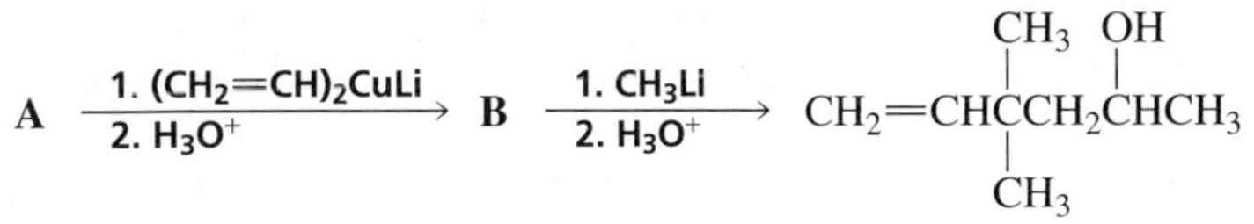

$$A \xrightarrow[\text{2. H}_3\text{O}^+]{\text{1. (CH}_2=\text{CH)}_2\text{CuLi}} B \xrightarrow[\text{2. H}_3\text{O}^+]{\text{1. CH}_3\text{Li}} CH_2=CHCCH_2CHCH_3$$

77. Propose a reasonable mechanism for each of the following reactions:

a.

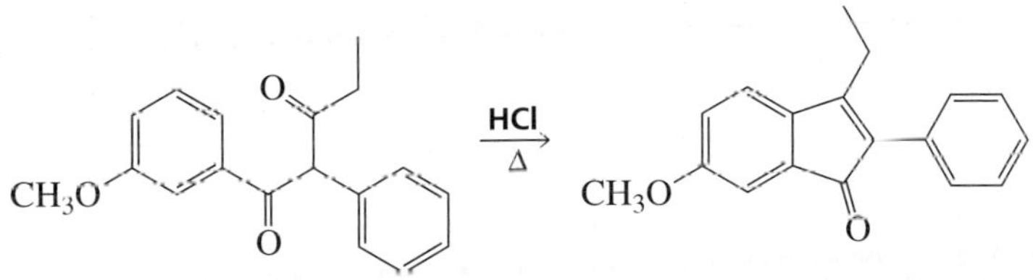

b.

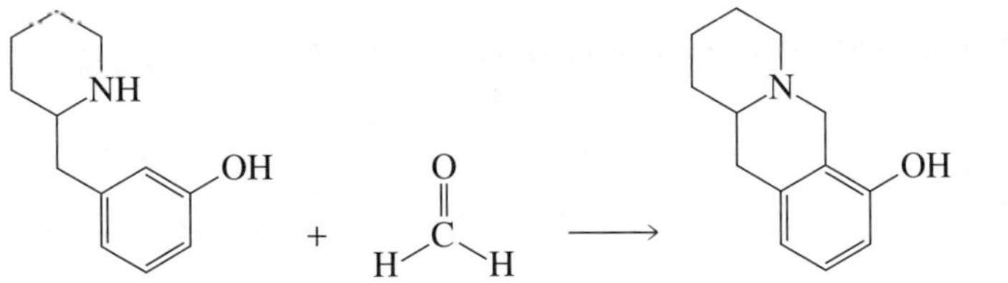

78. A compound reacts with methylmagnesium bromide followed by acidification to form the product with the following ^{1}H NMR spectrum. Identify the compound.

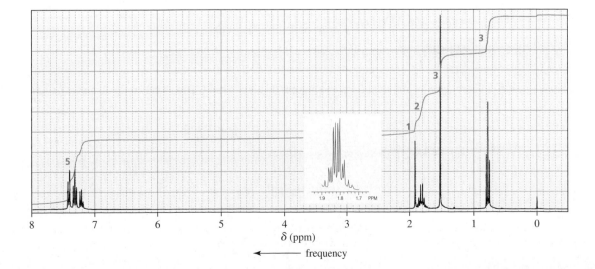

79. Show how each of the following compounds can be prepared from the given starting material. In each case, you will need to use a protecting group.

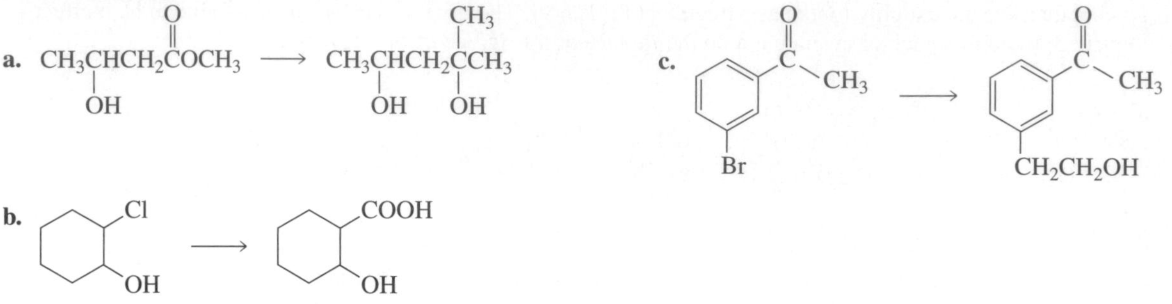

a. $\underset{\underset{OH}{|}}{CH_3CHCH_2}\overset{\overset{O}{\|}}{C}OCH_3 \longrightarrow \underset{\underset{OH}{|}}{CH_3CHCH_2}\underset{\underset{OH}{|}}{\overset{\overset{CH_3}{|}}{C}}CH_3$

c.

b.

80. When a cyclic ketone reacts with diazomethane, the next larger cyclic ketone is formed. This is called a ring expansion reaction. Provide a mechanism for this reaction.

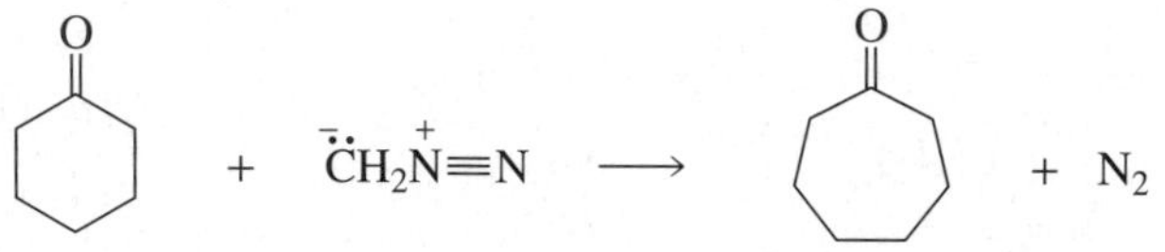

cyclohexanone   diazomethane   cycloheptanone

81. The pK_a values of oxaloacetic acid are 2.22 and 3.98.

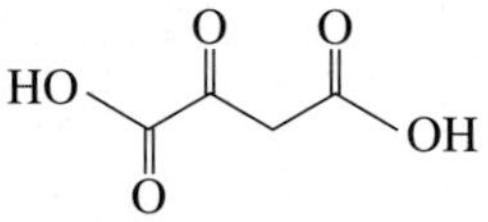

oxaloacetic acid

a. Which carboxyl group is more acidic?

b. The amount of hydrate present in an aqueous solution of oxaloacetic acid depends on the pH of the solution: 95% at pH = 0, 81% at pH = 1.3, 35% at pH = 3.1, 13% at pH = 4.7, 6% at pH = 6.7, and 6% at pH = 12.7. Explain this pH dependence.

82. The *Horner–Emmons modification* is a variation of a Wittig reaction in which a phosphonate-stabilized carbanion is used instead of a phosphonium ylide.

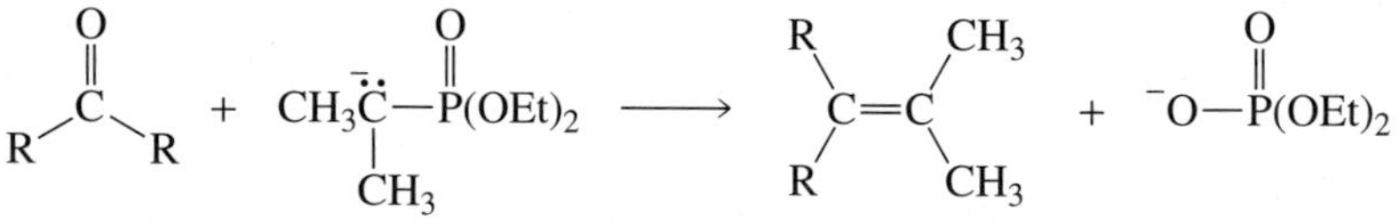

$Et = CH_3CH_2$

The phosphonate-stabilized carbanion is prepared from an appropriate alkyl halide. This is called the *Arbuzov reaction*.

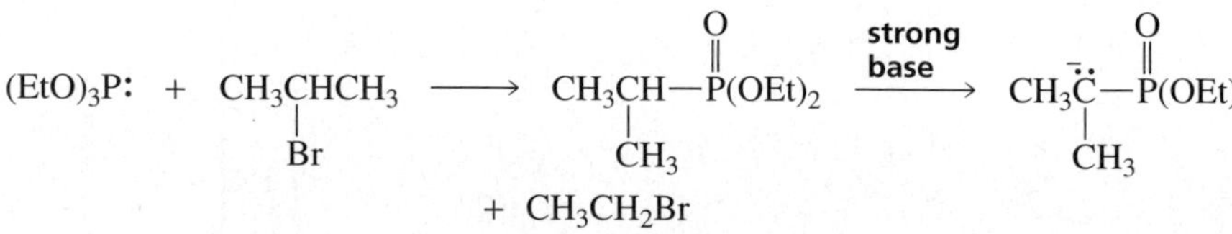

Because the Arbuzov reaction can be carried out with an α-bromo ketone or an α-bromo ester (in which case it is called a *Perkow reaction*), it provides a way to synthesize α,β-unsaturated ketones and esters.

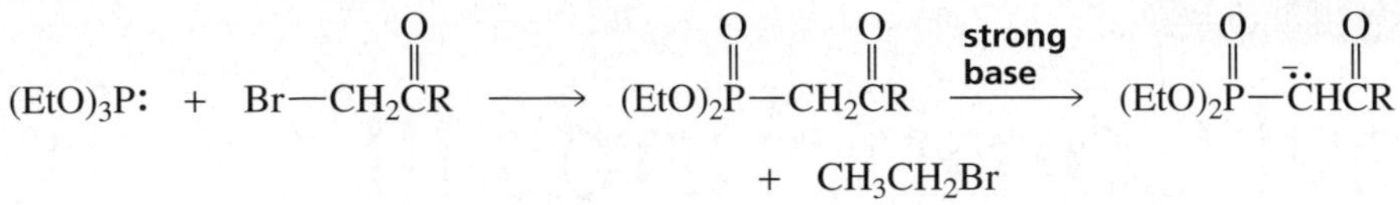

$$(EtO)_3P: \ + \ Br-CH_2\overset{O}{\overset{\|}{C}}R \ \longrightarrow \ (EtO)_2\overset{O}{\overset{\|}{P}}-CH_2\overset{O}{\overset{\|}{C}}R \ \xrightarrow[\text{base}]{\text{strong}} \ (EtO)_2\overset{O}{\overset{\|}{P}}-\overset{..}{\overset{-}{C}}HC R$$

$$+ \ CH_3CH_2Br$$

a. Propose a mechanism for the Arbuzov reaction.
b. Propose a mechanism for the Horner–Emmons modification.
c. Show how the following compounds can be prepared from the given starting material:

1. $CH_3CH_2\overset{O}{\overset{\|}{C}}H \ \longrightarrow \ CH_3CH_2CH{=}CH\overset{O}{\overset{\|}{C}}CH_3$

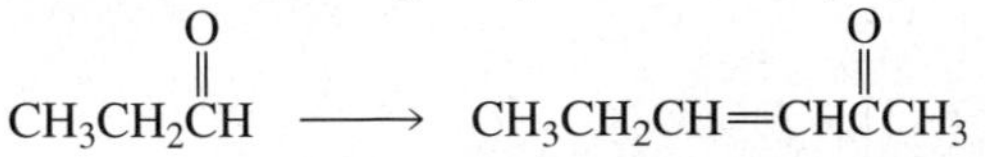

2.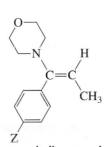

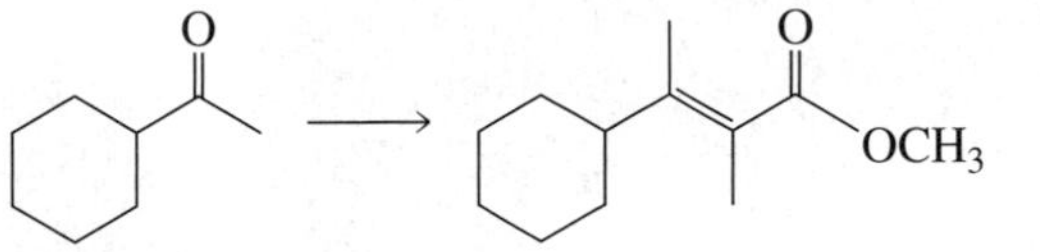

83. In order to solve this problem, you must read the description of the Hammett σ, ρ treatment given in Chapter 16, Problem 88. When the rate constants for the hydrolysis of several morpholine enamines of para-substituted propiophenones are determined at pH $= 4.7$, the ρ value is positive; however, when the rates of hydrolysis are determined at pH $= 10.4$, the ρ value is negative.
a. What is the rate-determining step of the hydrolysis reaction when it is carried out in a basic solution?
b. What is the rate-determining step of the reaction when it is carried out in an acidic solution?

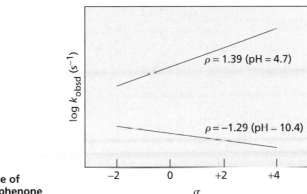

a morpholine enamine of
a para-substituted propiophenone

84. Propose a mechanism for each of the following reactions:

a.

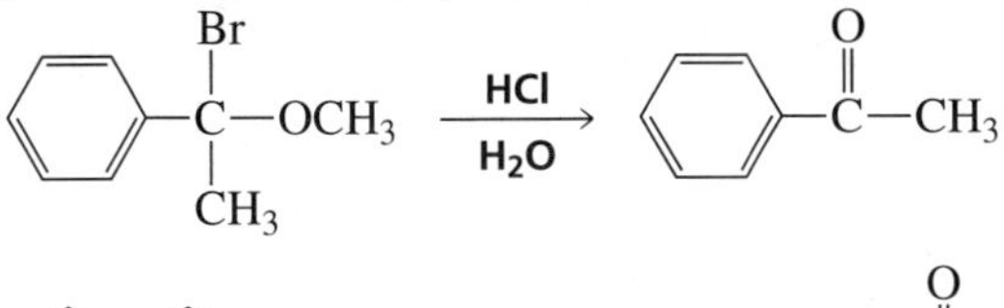

$$\underset{CH_3}{\overset{Br}{\underset{\|}{\underset{C}{\bigcirc}}}}{-}OCH_3 \ \xrightarrow[H_2O]{HCl} \ \bigcirc{-}\overset{O}{\overset{\|}{C}}{-}CH_3$$

b.

$$\bigcirc\!\!\bigcirc \ \xrightarrow[H_2O]{HCl} \ (HOCH_2CH_2CH_2)_2CH\overset{O}{\overset{\|}{C}}H$$

c.

$$\bigcirc\!\!\bigcirc \ + \ CH_3CH_2SH \ \longrightarrow \ \underset{OH}{\overset{OH}{\bigcirc\!\!\bigcirc}}{-}SCH_2CH_3$$

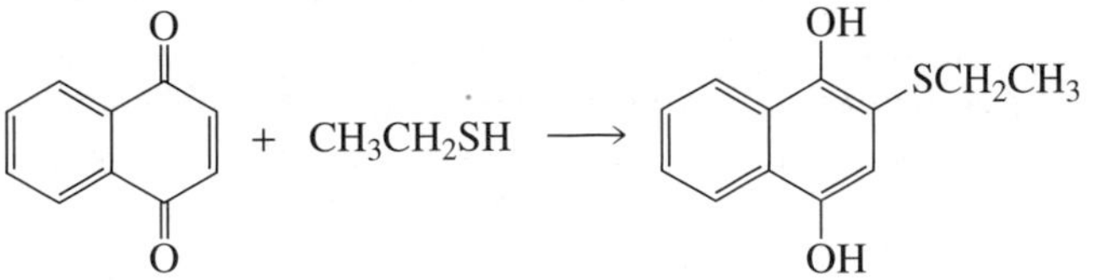

d. $CH_3CH{=}CH\overset{O}{\overset{\|}{C}}CH_3 \ + \ \bigcirc\!\!\!\overset{N:}{\bigcirc} \ \longrightarrow \ \overset{+N}{\bigcirc\!\!\!\bigcirc}{-}\underset{CHCH_2\overset{O}{\overset{\|}{C}}CH_3}{\overset{CH_3}{}} \ \xrightarrow[H_2O]{HCl} \ \overset{O}{\bigcirc}{-}\underset{CHCH_2\overset{O}{\overset{\|}{C}}CH_3}{\overset{CH_3}{}} \ + \ \underset{H \quad H}{\overset{+}{\bigcirc\!\!\!N}}$$

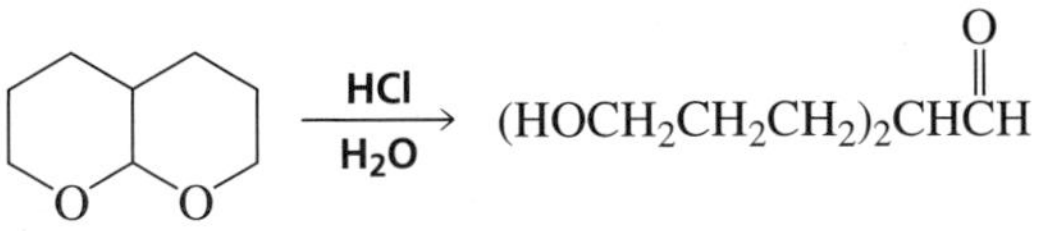

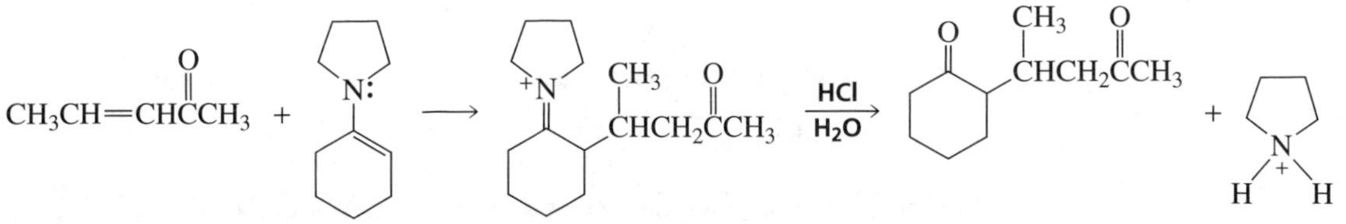

Carbonyl Compounds III
Reactions at the α-Carbon

acetyl-CoA

BUILDING ON FUNDAMENTALS

SECTION 18.1	Electron delocalization increases a compound's stability (7.6).
SECTION 18.2	When an acid is added to a reaction, the first thing that happens is the acid protonates the atom in the reactant that has the greatest electron density (10.4).
SECTION 18.8	Mild conditions favor the kinetically controlled product (the more rapidly formed product); more

vigorous conditions favor the thermodynamically controlled product (the more stable product) (7.11).

SECTION 18.12 Conjugated double bonds are more stable than isolated double bonds (7.7).

SECTION 18.14 Tetrahedral compounds are unstable if they contain a group that is a weak enough base to be expelled (16.5).

When we looked at the reactions of carbonyl compounds in Chapters 16 and 17, we saw that their site of reactivity is the partially positively charged carbonyl carbon, which is attacked by nucleophiles.

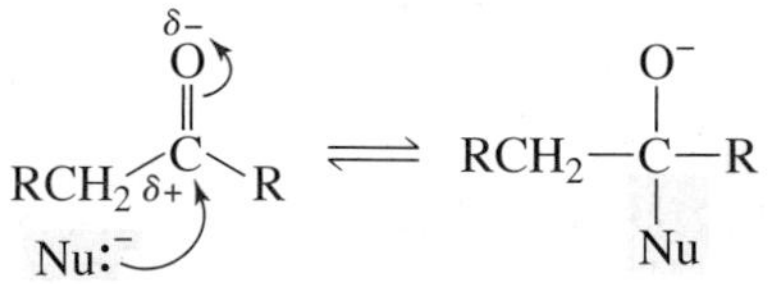

Aldehydes, ketones, esters, and *N,N*-disubstituted amides have a second site of reactivity. A hydrogen bonded to *a carbon adjacent to a carbonyl carbon* is sufficiently acidic to be removed by a strong base. The carbon adjacent to a carbonyl carbon is called an **α-carbon**. A hydrogen bonded to an α-carbon is therefore called an **α-hydrogen**.

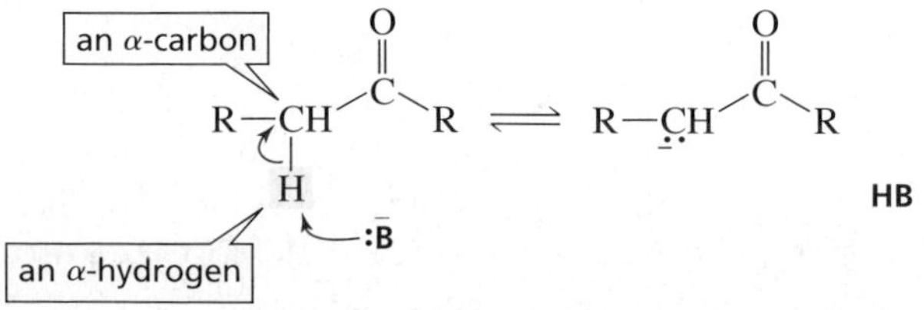

In Section 18.1, you will find out why a hydrogen bonded to a carbon adjacent to a carbonyl carbon is more acidic than hydrogens attached to other sp^3 hybridized carbons, and you will look at some reactions that result from this acidity. Later in the chapter, you will see that a proton is not the only substituent that can be removed from an α-carbon: a carboxyl group bonded to an α-carbon can be removed as CO_2. At the end of the chapter, you will be introduced to some synthetic schemes that rely on the ability to remove protons and carboxyl groups from α-carbons.

18.1 The Acidity of an α-Hydrogen

Hydrogen and carbon have similar electronegativities, which means that the electrons binding them together are shared almost equally by the two atoms. Consequently, a hydrogen bonded to a carbon is usually not acidic. This is particularly true for hydrogens bonded to sp^3 carbons, because these carbons are the most similar to hydrogen in electronegativity (Section 6.10). The high pK_a of ethane provides evidence for the low acidity of a hydrogen bonded to an sp^3 carbon.

$$CH_3CH_3 \quad pK_a > 60$$

A hydrogen bonded to an sp^3 carbon adjacent to a carbonyl carbon, however, is much more acidic than hydrogens bonded to other sp^3 carbons. For example, the pK_a value for dissociation of a proton from the α-carbon of an aldehyde or a ketone ranges from 16 to 20, and the pK_a value for dissociation of a proton from the α-carbon of an ester is about 25 (Table 18.1). Notice that, although an α-hydrogen is more acidic

Table 18.1 The pKₐ Values of Some Carbon Acids

Structure	pK_a	Structure	pK_a
$CH_2(H)C(O)N(CH_3)_2$	30	$N≡CCH(H)C≡N$	11.8
$CH_2(H)C(O)OCH_2CH_3$	25	$CH_3C(O)CH(H)C(O)OCH_2CH_3$	10.7
$CH_2(H)C≡N$	25	$C_6H_5C(O)CH(H)C(O)CH_3$	9.4
$CH_2(H)C(O)CH_3$	20	$CH_3C(O)CH(H)C(O)CH_3$	8.9
$CH_2(H)C(O)H$	17	$CH_3CH(H)NO_2$	8.6
$CH_3CH_2OC(O)CH(H)C(O)OCH_2CH_3$	13.3	$CH_3C(O)CH(H)C(O)H$	5.9
		$O_2NCH(H)NO_2$	3.6

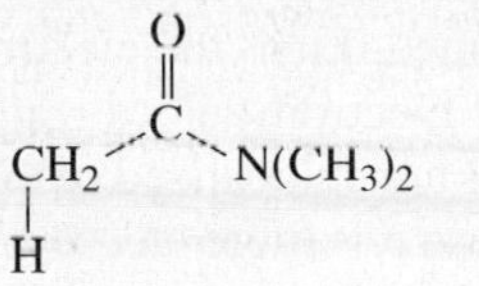

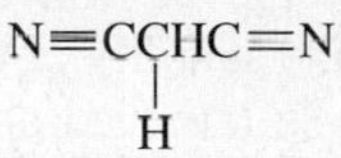

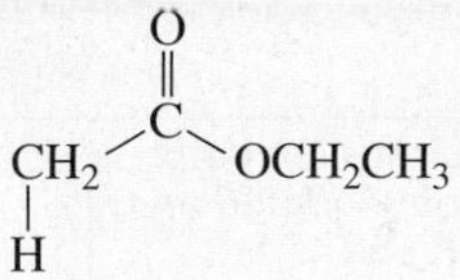

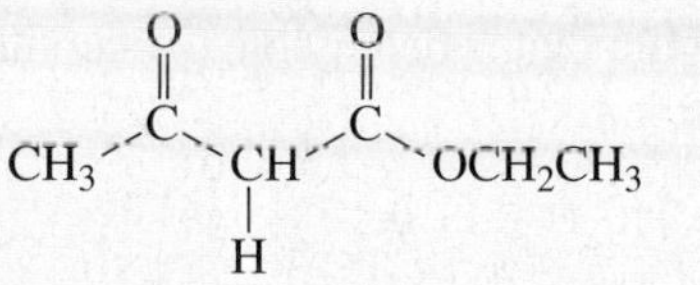

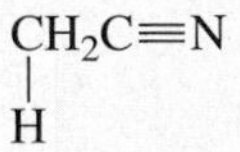

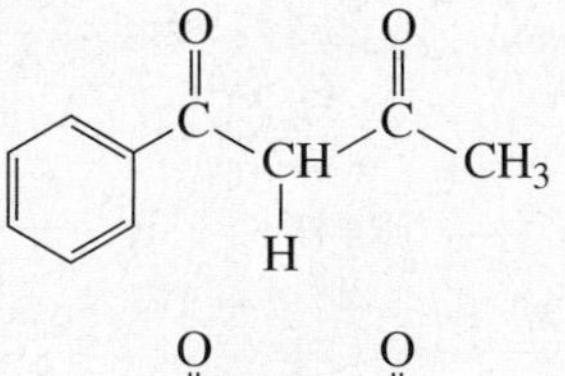

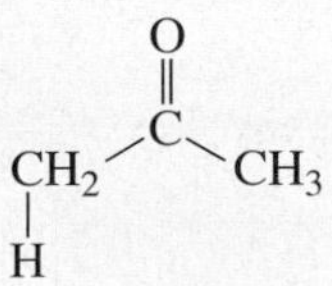

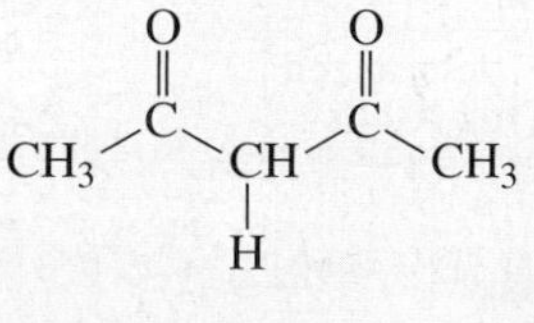

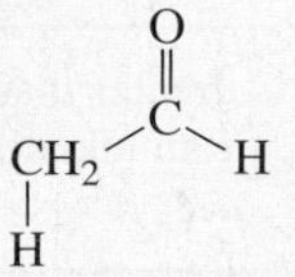

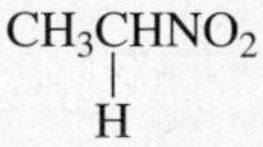

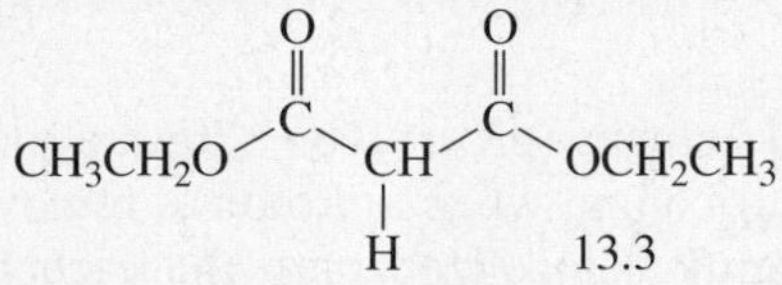

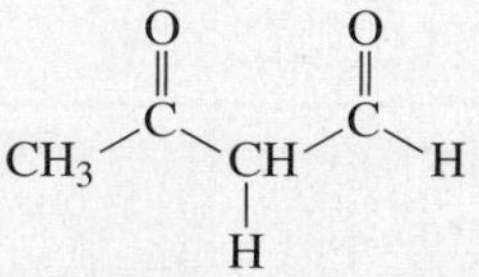

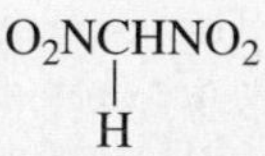

than most other carbon-bound hydrogens, it is less acidic than a hydrogen of water ($pK_a = 15.7$). A compound that contains a relatively acidic hydrogen bonded to an sp^3 carbon is called a **carbon acid**.

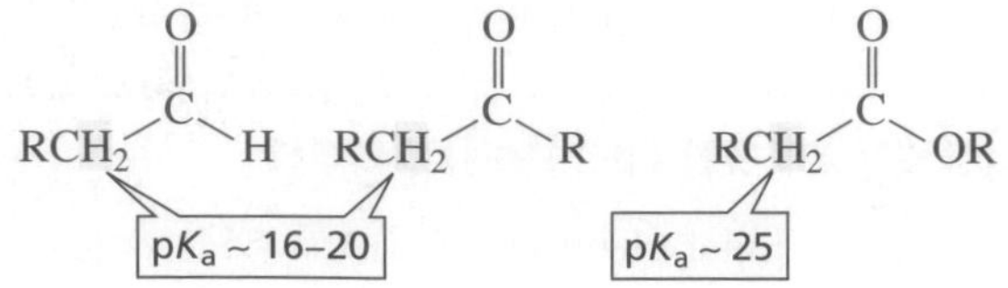

Why is a hydrogen bonded to an sp^3 carbon adjacent to a carbonyl carbon so much more acidic than hydrogens bonded to other sp^3 carbons? An α-hydrogen is more acidic because the base formed when a proton is removed from an α-carbon is more stable than a base formed when a proton is removed from other sp^3 carbons. As we have seen, the more stable the base, the stronger is its conjugate acid (Section 1.18).

Why is the base resulting from the removal of an α-hydrogen more stable? When a proton is removed from ethane, the electrons left behind reside solely on a carbon atom. Because carbon is not very electronegative, a carbanion is unstable. As a result, the pK_a of its conjugate acid is very high.

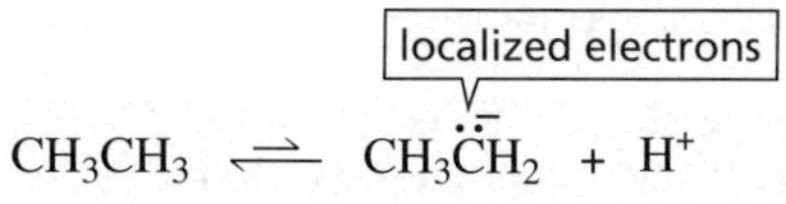

In contrast, when a proton is removed from a carbon adjacent to a carbonyl carbon, two factors combine to increase the stability of the base that is formed. First, the electrons left behind when the proton is removed are delocalized, and electron delocalization increases stability (Section 7.6). More importantly, the electrons are delocalized onto an oxygen, an atom that is better able to accommodate them because it is more electronegative than carbon.

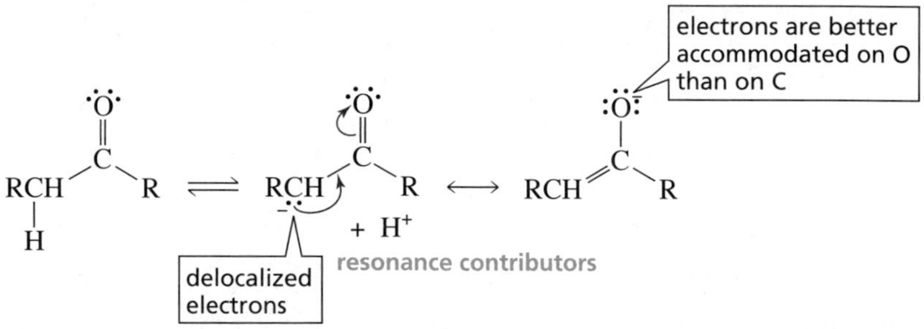

> **PROBLEM 1◆**
>
> The pK_a of a hydrogen bonded to the sp^3 carbon of propene is 42, which is greater than that of any of the carbon acids listed in Table 18.1 but less than the pK_a of an alkane. Explain.

Now we can understand why aldehydes and ketones ($pK_a = 16 - 20$) are more acidic than esters ($pK_a = 25$). The electrons left behind when a proton is removed from the α-carbon of an ester are not as readily delocalized onto the carbonyl oxygen as they would be in an aldehyde or a ketone. This is because the oxygen of the OR group of the ester also has a lone pair that can be delocalized onto the carbonyl oxygen. Thus, the two pairs of electrons compete for delocalization onto the same oxygen.

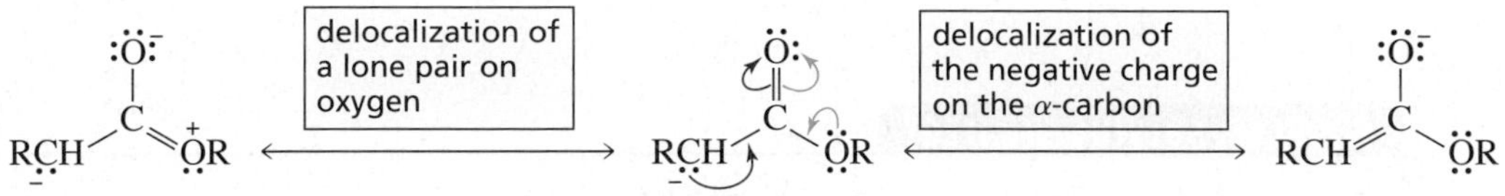

resonance contributors

Nitroalkanes, nitriles, and *N,N*-disubstituted amides also have a relatively acidic α-hydrogen (Table 18.1) because in each case the electrons left behind when the proton is removed can be delocalized onto an atom that is more electronegative than carbon.

> The α-hydrogen of a ketone or an aldehyde is more acidic than the α-hydrogen of an ester.

$CH_3CH_2NO_2$
nitroethane
$pK_a = 8.6$

$CH_3CH_2C{\equiv}N$
propanenitrile
$pK_a = 26$

$CH_3\overset{O}{\overset{\|}{C}}N(CH_3)_2$
N,N-dimethylacetamide
$pK_a = 30$

If the α-carbon is *between* two carbonyl groups, the acidity of an α hydrogen is even greater (Table 18.1). For example, the pK_a value for dissociation of a proton from the α-carbon of 2,4-pentanedione, a compound with an α-carbon between two ketone carbonyl groups, is 8.9; the pK_a value for dissociation of a proton from the α-carbon of ethyl 3-oxobutyrate, a compound with an α-carbon between a ketone carbonyl group and an ester carbonyl group, is 10.7. Ethyl 3-oxobutyrate is classified as a **β-keto ester** because the ester has a carbonyl group at the β-position; 2,4-pentanedione is a **β-diketone**.

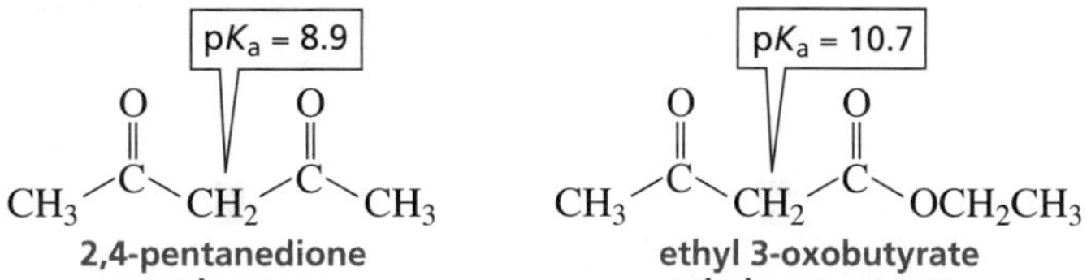

$pK_a = 8.9$

2,4-pentanedione
acetylacetone
a *β*-diketone

$pK_a = 10.7$

ethyl 3-oxobutyrate
ethyl acetoacetate
a *β*-keto ester

The acidity of α-hydrogens bonded to carbons flanked by two carbonyl groups increases because the electrons left behind when the proton is removed can be delocalized onto *two* oxygen atoms. β-Diketones have lower pK_a values than β-keto esters because, as we have seen, electrons are more readily delocalized onto ketone carbonyl groups than they are onto ester carbonyl groups.

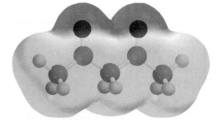

2,4-pentanedione

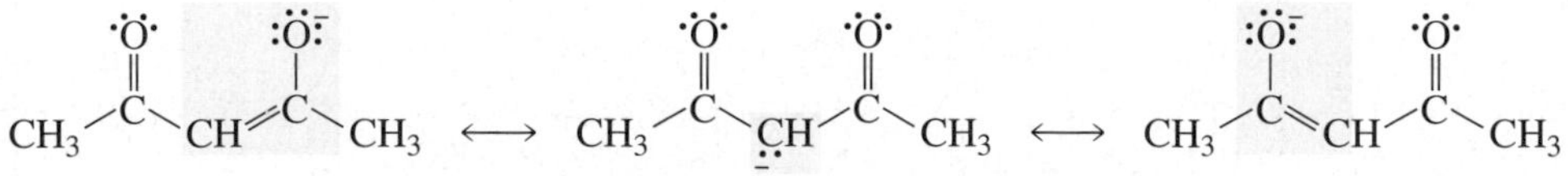

resonance contributors for the 2,4-pentanedione anion

PROBLEM 2◆

Give an example of

a. a β-keto nitrile.　　　　　　　**b.** a β-diester.

PROBLEM-SOLVING STRATEGY

The Acid–Base Behavior of a Carbonyl Compound

Explain why HO⁻ cannot remove a proton from the α-carbon of a carboxylic acid.

Finding that HO⁻ cannot remove a proton from the α-carbon of a carboxylic acid suggests that HO⁻ reacts with another portion of the molecule more rapidly. Because the proton on the carboxyl group is more acidic than the proton on the α-carbon, we can conclude that HO⁻ removes a proton from the carboxyl group rather than from the α-carbon.

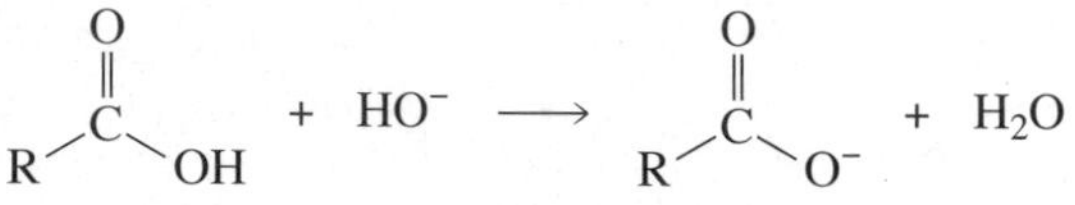

$$
\underset{R}{\overset{O}{\underset{}{\underset{\displaystyle\|}{C}}}}\text{—OH} \;+\; HO^- \;\longrightarrow\; \underset{R}{\overset{O}{\underset{}{\underset{\displaystyle\|}{C}}}}\text{—O}^- \;+\; H_2O
$$

Now continue on to Problem 3.

PROBLEM 3◆

Explain why a proton can be removed from the α-carbon of *N,N*-dimethylethanamide but not from the α-carbon of either *N*-methylethanamide or ethanamide.

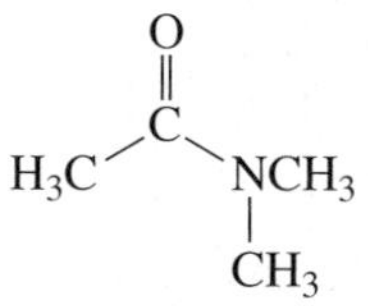

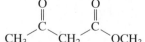

N,N-dimethylethanamide 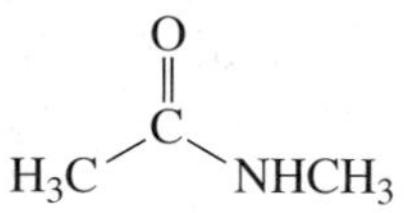　　　　**N-methylethanamide** 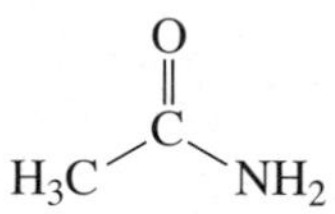　　　　**ethanamide**

PROBLEM 4◆

Explain why an *N,N*-disubstituted amide is less acidic than an ester.

PROBLEM 5◆

List the compounds in each of the following groups in order of decreasing acidity:

a. $CH_2{=}CH_2$　　　CH_3CH_3　　　$CH_3\overset{\displaystyle O}{\overset{\|}{C}}H$　　　$HC{\equiv}CH$

b.

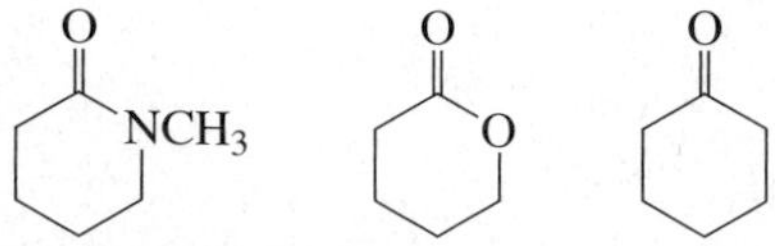

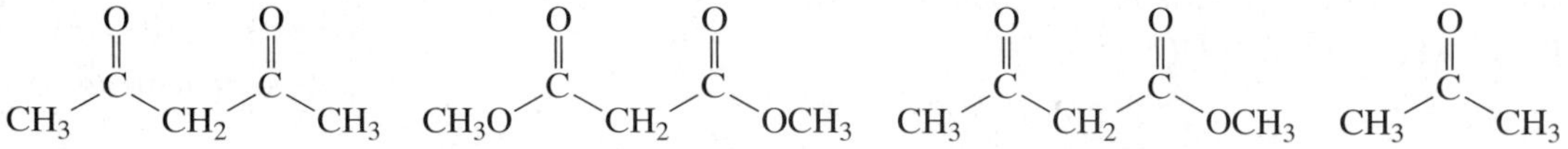

c.

18.2 Keto–Enol Tautomers

A ketone exists in equilibrium with its enol tautomer. Recall that **tautomers** are isomers that are in rapid equilibrium (Section 6.7). Keto–enol tautomers differ in the location of a double bond and a hydrogen.

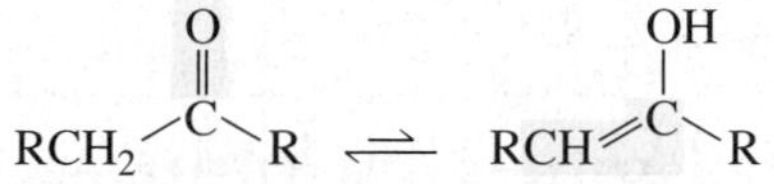

keto tautomer **enol tautomer**

For most ketones, the **enol tautomer** is much less stable than the **keto tautomer**. For example, an aqueous solution of acetone exists as an equilibrium mixture of more than 99.9% keto tautomer and less than 0.1% enol tautomer.

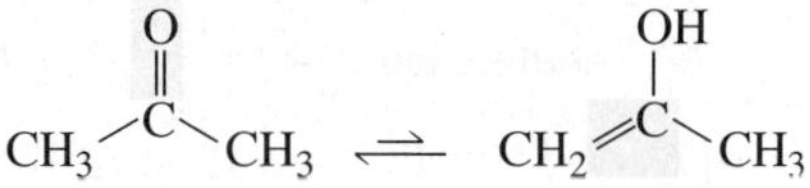

> 99.9% **keto tautomer** **< 0.1%** **enol tautomer**

The fraction of the enol tautomer in an aqueous solution is considerably greater for a β-diketone because the enol tautomer is stabilized by intramolecular hydrogen bonding and by conjugation of the carbon–carbon double bond with the second carbonyl group.

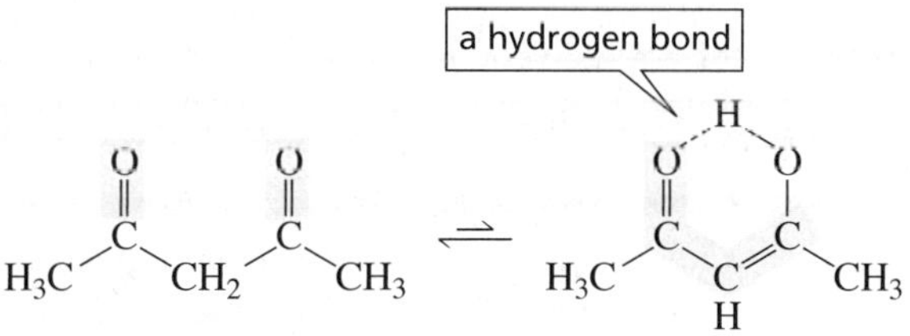

85% **keto tautomer** **15%** **enol tautomer**

Phenol is unusual in that its enol tautomer is *more* stable than its keto tautomer because the enol tautomer is aromatic, but the keto tautomer is not.

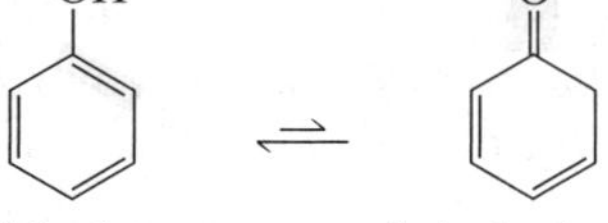

enol tautomer aromatic **keto tautomer** not aromatic

PROBLEM 6

Only 15% of 2,4-pentanedione exists as the enol tautomer in water, but 92% exists as the enol tautomer in hexane. Explain why this is so.

18.3 **Enolization**

Now that we know that a hydrogen bonded to a carbon adjacent to a carbonyl carbon is somewhat acidic, we can understand why keto and enol tautomers interconvert as we first saw in Chapter 6. The interconversion of the keto and enol tautomers is called **keto–enol tautomerization**, **keto–enol interconversion**, or **enolization**. The interconversion can be catalyzed by either a base or an acid. The mechanism for base-catalyzed keto–enol interconversion is shown below.

Mechanism for base-catalyzed keto–enol interconversion

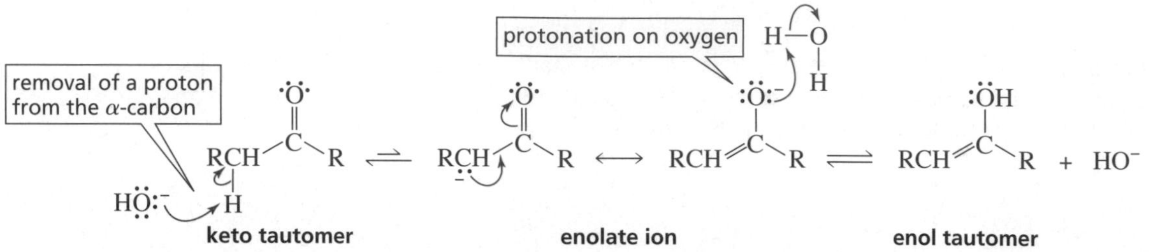

- Hydroxide ion removes a proton from the α-carbon of the keto tautomer, forming an anion called an enolate ion. The enolate ion has two resonance contributors.
- Protonation on oxygen forms the enol tautomer, whereas protonation on the α-carbon reforms the keto tautomer.

The mechanism for acid-catalyzed keto–enol interconversion follows.

Mechanism for acid-catalyzed keto–enol interconversion

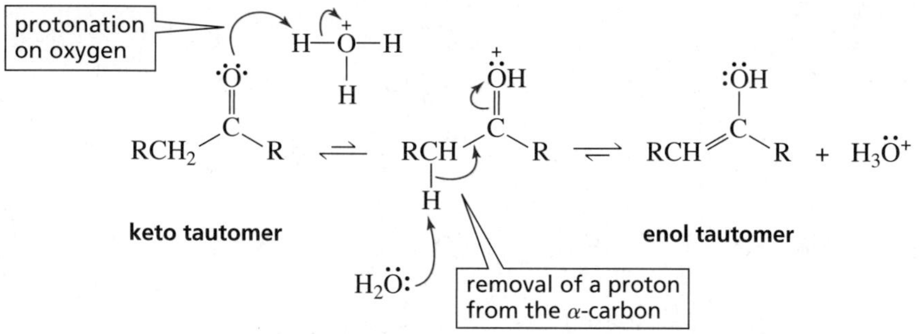

- The carbonyl oxygen of the keto tautomer is protonated.
- Water removes a proton from the α-carbon, forming the enol tautomer.

Notice that the steps are reversed in the base- and acid-catalyzed reactions. In the base-catalyzed reaction, the base removes a proton from an α-carbon in the first step and the oxygen is protonated in the second step. In the acid-catalyzed reaction, the oxygen is protonated in the first step and the proton is removed from the α-carbon in the second step. Notice also that the catalyst is regenerated in both the acid- and base-catalyzed reactions.

PROBLEM 7◆

Give the relative concentrations of enol and enolate ion in a solution of pH = 8, if the enol has a pK_a value of

a. 20

b. 16

c. 10

PROBLEM 8◆

Draw the enol tautomers for each of the following compounds. For compounds that have more than one enol tautomer, indicate which is more stable.

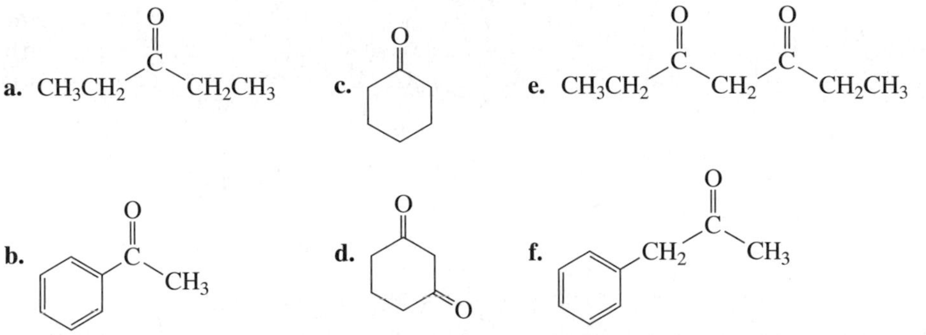

a. CH_3CH_2—C(=O)—CH_2CH_3

b. (phenyl)—C(=O)—CH_3

c. cyclohexanone

d. cyclohexane-1,3-dione

e. CH_3CH_2—C(=O)—CH_2—C(=O)—CH_2CH_3

f. (phenyl)—CH_2—C(=O)—CH_3

18.4 How Enols and Enolate Ions React

The carbon–carbon double bond of an enol suggests that it is a nucleophile—like an alkene. In fact, an enol is more electron rich than an alkene because the oxygen atom donates electrons by resonance. Therefore, an enol is a better nucleophile than an alkene.

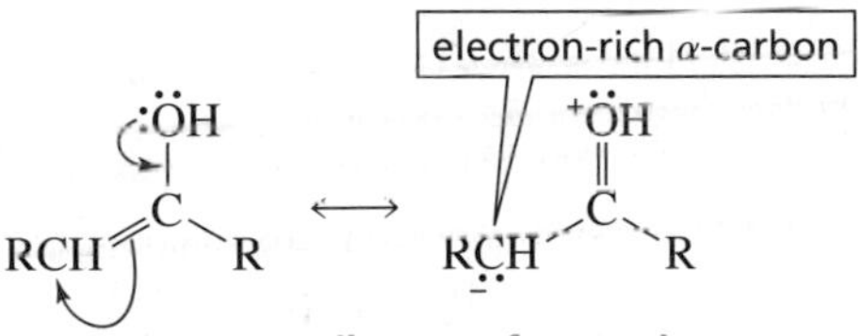

electron-rich α-carbon

resonance contributors of an enol

A carbonyl compound with an α-hydrogen can undergo a substitution reaction at the α-carbon. The substitution reaction can be catalyzed by a base or by an acid. The mechanism for base-catalyzed α-substitution is shown below.

Mechanism for base-catalyzed α-substitution

enolate ion

α-substituted product

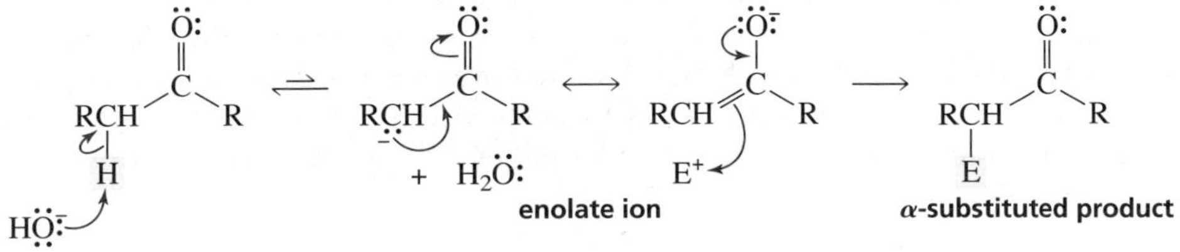

- A base removes a proton from the α-carbon, forming an enolate ion.
- The enolate ion then reacts with an electrophile. Because they are negatively charged, enolate ions are better nucleophiles than enols are.

The overall reaction is an **α-substitution reaction**; one electrophile (E^+) is substituted for another (H^+) at the α-carbon.

The mechanism for acid-catalyzed α-substitution follows.

Mechanism for acid-catalyzed α-substitution

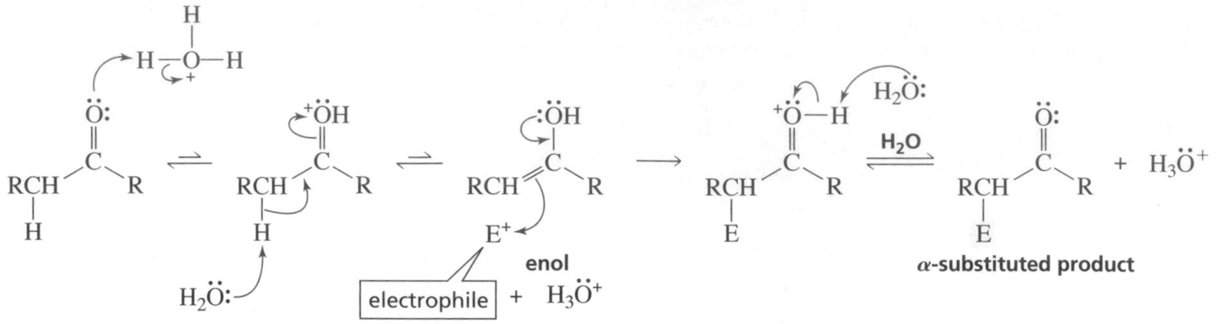

- The acid protonates the most electron-dense atom in the compound.
- Water removes a proton from the α-carbon.
- The enol then reacts with an electrophile.

The resonance contributors of the enolate ion show that it has two electron-rich sites: the α-carbon and the oxygen. The resonance contributor with the negatively charged oxygen makes the greater contribution to the hybrid. The enolate ion is an example of an *ambident nucleophile* (*ambi* is Latin for "both"; *dent* is Latin for "teeth"). An **ambident nucleophile** is a nucleophile with two nucleophilic sites ("two teeth").

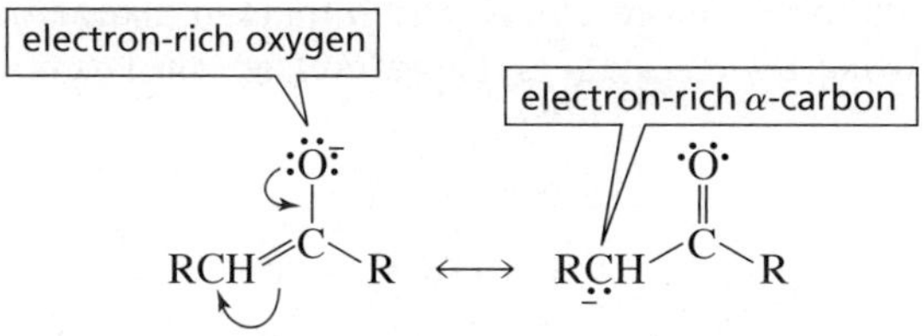

resonance contributors of an enolate ion

Which nucleophilic site (C or O) reacts with an electrophile depends on the electrophile and on the reaction conditions. Protonation occurs preferentially on oxygen (thus forming the kinetic product), because of the greater concentration of negative charge on the more electronegative oxygen atom; the reaction is reversible, however, so the ketone (the thermodynamic product) predominates at equilibrium. When the electrophile is something other than a proton, carbon is more likely to be the nucleophile because carbon is a better nucleophile than oxygen.

Notice the similarity between keto–enol interconversion and α-substitution. Actually, keto–enol interconversion is an α-substitution reaction in which hydrogen serves as both the electrophile that is removed from the α-carbon and the electrophile that is added to the α-carbon when the enol or enolate ion reverts back to the keto tautomer.

As various α-substitution reactions are discussed in this chapter, notice that they all follow basically the same mechanism: a base removes a proton from an α-carbon and the resulting enol or enolate ion reacts with an electrophile. The reactions differ only in the nature of the base and the electrophile—and in whether the reactions are carried out under acidic or basic conditions.

PROBLEM 9

When a dilute solution of acetaldehyde is shaken in NaOD in D_2O, explain why the methyl hydrogens are exchanged with deuterium but the aldehyde hydrogen (the one attached to the carbonyl carbon) is not.

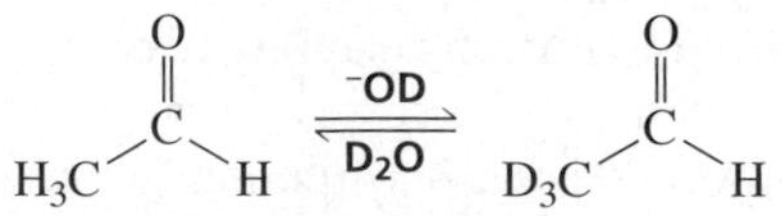

18.5 Halogenation of the α-Carbon of Aldehydes and Ketones

When Br_2, Cl_2, or I_2 is added to a solution of an aldehyde or a ketone, a halogen replaces *one or more* of the α-hydrogens of the carbonyl compound.

Acid-Catalyzed Halogenation

In the acid-catalyzed reaction, the halogen replaces *one* of the α-hydrogens:

Under acidic conditions, one α-hydrogen is substituted for a halogen.

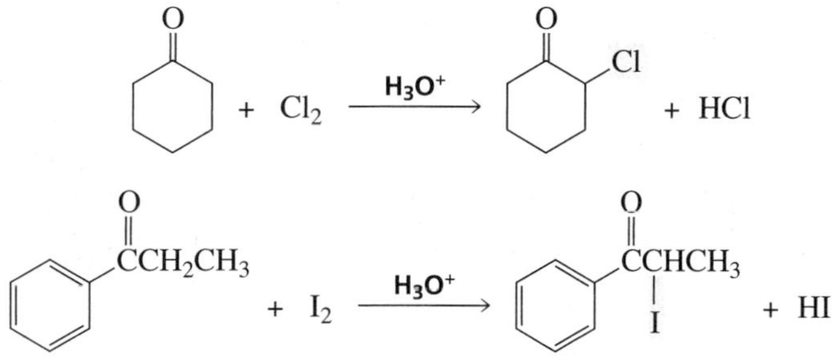

α-Halogenation-1

The mechanism for the reaction is shown below.

Mechanism for acid-catalyzed halogenation

α-Halogenation-2

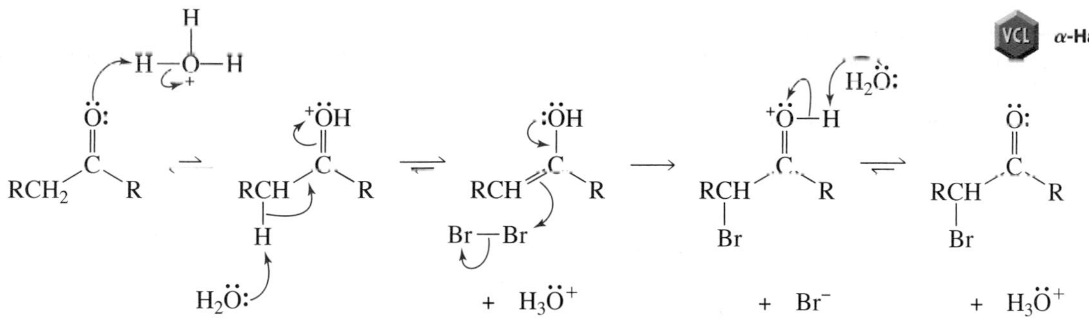

- The carbonyl oxygen is protonated.
- Water removes a proton from the α-carbon, forming an enol.
- The enol reacts with an electrophilic halogen.

Base-Promoted Halogenation

When excess Br_2, Cl_2, or I_2 is added to a *basic* solution of an aldehyde or a ketone, the halogen replaces *all* the α-hydrogens.

Under basic conditions, all the α-hydrogens are substituted for halogens.

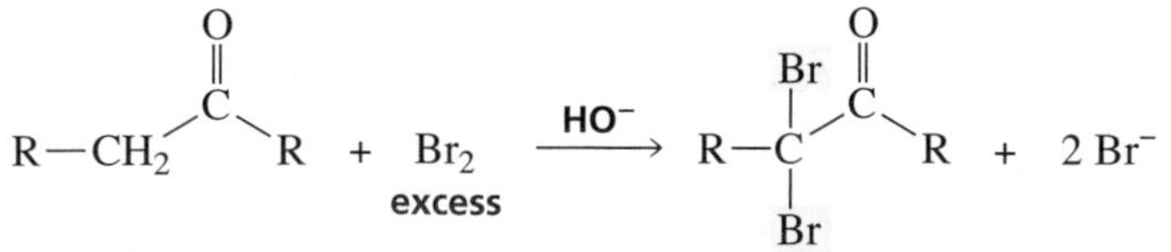

The mechanism for the reaction is shown below.

Mechanism for base-promoted halogenation

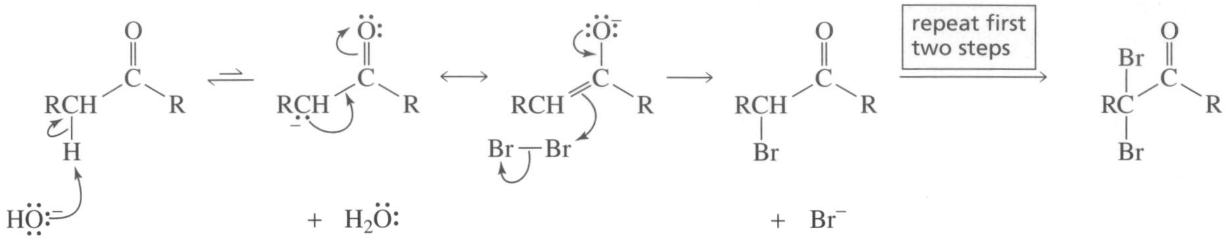

α-Halogenation-3

- Hydroxide ion removes a proton from the α-carbon
- The enolate ion reacts with the electrophilic bromine.

These two steps are repeated until all the α-hydrogens are replaced by the halogen. Each successive halogenation is *more rapid* than the previous one because the electron-withdrawing halogen—in this case, bromine—increases the acidity of the remaining α-hydrogens. This is why *all* the α-hydrogens are replaced by bromines.

Under acidic conditions, on the other hand, each successive halogenation is *slower* than the previous one because the electron-withdrawing bromine decreases the basicity of the carbonyl oxygen, thereby making protonation of the carbonyl oxygen less favorable.

The Haloform Reaction

In the presence of excess base and excess halogen, a methyl ketone is converted into a carboxylate ion.

Mechanism for the haloform reaction

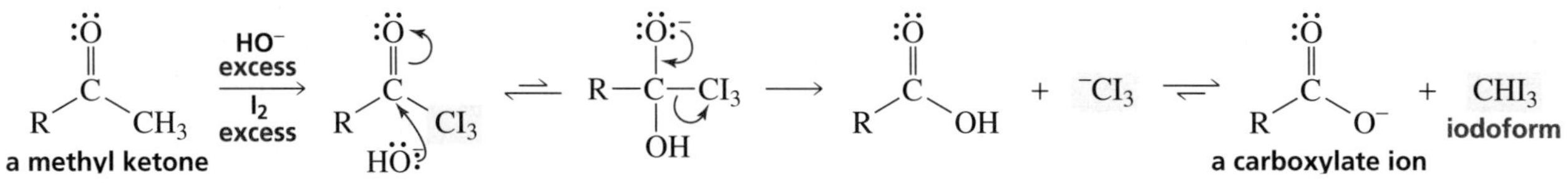

- First, all the hydrogens of the methyl group are replaced by halogens, forming a trihalo-substituted ketone.
- Hydroxide ion attacks the carbonyl carbon of the trihalo-substituted ketone.
- Because the trihalomethyl ion is a weaker base than hydroxide ion (the pK_a of CHI_3 is 14; the pK_a of H_2O is 15.7), the trihalomethyl ion is the group more easily expelled from the tetrahedral intermediate, forming a carboxylic acid.
- Proton transfer forms a carboxylate ion and iodoform.

The conversion of a methyl ketone to a carboxylate ion is called a **haloform reaction** because one of the products is haloform—either $CHCl_3$ (chloroform), $CHBr_3$ (bromoform), or CHI_3 (iodoform). Before spectroscopy became a routine analytical tool, the haloform reaction served as a test for methyl ketones: the formation of iodoform, a bright yellow compound, signaled that a methyl ketone was present.

PROBLEM 10◆

Why do only methyl ketones undergo the haloform reaction?

PROBLEM 11◆

A ketone undergoes acid-catalyzed bromination, acid-catalyzed chlorination, and acid-catalyzed deuterium exchange at the α-carbon, all at about the same rate. What does this tell you about the mechanism of these reactions?

18.6 Halogenation of the α-Carbon of Carboxylic Acids: The Hell–Volhard–Zelinski Reaction

Carboxylic acids cannot undergo substitution reactions at the α-carbon, because a base will remove a proton from the OH group instead of from the α-carbon, since the OH group is more acidic. If, however, a carboxylic acid is treated with PBr_3 and Br_2, the α-carbon can be brominated. (Red phosphorus can be used in place of PBr_3 since P and excess Br_2 react to form PBr_3.) This halogenation reaction is called the **Hell–Volhard–Zelinski reaction** or, more simply, the **HVZ reaction**.

the HVZ reaction

$$
\underset{\text{RCH}_2}{\overset{O}{\underset{\|}{C}}}\text{OH} \xrightarrow[\text{2. H}_2\text{O}]{\text{1. PBr}_3 \text{ (or P), Br}_2} \underset{\underset{\text{Br}}{\text{RCH}}}{\overset{O}{\underset{\|}{C}}}\text{OH}
$$

You will see, when you examine its mechanism, that α-substitution occurs because an acyl bromide, rather than a carboxylic acid, is the compound that undergoes α-substitution.

Mechanism for the Hell–Volhard–Zelinski reaction

$$
\underset{\substack{\text{RCH}_2 \\ \text{a carboxylic acid}}}{\overset{\ddot{O}:}{\underset{\|}{C}}}\text{OH} \xrightarrow{\text{PBr}_3} \underset{\substack{\text{RCH}_2 \\ \text{an acyl bromide}}}{\overset{\ddot{O}:}{\underset{\|}{C}}}\text{Br} \rightleftharpoons \underset{\substack{\text{RCH} \\ \text{enol}}}{\overset{:\ddot{O}\text{H}}{C}}\text{Br}
$$

Br—Br

$$
\underset{\substack{\underset{\text{Br}}{\text{RCH}} \\ \text{an } \alpha\text{-brominated} \\ \text{carboxylic acid}}}{\overset{\ddot{O}:}{\underset{\|}{C}}}\text{OH} \xleftarrow{\text{H}_2\text{O}} \underset{\substack{\underset{\text{Br}}{\text{RCH}} \\ \text{an } \alpha\text{-brominated} \\ \text{acyl bromide}}}{\overset{\ddot{O}:}{\underset{\|}{C}}}\text{Br} + \text{HBr} \rightleftharpoons \underset{\underset{\text{Br}}{\text{RCH}}}{\overset{+\ddot{O}\text{H}}{C}}\text{Br} + \text{Br}^-
$$

- PBr_3 converts the carboxylic acid into an acyl bromide by a mechanism similar to the one followed when PBr_3 converts an alcohol into an alkyl bromide (Section 10.2). Notice that in both reactions PBr_3 replaces an OH with a Br.
- The acyl bromide is in equilibrium with its enol.
- Bromination of the enol forms the α-brominated acyl bromide, which is hydrolyzed to the α-brominated carboxylic acid.

18.7 α-Halogenated Carbonyl Compounds Are Useful in Synthesis

Removing a proton from an α-carbon makes the α-carbon a nucleophile.

You have seen that when a base removes a proton from an α-carbon of an aldehyde or a ketone (Section 18.2), the α-carbon becomes *nucleophilic*—it reacts with electrophiles.

α-carbon reacts with electrophiles

base   E⁺

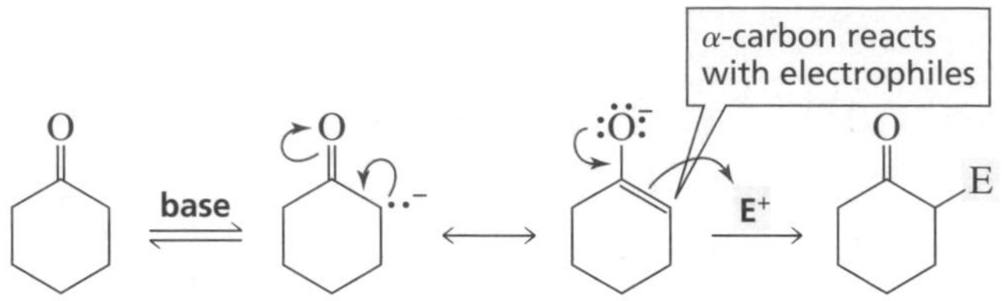

Substituting a bromine for a hydrogen bonded to an α-carbon makes the α-carbon an electrophile.

However, when the α-position is halogenated, the α-carbon becomes *electrophilic*—it reacts with nucleophiles. Therefore, both electrophiles and nucleophiles can be placed on α-carbons.

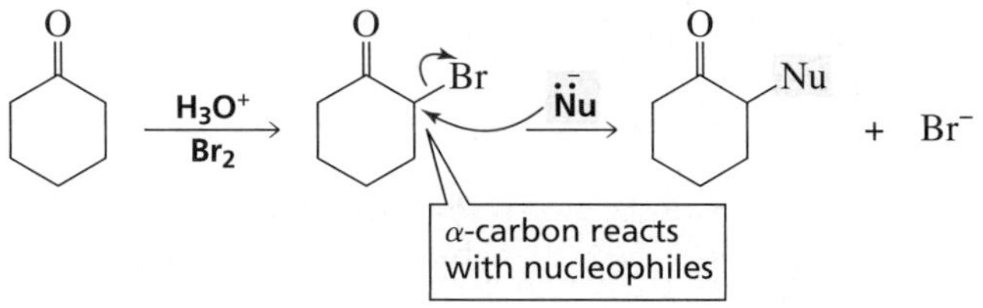

H₃O⁺
Br₂   Nu   + Br⁻

α-carbon reacts with nucleophiles

α-Brominated carbonyl compounds are also useful to synthetic chemists because once a bromine has been introduced into the α-position of a carbonyl compound, an α,β-unsaturated carbonyl compound can be prepared by means of an E2 elimination reaction, using a strong and bulky base to encourage elimination over substitution (Section 9.2).

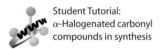

Student Tutorial:
α-Halogenated carbonyl compounds in synthesis

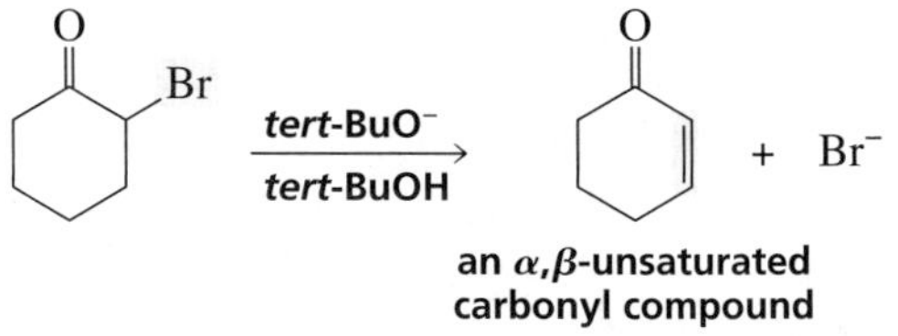

tert-BuO⁻
tert-BuOH   + Br⁻

an α,β-unsaturated
carbonyl compound

PROBLEM 12

How could you prepare the following compounds from the given starting materials?

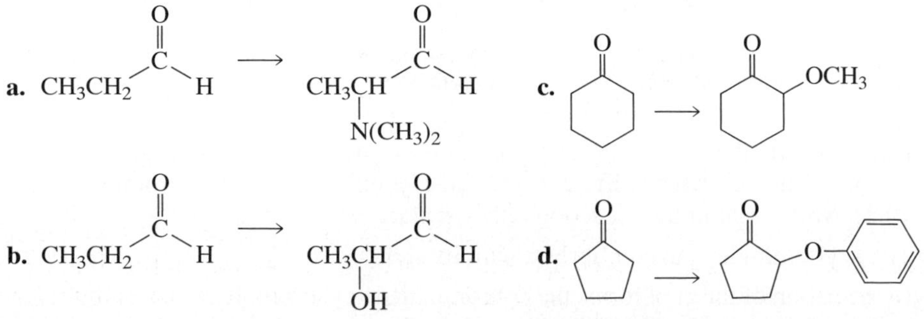

a. CH₃CH₂—C(=O)—H → CH₃CH(N(CH₃)₂)—C(=O)—H

b. CH₃CH₂—C(=O)—H → CH₃CH(OH)—C(=O)—H

c. [cyclohexanone] → [2-methoxycyclohexanone with OCH₃]

d. [cyclopentanone] → [2-phenoxycyclopentanone with O—phenyl]

PROBLEM 13

How could the following compounds be prepared from a carbonyl compound with no carbon–carbon double bonds?

a.

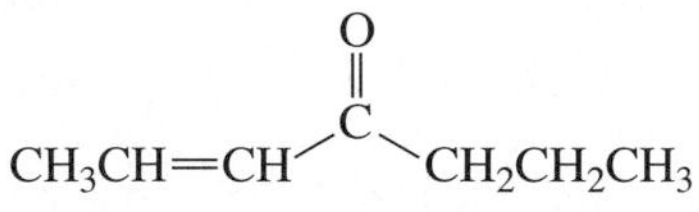

$$CH_3CH=CH-\overset{\overset{\displaystyle O}{\|}}{C}-CH_2CH_2CH_3$$

b.

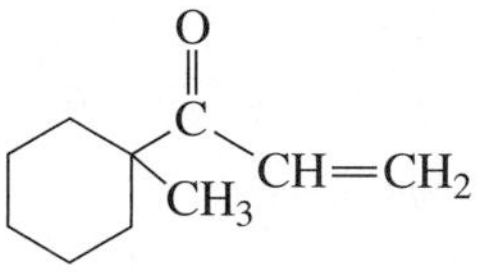

PROBLEM 14

How could the following compounds be prepared from cyclohexanone?

a.

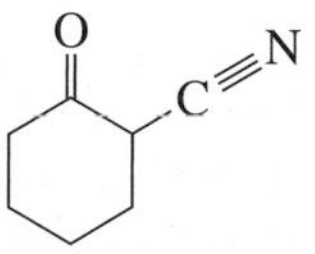

b.

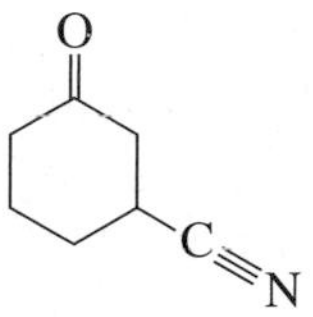

c.

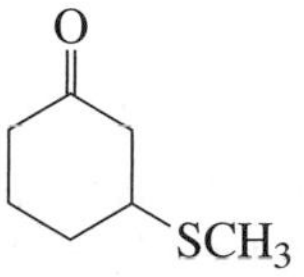

d.

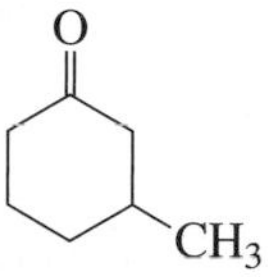

18.8 Using LDA to Form an Enolate Ion

The amount of carbonyl compound converted to an enolate ion depends on the pK_a of the carbonyl compound and the particular base used to remove the α-hydrogen. For example, when hydroxide ion (the pK_a of its conjugate acid is 15.7) is used to remove an α-hydrogen from cyclohexanone ($pK_a = 17$), only a small amount of the carbonyl compound is converted into the enolate ion because hydroxide ion is a weaker base than the base being formed. (Recall that the equilibrium of an acid–base reaction favors reaction of the strong acid and formation of the weak acid, Section 1.19.)

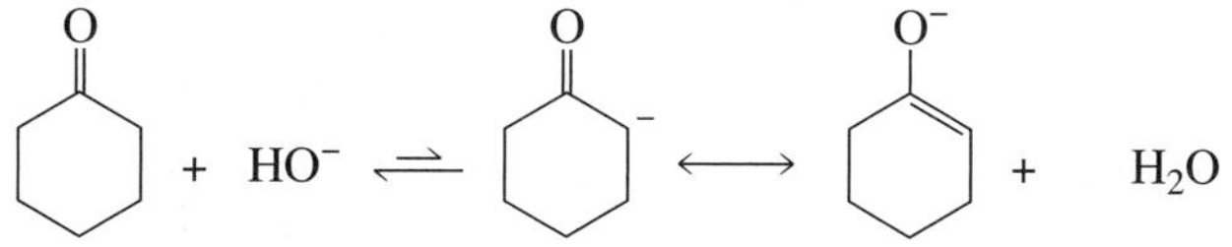

In contrast, when lithium diisopropylamide (LDA) is used to remove the α-hydrogen (the pK_a of LDA's conjugate acid is ~35), essentially all the carbonyl compound is converted to the enolate ion because LDA is a much stronger base than the base being formed. Therefore, LDA is the base of choice for those reactions that require the carbonyl compound to be completely converted to an enolate ion before it reacts with an electrophile.

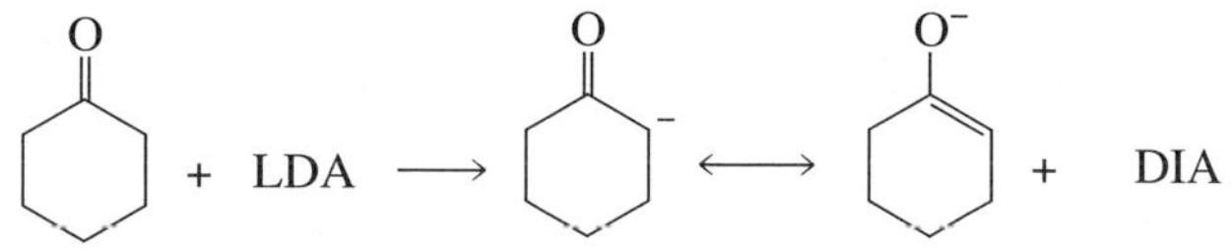

Using a nitrogen base to form an enolate ion can be a problem because a nitrogen base can also react as a nucleophile and attack the carbonyl carbon (Section 17.8). However, the two bulky alkyl substituents bonded to the nitrogen of LDA make it difficult for the nitrogen to get close enough to the carbonyl carbon to react with it. Consequently,

LDA is a strong base but a poor nucleophile: it removes an α-hydrogen much faster than it attacks a carbonyl carbon. LDA is easily prepared by adding butyllithium to diisopropylamine (DIA) in THF at $-78\,°C$.

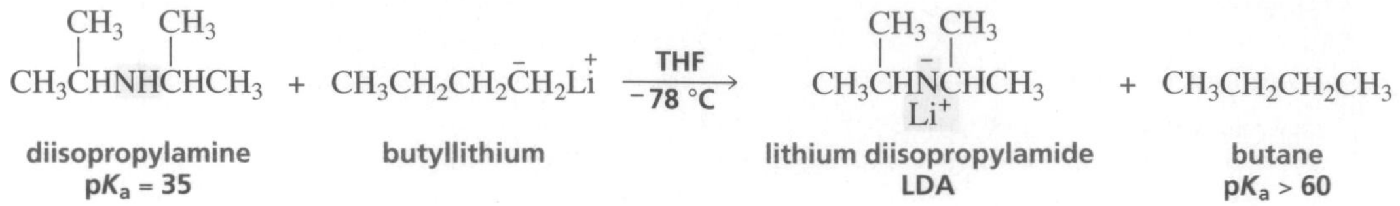

$$\underset{\substack{\text{diisopropylamine}\\ pK_a = 35}}{\overset{\substack{\text{CH}_3 \quad \text{CH}_3 \\ | \qquad |}}{\text{CH}_3\text{CHNHCHCH}_3}} + \underset{\text{butyllithium}}{\text{CH}_3\text{CH}_2\text{CH}_2\overset{-}{\text{C}}\text{H}_2\overset{+}{\text{Li}}} \xrightarrow[-78\,°C]{\text{THF}} \underset{\substack{\text{lithium diisopropylamide}\\ \textbf{LDA}}}{\overset{\substack{\text{CH}_3 \quad \text{CH}_3 \\ | \qquad |}}{\underset{\text{Li}^+}{\text{CH}_3\text{CHNCHCH}_3}}} + \underset{\substack{\text{butane}\\ pK_a > 60}}{\text{CH}_3\text{CH}_2\text{CH}_2\text{CH}_3}$$

18.9 Alkylation of the α-Carbon of Carbonyl Compounds

Enolate ions can be alkylated on the α-carbon.

Alkylation of the α-carbon of a carbonyl compound is an important reaction because it gives us another way to form a carbon–carbon bond. Alkylation is carried out by first removing a proton from the α-carbon with a strong base such as LDA and then adding the appropriate alkyl halide. Because the alkylation is an S_N2 reaction, it works best with methyl halides and primary alkyl halides (Section 8.2).

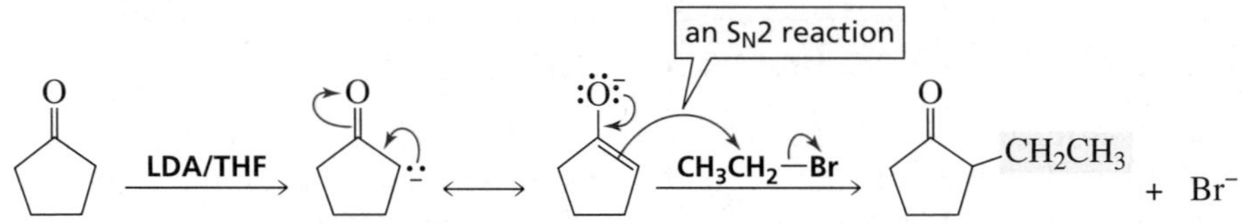

This method can be used to alkylate ketones, esters, and nitriles at the α-carbon. Aldehydes, however, give poor yields of α-alkylated products (Section 18.11).

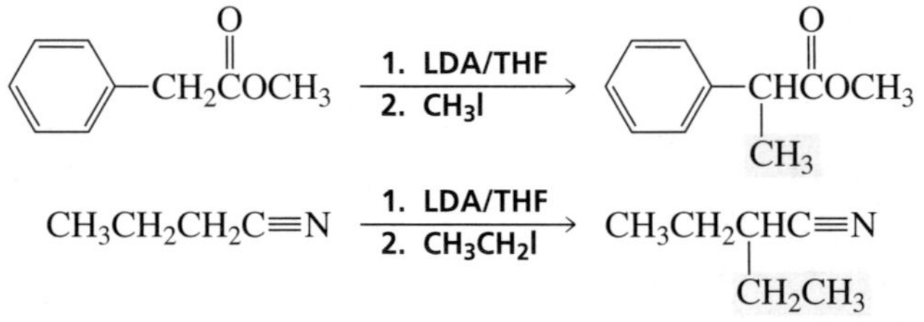

Two different products can be formed when the ketone is not symmetrical, because either α-carbon can be alkylated. For example, methylation of 2-methylcyclohexanone with one equivalent of methyl iodide forms both 2,6-dimethylcyclohexanone and 2,2-dimethylcyclohexanone. The relative amounts of the two products depend on the reaction conditions.

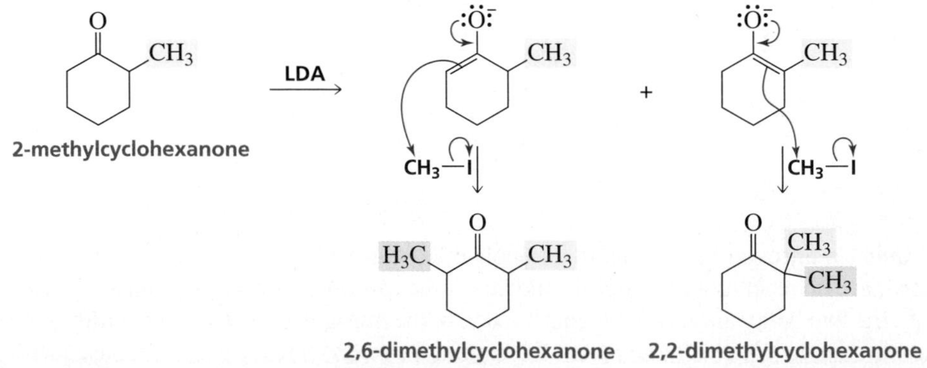

2-methylcyclohexanone

2,6-dimethylcyclohexanone   2,2-dimethylcyclohexanone

The enolate ion leading to 2,6-dimethylcyclohexanone is the *kinetic* enolate ion because the α-hydrogen that is removed to make this enolate ion is more accessible to the base (particularly if a hindered base like LDA is used) and is slightly more acidic. Therefore, 2,6-dimethylcyclohexanone is formed faster and is the major product if the reaction is carried out under conditions (−78 °C) that cause the reaction to be irreversible (Section 7.11).

The enolate leading to 2,2-dimethylcyclohexanone is the *thermodynamic* enolate ion, because it is the more stable enolate ion since it has the more substituted double bond. (Alkyl substitution increases enolate ion stability for the same reason it increases alkene stability; Section 4.11.) Therefore, 2,2-dimethylcyclohexanone is the major product if the reaction is carried out under conditions that cause enolate ion formation to be reversible, and if a less hindered base (KH) is used to cause steric differences in enolate ion formation to be less important.

The less substituted α-carbon can be alkylated—without having to control the conditions to make certain that the reaction does not become reversible—by first making the *N,N*-dimethylhydrazone of the ketone.

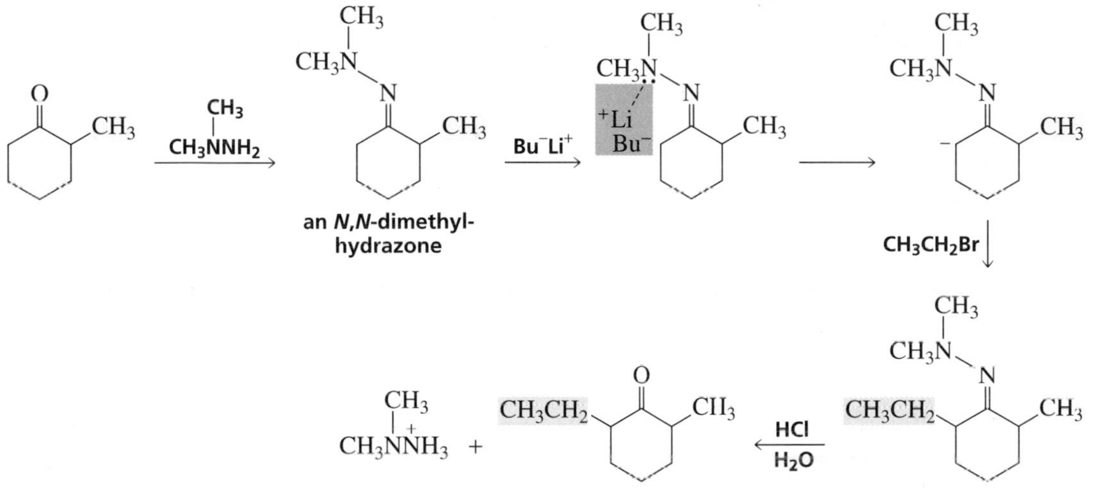

an *N,N*-dimethyl-
hydrazone

The *N,N*-dimethylhydrazone forms with the dimethylamino group pointing away from the more substituted α-carbon. The nitrogen of the dimethylamino group directs the base to the less substituted carbon by coordinating with the lithium ion of butyllithium (Bu⁻Li⁺), the base generally employed in this reaction. Hydrolysis of the hydrazone reforms the ketone (Section 17.8).

THE SYNTHESIS OF ASPIRIN

The first step in the industrial synthesis of aspirin is known as the **Kolbe–Schmitt carboxylation reaction**. In this step, the phenolate ion reacts with carbon dioxide under pressure to form *o*-hydroxybenzoic acid, also known as salicylic acid. Acetylation of salicylic acid with acetic anhydride forms acetylsalicylic acid (aspirin).

During World War I, the Bayer Company bought as much phenol as it could on the international market, knowing that eventually all the phenol could be used to manufacture aspirin. This left little phenol available for other countries to purchase for the synthesis of 2,4,6-trinitrophenol, a common explosive at that time.

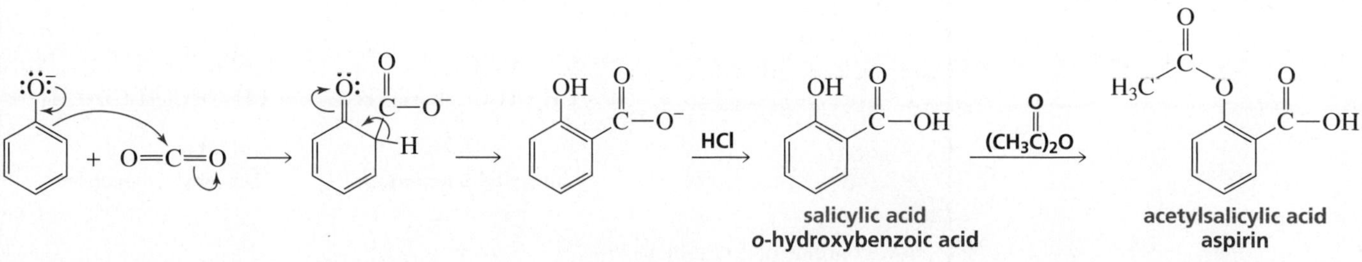

salicylic acid
o-hydroxybenzoic acid

acetylsalicylic acid
aspirin

PROBLEM 15◆

What compound is formed when a dilute solution of cyclohexanone is shaken with NaOD in D_2O for several hours?

PROBLEM 16

Explain why alkylation of an α-carbon works best if the alkyl halide used in the reaction is a primary alkyl halide, and why alkylation does not work at all if a tertiary alkyl halide is used.

PROBLEM-SOLVING STRATEGY

Alkylating a Carbonyl Compound

How could you prepare 4-methyl-3-hexanone from a ketone containing no more than six carbon atoms?

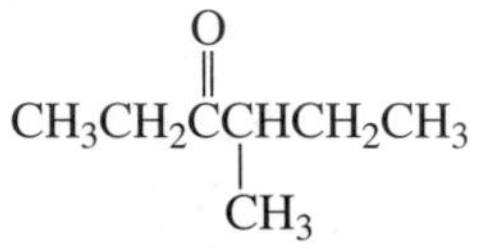

$$
\begin{array}{c}
\quad\quad O \\
\quad\quad \| \\
CH_3CH_2CCHCH_2CH_3 \\
\quad\quad\quad | \\
\quad\quad\quad CH_3
\end{array}
$$
4-methyl-3-hexanone

Either of two sets of ketone and alkyl halide could be used for the synthesis: one set is 3-hexanone and a methyl halide; the other is 3-pentanone and an ethyl halide.

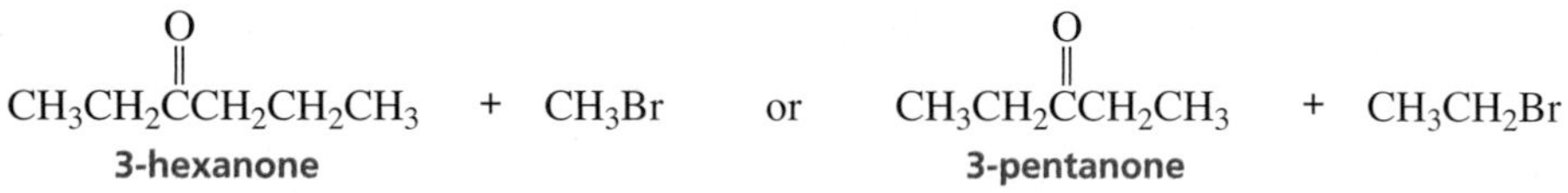

$$
\begin{array}{ccc}
\quad\quad O & & \quad\quad O \\
\quad\quad \| & & \quad\quad \| \\
CH_3CH_2CCH_2CH_2CH_3 + CH_3Br & \text{or} & CH_3CH_2CCH_2CH_3 + CH_3CH_2Br \\
\textbf{3-hexanone} & & \textbf{3-pentanone}
\end{array}
$$

3-Pentanone and an ethyl halide are the preferred starting materials. Because 3-pentanone is symmetrical, only one α-substituted ketone will be formed. In contrast, 3-hexanone can form two different enolate ions, so two α-substituted ketones can be formed.

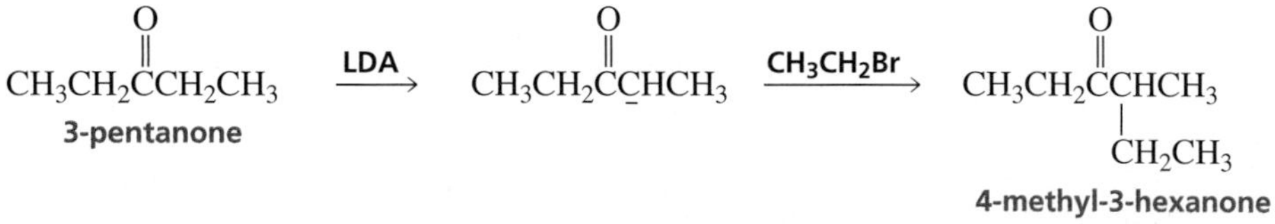

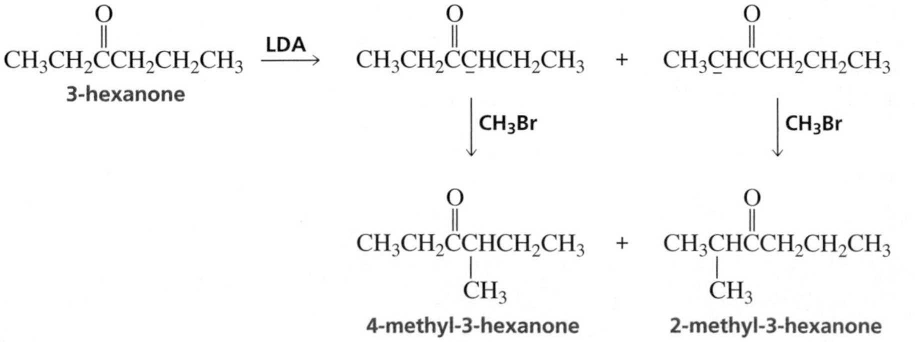

$$
\begin{array}{ccccc}
\quad O & & \quad O & & \quad O \\
\quad \| & & \quad \| & & \quad \| \\
CH_3CH_2CCH_2CH_3 & \xrightarrow{LDA} & CH_3CH_2CCHCH_3 & \xrightarrow{CH_3CH_2Br} & CH_3CH_2CCHCH_3 \\
\textbf{3-pentanone} & & & & \quad\quad\quad | \\
& & & & \quad\quad\quad CH_2CH_3
\end{array}
$$
4-methyl-3-hexanone

$$
\begin{array}{ccc}
\quad O & & \\
\quad \| & & \\
CH_3CH_2CCH_2CH_2CH_3 & \xrightarrow{LDA} &
\begin{array}{c}
\quad O \\
\quad \| \\
CH_3CH_2CCHCH_2CH_3
\end{array}
+
\begin{array}{c}
\quad O \\
\quad \| \\
CH_3CHCCH_2CH_2CH_3
\end{array} \\
\textbf{3-hexanone} & & \\
& & \downarrow CH_3Br \quad\quad\quad\quad\quad \downarrow CH_3Br
\end{array}
$$

$$
\begin{array}{cc}
\quad\quad O & \quad\quad O \\
\quad\quad \| & \quad\quad \| \\
CH_3CH_2CCHCH_2CH_3 + & CH_3CHCCH_2CH_2CH_3 \\
\quad\quad\quad | & \quad | \\
\quad\quad\quad CH_3 & \quad CH_3 \\
\textbf{4-methyl-3-hexanone} & \textbf{2-methyl-3-hexanone}
\end{array}
$$

Now continue on to Problem 17.

PROBLEM 17

How could each of the following compounds be prepared from a ketone and an alkyl halide?

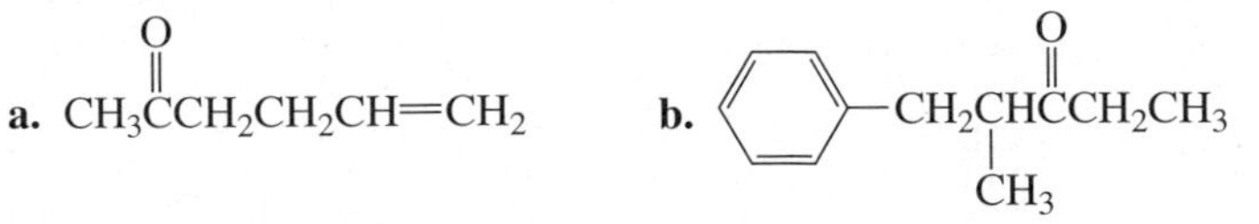

a. $CH_3\overset{O}{\overset{\|}{C}}CH_2CH_2CH{=}CH_2$ b.

18.10 Alkylation and Acylation of the α-Carbon Using an Enamine Intermediate

We have seen that an enamine is formed when an aldehyde or a ketone reacts with a secondary amine (Section 17.8).

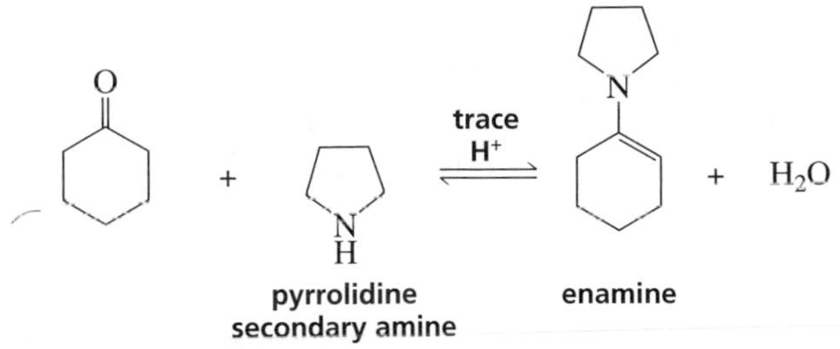

pyrrolidine
secondary amine

enamine

Enamines react with electrophiles in the same way that enolate ions do.

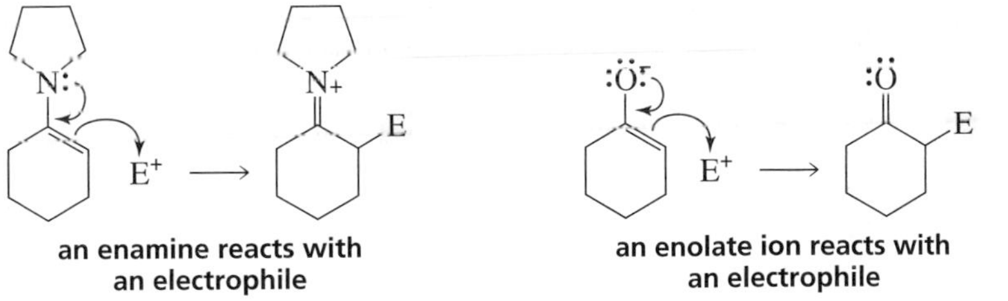

an enamine reacts with
an electrophile

an enolate ion reacts with
an electrophile

This means that electrophiles can be added to the α-carbon of an aldehyde or a ketone by

- first converting the carbonyl compound to an enamine (by treating the carbonyl compound with a secondary amine);
- then adding the electrophile;
- finally, hydrolyzing the imine back to the ketone.

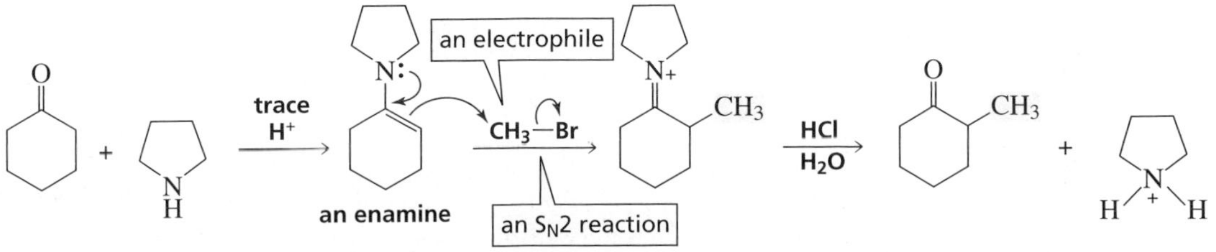

trace
H$^+$

an electrophile

an enamine

CH$_3$—Br

an S$_N$2 reaction

HCl
H$_2$O

Because the alkylation step is an S$_N$2 reaction, only primary alkyl halides or methyl halides should be used (Section 8.2).

One advantage to using an enamine intermediate to alkylate an aldehyde or a ketone is that only the monoalkylated product is formed.

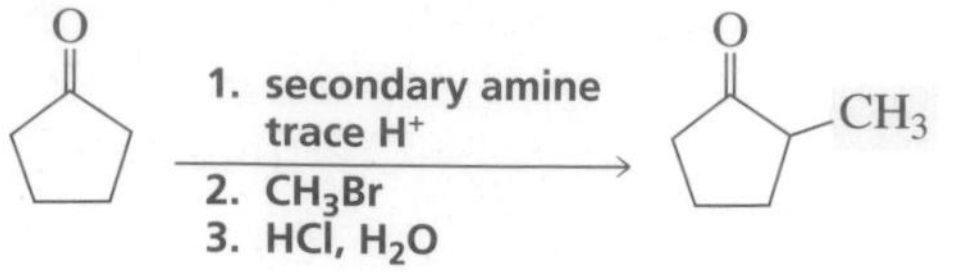

In contrast, when a carbonyl compound is alkylated directly, dialkylated and O-alkylated products can also be formed.

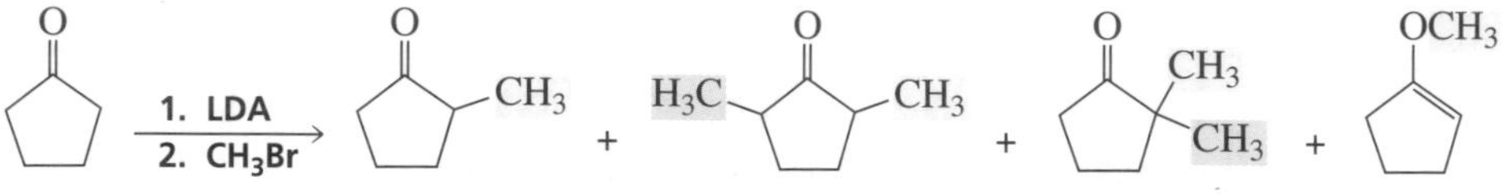

In addition to using enamine intermediates to *alkylate* aldehydes and ketones, we can also use them to *acylate* aldehydes and ketones.

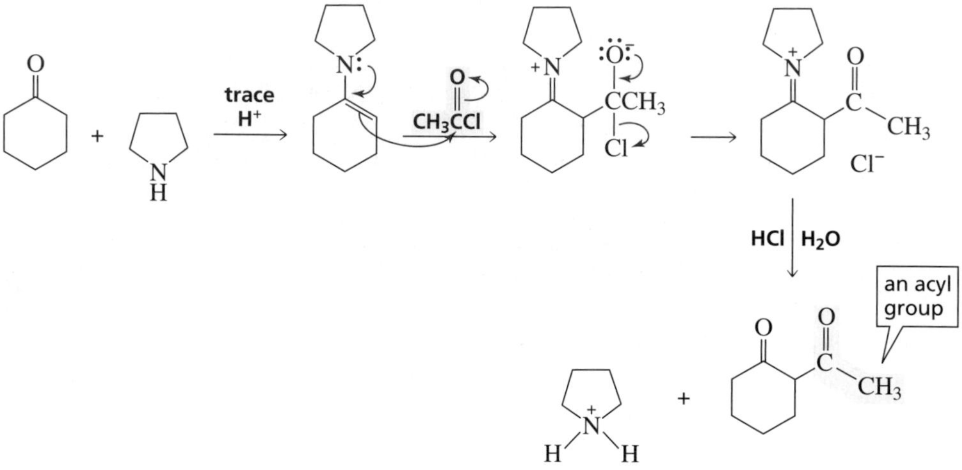

In summary, the α-carbon of an aldehyde or a ketone can be made to react with an *electrophile* either by first treating the carbonyl compound with LDA or by converting the carbonyl compound to an enamine.

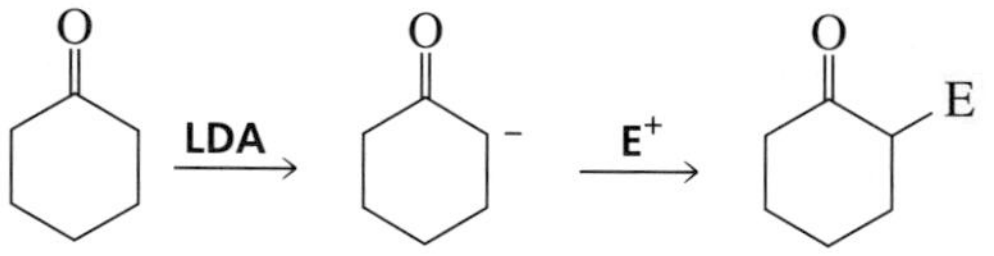

Alternatively, the α-carbon of an aldehyde or a ketone can be made to react with a *nucleophile* by first brominating the α-position of the carbonyl compound.

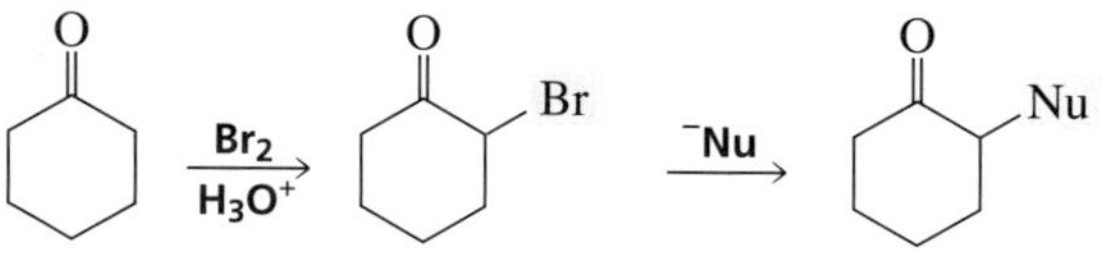

PROBLEM 18

Describe how the following compounds could be prepared using an enamine intermediate:

a.

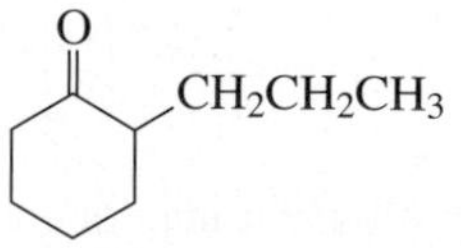

b.

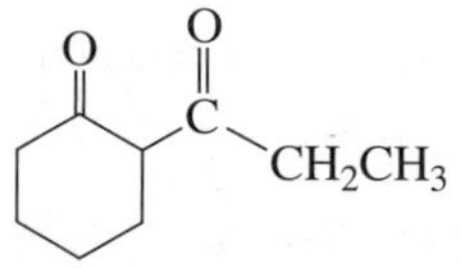

18.11 Alkylation of the β-Carbon: The Michael Reaction

In Section 17.16, you saw that nucleophiles react with α,β-unsaturated aldehydes and ketones, forming either direct addition products or conjugate addition products; in Section 17.17, you saw that nucleophiles react with α,β-unsaturated carboxylic acid derivatives, forming either nucleophilic acyl substitution products or conjugate addition products.

direct addition   conjugate addition

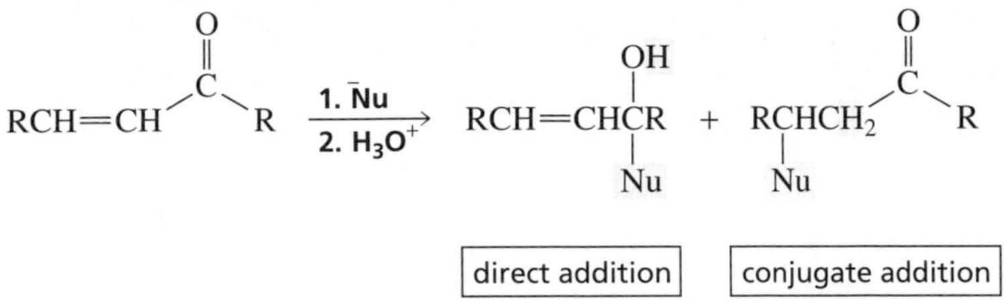

BIOGRAPHY

Arthur Michael (1853–1942) *was born in Buffalo, New York. He studied at the University of Heidelberg, the University of Berlin, and L'École de Médicine in Paris. He was a professor of chemistry at Tufts and Harvard Universities, retiring from Harvard when he was 83.*

When the nucleophile in these reactions is an enolate ion, the addition reaction has a special name: it is called a **Michael reaction**. The enolate ions that work best in Michael reactions are those that are flanked by two electron-withdrawing groups: enolate ions of β-diketones, β-diesters, β-keto esters, and β-keto nitriles. Because these enolate ions are relatively weak bases, addition occurs at the β-carbon of α,β-unsaturated aldehydes and ketones. The enolate ions also add to the β-carbon of α,β-unsaturated esters and amides because of the low reactivity of the carbonyl group. Notice that Michael reactions form 1,5-dicarbonyl compounds: if a carbonyl carbon of the enolate ion is given the 1-position, the carbonyl carbon of the other reactant is at the 5-position.

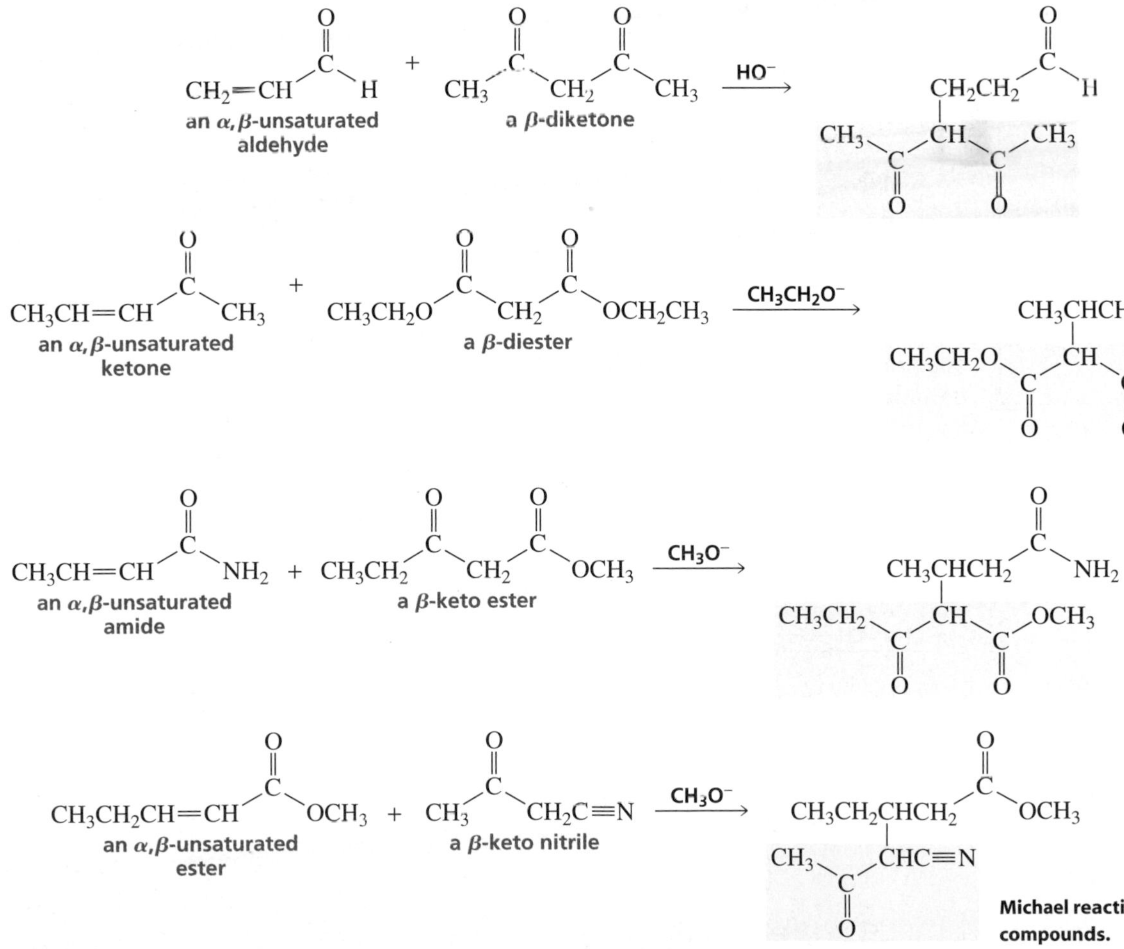

Michael reactions form 1,5-dicarbonyl compounds.

All these reactions take place by the same mechanism:

Mechanism for the Michael reaction

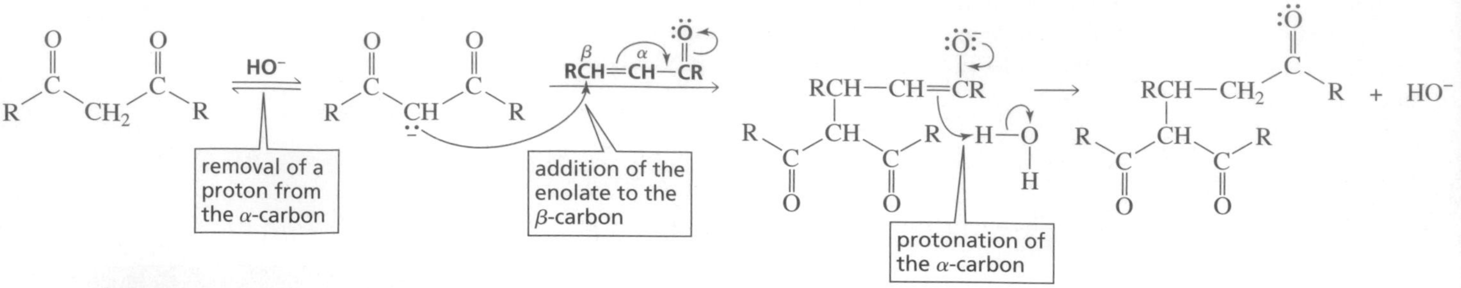

A base removes a proton from the α-carbon of the carbon acid.

[reaction mechanism scheme with labeled steps: "removal of a proton from the α-carbon", "addition of the enolate to the β-carbon", "protonation of the α-carbon"]

BIOGRAPHY

Gilbert Stork *was born in Belgium in 1921, graduated from the University of Florida, and received a Ph.D. from the University of Wisconsin. He was a professor of chemistry at Harvard University and has been a professor at Columbia University since 1953. Stork is credited with the development of many new synthetic procedures in addition to the one bearing his name.*

- A base removes a proton from the α-carbon of the carbon acid.
- The enolate ion adds to the β-carbon of an α,β-unsaturated carbonyl compound.
- The α-carbon obtains a proton from the solvent.

Notice that if either of the reactants in a Michael reaction has an ester group, the base used to remove the α-proton is the same as the leaving group of the ester. This is done because the base, in addition to being able to remove an α-proton, can react as a nucleophile and attack the carbonyl group of the ester; but if the nucleophile is identical to the OR group of the ester, nucleophilic attack on the carbonyl group will not change the reactant.

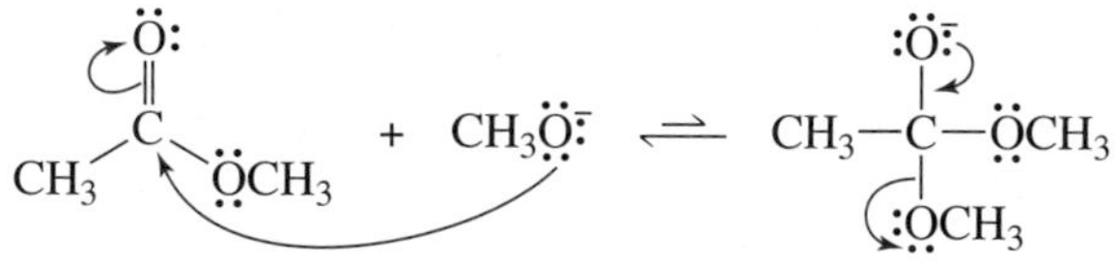

[reaction scheme]

Enamines can be used in place of enolate ions in Michael reactions. When an enamine is used as a nucleophile in a Michael reaction, the reaction is called a **Stork enamine reaction**.

Mechanism for the Stork enamine reaction

[reaction mechanism scheme showing pyrrolidine enamine of cyclohexanone reacting with CH$_2$=CH–CHO, followed by HCl, H$_2$O]

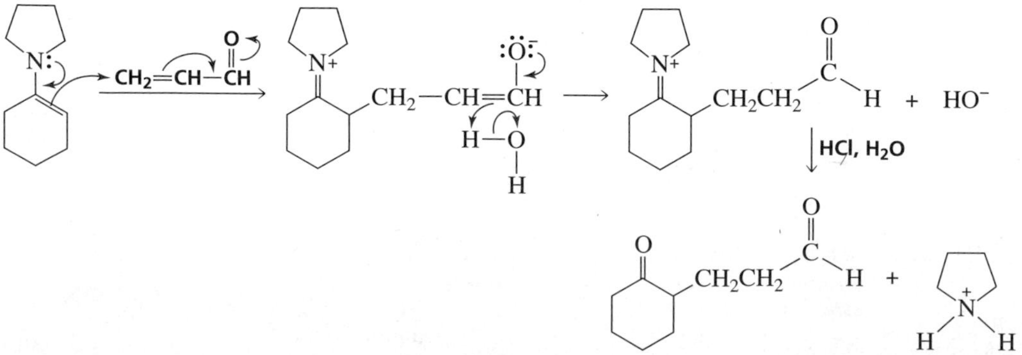

PROBLEM 19◆

What reagents would you use to prepare the following compounds?

a.

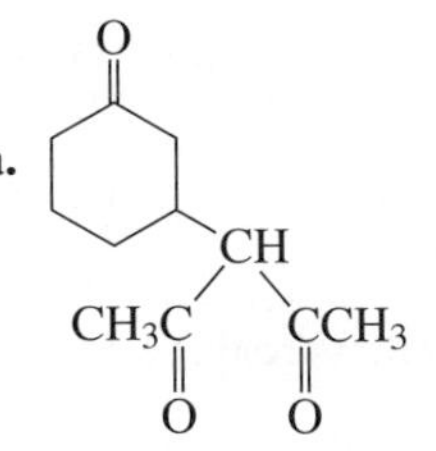

b.

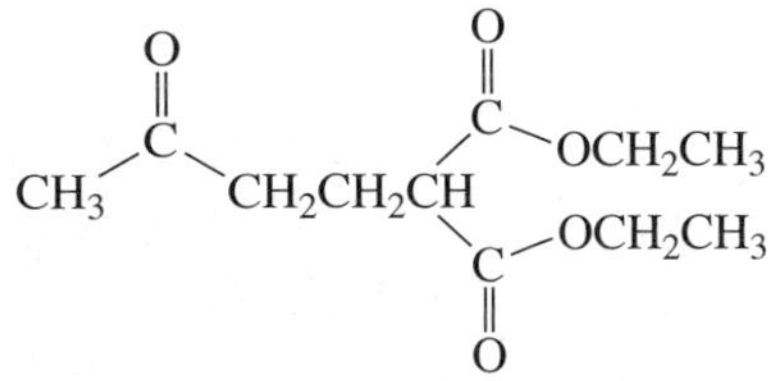

18.12 An Aldol Addition Forms β-Hydroxyaldehydes or β-Hydroxyketones

We saw in Chapter 17 that the carbonyl carbon of aldehydes and ketones is an electrophile. We have just seen that a proton can be removed from the α-carbon of an aldehyde or a ketone, converting the α-carbon into a nucleophile. An **aldol addition** is a reaction in which *both* of these activities are observed: one molecule of a carbonyl compound—after a proton is removed from an α-carbon—reacts as a *nucleophile* and attacks the *electrophilic* carbonyl carbon of a second molecule of the carbonyl compound.

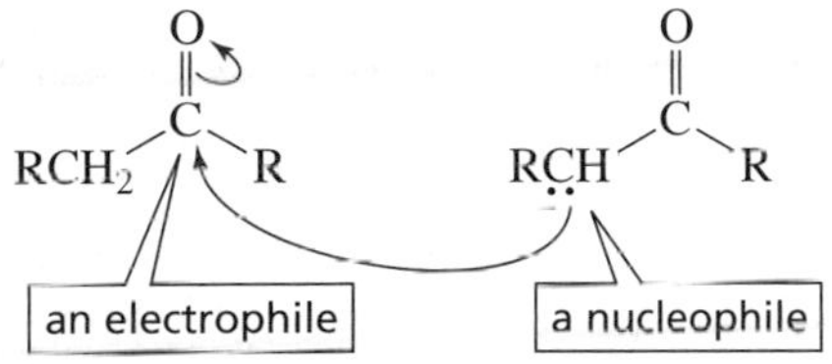

| an electrophile | a nucleophile |

Thus, an aldol addition is a reaction between two molecules of an *aldehyde* or two molecules of a *ketone*. When the reactant is an aldehyde, the product is a β-hydroxyaldehyde, which is why the reaction is called an aldol addition ("ald" for aldehyde, "ol" for alcohol). When the reactant is a ketone, the product is a β-hydroxyketone. Because the addition reaction is reversible, good yields of the addition product are obtained only if it is removed from the solution as it is formed.

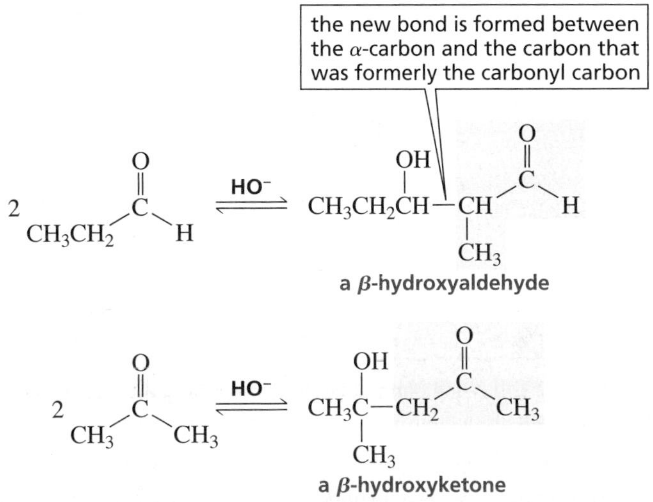

the new bond is formed between the α-carbon and the carbon that was formerly the carbonyl carbon

aldol additions

 Aldol-1

The new C—C bond formed in an aldol addition connects the α-carbon of one molecule and the carbon that formerly was the carbonyl carbon of the other molecule.

a β-hydroxyaldehyde

a β-hydroxyketone

The mechanism of the reaction is shown below.

Mechanism for the aldol addition

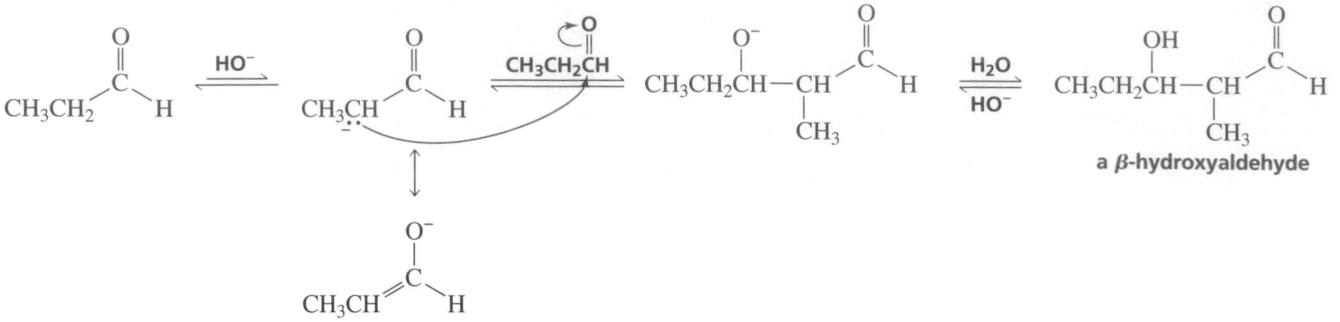

a β-hydroxyaldehyde

- A base removes a proton from the α-carbon, creating an enolate ion.
- The enolate ion adds to the carbonyl carbon of a second molecule of the carbonyl compound.
- The negatively charged oxygen is protonated by the solvent.

Because an aldol addition reaction occurs between two molecules of the same carbonyl compound, the product has twice as many carbons as the reacting aldehyde or ketone.

Ketones are less susceptible than aldehydes to attack by nucleophiles (Section 17.2), so aldol additions occur more slowly with ketones.

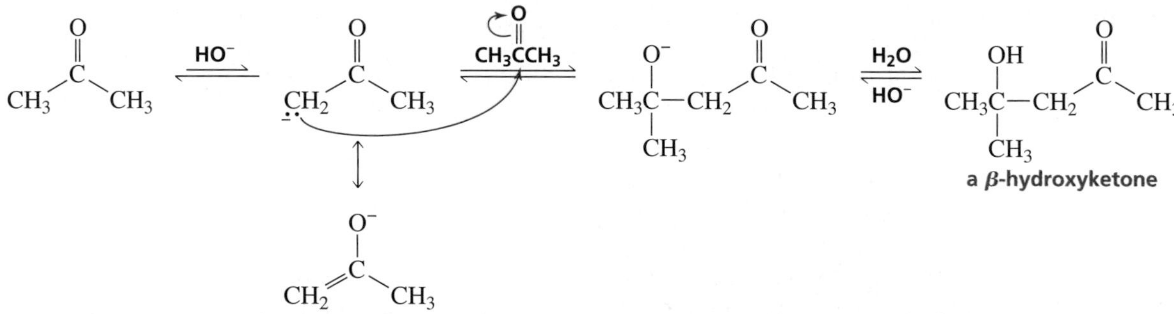

a β-hydroxyketone

The relatively high reactivity of aldehydes in competing aldol addition reactions is what causes them to give low yields of α-alkylation products (Section 18.9).

PROBLEM 20

Show the aldol addition product that would be formed from each of the following compounds:

a. 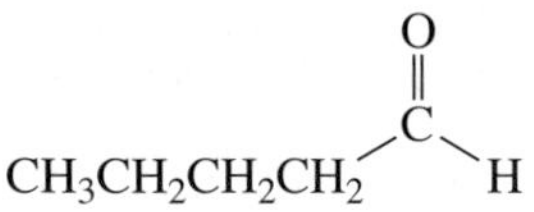 $CH_3CH_2CH_2CH_2$—C(=O)—H

c. CH_3CH_2—C(=O)—CH_2CH_3

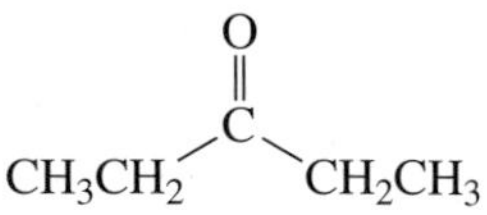

b. 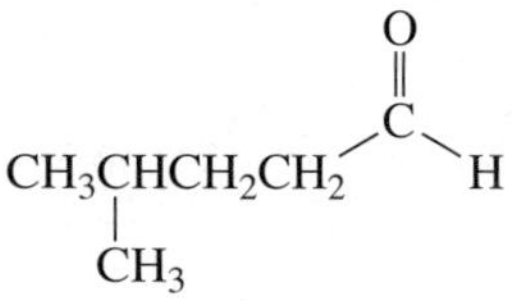 $CH_3CHCH_2CH_2$—C(=O)—H, with CH_3 substituent

d. 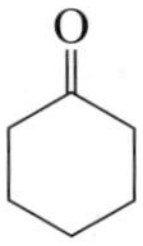 (cyclohexanone structure)

PROBLEM 21◆

For each of the following compounds, indicate the aldehyde or ketone from which it would be formed by an aldol addition:

a. 2-ethyl-3-hydroxyhexanal

b. 4-hydroxy-4-methyl-2-pentanone

c. 2,4-dicyclohexyl-3-hydroxybutanal

d. 5-ethyl-5-hydroxy-4-methyl-3-heptanone

> **PROBLEM 22**
>
> An aldol addition can be catalyzed by acids as well as by bases. Propose a mechanism for the acid-catalyzed aldol addition of propanal.

18.13 Dehydration of Aldol Addition Products Forms α,β-Unsaturated Aldehydes and Ketones

We have seen that alcohols undergo dehydration when they are heated with acid (Section 10.4). The β-hydroxyaldehyde and β-hydroxyketone products of aldol addition reactions are easier to dehydrate than many other alcohols because the double bond formed as the result of dehydration is conjugated with a carbonyl group. Conjugation increases the stability of the product (Section 7.7) and therefore makes it easier to form. If the product of an aldol addition undergoes dehydration, the overall reaction is called an **aldol condensation**. A **condensation reaction** is a reaction that combines two molecules by forming a new C—C bond while removing a small molecule (usually water or an alcohol). Notice that an aldol condensation forms an α,β-unsaturated aldehyde or an α,β-unsaturated ketone.

An aldol addition product loses water to form an aldol condensation product.

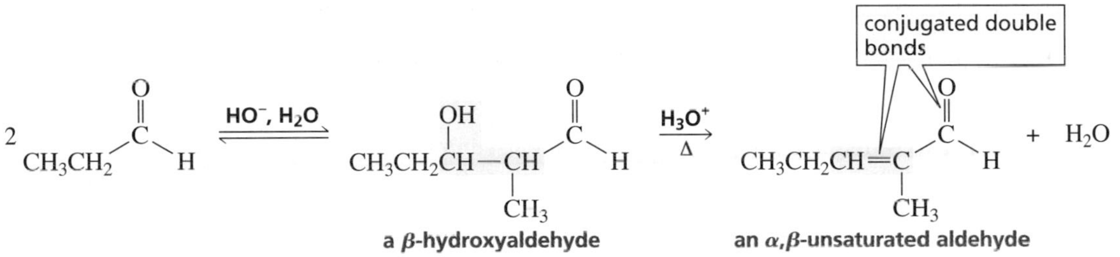

a β-hydroxyaldehyde   an α,β-unsaturated aldehyde

β-Hydroxyaldehydes and β-hydroxyketones can also undergo dehydration under basic conditions. Thus, heating the aldol addition product in either acid or base leads to dehydration. The product of dehydration is called an enone—"ene" for the double bond and "one" for the carbonyl group.

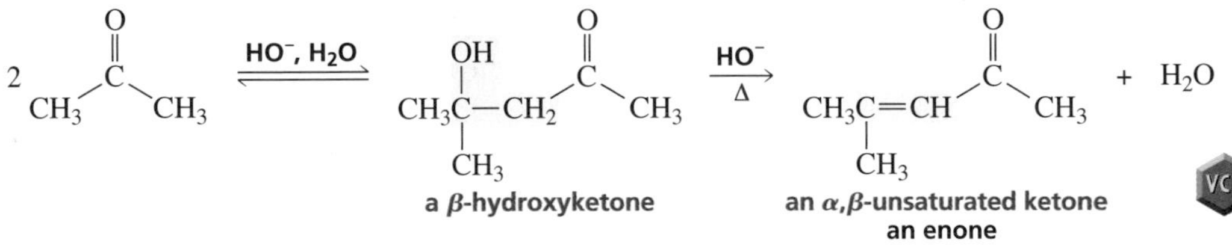

a β-hydroxyketone   an α,β-unsaturated ketone
                                    an enone

VCL Aldol-2

Dehydration sometimes occurs under the conditions in which the aldol addition is carried out, without requiring additional heat. In such cases, the β-hydroxycarbonyl compound is an intermediate and the enone is the final product of the reaction. For example, the β-hydroxyketone formed from the aldol addition of acetophenone loses water as soon as it is formed because the new double bond is conjugated not only with the carbonyl group but also with the benzene ring. Conjugation stabilizes the dehydrated product and therefore makes it relatively easy to form.

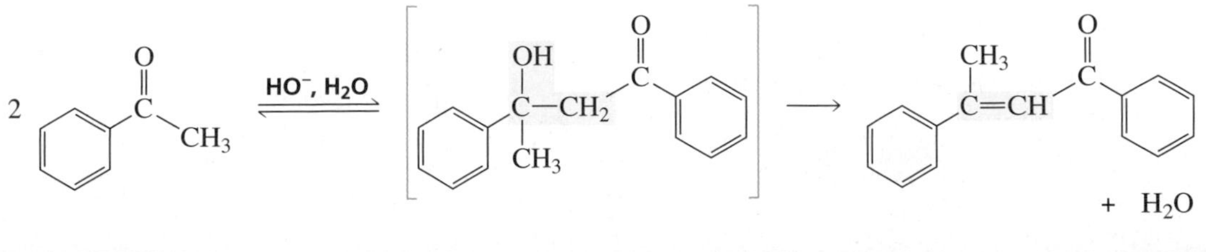

PROBLEM 23◆

What product is obtained from the aldol condensation of cyclohexanone?

PROBLEM 24 **SOLVED**

How could you prepare the following compounds using a starting material containing no more than three carbons?

a. 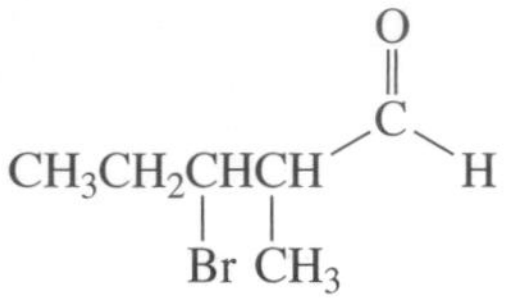 CH₃CH₂CHCH—C(=O)H with Br and CH₃ substituents

b. CH₃CH₂CH₂—C(=O)—CH₃

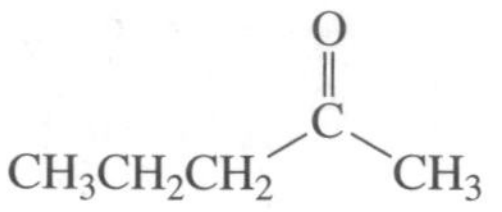

c. CH₃CH₂CH₂—C(=O)—OH

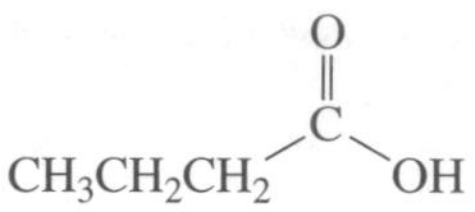

Solution to 24a A compound with the correct six-carbon skeleton can be obtained if a three-carbon aldehyde undergoes an aldol addition. Dehydration of the addition product forms an α,β-unsaturated aldehyde. Conjugate addition of HBr (Section 17.16) forms the target compound.

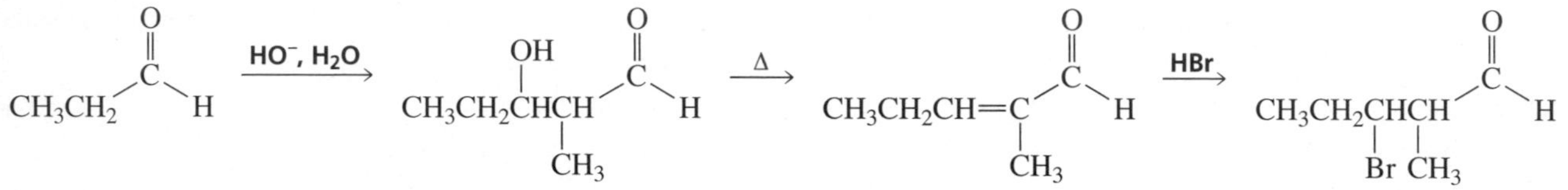

CH₃CH₂—C(=O)H →(HO⁻, H₂O)→ CH₃CH₂CHCH—C(=O)H (with OH and CH₃) →(Δ)→ CH₃CH₂CH=C—C(=O)H (with CH₃) →(HBr)→ CH₃CH₂CHCH—C(=O)H (with Br and CH₃)

18.14 The Mixed Aldol Addition

If two different carbonyl compounds are used in an aldol addition, four products can be formed because each enolate ion can react with either of the two carbonyl compounds. In the following example, both carbonyl compound A and carbonyl compound B can lose a proton from an α-carbon to form enolate ion A⁻ or B⁻; A⁻ can react with either A or B, and B⁻ can react with either A or B:

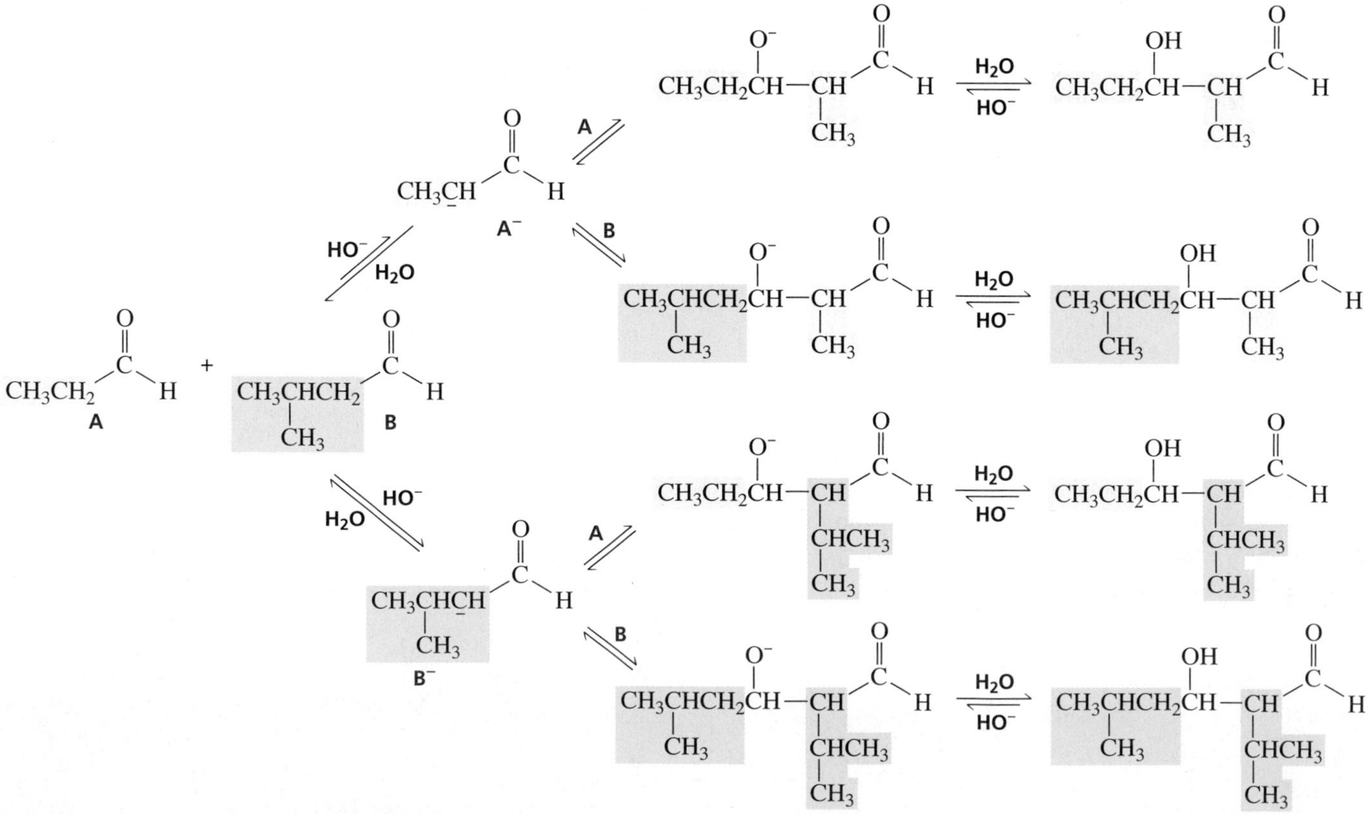

The preceding reaction is called a **mixed aldol addition** or a **crossed aldol addition**. The four products have similar physical properties, making them difficult to separate. Consequently, a mixed aldol addition that forms four products is not a synthetically useful reaction.

Under certain conditions, a mixed aldol addition can lead primarily to one product. If one of the carbonyl compounds does not have any α-hydrogens, it cannot form an enolate ion. This reduces the number of possible products from four to two. A greater amount of one of the two products will be formed if the compound *without* α-hydrogens is always present in excess, because then the enolate ion will be more likely to react with it, rather than with another molecule of the enolate ion's parent compound. Therefore, the compound with α-hydrogens should be added slowly to a basic solution of the compound without α-hydrogens.

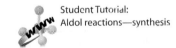

excess   **add slowly**

If both carbonyl compounds have α-hydrogens, primarily one aldol product can be obtained if LDA is used to remove the α-hydrogen that creates the enolate ion. Because LDA is a strong base (Section 18.8), all the carbonyl compound will be converted into an enolate ion, so there will be none of that carbonyl compound left for the enolate ion to react with in an aldol addition. An aldol addition will not be able to occur until the second carbonyl compound is added to the reaction mixture. If the second carbonyl compound is added slowly, the chance that it will form an enolate ion and then react with another molecule of its parent carbonyl compound will be minimized.

Student Tutorial:
Aldol reactions—synthesis

PROBLEM 25

Give the products obtained from mixed aldol additions of the following compounds:

a. $CH_3CH_2CH_2$—C(=O)—H + $CH_3CH_2CH_2CH_2$—C(=O)—H

b. CH_3—C(=O)—H + CH_3CH_2—C(=O)—CH_2CH_3

c. (cyclohexanone) + CH_3CH_2—C(=O)—H

d. H—C(=O)—H + CH_3CH_2—C(=O)—H
   excess

PROBLEM 26

Describe how the following compounds could be prepared using an aldol addition in the first step of the synthesis:

a. CH_3CH_2—C(=O)—CHCH$_2$OH
              |
             CH_3

b. (phenyl)—CH=CH—C(=O)—CH=CH—(phenyl)

c. (cyclohexanone with =CH—phenyl)

PROBLEM 27

Propose a mechanism for the following reaction:

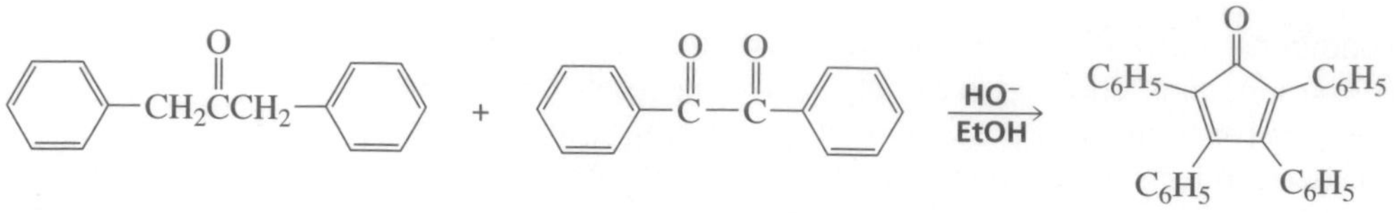

18.15 A Claisen Condensation Forms a β-Keto Ester

When two molecules of an *ester* undergo a condensation reaction, the reaction is called a **Claisen condensation**. The product of a Claisen condensation is a β-keto ester.

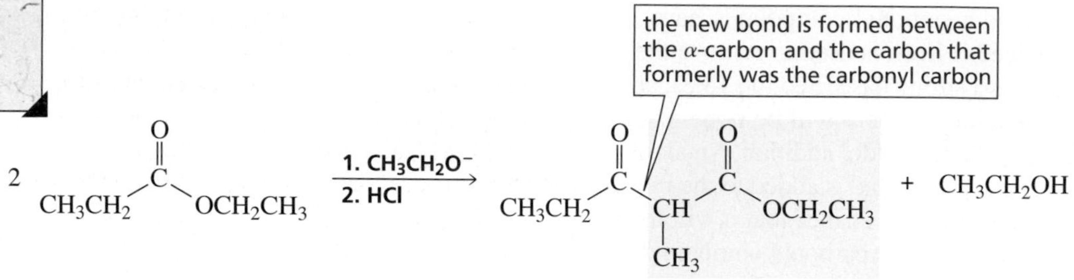

the new bond is formed between the α-carbon and the carbon that formerly was the carbonyl carbon

a β-keto ester

In a Claisen condensation, as in an aldol addition, one molecule of carbonyl compound is converted into an enolate ion when a strong base removes a proton from an α-carbon. The enolate ion attacks the carbonyl carbon of a second molecule of ester. The base employed corresponds to the leaving group of the ester so that the reactant will not change if the base were to act as a nucleophile and attack the carbonyl group (Section 18.11).

Mechanism for the Claisen condensation

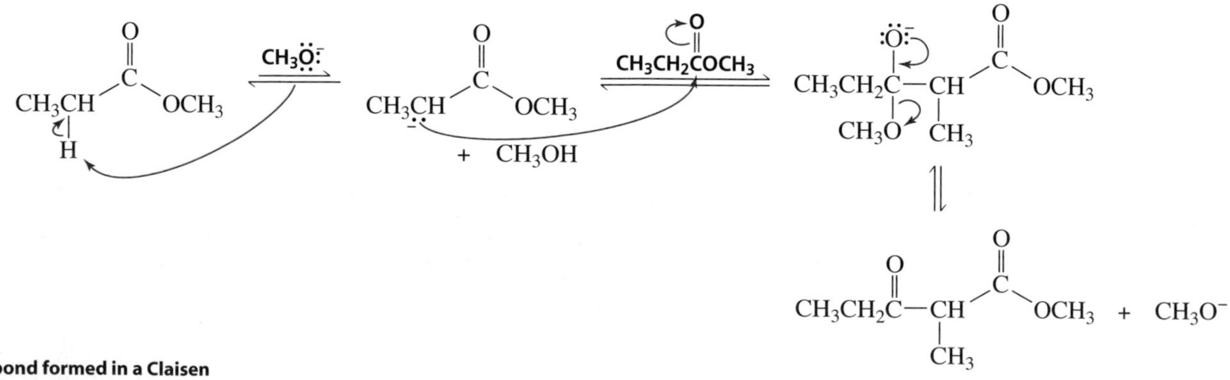

The new C—C bond formed in a Claisen condensation connects the α-carbon of one molecule and the carbon that formerly was the carbonyl carbon of the other molecule.

After nucleophilic attack, the Claisen condensation and the aldol addition differ. In the Claisen condensation, the negatively charged oxygen reforms the carbon–oxygen

π bond and expels the $^-$OR group. In the aldol addition, the negatively charged oxygen obtains a proton from the solvent.

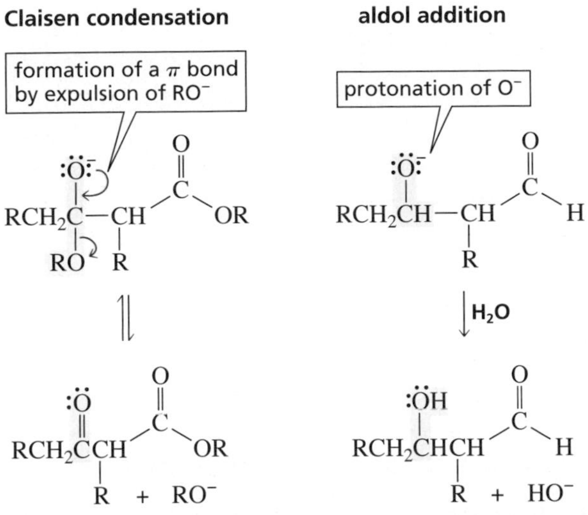

Claisen condensation   **aldol addition**

The difference between the last step of the Claisen condensation and the last step of the aldol addition arises from the difference between esters and aldehydes or ketones. With esters (class I carbonyl compounds), the carbon bonded to the negatively charged oxygen is also bonded to a group that can be expelled. With aldehydes or ketones (class II carbonyl compounds), the carbon bonded to the negatively charged oxygen is not bonded to a group that can be expelled. Thus, the Claisen condensation is a nucleophilic acyl substitution reaction, whereas the aldol addition is a nucleophilic addition reaction.

Expulsion of the alkoxide ion is reversible because the alkoxide ion can readily reform the tetrahedral intermediate by reacting with the β-keto ester. The condensation reaction can be driven to completion, however, if a proton is removed from the β-keto ester. Removing a proton prevents the reverse reaction from occurring, because the negatively charged alkoxide ion will not react with the negatively charged anion of the β-keto ester. A proton is easily removed from the β-keto ester because the central α-carbon is flanked by two carbonyl groups, making its α-hydrogen much more acidic than the α-hydrogens of the ester starting material.

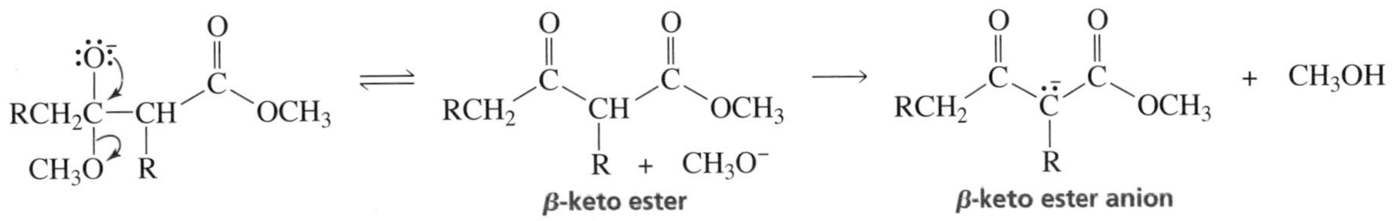

VCL **Claisen Condensation-1**

Consequently, a successful Claisen condensation requires an ester with two α-hydrogens and an equivalent amount of base rather than a catalytic amount of base. When the reaction is over, addition of acid to the reaction mixture reprotonates the β-keto ester anion. Any remaining alkoxide ion that could cause the reaction to reverse is also protonated.

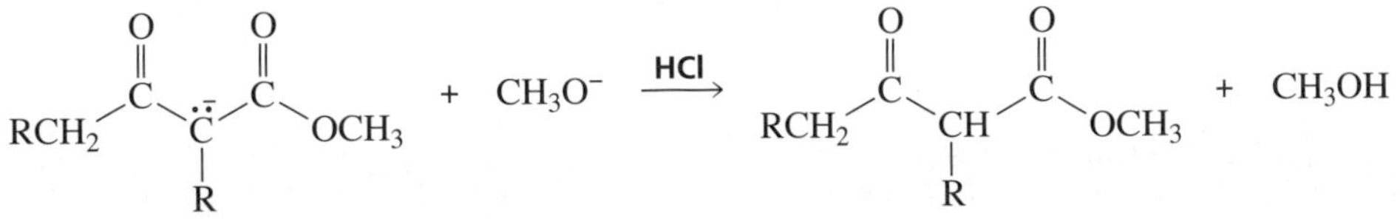

PROBLEM 28◆

Give the products of the following reactions:

a. CH₃CH₂CH₂COCH₃ $\xrightarrow{\text{1. CH}_3\text{O}^-}{\text{2. HCl}}$

b. CH₃CHCH₂COCH₂CH₃ $\xrightarrow{\text{1. CH}_3\text{CH}_2\text{O}^-}{\text{2. HCl}}$
      |
      CH₃

PROBLEM 29◆

Which of the following esters cannot undergo a Claisen condensation?

A CH₃CH=CHCOCH₃   **B** HCOCH₃   **C** CH₃COCH₃   **D** ⬡—COCH₃

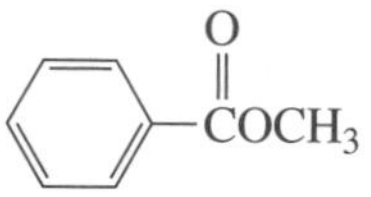

18.16 The Mixed Claisen Condensation

A **mixed Claisen condensation** is a condensation reaction between two different esters. Like a mixed aldol addition, a mixed Claisen condensation is a useful reaction only if it is carried out under conditions that foster the formation of primarily one product. Otherwise, the reaction will form a mixture of products that are difficult to separate. Primarily one product will be formed if one of the esters has no α-hydrogens (and therefore cannot form an enolate ion) and the other ester is added slowly so that the ester without α-hydrogens is always in excess.

a mixed Claisen condensation

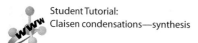

CH₃CH₂CH₂—C—OCH₂CH₃ (add slowly) + ⬡—C—OCH₂CH₃ (excess) $\xrightarrow{\text{1. CH}_3\text{CH}_2\text{O}^-}{\text{2. HCl}}$ ⬡—C—CH—C—OCH₂CH₃ (with CH₂CH₃) + CH₃CH₂OH

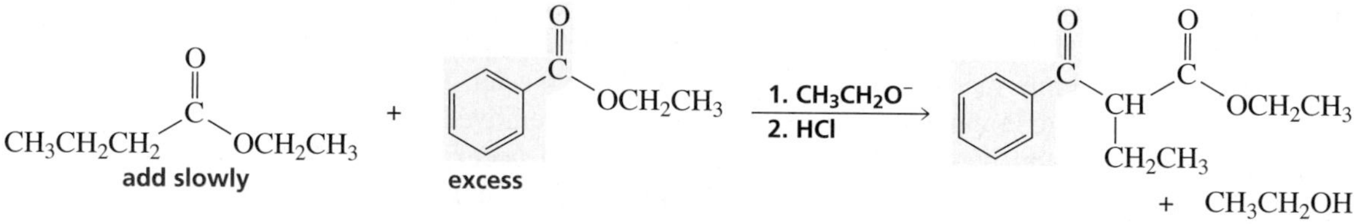

Student Tutorial:
Claisen condensations—synthesis

A reaction similar to a mixed Claisen condensation is the condensation of a ketone and an ester. Because the α-hydrogens of a ketone are more acidic than those of an ester, primarily one product is formed if the ketone and the base are both added slowly to the ester. The product is a β-diketone. Because of the difference in acidities of the α-hydrogens, primarily one condensation product is obtained even if both reagents have α-hydrogens.

condensation of a ketone and an ester

(cyclohexanone, add slowly) + CH₃—C—OCH₂CH₃ (ethyl acetate, excess) $\xrightarrow{\text{1. CH}_3\text{CH}_2\text{O}^-}{\text{2. HCl}}$ (a β-diketone) + CH₃CH₂OH

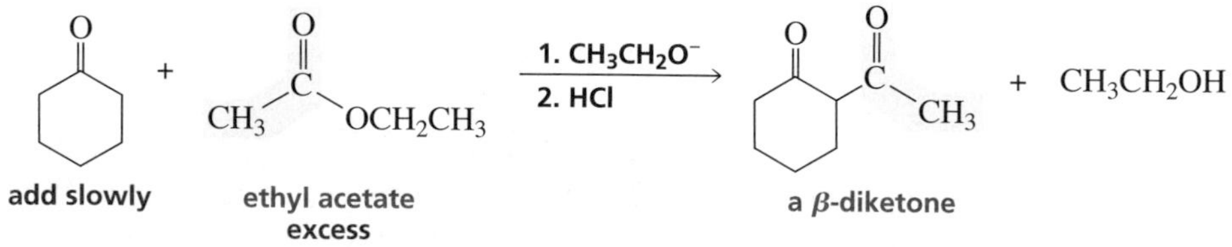

A β-keto aldehyde is formed when a ketone condenses with a formate ester.

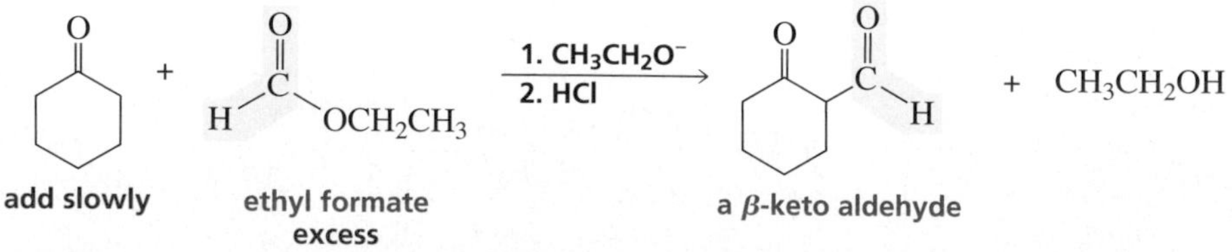

(cyclohexanone, add slowly) + H—C—OCH₂CH₃ (ethyl formate, excess) $\xrightarrow{\text{1. CH}_3\text{CH}_2\text{O}^-}{\text{2. HCl}}$ (a β-keto aldehyde) + CH₃CH₂OH

A β-keto ester is formed when a ketone condenses with diethyl carbonate.

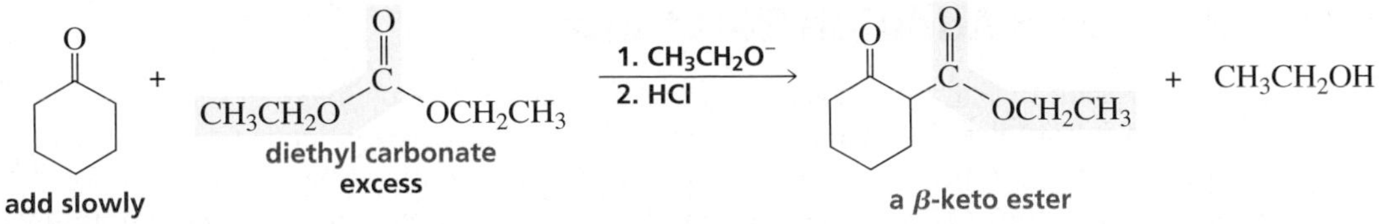

cyclohexanone + CH$_3$CH$_2$O—C(=O)—OCH$_2$CH$_3$ (diethyl carbonate, excess) $\xrightarrow[\text{2. HCl}]{\text{1. CH}_3\text{CH}_2\text{O}^-}$ β-keto ester + CH$_3$CH$_2$OH

diethyl carbonate
excess

add slowly

a β-keto ester

PROBLEM 30

Give the product of each of the following reactions:

 Claisen Condensation-2

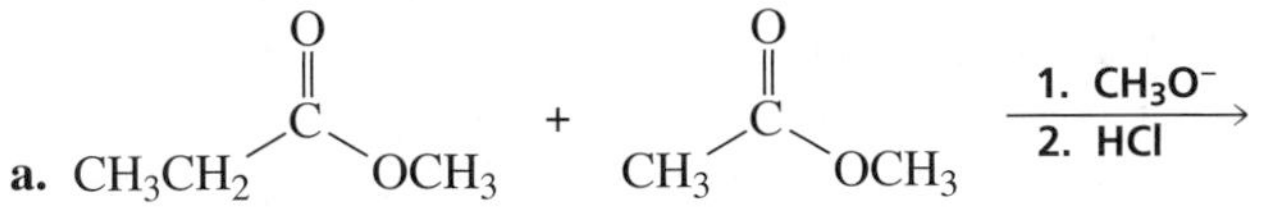

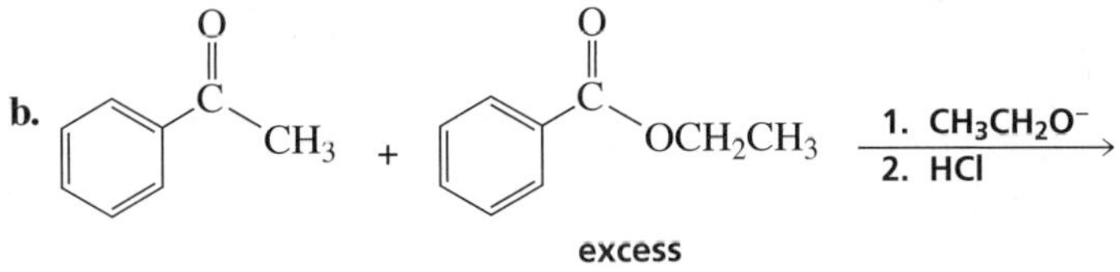

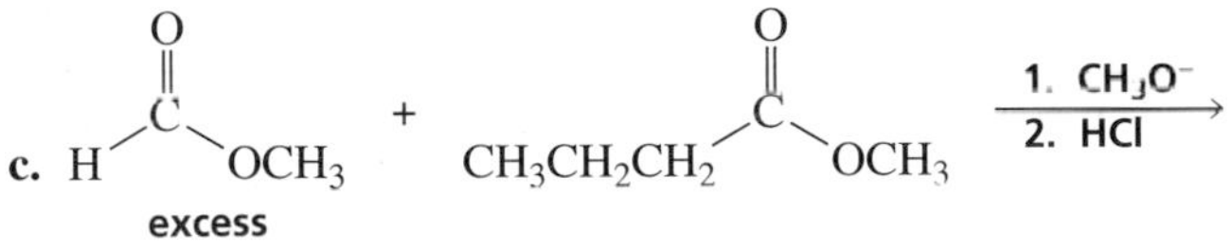

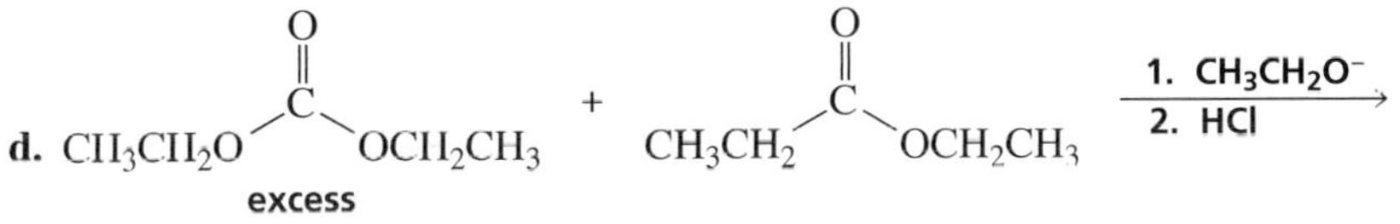

a. CH$_3$CH$_2$—C(=O)—OCH$_3$ + CH$_3$—C(=O)—OCH$_3$ $\xrightarrow[\text{2. HCl}]{\text{1. CH}_3\text{O}^-}$

b. (phenyl)C(=O)CH$_3$ + (phenyl)C(=O)OCH$_2$CH$_3$ (excess) $\xrightarrow[\text{2. HCl}]{\text{1. CH}_3\text{CH}_2\text{O}^-}$

c. H—C(=O)—OCH$_3$ (excess) + CH$_3$CH$_2$CH$_2$—C(=O)—OCH$_3$ $\xrightarrow[\text{2. HCl}]{\text{1. CH}_3\text{O}^-}$

 Claisen Condensation-3

d. CH$_3$CH$_2$O—C(=O)—OCH$_2$CH$_3$ (excess) + CH$_3$CH$_2$—C(=O)—OCH$_2$CH$_3$ $\xrightarrow[\text{2. HCl}]{\text{1. CH}_3\text{CH}_2\text{O}^-}$

PROBLEM 31 *SOLVED*

Show how the following compounds could be prepared from 3-thiomethylcyclohexanone:

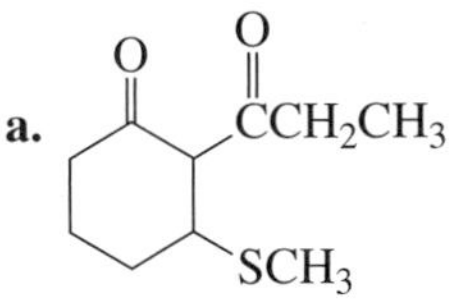

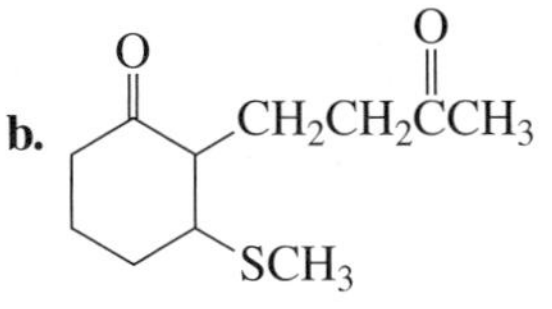

a. 2-(C(=O)CH$_2$CH$_3$)-3-(SCH$_3$)cyclohexanone

b. 2-(CH$_2$CH$_2$C(=O)CH$_3$)-3-(SCH$_3$)cyclohexanone

Solution to 31a Because the desired compound is a 1,3-dicarbonyl compound, it can be prepared by treating an enolate ion with an ester.

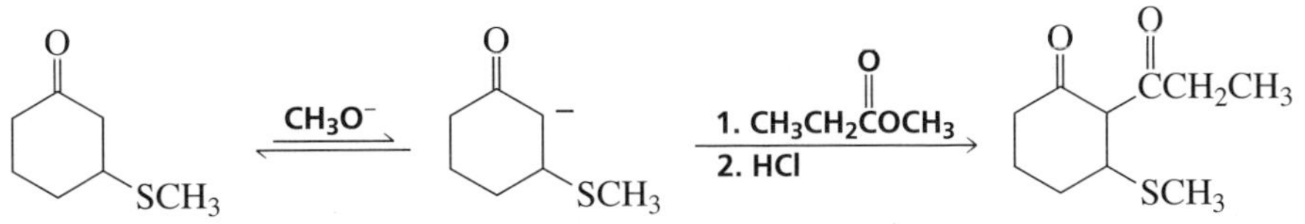

3-(SCH$_3$)cyclohexanone $\xrightarrow{\text{CH}_3\text{O}^-}$ enolate $\xrightarrow[\text{2. HCl}]{\text{1. CH}_3\text{CH}_2\text{C(=O)OCH}_3}$ 2-(C(=O)CH$_2$CH$_3$)-3-(SCH$_3$)cyclohexanone

Solution to 31b Because the desired compound is a 1,5-dicarbonyl compound, it can be prepared by treating an enolate ion with an α,β-unsaturated carbonyl compound (a Michael reaction).

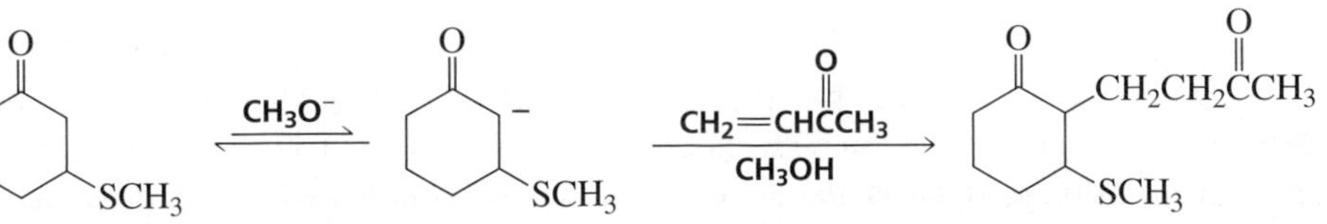

3-(SCH$_3$)cyclohexanone $\xrightarrow{\text{CH}_3\text{O}^-}$ enolate $\xrightarrow[\text{CH}_3\text{OH}]{\text{CH}_2=\text{CHC(=O)CH}_3}$ 2-(CH$_2$CH$_2$C(=O)CH$_3$)-3-(SCH$_3$)cyclohexanone

18.17 Intramolecular Condensation and Addition Reactions

We have seen that if a compound has two functional groups that can react with each other, an intramolecular reaction readily occurs if the reaction leads to the formation of five- or six-membered rings (Section 8.11). Consequently, if base is added to a compound that contains two carbonyl groups, an intramolecular reaction occurs if a product with a five- or six-membered ring can be formed. Thus, a compound with two ester groups would undergo an intramolecular Claisen condensation, and a compound with two aldehyde or ketone groups would undergo an intramolecular aldol addition.

Intramolecular Claisen Condensations

The addition of base to a 1,6-diester causes the diester to undergo an intramolecular Claisen condensation, thereby forming a five-membered ring β-keto ester. An intramolecular Claisen condensation is called a **Dieckmann condensation**.

BIOGRAPHY

Walter Dieckmann (1869–1925) *was born in Germany. He received a Ph.D. from the University of Munich and became a professor of chemistry there.*

1. CH_3O^-
2. HCl
+ CH_3OH

a 1,6-diester

a β-keto ester

VCL Dieckmann Reaction

A six-membered ring β-keto ester is formed from a Dieckmann condensation of a 1,7-diester.

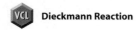

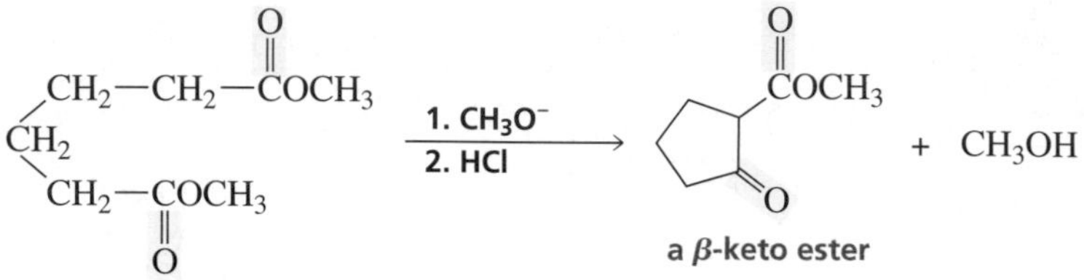

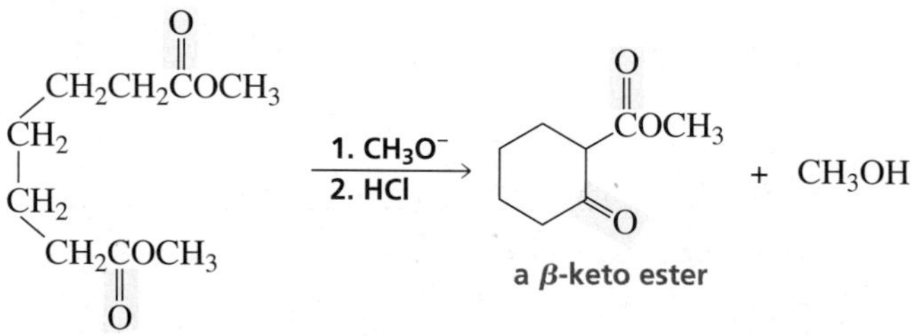

1. CH_3O^-
2. HCl
+ CH_3OH

a 1,7-diester

a β-keto ester

The mechanism of the Dieckmann condensation is the same as the mechanism of the Claisen condensation. The only difference between the two reactions is that the attacking enolate ion and the carbonyl group undergoing nucleophilic attack are in different molecules in the Claisen condensation but are in the same molecule in the Dieckmann condensation.

Mechanism for the Dieckmann condensation

CH_3O^-
+ CH_3OH
+ CH_3O^-

The Dieckmann condensation, like the Claisen condensation, can be driven to completion by carrying out the reaction with enough base to remove a proton from the α-carbon of the β-keto ester product. When the reaction is over, acid is added to reprotonate the condensation product.

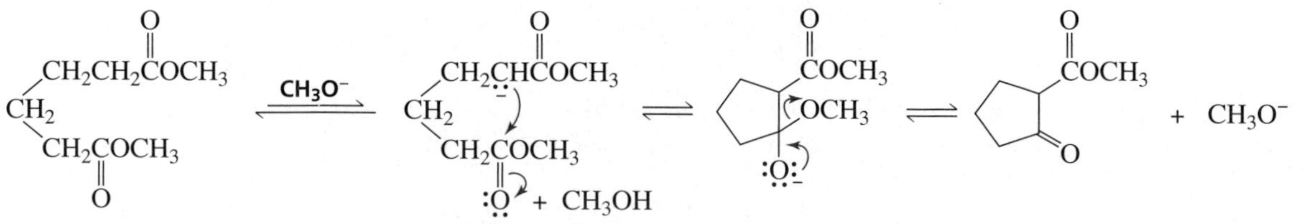

PROBLEM 32

Write the mechanism for the base-catalyzed formation of a cyclic β-keto ester from a 1,7-diester.

Intramolecular Aldol Additions

Because a 1,4-diketone has two different sets of α-hydrogens, two different intramolecular addition products can potentially form, one with a five-membered ring, the other with a three-membered ring. The greater stability of five- and six-membered rings causes them to be formed preferentially (Section 2.11). In fact, the five-membered ring product is the only product formed from the intramolecular aldol addition of a 1,4-diketone.

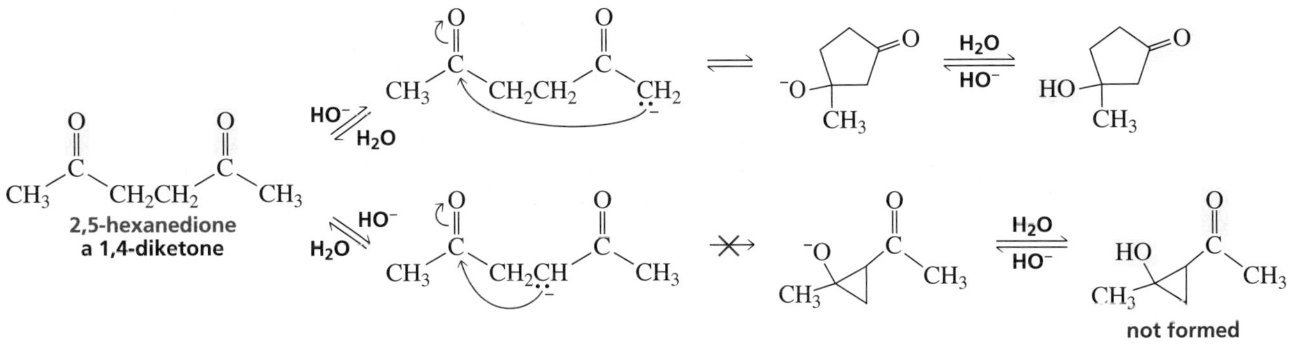

The intramolecular aldol addition of a 1,6-diketone can potentially lead to either a seven- or a five-membered ring product. Again, the more stable product—the one with the five-membered ring—is the only product of the reaction.

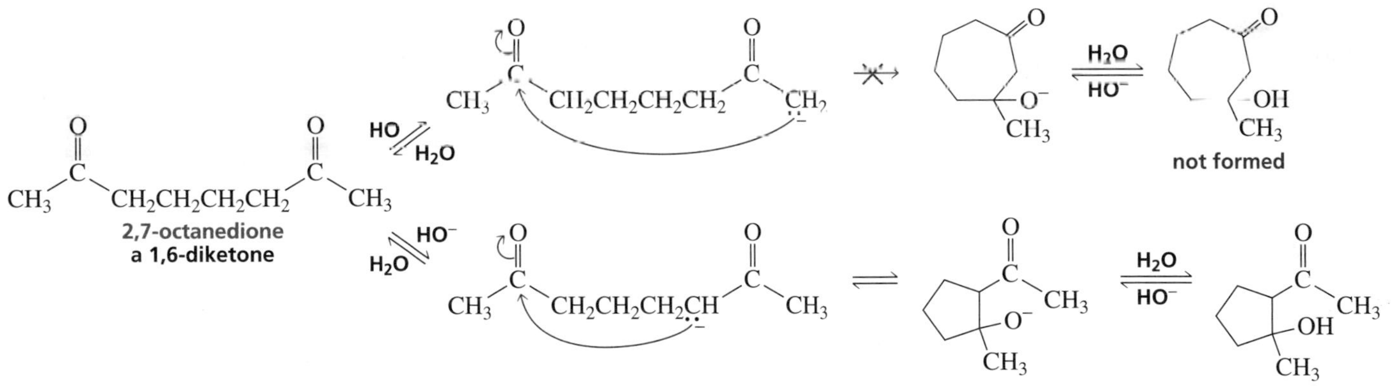

1,5-Diketones and 1,7-diketones undergo intramolecular aldol additions to form six-membered ring products.

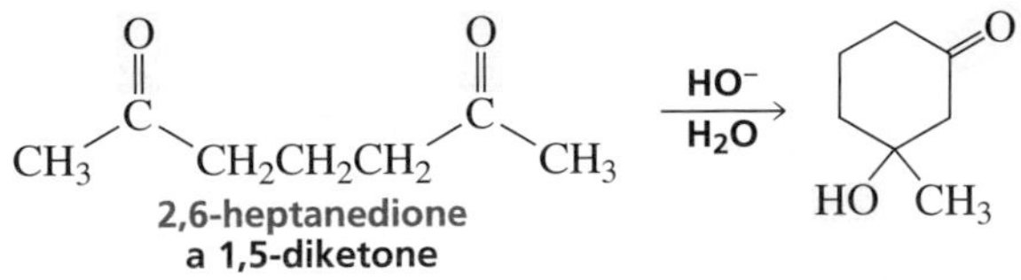

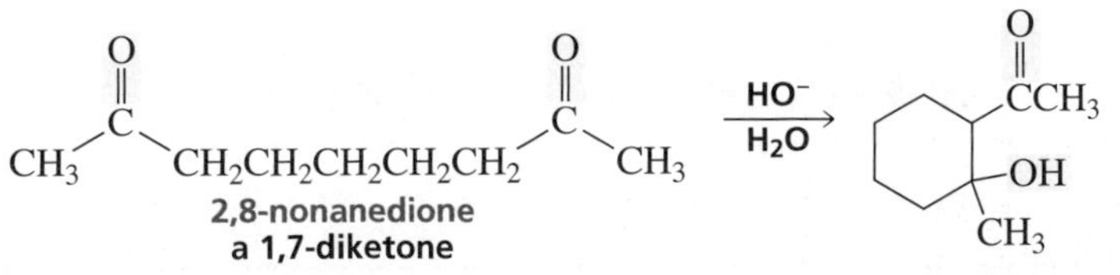

PROBLEM 33♦

If the preference for formation of a six-membered ring were not so great, what other cyclic product would be formed from the intramolecular aldol addition of

a. 2,6-heptanedione?   **b.** 2,8-nonanedione?

PROBLEM 34

Can 2,4-pentanedione undergo an intramolecular aldol addition? If so, why? If not, why not?

PROBLEM 35   *SOLVED*

What products can be obtained if the following keto aldehyde is treated with a base? Which would you expect to be the major product?

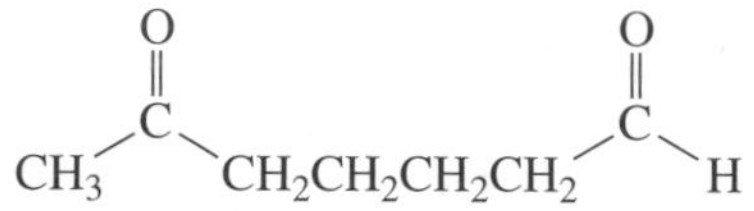

Solution   Three products are possible because there are three different sets of α-hydrogens. More B and C are formed than A because a five-membered ring is formed in preference to a seven-membered ring. The major product depends on the reaction conditions. B is the thermodynamic product because it is formed from the more stable enolate. C is the kinetic product because the α-hydrogen of the aldehyde is more acidic than the α-hydrogen of the ketone (Table 18.1).

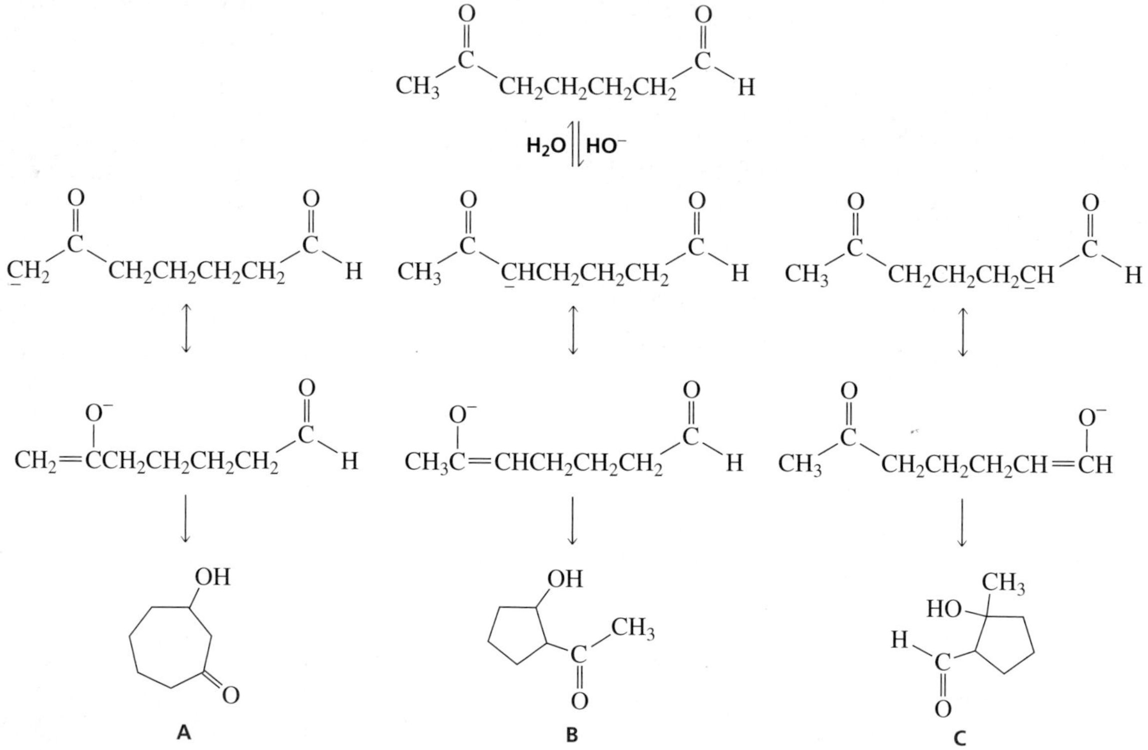

PROBLEM 36♦

Give the product of the reaction of each of the following compounds with a base:

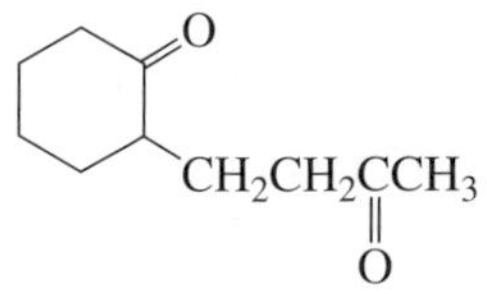

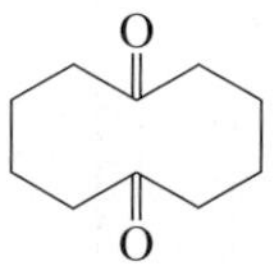

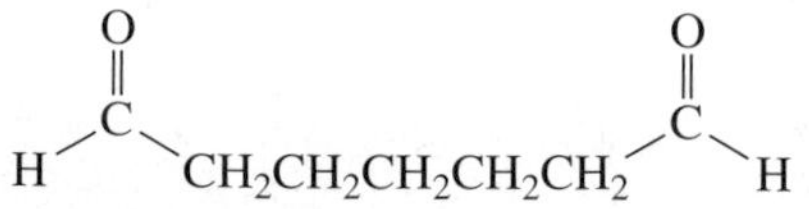

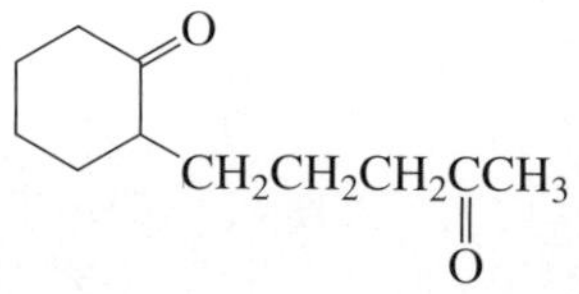

The Robinson Annulation

Reactions that form carbon–carbon bonds are important to synthetic chemists. Without such reactions, large organic molecules could not be prepared from smaller ones. We have seen that Michael reactions and aldol additions form carbon–carbon bonds. The **Robinson annulation** is a reaction that puts these two carbon–carbon bond-forming reactions together, providing a route to the synthesis of many complicated organic molecules. "Annulation" comes from *annulus,* Latin for "ring." Thus, an **annulation reaction** is a ring-forming reaction.

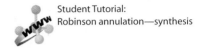

Student Tutorial:
Robinson annulation—synthesis

the Robinson annulation

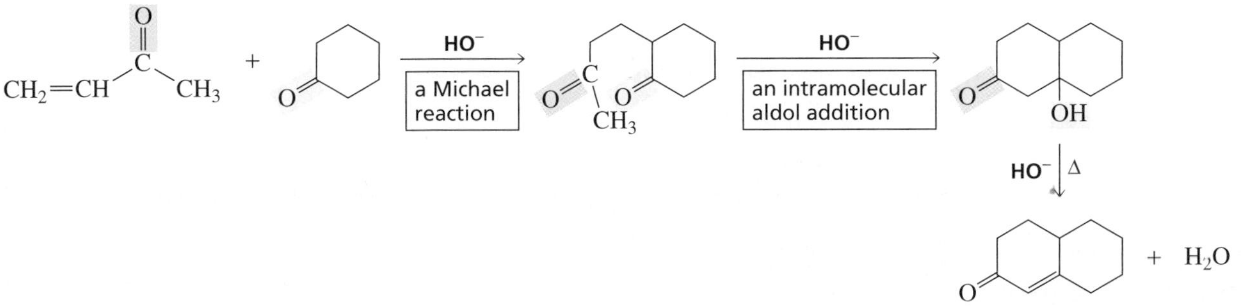

- The first stage of a Robinson annulation is a Michael reaction that forms a 1,5-diketone.
- The second stage is an intramolecular aldol addition.
- Heating the basic solution dehydrates the alcohol.

Notice that a Robinson annulation results in a product that has a 2-cyclohexenone ring.

PROBLEM-SOLVING STRATEGY

Analyzing a Robinson Annulation

Propose a synthesis for each of the following compounds, using a Robinson annulation:

a.

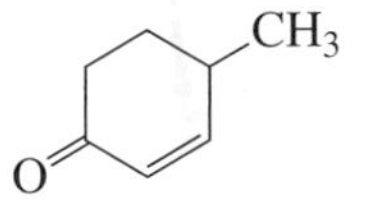

b.

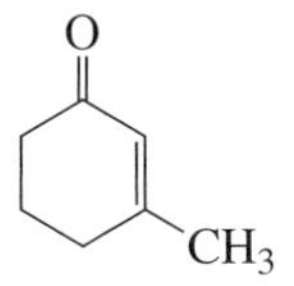

Careful examination of a Robinson annulation tell us which part of the molecule comes from which reactant. This knowledge, then, allows us to choose appropriate reactants for any other Robinson annulation. Analysis shows that the keto group of the cyclohexenone comes from the α,β-unsaturated carbonyl compound and the double bond results from attack of the enolate ion of the α,β-unsaturated carbonyl compound on the carbonyl group of the other reactant. Thus, we can arrive at the appropriate reactants by cutting through the double bond of the target compound and cutting between the β- and γ-carbons on the other side of the carbonyl group.

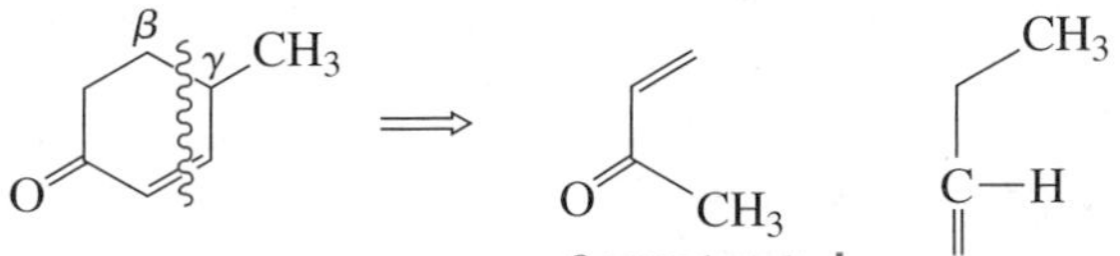

α,β-unsaturated
carbonyl compound

Therefore, the required reactants for part a are

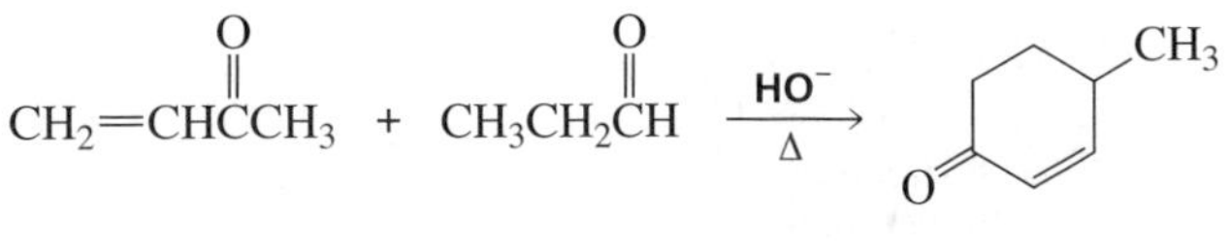

$$CH_2{=}CHCCH_3 \ + \ CH_3CH_2CH \xrightarrow[\Delta]{HO^-}$$

By cutting through the compound in part b, we can determine the required reactants for its synthesis:

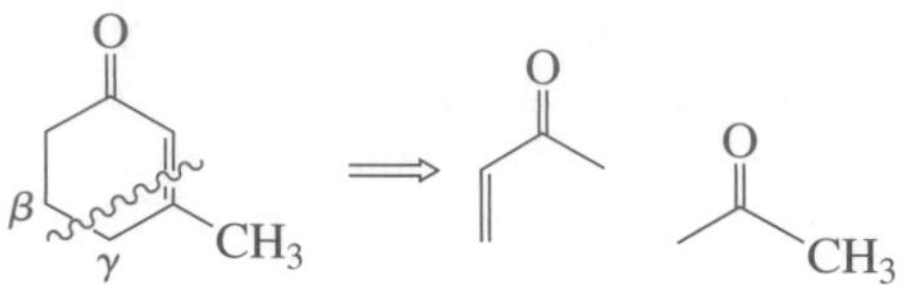

Therefore, the required reactants for part b are

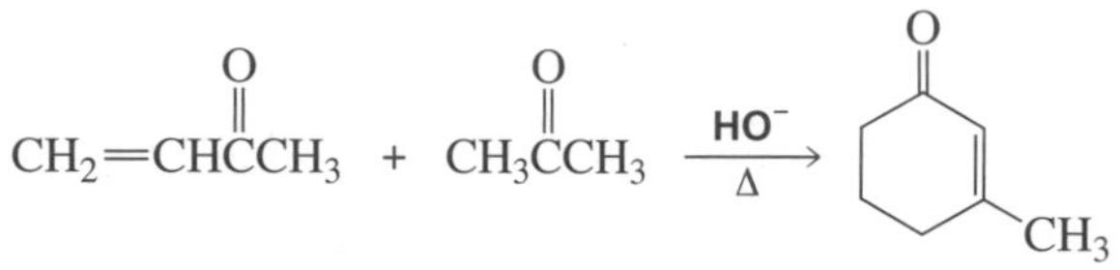

$$CH_2=CHCCH_3 \ + \ CH_3CCH_3 \ \xrightarrow[\Delta]{HO^-}$$

Now continue on to Problem 37.

PROBLEM 37

Propose a synthesis for each of the following compounds, using a Robinson annulation:

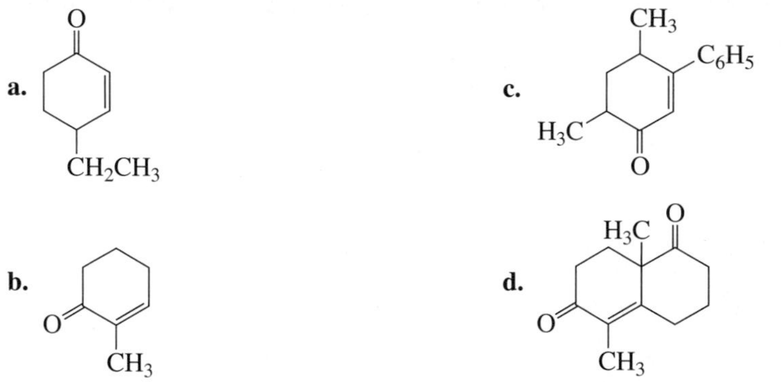

a.

CH₂CH₃

c.

b.

CH₃

d.

CH₃

18.18 3-Oxocarboxylic Acids Can Be Decarboxylated

Carboxylate ions do not lose CO_2, for the same reason that alkanes such as ethane do not lose a proton—because the leaving group would be a carbanion. Carbanions are very strong bases and therefore are very poor leaving groups.

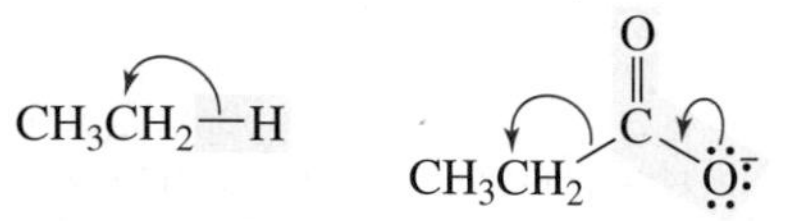

If, however, the CO_2 group is bonded to a carbon adjacent to a carbonyl carbon, the CO_2 group can be removed, because the electrons left behind can be delocalized onto the carbonyl oxygen. Consequently, 3-oxocarboxylate ions (carboxylate ions with a carbonyl group at the 3-position) lose CO_2 when they are heated. Loss of CO_2 from a molecule is called **decarboxylation**.

3-Oxocarboxylic acids decarboxylate when heated.

removing CO₂ from an α-carbon

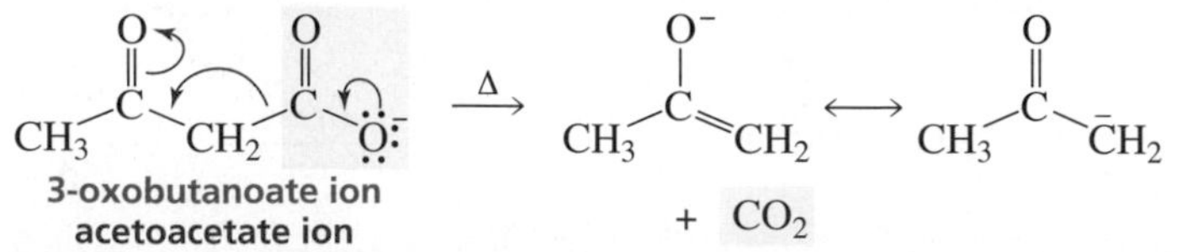

3-oxobutanoate ion
acetoacetate ion

$+ \ CO_2$

Notice the similarity between removal of CO_2 from a 3-oxocarboxylate ion and removal of a proton from an α-carbon. In both reactions, a substituent—CO_2 in one case, H^+ in the other—is removed from an α-carbon and its bonding electrons are delocalized onto an oxygen.

removing a proton from an α-carbon

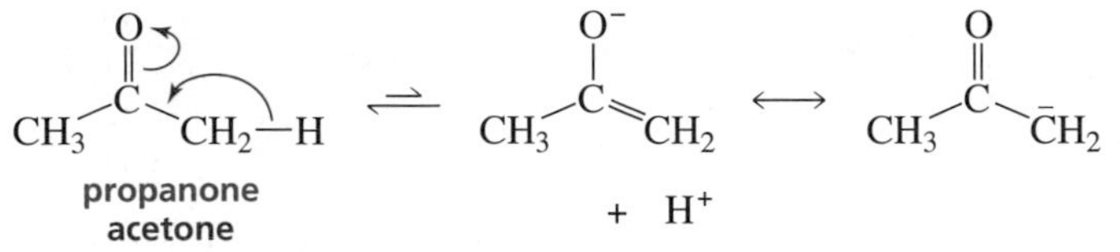

propanone
acetone

$+ H^+$

Decarboxylation is even easier if the reaction is carried out under acidic conditions, because the reaction is catalyzed by an intramolecular transfer of a proton from the carboxyl group to the carbonyl oxygen. The enol that is formed immediately tautomerizes to a ketone.

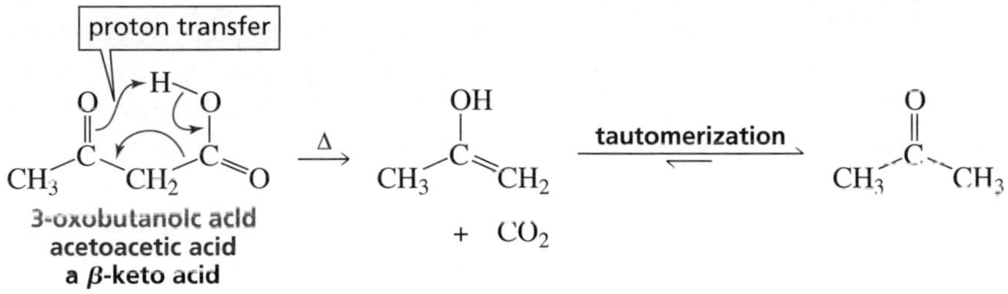

proton transfer

3-oxobutanoic acid
acetoacetic acid
a β-keto acid

$+ CO_2$

tautomerization

We saw in Section 18.1 that it is harder to remove a proton from an α-carbon if the electrons are delocalized onto the carbonyl group of an ester rather than onto the carbonyl group of a ketone. For the same reason, a higher temperature is required to decarboxylate a β-dicarboxylic acid such as malonic acid than to decarboxylate a β-keto acid.

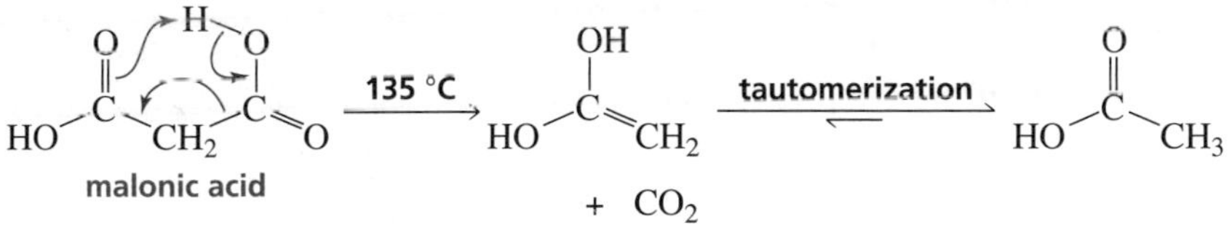

malonic acid

135 °C

$+ CO_2$

tautomerization

In summary, carboxylic acids with a carbonyl group at the 3-position (both β-ketocarboxylic acids and β-dicarboxylic acids) lose CO_2 when they are heated.

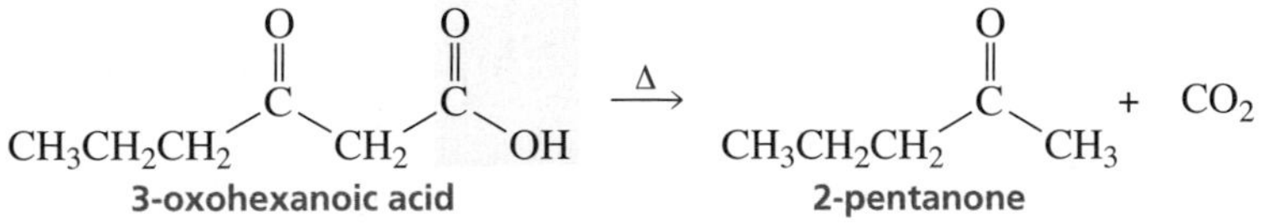

3-oxohexanoic acid

2-pentanone

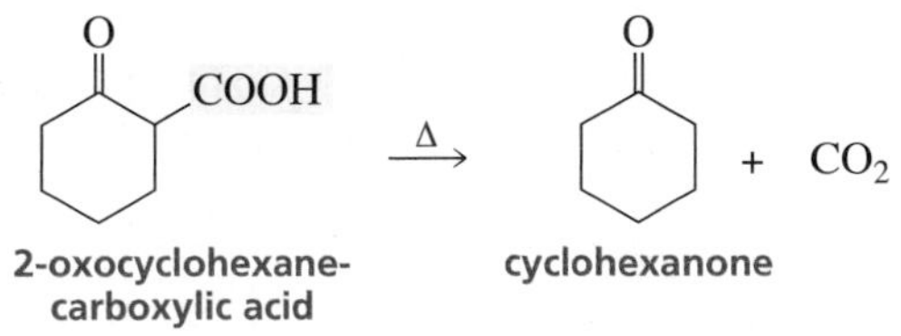

$+ CO_2$

2-oxocyclohexane-
carboxylic acid

cyclohexanone

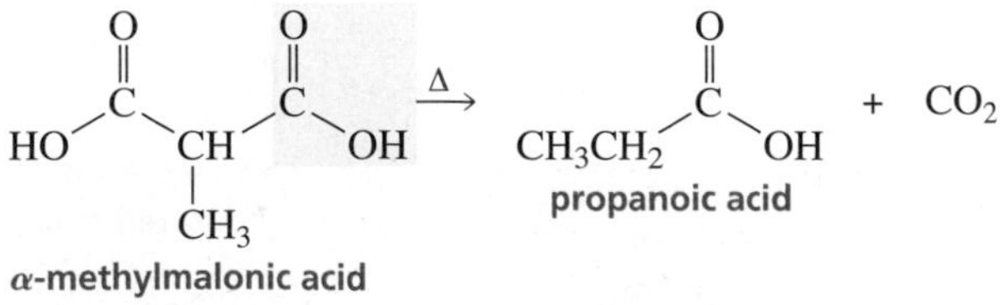

$+ CO_2$

α-methylmalonic acid

propanoic acid

$+ CO_2$

PROBLEM 38◆

Which of the following compounds would be expected to decarboxylate when heated?

A

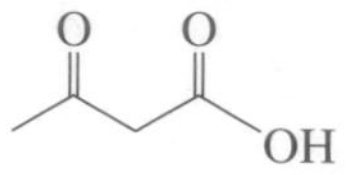

C

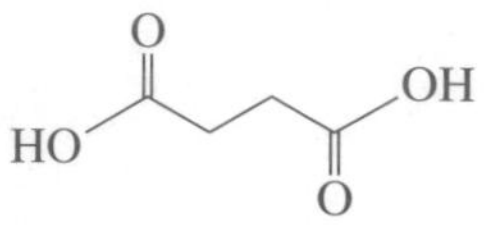

B

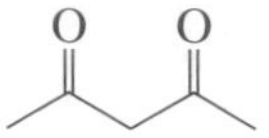

D

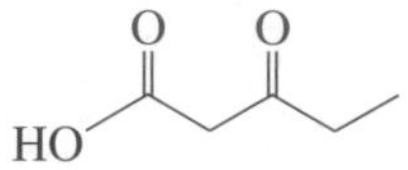

18.19 The Malonic Ester Synthesis: A Way to Synthesize a Carboxylic Acid

A malonic ester synthesis forms a carboxylic acid with two more carbon atoms than the alkyl halide.

A combination of two of the reactions discussed in this chapter—alkylation of an α-carbon and decarboxylation of a β-dicarboxylic acid—can be used to prepare carboxylic acids of any desired chain length. The procedure is called the **malonic ester synthesis** because the starting material for the synthesis is the diester of malonic acid. The first two carbons of the carboxylic acid being synthesized come from malonic ester, and the rest of the carboxylic acid comes from the alkyl halide used in the second step of the reaction.

Student Tutorial:
Malonic ester synthesis

malonic ester synthesis

$$\underset{\substack{\text{diethyl malonate}\\ \text{malonic ester}}}{C_2H_5O-\overset{O}{\overset{\|}{C}}-CH_2-\overset{O}{\overset{\|}{C}}-OC_2H_5} \xrightarrow[\substack{\text{2. RBr}\\ \text{3. HCl, H}_2\text{O, }\Delta}]{\text{1. CH}_3\text{CH}_2\text{O}^-} R-CH_2-\overset{O}{\overset{\|}{C}}-OH$$

from malonic ester

from the alkyl halide

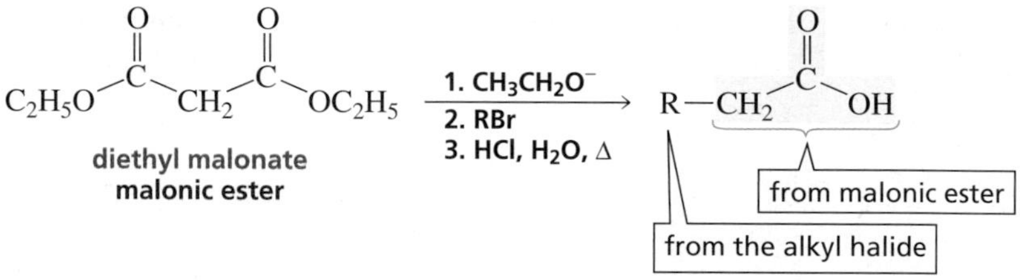

The mechanism for the malonic ester synthesis is shown below.

Mechanism for the malonic ester synthesis

akylation of the α-carbon

removal of a proton from the α-carbon

an α-substituted malonic ester

HCl, H₂O │ Δ ← hydrolysis

$$R-CH_2-\overset{O}{\overset{\|}{C}}-OH \; + \; CO_2 \; \overset{\Delta}{\longleftarrow} \; HO-\overset{O}{\overset{\|}{C}}-\underset{R}{CH}-\overset{O}{\overset{\|}{C}}-OH \; + \; 2\,CH_3CH_2OH$$

decarboxylation

an α-substituted malonic acid

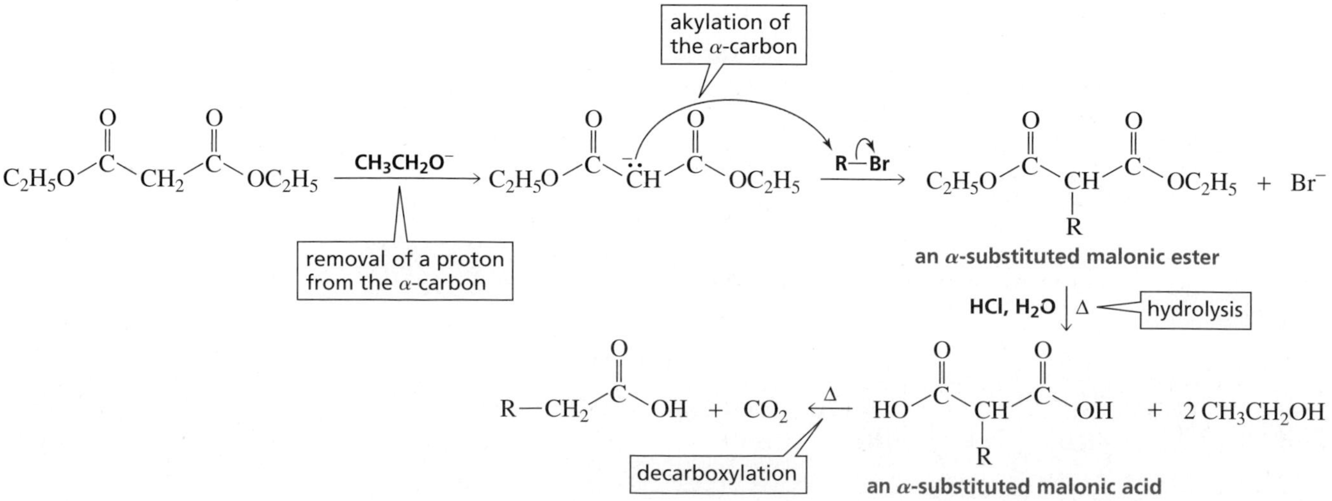

- A proton is easily removed from the α-carbon because it is flanked by two ester groups ($pK_a = 13$).
- The resulting α-carbanion reacts with an alkyl halide, forming an α-substituted malonic ester. Because alkylation is an S_N2 reaction, it works best with primary alkyl halides and methyl halides (Section 8.2).
- Heating the α-substituted malonic ester in an acidic aqueous solution hydrolyzes both ester groups to carboxylic acid groups, forming an α-substituted malonic acid.
- Further heating decarboxylates the 3-oxocarboxylic acid.

Carboxylic acids with two substituents bonded to the α-carbon can be prepared by carrying out two successive α-carbon alkylations.

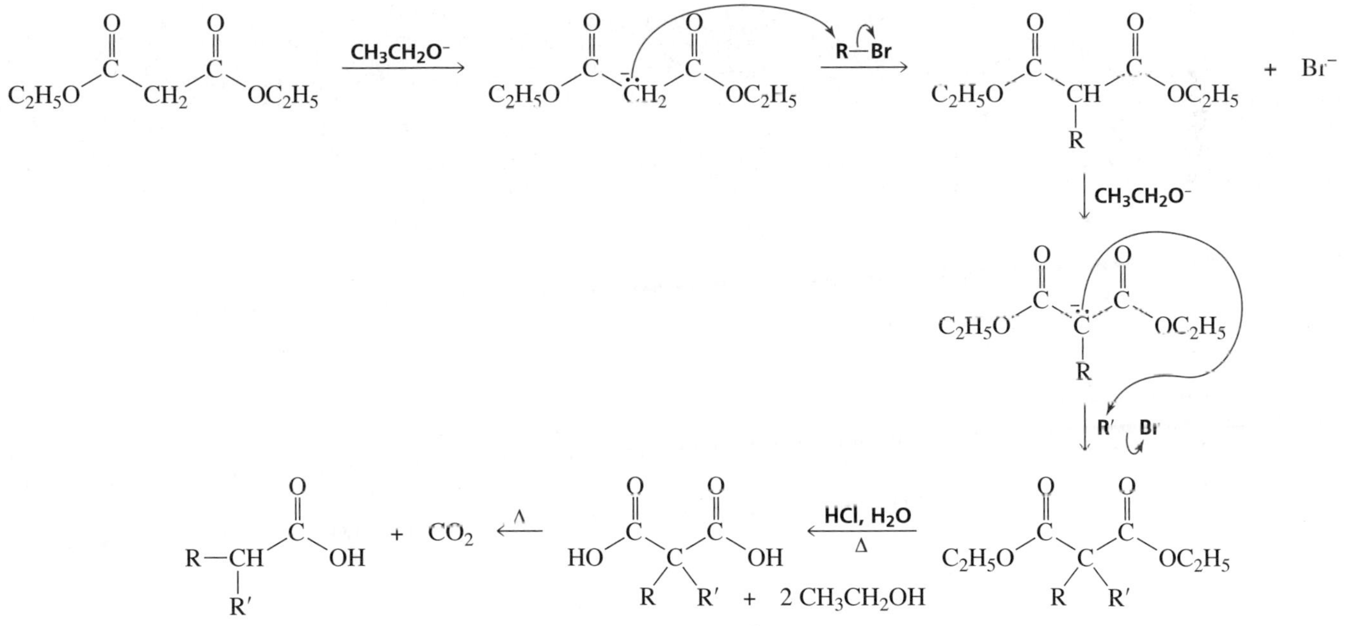

PROBLEM 39◆

What alkyl bromide(s) should be used in the malonic ester synthesis of each of the following carboxylic acids?

a. propanoic acid

b. 2-methylpropanoic acid

c. 3-phenylpropanoic acid

d. 4-methylpentanoic acid

PROBLEM 40

Explain why the following carboxylic acids cannot be prepared by the malonic ester synthesis:

a. 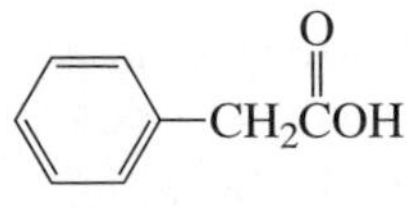 —CH₂COH

b.  CH₂=CHCH₂COH

c. 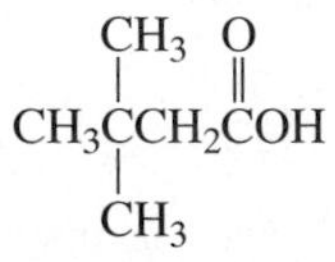 CH₃CCH₂COH with CH₃ groups

18.20 The Acetoacetic Ester Synthesis: A Way to Synthesize a Methyl Ketone

The only difference between the acetoacetic ester synthesis and the malonic ester synthesis is the use of acetoacetic ester rather than malonic ester as the starting material. The difference in starting material causes the product of the **acetoacetic ester synthesis** to be a *methyl ketone* rather than a *carboxylic acid*. The carbonyl group of the methyl ketone and the carbon atoms on either side of it come from acetoacetic ester; the rest of the ketone comes from the alkyl halide used in the second step of the reaction.

An acetoacetic ester synthesis forms a methyl ketone with two more carbon atoms than the alkyl halide.

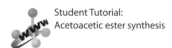
Student Tutorial:
Acetoacetic ester synthesis

acetoacetic ester synthesis

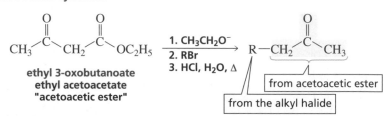

ethyl 3-oxobutanoate
ethyl acetoacetate
"acetoacetic ester"

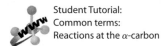
Student Tutorial:
Common terms:
Reactions at the α-carbon

The mechanisms for the acetoacetic ester synthesis and the malonic ester synthesis are similar. The last step in the acetoacetic ester synthesis is the decarboxylation of a substituted acetoacetic acid rather than a substituted malonic acid.

Mechanism for the acetoacetic ester synthesis

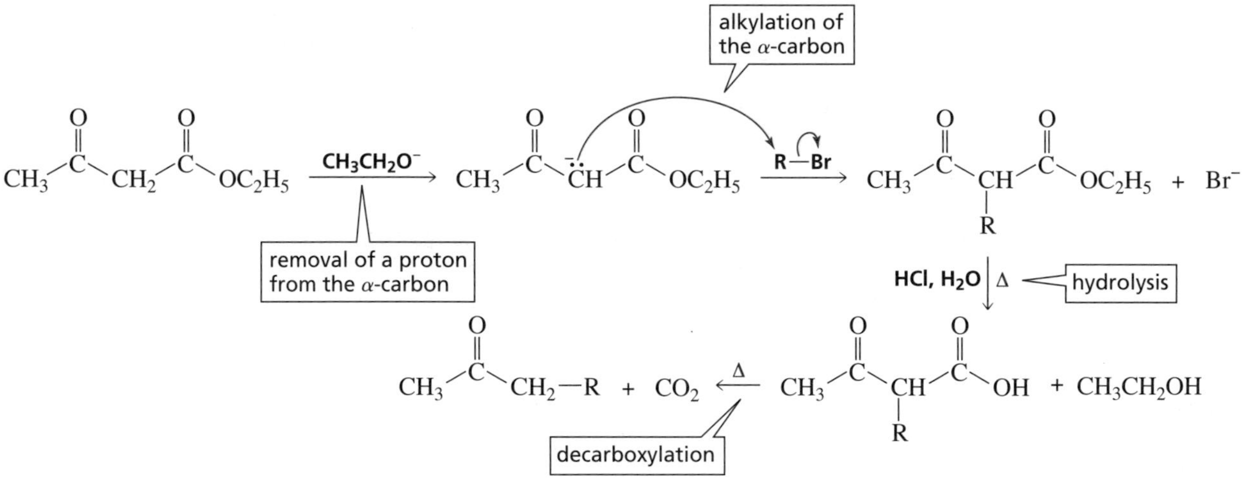

PROBLEM 41 **SOLVED**

Starting with methyl propanoate, how could you prepare 4-methyl-3-heptanone?

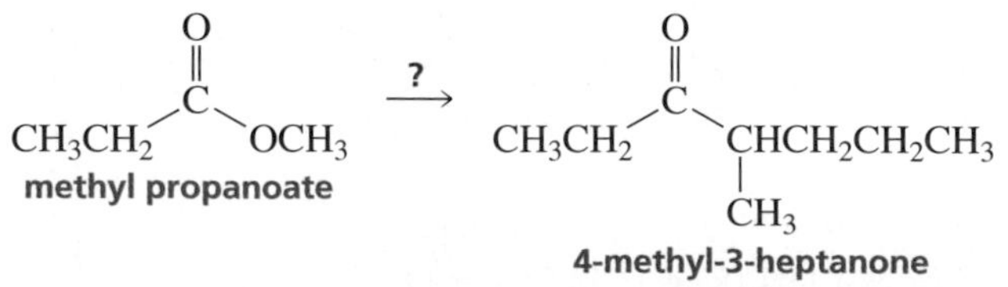

methyl propanoate

4-methyl-3-heptanone

Because the starting material is an ester and the target molecule has more carbons than the starting material, a Claisen condensation appears to be a good way to start this synthesis. The Claisen condensation forms a β-keto ester that can be easily alkylated at the desired carbon because it is flanked by two carbonyl groups. Acid-catalyzed hydrolysis will form a 3-oxocarboxylic acid that will decarboxylate when it is heated.

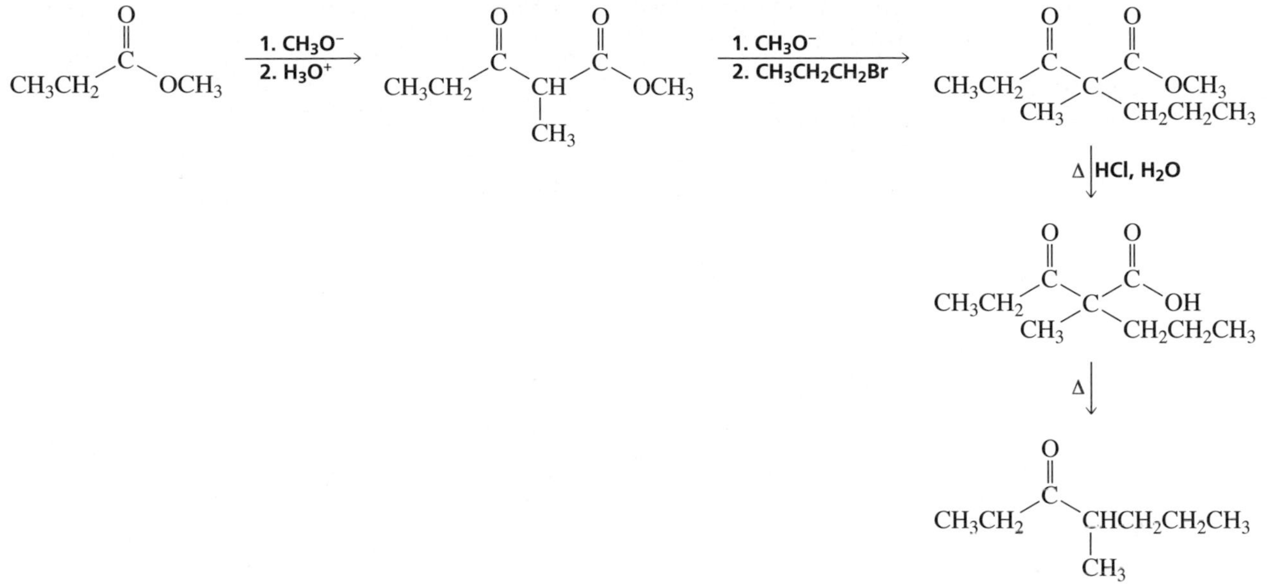

CH$_3$CH$_2$—C(=O)—OCH$_3$ $\xrightarrow[\text{2. H}_3\text{O}^+]{\text{1. CH}_3\text{O}^-}$ CH$_3$CH$_2$—C(=O)—CH(CH$_3$)—C(=O)—OCH$_3$ $\xrightarrow[\text{2. CH}_3\text{CH}_2\text{CH}_2\text{Br}]{\text{1. CH}_3\text{O}^-}$ CH$_3$CH$_2$—C(=O)—C(CH$_3$)(CH$_2$CH$_2$CH$_3$)—C(=O)—OCH$_3$

Δ | HCl, H$_2$O

CH$_3$CH$_2$—C(=O)—C(CH$_3$)(CH$_2$CH$_2$CH$_3$)—C(=O)—OH

$\Delta \downarrow$

CH$_3$CH$_2$—C(=O)—CH(CH$_3$)CH$_2$CH$_2$CH$_3$

PROBLEM 42◆

What alkyl bromide should be used in the acetoacetic ester synthesis of each of the following methyl ketones?

a. 2-pentanone **b.** 2-octanone **c.** 4-phenyl-2-butanone

18.21 Designing a Synthesis VII: Making New Carbon–Carbon Bonds

When you are planning the synthesis of a compound that requires the formation of a new carbon–carbon bond, first locate the new bond that must be made. For example, in the synthesis of the following β-diketone, the new bond is the one that makes the second five-membered ring:

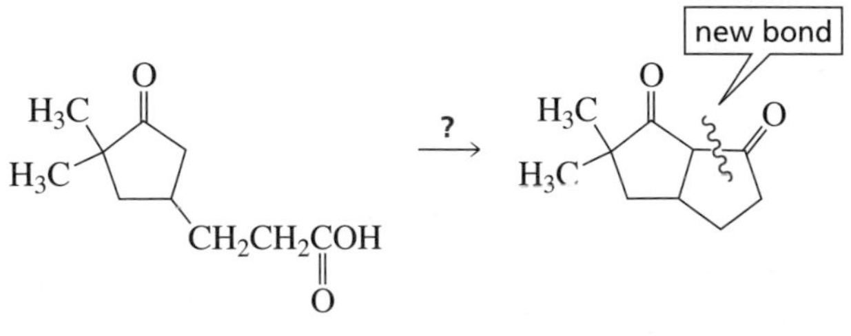

Next, determine which of the atoms that form the bond should be the nucleophile and which should be the electrophile. In this case, it is easy to choose between the two possibilities because you know that a carbonyl carbon is an electrophile.

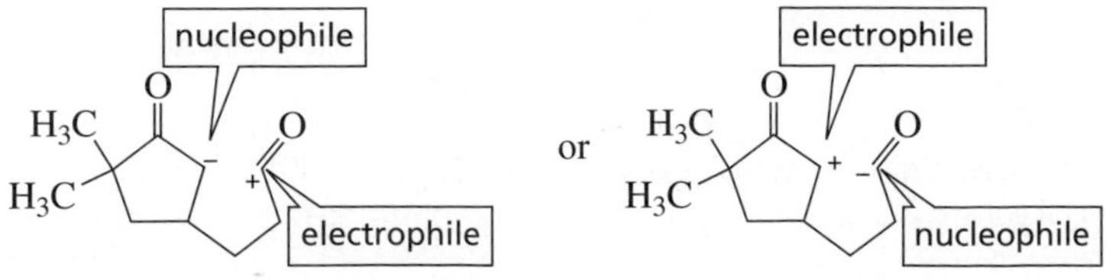

Now you need to determine what compound you could use that would give you the desired electrophilic and nucleophilic sites. If you are told what the starting material is, use it as a clue to arrive at the desired compound. For example, an ester carbonyl group would be a good electrophile for this synthesis because it has a group that would be eliminated. Moreover, the α-hydrogens of the ketone are more acidic than the α-hydrogens of the ester so the desired nucleophile would be easy to obtain. The ester can easily be prepared from the carboxylic acid starting material.

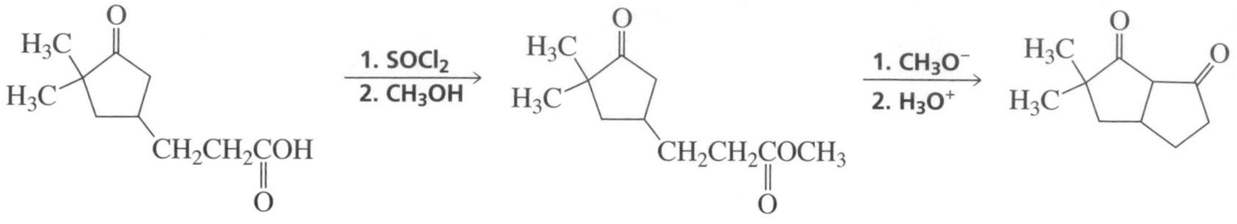

In the next synthesis, two new carbon–carbon bonds must be formed.

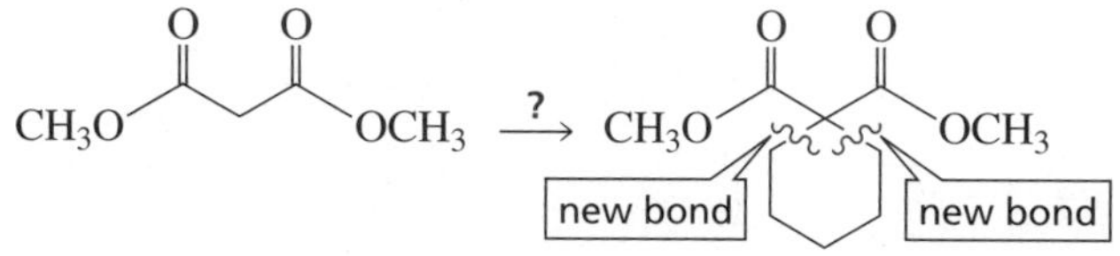

After identifying the electrophilic and nucleophilic sites, you can see that two successive alkylations of a diester of malonic acid, using 1,5-dibromopentane for the alkyl halide, will produce the desired compound.

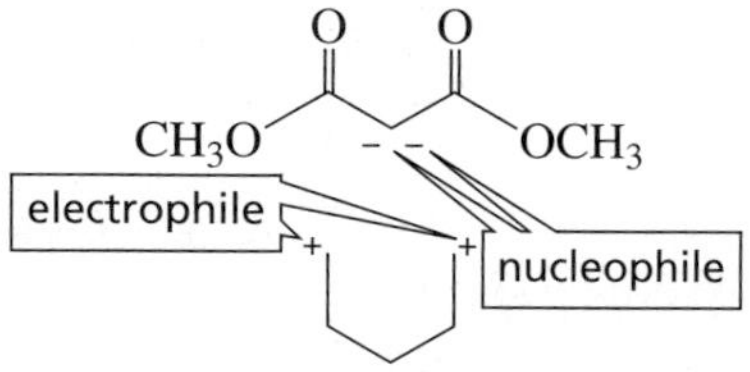

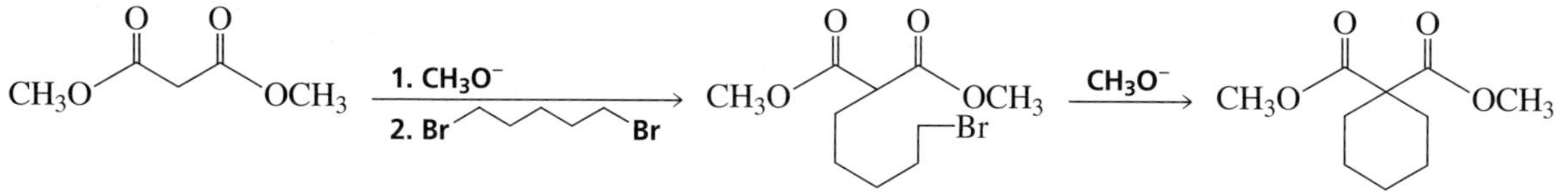

In planning the following synthesis, the diester given as the starting material suggests that you should use a Dieckmann condensation to obtain the cyclic compound:

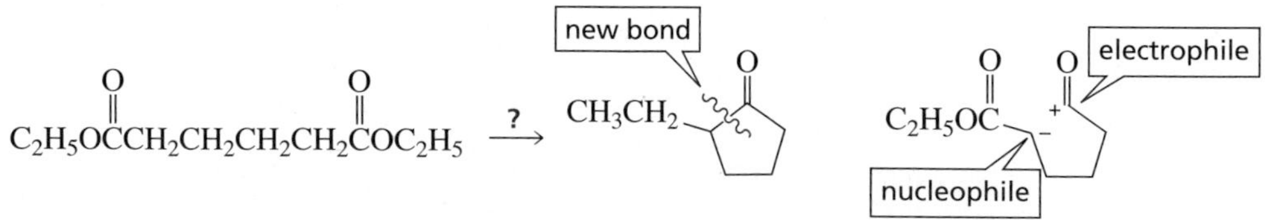

After the cyclopentanone ring is formed from a Dieckmann condensation of the starting material, alkylation of the α-carbon followed by hydrolysis of the β-keto ester and decarboxylation forms the desired product.

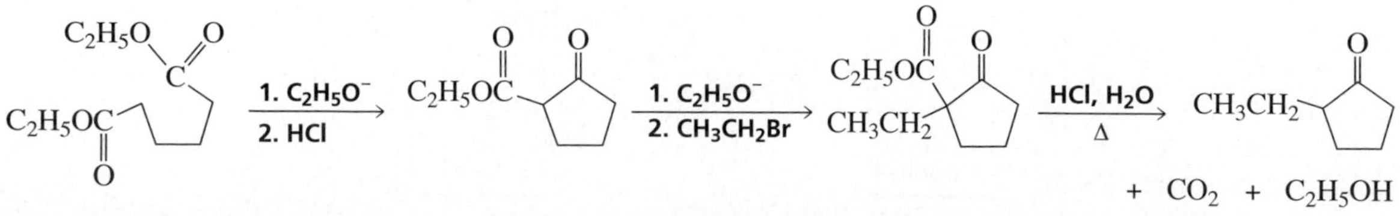

PROBLEM 43

Design a synthesis for each of the following compounds using the given starting material:

a.

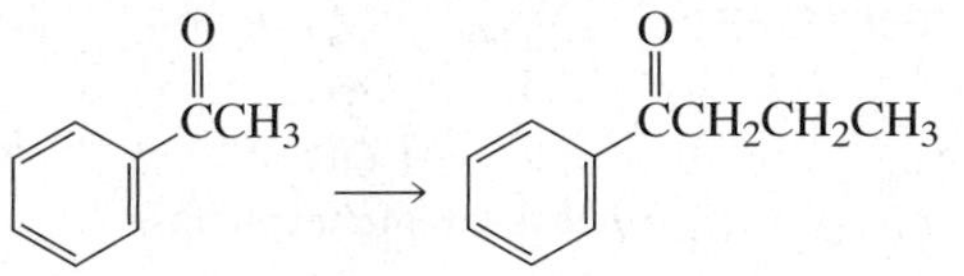

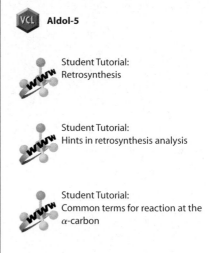

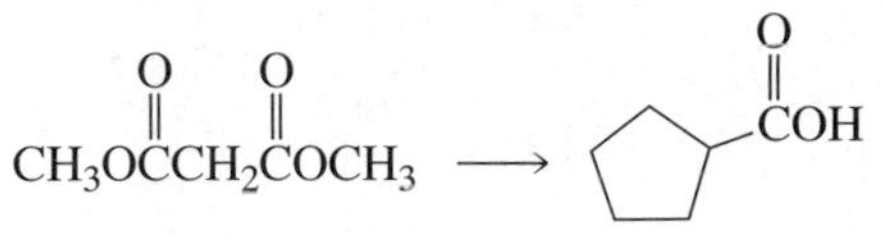

VCL **Aldol-5**

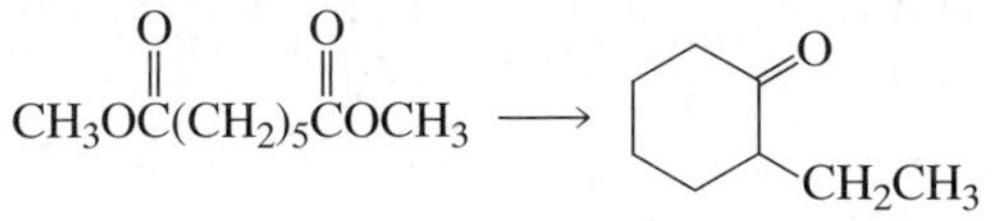

Student Tutorial:
Retrosynthesis

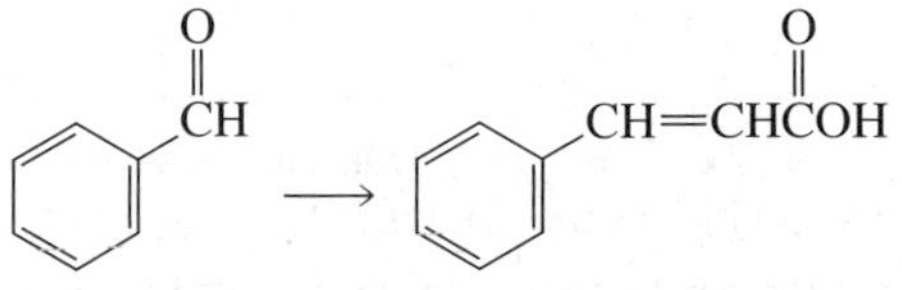

Student Tutorial:
Hints in retrosynthesis analysis

Student Tutorial:
Common terms for reaction at the α-carbon

b. $CH_3OC(CH_2)_5COCH_3 \longrightarrow$

c.

d. $CH_3OCCH_2COCH_3 \longrightarrow$

18.22 Reactions at the α-Carbon in Biological Systems

Many reactions that occur in biological systems involve reactions at the α-carbon—the kinds of reactions you have studied in this chapter. We will now look at a few examples.

A Biological Aldol Addition

Glucose, the most abundant sugar found in nature, is synthesized in biological systems from two molecules of pyruvate. The series of reactions that convert two molecules of pyruvate into glucose is called **gluconeogenesis**. The reverse process—the breakdown of glucose into two molecules of pyruvate—is called **glycolysis** (Section 25.7).

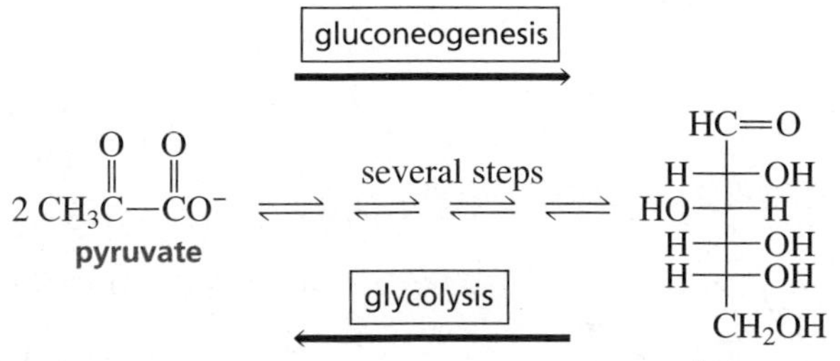

$$2 \ CH_3C-CO^- \rightleftharpoons \rightleftharpoons \rightleftharpoons \rightleftharpoons$$

pyruvate

gluconeogenesis

several steps

glycolysis

HC=O
H——OH
HO——H
H——OH
H——OH
CH$_2$OH

glucose

Because glucose has twice as many carbons as pyruvate, you should not be surprised to learn that one of the steps in the biosynthesis of glucose is an aldol addition. An enzyme called aldolase catalyzes an aldol addition between dihydroxyacetone phosphate and glyceraldehyde-3-phosphate. The product is fructose-1,6-diphosphate, which is subsequently converted to glucose. The mechanism of this reaction will be discussed in Section 23.9.

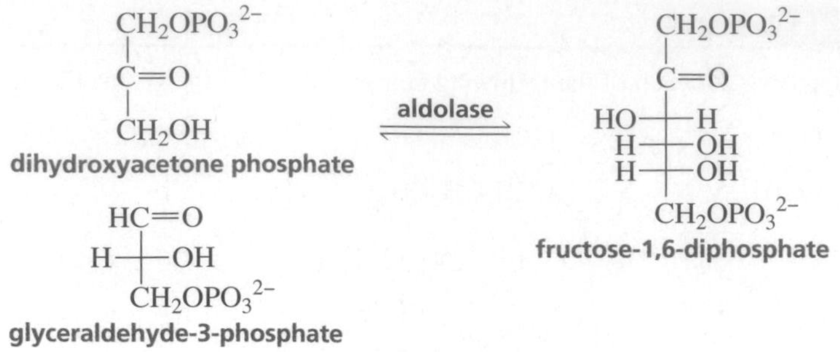

$$CH_2OPO_3^{2-}$$
$$C=O$$
$$CH_2OH$$
dihydroxyacetone phosphate

$$HC=O$$
$$H \rule[0.5ex]{2em}{0.1ex} OH$$
$$CH_2OPO_3^{2-}$$
glyceraldehyde-3-phosphate

$\xrightleftharpoons{\text{aldolase}}$

$$CH_2OPO_3^{2-}$$
$$C=O$$
$$HO \rule[0.5ex]{2em}{0.1ex} H$$
$$H \rule[0.5ex]{2em}{0.1ex} OH$$
$$H \rule[0.5ex]{2em}{0.1ex} OH$$
$$CH_2OPO_3^{2-}$$
fructose-1,6-diphosphate

PROBLEM 44

Propose a mechanism for the formation of fructose-1,6-diphosphate from dihydroxyacetone phosphate and glyceraldehyde-3-phosphate, using HO^- as the catalyst.

A Biological Aldol Condensation

Collagen is the most abundant protein in mammals, amounting to about one-fourth of the total protein. It is the major fibrous component of bone, teeth, skin, cartilage, and tendons. Individual collagen molecules—called tropocollagen—can be isolated only from tissues of young animals. As animals age, the individual molecules become cross-linked, which is why meat from older animals is tougher than meat from younger ones. Collagen cross-linking is an example of an aldol condensation.

Before collagen molecules can cross-link, the primary amino groups of the lysine residues of collagen must be converted to aldehyde groups. The enzyme that catalyzes this reaction is called lysyl oxidase. An aldol condensation between two aldehyde residues results in a cross-linked protein.

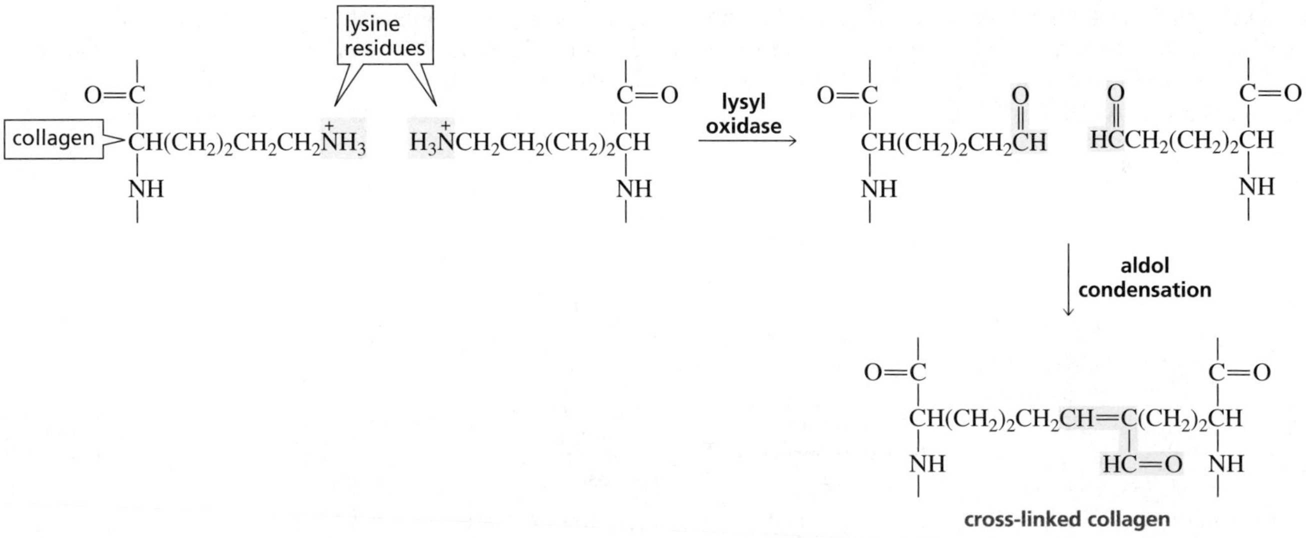

cross-linked collagen

A Biological Claisen Condensation

Fatty acids are long-chain, unbranched carboxylic acids (Sections 16.14 and 26.1). Most naturally occurring fatty acids contain an even number of carbons because they are synthesized from acetate, which has two carbons.

In Section 16.22, you saw that carboxylic acids can be activated in biological systems by being converted to thioesters of coenzyme A.

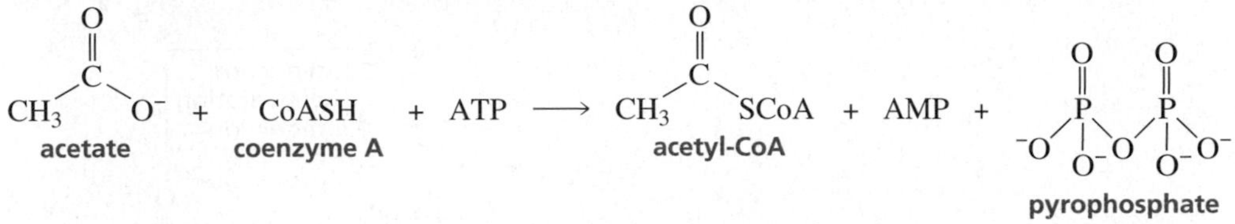

One of the necessary reactants for fatty acid synthesis is malonyl-CoA, which is obtained by carboxylation of acetyl-CoA. The mechanism for this reaction will be discussed in Section 24.5.

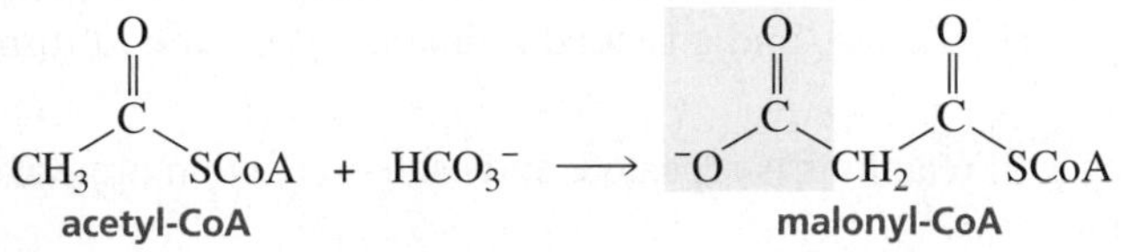

Before fatty acid synthesis can occur, the acyl groups of acetyl-CoA and malonyl-CoA are transferred to other thiols by means of a transesterification reaction.

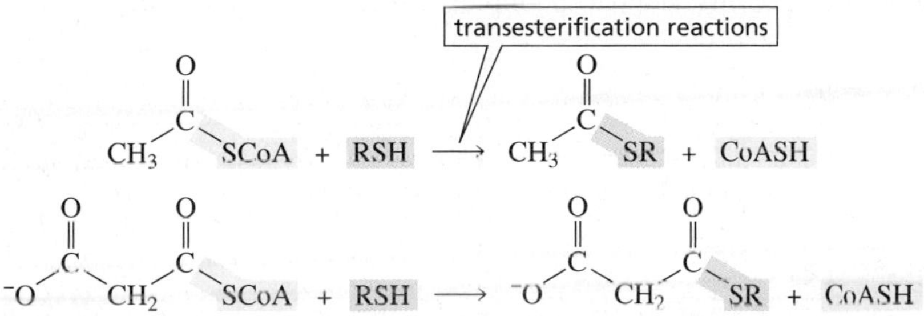

A molecule of acetyl thioester and a molecule of malonyl thioester are the reactants for the first round in the biosynthesis of a fatty acid.

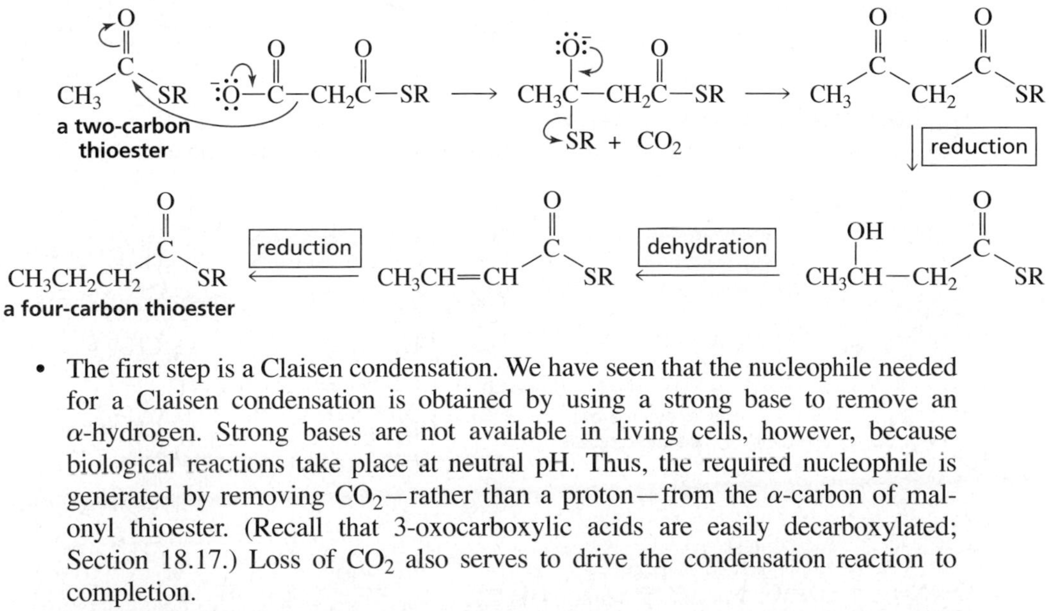

- The first step is a Claisen condensation. We have seen that the nucleophile needed for a Claisen condensation is obtained by using a strong base to remove an α-hydrogen. Strong bases are not available in living cells, however, because biological reactions take place at neutral pH. Thus, the required nucleophile is generated by removing CO_2—rather than a proton—from the α-carbon of malonyl thioester. (Recall that 3-oxocarboxylic acids are easily decarboxylated; Section 18.17.) Loss of CO_2 also serves to drive the condensation reaction to completion.

- The product of the condensation reaction undergoes a reduction, a dehydration, and a second reduction to give a four-carbon thioester.

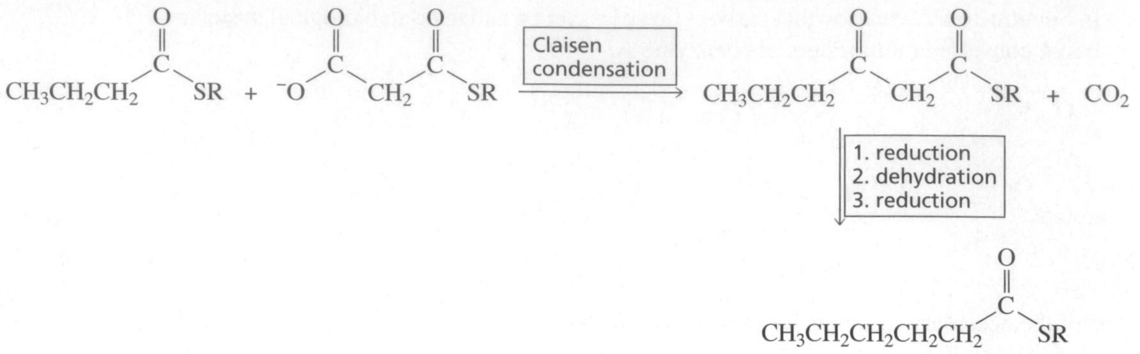

- The four-carbon thioester and a molecule of malonyl thioester are the reactants for the second round. Again, the product of the condensation reaction undergoes a reduction, a dehydration, and a second reduction, this time to form a six-carbon thioester.

- The sequence of reactions is repeated, and each time two more carbons are added to the chain.

This mechanism explains why naturally occurring fatty acids are unbranched and generally contain an even number of carbons. Once a thioester with the appropriate number of carbon is obtained, it undergoes a transesterification reaction with glycerol in order to form fats, oils, and phospholipids (Sections 26.3 and 26.4).

PROBLEM 45◆

Palmitic acid is a straight-chain saturated 16-carbon fatty acid. How many moles of malonyl-CoA are required for the synthesis of one mole of palmitic acid?

PROBLEM 46◆

a. If the biosynthesis of palmitic acid were carried out with CD_3COSR and nondeuterated malonyl thioester, how many deuterium atoms would be incorporated into palmitic acid?

b. If the biosynthesis of palmitic acid were carried out with $^-OOCCD_2COSR$ and non-deuterated acetyl thioester, how many deuterium atoms would be incorporated into palmitic acid?

A Biological Decarboxylation

An example of a decarboxylation that occurs in biological systems is the decarboxylation of acetoacetate.

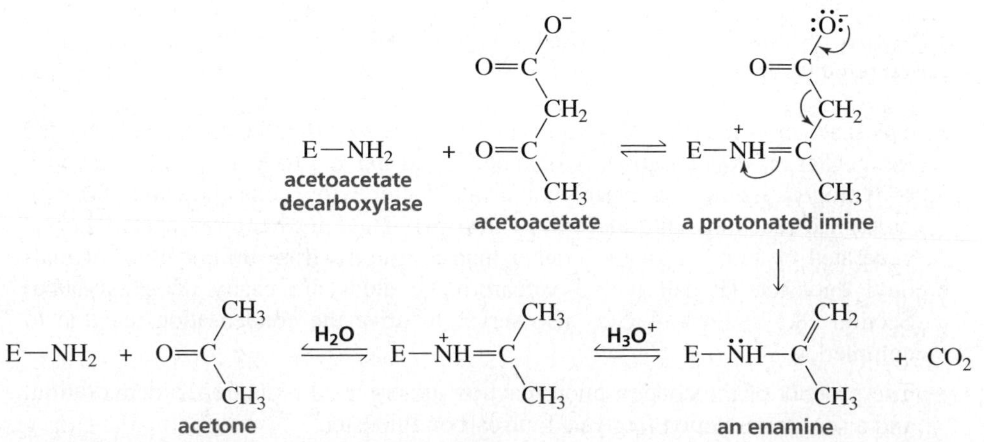

- Acetoacetate decarboxylase, the enzyme that catalyzes the reaction, forms an imine with acetoacetate.
- Under physiological conditions the imine is protonated and therefore readily accepts the pair of electrons left behind when the substrate loses CO_2.
- Decarboxylation forms an enamine.
- Hydrolysis of the enamine produces the decarboxylated product (acetone) and regenerates the enzyme (Section 17.8).

PROBLEM 47

When the enzymatic decarboxylation of acetoacetate is carried out in $H_2^{18}O$, the acetone that is formed contains ^{18}O. What does this tell you about the mechanism of the reaction?

SUMMARY

A hydrogen bonded to an α-carbon of an aldehyde, ketone, ester, or *N,N*-disubstituted amide is sufficiently acidic to be removed by a strong base because the base that is formed when the proton is removed is stabilized by delocalization of its negative change onto an oxygen. A **carbon acid** is a compound with a relatively acidic hydrogen bonded to an sp^3 hybridized carbon. Aldehydes and ketones ($pK_a \sim 16$ to 20) are more acidic than esters ($pK_a \sim 25$). **β-Diketones** ($pK_a \sim 9$) and β-keto esters ($pK_a = 11$) are even more acidic.

Keto–enol interconversion can be catalyzed by acids or by bases. Generally, the **keto tautomer** is more stable. When an **α-substitution reaction** takes place under acidic conditions, an enol reacts with an electrophile; when the reaction takes place under basic conditions, an enolate ion reacts with an electrophile. Whether C or O reacts with the electrophile depends on the electrophile and the reaction conditions.

Aldehydes and ketones react with Br_2, Cl_2, or I_2: under acidic conditions, a halogen replaces one of the α-hydrogens of the carbonyl compound; under basic conditions, halogens replace all the α-hydrogens. The **HVZ** reaction brominates the α-carbon of a carboxylic acid. When the α-position is halogenated, the α-carbon reacts with nucleophiles.

LDA is used to form an enolate ion in reactions that require the carbonyl compound to be completely converted to the enolate ion before it reacts with an electrophile. If the electrophile is an alkyl halide, the enolate ion is alkylated. The less substituted α-carbon is alkylated when the reaction is under kinetic control; the more substituted α-carbon is alkylated when the reaction is under thermodynamic control. Aldehydes and ketones can be alkylated or acylated via an enamine intermediate. Enolate ions of

β-diketones, β-diesters, β-keto esters, and β-keto nitriles undergo **Michael reactions** with α,β-unsaturated carbonyl compounds. Michael reactions form 1,5-dicarbonyl compounds.

In an **aldol addition**, the enolate ion of an aldehyde or a ketone reacts with the carbonyl carbon of a second molecule of aldehyde or ketone, forming a β-hydroxyaldehyde or a β-hydroxyketone. The new C—C bond forms between the α-carbon of one molecule and the carbon that formerly was the carbonyl carbon of the other molecule. The product of an aldol addition can be dehydrated to give an **aldol condensation** product. In a **Claisen condensation**, the enolate ion of an ester reacts with a second molecule of ester, eliminating an ^-OR group to form a β-keto ester. A **Dieckmann condensation** is an intramolecular Claisen condensation. A **Robinson annulation** is a ring-forming reaction in which a Michael reaction and an intramolecular aldol addition occur sequentially.

Carboxylic acids with a carbonyl group at the 3-position **decarboxylate** when they are heated. Carboxylic acids can be prepared by a **malonic ester synthesis**; the α-carbon of the diester is alkylated and the α-substituted malonic ester undergoes acid-catalyzed hydrolysis and decarboxylation; the resulting carboxylic acid has two more carbons than the alkyl halide. Similarly, methyl ketones can be prepared by an **acetoacetic ester synthesis**; the carbonyl group and the carbon on either side of it come from acetoacetic ester, and the rest of the methyl ketone comes from the alkyl halide.

When planning the synthesis of a compound that requires the formation of a new carbon–carbon bond, first locate the new bond that must be made, and then determine which of the atoms that form the bond should be the nucleophile and which should be the electrophile.

SUMMARY OF REACTIONS

1. Keto-enol interconversion (Section 18.3)

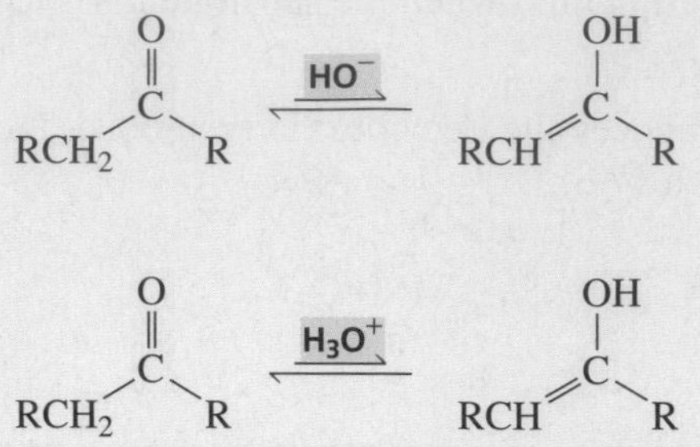

$$\underset{RCH_2}{\overset{O}{\underset{}{\parallel}}}\!\!\!C\!\!-\!\!R \quad \underset{HO^-}{\rightleftharpoons} \quad \underset{RCH}{\overset{OH}{\underset{}{\vert}}}\!\!\!C\!\!=\!\!R$$

$$\underset{RCH_2}{\overset{O}{\underset{}{\parallel}}}\!\!\!C\!\!-\!\!R \quad \underset{H_3O^+}{\rightleftharpoons} \quad \underset{RCH}{\overset{OH}{\underset{}{\vert}}}\!\!\!C\!\!=\!\!R$$

2. Halogenation of the α-carbon of aldehydes and ketones (Section 18.5)

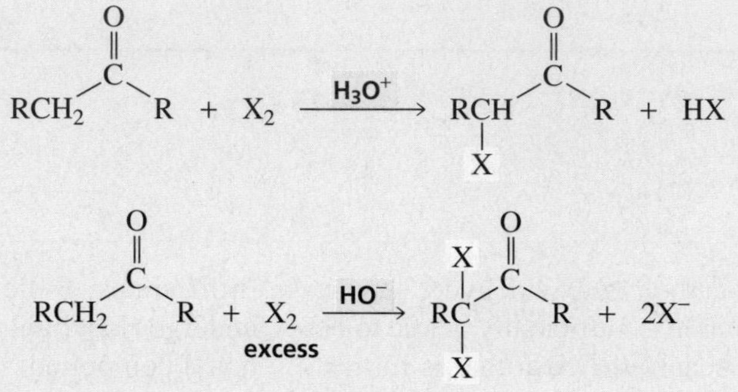

$$\underset{RCH_2}{\overset{O}{\parallel}}C{-}R \;+\; X_2 \;\xrightarrow{H_3O^+}\; \underset{\underset{X}{\vert}}{\underset{RCH}{\overset{O}{\parallel}}}C{-}R \;+\; HX$$

$$\underset{RCH_2}{\overset{O}{\parallel}}C{-}R \;+\; \underset{\textbf{excess}}{X_2} \;\xrightarrow{HO^-}\; \underset{\underset{X}{\vert}}{\overset{\overset{X}{\vert}}{\underset{RC}{\overset{O}{\parallel}}C{-}R}} \;+\; 2X^-$$

$$X_2 = Cl_2,\ Br_2,\ or\ I_2$$

3. Halogenation of the α-carbon of carboxylic acids: the Hell–Volhard–Zelinski reaction (Section 18.6)

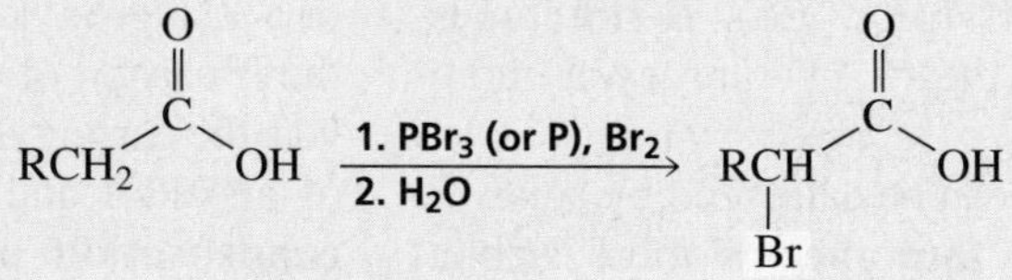

$$\underset{RCH_2}{\overset{O}{\parallel}}C{-}OH \;\xrightarrow[\textbf{2. H}_2\textbf{O}]{\textbf{1. PBr}_3\textbf{ (or P), Br}_2}\; \underset{\underset{Br}{\vert}}{\underset{RCH}{\overset{O}{\parallel}}}C{-}OH$$

4. When an α-hydrogen is removed, the α-carbon is a nucleophile and reacts with an electrophile (Section 18.7).

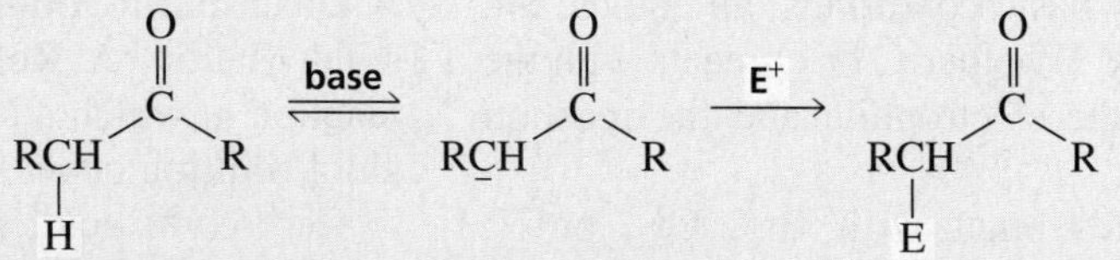

$$\underset{\underset{H}{\vert}}{\underset{RCH}{\overset{O}{\parallel}}}C{-}R \;\underset{\textbf{base}}{\rightleftharpoons}\; \underset{RCH}{\overset{O}{\parallel}}C{-}R \;\xrightarrow{E^+}\; \underset{\underset{E}{\vert}}{\underset{RCH}{\overset{O}{\parallel}}}C{-}R$$

5. When the α-carbon is halogenated, the α-carbon is an electrophile and reacts with a nucleophile (Section 18.7).

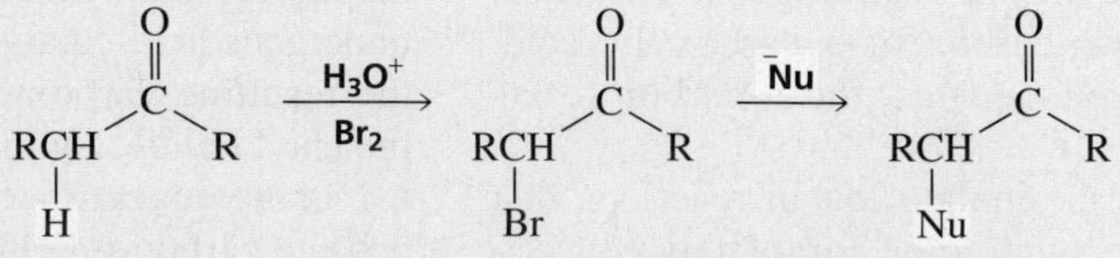

$$\underset{\underset{H}{\vert}}{\underset{RCH}{\overset{O}{\parallel}}}C{-}R \;\xrightarrow[\textbf{Br}_2]{\textbf{H}_3\textbf{O}^+}\; \underset{\underset{Br}{\vert}}{\underset{RCH}{\overset{O}{\parallel}}}C{-}R \;\xrightarrow{^-Nu}\; \underset{\underset{Nu}{\vert}}{\underset{RCH}{\overset{O}{\parallel}}}C{-}R$$

6. Compounds with halogenated α-carbons can form α,β-unsaturated carbonyl compounds (Section 18.7).

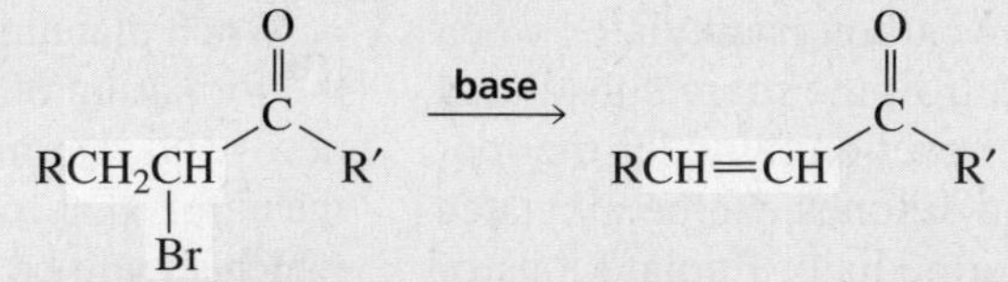

$$\underset{\underset{Br}{\vert}}{\underset{RCH_2CH}{\overset{O}{\parallel}}}C{-}R' \;\xrightarrow{\textbf{base}}\; \underset{RCH=CH}{\overset{O}{\parallel}}C{-}R'$$

7. Alkylation of the α-carbon of carbonyl compounds (Section 18.9)

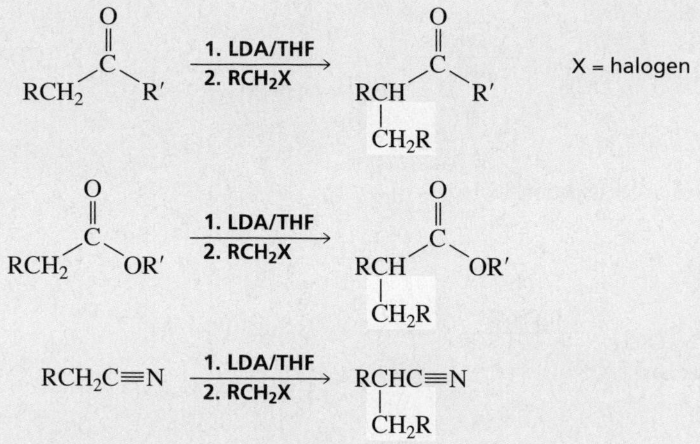

$$
\underset{RCH_2}{\overset{\overset{\displaystyle O}{\parallel}}{C}} R' \quad \xrightarrow[\text{2. RCH}_2X]{\text{1. LDA/THF}} \quad \underset{\underset{CH_2R}{\overset{|}{RCH}}}{\overset{\overset{\displaystyle O}{\parallel}}{C}} R' \qquad X = \text{halogen}
$$

$$
\underset{RCH_2}{\overset{\overset{\displaystyle O}{\parallel}}{C}} OR' \quad \xrightarrow[\text{2. RCH}_2X]{\text{1. LDA/THF}} \quad \underset{\underset{CH_2R}{\overset{|}{RCH}}}{\overset{\overset{\displaystyle O}{\parallel}}{C}} OR'
$$

$$
RCH_2C{\equiv}N \quad \xrightarrow[\text{2. RCH}_2X]{\text{1. LDA/THF}} \quad \underset{\overset{|}{CH_2R}}{RCHC{\equiv}N}
$$

8. Alkylation and acylation of the α-carbon of aldehydes and ketones by means of an enamine intermediate (Sections 18.10 and 18.11)

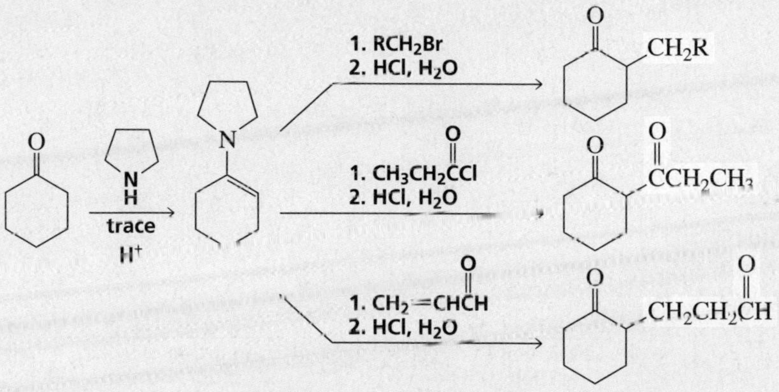

9. Michael reaction: attack of an enolate on an α,β-unsaturated carbonyl compound (Section 18.10)

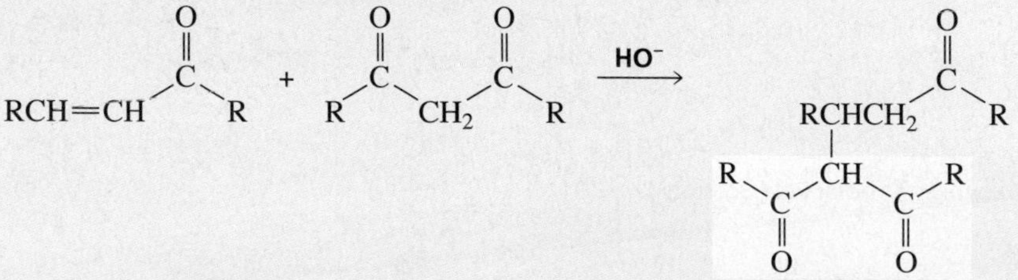

10. Aldol addition of two aldehydes, two ketones, or an aldehyde and a ketone (Section 18.12)

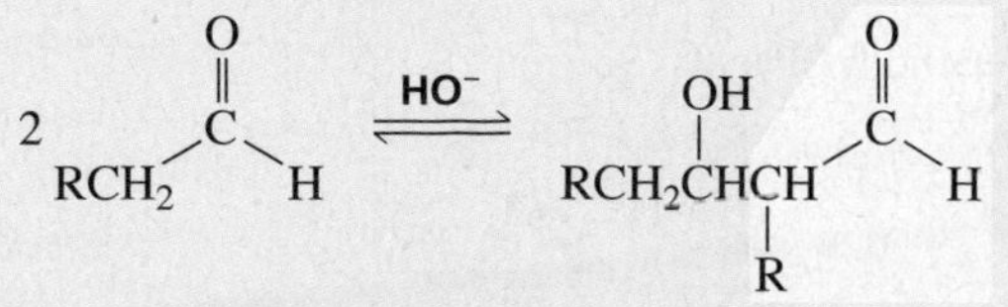

$$
2 \underset{RCH_2}{\overset{\overset{\displaystyle O}{\parallel}}{C}} H \quad \overset{HO^-}{\rightleftharpoons} \quad \underset{\underset{R}{\overset{|}{RCH_2CHCH}}}{\overset{OH}{}} \overset{\overset{\displaystyle O}{\parallel}}{C} H
$$

11. Aldol condensation: dehydration of the product of an aldol addition (Section 18.13)

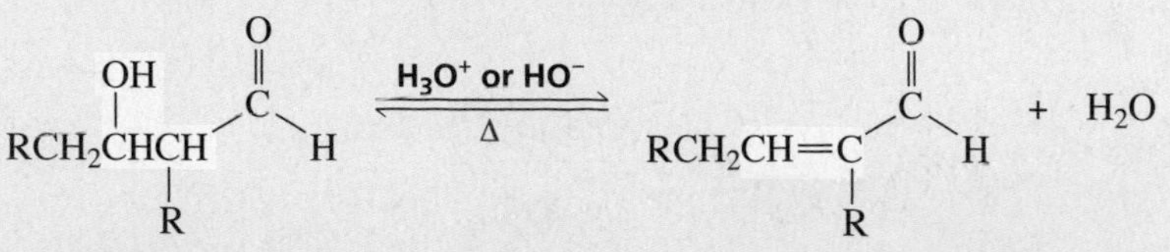

$$
\underset{\overset{|}{R}}{\overset{OH}{RCH_2CHCH}} \overset{\overset{\displaystyle O}{\parallel}}{C} H \quad \underset{\Delta}{\overset{H_3O^+ \text{ or } HO^-}{\rightleftharpoons}} \quad \underset{\overset{|}{R}}{RCH_2CH{=}C} \overset{\overset{\displaystyle O}{\parallel}}{C} H \quad + \quad H_2O
$$

12. Claisen condensation of two esters (Sections 18.15 and 18.16)

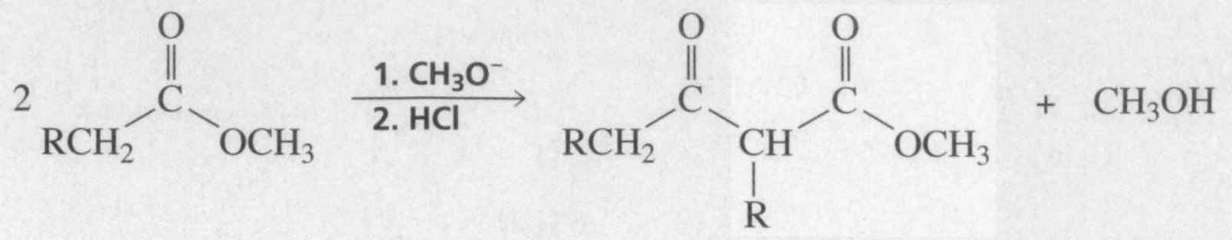

13. Condensation of a ketone and an ester (Section 18.16)

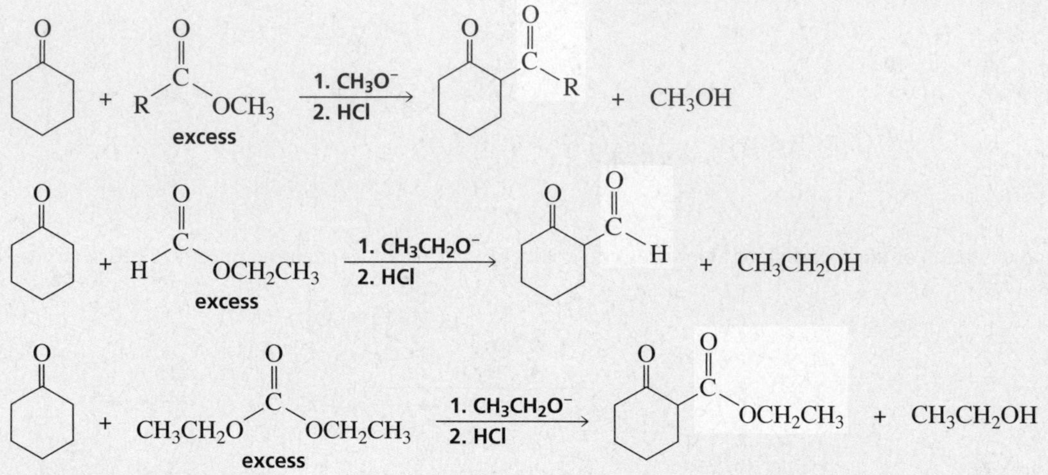

14. Robinson annulation (Section 18.17)

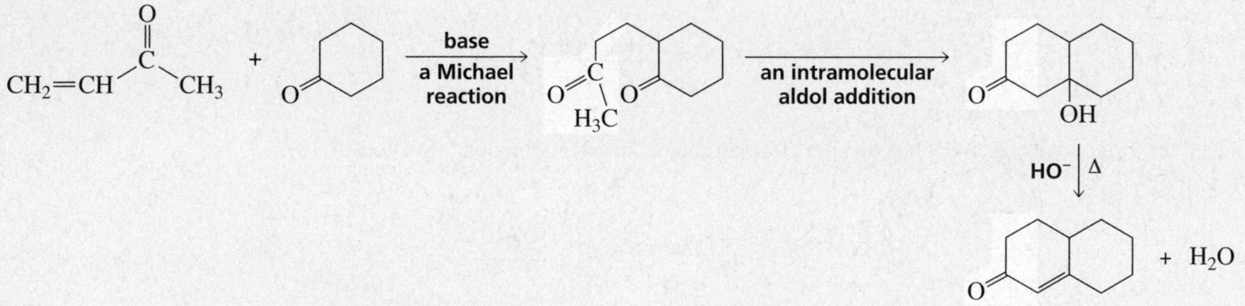

15. Decarboxylation of 3-oxocarboxylic acids (Section 18.18)

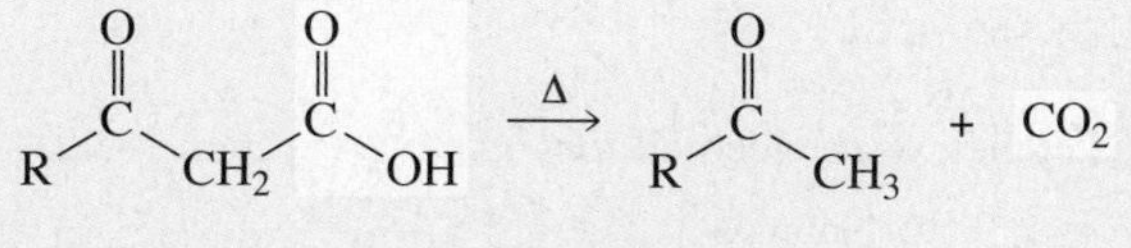

16. Malonic ester synthesis: preparation of carboxylic acids (Section 18.19)

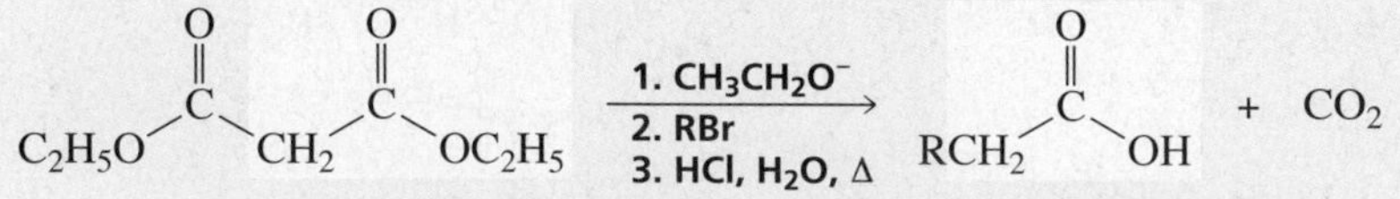

17. Acetoacetic ester synthesis: preparation of methyl ketones (Section 18.20)

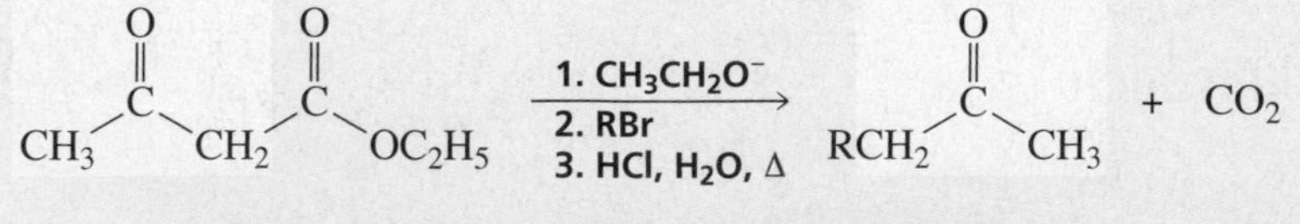

KEY TERMS

acetoacetic ester synthesis (p. 888)
aldol addition (p. 871)
aldol condensation (p. 873)
ambident nucleophile (p. 858)
annulation reaction (p. 883)
α-carbon (p. 850)
carbon acid (p. 852)
Claisen condensation (p. 876)
condensation reaction (p. 873)
crossed aldol addition (p. 875)
decarboxylation (p. 884)
Dieckmann condensation (p. 880)

β-diketone (p. 853)
enolization (p. 856)
enol tautomer (p. 855)
gluconeogenesis (p. 891)
glycolysis (p. 891)
haloform reaction (p. 860)
Hell–Volhard–Zelinski (HVZ)
   reaction (p. 861)
α-hydrogen (p. 850)
keto–enol interconversion (p. 856)
keto–enol tautomerization (p. 856)
β-keto ester (p. 853)

keto tautomer (p. 855)
Kolbe–Schmitt carboxylation
   reaction (p. 865)
malonic ester
   synthesis (p. 886)
Michael reaction (p. 869)
mixed aldol addition (p. 875)
mixed Claisen condensation (p. 878)
Robinson annulation (p. 883)
α-substitution reaction (p. 857)
Stork enamine reaction (p. 870)
tautomers (p. 855)

PROBLEMS

48. Write a structure for each of the following:
  a. ethyl acetoacetate
  b. α-methylmalonic acid
  c. a β-keto ester
  d. the enol tautomer of cyclopentanone
  e. the carboxylic acid obtained from the malonic ester synthesis when the alkyl halide is propyl bromide

49. Give the products of the following reactions:
  a. diethyl heptanedioate: (1) sodium ethoxide; (2) HCl
  b. pentanoic acid + PBr$_3$ + Br$_2$, followed by hydrolysis
  c. acetone + ethyl acetate: (1) sodium ethoxide; (2) HCl
  d. diethyl 2-ethylhexanedioate: (1) sodium ethoxide; (2) HCl
  e. diethyl malonate: (1) sodium ethoxide; (2) isobutyl bromide; (3) HCl, H$_2$O + Δ
  f. acetophenone + diethyl carbonate: (1) sodium ethoxide; (2) HCl
  g. 1,3-cyclohexanedione + allyl bromide + sodium hydroxide
  h. dibenzyl ketone + methyl vinyl ketone + excess sodium hydroxide
  i. cyclopentanone: (1) pyrrolidine + catalytic H$^+$; (2) ethyl bromide; (3) HCl, H$_2$O
  j. γ-butyrolactone + LDA in THF, followed by methyl iodide
  k. 2,7-octanedione + sodium hydroxide
  l. diethyl 1,2-benzenedicarboxylate + ethyl acetate: (1) excess sodium ethoxide; (2) HCl

50. Which compound decarboxylates at the lowest temperature?

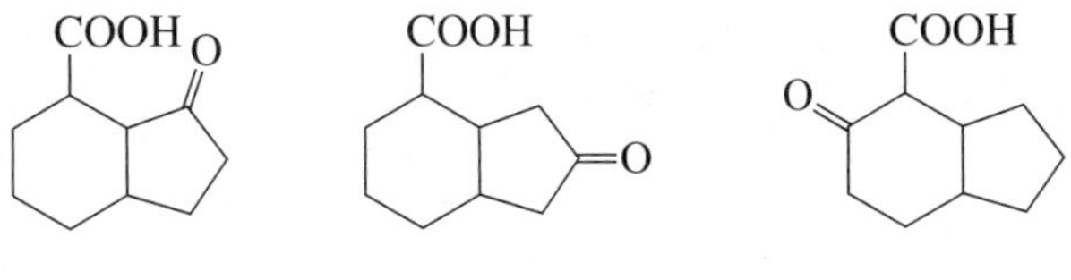

51. The chemical shifts of nitromethane, dinitromethane, and trinitromethane are at δ 6.10, δ 4.33, and δ 7.52. Match each chemical shift with the compound. Explain how chemical shift correlates with pK_a.

52. a. Explain why a racemic mixture of 2-methyl-1-phenyl-1-butanone is formed when (R)-2-methyl-1-phenyl-1-butanone is dissolved in an acidic or basic aqueous solution.
  b. Give an example of another ketone that would undergo acid- or base-catalyzed racemization.

53. What is the product of the following reaction?

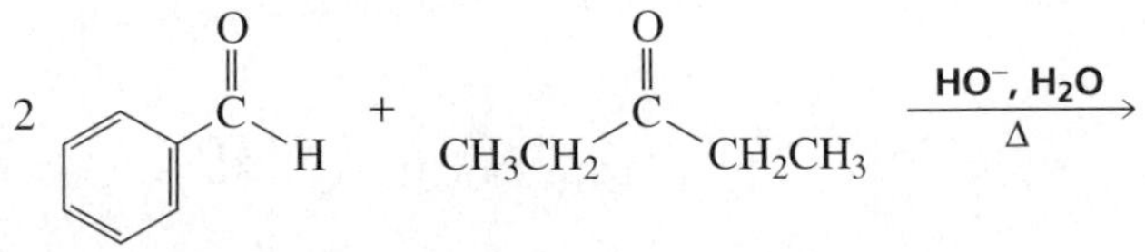

54. Identify A–L. (*Hint:* A shows three singlets in its ^{1}H NMR spectrum with integral ratios 3 : 2 : 3 and gives a positive iodoform test; see Section 18.4)

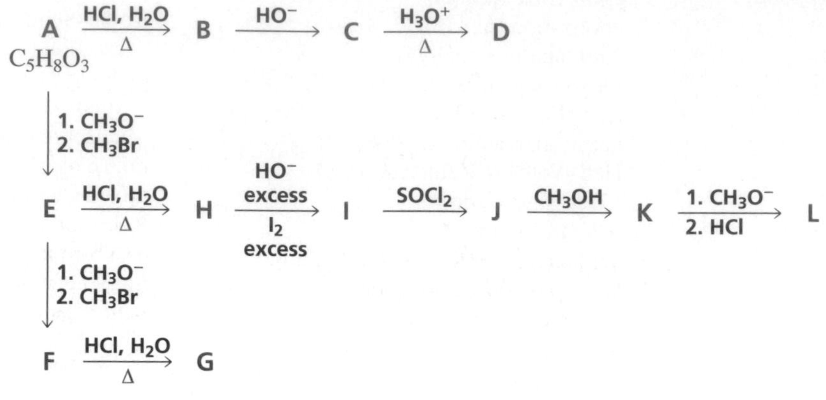

$$A \xrightarrow[\Delta]{HCl, H_2O} B \xrightarrow{HO^-} C \xrightarrow[\Delta]{H_3O^+} D$$
$$C_5H_8O_3$$

$$\downarrow \begin{array}{l} 1.\ CH_3O^- \\ 2.\ CH_3Br \end{array}$$

$$E \xrightarrow[\Delta]{HCl, H_2O} H \xrightarrow[\substack{I_2 \\ excess}]{HO^- \\ excess} I \xrightarrow{SOCl_2} J \xrightarrow{CH_3OH} K \xrightarrow[2.\ HCl]{1.\ CH_3O^-} L$$

$$\downarrow \begin{array}{l} 1.\ CH_3O^- \\ 2.\ CH_3Br \end{array}$$

$$F \xrightarrow[\Delta]{HCl, H_2O} G$$

55. Show how the following compounds could be prepared from cyclohexanone:

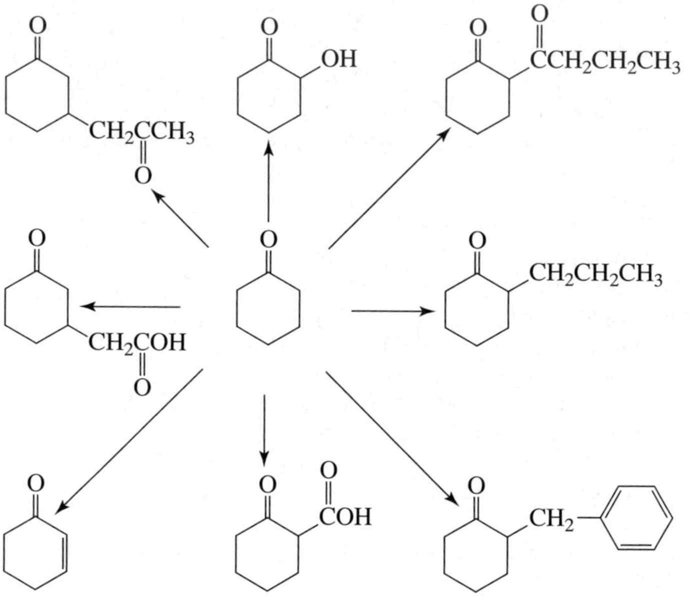

56. A β,γ-unsaturated carbonyl compound rearranges to a more stable conjugated α,β-unsaturated compound in the presence of either acid or base.
  a. Propose a mechanism for the base-catalyzed rearrangement.
  b. Propose a mechanism for the acid-catalyzed rearrangement.

a β,γ-unsaturated   an α,β-unsaturated
carbonyl compound   carbonyl compound

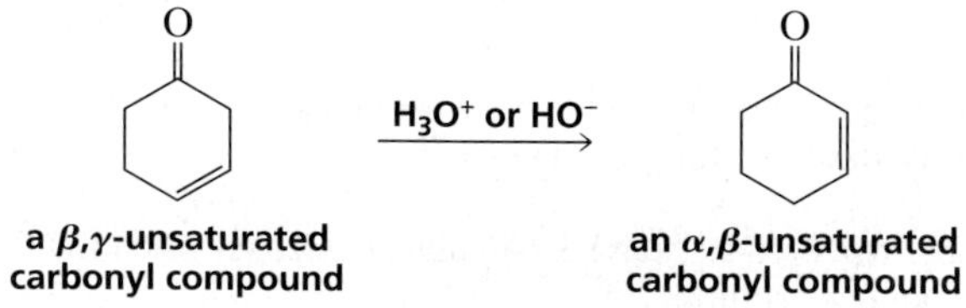

57. There are other condensation reactions similar to the aldol and Claisen condensations:
  a. The *Perkin condensation* is the condensation of an aromatic aldehyde and acetic anhydride. Give the product obtained from the following Perkin condensation:

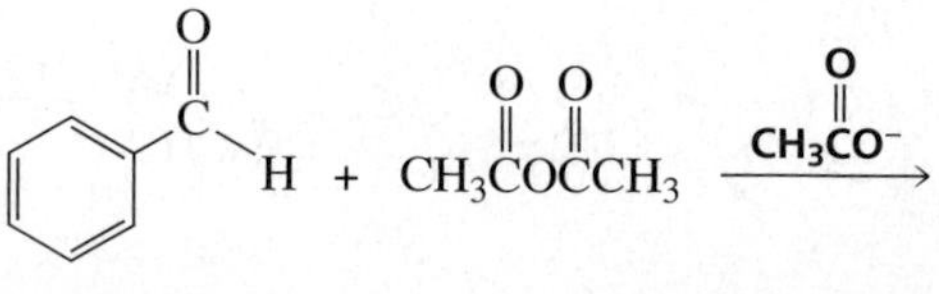

b. What compound would result if water were added to the product of a Perkin condensation?

c. The *Knoevenagel condensation* is the condensation of an aldehyde or a ketone that has no α-hydrogens and a compound such as diethyl malonate that has an α-carbon flanked by two electron-withdrawing groups. Give the product obtained from the following Knoevenagel condensation:

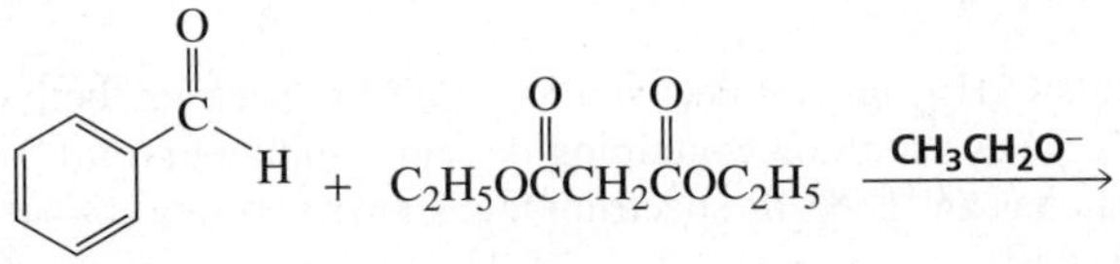

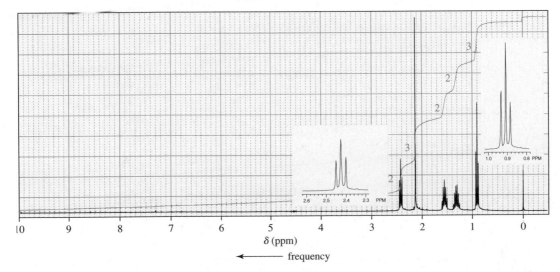

d. What product would be obtained if the product of a Knoevenagel condensation were heated in an aqueous acidic solution?

58. The *Reformatsky reaction* is an addition reaction in which an organozinc reagent is used instead of a Grignard reagent to attack the carbonyl group of an aldehyde or a ketone. Because the organozinc reagent is less reactive than a Grignard reagent, a nucleophilic addition to the ester group does not occur. The organozinc reagent is prepared by treating an *a*-bromo ester with zinc.

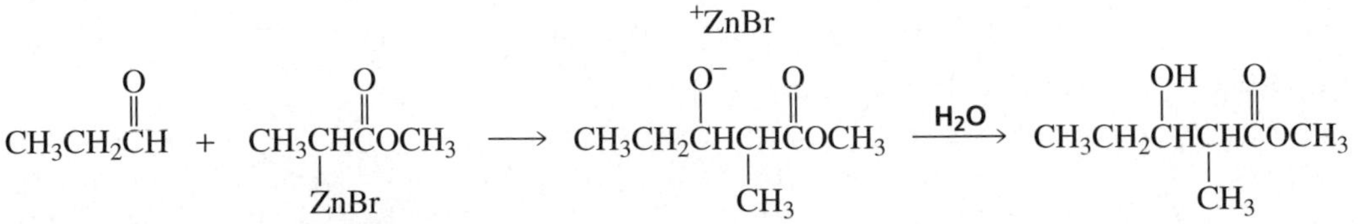

Describe how each of the following compounds could be prepared, using a Reformatsky reaction:

a. 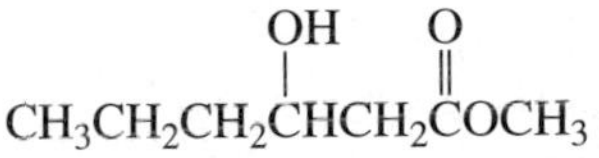 $CH_3CH_2CH_2CHCH_2COCH_3$ (OH, O)

b. 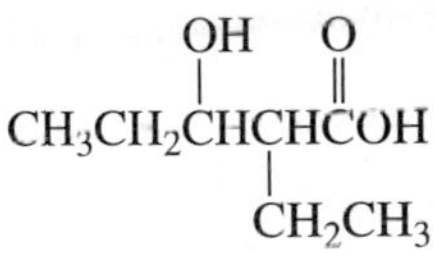 $CH_3CH_2CHCHCOH$ (OH, O; CH_2CH_3)

c. 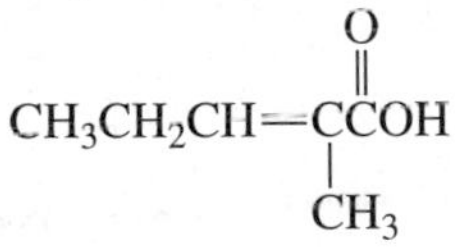 $CH_3CH_2CH=CCOH$ (O; CH_3)

d. 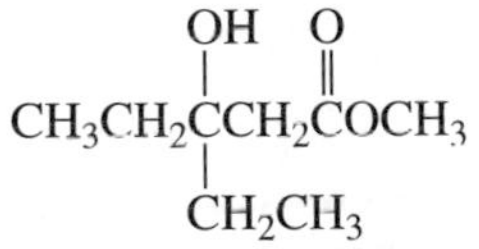 $CH_3CH_2CCH_2COCH_3$ (OH, O; CH_2CH_3)

59. The ketone whose ^1H NMR spectrum is shown here was obtained as the product of an acetoacetic ester synthesis. What alkyl halide was used in the synthesis?

δ (ppm)

← frequency

60. Indicate how the following compounds could be synthesized from cyclohexanone and any other necessary reagents:

a. 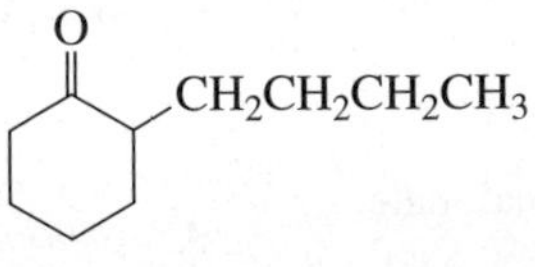 $CH_2CH_2CH_2CH_3$ (O)

b. 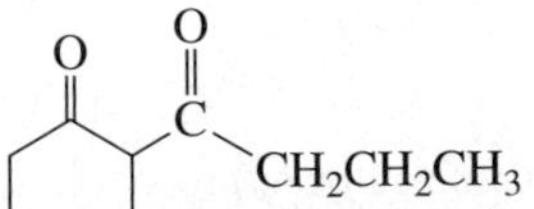 $CH_2CH_2CH_3$ (O, O) **(two ways)**

c. 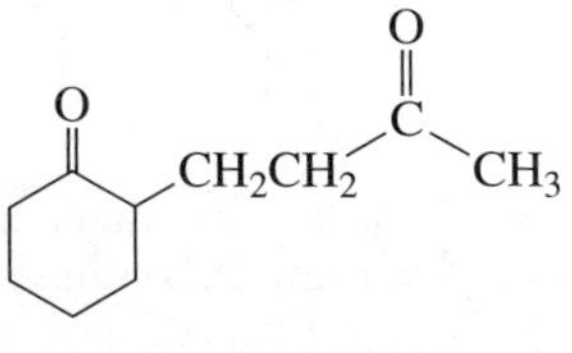 CH_2CH_2 CH_3 (O, O)

d. 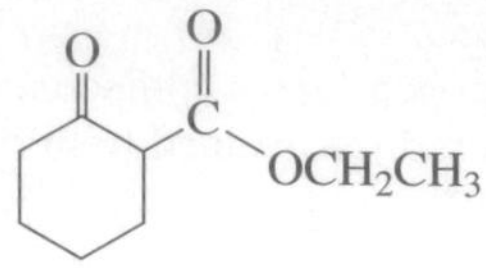 e. 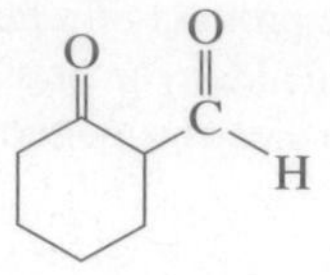 f.

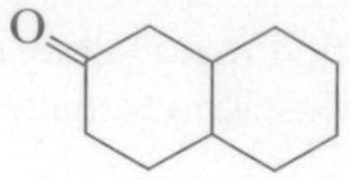

61. Compound A with molecular formula C_6H_{10} has two peaks in its ^1H NMR spectrum, both of which are singlets (with ratio 9 : 1). Compound A reacts with an acidic aqueous solution containing mercuric sulfate to form compound B, which gives a positive iodoform test (Section 18.4) and that has an ^1H NMR spectrum that shows two singlets (with ratio 3 : 1). Identify A and B.

62. Indicate how each of the following compounds could be synthesized from the given starting material and any other necessary reagents:

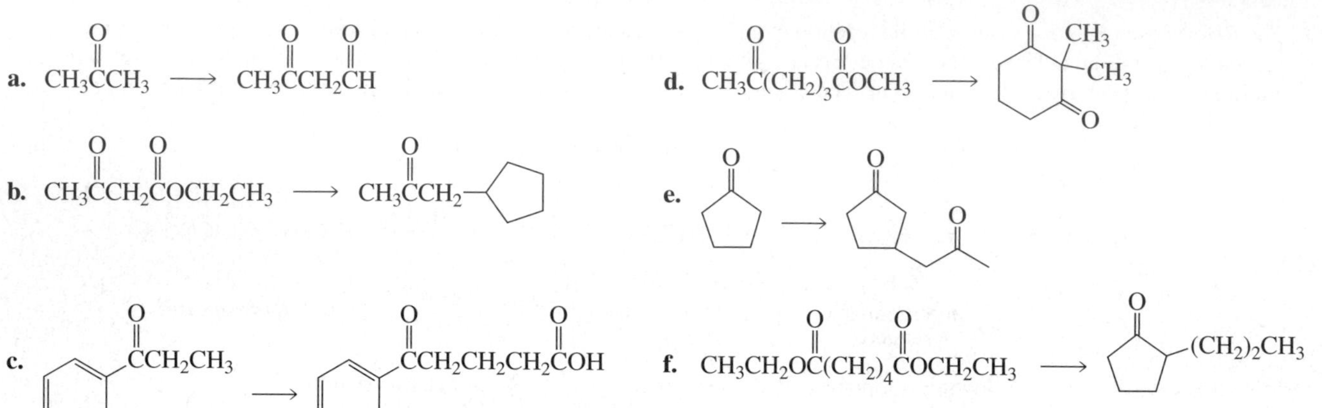

a. $CH_3\overset{O}{\overset{\|}{C}}CH_3 \longrightarrow CH_3\overset{O}{\overset{\|}{C}}CH_2\overset{O}{\overset{\|}{C}}H$

b. $CH_3\overset{O}{\overset{\|}{C}}CH_2\overset{O}{\overset{\|}{C}}OCH_2CH_3 \longrightarrow CH_3\overset{O}{\overset{\|}{C}}CH_2$—

c.

d. $CH_3\overset{O}{\overset{\|}{C}}(CH_2)_3\overset{O}{\overset{\|}{C}}OCH_3 \longrightarrow$

e.

f. $CH_3CH_2O\overset{O}{\overset{\|}{C}}(CH_2)_4\overset{O}{\overset{\|}{C}}OCH_2CH_3 \longrightarrow$

63. Bupropion hydrochloride is an antidepressant marketed under the trade name Wellbutrin. Propose a synthesis of bupropion hydrochloride, starting with benzene.

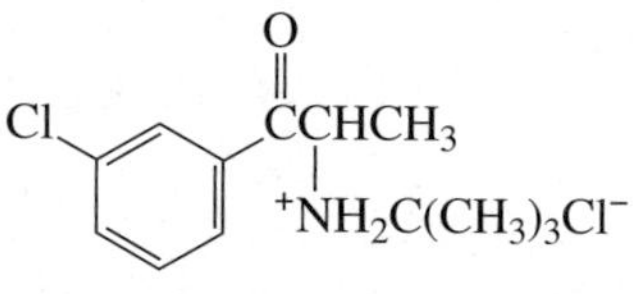

bupropion hydrochloride

64. What reagents would be required to carry out the following transformations?

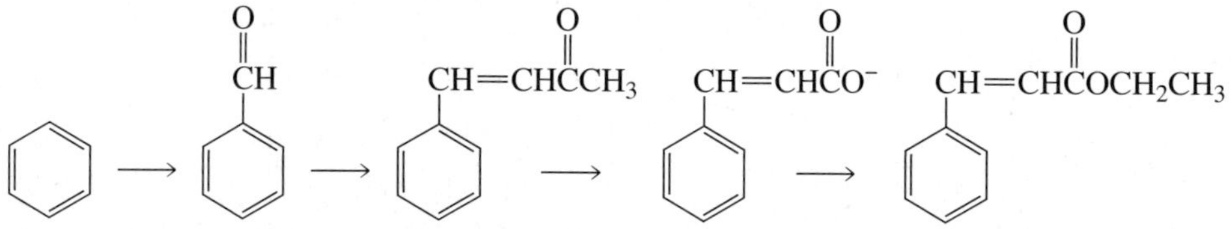

65. Give the products of the following reactions:

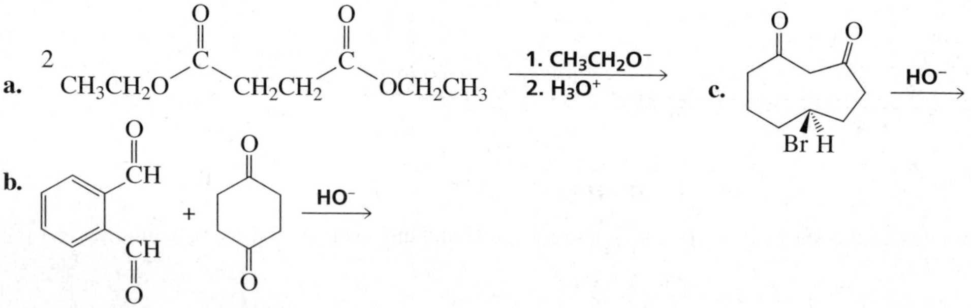

a. $2\ \ CH_3CH_2O\overset{O}{\overset{\|}{C}}CH_2CH_2\overset{O}{\overset{\|}{C}}OCH_2CH_3 \xrightarrow[\text{2. } H_3O^+]{\text{1. } CH_3CH_2O^-}$

b.

c.

66. a. Show how the amino acid alanine can be synthesized from propanoic acid.
b. Show how the amino acid glycine can be synthesized from phthalimide and diethyl 2-bromomalonate.

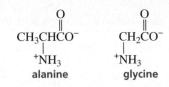

alanine   glycine

67. Cindy Synthon tried to prepare the following compounds using aldol condensations. Which of these compounds was she successful in synthesizing? Explain why the other syntheses were not successful.

a. 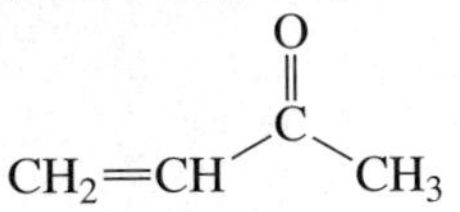 CH₂=CH—C(=O)—CH₃

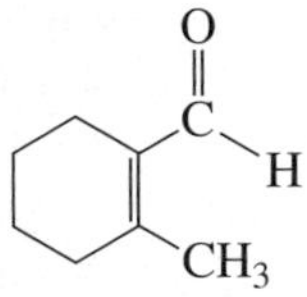

b. 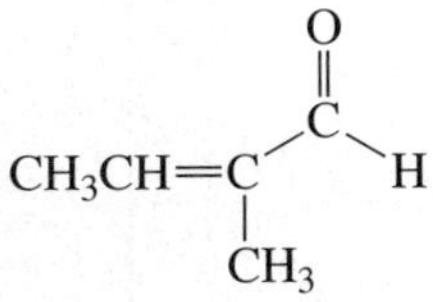 CH₃CH=C(CH₃)—C(=O)—H

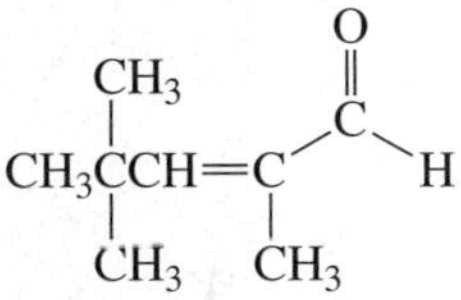

c. 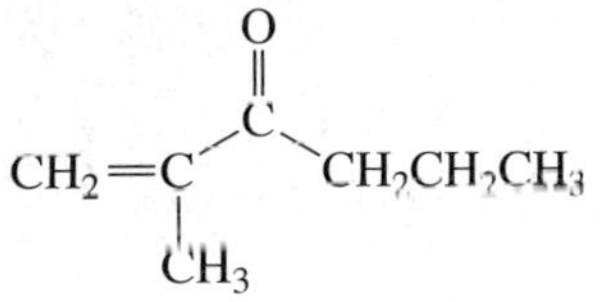 CH₂=C(CH₃)—C(=O)—CH₂CH₂CH₃

d. CH₃C(CH₃)=CH—C(=O)—H

e. (cyclohexenone)—CH₂CH₃

f. 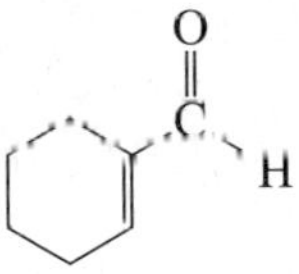 (cyclohexenyl)—C(=O)—H

g. (cyclohexenyl with CH₃)—C(=O)—H

h. CH₃C(CH₃)CH=C(CH₃)—C(=O)—H

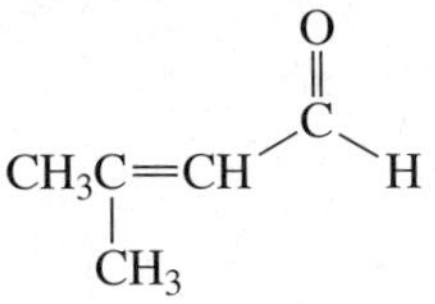

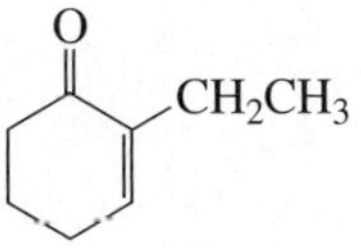

l. CH₃C(CH₃)CH₂CH=C(CH₃)—C(=O)—H

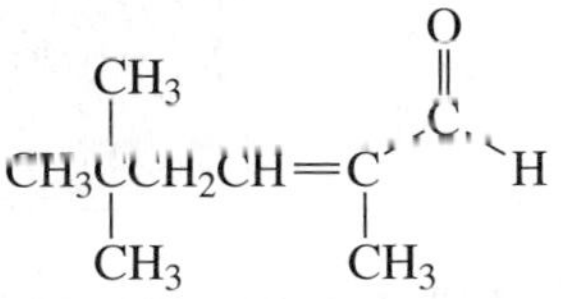

68. Show how the following compounds could be synthesized. The only carbon-containing compounds available to you for each synthesis are shown.

a. 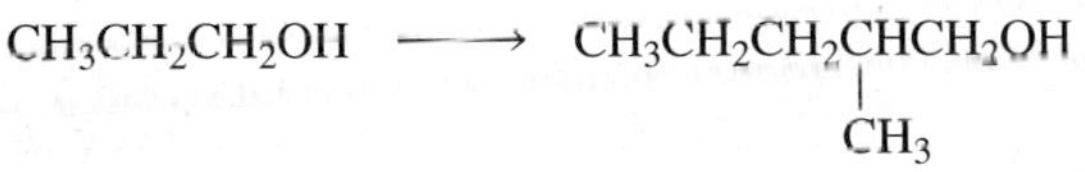 CH₃CH₂CH₂OH ⟶ CH₃CH₂CH₂CHCH₂OH
                                          |
                                         CH₃

b. 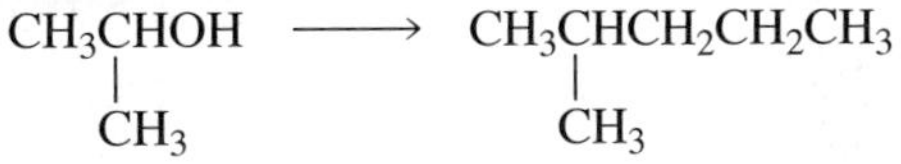 CH₃CHOH ⟶ CH₃CHCH₂CH₂CH₃
         | |
        CH₃   CH₃

c. 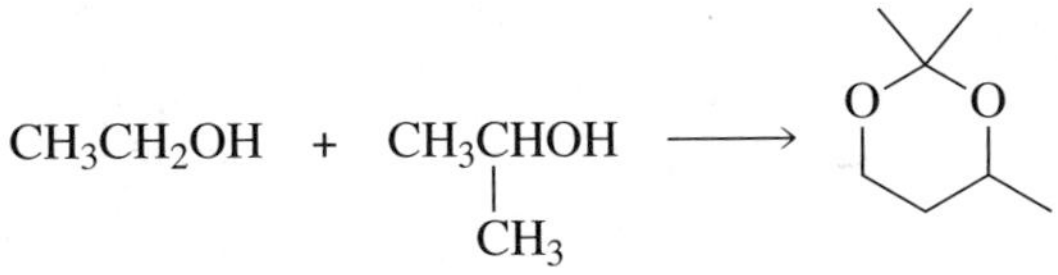 CH₃CH₂OH + CH₃CHOH ⟶ (dioxane ring structure)
                      |
                     CH₃

69. Explain why the following bromoketone forms different bicyclic compounds under different reaction conditions:

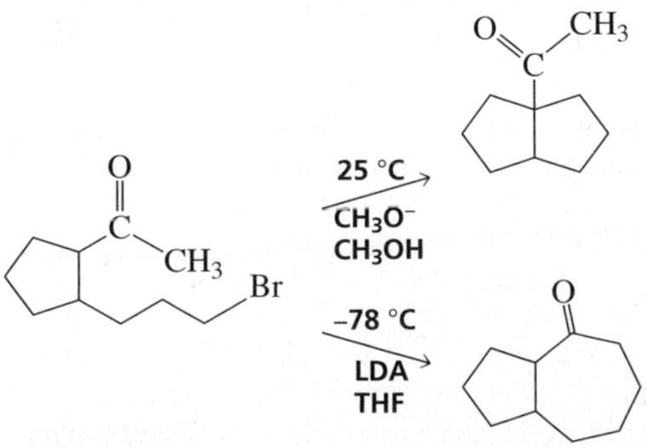

(bromoketone structure)

25 °C, CH₃O⁻, CH₃OH

−78 °C, LDA, THF

70. A *Mannich reaction* puts a $\diagdown$ NCH$_2$— group on the α-carbon of a carbon acid. Propose a mechanism for the reaction.

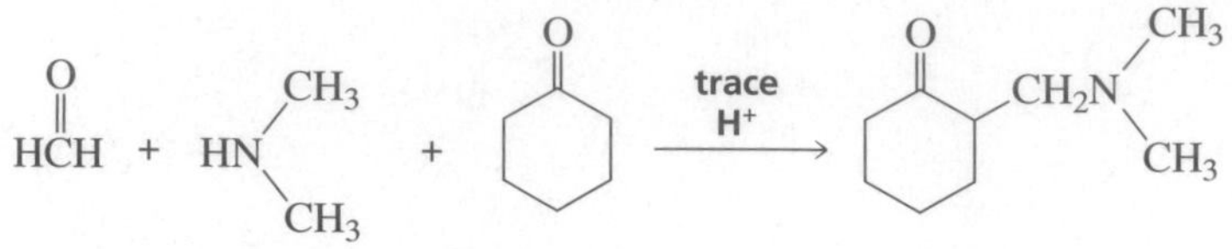

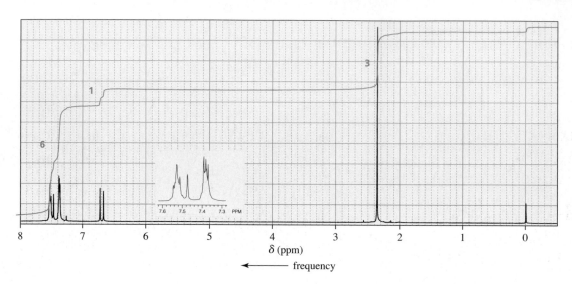

71. What carbonyl compounds are required to prepare a compound with molecular formula $C_{10}H_{10}O$ whose ^{1}H NMR spectrum is shown?

72. Ninhydrin reacts with an amino acid to form a purple-colored compound. Propose a mechanism to account for the formation of the colored compound.

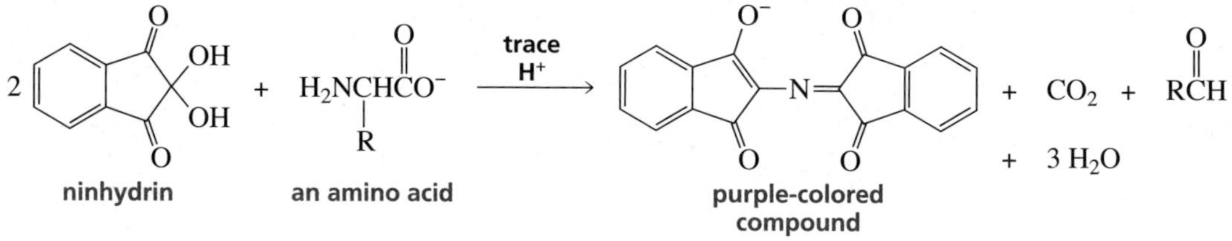

ninhydrin   an amino acid   purple-colored compound

73. A carboxylic acid is formed when an α-haloketone reacts with hydroxide ion. This reaction is called a *Favorskii reaction*. Propose a mechanism for the following Favorskii reaction. (*Hint:* In the first step, HO$^-$ removes a proton from the α-carbon that is not bonded to Br; a three-membered ring is formed in the second step; and HO$^-$ is a nucleophile in the third step.)

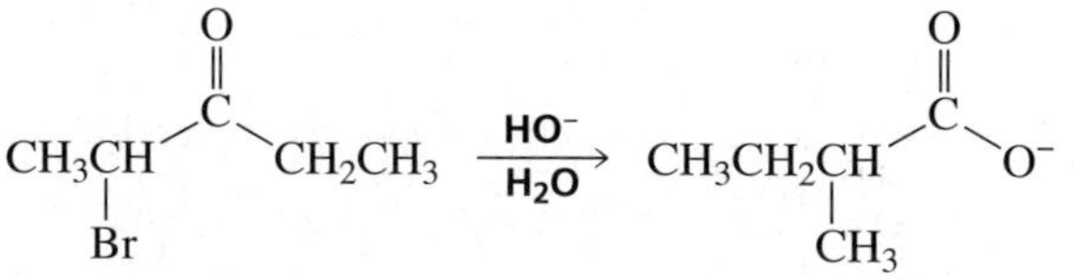

74. An α,β-unsaturated carbonyl compound can be prepared by a reaction known as a selenenylation–oxidation reaction. A selenoxide is formed as an intermediate. Propose a mechanism for the reaction.

1. LDA/THF
2. C$_6$H$_5$SeBr
3. H$_2$O$_2$

a selenoxide

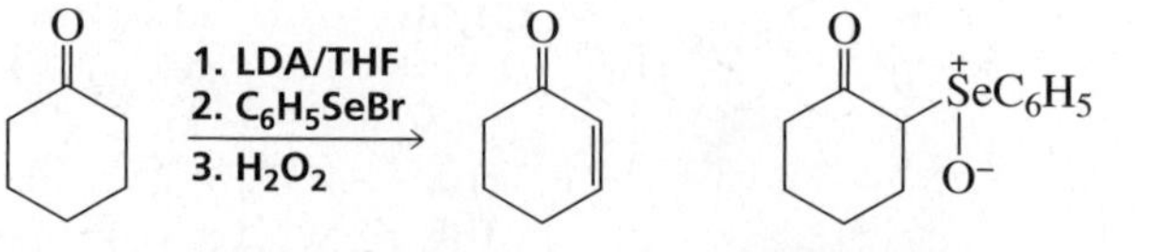

75. a. What carboxylic acid would be formed if the malonic ester synthesis were carried out with one equivalent of malonic ester, one equivalent of 1,5-dibromopentane, and two equivalents of base?

  b. What carboxylic acid would be formed if the malonic ester synthesis were carried out with two equivalents of malonic ester, one equivalent of 1,5-dibromopentane, and two equivalents of base?

76. A *Cannizzaro reaction* is the reaction of an aldehyde that has no α-hydrogens with concentrated aqueous sodium hydroxide. In this reaction, half the aldehyde is converted to a carboxylic acid and the other half is converted to an alcohol. Propose a reasonable mechanism for the following Cannizzaro reaction:

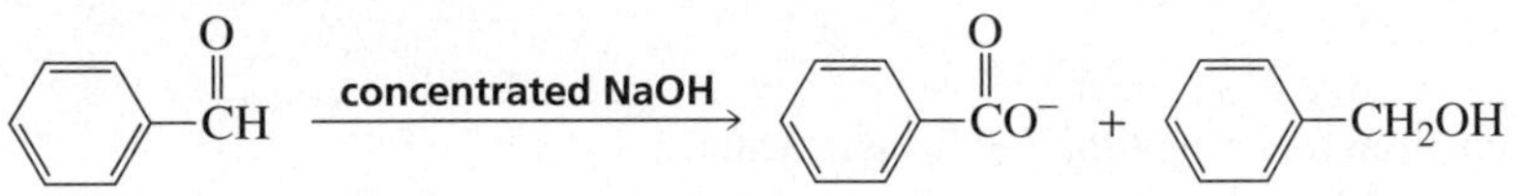

77. Propose a reasonable mechanism for each of the following reactions:

  a.

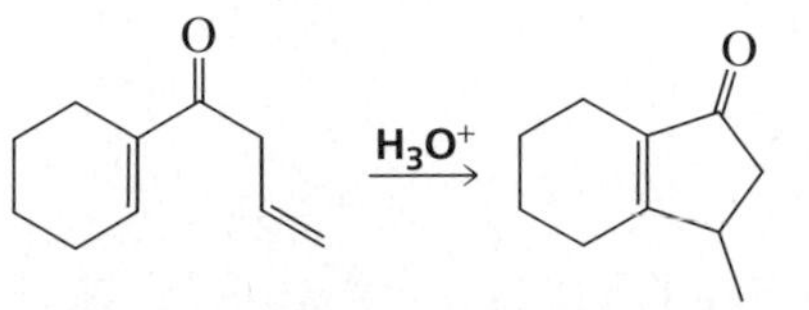

  b.

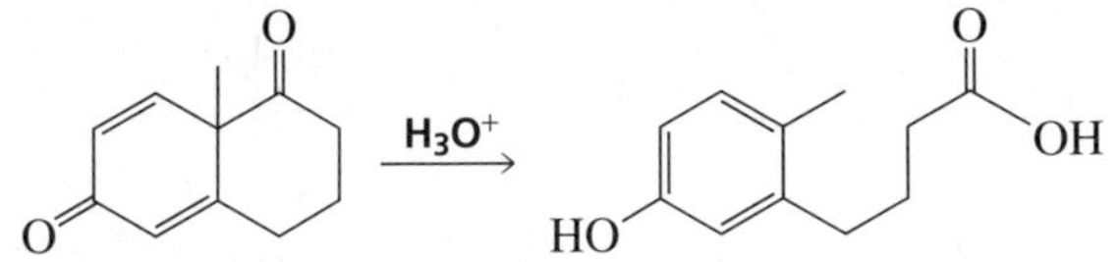

78. The following reaction is known as the *benzoin condensation*. The reaction will not take place if sodium hydroxide is used instead of sodium cyanide. Propose a mechanism for the reaction.

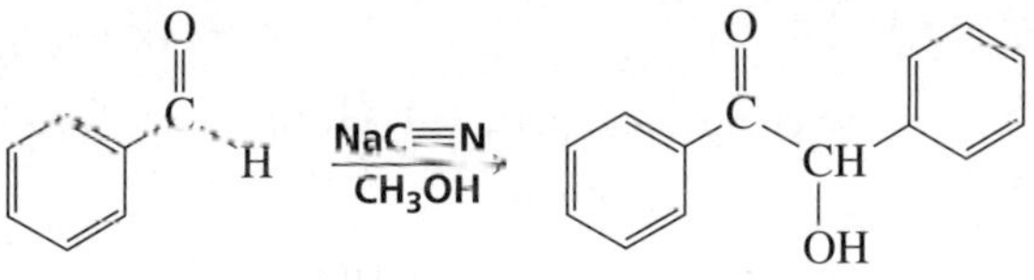

79. Orsellinic acid, a common constituent of lichens, is synthesized from the condensation of acetyl thioester and malonyl thioester. If a lichen were grown on a medium containing acetate that was radioactively labeled with ^{14}C at the carbonyl carbon, which carbons would be labeled in orsellinic acid?

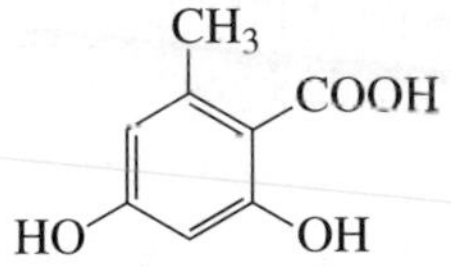

orsellinic acid

80. Propose a mechanism for the following reaction. (*Hint:* the intermediate has a cumulated double bond.)

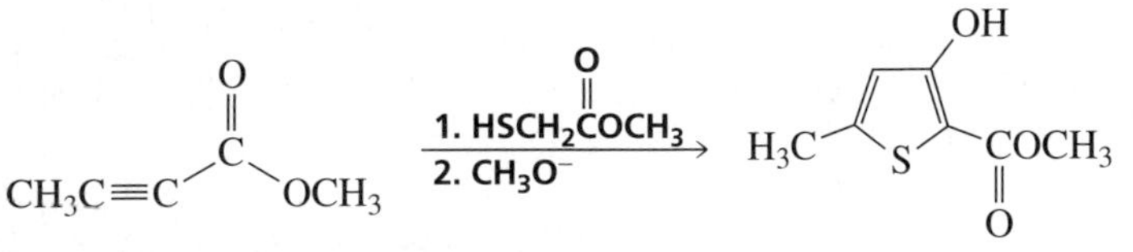

81. A compound known as *Hagemann's ester* can be prepared by treating a mixture of formaldehyde and ethyl acetoacetate first with base and then with acid and heat. Write the structure for the product of each of the steps.

  a. The first step is an aldol-like condensation.
  b. The second step is a Michael addition.
  c. The third step is an intramolecular aldol addtion.
  d. The final transformation includes a dehydration and a hydrolysis followed by a decarboxylation.

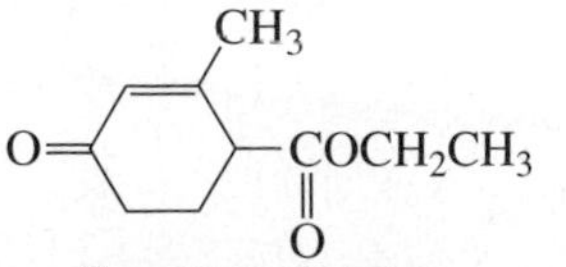

Hagemann's ester

82. Amobarbital is a sedative marketed under the trade name Amytal. Propose a synthesis of amobarbital, using diethyl malonate and urea (page 733) as two of the starting materials.

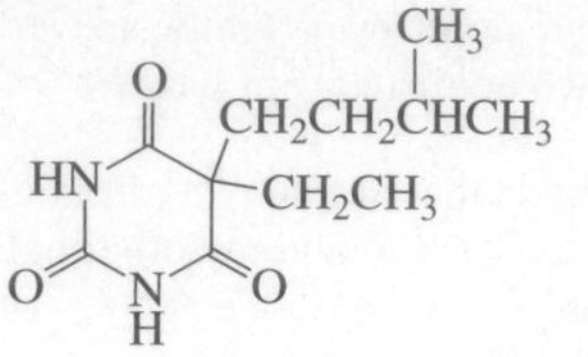

Amytal®

83. Propose a reasonable mechanism for each of the following reactions:

a.

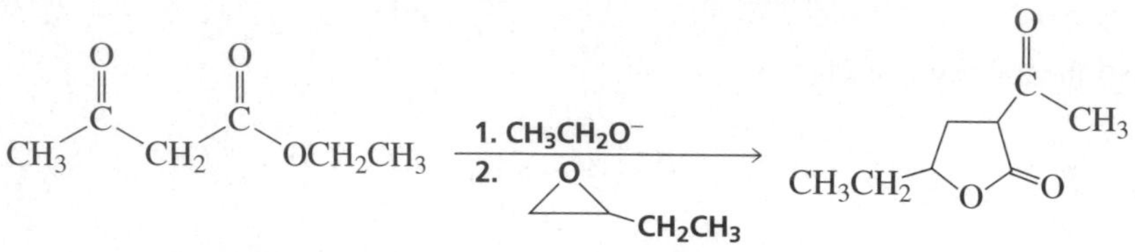

b.

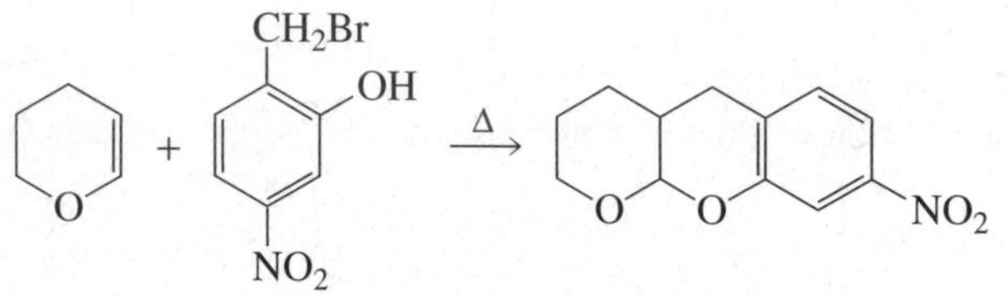

84. Tyramine is an alkaloid found in mistletoe, ripe cheese, and putrefied animal tissue. Dopamine is a neurotransmitter involved in the regulation of the central nervous system.

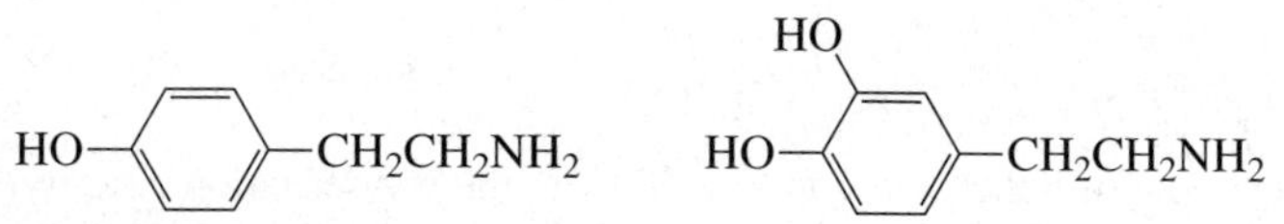

tyramine   dopamine

    a. Give two ways to prepare β-phenylethylamine from β-phenylethyl chloride.
    b. How can β-phenylethylamine be prepared from benzyl chloride?
    c. How can β-phenylethylamine be prepared from benzaldehyde?
    d. How can tyramine be prepared from β-phenylethylamine?
    e. How can dopamine be prepared from tyramine?

85. a. Ketoprofen, like ibuprofen, is an anti-inflammatory analgesic. How could ketoprofen be synthesized from the given starting material?

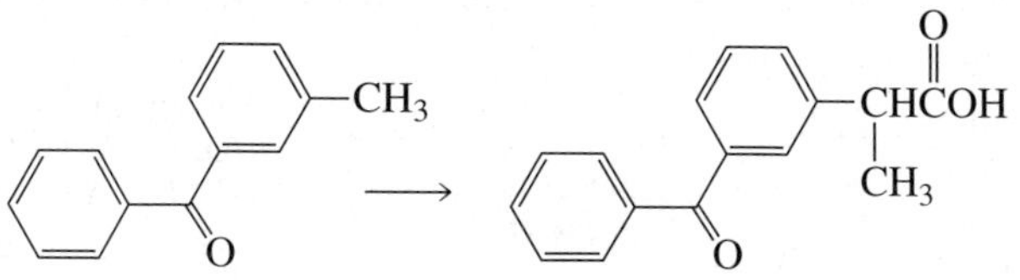

ketoprofen

    b. Ketoprofen and ibuprofen both have a propanoic acid substituent (see Problem 81 in Chapter 16). Explain why the identical subunits are synthesized in different ways.

More About Oxidation–Reduction Reactions and Amines

The two chapters in Part 7 discuss in greater detail two topics that were presented in preceding chapters—oxidation–reduction reactions and amines.

CHAPTER 19
More About Oxidation–Reduction Reactions

You first encountered oxidation–reduction reactions in Chapter 4. You learned that if the reaction increases the number of C—H bonds or decreases the number of C—O, C—N, or C—X bonds (where X denotes a halogen), the compound has been reduced. On the other hand, if the reaction decreases the number of C—H bonds or increases the number of C—O, C—N, or C—X bonds, the compound has been oxidized. **Chapter 19** reviews some of the redox reactions you have seen previously and introduces you to many more.

CHAPTER 20
More About Amines • Heterocyclic Compounds

Chapter 20 extends the coverage of amines and discusses the chemistry of heterocyclic compounds. Amines do not undergo addition, substitution, or elimination reactions. Their importance lies in how they react with other organic compounds. This reactivity is far too important to wait until the second half of the book to be discussed. You first encountered amines in Chapter 1 and you continued to encounter them in almost every chapter since.

CHAPTER 19

More About Oxidation–Reduction Reactions

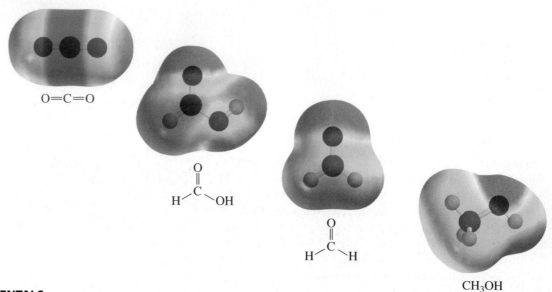

O=C=O

H—C(=O)—OH

H—C(=O)—H

CH_3OH

BUILDING ON FUNDAMENTALS

SECTION 19.0 If the reaction increases the number of C—H bonds or decreases the number of C—O, C—N, or C—X bonds (where X denotes a halogen), the compound has been reduced (4.8).

SECTION 19.0 If the reaction decreases the number of C—H bonds or increases the number of C—O, C—N, or C—X bonds, the compound has been oxidized (4.9).

SECTION 19.1 Alkenes can be reduced to alkanes (4.11) and alkynes can be reduced to alkenes or alkanes (6.9).

SECTION 19.1 Aldehydes, ketones, carboxylic acids, and esters are reduced to alcohols (17.6).

SECTION 19.1 Nitriles, amides, and imines are reduced to amines (16.19, 17.6, and 17.8).

SECTION 19.2 Secondary alcohols are oxidized to ketones and primary alcohols are oxidized to aldehydes and carboxylic acids (10.5).

SECTION 19.4 Alkenes can be oxidized to epoxides (4.9).

An important group of organic reactions consists of those that involve the transfer of electrons from one molecule to another. Organic chemists use these reactions—called **oxidation–reduction reactions** or **redox reactions**—to synthesize a large variety of compounds. Redox reactions are also important in biological systems because many of these reactions produce energy. These will be discussed in Chapters 24 and 25. You have already seen a number of oxidation and reduction reactions in other chapters, but discussing them as a group will give you the opportunity to compare them.

In an oxidation–reduction reaction, one species loses electrons and one gains electrons. The species that loses electrons is oxidized, and the one that gains electrons is reduced. One way to remember the difference between oxidation and reduction is with the phrase "LEO the lion says GER": *Loss of Electrons is Oxidation; Gain of Electrons is Reduction.*

The following is an example of an oxidation–reduction reaction involving inorganic reagents:

$$Cu^+ + Fe^{3+} \longrightarrow Cu^{2+} + Fe^{2+}$$

In this reaction, Cu^+ loses an electron. Thus, Cu^+ is oxidized. Fe^{3+} gains an electron and it, therefore, is reduced. This reaction demonstrates two important features of oxidation–reduction reactions. First, *oxidation is always coupled with reduction.* In other words, a species cannot gain electrons (be reduced) unless another species in the reaction simultaneously loses electrons (is oxidized). Second, the species that is oxidized (Cu^+) is called the **reducing agent** because it loses the electrons that are used to reduce the other species (Fe^{3+}). Similarly, the species that is reduced (Fe^{3+}) is called the **oxidizing agent** because it gains the electrons given up by the other species (Cu^+) when it is oxidized.

It is easy to tell whether an organic compound has been oxidized or reduced simply by looking at the change in the compound's structure. We will be looking primarily at reactions where **oxidation** or **reduction** has taken place on carbon: if the reaction increases the number of C—H bonds or decreases the number of C—O, C—N, or C—X bonds (where X denotes a halogen), the compound has been reduced. If the reaction decreases the number of C—H bonds or increases the number of C—O, C—N, or C—X bonds, the compound has been oxidized. Notice that the oxidation level of a carbon atom is equal to the total number of its C—O, C—N, and C—X bonds.

> **Reduction at carbon increases the number of C—H bonds or decreases the number of C—O, C—N, or C—X bonds.**

> **Oxidation at carbon decreases the number of C—H bonds or increases the number of C—O, C—N, or C—X bonds.**

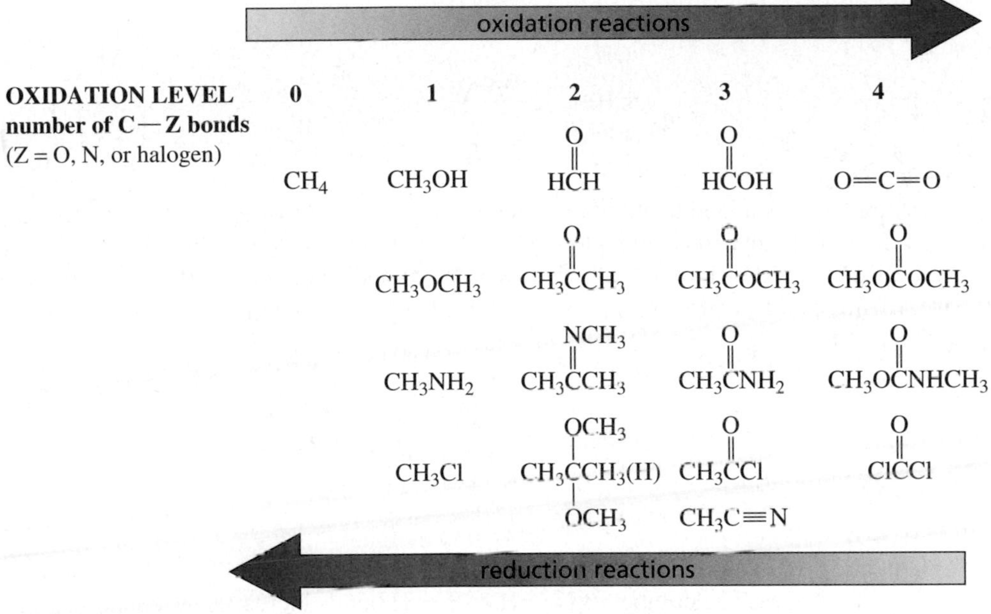

Student Tutorial:
Changes in oxidation levels

Let's now look at some examples of oxidation–reduction reactions that take place on carbon. Notice that in each of the following reactions, the product has more C—H bonds than the reactant has: the alkene, ketone, and aldehyde, therefore, have been reduced. Hydrogen, hydrazine, and sodium borohydride are the reducing agents. You have seen these reactions in previous chapters (Sections 4.11, 14.16, and 17.6).

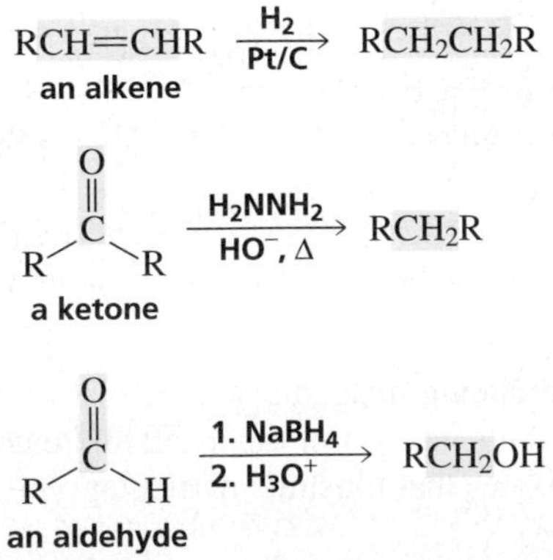

$$RCH\!=\!CHR \xrightarrow[\text{Pt/C}]{\text{H}_2} RCH_2CH_2R$$
an alkene

$$\underset{\textbf{a ketone}}{\overset{\displaystyle O}{\underset{R}{\overset{\|}{\text{C}}}R}} \xrightarrow[\text{HO}^-, \Delta]{\textbf{H}_2\textbf{NNH}_2} RCH_2R$$

$$\underset{\textbf{an aldehyde}}{\overset{\displaystyle O}{\underset{R}{\overset{\|}{\text{C}}}H}} \xrightarrow[\textbf{2. H}_3\textbf{O}^+]{\textbf{1. NaBH}_4} RCH_2OH$$

In the first reaction of the next group, the number of C—Br bonds increases. In the second and third reactions, the number of C—H bonds decreases and the number of C—O bonds increases. Thus, the alkene, aldehyde, and alcohol are being oxidized. Bromine and chromic acid (H_2CrO_4) are the oxidizing agents. Notice that the increase in the number of C—O bonds in the third reaction results from a carbon–oxygen single bond becoming a carbon–oxygen double bond. You have also seen these reactions in previous chapters (Sections 4.7 and 10.5).

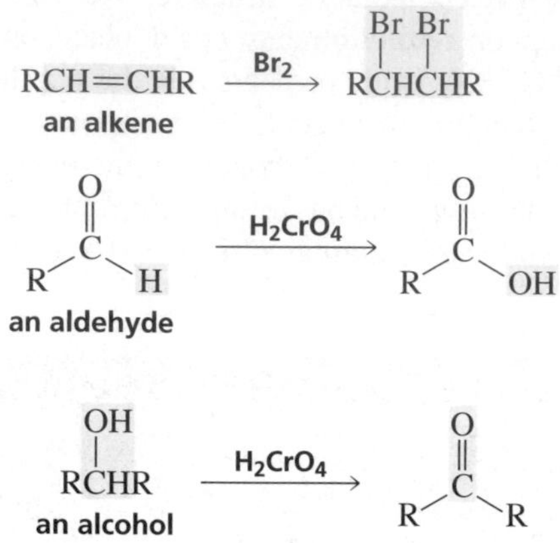

$$
\underset{\textbf{an alkene}}{RCH{=}CHR} \xrightarrow{Br_2} \underset{}{\overset{Br\ Br}{\underset{}{RCHCHR}}}
$$

$$
\underset{\textbf{an aldehyde}}{\overset{O}{\underset{R\ \ \ \ H}{\parallel C}}} \xrightarrow{H_2CrO_4} \underset{}{\overset{O}{\underset{R\ \ \ \ OH}{\parallel C}}}
$$

$$
\underset{\textbf{an alcohol}}{\overset{OH}{\underset{RCHR}{|}}} \xrightarrow{H_2CrO_4} \underset{R\ \ \ \ R}{\overset{O}{\parallel C}}
$$

When water is added to an alkene, the product has one more C—H bond than the reactant, but it also has one more C—O bond. In this reaction, one carbon is reduced and another is oxidized. The two processes cancel each other as far as the overall molecule is concerned, so the overall reaction is neither an oxidation nor a reduction.

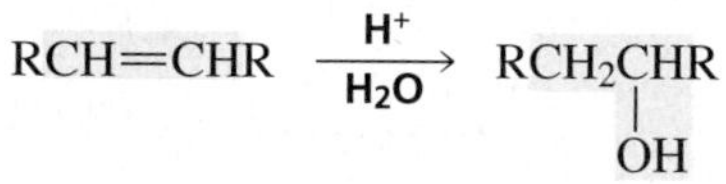

$$
RCH{=}CHR \xrightarrow[H_2O]{H^+} \underset{\underset{OH}{|}}{RCH_2CHR}
$$

Oxidation–reduction reactions that take place on nitrogen or sulfur produce similar structural changes. The number of N—H or S—H bonds increases in reduction reactions, and the number of N—O or S—O bonds increases in oxidation reactions. In the following reactions, nitrobenzene and the disulfide are reduced (Sections 14.19 and 22.7), and the thiol is oxidized to a sulfonic acid:

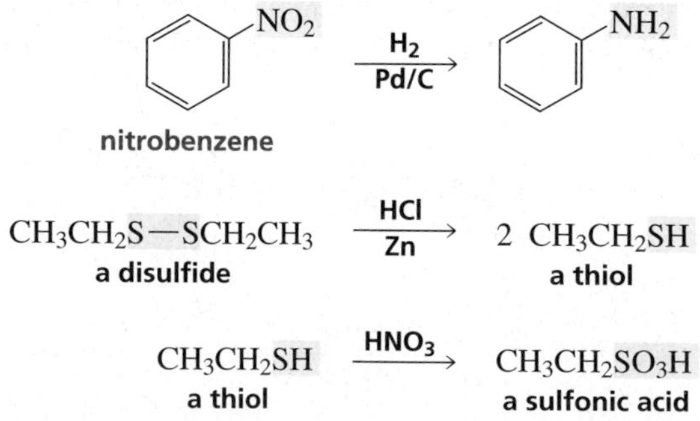

$$
\underset{\textbf{a disulfide}}{CH_3CH_2S{-}SCH_2CH_3} \xrightarrow[Zn]{HCl} \underset{\textbf{a thiol}}{2\ CH_3CH_2SH}
$$

$$
\underset{\textbf{a thiol}}{CH_3CH_2SH} \xrightarrow{HNO_3} \underset{\textbf{a sulfonic acid}}{CH_3CH_2SO_3H}
$$

Many oxidizing and reducing reagents are available to organic chemists. This chapter highlights only a small fraction of the available reagents. The ones selected are some of the more common ones that illustrate particular types of transformations caused by oxidation or reduction.

PROBLEM 1◆

Identify each of the following reactions as an oxidation reaction, a reduction reaction, or neither:

a. 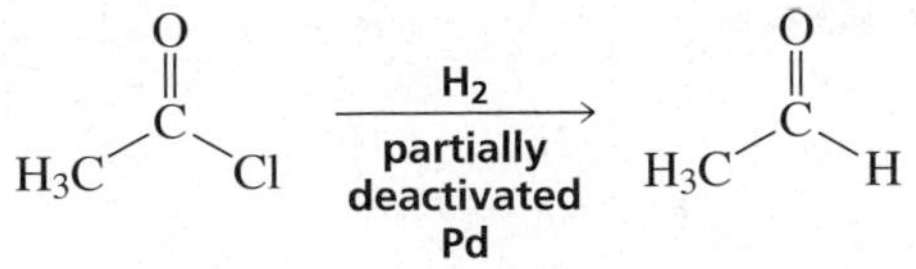 $\xrightarrow[\substack{\text{partially} \\ \text{deactivated} \\ \text{Pd}}]{H_2}$

b. $RCH{=}CHR \xrightarrow{HBr} RCH_2\overset{\displaystyle Br}{\underset{\displaystyle |}{C}}HR$

c. $\xrightarrow[h\nu]{Br_2}$

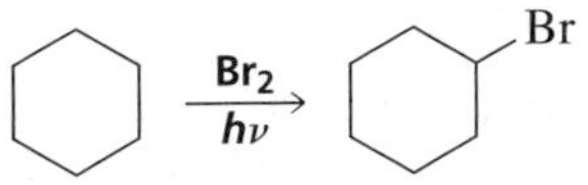

d. 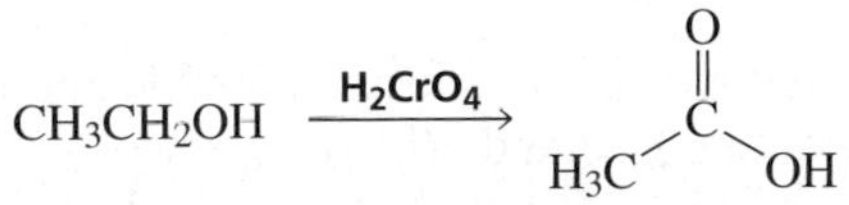 $CH_3CH_2OH \xrightarrow{H_2CrO_4}$

e. $CH_3C{\equiv}N \xrightarrow{\substack{H_2 \\ Pt/C}} CH_3CH_2NH_2$

f. $CH_3CH_2CH_2Br \xrightarrow{HO^-} CH_3CH_2CH_2OH$

19.1 Reduction Reactions

An organic compound is reduced when hydrogen (H_2) is added to it. A molecule of H_2 can be thought of as being composed of (1) two hydrogen atoms, (2) two electrons and two protons, or (3) a hydride ion and a proton. In the sections that follow, you will see that these three ways to describe H_2 correspond to the three mechanisms by which H_2 is added to an organic compound.

components of H:H

H· ·H	·⁻ H⁺ ·⁻ H⁺	H:⁻ H⁺
two hydrogen atoms	two electrons and two protons	a hydride ion and a proton

Reduction by Addition of Two Hydrogen Atoms

We have seen that hydrogen can be added to carbon–carbon double and triple bonds in the presence of a metal catalyst (Sections 4.11 and 6.9). These reactions, called **catalytic hydrogenations**, are reduction reactions because there are more C—H bonds in the products than in the reactants. Alkenes and alkynes are both reduced to alkanes.

Student Tutorial:
Catalytic hydrogenation of ethylene

$$CH_3CH_2CH{=}CH_2 \; + \; H_2 \xrightarrow{\text{Pt, Pd, or Ni}} CH_3CH_2CH_2CH_3$$
$$\underset{\text{1-butene}}{} \qquad\qquad\qquad\qquad\qquad \underset{\text{butane}}{}$$

$$CH_3CH_2CH_2C{\equiv}CH \; + \; 2\,H_2 \xrightarrow{\text{Pt, Pd, or Ni}} CH_3CH_2CH_2CH_2CH_3$$
$$\underset{\text{1-pentyne}}{} \qquad\qquad\qquad\qquad\qquad\quad \underset{\text{pentane}}{}$$

In a catalytic hydrogenation, the H—H bond breaks homolytically (Section 4.11). This means that reduction occurs through the addition of two hydrogen atoms to the organic molecule.

We have seen that the catalytic hydrogenation of an alkyne can be stopped at a cis alkene if a partially deactivated catalyst is used (Section 6.9).

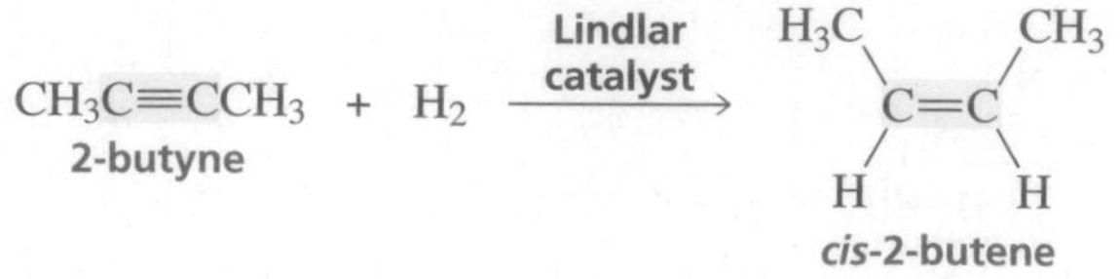

$$CH_3C{\equiv}CCH_3 \ + \ H_2 \ \xrightarrow{\textbf{Lindlar catalyst}} \ \underset{\textit{cis}\text{-2-butene}}{\overset{\displaystyle H_3C \qquad CH_3}{\underset{\displaystyle H \qquad\quad H}{C{=}C}}}$$

2-butyne

Only the alkene substituent is reduced in the following reaction. The very stable benzene ring can be reduced only under special conditions.

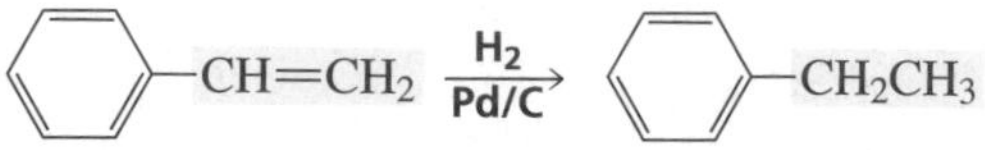

$$\text{C}_6\text{H}_5{-}CH{=}CH_2 \ \xrightarrow{\overset{\textbf{H}_2}{\textbf{Pd/C}}} \ \text{C}_6\text{H}_5{-}CH_2CH_3$$

Catalytic hydrogenation can also be used to reduce carbon–nitrogen double and triple bonds. The reaction products are amines (Sections 17.8 and 16.19).

$$CH_3CH_2CH{=}NCH_3 \ + \ H_2 \ \xrightarrow{\textbf{Pd/C}} \ \underset{\textbf{methylpropylamine}}{CH_3CH_2CH_2NHCH_3}$$

$$CH_3CH_2CH_2C{\equiv}N \ + \ 2\,H_2 \ \xrightarrow{\textbf{Pd/C}} \ \underset{\textbf{butylamine}}{CH_3CH_2CH_2CH_2NH_2}$$

The carbonyl group of ketones and aldehydes can be reduced by catalytic hydrogenation, with Raney nickel as the metal catalyst. (Raney nickel is finely dispersed nickel with adsorbed hydrogen, so an external source of H_2 is not needed.) Aldehydes are reduced to primary alcohols, and ketones are reduced to secondary alcohols.

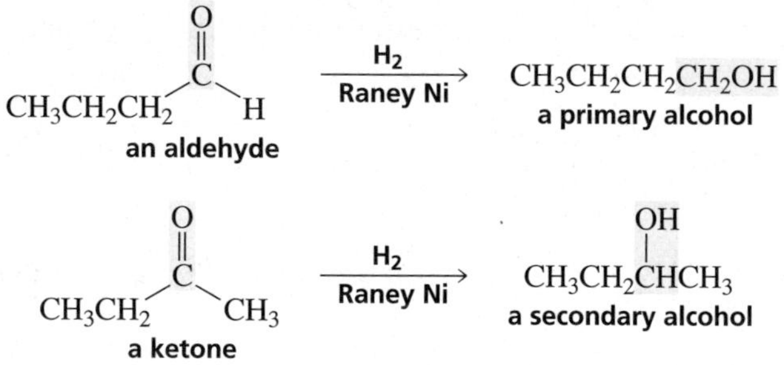

$$\underset{\textbf{an aldehyde}}{CH_3CH_2CH_2{-}\overset{\displaystyle O}{\overset{\|}{C}}{-}H} \ \xrightarrow{\overset{\textbf{H}_2}{\textbf{Raney Ni}}} \ \underset{\textbf{a primary alcohol}}{CH_3CH_2CH_2CH_2OH}$$

$$\underset{\textbf{a ketone}}{CH_3CH_2{-}\overset{\displaystyle O}{\overset{\|}{C}}{-}CH_3} \ \xrightarrow{\overset{\textbf{H}_2}{\textbf{Raney Ni}}} \ \underset{\textbf{a secondary alcohol}}{CH_3CH_2\overset{\displaystyle OH}{\overset{|}{C}H}CH_3}$$

The reduction of an acyl chloride can be stopped at an aldehyde if a partially deactivated catalyst is used. This reaction is known as the **Rosenmund reduction**. The catalyst for the Rosenmund reduction is similar to the partially deactivated palladium catalyst used to stop the reduction of an alkyne at a cis alkene (Section 6.9).

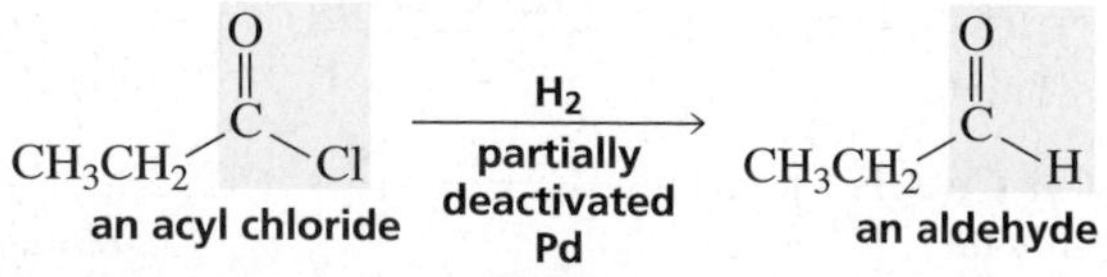

$$\underset{\textbf{an acyl chloride}}{CH_3CH_2{-}\overset{\displaystyle O}{\overset{\|}{C}}{-}Cl} \ \xrightarrow[\substack{\textbf{partially}\\\textbf{deactivated}\\\textbf{Pd}}]{\textbf{H}_2} \ \underset{\textbf{an aldehyde}}{CH_3CH_2{-}\overset{\displaystyle O}{\overset{\|}{C}}{-}H}$$

The carbonyl groups of carboxylic acids, esters, and amides are less reactive than the carbonyl groups of aldehydes and ketones and therefore are harder to reduce (Section 17.6). They cannot be reduced by catalytic hydrogenation (except under extreme conditions). They can, however, be reduced by a method we will discuss later in this section.

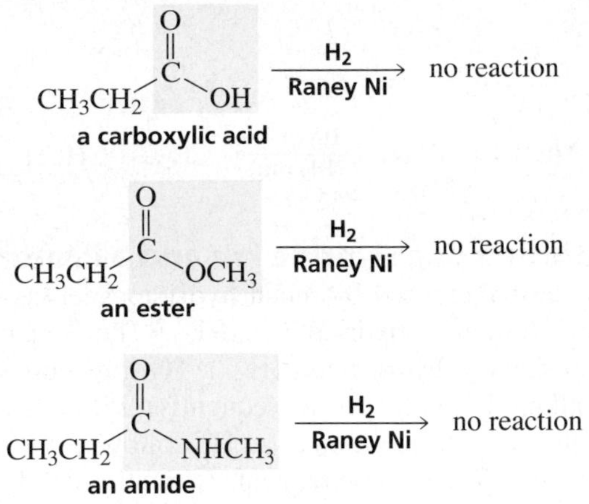

$$CH_3CH_2-\overset{\overset{\displaystyle O}{\|}}{C}-OH \xrightarrow[\text{Raney Ni}]{H_2} \text{no reaction}$$
a carboxylic acid

$$CH_3CH_2-\overset{\overset{\displaystyle O}{\|}}{C}-OCH_3 \xrightarrow[\text{Raney Ni}]{H_2} \text{no reaction}$$
an ester

$$CH_3CH_2-\overset{\overset{\displaystyle O}{\|}}{C}-NHCH_3 \xrightarrow[\text{Raney Ni}]{H_2} \text{no reaction}$$
an amide

PROBLEM 2◆

Give the products of the following reactions:

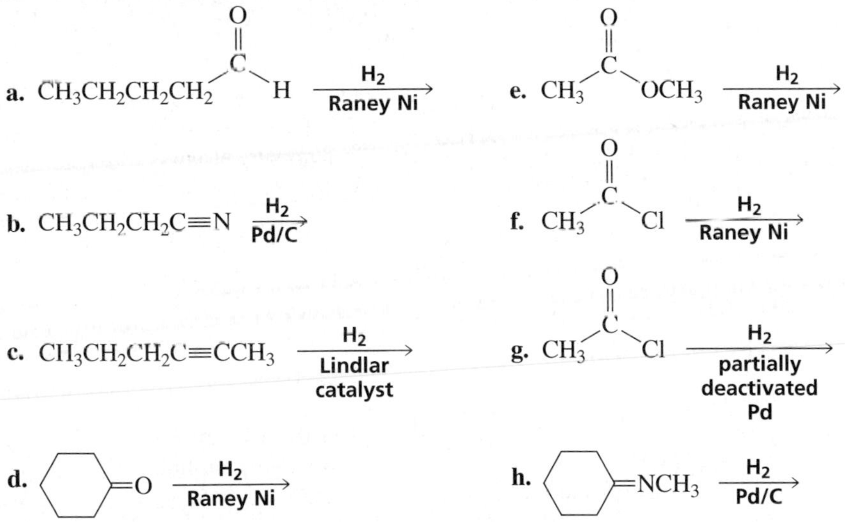

a. $CH_3CH_2CH_2CH_2-\overset{\overset{\displaystyle O}{\|}}{C}-H \xrightarrow[\text{Raney Ni}]{H_2}$

e. $CH_3-\overset{\overset{\displaystyle O}{\|}}{C}-OCH_3 \xrightarrow[\text{Raney Ni}]{H_2}$

b. $CH_3CH_2CH_2C\equiv N \xrightarrow[\text{Pd/C}]{H_2}$

f. $CH_3-\overset{\overset{\displaystyle O}{\|}}{C}-Cl \xrightarrow[\text{Raney Ni}]{H_2}$

c. $CH_3CH_2CH_2C\equiv CCH_3 \xrightarrow[\substack{\text{Lindlar}\\\text{catalyst}}]{H_2}$

g. $CH_3-\overset{\overset{\displaystyle O}{\|}}{C}-Cl \xrightarrow[\substack{\text{partially}\\\text{deactivated}\\\text{Pd}}]{H_2}$

d. $\bigcirc\!\!=\!\!O \xrightarrow[\text{Raney Ni}]{H_2}$

h. $\bigcirc\!\!=\!\!NCH_3 \xrightarrow[\text{Pd/C}]{H_2}$

Reduction by Addition of an Electron, a Proton, an Electron, and a Proton

When a compound is reduced using sodium in liquid ammonia, sodium donates an electron to the compound and ammonia donates a proton. This sequence is repeated, so the overall reaction adds two electrons and two protons to the compound. Such a reaction is known as a **dissolving-metal reduction**.

In Section 6.9, you saw the mechanism for the dissolving-metal reduction that converts an alkyne to a trans alkene.

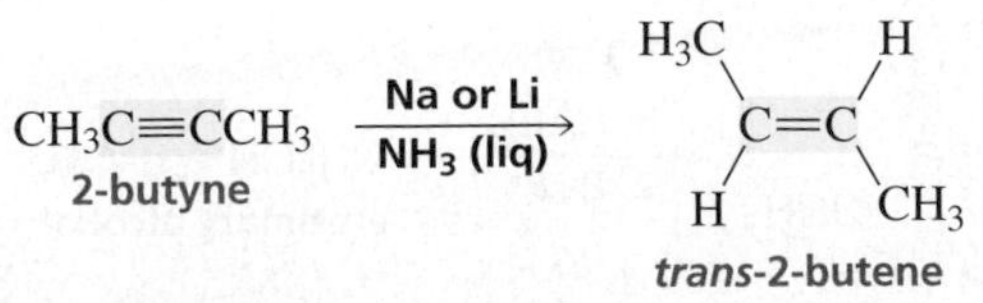

$$CH_3C\equiv CCH_3 \xrightarrow[\text{NH}_3 \text{ (liq)}]{\text{Na or Li}} \underset{\substack{\displaystyle H \quad\quad CH_3 \\ \textit{trans}\text{-2-butene}}}{\overset{\displaystyle H_3C \quad\quad H}{\underset{\displaystyle}{C=C}}}$$
2-butyne

Sodium (or lithium) in liquid ammonia cannot reduce a carbon–carbon double bond. This reagent is therefore useful for reducing a triple bond in a compound that also contains a double bond.

$$CH_3\overset{CH_3}{C}=CHCH_2C\equiv CCH_3 \xrightarrow[\text{NH}_3\text{ (liq)}]{\text{Na or Li}} CH_3\overset{CH_3}{C}=CHCH_2 \overset{H}{\underset{H}{C}}=\overset{CH_3}{\underset{H}{C}}$$

Reduction by Addition of a Hydride Ion and a Proton

Carbonyl groups are easily reduced by metal hydrides such as sodium borohydride ($NaBH_4$) or lithium aluminum hydride ($LiAlH_4$). The actual reducing agent in **metal-hydride reductions** is hydride ion ($H:^-$). Hydride ion adds to the carbonyl carbon, forming an alkoxide ion that is subsequently protonated. In other words, the carbonyl group is reduced by the addition of an H^- followed by the addition of an H^+. The mechanisms for reduction by these reagents were discussed in Section 17.6.

$$H:^- \quad \overset{O}{\underset{|}{C}} \longrightarrow \overset{O^-}{\underset{|}{-C-}} \xrightarrow{H_3O^+} \overset{OH}{\underset{|}{-C-}}$$

The numbers in front of the reagents above or below a reaction arrow indicate that the second reagent is not added until reaction with the first reagent is completed.

Aldehydes, ketones, and acyl halides can be reduced to alcohols by sodium borohydride.

$$CH_3CH_2CH_2\overset{O}{\underset{}{C}}H \xrightarrow[\text{2. H}_3\text{O}^+]{\text{1. NaBH}_4} CH_3CH_2CH_2CH_2OH$$
an aldehyde — a primary alcohol

$$CH_3CH_2CH_2\overset{O}{\underset{}{C}}CH_3 \xrightarrow[\text{2. H}_3\text{O}^+]{\text{1. NaBH}_4} CH_3CH_2CH_2\overset{OH}{\underset{}{C}HCH_3}$$
a ketone — a secondary alcohol

$$CH_3CH_2CH_2\overset{O}{\underset{}{C}}Cl \xrightarrow[\text{2. H}_3\text{O}^+]{\text{1. NaBH}_4} CH_3CH_2CH_2CH_2OH$$
an acyl chloride — a primary alcohol

The metal–hydrogen bonds in lithium aluminum hydride are more polar (that is, the difference in electronegativity between aluminum and hydrogen is larger) than those in sodium borohydride, so $LiAlH_4$ is a stronger reducing agent. Consequently, both $LiAlH_4$ and $NaBH_4$ reduce aldehydes, ketones, and acyl halides, but $LiAlH_4$ is not generally used for this purpose since $NaBH_4$ is safer and easier to use. $LiAlH_4$ is generally used to reduce only compounds—such as carboxylic acids, esters, and amides—that cannot be reduced by the milder reagent.

$$CH_3CH_2CH_2\overset{O}{\underset{}{C}}OH \xrightarrow[\text{2. H}_3\text{O}^+]{\text{1. LiAlH}_4} CH_3CH_2CH_2CH_2OH + H_2O$$
a carboxylic acid — a primary alcohol

$$CH_3CH_2\overset{O}{\underset{}{C}}OCH_3 \xrightarrow[\text{2. H}_3\text{O}^+]{\text{1. LiAlH}_4} CH_3CH_2CH_2OH + CH_3OH$$
an ester — a primary alcohol

If diisobutylaluminum hydride (DIBALH) is used as the hydride donor at a low temperature, the reduction of the ester can be stopped after the addition of one equivalent of hydride ion. Therefore, the final products of the reaction are an aldehyde and an alcohol (Section 17.6).

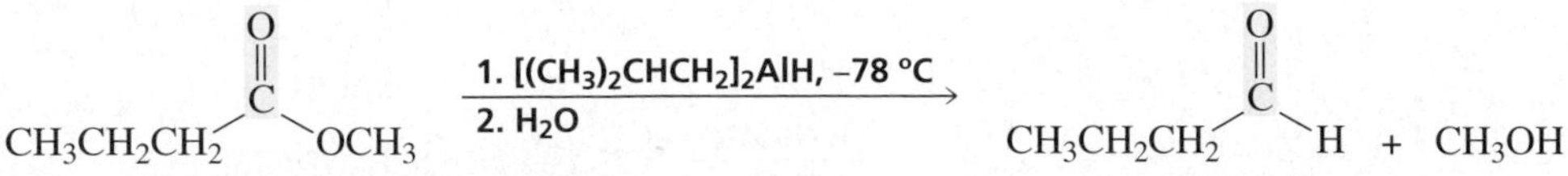

$$CH_3CH_2CH_2 \overset{\overset{\displaystyle O}{\|}}{C} OCH_3 \xrightarrow[\text{2. H}_2\text{O}]{\text{1. [(CH}_3\text{)}_2\text{CHCH}_2\text{]}_2\text{AlH, }-78\text{ °C}} CH_3CH_2CH_2 \overset{\overset{\displaystyle O}{\|}}{C} H + CH_3OH$$

an ester $\qquad\qquad\qquad\qquad\qquad\qquad$ an aldehyde

Replacing some of the hydrogens of $LiAlH_4$ with alkoxy (OR) groups decreases the reactivity of the metal hydride. For example, lithium tri-*tert*-butoxyaluminum hydride reduces an acyl chloride to an aldehyde, whereas $LiAlH_4$ reduces the acyl chloride all the way to an alcohol.

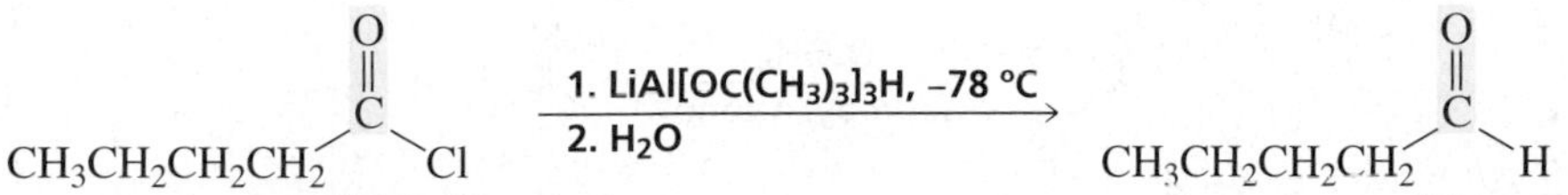

$$CH_3CH_2CH_2CH_2 \overset{\overset{\displaystyle O}{\|}}{C} Cl \xrightarrow[\text{2. H}_2\text{O}]{\text{1. LiAl[OC(CH}_3\text{)}_3\text{]}_3\text{H, }-78\text{ °C}} CH_3CH_2CH_2CH_2 \overset{\overset{\displaystyle O}{\|}}{C} H$$

an acyl chloride $\qquad\qquad\qquad\qquad\qquad\qquad$ an aldehyde

The carbonyl group of an amide is reduced to a methylene group (CH_2) by lithium aluminum hydride (Section 17.6). Primary, secondary, and tertiary amines are formed, depending on the number of substituents bonded to the nitrogen of the amide. To obtain the amine in its neutral basic form, acid is not used in the second step of the reaction.

$$CH_3CH_2CH_2 \overset{\overset{\displaystyle O}{\|}}{C} NH_2 \xrightarrow[\text{2. H}_2\text{O}]{\text{1. LiAlH}_4} CH_3CH_2CH_2CH_2NH_2$$

a primary amine

$$CH_3CH_2CH_2 \overset{\overset{\displaystyle O}{\|}}{C} NHCH_3 \xrightarrow[\text{2. H}_2\text{O}]{\text{1. LiAlH}_4} CH_3CH_2CH_2CH_2NHCH_3$$

a secondary amine

$$CH_3CH_2CH_2 \overset{\overset{\displaystyle O}{\|}}{C} \underset{\underset{\displaystyle CH_3}{|}}{N}CH_3 \xrightarrow[\text{2. H}_2\text{O}]{\text{1. LiAlH}_4} CH_3CH_2CH_2CH_2\underset{\underset{\displaystyle CH_3}{|}}{\overset{\overset{\displaystyle CH_3}{|}}{N}}CH_3$$

a tertiary amine

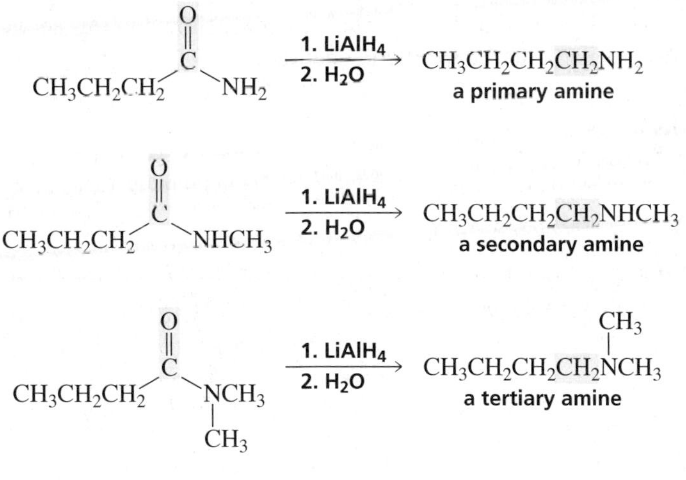

Because sodium borohydride cannot reduce an ester, amide, or carboxylic acid, it can be used to selectively reduce an aldehyde or a ketone group in a compound that also contains a less reactive group. Acid is not used in the second step of the reaction shown here to avoid hydrolyzing the ester:

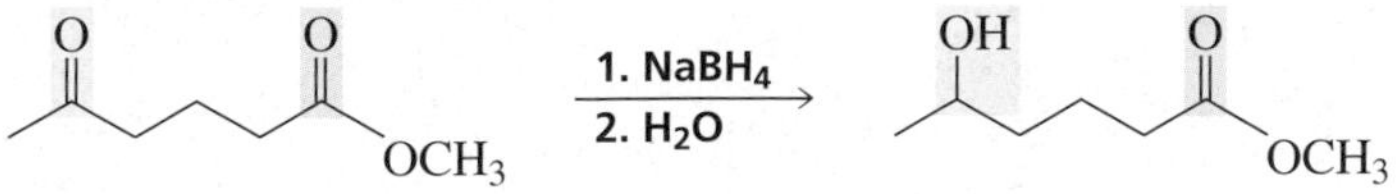

(structure with ketone and ester) $\xrightarrow[\text{2. H}_2\text{O}]{\text{1. NaBH}_4}$ (structure with OH and ester)

The multiply bonded carbon atoms of alkenes and alkynes do not possess a partial positive charge and therefore will not react with reagents that reduce compounds by donating a hydride ion.

$$CH_3CH_2CH{=}CH_2 \xrightarrow{\text{NaBH}_4} \text{no reduction reaction}$$

$$CH_3CH_2C{\equiv}CH \xrightarrow{\text{NaBH}_4} \text{no reduction reaction}$$

Because sodium borohydride cannot reduce carbon–carbon double bonds, a carbonyl group in a compound that also has an alkene functional group can be selectively reduced, as long as the double bonds are not conjugated (Section 17.16). Acid is not used in the second step of the reaction to avoid addition of the acid to the double bond.

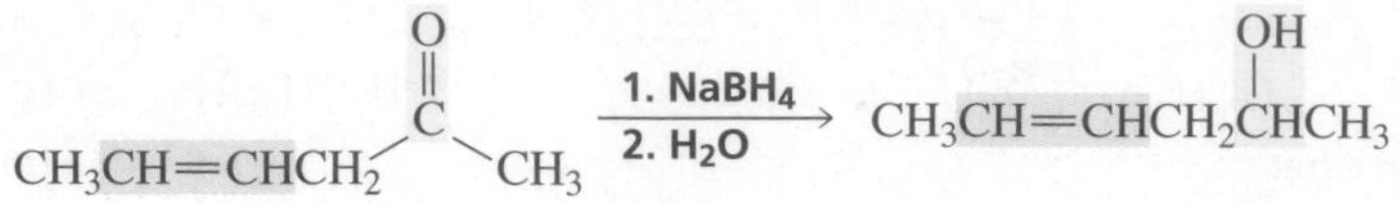

$$CH_3CH{=}CHCH_2\overset{O}{\underset{}{\overset{\|}{C}}}CH_3 \xrightarrow[\text{2. H}_2\text{O}]{\text{1. NaBH}_4} CH_3CH{=}CHCH_2\overset{OH}{\underset{}{\overset{|}{C}}}HCH_3$$

A **chemoselective reaction** is a reaction in which a reagent reacts with one functional group in preference to another. For example, NaBH$_4$ in isopropyl alcohol reduces aldehydes faster than it reduces ketones.

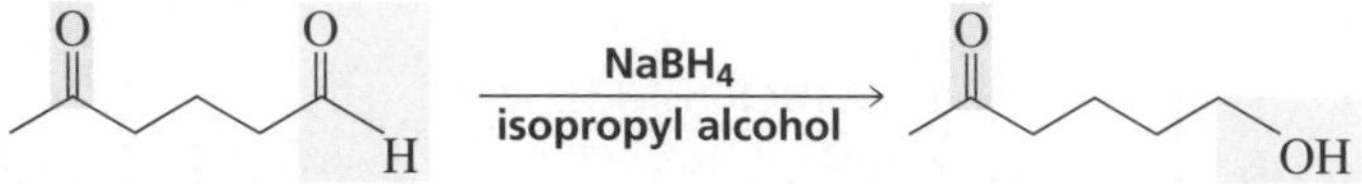

$$\xrightarrow[\text{isopropyl alcohol}]{\text{NaBH}_4}$$

In contrast, NaBH$_4$ in aqueous ethanol at $-15\ °C$ in the presence of cerium trichloride reduces ketones faster than it reduces aldehydes.

Student Tutorial:
Reductions

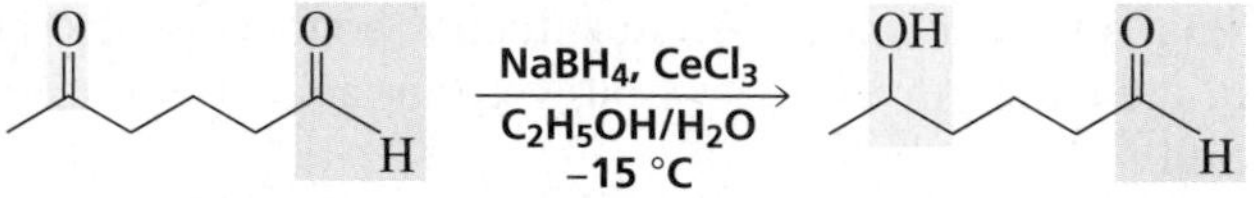

$$\xrightarrow[\substack{\text{C}_2\text{H}_5\text{OH/H}_2\text{O} \\ -15\ °C}]{\text{NaBH}_4,\ \text{CeCl}_3}$$

As we have already mentioned, the reducing agents shown in this section are only a fraction of those available to the synthetic chemist.

PROBLEM 3◆

Explain why terminal alkynes cannot be reduced by Na in liquid NH$_3$.

PROBLEM 4◆

Give the products of the following reactions:

a. 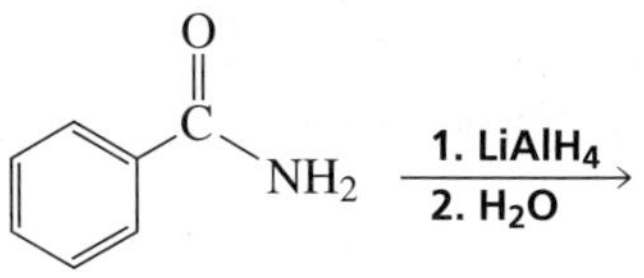 $\xrightarrow[\text{2. H}_2\text{O}]{\text{1. LiAlH}_4}$

d. 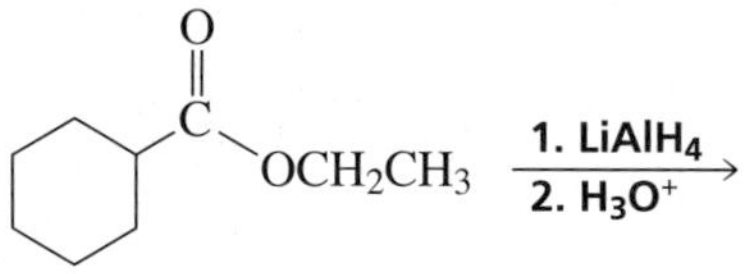 $\xrightarrow[\text{2. H}_3\text{O}^+]{\text{1. LiAlH}_4}$

b. 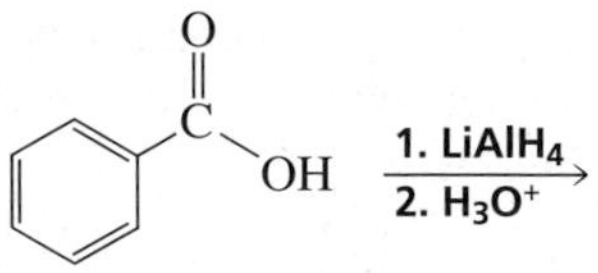 $\xrightarrow[\text{2. H}_3\text{O}^+]{\text{1. LiAlH}_4}$

e.  $CH_3CH_2\overset{O}{\underset{}{\overset{\|}{C}}}NHCH_2CH_3 \xrightarrow[\text{2. H}_2\text{O}]{\text{1. LiAlH}_4}$

c. $CH_3CH_2\overset{O}{\underset{}{\overset{\|}{C}}}CH_2CH_3 \xrightarrow[\text{2. H}_3\text{O}^+]{\text{1. NaBH}_4}$

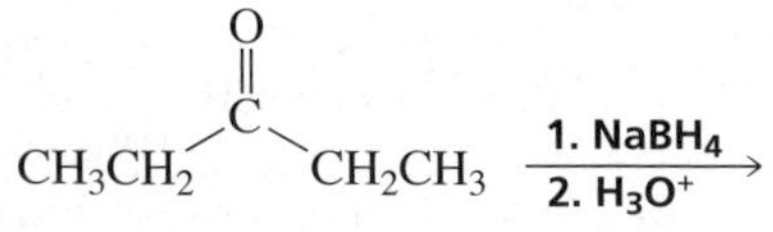

f. 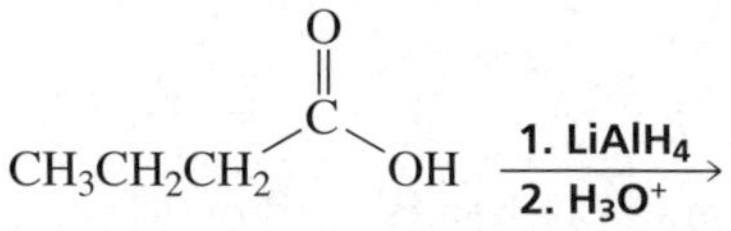 $CH_3CH_2CH_2\overset{O}{\underset{}{\overset{\|}{C}}}OH \xrightarrow[\text{2. H}_3\text{O}^+]{\text{1. LiAlH}_4}$

PROBLEM 5◆

Can carbon–nitrogen double and triple bonds be reduced by lithium aluminum hydride? Explain your answer.

PROBLEM 6

Give the products of the following reactions (assume that excess reducing agent is used in part d):

a.

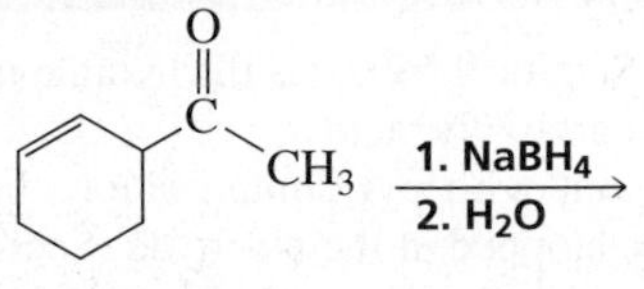

![chemical structure with cyclohexene ring bearing C(=O)CH₃] 1. NaBH₄
2. H₂O

c.

![chemical structure cyclohexanone with CH₂COCH₃] 1. NaBH₄
2. H₂O

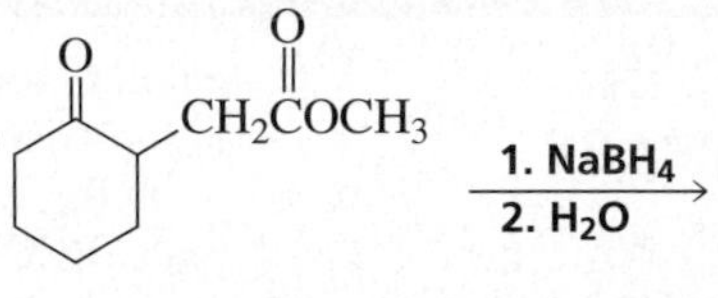

b.

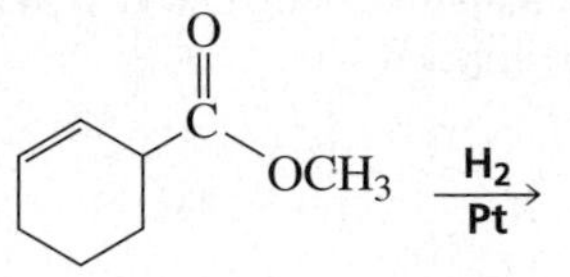

![chemical structure cyclohexene ring bearing C(=O)OCH₃] H₂ / Pt

d.

![chemical structure cyclohexanone with CH₂COCH₃] 1. LiAlH₄
2. H₂O

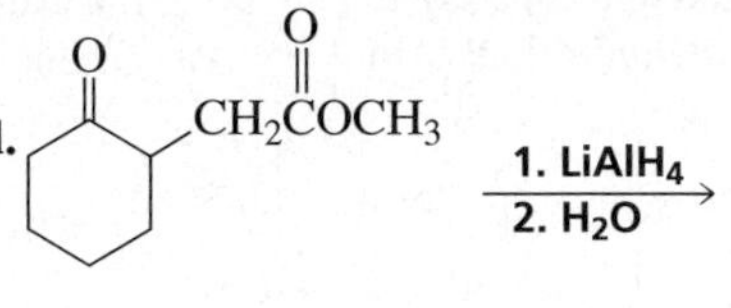

PROBLEM 7 | **SOLVED**

How could you synthesize the following compounds from starting materials containing no more than four carbons?

a.

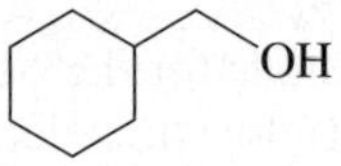

![cyclohexane with CH₂OH]

b.

![cyclohexane with ethyl group]

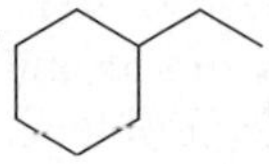

Solution to 7a The six-membered ring indicates that the compound can be synthesized by means of a Diels–Alder reaction (Section 7.12).

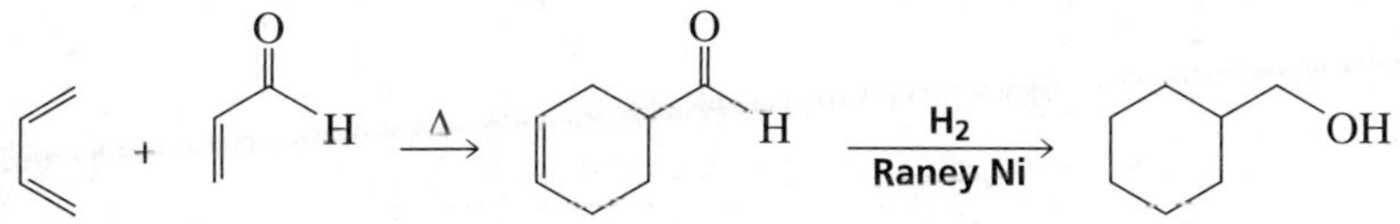

![Diels-Alder scheme: diene + acrolein →Δ→ cyclohexene carbaldehyde →H₂/Raney Ni→ cyclohexane-CH₂OH]

19.2 **Oxidation of Alcohols**

Student Tutorial:
Oxidation reaction of alcohols

Oxidation is the reverse of reduction. For example, a ketone is *reduced* to a secondary alcohol, and the reverse reaction is the *oxidation* of a secondary alcohol to a ketone.

$$\text{ketone} \underset{\text{oxidation}}{\overset{\text{reduction}}{\rightleftarrows}} \text{secondary alcohol}$$

We have seen that chromic acid (H_2CrO_4) is a reagent commonly used to oxidize secondary alcohols to ketones (Section 10.5).

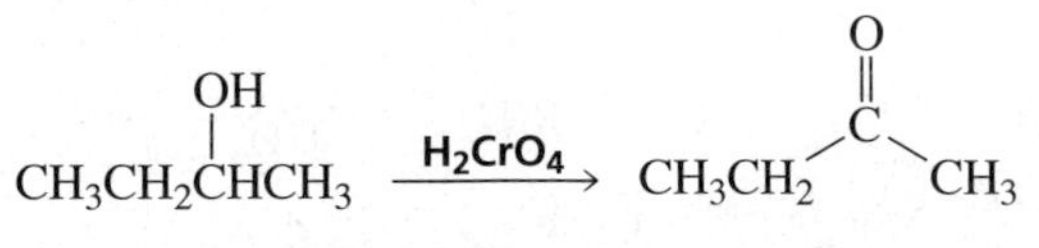

$$\underset{\text{secondary alcohol}}{CH_3CH_2\overset{\overset{\displaystyle OH}{|}}{C}HCH_3} \xrightarrow{H_2CrO_4} \underset{\text{ketone}}{CH_3CH_2\overset{\overset{\displaystyle O}{\|}}{C}CH_3}$$

Primary alcohols are initially oxidized to aldehydes by chromic acid and other chromium-containing reagents. The reaction does not stop at the aldehyde, however. Instead, the aldehyde is further oxidized to a carboxylic acid (Section 10.5).

VCL **Alcohol Oxidation-1**

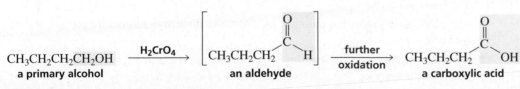

$$\underset{\text{a primary alcohol}}{CH_3CH_2CH_2CH_2OH} \xrightarrow{H_2CrO_4} \underset{\text{an aldehyde}}{\left[CH_3CH_2CH_2\overset{\overset{\displaystyle O}{\|}}{C}H \right]} \xrightarrow[\text{oxidation}]{\text{further}} \underset{\text{a carboxylic acid}}{CH_3CH_2CH_2\overset{\overset{\displaystyle O}{\|}}{C}OH}$$

These reactions are easily recognized as oxidations because the number of C—H bonds in the reactant decreases and the number of C—O bonds increases.

THE ROLE OF HYDRATES IN THE OXIDATION OF PRIMARY ALCOHOLS

When a primary alcohol is oxidized to a carboxylic acid, the alcohol is initially oxidized to an aldehyde that is in equilibrium with its hydrate (Section 17.9). It is the hydrate that is subsequently oxidized to a carboxylic acid.

If the reaction is carried out with pyridinium chlorochromate (PCC), oxidation can be stopped at the aldehyde (Section 10.5), because PCC is used in an anhydrous solvent. If water is not present, the hydrate cannot be formed.

$$CH_3CH_2OH \xrightarrow{H_2CrO_4} \underset{CH_3}{\overset{O}{\|}}\!\!-\!\!H \underset{H_2O}{\overset{H^+}{\rightleftharpoons}} CH_3\underset{OH}{\overset{OH}{CH}} \xrightarrow{H_2CrO_4} \underset{CH_3}{\overset{O}{\|}}\!\!-\!\!OH$$

Because of the toxicity of chromium-based reagents, other methods for the oxidation of alcohols have been developed. One of the most widely employed, called the **Swern oxidation**, uses dimethyl sulfoxide [$(CH_3)_2SO$], oxalyl chloride [$(COCl)_2$], and triethylamine. Since the reaction is *not* carried out in an aqueous solution, the oxidation of a primary alcohol stops at the aldehyde (as in PCC oxidation). Secondary alcohols are oxidized to ketones.

Alcohol Oxidation-2

$$CH_3CH_2CH_2OH \xrightarrow[\text{2. triethylamine}]{\text{1. }CH_3\overset{O}{\overset{\|}{S}}CH_3,\ Cl\!-\!\overset{O}{\overset{\|}{C}}\!-\!\overset{O}{\overset{\|}{C}}\!-\!Cl,\ -60\ °C} CH_3CH_2\overset{O}{\overset{\|}{C}}H$$

a primary alcohol → **an aldehyde**

$$\underset{\text{a secondary alcohol}}{CH_3CH_2\overset{OH}{\overset{|}{C}}HCH_3} \xrightarrow[\text{2. triethylamine}]{\text{1. }CH_3\overset{O}{\overset{\|}{S}}CH_3,\ Cl\!-\!\overset{O}{\overset{\|}{C}}\!-\!\overset{O}{\overset{\|}{C}}\!-\!Cl,\ -60\ °C} \underset{\text{a ketone}}{CH_3CH_2\overset{O}{\overset{\|}{C}}CH_3}$$

The actual oxidizing agent in the Swern oxidation is dimethylchlorosulfonium ion, which is formed from an S_N2 reaction between dimethyl sulfoxide and oxalyl chloride.

Mechanism of the Swern oxidation

an E2 reaction

an S_N2 reaction

alcohol   dimethylchlorosulfonium ion

$(CH_3CH_2)_3\ddot{N}$
triethylamine

aldehyde or ketone

+ Cl⁻   + HCl   + CH_3SCH_3

- The alcohol displaces the chloride ion from the dimethylchlorosulfonium ion in an S_N2 reaction.
- Like chromic acid oxidation, the Swern oxidation uses an E2 reaction to form the aldehyde or ketone.

To understand how dimethyl sulfoxide and oxalyl chloride react to form the dimethylchlorosulfonium ion, see Problem 60.

TREATING ALCOHOLISM WITH ANTABUSE

Disulfiram, most commonly known by one of its trade names, Antabuse, is used to treat alcoholism. After the drug is taken, it causes violently unpleasant effects if ethanol is consumed within the next two days.

$$CH_3CH_2 \quad S \qquad\qquad S \quad CH_2CH_3$$
$$N-C-S-S-C-N$$
$$CH_3CH_2 \qquad\qquad\qquad CH_2CH_3$$
Antabuse®

Antabuse inhibits aldehyde dehydrogenase, the enzyme responsible for oxidizing acetaldehyde (a product of ethanol metabolism) to acetic acid, resulting in a buildup of acetaldehyde. It is the acetaldehyde that causes the unpleasant physiological effects of intoxication: intense flushing, nausea, dizziness, sweating, throbbing headaches, decreased blood pressure, and, ultimately, shock. Consequently, Antabuse should be taken only under strict medical supervision. In some people, aldehyde dehydrogenase does not function properly even under normal circumstances. Their symptoms in response to ingesting alcohol are nearly the same as those of individuals who are medicated with Antabuse.

Antabuse® inhibits this enzyme

$$CH_3CH_2OH \xrightarrow[\text{dehydrogenase}]{\text{alcohol}} \underset{\text{acetaldehyde}}{CH_3-\overset{\displaystyle O}{C}-H} \xrightarrow[\text{dehydrogenase}]{\text{aldehyde}} \underset{\text{acetic acid}}{CH_3-\overset{\displaystyle O}{C}-OH}$$
ethanol

FETAL ALCOHOL SYNDROME

The damage done to a human fetus when the mother drinks alcohol during her pregnancy is known as *fetal alcohol syndrome*. The harmful effects— growth retardation, decreased mental functioning, and facial and limb abnormalities—are attributable to acetaldehyde, formed from the oxidation of ethanol, crossing the placenta and accumulating in the liver of the fetus.

PROBLEM 8

Give the product formed from the reaction of each of the following alcohols with

a. an acidic solution of sodium dichromate.

b. the reagents required for a Swern oxidation.

   1. 3-pentanol   **3.** 2-methyl-2-pentanol   **5.** cyclohexanol

   2. 1-pentanol   **4.** 2,4-hexanediol   **6.** 1,4-butanediol

VCL   **Alcohol Oxidation-3**

PROBLEM 9

Propose a mechanism for the chromic acid oxidation of 1-propanol to propanal.

19.3 Oxidation of Aldehydes and Ketones

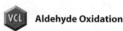

VCL   **Aldehyde Oxidation**

Aldehydes are oxidized to carboxylic acids. Because aldehydes are generally easier to oxidize than primary alcohols, any of the reagents described in the preceding section or in Section 10.5 for oxidizing primary alcohols to carboxylic acids can be used to oxidize aldehydes to carboxylic acids.

$$CH_3CH_2\overset{\overset{\displaystyle O}{\|}}{C}H \xrightarrow[\text{H}_2\text{SO}_4]{\text{Na}_2\text{Cr}_2\text{O}_7} CH_3CH_2\overset{\overset{\displaystyle O}{\|}}{C}OH$$

$$\text{(cyclohexyl)}CH \xrightarrow{\text{H}_2\text{CrO}_4} \text{(cyclohexyl)}C\text{OH}$$

aldehydes   **carboxylic acids**

BIOGRAPHY

Bernhard Tollens (1841–1918) *was born in Germany. He was a professor of chemistry at the University of Göttingen, the same university from which he received a Ph.D.*

BIOGRAPHY

Johann Friedrich Wilhelm Adolf von Baeyer (1835–1917) *started his study of chemistry under Bunsen and Kekulé (Section 7.2) at the University of Heidelberg and received a Ph.D. from the University of Berlin, where he studied under Hofmann (Section 20.4). (See also Section 2.11.)*

BIOGRAPHY

Victor Villiger (1868–1934) *was Baeyer's student. The two published the first paper on the Baeyer–Villiger oxidation in Chemische Berichte in 1899.*

Silver oxide is a mild oxidizing agent. *Tollens reagent*, a dilute solution of silver oxide in aqueous ammonia, oxidizes an aldehyde but it is too weak to oxidize an alcohol or any other functional group. An advantage to using Tollens reagent to oxidize an aldehyde is that the reaction occurs under basic conditions. Therefore, you do not have to worry about affecting other functional groups in the molecule that might undergo a reaction in an acidic solution.

$$CH_3CH_2\overset{\overset{\displaystyle O}{\|}}{C}H \xrightarrow[\text{2. H}_3\text{O}^+]{\text{1. Ag}_2\text{O, NH}_3} CH_3CH_2\overset{\overset{\displaystyle O}{\|}}{C}OH + \underset{\substack{\text{metallic}\\\text{silver}}}{\text{Ag}}$$

The specific oxidizing agent in Tollens reagent is Ag$^+$, which is reduced to metallic silver. This reaction is the basis of the **Tollens test**; if Tollens reagent is added to a small amount of an aldehyde in a test tube, the inside of the test tube becomes coated with a shiny mirror of metallic silver. Consequently, if a mirror is not formed when Tollens reagent is added to a compound, we can conclude that the compound does not contain an aldehyde functional group.

Ketones do not react with most of the reagents used to oxidize aldehydes. However, both aldehydes *and* ketones can be oxidized by a peroxyacid. Aldehydes are oxidized to carboxylic acids and ketones are oxidized to esters. A **peroxyacid** (also called a percarboxylic acid or an acyl hydroperoxide) contains one more oxygen than a carboxylic acid, and it is this oxygen that is inserted between the carbonyl carbon and the H of an aldehyde or the R of a ketone. The reaction is called a **Baeyer–Villiger oxidation**. A particularly good reagent for the Baeyer–Villiger reaction is peroxytrifluoroacetic acid.

Baeyer-Villiger oxidations

$$\underset{\text{an aldehyde}}{CH_3CH_2CH_2\overset{\overset{\displaystyle O}{\|}}{C}H} + \underset{\text{a peroxyacid}}{R\overset{\overset{\displaystyle O}{\|}}{C}OOH} \longrightarrow \underset{\text{a carboxylic acid}}{CH_3CH_2CH_2\overset{\overset{\displaystyle O}{\|}}{C}OH} + R\overset{\overset{\displaystyle O}{\|}}{C}OH$$

$$\underset{\text{a ketone}}{CH_3CH_2\overset{\overset{\displaystyle O}{\|}}{C}CH_2CH_3} + \underset{\text{a peroxyacid}}{R\overset{\overset{\displaystyle O}{\|}}{C}OOH} \longrightarrow \underset{\text{an ester}}{CH_3CH_2\overset{\overset{\displaystyle O}{\|}}{C}OCH_2CH_3} + R\overset{\overset{\displaystyle O}{\|}}{C}OH$$

If the two alkyl substituents attached to the carbonyl group of the ketone are not the same, on what side of the carbonyl carbon is the oxygen inserted? For example, does the oxidation of cyclohexyl methyl ketone form methyl cyclohexanecarboxylate or cyclohexyl acetate?

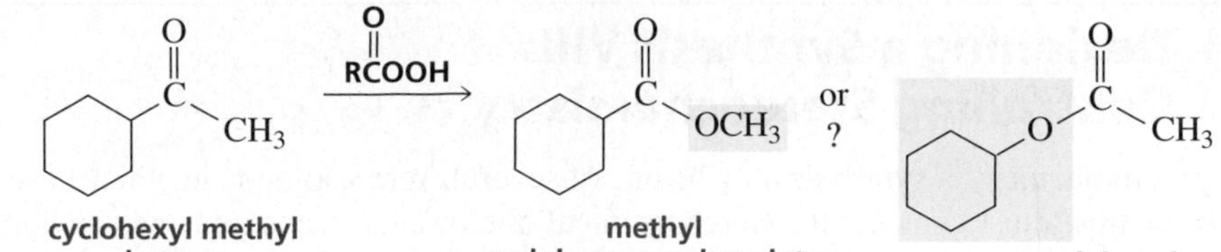

O
‖
RCOOH
———→

cyclohexyl methyl
ketone

methyl
cyclohexanecarboxylate

or ?

cyclohexyl
acetate

To answer this question, we must look at the mechanism of the reaction.

Mechanism of the Baeyer–Villiger oxidation

$$R-C(=O)-R' \ + \ CF_3COO:^- \ \rightleftharpoons \ R-\overset{:O:^-}{\underset{O-OCCF_3}{\overset{|}{C}}}-R' \ \longrightarrow \ R'-C(=O)-OR \ + \ CF_3CO^-$$

a weak O—O bond

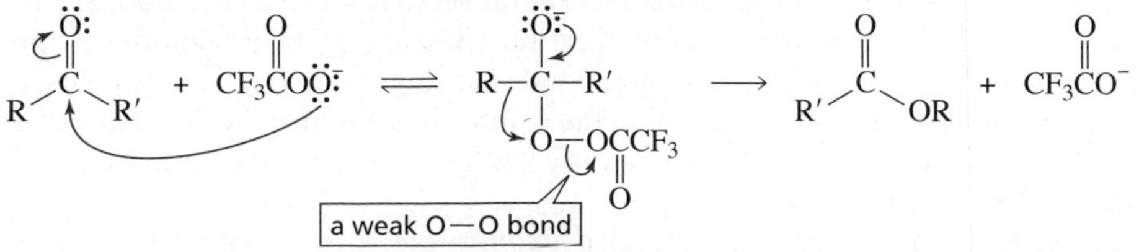

an unstable
intermediate

- The ketone and the peroxyacid react to form an unstable tetrahedral intermediate with a weak O—O bond.
- As the O—O bond breaks heterolytically, one of the alkyl groups migrates to an oxygen. This rearrangement is similar to the 1,2-shifts that occur when carbocations rearrange (Section 4.6).

Studies of the migration tendencies of different groups have established the following order:

relative migration tendencies

most likely
to migrate ⟶ H > *tert*-alkyl > *sec*-alkyl = phenyl > primary alkyl > methyl ⟵ least likely
to migrate

Therefore, the product of the Baeyer–Villiger oxidation of cyclohexyl methyl ketone will be cyclohexyl acetate because a secondary alkyl group (the cyclohexyl group) is more likely to migrate than a methyl group. Aldehydes are always oxidized to carboxylic acids, since H has the greatest tendency to migrate.

 **VCL** **Baeyer–Villiger Oxidation**

PROBLEM 10

Give the products of the following reactions:

a. 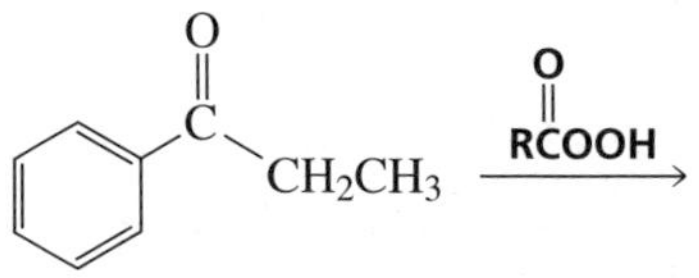
O
‖
C—CH₂CH₃
RCOOH
———→

d. CH₃CH(CH₃)—C(=O)—C(CH₃)CH₃ (with CH₃)
RCOOH
———→

b.
O
‖
C—H
RCOOH
———→

e. CH₃CH₂CH₂—C(=O)—H
1. Ag₂O, NH₃
2. H₃O⁺
———→

c.
O
‖ (cyclopentanone)
—CH₃
RCOOH
———→

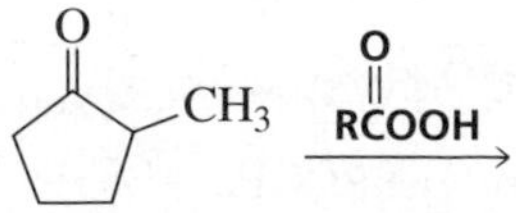

f. 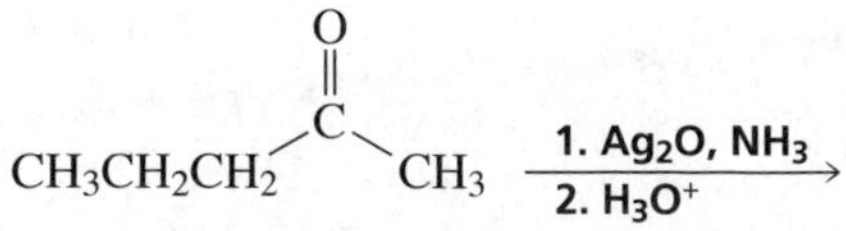 CH₃CH₂CH₂—C(=O)—CH₃
1. Ag₂O, NH₃
2. H₃O⁺
———→

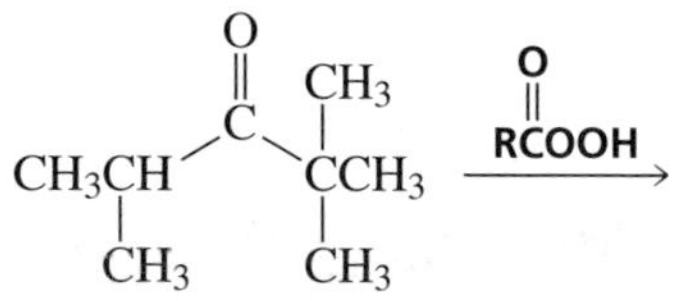

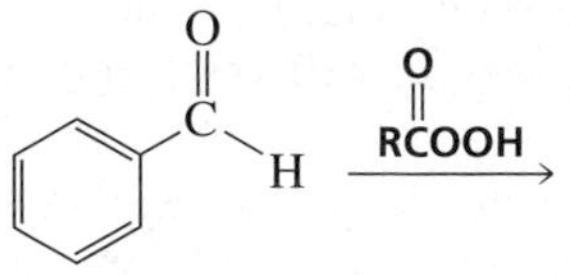

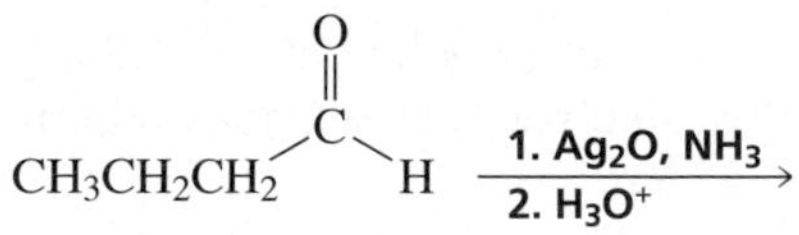

19.4 Designing a Synthesis VIII: Controlling Stereochemistry

The target molecule of a synthesis may be one of several stereoisomers, in which case an organic chemist must consider the stereochemical outcome of every step and use highly stereoselective reagents to achieve the desired configuration. If the stereochemistry of the reactions is not controlled, the resulting mixture of stereoisomers may be difficult or even impossible to separate.

The number of possible stereoisomers complicating a synthesis depends on the number of double bonds and asymmetric centers in the target molecule, because each double bond can exist in an *E* or *Z* configuration (Section 3.5) and each asymmetric center can have an *R* or *S* configuration (Section 5.7). In addition, if the target molecule has rings with a common bond, the rings can be either trans fused or cis fused (Section 26.10). In designing the synthesis, care must be taken to make sure that each double bond, each asymmetric center, and each ring fusion has the appropriate configuration.

Some stereoselective reactions are also *enantioselective*; an **enantioselective reaction** forms more of one enantiomer than of another. We have seen that an enantiomerically pure compound can be obtained if an enzyme is used to catalyze the reaction that forms the compound. Enzyme-catalyzed reactions result in the exclusive formation of one enantiomer since enzymes are chiral (Section 5.20). For example, ketones are enzymatically reduced to alcohols by enzymes called alcohol dehydrogenases. Whether the *R* or the *S* enantiomer is formed depends on the particular alcohol dehydrogenase used: alcohol dehydrogenase from the bacterium *Lactobacillus kefir* forms *R* alcohols, whereas alcohol dehydrogenases from yeast, horse liver, and the bacterium *Thermoanaerobium brocki* form *S* alcohols. The alcohol dehydrogenases use the biological reducing agent NADPH to carry out the reduction (Section 24.2).

$$ + \text{ NADPH } + \text{ H}^+ \xrightarrow[\text{alcohol dehydrogenase}]{\textit{Lactobacillus kefir}} + \text{ NADP}^+ $$

(R)-2,2-trifluoro-1-phenyl-1-ethanol

$$ + \text{ NADPH } + \text{ H}^+ \xrightarrow[\text{alcohol dehydrogenase}]{\textit{Thermoanaerobium brocki}} + \text{ NADP}^+ $$

(S)-2,2-trifluoro-1-phenyl-1-ethanol

Unfortunately, this method of controlling the configuration of a target is not universally applicable because enzymes require substrates of very specific size and shape (Section 23.8).

Alternatively, an enantiomerically pure catalyst that is not an enzyme can be used to obtain an enantiomerically pure compound. We have seen that an alkene can be oxidized to an epoxide by a peroxyacid (Section 4.9).

$$ \underset{\text{an alkene}}{\text{RCH=CH}_2} + \underset{\text{a peroxyacid}}{\text{RCOOH}} \longrightarrow \underset{\text{an epoxide}}{\text{RCH–CH}_2} + \underset{\text{a carboxylic acid}}{\text{RCOH}} $$

An enantiomerically pure epoxide of an allylic alcohol can be prepared by treating the alcohol with *tert*-butyl hydroperoxide, titanium isopropoxide, and enantiomerically pure diethyl tartrate (DET). The structure of the epoxide depends on which enantiomer of diethyl tartrate is used.

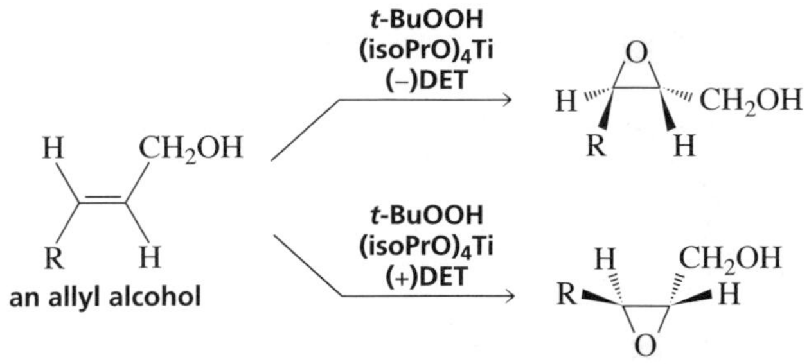

This method, developed in 1980 by Barry Sharpless, has proven to be useful for the synthesis of a wide variety of enantiomerically pure compounds, because an epoxide, being very susceptible to attack by nucleophiles, can easily be converted into a compound with two adjacent asymmetric centers. In the following example, an allylic alcohol is converted into an enantiomerically pure epoxide, which is then used to form an enantiomerically pure diol.

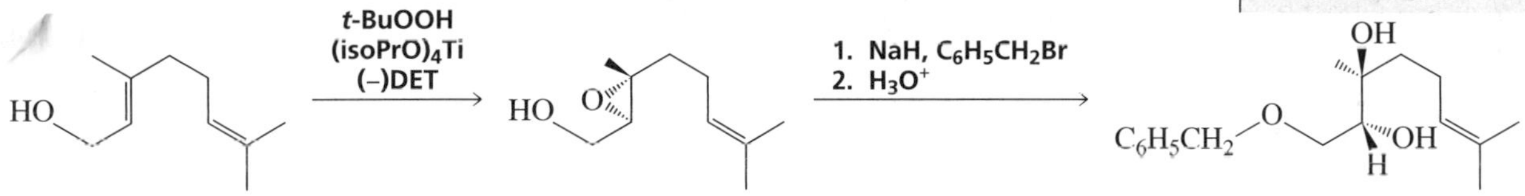

PROBLEM 11

What is the product of the reaction of methylmagnesium bromide with either of the enantiomerically pure epoxides that can be prepared from (*E*)-3-methyl-2-penten-1-ol by the method shown immediately above? Assign *R* or *S* configurations to the asymmetric centers of each product.

PROBLEM 12◆

What stereoisomers are formed from the reaction of cyclohexene oxide with methoxide ion in methanol?

PROBLEM 13◆

Is the addition of Br_2 to an alkene such as *trans*-2-pentene a stereoselective reaction? Is it a stereospecific reaction? Is it an enantioselective reaction?

19.5 Hydroxylation of Alkenes

An alkene can be oxidized to a 1,2-diol either by potassium permanganate ($KMnO_4$) in a cold basic solution or by osmium tetroxide (OsO_4). The solution of potassium permanganate must be basic, and the oxidation must be carried out at room temperature or below. If the solution is heated or if it is acidic, the diol will be oxidized further (Section 19.7). A diol is also called a **glycol**. The OH groups are on adjacent carbons in 1,2-diols, so 1,2-diols are also known as **vicinal diols** or **vicinal glycols**. (Recall that vicinal means the two OH groups are on adjacent carbons; Section 4.7.)

$$CH_3CH{=}CHCH_3 \xrightarrow[\text{cold}]{KMnO_4,\ HO^-,\ H_2O} CH_3\underset{|}{\overset{OH}{C}}H{-}\overset{OH}{\underset{|}{C}}HCH_3$$

a vicinal diol

BIOGRAPHY

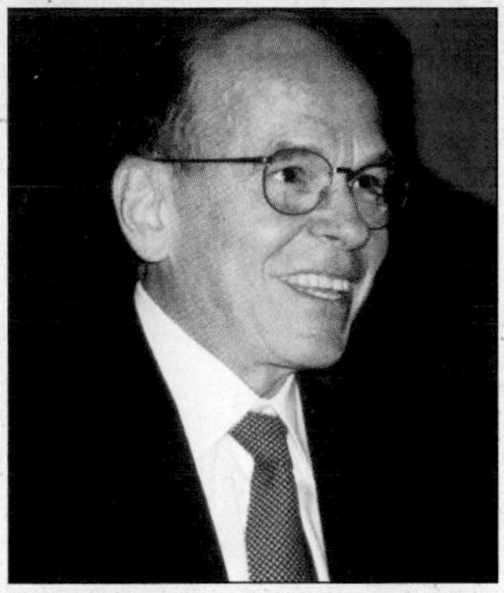

The 2001 Nobel Prize in chemistry was awarded to **K. Barry Sharpless, William S. Knowles,** *and* **Royji Noyroi** *for their work in the field of asymmetric catalysis.* **K. Barry Sharpless** *was born in Philadelphia in 1941. He received a B.A. from Dartmouth in 1963 and a Ph.D. in chemistry from Stanford in 1968. He served as a professor at MIT and Stanford. Currently, he is at the Scripps Research Institute in La Jolla, California. He received the 2001 Nobel Prize in chemistry for his work on chirally catalyzed oxidation reactions. (See also Section 24.24.)* **William Knowles,** *was born in 1917. He received a Ph.D. from Columbia University in 1942 and is a scientist at Monsanto Company, St. Louis, Missouri.* **Royji Noyroi** *was born in 1938 in Kobe, Japan. He received a Ph.D. from Kyoto University and is a Professor of materials science at Nagoya University in Nagoya, Japan. Knowles and Noyroi were cited for their work on chirally catalyzed hydrogenation.*

$$CH_3CH_2CH{=}CH_2 \xrightarrow[\text{2. H}_2\text{O}_2]{\text{1. OsO}_4} \overset{\overset{\displaystyle OH}{|}}{CH_3CH_2CHCH_2OH}$$
a vicinal diol

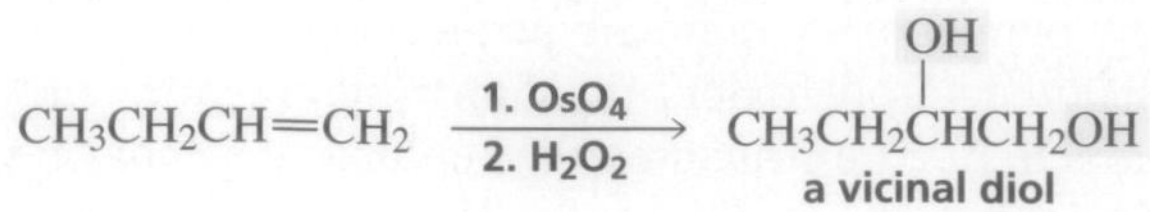

Both $KMnO_4$ and OsO_4 form a cyclic intermediate when they react with an alkene. The reactions occur because manganese and osmium are in a highly positive oxidation state and therefore attract electrons. (Manganese and osmium have oxidation states of $+7$ and $+8$, respectively.)

Mechanisms for cis glycol formation

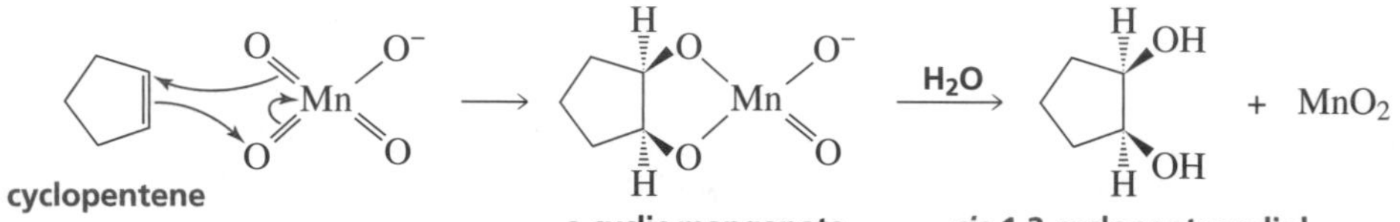

cyclopentene   →   a cyclic manganate intermediate   →   cis-1,2-cyclopentanediol $+$ MnO_2

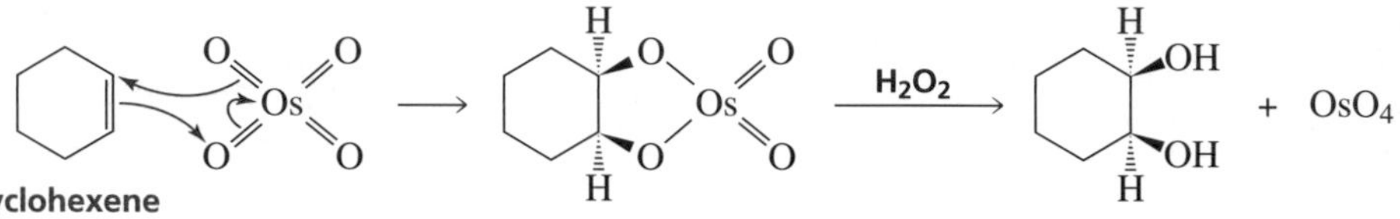

cyclohexene   →   a cyclic osmate intermediate   →   cis-1,2-cyclohexanediol $+$ OsO_4

Alkene Dihydroxylation

- Formation of the cyclic intermediate is a syn addition because both oxygens are delivered to the same side of the double bond. Therefore, the oxidation reaction is stereospecific—a cis cycloalkene forms only a cis diol.

- When hydrolyzed, the cyclic manganate intermediate opens to a cis diol. The cyclic osmate intermediate is hydrolyzed with hydrogen peroxide that reoxidizes the osmium reagent to osmium tetroxide. (Because osmium tetroxide is recycled, only a catalytic amount of this expensive and toxic oxidizing agent is needed.)

Higher yields of the diol are obtained with osmium tetroxide than with potassium permanganate because the cyclic osmate intermediate is less likely to undergo side reactions.

PROBLEM 14◆

Give the products that would be formed from the reaction of each of the following alkenes with OsO_4 followed by aqueous H_2O_2:

a. $CH_3C{=}CHCH_2CH_3$
        $|$
        CH_3

b. 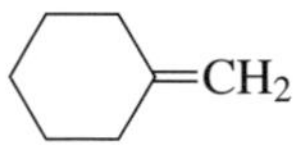 ⬡$={CH_2}$

PROBLEM 15

What stereoisomers would be formed from the reaction of each of the following alkenes with OsO_4 followed by H_2O_2?

a. *trans*-2-butene   **b.** *cis*-2-butene   **c.** *cis*-2-pentene   **d.** *trans*-2-pentene

19.6 Oxidative Cleavage of 1,2-Diols

1,2-Diols are oxidized to ketones or aldehydes or both by periodic acid (HIO_4). The reaction takes place because iodine is in a highly positive oxidation state (+7); it readily accepts electrons. The periodic acid reacts with the diol to form a cyclic intermediate. When the intermediate breaks down, the bond between the two carbons bonded to the oxygens breaks. If a carbon that was bonded to an OH group is also bonded to two R groups, the product will be a ketone; if the carbon is bonded to an R and an H, the product will be an aldehyde. Because this oxidation reaction cuts the reactant into two pieces, it is called an **oxidative cleavage**.

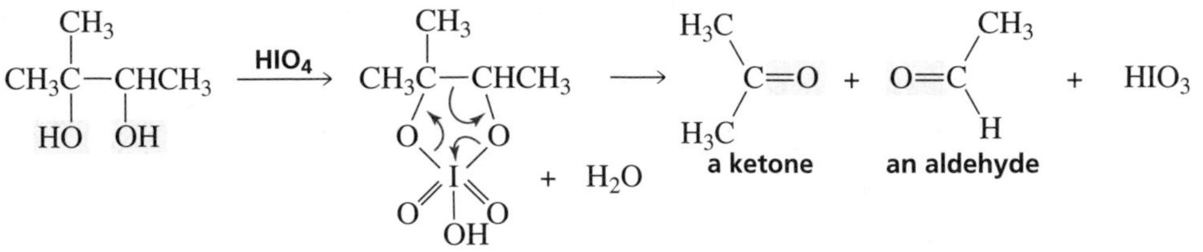

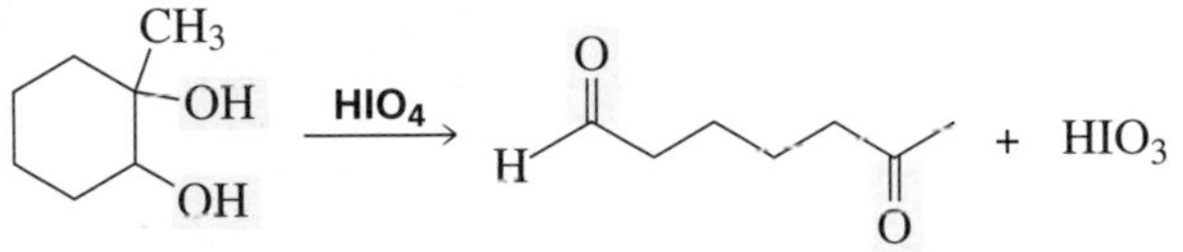

PROBLEM 16◆

An alkene is treated with OsO_4 followed by H_2O_2. When the resulting diol is treated with HIO_4, the only product obtained is an unsubstituted cyclic ketone with molecular formula $C_6H_{10}O$. What is the structure of the alkene?

PROBLEM-SOLVING STRATEGY

Predicting Oxidative Cleavage of 1,2-Diols

Of the following five compounds, explain why only D cannot be cleaved by periodic acid.

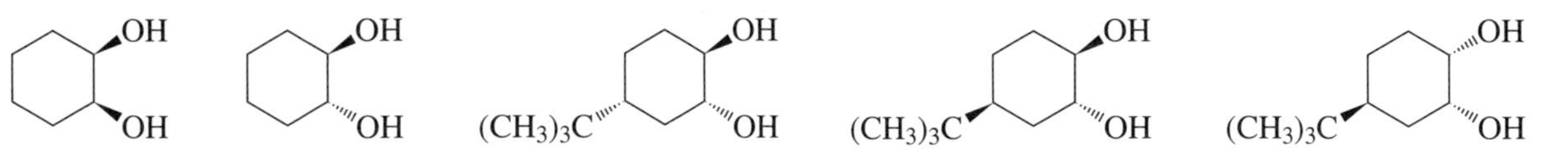

A B C D E

To figure out why one of a series of similar compounds is unreactive, we first need to consider what kinds of compounds undergo the reaction and any stereochemical requirements of the reaction. We know that periodic acid cleaves 1,2-diols. Because the reaction forms a cyclic intermediate, the two OH groups of the diol must be positioned so that they can form the intermediate.

The two OH groups of a 1,2-cyclohexanediol can both be equatorial, they can both be axial, or one can be equatorial and the other axial.

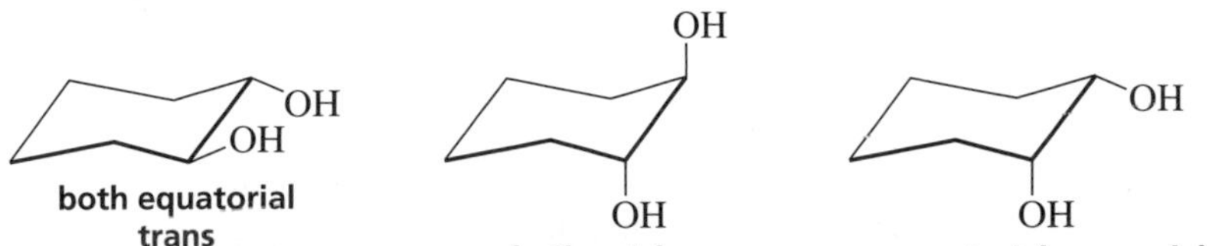

both equatorial trans

both axial trans

one equatorial, one axial cis

In a cis 1,2-cyclohexanediol, one OH is equatorial and the other is axial. Because both cis 1,2-diols (A and E) are cleaved, we know that the cyclic intermediate can be formed when the OH groups are in these positions. In a trans 1,2-diol, both OH groups are equatorial *or* both are axial (Section 2.14). Two of the trans 1,2-diols can be cleaved (B and C), and one cannot (D). We can conclude that the one that cannot be cleaved must have both OH groups in axial positions because they would be too far from each other to form a cyclic intermediate. Now we need to draw the most stable conformers of B, C, and D to see why only D has both OH groups in axial positions.

The most stable conformer of B is the one with both OH groups in equatorial positions. The steric requirements of the bulky *tert*-butyl group force it into an equatorial position, where there is more room for such a large substituent. This causes both OH groups in compound C to be in equatorial positions and both OH groups in compound D to be in axial positions. Therefore, C can be cleaved by periodic acid, but D cannot.

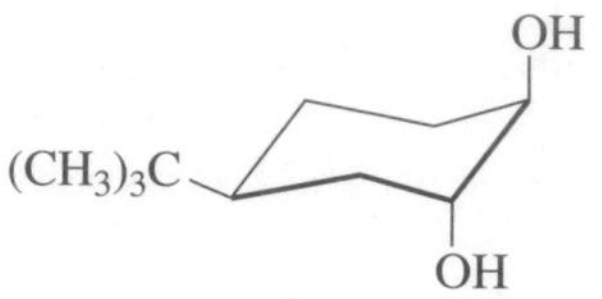

Now continue on to Problem 17.

PROBLEM 17◆

Which of each pair of diols is cleaved more rapidly by periodic acid?

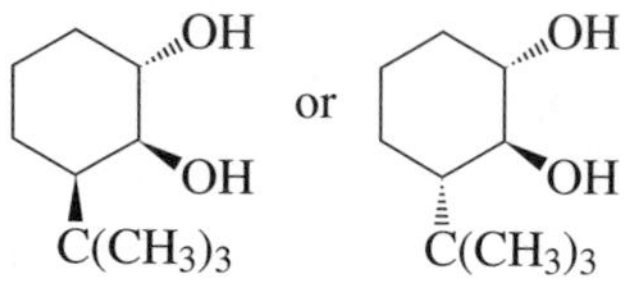

19.7 Oxidative Cleavage of Alkenes

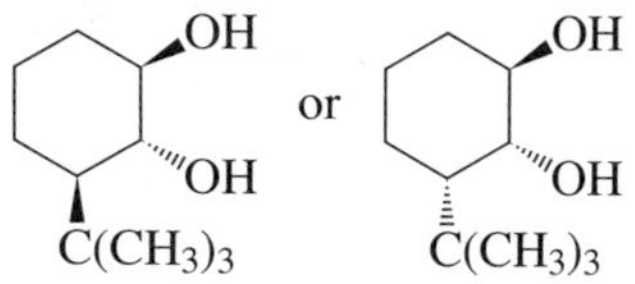

We have seen that alkenes can be oxidized to 1,2-diols and that 1,2-diols can be further oxidized to aldehydes and ketones (Sections 19.5 and 19.6, respectively). Alternatively, alkenes can be directly oxidized to carbonyl compounds by ozone (O_3) or by potassium permanganate.

Ozonolysis

When an alkene is treated with ozone at low temperatures, the double bond breaks and the carbons that were doubly bonded to each other are now doubly bonded to oxygens instead. This oxidation reaction is known as **ozonolysis**.

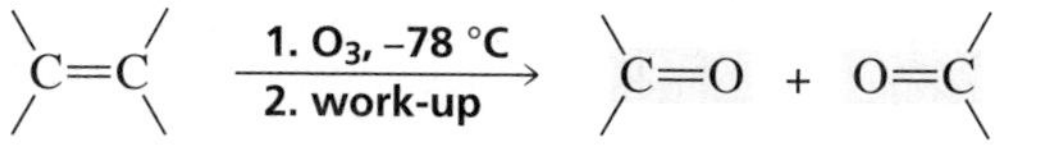

Ozone can be produced by passing oxygen gas through an electric discharge. The structure of ozone is represented by the following resonance contributors:

resonance contributors of ozone

Ozone and the alkene undergo a concerted cycloaddition reaction—the oxygen atoms add to the two sp^2 carbons in a single step. The addition of ozone to the alkene should remind you of the electrophilic addition reactions of alkenes discussed in Chapter 4.

Mechanism for ozonide formation

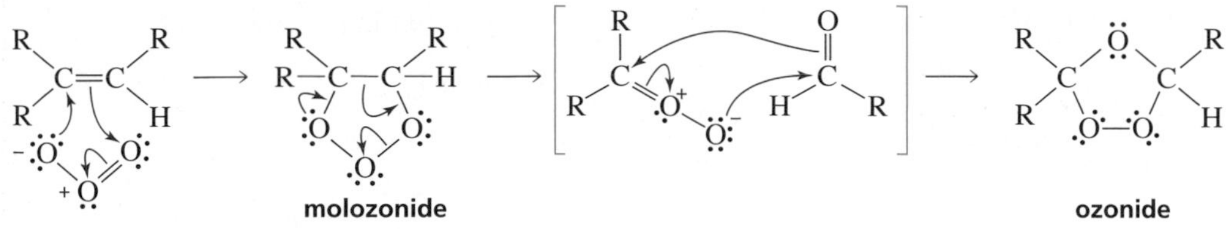

molozonide   **ozonide**

- An electrophile adds to one of the sp^2 carbons, and a nucleophile adds to the other. The electrophile is the oxygen at one end of the ozone molecule, and the nucleophile is the oxygen at the other end. The product of ozone addition to an alkene is a **molozonide**. (The name "molozonide" indicates that one mole of ozone has added to the alkene.)

- The molozonide is unstable because it has two O—O bonds; it immediately rearranges to a more stable **ozonide**.

Ozonides are seldom isolated, because they are explosive. In solution, they are easily cleaved to carbonyl compounds. When an ozonide is cleaved in the presence of a reducing agent, such as zinc or dimethyl sulfide, the products will be ketones, aldehydes, or both. (The product will be a ketone if the sp^2 carbon of the alkene is bonded to two carbon-containing substituents; the product will be an aldehyde if at least one of the substituents bonded to the sp^2 carbon is a hydrogen.) The reducing agent prevents aldehydes from being oxidized to carboxylic acids. Cleaving the ozonide in the presence of zinc or dimethyl sulfide is referred to as "working up the ozonide under reducing conditions."

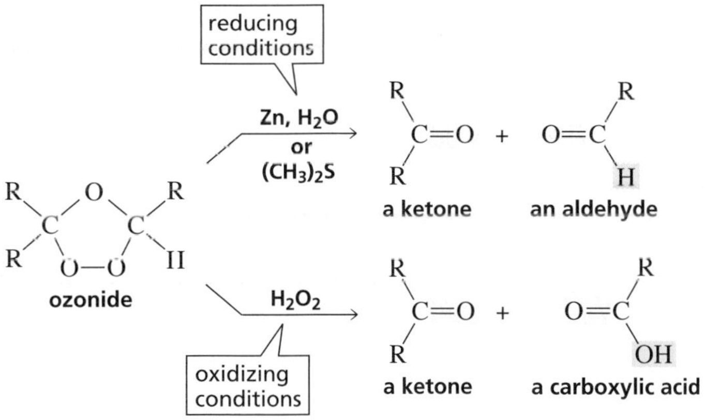

If the ozonide is cleaved in the presence of an oxidizing agent such as hydrogen peroxide (H_2O_2), the products will be ketones, carboxylic acids, or both. Carboxylic acids are formed instead of aldehydes because any aldehyde that is initially formed will be immediately oxidized to a carboxylic acid by hydrogen peroxide. Cleavage in the presence of H_2O_2 is referred to as "working up the ozonide under oxidizing conditions."

The following reactions are examples of the oxidative cleavage of alkenes by ozonolysis:

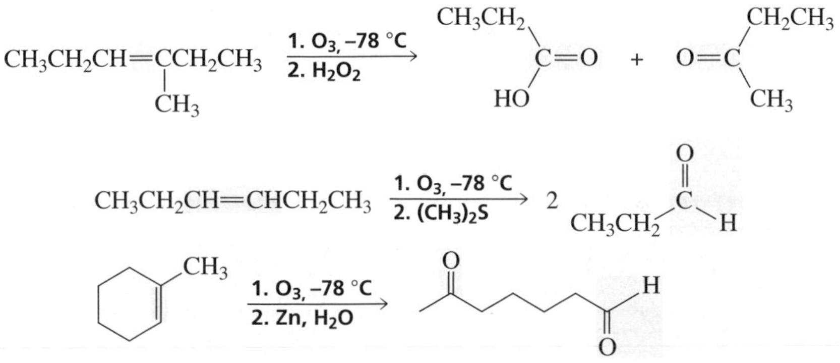

To determine the product of ozonolysis, replace C=C with C=O O=C. If work-up is done under oxidizing conditions, convert any aldehyde products to carboxylic acids.

The one-carbon fragment obtained from the reaction of a terminal alkene with ozone will be oxidized to formaldehyde if the ozonide is worked up under reducing conditions and to formic acid if it is worked up under oxidizing conditions.

$$CH_3CH_2CH_2CH=CH_2 \xrightarrow[\text{2. Zn, H}_2\text{O}]{\text{1. O}_3, -78\ ^\circ\text{C}}$$

$$\begin{array}{c} CH_3CH_2CH_2 \\ \diagdown \\ C=O \\ \diagup \\ H \end{array} + \begin{array}{c} H \\ \diagdown \\ O=C \\ \diagup \\ H \end{array}$$

$$CH_3CH_2CH_2CH=CH_2 \xrightarrow[\text{2. H}_2\text{O}_2]{\text{1. O}_3, -78\ ^\circ\text{C}}$$

$$\begin{array}{c} CH_3CH_2CH_2 \\ \diagdown \\ C=O \\ \diagup \\ HO \end{array} + \begin{array}{c} H \\ \diagdown \\ O=C \\ \diagup \\ OH \end{array}$$

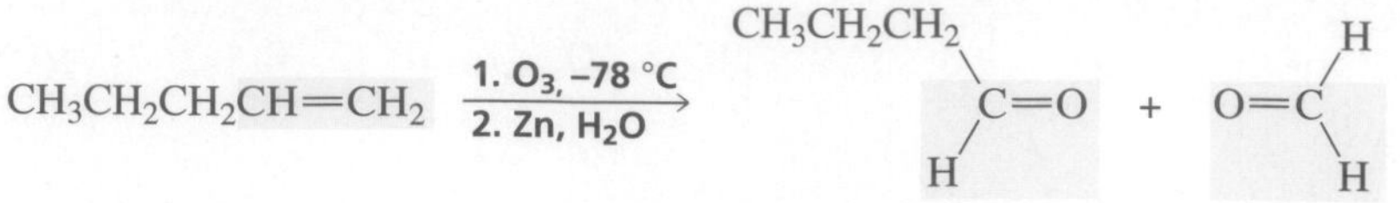

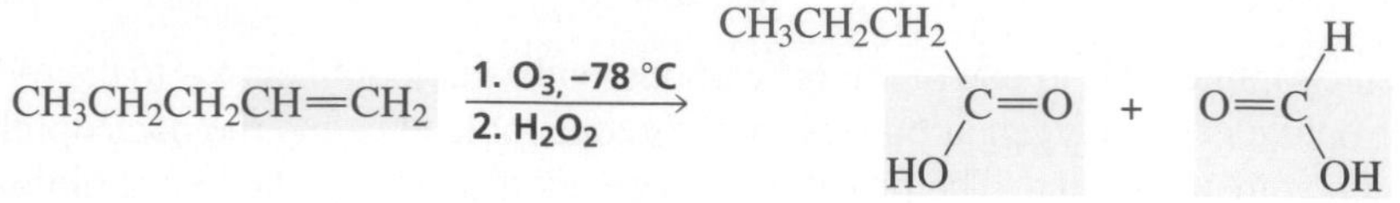

Only the side-chain double bond will be oxidized in the following reaction because the stable benzene ring is oxidized only under prolonged exposure to ozone.

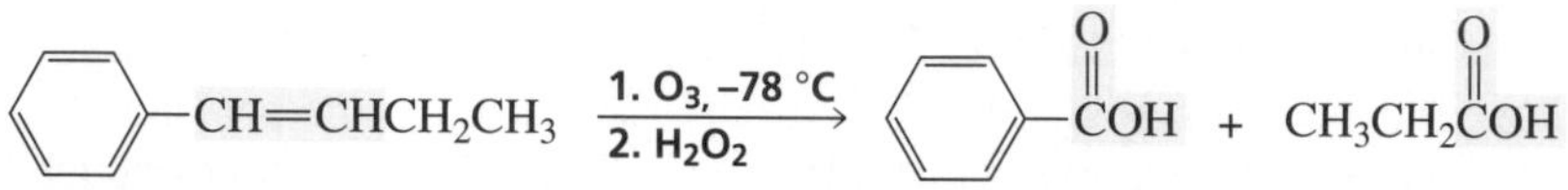

$$\text{C}_6\text{H}_5\text{—CH=CHCH}_2\text{CH}_3 \xrightarrow[\text{2. H}_2\text{O}_2]{\text{1. O}_3, -78\ ^\circ\text{C}} \text{C}_6\text{H}_5\text{—}\overset{\text{O}}{\overset{\|}{\text{C}}}\text{OH} + \text{CH}_3\text{CH}_2\overset{\text{O}}{\overset{\|}{\text{C}}}\text{OH}$$

PROBLEM 18◆

Give an example of an alkene that forms the same ozonolysis products, regardless of whether the ozonide is worked up under reducing conditions (Zn, H₂O) or oxidizing conditions (H₂O₂).

PROBLEM 19

Give the products you would expect to obtain when the following compounds are treated with ozone, followed by work-up with

a. Zn, H₂O:

1. $\overset{\text{CH}_3}{\overset{|}{\text{CH}_3\text{CH}_2\text{CH}_2\text{C}}}=\text{CHCH}_3$

2. $CH_2=CHCH_2CH_2CH_2CH_3$

3. 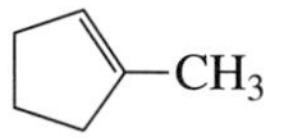 cyclopentene–CH₃

b. H₂O₂:

4. 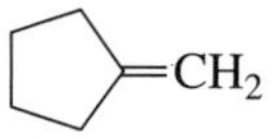 cyclopentane=CH₂

5. $CH_3CH_2CH_2CH=CHCH_2CH_2CH_3$

6. 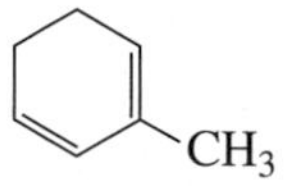 cyclohexadiene–CH₃

Ozonolysis can be used to determine the structure of an unknown alkene. If you know what carbonyl compounds are formed by ozonolysis, you can mentally work backward to deduce the structure of the alkene. For example, if ozonolysis of an alkene followed by a work-up under reducing conditions forms acetone and butanal as products, you can conclude that the alkene was 2-methyl-2-hexene.

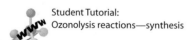
Student Tutorial:
Ozonolysis reactions—synthesis

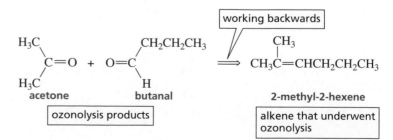

PROBLEM 20◆

a. What alkene would give only acetone as a product of ozonolysis?

b. What alkenes would give only butanal as a product of ozonolysis?

PROBLEM 21 | **SOLVED**

The following products were obtained from ozonolysis of a diene followed by work-up under reducing conditions. Give the structure of the diene.

$$
\underset{\text{HCCH}_2\text{CH}_2\text{CH}_2\text{CH}}{\overset{\text{O}\quad\quad\text{O}}{\parallel\quad\quad\parallel}} \quad + \quad \underset{\text{HCH}}{\overset{\text{O}}{\parallel}} \quad + \quad \underset{\text{CH}_3\text{CH}_2\text{CH}}{\overset{\text{O}}{\parallel}}
$$

Solution The five-carbon dicarbonyl compound indicates that the diene must contain five carbons flanked by two double bonds.

$$
\underset{\text{HCCH}_2\text{CH}_2\text{CH}_2\text{CH}}{\overset{\text{O}\quad\quad\text{O}}{\parallel\quad\quad\parallel}} \quad \Longrightarrow \quad =\text{CHCH}_2\text{CH}_2\text{CH}_2\text{CH}=
$$

One of the other two carbonyl compounds obtained from ozonolysis has one carbon atom, and the other has three carbon atoms. Therefore, one carbon has to be added to one end of the diene, and three carbons have to be added to the other end.

$$
\text{CH}_2=\text{CHCH}_2\text{CH}_2\text{CH}_2\text{CH}=\text{CHCH}_2\text{CH}_3
$$

PROBLEM 22◆

What aspect of the structure of the alkene does ozonolysis not tell you?

Permanganate Cleavage

We have seen that alkenes are oxidized to 1,2-diols by a basic solution of potassium permanganate at room temperature or below, and that the 1,2-diols can subsequently be cleaved by periodic acid to form aldehydes, ketones, or both (Sections 19.5 and 19.6). If, however, the basic solution of potassium permanganate is heated or if the solution is acidic, the reaction will not stop at the diol. Instead, the alkene will be cleaved, and the reaction products will be ketones and carboxylic acids. If the reaction is carried out under basic conditions, any carboxylic acid product will be in its basic form (RCOO^-); if the reaction is carried out under acidic conditions, any carboxylic acid product will be in its acidic form (RCOOH) (Section 1.24). Terminal alkenes form CO_2 as a product.

$$
\underset{\text{CH}_3\text{CH}_2\text{C}=\text{CHCH}_3}{\overset{\text{CH}_3}{\overset{\mid}{}}} \quad\xrightarrow[\Delta]{\textbf{KMnO}_4\text{, HO}^-}\quad \underset{\text{CH}_3\text{CH}_2}{\overset{\text{CH}_3}{}}\!\!\text{C}=\text{O} \;+\; \text{O}=\text{C}\!\!\underset{\text{O}^-}{\overset{\text{CH}_3}{}}
$$

$$
\text{CH}_3\text{CH}_2\text{CH}=\text{CH}_2 \quad\xrightarrow[\text{H}^+]{\textbf{KMnO}_4}\quad \underset{\text{HO}}{\overset{\text{CH}_3\text{CH}_2}{}}\!\!\text{C}=\text{O} \;+\; \text{CO}_2
$$

$=\text{CH}_2 \quad\xrightarrow[\Delta]{\textbf{KMnO}_4\text{, HO}^-}\quad$ $=\text{O} \;+\; \text{CO}_2$

A peroxyacid, OsO₄, and cold basic KMnO₄ break only the π bond of the alkene. Ozone and acidic (or hot basic) KMnO₄ break both the π bond and the σ bond.

The various methods used to oxidize an alkene are summarized in Table 19.1.

Table 19.1   Summary of the Methods Used to Oxidize an Alkene

---

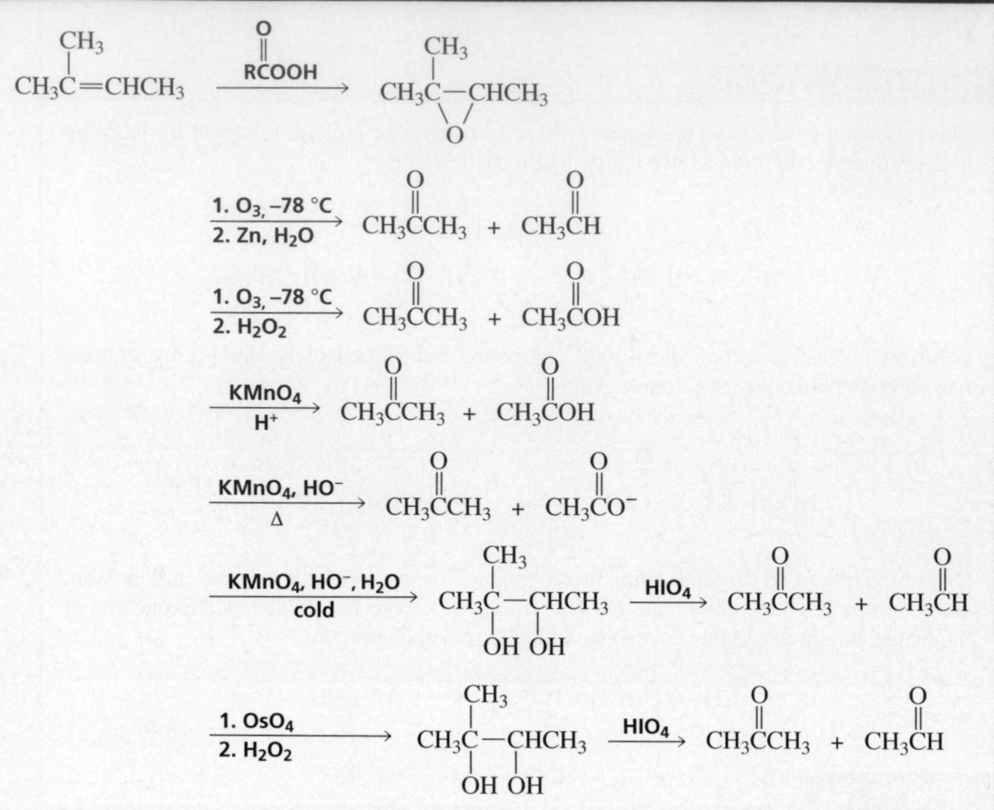

PROBLEM 23   SOLVED

Describe how the following compound could be prepared, using the given starting materials. (Perform a retrosynthetic analysis to help you arrive at your answer.)

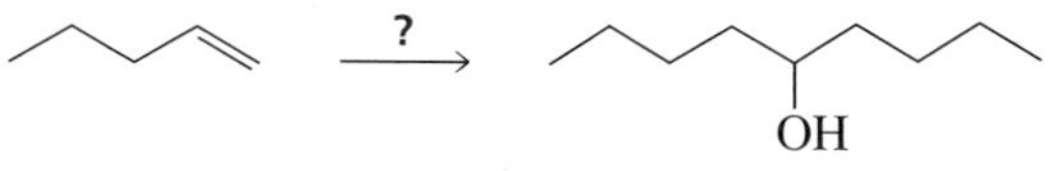

Solution   In a retrosynthetic analysis, the target molecule can be disconnected to give a five-carbon positively charged fragment and a four-carbon negatively charged fragment (Section 17.15). Pentanal and a butyl magnesium bromide are the synthetic equivalents for these two fragments. The five-carbon starting material can be converted to the required four-carbon compound by ozonolysis.

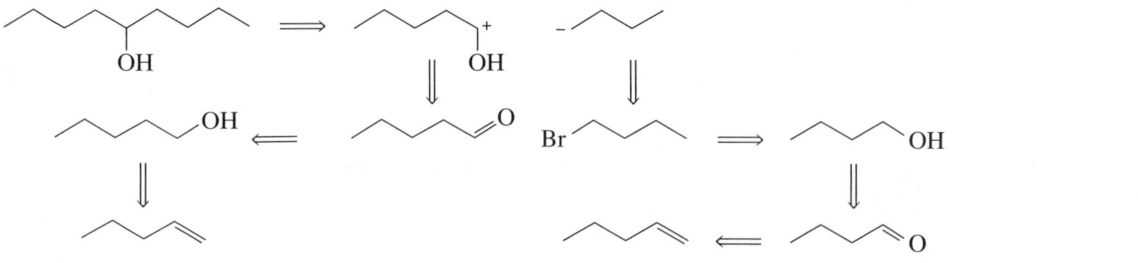

We can now write the syntheses of pentanal and butyl magnesium bromide from the given starting material in the forward direction, along with the necessary reagents.

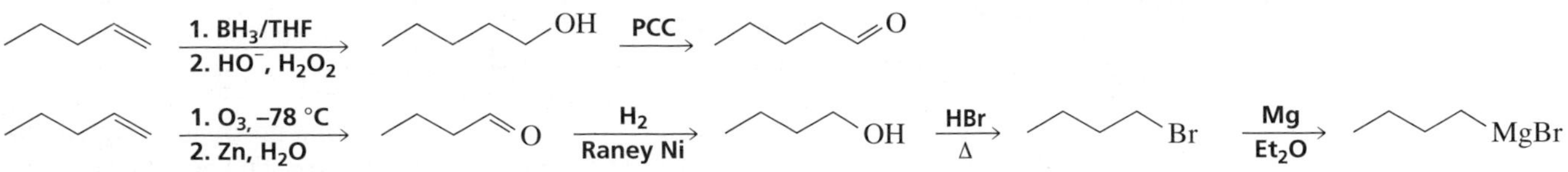

The reaction of pentanal with butyl magnesium bromide forms the target compound.

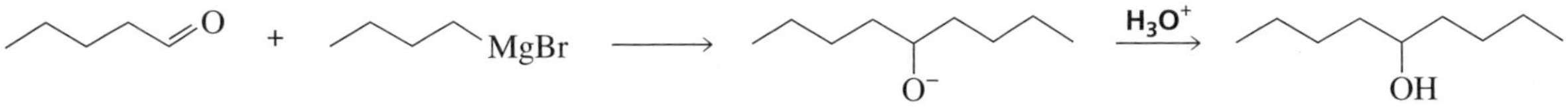

PROBLEM 24

a. How could you synthesize the following compound from starting materials containing no more than four carbons? (*Hint:* A 1,6-diketone can be synthesized by oxidative cleavage of a 1,2-disubstituted cyclohexene.)

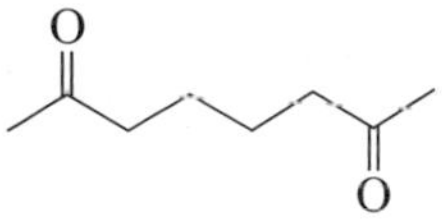

b. How could you synthesize the same compound in two steps from starting materials containing no more than six carbons?

19.8 Oxidative Cleavage of Alkynes

The same reagents that oxidize alkenes also oxidize alkynes. Alkynes are oxidized to diketones by a basic solution of $KMnO_4$ at room temperature and are cleaved by ozonolysis to carboxylic acids. Ozonolysis of alkynes requires neither oxidative nor reductive work-up; it is followed only by hydrolysis. Carbon dioxide is obtained from the CH group of a terminal alkyne.

$$CH_3C{\equiv}CCH_2CH_3 \xrightarrow[\text{HO}^-]{\text{KMnO}_4} CH_3\overset{O}{\underset{}{C}}-\overset{O}{\underset{}{C}}CH_2CH_3$$
2-pentyne

$$CH_3C{\equiv}CCH_2CH_3 \xrightarrow[\text{2. H}_2\text{O}]{\text{1. O}_3, -78\ °C} CH_3\overset{O}{\underset{}{C}}OH + CH_3CH_2\overset{O}{\underset{}{C}}OH$$
2-pentyne

$$CH_3CH_2CH_2C{\equiv}CH \xrightarrow[\text{2. H}_2\text{O}]{\text{1. O}_3, -78\ °C} CH_3CH_2CH_2\overset{O}{\underset{}{C}}OH + CO_2$$
1-pentyne

PROBLEM 25◆

What is the structure of the alkyne that gives each of the following sets of products upon ozonolysis followed by hydrolysis?

a.

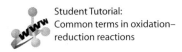

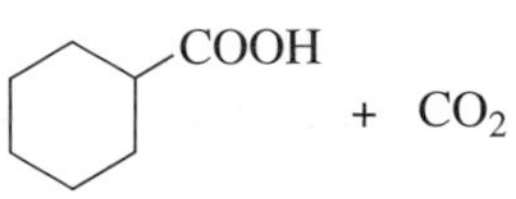

$+\ CO_2$

b.

$+\ 2$

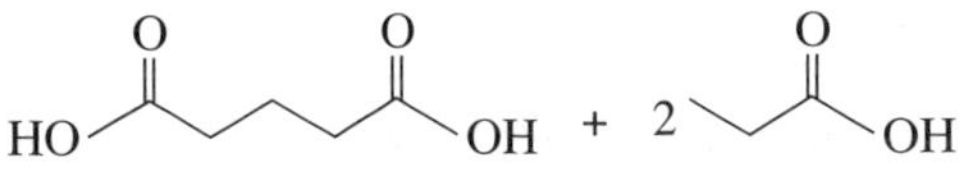

19.9 Designing a Synthesis IX: Functional Group Interconversion

Converting one functional group into another is called **functional group interconversion**. Our knowledge of oxidation–reduction reactions has greatly expanded our ability to carry out functional group interconversions. For example, an aldehyde can be converted into a primary alcohol, an alkene, a secondary alcohol, a ketone, a carboxylic acid, an acyl chloride, an ester, an amide, or an amine.

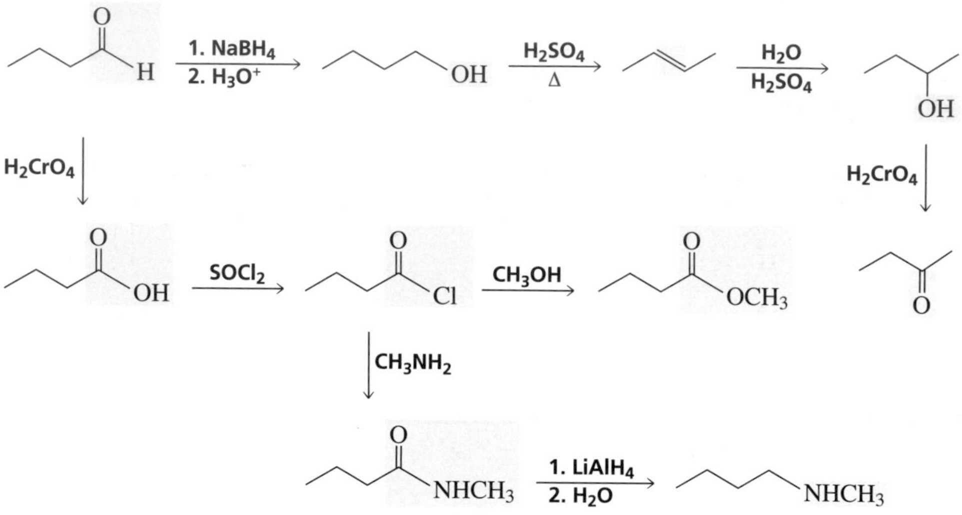

Student Tutorial:
Multistep synthesis

Quinone Reduction

A ketone can be converted into an ester or an alcohol.

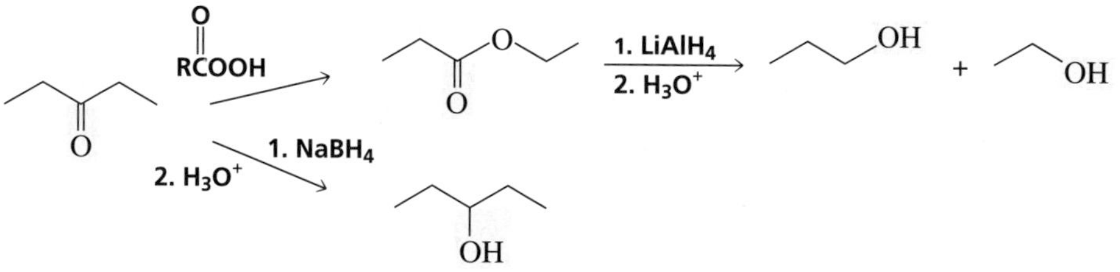

As the number of reactions with which you are familiar become more numerous, the number of functional group interconversions you can perform will also increase. As a result, you will find that you have more than one route available when you design a synthesis. The route you actually decide to use will depend on the availability and cost of the starting materials and on the ease with which the reactions in the synthetic pathway can be carried out.

PROBLEM 26

Add the necessary reagents over the reaction arrows.

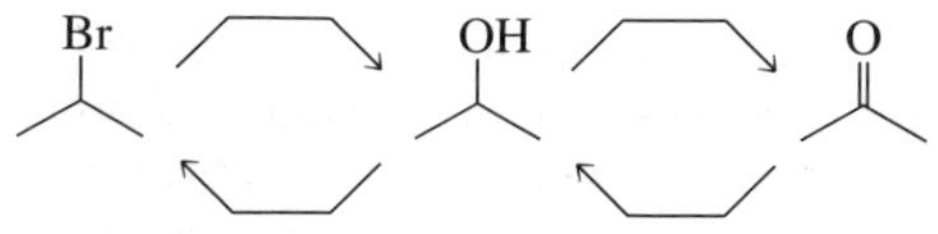

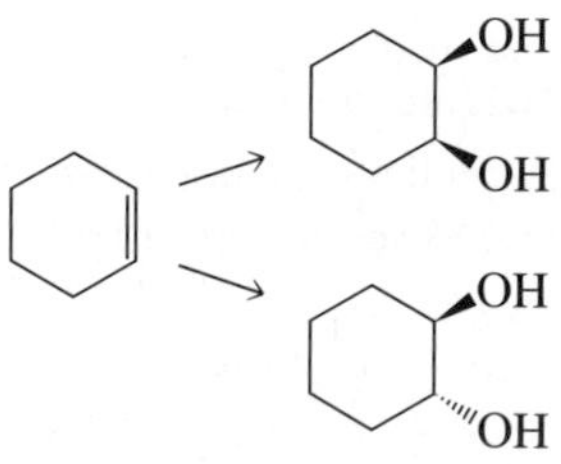

PROBLEM 27

a. Show two ways to convert an alkyl halide into an alcohol that contains one additional carbon atom.

b. Show how a primary alkyl halide can be converted into an amine that contains one additional carbon atom.

c. Show how a primary alkyl halide can be converted into an amine that contains one less carbon atom.

PROBLEM 28

Show how each of the following compounds could be synthesized from the given starting material:

a.

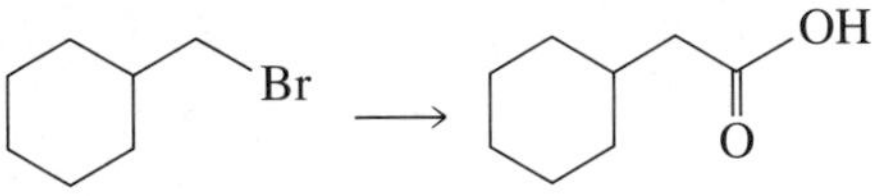

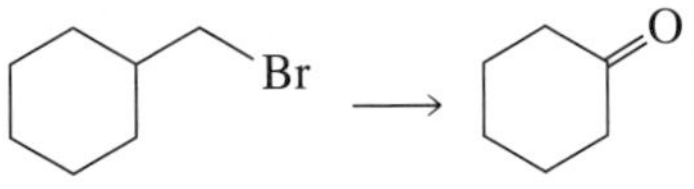

b.

c.

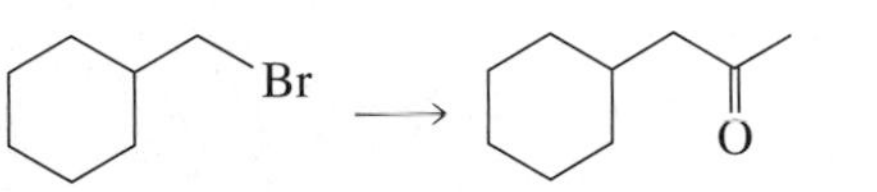

d.

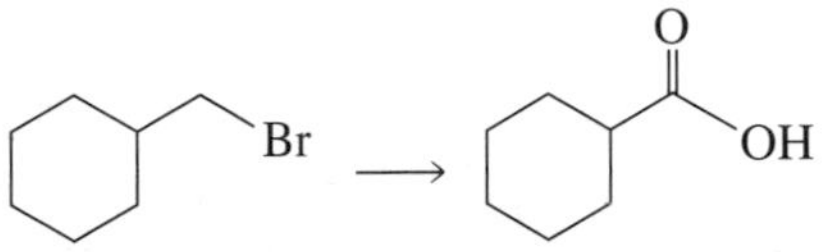

PROBLEM 29

How many different functional groups can you use to synthesize a primary alcohol?

SUMMARY

Oxidation is coupled with reduction: a **reducing agent** is oxidized and an **oxidizing agent** is reduced. If the reaction increases the number of C—H bonds or decreases the number of C—O, C—N, or C—X bonds (where X denotes a halogen), the compound has been reduced; if the reaction decreases the number of C—H bonds or increases the number of C—O, C—N, or C—X bonds, the compound has been oxidized. Similarly, reduction increases the number of N—H or S—H bonds, and oxidation increases the number of N—O or S—O bonds.

Reduction by the addition of H_2 to an organic compound occurs by one of three mechanisms: **catalytic hydrogenations** add two hydrogen atoms, **dissolving-metal reductions** add two electrons and two protons, and **metal-hydride reductions** consist of the addition of a hydride ion followed by a proton. Carbon–carbon, carbon–nitrogen, and some carbon–oxygen multiple bonds can be reduced by catalytic hydrogenation. An alkyne is reduced by sodium and liquid ammonia to a trans alkene. $NaBH_4$ is used to reduce aldehydes, ketones, and acyl halides; $LiAlH_4$, a stronger reducing agent, is used to reduce carboxylic acids, esters, and amides. Replacing some of the hydrogens of $LiAlH_4$ with OR groups decreases the reactivity of the metal hydride.

Multiply bonded carbon atoms cannot be reduced by metal hydrides.

Primary alcohols are oxidized to carboxylic acids by chromium-containing reagents and to aldehydes by PCC or a **Swern oxidation**. Secondary alcohols are oxidized to ketones. Tollens reagent oxidizes only aldehydes. A **peroxyacid** oxidizes an aldehyde to a carboxylic acid, a ketone to an ester (in a **Baeyer–Villiger oxidation**), and an alkene to an epoxide. Alkenes are oxidized to 1,2-diols by potassium permanganate ($KMnO_4$) in a cold basic solution or by osmium tetroxide (OsO_4).

1,2-Diols are **oxidatively cleaved** to ketones and/or aldehydes by periodic acid (HIO_4). Ozonolysis oxidatively cleaves alkenes to ketones, aldehydes, or both when worked up under reducing conditions and to ketones, carboxylic acids, or both when worked up under oxidizing conditions. Acidic solutions and hot basic solutions of potassium permanganate also oxidatively cleave alkenes to ketones, carboxylic acids, or both.

A **chemoselective reaction** is a reaction in which a reagent reacts with one functional group in preference to another. An **enantioselective reaction** forms more of one enantiomer than of another. Converting one functional group into another is called **functional group interconversion**.

SUMMARY OF REACTIONS

1. Catalytic hydrogenation of double and triple bonds (Section 19.1)

a. $RCH{=}CHR + H_2 \xrightarrow{\text{Pt, Pd, or Ni}} RCH_2CH_2R$

$RC{\equiv}CR + 2\,H_2 \xrightarrow{\text{Pt, Pd, or Ni}} RCH_2CH_2R$

$RCH{=}NR + H_2 \xrightarrow{\text{Pt, Pd, or Ni}} RCH_2NHR$

$RC{\equiv}N + 2\,H_2 \xrightarrow{\text{Pt, Pd, or Ni}} RCH_2NH_2$

b. $\underset{O}{RCH} + H_2 \xrightarrow{\text{Raney Ni}} RCH_2OH$

$\underset{O}{RCR} + H_2 \xrightarrow{\text{Raney Ni}} \underset{OH}{RCHR}$

c. $\underset{O}{RCCl} + H_2 \xrightarrow[\text{Pd}]{\text{partially deactivated}} \underset{O}{RCH}$

2. Reduction of alkynes to alkenes (Section 19.1)

$RC{\equiv}CR \xrightarrow[\text{Lindlar catalyst}]{H_2} \overset{H}{\underset{R}{\diagdown}}C{=}C\overset{H}{\underset{R}{\diagup}}$

$RC{\equiv}CR \xrightarrow[\text{NH}_3\text{ (liq)}]{\text{Na or Li}} \overset{H}{\underset{R}{\diagdown}}C{=}C\overset{R}{\underset{H}{\diagup}}$

3. Reduction of carbonyl compounds with reagents that donate hydride ion (Section 19.1)

a. $\underset{O}{RCH} \xrightarrow[\text{2. H}_3\text{O}^+]{\text{1. NaBH}_4} RCH_2OH$

b. $\underset{O}{RCR} \xrightarrow[\text{2. H}_3\text{O}^+]{\text{1. NaBH}_4} \underset{OH}{RCHR}$

c. $\underset{O}{RCCl} \xrightarrow[\text{2. H}_3\text{O}^+]{\text{1. NaBH}_4} RCH_2OH$

d. $\underset{O}{RCOH} \xrightarrow[\text{2. H}_3\text{O}^+]{\text{1. LiAlH}_4} RCH_2OH$

e. $\underset{O}{RCOR'} \xrightarrow[\text{2. H}_3\text{O}^+]{\text{1. LiAlH}_4} RCH_2OH + R'OH$

f. $\underset{O}{RCNHR'} \xrightarrow[\text{2. H}_2\text{O}]{\text{1. LiAlH}_4} RCH_2NHR'$

g. $\underset{O}{RCOR'} \xrightarrow[\text{2. H}_2\text{O}]{\text{1. [(CH}_3)_2\text{CHCH}_2]_2\text{AlH, }-78\,°\text{C}} \underset{O}{RCH} + R'OH$

h. $\underset{O}{RCCl} \xrightarrow[\text{2. H}_2\text{O}]{\text{1. LiAl[OC(CH}_3)_3]_3\text{H, }-78\,°\text{C}} \underset{O}{RCH}$

4. Oxidation of alcohols (Section 19.2)

primary alcohols $RCH_2OH \xrightarrow{\text{H}_2\text{CrO}_4} \left[\underset{O}{RCH}\right] \xrightarrow[\text{oxidation}]{\text{further}} \underset{O}{RCOH}$

$RCH_2OH \xrightarrow[\text{CH}_2\text{Cl}_2]{\text{PCC}} \underset{O}{RCH}$

$RCH_2OH \xrightarrow[\text{2. triethylamine}]{\text{1. CH}_3\text{SCH}_3,\ \text{ClC}{-}\text{CCl},\,-60\,°\text{C}} \underset{O}{RCH}$

secondary alcohols $\underset{OH}{RCHR} \xrightarrow{\text{H}_2\text{CrO}_4} \underset{O}{RCR}$

$\underset{OH}{RCHR} \xrightarrow[\text{2. triethylamine}]{\text{1. CH}_3\text{SCH}_3,\ \text{ClC}{-}\text{CCl},\,-60\,°\text{C}} \underset{O}{RCR}$

5. Oxidation of aldehydes and ketones (Section 19.3)

a. **aldehydes** $\underset{\text{RCH}}{\overset{O}{\|}} \xrightarrow{\text{H}_2\text{CrO}_4} \underset{\text{RCOH}}{\overset{O}{\|}}$

$\underset{\text{RCH}}{\overset{O}{\|}} \xrightarrow[\text{2. H}_3\text{O}^+]{\text{1. Ag}_2\text{O, NH}_3} \underset{\text{RCOH}}{\overset{O}{\|}} + \underset{\substack{\text{metallic} \\ \text{silver}}}{\text{Ag}}$

$\underset{\text{RCH}}{\overset{O}{\|}} \xrightarrow{\overset{O}{\overset{\|}{\text{R'COOH}}}} \underset{\text{RCOH}}{\overset{O}{\|}} + \underset{\text{R'COH}}{\overset{O}{\|}}$

b. **ketones** $\underset{\text{RCR}}{\overset{O}{\|}} \xrightarrow{\overset{O}{\overset{\|}{\text{R'COOH}}}} \underset{\text{RCOR}}{\overset{O}{\|}} + \underset{\text{R'COH}}{\overset{O}{\|}}$

6. Oxidation of alkenes (Sections 19.4, 19.5 and 19.7)

a. $\underset{\text{RC=CHR'}}{\overset{R}{|}} \xrightarrow[\substack{\text{2. Zn, H}_2\text{O} \\ \text{or} \\ \text{(CH}_3)_2\text{S}}]{\text{1. O}_3, -78\,°\text{C}} \underset{\text{RCR}}{\overset{O}{\|}} + \underset{\text{R'CH}}{\overset{O}{\|}}$

$\xrightarrow[\text{2. H}_2\text{O}_2]{\text{1. O}_3, -78\,°\text{C}} \underset{\text{RCR}}{\overset{O}{\|}} + \underset{\text{R'COH}}{\overset{O}{\|}}$

b. $\underset{\text{RC=CHR'}}{\overset{R}{|}} \xrightarrow{\text{KMnO}_4,\ \text{H}^+} \underset{\text{RCR}}{\overset{O}{\|}} + \underset{\text{R'COH}}{\overset{O}{\|}}$

c. $\underset{\text{RC=CHR'}}{\overset{R}{|}} \xrightarrow[\text{cold}]{\text{KMnO}_4,\ \text{HO}^-,\ \text{H}_2\text{O}} \underset{\underset{\text{OH OH}}{|\ \ |}}{\overset{R}{|}}\text{RC—CHR'} \xrightarrow{\text{HIO}_4} \underset{\text{RCR}}{\overset{O}{\|}} + \underset{\text{R'CH}}{\overset{O}{\|}}$

$\xrightarrow[\text{2. H}_2\text{O}_2]{\text{1. OsO}_4} \underset{\underset{\text{OH OH}}{|\ \ |}}{\overset{R}{|}}\text{RC—CHR'} \xrightarrow{\text{HIO}_4} \underset{\text{RCR}}{\overset{O}{\|}} + \underset{\text{R'CH}}{\overset{O}{\|}}$

$\xrightarrow{\overset{O}{\overset{\|}{\text{RCOOH}}}} \underset{\overset{\displaystyle |}{R}}{\text{RC——CHR'}} \overset{O}{\triangle}$

7. Oxidation of 1,2-diols (Section 19.6)

$\underset{\underset{\text{OH OH}}{|\ \ |}}{\overset{R}{|}}\text{RC—CHR'} \xrightarrow{\text{HIO}_4} \underset{\text{RCR}}{\overset{O}{\|}} + \underset{\text{R'CH}}{\overset{O}{\|}}$

8. Oxidation of alkynes (Section 19.8)

a. $\text{RC}\equiv\text{CR'} \xrightarrow[\text{HO}^-]{\text{KMnO}_4} \underset{\text{RC—CR'}}{\overset{O\ \ O}{\|\ \ \|}}$

c. $\text{RC}\equiv\text{CH} \xrightarrow[\text{2. H}_2\text{O}]{\text{1. O}_3, -78\,°\text{C}} \underset{\text{RCOH}}{\overset{O}{\|}} + \text{CO}_2$

b. $\text{RC}\equiv\text{CR'} \xrightarrow[\text{2. H}_2\text{O}]{\text{1. O}_3, -78\,°\text{C}} \underset{\text{RCOH}}{\overset{O}{\|}} + \underset{\text{R'COH}}{\overset{O}{\|}}$

KEY TERMS

Baeyer–Villiger oxidation (p. 920)
catalytic hydrogenation (p. 911)
chemoselective reaction (p. 916)
dissolving-metal reduction (p. 913)
enantioselective reaction (p. 922)
functional group interconversion (p. 932)
glycol (p. 923)
metal-hydride reduction (p. 914)

molozonide (p. 927)
oxidation (p. 909)
oxidation–reduction reaction (p. 908)
oxidative cleavage (p. 925)
oxidizing agent (p. 909)
ozonide (p. 927)
ozonolysis (p. 926)
peroxyacid (p. 920)

redox reaction (p. 908)
reducing agent (p. 909)
reduction (p. 909)
Rosenmund reduction (p. 912)
Swern oxidation (p. 918)
Tollens test (p. 920)
vicinal diol (p. 923)
vicinal glycol (p. 923)

PROBLEMS

30. Fill in the blank with "oxidized" or "reduced."
   a. Secondary alcohols are _____ to ketones.
   b. Acyl halides are _____ to aldehydes.
   c. Aldehydes are _____ to primary alcohols.
   d. Alkenes are _____ to aldehydes and/or ketones.
   e. Aldehydes are _____ to carboxylic acids.
   f. Alkenes are _____ to 1,2-diols.
   g. Alkenes are _____ to alkanes.

31. Give the products of the following reactions. Indicate whether each reaction is an oxidation or a reduction:

a. $CH_3CH_2CH_2CH_2CH_2OH \xrightarrow[H_2SO_4]{Na_2Cr_2O_7}$

b. 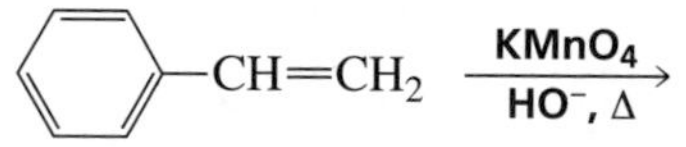 [benzene ring]–$CH=CH_2 \xrightarrow[HO^-, \Delta]{KMnO_4}$

c. $CH_3CH_2CH_2\overset{O}{\underset{\|}{C}}Cl \xrightarrow[\substack{\text{partially} \\ \text{deactivated Pd}}]{H_2}$

d. $CH_3CH_2C \equiv CH \xrightarrow[\substack{\text{1. disiamylborane} \\ \text{2. } H_2O_2, HO^-, H_2O \\ \text{3. } NaBH_4 \\ \text{4. } H_3O^+}]{}$

e. $CH_3CH_2CH = CHCH_2CH_3 \xrightarrow[\text{2. Zn, } H_2O]{\text{1. } O_3, -78\ °C}$

f. $CH_3CH_2CH_2\overset{O}{\underset{\|}{C}}NHCH_3 \xrightarrow[\text{2. } H_2O]{\text{1. LiAlH}_4}$

g. 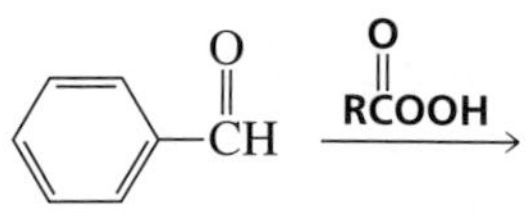 [benzene ring]–$\overset{O}{\underset{\|}{C}}H \xrightarrow{\overset{O}{\underset{\|}{R}COOH}}$

h. [benzene ring]–$\overset{O}{\underset{\|}{C}}OCHCH_3 \xrightarrow[\text{2. } H_3O^+]{\text{1. LiAlH}_4}$ with CH_3

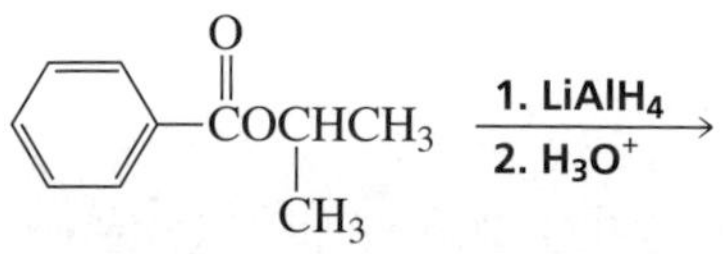

i. 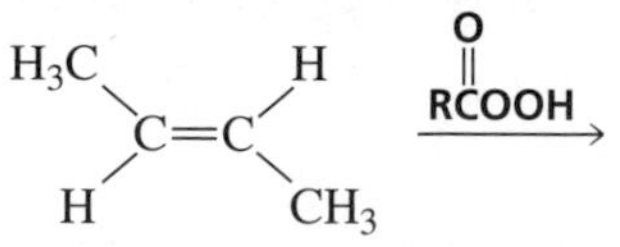 $\overset{H_3C}{\underset{H}{}}C = C\overset{H}{\underset{CH_3}{}} \xrightarrow{\overset{O}{\underset{\|}{R}COOH}}$

j. 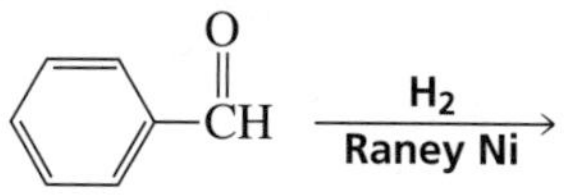 [benzene ring]–$\overset{O}{\underset{\|}{C}}H \xrightarrow[\text{Raney Ni}]{H_2}$

k. $CH_3CH_2CH_2C \equiv CCH_3 \xrightarrow[NH_3 \text{ (liq)}]{Na}$

l. $CH_3CH_2CH_2C \equiv CCH_3 \xrightarrow[\text{2. } H_2O]{\text{1. } O_3, -78\ °C}$

m. 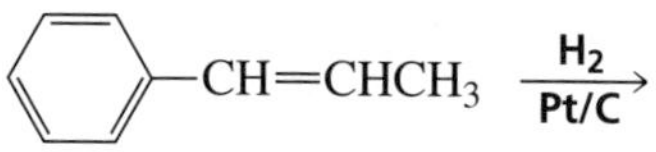 [benzene ring]–$CH = CHCH_3 \xrightarrow[Pt/C]{H_2}$

n. 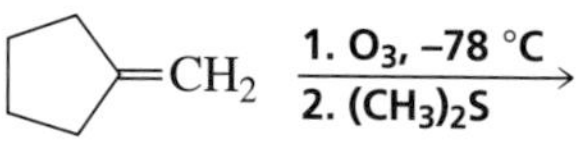 [cyclopentane ring]$=CH_2 \xrightarrow[\text{2. } (CH_3)_2S]{\text{1. } O_3, -78\ °C}$

o. [cyclohexene] $\xrightarrow[\substack{\text{2. } CH_3MgBr \\ \text{3. } H_3O^+}]{\text{1. } \overset{O}{\underset{\|}{R}COOH}}$

p. [cyclohexene] $\xrightarrow[HO^-, \text{cold}]{KMnO_4}$

q. [cyclohexene] $\xrightarrow[HO^-, \Delta]{KMnO_4}$

r. [cyclohexadiene] $\xrightarrow[\text{2. } H_2O_2]{\text{1. } O_3, -78\ °C}$

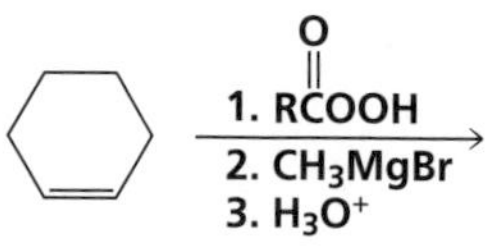

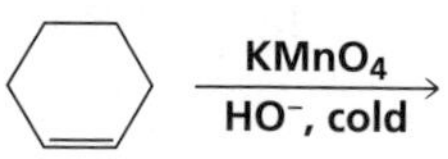

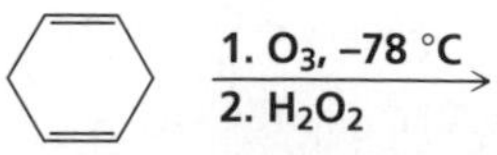

32. How could each of the following compounds be converted to $CH_3CH_2CH_2\overset{O}{\overset{\|}{C}}OH$?

HLC

a. $CH_3CH_2CH_2\overset{O}{\overset{\|}{C}}H$

b. $CH_3CH_2CH_2CH_2OH$

c. $CH_3CH_2CH_2CH_2Br$

d. $CH_3CH_2CH=CH_2$

33. Identify A–G:

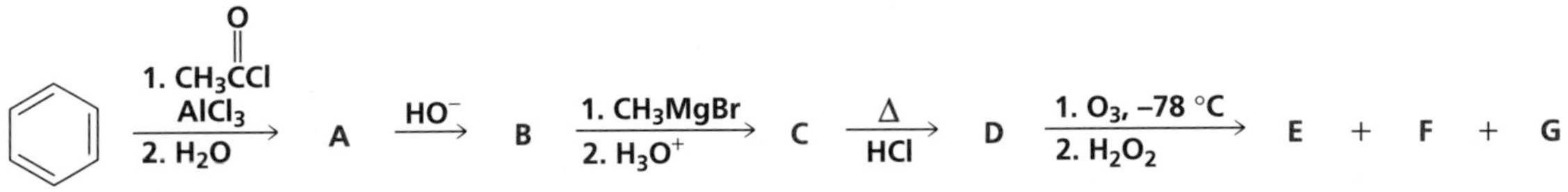

$$\text{benzene} \xrightarrow[\text{2. H}_2\text{O}]{\substack{\text{1. CH}_3\overset{O}{\overset{\|}{C}}\text{Cl} \\ \text{AlCl}_3}} \text{A} \xrightarrow{\text{HO}^-} \text{B} \xrightarrow[\text{2. H}_3\text{O}^+]{\text{1. CH}_3\text{MgBr}} \text{C} \xrightarrow[\text{HCl}]{\Delta} \text{D} \xrightarrow[\text{2. H}_2\text{O}_2]{\text{1. O}_3, -78\,°\text{C}} \text{E} + \text{F} + \text{G}$$

34. Identify the alkene that would give each of the following products upon ozonolysis followed by treatment with hydrogen peroxide:

a. $CH_3CH_2CH_2\overset{O}{\overset{\|}{C}}OH$ + $CH_3\overset{O}{\overset{\|}{C}}CH_3$

b. $CH_3\overset{O}{\overset{\|}{C}}CH_2CH_2CH_2CH_2\overset{O}{\overset{\|}{C}}CH_2CH_3$

c. $HO\overset{O}{\overset{\|}{C}}$ + [diketone-acid structure]

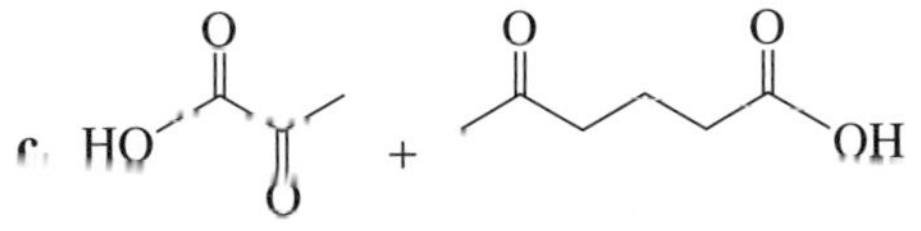

d. [cyclohexanone] + $CH_3CH_2\overset{O}{\overset{\|}{C}}OH$

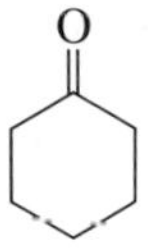

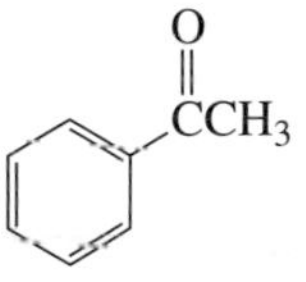

e. [phenyl-$\overset{O}{\overset{\|}{C}}CH_3$] + $H\overset{O}{\overset{\|}{C}}OH$

f. [cyclohexane diCOOH structure] + $HO-\overset{O}{\overset{\|}{C}}-\overset{O}{\overset{\|}{C}}-OH$

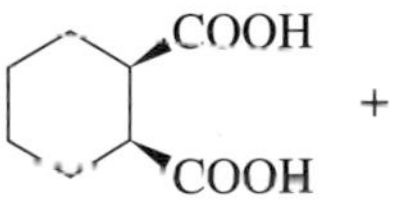

35. Fill in each box with the appropriate reagent:

a. $CH_3CH_2CH=CH_2 \xrightarrow[\text{2. }\Box]{\text{1. }\Box} CH_3CH_2CH_2CH_2OH \xrightarrow{\Box} CH_3CH=CHCH_3 \xrightarrow{\Box} CH_3\overset{O}{\overset{\|}{C}}OH$

b. $CH_3CH_2Br \xrightarrow{\Box} \Box \xrightarrow[\text{2. }\Box]{\text{1. }\Box} CH_3CH_2CH_2CH_2OH \xrightarrow{\Box} CH_3CH_2CH_2\overset{O}{\overset{\|}{C}}H$

c. [cyclohexane] $\xrightarrow{\Box}$ [cyclohexane-Br] $\xrightarrow{\Box}$ [cyclohexane-OH] + [cyclohexene] $\xrightarrow{\Box}$ [cyclohexanone] + $HO\overset{O}{\overset{\|}{C}}CH_2CH_2CH_2CH_2\overset{O}{\overset{\|}{C}}OH$

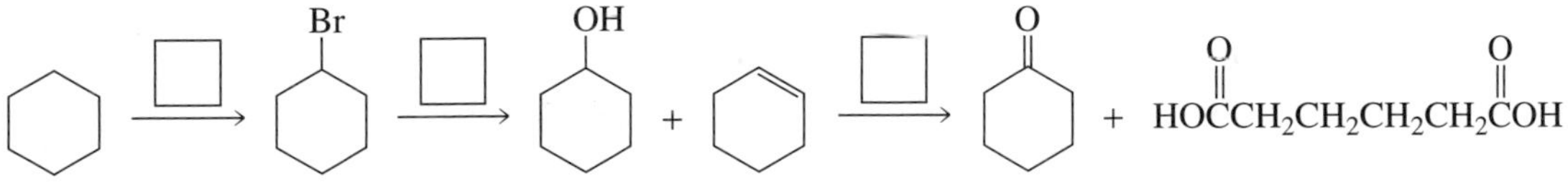

36. a. Give the products obtained from ozonolysis of each of the following compounds, followed by work-up under oxidizing conditions:

1. 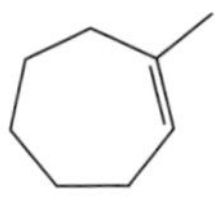   **2.**   **3.** 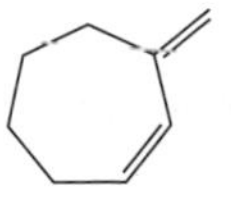   **4.** 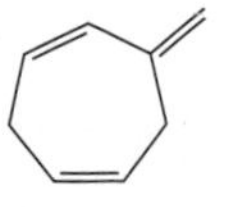   **5.**

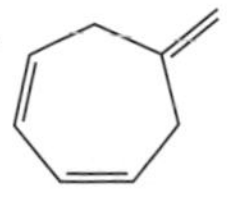

b. What hydrocarbon would form the following products upon reaction with ozone, followed by work-up under reducing conditions?

$$H\overset{O}{\overset{\|}{C}}CH_2CH_2\overset{O}{\overset{\|}{C}}-\overset{O}{\overset{\|}{C}}H + H-\overset{O}{\overset{\|}{C}}-\overset{O}{\overset{\|}{C}}-H + H\overset{O}{\overset{\|}{C}}H$$

37. Show how each of the following compounds can be prepared from cyclohexene:

a. b. 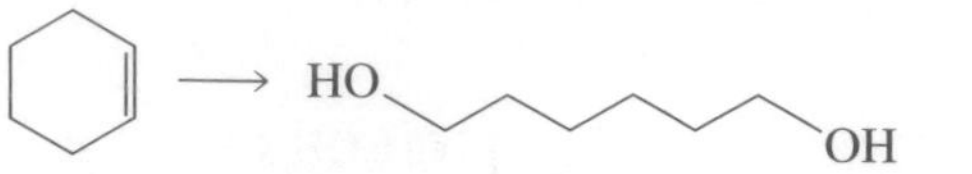 c.

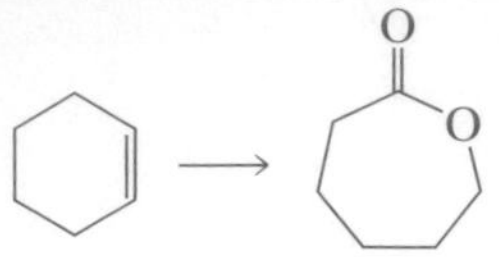

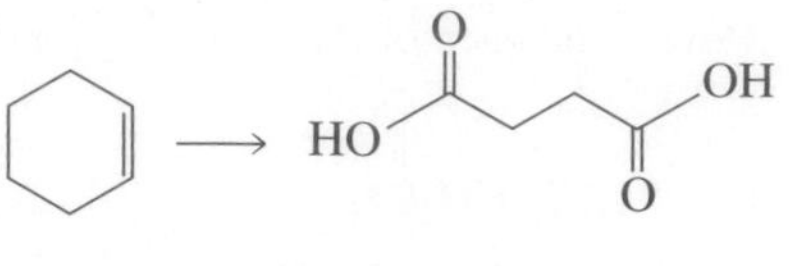

38. The ^{1}H NMR spectrum of the product obtained when an unknown alkene reacts with ozone and the ozonolysis product is worked up under oxidizing conditions is shown. Identify the alkene.

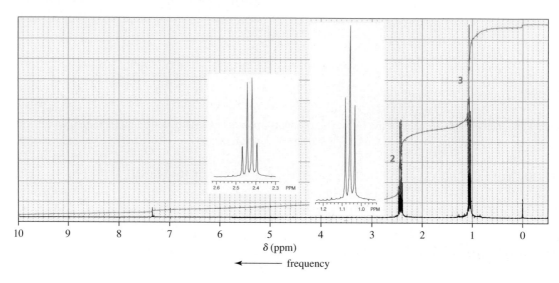

δ (ppm)

← frequency

39. Identify A–N:

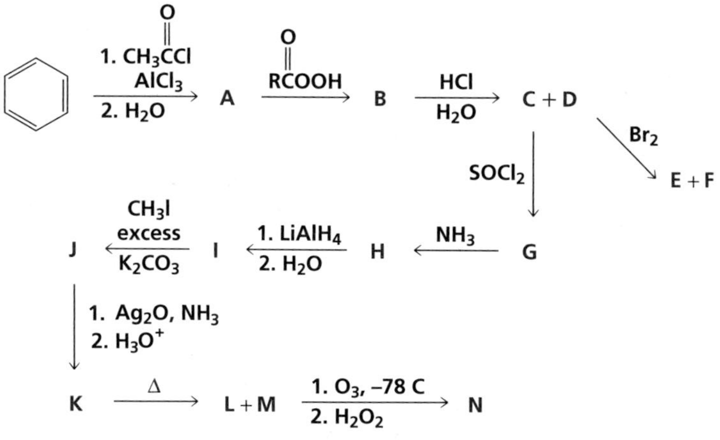

40. Chromic acid oxidizes 2-propanol six times faster than it oxidizes 2-deuterio-2-propanol. Explain. (*Hint:* See Sections 9.7 and 10.9.)

41. Fill in each box with the appropriate reagent:

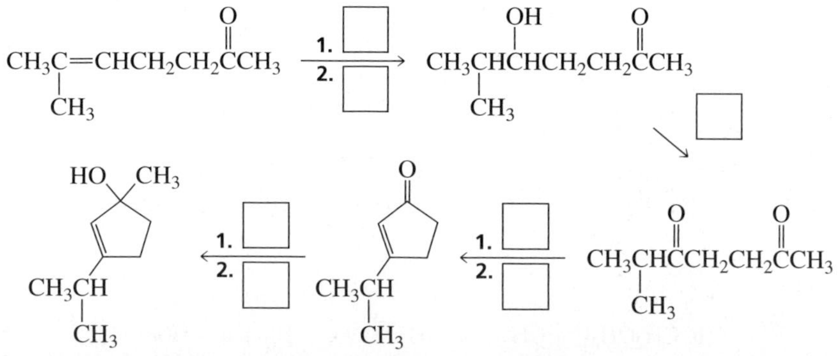

42. Show how each of the following compounds could be prepared, using the given starting material:

a. $CH_3CH_2\overset{\overset{O}{\|}}{C}H \longrightarrow CH_3CH_2\overset{\overset{O}{\|}}{C}OCH_2CH_2CH_3$

d. $\longrightarrow HO\overset{\overset{O}{\|}}{C}CH_2CH_2CH_2CH_2\overset{\overset{O}{\|}}{C}OH$

b. $CH_3CH_2CH_2CH_2OH \longrightarrow CH_3CH_2CH_2\overset{\overset{O}{\|}}{C}CH_2CH_3$

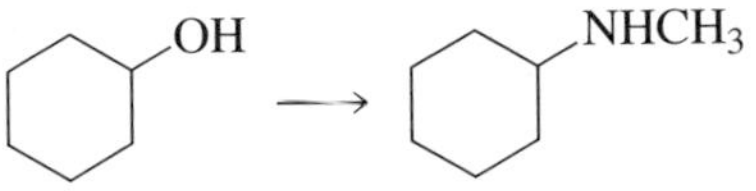

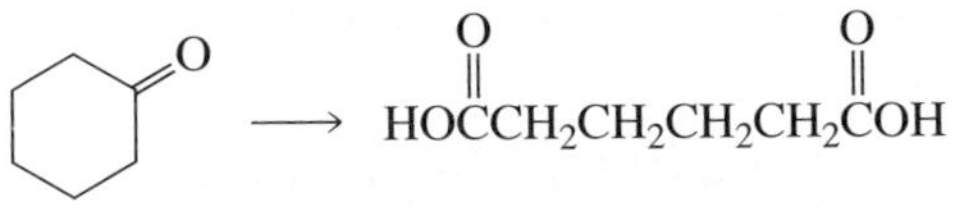

e. $\longrightarrow HO\overset{\overset{O}{\|}}{C}CH_2CH_2CH_2CH_2\overset{\overset{O}{\|}}{C}CH_3$

c.

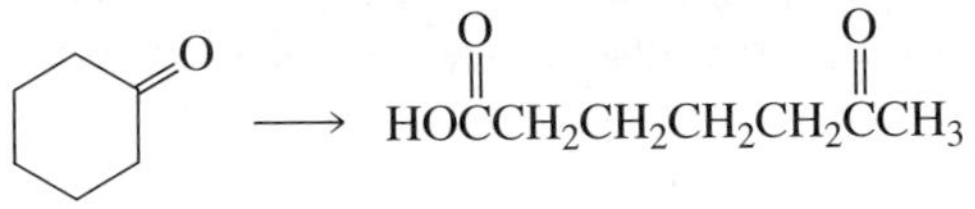

43. Treatment of an alkene with ozone followed by work-up with hydrogen peroxide forms formic acid and a compound that shows three signals (a singlet, a triplet, and a quartet) in its ^1H NMR spectrum. Identify the alkene.

44. Which of the following compounds would be more rapidly cleaved by HIO_4?

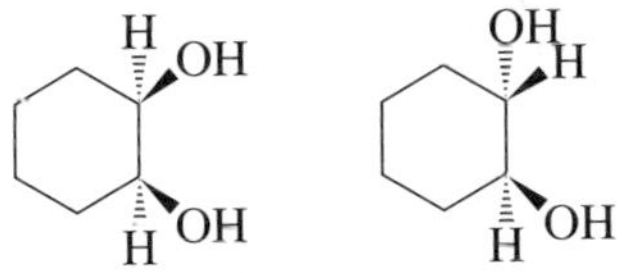

45. Show the reagents needed to carry out the following functional group interconversions:

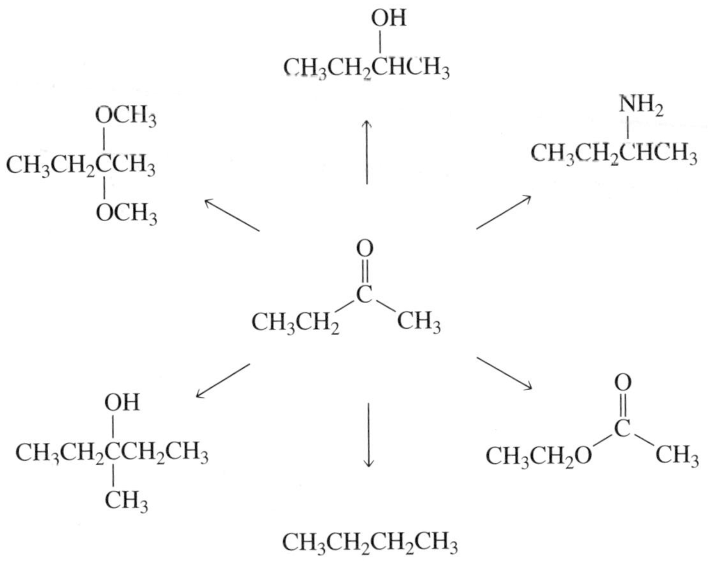

46. Show how cyclohexylacetylene can be converted into each of the following compounds:

a.

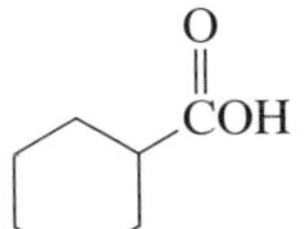

b.

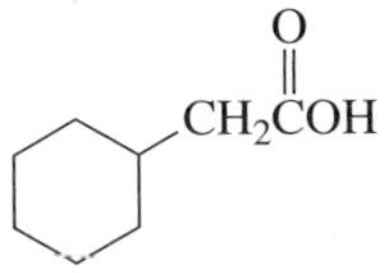

47. Show how the following compounds could be synthesized. The only carbon-containing reagents that are available for each synthesis are shown.

a. $CH_3CH_2CH_2OH \longrightarrow CH_3CH_2CH_2\underset{\underset{CH_3}{|}}{C}HCH_2OH$

c. $CH_3CH_2OH + CH_3\underset{\underset{CH_3}{|}}{C}HOH \longrightarrow$

b. $CH_3\underset{\underset{CH_3}{|}}{C}HOH \longrightarrow CH_3\underset{\underset{CH_3}{|}}{C}HCH_2CH_2CH_3$

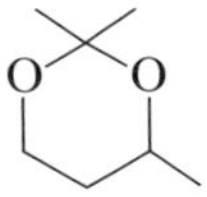

48. The catalytic hydrogenation of 0.5 g of a hydrocarbon at 25 °C consumed about 200 mL of H_2 under 1 atm of pressure. Reaction of the hydrocarbon with ozone, followed by treatment with hydrogen peroxide, gave one product, which was found to be a four-carbon carboxylic acid. Identify the hydrocarbon.

49. Tom Thumbs was asked to prepare the compounds shown below from the given starting materials. The reagents he chose to use for each synthesis are shown.
   a. Which of his syntheses were successful?
   b. What products did he obtain from the other syntheses?
   c. In his unsuccessful syntheses, what reagents should he have used to obtain the desired product?

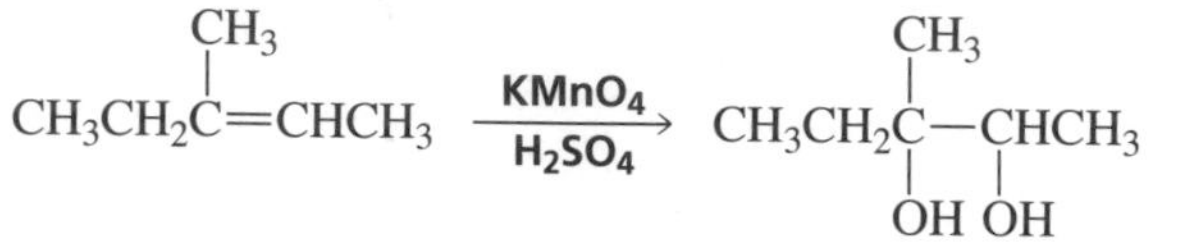

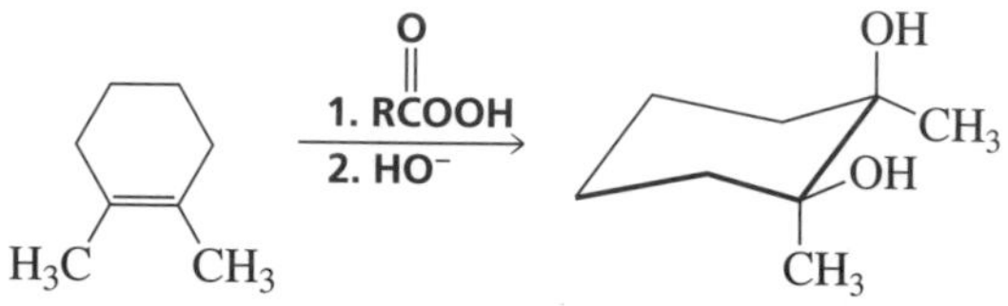

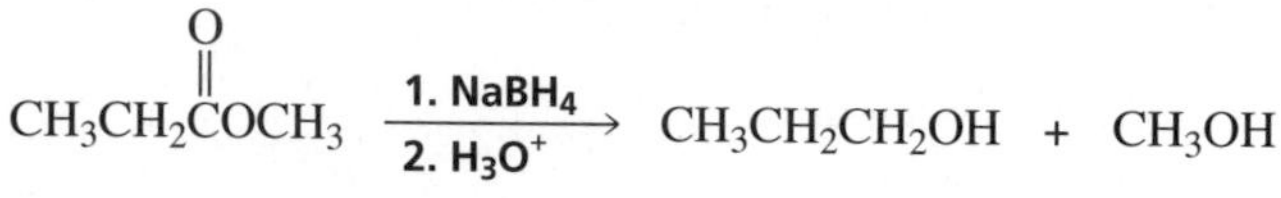

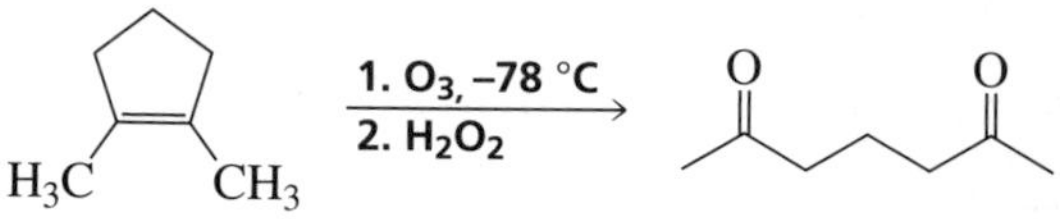

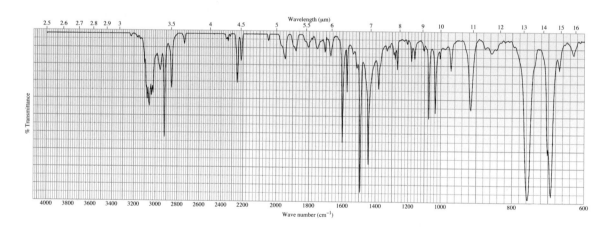

50. The catalytic hydrogenation of compound A formed compound B. The IR spectrum of compound A and the ^1H NMR spectrum of compound B are shown. Identify the compounds.

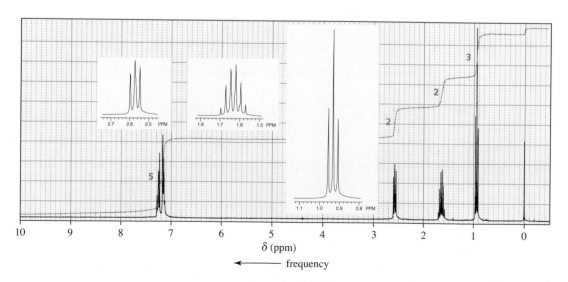

51. Diane Diol worked for several days to prepare the following compounds:

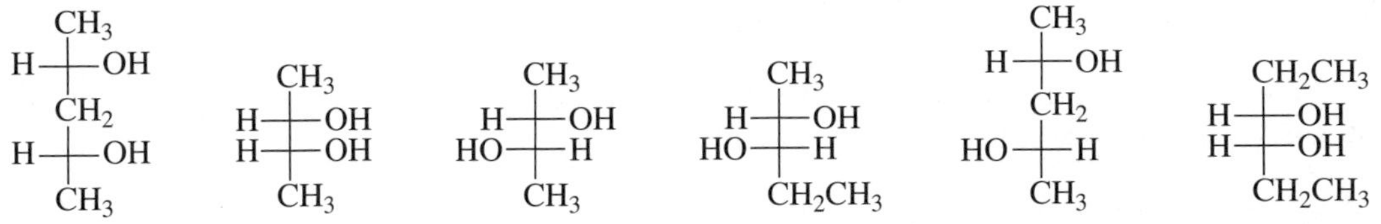

She labeled them carefully and went to lunch. To her horror, when she returned, she found that the labels had fallen off the bottles and onto the floor. Gladys Glycol, the student at the next bench, told her that the diols could be easily distinguished by two experiments. All Diane had to do was determine which ones were optically active and how many products were obtained when each was treated with periodic acid. Diane did what Gladys suggested and learned that:

1. Compounds A, E, and F are optically active, and B, C, and D are optically inactive.
2. One product is obtained from the reaction of A, B, and D with periodic acid.
3. Two products are obtained from the reaction of F with periodic acid.
4. C and E do not react with periodic acid.

Will Diane be able to distinguish between the six diols and label them from A to F with only the preceding information? Provide a systematic name for each structure.

52. Show how propyl propionate could be prepared, using allyl alcohol as the only source of carbon.

53. Compound A has a molecular formula of $C_5H_{12}O$ and is oxidized by an acidic solution of sodium dichromate to give compound B, whose molecular formula is $C_5H_{10}O$. When compound A is heated with H_2SO_4, C and D are obtained. Considerably more D is obtained than C. Reaction of compound C with O_3 followed by treatment with H_2O_2 gives two products: formic acid and compound E, whose molecular formula is C_4H_8O. Reaction of compound D with O_3 followed by treatment with H_2O_2 gives compound F, whose molecular formula is C_3H_6O, and compound G, whose molecular formula is $C_2H_4O_2$. What are the structures of compounds A through G?

54. A compound forms *cis*-1,2-dimethylcyclopropane when it is reduced with H_2 and Pd/C. The ^1H NMR spectrum of the compound shows only two singlets. What is the structure of the compound?

55. Show how you could convert
  a. maleic acid to (2R,3S)-tartaric acid.   **c.** maleic acid to (2R,3R)- and (2S,3S)-tartaric acid.
  b. fumaric acid to (2R,3S)-tartaric acid.   **d.** fumaric acid to (2R,3R)- and (2S,3S)-tartaric acid.

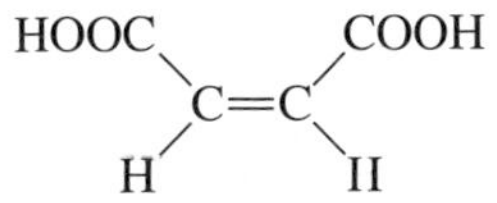

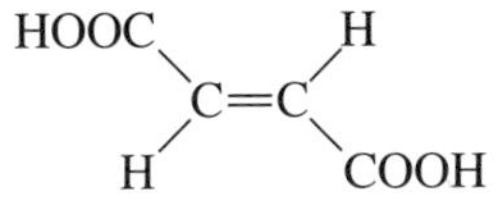

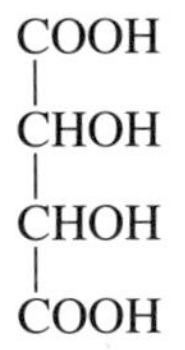

56. Identify A through O:

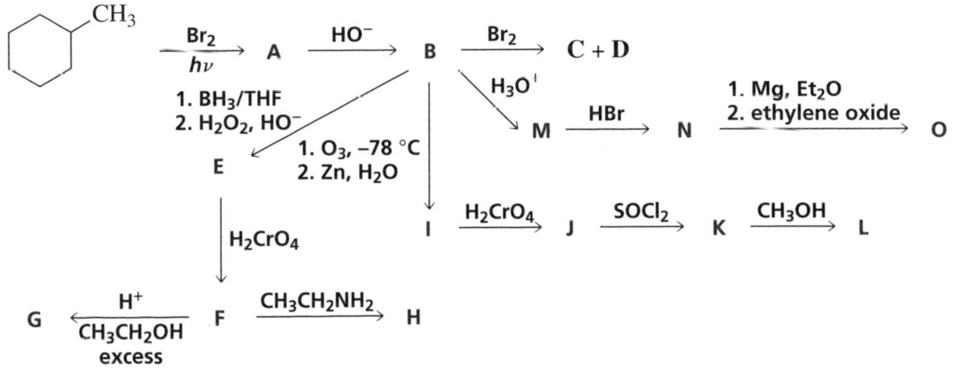

57. Show how the following compounds could be prepared using only the indicated starting material as the source of carbon:

a. $\overset{O}{\overset{||}{CH_3CCH_3}}$ from $\overset{CH_3}{\overset{|}{CH_3CHCH_3}}$

b. $\overset{CH_3}{\overset{|}{CH_3CH=CCH_3}}$ using propane as the only source of carbon

c. $\overset{O\ \ CH_3}{\overset{||\ \ \ |}{CH_3C-CHCH_3}}$ from propane and any molecule with two carbon atoms

d. $\overset{O}{\overset{||}{CH_3CH_2CH}}$ from two molecules of ethane

58. A primary alcohol can be oxidized only as far as the aldehyde stage if the alcohol is first treated with tosyl chloride (TsCl) and the resulting tosylate is allowed to react with dimethyl sulfoxide (DMSO). Propose a mechanism for this reaction. (*Hint:* See Section 19.2.)

$$CH_3CH_2CH_2CH_2OH \xrightarrow[\text{pyridine}]{\textbf{TsCl}} CH_3CH_2CH_2CH_2OTs \xrightarrow{\textbf{DMSO}} CH_3CH_2CH_2\overset{O}{\overset{||}{CH}}$$

59. Identify the alkene that gives each of the following products upon ozonolysis followed by treatment with dimethyl sulfide:

a.

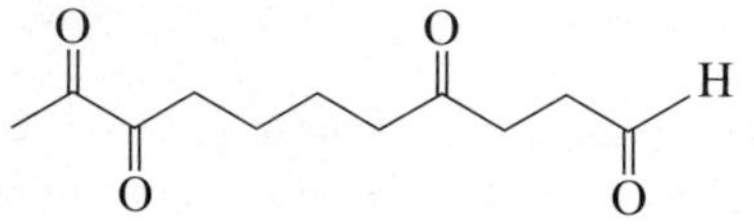

b. 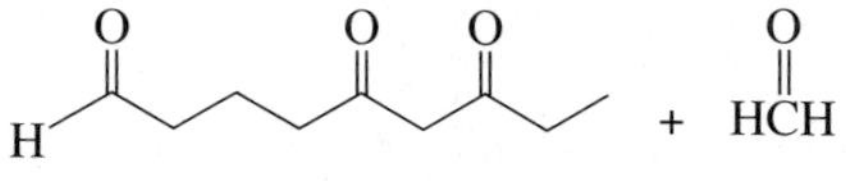   +   HCH

60. Propose a mechanism to explain how dimethyl sulfoxide and oxalyl chloride react to form the dimethylchlorosulfonium ion used as the oxidizing agent in the Swern oxidation.

$$\underset{\textbf{dimethyl sulfoxide}}{CH_3-\overset{O}{\overset{||}{S}}-CH_3} \quad + \quad \underset{\textbf{oxalyl chloride}}{Cl-\overset{O}{\overset{||}{C}}-\overset{O}{\overset{||}{C}}-Cl} \quad \longrightarrow \quad \underset{\substack{\textbf{dimethylchloro-}\\\textbf{sulfonium ion}}}{CH_3-\overset{Cl}{\overset{|}{\underset{+}{S}}}-CH_3} + CO_2 + CO + Cl^-$$

61. Show how the following compounds could be prepared using only the indicated starting material as the source of carbon:

a.

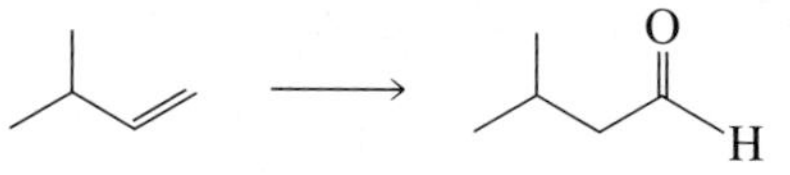

b.

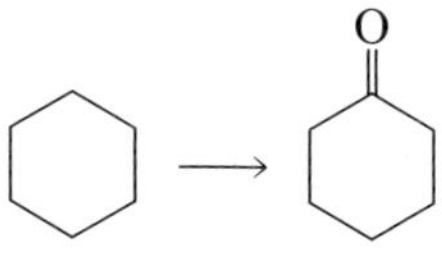

c.

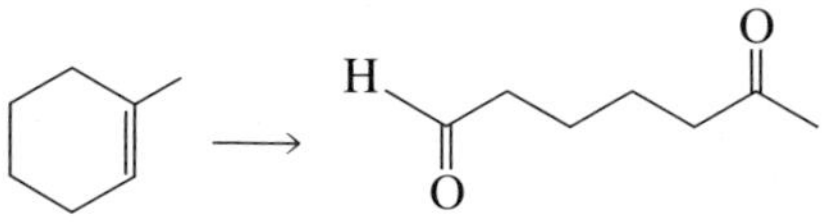

d.

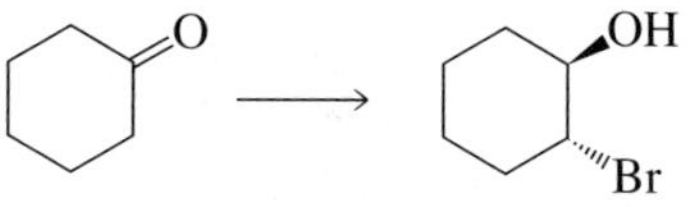

e.

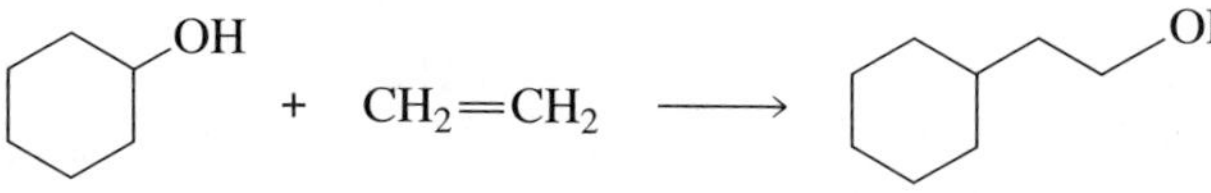

f.

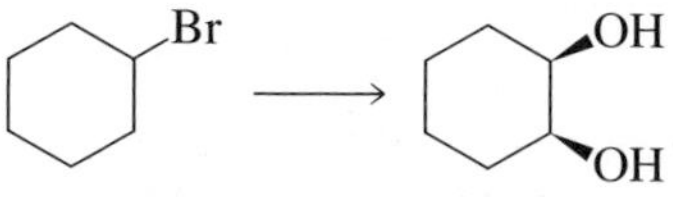

CHAPTER 20

More About Amines • Heterocyclic Compounds

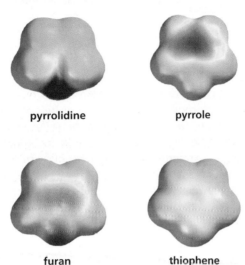

pyrrolidine   pyrrole

furan   thiophene

BUILDING ON FUNDAMENTALS

SECTIONS 20.4 AND 20.6 The Hoffman elimination and Cope elimination reactions violate Zaitsev's rule, just like the elimination reactions of alkyl fluorides do and for the same reason (9.2).

SECTION 20.4 Primary carbanions are more stable than secondary carbanions, which are more stable than tertiary carbanions (9.2).

SECTION 20.5 Quaternary ammonium salts can be employed as phase-transfer catalysts, just like crown ethers can be (10.10).

SECTION 20.8 An aromatic compound is cyclic and planar, every atom in the ring has a π orbital, and the π cloud contains an odd number of pairs of π electrons (14.2).

SECTION 20.8 Pyrrole, furan, and thiophene are aromatic compounds that undergo electrophilic aromatic substitution reactions just like benzene does (14.10).

SECTION 20.8 The more stable and the more nearly equivalent the resonance contributors, the greater is the delocalization (resonance) energy (7.6).

SECTIONS 20.8 AND 20.9 An sp hybridized atom is more electronegative than the same atom that is sp^2 hybridized, which is more electronegative than the same atom that is sp^3 hybridized (1.20).

SECTION 20.9 Pyridine is an aromatic compound that undergoes electrophilic aromatic substitution reactions more slowly than benzene (15.2) and nucleophilic aromatic substitution reactions more rapidly than benzene (15.12).

Amines, compounds in which one or more of the hydrogens of ammonia (NH_3) have been replaced by an alkyl group, are among some of the most abundant compounds in the biological world. We will come to appreciate their biological importance as we explore the structures and properties of amino acids and proteins in Chapter 22; study how enzymes catalyze chemical reactions in Chapter 23; investigate the ways in which coenzymes—compounds derived from vitamins—help enzymes catalyze chemical reactions in Chapter 24; learn about nucleic acids (DNA and RNA) in Chapter 27; and look at how drugs are discovered and designed in Chapter 30.

Amines are also exceedingly important compounds to organic chemists—far too important to leave until the end of a course in organic chemistry. We have, therefore, already studied many aspects of amines and their chemistry. For example, we have seen that the nitrogen in amines is sp^3 hybridized with the lone pair residing in an sp^3 orbital (Section 2.8), and we know that amines invert rapidly at room temperature through a transition state in which the sp^3 nitrogen becomes an sp^2 nitrogen (Section 5.17).

We also have examined the physical properties of amines—their hydrogen-bonding properties, boiling points, and solubilities (Section 2.9)—and we learned how amines are named (Section 2.7). Most important, we have seen that the lone-pair electrons of the nitrogen atom cause amines to react as bases, sharing the lone pair with a proton, and as nucleophiles, sharing the lone pair with an atom other than a proton.

an amine is a base:

$$R-\ddot{N}H_2 \;+\; H-Br \longrightarrow R-\overset{+}{N}H_3 \;+\; Br^-$$

an amine is a nucleophile:

$$R-\ddot{N}H_2 \;+\; CH_3-Br \longrightarrow R-\overset{+}{N}H_2-CH_3 \;+\; Br^-$$

In this chapter, we will revisit some of these topics and look at other aspects of amines and their chemistry that we have not considered previously.

Some amines are **heterocyclic compounds** (or **heterocycles**)—cyclic compounds in which one or more of the atoms of the ring are **heteroatoms** (Section 14.4). A variety of atoms, such as N, O, S, Se, P, Si, B, and As, can be incorporated into ring structures.

Heterocycles are an extraordinarily important class of compounds, making up more than half of all known organic compounds. Almost all the compounds we know as drugs (Chapter 30), most vitamins (Chapter 24), and many other natural products are heterocycles. In this chapter, we will consider the most prevalent heterocyclic compounds—the ones containing the heteroatom N, O, or S.

20.1 More About Amine Nomenclature

In Section 2.7, we saw that amines are classified as primary, secondary, or tertiary, depending on whether one, two, or three hydrogens of ammonia, respectively, have been replaced by an alkyl group. We also saw that amines have both common and systematic names. Common names are obtained by citing the names of the alkyl substituents (in alphabetical order) that have replaced the hydrogens of ammonia. Systematic names employ "amine" as a functional group suffix.

$CH_3CH_2CH_2CH_2CH_2NH_2$	$CH_3CH_2CH_2CH_2NHCH_2CH_3$	$CH_3CH_2CH_2\overset{\underset{\displaystyle |}{CH_3}}{N}CH_2CH_3$
a primary amine	**a secondary amine**	**a tertiary amine**
1-pentanamine	*N*-ethyl-1-butanamine	*N*-ethyl-*N*-methyl-1-propanamine
pentylamine	butylethylamine	ethylmethylpropylamine

systematic name:
common name:

A saturated cyclic amine—a cyclic amine without any double bonds—can be named as a cycloalkane, using the prefix "aza" to denote the nitrogen atom. There are, however, other acceptable names. Some of the more commonly used names are shown here. Notice that heterocyclic rings are numbered so that the heteroatom has the lowest possible number.

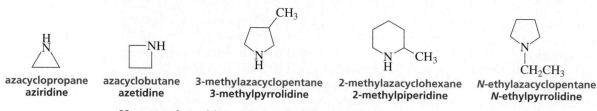

azacyclopropane	**azacyclobutane**	**3-methylazacyclopentane**	**2-methylazacyclohexane**	***N*-ethylazacyclopentane**
aziridine	**azetidine**	**3-methylpyrrolidine**	**2-methylpiperidine**	***N*-ethylpyrrolidine**

Heterocycles with oxygen and sulfur heteroatoms are named similarly. The prefix for oxygen is "oxa" and that for sulfur is "thia."

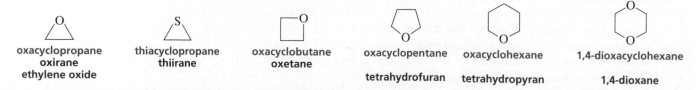

oxacyclopropane	**thiacyclopropane**	**oxacyclobutane**	**oxacyclopentane**	**oxacyclohexane**	**1,4-dioxacyclohexane**
oxirane	**thiirane**	**oxetane**			
ethylene oxide			**tetrahydrofuran**	**tetrahydropyran**	**1,4-dioxane**

Name the following compounds:

a.

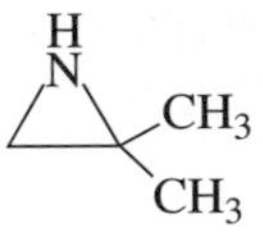

c.

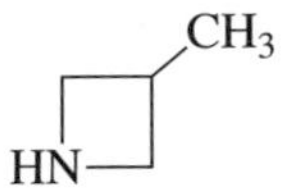

e.

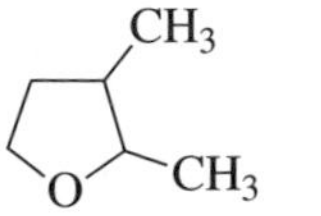

b.

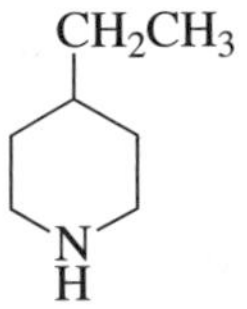

d.

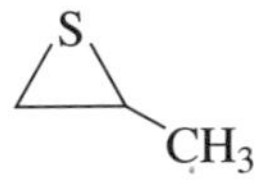

f.

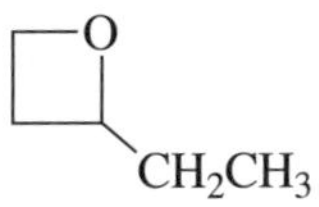

20.2 **More About the Acid–Base Properties of Amines**

Amines are the most common organic bases. We have seen that ammonium ions have pK_a values of about 11 (Sections 1.18 and 10.6) and anilinium ions have pK_a values of about 5 (Sections 7.9 and 15.4). The greater acidity of anilinium ions compared with ammonium ions is due to the greater stability of the conjugate bases of the anilinium ions as a result of electron delocalization. Amines have very high pK_a values. For example, the pK_a of methylamine is 40.

$$CH_3CH_2CH_2\overset{+}{N}H_3$$

an ammonium ion
pK_a = 10.8

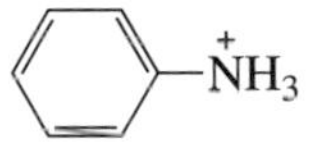

$-\overset{+}{N}H_3$

an anilinium ion
pK_a = 4.58

CH_3NH_2

an amine
pK_a = 40

Saturated amine heterocycles containing five or more atoms have physical and chemical properties typical of the acyclic amines. For example, pyrrolidine, piperidine, and morpholine are typical secondary amines, and N-methylpyrrolidine and quinuclidine are typical tertiary amines. The conjugate acids of these amines have pK_a values expected for ammonium ions. We have seen that the basicity of amines allows them to be easily separated from other organic compounds (Chapter 1, Problems 95 and 96).

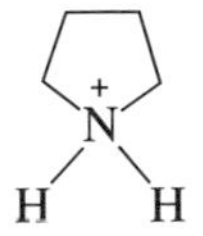

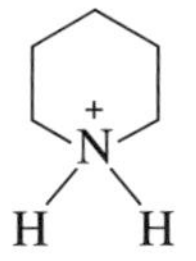

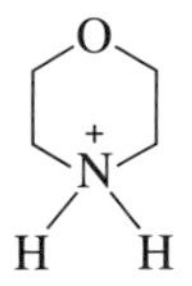

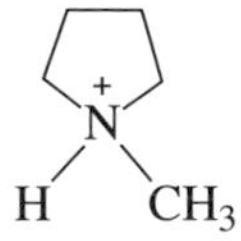

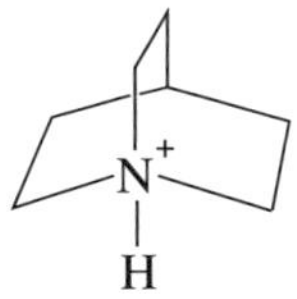

the ammonium ions of:

pyrrolidine
pK_a = 11.27

piperidine
pK_a = 11.12

morpholine
pK_a = 9.28

N-methylpyrrolidine
pK_a = 10.32

quinuclidine
pK_a = 11.38

Why is the pK_a of the conjugate acid of morpholine significantly lower than the pK_a of the conjugate acid of piperidine?

a. Draw the structure of 3-quinuclidinone.
b. What is the approximate pK_a of its conjugate acid?
c. Which has a lower pK_a, the conjugate acid of 3-bromoquinuclidine or the conjugate acid of 3-chloroquinuclidine?

20.3 Amines React as Bases and as Nucleophiles

We have seen that the leaving group of an amine ($^-NH_2$) is such a strong base that amines cannot undergo the substitution and elimination reactions that alkyl halides, alcohols, and ethers undergo (Section 10.6). The relative reactivities of these compounds—each with an electron-withdrawing group bonded to an sp^3 carbon—can be appreciated by comparing the pK_a values of the conjugate acids of their leaving groups, keeping in mind that the weaker the acid, the stronger its conjugate base and the poorer the base is as a leaving group.

relative reactivities

most reactive $\rightarrow$ RCH_2F	> RCH_2OH	~ RCH_2OCH_3	> RCH_2NH_2 $\leftarrow$ least reactive
strongest acid, weakest conjugate base $\rightarrow$ HF $pK_a = 3.2$	H_2O $pK_a = 15.7$	RCH_2OH $pK_a = 15.5$	NH_3 $pK_a = 36$ $\leftarrow$ weakest acid, strongest conjugate base

The lone pair on the nitrogen of an amine causes it to be nucleophilic as well as basic. We have seen that amines react as bases in proton-transfer reactions and in elimination reactions (Sections 1.18, 9.10, and 10.2). We have also seen that they react as nucleophiles in a number of different reactions: in nucleophilic substitution reactions—reactions that *alkylate* the amine (Section 8.4)—such as

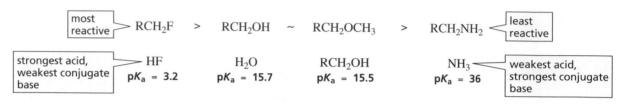

$$CH_3CH_2Br \ + \ CH_3NH_2 \ \longrightarrow \ CH_3CH_2\overset{+}{N}H_2CH_3 \ \rightleftharpoons \ CH_3CH_2NHCH_3 \ + \ HBr$$

methylamine Br^- ethylmethylamine

and in nucleophilic acyl substitution reactions—reactions that *acylate* the amine (Sections 16.8, 16.9, and 16.10)—such as

and in nucleophilic addition–elimination reactions—the reactions of aldehydes and ketones with primary amines to form imines and with secondary amines to form enamines (Section 17.8)—such as

and in conjugate addition reactions (Section 17.16)—such as,

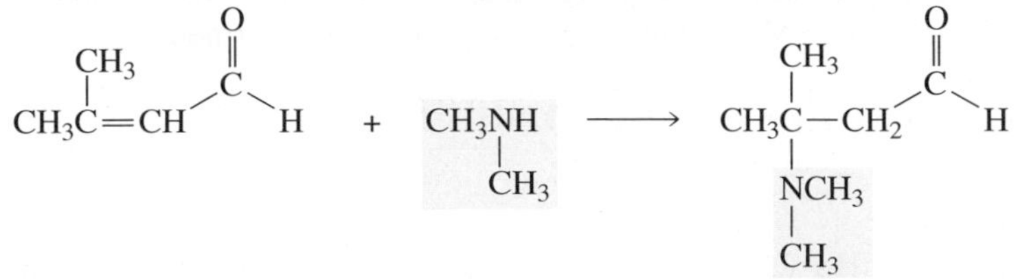

$$CH_3\overset{CH_3}{C}=CH\overset{O}{\underset{}{C}}H \;+\; CH_3\underset{CH_3}{NH} \longrightarrow CH_3\overset{CH_3}{\underset{\underset{CH_3}{NCH_3}}{C}}-CH_2\overset{O}{\underset{}{C}}H$$

We have seen that primary arylamines react with nitrous acid to form stable arenediazonium salts (Section 15.11). Arenediazonium salts are useful to synthetic chemists because the diazonium group can be replaced by a wide variety of nucleophiles. This reaction allows a wider variety of substituted benzenes to be prepared than can be prepared solely from electrophilic aromatic substitution reactions.

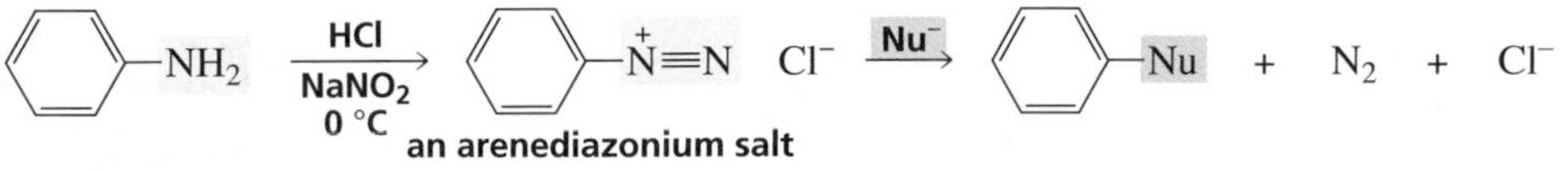

$$\text{C}_6\text{H}_5-NH_2 \xrightarrow[\substack{NaNO_2 \\ 0\,°C}]{HCl} \text{C}_6\text{H}_5-\overset{+}{N}\equiv N \;\; Cl^- \xrightarrow{Nu^-} \text{C}_6\text{H}_5-Nu \;+\; N_2 \;+\; Cl^-$$

an arenediazonium salt

PROBLEM 4

Give the product of each of the following reactions:

a.

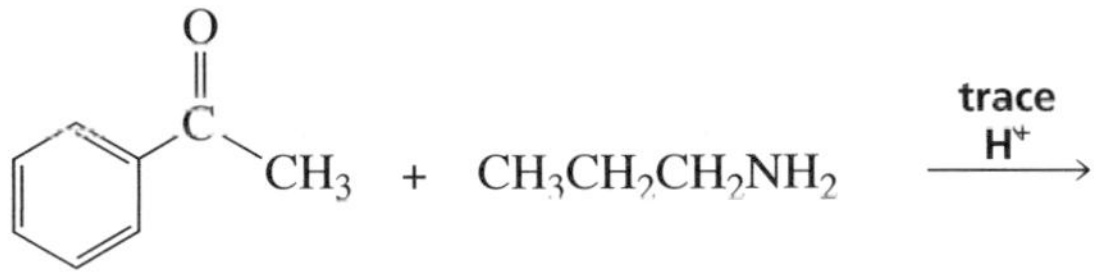

$$\text{C}_6\text{H}_5\overset{O}{\underset{}{C}}CH_3 \;+\; CH_3CH_2CH_2NH_2 \xrightarrow{\substack{trace \\ H^+}}$$

b.

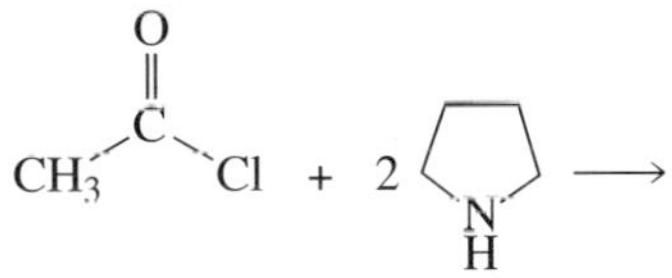

$$CH_3\overset{O}{\underset{}{C}}Cl \;+\; 2\;\text{(pyrrolidine, N-H)} \longrightarrow$$

c.

$$\text{C}_6\text{H}_5-NH_2 \xrightarrow[\text{2. H}_2\text{O, Cu}_2\text{O, Cu(NO}_3)_2]{\text{1. HCl, NaNO}_2,\ 0\ °C}$$

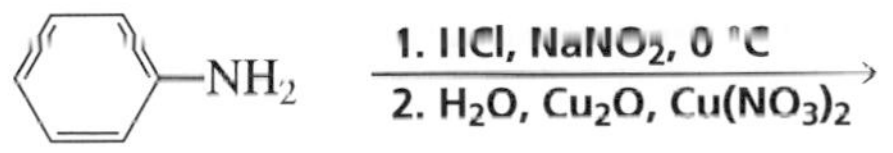

d.

$$\text{C}_6\text{H}_5\overset{O}{\underset{}{C}}CH_3 \;+\; CH_3CH_2NHCH_2CH_3 \xrightarrow{\substack{trace \\ H^+}}$$

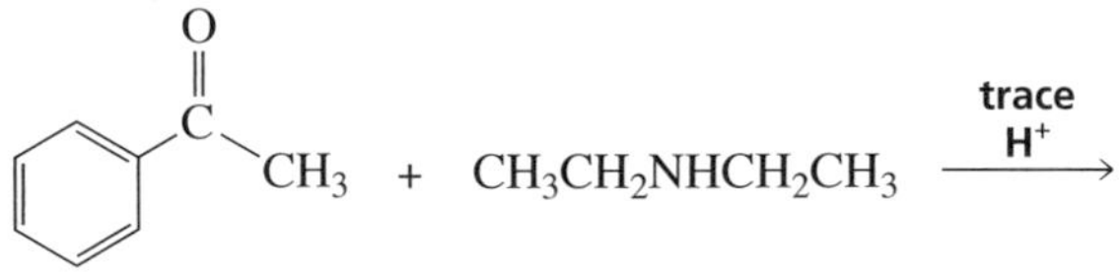

20.4 Quaternary Ammonium Hydroxides Undergo Elimination Reactions

The leaving group of a **quaternary ammonium ion** has about the same leaving tendency as a protonated amino group, but it does not have an acidic hydrogen that would protonate a basic reactant. A quaternary ammonium ion, therefore, can undergo a reaction with a strong base. The reaction of a quaternary ammonium ion with hydroxide ion is known as a **Hofmann elimination reaction**. The leaving group in a Hofmann elimination reaction is a tertiary amine. Because a tertiary amine is a relatively poor leaving group, the reaction requires heat.

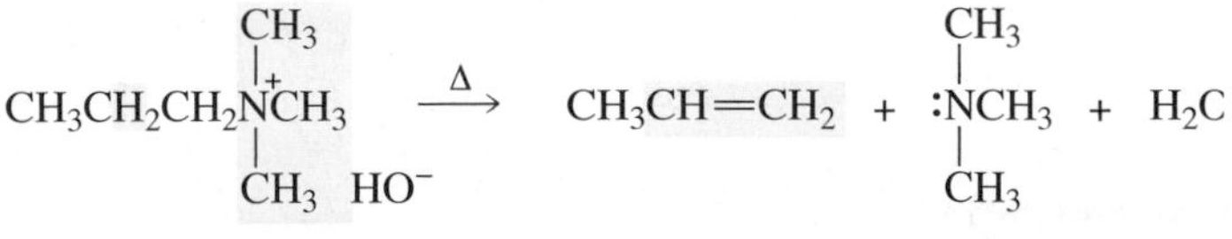

$$CH_3CH_2CH_2\overset{\overset{CH_3}{|}}{\underset{\underset{CH_3\;\;HO^-}{|}}{\overset{+}{N}}}CH_3 \xrightarrow{\Delta} CH_3CH=CH_2 \;+\; :\overset{CH_3}{\underset{CH_3}{N}}CH_3 \;+\; H_2C$$

A Hofmann elimination reaction is an E2 reaction, which, as you will recall, is a concerted, one-step reaction: the proton and the tertiary amine are removed in the same step (Section 9.1). Very little substitution product is formed.

Student Tutorial:
Hofmann elimination reaction

Mechanism of the Hofmann elimination

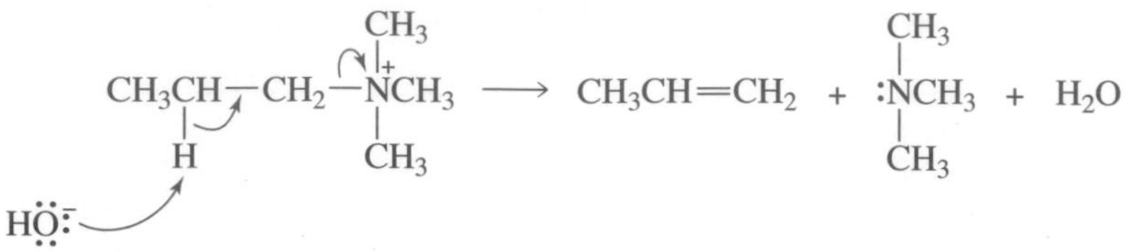

$$CH_3CH{-}CH_2{-}\overset{\overset{CH_3}{|}}{\underset{\underset{CH_3}{|}}{\overset{+}{N}}}CH_3 \longrightarrow CH_3CH{=}CH_2 + :\overset{\overset{CH_3}{|}}{\underset{\underset{CH_3}{|}}{N}}CH_3 + H_2O$$

$$\underset{H}{|}$$

$$H\ddot{O}:^-$$

PROBLEM 5◆

What is the difference between the reaction that occurs when isopropyltrimethylammonium hydroxide is heated and the reaction that occurs when 2-bromopropane is treated with hydroxide ion?

The carbon to which the tertiary amine is attached is designated as the α-carbon, so the adjacent carbon, from which the proton is removed, is called the β-carbon. (Recall that E2 reactions are also called β-elimination reactions, since elimination is initiated by removing a proton from a β-carbon; Section 9.1.) If the quaternary ammonium ion has more than one β-carbon, the major alkene product is the one obtained by removing a proton from the β-carbon bonded to the greater number of hydrogens. In the following reaction, the major alkene product is obtained by removing a proton from the β-carbon bonded to three hydrogens, and the minor alkene product results from removing a proton from the β-carbon bonded to two hydrogens.

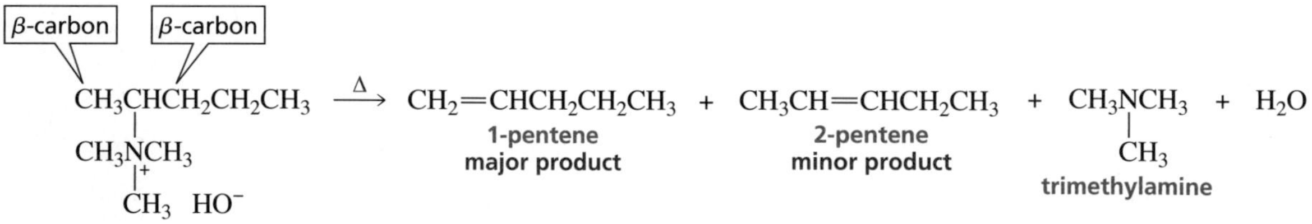

$\boxed{\beta\text{-carbon}}$ $\boxed{\beta\text{-carbon}}$

$$CH_3CHCH_2CH_2CH_3 \xrightarrow{\Delta} CH_2{=}CHCH_2CH_2CH_3 + CH_3CH{=}CHCH_2CH_3 + CH_3NCH_3 + H_2O$$

$$\underset{\underset{\underset{CH_3\ HO^-}{|}}{\overset{+}{CH_3NCH_3}}}{|}$$

1-pentene
major product

2-pentene
minor product

$$\underset{CH_3}{|}$$
trimethylamine

The major alkene product of the following reaction comes from removing a proton from the β-carbon bonded to two hydrogens, because the other β-carbon is bonded to only one hydrogen.

In a Hofmann elimination reaction, the proton is removed from the β-carbon bonded to the most hydrogens.

$\boxed{\beta\text{-carbon}}$ $\boxed{\beta\text{-carbon}}$

$$CH_3CHCH_2\overset{\overset{CH_3}{|}}{\underset{\underset{CH_3\ HO^-}{|}}{\overset{+}{N}}}CH_2CH_2CH_3 \xrightarrow{\Delta} CH_3CHCH_2\overset{\overset{CH_3}{|}}{\underset{\underset{CH_3}{|}}{N}} + CH_2{=}CHCH_3 + H_2O$$

isobutyldimethylamine

propene

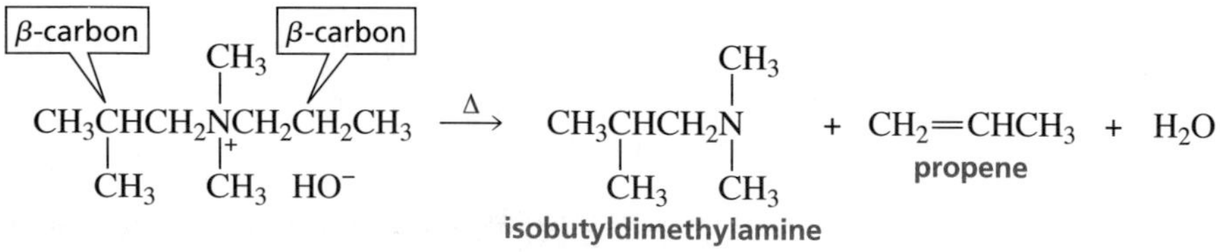

PROBLEM 6◆

What are the minor products in the preceding Hofmann elimination reaction?

We have seen that in an E2 reaction of an alkyl chloride, alkyl bromide, or alkyl iodide, the proton is removed from the β-carbon bonded to the *fewest* hydrogens (*Zaitsev's rule*; Section 9.2). Now, however, we see that in an E2 reaction of a quaternary ammonium ion, the proton is removed from the β-carbon bonded to the *most* hydrogens (*anti-Zaitsev elimination*).

Why do alkyl halides follow Zaitsev's rule while quaternary amines violate the rule? Quaternary amines violate the rule for the same reason that alkyl fluorides violate it (Section 9.2). Alkyl halides, other than alkyl fluorides, have relatively good leaving groups. Therefore, when hydroxide ion starts to remove a proton in an E2 reaction, the halide ion immediately begins to depart, forming a transition state with an *alkene-like* structure. The proton is removed from the β-carbon bonded to the fewest hydrogens in order to achieve the most stable alkene-like transition state.

Zaitsev elimination alkene-like transition state		anti-Zaitsev elimination carbanion-like transition state	

$$
\begin{array}{cccc}
\overset{\delta-}{\text{OH}} & \overset{\delta-}{\text{OH}} & \overset{\delta-}{\text{OH}} & \overset{\delta-}{\text{OH}} \\
\vdots & \vdots & \vdots & \vdots \\
\text{H} & \text{H} & \text{H} & \text{H} \\
\vdots & \vdots & \vdots & \vdots \\
\text{CH}_3\text{CH}\!=\!\!\text{CHCH}_3 & \text{CH}_3\text{CH}_2\text{C}\!=\!\!\text{CH}_2 & \overset{}{\underset{\delta-}{\text{CH}_2}}\text{CHCH}_2\text{CH}_2\text{CH}_3 & \text{CH}_3\text{CHCHCH}_2\text{CH}_3 \\
\underset{\text{Br}}{\overset{\delta-}{|}} & \underset{\text{Br}}{\overset{\delta-}{|}} & \underset{{}^+\text{N(CH}_3)_3}{|} & \underset{{}^+\text{N(CH}_3)_3}{\overset{\delta-}{|}} \\
\textbf{more stable} & \textbf{less stable} & \textbf{more stable} & \textbf{less stable}
\end{array}
$$

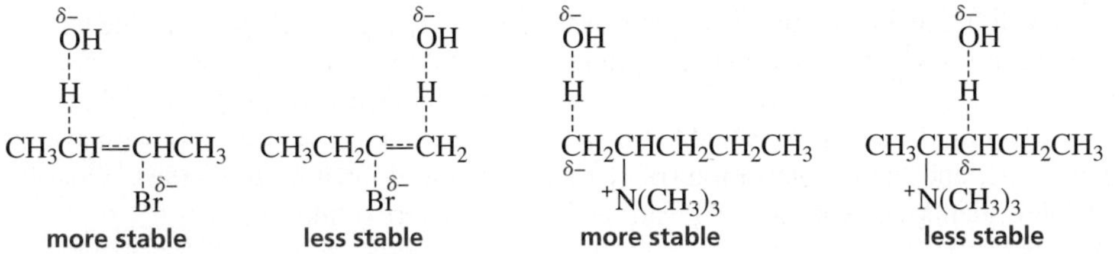

Quaternary ammonium ions, like alkyl fluorides, have more basic (poorer) leaving groups. As a result, when hydroxide ion starts to remove a proton in an E2 reaction, the leaving group does not start to leave, so a partial negative charge builds up on the carbon from which the proton is being removed. This gives the transition state a *carbanion-like* structure rather than an alkene-like structure. By removing a proton from the β-carbon bonded to the most hydrogens, the most stable carbanion-like transition state is achieved. (Recall from Section 9.2 that primary carbanions are more stable than secondary carbanions, which are more stable than tertiary carbanions.) Steric factors in the Hofmann reaction also favor anti-Zaitsev elimination. Because the Hofmann elimination reaction occurs in an anti-Zaitsev manner, *anti-Zaitsev elimination* is also referred to as *Hofmann elimination*.

PROBLEM 7♦

Give the major products of each of the following reactions:

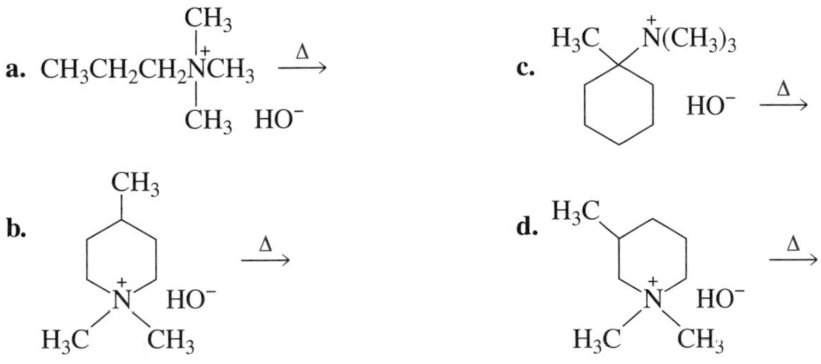

a. $\underset{\underset{\text{CH}_3}{|}}{\overset{\overset{\text{CH}_3}{|}}{\text{CH}_3\text{CH}_2\text{CH}_2\overset{+}{\text{N}}\text{CH}_3}}$ $\xrightarrow{\Delta}$ HO⁻

b. (4-methyl ring with N⁺, H₃C and CH₃ substituents, HO⁻) $\xrightarrow{\Delta}$

c. (cyclohexane ring with H₃C and ⁺N(CH₃)₃) HO⁻ $\xrightarrow{\Delta}$

d. (ring with H₃C and N⁺, H₃C and CH₃ substituents, HO⁻) $\xrightarrow{\Delta}$

For a quaternary ammonium ion to undergo an elimination reaction, the counterion must be hydroxide ion, because a strong base is needed to remove the proton from the β-carbon. Halide ions, in contrast, are weak bases, so quaternary ammonium *halides* cannot undergo a Hofmann elimination reaction. However, a quaternary ammonium *halide* can be converted into a quaternary ammonium *hydroxide* by treatment with silver oxide and water. The silver halide precipitates, and the halide ion is replaced by hydroxide ion. The compound can now undergo an elimination reaction.

$$
2\ \text{R}\!-\!\overset{\overset{\text{R}}{|}}{\underset{\underset{\text{R}\ \ \text{I}^-}{|}}{\overset{+}{\text{N}}}}\!-\!\text{R}\ +\ \text{Ag}_2\text{O}\ +\ \text{H}_2\text{O}\ \longrightarrow\ 2\ \text{R}\!-\!\overset{\overset{\text{R}}{|}}{\underset{\underset{\text{R}\ \ \text{HO}^-}{|}}{\overset{+}{\text{N}}}}\!-\!\text{R}\ +\ 2\ \text{AgI} \downarrow
$$

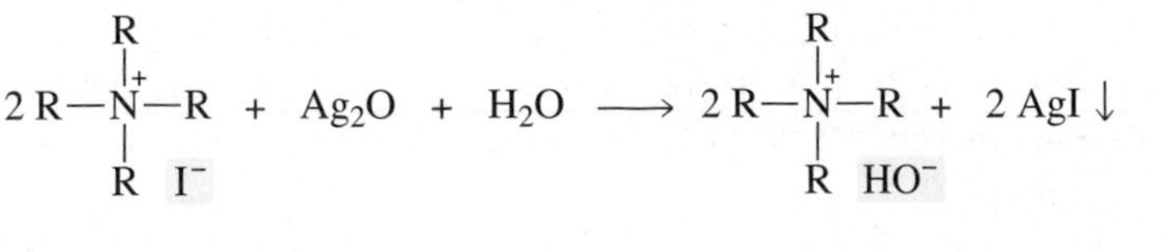

The reaction of an amine with sufficient methyl iodide to convert the amine into a quaternary ammonium iodide is called **exhaustive methylation** (see Chapter 8, Problem 12). The reaction is carried out in a basic solution of potassium carbonate, so that the amines formed as intermediates will be predominantly in their basic forms.

 **Amine Formation**

| exhaustive methylation |

$$CH_3CH_2CH_2NH_2 \ + \ CH_3I \xrightarrow[\text{excess}]{K_2CO_3} \ CH_3CH_2CH_2\overset{+}{N}CH_3 \quad I^-$$

(with CH_3 above and CH_3 below the N)

The Hofmann elimination reaction was used by early organic chemists as the last step of a process known as a Hofmann degradation—a method used to identify amines. In a *Hofmann degradation*, an amine is exhaustively methylated with methyl iodide, treated with silver oxide to convert the quaternary ammonium iodide to a quaternary ammonium hydroxide, and then heated to allow it to undergo a Hofmann elimination. Once the alkene is identified, working backward gives the structure of the amine.

A USEFUL BAD-TASTING COMPOUND

Several practical uses have been found for Bitrex, a quaternary ammonium salt, because it is one of the most bitter-tasting substances known and is nontoxic. Bitrex is used, for example, to encourage deer to look elsewhere for food; it is put on the backs of animals to keep them from biting one another; it is put on children's' fingers to persuade them to stop sucking their thumbs or biting their fingernails; and it is added to toxic substances to keep them from being ingested accidentally.

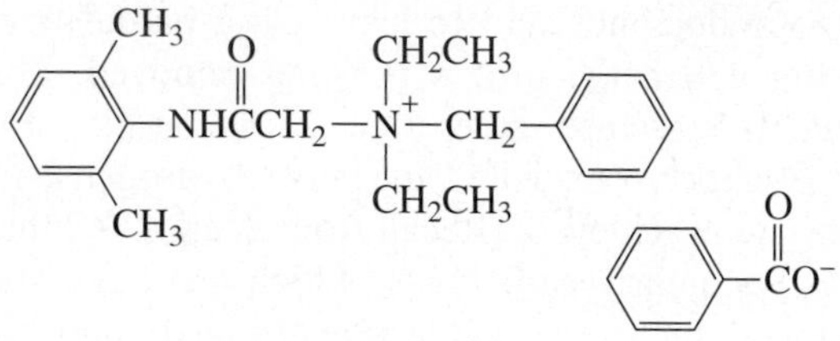

Bitrex®

PROBLEM 8◆

Identify the amine in each case:

a. 4-Methyl-2-pentene is obtained from the Hofmann degradation of a primary amine.

b. 2-Methyl-1-3-butadiene is obtained from two successive Hofmann degradations of a secondary amine.

PROBLEM 9 | SOLVED

Describe a synthesis for each of the following compounds, using the given starting material and any necessary reagents:

a. $CH_3CH_2CH_2CH_2NH_2 \longrightarrow CH_3CH_2CH=CH_2$

c. (pyrrolidine) $\longrightarrow CH_2=CH-CH=CH_2$

b. $CH_3CH_2CH_2\underset{|}{CH}CH_3 \longrightarrow CH_3CH_2CH_2CH=CH_2$
(Br on the CH)

Solution to 9a Although an amine cannot undergo an elimination reaction, a quaternary ammonium hydroxide can. The amine, therefore, must first be converted into a quaternary ammonium iodide by reacting with excess methyl iodide. Treatment with aqueous silver oxide forms the quaternary ammonium hydroxide. Heat is required for the elimination reaction.

$$CH_3CH_2CH_2CH_2NH_2 \xrightarrow[\substack{\text{excess} \\ K_2CO_3}]{CH_3I} CH_3CH_2CH_2CH_2\overset{+}{N}(CH_3)_3 \xrightarrow[H_2O]{Ag_2O} CH_3CH_2CH_2CH_2\overset{+}{N}(CH_3)_3 \xrightarrow{\Delta} CH_3CH_2CH=CH_2 \ + \ H_2O$$
(I^- under the first product; HO^- under the second product)

20.5 **Phase-Transfer Catalysis**

One problem that often arises in the laboratory is finding a solvent that will dissolve all the reactants needed for a given reaction. For example, if we want cyanide ion to react with 1-bromohexane, we must find a way of mixing sodium cyanide, an ionic compound soluble only in water, with the alkyl halide, which is insoluble in water. If we mix an aqueous solution of sodium cyanide with a solution of 1-bromohexane in a nonpolar solvent, there will be two distinct phases—an aqueous phase and an organic phase—because the two solutions are immiscible. How, then, can a reaction between sodium cyanide and the alkyl halide take place?

$$CH_3CH_2CH_2CH_2CH_2CH_2Br \ + \ ^-C{\equiv}N \ \xrightarrow{\ ?\ } \ CH_3CH_2CH_2CH_2CH_2CH_2C{\equiv}N \ + \ Br^-$$
1-bromohexane

The two compounds will be able to react with each other if a catalytic amount of a **phase-transfer catalyst** is added to the reaction mixture.

$$CH_3CH_2CH_2CH_2CH_2CH_2Br \ + \ ^-C{\equiv}N \ \xrightarrow[\quad R_4\overset{+}{N}\ HSO_4^-\quad]{\text{phase-transfer catalyst}} \ CH_3CH_2CH_2CH_2CH_2CH_2C{\equiv}N \ + \ Br^-$$

Quaternary ammonium salts are the most common phase-transfer catalysts. However, we saw in Section 10.10 that crown ethers can also be used as phase-transfer catalysts.

phase-transfer catalysts

$CH_2CH_2CH_2CH_3$	CH_3	CH_2CH_3
$CH_3CH_2CH_2CH_2\overset{+}{N}CH_2CH_2CH_2CH_3$	$CH_3(CH_2)_{14}CH_2\overset{+}{N}CH_3$	—$CH_2\overset{+}{N}CH_2CH_3$
$HSO_4^-\ CH_2CH_2CH_2CH_3$	$HSO_4^-\ CH_3$	$HSO_4^-\ CH_2CH_3$
tetrabutylammonium hydrogen sulfate	**hexadecyltrimethylammonium hydrogen sulfate**	**benzyltriethylammonium hydrogen sulfate**

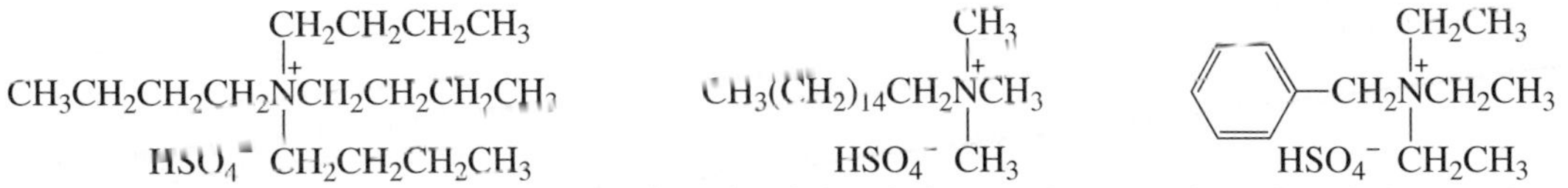

How does adding a phase-transfer catalyst enable the reaction of cyanide ion with 1-bromohexane to take place? The answer is that, because of its nonpolar alkyl groups, the quaternary ammonium salt is soluble in nonpolar solvents, and because of its charge, it is also soluble in water. Consequently, it can act as an intermediary between the two immiscible phases. When a phase-transfer catalyst such as tetrabutylammonium hydrogen sulfate passes into the nonpolar, organic phase, it must carry a counterion with it to balance its positive charge. The counterion can be either its original counterion (hydrogen sulfate) or another ion that is present in the solution (in the example under discussion, it could be cyanide ion or Br⁻). Because there is more cyanide ion than hydrogen sulfate or Br⁻ in the aqueous phase, the accompanying ion is more often going to be a cyanide ion. Once a cyanide ion has been carried into the organic phase, it can react with the alkyl halide. (When hydrogen sulfate is transported into the organic phase, it is unreactive because it is both a weak base and a poor nucleophile.) The quaternary ammonium ion will then pass back into the aqueous phase, carrying with it either hydrogen sulfate or bromide ion as a counterion. The reaction continues with the phase-transfer catalyst shuttling back and forth between the two phases. **Phase-transfer catalysis** has been successfully used in a wide variety of organic reactions.

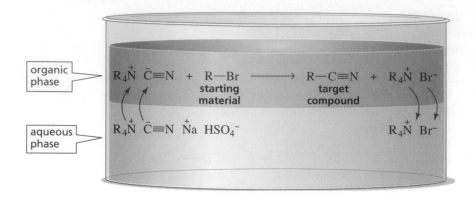

20.6 Oxidation of Amines • The Cope Elimination Reaction

Amines are easily oxidized, sometimes just by being exposed to air. They are therefore stored as salts (for example, as amine hydrochlorides), which is why drugs that contain amino groups are often sold as salts; the salts also have greater solubility in the bloodstream.

Primary amines are oxidized to hydroxylamines, which in turn are oxidized to nitroso compounds, which are oxidized to nitro compounds. Hydrogen peroxide, peroxyacids, and other common oxidizing agents are used to oxidize amines. The oxidation reactions appear to involve radicals but are not well understood.

$$R-NH_2 \xrightarrow{\text{oxidation}} R-NH-OH \xrightarrow{\text{oxidation}} R-N=O \xrightarrow{\text{oxidation}} R-\overset{\overset{\displaystyle O}{\|}}{\underset{\underset{\displaystyle O^-}{}}{N^+}}$$

a primary amine　　**a hydroxylamine**　　**a nitroso compound**　　**a nitro compound**

Secondary amines are oxidized to secondary hydroxylamines, and tertiary amines are oxidized to tertiary amine oxides.

$$\underset{\text{a secondary amine}}{R-\overset{\displaystyle R}{\underset{\displaystyle \cdot\cdot}{N}}H} + HO-OH \longrightarrow R-\overset{\displaystyle R}{\underset{\displaystyle OH}{\overset{+}{N}H}} + HO^- \longrightarrow \underset{\text{a secondary hydroxylamine}}{R-\overset{\displaystyle R}{\underset{\displaystyle OH}{N\!:}}} + H_2O$$

$$\underset{\text{a tertiary amine}}{R-\overset{\displaystyle R}{\underset{\displaystyle \cdot\cdot}{N}}-R} + HO-OH \longrightarrow R-\overset{\displaystyle R}{\underset{\displaystyle OH}{\overset{+}{N}}}-R + HO^- \longrightarrow \underset{\text{an tertiary amine oxide}}{R-\overset{\displaystyle R}{\underset{\displaystyle O^-}{\overset{+}{N}}}-R} + H_2O$$

Tertiary amine oxides undergo a reaction similar to the Hofmann elimination reaction, called a **Cope elimination reaction**. In this reaction, a tertiary amine oxide, rather than a quaternary ammonium ion, undergoes elimination.

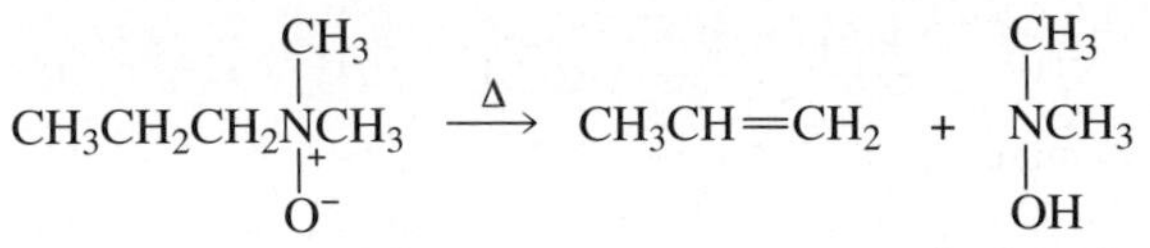

$$CH_3CH_2CH_2\overset{\overset{\displaystyle CH_3}{|}}{\underset{\underset{\displaystyle O^-}{|}}{N^+}}CH_3 \xrightarrow{\Delta} CH_3CH=CH_2 \ + \ \overset{\overset{\displaystyle CH_3}{|}}{\underset{\underset{\displaystyle OH}{|}}{N}}CH_3$$

a tertiary amine oxide   **a hydroxylamine**

A strong base is not needed for a Cope elimination because the amine oxide acts as its own base. Therefore, a Cope elimination reaction occurs under milder conditions than does a Hofmann elimination reaction. The Cope elimination is an intramolecular E2 reaction involving syn elimination.

Mechanism of the Cope elimination reaction

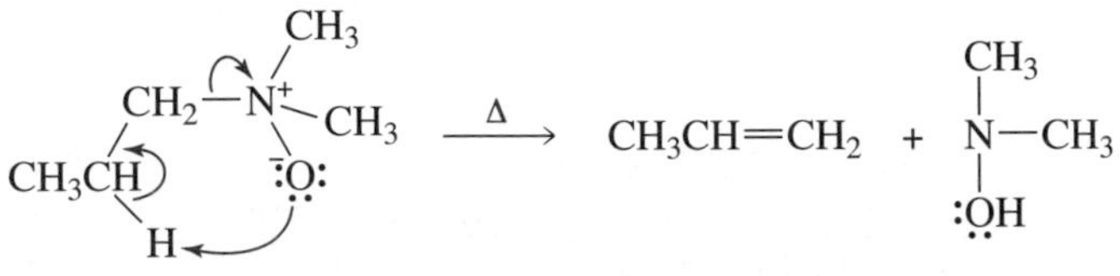

The major product of the Cope elimination, like that of the Hofmann elimination, is the one obtained by removing a proton from the β-carbon bonded to the greater number of hydrogens.

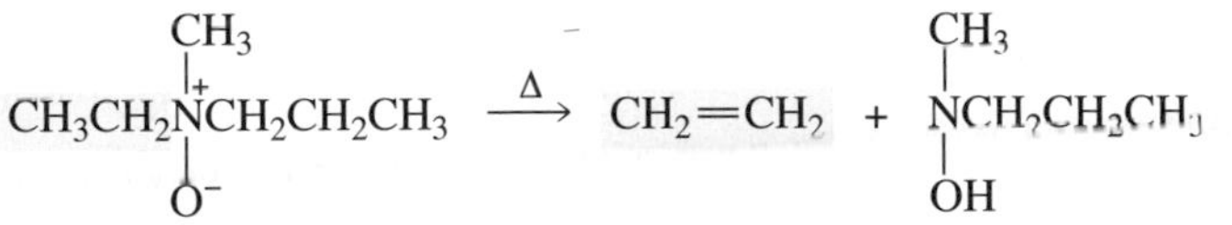

$$CH_3CH_2\overset{\overset{\displaystyle CH_3}{|}}{\underset{\underset{\displaystyle O^-}{|}}{N^+}}CH_2CH_2CH_3 \xrightarrow{\Delta} CH_2=CH_2 \ + \ \overset{\overset{\displaystyle CH_3}{|}}{\underset{\underset{\displaystyle OH}{|}}{N}}CH_2CH_2CH_3$$

In a Cope elimination, the proton is removed from the β-carbon bonded to the most hydrogens.

PROBLEM 10◆

Does the Cope elimination have an alkene-like transition state or a carbanion-like transition state?

PROBLEM 11◆

Give the products that would be obtained by treating the following tertiary amines with hydrogen peroxide followed by heat:

a. 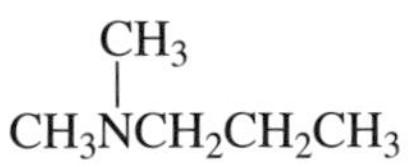 $CH_3\overset{\overset{\displaystyle CH_3}{|}}{N}CH_2CH_2CH_3$

c. 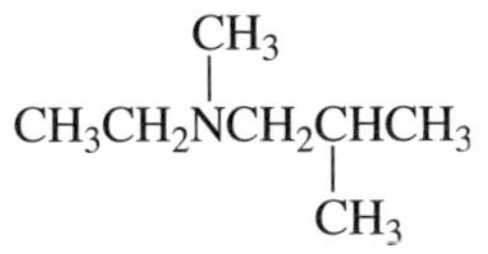 $CH_3CH_2\overset{\overset{\displaystyle CH_3}{|}}{N}CH_2\overset{\underset{\underset{\displaystyle CH_3}{|}}{}}{C}HCH_3$

b. 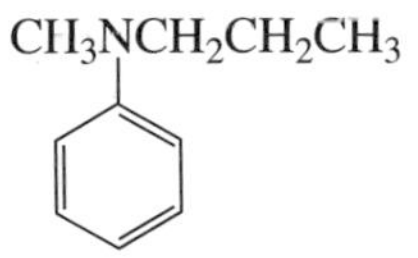 $CH_3\overset{}{N}CH_2CH_2CH_3$ (on a benzene ring)

d. 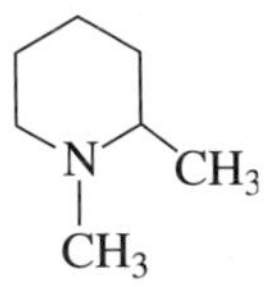 (piperidine ring with N—CH$_3$ and 2-CH$_3$)

20.7 Synthesis of Amines

Because ammonia and amines are good nucleophiles, they readily undergo S_N2 reactions with alkyl halides (here X denotes a halogen):

$$\overset{..}{N}H_3 \xrightarrow{RCH_2-X} RCH_2-\overset{+}{N}H_3 \rightleftharpoons \underset{\substack{\text{a primary amine} \\ + \; HX}}{RCH_2-\overset{..}{N}H_2} \xrightarrow{RCH_2-X} RCH_2-\underset{\substack{| \\ RCH_2}}{\overset{+}{N}H_2} \rightleftharpoons \underset{\substack{a \; secondary \; amine}}{RCH_2-\underset{\substack{| \\ RCH_2}}{\overset{..}{N}H}} + HX$$

$$\overset{RCH_2-X}{\Big\downarrow}$$

$$\underset{\substack{\text{a quaternary} \\ \text{ammonium salt}}}{RCH_2-\underset{\substack{| \\ RCH_2 \;\; X^-}}{\overset{\overset{| \\ RCH_2}}{\overset{+}{N}}}-CH_2R} \xleftarrow{RCH_2-X} \underset{\substack{a \; tertiary \; amine \\ + \; HX}}{RCH_2-\underset{\substack{| \\ RCH_2}}{\overset{\overset{RCH_2}{|}}{N:}}} \rightleftharpoons RCH_2-\underset{\substack{| \\ RCH_2}}{\overset{\overset{RCH_2}{|}}{\overset{+}{N}H}}$$

Although these S_N2 reactions can be used to synthesize amines, the yields are poor because it is difficult to stop the reaction when the desired number of alkyl substituents has been placed on the nitrogen since ammonia and primary, secondary, and tertiary amines have similar reactivities.

A much better way to prepare a primary amine is by means of a Gabriel synthesis (Section 16.18). This reaction involves alkylating phthalimide and then hydrolyzing the N-substituted phthalimide.

Gabriel synthesis

Primary amines also can also be prepared in good yields by the use of azide ion ($^-N_3$) as the nucleophile in an S_N2 reaction. The product of the reaction is an alkyl azide, which can be reduced to a primary amine. (See Chapter 8, Problem 13.)

$$\underset{\text{butyl bromide}}{CH_3CH_2CH_2CH_2Br} \xrightarrow{^-N_3} \underset{\text{butyl azide}}{CH_3CH_2CH_2CH_2\overset{+}{N}=\overset{..}{N}=\overset{..}{N}^-} \xrightarrow[Pd/C]{H_2} \underset{\text{butylamine}}{CH_3CH_2CH_2CH_2NH_2}$$

The catalytic reduction of a nitrile is yet another way to form a primary amine (Section 16.19). (Recall that a nitrile can be obtained from the reaction of cyanide ion with an alkyl halide.)

$$\underset{\text{butyl bromide}}{CH_3CH_2CH_2CH_2Br} \xrightarrow[HCl]{NaC\equiv N} \underset{\text{pentanenitrile}}{CH_3CH_2CH_2CH_2C\equiv N} \xrightarrow[Pd/C]{H_2} \underset{\text{pentylamine}}{CH_3CH_2CH_2CH_2CH_2NH_2}$$

A primary amine is obtained from the reduction of a nitroalkane, and an arylamine is obtained from the reduction of nitrobenzene.

$$\underset{\text{nitroethane}}{CH_3CH_2NO_2} + H_2 \xrightarrow{Pd/C} \underset{\text{ethylamine}}{CH_3CH_2NH_2}$$

Primary, secondary, and tertiary amines can be formed from the reduction of an amide with $LiAlH_4$ (Sections 17.6 and 19.1). The class of amine obtained depends on the number of substituents on the nitrogen atom of the amide.

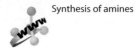
Synthesis of amines

$$\underset{R}{\overset{O}{\underset{\|}{C}}}{}_{NH_2} \xrightarrow[\text{2. } H_2O]{\text{1. } LiAlH_4} RCH_2NH_2$$
a primary amine

$$\underset{R}{\overset{O}{\underset{\|}{C}}}{}_{NHCH_3} \xrightarrow[\text{2. } H_2O]{\text{1. } LiAlH_4} RCH_2NHCH_3$$
a secondary amine

$$\underset{R}{\overset{O}{\underset{\|}{C}}}{}_{\underset{CH_3}{N CH_3}} \xrightarrow[\text{2. } H_2O]{\text{1. } LiAlH_4} RCH_2NCH_3 \mid CH_3$$
a tertiary amine

Primary, secondary, and tertiary amines can also be prepared by reductive amination (Section 17.8).

20.8 Aromatic Five-Membered-Ring Heterocycles

Now we will look at heterocycles that have a five-membered aromatic ring.

Pyrrole, Furan, and Thiophene

Pyrrole, **furan**, and **thiophene** are five-membered-ring heterocycles. Each has three pairs of delocalized π electrons, two of which are shown as π bonds, and one of which is shown as a lone pair on the heteroatom. Furan and thiophene have a second lone pair that is not part of the π cloud. These electrons are in an sp^2 orbital perpendicular to the p orbitals. Pyrrole, furan, and thiophene are aromatic because they are cyclic and planar, every carbon in the ring has a p orbital, and the π cloud contains *three* pairs of π electrons (Sections 14.1 and 14.4).

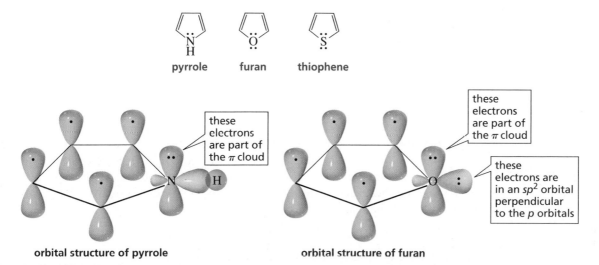

pyrrole   furan   thiophene

these electrons are part of the π cloud

these electrons are part of the π cloud

these electrons are in an sp^2 orbital perpendicular to the p orbitals

orbital structure of pyrrole   **orbital structure of furan**

Pyrrole is an extremely weak base because the electrons shown as a lone pair are part of the π cloud. Therefore, when pyrrole is protonated, its aromaticity is destroyed. Consequently, the conjugate acid of pyrrole is a very strong acid ($pK_a = -3.8$); that is, it has a strong tendency to lose a proton.

The resonance contributors of pyrrole show that nitrogen donates the electrons depicted as a lone pair into the five-membered ring.

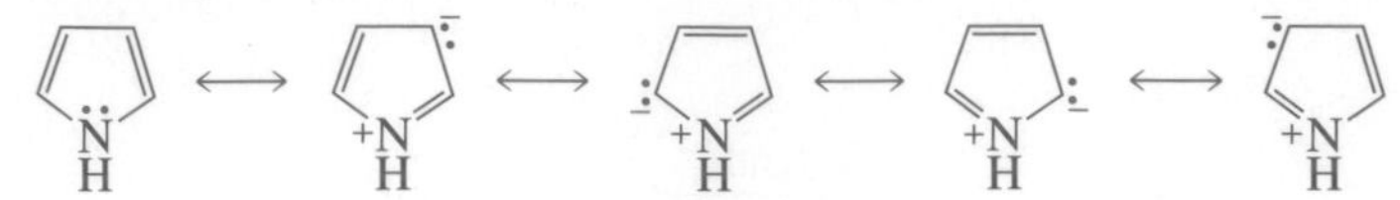

resonance contributors of pyrrole

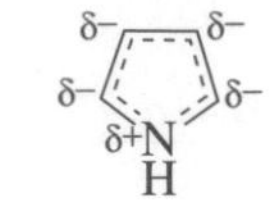

resonance hybrid

PROBLEM 12

Draw arrows to show the movement of electrons in going from one resonance contributor to the next in pyrrole.

Pyrrole—an unsaturated five-membered-ring heterocyclic amine—has a dipole moment of 1.80 D (Section 1.15). Pyrrolidine—a saturated five-membered-ring heterocyclic amine—has a slightly smaller dipole moment of 1.57 D, but as we see from the electrostatic potential maps, the two dipole moments are in opposite directions. (The red areas are on opposite sides of the two molecules.) The dipole moment in pyrrolidine is due to inductive electron withdrawal by the nitrogen atom. Apparently, the ability of pyrrole's nitrogen to donate electrons into the ring by resonance more than makes up for its inductive electron withdrawal (Section 15.2).

pyrrolidine $\mu = 1.57$ **D** $\mu = 1.80$ **D**   **pyrrole**

In Section 7.6, we saw that the more stable and more nearly equivalent the resonance contributors, the greater is the compound's resonance energy. The resonance energies of pyrrole, furan, and thiophene are not as great as the resonance energies of benzene and the cyclopentadienyl anion, each a compound for which the resonance contributors are all equivalent. Thiophene, with the least electronegative heteroatom, has the greatest resonance energy of the three five-membered ring heterocycles; and furan, with the most electronegative heteroatom, has the smallest resonance energy, because the resonance contributors with a positive charge on the heteroatom are the least stable for the compound with the most electronegative heteroatom.

relative resonance energies of some aromatic compounds

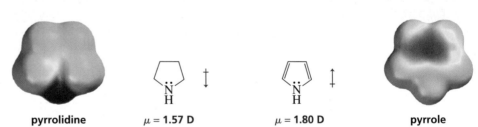

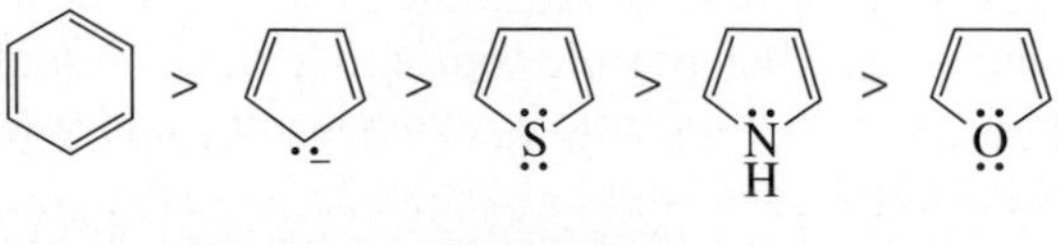

Because pyrrole, furan, and thiophene are aromatic, they undergo electrophilic aromatic substitution reactions.

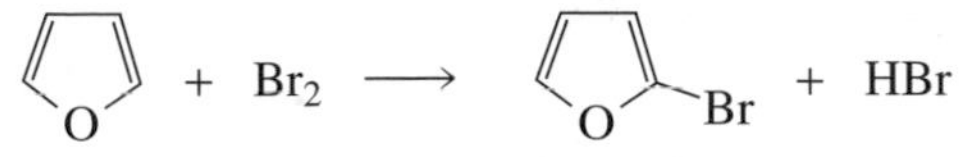

2-bromofuran

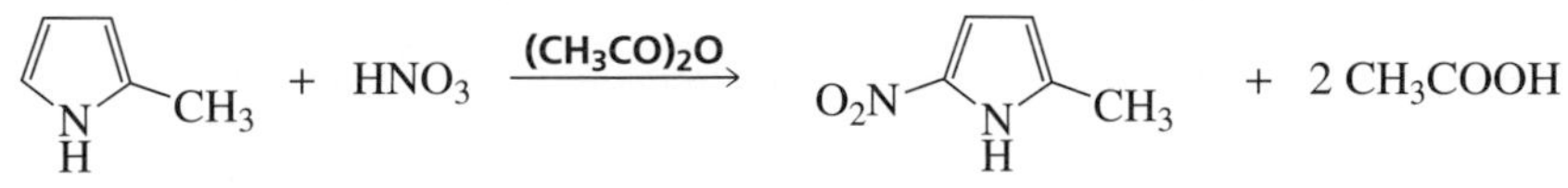

2-methyl-5-nitropyrrole

Mechanism for electrophilic aromatic substitution

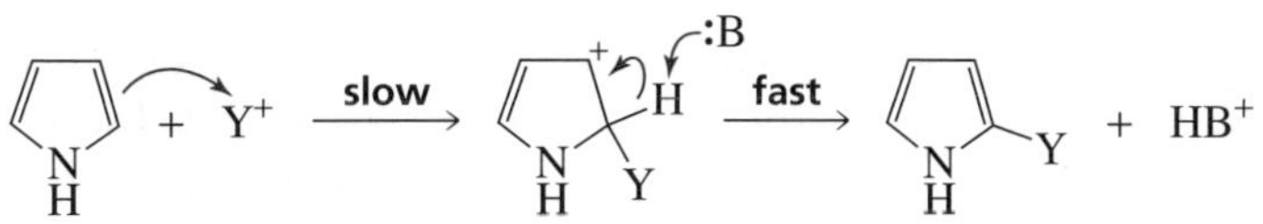

Where :B is any base in the solution.

Substitution occurs preferentially at C-2 because the intermediate obtained by attaching a substituent at this position is more stable than the intermediate obtained by attaching a substituent at C-3 (Figure 20.1). Both intermediates have a relatively stable resonance contributor in which all the atoms (except H) have complete octets. The intermediate resulting from C-2 substitution of pyrrole has *two* additional resonance contributors, each with a positive charge on a *secondary allylic* carbon. The intermediate resulting from C-3 substitution, however, has only *one* additional resonance contributor, which has a positive charge on a *secondary* carbon. This resonance contributor is further destabilized by being adjacent to an electron-withdrawing nitrogen atom.

Pyrrole, furan, and thiophene undergo electrophilic substitution preferentially at C-2.

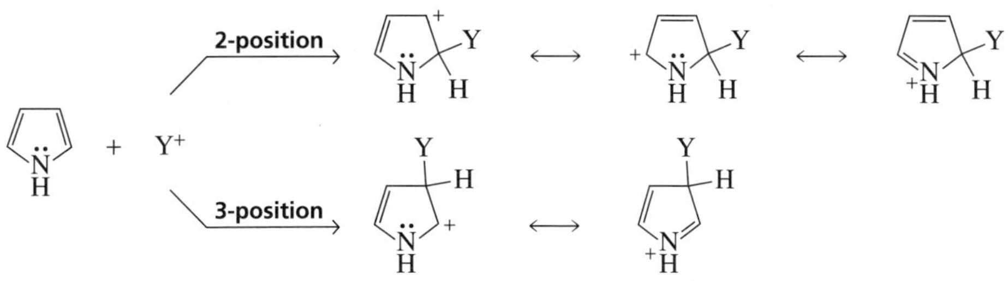

◀ **Figure 20.1**
Structures of the intermediates that can be formed from the reaction of an electrophile with pyrrole at C-2 and C-3.

If both positions adjacent to the heteroatom are occupied, electrophilic substitution will take place at C-3.

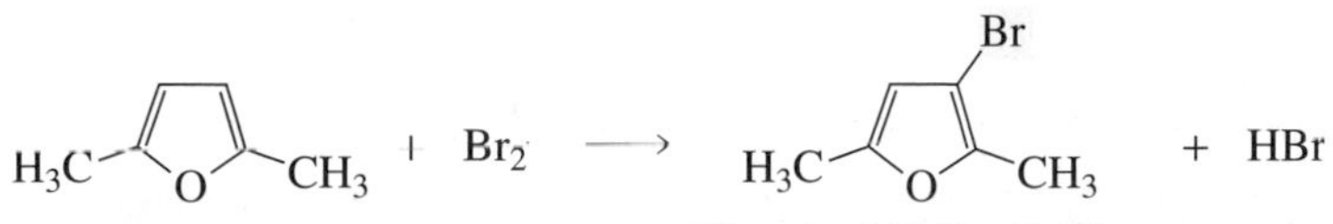

3-bromo-2,5-dimethylfuran

Pyrrole, furan, and thiophene are all more reactive than benzene toward electrophilic aromatic substitution because they are better able to stabilize the positive charge on the carbocation intermediate, since the lone pair on the heteroatom can donate electrons into the ring by resonance (Figure 20.1).

relative reactivity toward electrophilic aromatic substitution

pyrrole furan thiophene benzene

Pyrrole, furan, and thiophene are more reactive than benzene toward electrophilic aromatic substitution.

Furan is not as reactive as pyrrole in electrophilic aromatic substitution reactions. The oxygen of furan is more electronegative than the nitrogen of pyrrole, so oxygen is not as effective as nitrogen in stabilizing the carbocation. Thiophene is less reactive than furan because sulfur's π electrons are in a $3p$ orbital, which overlaps less effectively than the $2p$ orbital of nitrogen or oxygen with the $2p$ orbital of carbon. The electrostatic potential maps illustrate the different electron densities of the three rings.

The relative reactivities of the five-membered-ring heterocycles are reflected in the Lewis acid required to catalyze a Friedel–Crafts acylation reaction (Section 14.14). Benzene requires $AlCl_3$, a relatively strong Lewis acid. Thiophene is more reactive than benzene, so it can undergo a Friedel–Crafts reaction using $SnCl_4$, a weaker Lewis acid. An even weaker Lewis acid, BF_3, can be used when the reactant is furan. Pyrrole is so reactive that an anhydride is used instead of a more reactive acyl chloride, and no catalyst is necessary.

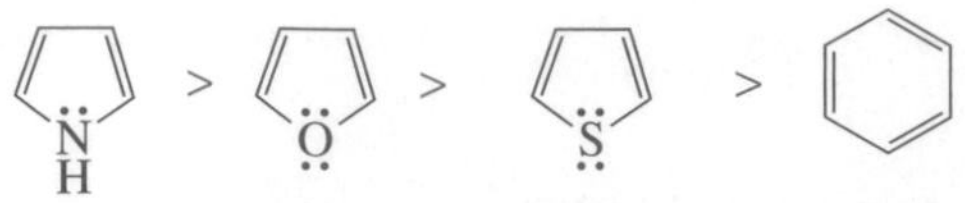

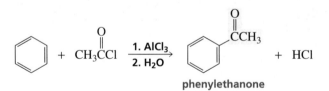

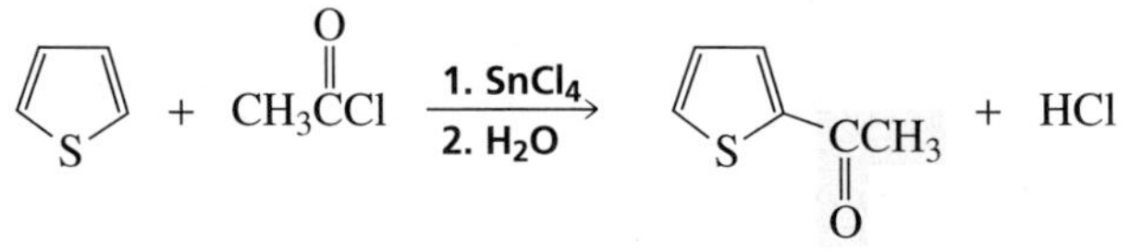

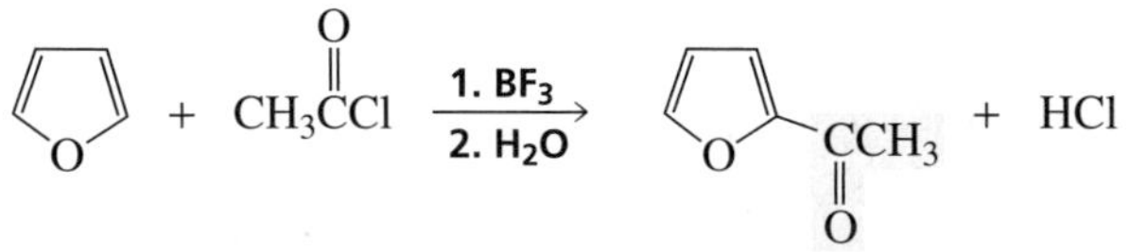

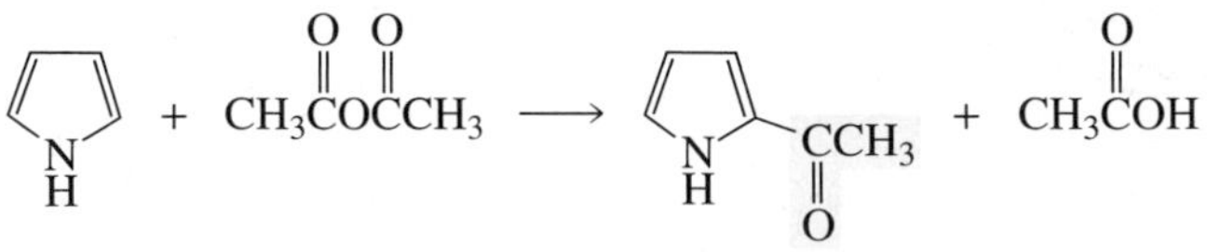

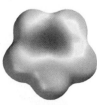

phenylethanone

2-acetylthiophene

2-acetylfuran

2-acetylpyrrole

pyrrole

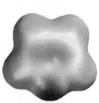

furan

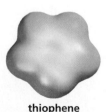

thiophene

The resonance hybrid of pyrrole indicates that there is a partial positive charge on the nitrogen. Therefore, pyrrole is protonated on C-2 rather than on nitrogen. Remember, a proton is an electrophile and, like other electrophiles, attaches to the C-2 position of pyrrole.

$$pK_a = -3.8$$

Pyrrole is unstable in strongly acidic solutions because, once protonated, it can readily polymerize.

$$\longrightarrow \quad \longrightarrow \quad \longrightarrow \quad \textbf{polymer}$$

Student Tutorial:
Basic sites in nitrogen heterocycles

The nitrogen in pyrrole is sp^2 hybridized and is therefore more electronegative than the sp^3 nitrogen of a saturated amine (Section 1.20). As a result, pyrrole ($pK_a = \sim17$) is more acidic (Table 20.1) than the analogous saturated amine ($pK_a = \sim36$). The partial positive charge on the nitrogen atom (that is apparent in the resonance hybrid) also contributes to pyrrole's increased acidity.

$$pK_a = \sim17 \qquad\qquad pK_a = \sim36$$

Table 20.1 The pK_a Values of Several Nitrogen Heterocycles

$pK_a = -3.8$	$pK_a = -2.4$	$pK_a = 1.0$	$pK_a - 2.5$	$pK_a - 4.85$	$pK_a = 5.16$
$pK_a = 6.8$	$pK_a = 8.0$	$pK_a = 11.1$	$pK_a = 14.4$	$pK_a = \sim17$	$pK_a = \sim36$

PROBLEM 13

When pyrrole is added to a dilute solution of D_2SO_4 in D_2O, 2-deuteropyrrole is formed. Propose a mechanism to account for the formation of this compound.

PROBLEM 14

Use resonance contributors to explain why pyrrole is protonated on C-2 rather than on nitrogen.

PROBLEM 15◆

Explain why pyrrole ($pK_a \sim 17$) is less acidic than cyclopentadiene ($pK_a = 15$), even though nitrogen is considerably more electronegative than carbon.

Indole, Benzofuran, and Benzothiophene

Indole, benzofuran, and benzothiophene all contain a five-membered aromatic ring fused to a benzene ring. The atoms in the rings are numbered so as to give the heteroatom the lowest possible number. Indole, benzofuran, and benzothiophene are aromatic because they are cyclic and planar, every carbon in the ring has a *p* orbital, and the π cloud of each compound contains *five* pairs of π electrons (Section 14.2). Notice that the electrons shown as a lone pair on the indole nitrogen are part of the π cloud; therefore, the conjugate acid of indole, like the conjugate acid of pyrrole, is a strong acid ($pK_a = -2.4$). In other words, indole is an extremely weak base.

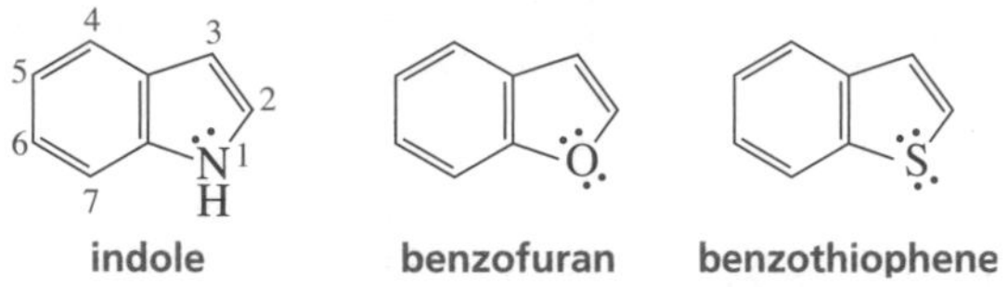

indole   benzofuran   benzothiophene

20.9 Aromatic Six-Membered-Ring Heterocycles

Next, we will look at heterocycles that have a six-membered aromatic ring.

Pyridine

When one of the carbons of a benzene ring is replaced by a nitrogen, the resulting compound is called **pyridine**.

these electrons are in an sp^2 orbital perpendicular to the *p* orbitals

pyridine   orbital structure of pyridine

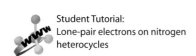

Student Tutorial:
Lone-pair electrons on nitrogen heterocycles

The pyridinium ion is a stronger acid than a typical ammonium ion because the acidic hydrogen of a pyridinium ion is attached to an sp^2 nitrogen, which is more electronegative than an sp^3 nitrogen (Section 1.20).

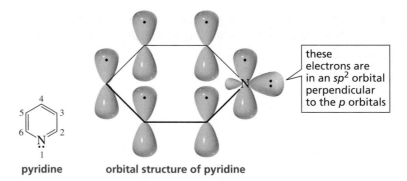

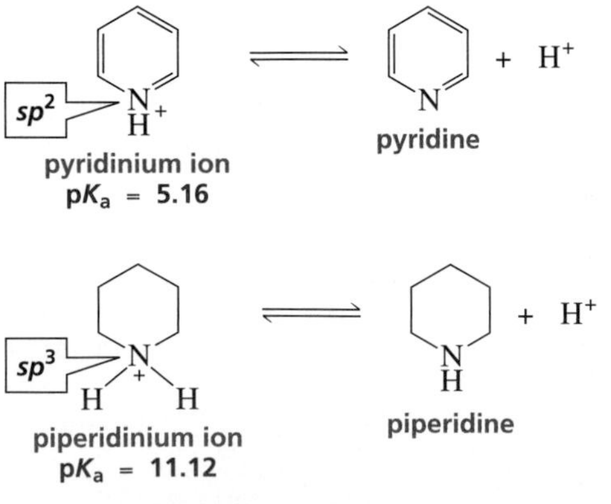

pyridinium ion
$pK_a = 5.16$

pyridine

piperidinium ion
$pK_a = 11.12$

piperidine

Pyridine undergoes reactions characteristic of tertiary amines. For example, pyridine undergoes S_N2 reactions with alkyl halides (Section 8.4), and it reacts with hydrogen peroxide to form an *N*-oxide (Section 20.6).

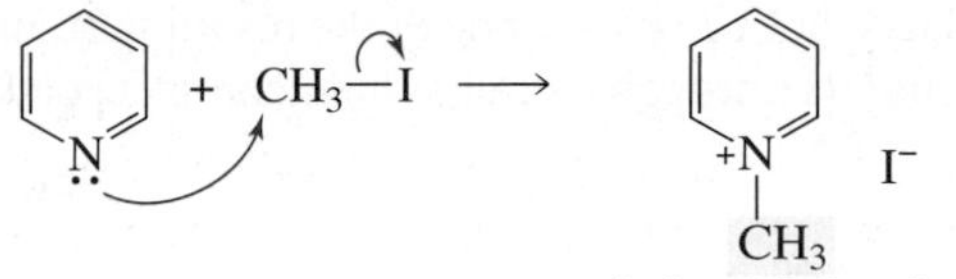

N-methylpyridinium iodide

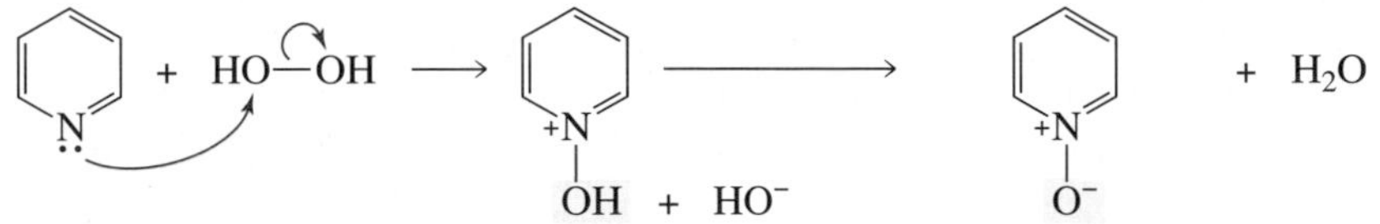

pyridine-*N*-oxide

$pK_a = 0.79$

PROBLEM 16 *SOLVED*

Will an amide be formed from the reaction of an acyl chloride with an aqueous solution of pyridine? Explain your answer.

Solution An amide will *not* be formed because the positively charged nitrogen causes pyridine to be an excellent leaving group. As a result, the final product of the reaction will be a carboxylic acid. (If the final pH of the solution is greater than the pK_a of the carboxylic acid, the carboxylic acid will be predominantly in its basic form.)

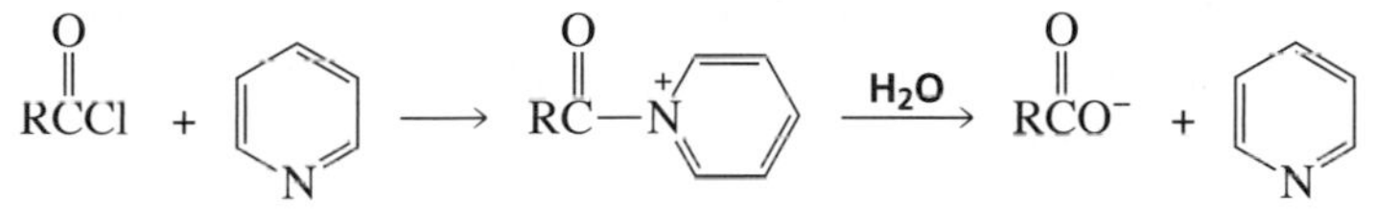

Pyridine is aromatic. Like benzene, it has two uncharged resonance contributors. Because of the electron-withdrawing nitrogen, pyridine has three charged resonance contributors that benzene does not have.

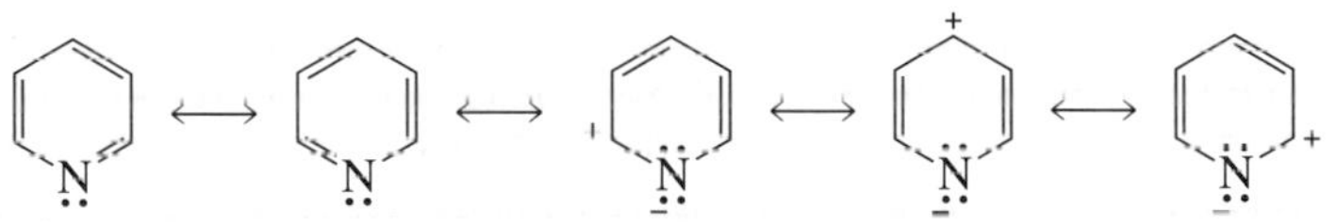

resonance contributors of pyridine

The dipole moment of pyridine is 1.57 D. As the resonance contributors and the electrostatic potential map indicate, the electron-withdrawing nitrogen is the negative end of the dipole.

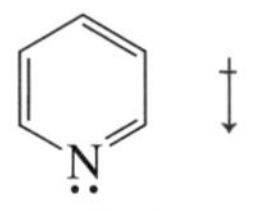

$\mu = 1.57 D$

Being aromatic, pyridine (like benzene) undergoes electrophilic aromatic substitution reactions.

Mechanism for electrophilic aromatic substitution

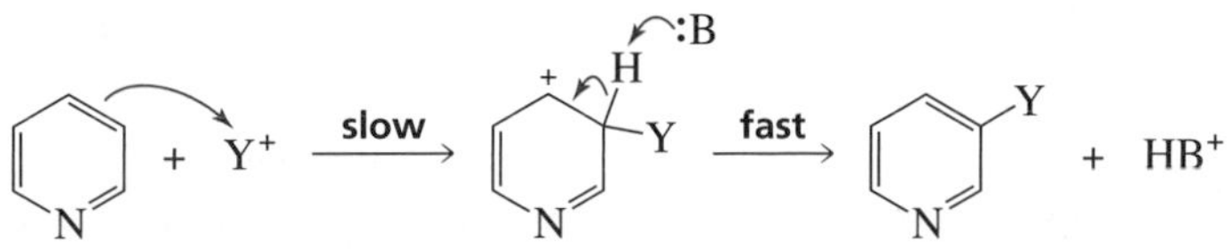

where :B is any base in the solution.

Electrophilic aromatic substitution of pyridine takes place at C-3 because the most stable intermediate is obtained by placing an electrophilic substituent at that position

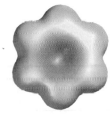
benzene

pyridine

Pyridine undergoes electrophilic aromatic substitution at C-3.

(Figure 20.2). When the substituent is placed at C-2 or C-4, one of the resulting resonance contributors is particularly unstable because its nitrogen atom has an incomplete octet *and* a positive charge.

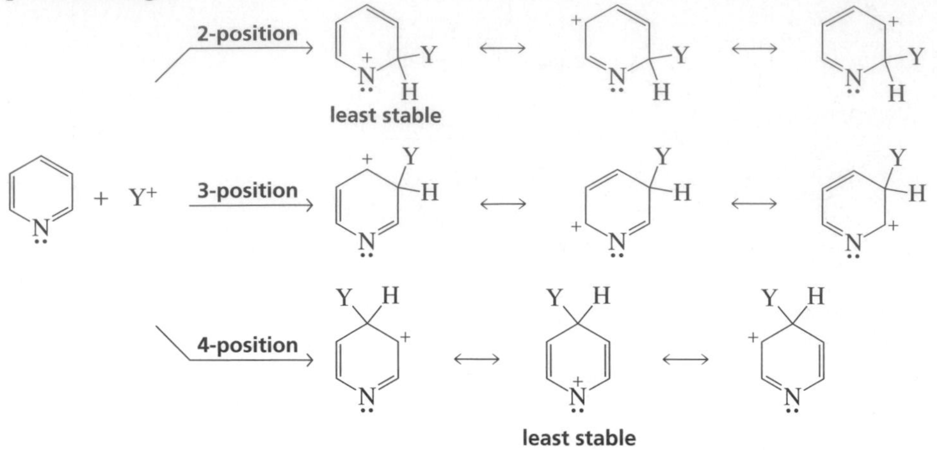

Figure 20.2 ▶ Structures of the intermediates that can be formed from the reaction of an electrophile with pyridine.

The electron-withdrawing nitrogen atom makes the intermediate obtained from electrophilic aromatic substitution of pyridine less stable than the carbocation intermediate obtained from electrophilic aromatic substitution of benzene. Pyridine, therefore, is less reactive than benzene. Indeed, it is even less reactive than nitrobenzene. (Recall from Section 15.2 that an electron-withdrawing nitro group strongly deactivates a benzene ring toward electrophilic aromatic substitution.)

relative reactivity toward electrophilic aromatic substitution

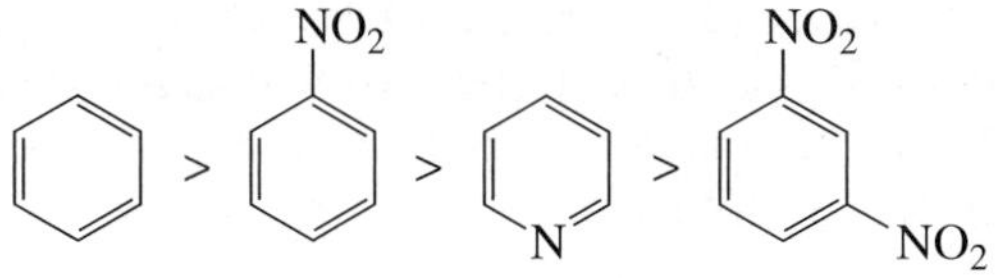

Pyridine, therefore, undergoes electrophilic aromatic substitution reactions only under vigorous conditions, and the yields of these reactions are often quite low. If the nitrogen becomes protonated under the reaction conditions, the reactivity decreases further because a positively charged nitrogen would make the carbocation intermediate even less stable.

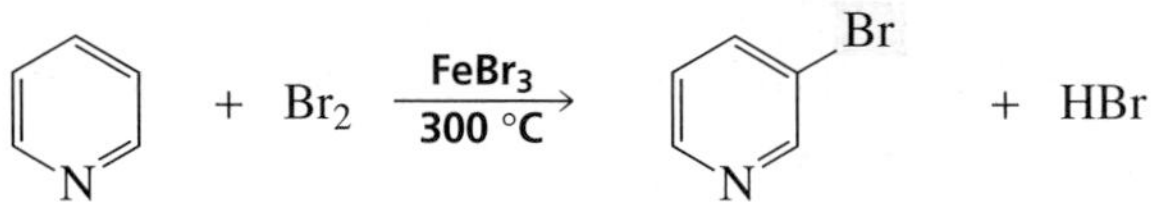

3-bromopyridine
30%

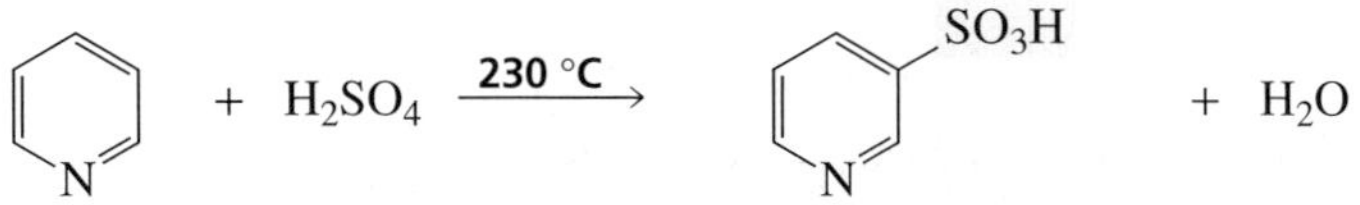

pyridine-3-sulfonic acid
71%

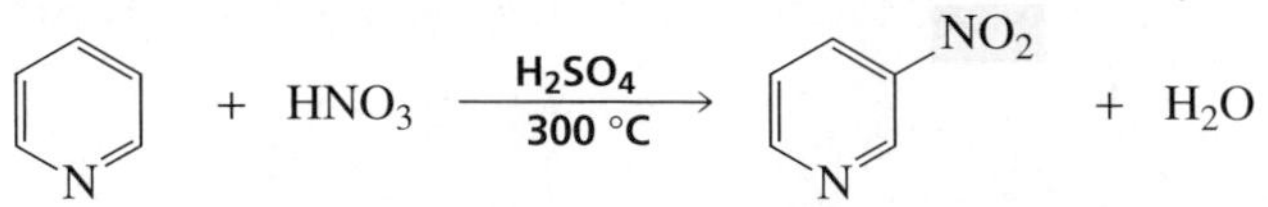

3-nitropyridine
22%

We have seen that highly deactivated benzene rings do not undergo Friedel–Crafts alkylation or acylation reactions. Therefore, pyridine, whose reactivity is similar to that of a highly deactivated benzene, does not undergo these reactions either.

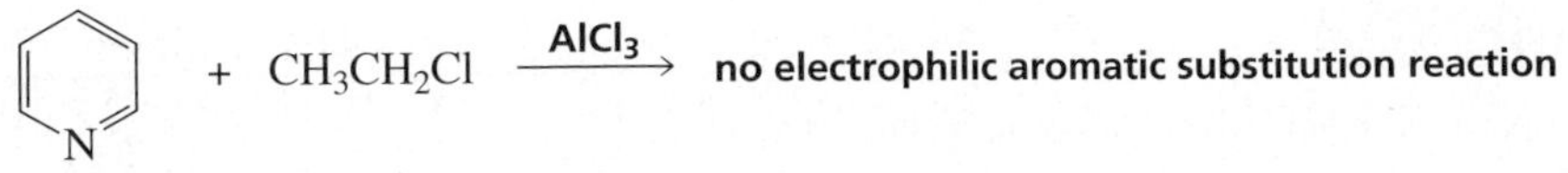

$$\text{pyridine} + CH_3CH_2Cl \xrightarrow{\text{AlCl}_3} \text{no electrophilic aromatic substitution reaction}$$

PROBLEM 17◆

Give the product of the following reaction:

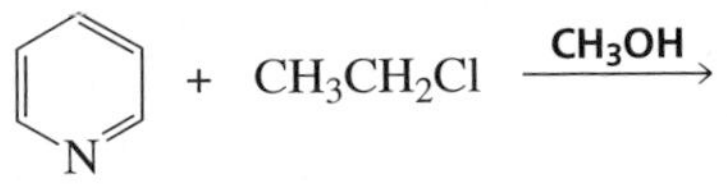

$$\text{pyridine} + CH_3CH_2Cl \xrightarrow{\text{CH}_3\text{OH}}$$

Since pyridine is *less* reactive than benzene in *electrophilic* aromatic substitution reactions, it is not surprising that pyridine is *more* reactive than benzene in *nucleophilic* aromatic substitution reactions. The electron-withdrawing nitrogen atom that destabilizes the intermediate in electrophilic aromatic substitution stabilizes it in nucleophilic aromatic substitution.

> **Pyridine is *less* reactive than benzene toward electrophilic aromatic substitution and *more* reactive than benzene toward nucleophilic aromatic substitution.**

Mechanism for nucleophilic aromatic substitution

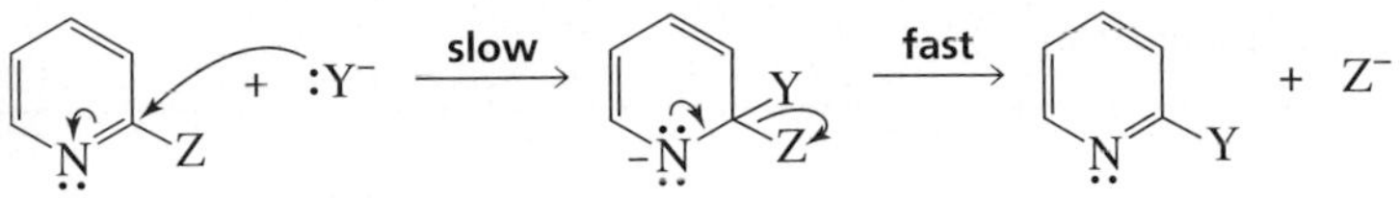

Nucleophilic aromatic substitution of pyridine takes place at C-2 and C-4, because attack at these positions leads to the most stable intermediate. Only when nucleophilic attack occurs at these positions is a resonance contributor obtained that has the greatest electron density on nitrogen, the most electronegative of the ring atoms (Figure 20.3).

> **Pyridine undergoes nucleophilic aromatic substitution at C-2 and C-4.**

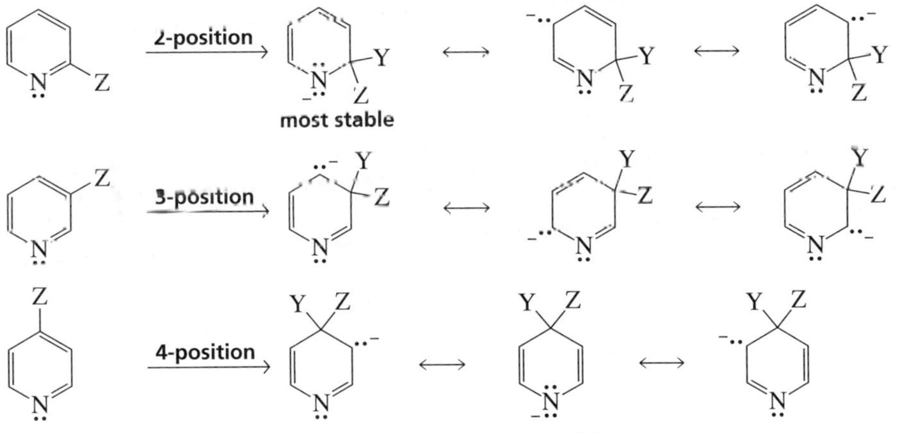

◄ **Figure 20.3**
Structures of the intermediates that can be formed from the reaction of a nucleophile with pyridine.

If the leaving groups at C-2 and C-4 are different, the incoming nucleophile will preferentially substitute for the weaker base (the better leaving group).

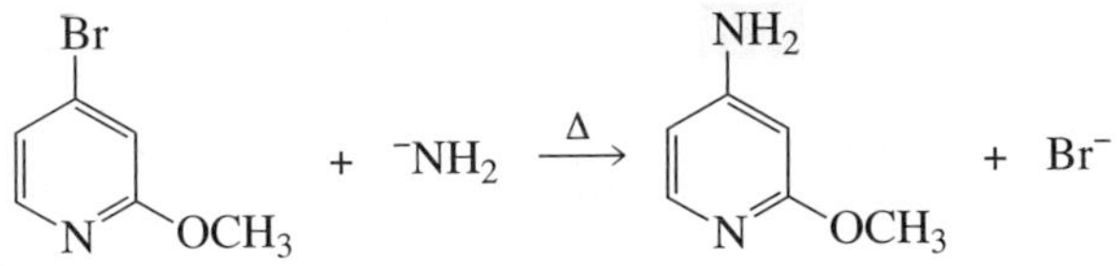

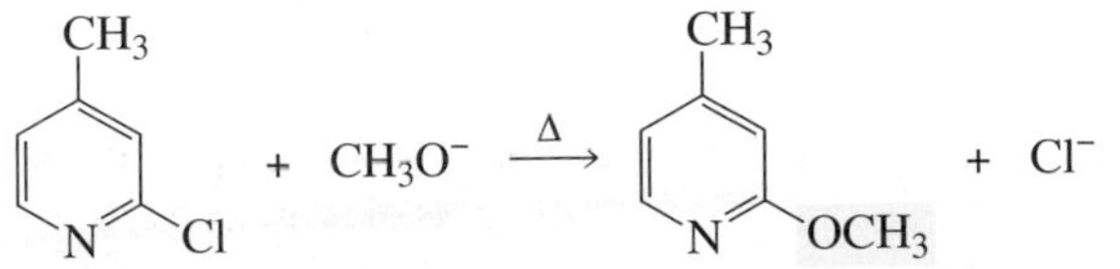

PROBLEM 18

Compare the mechanisms of the following reactions:

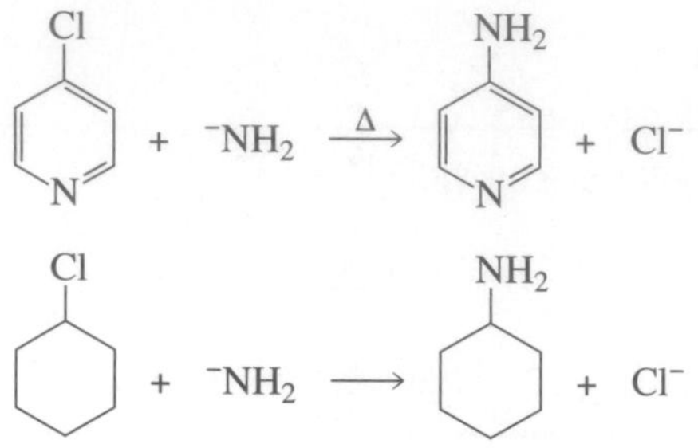

PROBLEM 19

a. Propose a mechanism for the following reaction:

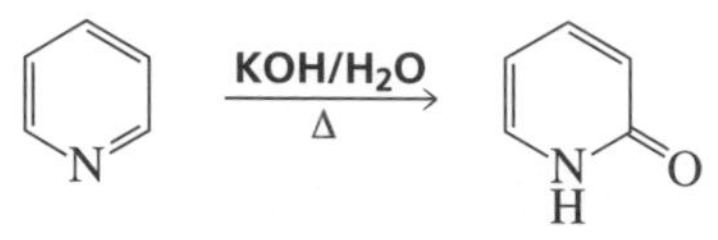

b. What other product is formed?

Substituted pyridines undergo many of the side-chain reactions that substituted benzenes undergo. For example, alkyl-substituted pyridines can be brominated and oxidized.

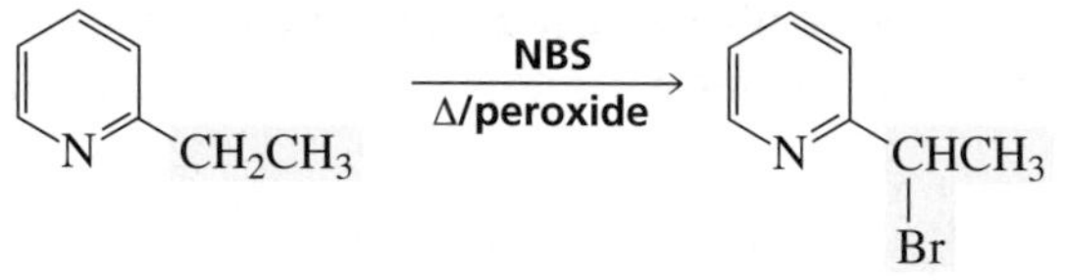

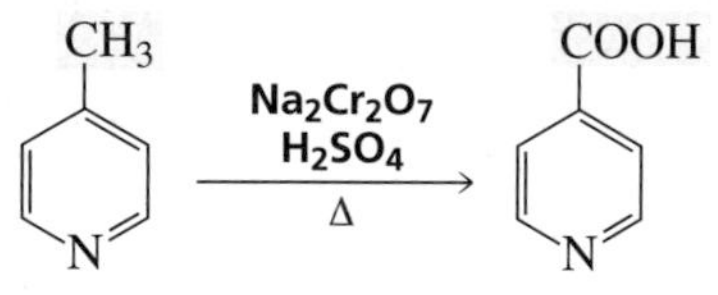

When 2- or 4-aminopyridine is diazotized, α-pyridone or γ-pyridone is formed. Apparently, the diazonium salt reacts immediately with water to form a hydroxypyridine (Section 15.9). The product of the reaction is a pyridone because the keto form of a hydroxypyridine is more stable than the enol form. (The mechanism for the conversion of a primary amino group into a diazonium group is shown in Section 15.11.)

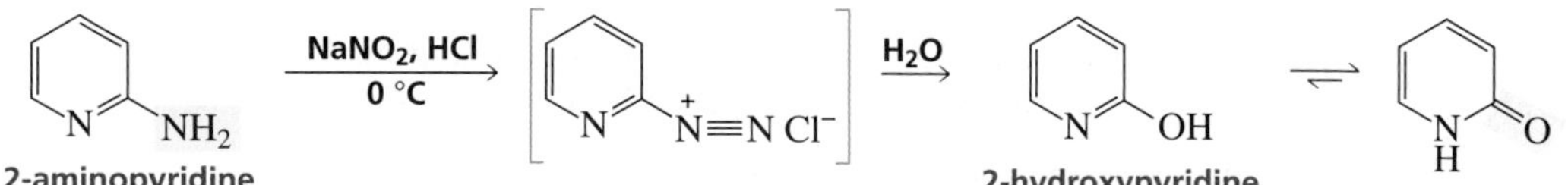

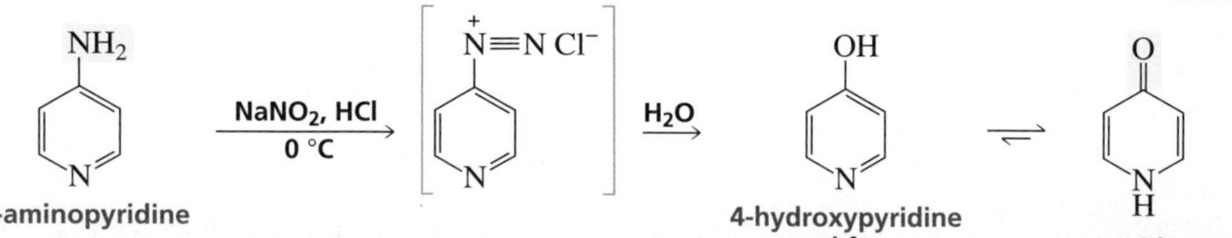

The electron-withdrawing nitrogen causes the α-hydrogens of alkyl groups attached to the 2- and 4-positions of the pyridine ring to have about the same acidity as the α-hydrogens of ketones (Section 18.1).

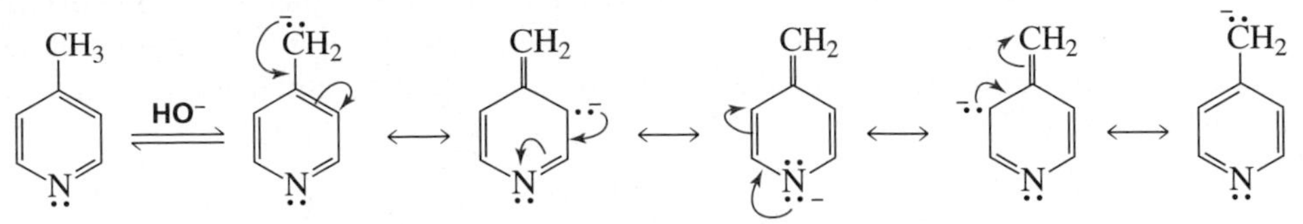

Consequently, the α-hydrogens of alkyl substituents can be removed by base, and the resulting carbanions can react as nucleophiles.

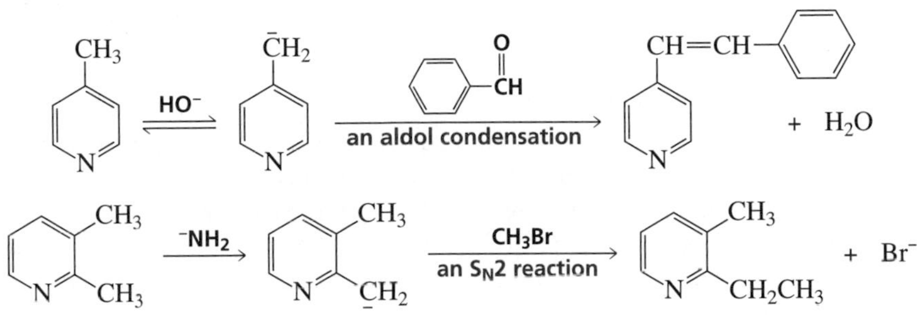

PROBLEM 20◆

Rank the following compounds in order of decreasing ease of removing a proton from a methyl group.

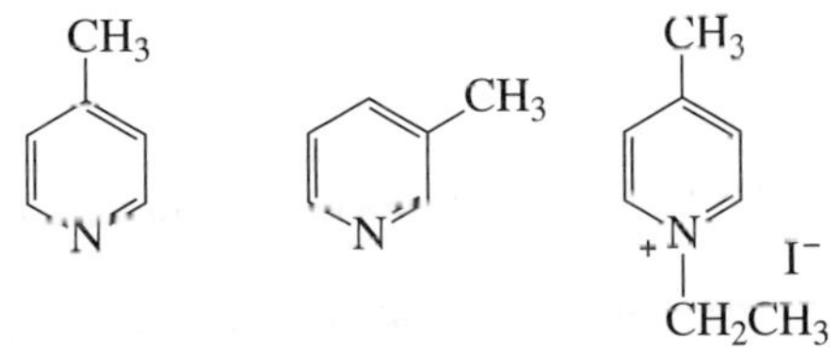

Quinoline and Isoquinoline

Quinoline and isoquinoline are known as *benzopyridines* because they have both a benzene ring and a pyridine ring. Like benzene and pyridine, they are aromatic compounds. The pK_a values of their conjugate acids are similar to the pK_a value of the conjugate acid of pyridine.

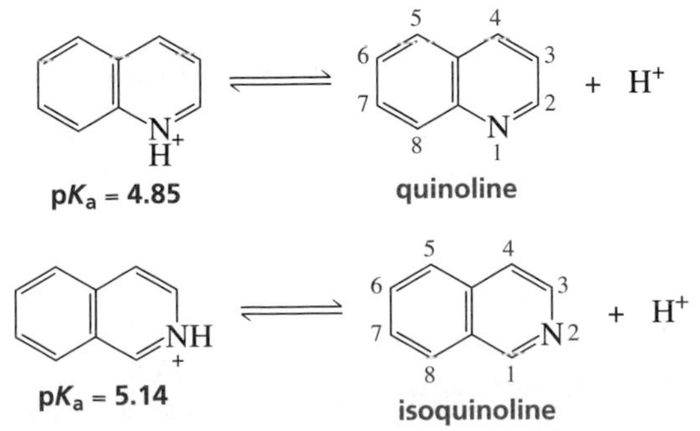

pKa = 4.85

quinoline

pKa = 5.14

isoquinoline

Note that in order for the carbons in quinoline and isoquinoline to have the same numbers, the nitrogen in isoquinoline is assigned the 2-position, not the lowest possible number.

20.10 Amine Heterocycles Have Important Roles in Nature

Proteins are naturally occurring polymers of α-amino acids (Chapter 22). Three of the 20 most common naturally occurring amino acids contain heterocyclic rings: proline contains a pyrrolidine ring, tryptophan contains an indole ring, and histidine contains an imidazole ring.

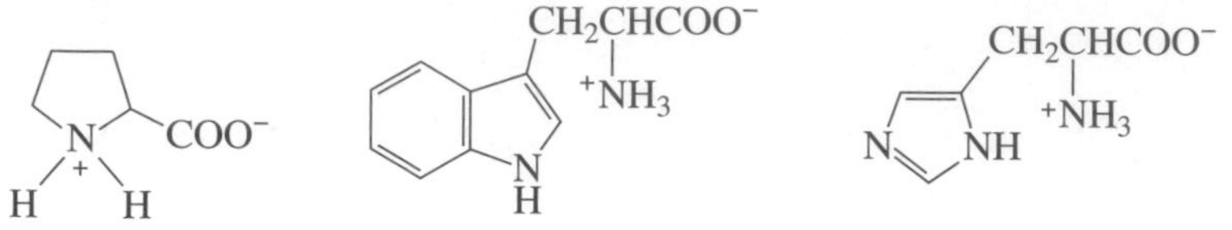

proline   **tryptophan**   **histidine**

Imidazole

Imidazole, the heterocyclic ring of histidine, is the first heterocyclic compound we have encountered that has two heteroatoms. Imidazole is an aromatic compound because it is cyclic and planar, every carbon in the ring has a p orbital, and the π cloud contains *three* pairs of π electrons (Section 14.2). The electrons drawn as lone-pair electrons on N-1 (see below) are part of the π cloud because they are in a p orbital, whereas the lone-pair electrons on N-3 are not part of the π cloud because they are in an sp^2 orbital, perpendicular to the p orbitals.

these electrons are in an sp^2 orbital perpendicular to the p orbitals

these electrons are part of the π cloud

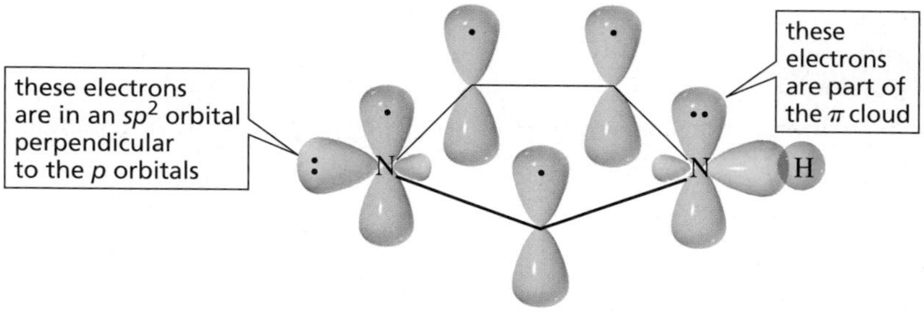

orbital structure of imidazole

The resonance energy of imidazole is 14 kcal/mol (59 kJ/mol), significantly less than the resonance energy of benzene (36 kcal/mol or 151 kJ/mol).

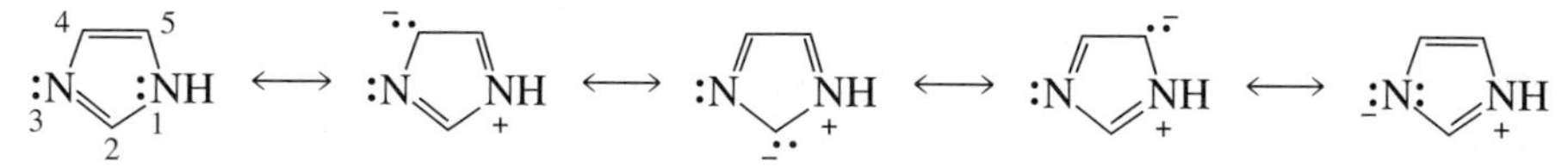

resonance contributors of imidazole

Unlike pyrrole, imidazole is protonated in acidic solutions because the lone-pair electrons in the sp^2 orbital are not part of the π cloud. Since the conjugate acid of imidazole has a pK_a of 6.8, imidazole exists in both the protonated and unprotonated forms at physiological pH (7.3). This is one of the reasons that histidine, the imidazole-containing amino acid, is an important catalytic component of many enzymes (Section 23.9).

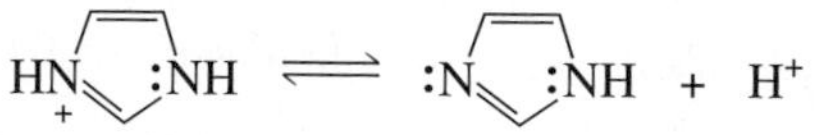

pK_a = 6.8

Neutral imidazole is a stronger acid (pK_a = 14.4) than neutral pyrrole (pK_a ~17) because of the second nitrogen in the ring that is electron withdrawing.

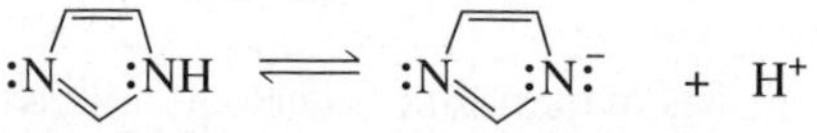

pK_a = 14.4

   Notice that both protonated imidazole and the imidazole anion have two equivalent resonance contributors. Therefore, the two nitrogens become equivalent when imidazole is either protonated or deprotonated.

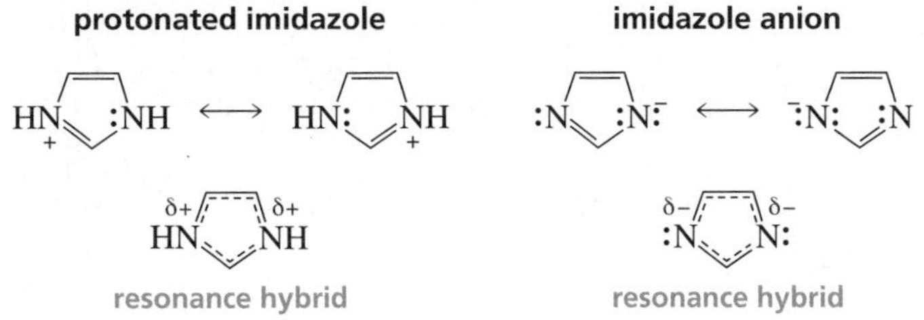

protonated imidazole   **imidazole anion**

resonance hybrid   resonance hybrid

PROBLEM 21◆

Give the major product of the following reaction:

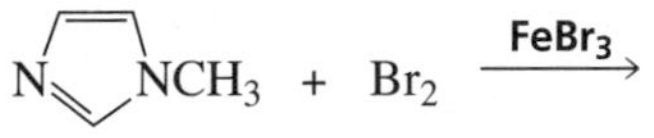

PROBLEM 22◆

List imidazole, pyrrole, and benzene in order of decreasing reactivity toward electrophilic aromatic substitution.

PROBLEM 23◆

Imidazole boils at 257 °C, whereas N-methylimidazole boils at 199 °C. Explain this difference in boiling points.

PROBLEM 24◆

What percent of imidazole will be protonated at physiological pH (7.3)?

Purine and Pyrimidine

Nucleic acids (DNA and RNA) contain substituted **purines** and substituted **pyrimidines** (Section 27.1); DNA contains adenine, guanine, cytosine, and thymine (abbreviated A, G, C, and T), and RNA contains adenine, guanine, cytosine, and uracil (A, G, C, and U). Why DNA contains T instead of U is explained in Section 27.12. Unsubstituted purine and pyrimidine are not found in nature. Notice that hydroxypurines and hydroxypyrimidines are more stable in the keto form (see page 964). We will see that the preference for the keto form is crucial for proper base pairing in DNA (Section 27.3).

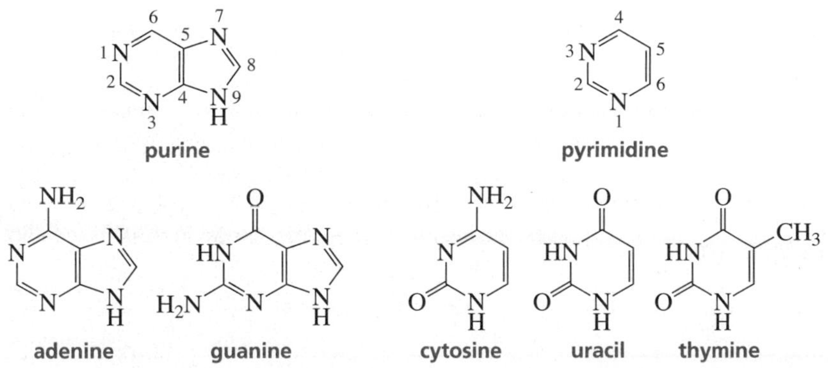

purine   pyrimidine

adenine   guanine   cytosine   uracil   thymine

Porphyrin

Substituted *porphyrins* are another group of important naturally occurring heterocyclic compounds. A **porphyrin ring system** consists of four pyrrole rings joined by one-carbon bridges. Heme, which is found in hemoglobin and myoglobin, contains an iron ion (Fe^{2+}) ligated by the four nitrogens of a porphyrin ring system. **Ligation** is the sharing of nonbonding electrons with a metal ion. The porphyrin ring system of heme is known as **protoporphyrin IX**; the ring system plus the iron atom is called *iron protoporphyrin IX*.

a porphyrin ring system

iron protoporphyrin IX
heme

Hemoglobin is responsible for transporting oxygen to cells and carbon dioxide away from cells, whereas myoglobin is responsible for storing oxygen in cells. Hemoglobin has four polypeptide chains and four heme groups; myoglobin has one polypeptide chain and one heme group. The iron atoms in hemoglobin and myoglobin, in addition to being ligated to the four nitrogens of the porphyrin ring, are also ligated to a histidine of the protein component (globin), and the sixth ligand is oxygen or carbon dioxide. Carbon monoxide is about the same size and shape as O_2, but CO binds more tightly than O_2 to Fe^{2+}. Consequently, breathing carbon monoxide can be fatal because it binds to hemoglobin more tightly than oxygen does and interferes with the transport of oxygen through the bloodstream.

The extensive conjugated system of porphyrin gives blood its characteristic red color. Its high molar absorptivity (about $160,000 \ M^{-1}cm^{-1}$) allows concentrations as low as $1 \times 10^{-8} \ M$ to be detected by UV spectroscopy (Section 12.17).

The ring system in chlorophyll *a*, the substance responsible for the green color of plants (Section 12.19), is similar to porphyrin but contains a cyclopentanone ring, and one of its pyrrole rings is partially reduced. The metal ion in chlorophyll *a* is magnesium (Mg^{2+}).

Vitamin B_{12} also has a ring system similar to porphyrin, but in this case, the metal ion is cobalt (Co^{3+}). Other aspects of vitamin B_{12}'s structure and chemistry are discussed in Section 24.7.

PROBLEM 25◆

Is porphyrin aromatic?

PORPHYRIN, BILIRUBIN, AND JAUNDICE

The average human body breaks down about 6 g of hemoglobin each day. The protein portion (globin) and the iron are reutilized, but the porphyrin ring is broken down, being reduced first to biliverdin, a green compound, and then to bilirubin, a yellow compound. If more bilirubin is formed than can be excreted by the liver, it accumulates in the blood. When its concentration there reaches a certain level, it diffuses into the tissues, giving them a yellow appearance. This condition is known as jaundice.

SUMMARY

Amines are classified as primary, secondary, or tertiary, depending on whether one, two, or three hydrogens of ammonia have been replaced by alkyl groups. Some amines are **heterocyclic compounds**—cyclic compounds in which one or more of the atoms of the ring is an atom other than carbon. Heterocyclic rings are numbered so that the **heteroatom** has the lowest possible number.

The lone pair on the nitrogen causes amines to be both bases and nucleophiles. Amines react as nucleophiles in nucleophilic substitution reactions, in nucleophilic acyl substitution reactions, in nucleophilic addition–elimination reactions, and in conjugate addition reactions. Amines react as bases in acid–base reactions and in elimination reactions.

Amines cannot undergo the substitution and elimination reactions that alkyl halides undergo, because the leaving groups of amines are too basic. Amines are easily oxidized. Quaternary ammonium hydroxides and amine oxides undergo E2 elimination reactions known as Hofmann elimination reactions and Cope elimination reactions, respectively. In both reactions, the proton from the β-carbon bonded to the greater number of hydrogens is removed. Quaternary ammonium salts are the most common **phase-transfer catalysts**.

Amines can be synthesized by means of a Gabriel synthesis, by reduction of an amide, an alkyl azide, a nitrile, or a nitro-substituted compound, or by reductive amination.

Saturated heterocycles containing five or more atoms have physical and chemical properties typical of acyclic compounds that contain the same heteroatom. **Pyrrole, furan,** and **thiophene** are heterocyclic aromatic compounds that undergo electrophilic aromatic substitution reactions preferentially at C-2. They are more reactive than benzene toward electrophilic aromatic substitution. When pyrrole is protonated, its aromaticity is destroyed. Pyrrole polymerizes in strongly acidic solutions. Indole, benzofuran, and benzothiophene are heterocyclic aromatic compounds that contain a five-membered aromatic ring fused to a benzene ring.

Replacing one of benzene's carbons with a nitrogen forms **pyridine**, a heterocyclic aromatic compound that undergoes electrophilic aromatic substitution reactions at C-3 and nucleophilic aromatic substitution reactions at C-2 and C-4. Pyridine is less reactive than benzene in electrophilic aromatic substitution reactions and more reactive in nucleophilic aromatic substitution reactions. Quinoline and isoquinoline are heterocyclic aromatic compounds with both a benzene ring and a pyridine ring.

Imidazole is the heterocyclic ring of the amino acid histidine. The conjugate acid of imidazole has a pK_a of 6.8, allowing it to exist in both the protonated and unprotonated forms at physiological pH (pH = 7.3) Nucleic acids (DNA and RNA) contain substituted **purines** and substituted **pyrimidines**. Hydroxypurines and hydroxypyrimidines are more stable in the keto form. A **porphyrin ring system** consists of four pyrrole rings joined by one-carbon bridges; in hemoglobin and myoglobin, the four nitrogen atoms are ligated to Fe^{2+}. The metal ion in chlorophyll a is Mg^{2+} and the metal ion in vitamin B_{12} is Co^{2+}.

SUMMARY OF REACTIONS

1. Reactions of amines as nucleophiles (Section 20.3)

a. In alkylation reactions:

b. In acylation reactions:

c. In nucleophilic addition–elimination reactions:

    i. Reaction of a primary amine with an aldehyde or ketone to form an imine:

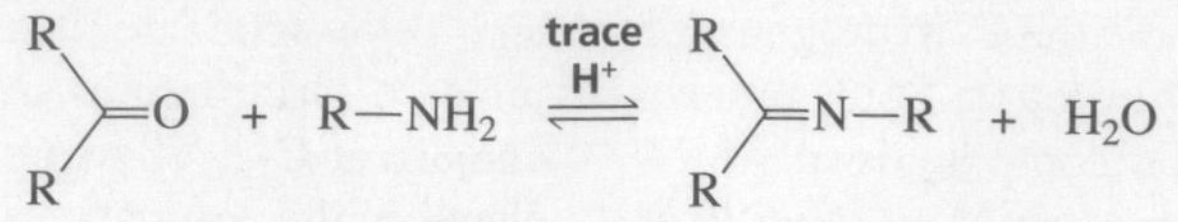

    ii. Reaction of a secondary amine with an aldehyde or ketone to form an enamine:

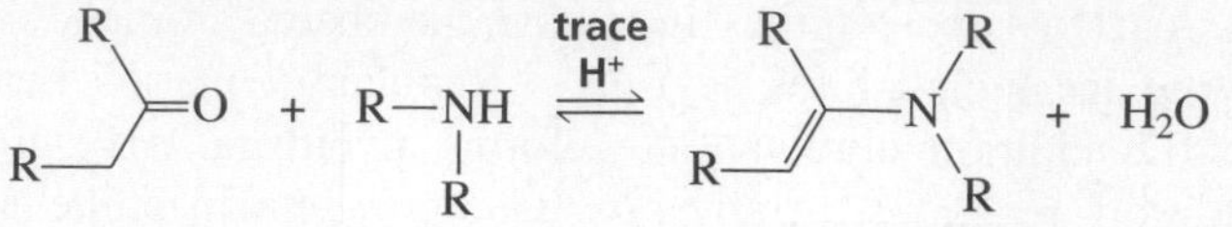

d. In conjugate addition reactions:

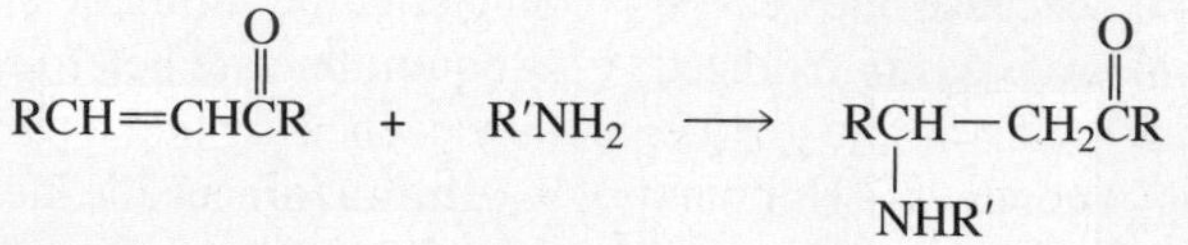

2. Primary arylamines react with nitrous acid to form stable arenediazonium salts (Section 20.3).

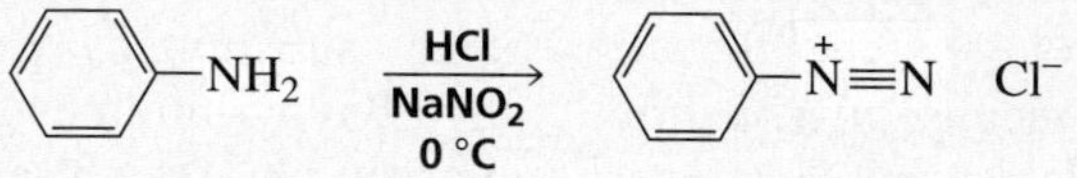

3. Oxidation of amines: primary amines are oxidized to nitro compounds, secondary amines to hydroxylamines, and tertiary amines to amine oxides (Section 20.6).

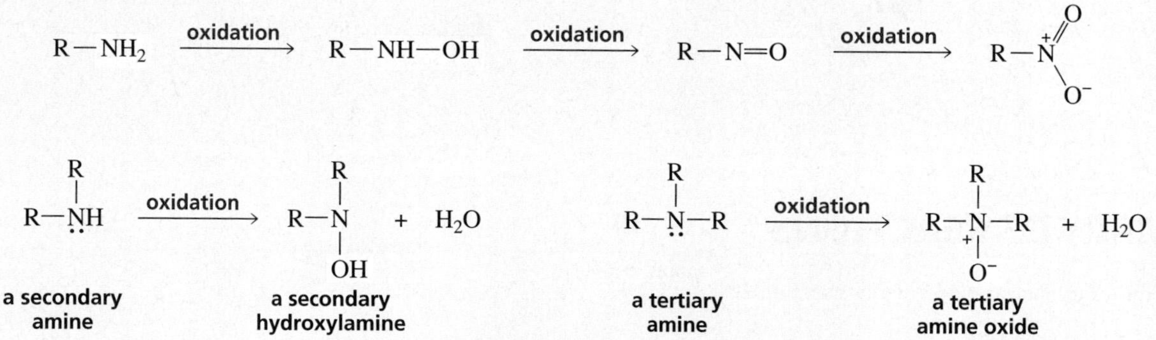

$$R-NH_2 \xrightarrow{\text{oxidation}} R-NH-OH \xrightarrow{\text{oxidation}} R-N=O \xrightarrow{\text{oxidation}} R-\overset{+}{N}\underset{O^-}{\overset{O}{\|}}$$

  a secondary   a secondary   a tertiary   a tertiary
    amine   hydroxylamine   amine   amine oxide

4. Elimination reactions of *quaternary ammonium hydroxides* or *tertiary amine oxides* (Sections 20.4 and 20.6)

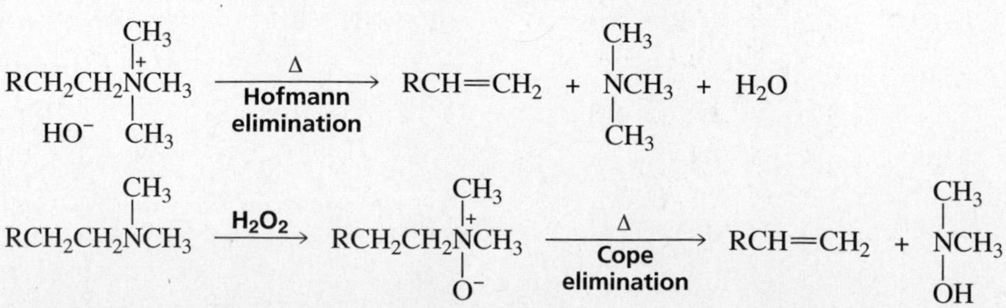

in both eliminations, the proton is removed from the β-carbon bonded to the most hydrogens

5. Synthesis of amines (Section 20.7)

a. Gabriel synthesis of primary amines:

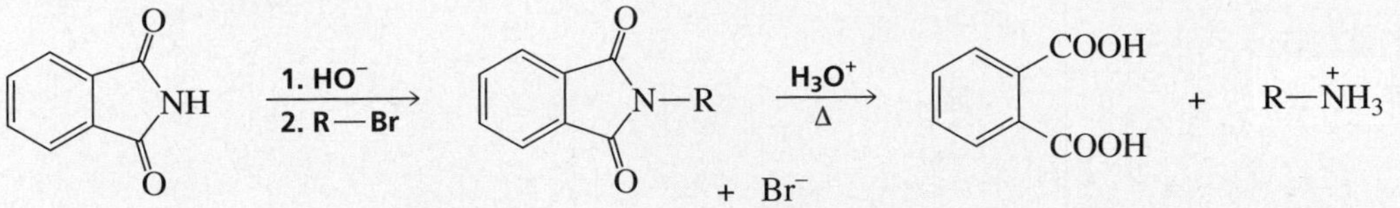

b. Reduction of an alkyl azide or a nitrile:

$$R-Br \xrightarrow{\ ^-N_3\ } R-N=\overset{+}{N}=N^- \xrightarrow[Pd/C]{H_2} R-NH_2$$

$$R-C\equiv N \xrightarrow[Pd/C]{H_2} R-CH_2NH_2$$

c. Reduction of a nitroalkane or nitrobenzene:

$$CH_3CH_2CH_2NO_2 \ + \ H_2 \xrightarrow{Pd/C} CH_3CH_2CH_2NH_2$$

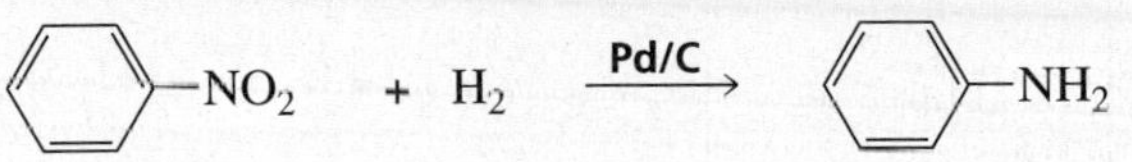

6. Electrophilic aromatic substitution reactions

a. Pyrrole, furan, and thiophene (Section 20.8):

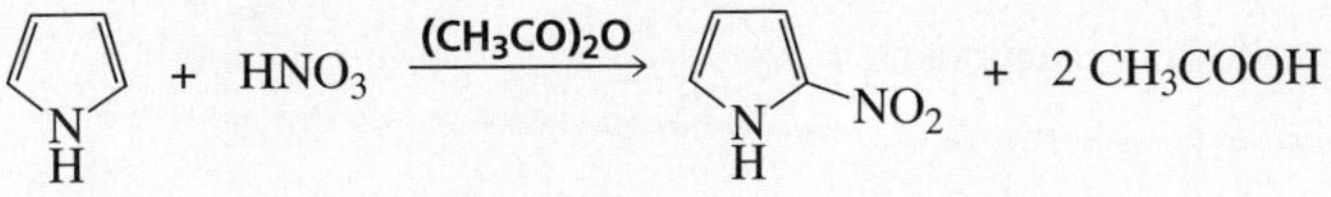

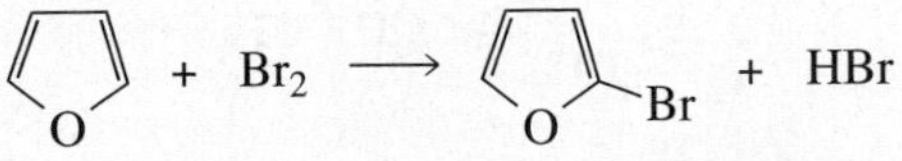

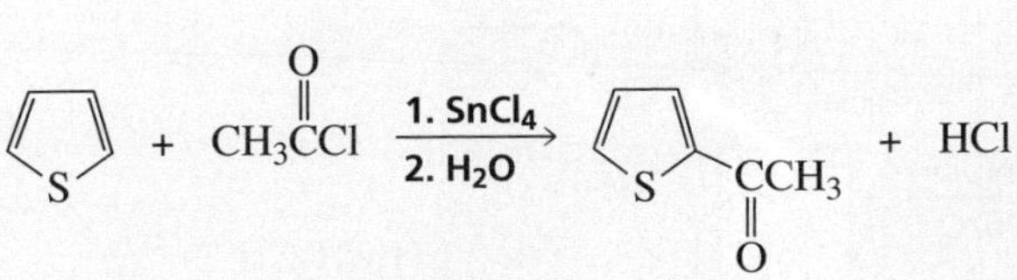

b. Pyridine (Section 20.9):

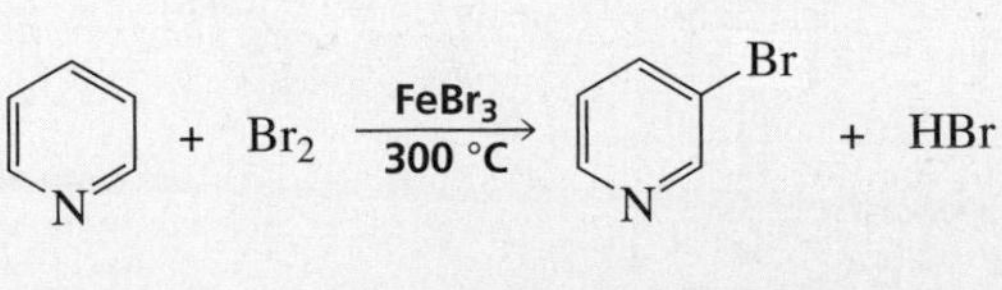

7. Nucleophilic aromatic substitution reactions of pyridine (Section 20.9)

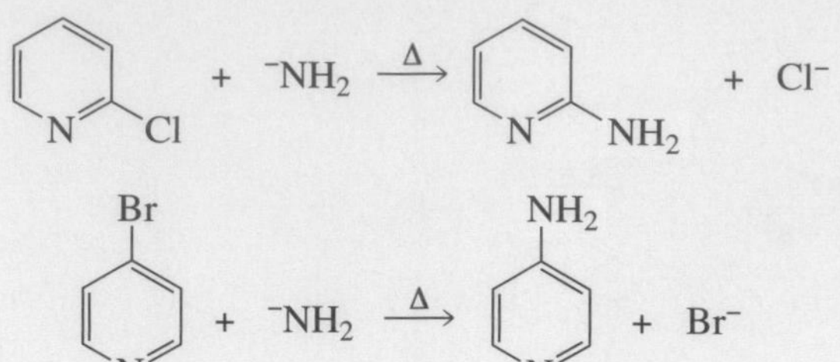

KEY TERMS

amines (p. 943)
Cope elimination reaction (p. 952)
exhaustive methylation (p. 950)
furan (p. 955)
heteroatom (p. 944)
heterocycle (p. 944)
heterocyclic compound (p. 944)

Hofmann elimination
   reaction (p. 947)
imidazole (p. 966)
ligation (p. 968)
phase-transfer catalysis (p. 952)
phase-transfer catalyst (p. 951)
porphyrin ring system (p. 968)

purine (p. 967)
pyridine (p. 960)
pyrimidine (p. 967)
pyrrole (p. 955)
quaternary ammonium
   ion (p. 947)
thiophene (p. 955)

PROBLEMS

26. Name the following compounds:

a. 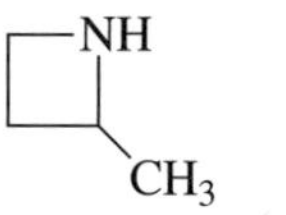 **b.** 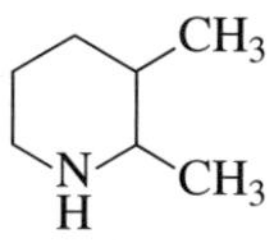 **c.** 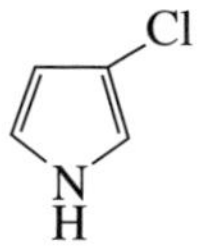 **d.**

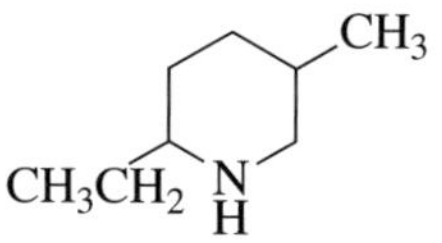

27. Give the product of each of the following reactions:

a. 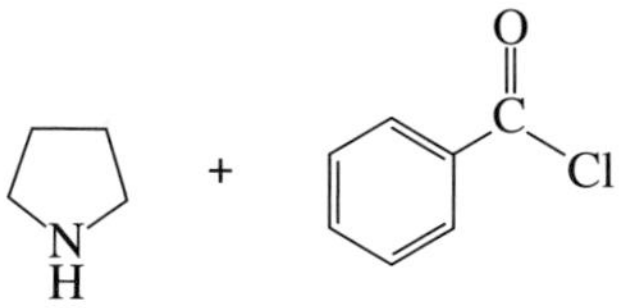 **d.** **g.**

b. 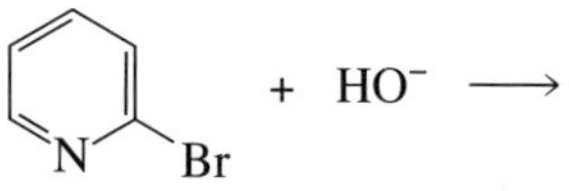 **e.** **h.**

c. CH₃CH₂CH₂CH₂Br $\xrightarrow[\text{2. H}_2, \text{ Pd/C}]{\text{1. } \bar{\text{C}} \equiv \text{N}}$ **f.** 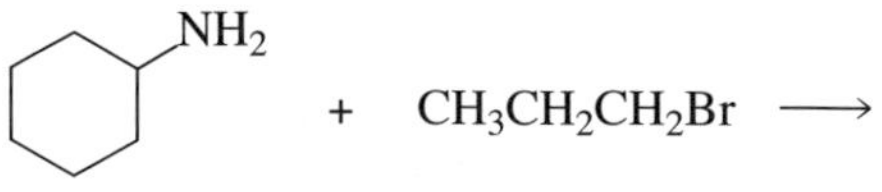 **i.**

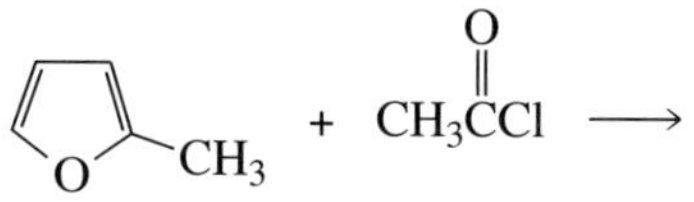

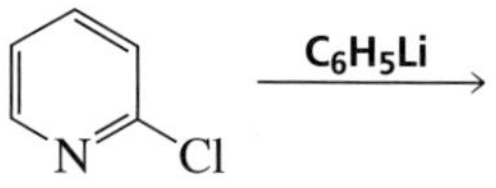

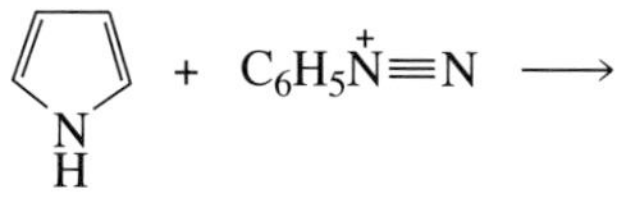

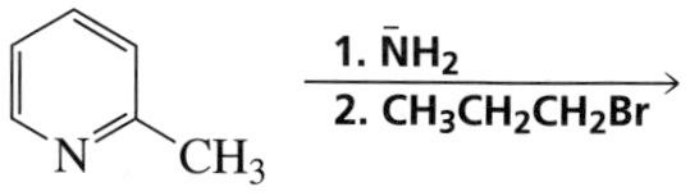

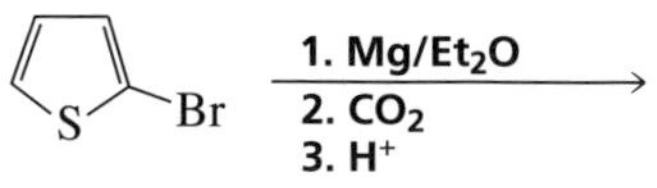

28. List the following compounds in order of decreasing acidity:

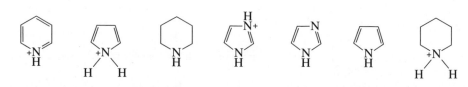

29. Which of the following compounds is easier to decarboxylate?

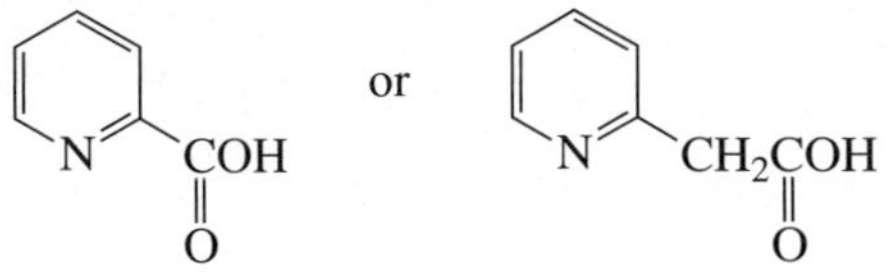

or

30. Rank the following compounds in order of decreasing reactivity in an electrophilic aromatic substitution reaction:

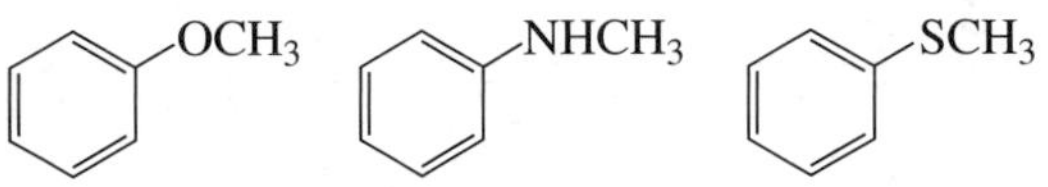

31. One of the following compounds undergoes electrophilic aromatic substitution predominantly at C-3, and one undergoes electrophilic aromatic substitution predominantly at C-4. Which is which?

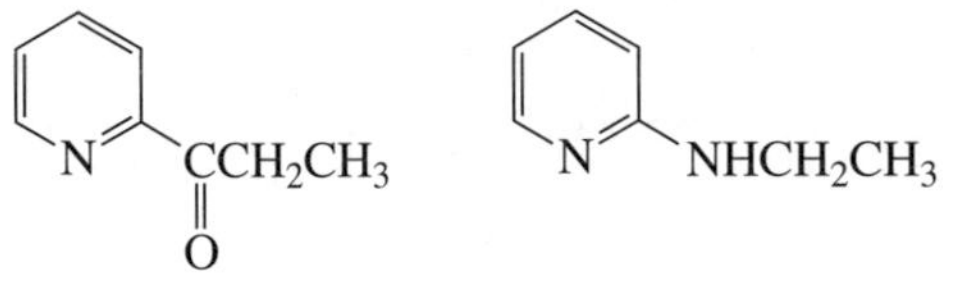

32. Benzene undergoes electrophilic aromatic substitution reactions with aziridines in the presence of a Lewis acid such as $AlCl_3$.
   a. What are the major and minor products of the following reaction?

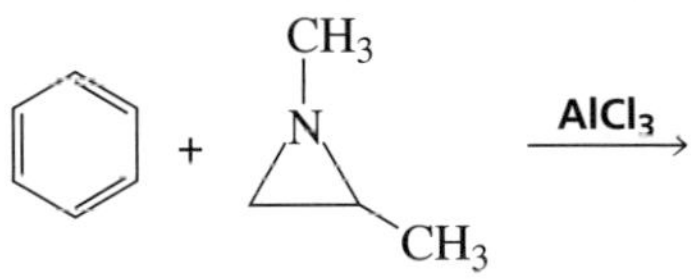

   b. Would you expect epoxides to undergo similar reactions?

33. A Hofmann degradation of a primary amine forms an alkene that gives butanal and 2-methylpropanal upon ozonolysis and work-up under reducing conditions. Identify the amine.

34. The dipole moments of furan and tetrahydrofuran are in the same direction. One compound has a dipole moment of 0.70 D, and the other has a dipole moment of 1.73 D. Which is which?

35. Show how the vitamin niacin can be synthesized from nicotine.

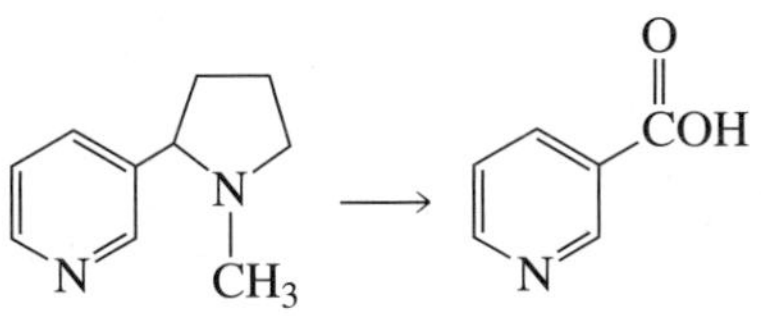

36. The chemical shifts of the C-2 hydrogen in the spectra of pyrrole, pyridine, and pyrrolidine are δ 2.82, δ 6.42, and δ 8.50. Match each chemical shift with its heterocycle.

37. Explain why protonation of aniline has a dramatic effect on the compound's UV spectrum, whereas protonation of pyridine has only a small effect on that compound's UV spectrum.

38. Explain why pyrrole ($pK_a \sim 17$) is a much stronger acid than ammonia ($pK_a = 36$).

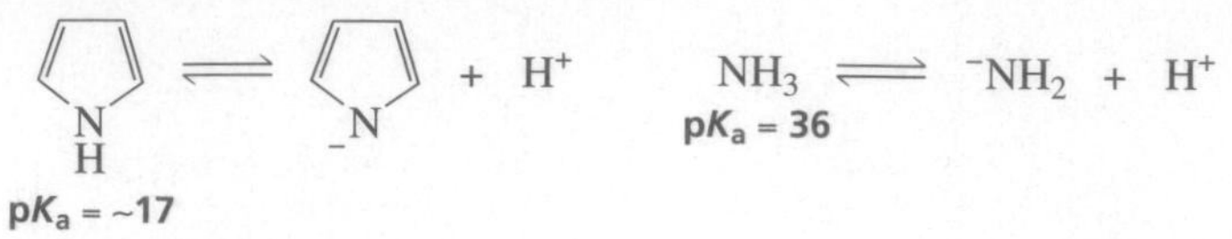

$$\text{pyrrole} \rightleftharpoons \text{pyrrolide} + H^+ \qquad NH_3 \rightleftharpoons {}^-NH_2 + H^+$$
$$pK_a = 36$$
$$pK_a = \sim 17$$

39. Propose a mechanism for the following reaction:

$$2 \text{ pyrrole} + H_2C{=}O \xrightarrow{\text{trace } H^+} \text{dipyrromethane}$$

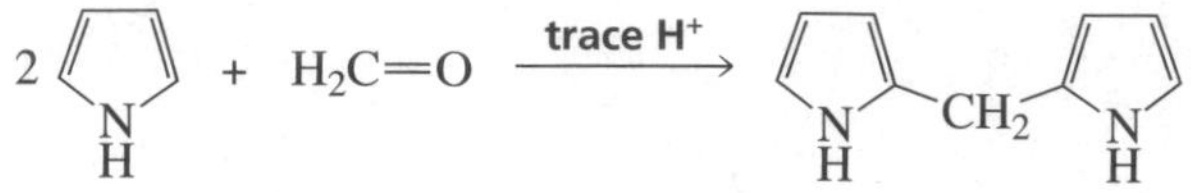

40. Quinolines are commonly synthesized by a method known as the Skraup synthesis, in which aniline reacts with glycerol under acidic conditions. Nitrobenzene is added to the reaction mixture to serve as an oxidizing agent. The first step in the synthesis is the dehydration of glycerol to propenal.

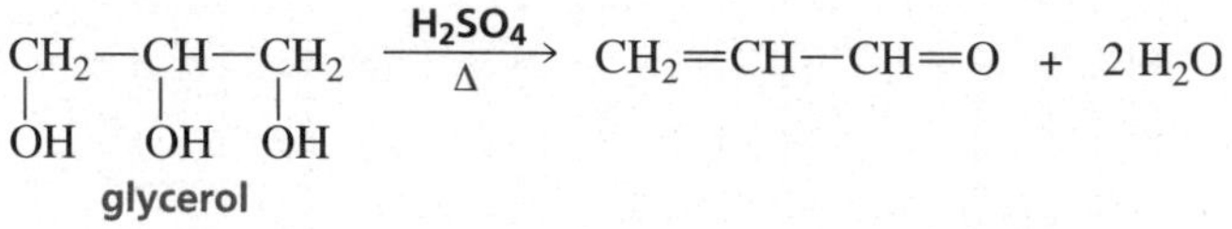

$$\underset{\underset{\text{glycerol}}{\overset{|}{\underset{OH}{}} \quad \overset{|}{\underset{OH}{}} \quad \overset{|}{\underset{OH}{}}}{CH_2{-}CH{-}CH_2} \xrightarrow[\Delta]{H_2SO_4} CH_2{=}CH{-}CH{=}O + 2 H_2O$$

  a. What product would be obtained if *para*-ethylaniline were used instead of aniline?
  b. What product would be obtained if 3-hexen-2-one were used instead of glycerol?
  c. What starting materials are needed for the synthesis of 2,7-diethyl-3-methylquinoline?

41. Propose a mechanism for each of the following reactions:

  a. 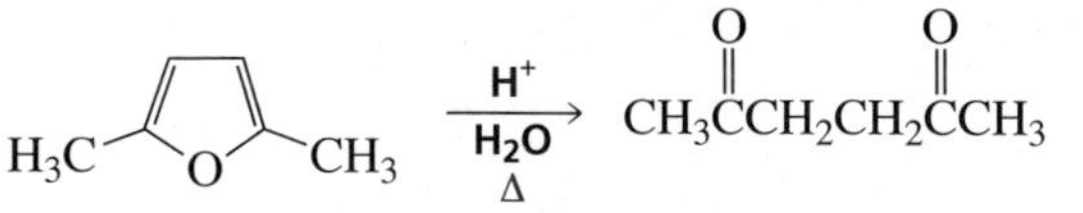
$$\text{2,5-dimethylfuran} \xrightarrow[\underset{\Delta}{H_2O}]{H^+} \overset{O}{\underset{}{CH_3\overset{\|}{C}CH_2CH_2\overset{\|}{C}CH_3}}$$

  b. 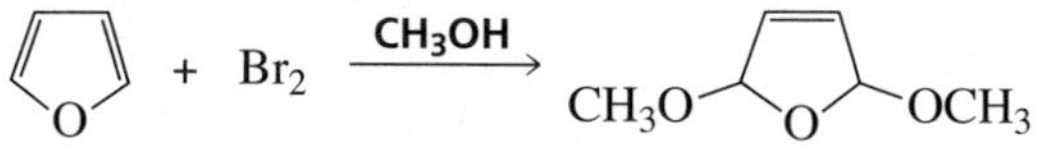
$$\text{furan} + Br_2 \xrightarrow{CH_3OH} \text{2,5-dimethoxy product}$$

42. Give the major product of each of the following reactions:

  a. 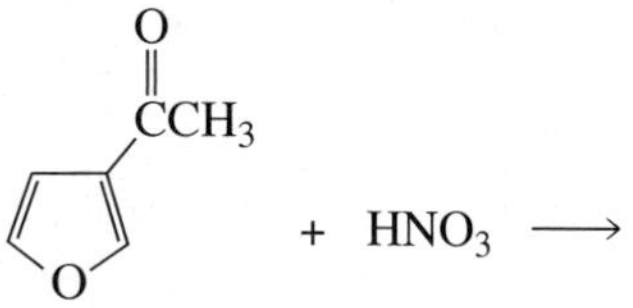 3-acetylfuran $+ HNO_3 \longrightarrow$

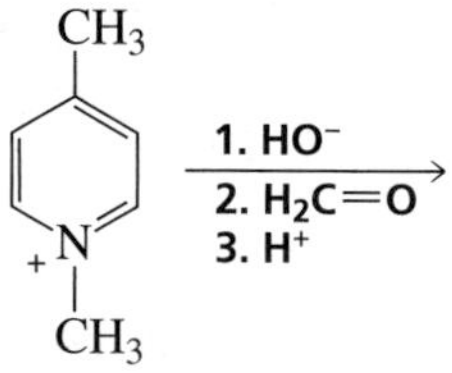

  b. 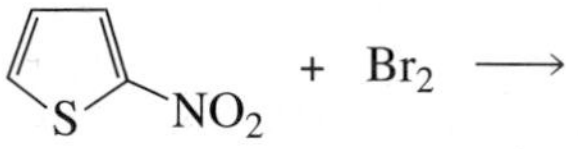 2-nitrothiophene $+ Br_2 \longrightarrow$

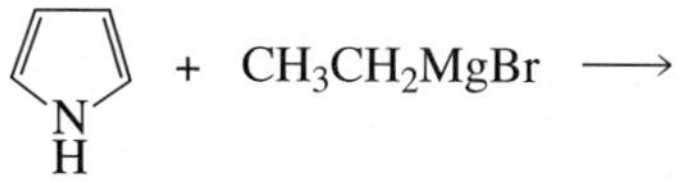

  c. 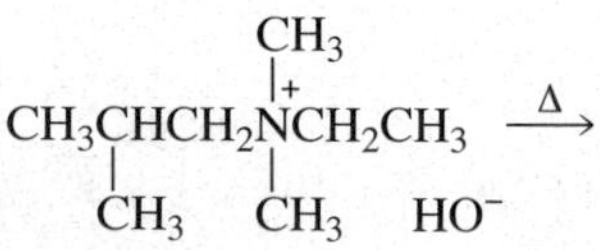 $\underset{\underset{CH_3}{|}}{CH_3CHCH_2}\overset{\overset{CH_3}{|}}{\underset{\underset{HO^-}{}}{\overset{+}{N}CH_2CH_3}} \xrightarrow{\Delta}$
      $\underset{CH_3}{}$

  d. 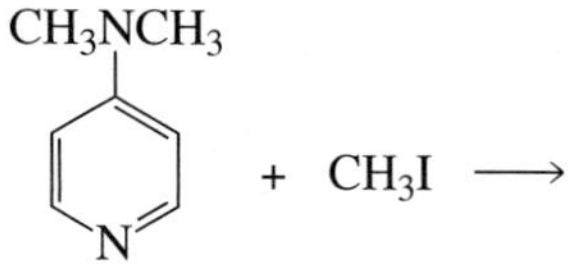 4-(dimethylamino)pyridine $+ CH_3I \longrightarrow$

  e. 2-pyridone $+ PCl_3 \longrightarrow$

  f. 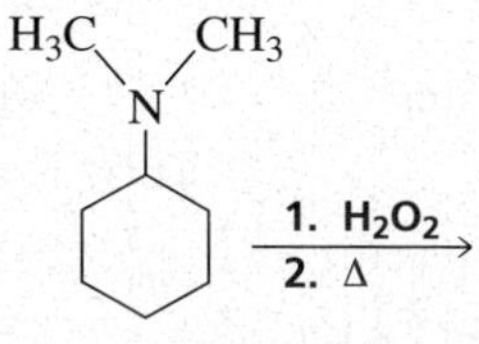 N,N-dimethylcyclohexylamine $\xrightarrow[\text{2. } \Delta]{\text{1. } H_2O_2}$

  g. 1,4-dimethylpyridinium $\xrightarrow[\text{3. } H^+]{\substack{\text{1. } HO^- \\ \text{2. } H_2C{=}O}}$

  h. pyrrole $+ CH_3CH_2MgBr \longrightarrow$

  i. 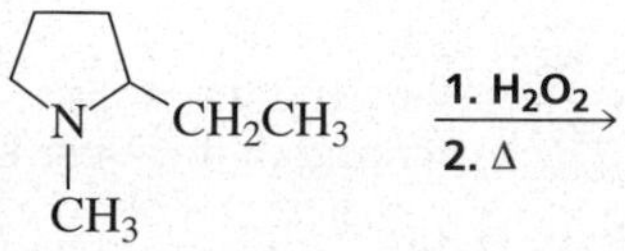 1-methyl-2-ethylpyrrolidine $\xrightarrow[\text{2. } \Delta]{\text{1. } H_2O_2}$

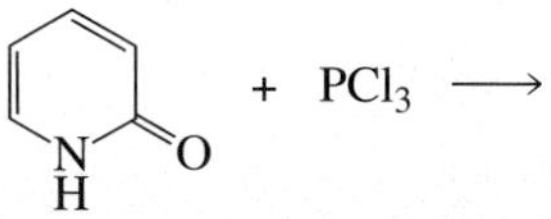

43. When piperidine undergoes the series of reactions indicated below, 1,4-pentadiene is obtained as the product. When the four different methyl-substituted piperidines undergo the same series of reactions, each forms a different diene: 1,5-hexadiene, 1,4-pentadiene, 2-methyl-1,4-pentadiene, and 3-methyl-1,4-pentadiene. Which methyl-substituted piperidine yields which diene?

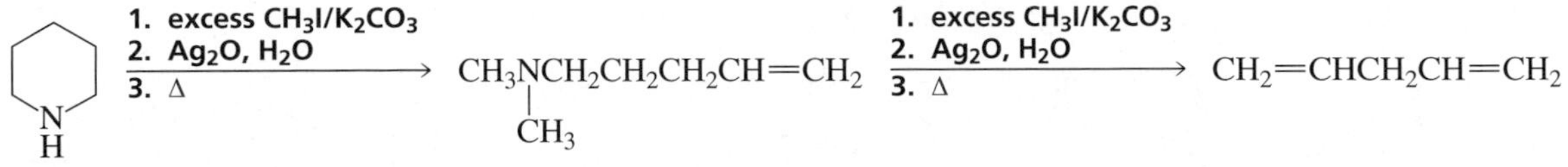

piperidine

44. a. Draw resonance contributors to show why pyridine-N-oxide is more reactive than pyridine toward electrophilic aromatic substitution

b. At what position does pyridine-N-oxide undergo electrophilic aromatic substitution?

45. Propose a mechanism for the following reaction:

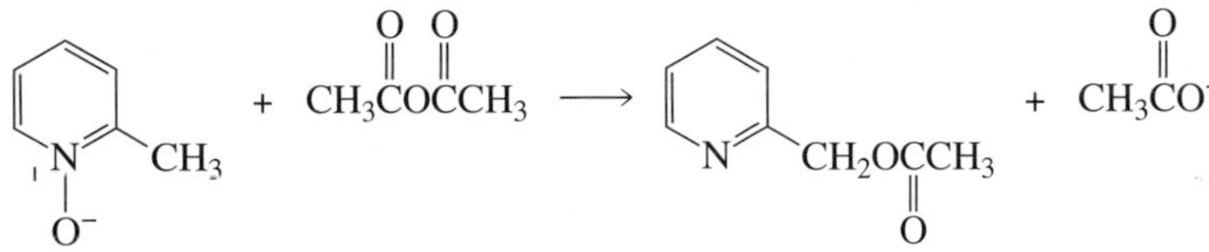

46. Explain why the aziridinium ion has a considerably lower pK_a (8.0) than that of a typical secondary ammonium ion ($\sim$11). (*Hint:* Recall that the larger the bond angle, the greater the s character, and the greater the s character, the more electronegative the atom.)

aziridinium ion
pK_a - 8.04

47. Pyrrole reacts with excess *para*-(N,N-dimethylamino)benzaldehyde to form a highly colored compound. Draw the structure of the colored compound.

48. 2-Phenylindole is prepared from the reaction of acetophenone and phenylhydrazine, a method known as the Fischer indole synthesis. Propose a mechanism for this reaction. (*Hint:* The reactive species is the enamine tautomer of the phenylhydrazone.)

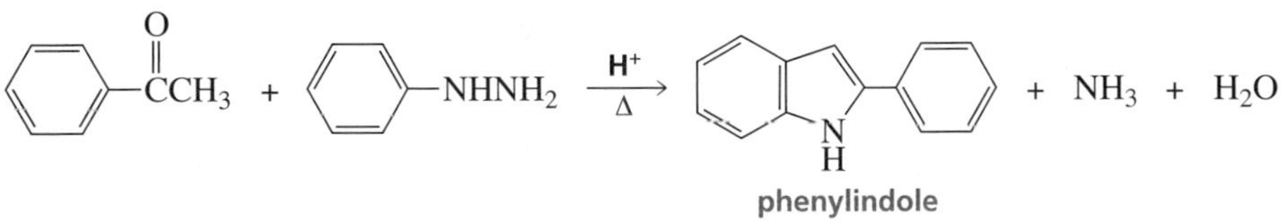

phenylindole

49. What starting materials are required to synthesize the following compounds, using the Fischer indole synthesis? (*Hint:* See Problem 48.)

a.

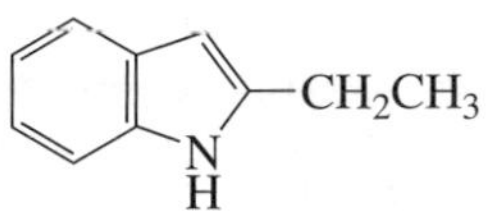

b.

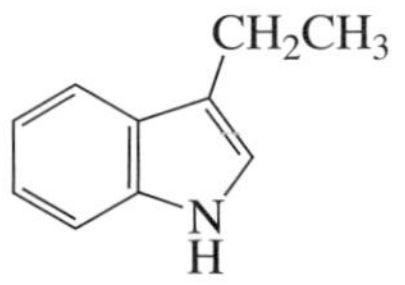

c.

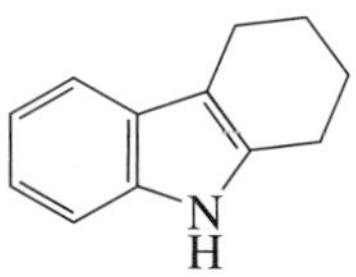

50. Organic chemists work with tetraphenylporphyrins rather than with porphyrins because tetraphenylporphyrins are much more resistant to air oxidation. Tetraphenylporphyrin can be prepared by the reaction of benzaldehyde with pyrrole. Propose a mechanism for the formation of the ring system shown here:

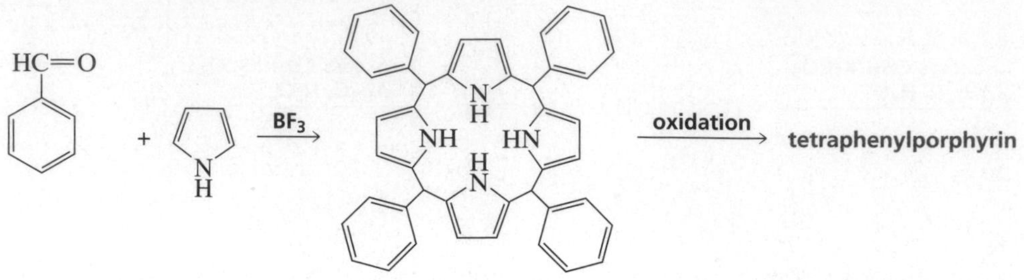

PART 8

Bioorganic Compounds

Chapters 21 through 27 discuss the chemistry of organic compounds found in biological systems. Many of these compounds are larger than the organic compounds you have seen up to this point, and they may have more than one functional group, but the principles that govern their structure and reactivity are essentially the same as those that govern the structure and reactivity of the compounds that you have been studying. These chapters, therefore, will give you the opportunity to review much of the organic chemistry you have learned and apply it to compounds found in the biological world.

CHAPTER 21
Carbohydrates

Chapter 21 introduces you to the chemistry of carbohydrates, the most abundant class of compounds in the biological world. First you will learn about the structures and reactions of monosaccharides. Then you will see how they are linked to form disaccharides and polysaccharides. Many examples of carbohydrates found in nature will be discussed.

CHAPTER 22
Amino Acids, Peptides, and Proteins

Chapter 22 starts by looking at the physical properties of amino acids. Then you will see how amino acids are linked to form peptides and proteins. You will also see how proteins are made in the laboratory. Later, when you read Chapter 27, you will be able to compare this with how they are synthesized in nature. What you learn about protein structure in this chapter will prepare you for understanding how enzymes catalyze chemical reactions, which is discussed in Chapter 23.

CHAPTER 23
Catalysis

Chapter 23 first describes the various ways that organic reactions can be catalyzed and then shows how enzymes employ these same methods to catalyze reactions in biological systems.

CHAPTER 24
The Organic Mechanisms of the Coenzymes

Chapter 24 describes the chemistry of the coenzymes—organic compounds that some enzymes need to catalyze biological reactions. Coenzymes play a variety of chemical roles: some function as oxidizing and reducing agents, some allow electrons to be delocalized, some activate groups for further reaction, and some provide good nucleophiles or strong bases needed for reactions. Because coenzymes are derived from vitamins, you will see why vitamins are necessary for many of the organic reactions that occur in biological systems.

CHAPTER 25
Metabolism

Chapter 25 looks at the organic reactions living organisms carry out to obtain the energy they need and to synthesize the compounds they require.

CHAPTER 26
Lipids

Chapter 26 discusses the chemistry of lipids. Lipids are water-insoluble compounds found in animals and plants. First you will study the structure and function of different kinds of lipids. You will then be able to understand such things as how aspirin prevents inflammation, what causes butter to become rancid, and how cholesterol and other terpenes are synthesized in nature.

CHAPTER 27
Nucleosides, Nucleotides, and Nucleic Acids

Chapter 27 covers the chemistry and structures of nucleosides, nucleotides, and nucleic acids (RNA and DNA). You will see, from a mechanistic standpoint, why ATP is the universal carrier of chemical energy, how nucleotides are linked to form nucleic acids, why DNA contains thymine instead of uracil, and how the genetic messages encoded in DNA are transcribed into mRNA and then translated into proteins. Also explained are how the sequence of bases in DNA is determined and how DNA with specific base sequences can be synthesized.

CHAPTER 21

Carbohydrates

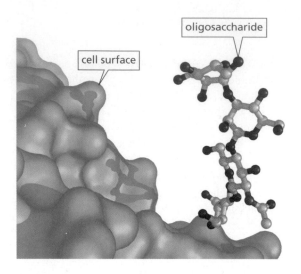

oligosaccharide

cell surface

BUILDING ON FUNDAMENTALS

SECTION 21.5 A base can remove a proton from the α-carbon of an aldehyde or a ketone, forming an enolate ion (18.3).

SECTION 21.6 Aldehydes are reduced to primary alcohols and ketones are reduced to secondary alcohols (19.1).

SECTION 21.6 Aldehydes and primary alcohols are oxidized to carboxylic acids (10.5).

SECTION 21.7 An aldehyde (or a ketone) reacts with a derivative of ammonia to form a compound with an imine linkage (17.8).

SECTION 21.8 An imine is hydrolyzed under acidic conditions to an aldehyde (or a ketone) and ammonia (or an amine) (17.8).

SECTION 21.9 Nitriles are unstable in basic solutions (17.7).

SECTIONS 21.11 AND 21.14 Aldehydes and ketones react with alcohols to form hemiacetals and hemiketals, which react with alcohols to from acetals and ketals (17.10). These reactions are fundamental to the structure and behavior of carbohydrates.

D-glucose

D-fructose

Bioorganic compounds are organic compounds found in biological systems. Their structures can be quite complex, and yet their reactivity is governed by the same principles as the comparatively simple organic molecules we have discussed so far. The organic reactions that chemists carry out in the laboratory are in many ways just the same as those performed by nature inside a living cell. In other words, bioorganic reactions can be thought of as organic reactions that take place in tiny flasks called cells.

Most bioorganic compounds have more complicated structures than those of the organic compounds you are now used to seeing, but do not let the structures fool you into thinking that their chemistry must be equally complicated. One reason the structures of bioorganic compounds are more complicated is that bioorganic compounds must be able to recognize each other. Much of their structure is for that very purpose, a function called **molecular recognition**.

The first group of bioorganic compounds we will look at are *carbohydrates*—the most abundant class of compounds in the biological world, making up more than 50% of the dry weight of the Earth's biomass. Carbohydrates are important constituents of all living organisms and have a variety of different functions. Some are important structural components of cells; others act as recognition sites on cell surfaces. For example, the first event in any of our lives was a sperm recognizing a carbohydrate on the outer surface of an egg. Other carbohydrates serve as a major source of metabolic energy. For example, the leaves, fruits, seeds, stems, and roots of plants contain carbohydrates that plants use for their own metabolic needs but that can also serve the metabolic needs of the animals that eat the plants.

Early chemists noted that carbohydrates have molecular formulas that make them appear to be hydrates of carbon, $C_n(H_2O)_n$—hence the name. Later structural studies revealed, however, that these compounds are *not* hydrates because they do not contain intact water molecules; nevertheless, the term "carbohydrate" persists.

Carbohydrates are polyhydroxy aldehydes such as D-glucose, polyhydroxy ketones such as D-fructose, and compounds such as sucrose formed by linking polyhydroxy aldehydes or polyhydroxy ketones together (Section 21.17). The chemical structures of carbohydrates are commonly represented by wedge-and-dash structures or by Fischer projections. Notice that both D-glucose and D-fructose have the molecular formula $C_6H_{12}O_6$, consistent with the general formula $C_6(H_2O)_6$ that made early chemists think that those compounds were hydrates of carbon.

> Recall from Section 5.7 that horizontal bonds point toward the viewer and vertical bonds point away from the viewer in Fischer projections.

| wedge-and-dash structure | Fischer projection | wedge-and-dash structure | Fischer projection |

D-glucose
a polyhydroxy aldehyde

D-fructose
a polyhydroxy ketone

The most abundant carbohydrate in nature is glucose. Living cells oxidize glucose in the first of a series of processes that provide them with energy. When animals have more glucose than they need for energy, they convert excess glucose into a polymer called glycogen (Section 21.18). Later, when an animal needs energy, glycogen is broken down into individual glucose molecules. Plants convert excess glucose into a polymer known as starch. Cellulose—the major structural component of plants—is another polymer of glucose. Chitin, a carbohydrate similar to cellulose, makes up the exoskeletons of crustaceans, insects, and other arthropods and is also the structural material of fungi.

Animals obtain glucose from food that contains glucose, such as plants. Plants produce glucose by *photosynthesis*. During photosynthesis, plants take up water through their roots and use carbon dioxide from the air to synthesize glucose and oxygen. Because photosynthesis is the reverse of the process used by organisms to obtain energy—specifically, the oxidation of glucose to carbon dioxide and water—plants require energy to carry out photosynthesis. Plants obtain the energy they need for photosynthesis from sunlight, captured by chlorophyll molecules in green plants. Photosynthesis uses the CO_2 that animals exhale as waste and generates the O_2 that animals inhale to sustain life. Nearly all the oxygen in the atmosphere has been released by photosynthetic processes.

$$C_6H_{12}O_6 \; + \; 6\,O_2 \; \underset{\text{photosynthesis}}{\overset{\text{oxidation}}{\rightleftharpoons}} \; 6\,CO_2 \; + \; 6\,H_2O$$
$$\text{glucose}$$

21.1 Classification of Carbohydrates

The terms "carbohydrate," "saccharide," and "sugar" are used interchangeably. "Saccharide" comes from the word for "sugar" in several early languages (*sarkara* in Sanskrit, *sakcharon* in Greek, and *saccharum* in Latin).

There are two classes of carbohydrates, simple and complex. **Simple carbohydrates** are **monosaccharides** (single sugars), whereas **complex carbohydrates** contain two or more monosaccharides linked together. **Disaccharides** have two monosaccharides linked together, **oligosaccharides** have three to 10 (*oligos* is Greek for "few"), and **polysaccharides** have more than 10. Disaccharides, oligosaccharides, and polysaccharides can be broken down to monosaccharides by hydrolysis.

a monosaccharide subunit

$$—M—M—M—M—M—M—M—M—M— \xrightarrow{\text{hydrolysis}} \text{x M}$$

polysaccharide   monosaccharide

A *monosaccharide* can be a polyhydroxy aldehyde such as D-glucose or a polyhydroxy ketone such as D-fructose. Polyhydroxy aldehydes are called **aldoses** ("ald" is for aldehyde; "ose" is the suffix for a sugar), whereas polyhydroxy ketones are called **ketoses**. Monosaccharides are also classified according to the number of carbons they contain: those with three carbons are **trioses**, those with four carbons are **tetroses**, those with five carbons are **pentoses**, and those with six and seven carbons are **hexoses** and **heptoses**, respectively. Therefore, a six-carbon polyhydroxy aldehyde such as D-glucose is an aldohexose, whereas a six-carbon polyhydroxy ketone such as D-fructose is a ketohexose.

PROBLEM 1◆

Classify the following monosaccharides:

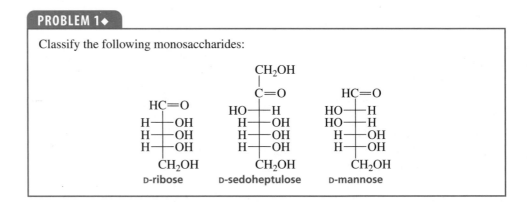

D-ribose   D-sedoheptulose   D-mannose

21.2 The D and L Notation

The smallest aldose, and the only one whose name does not end in "ose," is glyceraldehyde, an aldotriose.

A carbon to which four different groups are attached is an asymmetric center.

asymmetric center

$$\text{HOCH}_2\text{CHCH}$$
with O double bond and OH

glyceraldehyde

Because glyceraldehyde has an asymmetric center, it can exist as a pair of enantiomers. The *R* and *S* isomers (Section 5.7) are

clockwise is *R*

(*R*)-(+)-glyceraldehyde   (*S*)-(−)-glyceraldehyde

perspective formulas

H is on a horizontal bond so counterclockwise is *R*

(*R*)-(+)-glyceraldehyde   (*S*)-(−)-glyceraldehyde

Fischer projections

The notations D and L are used to describe the configurations of carbohydrates and amino acids (Section 22.2). In a Fischer projection of a monosaccharide, the carbonyl group is always placed on top (in the case of aldoses) or as close to the top as possible (in the case of ketoses). Examine the Fischer projection of galactose shown below and note that the compound has four asymmetric centers (C-2, C-3, C-4, and C-5). *If the OH group attached to the bottom-most asymmetric center (the carbon second from the bottom) is on the right, the compound is a D-sugar. If that OH group is on the left, the compound is an L-sugar.* Almost all sugars found in nature are D-sugars. Notice that the mirror image of a D-sugar is an L-sugar.

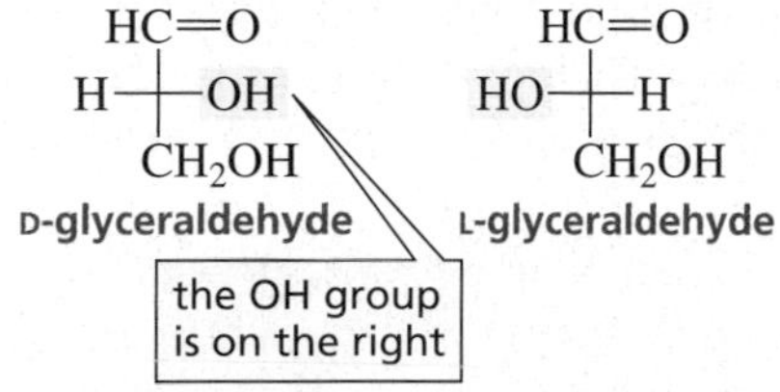

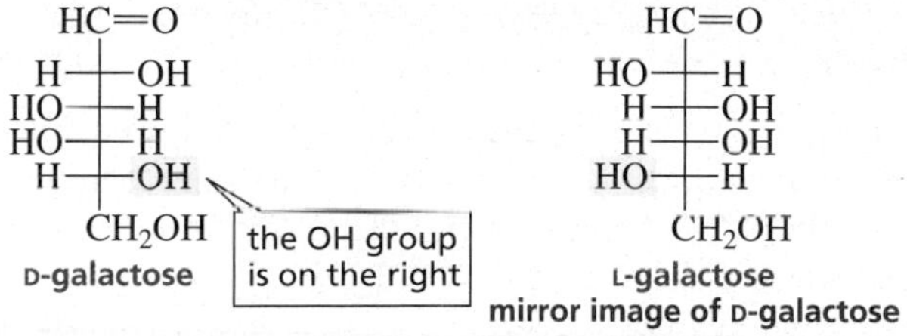

$$
\begin{array}{cc}
\text{HC}{=}\text{O} & \text{HC}{=}\text{O} \\
\text{H}{-}\!\!-\!\!{-}\text{OH} & \text{HO}{-}\!\!-\!\!{-}\text{H} \\
\text{CH}_2\text{OH} & \text{CH}_2\text{OH} \\
\text{D-glyceraldehyde} & \text{L-glyceraldehyde}
\end{array}
$$

the OH group is on the right

$$
\begin{array}{cc}
\text{HC}{=}\text{O} & \text{HC}{=}\text{O} \\
\text{H}{-}\text{OH} & \text{HO}{-}\text{H} \\
\text{HO}{-}\text{H} & \text{H}{-}\text{OH} \\
\text{HO}{-}\text{H} & \text{H}{-}\text{OH} \\
\text{H}{-}\text{OH} & \text{HO}{-}\text{H} \\
\text{CH}_2\text{OH} & \text{CH}_2\text{OH} \\
\text{D-galactose} & \text{L-galactose}
\end{array}
$$

the OH group is on the right

mirror image of D-galactose

Emil Fischer and his colleagues studied carbohydrates in the late nineteenth century, when techniques for determining the absolute configurations of compounds were not available. Fischer arbitrarily assigned the *R*-configuration to the dextrorotatory isomer of glyceraldehyde that we call D-glyceraldehyde. He turned out to be correct: D-glyceraldehyde is (*R*)-(+) glyceraldehyde, and L-glyceraldehyde is (*S*)-(−)-glyceraldehyde (Section 5.15).

Like *R* and *S*, the symbols D and L indicate the configuration of an asymmetric center, but they do not indicate whether the compound rotates polarized light to the right (+) or to the left (−) (Section 5.7). For example, D-glyceraldehyde is dextrorotatory, whereas D-lactic acid is levorotatory. In other words, optical rotation, like melting or boiling points, is a physical property of a compound, whereas "*R, S,* D, and L" are conventions humans use to indicate the configuration about an asymmetric center.

Student Tutorial:
D and L Notation

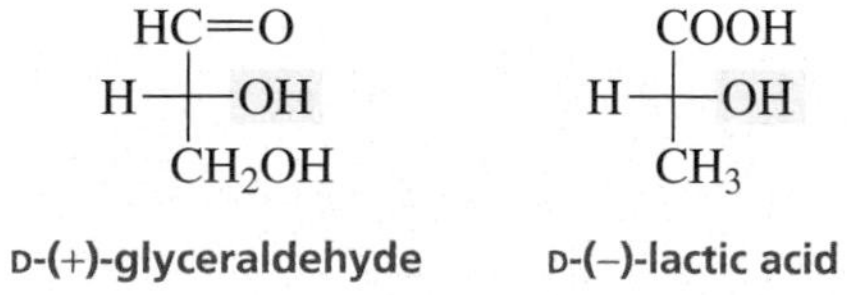

$$
\begin{array}{cc}
\text{HC}{=}\text{O} & \text{COOH} \\
\text{H}{-}\!\!-\!\!{-}\text{OH} & \text{H}{-}\!\!-\!\!{-}\text{OH} \\
\text{CH}_2\text{OH} & \text{CH}_3 \\
\text{D-(+)-glyceraldehyde} & \text{D-(−)-lactic acid}
\end{array}
$$

The common name of the monosaccharide, together with the D or L designation, completely defines its structure, because the configurations of all the asymmetric centers are implicit in that name.

PROBLEM 2

Draw Fischer projections of L-glucose and L-fructose.

PROBLEM 3◆

Indicate whether each of the following is D-glyceraldehyde or L-glyceraldehyde, assuming that the horizontal bonds point toward you and the vertical bonds point away from you (Section 5.6):

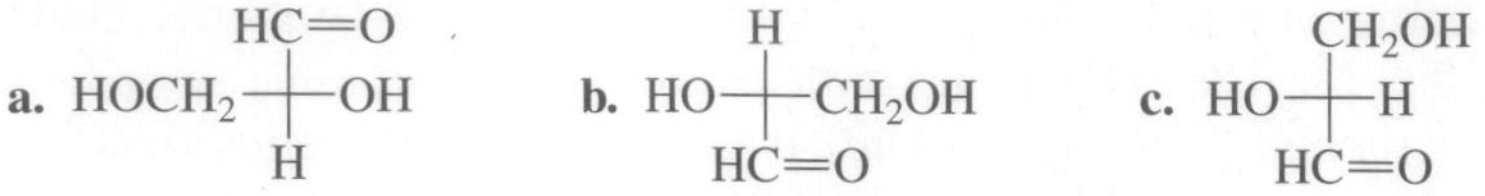

a.
$$HC=O$$
$$HOCH_2 \text{---} OH$$
$$H$$

b.
$$H$$
$$HO \text{---} CH_2OH$$
$$HC=O$$

c.
$$CH_2OH$$
$$HO \text{---} H$$
$$HC=O$$

21.3 The Configurations of Aldoses

Aldotetroses have two asymmetric centers and therefore four stereoisomers. Two of the stereoisomers are D-sugars and two are L-sugars. The names of the aldotetroses—erythrose and threose—were used to name the erythro and threo pairs of enantiomers described in Section 5.11.

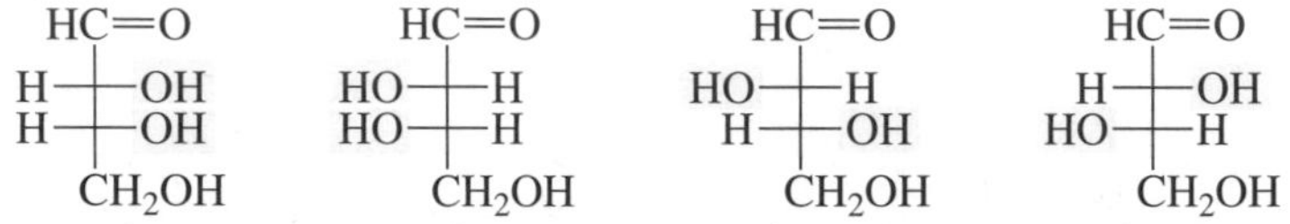

$$\begin{array}{cccc}
HC=O & HC=O & HC=O & HC=O \\
H \text{---} OH & HO \text{---} H & HO \text{---} H & H \text{---} OH \\
H \text{---} OH & HO \text{---} H & H \text{---} OH & HO \text{---} H \\
CH_2OH & CH_2OH & CH_2OH & CH_2OH \\
\text{D-erythrose} & \text{L-erythrose} & \text{D-threose} & \text{L-threose}
\end{array}$$

Aldopentoses have three asymmetric centers and therefore eight stereoisomers (four pairs of enantiomers); and aldohexoses have four asymmetric centers and 16 stereoisomers (eight pairs of enantiomers). The four D-aldopentoses and the eight D-aldohexoses are shown in Table 21.1.

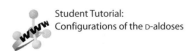

Student Tutorial:
Configurations of the D-aldoses

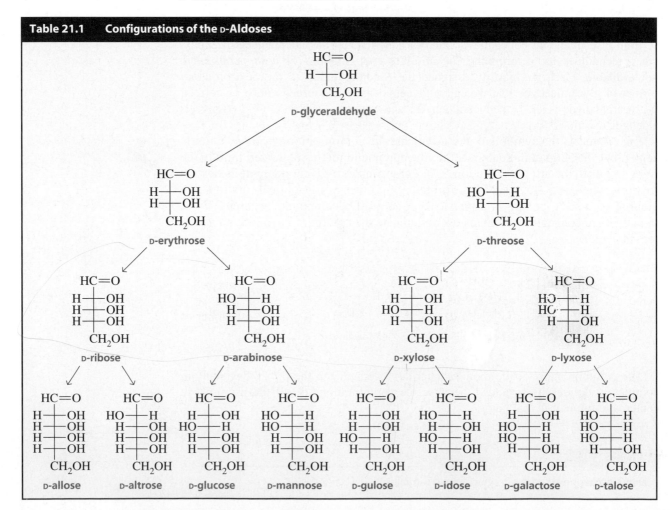

Table 21.1   Configurations of the D-Aldoses

Diastereomers that differ in configuration at only one asymmetric center are called **epimers**. For example, D-ribose and D-arabinose are C-2 epimers because they differ in configuration only at C-2; D-idose and D-talose are C-3 epimers.

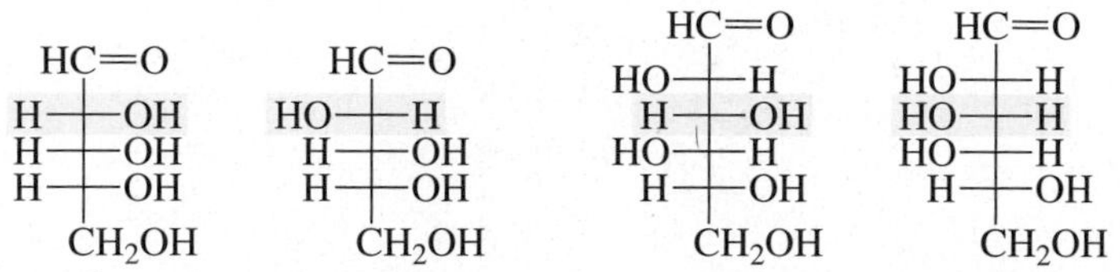

D-ribose   D-arabinose   D-idose   D-talose

C-2 epimers   **C-3 epimers**

D-Glucose, D-mannose, and D-galactose are the most common aldohexoses in biological systems. An easy way to learn their structures is to memorize the structure of D-glucose and then remember that D-mannose is the C-2 epimer of D-glucose and D-galactose is the C-4 epimer of D-glucose. Sugars such as D-glucose and D-galactose are also diastereomers because they are stereoisomers that are not enantiomers (Section 5.11). An epimer is a particular kind of diastereomer.

D-Mannose is the C-2 epimer of D-glucose.

D-Galactose is the C-4 epimer of D-glucose.

Diastereomers are stereoisomers that are not enantiomers.

PROBLEM 4◆

a. Are D-erythrose and L-erythrose enantiomers or diastereomers?
b. Are L-erythrose and L-threose enantiomers or diastereomers?

PROBLEM 5◆

a. What sugar is the C-3 epimer of D-xylose?
b. What sugar is the C-5 epimer of D-allose?
c. What sugar is the C-4 epimer of L-gulose?

PROBLEM 6◆

Give systematic names to the following compounds. Indicate the configuration (*R* or *S*) of each asymmetric center:

a. D-glucose   **b.** D-mannose   **c.** D-galactose   **d.** L-glucose

---

21.4 The Configurations of Ketoses

Naturally occurring ketoses have the ketone group in the 2-position. The configurations of the D-2-ketoses are shown in Table 21.2. A ketose has one fewer asymmetric center than does an aldose with the same number of carbon atoms. Therefore, a ketose has only half as many stereoisomers as an aldose with the same number of carbon atoms.

PROBLEM 7◆

What sugar is the C-3 epimer of D-fructose?

PROBLEM 8◆

How many stereoisomers are possible for

a. a 2-ketoheptose?   **b.** an aldoheptose?   **c.** a ketotriose?

Table 21.2   Configurations of the D-Ketoses

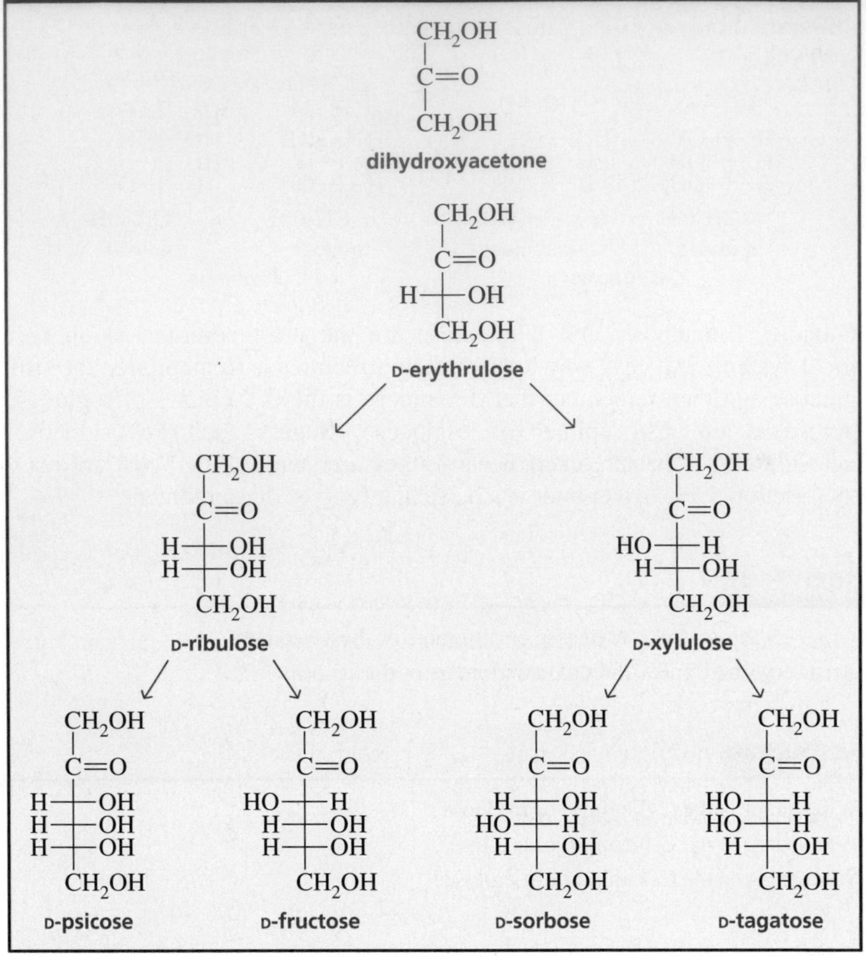

$$CH_2OH$$
$$C=O$$
$$CH_2OH$$

dihydroxyacetone

$$CH_2OH$$
$$C=O$$
$$H{-}{-}OH$$
$$CH_2OH$$

D-erythrulose

D-ribulose

$$CH_2OH$$
$$C=O$$
$$H{-}OH$$
$$H{-}OH$$
$$CH_2OH$$

D-xylulose

$$CH_2OH$$
$$C=O$$
$$HO{-}H$$
$$H{-}OH$$
$$CH_2OH$$

D-psicose, D-fructose, D-sorbose, D-tagatose

21.5 Reactions of Monosaccharides in Basic Solutions

Monosaccharides cannot undergo reactions with basic reagents, because in a basic solution, a monosaccharide is converted to a complex mixture of polyhydroxy aldehydes and polyhydroxyketones. Let's first look at what happens to D-glucose in a basic solution, beginning with its conversion to its C-2 epimer.

Mechanism for the base-catalyzed epimerization of a monosaccharide

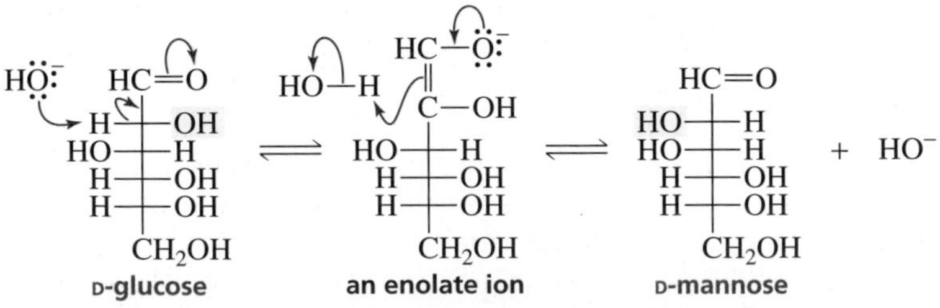

- The base removes a proton from an α-carbon, forming an enolate ion (Section 18.3).
- In the enolate ion, C-2 is no longer an asymmetric center.
- When C-2 is reprotonated, the proton can come from the top or the bottom of the planar sp^2 carbon, forming both D-glucose and D-mannose.

Because the reaction forms a pair of C-2 epimers, it is called epimerization. **Epimerization** changes the configuration of a carbon by removing a proton and then reprotanating it.

In a basic solution, in addition to forming its C-2 epimer, D-glucose can also undergo a reaction called an enediol rearrangement, which results in the formation of D-fructose and other ketohexoses.

In a basic solution an aldose forms a C-2 epimer and one or more ketoses.

Mechanism for the base-catalyzed enediol rearrangement of a monosaccharide

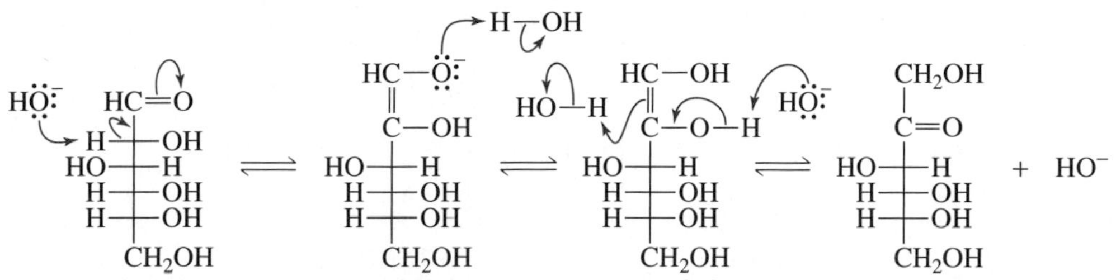

| D-glucose | an enolate ion | an enediol | D-fructose |

- The base removes a proton from an α-carbon, forming an enolate ion (Section 18.3).
- The enolate ion can be protonated to form an enediol.
- The enediol has two OH groups that can ketonize (form a carbonyl group). Ketonizing the OH at C-1 reforms D-glucose; ketonizing the OH group at C-2 forms D-fructose.

Another enediol rearrangement, initiated by a base removing a proton from C-3 of D-fructose, forms a ketose with the carbonyl group at C-3. Thus, the carbonyl group can be moved up and down the chain.

PROBLEM 9

Show how an enediol rearrangement can move the carbonyl carbon of fructose from C-2 to C-3.

PROBLEM 10

Write the mechanism for the base-catalyzed conversion of D-fructose into D-glucose and D-mannose.

PROBLEM 11◆

When D-tagatose is added to a basic aqueous solution, an equilibrium mixture of three monosaccharides is obtained. What are these monosaccharides?

21.6 Redox Reactions of Monosaccharides

Because they contain *alcohol* functional groups and *aldehyde* or *ketone* functional groups, the reactions of monosaccharides are an extension of what you have already learned about the reactions of alcohols, aldehydes, and ketones. For example, an aldehyde group in a monosaccharide can be oxidized or reduced and can react with nucleophiles to form imines, hemiacetals, and acetals. As you read this section and those that follow, dealing with the reactions of monosaccharides, you will find cross-references to earlier discussions of simpler organic compounds undergoing the same reactions. Go back and look at these earlier discussions when they are mentioned; they will make learning about carbohydrates a lot easier.

Reduction

The carbonyl group in aldoses and ketoses can be reduced by the usual carbonyl-group reducing agents (such as $NaBH_4$; Section 19.1). The product of the reduction is a polyalcohol, known as an **alditol**. Reduction of an aldose forms one alditol. Reduction of a ketose forms two alditols because the reaction creates a new asymmetric center in the product. D-Mannitol, the alditol formed from the reduction of D-mannose, is found in mushrooms, olives, and onions. The reduction of D-fructose forms D-mannitol and D-glucitol, the C-2 epimer of D-mannitol. D-Glucitol—also called sorbitol—is about 60% as sweet as sucrose. It is found in plums, pears, cherries, and berries and is used as a sugar substitute in the manufacture of candy.

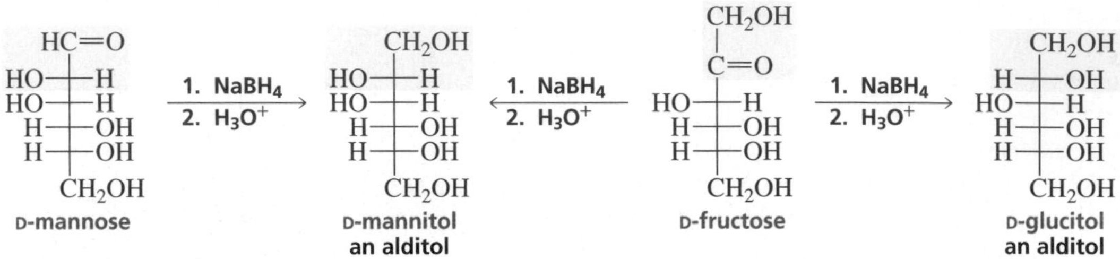

D-Glucitol is also obtained from the reduction of either D-glucose or L-gulose.

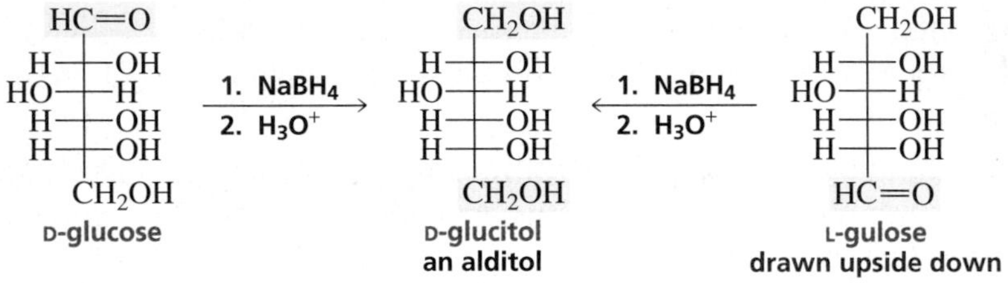

D-Xylitol—obtained from the reduction of D-xylose—is used as a sweetening agent in cereals and in "sugarless" gum.

PROBLEM 12◆

What products are obtained from the reduction of

a. D-idose? **b.** D-sorbose?

PROBLEM 13◆

a. What other monosaccharide is reduced only to the alditol obtained from the reduction of
   1. D-talose?
   2. D-galactose?
b. What monosaccharide is reduced to two alditols, one of which is the alditol obtained from the reduction of
   1. D-talose?
   2. D-allose?

Oxidation

Aldoses can be distinguished from ketoses by observing what happens to the color of an aqueous solution of Br_2 when it is added to the sugar. Br_2 is a mild oxidizing agent and easily oxidizes the aldehyde group, but it cannot oxidize ketones or alcohols. Consequently, if a small amount of an aqueous solution of Br_2 is added to an unknown monosaccharide, the reddish-brown color of Br_2 will disappear if the monosaccharide is an aldose because Br_2 will be reduced to Br^-, which is colorless. If the red color persists, indicating no reaction with Br_2, the monosaccharide is a ketose, which does not react with Br_2. The product of the oxidation reaction is an **aldonic acid**.

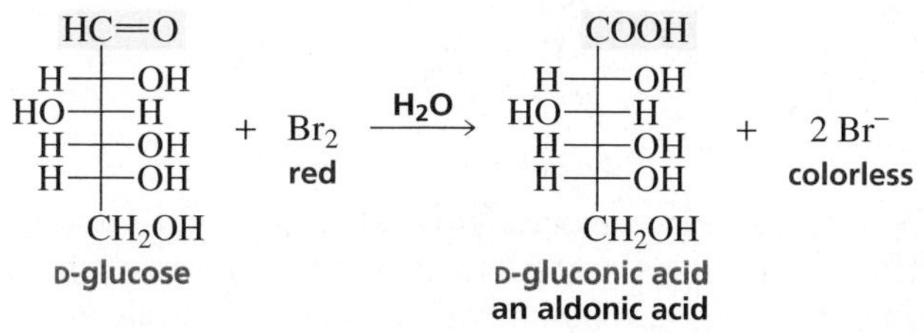

HC=O   H—OH   HO—H   H—OH   H—OH CH_2OH
D-glucose

$+ Br_2$ (red) $\xrightarrow{H_2O}$

COOH   H—OH   HO—H   H—OH   H—OH CH_2OH
D-gluconic acid
an aldonic acid

$+ 2 Br^-$ (colorless)

Both aldoses and ketoses are oxidized to aldonic acids by Tollens reagent (Ag^+, NH_3, HO^-), which therefore cannot be used to distinguish them. You saw, however, in Section 19.3, that Tollens reagent oxidizes aldehydes but not ketones. Why, then, are ketoses oxidized by Tollens reagent, if ketones are not? Ketoses are oxidized because the oxidation reaction is carried out under basic conditions, and in a basic solution, a ketose can be converted into an aldose by an enediol rearrangement (Section 21.5).

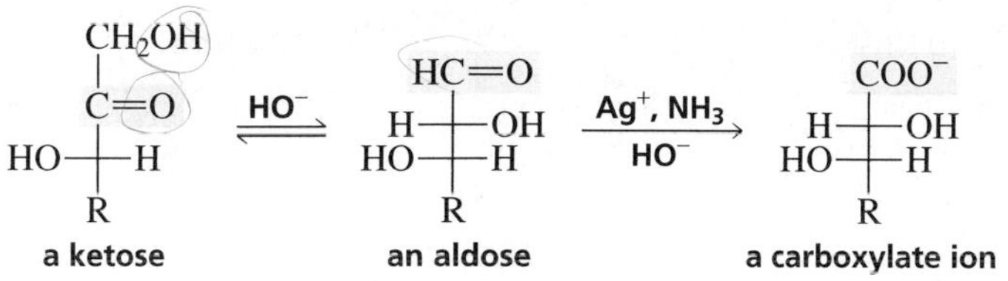

CH_2OH C=O   HO—H   R
a ketose

$\xrightleftharpoons{HO^-}$

HC=O   H—OH   HO—H   R
an aldose

$\xrightarrow[HO^-]{Ag^+, NH_3}$

COO^- H—OH   HO—H   R
a carboxylate ion

In the presence of an oxidizing agent stronger than the ones discussed above (such as HNO_3), one or more of the alcohol groups can be oxidized in addition to the aldehyde group. A primary alcohol is the one most easily oxidized. The product obtained when both the aldehyde and the primary alcohol groups of an aldose are oxidized is called an **aldaric acid**. (In an ald*on*ic acid, *on*e end is oxidized. In an ald*ar*ic acid, both ends *ar*e oxidized.)

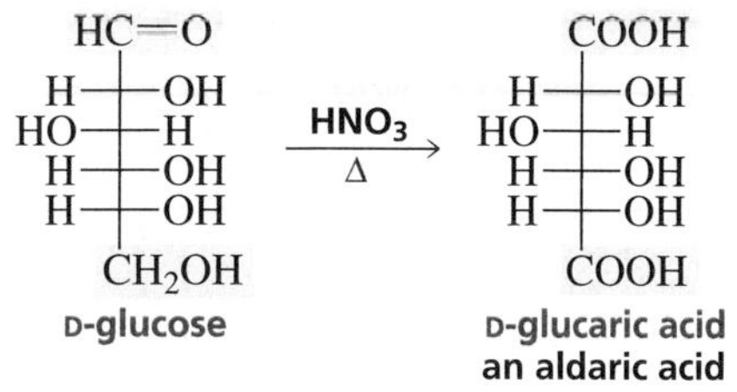

HC=O   H—OH   HO—H   H—OH   H—OH CH_2OH
D-glucose

$\xrightarrow[\Delta]{HNO_3}$

COOH   H—OH   HO—H   H—OH   H—OH   COOH
D-glucaric acid
an aldaric acid

PROBLEM 14♦

a. Name an aldohexose other than D-glucose that is oxidized to D-glucaric acid by nitric acid.

b. What is another name for D-glucaric acid?

c. Name another pair of aldohexoses that are oxidized to identical aldaric acids.

21.7 Monosaccharides Form Crystalline Osazones

The tendency of monosaccharides to form syrups that do not crystallize made the purification and isolation of monosaccharides difficult. However, Emil Fischer found that when phenylhydrazine is added to an aldose or a ketose, a yellow crystalline solid that is insoluble in water is formed. He called this derivative an **osazone** ("ose" for sugar; "azone" for hydrazone). Osazones are easily isolated and purified and were once used extensively to identify monosaccharides.

$$
\begin{array}{c}
\text{HC}{=}\text{O} \\
\text{H}{-}\text{OH} \\
\text{HO}{-}\text{H} \\
\text{H}{-}\text{OH} \\
\text{H}{-}\text{OH} \\
\text{CH}_2\text{OH}
\end{array}
\quad + \quad 3\ \text{NH}_2\text{NH}{-}\bigcirc
\quad\xrightarrow[\text{H}^+]{\text{trace}}\quad
\begin{array}{c}
\text{HC}{=}\text{NNHC}_6\text{H}_5 \\
\text{C}{=}\text{NNHC}_6\text{H}_5 \\
\text{HO}{-}\text{H} \\
\text{H}{-}\text{OH} \\
\text{H}{-}\text{OH} \\
\text{CH}_2\text{OH}
\end{array}
\quad + \quad \bigcirc\text{NH}_2 \quad + \quad \text{NH}_3 \quad + \quad 2\ \text{H}_2\text{O}
$$

D-glucose　　　　　　　　　　　　　　　**the osazone of D-glucose**

Aldehydes and ketones react with one equivalent of phenylhydrazine, forming phenyl-hydrazones (Section 17.8). Aldoses and ketoses, in contrast, react with three equivalents of phenylhydrazine, forming osazones. With aldoses and ketoses, one equivalent functions as an oxidizing agent and is reduced to aniline and ammonia. Two equivalents form imines with carbonyl groups. The reaction stops at this point, regardless of how much phenylhydrazine is present.

Because the configuration of the number-2 carbon is lost during osazone formation, C-2 epimers form identical osazones. For example, D-idose and D-gulose, which are C-2 epimers, both form the same osazone.

C-2 epimers form identical osazones.

$$
\begin{array}{c}
\text{HC}{=}\text{O} \\
\text{HO}{-}\text{H} \\
\text{H}{-}\text{OH} \\
\text{HO}{-}\text{H} \\
\text{H}{-}\text{OH} \\
\text{CH}_2\text{OH}
\end{array}
\xrightarrow[\text{H}^+]{\substack{\textbf{3 NH}_2\textbf{NH}{-}\bigcirc \\ \text{trace}}}
\begin{array}{c}
\text{HC}{=}\text{NNHC}_6\text{H}_5 \\
\text{C}{=}\text{NNHC}_6\text{H}_5 \\
\text{H}{-}\text{OH} \\
\text{HO}{-}\text{H} \\
\text{H}{-}\text{OH} \\
\text{CH}_2\text{OH}
\end{array}
\xleftarrow[\text{H}^+]{\substack{\textbf{3 NH}_2\textbf{NH}{-}\bigcirc \\ \text{trace}}}
\begin{array}{c}
\text{HC}{=}\text{O} \\
\text{H}{-}\text{OH} \\
\text{H}{-}\text{OH} \\
\text{HO}{-}\text{H} \\
\text{H}{-}\text{OH} \\
\text{CH}_2\text{OH}
\end{array}
$$

D-idose　　　　　　　**the osazone of**　　　　　　　**D-gulose**
　　　　　　　　　　D-idose and of D-gulose

The number-1 and number-2 carbons of ketoses react with phenylhydrazine, too. Consequently, D-fructose, D-glucose, and D-mannose all form the same osazone.

$$
\begin{array}{c}
\text{HC}{=}\text{O} \\
\text{H}{-}\text{OH} \\
\text{HO}{-}\text{H} \\
\text{H}{-}\text{OH} \\
\text{H}{-}\text{OH} \\
\text{CH}_2\text{OH}
\end{array}
\xrightarrow[\text{H}^+]{\substack{\textbf{3 H}_2\textbf{NNH}{-}\bigcirc \\ \text{trace}}}
\begin{array}{c}
\text{HC}{=}\text{NNHC}_6\text{H}_5 \\
\text{C}{=}\text{NNHC}_6\text{H}_5 \\
\text{HO}{-}\text{H} \\
\text{H}{-}\text{OH} \\
\text{H}{-}\text{OH} \\
\text{CH}_2\text{OH}
\end{array}
\xleftarrow[\text{H}^+]{\substack{\textbf{3 H}_2\textbf{NNH}{-}\bigcirc \\ \text{trace}}}
\begin{array}{c}
\text{CH}_2\text{OH} \\
\text{C}{=}\text{O} \\
\text{HO}{-}\text{H} \\
\text{H}{-}\text{OH} \\
\text{H}{-}\text{OH} \\
\text{CH}_2\text{OH}
\end{array}
$$

D-glucose　　　　　　　　**the osazone of**　　　　　　　**D-fructose**
　　　　　　　D-glucose, D-mannose, and D-fructose

MEASURING THE BLOOD GLUCOSE LEVELS IN DIABETES

Glucose in the bloodstream reacts with an NH$_2$ group of hemoglobin to form an imine (Section 17.8) that subsequently undergoes an irreversible rearrangement to a more stable α-aminoketone known as hemoglobin-A$_{Ic}$.

$$
\begin{array}{c}
\text{HC}{=}\text{O} \\
\text{H}{-}\text{OH} \\
\text{HO}{-}\text{H} \\
\text{H}{-}\text{OH} \\
\text{H}{-}\text{OH} \\
\text{CH}_2\text{OH}
\end{array}
\xrightarrow[\text{H}^+]{\substack{\textbf{NH}_2\textbf{-hemoglobin} \\ \text{trace}}}
\begin{array}{c}
\text{HC}{=}\text{N}{-}\text{hemoglobin} \\
\text{H}{-}\text{OH} \\
\text{HO}{-}\text{H} \\
\text{H}{-}\text{OH} \\
\text{H}{-}\text{OH} \\
\text{CH}_2\text{OH}
\end{array}
\xrightarrow{\text{rearrangement}}
\begin{array}{c}
\text{CH}_2\text{NH}{-}\text{hemoglobin} \\
\text{C}{=}\text{O} \\
\text{HO}{-}\text{H} \\
\text{H}{-}\text{OH} \\
\text{H}{-}\text{OH} \\
\text{CH}_2\text{OH}
\end{array}
$$

D-glucose　　　　　　　　　　　　　　　　　　　　　　　　　**hemoglobin-A$_{Ic}$**

Insulin is the hormone that regulates the level of glucose—and thus the amount of hemoglobin-A$_{Ic}$—in the blood. Diabetes is a condition in which the body does not produce sufficient insulin, or in which the insulin it produces does not function properly. Because people with untreated diabetes have increased blood glucose levels, they also have a higher concentration of hemoglobin-A$_{Ic}$ than people without diabetes. Thus, measuring the hemoglobin-A$_{Ic}$ level is a way to determine whether the blood glucose level of a diabetic patient is being controlled.

Cataracts, a common complication in diabetes, are caused by the reaction of glucose with the NH$_2$ group of proteins in the lens of the eye. Some think the arterial rigidity common in old age may be attributable to a similar reaction of glucose with the NH$_2$ group of proteins.

PROBLEM 15◆

Name a ketose and another aldose that form the same osazone as

a. D-ribose
b. D-altrose
c. L-idose
d. D-galactose

PROBLEM 16◆

What monosaccharides form the same osazone as D-sorbose?

21.8 Lengthening the Chain: The Kiliani–Fischer Synthesis

BIOGRAPHY

Heinrich Kiliani (1855–1945) *was born in Germany. He received a Ph.D. from the University of Munich, studying under Professor Emil Erlenmeyer. Kiliani became a professor of chemistry at the University of Freiburg.*

The carbon chain of an aldose can be increased by one carbon in a **Kiliani–Fischer synthesis**. In other words, tetroses can be converted into pentoses, and pentoses can be converted into hexoses.

the modified Kiliani–Fischer synthesis

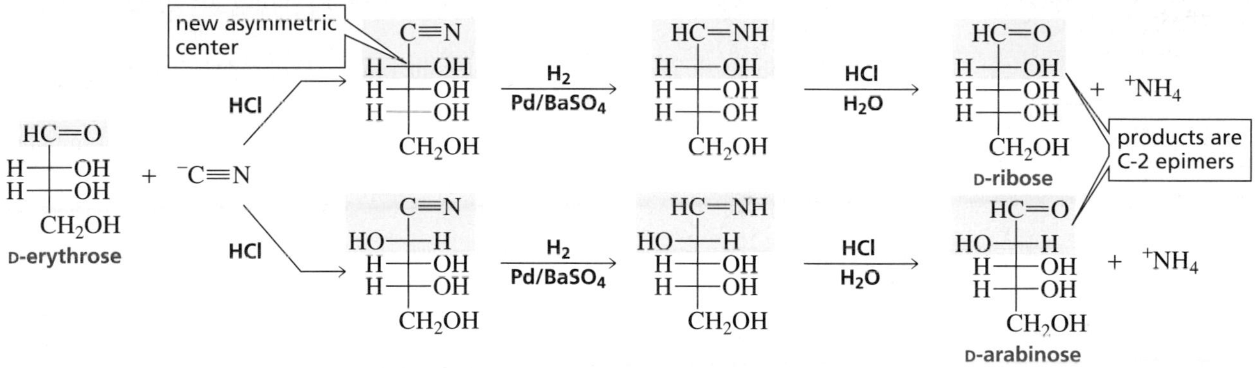

- In the first step of the synthesis, cyanide ion adds to the carbonyl group. This reaction converts the carbonyl carbon in the starting material to an asymmetric center. (The OH bonded to C-2 in the product of this first step can therefore be on the right or on the left in the Fischer projection.) Consequently, two products are formed that differ only in their configuration at C-2. The configurations of the other asymmetric centers do not change, because no bond to any of the asymmetric centers is broken during the course of the reaction.
- The C≡N bond is reduced to an imine, using a partially deactivated palladium catalyst so that the imine is not further reduced to an amine (Section 6.9).
- The two imines are hydrolyzed to aldoses (Section 17.8).

Notice that the synthesis leads to a pair of C-2 epimers because the first step of the reaction converts the carbonyl carbon in the starting material to an asymmetric center. The two epimers are not obtained in equal amounts, however, because the first step of the reaction produces a pair of diastereomers and diastereomers are generally formed in unequal amounts (Section 5.19).

The Kiliani–Fischer synthesis leads to a pair of C-2 epimers.

PROBLEM 17◆

What monosaccharides would be formed in a Kiliani–Fischer synthesis starting with each of these compounds?

a. D-xylose
b. L-threose

21.9 Shortening the Chain: The Wohl Degradation

The **Wohl degradation**—the opposite of the Kiliani–Fischer synthesis—shortens an aldose chain by one carbon: hexoses are converted into pentoses, and pentoses are converted into tetroses.

the Wohl degradation

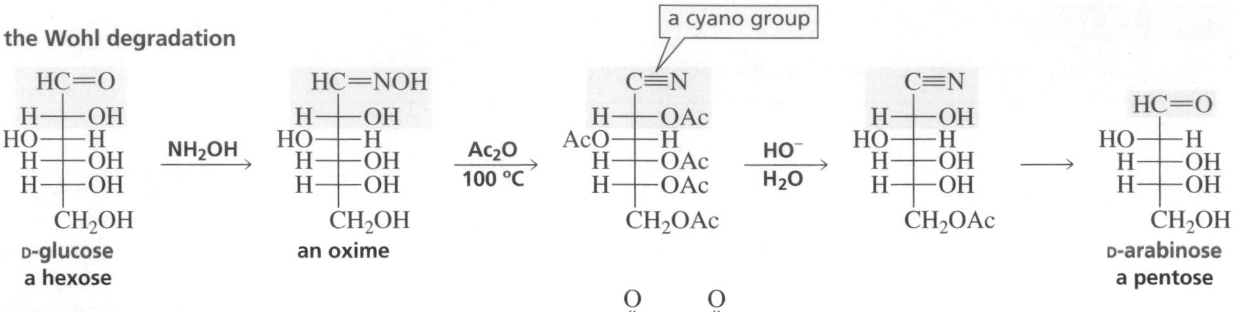

a cyano group

D-glucose
a hexose

an oxime

D-arabinose
a pentose

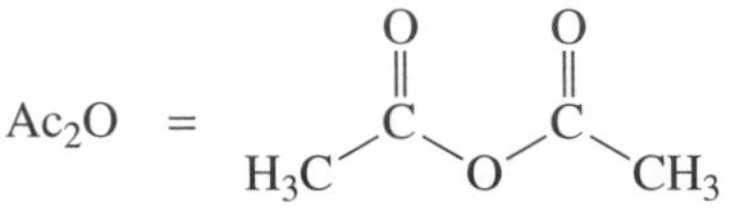

$$Ac_2O = \underset{H_3C}{\overset{O}{\underset{}{\parallel}}}C-O-\underset{CH_3}{\overset{O}{\underset{}{\parallel}}}C$$

- In the first step, the aldehyde reacts with hydroxylamine to form an oxime (Section 17.8).
- Heating with acetic anhydride dehydrates the oxime, forming a nitrile; all the OH groups are converted to esters.
- Treatment with base hydrolyzes the ester groups (Section 16.12) and eliminates the cyano group (Section 17.7).

PROBLEM 18◆

What two monosaccharides can be degraded to

a. D-ribose   **b.** D-arabinose?   **c.** L-ribose?

21.10 Stereochemistry of Glucose: The Fischer Proof

In 1891, Emil Fischer (Section 5.4) determined the stereochemistry of glucose using one of the most brilliant examples of reasoning in the history of chemistry. He chose (+)-glucose for his study because it is the most common monosaccharide found in nature.

Fischer knew that (+)-glucose is an aldohexose, but 16 different structures can be written for an aldohexose. Which of them represents the structure of (+)-glucose? The 16 stereoisomers of an aldohexose are actually eight pairs of enantiomers, so if you know the structures of one set of eight, you automatically know the structures of the other set. Therefore, Fischer needed to consider only one set of eight. He considered the eight stereoisomers that had their C-5 OH group on the right in the Fischer projection (the stereoisomers shown below that we now call the D-sugars).

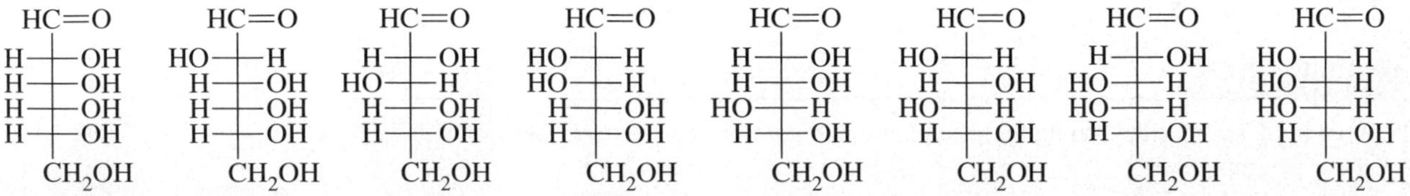

D-allose D-altrose D-glucose D-mannose D-gulose D-idose D-galactose D-talose
1   2   3   4   5   6   7   8

One of these is (+)-glucose, and its mirror image is (−)-glucose. It would not be possible to determine whether (+)-glucose was D-glucose or L-glucose until 1951. Fischer used the following information to determine glucose's stereochemistry—that is, to determine the configuration of each of its asymmetric centers.

1. When the Kiliani–Fischer synthesis is performed on the sugar known as (−)-arabinose, the two sugars known as (+)-glucose, and (+)-mannose are obtained. This means that (+)-glucose and (+)-mannose are C-2 epimers; in other words, they have the same configuration at C-3, C-4, and C-5. Consequently, (+)-glucose and (+)-mannose have to be one of the following pairs: sugars 1 and 2, 3 and 4, 5 and 6, or 7 and 8.

2. (+)-Glucose and (+)-mannose are both oxidized by nitric acid to optically active aldaric acids. The aldaric acids of sugars 1 and 7 would not be optically active because each has a plane of symmetry. (A compound containing a plane of symmetry is achiral, meaning it has a superimposable mirror image; Section 5.12.) Excluding sugars 1 and 7 means that (+)-glucose and (+)-mannose must be sugars 3 and 4 or 5 and 6.

3. Since (+)-glucose and (+)-mannose are the products obtained when the Kiliani–Fischer synthesis is carried out on (−)-arabinose, Fischer knew that if (−)-arabinose has the structure shown below on the left, (+)-glucose and (+)-mannose are sugars 3 and 4. On the other hand, if (−)-arabinose has the structure shown on the right, (+)-glucose and (+)-mannose are sugars 5 and 6:

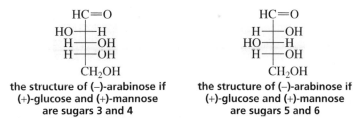

the structure of (–)-arabinose if
(+)-glucose and (+)-mannose
are sugars 3 and 4

the structure of (–)-arabinose if
(+)-glucose and (+)-mannose
are sugars 5 and 6

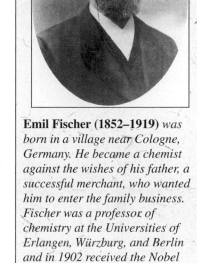
When (−)-arabinose is oxidized with nitric acid, the aldaric acid that results is optically active. This means that the aldaric acid does *not* have a plane of symmetry. Therefore, (−)-arabinose must have the structure shown on the left because the aldaric acid of the sugar on the right has a plane of symmetry. Thus, (+)-glucose and (+)-mannose are represented by sugars 3 and 4.

4. Now the only question remaining was, is (+)-glucose sugar 3 or sugar 4? To answer this question, Fischer had to develop a chemical method for interchanging the aldehyde and hydroxymethyl groups of an aldohexose. When he chemically interchanged those groups on the sugar known as (+)-glucose, he obtained an aldohexose that was different from (+)-glucose. When he chemically interchanged the aldehyde and hydroxymethyl groups of (+)-mannose, he still had (+)-mannose. Therefore, he concluded that (+)-glucose is sugar 3 because interchanging the groups of sugar 3 leads to a different sugar (L-gulose).

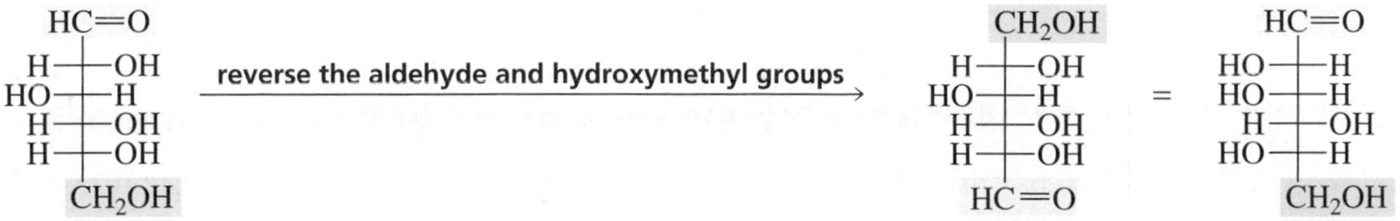

HC=O
H——OH
HO——H
H——OH
H——OH
CH₂OH
D-glucose

reverse the aldehyde and hydroxymethyl groups →

CH₂OH
H——OH
HO——H
H——OH
H——OH
HC=O
L-gulose
drawn upside down

=

HC=O
HO——H
HO——H
H——OH
HO——H
CH₂OH
L-gulose

If (+)-glucose is sugar 3, (+)-mannose must be sugar 4. As predicted, when the aldehyde and hydroxymethyl groups of sugar 4 are interchanged, the same sugar is obtained.

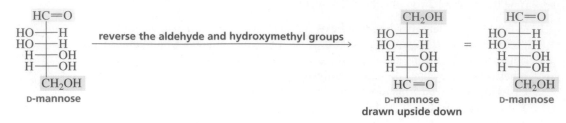

Using similar reasoning, Fischer went on to determine the stereochemistry of 14 of the 16 aldohexoses. He received the Nobel Prize in chemistry in 1902 for this achievement. His original guess that (+)-glucose is a D-sugar was later shown to be correct, so all of his structures are correct (Section 5.15). If he had been wrong and (+)-glucose had been an L-sugar, his contribution to the stereochemistry of aldoses would still have had the same importance, but all his stereochemical assignments would have had to be reversed.

GLUCOSE/DEXTROSE

André Dumas first used the term "glucose" in 1838 to refer to the sweet compound that comes from honey and grapes. Later, Kekulé (Section 7.1) decided that it should be called dextrose because it was dextrorotatory. When Fischer studied the sugar, he called it glucose, and chemists have called it glucose ever since, although dextrose is often found on food labels.

PROBLEM 19 SOLVED

Aldohexoses A and B form the same osazone. Compound A is oxidized by nitric acid to an optically active aldaric acid, and compound B is oxidized to an optically inactive aldaric acid. Wohl degradation of either A or B forms aldopentose C, which is oxidized by nitric acid to an optically active aldaric acid. Wohl degradation of C forms D, which is oxidized by nitric acid to an optically active aldaric acid. Wohl degradation of D forms (+)-glyceraldehyde. Identify A, B, C, and D.

Solution This is the kind of problem that should be solved by working backward. The bottom-most asymmetric center in D must have the OH group on the right because D is degraded to (+)-glyceraldehyde. Since D is oxidized to an optically active aldaric acid, D must be D-threose. The two bottom-most asymmetric centers in C and D have the same configuration because C is degraded to D. Since C is oxidized to an optically active aldaric acid, C must be D-lyxose. Compounds A and B, therefore, must be D-galactose and D-talose. Because A is oxidized to an optically active aldaric acid, it must be D-talose and B must be D-galactose.

PROBLEM 20◆

Identify A, B, C, and D in the preceding problem if D is oxidized to an optically *inactive* aldaric acid, A, B, and C are oxidized to optically active aldaric acids, and if interchanging the aldehyde and alcohol groups of A leads to a different sugar.

21.11 Monosaccharides Form Cyclic Hemiacetals

D-Glucose exists in three different forms: the open-chain form of D-glucose that we have been discussing and two cyclic forms—α-D-glucose and β-D-glucose. We know that the two cyclic forms are different because they have different physical properties.

For example, α-D-glucose melts at 146 °C, whereas β-D-glucose melts at 150 °C, and α-D-glucose has a specific rotation of +112.2, whereas β-D-glucose has a specific rotation of 18.7.

How can D-glucose exist in a cyclic form? In Section 17.10, we saw that an aldehyde reacts with an equivalent of an alcohol to form a hemiacetal. A monosaccharide such as D-glucose has an aldehyde group and several alcohol groups. The alcohol group bonded to C-5 of D-glucose reacts with the aldehyde group. The reaction forms two cyclic (six-membered ring) hemiacetals. At equilibrium, there is almost twice as much β-D-glucose (64%) as α-D-glucose (36%).

To see that C-5 is in the proper position to attack the aldehyde group, we need to convert the Fischer projection of D-glucose to a flat ring structure. To do this, draw the hydroxymethyl group *up* from the back left-hand corner. Groups on the *right* in a Fischer projection are *down* in the cyclic structure, whereas groups on the *left* in a Fischer projection are *up* in the cyclic structure.

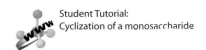
Student Tutorial: Cyclization of a monosaccharide

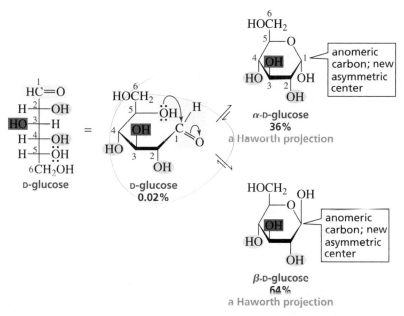

The six-membered ring hemiacetals shown here are drawn as Haworth projections. In a **Haworth projection**, the six-membered ring is represented as flat and is viewed edge-on. The ring oxygen is always placed in the back right-hand corner of the ring, with the anomeric carbon (C-1) on the right-hand side and the hydroxymethyl group drawn *up* from the back left-hand corner (C-5).

Two different hemiacetals are formed because the carbonyl carbon of the open-chain aldehyde becomes a new asymmetric center in the cyclic hemiacetal. If the OH group bonded to the new asymmetric center is down (trans to the hydroxymethyl group at C-5), the hemiacetal is α-D-glucose; if the OH group is up (cis to the hydroxymethyl group at C-5), the hemiacetal is β-D-glucose. The mechanism for cyclic hemiacetal formation is the same as the mechanism for hemiacetal formation between individual aldehyde and alcohol molecules (Section 17.10).

α-D-Glucose and β-D-glucose are anomers. **Anomers** are two sugars that differ in configuration only at the carbon that was the carbonyl carbon in the open-chain form. This carbon is called the **anomeric carbon**. (*Ano* is Greek for "upper"; thus, anomers differ in configuration at the asymmetric center that would be uppermost in a Fischer projection.) The prefixes α- and β- denote the configuration about the anomeric carbon. Because anomers, like epimers, differ in configuration at only one carbon atom, they too are a particular kind of diastereomers. Notice that the anomeric carbon is the only carbon in the molecule that is bonded to two oxygens.

In an aqueous solution, the open-chain form of D-glucose is in equilibrium with the two cyclic hemiacetals. However, because formation of the cyclic hemiacetals

Groups on the *right* in a Fischer projection are *down* in a Haworth projection.

Groups on the *left* in a Fischer projection are *up* in a Haworth projection.

BIOGRAPHY

Sir Walter Norman Haworth (1883–1950) *was born in England, received a Ph.D. in Germany from the University of Göttingen, and later was a professor of chemistry at the Universities of Durham and Birmingham in Great Britain. He was the first person to synthesize vitamin C, which he named ascorbic acid. During World War II, he worked on the atomic bomb project. Haworth received the Nobel Prize in chemistry in 1937 and was knighted in 1947.*

proceeds nearly to completion (unlike formation of acyclic hemiacetals), very little of the glucose is in the open-chain form (about 0.02%). Even so, the sugar still undergoes the reactions discussed in previous sections (oxidation, reduction, and osazone formation) because the reagents react with the small amount of open-chain aldehyde that is present. As the aldehyde reacts, the equilibrium shifts to produce more open-chain aldehyde, which can then undergo reaction. Eventually, all the glucose molecules react by way of the open-chain aldehyde.

When crystals of pure α-D-glucose are dissolved in water, the specific rotation gradually changes from +112.2 to +52.7 When crystals of pure β-D-glucose are dissolved in water, the specific rotation gradually changes from +18.7 to +52.7. This change in rotation occurs because in water, the hemiacetal opens to form the aldehyde and when the aldehyde recyclizes, both α-D-glucose and β-D-glucose can be formed. Eventually, the three forms of glucose reach equilibrium concentrations. The specific rotation of the equilibrium mixture is +52.7. This is why the same specific rotation results whether the crystals originally dissolved in water are all α-D-glucose or all β-D-glucose or any mixture of the two. A slow change in optical rotation to an equilibrium value is called **mutarotation**.

If an aldose can form a five- or a six-membered-ring, it will exist predominantly as a cyclic hemiacetal in solution. Whether a five- or a six-membered-ring is formed depends on their relative stabilities. D-Ribose is an example of an aldose that forms five-membered-ring hemiacetals: α-D-ribose and β-D-ribose. The Haworth projection of a five-membered-ring sugar is viewed edge-on, with the ring oxygen pointing away from the viewer. The anomeric carbon is again on the right-hand side of the molecule, and the hydroxymethyl group is drawn up from the back left-hand corner.

Six-membered-ring sugars are called **pyranoses**, and five-membered-ring sugars are called **furanoses**. These terms come from *pyran* and *furan,* the names of the five- and six-membered-ring cyclic ethers shown in the margin. Consequently, α-D-glucose is also called α-D-glucopyranose. The prefix α- indicates the configuration about the anomeric carbon, and "pyranose" indicates that the sugar exists as a six-membered-ring cyclic hemiacetal.

pyran

furan

Ketoses also exist in solution predominantly in cyclic forms. For example, D-fructose forms a five-membered-ring hemiketal as a result of its C-5 OH group reacting with its ketone carbonyl group. If the new asymmetric center has its OH group trans to the hydroxymethyl group, the compound is α-D-fructose; if its OH group is cis to the hydroxymethyl group, the compound is β-D-fructose. These sugars can also be called α-D-fructofuranose and β-D-fructofuranose. Notice that in fructose the anomeric carbon is C-2, not C-1 as in aldoses. D-Fructose can also form a six-membered ring by using the C-6 OH group. The pyranose form predominates in the monosaccharide, whereas the furanose form predominates when the sugar is part of a disaccharide. (See the structure of sucrose in Section 21.17.)

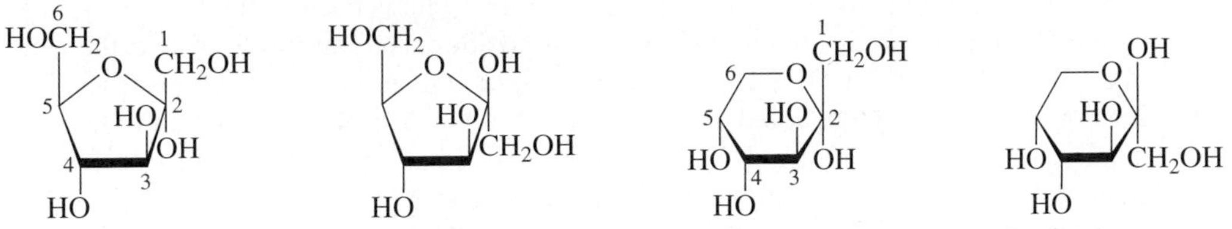

α-D-fructofuranose   β-D-fructofuranose   α-D-fructopyranose   β-D-fructopyranose

Haworth projections are useful because they show clearly whether the OH groups on the ring are cis or trans to each other. Five-membered rings are nearly planar, so furanoses are represented fairly accurately by Haworth projections. Haworth projections, however, are structurally misleading for pyranoses because a six-membered ring is not flat but exists preferentially in a chair conformation (Section 2.12).

PROBLEM 21 *SOLVED*

4-Hydroxy- and 5-hydroxyaldehydes exist primarily as cyclic hemiacetals. Draw the structure of the cyclic hemiacetal formed by each of the following:

a. 4-hydroxybutanal

c. 4-hydroxypentanal

b. 5-hydroxypentanal

d. 4-hydroxyheptanal

Solution to 21a   Draw the reactant with its alcohol and carbonyl groups on the same side of the molecule. Then look to see what size ring will form. Two cyclic products are obtained because the carbonyl carbon of the reactant has been converted into a new asymmetric center in the product.

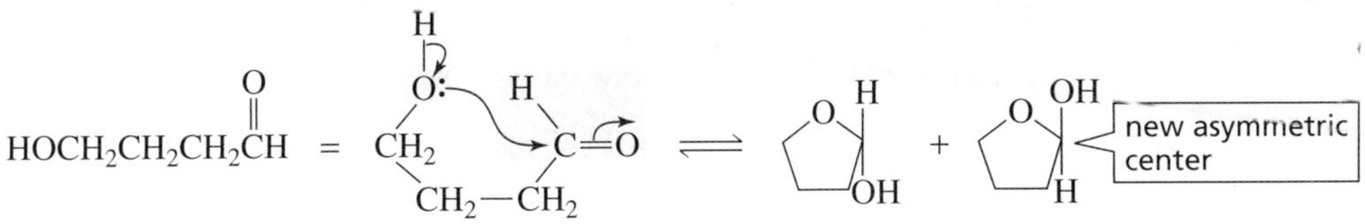

PROBLEM 22

Draw the following sugars using Haworth projections:

a. β-D-galactopyranose   **b.** α-D-tagatopyranose   **c.** α-L-glucopyranose

---

PROBLEM 23

D-Glucose most often exists as a pyranose, but it can also exist as a furanose. Draw the Haworth projection of α-D-glucofuranose.

21.12 **Glucose Is the Most Stable Aldohexose**

Drawing glucose in its chair conformation shows why it is the most common aldohexose in nature. To convert the Haworth projection of D-glucose into a chair conformation, start by drawing the chair so that the backrest is on the left and the footrest is on the right. Then place the ring oxygen at the back right-hand corner and the primary alcohol group in the equatorial position. (It would be helpful here to build a molecular model.) The primary alcohol group is the largest of all the substituents, and large substituents are more stable in the equatorial position because there is less steric strain in that position (Section 2.13). Because the OH group bonded to C-4 is trans to the primary alcohol group (this is easily seen in the Haworth projection), the C-4 OH group is also in the equatorial position. (Recall from Section 2.14 that 1,2-diequatorial substituents are trans to one another.) The C-3 OH group is trans to the C-4 OH group, so the C-3 OH group is also in the equatorial position. As you move around the ring, you will find that all the OH substituents in β-D-glucose are in equatorial positions. The axial positions are all occupied by hydrogens, which require little space and therefore experience little steric strain. No other aldohexose exists in such a strain-free conformation. This means that β-D-glucose is the most stable of all the aldohexoses, so we should not be surprised that it is the most prevalent aldohexose in nature.

The α-position is down in a Haworth projection and axial in a chair conformation.

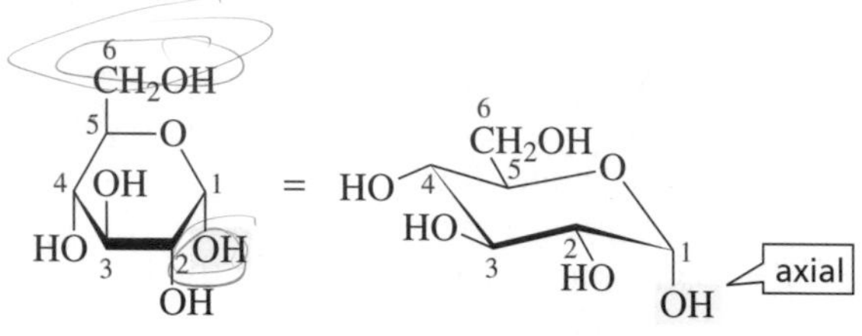

α-D-glucose
chair conformation

The β-position is up in a Haworth projection and equatorial in a chair conformation.

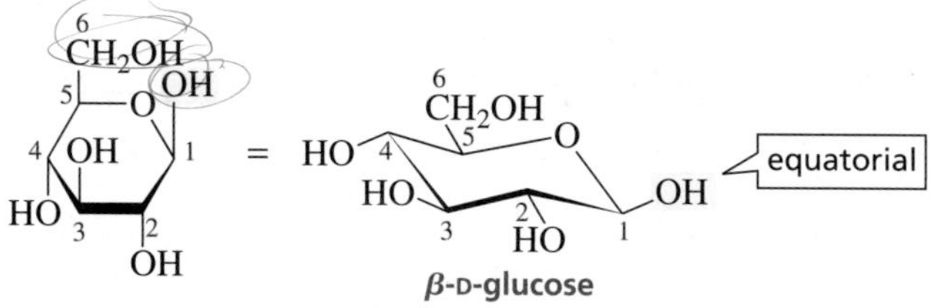

β-D-glucose
chair conformation

The chair conformations show why there is more β-D-glucose than α-D-glucose in an aqueous solution at equilibrium. The OH group bonded to the anomeric carbon is in the equatorial position in β-D-glucose, whereas it is in the axial position in α-D-glucose. Therefore, β-D-glucose is more stable than α-D-glucose, so β-D-glucose predominates at equilibrium in an aqueous solution.

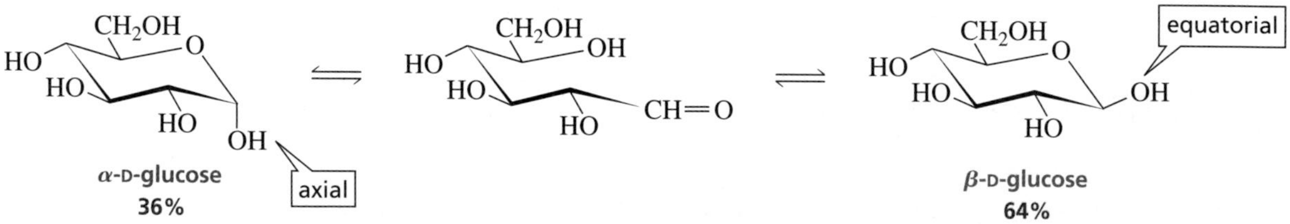

α-D-glucose
36%

β-D-glucose
64%

If you remember that all the OH groups in β-D-glucose are in equatorial positions, you will find it easy to draw the chair conformation of any other pyranose. For example, if you want to draw α-D-galactose, you would put all the OH groups in equatorial positions except the OH groups at C-4 (because galactose is a C-4 epimer of glucose) and at C-1 (because it is the α-anomer). You would put these two OH groups in axial positions.

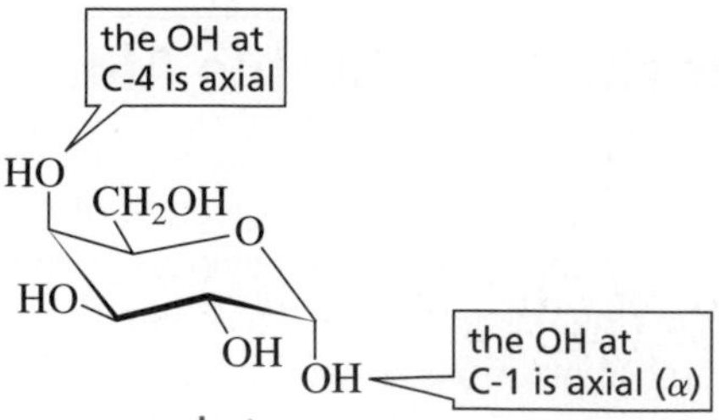

the OH at
C-4 is axial

the OH at
C-1 is axial (α)

α-D-galactose

To draw an L-pyranose, draw the D-pyranose first, and then draw its mirror image. For example, to draw β-L-gulose, first draw β-D-gulose. (Gulose differs from glucose at C-3 and C-4, so the OH groups at these positions are in axial positions.) Then draw the mirror image of β-D-gulose to get β-L-gulose.

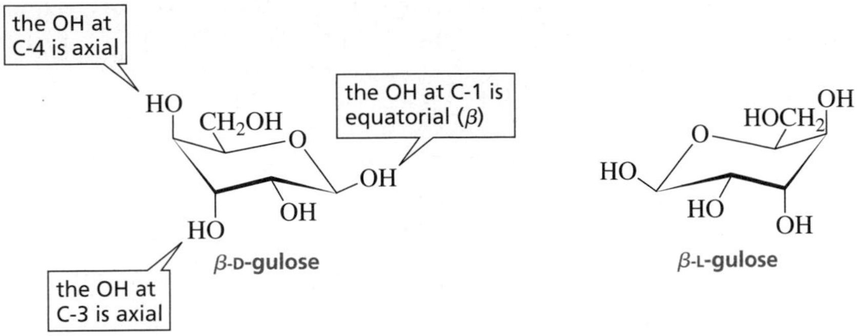

β-D-gulose   β-L-gulose

PROBLEM 24◆ *SOLVED*

Which OH groups are in the axial position in

a. β-D-mannopyranose?   **b.** β-D-idopyranose?   **c.** α-D-allopyranose?

Solution to 24a   All the OH groups in β-D-glucose are in equatorial positions. Because β-D-mannose is a C-2 epimer of β-D-glucose, only the C-2 OH group of β-D-mannose will be in the axial position.

21.13 **Formation of Glycosides**

In the same way that a hemiacetal (or hemiketal) reacts with an alcohol to form an acetal (or ketal) (Section 17.10), the cyclic hemiacetal (or hemiketal) formed by a monosaccharide can react with an alcohol to form an acetal (or ketal). The acetal or ketal of a sugar is called a **glycoside**, and the bond between the anomeric carbon and the alkoxy oxygen is called a **glycosidic bond**. Glycosides are named by replacing the "e" ending of the sugar's name with "ide." Thus, a glycoside of glucose is a glucoside, a glycoside of galactose is a galactoside, and so on. If the pyranose or furanose name is used, the acetal (or ketal) is called a **pyranoside** or a **furanoside**.

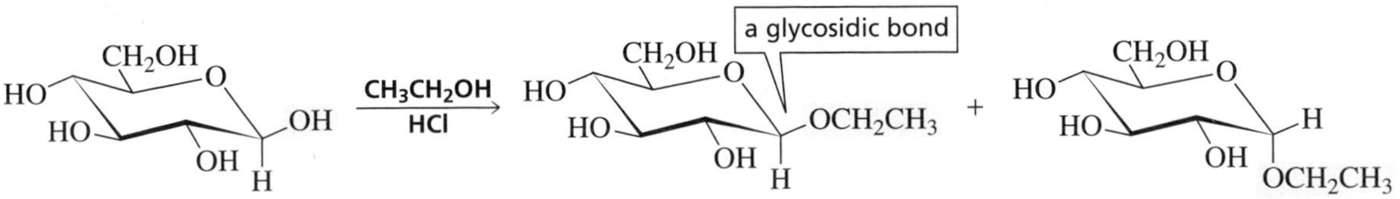

β-D-glucose
β-D-glucopyranose

ethyl β-D-glucoside
ethyl β-D-glucopyranoside

ethyl α-D-glucoside
ethyl α-D-glucopyranoside

Notice that the reaction of a single anomer with an alcohol leads to the formation of both the α- and β-glycosides. The mechanism of the reaction shows why both glycosides are formed.

Mechanism for glycoside formation

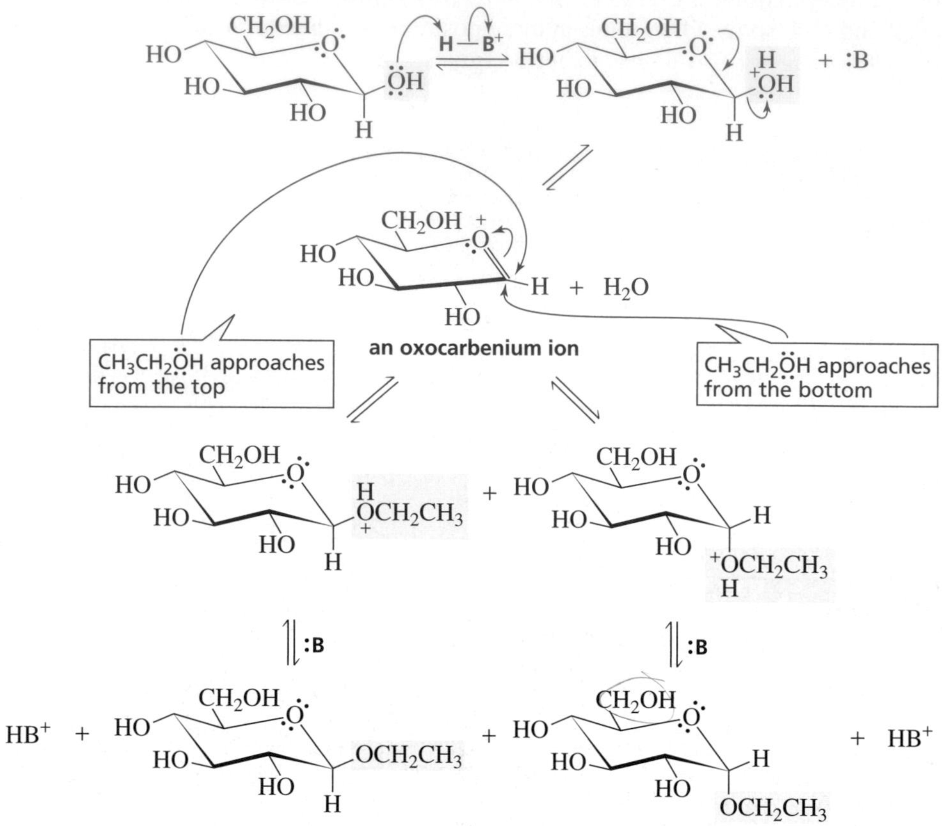

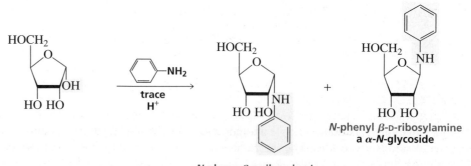

an oxocarbenium ion

CH₃CH₂ÖH approaches from the top

CH₃CH₂ÖH approaches from the bottom

a β-glycoside

an α-glycoside
major product

- The OH group bonded to the anomeric carbon becomes protonated in the acidic solution.

- A lone pair on the ring oxygen helps expel a molecule of water. The anomeric carbon in the resulting oxocarbenium ion is sp^2 hybridized, so that part of the molecule is planar. (An **oxocarbenium ion** has a positive charge that is shared by a carbon and an oxygen.)

- When the alcohol approaches from the top of the plane, the β-glycoside is formed; when it appraoches from the bottom of the plane, the α-glycoside is formed.

Notice that the mechanism is the same as that shown for acetal formation in Section 17.10. Surprisingly, D-glucose forms more of the α-glycoside than the β-glycoside. The reason for this is explained in the next section.

Similar to the reaction of a monosaccharide with an alcohol is the reaction of a monosaccharide with an amine in the presence of a trace amount of acid. The product of the reaction is an *N*-glycoside. An **N-glycoside** has a nitrogen in place of the oxygen at the glycosidic linkage. The subunits of DNA and RNA are β-*N*-glycosides (Section 27.1).

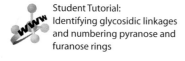

Student Tutorial:
Identifying glycosidic linkages and numbering pyranose and furanose rings

N-pheny β-D-ribosylamine
an α-*N*-glycoside

N-phenyl β-D-ribosylamine
a α-*N*-glycoside

PROBLEM 25◆

Why is only a trace amount of acid used in the formation of an *N*-glycoside?

21.14 **The Anomeric Effect**

We have seen that β-D-glucose is more stable than α-D-glucose because there is more room for a substituent in the equatorial position. The relative amounts of β-D-glucose and α-D-glucose are only $2:1$ (Section 21.10), however, so the preference of the OH group for the equatorial position is not as large as one might expect. Compare this, for example, with the preference of the OH group for the equatorial position in cyclohexanol, where the ratio is $5.4:1$ (Table 2.10 on page 112).

When glucose reacts with an alcohol to form a glucoside, the major product is the α-glucoside. Because acetal formation is reversible, the α-glucoside must be more stable than the β-glucoside. The preference of certain substituents bonded to the anomeric carbon for the axial position is called the **anomeric effect**.

What is responsible for the anomeric effect? The C—Z bond has a σ^* antibonding orbital. If one of the ring oxygen's lone pairs is in an orbital that is parallel to the σ^* antibonding orbital, the molecule can be stabilized by some of the electron density moving from oxygen into the σ^* antibonding orbital. The orbital containing the axial lone pair of the ring oxygen can overlap the σ^* antibonding orbital only if the substituent is axial. If the substituent is equatorial, neither of the orbitals that contain a lone pair is aligned correctly for overlap. As a result of overlap between the lone pair and the σ^* antibonding orbital, the C—Z bond is longer and weaker than normal, and the C—O bond within the ring is shorter and stronger than normal.

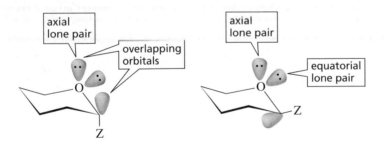

21.15 **Reducing and Nonreducing Sugars**

Because glycosides are acetals (or ketals), they are not in equilibrium with the open-chain aldehyde (or ketone) in neutral or basic aqueous solutions. Without being in equilibrium with a compound that has a carbonyl group, they cannot be oxidized by reagents such as Ag^+ or Br_2. Glycosides, therefore, are nonreducing sugars—they cannot reduce Ag^+ or Br_2.

In contrast, hemiacetals (or hemiketals) are in equilibrium with the open-chain sugars in aqueous solution, so they can reduce Ag^+ or Br_2. In summary, as long as a sugar has an aldehyde, a ketone, a hemiacetal, or a hemiketal group, it is able to reduce an oxidizing agent and therefore is classified as a **reducing sugar**. Without one of these groups, it is a **nonreducing sugar**.

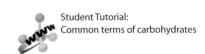
Student Tutorial:
Common terms of carbohydrates

A sugar with an aldehyde, a ketone, a hemiacetal, or a hemiketal group is a reducing sugar. A sugar without one of these groups is a nonreducing sugar.

Name the following compounds, and indicate whether each is a reducing sugar or a nonreducing sugar:

a.

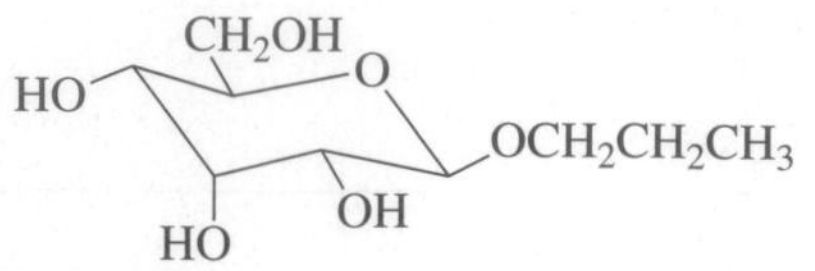

c.

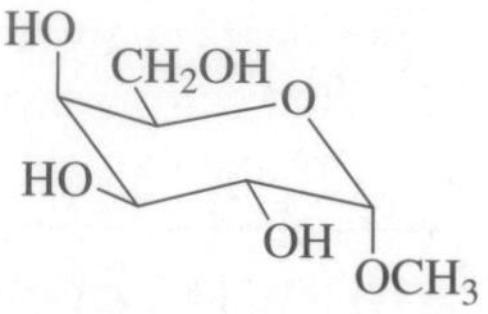

b.

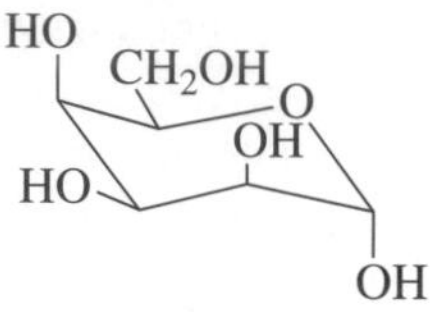

d.

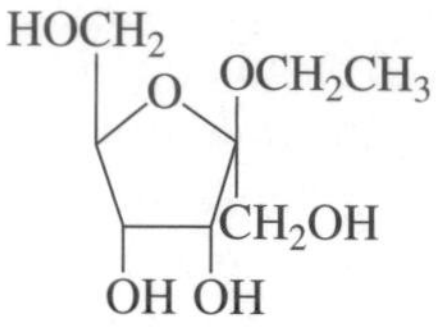

Solution to 26a The only OH group in an axial position in part a is the one at C-3. Therefore, this sugar is the C-3 epimer of D-glucose, which is D-allose. The substituent at the anomeric carbon is in the β-position. Thus, the sugar's name is propyl β-D-alloside or propyl β-D-allopyranoside. Because the sugar is an acetal, it is a nonreducing sugar.

21.16 **Disaccharides**

If the hemiacetal group of a monosaccharide forms an acetal by reacting with an alcohol group of another monosaccharide, the glycoside that is formed is a disaccharide. **Disaccharides** are compounds consisting of two monosaccharide subunits hooked together by an acetal linkage. For example, maltose is a disaccharide obtained from the hydrolysis of starch. It contains two D-glucose subunits connected by an acetal linkage. This particular acetal linkage is called an **α-1,4′-glycosidic linkage** because the linkage is between C-1 of one sugar subunit and C-4 of the other, and the oxygen bonded to the anomeric carbon in the glycosidic linkage is in the α-position. The "prime" superscript indicates that C-4 is not in the same ring as C-1. *Remember that the α-position is axial and the β-position is equatorial when a sugar is shown in a chair conformation.*

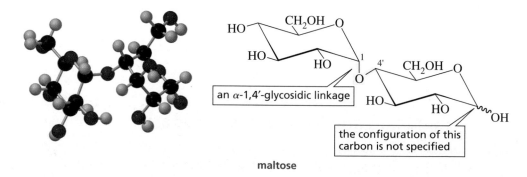

an α-1,4′-glycosidic linkage

the configuration of this carbon is not specified

maltose

Note that the structure of maltose does not specify the configuration of the anomeric carbon that is not an acetal (the anomeric carbon of the subunit on the right marked with a wavy line), because maltose can exist in both the α and β forms. In α-maltose, the OH group bonded to this anomeric carbon is in the axial position. In β-maltose, the OH group is in the equatorial position. Because maltose can exist in both α and β forms, mutarotation occurs when crystals of one form are dissolved in a solvent. Maltose is a reducing sugar because the right-hand subunit is a hemiacetal and therefore is in equilibrium with the open-chain aldehyde that is easily oxidized.

Cellobiose, a disaccharide obtained from the hydrolysis of cellulose, also contains two D-glucose subunits. Cellobiose differs from maltose in that the two glucose subunits are hooked together by a **β-1,4′-glycosidic linkage**. Thus, the only difference in the structures of maltose and cellobiose is the configuration of the glycosidic linkage. Like maltose, cellobiose exists in both α and β forms because the OH group bonded to the anomeric carbon not involved in acetal formation can be in either the axial position (in α-cellobiose) or the equatorial position (in β-cellobiose). Cellobiose is a reducing sugar because the subunit on the right is a hemiacetal.

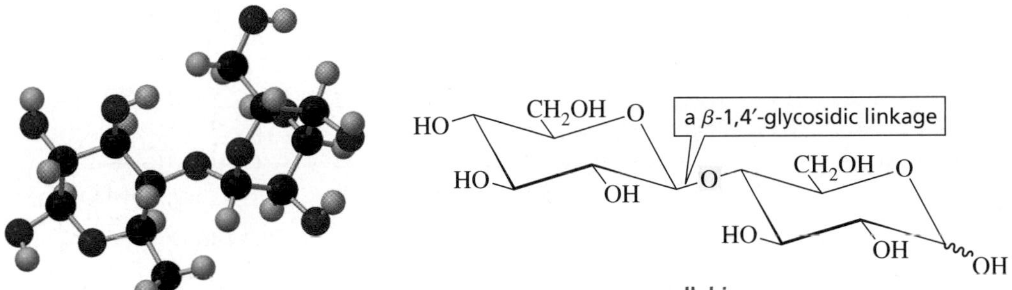

cellobiose

Lactose is a disaccharide found in milk. It constitutes 4.5% of cow's milk by weight and 6.5% of human milk. One of the subunits of lactose is D-galactose, and the other is D-glucose. The D-galactose subunit is an acetal, and the D-glucose subunit is a hemiacetal. The subunits are joined by a β-1,4′-glycosidic linkage. Because one of the subunits is a hemiacetal, lactose is a reducing sugar and undergoes mutarotation.

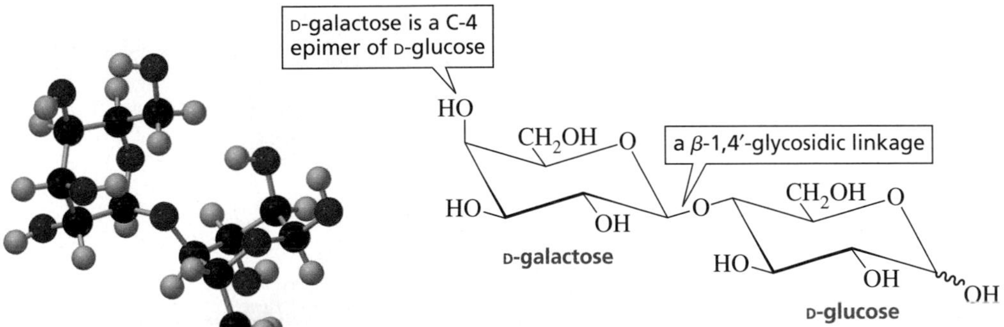

lactose

A simple experiment can prove that the hemiacetal linkage in lactose belongs to the glucose residue rather than to the galactose residue. The disaccharide is treated with excess methyl iodide in the presence of Ag_2O, a reaction that methylates all the OH groups. Silver oxide is used to increase the leaving tendency of the iodide ion in the S_N2 reaction since the OH group is a relatively poor nucleophile. The product is then hydrolyzed under acidic conditions. This treatment causes the two acetal linkages to be hydrolyzed, but all the ether linkages formed by methylating the OH groups are untouched. Identification of the products shows that the galactose residue contained the acetal linkage in the disaccharide, because the C-4 OH of galactose was able to react with methyl iodide. The C-4 OH of glucose, on the other hand, was unable to react with methyl iodide because it was used to form the acetal with galactose.

$$\text{(lactose structure)} \xrightarrow[\textbf{Ag}_2\textbf{O}]{\begin{array}{c}\textbf{CH}_3\textbf{I}\\\text{excess}\end{array}} \text{(permethylated structure)}$$

acetal linkage

hemiacetal linkage

acetal is hydrolyzed

acetal is hydrolyzed

$$\downarrow \text{H}_2\text{O}$$

reacted with CH₃I

did not react with CH₃I

2,3,4,6-tetra-*O*-methylgalactose

2,3,6-tri-*O*-methylglucose

LACTOSE INTOLERANCE

Lactase is an enzyme that specifically breaks the β-1,4′-glycosidic linkage of lactose. Cats and dogs lose their intestinal lactase when they become adults; they are then no longer able to digest lactose. Consequently, when they are fed milk or milk products, the undegraded lactose causes digestive problems such as bloating, abdominal pain, and diarrhea. These problems occur because only monosaccharides can pass into the bloodstream, so lactose has to pass undigested into the large intestine. When humans have stomach flu or other intestinal disturbances, they can temporarily lose their lactase, thereby becoming lactose intolerant. Some humans lose their lactase permanently as they mature. Approximately 10% of the adult Caucasian population of the United States has lost its lactase. Lactose intolerance is much more common in people whose ancestors came from nondairy-producing countries. For example, only 3% of Danes are lactose intolerant, compared with 90% of all Chinese and Japanese and 97% of Thais. This is why you are not likely to find dairy items on Chinese menus.

GALACTOSEMIA

After lactose is degraded into glucose and galactose, the galactose must be converted into glucose before it can be used by cells. Individuals who do not have the enzyme that converts galactose into glucose have the genetic disease known as galactosemia. Without this enzyme, galactose accumulates in the bloodstream. This condition can cause mental retardation and even death in infants. Galactosemia is treated by excluding galactose from the diet.

The most common disaccharide, sucrose, is the substance we know as table sugar. Obtained from sugar beets and sugarcane, sucrose consists of a D-glucose subunit and a D-fructose subunit linked by a glycosidic bond between C-1 of glucose (in the α-position) and C-2 of fructose (in the β-position). About 90 million tons of sucrose are produced commercially throughout the world each year.

Unlike the other disaccharides that have been discussed, sucrose is not a reducing sugar and does not exhibit mutarotation, because its glycosidic bond is between the anomeric carbon of glucose and the anomeric carbon of fructose. Sucrose, therefore, does not have a hemiacetal or hemiketal group, so it is not in equilibrium with the readily oxidized open-chain aldehyde or ketone form in aqueous solution.

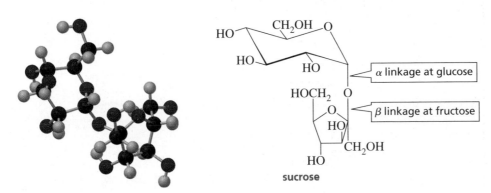

sucrose

Sucrose has a specific rotation of $+66.5$. When it is hydrolyzed, the resulting $1:1$ mixture of glucose and fructose has a specific rotation of -22.0. Because the sign of the rotation changes when sucrose is hydrolyzed, a $1:1$ mixture of glucose and fructose is called *invert sugar*. The enzyme that catalyzes the hydrolysis of sucrose is called *invertase*. Honeybees have invertase, so the honey they produce is a mixture of sucrose, glucose, and fructose. Because fructose is sweeter than sucrose, invert sugar is sweeter than sucrose. Some foods are advertised as containing fructose instead of sucrose, which means that they achieve the same level of sweetness with a lower sugar (lower calorie) content.

21.17 Polysaccharides

Polysaccharides contain as few as 10 or as many as several thousand monosaccharide units joined together by glycosidic linkages. The molecular weight of the individual polysaccharide chains is variable. The most common polysaccharides are starch and cellulose.

Starch is the major component of flour, potatoes, rice, beans, corn, and peas. It is a mixture of two different polysaccharides: amylose (about 20%) and amylopectin (about 80%). Amylose is composed of unbranched chains of D-glucose units joined by α-1,4′-glycosidic linkages.

three subunits of amylose

Amylopectin is a branched polysaccharide. Like amylose, it is composed of chains of D-glucose units joined by α-1,4′-glycosidic linkages. Unlike amylose, however, amylopectin also contains **α-1,6′-glycosidic linkages**. These linkages create the branches in the polysaccharide (Figure 21.1). Amylopectin can contain up to 10^6 glucose units, making it one of the largest molecules found in nature.

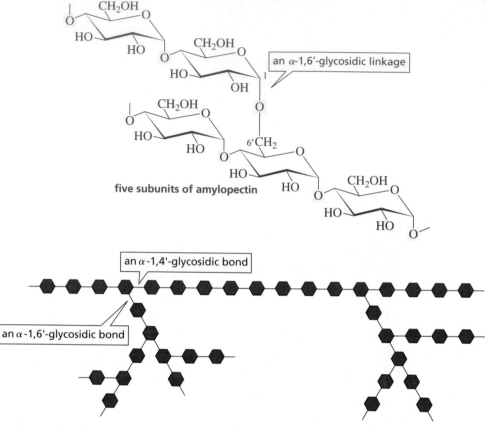

five subunits of amylopectin

an α-1,6′-glycosidic linkage

an α-1,4′-glycosidic bond

an α-1,6′-glycosidic bond

Figure 21.1 ▶
Branching in amylopectin. The hexagons represent glucose units. They are joined by α-1,4′- and α-1,6′-glyclosidic bonds.

Living cells oxidize D-glucose in the first of a series of processes that provide them with energy (Section 25.7). When animals have more D-glucose than they need for energy, they convert the excess D-glucose into a polymer called glycogen. Glycogen has a structure similar to that of amylopectin, but glycogen has more branches (Figure 21.2). The branch points in glycogen occur about every 10 residues, whereas those in amylopectin occur about every 25 residues. The high degree of branching in glycogen has important physiological consequences. When an animal needs energy, many individual glucose units can be simultaneously removed from the ends of many branches. Plants convert excess D-glucose into starch.

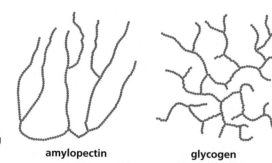

Figure 21.2 ▶
A comparison of the branching in amylopectin and glycogen.

amylopectin   **glycogen**

WHY THE DENTIST IS RIGHT

Bacteria found in the mouth have an enzyme that converts sucrose into a polysaccharide called dextran. Dextran is made up of glucose units joined mainly through α-1,3′- and α-1,6′-glycosidic linkages. About 10% of dental plaque is composed of dextran. This is the chemical basis of why your dentist cautions you not to eat candy.

Cellulose is the structural material of higher plants. Cotton, for example, is composed of about 90% cellulose, and wood is about 50% cellulose. Like amylose, cellulose is composed of unbranched chains of D-glucose units. Unlike amylose, however, the glucose units in cellulose are joined by β-1,4'-glycosidic linkages rather than by α-1,4'-glycosidic linkages.

▲ **Figure 21.3**
The α-1,4'-glycosidic linkages in amylose cause it to form a left-handed helix. Many of its OH groups form hydrogen bonds with water molecules.

Strands of cellulose in a plant fiber.

α-1,4'-Glycosidic linkages are easier to hydrolyze than β-1,4'-glycosidic linkages because of the anomeric effect that weakens the bond to the anomeric carbon (Section 21.14). All mammals have the enzyme (α-glucosidase) that hydrolyzes the α-1,4'-glycosidic linkages that join glucose units in amylose, amylopectin, and glycogen, but they do not have the enzyme (β-glucosidase) that hydrolyzes β-1,4'-glycosidic linkages. (It is also why we need lactase to metabolize lactose.) As a result, mammals *cannot* obtain the glucose they need by eating cellulose. However, bacteria that possess β-glucosidase inhabit the digestive tracts of grazing animals, so cows can eat grass and horses can eat hay to meet their nutritional requirements for glucose. Termites also harbor bacteria that break down the cellulose in the wood they eat.

The different glycosidic linkages in starch and cellulose give these compounds very different physical properties. The α-linkages in starch cause amylose to form a helix that promotes hydrogen bonding of its OH groups to water molecules (Figure 21.3). As a result, starch is soluble in water.

On the other hand, the β-linkages in cellulose promote the formation of intramolecular hydrogen bonds. Consequently, these molecules arrange themselves in linear arrays (Figure 21.4), held together by intermolecular hydrogen bonds between adjacent chains. These large aggregates cause cellulose to be insoluble in water. The strength of these bundles of polymer chains makes cellulose an effective structural material. Processed cellulose is also used for the production of paper and cellophane.

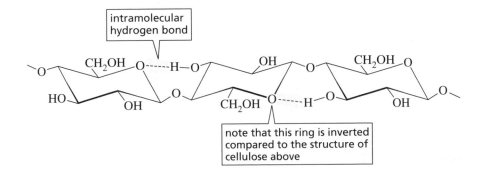

◀ **Figure 21.4**
The β-1,4'-glycosidic linkages in cellulose form intramolecular hydrogen bonds, which cause the molecules to assemble in linear arrays.

Chitin is a polysaccharide that is structurally similar to cellulose. It is the major structural component of the shells of crustaceans (such as lobsters, crabs, and shrimps) and the exoskeletons of insects and other arthropods. Like cellulose, chitin has β-1,4'-glycosidic linkages. It differs from cellulose in that it has an *N*-acetylamino group instead of an OH group at the C-2 position. The β-1,4'-glycosidic linkages give chitin its structural rigidity.

The shell of this bright orange crab from Australia is composed largely of chitin.

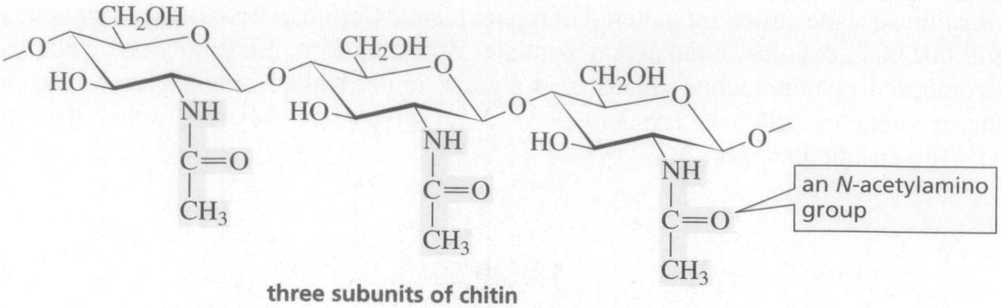

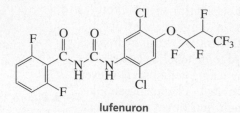

three subunits of chitin

an *N*-acetylamino group

CONTROLLING FLEAS

Several different drugs have been developed to help pet owners control fleas. One of these drugs is lufenuron, the active ingredient in Program.

Lufenuron interferes with the flea's production of chitin. The consequences are fatal for the flea because its exoskeleton is composed primarily of chitin.

lufenuron

PROBLEM 27

What is the main structural difference between

a. amylose and cellulose?

b. amylose and amylopectin?

c. amylopectin and glycogen?

d. cellulose and chitin?

21.18 Some Naturally Occurring Products Derived from Carbohydrates

Deoxy sugars are sugars in which one of the OH groups is replaced by a hydrogen (*deoxy* means "without oxygen"). 2-Deoxyribose, an important example of a deoxy sugar, is missing the oxygen at the C-2 position. D-Ribose is the sugar component of ribonucleic acid (RNA), whereas 2-deoxyribose is the sugar component of deoxyribonucleic acid (DNA); (see Section 27.1).

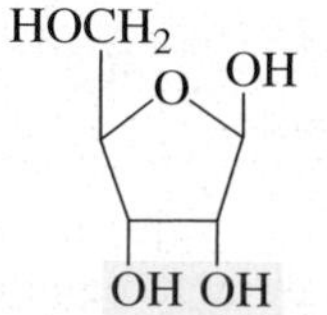

β-D-**ribose** 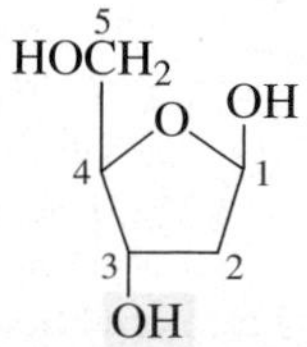   β-D-**2-deoxyribose**

In **amino sugars**, one of the OH groups is replaced by an amino group. *N*-Acetylglucosamine—the subunit of chitin and one of the subunits of certain bacterial cell walls—is an example of an amino sugar (Sections 21.17 and 23.9). Some important antibiotics contain amino sugars. For example, the three subunits of the antibiotic gentamicin are deoxyamino sugars. Notice that the middle subunit is missing the ring oxygen, so it is not really a sugar.

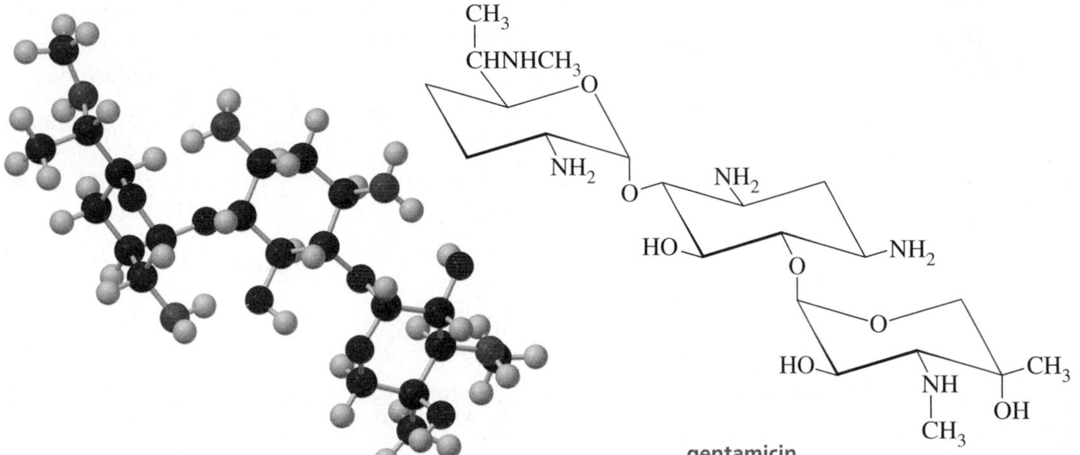

**gentamicin
an antibiotic**

HEPARIN

Heparin is a natural anticoagulant that is released when an injury occurs for the purpose of preventing excessive blood clot formation. This polysaccharide is made up of glucosamine, glucuronic acid, and iduronic acid subunits. The C-6 OH groups of the glucosamine subunits and the C-2 OH groups of the iduronic acid subunits are sulfonated. Some of the amino groups are sulfonated as well and some are acetylated. As a result, heparin is a highly negatively charged molecule. It is found principally in cells that line arterial walls. Heparin is widely used clinically as an anticoagulant.

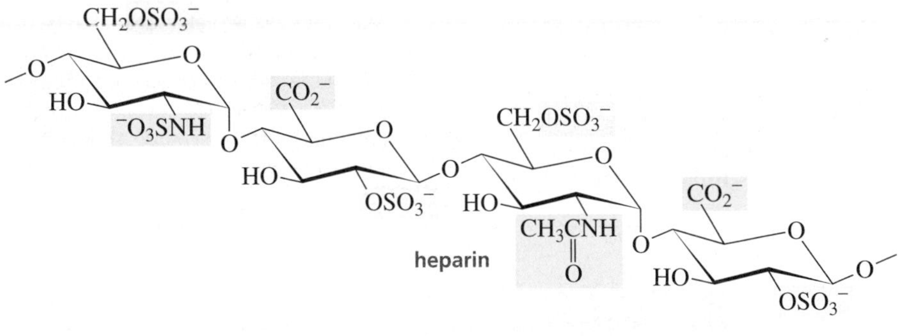

heparin

L-Ascorbic acid (vitamin C) is synthesized from D-glucose in plants and in the livers of most vertebrates. Humans, monkeys, and guinea pigs do not have the enzymes necessary for the biosynthesis of vitamin C, so they must obtain the vitamin in their diets. The biosynthesis of vitamin C involves the enzymatic conversion of D-glucose into L-gulonic acid, reminiscent of the last step in the Fischer proof. L-Gulonic acid is converted into a γ-lactone by the enzyme lactonase, and then an enzyme called oxidase oxidizes the lactone to L-ascorbic acid. The L-designation of ascorbic acid refers to the configuration at C-5, which was C-2 in D-glucose.

the synthesis of L-ascorbic acid

D-glucose → (oxidizing enzyme) → (reducing enzyme) → rotate 180° → L-gulonic acid

↓ lactonase

a γ-lactone

L-ascorbic acid vitamin C ← (oxidase) ←

$pK_a = 4.17$

L-configuration

L-dehydroascorbic acid ← (oxidation) ← L-ascorbic acid vitamin C

An English sailor circa 1829.

Although L-ascorbic acid does not have a carboxylic acid group, it is an acidic compound because the pK_a of the C-3 OH group is 4.17. L-Ascorbic acid is readily oxidized to L-dehydroascorbic acid, which is also physiologically active. If the lactone ring is opened by hydrolysis, all vitamin C activity is lost. Therefore, not much intact vitamin C survives in food that has been thoroughly cooked. Worse, if the food is cooked in water and then drained, the water-soluble vitamin is thrown out with the water!

VITAMIN C

Vitamin C traps radicals formed in aqueous environments (Section 11.10), preventing harmful oxidation reactions the radicals would cause. Not all the physiological functions of vitamin C are known; however, we do know it is required for the synthesis of collagen, which is the structural protein of skin, tendons, connective tissue, and bone. Vitamin C is abundant in citrus fruits and tomatoes, but when it is not present in the diet, lesions appear on the skin, severe bleeding occurs about the gums, in the joints, and under the skin, and any wounds heal slowly. The condition, known as *scurvy*, was the first disease to be treated by adjusting the diet. British sailors who shipped out to sea after the late 1700s were required to eat limes to prevent it (which is how they came to be called "limeys"). *Scorbutus* is Latin for "scurvy"; *ascorbic*, therefore, means "no scurvy."

PROBLEM 28

Explain why the C-3 OH group of vitamin C is more acidic than the C-2 OH group.

21.19 Carbohydrates on Cell Surfaces

Many cells have short oligosaccharide chains on their surface that enable the cells to recognize and interact with other cells and with invading viruses and bacteria. These oligosaccharides are linked to the surface of the cell by the reaction of an OH or an NH_2 group of a cell-membrane protein with the anomeric carbon of a cyclic sugar. Proteins bonded to oligosaccharides are called **glycoproteins**. The percentage of carbohydrate in glycoproteins is variable; some glycoproteins contain as little as 1% carbohydrate by weight, whereas others contain as much as 80%.

glycoproteins

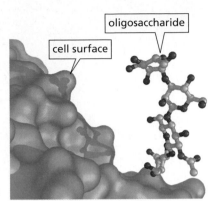

cell surface

oligosaccharide

There are many different types of glycoproteins, including structural proteins such as collagen, proteins found in mucous secretions, immunoglobulins, follicle-stimulating hormone and thyroid-stimulating hormone, interferon (an antiviral protein), and blood plasma proteins. One of the functions of the oligosaccharide chains of glycoproteins is to act as receptor sites on the cell surface, transmitting signals from hormones and other molecules into the cell. The carbohydrates on cell surfaces also serve as points of attachment for other cells, viruses, and toxins.

Carbohydrates on the surfaces of cells provide a way for cells to recognize one another. Such interactions between surface carbohydrates have been found to play a role in activities as diverse as infection, the prevention of infection, fertilization, inflammatory diseases like rheumatoid arthritis and septic shock, and blood clotting. Thus, the goal of the HIV protease inhibitor drugs, for example, is to prevent HIV from recognizing cells by means of their surface oligosaccharides and penetrating them. The fact that several known antibiotics contain amino sugars (Section 21.18) suggests that they function by recognizing target cells. Carbohydrate interactions also are involved in the regulation of cell growth, so changes in membrane glycoproteins are thought to be correlated with malignant transformations.

Differences in blood type (A, B, or O) are actually differences in the sugars bound to the surfaces of red blood cells. Each type of blood is associated with a different carbohydrate structure (Figure 21.5). Type AB blood is a mixture of both type A and type B carbohydrates.

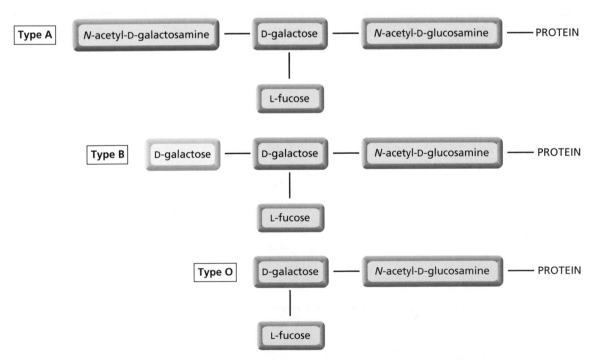

▲ **Figure 21.5**
Blood type is determined by the nature of the sugar on the surfaces of red blood cells. Fucose is 6-deoxygalactose.

Antibodies are proteins that are synthesized by the body in response to foreign substances called *antigens*. Interaction with the antibody causes the antigen to either precipitate it or flag it for destruction by immune system cells. This is why, for example, blood cannot be transferred from one person to another unless the blood types of the donor and acceptor are compatible. Otherwise the donated blood will be considered a foreign substance and provoke an immune response.

Looking at Figure 21.5, we can see why the immune system of type A people recognizes type B blood as foreign and vice versa. The immune system of people with type A, B, or AB blood does not, however, recognize type O blood as foreign, because the carbohydrate in type O blood is also a component of types A, B, and AB. Thus, anyone can accept type O blood, so people with type O blood are called universal donors. People with type AB blood can accept types AB, A, B, and O blood, so they are referred to as universal acceptors.

> **PROBLEM 29◆**
>
> Refer to Figure 21.5 to answer the following questions:
>
> **a.** People with type O blood can donate blood to anyone, but they cannot receive blood from everyone. From whom can they *not* receive blood?
>
> **b.** People with type AB blood can receive blood from anyone, but they cannot give blood to everyone. To whom can they *not* give blood?

21.20 Synthetic Sweeteners

For a molecule to taste sweet, it must bind to a receptor on a taste bud cell of the tongue. The binding of this molecule causes a nerve impulse to pass from the taste bud to the brain, where the molecule is interpreted as being sweet. Sugars differ in their degree of "sweetness." Compared with the sweetness of glucose, which is assigned a relative value of 1.00, the sweetness of sucrose is 1.45, and that of fructose, the sweetest of all sugars, is 1.65.

Developers of synthetic sweeteners must evaluate potential products in terms of several factors—such as toxicity, stability, and cost—in addition to taste. Saccharin (Sweet'N Low), the first synthetic sweetener, was discovered accidentally by Ira Remsen and his student Constantine Fahlberg at Johns Hopkins University in 1878. Fahlberg was studying the oxidation of ortho-substituted toluenes in Remsen's laboratory when he found that one of his newly synthesized compounds had an extremely sweet taste. (As strange as it may seem today, at one time it was common for chemists to taste compounds in order to characterize them.) He called this compound saccharin; it was eventually found to be about 300 times sweeter than glucose. Notice that, in spite of its name, saccharin is *not* a saccharide.

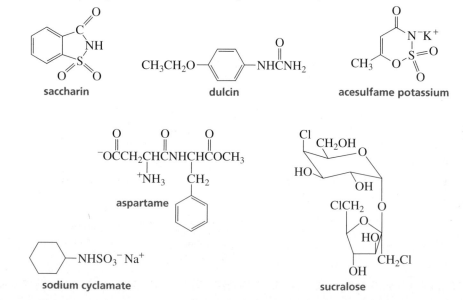

saccharin   dulcin   acesulfame potassium

aspartame   sucralose

sodium cyclamate

Because it has little caloric value, saccharin became an important substitute for sucrose after becoming commercially available in 1885. The chief nutritional problem in the West was—and still is—the overconsumption of sugar and its consequences: obesity, heart disease, and dental decay. Saccharin is also a boon to people with diabetes, who must limit their consumption of sucrose and glucose. Although the toxicity of saccharin had not been studied carefully when the compound was first marketed (our current concern with toxicity is a fairly recent development), extensive studies since then have shown saccharin to be harmless. In 1912, saccharin was temporarily banned in the United States, not because of any concern about its toxicity, but because of a concern that people would miss out on the nutritional benefits of sugar.

THE WONDER OF DISCOVERY

Ira Remsen gave the following account of why he became a scientist.* He was working as a physician and came across the statement "Nitric acid acts upon copper" in a chemistry book. He decided to see what "acts on" meant. He poured nitric acid on a penny sitting on a table. "But what was this wonderful thing which I beheld? The cent had already changed and it was not small change either. A greenish blue liquid foamed and fumed over the cent and over the table. The air in the neighborhood of the performance became dark red. A great colored cloud arose. This was disagreeable and suffocating—how should I stop this? I tried to get rid of the objectionable mess by picking it up and throwing it out of the window, which

I had meanwhile opened. I learned another fact—nitric acid not only acts upon copper but it acts upon my fingers. The pain led to another unpremeditated experiment. I drew my fingers across my trousers and another fact was discovered. Nitric acid also acts upon trousers. Taking everything else into consideration, that was the most impressive experiment, and, relatively, probably the most expensive experiment I ever performed. I tell of it even now with interest. It was a revelation to me. It resulted in a desire on my part to learn even more about that remarkable kind of action. Plainly the only way to learn about it was to see its results, to experiment, to work in a laboratory."

* L. R. Summerlin, C. L. Bordford, and J. B. Ealy, *Chemical Demonstrations,* 2nd ed. (Washington, DC: American Chemical Society, 1988).

Dulcin was the second synthetic sweetener to be discovered (in 1884). Even though it did not have the bitter, metallic aftertaste associated with saccharin, it never achieved much popularity. Dulcin was taken off the market in 1951 in response to concerns about its toxicity.

Sodium cyclamate became a widely used nonnutritive sweetener in the 1950s, but was banned in the United States some 20 years later in response to two studies that appeared to show that large amounts of sodium cyclamate cause liver cancer in mice.

Aspartame (NutraSweet, Equal), about 200 times sweeter than sucrose, was approved by the U.S. Food and Drug Administration (FDA) in 1981. Because aspartame contains a phenylalanine subunit, it should not be used by people with the genetic disease known as phenylketonuria (PKU) (Section 25.9).

Acesulfame potassium (Sweet and Safe, Sunette, Sweet One) was approved in 1988. Also called ascesulfame-K, it too is about 200 times sweeter than glucose. It has less aftertaste than saccharine and is more stable than aspartame at high temperatures.

Sucralose (Splenda), 600 times sweeter than glucose, is the most recently approved (1991) synthetic sweeter. It maintains its sweetness in foods stored for long periods and at temperatures used in baking. Sucralose is made from sucrose by selectively replacing three of sucrose's OH groups with chlorines. During the chlorination, the 4-position of the glucose ring becomes inverted, so sucralose is a galactopyranoside, not a glucopyranoside. The body does not recognize sucralose as a carbohydrate, so instead of being metabolized it is eliminated from the body unchanged.

The fact that these synthetic sweeteners have such different structures shows that the sensation of sweetness is not induced by a single molecular shape.

ACCEPTABLE DAILY INTAKE

The FDA has established an acceptable daily intake (ADI) value for many of the food ingredients it clears for use. The ADI is the amount of the substance a person can consume safely, each day of his or her life.

For example, the ADI for acesulfame-K is 15 mg/kg/day. This means that each day a 132-lb person can consume the amount of acesulfame-K that would be found in two gallons of an artificially sweetened beverage. The ADI for sucralose is also 15 mg/kg/day.

SUMMARY

Bioorganic compounds—organic compounds found in biological systems—obey the same chemical principles that smaller organic molecules do. Much of the structure of bioorganic compounds exists for the purpose of **molecular recognition**.

Carbohydrates are the most abundant class of compounds in the biological world. They are polyhydroxy aldehydes (**aldoses**) and polyhydroxy ketones (**ketoses**) or compounds formed by linking up aldoses and ketoses. The notations D and L describe the configuration of the bottommost asymmetric center of a **monosaccharide** in a Fischer projection; the configurations of the other carbons are implicit in the common name. Most naturally occurring sugars are D-sugars. Naturally occurring ketoses have the ketone group in the 2-position. **Epimers** differ in configuration at only one asymmetric center: D-mannose is the C-2 epimer of D-glucose and D-galactose is the C-4 epimer of D-glucose.

In a basic solution, a monosaccharide is converted to a complex mixture of polyhydroxyaldehydes and polyhydroxyketones. Reduction of an aldose forms one **alditol**; reduction of a ketose forms two alditols. Br_2 oxidizes aldoses but not ketoses; Tollens reagent oxidizes both. Aldoses are oxidized to **aldonic acids** or to **aldaric acids**. Aldoses and ketoses react with three equivalents of phenylhydrazine, forming **osazones**. C-2 epimers form identical osazones. The **Kiliani–Fischer synthesis** increases the carbon chain of an aldose by one carbon; it forms C-2 epimers. The **Wohl degradation** decreases the carbon chain by one carbon. The OH groups of monosaccharides react with methyl iodide/silver oxide to form ethers.

The aldehyde or keto group of a **monosaccharide** reacts with one of its OH groups to form cyclic hemiacetals or hemiketals: glucose forms α-D-glucose and β-D-glucose. The α-position is axial when a sugar is shown in a chair conformation and down when the sugar is shown in a Haworth projection; the β-position is equatorial when a sugar is shown in a chair conformation and up when the sugar is shown in a Haworth projection. At equilibrium, more β-D-glucose is present than α-D-glucose. α-D-Glucose and β-D-glucose are

anomers—they differ in configuration only at the **anomeric carbon**, which is the carbon that was the carbonyl carbon in the open-chain form. Anomers have different physical properties. Six-membered-ring sugars are **pyranoses**; five-membered-ring sugars are **furanoses**. The most abundant monosaccharide in nature is glucose. All the OH groups in β-D-glucose are in equatorial positions. A slow change in optical rotation to an equilibrium value is called **mutarotation**.

The cyclic hemiacetal (or hemiketal) can react with an alcohol to form an acetal (or ketal), called a **glycoside**. If the name "pyranose" or "furanose" is used, the acetal is called a **pyranoside** or a **furanoside**. The bond between the anomeric carbon and the alkoxy oxygen is called a **glycosidic bond**. The preference for the axial position by certain substituents bonded to the anomeric carbon is called the **anomeric effect**. If a sugar has an aldehyde, ketone, hemiacetal, or hemiketal group, it is a reducing sugar.

Disaccharides consist of two monosaccharides hooked together by an acetal linkage. Maltose has an **α-1,4'-glycosidic linkage**; cellobiose has a **β-1,4'-glycosidic linkage**. The most common disaccharide is sucrose; it has a D-glucose subunit and a D-fructose subunit linked by their anomeric carbons.

Polysaccharides contain as few as 10 or as many as several thousand monosaccharides joined together by glycosidic linkages. Starch is composed of amylose and amylopectin. Amylose has unbranched chains of D-glucose units joined by α-1,4'-glycosidic linkages. Amylopectin, too, has chains of D-glucose units joined by α-1,4'-glycosidic linkages, but it also has α-1,6'-glycosidic linkages that create branches. Glycogen is similar to amylopectin but has more branches. Cellulose has unbranched chains of D-glucose units joined by β-1,4'-glycosidic linkages. The α-linkages cause amylose to form a helix; the β-linkages allow the molecules of cellulose to form intramolecular hydrogen bonds.

The surfaces of many cells contain short oligosaccharide chains that allow the cells to interact with each other. These oligosaccharides are linked to the cell surface by protein groups. Proteins bonded to oligosaccharides are called **glycoproteins**.

SUMMARY OF REACTIONS

1. Epimerization (Section 21.5)

$$
\begin{array}{ccc}
\text{HC=O} & & \text{HC=O} \\
\text{H}\!-\!\text{OH} & \xrightleftharpoons[\text{H}_2\text{O}]{\text{HO}^-} & \text{HO}\!-\!\text{H} \\
\text{(CHOH)}_n & & \text{(CHOH)}_n \\
\text{CH}_2\text{OH} & & \text{CH}_2\text{OH}
\end{array}
$$

2. Enediol rearrangement (Section 21.5)

$$
\begin{array}{ccc}
\text{HC}=\text{O} & & \text{HC}-\text{OH} & & \text{CH}_2\text{OH}\\
|& & \parallel & & |\\
\text{CHOH} & \xrightleftharpoons[\text{H}_2\text{O}]{\text{HO}^-} & \text{C}-\text{OH} & \xrightleftharpoons[\text{HO}^-]{\text{H}_2\text{O}} & \text{C}=\text{O}\\
|& & |& & |\\
(\text{CHOH})_n & & (\text{CHOH})_n & & (\text{CHOH})_n\\
|& & |& & |\\
\text{CH}_2\text{OH} & & \text{CH}_2\text{OH} & & \text{CH}_2\text{OH}
\end{array}
$$

3. Reduction (Section 21.6)

$$
\begin{array}{ccc}
\text{HC}=\text{O} & & \text{CH}_2\text{OH}\\
|& \xrightarrow[\text{2. H}_3\text{O}^+]{\text{1. NaBH}_4} & |\\
(\text{CHOH})_n & & (\text{CHOH})_n\\
|& & |\\
\text{CH}_2\text{OH} & & \text{CH}_2\text{OH}
\end{array}
\qquad
\begin{array}{ccc}
\text{CH}_2\text{OH} & & \text{CH}_2\text{OH}\\
|& & |\\
\text{C}=\text{O} & \xrightarrow[\text{2. H}_3\text{O}^+]{\text{1. NaBH}_4} & \text{CHOH}\\
|& & |\\
(\text{CHOH})_n & & (\text{CHOH})_n\\
|& & |\\
\text{CH}_2\text{OH} & & \text{CH}_2\text{OH}
\end{array}
$$

4. Oxidation (Section 21.6)

a.
$$
\begin{array}{ccc}
\text{HC}=\text{O} & & \text{COO}^-\\
|& \xrightarrow[\text{HO}^-]{\text{Ag}^+,\ \text{NH}_3} & |\\
(\text{CHOH})_n & & (\text{CHOH})_n \quad +\quad \text{Ag}\\
|& & |\\
\text{CH}_2\text{OH} & & \text{CH}_2\text{OH}
\end{array}
$$

c.
$$
\begin{array}{ccc}
\text{HC}=\text{O} & & \text{COOH}\\
|& \xrightarrow[\text{H}_2\text{O}]{\text{Br}_2} & |\\
(\text{CHOH})_n & & (\text{CHOH})_n \quad +\quad 2\ \text{Br}^-\\
|& & |\\
\text{CH}_2\text{OH} & & \text{CH}_2\text{OH}
\end{array}
$$

b.
$$
\begin{array}{ccc}
\text{CH}_2\text{OH} & & \text{COO}^-\\
|& & |\\
\text{C}=\text{O} & \xrightarrow[\text{HO}^-]{\text{Ag}^+,\ \text{NH}_3} & \text{CHOH}\\
|& & |\\
(\text{CHOH})_n & & (\text{CHOH})_n \quad +\quad \text{Ag}\\
|& & |\\
\text{CH}_2\text{OH} & & \text{CH}_2\text{OH}
\end{array}
$$

d.
$$
\begin{array}{ccc}
\text{HC}=\text{O} & & \text{COOH}\\
|& \xrightarrow[\Delta]{\text{HNO}_3} & |\\
(\text{CHOH})_n & & (\text{CHOH})_n\\
|& & |\\
\text{CH}_2\text{OH} & & \text{COOH}
\end{array}
$$

5. Osazone formation (Section 21.7)

$$
\begin{array}{c}
\text{HC}=\text{O}\\
|\\
\text{CHOH}\\
|\\
(\text{CHOH})_n\\
|\\
\text{CH}_2\text{OH}
\end{array}
\ +\ 3\ \text{NH}_2\text{NH}{-}\bigcirc
\xrightarrow[\text{H}^+]{\text{trace}}
\begin{array}{c}
\text{HC}=\text{NNHC}_6\text{H}_5\\
|\\
\text{C}=\text{NNHC}_6\text{H}_5\\
|\\
(\text{CHOH})_n\\
|\\
\text{CH}_2\text{OH}
\end{array}
\ +\ \bigcirc\!{-}\text{NH}_2\ +\ \text{NH}_3\ +\ 2\ \text{H}_2\text{O}
$$

6. Chain elongation: the Kiliani–Fischer synthesis (Section 21.8)

$$
\begin{array}{ccc}
\text{HC}=\text{O} & & \text{HC}=\text{O}\\
|& \xrightarrow[\text{3. H}_3\text{O}^+]{\substack{\text{1. NaC}\equiv\text{N/HCl}\\ \text{2. H}_2,\ \text{Pd/BaSO}_4}} & |\\
(\text{CHOH})_n & & (\text{CHOH})_{n+1}\\
|& & |\\
\text{CH}_2\text{OH} & & \text{CH}_2\text{OH}
\end{array}
$$

7. Chain shortening: the Wohl degradation (Section 21.9)

$$
\begin{array}{ccc}
\text{HC}=\text{O} & & \text{HC}=\text{O}\\
|& \xrightarrow[\text{3. HO}^-,\ \text{H}_2\text{O}]{\substack{\text{1. NH}_2\text{OH}\\ \text{2. Ac}_2\text{O, 100 °C}}} & |\\
(\text{CHOH})_n & & (\text{CHOH})_{n-1}\\
|& & |\\
\text{CH}_2\text{OH} & & \text{CH}_2\text{OH}
\end{array}
$$

8. Acetal (and ketal) formation (Section 21.13)

KEY TERMS

aldaric acid (p. 987)
alditol (p. 986)
aldonic acid (p. 987)
aldose (p. 980)
amino sugar (p. 1006)
anomeric carbon (p. 993)
anomeric effect (p. 999)
anomers (p. 993)
bioorganic compound (p. 978)
carbohydrate (p. 979)
complex carbohydrate (p. 980)
deoxy sugar (p. 1006)
disaccharide (p. 980)
epimerization (p. 985)
epimers (p. 983)

furanose (p. 994)
furanoside (p. 997)
glycoprotein (p. 1008)
glycoside (p. 997)
N-glycoside (p. 998)
glycosidic bond (p. 997)
α-1,4′-glycosidic linkage (p. 1000)
α-1,6′-glycosidic linkage (p. 1004)
β-1,4′-glycosidic linkage (p. 1001)
Haworth projection (p. 993)
heptose (p. 980)
hexose (p. 980)
ketose (p. 980)
Kiliani–Fischer synthesis (p. 989)
molecular recognition (p. 978)

monosaccharide (p. 980)
mutarotation (p. 994)
nonreducing sugar (p. 999)
oligosaccharide (p. 980)
osazone (p. 987)
oxocarbenium ion (p. 998)
pentose (p. 980)
polysaccharide (p. 980)
pyranose (p. 994)
pyranoside (p. 997)
reducing sugar (p. 999)
simple carbohydrate (p. 980)
tetrose (p. 980)
triose (p. 980)
Wohl degradation (p. 990)

PROBLEMS

30. Give the product or products that are obtained when D-galactose reacts with each of the following substances:
  a. nitric acid
  b. Tollens reagent
  c. $NaBH_4$, followed by H_3O^+
  d. three equivalents of phenylhydrazine + trace acid
  e. Br_2 in water
  f. ethanol + HCl
  g. 1. hydroxylamine, 2. acetic anhydride 3. HO^-

31. Identify the sugar in each description:
  a. An aldopentose that is not D-arabinose forms D-arabinitol when it is reduced with $NaBH_4$.
  b. A sugar forms the same osazone as D-galactose with phenylhydrazine, but it is not oxidized by an aqueous solution of Br_2.
  c. A sugar that is not D-altrose forms D-altraric acid when it reacts with nitric acid.
  d. A ketose that, when reduced with $NaBH_4$, forms D-altritol and D-allitol.

32. D-Xylose and D-lyxose are formed when D-threose undergoes a Kiliani–Fischer synthesis. D-Xylose is oxidized to an optically inactive aldaric acid, whereas D-lyxose forms an optically active aldaric acid. What are the structures of D-xylose and D-lyxose?

33. Answer the following questions about the eight aldopentoses:
  a. Which are enantiomers?
  b. Which give identical osazones?
  c. Which form an optically active compound when oxidized with nitric acid?

34. The reaction of D-ribose with one equivalent of methanol plus HCl forms four products. Give the structures of the products.

35. Compounds A, B, and C are three different D-aldohexoses. Compounds A and B are reduced to the same optically active alditol, but form different osazones when treated with phenylhydrazine. Compounds B and C form the same osazone but are reduced to different alditols. Give the structures of A, B, and C.

36. Determine the structure of D-galactose, using arguments similar to those used by Fischer to prove the structure of D-glucose.

37. Dr. Isent T. Sweet isolated a monosaccharide and determined that it had a molecular weight of 150. Much to his surprise, he found that it was not optically active. What is the structure of the monosaccharide?

38. The ^1H NMR spectrum of D-glucose in D_2O exhibits two high-frequency (low-field) doublets. What is responsible for these doublets?

39. Treatment with sodium borohydride converts aldose A into an optically inactive alditol. Wohl degradation of A forms B, whose alditol is optically inactive. Wohl degradation of B forms D-glyceraldehyde. Identify A and B.

40. A hexose was obtained after (+)-glyceraldehyde underwent three successive Kiliani–Fischer syntheses. Identify the hexose from the following experimental information:
  a. Oxidation with nitric acid forms an optically active aldaric acid.
  b. A Wohl degredation followed by oxidation with nitric acid forms an optically inactive aldaric acid.
  c. A second Wohl degradation forms erythrose.

41. D-Glucuronic acid is found widely in plants and animals. One of its functions is to detoxify poisonous HO-containing compounds by reacting with them in the liver to form glucuronides. Glucuronides are water soluble and therefore readily excreted. After ingestion of a poison such as turpentine, morphine, or phenol, the glucuronides of these compounds are found in the urine. Give the structure of the glucuronide formed by the reaction of β-D-glucuronic acid and phenol.

β-D-glucuronic acid

42. A D-aldopentose is oxidized by nitric acid to an optically active aldaric acid. A Wohl degradation of the aldopentose leads to a monosaccharide that is oxidized by nitric acid to an optically inactive aldaric acid. Identify the D-aldopentose.

43. Hyaluronic acid, a component of connective tissue, is the fluid that lubricates the joints. It is a polymer of alternating N-acetyl-D-glucosamine and D-glucuronic acid subunits joined by β-1,3'-glycosidic linkages. Draw a short segment of hyaluronic acid.

44. In order to synthesize D-galactose, Professor Amy Losse went to the stockroom to get some D-lyxose to use as a starting material. She found that the labels had fallen off the bottles containing D-lyxose and D-xylose. How could she determine which bottle contains D-lyxose?

45. A hexose was obtained when the residue of a shrub *Sterculia setigeria* undergoes acid hydrolysis. Identify the hexose from the following experimental information:
   a. It undergoes mutarotation.
   b. It does not react with Br_2.
   c. When it reacts with Tollens reagent, D-galactonic acid and D-talonic acid are formed.

46. When D-fructose is dissolved in D_2O and the solution is made basic, the D-fructose recovered from the solution has an average of 1.7 deuterium atoms attached to the C-1 carbon per molecule. Show the mechanism that accounts for the incorporation of these deuterium atoms into D-fructose.

47. How many aldaric acids are obtained from the 16 aldohexoses?

48. Calculate the percentages of α-D-glucose and β-D-glucose present at equilibrium from the specific rotations of α-D-glucose, β-D-glucose, and the equilibrium mixture. Compare your values with those given in Section 21.11. (*Hint:* The specific rotation of the mixture equals the specific rotation of α-D-glucose times the fraction of glucose present in the α form plus the specific rotation of β-D-glucose times the fraction of glucose present in the β-form.)

49. An unknown disaccharide gives a positive Tollens test. A β-1,4'-glycosidase hydrolyzes it to D-galactose and D-mannose. When the disaccharide is treated with methyl iodide and Ag_2O (Section 21.16) and then hydrolyzed with dilute HCl, the products are 2,3,4,6-tetra-O-methylgalactose and 2,3,4-tri-O-methylmannose. Propose a structure for the disaccharide.

50. Predict whether D-altrose exists preferentially as a pyranose or a furanose. (*Hint:* In the most stable arrangement for a five-membered ring, all the adjacent substituents are trans.)

51. Trehalose, $C_{12}H_{22}O_{11}$, is a nonreducing sugar that is only 45% as sweet as sugar but does not attract moisture and so remains free-flowing and dry. When hydrolyzed by aqueous acid or the enzyme maltase, it forms only D-glucose. When it is treated with excess methyl iodide in the presence of Ag_2O and then hydrolyzed with water under acidic conditions, only 2,3,4,6-tetra-O-methyl-D-glucose is formed.
   a. Draw the structure for trehalose.
   b. What is the function of Ag_2O?

52. Propose a mechanism for the rearrangement that converts an α-hydroxyimine into an α-aminoketone in the presence of a trace amount of acid (Section 21.7).

53. A disaccharide forms a silver mirror with Tollens reagent and is hydrolyzed by a β-glycosidase. When the disaccharide is treated with excess methyl iodide in the presence of Ag_2O and then hydrolyzed with water under acidic conditions, 2,3,4-tri-*O*-methylmannose and 2,3,4,6-tetra-*O*-methylgalactose are formed. Draw the structure of the disaccharide.

54. All the glucose units in dextran have six-membered rings. When a sample of dextran is treated with methyl iodide and Ag_2O and the product is hydrolyzed under acidic conditions, the final products are 2,3,4,6-tetra-*O*-methyl-D-glucose, 2,4,6-tri-*O*-methyl-D-glucose, 2,3,4-tri-*O*-methyl-D-glucose, and 2,4-di-*O*-methyl-D-glucose. Draw a short segment of dextran.

55. When a pyranose is in the chair conformation in which the CH_2OH group and the C-1 OH group are both in axial positions, the two groups can react to form an acetal. This is called the anhydro form of the sugar (it has "lost water"). The anhydro form of D-idose is shown here. In an aqueous solution at 100 °C, about 80% of D-idose exists in the anhydro form. Under the same conditions, only about 0.1% of D-glucose exists in the anhydro form. Explain.

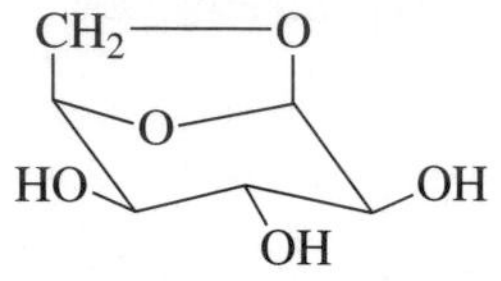

anhydro form of D-idose

56. Devise a method to convert D-glucose into D-allose.

CHAPTER 22

Amino Acids, Peptides, and Proteins

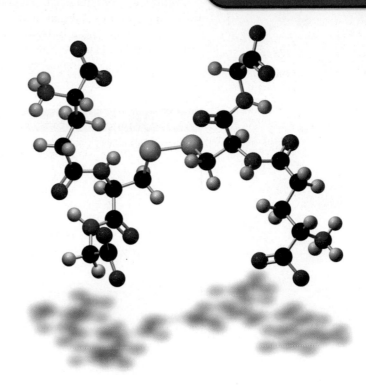

oxidized glutathione

BUILDING ON FUNDAMENTALS

SECTION 22.3	The acidic form of an acid predominates if the pH of the solution is less than the pK_a value of the compound, and the basic form predominates if the pH of the solution is greater than the pK_a value of the compound (1.24).
SECTION 22.3	When pH = pK_a, half of an ionizable group will be in its acidic form and half will be in its basic form (1.24).
SECTION 22.5	Hydrates dehydrate to ketones (or aldehydes) (17.9).
SECTION 22.5	A ketone reacts with a primary amine to form an imine (17.8).
SECTION 22.6	Another way to separate enantiomers is described (5.16).
SECTION 22.7	Oxidation decreases the number of C—H, S—H, or N—H bonds; reduction increases the number of C—H, S—H, or N—H bonds (19.0).
SECTION 22.12	A thiol is a good nucleophile (10.11); Br⁻ is a good leaving group (8.3).
SECTION 22.12	Amides are hydrolyzed in an acidic solution (16.17).

The three kinds of polymers prevalent in nature are polysaccharides, proteins, and nucleic acids. You have already learned about polysaccharides, which are naturally occurring polymers of monosaccharide subunits (Section 21.17), and nucleic acids are covered in Chapter 27. We will now look at proteins and the structurally similar, but shorter, peptides.

Peptides and **proteins** are polymers of amino acids. The amino acids are linked together by amide bonds. An **amino acid** is a carboxylic acid with an ammonium group on the α-carbon. The repeating units are called **amino acid residues**.

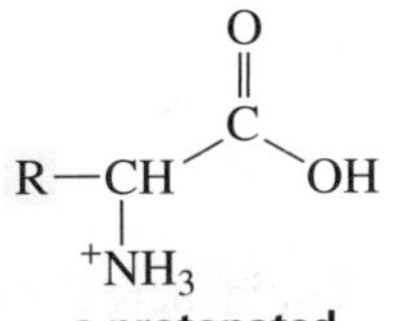

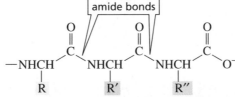

a protonated
α-aminocarboxylic acid
an amino acid

amino acids are linked together by amide bonds

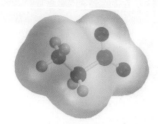

glycine

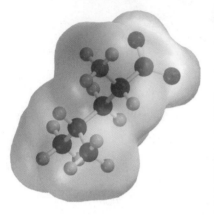

leucine

aspartate

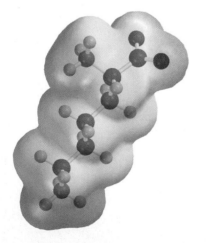

lysine

Amino acid polymers can be composed of any number of amino acid residues. A **dipeptide** contains 2 amino acid residues, a **tripeptide** contains 3, an **oligopeptide** contains 3 to 10, and a **polypeptide** contains many. Proteins are naturally occurring polypeptides made up of 40 to 4000 amino acid residues. Proteins and peptides serve many functions in biological systems (Table 22.1).

Table 22.1	Examples of the Many Functions of Proteins in Biological Systems
Structural proteins	These proteins impart strength to biological structures or protect organisms from their environment. For example, collagen is the major component of bones, muscles, and tendons; keratin is the major component of hair, hooves, feathers, fur, and the outer layer of skin.
Protective proteins	Snake venoms and plant toxins protect their owners from predators. Blood-clotting proteins protect the vascular system when it is injured. Antibodies and peptide antibiotics protect us from disease.
Enzymes	Enzymes are proteins that catalyze the reactions that occur in living systems.
Hormones	Some of the hormones, such as insulin, that regulate the reactions that occur in living systems are proteins.
Proteins with physiological functions	These proteins are responsible for physiological functions such as the transport and storage of oxygen in the body, the storage of oxygen in the muscles, and the contraction of muscles.

Proteins can be divided roughly into two classes. **Fibrous proteins** contain long chains of polypeptides arranged in bundles; these proteins are insoluble in water. All structural proteins are fibrous proteins. **Globular proteins** tend to have roughly spherical shapes and are soluble in water. Essentially all enzymes are globular proteins.

22.1 Classification and Nomenclature of Amino Acids

The structures of the 20 most common naturally occurring amino acids and the frequency with which each occurs in proteins are shown in Table 22.2. Other amino acids occur in nature, but only infrequently. Note that the amino acids differ only in the substituent (R) attached to the α-carbon. The wide variation in these substituents (called side chains) is what gives proteins their great structural diversity and, as a consequence, their great functional diversity. Notice too that all amino acids except proline contain a primary amino group. Proline contains a secondary amino group incorporated into a five-membered ring.

The amino acids are almost always called by their common names. Often, the name tells you something about the amino acid. For example, glycine got its name as a result of its sweet taste (*glykos* is Greek for "sweet"), and valine, like valeric acid, has five carbon atoms. Asparagine was first found in asparagus, and tyrosine was isolated from cheese (*tyros* is Greek for "cheese").

Dividing the amino acids into classes, as in Table 22.2, makes them easier to learn. The aliphatic side chain amino acids include glycine, the amino acid in which R=H, and four amino acids with alkyl side chains. Alanine is the amino acid with a methyl side chain, and valine has an isopropyl side chain. Notice that, in spite of its name, isoleucine does not have an isobutyl substituent; it has a *sec*-butyl substituent. Leucine is the amino acid that has an isobutyl substituent. Each of the amino acids has both a three-letter abbreviation (the first three letters of the name, in most cases) and a single-letter abbreviation.

Table 22.2 The Most Common Naturally Occurring Amino Acids. *The amino acids are shown in the form that predominates at physiological pH (7.3).*

	Formula	Name	Abbreviations		Average relative abundance in proteins
Aliphatic side chain amino acids	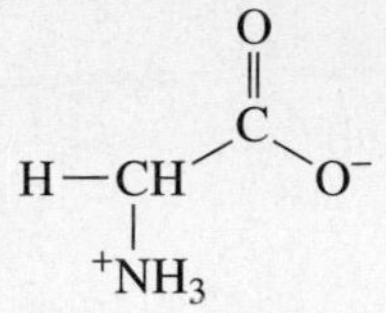 H—CH($^+$NH$_3$)—C(=O)O$^-$	Glycine	Gly	G	7.5%
	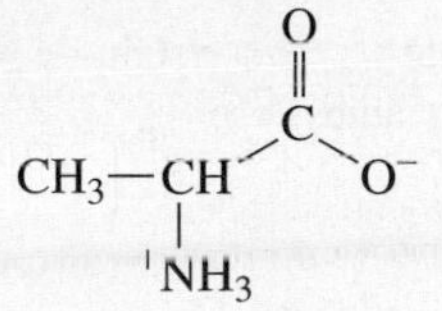 CH$_3$—CH(NH$_3$)—C(=O)O$^-$	Alanine	Ala	A	9.0%
	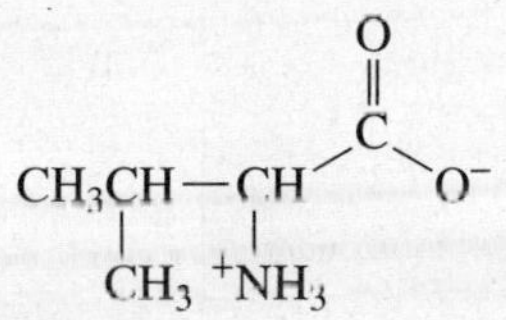 CH$_3$CH(CH$_3$)—CH($^+$NH$_3$)—C(=O)O$^-$	Valine*	Val	V	6.9%
	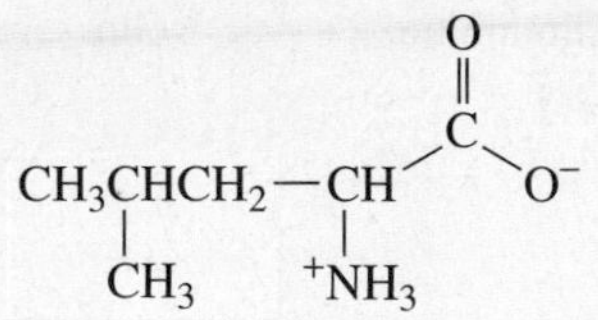 CH$_3$CHCH$_2$(CH$_3$)—CH($^+$NH$_3$)—C(=O)O$^-$	Leucine*	Leu	L	7.5%
	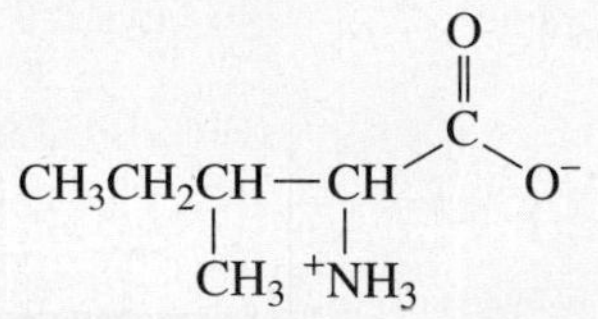 CH$_3$CH$_2$CH(CH$_3$)—CH($^+$NH$_3$)—C(=O)O$^-$	Isoleucine*	Ile	I	4.6%
Hydroxy-containing amino acids	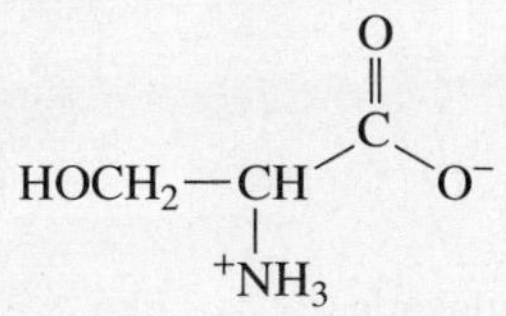 HOCH$_2$—CH($^+$NH$_3$)—C(=O)O$^-$	Serine	Ser	S	7.1%
	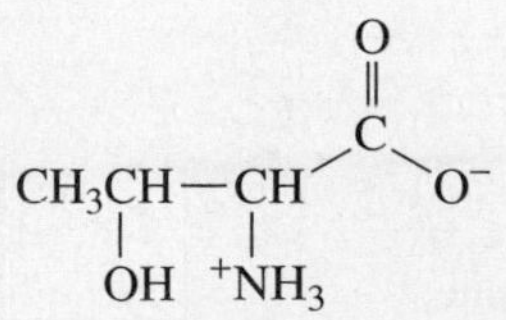 CH$_3$CH(OH)—CH($^+$NH$_3$)—C(=O)O$^-$	Threonine*	Thr	T	6.0%

(Continued)

Table 22.2 Continued

	Formula	Name	Abbreviations		Average relative abundance in proteins
Sulfur-containing amino acids	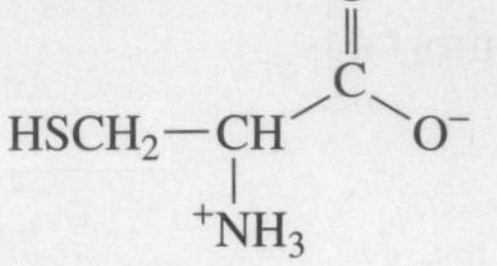 $HSCH_2-CH$, $\overset{+}{N}H_3$, $C=O$, O^-	Cysteine	Cys	C	2.8%
	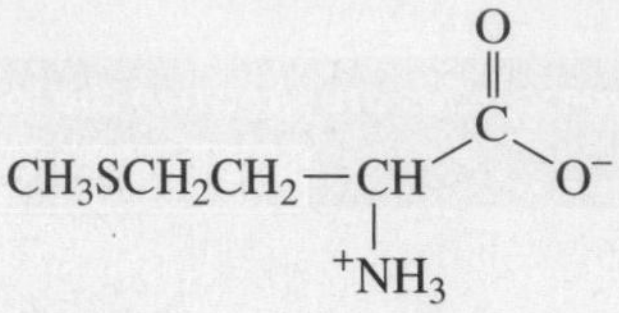 $CH_3SCH_2CH_2-CH$, $\overset{+}{N}H_3$, $C=O$, O^-	Methionine*	Met	M	1.7%
Acidic amino acids	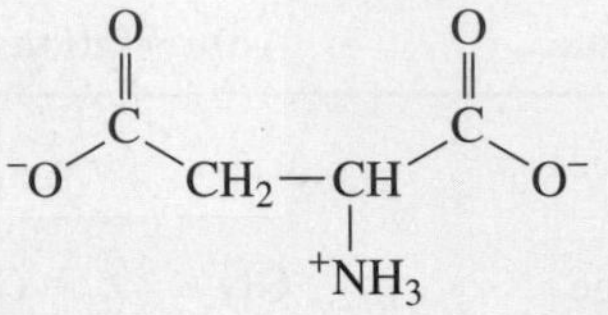 $^-O-C(=O)-CH_2-CH$, $\overset{+}{N}H_3$, $C=O$, O^-	Aspartate (aspartic acid)	Asp	D	5.5%
	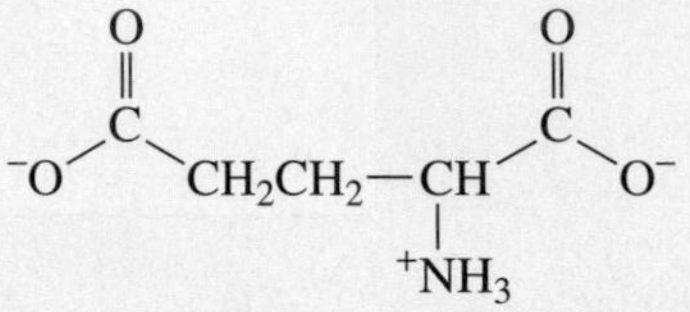 $^-O-C(=O)-CH_2CH_2-CH$, $\overset{+}{N}H_3$, $C=O$, O^-	Glutamate (glutamic acid)	Glu	E	6.2%
Amides of acidic amino acids	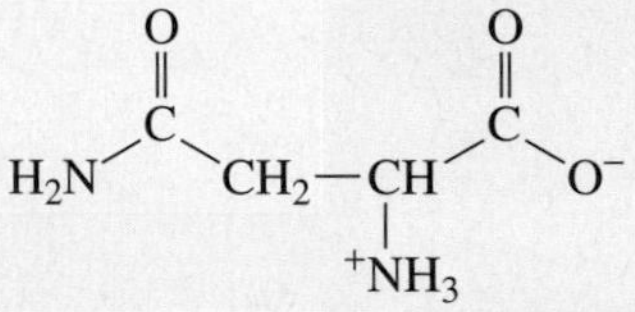 $H_2N-C(=O)-CH_2-CH$, $\overset{+}{N}H_3$, $C=O$, O^-	Asparagine	Asn	N	4.4%
	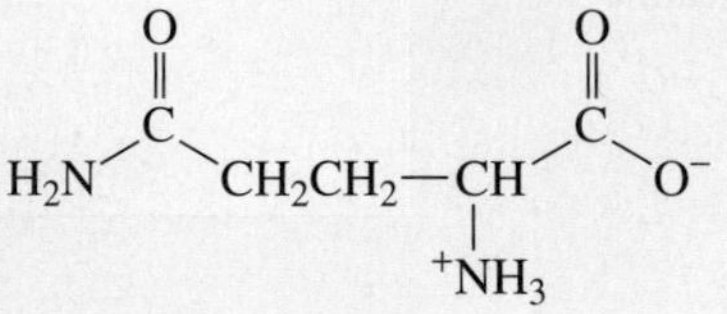 $H_2N-C(=O)-CH_2CH_2-CH$, $\overset{+}{N}H_3$, $C=O$, O^-	Glutamine	Gln	Q	3.9%
Basic amino acids	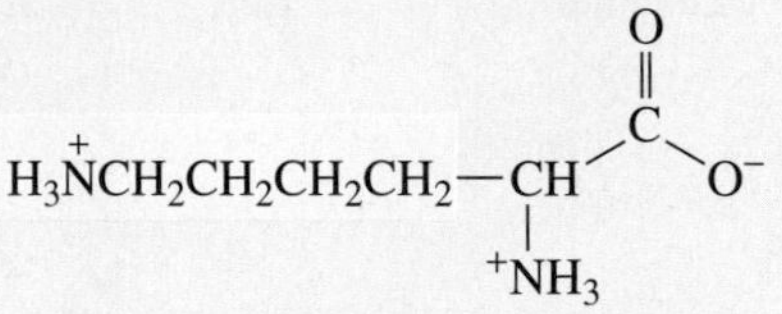 $\overset{+}{H_3N}CH_2CH_2CH_2CH_2-CH$, $\overset{+}{N}H_3$, $C=O$, O^-	Lysine*	Lys	K	7.0%
	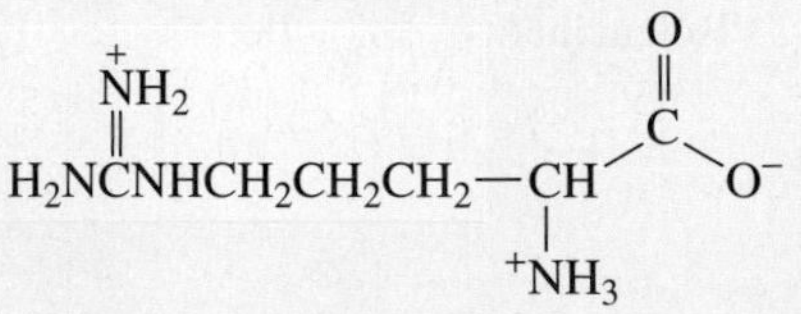 $H_2NCNHCH_2CH_2CH_2-CH$, $\overset{+}{N}H_2$, $\overset{+}{N}H_3$, $C=O$, O^-	Arginine*	Arg	R	4.7%
Benzene-containing amino acids	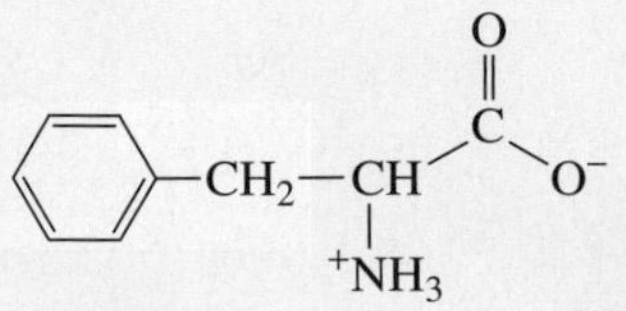 $C_6H_5-CH_2-CH$, $\overset{+}{N}H_3$, $C=O$, O^-	Phenylalanine*	Phe	F	3.5%
	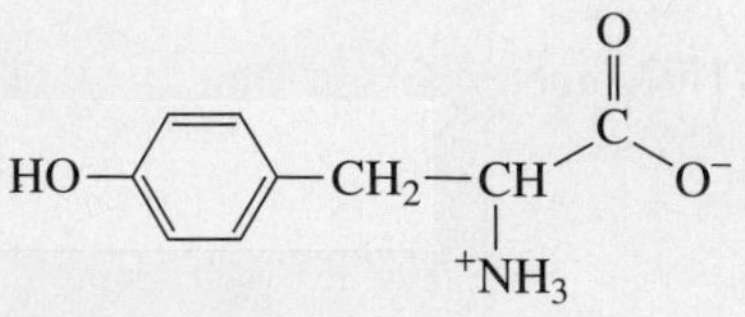 $HO-C_6H_4-CH_2-CH$, $\overset{+}{N}H_3$, $C=O$, O^-	Tyrosine	Tyr	Y	3.5%

	Formula	Name	Abbreviations		Average relative abundance in proteins
Heterocylic amino acids	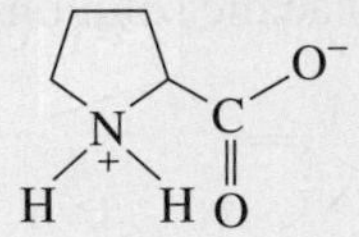	Proline	Pro	P	4.6%
	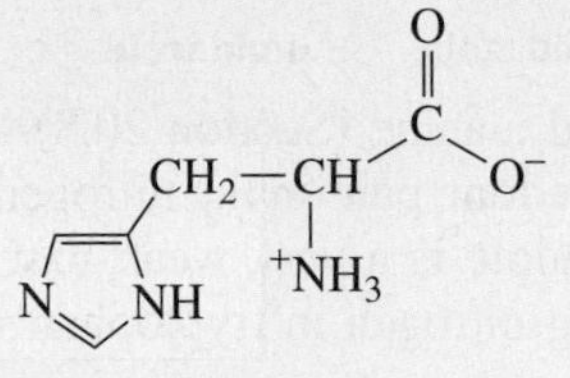	Histidine*	His	H	2.1%
	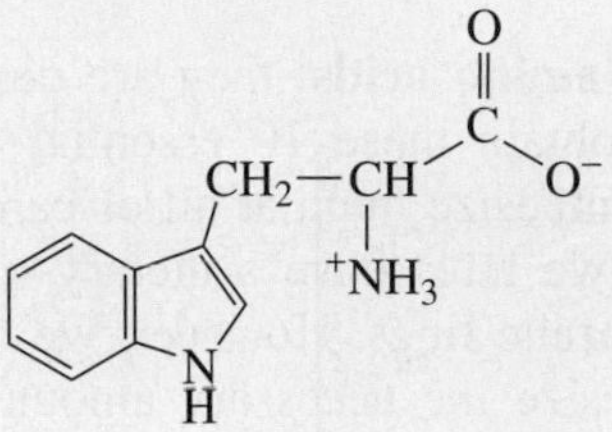	Tryptophan*	Trp	W	1.1%

Table 22.2 Continued

* Essential amino acids

Two amino acid side chains—serine and threonine—contain alcohol groups. Serine is an HO-substituted alanine and threonine has a branched ethanol substituent. There are also two sulfur-containing amino acids: cysteine is an HS-substituted alanine and methionine has a 2-(methylthio)ethyl substituent.

There are two acidic amino acids (amino acids with two carboxylic acid groups): aspartate and glutamate. Aspartate is a carboxy-substituted alanine and glutamate has one more methylene group than aspartate. (If their carboxyl groups are protonated, they are called aspartic acid and glutamic acid, respectively.) Two amino acids— asparagine and glutamine—are amides of the acidic amino acids; asparagine is the amide of aspartate and glutamine is the amide of glutamate. Notice that the obvious one-letter abbreviations cannot be used for these four amino acids because A and G are used for alanine and glycine. Aspartate and glutamate are abbreviated D and E; asparagine and glutamine are abbreviated N and Q.

There are two basic amino acids (amino acids with two basic nitrogen-containing groups): lysine and arginine. Lysine has an ϵ-amino group and arginine has a δ-guanidino group. At physiological pH, these groups are protonated. The ϵ and δ can remind you how many methylene groups each amino acid has.

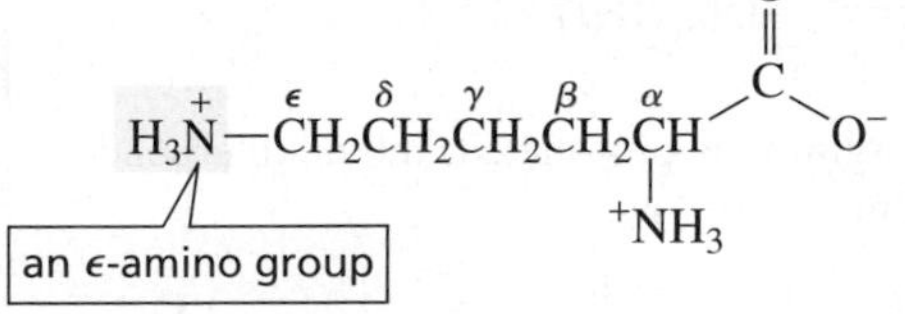

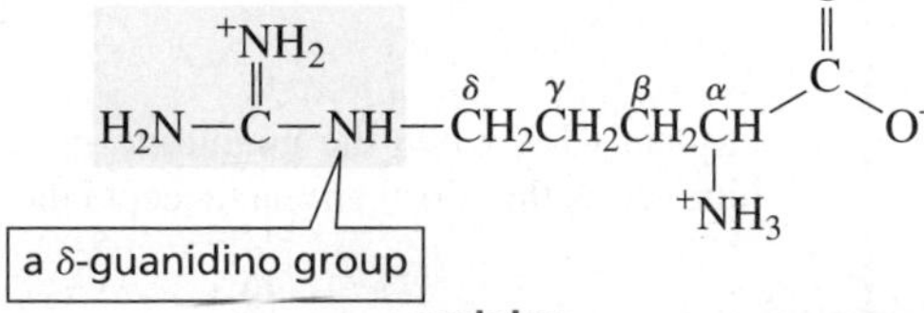

Two amino acids—phenylalanine and tyrosine—contain benzene rings. As its name indicates, phenylalanine is phenyl-substituted alanine. Tyrosine is phenylalanine with a *para*-hydroxy substituent.

Proline, histidine, and tryptophan are heterocyclic amino acids. We have noted that proline, with its nitrogen incorporated into a five-membered ring, is the only amino acid that contains a secondary amino group. Histidine is an imidazole-substituted alanine. Imidazole is an aromatic compound because it is cyclic and planar and has three pairs of delocalized π electrons (Section 22.10). The pK_a of a protonated imidazole ring is 6.0, so the ring will be protonated in acidic solutions and nonprotonated in basic solutions (see page 1026).

$$HN\overset{+}{\diagup}:NH \quad \rightleftharpoons \quad :N\diagup:NH \;+\; H^+$$

protonated imidazole   **imidazole**

indole

Tryptophan is an indole-substituted alanine (Section 20.8). Like imidazole, indole is an aromatic compound. Because the lone pair on the nitrogen atom of indole is needed for the compound's aromaticity, indole is a very weak base. (The pK_a of protonated indole is -2.4.) Therefore, the ring nitrogen in tryptophan is never protonated under physiological conditions.

Ten amino acids are **essential amino acids**; they are denoted in Table 22.2 by asterisks (*). We humans must obtain these 10 essential amino acids from our diets because we either cannot synthesize them at all or cannot synthesize them in adequate amounts. For example, we must have a dietary source of phenylalanine because we cannot synthesize benzene rings. However, we do not need tyrosine in our diets, because we can synthesize the necessary amounts from phenylalanine. Although humans can synthesize arginine, it is needed for growth in greater amounts than can be synthesized. So arginine is considered an essential amino acid for children but a nonessential amino acid for adults.

PROTEINS AND NUTRITION

Proteins are an important component of our diets. Dietary protein is hydrolyzed in the body to individual amino acids. Some of these are used to synthesize proteins needed by the body, some are broken down further to supply energy to the body, and some are used as starting materials for the synthesis of nonprotein compounds the body needs, such as thyroxine (Section 14.11), adrenaline, and melanin (Section 25.9).

Not all proteins contain the same amino acids. Most proteins from meat and dairy products contain all the amino acids needed by the body. However, most proteins from vegetable sources are *incomplete* proteins; they contain too little of one or more essential amino acids to support human growth. For example, bean protein is deficient in methionine and wheat protein is deficient in lysine. Therefore, a balanced diet must include proteins from different sources.

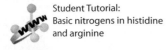
Student Tutorial:
Basic nitrogens in histidine and arginine

PROBLEM 1

a. Explain why, when the imidazole ring of histidine is protonated, the double-bonded nitrogen is the nitrogen that accepts the proton.

b. Explain why, when the guanidino group of arginine is protonated, the double-bonded nitrogen is the nitrogen that accepts the proton.

22.2 **Configuration of the Amino Acids**

The α-carbon of all the naturally occurring amino acids except glycine is an asymmetric center. Therefore, 19 of the 20 amino acids listed in Table 22.2 can exist as enantiomers. The D and L notation used for monosaccharides (Section 21.2) is also used for amino acids. An amino acid drawn in a Fischer projection with the carboxyl group on the top and the R group on the bottom of the vertical axis is a **D-amino acid** if the amino group is on the right, and an **L-amino acid** if the amino group is on the left. Unlike monosaccharides, where the D isomer is the one found in nature, most amino acids found in nature have the L configuration. To date, D-amino acid residues have been found only in a few peptide antibiotics and in some small peptides attached to the cell walls of bacteria.

alanine
an amino acid

D-glyceraldehyde   L-glyceraldehyde

Naturally occurring monosaccharides have the D-configuration.

D-amino acid   L-amino acid

Naturally occurring amino acids have the L-configuration.

Why D-sugars and L-amino acids? While it makes no difference which isomer nature "selected" to be synthesized, it is important that the same isomer be synthesized by all organisms. For example, since mammals ended up having L-amino acids, L-amino acids must be the isomers synthesized by the organisms upon which mammals depend for food.

AMINO ACIDS AND DISEASE

The Chamorro people of Guam have a high incidence of a syndrome that resembles amyotrophic lateral sclerosis (ALS or Lou Gehrig's disease) with elements of Parkinson's disease and dementia. This syndrome developed during World War II when, as a result of food shortages, the tribe ate large quantities of *Cycas circinalis* seeds. These seeds contain β-methylamino-L-alanine, an amino acid that binds to glutamate receptors. When monkeys are given β-methylamino-L-alanine, they develop some of the features of this syndrome. There is hope that, by studying the mechanism of action of β-methylamino-L-alanine, we may gain an understanding of how ALS and Parkinson's disease arise.

L-alanine

β-methylamino-L-alanine

PROBLEM 2◆

a. Which isomer—(R)-alanine or (S)-alanine—is D-alanine?

b. Which isomer—(R)-aspartate or (S)-aspartate—is D-aspartate?

c. Can a general statement be made relating R and S to D and L?

A PEPTIDE ANTIBIOTIC

Gramicidin S, an antibiotic produced by a strain of bacteria, is a cyclic decapeptide. Notice that one of its residues is ornithine, an amino acid not listed in Table 22.2 because it occurs rarely in nature. Ornithine resembles lysine but has one less methylene group in its side chain. Notice also that the antibiotic contains two D-amino acids.

L-Val
L-Pro   L-Orn
L-Phe   L-Leu
L-Leu   D-Phe
D-Orn   L-Pro
L-Val

gramicidin S

$$H_3\overset{+}{N}CH_2CH_2CH_2\underset{\underset{+NH_3}{|}}{CH}\overset{\overset{O}{\|}}{C}O^-$$

ornithine

PROBLEM 3 — SOLVED

Threonine has two asymmetric centers and therefore has four stereoisomers.

COO⁻	COO⁻	COO⁻	COO⁻
H — ⁺NH₃	H₃⁺N — H	H — ⁺NH₃	H₃⁺N — H
H — OH	HO — H	HO — H	H — OH
CH₃	CH₃	CH₃	CH₃
1	**2**	**3**	**4**

Naturally occurring L-threonine is (2S,3R)-threonine. Which of the stereoisomers is L-threonine?

Solution   Stereoisomer number 1 has the *R* configuration at both C-2 and C-3, because in both cases the arrow drawn from the highest to the next-highest-priority substituent is counterclockwise. In both cases, counterclockwise signifies *R* because the lowest-priority substituent (H) is on a horizontal bond. Therefore, the configuration of (2S,3R)-threonine is the opposite of that in stereoisomer number 1 at C-2 and the same as that in stereoisomer number 1 at C-3. Thus, L-threonine is stereoisomer number 4. Notice that the ⁺NH₃ group is on the left, just as we would expect for the Fischer projection of an L-amino acid.

PROBLEM 4◆

Do any other amino acids in Table 22.2 have more than one asymmetric center?

22.3 Acid–Base Properties of Amino Acids

Every amino acid has a carboxyl group and an amino group, and each group can exist in an acidic form or a basic form, depending on the pH of the solution in which the amino acid is dissolved.

We have seen that compounds exist primarily in their acidic forms (that is, with their protons attached) in solutions that are more acidic than their pK_a values and primarily in their basic forms (that is, without their protons) in solutions that are more basic than their pK_a values (Section 1.24). The carboxyl groups of the amino acids have pK_a values of approximately 2, and the protonated amino groups have pK_a values near 9 (Table 22.3). Both groups, therefore, will be in their acidic forms

Table 22.3   The pK_a Values of Amino Acids

Amino acid	pK_a $\alpha\text{-COOH}$	pK_a $\alpha\text{-NH}_3^+$	pK_a side chain
Alanine	2.34	9.69	—
Arginine	2.17	9.04	12.48
Asparagine	2.02	8.84	—
Aspartic acid	2.09	9.82	3.86
Cysteine	1.92	10.46	8.35
Glutamic acid	2.19	9.67	4.25
Glutamine	2.17	9.13	—
Glycine	2.34	9.60	—
Histidine	1.82	9.17	6.04
Isoleucine	2.36	9.68	—
Leucine	2.36	9.60	—
Lysine	2.18	8.95	10.79
Methionine	2.28	9.21	—
Phenylalanine	2.16	9.18	—
Proline	1.99	10.60	—
Serine	2.21	9.15	—
Threonine	2.63	9.10	—
Tryptophan	2.38	9.39	—
Tyrosine	2.20	9.11	10.07
Valine	2.32	9.62	—

in a very acidic solution (pH ~ 0). At pH = 7, the pH of the solution is greater than the pK_a of the carboxyl group, but less than the pK_a of the protonated amino group. The carboxyl group, therefore, will be in its basic form and the amino group will be in its acidic form. In a strongly basic solution (pH ~ 11), both groups will be in their basic forms.

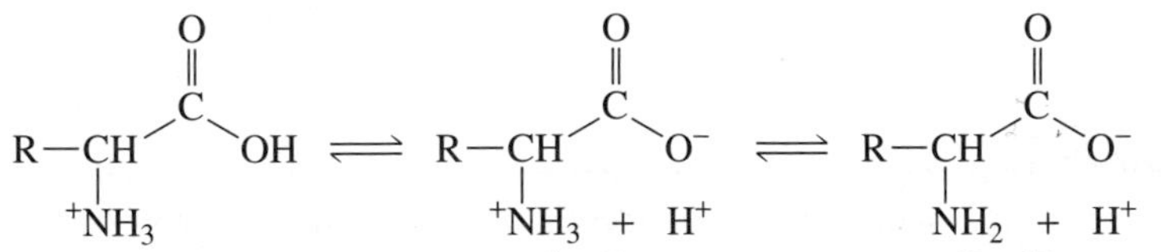

Recall from the Henderson–Hasselbalch equation that the acidic form predominates if the pH of the solution is less than the pK_a of the ionizable group and the basic form predominates if the pH of the solution is greater than the pK_a of the ionizable group.

Notice that an amino acid can never exist as an uncharged compound, regardless of the pH of the solution. To be uncharged, an amino acid would have to lose a proton from an $^+NH_3$ group with a pK_a of about 9 before it would lose a proton from a COOH group with a pK_a of about 2. This clearly is impossible: a weak acid ($pK_a = 9$) cannot be more acidic than a strong acid ($pK_a = 2$). Therefore, at physiological pH (7.3) an amino acid such as alanine exists as a dipolar ion, called a zwitterion. A **zwitterion** is a compound that has a negative charge on one atom and a positive charge on a nonadjacent atom. (The name comes from *zwitter*, German for "hermaphrodite" or "hybrid.")

A few amino acids have side chains with ionizable hydrogens (Table 22.3). The protonated imidazole side chain of histidine, for example, has a pK_a of 6.04.

Histidine, therefore, can exist in four different forms, and the form that predominates depends on the pH of the solution.

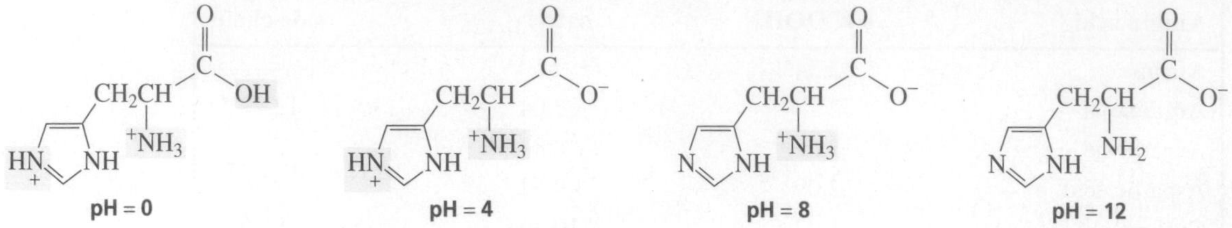

histidine

PROBLEM 5♦

Why are the carboxylic acid groups of the amino acids so much more acidic ($pK_a \sim 2$) than a carboxylic acid such as acetic acid ($pK_a = 4.76$)?

PROBLEM 6 **SOLVED**

Draw the predominant form for each of the following amino acids at physiological pH (7.3):

a. aspartate **c.** glutamine **e.** arginine
b. histidine **d.** lysine **f.** tyrosine

Solution to 6a Both carboxyl groups are in their basic forms because the pH of the solution is greater than their pK_a values. The protonated amino group is in its acidic form because the pH of the solution is less than its pK_a value.

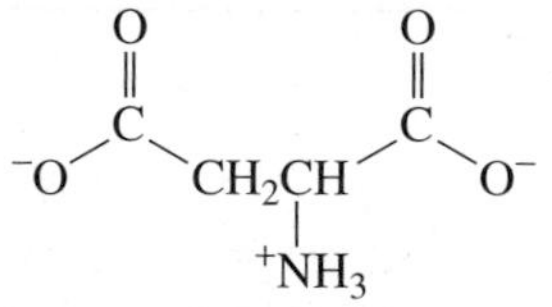

PROBLEM 7♦

Draw the predominant form for glutamate in a solution with the following pH:

a. 0 **b.** 3 **c.** 6 **d.** 11

PROBLEM 8

a. Why is the pK_a of the glutamate side chain greater than the pK_a of the aspartate side chain?
b. Why is the pK_a of the arginine side chain greater than the pK_a of the lysine side chain?

22.4 The Isoelectric Point

The **isoelectric point** (pI) of an amino acid is the pH at which it has no net charge. In other words, it is the pH at which the amount of positive charge on an amino acid exactly balances the amount of negative charge:

pI = pH at which there is no net charge

The pI of an amino acid that does *not* have an ionizable side chain—such as alanine—is midway between its two pK_a values. This is because at pH = 2.34, half the molecules have a negatively charged carboxyl group and half have an uncharged carboxyl group, and at pH = 9.69, half the molecules have a positively charged amino group and half have an uncharged amino group. As the pH increases from 2.34, the carboxyl group of more molecules becomes negatively charged; as the pH decreases from 9.69, the amino group of more molecules becomes positively charged. Therefore, at the average of the two pK_a values, the number of negatively charged groups equals the number of positively charged groups.

> **Recall from the Henderson–Hasselbalch equation that when pH = pK_a, half the group is in its acidic form and half is in its basic form.**

> **An amino acid will be positively charged if the pH of the solution is less than the pI of the amino acid and will be negatively charged if the pH of the solution is greater than the pI of the amino acid.**

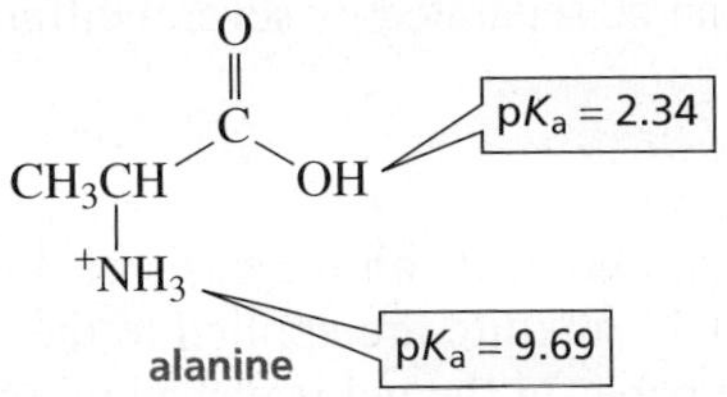

$$pI = \frac{2.34 + 9.69}{2} = \frac{12.03}{2} = 6.02$$

The pI of most amino acids that *have* an ionizable side chain (see Problem 12) is the average of the pK_a values of the similarly ionizing groups (either positively charged groups ionizing to uncharged groups or uncharged groups ionizing to negatively charged groups). For example, the pI of lysine is the average of the pK_a values of the two groups that are positively charged in their acidic form and uncharged in their basic form. The pI of glutamic acid, on the other hand, is the average of the pK_a values of the two groups that are uncharged in their acidic form and negatively charged in their basic form.

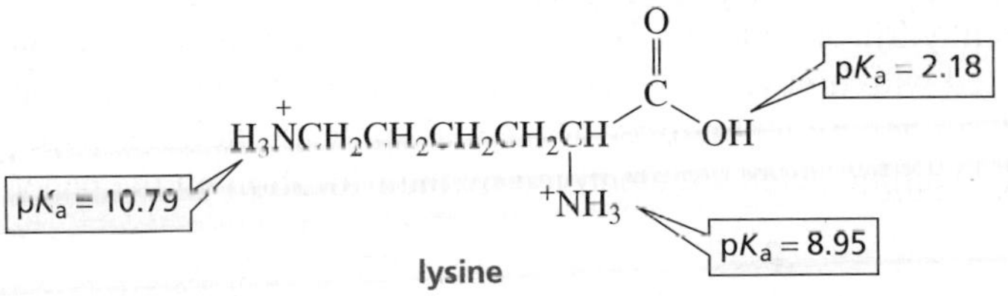

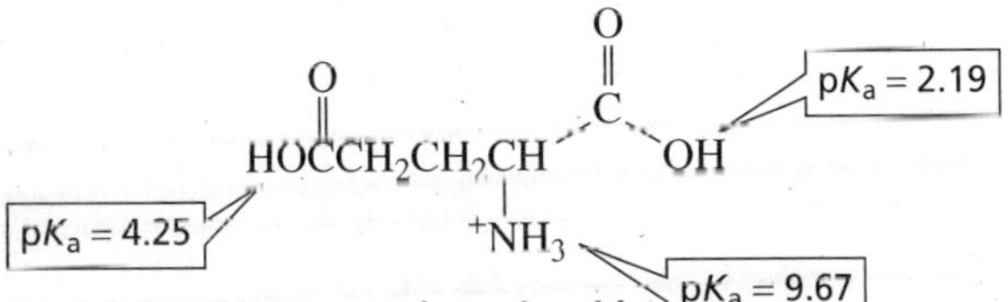

$$pI = \frac{8.95 + 10.79}{2} = \frac{19.74}{2} = 9.87$$

$$pI = \frac{2.19 + 4.25}{2} = \frac{6.44}{2} = 3.22$$

PROBLEM 9

Explain why the pI of lysine is the average of the pK_a values of its two protonated amino groups.

PROBLEM 10◆

Calculate the pI of each of the following amino acids:

a. asparagine   b. arginine   c. serine   d. aspartate

PROBLEM 11◆

a. Which amino acid has the lowest pI value?
b. Which amino acid has the highest pI value?
c. Which amino acid has the greatest amount of negative charge at pH = 6.20?
d. Which amino acid has a greater negative charge at pH = 6.20, glycine or methionine?

PROBLEM 12

Explain why the pI values of tyrosine and cysteine cannot be determined by the method just described.

22.5 Separation of Amino Acids

A mixture of amino acids can be separated by several different techniques.

Electrophoresis

Student Tutorial:
Electrophoresis and pI

Electrophoresis separates amino acids on the basis of their pI values. A few drops of a solution of an amino acid mixture are applied to the middle of a piece of filter paper or to a gel. When the paper or the gel is placed in a buffered solution between two electrodes and an electric field is applied (Figure 22.1), an amino acid with a pI greater than the pH of the solution will have an overall positive charge and will migrate toward the cathode (the negative electrode). The farther the amino acid's pI is from the pH of the buffer, the more positive the amino acid will be and the farther it will migrate toward the cathode in a given amount of time. An amino acid with a pI less than the pH of the buffer will have an overall negative charge and will migrate toward the anode (the positive electrode). If two molecules have the same charge, the larger one will move more slowly during electrophoresis because the same charge has to move a greater mass.

Considering that amino acids are colorless, how can we detect them after they have been separated? After the amino acids have been separated by electrophoresis, the filter paper is sprayed with ninhydrin and dried in a warm oven. Most amino acids form a purple product when heated with ninhydrin. The number of different kinds of amino acids in the mixture is determined by the number of colored spots on the filter paper (Figure 22.1). The individual amino acids are identified by their location on the paper compared with a standard.

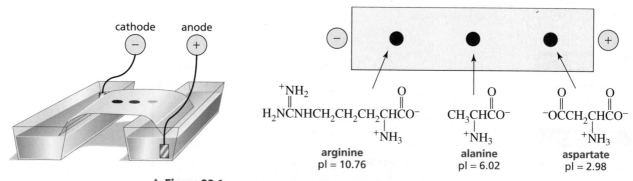

▲ **Figure 22.1**
Arginine, alanine, and aspartate separated by electrophoresis at pH = 5.

The mechanism for formation of the colored product is shown below, omitting the mechanisms for the steps involving dehydration, imine formation, and imine hydrolysis. (These mechanisms are shown in Sections 17.8 and 17.9.)

Mechanism for the reaction of an amino acid with ninhydrin to form a colored product

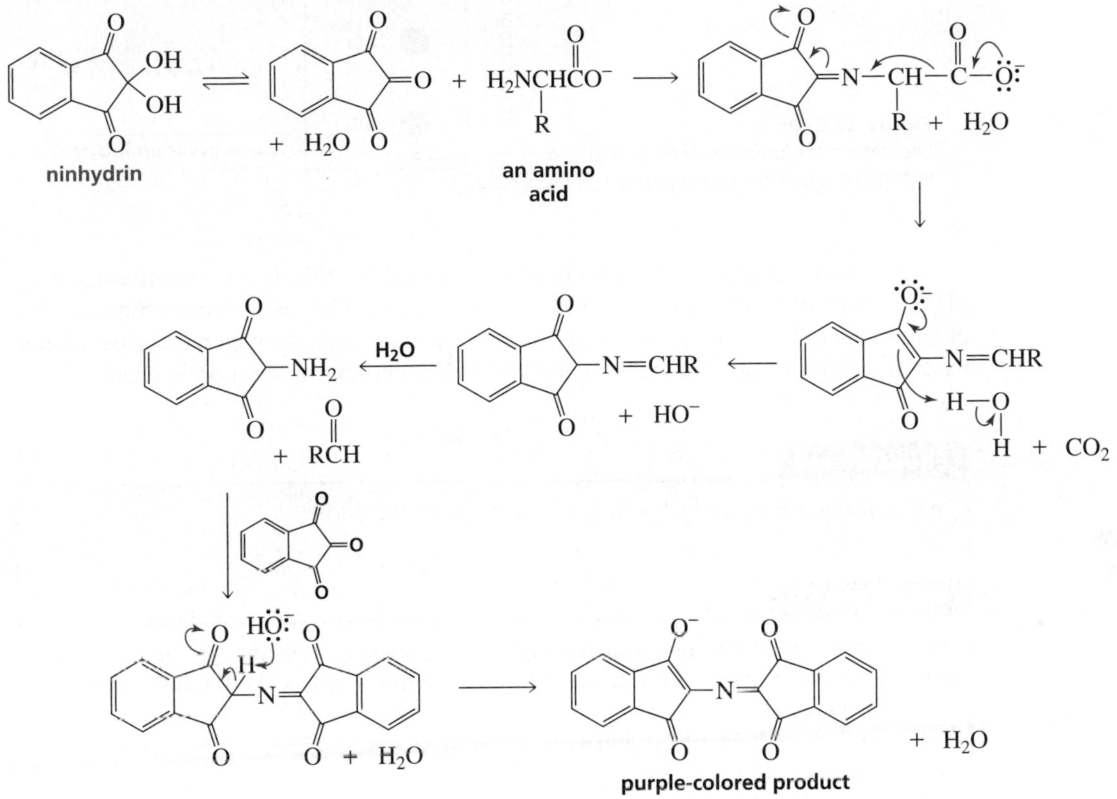

ninhydrin

an amino acid

purple-colored product

Paper Chromatography and Thin-Layer Chromatography

Paper chromatography once played an important role in biochemical analysis because it provided a method for separating amino acids using very simple equipment. Although more modern techniques are generally employed today, we will describe the principles behind paper chromatography because many of the same principles are employed in modern separation techniques.

Paper chromatography separates amino acids on the basis of polarity. A few drops of a solution of an amino acid mixture are applied to the bottom of a strip of filter paper. The edge of the paper is then placed in a solvent (typically a mixture of water, acetic acid, and butanol). The solvent moves up the paper by capillary action, carrying the amino acids with it. Depending on their polarities, the amino acids have different affinities for the mobile (solvent) and stationary (paper) phases and therefore travel up the paper at different rates. The more polar the amino acid, the more strongly it is adsorbed onto the relatively polar paper. The less polar amino acids travel up the paper more rapidly, since they have a greater affinity for the mobile phase. Therefore, when the paper is developed with ninhydrin, the colored spot closest to the origin is the most polar amino acid and the spot farthest away from the origin is the least polar amino acid (Figure 22.2).

Less polar amino acids travel up the paper more rapidly.

The most polar amino acids are those with charged side chains, the next most polar are those with side chains that can form hydrogen bonds, and the least polar are those with hydrocarbon side chains. For amino acids with hydrocarbon side chains, the larger the alkyl group, the less polar the amino acid. In other words, leucine is less polar than valine.

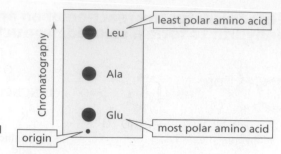

Figure 22.2 ▶
Separation of glutamate, alanine, and leucine by paper chromatography.

Paper chromatography has largely been replaced by **thin-layer chromatography** (TLC), which differs from paper chromatography in that TLC uses a plate with a coating of solid material instead of filter paper. The physical property on which the separation is based depends on the solid material and the solvent chosen for the mobile phase.

PROBLEM 13♦

What aldehyde is formed when valine is treated with ninhydrin?

PROBLEM 14♦

A mixture of seven amino acids (glycine, glutamate, leucine, lysine, alanine, isoleucine, and aspartate) is separated by TLC. Explain why only six spots show up when the chromatographic plate is sprayed with ninhydrin and heated.

Ion-Exchange Chromatography

Electrophoresis and thin-layer chromatography are analytical separations—small amounts of amino acids are separated for analysis. Preparative separation, in which larger amounts of amino acids are separated for use in subsequent processes, can be achieved using **ion-exchange chromatography**. This technique employs a column packed with an insoluble resin. A solution of a mixture of amino acids is loaded onto the top of the column, and a series of buffer solutions of increasing pH are poured through the column. The amino acids separate because they flow through the column at different rates, as explained below.

The resin is a chemically inert material with charged side chains. The structure of a commonly used resin is shown in Figure 22.3. If a mixture of lysine and glutamate in

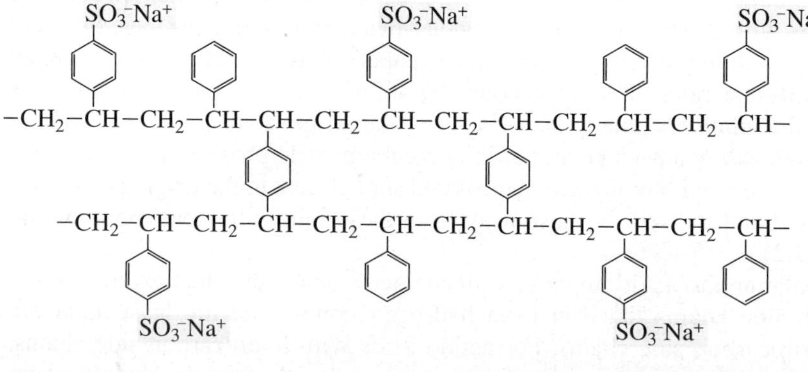

Figure 22.3 ▶
A section of a cation-exchange resin. This particular resin is called Dowex 50.

a solution with a pH of 6 were to be loaded onto the column, glutamate would travel down the column rapidly because its negatively charged side chain would be repelled by the negatively charged sulfonic acid groups of the resin. The positively charged side chain of lysine, on the other hand, would cause that amino acid to be retained on the column. This kind of resin is called a **cation-exchange resin** because it exchanges the Na^+ counterions of the SO_3^- groups for the positively charged species traveling through the column. In addition, the relatively nonpolar nature of the column causes it to retain nonpolar amino acids longer than polar amino acids.

Cations bind most strongly to cation-exchange resins.

Resins with positively charged groups are called **anion-exchange resins** because they impede the flow of anions by exchanging their negatively charged counterions for negatively charged species traveling through the column. A common anion-exchange resin, Dowex 1, has $CH_2N^+(CH_3)_3Cl^-$ groups in place of the $SO_3^-Na^+$ groups in Figure 22.3.

Anions bind most strongly to anion-exchange resins.

An **amino acid analyzer** is an instrument that automates ion-exchange chromatography. When a solution of an amino acid mixture passes through the column of an amino acid analyzer containing a cation-exchange resin, the amino acids move through the column at different rates, depending on their overall charge. The solution that flows out of the column (the effluent) is collected in fractions, and these are collected often enough that a different amino acid ends up in each one (Figure 22.4).

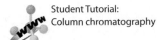

Student Tutorial:
Column chromatography

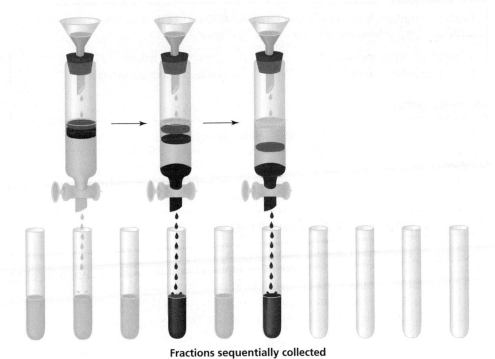

Fractions sequentially collected

◀ **Figure 22.4**
Separation of amino acids by ion-exchange chromatography.

If ninhydrin is added to each of the fractions, the concentration of amino acid in each fraction can be determined by the amount of absorbance at 570 nm, because the colored compound formed by the reaction of an amino acid with ninhydrin has a λ_{max} of 570 (Section 12.17). This information combined with each fraction's rate of passage through the column allow the identity and relative amount of each amino acid to be determined (Figure 22.5).

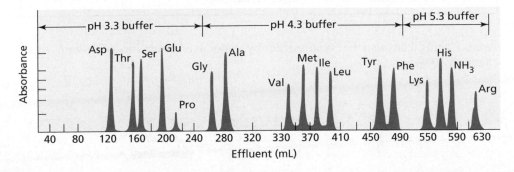

◀ **Figure 22.5**
A typical chromatogram obtained from the separation of a mixture of amino acids using an automated amino acid analyzer.

WATER SOFTENERS: EXAMPLES OF CATION-EXCHANGE CHROMATOGRAPHY

Water-softening systems contain a column packed with a cation-exchange resin that has been flushed with concentrated sodium chloride. When "hard water" (water with high levels of calcium and magnesium ions; Section 16.14) passes through the column, the resin binds the magnesium and calcium ions more tightly than it binds sodium ions. It thus removes magnesium and calcium ions from the water and replaces them with sodium ions. The resin must be recharged from time to time by being flushed with concentrated sodium chloride, to replace the bound magnesium and calcium ions with sodium ions.

PROBLEM 15

Why are buffer solutions of increasingly higher pH used to elute the column that generates the chromatogram shown in Figure 22.5? (*Elute* means wash out with a solvent.)

PROBLEM 16

Explain the order of elution (with a buffer of pH 4) of each of the following pairs of amino acids through a column packed with Dowex 50 (Figure 22.3):

a. aspartate before serine

b. glycine before alanine

c. valine before leucine

d. tyrosine before phenylalanine

PROBLEM 17◆

Give the order in which the following amino acids would be eluted with a buffer of pH 4 from a column containing an anion-exchange resin (Dowex 1): histidine, serine, aspartate, valine.

22.6 Synthesis of Amino Acids

Chemists do not have to rely on nature to produce amino acids; they can synthesize them in the laboratory, using a variety of methods. One of the oldest methods replaces an α-hydrogen of a carboxylic acid with a bromine in a Hell–Volhard–Zelinski reaction (Section 18.6). The resulting α-bromocarboxylic acid then undergoes an S_N2 reaction with ammonia to form the amino acid (Section 8.4).

$$\underset{\substack{\text{a carboxylic}\\\text{acid}}}{RCH_2\overset{\overset{\displaystyle O}{\|}}{C}OH} \xrightarrow[\text{2. H}_2\text{O}]{\text{1. Br}_2\text{, PBr}_3} \underset{\underset{\displaystyle Br}{|}}{RCH\overset{\overset{\displaystyle O}{\|}}{C}OH} \xrightarrow{\substack{\text{excess}\\\text{NH}_3}} \underset{\substack{\\\overset{+}{NH_3}\\\text{an amino acid}}}{RCH\overset{\overset{\displaystyle O}{\|}}{C}O^-} + \overset{+}{N}H_4Br^-$$

PROBLEM 18◆

Why is excess ammonia used in the preceding reaction?

Amino acids can also be synthesized by reductive amination of α-keto acids (Section 17.8).

$$\underset{\underset{\displaystyle O}{\|}}{RC}\overset{\overset{\displaystyle O}{\|}}{C}OH \xrightarrow[\substack{\|\\O\\\text{NaBH(OCCH}_3)_3}]{\text{excess ammonia}} \underset{\overset{+}{NH_3}}{RCH}\overset{\overset{\displaystyle O}{\|}}{C}O^-$$

PROBLEM 19◆

Biological organisms can also convert α-keto acids into amino acids, but because the reagents organic chemists use for this reaction are not available in living cells, cells carry out this reaction by a different mechanism (Section 24.6).

a. What amino acid is obtained from the reductive amination of each of the following metabolic intermediates in the cell?

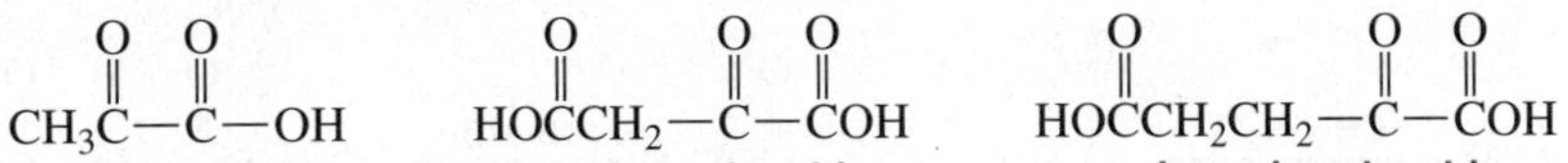

pyruvic acid oxaloacetic acid α-ketoglutaric acid

b. What amino acids are obtained from the same metabolic intermediates when the amino acids are synthesized in the laboratory?

Amino acids can be synthesized in much better yields than those obtained in the previous two reactions by the *N*-phthalimidomalonic ester synthesis, a method that combines the malonic ester synthesis (Section 18.19) and the Gabriel synthesis (Section 16.18). The mechanism of the reaction is shown below.

Mechanism for the *N*-phthalimidomalonic ester synthesis

α-bromomalonic ester potassium phthalimide *N*-phthalimidomalonic ester

phthalic acid an amino acid

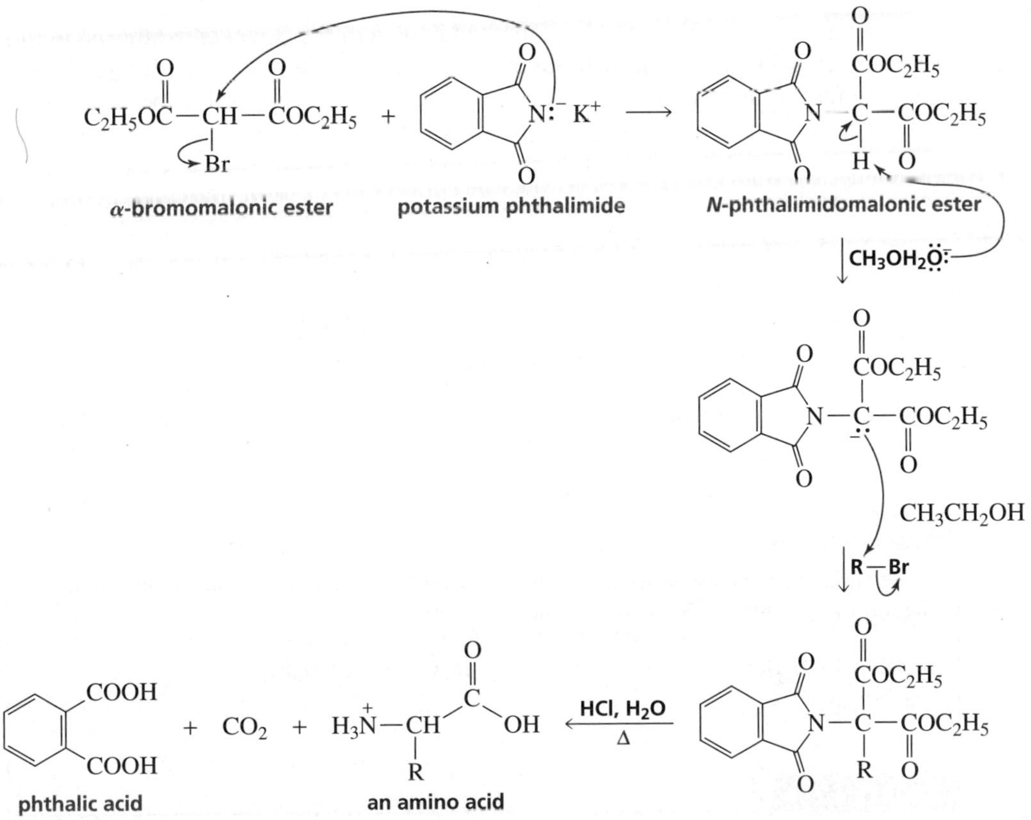

- α-Bromomalonic ester and potassium phthalimide undergo an S_N2 reaction.
- A proton is easily removed from the α-carbon of *N*-phthalimidomalonic ester since it is flanked by two ester groups.

- The resulting carbanion undergoes an S_N2 reaction with an alkyl halide.
- Heating in an acidic aqueous solution hydrolyzes both ester groups and both amide groups and decarboxylates the 3-oxocarboxylic acid.

A variation of the *N*-phthalimidomalonic ester synthesis uses acetamidomalonic ester in place of *N*-phthalimidomalonic ester.

acetamidomalonic ester

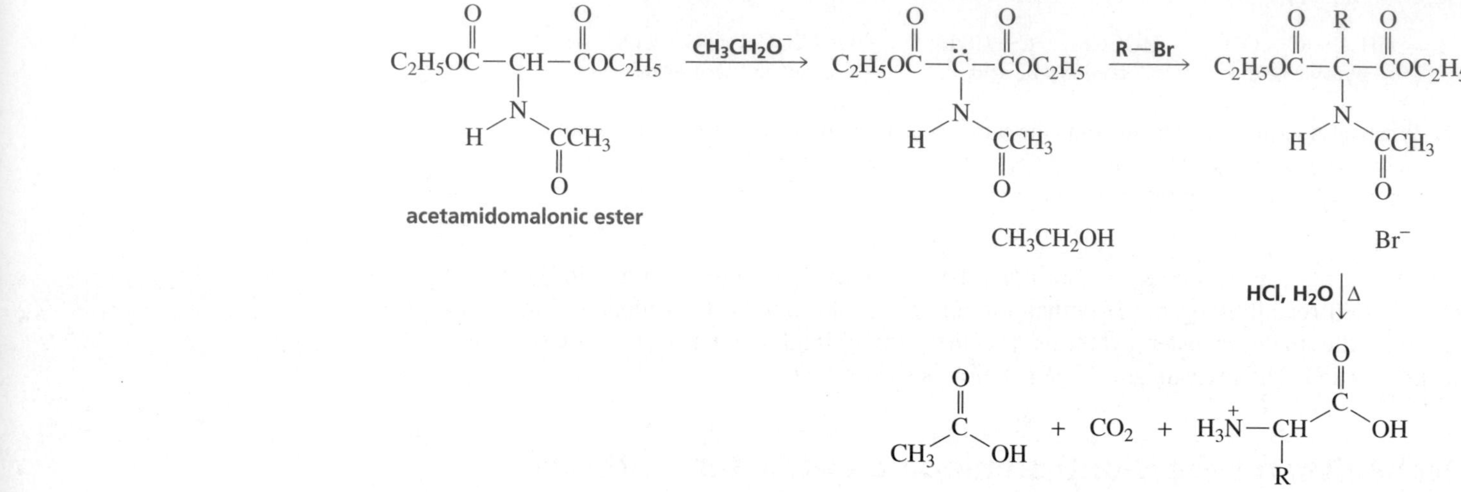

Amino acids can be prepared by a method known as the Strecker synthesis. In this synthesis, an aldehyde reacts with ammonia forming an imine. An addition reaction with cyanide ion forms an intermediate, which, when hydrolyzed, forms the amino acid (Section 16.19). Compare this reaction with the Kiliani–Fischer synthesis of aldoses in Section 21.8.

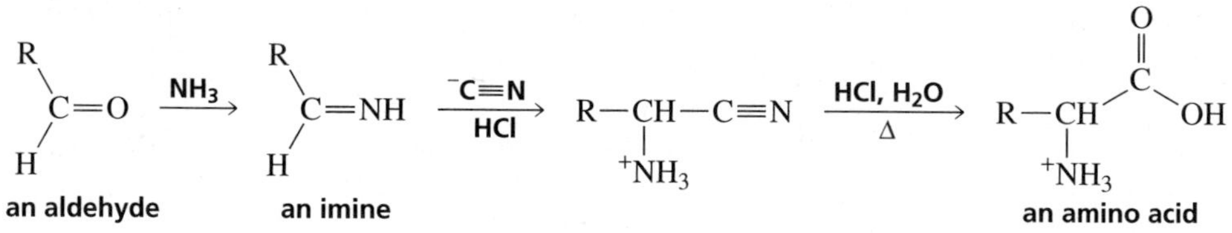

an aldehyde   **an imine**   **an amino acid**

PROBLEM 20◆

What amino acid would be formed using the *N*-phthalimidomalonic ester synthesis when the following compounds are used in the third step?

a. CH₃CHCH₂Br
      |
      CH₃

b. CH₃SCH₂CH₂Br

PROBLEM 21◆

What alkyl halide would you use in the acetamidomalonic ester synthesis to prepare

a. lysine?

b. phenylalanine?

What amino acid would be formed when the aldehyde used in the Strecker synthesis is

a. acetaldehyde?　　　　**b.** 2-methylbutanal?　　　　**c.** 3-methylbutanal?

22.7 Resolution of Racemic Mixtures of Amino Acids

When amino acids are synthesized in nature, only the L-enantiomer is formed (Section 5.20). However, when amino acids are synthesized in the laboratory, the product is usually a racemic mixture of D and L enantiomers. If only one isomer is desired, the enantiomers must be separated, which can be accomplished by means of an enzyme-catalyzed reaction. Because an enzyme is chiral, it will react at a different rate with each of the enantiomers (Section 5.21). For example, pig kidney aminoacylase is an enzyme that catalyzes the hydrolysis of N-acetyl-L-amino acids, but not N-acetyl-D-amino acids. Therefore, if the racemic amino acid is converted into a pair of N-acetylamino acids and the N-acetylated mixture is hydrolyzed with pig kidney aminoacylase, the products will be the L-amino acid and N-acetyl-D-amino acid, which are easily separated. Because the resolution (separation) of the enantiomers depends on the difference in the rates of reaction of the enzyme with the two N-acetylated compounds, this technique is known as a **kinetic resolution**.

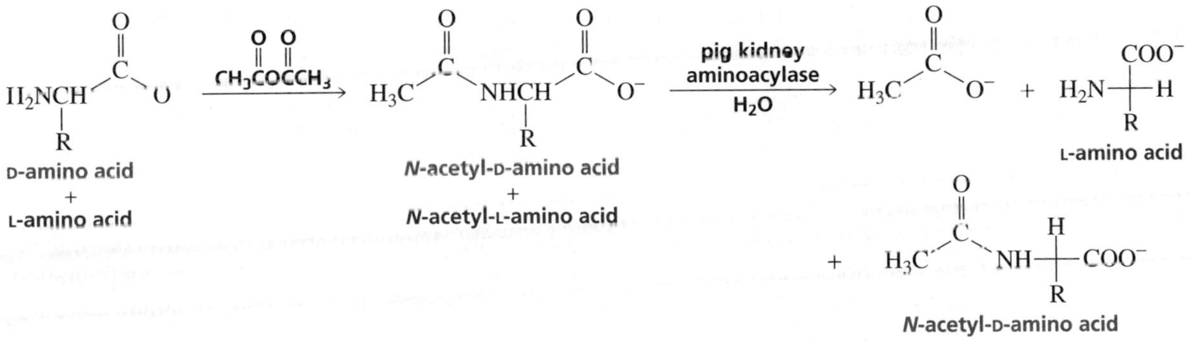

Pig liver esterase is an enzyme that catalyzes the hydrolysis of esters. It hydrolyzes esters of L-amino acids more rapidly than esters of D-amino acids. How can this enzyme be used to separate a racemic mixture of amino acids?

22.8 Peptide Bonds and Disulfide Bonds

Peptide bonds and disulfide bonds are the only covalent bonds that join amino acid residues together in a peptide or a protein.

Peptide Bonds

The amide bonds that link amino acid residues are called **peptide bonds**. By convention, peptides and proteins are drawn with the free amino group (of the **N-terminal amino acid**) on the left and the free carboxyl group (of the **C-terminal amino acid**) on the right.

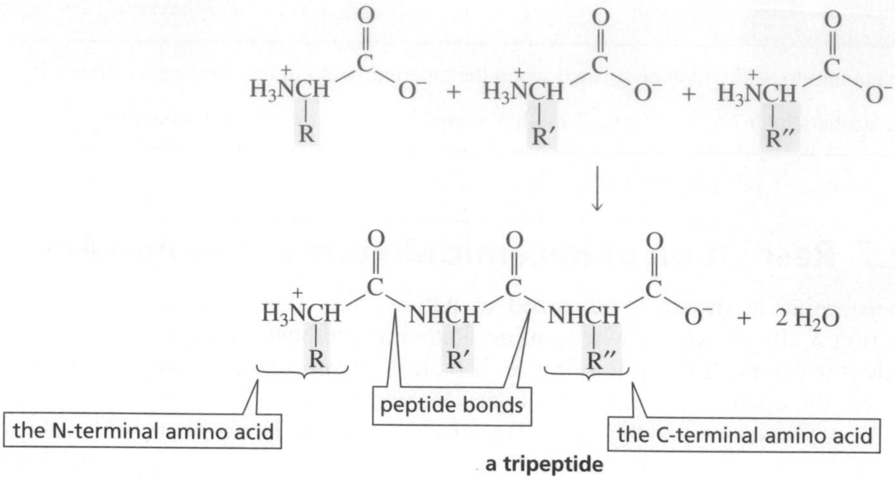

a tripeptide

When the identities of the amino acids in a peptide are known but their sequence is not known, the amino acids are written separated by commas. When the sequence of amino acids is known, the amino acids are written connected by hyphens. In the pentapeptide represented on the right, valine is the N-terminal amino acid and histidine is the C-terminal amino acid. The amino acids are numbered starting with the N-terminal end. The glutamate residue is therefore referred to as Glu 4, because it is the fourth amino acid from the N-terminal end. In naming the peptide, adjective names (ending in "yl") are used for all the amino acids except the C-terminal amino acid. Thus, this pentapeptide is named valylcysteylalanylglutamylhistidine. It is assumed that the configuration of each amino acid is L unless otherwise specified.

<table>
<tr><td>Glu, Cys, His, Val, Ala</td><td>Val-Cys-Ala-Glu-His</td></tr>
<tr><td>the pentapeptide contains the indicated amino acids, but their sequence is not known</td><td>the amino acids in the pentapeptide have the indicated sequence</td></tr>
</table>

A peptide bond has about 40% double-bond character because of electron delocalization (Section 16.2). Steric hindrance in the cis configuration causes the trans configuration about the amide linkage to be more stable, so the α-carbons of adjacent amino acids are trans to each other (Section 4.11).

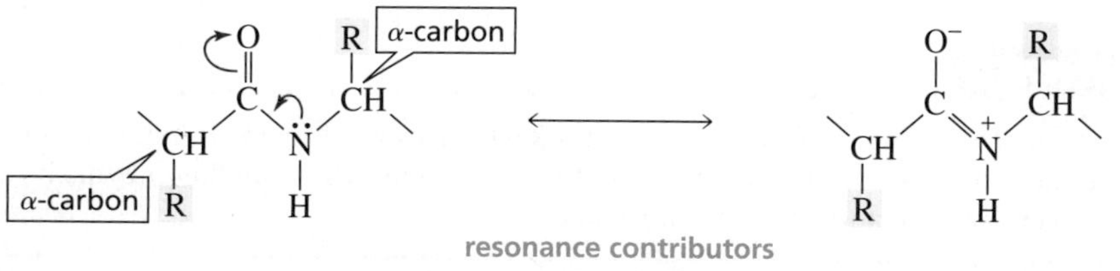

resonance contributors

The partial double-bond character prevents free rotation about the peptide bond, so the carbon and nitrogen atoms of the peptide bond and the two atoms to which each is attached are held rigidly in a plane (Figure 22.6). This regional planarity affects the way

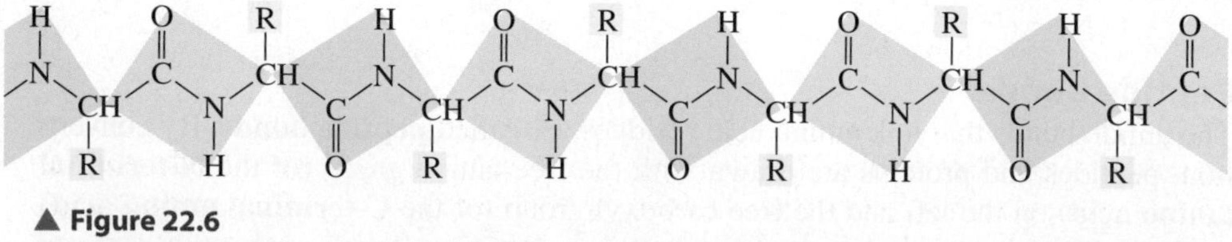

▲ **Figure 22.6**
A segment of a polypeptide chain. Colored squares indicate the plane defined by each peptide bond. Notice that the R groups bonded to the α-carbons are on alternate sides of the peptide backbone.

a chain of amino acids can fold, so electron delocalization has important implications for the three-dimensional shapes of peptides and proteins (Section 22.13).

PROBLEM 24

Draw the tetrapeptide Ala-Thr-Asp-Asn and indicate the peptide bonds.

PROBLEM 25♦

Using the three-letter abbreviations, write the six tripeptides consisting of Ala, Gly, and Met.

PROBLEM 26

Draw a peptide bond in the cis configuration.

PROBLEM 27♦

Which bonds in the backbone of a peptide can rotate freely?

Disulfide Bonds

When thiols are oxidized under mild conditions, they form disulfides. A **disulfide** is a compound with an S—S bond. (Recall that the number of S—H bonds decreases in an oxidation reaction and increases in a reduction reaction.)

$$2\ R\!-\!SH \xrightarrow{\text{mild oxidation}} RS\!-\!SR$$
$$\text{a thiol} \qquad\qquad\qquad \text{a disulfide}$$

An oxidizing agent commonly used for this reaction is Br_2 (or I_2) in a basic solution.

Mechanism for oxidation of a thiol to a disulfide

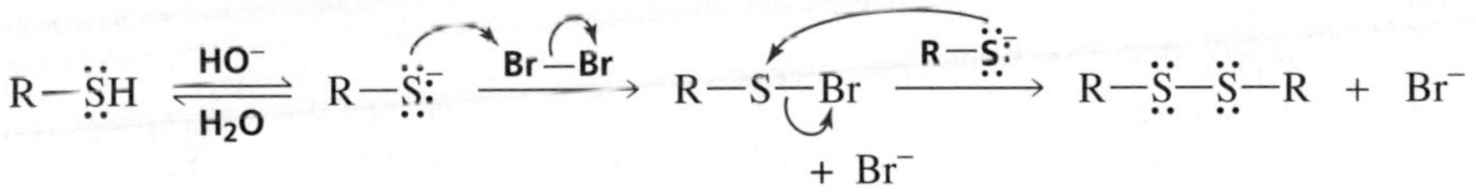

$$R\!-\!\overset{..}{\underset{..}{S}}H \underset{H_2O}{\overset{HO^-}{\rightleftharpoons}} R\!-\!\overset{..}{\underset{..}{S}}\!:^{-} \xrightarrow{Br\!-\!Br} R\!-\!S\!-\!Br \xrightarrow{R\!-\!\overset{..}{S}\!:^{-}} R\!-\!\overset{..}{\underset{..}{S}}\!-\!\overset{..}{\underset{..}{S}}\!-\!R\ +\ Br^-$$
$$+\ Br^-$$

Because thiols can be oxidized to disulfides, disulfides can be reduced to thiols.

$$RS\!-\!SR \xrightarrow{\text{reduction}} 2\ R\!-\!SH$$
$$\text{a disulfide} \qquad\qquad \text{a thiol}$$

Cysteine is an amino acid that contains a thiol group. Two cysteine molecules therefore can be oxidized to a disulfide. This disulfide is called cystine.

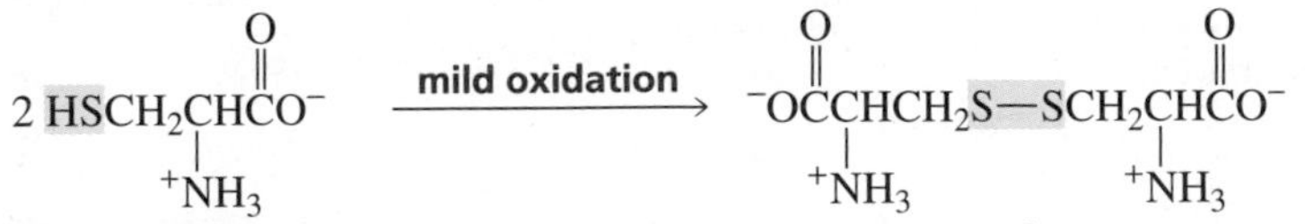

$$2\ \underset{\underset{\text{cysteine}}{\overset{|}{{}^+NH_3}}}{HSCH_2CHCO^-} \xrightarrow{\text{mild oxidation}} \underset{\underset{\text{cystine}}{\overset{|}{{}^+NH_3}\qquad\qquad\overset{|}{{}^+NH_3}}}{{}^-OCCHCH_2S\!-\!SCH_2CHCO^-}$$

Two cysteine residues in a protein can be oxidized to a disulfide, creating a bond known as a **disulfide bridge**. Disulfide bridges are the only covalent bonds that are found between nonadjacent amino acids in peptides and proteins. They contribute to the overall shape of a protein by linking cysteine residues found in different parts of the peptide backbone, as shown in Figure 22.7.

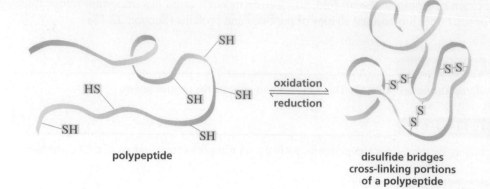

Figure 22.7 ▶
Disulfide bridges cross-linking portions of a peptide.

polypeptide

disulfide bridges cross-linking portions of a polypeptide

The hormone insulin, secreted by the pancreas, controls the level of glucose in the blood by regulating glucose metabolism. Insulin is a polypeptide with two peptide chains. The short chain (the A-chain) contains 21 amino acids, and the long chain (the B-chain) contains 30 amino acids. The A- and B-chains are connected to each other by two disulfide bridges. These are **interchain disulfide bridges** (between the A- and B-chains). Insulin also has an **intrachain disulfide bridge** (within the A-chain).

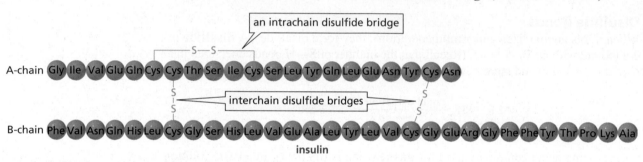

an intrachain disulfide bridge

A-chain Gly Ile Val Glu Gln Cys Cys Thr Ser Ile Cys Ser Leu Tyr Gln Leu Glu Asn Tyr Cys Asn

interchain disulfide bridges

B-chain Phe Val Asn Gln His Leu Cys Gly Ser His Leu Val Glu Ala Leu Tyr Leu Val Cys Gly Glu Arg Gly Phe Phe Tyr Thr Pro Lys Ala

insulin

HAIR: STRAIGHT OR CURLY?

Hair is made up of a protein called keratin that contains an unusually large number of cysteine residues (about 8% of the amino acids compared with an average of 2.8% for other proteins). These furnish keratin with many disulfide bridges that preserve its three-dimensional structure. People can alter the structure of their hair (if they think it is either too straight or too curly) by changing the location of these disulfide bridges. This is accomplished by first applying a reducing agent to the hair to reduce all the disulfide bridges in the protein strands. Then, after rearranging the hair into the desired shape (using curlers to curl it or combing it straight to uncurl it), an oxidizing agent is applied to form new disulfide bridges. The new disulfide bridges hold the hair in its new shape. When this treatment is applied to straight hair, it is called a "permanent." When it is applied to curly hair, it is called "hair straightening."

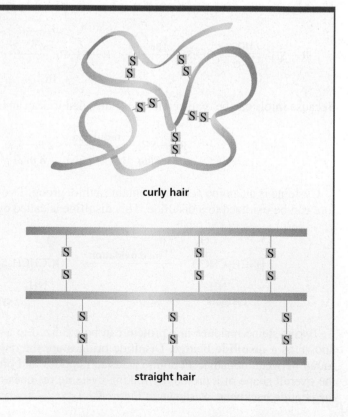

curly hair

straight hair

22.9 **Some Interesting Peptides**

Enkephalins are pentapeptides synthesized by the body to control pain. They decrease the body's sensitivity to pain by binding to receptors in certain brain cells. Part of the three-dimensional structures of enkephalins must be similar to those of morphine and related painkillers such as Demerol because they bind to the same receptors (Sections 30.3 and 30.6).

Tyr-Gly-Gly-Phe-Leu   Tyr-Gly-Gly-Phe-Met
leucine enkephalin   methionine enkephalin

Bradykinin, vasopressin, and oxytocin are peptide hormones. They are all nonapeptides. Bradykinin inhibits the inflammation of tissues. Vasopressin controls blood pressure by regulating the contraction of smooth muscle; it is also an antidiuretic. Oxytocin induces labor in pregnant women by stimulating the uterine muscle to contract, and it also stimulates milk production in nursing mothers. Vasopressin and oxytocin both have an intrachain disulfide bond, and their C-terminal amino acids contain amide rather than carboxyl groups. Notice that the C-terminal amide group is indicated by writing "NH$_2$" after the name of the C-terminal amino acid. Vasopressin and oxytocin also act on the brain. Vasopressin is a "fight-or-flight" hormone, whereas oxytocin has the opposite effect; it calms the body and promotes social bonding. In spite of their very different physiological effects, vasopressin and oxytocin differ only by two amino acids.

BIOGRAPHY

Oxytocin was the first small peptide to be synthesized. This was accomplished in 1953 by **Vincent du Vigneaud (1901–1978)**, *who later synthesized vasopressin. Du Vigneaud was born in Chicago and was a professor at George Washington University Medical School and later at Cornell University Medical College. For synthesizing these nonapeptides, he received the Nobel Prize in chemistry in 1955.*

bradykinin   Arg-Pro-Pro-Gly-Phe-Ser-Pro-Phe-Arg

vasopressin   Cys-Tyr-Phe-Gln-Asn-Cys-Pro-Arg-Gly-NH$_2$
                S——————————S

oxytocin   Cys-Tyr-Ile-Gln-Asn-Cys-Pro-Leu-Gly-NH$_2$
                S——————————S

The synthetic sweetener aspartame, or NutraSweet (Section 21.20), is the methyl ester of a dipeptide of L-aspartate and L-phenylalanine. Aspartame is about 200 times sweeter than sucrose. The ethyl ester of the same dipeptide is not sweet. If a D-amino acid is substituted for either of the L-amino acids of aspartame, the resulting dipeptide is bitter rather than sweet.

aspartame
NutraSweet®

Glutathione is a tripeptide consisting of glutamate, cysteine, and glycine. Its function is to destroy harmful oxidizing agents in the body. Oxidizing agents are thought to be responsible for some of the effects of aging and to play a causative role in cancer. Glutathione removes oxidizing agents by reducing them, and as a result undergoing oxidation that forms a disulfide bond between two glutathione molecules (see pages 1017 and 1040). An enzyme subsequently reduces the disulfide bond, returning glutathione to its original condition so it can react with more oxidizing agents.

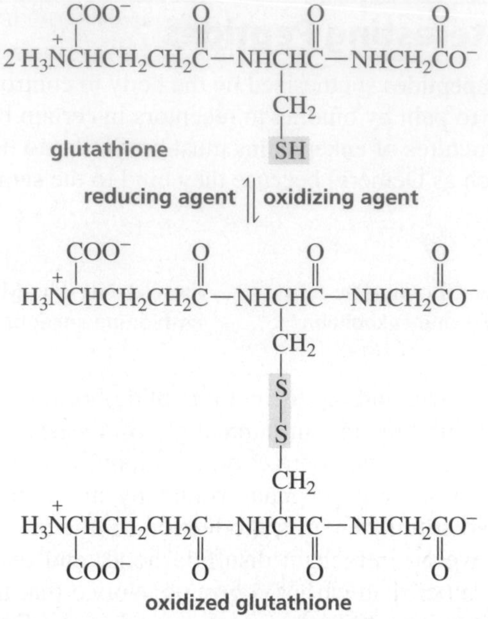

$$2 \ \overset{+}{\text{H}_3}\text{NCHCH}_2\text{CH}_2\overset{\text{COO}^-}{\underset{}{\overset{|}{\text{C}}}}-\text{NHCHC}-\text{NHCH}_2\text{CO}^-$$

glutathione

reducing agent ‖ oxidizing agent

oxidized glutathione

What is unusual about glutathione's structure? (If you have trouble answering this question, draw the structure you would expect for Glu-Cys-Gly, and compare your structure with the structure of glutathione.)

22.10 Strategy of Peptide Bond Synthesis: N-Protection and C-Activation

One difficulty in synthesizing a polypeptide once its structure is known, is that the amino acids have two functional groups that enable them to combine in various ways. For example, suppose you wanted to make the dipeptide Gly-Ala. That dipeptide is only one of four possible dipeptides that could be formed from a mixture of alanine and glycine.

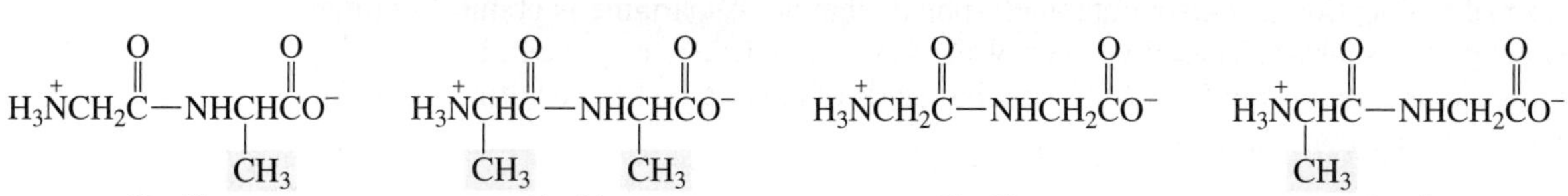

Gly-Ala   **Ala-Ala**   **Gly-Gly**   **Ala-Gly**

If the amino group of the amino acid that is to be on the N-terminal end (in this case, Gly) is protected (Section 17.11), it will not be available to form a peptide bond. If the carboxyl group of this same amino acid is activated before the second amino acid is added, the amino group of the added amino acid (in this case, Ala) will react with the activated carboxyl group of glycine in preference to reacting with a nonactivated carboxyl group of another alanine molecule.

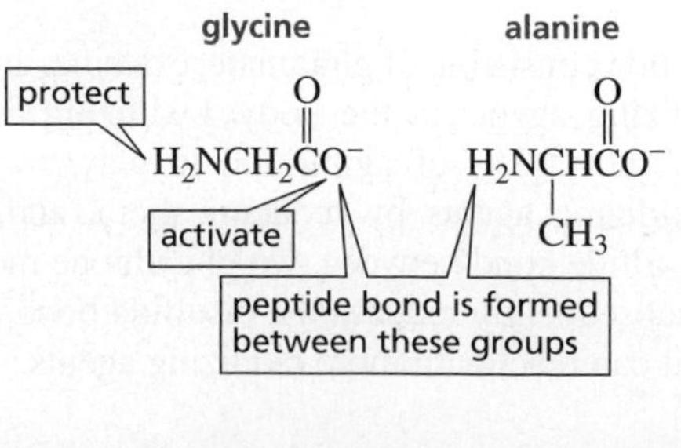

glycine   alanine

protect

activate

peptide bond is formed between these groups

The reagent most often used to protect the amino group of an amino acid is di-*tert*-butyl dicarbonate. The protecting group is known by the acronym *t*-BOC (pronounced "tee-bok"). The popularity of this reagent is due to the ease with which the protecting group can be removed when the need for protection is over.

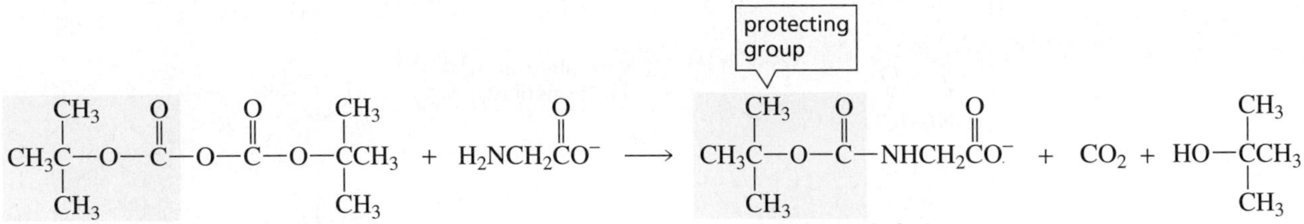

di-*tert*-butyl dicarbonate   glycine   N-protected glycine

Carboxylic acids are generally activated by being converted into acyl chlorides (Section 16.21). Acyl chlorides, however, are so reactive that they can readily react with the side chains of some of the amino acids during peptide synthesis, creating unwanted products. The preferred method for activating the carboxyl group of an N-protected amino acid is to convert it into an imidate using dicyclohexylcarbodiimide (DCC). (By now, you have probably noticed that biochemists are even fonder of acronyms than organic chemists are.) DCC activates a carboxyl group by putting a good leaving group on the carbonyl carbon.

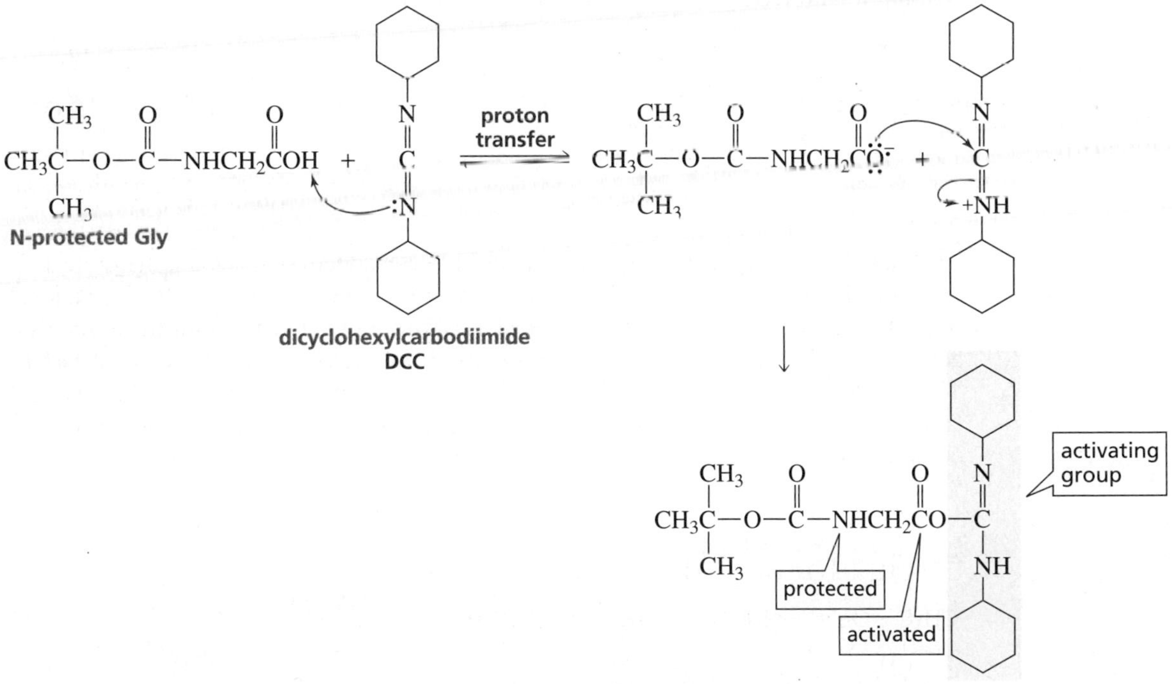

After the amino acid's N-terminal group is protected and its C-terminal group is activated, the second amino acid is added. The unprotected amino group of the second amino acid attacks the activated carboxyl group, forming a tetrahedral intermediate. The C—O bond of the tetrahedral intermediate is easily broken because the bonding electrons are delocalized; breaking this bond forms dicyclohexylurea, a stable diamide. Recall that the weaker (more stable) the base, the better a leaving group it is; see Section 16.6.

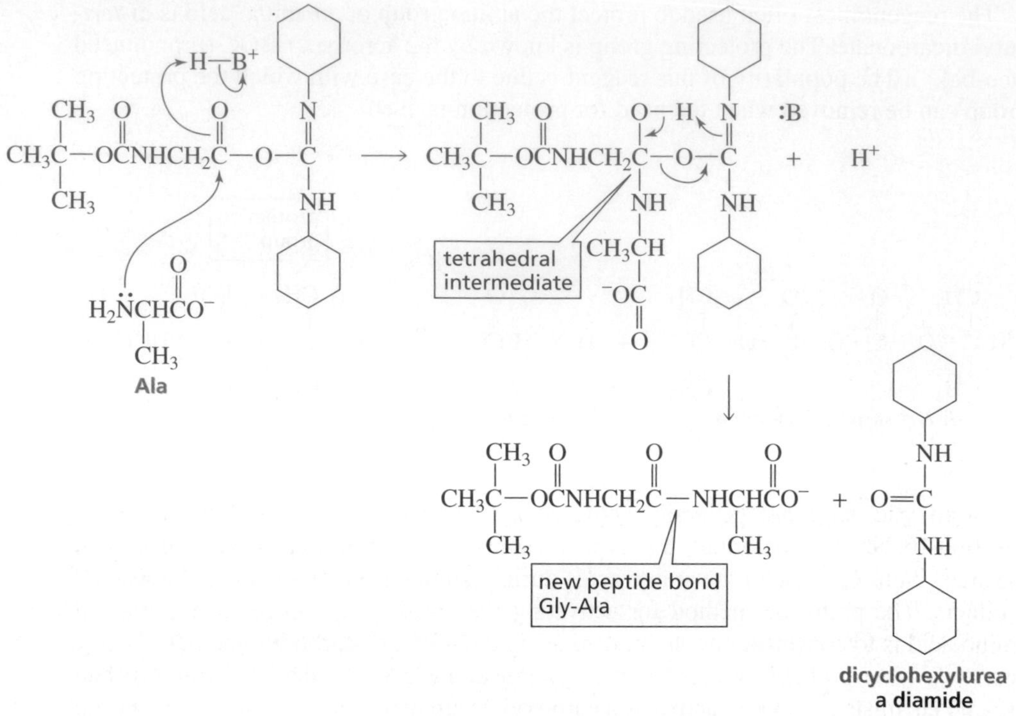

Amino acids can be added to the growing C-terminal end by repeating the same two steps: activating the carboxyl group of the C-terminal amino acid of the peptide by treating it with DCC and then adding a new amino acid.

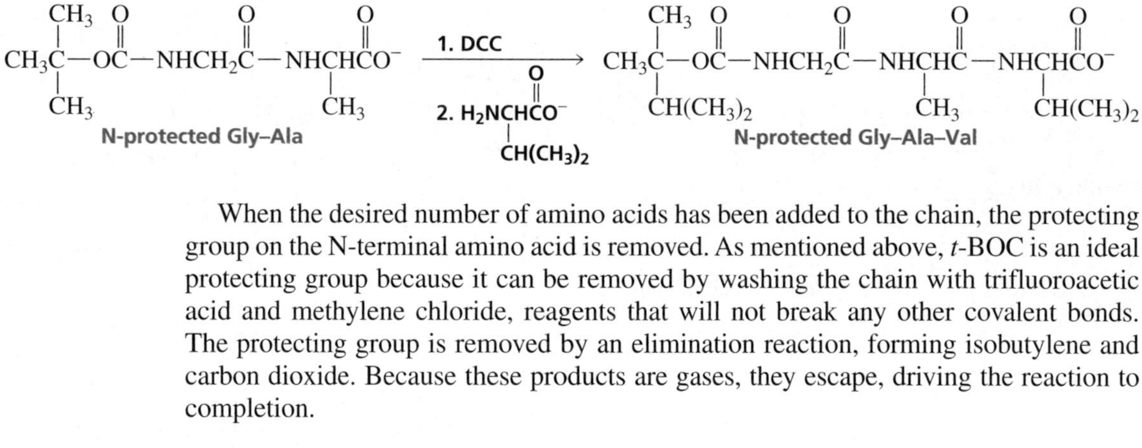

When the desired number of amino acids has been added to the chain, the protecting group on the N-terminal amino acid is removed. As mentioned above, *t*-BOC is an ideal protecting group because it can be removed by washing the chain with trifluoroacetic acid and methylene chloride, reagents that will not break any other covalent bonds. The protecting group is removed by an elimination reaction, forming isobutylene and carbon dioxide. Because these products are gases, they escape, driving the reaction to completion.

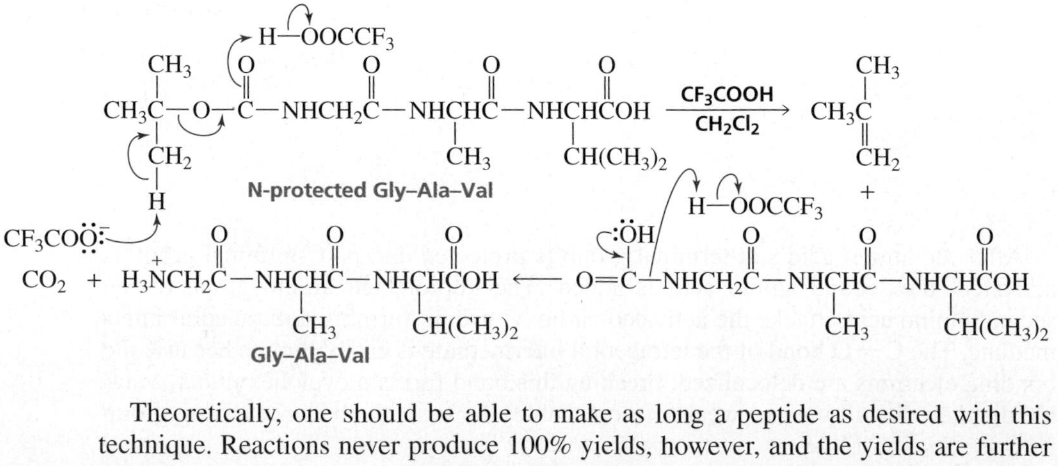

Theoretically, one should be able to make as long a peptide as desired with this technique. Reactions never produce 100% yields, however, and the yields are further decreased during the purification process. (In addition, after each step of the synthesis, the peptide must be purified to prevent subsequent unwanted reactions with leftover reagents.) Assuming that each amino acid can be added to the growing end of the peptide chain with an 80% yield (a relatively high yield, as you can probably appreciate from your own experience in the laboratory), the overall yield of the synthesis of a nonapeptide such as bradykinin would be only 17%. It is clear that large polypeptides could never be efficiently synthesized in this way.

Number of amino acids

	2	3	4	5	6	7	8	9
Overall yield	80%	64%	51%	41%	33%	26%	21%	17%

PROBLEM 29◆

What dipeptides would be formed by heating a mixture of valine and N-protected leucine?

PROBLEM 30

Suppose you are trying to synthesize the dipeptide Val-Ser. Compare the product that would be obtained if thionyl chloride were used to activate the carboxyl group of N-protected valine with the product that would be obtained if the carboxyl group were activated with DCC.

PROBLEM 31

Show the steps in the synthesis of the tetrapeptide Leu-Phe-Lys-Val.

PROBLEM 32◆

a. Calculate the overall yield of bradykinin when the yield for the addition of each amino acid to the chain is 70%.
b. What would be the overall yield of a peptide containing 15 amino acid residues if the yield for the incorporation of each is 80%?

22.11 **Automated Peptide Synthesis**

In addition to producing low overall yields, the method of peptide synthesis described in Section 22.10 is extremely time-consuming because the product must be purified at each step of the synthesis. In 1969, Bruce Merrifield described a method that revolutionized the synthesis of peptides because it provided a much faster way to produce peptides in much higher yields. Furthermore, because it is automated, the synthesis requires fewer hours of direct attention. With this technique, bradykinin was synthesized in 27 hours with an 85% yield. Subsequent refinements in the technique now allow a reasonable yield of a peptide containing 100 amino acids to be synthesized in four days.

In the Merrifield method, the C-terminal amino acid is covalently attached to a solid support in a column. Each N-terminal-protected amino acid is then added one at a time, along with other needed reagents, so the protein is synthesized from the C-terminal end to the N-terminal end. Notice that this is opposite to the way proteins are synthesized in nature (from the N-terminal end to the C-terminal end; Section 27.8). Because his process uses a solid support and is automated, Merrifield's method of protein synthesis is called **automated solid-phase peptide synthesis**.

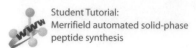

Merrifield automated solid-phase synthesis of a tripeptide

$$CH_3C \begin{matrix} CH_3 \\ | \\ C \\ || \\ CH_2 \end{matrix} + CO_2 + \underset{R}{H_2NCHC} \overset{O}{\underset{||}{C}} - \underset{R}{NHCHC} \overset{O}{\underset{||}{C}} - \underset{R}{NHCHCO} - CH_2 - \bigcirc - \bullet$$

$$\downarrow HF$$

$$\underset{R}{H_3NCHC} \overset{O}{\underset{||}{C}} - \underset{R}{NHCHC} \overset{O}{\underset{||}{C}} - \underset{R}{NHCHCOH} + HOCH_2 - \bigcirc - \bullet$$

The solid support to which the C-terminal amino acid is attached is a resin similar to the one used in ion-exchange chromatography (Section 22.5), except that the benzene rings have chloromethyl substituents instead of sulfonic acid substituents. Before the C-terminal amino acid is attached to the resin, its amino group is protected with *t*-BOC to prevent the amino group from reacting with the resin. The C-terminal amino acid is attached to the resin by means of an S_N2 reaction: its carboxyl group attacks a benzyl carbon of the resin, displacing a chloride ion (Section 8.4).

After the C-terminal amino acid is attached to the resin, the *t*-BOC protecting group is removed (Section 22.9). The next amino acid, with its amino group protected with *t*-BOC and its carboxyl group activated with DCC, is added to the column.

A huge advantage of the Merrifield method of peptide synthesis is that the growing peptide can be purified by washing the column with an appropriate solvent after each step of the procedure. The impurities are washed out of the column because they are not attached to the solid support. Since the peptide is covalently attached to the resin, none of it is lost in the purification step, leading to high yields of purified product.

After the required amino acids have been added one by one, the peptide can be removed from the resin by treatment with HF under mild conditions that do not break the peptide bonds.

Merrifield's technique is constantly being improved so that peptides can be made more rapidly and more efficiently. However, it still cannot begin to compare with nature. A bacterial cell is able to synthesize a protein containing thousands of amino acids in seconds and can simultaneously synthesize thousands of different proteins with no mistakes.

Since the early 1980s, it has been possible to synthesize proteins by genetic engineering techniques. Strands of DNA, introduced into bacterial cells, will cause the cells to produce large amounts of a desired protein (Section 27.12). Until genetic engineering techniques became available, cattle and pigs were the sources of insulin for people with diabetes. The insulin was effective, but there were concerns about whether enough could be obtained over the long term for the growing population of diabetics. In addition, this insulin did not have exactly the same primary structure as human insulin, so there were concerns about allergic reactions. Now, however, mass quantities of synthetic insulin, chemically identical to human insulin, are produced from genetically modified *E. coli*. Genetic engineering techniques also have been useful in synthesizing proteins that differ in one or a few amino acids from a natural protein. Such synthetic proteins have been used, for example, to learn how a change in a single amino acid affects the properties of a protein (Section 23.9).

PROBLEM 33

Show the steps in the synthesis of the tetrapeptide in Problem 31, using Merrifield's method.

22.12 **An Introduction to Protein Structure**

Proteins are described by four levels of structure. The **primary structure** of a protein is the sequence of amino acids in the chain and the location of all the disulfide bridges. **Secondary structures** are regular conformations assumed by segments of the protein's backbone when it folds. The **tertiary structure** is the three-dimensional shape of the entire polypeptide. If a protein has more than one polypeptide chain, it also has quaternary structure. The **quaternary structure** is the way the individual polypeptide chains are arranged with respect to each other.

PRIMARY STRUCTURE AND EVOLUTION

When scientists examine the primary structures of proteins that carry out the same function in different organisms, they can correlate the number of amino acid differences in the proteins to the closeness of the taxonomic relationship between the species. For example, cytochrome *c*, a protein that transfers electrons in biological oxidations, has about 100 amino acid residues. Yeast cytochrome *c* differs by 48 amino acids from horse cytochrome *c*, whereas duck cytochrome *c* differs by only two amino acids from chicken cytochrome *c*. Ducks and chickens have a much closer taxonomic relationship than horses and yeast. Likewise, the cytochrome *c* in chickens and turkeys have identical primary structures. Humans and chimpanzees also have identical cytochrome *c*'s, differing by one amino acid from the cytochrome *c* of the rhesus monkey.

BIOGRAPHY

Insulin was the first protein to have its primary sequence determined. This was done in 1953 by **Frederick Sanger**, *who was awarded the 1958 Nobel Prize in chemistry for his work. Sanger was born in England in 1918 and received a Ph.D. from Cambridge University, where he has worked for his entire career. He also received a share of the 1980 Nobel Prize in chemistry (Section 27.10) for being the first to sequence a DNA molecule (with 5375 nucleotide pairs).*

22.13 How to Determine the Primary Structure of a Peptide or a Protein

The first step in determining the sequence of amino acids in a peptide or a protein is to reduce any disulfide bridges in the peptide or protein. A commonly used reducing agent is 2-mercaptoethanol, which is oxidized to a disulfide in this reaction. Reaction of the protein thiol groups with iodoacetic acid prevents the disulfide bridges from reforming as a result of oxidation by O_2.

cleaving disulfide bridges

$$+ \ 2 \ HSCH_2CH_2OH \longrightarrow$$
2-mercaptoethanol

$$+ \ \begin{array}{c} SCH_2CH_2OH \\ | \\ SCH_2CH_2OH \end{array}$$

$$\underset{\text{iodoacetic acid}}{ICH_2COH}$$

$$+ \ 2 \ HI$$

PROBLEM 34

Write the mechanism for the reaction of a cysteine residue with iodoacetic acid.

The next step is to determine the number and kinds of amino acids in the peptide or protein. To do this, a sample of the peptide or protein is dissolved in 6 M HCl and heated at 100 °C for 24 hours. This treatment hydrolyzes all the amide bonds in the protein, including the amide bonds in the side chains of asparagine and glutamine.

$$\text{protein} \xrightarrow[\substack{100 \text{ °C} \\ 24 \text{ h}}]{6 \text{ M HCl}} \text{amino acids}$$

The mixture of amino acids is then passed through an amino acid analyzer to determine the number and kind of each amino acid in the peptide or protein (Section 22.5).

Because all the asparagine and glutamine residues have been hydrolyzed to aspartate and glutamate residues, the number of aspartate or glutamate residues in the amino acid mixture tells us the number of aspartate plus asparagine or glutamate plus glutamine residues in the original protein. A separate test must be conducted to distinguish between aspartate and asparagine or between glutamate and glutamine in the original protein.

The strongly acidic conditions used for hydrolysis destroy all the tryptophan residues, because the indole ring is unstable in acid (Section 20.8). However, the tryptophan content can be determined by hydroxide-ion-promoted hydrolysis of the protein. This is not a general method for peptide bond hydrolysis because the strongly basic conditions destroy several other amino acid residues.

One of the most widely used methods to identify the N-terminal amino acid of a peptide or protein is to treat the protein with phenyl isothiocyanate (PITC), more commonly known as **Edman's reagent**. This reagent reacts with the N-terminal amino group, and the resulting thiazolinone derivative is cleaved from the protein under mildly acidic conditions, leaving behind a peptide with one fewer amino acids. The thiazolinone derivative is extracted into an organic solvent and in the presence of acid, rearranges to a more stable phenylthiohydantoin (PTH).

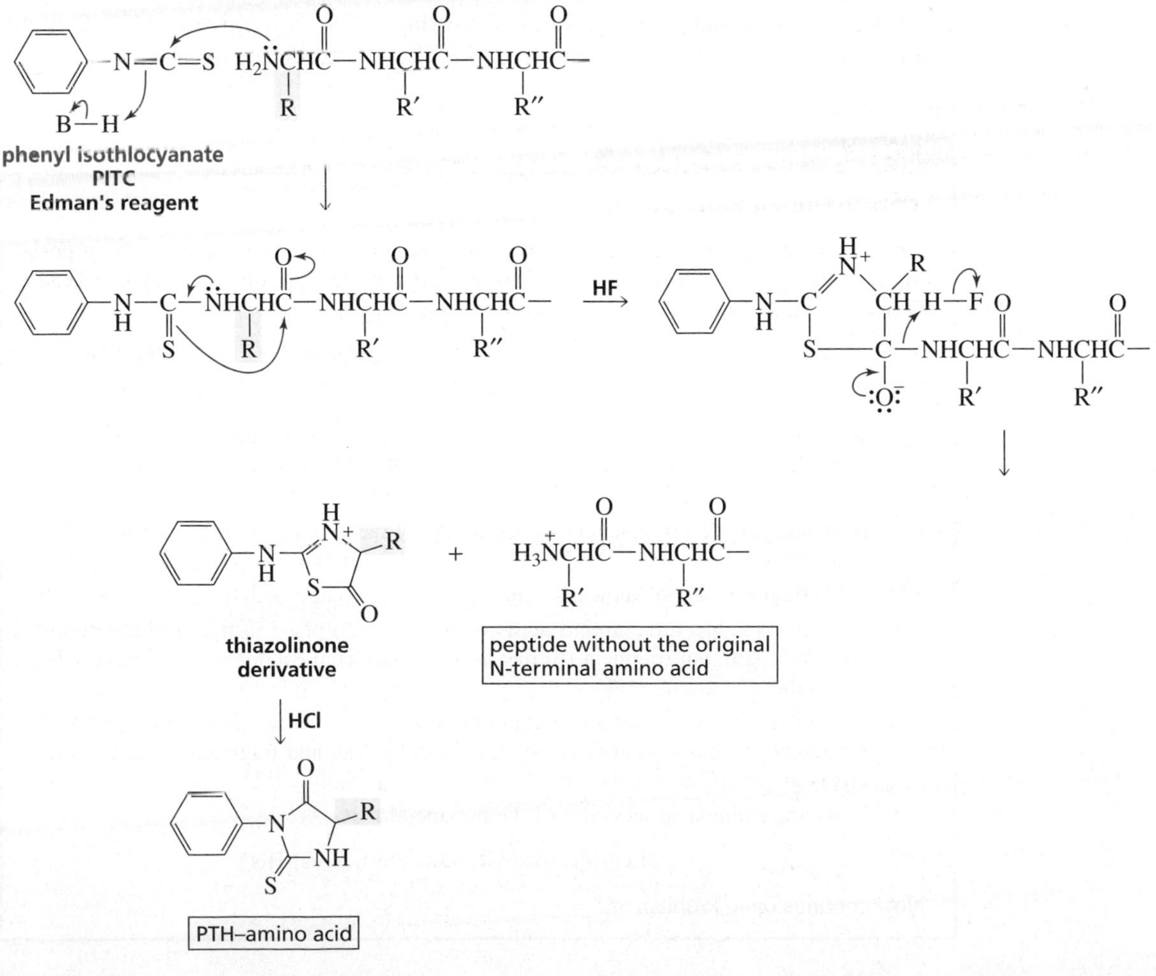

Because each amino acid has a different substituent (R), each amino acid forms a different PTH–amino acid. The particular PTH–amino acid can be identified by chromatography using known standards. Several successive Edman degradations can be carried out on a protein. An automated instrument known as a *sequenator* allows about 50 successive Edman degradations to be performed. The entire primary sequence cannot be determined in this way, however, because side products accumulate that interfere with the results.

The C-terminal amino acid of the peptide or protein can be identified by treating the protein with a peptidase called carboxypeptidase A. A **peptidase** is an enzyme that catalyzes the hydrolysis of a peptide bond. Carboxypeptidase A catalyzes the hydrolysis of the C-terminal peptide bond, cleaving off the C-terminal amino acid, as long as it is *not* arginine or lysine (Section 23.9). On the other hand, carboxypeptidase B cleaves off the C-terminal amino acid *only* if it is arginine or lysine. Carboxypeptidase A and B are more specifically called exopeptidases; an **exopeptidase** is an enzyme that catalyzes the hydrolysis of a peptide bond at the end of a peptide chain.

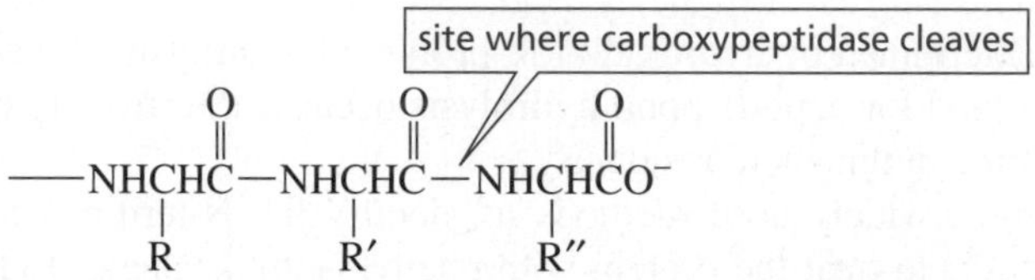

$$\overset{\text{O}}{\overset{\|}{\text{——NHCHC}}}-\overset{\text{O}}{\overset{\|}{\underset{\text{R}'}{\text{NHCHC}}}}-\overset{\text{O}}{\overset{\|}{\underset{\text{R}''}{\text{NHCHCO}^-}}}$$

site where carboxypeptidase cleaves

Once the N-terminal and C-terminal amino acids have been identified, a sample of the protein is hydrolyzed with dilute acid. This treatment, called **partial hydrolysis**, hydrolyzes only some of the peptide bonds. The resulting fragments are separated and the amino acid composition of each determined by using electrophoresis on an amino acid analyzer. The sequence of the original protein can then be deduced by lining up the peptides and looking for regions of overlap. (The N-terminal and C-terminal amino acids of each fragment can also be identified, if needed.)

PROBLEM-SOLVING STRATEGY

Sequencing an Oligopeptide

A nonapeptide undergoes partial hydrolysis to give peptides whose amino acid compositions are shown below. Reaction of the intact nonapeptide with Edman's reagent releases PTH-Leu. What is the sequence of the nonapeptide?

1. Pro, Ser	**3.** Met, Ala, Leu	**5.** Glu, Ser, Val, Pro	**7.** Met, Leu
2. Gly, Glu	**4.** Gly, Ala	**6.** Glu, Pro, Gly	**8.** His, Val

- Because we know that the N-terminal amino acid is Leu, we need to look for a fragment that contains Leu. Fragment (7) tells us that Met is next to Leu and fragment (3) tells us that Ala is next to Met.
- Now we look for a fragment that contains Ala. Fragment (4) contains Ala and tells us that Gly is next to Ala.
- From fragment (2), we know that Glu comes next; Glu is in both fragments (5) and (6).
- Fragment (5) has three amino acids we have yet to place in the growing peptide (Ser, Val, Pro), but fragment (6) has only one, so from fragment (6), we know that Pro is the next amino acid.
- Fragment (1) tells us that the next amino acid is Ser; now we can use fragment (5). Fragment (5) tells us that the next amino acid is Val, and fragment (8) tells us that His is the last (C-terminal) amino acid.
- Thus, the amino acid sequence of the nonapeptide is

  Leu-Met-Ala-Gly-Glu-Pro-Ser-Val-His

Now continue on to Problem 35.

PROBLEM 35◆

A decapeptide undergoes partial hydrolysis to give peptides whose amino acid compositions are shown. Reaction of the intact decapeptide with Edman's reagent releases PTH-Gly. What is the sequence of the decapeptide?

1. Ala, Trp	**3.** Pro, Val	**5.** Trp, Ala, Arg	**7.** Glu, Ala, Leu
2. Val, Pro, Asp	**4.** Ala, Glu	**6.** Arg, Gly	**8.** Met, Pro, Leu, Glu

The peptide or protein can also be partially hydrolyzed using endopeptidases. An **endopeptidase** is an enzyme that catalyzes the hydrolysis of a peptide bond that is not at the end of a peptide chain. Trypsin, chymotrypsin, and elastase are endopeptidases that catalyze the hydrolysis of only the specific peptide bonds listed in Table 22.4. Trypsin, for example, catalyzes the hydrolysis of the peptide bond on the C-side of only arginine or lysine residues.

C-side of lysine
C-side of arginine
lysine
arginine

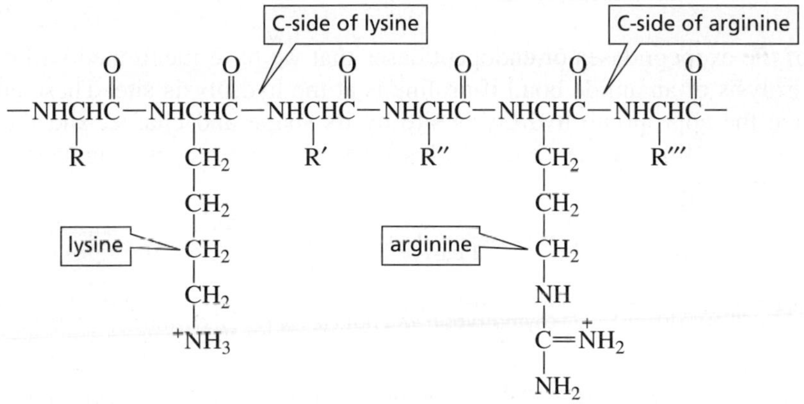

Table 22.4 Specificity of Peptide or Protein Cleavage

Reagent	Specificity
Chemical reagents	
Edman's reagent	removes the N-terminal amino acid
Cyanogen bromide	hydrolyzes on the C-side of Met
Exopeptidases*	
Carboxypeptidase A	removes the C-terminal amino acid (not Arg or Lys)
Carboxypeptidase B	removes the C-terminal amino acid (only Arg or Lys)
Endopeptidases*	
Trypsin	hydrolyzes on the C-side of Arg and Lys
Chymotrypsin	hydrolyzes on the C-side of amino acids that contain aromatic six-membered rings (Phe, Tyr, Trp)
Elastase	hydrolyzes on the C-side of small amino acids (Gly and Ala)

* Cleavage will not occur if Pro is on either side of the bond to be hydrolyzed.

Thus, trypsin will catalyze the hydrolysis of three peptide bonds in the following peptide, creating a hexapeptide, a dipeptide, and two tripeptides.

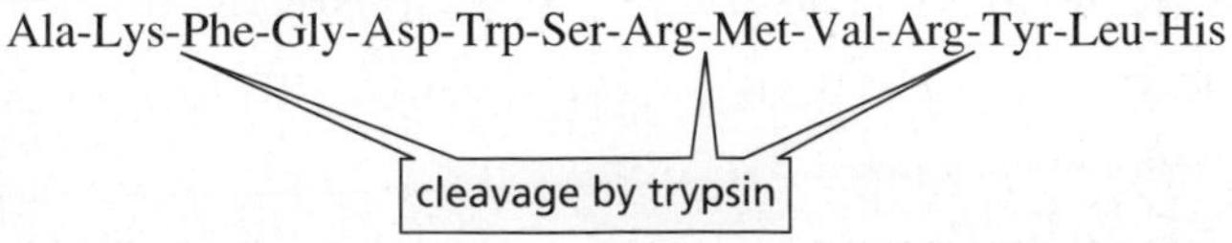

Ala-Lys-Phe-Gly-Asp-Trp-Ser-Arg-Met-Val-Arg-Tyr-Leu-His
cleavage by trypsin

Chymotrypsin catalyzes the hydrolysis of the peptide bond on the C-side of amino acids that contain aromatic six-membered rings (Phe, Tyr, Trp).

Ala-Lys-Phe-Gly-Asp-Trp-Ser-Arg-Met-Val-Arg-Tyr-Leu-His

cleavage by chymotrypsin

Elastase catalyzes the hydrolysis of peptide bonds on the C-side of the two smallest amino acids (Gly, Ala). Chymotrypsin and elastase are much less specific than trypsin. (An explanation for the specificity of these enzymes is given in Section 23.9.)

Ala-Lys-Phe-Gly-Asp-Trp-Ser-Arg-Met-Val-Arg-Tyr-Leu-His

cleavage by elastase

None of the exopeptidases or endopeptidases that we have mentioned will catalyze the hydrolysis of an amide bond if proline is at the hydrolysis site. These enzymes recognize the appropriate hydrolysis site by its shape and charge, and the cyclic structure of proline causes the hydrolysis site to have an unrecognizable three-dimensional shape.

Ala-Lys-Pro   Leu-Phe-Pro   Pro-Phe-Val

trypsin will not cleave   chymotrypsin will not cleave   chymotrypsin will cleave

Cyanogen bromide (BrC≡N) causes the hydrolysis of the peptide bond on the C-side of a methionine residue. Cyanogen bromide is more specific than the endopeptidases about what peptide bonds it cleaves, so it provides more reliable information about the primary structure (the sequence of amino acids). Cyanogen bromide is not a protein and therefore does not recognize the substrate by its shape, so it will still cleave the peptide bond if proline is at the cleavage site.

Ala-Lys-Phe-Gly-Met-Pro-Ser-Arg-Met-Val-Arg-Tyr-Leu-His

cleavage by cyanogen bromide

The mechanism for cleavage of a peptide bond by cyanogen bromide is shown below.

Mechanism for the cleavage of a peptide bond by cyanogen bromide

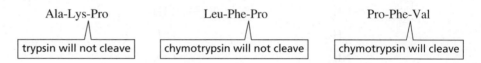

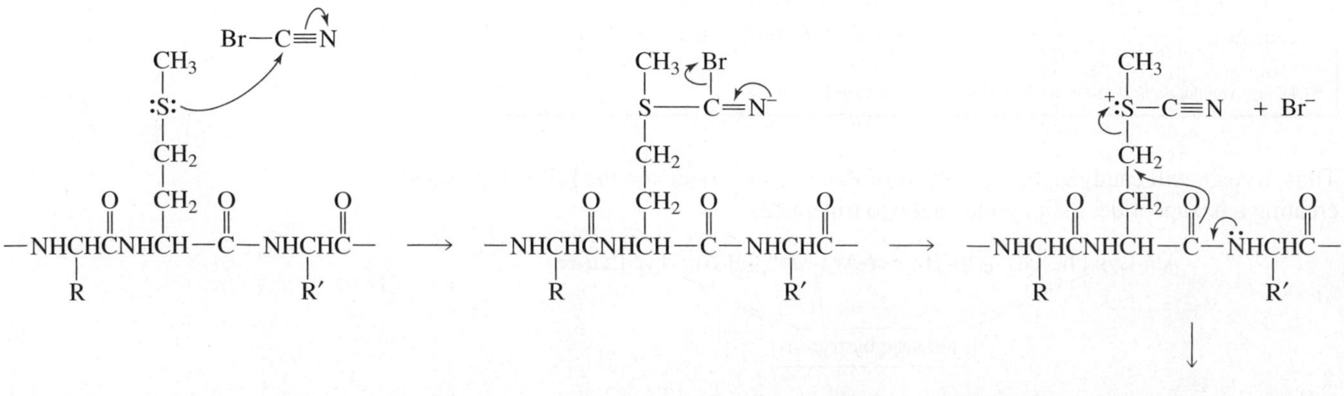

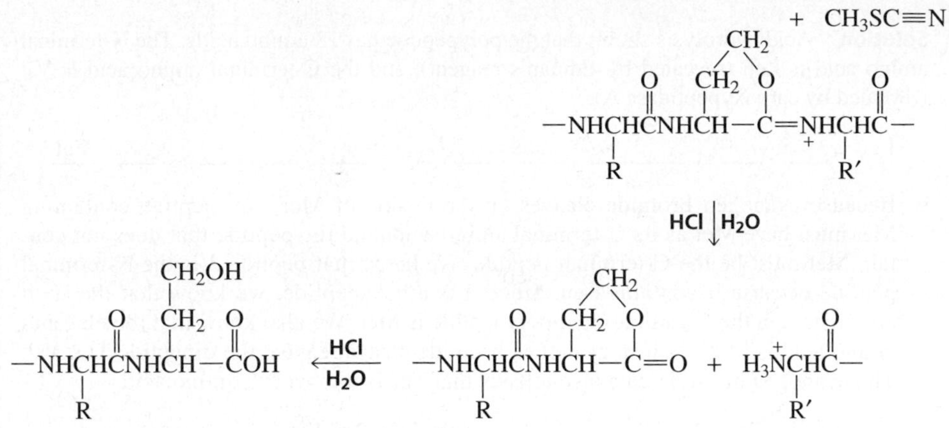

$$+ \quad CH_3SC{\equiv}N$$

$$\underset{\begin{array}{c}\\ \\ R \end{array}}{-NHCHC}\overset{O}{\underset{}{\|}}\underset{}{NHCH}\overset{\underset{}{CH_2}}{\underset{}{\overset{CH_2}{|}}}\overset{O}{\underset{}{\|}}C{=}\underset{+}{NHCHC}\overset{O}{\underset{}{\|}}\underset{R'}{-}$$

$$HCl \bigg\downarrow H_2O$$

$$\underset{R}{-NHCHCNHCH-COH} \overset{HCl}{\underset{H_2O}{\longleftarrow}} \underset{R}{-NHCHCNHCH-C}{=}O \quad + \quad \underset{R'}{H_3\overset{+}{N}CHC}-$$

(with substituents CH_2OH, CH_2, CH_2 as shown)

- The highly nucleophilic sulfur of methionine attacks the carbon of cyanogen bromide.
- Nucleophilic attack by the oxygen on the methylene group, resulting in departure of the weakly basic leaving group, forms a five-membered ring.
- Acid-catalyzed hydrolysis of the imine cleaves the protein (Section 17.8).
- Further hydrolysis causes the lactone (a cyclic ester) to open to a carboxyl group and an alcohol group (Section 16.11).

The last step in determining the primary structure of a protein is to figure out the location of any disulfide bonds. This is done by hydrolyzing a sample of the protein that has intact disulfide bonds. From a determination of the amino acids in the cysteine-containing fragments, the locations of the disulfide bonds in the protein can be established (Problem 52).

PROBLEM 36◆

Indicate the peptides that would result from cleavage by the indicated reagent:

a. His-Lys-Leu-Val-Glu-Pro-Arg-Ala-Gly-Ala by trypsin
b. Leu-Gly-Ser-Met-Phe-Pro-Tyr-Gly-Val by chymotrypsin

PROBLEM 37

Why does cyanogen bromide not cleave at cysteine residues?

PROBLEM 38 **SOLVED**

Determine the amino acid sequence of a polypeptide from the following data:

Acid hydrolysis gives Ala, Arg, His, 2 Lys, Leu, 2 Met, Pro, 2 Ser, Thr, Val.

Carboxypeptidase A releases Val.

Edman's reagent releases PTH-Leu.

Cleavage with cyanogen bromide gives three peptides with the following amino acid compositions:

   1. His, Lys, Met, Pro, Ser   **3.** Ala, Arg, Leu, Lys, Met, Ser
   2. Thr, Val

Trypsin-catalyzed hydrolysis gives three peptides and a single amino acid:

   1. Arg, Leu, Ser   **3.** Lys
   2. Met, Pro, Ser, Thr, Val   **4.** Ala, His, Lys, Met

Solution   Acid hydrolysis shows that the polypeptide has 13 amino acids. The N-terminal amino acid is Leu (revealed by Edman's reagent), and the C-terminal amino acid is Val (revealed by carboxypeptidase A).

Leu __ __ __ __ __ __ __ __ __ __ __ Val

- Because cyanogen bromide cleaves on the C-side of Met, any peptide containing Met must have Met as its C-terminal amino acid, and the peptide that does not contain Met must be the C-terminal peptide. We know that peptide 3 is the N-terminal peptide because it contains Leu. Since it is a hexapeptide, we know that the sixth amino acid in the 13-amino-acid polypeptide is Met. We also know that the eleventh amino acid is Met because cyanogen bromide cleavage gave the dipeptide Thr, Val. The cyanogen bromide data also tells us that Thr is the twelfth amino acid.

Ala, Arg, Lys, Ser   His, Lys, Pro, Ser

Leu __ __ __ __ Met __ __ __ __ Met Thr Val

- Because trypsin cleaves on the C-side of Arg and Lys, any peptide containing Arg or Lys must have that amino acid as its C-terminal amino acid. Therefore, Arg is the C-terminal amino acid of peptide 1, so we now know that the first three amino acids are Leu-Ser-Arg. We also know that the next two are Lys-Ala because if they were Ala-Lys, trypsin cleavage would give an Ala, Lys dipeptide. The trypsin data also identify the positions of His and Lys.

Pro, Ser

Leu Ser Arg Lys Ala Met His Lys __ __ Met Thr Val

- Finally, because trypsin successfully cleaves on the C-side of Lys, Pro cannot be adjacent to Lys. Thus, the amino acid sequence of the polypeptide is

Leu Ser Arg Lys Ala Met His Lys Ser Pro Met Thr Val

PROBLEM 39◆

Determine the primary structure of an octapeptide from the following data:

Acid hydrolysis gives 2 Arg, Leu, Lys, Met, Phe, Ser, Tyr.

Carboxypeptidase A releases Ser.

Edman's reagent releases Leu.

Cyanogen bromide forms two peptides with the following amino acid compositions:
   1. Arg, Phe, Ser   **2.** Arg, Leu, Lys, Met, Tyr

Trypsin forms the following two amino acids and two peptides:
   1. Arg   **3.** Arg, Met, Phe
   2. Ser   **4.** Leu, Lys, Tyr

PROBLEM 40◆

While determining the primary structure of insulin, what would lead you to conclude that insulin had more than one polypeptide chain?

22.14 Secondary Structure of Proteins

Secondary structure describes the repetitive conformations assumed by segments of the backbone chain of a peptide or protein. In other words, the secondary structure describes how segments of the backbone fold. Three factors determine the secondary structure of a stretch of protein:

- the regional planarity about each peptide bond (as a result of the partial double-bond character of the amide bond), which limits the possible conformations of the peptide chain (Section 22.7)
- the maximizing of the number of peptide groups that engage in hydrogen bonding to minimize energy (that is, hydrogen bonding between the carbonyl oxygen of one amino acid residue and the amide hydrogen of another)
- the need for adequate separation between neighboring R groups to avoid steric strain and repulsion of like charges

α-Helix

One type of secondary structure is the **α-helix**. In an α-helix, the backbone of the polypeptide coils around the long axis of the protein molecule. The substituents on the α-carbons of the amino acids protrude outward from the helix, thereby minimizing steric strain (Figure 22.8a). The helix is stabilized by hydrogen bonds (Figure 22.8b): each hydrogen attached to an amide nitrogen is hydrogen bonded to a carbonyl oxygen of an amino acid four amino acids away. Because the amino acids have the L-configuration, the α-helix is a right-handed helix; that is, it rotates in a clockwise direction as it spirals down (Figure 22.8c). Each turn of the helix contains 3.6 amino acid residues, and the repeat distance of the helix is 5.4 Å.

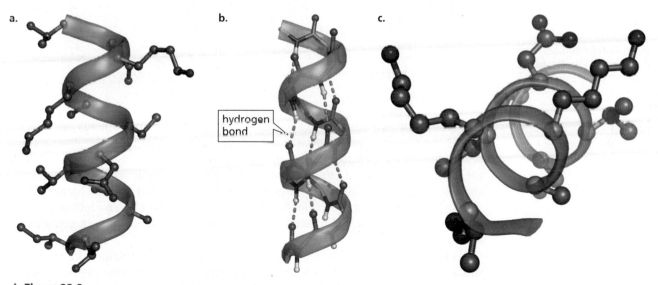

▲ **Figure 22.8**
(a) A segment of a protein in an α-helix. (b) The helix is stabilized by hydrogen bonding between peptide groups. (c) Looking at the longitudinal axis of an α-helix.

Not all amino acids are able to fit into an α-helix. A proline residue, for example, causes a distortion in a helix because the bond between the proline nitrogen and the α-carbon cannot rotate to let proline fit into a helix properly. Similarly, two adjacent amino acids that have more than one substituent on a β-carbon (valine, isoleucine, or threonine) cannot fit into a helix because of steric crowding between the R groups. Finally, two adjacent amino acids with like-charged substituents cannot fit into a helix because of electrostatic repulsion between the R groups. The percentage of amino acid residues coiled into an α-helix varies from protein to protein, but, on average, about 25% of the residues in globular proteins are in α-helices.

β-Pleated Sheet

The second type of secondary structure is the **β-pleated sheet**. In a β-pleated sheet, the polypeptide backbone is extended in a zigzag structure resembling a series of pleats. A β-pleated sheet is almost fully extended—the average two-residue repeat distance is 7.0 Å. The hydrogen bonding in a β-pleated sheet occurs between neighboring peptide chains, and these chains can run in the same direction or in opposite directions. In a **parallel β-pleated sheet**, the adjacent chains run in the same direction. In an **antiparallel β-pleated sheet**, the adjacent chains run in opposite directions (Figure 22.9).

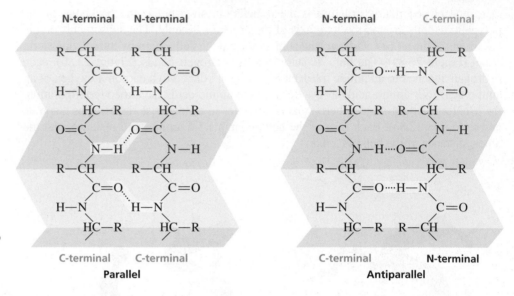

Figure 22.9 ▶
Segments of β-pleated sheets drawn to illustrate their pleated character. Note that the first is parallel and the second is antiparallel.

Because the substituents (R) on the α-carbons of the amino acids on adjacent chains are close to each other, the substituents must be small if the chains are to nestle closely enough together to maximize hydrogen-bonding interactions. Silk, for example, contains a large proportion of relatively small amino acids (glycine and alanine) and therefore has large segments of β-pleated sheet. The number of side-by-side strands in a β-pleated sheet ranges from 2 to 15 in a globular protein. The average strand in a β-pleated sheet section of a globular protein contains six amino acid residues.

Wool and the fibrous protein of muscle are examples of proteins with secondary structures that are almost all α-helices. Consequently, these proteins can be stretched. In contrast, proteins with secondary structures that are predominantly β-pleated sheets, such as silk and spider webs, cannot be stretched because a β-pleated sheet is almost fully extended.

Coil Conformation

Generally, less than half of the backbone of a globular protein is arranged in a defined secondary structure, whether either an α-helix or a β-pleated sheet (Figure 22.10). Most of the rest of the protein, however, although highly ordered, is difficult to describe. Many of these polypeptide fragments are said to be in **coil** or **loop conformations**.

PROBLEM 41◆

How long is an α-helix that contains 74 amino acids? Compare the length of this α-helix with the length of a fully extended peptide chain containing the same number of amino acids. (The distance between consecutive amino acids in a fully extended chain is 3.5 Å.)

◀ **Figure 22.10**
The backbone structure of carboxypeptidase A: α-helical segments are purple; β-pleated sheets are indicated by flat green arrows pointing in the N → C direction.

β-PEPTIDES: AN ATTEMPT TO IMPROVE ON NATURE

Chemists are currently studying β-peptides, which are polymers of β-amino acids. These peptides have backbones one carbon longer than the peptides nature synthesizes using α-amino acids. Therefore, each β-amino acid residue has two carbons to which side chains can be attached.

Like α-polypeptides, β-polypeptides fold into relatively stable helical and pleated sheet conformations, causing scientists to wonder whether biological activity might be possible with such peptides. Recently, a β-peptide with biological activity has been synthesized that mimics the activity of the hormone somatostatin. There is hope that β-polypeptides will provide a source of new drugs and catalysts. Surprisingly, the peptide bonds in β-polypeptides are resistant to the enzymes that catalyze the hydrolysis of peptide bonds in α-polypeptides. This resistance to hydrolysis means that a β-polypeptide drug would have a longer duration of action in the bloodstream.

22.15 Tertiary Structure of Proteins

The *tertiary structure* of a protein is the three-dimensional arrangement of all the atoms in the protein. Proteins fold spontaneously in solution to maximize their stability. Every time there is a stabilizing interaction between two atoms, free energy is released. The more free energy released (the more negative the $\Delta G°$), the more stable the protein. Consequently, a protein tends to fold in a way that maximizes the number of stabilizing interactions (Figure 22.11).

The stabilizing interactions in a protein include disulfide bonds, hydrogen bonds, electrostatic attractions (attractions between opposite charges), and hydrophobic (van der Waals) interactions (Figure 22.12). Stabilizing interactions can occur between peptide groups (atoms in the backbone of the protein), between side-chain groups (α-substituents), and between peptide and side-chain groups. Because the side-chain groups help determine how a protein folds, the tertiary structure of a protein is determined by its primary structure.

Disulfide bonds are the only covalent bonds that can form when a protein folds. The other bonding interactions that occur in folding are much weaker, but because there are so many of them, they are the important interactions in determining how a protein folds.

John Kendrew (1917–1997) *was born in England and educated at Cambridge University, where Max Perutz was working on the structure of hemoglobin (completed in 1959). Perutz assigned the work on myoglobin, a smaller protein, to Kendrew, who completed it in 1957.*

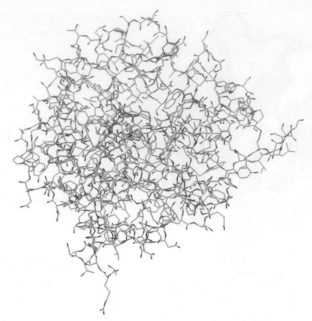

▲ **Figure 22.11**
The three-dimensional structure of carboxypeptidase A.

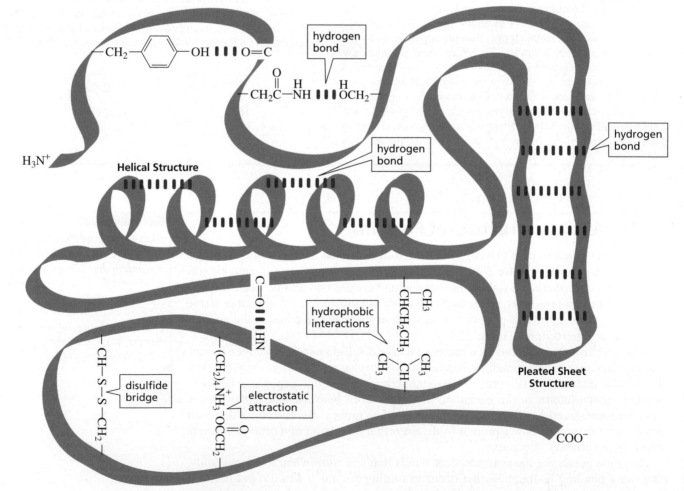

▲ **Figure 22.12**
Stabilizing interactions responsible for the tertiary structure of a protein.

Most proteins exist in aqueous environments. Therefore, they tend to fold in a way that exposes the maximum number of polar groups to the surrounding water and that buries the nonpolar groups in the protein's interior, away from water.

The **hydrophobic interactions** between nonpolar groups in the protein increase its stability by increasing the entropy of water molecules. Water molecules that surround nonpolar groups are highly structured. When two nonpolar groups come together, the surface area in contact with water decreases, decreasing the amount of structured water. Decreasing structure increases entropy, which in turn decreases the free energy, which increases the stability of the protein. (Recall that $\Delta G° = \Delta H° − T\Delta S°$.)

PROBLEM 42◆

How would a protein that resides in the interior of a membrane fold, compared with the water-soluble protein just discussed? (*Hint:* See Section 26.4.)

22.16 **Quaternary Structure of Proteins**

Proteins that have more than one peptide chain are called **oligomers**. The individual chains are called **subunits**. A protein with a single subunit is called a *monomer*; one with two subunits is called a *dimer*; one with three subunits is called a *trimer*; and one with four subunits is called a *tetramer.* The quaternary structure of a protein describes the way the subunits are arranged in space. Some of the possible arrangements of the six subunits of a hexamer, for example, are

possible quaternary structures for a hexamer

The subunits are held together by the same kinds of interactions that hold the individual protein chains in a particular three-dimensional conformation: hydrophobic interactions, hydrogen bonding, and electrostatic attractions. Hemoglobin is an example of a tetramer. It has two different kinds of subunits and two of each kind. The quaternary structure of hemoglobin is shown in Figure 22.13.

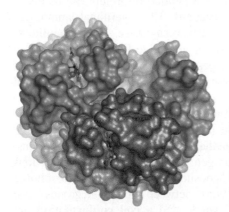

◀ **Figure 22.13**
A representation of the quaternary structure of hemoglobin generated by computer graphics. The orange and green represent the polypeptide chains; there are two identical orange subunits and two identical green subunits. Two of the porphyrin rings (gray beads; Section 20.10) are visible, ligated to iron (pink) and bonded to oxygen (red).

PROBLEM 43◆

a. Which water-soluble protein or subunit would have the greatest percentage of polar amino acids, a spherical protein, a cigar-shaped protein, or a subunit of a hexamer?

b. Which would have the smallest percentage of polar amino acids?

22.17 **Protein Denaturation**

Destroying the highly organized tertiary structure of a protein is called **denaturation**. Anything that breaks the bonds maintaining the three-dimensional shape of the protein will cause the protein to denature (unfold). Because these bonds are weak, proteins are easily denatured. The totally random conformation of a denatured protein is called a **random coil**. The following are some of the ways that proteins can be denatured:

- Changing the pH denatures proteins because it changes the charges on many of the side chains. This disrupts electrostatic attractions and hydrogen bonds.

- Certain reagents such as urea and guanidine hydrochloride denature proteins by forming hydrogen bonds to the protein groups that are stronger than the hydrogen bonds formed between the groups.

- Detergents such as sodium dodecyl sulfate denature proteins by associating with the nonpolar groups of the protein, thus interfering with the normal hydrophobic interactions.

- Organic solvents denature proteins by disrupting hydrophobic interactions.

- Proteins can also be denatured by heat or by agitation. Both increase molecular motion, which can disrupt the attractive forces. A well-known example is the change that occurs to the white of an egg when it is heated or whipped.

SUMMARY

Peptides and **proteins** are polymers of **amino acids** linked together by **peptide** (amide) bonds. A **dipeptide** contains 2 amino acid residues, a **tripeptide** contains 3, an **oligopeptide** contains 3 to 10, and a **polypeptide** contains many amino acid residues. Proteins have 40 to 4000 amino acid residues. The **amino acids** differ only in the substituent attached to the α-carbon. Almost all amino acids found in nature have the L configuration.

The carboxyl groups of the amino acids have pK_a values of ~2, and the protonated amino groups have pK_a values of ~9. At physiological pH, an amino acid exists as a **zwitterion**. A few amino acids have side chains with ionizable hydrogens. The **isoelectric point** (pI) of an amino acid is the pH at which the amino acid has no net charge. A mixture of amino acids can be separated based on their pI's by **electrophoresis** or based on their polarities by **paper chromatography** or **thin-layer chromatography**. Preparative separation can be achieved using **ion-exchange chromatography** employing an **anion-** or a **cation-exchange resin**. An **amino acid analyzer** is an instrument that automates ion-exchange chromatography. A racemic mixture of amino acids can be separated by a **kinetic resolution** using an enzyme-catalyzed reaction.

A peptide bond has restricted rotation because it has about 40% double-bond character. Two cysteine residues can be oxidized to a **disulfide bridge**, the only kind of covalent bond that is found between nonadjacent amino acids. By convention, peptides and proteins are written with the free amino group (the **N-terminal amino acid**) on the left and the free carboxyl group (the **C-terminal amino acid**) on the right.

To synthesize a peptide bond, the amino group of the N-terminal amino acid must be protected (for example, by t-BOC) and its carboxyl group activated (with DCC). The second amino acid is added to form a dipeptide. Amino acids can be added to the growing C-terminal end by repeating the same two steps: activating the carboxyl group of the C-terminal amino acid with DCC and adding a new amino acid. **Automated solid-phase peptide synthesis** allows peptides to be made more rapidly and in higher yields.

The **primary structure** of a protein is the sequence of its amino acids and the location of all its disulfide bridges. The N-terminal amino acid of a peptide or protein can be determined with **Edman's reagent**. The C-terminal amino acid can be identified with carboxypeptidase. An **exopeptidase** catalyzes the hydrolysis of a peptide bond at the end of a peptide chain. An **endopeptidase** catalyzes the hydrolysis of a peptide bond that is not at the end of a peptide chain. **Partial hydrolysis** hydrolyzes only some of the peptide bonds.

The **secondary structure** of a protein describes how local segments of the protein's backbone folds. A protein folds so as to maximize the number of stabilizing interactions: disulfide bonds, hydrogen bonds, electrostatic attractions (attraction between opposite charges), and **hydrophobic interactions** (interactions between nonpolar groups). An α-helix, a β-pleated sheet, and a **coil conformation** are types of secondary structure. The **tertiary structure** of a protein is the three-dimensional arrangement of all the atoms in the protein. Proteins with more than one peptide chain are called **oligomers**. The individual chains are called **subunits**. The **quaternary structure** of a protein describes the way the subunits are arranged with respect to each other in space.

KEY TERMS

amino acid (p. 1017)
D-amino acid (p. 1023)
L-amino acid (p. 1023)
amino acid analyzer (p. 1031)
amino acid residue (p. 1017)
anion-exchange resin (p. 1031)
antiparallel β-pleated sheet (p. 1054)
automated solid-phase peptide
  synthesis (p. 1043)
cation-exchange resin (p. 1031)
coil conformation (p. 1054)
C-terminal amino acid (p. 1035)
denaturation (p. 1058)
dipeptide (p. 1018)
disulfide (p. 1037)
disulfide bridge (p. 1037)
Edman's reagent (p. 1047)
electrophoresis (p. 1028)

endopeptidase (p. 1049)
essential amino acid (p. 1022)
exopeptidase (p. 1048)
fibrous protein (p. 1018)
globular protein (p. 1018)
α-helix (p. 1053)
hydrophobic interactions (p. 1057)
interchain disulfide bridge (p. 1038)
intrachain disulfide bridge (p. 1038)
ion-exchange chromatography
  (p. 1030)
isoelectric point (p. 1026)
kinetic resolution (p. 1035)
loop conformation (p. 1054)
N-terminal amino acid (p. 1035)
oligomer (p. 1057)
oligopeptide (p. 1018)
paper chromatography (p. 1029)

parallel β-pleated sheet (p. 1054)
partial hydrolysis (p. 1048)
peptidase (p. 1048)
peptide (p. 1017)
peptide bond (p. 1035)
β-pleated sheet (p. 1054)
polypeptide (p. 1018)
primary structure (p. 1045)
protein (p. 1017)
quaternary structure (p. 1046)
random coil (p. 1058)
secondary structure (p. 1045)
subunit (p. 1057)
tertiary structure (p. 1046)
thin-layer chromatography (p. 1030)
tripeptide (p. 1018)
zwitterion (p. 1025)

PROBLEMS

44. Explain why, unlike most amines and carboxylic acids, amino acids are insoluble in diethyl ether.

45. Show the peptides that would result from cleavage by the indicated reagent:
  a. Val-Arg-Gly-Met-Arg-Ala-Ser by carboxypeptidase A
  b. Ser-Phe-Lys-Met-Pro-Ser-Ala-Asp by cyanogen bromide
  c. Arg-Ser Pro-Lys-Lys-Ser-Glu-Gly by trypsin

46. Aspartame has a pI of 5.9. Draw its prevailing form at physiological pH.

47. Draw the form of aspartic acid that predominates at
  a. pH − 1.0   **b.** pH = 2.6   **c.** pH = 6.0   **d.** pH = 11.0

48. Dr. Kim S. Tree was preparing a manuscript for publication in which she reported that the pI of the tripeptide Lys-Lys-Lys was 10.6. One of her students pointed out that there must be an error in her calculations because the pK_a of the ε-amino group of lysine is 10.8 and the pI of the tripeptide has to be greater than any of its individual pK_a values. Was the student correct?

49. A mixture of amino acids that do not separate sufficiently when a single technique is used can often be separated by two-dimensional chromatography. In this technique, the mixture of amino acids is applied to a piece of filter paper and separated by chromatographic techniques. The paper is then rotated 90°, and the amino acids are further separated by electrophoresis, producing a type of chromatogram called a *fingerprint*. Identify the spots in the fingerprint obtained from a mixture of Ser, Glu, Leu, His, Met, and Thr.

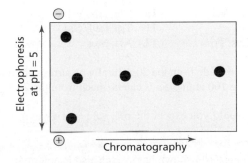

50. Explain the difference in the pK_a values of the carboxyl groups of alanine, serine, and cysteine.

51. Which would be a more effective buffer at physiological pH, a solution of 0.1 M glycylglycylglycylglycine or a solution of 0.2 M glycine?

52. Identify the location and type of charge on the hexapeptide Lys-Ser-Asp-Cys-His-Tyr at
 a. pH = 7. b. pH = 5. c. pH = 9.

53. Give the product obtained when a lysine residue in a polypeptide reacts with maleic anhydride.

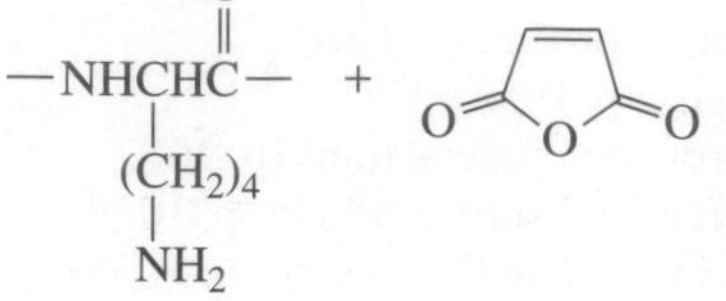

54. The following polypeptide was treated with 2-mercaptoethanol and then with iodoacetic acid. After reacting with maleic anhydride, the peptide was hydrolyzed by trypsin. (After a peptide is treated with maleic anhydride, trypsin will cleave it at arginine residues only.)

Gly-Ser-Asp-Ala-Leu-Pro-Gly-Ile-Thr-Ser-Arg-Asp-Val-Ser-Lys-Val-Glu-Tyr-Phe-
Glu-Ala-Gly-Arg-Ser-Glu-Phe-Lys-Glu-Pro-Arg-Leu-Tyr-Met-Lys-Val-Glu-Gly-
Arg-Pro-Val-Ser-Ala-Gly-Leu-Trp

 a. After a peptide is treated with maleic anhydride, why does trypsin no longer cleave it at lysine residues?
 b. How many fragments are obtained from the peptide?
 c. In what order would the fragments be eluted from an anion-exchange column using a buffer of pH = 5.

55. Treatment of a polypeptide with 2-mercaptoethanol yields two polypeptides with the following primary sequences:

Val-Met-Tyr-Ala-Cys-Ser-Phe-Ala-Glu-Ser

Ser-Cys-Phe-Lys-Cys-Trp-Lys-Tyr-Cys-Phe-Arg-Cys-Ser

Treatment of the original intact polypeptide with chymotrypsin yields the following peptides:
 1. Ala, Glu, Ser 3. Tyr, Val, Met 5. Ser, Phe, 2 Cys, Lys, Ala, Trp
 2. 2 Phe, 2 Cys, Ser 4. Arg, Ser, Cys 6. Tyr, Lys

Determine the positions of the disulfide bridges in the original polypeptide.

56. Show how aspartame can be synthesized using DCC.

57. Show how valine can be prepared by
 a. a Hell–Volhard–Zelinski reaction.
 b. a Strecker synthesis.
 c. a reductive amination.
 d. a N-phthalimidomalonic ester synthesis.
 e. an acetamidomalonic ester synthesis.

58. Reaction of a polypeptide with carboxypeptidase A releases Met. The polypeptide undergoes partial hydrolysis to give the following peptides. What is the sequence of the polypeptide?
 1. Ser, Lys, Trp 5. Met, Ala, Gly 9. Lys, Ser
 2. Gly, His, Ala 6. Ser, Lys, Val 10. Glu, His, Val
 3. Glu, Val, Ser 7. Glu, His 11. Trp, Leu, Glu
 4. Leu, Glu, Ser 8. Leu, Lys, Trp 12. Ala, Met

59. a. How many different octapeptides can be made from the 20 naturally occurring amino acids?
 b. How many different proteins containing 100 amino acids can be made from the 20 naturally occurring amino acids?

60. Glycine has pK_a values of 2.3 and 9.6. Would you expect the pK_a values of glycylglycine to be higher or lower than these values?

61. A mixture of 15 amino acids gave the fingerprint shown below (see also Problem 49). Identify the spots. (*Hint 1:* Pro reacts with ninhydrin to produce a yellow color; Phe and Tyr produce a yellow-green color. *Hint 2:* Count the number of spots before you start.)

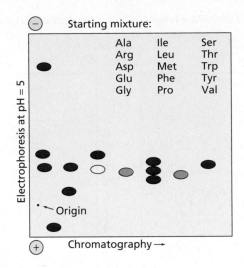

62. Dithiothreitol reacts with disulfide bridges in the same way that 2-mercaptoethanol does. With dithiothreitol, however, the equilibrium lies much more to the right. Explain.

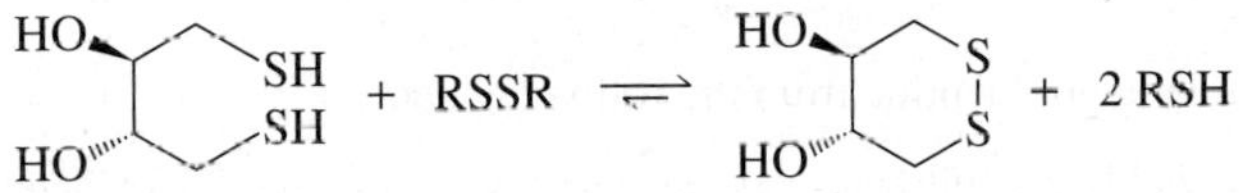

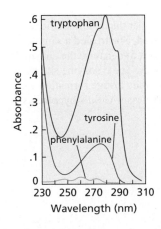

63. α-Amino acids can be prepared by treating an aldehyde with ammonia and hydrogen cyanide, followed by acid-catalyzed hydrolysis.
   a. Give the structures of the two intermediates formed in this reaction.
   b. What amino acid is formed when the aldehyde that is used is 3-methylbutanal?
   c. What aldehyde would be needed to prepare valine?

64. The UV spectra of tryptophan, tyrosine, and phenylalanine are shown here. Each spectrum is that of a 1×10^{-3} M solution of the amino acid, buffered at pH = 6.0, in a cell with a light path of 1 cm. Calculate the approximate molar absorptivity of each of the three amino acids at 280 nm.

65. A normal polypeptide and a mutant of the polypeptide were hydrolyzed by an endopeptidase under the same conditions. The normal and mutant polypeptide differ by one amino acid residue. The fingerprints of the peptides obtained from the two polypeptides are shown below. What kind of amino acid substitution occurred as a result of the mutation? (That is, is the substituted amino acid more or less polar than the original amino acid? Is its pI lower or higher?)

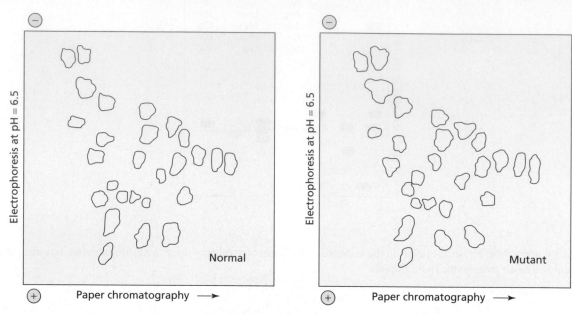

66. Determine the amino acid sequence of a polypeptide from the following data:

Complete hydrolysis of the peptide yields the following amino acids: Ala, Arg, Gly, 2 Lys, Met, Phe, Pro, 2 Ser, Tyr, Val.

Treatment with Edman's reagent gives PTH-Val.

Carboxypeptidase A releases Ala.

Treatment with cyanogen bromide yields the following two peptides:
 1. Ala, 2 Lys, Phe, Pro, Ser, Tyr
 2. Arg, Gly, Met, Ser, Val

Treatment with trypsin yields the following three peptides:
 1. Gly, Lys, Met, Tyr
 2. Ala, Lys, Phe, Pro, Ser
 3. Arg, Ser, Val

Treatment with chymotrypsin yields the following three peptides:
 1. 2 Lys, Phe, Pro
 2. Arg, Gly, Met, Ser, Tyr, Val
 3. Ala, Ser

67. Professor Mary Gold wanted to test her hypothesis that the disulfide bridges that form in many proteins do so after the minimum energy conformation of the protein has been achieved. She treated a sample of lysozyme, an enzyme containing four disulfide bridges, with 2-mercaptoethanol and then added urea to denature the enzyme. She slowly removed these reagents so that the enzyme could refold and reform the disulfide bridges. The lysozyme she recovered had 80% of its original activity. What would be the percent activity in the recovered enzyme if disulfide bridge formation were entirely random rather than determined by the tertiary structure? Does this experiment support Professor Gold's hypothesis?

CHAPTER 23

Catalysis

an RNA catalyst

BUILDING ON FUNDAMENTALS

SECTION 23.0 A catalyst increases the rate of a chemical reaction without itself being consumed or changed in the reaction (4.5).

SECTION 23.2 An acid catalyst donates a proton to a reactant (4.5).

SECTION 23.3 A base catalyst removes a proton from a reactant (18.3).

SECTION 23.4 A nucleophilic catalyst forms an intermediate by forming a covalent bond with a reactant (16.12).

SECTION 23.6 An intramolecular reaction that forms a five- or a six-membered ring occurs more readily than the analogous intermolecular reaction (8.11).

SECTION 23.9 An enzyme is specific for the reactant whose reaction it catalyzes (5.20).

A **catalyst** is a substance that *increases the rate of a chemical reaction without itself being consumed or changed in the reaction* (Section 4.5). We have seen that the rate of a chemical reaction depends on the energy barrier that must be overcome in the process of converting reactants into products (Section 3.7). The height of the "energy hill" is indicated by the free energy of activation ($\Delta G^{\ddagger}$). A catalyst increases the rate of a chemical reaction by providing a pathway with a lower $\Delta G^{\ddagger}$.

A catalyst can decrease $\Delta G^{\ddagger}$ in one of three ways:

1. The catalyzed and uncatalyzed reactions can have different, but similar, mechanisms, with the catalyst providing a way to convert the reactant into a *less stable species* (Figure 23.1a).

2. The catalyzed and uncatalyzed reactions can have different, but similar, mechanisms, with the catalyst providing a way to make *the transition state more stable* (Figure 23.1b).

3. The catalyst can completely *change the mechanism* of the reaction, providing an alternative pathway with a smaller $\Delta G^{\ddagger}$ than that of the uncatalyzed reaction (Figure 23.2).

When we say that a catalyst is neither consumed nor changed by a reaction, we do not mean that it does not participate in the reaction. A catalyst *must* participate in the reaction if it is going to make it go faster. What we mean is that a catalyst has the same form after the reaction as it had before the reaction. Because the catalyst is not used up during

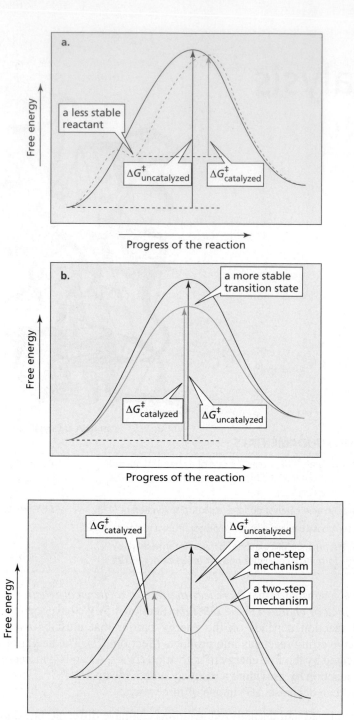

Figure 23.1 ▶
Reaction coordinate diagrams for an uncatalyzed reaction and for a catalyzed reaction. (a) The catalyst converts the reactant to a less stable species. (b) The catalyst stabilizes the transition state.

Figure 23.2 ▶
Reaction coordinate diagrams for an uncatalyzed reaction and for a catalyzed reaction. The catalyzed reaction takes place by an alternative and energetically more favorable pathway.

the reaction, only a small amount of the catalyst is needed. (If a catalyst is used up in one step of the reaction, it must be regenerated in a subsequent step.) Therefore, a catalyst is added to a reaction mixture in small *catalytic* amounts, much smaller than the number of moles of reactant (typically 1% to 10% of the number of moles of reactant).

Notice that the stability of the original reactants and final products is the same in both the catalyzed and corresponding uncatalyzed reactions. In other words, the catalyst does not change the equilibrium constant of the reaction. (Observe that $\Delta G°$ is the same for the catalyzed and uncatalyzed reactions in Figures 23.1a, 23.1b, and 23.2.) Because the catalyst does not change the equilibrium constant, it does not change the *amount* of product formed when the reaction has reached equilibrium. It changes only the *rate* at which the product is formed.

PROBLEM 1♦

Which of the following parameters would be different for a reaction carried out in the presence of a catalyst, compared with the same reaction carried out in the absence of a catalyst? (*Hint:* See Section 3.7.)

$$\Delta G^\circ,\ \Delta H^\ddagger,\ E_a,\ \Delta S^\ddagger,\ \Delta H^\circ,\ K_{eq},\ \Delta G^\ddagger,\ \Delta S^\circ,\ k_{rate}$$

23.1 Catalysis in Organic Reactions

There are several ways a catalyst can provide a more favorable pathway for an organic reaction:

- It can increase the susceptibility of an electrophile to nucleophilic attack.
- It can increase the reactivity of a nucleophile.
- It can increase the leaving ability of a group by converting it into a weaker base.
- It can increase the stability of a transition state.

In this chapter, we will look at some of the most common catalysts—acid catalysts, base catalysts, nucleophilic catalysts, and metal-ion catalysts—and the ways in which they provide an energetically more favorable pathway for an organic reaction. We will then see how the same modes of catalysis are used in enzyme-catalyzed reactions.

THE NOBEL PRIZE

Throughout this text, you have seen biographical sketches that give you some information about the men and women who created the science you are studying. You have found that many of these people are Nobel Prize winners. The Nobel Prize is considered by many to be the most coveted award a scientist can receive. These awards were established by **Alfred Bernhard Nobel (1833–1896)** and were first awarded in 1901.

Nobel was born in Stockholm, Sweden. When he was nine, he moved with his parents to St. Petersburg, where his father manufactured torpedoes and submarine mines he had invented for the Russian government. As a young man, Alfred did research on explosives in a factory his father owned near Stockholm. In 1864, an explosion in the factory killed his younger brother, causing Alfred to look for ways to make explosives easier to handle and transport. The Swedish government would not allow the factory to be rebuilt because of the many accidents that had occurred there. Nobel, therefore, established an explosives factory in Germany, where, in 1867, he discovered that if nitroglycerin is mixed with diatomaceous earth, the mixture can be molded into sticks that cannot be set off without a detonating cap. Thus, Nobel invented dynamite. He also invented blasting gelatin and smokeless powder. Although he was the inventor of explosives used by the military, he was a strong supporter of peace movements.

The 355 patents Nobel held made him a wealthy man. He never married, and when he died, his will stipulated that the bulk of his estate ($9,200,000) be used to establish prizes to be awarded to those who "have conferred the greatest benefit on mankind." He instructed that the money be invested and the interest earned each year be divided into five equal portions "to be awarded to the persons having made the most important contributions in the fields of chemistry, physics, physiology or medicine, literature, and to the one who had done the most or the best work toward fostering fraternity among nations, the abolition of standing armies, and the holding and promotion of peace congresses." Nobel also stipulated that in awarding the prizes, no consideration be given to the nationality of the candidate, that each prize be shared by no more than three persons, and that no prize be awarded posthumously.

Nobel gave instructions that the prizes for chemistry and physics were to be awarded by the Royal Swedish Academy of Sciences, for physiology or medicine by the Karolinska Institute in Stockholm, for literature by the Swedish Academy, and for peace by a five-person committee appointed by the Norwegian Parliament. The deliberations are secret, and the decisions cannot be appealed. In 1969, the Swedish Central Bank established a prize in economics in Nobel's honor. The recipient of this prize is selected by the Royal Swedish Academy of Sciences. On December 10—the anniversary of Nobel's death—the prizes are awarded in Stockholm, except for the peace prize, which is awarded in Oslo.

Alfred Bernhard Nobel

23.2 **Acid Catalysis**

A proton is donated to the reactant in an acid-catalyzed reaction.

An **acid catalyst** increases the rate of a reaction by donating a proton to a reactant. In the preceding chapters we have seen many examples of acid catalysis. For example, we saw that an acid provides the electrophilic proton needed for the addition of water or an alcohol to an alkene (Section 4.5). We also saw that an alcohol cannot undergo substitution and elimination reactions unless an acid is present to protonate the OH group, making the leaving group a weaker base and therefore a better leaving group (Section 10.1).

To review some of the important ways an acid can catalyze a reaction, let's look at the mechanism for the acid-catalyzed hydrolysis of an ester (Section 16.11). The reaction has two slow steps: formation of the tetrahedral intermediate and collapse of the tetrahedral intermediate. Donation of a proton to, and removal of a proton from, an electronegative atom such as oxygen are fast steps.

Mechanism for acid-catalyzed ester hydrolysis

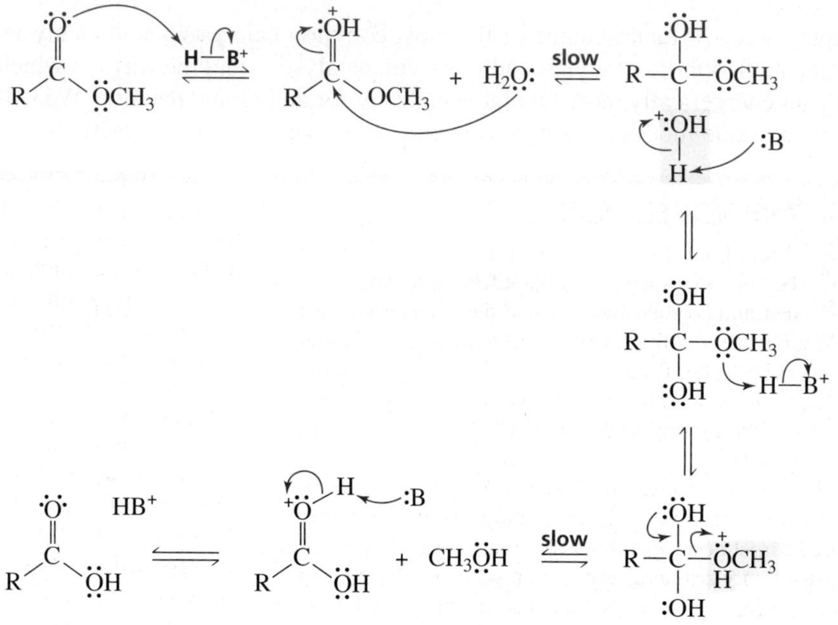

A catalyst must increase the rate of a slow step because increasing the rate of a fast step will not increase the rate of the overall reaction. The acid increases the rates of both slow steps of this reaction. It increases the rate of formation of the tetrahedral intermediate by protonating the carbonyl oxygen, thereby making the carbonyl group more susceptible to nucleophilic attack than an unprotonated carbonyl group would be. Increasing the reactivity of the carbonyl group by protonating it is an example of providing a way to convert the reactant into a less stable (more reactive) species (Figure 23.1a).

A catalyst must increase the rate of a slow step. Increasing the rate of a fast step will not increase the rate of the overall reaction.

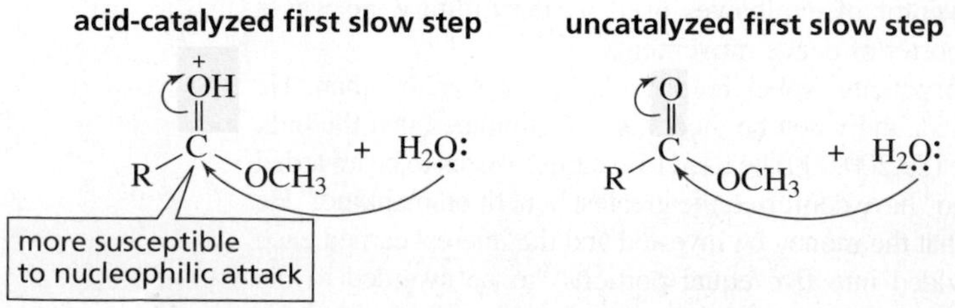

acid-catalyzed first slow step   **uncatalyzed first slow step**

The acid increases the rate of the second slow step by changing the basicity of the group that is expelled when the tetrahedral intermediate collapses. In the presence of an acid, methanol (pK_a of $\overset{+}{CH_3OH_2}$ = −2.5) is expelled; in the absence of an acid, methoxide ion (pK_a of CH_3OH = 15.7) is expelled. Methanol is a weaker base than methoxide ion, so it is more easily eliminated.

acid-catalyzed second slow step   **uncatalyzed second slow step**

CH₃OH is the leaving group

CH₃O⁻ is the leaving group

The mechanism for the acid-catalyzed hydrolysis of an ester shows that the reaction can be divided into two distinct parts: formation of a tetrahedral intermediate and collapse of a tetrahedral intermediate. There are three steps in each part. Notice that in each part, the first step is a fast protonation step, the second step is a slow catalyzed step that involves either breaking a π bond or forming a π bond, and the last step is a fast deprotonation step (to regenerate the catalyst).

PROBLEM 2

Compare each of the mechanisms listed below with the mechanism for each part of the acid-catalyzed hydrolysis of an ester, indicating

a. similarities.   **b.** differences.

1. acid-catalyzed formation of a hydrate (Section 17.9)
2. acid-catalyzed conversion of an aldehyde into a hemiacetal (Section 17.10)
3. acid-catalyzed conversion of a hemiacetal into an acetal (Section 17.10)
4. acid-catalyzed hydrolysis of an amide (Section 16.17)

There are two types of acid catalysis: specific-acid catalysis and general-acid catalysis. In **specific-acid catalysis**, the proton is fully transferred to the reactant *before* the slow step of the reaction begins (Figure 23.3a). In **general-acid catalysis**, the proton is transferred to the reactant *during* the slow step of the reaction (Figure 23.3b). Specific-acid and

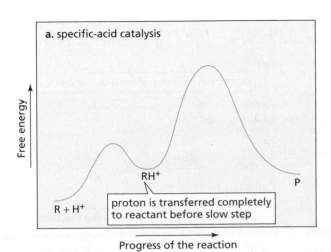

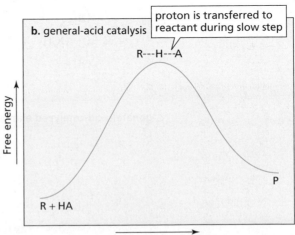

▲ **Figure 23.3**
(a) Reaction coordinate diagram for a specific-acid-catalyzed reaction. The proton is transferred completely to the reactant before the slow step of the reaction begins. (b) Reaction coordinate diagram for a general-acid-catalyzed reaction. The proton is transferred partially to the reactant in the transition state of the slow step of the reaction.

general-acid catalysis increase the rate of a reaction in the same way—by donating a proton in order to make either bond making or bond breaking easier. The two types of acid catalysis differ only in the extent to which the proton is transferred in the transition state of the slow step of the reaction.

In the examples that follow, notice the difference in the extent to which the proton has been transfered when the nucleophile attacks the reactant.

- In specific-acid-catalyzed attack by water on a carbonyl group, the nucleophile attacks a fully protonated carbonyl group. In general-acid-catalyzed attack by water on a carbonyl group, the carbonyl group becomes protonated as the nucleophile attacks it.

specific-acid-catalyzed attack by water

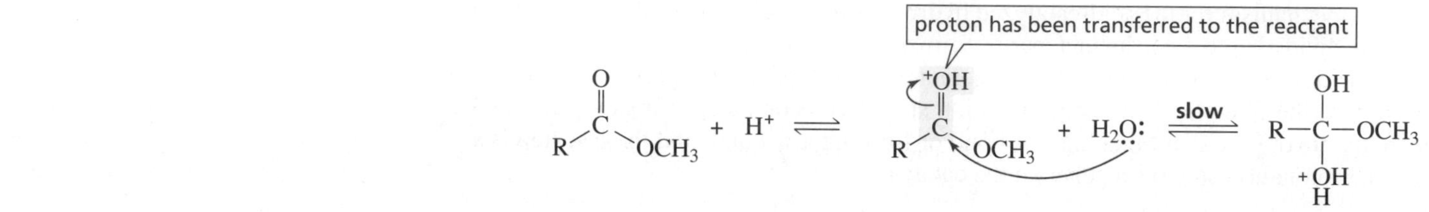

general-acid-catalyzed attack by water

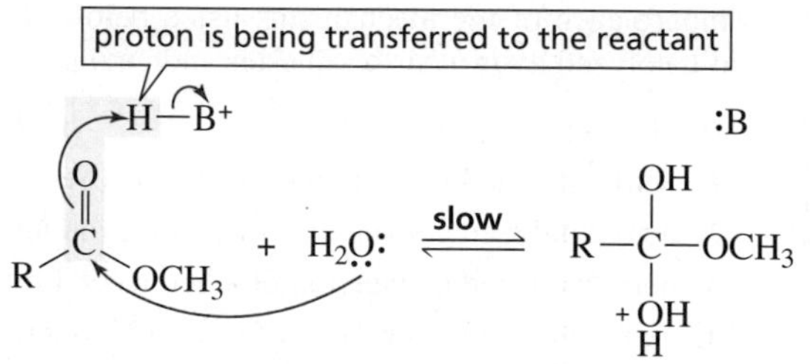

- In specific-acid-catalyzed collapse of a tetrahedral intermediate, a fully protonated leaving group is eliminated, whereas in general-acid-catalyzed collapse of a tetrahedral intermediate, the leaving group picks up a proton as the group is eliminated.

specific-acid-catalyzed elimination of the leaving group

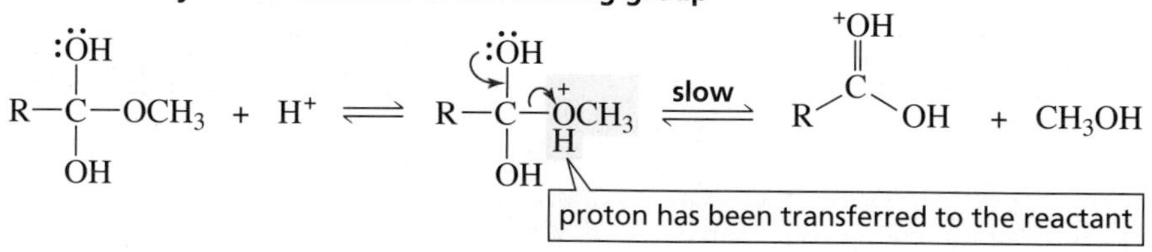

general-acid-catalyzed elimination of the leaving group

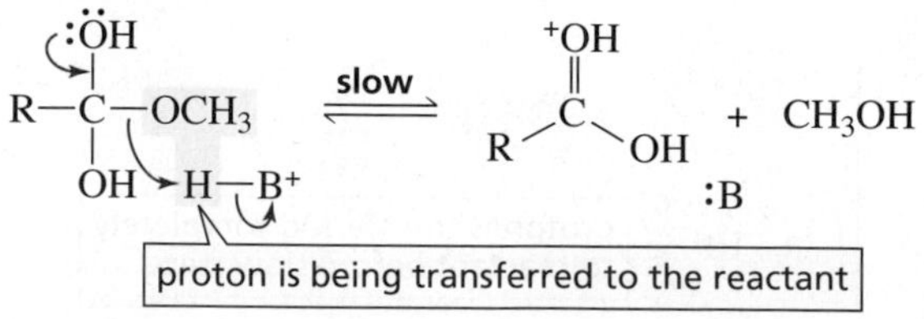

The proton is donated to the reactant *before* the slow step in a specific-acid-catalyzed reaction, and *during* the slow step in a general-acid-catalyzed reaction.

A specific-acid catalyst must be an acid that is strong enough to protonate the reactant fully before the slow step begins. A general-acid catalyst can be a weaker acid because it only partially transfers a proton in the transition state of the slow step. A list of acids and their pK_a values is given in Appendix II.

Are the slow steps for the acid-catalyzed hydrolysis of an ester on page 1066 general-acid catalyzed or specific-acid catalyzed?

PROBLEM 4

The following reaction occurs by a general-acid-catalyzed mechanism:

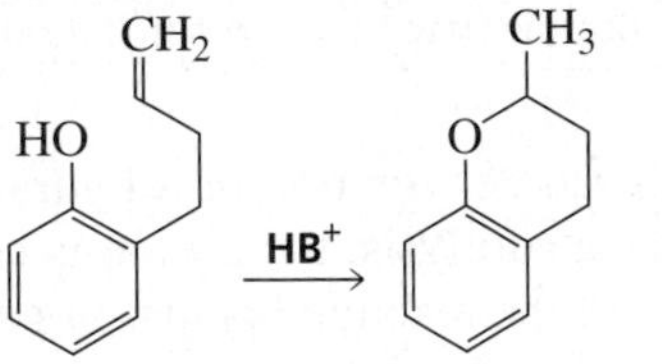

Propose a mechanism for this reaction.

PROBLEM 5   SOLVED

An alcohol will not react with aziridine unless an acid is present. Why is the acid necessary?

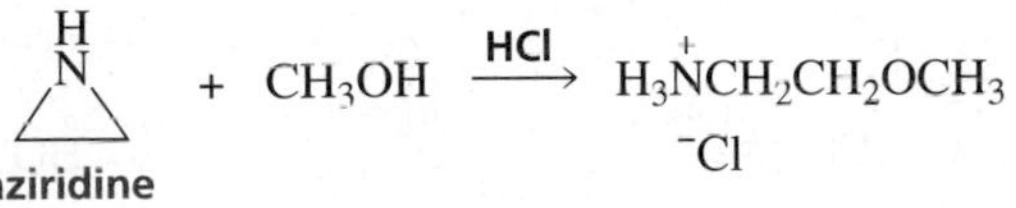

Solution Although relief of ring strain is sufficient by itself to cause an epoxide to undergo a ring-opening reaction (Section 10.8), it is not sufficient to cause an aziridine to undergo a ring-opening reaction. A negatively charged nitrogen is a stronger base, and therefore a poorer leaving group, than a negatively charged oxygen. An acid, therefore, is needed to make the ring nitrogen a better leaving group by protonating it.

23.3 Base Catalysis

We have also seen several base-catalyzed reactions such as the enediol rearrangement (Section 21.5) and the interconversion of keto and enol tautomers (Section 18.3). A **base catalyst** increases the rate of a reaction by removing a proton from the reactant. For example, dehydration of a hydrate in the presence of hydroxide ion is a base-catalyzed reaction. Hydroxide ion (the base) increases the rate of the reaction by removing a proton from the neutral hydrate.

specific-base-catalyzed dehydration

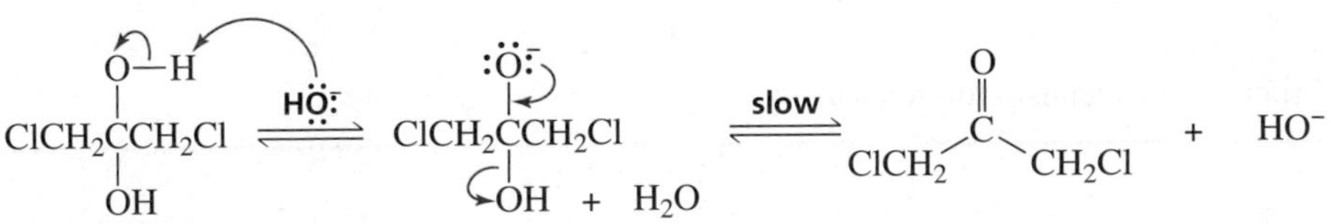

Removing a proton from the hydrate increases the rate of dehydration by providing a pathway with a more stable transition state. The transition state for elimination of HO^- from a negatively charged tetrahedral intermediate is more stable because a positive charge does not develop on the electronegative oxygen atom,

A proton is removed from the reactant in a base-catalyzed reaction.

as it does in the transition state for elimination of HO⁻ from a neutral tetrahedral intermediate.

$$\delta-\text{O}$$

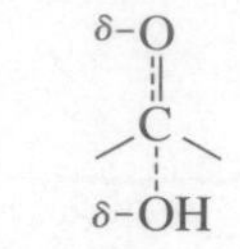

transition state for elimination of HO⁻ from a negatively charged tetrahedral intermediate	transition state for elimination of HO⁻ from a neutral tetrahedral intermediate

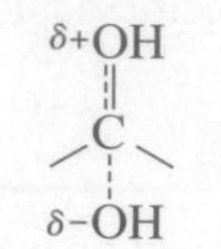

The foregoing base-catalyzed dehydration of a hydrate is an example of specific-base catalysis. In **specific-base catalysis**, the proton is completely removed from the reactant *before* the slow step of the reaction begins. In **general-base catalysis**, on the other hand, the proton is removed from the reactant *during* the slow step of the reaction. Compare the extent of proton transfer in the slow step of the preceding specific-base-catalyzed dehydration with the extent of proton transfer in the slow step of the following general-base-catalyzed dehydration:

general-base-catalyzed dehydration

a hydrate

> **The proton is removed from the reactant before the slow step in a specific-base-catalyzed reaction, and during the slow step in a general-base-catalyzed reaction.**

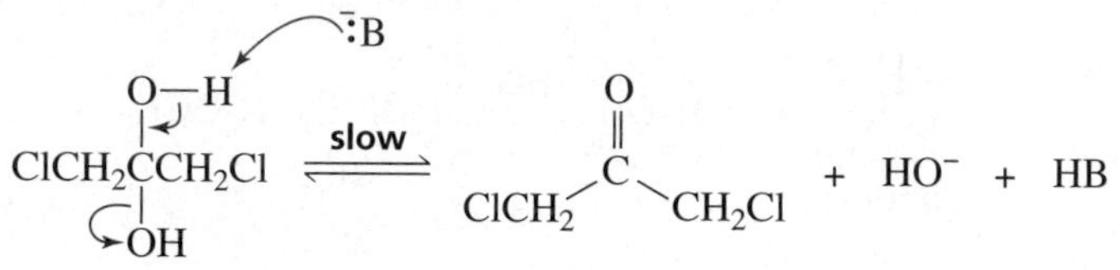

In specific-base catalysis, the base has to be strong enough to remove a proton from the reactant completely before the slow step begins. In general-base catalysis, the base can be weaker because the proton is only partially transferred to the base in the transition state of the slow step. We will see that enzymes catalyze reactions using general-acid and general-base catalytic groups because at physiological pH (7.3), too small a concentration ($\sim 1 \times 10^{-7}$ M) of H⁺ for specific-acid catalysis or HO⁻ for specific-base catalysis is available.

PROBLEM 6

The following reaction occurs by a mechanism involving general-base catalysis:

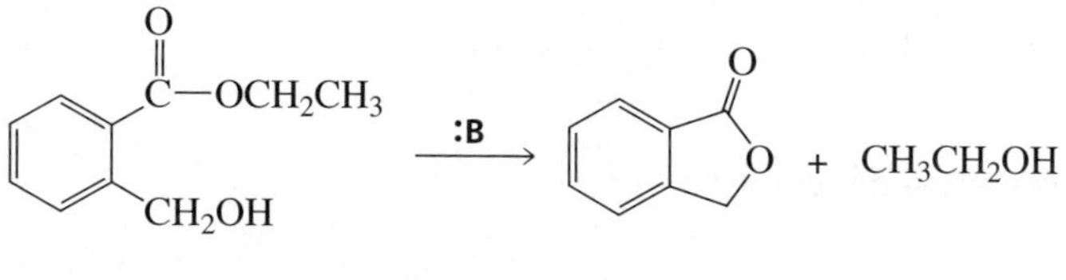

Propose a mechanism for this reaction.

23.4 Nucleophilic Catalysis

A **nucleophilic catalyst** increases the rate of a reaction by acting as a nucleophile; it forms an intermediate by forming a covalent bond with the reactant. **Nucleophilic catalysis**, therefore, is also called **covalent catalysis**. A nucleophilic catalyst increases the reaction rate by completely changing the mechanism of the reaction.

In the following reaction, iodide ion increases the rate of conversion of ethyl chloride into ethyl alcohol by acting as a nucleophilic catalyst:

a nucleophilic catalyst

$$CH_3CH_2Cl \ + \ HO^- \xrightarrow[H_2O]{I^-} \ CH_3CH_2OH \ + \ Cl^-$$

To understand how iodide ion catalyzes this reaction, we need to look at the mechanism of the reaction both with and without the catalyst. In the absence of iodide ion, ethyl chloride is converted into ethyl alcohol in a one-step S_N2 reaction.

Mechanism of the uncatalyzed reaction

$$H\ddot{O}^- \ + \ CH_3CH_2{-}Cl \ \longrightarrow \ CH_3CH_2OH \ + \ Cl^-$$

If iodide ion is present in the reaction mixture, the reaction takes place by two successive S_N2 reactions.

Mechanism of the iodide-ion-catalyzed reaction

$$:\ddot{I}^- \ + \ CH_3CH_2{-}Cl \ \longrightarrow \ CH_3CH_2I \ + \ Cl^-$$
$$H\ddot{O}^- \ + \ CH_3CH_2{-}I \ \longrightarrow \ CH_3CH_2OH \ + \ I^-$$

A nucleophilic catalyst forms a covalent bond with the reactant.

The first S_N2 reaction in the catalyzed reaction is faster than the uncatalyzed reaction because in a protic solvent, iodide ion is a better nucleophile than hydroxide ion (Section 8.3), which is the nucleophile in the uncatalyzed reaction. The second S_N2 reaction in the catalyzed reaction is also faster than the uncatalyzed reaction because iodide ion is a weaker base and therefore a better leaving group than chloride ion, the leaving group in the uncatalyzed reaction. Thus, iodide ion increases the rate of formation of ethanol by changing a relatively slow one-step reaction into a reaction with two relatively fast steps (Figure 23.2).

Iodide ion is a nucleophilic catalyst because it reacts as a nucleophile, forming a covalent bond with the reactant. The iodide ion that is consumed in the first reaction is regenerated in the second, so it comes out of the reaction unchanged.

Another reaction in which a nucleophilic catalyst provides a more favorable pathway by changing the mechanism of the reaction is the imidazole-catalyzed hydrolysis of an ester.

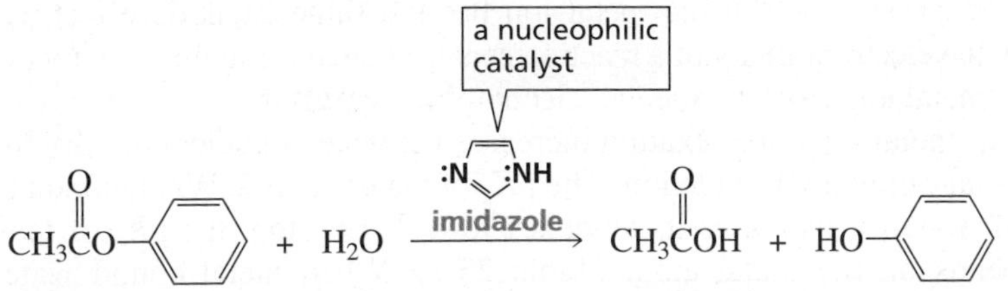

a nucleophilic catalyst

$$\underset{\text{phenyl acetate}}{CH_3\overset{O}{\overset{\|}{C}}O{-}\phi} \ + \ H_2O \ \xrightarrow[\text{imidazole}]{:N \quad :NH} \ \underset{\text{acetic acid}}{CH_3\overset{O}{\overset{\|}{C}}OH} \ + \ \underset{\text{phenol}}{HO{-}\phi}$$

Imidazole is a better nucleophile than water, so imidazole reacts faster with the ester than water does. The acyl imidazole that is formed is particularly reactive because the positively charged nitrogen makes imidazole a very good leaving group. Therefore, it is hydrolyzed much more rapidly than the ester would have been. Because formation of

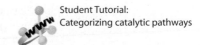

Student Tutorial:
Categorizing catalytic pathways

the acyl imidazole and its subsequent hydrolysis are both faster than ester hydrolysis, imidazole increases the rate of ester hydrolysis.

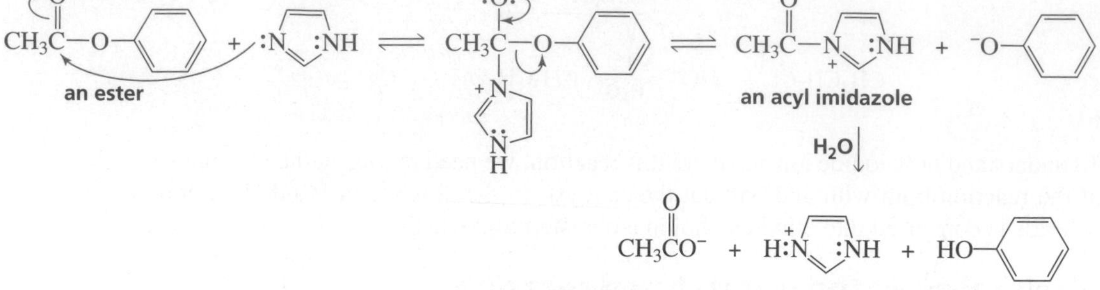

23.5 **Metal-Ion Catalysis**

Metal ions exert their catalytic effect by coordinating (complexing) with atoms that have lone-pair electrons. In other words, metal ions are Lewis acids (Section 1.26). A *metal ion* can increase the rate of a reaction in several ways.

- It can make a reaction center more susceptible to receiving electrons, as in A in the following diagram:

Metal atoms are Lewis acids.

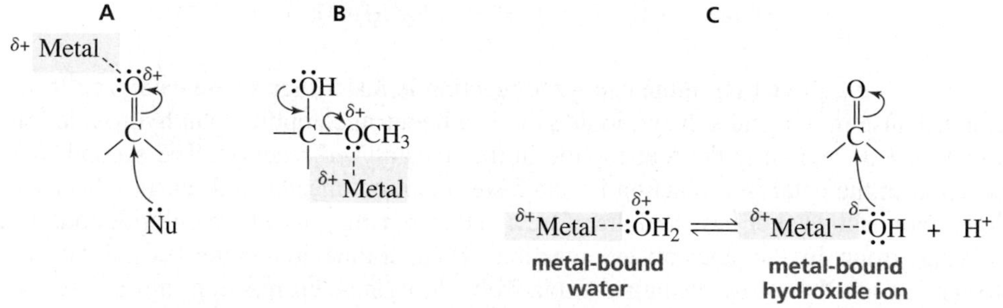

- It can make a leaving group a weaker base, and therefore a better leaving group, as in B.
- It can increase the rate of a hydrolysis reaction by increasing the nucleophilicity of water, as in C.

In cases A and B, the metal ion exerts the same kind of catalytic effect as a proton does. In a reaction in which the metal ion has the same catalytic effect as a proton (increasing the electrophilicity of a reaction center or decreasing the basicity of a leaving group), the metal ion is often called an **electrophilic catalyst**.

In case C, metal-ion complexation increases the water's nucleophilicity by converting it to metal-bound hydroxide ion. The pK_a of water is 15.7. When a metal ion complexes with water, it increases its tendency to lose a proton: the pK_a of metal-bound water depends on the metal atom (Table 23.1). When metal-bound water loses a proton, metal-bound hydroxide ion is formed. Metal-bound hydroxide ion, while not as good a nucleophile as hydroxide ion, is a better nucleophile than water. Metal-ion catalysis is important in biological systems because hydroxide ion itself would not be available at physiological pH (7.3).

Table 23.1	The pK_a of Metal-Bound Water		
M^{2+}	**pK_a**	**M^{2+}**	**pK_a**
Ca^{2+}	12.7	Co^{2+}	8.9
Mg^{2+}	11.8	Zn^{2+}	8.7
Cd^{2+}	11.6	Fe^{2+}	7.2
Mn^{2+}	10.6	Cu^{2+}	6.8
Ni^{2+}	9.4	Be^{2+}	5.7

Now we will look at some examples of metal-ion-catalyzed reactions. The decarboxylation of dimethyloxaloacetate can be catalyzed by either Cu^{2+} or Al^{3+}.

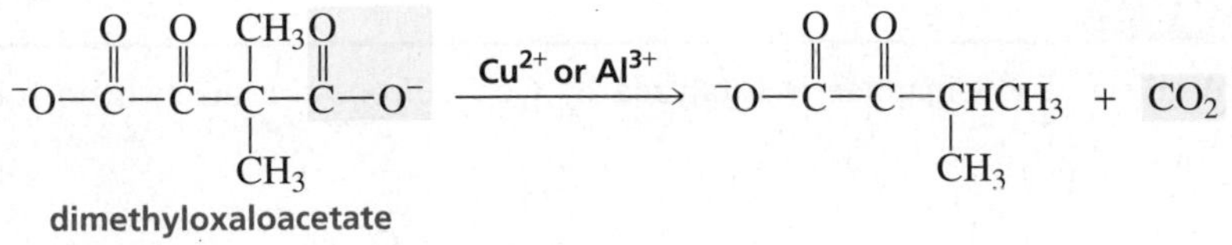

dimethyloxaloacetate

In this reaction, the metal ion complexes with two oxygen atoms of the reactant. Complexation increases the rate of decarboxylation by making the carbonyl group more susceptible to receiving the electrons left behind when CO$_2$ is eliminated.

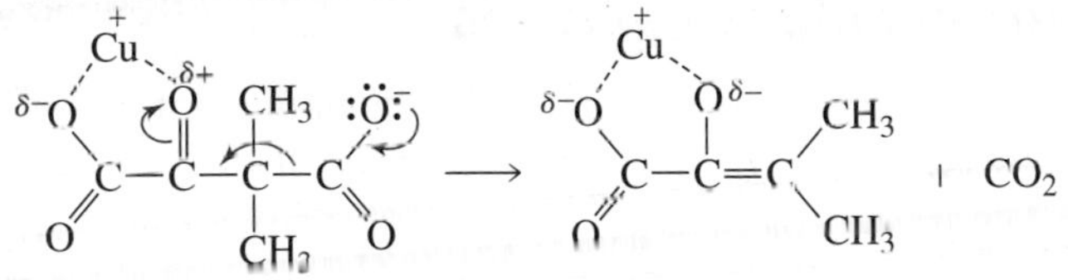

The hydrolysis of methyl trifluoroacetate has two slow steps. Zn^{2+} increases the rate of the first slow step by providing metal-bound hydroxide ion, a better nucleophile than water. Zn^{2+} increases the rate of the second slow step by decreasing the basicity of the group that is eliminated from the tetrahedral intermediate.

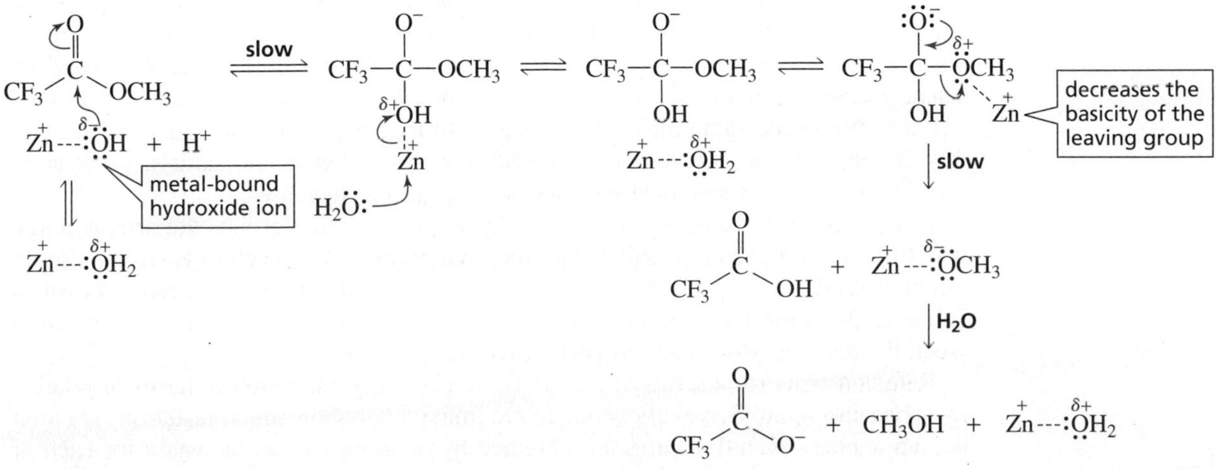

PROBLEM 7♦

Although metal ions increase the rate of decarboxylation of dimethyloxaloacetate, they have no effect on the rate of decarboxylation either of the monoethyl ester of dimethyloxaloacetate or of acetoacetate. Explain why this is so.

$$^{-}O-\overset{\overset{\displaystyle O}{\|}}{C}-\overset{\overset{\displaystyle O}{\|}}{C}-\overset{\overset{\displaystyle CH_3}{|}}{\underset{\underset{\displaystyle CH_3}{|}}{C}}-\overset{\overset{\displaystyle O}{\|}}{C}-O^{-}$$

dimethyloxaloacetate

$$CH_3CH_2O-\overset{\overset{\displaystyle O}{\|}}{C}-\overset{\overset{\displaystyle O}{\|}}{C}-\overset{\overset{\displaystyle CH_3}{|}}{\underset{\underset{\displaystyle CH_3}{|}}{C}}-\overset{\overset{\displaystyle O}{\|}}{C}-O^{-}$$

**monoethyl ester of
dimethyloxaloacetate**

$$CH_3-\overset{\overset{\displaystyle O}{\|}}{C}-CH_2-\overset{\overset{\displaystyle O}{\|}}{C}-O^{-}$$

acetoacetate

PROBLEM 8

The hydrolysis of glycinamide is catalyzed by Co^{2+}. Propose a mechanism for this reaction.

$$H_2NCH_2\overset{\overset{\displaystyle O}{\|}}{C}NH_2 \;+\; H_2O \;\xrightarrow{\;Co^{2+}\;}\; H_2NCH_2\overset{\overset{\displaystyle O}{\|}}{C}O^{-} \;+\; {}^{+}NH_4$$

23.6 Intramolecular Reactions

The rate of a chemical reaction is determined by the number of molecular collisions with sufficient energy *and* with the proper orientation in a given period of time (Section 3.7):

$$\text{rate of reaction} \;=\; \frac{\text{number of collisions}}{\text{unit of time}} \;\times\; \frac{\text{fraction with}}{\text{sufficient energy}} \;\times\; \frac{\text{fraction with}}{\text{proper orientation}}$$

Because a catalyst decreases the energy barrier of a reaction, it increases the fraction of collisions that occur with sufficient energy to overcome the barrier.

The rate of a reaction can also be increased by increasing the frequency of the collisions. Increasing the concentration of the reactants increases the frequency of the collisions. In addition, we have seen that an **intramolecular reaction** that forms a five- or a six-membered ring occurs more readily than the analogous **intermolecular reaction**. This is because an intramolecular reaction has the advantage that the reacting groups are tied together in the same molecule, giving them a better chance of finding each other than if they were in two different molecules in a solution of the same concentration (Section 8.11). As a result, the frequency of the collisions increases.

If, in addition to being in the same molecule, the reacting groups are arranged in a way that increases the probability that they will collide with each other in the proper orientation, the rate of the reaction is further increased. The relative rates shown in Table 23.2 demonstrate the enormous increase that occurs in the rate of a reaction when the reacting groups are properly oriented.

Rate constants for a series of reactions are generally compared in terms of relative rates because relative rates allow one to see immediately how much faster one reaction is than another. **Relative rates** are obtained by dividing the rate constant for each of the reactions by the rate constant of the slowest reaction in the series. The slowest

Table 23.2 Relative Rates of an Intermolecular Reaction and Five Intramolecular Reactions

Reaction	Relative rate
A	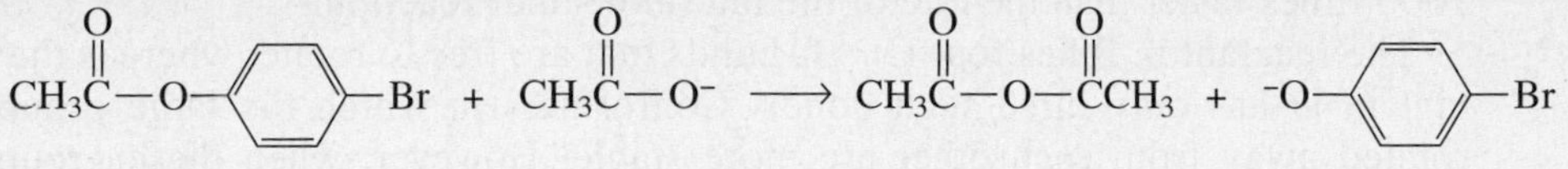 1.0
B	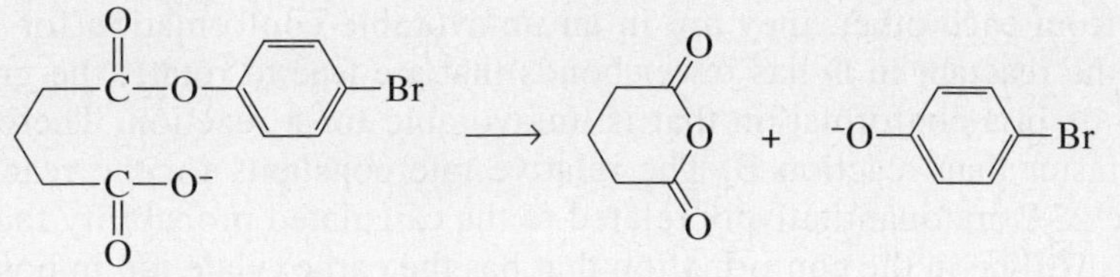 1×10^3 M
C	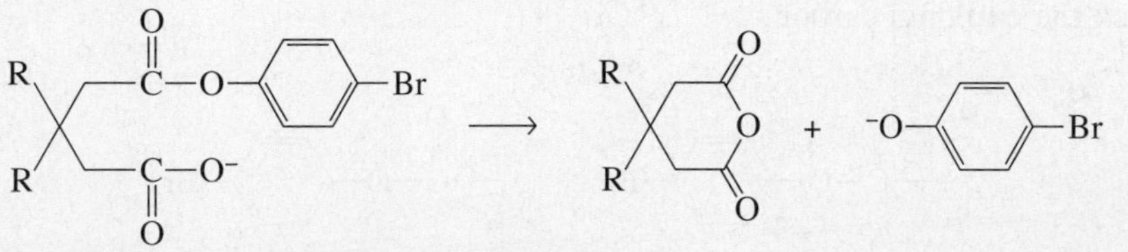 R = CH_3 2.3×10^4 M R = $(CH_3)_2CH$ 1.3×10^6 M
D	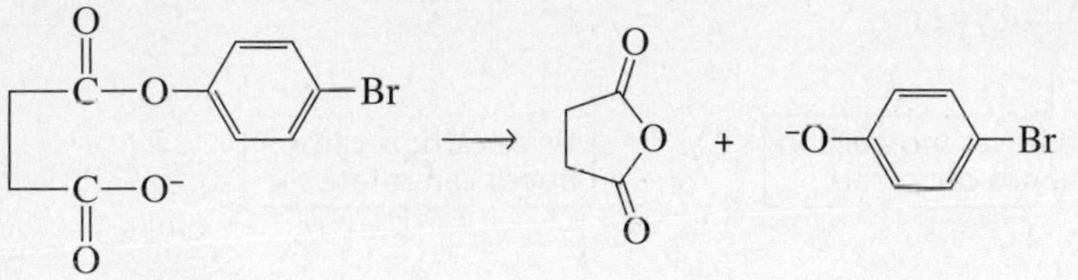 2.2×10^5 M
E	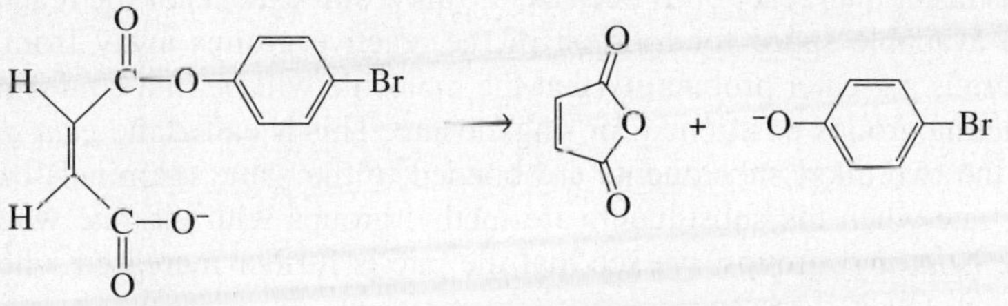 1×10^7 M
F	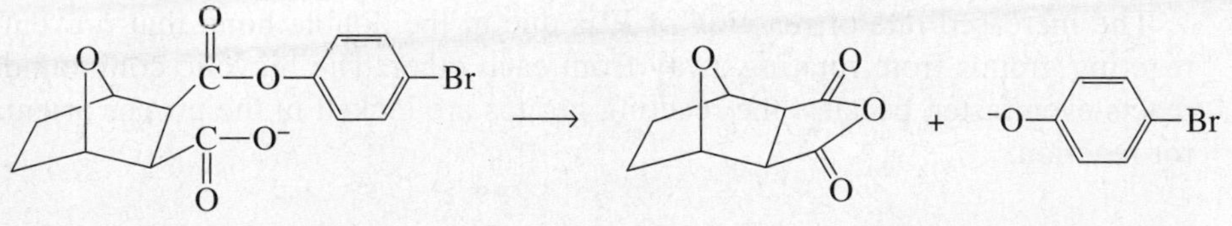 5×10^7 M

reaction in Table 23.2 is an intermolecular reaction; all the others are intramolecular reactions. Because an intramolecular reaction is a first-order reaction (it has units of $time^{-1}$) and an intermolecular reaction is a second-order reaction (it has units of $time^{-1}M^{-1}$), the relative rates in Table 23.2 have units of molarity (M) (Section 3.7).

$$\text{relative rate} = \frac{\textbf{first-order rate constant}}{\textbf{second-order rate constant}} = \frac{time^{-1}}{time^{-1}M^{-1}} = M$$

The relative rates shown in Table 23.2 are also called *effective molarities*. **Effective molarity** is the concentration of the reactant that would be required in an *intermolecular* reaction for it to have the same rate as the *intramolecular* reaction. In other words, the effective molarity is the advantage given to a reaction by having the reacting groups in the same molecule. In some cases, juxtaposing the reacting groups provides such an enormous increase in rate that the effective molarity is greater than the concentration of the reactant in its solid state!

The first reaction shown in Table 23.2, reaction A, is an intermolecular reaction between an ester and a carboxylate ion. The second reaction, B, has those same two reacting groups in a single molecule. The rate of the intramolecular reaction is 1000 times faster than the rate of the intermolecular reaction.

The reactant in B has four C—C bonds that are free to rotate, whereas the reactant in D has only three such bonds. Conformers in which the large groups are rotated away from each other are more stable. However, when these groups are pointed away from each other, they are in an unfavorable conformation for reaction. Because the reactant in D has fewer bonds that are free to rotate, the groups are less apt to be in a conformation that is unfavorable for a reaction. Therefore, reaction D is faster than reaction B. The relative rate constants for the reactions shown in Table 23.2 are quantitatively related to the calculated probability that the reacting groups will be in the conformation that has the carboxylate ion in position to attack the carbonyl carbon.

Reaction C is faster than reaction B because the alkyl substituents of the reactant in C decrease the available space for rotation of the reactive groups away from each other. Thus, there is a greater probability that the molecule will be in a conformation that has the reacting groups positioned for ring closure. This is called the *gem-dialkyl effect* because the two alkyl substituents are bonded to the same (geminal) carbon. Comparing the rate when the substituents are methyl groups with the rate when the substituents are isopropyl groups, we see that the rate is further increased when the size of the alkyl groups is increased.

The increased rate of reaction of E is due to the double bond that prevents the reacting groups from rotating away from each other. The bicyclic compound in F reacts even faster, because the reacting groups are locked in the proper orientation for reaction.

PROBLEM 9◆

The relative rate of reaction of the cis alkene (E) is given in Table 23.2. What would you expect the relative rate of reaction of the trans isomer to be?

23.7 Intramolecular Catalysis

Just as having two reacting groups in the same molecule increases the rate of a reaction compared with having the groups in separate molecules, having a *reacting group* and a *catalyst* in the same molecule increases the rate of a reaction compared with having them in separate molecules. When a catalyst is part of the reacting molecule, the catalysis is called **intramolecular catalysis**. Intramolecular nucleophilic catalysis, intramolecular general-acid or general-base catalysis, and intramolecular metal-ion catalysis are all possible. Intramolecular catalysis is also known as *anchimeric assistance* (*anchimeric* means "adjacent parts" in Greek). Let's now look at some examples of intramolecular catalysis.

When chlorocyclohexane reacts with an aqueous solution of ethanol, an alcohol and an ether are formed. Two products are formed because there are two nucleophiles (H_2O and CH_3CH_2OH) in the solution.

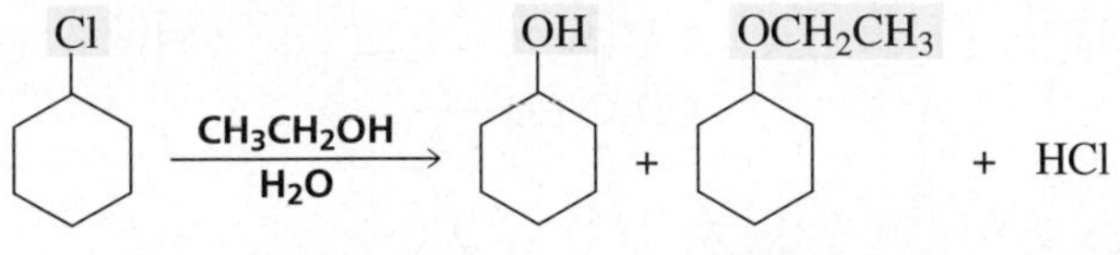

A 2-thio-substituted chlorocyclohexane undergoes the same reaction. However, the rate of the reaction depends on whether the thio substituent is cis or trans to the chloro substituent. If it is trans, the 2-thio-substituted compound reacts about 70,000 times faster than the unsubstituted compound. But if it is cis, the 2-thio-substituted compound reacts a little more slowly than the unsubstituted compound.

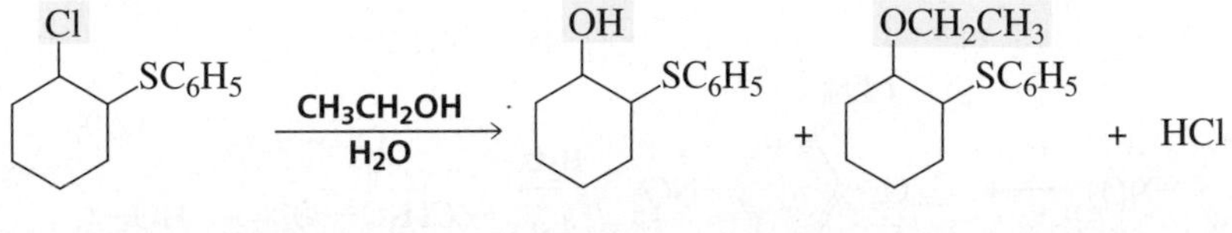

What accounts for the much faster reaction of the trans-substituted compound? In this reaction, the thio substituent is an intramolecular nucleophilic catalyst. It displaces the chloro substituent by attacking the back side of the carbon to which the chloro substituent is attached. Back-side attack requires both substituents to be in axial positions, and only the trans isomer can have both of its substituents in axial positions (Section 2.14). Subsequent attack by water or ethanol on the sulfonium ion is rapid because the positively charged sulfur is an excellent leaving group and because breaking the three-membered ring releases strain.

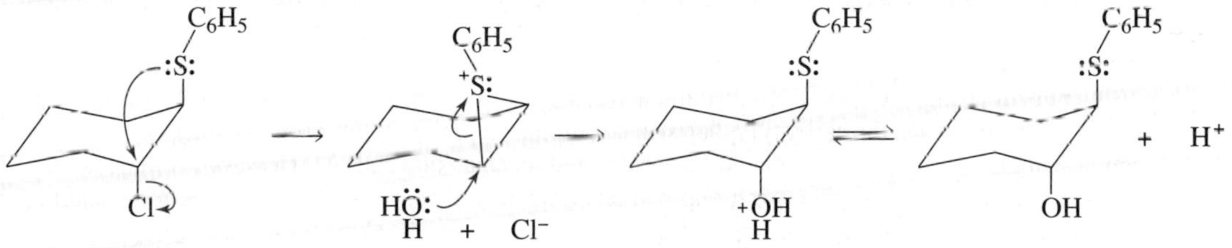

PROBLEM 10◆

Show all the products, and their configurations, that would be obtained from solvolysis of the trans-substituted compound illustrated in the preceding diagram.

The rate of hydrolysis of phenyl acetate is increased about 150-fold at neutral pH by the presence of a carboxylate ion in the ortho position. The *ortho*-carboxyl-substituted ester is commonly known as aspirin (Section 18.9). In the following reactions, each reactant and product is shown in the form that predominates at physiological pH (7.3).

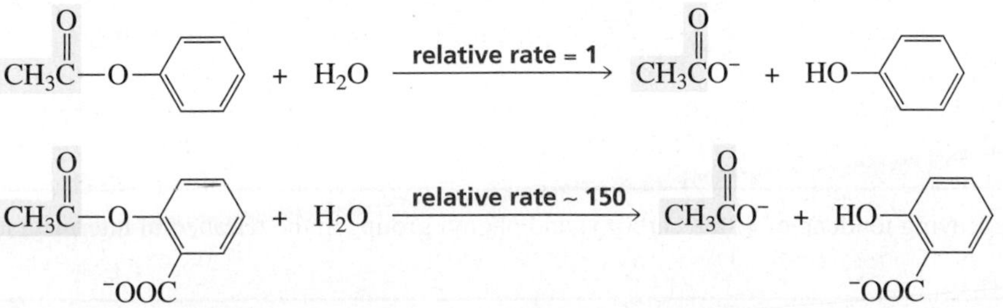

The *ortho*-carboxylate group is an intramolecular general-base catalyst that increases the nucleophilicity of water, thereby increasing the rate of formation of the tetrahedral intermediate.

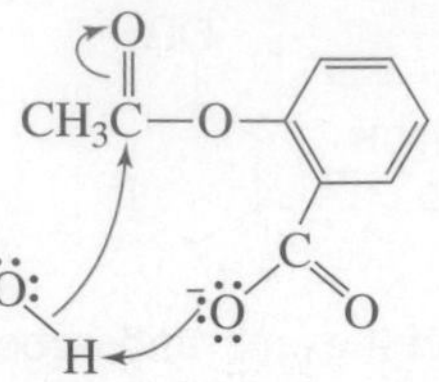

If nitro groups are put on the benzene ring, the *ortho*-carboxyl substituent acts as an intramolecular *nucleophilic catalyst* instead of an intramolecular *general-base catalyst*. It increases the rate of the hydrolysis reaction by converting the ester into an anhydride, which is more rapidly hydrolyzed than an ester (Section 16.6).

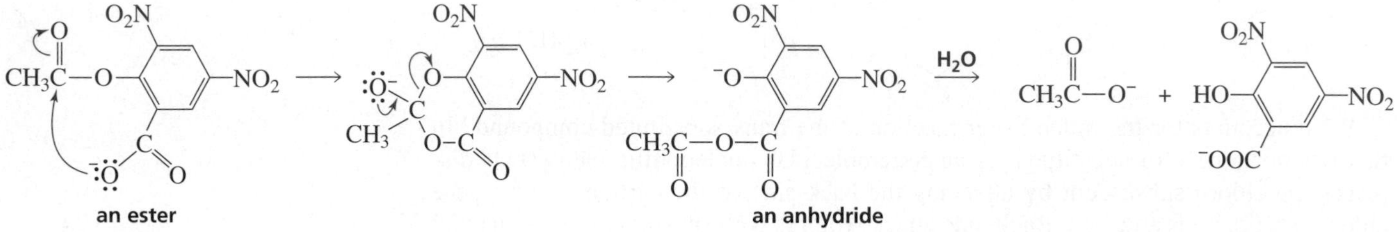

PROBLEM 11 **SOLVED**

What causes the mode of catalysis to change from general base to nucleophilic in the hydrolysis of an *ortho*-carboxyl-substituted phenyl acetate?

Solution The *ortho*-carboxyl substituent is in position to form a tetrahedral intermediate. If the carboxyl group in the tetrahedral intermediate is a better leaving group than the phenoxy group, the carboxyl group will be eliminated preferentially from the intermediate. This will reform the starting material, which will be hydrolyzed by a general-base-catalyzed mechanism (path A). However, if the phenoxy group is a better leaving group than the carboxyl group, the phenoxy group will be eliminated, thereby forming the anhydride, and the reaction will have occurred by a mechanism involving nucleophilic catalysis (path B).

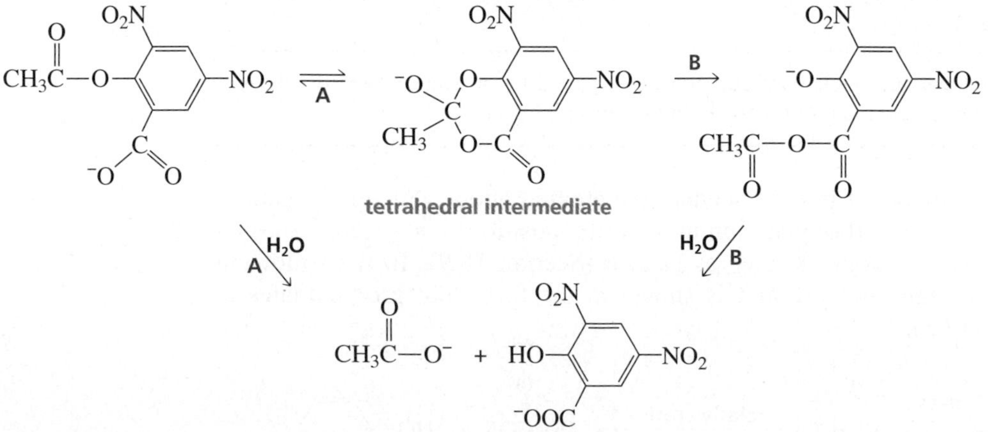

tetrahedral intermediate

PROBLEM 12◆

Why do the nitro groups change the relative leaving tendencies of the carboxyl and phenyl groups in the tetrahedral intermediate in Problem 11?

PROBLEM 13

Whether the *ortho*-carboxyl substituent acts as an intramolecular general-base catalyst or as an intramolecular nucleophilic catalyst can be determined by carrying out the hydrolysis of aspirin with ^{18}O-labeled water and determining whether ^{18}O is incorporated into *ortho*-carboxyl-substituted phenol. Explain the results that would be obtained with the two types of catalysis.

The following reaction, in which Ni^{2+} catalyzes the hydrolysis of the ester, is an example of intramolecular metal-ion catalysis:

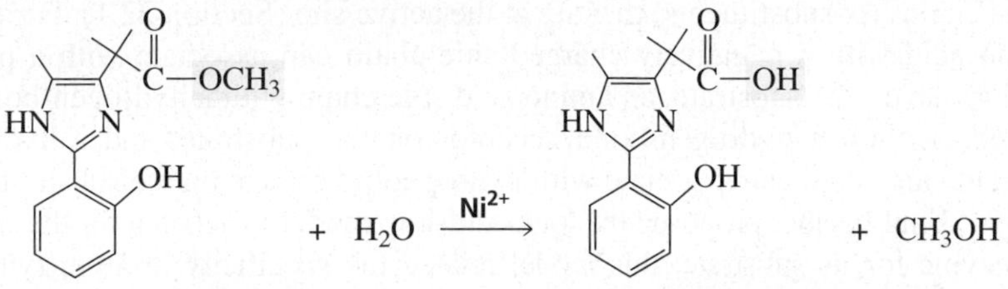

VCL Epoxidation-3

The metal ion complexes with an oxygen and a nitrogen of the reactant, as well as with a molecule of water. The metal ion increases the rate of the reaction by positioning the water molecule and converting it to metal-bound hydroxide, thus increasing its nucleophilicity.

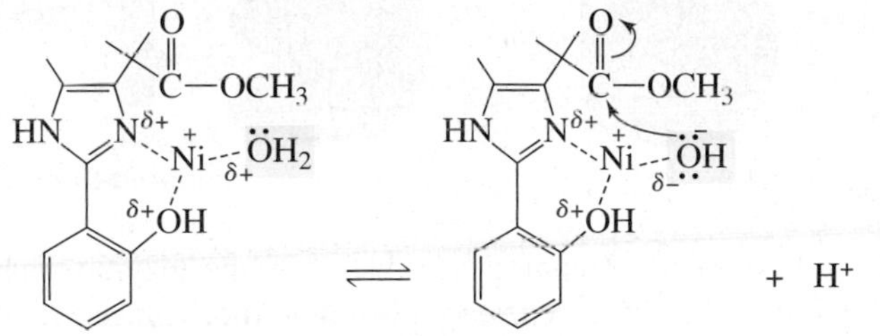

23.8 Catalysis in Biological Reactions

Essentially all organic reactions that occur in biological systems require a catalyst. Most biological catalysts are **enzymes**, which are globular proteins (Section 22.1), although some RNA catalysts are known (page 1063). Each biological reaction is catalyzed by a different enzyme. Enzymes are extraordinarily good catalysts—they can increase the rate of an intermolecular reaction by as much as 10^{16}. In contrast, rate enhancements achieved by nonbiological catalysts in intermolecular reactions are seldom greater than 10,000-fold.

The reactant of an enzyme-catalyzed reaction is called a **substrate**. The enzyme has a pocket or cleft known as an **active site**. The substrate specifically fits and binds to the active site (Figure 23.4 on page 1080).

$$\text{substrate} \xrightarrow{\text{enzyme}} \text{product}$$

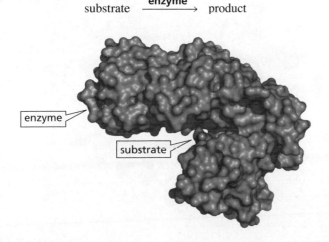

All the bond-making and bond-breaking steps of the reaction occur while the substrate is bound to the active site. Enzymes differ from nonbiological catalysts in that they are specific for the substrate whose reaction they catalyze (Section 5.21). All enzymes, however, do not have the same degree of specificity. Some are specific for a single compound and will not tolerate even the slightest variation in structure, whereas some catalyze the reaction of a family of compounds with related structures. The specificity of an enzyme for its substrate is an example of the phenomenon known as **molecular recognition**—the ability of one molecule to recognize another (Section 21.0).

The specificity of an enzyme results from its conformation and the particular amino acid side chains (α-substituents) that are at the active site (Section 22.1). For example, an amino acid with a negatively charged side chain can associate with a positively charged group on the substrate, an amino acid side chain with a hydrogen-bond donor can associate with a hydrogen-bond acceptor on the substrate, and a hydrophobic amino acid side chain can associate with hydrophobic groups on the substrate.

In 1894, Emil Fischer proposed the **lock-and-key model** to account for the specificity of an enzyme for its substrate: this model related the specificity of an enzyme for its substrate to the specificity of a lock for a correctly shaped key.

lock-and-key model   induced-fit model

The energy released as a result of binding the substrate to the enzyme can be used to induce a change in the conformation of the enzyme, leading to more precise binding between the substrate and the active site. This change in conformation of the enzyme is known as induced fit. In the **induced-fit model**, the shape of the active site does not become completely complementary to the shape of the substrate until the enzyme has bound the substrate. An example of induced fit is shown in Figure 23.4. The three-dimensional structure of the enzyme hexokinase is shown before and after binding glucose, its substrate. Notice the change in conformation that occurs upon binding the substrate.

Figure 23.4 ▶
The structure of hexokinase before binding its substrate is shown in red. The structure of hexokinase after binding its substrate is shown in green.

Various factors contribute to the remarkable catalytic ability of enzymes. Some of the most important are:

- Reacting groups are brought together at the active site in the proper orientation for reaction. This is analogous to the way proper positioning of reacting groups increases the rate of intramolecular reactions (Section 23.6).

- Some of the amino acid side chains of the enzyme serve as catalysts, and many enzymes also have metal ions at their active site that act as catalysts. These species are positioned relative to the substrate precisely where they are needed for catalysis. This factor is analogous to the way intramolecular catalysis by acids, bases, and metal ions is able to enhance reaction rates (Section 23.7).

- Amino acid side chains can stabilize transition states and intermediates—by van der Waals interactions, electrostatic interactions, and hydrogen bonding—which makes them easier to form (Figure 23.1b).

As we look at some examples of enzyme-catalyzed reactions, notice that the functional groups on the enzyme side chains are the same functional groups you are used to seeing in simple organic compounds, and the modes of catalysis used by enzymes are the same as the modes of catalysis used in organic reactions. The remarkable catalytic ability of enzymes stems in part from their ability to use several modes of catalysis in the same reaction. Factors other than those listed above can contribute to the increased rate of enzyme-catalyzed reactions, but not all factors are employed by every enzyme. We will consider some of these rate-enhancing factors when we discuss individual enzymes.

23.9 Enzyme-Catalyzed Reactions

Now we will look at the mechanisms of five enzyme-catalyzed reactions in order to understand how the amino acid side chains at the active site act as catalytic groups. In examining these reactions, notice that they are similar to reactions that you have seen organic compounds undergo. If you refer back to sections referenced throughout this chapter, you will be able to see that much of the organic chemistry you have learned applies to the reactions of compounds found in the biological world.

Mechanism for carboxypeptidase A

The names of most enzymes end in "ase," and the enzyme's name tells you the reaction it catalyzes. For example, carboxypeptidase A catalyzes the hydrolysis of the C-terminal (carboxy-terminal) peptide bond in peptides and proteins, releasing the C-terminal amino acid (Section 22.13).

Carboxypeptidase A is a *metalloenzyme,* an enzyme that contains a tightly bound metal ion. The metal ion in carboxypeptidase A is Zn^{2+}. Carboxypeptidase A is one of several hundred enzymes known to contain zinc. In bovine pancreatic carboxypeptidase A, Zn^{2+} is bound to the enzyme at its active site by forming a complex with Glu 72, His 196, and His 69, as well as with a water molecule (Figure 23.5). (The source of the enzyme is specified because, although carboxypeptidase A's from different sources follow the same mechanism, they have slightly different primary structures.)

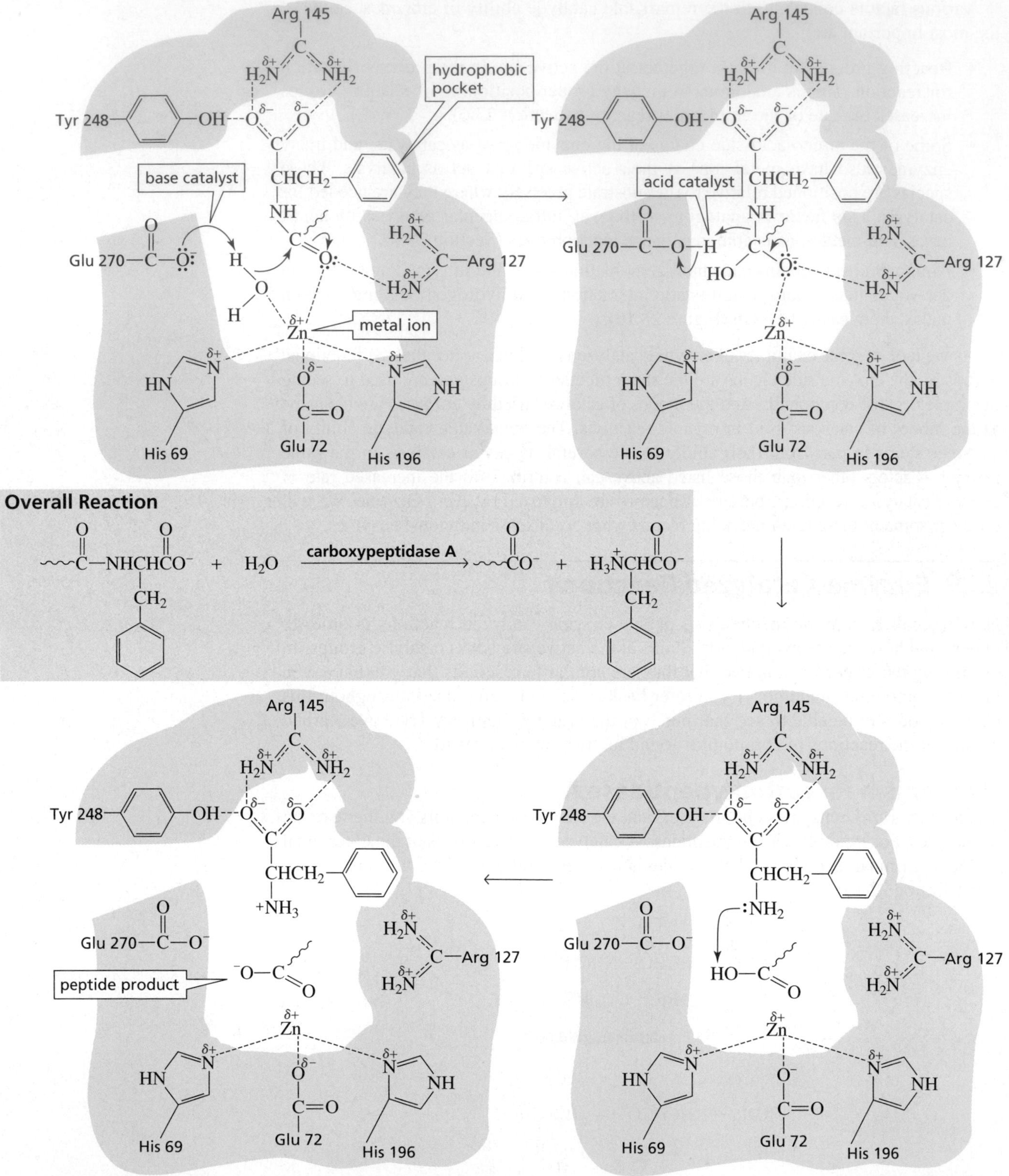

Overall Reaction

▲ Figure 23.5
Proposed mechanism for the carboxypeptidase A–catalyzed hydrolysis of a peptide bond.

Student Tutorial:
Mechanism: Carboxypeptidase A

Several groups at the active site of carboxypeptidase A participate in binding the substrate in the optimal position for reaction (Figure 23.5). Arg 145 forms two hydrogen bonds and Tyr 248 forms one hydrogen bond with the C-terminal carboxyl group of the substrate. (In this example, the C-terminal amino acid is phenylalanine.) The side chain of the C-terminal amino acid is positioned in a hydrophobic pocket, which is why carboxypeptidase A is not active if the C-terminal amino acid is arginine or lysine. Apparently, the long, positively charged side chains of these amino acid residues (Table 22.2) cannot fit into the nonpolar pocket. The reaction takes place as follows:

- When the substrate binds to the active site, Zn^{2+} partially complexes with the oxygen of the carbonyl group of the amide that will be hydrolyzed (Figure 23.5). Zn^{2+} polarizes the carbon–oxygen double bond, making the carbonyl carbon more susceptible to nucleophilic attack and stabilizing the negative charge that develops on the oxygen atom in the transition state that leads to the tetrahedral intermediate. Arg 127 also increases the carbonyl group's electrophilicity and stabilizes the developing negative charge on the oxygen atom in the transition state. Zn^{2+} also complexes with water, thereby making it a better nucleophile. Glu 270 functions as a base catalyst, further increasing water's nucleophilicity.

- In the second step of the reaction, Glu 270 functions as an acid catalyst, increasing the leaving tendency of the amino group. When the reaction is over, the amino acid (phenylalanine in this example) and the peptide with one less amino acid residue dissociate from the enzyme, and another molecule of substrate binds to the active site. It has been suggested that the unfavorable electrostatic interaction between the negatively charged carboxyl group of the peptide product and the negatively charged Glu 270 residue facilitates the release of the product from the enzyme.

Notice that in these enzyme-catalyzed reactions, the acid and base catalysis that occurs is general-acid and general-base catalysis (Sections 23.2 and 23.3). The protons are being removed and donated during (rather than before) the other bond-making and bond-breaking processes. At physiological pH (7.3), too small a concentration ($\sim 1 \times 10^{-7}$ M) of H^+ for specific-acid catalysis or HO^- for specific-base catalysis is available.

PROBLEM 14 *SOLVED*

Which of the following amino acid side chains can aid the departure of a leaving group by protonating it?

$$-CH_2CH_2SCH_3 \qquad -CH(CH_3)_2 \qquad -CH_2\overset{+}{-}\!\!\!\diagup\!\!\!\diagdown\!NH \qquad -CH_2\overset{O}{\overset{\|}{C}}OH$$

      1   **2**   **3**   **4**

Solution Side chains **1** and **2** do not have an acidic proton, so they cannot aid the departure of a leaving group. Side chains **3** and **4** each have an acidic proton, so they can aid the departure of a leaving group.

PROBLEM 15◆

Which of the following amino acid side chains can help remove a proton from the α-carbon of an aldehyde?

$$-CH_2\overset{O}{\overset{\|}{C}}NH_2 \qquad -\!\!\diagup\!\!\diagdown\!\!-O^- \qquad -CH_2\!\!-\!\!\diagup\!\!\diagdown\!N \qquad -CH_2\overset{O}{\overset{\|}{C}}O^-$$

      1   **2**   **3**   **4**

Which of the following C-terminal peptide bonds would be more readily cleaved by carboxypeptidase A?

Ser-Ala-Phe   or   Ser-Ala-Asp

Explain your choice.

Carboxypeptidase A has esterase activity as well as peptidase activity. In other words, the compound can hydrolyze ester bonds as well as peptide bonds. When carboxypeptidase A hydrolyzes ester bonds, Glu 270 acts as a nucleophilic catalyst instead of a general-base catalyst. Propose a mechanism for the carboxypeptidase A–catalyzed hydrolysis of an ester bond.

Mechanism for the serine proteases

Trypsin, chymotrypsin, and elastase are members of a large group of *endopeptidases* known collectively as serine proteases. Recall that an endopeptidase cleaves a peptide bond that is not at the end of a peptide chain (Section 22.13). They are called *proteases* because they catalyze the hydrolysis of protein peptide bonds. They are called *serine proteases* because they all have a serine residue at the active site that participates in the catalysis.

The various serine proteases have similar primary structures, suggesting that they are evolutionarily related. They all have the same three catalytic residues at the active site: an aspartate, a histidine, and a serine. But they have one important difference—the composition of the pocket at the active site that binds the side chain of the amino acid residue undergoing hydrolysis (Figure 23.6). This pocket is what gives the serine proteases their different specificities (Section 22.13).

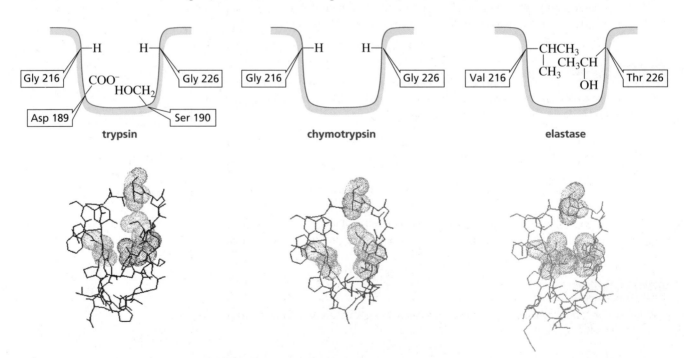

▲ **Figure 23.6**
The binding pockets in trypsin, chymotrypsin, and elastase. The negatively charged aspartate is shown in red, and the relatively nonpolar amino acids are shown in green. The structures of the binding pockets explain why trypsin binds long, positively charged amino acids; chymotrypsin binds flat, nonpolar amino acids; and elastase binds only small amino acids.

The pocket in trypsin is narrow and has a serine and a negatively charged aspartate carboxyl group at its bottom. The shape and charge of the binding pocket cause it to bind long, positively charged amino acid side chains (Lys and Arg). This is why trypsin hydrolyzes peptide bonds on the C-side of arginine and lysine residues. The pocket in chymotrypsin is narrow and is lined with nonpolar amino acids, so chymotrypsin cleaves on the C-side of amino acids with flat, nonpolar side chains (Phe, Tyr, Trp). In elastase, two glycines on the sides of the pocket in trypsin and in chymotrypsin are replaced by relatively bulky valine and threonine residues. Consequently, only small amino acids can fit into the pocket. Elastase, therefore, hydrolyzes peptide bonds on the C-side of small amino acids (Gly and Ala).

The mechanism for bovine chymotrypsin-catalyzed hydrolysis of a peptide bond is shown in Figure 23.7. The other serine proteases follow the same mechanism. The reaction proceeds as follows:

- As a consequence of binding the flat, nonpolar side chain in the pocket, the amide linkage that is to be hydrolyzed is positioned very close to Ser 195. His 57 functions as a base catalyst, increasing the nucleophilicity of serine, which attacks the carbonyl group. This process is helped by Asp 102, which uses its negative charge to stabilize the resulting positive charge on His 57 and to position the five-membered ring so that its basic N atom is close to the OH of serine. The stabilization of a charge by an opposite charge is called **electrostatic catalysis**. Formation of the tetrahedral intermediate causes a slight change in the conformation of the protein that allows the negatively charged oxygen to slip into a previously unoccupied area of the active site known as the *oxyanion hole*. Once in the oxyanion hole, the negatively charged oxygen can hydrogen bond with two peptide groups (Gly 193 and Scr 195), which stabilizes the tetrahedral intermediate.

- In the next step, the tetrahedral intermediate collapses, expelling the amino group. This is a strongly basic group that cannot be expelled without the participation of His 57, which acts as an acid catalyst. The product of the second step is an **acyl-enzyme intermediate** because the serine group of the enzyme has been acylated. (An acyl group has been put on it.)

- The third step is just like the first step, except that water instead of serine is the nucleophile. Water attacks the acyl group of the acyl-enzyme intermediate, with His 57 functioning as a base catalyst to increase water's nucleophilicity and Asp 102 stabilizing the positively charged histidine residue.

- In the final step of the reaction, the tetrahedral intermediate collapses, expelling serine. His 57 functions as an acid catalyst in this step, increasing serine's leaving ability.

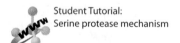 Student Tutorial: Serine protease mechanism

The mechanism for chymotrypsin-catalyzed hydrolysis shows the importance of histidine as a catalytic group. Because the pK_a of the imidazole ring of histidine ($pK_a = 6.0$) is close to neutrality, histidine can act both as an acid catalyst and as a base catalyst at physiological pH.

Much information about the relationship between the structure of a protein and its function has been determined by **site-specific mutagenesis**, a technique that replaces one amino acid of a protein with another. For example, when Asp 102 of chymotrypsin is replaced with Asn 102, the enzyme's ability to bind the substrate is unchanged, but its ability to catalyze the reaction decreases to less than 0.05% of the value for the native enzyme. Clearly, Asp 102 must be involved in the catalytic process. We just saw that its role is to position histidine and use its negative charge to stabilize histidine's positive charge.

side chain of an aspartate (Asp) residue   side chain of an asparagine (Asn) residue

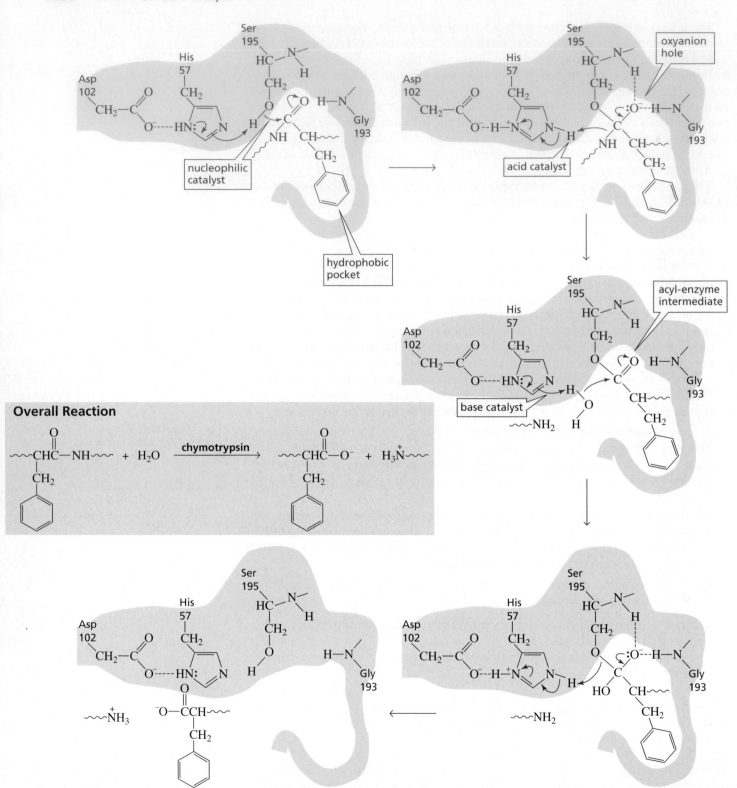

▲ **Figure 23.7**
Proposed mechanism for the chymotrypsin-catalyzed hydrolysis of a peptide bond.

PROBLEM 18◆

Arginine and lysine side chains fit into trypsin's binding pocket. One of these side chains forms a direct hydrogen bond with serine and an indirect hydrogen bond (mediated through a water molecule) with aspartate. The other side chain forms direct hydrogen bonds with both serine and aspartate. Which is which?

PROBLEM 19

Serine proteases do not catalyze hydrolysis if the amino acid at the hydrolysis site is a D-amino acid. Trypsin, for example, cleaves on the C-side of L-Arg and L-Lys, but not on the C-side of D-Arg and D-Lys. Explain.

Mechanism for lysozyme

Lysozyme is an enzyme that destroys bacterial cell walls. These cell walls are composed of alternating *N*-acetylmuramic acid (NAM) and *N*-acetylglucosamine (NAG) units linked by β-1,4'-glycosidic linkages (Section 21.17). Lysozyme destroys the cell wall by catalyzing the hydrolysis of the NAM–NAG bond.

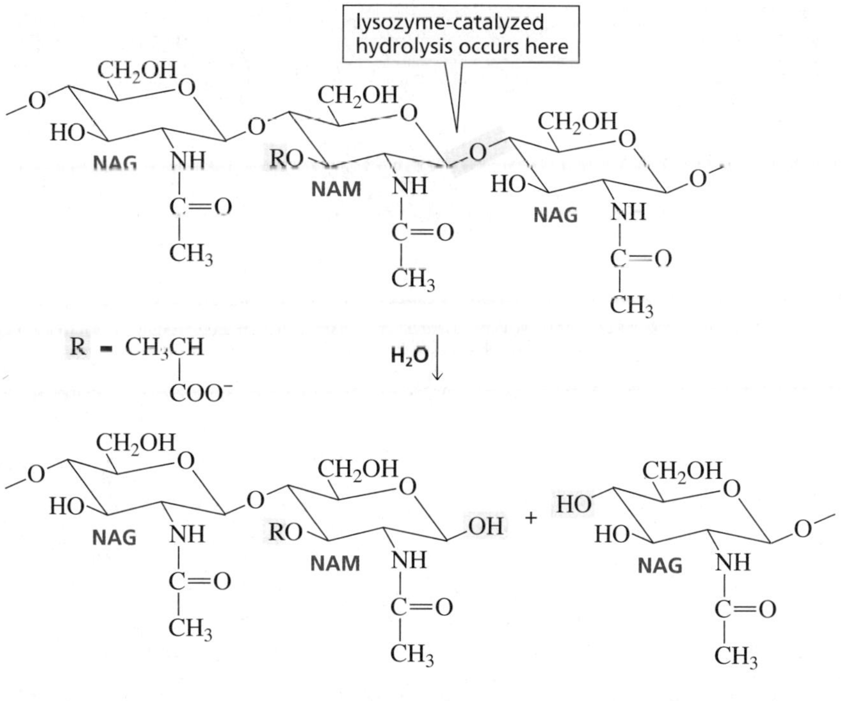

The active site of hen egg-white lysozyme binds six sugar residues of the substrate. The many amino acid residues involved in binding the substrate in the correct position in the active site are shown in Figure 23.8. The six sugar residues are labeled A, B, C, D, E, and F. The carboxylic acid substituent of the RO group of NAM cannot fit into the binding site for C or E. This means that NAM units must bind at the sites for B, D, and F. Hydrolysis occurs between D and E.

Lysozyme has two catalytic groups at the active site: Glu 35 and Asp 52 (Figure 23.9). The discovery that the enzyme-catalyzed reaction takes place with retention of configuration at the anomeric carbon led to the conclusion that it cannot be a one-step S_N2 reaction; the reaction must involve either two sequential S_N2 reactions or an S_N1 reaction with the enzyme blocking one face of the oxocarbenium ion intermediate from nucleophilic attack. Although lysozyme was the first enzyme to have its mechanism

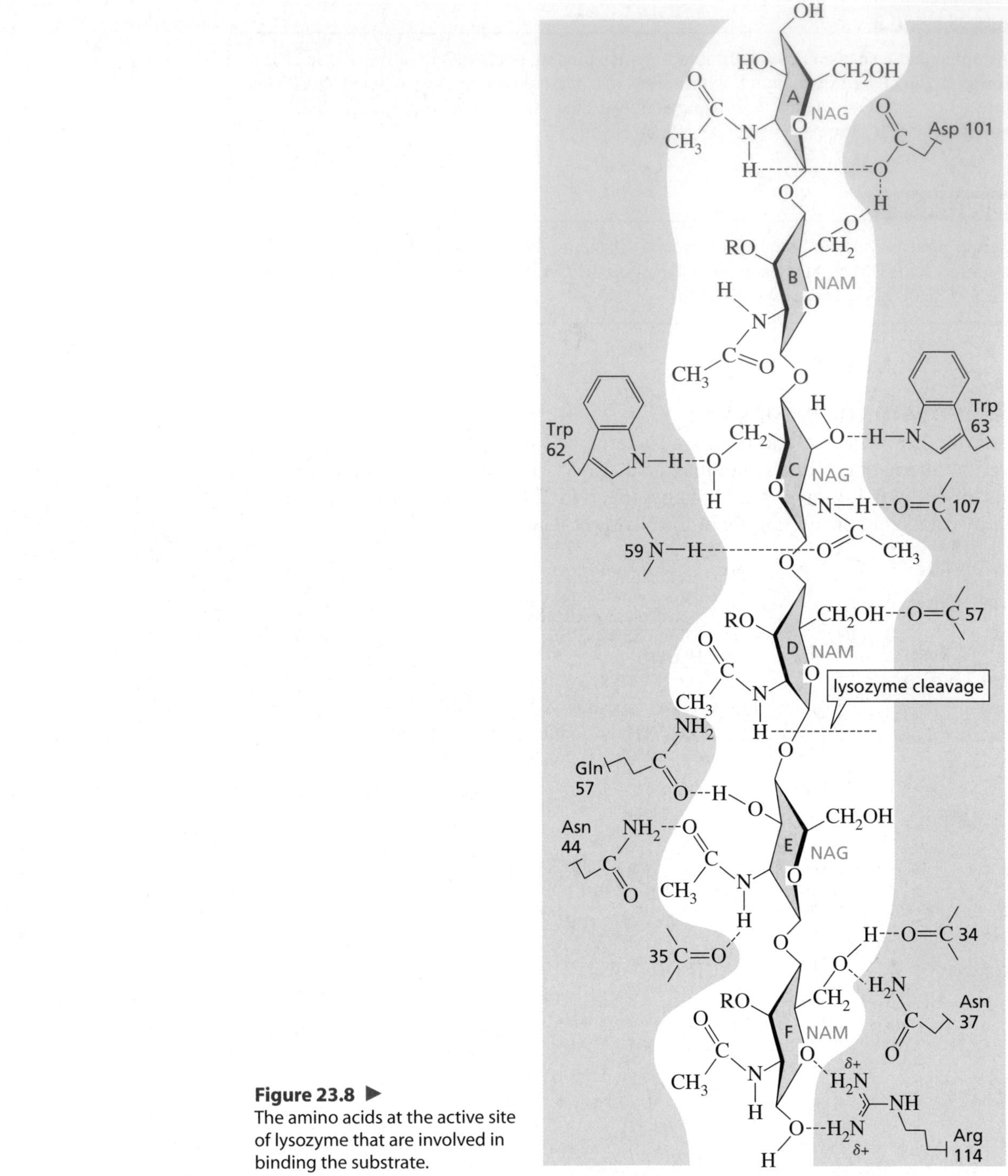

Figure 23.8 ▶
The amino acids at the active site of lysozyme that are involved in binding the substrate.

studied—and it has been studied extensively for almost 40 years—only recently have data been obtained that support the mechanism involving two sequential S_N2 reactions shown in Figure 23.9:

- In the first step of the reaction, Asp 52 acts as a nucleophilic catalyst and attacks the anomeric carbon (C-1) of the NAM residue, displacing the leaving group. Glu 35 acts as an acid catalyst, protonating the leaving group and thereby making it a weaker base and a better leaving group. Site-specific mutagenesis studies show that when Glu 35 is replaced by Asp, the enzyme has only weak activity. Apparently, Asp does not lie at the optimal distance from and angle to the oxygen atom to

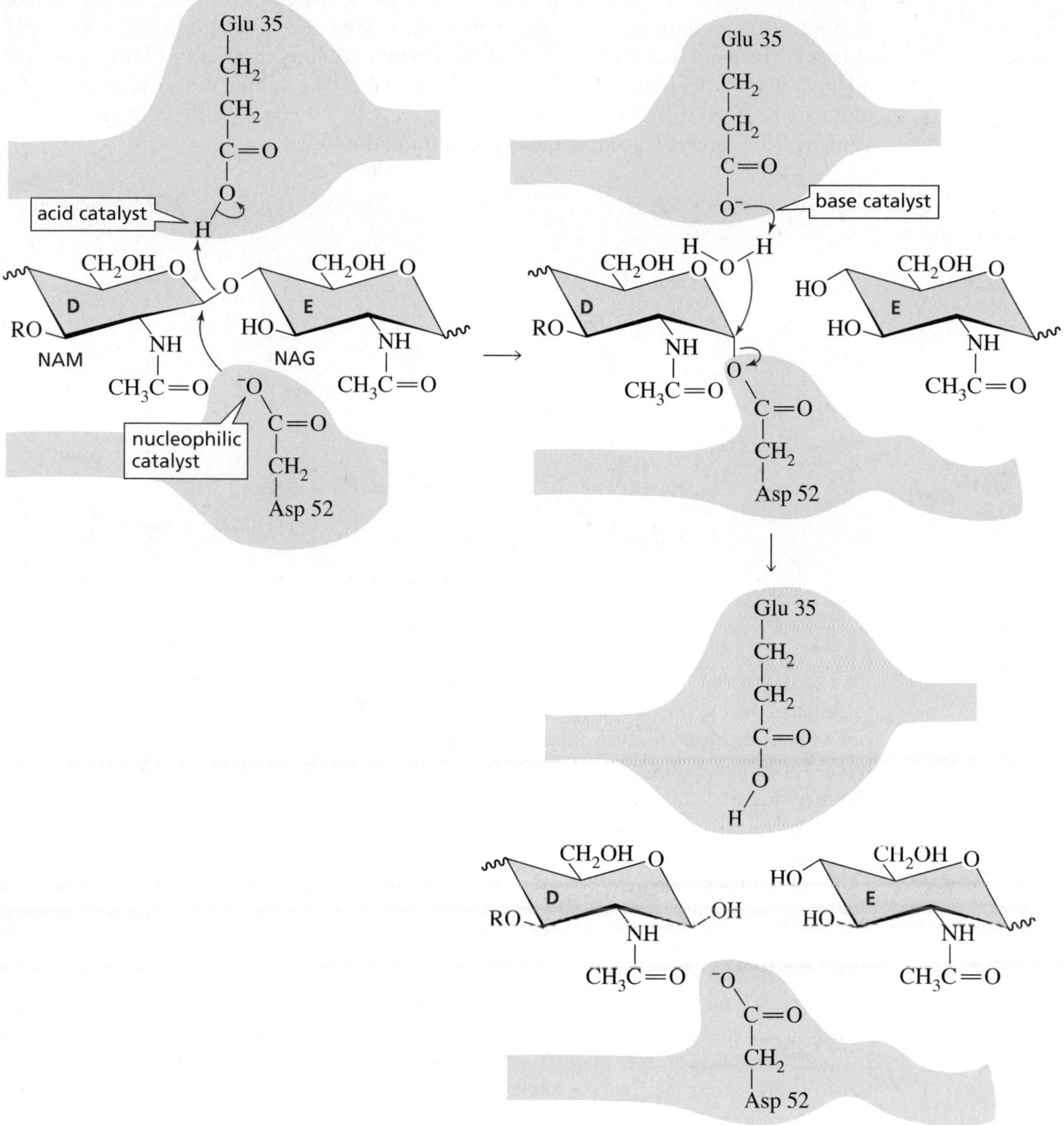

▲ **Figure 23.9**
Proposed mechanism for the lysozyme-catalyzed hydrolysis of a cell wall.

protonate it. When Glu 35 is replaced by Ala, an amino acid that cannot act as an acid catalyst, the activity of the enzyme is completely lost.

- In the second step of the reaction, Glu 35 acts as a base catalyst to increase water's nucleophilicity.

PROBLEM 20◆

If H_2O^{18} were used to hydrolyze lysozyme, which ring would contain the label, NAM or NAG?

A plot of the activity of an enzyme as a function of the pH of the reaction mixture is called a **pH–activity profile** or a **pH–rate profile** (Section 17.8). The pH–activity profile for lysozyme is shown in Figure 23.10. It is a bell-shaped curve with the maximum rate occurring at about pH 5.3. The pH at which the enzyme is 50% active is 3.8 on the ascending leg of the curve and 6.7 on the descending leg. These pH values correspond to the pK_a values of the enzyme's catalytic groups. (This is true for all bell-shaped pH–rate profiles, provided that the pK_a values are at least two pK_a units apart. If the difference between them is less than two pK_a units, the precise pK_a values of the catalytic groups must be determined in other ways.)

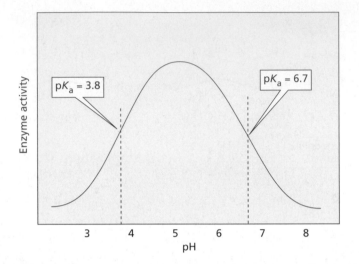

Figure 23.10 ▶
Dependence of lysozyme activity on the pH of the reaction mixture.

The pK_a given by the ascending leg is the pK_a of a group that is catalytically active in its basic form. When that group is fully protonated, the enzyme is not active. As the pH of the reaction mixture increases, a larger fraction of the group is present in its basic form, and as a result, the enzyme shows increasing activity. Similarly, the pK_a given by the descending leg is the pK_a of a group that is catalytically active in its acidic form. Maximum catalytic activity occurs when the group is fully protonated; activity decreases with increasing pH because a greater fraction of the group lacks a proton.

From the lysozyme mechanism shown in Figure 23.9, we can conclude that Asp 52 is the group with a pK_a of 3.8 and Glu 35 is the group with a pK_a of 6.7. The pH–activity profile indicates that lysozyme is maximally active when Asp 52 is in its basic form and Glu 35 is in its acidic form.

Table 23.2 on page 1075 shows that the pK_a of aspartic acid is 3.86 and the pK_a of glutamic acid is 4.25. The pK_a of Asp 52 agrees with the pK_a of aspartic acid, but the pK_a of Glu 35 is much greater than the pK_a of glutamic acid. Why is the pK_a of the glutamic acid residue at the active site of the enzyme so much greater than the pK_a given for glutamic acid in the table? The pK_a values in the table were determined in water. In the enzyme, Asp 52 is surrounded by polar groups, which means that its pK_a should be close to the pK_a determined in water, a polar solvent. Glu 35, however, is in a predominantly nonpolar microenvironment, so its pK_a should be greater than the pK_a determined in water. We have seen that the pK_a of a carboxylic acid is greater in a nonpolar solvent because there is less tendency to form charged species in nonpolar solvents (Section 8.10).

Part of the catalytic efficiency of lysozyme results from its ability to provide different solvent environments at the active site. This allows one catalytic group to exist in its acidic form at the same surrounding pH at which a second catalytic group exists in its basic form. This property is unique to enzymes; chemists cannot provide different solvent environments for different parts of nonenzymatic systems.

PROBLEM 21◆

When apples that have been cut are exposed to oxygen, an enzyme-catalyzed reaction causes them to turn brown. Coating them with lemon juice (pH ~ 3.5) as soon as they are cut prevents the color change. Explain why this is so.

Mechanism for glucose-6-phosphate isomerase

Glycolysis is the name given to the series of enzyme-catalyzed reactions responsible for converting glucose into two molecules of pyruvate (Sections 18.22 and 25.7). The second reaction in glycolysis is an isomerization reaction that converts glucose-6-phosphate to fructose-6-phosphate. Recall that the open-chain form of glucose is an aldohexose, whereas the open-chain form of fructose is a ketohexose. Therefore, the enzyme that catalyzes this reaction—glucose-6-phosphate isomerase—converts an aldose to a ketose (Section 21.5). In solution, the sugars exist predominantly in their cyclic forms, so the enzyme must open the six-membered-ring sugar and convert it to the five-membered-ring sugar. Glucose-6-phosphate isomerase is known to have at least three catalytic groups at its active site, one functioning as an acid catalyst and two acting as base catalysts (Figure 23.11). The reaction takes place as follows:

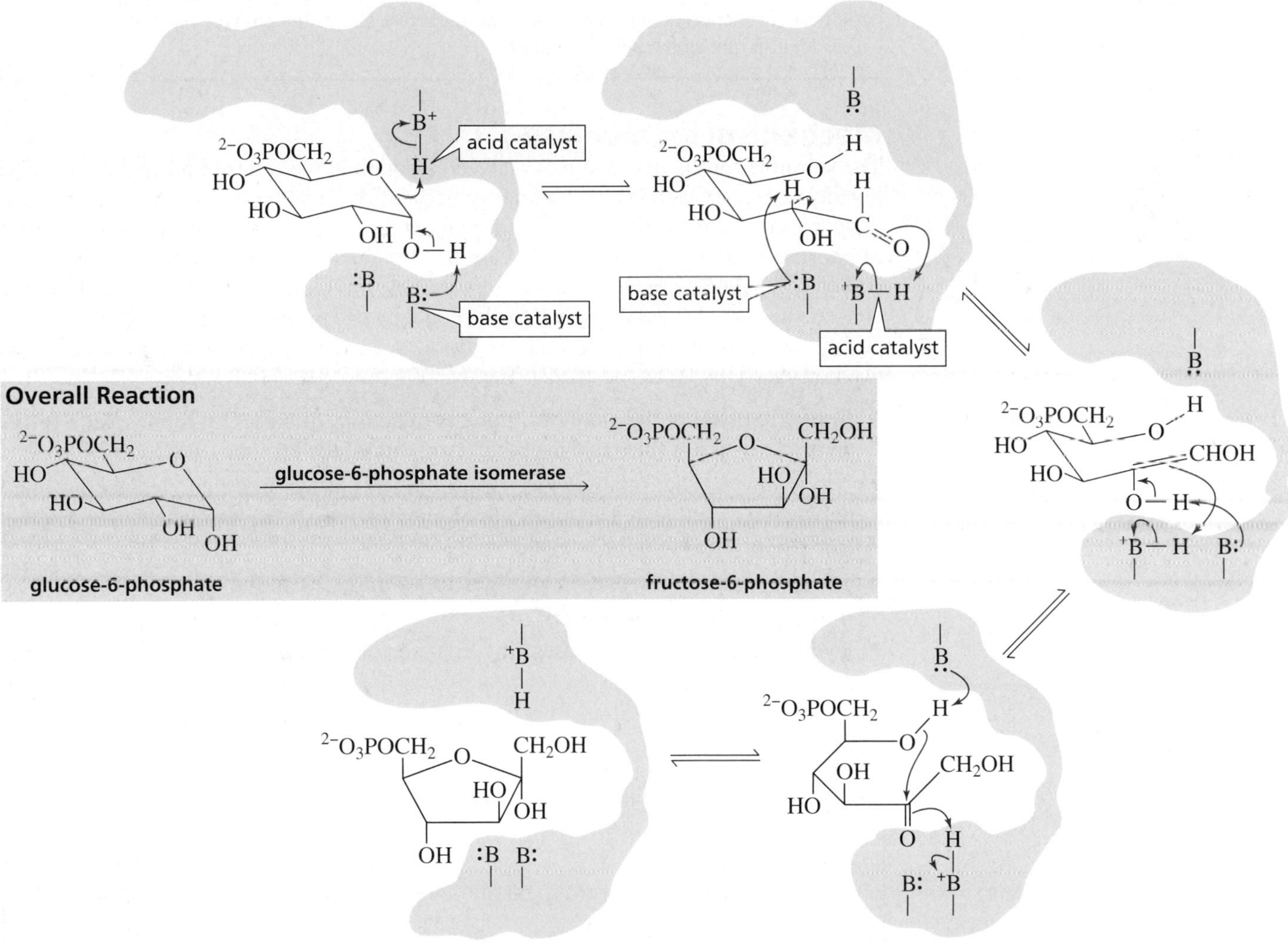

▲ Figure 23.11
Proposed mechanism for the isomerization of glucose-6-phosphate to fructose-6-diphosphate.

- The first step is a ring-opening reaction. A base catalyst (presumably a histidine side chain) removes a proton from the OH group, and an acid catalyst (a protonated lysine side chain) aids the departure of the leaving group by protonating it, thereby making it a weaker base and therefore a better leaving group (Section 17.10).

- In the second step of the reaction, a base catalyst (a glutamate side chain) removes a proton from the α-carbon of the aldehyde. Recall that α-hydrogens are relatively acidic (Section 18.1).

- In the next step, the enol is converted to a ketone (Section 21.5).
- In the final step of the reaction, the conjugate base of the acid catalyst employed in the first step and the conjugate acid of the base catalyst employed in the first step catalyze ring closure.

PROBLEM 22

When glucose undergoes base-catalyzed isomerization in the absence of the enzyme, three products result: glucose, fructose, and mannose (Section 21.5). Why is mannose not formed in the enzyme-catalyzed reaction?

PROBLEM 23◆

The descending leg of the pH–rate profile for glucose-6-phosphate isomerase indicates that one of the amino acid side chains at the active site of the enzyme has a pK_a value of 9.3. Identify the amino acid side chain.

Mechanism for aldolase

The substrate for the first enzyme-catalyzed reaction in the series of reactions known as glycolysis is a six-carbon compound (D-glucose). The final product of glycolysis is two molecules of a three-carbon compound (pyruvate). Therefore, at some point in the series of enzyme-catalyzed reactions, a six-carbon compound must be cleaved into two three-carbon compounds. The enzyme *aldolase* catalyzes this cleavage (Figure 23.12). Aldolase converts fructose-1,6-diphosphate into glyceraldehyde-3-phosphate and dihydroxyacetone phosphate. The enzyme is called aldolase because the reverse reaction is an aldol addition reaction (Sections 18.12 and 18.22). The reaction proceeds as follows:

- In the first step of the aldolase-catalyzed reaction, fructose-1,6-diphosphate forms an imine with a lysine residue at the active site of the enzyme (Section 17.8).
- A tyrosine residue functions as a base catalyst in the step that cleaves the bond between C-3 and C-4. The molecule of glyceraldehyde-3-phosphate formed in this step dissociates from the enzyme.
- The enamine intermediate rearranges to an imine, with the tyrosine residue now functioning as an acid catalyst.
- Hydrolysis of the imine releases dihydroxyacetone phosphate, the other three-carbon product.

PROBLEM 24◆

Which of the following amino acid side chains can form an imine with the substrate?

$$\underset{\textbf{1}}{-CH_2\overset{\overset{\displaystyle O}{\|}}{C}NH_2} \qquad \underset{\textbf{2}}{-(CH_2)_4NH_2} \qquad \underset{\textbf{3}}{-(CH_2)_3NH\overset{\overset{\displaystyle NH}{\|}}{C}NH_2} \qquad \underset{\textbf{4}}{-CH_2OH}$$

PROBLEM 25◆

Propose a mechanism for the aldolase-catalyzed cleavage of fructose-1,6-diphosphate if it did not form an imine with the substrate. What is the advantage gained by imine formation?

PROBLEM 26

In glycolysis, why must glucose-6-phosphate isomerize to fructose-6-phosphate before the cleavage reaction with aldolase occurs? (See page 1150.)

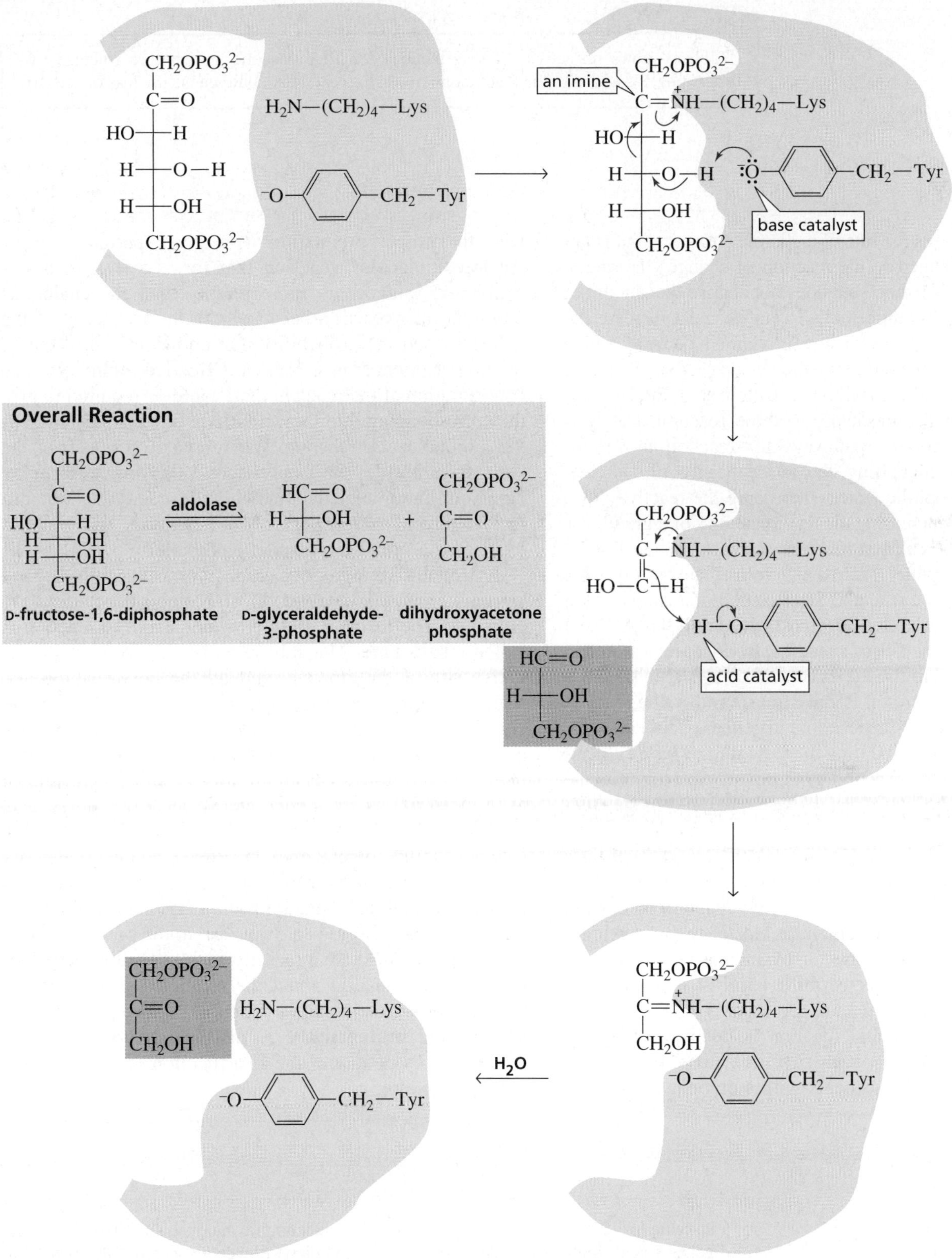

Overall Reaction

▲ **Figure 23.12**
Proposed mechanism for the aldolase-catalyzed cleavage of fructose-1,6-diphosphate to form glyceraldehyde-3-phosphate and dihydroxyacetone phosphate.

PROBLEM 27◆

Aldolase shows no activity if it is incubated with iodoacetic acid before fructose-1,6-diphosphate is added to the reaction mixture. Suggest what could cause the loss of activity.

SUMMARY

A **catalyst** increases the rate of a chemical reaction but is not consumed or changed in the reaction; it changes the rate at which a product is formed but does not change the amount of product formed at equilibrium. A catalyst must increase the rate of a slow step. It does this by providing a pathway with a lower $\Delta G^{\ddagger}$, either by converting the reactant into a less stable species, by making the transition state more stable, or by completely changing the mechanism of the reaction. Catalysts provide more favorable pathways for reaction in various ways, including increasing the susceptibility of an electrophile to nucleophilic attack, increasing the reactivity of a nucleophile, and increasing the leaving ability of a group.

A **nucleophilic catalyst** increases the rate of a reaction by acting as a nucleophile: it forms an intermediate by forming a covalent bond with a reactant. Stabilization of a charge by an opposite charge is called **electrostatic catalysis**. An acid catalyst increases the rate of a reaction by donating a proton to a reactant. There are two types of acid catalysis: in **specific-acid catalysis**, the proton is fully transferred to the reactant before the slow step of the reaction; in **general-acid catalysis**, the proton is transferred during the slow step. A **base catalyst** increases the rate of a reaction by removing a proton from the reactant. There are two types of base catalysis: in **specific-base catalysis**, the proton is completely removed from the reactant before the slow step of the reaction; in **general-base catalysis**, the proton is removed during the slow step.

A **metal ion** can increase the rate of a reaction by making a reaction center more receptive to electrons, by making a leaving group a weaker base, or by increasing the nucleophilicity of water. An **electrophilic catalyst** is a metal ion with the same catalytic effect as a proton.

The rate of a chemical reaction is determined by the number of collisions between two molecules, or between two intramolecular constituents with sufficient energy and with the proper orientation in a given period of time. An **intramolecular reaction** that forms a five- or a six-membered ring occurs more readily than the analogous intermolecular reaction because both the frequency of the collisions and the probability that collisions will occur in the proper orientation increases. **Effective molarity** is the concentration of the reactant that would be required to give the corresponding intermolecular reaction the same rate as the intramolecular reaction. When a catalyst is part of the reacting molecule, the catalysis is called **intramolecular catalysis**; intramolecular nucleophilic catalysis, intramolecular general-acid or general-base catalysis, and intramolecular metal-ion catalysis all are possible.

Essentially all organic reactions that occur in biological systems require a catalyst. Most biological catalysts are **enzymes**. The reactant of an enzyme-catalyzed reaction is called a **substrate**. The substrate specifically binds to the **active site** of the enzyme, and all the bond-making and bond-breaking steps of the reaction occur while it is at that site. The specificity of an enzyme for its substrate is an example of **molecular recognition**. The change in conformation of the enzyme when it binds the substrate is known as **induced fit**.

Two important factors contributing to the remarkable catalytic ability of enzymes are that reacting groups are brought together at the active site in the proper orientation for reaction and that the amino acid side chains and a metal ion in the case of some enzymes are in the proper position relative to the substrate, precisely where they are needed for catalysis. Information about the relationship between the structure of a protein and its function has been obtained by **site-specific mutagenesis**. A **pH–rate profile** is a plot of the activity of an enzyme as a function of the pH of the reaction mixture.

KEY TERMS

acid catalyst (p. 1066)
active site (p. 1079)
acyl-enzyme intermediate (p. 1085)
base catalyst (p. 1069)
catalyst (p. 1063)
covalent catalysis (p. 1070)
effective molarity (p. 1075)
electrophilic catalyst (p. 1072)
electrostatic catalysis (p. 1085)

enzyme (p. 1079)
general-acid catalysis (p. 1067)
general-base catalysis (p. 1070)
induced-fit model (p. 1080)
intramolecular
  catalysis (p. 1076)
lock-and-key model (p. 1080)
metal-ion catalysis (p. 1072)
molecular recognition (p. 1080)

nucleophilic catalysis (p. 1070)
nucleophilic catalyst (p. 1070)
pH–activity profile (p. 1089)
pH–rate profile (p. 1089)
relative rate (p. 1074)
site-specific mutagenesis (p. 1085)
specific-acid catalysis (p. 1067)
specific-base catalysis (p. 1070)
substrate (p. 1079)

PROBLEMS

28. Which of the following two compounds would eliminate HBr more rapidly in basic solutions?

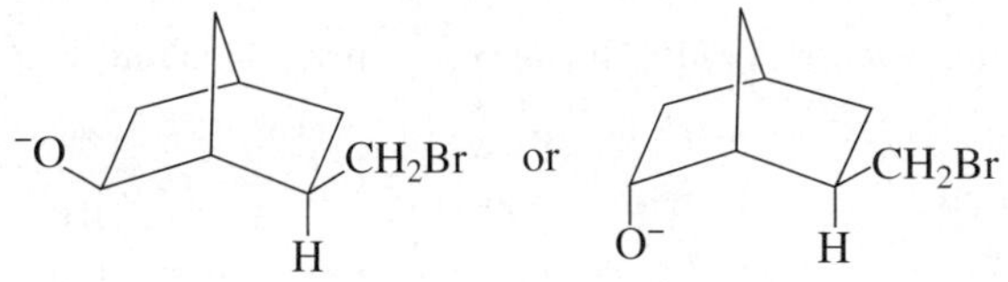

29. Which compound would form a lactone more rapidly?

a.   or

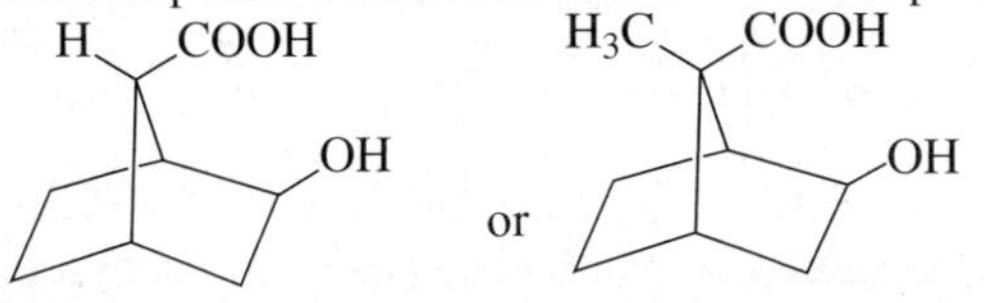

b.   or

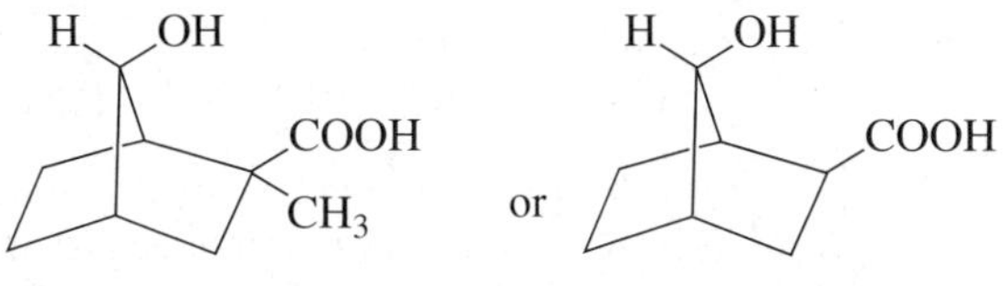

30. Which compound would form an anhydride more rapidly?

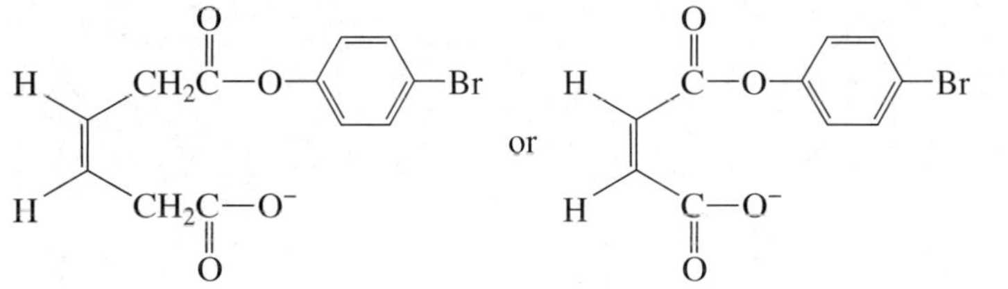

or

31. Which compound has the greatest rate of hydrolysis: benzamide, *o*-carboxybenzamide, *o*-formylbenzamide, or *o*-hydroxybenzamide?

32. Indicate the type of catalysis that is occurring in the slow step in each of the following reaction sequences:

a. $CH_3CH_2SCH_2CH_2Cl$ $\xrightarrow{\text{slow}}$ $\xrightarrow{HO^-}$ $CH_3CH_2SCH_2CH_2OH$
$+ Cl^-$

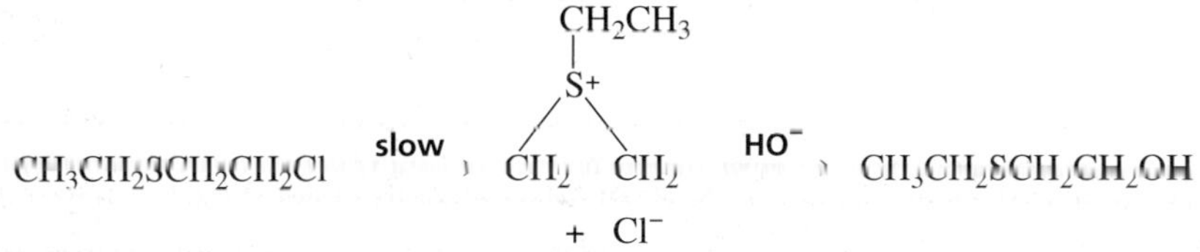

b.

$\xrightarrow{\text{slow}}$ $+$ $\longrightarrow$ $+$

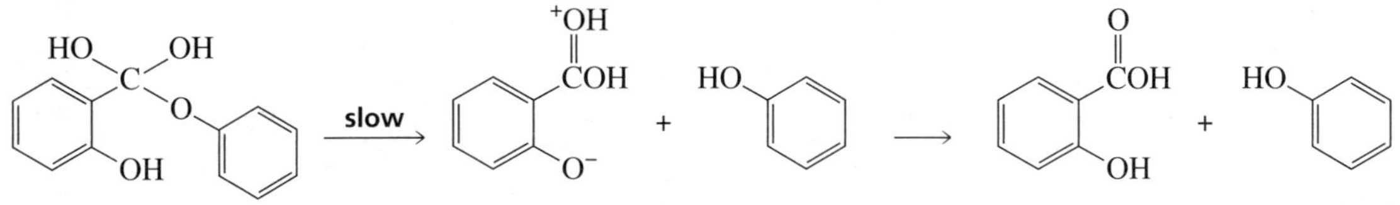

33. The deuterium kinetic isotope effect (k_{H_2O}/k_{D_2O}) for the hydrolysis of aspirin is 2.2. What does this tell you about the kind of catalysis exerted by the *ortho*-carboxyl substituent? (*Hint:* It is easier to break an O—H bond than an O—D bond.)

34. The rate constant for the uncatalyzed reaction of two molecules of glycine ethyl ester to form glycylglycine ethyl ester is $0.6 \text{ s}^{-1} \text{ M}^{-1}$. In the presence of Co^{2+}, the rate constant is $1.5 \times 10^6 \text{ s}^{-1} \text{ M}^{-1}$. What rate enhancement does the catalyst provide?

35. Draw the pH–activity profile for an enzyme that has one catalytic group at the active site. The catalytic group is an acid catalyst with a pK_a of 5.6.

36. A Co^{2+} complex catalyzes the hydrolysis of the lactam shown:

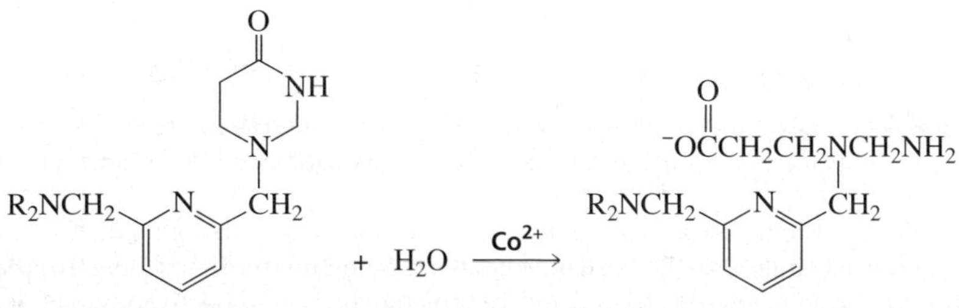

$+ \text{ H}_2\text{O} \xrightarrow{Co^{2+}}$

Propose a mechanism for the metal-ion catalyzed reaction.

37. There are two kinds of aldolases. Class I aldolases are found in animals and plants; class II aldolases are found in fungi, algae, and some bacteria. Only class I aldolases form an imine. Class II aldolases have a metal ion (Zn^{2+}) at the active site. The mechanism for catalysis by class I aldolases was given in Section 23.9. Propose a mechanism for catalysis by class II aldolases.

38. Propose a mechanism for the following reaction. (*Hint:* The rate of the reaction is much slower if the nitrogen atom is replaced by CH.)

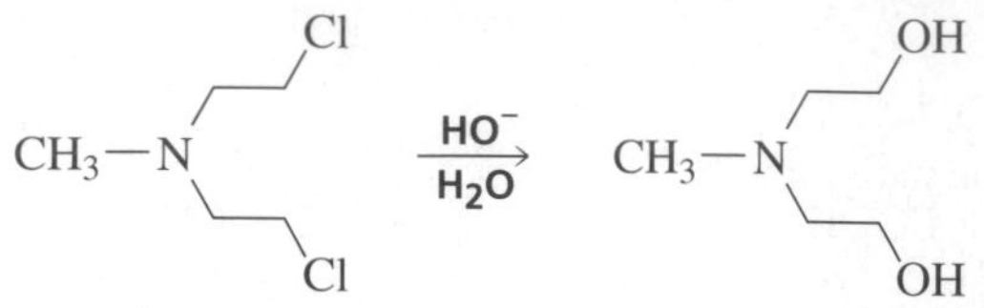

39. The hydrolysis of the ester shown here is catalyzed by morpholine, a secondary amine. Propose a mechanism for this reaction. (*Hint:* The pK_a of the conjugate acid of morpholine is 9.3, so morpholine is too weak a base to function as a base catalyst in this reaction.)

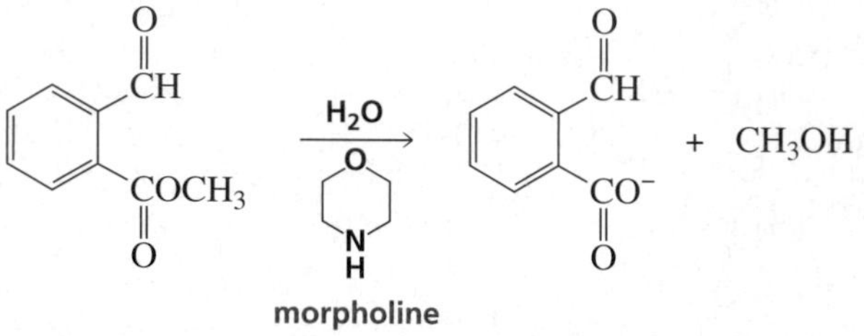

morpholine

40. The enzyme carbonic anhydrase catalyzes the conversion of carbon dioxide into bicarbonate ion (Section 1.17). It is a metalloenzyme, with Zn^{2+} coordinated at the active site by three histidine residues. Propose a mechanism for this reaction.

$$CO_2 + H_2O \xrightarrow{\text{carbonic anhydrase}} HCO_3^- + H^+$$

41. At pH = 12, the rate of hydrolysis of ester A is greater than the rate of hydrolysis of ester B. At pH = 8, the relative rates reverse. (Ester B hydrolyzes faster than ester A.) Explain these observations.

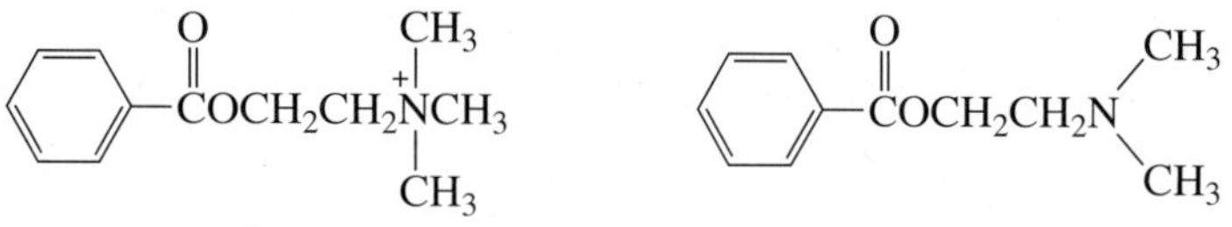

A   B

42. 2-Acetoxycyclohexyl tosylate reacts with acetate ion to form 1,2-cyclohexanediol diacetate. The reaction is stereospecific; the stereoisomers obtained as products depend on the stereoisomer used as a reactant. Explain the following observations:
  a. Both cis reactants form an optically active trans product, but each cis reactant forms a different trans product.
  b. Both trans reactants form the same racemic mixture.
  c. A trans reactant is more reactive than a cis reactant.

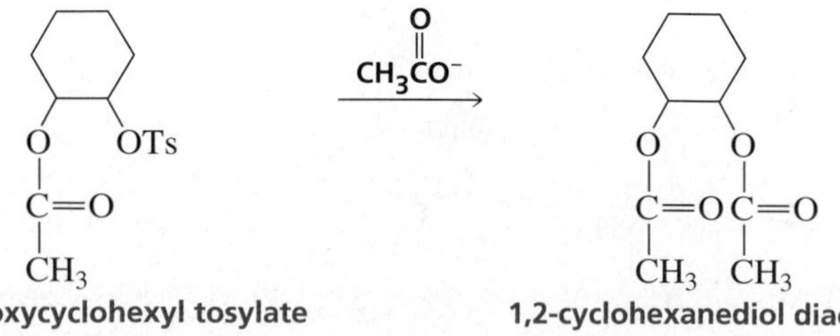

2-acetoxycyclohexyl tosylate   1,2-cyclohexanediol diacetate

43. Proof that an imine was formed between aldolase and its substrate was obtained by using D-fructose-1,6-diphosphate, labeled at the C-2 position with ^{14}C, as the substrate. $NaBH_4$ was added to the reaction mixture. A radioactive product was isolated from the reaction mixture and hydrolyzed in an acidic solution. Draw the structure of the radioactive product obtained from the acidic solution. (*Hint:* $NaBH_4$ reduces an imine linkage.)

44. 3-Amino-2-oxindole catalyzes the decarboxylation of α-keto acids.
  a. Propose a mechanism for the catalyzed reaction.
  b. Would 3-aminoindole be equally effective as a catalyst?

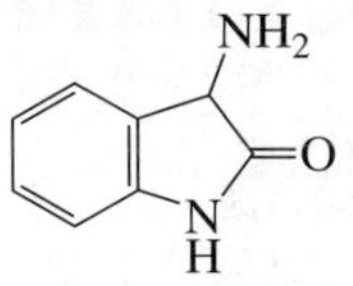

3-amino-2-oxindole

45. a. Explain why the alkyl halide shown here reacts much more rapidly than do primary alkyl halides, such as butyl chloride and pentyl chloride, with guanine residues.

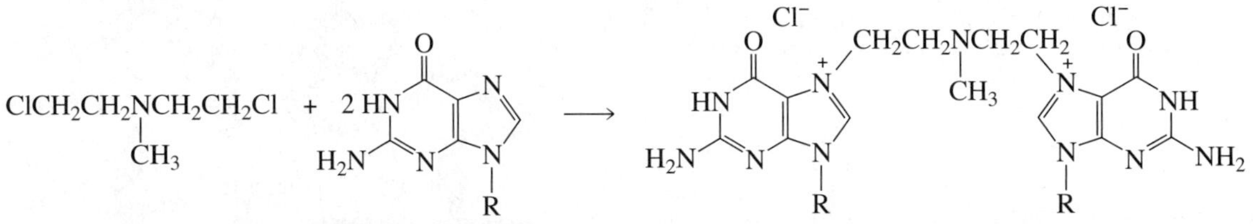

  b. The alkyl halide can react with two guanine residues in two different chains, thereby cross-linking the chains. Propose a mechanism for the cross-linking reaction.

46. Triosephosphate isomerase catalyzes the conversion of dihydroxyacetone phosphate to glyceraldehyde-3-phosphate. The enzyme's catalytic groups are Glu 165 and His 95. In the first step of the reaction, these catalytic groups function as a base and an acid catalyst, respectively. Propose a mechanism for the reaction.

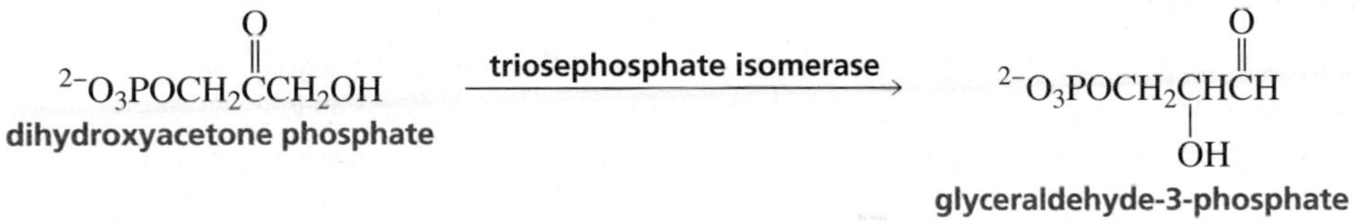

$$^{2-}O_3POCH_2CCH_2OH \xrightarrow{\text{triosephosphate isomerase}} {}^{2-}O_3POCH_2CHCH$$

dihydroxyacetone phosphate   **glyceraldehyde-3-phosphate**

CHAPTER 24

The Organic Mechanisms of the Coenzymes

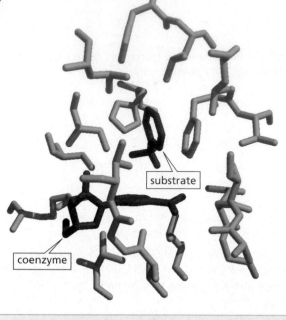

substrate

coenzyme

BUILDING ON FUNDAMENTALS

SECTION 24.2 Many reducing agents reduce their substrates by adding a hydride ion (17.6).

SECTION 24.4 A decarboxylation reaction can occur if the electrons left behind after CO_2 is removed can be delocalized (18.18).

SECTION 24.5 Enolate ions are good nucleophiles (18.4).

SECTION 24.6 The reaction of an aldehyde with a primary amine forms an imine (17.8).

SECTION 24.6 If a proton is removed from an asymmetric center, both the *R* and *S* isomers will form upon reprotonation (21.5).

SECTION 24.9 Disulfides are reduced to thiols; under mild conditions, thiols are oxidized to disulfides (22.8).

Many enzymes cannot catalyze a reaction without the help of a cofactor. **Cofactors** are substances that assist enzymes in catalyzing certain reactions that cannot be catalyzed by the amino acid side chains of the protein alone. Some cofactors are *metal ions,* while others are *organic molecules.*

A metal-ion cofactor acts as a Lewis acid in a variety of ways to help an enzyme catalyze a reaction. It can coordinate with groups on the enzyme, causing them to align in an arrangement that facilitates the reaction; it can help bind the substrate to the active site of the enzyme; it can form a coordination complex with the substrate to increase its reactivity; or it can increase the nucleophilicity of water at the active site (Section 24.5). An enzyme that has a tightly bound metal ion is called a **metalloenzyme**. We have seen how important Zn^{2+} is to the mechanism of the reaction catalyzed by carboxypeptidase A, a metalloenzyme (see page 1082).

> **PROBLEM 1♦**
>
> How does the metal ion in carboxypeptidase A (Section 23.9) increase the enzyme's catalytic activity?

Cofactors that are organic molecules are called **coenzymes**. Coenzymes are derived from organic compounds commonly known as *vitamins*. A **vitamin** is a substance needed in small amounts for normal body function that the body cannot synthesize. Table 24.1 lists the vitamins and their biochemically active coenzyme forms.

Table 24.1   The Vitamins, the Coenzymes for Which They Are Precursors, and the Chemical Functions of the Coenzymes

Vitamin	Coenzyme	Reaction catalyzed	Human deficiency disease
Water-Soluble Vitamins			
Niacin (niacinamide)	NAD^+, $NADP^+$ NADH, NADPH	Oxidation Reduction	Pellagra
Riboflavin (vitamin B_2)	FAD, FMN $FADH_2$, $FMNH_2$	Oxidation Reduction	Skin inflammation
Thiamine (vitamin B_1)	Thiamine pyrophosphate (TPP)	Two-carbon transfer	Beriberi
Lipoic acid (lipoate)	Lipoate Dihydrolipoate	Oxidation Reduction	—
Pantothenic acid (pantothenate)	Coenzyme A (CoASH)	Acyl transfer	—
Biotin (vitamin H)	Biotin	Carboxylation	—
Pyridoxine (vitamin B_6)	Pyridoxal phosphate (PLP)	Decarboxylation Transamination Racemization C_α—C_β bond cleavage α,β-Elimination β-Substitution	Anemia
Vitamin B_{12}	Coenzyme B_{12}	Isomerization	Pernicious anemia
Folic acid (folate)	Tetrahydrofolate (THF)	One-carbon transfer	Megaloblastic anemia
Ascorbic acid (vitamin C)	—	—	Scurvy
Water-Insoluble (lipid-soluble) Vitamins			
Vitamin A	—	—	—
Vitamin D	—	—	Rickets
Vitamin E	—	—	—
Vitamin K	Vitamin KH_2	Carboxylation	—

We have seen that enzymes catalyze reactions following the principles of organic chemistry (Section 23.9). Coenzymes use these same principles. We will see that coenzymes play a variety of chemical roles that the amino acid side chains of enzymes cannot play. Some coenzymes function as oxidizing and reducing agents, some allow electrons to be delocalized, some activate groups for further reaction, and some provide good nucleophiles or strong bases needed for a reaction. Because it would be highly inefficient for the body to use a compound only once and then discard it, coenzymes are recycled. Therefore, we will see that any coenzyme that undergoes a change during the course of a reaction is subsequently converted back to its original form.

An enzyme plus the cofactor it requires to catalyze a reaction is called a **holoenzyme**. An enzyme that has had its cofactor removed is called an **apoenzyme**. Holoenzymes are catalytically active, whereas apoenzymes are catalytically inactive because they have lost their cofactors.

Early nutritional studies divided vitamins into two classes: water-soluble vitamins and water-insoluble vitamins (Table 24.1). Vitamins A, D, E, and K are water insoluble. Vitamin K is the only water-insoluble vitamin currently known to be a precursor for a coenzyme. Vitamin A is required for proper vision, vitamin D regulates calcium and phosphate metabolism, and vitamin E is an antioxidant. Because they do not function as coenzymes, vitamins A, D, and E are not discussed in this chapter. Vitamins A and E are discussed in Sections 11.10 and 26.7, and vitamin D is discussed in Section 29.6.

All the water-soluble vitamins except vitamin C are precursors for coenzymes. In spite of its name, vitamin C is not actually a vitamin because it is required in fairly high amounts and most mammals are able to synthesize it (Section 21.18). Humans and guinea pigs cannot synthesize it, however, so it must be included in their diets. We have seen that vitamin C and vitamin E are radical inhibitors and therefore are antioxidants. Vitamin C traps radicals formed in aqueous environments, whereas vitamin E traps radicals formed in nonpolar environments (Section 11.10).

It is hard to overdose on water-soluble vitamins because the body can generally eliminate any excess. One can, however, overdose on water-insoluble vitamins because they are *not* easily eliminated by the body and can accumulate in cell membranes and other nonpolar components of the body. For example, excess vitamin D causes calcification of soft tissues. The kidneys are particularly susceptible to calcification, which eventually leads to kidney failure. Vitamin D is formed in the skin as a result of a photochemical reaction caused by the ultraviolet rays from the sun (Section 29.6).

VITAMIN B₁

Christiaan Eijkman (1858–1930) was a member of a medical team that was sent to the East Indies to study beriberi in 1886. At that time, all diseases were thought to be caused by microorganisms. When the microorganism that caused beriberi could not be found, the team left the East Indies, but Eijkman stayed behind to become the director of a new bacteriological laboratory. In 1896, he accidentally discovered the cause of beriberi when he noticed that chickens used in the laboratory had developed symptoms characteristic of the disease. He found that the symptoms had developed when a cook had started feeding the chickens rice meant for hospital patients. The symptoms disappeared when a new cook resumed giving chicken feed to the chickens. Later it was recognized that thiamine (vitamin B₁) is present in rice husks but not in polished rice.

For these advances, Eijkman shared the 1929 Nobel Prize in physiology or medicine with Frederick Hopkins.

Christiaan Eijkman

"VITAMINE"—AN AMINE REQUIRED FOR LIFE

Sir Frederick Hopkins was the first to suggest that diseases such as rickets and scurvy might result from the absence of substances in the diet that are needed only in very small quantities. The first such compound recognized to be essential in the diet was an amine, which led Casimir Funk to conclude incorrectly that all such compounds were amines. He, therefore, called them vitamines ("amines required for life"). The *e* was later dropped from the name.

Sir Frederick G. Hopkins (1861–1947) *was born in England. His finding that one sample of a protein supported life while another did not led him to conclude that the former contained a trace amount of a substance essential to life. His hypothesis later became known as the "vitamin concept," for which he received a share of the 1929 Nobel Prize in physiology or medicine. He also originated the concept of essential amino acids.*

Casimir Funk (1884–1967) *was born in Poland, received his medical degree from the University of Bern, and became a U.S. citizen in 1920. In 1923, he returned to Poland to direct the State Institute of Hygiene. He returned to the United States permanently when World War II broke out.*

Sir Frederick Hopkins

24.1 An Introduction to Metabolism

The reactions that living organisms carry out to obtain the energy they need and to synthesize the compounds they require are collectively known as **metabolism**. Metabolism can be divided into two parts: *catabolism* and *anabolism*.

- **Catabolic reactions** break down complex nutrient molecules to provide energy and to obtain simple precursor molecules for synthesis.
- **Anabolic reactions** require energy and result in the synthesis of complex biomolecules from simpler precursor molecules.

We will take a closer look at metabolism in Chapter 25. Then you will see how many of the reactions in this chapter fit into the overall metabolic scheme.

24.2 The Vitamin Needed for Many Redox Reactions: Vitamin B₃

Any enzyme that catalyzes an oxidation reaction or a reduction reaction requires a coenzyme because none of the amino acid side chains are oxidizing or reducing agents. The coenzyme serves as the oxidizing or reducing agent. The enzyme's role is to hold the substrate and coenzyme together so that the oxidation or reduction reaction can take place (see the model on page 1098).

The Pyridine Nucleotide Coenzymes

The coenzymes most commonly used by enzymes to catalyze oxidation reactions are **nicotinamide adenine dinucleotide (NAD⁺)** and **nicotinamide adenine dinucleotide phosphate (NADP⁺)**.

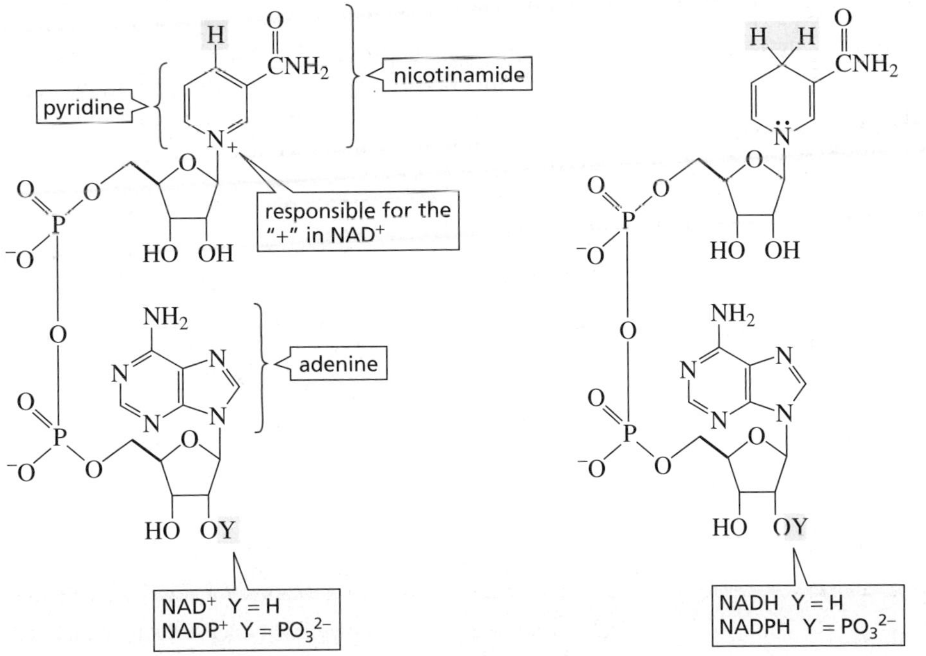

NAD⁺ and NADP⁺ are oxidizing agents.

NADH and NADPH are reducing agents.

When NAD⁺ (or NADP⁺) oxidizes a substrate, the coenzyme is reduced to NADH (or NADPH). NADH and NADPH are reducing agents; they are used as coenzymes by enzymes that catalyze reduction reactions. Enzymes that catalyze oxidation reactions bind NAD⁺ (or NADP⁺) more tightly than they bind NADH (or NADPH). When the oxidation reaction is over, the relatively loosely bound NADH (or NADPH) dissociates from the enzyme. Likewise, enzymes that catalyze reduction reactions bind **NADH (or NADPH)** more tightly than they bind NAD⁺ (or NADP⁺). When the reduction reaction is over, the relatively loosely bound NAD^+ (or $NADP^+$) dissociates from the enzyme.

$$\text{substrate}_{\text{reduced}} + NAD^+ \xrightleftharpoons{\text{enzyme}} \text{substrate}_{\text{oxidized}} + NADH + H^+$$

$$\text{substrate}_{\text{reduced}} + NADP^+ \xrightleftharpoons{\text{enzyme}} \text{substrate}_{\text{oxidized}} + NADPH + H^+$$

NAD^+ is composed of two nucleotides linked together through their phosphate groups. A **nucleotide** consists of a heterocyclic compound attached, in a β-configuration, to C-1 of a phosphorylated ribose (Section 27.1). In a **heterocyclic compound**, one or more of the ring atoms is an atom other than carbon (Sections 14.4, 20.8, and 20.9).

a nucleotide · adenine · niacinamide nicotinamide · niacin nicotinic acid

The heterocyclic component of one of the nucleotides of NAD^+ is nicotinamide, and the heterocyclic component of the other is adenine. This accounts for the coenzyme's name (**n**icotinamide **a**denine **d**inucleotide). The positive charge in the NAD^+ abbreviation indicates the positively charged nitrogen of the substituted pyridine ring.

The only way in which $NADP^+$ differs structurally from NAD^+ is in the phosphate group bonded to the 2'-OH group of the ribose of the adenine nucleotide; this explains the addition of "P" to the name. NAD^+ and NADH are generally used as coenzymes in catabolic reactions; $NADP^+$ and NADPH are generally used as coenzymes in anabolic reactions.

The adenine nucleotide for the coenzymes is provided by ATP. Niacin (vitamin B_3) is the portion of the coenzyme that the body cannot synthesize and must acquire through the diet. (Actually, humans can synthesize a small amount of niacin from the amino acid tryptophan but not in sufficient quantity to meet the body's metabolic needs.)

adenosine triphosphate
ATP

NIACIN DEFICIENCY

A deficiency in niacin causes pellagra, a disease that begins with dermatitis and ultimately causes insanity and death. More than 120,000 cases of pellagra were reported in the United States in 1927, mainly among poor people with unvaried diets. A factor known to be present in preparations of vitamin B prevented pellagra, but it was not until 1937 that the factor was identified as nicotinic acid. Mild deficiencies slow down metabolism, which is a potential contributing factor in obesity.

When bread companies started adding nicotinic acid to their bread, they insisted that its name be changed to niacin because nicotinic acid sounded too much like nicotine and they did not want their vitamin-enriched bread to be associated with a harmful substance. Niacinamide is a nutritionally equivalent form of niacin.

Malate dehydrogenase is the enzyme that catalyzes the oxidation of the *secondary alcohol group* of malate to a *ketone group*. (We will see that it is one of the reactions in the catabolic pathway known as the citric acid cycle; Section 25.10). The oxidizing agent in this reaction is NAD$^+$. Many enzymes that catalyze oxidation reactions are called **dehydrogenases**. Recall that the number of C—H bonds decreases in an oxidation reaction (Section 10.5). In other words, dehydrogenases remove hydrogen.

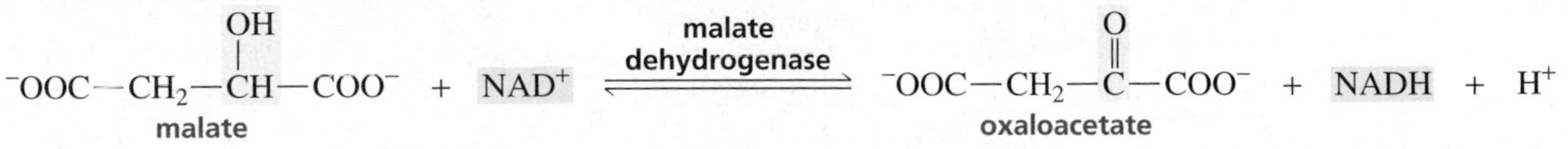

$$^-OOC-CH_2-\underset{\underset{\textbf{malate}}{|}}{\overset{\overset{OH}{|}}{CH}}-COO^- + NAD^+ \xrightleftharpoons{\overset{\textbf{malate}}{\textbf{dehydrogenase}}} {}^-OOC-CH_2-\underset{\underset{\textbf{oxaloacetate}}{}}{\overset{\overset{O}{\parallel}}{C}}-COO^- + NADH + H^+$$

β-Aspartate-semialdehyde is reduced to homoserine in an anabolic pathway, with NADPH as the reducing agent.

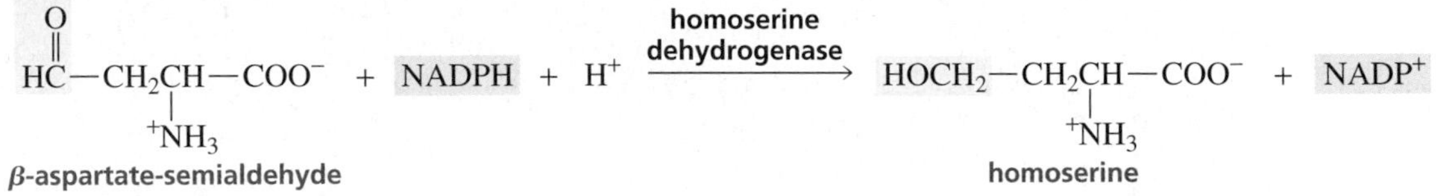

$$\underset{\underset{\textbf{β-aspartate-semialdehyde}}{}}{\overset{\overset{O}{\parallel}}{HC}}-CH_2\underset{\underset{{}^+NH_3}{|}}{CH}-COO^- + NADPH + H^+ \xrightarrow{\overset{\textbf{homoserine}}{\textbf{dehydrogenase}}} HOCH_2-CH_2\underset{\underset{{}^+NH_3}{|}}{\underset{\underset{\textbf{homoserine}}{}}{CH}}-COO^- + NADP^+$$

The differentiation between the coenzymes used in catabolism and those used in anabolism results from the strong specificity each of the enzymes that catalyze these oxidation–reduction reactions exhibits for a particular coenzyme. For example, an enzyme that catalyzes an oxidation reaction can readily tell the difference between NAD$^+$ and NADP$^+$; if the enzyme is in a catabolic pathway, it will bind NAD$^+$, but not NADP$^+$. The relative concentrations of the coenzymes in a cell also encourage binding of the appropriate coenzyme. For example, because NAD$^+$ and NADH are catabolic coenzymes and catabolic reactions are most often oxidation reactions, the NAD$^+$ concentration in a cell is much greater than the NADH concentration. (The cell maintains its [NAD$^+$]/[NADH] ratio near 1000.) Because NADP$^+$ and NADPH are anabolic coenzymes and anabolic pathways are predominantly reduction reactions, the concentration of NADPH in the cell is greater than the concentration of NADP$^+$. (The ratio of [NADP$^+$]/[NADH] is maintained at about 0.01.)

Mechanisms for the pyridine nucleotide coenzymes

How do these oxidation–reduction reactions take place? All the chemistry of the pyridine nucleotide coenzymes (NAD$^+$, NADP$^+$, NADH, and NADPH) takes place at the 4-position of the pyridine ring. The rest of the molecule has the job of binding the coenzyme to the proper site on the enzyme. A substrate that is being *oxidized* donates a hydride ion (H$^-$) to the 4-position of the pyridine ring. In the following reaction, for example, the primary alcohol is oxidized to an aldehyde. A basic amino acid side chain of the enzyme can help the reaction by removing a proton from the oxygen atom of the substrate.

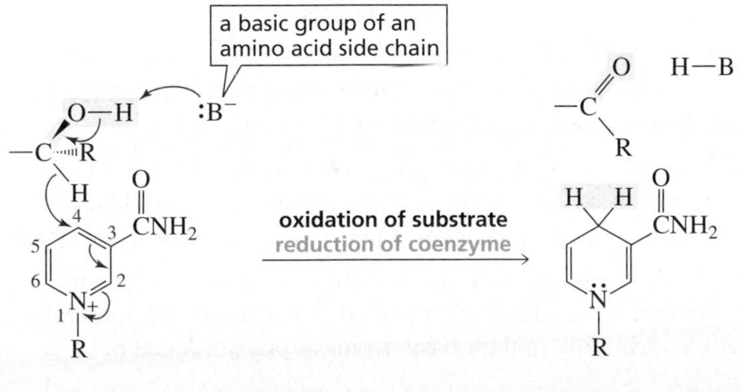

Glyceraldehyde-3-phosphate dehydrogenase is another example of an enzyme that uses NAD^+ as an oxidizing coenzyme. The enzyme catalyzes the oxidation of the aldehyde group of glyceraldehyde-3-phosphate to an anhydride of a carboxylic acid and phosphoric acid. We will see that this is a reaction that occurs in glycolysis (Section 25.7).

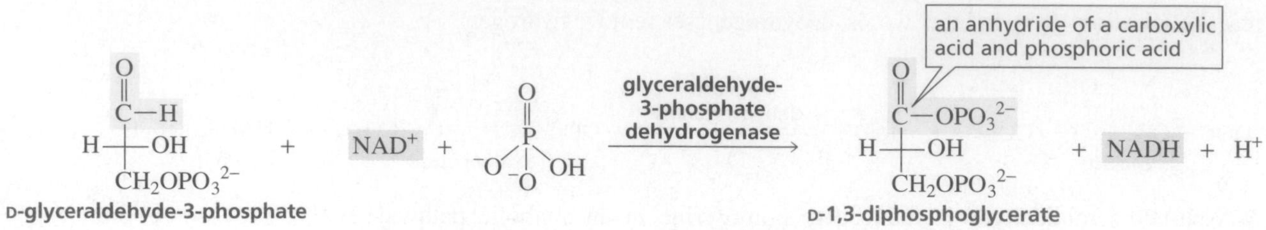

D-glyceraldehyde-3-phosphate $+$ NAD^+ $+$ [phosphate] $\xrightarrow{\text{glyceraldehyde-3-phosphate dehydrogenase}}$ D-1,3-diphosphoglycerate $+$ $NADH$ $+$ H^+

an anhydride of a carboxylic acid and phosphoric acid

The mechanism of the reaction is shown below.

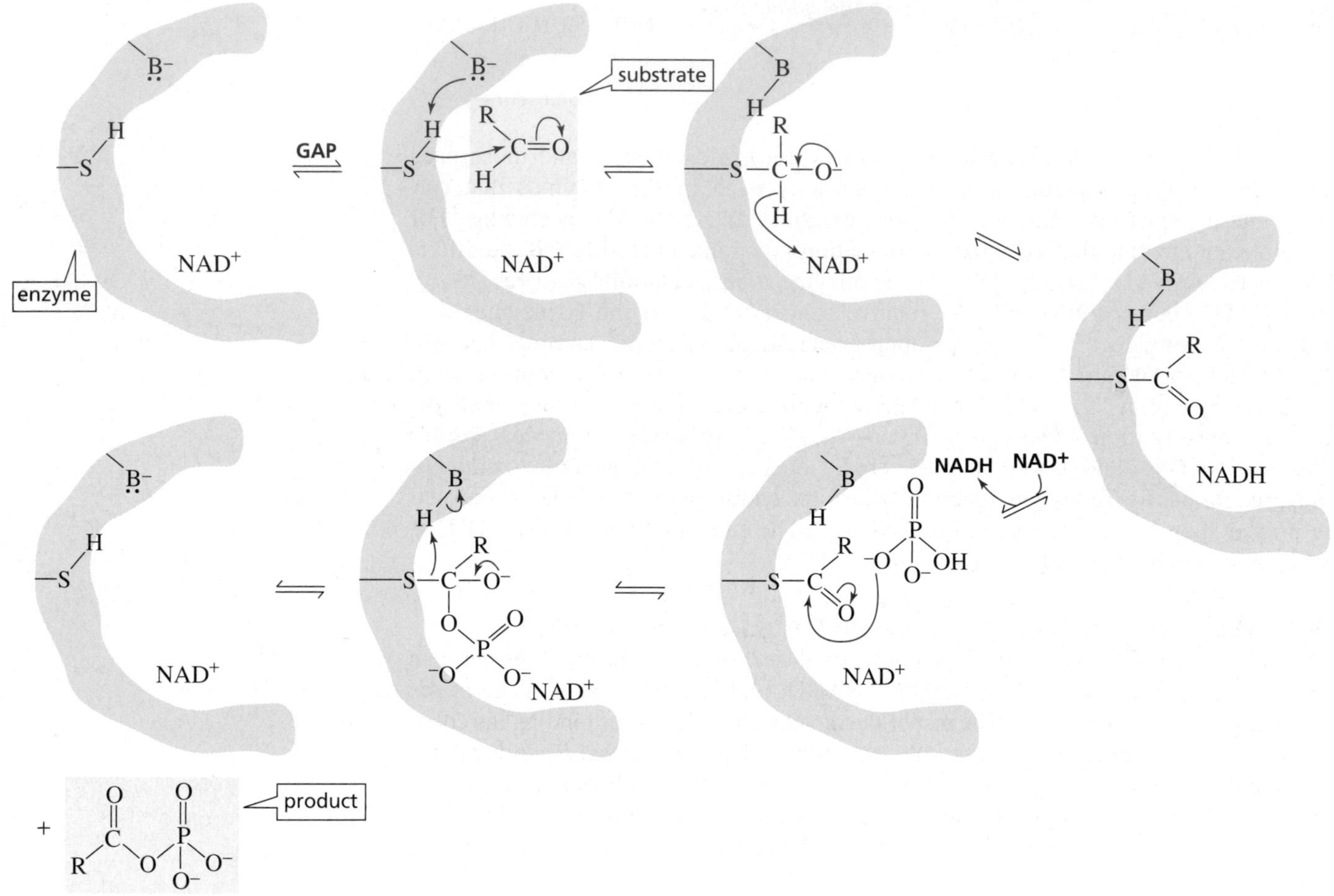

- The enzyme binds the substrate at its active site.
- An SH group of a cysteine side chain at the active site of the enzyme reacts with glyceraldehyde-3-phosphate to form a tetrahedral intermediate. A side chain of the enzyme increases cysteine's nucleophilicity by acting as a base catalyst (Section 23.3).
- The tetrahedral intermediate expels a hydride ion, transferring it to the 4-position of the pyridine ring of an NAD^+ that is bonded to the enzyme at an adjacent site.
- NADH dissociates from the enzyme, and the enzyme binds a new NAD^+.
- Phosphate reacts with the thioester, forming the anhydride product and releasing cysteine. (Phosphoric acid has pK_a values of 1.9, 6.7, and 12.4; therefore, two of the groups will be primarily in their basic forms at physiological pH.)

Notice that at the end of the reaction the holoenzyme is exactly as it was at the beginning, so the catalytic cycle can be repeated.

PROBLEM 2◆

What is the product of the following reaction?

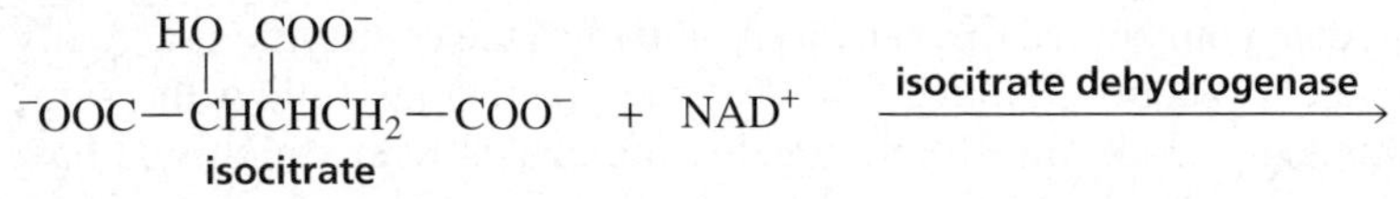

$$\begin{matrix} \text{HO} & \text{COO}^- \\ | & | \\ {}^-\text{OOC}-\text{CHCHCH}_2-\text{COO}^- & + & \text{NAD}^+ \end{matrix} \xrightarrow{\text{isocitrate dehydrogenase}}$$

isocitrate

The mechanism for reduction by NADH (or by NADPH) is the reverse of the mechanism for oxidation by NAD$^+$ (or by NADP$^+$). If a substrate is being *reduced*, the dihydropyridine ring of NADH (or NADPH) donates a hydride ion from its 4-position to the substrate. An acidic amino acid side chain of the enzyme aids the reaction by donating a proton to the substrate.

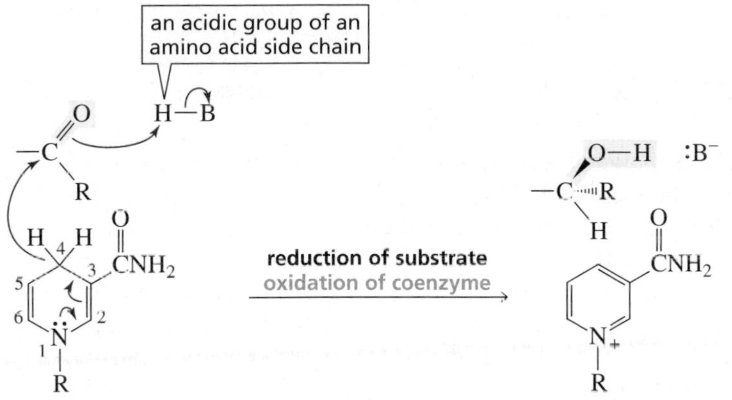

an acidic group of an amino acid side chain

reduction of substrate
oxidation of coenzyme

Because NADH and NADPH reduce compounds by donating a hydride ion, they can be considered biological equivalents of NaBH$_4$ or LiAlH$_4$, the hydride donors we have seen used as reducing reagents in nonbiological reactions (Sections 17.6 and 19.1).

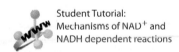

Student Tutorial:
Mechanisms of NAD$^+$ and NADH dependent reactions

Why are the structures of biological redox (reducing and oxidizing) reagents so much more complicated than the structures of the redox reagents used to carry out the same reactions in the laboratory? NADH is certainly more complicated than LiAlH$_4$, although both reagents reduce compounds by donating a hydride ion. Much of the structural complexity of a coenzyme is for molecular recognition—to allow it to be recognized by the enzyme. **Molecular recognition** allows the enzyme to bind the substrate and the coenzyme in the proper orientation for reaction.

Another reason for the difference in complexity is that a redox reagent found in a biological system must be highly selective and therefore less reactive than a laboratory redox reagent. For example, a biological reducing agent cannot reduce just any reducible compound with which it comes into contact. Biological reactions have to be much more carefully controlled than that. Because the coenzymes are relatively unreactive compared with nonbiological redox agents, the reaction between the substrate and the coenzyme does not occur at all or takes place very slowly without the enzyme. For example, NADH will not reduce an aldehyde or a ketone unless an enzyme is present. NaBH$_4$ and LiAlH$_4$ are more reactive hydride donors—in fact, much too reactive to even exist in the aqueous environment of the cell. Similarly, NAD$^+$ is a much more selective oxidizing agent than the typical oxidizing agents used in the laboratory; for example, NAD$^+$ will oxidize an alcohol only in the presence of an enzyme.

Because a biological reducing agent must be recycled (rather than being discarded in its oxidized form, as is the fate of a reducing agent used in a laboratory), the equilibrium constant for its oxidized and reduced forms is generally close to unity. Therefore, biological redox reactions are not highly exergonic; rather, they are equilibrium reactions driven in the appropriate direction by the removal of reaction products as a result of their participation in subsequent reactions.

As you study the coenzymes in this chapter, do not let the complexity of their structures intimidate you. Notice that only a small part of the coenzyme is actually involved in the chemical reaction. Notice also that the coenzymes follow the same rules of organic chemistry as do the simple organic molecules with which you have become familiar.

An oxidizing enzyme can distinguish, as a laboratory reagent cannot, between the two hydrogens on the carbon from which it catalyzes the removal of a hydride ion. For example, alcohol dehydrogenase removes only the pro-R hydrogen (H_a) of ethanol. (It is called the pro-R hydrogen because if it were replaced by deuterium, the asymmetric center would have the R configuration; H_b is the pro-S hydrogen.)

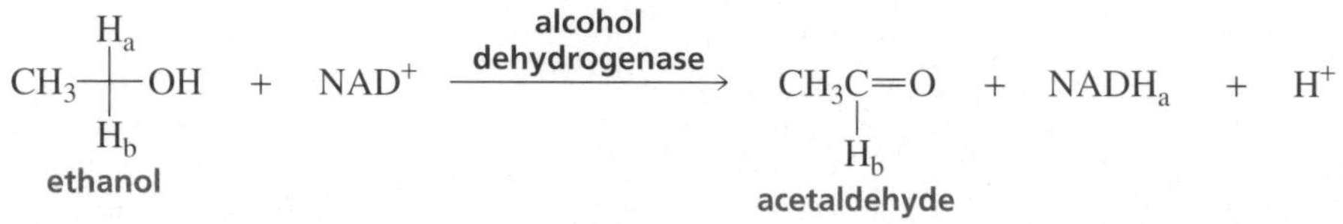

$$CH_3 \overset{H_a}{\underset{H_b}{\vert}} OH \; + \; NAD^+ \; \xrightarrow{\text{alcohol dehydrogenase}} \; CH_3C{=}O \; + \; NADH_a \; + \; H^+$$

ethanol   acetaldehyde (H_b)

Similarly, a reducing enzyme can distinguish between the two hydrogens at the 4-position of the nicotinamide ring of NADH. An enzyme has a specific binding site for the coenzyme, and when it binds the coenzyme, it blocks one of its sides. If the enzyme blocks the B-side of NADH, the substrate will bind to the A-side and the H_a hydride ion will be transferred to the substrate. If the enzyme blocks the A-side of the coenzyme, the substrate will have to bind to the B-side, and the H_b hydride ion will be transferred. Currently, 156 dehydrogenases are known to transfer H_a, and 121 are known to transfer H_b.

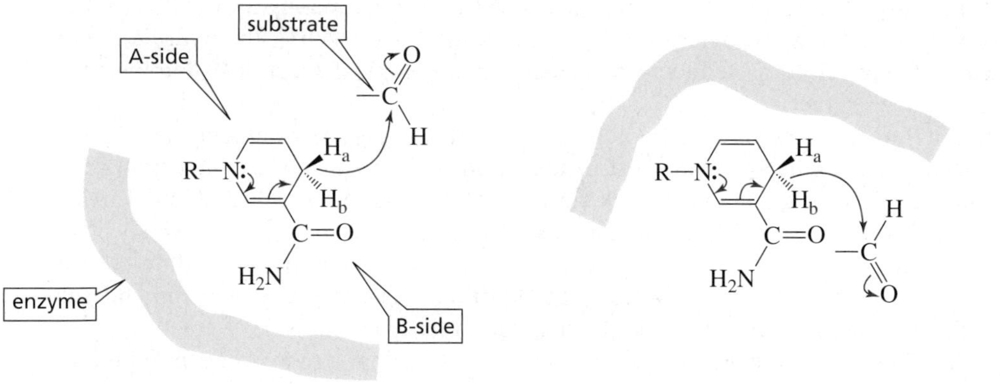

The enzyme blocks the B-side of the coenzyme, so the substrate binds to the A-side.

The enzyme blocks the A-side of the coenzyme, so the substrate binds to the B-side.

PROBLEM 3◆

What is the product of the following reaction?

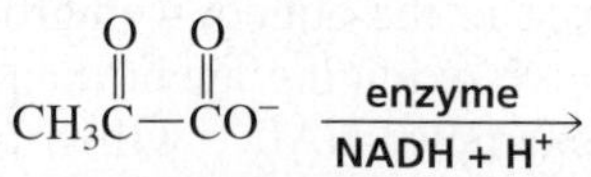

$$CH_3\overset{O}{\overset{\|}{C}}{-}\overset{O}{\overset{\|}{C}}O^- \; \xrightarrow[\text{NADH + H}^+]{\text{enzyme}}$$

24.3 Flavin Adenine Dinucleotide and Flavin Mononucleotide: Vitamin B$_2$

Flavin adenine dinucleotide (FAD) and **flavin mononucleotide (FMN)**, like NAD^+ and $NADP^+$, are coenzymes used to oxidize substrates.

The Flavin Nucleotide Coenzymes

As its name indicates, FAD is a dinucleotide in which one of the heterocyclic components is flavin and the other is adenine. FMN contains flavin but not adenine and so it is a mononucleotide. Notice that instead of ribose, the flavin nucleotide has a reduced ribose (a ribitol group). Flavin plus ribitol is the vitamin known as *riboflavin* or *vitamin B$_2$*. (Flavin is a bright yellow compound; *flavus* is Latin for "yellow.") A vitamin B$_2$ deficiency causes inflammation of the skin.

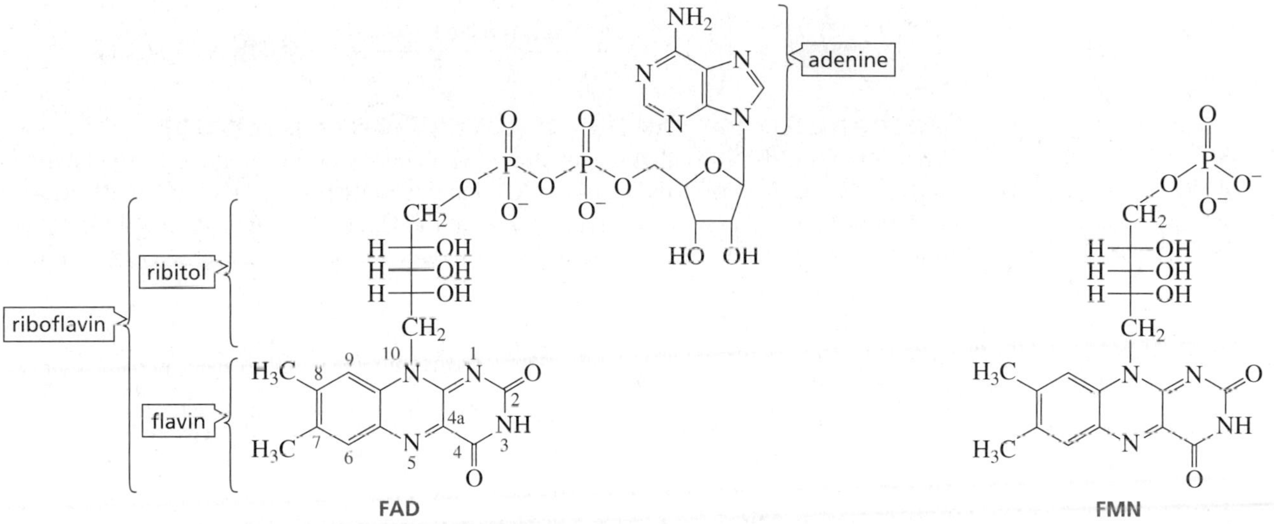

FAD

FMN

A *flavoprotein* is an enzyme that contains either FAD or FMN. In most flavoproteins, FAD (or FMN) is bound quite tightly. Tight binding allows the enzyme to control the oxidation potential of the coenzyme. (The more positive the oxidation potential, the stronger is the oxidizing agent.) Consequently, some flavoproteins are stronger oxidizing agents than others.

PROBLEM 4◆

FAD is obtained by an enzyme-catalyzed reaction that uses FMN and ATP as substrates. What is the other product of this reaction?

How can we tell which enzymes use FAD (or FMN) and which use NAD^+ (or $NADP^+$) as the oxidizing coenzyme? A rough guideline is that NAD^+ and $NADP^+$ are the coenzymes used in enzyme-catalyzed oxidation reactions involving carbonyl compounds (alcohols being oxidized to ketones, aldehydes, or carboxylic acids), while FAD and FMN are the coenzymes used in other types of oxidations. For example, in the following reactions, FAD oxidizes a dithiol to a disulfide, an amine to an imine, and a saturated alkyl group to an unsaturated alkene, and FMN oxidizes NADH to NAD^+. (This is only an approximate guideline, however, because FAD is used in some oxidations that involve carbonyl compounds, and NAD^+ and $NADP^+$ are used in some oxidations that do not involve carbonyl compounds.)

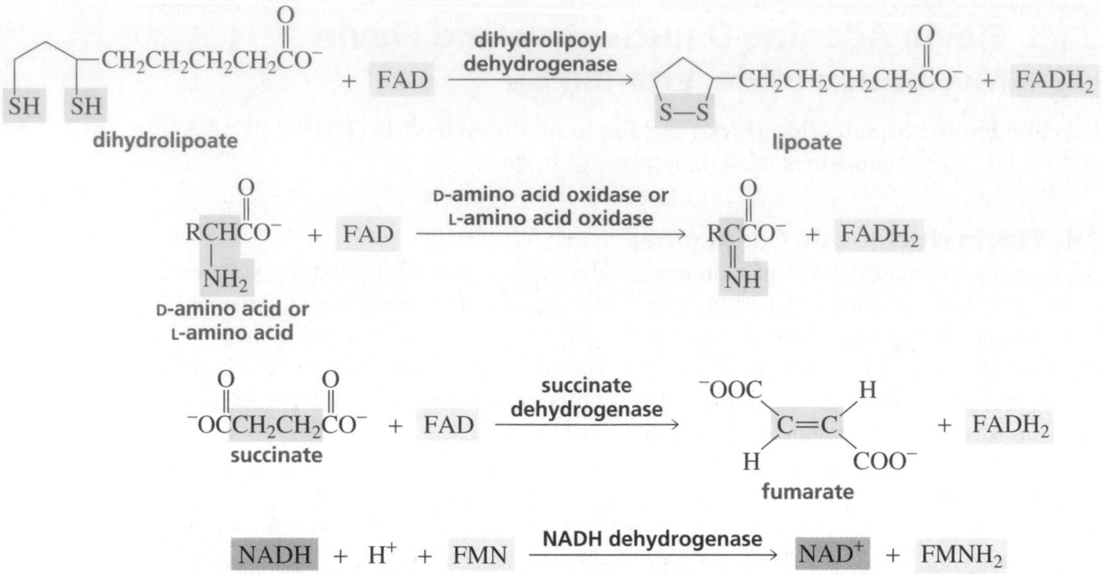

dihydrolipoate + FAD → lipoate + FADH₂ (dihydrolipoyl dehydrogenase)

D-amino acid or L-amino acid + FAD → (D-amino acid oxidase or L-amino acid oxidase) + FADH₂

succinate + FAD → fumarate + FADH₂ (succinate dehydrogenase)

$$\text{NADH} + \text{H}^+ + \text{FMN} \xrightarrow{\text{NADH dehydrogenase}} \text{NAD}^+ + \text{FMNH}_2$$

Mechanisms for the flavin nucleotide coenzymes

When FAD (or FMN) oxidizes a substrate (S), the coenzyme is reduced to FADH₂ (or FMNH₂). FADH₂ and FMNH₂, like NADH and NADPH, are reducing agents. All the oxidation–reduction chemistry takes place on the flavin ring. Reduction of the flavin ring disrupts the conjugated system, so the reduced coenzymes are less colored than their oxidized forms.

FAD and FMN are oxidizing agents.

FADH₂ and FMNH₂ are reducing agents.

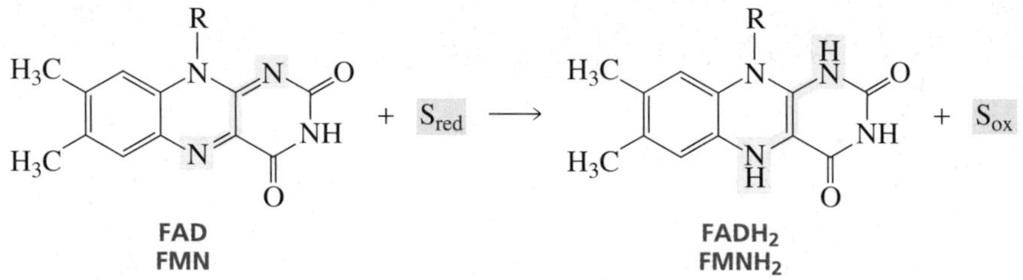

$$\text{FAD / FMN} + S_{red} \longrightarrow \text{FADH}_2 / \text{FMNH}_2 + S_{ox}$$

PROBLEM 5◆

How many conjugated double bonds are there in

a. FAD?   **b.** FADH₂?

The mechanism for the FAD-catalyzed oxidation of dihydrolipoate to lipoate is shown below.

Mechanism for dihydrolipoyl dehydrogenase

dihydrolipoate   lipoate

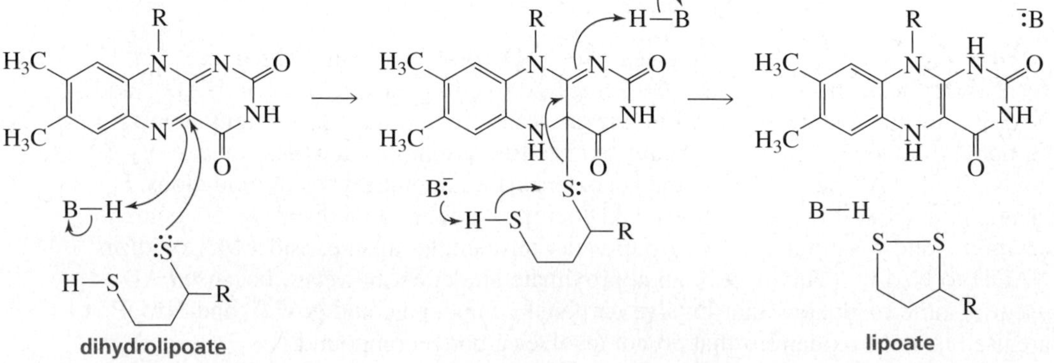

- The thiolate ion attacks the C-4a position of the flavin ring. This is an acid-catalyzed reaction: as the thiolate ion attacks the ring, a proton is donated to the N-5 nitrogen (Section 23.2).

- A second nucleophilic attack by a thiolate ion, this time on the sulfur that is covalently attached to the coenzyme, generates the oxidized product and FADH$_2$.

The mechanism for the FAD-catalyzed oxidation of an amino acid to an imino acid is quite different from the preceding FAD-catalyzed reaction.

Mechanism for D- or L-amino acid oxidase

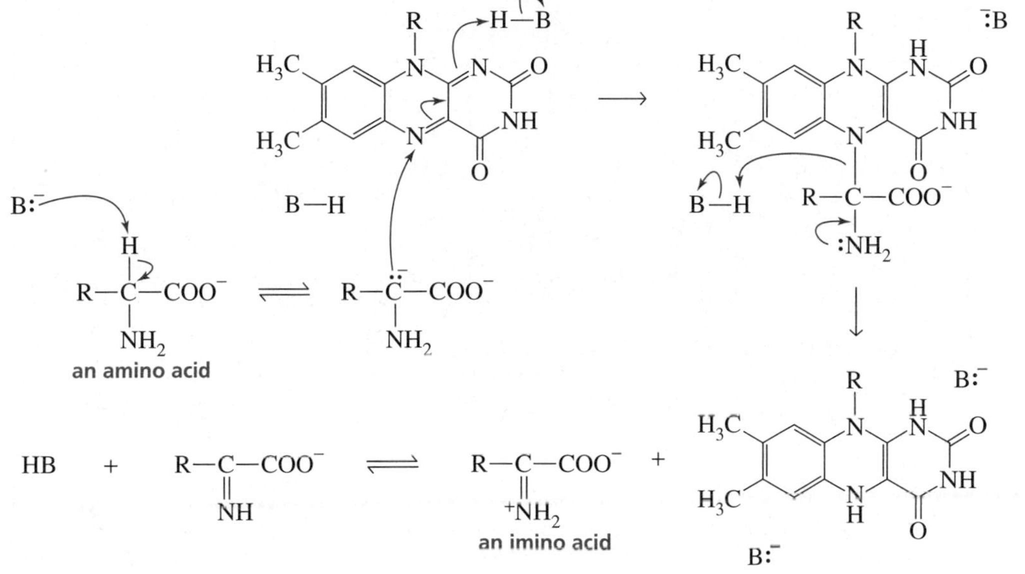

- A basic amino acid side chain at the active site of the enzyme removes a proton from the α-carbon of the amino acid, forming a carbanion.

- The carbanion attacks the N-5 position of the flavin ring.

- Collapse of the resulting tetrahedral intermediate yields the oxidized amino acid (an imino acid) and the reduced coenzyme (FADH$_2$).

These two mechanisms show that FAD is a more versatile coenzyme than NAD$^+$. Unlike NAD$^+$, which always uses the same mechanism, flavin coenzymes can use several different mechanisms to carry out oxidation reactions. For example, we have just seen that when FAD oxidizes dihydrolipoate, nucleophilic attack occurs on the C-4a position of the flavin ring, but when it oxidizes an amino acid, nucleophilic attack occurs on the N-5 position.

Cells contain very low concentrations of FAD and much higher concentrations of NAD$^+$. This difference in concentration is responsible for a significant difference in the enzymes (**E**) that use NAD$^+$ as an oxidizing agent and those that use FAD. Generally, FAD is covalently bound to its enzyme and remains bound after being reduced to FADH$_2$. Unlike NADH, FADH$_2$ does not dissociate from the enzyme. It, therefore, must be reoxidized to FAD before the enzyme can begin another round of catalysis. The oxidizing agent used for this reaction is NAD$^+$ or O$_2$. Therefore, an enzyme that uses an oxidizing coenzyme other than NAD$^+$ may still require NAD$^+$ to reoxidize the reduced coenzyme. For this reason, NAD$^+$ has been called the "common currency" of biological oxidation–reduction reactions.

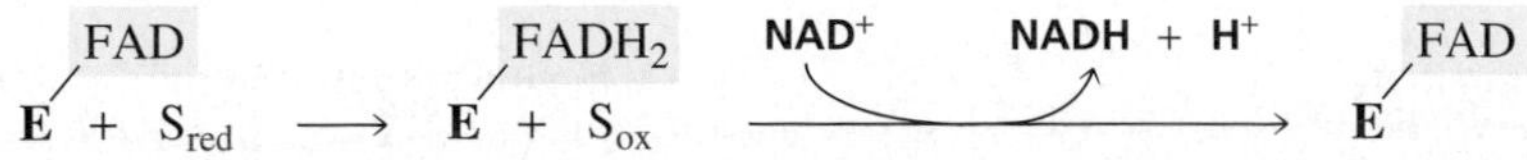

PROBLEM 6

Propose a mechanism for the reduction of lipoate by $FADH_2$.

PROBLEM 7 **SOLVED**

A common way for FAD to become covalently bound to its enzyme is by having a proton removed from the C-8 methyl group and a proton donated to N-1. Then a cysteine or other nucleophilic amino acid side chain of the enzyme attacks the methylene carbon at C-8 as a proton is donated to N-5. Describe these events mechanistically.

Solution

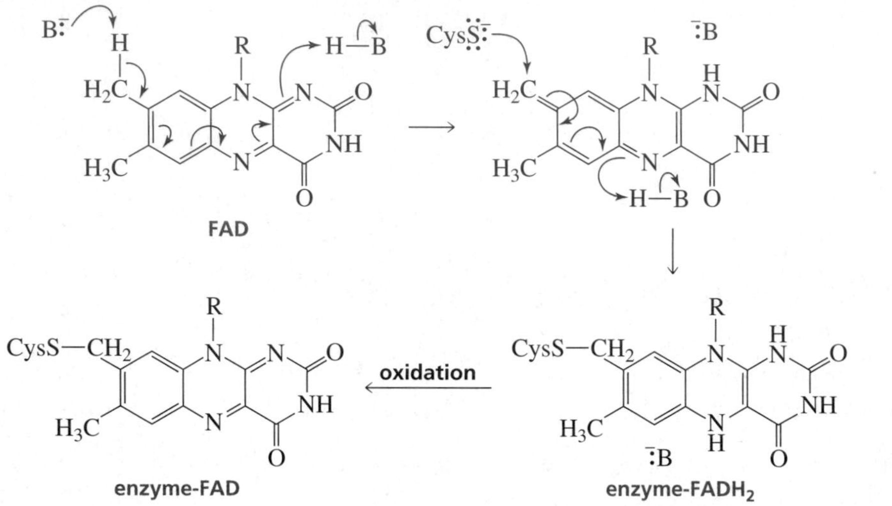

Notice that during the process of being attached to the enzyme, FAD is reduced to $FADH_2$. It is subsequently oxidized back to FAD. Once the coenzyme is attached to the enzyme, it does not come off.

PROBLEM 8

Explain why the hydrogens of the methyl group bonded to flavin at C-8 are more acidic than those of the methyl group bonded at C-7.

24.4 Thiamine Pyrophosphate: Vitamin B₁

Thiamine was the first of the B vitamins to be identified, so it became known as vitamin B_1. The absence of thiamine in the diet causes a disease called beriberi, which damages the heart, impairs nerve reflexes, and in extreme cases causes paralysis. One major dietary source of vitamin B_1 is the hulls of rice kernels (page 1100). A deficiency is therefore most likely to occur when highly polished rice is a major component of the diet. A deficiency is also seen in alcoholics who are severely malnourished.

Vitamin B_1 is used to form the coenzyme **thiamine pyrophosphate (TPP)**. TPP is the coenzyme required by enzymes that catalyze the transfer of a two-carbon fragment from one species to another.

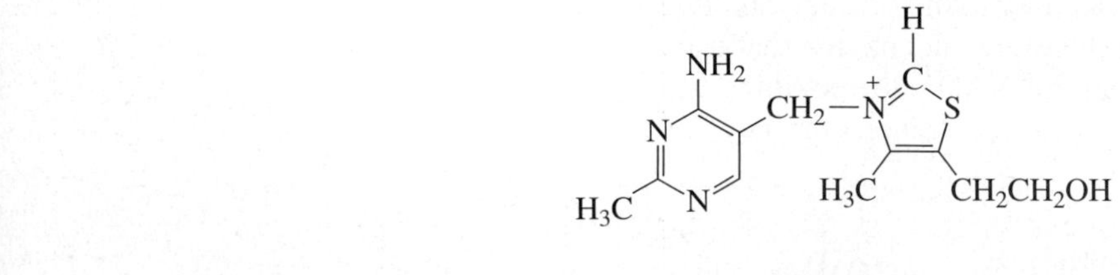

vitamin B₁

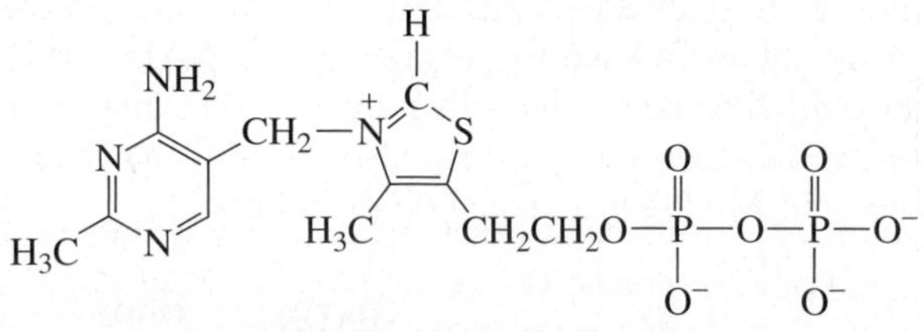

thiamine pyrophosphate
TPP

Pyruvate decarboxylase is an example of an enzyme that requires thiamine pyrophosphate. This enzyme catalyzes the decarboxylation of pyruvate and transfers the remaining two-carbon fragment to a proton, resulting in the formation of acetaldehyde.

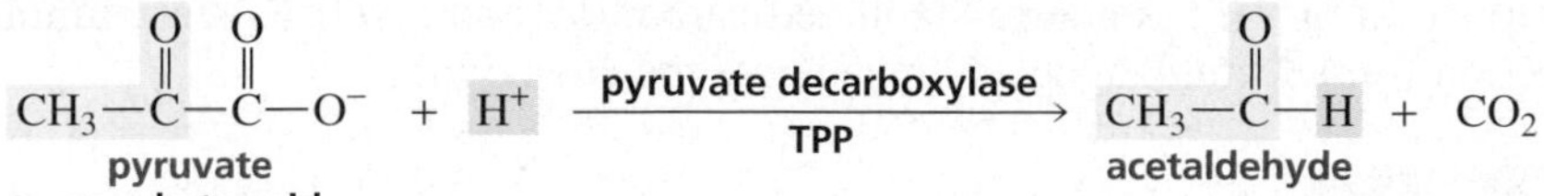

$$CH_3-\overset{O}{\underset{}{C}}-\overset{O}{\underset{}{C}}-O^- + H^+ \xrightarrow[\text{TPP}]{\text{pyruvate decarboxylase}} CH_3-\overset{O}{\underset{}{C}}-H + CO_2$$

pyruvate
an α-keto acid → acetaldehyde

You may wonder why an α-keto acid such as pyruvate can be decarboxylated, since the electrons left behind when CO_2 is removed cannot be delocalized onto the carbonyl oxygen. We will see that the thiazolium ring of the coenzyme provides a site for the delocalization of the electrons. A site to which electrons can be delocalized is called an **electron sink**.

Mechanism for pyruvate decarboxylase

The hydrogen bonded to the imine carbon of TPP is relatively acidic ($pK_a = 12.7$) because the ylide carbanion formed when the proton is removed is stabilized by the adjacent positively charged nitrogen. The ylide carbanion is a good nucleophile (Section 17.13).

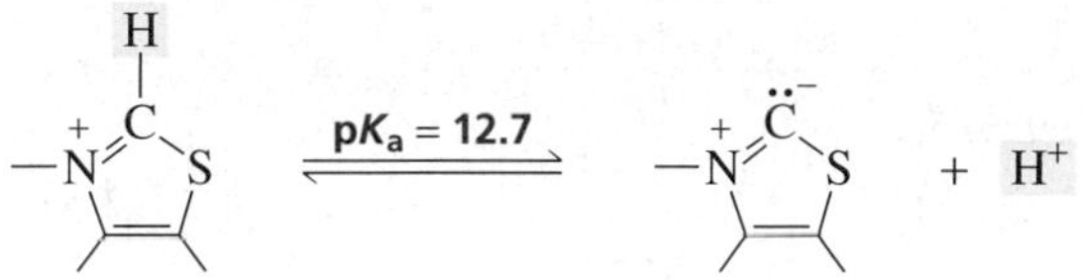

$$-\overset{+}{N}\underset{}{\overset{\overset{H}{|}}{\underset{}{C}}}S \underset{}{\overset{pK_a = 12.7}{\rightleftharpoons}} -\overset{+}{N}\overset{\ddot{\overset{..}{C}}{}^-}{\underset{}{}}S + H^+$$

thiazolium ring   ylide carbanion

A decarboxylase is an enzyme that catalyzes the removal of CO_2 from the substrate. The mechanism for the reaction catalyzed by pyruvate decarboxylase is shown below.

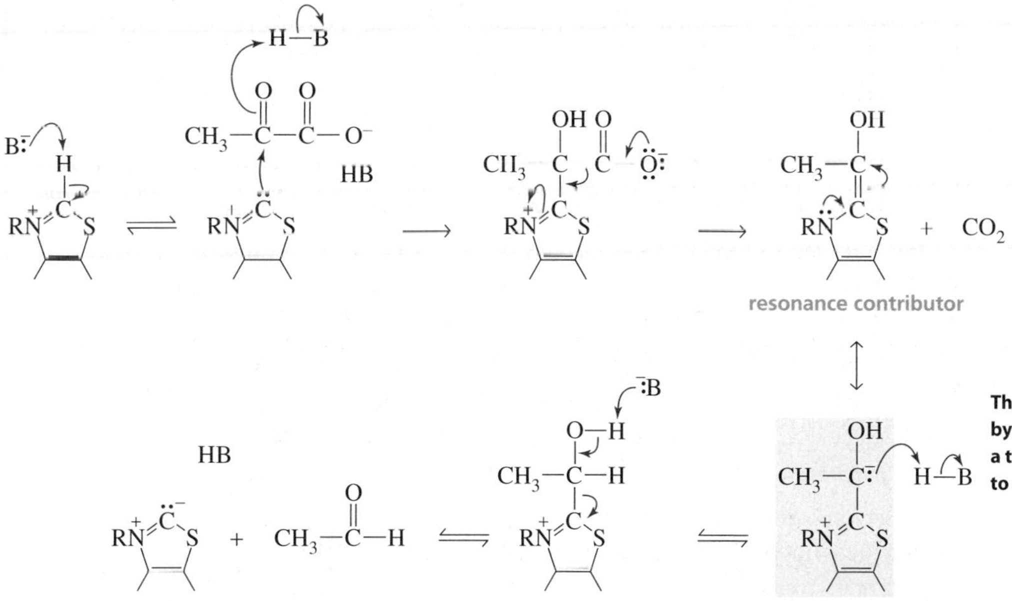

resonance contributor

Thiamine pyrophosphate (TPP) is required by enzymes that catalyze the transfer of a two-carbon fragment from one species to another.

resonance contributor
resonance-stabilized carbanion

- The nucleophilic ylide carbanion attacks the electrophilic ketone group of the α-keto acid.

- The intermediate thus formed can easily undergo decarboxylation because the electrons left behind when CO_2 is removed can be delocalized onto the positively charged nitrogen. The positively charged nitrogen of TPP is a more effective electron sink than the β-keto group of a β-keto acid, a class of compounds that are fairly easily decarboxylated (Section 18.18).

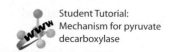

Student Tutorial:
Mechanism for pyruvate
decarboxylase

- The decarboxylated product is stabilized by electron delocalization. One of the resonance contributors is neutral and the other has separated charges. (We will call the decarboxylated product a *resonance-stabilized carbanion.*)

- Protonation of the resonance-stabilized carbanion and a subsequent elimination reaction form acetaldehyde and regenerate the coenzyme.

PROBLEM 9

Draw structures that show the similarity between decarboxylation of the pyruvate–TPP intermediate and decarboxylation of a β-keto acid.

PROBLEM 10

Acetolactate synthase is another TPP-requiring enzyme. It too catalyzes the decarboxylation of pyruvate, but it transfers the resulting two-carbon fragment to another molecule of pyruvate, forming acetolactate. This is the first step in the biosynthesis of the amino acids valine and leucine. Propose a mechanism for acetolactate synthase.

$$2\ CH_3-\overset{O}{\overset{\|}{C}}-\overset{O}{\overset{\|}{C}}-O^- \xrightarrow[\text{TPP}]{\text{acetolactate synthase}} CH_3-\overset{O}{\overset{\|}{C}}-\underset{\underset{CH_3}{|}}{\overset{OH}{\overset{|}{C}}}-\overset{O}{\overset{\|}{C}}-O^- + CO_2$$

pyruvate   acetolactate

PROBLEM 11

Acetolactate synthase can also transfer the two-carbon fragment from pyruvate to α-ketobutyrate, forming α-aceto-α-hydroxybutyrate. This is the first step in the formation of isoleucine. Propose a mechanism for this reaction.

$$CH_3\overset{O}{\overset{\|}{C}}-\overset{O}{\overset{\|}{C}}O^- + CH_3CH_2\overset{O}{\overset{\|}{C}}-\overset{O}{\overset{\|}{C}}O^- \xrightarrow[\text{TPP}]{\text{acetolactate synthase}} CH_3\overset{O}{\overset{\|}{C}}-\underset{\underset{CH_2CH_3}{|}}{\overset{OH}{\overset{|}{C}}}COO^- + CO_2$$

     α-ketobutyrate α-aceto-α-hydroxybutyrate

We will see in Chapter 25 that the final product of carbohydrate metabolism is pyruvate. For pyruvate to be further metabolized, it must be converted to acetyl-CoA. The *pyruvate dehydrogenase system* is a group of three enzymes responsible for the conversion of pyruvate to acetyl-CoA.

$$CH_3-\overset{O}{\overset{\|}{C}}-\overset{O}{\overset{\|}{C}}-O^- + CoASH \xrightarrow[\text{system}]{\text{pyruvate dehydrogenase}} CH_3-\overset{O}{\overset{\|}{C}}-SCoA + CO_2$$

pyruvate   acetyl-CoA

The mechanism for the pyruvate dehydrogenase system requires TPP and four other coenzymes: lipoate, coenzyme A, FAD, and NAD$^+$.

Mechanism for the pyruvate dehydrogenase system

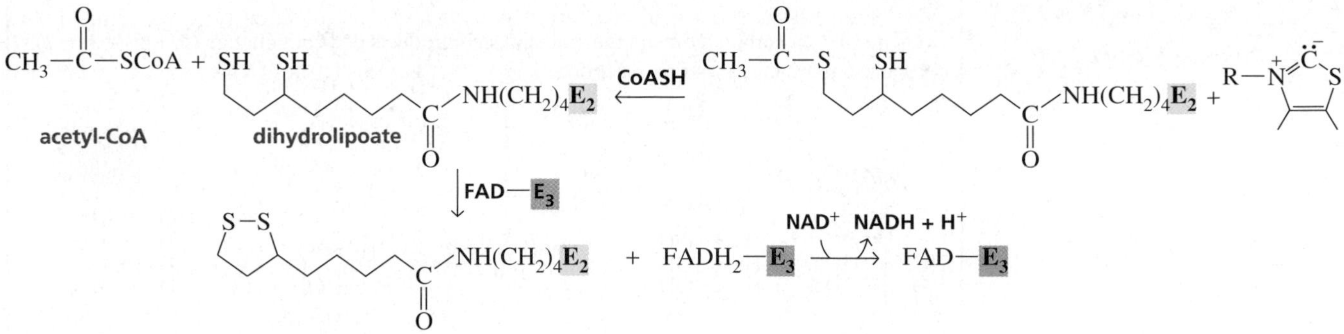

- The first enzyme in the system catalyzes the reaction of TPP with pyruvate to form the same resonance-stabilized carbanion formed by pyruvate decarboxylase and by the enzyme in Problems 10 and 11.

- The second enzyme of the system (**E$_2$**) is attached to the coenzyme **lipoate**, as a result of using the amino group of a lysine side chain to form an amide with the coenzyme. The disulfide linkage of lipoate is cleaved when it undergoes nucleophilic attack by the resonance-stabilized carbanion.

- The TPP ylide carbanion is eliminated from the tetrahedral intermediate.

- **Coenzyme A (CoASH)** reacts with the thioester in a transthioesterification reaction (one thioester is converted into another), substituting coenzyme A for dihydrolipoate. At this point, the final reaction product (acetyl-CoA) has been formed.

- Before another catalytic cycle can occur, however, dihydrolipoate must be oxidized back to lipoate. This is done by the third enzyme (**E$_3$**), an FAD-requiring enzyme (Section 24.3). Oxidation of dihydrolipoate by FAD forms enzyme-bound FADH$_2$.

- NAD$^+$ then oxidizes FADH$_2$ back to FAD.

The vitamin needed to make **coenzyme A** is pantothenate. We have seen that CoASH is used in biological systems to activate carboxylic acids by converting them into thioesters, which are much more reactive toward nucleophilic acyl substitution reactions than are carboxylic acids (Section 16.22). At physiological pH (7.3), a carboxylic acid would be present in its negatively charged basic form, which could not be approached by a nucleophile.

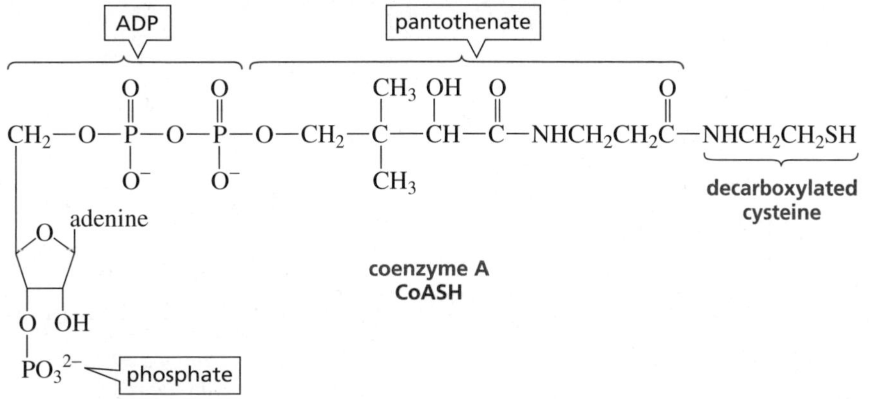

coenzyme A
CoASH

PROBLEM 12 **SOLVED**

TPP is a coenzyme for transketolase, the enzyme that catalyzes the conversion of a ketopentose (xylulose-5-phosphate) and an aldopentose (ribose-5-phosphate) into an aldotriose (glyceraldehyde-3-phosphate) and a ketoheptose (sedoheptulose-7-phosphate). Notice that the total number of carbon atoms in the reactants and products does not change $(5 + 5 = 3 + 7)$. Propose a mechanism for this reaction.

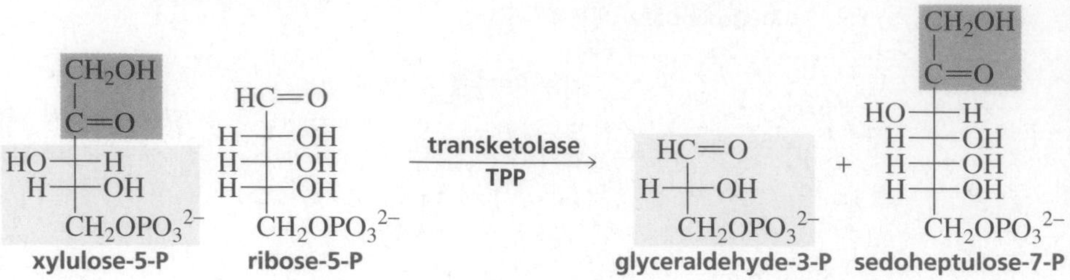

CH$_2$OH | C=O | HO—H | H—OH | CH$_2$OPO$_3{}^{2-}$ — **xylulose-5-P**   HC=O | H—OH | H—OH | H—OH | CH$_2$OPO$_3{}^{2-}$ — **ribose-5-P**   →(transketolase / TPP)→   HC=O | H—OH | CH$_2$OPO$_3{}^{2-}$ — **glyceraldehyde-3-P**   +   CH$_2$OH | C=O | HO—H | H—OH | H—OH | H—OH | CH$_2$OPO$_3{}^{2-}$ — **sedoheptulose-7-P**

Solution   The reaction shows that a two-carbon fragment is transferred from xylulose-5-phosphate to ribose-5-phosphate. Because TPP transfers two-carbon fragments, we know that TPP must remove the two-carbon fragment that is to be transferred from xylulose-5-phosphate. Thus, the reaction must start by TPP attacking the carbonyl group of xylulose-5-phosphate. We can add an acid group to accept the electrons from the carbonyl group and a basic group to aid in the removal of the two-carbon fragment. The two-carbon fragment that becomes attached to TPP is a resonance-stabilized carbanion that adds to the carbonyl group of ribose-5-phosphate. Again, an acid group accepts the electrons from the carbonyl group, and a basic group aids in the elimination of TPP.

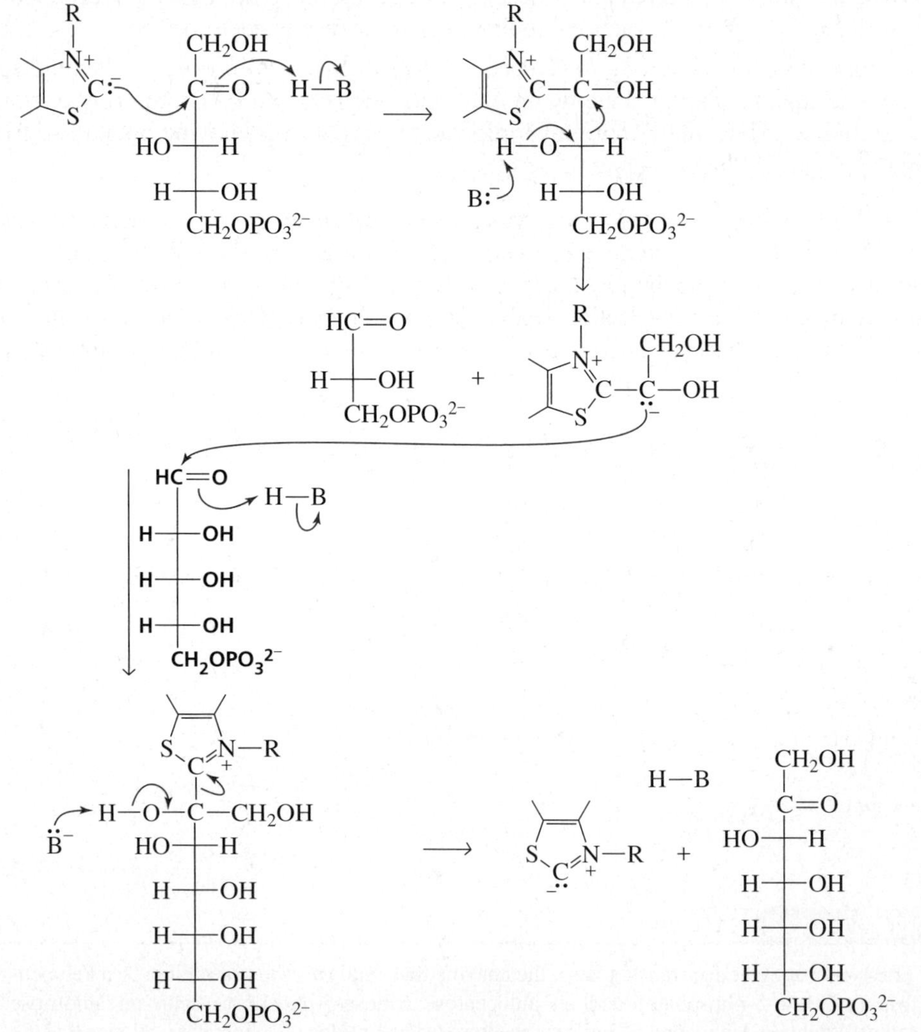

Notice the similar function of TPP in all TPP-requiring enzymes. In each reaction, the nucleophilic coenzyme attacks a carbonyl group of the substrate and allows a bond to that carbonyl group to be broken because the electrons left behind can be delocalized into the thiazolium ring. The resulting two-carbon fragment is then transferred—to a proton in the case of pyruvate decarboxylase, to coenzyme A (via lipoate) in the pyruvate dehydrogenase system, and to a carbonyl group in Problems 10, 11, and 12.

PROBLEM 13

An unfortunate effect of drinking too much alcohol, known as a hangover, is attributable to the acetaldehyde formed when ethanol is oxidized. There is some evidence that vitamin B_1 can cure a hangover. How can the vitamin do this?

24.5 Biotin: Vitamin H

Biotin (vitamin H) is an unusual vitamin in that it is synthesized by bacteria that live in the intestine. Consequently, biotin does not have to be included in our diet and deficiencies are rare. Biotin deficiencies, however, can be found in people who maintain a diet high in raw eggs. Egg whites contain a protein (avidin) that binds biotin tightly and thereby prevents it from acting as a coenzyme. When eggs are cooked, avidin is denatured, and the denatured protein does not bind biotin. Biotin is attached to its enzyme (**E**) by forming an amide with the amino group of a lysine side chain.

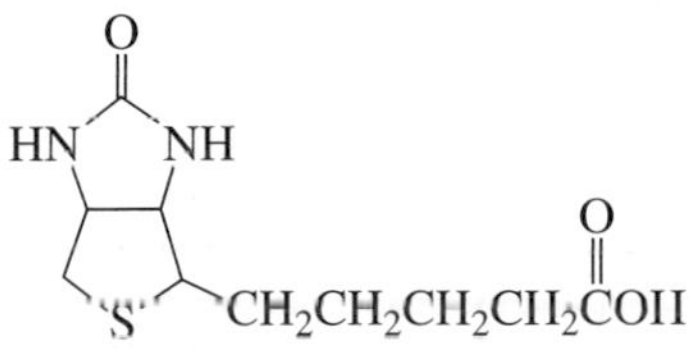

biotin 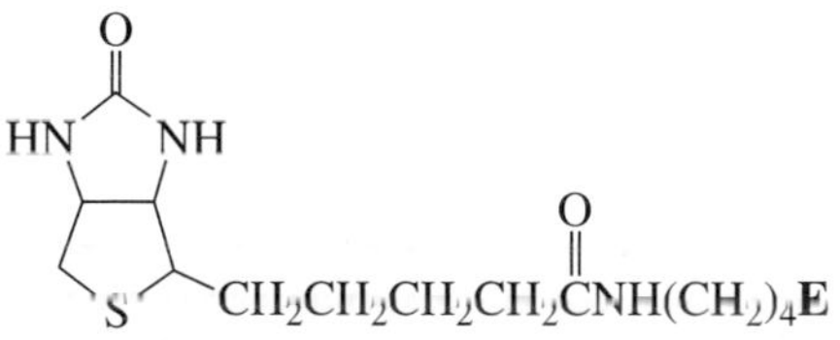   enzyme-bound biotin

Biotin is the coenzyme required by enzymes that catalyze carboxylation of an α-carbon (a carbon adjacent to a carbonyl group). Therefore, the enzymes that require biotin as a coenzyme are called carboxylases. For example, pyruvate carboxylase converts pyruvate to oxaloacetate. Acetyl-CoA carboxylase converts acetyl-CoA into malonyl-CoA. Biotin-requiring enzymes use bicarbonate (HCO_3^-) for the source of the carboxyl group that becomes attached to the substrate.

Biotin is required by enzymes that catalyze the carboxylation of a carbon adjacent to a carbonyl group.

a carboxyl group has been added to the α-carbon

a carboxyl group has been added to the α-carbon

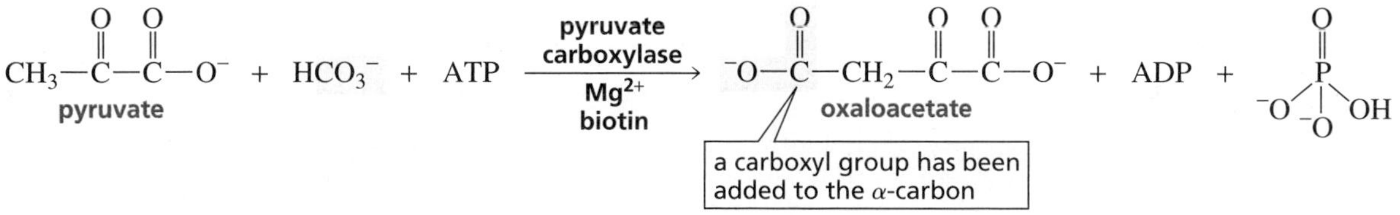

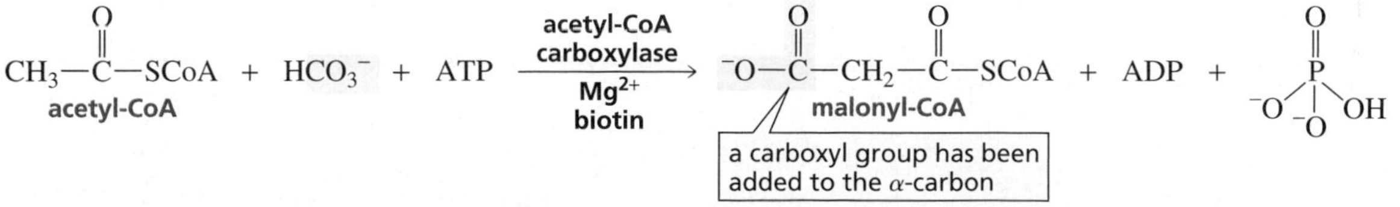

In addition to requiring bicarbonate, biotin-requiring enzymes require Mg^{2+} and ATP. The function of Mg^{2+} is to decrease the overall negative charge on ATP by complexing with two of its negatively charged oxygens. Unless its negative charge is reduced, ATP cannot be approached by a nucleophile (Section 25.5). The function of ATP is to increase the reactivity of bicarbonate by converting it to "activated bicarbonate," a compound with a good leaving group (Section 25.3). Notice that "activated bicarbonate" is a mixed anhydride of carbonic acid and phosphoric acid (Section 16.1).

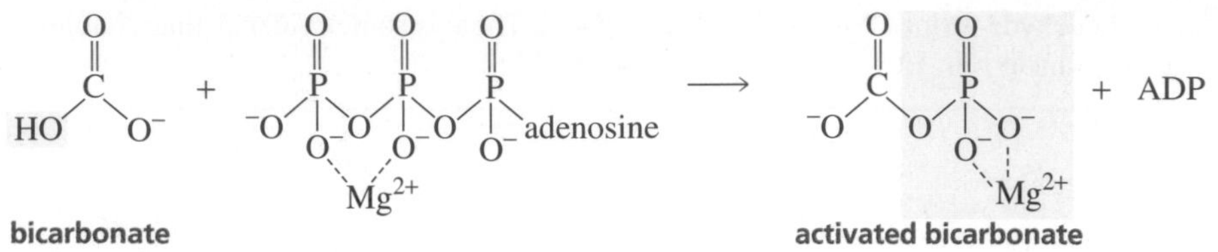

Once bicarbonate has been activated, the catalytic reaction can begin. The mechanism for carboxylation of acetyl CoA by acetyl-CoA carboxylase is shown below.

Mechanism for carboxylation of acetyl-CoA by acetyl-CoA carboxylase

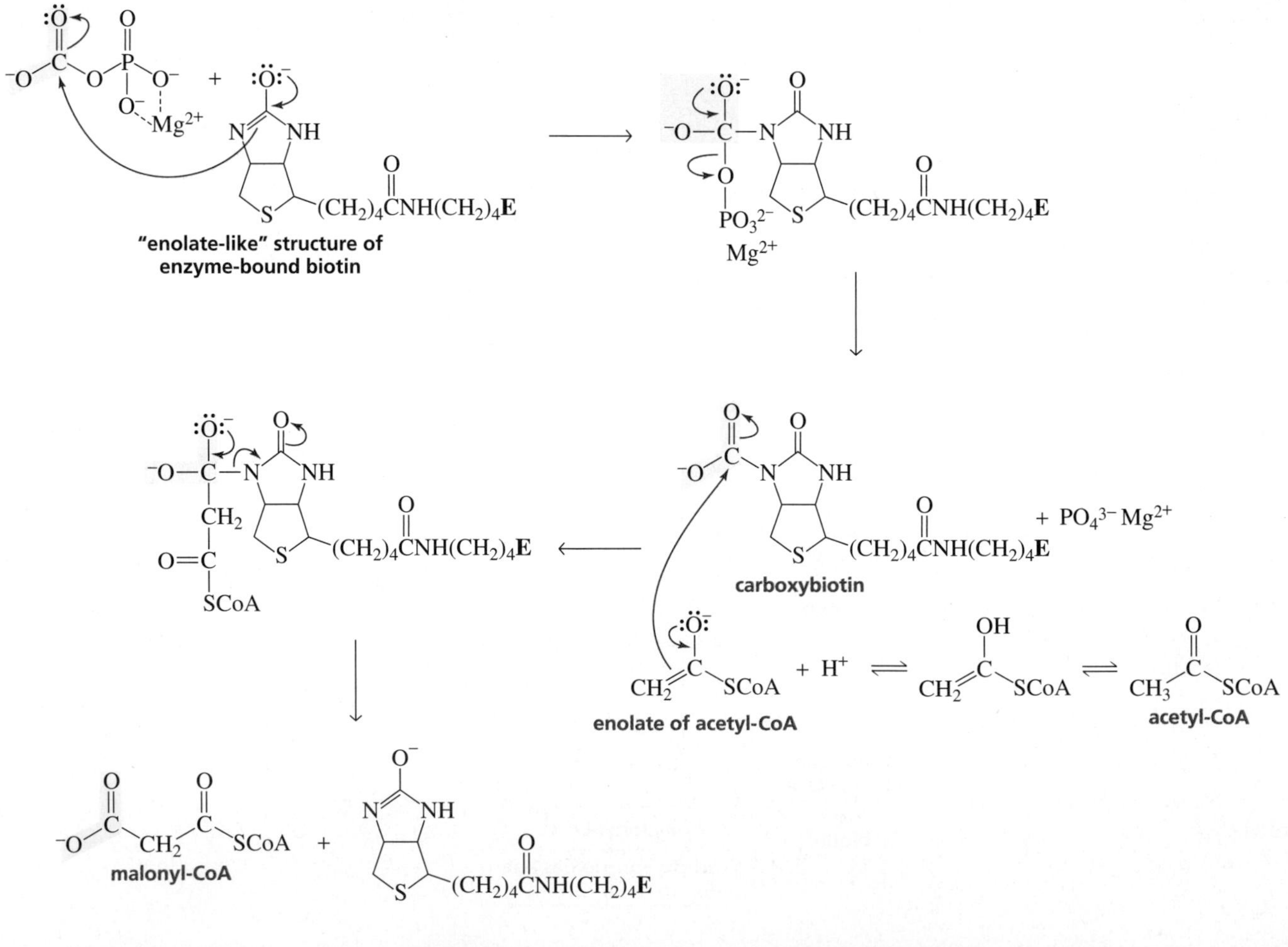

- Nucleophilic attack by biotin on activated bicarbonate forms carboxybiotin. Because the nitrogen of an amide is not nucleophilic, it is likely that the active form of biotin has an enolate-like structure.

- Nucleophilic attack by the substrate (in this case, the enolate of acetyl-CoA) on carboxybiotin transfers the carboxyl group from biotin to the substrate.

All biotin-requiring enzymes follow the same three steps: activation of bicarbonate by ATP, reaction of activated bicarbonate with biotin to form carboxybiotin, and transfer of the carboxyl group from carboxybiotin to the substrate.

24.6 Pyridoxal Phosphate: Vitamin B$_6$

The coenzyme **pyridoxal phosphate (PLP)** is derived from the vitamin known as pyridoxine or vitamin B$_6$. (Pyridoxal's "al" suffix indicates the coenzyme is an aldehyde.) A deficiency in vitamin B$_6$ causes anemia; severe deficiencies can cause seizures and death.

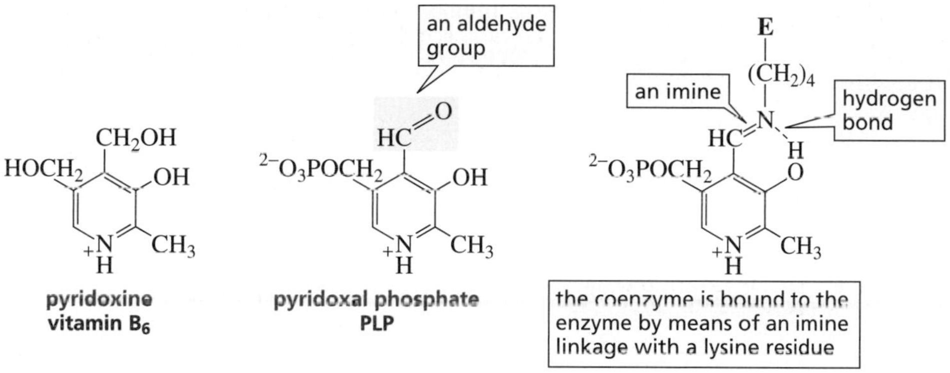

pyridoxine
vitamin B$_6$

pyridoxal phosphate
PLP

the coenzyme is bound to the enzyme by means of an imine linkage with a lysine residue

PLP is required by enzymes that catalyze certain transformations of amino acids. The most common are decarboxylation, transamination, racemization (the interconversion of L- and D-amino acids), C$_\alpha$—C$_\beta$ bond cleavage, and α,β-elimination.

decarboxylation

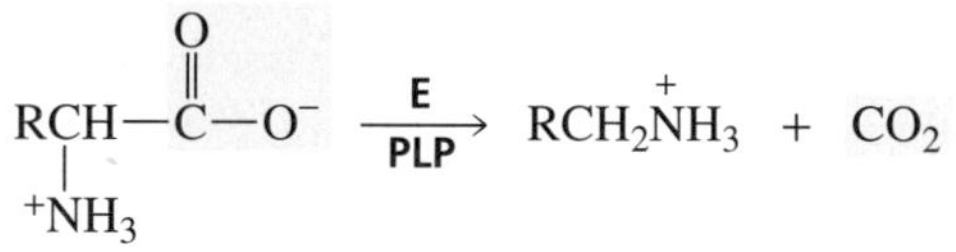

transamination

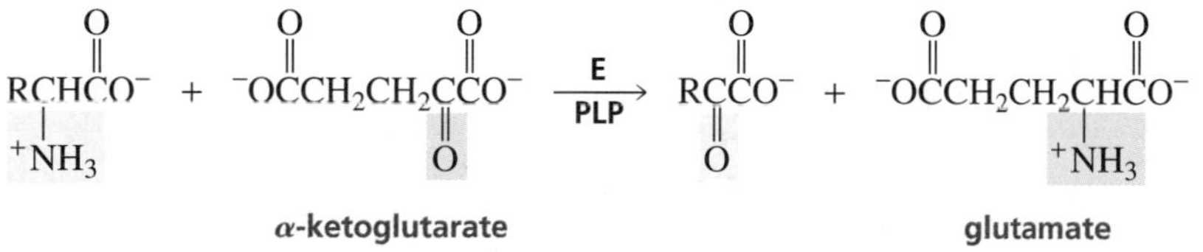

α-ketoglutarate

glutamate

Pyridoxal phosphate (PLP) is required by enzymes that catalyze certain transformations of amino acids.

racemization

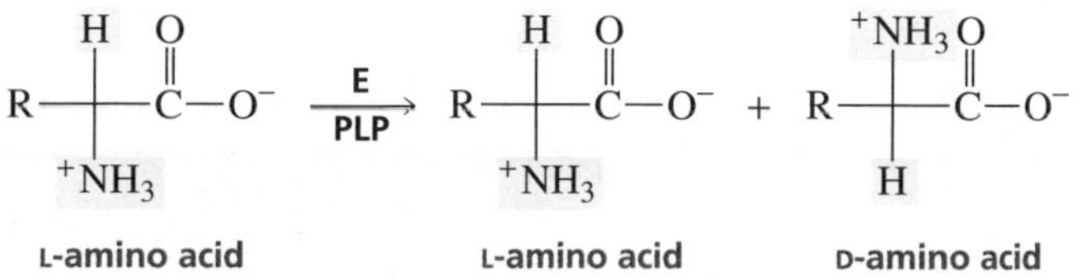

L-amino acid

L-amino acid

D-amino acid

$C_\alpha - C_\beta$ **bond cleavage**

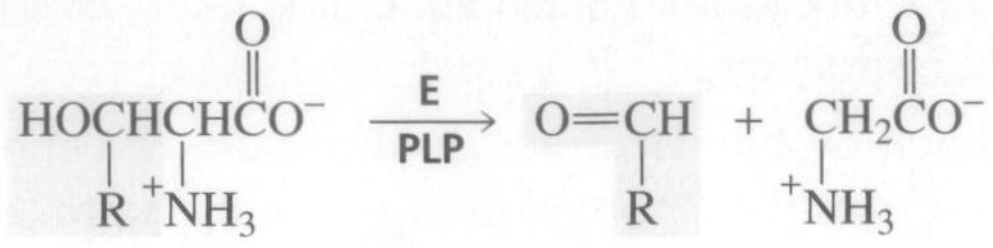

$$HOCHCHCO^- \xrightarrow[\text{PLP}]{\text{E}} O{=}CH + CH_2CO^-$$

(with R and $^+NH_3$ substituents on the left; R on the middle; $^+NH_3$ on the right; carbonyl oxygens shown)

α,β-**elimination**

$$XCH_2CHCO^- \xrightarrow[\text{PLP}]{\text{E}} CH_3CCO^- + X^- + \overset{+}{N}H_4$$

(with $^+NH_3$ substituent on the left; carbonyl oxygens shown)

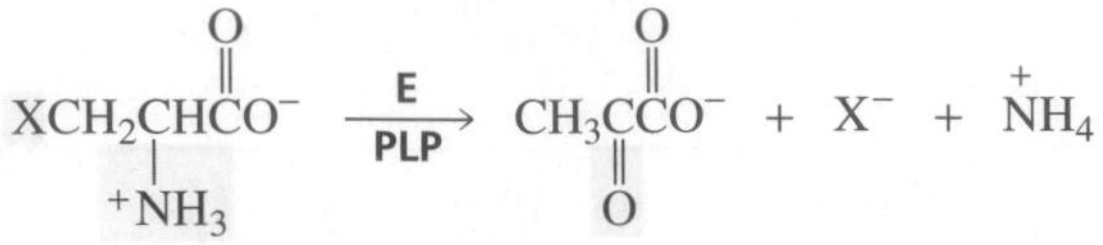

In each of these transformations, one of the bonds to the α-carbon of the amino acid substrate is broken in the first step of the reaction. Decarboxylation breaks the bond joining the carboxyl group to the α-carbon; transamination, racemization, and α,β-elimination break the bond joining the hydrogen to the α-carbon; and $C_\alpha - C_\beta$ bond cleavage breaks the bond joining the R group to the α-carbon.

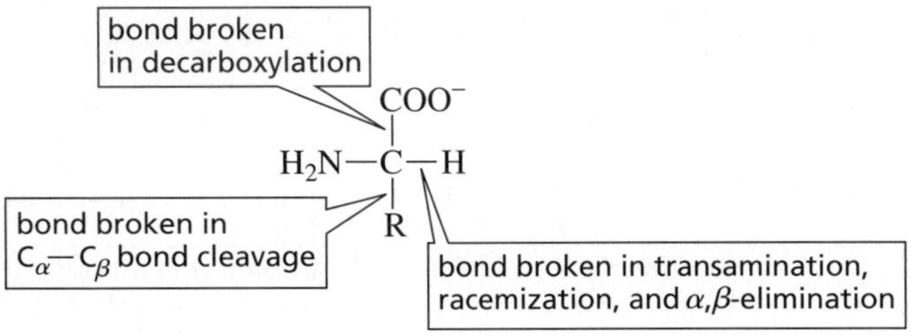

bond broken in decarboxylation

$$H_2N{-}\overset{\displaystyle COO^-}{\underset{\displaystyle R}{C}}{-}H$$

bond broken in $C_\alpha - C_\beta$ bond cleavage

bond broken in transamination, racemization, and α,β-elimination

PLP becomes attached to its enzyme by forming an imine with the amino group of a lysine side chain. The first step in all PLP-requiring enzymes is a transimination reaction. A **transimination** reaction converts one imine into another imine.

Mechanism for transimination

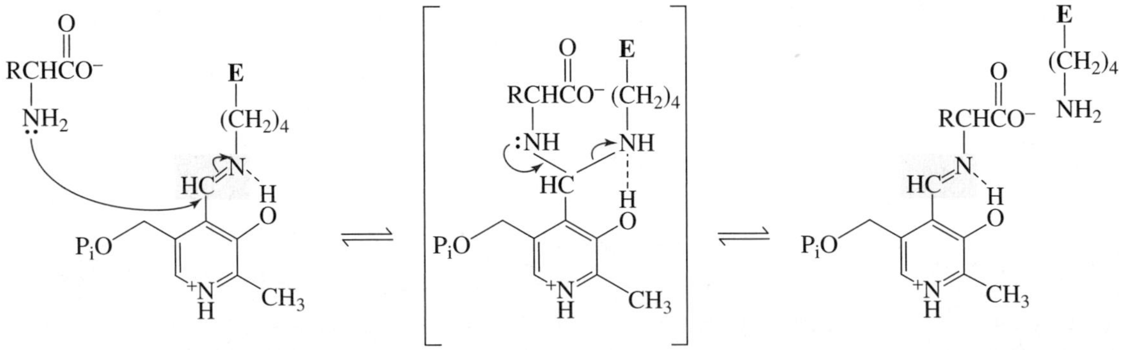

enzyme-bound PLP   **amino acid–bound PLP**

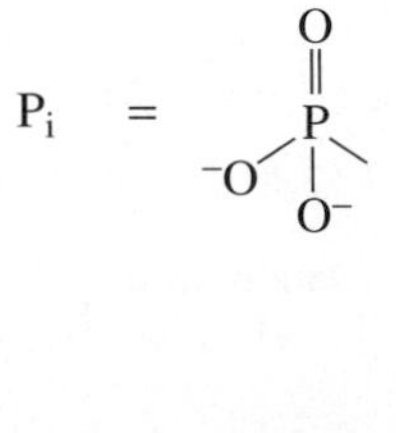

$$P_i = \overset{\displaystyle O}{\underset{\displaystyle O^-}{\overset{\|}{\underset{|}{-O{-}P{-}}}}}$$

- In the transimination reaction, the amino acid substrate reacts with the imine formed by *PLP and the enzyme,* forming a tetrahedral intermediate.
- The lysine group of the enzyme is expelled, forming a new imine between *PLP and the amino acid.*

Once the amino acid has formed an imine with PLP, a bond to the α-carbon can be broken because the electrons left behind when the bond breaks can be delocalized onto the positively charged protonated nitrogen of the pyridine ring. In other words, the protonated nitrogen of the pyridine ring is an electron sink. If the OH group is removed from the pyridine ring, the cofactor loses much of its activity. Apparently, the hydrogen bond formed by the OH group helps weaken the bond to the α-carbon.

Decarboxylation

If the PLP-catalyzed reaction is a decarboxylation, the first step is removal of the carboxyl group from the α-carbon of the amino acid.

Mechanism for PLP-catalyzed decarboxylation of an amino acid

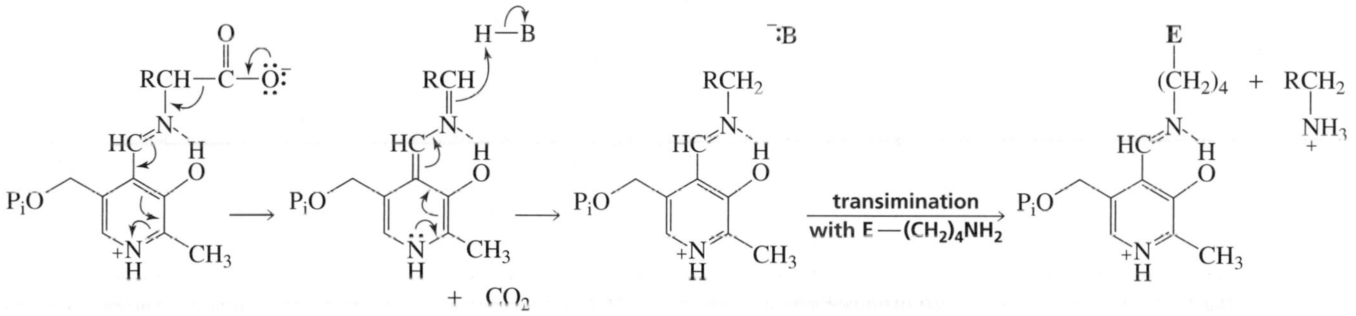

- After the carboxyl group is removed, electron rearrangement and protonation of the α-carbon of the decarboxylated intermediate by a protonated amino group of a lysine side chain or by some other acid group reestablishes the aromaticity of the pyridine ring.
- The last step in all PLP-requiring enzymes is a transimination reaction with a lysine side chain, in order to release the product of the enzyme-catalyzed reaction and regenerate enzyme-bound PLP.

Transamination

The first reaction in the catabolism of most amino acids is replacement of the amino group of the amino acid by a ketone group. This is called a **transamination** reaction because the amino group removed from the amino acid is not lost, but is *transferred* to the ketone group of α-ketoglutarate, thereby forming glutamate. The enzymes that catalyze transamination reactions are called *aminotransferases*. Transamination allows the amino groups of the various amino acids to be collected into a single amino acid (glutamate) so that excess nitrogen can be easily excreted. (Do not confuse *transamination* with *transimination,* discussed previously.)

Mechanism for PLP-catalyzed transamination of an amino acid

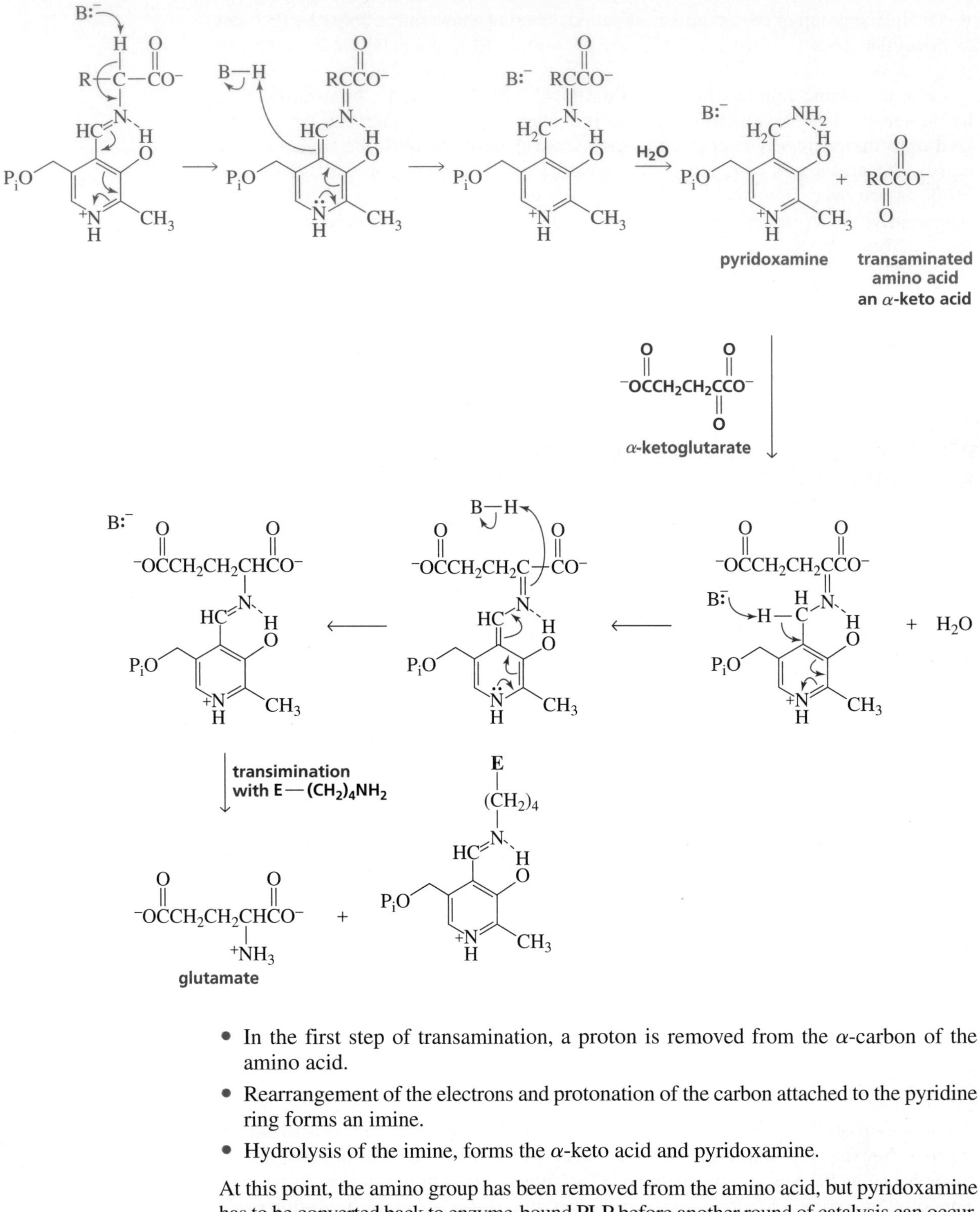

pyridoxamine | transaminated amino acid an α-keto acid

α-ketoglutarate

glutamate

transimination with E—$(CH_2)_4NH_2$

- In the first step of transamination, a proton is removed from the α-carbon of the amino acid.

- Rearrangement of the electrons and protonation of the carbon attached to the pyridine ring forms an imine.

- Hydrolysis of the imine, forms the α-keto acid and pyridoxamine.

At this point, the amino group has been removed from the amino acid, but pyridoxamine has to be converted back to enzyme-bound PLP before another round of catalysis can occur.

- Pyridoxamine forms an imine with α-ketoglutarate, the second substrate of the reaction.

- Removal of a proton from the carbon attached to the pyridine ring, followed by rearrangement of the electrons and donation of a proton to the α-carbon of the substrate forms an imine that, when transiminated with a lysine side chain, releases glutamate and reforms enzyme-bound PLP.

Notice that the proton transfer steps are reversed in the two phases of the reaction. Transfer of the amino group of the amino acid to pyridoxal requires removal of the proton from the α-carbon of the amino acid and donation of a proton to the carbon bonded to the pyridine ring. Transfer of the amino group of pyridoxamine to α-ketoglutarate requires removal of the proton from the carbon bonded to the pyridine ring and donation of a proton to the α-carbon of α-ketoglutarate.

ASSESSING THE DAMAGE AFTER A HEART ATTACK

After a heart attack, aminotransferases and other enzymes leak from the damaged cells of the heart into the bloodstream. The severity of damage done to the heart can be determined from the concentrations of alanine aminotransferase and aspartate aminotransferase in the blood.

PROBLEM 14◆

α-Keto acids other than α-ketoglutarate can accept the amino group from pyridoxamine in enzyme-catalyzed transaminations. What amino acids are formed from the following α-keto acids?

a. CH$_3$C—CO$^-$
  pyruvate

b. $^-$OCCH$_2$C—CO$^-$
  oxaloacetate

Racemization

The mechanism for the PLP-catalyzed racemization of an L-amino acid is shown below.

Mechanism for PLP-catalyzed racemization of an L-amino acid

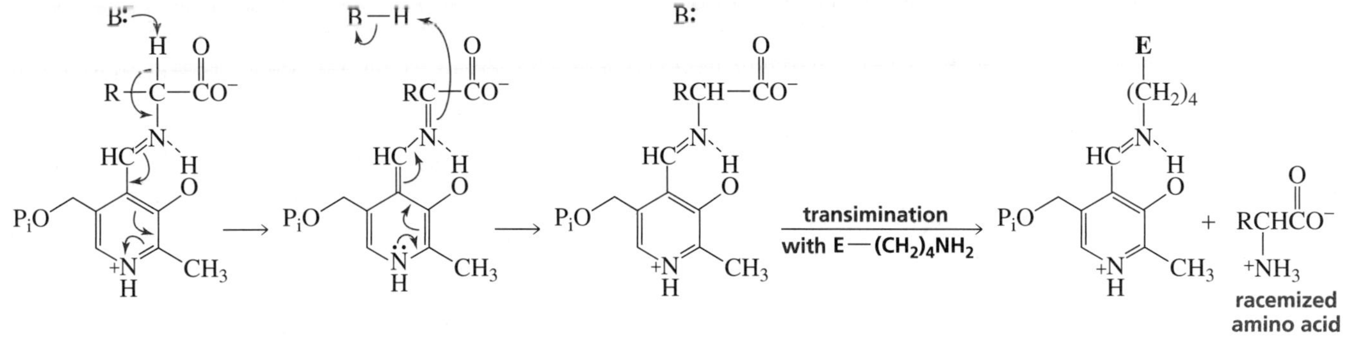

- The first step is the same as the first step in the PLP-catalyzed transamination of an amino acid—removal of a proton from the α-carbon of the amino acid bound to PLP.

- Unlike transamination, where reprotonation occurs on the carbon attached to the pyridine ring, reprotonation in racemization occurs on the α-carbon of the amino acid.

The proton can be donated to the sp^2 hybridized α-carbon from either side of the plane defined by the double bond. Consequently, both D- and L-amino acids are formed. In other words, the L-amino acid is racemized.

Compare the second step in a PLP-catalyzed transamination with the second step in a PLP-catalyzed racemization. In an enzyme that catalyzes transamination, an acidic group at the active site of the enzyme is in position to donate a proton to the carbon attached to the pyridine ring. The enzyme that catalyzes racemization does not have this acidic group, so the substrate is reprotonated at the α-carbon. In other words, the *coenzyme* carries out the chemical reaction, but the *enzyme* determines the course of the reaction.

C_{α}—C_{β} Bond Cleavage

In the first step of the mechanism for PLP-catalyzed C_{α}—C_{β} bond cleavage, a basic group at the active site of the enzyme removes a proton from an OH group bonded to the β-carbon of the amino acid.

Mechanism for PLP-catalyzed C_{α}—C_{β} bond cleavage

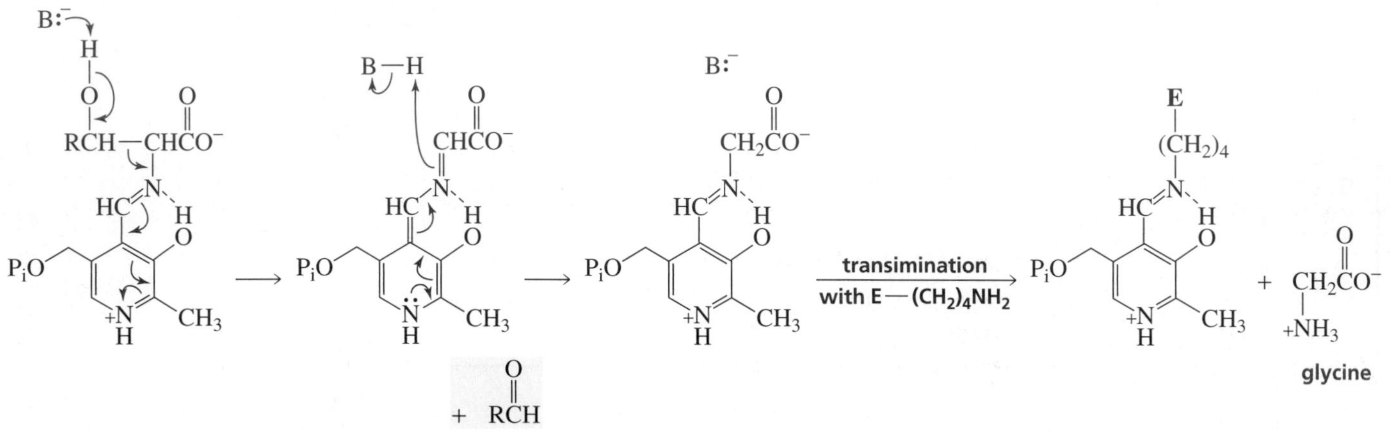

- Removing the proton from the OH group causes the C_{α}—C_{β} bond to be cleaved. Serine and threonine are the only two amino acids that can serve as substrates for the reaction because they are the only amino acids with an OH group bonded to their β-carbon. When serine is the substrate, the cleavage step releases formaldehyde (R = H); when threonine is the substrate, the cleavage step releases acetaldehyde (R = CH$_3$).

- Electron rearrangement and protonation of the α-carbon of the amino acid, followed by transimination with a lysine side chain, releases glycine.

The formaldehyde formed when serine undergoes C_{α}—C_{β} bond cleavage never leaves the active site of the enzyme; it is immediately transferred to tetrahydrofolate (Section 24.8).

PROBLEM 15

Propose a mechanism for a PLP-catalyzed α,β-elimination.

Choosing the Bond to Be Cleaved

If all PLP-requiring enzymes start with the same substrate—an amino acid bound to pyridoxal phosphate by an imine linkage—how can three different bonds be cleaved in the first step of the reaction? The bond cleaved in the first step depends on the conformation of the amino acid that the enzyme binds. There is free rotation about the C_{α}—N bond of the amino acid, and an enzyme can bind any of the possible conformations about this bond. The enzyme will bind the conformation in which the overlapping orbitals of the bond to be broken in the first step of the reaction lie parallel to the *p* orbitals of the conjugated system. As a result, the orbital containing the electrons left behind when the bond is broken can overlap with the orbitals of the conjugated system. If such overlap could not occur, the electrons could not be delocalized into the conjugated system and the carbanion intermediate could not be stabilized.

Student Tutorial:
Mechanism of PLP-dependent reactions

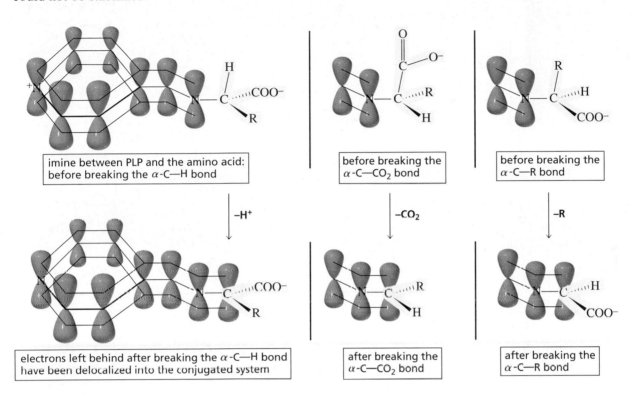

imine between PLP and the amino acid: before breaking the α-C—H bond

before breaking the α-C—CO₂ bond

before breaking the α-C—R bond

−H⁺

−CO₂

−R

electrons left behind after breaking the α-C—H bond have been delocalized into the conjugated system

after breaking the α-C—CO₂ bond

after breaking the α-C—R bond

PROBLEM 16◆

Which of the following compounds is more easily decarboxylated?

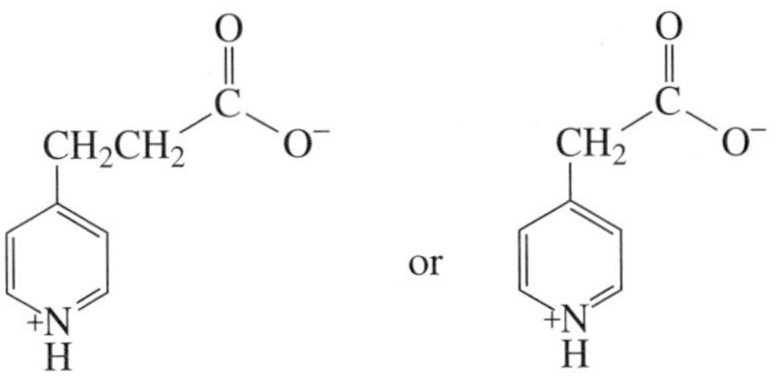

or

PROBLEM 17

Explain why the ability of PLP to catalyze an amino acid transformation is greatly reduced if a PLP-requiring enzymatic reaction is carried out at a pH at which the pyridine nitrogen is not protonated.

PROBLEM 18

Explain why the ability of PLP to catalyze an amino acid transformation is greatly reduced if the OH substituent of pyridoxal phosphate is replaced by an OCH₃.

Dorothy Crowfoot Hodgkin (1910–1994) *was born in Egypt to English parents. She received an undergraduate degree from Somerville College at Oxford University and earned a Ph.D. from Cambridge University. She performed the first three-dimensional calculations in crystallography, and she was the first to use computers to determine the structures of compounds, successfully determining the structures of penicillin, insulin, and vitamin B_{12}. For her work on vitamin B_{12}, she received the 1964 Nobel Prize in chemistry. Hodgkin was a professor of chemistry at Somerville, where one of her graduate students was former Prime Minister of England Margaret Thatcher. She was also a founding member of Pugwash, an organization whose purpose was to further communication between scientists on both sides of the Iron Curtain.*

24.7 Coenzyme B_{12}: Vitamin B_{12}

Enzymes that catalyze certain rearrangement reactions require **coenzyme B_{12}**, a coenzyme derived from vitamin B_{12}. The structure of vitamin B_{12} was determined by Dorothy Crowfoot Hodgkin, using X-ray crystallography. The vitamin has a cyano group (or HO^- or H_2O) coordinated with cobalt (Section 20.10). In coenzyme B_{12}, this group is replaced by a 5′-deoxyadenosyl group.

Animals and plants cannot synthesize vitamin B_{12}. In fact, only a few microorganisms can synthesize it. Humans must obtain all their vitamin B_{12} from their diet, particularly from meat. A deficiency causes pernicious anemia. Because vitamin B_{12} is needed in only very small amounts, deficiencies caused by consumption of insufficient amounts of the vitamin are rare but have been found in vegetarians who eat no animal products. Most deficiencies are caused by an inability to absorb the vitamin in the intestine.

The following are examples of enzyme-catalyzed reactions that require coenzyme B$_{12}$:

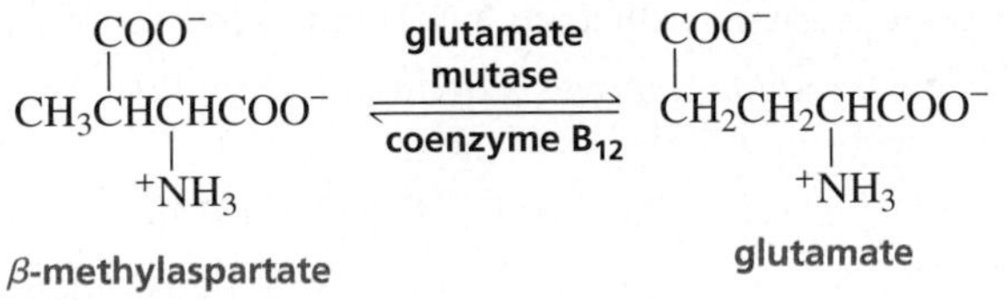

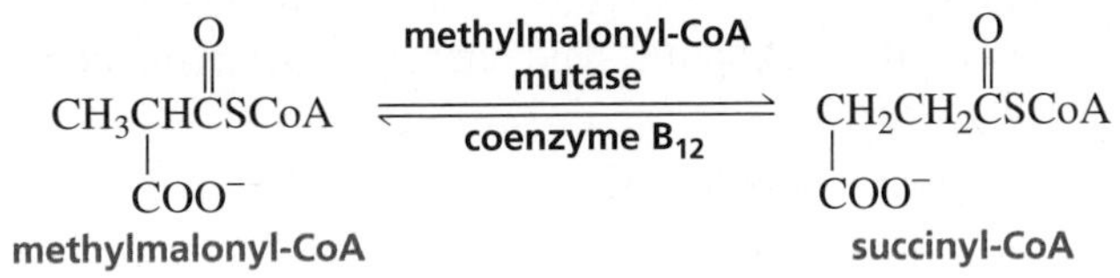

$$\underset{\substack{\text{β-methylaspartate}}}{\underset{\underset{{}^{+}\text{NH}_3}{|}}{\text{CH}_3\text{CHCHCOO}^-}} \;\underset{\text{coenzyme B}_{12}}{\overset{\substack{\text{glutamate}\\\text{mutase}}}{\rightleftharpoons}}\; \underset{\substack{\text{glutamate}}}{\underset{\underset{{}^{+}\text{NH}_3}{|}}{\text{CH}_2\text{CH}_2\text{CHCOO}^-}}$$

$$\underset{\substack{\text{methylmalonyl-CoA}}}{\underset{\underset{\text{COO}^-}{|}}{\overset{\overset{\text{O}}{\|}}{\text{CH}_3\text{CHCSCoA}}}} \;\underset{\text{coenzyme B}_{12}}{\overset{\substack{\text{methylmalonyl-CoA}\\\text{mutase}}}{\rightleftharpoons}}\; \underset{\substack{\text{succinyl-CoA}}}{\underset{\underset{\text{COO}^-}{|}}{\overset{\overset{\text{O}}{\|}}{\text{CH}_2\text{CH}_2\text{CSCoA}}}}$$

Coenzyme B$_{12}$ is required by enzymes that catalyze the exchange of a hydrogen bonded to one carbon with a group bonded to an adjacent carbon.

$$\underset{\substack{\text{1,2-propanediol}}}{\underset{\underset{\text{OH}}{|}}{\text{CH}_3\text{CHCH}_2\text{OH}}} \;\underset{\text{coenzyme B}_{12}}{\overset{\text{dioldehydrase}}{\longrightarrow}}\; \underset{\substack{\text{a hydrate}}}{\left[\underset{\underset{\text{OH}}{|}}{\text{CH}_3\text{CH}_2\text{CHOH}}\right]} \longrightarrow \underset{\substack{\text{propanal}}}{\overset{\overset{\text{O}}{\|}}{\text{CH}_3\text{CH}_2\text{CH}}} + \text{H}_2\text{O}$$

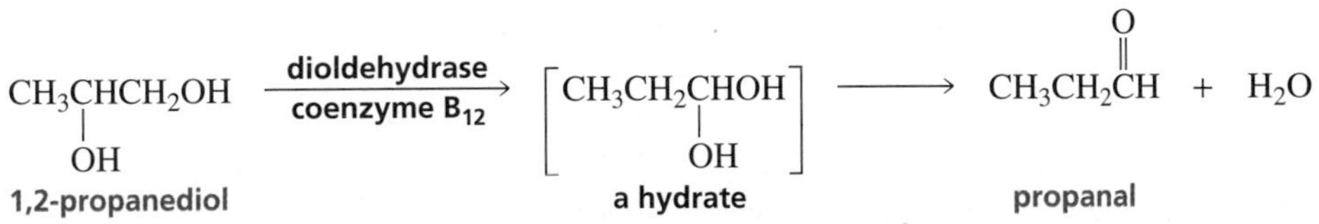

In each of these coenzyme B$_{12}$–requiring reactions, a group (Y) bonded to one carbon changes places with a hydrogen bonded to an adjacent carbon.

$$\underset{\underset{\text{H}}{|}}{-\text{C}}\underset{\underset{\text{Y}}{|}}{\text{C}}- \;\overset{\substack{\text{a coenzyme B}_{12}-\\\text{requiring enzyme}}}{\rightleftharpoons}\; \underset{\underset{\text{Y}}{|}}{-\text{C}}\underset{\underset{\text{H}}{|}}{\text{C}}-$$

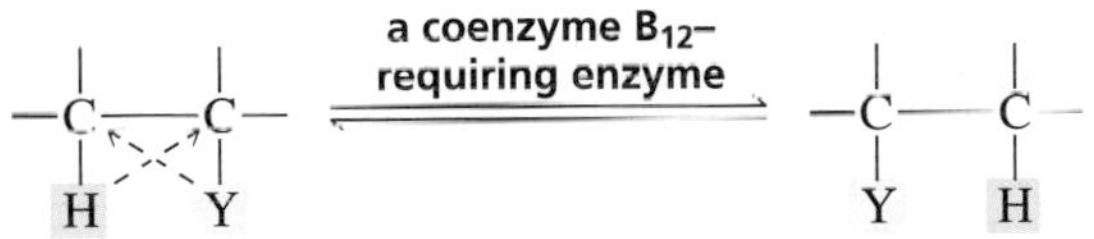

For example, glutamate mutase and methylmalonyl-CoA mutase both catalyze a reaction in which the COO$^-$ group bonded to one carbon changes places with a hydrogen of an adjacent methyl group. In the reaction catalyzed by dioldehydrase, an OH group changes places with a methylene hydrogen. The resulting product is a hydrate that loses water to form propanal.

The chemistry of coenzyme B$_{12}$ takes place at the bond joining the cobalt and the 5'-deoxyadenosyl group. The currently accepted mechanism for dioldehydrase is shown below.

The mechanism for a coenzyme B$_{12}$-requiring enzyme-catalyzed reaction

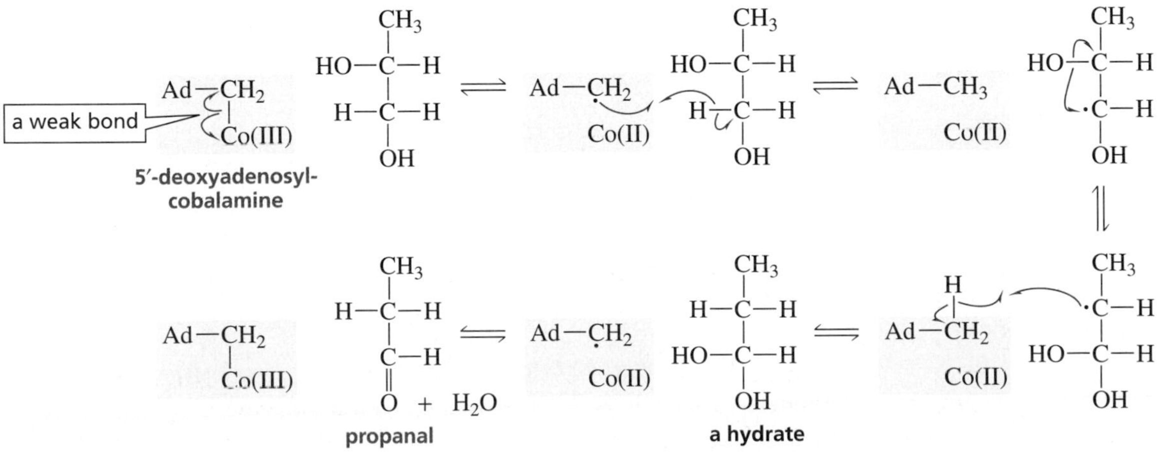

- The unusually weak Co—C bond (26 kcal/mol or 109 kJ/mol, compared with 99 kcal/mol or 414 kJ/mol for a C—H bond) breaks homolytically, forming a 5′-deoxyadenosyl radical and reducing Co(III) to Co(II).
- The 5′-deoxyadenosyl radical removes a hydrogen atom from the C-1 carbon of the substrate, thereby becoming 5′-deoxyadenosine.
- A hydroxyl radical (·OH) migrates from C-2 to C-1, creating a radical at C-2.
- The radical at C-2 removes a hydrogen atom from 5′-deoxyadenosine, forming the rearranged product and regenerating the 5′-deoxyadenosyl radical, which recombines with Co(II) to regenerate the coenzyme. The enzyme–coenzyme complex is then ready for another catalytic cycle.
- The initial product is a hydrate that loses water to form propanal, the final product of the reaction.

It is likely that all coenzyme B_{12}–requiring enzymes catalyze reactions by means of the same general mechanism. The role of the coenzyme is to provide a way to remove a hydrogen atom from the substrate. Once the hydrogen atom has been removed, an adjacent group can migrate to take its place. The coenzyme then gives back the hydrogen atom, delivering it to the carbon that lost the migrating group.

PROBLEM 19

Ethanolamine ammonia lyase, a coenzyme B_{12}–requiring enzyme, catalyzes the following reaction:

$$HOCH_2CH_2NH_2 \xrightarrow[\text{ammonia lyase}]{\text{ethanolamine}} CH_3\overset{\overset{\displaystyle O}{\|}}{C}H + NH_3$$

Propose a mechanism for this reaction.

PROBLEM 20◆

A fatty acid with an even number of carbon atoms is metabolized to acetyl-CoA, which can enter the citric acid cycle. A fatty acid with an odd number of carbon atoms is metabolized to acetyl-CoA and one equivalent of propionyl-CoA. Two coenzyme-requiring enzymes are needed to convert propionyl-CoA into succinyl-CoA, a citric acid cycle intermediate. Write the two enzyme-catalyzed reactions and include the names of the required coenzymes.

24.8 Tetrahydrofolate: Folic Acid

Tetrahydrofolate (THF) is the coenzyme used by enzymes that catalyze reactions that transfer a group containing a single carbon to their substrates. The one-carbon group can be a methyl group (CH_3), a methylene group (CH_2), or a formyl group ($HC=O$). Tetrahydrofolate is produced by the reduction of two double bonds of folic acid (folate), its precursor vitamin. Bacteria synthesize folate, but mammals cannot.

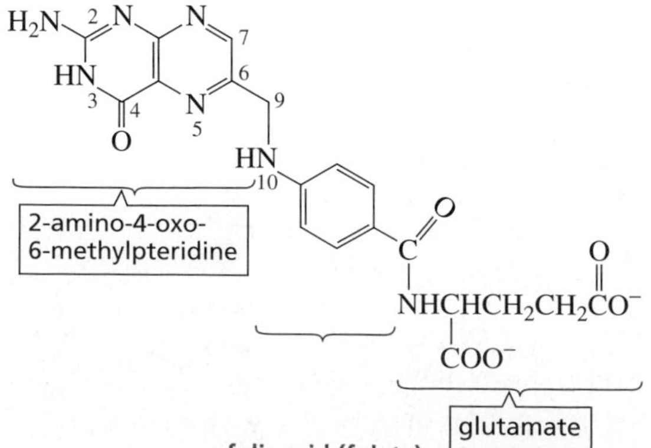

folic acid (folate)

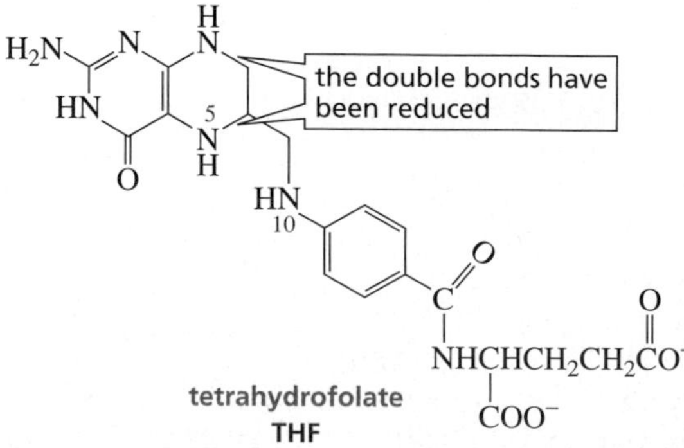

tetrahydrofolate
THF

There are actually six different THF-coenzymes. N^5-Methyl-THF transfers a methyl group (CH_3), N^5,N^{10}-methylene-THF transfers a methylene group (CH_2), and the others transfer a formyl group ($HC=O$).

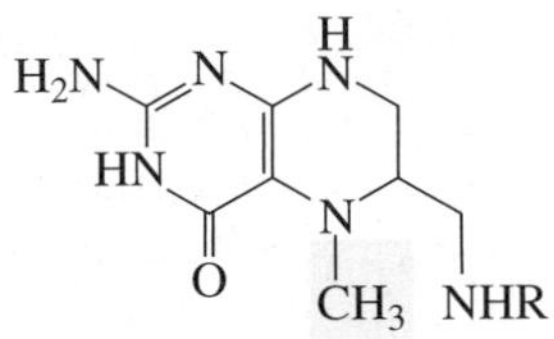

N^5-methyl-THF

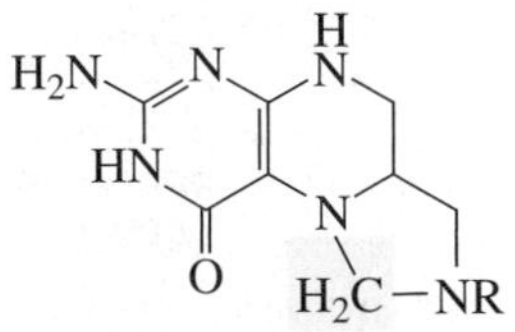

N^5,N^{10}-methylene-THF

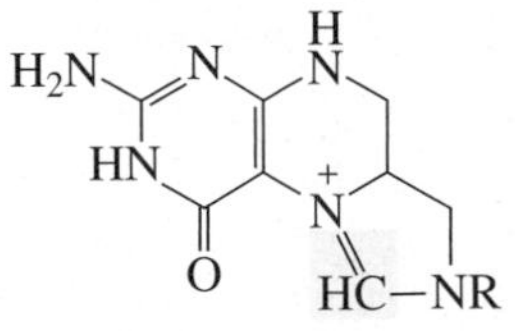

N^5,N^{10}-methenyl-THF

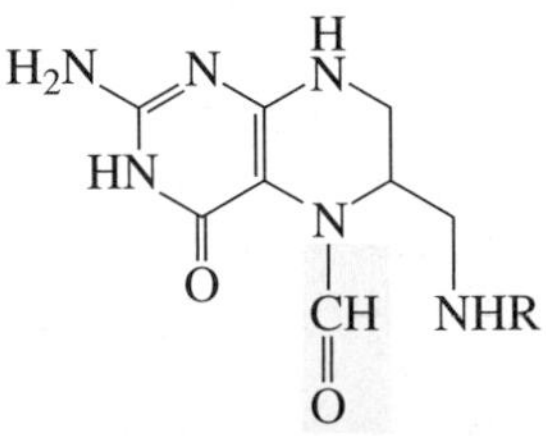

N^5-formyl-THF

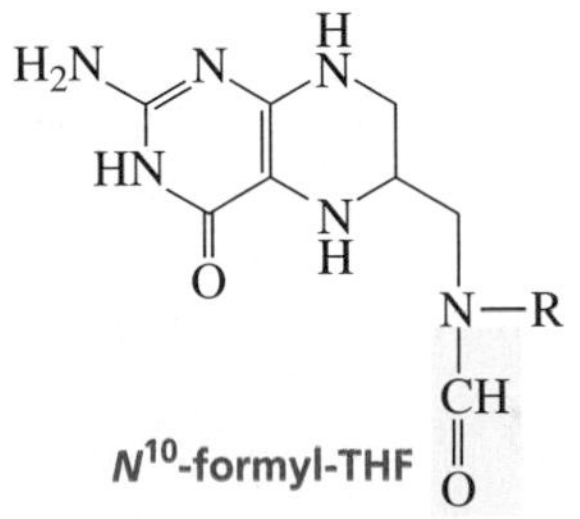

N^{10}-formyl-THF

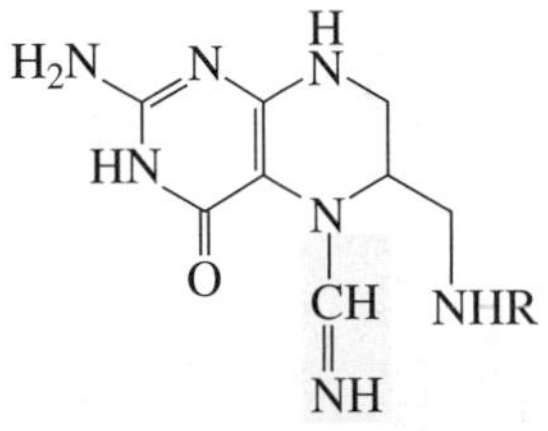

N^5-formimino-THF

Glycinamide ribonucleotide (GAR) transformylase is an example of an enzyme that requires a THF-coenzyme. The formyl group given to the substrate eventually ends up being the C-8 carbon of the purine nucleotides (Section 27.1).

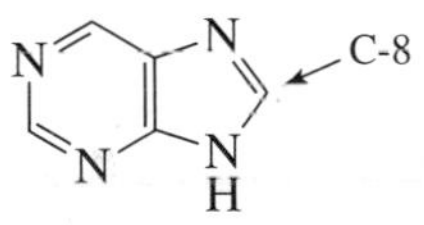

ribose-5-phosphate $\quad + \quad N^{10}$-formyl-THF $\quad \xrightarrow{\text{GAR transformylase}} \quad$ ribose-5-phosphate $\quad + \quad$ THF

purine

Homocysteine methyl transferase, an enzyme required for the synthesis of methionine, also requires a THF-coenzyme.

Tetrahydrofolate (THF) is the coenzyme required by enzymes that catalyze the transfer of a group containing one carbon to their substrates.

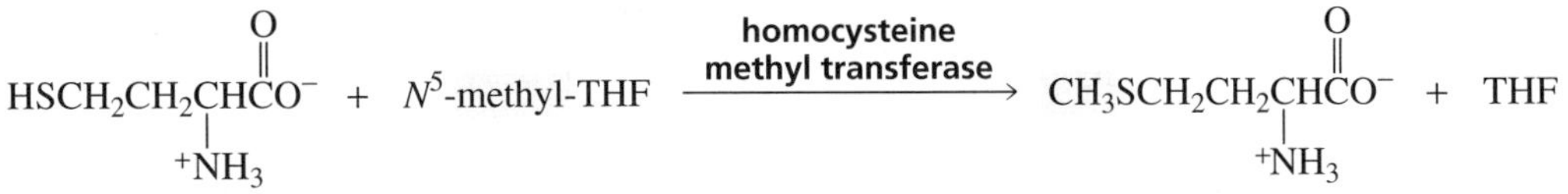

$HSCH_2CH_2CHCO^-$ $\quad + \quad N^5$-methyl-THF $\quad \xrightarrow[\text{methyl transferase}]{\text{homocysteine}} \quad CH_3SCH_2CH_2CHCO^-$ $\quad + \quad$ THF

homocysteine

methionine

Thymidylate Synthase: The Enzyme That Converts U's into T's

The heterocyclic bases in RNA are adenine, guanine, cytosine, and uracil (A, G, C, and U); the heterocyclic bases in DNA are adenine, guanine, cytosine, and thymine (A, G, C, and T; these are described in Section 27.1). In other words, the heterocyclic bases in RNA and DNA are the same, except RNA contains U's, whereas DNA contains T's. Why DNA contains T's instead of U's is explained in Section 27.9.

The T's used for the biosynthesis of DNA are synthesized from U's by thymidylate synthase, an enzyme that requires N^5,N^{10}-methylene-THF as a coenzyme. Even though the only structural difference between a T and a U is a *methyl* group, a T is synthesized by first transferring a *methylene* group to a U. The methylene group is then reduced to a methyl group.

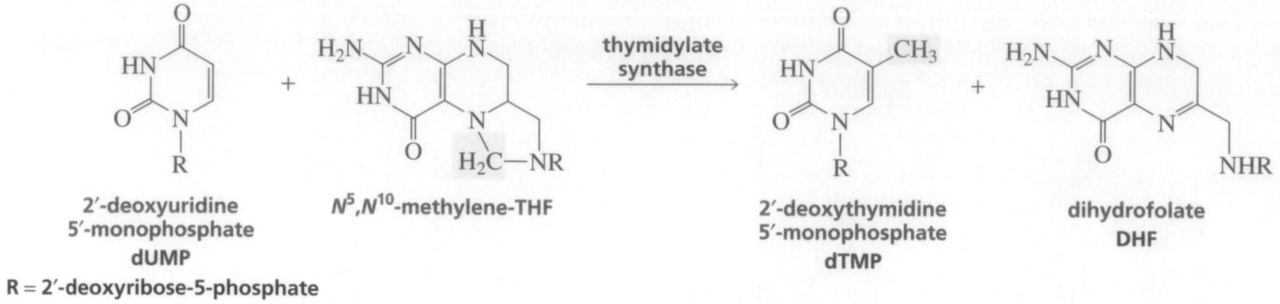

**2′-deoxyuridine
5′-monophosphate
dUMP**

R = **2′-deoxyribose-5-phosphate**

N^5,N^{10}-**methylene-THF**

**2′-deoxythymidine
5′-monophosphate
dTMP**

**dihydrofolate
DHF**

The mechanism of the reaction catalyzed by thymidylate synthase is shown below.

Mechanism for catalysis by thymidylate synthase

an E2 reaction

oxidized
coenzyme

Student Tutorial:
Mechanism for catalysis
by thymidylate synthase

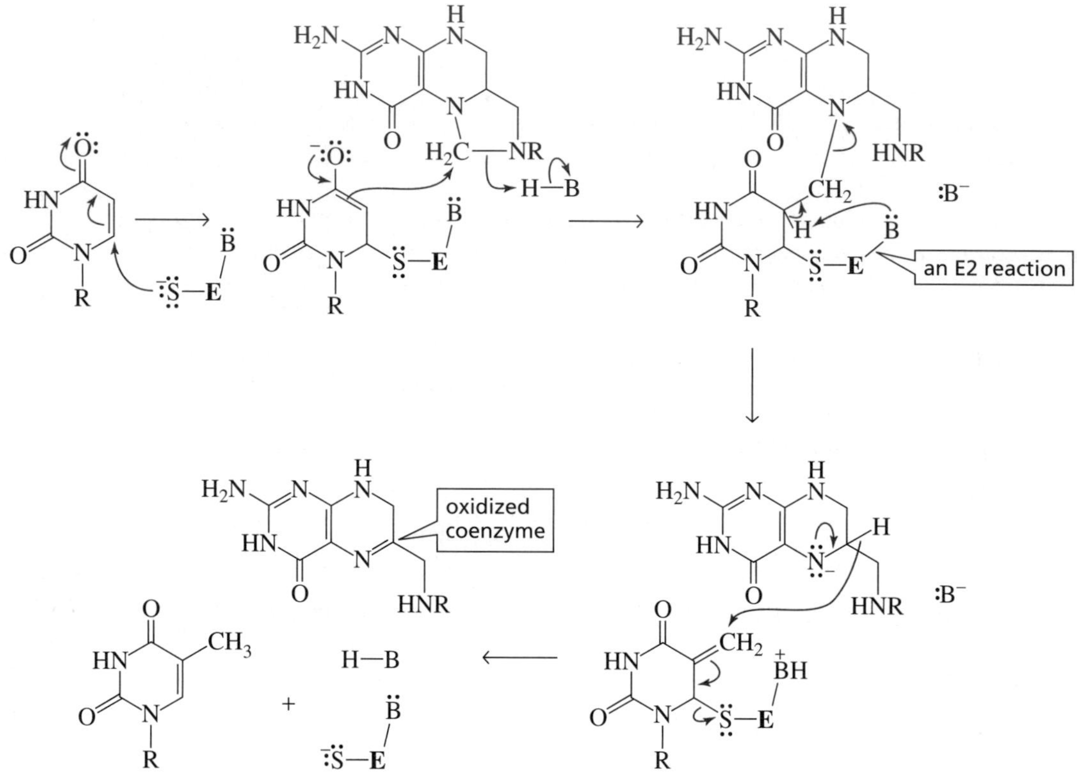

- A nucleophilic cysteine group at the active site of the enzyme attacks the β-carbon of uridine. (This is an example of conjugate addition; see Section 17.17.)
- Nucleophilic attack by the α-carbon of uridine on the methylene group of N^5,N^{10}-methylene-THF forms a covalent bond between uridine and the coenzyme.
- A proton on the α-carbon of uridine is removed with the help of a basic group of an amino acid side chain at the active site of the enzyme, eliminating the coenzyme.
- Transfer of a hydride ion from the coenzyme to the methylene group, followed by elimination of the enzyme, forms thymidine and dihydrofolate (DHF).

Notice that the coenzyme that transfers the methylene group to the substrate is also the reducing agent that subsequently reduces the methylene group to a methyl group. Because the coenzyme is the reducing agent, it is simultaneously oxidized. The oxidized coenzyme is dihydrofolate.

When the reaction is over, dihydrofolate must be converted back to N^5,N^{10}-methylene-THF so that the coenzyme can undergo another catalytic cycle. Dihydrofolate is first reduced to tetrahydrofolate. Then serine hydroxymethyl transferase—the PLP-requiring enzyme that cleaves the C_α—C_β bond of serine to form glycine and formaldehyde—transfers formaldehyde to the coenzyme (Section 24.6). In other words, the formaldehyde that is cleaved off serine is immediately transferred to THF to form N^5,N^{10}-methylene-THF, which is fortunate because formaldehyde is cytotoxic. (It kills cells.)

$$\text{dihydrofolate} + \text{NADPH} + \text{H}^+ \xrightarrow{\text{dihydrofolate reductase}} \text{tetrahydrofolate} + \text{NADP}^+$$

$$\text{tetrahydrofolate} + \underset{\substack{| \\ {}^+\text{NH}_3 \\ \textbf{serine}}}{\text{HOCH}_2\text{CHCOO}^-} \xrightarrow[\text{PLP}]{\text{serine hydroxymethyl transferase}} N^5,N^{10}\text{-methylene-THF} + \underset{\substack{| \\ {}^+\text{NH}_3 \\ \textbf{glycine}}}{\text{CH}_2\text{COO}^-}$$

Cancer Chemotherapy

Cancer is the abnormal (uncontrolled) growth and proliferation of cells. Because cells cannot multiply if they cannot synthesize DNA, several of the cancer chemotherapeutic agents that have been developed are designed to inhibit thymidylate synthase and dihydrofolate reductase. If a cell cannot make thymidine (T), it cannot synthesize DNA. Inhibiting dihydrofolate reductase also prevents the synthesis of thymidine because cells have a limited amount of tetrahydrofolate. If they cannot convert dihydrofolate back to tetrahydrofolate, they cannot continue to synthesize thymidine.

A common anticancer drug that inhibits thymidylate synthase is 5-fluorouracil. 5-Fluorouracil and uracil react with thymidylate synthase in the same way. However, the fluorine of 5-fluorouracil cannot be removed by the base in the third step of the reaction because fluorine is too electronegative to come off as F^+. As a consequence, the reaction stops, leaving the enzyme permanently attached to the substrate. The active site of the enzyme is now blocked with 5-fluorouracil and cannot bind dUMP. Therefore, dTMP can no longer be synthesized and, without dTMP, DNA cannot be synthesized. Unfortunately, most anticancer drugs cannot discriminate between diseased and normal cells, so cancer chemotherapy is accompanied by debilitating side effects. However, because cancer cells, with their uncontrolled cell division, are dividing more rapidly than normal cells, they are harder hit by cancer-fighting chemotherapeutic agents than normal cells are.

5-fluorouracil
5-FU

the enzyme has become irreversibly attached to the substrate

5-Fluorouracil is a **mechanism-based inhibitor**: it inactivates the enzyme by taking part in the normal catalytic mechanism. It is also called a **suicide inhibitor** because the enzyme effectively "commits suicide" by reacting with it. The therapeutic use of

5-fluorouracil illustrates the importance of knowing the mechanism for enzyme-catalyzed reactions. If scientists know the mechanism for a reaction, they may be able to design an inhibitor to turn the reaction off at a certain step.

Aminopterin and methotrexate are anticancer drugs that are inhibitors of dihydro-folate reductase. Because their structures are similar to that of dihydrofolate, they compete with it for binding at the active site of the enzyme. Since they bind 1000 times more tightly to the enzyme than does dihydrofolate, they inhibit the enzyme's activity. These two compounds are examples of **competitive inhibitors**.

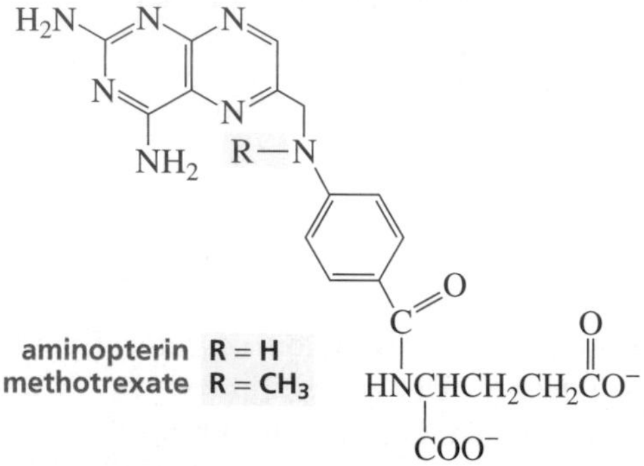

aminopterin **R = H**
methotrexate **R = CH₃**

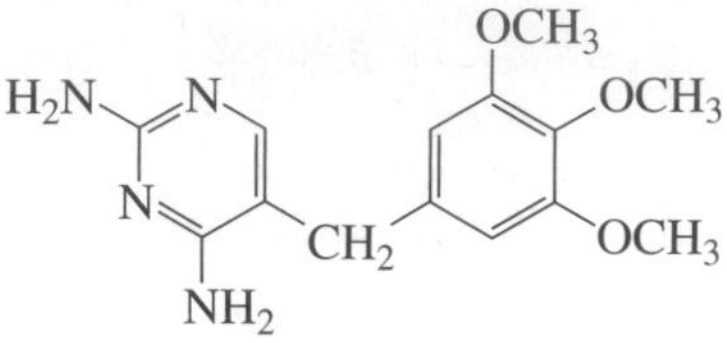

trimethoprim

Because aminopterin and methotrexate inhibit the synthesis of THF, they interfere with the synthesis of any compound that requires a THF-coenzyme in one of the steps of its synthesis. Thus, not only do they prevent the synthesis of thymidine but they also inhibit the synthesis of adenine and guanine—other heterocyclic compounds needed for the synthesis of DNA—because their synthesis also requires a THF-coenzyme. One clinical technique used in chemotherapy to fight cancer calls for the patient to be given a lethal dose of methotrexate and then, after cancer cells have died, to "save" him or her by administering N^5-formyl-THF.

Trimethoprim is used as an antibiotic because it binds to bacterial dihydrofolate reductase much more tightly than to mammalian dihydrofolate reductase.

PROBLEM 21◆

What is the source of the methyl group in thymidine?

THE FIRST ANTIBIOTICS

Sulfonamides—commonly known as sulfa drugs—were introduced clinically in 1934 as the first effective antibiotics (Section 30.4). Donald Woods, a British bacteriologist, noticed that sulfanilamide, initially the most widely used sulfonamide, was structurally similar to *p*-aminobenzoic acid, a compound necessary for bacterial growth. He proposed that sulfanilamide's antibacterial properties were due to its being able to block the normal utilization of *p*-aminobenzoic acid.

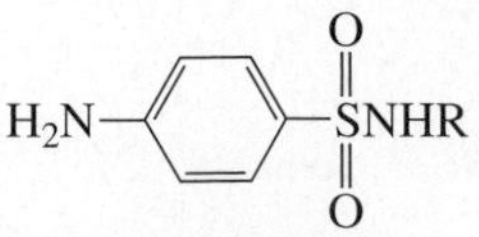

a sulfonamide 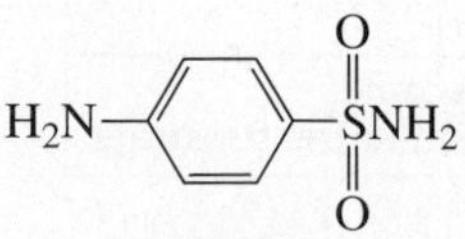   sulfanilamide 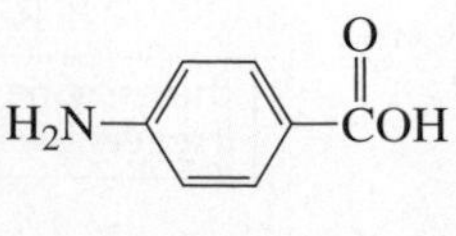   *p*-aminobenzoic acid

Woods and Paul Flores suggested that sulfanilamide acts by inhibiting the enzyme that incorporates *p*-aminobenzoic acid into folic acid. Because the enzyme cannot tell the difference between sulfanilamide and *p*-aminobenzoic acid, both compounds compete for the active site of the enzyme. Humans are not adversely affected by the drug because they do not synthesize folate instead, they get all their folate from their diets.

Donald D. Woods (1912–1964) *was born in Ipswich, England, and received a B.A. and a Ph.D. from Cambridge University. He worked in Paul Flores's laboratory at the London Hospital Medical College.*

Paul B. Flores (1882–1971) *was born in London. He moved his laboratory to Middlesex Hospital Medical School when a bacterial chemistry unit was established there. He was knighted in 1946.*

24.9 Vitamin KH₂: Vitamin K

Vitamin K is required for proper clotting of blood. The letter K comes from *koagulation,* which is German for "clotting." A series of reactions utilizing six proteins causes blood to clot. The process requires these proteins to bind Ca^{2+}. Vitamin K is required for proper Ca^{2+} binding. Vitamin K is found in the leaves of green plants. Vitamin K deficiencies are rare because the vitamin is also synthesized by intestinal bacteria. **Vitamin KH₂** (the hydroquinone of vitamin K) is the coenzyme form of the vitamin.

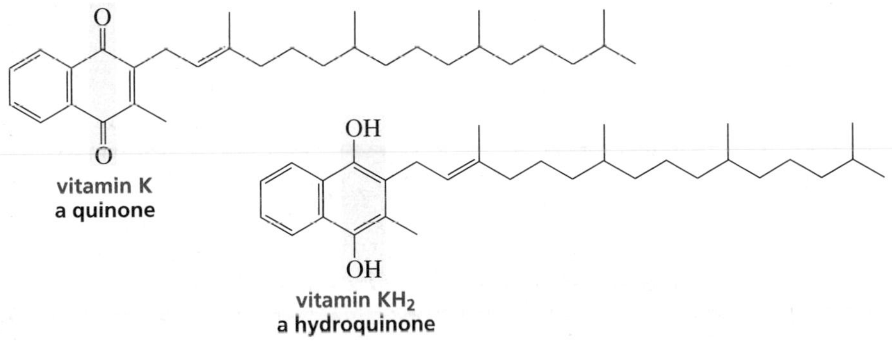

vitamin K
a quinone

vitamin KH₂
a hydroquinone

Vitamin KH₂ is the coenzyme for the enzyme that catalyzes the carboxylation of the γ-carbon of glutamate side chains in proteins, forming γ-carboxyglutamates. γ-Carboxyglutamates complex Ca^{2+} much more effectively than glutamates do. The enzyme uses CO_2 for the carboxyl group it puts on the glutamate side chains. The proteins responsible for blood clotting all have several glutamates near their N-terminal ends. For example prothrombin, a blood-clotting protein, has glutamates at positions 7, 8, 15, 17, 20, 21, 26, 27, 30, and 33.

Vitamin KH₂ is required by the enzyme that catalyzes the carboxylation of the γ-carbon of a glutamate side chain in a protein.

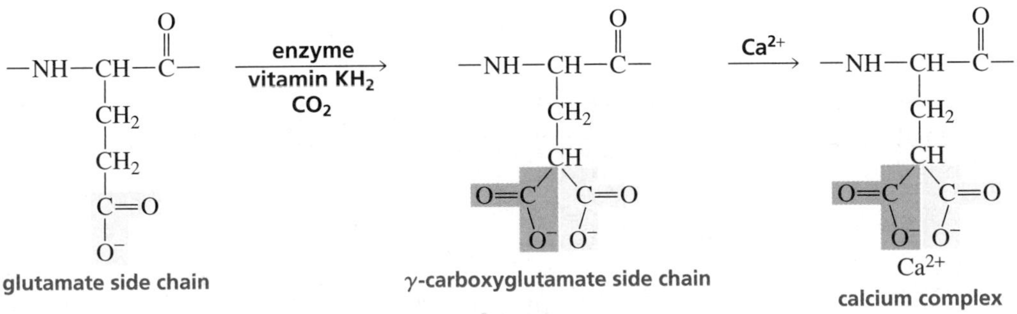

glutamate side chain

γ-carboxyglutamate side chain

calcium complex

The mechanism for the vitamin KH₂–catalyzed carboxylation of glutamate had puzzled chemists because the γ-proton that must be removed from glutamate before it can attack CO_2 is not very acidic. The mechanism, therefore, must involve the creation of a strong base. The mechanism shown below was proposed by Paul Dowd.

Mechanism for the vitamin KH_2-dependent carboxylation of glutamate

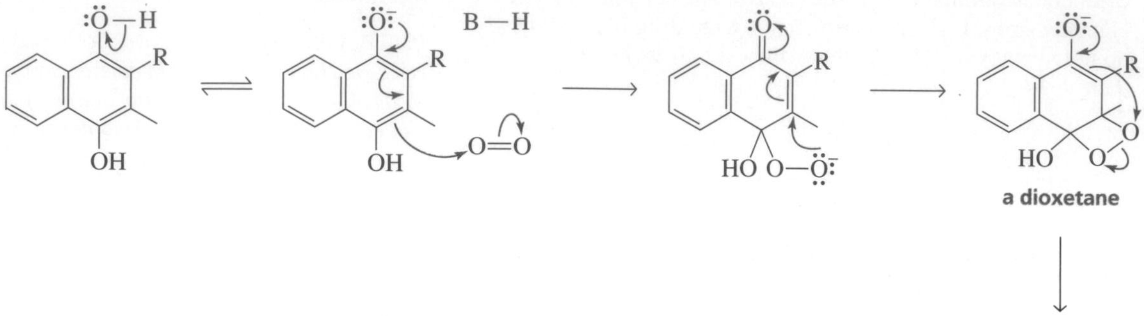

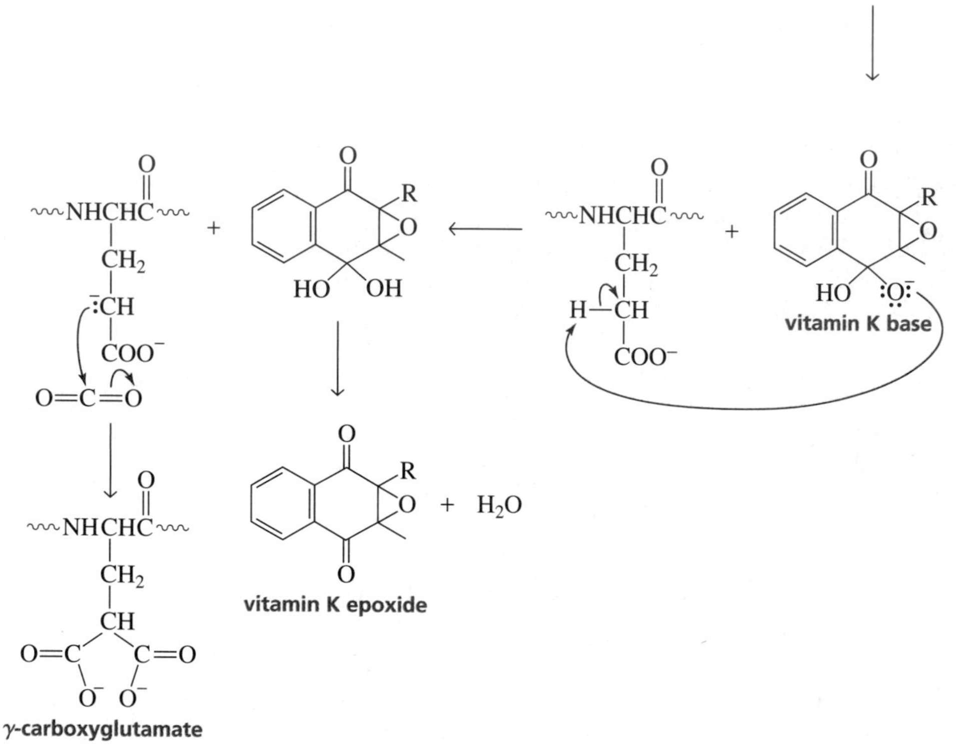

a dioxetane

vitamin K base

γ-carboxyglutamate

vitamin K epoxide

- The vitamin loses a proton from a phenolic OH group.
- The base thus formed attacks molecular oxygen.
- A dioxetane is formed, which collapses to form a vitamin K base.
- The vitamin K base is strong enough to remove a proton from the γ-carbon of glutamate.
- The glutamate carbanion attacks CO_2 to form γ-carboxyglutamate, and the protonated vitamin K base (a hydrate) loses water, forming vitamin K epoxide.

Vitamin K epoxide is reduced back to vitamin KH_2 by an enzyme that uses the coenzyme dihydrolipoate as the reducing agent. The epoxide is first reduced to vitamin K, which is then further reduced to vitamin KH_2.

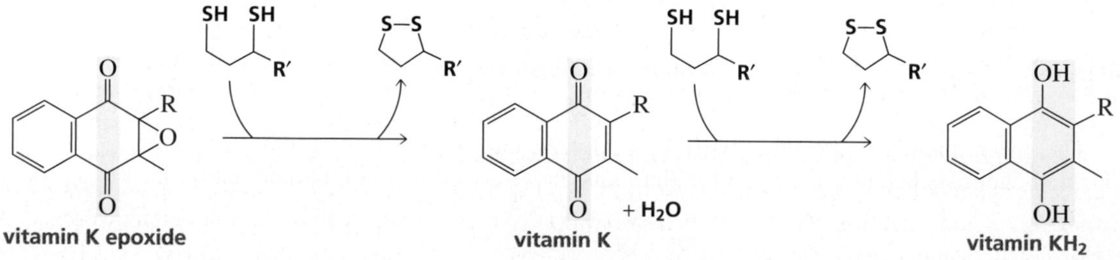

vitamin K epoxide   vitamin K   vitamin KH_2

Warfarin and dicoumarol are used clinically as anticoagulants. These prevent clotting by inhibiting the enzyme that synthesizes vitamin KH_2 from vitamin K epoxide by binding to the enzyme's active site. The enzyme cannot tell the difference between these two compounds and vitamin K epoxide, so the compounds act as *competitive inhibitors*. Warfarin is also a common rat poison, causing death by internal bleeding.

warfarin   dicoumarol

Vitamin E too has recently been found to be an anticoagulant. It directly inhibits the enzyme that carboxylates glutamate residues.

TOO MUCH BROCCOLI

An article describing two women with diseases characterized by abnormal blood clotting reported that they did not improve when they were given warfarin. When questioned about their diets, one woman said that she ate at least a pound (0.45 kg) of broccoli every day, and the other ate broccoli soup and a broccoli salad every day. When broccoli was removed from their diets, warfarin became effective in preventing the abnormal clotting of their blood. Because broccoli is high in vitamin K, these patients had been getting enough dietary vitamin K to compete with the drug for the enzyme's active site, thereby making the drug ineffective.

PROBLEM 22

Thiols such as ethanethiol and propanethiol can be used to reduce vitamin K epoxide back to vitamin KH_2, but they react much more slowly than dihydrolipoate. Explain.

SUMMARY

Cofactors assist enzymes in catalyzing a variety of reactions that cannot be catalyzed solely by the enzymes' amino acid side chains. Cofactors can be metal ions or organic molecules. An enzyme with a tightly bound metal ion is called a **metalloenzyme**. Cofactors that are organic molecules are called **coenzymes**; coenzymes are derived from **vitamins**, which are substances needed in small amounts for normal body function that the body cannot synthesize. All the water-soluble vitamins except vitamin C are precursors for coenzymes. Vitamin K is the only water-insoluble vitamin that is a precursor for a coenzyme.

Coenzymes play a variety of chemical roles that the amino acid side chains of enzymes cannot perform. Some function as oxidizing and reducing agents; some allow electrons to be delocalized; some activate groups for further reaction; and some provide good nucleophiles or strong bases needed for a reaction. **Molecular recognition** allows the enzyme to bind the substrate and the coenzyme in the proper orientation for reaction. Coenzymes are recycled. An enzyme plus its cofactor is called a **holoenzyme**. An enzyme that has had its cofactor removed is called an **apoenzyme**.

Metabolism—the set of reactions that living organisms carry out to obtain energy and to synthesize the compounds they require—can be divided into catabolism and anabolism. **Catabolic reactions** break down complex molecules to provide energy and simple molecules. **Anabolic reactions** require energy and lead to the synthesis of complex biomolecules.

NAD^+, $NADP^+$, **FAD**, and **FMN** are coenzymes used to catalyze oxidation reactions; those used to catalyze reduction reactions are **NADH**, **NADPH**, $FADH_2$, and $FMNH_2$.

Many enzymes that catalyze oxidation reactions are called **dehydrogenases**. All the redox chemistry of the **pyridine nucleotide coenzymes** takes place at the 4-position of the pyridine ring. All the redox chemistry of the **flavin coenzymes** takes place on the flavin ring.

Thiamine pyrophosphate (TPP) is the coenzyme required by enzymes that catalyze the transfer of a two-carbon fragment. **Biotin** is the coenzyme required by enzymes that catalyze carboxylation of a carbon adjacent to a carbonyl group. **Pyridoxal phosphate (PLP)** is the coenzyme required by enzymes that catalyze certain transformations of amino acids: decarboxylation, transamination, racemization, C_α—C_β bond cleavage, and α,β-elimination. In a **transimination reaction**, one imine is converted into another

imine; in a **transamination reaction**, the amino group is removed from a substrate, leaving a keto group in its place, and transferred to another molecule.

In a **coenzyme B$_{12}$**–requiring reaction, a group bonded to one carbon changes places with a hydrogen bonded to an adjacent carbon. **Tetrahydrofolate (THF)** is the coenzyme used by enzymes catalyzing reactions that transfer a group containing a single carbon—methyl, methylene, or formyl—to their substrates. **Vitamin KH$_2$** is the coenzyme for the enzyme that catalyzes the carboxylation of the γ-carbon of glutamate side chains, a reaction required for blood clotting. A **suicide inhibitor** inactivates an enzyme by taking part in the normal catalytic mechanism. **Competitive inhibitors** compete with the substrate for binding at the active site of the enzyme.

KEY TERMS

anabolic reaction (p. 1101)
apoenzyme (p. 1099)
biotin (p. 1115)
catabolic reaction (p. 1101)
coenzyme (p. 1098)
coenzyme A (CoASH) (p. 1113)
coenzyme B$_{12}$ (p. 1124)
cofactor (p. 1098)
competitive inhibitor (p. 1130)
dehydrogenase (p. 1103)
electron sink (p. 1111)

flavin adenine dinucleotide
  (FAD) (p. 1107)
flavin mononucleotide (FMN) (p. 1107)
heterocyclic compound (p. 1102)
holoenzyme (p. 1099)
lipoate (p. 1113)
mechanism-based inhibitor (p. 1129)
metabolism (p. 1101)
metalloenzyme (p. 1098)
nicotinamide adenine dinucleotide
  (NAD$^+$) (p. 1101)

nicotinamide adenine dinucleotide
  phosphate (NADP$^+$) (p. 1101)
nucleotide (p. 1102)
pyridoxal phosphate (PLP) (p. 1117)
suicide inhibitor (p. 1129)
tetrahydrofolate (THF) (p. 1126)
thiamine pyrophosphate (TPP) (p. 1110)
transamination (p. 1119)
transimination (p. 1118)
vitamin (p. 1098)
vitamin KH$_2$ (p. 1131)

PROBLEMS

23. Answer the following questions:
   a. What six cofactors act as oxidizing agents?
   b. What are the cofactors that donate one-carbon groups?
   c. What three one-carbon groups do various tetrahydrofolates donate to substrates?
   d. What is the function of FAD in the pyruvate dehydrogenase complex?
   e. What is the function of NAD$^+$ in the pyruvate dehydrogenase complex?
   f. What reaction necessary for proper blood clotting is catalyzed by vitamin KH$_2$?
   g. What coenzymes are used for decarboxylation reactions?
   h. What kinds of substrates do the decarboxylating coenzymes work on?
   i. What coenzymes are used for carboxylation reactions?
   j. What kinds of substrates do the carboxylating coenzymes work on?

24. Name the coenzymes that
   a. allow electrons to be delocalized.
   b. activate groups for further reaction.
   c. provide a good nucleophile.
   d. provide a strong base.

25. For each of the following reactions, name the enzyme that catalyzes the reaction and name the required coenzyme:

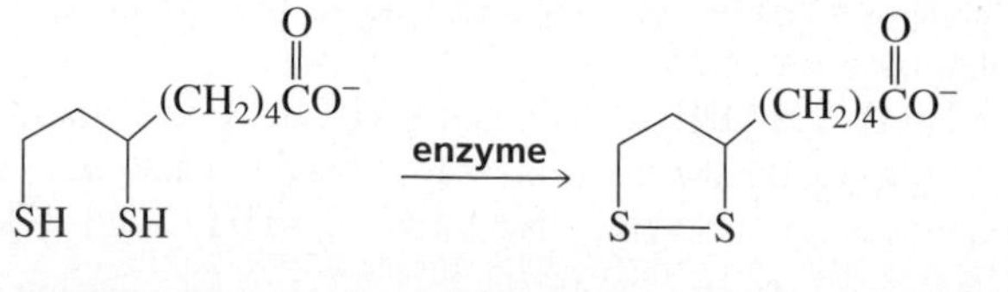

c. 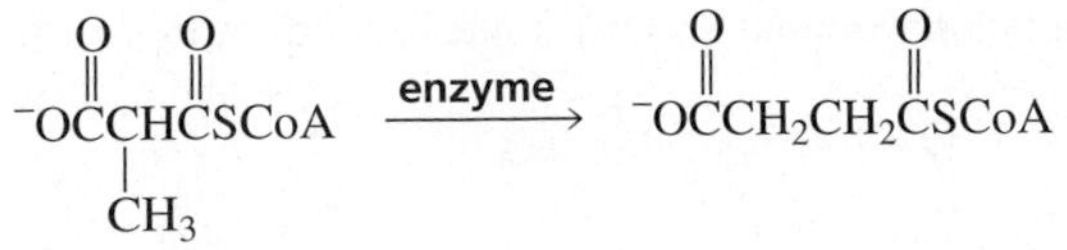
$$\underset{\underset{CH_3}{|}}{\overset{O\;\;O}{\overset{\|\;\;\|}{^-OCCHCSCoA}}} \xrightarrow{\text{enzyme}} \overset{O\quad\quad O}{\overset{\|\quad\quad\|}{^-OCCH_2CH_2CSCoA}}$$

d. 
$$\overset{O\;\;O}{\overset{\|\;\;\|}{CH_3C-CO^-}} \xrightarrow[\text{(a catabolic reaction)}]{\text{enzyme}} \underset{}{\overset{OH\;\;O}{\overset{|\quad\;\|}{CH_3CH-CO^-}}}$$

e. 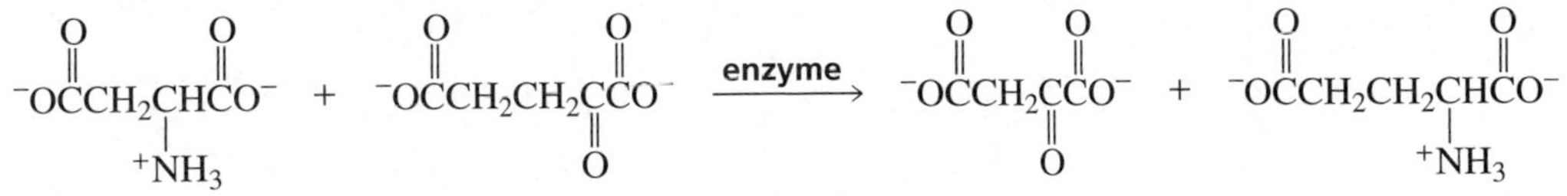
$$\underset{\underset{^+NH_3}{|}}{\overset{O\;\;\;\;O}{\overset{\|\;\;\;\;\|}{^-OCCH_2CHCO^-}}} + \underset{\underset{O}{\|}}{\overset{O\quad\;\;O}{\overset{\|\quad\;\;\|}{^-OCCH_2CH_2CCO^-}}} \xrightarrow{\text{enzyme}} \underset{\underset{O}{\|}}{\overset{O\quad O}{\overset{\|\quad\|}{^-OCCH_2CCO^-}}} + \underset{\underset{^+NH_3}{|}}{\overset{O\quad\quad\;\;O}{\overset{\|\quad\quad\;\;\|}{^-OCCH_2CH_2CHCO^-}}}$$

f. 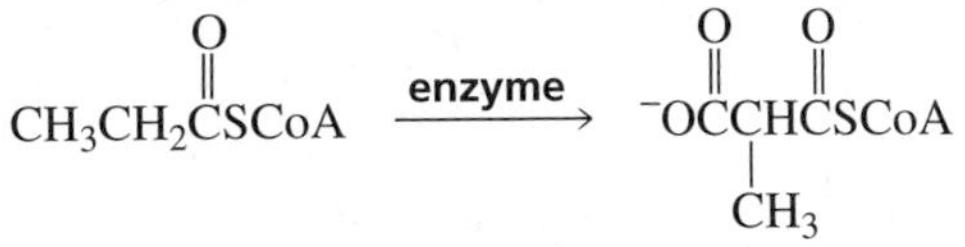
$$\overset{O}{\overset{\|}{CH_3CH_2CSCoA}} \xrightarrow{\text{enzyme}} \underset{\underset{CH_3}{|}}{\overset{O\;\;O}{\overset{\|\;\;\|}{^-OCCHCSCoA}}}$$

26. *S*-Adenosylmethionine (SAM) is formed from the reaction between ATP (Section 8.12) and methionine. The other product of the reaction is triphosphate. Propose a mechanism for this reaction.

27. Five coenzymes are required by α-ketoglutarate dehydrogenase, the enzyme in the citric acid cycle that converts α-ketoglutarate to succinyl-CoA.
    a. Identify the coenzymes.
    b. Propose a mechanism for this reaction.

$$\underset{\textbf{α-ketoglutarate}}{\overset{O\quad\quad\;\;O\;\;O}{\overset{\|\quad\quad\;\;\|\;\;\|}{^-OCCH_2CH_2C-CO^-}}} \xrightarrow{\textbf{α-ketoglutarate dehydrogenase}} \underset{\textbf{succinyl-CoA}}{\overset{O\quad\quad\;\;\;O}{\overset{\|\quad\quad\;\;\;\|}{^-OCCH_2CH_2CSCoA}}} + CO_2$$

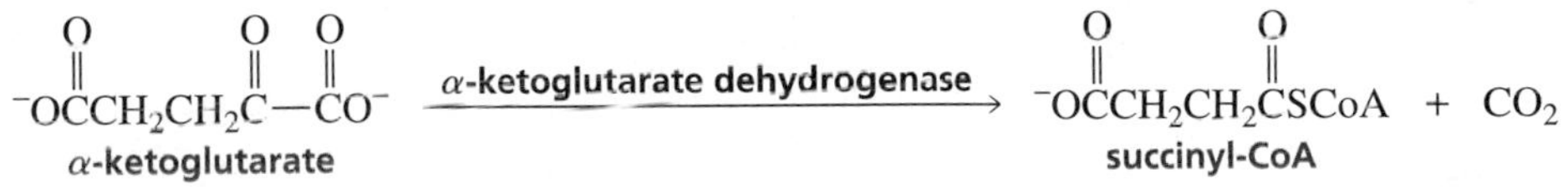

28. Give the products of the following reaction, where T is tritium:

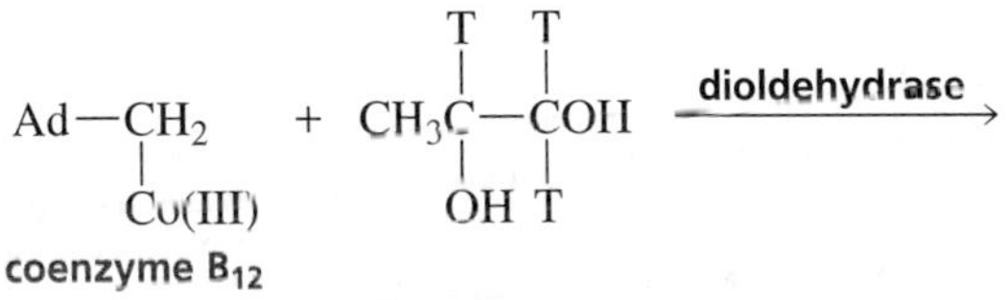

$$\underset{\underset{\textbf{coenzyme B}_{12}}{\overset{Co(III)}{|}}}{\overset{}{Ad-CH_2}} + \underset{\underset{OH\;T}{|\;\;|}}{\overset{T\;\;T}{\overset{|\;\;|}{CH_3C-COH}}} \xrightarrow{\textbf{dioldehydrase}}$$

(*Hint:* Tritium is a hydrogen atom with two neutrons. Although a C—T bond breaks four times more slowly than a C—H bond, it is still the first bond in the substrate that breaks.)

29. Propose a mechanism for methylmalonyl-CoA mutase, the enzyme that converts methylmalonyl-CoA into succinyl-CoA.

30. When transaminated, the three branched-chain amino acids (valine, leucine, and isoleucine) form compounds that have the characteristic odor of maple syrup. An enzyme known as branched-chain α-keto acid dehydrogenase converts these compounds into CoA esters. People who do not have this enzyme have the genetic disease known as maple syrup urine disease, so-called because their urine smells like maple syrup.
    a. Give the structures of the compounds that smell like maple syrup.
    b. Give the structures of the CoA esters.
    c. Branched-chain α-keto acid dehydrogenase has five coenzymes. Identify them.
    d. Suggest a way to treat maple syrup urine disease.

31. When UMP is dissolved in T_2O (T = tritium; see Problem 28), exchange of T for H occurs at the 5-position. Propose a mechanism for this exchange.

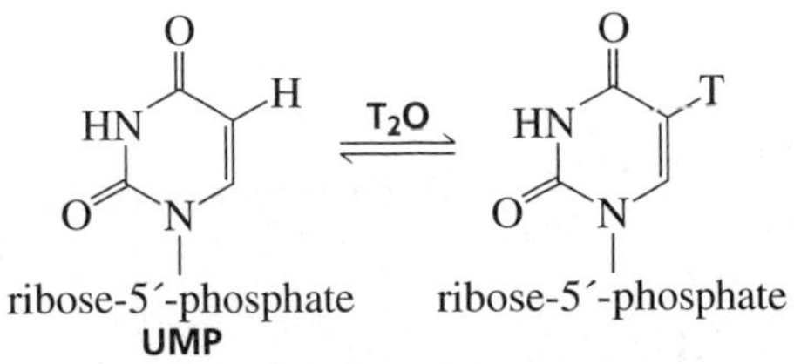

ribose-5´-phosphate   ribose-5´-phosphate
UMP

32. Dehydratase is a pyridoxal-requiring enzyme that catalyzes an α,β-elimination reaction. Propose a mechanism for this reaction.

$$\text{HOCH}_2\underset{\overset{|}{^+\text{NH}_3}}{\text{CH}}\overset{\overset{\text{O}}{\|}}{\text{C}}\text{O}^- \xrightarrow[\text{PLP}]{\text{dehydratase}} \text{CH}_3\overset{\overset{\text{O}}{\|}}{\underset{\overset{\|}{\text{O}}}{\text{C}}}\text{CO}^- + \overset{+}{\text{NH}_4}$$

33. In addition to the reactions mentioned in Section 24.6, PLP can catalyze β-substitution reactions. Propose a mechanism for the following PLP-catalyzed β-substitution reaction:

$$\text{XCH}_2\underset{\overset{|}{^+\text{NH}_3}}{\text{CH}}\overset{\overset{\text{O}}{\|}}{\text{C}}\text{O}^- + \text{Y}^- \xrightarrow[\text{PLP}]{\text{enzyme}} \text{YCH}_2\underset{\overset{|}{^+\text{NH}_3}}{\text{CH}}\overset{\overset{\text{O}}{\|}}{\text{C}}\text{O}^- + \text{X}^-$$

34. PLP can catalyze both α,β-elimination reactions (Problem 32) and β,γ-elimination reactions. Propose a mechanism for the following PLP-catalyzed β,γ-elimination:

$$\text{XCH}_2\text{CH}_2\underset{\overset{|}{^+\text{NH}_3}}{\text{CH}}\overset{\overset{\text{O}}{\|}}{\text{C}}\text{O}^- \xrightarrow[\text{PLP}]{\text{enzyme}} \text{CH}_3\text{CH}_2\overset{\overset{\text{O}}{\|}}{\underset{\overset{\|}{\text{O}}}{\text{C}}}\text{CO}^- + \text{X}^- + \overset{+}{\text{NH}_4}$$

35. The glycine cleavage system is a group of four enzymes that together catalyze the following reaction:

$$\text{glycine} + \text{THF} \xrightarrow{\textbf{glycine cleavage system}} N^5,N^{10}\text{-methylene-THF} + \text{CO}_2$$

Use the following information to determine the sequence of reactions carried out by the glycine cleavage system:
a. The first enzyme is a PLP-requiring decarboxylase.
b. The second enzyme is aminomethyltransferase. This enzyme has a lipoate coenzyme.
c. The third enzyme is an N^5,N^{10}-methylene-THF synthesizing enzyme. It catalyzes a reaction that forms $^+\text{NH}_4$ as one of the products.
d. The fourth enzyme is an FAD-requiring enzyme.
e. The cleavage system also requires NAD$^+$.

36. Nonenzyme-bound FAD is a stronger oxidizing agent than NAD$^+$. How, then, can NAD$^+$ oxidize the reduced flavoenzyme in the pyruvate dehydrogenase system?

37. FADH$_2$ reduces α,β-unsaturated thioesters to saturated thioesters. The reaction is thought to take place by a mechanism that involves radicals. Propose a mechanism for this reaction.

$$\text{RCH}{=}\text{CH}\overset{\overset{\text{O}}{\|}}{\text{C}}\text{SR} + \text{FADH}_2 \longrightarrow \text{RCH}_2\text{CH}_2\overset{\overset{\text{O}}{\|}}{\text{C}}\text{SR} + \text{FAD}$$

CHAPTER 25

The Chemistry of Metabolism

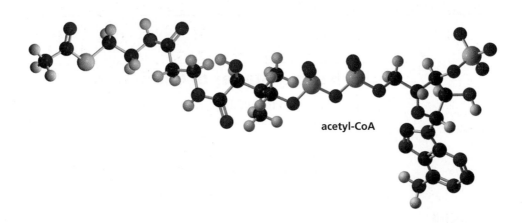

acetyl-CoA

BUILDING ON FUNDAMENTALS

SECTION 25.2	Strong bases are poor leaving groups (10.1). We will see that ATP provides reaction pathways for certain biochemical reactions that would otherwise not occur because of poor leaving groups.
SECTION 25.3	Carboxylic acids in biological systems are activated by being converted into acyl phosphates, acyl pyrophosphates, and acyl adenylates (16.22).
SECTION 25.4	Charged species are stabilized in an aqueous solution by solvation (2.9).
SECTION 25.4	Molecules are stabilized by electron delocalization (7.6).
SECTION 25.7	The enediol rearrangement interconverts aldoses and ketoses (21.5).
SECTION 25.10	A hydrogen bonded to a carbon adjacent to a carbonyl carbon is more acidic than hydrogens attached to other sp^3 carbons (18.1).
SECTION 25.10	If a carboxyl group is bonded to a carbon adjacent to a carbonyl carbon, the carboxyl group can be removed as CO_2 (18.18).
SECTION 25.10	α,β-Unsaturated carbonyl compounds can undergo conjugate addition reactions with nucleophiles (17.17).

The reactions that living organisms carry out to obtain the energy they need and to synthesize the compounds they require are collectively known as **metabolism**. Metabolism can be divided into two parts: catabolism and anabolism. *Catabolic reactions* break down complex nutrient molecules to produce energy and simple molecules that can be used for synthesis. *Anabolic reactions* synthesize complex biomolecules from simpler precursor molecules; these reactions require energy.

> **catabolism:** complex molecules $\longrightarrow$ simple molecules + energy
> **anabolism:** simple molecules + energy $\longrightarrow$ complex molecules

It is important to remember that almost every reaction that occurs in a living system is catalyzed by an enzyme. The enzyme holds the reactants and any necessary coenzymes in place, orienting the reacting functional groups and the catalyzing amino acid side chains in such a way that the enzyme-catalyzed reaction is able to take place (Section 23.8).

Most of the reactions described in this chapter are reactions you have studied in previous chapters. If you take the time to go back to the sections referred to and review these reactions, you will see that many of the organic reactions done by cells are the same as the organic reactions done by chemists.

DIFFERENCES IN METABOLISM

Humans do not necessarily metabolize compounds in the same way as other species do. This becomes a significant problem when drugs are tested on animals (Section 30.4). For example, chocolate is metabolized to different compounds in humans and in dogs; the metabolites produced in humans are nontoxic, whereas those produced in dogs can be highly toxic. Differences in metabolism have been found even within the same species. For example, isoniazid—an antituberculosis drug—is metabolized by Eskimos much faster than by Egyptians. Current research is showing that men and women metabolize certain drugs differently. For example, kappa opioids—a class of painkillers—have been found to be about twice as effective in women as they are in men.

25.1 The Four Stages of Catabolism

The reactants required for all life processes ultimately come from our diet. In that sense, we really are what we eat. Catabolism can be divided into four stages (Figure 25.1). The *first stage of catabolism* is called digestion. In this stage, fats, carbohydrates, and proteins are hydrolyzed to fatty acids, monosaccharides, and amino acids, respectively. These reactions occur in the mouth, stomach, and small intestine.

> In the first stage of catabolism, fats, carbohydrates, and proteins are hydrolyzed to fatty acids, monosaccharides, and amino acids.

You are what you eat.

Figure 25.1 ▶
The four stages of catabolism:
1. digestion 2. conversion of fatty acids, monosaccharides, and amino acids to compounds that can enter the citric acid cycle 3. the citric acid cycle 4. oxidative phosphorylation

> Compounds that can enter the citric acid cycle are citric acid cycle intermediates, acetyl-CoA, and pyruvate.

In the *second stage of catabolism*, the products obtained from the first stage—fatty acids, monosaccharides, and amino acids—are converted to compounds that can enter the citric acid cycle. In order to enter the citric acid cycle, a compound must be either one of the compounds in the cycle itself—these are called citric acid cycle intermediates—or it must be acetyl-CoA or pyruvate.

Acetyl CoA is the only non–citric acid cycle intermediate that can enter the cycle; it enters by being converted to citrate, a citric acid cycle intermediate. Pyruvate enters the citric acid cycle by being converted to acetyl-CoA; this is the reaction catalyzed by the pyruvate dehydrogenase system discussed in Section 24.4.

The *third stage of catabolism* is the citric acid cycle. In the citric acid cycle, the acetyl group of each molecule of acetyl-CoA is converted to two molecules of CO_2.

The citric acid cycle is the third stage of catabolism.

Metabolic energy is measured in terms of adenosine 5'-triphosphate (ATP). Cells get the energy they need by using nutrient molecules to make ATP. Only a small amount of ATP is formed in the first three stages of catabolism. Most ATP is formed in the fourth stage of catabolism.

Metabolic energy is measured in terms of adenosine triphosphate (ATP).

We will see that many catabolic reactions are oxidation reactions. In the *fourth stage of catabolism*, every molecule of NADH formed in one of the earlier stages of catabolism (from NAD^+ being used to carry out an oxidation reaction) is converted into three molecules of ATP in a process called *oxidative phosphorylation*. Oxidative phosphorylation also converts every molecule of $FADH_2$ formed earlier (as a result of FAD carrying out an oxidation reaction) into two molecules of ATP. Thus, most of the energy (ATP) provided by fats, carbohydrates, and proteins is obtained in the fourth stage of catabolism.

Oxidative phosphorylation is the fourth stage of catabolism.

25.2 ATP Is the Carrier of Chemical Energy

All cells require energy to live and reproduce. They get the energy they need from nutrients that they convert into a chemically useful form. The most important form of chemical energy is **adenosine 5'-triphosphate (ATP)**. The importance of ATP to biological reactions is reflected in its turnover rate in humans: each day, a person uses an amount of ATP equivalent to his or her body weight.

When phosphoric acid is heated with P_2O_5 it loses water, forming a phospho-anhydride called pyrophosphoric acid. Its name comes from *pyr,* the Greek word for "fire," because pyrophosphoric acid is prepared by "fire"—that is, by heating. Triphosphoric acid and higher polyphosphoric acids are also formed. ATP is a monoester of triphosphoric acid.

ATP is the universal carrier of chemical energy because, as is commonly stated, "the energy released from hydrolysis of ATP converts endergonic reactions into exergonic reactions." In other words, the ability of ATP to enable otherwise unfavorable reactions to occur is attributed to the large amount of energy released when ATP is hydrolyzed, which can be used to drive endergonic reactions. For example, the reaction of glucose with hydrogen phosphate to form glucose-6-phosphate is endergonic ($\Delta G^{\circ\prime} = +3.3$ kcal/mol or $+13.8$ kJ/mol).* The hydrolysis of ATP, on the other hand, is highly exergonic ($\Delta G^{\circ\prime} = -7.3$ kcal/mol or -30.5 kJ/mol). When the two reactions are added together (the species occurring on both sides of the reaction arrow cancel), the net reaction is exergonic ($\Delta G^{\circ\prime} = -4.0$ kcal/mol or -16.7 kJ/mol). Thus, the

---

* The prime in $\Delta G^{\circ\prime}$ indicates that two additional parameters have been added to the ΔG° defined in Section 3.7: the reaction occurs in aqueous solution at pH = 7 and the concentration of water is assumed to be constant.

energy released from the hydrolysis of ATP is more than enough to drive the phosphorylation of glucose. Two reactions in which the energy of one is said to drive the other are called *coupled reactions*.

$$\Delta G°'$$

glucose + hydrogen phosphate $\longrightarrow$ glucose-6-phosphate + H_2O	+ 3.3 kcal/mol	or	+ 13.8 kJ/mol
ATP + H_2O $\longrightarrow$ ADP + hydrogen phosphate	− 7.3 kcal/mol	or	− 30.5 kJ/mol
glucose + ATP $\longrightarrow$ glucose-6-phosphate + ADP	− 4.0 kcal/mol	or	− 16.7 kJ/mol

This nonmechanistic description of ATP's power makes ATP seem like a bullet of energy that can be shot at an endothermic reaction. Let's look at the mechanism of the reaction to see what really happens. The reaction is a simple one-step nucleophilic substitution reaction. The 6-OH group of glucose attacks the terminal phosphate of ATP, breaking a **phosphoanhydride bond** without forming an intermediate. Essentially, it is a substitution reaction with an adenosine 5′-diphosphate (ADP) leaving group.

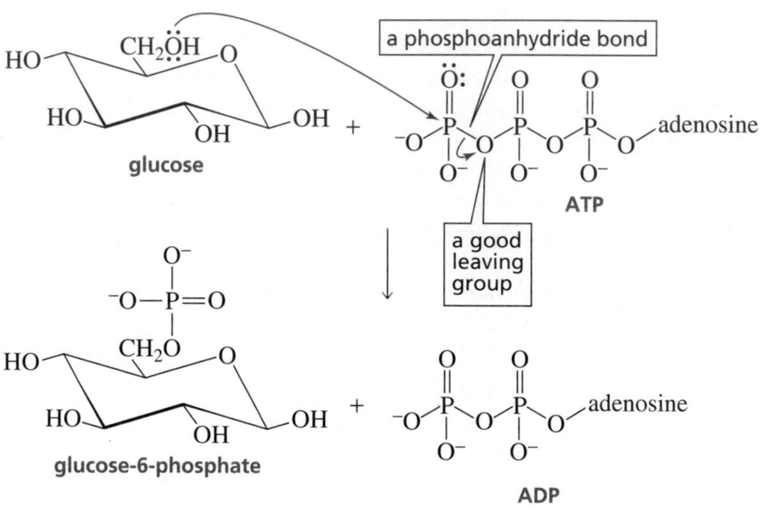

Now we have a chemical understanding of why the phosphorylation of glucose requires ATP. Without ATP, the 6-OH group of glucose would have to displace a very basic HO^- group from hydrogen phosphate. With ATP, the 6-OH group of glucose displaces the weakly basic ADP.

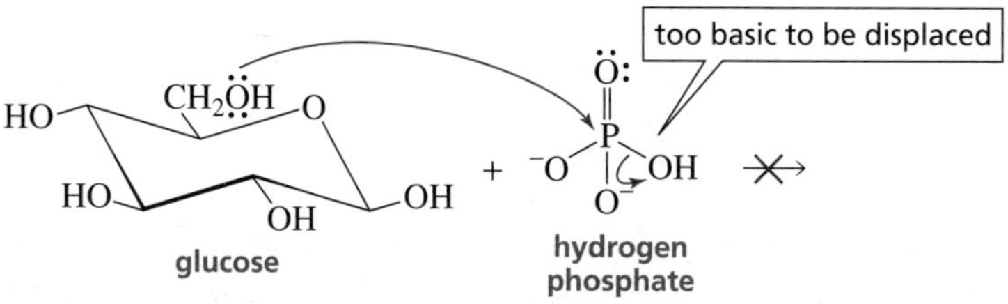

Although the reaction of glucose with hydrogen phosphate is described above as being driven by the "hydrolysis" of ATP, you can see from the mechanism that glucose does not react with hydrogen phosphate and ATP is not hydrolyzed because it does not react with water. In other words, neither of the coupled reactions actually occurs. Instead, the phosphoryl group of ATP is transferred directly to glucose.

The transfer of a phosphoryl group from ATP to glucose is an example of a **phosphoryl transfer reaction**. There are many phosphoryl transfer reactions in biological systems. In all of these reactions, an electrophilic phosphoryl group is transferred to a nucleophile as a result of breaking a phosphoanhydride bond. The above example of a phosphoryl transfer reaction demonstrates the actual chemical function of ATP: *it provides a reaction pathway involving a good leaving group for a reaction that cannot occur (or would occur very slowly) because of a poor leaving group.*

ATP provides a reaction pathway involving a good leaving group for a reaction that cannot occur because of a poor leaving group.

PROBLEM 1 **SOLVED**

The hydrolysis of phosphoenolpyruvate is so highly exergonic ($\Delta G^{\circ\prime} = -14.8$ kcal/mol or -61.9 kJ/mol) that it can be used to "drive the formation" of ATP from ADP and hydrogen phosphate ($\Delta G^{\circ\prime} = +7.3$ kcal/mol or $+30.5$ kJ/mol). Propose a mechanism for this reaction.

Solution As we saw in the example of phosphorylation of glucose by ATP, neither of the coupled reactions actually occurs: phosphoenolpyruvate does not react with water and ADP does not react with hydrogen phosphate. Just as ATP "drives the formation" of glucose-6-phosphate by supplying glucose (a nucleophile) with a compound that has a phosphoryl group attached to a good leaving group (ADP), phosphoenolpyruvate "drives the formation" of ATP by supplying ADP (a nucleophile) with a compound that has a phosphoryl group attached to a good leaving group (pyruvate).

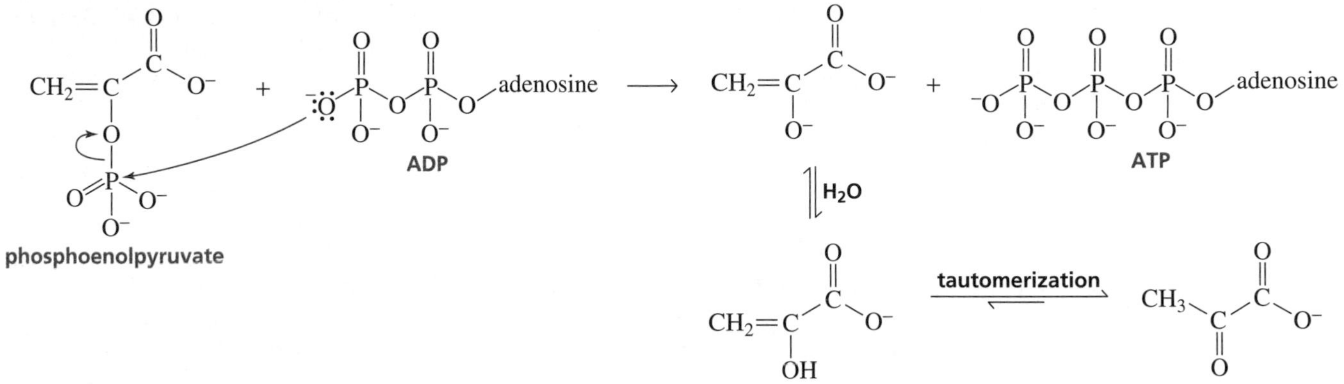

PROBLEM 2◆

Why is pyruvate a good leaving group?

PROBLEM 3◆

Several important biomolecules and the $\Delta G^{\circ\prime}$ values for their hydrolysis are listed here. Which of them "hydrolyzes" with sufficient energy to "drive the formation" of ATP?

glycerol-1-phosphate:
-2.2 kcal/mol (-9.2 kJ/mol)

phosphocreatine.
-11.8 kcal/mol (-49.4 kJ/mol)

fructose-6-phosphate:
-3.8 kcal/mol (-15.9 kJ/mol)

glucose-6-phosphate:
-3.3 kcal/mol (-13.8 kJ/mol)

25.3 **There Are Three Mechanisms for Phosphoryl Transfer Reactions**

There are three possible mechanisms for a phosphoryl transfer reaction. We will illustrate them using the following nucleophilic acyl substitution reaction:

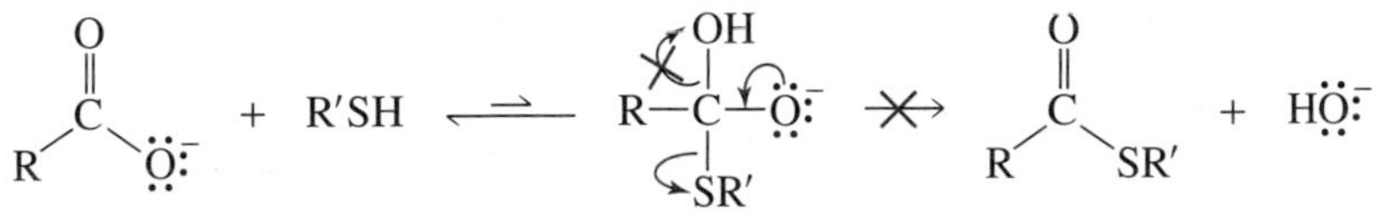

The above reaction does not occur because the negatively charged carboxylate ion resists nucleophilic attack, and even if the tetrahedral intermediate could be formed, the incoming nucleophile is a weaker base than the base that would have to be expelled from the tetrahedral intermediate to form the thioester. In other words, the thiol would be expelled from the tetrahedral intermediate, reforming the carboxylate ion (Section 16.5).

If ATP is added to the reaction mixture, the reaction occurs. The carboxylate ion attacks one of the phosphorus atoms of ATP, breaking a phosphoanhydride bond. This puts a leaving group on the carboxyl group that can be displaced by the thiol. There are three possible mechanisms for the reaction of a nucleophile with ATP because each of the three phosphorus atoms of ATP can undergo nucleophilic attack and each mechanism puts a different phosphoryl leaving group on the nucleophile.

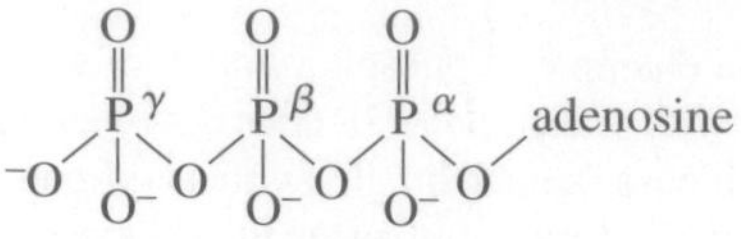

ATP

If the carboxylate ion attacks the γ-phosphorus of ATP, an **acyl phosphate** is formed. The acyl phosphate then reacts with the thiol in a nucleophilic acyl substitution reaction (Section 16.7) to form the thioester.

nucleophilic attack on the γ-phosphorus

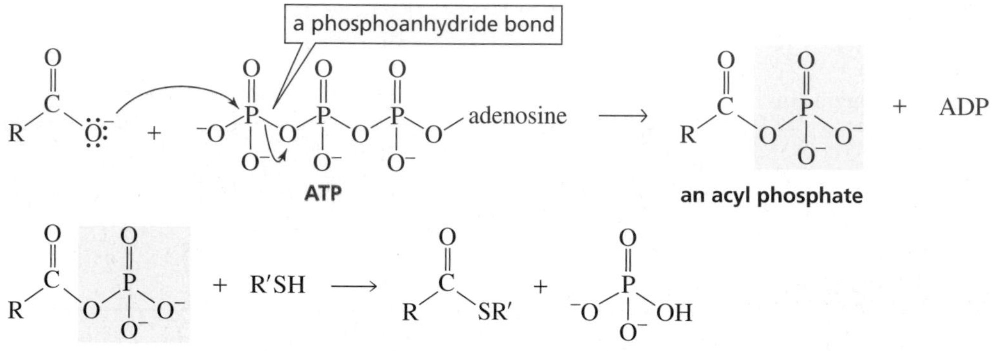

overall reaction

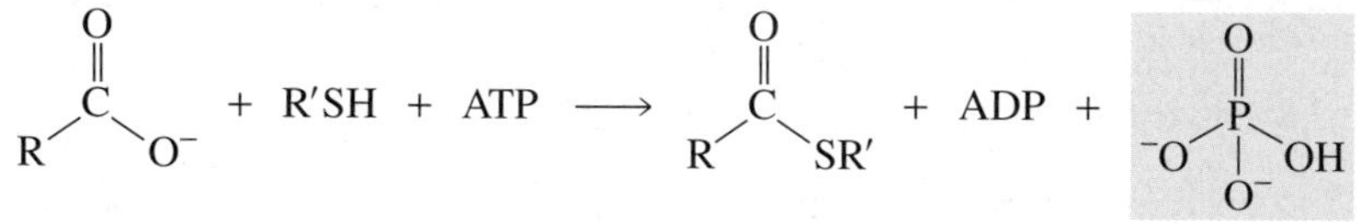

If the carboxylate ion attacks the β-phosphorus of ATP, an **acyl pyrophosphate** is formed (see Problem 4). The acyl pyrophosphate then reacts with the thiol in a nucleophilic acyl substitution reaction to form the thioester.

nucleophilic attack on the β-phosphorus

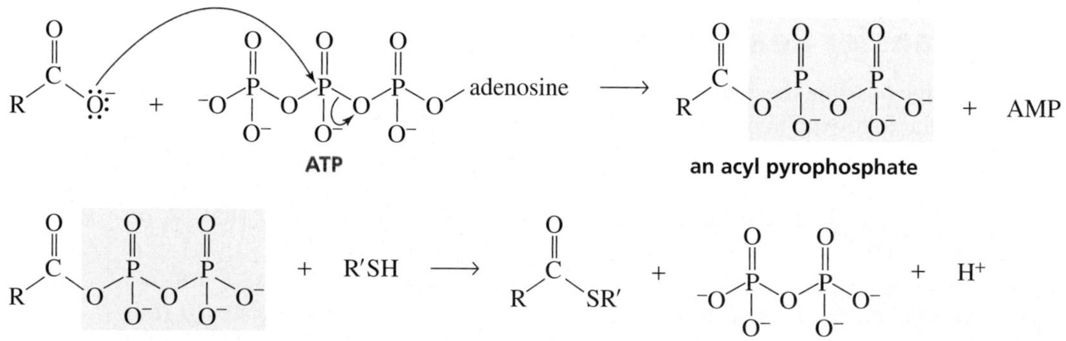

overall reaction

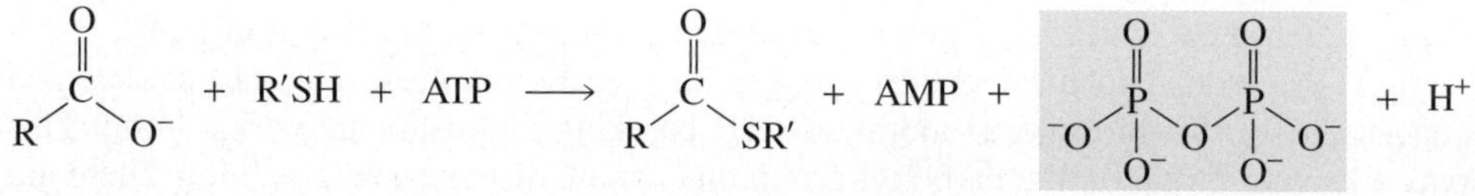

In the third possible mechanism, the carboxylate ion attacks the α-phosphorus of ATP, forming an **acyl adenylate**. The acyl adenylate then reacts with the thiol in a nucleophilic acyl substitution reaction to form the thioester.

nucleophilic attack on the α-phosphorus

an acyl adenylate

overall reaction

Notice that each of the three mechanisms puts a leaving group on the carboxylate ion that can easily be displaced by a nucleophile. The only difference in the three mechanisms is the particular phosphate atom that is attacked by the nucleophile (the carboxylate ion) and the nature of the intermediate that is formed.

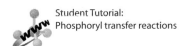

Student Tutorial:
Phosphoryl transfer reactions

Many different nucleophiles react with ATP in biological systems. Whether nucleophilic attack occurs on the α-, β-, or γ-phosphorus in any particular reaction depends on the enzyme catalyzing the reaction. Mechanisms involving nucleophilic attack on the γ-phosphorus form ADP and phosphate as side products, whereas mechanisms involving nucleophilic attack on the α- or β-phosphorus form adenosine 5'-monophosphate (AMP) and pyrophosphate as side products.

When pyrophosphate is one of the side products, it is subsequently hydrolyzed to two equivalents of phosphate. The hydrolysis drives the reaction to the right, ensuring its irreversibility.

pyrophosphate   phosphate

Therefore, enzyme-catalyzed reactions in which irreversibility is important take place by one of the mechanisms that form pyrophosphate as a product (attack on the α- or β-phosphorus of ATP). For example, both the reaction that links nucleotide subunits to form nucleic acids (Section 27.3) and the reaction that binds an amino acid to a tRNA (the first step in translating RNA into a protein; Section 27.8) involve nucleophilic attack on the α-phosphorus of ATP.

PROBLEM 4

The β-phosphorus of ATP has two phosphoanhydride linkages, but generally only the one linking the β-phosphorus to the α-phosphorus is broken in phosphoryl transfer reactions. Explain why the one linking the β-phosphorus to the γ-phosphorus is almost never broken.

25.4 The "High-Energy" Character of Phosphoanhydride Bonds

Because the hydrolysis of a phosphoanhydride bond is a highly exergonic reaction, phosphoanhydride bonds are called **high-energy bonds**. The term "high-energy" in this context means that a lot of energy is released when the bond breaks. Do not confuse it with "bond energy," the term chemists use to describe how difficult it is to break a bond. A bond with a *high bond energy* is hard to break, whereas a *high-energy bond* breaks readily (Section 3.7).

Why is the hydrolysis of a phosphoanhydride bond so exergonic? In other words, why is the $\Delta G^{\circ\prime}$ value for its hydrolysis large and negative? A large negative $\Delta G^{\circ\prime}$ means that the products of the reaction are much more stable than the reactants. Let's look at ATP and its hydrolysis products, phosphate and ADP, to see why this is so.

Three factors contribute to the greater stability of ADP and phosphate compared to ATP:

1. **Greater electrostatic repulsion in ATP.** At physiological pH (pH = 7.3), ATP has 3.3 negative charges (see Problem 5a), ADP has 2.8 negative charges, and phosphate has 1.8 negative charges (Section 1.24). Because of ATP's greater negative charge, more electrostatic repulsions are present in ATP than in ADP or phosphate. Electrostatic repulsions destabilize a molecule.

2. **More solvation stabilization in the products.** Negatively charged ions are stabilized in an aqueous solution by solvation (Section 2.9). Because the reactant has 3.3 negative charges, while the sum of the negative charges on the products is 3.9 (2.8 + 1.1), there is more solvation stabilization in the products than in the reactant.

3. **Greater electron delocalization in the products.** A lone pair on the oxygen joining the two phosphorus atoms is not effectively delocalized, because delocalization would put a positive charge on an oxygen that is next to a partially positively charged phosphorus atom. When the phosphoanhydride bond breaks, one additional lone pair can be effectively delocalized. Electron delocalization stabilizes a molecule (Section 7.6).

Similar factors explain the large negative $\Delta G^{\circ\prime}$ when ATP is hydrolyzed to AMP and pyrophosphate, and when pyrophosphate is hydrolyzed to two equivalents of phosphate.

PROBLEM 5 **SOLVED**

ATP has pK_a values of 0.9, 1.5, 2.3, and 7.7; ADP has pK_a values of 0.9, 2.8, and 6.8; and phosphoric acid has pK_a values of 1.9, 6.7, and 12.4.
Do the calculation showing that at pH 7.3

a. the charge on ATP is −3.3.

b. the charge on ADP is −2.8.

c. the charge on phosphate is −1.8.

Solution to 5a   Because pH 7.3 is much more basic than the pK_a values of the first three ionizations of ATP, we know that these three groups will be entirely in their basic forms at that pH, giving ATP three negative charges. We need to determine what fraction of the group with pK_a 7.7 will be in its basic form at pH 7.3.

$$\frac{\text{concentration of the basic form}}{\text{total concentration}} = \frac{[\text{A}^-]}{[\text{A}^-] + [\text{HA}]}$$

$$[\text{A}^-] = \text{concentration of the basic form}$$

$$[\text{HA}] = \text{concentration of the acidic form}$$

Because this equation has two unknowns, one of the unknowns must be expressed in terms of the other unknown. Using the definition of the acid dissociation constant (K_a), we can define [HA] in terms of [A$^-$], K_a, and [H$^+$].

$$K_a = \frac{[\text{A}^-][\text{H}^+]}{[\text{HA}]}$$

$$[\text{HA}] = \frac{[\text{A}^-][\text{H}^+]}{K_a}$$

$$\frac{[\text{A}^-]}{[\text{A}^-] + [\text{HA}]} = \frac{[\text{A}^-]}{[\text{A}^-] + \dfrac{[\text{A}^-][\text{H}^+]}{K_a}} = \frac{K_a}{K_a + [\text{H}^+]}$$

Now we can calculate the fraction of the group with pK_a 7.7 that will be in the basic form. (Note that K_a is calculated from pK_a and [H$^+$] is calculated from pH.)

$$\frac{K_a}{K_a + [\text{H}^+]} = \frac{2.0 \times 10^{-8}}{2.0 \times 10^{-8} + 5.0 \times 10^{-8}} = 0.3$$

Thus the total negative charge on ATP = 3.0 + 0.3 = 3.3.

25.5   **Why ATP Is Kinetically Stable in a Cell**

Although ATP reacts readily in enzyme-catalyzed reactions, it reacts quite slowly in the absence of an enzyme. For example, a carboxylic acid anhydride hydrolyzes in a matter of minutes, but ATP (a phosphoric acid anhydride) takes several weeks to hydrolyze. The low rate of ATP hydrolysis is important because it allows ATP to exist in the cell until it is needed for an enzyme-catalyzed reaction.

The negative charges on ATP are what make it relatively unreactive. These negative charges repel the approach of nucleophiles. When ATP is bound at an active site of an enzyme, it complexes with magnesium (Mg^{2+}), which decreases the overall negative charge on ATP. (This is why ATP-requiring enzymes also require Mg^{2+}; Section 24.5.) The other two negative charges can be stabilized by positively charged groups such as arginine or lysine side chains at the active site, as shown in Figure 25.2. In this form, ATP is readily approached by nucleophiles, so ATP reacts rapidly in an enzyme-catalyzed reaction, but only very slowly in the absence of an enzyme.

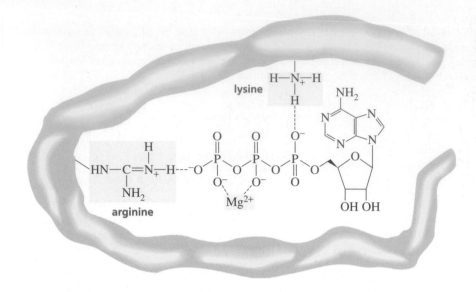

Figure 25.2 ▶
The interactions between ATP, Mg^{2+}, and arginine and lysine residues at the active site of an enzyme.

25.6 The Catabolism of Fats

We have seen that in the first two stages of catabolism, fats, carbohydrates, and proteins are converted into compounds that can enter the citric acid cycle (Section 25.1). Now we will look at the reactions that allow fats to enter the cycle.

In the first stage of fat catabolism, the fat's three ester groups are hydrolyzed by enzymes to glycerol and three fatty acid molecules.

$$\underset{\text{a fat}}{\begin{array}{c} CH_2O-\overset{\displaystyle O}{\overset{\|}{C}}-R^1 \\ | \\ CHO-\overset{\displaystyle O}{\overset{\|}{C}}-R^2 \\ | \\ CH_2O-\overset{\displaystyle O}{\overset{\|}{C}}-R^3 \end{array}} + 3\,H_2O \longrightarrow \underset{\text{glycerol}}{\begin{array}{c} CH_2OH \\ | \\ CHOH \\ | \\ CH_2OH \end{array}} + \underset{\text{fatty acids}}{\begin{array}{c} R^1-\overset{\displaystyle O}{\overset{\|}{C}}-OH \\ R^2-\overset{\displaystyle O}{\overset{\|}{C}}-OH \\ R^3-\overset{\displaystyle O}{\overset{\|}{C}}-OH \end{array}}$$

Glycerol reacts with ATP to form glycerol-3-phosphate in the same way that glucose reacts with ATP to form glucose-6-phosphate (Section 25.2). The enzyme that catalyzes this reaction is called glycerol kinase. A **kinase** is an enzyme that puts a phosphoryl group on its substrate; thus, glycerol kinase puts a phosphoryl group on glycerol. Notice that this ATP-requiring enzyme also requires Mg^{2+} (Section 25.5).

$$\underset{\text{glycerol}}{\begin{array}{c} CH_2OH \\ | \\ CHOH \\ | \\ CH_2OH \end{array}} \quad \overset{\textbf{ATP, Mg}^{2+} \quad \textbf{ADP}}{\underset{\substack{\textbf{glycerol} \\ \textbf{kinase}}}{\longrightarrow}} \quad \underset{\text{glycerol-3-phosphate}}{\begin{array}{c} CH_2OH \\ | \\ CHOH \\ | \\ CH_2OPO_3{}^{2-} \end{array}} \quad \overset{\textbf{NAD}^+\textbf{, Zn}^{2+} \quad \textbf{NADH, H}^+}{\underset{\substack{\textbf{glycerol} \\ \textbf{phosphate} \\ \textbf{dehydrogenase}}}{\longrightarrow}} \quad \underset{\substack{\text{dihydroxyacetone} \\ \text{phosphate}}}{\begin{array}{c} CH_2OH \\ | \\ C=O \\ | \\ CH_2OPO_3{}^{2-} \end{array}}$$

a phosphoryl group

The secondary alcohol group of glycerol-3-phosphate is then oxidized by NAD^+ to a ketone. The enzyme that catalyzes this reaction is called glycerol phosphate dehydrogenase. Recall that a **dehydrogenase** is an enzyme that oxidizes its substrate (Section 24.2). We have seen that when a substrate is oxidized by NAD^+, the substrate donates a hydride ion to the 4-position of NAD^+'s pyridine ring (Section 24.2). Zn^{2+} is a cofactor for the reaction; it increases the acidity of the secondary alcohol's proton by coordinating to the oxygen (Section 23.5).

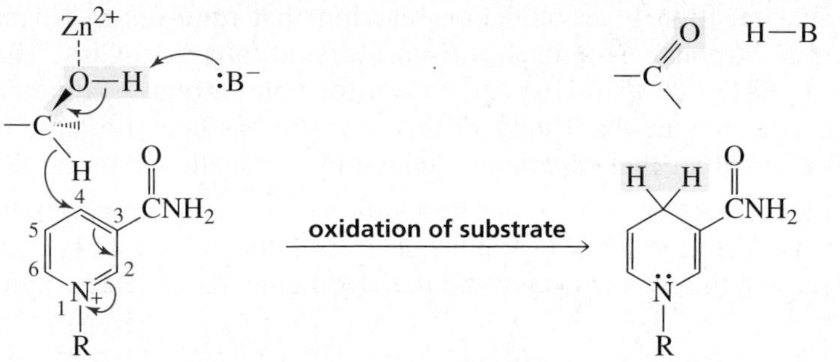

The product of the reaction, dihydroxyacetone phosphate, is one of the compounds in the glycolytic pathway, so it can enter that pathway and be broken down further (Section 25.7).

Notice that when biochemical reactions are written, the only structures shown are those of the primary reactant and primary product. The names of other reactants and products are abbreviated and placed on a curved arrow that intersects the reaction arrow.

PROBLEM 6

Show the mechanism for the reaction of glycerol with ATP to form glycerol-3-phosphate.

PROBLEM 7

The asymmetric center of glycerol-3-phosphate has the R configuration. Draw the structure of (R)-glycerol-3-phosphate.

Before a fatty acid can be metabolized, it must be activated. We have seen that one way a carboxylic acid is activated in biological systems is by being converted into a thioester (Section 16.22). Here the *fatty acid* is activated by being converted into a *fatty acyl-CoA.*

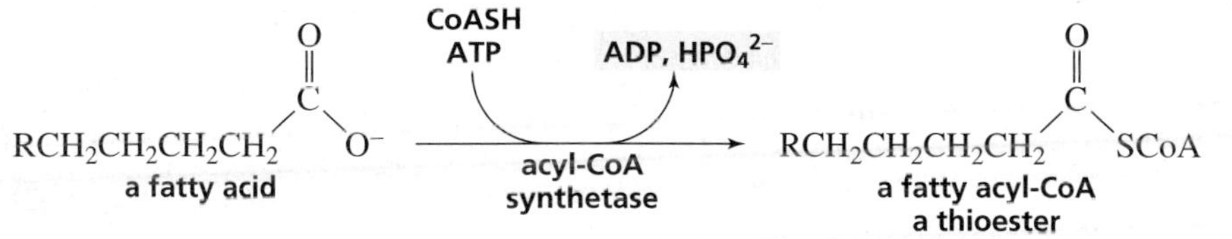

The fatty acyl-CoA is converted to acetyl-CoA in a repeating pathway called **β-oxidation**—a series of four reactions. Each passage through the four reactions removes two carbons from the fatty acyl-CoA by converting them into acetyl-CoA (Figure 25.3). Each of the four reactions is catalyzed by a different enzyme.

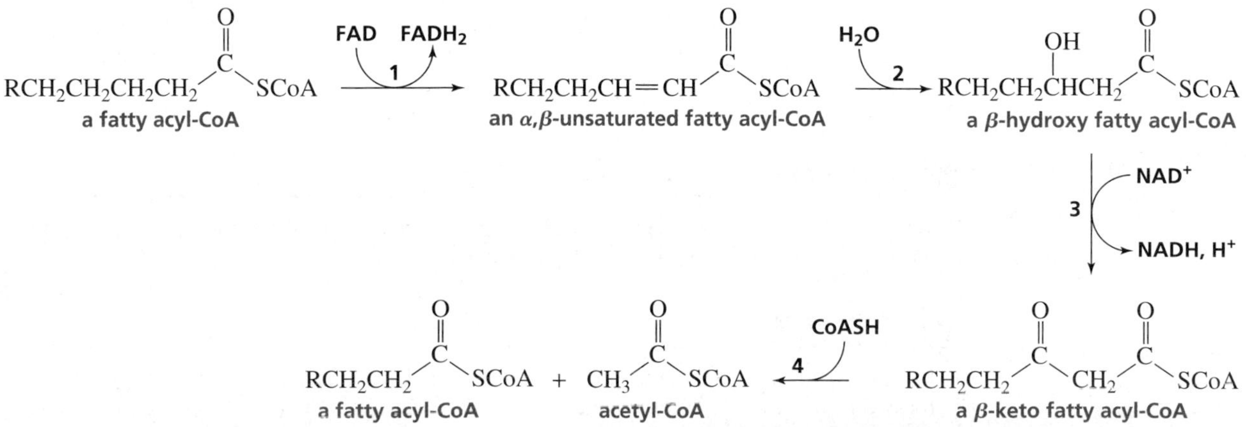

▲ **Figure 25.3**
In **β-oxidation** a series of four enzyme-catalyzed reactions is repeated until the entire fatty acyl-CoA molecule has been converted to acetyl-CoA molecules. The enzymes that catalyze the reactions are:
1. acyl-CoA dehydrogenase
2. enoyl-CoA hydratase
3. 3-L-hydroxyacyl-CoA dehydrogenase
4. β-ketoacyl-CoA thiolase

1. The first reaction is an oxidation reaction that removes hydrogen from the α- and β-carbons, forming an α,β-unsaturated fatty acyl-CoA. The oxidizing agent is FAD, the oxidizing agent used for non-carbonyl oxidations (Section 24.2). The enzyme that catalyzes this reaction has been found to be deficient in 10% of babies that experience sudden infant death syndrome (SIDS).

2. The second reaction is the conjugate addition of water to the α,β-unsaturated fatty acyl-CoA (Section 17.18). A glutamate side chain (Glu 144) acts as a base catalyst, and the enolate ion is protonated by a glutamic acid side chain (Glu 141).

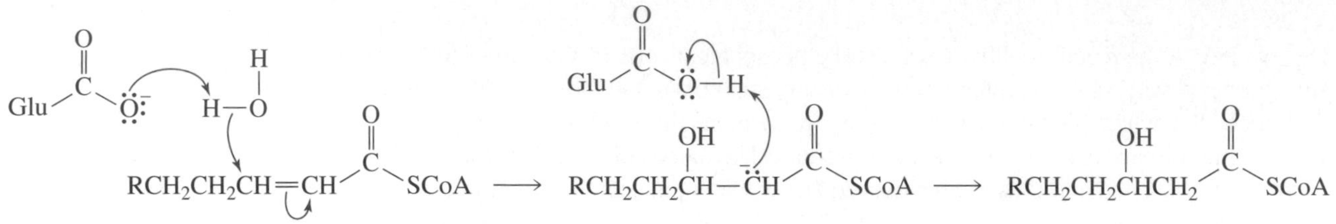

3. The third reaction is another oxidation reaction: NAD^+ oxidizes the secondary alcohol to a ketone.

4. The fourth reaction is the reverse of a Claisen condensation (Section 18.15), followed by conversion of the enol tautomer to the keto tautomer (Section 18.3). The final product is acetyl-CoA and a fatty acyl-CoA with two fewer carbons than the starting fatty acyl-CoA. The mechanism for this reaction is shown below.

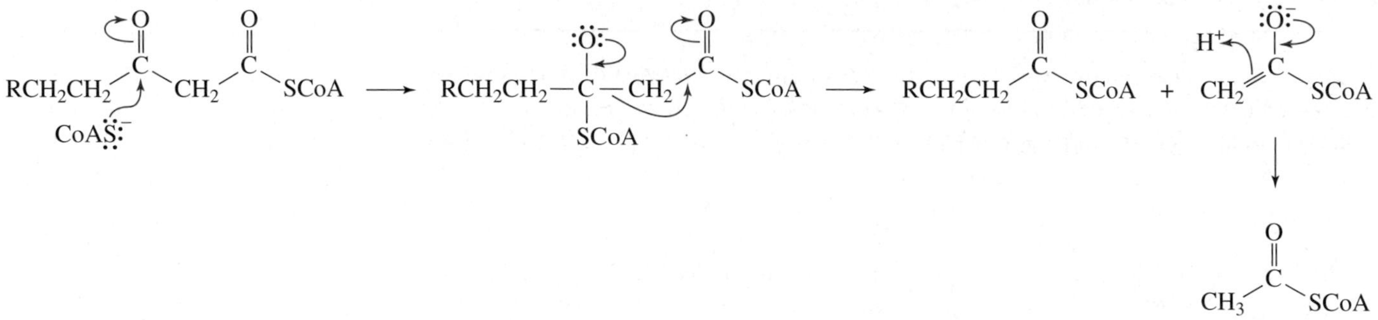

Fatty acids are converted to molecules of acetyl-CoA.

The four reactions are repeated, forming another molecule of acetyl-CoA and a fatty acyl-CoA that is now four carbons shorter than it was originally. Each time the series of four reactions is repeated, two more carbons are removed (as acetyl-CoA) from the fatty acyl-CoA. The series of reactions is repeated until the entire fatty acid has been converted into acetyl-CoA molecules.

We will see that acetyl-CoA enters the citric acid cycle by reacting with oxaloacetate (a citric acid cycle intermediate) to form citrate, another citric acid cycle intermediate (Section 25.6).

PROBLEM 8◆

Why does the OH group add to the β-carbon rather than to the α-carbon in the second reaction in the catabolism of fats? (*Hint:* See Section 17.17.)

PROBLEM 9◆

Palmitic acid is a 16-carbon saturated fatty acid. How many moles of acetyl-CoA are formed from 1 mol of palmitic acid?

PROBLEM 10◆

How many moles of NADH are formed from the β-oxidation of 1 mol of palmitic acid?

25.7 The Catabolism of Carbohydrates

In the first stage of carbohydrate catabolism, the acetal groups that hold glucose sub-units together are hydrolyzed, forming individual glucose molecules (Section 21.17).

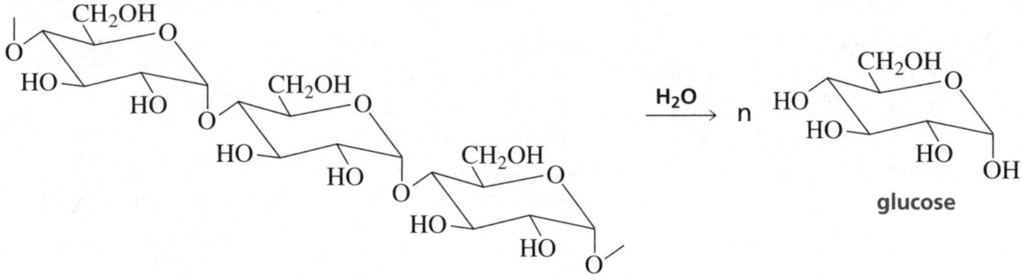

glucose

Each glucose molecule is converted to two molecules of pyruvate in a series of 10 reactions known as **glycolysis** or the *glycolytic pathway* (Figure 25.4).

Glucose is converted to two molecules of pyruvate.

1. In the first reaction, glucose is converted to glucose-6-phosphate, a reaction we just looked at in Section 25.2.

2. Glucose-6-phosphate then isomerizes to fructose-6-phosphate, a reaction whose mechanism we examined in Section 23.9.

3. In the third reaction, ATP puts a second phosphoryl group on fructose-6-phosphate. The product of the reaction is fructose-1,6-diphosphate.

4. The fourth reaction is the reverse of an aldol addition reaction. We looked at the mechanism of this reaction in Section 23.9.

5. Dihydroxyacetone phosphate, produced in the fourth reaction, is converted into glyceraldehyde-3-phosphate by forming an enediol that then forms glyceraldehyde-3-phosphate (if the OH group at C-1 ketonizes) or reforms dihydroxyacetone phosphate (if the OH group at C-2 ketonizes).

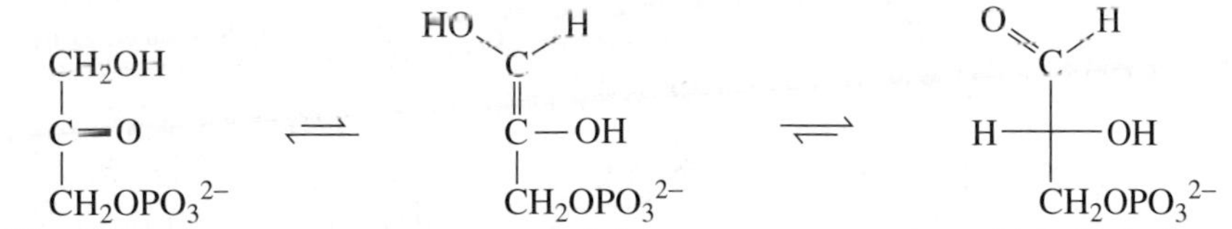

dihydroxyacetone phosphate   an enediol   glyceraldehyde-3-phosphate

A glutamate side chain is the base that removes a proton from the α-carbon, and a protonated histidine donates a proton to the carbonyl oxygen. In the second step, the histidine removes a proton from the C-1 OH group, and a glutamic acid protonates the carbon. Compare this mechanism with the enediol rearrangement shown in Section 21.5.

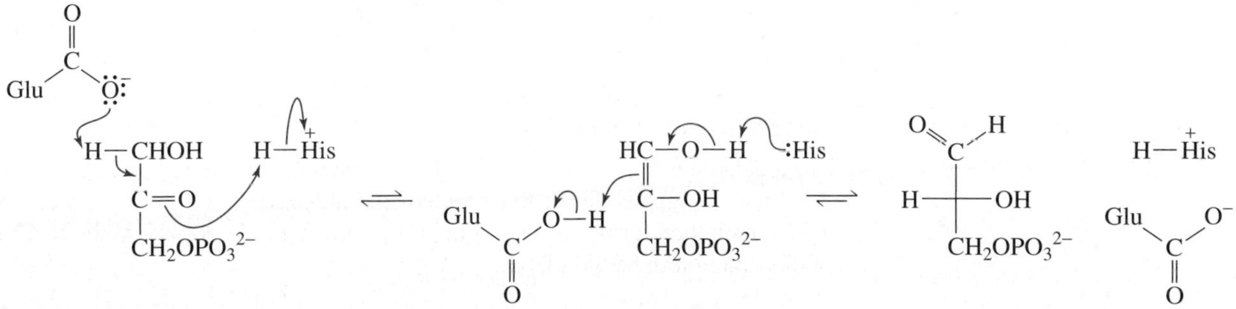

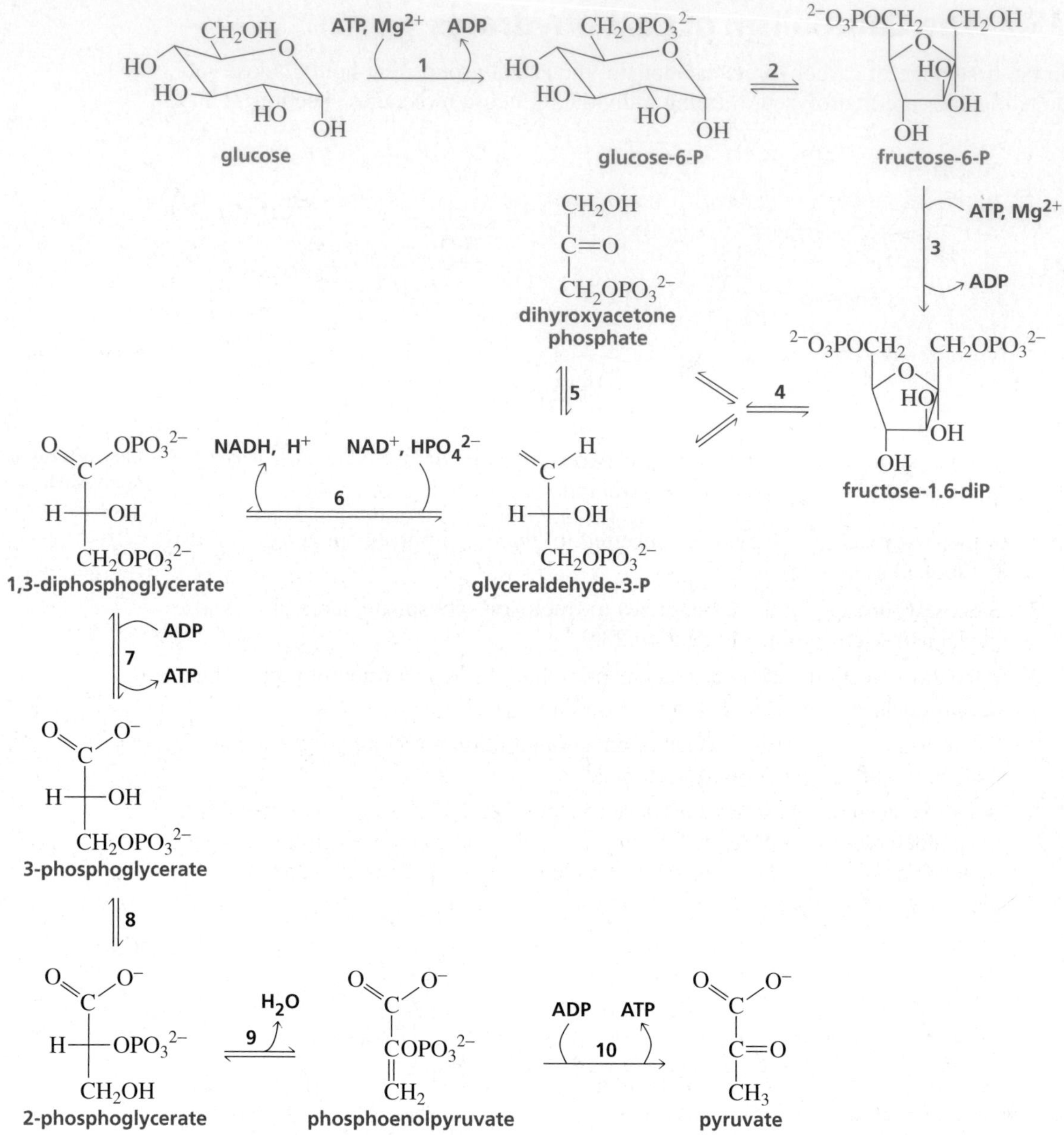

▲ **Figure 25.4**

Glycolysis, the series of enzyme-catalyzed reactions responsible for converting 1 mol of glucose into 2 mol of pyruvate. The enzymes that catalyze the reactions are:

1. hexokinase
2. phosphoglucose isomerase
3. phosphofructokinase
4. aldolase
5. triosephosphate isomerase
6. glyceraldehyde-3-phosphate dehydrogenase
7. phosphoglycerate kinase
8. phosphoglycerate mutase
9. enolase
10. pyruvate kinase

Thus, overall, each molecule of glucose is converted to two molecules of glyceraldehyde-3-phosphate.

6. The aldehyde group of glyceraldehyde-3-phosphate is oxidized by NAD^+, forming 1,3-diphosphoglycerate. In this reaction the aldehyde is oxidized to a carboxylic acid, which then forms an ester with phosphoric acid. We looked at the mechanism of this reaction in Section 24.2.

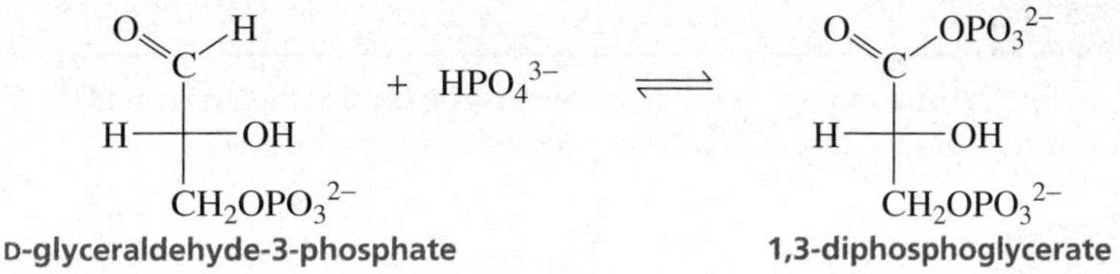

D-glyceraldehyde-3-phosphate + HPO_4^{3-} ⇌ 1,3-diphosphoglycerate

7. In the seventh reaction, 1,3-diphosphoglycerate transfers a phosphoryl group to ADP, thereby forming ATP and 3-phosphoglycerate.

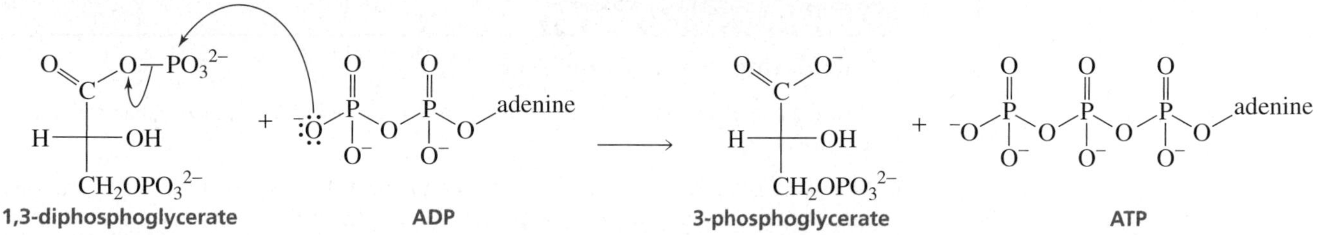

1,3-diphosphoglycerate + ADP → 3-phosphoglycerate + ATP

8. The eighth reaction is an isomerization: 3-phosphoglycerate is converted to 2-phosphoglycerate. The enzyme that catalyzes this reaction has a phosphoryl group on one of its amino acid side chains that it transfers to the 2-position of 3-phosphoglycerate to form an intermediate with two phosphoryl groups. The intermediate transfers the phosphoryl group on its 3-position back to the enzyme.

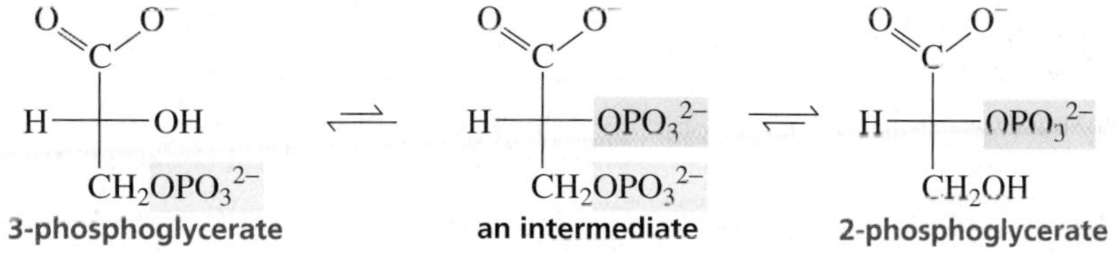

3-phosphoglycerate ⇌ an intermediate ⇌ 2-phosphoglycerate

9. The ninth reaction is a dehydration reaction (an E2 elimination reaction) that forms phosphoenolpyruvate. A lysine side chain at the active site of the enzyme is the base that removes a proton from the α-carbon (Section 18.1). The HO$^-$ group is protonated by a glutamic acid side chain, which makes it a better leaving group (Section 10.1).

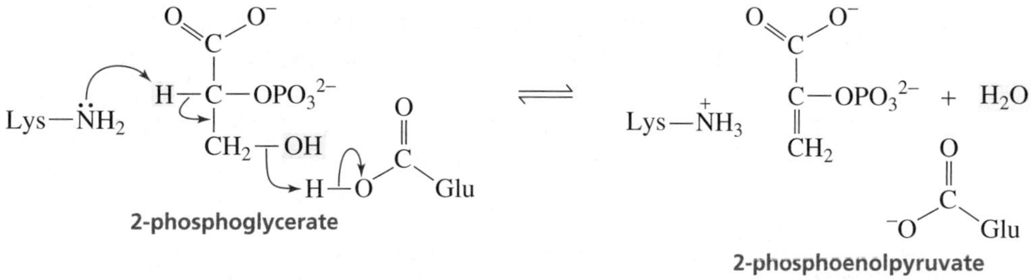

2-phosphoglycerate ⇌ 2-phosphoenolpyruvate + H_2O

10. In the last reaction of the glycolytic pathway, phosphoenolpyruvate transfers its phosphoryl group to ADP, forming ATP and pyruvate (see Problem 1 on page 1141).

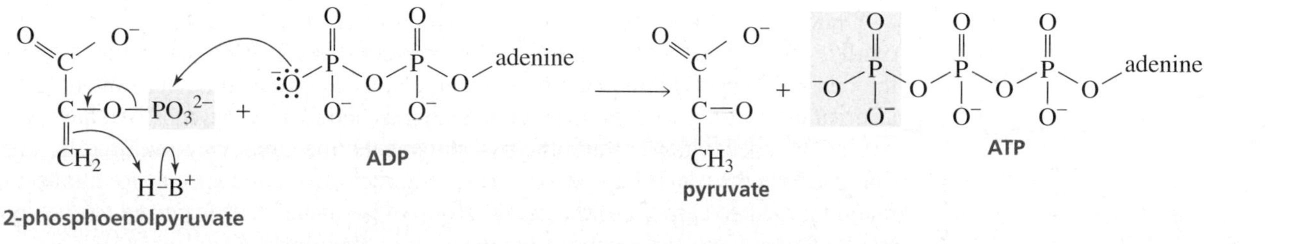

2-phosphoenolpyruvate + ADP → pyruvate + ATP

PROBLEM 11

Propose a mechanism for the third reaction in glycolysis: the formation of fructose-1,6-diphosphate from the reaction of fructose-6-phosphate with ATP.

PROBLEM 12◆

a. Which steps in glycolysis require ATP?   **b.** Which steps in glycolysis produce ATP?

PROBLEM-SOLVING STRATEGY

Calculating Production of ATP

How many molecules of ATP are obtained from each molecule of glucose that is metabolized to pyruvate?

To solve this kind of problem, we first need to count the number of ATPs that are used (in this case, to convert glucose to pyruvate). We see that two are used: one to form glucose-1-phosphate, and the other to form fructose-1,6-diphosphate. Next, we need to know many ATPs are formed. Each glyceraldehyde-3-phosphate that is metabolized to pyruvate forms two ATPs. Because each glucose molecule forms two molecules of glyceraldehyde-3-phosphate, 4 ATPs are formed from each molecule of glucose. Subtracting the molecules used, we find that each molecule of glucose metabolized to pyruvate forms two molecules of ATP.

Now continue on to Problem 13.

PROBLEM 13◆

How many moles of NAD^+ are required to convert 1 mol of glucose to pyruvate?

25.8 The Fates of Pyruvate

We have just seen that NAD^+ is used as an oxidizing agent in glycolysis. If glycolysis is to continue, the NADH produced as a result of that oxidation reaction has to be oxidized back to NAD^+ so NAD^+ will continue to be available as an oxidizing agent.

If oxygen is present, it is the oxidizing agent used to oxidize NADH back to NAD^+; this happens *in the fourth stage of catabolism*. If oxygen is not present—as occurs, for example in muscle cells when intense muscle activity causes all the oxygen to be depleted—pyruvate (the product of glycolysis) is used to oxidize NADH back to NAD^+. In the process, pyruvate is reduced to lactate (lactic acid). The acidic conditions caused by a buildup of lactic acid in muscles are responsible for the burning sensation you may have experienced when exercising.

pyruvate → lactate
(NADH, H⁺ → NAD⁺, lactate dehydrogenase)

Under normal (aerobic) conditions, when oxygen rather than pyruvate is used to oxidize NADH to NAD^+, pyruvate is converted to acetyl-CoA, which then enters the citric acid cycle. This reaction is catalyzed by the pyruvate dehydrogenase system, a series of reactions that requires three enzymes and five coenzymes. We have seen that one of the coenzymes is thiamine pyrophosphate, the coenzyme required by enzymes that catalyze the transfer of a two-carbon fragment from one species to another; in this reaction, the two carbons of the acetyl group of pyruvate are transferred to coenzyme A. We looked at the mechanism of this reaction in Section 24.4.

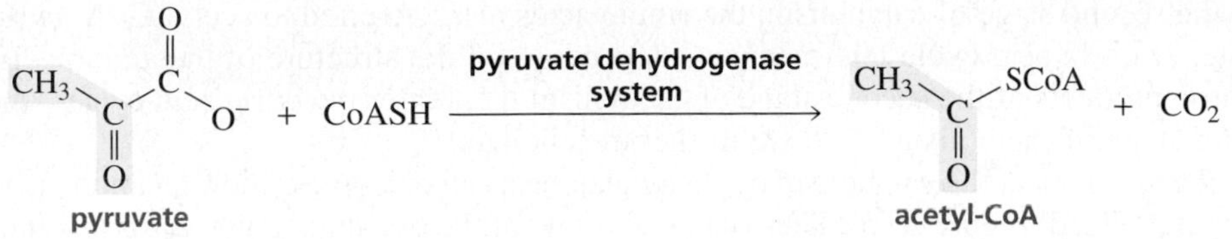

We have just seen that under anaerobic (no oxygen) conditions, pyruvate is reduced to lactate. Under these same conditions in yeast, however, pyruvate has a different fate: it is decarboxylated to acetaldehyde by pyruvate decarboxylase, a TPP-dependent (recall that TPP = thiamine pyrophosphate; Section 24.4) enzyme (that is, an enzyme that requires thiamine pyrophosphate) whose mechanism we looked at in Section 24.4. In this case, acetaldehyde is the compound that oxidizes NADH back to NAD^+ and in the process is reduced to ethanol—a reaction that has been used by humankind for thousands of years to produce wine, beer, and other fermented drinks.

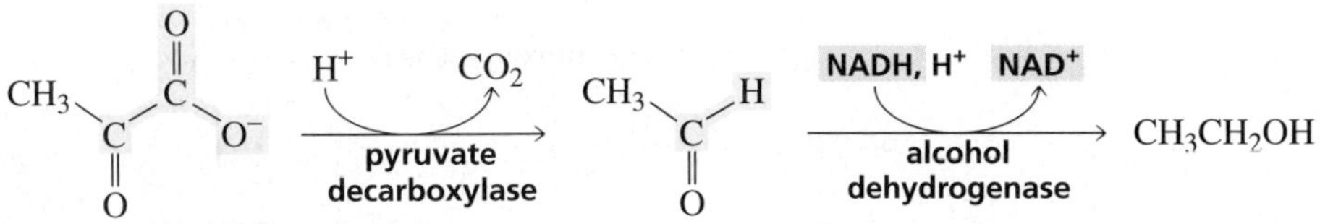

PROBLEM 14◆

What functional group of pyruvate is reduced when pyruvate is converted to lactate?

PROBLEM 15◆

What coenzyme is required to convert pyruvate to acetaldehyde?

PROBLEM 16◆

What functional group of acetaldehyde is reduced when acetaldehyde is converted to ethanol?

PROBLEM 17

Propose a mechanism for the reduction of acetaldehyde by NADH to ethanol. (*Hint:* See Section 24.2.)

25.9 The Catabolism of Proteins

In the first stage of protein catabolism, proteins are hydrolyzed to amino acids.

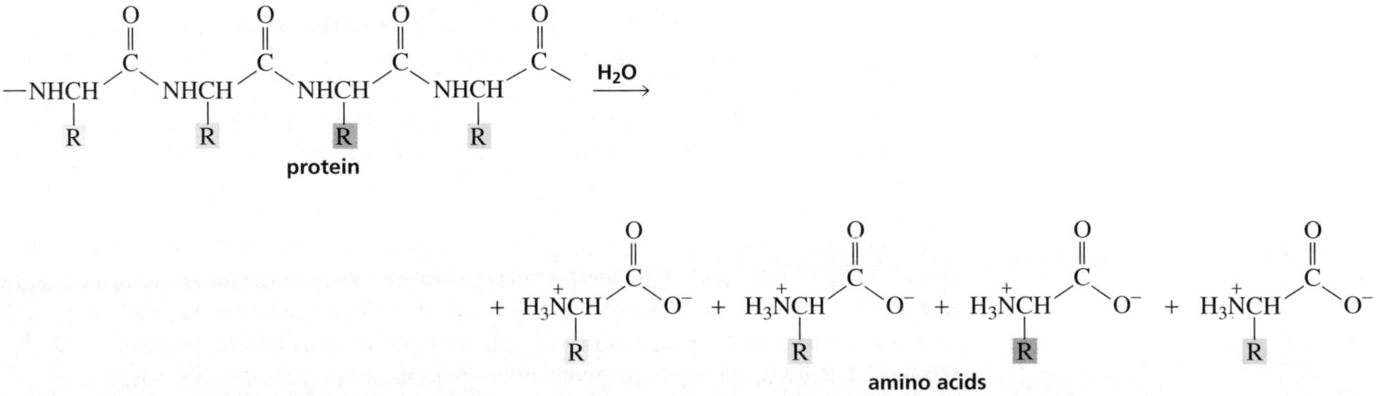

Amino acids are converted to acetyl-CoA, pyruvate, or citric acid cycle intermediates.

In the second stage of catabolism, the amino acids are converted to acetyl-CoA, pyruvate, or citric acid cycle intermediates, depending on the structure of the amino acid. These products of the second stage of catabolism then enter the citric acid cycle—the third stage of catabolism—and are further metabolized.

We will look at the catabolism of phenylalanine as an example of how an amino acid is metabolized (Figure 25.5). Phenylalanine is one of the essential amino acids and thus must be included in our diet (Section 22.1). The enzyme phenylalanine hydroxylase converts phenylalanine into tyrosine. Thus, tyrosine is not technically an essential amino acid, but if our diet lacks phenylalanine, we will also be deficient in tyrosine.

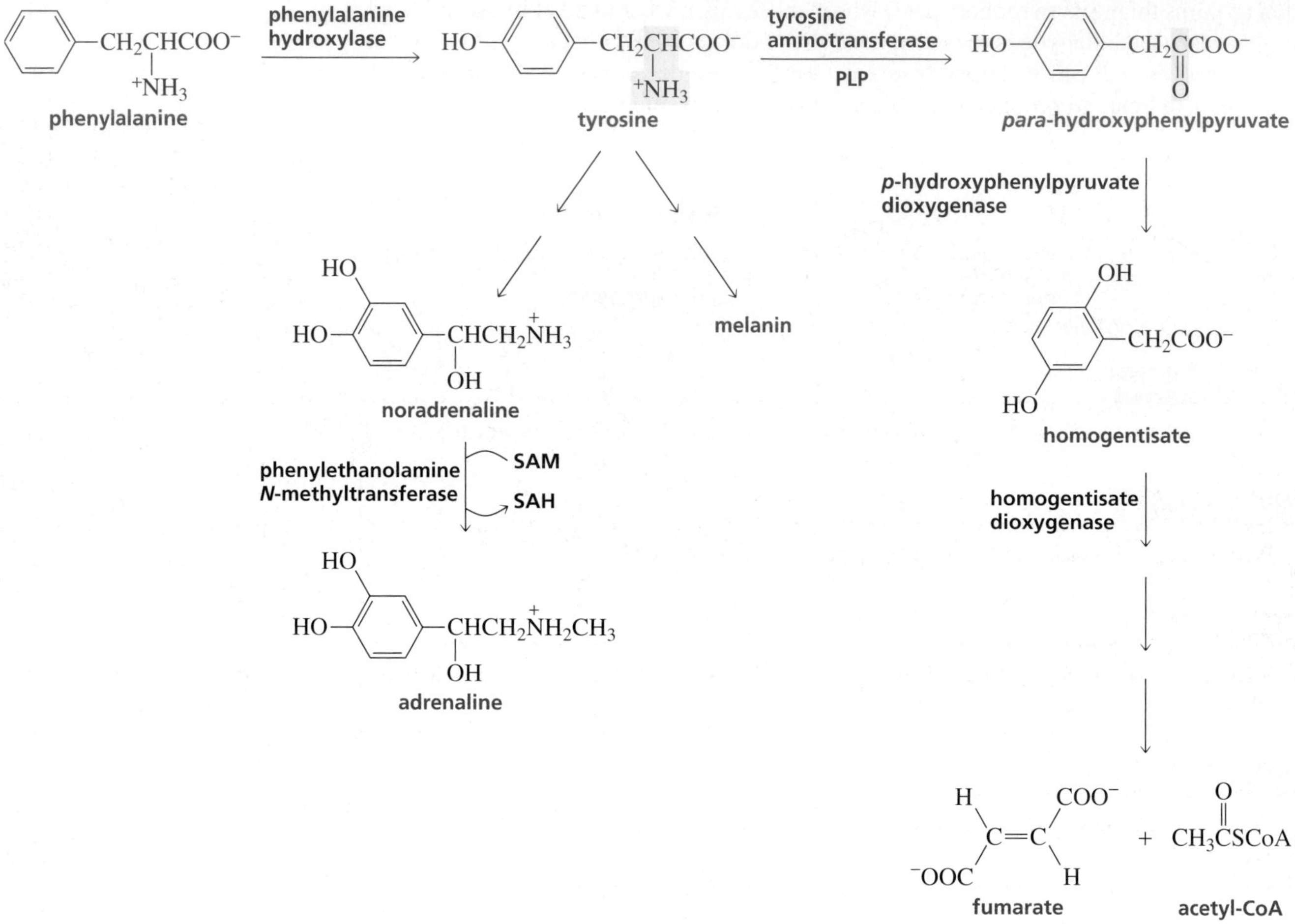

▲ **Figure 25.5**
The catabolism of phenylalanine.

The first reaction in the catabolism of most amino acids is transamination, a PLP-dependent reaction (recall PLP = pyridoxal pyrophosphate; Section 24.6) that replaces the amino group of the amino acid with a ketone group. *para*-Hydroxyphenylpyruvate, the product of transamination of tyrosine, is converted by a series of reactions to fumarate and acetyl-CoA. Fumarate is a citric acid cycle intermediate, so it can enter the citric acid cycle directly, and acetyl-CoA gets into the cycle by reacting with oxaloacetate to form citrate (Section 25.10). Each of the reactions in this catabolic pathway is catalyzed by a different enzyme.

In addition to being used for energy, the amino acids we ingest are also used for the synthesis of proteins and other compounds the body needs. For example, tyrosine is used to synthesize neurotransmitters (dopamine and adrenaline) and melanin, the compound responsible for skin pigmentation. Recall that SAM (*S*-adenosylmethionine) is the biological methylating agent that converts noradrenalin to adrenalin; Section 8.12.

PHENYLKETONURIA: AN INBORN ERROR OF METABOLISM

About one in every 20,000 babies is born without phenylalanine hydroxylase, the enzyme that converts phenylalanine into tyrosine. This genetic disease is called phenylketonuria (PKU). Without phenylalanine hydroxylase, the level of phenylalanine builds up; when it reaches a high concentration, it is transaminated to phenylpyruvate. The high level of phenylpyruvate found in urine gives the disease its name.

$$\text{phenylalanine} \quad \overset{\text{transamination}}{\longrightarrow} \quad \text{phenylpyruvate}$$

Within 24-hours after birth, all babies born in the United States are tested for high serum phenylalanine levels, which indicate a buildup of phenylalanine caused by an absence of phenylalanine hydroxylase. Babies with high levels are immediately put on a diet low in phenylalanine and high in tyrosine. As long as the phenylalanine level is kept under careful control for the first 5 to 10 years of life, the child will experience no adverse effects. You may have noticed the warning on containers of foods that contain Nutrasweet, announcing that it contains phenylalanine. (Recall that this sweetner is a methyl ester of a dipeptide of L-aspartate and L-phenylalanine; Section 22.9).

If the diet is not controlled, however, the baby will be severely mentally retarded by the time he or she is a few months old. Untreated children have paler skin and fairer hair than other members of their family because, without tyrosine, they cannot synthesize melanin, the black skin pigment. Half of untreated phenylketonurics are dead by age 20. When a woman with PKU becomes pregnant, she must return to the low phenylalanine diet she had as a child, because a high level of phenylalanine can cause abnormal development of the fetus.

ALCAPTONURIA

Another genetic disease that results from a deficiency of an enzyme in the pathway for phenylalanine degradation is alcaptonuria, which is caused by lack of homogentisate dioxygenase. The only ill effect of this enzyme deficiency is black urine. The reason the urine of those afflicted with alcaptonuria turns black is that the homogentisate they excrete immediately oxidizes in the air, forming a black compound.

PROBLEM 18◆

What coenzyme is required for transamination?

PROBLEM 19◆

When the amino acid alanine undergoes transamination, what compound is formed?

25.10 The Citric Acid Cycle

The **citric acid cycle** is the third stage of catabolism. In this series of eight reactions, the acetyl group of each molecule of acetyl-CoA—formed by the catabolism of fats, carbohydrates, and amino acids—is converted into two molecules of CO_2 (Figure 25.6).

$$\longrightarrow \quad 2\,CO_2 \ + \ CoASH$$

The acetyl group of each molecule of acetyl-CoA that enters the citric acid cycle is converted to two molecules of CO_2.

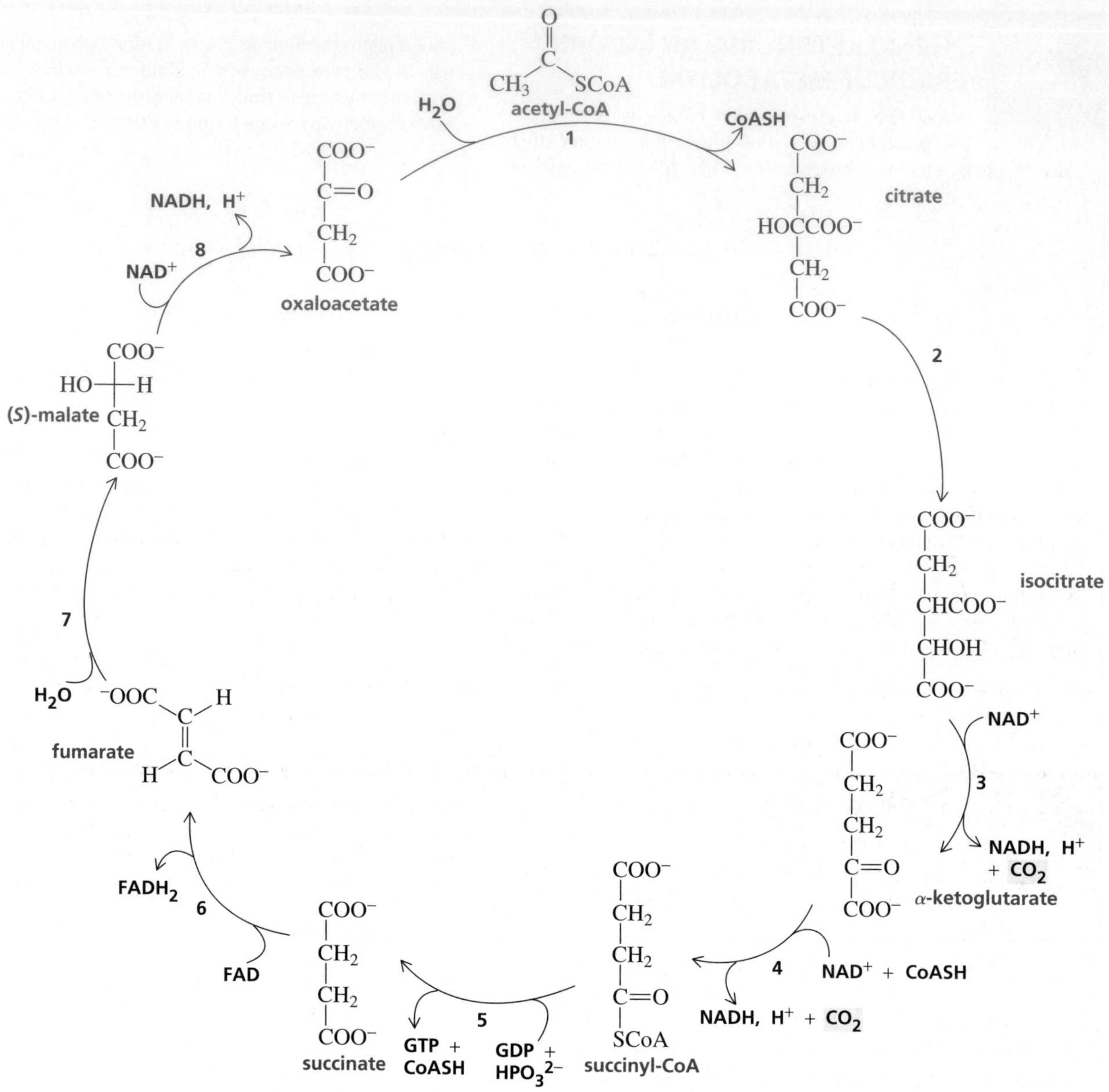

▲ **Figure 25.6**
The citric acid cycle, the series of enzyme-catalyzed reactions responsible for the oxidation of the acetyl group of acetyl-CoA to 2 molecules of CO_2. The enzymes that catalyze the reactions are:

1. citrate synthase
2. aconitase
3. isocitrate dehydrogenase
4. α-ketoglutarate dehydrogenase
5. succinyl-CoA synthetase
6. succinate dehydrogenase
7. fumarase
8. malate dehydrogenase

The series of reactions is called a *cycle* because they comprise a closed loop in which the product of the eighth reaction (oxaloacetate) is the reactant for the first reaction.

1. In the first reaction, acetyl-CoA reacts with oxaloacetate to form citrate. The mechanism of the reaction shows that an aspartate side chain of the enzyme removes a proton from the α-carbon of acetyl-CoA, creating a nucleophile that attacks the keto carbonyl carbon of oxaloacetate. The carbonyl oxygen picks up a proton from a histidine side chain. The intermediate that results is hydrolyzed to citrate in a nucleophilic acyl substitution reaction (Section 16.7).

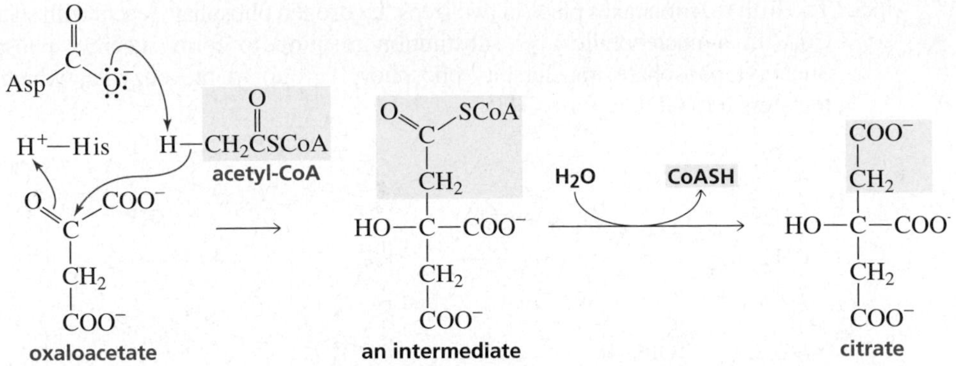

2. In the second reaction, citrate is converted to isocitrate, its isomer. The reaction takes place in two steps. The first step is a dehydration reaction (Section 10.4); a serine side chain of the enzyme removes a proton and the OH leaving group is protonated by a histidine side chain to make it a weaker base (H_2O) and therefore a better leaving group. In the second step, conjugate addition of water to the intermediate forms isocitrate (Section 17.17).

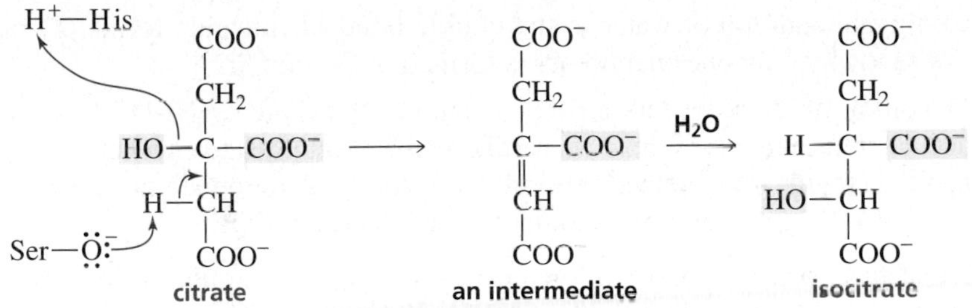

3. The third reaction also has two steps. In the first step, the secondary alcohol group of isocitrate is oxidized to a ketone by NAD^+. In the second step, the ketone loses CO_2. We have seen that a COO^- group that is bonded to a carbon adjacent to a carbonyl carbon can be removed because the electrons left behind can be delocalized onto the carbonyl oxygen (Section 18.18).

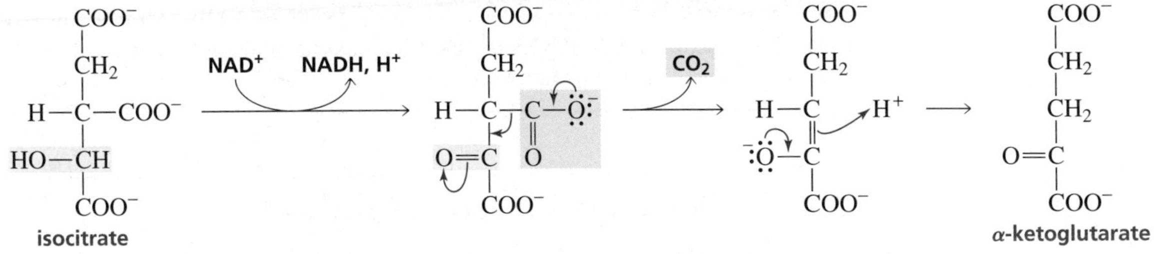

4. In the fourth reaction, NAD^+ is again the oxidizing agent. This is the reaction that releases the second molecule of CO_2. It requires a group of enzymes and the same five coenzymes required by the pyruvate dehydrogenase system that forms acetyl-CoA (Section 24.4). The product of the reaction is succinyl-CoA.

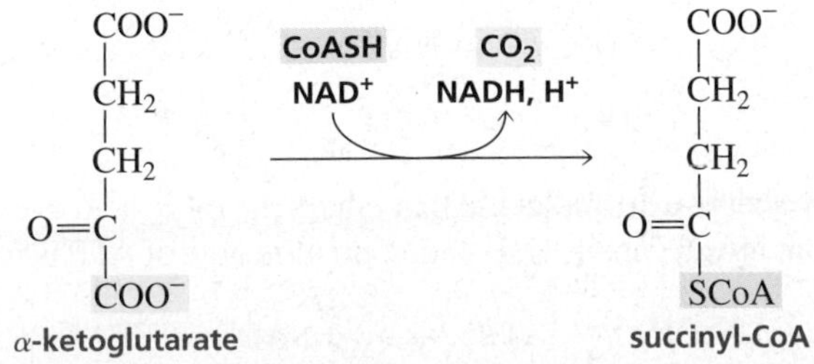

5. The fifth reaction takes place in two steps. Hydrogen phosphate reacts with succinyl-CoA in a nucleophilic acyl substitution reaction to form succinyl phosphate. Succinyl phosphate transfers its phosphoryl group to the enzyme, which then transfers it to GDP to form GTP.

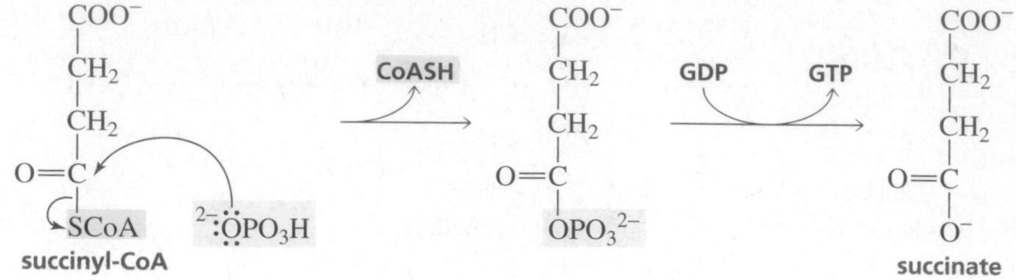

GTP transfers a phosphoryl group to ADP to from ATP.

$$\text{GTP} + \text{ADP} \rightleftharpoons \text{GDP} + \text{ATP}$$

6. In the sixth reaction, FAD oxidizes succinate to fumarate. We looked at the mechanism of this reaction in Section 24.3.

7. Conjugate addition of water to the double bond of fumarate forms (S)-malate. We saw why only one enantiomer is formed in Section 5.20.

8. Oxidation of the secondary alcohol group of (S)-malate by NAD^+ forms oxaloacetate, returning the cycle to its starting point. Oxaloacetate now begins the cycle again, reacting with another molecule of acetyl-CoA to initiate the conversion of acetyl-CoA's acetyl group to another two molecules of CO_2.

Notice that reactions 6, 7, and 8 in the citric acid cycle are similar to reactions 1, 2, and 3 in the β-oxidation of fatty acids (Section 25.6).

PROBLEM 20◆

What functional group of isocitrate is oxidized in the third reaction of the citric acid cycle?

PROBLEM 21◆

The citric acid cycle is also called the tricarboxylic acid cycle. Which of the citric acid cycle intermediates are tricarboxylic acids?

25.11 Oxidative Phosphorylation

Each round of the citric acid cycle forms 3 molecules of NADH, 1 molecule of $FADH_2$ and 1 molecule of ATP. The NADH and $FADH_2$ molecules undergo **oxidative phosphorylation**—the fourth stage of catabolism—which oxidizes them back to NAD^+ and FAD. For each NADH that undergoes oxidative phosphorylation, 3 molecules of ATP are formed, and for each $FADH_2$ that undergoes oxidative phosphorylation, 2 molecules of ATP are formed.

In oxidative phosphorylation, each molecule of NADH is converted to three molecules of ATP and each molecule of $FADH_2$ is converted to 2 molecules of ATP.

$$\text{NADH} \longrightarrow \text{NAD}^+ + 3 \text{ ATP}$$

$$\text{FADH}_2 \longrightarrow \text{FAD} + 2 \text{ ATP}$$

Therefore, for every acetyl-CoA molecule that enters the citric acid cycle, 11 molecules of ATP are formed from NADH and $FADH_2$ and one molecule of ATP is formed in the cycle.

$$3 \text{ NADH} + \text{FADH}_2 + \text{ATP} \longrightarrow 3 \text{ NAD}^+ + \text{FAD} + 12 \text{ ATP}$$

How many molecules of ATP are obtained when the NADH and $FADH_2$ formed during the metabolism of one molecule of acetyl-CoA to CO_2 undergo oxidative phosphorylation?

25.12 Anabolism

Anabolism can be thought of as the reverse of catabolism. In anabolism, acetyl-CoA, pyruvate, and citric acid cycle intermediates are the starting materials for the synthesis of fatty acids, monosaccharides, and amino acids. These compounds are then used to form fats, carbohydrates, and proteins. The mechanisms utilized by biological systems to synthesize fats and proteins are discussed in Sections 18.22 and 27.8.

BASAL METABOLIC RATE

Your basal metabolic rate (BMR) is the number of calories you would burn if you stayed in bed all day. A person's BMR is affected by gender, age, and genetics: it is greater for men than for women, it is greater for young people than for old people, and some people are born with a faster metabolic rate than others. The BMR is also affected by percentage of body fat: the higher the percentage, the lower the BMR. For humans, the average BMR is about 1600 kcal/day.

In addition to consuming sufficient calories to sustain your basal metabolism, you must also obtain energy for physical activities. The more active you are, the more calories you must consume in order to maintain your current weight. People who consume more calories than required by their BMR plus their level of physical activity will gain weight; if they consume fewer calories, they will lose weight.

SUMMARY

Metabolism is the set of reactions living organisms carry out to obtain energy and to synthesize the compounds they require. Metabolism can be divided into **catabolism** and **anabolism**. Catabolic reactions break down complex molecules to provide energy and simple molecules. Anabolic reactions require energy and lead to the synthesis of complex biomolecules from simple molecules.

ATP is a cell's most important source of chemical energy; ATP provides a reaction pathway involving a good leaving group for a reaction that would not otherwise occur because of a poor leaving group. This occurs by way of a **phosphoryl transfer reaction** in which a phosphoryl group of ATP is transferred to a nucleophile as a result of breaking a **phosphoanhydride bond**. The reaction involves one of three intermediates: an **acyl phosphate**, an **acyl pyrophosphate**, or an **acyl adenylate**. Cleavage of a phosphoanhydride bond is highly exergonic because of electrostatic repulsions, solvation, and electron delocalization.

Catabolism can be divided into four stages. In the *first stage*, fats, carbohydrates, and proteins are hydrolyzed to fatty acids, monosaccharides, and amino acids, respectively. In the *second stage*, the products obtained from the first stage are converted to compounds that can enter the citric acid cycle (the third stage of catabolism). In order to enter

the citric acid cycle, a compound must be either a citric acid cycle intermediate, acetyl-CoA, or pyruvate.

In the second stage of catabolism, a fatty acyl-CoA is converted to acetyl-CoA in a pathway called **β-oxidation**. The series of four reactions is repeated until the entire fatty acid has been converted to acetyl-CoA molecules. Glucose, in the second stage of catabolism, is converted to two molecules of pyruvate in a series of 10 reactions known as **glycolysis**. Under normal (aerobic) conditions, pyruvate is converted to acetyl-CoA, which then enters the citric acid cycle. Amino acids are metabolized to acetyl-CoA, pyruvate, or citric acid cycle intermediates in the second stage of catabolism, depending on the amino acid. The amino acids we ingest are used for energy and for the synthesis of proteins and other compounds needed by the body.

The **citric acid cycle** is the *third stage* of catabolism, it is a series of eight reactions that converts the acetyl group of each molecule of acetyl-CoA that enters the cycle into two molecules of CO_2.

Metabolic energy is measured in terms of ATP. In the *fourth stage* of catabolism, called **oxidative phosphorylation**, each molecule of NADH and $FADH_2$ formed in oxidation reactions in the second and third stages of catabolism is converted into three molecules of ATP and two molecules of ATP, respectively.

KEY TERMS

acyl adenylate (p. 1143)
acyl phosphate (p. 1142)
acyl pyrophosphate (p. 1142)
adenosine 5′-triphosphate
   (ATP) (p. 1139)
anabolism (p. 1137)

catabolism (p. 1137)
citric acid cycle (p. 1155)
dehydrogenase (p. 1146)
glycolysis (p. 1149)
high-energy bond (p. 1144)
kinase (p. 1146)

metabolism (p. 1137)
β-oxidation (p. 1147)
oxidative phosphorylation (p. 1139)
phosphoanhydride bond (p. 1140)
phosphoryl transfer reaction (p. 1140)

PROBLEMS

23. Indicate whether an anabolic pathway or a catabolic pathway does the following:
   a. produces energy in the form of ATP
   b. involves primarily oxidation reactions

24. Galactose can enter the glycolytic cycle but it must first react with ATP to form galactose-1-phosphate. Propose a mechanism for this reaction.

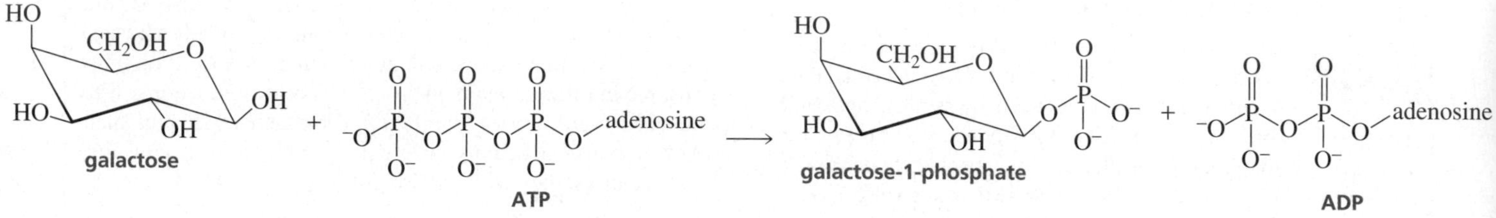

25. Pyruvate is the leaving group in the tenth reaction of glycolysis. Why is it a good leaving group?

26. When pyruvate is reduced by NADH to lactate, which hydrogen in pyruvate comes from NADH?

27. What reactions in the citric acid cycle form a product with a new asymmetric center?

28. If the phosphorus atom in 3-phosphoglycerate is radioactively labeled, where will the radioactive label be when the reaction that forms 2-phosphoglycerate is over?

29. What carbon atoms of glucose end up as a carboxyl group in pyruvate?

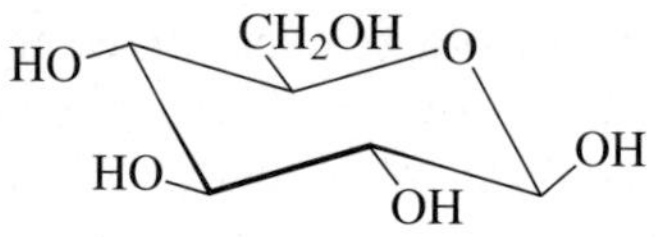

glucose

30. What carbon atoms of glucose end up in ethanol under anaerobic conditions in yeast?

31. How many molecules of acetyl-CoA are obtained from β-oxidation of one molecule of a 16-carbon saturated fatty acyl-CoA?

32. How many molecules of CO_2 are obtained from the complete metabolism of one molecule of a 16-carbon saturated fatty acyl-CoA?

33. How many molecules of ATP are obtained from β-oxidation of one molecule of a 16-carbon saturated fatty acyl-CoA?

34. How many molecules of NADH and $FADH_2$ are obtained from β-oxidation of one molecule of a 16-carbon saturated fatty acyl-CoA?

35. How many molecules of ATP are obtained from the NADH and $FADH_2$ formed in β-oxidation of one molecule of a 16-carbon saturated fatty acyl-CoA?

36. How many molecules of ATP are obtained from complete (including the fourth stage of catabolism) metabolism of one molecule of a 16-carbon saturated fatty acyl-CoA?

37. How many molecules of ATP are obtained from complete (including the fourth stage of catabolism) metabolism of one molecule of glucose?

38. Conjugate addition, like direct addition, can occur at the compound's *Re* face or its *Si* face (Section 17.14). Does attack of water occur at the *Re* or *Si* face of fumarate?

39. Most fatty acids have an even number of carbon atoms and therefore are completely metabolized to acetyl-CoA. A fatty acid with an odd number of carbon atoms is metabolized to acetyl-CoA and one equivalent of propionyl-CoA. The following two reactions convert propionyl-CoA into succinyl-CoA, a citric acid cycle intermediate, so it can be further metabolized. Each of the reactions requires a coenzyme. Identify the coenzyme for each step. From what vitamins are the coenzymes derived? (*Hint:* see Chapter 24.)

propionyl-CoA   **methylmalonyl-CoA**   **succinyl-CoA**

40. If glucose is labeled with ^{14}C in the indicated position, where will the label be in pyruvate?
  a. glucose-1-^{14}C **c.** glucose-3-^{14}C **e.** glucose-5-^{14}C
  b. glucose-2-^{14}C **d.** glucose-4-^{14}C **f.** glucose-6-^{14}C

41. Under conditions of starvation, acetyl-CoA, instead of being degraded in the citric acid cycle, is converted into acetone and 3-hydroxybutyrate, compounds called ketone bodies that can be used by the brain as a temporary fuel. Propose a mechanism for their formation.

                     acetone   **3-hydroxybutyrate**

42. Acyl-CoA synthase is the enzyme that activates a fatty acid by converting it to a fatty acyl-CoA (Section 25.2) in a series of two reactions. In the first reaction, the fatty acid reacts with ATP, and one of the products formed is ADP. The other product reacts in a second reaction with CoASH to form the fatty acyl-CoA. Propose a mechanism for each of the reactions.

43. a. UDP-Galactose-4-epimerase converts UDP-galactose to UDP-glucose. The reaction requires NAD^+ as a coenzyme. Propose a mechanism for the reaction.
   b. Why is the enzyme called an epimerase?

44. Professor Anna Bol is trying to determine the mechanism of a reaction in which ATP activates a carboxylate ion. Whether the carboxylate ion attacks the α-phosphorus or the β-phosphorus of ATP cannot be determined from the reaction products because AMP and pyrophosphate are products in both reactions. However, the mechanisms can be distinguished by a labeling experiment in which the enzyme, the carboxylate ion, ATP, and radioactively labeled pyrophosphate are incubated, and the ATP is isolated. If the isolated ATP is radioactive, the attack occurred on the α-phosphorus. If it is not radioactive, the attack occurred on the β-phosphorus. Explain these conclusions.

45. What would be the results of the experiment in Problem 44 if radioactive AMP were added to the incubation mixture instead of radioactive pyrophosphate?

CHAPTER 26

Lipids

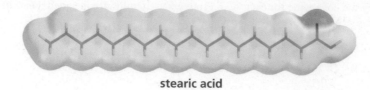

stearic acid

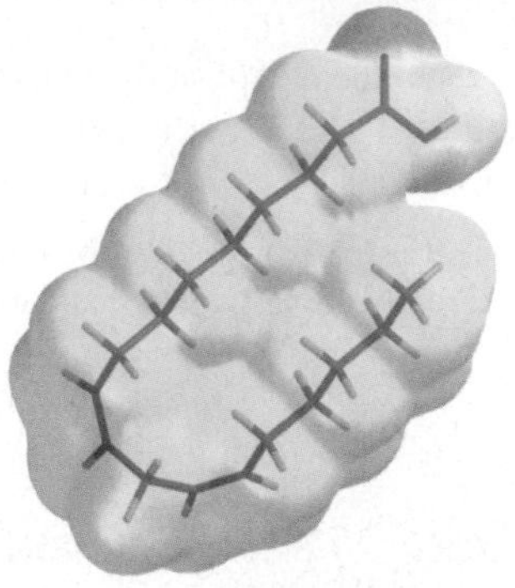

linoleic acid

BUILDING ON FUNDAMENTALS

SECTION 26.3 Electron delocalization increases the stability of a compound (7.6).

SECTION 26.3 Double bonds are reduced by catalytic hydrogenation (4.11).

SECTION 26.4 An alcohol reacts with a carboxylic acid to form an ester; an alcohol reacts with phosphoric acid to form a phosphate ester (16.15).

SECTION 26.8 Weak bases are good leaving groups (8.3).

SECTION 26.8 We will see a biosynthesis that involves a Claisen condensation (18.15), an aldol addition (18.12), thioester hydrolysis (16.22), and decarboxylation (18.18).

SECTION 26.8 A thioester, like an oxygen ester, is reduced to an alcohol by two equivalents of hydride ion (17.6).

SECTION 26.9 Trans fused rings have both substituents in equatorial positions, so they are more stable than cis fused rings (2.14).

Lipids are organic compounds, found in living organisms, that are soluble in nonpolar solvents. Because compounds are classified as lipids on the basis of a physical property—their solubility—rather than on the basis of their structures, lipids have a variety of structures and functions, as the following examples illustrate:

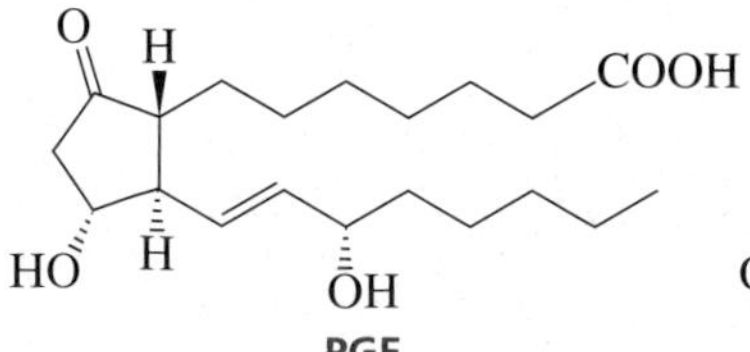

PGE$_1$
a vasodilator

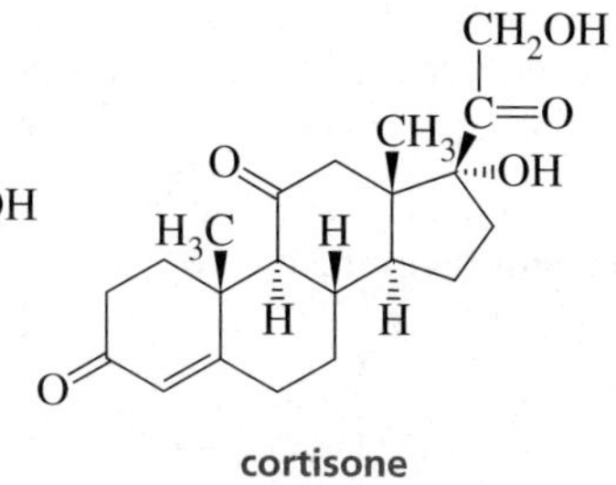

cortisone
a hormone

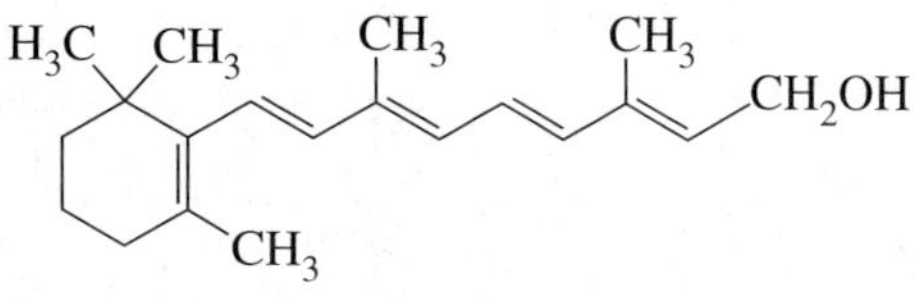

vitamin A
a vitamin

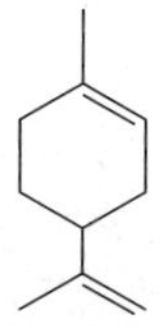

limonene
in orange and
lemon oils

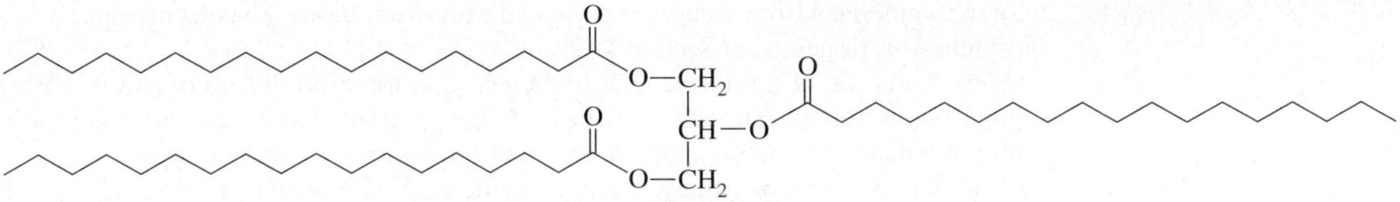

tristearin
a fat

The ability of lipids to dissolve in nonpolar organic solvents results from their significant hydrocarbon component—the part of the molecule responsible for its "oiliness" or "fattiness." The word *lipid* comes from the Greek *lipos,* which means "fat."

26.1 Fatty Acids Are Long-Chain Carboxylic Acids

Fatty acids, one major group of lipids, are carboxylic acids with long hydrocarbon chains. The fatty acids most frequently found in nature are shown in Table 26.1. Most naturally occurring fatty acids contain an even number of carbon atoms and are unbranched because

Table 26.1   Common Naturally Occurring Fatty Acids

Number of carbons	Common name	Systematic name	Structure	Melting point °C
Saturated				
12	lauric acid	dodecanoic acid	 COOH	44
14	myristic acid	tetradecanoic acid	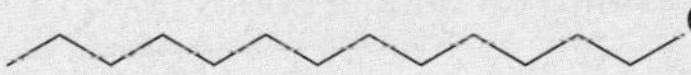 COOH	58
16	palmitic acid	hexadecanoic acid	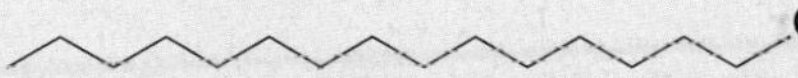 COOH	63
18	stearic acid	octadecanoic acid	 COOH	69
20	arachidic acid	eicosanoic acid	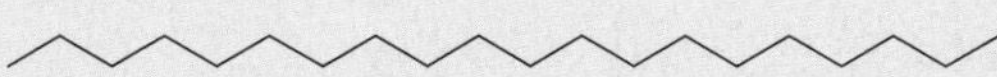 COOH	77
Unsaturated				
16	palmitoleic acid	(9Z)-hexadecenoic acid	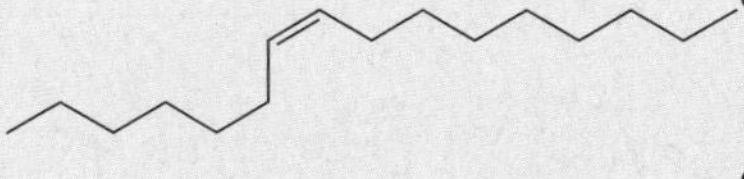 COOH	0
18	oleic acid	(9Z)-octadecenoic acid	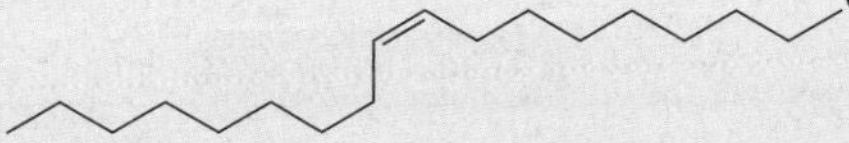 COOH	13
18	linoleic acid	(9Z,12Z)-octadecadienoic acid	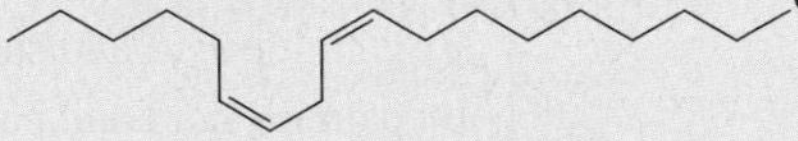 COOH	−5
18	linolenic acid	(9Z,12Z,15Z)-octadecatrienoic acid	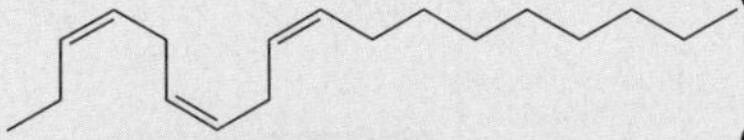 COOH	−11
20	arachidonic acid	(5Z,8Z,11Z,14Z)-eicosatetraenoic acid	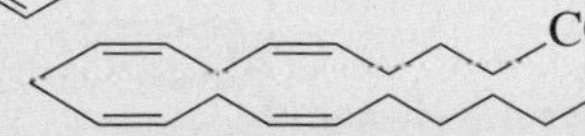 COOH	−50
20	EPA	(5Z,8Z,11Z,14Z,17Z)-eicosapentaenoic acid	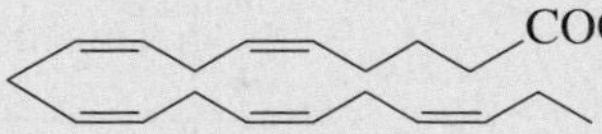 COOH	−50

they are synthesized from acetate, a compound with two carbons. The mechanism for their biosynthesis is discussed in Section 18.22.

Fatty acids can be saturated with hydrogen (and therefore have no carbon–carbon double bonds) or unsaturated (and have carbon–carbon double bonds). Fatty acids with more than one double bond are called **polyunsaturated fatty acids**.

The melting points of saturated fatty acids increase with increasing molecular weight because of increased van der Waals interactions between the molecules (Section 2.9). The melting points of unsaturated fatty acids also increase with increasing molecular weight, but are lower than those of saturated fatty acids with comparable molecular weights (Table 26.1).

The double bonds in naturally occurring unsaturated fatty acids have the cis configuration and are always separated by one CH_2 group. The cis double bond produces a bend in the molecule, which prevents them from packing together as tightly as saturated fatty acids. As a result, unsaturated fatty acids have fewer intermolecular interactions and therefore lower melting points than saturated fatty acids with comparable molecular weights (Table 26.1). The melting points of the unsaturated fatty acids decrease as the number of double bonds increases. For example, an 18-carbon fatty acid melts at 69 °C if it is saturated, at 13 °C if it has one double bond, at −5 °C if it has two double bonds, and at −11 °C if it has three double bonds.

Unsaturated fatty acids have lower melting points than saturated fatty acids.

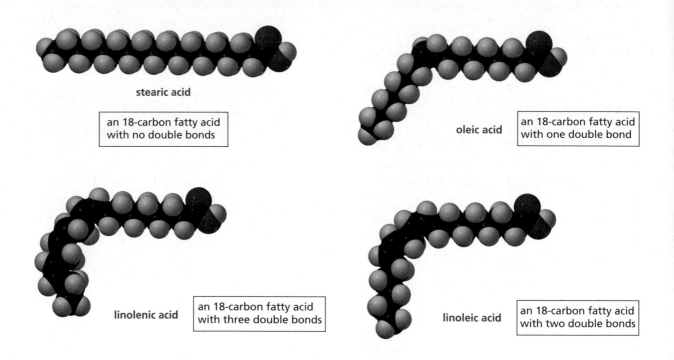

stearic acid

an 18-carbon fatty acid with no double bonds

oleic acid

an 18-carbon fatty acid with one double bond

linolenic acid

an 18-carbon fatty acid with three double bonds

linoleic acid

an 18-carbon fatty acid with two double bonds

PROBLEM 1

Explain the difference in the melting points of the following fatty acids:

a. palmitic acid and stearic acid   **c.** oleic acid and linoleic acid

b. palmitic acid and palmitoleic acid

PROBLEM 2◆

What products are formed when arachidonic acid reacts with excess ozone followed by treatment with H_2O_2? (*Hint:* See Section 19.7.)

OMEGA FATTY ACIDS

Omega is a term used to indicate the position of the first double bond, counting from the methyl end in an unsaturated fatty acid. For example, linoleic acid is called omega-6 fatty acid because its first double bond is after the sixth carbon and linolenic acid is called omega-3 fatty acid because its first double bond is after the third carbon. Mammals lack the enzyme that introduces a double bond beyond C-9 (the carboxyl carbon is C-1). Linoleic acid and linolenic acids, therefore, are essential fatty acids for mammals: we cannot synthesize them, yet we require them for normal body function, so they must be included in our diets.

Linoleic and linolenic acids are essential fatty acids for mammals.

omega-6 fatty acid / linoleic acid

omega-3 fatty acid / linolenic acid / COOH

26.2 Waxes Are High-Molecular Weight Esters

Waxes are esters formed from long-chain carboxylic acids and long-chain alcohols. For example, beeswax, the structural material of beehives, has a 26-carbon carboxylic acid component and a 30-carbon alcohol component. The word *wax* comes from the Old English *weax*, meaning "material of the honeycomb." Carnauba wax is a particularly hard wax because of its relatively high molecular weight; it has a 32-carbon carboxylic acid component and a 34-carbon alcohol component. Carnauba wax is widely used as a car wax and in floor polishes.

Layers of honeycomb in a beehive.

$CH_3(CH_2)_{24}CO(CH_2)_{29}CH_3$
a major component of beeswax structural material of beehives

$CH_3(CH_2)_{30}CO(CH_2)_{33}CH_3$
a major component of carnauba wax coating on the leaves of a Brazilian palm

$CH_3(CH_2)_{14}CO(CH_2)_{15}CH_3$
a major component of spermaceti wax from the heads of sperm whales

Waxes are common in living organisms. The feathers of birds are coated with wax to make them water repellent. Some vertebrates secrete wax in order to keep their fur lubricated and water repellent. Insects secrete a waterproof, waxy layer on the outside of their exoskeletons. Wax is also found on the surfaces of certain leaves and fruits, where it serves as a protectant against parasites and minimizes the evaporation of water.

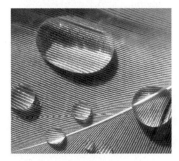

Raindrops on a feather.

26.3 Fats and Oils

Triacylglycerols, also called triglycerides, are compounds in which each of the three OH groups of glycerol has formed an ester with a fatty acid. If the three fatty acid components of a triacylglycerol are the same, the compound is called a **simple triacylglycerol**. **Mixed triacylglycerols** contain two or three different fatty acid components and are more common than simple triacylglycerols. Not all triacylglycerol molecules from a single source are necessarily identical; substances such as lard and olive oil, for example, are mixtures of several different triacylglycerols (Table 26.2).

$$
\begin{array}{ccc}
\text{CH}_2\text{—OH} & \text{R}^1\text{—}\overset{\overset{\displaystyle O}{\|}}{\text{C}}\text{—OH} & \text{CH}_2\text{—O}\text{—}\overset{\overset{\displaystyle O}{\|}}{\text{C}}\text{—R}^1 \\
| & & | \\
\text{CH—OH} & \text{R}^2\text{—}\overset{\overset{\displaystyle O}{\|}}{\text{C}}\text{—OH} & \text{CH—O}\text{—}\overset{\overset{\displaystyle O}{\|}}{\text{C}}\text{—R}^2 \\
| & & | \\
\text{CH}_2\text{—OH} & \text{R}^3\text{—}\overset{\overset{\displaystyle O}{\|}}{\text{C}}\text{—OH} & \text{CH}_2\text{—O}\text{—}\overset{\overset{\displaystyle O}{\|}}{\text{C}}\text{—R}^3 \\
\textbf{glycerol} & \textbf{fatty acids} & \textbf{a triacylglycerol} \\
& & \textbf{a fat or an oil}
\end{array}
$$

Table 26.2   Approximate Percentage of Fatty Acids in Some Common Fats and Oils

| | mp (°C) | Saturated fatty acids | | | | Unsaturated fatty acids | | |
		lauric C_{12}	myristic C_{14}	palmitic C_{16}	stearic C_{18}	oleic C_{18}	linoleic C_{18}	linolenic C_{18}
Animal fats								
Butter	32	2	11	29	9	27	4	—
Lard	30	—	1	28	12	48	6	—
Human fat	15	1	3	25	8	46	10	—
Whale blubber	24	—	8	12	3	35	10	—
Plant oils								
Corn	20	—	1	10	3	50	34	—
Cottonseed	−1	—	1	23	1	23	48	—
Linseed	−24	—	—	6	3	19	24	47
Olive	−6	—	—	7	2	84	5	—
Peanut	3	—	—	8	3	56	26	—
Safflower	−15	—	—	3	3	19	70	3
Sesame	−6	—	—	10	4	45	40	—
Soybean	−16	—	—	10	2	29	51	7

The percentages of fatty acids do not add up to 100 because the fats and oils also contain fatty acids other than those listed in the table.

Triacylglycerols that are solids or semisolids at room temperature are called **fats**. Most fats are obtained from animals and are composed largely of triacylglycerols with fatty acid components that are either saturated or have only one double bond. The saturated fatty acid tails pack closely together, giving these triacylglycerols relatively high melting points, causing them to be solids at room temperature.

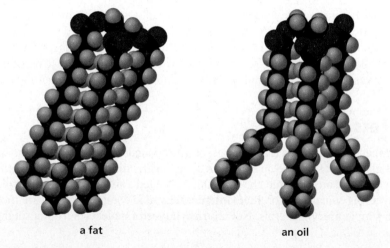

a fat   an oil

Liquid triacylglycerols are called **oils**. Oils typically come from plant products such as corn, soybeans, olives, and peanuts. They are composed primarily of triacylglycerols with fatty acid components that are unsaturated and therefore cannot pack tightly together. Consequently, they have relatively low melting points, causing them to be liquids at room temperature. Compare the approximate fatty acid compositions of the common fats and oils shown in Table 26.2.

Some or all of the double bonds of polyunsaturated oils can be reduced by catalytic hydrogenation (Section 4.11). Margarine and shortening are prepared by hydrogenating vegetable oils, such as soybean oil or safflower oil, until they have the desired consistency. The hydrogenation reaction must be carefully controlled, however, because reducing all the carbon–carbon double bonds would produce a hard fat with the consistency of beef tallow.

$$RCH{=}CHCH_2CH{=}CHCH_2CH{=}CH{-} \quad \xrightarrow[\text{Pt}]{\text{H}_2} \quad RCH_2CH_2CH_2CH{=}CHCH_2CH_2CH_2{-}$$

Vegetable oils have become popular in food preparation because some studies have linked the consumption of saturated fats with heart disease. Recent studies have shown that *un*saturated fats may also be implicated in heart disease. However, one unsaturated fatty acid—a 20-carbon fatty acid with five double bonds, known as EPA and found in high concentrations in fish oils—is thought to lower the chance of developing certain forms of heart disease. Once consumed, dietary fat is hydrolyzed in the intestine, releasing glycerol and fatty acids (Section 16.11). We have seen that the hydrolysis of fats under basic conditions forms glycerol and salts of fatty acids that are commonly known as *soap* (Section 16.14).

This puffin's diet is high in fish oil.

PROBLEM 3♦

Which has a higher melting point, glyceryl tripalmitoleate or glyceryl tripalmitate?

PROBLEM 4

Draw the structure of an optically inactive fat that when hydrolyzed gives glycerol, one equivalent of lauric acid, and two equivalents of stearic acid.

PROBLEM 5

Draw the structure of an optically active fat that when hydrolyzed gives the same products as the fat in Problem 4.

Animals have a subcutaneous layer of fat cells that serves as both an energy source and an insulator. The fat content of the average man is about 21%, whereas the fat content of the average woman is about 25%. A fat provides about six times as much metabolic energy as an equal weight of hydrated glycogen because fats are less oxidized than carbohydrates and because fats are nonpolar, they do not bind water. In contrast, two-thirds of the weight of stored glycogen is water (Section 21.17).

Humans can store sufficient fat to provide for the body's metabolic needs for two to three months, but can store only enough carbohydrate to provide for its metabolic needs for less than 24 hours. Carbohydrates, therefore, are used primarily as a quick, short-term energy source.

Polyunsaturated fats and oils are easily oxidized by O_2 by means of a radical chain reaction. In the initiation step, a radical removes a hydrogen from a methylene group that is flanked by two double bonds. This hydrogen is the one most easily removed because the unpaired electron is shared by three carbons. This radical reacts with O_2, forming a peroxy radical with conjugated double bonds. The peroxy radical removes a hydrogen from a methylene group of another molecule of fatty acid, forming an alkyl hydroperoxide. The two propagating steps are repeated over and over.

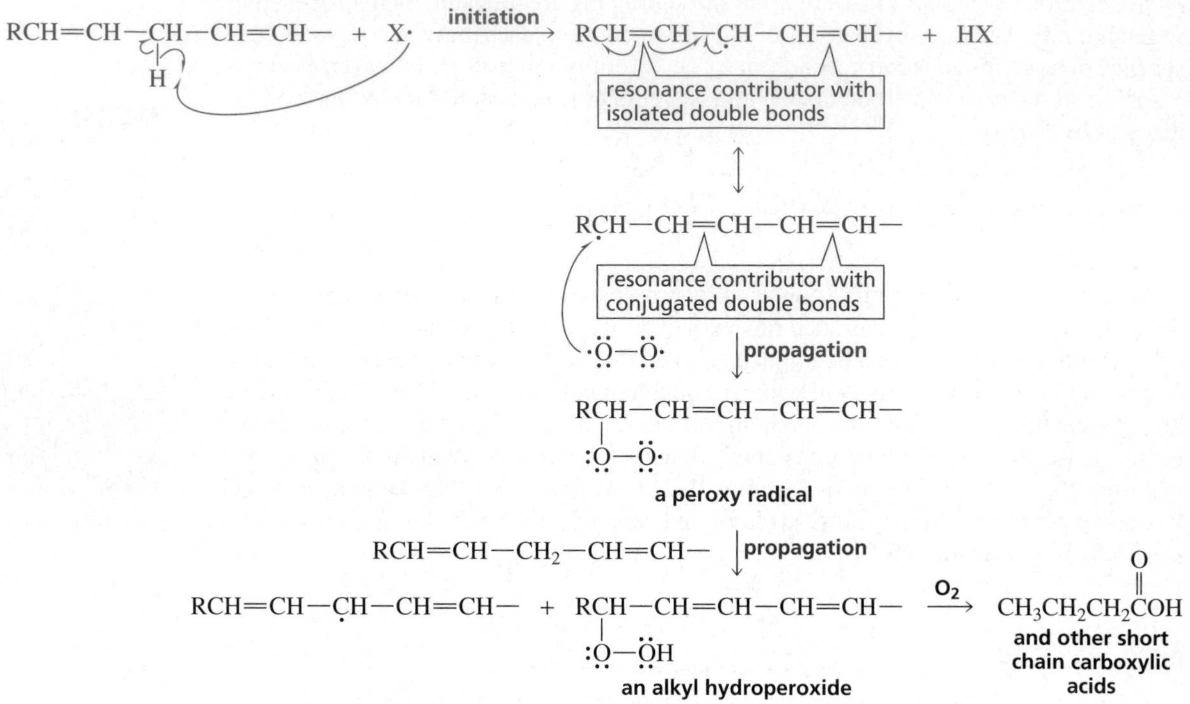

RCH=CH—CH—CH=CH— + X· $\xrightarrow{\text{initiation}}$ RCH=CH—ĊH—CH=CH— + HX

|H

resonance contributor with isolated double bonds

RCH—CH=CH—CH=CH—
·

resonance contributor with conjugated double bonds

·Ö—Ö· propagation

RCH—CH=CH—CH=CH—
|
:Ö—Ö·

a peroxy radical

RCH=CH—CH₂—CH=CH— propagation

RCH=CH—ĊH—CH=CH— + RCH—CH=CH—CH=CH— $\xrightarrow{O_2}$ CH₃CH₂CH₂COH
                                          |
                                          :Ö—ÖH

and other short chain carboxylic acids

an alkyl hydroperoxide

The reaction of fatty acids with O_2 causes them to become rancid. The unpleasant taste and smell associated with rancidity result from the further oxidation of the alkyl hydroperoxide to short chain carboxylic acids such as butyric acid that have strong odors. The same process contributes to the odor of sour milk.

PROBLEM 6   SOLVED

An oil obtained from coconuts is unusual in that all three fatty acid components are identical. The molecular formula of the oil is $C_{45}H_{86}O_6$. What is the molecular formula of the carboxylate ion obtained when the oil is saponified?

Solution   When the oil is saponified, it forms glycerol and three equivalents of carboxylate ion. In losing glycerol, the oil loses three carbons and five hydrogens. Thus, the three equivalents of carboxylate ion have a combined molecular formula of $C_{42}H_{81}O_6$. Dividing by three gives a molecular formula of $C_{14}H_{27}O_2$ for the carboxylate ion.

PROBLEM 7

Draw the resonance contributors for the radical formed when a hydrogen atom is removed from C-10 of arachidonic acid.

OLESTRA: NONFAT WITH FLAVOR

Chemists have been searching for ways to reduce the caloric content of foods without decreasing their flavor. Many people who believe that "no fat" is synonymous with "no flavor" think this is a worthy endeavor. Procter and Gamble spent 30 years and more than $2 billion to develop a fat substitute they named Olestra. After reviewing the results of more than 150 studies, the Federal Food and Drug Administration (Section 30.13) approved the limited use of Olestra in snack foods.

Olestra is a semisynthetic compound. That is, Olestra itself does not exist in nature, but its components do. Developing a compound that can be made from units that are a normal part of our diet decreases the likelihood that the new compound will be toxic. Olestra is made by esterifying all the OH groups of sucrose with fatty acids obtained from cottonseed oil and soybean oil. Therefore, its component parts are table sugar and vegetable oil. Because its ester linkages are too sterically hindered to be hydrolyzed by digestive enzymes, Olestra tastes like fat but it cannot be digested and so it has no caloric value.

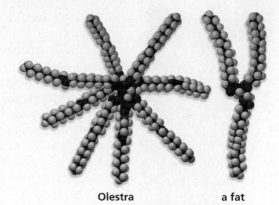

Olestra   **a fat**

Courtesy of Procter & Gamble Company.

ester groups are too hindered to be hydrolyzed

glucose

fructose

Olestra

WHALES AND ECHOLOCATION

Whales have enormous heads, accounting for 33% of their total weight. They have large deposits of fat in their heads and lower jaws. This fat is very different from both the whale's normal body fat and its dietary fat. Because major anatomical modifications were necessary to accommodate this fat, it must have some important function for the animal. It is now believed that the fat is used for echolocation—the emitting of sounds in pulses in order to gain information by analyzing the returning echoes. The fat in the whale's head focuses the emitted sound waves in a directional beam, and the echoes are received by the fat organ in the lower jaw. This organ transmits the sound to the brain for processing and interpretation, providing the whale with information about the depth of the water, changes in the sea floor, and the location of the coastline. The fat deposits in the whale's head and jaw therefore give the animal a unique acoustic sensory system and allow it to compete successfully for survival with the shark, which also has a well-developed sense of sound direction.

Humpback whale in Alaska.

26.4 **Phospholipids and Sphingolipids Are Components of Membranes**

For organisms to operate properly, some of their parts must be separated from other parts. On a cellular level, for example, the outside of the cell must be separated from the inside. "Greasy" lipid **membranes** serve as the barrier. In addition to isolating the cell's contents, membranes allow the selective transport of ions and organic molecules into and out of the cell.

Phosphoacylglycerols (also called **phosphoglycerides**), the major components of cell membranes, belong to a class of compounds called **phospholipids**—lipids that contain a phosphate group. Phosphoacylglycerols are similar to triacylglycerols except that a terminal OH group of glycerol is esterified with phosphoric acid rather than with a fatty acid, forming a **phosphatidic acid**. The C-2 carbon of glycerol in phosphoacylglycerols has the *R* configuration.

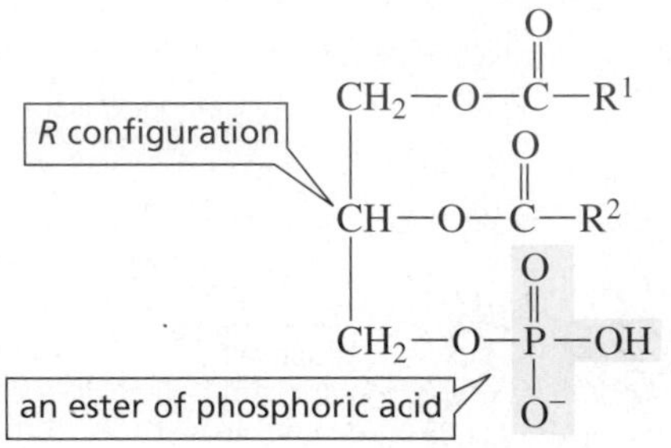

phosphatidylserine
a phosphoacylglycerol

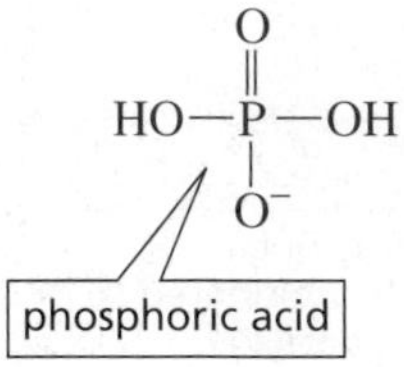

Phosphatidic acids are the simplest phosphoacylglycerols and are present only in small amounts in membranes. The most common phosphoacylglycerols in membranes have a second phosphate ester linkage—they are phosphodiesters.

phosphoacylglycerides

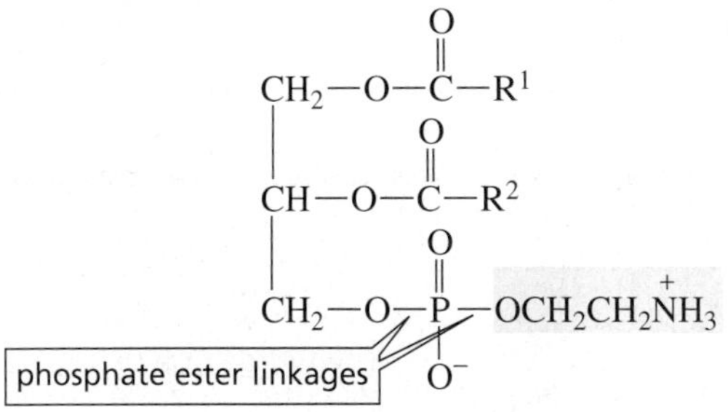

a phosphatidylethanolamine
a cephalin

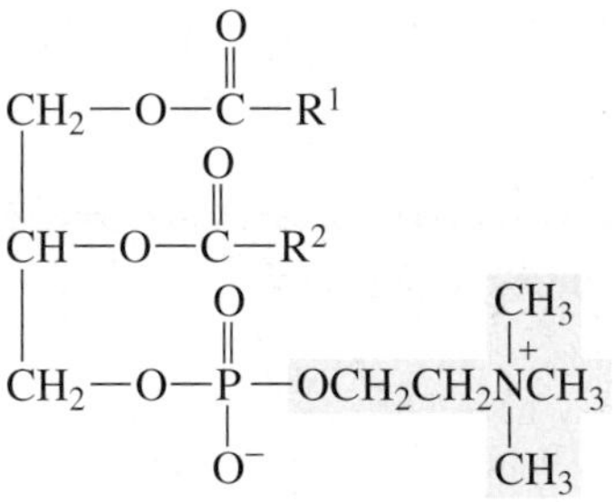

a phosphatidylcholine
a lecithin

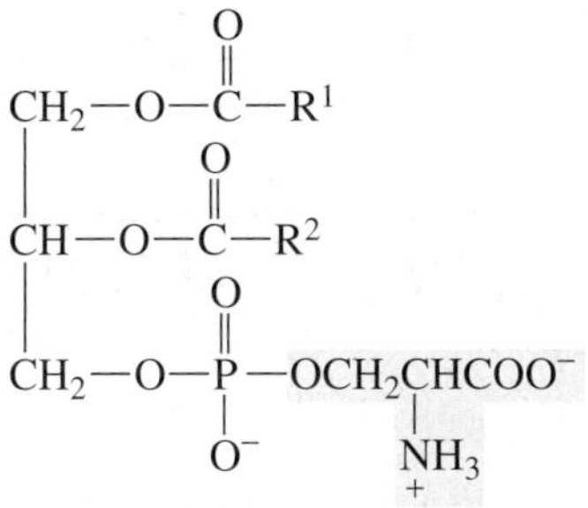

a phosphatidylserine

The most common phosphoacylglycerols are phosphodiesters.

The alcohols most commonly used to form the second ester group are ethanolamine, choline, and serine. Phosphatidylethanolamines are also called *cephalins*, and phosphatidylcholines are called *lecithins*. Lecithins are added to foods such as mayonnaise to prevent the aqueous and fat components from separating.

PROBLEM 8◆

Do the identities of R^1 and R^2 in phosphatidic acid change the configuration of the asymmetric center?

Phosphoacylglycerols form **membranes** by arranging themselves in a **lipid bilayer**. The polar heads of the phosphoacylglycerols are on both surfaces of the bilayer, and the fatty acid chains form the interior of the bilayer. Cholesterol, a membrane lipid discussed in Section 26.9, is also found in the interior of the bilayer (Figure 26.1). A typical bilayer is about 50 Å thick. (Compare the bilayer structure with that of the micelles formed by soap in aqueous solution, described in Section 16.14.)

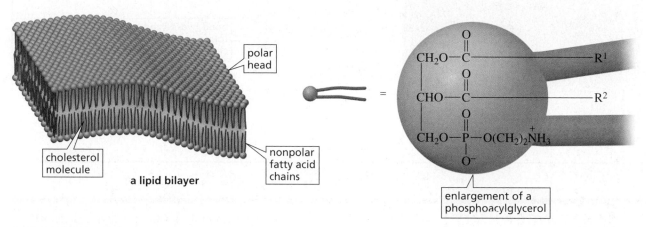

▲ **Figure 26.1**
Anatomy of a lipid bilayer.

The fluidity of a membrane—the ease with which membrane components are able to change position within the membrane—is controlled by the fatty acid components of the phosphoacylglycerols. Saturated fatty acids decrease membrane fluidity because their hydrocarbon chains pack closely together. Unsaturated fatty acids increase fluidity because they pack less closely together. Cholesterol also decreases fluidity (Section 26.9). Only animal membranes contain cholesterol, so they are more rigid than plant membranes.

The unsaturated fatty acid chains of phosphoacylglycerols are susceptible to reaction with O_2, similar to the reaction described on page 1168 for fats and oils. This oxidation reaction leads to the degradation of membranes. Vitamin E is an important antioxidant that protects fatty acid chains from degradation through oxidation. Vitamin E, also called α-tocopherol, is classified as a lipid because it is soluble in nonpolar solvents. It is therefore able to enter biological membranes; once there, it reacts more rapidly with oxygen than the bilayer triacylglycerols do, thus preventing them from reacting with oxygen (Section 11.10). There are some who believe that vitamin E slows the aging process. The ability of vitamin E to react with oxygen more rapidly than fats do is the reason it is added as a preservative to many fat-containing foods.

α-**tocopherol**
vitamin E

SNAKE VENOM

The venom of some poisonous snakes contains a phospholipase, an enzyme that hydrolyzes an ester group of a phosphoglyceride. For example, both the eastern diamondback rattlesnake and the Indian cobra contain a phospholipase that hydrolyzes an ester bond of cephalins, causing the membranes of red blood cells to rupture.

An eastern diamondback rattlesnake.

$$
\begin{array}{l}
CH_2O-\overset{\displaystyle O}{\overset{\|}{C}}-R \\[2mm]
CHO-\overset{\displaystyle O}{\overset{\|}{C}}-R \\[2mm]
CH_2O-\overset{\displaystyle }{\underset{O^-}{\overset{\|}{P}}}-OCH_2CH_2\overset{+}{N}H_3
\end{array}
$$

bond hydrolyzed by the phospholipase found in the Indian cobra and the eastern diamondback rattlesnake

IS CHOCOLATE A HEALTH FOOD?

We have long been told that our diets should include lots of fruits and vegetables because they are good sources of antioxidants. Antioxidants protect against cardiovascular disease, cancer, and cataracts, and they are thought to slow the effects of aging. Recent studies show that chocolate also has high levels of antioxidants. On a weight basis, the concentration of antioxidants in chocolate is higher than the concentration in red wine or green tea and 20 times higher than the concentration in tomatoes. Another piece of good news for chocolate lovers is that stearic acid, the main fatty acid in chocolate, does not appear to raise blood cholesterol levels the way other saturated fatty acids do. Dark chocolate contains more than twice the level of antioxidants as milk chocolate. Unfortunately, white chocolate contains no antioxidants.

PROBLEM 9◆

Membranes contain proteins. Integral membrane proteins extend partly or completely through the membrane, whereas peripheral membrane proteins are found on the inner or outer surface of the membrane. What is the likely difference in the amino acid composition of integral and peripheral membrane proteins?

PROBLEM 10◆

A colony of bacteria accustomed to an environment with a temperature of 25 °C was moved to an identical environment, except that its temperature was 35 °C. The higher temperature increased the fluidity of the bacterial membranes. What could the bacteria do to regain their original membrane fluidity?

Sphingolipids are another kind of lipid found in membranes. They are the major lipid components in the myelin sheaths of nerve fibers. Sphingolipids contain an amino alcohol called sphingosine instead of glycerol. In sphingolipids, the amino group of sphingosine is bonded to the acyl group of a fatty acid. Both asymmetric centers in sphingosine have the *S* configuration.

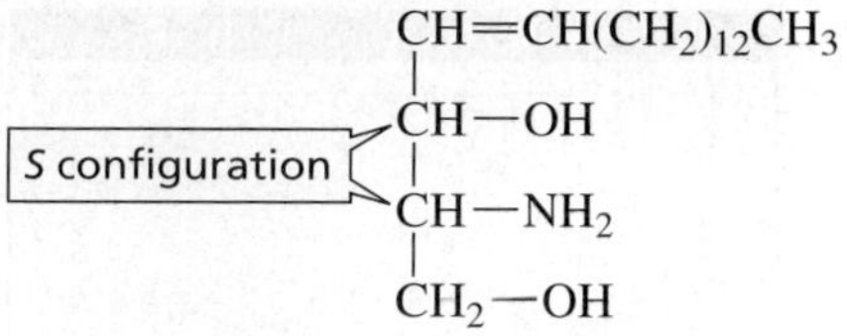

Two of the most common kinds of sphingolipids are *sphingomyelins* and *cerebrosides*. In sphingomyelins, the primary OH group of sphingosine is bonded to phosphocholine or phosphoethanolamine, in a manner similar to the bonding in lecithins and cephalins. In cerebrosides, the primary OH group of sphingosine is bonded to a sugar residue through a β-glycosidic linkage (Section 21.13). Sphingomyelins are phospholipids because they contain a phosphate group. Cerebrosides, on the other hand, are not phospholipids.

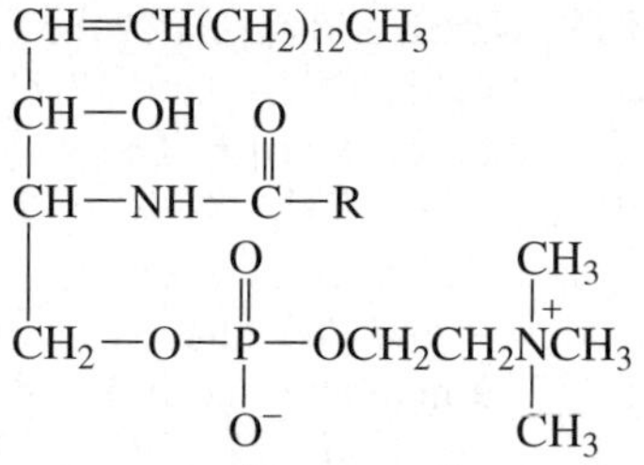

$$CH{=}CH(CH_2)_{12}CH_3$$
$$CH{-}OH$$
$$S\ configuration$$
$$CH{-}NH_2$$
$$CH_2{-}OH$$
sphingosine

a sphingomyelin

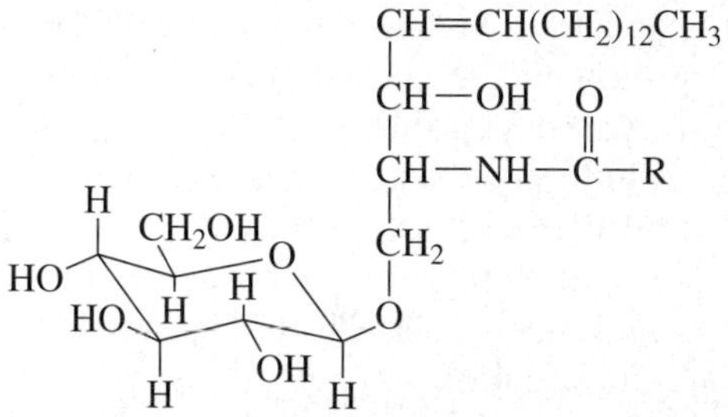

a glucocerebroside

MULTIPLE SCLEROSIS AND THE MYELIN SHEATH

The myelin sheath is a lipid-rich covering that is wrapped around the axons of nerve cells. Composed largely of sphingomyelins and cerebrosides, the sheath increases the velocity of nerve impulses. Multiple sclerosis is a disease characterized by loss of the myelin sheath, a consequent slowing of nerve impulses, and eventual paralysis.

PROBLEM 11

a. Draw the structures of three different sphingomyelins.
b. Draw the structure of a galactocerebroside.

PROBLEM 12

The membrane phospholipids in deer and elk have a higher degree of unsaturation in cells closer to the hoof than in cells closer to the body. Explain how this trait can be important for survival.

26.5 Prostaglandins Regulate Physiological Responses

Prostaglandins are found in all body tissues and are responsible for regulating a variety of physiological responses, such as inflammation, blood pressure, blood clotting, fever, pain, the induction of labor, and the sleep–wake cycle. All prostaglandins have a five-membered ring with a seven-carbon carboxylic acid

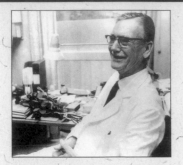

substituent and an eight-carbon hydrocarbon substituent. The two substituents are trans to each other.

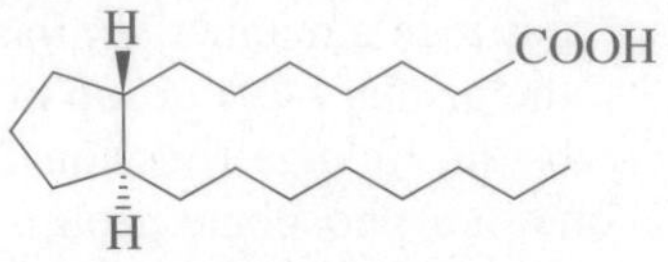

prostaglandin skeleton

Prostaglandins are classified using the formula PGX, where X designates the functional groups of the compound's five-membered ring. PGAs, PGBs, and PGCs all contain a carbonyl group and a double bond in the five-membered ring. The location of the double bond determines whether a prostaglandin is a PGA, PGB, or PGC. PGDs and PGEs are β-hydroxy ketones, and PGFs are 1,3-diols. A subscript indicates the total number of double bonds in the side chains, and "α" and "β" specify the configuration of the two OH groups in a PGF: "α" indicates a cis diol and "β" indicates a trans diol.

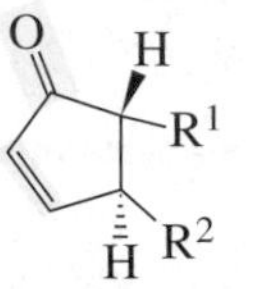

PGAs

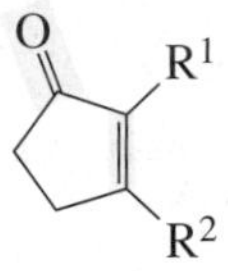

PGBs

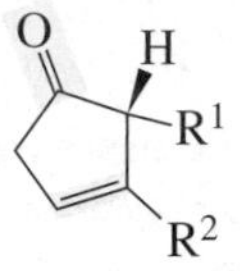

PGCs

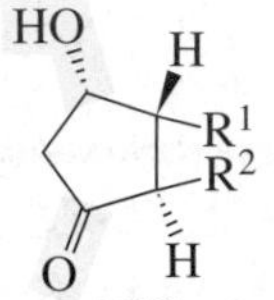

PGDs

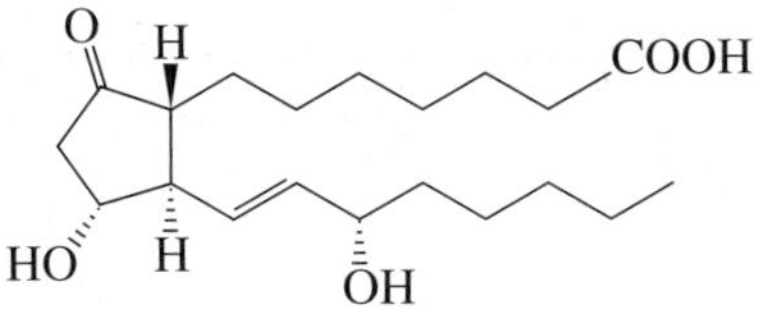

PGE_1

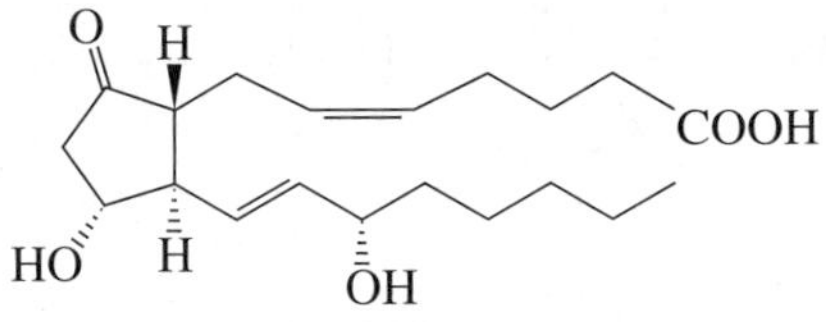

PGE_2

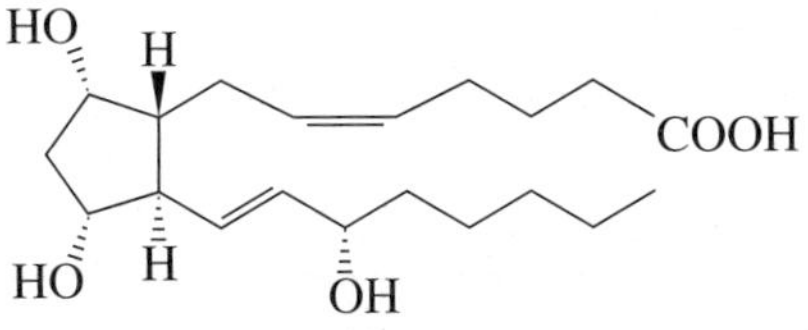

$PGF_{2\alpha}$

Prostaglandins are synthesized from arachidonic acid, a 20-carbon fatty acid with four cis double bonds. In the cell, arachidonic acid is found esterified to the 2-position of glycerol in many phospholipids. Arachidonic acid is synthesized from linoleic acid. Because linoleic acid cannot be synthesized by mammals, it must be acquired through the diet.

An enzyme called prostaglandin endoperoxide synthase catalyzes the conversion of arachidonic acid to PGH_2, the precursor of all prostaglandins. There are two forms of this enzyme; one carries out the normal physiological production of prostaglandin, and the other synthesizes additional prostaglandin in response to inflammation. The enzyme has two activities: a *cyclooxygenase activity* and a *hydroperoxidase activity.* It uses its cyclooxygenase activity to form the five-membered ring.

BIOGRAPHY

Ulf Svante von Euler (1905–1983) *first identified prostaglandins—from semen—in the early 1930s. He named them for their source, the prostate gland. By the time scientists realized that all cells except red blood cells synthesize prostaglandins, the name had become entrenched. Von Euler was born in Stockholm and received an M.D. from the Karolinska Institute, where he remained as a member of the faculty. He also discovered noradrenaline and identified its function as a chemical intermediate in nerve transmission. For this work, he shared the 1970 Nobel Prize in physiology or medicine with Julius Axelrod and Sir Bernard Katz.*

BIOGRAPHY

For their work on prostaglandins, **Sune Bergström, Bengt Ingemar Samuelsson,** *and* **John Robert Vane** *shared the 1982 Nobel Prize in physiology or medicine. Bergström and Samuelsson were born in Sweden—Bergström in 1916 and Samuelsson in 1934. They are both at the Karolinska Institute. Vane was born in England in 1927 and is at the Wellcome Foundation in Beckenham, England.*

biosynthesis of prostaglandins, thromboxanes, and prostacyclins

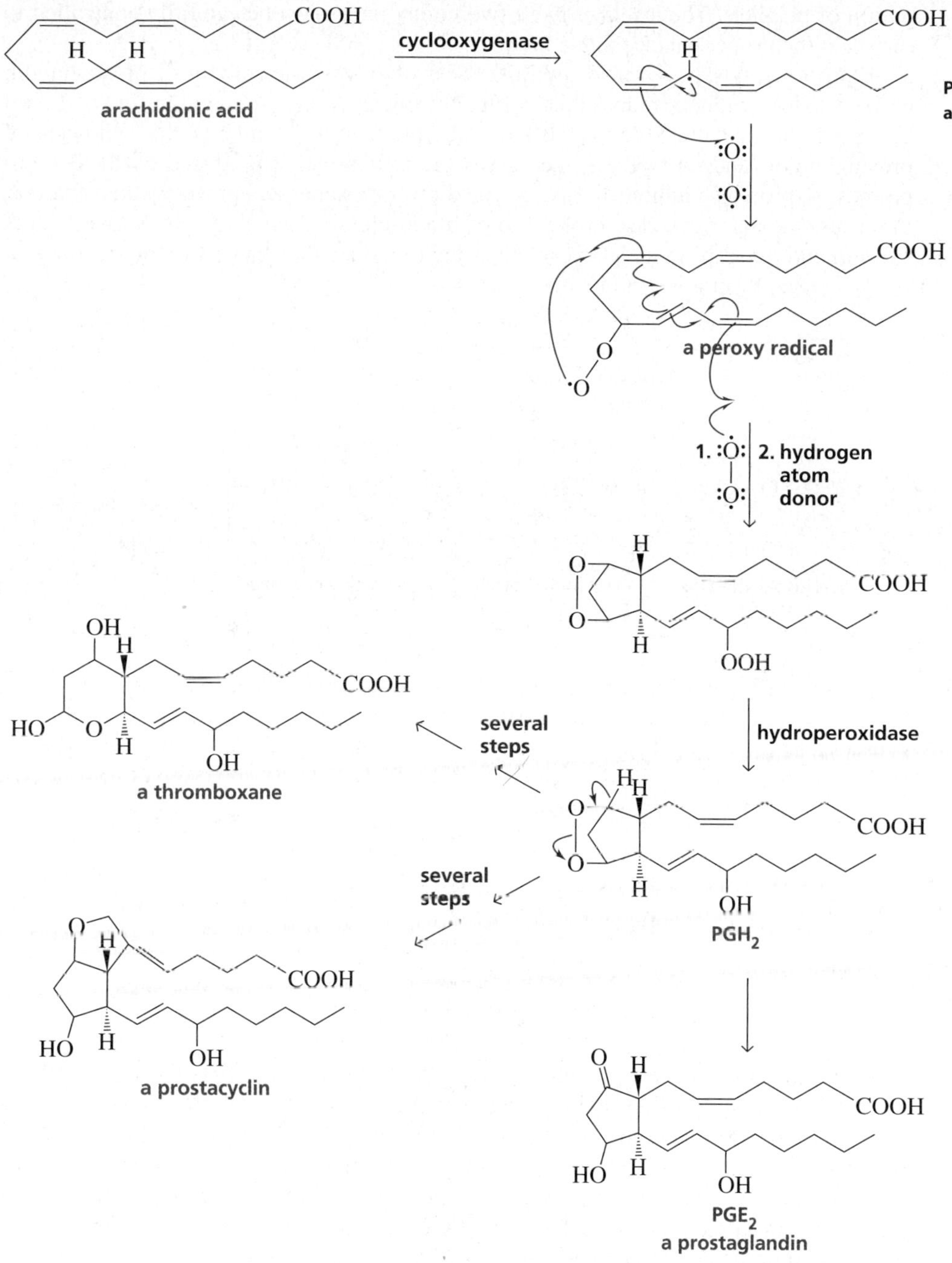

Prostaglandins are synthesized from arachidonic acid.

- In the first step, a hydrogen atom is removed from a carbon flanked by two double bonds. This hydrogen is removed relatively easily because the resulting radical is stabilized by electron delocalization.

- The radical reacts with oxygen to form a peroxy radical. Notice that these two steps are the same as the first two steps in the reaction that causes fats to become rancid (Section 26.3).

- The peroxy radical rearranges and reacts with a second molecule of oxygen.

- The enzyme then uses its hydroperoxidase activity to convert the OOH group into an OH group, forming PGH_2, which rearranges to form PGE_2, a prostaglandin.

In addition to serving as a precursor for the synthesis of prostaglandins, PGH_2 is a precursor for the synthesis of *thromboxanes* and *prostacyclins*. Thromboxanes constrict blood vessels and stimulate the aggregation of platelets, the first step in blood clotting. Prostacyclins have the opposite effect, dilating blood vessels and inhibiting the aggregation of platelets. The levels of these two compounds must be carefully controlled to maintain the proper balance in the blood.

Aspirin (acetylsalicylic acid) inhibits the cyclooxygenase activity of prostaglandin endoperoxide synthase. It does this by transferring an acetyl group to a serine hydroxyl group of the enzyme (Section 16.10). Aspirin, therefore, inhibits the synthesis of prostaglandins and, in that way, decreases the inflammation produced by these compounds. Aspirin also inhibits the synthesis of thromboxanes and prostacyclins. Overall, this causes a slight decrease in the rate of blood clotting, which is why some doctors recommend one aspirin tablet every other day to reduce the chance of a heart attack or stroke caused by clotting in blood vessels.

Other anti-inflammatory drugs, such as ibuprofen (the active ingredient in Advil, Motrin, and Nuprin) and naproxen (the active ingredient in Aleve), also inhibit the synthesis of prostaglandins. They compete with either arachidonic acid or the peroxy radical for the enzyme's binding site.

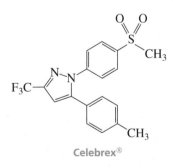

Celebrex®

Both aspirin and these other nonsteroidal anti-inflammatory drugs (NSAIDs) inhibit the synthesis of all prostaglandins—those produced under normal physiological conditions and those produced in response to inflammation. One prostaglandin regulates the production of acid in the stomach, so when prostaglandin synthesis stops, the acidity of the stomach can rise above normal levels. Celebrex, a relatively new drug, inhibits only the enzyme that produces prostaglandin in response to inflammation. Thus, inflammatory conditions now can be treated without some of the harmful side effects.

Arachidonic acid can also be converted to a *leukotriene*. Because they induce contraction of the muscle that lines the airways to the lungs, leukotrienes are implicated in allergic reactions, inflammatory reactions, and heart attacks. They also bring on the symptoms of asthma and contribute to the onset of anaphylactic shock, a potentially fatal allergic reaction. Several antileukotriene agents are available for the treatment of asthma.

arachidonic acid → **a leukotriene**

PROBLEM 13

Treating PGA$_2$ with a strong base such as sodium *tert*-butoxide followed by addition of acid converts it to PGC$_2$. Propose a mechanism for this reaction.

26.6   **Terpenes Contain Carbon Atoms in Multiples of Five**

Terpenes are a diverse class of lipids. More than 20,000 terpenes, many found in the oils extracted from fragrant plants, are known. They can be hydrocarbons, or they can contain oxygen and be alcohols, ketones, or aldehydes. Oxygen-containing terpenes are sometimes called **terpenoids**. Certain terpenes and terpenoids have been used as spices, perfumes, and medicines for thousands of years.

BIOGRAPHY

Leopold Stephen Ružička (1887–1976) *was the first to recognize that many organic compounds contain multiples of five carbons. A Croatian, Ružička attended college in Switzerland and became a Swiss citizen in 1917. He was a professor of chemistry at the University of Utrecht in the Netherlands and later at the Federal Institute of Technology in Zürich. For his work on terpenes, he shared the 1939 Nobel Prize in chemistry with Adolph Butenandt (page 1188).*

menthol
peppermint oil

geraniol
geranium oil

zingiberene
oil of ginger

β-selinene
oil of celery

After analyzing a large number of terpenes, organic chemists realized that they contained carbon atoms in multiples of 5. These naturally occurring compounds contain 10, 15, 20, 25, 30, and 40 carbon atoms, which suggests that there is a compound with five carbon atoms that serves as their building block. Further investigation showed that their structures are consistent with their being made by joining together isoprene units, usually in a head-to-tail fashion. (The branched end of isoprene is called the head, and the unbranched end is called the tail.) Isoprene is the common name for 2-methyl-1,3-butadiene, a compound containing five carbon atoms.

That isoprene units are linked in a head-to-tail fashion to form terpenes is known as the **isoprene rule**.

tail

tail

head

head

**carbon skeleton of two isoprene units with a bond
between the tail of one and the head of another**

In the case of cyclic compounds, the linkage of the head of one isoprene unit to the tail of another is followed by an additional linkage to form the ring. The second linkage is not necessarily head-to-tail but is whatever linkage is necessary to form a stable five- or six-membered ring.

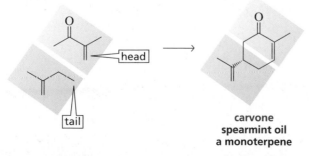

head

tail

**carvone
spearmint oil
a monoterpene**

In Section 26.8, we will see that the compound actually used in the biosynthesis of terpenes is not isoprene, but isopentenyl pyrophosphate, a compound that has the same carbon skeleton as isoprene. We will also look at the mechanism by which isopentenyl pyrophosphate units are joined together in a head-to-tail fashion.

Terpenes are classified according to the number of carbons they contain (Table 26.3). **Monoterpenes** are composed of two isoprene units, so they have 10 carbons.

A monoterpene has ten carbon atoms.

Table 26.3	Classification of Terpenes
Carbon atoms	**Classification**
10	monoterpenes
15	sesquiterpenes
20	diterpenes
25	sesterpenes
30	triterpenes
40	tetraterpenes

Sesquiterpenes, with 15 carbons, are composed of three isoprene units. Many fragrances and flavorings found in plants are monoterpenes and sesquiterpenes. These compounds are known as *essential oils*.

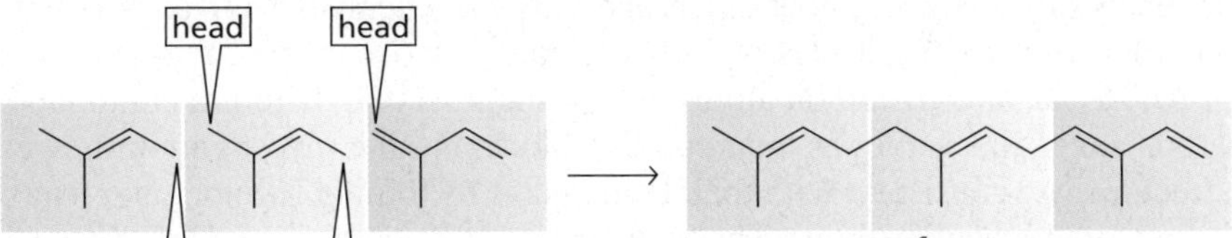

α-farnesene
a sesquiterpene found in the waxy coating on apple skins

Triterpenes (six isoprene units) and **tetraterpenes** (eight isoprene units) have important biological roles. For example, **squalene**, a triterpene, is the precursor of cholesterol, which is the precursor of all the other steroid molecules (Section 26.9).

Student Tutorial:
Isoprene units in terpenes

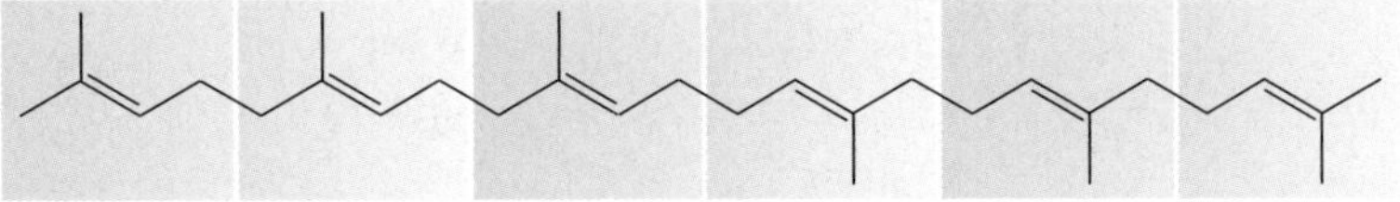

squalene

Carotenoids are tetraterpenes. Lycopene, the compound responsible for the red coloring of tomatoes and watermelon, and *β*-carotene, the compound that causes carrots and apricots to be orange, are examples of carotenoids. Some of the properties of carotenoids were discussed in Section 12.19.

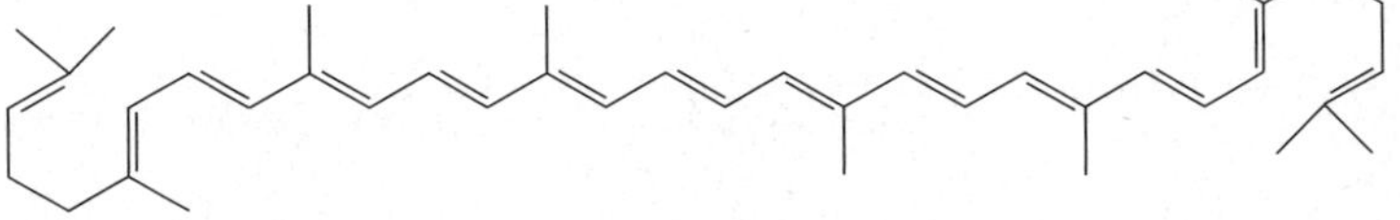

lycopene

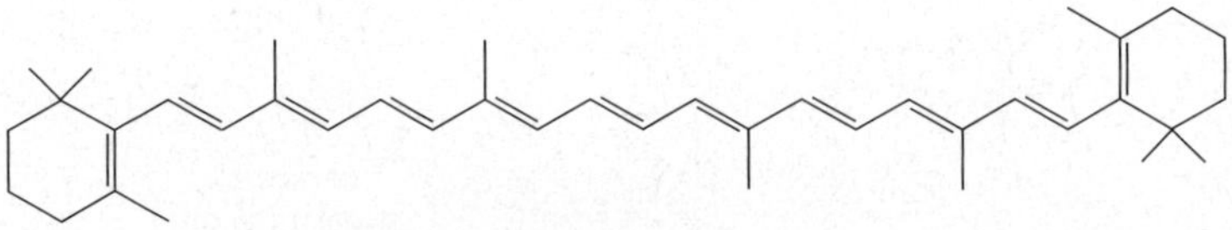

β-carotene

PROBLEM 14◆

Limonene has an asymmetric center so it exists as two different stereoisomers. The *R* isomer is found in oranges and the *S* isomer is found in lemons. Which of the following structures is found in oranges?

(+)-limonene   (−)-limonene

PROBLEM 15 SOLVED

Mark off the isoprene units in menthol, zingiberene, β-selinene, and squalene.

Solution   In zingiberene we find

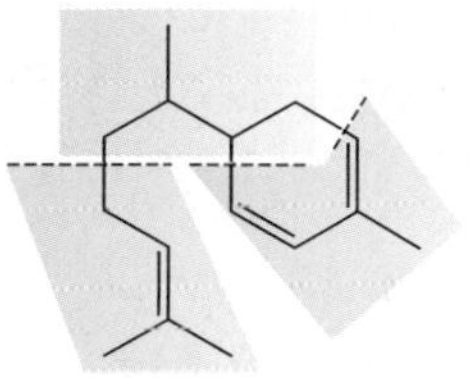

PROBLEM 16◆

One of the linkages in squalene is tail-to-tail, not head-to-tail. What does this suggest about how squalene is synthesized in nature? (*Hint:* Locate the position of the tail-to-tail linkage.)

PROBLEM 17

Mark off the isoprene units in lycopene and β-carotene (see page 1178). Can you detect a similarity in the way in which squalene, lycopene, and β-carotene are biosynthesized?

26.7 Vitamin A Is a Diterpene

Of the vitamins that are lipids—vitamins A, D, E, and K (Sections 11.10, 24.9, and 29.6)—vitamin A, a diterpene, is the only one we have not already discussed. Cleaving β-carotene, the major dietary source of the vitamin, forms two molecules of vitamin A. Also called retinol, vitamin A plays an important role in vision.

The retina of the eye contains cone cells and rod cells. The cone cells are responsible for color vision and for vision in bright light. The rod cells are responsible for vision in dim light. In rod cells, vitamin A is oxidized to an aldehyde and the trans double bond at C-11 is isomerized to a cis double bond. The mechanism for the enzyme-catalyzed interconversion of cis and trans double bonds is discussed in Section 17.18. The protein *opsin* uses a lysine side chain (Lys 216) to form an imine with (11*Z*)-retinal, resulting in a complex known as *rhodopsin.* When rhodopsin absorbs visible light, the cis double bond isomerizes to the trans isomer. This change in molecular geometry causes an electrical signal to be sent to the brain, where it is perceived as a visual image. The trans isomer of rhodopsin is not stable and is hydrolyzed to (11*E*)-retinal and opsin in a reaction referred to as bleaching of the visual pigment. (11*E*)-Retinal is then converted back to (11*Z*)-retinal to complete the vision cycle.

the chemistry of vision

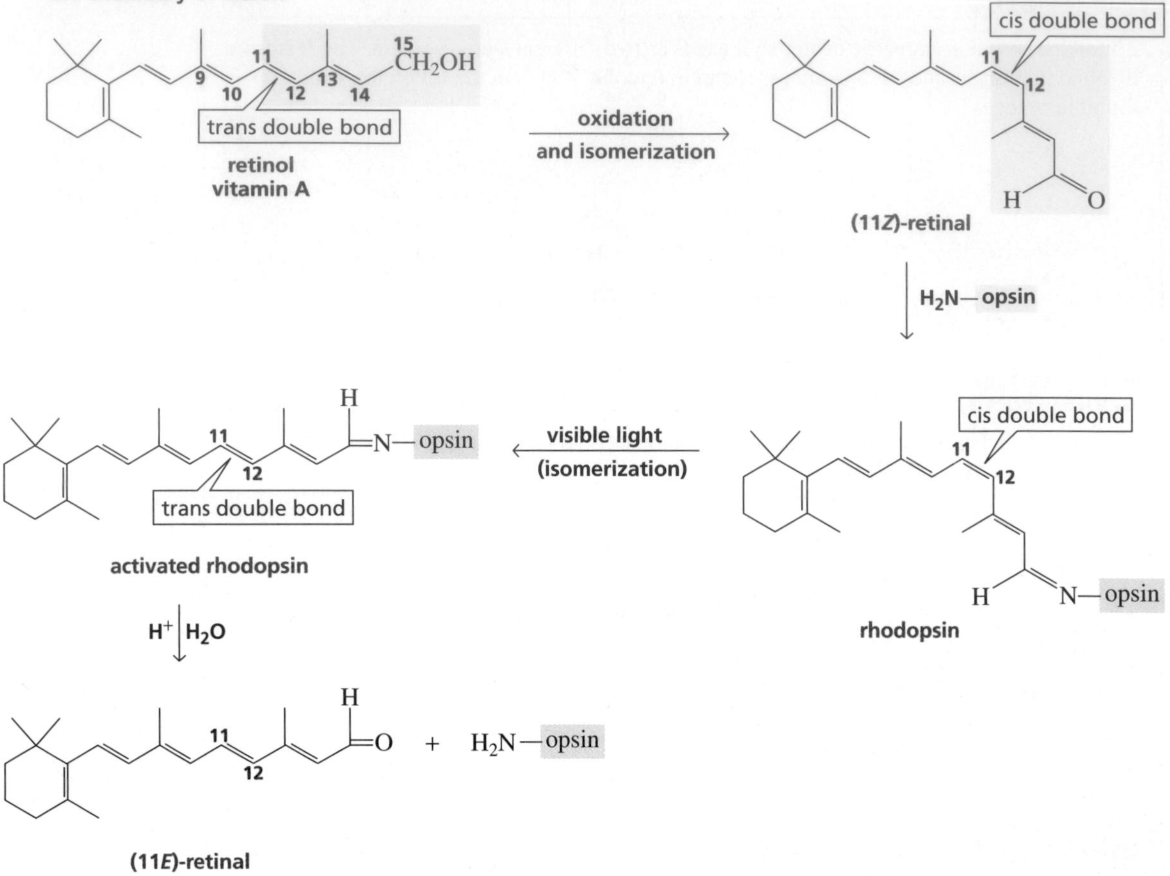
15
CH_2OH
9
10
11
12
13
14
trans double bond
retinol
vitamin A
oxidation
and isomerization
cis double bond
11
12
H
O
(11Z)-retinal
H_2N—opsin
cis double bond
11
12
H
N—opsin
rhodopsin
visible light
(isomerization)
H
N—opsin
11
12
trans double bond
activated rhodopsin
H^+
H_2O
H
O
11
12
+ H_2N—opsin
(11E)-retinal

The details of how the foregoing sequence of reactions creates a visual image are not clearly understood. The fact that a simple change in configuration can be responsible for initiating a process as complicated as vision, though, is remarkable.

26.8 How Terpenes Are Biosynthesized

The five-carbon compound used for the biosynthesis of terpenes is 3-methyl-3-butenylpyrophosphate, loosely called isopentenyl pyrophosphate by biochemists.

Biosynthesis of Isopentenyl Pyrophosphate

Each step in the biosynthesis of isopentenyl pyrophosphate is catalyzed by a different enzyme.

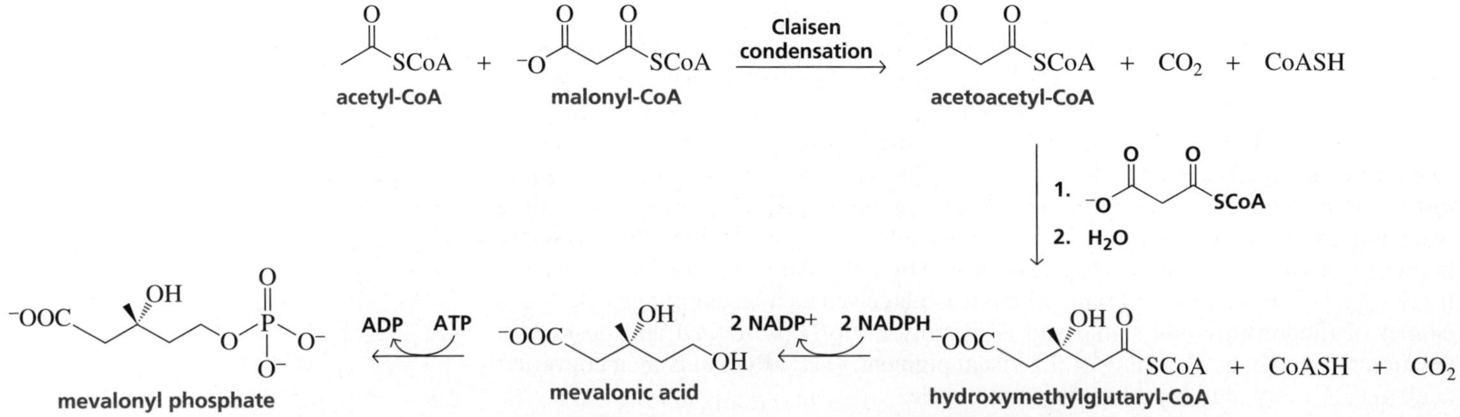
O
SCoA
+
O
O
$^-$O
SCoA
Claisen
condensation
O
O
SCoA
+ CO_2 + CoASH
acetyl-CoA
malonyl-CoA
acetoacetyl-CoA
1.
O
O
$^-$O
SCoA
2. H_2O
$^-$OOC
OH
O
SCoA
+ CoASH + CO_2
hydroxymethylglutaryl-CoA
2 NADP+
2 NADPH
$^-$OOC
OH
OH
mevalonic acid
ADP
ATP
$^-$OOC
OH
O
P
O$^-$
O$^-$
mevalonyl phosphate

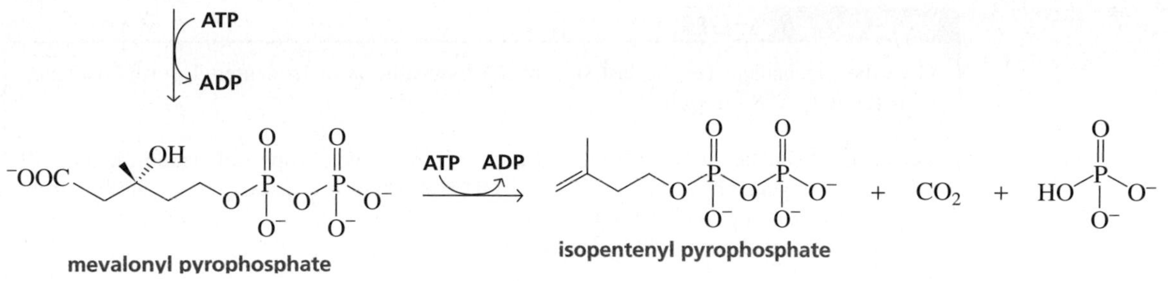

mevalonyl pyrophosphate → isopentenyl pyrophosphate + CO_2 + phosphate

- The first step is the same Claisen condensation that occurs in the first step of the biosynthesis of fatty acids, except that the acetyl and malonyl groups remain attached to coenzyme A rather than being transferred to other thiols (Section 18.22).
- The Claisen condensation is followed by an aldol addition with a second molecule of malonyl-CoA, which is followed by hydrolysis of one of the thioester groups and decarboxylation (Sections 18.12 and 18.18).
- The thioester is reduced with two equivalents of NADPH to form mevalonic acid (Sections 17.6 and 24.2).
- A pyrophosphate group is added by means of two successive phosphorylations with ATP (Section 25.2).
- The OH group is phosphorylated; decarboxylation and loss of the phosphate group result in isopentenyl pyrophosphate (see Problem 18).

The mechanism for converting mevalonic acid into mevalonyl phosphate is essentially an S_N2 reaction between the alcohol group and ATP; adenosyl diphosphate (ADP) is the leaving group (Section 25.2). A second S_N2 reaction converts mevalonyl phosphate to mevalonyl pyrophosphate. ATP is an excellent phosphorylating reagent for nucleophiles because its phosphoanhydride bonds are easily broken. The reason that phosphoanhydride bonds are so easily broken is discussed in Section 25.4.

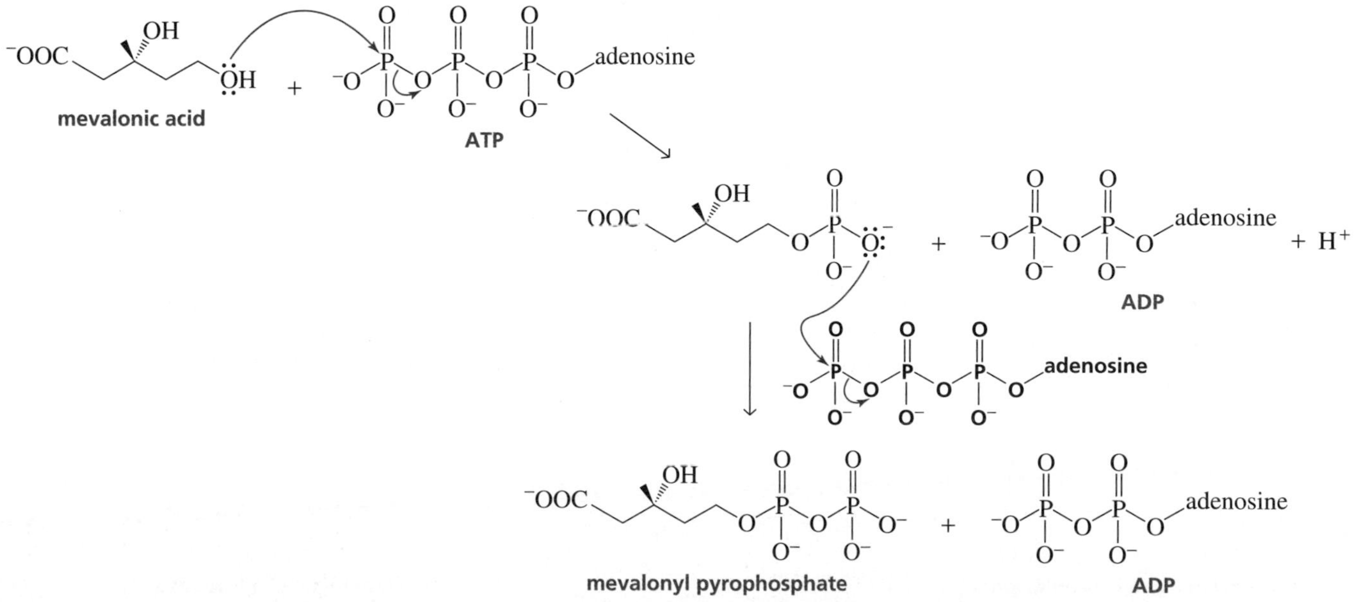

PROBLEM 18 *SOLVED*

Give the mechanism for the last step in the biosynthesis of isopentenyl pyrophosphate, showing why ATP is required.

Solution In the last step of the biosynthesis of isopentenyl pyrophosphate, elimination of CO_2 is accompanied by elimination of an ^-OH group, which is a strong base and therefore a poor leaving group. ATP is used to convert the OH group into a phosphate group, which is easily eliminated because it is a weak base.

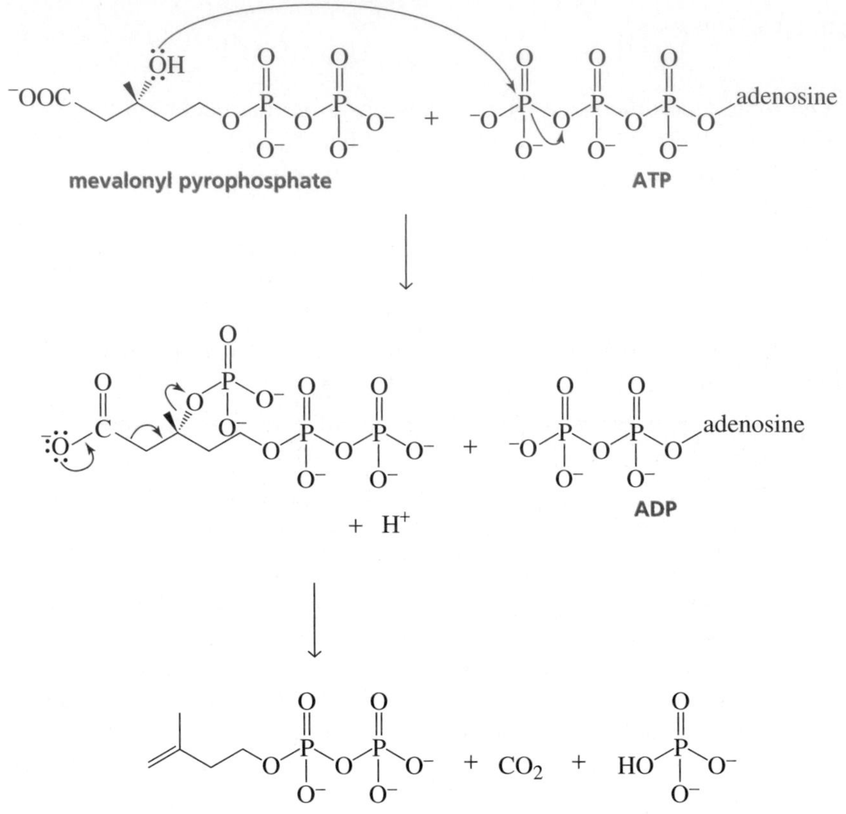

mevalonyl pyrophosphate   ATP

ADP

$+ H^+$

$+ CO_2 + $

PROBLEM 19

Give the mechanisms for the Claisen condensation and aldol addition that occur in the first two steps of the biosynthesis of isopentenyl pyrophosphate.

Biosynthesis of Dimethylallyl Pyrophosphate

Both **isopentenyl pyrophosphate** and **dimethylallyl pyrophosphate** are needed for the biosynthesis of terpenes. Therefore, some isopentenyl pyrophosphate is converted to dimethylallyl pyrophosphate by an enzyme-catalyzed isomerization reaction. The isomerization involves addition of a proton to the sp^2 carbon of isopentenyl pyrophosphate that is bonded to the greater number of hydrogens (Section 4.4) and elimination of a proton from the carbocation intermediate in accordance with Zaitsev's rule (Section 9.2).

Addition of a proton and loss of a proton convert isopentenyl pyrophosphate into dimethylallyl pyrophosphate.

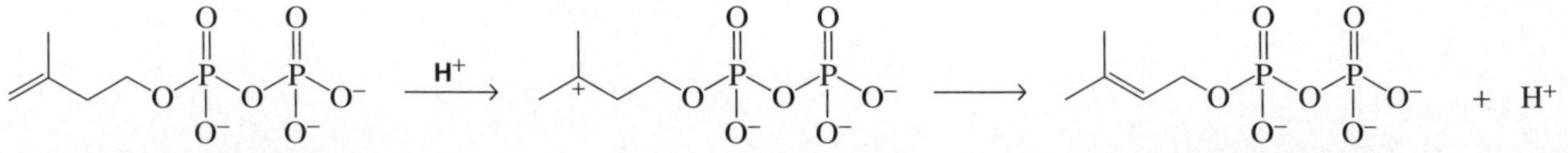

isopentenyl pyrophosphate   dimethylallyl pyrophosphate

Dimethylallyl Pyrophosphate and Isopentenyl Pyrophosphate Are the Starting Materials for the Biosynthesis of Terpenes

The reaction of dimethylallyl pyrophosphate with isopentenyl pyrophosphate forms geranyl pyrophosphate, a 10-carbon compound. In the first step of the reaction, isopentenyl pyrophosphate acts as a nucleophile and displaces a pyrophosphate group from dimethylallyl pyrophosphate. Pyrophosphate is an excellent leaving group: its four OH groups have pK_a values of 0.9, 2.0, 6.6, and 9.4. Therefore, three of the four groups will be primarily in their basic forms at physiological pH (pH = 7.3). A proton is removed in the next step, resulting in the formation of geranyl pyrophosphate.

Terpenes are biosynthesized from isopentenyl pyrophosphate and dimethylallyl pyrophosphate.

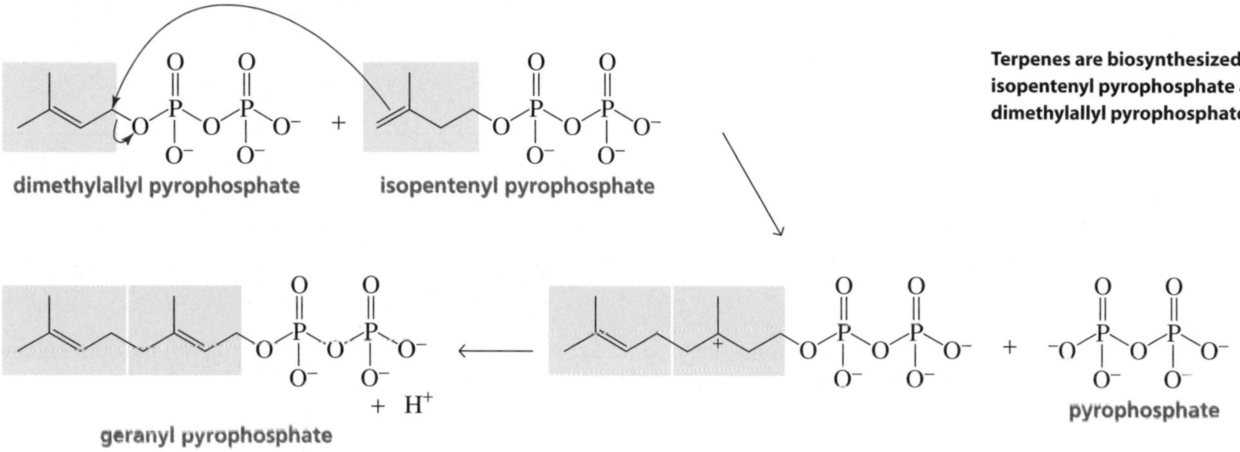
dimethylallyl pyrophosphate
isopentenyl pyrophosphate
+ H+
geranyl pyrophosphate
pyrophosphate

The following scheme shows how some of the many monoterpenes could be synthesized from geranyl pyrophosphate:

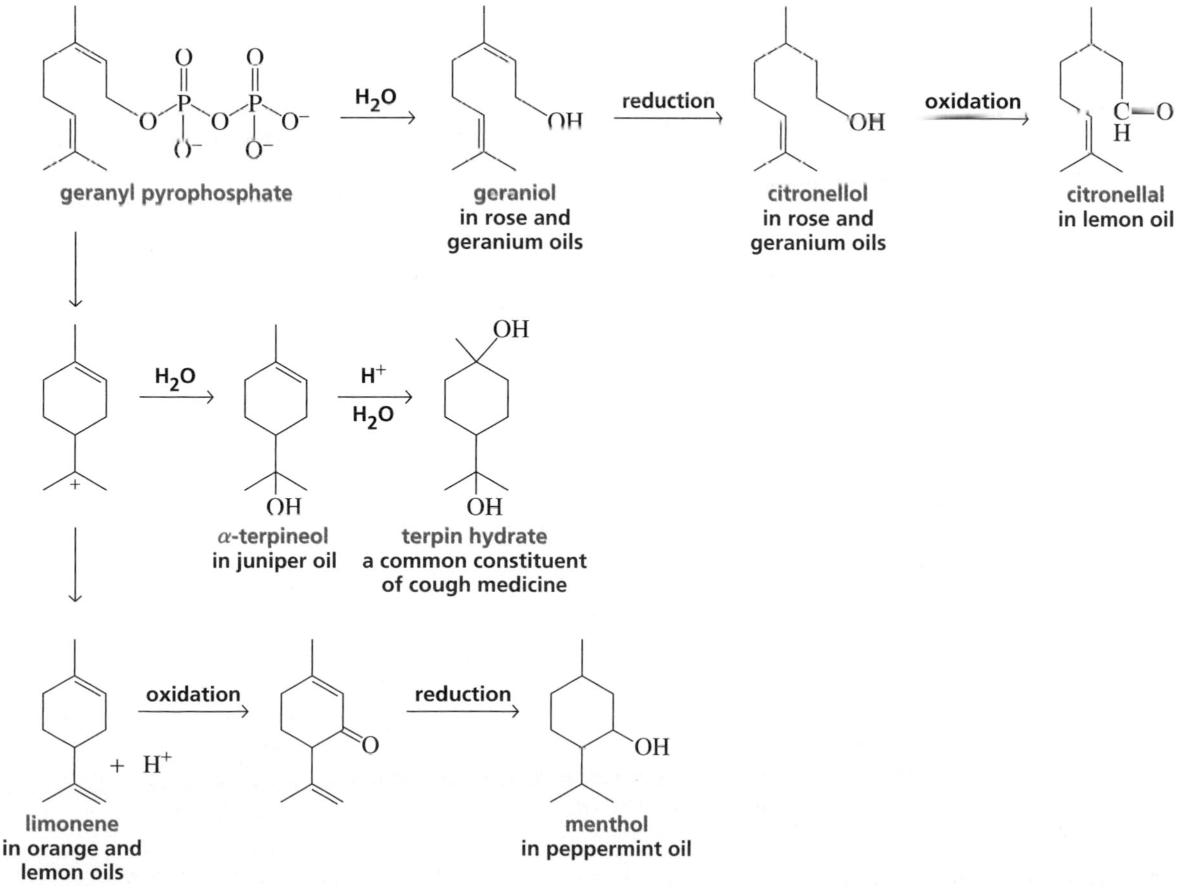
H2O
reduction
oxidation
geranyl pyrophosphate
geraniol
in rose and
geranium oils
citronellol
in rose and
geranium oils
citronellal
in lemon oil
H2O
H+
H2O
α-terpineol
in juniper oil
terpin hydrate
a common constituent
of cough medicine
oxidation
reduction
+ H+
limonene
in orange and
lemon oils
menthol
in peppermint oil

PROBLEM 20

Propose a mechanism for the conversion of the *E* isomer of geranyl pyrophosphate to the *Z* isomer.

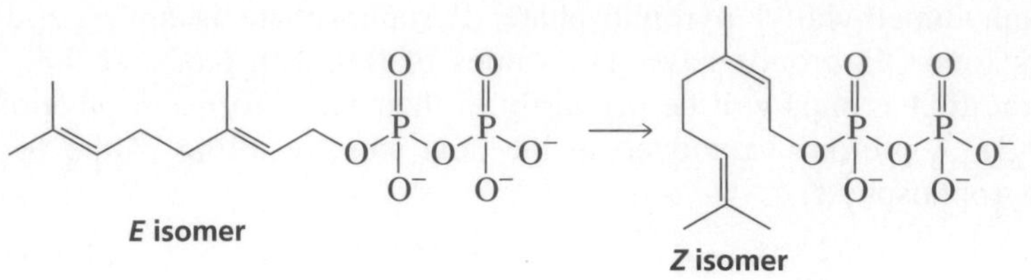

E isomer

Z isomer

PROBLEM 21

Propose mechanisms for the formation of α-terpineol and limonene from geranyl pyrophosphate.

Geranyl pyrophosphate can react with another molecule of isopentenyl pyrophosphate to form farnesyl pyrophosphate, a 15-carbon compound.

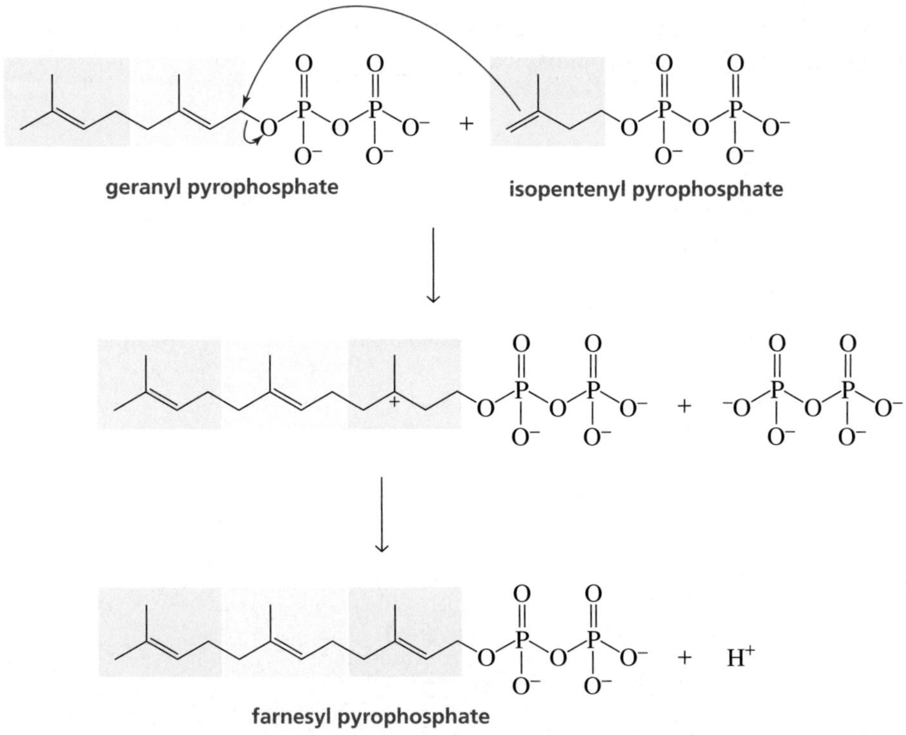

geranyl pyrophosphate

isopentenyl pyrophosphate

farnesyl pyrophosphate

Two molecules of farnesyl pyrophosphate form squalene, a 30-carbon compound. The reaction is catalyzed by the enzyme squalene synthase, which joins the two molecules in a tail-to-tail linkage. As we noted earlier, squalene is the precursor of cholesterol, and cholesterol is the precursor of all other steroids.

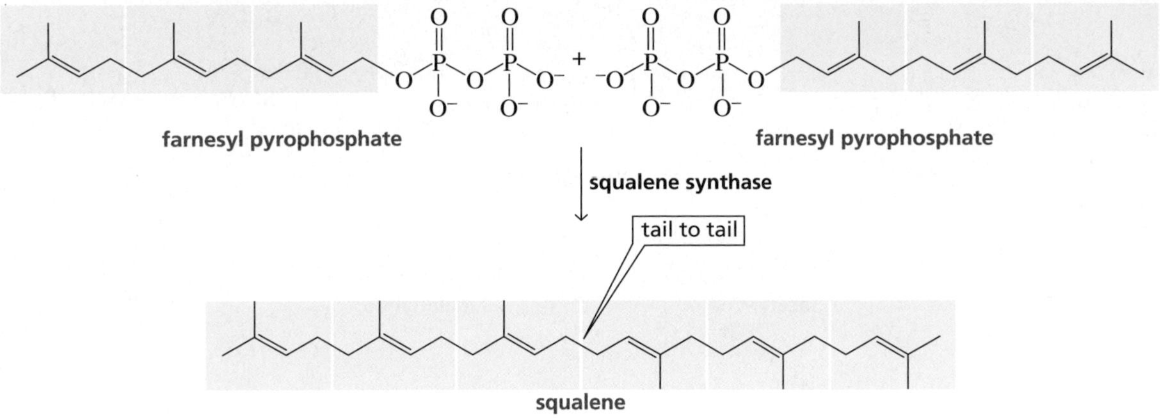

farnesyl pyrophosphate + farnesyl pyrophosphate

squalene synthase

tail to tail

squalene

Farnesyl pyrophosphate can react with another molecule of isopentenyl pyrophosphate to form geranylgeranyl pyrophosphate, a 20-carbon compound. Two geranylgeranyl pyrophosphates can join to form phytoene, a 40-carbon compound. Phytoene is the precursor of the carotenoid (tetraterpene) pigments in plants.

PROBLEM 22

In aqueous acidic solution, farnesyl pyrophosphate forms the following sesquiterpene:

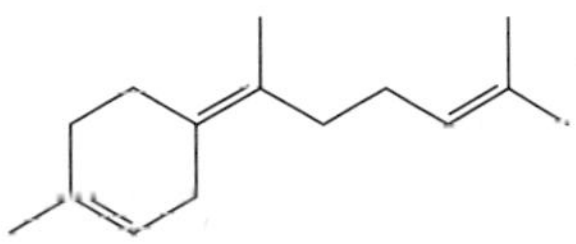

Propose a mechanism for this reaction.

PROBLEM 23 **SOLVED**

If squalene were synthesized in a medium containing acetate whose carbonyl carbon was labeled with radioactive ^{14}C, which carbons in squalene would be labeled?

Solution Acetate reacts with ATP to form acetyl adenylate, which then reacts with CoASH to form acetyl-CoA (Section 16.22). Because malonyl-CoA is prepared from acetyl-CoA, the thioester carbonyl carbon of malonyl-CoA will also be labeled. Examining each step of the mechanism for the biosynthesis of isopentenyl pyrophosphate from acetyl-CoA and malonyl-CoA allows you to determine the locations of the radioactively labeled carbons in isopentenyl pyrophosphate. Similarly, the locations of the radioactively labeled carbons in geranyl pyrophosphate can be determined from the mechanism for its biosynthesis from isopentenyl pyrophosphate. And the locations of the radioactively labeled carbons in farnesyl pyrophosphate can be determined from the mechanism for its biosynthesis from geranyl pyrophosphate. Knowing that squalene is obtained from a tail-to-tail linkage of two farnesyl pyrophosphates tells you which carbons in squalene will be labeled.

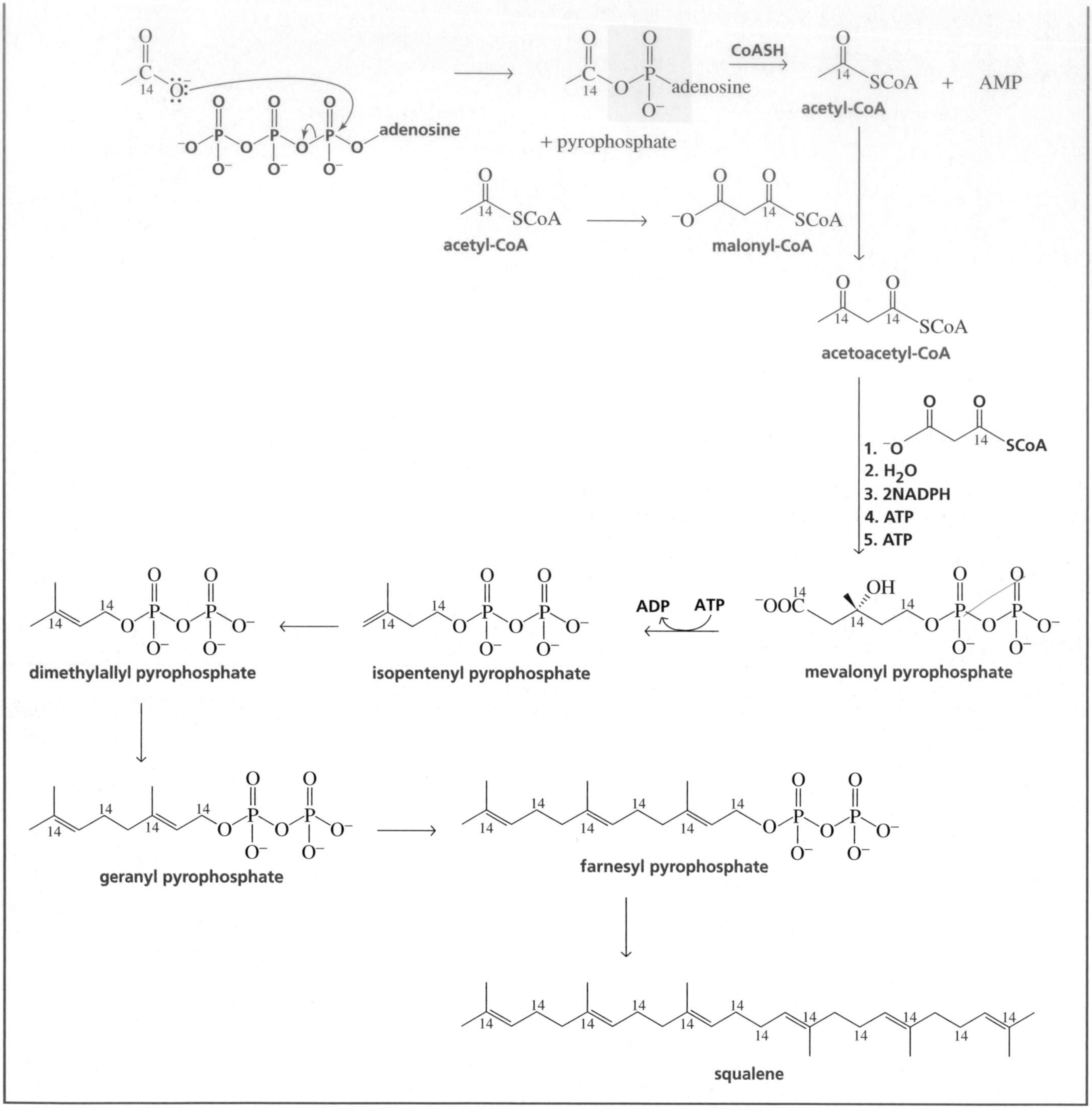

26.9 Steroids Are Chemical Messengers

Hormones are chemical messengers—organic compounds synthesized in glands and delivered by the bloodstream to target tissues in order to stimulate or inhibit some process. Many hormones are **steroids**. Because steroids are nonpolar compounds, they too are lipids. Their nonpolar character allows them to cross cell membranes, so they can leave the cells in which they are synthesized and enter their target cells.

All steroids contain a tetracyclic ring system. The four rings are designated A, B, C, and D. Rings A, B, and C are six-membered rings, and D is a five-membered ring. The carbons in the steroid ring system are numbered as shown in the margin. Rings can be **trans fused** or **cis fused**; trans fused rings are more stable.

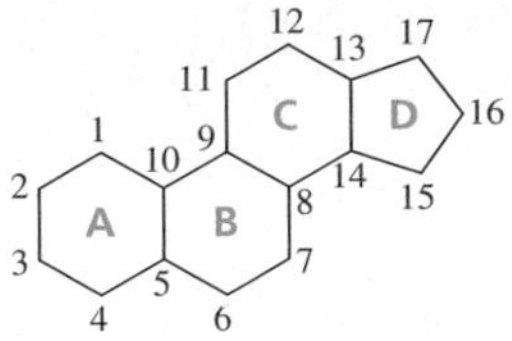

the steroid ring system

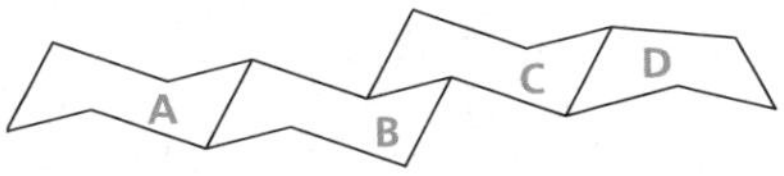

all the rings are trans fused

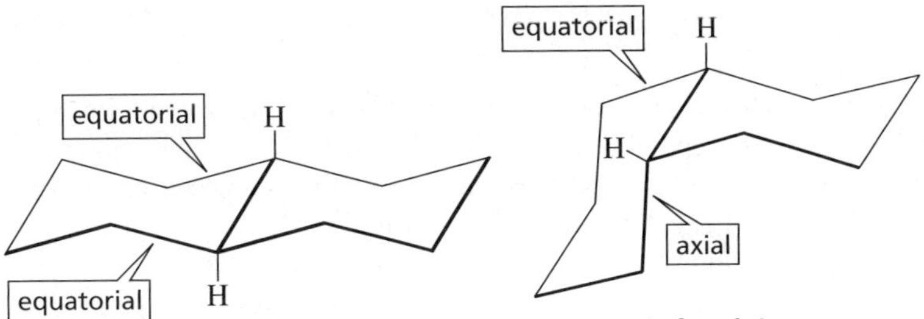

trans-fused rings
more stable

cis-fused rings
less stable

In steroids, the B, C, and D rings are all trans fused. In most naturally occurring steroids, the A and B rings are also trans fused.

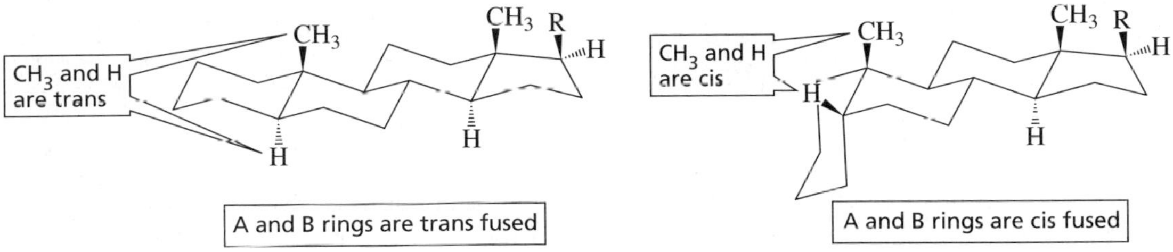

A and B rings are trans fused

A and B rings are cis fused

Many steroids have methyl groups at the 10- and 13-positions. These are called **angular methyl groups**. When steroids are drawn, both angular methyl groups are shown to be above the plane of the steroid ring system. Substituents on the same side of the steroid ring system as the angular methyl groups are designated β-substituents (indicated by a solid wedge). Those on the opposite side of the plane of the ring system are α-substituents (indicated by a hatched wedge).

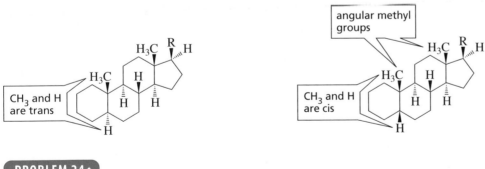

CH₃ and H are trans

angular methyl groups

CH₃ and H are cis

PROBLEM 24◆

A β-hydrogen at C-5 means that the A and B rings are _____ fused; an α-hydrogen at C-5 means that they are _____ fused.

The most abundant member of the steroid family in animals is **cholesterol**, the precursor of all other steroids. Cholesterol is biosynthesized from squalene, a triterpene (Section 26.6), and is an important component of cell membranes (Figure 26.1). Its ring structure makes it more rigid than other membrane lipids. Because cholesterol has eight asymmetric centers, 256 stereoisomers are possible, but only one exists in nature (Chapter 5, Problem 24).

cholesterol

Cholesterol is the precursor of all other steroids.

Student Tutorial: Steroids

The steroid hormones can be divided into five classes: glucocorticoids, mineralocorticoids, androgens, estrogens, and progestins. Glucocorticoids and mineralocorticoids are synthesized in the adrenal cortex and are collectively known as *adrenal cortical steroids*. All adrenal cortical steroids have an oxygen at C-11.

Glucocorticoids, as their name suggests, are involved in glucose metabolism; they also participate in the metabolism of proteins and fatty acids. Cortisone is an example of a glucocorticoid. Because of its anti-inflammatory effect, it is used clinically to treat arthritis and other inflammatory conditions.

cortisone

aldosterone

Mineralocorticoids cause increased reabsorption of Na^+, Cl^-, and HCO_3^- by the kidneys, leading to an increase in blood pressure. Aldosterone is an example of a mineralocorticoid.

PROBLEM 25◆

Is the OH substituent of the A ring of cholesterol an α-substituent or a β-substituent?

PROBLEM 26

Aldosterone is in equilibrium with its cyclic hemiacetal. Draw the hemiacetal form of aldosterone.

Sex hormones, known as *androgens*, are secreted primarily by the testes. They are responsible for the development of male secondary sex characteristics during puberty, including muscle growth. Testosterone and 5α-dihydrotestosterone are androgens.

5α-dihydrotestosterone

testosterone

Estradiol and estrone are sex hormones known as *estrogens*. They are secreted primarily by the ovaries and are responsible for the development of female secondary sex characteristics. They also regulate the menstrual cycle. Progesterone is the most important member of a group of hormones called *progestins*: it is the hormone that prepares the lining of the uterus for implantation of an ovum and is essential for the maintenance of pregnancy. It also prevents ovulation during pregnancy.

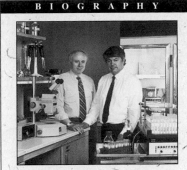

B I O G R A P H Y

Michael S. Brown *and* **Joseph Leonard Goldstein** *shared the 1985 Nobel Prize in physiology or medicine for their work on the regulation of cholesterol metabolism and the treatment of disease caused by elevated cholesterol levels in the blood. Brown was born in New York in 1941, Goldstein in South Carolina in 1940. They are both professors of medicine at the University of Texas Southwestern Medical Center.*

estradiol   estrone   progesterone

Although the various steroid hormones have remarkably different physiological effects, their structures are quite similar. For example, the only difference between testosterone and progesterone is the substituent at C-17, and the only difference between 5α-dihydrotestosterone and estradiol is a methyl group and three hydrogens, but these compounds make the difference between being male and being female. These examples illustrate the extreme specificity of biochemical reactions.

In addition to being the precursor of all the steroid hormones, cholesterol is the precursor of the *bile acids*. In fact, the word *cholesterol* is derived from the Greek words *chole* meaning "bile" and *stereos* meaning "solid." The bile acids— cholic acid and chenodeoxycholic acid—are synthesized in the liver, stored in the gallbladder, and secreted into the small intestine, where they act as emulsifying agents so that fats and oils can be digested by water-soluble digestive enzymes. Cholesterol is also the precursor of vitamin D (Section 29.6).

cholic acid   chenodeoxycholic acid

CHOLESTEROL AND HEART DISEASE

Cholesterol is probably the best-known lipid because of the widely publicized correlation between cholesterol levels in the blood and heart disease. Cholesterol is synthesized in the liver and is also found in almost all body tissues. Cholesterol is also found in many foods, but we do not require it in our diet because the body can synthesize all we need. A diet high in cholesterol can lead to high levels of cholesterol in the bloodstream, and the excess can accumulate on the walls of arteries, restricting the flow of blood. This disease of the circulatory system is known as *atherosclerosis* and is a primary cause of heart disease. Cholesterol travels through the bloodstream packaged in particles that also contain cholesterol esters, phospholipids, and proteins. The particles are classified according to their density. Low-density lipoprotein (LDL) particles transport cholesterol from the liver to other tissues. Receptors on the surfaces of cells bind LDL particles, allowing them to be brought into the cell so that it can use the cholesterol. High-density lipoprotein (HDL) is a cholesterol scavenger, removing cholesterol from the surfaces of membranes and delivering it back to the liver, where it is converted into bile acids. LDL is the so-called bad cholesterol, whereas HDL is the "good" cholesterol. The more cholesterol we eat, the less the body synthesizes. But this does not mean that the presence of dietary cholesterol has no effect on the total amount of cholesterol in the bloodstream, because dietary cholesterol also inhibits the synthesis of the LDL receptors. So the more cholesterol we eat, the less the body synthesizes, but also, the less the body can get rid of by bringing it into target cells.

CLINICAL TREATMENT OF HIGH CHOLESTEROL

Statins are the newest class of cholesterol-reducing drugs. Statins reduce serum cholesterol levels by inhibiting the enzyme that catalyzes the reduction of hydroxymethylglutaryl-CoA to mevalonic acid (Section 26.8). Decreasing the mevalonic acid concentration decreases the isopentenyl pyrophosphate concentration, so the biosynthesis of all terpenes, including cholesterol, is diminished. As a consequence of diminished cholesterol synthesis in the liver, the liver forms more LDL receptors—the receptors that help clear LDL from the bloodstream. Studies show that for every 10% that cholesterol is reduced, deaths from coronary heart disease are reduced by 15% and total death risk is reduced by 11%.

Simvastatin and lovastatin are natural statins used clinically under the trade names Zocor and Mevacor. Atorvastatin (Lipitor), a synthetic statin, is now the most popular statin. Lipitor has greater potency and a longer half-life than natural statins have, because its metabolites are as active as the parent drug in reducing cholesterol levels. Therefore, smaller doses of the drug may be administered. The required dose is reduced further because Lipitor is marketed as a single enantiomer. In addition, it is more lipophilic than compactin and lovastatin, so it has a greater tendency to remain in the endoplasmic reticulum of the liver cells, where it is needed. Lipitor was the second most widely prescribed drug in the United States in 2004 (Section 30.0).

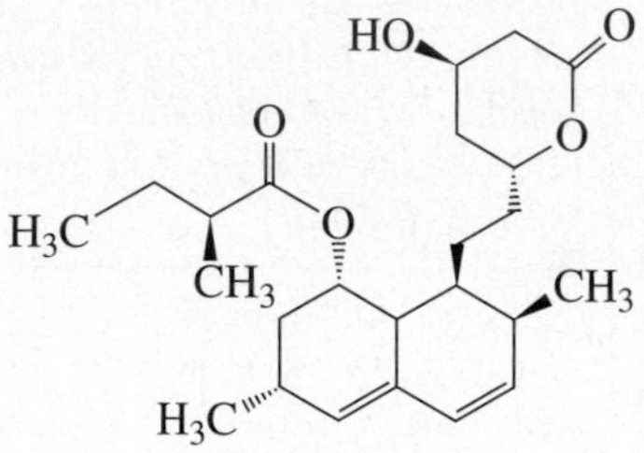

lovastatin
Mevacor®

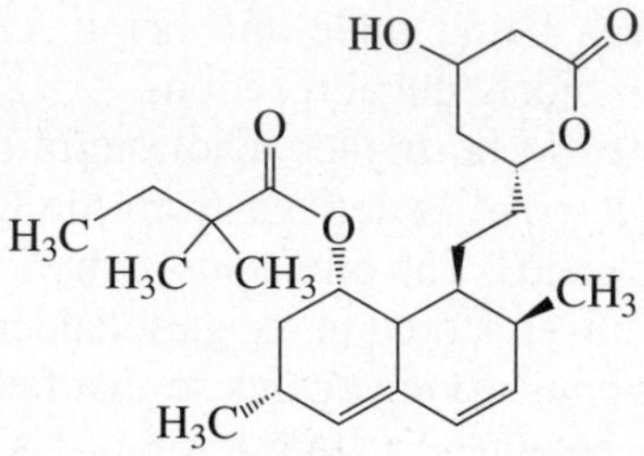

simvastatin
Zocor®

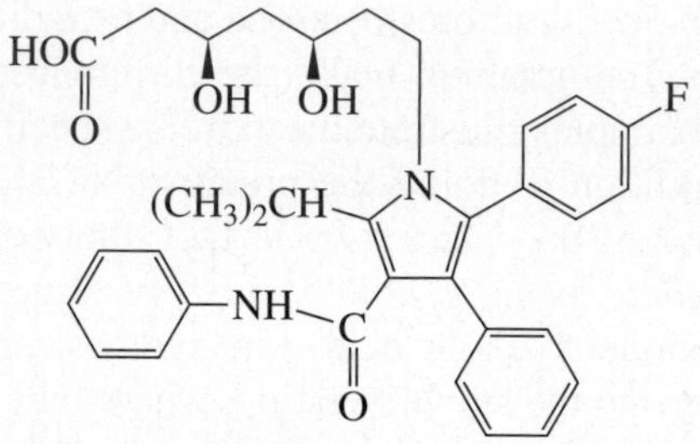

atorvastatin
Lipitor®

PROBLEM 27

The acid component of a cholesterol ester is a fatty acid such as linoleic acid. Draw the structure of a cholesterol ester.

PROBLEM 28◆

Are the three OH groups of cholic acid axial or equatorial?

26.10 How Nature Synthesizes Cholesterol

How is cholesterol, the precursor of all the steroid hormones, biosynthesized? The starting material for the biosynthesis is the triterpene squalene, which must first be converted to lanosterol. Lanosterol is converted to cholesterol in a series of 19 steps.

biosynthesis of lanosterol and cholesterol

squalene

$\xrightarrow{\text{squalene epoxidase}}$
O_2

squalene oxide

Cholesterol is synthesized from squalene.

H^+

acid-catalyzed opening of the epoxide

lanosterol

$+ \; H^+$

protosterol cation

the C-9 proton

19 steps

cholesterol

- The first step is epoxidation of the 2,3-double bond of squalene.
- Acid-catalyzed opening of the epoxide initiates a series of cyclizations resulting in the protosterol cation.
- Elimination of a C-9 proton from the cation initiates a series of 1,2-hydride and 1,2-methyl shifts, resulting in lanosterol.

Converting lanosterol to cholesterol requires removing three methyl groups from lanosterol, in addition to reducing two double bonds and creating a new double bond. Removing methyl groups from carbon atoms is not easy, and many different enzymes are required to carry out the 19 steps. So why does nature bother? Why not just use lanosterol instead of cholesterol? Konrad Bloch answered that question when he found that membranes containing lanosterol instead of cholesterol are much more permeable. Small molecules are able to pass easily through lanosterol-containing membranes. As each methyl group is removed from lanosterol, the membrane becomes less and less permeable.

B I O G R A P H Y

Konrad Bloch *and* **Feodor Lynen** *shared the 1964 Nobel Prize in physiology or medicine.* **Konrad Emil Bloch (1912–2000)** *was born in Upper Silesia (then a part of Germany), left Nazi Germany for Switzerland in 1934, and came to the United States in 1936, becoming a U.S. citizen in 1944. He received a Ph.D. from Columbia in 1938, taught at the University of Chicago, and became a professor of biochemistry at Harvard in 1954. Bloch showed how fatty acids and cholesterol are biosynthesized from acetate, and for this shared the 1964 Nobel Prize in physiology or medicine with Feodor Lynen.*

B I O G R A P H Y

Feodor Lynen (1911–1979) *was born in Germany, received a Ph.D. under Heinrich Wieland, and married Wieland's daughter. He was head of the Institute of Cell Chemistry at the University of Munich. Because he showed that the two-carbon acetate unit is actually acetyl-CoA, and also determined the structure of coenzyme A, Lynen shared he 1964 Nobel Prize in physiology or medicine with Konrad Bloch.*

PROBLEM 29

Draw the individual 1,2-hydride and 1,2-methyl shifts responsible for conversion of the protosterol cation to lanosterol. How many hydride shifts are involved? How many methyl shifts?

26.11 **Synthetic Steroids**

The potent physiological effects of steroids led scientists, in their search for new drugs, to synthesize steroids that are not available in nature and to investigate their physiological effects. Stanozolol and Dianabol are two drugs developed in this way that have the same muscle-building effect as testosterone. Steroids that aid in the development of muscle are called *anabolic steroids*. These drugs are available by prescription and are used to treat people suffering from traumas accompanied by muscle deterioration. The same drugs have been administered to athletes and racehorses to increase their muscle mass. Stanozolol was the drug detected in several athletes in the 1988 Olympics. Anabolic steroids, when taken in relatively high dosages, have been found to cause liver tumors, personality disorders, and testicular atrophy.

stanozolol

Dianabol®

Many synthetic steroids have been found to be much more potent than natural steroids. Norethindrone, for example, is better than progesterone in arresting ovulation. Another synthetic steroid, RU 486, when taken along with prostaglandins, terminates pregnancy within the first nine weeks of gestation. Its name comes from Roussel-Uclaf, the French pharmaceutical company where it was first synthesized in 1980, and an arbitrary lab serial number. Notice that these compounds have structures similar to that of progesterone.

norethindrone

mefipristone
RU 486

SUMMARY

Lipids are bioorganic compounds that are soluble in nonpolar solvents. One class of lipids, **fatty acids**, are carboxylic acids with long unbranched hydrocarbon chains. Double bonds in fatty acids have the cis configuration. Fatty acids with more than one double bond are called **polyunsaturated fatty acids**. Double bonds in naturally occurring unsaturated fatty acids are separated by one methylene group. **Waxes** are esters formed from long-chain carboxylic acids and long-chain alcohols. **Prostaglandins** are synthesized from arachidonic acid and are responsible for regulating a variety of physiological responses.

Triacylglycerols (triglycerides) are compounds in which the three OH groups of glycerol are esterified with fatty acids. Triacylglycerols that are solids or semisolids

at room temperature are called **fats**; liquid triacylglycerols are called **oils**. Some or all of the double bonds of polyunsaturated oils can be reduced by catalytic hydrogenation. **Phosphoacylglycerols** differ from triacylglycerols in that the terminal OH group of glycerol is esterified with phosphoric acid instead of a fatty acid. Phosphoacylglycerols form membranes by arranging themselves in a **lipid bilayer**. **Phospholipids** are lipids that contain a phosphate group. **Sphingolipids**, also found in membranes, contain sphingosine (an amino alcohol) instead of glycerol.

Terpenes contain carbon atoms in multiples of 5. They are made by joining together five-carbon units, usually in a head-to-tail fashion; this is called the **isoprene rule**.

Monoterpenes—terpenes with two isoprene units—have 10 carbons; **sesquiterpenes** have 15. Squalene, a **triterpene** (a terpene with six isoprene units), is a precursor of steroid molecules. Lycopene and β-carotene are **tetraterpenes** called **carotenoids**. β-Carotene is cleaved to form two molecules of vitamin A.

The five-carbon compound used for the synthesis of terpenes is isopentenyl pyrophosphate. The reaction of **dimethylallyl pyrophosphate** (formed from isopentenyl pyrophosphate) with **isopentenyl pyrophosphate** forms geranyl pyrophosphate, a 10-carbon compound. Geranyl pyrophosphate can react with another molecule of isopentenyl pyrophosphate to form farnesyl pyrophosphate, a 15-carbon compound. Two molecules of farnesyl pyrophosphate form **squalene**, a 30-carbon compound. Squalene is the precursor of **cholesterol**. Farnesyl pyrophosphate can react with another molecule of isopentenyl pyrophosphate to form geranylgeranyl pyrophosphate, a 20-carbon compound. Two geranylgeranyl pyrophosphates join to form phytoene, a 40-carbon compound. Phytoene is the precursor of the **carotenoids**.

Hormones are chemical messengers. Many hormones are **steroids**. All steroids contain a tetracyclic ring system. The B, C, and D rings are **trans fused**. In most naturally occurring steroids, the A and B rings are also trans fused. Methyl groups at C-10 and C-13 are called **angular methyl groups**. β-Substituents are on the same side of the steroid ring system as the angular methyl groups; α-substituents are on the opposite side. Synthetic steroids are steroids that are not found in nature.

The most abundant member of the steroid family in animals is **cholesterol**, the precursor of all other steroids. Cholesterol is an important component of cell membranes; its ring structure causes it to be more rigid than other membrane lipids. In the biosynthesis of cholesterol, squalene is converted to lanosterol, which is converted to cholesterol.

KEY TERMS

angular methyl group (p. 1187)
carotenoid (p. 1178)
cholesterol (p. 1187)
cis fused (p. 1187)
dimethylallyl pyrophos-
  phate (p. 1182)
fat (p. 1166)
fatty acid (p. 1163)
hormone (p. 1186)
isopentenyl pyrophosphate (p. 1182)
isoprene rule (p. 1177)
lipid (p. 1162)

lipid bilayer (p. 1171)
membrane (p. 1171)
mixed triacylglycerol (p. 1165)
monoterpene (p. 1178)
oil (p. 1167)
phosphatidic acid (p. 1170)
phosphoacylglycerol (p. 1170)
phosphoglycerides (p. 1170)
phospholipid (p. 1170)
polyunsaturated fatty
  acid (p. 1164)
prostaglandin (p. 1173)

sesquiterpene (p. 1178)
simple triacylglycerol (p. 1165)
sphingolipid (p. 1172)
squalene (p. 1178)
steroid (p. 1186)
terpene (p. 1177)
terpenoid (p. 1177)
tetraterpene (p. 1178)
trans fused (p. 1187)
triacylglycerol (p. 1165)
triterpene (p. 1178)
wax (p. 1165)

PROBLEMS

30. An optically active fat, when completely hydrolyzed, yields twice as much stearic acid as palmitic acid. Draw the structure of the fat.

31. Do all triacylglycerols have the same number of asymmetric centers?

32. a. How many different triacylglycerols are there in which one of the fatty acid components is lauric acid and two are myristic acid?
  b. How many different triacylglycerols are there in which one of the fatty acid components is lauric acid, one is myristic acid, and one is palmitic acid?

33. Cardiolipins are found in heart muscles. Give the products formed when a cardiolipin undergoes complete acid-catalyzed hydrolysis.

$$
\begin{array}{ccc}
& O & & O \\
& \parallel & & \parallel \\
CH_2O-CR^1 & & R^3C-OCH_2 \\
| \quad O & & O \quad | \\
\quad \parallel & & \parallel \\
CHO-CR^2 & & R^4C-OCH \\
| \quad O & & O \quad | \\
\quad \parallel & & \parallel \\
CH_2O-P-OCH_2CHCH_2O-P-OCH_2 \\
\quad | \qquad\qquad | \qquad\qquad | \\
\quad O^- \qquad\quad OH \qquad\quad O^-
\end{array}
$$

a cardiolipin

34. Nutmeg contains a simple, fully saturated triacylglycerol with a molecular weight of 722. Draw its structure.

35. Give the product that would be obtained from the reaction of cholesterol with each of the following reagents:
(*Hint:* Because of steric hindrance from the angular methyl groups, the α-face is more susceptible to attack by reagents than the β-face.)
  a. H_2O, H^+
  b. BH_3 in THF, followed by H_2O_2 + HO^-
  c. H_2, Pd/C
  d. Br_2 + H_2O
  e. peroxyacetic acid
  f. the product of part e + CH_3O^-

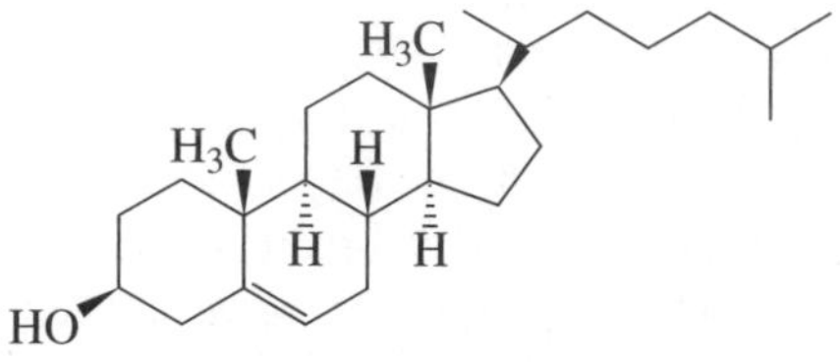

cholesterol

36. Dr. Cole S. Terol synthesized the following samples of mevalonic acid and fed them to a group of lemon trees:

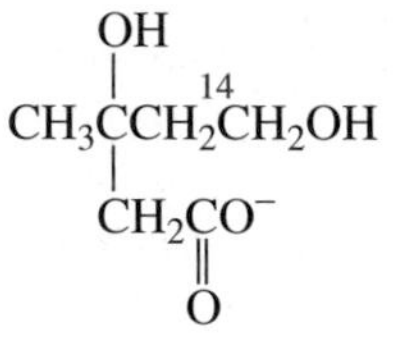

sample A 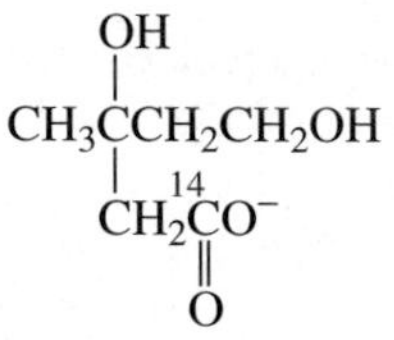   sample B 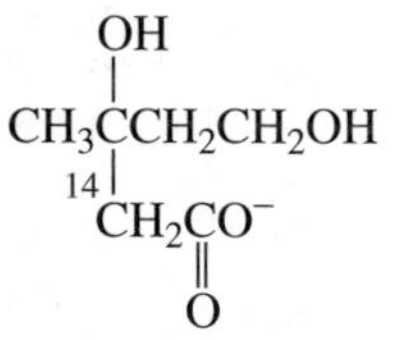   sample C

Which carbons in citronellal (page 1183), which is isolated from lemon oil, will be labeled in trees that were fed the following?
  a. sample A   **b.** sample B   **c.** sample C

37. An optically active monoterpene (compound A) with molecular formula $C_{10}H_{18}O$ undergoes catalytic hydrogenation to form an optically inactive compound with molecular formula $C_{10}H_{20}O$ (compound B). When compound B is heated with acid, followed by reaction with O_3, and workup under reducing conditions (Zn, H_2O), one of the products obtained is 4-methylcyclohexanone. Give possible structures for compound A.

38. Eudesmol is a sesquiterpene found in eucalyptus. Propose a mechanism for its biosynthesis from isopentenyl pyrophosphate.

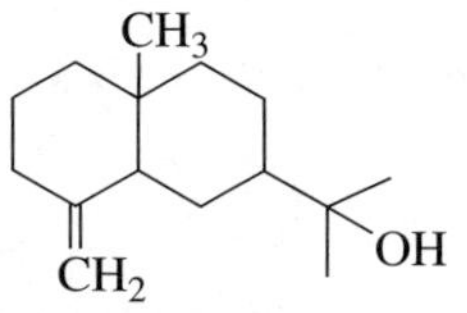

eudesmol

39. If junipers were allowed to grow in a medium containing acetate in which the methyl carbon was labeled with ^{14}C, which carbons in α-terpineol would be labeled?

40. a. Propose a mechanism for the following reaction:

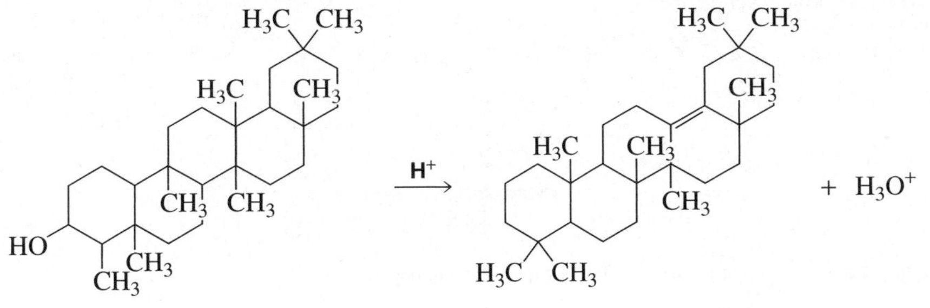

b. To what class of terpene does the starting material belong? Mark off the isoprene units in the starting material.

41. 5-Androstene-3,17-dione is isomerized to 4-androstene-3,17-dione by hydroxide ion. Propose a mechanism for this reaction.

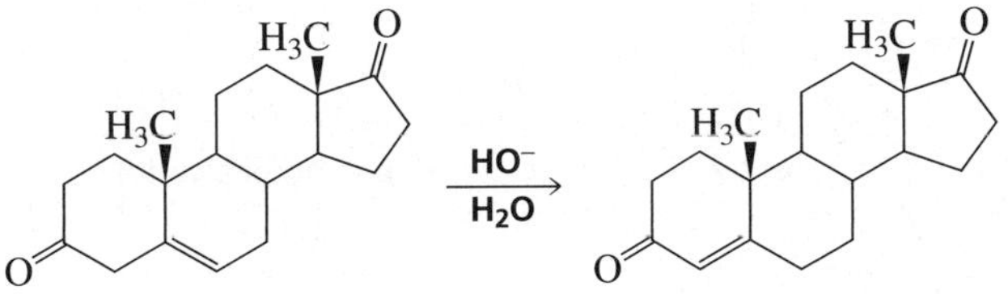

5-androstene-3,17-dione **4-androstene-3,17-dione**

42. Both OH groups of one of the following steroid diols react with excess ethyl chloroformate, but only one OH group of the other steroid diol reacts under the same conditions:

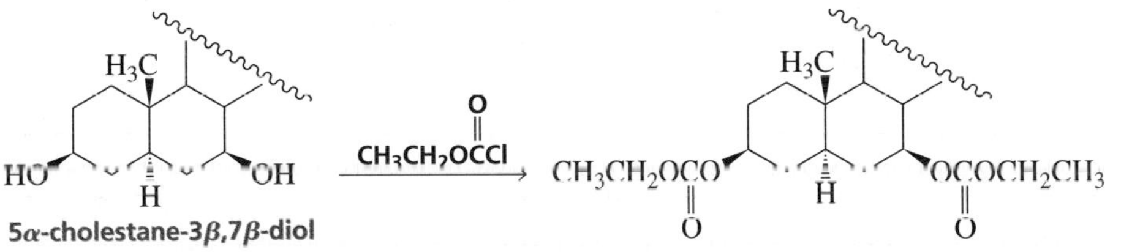

5α-cholestane-3β,7β-diol

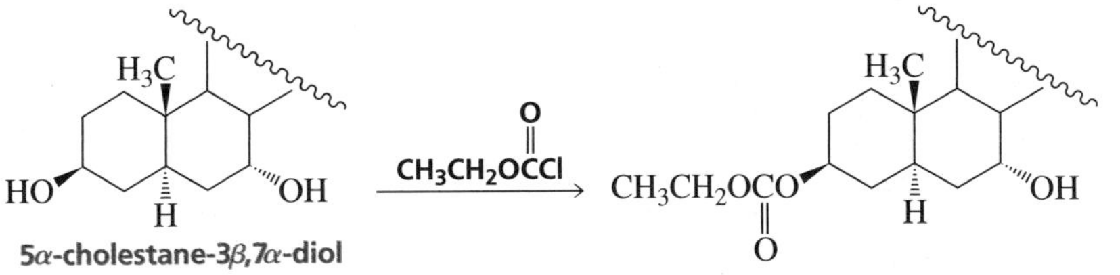

5α-cholestane-3β,7α-diol

Explain the difference in reactivity.

43. The acid-catalyzed dehydration of an alcohol to a rearranged alkene is known as a Wagner–Meerwein rearrangement. Propose a mechanism for the following Wagner–Meerwein rearrangement:

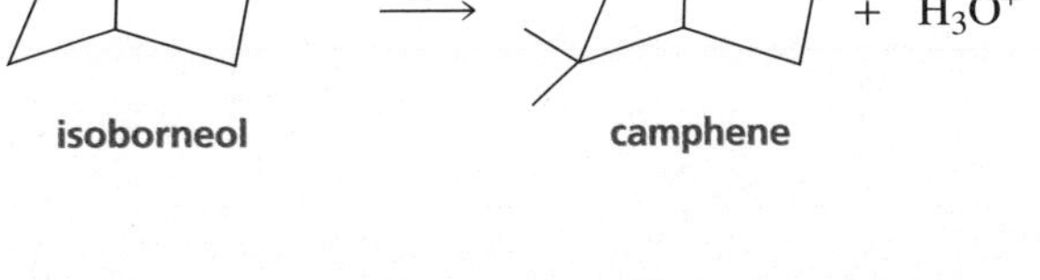

isoborneol **camphene**

44. Diethylstilbestrol (DES) was given to pregnant women to prevent miscarriage, until it was found that the drug caused cancer in both the mothers and their female children. DES has estradiol activity even though it is not a steroid. Draw DES in a way that shows that it is structurally similar to estradiol.

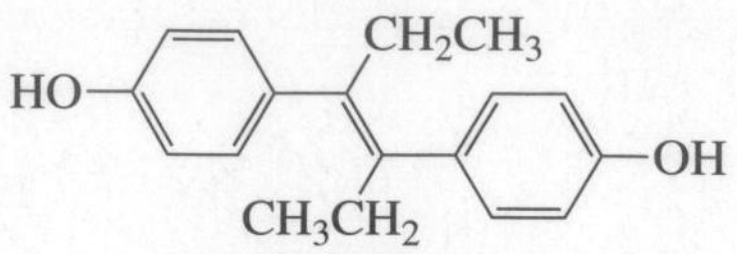

diethylstilbestrol
DES

45. Cedrol and patchouli alcohol are terpenes that are isolated from essential oils.
   a. What class of terpene are they?
   b. Mark off the isoprene units in each.

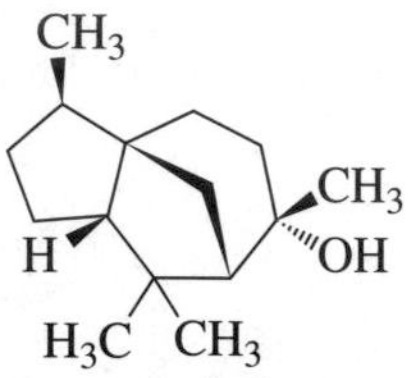

cedrol 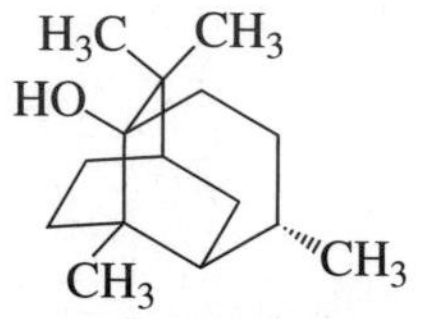   **patchouli alcohol**

CHAPTER 27

Nucleosides, Nucleotides, and Nucleic Acids

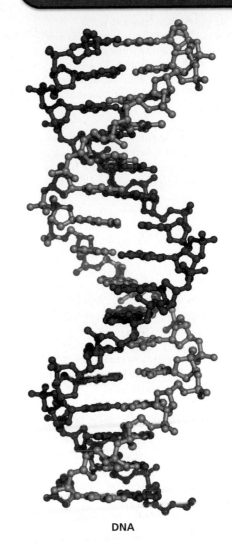

DNA

BUILDING ON FUNDAMENTALS

SECTION 27.3	A hydrogen bonded to an O, N, or F forms a hydrogen bond with an O, N, or F in another molecule (2.9).	**SECTION 27.8**	Hydrolysis of an imine forms a carbonyl compound and ammonia or a primary amine (17.8).
SECTION 27.7	A nucleophilic acyl substitution reaction attaches an amino acid to a tRNA (16.7).	**SECTION 27.12**	A hydrogen bonded to a carbon adjacent to a cyano group is relatively acidic (18.1).

In previous chapters, we studied two of the three major kinds of biopolymers: polysaccharides and proteins. Now we will look at the third—nucleic acids. There are two types of nucleic acids: **deoxyribonucleic acid (DNA)** and **ribonucleic acid (RNA)**. DNA encodes an organism's entire hereditary information and controls the growth and division of cells. In all organisms (except certain viruses), the genetic information stored in DNA is transcribed into RNA. This information can then be translated for the synthesis of all the proteins needed for cellular structure and function.

DNA was first isolated in 1869 from the nuclei of white blood cells. Because it was found in the nucleus and was acidic, it was called *nucleic acid*. Eventually, scientists found that the nuclei of all cells contain DNA, but until experiments were reported in 1944 in which DNA was transferred from one species to another, along with inheritable traits, they did not realize that nucleic acids are the carriers of genetic information. In 1953, James Watson and Francis Crick described the three-dimensional structure of DNA—the famed double helix.

BIOGRAPHY

Studies that determined the structures of the nucleic acids and paved the way for the discovery of the DNA double helix were carried out by ***Phoebus Levene*** *and elaborated by* ***Sir Alexander Todd****.*

Phoebus Aaron Theodor Levene (1869–1940) *was born in Russia. When he immigrated to the United States with his family in 1891, his Russian name, Fishel, was changed to Phoebus. His medical school education had been interrupted, so he returned to Russia to complete his studies. When he returned to the United States, he took chemistry courses at Columbia University. Deciding to forgo medicine for a career in chemistry, he went to Germany to study under Emil Fischer. He was a professor of chemistry at the Rockefeller Institute (now Rockefeller University).*

BIOGRAPHY

Alexander R. Todd (1907–1997) *was born in Scotland. He received two Ph.D. degrees, one from Johann Wolfgang Goethe University in Frankfurt (1931) and one from Oxford University (1933). He was a professor of chemistry at the University of Edinburgh, then at the University of Manchester, and from 1944 to 1971 at Cambridge University. He was knighted in 1954 and was made a baron in 1962 (Baron Todd of Trumpington). For his work on nucleotides, he was awarded the 1957 Nobel Prize in chemistry.*

27.1 Nucleosides and Nucleotides

Nucleic acids are chains of five-membered-ring sugars linked by phosphate groups (Figure 27.1). The anomeric carbon of each sugar is bonded to a nitrogen of a heterocyclic compound in a β-glycosidic linkage. (Recall from Section 21.11 that a β-linkage is one in which the substituents at C-1 and C-4 are on the same side of the furanose ring.) Because the heterocyclic compounds are amines, they are commonly referred to as **bases**. In RNA the five-membered-ring sugar is D-ribose. In DNA it is 2′-deoxy-D-ribose (D-ribose without an OH group in the 2′-position).

Phosphoric acid, the parent compound of the phosphate groups in RNA and DNA, links the sugars in both RNA and DNA. Each of the OH groups of phosphoric acid can react with an alcohol to form a *phosphomonoester,* a *phosphodiester,* or a *phosphotriester,* depending on the number of OH groups that form ester bonds. In nucleic acids the phosphate group is a **phosphodiester**.

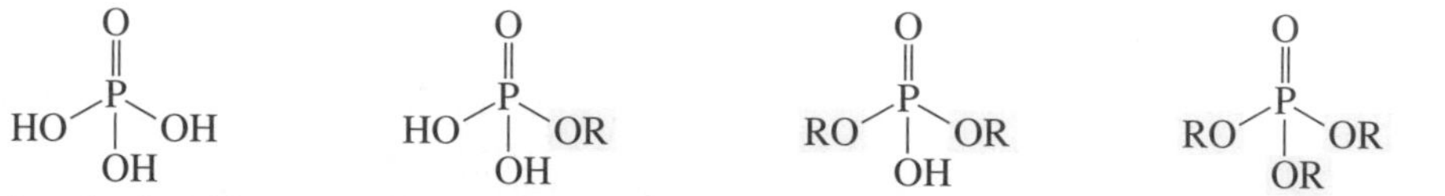

phosphoric acid **a phosphomonoester** **a phosphodiester** **a phosphotriester**

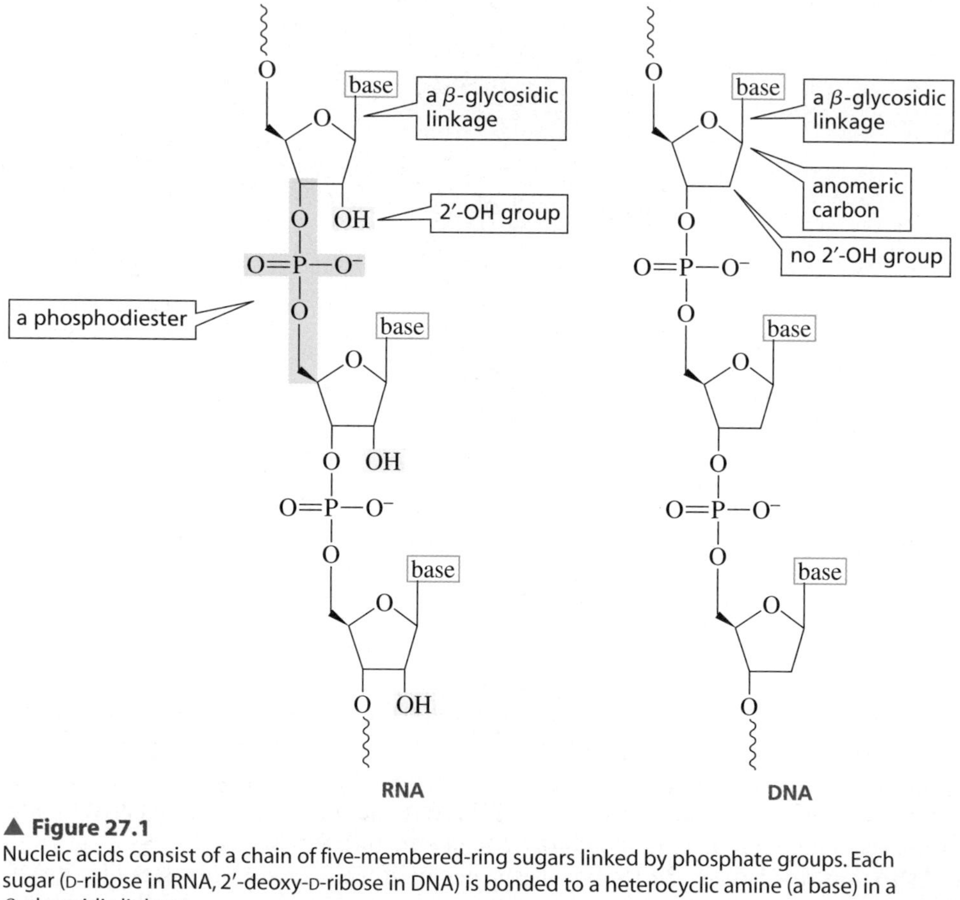

▲ **Figure 27.1**
Nucleic acids consist of a chain of five-membered-ring sugars linked by phosphate groups. Each sugar (D-ribose in RNA, 2′-deoxy-D-ribose in DNA) is bonded to a heterocyclic amine (a base) in a β-glycosidic linkage.

The vast differences in heredity between different species and between different members of the same species are determined by the sequence of the bases in DNA.

THE STRUCTURE OF DNA: WATSON, CRICK, FRANKLIN, AND WILKINS

James D. Watson was born in Chicago in 1928. He graduated from the University of Chicago at the age of 19 and received a Ph.D. three years later from Indiana University. In 1951, as a postdoctoral fellow at Cambridge University, Watson worked on determining the three-dimensional structure of DNA.

Francis H. C. Crick (1916–2004) was born in Northampton, England. Originally trained as a physicist, Crick did research on radar during World War II. After the war, deciding that the most interesting problem in science was the attempt to understand the physical basis of life, he entered Cambridge University to study the structure of biological molecules by X-ray analysis. He was a graduate student when he carried out his portion of the work that led to the proposal of the double helical structure of DNA. He received a Ph.D. in chemistry in 1953.

Rosalind Franklin (1920–1958) was born in London. She graduated from Cambridge University and in 1942 accepted a position as a research officer in the British Coal Utilisation Research Association. After the war, she studied X-ray diffraction techniques in Paris. In 1951 she returned to England and accepted

a position to develop an X-ray diffraction unit in the biophysics department at King's College. Her X-ray studies showed that DNA was a helix with phosphate groups on the outside of the molecule. Franklin died without knowing the role her work had played in determining the structure of DNA and without being recognized for her contribution.

Watson and Crick shared the 1962 Nobel Prize in medicine or physiology with Maurice Wilkins for determining the double helical structure of DNA. Wilkins (1916–2004), who contributed X-ray studies that confirmed the double helical structure, was born in New Zealand and moved to England six years later with his parents. During World War II he joined other British scientists who were working with American scientists on the development of the atomic bomb.

Francis Crick (*left*) and James Watson (*right*)

Rosalind Franklin (By courtesy of the National Portrait Gallery, London)

Surprisingly, there are only four bases in DNA: two are substituted purines (adenine and guanine), and two are substituted pyrimidines (cytosine and thymine).

purine

pyrimidine

adenine

guanine

cytosine

uracil

thymine

RNA also contains only four bases. Three (adenine, guanine, and cytosine) are the same as those in DNA, but the fourth base in RNA is uracil instead of thymine. Notice that thymine and uracil differ only by a methyl group. (Thymine is 5-methyluracil.) The reason DNA contains thymine instead of uracil is explained in Section 27.9.

The purines and pyrimidines are bonded to the anomeric carbon of the furanose ring—purines at N-9 and pyrimidines at N-1—in a β-glycosidic linkage. A compound containing a base bonded to D-ribose or to 2′-deoxy-D-ribose is called a **nucleoside**. The ring positions of the sugar component of a nucleoside are indicated by primed numbers to distinguish them from the ring positions of the base. This is why the sugar component of DNA is referred to as 2′-deoxy-D-ribose.

Notice the difference in the base names and their corresponding nucleoside names in Table 27.1. For example, adenine is the base, whereas adenosine is the nucleoside; similarly, cytosine is the base, whereas cytidine is the nucleoside, and so forth. Because uracil is found only in RNA, it is shown attached to D-ribose but not to 2′-deoxy-D-ribose; because thymine is found only in DNA, it is shown attached to 2′-deoxy-D-ribose but not to D-ribose.

Table 27.1   The Names of the Bases, the Nucleosides, and the Nucleotides

Base	Ribonucleoside	Deoxyribonucleoside	Ribonucleotide	Deoxyribonucleotide
Adenine	Adenosine	2′-Deoxyadenosine	Adenosine 5′-phosphate	2′-Deoxyadenosine 5′-phosphate
Guanine	Guanosine	2′-Deoxyguanosine	Guanosine 5′-phosphate	2′-Deoxyguanosine 5′-phosphate
Cytosine	Cytidine	2′-Deoxycytidine	Cytidine 5′-phosphate	2′-Deoxycytidine 5′-phosphate
Thymine	—	Thymidine	—	Thymidine 5′-phosphate
Uracil	Uridine	—	Uridine 5′-phosphate	

nucleosides

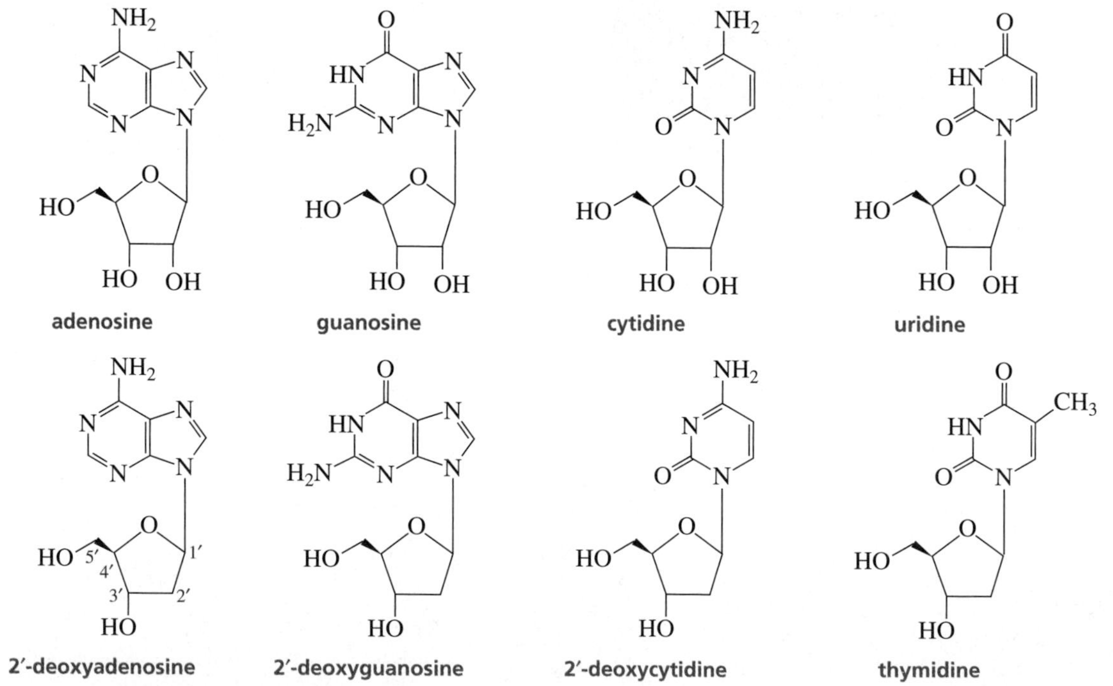

adenosine   guanosine   cytidine   uridine

2′-deoxyadenosine   2′-deoxyguanosine   2′-deoxycytidine   thymidine

PROBLEM 1

In acidic solutions, nucleosides are hydrolyzed to a sugar and a heterocyclic base. Propose a mechanism for this reaction.

A **nucleotide** is a nucleoside with either the 5′- or the 3′-OH group bonded in an ester linkage to phosphoric acid. The nucleotides of RNA—where the sugar is D-ribose—are more precisely called **ribonucleotides**, whereas the nucleotides of DNA—where the sugar is 2′-deoxy-D-ribose—are called **deoxyribonucleotides**.

nucleoside = base + sugar

nucleotide = base + sugar + phosphate

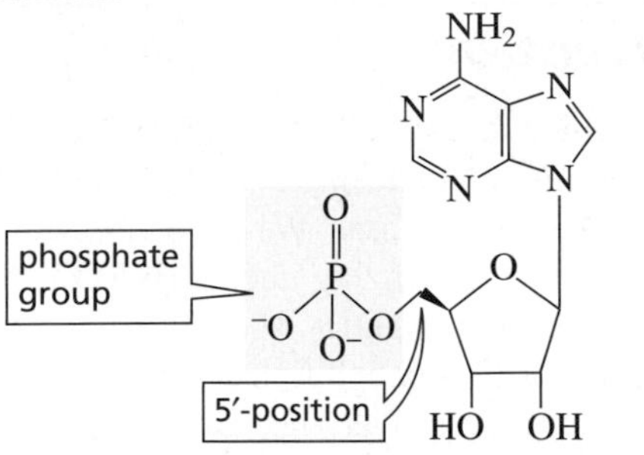

adenosine 5′-monophosphate
a ribonucleotide

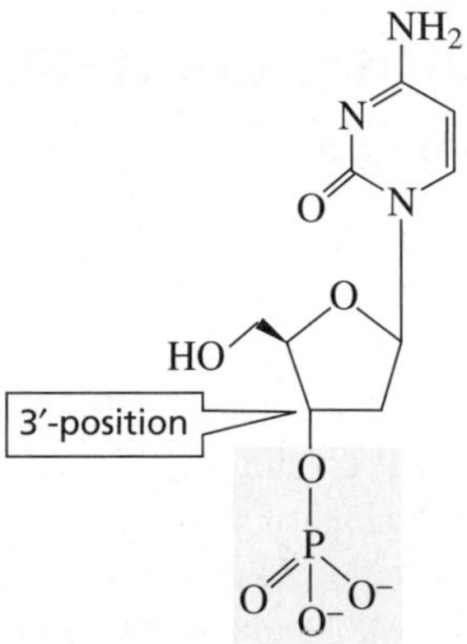

2′-deoxycytidine 3′-monophosphate
a deoxyribonucleotide

Because phosphoric acid can form an anhydride (Section 25.2), nucleotides can exist as monophosphates, diphosphates, and triphosphates. They are named by adding *monophosphate* or *diphosphate* or *triphosphate* to the name of the nucleoside.

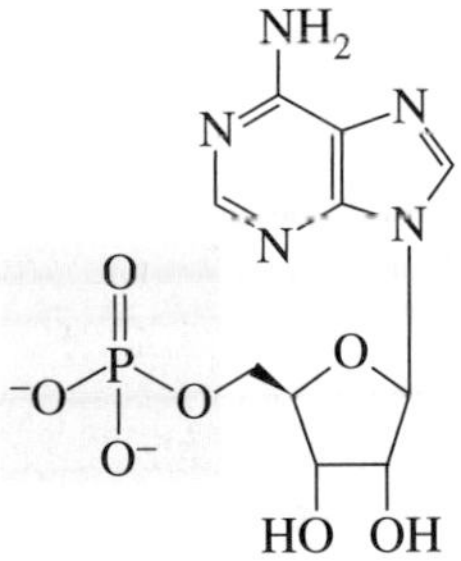

adenosine
5′-monophosphate
AMP

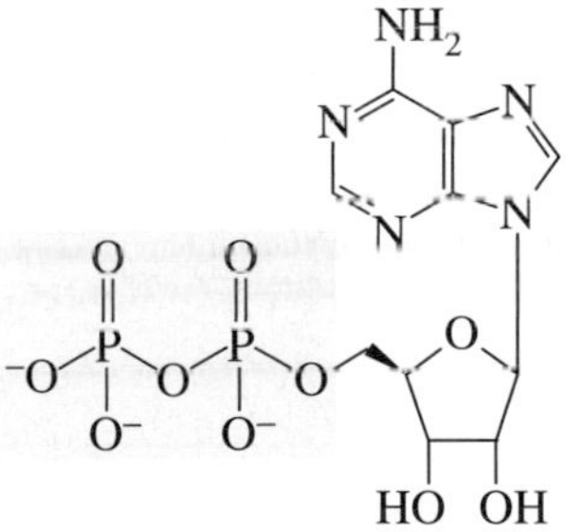

adenosine
5′-diphosphate
ADP

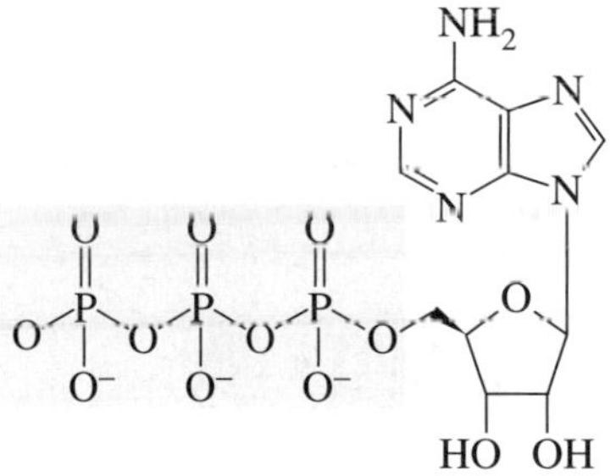

adenosine
5′-triphosphate
ATP

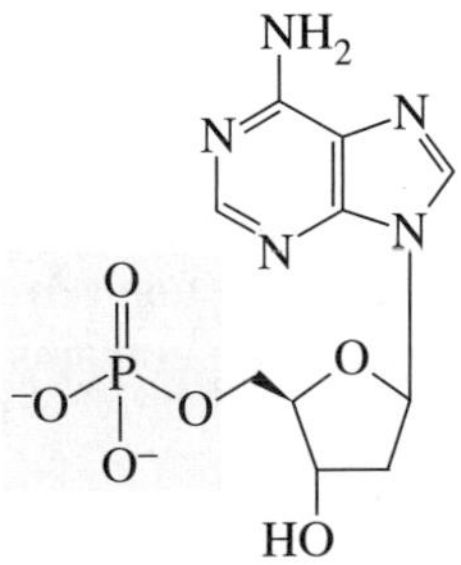

2′-deoxyadenosine
5′-monophosphate
dAMP

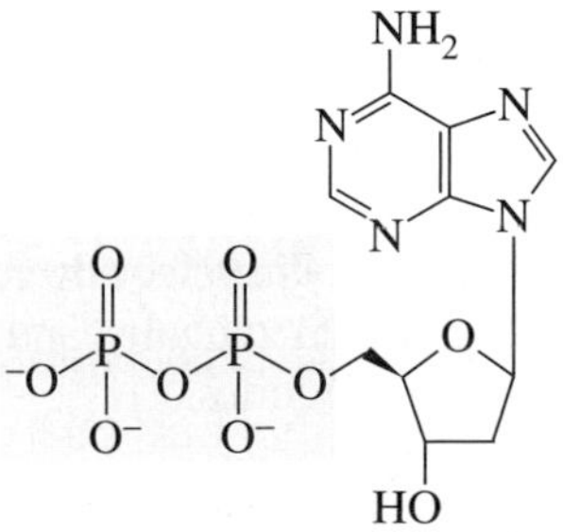

2′-deoxyadenosine
5′-diphosphate
dADP

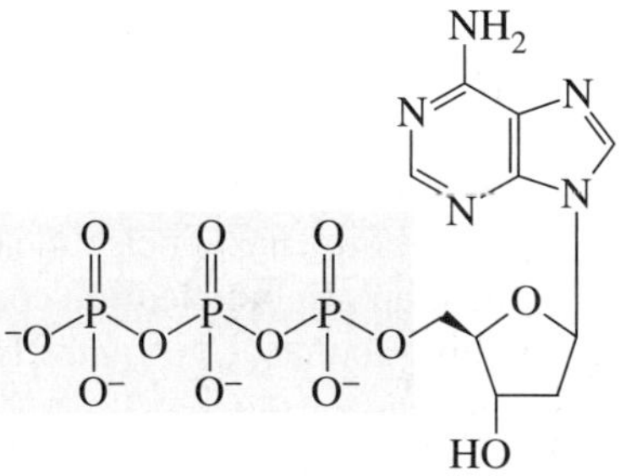

2′-deoxyadenosine
5′-triphosphate
dATP

Notice that nucleotide names are abbreviated (A, G, C, T, U—followed by MP, DP, or TP, depending on whether it is a monophosphate, diphosphate, or triphosphate—with a d in front if it contains a 2′-deoxy-D-ribose).

27.2 Other Important Nucleotides

We have seen that ATP is the carrier of chemical energy (Section 25.2). ATP, however, is not the only biologically important nucleotide. Guanosine 5′-triphosphate (GTP) is used in place of ATP in some phosphoryl transfer reactions. We have also seen in Sections 24.2 and 24.3 that dinucleotides are used as oxidizing agents (NAD$^+$, NADP$^+$, FAD, FMN) and reducing agents (NADH, NADPH, FADH$_2$, FMNH$_2$).

Another important nucleotide is adenosine 3′,5′-monophosphate, commonly known as cyclic AMP. Cyclic AMP is called a "second messenger" because it serves as a link between several hormones (the first messengers) and certain enzymes that regulate cellular function. Secretion of certain hormones, such as adrenaline, activates adenylate cyclase, the enzyme responsible for the synthesis of cyclic AMP from ATP. Cyclic AMP then activates an enzyme, generally by phosphorylating it. Cyclic nucleotides are so important in regulating cellular reactions that an entire scientific journal is devoted to these processes.

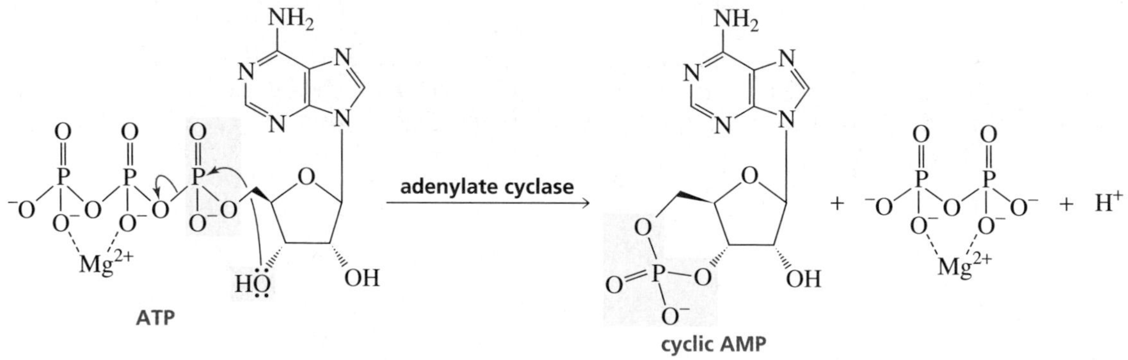

ATP → adenylate cyclase → cyclic AMP

27.3 Nucleic Acids Are Composed of Nucleotide Subunits

We have seen that **nucleic acids** are composed of long strands of nucleotide subunits linked by phosphodiester bonds that join the 3′-OH group of one nucleotide to the 5′-OH group of the next nucleotide (Figure 27.1). A **dinucleotide** contains two nucleotide subunits, an **oligonucleotide** contains three to ten subunits, and a **polynucleotide** contains many subunits. DNA and RNA are polynucleotides.

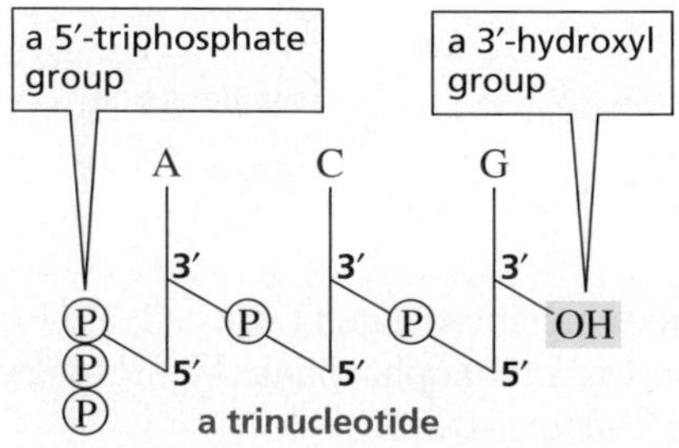

a trinucleotide

Nucleotide triphosphates are the starting materials for the biosynthesis of nucleic acids. DNA is synthesized by enzymes called *DNA polymerases*; RNA is synthesized by enzymes called *RNA polymerases*. The nucleotides are linked as a result of nucleophilic attack by a 3'-OH group of one nucleotide triphosphate on the α-phosphorus of another nucleotide triphosphate, breaking a phosphoanhydride bond and eliminating pyrophosphate (Figure 27.2). This means that the growing polymer is synthesized in the $5' \longrightarrow 3'$ direction; in other words, new nucleotides are added to the 3'-end. Pyrophosphate is subsequently hydrolyzed, which makes the reaction irreversible (Section 25.3). RNA strands are biosynthesized in the same way, using ribonucleotides instead of 2'-deoxyribonucleotides.

DNA is synthesized in the $5' \longrightarrow 3'$ direction.

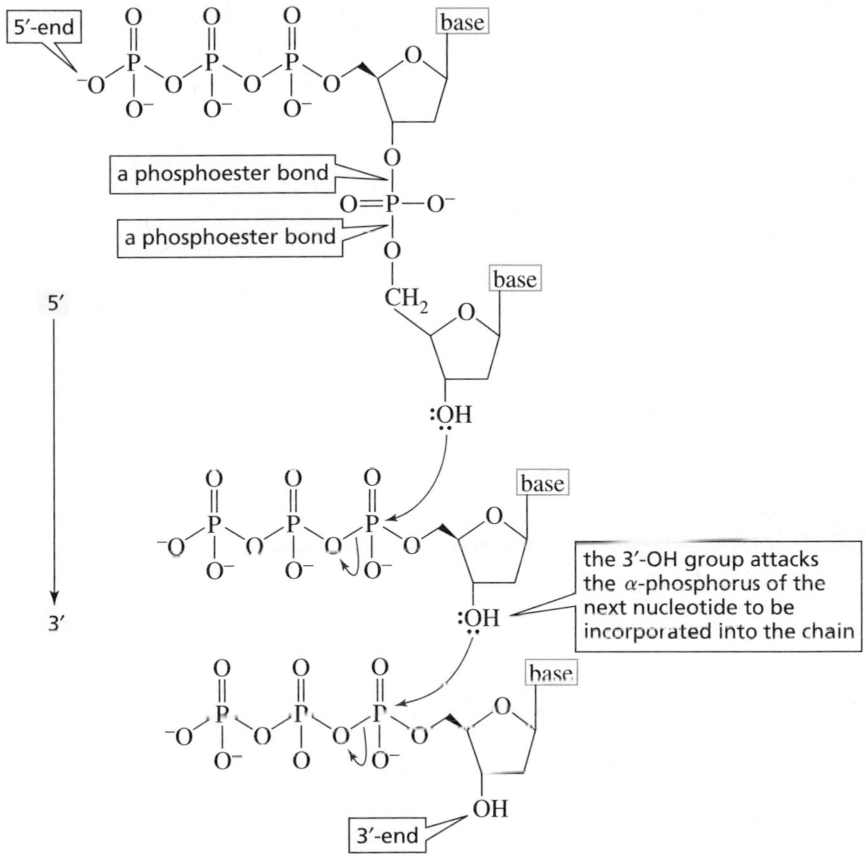

◄ **Figure 27.2**
Addition of nucleotides to a growing strand of DNA. Biosynthesis occurs in the $5' \longrightarrow 3'$ direction.

The **primary structure** of a nucleic acid is the sequence of bases in the strand. By convention, the sequence of bases in a polynucleotide is written in the $5' \longrightarrow 3'$ direction (the 5'-end is on the left). Remember that the nucleotide at the 5'-end of the strand has an unlinked 5'-triphosphate group, and the nucleotide at the 3'-end has an unlinked 3'-hydroxyl group.

Watson and Crick concluded that DNA consists of two strands of nucleic acids with the sugar–phosphate backbone on the outside and the bases on the inside. The strands are antiparallel (they run in opposite directions) and are held together by hydrogen bonds between the bases on one strand and the bases on the other strand (Figures 27.3 and 27.4). The width of the double-stranded molecule is relatively constant, so a purine must pair with a pyrimidine. If the larger purines paired, the strands would bulge; if the smaller pyrimidines paired, the strands would have to pull in to bring the two pyrimidines close enough to form hydrogen bonds.

Experiments carried out by Erwin Chargaff were critical to Watson and Crick's proposal for the secondary structure of DNA. These experiments showed that the number of adenines in DNA equals the number of thymines and the number of

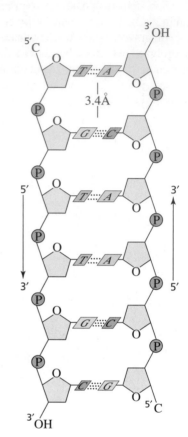

Figure 27.3 ▶
Complementary base pairing in DNA. Adenine (a purine) always pairs with thymine (a pyrimidine); guanine (a purine) always pairs with cytosine (a pyrimidine). Therefore,
[A] = [T]
[G] = [C]

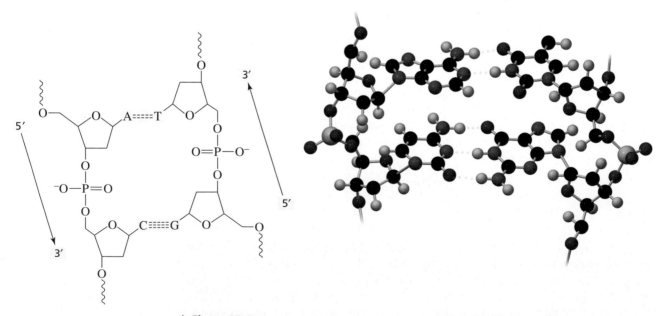

▲ **Figure 27.4**
The sugar–phosphate backbone of DNA is on the outside, and the bases are on the inside, with A's pairing with T's and G's pairing with C's. The two strands are antiparallel—they run in opposite directions.

guanines equals the number of cytosines. Chargaff also noted that the number of adenines and thymines relative to the number of guanines and cytosines is characteristic of a given species but varies from species to species. In human DNA, for example, 60.4% of the bases are adenines and thymines, whereas 74.2% of them are adenines and thymines in the DNA of the bacterium *Sarcina lutea.*

Chargaff's data showing that [adenine] = [thymine] and [guanine] = [cytosine] could be explained if adenine (A) always paired with thymine (T) and guanine (G) always paired with cytosine (C). This means the two strands are *complementary*: where there is an A in one strand, there is a T in the opposing strand, and where there is a G in one strand there is a C in the other strand (Figure 27.3). Thus, if you know the sequence of bases in one strand, you can figure out the sequence of bases in the other strand.

What causes adenine to pair with thymine rather than with cytosine (the other pyrimidine)? The base pairing is dictated by hydrogen bonding. Learning that the bases exist in the keto form (Section 18.2) allowed Watson to explain the pairing.* Adenine forms two hydrogen bonds with thymine but would form only one hydrogen bond with cytosine. Guanine forms three hydrogen bonds with cytosine but would form only one hydrogen bond with thymine (Figure 27.5). The N—H----N and N—H----O bonds that hold the bases together are all about the same length (2.9 ± 0.1 Å).

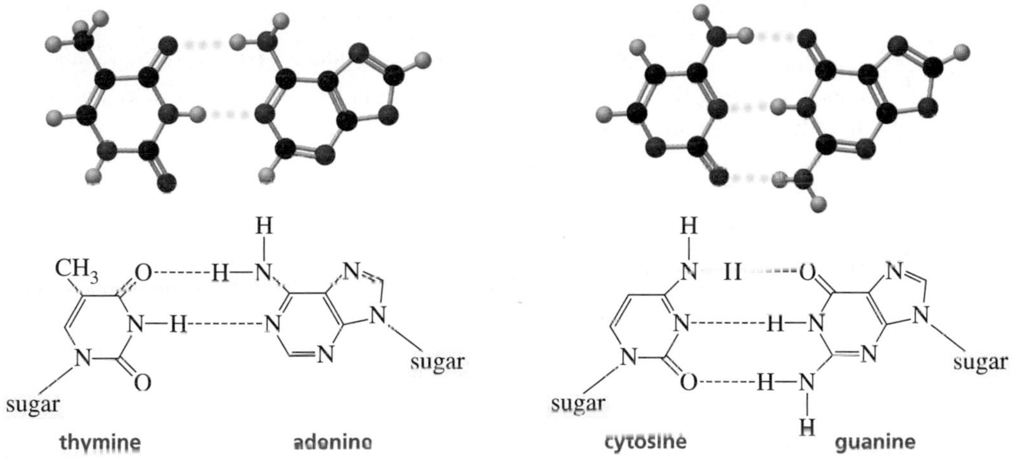

▲ Figure 27.5
Base pairing in DNA: adenine and thymine form two hydrogen bonds; cytosine and guanine form three hydrogen bonds.

The DNA strands are not linear but are twisted into a helix around a common axis (see Figure 27.6a). The base pairs are planar and parallel to each other on the inside of the helix (Figures 27.6b and c). The secondary structure is therefore known as a **double helix**. The double helix resembles a ladder (the base pairs are the rungs) twisted around an axis running down through its rungs (Figure 27.6c). The sugar–phosphate backbone is wrapped around the bases. The phosphate OH group has a pK_a of about 2, so it is in its basic form (negatively charged) at physiological pH. The negatively charged backbone repels nucleophiles, thereby preventing cleavage of the phosphodiester bonds.

Hydrogen bonding between base pairs is just one of the forces holding the two strands of the DNA double helix together. The bases are planar aromatic molecules that stack on top of one another, each pair slightly rotated with respect to the next pair, like a partially spread-out hand of cards. In this arrangement, there are favorable van der Waals interactions between the mutually induced dipoles of adjacent pairs of bases. These interactions, known as **stacking interactions**, are weak attractive forces, but when added together they contribute significantly to the stability of the double helix.

---

* Watson was having difficulty understanding the base pairing in DNA because he thought the bases existed in the enol form (see Problem 8). When Jerry Donohue, an American crystallographer, informed him that the bases more likely existed in the keto form, Chargaff's data could easily be explained by hydrogen bonding between adenine and thymine and between guanine and cytosine.

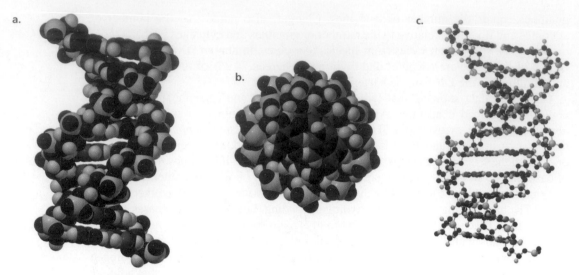

▲ **Figure 27.6**
(a) The DNA double helix. (b) View looking down the long axis of the helix. (c) The bases are planar and parallel on the inside of the helix.

Stacking interactions are strongest between two purines and weakest between two pyrimidines. Confining the bases to the inside of the helix has an additional stabilizing effect—it reduces the surface area of the relatively nonpolar residues' exposure to water. This increases the entropy of the surrounding water molecules (Section 22.15).

There are two different alternating grooves in a DNA helix; a **major groove** and a narrower **minor groove**. Cross sections of the double helix show that one side of each base pair faces into the major groove and the other side faces into the minor groove. Proteins and other molecules can bind to the grooves. The hydrogen-bonding properties of the functional groups facing into each groove determine what kind of molecules will bind to the groove. For example, netropsin is an antibiotic that works by binding to the minor groove of DNA (Section 30.10).

PROBLEM 4

Indicate whether each functional group of the five heterocyclic bases in nucleic acids can function as a hydrogen bond acceptor (A), a hydrogen bond donor (D), or both (D/A).

PROBLEM 5

Using the D, A, and D/A designations in Problem 4, indicate how base pairing would be affected if the bases existed in the enol form.

PROBLEM 6

The 2′,3′-cyclic phosphodiester that is formed when RNA is hydrolyzed (Figure 27.7), reacts with water, forming a mixture of nucleotide 2′- and 3′-phosphates. Propose a mechanism for this reaction.

PROBLEM 7◆

If one of the strands of DNA has the following sequence of bases running in the 5′ ⟶ 3′ direction,

$$5'-G-G-A-C-A-A-T-C-T-G-C-3'$$

a. what is the sequence of bases in the complementary strand?
b. what base is closest to the 5′-end in the complementary strand?

PROBLEM 8

5-Bromouracil, a highly mutagenic compound (that is, a compound that causes changes in DNA), is used in cancer chemotherapy. When administered to a patient, it is converted to the triphosphate and incorporated into DNA in place of thymine, which it resembles sterically. Why does it cause mutations? (*Hint:* The bromo substituent increases the stability of the enol tautomer.)

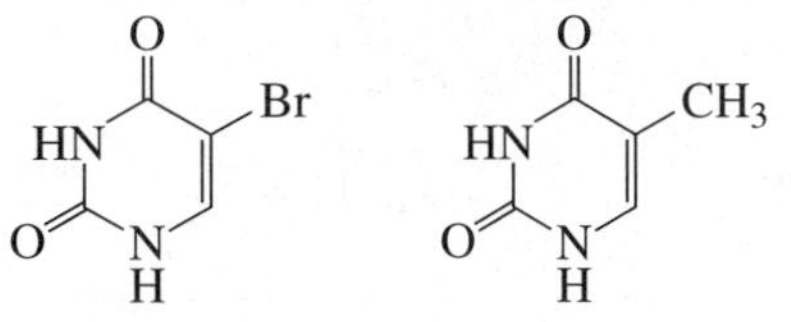

5-bromouracil   thymine

27.4 DNA Is Stable But RNA Is Easily Cleaved

Unlike DNA, RNA is easily cleaved because the 2'-OH group of ribose can act as the nucleophile that cleaves the strand (Figure 27.7). This explains why the 2'-OH group is absent in DNA. DNA must remain intact throughout the life span of a cell to preserve the genetic information. Easy cleavage of DNA would have disastrous consequences for the cell and for life itself. RNA, in contrast, is synthesized as it is needed and is degraded once it has served its purpose.

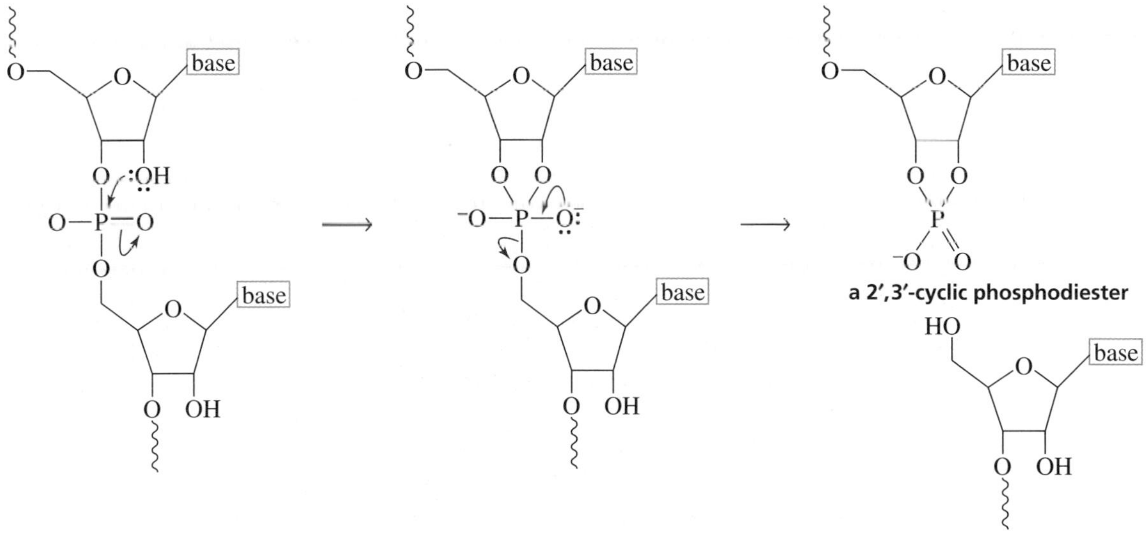

a 2',3'-cyclic phosphodiester

▲ **Figure 27.7**
Cleavage of RNA; the 2'-OH group is an intramolecular nucleophilic catalyst. RNA is thought to undergo cleavage 3 billion times faster than DNA.

27.5 Biosynthesis of DNA Is Called Replication

The genetic information of a human cell is contained in 23 pairs of chromosomes. Each chromosome is composed of several thousand **genes** (segments of DNA). The total DNA from a human cell—the **human genome**—contains 3.1 billion base pairs.

Part of the excitement created by Watson and Crick's proposed structure for DNA was due to the fact that the structure immediately suggested how DNA is able to pass on genetic information along to succeeding generations. Because the two strands are complementary, both carry the same genetic information. Both strands serve as

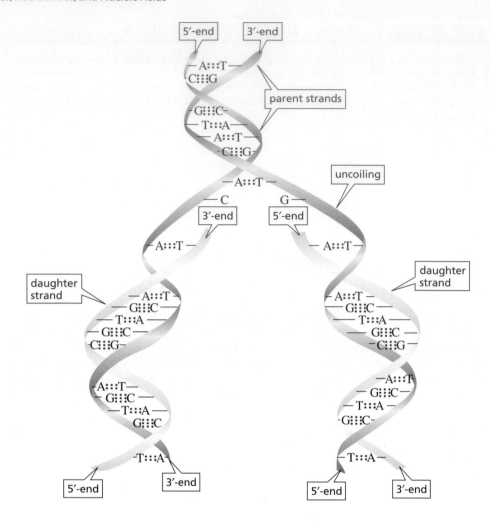

Figure 27.8 ▶
Replication of DNA. The daughter strand on the left is synthesized continuously in the 5′ ⟶ 3′ direction; the daughter strand on the right is synthesized discontinuously in the 5′ ⟶ 3′ direction.

templates for the synthesis of complementary new strands (Figure 27.8). The new (daughter) DNA molecules are identical to the original (parent) molecule—they contain all the original genetic information. The synthesis of identical copies of DNA is called **replication**.

All reactions involved in nucleic acid synthesis are catalyzed by enzymes. The synthesis of DNA takes place in a region of the molecule where the strands have started to separate, called a **replication fork**. Because a nucleic acid can be synthesized only in the 5′ ⟶ 3′ direction, only the daughter strand on the left in Figure 27.8 is synthesized continuously in a single piece (because it is synthesized in the 5′ ⟶ 3′ direction). The other daughter strand needs to grow in a 3′ ⟶ 5′ direction, so it is synthesized discontinuously in small pieces. Each piece is synthesized in the 5′ ⟶ 3′ direction and the fragments are joined together by an enzyme called DNA ligase. Each of the two resulting daughter molecules of the replicating DNA contains one of the original parent strands (blue strand in Figure 27.8) plus a newly synthesized strand (green strand). This process is called **semiconservative replication**.

PROBLEM 9

Using a dark line for the original parental DNA and a wavy line for DNA synthesized from parental DNA, show what the population of DNA molecules would look like in the fourth generation of replicated molecules.

27.6 Biosynthesis of RNA Is Called Transcription

If DNA contains hereditary information, there must be a method to decode that information. The decoding occurs in two steps.

1. The sequence of bases in DNA provides a blueprint for the synthesis of RNA; the synthesis of RNA from a DNA blueprint, called **transcription**, takes place in the nucleus of the cell. Then the newly synthesized RNA leaves the nucleus, carrying the genetic information into the cytoplasm (the cell material outside the nucleus).
2. The sequence of bases in RNA determines the sequence of amino acids in a protein; the synthesis of a protein from an RNA blueprint is called **translation** (Section 27.8).

First, we will look at transcription. DNA contains certain sequences of bases known as **promoter sites** that mark the beginning of genes. An enzyme recognizes a promoter site and binds to it, initiating RNA synthesis. The DNA at a promoter site unwinds into two single strands with their bases exposed. One of the strands is called the **sense strand**, or **informational strand**. To synthesize RNA, the complementary strand is called the **template strand**, or **antisense strand**. To synthesize RNA, the template strand is read in the $3' \longrightarrow 5'$ direction, so that RNA can be synthesized in the $5' \longrightarrow 3'$ direction (Figure 27.9). The bases in the template strand specify the bases that need to be incorporated into RNA, following the same base-pairing principle that allows replication of DNA. For example, each guanine in the template strand specifies the incorporation of a cytosine into RNA, and each adenine in the template strand specifies the incorporation of a uracil into RNA. (Recall that in RNA, uracil is used instead of thymine.) Because both RNA and the sense strand of DNA are complementary to the template strand, RNA and the sense strand of DNA have the same base sequence, except that RNA has a uracil wherever the sense strand has a thymine. Just as there are promoter sites that signal the places to start RNA synthesis, there are sites in DNA that signal that no more bases should be added to the growing strand of RNA, at which point synthesis stops.

BIOGRAPHY

Severo Ochoa (1905–1993) *was born in Spain. He graduated from the University of Malaga in 1921 and received an M.D. from the University of Madrid. He spent the next four years studying in Germany and England and then joined the faculty at New York University College of Medicine. He became a U.S. citizen in 1956.* **Ochoa** *was the first to prepare synthetic strands of RNA by incubating nucleotides in the presence of enzymes that are catalyze the biosynthesis of RNA.*

BIOGRAPHY

Arthur Kornberg *was born in New York in 1918. He graduated from the College of the City of New York and received an M.D. from the University of Rochester. He is a professor of biochemistry at Stanford University.* **Kornberg** *prepared synthetic strands of DNA in a manner similar to that done by Ochoa. For this work, they shared the 1959 Nobel Prize in physiology or medicine.*

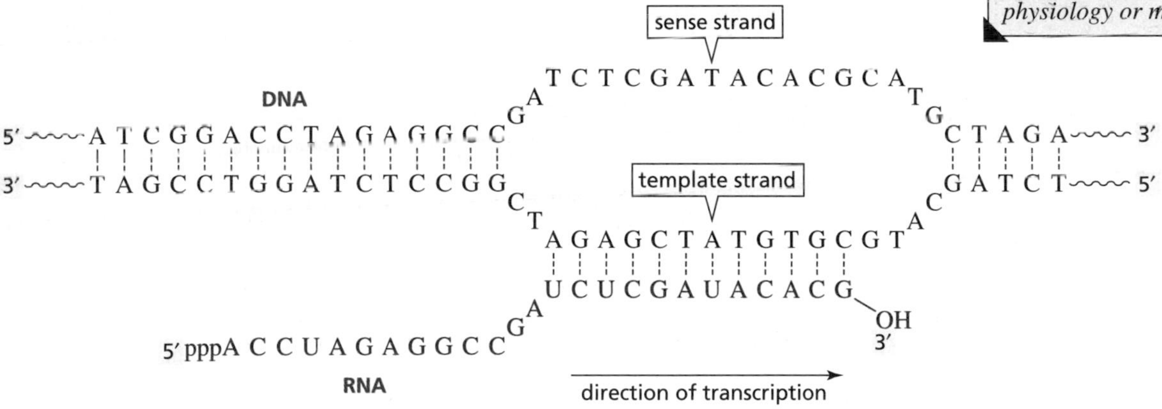

▲ Figure 27.9
Transcription: using DNA as a blueprint for the synthesis of RNA.

Surprisingly, a gene is not necessarily a continuous sequence of bases. Often the bases of a gene are interrupted by bases that appear to have no informational content. A stretch of bases representing a portion of a gene is called an **exon**, while a stretch of bases that contains no genetic information is called an **intron**. The RNA that is synthesized is complementary to the entire sequence of DNA bases—exons and introns—between the promoter site and the stop signal. So after the RNA is synthesized, but before it leaves the nucleus, the so-called nonsense bases (encoded by the introns) are cut out and the informational fragments are spliced together, resulting in a much shorter RNA molecule. This RNA-processing step is known as **RNA splicing**. Scientists have found that only about 2% of DNA contains genetic information, while 98% consists of introns.

Transcription: DNA ⟶ RNA

Translation: mRNA ⟶ protein

RNA is synthesized in the $5' \longrightarrow 3'$ direction.

It has been suggested that the purpose of introns is to make RNA more versatile. The originally synthesized long strand of RNA can be spliced in different ways to create a variety of shorter RNAs.

PROBLEM 10

Why do both thymine and uracil specify the incorporation of adenine?

27.7 There Are Three Kinds of RNA

RNA molecules are much shorter than DNA molecules and are generally single-stranded. Although DNA molecules can have billions of base pairs, RNA molecules rarely have more than 10,000 nucleotides. There are three kinds of RNA:

- **messenger RNA (mRNA)**, whose sequence of bases determines the sequence of amino acids in a protein
- **ribosomal RNA (rRNA)**, a structural component of ribosomes, which are the particles on which the biosynthesis of proteins takes place
- **transfer RNA (tRNA)**, the carrier of amino acids for protein synthesis.

tRNA molecules are much smaller than mRNA or rRNA molecules; a tRNA contains only 70 to 90 nucleotides. The single strand of tRNA is folded into a characteristic cloverleaf structure, with three loops and a little bulge next to the right-hand loop (Figure 27.10a). There are at least four regions with complementary base pairing. All tRNAs have a CCA sequence at the 3'-end. The three bases at the bottom of the loop directly opposite the 5'-and 3'-ends are called an **anticodon** (Figures 27.10a and b).

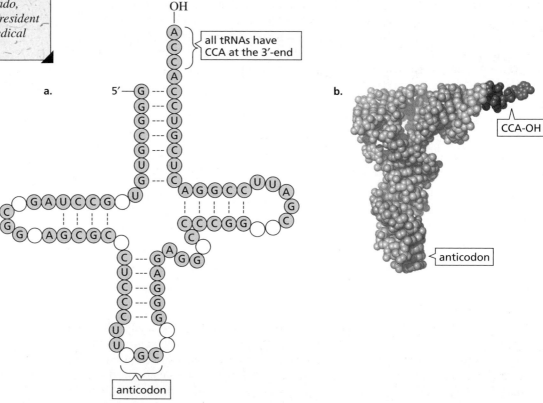

▲ **Figure 27.10**

(a) tRNA^{Ala}, a transfer RNA that carries alanine. Compared with other RNAs, tRNA contains a high percentage of unusual bases (shown as empty circles). These bases result from enzymatic modification of the four normal bases. (b) tRNA^{Phe}: the anticodon is green; the CCA at the 3'-end is red.

Each tRNA can carry an amino acid bound as an ester to its terminal 3'-OH group. The amino acid will be inserted into a protein during protein biosynthesis. Each tRNA can carry only one particular amino acid. A tRNA that carries alanine is designated as tRNA^{Ala}.

Attachment of an amino acid to a tRNA is catalyzed by an enzyme called aminoacyl-tRNA synthetase. The mechanism for the reaction is shown in Figure 27.11.

▲ **Figure 27.11**
The proposed mechanism for aminoacyl-tRNA synthetase—the enzyme that catalyzes the attachment of an amino acid to a tRNA.

- The carboxylate group of the amino acid is activated by attacking the α-phosphorus of ATP, forming an acyl adenylate.
- The pyrophosphate that is expelled is subsequently hydrolyzed, ensuring the irreversibility of the phosphoryl transfer reaction (Section 25.3).
- A nucleophilic acyl substitution reaction occurs in which the 3'-OH group of tRNA attacks the carbonyl carbon of the acyl adenylate, forming a tetrahedral intermediate.
- The amino acyl tRNA is formed when AMP is expelled from the tetrahedral intermediate.

All the steps take place at the active site of the enzyme. Each amino acid has its own aminoacyl-tRNA synthetase. Each synthetase has two specific binding sites, one for the amino acid and one for the tRNA that carries that amino acid (Figure 27.12).

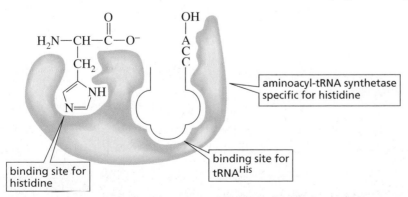

▲ **Figure 27.12**
An aminoacyl-tRNA synthetase has a binding site for tRNA and a binding site for the particular amino acid that is to be attached to that tRNA. Histidine is the amino acid and tRNA^{His} is the tRNA molecule in this example.

It is critical that the correct amino acid be attached to the tRNA. Otherwise, the protein will not be synthesized correctly. Fortunately, the synthetases correct their own mistakes. For example, valine and threonine are approximately the same size but threonine has an OH group in place of a CH_3 group of valine. Both amino acids, therefore, can bind at the amino acid binding site of the aminoacyl-tRNA synthetase for valine, and both can then be activated by reacting with ATP to form an acyl adenylate. The aminoacyl-tRNA synthetase for valine has two adjacent catalytic sites, one for attaching the acyl adenylate to tRNA and one for hydrolyzing the acyl adenylate.

The acylation site is hydrophobic, so valine is preferred over threonine for the tRNA acylation reaction. The hydrolytic site is polar, so threonine is preferred over valine for the hydrolysis reaction. Thus, if threonine is activated by the aminoacyl-tRNA synthetase for valine, it will be hydrolyzed rather than transferred to the tRNA.

27.8 Biosynthesis of Proteins Is Called Translation

A protein is synthesized from its N-terminal end to its C-terminal end by a process that reads the bases along the mRNA strand in the $5' \longrightarrow 3'$ direction. Each amino acid that is to be incorporated into a protein is specified by one or more three-base sequences called **codons**. The bases are read consecutively and are never skipped. The three-base sequences and the amino acid each sequence codes for are known as the **genetic code** (Table 27.2). A codon is written with the 5′-nucleotide on the left. For example, the codon UCA on mRNA codes for the amino acid serine, whereas CAG codes for glutamine.

Table 27.2	The Genetic Code				
5′-Position		**Middle position**			**3′-Position**
	U	**C**	**A**	**G**	
U	Phe	Ser	Tyr	Cys	U
	Phe	Ser	Tyr	Cys	C
	Leu	Ser	Stop	Stop	A
	Leu	Ser	Stop	Trp	G
C	Leu	Pro	His	Arg	U
	Leu	Pro	His	Arg	C
	Leu	Pro	Gln	Arg	A
	Leu	Pro	Gln	Arg	G
A	Ile	Thr	Asn	Ser	U
	Ile	Thr	Asn	Ser	C
	Ile	Thr	Lys	Arg	A
	Met	Thr	Lys	Arg	G
G	Val	Ala	Asp	Gly	U
	Val	Ala	Asp	Gly	C
	Val	Ala	Glu	Gly	A
	Val	Ala	Glu	Gly	G

Because there are four bases and the codons are triplets, $4^3 = 64$ different codons are possible. This is many more than are needed to specify the 20 different amino acids, so all the amino acids—except methionine and tryptophan—have more than one codon. It is not surprising, therefore, that methionine and tryptophan are the least abundant amino acids in proteins. Actually, 61 of the codons specify amino acids, and three codons are stop codons. **Stop codons** tell the cell to "stop protein synthesis here."

Translation is the process by which the genetic message in DNA that has been passed to mRNA is decoded and used to build proteins. Each of the approximately 100,000 proteins in the human body is synthesized from a different mRNA. Don't confuse transcription and translation: these words are used just as they are used in English. Transcription (DNA to RNA) is copying *within the same language,* in this case the language is nucleotides. Translation (RNA to protein) is *changing to another language*—the language of amino acids.

How the information in mRNA is translated into a polypeptide is shown in Figure 27.13. In this figure, serine was the last amino acid incorporated into the growing polypeptide chain. Serine was specified by the AGC codon in the mRNA because the anticodon of the tRNA that carries serine is GCU (3'-UCG-5'). (Remember that a base sequence is read in the 5' $\longrightarrow$ 3' direction, so the sequence of bases in an anticodon must be read from right to left.)

- The next codon, CUU, signals for a tRNA with an anticodon of AAG (3'-GAA-5'). That particular tRNA carries leucine. The amino group of leucine reacts in an enzyme-catalyzed nucleophilic acyl substitution reaction (Section 16.7) with the ester on the adjacent serine tRNA, displacing that tRNA.

- The next codon (GCC) specifies a tRNA carrying alanine. The amino group of alanine displaces the tRNA that brought in leucine.

Subsequent amino acids are brought in one at a time in the same way, with the codon in mRNA specifying the amino acid to be incorporated by complementary base pairing with the anticodon of the tRNA that carries that amino acid.

Protein synthesis takes place on the ribosomes (Figure 27.14). The smaller of the two subunits in the ribosome has three binding sites for RNA molecules. It binds the mRNA whose base sequence is to be read, the tRNA carrying the growing peptide chain, and the tRNA carrying the next amino acid to be incorporated into the protein. The larger subunit of the ribosome catalyzes peptide bond formation.

A sculpture done by Robert Holley.

Robert W. Holley (1922–1993) *was born in Illinois and received a bachelor's degree from the University of Illinois and a Ph.D. from Cornell University. During World War II he worked on the synthesis of penicillin at Cornell Medical School. He was a professor at Cornell and later at the University of California, San Diego. He was also a noted sculptor.*

Student Tutorial: Translation

A protein is synthesized in the N-terminal $\longrightarrow$ C-terminal direction.

PROBLEM 11◆

If methionine is the first amino acid incorporated into a polypeptide, what polypeptide is encoded for by the following stretch of mRNA?

5'—G—C—A—U—G—G—A—C—C—C—C—G—U—U—A—U—
U—A—A—A—C—A—C—3'

PROBLEM 12◆

Four C's occur in a row in the segment of mRNA in Problem 11. What polypeptide would be formed from the mRNA if one of the four C's were cut out of the strand?

PROBLEM 13

UAA is a stop codon. Why does the UAA sequence in mRNA in Problem 11 not cause protein synthesis to stop?

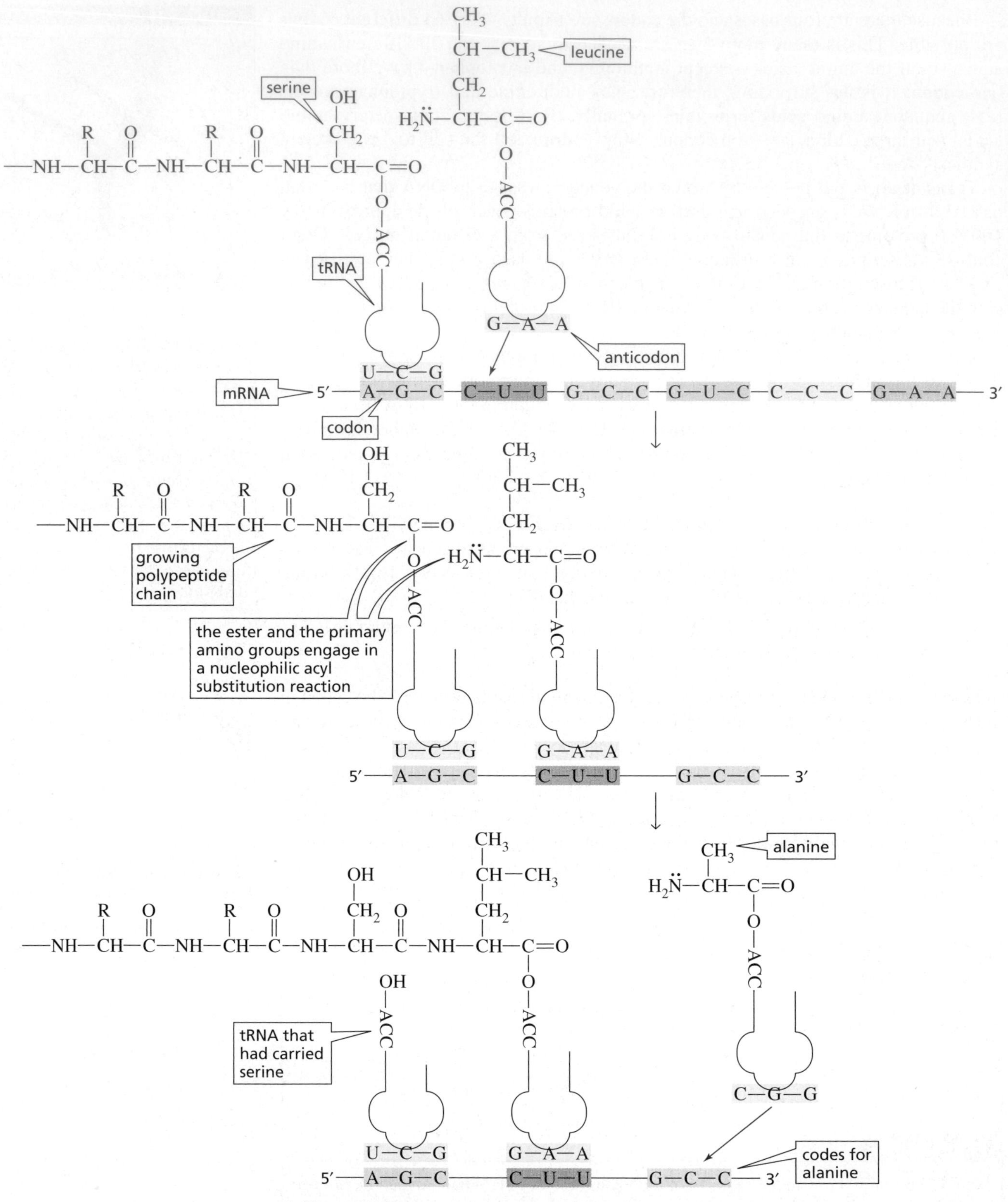

▲ Figure 27.13
Translation. The sequence of bases in mRNA determines the sequence of amino acids in a protein.

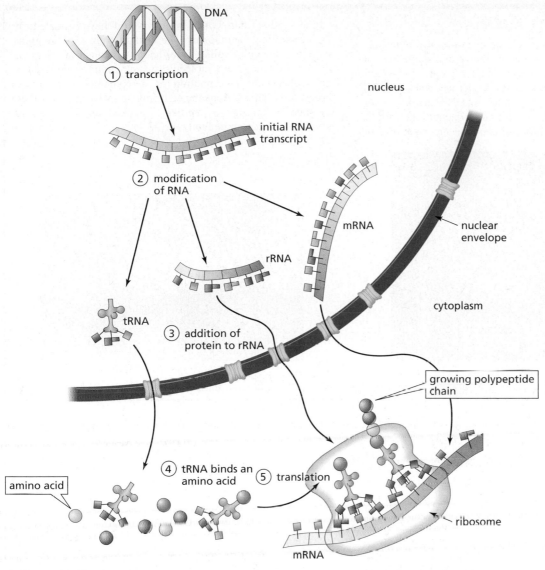

▲ **Figure 27.14**
Transcription and translation. 1. Transcription of DNA occurs in the nucleus of the cell. The initial RNA transcript is the precursor of all RNA: tRNA, rRNA, and mRNA. 2. The initially formed RNA often must be chemically modified before it acquires biological activity. Modification can entail removing nucleotide segments, adding nucleotides to the 5'- or 3'-ends, or chemically altering certain nucleotides. 3. Proteins are added to rRNA to form ribosomal subunits. tRNA, mRNA, and ribosomal subunits leave the nucleus. 4. Each tRNA binds the appropriate amino acid. 5. tRNA, mRNA, and a ribosome work together to translate the mRNA information into a protein.

PROBLEM 14◆

Write the sequences of bases in the sense strand of DNA that resulted in the mRNA in Problem 11.

PROBLEM 15

List the possible codons on mRNA that specify each amino acid in Problem 11 and the anticodon on the tRNA that carries that amino acid.

SICKLE CELL ANEMIA

Sickle cell anemia is an example of the damage that can be caused by a change in a single base of DNA (Problem 65 in Chapter 22). It is a hereditary disease caused when a GAG triplet becomes a GTG triplet in the sense strand of a section of DNA that codes for the β-subunit of hemoglobin. As a consequence, the mRNA codon becomes GUG—which signals for incorporation of valine—rather than GAG, which would have signaled for incorporation of glutamate. The change from a polar glutamate to a nonpolar valine is sufficient to change the shape of the deoxyhemoglobin molecule, which induces aggregation and subsequent precipitation of deoxyhemoglobin in red blood cells. This stiffens the cells, making it difficult for them to squeeze through a capillary. Blocked capillaries cause severe pain and can be fatal.

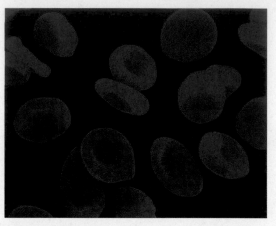

Normal red blood cells

Sickle red blood cells

ANTIBIOTICS THAT ACT BY INHIBITING TRANSLATION

Puromycin is a naturally occurring antibiotic, one of several that acts by inhibiting translation. It does so by mimicking the 3′-CCA-aminoacyl portion of a tRNA, fooling the enzyme into transferring the growing peptide chain to the amino group of puromycin rather than to the amino group of the incoming 3′-CCA-aminoacyl tRNA. As a result, protein synthesis stops. Because puromycin blocks protein synthesis in eukaryotes as well as in prokaryotes, it is poisonous to humans and therefore is not a clinically useful antibiotic. To be clinically useful, an antibiotic must affect protein synthesis only in prokaryotic cells.

puromycin

Clinically useful antibiotics	Mode of action
Tetracycline	Prevents the aminoacyl-tRNA from binding to the ribosome
Erythromycin	Prevents the incorporation of new amino acids into the protein
Streptomycin	Inhibits the initiation of protein synthesis
Chloramphenicol	Prevents the new peptide bond from being formed

27.9 Why DNA Contains Thymine Instead of Uracil

In Section 24.8 we saw that dTMP is formed when dUMP is methylated, with coenzyme N^5,N^{10}-methylenetetrahydrofolate supplying the methyl group. Because the incorporation of the methyl group into uracil oxidizes tetrahydrofolate to dihydrofolate, dihydrofolate must be reduced back to tetrahydrofolate to prepare the coenzyme for another catalytic cycle. The reducing agent is NADPH. Every NADPH formed in a biological organism can drive the formation of three ATPs (Section 25.11), so using an NADPH to reduce dihydrofolate comes at the expense of ATP. This means that the synthesis of thymine is energetically expensive, so there must be a good reason for DNA to contain thymine instead of uracil.

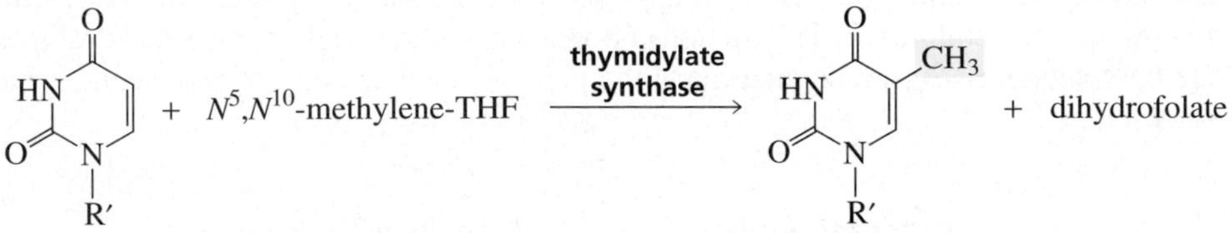

$R' = $ 2′-deoxyribose-5-P

$$\text{dihydrofolate} + \text{NADPH} + \text{H}^+ \xrightarrow{\text{dihydrofolate reductase}} \text{tetrahydrofolate} + \text{NADP}^+$$

The presence of thymine instead of uracil in DNA prevents potentially lethal mutations. Cytosine can tautomerize to form an imine, which can be hydrolyzed to uracil (Section 17.8). The overall reaction is called a **deamination** because it removes an amino group.

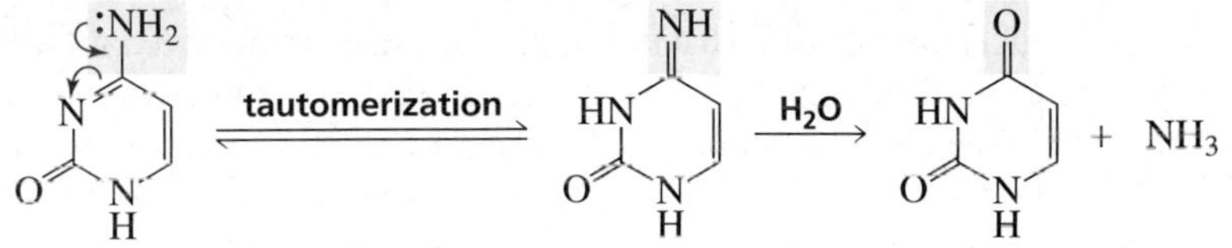

If a cytosine in DNA is deaminated to a uracil, uracil will specify incorporation of an adenine into the daughter strand during replication instead of the guanine that would have been specified by cytosine. Fortunately, a U in DNA is recognized as a "mistake" by an enzyme before an incorrect base can be inserted into the daughter strand. The enzyme cuts out the U and replaces it with a C. If U's were normally found in DNA, the enzyme could not distinguish between a normal U and a U formed by deamination of cytosine. Having T's in place of U's in DNA allows the U's that are found in DNA to be recognized as mistakes.

Unlike DNA, which replicates itself, any mistake in RNA does not survive for long because RNA is constantly being degraded and then resynthesized from the DNA template. Therefore, it is not worth spending the extra energy to incorporate T's into RNA.

PROBLEM 16◆

Adenine can be deaminated to hypoxanthine, and guanine can be deaminated to xanthine. Draw structures for hypoxanthine and xanthine.

PROBLEM 17

Explain why thymine cannot be deaminated.

27.10 **How the Base Sequence of DNA Is Determined**

In June 2000, two teams of scientists (one from a private biotechnology company and one from the publicly funded Human Genome Project) announced that they had completed the first draft of the sequence of the 3.1 billion base pairs in human DNA. This is an enormous accomplishment. Consider, for example, that if the sequence of

1 million base pairs were determined each day, it would take more than 10 years to complete the sequence of the human genome.

DNA molecules are too large to sequence as a unit, so DNA is first cleaved at specific base sequences and the resulting DNA fragments are then sequenced individually. The enzymes that cleave DNA at specific base sequences are called **restriction endonucleases**, and the DNA fragments they produce are called **restriction fragments**. Several hundred restriction endonucleases are now known; a few examples, the base sequence each recognizes, and the point of cleavage in that base sequence are shown here.

restriction enzyme	recognition sequence
*Alu*I	AG|CT TC|GA
*Fnu*DI	GG|CC CC|GG
*Pst*I	CTGCA|G G|ACGTC

The base sequences that most restriction endonucleases recognize are *palindromes.* A palindrome is a word or a group of words that reads the same forward and backward. "Toot" and "race car" are examples of palindromes.* A restriction endonuclease recognizes a piece of DNA in which *the template strand is a palindrome of the sense strand.* In other words, the sequence of bases in the template strand (reading from right to left) is identical to the sequence of bases in the sense strand (reading from left to right).

BIOGRAPHY

Frederick Sanger *(Section 22.13) and* **Walter Gilbert** *shared half of the 1980 Nobel Prize in chemistry for their work on DNA sequencing. The other half went to* **Paul Berg**, *who had developed a method of cutting nucleic acids at specific sites and recombining the fragments in new ways, a technique known as recombinant DNA technology.*

BIOGRAPHY

Walter Gilbert *was born in Boston in 1932. He received a master's degree in physics from Harvard and a Ph.D. in mathematics from Cambridge University. In 1958 he joined the faculty at Harvard, where he became interested in molecular biology.*

BIOGRAPHY

Paul Berg *was born in New York in 1926. He received a Ph.D. from Western Reserve University (now Case Western Reserve University). He joined the faculty at Washington University in St. Louis in 1955 and became a professor of biochemistry at Stanford in 1959.*

PROBLEM 18◆

Which of the following base sequences would most likely be recognized by a restriction endonuclease?

a. ACGCGT
b. ACGGGT
c. ACGGCA
d. ACACGT
e. ACATCGT
f. CCAACC

The restriction fragments can be sequenced using a procedure developed by Frederick Sanger, known as the **dideoxy method**. This method involves generating fragments whose length depends on the last base added to the fragment. Because of its simplicity, it has superceded alternative methods.

In the dideoxy method a small piece of DNA called a primer, labeled at the 5′-end with ^{32}P, is added to the restriction fragment whose sequence is to be determined. Next, the four 2′-deoxyribonucleoside triphosphates are added, and then DNA polymerase—the enzyme that adds nucleotides to a strand of DNA—is added as well. In addition, a small amount of the 2′,3′-dideoxynucleoside triphosphate of one of the bases is added to the reaction mixture. A 2′,3′-dideoxynucleoside triphosphate has no OH group at either the 2′- or 3′-position.

* Some other palindromes are "Mom," "Dad," "Bob," "Lil," "radar," "noon," "wow," "poor Dan in a droop," "a man, a plan, a canal, Panama," "Sex at noon taxes," and "He lived as a devil, eh?"

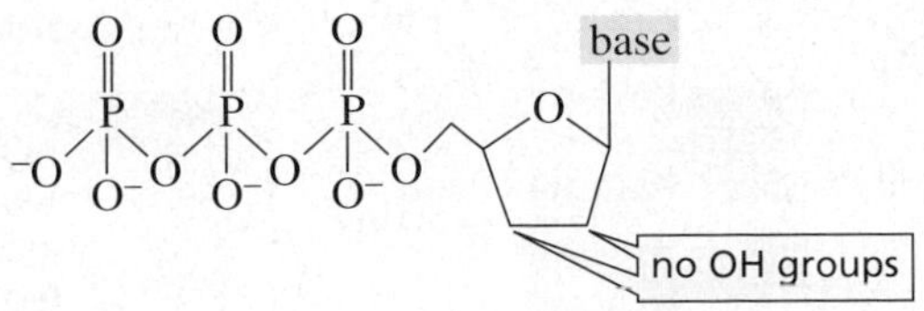

Nucleotides will be added to the primer by base pairing with the restriction fragment. Synthesis will stop if the 2′,3′-dideoxy analog of dATP is added instead of dATP, because the 2′,3′-dideoxy analog does not have a 3′-OH to which additional nucleotides can be added. Therefore, three different chain-terminated fragments will be obtained from the DNA restriction fragment shown here.

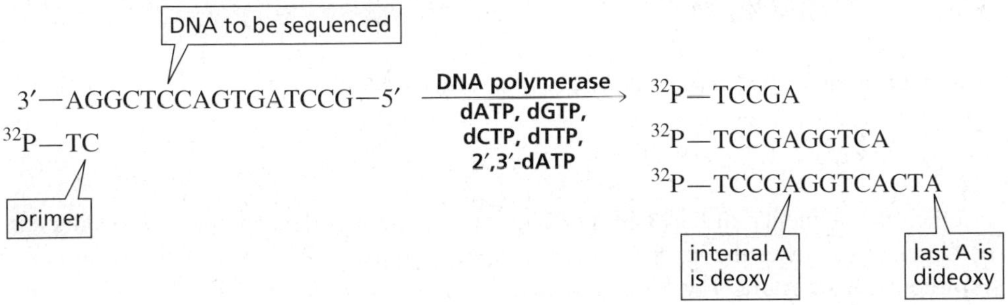

The procedure is repeated three more times using a 2′,3′-dideoxy analog of dGTP, then a 2′,3′-dideoxy analog of dCTP, and then a 2′,3′-dideoxy analog of dTTP.

PROBLEM 19

What labeled fragments would be obtained from the segment of DNA shown above if a 2′,3′-dideoxy analog of dGTP had been added to the reaction mixture instead of a 2′,3′-dideoxy analog of dATP?

The chain-terminated fragments obtained from each of the four experiments are loaded onto separate lanes of a buffered polyacrylamide gel: the fragments obtained from using a 2′,3′-dideoxy analog of dATP are loaded onto one lane; the fragments obtained from using a 2′,3′-dideoxy analog of dGTP onto another lane; and so on (see Figure 27.15 on page 1220). An electric field is applied across the ends of the gel, causing the negatively charged fragments to travel toward the positively charged electrode (the anode). The smaller fragments fit through the spaces in the gel relatively easily and therefore travel through the gel faster, while the larger fragments pass through the gel more slowly.

After the fragments have been separated, the gel is placed in contact with a photographic plate. Radiation from ^{32}P causes a dark spot to appear on the plate opposite he location of each labeled fragment in the gel. This technique is called autoradiography, and the exposed photographic plate is known as an **autoradiograph** (Figure 27.15).

The sequence of bases in the original restriction fragment can be read directly from the autoradiograph. The identity of each base is determined by noting the column where each successive dark spot (larger piece of labeled fragment) appears, starting at the bottom of the gel to identify the base at the 5′-end. The sequence of the fragment of DNA responsible for the autoradiograph in Figure 27.15a is shown on the left-hand side of the figure.

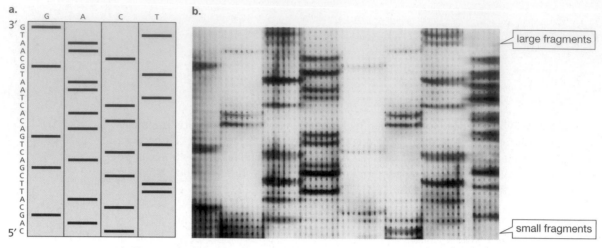

▲ **Figure 27.15**
(a) A schematic drawing of an autoradiograph. (b) An actual autoradiograph.

Once the sequence of bases in a restriction fragment is determined, the results can be checked by using the same process to obtain the base sequence of the fragment's complementary strand. The base sequence in the original piece of DNA can be determined by repeating the entire procedure with a different restriction endonuclease and noting overlapping fragments.

27.11 **Polymerase Chain Reaction (PCR)**

The **polymerase chain reaction (PCR)**, developed in 1983, is a technique that allows scientists to amplify DNA—making billions of copies of a segment of DNA in a very short time. With PCR, sufficient DNA for analysis can be obtained from a single hair or sperm.

PCR is done by adding the following to a solution containing the segment of DNA to be amplified (the target DNA):

- a large excess of primers (short pieces of DNA) that base pair with (anneal with) the DNA that flanks the target DNA
- the four deoxyribonucleotide triphosphates (dATP, dGTP, dCTP, dTTP)
- a heat-stable DNA polymerase

The following three steps are then carried out (Figure 27.16):

- **Strand separation.** The solution is heated to 95 °C, causing double-stranded DNA to separate into two single strands.
- **Base-pairing of the primers.** The solution is cooled to 54 °C, allowing the primers to pair with the bases in the DNA that flank the 3′-end of the target DNA.
- **DNA synthesis.** The solution is heated to 72 °C, a temperature at which DNA polymerase catalyzes the addition of nucleotides to the primer. Notice that because the primers attach to the 3′-end of target DNA, copies of target DNA are synthesized in the required 5′ ⟶ 3′ direction.

The solution is then heated to 95 °C to begin a second cycle; the second cycle produces four copies of double-stranded DNA. The third cycle, therefore, starts with eight single strands of DNA and produces 16 single strands. It takes about an hour to complete 30 cycles, at which point DNA has been amplified a billionfold.

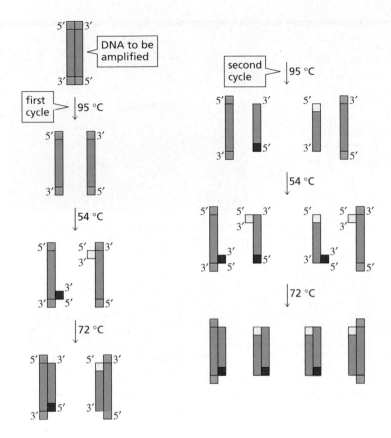

The DNA from a forty-million-year-old leaf preserved in amber was amplified by PCR and then sequenced.

◀ **Figure 27.16**
Two cycles of the polymerase chain reaction.

PCR has a wide range of clinical uses. It can be used to detect mutations that lead to cancer, to diagnose genetic diseases, to reveal the presence of HIV that would be missed by an antibody assay, to monitor cancer chemotherapy, and to rapidly identify an infectious disease.

DNA FINGERPRINTING

PCR is used by forensic chemists to compare DNA samples collected at the scene of a crime with the DNA of the suspected perpetrator. The base sequences of segments of noncoding DNA vary from individual to individual. Forensic laboratories have identified 13 of these segments that are the most accurate for identification purposes. If the sequence of bases from two DNA samples are the same, the chance is about 80 billion to one that they are from the same individual. DNA fingerprinting is also used to establish paternity, accounting for about 100,000 DNA profiles a year.

27.12 **Genetic Engineering**

Recombinant DNA molecules are DNA molecules (natural or synthetic) that have been incorporated into the DNA of a compatible host cell. Recombinant DNA technology—also known as **genetic engineering**—has many practical applications. For example, the DNA for a particular protein can be inserted into a vector (a microorganism capable of replicating DNA) that produces millions of copies, leading to the synthesis of large amounts of protein. Human insulin is made in this way.

Agriculture is benefiting from genetic engineering as crops are being produced with new genes that increase their resistance to drought and insects. For example,

RESISTING HERBICIDES

Glyphosate, an herbicide known as Roundup, kills weeds by inhibiting an enzyme that plants need to synthesize phenylalanine and tryptophan, amino acids required for growth. Studies are being done with corn and cotton that have been genetically engineered to tolerate the herbicide. These crops have been given a gene for an enzyme that makes glyphosate inactive by acetylating it with acetyl-CoA. When fields of these plants are sprayed with glyphosate, the weeds are killed but not the crops.

Corn genetically engineered to resist the herbicide glyphosphate by acetylating it.

glyphosate
an herbicide

+ enzyme →

N-acetylglyphosate
harmless to plants

+ CoASH

genetically engineered cotton crops are resistant to the cotton bollworm and genetically engineered corn is resistant to the corn rootworm. With crops like these, growers no longer have to use herbicides, which threaten groundwater supplies. Genetically modified organisms (GMOs) have been responsible for a nearly 50% reduction in agricultural chemical sales in the United States. On the other hand, many people worry about the potential consequences of changing the genes of any species.

27.13 Laboratory Synthesis of DNA Strands

There are many reasons why scientists are working to develop practical methods for synthesizing oligonucleotides with specific base sequences. One is that it would allow them to synthesize a gene that could be inserted into the DNA of microorganisms, causing the organisms to synthesize a particular protein. Alternatively, a synthetic gene could be inserted into the DNA of an organism defective in that gene, thus correcting the defect; this treatment strategy is known as **gene therapy**.

Synthesizing an oligonucleotide with a particular base sequence is an even more challenging task than synthesizing a polypeptide with a specific amino acid sequence, because each nucleotide has several groups that must be protected and then deprotected at the proper times. Oligonucleotide synthesis is carried out in an automated method similar to automated peptide synthesis (Section 22.11); the growing nucleotide is attached to a solid support so that purification can be carried out by flushing the reaction container with an appropriate solvent. With this technique, none of the synthesized product is lost during purification.

Phosphoramidite Monomers

One current method synthesizes oligonucleotides using phosphoramidite monomers. The 5'-OH group of each phosphoramidite monomer is attached to a *para*-dimethoxytrityl (DMTr) protecting group. The particular group (Pr) used to protect the base depends on the base.

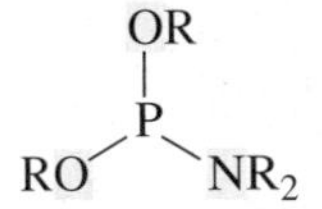

a phosphoramidite

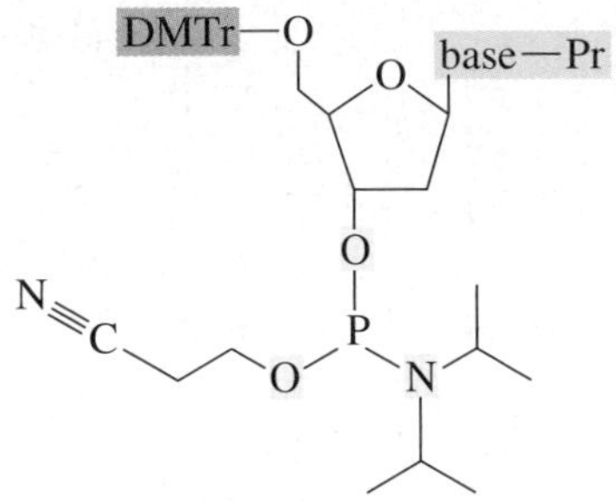

a phosphoramidite monomer
used for
oligonucleotide synthesis

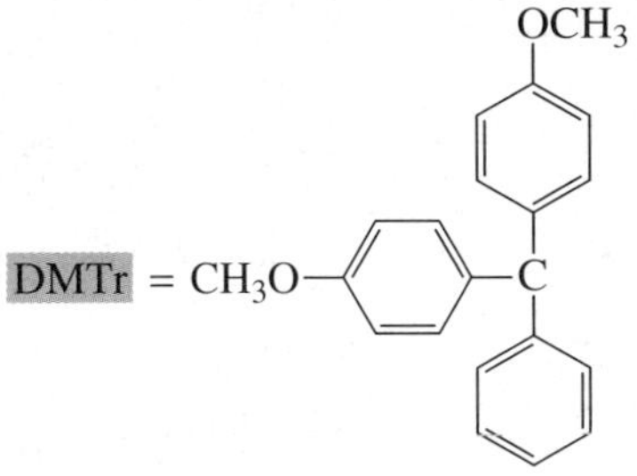

DMTr = CH$_3$O—

The 3'-nucleoside of the oligonucleotide to be synthesized is attached to a controlled-pore glass solid support and the oligonucleotide is synthesized from the 3'-end. When a monomer is added to the nucleoside attached to the solid support, the only nucleophile in the reaction mixture is the 5'-OH group of the sugar bonded to the solid support. This nucleophile attacks the phosphorus of the phosphoramidite, displacing the amine and forming a phosphite. The amine is too strong a base to be expelled without being protonated. Protonated tetrazole is the acid used for protonation because it is strong enough to protonate the diisopropylamine leaving group, but not strong enough to remove the DMTr protecting group. The phosphite is oxidized to a phosphate using I$_2$ or *tert*-butylhydroperoxide. The DMTr protecting group on the 5'-end of the dinucleotide is removed with mild acid. The cycle of (1) monomer addition, (2) oxidation, and (3) deprotection with acid is repeated over and over until a polymer of the desired length and sequence is obtained. Notice that the DNA polymer is synthesized in the 3' ⟶ 5' direction, opposite to the direction (5' ⟶ 3') in which DNA is synthesized in nature.

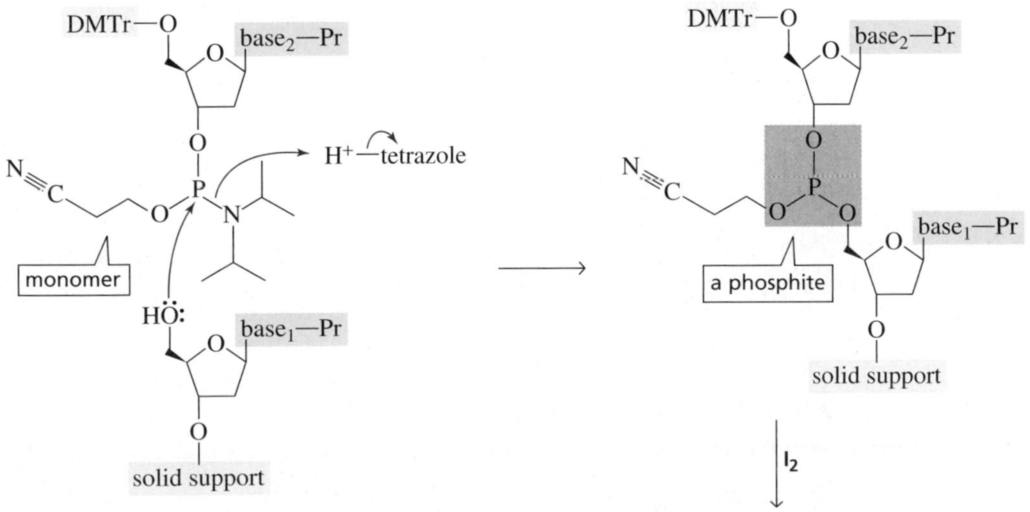

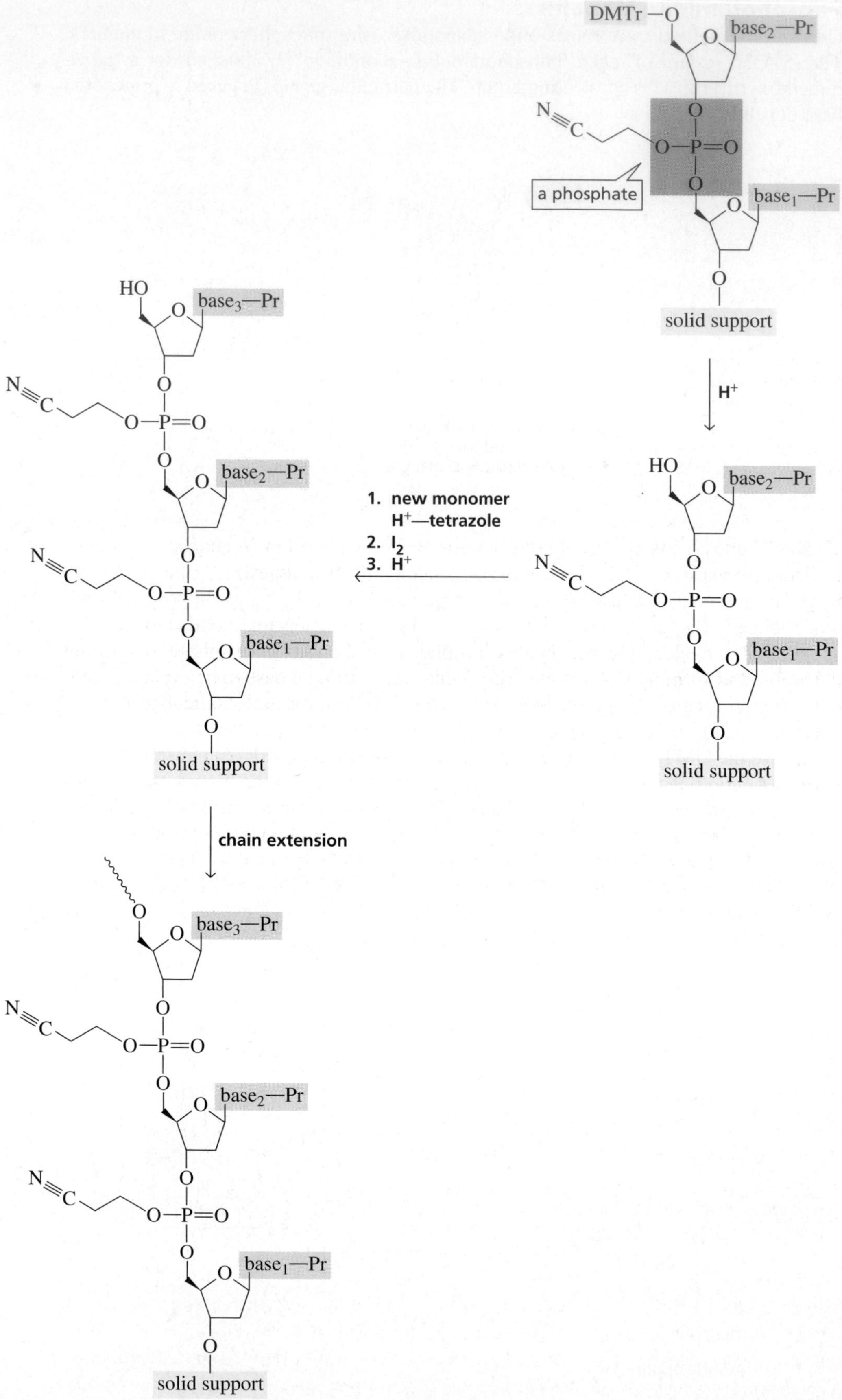
DMTr—O
base2—Pr
N≡C
O—P=O
a phosphate
base1—Pr
solid support
H+
HO
base2—Pr
1. new monomer
H+—tetrazole
2. I2
3. H+
HO
base3—Pr
chain extension

The NH$_2$ groups of cytosine, adenine, and guanine are nucleophiles and therefore must be protected to prevent them from reacting with a newly added monomer; they are protected as amides. Thymine does not contain any nucleophilic groups, so it does not have to be protected.

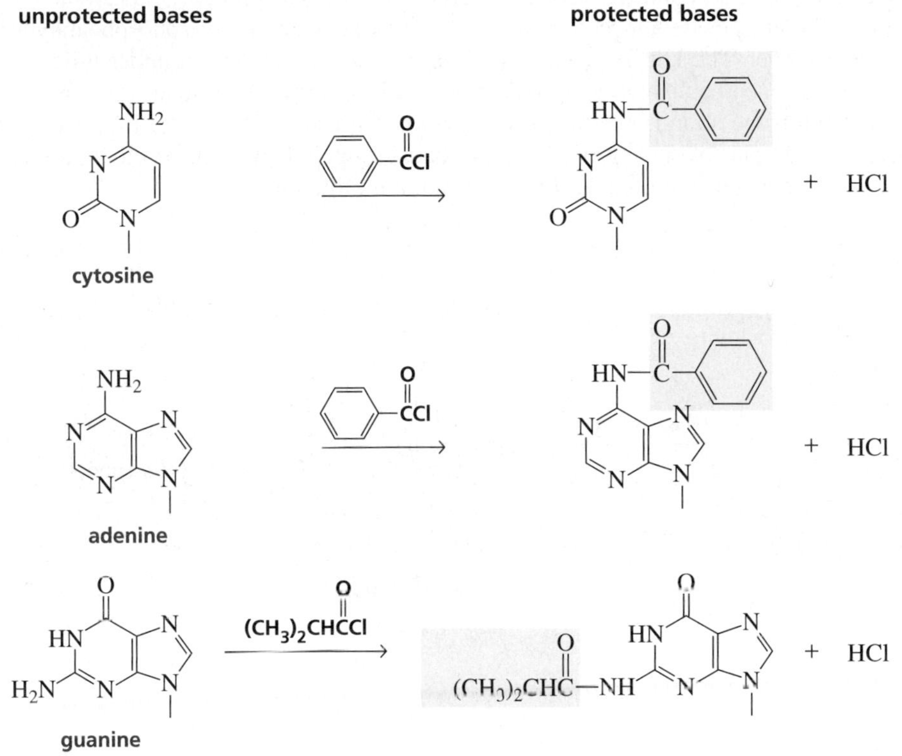

unprotected bases   **protected bases**

cytosine

adenine

guanine

After the oligonucleotide is synthesized, the protecting groups on the phosphates and the protecting groups on the bases must be removed, and the oligonucleotide has to be detached from the solid support. This can all be done in a single step using aqueous ammonia. Because a hydrogen bonded to a carbon adjacent to a cyano group is relatively acidic (Section 18.1), ammonia can be used as the base in the elimination reaction that removes the phosphate protecting group.

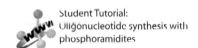

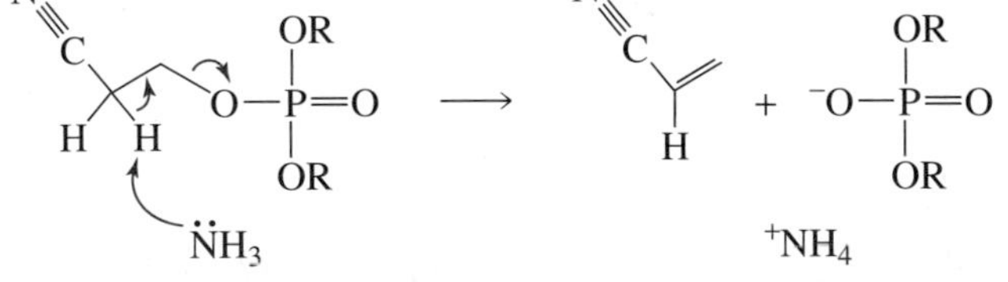

Student Tutorial:
Oligonucleotide synthesis with phosphoramidites

Currently, automated DNA synthesizers can synthesize nucleic acids containing as many as 130 nucleotides in acceptable yields, adding one nucleotide every 2 to 3 minutes. (Imagine doing ~390 reactions—130 couplings, 130 oxidations, and 130 DMT removals— plus final deprotections, and having to isolate the product after each one!) Longer nucleic acids can be prepared by splicing together two or more individually prepared strands. To ensure a good overall yield of oligonucleotide, the addition of each monomer must occur in high yield (>98%). This can be accomplished by using a large excess of monomer. This, however, makes oligonucleotide synthesis very expensive because the unreacted monomers are wasted.

H-Phosphonate Monomers

A second method, one using H-phosphonate monomers for synthesizing oligonucleotides that have specific base sequences, improves on the phosphoramidite method in that the monomers are easier to handle and phosphate protecting groups are not needed. However, the yields are not as good. The H-phosphonate monomers are activated by reacting with an acyl chloride, which converts them into phosphoanhydrides. The 5'-OH group of the nucleoside attached to the solid support reacts with the phosphoanhydride, forming a dimer. The DMTr protecting group is removed with acid under mild conditions, and a second activated monomer is added. Monomers are added one at a time in this way until the strand is complete. Oxidation with I_2 (or *tert*-butyl hydroperoxide) converts the H-phosphonate groups to phosphate groups. The base-protecting groups are removed by aqueous ammonia as in the phosphoramidite method.

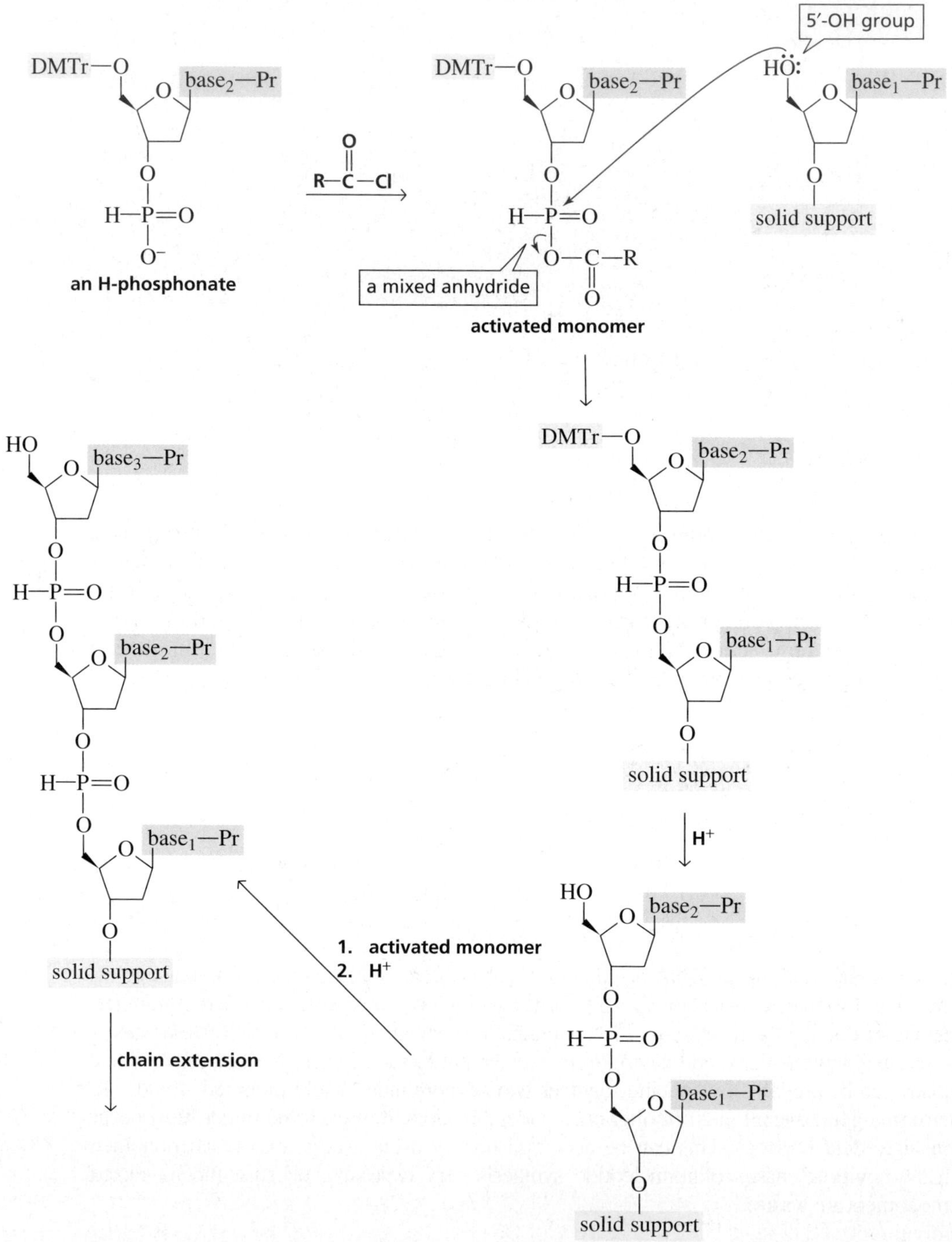

Notice that in the phosphoramidite method, an oxidation takes place each time a monomer is added, whereas in the H-phosphonate method, a single oxidation is carried out after the entire strand has been synthesized.

PROBLEM 20

Propose a mechanism for removal of the DMTr protecting group by treatment with acid.

SUMMARY

There are two types of nucleic acids: **deoxyribonucleic acid (DNA)** and **ribonucleic acid (RNA)**. DNA encodes an organism's hereditary information and controls the growth and division of cells. In most organisms, the genetic information stored in DNA is **transcribed** into RNA. The information thus contained in the RNA strand can then be **translated** for the synthesis of all the proteins needed for cellular structure and function.

A **nucleoside** contains a base bonded to D-ribose or to 2-deoxy-D-ribose. A **nucleotide** is a nucleoside with either the 5'- or the 3'-OH group bonded to phosphoric acid by an ester linkage. **Nucleic acids** are composed of long strands of nucleotide subunits linked by phosphodiester bonds. These linkages join the 3'-OH group of one nucleotide to the 5'-OH group of the next nucleotide. A **dinucleotide** contains two nucleotide subunits, an **oligonucleotide** contains three to 10 subunits, and a **polynucleotide** contains many subunits. DNA contains 2'-deoxy-D-ribose while RNA contains D-ribose. The difference in the sugars causes DNA to be stable and RNA to be easily cleaved.

The **primary structure** of a nucleic acid is the sequence of bases in its strand. DNA contains **A**, **G**, **C**, and **T**; RNA contains **A**, **G**, **C**, and **U**. The presence of thymine instead

of uracil in DNA prevents mutations caused by imine hydrolysis of C to form U. DNA is double-stranded. The strands run in opposite directions and are twisted into a helix, giving DNA a major groove and a minor groove. The bases are confined to the inside of the helix and the sugar and phosphate groups are on the outside. The strands are held together by hydrogen bonds between bases of opposing strands as well as by **stacking interactions**; these are van der Waals attractions between adjacent bases on the same strand. The two strands—one is called a **sense strand** and the other a **template strand**—are complementary: **A** pairs with **T**, and **G** pairs with **C**. DNA is synthesized in the 5' ⟶ 3' direction by a process called **semiconservative replication**.

The sequence of bases in DNA provides the blueprint for the synthesis (**transcription**) of RNA. RNA is synthesized in the 5' ⟶ 3' direction by reading the bases along the DNA template strand in the 3' ⟶ 5' direction. There are three kinds of RNA: messenger RNA, ribosomal RNA, and transfer RNA. Protein synthesis (**translation**) proceeds from the N-terminal end to the C-terminal end by a process that reads the bases along the mRNA strand in the 5' ⟶ 3' direction. Each three-base sequence—a **codon**—specifies the particular amino acid to be incorporated into a protein. A tRNA carries the amino acid bound as an ester to its 3'-terminal position. The codons and the amino acids they specify are known as the **genetic code**.

Restriction endonucleases cleave DNA at specific palindromes forming **restriction fragments**. The **dideoxy** **method** is the method preferred to determine the sequence of bases in the restriction fragments. The **polymerase chain reaction (PCR)** method amplifies segments of DNA, making billions of copies in a very short time. Oligonucleotides with specific base sequences can be synthesized using phosphoramidite monomers or H-phosphonate monomers.

KEY TERMS

anticodon (p. 1210)
antisense strand (p. 1209)
autoradiograph (p. 1219)
base (p. 1198)
codon (p. 1212)
deamination (p. 1217)
deoxyribonucleic acid (DNA) (p. 1197)
deoxyribonucleotide (p. 1201)
dideoxy method (p. 1218)
dinucleotide (p. 1201)
double helix (p. 1205)
exon (p. 1209)
gene (p. 1207)
gene therapy (p. 1222)
genetic code (p. 1212)
genetic engineering (p. 1221)

human genome (p. 1207)
informational strand (p. 1209)
intron (p. 1209)
major groove (p. 1206)
messenger RNA (mRNA) (p. 1210)
minor groove (p. 1206)
nucleic acid (p. 1198)
nucleoside (p. 1200)
nucleotide (p. 1201)
oligonucleotide (p. 1202)
phosphodiester (p. 1198)
polymerase chain reaction (PCR) (p. 1220)
polynucleotide (p. 1202)
primary structure of DNA (p. 1203)
promoter site (p. 1209)
recombinant DNA (p. 1221)

replication (p. 1208)
replication fork (p. 1208)
restriction endonuclease (p. 1218)
restriction fragment (p. 1218)
ribonucleic acid (RNA) (p. 1197)
ribonucleotide (p. 1201)
ribosomal RNA (rRNA) (p. 1210)
RNA splicing (p. 1209)
semiconservative replication (p. 1208)
sense strand (p. 1209)
stacking interactions (p. 1205)
stop codon (p. 1213)
template strand (p. 1209)
transcription (p. 1209)
transfer RNA (tRNA) (p. 1210)
translation (p. 1209)

PROBLEMS

21. Name the following compounds:

a.

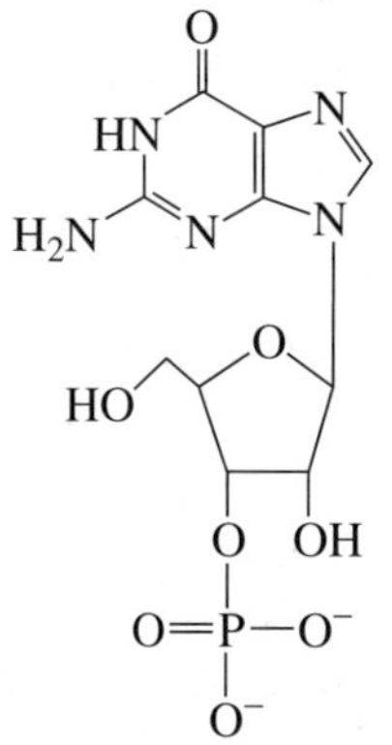

b.

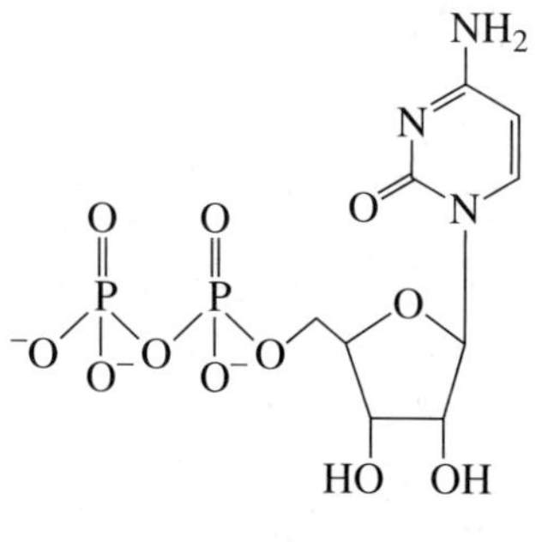

c.

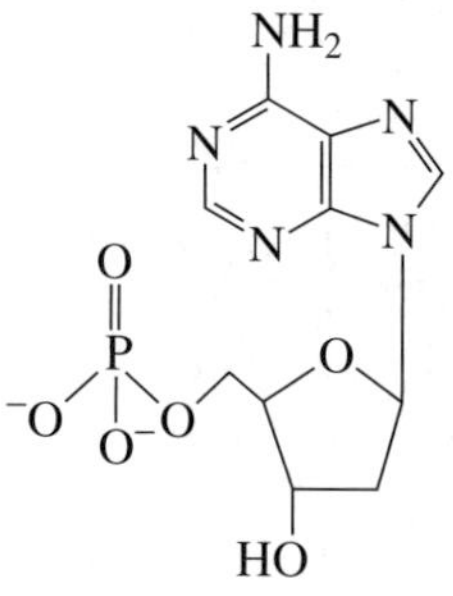

d.

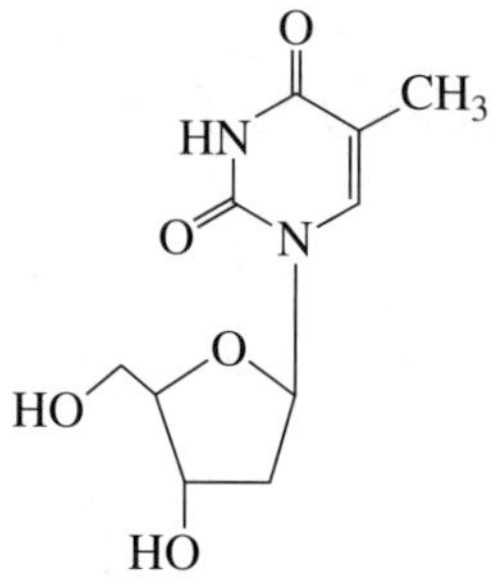

22. What nonapeptide is coded for by the following piece of mRNA?

$$5' - AAA - GUU - GGC - UAC - CCC - GGA - AUG - GUG - GUC - 3'$$

23. What is the sequence of bases in the template strand of DNA that codes for the mRNA in Problem 22?

24. What is the sequence of bases in the sense strand of DNA that codes for the mRNA in Problem 22?

25. What would be the C-terminal amino acid if the codon at the 3'-end of the mRNA in Problem 22 underwent the following mutation:
a. if the first base was changed to A?
b. if the second base was changed to A?
c. if the third base was changed to A?
d. if the third base was changed to G?

26. What would be the base sequence of the segment of DNA that is responsible for the biosynthesis of the following hexapeptide?

Gly-Ser-Arg-Val-His-Glu

27. Propose a mechanism for the following reaction:

$$^-\text{O}\overset{\text{O}}{\overset{\|}{\text{C}}}\text{CH}_2\text{CH}_2\overset{\text{O}}{\underset{^+\text{NH}_3}{\text{CH}}}\overset{\|}{\text{C}}\text{O}^- + \text{NH}_3 + \text{ATP} \longrightarrow \text{H}_2\text{N}\overset{\text{O}}{\overset{\|}{\text{C}}}\text{CH}_2\text{CH}_2\overset{\text{O}}{\underset{^+\text{NH}_3}{\text{CH}}}\overset{\|}{\text{C}}\text{O}^- + \text{ADP} + \text{ }^-\text{O}\overset{\text{O}}{\overset{\|}{\underset{\text{OH}}{\text{P}}}}\text{O}^-$$

28. Match the codon with the anticodon:

Codon	Anticodon
AAA	ACC
GCA	CCU
CUU	UUU
AGG	AGG
CCU	UGA
GGU	AAG
UCA	GUC
GAC	UGC

29. Using the single-letter abbreviations for the amino acids in Table 22.1, write the sequence of amino acids in a tetrapeptide represented by the first four different letters in your first name. Do not use any letter twice. (Because not all letters are assigned to amino acids, you might have to use one or two letters in your last name.) Write one of the sequences of bases in mRNA that would result in the synthesis of that polypeptide. Write the sequence of bases in the sense strand of DNA that would result in formation of that fragment of mRNA.

30. Which of the following pairs of dinucleotides are present in equal amounts in DNA?
a. CC and GG   c. CA and TG   e. GT and CA
b. CG and GT   d. CG and AT   f. TA and AT

31. Human immunodeficiency virus (HIV) is the retrovirus that causes AIDS. AZT is the best known of the drugs designed to interfere with retroviral DNA synthesis. When cells take up AZT, they convert it to AZT-triphosphate. Explain how AZT interferes with DNA synthesis.

3′-azido-2′-deoxythymidine
AZT

32. Why is the codon a triplet rather than a doublet or a quartet?

33. RNAase, the enzyme that catalyzes the hydrolysis of RNA, has two catalytically active histidine residues at its active site. One of the histidine residues is catalytically active in its acidic form and the other is catalytically active in its basic form. Propose a mechanism for RNAase.

34. The amino acid sequences of peptide fragments obtained from a normal protein were compared with those obtained from the same protein synthesized by a defective gene. They were found to differ in only one peptide fragment. The primary sequences of the fragments that differed are shown here.

Normal: Gln-Tyr-Gly-Thr-Arg-Tyr-Val
Mutant: Gln-Ser-Glu-Pro-Gly-Thr

a. What is the defect in DNA?
b. It was later determined that the normal peptide fragment is an octapeptide with a C-terminal Val-Leu. What is the C-terminal amino acid of the mutant peptide?

35. Which cytosine in the following sense strand of DNA could cause the most damage to the organism if it were deaminated?

$$5'—A—T—G—T—C—G—C—T—A—A—T—C—3'$$

36. Sodium nitrite, a common food preservative (page 706), is capable of causing mutations in an acidic environment by converting cytosines to uracils. Explain how this occurs.

37. *Staphylococcus* nuclease is an enzyme that catalyzes the hydrolysis of DNA. The overall hydrolysis reaction is shown below:

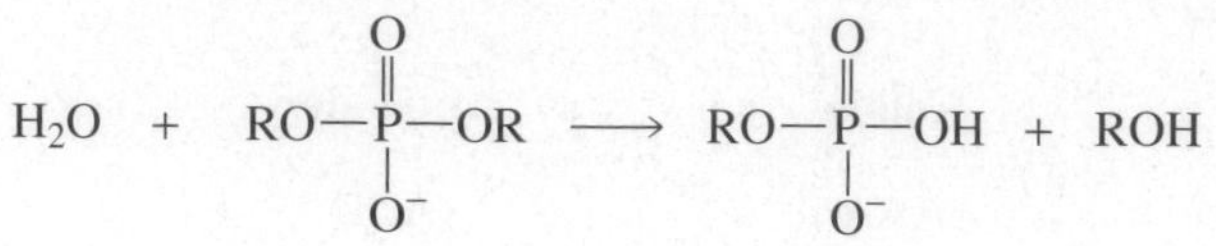

$$H_2O + RO—\underset{\underset{O^-}{|}}{\overset{\overset{O}{\|}}{P}}—OR \longrightarrow RO—\underset{\underset{O^-}{|}}{\overset{\overset{O}{\|}}{P}}—OH + ROH$$

The reaction is catalyzed by Ca^{2+}, Glu 43, and Arg 87. Propose a mechanism for this reaction. Recall that the nucleotides in DNA have phosphodiester linkages.

38. The first amino acid incorporated into a polypeptide chain during its biosynthesis in prokaryotes is *N*-formylmethionine. Explain the purpose of the formyl group. (*Hint:* The risosome has a binding site for the growing peptide chain and a binding site for the incoming amino acid.)

39. Why does DNA not unravel completely before replication begins?

PART 9

Special Topics in Organic Chemistry

CHAPTER 28
Synthetic Polymers

Previous chapters discussed the polymers synthesized by biological systems—proteins, carbohydrates, and nucleic acids. **Chapter 28** discusses polymers synthesized by chemists. These synthetic polymers have physical properties that make them useful components of fabrics, bottles, food wrap, automobile parts, compact discs, and thousands of other objects and materials that pervade our lives.

CHAPTER 29
Pericyclic Reactions

Chapter 29 discusses pericyclic reactions. These are reactions that occur as a result of a cyclic reorganization of electrons. In this chapter, you will learn how the conservation of orbital symmetry theory explains the relationships between reactant, product, and reaction conditions in a pericyclic reaction.

CHAPTER 30
The Organic Chemistry of Drugs:
Discovery and Design

Chapter 30 introduces you to medicinal chemistry. Here you will see how many of our commonly used drugs were discovered, and you will learn about some of the techniques used to develop new ones.

CHAPTER 28

Synthetic Polymers

Super Glue

BUILDING ON FUNDAMENTALS

SECTION 28.2 Some polymers are produced by chain reactions, which earlier discussions described as having initiation, propagation, and termination steps (11.6).

SECTION 28.2 A carbocation will rearrange if rearrangement leads to a more stable carbocation (4.6).

SECTION 28.2 A radical initiator has a weak bond that readily undergoes homolytic cleavage (11.6).

SECTION 28.2 In an electrophilic addition reaction, the electrophile adds to the sp^2 carbon bonded to the greater number of hydrogens (4.4).

SECTION 28.2 Under basic conditions, a nucleophile attacks the less sterically hindered carbon of an epoxide; under acidic conditions, it attacks the more substituted carbon of an epoxide (10.8).

Probably no group of synthetic compounds is more important to modern life than synthetic polymers. A **polymer** is a large molecule made by linking together repeating units of small molecules called **monomers**. The process of linking them together is called **polymerization**.

$$n\text{M} \xrightarrow{\text{polymerization}} \text{—M—M—M—M—M—M—M—M—M—}$$

monomer   polymer

ethylene monomers   polyethylene

Unlike small organic molecules, which are of interest because of their *chemical* properties, these giant molecules—with molecular weights ranging from thousands to millions—are interesting primarily because of their *physical* properties that make them useful in everyday life. Some synthetic polymers resemble natural substances, but most are quite different from materials found in nature. Such diverse products as photographic film, compact discs, rugs, food wrap, artificial joints, Super Glue, toys, plastic bottles, weather stripping, automobile body parts, and shoe soles are made of synthetic polymers.

Polymers can be divided into two broad groups: **synthetic polymers** and **biopolymers** (natural polymers). Synthetic polymers are synthesized by scientists, whereas biopolymers are synthesized by organisms. Examples of biopolymers are DNA, which is the storage molecule for genetic information; RNA and proteins, the molecules that induce biochemical transformations; and polysaccharides, which store energy and also function as structural materials. The structures and properties of these biopoylmers are presented in other chapters. In this chapter, we will explore synthetic polymers.

Humans first relied on *biopolymers* for clothing, wrapping themselves in animal skins and furs. Later, they learned to spin natural fibers into thread and to weave the thread into cloth. Today, much of our clothing is made of *synthetic polymers* (such as nylon, polyester, and polyacrylonitrile). Many people prefer clothing made of natural polymers (like cotton, wool, and silk), but it has been estimated that if synthetic polymers were not available, all the arable land in the United States would have to be used for the production of cotton and wool for clothing.

A **plastic** is a polymer capable of being molded. The first commercial plastic was celluloid. Invented in 1856 by Alexander Parke, it was a mixture of nitrocellulose and camphor. Celluloid was used in the manufacture of billiard balls and piano keys, replacing scarce ivory and providing a reprieve for many elephants. Celluloid was also used for motion picture film until it was replaced by cellulose acetate, a more stable polymer.

The first synthetic fiber was rayon. In 1865, the French silk industry was threatened by an epidemic that killed many silkworms, highlighting the need for an artificial silk substitute. Louis Chardonnet accidentally discovered the starting material for a synthetic fiber when, while wiping up some spilled nitrocellulose from a table, he noticed long silklike strands adhering to both the cloth and the table. "Chardonnet silk" was introduced at the Paris Exposition in 1891. It was called *rayon* because it was so shiny that it appeared to give off rays of light. The "rayon" used today does not contain any nitro groups.

The first synthetic rubber was synthesized by German chemists in 1917. Their efforts were in response to a severe shortage of raw materials as a result of blockading during World War I.

Hermann Staudinger was the first to recognize that the various polymers being produced were not disorderly conglomerates of monomers, but instead were made up of chains of monomers joined together. Today, the synthesis of polymers has grown from a process carried out with little chemical understanding to a sophisticated science in which molecules are engineered to predetermined specifications in order to produce new materials tailored to fit human needs. Examples include Lycra, a fabric with elastic properties, and Dyneema, the strongest fabric commercially available.

Polymer chemistry is now part of the larger discipline of **materials science**, which involves the creation of new materials to replace metals, glass, fabrics, wood, cardboard, and paper. Polymer chemistry has evolved into a multibillion-dollar industry. More than 2.5×10^{13} kilograms of synthetic polymers are produced in the United States each year, and approximately 30,000 polymer patents are currently in force. We can expect many more new materials to be developed by scientists in the years to come.

> **BIOGRAPHY**
>
> **Alexander Parke (1813–1890)** *was born in Birmingham, England. He called the polymer that he invented "pyroxylin" but was unable to market it.*

> **BIOGRAPHY**
>
> *The inventor* **John Wesley Hyatt (1837–1920)** *was born in New York. When a New York firm offered a prize of $10,000 for a substitute for ivory billiard balls, Hyatt improved the synthesis of pyroxylin, changed its name to "celluloid," and patented a method for using it to make billiard balls. He did not, however, win the prize.*

> **BIOGRAPHY**
>
> **Louis-Marie-Hilaire Bernigaud, Comte de Chardonnet (1839–1924)** *was born in France. In the early stages of his career, he was an assistant to Louis Pasteur. Because the rayon he initially produced was made from nitrocellulose, it was dangerously flammable. Eventually, chemists learned to remove some of the nitro groups after the fiber was formed, which made the fiber much less flammable but not as strong.*

> **BIOGRAPHY**
>
> **Hermann Staudinger (1881–1965)**, *the son of a professor, was born in Germany. He became a professor at the Technical Institute of Karlsruhe and at the University of Freiburg. He received the Nobel Prize in chemistry in 1953 for his contributions to polymer chemistry.*

28.1 There Are Two Major Classes of Synthetic Polymers

Synthetic polymers can be divided into two major classes, depending on their method of preparation. **Chain-growth polymers**, also known as **addition polymers**, are made by **chain reactions**—the addition of monomers to the end of a growing chain. The end of the chain is reactive because it is a radical, a cation, or an anion. Polystyrene—used for disposable food containers, insulation, and toothbrush handles, among other things—is

Chain-growth polymers are also called addition polymers.

an example of a chain-growth polymer. Polystyrene is pumped full of air to produce the material used as insulation in house construction.

$$CH_2=CH \quad CH_2=CH \quad CH_2=CH \quad \longrightarrow \quad -CH_2-CH-CH_2-CH+CH_2-CH-$$

styrene

polystyrene
a chain-growth polymer

repeating unit

Step-growth polymers, also called **condensation polymers**, are made by combining two molecules while, in most cases, removing a small molecule, generally water or an alcohol. The reacting molecules have reactive functional groups at each end. Unlike chain-growth polymerization, which requires the individual molecules to add to the end of a growing chain, step-growth polymerization allows any two reactive molecules to combine. Dacron is an example of a step-growth polymer.

repeating unit

$$CH_3O-\overset{O}{\underset{\|}{C}}-\overset{O}{\underset{\|}{C}}-OCH_3 \;+\; HOCH_2CH_2OH \;\overset{\Delta}{\longrightarrow}\; -OCH_2CH_2O-\overset{O}{\underset{\|}{C}}-\overset{O}{\underset{\|}{C}}-OCH_2CH_2O- \;+\; 2n\,CH_3OH$$

dimethyl terephthalate   1,2-ethanediol   poly(ethylene terephthalate)
Dacron®
a step-growth polymer

Dacron is the most common of the group of polymers known as **polyesters**, which are polymers containing many ester groups. Polyesters are used for clothing and are responsible for the wrinkle-resistant behavior of many fabrics. Polyester is also used to make the plastic film called Mylar, used in the manufacture of magnetic recording tape. This film is tear resistant and, when processed, has a tensile strength nearly as great as that of steel. Aluminized Mylar was used to make the Echo satellite that was put into orbit around the Earth as a giant reflector. The polymer in soft drink bottles is also a polyester.

28.2 Chain-Growth Polymers

The monomers used most commonly in chain-growth polymerization are ethylene (ethene) and substituted ethylenes. In the chemical industry, monosubstituted ethylenes are known as *alpha olefins*. Polymers formed from ethylene or substituted ethylenes are called **vinyl polymers**. Some of the many vinyl polymers synthesized by chain-growth polymerization are listed in Table 28.1.

Chain-growth polymerization proceeds by one of three mechanisms: **radical polymerization**, **cationic polymerization**, or **anionic polymerization**. Each mechanism has three distinct phases: an *initiation step* that starts the polymerization, *propagation steps* that allow the chain to grow, and *termination steps* that stop the growth of the chain. We will see that the mechanism for a given chain-growth reaction depends on the structure of the monomer *and* the initiator used to activate the monomer.

Radical Polymerization

For chain-growth polymerization to occur by a radical mechanism, a radical initiator must be added to the monomer to convert some of the monomer molecules into radicals.

Table 28.1 Some Important Chain-Growth Polymers and Their Uses

Monomer	Repeating unit	Polymer name	Uses
$CH_2{=}CH_2$	$-CH_2-CH_2-$	polyethylene	film, toys, bottles, plastic bags

$CH_2{=}CH$
    │
    Cl
	$-CH_2-CH-$
	    │
	    Cl
		poly(vinyl chloride)	"squeeze" bottles, pipe, siding, flooring

$CH_2{=}CH-CH_3$	$-CH_2-CH-$
	    │
	    CH_3
		polypropylene	molded caps, margarine tubs, indoor/outdoor carpeting, upholstery

$CH_2{=}CH$
    │
    (phenyl)
	$-CH_2-CH-$
	    │
	    (phenyl)
		polystyrene	packaging, toys, clear cups, egg cartons, hot drink cups

$CF_2{=}CF_2$	$-CF_2-CF_2-$	poly(tetrafluoroethylene) Teflon	nonsticking surfaces, liners, cable insulation

$CH_2{=}CH$
    │
    $C{\equiv}N$
	$-CH_2-CH-$
	    │
	    $C{\equiv}N$
		poly(acrylonitrile) Orlon, Acrilan	rugs, blankets, yarn, apparel, simulated fur

$CH_2{=}C-CH_3$
    │
    $COCH_3$
    ‖
    O
	CH_3
	    │
	$-CH_2-C-$
	    │
	    $COCH_3$
	    ‖
	    O
		poly(methyl methacrylate) Plexiglas, Lucite	lighting fixtures, signs, solar panels, skylights

$CH_2{=}CH$
    │
    $OCCH_3$
    ‖
    O
	$-CH_2-CH-$
	    │
	    $OCCH_3$
	    ‖
	    O
		poly(vinyl acetate)	latex paints, adhesives

chain initiating steps in radical polymerization

$$RO-OR \xrightarrow{\Delta} 2\,RO\cdot$$

a radical initiator → radicals

$$RO\cdot + CH_2{=}CH \longrightarrow ROCH_2\overset{\cdot}{C}H$$
$$\quad\quad\quad\quad | \quad\quad\quad\quad\quad |$$
$$\quad\quad\quad\quad Z \quad\quad\quad\quad\quad Z$$

the alkene monomer reacts with a radical

chain-propagating steps

propagating sites

$$ROCH_2\overset{\cdot}{C}H + CH_2{=}CH \longrightarrow ROCH_2CHCH_2\overset{\cdot}{C}H$$
$$\quad\quad | \quad\quad\quad\quad | \quad\quad\quad\quad\quad\quad | \quad\quad |$$
$$\quad\quad Z \quad\quad\quad\quad Z \quad\quad\quad\quad\quad\quad Z \quad\quad Z$$

$$ROCH_2CHCH_2\overset{\cdot}{C}H + CH_2{=}CH \longrightarrow ROCH_2CHCH_2CHCH_2\overset{\cdot}{C}H \xrightarrow{\text{etc.}}$$
$$\quad\quad | \quad\quad | \quad\quad\quad\quad | \quad\quad\quad\quad\quad\quad\quad | \quad\quad | \quad\quad |$$
$$\quad\quad Z \quad\quad Z \quad\quad\quad\quad Z \quad\quad\quad\quad\quad\quad\quad Z \quad\quad Z \quad\quad Z$$

- The initiator breaks homolytically into radicals, and each radical adds to an alkene monomer, converting it into a radical.
- This radical reacts with another monomer, adding a new subunit that propagates the chain. The radical site is now at the end of the most recent unit added to the end of the chain. This is called the **propagating site**.

This process is repeated over and over. Hundreds or even thousands of alkene monomers can add one at a time to the growing chain. Eventually, the chain reaction stops because the propagating sites are destroyed. Propagating sites can be destroyed when

- two chains combine at their propagating sites;
- two chains undergo *disproportionation,* with one chain being oxidized to an alkene and the other being reduced to an alkane as a result of hydrogen atom transfer;
- a chain reacts with an impurity that consumes the radical.

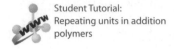

Student Tutorial:
Repeating units in addition polymers

three ways to terminate the chain

chain combination

$$2\ RO{-}[CH_2CH]_n{-}CH_2\dot{C}H \longrightarrow RO{-}[CH_2CH]_n{-}CH_2CHCHCH_2{-}[CHCH_2]_n{-}OR$$

Z · Z · Z · Z · Z

disproportionation

$$2\ RO{-}[CH_2CH]_n{-}CH_2\dot{C}H \longrightarrow RO{-}[CH_2CH]_n{-}CH{=}CH\ +\ RO{-}[CH_2CH]_n{-}CH_2CH_2$$

Z · Z · Z · Z · Z · Z

reaction with an impurity

$$RO{-}[CH_2CH]_n{-}CH_2\dot{C}H\ +\ impurity \longrightarrow RO{-}[CH_2CH]_n{-}CH_2CH{-}impurity$$

Z · Z · Z · Z

The above three steps are called termination steps. Thus, *radical polymerizations* have chain-initiating, chain-propagating, and chain-terminating steps similar to the steps that take place in the radical reactions discussed in Sections 11.2 and 11.6.

As long as the polymer has a high molecular weight, the groups (RO) at the ends of the polymer are relatively unimportant in determining its physical properties and are generally not even specified; it is the rest of the molecule that determines the properties of the polymer.

The molecular weight of the polymer can be controlled by a process known as **chain transfer**. In chain transfer, the growing chain reacts with a molecule XY in a manner that allows X· to terminate the chain, leaving behind Y· to initiate a new chain. Molecule XY can be a solvent, a radical initiator, or any molecule with a bond that can be cleaved homolytically.

$$-CH_2{-}[CH_2CH]_n{-}CH_2\dot{C}H\ +\ XY \longrightarrow -CH_2{-}[CH_2CH]_n{-}CH_2CHX\ +\ Y\cdot$$

Z · Z · Z · Z

Chain-growth polymerization of monosubstituted ethylenes exhibits a marked preference for **head-to-tail addition**, where the head of one monomer is attached to the tail of another.

tail · head

$$CH_2{=}CH$$

Z

$$-CH_2CHCH_2CH- \qquad -CH_2CHCHCH_2- \qquad -CHCH_2CH_2CH-$$

Z · Z · Z · Z · Z · Z

head-to-tail · **head-to-head** · **tail-to-tail**

Head-to-tail addition of a substituted ethylene results in a polymer in which every other carbon bears a substituent.

$$CH_2\!=\!CH \longrightarrow -CH_2CHCH_2CHCH_2CHCH_2CHCH_2CHCH_2CH-$$

Cl	Cl Cl Cl Cl Cl Cl
vinyl chloride	**poly(vinyl chloride)**

Head-to-tail addition is favored for steric reasons. The propagating site experiences less steric hindrance at the unsubstituted sp^2 carbon of the alkene and therefore attacks it preferentially. Groups that stabilize radicals also promote head-to-tail addition. For example, when Z (in the species on page 1236) is a phenyl substituent, the benzene ring stabilizes the radical by electron delocalization, so the propagating site is the carbon that bears the phenyl substituent.

In cases where Z is small (so steric considerations are less important) and less able to stabilize the growing end of the chain by electron delocalization, some head-to-head addition and some tail-to-tail addition also occur. This has been observed primarily in situations where Z is fluorine. Abnormal addition, however, has never been found to constitute more than 10% of the overall chain.

Monomers that most readily undergo chain-growth polymerization by a radical mechanism are those in which the substituent Z is an electron-withdrawing group or is a group able to stabilize the growing radical species by electron delocalization. Examples of monomers that undergo radical polymerization are shown in Table 28.2.

Table 28.2	**Examples of Alkenes That Undergo Radical Polymerization**	
$CH_2\!=\!CH$ *styrene*	$CH_2\!=\!CH$ | $OCCH_3$ ‖ O *vinyl acetate*	$CH_2\!=\!CCH_3$ | $COCH_3$ ‖ O *methyl methacrylate*
$CH_2\!=\!CH$ | Cl *vinyl chloride*	$CH_2\!=\!CH$ | $C\!\equiv\!N$ *acrylonitrile*	$CH_2\!=\!CH$ | $CH\!=\!CH_2$ *1,3-butadiene*

Any compound that readily undergoes homolytic cleavage to form radicals that are sufficiently energetic to convert an alkene into a radical can serve as an initiator for radical polymerization. Several radical initiators are shown in Table 28.3.

A common feature of all radical initiators is a relatively weak bond that readily undergoes homolytic cleavage (Section 11.6). In all but one of the radical initiators shown in Table 28.3, the weak bond is an oxygen–oxygen bond. Two factors enter into the choice of radical initiator for a particular chain-growth polymerization. The first is the desired solubility of the initiator. For example, potassium persulfate is often used if the initiator needs to be soluble in water, whereas an initiator with several carbons is chosen if the initiator must be soluble in a nonpolar solvent. The second factor is the temperature at which the polymerization reaction is to be carried out. For example, a *tert*-butoxy radical is relatively stable, so an initiator that forms a *tert*-butoxy radical is used for polymerizations carried out at relatively high temperatures.

Table 28.3   Some Radical Initiators

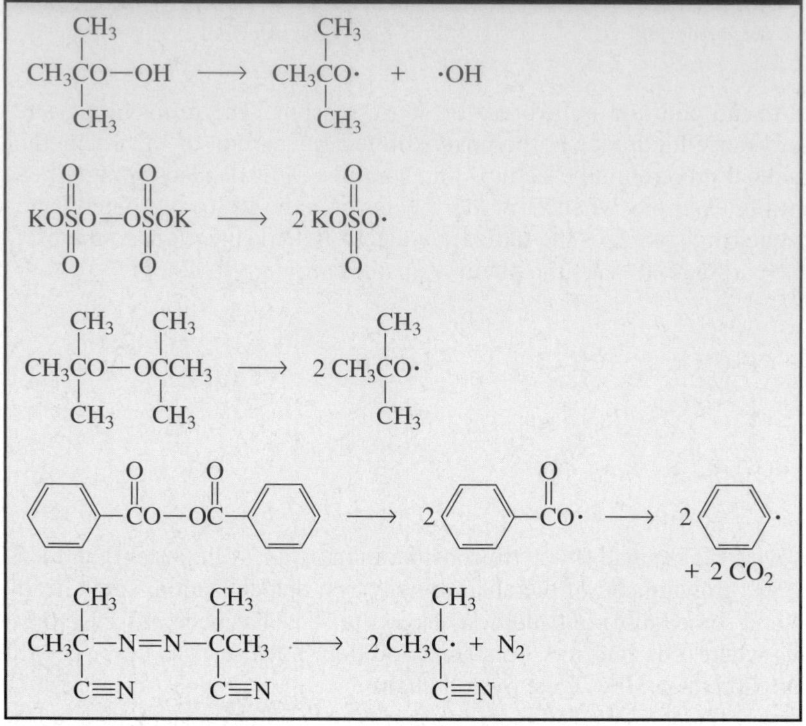

$$CH_3\underset{CH_3}{\overset{CH_3}{CO}}-OH \longrightarrow CH_3\underset{CH_3}{\overset{CH_3}{CO}}\cdot + \cdot OH$$

$$KOSO-OSOK \longrightarrow 2\,KOSO\cdot$$

$$CH_3\underset{CH_3}{\overset{CH_3}{CO}}-\underset{CH_3}{\overset{CH_3}{OCCH_3}} \longrightarrow 2\,CH_3\underset{CH_3}{\overset{CH_3}{CO}}\cdot$$

$$CH_3\underset{C\equiv N}{\overset{CH_3}{C}}-N=N-\underset{C\equiv N}{\overset{CH_3}{CCH_3}} \longrightarrow 2\,CH_3\underset{C\equiv N}{\overset{CH_3}{C}}\cdot + N_2$$

PROBLEM 1◆

What monomer would you use to form each of the following polymers?

a. —CH$_2$CHCH$_2$CHCH$_2$CHCH$_2$CHCH$_2$CH—
        | | | | |
        Cl   Cl   Cl   Cl   Cl

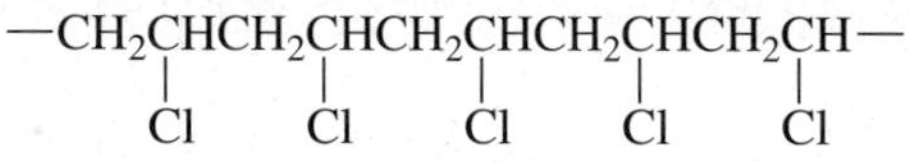

b.
    CH$_3$ CH$_3$ CH$_3$ CH$_3$ CH$_3$ CH$_3$
—CH$_2$CCH$_2$CCH$_2$CCH$_2$CCH$_2$CCH$_2$C—
    C=O C=O C=O C=O C=O C=O
    O   O   O   O   O   O
    CH$_3$ CH$_3$ CH$_3$ CH$_3$ CH$_3$ CH$_3$

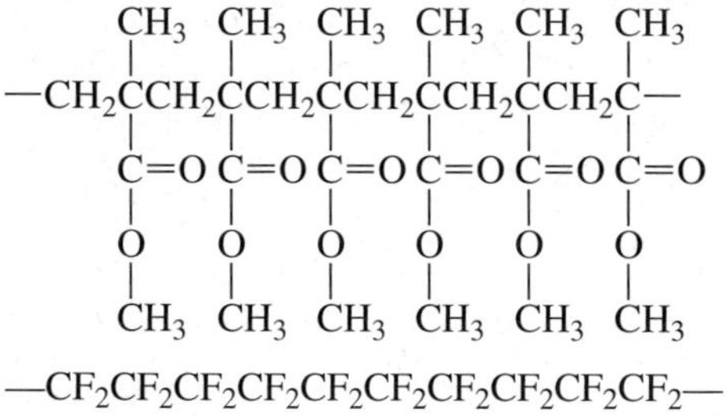

c. —CF$_2$CF$_2$CF$_2$CF$_2$CF$_2$CF$_2$CF$_2$CF$_2$CF$_2$CF$_2$—

PROBLEM 2◆

Which polymer would be more apt to contain abnormal head-to-head linkages: poly(vinyl chloride) or polystyrene?

PROBLEM 3

Draw a segment of polystyrene that contains abnormal head-to-head and tail-to-tail linkages.

PROBLEM 4

Show the mechanism for the formation of a segment of poly(vinyl chloride) containing three units of vinyl chloride and initiated by hydrogen peroxide.

Branching of the Polymer Chain

If the propagating site abstracts a hydrogen atom from a chain, a branch can grow off the chain at that point.

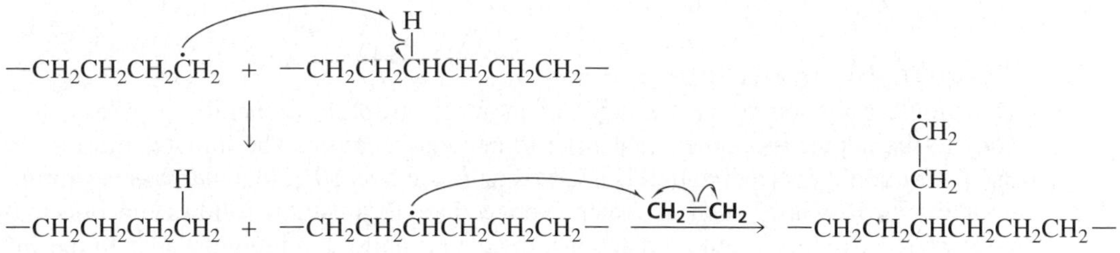

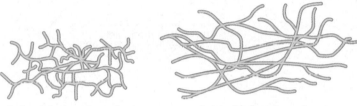

Removing a hydrogen atom from a carbon near the end of a chain leads to short branches, whereas removing a hydrogen atom from a carbon near the middle of a chain results in long branches. Short branches are more likely to be formed than long ones because the ends of the chain are more accessible.

chain with short branches   chain with long branches

Branching greatly affects the physical properties of the polymer. Unbranched chains can pack together more closely than branched chains can. Consequently, linear polyethylene (known as high-density polyethylene) is a relatively hard plastic, used for the production of such things as artificial hip joints, whereas branched polyethylene (low-density polyethylene) is a much more flexible polymer, used for trash bags and dry-cleaning bags.

Branched polymers are more flexible.

RECYCLING SYMBOLS

When plastics are recycled, the various types must be separated from one another. To aid in the separation, many states require manufacturers to place a recycling symbol on their products to indicate the type of plastic it is. You are probably familiar with these symbols, which are often embossed on the bottom of plastic containers. The symbols consist of three arrows around one of seven numbers; an abbreviation below the symbol indicates the type of polymer from which the container is made. The lower the number in the middle of the symbol, the greater is the ease with which the material can be recycled: 1 (PET) stands for poly(ethylene terephthalate), 2 (HDPE) for high-density polyethylene, 3 (V) for poly(vinyl chloride), 4 (LDPE) for low-density polyethylene, 5 (PP) for polypropylene, 6 (PS) for polystyrene, and 7 for all other plastics.

Recycling labels

PROBLEM 5◆

Polyethylene can be used for the production of beach chairs as well as beach balls. Which of these items is made from more highly branched polyethylene?

PROBLEM 6

Draw a short segment of branched polystyrene that shows the linkages at the branch point.

Cationic Polymerization

In cationic polymerization, the initiator is an electrophile (generally a proton) that adds to the alkene monomer, causing it to become a cation. The initiator most often used in cationic polymerization is a Lewis acid, such as BF_3, together with a proton-donating Lewis base, such as water. Notice that the reaction follows the rule that governs electrophilic addition reactions: the electrophile (the initiator) adds to the sp^2 carbon bonded to the greater number of hydrogens (Section 4.4).

chain-initiating step

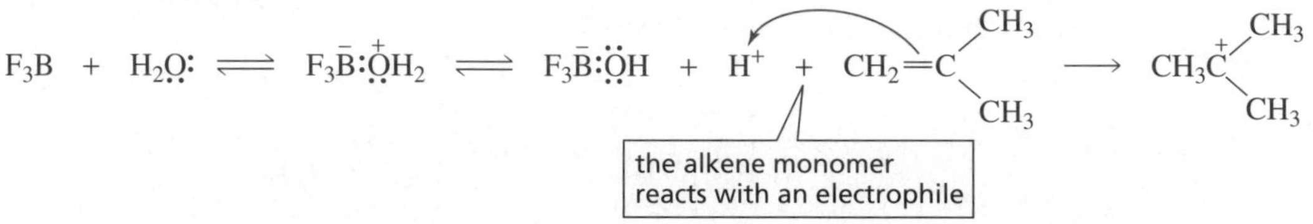

the alkene monomer reacts with an electrophile

chain-propagating steps in cationic polymerization

propagating sites

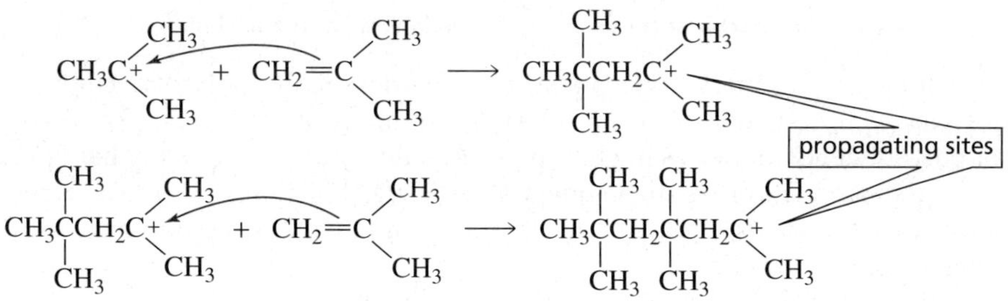

- The cation formed in the initiation step reacts with a second monomer, forming a new cation that reacts in turn with a third monomer. As each subsequent monomer adds to the chain, the new positively charged propagating site is at the end of the most recently added unit.

Cationic polymerization can be terminated by

- loss of a proton;
- addition of a nucleophile to the propagating site;
- a chain-transfer reaction with the solvent (YZ).

three ways to terminate the chain

loss of a proton

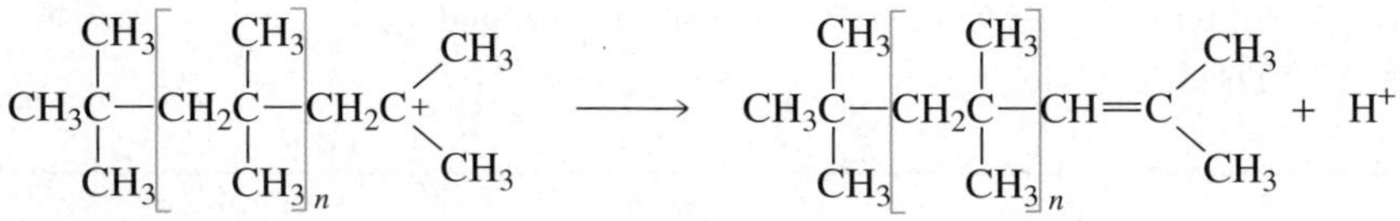

reaction with a nucleophile

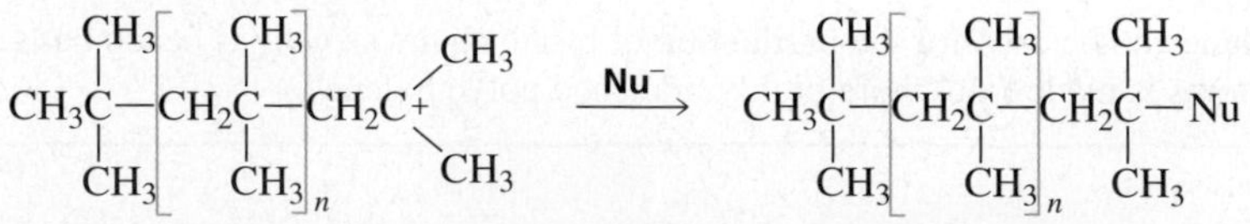

chain-transfer reaction with the solvent

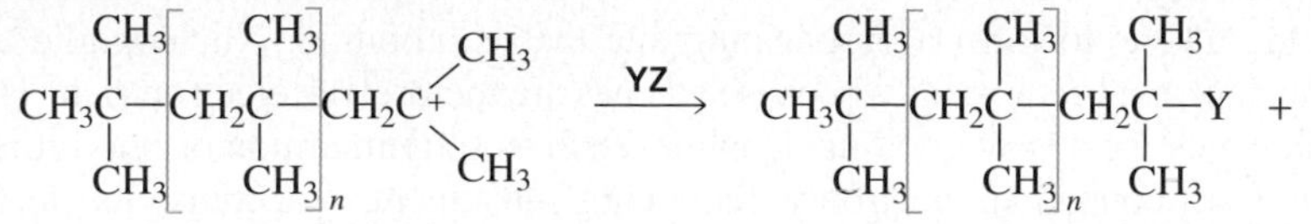

$$CH_3C-[CH_2C-]_n CH_2C^+ \xrightarrow{\text{YZ}} CH_3C-[CH_2C-]_n CH_2C-Y \ +$$

The carbocation intermediates formed during cationic polymerization, like any other carbocations, can undergo rearrangement by either a 1,2-hydride shift or a 1,2-methyl shift if rearrangement leads to a more stable carbocation (Section 4.6). For example, the polymer formed from the cationic polymerization of 3-methyl-1-butene contains both unrearranged and rearranged units. The unrearranged propagating site is a secondary carbocation, whereas the rearranged propagating site, obtained by a 1,2-hydride shift, is a more stable tertiary carbocation. The extent of rearrangement depends on the reaction temperature.

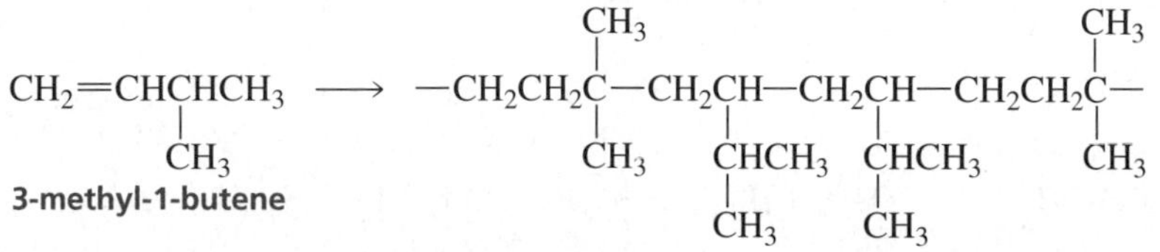

$$CH_2=CHCHCH_3 \longrightarrow$$

3-methyl-1-butene

unrearranged propagating site

rearranged propagating site

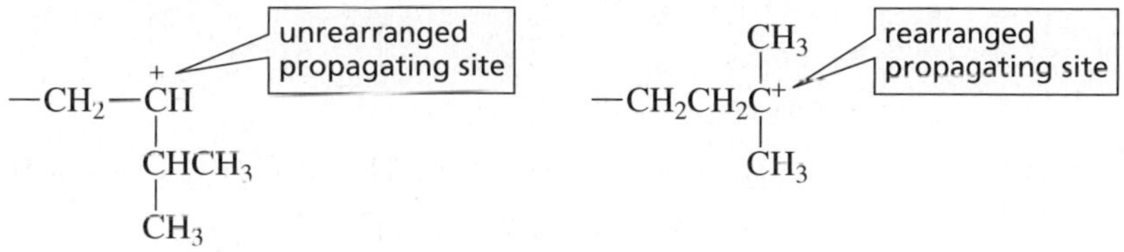

Monomers that are best able to undergo polymerization by a cationic mechanism are those with substituents that can stabilize the positive charge at the propagating site by donating electrons by hyperconjugation or by resonance (Section 15.2). Examples of monomers that undergo cationic polymerization are given in Table 28.4.

Table 28.4	Examples of Alkenes That Undergo Cationic Polymerization	
$CH_2=CCH_3$ CH_3 isobutylene	$CH_2=CH$ OCH_3 methyl vinyl ether	$CH_2=CH$ styrene

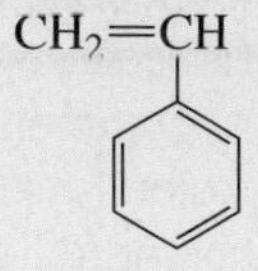

PROBLEM 7◆

List the following groups of monomers in order of decreasing ability to undergo cationic polymerization:

a. $CH_2=CH$ (with NO_2) $CH_2=CH$ (with CH_3) $CH_2=CH$ (with OCH_3)

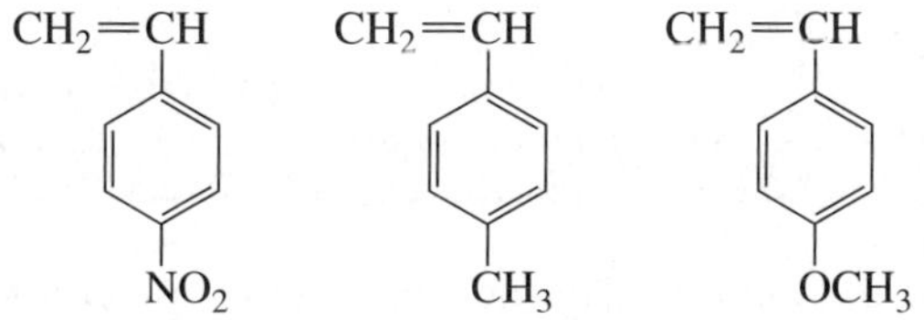

b. $CH_2=CCH_3$ with CH_3 $CH_2=CHOCH_3$ $CH_2=CHCOCH_3$ (with O)

c. $CH_2=CH$ $CH_2=CCH_3$

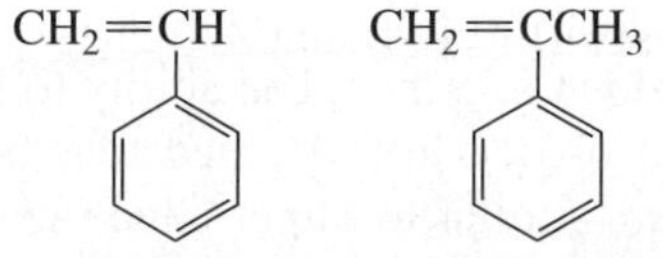

Anionic Polymerization

In anionic polymerization, the initiator is a nucleophile (generally a strong base) that reacts with the alkene to form a propagating site that is an anion. Nucleophilic attack on an alkene does not occur readily because alkenes are themselves electron rich. Therefore, the initiator must be a very good nucleophile, such as sodium amide or butyllithium, and the alkene must contain an electron-withdrawing substituent to decrease its electron density.

chain-initiating step in anionic polymerization

the alkene monomer reacts with a nucleophile

chain-propagating steps

propagating sites

The chain can be terminated by a chain transfer reaction with the solvent or by reaction with an impurity in the reaction mixture. If the solvent cannot donate a proton to terminate the chain and if all impurities that can react with a carbanion are rigorously excluded, chain propagation will continue until all the monomer has been consumed. At this point, the propagating site will still be active, so the polymerization reaction will continue if more monomer is added to the system. Such nonterminated chains are called **living polymers** because the chains remain active until they are "killed." Living polymers result most commonly from anionic polymerization because the chains cannot be terminated by proton loss from the polymer, as they can in cationic polymerization, or by disproportionation or radical recombination, as they can in radical polymerization. Some alkenes that undergo polymerization by an anionic mechanism are shown in Table 28.5.

Table 28.5	Examples of Alkenes That Undergo Anionic Polymerization		
$CH_2{=}CH$ $\|$ CH $\|\|$ O **acrolein**	$CH_2{=}CH$ $\|$ CNH_2 $\|\|$ O **acrylamide**	$CH_2{=}CCH_3$ $\|$ $COCH_3$ $\|\|$ O **methyl methacrylate**	$CH_2{=}CH$ **styrene**

Super Glue is a polymer of methyl α-cyanoacrylate. Because the monomer has two electron-withdrawing groups, it requires only a moderately good nucleophile to initiate anionic polymerization, such as surface-absorbed water. You may well have experienced this reaction if you have ever spilled a drop of Super Glue on your fingers. A nucleophilic group on the surface of the skin initiates the polymerization reaction, with the result that two fingers can become firmly glued together. The ability to form covalent bonds with groups on the surfaces of the objects to be glued together is what gives Super Glue its amazing strength. Polymers similar to Super Glue (they are butyl, isobutyl, or octyl esters rather than methyl esters) are used by surgeons to close wounds.

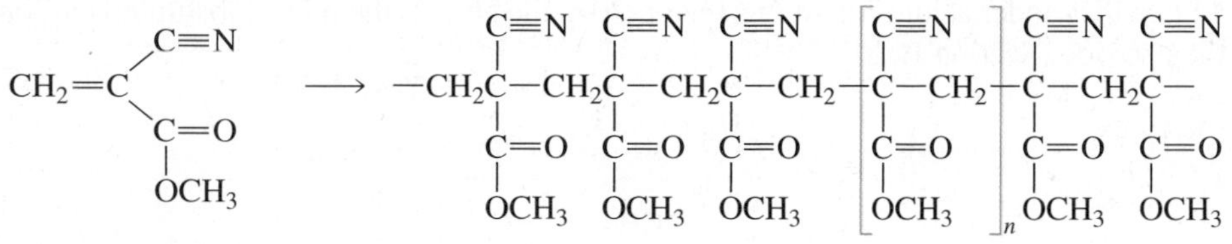

methyl α-cyanoacrylate → **Super Glue®**

PROBLEM 8◆

List the following groups of monomers in order of decreasing ability to undergo anionic polymerization:

a.

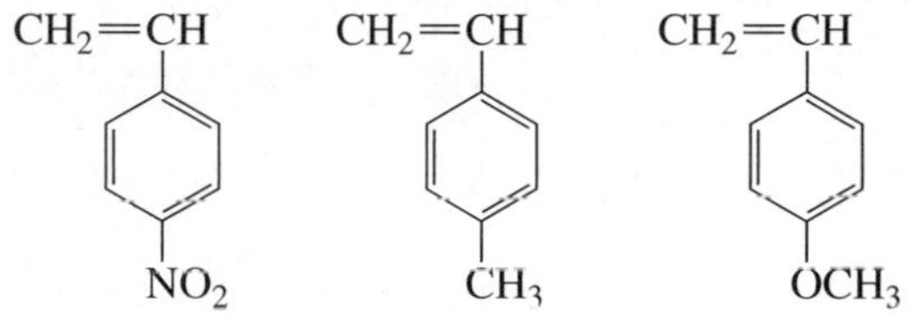

b. CH_2—$CHCH_3$ CH_2=$CHCl$ CH_2=CHC≡N

What Determines the Mechanism?

We have seen that the substituent on the alkene determines the best mechanism for chain-growth polymerization. Alkenes with substituents that can stabilize radicals readily undergo radical polymerization, alkenes with electron-donating substituents that can stabilize cations undergo cationic polymerization, and alkenes with electron-withdrawing substituents that can stabilize anions undergo anionic polymerizations.

Some alkenes undergo polymerization by more than one mechanism. For example, styrene can undergo polymerization by radical, cationic, and anionic mechanisms because the phenyl group can stabilize benzylic radicals, benzylic cations, and benzylic anions. The mechanism followed for its polymerization depends on the nature of the initiator chosen to start the reaction.

Ring-Opening Polymerizations

Although ethylene and substituted ethylenes are the monomers most commonly used for chain-growth polymerization reactions, other compounds can polymerize as well. For example, epoxides undergo chain-growth polymerization reactions. If the initiator is a nucleophile such as HO⁻ or RO⁻, polymerization occurs by an anionic mechanism. Notice that the nucleophile attacks the less sterically hindered carbon of the epoxide (Section 10.8).

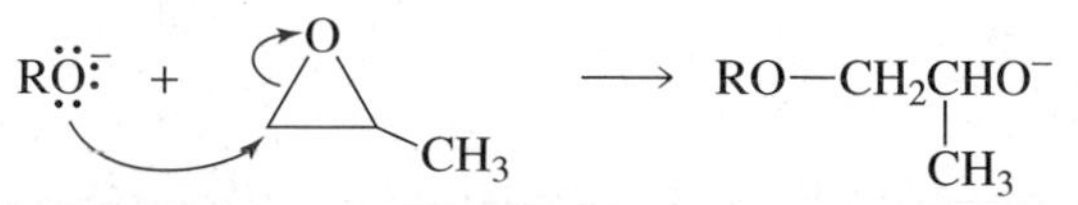

propylene oxide

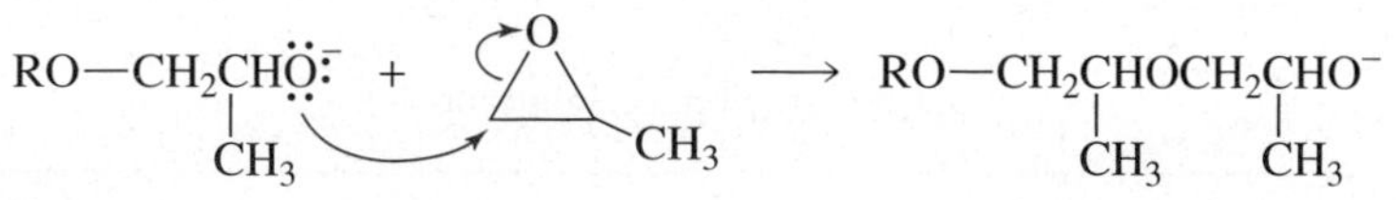

If the initiator is a Lewis acid or a proton-donating acid, epoxides are polymerized by a cationic mechanism. Polymerization reactions that involve ring-opening reactions, such as the polymerization of propylene oxide, are called **ring-opening polymerizations**. Notice that under acidic conditions the nucleophile attacks the more substituted carbon of the epoxide (Section 10.8).

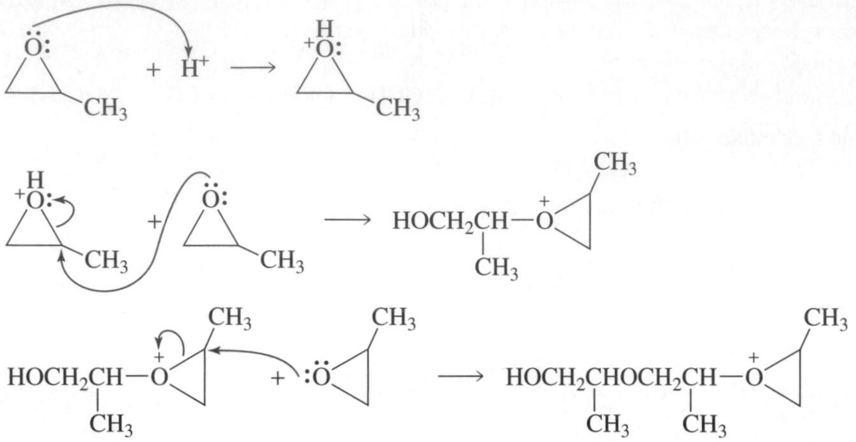

Explain why, when propylene oxide undergoes anionic polymerization, nucleophilic attack occurs at the less substituted carbon of the epoxide, but when it undergoes cationic polymerization, nucleophilic attack occurs at the more substituted carbon.

PROBLEM 10

Describe the polymerization of 2,2-dimethyloxirane by

a. an anionic mechanism.
b. a cationic mechanism.

PROBLEM 11◆

Which monomer and which type of initiator would you use to synthesize each of the following polymers?

a.

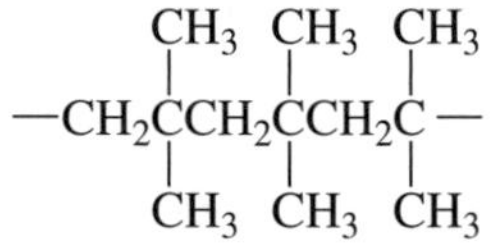

$$-CH_2\overset{\underset{\displaystyle CH_3}{|}}{C}CH_2\overset{\underset{\displaystyle CH_3}{|}}{C}CH_2\overset{\underset{\displaystyle CH_3}{|}}{C}-$$

with CH₃ groups above each C

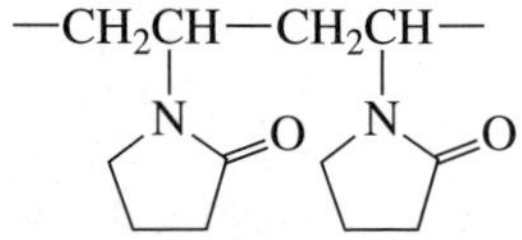

c. $-CH_2CH_2OCH_2CH_2O-$

b. $-CH_2CH-CH_2CH-$
with N-pyrrolidinone (lactam) groups

d. 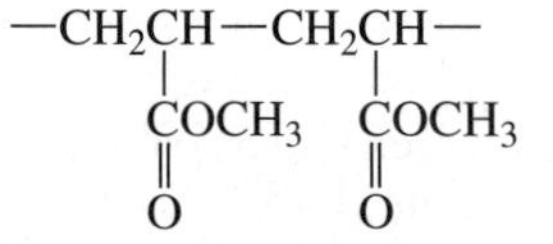 $-CH_2CH-CH_2CH-$
with COCH₃ (=O) groups

PROBLEM 12◆

Draw a short segment of the polymer formed from cationic polymerization of 3,3-dimethyloxacyclobutane.

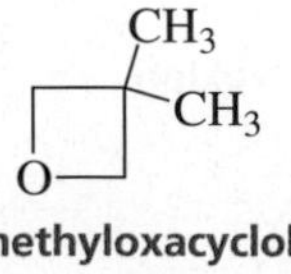

3,3-dimethyloxacyclobutane

28.3 Stereochemistry of Polymerization · Ziegler–Natta Catalysts

Polymers formed from monosubstituted ethylenes can exist in three configurations: isotactic, syndiotactic, and atactic. An **isotactic polymer** has all of its substituents on the same side of the fully extended carbon chain. (*Iso* and *taxis* are Greek for "the same" and "order," respectively.) In a **syndiotactic polymer** (*syndio* means "alternating"), the substituents regularly alternate on both sides of the carbon chain. The substituents in an **atactic polymer** are randomly oriented.

isotactic configuration (same side)

H CH₃   H CH₃   H CH₃   H CH₃   H CH₃   H CH₃

syndiotactic configuration (both sides)

H CH₃ H₃C H   H CH₃ H₃C H   H CH₃ H₃C H

The configuration of the polymer affects its physical properties. Polymers in the isotactic or syndiotactic configuration are more likely to be crystalline solids, because positioning the substituents in a regular order allows for a more regular packing arrangement. Polymers in the atactic configuration are more disordered and cannot pack together as well, so these polymers are less rigid and therefore softer.

The configuration of the polymer depends on the mechanism by which polymerization occurs. In general, radical polymerization leads primarily to branched polymers in the atactic configuration. Anionic polymerization can produce polymers with the most stereoregularity. The percentage of chains in the isotactic or syndiotactic configuration increases as the polymerization temperature decreases and the solvent polarity decreases.

In 1953, Karl Ziegler and Giulio Natta found that the structure of a polymer could be controlled if the growing end of the chain and the incoming monomer were coordinated with an aluminum–titanium initiator. These initiators are now called **Ziegler–Natta catalysts**. Long, unbranched polymers with either the isotactic or the syndiotactic configuration can be prepared using Ziegler–Natta catalysts. Whether the chain is isotactic or syndiotactic depends on the particular Ziegler–Natta catalyst used. These catalysts revolutionized the field of polymer chemistry because they allow the synthesis of stronger and stiffer polymers that have greater resistance to cracking and heat. High-density polyethylene is prepared using a Ziegler–Natta process.

A proposed mechanism for Ziegler–Natta-catalyzed polymerization of a substituted ethylene is shown in Figure 28.1. The monomer forms a π complex (Section 6.6) with titanium at an open coordination site (a site available to accept electrons) and the coordinated alkene is inserted between the titanium and the growing polymer, thereby extending the polymer chain. Because a new coordination site opens up during insertion of the monomer, the process can be repeated over and over.

Polyacetylene is another polymer prepared by a Ziegler–Natta process. It can be converted to a **conducting polymer** because the conjugated double bonds in polyacetylene enable electricity to travel down its backbone after several electrons are removed from or added to the backbone (called "doping").

HC≡CH $\xrightarrow{\text{a Ziegler–Natta catalyst}}$ —CH=CH $\left[\text{CH=CH}\right]_n$ CH=CH—

acetylene   **polyacetylene**

▲ Figure 28.1
The mechanism of the Ziegler–Natta-catalyzed polymerization of a substituted ethylene. A monomer forms a π complex with an open coordination site of titanium and then is inserted between the titanium and the growing polymer.

28.4 Polymerization of Dienes • The Manufacture of Rubber

When the bark of a rubber tree is cut, a sticky white liquid oozes out. This same liquid is found inside the stalks of dandelions and milkweed. The sticky material is *latex,* a suspension of rubber particles in water. Its biological function is to protect the tree after an injury by covering the wound like a bandage.

Natural rubber is a polymer of 2-methyl-1,3-butadiene, also called isoprene. On average, a molecule of rubber contains 5000 isoprene units. All the double bonds in natural rubber are cis. Rubber is a waterproof material because its tangled hydrocarbon chains have no affinity for water. Charles Macintosh, a Scotsman, was the first to use rubber as a waterproof coating for raincoats. As in the case of other naturally occurring terpenes, the five-carbon compound actually used in the biosynthesis is isopentenylpyrophosphate (Section 26.8).

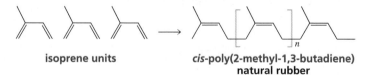

isoprene units → ***cis*-poly(2-methyl-1,3-butadiene)**
natural rubber

Gutta-percha (from the Malaysian words *getah,* meaning "gum," and *percha,* meaning "tree") is a naturally occurring isomer of rubber in which all the double bonds are trans. Like rubber, gutta-percha is exuded by certain trees, but it is much less common. It is also harder and more brittle than rubber. Gutta-percha is the filling material that dentists use in root canals. In the past, it was used for the casing of golf balls.

PROBLEM 13

Draw a short segment of gutta-percha.

By mimicking nature, scientists have learned to make synthetic rubbers with properties tailored to meet human needs. These materials have some of the properties of natural rubber, including being waterproof and elastic, but they have some improved properties as well; they are tougher, more flexible, and more durable than natural rubber.

Latex being collected from a rubber tree.

Synthetic rubbers have been made by emulsion polymerization of dienes other than isoprene. Emulsion polymerization involves radical polymerization of the monomers in an emulsion, which makes it easier to control the reaction. One synthetic rubber is a polymer, formed by 1,4-polymerization of 1,3-butadiene, in which all the double bonds are cis. There are two reasons that cause 1,4-polymerization to predominate over 1,2-polymerization. First, there is less steric hindrance at the 4-position. Second, the product of 1,4-polymerization is more stable since it has alkene groups with two substituents, whereas the product of 1,2-polymerization has alkene groups with only one substituent (Section 4.11).

1,3-butadiene monomers → ***cis*-poly(1,3-butadiene)**
a synthetic rubber

Neoprene is a synthetic rubber made by polymerizing 2-chloro-1,3-butadiene. It is used to make wet suits, shoe soles, tires, hoses, and coated fabrics.

$$CH_2=CCH=CH_2$$
2-chloro-1,3-butadiene
chloroprene

neoprene

Charles Goodyear (1800–1860), *the son of an inventor of farm implements, was born in Connecticut. He patented the process of vulcanization in 1844. The process was so simple, however, that it could easily be copied, so he spent many years contesting infringements on his patent. In 1852, with Daniel Webster as his lawyer, he obtained the right to the patent.*

A problem common to both natural and most synthetic rubbers is that the polymers are very soft and sticky. They can be hardened by a process known as *vulcanization*. Charles Goodyear discovered this process while looking for ways to improve the properties of rubber. He accidentally spilled a mixture of rubber and sulfur on a hot stove. To his surprise, the mixture became hard but flexible. He called the heating of rubber with sulfur vulcanization, after Vulcan, the Roman god of fire.

Heating rubber with sulfur causes **cross-linking** of separate polymer chains through disulfide bonds (Figure 28.2; see also Section 22.8). Thus, instead of the chains being tangled together, the vulcanized chains are covalently bonded to each other in one giant molecule. Because the polymer has double bonds, the chains have bends and kinks that prevent them from forming a tightly packed crystalline polymer. When the rubber is stretched, the chains straighten out in the direction of the pull. The cross-linking prevents rubber from being torn when it is stretched; moreover, the cross-links provide a reference framework for the material to return to when the stretching force is removed.

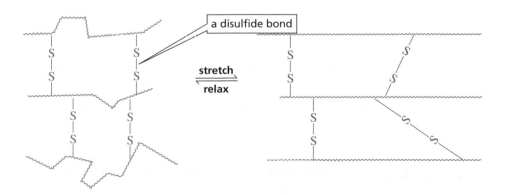

a disulfide bond

stretch / relax

◀ **Figure 28.2**
The rigidity of rubber is increased by cross-linking the polymer chains with disulfide bonds. When rubber is stretched, the randomly coiled chains straighten out and orient themselves along the direction of the stretch.

The physical properties of rubber can be controlled by regulating the amount of sulfur used in the vulcanization process. Rubber made with 1%–3% sulfur is soft and stretchy and is used to make rubber bands. Rubber made with 3%–10% sulfur is more

The greater the degree of cross-linking, the more rigid is the polymer.

rigid and is used in the manufacture of tires. Goodyear's name can be found on many tires sold today. The story of rubber is an example of a scientist taking a natural material and finding ways to improve its useful properties.

> **PROBLEM 14**
>
> Draw a segment of the polymer that would be formed from 1,2-polymerization of 1,3-butadiene.

28.5 Copolymers

The polymers we have discussed so far are formed from only one type of monomer and are called **homopolymers**. Often, two or more different monomers are used to form a polymer. The resulting product is called a **copolymer**. Increasing the number of different monomers used to form the copolymer dramatically increases the number of different copolymers those monomers can form. Even if only two kinds of monomers are used, copolymers with very different properties can be prepared by varying the amounts of each monomer. Both chain-growth polymers and step-growth polymers can be copolymers. Many of the synthetic polymers used today are copolymers. Table 28.6 shows some common copolymers and the monomers from which they are synthesized.

Table 28.6   Some Examples of Copolymers and Their Uses

Monomer	Copolymer name	Uses
$CH_2{=}CH$ (Cl) **vinyl chloride** $+$ $CH_2{=}CCl$ (Cl) **vinylidene chloride**	Saran	film for wrapping food
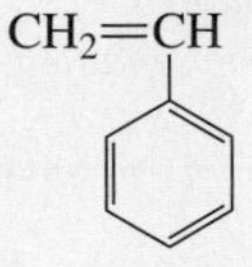 $CH_2{=}CH$ (phenyl) **styrene** $+$ $CH_2{=}CH$ ($C{\equiv}N$) **acrylonitrile**	SAN	dishwasher-safe objects, vaccum cleaner parts
$CH_2{=}CH$ ($C{\equiv}N$) **acrylonitrile** $+$ $CH_2{=}CH$ ($CH{=}CH_2$) **1,3-butadiene** 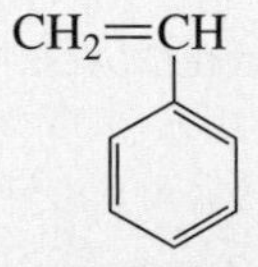 $+$ $CH_2{=}CH$ (phenyl) **styrene**	ABS	bumpers, crash helmets, telephones, luggage
$CH_2{=}CCH_3$ (CH_3) **isobutylene** $+$ $CH_2{=}CHC{=}CH_2$ (CH_3) **isoprene**	butyl rubber	inner tubes, balls, inflatable sporting goods

There are four types of copolymers. In an **alternating copolymer**, the two monomers alternate. A **block copolymer** consists of blocks of each kind of monomer. In a **random copolymer**, the distribution of monomers is random. A **graft copolymer** contains branches derived from one monomer grafted onto a backbone derived from another monomer. These structural differences extend the range of physical properties available to the scientist designing the copolymer.

an alternating copolymer	ABABABABABABABABABABABA
a block copolymer	AAAAABBBBBAAAAABBBBBAAA
a random copolymer	AABABABBBABAABBABABBBAAAB
a graft copolymer	AAAAAAAAAAAAAAAAAAAAAAAA

a graft copolymer columns (B chains hanging from the A backbone):

```
AAAAAAAAAAAAAAAAAAAAAAAA
   B         B         B
   B         B         B
   B         B         B
   B         B         B
   B         B         B
   B         B         B
```

28.6 Step-Growth Polymers

Step-growth polymers are formed by the intermolecular reaction of bifunctional molecules (molecules with two functional groups). When the functional groups react, in most cases a small molecule such as H_2O, alcohol, or HCl is lost. This is why these polymers are also called *condensation polymers*.

There are two types of step-growth polymers. One type is formed by the reaction of a single compound that possesses two different functional groups, which we will call A and B. Functional group A of one molecule of the compound reacts with functional group B of another molecule to form the monomer (A—X—B) that undergoes polymerization.

$$A—B \quad A—B \quad \longrightarrow \quad A—X—B$$

The other type of step-growth polymer is formed by the reaction of two different bifunctional compounds. One compound contains two A functional groups and the other compound contains two B functional groups. Functional group A of one compound reacts with functional group B of the other compound to form the monomer (A—X—B).

$$A—A \quad B—B \quad \longrightarrow \quad A—X—B$$

Step-growth polymers are made by combining molecules with reactive groups at each end.

The formation of step-growth polymers, unlike the formation of chain-growth polymers, does not occur through chain reactions. Any two monomers (or short chains) can react. The progress of a typical step-growth polymerization is shown schematically in Figure 28.3. When the reaction is 50% complete (12 bonds have formed between 25 monomers), the reaction products are primarily dimers and trimers. Even at 75% completion, no long chains have been formed. This means that if step-growth polymerization is to lead to long-chain polymers, very high yields must be achieved.

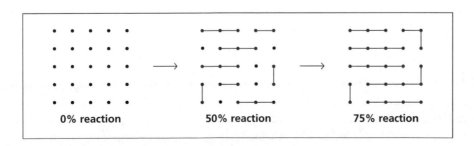

0% reaction	50% reaction	75% reaction

◀ **Figure 28.3**
Progress of a step-growth polymerization.

Polyamides

Nylon is the common name of a synthetic **polyamide**. Originally named nylon 6, it is an example of a step-growth polymer formed by a monomer with two different functional groups. The carboxylic acid group of one monomer reacts with the amino group of another monomer to form amide groups. Structurally, the reaction is similar to the polymerization of α-amino acids to form proteins (Section 22.8). This particular nylon was called nylon 6 because it is formed from the polymerization of 6-aminohexanoic acid, a compound that contains six carbons.

$$\underset{\textbf{6-aminohexanoic acid}}{H_3\overset{+}{N}(CH_2)_5\overset{O}{\overset{\|}{C}}O^-} \xrightarrow[-H_2O]{\Delta} \underset{\substack{\textbf{nylon 6} \\ \textbf{a polyamide}}}{-NH(CH_2)_5\overset{O}{\overset{\|}{C}}\left[NH(CH_2)_5\overset{O}{\overset{\|}{C}}\right]_n NH(CH_2)_5\overset{O}{\overset{\|}{C}}-}$$

The starting material for the commericial synthesis of nylon 6 is ε-caprolactam. The lactam is opened by a base.

$$\underset{\textbf{ε-caprolactam}}{\text{[structure]}} \xrightarrow[\Delta]{HO^-} \underset{\textbf{nylon 6}}{-NH(CH_2)_5\overset{O}{\overset{\|}{C}}\left[NH(CH_2)_5\overset{O}{\overset{\|}{C}}\right]_n NH(CH_2)_5\overset{O}{\overset{\|}{C}}-}$$

A related polyamide, nylon 66 is an example of a step-growth polymer formed by two different bifunctional monomers: adipic acid and 1,6-hexanediamine. It is called nylon 66 because it is formed from a six-carbon diacid and a six-carbon diamine.

$$\underset{\textbf{adipic acid}}{HO\overset{O}{\overset{\|}{C}}(CH_2)_4\overset{O}{\overset{\|}{C}}OH} + \underset{\textbf{1,6-hexanediamine}}{H_2N(CH_2)_6NH_2} \xrightarrow[-H_2O]{\Delta} \underset{\textbf{nylon 66}}{-\overset{O}{\overset{\|}{C}}(CH_2)_4\overset{O}{\overset{\|}{C}}\left[NH(CH_2)_6NH\overset{O}{\overset{\|}{C}}(CH_2)_4\overset{O}{\overset{\|}{C}}\right]_n NH(CH_2)_6NH-}$$

Nylon first found wide use in textiles and carpets. Because it is resistant to stress, it is also used in such applications as mountaineering ropes, tire cords, and fishing lines, and as a substitute for metal in bearings and gears. The usefulness of nylon precipitated a search for yet more new "super fibers" with super strength and super heat resistance.

Nylon is pulled from a beaker of adipoyl chloride and 1,6-hexanediamine.

PROBLEM 15◆

a. Draw a short segment of nylon 4.
b. Draw a short segment of nylon 44.

PROBLEM 16

Write an equation that explains what will happen if a scientist working in the laboratory spills sulfuric acid on her nylon 66 hose.

PROBLEM 17

Propose a mechanism for the base-catalyzed polymerization of ε-caprolactam.

One super fiber is Kevlar, a polymer of 1,4-benzenedicarboxylic acid and 1,4-diaminobenzene. Aromatic polyamides are called **aramides**. The incorporation of aromatic rings into polymers has been found to result in polymers with great physical strength; Kevlar is five times stronger than steel on an equal weight basis. Army helmets are made of Kevlar, which is also used for lightweight bullet-resistant vests, automobile parts, high-performance skis, the ropes used on the Mars Pathfinder, and high-performance sails used in the Americas Cup. Because it is stable at very high temperatures, it is used in the protective clothing worn by firefighters.

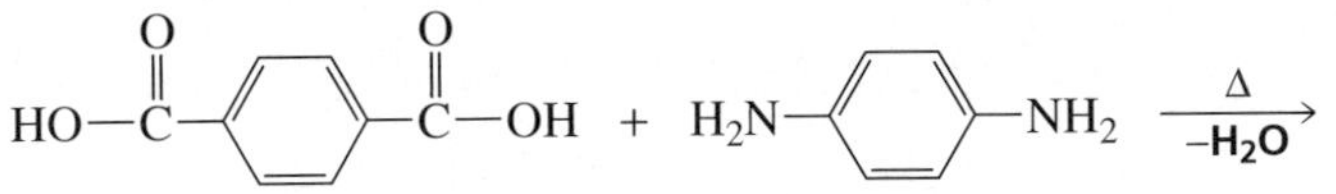

1,4-benzenedicarboxylic acid **1,4-diaminobenzene**

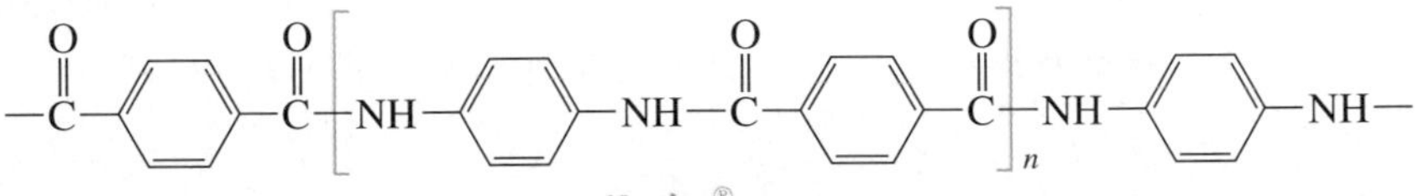

Kevlar®
an aramide

Kevlar owes its strength to the way in which the individual polymer chains interact with each other. The chains are hydrogen bonded, forming a sheetlike structure.

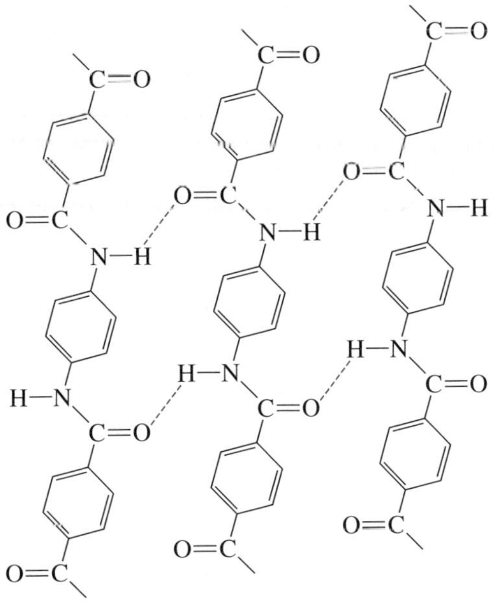

Polyesters

Polyesters are step-growth polymers in which the monomer units are joined together by ester groups. They have found wide commercial use as fibers, plastics, and coatings. The most common polyester is known by the trade name Dacron and is made by the transesterification (Section 16.10) of dimethyl terephthalate with ethylene glycol. High levels of resilience, durability, and moisture resistance are the properties of this polymer that contribute to its "wash-and-wear" characteristics.

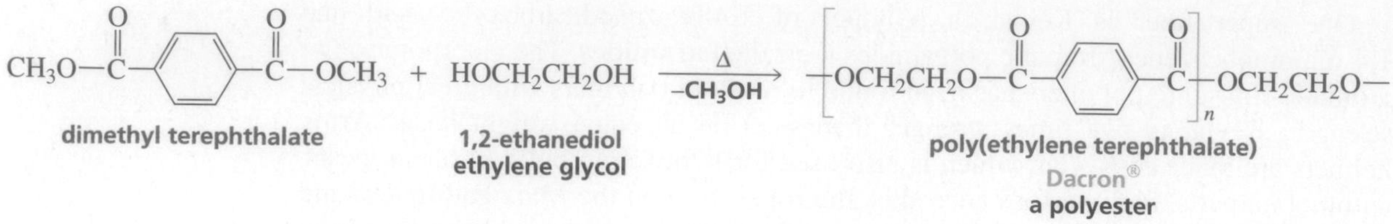

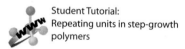

Kodel polyester is formed by the transesterification of dimethyl terephthalate with 1,4-di(hydroxymethyl)cyclohexane. The stiff polyester chain causes the fiber to have a harsh feel that can be softened by blending the fiber with wool or cotton.

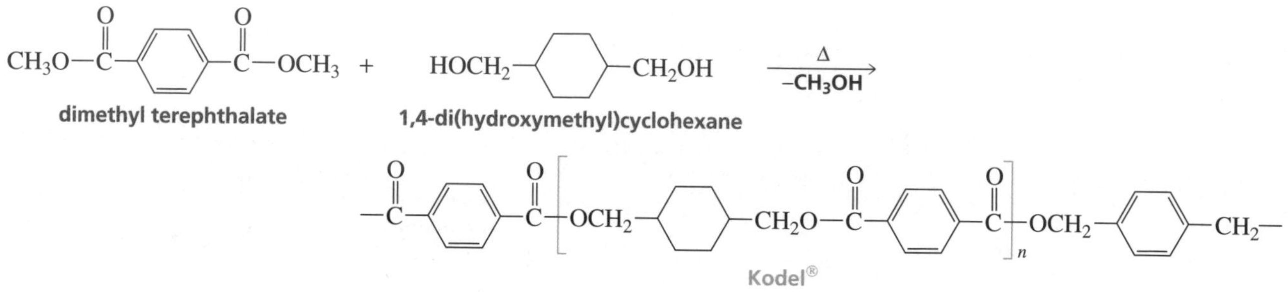

dimethyl terephthalate + **1,4-di(hydroxymethyl)cyclohexane** $\xrightarrow[-CH_3OH]{\Delta}$

Kodel®

PROBLEM 18

What happens to polyester slacks if aqueous NaOH is spilled on them?

Student Tutorial:
Repeating units in step-growth polymers

Polyesters with two ester groups bonded to the same carbon are known as **polycarbonates**. Lexan, a polycarbonate produced by the transesterification reaction of diphenyl carbonate with bisphenol A, is a strong, transparent polymer used for bulletproof windows and traffic-light lenses. In recent years, polycarbonates have become important in the automobile industry as well as in the manufacture of compact discs.

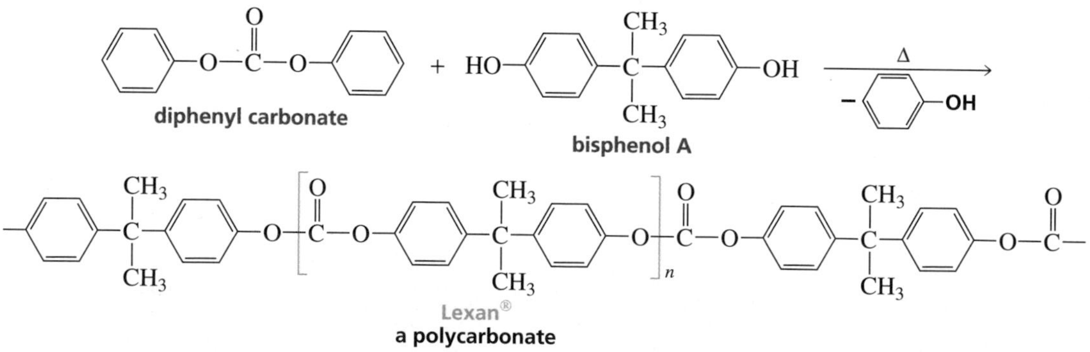

diphenyl carbonate + **bisphenol A** $\xrightarrow[-]{\Delta}$

Lexan®
a polycarbonate

Epoxy Resins

Epoxy resins are the strongest adhesives known. They can adhere to almost any kind of surface and are resistant to solvents and to extremes of temperature. Epoxy cement is purchased as a kit consisting of a low-molecular-weight *prepolymer* (the most common is a polymer of bisphenol A and epichlorohydrin) and a *hardener* that react when mixed to form a cross-linked polymer.

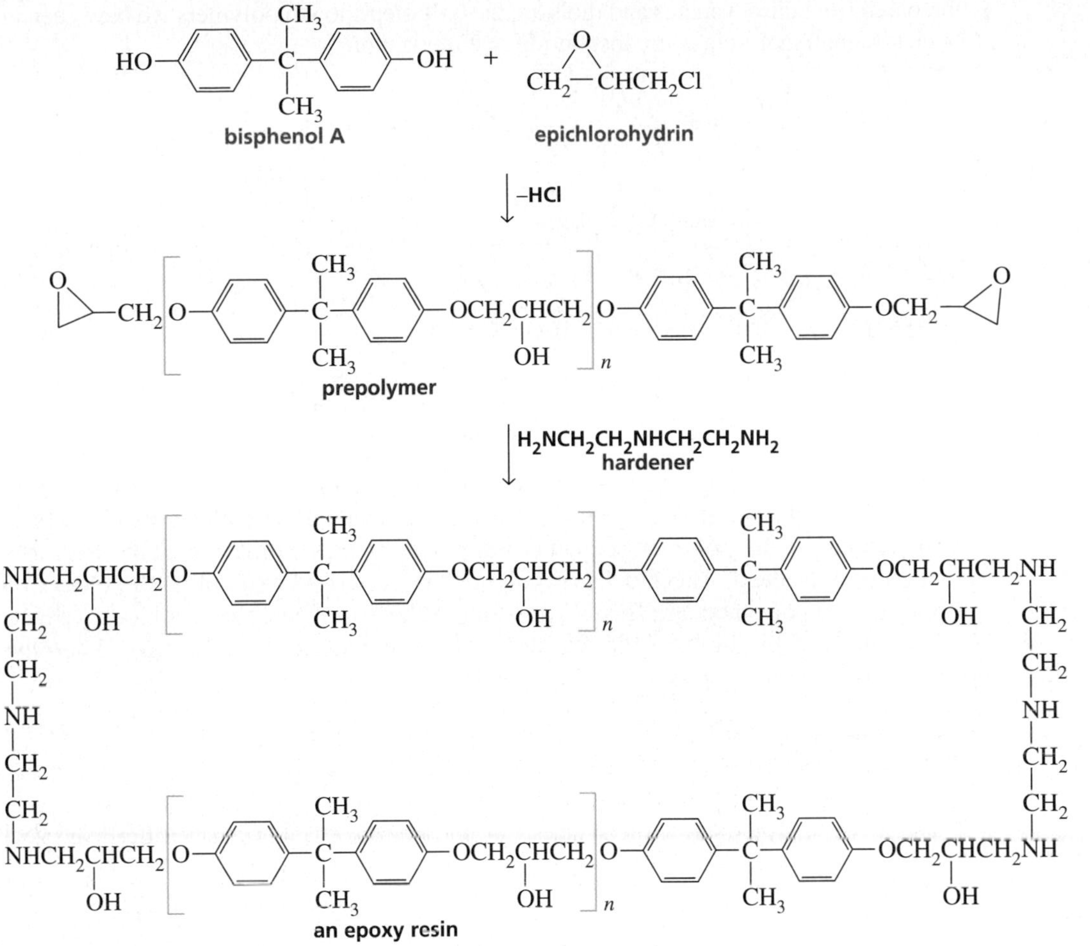

bisphenol A   epichlorohydrin

↓ −HCl

prepolymer

↓ H₂NCH₂CH₂NHCH₂CH₂NH₂ hardener

an epoxy resin

PROBLEM 19

a. Propose a mechanism for the formation of the prepolymer formed by bisphenol A and epichlorohydrin.

b. Propose a mechanism for the reaction of the prepolymer with the hardener.

Polyurethanes

A **urethane**—also called a carbamate—is a compound that has an OR group and an NHR group bonded to the same carbonyl carbon. Urethanes can be prepared by treating an isocyanate with an alcohol, in the presence of a catalyst such as a tertiary amine.

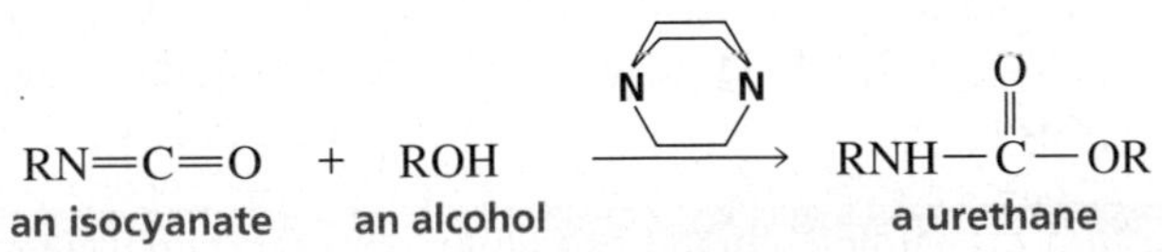

$$RN{=}C{=}O \ + \ ROH \ \longrightarrow \ RNH{-}\overset{\displaystyle O}{\overset{\|}{C}}{-}OR$$

an isocyanate   an alcohol   a urethane

Polyurethanes are polymers that contain urethane groups. One of the most common polyurethanes is prepared by the polymerization of toluene-2,6-diisocyanate and ethylene glycol. If the reaction is carried out in the presence of a blowing agent, the product is a polyurethane foam. Blowing agents are gases such as nitrogen or carbon dioxide. At one time, chlorofluorocarbons—low-boiling liquids that vaporize on heating—were used, but they have been banned for environmental reasons (Section 11.11). Polyurethane foams are used for furniture stuffing, carpet backings, and insulation. Notice that polyurethanes prepared from diisocyanates and diols are the only step-growth polymers we have seen in which a small molecule is *not* lost during polymerization.

$$O=C=N \underset{\text{toluene-2,6-diisocyanate}}{\overset{CH_3}{\underset{}{\diagdown}}} N=C=O \quad + \quad \underset{\text{ethylene glycol}}{HOCH_2CH_2OH} \longrightarrow$$

$$\underset{\text{a polyurethane}}{-\overset{O}{\overset{\|}{C}}-NH \overset{CH_3}{\diagdown} NH-\overset{O}{\overset{\|}{C}} \left[OCH_2CH_2O-\overset{O}{\overset{\|}{C}}-NH \overset{CH_3}{\diagdown} NH-\overset{O}{\overset{\|}{C}} \right]_n OCH_2CH_2O-\overset{O}{\overset{\|}{C}}-}$$

One of the most important uses of polyurethanes is in fabrics with elastic properties, such as Spandex (trade name is Lycra). These materials are block copolymers in which some of the polymer segments are polyurethanes, some are polyesters, and some are polyethers. The blocks of polyurethane are rigid and short; the blocks of polyesters of polyethers are flexible and long. When stretched, the soft blocks, which are crosslinked by the hard blocks, becomes crystalline (highly ordered). When the tension is released, they revert to their previous state (Section 28.7).

PROBLEM 20

If a small amount of glycerol is added to the reaction mixture of toluene-2,6-diisocyanate and ethylene glycol during the synthesis of polyurethane foam, a much stiffer foam is obtained. Explain.

$$\underset{\overset{|}{OH} \quad \overset{|}{OH} \quad \overset{|}{OH}}{CH_2-CH-CH_2}$$
$$\text{glycerol}$$

28.7 Physical Properties of Polymers

The individual chains of a polymer such as polyethylene are held together by van der Waals forces. Because these forces operate only at small distances, they are strongest if the polymer chains can line up in an ordered, closely packed array. Regions of the polymer in which the chains are highly ordered with respect to one another are called **crystallites** (Figure 28.4). Between the crystallites are amorphous, noncrystalline regions in which the chains are randomly oriented. The more crystalline—the more ordered—the polymer is, the denser, harder, and more resistant it is to heat (Table 28.7). If the polymer chains possess substituents (as does poly[methyl methacrylate], for example) or have branches that prevent them from packing closely together, the density of the polymer is reduced.

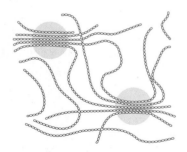

▲ **Figure 28.4**
In the regions called crystallites (indicated by circles), the polymer chains are highly ordered, much the way crystals are. Between the crystallites are noncrystalline (amorphous) regions in which the polymer chains are randomly oriented.

Table 28.7 Properties of Polyethylene as a Function of Crystallinity				
Crystallinity (%) 55	62	70	77	85
Density (g/cm^3) 0.92	0.93	0.94	0.95	0.96
Melting point (°C) 109	116	125	130	133

Thermoplastic Polymers

Plastics can be classified according to the physical properties they acquire from the way in which their individual chains are arranged. **Thermoplastic polymers** have both ordered crystalline regions and amorphous noncrystalline regions. These plastics are hard at room temperature but soft enough to be molded when heated, because the individual chains can slip past one another at elevated temperatures. Thermoplastic polymers are the plastics we encounter most often in our daily lives—in combs, toys, switch plates, and telephone casings. They are the plastics that are easily cracked.

Thermosetting Polymers

Very rigid materials can be obtained if polymer chains are cross-linked. The greater the degree of cross-linking, the more rigid is the polymer. Such cross-linked polymers are called **thermosetting polymers**. After they are hardened, they cannot be remelted by heating, because the cross-links are covalent bonds, not intermolecular van der Waals forces. Cross-linking reduces the mobility of the polymer chains, causing the polymer to be relatively brittle. Because thermosetting polymers do not have the wide range of properties characteristic of thermoplastic polymers, they are less widely used.

Melmac, a highly cross-linked thermosetting polymer of melamine and formaldehyde, is a hard, moisture-resistant material. Because it is colorless, Melmac can be made into materials with pastel colors. It is used to make counter surfaces and lightweight dishes.

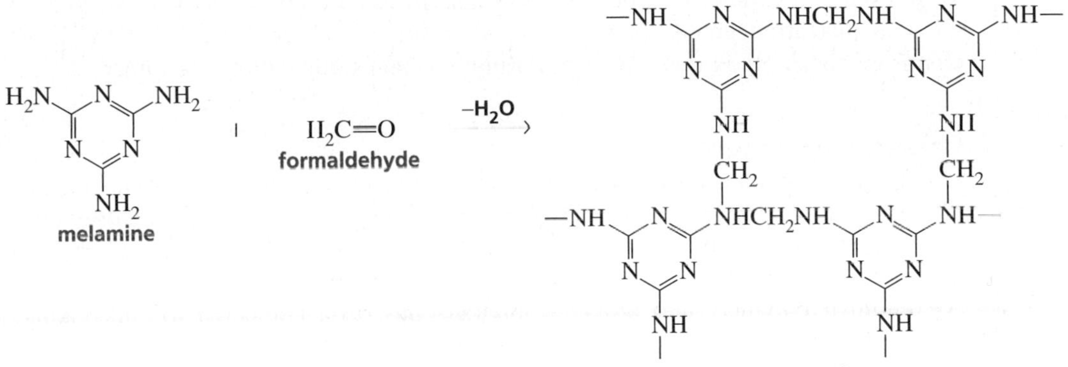

DESIGNING A POLYMER

Today, polymers are being designed to meet ever more exacting and specific needs. A polymer used for making dental impressions, for example, must be soft enough initially to be molded around the teeth but must become hard enough later to maintain a fixed shape. The polymer commonly used for dental impressions contains three-membered aziridine rings that react to form cross-links between the chains. Because aziridine rings are not very reactive, cross-linking occurs relatively slowly, so most of the hardening of the polymer does not occur until the polymer is removed from the patient's mouth.

A polymer used for making contact lenses must be sufficiently hydrophilic to allow lubrication of the eye. Such a polymer, therefore, has many OH groups.

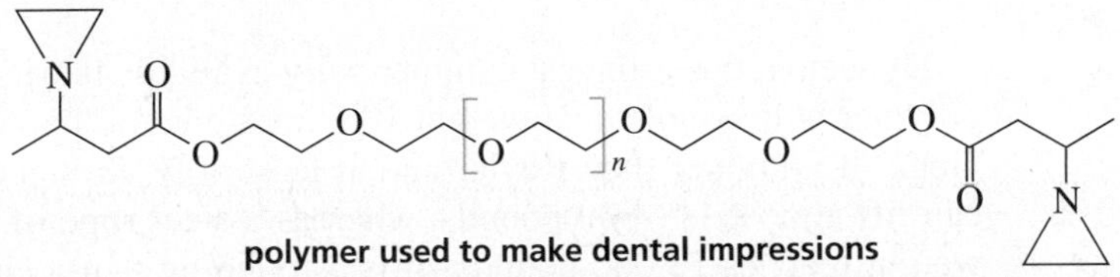

polymer used to make dental impressions

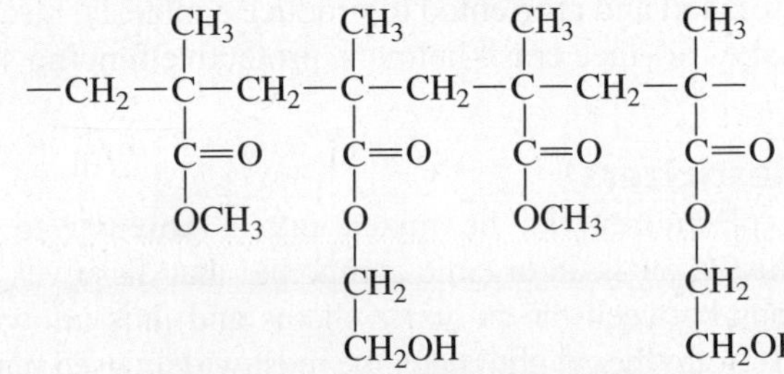

polymer used to make contact lenses

PROBLEM 21

Propose a mechanism for the formation of Melmac.

PROBLEM 22

Bakelite, a durable plastic used for the casing of early radios and TVs, was the first of the thermosetting polymers. It is a highly cross-linked polymer formed from the acid-catalyzed polymerization of phenol and formaldehyde. Because it is much darker than Melmac, Bakelite's range of colors is limited. Propose a structure for Bakelite.

Elastomers

An **elastomer** is a polymer that stretches and then reverts to its original shape. It is a randomly oriented amorphous polymer, but it must have some cross-linking so that the chains do not slip over one another. When elastomers are pulled, the random chains stretch out and crystallize. The van der Waals forces are not strong enough to keep them in that arrangement, however, so when the stretching force is removed, the chains go back to their random shapes. Rubber is an example of an elastomer.

Oriented Polymers

Polymer chains obtained by conventional polymerization can be stretched out and then packed back together in a more ordered, parallel arrangement than they had originally, resulting in polymers that are stronger than steel or that conduct electricity almost as well as copper (Figure 28.5). Such polymers are called **oriented polymers**. Converting conventional polymers into oriented polymers has been compared to "uncooking" spaghetti; the conventional polymer is analogous to disordered, cooked spaghetti, whereas the oriented polymer is like ordered, raw spaghetti.

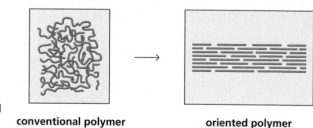

Figure 28.5 ▶
The creation of an oriented polymer.

conventional polymer   **oriented polymer**

Dyneema, the strongest commercially available fabric, is an oriented polyethylene polymer with a molecular weight 100 times greater than that of high-density polyethylene. It is lighter than Kevlar and at least 40% stronger. A rope made of Dyneema can lift almost 119,000 pounds, whereas a steel rope of similar size fails before the weight reaches 13,000 pounds! It is astounding that a chain of carbon atoms can be stretched and reoriented to produce a material stronger than steel. Dyneema is used to make full-face crash helmets, protective fencing suits, and hang gliders.

Plasticizers

A plasticizer can be mixed into a polymer to make the latter more flexible. A **plasticizer** is an organic compound that dissolves in the polymer, lowering the attractions between the polymer chains and thus allowing them to slide past one another. Di-2-ethylhexyl phthalate, the most widely used plasticizer, is added to poly(vinyl chloride)—normally a brittle polymer—to make products such as vinyl raincoats, shower curtains, and garden hoses.

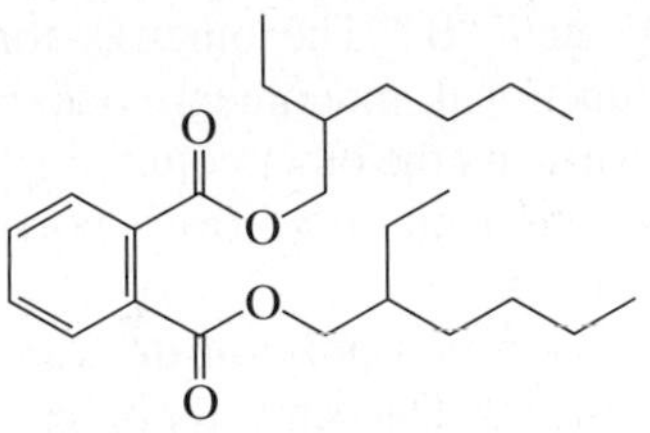

di-2-ethylhexyl phthalate
a plasticizer

An important property to consider in choosing a plasticizer is its permanence—how well the plasticizer remains in the polymer. The "new-car smell" appreciated by car owners is the odor of the plasticizer that has vaporized from the vinyl upholstery. When a significant amount of the plasticizer has evaporated, the upholstery becomes brittle and cracks.

Biodegradable Polymers

Biodegradable polymers are polymers that can be broken into small segments by enzyme-catalyzed reactions. The enzymes are produced by microorganisms. The carbon–carbon bonds of chain-growth polymers are inert to enzyme-catalyzed reactions, so the polymers are nonbiodegradable unless bonds that *can* be broken by enzymes are inserted into them. Then, when the polymer is buried as waste, microorganisms in the ground can degrade the polymer. One method proposed to make a polymer biodegradable inserts hydrolyzable ester groups into it. For example, if the acetal shown below is added to an alkene undergoing radical polymerization, ester groups will be inserted into the polymer, forming "weak links" that are susceptible to enzyme-catalyzed hydrolysis.

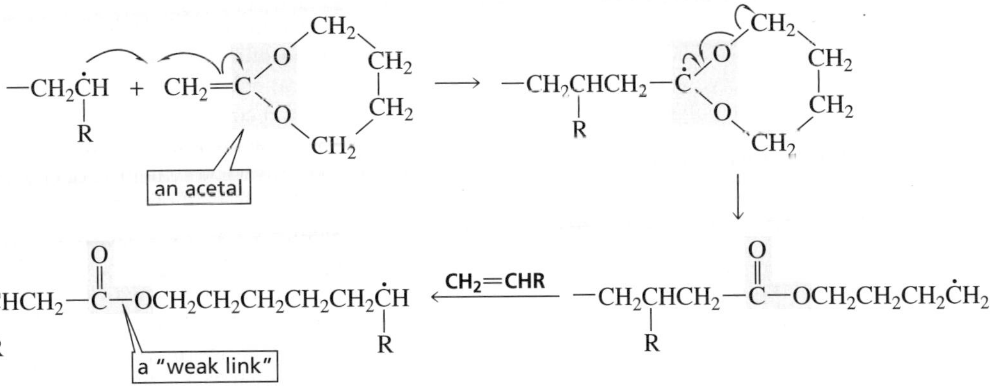

SUMMARY

A **polymer** is a giant molecule made by linking together repeating units of small molecules called **monomers**. The process of linking them is called **polymerization**. **Polymer chemistry** is part of the larger discipline of **materials science**. Polymers can be divided into two groups: **synthetic polymers**, which are synthesized by scientists, and **biopolymers**, which are synthesized by organisms. Synthetic polymers can be divided into two classes, depending on their method of preparation: **chain-growth polymers**, also called **addition polymers**; and **step-growth polymers**, also known as **condensation polymers**.

Chain-growth polymers are made by **chain reactions**, which add monomers to the end of a growing chain. These reactions take place by one of three mechanisms: **radical polymerization**, **cationic polymerization**, or **anionic polymerization**. Each mechanism has an initiation step that starts the polymerization, propagation steps that allow the chain to grow at the **propagating site**, and termination steps that stop the growth of the chain. The choice of mechanism depends on the structure of the **monomer** and the initiator used to activate the monomer. In radical polymerization, the initiator is a radical; in cationic polymerization, it is an electrophile; and in anionic polymerization, it is a nucleophile. Nonterminated polymer chains are called **living polymers**.

Chain-growth polymerization exhibits a preference for **head-to-tail addition**. Branching affects the physical properties of the polymer because linear unbranched chains can pack together more closely than branched chains can. The substituents are on the same side of the carbon chain in an **isotactic polymer**, alternate on both sides of the chain in a **syndiotactic polymer**, and are randomly oriented in an **atactic polymer**. The structure of a polymer can be controlled with **Ziegler–Natta catalysts**. Natural rubber is a polymer of 2-methyl-1,3-butadiene. Synthetic rubbers have been made by polymerizing dienes other than isoprene. Heating rubber with sulfur to cross-link the chains is called vulcanization.

Homopolymers are made of one kind of monomer, whereas **copolymers** are made of more than one kind. In an **alternating copolymer**, the two monomers alternate. In a **block copolymer**, there are blocks of each kind of monomer. In a **random copolymer**, the distribution of monomers is random. A **graft copolymer** contains branches derived from one monomer grafted onto a backbone derived from another.

Step-growth polymers are made by combining two molecules with reactive functional groups at each end. There are two types of step-growth polymers. One type is formed by using a single monomer with two different functional groups,

"A" and "B." The other is formed by using two different bifunctional monomers, one containing two A functional groups and the other containing two B functional groups. The formation of step-growth polymers does not involve chain reactions.

Nylon is a **polyamide**. Aromatic polyamides are called **aramides**. Dacron is a polyester. **Polycarbonates** are polyesters with two alkoxy groups bonded to the same carbonyl carbon. A **urethane** is a compound that has an ester and an amide group bonded to the same carbonyl carbon.

Crystallites are highly ordered regions of a polymer. The more crystalline the polymer is, the denser, harder, and more resistant to heat it is. **Thermoplastic polymers** have crystalline and noncrystalline regions. **Thermosetting polymers** have cross-linked polymer chains. The greater the degree of cross-linking, the more rigid is the polymer. An **elastomer** is a plastic that stretches and then reverts to its original shape. A **plasticizer** is an organic compound that dissolves in the polymer and allows the chains to slide past one another. **Biodegradable polymers** can be broken into small segments by enzyme-catalyzed reactions.

KEY TERMS

addition polymer (p. 1233)
alternating copolymer (p. 1248)
anionic polymerization (p. 1234)
aramide (p. 1251)
atactic polymer (p. 1245)
biodegradable polymer (p. 1257)
biopolymer (p. 1233)
block copolymer (p. 1248)
cationic polymerization (p. 1234)
chain-growth polymer (p. 1233)
chain reaction (p. 1233)
chain transfer (p. 1236)
condensation polymer (p. 1234)
conducting polymer (p. 1245)
copolymer (p. 1248)
cross-linking (p. 1247)

crystallites (p. 1254)
elastomer (p. 1256)
epoxy resin (p. 1252)
graft copolymer (p. 1248)
head-to-tail addition (p. 1236)
homopolymer (p. 1248)
isotactic polymer (p. 1245)
living polymer (p. 1242)
materials science (p. 1233)
monomer (p. 1232)
oriented polymer (p. 1256)
plastic (p. 1233)
plasticizer (p. 1256)
polyamide (p. 1250)
polycarbonate (p. 1252)
polyester (p. 1251)

polymer (p. 1232)
polymer chemistry (p. 1233)
polymerization (p. 1232)
polyurethane (p. 1253)
propagating site (p. 1235)
radical polymerization (p. 1234)
random copolymer (p. 1248)
ring-opening polymerization (p. 1244)
step-growth polymer (p. 1249)
syndiotactic polymer (p. 1245)
synthetic polymer (p. 1233)
thermoplastic polymer (p. 1254)
thermosetting polymer (p. 1255)
urethane (p. 1253)
vinyl polymer (p. 1234)
Ziegler–Natta catalyst (p. 1245)

PROBLEMS

23. Draw short segments of the polymers obtained from the following monomers:

a. $CH_2{=}CHF$

b. $CH_2{=}CHCO_2H$

c. $HO(CH_2)_5\overset{O}{\overset{\|}{C}}OH$

d. $Cl\overset{O}{\overset{\|}{C}}(CH_2)_5\overset{O}{\overset{\|}{C}}Cl \ + \ H_2N(CH_2)_5NH_2$

e. a benzene ring with CH₃, NCO, and OCN substituents + $HOCH_2CH_2OH$

In each case, indicate whether the polymerization is a chain-growth or a step-growth polymerization.

24. Draw the repeating unit of the step-growth polymer that will be formed from each of the following reactions:

a. 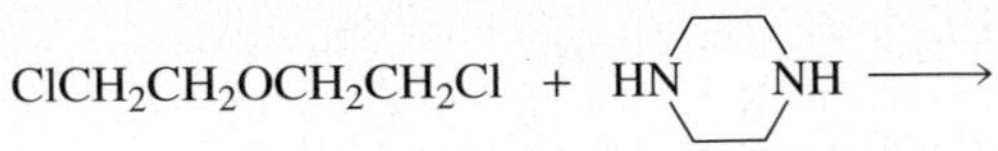 $ClCH_2CH_2OCH_2CH_2Cl$ + HN⎤⎣NH ⟶

b. 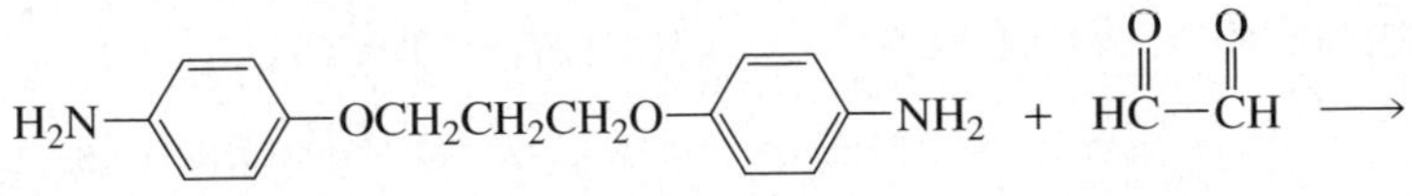 H_2N—⟨⟩—$OCH_2CH_2CH_2O$—⟨⟩—NH_2 + HC—CH ⟶ (with two C=O groups)

c. O=⟨⟩=O + $(C_6H_5)_3P$=CH—⟨⟩—CH=$P(C_6H_5)_3$ ⟶

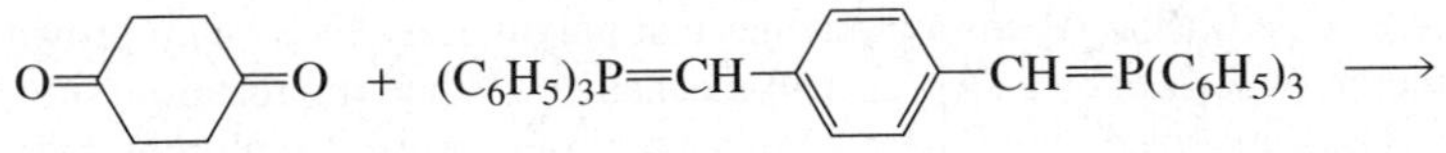

25. Draw the structure of the monomer or monomers used to synthesize the following polymers:

a. $-CH_2CH-$
  $\quad\quad CH_2CH_3$

b. $-CH_2CHO-$
  $\quad\quad CH_3$

c. 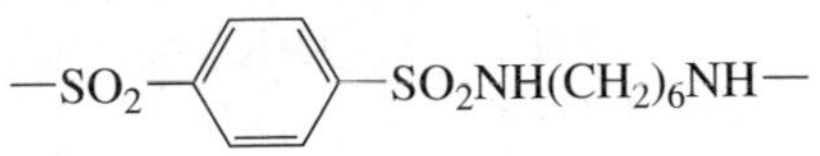 $-SO_2-$⟨⟩$-SO_2NH(CH_2)_6NH-$

d. 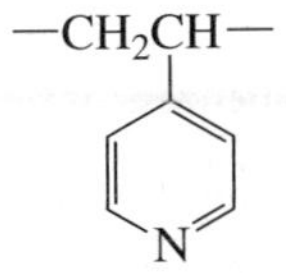 $-CH_2CH-$ (with pyridine ring, N)

e. $-CH_2C$=$CHCH_2-$
  $\quad\quad CH_3$

f. $-CH_2CH_2CH_2CH_2CH_2CO-$ (with C=O)

g. 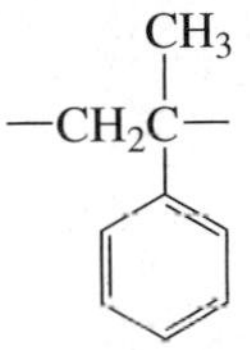 $-CH_2C-$
  $\quad\quad CH_3$ (with benzene ring)

h. 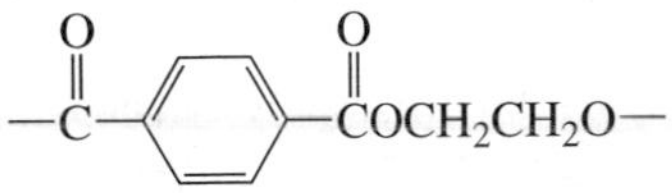 $-C$—⟨⟩—$COCH_2CH_2O-$ (with two C=O)

Indicate whether each polymer is a chain-growth polymer or a step-growth polymer.

26. Explain why the configuration of a polymer of isobutylene is not isotactic, syndiotactic, or atactic.

27. Draw short segments of the polymers obtained from the following compounds under the given reaction conditions:

a.  H_2C—$CHCH_3$ (epoxide) $\xrightarrow{CH_3O^-}$

b. 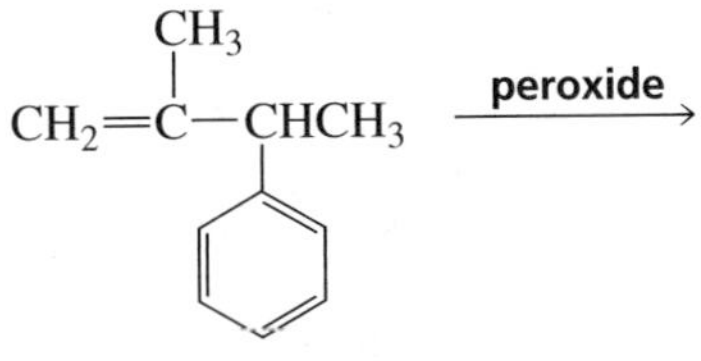 CH_2=C—$CHCH_3$ $\xrightarrow{peroxide}$
  $\quad\quad CH_3$ (with benzene ring)

c. CH_2=CH $\xrightarrow{CH_3CH_2CH_2CH_2Li}$
  $\quad CNH_2$
  $\quad \| $
  $\quad O$

d. CH_2=C—C=CH_2 $\xrightarrow{peroxide}$
  $\quad\quad CH_3 CH_3$

e. CH_2=$CHOCH_3$ $\xrightarrow{BF_3, H_2O}$

28. Quiana is a synthetic fabric that feels very much like silk.
  a. Is Quiana a nylon or a polyester?
  b. What monomers are used to synthesize Quiana?

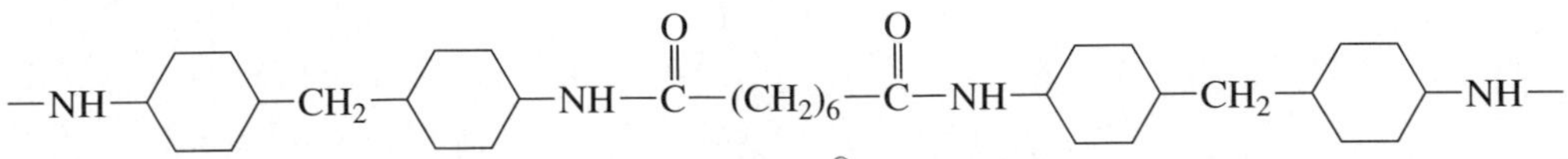

$-NH$—⟨⟩—CH_2—⟨⟩—NH—C—$(CH_2)_6$—C—NH—⟨⟩—CH_2—⟨⟩—$NH-$
(with two C=O groups)

Quiana®

29. Explain why a random copolymer is obtained when 3,3-dimethyl-1-butene undergoes cationic polymerization.

$$CH_2=CH-\underset{\underset{CH_3}{|}}{\overset{\overset{CH_3}{|}}{C}}-CH_3 \longrightarrow -CH_2-CH-CH_2-CH-\overset{\overset{CH_3}{|}}{\underset{\underset{CH_3}{|}}{C}}-CH_2-CH-\overset{\overset{CH_3}{|}}{\underset{\underset{CH_3}{|}}{C}}-CH_2-CH-$$

30. Polly Propylene has started two polymerization reactions. One flask contains a monomer that polymerizes by a chain-growth mechanism, and the other flask contains a monomer that polymerizes by a step-growth mechanism. When the reactions are terminated early in the process and the contents of the flasks analyzed, one flask contains a high-molecular-weight polymer and very little material of intermediate molecular weight. The other flask contains mainly material of intermediate molecular weight and very little high-molecular-weight material. Which flask contains which product? Explain.

31. Poly(vinyl alcohol) is a polymer used to make fibers and adhesives. It is synthesized by hydrolysis or alcoholysis of the polymer obtained from polymerization of vinyl acetate (as shown below).
   a. Why is poly(vinyl alcohol) not prepared by polymerizing vinyl alcohol?
   b. Is poly(vinyl acetate) a polyester?

$$-CH_2-CH-CH_2-CH-CH_2-CH- \xrightarrow[\Delta]{H_2O} -CH_2-CH-CH_2-CH-CH_2-CH-$$

poly(vinyl acetate)   poly(vinyl alcohol)

32. Five different repeating units are found in the polymer obtained by cationic polymerization of 4-methyl-1-pentene. Identify these repeating units.

33. If a peroxide is added to styrene, the polymer known as polystyrene is formed. If a small amount of 1,4-divinylbenzene is added to the reaction mixture, a stronger and more rigid polymer is formed. Draw a short section of this more rigid polymer.

$$CH_2=CH-\underset{\textbf{1,4-divinylbenzene}}{\boxed{}}-CH=CH_2$$

34. A particularly strong and rigid polyester used for electronic parts is marketed under the trade name Glyptal. It is a polymer of terephthalic acid and glycerol. Draw a segment of the polymer and explain why it is so strong.

35. The following two compounds form a 1 : 1 alternating copolymer. No initiator is needed for the polymerization. Propose a mechanism for formation of the copolymer.

36. Which monomer would give a greater yield of polymer, 5-hydroxypentanoic acid or 6-hydroxyhexanoic acid? Explain your choice.

37. When rubber balls and other objects made of natural rubber are exposed to the air for long periods, they turn brittle and crack. This happens more slowly to objects made of polyethylene. Explain.

38. When acrolein undergoes anionic polymerization, a polymer with two types of repeating units is obtained. Give the structures of the repeating units.

$$CH_2=CHCH\overset{\overset{O}{\|}}{}$$
acrolein

39. Why do vinyl raincoats become brittle as they get old, even if they are not exposed to air or to any pollutants?

40. The polymer shown here is synthesized by hydroxide ion–promoted hydrolysis of a copolymer of *para*-nitrophenyl methacrylate and acrylate.

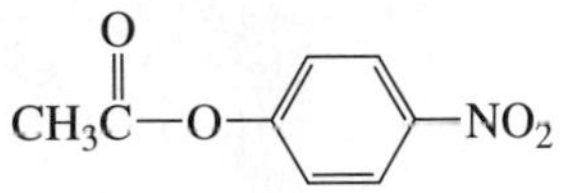

   a. Propose a mechanism for the formation of the copolymer.

   b. Explain why hydrolysis of the copolymer to form the polymer occurs much more rapidly than hydrolysis of *para*-nitrophenyl acetate.

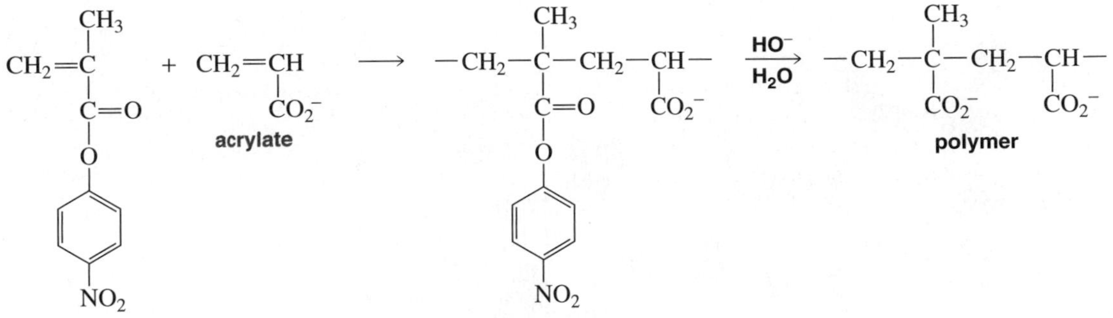

41. An alternating copolymer of styrene and vinyl acetate can be turned into a graft copolymer by hydrolyzing it and then adding ethylene oxide. Draw the structure of the graft copolymer.

42. How could head-to-head poly(vinyl bromide) be synthesized?

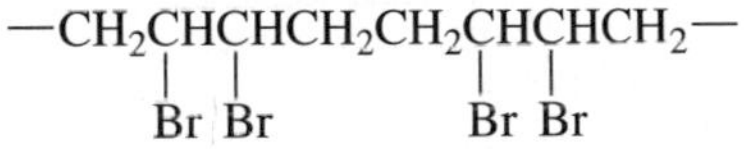

$$-CH_2CHCHCH_2CH_2CHCHCH_2-$$
$$Br\ BrBr\ Br$$

head-to-head poly(vinyl bromide)

43. Delrin (polyoxymethylene) is a tough self-lubricating polymer used in gear wheels. It is made by polymerizing formaldehyde in the presence of an acid catalyst.

   a. Propose a mechanism for formation of a segment of the polymer.

   b. Is Delrin a chain-growth polymer or a step-growth polymer?

CHAPTER 29

Pericyclic Reactions

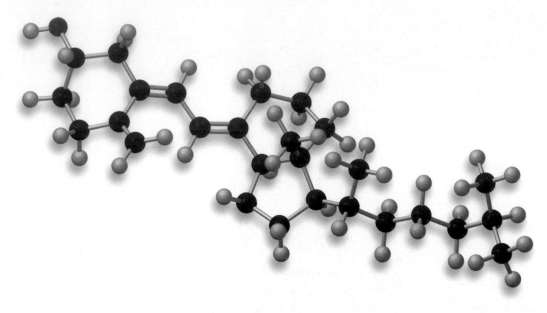

Vitamin D₃

BUILDING ON FUNDAMENTALS

SECTION 29.2 An electron goes into the available molecular orbital with the lowest energy; only two electrons can occupy a particular molecular orbital (1.6).

SECTION 29.2 The bonding MOs are lower in energy and the antibonding MOs higher in energy than the p atomic orbitals (1.6).

SECTION 29.2 Conjugation raises the energy of the HOMO and lowers the energy of the LUMO (7.8, 12.18).

SECTION 29.2 As the molecular orbitals increase in energy, they alternate in being symmetric and antisymmetric. Therefore, the ground-state HOMO and the

excited-state HOMO always have opposite symmetries: one is symmetric and the other is antisymmetric (7.8).

SECTION 29.3 In a stereospecific reaction, each stereoisomeric reactant leads to a different stereoisomeric product of a different set of streoisomeric products (5.18).

SECTION 29.4 The new σ bonds in the product of a cycloaddition reaction are formed by donation of electron density from one reactant to the other reactant. Since only an empty orbital can accept electrons, the electrons must be moving from the HOMO of one of the molecules to the LUMO of the other (7.12).

Reactions of organic compounds can be divided into three classes—polar reactions, radical reactions, and pericyclic reactions. The most common, and the ones most familiar to you, are polar reactions. A **polar reaction** is one in which a nucleophile reacts with an electrophile. Both electrons in the new bond come from the nucleophile.

a polar reaction

$$H:\ddot{O}:^- + \overset{\delta+}{CH_3} - \overset{\delta-}{Br} \longrightarrow CH_3 - OH + Br^-$$

A **radical reaction** is one in which a new bond is formed using one electron from each of the reactants.

a radical reaction

$$CH_3\dot{C}H_2 + Cl-Cl \longrightarrow CH_3CH_2-Cl + \cdot Cl$$

A **pericyclic reaction** occurs when the electrons in one or more reactants are reorganized in a cyclic manner. Pericyclic reactions are concerted and highly selective. In this chapter we will look at the three most common types of pericyclic reactions—electrocyclic reactions, cycloaddition reactions, and sigmatropic rearrangements.

29.1 Three Kinds of Pericyclic Reactions

An **electrocyclic reaction** is an intramolecular reaction in which a new σ (sigma) bond is formed between the ends of a conjugated π (pi) system. This reaction is easy to recognize—the product is a *cyclic* compound that has one more ring and one fewer π bond than the reactant.

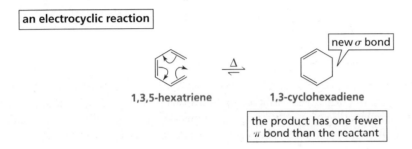

an electrocyclic reaction

1,3,5-hexatriene   1,3-cyclohexadiene

new σ bond

the product has one fewer π bond than the reactant

Electrocyclic reactions are reversible. In the reverse direction, an electrocyclic reaction is one in which a σ bond in a cyclic compound breaks, forming a conjugated π system that has one more π bond than the cyclic compound.

σ bond breaks

cyclobutene   1,3-butadiene

the product has one more π bond than the reactant

In a **cycloaddition reaction**, two different π bond–containing molecules react to form a cyclic compound. Each of the reactants loses a π bond, and the resulting cyclic product has two new σ bonds. The Diels–Alder reaction is a familiar example of a cycloaddition reaction (Section 7.12).

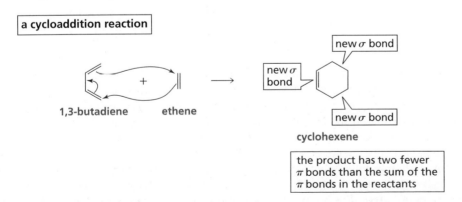

a cycloaddition reaction

1,3-butadiene   ethene

new σ bond

new σ bond

new σ bond

cyclohexene

the product has two fewer π bonds than the sum of the π bonds in the reactants

In a **sigmatropic rearrangement**, a σ bond is broken in the reactant, a new σ bond is formed in the product, and the π bonds rearrange. The number of π bonds does not change (the reactant and the product have the same number of π bonds). The σ bond that is broken can be in the middle of the π system or at the end of the π system. The π system consists of the double-bonded carbons and the carbons immediately adjacent to them.

sigmatropic rearrangements

σ bond is broken in the middle of the π system

σ bond is formed

product and reactant have the same number of π bonds, but their positions are changed

σ bond is broken at the end of the π system

σ bond is formed

Notice that electrocyclic reactions and sigmatropic rearrangements occur within a single π system—they are *intra*molecular reactions. In contrast, cycloaddition reactions involve the interaction of two different π systems—they are usually *inter*molecular reactions. The three kinds of pericyclic reactions share the following features:

• They are all concerted reactions. This means that all the electron reorganization takes place in a single step. Therefore, there is one cyclic transition state and no intermediate.

• Because the reactions are concerted, they are highly stereoselective.

• The reactions are generally not affected by catalysts or by changes in the solvent.

We will see that the configuration of the product formed in a pericyclic reaction depends on

• the configuration of the reactant;

• the number of conjugated double bonds or pairs of electrons in the reacting system;

• whether the reaction is a thermal reaction or a photochemical reaction.

A **photochemical reaction** is one that takes place when a reactant absorbs light. A **thermal reaction** takes place *without* the absorption of light. Despite its name, a thermal reaction does not necessarily require more heat than what is available at room temperature. Some thermal reactions do require additional heat in order to take place at a reasonable rate, but others readily occur at, or even below, room temperature.

For many years, pericyclic reactions puzzled chemists. Why did some pericyclic reactions take place only under thermal conditions, whereas others took place only under photochemical conditions, and others were successfully carried out under both thermal and photochemical conditions? The configurations of the products were also puzzling. After many pericyclic reactions had been investigated, chemists observed that if a pericyclic reaction could take place under both thermal and photochemical conditions, the configuration of the product formed under one set of conditions was different from the configuration of the product formed under the other set of conditions. For example, if the cis isomer was obtained under thermal conditions, the trans isomer was obtained under photochemical conditions and vice versa.

It took two very talented chemists, each bringing his own expertise to the problem, to explain the puzzling behavior of pericyclic reactions. In 1965, R. B. Woodward, an experimentalist, and Roald Hoffmann, a theorist, developed the **conservation of orbital symmetry theory** to explain the relationships between the structure and configuration of the reactant, the conditions (thermal, photochemical, or both) under which the reaction takes place, and the configuration of the product. Because the behavior of pericyclic reactions is so precise, it is not surprising that everything about their behavior can be explained by one simple theory. The difficult part was having the insight that led to the theory.

The conservation of orbital symmetry theory states that *in-phase orbitals overlap during the course of a pericyclic reaction.* The conservation of orbital symmetry theory was based on the **frontier orbital theory** put forth by Kenichi Fukui in 1954. Although Fukui's theory was more than 10 years old when the conservation of orbital symmetry theory was being developed, it had been overlooked because of its mathematical complexity and Fukui's failure to apply it to stereoselective reactions.

According to the conservation of orbital symmetry theory, the symmetry of a molecular orbital controls both the conditions under which a pericyclic reaction takes place and the configuration of the product that is formed. To understand pericyclic reactions, therefore, we must now review molecular orbital theory.

PROBLEM 1◆

Examine the following pericyclic reactions. For each reaction, indicate whether it is an electrocyclic reaction, a cycloaddition reaction, or a sigmatropic rearrangement.

a.

b.

c.

d.

29.2 **Molecular Orbitals and Orbital Symmetry**

The overlap of p atomic orbitals to form π molecular orbitals can be described mathematically using quantum mechanics. The result of the mathematical treatment can be described simply in nonmathematical terms by **molecular orbital (MO) theory**. You were introduced to molecular orbital theory in Sections 1.6 and 7.8. Take a few minutes to review the following key points raised in those sections.

- The two lobes of a p orbital have opposite phases. When two in-phase atomic orbitals interact, a covalent bond is formed. When two out-of-phase atomic orbitals interact, a node is created between the two nuclei.

- Electrons fill molecular orbitals according to the same rules—the aufbau principle, the Pauli exclusion principle, Hund's rule—that govern how they fill atomic orbitals: an electron goes into the available molecular orbital with the lowest energy; only two electrons can occupy a particular molecular orbital and they must be of opposite spin; and an electron will occupy an empty degenerate orbital before it will pair up (Section 1.2).

- Because the π-bonding portions of a molecule are perpendicular to the axes of the σ bonds, the π bonds can be treated independently. Each carbon atom that forms a π bond has a p atomic orbital, and the p atomic orbitals of the carbon atoms combine to produce a π molecular orbital. Thus, a molecular orbital can be described by the **linear combination of atomic orbitals (LCAO)**. In a π molecular orbital, each electron that previously occupied a *p* atomic orbital belonging to an individual carbon nucleus now occupies the entire part of the molecule that is encompassed by the interacting *p* orbitals.

A molecular orbital description of ethene is shown in Figure 29.1. (To show the different phases of the two lobes of a *p* orbital, one lobe is colored blue and the other is colored green.* Because ethene has one π bond, it has two *p* atomic orbitals that combine to produce two π molecular orbitals. The in-phase interaction of the two *p* atomic orbitals gives a bonding π molecular orbital, designated by ψ_1 (ψ is the Greek letter psi). The bonding molecular orbital is of lower energy than the isolated *p* atomic orbitals. The two *p* atomic orbitals of ethene can also interact out-of-phase. Interaction of out-of-phase orbitals gives an antibonding π^* molecular orbital, ψ_2, which is of higher energy than the *p* atomic orbitals. The bonding molecular orbital results from additive interaction of the atomic orbitals, whereas the antibonding molecular orbital results from subtractive interaction. In other words, the interaction of in-phase orbitals holds atoms together, while the interaction of out-of-phase orbitals pushes atoms apart. Because electrons reside in the available molecular orbitals with the lowest energy and two electrons can occupy a molecular orbital, the two π electrons of ethene reside in the bonding π molecular orbital. This molecular orbital picture describes all molecules with one carbon–carbon double bond.

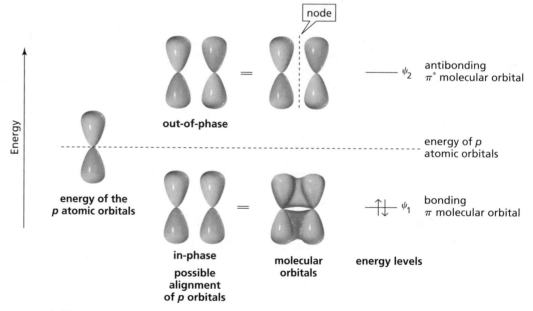

▲ **Figure 29.1**
The interaction of in-phase *p* atomic orbitals produces a bonding π molecular orbital that is lower in energy than the *p* atomic orbitals. The interaction of out-of-phase *p* atomic orbitals produces an antibonding π^* molecular orbital that is higher in energy than the *p* atomic orbitals.

1,3-Butadiene has two conjugated π bonds, so it has four *p* atomic orbitals (Figure 29.2). Four atomic orbitals can combine linearly in four different ways. Consequently, there are four π molecular orbitals: ψ_1, ψ_2, ψ_3, and ψ_4. Notice that orbitals are conserved: four atomic orbitals combine to produce four molecular orbitals. Half are bonding molecular orbitals (ψ_1 and ψ_2) and the other half are antibonding molecular orbitals (ψ_3 and ψ_4). Also notice that the bonding MOs are lower in energy and the antibonding MOs are higher in energy than the *p* atomic orbitals. Because the four π electrons will reside in the available molecular orbitals with the lowest energy,

---

* Some chemists represent the different phases by a $(+)$ and a $(-)$.

two electrons are in ψ_1 and two are in ψ_2. Remember that although the molecular orbitals have different energies, they all coexist. This molecular orbital picture describes all molecules with two conjugated carbon–carbon double bonds.

If you examine the interacting orbitals in Figure 29.2, you will see that in-phase orbitals interact to give a bonding interaction and out-of-phase orbitals interact to create a node. Recall that a node is a place in which there is zero probability of finding an electron (Section 1.5). You will also see that as the energy of the molecular orbital increases, the number of bonding interactions decreases and the number of nodes *between* the nuclei increases. For example, ψ_1 has three bonding interactions and zero nodes between the nuclei, ψ_2 has two bonding interactions and one node between the nuclei, ψ_3 has one bonding interaction and two nodes between the nuclei, and ψ_4 has zero bonding interactions and three nodes between the nuclei. *Notice that a molecular orbital is bonding if the number of bonding interactions is greater than the number of nodes between the nuclei, and a molecular orbital is antibonding if the number of bonding interactions is fewer than the number of nodes between the nuclei.*

Orbitals are conserved: two atomic orbitals combine to produce two molecular orbitals, four atomic orbitals combine to produce four molecular orbitals, six atomic orbitals combine to produce six molecular orbitals, and so on.

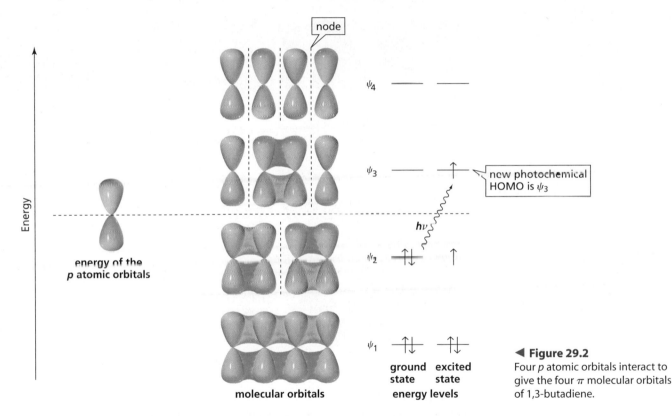

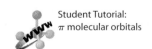
Figure 29.2
Four *p* atomic orbitals interact to give the four π molecular orbitals of 1,3-butadiene.

The normal electronic configuration of a molecule is known as its **ground state**. In the ground state of 1,3-butadiene, the **highest occupied molecular orbital (HOMO)** is ψ_2, and the **lowest unoccupied molecular orbital (LUMO)** is ψ_3. If a molecule absorbs light of an appropriate wavelength, the light will promote an electron from its ground-state HOMO to its LUMO (from ψ_2 to ψ_3). The molecule is then in an **excited state**. In the excited state, the HOMO is ψ_3 and the LUMO is ψ_4. *In a thermal reaction the reactant is in its ground state; in a photochemical reaction the reactant is in an excited state.*

Some molecular orbitals are *symmetric* and some are *antisymmetric* (they do not have a mirror plane, but would have one if one-half of the MO were turned upside down); they are easy to distinguish. If the *p* orbitals at the ends of the molecular orbital are in-phase (both have blue lobes on the top and green lobes on the bottom), the molecular orbital is symmetric. If the two end *p* orbitals are out-of-phase, the molecular orbital is antisymmetric. In Figure 29.2, ψ_1 and ψ_3 are **symmetric molecular orbitals** and ψ_2 and ψ_4 are **antisymmetric molecular orbitals**. Notice that as the molecular orbitals increase in energy, they alternate in being symmetric and antisymmetric. Therefore, *the ground-state*

Student Tutorial:
π molecular orbitals

The ground-state HOMO and the excited-state HOMO have opposite symmetries.

HOMO and the excited-state HOMO always have opposite symmetries: one is symmetric and the other is antisymmetric. A molecular orbital description of 1,3,5-hexatriene, a compound with three conjugated double bonds, is shown in Figure 29.3. As a review, examine the figure and note:

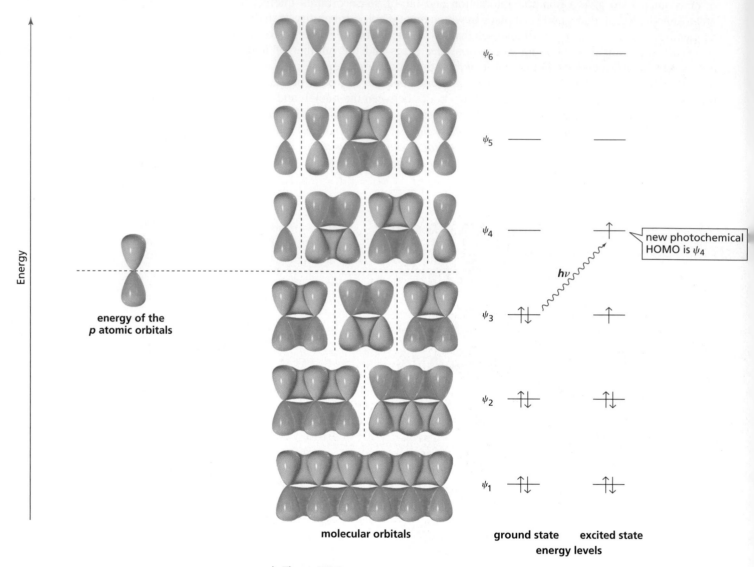

▲ **Figure 29.3**
Six p atomic orbitals interact to give the six π molecular orbitals of 1,3,5-hexatriene.

- the distribution of electrons in the ground and excited states;
- that the number of bonding interactions decreases and the number of nodes increases as the molecular orbitals increase in energy;
- that the molecular orbitals alternate from symmetric to antisymmetric as the molecular orbitals increase in energy;
- compared to the ground state, the excited state has a new HOMO and LUMO.

Although the chemistry of a compound is determined by all its molecular orbitals, we can learn a great deal about that chemistry by looking at only the **HOMO** and the **LUMO**. These two molecular orbitals are known as the **frontier orbitals**. We will now see that simply by evaluating *one* of the frontier orbitals of the reactant or reactants in a pericyclic reaction, we can predict the conditions under which the reaction will occur (thermal or photochemical, or both) and the products that will be formed.

PROBLEM 2◆

Answer the following questions for the π molecular orbitals of 1,3,5-hexatriene:

a. Which are the bonding orbitals, and which are the antibonding orbitals?

b. Which orbitals are the HOMO and the LUMO in the ground state?

c. Which orbitals are the HOMO and the LUMO in the excited state?

d. Which orbitals are symmetric, and which are antisymmetric?

e. What is the relationship between HOMO and LUMO and symmetric and antisymmetric orbitals?

PROBLEM 3◆

a. How many π molecular orbitals does 1,3,5,7-octatetraene have?

b. What is the designation of its HOMO (ψ_1, ψ_2, etc.)?

c. How many nodes does its highest-energy π molecular orbital have between the nuclei?

PROBLEM 4

Give a molecular orbital description for each of the following:

a. 1,3-pentadiene   **c.** 1,3,5-heptatriene

b. 1,4-pentadiene   **d.** 1,3,5,8-nonatetraene

29.3 **Electrocyclic Reactions**

An *electrocyclic reaction* is an intramolecular reaction in which the rearrangement of π electrons leads to a cyclic product that has one fewer π bond than the reactant. An electrocyclic reaction is completely stereoselective, preferentially forming one stereoisomer; an electrocyclic reaction is also stereospecific. For example, when (2E,4Z,6E)-octatriene undergoes an electrocyclic reaction under thermal conditions, only the cis product is formed; note that the cis isomer is a meso compound (Section 5.12). In contrast, when (2E,4Z,6Z)-octatriene undergoes an electrocyclic reaction under thermal conditions, only the trans product is formed; the trans isomer is a pair of enantiomers (Section 5.4). Recall that E means the high-priority groups are on opposite sides of the double bond, and Z means the high-priority groups are on the same side of the double bond (Section 3.5).

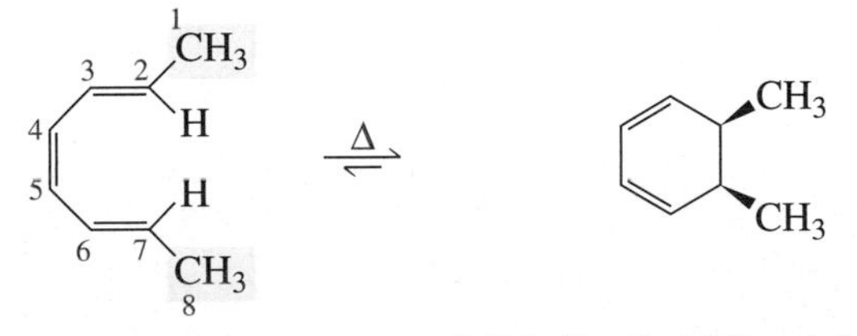

(2E,4Z,6E)-octatriene   *cis*-5,6-dimethyl-1,3-cyclohexadiene

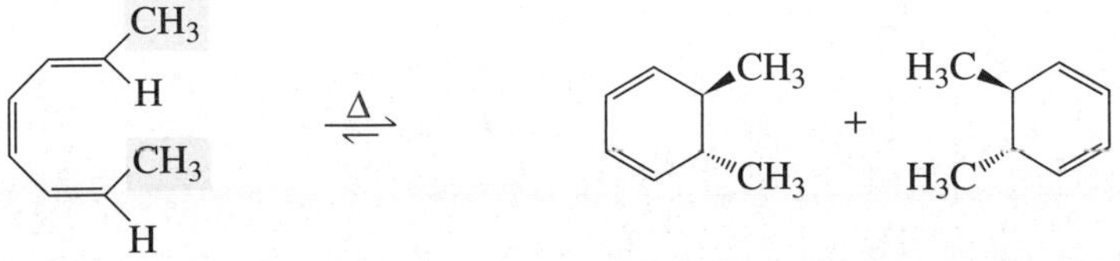

(2E,4Z,6Z)-octatriene   *trans*-5,6-dimethyl-1,3-cyclohexadiene

However, when the reactions are carried out under photochemical conditions, the products have opposite configurations: the compound that forms the cis isomer under thermal conditions forms the trans isomer under photochemical conditions, and the compound that forms the trans isomer under thermal conditions forms the cis isomer under photochemical conditions.

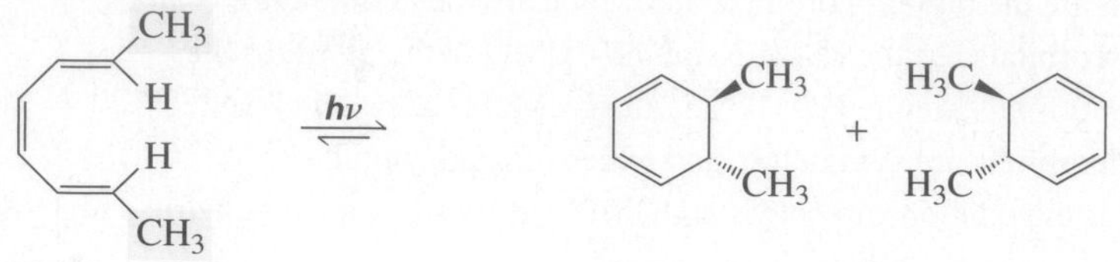

(2E,4Z,6E)-octatriene   *trans*-5,6-dimethyl-1,3-cyclohexadiene

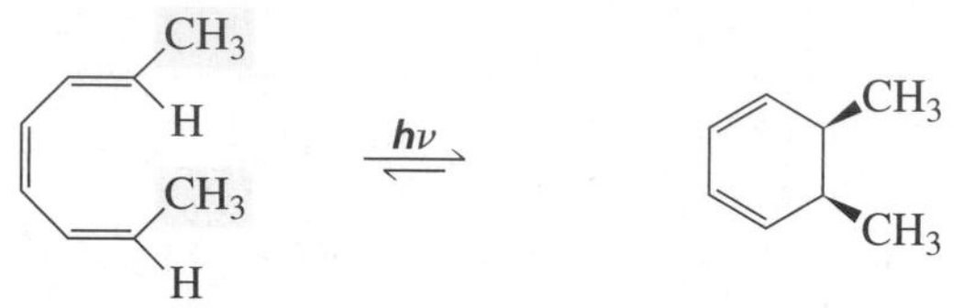

(2E,4Z,6Z)-octatriene   *cis*-5,6-dimethyl-1,3-cyclohexadiene

Under thermal conditions, (2E,4Z)-hexadiene cyclizes to *cis*-3,4-dimethylcyclobutene, and (2E,4E)-hexadiene cyclizes to *trans*-3,4-dimethylcyclobutene.

(2E,4Z)-hexadiene   *cis*-3,4-dimethylcyclobutene

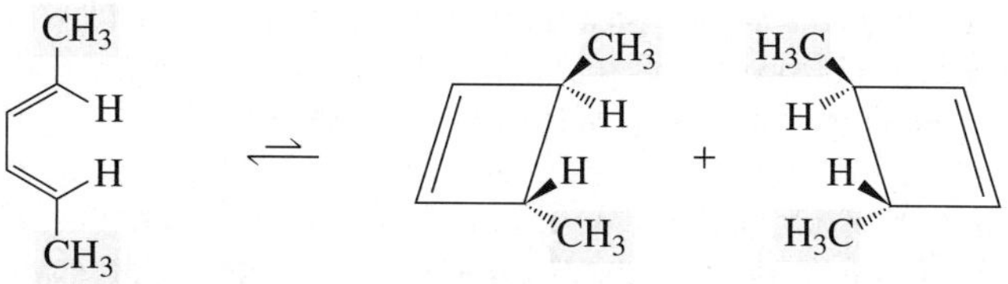

(2E,4E)-hexadiene   *trans*-3,4-dimethylcyclobutene

As we saw with the octatrienes, the configuration of the product changes if the reactions are carried out under photochemical conditions: the trans isomer is obtained from (2E,4Z)-hexadiene instead of the cis isomer; the cis isomer is obtained from (2E,4E)-hexadiene instead of the trans isomer.

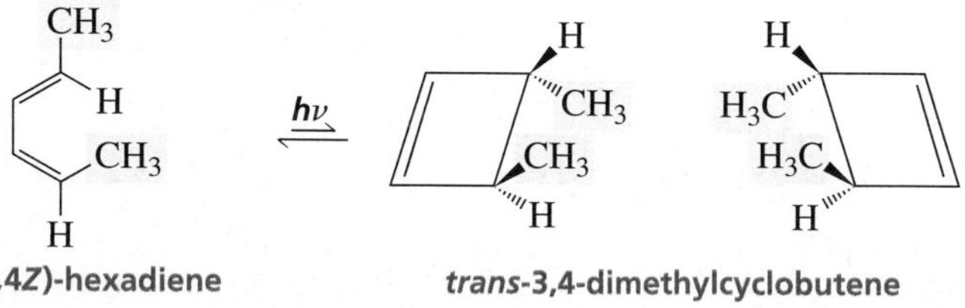

(2E,4Z)-hexadiene   *trans*-3,4-dimethylcyclobutene

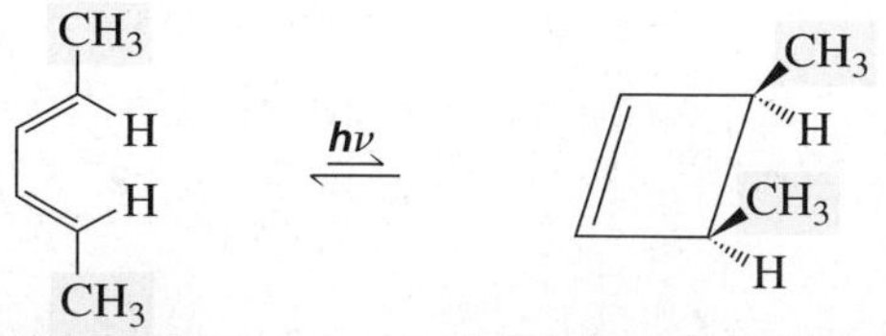

(2E,4E)-hexadiene   *cis*-3,4-dimethylcyclobutene

Electrocyclic reactions, as noted before, are reversible. The cyclic compound is favored for electrocyclic reactions that form six-membered rings, whereas the open-chain compound is favored for electrocyclic reactions that form four-membered rings because of the angle strain associated with four-membered rings (Section 2.11).

Now we will use what we have learned about molecular orbitals to explain the configuration of the products of the preceding reactions. We will then be able to predict the configuration of the product of any other electrocyclic reaction.

The product of an electrocyclic reaction results from the formation of a new σ bond. For this bond to form, the p orbitals at the ends of the conjugated system must rotate so they overlap head-to-head and rehybridize to sp^3. Rotation can occur in two ways. If both orbitals rotate in the same direction (both clockwise or both counterclockwise), ring closure is **conrotatory**.

If the orbitals rotate in opposite directions (one clockwise, the other counterclockwise), ring closure is **disrotatory**.

The mode of ring closure depends on the symmetry of the HOMO of the compound undergoing ring closure. Only the symmetry of the HOMO is important in determining the course of the reaction because this is where the highest energy electrons are. These are the most loosely held electrons and therefore the ones most easily moved during a reaction.

To form the new σ bond, the orbitals must rotate so that in-phase p orbitals overlap, because in-phase overlap is a bonding interaction. Out-of-phase overlap would be an antibonding interaction. If the HOMO is symmetric (the end orbitals are identical), rotation will have to be disrotatory to achieve in-phase overlap. In other words, disrotatory ring closure is symmetry-allowed, whereas conrotatory ring closure is symmetry-forbidden.

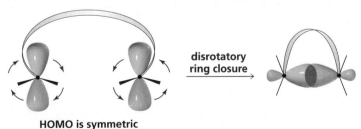

HOMO is symmetric

If the HOMO is antisymmetric, rotation has to be conrotatory in order to achieve in-phase overlap. In other words, conrotatory ring closure is symmetry-allowed, whereas disrotatory ring closure is symmetry-forbidden.

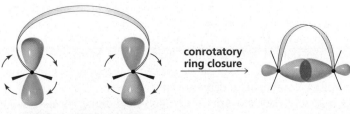

HOMO is antisymmetric

A symmetry-allowed pathway requires in-phase orbital overlap.

Notice that a **symmetry-allowed pathway** is one in which in-phase orbitals overlap; a **symmetry-forbidden pathway** is one in which out-of-phase orbitals would overlap. A symmetry-allowed reaction can take place under relatively mild conditions. If a reaction is symmetry-forbidden, it cannot take place by a concerted pathway. If a symmetry-forbidden reaction takes place at all, it must do so by a nonconcerted mechanism.

Now we are ready to learn why the electrocyclic reactions discussed at the beginning of this section form the indicated products, and why the configuration of the product changes if the reaction is carried out under photochemical conditions.

The ground-state HOMO (ψ_3) of a compound with three conjugated π bonds, such as (2*E*,4*Z*,6*E*)-octatriene, is symmetric (Figure 29.3). This means that ring closure under *thermal conditions* is disrotatory. In disrotatory ring closure of (2*E*,4*Z*,6*E*)-octatriene, the methyl groups are both pushed up (or down), which results in formation of the cis product.

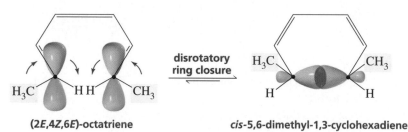

(2E,4Z,6E)-octatriene   ***cis*-5,6-dimethyl-1,3-cyclohexadiene**

In disrotatory ring closure of (2*E*,4*Z*,6*Z*)-octatriene, one methyl group is pushed up and the other is pushed down, which results in formation of the trans product. The enantiomer is obtained by reversing the groups that are pushed up and down.

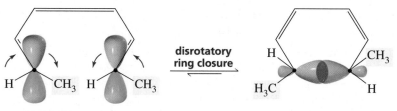

(2E,4Z,6Z)-octatriene   ***trans*-5,6-dimethyl-1,3-cyclohexadiene**

If the reaction takes place under *photochemical conditions,* we must consider the excited-state HOMO rather than the ground-state HOMO. The excited-state HOMO (ψ_4) of a compound with three π bonds is antisymmetric (Figure 29.3). Therefore, under photochemical conditions, (2*E*,4*Z*,6*Z*)-octatriene undergoes conrotatory ring closure, so both methyl groups are pushed down (or up) and the cis product is formed.

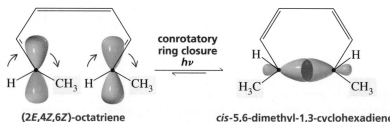

(2E,4Z,6Z)-octatriene   ***cis*-5,6-dimethyl-1,3-cyclohexadiene**

The symmetry of the HOMO of the compound undergoing ring closure controls the stereochemical outcome of an electrocyclic reaction.

We have just seen why the configuration of the product formed under photochemical conditions is the opposite of the configuration of the product formed under thermal conditions: the ground-state HOMO is symmetric, so disrotatory ring closure occurs, whereas the excited-state HOMO is antisymmetric, so conrotatory ring closure occurs. Thus, the stereochemical outcome of an electrocyclic reaction depends on the symmetry of the HOMO of the compound undergoing ring closure.

Now let's see why ring closure of (2*E*,4*Z*)-hexadiene forms *cis*-3,4-dimethyl-cyclobutene. The compound undergoing ring closure has two conjugated π bonds.

The ground-state HOMO of a compound with two conjugated π bonds is antisymmetric (Figure 29.2), so ring closure is conrotatory. Conrotatory ring closure of (2*E*,4*Z*)-hexadiene leads to the cis product.

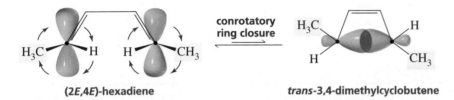

(2E,4Z)-hexadiene → **conrotatory ring closure** → ***cis*-3,4-dimethylcyclobutene**

Similarly, conrotatory ring closure of (2*E*,4*E*)-hexadiene leads to the trans product.

(2E,4E)-hexadiene → **conrotatory ring closure** → ***trans*-3,4-dimethylcyclobutene**

If the reaction is carried out under photochemical conditions, however, the excited-state HOMO of a compound with two conjugated π bonds is symmetric. (Recall that the ground-state HOMO and the excited-state HOMO have opposite symmetries.) So (2*E*,4*Z*)-hexadiene will undergo disrotatory ring closure, resulting in the trans product, whereas (2*E*,4*E*)-hexadiene will undergo disrotatory ring closure and form the cis product.

We have seen that the ground-state HOMO of a compound with two conjugated double bonds is antisymmetric, whereas the ground-state HOMO of a compound with three conjugated double bonds is symmetric. If we examine molecular orbital diagrams for compounds with four, five, six, and more conjugated double bonds, we can conclude that *the ground-state HOMO of a compound with an even number of conjugated double bonds is antisymmetric, whereas the ground-state HOMO of a compound with an odd number of conjugated double bonds is symmetric.* Therefore, from the number of conjugated double bonds in a compound, we can immediately tell whether ring closure will be conrotatory (an even number of conjugated double bonds) or disrotatory (an odd number of conjugated double bonds) under thermal conditions. However, if the reaction takes place under photochemical conditions, everything is reversed, since the ground-state and excited-state HOMOs have opposite symmetries; if the ground-state HOMO is symmetric, the excited-state HOMO is antisymmetric and vice versa.

We have seen that the stereochemistry of an electrocyclic reaction depends on the mode of ring closure, and the mode of ring closure depends on the number of conjugated π bonds in the reactant *and* on whether the reaction is carried out under thermal or photochemical conditions. What we have learned about electrocyclic reactions can be summarized by the **selection rules** listed in Table 29.1. These are also known as the **Woodward–Hoffmann rules** for electrocyclic reactions.

> The ground-state HOMO of a compound with an even number of conjugated double bonds is antisymmetric.

> The ground-state HOMO of a compound with an odd number of conjugated double bonds is symmetric.

Table 29.1 Woodward–Hoffmann Rules for Electrocyclic Reactions		
Number of conjugated π bonds	**Reaction conditions**	**Allowed mode of ring closure**
Even number	Thermal	Conrotatory
	Photochemical	Disrotatory
Odd number	Thermal	Disrotatory
	Photochemical	Conrotatory

The rules in Table 29.1 are for determining whether a given electrocyclic reaction is "allowed by orbital symmetry." There are also selection rules to determine whether cycloaddition reactions (Table 29.3 on page 1278) and sigmatropic rearrangements (Table 29.4 on page 1281) are "allowed by orbital symmetry." It can be rather burdensome to memorize these rules (and worrisome if they are forgotten during an exam), but all the rules can be summarized by the mnemonic "TE-AC." How to use "TE-AC" is explained in Section 29.7.

PROBLEM 5

a. For conjugated systems with two, three, four, five, six, and seven conjugated π bonds, construct quick molecular orbitals (just draw the p orbitals at the ends of the conjugated system as they are drawn on pages 1272 and 1273) to show whether the HOMO is symmetric or antisymmetric.

b. Using these drawings, convince yourself that the Woodward–Hoffmann rules in Table 29.1 are valid.

PROBLEM 6◆

a. Under thermal conditions, will ring closure of (2E,4Z,6Z,8E)-decatetraene be conrotatory or disrotatory?

b. Will the product have the cis or the trans configuration?

c. Under photochemical conditions, will ring closure be conrotatory or disrotatory?

d. Will the product have the cis or the trans configuration?

The series of reactions in Figure 29.4 illustrates just how easy it is to determine the mode of ring closure and therefore the product of an electrocyclic reaction. The reactant of the first reaction has three conjugated double bonds and is undergoing ring closure under thermal conditions. Ring closure, therefore, is disrotatory (Table 29.1). Disrotatory ring closure of this reactant causes the hydrogens to be cis in the ring-closed product. To determine the relative positions of the hydrogens, draw them in the reactant and then draw arrows showing disrotatory ring closure (Figure 29.4a).

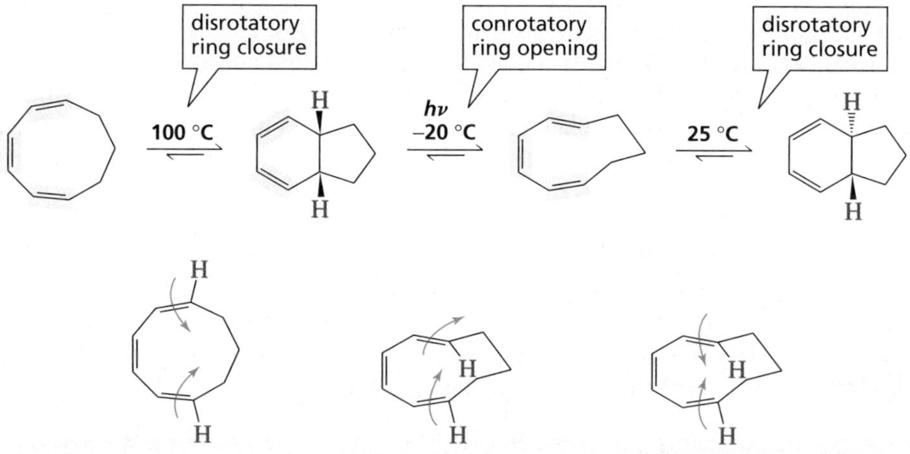

Figure 29.4 ▶
Determining the stereochemistry of the product of an electrocyclic reaction.

The second step in Figure 29.4 is a ring-opening electrocyclic reaction that takes place under photochemical conditions. Because of the principle of microscopic reversibility (Section 14.13), the orbital symmetry rules used for a ring-closure reaction also apply to the reverse ring-opening reaction. The compound undergoing ring closure has three conjugated double bonds. The reaction occurs under photochemical conditions, so both ring closure and the reverse ring opening are conrotatory. (Notice that the number of conjugated double bonds we use to determine the mode of ring opening or closure in reversible electrocyclic reactions is the number in the compound that would be undergoing ring closure.) If conrotatory rotation is to result in a product with cis hydrogens, the hydrogens in the compound undergoing ring closure must point in the same direction (Figure 29.4b).

The third step in Figure 29.4 is a thermal ring closure of a compound with three conjugated double bonds, so ring closure is disrotatory. Drawing the hydrogens and the arrows (Figure 29.4c) allows you to determine the relative positions of the hydrogens in the ring-closed product.

Notice that in all these electrocyclic reactions, if the bonds to the substituents (in this case, hydrogens) in the reactant point in *opposite directions* (as in Figure 29.4a), the substituents will be cis in the product if ring closure is disrotatory and trans if ring closure is conrotatory. On the other hand, if they point in the *same direction* (as in Figure 29.4b or 29.4c), they will be trans in the product if ring closure is disrotatory and cis if ring closure is conrotatory (Table 29.2).

Student Tutorial:
Electrocyclic reactions

Table 29.2 Configuration of the Product of an Electrocyclic Reaction		
Substituents in the reactant	**Mode of ring closure**	**Configuration of the product**
Point in opposite directions	Disrotatory	cis
	Conrotatory	trans
Point in the same direction	Disrotatory	trans
	Conrotatory	cis

PROBLEM 7◆

Which of the following are correct? Correct any false statements.

a. A conjugated diene with an even number of double bonds undergoes conrotatory ring closure under thermal conditions.

b. A conjugated diene with an antisymmetric HOMO undergoes conrotatory ring closure under thermal conditions.

c. A conjugated diene with an odd number of double bonds has a symmetric HOMO.

PROBLEM 8◆

a. Identify the mode of ring closure for each of the following electrocyclic reactions.

b. Are the indicated hydrogens cis or trans?

29.4 Cycloaddition Reactions

In a *cycloaddition reaction,* two different π bond–containing molecules react to form a cyclic molecule by rearrangment of the π electrons and formation of two new σ bonds. The Diels–Alder reaction is one of the best known examples of a cycloaddition reaction (Section 7.12). Cycloaddition reactions are classified according to the number of π electrons that interact to produce the product. The Diels–Alder reaction is a [4 + 2] cycloaddition reaction because one reactant has four interacting π electrons and the other reactant has two interacting π electrons. Only the π electrons participating in electron rearrangement are counted.

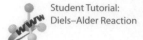
[4 + 2] cycloaddition (a Diels–Alder reaction)

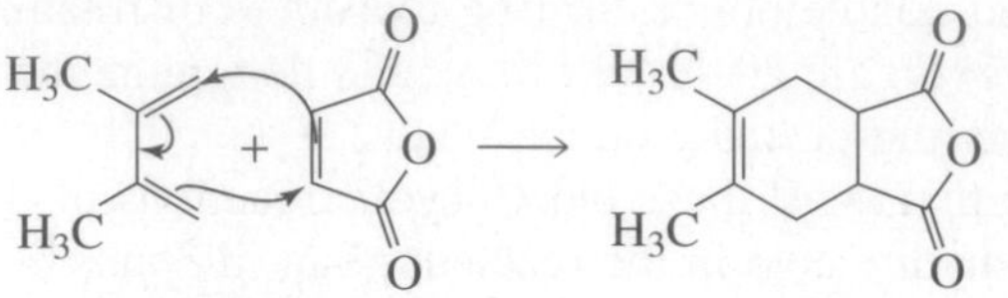

[2 + 2] cycloaddition

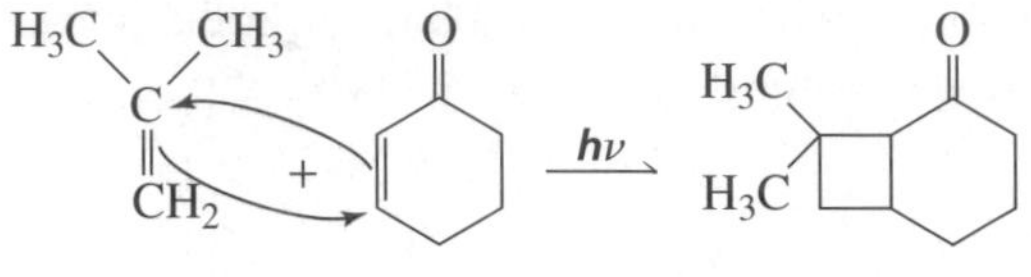

[8 + 2] cycloaddition

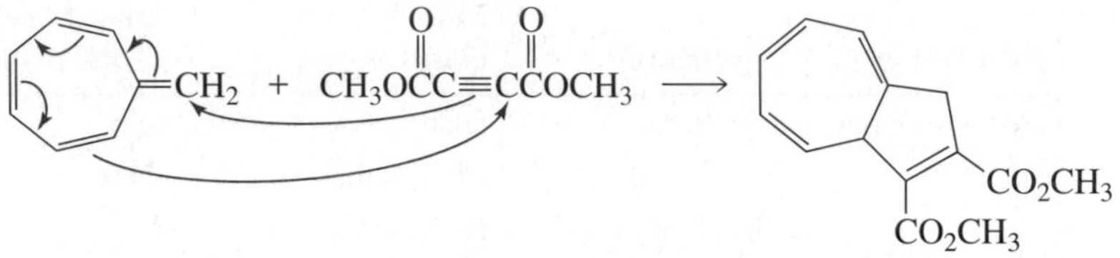

In a cycloaddition reaction, the orbitals of one molecule must overlap with the orbitals of the second molecule. Therefore, the frontier orbitals of both reactants must be evaluated to determine the outcome of the reaction. Because the new σ bonds in the product are formed by donation of electron density from one reactant to the other reactant, and because only an empty orbital can accept electrons, we must consider the HOMO of one of the molecules and the LUMO of the other. It does not matter which reacting molecule's HOMO is used as long as the donation occurs between the HOMO of one and the LUMO of the other.

There are two modes of orbital overlap for the simultaneous formation of two σ bonds, suprafacial and antarafacial. Bond formation is **suprafacial** if both σ bonds form on the same side of the π system. Bond formation is **antarafacial** if the two σ bonds form on opposite sides of the π system. Suprafacial bond formation is similar to syn addition, whereas antarafacial bond formation resembles anti addition (Section 5.19).

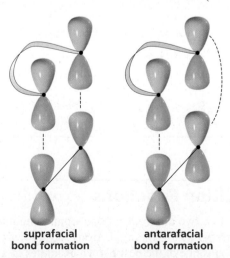

suprafacial bond formation   **antarafacial bond formation**

A cycloaddition reaction that forms a four-, five-, or six-membered ring must occur by suprafacial bond formation. The geometric constraints of these small rings make the antarafacial approach highly unlikely even if it is symmetry-allowed. (Remember that symmetry-allowed means the overlapping orbitals are in-phase.) Antarafacial bond formation is more likely in cycloaddition reactions that form larger rings.

Frontier orbital analysis of a [4 + 2] cycloaddition reaction shows that overlap of in-phase orbitals to form the two new σ bonds requires suprafacial orbital overlap (Figure 29.5). This is true whether we use the LUMO of the dienophile (a system with one π bond; Figure 29.1) and the HOMO of the diene (a system with two conjugated π bonds; Figure 29.2) or the HOMO of the dienophile and the LUMO of the diene to explain the reaction. Now we can understand why Diels–Alder reactions occur with relative ease (Section 7.12).

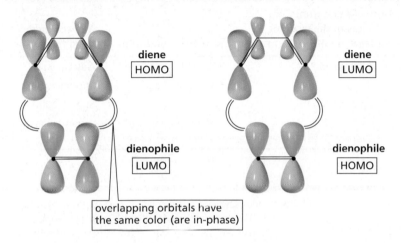

overlapping orbitals have the same color (are in-phase)

◀ **Figure 29.5**
Frontier molecular orbital analysis of a [4 + 2] cycloaddition reaction. The HOMO of either of the reactants can be used with the LUMO of the other. Both situations require suprafacial overlap for bond formation.

A [2 + 2] cycloaddition reaction does not occur under thermal conditions but does take place under photochemical conditions.

The frontier molecular orbitals in Figure 29.6 show why this is so. Under thermal conditions, suprafacial overlap is not symmetry-allowed (the overlapping orbitals are out-of-phase). Antarafacial overlap is symmetry-allowed but is not possible because of the small size of the ring. Under photochemical conditions, however, the reaction can take place because the symmetry of the excited-state HOMO is the opposite of that of

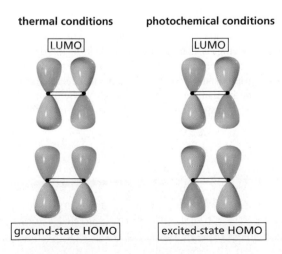

◀ **Figure 29.6**
Frontier molecular orbital analysis of a [2 + 2] cycloaddition reaction under thermal and photochemical conditions.

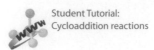

Student Tutorial:
Cycloaddition reactions

the ground-state HOMO. Therefore, overlap of the excited-state HOMO of one alkene with the LUMO of the second alkene involves symmetry-allowed suprafacial bond formation. Notice in the photochemical reaction that only one of the reactants is in an excited state. Because of the very short lifetimes of excited states, it is unlikely that two reactants in their excited states would find one another to interact. The selection rules for cycloaddition reactions are summarized in Table 29.3.

Table 29.3 Woodward–Hoffmann Rules for Cycloaddition Reactions		
Sum of the number of π bonds in the reacting systems of both reagents	**Reaction conditions**	**Allowed mode of bond formation**
Even number	Thermal	Antarafacial[a]
	Photochemical	Suprafacial
Odd number	Thermal	Suprafacial
	Photochemical	Antarafacial[a]
[a]Although antarafacial ring closure is symmetry-allowed, it can occur only with large rings.		

LUMINESCENCE

A reverse [2 + 2] cycloaddition reaction is responsible for the luminescence (also called cold light) given off by light sticks. A light stick contains a thin glass vial that holds a mixture of sodium hydroxide and hydrogen peroxide. The vial is suspended in a solution of diphenyloxalate and a dye. When the vial breaks, two nucleophilic acyl substitution reactions occur that form a compound with an unstable four-membered ring. Recall that the phenolate ion is a relatively good leaving group (Section 16.10).

Suprafacial overlap to form a four-membered ring can take place only under photochemical conditions, so one of the reactants must be in an excited state. Therefore, one of the two carbon dioxide molecules formed when the four-membered ring breaks is in an excited state (indicated by an asterisk in the following mechanism). When the electron in the excited state drops down to the ground state, a photon of ultraviolet light is released that is *not* visible to the human eye. However, when a dye is present, the excited carbon dioxide molecule can transfer some of its energy to the dye molecule, which causes an electron in the dye to be promoted to an excited state. When the electron of the dye drops down to the ground state, a photon of visible light is released that *is* visible to the human eye. In Section 29.6 you will see that a similar reaction is responsible for the light given off by fireflies.

PROBLEM 9

Explain why maleic anhydride reacts rapidly with 1,3-butadiene but does not react at all with ethene under thermal conditions.

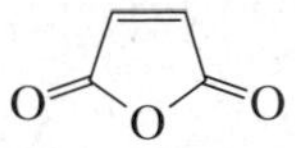

maleic anhydride

PROBLEM 10   *SOLVED*

Compare the reaction between 2,4,6-cycloheptatrienone and cyclopentadiene to the reaction between the former and ethene. Why does 2,4,6-cycloheptatrienone use two π electrons in one reaction and four π electrons in the other?

a.

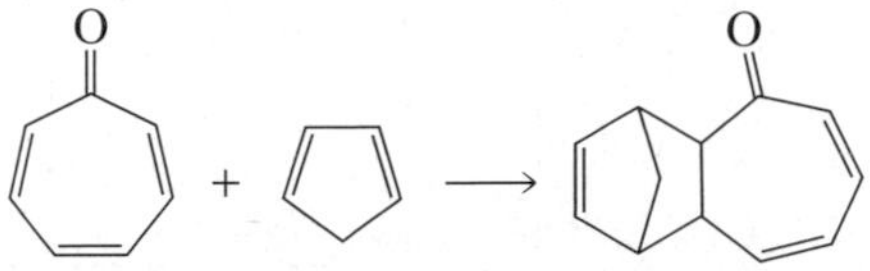

b.

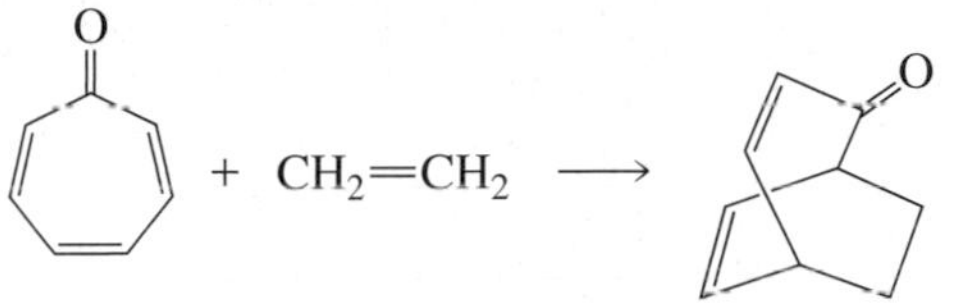

Solution   Both reactions are [4 + 2] cycloaddition reactions. When 2,4,6-cyclohepta-trienone reacts with cyclopentadiene, it uses two of its π electrons because cyclopenta-diene is the four-π-electron reactant. When 2,4,6-cycloheptatrienone reacts with ethene, it uses four of its π electrons because ethene is the two-π-electron reactant.

a.

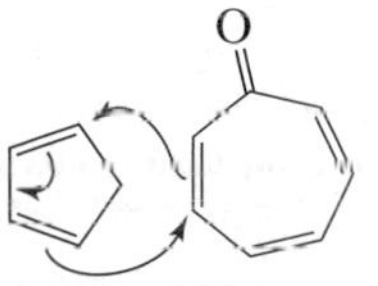

b.

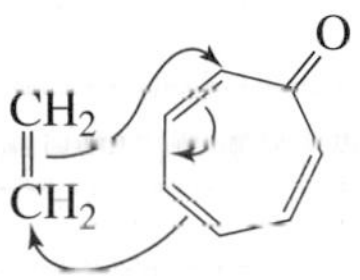

PROBLEM 11◆

Will a concerted reaction take place between 1,3-butadiene and 2-cyclohexenone in the presence of ultraviolet light?

29.5 **Sigmatropic Rearrangements**

The last class of concerted pericyclic reactions that we will consider is the group of reactions known as *sigmatropic rearrangements.* In a sigmatropic rearrangement, a σ bond in the reactant is broken, a new σ bond is formed, and the π electrons rearrange. The σ bond that breaks is a bond to an allylic carbon. It can be a σ bond between a carbon and a hydrogen, between a carbon and another carbon, or between a carbon and an oxygen, nitrogen, or sulfur. "Sigmatropic" comes from the Greek word *tropos,* which means "change," so sigmatropic means "sigma-change."

The numbering system used to describe a sigmatropic rearrangement differs from any numbering system you have seen previously. First, mentally break the σ bond in the reactant and give a number-1 label to both atoms that were attached by the bond. Then look at the new σ bond in the product. Count the number of atoms in each of the fragments that connect the broken σ bond and the new σ bond. The two numbers are put in brackets with the smaller number written first. The following is therefore a [2,3] sigmatropic rearrangement. Two atoms (N, N) connect the old and new σ bonds in one fragment, and three atoms (C, C, C) connect the old and new σ bonds in the other fragment.

a [2,3] sigmatropic rearrangement

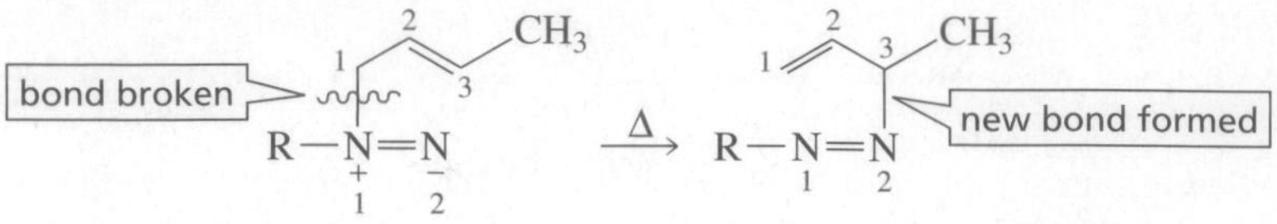

a [1,5] sigmatropic rearrangement

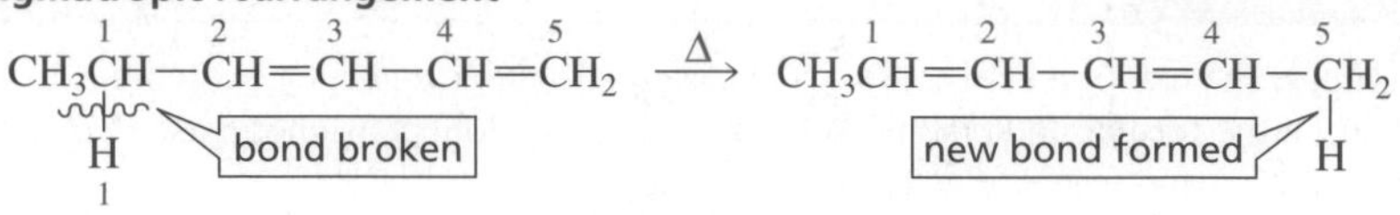

a [1,3] sigmatropic rearrangement

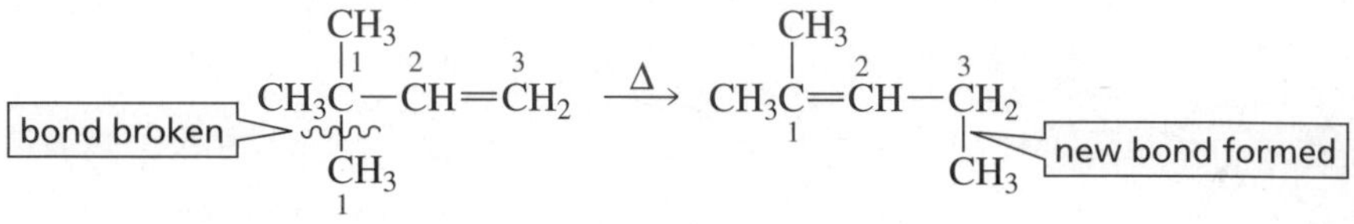

a [3,3] sigmatropic rearrangement

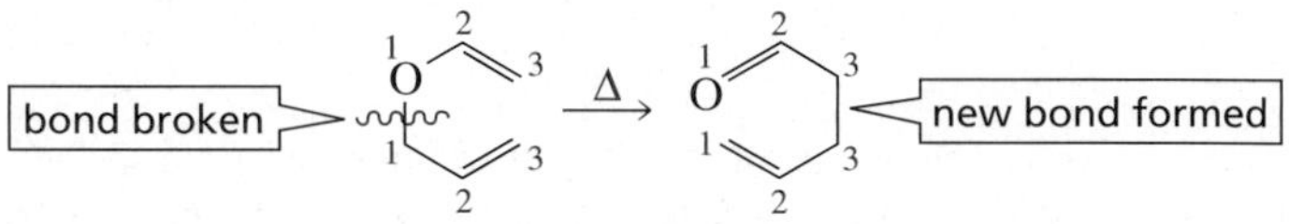

a. Name the kind of sigmatropic rearrangement that occurs in each of the following reactions.

b. Using arrows, show the electron rearrangement that takes place in each reaction.

1.

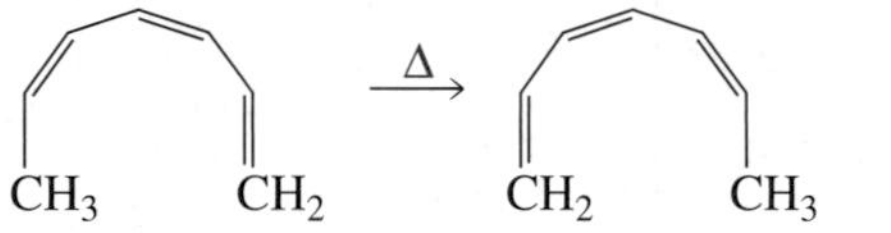

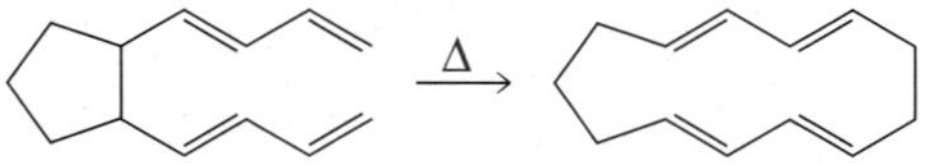

2.

3.

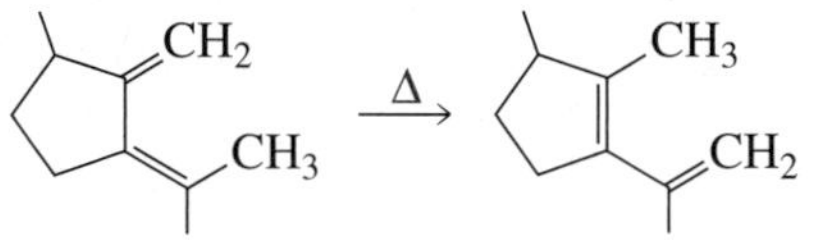

4.

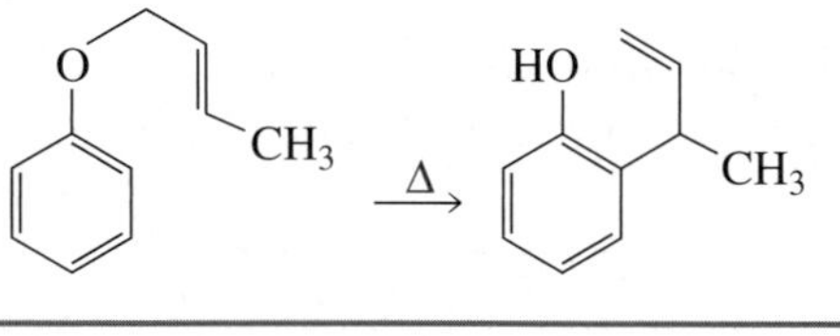

In the transition state of a sigmatropic rearrangement, the group that migrates is partially bonded to the migration origin and partially bonded to the migration terminus. There are two possible modes for rearrangement, analogous to the ones seen in cycloaddition reactions. If the migrating group remains on the same face of the π system, the rearrangement is *suprafacial*. If the migrating group moves to the opposite face of the π system, the rearrangement is *antarafacial*.

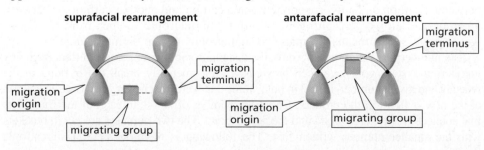

suprafacial rearrangement

antarafacial rearrangement

Sigmatropic rearrangements have cyclic transition states. If the transition state has six or fewer atoms in the ring, rearrangement must be suprafacial because of the geometric constraints of small rings.

We can describe a [1,3] sigmatropic rearrangement as involving a π bond and a pair of σ electrons, or we can say that it involves two pairs of electrons. A [1,5] sigmatropic rearrangement involves two π bonds and a pair of σ electrons (three pairs of electrons), and a [1,7] sigmatropic rearrangement involves four pairs of electrons. The symmetry rules for sigmatropic rearrangements are nearly the same as those for cycloaddition reactions. The only difference is that we count the number of pairs of electrons rather than the number of π bonds. (Compare Tables 29.3 and 29.4.) *Recall that the ground-state HOMO of a compound with an even number of conjugated double bonds is antisymmetric, whereas the ground-state HOMO of a compound with an odd number of conjugated double bonds is symmetric.*

Table 29.4 Woodward–Hoffmann Rules for Sigmatropic Rearrangements		
Number of pairs of electrons in the reacting system	**Reaction conditions**	**Allowed mode of rearrangement**
Even number	Thermal	Antarafacial[a]
	Photochemical	Suprafacial
Odd number	Thermal	Suprafacial
	Photochemical	Antarafacial[a]
[a]Although antarafacial ring closure is symmetry-allowed, it can occur only with large rings.		

A **Cope* rearrangement** is a [3,3] sigmatropic rearrangement of a 1,5-diene. A **Claisen† rearrangement** is a [3,3] sigmatropic rearrangement of an allyl vinyl ether. Both rearrangements form six-membered-ring transition states. The reactions, therefore, must be able to take place by a suprafacial pathway. Whether or not a suprafacial pathway is symmetry-allowed depends on the number of pairs of electrons involved in the rearrangement (Table 29.4). Because [3,3] sigmatropic rearrangements involve three pairs of electrons, they occur by a suprafacial pathway under thermal conditions. Therefore, both Cope and Claisen rearrangements readily take place under thermal conditions.

a Cope rearrangement

a Claisen rearrangement

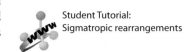

The Ireland-Claisen rearrangement uses an allyl ester instead of the allyl ether used in the Claisen rearrangement. A base removes a proton from the α-carbon of the ester and the enolate ion is trapped as a trimethylsilyl ether (Section 17.11). Mild heating results in a Claisen condensation.

Student Tutorial:
Sigmatropic rearrangements

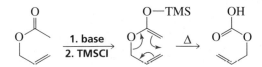

---

*After Arthur C. Cope, who also discovered the Cope elimination (page 952).
†After Ludwing Claisen, who also discovered the Claisen condensation (page 876).

PROBLEM 13♦

a. Give the product of the following reaction:

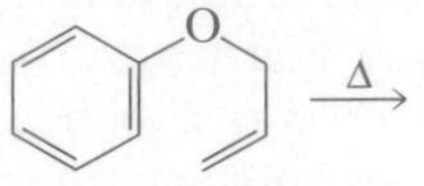

$$\text{(phenyl allyl ether)} \xrightarrow{\Delta}$$

b. If the terminal sp^2 carbon of the substituent bonded to the benzene ring is labeled with ^{14}C, where will the label be in the product?

Migration of Hydrogen

When a hydrogen migrates in a sigmatropic rearrangement, the hydrogen's s orbital is partially bonded to both the migration origin and the migration terminus in the transition state. Therefore, a [1,3] sigmatropic migration of hydrogen has a four-membered-ring transition state. Because two pairs of electrons are involved, the HOMO is antisymmetric. The selection rules, therefore, require an antarafacial rearrangement for a 1,3-hydrogen shift under thermal conditions (Table 29.4). Consequently, 1,3-hydrogen shifts do not occur under thermal conditions because the four-membered-ring transition state does not allow the required antarafacial rearrangement.

migration of hydrogen

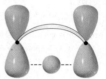

suprafacial rearrangement   **antarafacial rearrangement**

1,3-Hydrogen shifts can take place if the reaction is carried out under photochemical conditions because the HOMO is symmetric under photochemical conditions, allowing hydrogen to migrate by a suprafacial pathway (Table 29.4).

1,3-hydrogen shifts

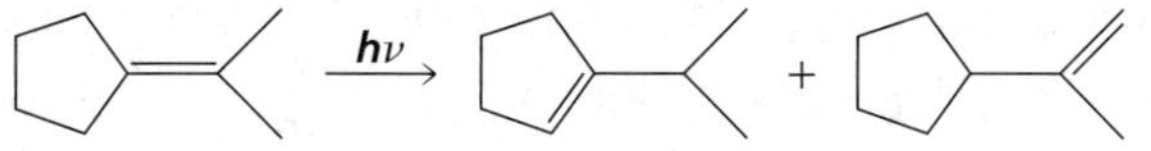

Two products are obtained in the preceding reaction because two different allylic hydrogens can undergo a 1,3-hydrogen shift.

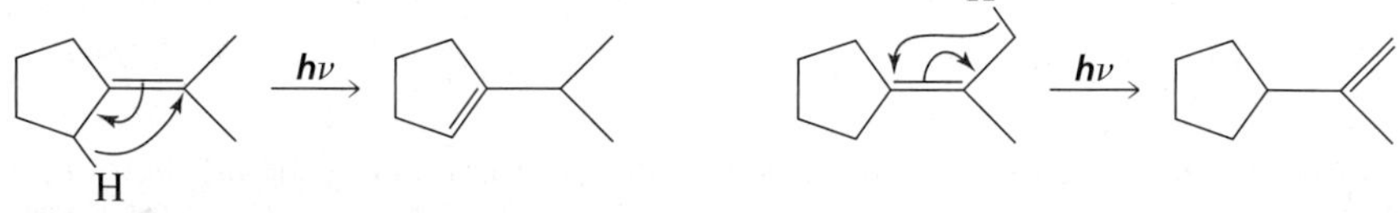

[1,5] Sigmatropic migrations of hydrogen are well known. They involve three pairs of electrons, so they take place by a suprafacial pathway under thermal conditions.

1,5-hydrogen shifts

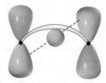

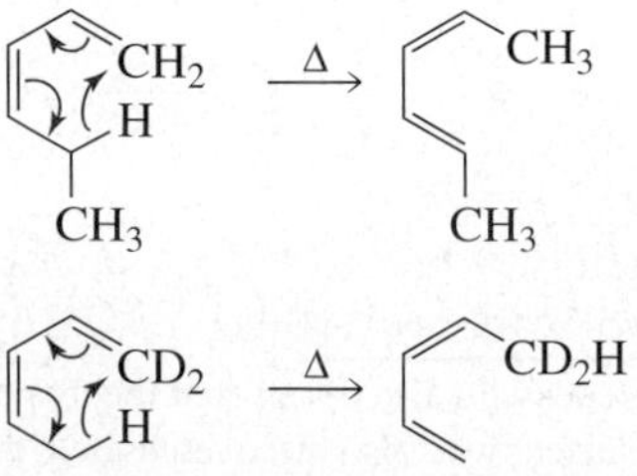

PROBLEM 14

Why was a deuterated compound used in the preceding example?

PROBLEM 15

Account for the difference in the products obtained under photochemical and thermal conditions:

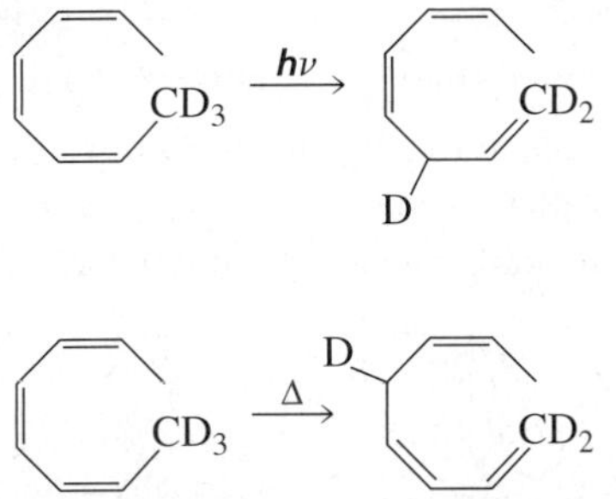

[1,7] Sigmatropic hydrogen migrations involve four pairs of electrons. They can take place under thermal conditions because the eight-membered-ring transition state allows the required antarafacial rearrangement.

1,7-hydrogen shift

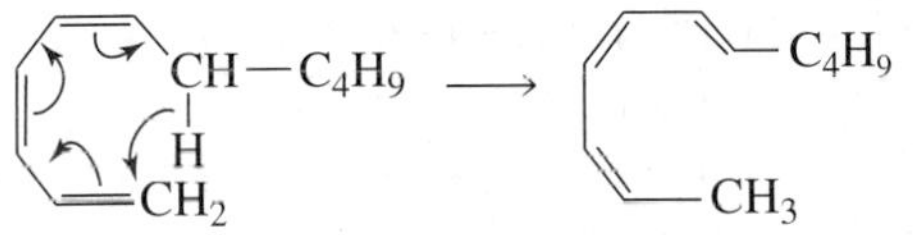

PROBLEM 16 SOLVED

5-Methyl-1,3-cyclopentadiene rearranges to give a mixture of 5-methyl-1,3-cyclopentadiene, 1-methyl-1,3-cyclopentadiene, and 2-methyl-1,3-cyclopentadiene. Show how these products are formed.

Solution   Notice that both equilibria involve [1,5] sigmatropic rearrangements. Although a hydrogen moves from one carbon to an adjacent carbon, the rearrangements are not considered 1,2-shifts because these would not account for all the atoms involved in the rearranged π electron system.

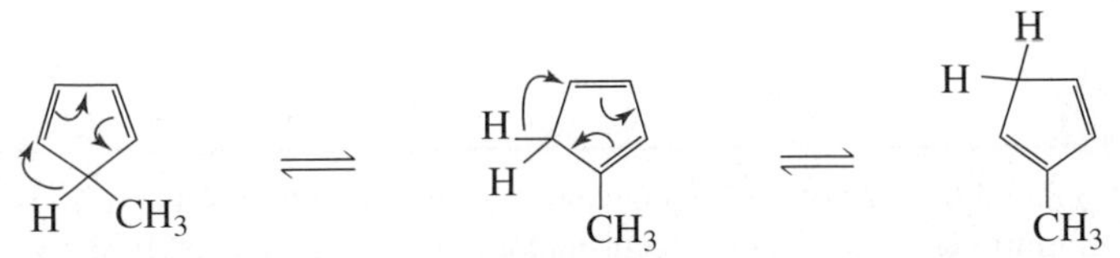

5-methyl-1,3-cyclopentadiene   **1-methyl-1,3-cyclopentadiene**   **2-methyl-1,3-cyclopentadiene**

Migration of Carbon

Unlike hydrogen, which can migrate in only one way because of its spherical s orbital, carbon has two ways to migrate because it has a two-lobed p orbital. Carbon can simultaneously interact with the migration origin and the migration terminus using one of its lobes.

carbon migrating with one of its lobes interacting

suprafacial rearrangement   **antarafacial rearrangement**

Carbon can also simultaneously interact with the migration source and the migration terminus using both lobes of its p orbital.

carbon migrating with both of its lobes interacting

 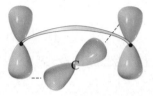

suprafacial rearrangement   **antarafacial rearrangement**

If the reaction requires a suprafacial rearrangement, carbon will migrate using one of its lobes if the HOMO is symmetric and will migrate using both of its lobes if the HOMO is antisymmetric.

When carbon migrates with only one of its p lobes interacting with the migration source and migration terminus, the migrating group retains its configuration because bonding is always to the same lobe. When carbon migrates with both of its p lobes interacting, bonding in the reactant and bonding in the product involve different lobes. Therefore, migration occurs with inversion of configuration.

The following [1,3] sigmatropic rearrangement has a four-membered-ring transition state that requires a suprafacial pathway. The reacting system has two pairs of electrons, so its HOMO is antisymmetric. Therefore, the migrating carbon interacts with the migration source and the migration terminus using both of its lobes and, as a result, carbon undergoes inversion of configuration.

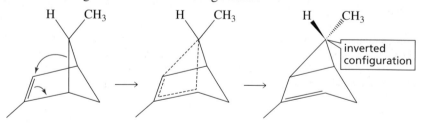

PROBLEM 17

[1,3] Sigmatropic migrations of hydrogen cannot occur under thermal conditions, but [1,3] sigmatropic migrations of carbon can occur under thermal conditions. Explain.

PROBLEM 18◆

a. Will thermal 1,3-migrations of carbon occur with retention or inversion of configuration?
b. Will thermal 1,5-migrations of carbon occur with retention or inversion of configuration?

29.6 Pericyclic Reactions in Biological Systems

We will now look at some pericyclic reactions that occur in biological systems.

Biological Cycloaddition Reactions

Exposure to ultraviolet light may cause skin cancer. This is one of the reasons many scientists are concerned about the thinning ozone layer. The ozone layer absorbs ultraviolet radiation high in the atmosphere, protecting organisms on Earth's surface (Section 11.11). One cause of skin cancer is the formation of *thymine dimers*. At any point in DNA where there are two adjacent thymine residues (Section 27.1), a [2 + 2] cycloaddition reaction can occur, resulting in the formation of a thymine dimer. Because [2 + 2] cycloaddition reactions take place only under photochemical conditions, the reaction takes place only in the presence of ultraviolet light.

two adjacent thymine residues in DNA $\xrightarrow{h\nu}$ **mutation-causing thymine dimer**

Thymine dimers can cause cancer because they interfere with the structural integrity of DNA. Any modification of DNA structure can lead to mutations and possibly to cancer.

Fortunately, there are enzymes that repair damaged DNA. When a repair enzyme recognizes a thymine dimer, it reverses the [2 + 2] cycloaddition reaction to regenerate the original T–T sequence. Repair enzymes, however, are not perfect, and some damage always remains uncorrected. People who do not have the repair enzyme (called DNA photolyase) that reverses thymine dimer formation do not often live beyond the age of 20. Fortunately, this genetic defect is rare.

Fireflies are one of several species that luminesce (emit cold light) as a result of a retro (reverse) [2 + 2] cycloaddition reaction, similar to the reaction that produces cold light in light sticks (Section 29.4). Fireflies have an enzyme (luciferase) that catalyzes the reaction between luciferin, ATP, and molecular oxygen to form a compound with an unstable four-membered ring. The purpose of ATP is to activate the carboxylate group by putting a group that can easily be expelled on the carbonyl carbon. (Notice that the carboxylate group attacks the α-phosphorus of ATP, expelling pyrophosphate; see Section 25.3.) When the four-membered ring breaks, an electron in oxyluciferin is promoted to an excited state because suprafacial overlap can occur only under photochemical conditions. When the electron in the excited state drops down to the ground state, a photon of light is released. In this example, the luciferin molecule is both the source of the unstable four-membered ring and the dye molecule that had to be added to the cold-light reaction.

A firefly with its stomach aglow.

luciferin $\xrightarrow[\text{luciferase}]{\text{ATP}}$ + pyrophosphate

an unstable
four-membered ring

+ AMP

reverse 2 + 2 cycloaddition

has an electron in an excited state

+ CO_2 $\longrightarrow$ + light

oxyluciferin

A Biological Reaction Involving an Electrocyclic Reaction and a Sigmatropic Rearrangement

Vitamin D is a general name for vitamins D_3 and D_2. Their only structural difference is vitamin D_2 has a double bond, that vitamin D_3 does not have, in the hydrocarbon chain attached to the five-membered ring. Vitamin D_3 is formed from 7-dehydrocholesterol (and vitamin D_2 is formed from ergostrol) by two pericyclic reactions. The first is an electrocyclic reaction that opens one of the six-membered rings to form provitamin D_3 (or provitamin D_2). This reaction occurs under photochemical conditions. The provitamin then undergoes a [1,7] sigmatropic rearrangement to form vitamin D_3 (or vitamin D_2). The sigmatropic rearrangement takes place under thermal conditions and is slower than the electrocyclic reaction, so the vitamins continue to be synthesized for several days after exposure to sunlight. The active form of the vitamin requires two successive hydroxylations of vitamins D_3 and D_2. The first occurs in the liver and the second in the kidneys.

ergosterol has a double bond here

an electrocyclic reaction
$h\nu$

provitamin D_2 has a double bond here

7-dehydrocholesterol
ergosterol

provitamin D_3
provitamin D_2

a [1,7] sigmatropic rearrangement

vitamin D_2 has a double bond here

cholecalciferol (vitamin D_3)
ergocalciferol (vitamin D_2)

THE SUNSHINE VITAMIN

Only the precursor molecules, not the vitamins themselves, are found in food. 7-Dehydrocholesterol comes from dairy products and fatty fish; ergosterol comes from some vegetables. Sunlight converts the precursor molecules to vitamins D_3 and D_2. All milk sold in the United States is enriched with vitamin D as a result of shining ultraviolet light on the milk to convert 7-dehydrocholesterol to vitamin D_3. Vitamin D controls calcium metabolism. In the presence of vitamin D, 30% of ingested calcium is absorbed; in its absence only 10% is absorbed.

A deficiency in vitamin D, which can be prevented by getting enough sun, causes a disease known as rickets. Rickets is characterized by deformed bones and stunted growth. Too much vitamin D is also harmful because it causes the calcification of soft tissues. It is thought that skin pigmentation evolved to protect the skin from the sun's ultraviolet rays in order to prevent the synthesis of too much vitamin D_3. This agrees with the observation that peoples indigenous to countries close to the equator have greater skin pigmentation.

PROBLEM 19◆

Does the [1,7] sigmatropic rearrangement that converts provitamin D_3 to vitamin D_3 involve suprafacial or antarafacial rearrangement?

PROBLEM 20◆

Explain why the hydrogen and the methyl substituent are trans to one another after photochemical ring closure of provitamin D_3 to form 7-dehydrocholesterol.

PROBLEM 21

Chorismate mutase is an enzyme that promotes a pericyclic reaction by forcing the substrate to assume the conformation needed for the reaction. The product of the pericyclic reaction is prephenate, which is subsequently converted into the amino acids phenylalanine and tyrosine. What kind of a pericyclic reaction does chorismate mutase catalyze?

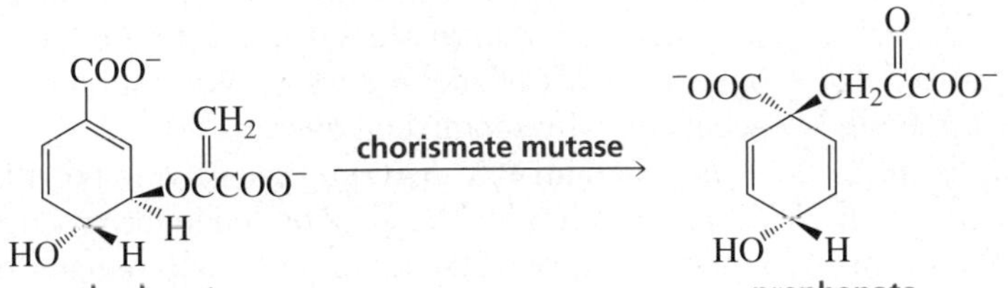

29.7 Summary of the Selection Rules for Pericyclic Reactions

The selection rules that determine the outcome of electrocyclic reactions, cycloaddition reactions, and sigmatropic rearrangements are summarized in Tables 29.1, 29.3, and 29.4, respectively. This is still a lot to remember. Fortunately, the selection rules for all pericyclic reactions can be summarized by "TE-AC."

- If TE (Thermal/Even) describes the reaction, the outcome *is given* by AC (Antarafacial or Conrotatory).

- If *one* of the letters of TE is different (the reaction is not Thermal/Even but is Thermal/Odd or Photochemical/Even), the outcome *is not given* by AC (the outcome is Suprafacial or Disrotatory).

- If *both* of the letters of TE are different (Photochemical/Odd), the outcome *is given* by AC (Antarafacial or Conrotatory), because "two negatives make a positive."

PROBLEM 22

Convince yourself that the "TE-AC" shortcut method for learning the information in Tables 29.1, 29.3, and 29.4 is valid.

SUMMARY

A **pericyclic reaction** is one in which the electrons in the reactant(s) are reorganized in a cyclic manner. Pericyclic reactions are concerted, highly stereoselective reactions that are generally not affected by catalysts or by a change in solvent. The three most common types of pericyclic reactions are *electrocyclic reactions, cycloaddition reactions,* and *sigmatropic rearrangements.*

The configuration of the product of a pericyclic reaction depends on the configuration of the reactant, the number of conjugated double bonds or pairs of electrons in the reacting system, and whether the reaction is **thermal** or **photochemical**. The outcome of pericyclic reactions is given by a set of **selection rules**, which can be summarized by **TE-AC**.

The two lobes of a p orbital have opposite phases. When two in-phase atomic orbitals interact, a covalent bond is formed; two out-of-phase orbitals interact to create a node. The conservation of orbital symmetry theory states that in-phase orbitals overlap during the course of a pericyclic reaction. In other words, a **symmetry-allowed pathway** is one in which in-phase orbitals overlap. If the p orbitals at the ends of the molecular orbital are in-phase, the molecular orbital is **symmetric**. If the two end p orbitals are out-of-phase, the molecular orbital is **antisymmetric**.

The normal electronic configuration of a molecule is known as its **ground state**. The ground-state HOMO of a compound with an even number of conjugated double bonds or an even number of pairs of electrons is antisymmetric; the ground-state HOMO of a compound with an odd number of conjugated double bonds or an odd number of pairs of electrons is symmetric. If a molecule absorbs light of an appropriate wavelength, the light will promote an electron from its ground-state **HOMO** to its **LUMO**. The molecule is then in an **excited state**. In a thermal reaction the reactant is in its ground state; in a photochemical reaction the reactant is in an excited state. The excited-state will have a new HOMO compared to the ground state, and the ground-state HOMO and the excited-state HOMO have opposite symmetries.

An **electrocyclic reaction** is an intramolecular reaction in which a new σ (sigma) bond is formed between the ends of a conjugated π (pi) system. To form this new σ bond, p orbitals at the ends of the conjugated system rotate so they can engage in in-phase overlap. If both orbitals rotate in the same direction, ring closure is **conrotatory**; if they rotate in opposite directions, it is **disrotatory**. If the HOMO is antisymmetric, conrotatory ring closure occurs: if it is symmetric, disrotatory ring closure occurs.

In a **cycloaddition reaction** two different π bond–containing molecules react to form a cyclic compound by rearranging the π electrons and forming two new σ bonds. Because a cycloaddition reaction involves two molecules, an explanation of the reaction must consider the HOMO of one molecule and the LUMO of the other. Bond formation is **suprafacial** if both σ bonds form on the same side of the π system; it is **antarafacial** if the two σ bonds form on opposite sides of the π system. Formation of small rings occurs only by suprafacial overlap.

In a **sigmatropic rearrangement**, a σ bond is broken in the reactant, a new σ bond is formed in the product, and the π bonds rearrange. If the migrating group remains on the same face of the π system, the rearrangement is **suprafacial**; if it moves to the opposite face of the π system, it is **antarafacial**.

KEY TERMS

antarafacial bond formation (p. 1276)
antarafacial rearrangement (p. 1280)
antisymmetric molecular orbital (p. 1267)
Claisen rearrangement (p. 1281)
conrotatory ring closure (p. 1271)
conservation of orbital symmetry theory (p. 1264)
Cope rearrangement (p. 1281)
cycloaddition reaction (p. 1263)
disrotatory ring closure (p. 1274)
electrocyclic reaction (p. 1263)
excited state (p. 1267)

frontier orbital analysis (p. 1277)
frontier orbitals (p. 1268)
frontier orbital theory (p. 1265)
ground state (p. 1267)
highest occupied molecular orbital (HOMO) (p. 1267)
linear combination of atomic orbitals (LCAO) (p. 1267)
lowest unoccupied molecular orbital (LUMO) (p. 1267)
molecular orbital (MO) theory (p. 1265)
pericyclic reaction (p. 1263)

photochemical reaction (p. 1264)
polar reaction (p. 1262)
radical reaction (p. 1262)
selection rules (p. 1273)
sigmatropic rearrangement (p. 1263)
suprafacial bond formation (p. 1276)
suprafacial rearrangement (p. 1280)
symmetric molecular orbital (p. 1267)
symmetry-allowed pathway (p. 1272)
symmetry-forbidden pathway (p. 1272)
thermal reaction (p. 1264)
Woodward–Hoffmann rules (p. 1273)

PROBLEMS

23. Give the product of each of the following reactions:

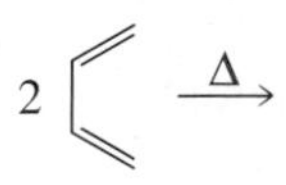

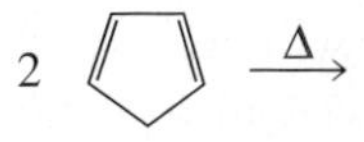

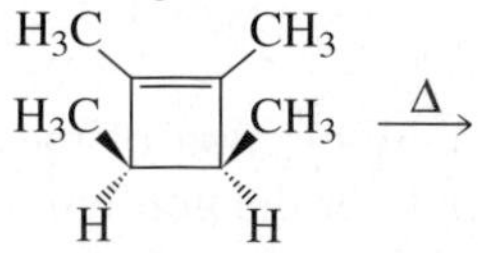

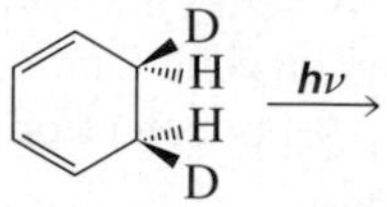

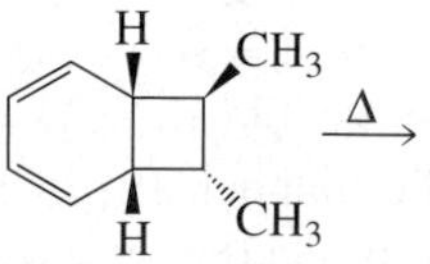

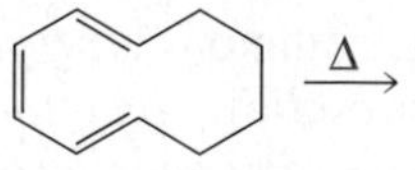

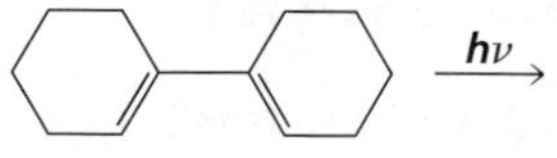

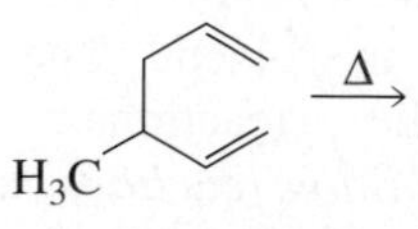

24. Give the product of each of the following reactions:

a.

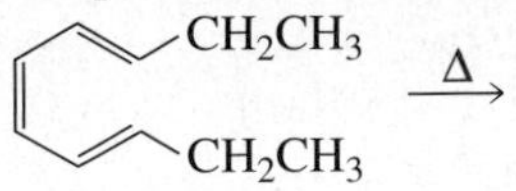

c.

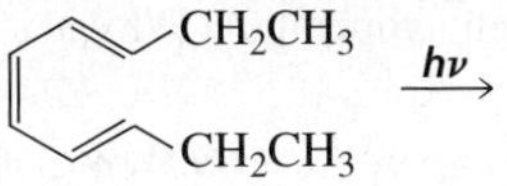

b.

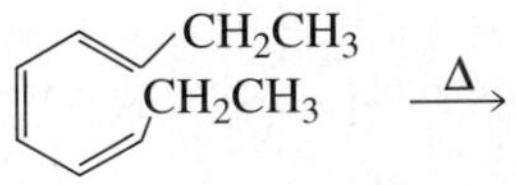

d.

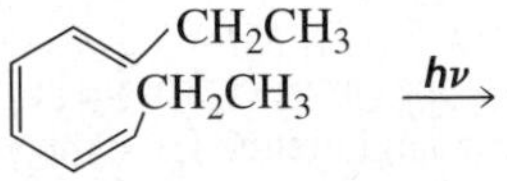

25. Account for the difference in the products of the following reactions:

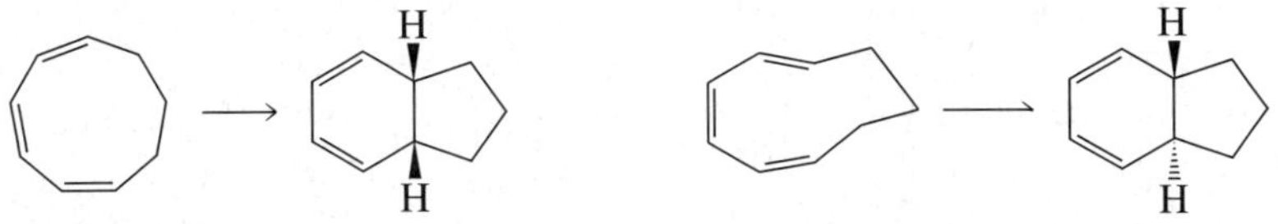

26. Show how norbornane could be prepared from cyclopentadiene.

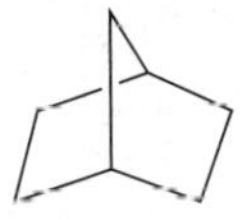

norbornane

27. Give the product formed when each of the following compounds undergoes an electrocyclic reaction

a. under thermal conditions.

b. under photochemical conditions.

1.

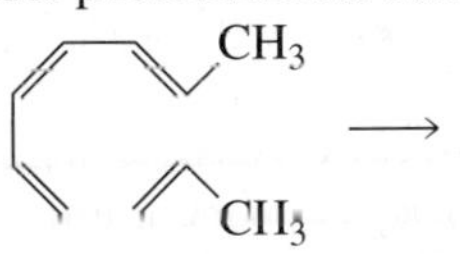

2.

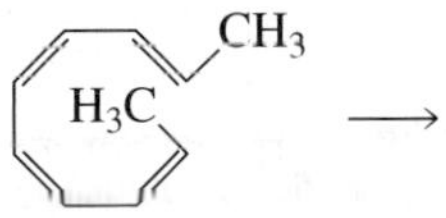

28. Give the product of each of the following reactions:

a.

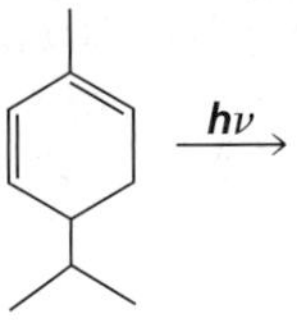

c.

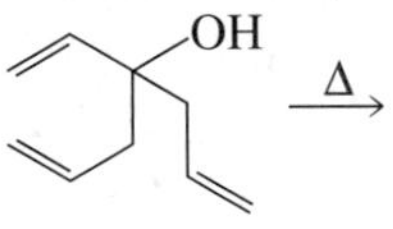

e.

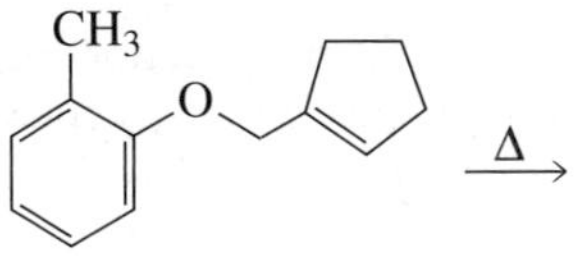

b.

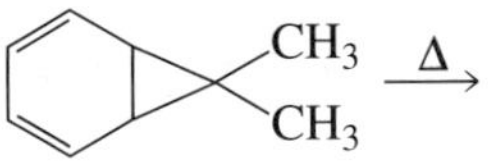

d.

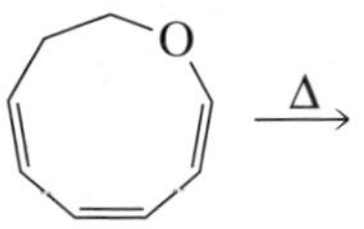

29. Which is the product of the following [1,3] sigmatropic rearrangement, A or B?

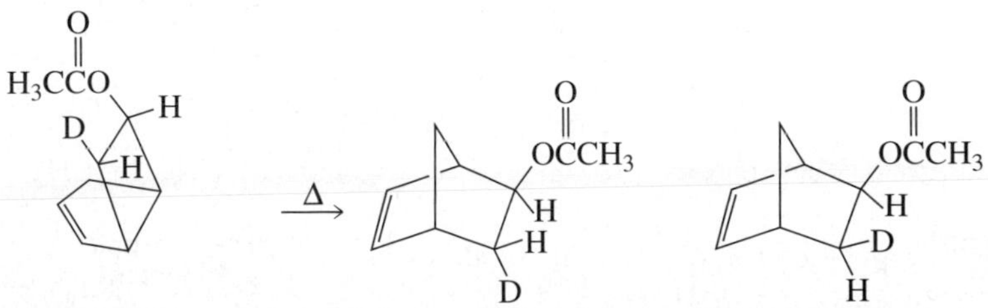

A   B

30. Dewar benzene is a highly strained isomer of benzene. In spite of its thermodynamic instability, it is very stable kinetically. It will rearrange to benzene, but only if heated to a very high temperature. Why is it kinetically stable?

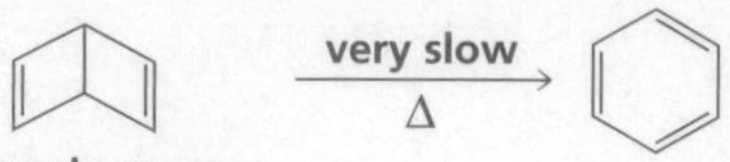

Dewar benzene

31. If the following compounds are heated, one will form one product from a [1,3] sigmatropic rearrangement and the other will form two products from two different [1,3] sigmatropic rearrangements. Give the products of the reactions.

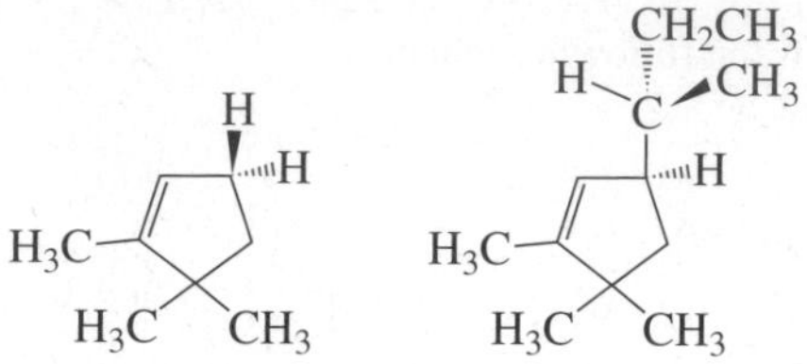

32. When the following compound is heated, a product is formed that shows an infrared absorption band at 1715 cm^{-1}. Draw the structure of the product.

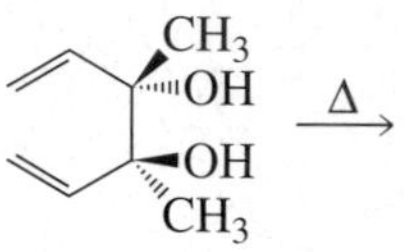

33. Two products are formed in the following [1,7] sigmatropic rearrangement, one due to hydrogen migration and the other to deuterium migration. Show the configuration of the products by replacing A and B with the appropriate atoms (H or D).

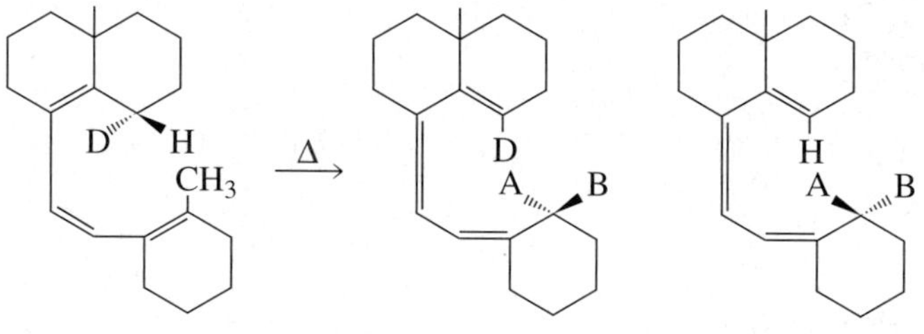

34. a. Propose a mechanism for the following reaction. (*Hint:* An electrocyclic reaction is followed by a Diels–Alder reaction.)
   b. What would be the reaction product if *trans*-2-butene were used instead of ethene?

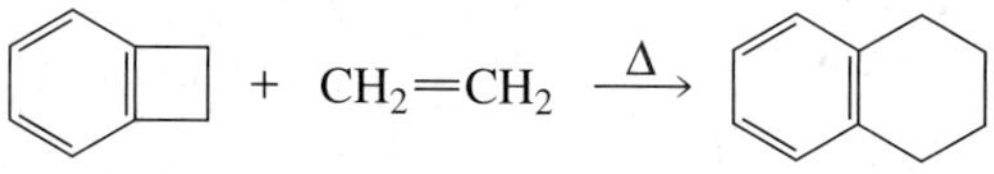

35. Explain why two different products are formed from disrotatory ring closure of (2E,4Z,6Z)-octatriene, but only one product is formed from disrotatory ring closure of (2E,4Z,6E)-octatriene.

36. Give the product of each of the following sigmatropic rearrangements:

a.   [3,3] sigmatropic rearrangement Δ

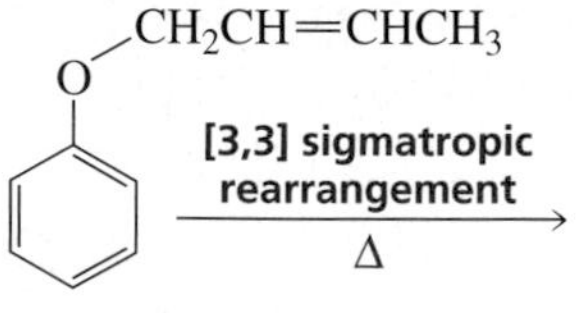

c.   [5,5] sigmatropic rearrangement Δ

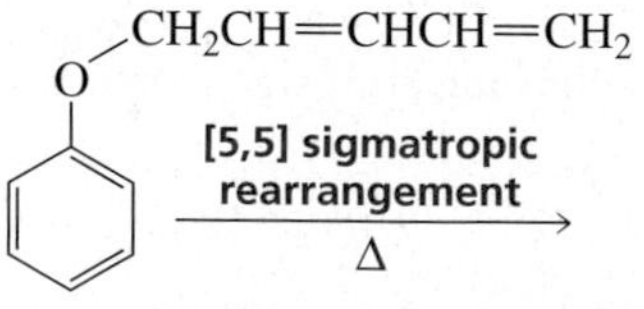

b.   [3,3] sigmatropic rearrangement Δ

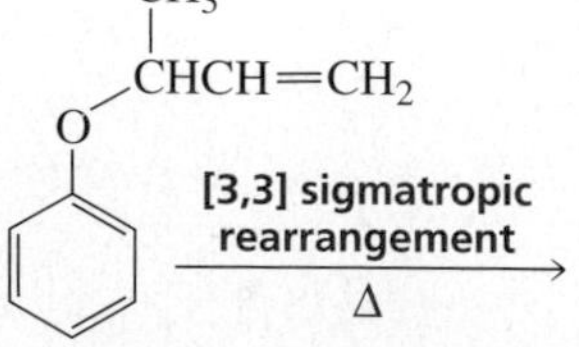

d.   [5,5] sigmatropic rearrangement Δ

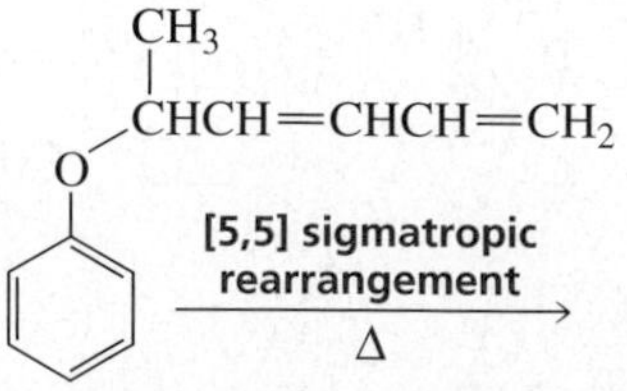

37. *cis*-3,4-Dimethylcyclobutene undergoes thermal ring opening to form the two products shown. One of the products is formed in 99% yield, the other in 1% yield. Which is which?

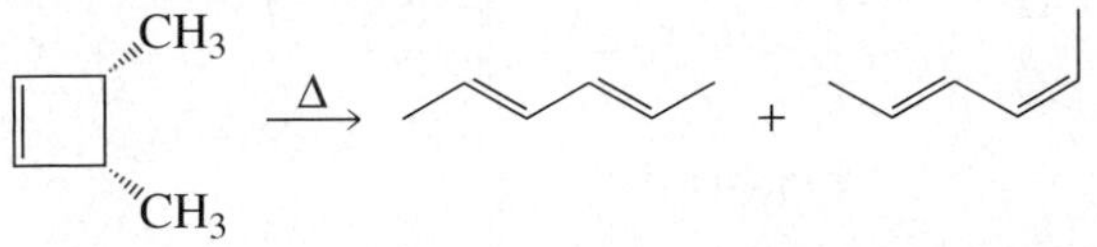

38. If isomer A is heated to about 100 °C, a mixture of isomers A and B is formed. Explain why there is no trace of isomer C or D.

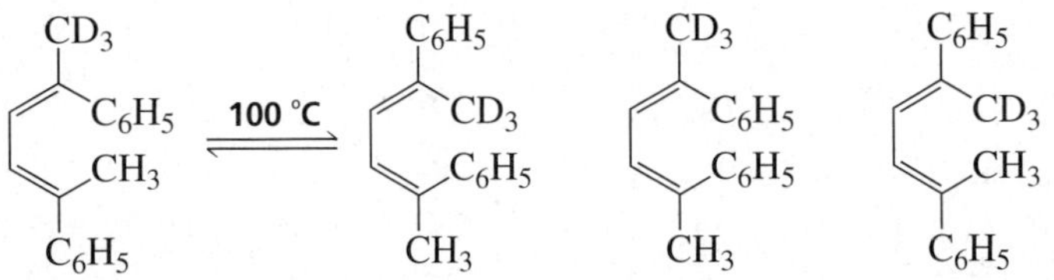

39. Propose a mechanism for the following reaction:

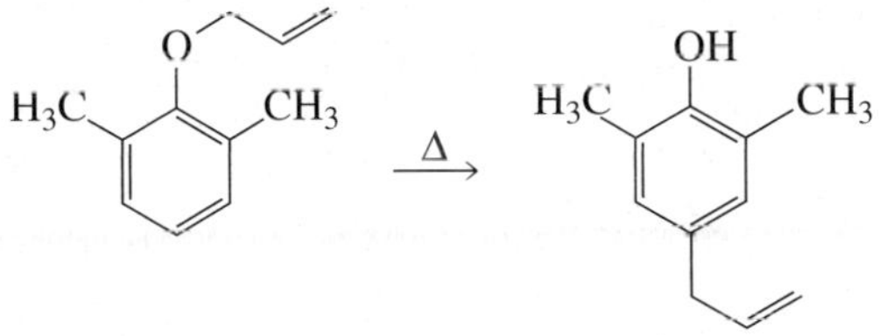

40. Compound A will not undergo a ring-opening reaction under thermal conditions, but compound B will. Explain.

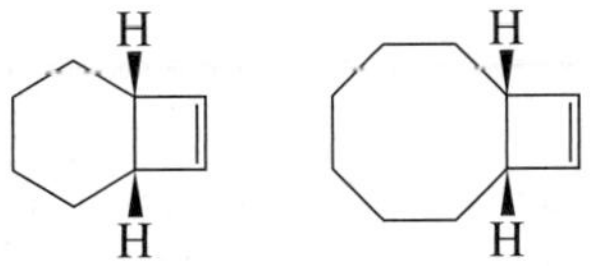

41. Professor Perry C. Click found that heating any one of the following isomers resulted in scrambling of the deuterium to all three positions on the five-membered ring. Propose a mechanism to account for this observation.

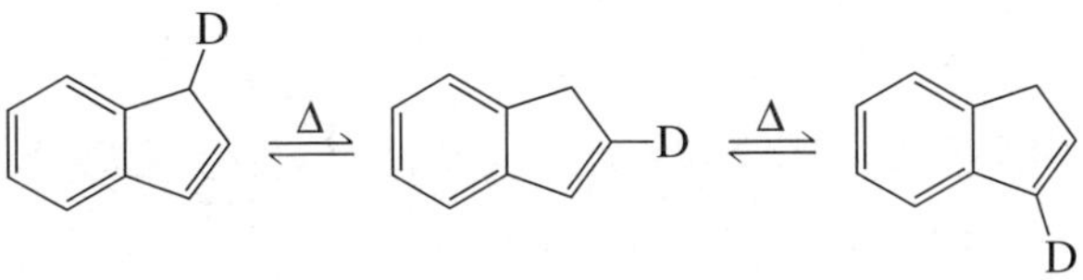

42. How could the following transformation be carried out using only heat or light?

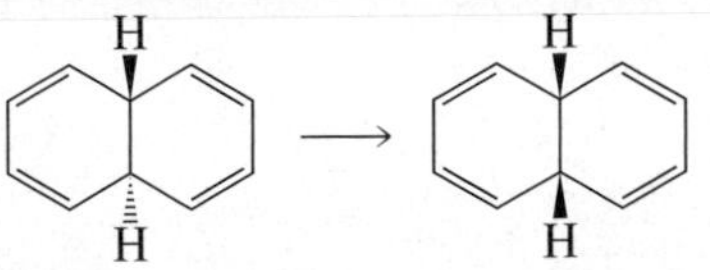

43. Show the steps involved in the following reaction:

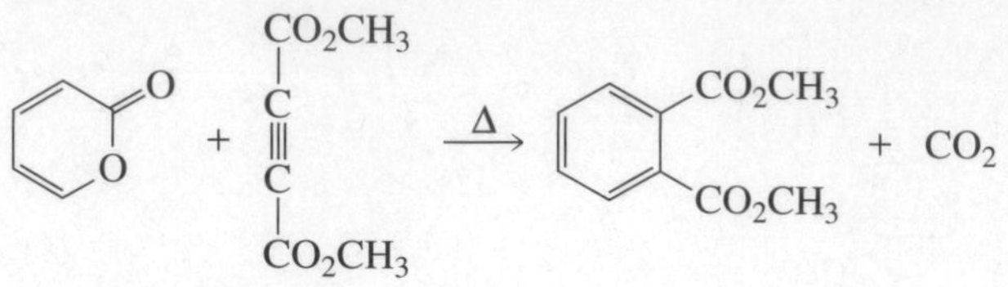

44. Propose a mechanism for the following reaction:

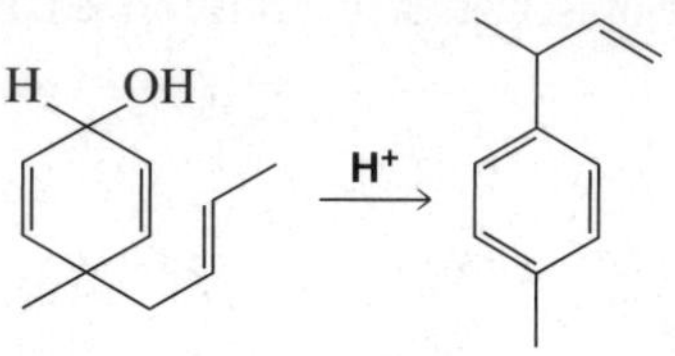

CHAPTER 30

The Organic Chemistry of Drugs
Discovery and Design

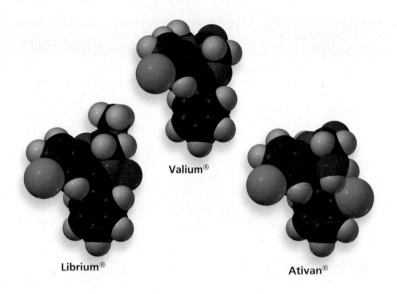

Valium®

Librium®

Ativan®

BUILDING ON FUNDAMENTALS

SECTION 30.2	Amides are less easily hydrolyzed than are esters (16.6).
SECTION 30.2	The rate of a reaction can be decreased by steric hindrance at the reaction site (8.2).
SECTION 30.6	The interactions of drugs with their receptors are the same as the interactions seen in other examples of molecular recognition: hydrogen bonding, electrostatic attractions, and hydrophobic (van der Waals) interactions (2.9, 22.15).
SECTION 30.6	Receptors typically recognize only one enantiomer (5.21).
SECTION 30.7	The ring strain in four-membered lactams causes these compounds to be more reactive than other amides (16.17).
SECTION 30.12	In DNA replication, the 3'-OH group of the last-incorporated nucleotide attacks the α-phosphorus of the next nucleotide to be incorporated into the chain (27.3).

A **drug** is any substance that is absorbed by the body and then changes or enhances a physical or psychological function. A drug can be a gas, a liquid, or a solid: it can have a simple structure or a complicated one. Drugs have been used by humans for thousands of years to alleviate pain and illness. By trial and error, people learned which herbs, berries, roots, and bark could be used for medicinal purposes. The knowledge about natural medicines was passed down from generation to generation without any understanding of how the drugs actually worked. Those who dispensed the drugs—medicine men and women, shamans, and witch doctors—were important members of every civilization. However, the drugs available to them were just a small fraction of the drugs available to us today.

Even at the beginning of the twentieth century, there were no drugs for the dozens of functional, degenerative, neurological, and psychiatric disorders from which people suffer; no hormone therapies; no vitamins; and—most significantly—no effective drug for the cure of any infectious disease. Local anesthetics had just been discovered, and there were only two analgesics to relieve major pain. One reason that families had many children was because some of the children were bound to succumb to childhood diseases. Life spans were generally short. In 1900, for example, the average life expectancy in the United States was 46 years for a man and 48 years for a woman. In 1920, about 80 of every 100,000 children died before their fifteenth birthday—most as a result of infections in their first year of life. Now there is a drug for almost every disease, and the effectiveness of this medical arsenal is reflected in current life expectancies: 74 years for a man and 79 years for a woman. Now only about four of every 100,000 children die before the age of 15, mainly from cancer, accidents, and inherited diseases.

The shelves of a typical modern pharmacy are stocked with almost 2000 preparations, most of which contain a single active ingredient, usually an organic compound. These medicines can be swallowed, injected, inhaled, or absorbed through the skin. In 2004, more than 3.1 billion prescriptions were dispensed in the United States. The most widely *prescribed drugs* in the United States are listed in Table 30.1, in order

Table 30.1 The Most Widely Prescribed Drugs in the United States in 2004 in Decreasing Order of the Number of Prescriptions Written

Brand name	Generic name	Structure	Use
Hydrocodone with APAP	hydrocodone with APAP	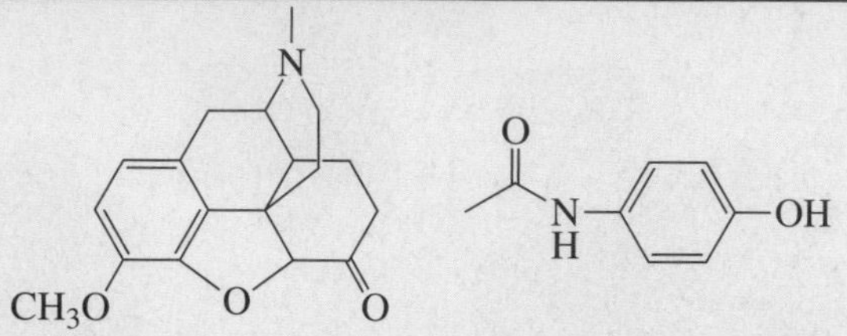	analgesic
Lipitor®	atorvastatin	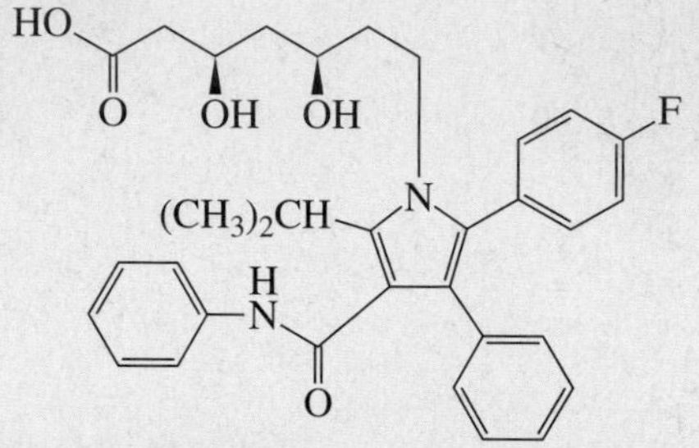	statin (cholesterol-reducing drug)
Lisinopril®	lisinopril	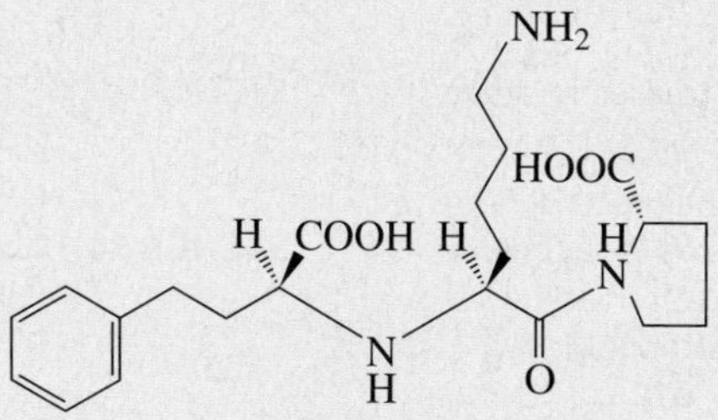	antihypertensive
Tenormin®	atenolol	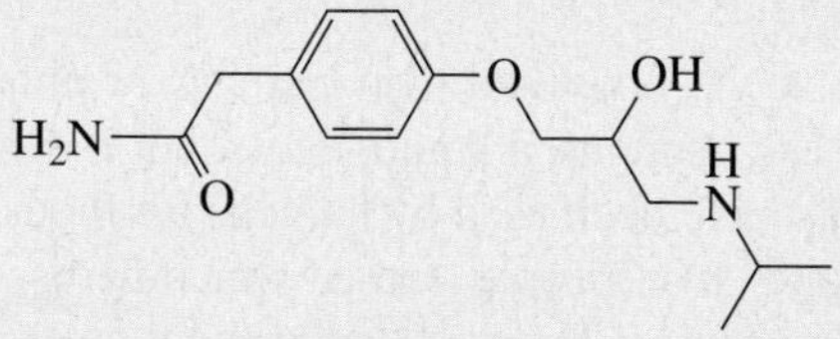	β-adrenergic blocking agent (antiarrhythmic, antihypertensive)
Synthroid®	levothyroxine	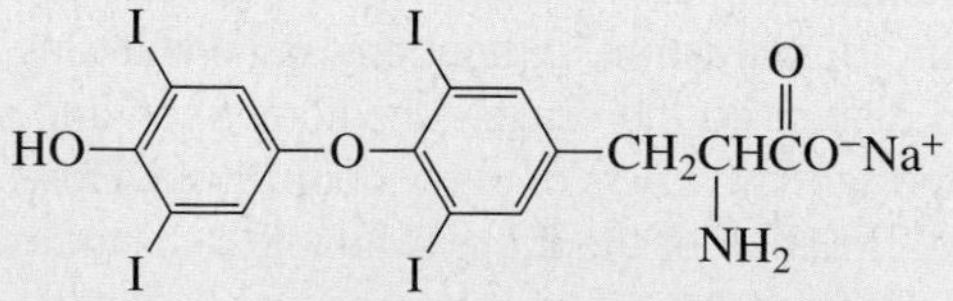	treatment of hypothyroidism
Trimox®	amoxicillin	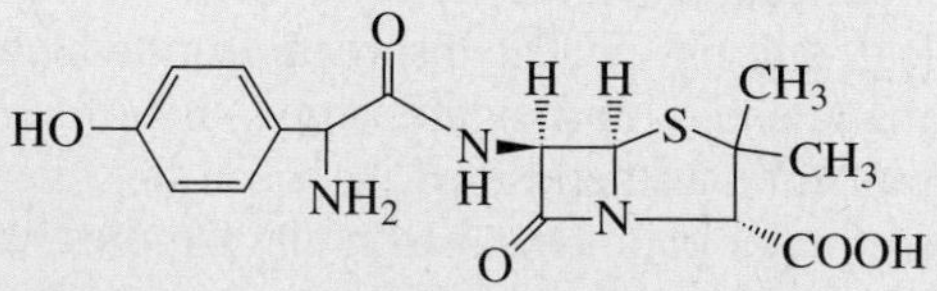	antibiotic
Hydrodiuril®	hydrochlorothiazide (HCTZ)	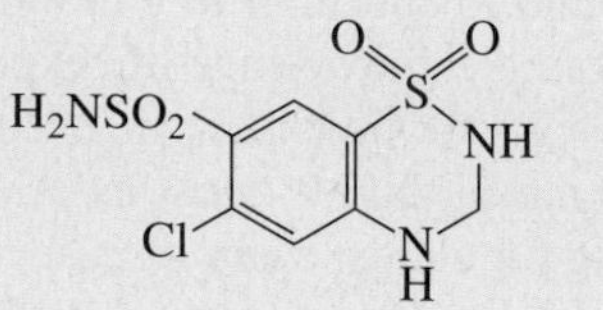	diuretic

Table 30.1 Continued

Brand name	Generic name	Structure	Use
Zithromax®	azithromycin	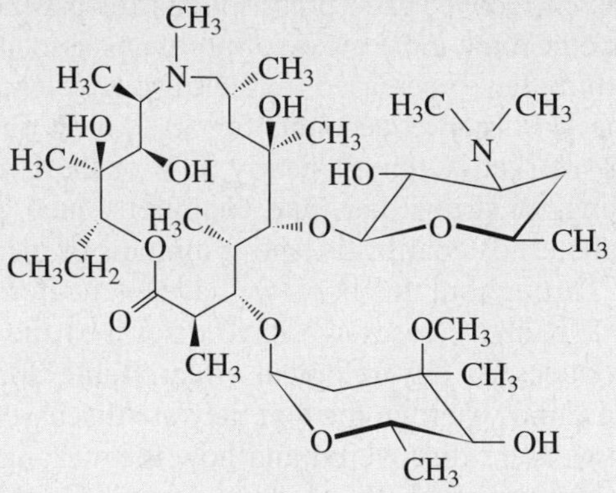	antibiotic
Lasix®	furosemide	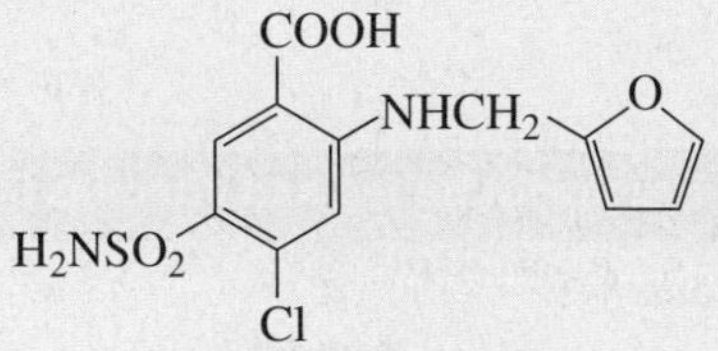	diuretic
Norvasc®	amlodipine	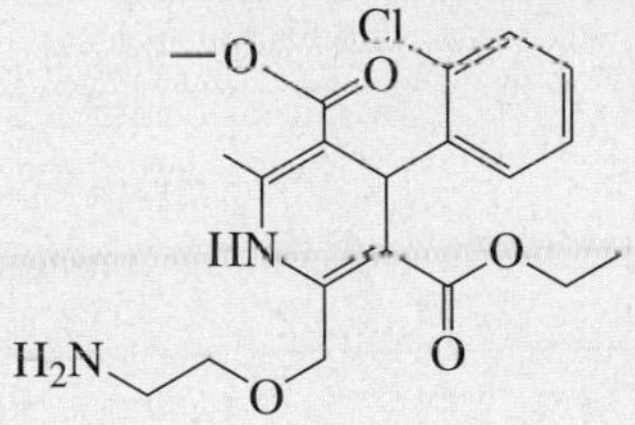	calcium channel blocker (antihypertensive)
Toprol-XL®	metoprolol	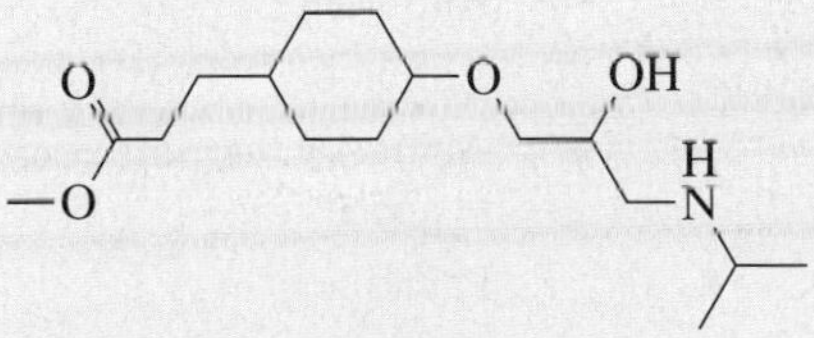	β-adrenergic blocking agent (antiarrhythmic, antihypertensive)
Xanax®	alprazolam	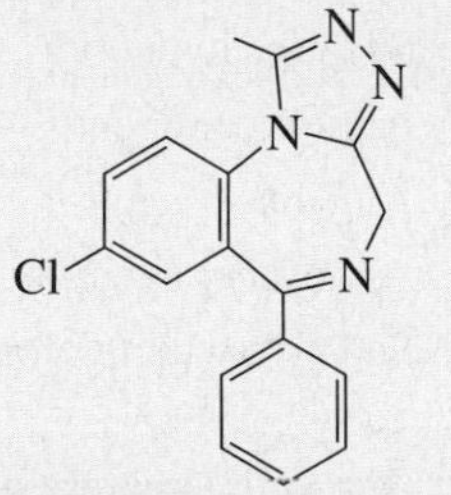	tranquilizer
Albuterol®	albuterol	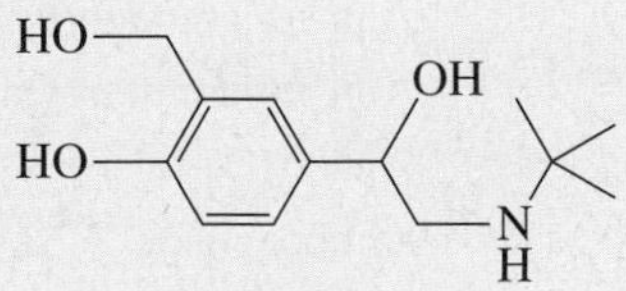	bronchodilator
Zoloft®	sertraline	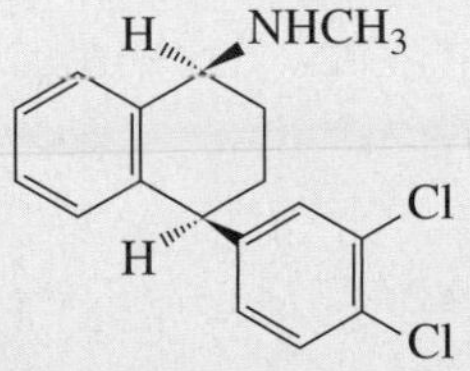	antidepressant

of the number of prescriptions written. Worldwide, antibiotics are the most widely prescribed class of drugs. In the developed world, heart drugs are the most widely prescribed, partly because they are generally taken for the remainder of the patient's life. In recent years, prescriptions for psychotropic drugs have decreased as doctors have become more aware of problems associated with addiction, and prescriptions for asthma have increased, reflecting a greater incidence (or awareness) of the disease. The U.S. market accounts for 50% of all pharmaceutical sales worldwide. Now that this market is aging—nearly 30% of the U.S. population is over 50—the demand is rising for drugs that treat conditions such as high cholesterol levels, hypertension, diabetes, osteoarthritis, and symptoms of menopause.

Throughout this book we have encountered many drugs, vitamins, and hormones, and in many cases we have discussed the mechanism by which each compound produces its physiological effect. Table 30.2 lists some of these compounds, their uses, and where in the text they are discussed. Now we will take a look at how some drugs were discovered and how they are named, and we will examine some of the techniques currently used by scientists in their search for new drugs.

Table 30.2 Drugs, Hormones, and Vitamins Discussed in Earlier Chapters		
Drug, hormone, or vitamin	**Remark**	**Section or chapter**
Sulfa drugs	The first antibiotics	Section 24.8
Tetracycline	Broad-spectrum antibiotic	Section 5.3
Puromycin	Broad-spectrum antibiotic	Section 27.8
Nonactin	Ionophorous antibiotic	Section 10.10
Penicillin	Antibiotic	Sections 16.4 and 16.17
Gentamicin	Antibiotic	Section 21.18
Gramicidin S	Antibiotic	Section 22.9
Aspirin	Analgesic, anti-inflammatory agent	Sections 18.10 and 26.5
Enkephalins	Painkillers	Section 22.9
Diethyl ether	Anesthetic	Section 10.7
Sodium pentothal	Sedative hypnotic	Section 10.7
5-Fluorouracil	Anticancer agent	Section 24.8
Methotrexate	Anticancer agent	Section 24.8
Taxol	Anticancer agent	Section 17.15
Thalidomide	Sedative with teratogenic side effects	Section 5.21
Ibuprofen	Anti-inflammatory agent	Section 26.5
Naproxen	Anti-inflammatory agent	Section 26.5
Celebrex	Anti-inflammatory agent	Section 26.5
AZT*	Anti-AIDS agent	page 1229
Warfarin	Anticoagulant	Section 24.9
Epinephrine	Vasoconstrictor, bronchodilator	Section 8.12
Vitamins		Chapter 24, Sections 11.10, 18.11, 21.8, and 26.7
Hormones		Sections 16.16, 22.9, and 26.9–26.11

* 3'-Azido-2'-deoxythymidine

30.1 **Naming Drugs**

The most accurate names for drugs are the chemical names that define their structures. However, these names are too long and complicated to appeal to the general public, or even to their physicians. The pharmaceutical company therefore chooses a brand name for any drug it develops. A **brand name** identifies the drug as a commercial product and distinguishes it from other products. Only the company that holds the patent can use the brand name for marketing the product. It is in the company's best interest to choose a name that is easy to remember and pronounce so that when the patent expires, the public will continue to request the drug by that name.

Each drug is also given a **generic name** that any pharmaceutical company can use to identify the product. The pharmaceutical company that developed the drug is allowed to choose the generic name from a list of 10 names provided by an independent group. It is the company's best interest to choose the generic name that is hardest to pronounce and least likely to be remembered, so that physicians and consumers will continue to use the familiar brand name. Brand names must always be capitalized, whereas generic names are not capitalized.

Drug manufacturers are permitted to patent and retain exclusive rights to the drugs they develop. A patent is valid for 20 years. Once the patent expires, other drug companies can market the drug under the generic name or under their own brand name, which is called a branded generic name and is a generic name that only they can use. For example, the antibiotic ampicillin is sold as Penbritin by the company that held the original patent. Now that the patent has expired, the drug is sold by other companies as Ampicin, Ampilar, Amplital, Binotal, Nuvapen, Pentrex, Ultrabion, Viccillin, and 30 other branded generic names.

The over-the-counter drugs that line the shelves of drugstores are available without prescription. They are often mixtures containing one or more active ingredients (generic or brand-name drugs), plus sweeteners and inert fillers. For example, the preparations called Advil (Whitehall Laboratories), Motrin (Upjohn), and Nuprin (Bristol-Meyers Squibb) all contain ibuprofen, a mild analgesic and anti-inflammatory drug. Ibuprofen was patented in Britain in 1964 by Boots, Inc., and the U.S. Food and Drug Administration (FDA) approved its use as a nonprescription drug in 1984.

30.2 **Lead Compounds**

The goal of the medicinal chemist is to find compounds that have potent effects on specific diseases while producing no or minimal side effects. In other words, a drug must react selectively with its target and have few if any negative effects. A drug must get to the right place in the body, at the right concentration, and at the right time. Therefore, a drug must have the appropriate solubility and other physical and chemical properties to allow it to be transported to the target cell. For example, if a drug is to be taken orally, it must be insensitive to the acid conditions of the stomach, and it also must resist enzymatic degradation by the liver before it reaches its target. Finally, it must eventually be excreted as is or be degraded to harmless compounds that can be excreted.

Medicinal agents used by humans since ancient times provided the starting point for the development of our current arsenal of drugs. The active ingredients were isolated from the herbs, berries, roots, and bark used in traditional medicine. Foxglove, for instance, furnished digitoxin, a cardiac stimulant. The bark of the cinchona tree yielded quinine for relief from malaria. Willow bark contains salicylates used to control fever and pain. A milky fluid obtained from the oriental opium poppy provided morphine for severe pain and codeine for the control of a cough. By 1882, more than 50 different herbs were commonly used to make medicines. Many were grown in the gardens of religious establishments that treated the sick.

Foxglove

Scientists still search the world for plants and berries and the oceans for flora and fauna that might yield new medicinal compounds. Taxol, a compound isolated from the bark of the Pacific yew tree, is a relatively recently recognized anticancer agent (Section 17.15). Almost half of the new drugs approved in recent years were natural products or derived from natural products.

Once a naturally occurring drug is isolated and its structure determined, it can serve as a prototype in a search for other biologically active compounds. The prototype is called a **lead compound** (that is, it plays a leading role in the search). Analogs of the lead compound are synthesized and tested to see if they are more effective than the lead compound or have fewer side effects. An analog may have a different substituent than the lead compound, a branched chain instead of a straight chain, a different functional group, or a different ring system.

Coca leaves

30.3 Molecular Modification

Producing analogs by changing the structure of a lead compound is called **molecular modification**. In a classic example of this process, a number of synthetic local anesthetics were developed from the lead compound, cocaine. Cocaine comes from the leaves of *Erythroxylon coca*, a bush native to the highlands of the South American Andes. Cocaine is a highly effective local anesthetic, but it produces undesireable effects on the central nervous system (CNS), ranging from initial euphoria to severe depression. By dissecting the cocaine molecule step by step—removing the methoxycarbonyl group and cleaving the seven-membered-ring system—scientists identified the portion of the molecule that carries the local anesthetic activity but does not induce the damaging CNS effects. This knowledge provided an improved lead compound— an ester of benzoic acid, with a terminal tertiary amino group on the alcohol component of the ester.

cocaine
lead compound

improved lead compound

Hundreds of related esters were then synthesized: esters with substituents on the aromatic ring, esters with alkyl groups bonded to the nitrogen, and esters with the length of the connecting alkyl chain modified. Successful anesthetics obtained through this process of molecular modification were benzocaine, a topical anesthetic, and procaine, commonly known by the brand name Novocain.

anesthetics

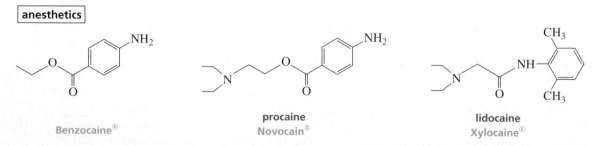

Benzocaine®

procaine
Novocain®

lidocaine
Xylocaine®

Because the ester group of procaine is hydrolyzed relatively rapidly by enzymes that catalyze ester hydrolysis, procaine has a short half-life. Researchers therefore focused next on synthesizing compounds with less easily hydrolyzed amide groups (Section 16.16). In this way, lidocaine, one of the most widely used injectable anesthetics, was discovered. The rate at which lidocaine is hydrolyzed is further decreased by its two *ortho*-methyl substituents, which sterically hinder the reactive carbonyl group.

Later, physicians recognized that the action of an anesthetic administered in vivo (in a living organism) could be lengthened considerably if it were administered along with epinephrine. Because epinephrine is a vasoconstrictor, it reduces the circulation of blood, allowing the drug to remain at its targeted site for a longer period.

In screening the structurally modified cocaine analogs for biological activity, scientists were surprised to find that replacing the ester linkage of procaine with an amide linkage led to a compound—procainamide hydrochloride—that had activity as a cardiac depressant as well as activity as a local anesthetic. Procainamide hydrochloride is currently used clinically as an antiarrhythmic.

procainamide hydrochloride

Morphine, the most widely used analgesic for severe pain, is the standard by which other painkilling medications are measured. Although scientists have learned how to synthesize morphine, all commercial morphine is obtained from opium, a milky fluid exuded by a species of poppy (see page 3). Morphine occurs in opium at concentrations as high as 10%. Opium was used for its analgesic properties as early as 4000 B.C. In Roman times both opium use and opium addiction were widespread. Methylating one of the OH groups of morphine produces codeine, which has one-tenth the analgesic activity of morphine but profoundly inhibits the cough reflex. Although 3% of opium is codeine, most commercial codeine is obtained by methylating morphine. Putting an acyl group on (acetylating) one of the OH groups of morphine produces a compound with a similar reduced potency. Acetylating both OH groups forms heroin, which is much more potent than morphine. It is less polar than morphine and therefore crosses the blood–brain barrier faster, resulting in a more rapid "high." Heroin has been banned in most countries because it is widely abused. It is synthesized by using acetic anhydride to acetylate morphine. Therefore, both heroin and acetic acid are formed as products. Drug enforcement agencies use dogs trained to recognize the pungent odor of acetic acid.

analgesics

morphine   codeine   heroin

Molecular modification of codeine led to dextromethorphan, the active ingredient in most cough medicines. Etorphine was synthesized as a result of scientists realizing that analgesic potency was related to the drug's ability to bind hydrophobically to the opiate receptor (Section 30.6). Etorphine is about 2000 times more potent than morphine, but it is not safe for use by humans. It has been used to tranquilize elephants and other large animals. Pentazocine is useful in obstetrics because it dulls the pain of labor but does not depress the respiration of the infant as morphine does.

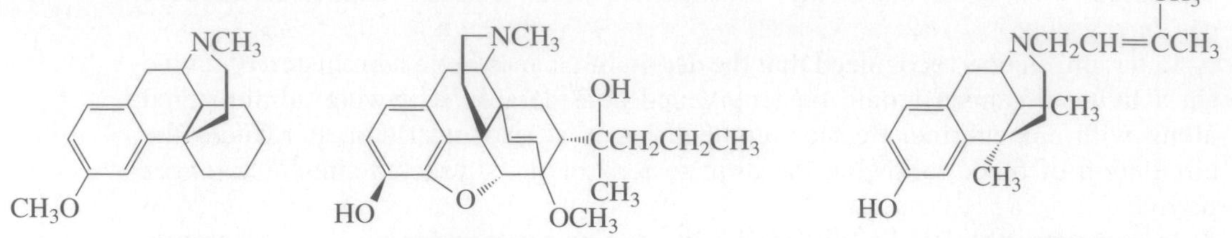

dextromethorphan   **etorphine**   **pentazocine**

Methadone was synthesized by German scientists in 1944 in an attempt to find a drug to treat muscle spasms. (It was originally called "Adolphile" in honor of Adolph Hitler.) No one recognized until 10 years later—after building molecular models—that methadone and morphine have similar shapes. Unlike morphine, however, methadone can be administered orally. Methadone also has a considerably longer half-life (24–26 hours) than morphine (2–4 hours). Repeated doses of methadone have cumulative effects, so it can be used in lower doses and at longer intervals. Because of these properties, methadone is used to treat chronic pain and the withdrawal symptoms of heroin addiction. Reducing the carbonyl group of methadone and acetylating it forms α-acetylmethadol. The levo ($-$) isomer of this compound can suppress withdrawal symptoms for 72 hours (Section 5.21). When the methadone analog Darvon (isomethadone) was introduced, it was initially thought to be the long-sought-after nonaddicting painkiller. However, it was later found to have no therapeutic advantage over less toxic and more effective analgesics.

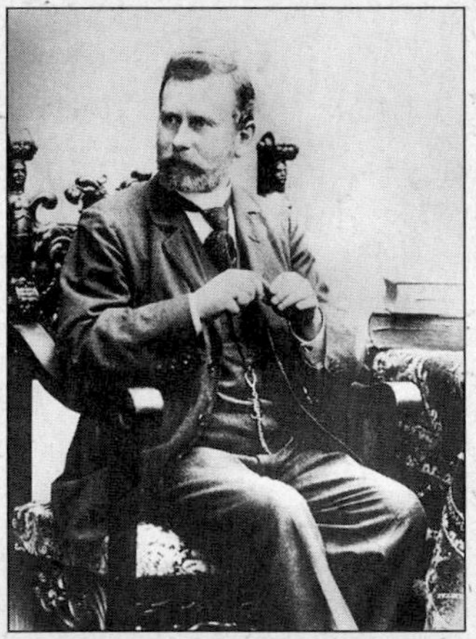

methadone α-**acetylmethadol**   **isomethadone**
Darvon®

Notice that morphine and all the compounds prepared by molecular modification of morphine have a structural feature in common: an aromatic ring attached to a quaternary carbon that is attached to a tertiary amine two carbons away.

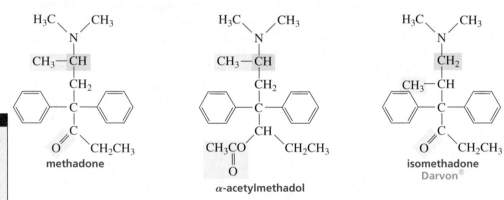

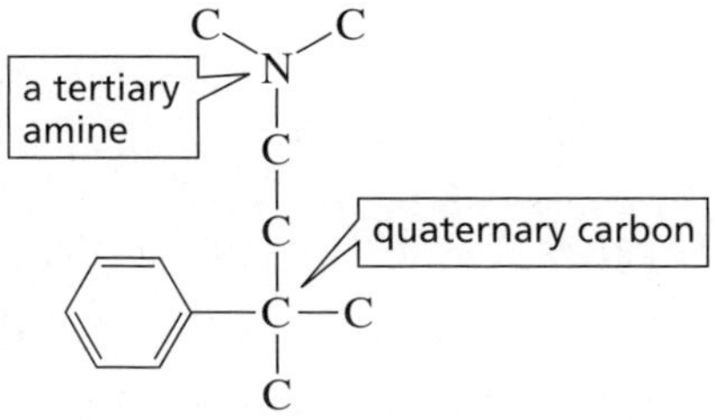

a tertiary amine

quaternary carbon

30.4 Random Screening

Most lead compounds are found by screening thousands of compounds randomly. A **random screen**, also known as a **blind screen**, is a search for a pharmacologically active compound done without the benefit of any information about what chemical structures might show activity. The first blind screen was carried out by Paul Ehrlich, who was searching for a "magic bullet" that would kill trypanosomes—the microorganisms that cause African sleeping sickness—without harming their human host. After testing more than 900 compounds against trypanosomes, Ehrlich tested some of his compounds against other bacteria. Compound 606 (salvarsan) was found to be dramatically effective against the microorganisms that cause syphilis, an often deadly disease and incurable at the time, that was exacting a toll on public health, much in the way HIV is doing today. Salvarsan was the most effective drug against syphilis from its discovery in 1909 until penicillin became available in the 1940s.

An important part of random screening is recognizing an effective compound. This requires the development of an assay for the desired biological activity. Some assays can be done in vitro ("in glass," that is, in a test tube or flask)—for example, searching for a compound that will inhibit a particular enzyme. Others are done in vivo (in a living organism)—for instance, searching for a compound that will save a mouse from a lethal dose of a virus. One problem with in vivo assays is that drugs can be metabolized differently by different animals (Section 11.10). Thus, an effective drug in a mouse may be less effective or even useless in a human. Another problem is the guesswork involved in regulating the dosages of both the virus and the drug. If the dosage of the virus is too high, it might kill the mouse in spite of the presence of a biologically active compound that could save the animal. If the dosage of the potential drug is too high, the drug might kill the mouse even though a lower dosage would have saved it.

The observation that azo dyes effectively dyed wool fibers (animal protein) gave scientists the idea that such dyes might selectively bind to bacterial proteins, too, and in the process perhaps harm the bacteria. Well over 10,000 dyes were screened in vitro in antibacterial tests, but none showed any antibiotic activity. At that point, some scientists suggested screening the dyes in vivo, saying that what physicians really needed were antibacterial agents that would cure infections in humans and animals, not in test tubes.

In vivo studies were therefore done in mice that had been infected with a bacterial culture. Now the luck of the investigators improved. Several dyes turned out to counteract gram-positive infections. The least toxic of these, Prontosil (a bright red dye), became the first drug to treat bacterial infections (Section 24.8).

Gram-negative bacteria have an outer membrane that covers their cell walls; gram-positive bacteria do not have an outer membrane, but tend to have thicker and more rigid cell walls. The two types can be distinguished by a stain, invented by Hans Christian Gram, that turns gram-negative bacteria pink and gram-positive bacteria purple.

Prontosil®

The fact that Prontosil was inactive in vitro but active in vivo should have been recognized as a sign that the dye was converted to an active compound by the mammalian organism, but this did not occur to the bacteriologists, who were content to have found a useful antibiotic. When scientists at the Pasteur Institute later investigated Prontosil, they noted that mice given the drug did not excrete a red compound. Urine analysis showed that the mice excreted *para*-acetamidobenzenesulfonamide, a colorless compound, instead. Chemists knew that anilines are acetylated in vivo, so they prepared the nonacetylated compound (sulfanilamide). When sulfanilamide was tested in mice infected with streptococcus, all the mice were cured, whereas untreated control mice died. Protonsil is an example of a **prodrug**, a compound that becomes an effective drug only after it undergoes a reaction in the body. Sulfanilamide was the first of the sulfa drugs—the first class of antibiotics (see page 732).

para-acetamidobenzenesulfonamide

para-aminobenzenesulfonamide
sulfanilamide

BIOGRAPHY

Gerhard Domagk (1895–1964) *was a research scientist at I. G. Farbenindustrie, a German manufacturer of dyes and other chemicals. He carried out studies that showed Prontosil to be an effective antibacterial agent. His daughter, who was dying of a streptococcal infection as a result of cutting her finger, was the first patient to receive the drug and be cured by it (1935). Prontosil received wider fame when it was used in 1936 to save the life of twenty-two-year-old Franklin D. Roosevelt Jr., son of the U.S. president. Domagk received the Nobel Prize in physiology or medicine in 1939, but Hitler did not allow Germans to accept Nobel Prizes because Carl von Ossietsky, a German who was in a concentration camp, had been awarded the Nobel Prize for peace in 1935. Domagk was eventually able to accept the prize in 1947 but, because of the time that had elapsed, was not given the monetary award.*

Sulfanilamide acts by inhibiting the bacterial enzyme that incorporates *para*-aminobenzoic acid into folic acid (Section 24.8). Thus, sulfanilamide is a **bacteriostatic drug**, a drug that inhibits the further growth of bacteria, rather than a **bactericidal drug**, a drug that kills the bacteria. Sulfanilamide inhibits the enzyme because the sulfonamide and carboxylic acid groups have similar sizes. Many other successful drugs have since been designed by this strategy of isosteric (like-size) replacements (Sections 24.8 and 24.9).

30.5 Serendipity in Drug Development

Many drugs have been discovered accidentally. Nitroglycerin, the drug used to relieve the symptoms of angina pectoris (heart pain), was discovered when workers handling nitroglycerin in the explosives industry experienced severe headaches. Investigation revealed that the headaches were caused by nitroglycerin's ability to produce a marked dilation of blood vessels. The pain associated with an angina attack results from the inability of the blood vessels to supply the heart adequately with blood. Nitroglycerin relieves the discomfort by dilating cardiac blood vessels.

$$CH_2-ONO_2$$
$$CH-ONO_2$$
$$CH_2-ONO_2$$
nitroglycerin

The tranquilizer Librium is another drug that was discovered accidentally. Leo Sternbach synthesized a series of quinazoline 3-oxides, but none of them showed any pharmacological activity. One of the compounds was not submitted for testing because it was not the quinazoline 3-oxide he had set out to synthesize. Two years after the project was abandoned, a laboratory worker came across this compound while cleaning up the lab, and Sternbach decided that he might as well submit it for testing before it was thrown away. The compound was shown to have tranquilizing properties and, when its structure was investigated, was found to be a benzodiazepine 4-oxide. Methylamine, instead of displacing the chloro substituent in a substitution reaction to form a quinazoline 3-oxide, had added to the imine group of the six-membered ring, causing the ring to open and then reclose to form a seven-membered ring. The compound was given the brand name Librium when it was put into clinical use in 1960.

a quinazoline
3-oxide

a benzodiazepine 4-oxide
chlordiazepoxide
Librium® (1960)

Librium was structurally modified in an attempt to find other tranquilizers. One successful modification produced Valium, a tranquilizer almost 10 times more potent than Librium. Currently, there are eight benzodiazepines in clinical use as tranquilizers in the United States and some 15 others abroad. Xanax is one of the most widely prescribed medications of any kind (Table 30.1); Rohypnol is one of the so-called date-rape drugs.

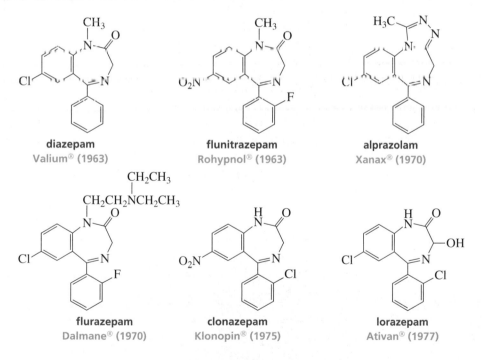

diazepam
Valium® (1963)

flunitrazepam
Rohypnol® (1963)

alprazolam
Xanax® (1970)

flurazepam
Dalmane® (1970)

clonazepam
Klonopin® (1975)

lorazepam
Ativan® (1977)

Viagra is a recent example of serendipity in drug development. Viagra was in clinical trials as a drug for heart ailments. When the clinical trials were canceled because Viagra turned out to be ineffective as a heart drug, those enrolled in the trials refused to return the remaining tablets. The pharmaceutical company then realized that the drug had other marketable effects.

30.6 **Receptors**

Many drugs exert their physiological effects by binding to a specific cellular binding site called a **receptor**, whose role is to trigger a response in the cell. That is why a small amount of a drug can bring about a measurable physiological effect. Because most receptors are chiral, different enantiomers of a drug can have different effects (Section 5.21). Drug receptors are often glycoproteins or lipoproteins. Some receptors are part of cell membranes, while others are found in the cytoplasm—the cell contents outside the nucleus. Nucleic acids—particularly DNA—also act as receptors for certain kinds of drugs. Because not all cells have the same receptors, drugs have considerable specificity. For example, epinephrine has intense effects on cardiac muscle, but almost no effect on muscle in other parts of the body.

A drug interacts with its receptor by means of the same kinds of bonding interactions—hydrogen bonding, electrostatic attractions, and hydrophobic (van der Waals) interactions—that we encountered in other examples of molecular recognition (Section 2.9 and 22.15). The most important factor in the interaction between a drug and a receptor is a snug fit: the greater the affinity of a drug for its binding site, the higher is the drug's potential biological activity. Two drugs for which DNA is a receptor are chloroquine (an antimalarial) and 3,6-diaminoacridine (an antibacterial). These flat cyclic compounds can slide into the DNA double helix between base pairs—like a card being inserted into a deck of playing cards—and interfere with the normal replication of DNA.

chloroquine

3,6-diaminoacridine

When scientists know something about the molecular basis of drug action—such as how a particular drug interacts with a receptor—they can design and synthesize compounds that might have a desired biological activity. For example, when excess histamine is produced by the body, it causes the symptoms associated with the common cold and allergic responses. This is thought to be the result of the protonated ethylamino group anchoring the histamine molecule to a negatively charged portion of the histamine receptor.

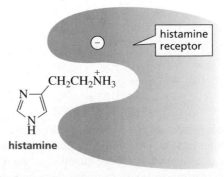

Drugs that interfere with the natural action of histamine—called antihistamines— bind to the histamine receptor but do not trigger the same response as histamine.

Like histamine, these drugs have a protonated amino group that binds to the receptor. The drugs also have bulky groups that keep the histamine molecule from approaching the receptor.

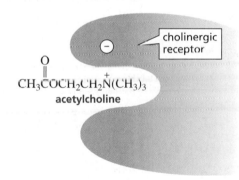

antihistamines

diphenhydramine
Benadryl®

promethazine
Promine®

promazine
Talofen®

Acetylcholine is a neurotransmitter that enhances peristalsis (the wavelike contractions of the digestive organs), wakefulness, and memory and is essential for nerve transmission. A deficiency of brain cell receptors that bind acetylcholine—cholinergic receptors—contributes to the characteristic loss of memory in Alzheimer's disease. Binding sites on cholinergic receptors are structurally similar to those that bind histamine. Therefore, antihistamines and cholinergic agents show overlapping activities. As a result, the antihistamine diphenhydramine has been used to treat insomnia and to combat motion sickness.

cholinergic receptor

$$CH_3\overset{O}{\overset{\|}{C}}OCH_2CH_2\overset{+}{N}(CH_3)_3$$
acetylcholine

Excess histamine production by the body also causes the hypersecretion of stomach acid by the cells of the stomach lining, leading to the development of ulcers. The antihistamines that block the histamine receptors, and thereby preventing the allergic responses associated with excess histamine production, have no effect on HCl production. This fact led scientists to conclude that a second kind of histamine receptor—a histamine H_2-receptor—triggers the release of acid into the stomach.

Because 4-methylhistamine was found to cause weak inhibition of HCl secretion, it was used as a lead compound. About 500 molecular modifications were performed over a 10-year period before a compound with a clinical effect on the histamine H_2-receptor was found. The first was Tagamet, followed by Zantac. Introduced in 1976, Tagamet was the first drug for the treatment of peptic ulcers. Previously, the only treatment was extensive bed rest, a bland diet, and antacids. Notice that steric blocking of the receptor site is not a factor in these compounds. Compared with the antihistamines, the effective antiulcer drugs have more polar rings and longer side chains.

BIOGRAPHY

Sir James W. Black *was born in Great Britain in 1924 and is a professor of physiology at King's College Hospital Medical School of the University of London. He led the project that discovered the H_2-receptor antagonists. For this and the development of β-blockers, he received the Nobel Prize in physiology and medicine in 1988.*

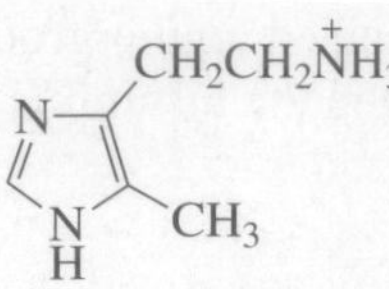

4-methylhistamine

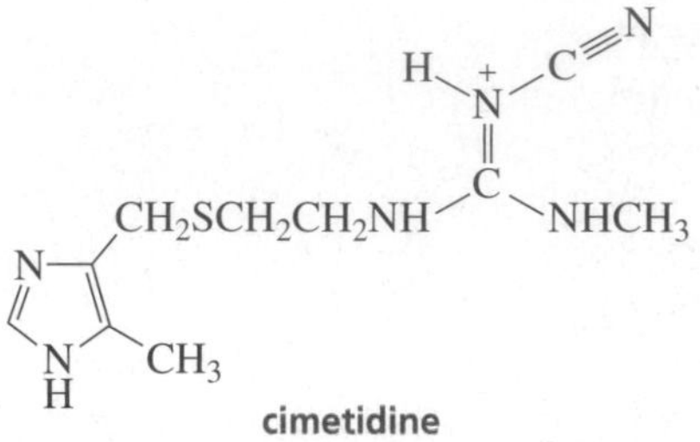

cimetidine
Tagamet®

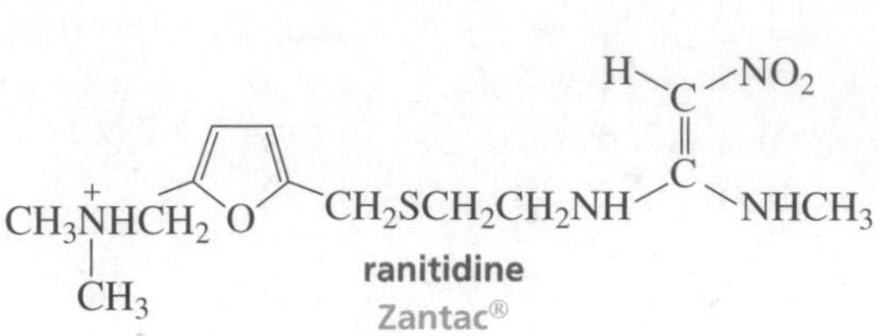

ranitidine
Zantac®

Tagamet has the same imidazole ring as 4-methylhistidine, but it has a sulfur atom and a functional group based on guanidine (Section 22.1). Zantac has a different heterocyclic ring and although its side chain is similar to that of Tagamet, it does not contain a guanidino group.

Research that implicated serotonin (a neurotransmitter like acetylcholine) in the generation of migraine attacks led to the development of drugs that bind to serotonin receptors. Sumatriptan, introduced in 1991, relieved not only the pain associated with migraines, but also many of migraine's other symptoms, including nausea and sensitivity to light and sound.

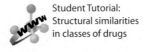

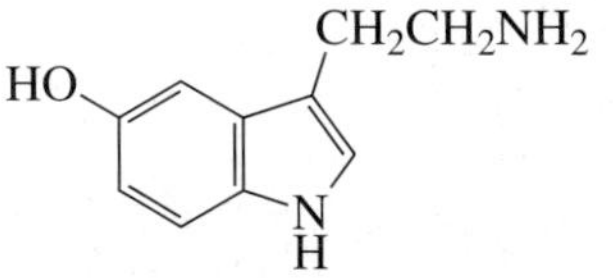

Student Tutorial:
Structural similarities
in classes of drugs

serotonin

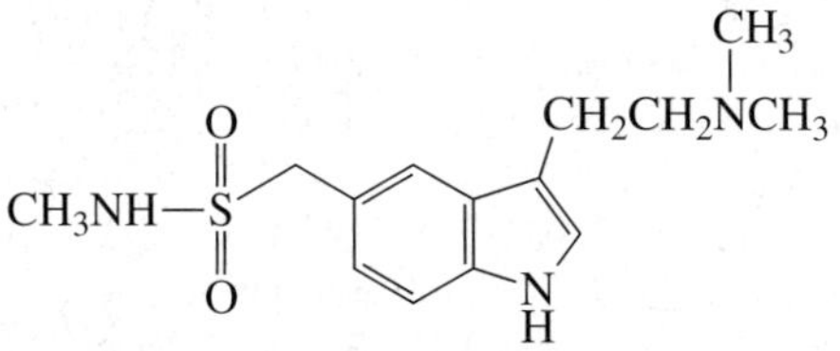

sumatriptan
Imitrex®

The success of sumatriptan spurred a search for other antimigraine agents by molecular modification, and three new triptans were introduced in 1997 and 1998. These second-generation triptans showed some improvements over sumatriptan—specifically, a longer half-life, reduced cardiac side effects, and improved hydrophobicity for CNS penetration.

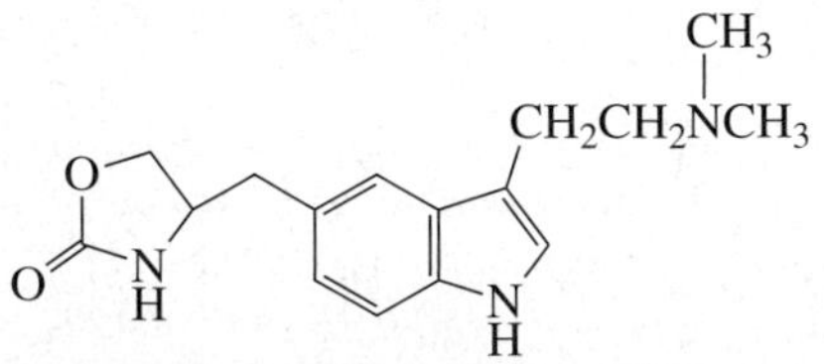

zolmitriptan
Zomig®

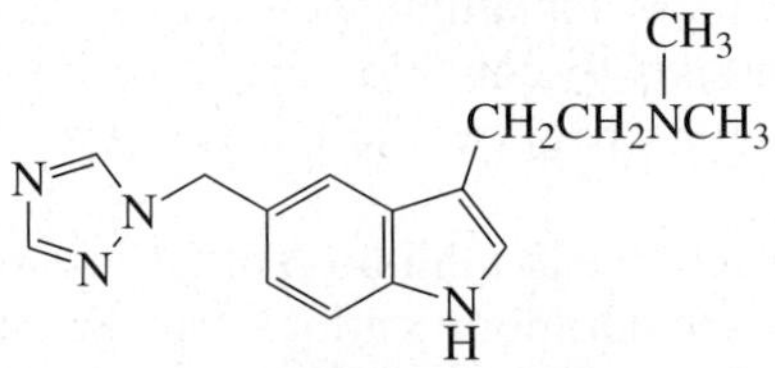

rizatriptan
Maxalt®

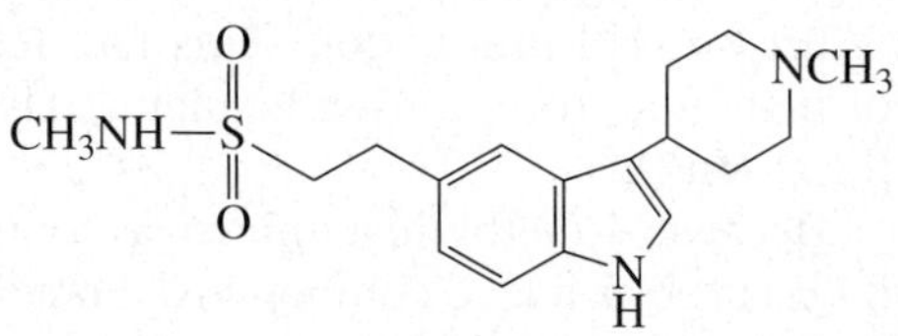

naratriptan
Amerge®

In screening modified compounds, it is not unusual to find a compound with a completely different pharmacological activity than the lead compound. For example, a molecular modification of a sulfonamide, an antibiotic, led to the discovery of tolbutamide, a drug with hypoglycemic activity (Section 25.8).

Molecular modification of promethazine, an antihistamine, led to chlorpromazine, a drug that, in addition to antihistamine activity, was found to lower body temperature. This drug found clinical use in chest surgery, where patients previously had to be cooled by being wrapped in cold, wet sheets. Because wrapping in cold, wet sheets had been used in the past to calm psychotic patients, a French psychiatrist tried the drug on some of his psychiatric patients. He found that chlorpromazine was able to suppress psychotic symptoms to the point that the patients assumed almost normal behavioral characteristics. Schizophrenic hallucinations and delusions went into remission, allowing many institutionalized patients to return to society. Before the development of this first antipsychotic drug, the only treatments offered to psychotic patients were lobotomy, electric shock, and insulin-coma therapy. The excitement over chlorpromazine was short-lived, however, as patients on the drug developed uncoordinated, involuntary movements, similar to those seen in Parkinson's disease. Research revealed that the drug blocks dopamine receptors in the brain. After thousands of molecular modifications, thioridazine was found to have the appropriate calming effect with less problematic side effects. It is now in clinical use as an antipsychotic.

chlorpromazine
Thorazine®

thioridazine
Mellaril®

fluoxetine
Prozac®

Molecular modification of the antihistamine benadryl (page 1305), led to the development of Prozac. Prozac was originally marketed as an antidepressant but is now used for a wide range of psychiatric disorders. Depression is thought to be caused, at least partly, by problems in the regulation of certain neurotransmitters, such as serotonin. Prozac did little else other than block the uptake of serotonin, so it lacked many of the undesirable side effects of its predecessors.

Sometimes a drug initially developed for one purpose is later found to have properties that make it better suited for a different purpose. Beta-blockers were originally intended to be used to alleviate the pain associated with angina by reducing the amount of work done by the heart. Later, they were found to have antihypertensive properties, so now they are used primarily to manage hypertension, a disease that is prevalent in the Western world. Notice the similar structures of the two β-blockers in Table 30.1, Tenormin and Tropol XL.

30.7 Drugs as Enzyme Inhibitors

In earlier chapters, we discussed several drugs that act by inhibiting enzymes (Sections 24.8 and 24.9). Penicillin, for example, destroys bacteria by inhibiting the enzyme that synthesizes bacterial cell walls (Section 16.17).

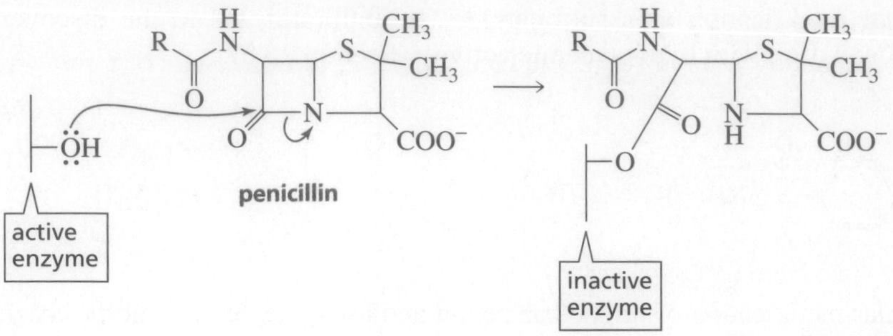

penicillin

active enzyme → inactive enzyme

Bacteria develop resistance to penicillin by secreting penicillinase, an enzyme that destroys penicillin by hydrolyzing its β-lactam ring before the drug can interfere with bacterial cell wall synthesis. (As the 4-membered ring opens, the nitrogen can obtain a proton from an acidic group at the active site of the enzyme, and the oxygen of the nucleophilic serine residue will give up a proton.)

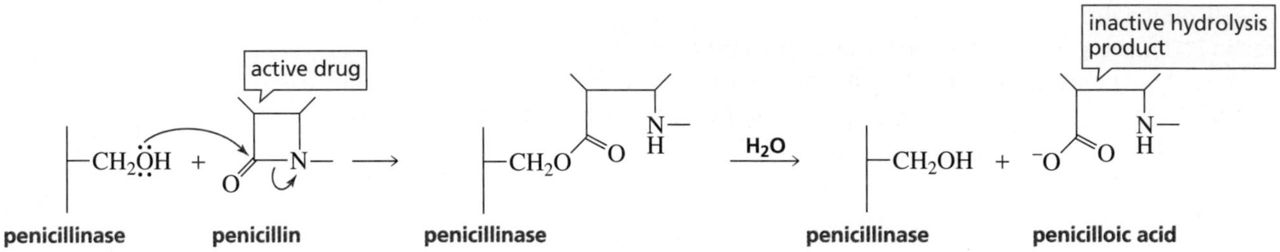

penicillinase   penicillin   penicillinase   penicillinase   penicilloic acid

Chemists have developed drugs that inhibit penicillinase. If one of these drugs is given to a patient along with penicillin, the antibiotic is not destroyed. The drug that inhibits penicillinase has no therapeutic effect itself, but acts by protecting a therapeutic drug.

One penicillinase inhibitor is a sulfone, which is easily prepared from penicillin by oxidizing the sulfur atom with a peroxyacid (Section 4.9).

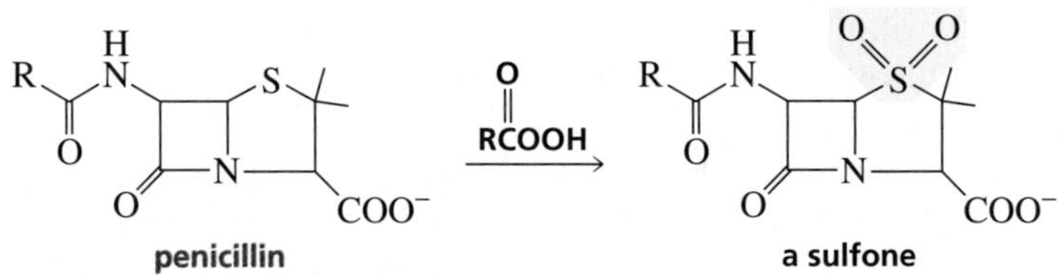

penicillin   **a sulfone**

Because the sulfone looks like the original antibiotic, penicillinase accepts it as a substrate, forming an ester, as it does with penicillin. If the ester were then hydrolyzed, penicillinase would be liberated and could react with penicillin. However, the electron-withdrawing sulfone provides an alternative pathway to hydrolysis that forms a stable imine. Because imines are susceptible to nucleophilic attack, an amino group at the active site of penicillinase reacts with the imine, forming a second covalent bond between the enzyme and the inhibitor. As a result of forming covalent bonds with the inhibitor, penicillinase is inactivated, thereby wiping out the resistance to penicillin.

The sulfone is another example of a mechanism-based **suicide inhibitor** (Section 24.8). The effectiveness of an inhibitor is indicated by its IC_{50} value, the concentration that produces 50% enzyme inhibition.

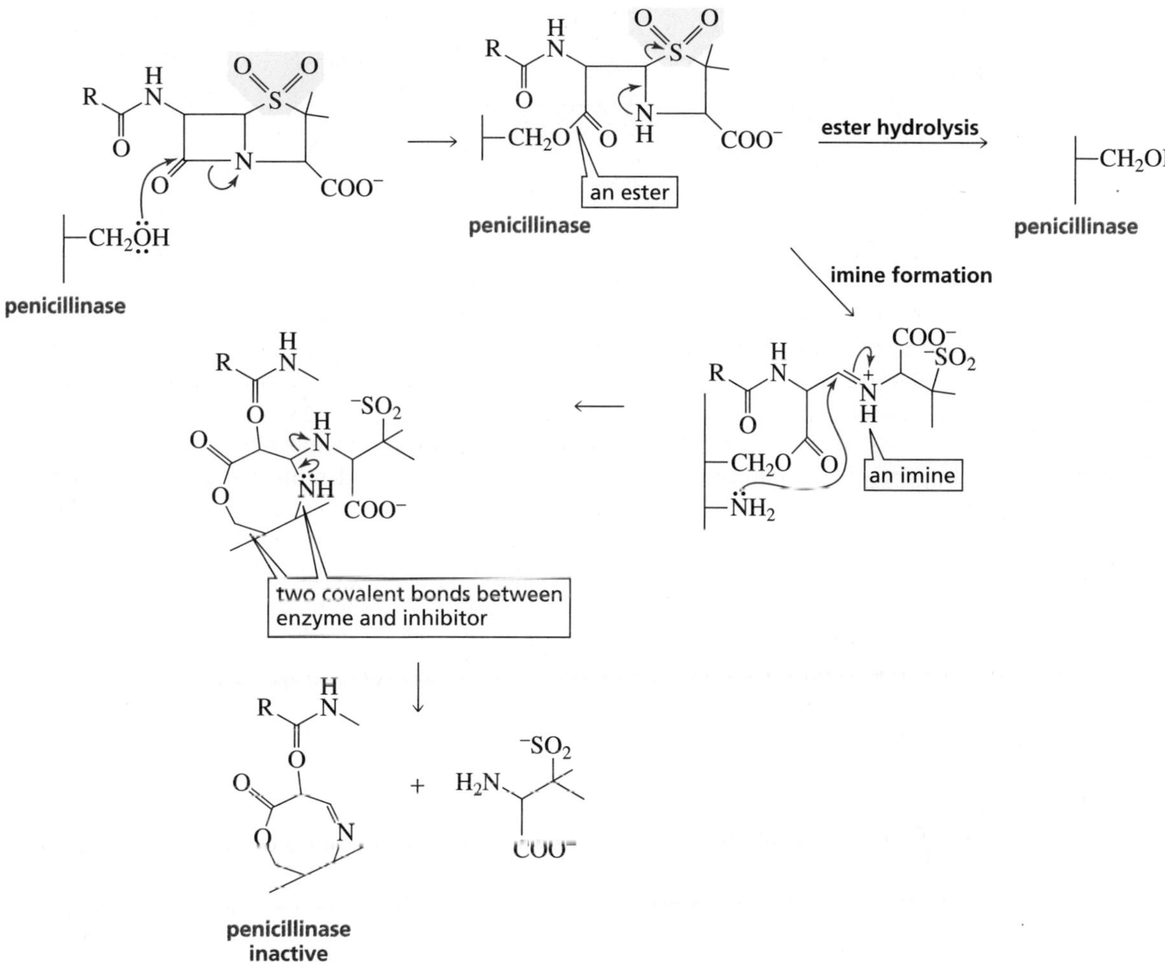

What makes the sulfone such an effective inhibitor is that the reactive imine group does not appear until after the sulfone has been bound to the enzyme that is to be inactivated. In this way, the inhibitor acts on a specific target. In contrast, if an imine were administered directly to a patient, it would react with whatever nucleophile it first encountered, and thus its activity would be nonspecific.

When two drugs are given simultaneously to a patient, their combined effect can be either additive, antagonistic, or synergistic. That is, the effect of two drugs used in combination can be equal to, less than, or greater than the sum of the effects obtained when the drugs are administered individually. Administering penicillin and the sulfone in combination results in **drug synergism**: the sulfone inhibits the penicillinase, so penicillin won't be destroyed and will be able to inhibit the enzyme that synthesizes bacterial cell walls.

Sometimes two drugs are administered in combination because some bacteria are resistant to one of them and the second drug minimizes the chance that the resistant baceria will proliferate. For example, two antimicrobial agents, isoniazid and rifampin, are given in combination to treat tuberculosis.

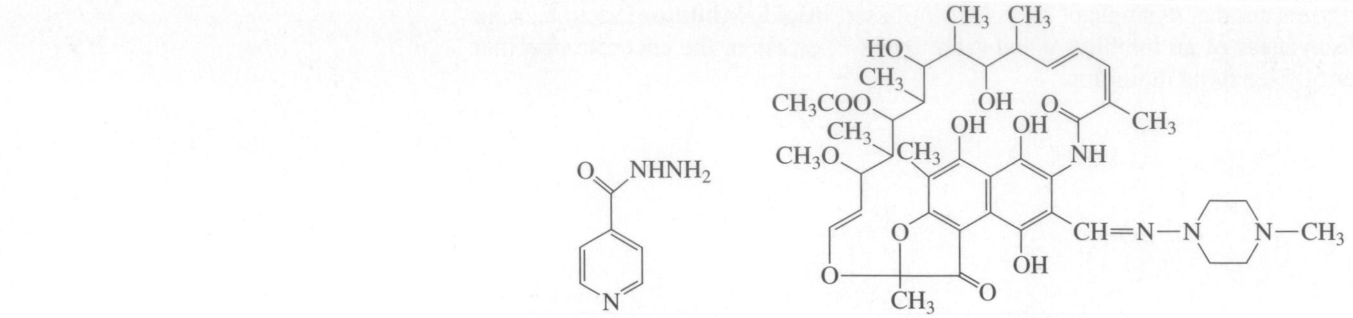

isoniazid
Nydrazid®

rifampin
Rifadin®

Typically, a bacterial strain takes 15 to 20 years to become resistant to an antibiotic. For example, penicillin became widely available in 1944. By 1952, however, 60% of all *Staphylococcus aureus* infections were penicillin resistant. The fluoroquinolones, the last class of antibiotics to be discovered until very recently, were discovered more than 30 years ago, so **drug resistance** has become an increasingly important problem in medicinal chemistry. More and more bacteria have become resistant to all antibiotics. Until recently, vancomycin was considered to be the antibiotic of last resort. Before 1989, there were no reported cases of vancomycin-resistant enterococci bacteria; now about 30% of these bacteria are resistant to vancomycin.

The antibiotic activity of the fluoroquinolones results from their ability to inhibit DNA gyrase, an enzyme required for transcription (Section 27.6). Humans are not harmed by the drug because the bacterial and mammalian forms of the enzyme are sufficiently different that the fluoroquinolones inhibit only the bacterial enzyme.

There are many different fluoroquinolones. All have fluorine substituents, which increase the lipophilicity of the drug, enabling it to penetrate into tissues and cells. If either the carboxyl group or the double bond in the 4-pyridinone ring is removed, all activity is lost. Changing the substituents on the piperazine ring can cause the drug to be excreted by the kidney rather than by the liver, which is useful to patients with impaired liver function. The substituents on the piperazine ring also affect the half-life (the time it takes for half of the drug to lose its reactivity) of the drug.

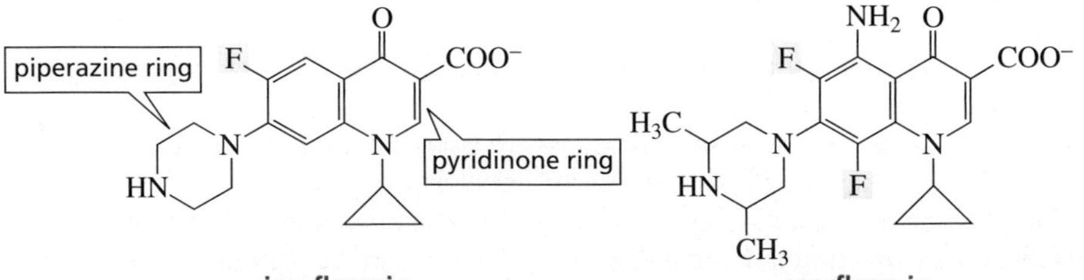

ciprofloxacin
Cipro®
active against gram-negative bacteria

sparfloxacin
Zagam®
**active against gram-negative bacteria
and gram-positive bacteria**

The approval of Zyvox by the FDA in April 2000 was met with great relief by the medical community. Zyvox is the first in a new family of antibiotics: the oxazolidinones. In clinical trials, Zyvox was found to cure 75% of the patients infected with bacteria that had become resistant to all other antibiotics. Another new class of antibiotic became available in 2005 when the FDA approved Cubicin, the first of the cyclic lipopeptide antibiotics.

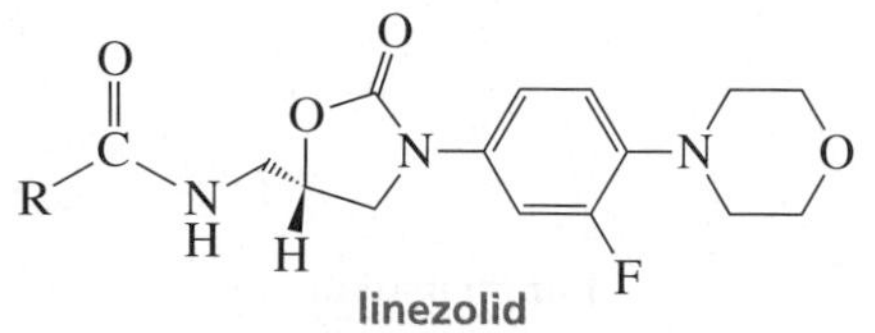

linezolid
Zyvox®

Zyvox is a synthetic compound designed by scientists to inhibit bacterial growth at a point different from that at which any other antibiotic exerts its effect (Section 27.8). Zyvox inhibits the initiation of protein synthesis by preventing the formation of the complex consisting of the first amino acid–bearing tRNA, mRNA, and the 30S ribosome (Section 27.8). Because of the drug's new mode of activity, resistance is expected to be rare at first and, hopefully, slow to emerge.

30.8 Designing a Suicide Substrate

A drug must be administered in sufficient quantity to achieve a therapeutic effect; the therapeutic ED_{50} is the dose that produces the maximum therapeutic effect in 50% of the test animals. Too much of almost any drug can be lethal. The **therapeutic index** of a drug is the ratio of the dose that causes a toxic effect to the therapeutic dose. The higher the therapeutic index, the greater is the margin of safety of the drug. Thus, the goal of pharmaceutical chemists is to design drugs with as few side effects and as high a therapeutic index as possible. A low therapeutic index may be tolerable for lethal diseases such as cancer, particularly if no other treatment is available.

Penicillin is an effective antibiotic that has a high therapeutic index. Both properties can be attributed to the fact that the drug works by interfering with cell wall synthesis (bacterial cells have walls, but human cells do not). What else is characteristic about cell walls that could lead to the design of an antibiotic? We know that enzymes and other proteins are polymers of L-amino acids. Cell walls, however, contain both L-amino acids and D-amino acids. Therefore, if the racemization of naturally occurring L-amino acids to mixtures of L- and D-amino acids could be prevented, D-amino acids would not be available for incorporation into cell walls, and bacterial cell wall synthesis could be stopped.

We have seen that amino acid racemization is catalyzed by an enzyme that requires pyridoxal phosphate as a coenzyme (Section 24.6). What we need, then, is a compound that will inhibit this enzyme. Because the natural substrate for the enzyme is an amino acid, an amino acid analog should be a good inhibitor.

The first step in racemization is removal of the α-hydrogen of the amino acid. If the inhibitor has a leaving group on the β-carbon, the electrons left behind when the proton is removed can displace the leaving group instead of being delocalized into the pyridine ring. (Compare the mechanism shown here with that shown for racemization in Section 24.6.) Transimination with the enzyme forms an α,β-unsaturated amino acid that reacts irreversibly with the imine formed by the enzyme and the coenzyme. Because the enzyme is now bound to the coenzyme in an amine linkage rather than in an imine linkage, the enzyme can no longer undergo a transimination reaction with its amino acid substrate and therefore has thus been irreversibly inactivated. This is another example of an inhibitor that does not become chemically active until it is at the active site of the targeted enzyme.

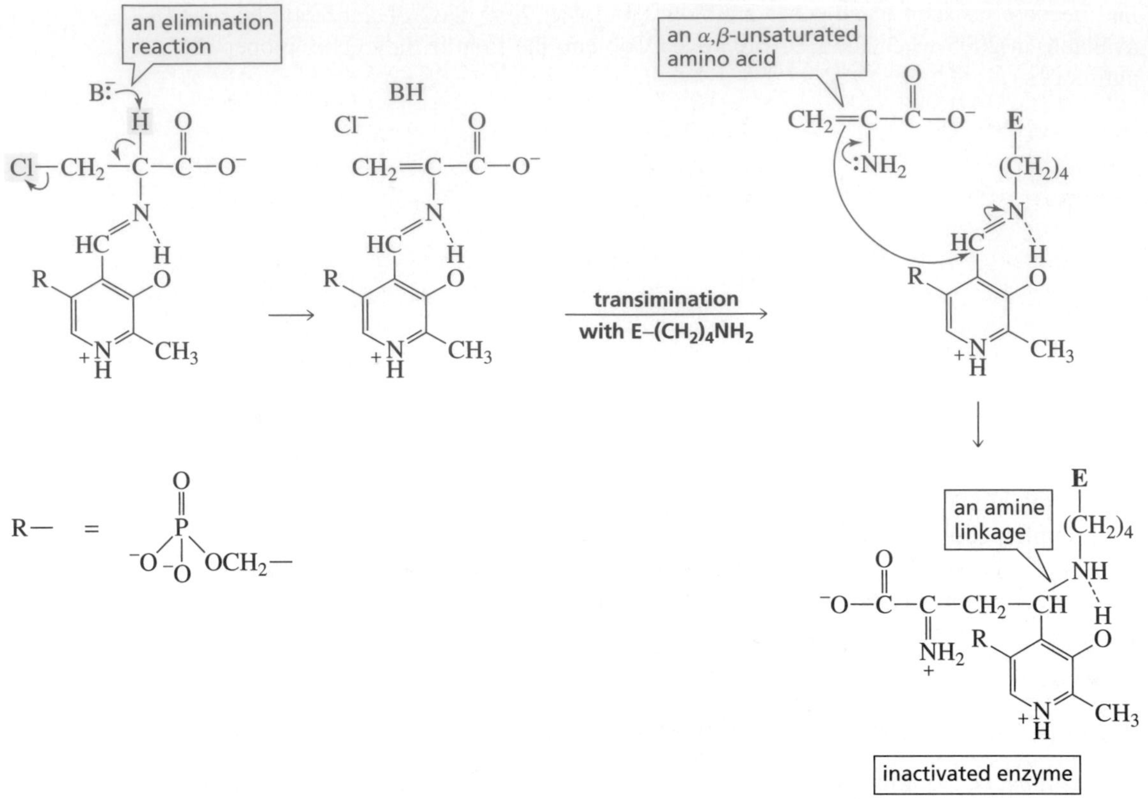

an elimination reaction

an α,β-unsaturated amino acid

transimination with E–(CH₂)₄NH₂

R— =

an amine linkage

inactivated enzyme

30.9 Quantitative Structure–Activity Relationships (QSAR)

The enormous cost of synthesizing and testing thousands of modified compounds in the attempt to find an active drug led scientists to develop a more rational approach—called **rational drug design**—to the creation of biologically active molecules. They realized that if a physical or chemical property of a series of drugs could be correlated with biological activity, they would know what property of the drug was related to that particular activity and could then design compounds that would have a good chance of exhibiting the desired activity. This strategy would be a great improvement over the more random approach to molecular modification that traditionally had been employed.

The first hint that a physical property of a drug could be related to biological activity dates back almost 100 years to when scientists recognized that chloroform ($CHCl_3$), diethyl ether, cyclopropane, and nitrous oxide (N_2O) were all useful general anesthetics. Clearly, the chemical structures of these diverse compounds could not account for their similar pharmacological effects. Instead, some physical property must explain the similarity of their biological activities.

In the early 1960s, Corwin Hansch postulated that the **biological activity** of a drug depended on two processes. The first is *distribution*: a drug must be able to get from the point where it enters the body to the receptor where it exerts its effect. For example, an anesthetic must be able to cross the aqueous milieu (blood) and penetrate the lipid barrier of nerve cell membranes. The second process is *binding*: when a drug reaches its receptor, it must interact properly with it.

Chloroform, diethyl ether, cyclopropane, and nitrous oxide were each put into a mixture of 1-octanol and water. 1-Octanol was chosen as the nonpolar solvent because, with its long chain and polar head group, it is a good model of a biological membrane. When the amount of drug dissolving in each of the layers was measured, it turned out that they all had a similar **distribution coefficient**, the ratio of the amount dissolving in

1-octanol to the amount dissolving in water. In other words, the distribution coefficient could be related to biological activity. Compounds with lower distribution coefficients—more polar compounds—could not penetrate the nonpolar cell membrane; compounds with greater distribution coefficients—more nonpolar compounds—could not cross the aqueous phase. This meant that the distribution coefficient of a compound could be used to determine whether that compound should be tested in vivo. The technique of relating a property of a series of compounds to biological activity is known as a **quantitative structure–activity relationship (QSAR)**.

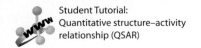

Student Tutorial:
Quantitative structure–activity
relationship (QSAR)

Determining the physical property of a drug cannot take the place of in vivo testing because a drug's behavior, once it reaches a suitable receptor, cannot be predicted by the distribution coefficient alone. Nevertheless, QSAR analysis provides a way for chemists to identify compounds that have the greatest probability of having a desired biological activity. In this way, chemists can avoid wasting resources on molecular modifications of compounds that are unsuitable for development as drugs.

In the following example, QSAR was useful not only in determining the structure of a potentially active drug but also in determining something about the structure of its receptor site. A series of substituted 2,4-diaminopyrimidines, which were being used as inhibitors of dihydrofolate reductase (Section 25.8), was investigated.

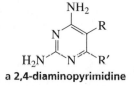

a 2,4-diaminopyrimidine

The potency of the inhibitors could be described by the equation

$$\text{potency} = 0.80\pi - 7.34\sigma - 8.14$$

where σ and π are substituent parameters.

The σ parameter is a measure of the electron-donating or electron-withdrawing ability of the substituents R and R′. The negative coefficient of σ indicates that potency is increased by electron donation (Chapter 16, Problem 85). The fact that increasing the basicity of the drug increases its potency suggests that the protonated drug is more active than the nonprotonated drug.

The π parameter is a measure of the hydrophobicity of the substituents. Potency was found to be better related to π when the π value for the more hydrophobic of the two substituents was used, rather than the sum of the π values for both substituents. This finding suggests that the receptor has a hydrophobic pocket that can accommodate one, but not both, substituents.

In another example, the potency of an analgesic was found to be described by the following equation, where *HA* indicates whether R is a hydrogen bond acceptor and where *B* is a steric factor:

potency $= -4.45 - 0.73HA + 6.5B - 1.55(B)^2$

In this case, analysis indicated that a vinyl substituted compound should be prepared.

In a search for a drug to be used to treat leukemia, a QSAR analysis showed that the antileukemic activity of a series of substituted triazines was related to the electron-donating ability of the substituent. When, however, another QSAR analysis showed that the toxicity of these compounds was also related to the electron-donating ability of the substituent, the researchers decided that it would be fruitless to continue synthesizing and testing this class of compounds.

In addition to solubility and substituent parameters, some of the properties that have been correlated with biological activity are oxidation–reduction potential, molecular size, interatomic distance between functional groups, degree of ionization, and molecular configuration.

30.10 Molecular Modeling

Because the shape of a molecule determines whether it will be recognized by a receptor and, therefore, whether it will exhibit biological activity, compounds with similar biological activity often have similar structures. Because computers can draw molecular models of compounds on a video display and move them around to assume different conformations, computer **molecular modeling** allows more rational drug design. There are computer programs that allow chemists to scan existing collections of thousands of compounds to find those with the appropriate structural and conformational properties.

Any compound that shows promise can be drawn on a computer display, along with the three-dimensional image of a receptor site. For example, the binding of netropsin, an antibiotic with a wide range of antimicrobial activity, to the minor groove of DNA is shown in Figure 30.1. Retonavir, a drug used to treat the AIDS virus, inactivates HIV protease, an enzyme essential for the maturation of the virus, by binding to its active site (Figure 30.2).

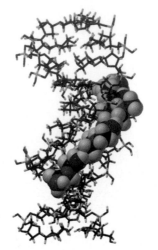

▲ **Figure 30.1**
The antibiotic netropsin bound in the minor groove of DNA.

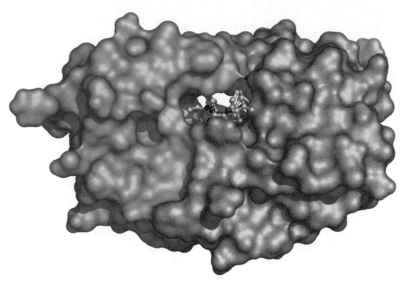

Figure 30.2 ▶
Ritonavir, a drug used to treat the AIDS virus, binds to the active site of HIV protease.

Being able to visualize the fit between the compound and the receptor may suggest modifications that can be made to the compound that might result in more favorable binding. In this way, the selection of compounds to be synthesized to screen for biological activity can be more rational, leading to faster discovery of pharmacologically active compounds. Molecular modeling will become more valuable as scientists learn more about receptor sites.

30.11 Combinatorial Organic Synthesis

The need for large collections of compounds to be screened for biological activity in the constant search for new drugs has led organic chemists to a synthetic strategy that employs the concept of mass production. In this strategy, called **combinatorial organic synthesis**, scientists synthesize a large group of related compounds—known as a library—by covalently connecting sets of building blocks of various structures. For example, if a compound can be synthesized by connecting three different kinds of building blocks from a "kit" containing 10 alternatives for each of the three kinds of blocks, then 1000 ($10 \times 10 \times 10$) different compounds can be prepared. This approach clearly mimics nature, which uses amino acids and nucleic acids as building blocks to synthesize an enormous number of different proteins and nucleic acids.

The first requirement in combinatorial synthesis is an assortment of reactive small molecules to be used as building blocks. Because of the ready availability of amino acids, combinatorial synthesis made its first appearance in the creation of peptide libraries. Peptides, however, have limited use as therapeutic agents because they generally cannot be taken orally and are rapidly metabolized. Currently, organic chemists create libraries of small organic molecules that can be used to modify *lead compounds* or to assist in *rational drug design*.

Student Tutorial:
Combinatorial synthesis

One example of combinatorial synthesis is its use to create a library of benzodiazepines (Section 30.5). These compounds can be thought of as originating from three different sets of building blocks: a substituted 2-aminobenzophenone, an amino acid, and an alkylating agent.

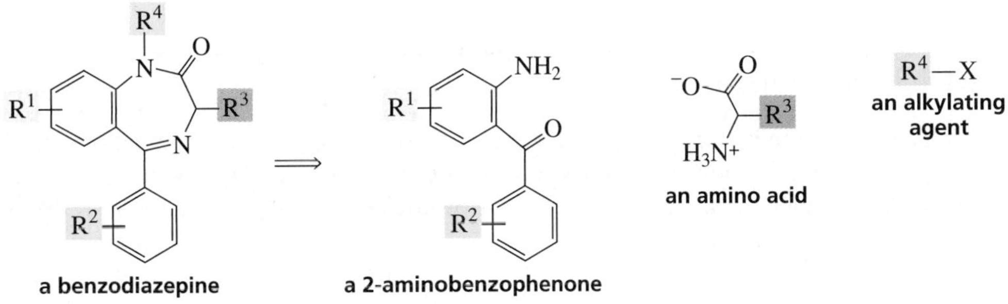

a benzodiazepine   a 2-aminobenzophenone

The 2-aminobenzophenone is attached to a solid support (Section 22.11) in a manner that allows it to be readily removed later by acid hydrolysis (Figure 30.3).

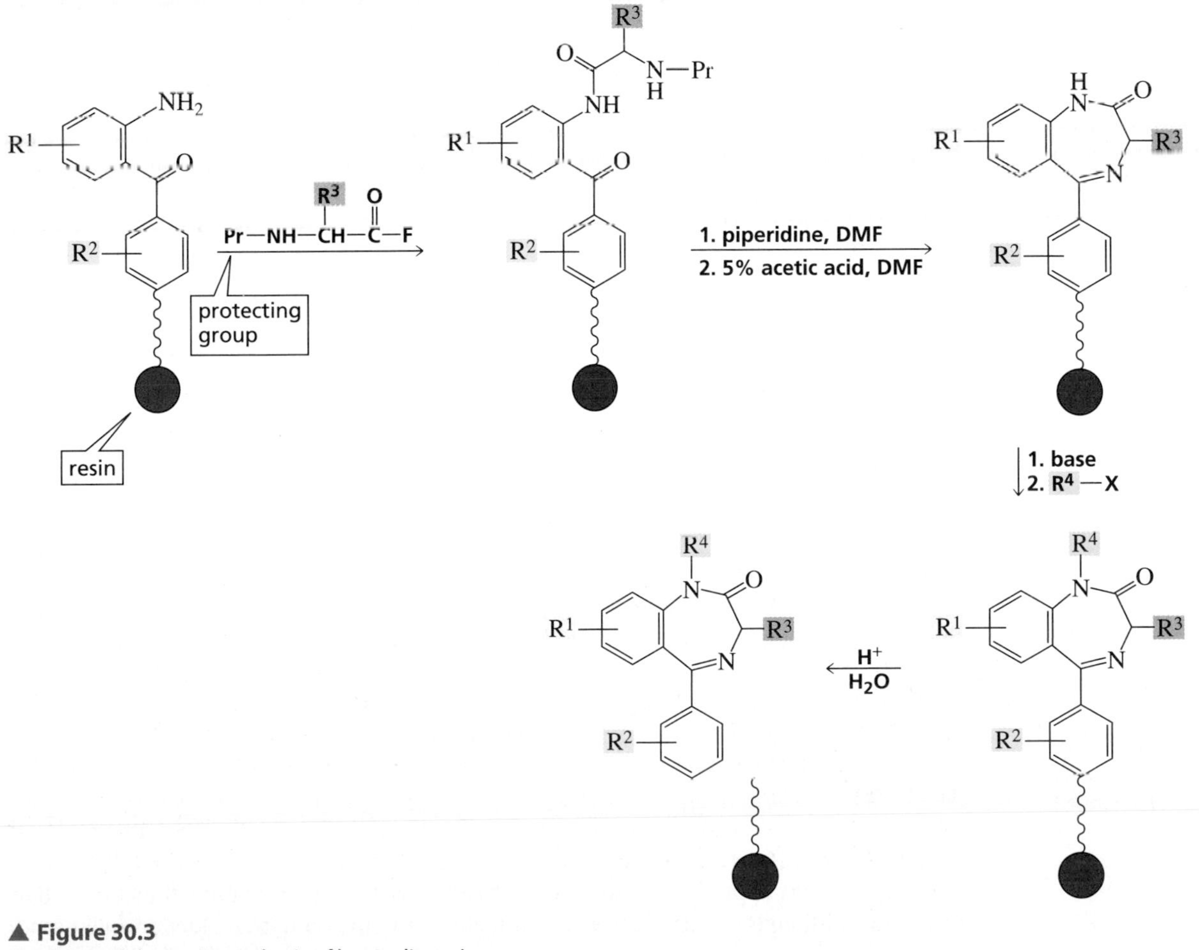

▲ **Figure 30.3**
Combinatorial organic synthesis of benzodiazepines.

The amino acid—*N*-protected and activated by being converted into an acyl fluoride—is then added. After the amide is formed, the protecting group is removed and the seven-membered ring is created through imine formation. Next, a base is added to remove the amide hydrogen and thus form a nucleophile that reacts with the added alkylating agent. The final product is then removed from the solid support.

In order to synthesize a library of these compounds, the solid support containing the 2-aminobenzophenone can be divided into several portions, and a different amino acid can then be added to each. Each ring-closed product can also be divided into several portions, and a different alkylating agent can be added to each of them. In this way, many different benzodiazepines can be prepared at the same time. Combinatorial syntheses do not have to have one of the reactants anchored to a solid support, but such a support tends to improve the yield because it prevents the loss of product during the purification step.

30.12 Antiviral Drugs

Relatively few clinically useful drugs have been developed for viral infections. The slow progress of this endeavor is due to the nature of viruses and the way they replicate. Viruses are smaller than bacteria and consist of nucleic acid—either DNA or RNA—surrounded by a coat of protein. Some viruses penetrate the host cell; others merely inject their nucleic acid into the cell. In either case, the nucleic acid is transcribed by the host and is integrated into the host genome.

Most **antiviral drugs** are analogs of nucleosides, interfering with the virus's DNA or RNA synthesis. In this way, they prevent the virus from replicating. For example, acyclovir, the drug used against herpes viruses, has a three-dimensional shape similar to guanine. Acyclovir can therefore fool the virus into incorporating the drug instead of guanine into the virus's DNA. Once this happens, the DNA strand can no longer grow, because acyclovir lacks a 3′-OH group. The terminated DNA binds to DNA polymerase and inactivates it irreversibly (Section 27.3).

acyclovir	cytarabine	ribavirin	idoxuridine
Aclovir®	Cytosar®	Viramid®	Herplex®
used against herpes simplex infections	**used against acute myelocytic leukemia**	**a broad-spectrum antiviral agent**	**approved for topical ophthalmic use**

Cytarabine, used for acute myelocytic leukemia, competes with cytosine for incorporation into viral DNA. Cytarabine contains an arabinose rather than a ribose (Table 21.1). Because the 2′-OH group is in the β-position (remember that the 2′-OH group of a ribonucleoside is in the α-position), the bases in DNA are not able to stack properly.

Ribavirin is a broad-spectrum antiviral agent that inhibits viral mRNA synthesis. A step in the metabolic pathway responsible for the synthesis of guanosine triphosphate (GTP) converts inosine monophosphate (IMP) into xanthosine monophosphate (XMP). Ribarvirin is a competitive inhibitor of the enzyme that catalyzes that step. Thus, ribarvirin interferes with the synthesis of GTP and, therefore, with the synthesis of all nucleic acids. It is used to treat children with chronic hepatitis C.

Idoxuridine is approved in the United States only for the topical treatment of ocular infections, although it is used for herpes infections in other countries. Idoxuridine has an iodo group in place of the methyl group of thymine and is incorporated into DNA in place of thymine. Chain elongation can continue because idoxuridine has a 3′-OH group but the resulting DNA is more easily broken and is also not transcribed properly. (Also see AZT on page 1229.)

30.13 **Economics of Drugs • Governmental Regulations**

The average cost of launching a new drug is $100 to $500 million. This cost has to be recovered quickly by the manufacturer because the starting date of a patent is the date the drug is first discovered. A patent is good for 20 years from the date it is applied for, but because it takes an average of 12 years to market a drug after its initial discovery, the patent protects the discoverer of the drug for an average of only 8 years. It is only during the 8 years of patent protection that drug sales can provide enough income to cover the initial costs and pay for research on new drugs. In addition, the average commercial lifetime of a drug is only 15 to 20 years. After that, the drug is generally replaced by a new and improved drug. Only about 1 in 3 drugs actually makes a profit for the company.

Why does it cost so much to develop a new drug? First of all, the Food and Drug Administration (FDA) has high standards that must be met before a drug is approved for a particular use (Section 30.4). An important factor leading to the high price of many drugs is the low rate of success in progressing from the initial concept to an approved product: only 1 or 2 of every 100 compounds tested become lead compounds; out of 100 structural modifications of a lead compound, only 1 is worthy of further study. For every 10,000 compounds evaluated in animal studies, only 10 will get to clinical trials. Clinical trials consist of three phases: phase I evaluates the effectiveness, safety, side effects, and dosage levels in up to 100 healthy volunteers; phase II investigates the effectiveness, safety, and side effects in 100 to 500 volunteers who have the condition the drug is meant to treat; and phase III establishes the effectiveness and appropriate dosage of the drug and monitors adverse reactions in several thousand volunteer patients. For every 10 compounds that enter clinical trials, only one satisfies the increasingly stringent requirements to become a marketable drug.

ORPHAN DRUGS

Because of the high cost of developing a drug, pharmaceutical companies are reluctant to carry out research on drugs for rare diseases. Even if a company were to find an effective drug for a disease, there would be no way to recoup the expenditure, because of the limited demand. Fortunately for people suffering from such diseases, the U.S. Congress, in 1983, passed the Orphan Drugs Act. The act creates public subsidies to fund research and provides tax credits for the development and marketing of drugs—called **orphan drugs**—for diseases or conditions that affect fewer than 200,000 people. In addition, the act stipulates that if the drug is not patentable, the company that develops it has four years of exclusive marketing rights. In the 10 years prior to the passage of this act, fewer than 10 orphan drugs were developed. Today, more than 100 are available, and some 600 others are in development.

Drugs originally developed as orphan drugs include AZT (to treat AIDS), Taxol (to treat ovarian cancer), Exosurf Neonatal (to treat respiratory distress syndrome in infants), and Opticrom (to treat corneal swelling).

SUMMARY

A **drug** is a compound that interacts with a biological molecule to trigger a physiological response. Each drug has a **brand name** that can be used only by the holder of the patent, which is valid for 20 years. Once a patent expires, other drug companies can market the drug under a **generic name** that can be used by any pharmaceutical company. Drugs no one wants to develop because they would be used for diseases or conditions that affect fewer than 200,000 people are called **orphan drugs**. The Food and Drug Administration (FDA) sets high standards that must be met before it approves a drug for a particular use. The average cost of launching a new drug is about $230 million.

The prototype for a new drug is called a **lead compound**. Changing the structure of a lead compound is called **molecular modification**. A **random screen**, or **blind screen**, is a search for a pharmacologically active lead compound without the benefit of any information about what structures might show activity. The technique of relating a property of a series of compounds to biological activity is known as a **quantitative structure–activity relationship (QSAR)**.

Many drugs exert their physiological effects by binding to a specific binding site called a **receptor**. Some drugs act by inhibiting enzymes or by binding to nucleic acids. Most **antiviral drugs** are analogs of nucleosides, interfering with DNA or RNA synthesis and thereby preventing the virus from replicating.

A **bacteriostatic drug** inhibits the further growth of bacteria; a **bactericidal drug** kills the bacteria. In recent years,

many bacteria have become resistant to antibiotics, so **drug resistance** is an increasingly important problem in medicinal chemistry.

The therapeutic index of a drug is the ratio of the lethal dose to the therapeutic dose. The higher the **therapeutic index**, the greater is the margin of safety of the drug. In **drug synergism**, the effect of two drugs used in combination is greater that the sum of the drugs' individual effects.

Large collections of compounds that can be screened for biological activity are prepared by **combinatorial organic synthesis**—the synthesis of a group of related compounds by covalently connecting sets of building blocks.

KEY TERMS

antiviral drug (p. 1316)
bactericidal drug (p. 1302)
bacteriostatic drug (p. 1302)
biological activity (p. 1312)
blind screen (p. 1300)
brand name (p. 1297)
combinatorial organic synthesis (p. 1314)
distribution coefficient (p. 1312)

drug (p. 1293)
drug resistance (p. 1310)
drug synergism (p. 1309)
generic name (p. 1297)
lead compound (p. 1298)
molecular modeling (p. 1314)
molecular modification (p. 1298)
orphan drug (p. 1317)

prodrug (p. 1301)
quantitative structure–activity relationship (QSAR) (p. 1313)
random screen (p. 1300)
rational drug design (p. 1312)
receptor (p. 1304)
suicide inhibitor (p. 1309)
therapeutic index (p. 1311)

PROBLEMS

1. What is the chemical name of each of the following drugs?
   a. benzocaine   **b.** procaine   **c.** idoxuridine

2. Based on the lead compound for the development of procaine and lidocaine, propose structures for other compounds that you would like to see tested for use as anesthetics.

3. Which of the following compounds is more likely to exhibit activity as a tranquilizer?

4. Which compound is more likely to be a general anesthetic?

$$CH_3CH_2CH_2OH \quad \text{or} \quad CH_3OCH_2CH_3$$

5. What accounts for the ease of imine formation between penicillinase and the sulfone antibiotic that counteracts penicillin resistance?

6. For each of the following pairs of compounds, indicate the compound that you would expect to be a more potent inhibitor of dihydrofolate reductase:

a.

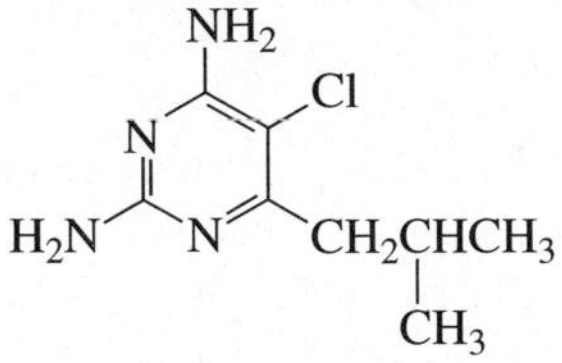

or

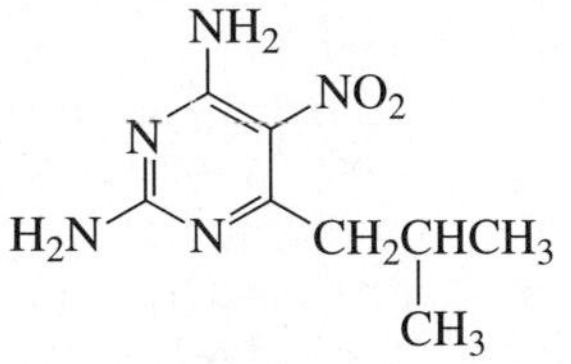

b.

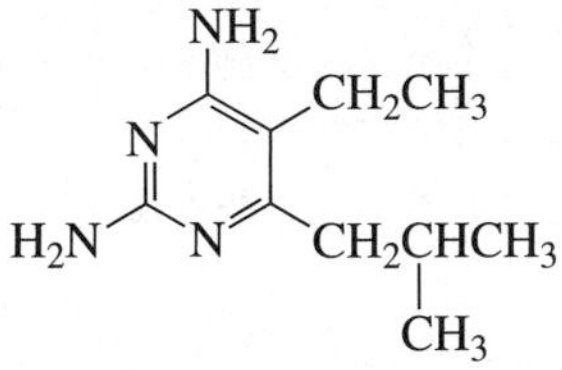

or

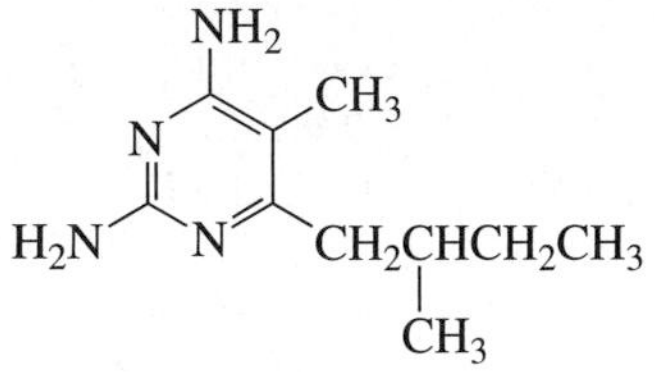

7. The lethal dose of tetrahydrocannabinol in mice is 2.0 g/kg, and the therapeutic dose in mice is 20 mg/kg. The lethal dose of sodium pentothal in mice is 100 mg/kg, and the therapeutic dose is 30 mg/kg. Which is the safer drug?

8. The following compound is a suicide inhibitor of the enzyme that catalyzes amino acid racemization:

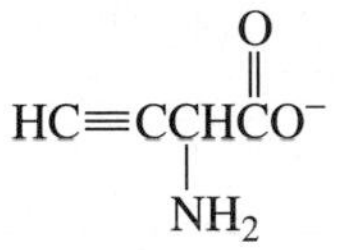

$$HC\equiv CCHCO^-$$

Propose a mechanism that explains how this compound irreversibly inactivates the enzyme.

9. Explain how each of the antiviral drugs shown in Section 30.12 differs from the naturally occurring nucleoside that it most closely resembles.

10. Show a mechanism for the formation of a benzodiazepine 4-oxide from the reaction of a quinazoline 3-oxide with methylamine (Section 30.5).

11. Show how Valium could be synthesized from benzoyl chloride, *para*-chloroaniline, methyl iodide, and the ethyl ester of glycine.

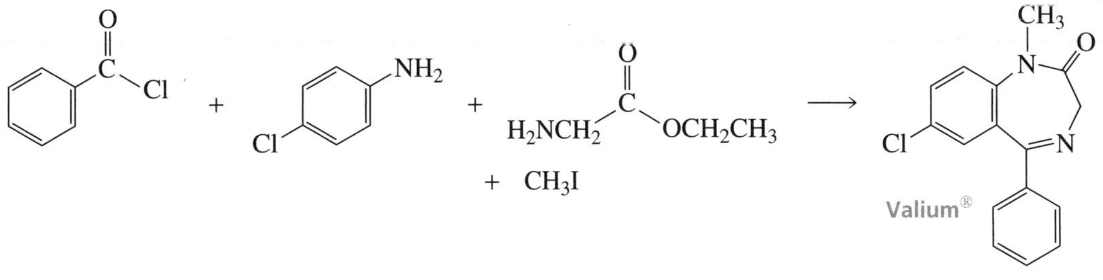

+   +

+ CH_3I

Valium®

12. Show how Tagamet could be synthesized from the indicated starting materials.

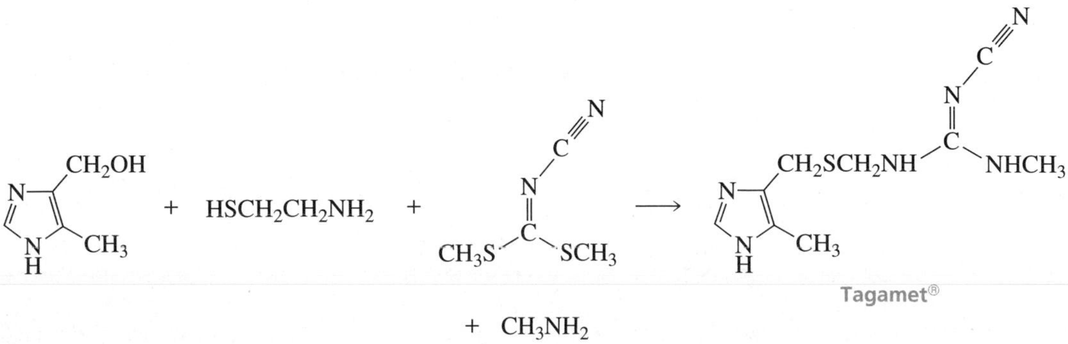

+ $HSCH_2CH_2NH_2$ +   ⟶

Tagamet®

+ CH_3NH_2

13. Show how lidocaine (page 1298) could be synthesized using chloroacetyl chloride, 2,6-dimethylaniline, and diethylamine.

Appendix I

Physical Properties of Organic Compounds

Physical Properties of Alkenes

Name	Structure	mp (°C)	bp (°C)	Density (g/mL)
Ethene	$CH_2\!=\!CH_2$	−169	−104	
Propene	$CH_2\!=\!CHCH_3$	−185	−47	
1-Butene	$CH_2\!=\!CHCH_2CH_3$	−185	−6.3	
1-Pentene	$CH_2\!=\!CH(CH_2)_2CH_3$		30	0.641
1-Hexene	$CH_2\!=\!CH(CH_2)_3CH_3$	−138	64	0.673
1-Heptene	$CH_2\!=\!CH(CH_2)_4CH_3$	−119	94	0.697
1-Octene	$CH_2\!=\!CH(CH_2)_5CH_3$	−101	122	0.715
1-Nonene	$CH_2\!=\!CH(CH_2)_6CH_3$	−81	146	0.730
1-Decene	$CH_2\!=\!CH(CH_2)_7CH_3$	−66	171	0.741
cis-2-Butene	cis-$CH_3CH\!=\!CHCH_3$	−180	37	0.650
trans-2-Butene	trans-$CH_3CH\!=\!CHCH_3$	−140	37	0.649
Methylpropene	$CH_2\!=\!C(CH_3)_2$	−140	−6.9	0.594
cis-2-Pentene	cis-$CH_3CH\!=\!CHCH_2CH_3$	−180	37	0.650
trans-2-Pentene	trans-$CH_3CH\!=\!CHCH_2CH_3$	−140	37	0.649
Cyclohexene		−104	83	0.811

Physical Properties of Alkynes

Name	Structure	mp (°C)	bp (°C)	Density (g/mL)
Ethyne	$HC\!\equiv\!CH$	−82	−84.0	
Propyne	$HC\!\equiv\!CCH_3$	−101.5	−23.2	
1-Butyne	$HC\!\equiv\!CCH_2CH_3$	−122	8.1	
2-Butyne	$CH_3C\!\equiv\!CCH_3$	−24	27	0.694
1-Pentyne	$HC\!\equiv\!C(CH_2)_2CH_3$	−98	39.3	0.695
2-Pentyne	$CH_3C\!\equiv\!CCH_2CH_3$	−101	55.5	0.714
3-Methyl-1-butyne	$HC\!\equiv\!CCH(CH_3)_2$		29	0.665
1-Hexyne	$HC\!\equiv\!C(CH_2)_3CH_3$	−132	71	0.715
2-Hexyne	$CH_3C\!\equiv\!C(CH_2)_2CH_3$	−92	84	0.731
3-Hexyne	$CH_3CH_2C\!\equiv\!CCH_2CH_3$	−101	81	0.725
1-Heptyne	$HC\!\equiv\!C(CH_2)_4CH_3$	−81	100	0.733
1-Octyne	$HC\!\equiv\!C(CH_2)_5CH_3$	−80	127	0.747
1-Nonyne	$HC\!\equiv\!C(CH_2)_6CH_3$	−50	151	0.757
1-Decyne	$HC\!\equiv\!C(CH_2)_7CH_3$	−44	174	0.766

Physical Properties of Cyclic Saturated Alkanes

Name	mp (°C)	bp (°C)	Density (g/mL)
Cyclopropane	−128	−33	
Cyclobutane	−80	−12	
Cyclopentane	−94	50	0.751
Cyclohexane	6.5	81	0.779
Cycloheptane	−12	118	0.811
Cyclooctane	14	149	0.834
Methylcyclopentane	−142	72	0.749
Methylcyclohexane	−126	100	0.769
cis-1,2-Dimethylcyclopentane	−62	99	0.772
trans-1,2-Dimethylcyclopentane	−120	92	0.750

Physical Properties of Ethers

Name	Structure	mp (°C)	bp (°C)	Density (g/mL)
Dimethyl ether	CH_3OCH_3	−141	−24.8	
Diethyl ether	$CH_3CH_2OCH_2CH_3$	−116	34.6	0.706
Dipropyl ether	$CH_3(CH_2)_2O(CH_2)_2CH_3$	−123	88	0.736
Diisopropyl ether	$(CH_3)_2CHOCH(CH_3)_2$	−86	69	0.725
Dibutyl ether	$CH_3(CH_2)_3O(CH_2)_3CH_3$	−98	142	0.764
Divinyl ether	$CH_2{=}CHOCH{=}CH_2$		35	
Diallyl ether	$CH_2{=}CHCH_2OCH_2CH{=}CH_2$		94	0.830
Tetrahydrofuran		−108	66	0.889
Dioxane	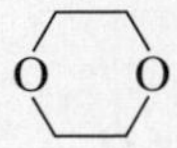	12	101	1.034

Physical Properties of Alcohols

Name	Structure	mp (°C)	bp (°C)	Solubility (g/100 g H$_2$O at 25 °C)
Methanol	CH$_3$OH	−97.8	64	∞
Ethanol	CH$_3$CH$_2$OH	−114.7	78	∞
1-Propanol	CH$_3$(CH$_2$)$_2$OH	−127	97.4	∞
1-Butanol	CH$_3$(CH$_2$)$_3$OH	−90	118	7.9
1-Pentanol	CH$_3$(CH$_2$)$_4$OH	−78	138	2.3
1-Hexanol	CH$_3$(CH$_2$)$_5$OH	−52	157	0.6
1-Heptanol	CH$_3$(CH$_2$)$_6$OH	−36	176	0.2
1-Octanol	CH$_3$(CH$_2$)$_7$OH	−15	196	0.05
2-Propanol	CH$_3$CHOHCH$_3$	−89.5	82	∞
2-Butanol	CH$_3$CHOHCH$_2$CH$_3$	−115	99.5	12.5
2-Methyl-1-propanol	(CH$_3$)$_2$CHCH$_2$OH	−108	108	10.0
2-Methyl-2-propanol	(CH$_3$)$_3$COH	25.5	83	∞
3-Methyl-1-butanol	(CH$_3$)$_2$CH(CH$_2$)$_2$OH	−117	130	2
2-Methyl-2-butanol	(CH$_3$)$_2$COHCH$_2$CH$_3$	−12	102	12.5
2,2-Dimethyl-1-propanol	(CH$_3$)$_3$CCH$_2$OH	55	114	∞
Allyl alcohol	CH$_2$=CHCH$_2$OH	129	97	∞
Cyclopentanol	C$_5$H$_9$OH	−19	140	s. sol.
Cyclohexanol	C$_6$H$_{11}$OH	24	161	s. sol.
Benzyl alcohol	C$_6$H$_5$CH$_2$OH	−15	205	4

Physical Properties of Alkyl Halides

Name	bp (°C)			
	Fluoride	*Chloride*	*Bromide*	*Iodide*
Methyl	−78.4	−24.2	3.6	42.4
Ethyl	−37.7	12.3	38.4	72.3
Propyl	−2.5	46.6	71.0	102.5
Isopropyl	−9.4	34.8	59.4	89.5
Butyl	32.5	78.4	100	130.5
Isobutyl		68.8	90	120
sec-Butyl		68.3	91.2	120.0
tert-Butyl		50.2	73.1	dec.
Pentyl	62.8	108	130	157.0
Hexyl	92	133	154	179

Physical Properties of Amines

Name	Structure	mp (°C)	bp (°C)	Solubility (g/100 g H_2O at 25 °C)
Primary Amines				
Methylamine	CH_3NH_2	−93	−6.3	v. sol.
Ethylamine	$CH_3CH_2NH_2$	−81	17	∞
Propylamine	$CH_3(CH_2)_2NH_2$	−83	48	∞
Isopropylamine	$(CH_3)_2CHNH_2$	−95	33	∞
Butylamine	$CH_3(CH_2)_3NH_2$	−49	78	v. sol.
Isobutylamine	$(CH_3)_2CHCH_2NH_2$	−85	68	∞
sec-Butylamine	$CH_3CH_2CH(CH_3)NH_2$	−72	63	∞
tert-Butylamine	$(CH_3)_3CNH_2$	−67	46	∞
Cyclohexylamine	$C_6H_{11}NH_2$	−18	134	s. sol.
Secondary Amines				
Dimethylamine	$(CH_3)_2NH$	−93	7.4	v. sol.
Diethylamine	$(CH_3CH_2)_2NH$	−50	55	10.0
Dipropylamine	$(CH_3CH_2CH_2)_2NH$	−63	110	10.0
Dibutylamine	$(CH_3CH_2CH_2CH_2)_2NH$	−62	159	s. sol.
Tertiary Amines				
Trimethylamine	$(CH_3)_3N$	−115	2.9	91
Triethylamine	$(CH_3CH_2)_3N$	−114	89	14
Tripropylamine	$(CH_3CH_2CH_2)_3N$	−93	157	s. sol.

Physical Properties of Benzene and Substituted Benzenes

Name	Structure	mp (°C)	bp (°C)	Solubility (g/100 g H_2O at 25 °C)
Aniline	$C_6H_5NH_2$	−6	184	3.7
Benzene	C_6H_6	5.5	80.1	s. sol.
Benzaldehyde	C_6H_5CHO	−26	178	s. sol.
Benzamide	$C_6H_5CONH_2$	132	290	s. sol.
Benzoic acid	C_6H_5COOH	122	249	0.34
Bromobenzene	C_6H_5Br	−30.8	156	insol.
Chlorobenzene	C_6H_5Cl	−45.6	132	insol.
Nitrobenzene	$C_6H_5NO_2$	5.7	210.8	s. sol.
Phenol	C_6H_5OH	43	182	s. sol.
Styrene	$C_6H_5CH{=}CH_2$	−30.6	145.2	insol.
Toluene	$C_6H_5CH_3$	−95	110.6	insol.

Physical Properties of Carboxylic Acids

Name	Structure	mp (°C)	bp (°C)	Solubility (g/100 g H_2O at 25 °C)
Formic acid	HCOOH	8.4	101	∞
Acetic acid	CH_3COOH	16.6	118	∞
Propionic acid	CH_3CH_2COOH	−21	141	∞
Butanoic acid	$CH_3(CH_2)_2COOH$	−5	162	∞
Pentanoic acid	$CH_3(CH_2)_3COOH$	−34	186	4.97
Hexanoic acid	$CH_3(CH_2)_4COOH$	−4	202	0.97
Heptanoic acid	$CH_3(CH_2)_5COOH$	−8	223	0.24
Octanoic acid	$CH_3(CH_2)_6COOH$	17	237	0.068
Nonanoic acid	$CH_3(CH_2)_7COOH$	15	255	0.026
Decanoic acid	$CH_3(CH_2)_8COOH$	32	270	0.015

Physical Properties of Dicarboxylic Acids

Name	Structure	mp (°C)	Solubility (g/100 g H_2O at 25 °C)
Oxalic acid	HOOCCOOH	189	S
Malonic acid	$HOOCCH_2COOH$	136	v. sol.
Succinic acid	$HOOC(CH_2)_2COOH$	185	s. sol.
Glutaric acid	$HOOC(CH_2)_3COOH$	98	v. sol.
Adipic acid	$HOOC(CH_2)_4COOH$	151	s. sol.
Pimelic acid	$HOOC(CH_2)_5COOH$	106	s. sol.
Phthalic acid	$1,2\text{-}C_6H_4(COOH)_2$	231	s. sol.
Maleic acid	*cis*-HOOCCH=CHCOOH	130.5	v. sol.
Fumaric acid	*trans*-HOOCCH=CHCOOH	302	s. sol.

Physical Properties of Acyl Chlorides and Acid Anhydrides

Name	Structure	mp (°C)	bp (°C)
Acetyl chloride	CH_3COCl	−112	51
Propionyl chloride	CH_3CH_2COCl	−94	80
Butyryl chloride	$CH_3(CH_2)_2COCl$	−89	102
Valeryl chloride	$CH_3(CH_2)_3COCl$	−110	128
Acetic anhydride	$CH_3(CO)O(CO)CH_3$	−73	140
Succinic anhydride	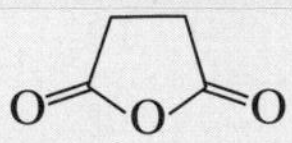		120

Physical Properties of Esters

Name	Structure	mp (°C)	bp (°C)
Methyl formate	$HCOOCH_3$	−100	32
Ethyl formate	$HCOOCH_2CH_3$	−80	54
Methyl acetate	CH_3COOCH_3	−98	57.5
Ethyl acetate	$CH_3COOCH_2CH_3$	−84	77
Propyl acetate	$CH_3COO(CH_2)_2CH_3$	−92	102
Methyl propionate	$CH_3CH_2COOCH_3$	−87.5	80
Ethyl propionate	$CH_3CH_2COOCH_2CH_3$	−74	99
Methyl butyrate	$CH_3CH_2CH_2COOCH_3$	−84.8	102.3
Ethyl butyrate	$CH_3CH_2CH_2COOCH_2CH_3$	−93	121

Physical Properties of Amides

Name	Structure	mp (°C)	bp (°C)
Formamide	$HCONH_2$	3	200 d*
Acetamide	CH_3CONH_2	82	221
Propanamide	$CH_3CH_2CONH_2$	80	213
Butanamide	$CH_3(CH_2)_2CONH_2$	116	216
Pentanamide	$CH_3(CH_2)_3CONH_2$	106	232

* d means the substance decomposes.

Physical Properties of Aldehydes

Name	Structure	mp (°C)	bp (°C)	Solubility (g/100 g H_2O at 25 °C)
Formaldehyde	$HCHO$	−92	−21	v. sol.
Acetaldehyde	CH_3CHO	−121	21	∞
Propionaldehyde	CH_3CH_2CHO	−81	49	16
Butyraldehyde	$CH_3(CH_2)_2CHO$	−96	75	7
Pentanal	$CH_3(CH_2)_3CHO$	−92	103	s. sol.
Hexanal	$CH_3(CH_2)_4CHO$	−56	131	s. sol.
Heptanal	$CH_3(CH_2)_5CHO$	−43	153	0.1
Octanal	$CH_3(CH_2)_6CHO$		171	insol.
Nonanal	$CH_3(CH_2)_7CHO$		192	insol.
Decanal	$CH_3(CH_2)_8CHO$	−5	209	insol.
Benzaldehyde	C_6H_5CHO	−26	178	0.3

Physical Properties of Ketones

Name	Structure	mp (°C)	bp (°C)	Solubility (g/100 g H_2O at 25 °C)
Acetone	CH_3COCH_3	−95	56	∞
2-Butanone	$CH_3COCH_2CH_3$	−86	80	25.6
2-Pentanone	$CH_3CO(CH_2)_2CH_3$	−78	102	5.5
2-Hexanone	$CH_3CO(CH_2)_3CH_3$	−57	127	1.6
2-Heptanone	$CH_3CO(CH_2)_4CH_3$	−36	151	0.4
2-Octanone	$CH_3CO(CH_2)_5CH_3$	−16	173	insol.
2-Nonanone	$CH_3CO(CH_2)_6CH_3$	−7	195	insol.
2-Decanone	$CH_3CO(CH_2)_7CH_3$	14	210	insol.
3-Pentanone	$CH_3CH_2COCH_2CH_3$	−40	102	4.8
3-Hexanone	$CH_3CH_2CO(CH_2)_2CH_3$		123	1.5
3-Heptanone	$CH_3CH_2CO(CH_2)_3CH_3$	−39	149	0.3
Acetophenone	$CH_3COC_6H_5$	19	202	insol.
Propiophenone	$CH_3CH_2COC_6H_5$	18	218	insol.

Appendix II

pK_a Values

Compound	pK_a	Compound	pK_a	Compound	pK_a
$CH_3C{\equiv}\overset{+}{N}H$	−10.1	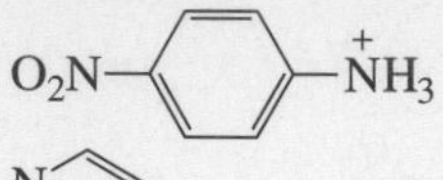 $O_2N{-}\text{<benzene>}{-}\overset{+}{N}H_3$	1.0	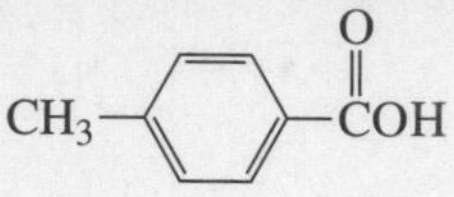 $CH_3{-}\text{<benzene>}{-}COOH$ (with C=O)	4.3
HI	−10	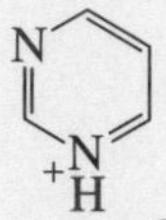 <pyrimidine, $\overset{+}{N}H$>	1.0	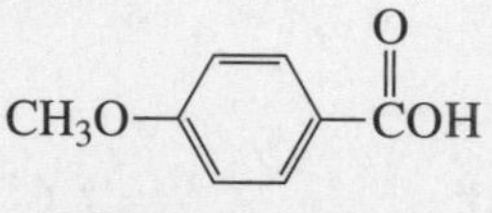 $CH_3O{-}\text{<benzene>}{-}COOH$ (with C=O)	4.5
HBr	−9	Cl_2CHCOH (with C=O)	1.3	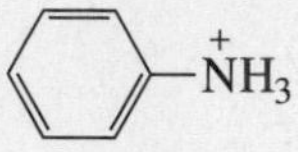 $\text{<benzene>}{-}\overset{+}{N}H_3$	4.6
$CH_3\overset{+OH}{CH}$	−8	HSO_4^-	2.0	CH_3COH (with C=O)	4.8
$CH_3\overset{+OH}{C}CH_3$	−7.3	H_3PO_4	2.1	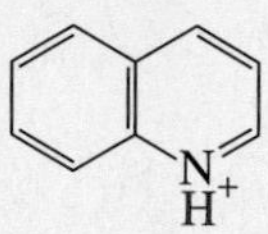 <quinoline, $\overset{+}{N}H$>	4.9
HCl	−7	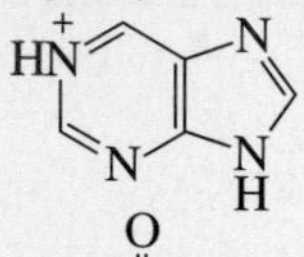 <purine, $H\overset{+}{N}$>	2.5	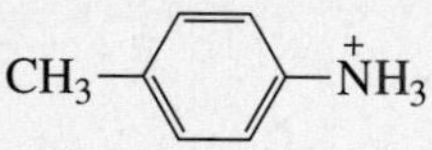	
$\text{<benzene>}{-}SO_3H$	−6.5	FCH_2COH (with C=O)	2.7	$CH_3{-}\text{<benzene>}{-}\overset{+}{N}H_3$	5.1
$CH_3\overset{+OH}{\underset{\parallel}{C}}OCH_3$	−6.5	$ClCH_2COH$ (with C=O)	2.8	<pyridine, $\overset{+}{N}H$>	5.2
$CH_3\overset{+OH}{\underset{\parallel}{C}}OH$	−6.1	$BrCH_2COH$ (with C=O)	2.9	$CH_3O{-}\text{<benzene>}{-}\overset{+}{N}H_3$	5.3
H_2SO_4	−5	ICH_2COH (with C=O)	3.2	$CH_3C{=}\overset{+}{N}HCH_3$ with CH_3	5.5
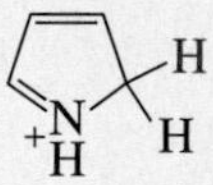 <pyrrolidinium, $\overset{+}{N}H$, H>	−3.8	HF	3.2		
$CH_3CH_2\overset{H}{\underset{+}{O}}CH_2CH_3$	−3.6	HNO_2	3.4	$CH_3\overset{O}{\underset{\parallel}{C}}CH_2\overset{O}{\underset{\parallel}{C}}H$	5.9
$CH_3CH_2\overset{H}{\underset{+}{O}}H$	−2.4	$O_2N{-}\text{<benzene>}{-}COH$ (with C=O)	3.4	$HO\overset{+}{N}H_3$	6.0
$CH_3\overset{H}{\underset{+}{O}}H$	−2.5	$HCOH$ (with C=O)	3.8	H_2CO_3	6.4
H_3O^+	−1.7			<imidazolium, $HN\overset{+}{N}H$>	6.8
HNO_3	−1.3	$Br{-}\text{<benzene>}{-}\overset{+}{N}H_3$	3.9	H_2S	7.0
CH_3SO_3H	−1.2	$Br{-}\text{<benzene>}{-}COH$ (with C=O)	4.0	$O_2N{-}\text{<benzene>}{-}OH$	7.1
$CH_3\overset{+OH}{\underset{\parallel}{C}}NH_2$	0.0	<pyridine-COH, with C=O>	4.2	$H_2PO_4^-$	7.2
F_3CCOH (with C=O)	0.2			$\text{<benzene>}{-}SH$	7.8
Cl_3CCOH (with C=O)	0.64				
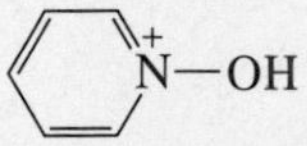 <pyridinium, $\overset{+}{N}{-}OH$>	0.79				

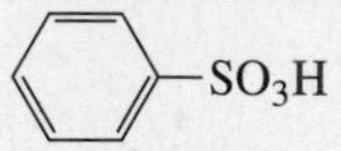

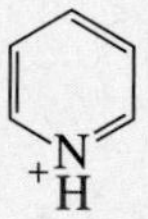

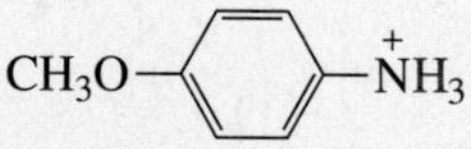

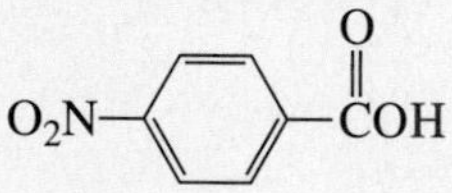

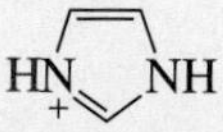

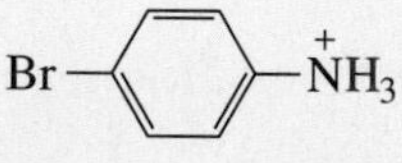

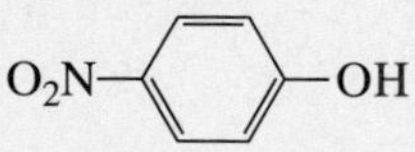

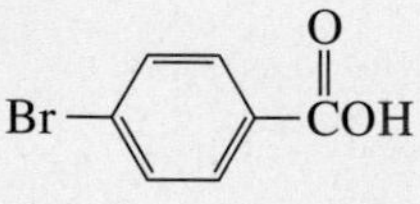

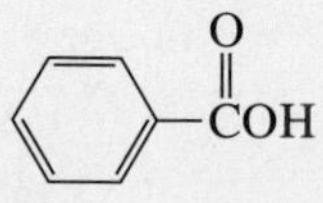

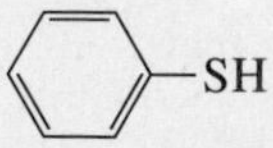

a pK_a values are for the red H in each structure

pKa Values (continued)

Compound	pKa	Compound	pKa	Compound	pKa
(aziridinium, $\overset{+}{N}H_2$)	8.0	(cyclohexyl-$\overset{+}{N}H_3$)	10.7	$CH_3\overset{\displaystyle O}{\overset{\|}{C}}H$	17
$H_2N\overset{+}{N}H_3$	8.1	$(CH_3)_2\overset{+}{N}H_2$	10.7	$(CH_3)_3COH$	18
$CH_3\overset{\displaystyle O}{\overset{\|}{C}}OOH$	8.2	(piperidinium, $\overset{+}{N}H_2$)	11.1	$CH_3\overset{\displaystyle O}{\overset{\|}{C}}CH_3$	20
$CH_3CH_2NO_2$	8.6	$CH_3CH_2\overset{+}{N}H_3$	11.0	$CH_3\overset{\displaystyle O}{\overset{\|}{C}}OCH_2CH_3$	24.5
$CH_3\overset{\displaystyle O}{\overset{\|}{C}}CH_2\overset{\displaystyle O}{\overset{\|}{C}}CH_3$	8.9	(pyrrolidinium, $\overset{+}{N}H_2$)	11.3	$HC\equiv CH$	25
$HC\equiv N$	9.1			$CH_3C\equiv N$	25
		$HOOH$	11.6	$CH_3\overset{\displaystyle O}{\overset{\|}{C}}N(CH_3)_2$	30
(morpholinium, $\overset{+}{N}H_2$)	9.3	HPO_4^{2-}	12.3	NH_3	36
		CF_3CH_2OH	12.4	(pyrrolidine, N–H)	36
$Cl-\!\!\langle\text{}\rangle\!\!-OH$	9.4	$CH_3CH_2O\overset{\displaystyle O}{\overset{\|}{C}}CH_2\overset{\displaystyle O}{\overset{\|}{C}}OCH_2CH_3$	13.3	CH_3NH_2	40
$\overset{+}{N}H_4$	9.4	$HC\equiv CCH_2OH$	13.5	(toluene, $-CH_3$)	41
$HOCH_2CH_2\overset{+}{N}H_3$	9.5	$H_2N\overset{\displaystyle O}{\overset{\|}{C}}NH_2$	13.7	(benzene)	43
$H_3\overset{+}{N}CH_2\overset{\displaystyle O}{\overset{\|}{C}}O^-$	9.8	$CH_3\overset{\displaystyle CH_3}{\overset{\|}{N}}CH_2CH_2OH$ $\underset{CH_3}{}$	13.9	$CH_2{=}CHCH_3$	43
(phenol, $-OH$)	10.0			$CH_2{=}CH_2$	44
$CH_3-\!\!\langle\text{}\rangle\!\!-OH$	10.2	(imidazole, $N{=}/NH$)	14.4	(cyclopropane)	46
HCO_3^-	10.2	CH_3OH	15.5	CH_4	60
CH_3NO_2	10.2	H_2O	15.7	CH_3CH_3	> 60
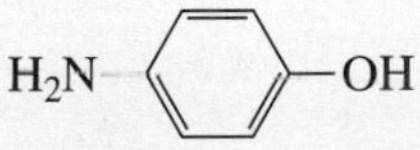 $H_2N-\!\!\langle\text{}\rangle\!\!-OH$	10.3	CH_3CH_2OH	16.0		
CH_3CH_2SH	10.5	$CH_3\overset{\displaystyle O}{\overset{\|}{C}}NH_2$	16		
$(CH_3)_3\overset{+}{N}H$	10.6	(cyclohexenyl $\overset{\displaystyle O}{\overset{\|}{C}}CH_3$)	16.0		
$CH_3\overset{\displaystyle O}{\overset{\|}{C}}CH_2\overset{\displaystyle O}{\overset{\|}{C}}OCH_2CH_3$	10.7	(pyrrole, N–H)	~17		
$CH_3\overset{+}{N}H_3$	10.7				

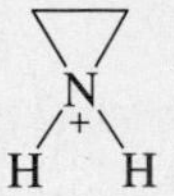

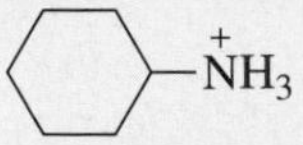

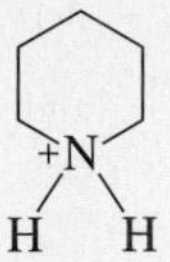

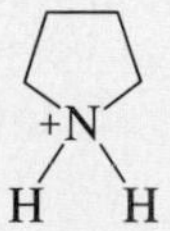

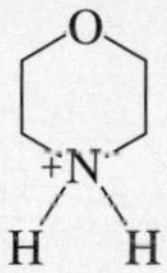

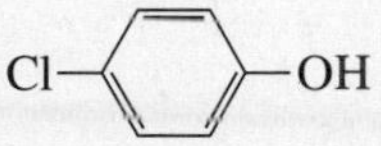

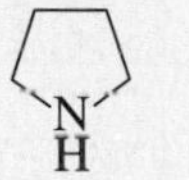

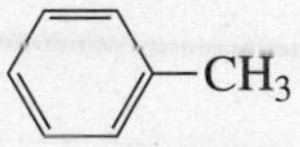

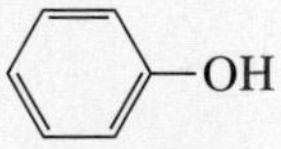

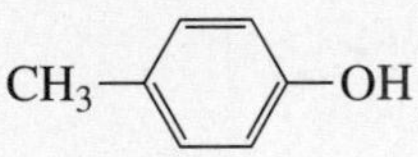

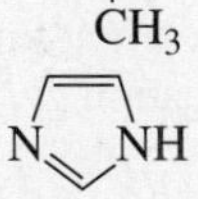

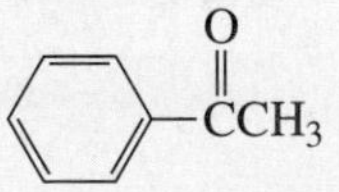

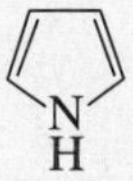

Appendix III

Derivations of Rate Laws

How to Determine Rate Constants

A **reaction mechanism** is a detailed analysis of how the chemical bonds (or the electrons) in the reactants rearrange to form the products. The mechanism for a given reaction must obey the observed rate law for the reaction. A **rate law** tells how the rate of a reaction depends on the concentration of the species involved in the reaction.

First-Order Reaction

The rate is proportional to the concentration of one reactant:

$$A \xrightarrow{k_1} products$$

Rate law: $\qquad\qquad rate = k_1[A]$

To determine the first-order rate constant (k_1),

Change in the concentration of A with respect to time:

$$\frac{-d[A]}{dt} = k_1[A]$$

Let a = the initial concentration of A;
let x = concentration of A that has reacted up to time t.
Therefore, the concentration of A left at time t is $(a - x)$
Substituting into the previous equation gives

$$\frac{-d(a - x)}{dt} = k_1(a - x)$$

$$\frac{-da}{dt} + \frac{dx}{dt} = k_1(a - x)$$

$$0 + \frac{dx}{dt} = k_1(a - x)$$

$$\frac{dx}{(a - x)} = k_1 dt$$

Integrating the previous equation yields

$$-\ln(a - x) = k_1 t + constant$$

At $t = 0$, $x = 0$; therefore,

$$constant = -\ln a$$

$$-\ln(a - x) = k_1 t - \ln a$$

$$\ln\frac{a}{a - x} = k_1 t$$

$$\ln\frac{a - x}{a} = -k_1 t$$

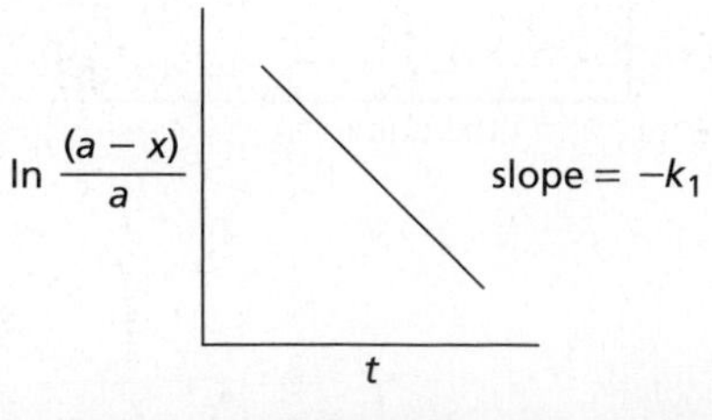

$\ln\frac{(a - x)}{a}$ versus t, slope = $-k_1$

Half-Life of a First-Order Reaction

The **half-life ($t_{1/2}$)** of a reaction is the time it takes for half the reactant to react (or for half the product to form). To derive the half-life of a reactant in a first-order reaction, we begin with the equation

$$\ln \frac{a}{(a-x)} = k_1 t$$

At $t_{1/2}$, $x = \dfrac{a}{2}$; therefore,

$$\ln \frac{a}{\left(a - \dfrac{a}{2}\right)} = k_1 t_{1/2}$$

$$\ln \frac{a}{\dfrac{a}{2}} = k_1 t_{1/2}$$

$$\ln 2 = k_1 t_{1/2}$$

$$0.693 = k_1 t_{1/2}$$

$$t_{1/2} = \frac{0.693}{k_1}$$

Notice that the half-life of a first-order reaction is independent of the concentration of the reactant.

Second-Order Reaction

The rate is proportional to the concentration of two reactants:

$$A + B \xrightarrow{k_2} \text{products}$$

Rate law: $\qquad\qquad$ rate $= k_2[A][B]$ $\qquad\qquad$ (k_2 is the rate constant)

To determine the second-order rate constant (k_2),
Change in the concentration of A with respect to time:

$$\frac{-d[A]}{dt} = k_2[A][B]$$

Let a = the initial concentration of A;
let b = the initial concentration of B;
let x = the concentration of A that has reacted at time t.

Therefore, the concentration of A left at time $t = (a - x)$, and the concentration of B left at time $t = (b - x)$.
Substitution gives

$$\frac{dx}{dt} = k_2(a - x)(b - x)$$

For the case where $a = b$ (this condition can be arranged experimentally),

$$\frac{dx}{dt} = k_2(a - x)^2$$

$$\frac{dx}{(a - x)^2} = k_2 \, dt$$

Integrating the equation gives

$$\frac{1}{(a - x)} = k_2 t + \text{constant}$$

At $t = 0$, $x = 0$; therefore,

$$\text{constant} = \frac{1}{a}$$

$$\frac{1}{(a-x)} - \frac{1}{a} = k_2 t$$

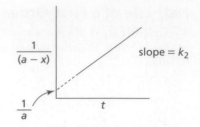

Half-Life of a Second-Order Reaction

$$\frac{1}{(a-x)} - \frac{1}{a} = k_2 t$$

At $t_{1/2}$, $x = \frac{a}{2}$; therefore,

$$\frac{1}{a} = k_2 t_{1/2}$$

$$t_{1/2} = \frac{1}{k_2 a}$$

Pseudo-First-Order Reaction

It is easier to determine a first-order rate constant than a second-order rate constant because the kinetic behavior of a first-order reaction is independent of the initial concentration of the reactant. Therefore, a first-order rate constant can be determined without knowing the initial concentration of the reactant. The determination of a second-order rate constant requires not only that the initial concentration of the reactants be known but also that the initial concentrations of the two reactants be identical in order to simplify the kinetic equation.

However, if the concentration of one of the reactants in a second-order reaction is much greater than the concentration of the other, the reaction can be treated as a first-order reaction. Such a reaction is known as a **pseudo-first-order reaction** and is given by

$$\frac{-d[A]}{dt} = k_2[A][B]$$

If $[B] \gg [A]$, then

$$\frac{-d[A]}{dt} = k_2'[A]$$

The rate constant obtained for a pseudo-first-order reaction (k_2') includes the concentration of B, but k_2 can be determined by carrying out the reaction at several different concentrations of B and determining the slope of a plot of the observed rate versus [B].

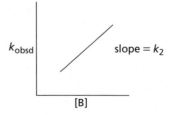

Appendix IV

Summary of Methods Used to Synthesize a Particular Functional Group

SYNTHESIS OF ACETALS

1. Acid-catalyzed reaction of an aldehyde with two equivalents of an alcohol (17.10).

SYNTHESIS OF ACID ANHYDRIDES

1. Reaction of an acyl halide with a carboxylate ion (16.8).
2. Preparation of a cyclic anhydride by heating a dicarboxylic acid (16.23).

SYNTHESIS OF ACYL CHLORIDES OR ACYL BROMIDES

1. Reaction of a carboxylic acid with $SOCl_2$, PCl_3 or PBr_3 (16.21).

SYNTHESIS OF ALCOHOLS

1. Acid-catalyzed hydration of an alkene (4.5).
2. Oxymercuration–demercuration of an alkene (4.8).
3. Hydroboration–oxidation of an alkene (4.10).
4. Reaction of an alkyl halide with HO^- (8.2, 8.5).
5. Reaction of a Grignard reagent with an epoxide (10.12).
6. Reduction of an aldehyde, a ketone, an acyl chloride, an anhydride, an ester, or a carboxylic acid (17.6, 19.1).
7. Reaction of a Grignard reagent with an aldehyde, a ketone, an acyl chloride, or an ester (17.4).
8. Reduction of a ketone with $NaBH_4$ in cold aqueous ethanol in the presence of cerium trichloride (19.1).
9. Cleavage of an ether with HI or HBr (10.7).
10. Reaction of an organozinc reagent with an aldehyde or a ketone (p. 901).

SYNTHESIS OF ALDEHYDES

1. Hydroboration–oxidation of a terminal alkyne with disiamylborane followed by $H_2O_2 + HO^-$ (6.8).
2. Oxidation of a primary alcohol with pyridinium chlorochromate (10.5, 19.2).
3. Swern oxidation of a primary alcohol with dimethyl sulfoxide, oxalyl chloride, and triethylamine (19.2).
4. Rosenmund reduction: catalytic hydrogenation of an acyl chloride (19.1).
5. Reaction of an acyl chloride with lithium tri(*tert*-butoxy)aluminum hydride (19.1).
6. Reaction of an ester with diisobutylaluminum hydride (DIBALH) (19.1).
7. Cleavage of a 1,2-diol with periodic acid (19.6).
8. Ozonolysis of an alkene, followed by workup under reducing conditions (19.7).

SYNTHESIS OF ALKANES

1. Catalytic hydrogenation of an alkene or an alkyne (4.11, 6.9, 19.1).
2. Reaction of a Grignard reagent with a source of protons (10.13).
3. Wolff–Kishner or Clemmensen reduction of an aldehyde or a ketone (14.18, 16.8).
4. Reduction of a thioacetal or thioketal with H_2 and Raney nickel (17.12).
5. Reaction of a Gilman reagent with an alkyl halide (10.13).
6. Preparation of a cyclopropane by the reaction of an alkene with a carbene (4.9).

SYNTHESIS OF ALKENES

1. Elimination of hydrogen halide from an alkyl halide (9.1, 9.2, 9.3).
2. Acid-catalyzed dehydration of an alcohol (10.4).
3. Hofmann elimination reaction: elimination of a proton and a tertiary amine from a quaternary ammonium hydroxide (20.4).
4. Exhaustive methylation of an amine, followed by a Hofmann elimination reaction (20.4).

5. Hydrogenation of an alkyne with Lindlar catalyst to form a cis alkene (6.9, 19.1).
6. Reduction of an alkyne with Na (or Li) and liquid ammonia to form a trans alkene (6.9, 19.1).
7. Formation of a cyclic alkene using a Diels–Alder reaction (7.12, 29.4).
8. Wittig reaction: reaction of an aldehyde or a ketone with a phosphonium ylide (17.13).
9. Reaction of a Gilman reagent with a halogenated alkene (10.13).
10. Heck reaction couples a vinyl halide with an alkene in a basic solution in the presence of $Pd(PPh_3)_4$ (10.13).
11. Stille reaction couples a vinyl halide with a stannane in the presence of $Pd(PPh_3)_4$ (10.13).
12. Suzuki reaction couples a vinyl halide with an organoborane in the presence of $Pd(PPh_3)_4$ (10.13).

SYNTHESIS OF ALKYL HALIDES
1. Addition of hydrogen halide (HX) to an alkene (4.1).
2. Addition of HBr + peroxide to an alkene (11.6).
3. Addition of halogen to an alkene (4.7).
4. Addition of hydrogen halide or a halogen to an alkyne (6.6).
5. Radical halogenation of an alkane, an alkene, or an alkyl benzene (11.2, 11.8).
6. Reaction of an alcohol with hydrogen halide, $SOCl_2$, PCl_3, or PBr_3 (10.1, 10.2).
7. Reaction of a sulfonate ester with halide ion (10.3).
8. Cleavage of an ether with HI or HBr (10.7).
9. Halogenation of an α-carbon of an aldehyde, a ketone, or a carboxylic acid (18.5, 18.6).

SYNTHESIS OF ALKYNES
1. Elimination of hydrogen halide from a vinyl halide (9.10).
2. Two successive eliminations of hydrogen halide from a vicinal dihalide or a geminal dihalide (9.10).
3. Reaction of an acetylide ion (formed by removing a proton from a terminal alkyne) with an alkyl halide (6.11).

SYNTHESIS OF AMIDES
1. Reaction of an acyl chloride, an acid anhydride, or an ester with ammonia or with an amine (16.8, 16.9, 16.10).
2. Reaction of a carboxylic acid and an amine with dicyclohexylcarbodiimide (22.10).
3. Reaction of a nitrile with a secondary or tertiary alcohol (p. 785).

SYNTHESIS OF AMINES
1. Reaction of an alkyl halide with NH_3, RNH_2, or R_2NH (8.4).
2. Reaction of an alkyl halide with azide ion, followed by reduction of the alkyl azide (8.4).
3. Reduction of an imine, a nitrile, or an amide (19.1).
4. Reductive amination of an aldehyde or a ketone (17.8).
5. Gabriel synthesis of primary amines: reaction of a primary alkyl halide with potassium phthalimide (16.18, 20.8).
6. Reduction of a nitro compound (14.19).
7. Condensation of a secondary amine and formaldehyde with a carbon acid (p. 904).

SYNTHESIS OF AMINO ACIDS
1. Hell–Volhard–Zelinski reaction: halogenation of a carboxylic acid, followed by treatment with excess NH_3 (18.6).
2. Reductive amination of an α-keto acid (22.6).
3. The N-phthalimidomalonic ester synthesis (Section 22.6).
4. The acetamidomalonic ester synthesis (Section 22.6).
5. The Strecker synthesis: reaction of an aldehyde with ammonia, followed by addition of cyanide ion and hydrolysis (Section 22.6).

SYNTHESIS OF CARBOXYLIC ACIDS
1. Oxidation of a primary alcohol (10.5, 19.2).
2. Oxidation of an aldehyde (19.3).
3. Ozonolysis of a monosubstituted alkene or a 1,2-disubstituted alkene, followed by workup under oxidizing conditions (19.7).
4. Ozonolysis of an alkyne (19.8).
5. Oxidation of an alkyl benzene (14.19).
6. Hydrolysis of an acyl halide, an acid anhydride, an ester, an amide, or a nitrile (16.8, 16.9, 16.10, 16.16, 16.19).

7. Haloform reaction: reaction of a methyl ketone with excess Br_2 (or Cl_2 or I_2) + HO^- (18.5).
8. Reaction of a Grignard reagent with CO_2 (17.4).
9. Malonic ester synthesis (18.19).
10. Favorskii reaction: reaction of an α-haloketone with hydroxide ion (p. 904).

SYNTHESIS OF CYANOHYDRINS
1. Reaction of an aldehyde or a ketone with sodium cyanide and HCl (17.7).

SYNTHESIS OF 1,2-DIOLS
1. Reaction of an epoxide with water (10.8).
2. Reaction of an alkene with osmium tetroxide or potassium permanganate (19.5).

SYNTHESIS OF DISULFIDES
1. Mild oxidation of a thiol (22.8).

SYNTHESIS OF ENAMINES
1. Reaction of an aldehyde or a ketone with a secondary amine (17.8).

SYNTHESIS OF EPOXIDES
1. Reaction of an alkene with a peroxyacid (4.9).
2. Reaction of a halohydrin with hydroxide ion (p. 480).
3. Reaction of an aldehyde or a ketone with a sulfonium ylide (p. 845).

SYNTHESIS OF ESTERS
1. Reaction of an acyl halide or an acid anhydride with an alcohol (16.8, 16.9).
2. Acid-catalyzed reaction of an ester or a carboxylic acid with an alcohol (16.10, 16.15).
3. Reaction of an alkyl halide with a carboxylate ion (8.4).
4. Reaction of a sulfonate ester with a carboxylate ion (10.3).
5. Oxidation of a ketone (19.3).
6. Preparation of a methyl ester by the reaction of a carboxylate ion with diazomethane (15.11).

SYNTHESIS OF ETHERS
1. Acid-catalyzed addition of an alcohol to an alkene (4.5).
2. Alkoxymercuration–demercuration of an alkene (4.8).
3. Williamson ether synthesis: reaction of an alkoxide ion with an alkyl halide (9.9).
4. Formation of symmetrical ethers by heating an acidic solution of a primary alcohol (10.4).

SYNTHESIS OF HALOHYDRINS
1. Reaction of an alkene with Br_2 (or Cl_2) and H_2O (4.7).
2. Reaction of an epoxide with a hydrogen halide (10.8).

SYNTHESIS OF IMINES
1. Reaction of an aldehyde or a ketone with a primary amine (17.8).

SYNTHESIS OF KETALS
1. Acid-catalyzed reaction of a ketone with two equivalents of an alcohol (17.10).

SYNTHESIS OF KETONES
1. Addition of water to an alkyne (6.7).
2. Hydroboration–oxidation of an alkyne (6.8).
3. Oxidation of a secondary alcohol (10.5, 19.2).
4. Cleavage of a 1,2-diol with periodic acid (19.6).
5. Ozonolysis of an alkene (19.7).
6. Friedel–Crafts acylation of an aromatic ring (14.14).
7. Preparation of a methyl ketone by the acetoacetic ester synthesis (18.20).
8. Reaction of a Gilman reagent with an acyl chloride (10.13).
9. Preparation of a cyclic ketone by the reaction of the next-size-smaller cyclic ketone with diazomethane (p. 848).

SYNTHESIS OF α,β-UNSATURATED KETONES
1. Elimination from an α-haloketone (18.7).
2. Selenenylation of a ketone, followed by oxidative elimination (p. 904).

SYNTHESIS OF NITRILES
1. Reaction of an alkyl halide with cyanide ion (8.4).
2. Reaction of a ketone with hydroxylamine followed by dehydration of the resulting oxime with acetic anhydride (21.9).

SYNTHESIS OF SUBSTITUTED BENZENES

1. Halogenation with Br_2 or Cl_2 and a Lewis acid (14.11).
2. Nitration with HNO_3 + H_2SO_4 (14.12).
3. Sulfonation: reaction with H_2SO_4 (14.13).
4. Friedel–Crafts acylation (14.14).
5. Friedel–Crafts alkylation (14.15, 14.16).
6. Sandmeyer reaction: reaction of an arenediazonium salt with CuBr, CuCl, or CuCN (15.9).
7. Formation of a phenol by reaction of an arenediazonium salt with water (15.9).
8. Formation of an aniline by reaction of a benzyne intermediate with $^-NH_2$ (15.13).
9. Reaction of a Gilman reagent with an aryl halide (10.13).
10. Heck reaction: couples a benzyl halide or an aryl halide or a triflate with an alkene in a basic solution in the presence of $Pd(PPh_3)_4$ (10.13).
11. Stille reaction: couples a benzyl halide or an aryl halide or a triflate with a stannane in the presence of $Pd(PPh_3)_4$ (10.13).
12. Suzuki reaction: couples a benzyl or aryl halide with an organoborane in the presence of $Pd(PPh_3)_4$ (10.13).

SYNTHESIS OF SULFIDES

1. Reaction of a thiol with an alkyl halide (8.4).
2. Reaction of a thiol with a sulfonate ester (10.3).

SYNTHESIS OF THIOLS

1. Reaction of an alkyl halide with hydrogen sulfide (8.4).
2. Catalytic hydrogenation of a disulfide (22.8).

Appendix V

Summary of Methods Employed to Form Carbon–Carbon Bonds

1. Reaction of an acetylide ion with an alkyl halide or a sulfonate ester (6.11, 8.4, 10.3).
2. Diels–Alder and other cycloaddition reactions (7.12, 29.4).
3. Reaction of a Grignard reagent with an epoxide (10.8).
4. Friedel–Crafts alkylation and acylation (14.14–14.16).
5. Reaction of a cyanide ion with an alkyl halide or a sulfonate ester (8.4, 10.3).
6. Reaction of a cyanide ion with an aldehyde or a ketone (17.7).
7. Reaction of a Grignard reagent with an epoxide, an aldehyde, a ketone, an ester, an amide, or CO_2 (10.12, 17.4).
8. Reaction of an organozinc reagent with an aldehyde or a ketone (p. 901).
9. Reaction of an alkene with a carbene (4.9).
10. Reaction of a Giman reagent (lithium dialkylcuprate) with an α,β-unsaturated ketone or an α,β-unsaturated aldehyde (17.16).
11. Aldol addition (18.12–18.14, 18.17).
12. Claisen condensation (18.15–18.17).
13. Perkin condensation (p. 900).
14. Knoevenagel condensation (p. 901).
15. Malonic ester synthesis and acetoacetic ester synthesis (18.19, 18.20).
16. Michael addition reaction (18.11).
17. Alkylation of an enamine (18.10).
18. Alkylation of the α-carbon of a carbonyl compound (18.9).
19. Reaction of a Gilman reagent with an aryl halide or a halogenated alkene (10.13).
20. Heck reaction: couples a vinyl, a benzyl, or an aryl halide, or a triflate with an alkene in a basic solution in the presence of $Pd(PPh_3)_4$ (10.13).
21. Stille reaction: couples a vinyl, a benzyl, or an aryl halide, or a triflate with a stannane in the presence of $Pd(PPh_3)_4$ (10.13).
22. Suzuki reaction: couples a vinyl, a benzyl, or an aryl halide with an organoborane in the presence of $Pd(PPh_3)_4$ (10.13).

Appendix VI
Spectroscopy Tables

Mass Spectrometry			
Common fragment ions*			
m/z	**Ion**	*m/z*	**Ion**
14	CH_2	46	NO_2
15	CH_3	47	CH_2SH, CH_3S
16	O	48	$CH_3S + H$
17	OH	49	CH_2Cl
18	H_2O, NH_4	51	CHF_2
19	F, H_3O	53	C_4H_5
26	$C\equiv N$	54	$CH_2CH_2C\equiv N$
27	C_2H_3	55	C_4H_7, $CH_2=CHC=O$
28	C_2H_4, CO, N_2, $CH=NH$	56	C_4H_8
29	C_2H_5, CHO	57	C_4H_9, $C_2H_5C=O$
30	CH_2NH_2, NO		
31	CH_2OH, OCH_3	58	$CH_3\overset{\displaystyle O}{\overset{\|}{C}}CH_2 + H$, $C_2H_5CHNH_2$, $(CH_3)_2NCH_2$,
32	O_2 (air)		$C_2H_5NHCH_2$, C_2H_2S
33	SH, CH_2F		
34	H_2S	59	$(CH_3)_2COH$, $CH_2OC_2H_5$, $\overset{\displaystyle O}{\overset{\|}{C}}OCH_3$,
35	Cl		$CH_2C=O + H$, CH_3OCHCH_3,
36	HCl		$\overset{\|}{NH_2}$
39	C_3H_3		
40	$CH_2C\equiv N$	60	CH_3CHCH_2OH
41	C_3H_5, $CH_2C\equiv N + H$, C_2H_2NH		$CH_2COOH + H$, CH_2ONO
42	C_3H_6		
43	C_3H_7, $CH_3C=O$, C_2H_5N		
44	$CH_2CH=O + H$, CH_3CHNH_2, CO_2, $NH_2C=O$, $(CH_3)_2N$		
45	CH_3CHOH, CH_2CH_2OH, CH_2OCH_3, COOH, $CH_3CHO + H$		

* All of these ions have a single positive charge.

Mass Spectrometry

Common fragment lost

Molecular ion minus	Fragment lost		Molecular ion minus	Fragment lost
1	H			
15	CH_3		43	C_3H_7, $CH_3\overset{O}{\overset{\|}{C}}$, $CH_2=CHO$, HCNO, $CH_3 + CH_2=CH_2$
17	HO			
18	H_2O		44	$CH_2=CHOH$, CO_2, N_2O, $CONH_2$, $NHCH_2CH_3$
19	F		45	CH_3CHOH, CH_3CH_2O, CO_2H, $CH_3CH_2NH_2$
20	HF		46	$H_2O + CH_2=CH_2$, CH_3CH_2OH, NO_2
26	$CH\equiv CH$, $C\equiv N$		47	CH_3S
27	$CH_2=CH$, $HC\equiv N$		48	CH_3SH, SO, O_3
28	$CH_2=CH_2$, CO, (HCN + H)		49	CH_2Cl
29	CH_3CH_2, CHO		51	CHF_2
30	NH_2CH_2, CH_2O, NO		52	C_4H_4, C_2N_2
31	OCH_3, CH_2OH, CH_3NH_2		53	C_4H_5
32	CH_3OH, S		54	$CH_2=CHCH=CH_2$
33	HS, (CH_3 and H_2O)		55	$CH_2=CHCHCH_3$
34	H_2S		56	$CH_2=CHCH_2CH_3$, $CH_3CH=CHCH_3$
35	Cl		57	C_4H_9
36	HCl, 2 H_2O		58	NCS, NO + CO, CH_3COCH_3
37	HCl + H			
38	C_3H_2, C_2N, F_2		59	$CH_3O\overset{O}{\overset{\|}{C}}$, $CH_3\overset{O}{\overset{\|}{C}}NH_2$
39	C_3H_3, HC_2N		60	C_3H_7OH
40	$CH_3C\equiv CH$			
41	$CH_2=CHCH_2$			
42	$CH_2=CHCH_3$, $CH_2=C=O$, $\underset{CH_2-CH_2}{\overset{CH_2}{\diagup \diagdown}}$, NCO			

¹H NMR Chemical Shifts

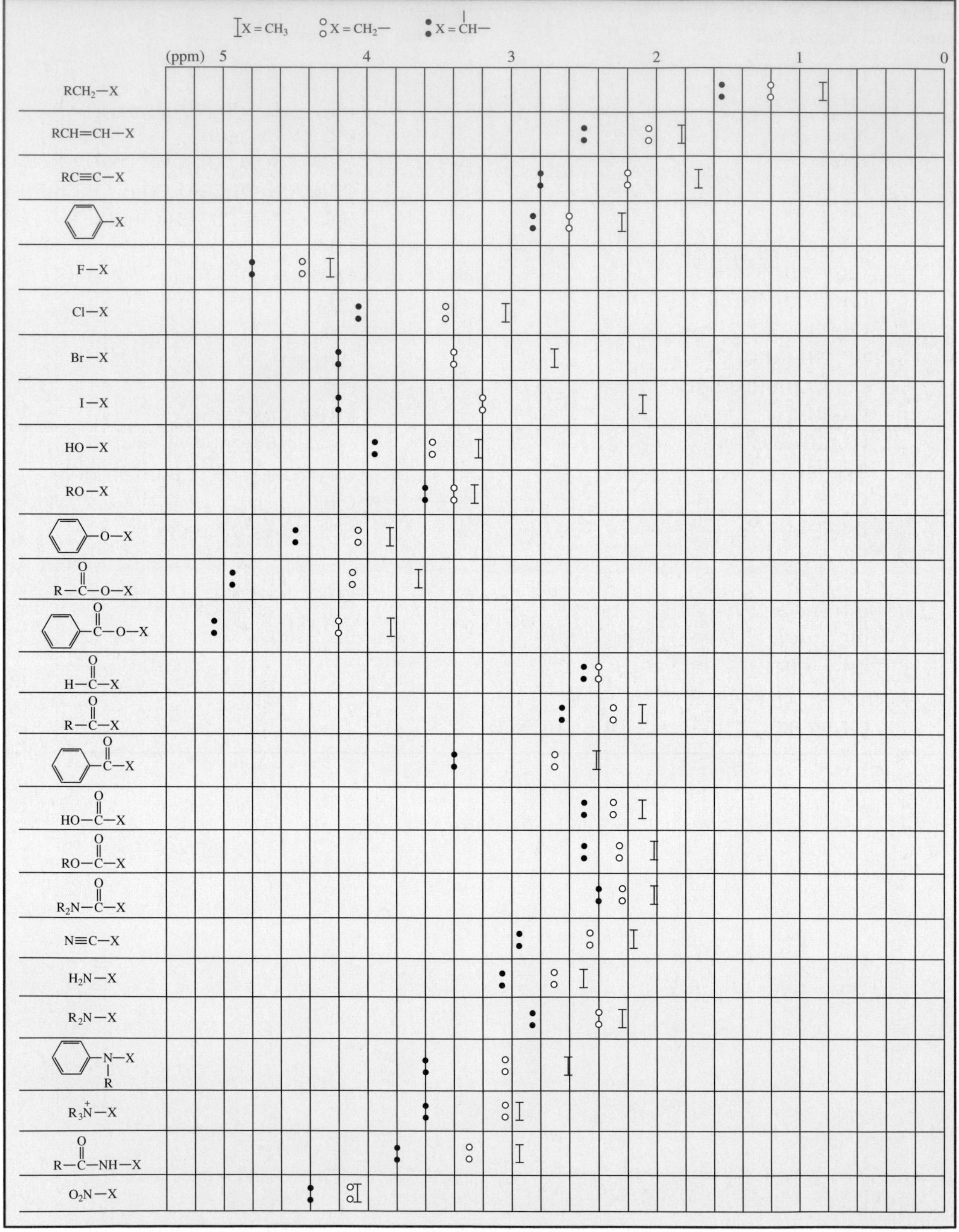

$\text{X} = CH_3$ $\text{X} = CH_2-$ $\text{X} = \overset{|}{C}H-$

(ppm) 5   4   3   2   1   0

RCH_2-X	
$RCH=CH-X$	
$RC\equiv C-X$	
⬡—X	
$F-X$	
$Cl-X$	
$Br-X$	
$I-X$	
$HO-X$	
$RO-X$	
⬡—O—X	
$R-\overset{O}{\overset{\|}{C}}-O-X$	
⬡—$\overset{O}{\overset{\|}{C}}$—O—X	
$H-\overset{O}{\overset{\|}{C}}-X$	
$R-\overset{O}{\overset{\|}{C}}-X$	
⬡—$\overset{O}{\overset{\|}{C}}$—X	
$HO-\overset{O}{\overset{\|}{C}}-X$	
$RO-\overset{O}{\overset{\|}{C}}-X$	
$R_2N-\overset{O}{\overset{\|}{C}}-X$	
$N\equiv C-X$	
H_2N-X	
R_2N-X	
⬡—$\overset{|}{\underset{R}{N}}$—X	
$R_3\overset{+}{N}-X$	
$R-\overset{O}{\overset{\|}{C}}-NH-X$	
O_2N-X	

Characteristic Infrared Group Frequencies (S = strong, M = medium, W = weak). (Courtesy of N.B. Colthup, Stamford Research Laboratories, American Cyanamid Company, and the editor of the *Journal of the Optical Society*.) Overtone bands are marked 2ν.

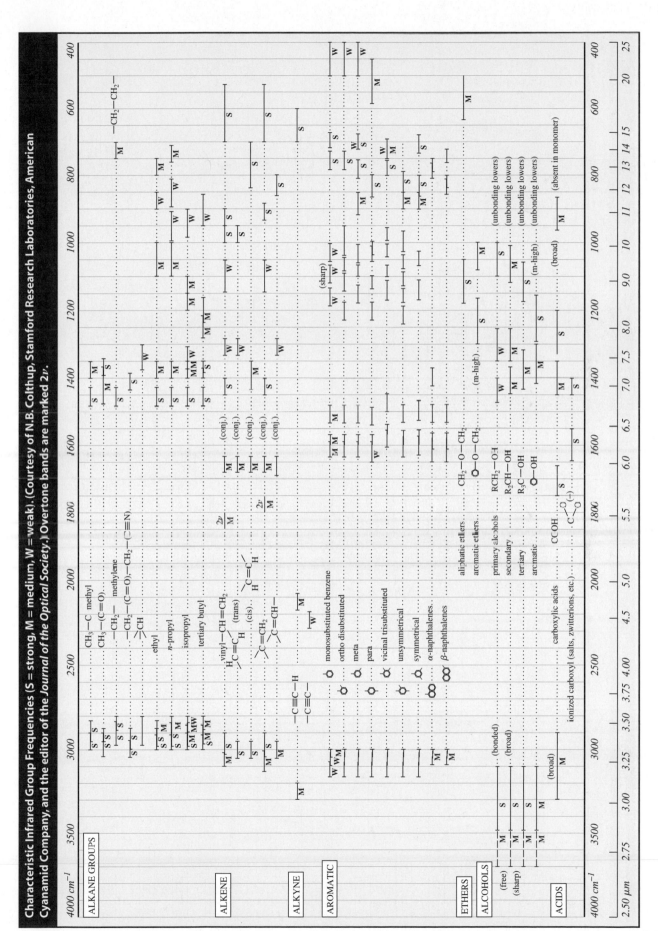

A-21

Characteristic Infrared Group Frequencies (continued)

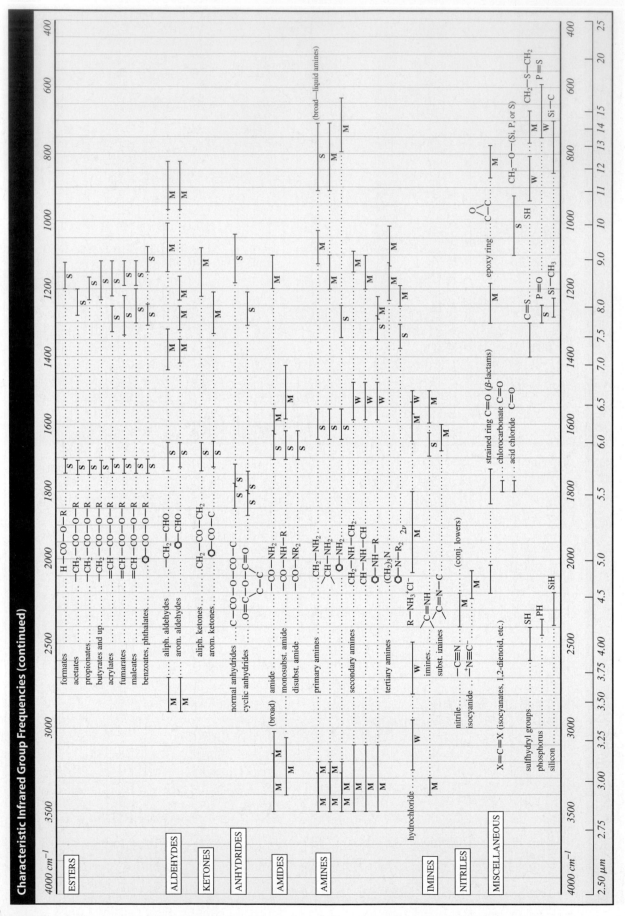

A-22

Characteristic Infrared Group Frequencies (continued)

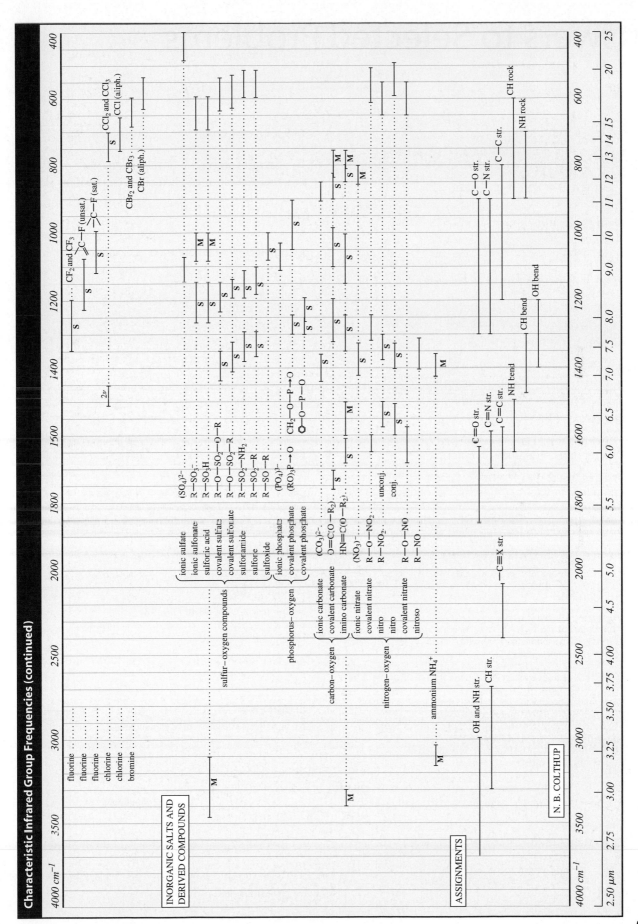

INORGANIC SALTS AND DERIVED COMPOUNDS

ASSIGNMENTS

N. B. COLTHUP

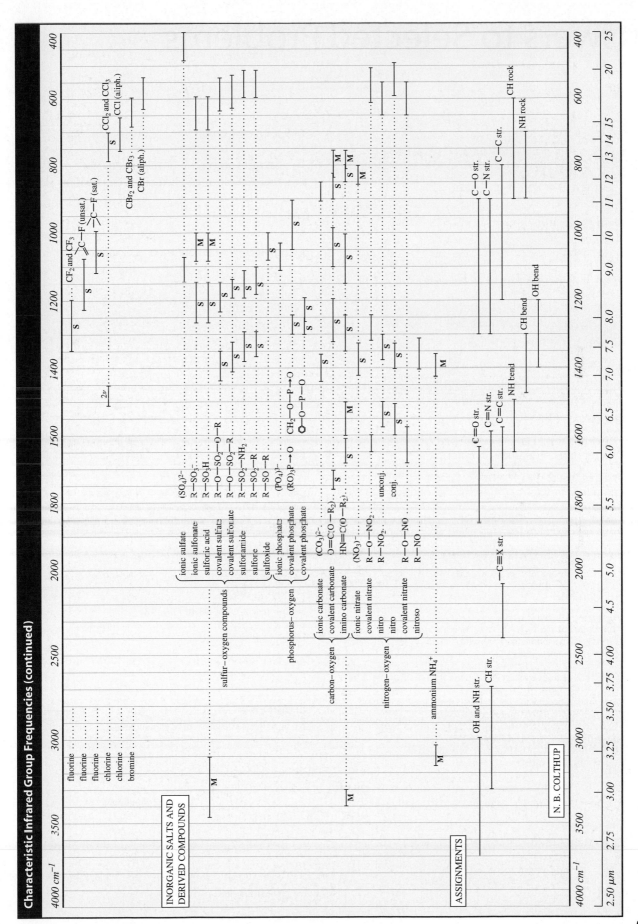

A-23

Answers to Selected Problems

CHAPTER 1

1-1. $8 + 8, 8 + 9, 8 + 10$ **1-2. a.** 3 **b.** 5 **c.** 6 **d.** 7 **1-3. a.** 1 **b.** 4s

1-4. a. Cl $1s^2 2s^2 2p^6 3s^2 3p^5$; Br $1s^2 2s^2 2p^6 3s^2 3p^6 4s^2 3d^{10} 4p^5$;
I $1s^2 2s^2 2p^6 3s^2 3p^6 4s^2 3d^{10} 4p^6 5s^2 4d^{10} 5p^5$ **b.** 7

1-6. a. Cl—CH$_3$ **b.** H—OH **c.** H—F **d.** Cl—CH$_3$

1-7. a. KCl **b.** Cl$_2$

1-9. a. $\overset{\delta-}{\text{HO}}$—$\overset{\delta+}{\text{H}}$ **c.** $\text{H}_3\overset{\delta+}{\text{C}}$—$\overset{\delta-}{\text{NH}_2}$ **e.** $\overset{\delta-}{\text{HO}}$—$\overset{\delta+}{\text{Br}}$ **g.** $\overset{\delta+}{\text{I}}$—$\overset{\delta-}{\text{Cl}}$

b. $\overset{\delta-}{\text{F}}$—$\overset{\delta+}{\text{Br}}$ **d.** $\text{H}_3\overset{\delta+}{\text{C}}$—$\overset{\delta-}{\text{Cl}}$ **f.** $\text{H}_3\overset{\delta-}{\text{C}}$—$\overset{\delta+}{\text{MgBr}}$ **h.** $\text{H}_2\overset{\delta+}{\text{N}}$—$\overset{\delta-}{\text{OH}}$

1-10. a. LiH and HF **b.** Its hydrogen has the greatest electron density. **c.** HF

1-11. a. oxygen **b.** oxygen **c.** oxygen **d.** hydrogen

1-12. a. $\text{CH}_3-\overset{\cdot\cdot}{\underset{\text{H}}{\overset{+}{\text{O}}}}-\text{CH}_3$ **c.** $\text{CH}_3-\overset{\text{CH}_3}{\underset{\text{CH}_3}{\overset{+}{\text{N}}}}-\text{CH}_3$

b. $\text{H}-\overset{-}{\underset{\text{H}}{\overset{\cdot\cdot}{\text{C}}}}-\text{H}$ **d.** $\text{H}-\overset{\text{H}\ \ \text{H}}{\underset{\text{H}\ \ \text{H}}{\overset{+}{\text{N}}-\overset{-}{\text{B}}}}-\text{H}$

1-14. a. $\text{CH}_3\text{CH}_2\text{OH}$ CH_3OCH_3

b. $\text{CH}_3\text{CH}_2\text{CH}_2\text{OH}$ $\text{CH}_3\overset{}{\underset{\text{OH}}{\text{CHCH}_3}}$ $\text{CH}_3\text{CH}_2\text{OCH}_3$

1-15. a. $\text{CH}_3\text{CH}_2\overset{\cdot\cdot}{\text{N}}\text{H}_2$ **c.** $\text{CH}_3\text{CH}_2\overset{\cdot\cdot}{\underset{\cdot\cdot}{\text{O}}}\text{H}$ **e.** $\text{CH}_3\text{CH}_2\overset{\cdot\cdot}{\underset{\cdot\cdot}{\text{Cl}}}:$

b. $\text{CH}_3\overset{\cdot\cdot}{\text{N}}\text{HCH}_3$ **d.** $\text{CH}_3\overset{\cdot\cdot}{\underset{\cdot\cdot}{\text{O}}}\text{CH}_3$ **f.** $\text{H}\overset{\cdot\cdot}{\underset{\cdot\cdot}{\text{O}}}\overset{\cdot\cdot}{\text{N}}\text{H}_2$

1-16. a. $\text{CH}_3\text{CH}_2\text{CH}_2\text{Cl}$ **c.** $\text{CH}_3\text{CH}_2\overset{\text{O}}{\overset{\|}{\text{C}}}\text{NCH}_2\text{CH}_3$ with CH_3 below N

b. $\text{CH}_3\overset{\text{O}}{\overset{\|}{\text{C}}}\text{OCH}_2\text{CH}_3$ **d.** $\text{CH}_3\text{CH}_2\text{C}\equiv\text{N}$

1-17. a. Cl **b.** O **c.** N **d.** C and H **1-19.** yes

1-20. a. π^* **b.** σ^* **c.** σ^* **d.** σ **1-21.** The C—C bonds are formed by sp^3–sp^3 overlap; the C—H bonds are formed by sp^3–s overlap. **1-23.** $> 104.5° < 109.5°$

1-24. hydrogens **1-25.** most = water least = methane **1-26.** $107.3°$

1-27. a. relative lengths: $\text{Br}_2 > \text{Cl}_2$; relative strengths: $\text{Cl}_2 > \text{Br}_2$ **b.** relative lengths; $\text{CH}_3\text{Br} > \text{CH}_3\text{Cl} > \text{CH}_3\text{F}$; relative strengths: $\text{CH}_3\text{F} > \text{CH}_3\text{Cl} > \text{CH}_3\text{Br}$

1-28. a. 1. C-I **2.** C-Cl **b. 1.** C-Cl **2.** C-C **3.** H-H **1-29.** σ

1-30. sp^2–sp^2 is stronger because the more s character, the stronger the bond

1-32. a. $109.5°$ **b.** $109.5°$ **c.** $109.5°$ **d.** $109.5°$ **1-35.** a, e, g, h

1-36. a. 1. $^+\text{NH}_4$ **2.** HCl **3.** H_2O **4.** H_3O^+ **b. 1.** $^-\text{NH}_2$ **2.** Br^- **3.** NO_3^-
4. HO^- **1-37. a.** 5.2 **b.** 3.4×10^{-3} **1-38.** 8.16×10^{-8}

1-39. $K_a = 1.51 \times 10^{-5}$; weaker **1-41. a.** basic **b.** acidic **c.** basic

1-42. a. CH_3COO^- **b.** $^-\text{NH}_2$ **c.** H_2O

1-43. $\text{CH}_3\text{NH}^- > \text{CH}_3\text{O}^- > \text{CH}_3\text{NH}_2 > \text{CH}_3\overset{\text{O}}{\overset{\|}{\text{C}}}\text{O}^- > \text{CH}_3\text{OH}$

1-47. a. 2.0×10^{-5} **b.** 3.2×10^{-7} **c.** 1.0×10^{-5} **d.** 4.0×10^{-13}

1-48. a. HBr **b.** $\text{CH}_3\text{CH}_2\text{CH}_2\overset{+}{\text{O}}\text{H}_2$ **c.** the one on the right **d.** $\text{H}_3\text{C}\overset{\text{O}}{\overset{\|}{\text{C}}}\text{SH}$

1-49. a. F^- **b.** I^- **1-50. a.** oxygen **b.** H_2S **c.** CH_3SH

1-51. a. HO^- **b.** NH_3 **c.** CH_3O^- **d.** CH_3O^-

1-52. a. $\text{CH}_3\text{OCH}_2\text{CH}_2\text{OH}$ **c.** $\text{CH}_3\text{CH}_2\text{OCH}_2\text{CH}_2\text{OH}$

b. $\text{CH}_3\text{CH}_2\text{CH}_2\overset{+}{\text{O}}\text{H}_2$ **d.** $\text{CH}_3\text{CH}_2\overset{\text{O}}{\overset{\|}{\text{C}}}\text{OH}$

1-53. $\text{CH}_3\overset{}{\underset{\text{F}}{\text{CHCH}_2\text{OH}}} > \text{CH}_3\overset{}{\underset{\text{Cl}}{\text{CHCH}_2\text{OH}}} > \text{CH}_2\overset{}{\underset{\text{Cl}}{\text{CH}_2\text{CH}_2\text{OH}}} > \text{CH}_3\text{CH}_2\text{CH}_2\text{OH}$

1-54.

a. $\text{CH}_3\overset{\text{O}}{\overset{\|}{\underset{\text{Br}}{\text{CHCO}^-}}}$ **b.** $\text{CH}_3\overset{\text{O}}{\overset{\|}{\underset{\text{Cl}}{\text{CHCH}_2\text{CO}^-}}}$ **c.** $\text{CH}_3\text{CH}_2\overset{\text{O}}{\overset{\|}{\text{CO}^-}}$ **d.** $\text{CH}_3\overset{\text{O}}{\overset{\|}{\text{C}}}\text{CH}_2\text{CH}_2\text{O}^-$

1-56. $\text{CH}_3\overset{\text{O}}{\underset{\text{O}}{\overset{\|}{\underset{\|}{\text{S}}}}}\text{—OH}$

1-57. a. $^-\text{O}\overset{\text{O}}{\overset{\|}{\text{C}}}\text{O}^- \longleftrightarrow ^-\text{O}\overset{\text{O}^-}{\overset{|}{\text{C}}}\text{O} \longleftrightarrow \text{O}\overset{\text{O}^-}{\overset{|}{\text{C}}}\text{O}^-$

b. $^-\text{O}\overset{\text{O}}{\overset{\|}{\overset{+}{\text{N}}}}\text{O}^- \longleftrightarrow ^-\text{O}\overset{\text{O}^-}{\overset{|}{\overset{+}{\text{N}}}}\text{O} \longleftrightarrow \text{O}\overset{\text{O}^-}{\overset{|}{\overset{+}{\text{N}}}}\text{O}^-$

1-58. a. $\text{CH}_3\text{C}\equiv\overset{+}{\text{N}}\text{H}$ **b.** CH_3CH_3 **c.** F_3CCOOH **d.** sp^2

e. $\text{HC}\equiv\text{CH} > \text{CH}_2\text{=CH}_2 > \text{CH}_3\text{CH}_3$ **f.** HNO_3

1-59. a. CH_3COO^- **b.** $\text{CH}_3\text{CH}_2\overset{+}{\text{N}}\text{H}_3$ **c.** H_2O **d.** Br^- **e.** $^+\text{NH}_4$ **f.** $\text{HC}\equiv\text{N}$
g. NO_2^- **h.** NO_3^- **1-60. a. 1.** neutral **2.** neutral **3.** charged **4.** charged
5. charged **6.** charged **b. 1.** charged **2.** charged **3.** charged
4. charged **5.** neutral **6.** neutral **c. 1.** neutral **2.** neutral **3.** neutral **4.** neutral
5. neutral **6.** neutral

1-61. a. $\text{CH}_3\overset{\text{O}}{\overset{\|}{\underset{^+\text{NH}_3}{\text{CHCO}^-}}}$ **b.** no **c.** 6.02

1-62. a. 10.4 **b.** 2.7 **c.** 6.4 **d.** 7.3 **e.** 5.6

1-63. a. 1. 4.9 **2.** 10.7 **b. 1.** >6.9 **2.** <8.7 **1-64.** 10.4

1-65. a. $\text{CH}_3\text{COO}^- + \text{H}^+ \rightleftharpoons \text{CH}_3\text{COOH}$

b. $\text{CH}_3\text{COOH} + \text{HO}^- \rightleftharpoons \text{CH}_3\text{COO}^- + \text{H}_2\text{O}$

CHAPTER 2

2-1. a. n-propyl alcohol or propyl alcohol **b.** dimethyl ether **c.** n-propylamine or propylamine

2-2. a. $\text{CH}_3\overset{\text{CH}_3}{\overset{|}{\text{CHCH}_2\text{CH}_3}}$ **b.** $\text{CH}_3\overset{\text{CH}_3}{\underset{\text{CH}_3}{\overset{|}{\underset{|}{\text{C}}}\text{CH}_3}}$
2-methylbutane   **2,2-dimethylpropane**

2-3.

1. $\text{CH}_3\text{CH}_2\text{CH}_2\text{CH}_2\text{Br}$ $\text{CH}_3\overset{}{\underset{\text{CH}_3}{\text{CHCH}_2\text{Br}}}$ $\text{CH}_3\text{CH}_2\overset{}{\underset{\text{Br}}{\text{CHCH}_3}}$ $\text{CH}_3\overset{\text{CH}_3}{\underset{\text{CH}_3}{\overset{|}{\underset{|}{\text{C}}}\text{Br}}}$

butyl bromide   **isobutyl bromide**   **sec-butyl bromide**   **tert-butyl bromide**
or
n-butyl bromide

2-4. c

2-5. a. 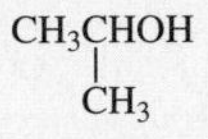 CH_3CHOH
CH_3 (below)

d. 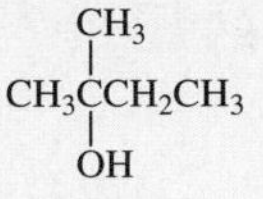 $CH_3CCH_2CH_3$ with CH_3 above and OH below

b. 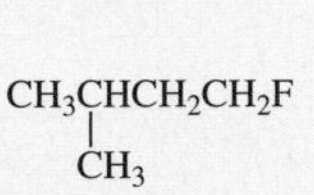 $CH_3CHCH_2CH_2F$ with CH_3 below

e. 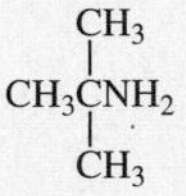 CH_3CNH_2 with CH_3 above and CH_3 below

c. 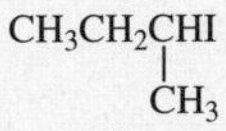 CH_3CH_2CHI with CH_3 below

f. $CH_3CH_2CH_2CH_2CH_2CH_2CH_2CH_2Br$

2-6. a. ethyl methyl ether **b.** methyl propyl ether **c.** *sec*-butylamine **d.** *n*-butyl alcohol **e.** isobutyl bromide **f.** *sec*-butyl chloride

2-7. a. 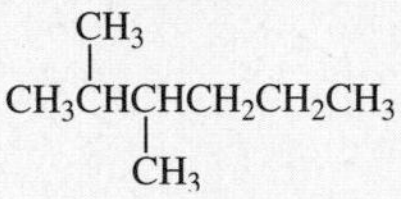 $CH_3CHCHCH_2CH_2CH_3$ with CH_3 below

d. 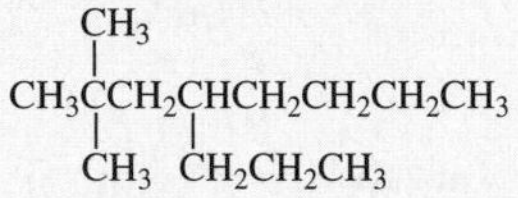 $CH_3CCH_2CHCH_2CH_2CH_3$ with CH_3 above, CH_3 and $CH_2CH_2CH_3$ below

b. 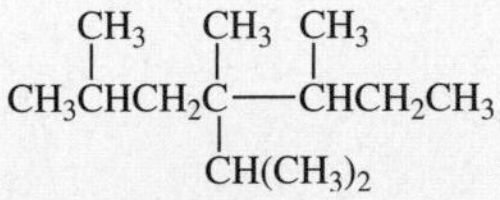 CH_3CHCH_2C—$CHCH_2CH_3$ with CH_3, CH_3, CH_3 above and $CH(CH_3)_2$ below

e. 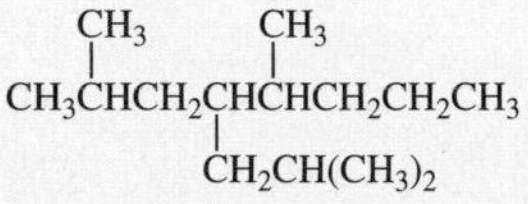 $CH_3CHCH_2CHCHCH_2CH_2CH_3$ with CH_3, CH_3 above and $CH_2CH(CH_3)_2$ below

c. 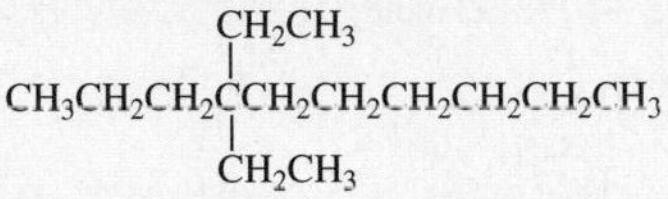 $CH_3CH_2CH_2CCH_2CH_2CH_2CH_2CH_3$ with CH_2CH_3 above and CH_2CH_3 below

f. 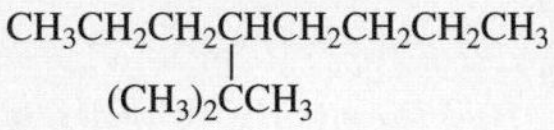 $CH_3CH_2CH_2CHCH_2CH_2CH_3$ with $(CH_3)_2CCH_3$ below

2-9. a. 2,2,4-trimethylhexane **b.** 2,2-dimethylbutane **c.** 3-methyl-4-propylheptane **d.** 2,2,5-trimethylhexane **e.** 3,3-diethyl-4-methyl-5-propyloctane **f.** 5-ethyl-4,4-dimethyloctane **g.** 3,3-diethylhexane **h.** 4-isopropyloctane **i.** 2,5-dimethylheptane

2-10. a. $CH_3CH_2CH_2CH_2CH_3$
pentane

c. 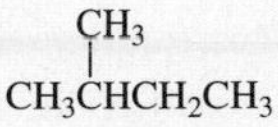 $CH_3CHCH_2CH_3$ with CH_3 above
2-methylbutane

b. 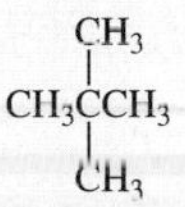 CH_3CCH_3 with CH_3 above and CH_3 below
2,2-dimethylpropane

d. 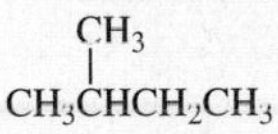 $CH_3CHCH_2CH_3$ with CH_3 above
2-methylbutane

2-12. a.  (structure) OH **d.** 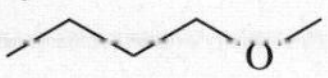 (structure) O

b. 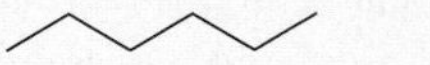 (structure)

e. 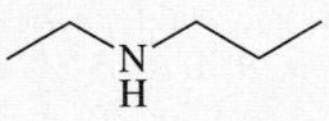 (structure) N–H

c. 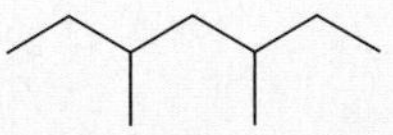 (structure)

f. 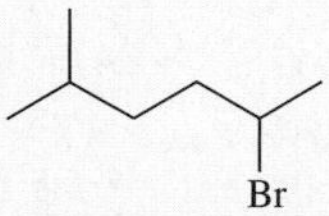 (structure) Br

2-14. a. 1-ethyl-2-methylcyclopentane **b.** ethylcyclobutane **c.** 4-ethyl-1,2-dimethylcyclohexane **d.** 3,6-dimethyldecane **e.** 2-cyclopropylpentane **f.** 1-ethyl-3-isobutylcyclohexane **g.** 5-isopropylnonane **h.** 1-*sec*-butyl-4-isopropylcyclohexane **2-15. a.** *sec*-butyl chloride, 2-chlorobutane, **secondary b.** isohexyl chloride, 1-chloro-4-methylpentane, **primary c.** cyclohexyl bromide, bromocyclohexane, **secondary d.** isopropyl fluoride, 2-fluoropropane, **secondary 2-17. a. 1.** methoxyethane **2.** ethoxyethane **3.** 4-methoxyoctane **4.** 1-isopropoxy-3-methylbutane **5.** 1-propoxybutane **6.** 2-isopropoxypentane **b.** no **c. 1.** ethyl methyl ether **2.** diethyl ether **4.** isopentyl isopropyl ether **5.** butyl propyl ether **2-19. a.** 1-pentanol, **primary b.** 4-methylcyclohexanol, **secondary c.** 5-chloro-2-methyl-2-pentanol, **tertiary d.** 5-methyl-3-hexanol, **secondary e.** 4-chloro-3-ethylcyclohexanol, **secondary f.** 2,6-dimethyl-4-octanol, **secondary**

2-20. $CH_3CCH_2CH_2CH_3$ with CH_3 above and OH below
2-methyl-2-pentanol

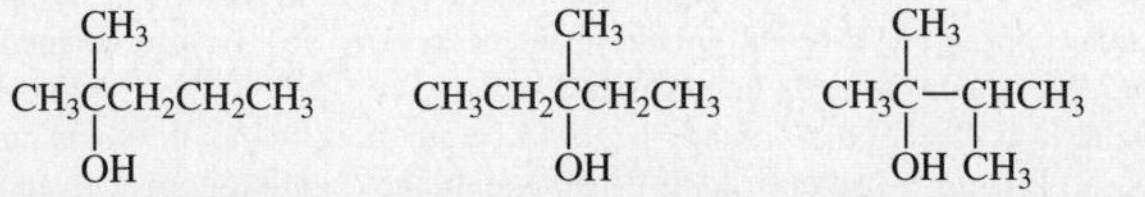

$CH_3CH_2CCH_2CH_3$ with CH_3 above and OH below
3-methyl-3-pentanol

CH_3C—$CHCH_3$ with CH_3 above and OH CH_3 below
2,3-dimethyl-2-butanol

2-21. a. tertiary **b.** tertiary **c.** primary **2-22. a.** hexylamine, 1-hexanamine, **primary b.** *sec*-butylisobutylamine, *N*-isobutyl-2-butanamine, **secondary c.** cyclohexylamine, cyclohexanamine, **primary d.** butylpropylamine, *N*-propyl-1-butanamine, **secondary e.** ethylmethylpropylamine, *N*-ethyl-*N*-methyl-1-propanamine, **tertiary f.** no common name, *N*-ethyl-3-methylcyclopentanamine, **secondary**

2-23. a.  $CH_3CHCH_2NHCH_2CH_3$ with CH_3 above **b.** $CH_3CH_2NHCH_2CH_3$

c. 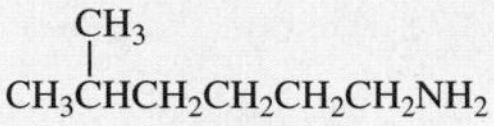 $CH_3CHCH_2CH_2CH_2NH_2$ with CH_3 above **d.** 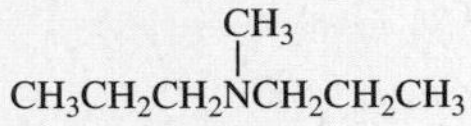 $CH_3CH_2CH_2NCH_2CH_3$ with CH_3 above

e. 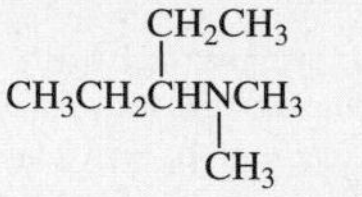 $CH_3CH_2CHNCH_3$ with CH_2CH_3 above and CH_3 below **f.** 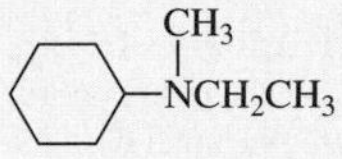 (cyclohexyl)–NCH_2CH_3 with CH_3 above

2-24. a. 6-methyl-1-heptanamine, isooctylamine, **primary b.** 3-methyl-*N*-propyl-1-butanamine, isopentylpropylamine, **secondary c.** *N*-ethyl-*N*-methylethanamine, diethylmethylamine, **tertiary d.** 2,5-dimethylcyclohexanamine, no common name, **primary**
2-25. a. 104.5° **b.** 107.3° **c.** 104.5° **d.** 109.5° **2-26. a.** 1, 4, 5 **b.** 1, 2, 4, 5, 6
2-28.

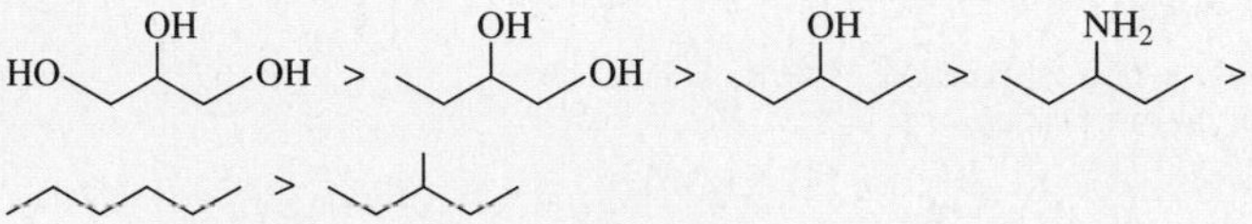

HO—(structure)—OH > (structure)—OH > (structure)—OH > (structure)—NH₂ >
(structure) > (structure)

2-30. a. $HOCH_2CH_2CH_2OH > CH_3CH_2CH_2OH >$
$CH_3CH_2CH_2CH_2OH > CH_3CH_2CH_2CH_2Cl$

b. 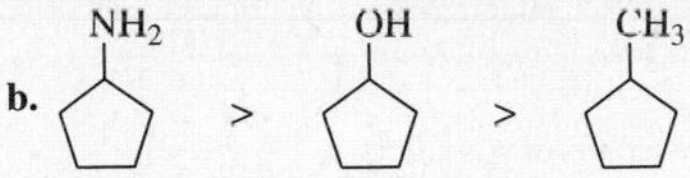 (cyclopentane with NH₂) > (cyclopentane with OH) > (cyclopentane with CH₃)

2-31. ethanol
2-33. a. 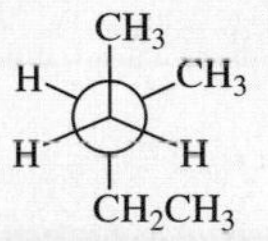 (Newman projection with CH_3, CH_3, H, H, H, H, CH_2CH_3) **b.** 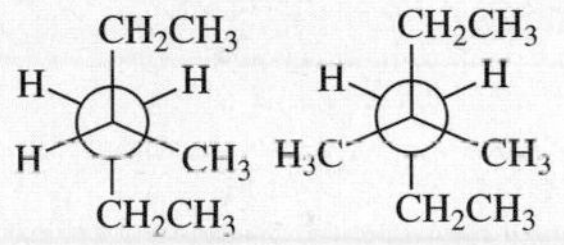 (Newman projection CH_2CH_3, H, H, H, CH_3, CH_2CH_3) (third Newman projection CH_2CH_3, H, H_3C, CH_3, CH_2CH_3)

2-34. a. 135° **b.** 140° **2-35.** hexothal **2-37.** 6.2 kcal/mol
2-38. 0.13 kcal/mol **2-39.** 84% **2-40. a.** cis **b.** cis **c.** cis **d.** trans **e.** trans **f.** trans **2-41.** *cis*-1-*tert*-butyl-3-methylcyclohexane **2-43. a.** one equatorial and one axial **b.** both equatorial and both axial **c.** both equatorial and both axial **d.** one equatorial and one axial **e.** one equatorial and one axial **f.** both equatorial and both axial **2-44. a.** 3.6 kcal/mol **b.** 0

CHAPTER 3

3-1. a. C_5H_8 **b.** C_4H_6 **c.** $C_{10}H_{16}$ **d.** C_8H_{10} **3-2. a.** 3 **b.** 4 **c.** 1 **d.** 3 **e.** 13

3-4. a. 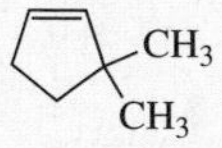 (cyclopentene with CH_3, CH_3) **c.** CH_3CH_2OCH=CH_2

b. 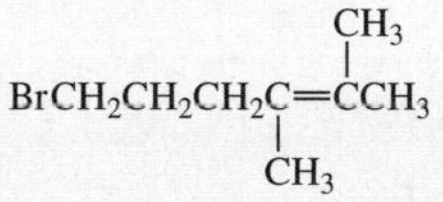 $BrCH_2CH_2CH_2C$=CCH_3 with CH_3 above and CH_3 below **d.** CH_2—$CHCH_2OH$

3-5. a. 4-methyl-2-pentene **b.** 2-chloro-3,4-dimethyl-3-hexene **c.** 1-bromocyclopentene **d.** 1-bromo-4-methyl-3-hexene **e.** 1,5-dimethylcyclohexene **f.** 1-butoxy-1-propene
3-6. a. 5 **b.** 4 **c.** 4 **d.** 6
3-7. a. 1 and 3

b. 1. (structure) H_3C and $CH_2CH_2CH_3$ / H and H, C=C, cis; H_3C and H / H and $CH_2CH_2CH_3$, C=C, trans

3. 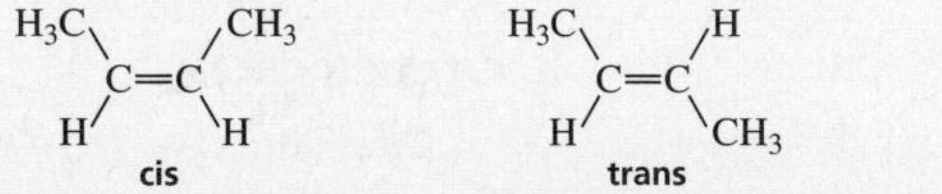 H_3C and CH_3 / H and H, C=C, cis; H_3C and H / H and CH_3, C=C, trans

3-8. $CH_3CH_2CH_2CH=CH_2$ $CH_3CH=CCH_3$ $CH_3CHCH=CH_2$
    (with CH_3 substituents below)

3-9. C **3-10. a.** $-I > -Br > -OH > -CH_3$ **b.** $-OH > -CH_2Cl >$
$-CH=CH_2 > -CH_2CH_2OH$ **3-12. a.** (*E*)-2-heptene **b.** (*Z*)-3,4-dimethyl-
2-penttene **c.** (*Z*)-1-chloro-3-ethyl-4-methyl-3-hexene

3-16. electrophiles: $CH_3\overset{+}{C}HCH_3$

     nucleophiles: H^- CH_3O^- $CH_3C\equiv CH$ NH_3

3-20. a. all **b.** *tert*-butyl **c.** *tert*-butyl **d.** -1.7 kcal/mol or -7.1 kJ/mol

3-22. a. 1. $A + B \rightleftharpoons C$ **b.** none
     2. $A + B \rightleftharpoons C$

3-23. a. 1. $\Delta G° = -1.5$; $K_{eq} = 7.2 \times 10^{10}$ **2.** $\Delta G° = -16$; $K_{eq} = 1.8 \times 10^8$
b. the greater the temperature, the more negative is the $\Delta G°$ **c.** the greater the
temperature, the smaller is the K_{eq} **3-24. a.** -21 kcal/mol **b.** -36 kcal/mol
c. exothermic **d.** exergonic **3-25. a.** a and b **b.** b **c.** c **3-28.** decreasing;
increasing **3-29. a.** it will decrease it **b.** it will increase it **3-30. a.** the first
reaction **b.** the first reaction **3-32. a.** first step **b.** revert to reactants
c. second step **3-33. a.** 1 **b.** 2 **c.** see the *Solutions Manual* **d.** k_{-1} **e.** k_{-1}
f. $B \longrightarrow C$ **g.** $C \longrightarrow B$

CHAPTER 4

4-1. a. 0 **b.** ethyl cation **4-2. a. 1.** 3 **2.** 3 **3.** 6
b. *sec*-butyl cation

4-3. a. $CH_3CH_2\overset{+}{C}CH_3 > CH_3CH_2\overset{+}{C}HCH_3 > CH_3CH_2CH_2\overset{+}{C}H_2$ (with CH_3 above first)

b. $CH_3\overset{+}{C}HCH_2\overset{+}{C}H_2 > CH_3CHCH_2\overset{+}{C}H_2 > CH_3CHCH_2\overset{+}{C}H_2$ (with CH_3, Cl, F substituents)

4-4. a. products **b.** reactants **c.** reactants **d.** products

4-5. a. $CH_3CH_2CHCH_3$ (Br below) **c.** [cyclopentane with CH₃ and Br] **e.** [cyclohexane with CH₃ and Br]

b. $CH_3CH_2CCH_3$ (CH₃ above, Br below) **d.** $CH_3CCH_2CH_2CH_3$ (CH₃ above, Br below) **f.** $CH_3CH_2CHCH_3$ (Br below)

4-6. a. $CH_2=CCH_3$ (CH₃ above) **c.** [cyclohexane]$-C=CH_2$ (CH₃ above)

b. [cyclohexane]$-CH_2CH=CH_2$ **d.** [cyclohexane]$=CHCH_3$ or [cyclohexene]$-CH_2CH_3$

4-7. a. $CH_3CH_2C=CH_2$ (CH₃ above) **b.** [cyclohexane with =CH₂]

4-8. greater than -2.5 and less than 15 **4-9. a.** 3 **b.** 2 **c.** second step
4-10.

a. $CH_3CH_2CH_2CHCH_3$ (OH below) **c.** $CH_3CH_2CH_2CHCH_2CH_3$ (OH below) **d.** [cyclohexane with CH₃ and OH]

                              and

b. [cyclohexane with OH] $CH_3CH_2CH_2CH_2CHCH_3$ (OH below)

4-15. a. $CH_3CCH_2CH_3$ (CH₃ above, Br below) **c.** [cyclohexane with CH₃ and Br]

b. [cyclohexane with CH₃ and Br] **d.** $CH_3CHCH_2CHCH_3$ (CH₃ and Br below)

e. $CH_3CH-CCH_3$ (CH₃ above, CH₃ and Br below) **f.** [methylcyclohexane with Br] and [methylcyclohexane with Br]

4-16. $CH_3CCH_2CH_3$ (CH₃ above, Br below) **4-20.** $CH_3CH_2CHCH_2I$ (Cl below)

4-21. a. $CH_2CHCH_2CH_3$ (Br Br below) **c.** $CH_2CHCH_2CH_3$ (Br OCH₂CH₃ below)

b. $CH_2CHCH_2CH_3$ (Br OH below) **d.** $CH_2CHCH_2CH_3$ (Br OCH₃ below)

4-24. a. [epoxide]$-CH_2CH_2CH_3$ **b.** [cyclohexene oxide]

c. [dimethyl epoxide] **d.** [dimethyl epoxide with CH₂CH₃]

4-25. a. cyclohexene **b.** 1-butene **4-27.** 2/3 mole

4-28. a. $CH_3CHCHCH_3$ (OH below) **b.** [cyclohexane with CH₃ and OH]

4-31. A

4-32. a. [cyclohexene with two CH₂CH₃] **b.** [cyclohexene with two CH₂CH₃] **c.** [cyclohexene with two CH₂CH₃]

4-33. *cis*-3,4-dimethyl-3-hexene > *trans*-3-hexene > *cis*-3-hexene >
cis-2,5-dimethyl-3-hexene
4-35. 3-Methylcyclohexanol would also be formed.

CHAPTER 5

5-1. a. $CH_3CH_2CH_2OH$ CH_3CHOH (CH₃ below) $CH_3CH_2OCH_3$ **b.** 7

5-3. a. , b. , c. , f. , h. **5-4. a.** P, F, J, L, G, R, Q, N, Z **b.** T, M, O, A, U, V,
H, I, X, Y **5-5.** a, c, and f **5-7.** a, c, and f **5-9.** A, B, and C
5-10. a. *R* **b.** *R* **c.** *R* **d.** *R* **5-11. a.** *S* **b.** *R* **c.** *S* **d.** *S* **5-12. a.** identical
b. enantiomers **c.** enantiomers **d.** enantiomers

5-14. a. — **1** $-CH_2OH$ **3** $-CH_3$ **2** $-CH_2CH_2OH$ **4** $-H$

b. **2** $-CH=O$ **1** $-OH$ **4** $-CH_3$ **3** $-CH_2OH$

c. **2** $-CH(CH_3)_2$ **3** $-CH_2CH_2Br$ **1** $-Cl$ **4** $-CH_2CH_2CH_2Br$

d. **2** $-CH=CH_2$ **3** $-CH_2CH_3$ **1** [phenyl] **4** $-CH_3$

5-15. a. levorotatory **b.** dextrorotatory **5-16.** $+168$ **5-17. a.** -24 **b.** 0
5-18. a. $+79$ **b.** 0 **c.** -79 **5-19. a.** do not know **b.** 98.5% dextrorotatory; 1.5%
levorotatory **5-22.** no **5-23. a.** enantiomers **b.** identical **c.** diastereomers
5-24. a. 8 **b.** $2^8 = 256$ **5-28.** 1-chloro-1-methylcyclooctane, *cis*-1-chloro-
5-methylcyclooctane, *trans*-1-chloro-5-methylcyclooctane **5-30.** b, d, and f
5-34. left $= R$; right $= R$ **5-36.** the second one **5-37. a.** (2*R*,3*R*)-2,3-dichlo-
ropentane **b.** (2*R*,3*R*)-2-bromo-3-chloropentane **c.** (1*R*,3*S*)-1,3-cyclopentanediol
d. (3*R*,4*S*)-3-chloro-4-methylhexane **5-39.** *S*
5-40. a. *R* **b.** *R* **c.** *S* **d.** *S* **5-41.** b **5-43. a.** no **b.** no **c.** no **d.** yes **e.** no **f.** no
5-46. a. 1. *trans*-3-heptene **2.** *cis*-3-heptene **b.** the enantiomer of each epoxide
would also be formed

5-47. a.

b.

c.

d.

5-54. a. 1-bromo-2-chloropropane **b.** equal amounts of R and S
5-55. a. (R)-malate and (S)-malate **b.** (R)-malate and (S)-malate **5-56.** >99%
5-57. the one on the left

CHAPTER 6

6-1. $C_{14}H_{20}$

6-2. a. $ClCH_2CH_2C{=}CCH_2CH_3$ **c.** $CH_3CHC{=}CH$ **e.** $HC{\equiv}CCH_2CCH_3$

b.

d. $HC{=}CCH_2Cl$ **f.** $CH_3C{\equiv}CCH_3$

6-3. a. 1-bromo-1,3-pentadiene **b.** 1-hepten-5-yne **c.** 4-hepten-1-yne
6-4.

$HC{\equiv}CCH_2CH_2CH_2CH_3$ $CH_3C{\equiv}CCH_2CH_2CH_3$ $CH_3CH_2C{\equiv}CCH_2CH_3$
    1-hexyne   **2-hexyne**   **3-hexyne**
  butylacetylene   **methylpropylacetylene**   **diethylacetylene**

methyl-1-pentyne   **4-methyl-1-pentyne**   **4-methyl-2-pentyne**   **3,3-dimethyl-1-butyne**
sec-butylacetylene   **isobutylacetylene**   **isopropylmethylacetylene**   *tert*-**butylacetylene**

6-5. a. 5-bromo-2-pentyne **b.** 6-bromo-2-chloro-4-octyne **c.** 1-methoxy-2-pentyne
d. 3-ethyl-1-hexyne **6-6. a.** 1-hepten-4-yne **b.** 4-methyl-1,4-hexadiene
c. 5-vinyl-5-octen-1-yne **d.** 3-butyn-1-ol **e.** 1,3,5-heptatriene **f.** 2,4-dimethyl-
4-hexen-1-ol **6-7.** pentane, 1-pentene, 1-pentyne **6-8. a.** $sp^2{-}sp^2$ **b.** $sp^2{-}sp^3$
c. $sp{-}sp^2$ **d.** $sp{-}sp^3$ **e.** $sp{-}sp$ **f.** $sp^2{-}sp^2$ **g.** $sp^2{-}sp^3$ **h.** $sp{-}sp^3$ **i.** $sp^2{-}sp$
6-9. If the less stable reactant has the more stable transition state or if the less
stable reactant has the less stable transition state and the difference in stabilities
of the reactants is greater than the difference in the stabilities of the transition
states.

6-10. a. $CH_2{=}CCH_3$ **b.** CH_3CCH_3 **c.** $CH_3C{=}CCH_3$ **d.** $HC{-}CCH_3$

e. $CH_3CH_2CCH_3$ **f.** $CH_3CCH_2CH_2CH_3$ + $CH_3CH_2CCH_2CH_3$

6-11.

6-12. $CH_3CH_2CCH_2CH_2CH_2CH_3$ and $CH_3CH_2CH_2CCH_2CH_2CH_3$

6-13. a. $CH_3C{\equiv}CH$ **b.** $CH_3CH_2C{\equiv}CCH_2CH_3$ **c.** $HC{\equiv}C{-}$

6-14. a. $CH_2{=}CCH_3$

b. $CH_3CH{=}CCH_2CH_2CH_3$ and $CH_3CH_2C{=}CHCH_2CH_3$

---

c. $CH_2{=}C$

and CH_3C

6-15. a. 1. $CH_3CH_2CCH_3$ **a. 2.** $CH_3CH_2CH_2CH$

b. 1. $CH_3CH_2CCH_3$ **b. 2.** $CH_3CH_2CCH_3$

c. 1. $CH_3CH_2CH_2CCH_3$ and $CH_3CH_2CCH_2CH_3$

c. 2. $CH_3CH_2CH_2CCH_3$ and $CH_3CH_2CCH_2CH_3$

6-16. ethyne (acetylene)

6-17. a. $CH_3CH_2CH_2C{\equiv}CH$ or $CH_3CH_2C{\equiv}CCH_3$ $\xrightarrow[\text{Pt}]{\text{H}_2}$

b. $CH_3C{\equiv}CCH_3$ $\xrightarrow[\substack{\text{Lindlar} \\ \text{catalyst}}]{\text{H}_2}$

c. $CH_3CH_2C{\equiv}CCH_3$ $\xrightarrow[\text{NH}_3]{\text{Na}}$

d. $CH_3CH_2CH_2CH_2C{\equiv}CH$ $\xrightarrow[\substack{\text{Lindlar} \\ \text{catalyst}}]{\text{H}_2}$ or $\xrightarrow[\text{NH}_3]{\text{Na}}$

6-18. 25 **6-19. a.** $CH_3\overset{+}{CH_2}$ **b.** $H_2C{=}\overset{+}{CH}$ **6-20.** The carbanion that would
be formed is a stronger base than the amide ion. **6-21. a.** $CH_3CH_2CH_2\overset{-}{C}H_2 >$
$CH_3CH_2CH{=}\overset{-}{C}H > CH_3CH_2C{\equiv}\overset{-}{C}$ **b.** $^-NH_2 > CH_3C{\equiv}\overset{-}{C} > CH_3CH_2O^- > F^-$

CHAPTER 7

7-1. a. 1. 4 **2.** 2 **b. 1.** 5 **2.** 5 **7-3. a.** all are the same length **b.** 2/3 of a
negative charge

7-7. a. $CH_3CH_2CH{=}\overset{+}{CH}CH_2$ **c.** $CH_3\overset{\underset{\displaystyle O^-}{|}}{C}{=}CHCH_3$

b. $CH_3CCH{=}CHCH_3$ **d.** $CH_3{-}\overset{\underset{\displaystyle NH_2}{|}}{\overset{\overset{\displaystyle +NH_2}{|}}{C}}{-}NH_2$

7-8. the dianion **7-9.** conjugated > isolated > cumulated

7-10. $CH_3\overset{\underset{\displaystyle CH_3}{|}}{C}{=}CHCH{=}\overset{\underset{\displaystyle CH_3}{|}}{C}CH_3 > CH_3CH{=}CHCH{=}CHCH_3 >$
    2,5-dimethyl-2,4-hexadiene   **2,4-hexadiene**

$CH_3CH{=}CHCH{=}CH_2 > CH_2{=}CHCH_2CH{=}CH_2$
    1,3-pentadiene   **1,4-pentadiene**

7-11. a.

b. $CH_3\overset{+}{N}HCH_2$ **c.**

7-12. $\psi_3 = 3$; $\psi_4 = 4$ **7-13. a.** bonding $= \psi_1$ and ψ_2; antibonding $= \psi_3$ and ψ_4
b. symmetric $= \psi_1$ and ψ_3; antisymmetric ψ_2 and ψ_4 **c.** HOMO $= \psi_2$;
LUMO $= \psi_3$; **d.** HOMO $= \psi_3$; LUMO $= \psi_4$ **e.** If the HOMO is symmetric,
the LUMO is antisymmetric and vice versa. **7-14. a.** bonding $= \psi_1, \psi_2$,
and ψ_3; antibonding $= \psi_4, \psi_5$, and ψ_6 **b.** symmetric $= \psi_1, \psi_3$, and ψ_5;
antisymmetric ψ_2, ψ_4, and ψ_6 **c.** HOMO $= \psi_3$; LUMO $= \psi_4$
d. HOMO $= \psi_4$; LUMO $= \psi_5$ **e.** If the HOMO is symmetric, the LUMO
is antisymmetric and vice versa. **7-15. a.** $\psi_1 = 3$; $\psi_2 = 2$ **b.** $\psi_1 = 7$; $\psi_2 = 6$
7-16. a. $CH_3CH{=}CHOH$

b. CH_3COH **c.** $CH_3CH{=}CHOH$ **d.** $CH_3CH{=}\overset{+}{CH}NH_3$

7-17. a. ethylamine **b.** ethoxide ion **c.** ethoxide ion

7-18.

${-}COOH > $ ${-}OH > $ ${-}CH_2OH$

7-20. resonance electron donation **7-21.** 1-bromo-1-phenylpropane

7-24.

7-25. a. $CH_3CHCHCH=CHCH_3$ + $CH_3CHCH=CHCHCH_3$
| | |
Cl Cl   Cl   Cl

b.
     Br
     |
$CH_3CH_2C-C=CHCH_3$ + $CH_3CH_2C=CCHCH_3$
| |
CH_3 CH_3 CH_3CH_3 Br

c.

7-29. a. Addition at C-1 forms the more stable carbocation. **b.** To cause the 1,2- and, 1,4-products to be different. **7-30. a.** formation of the carbocation **b.** reaction of the carbocation with the nucleophile

7-33. a.

c.

b.

d.

7-34.

7-35. a.

b.

7-36. a and d **7-38. a.** The product is a meso compound. **b.** The product is a racemic mixture.

7-39. a.

b.

c.

d.

e.

f.

CHAPTER 8

8-1. It will be 0.05 of the original rate. **8-2.** decrease

8-3.

           CH_3 CH_3 CH_3
           | | |
$CH_3CH_2CH_2CH_2CH_2Br$ > $CH_3CHCH_2CH_2Br$ > $CH_3CH_2CHCH_2Br$ > CH_3CH_2CBr
                                                                          |
                                                                          CH_3

8-4. b. (S)-2-butanol **c.** (R)-2-hexanol **d.** 3-pentanol **8-5. a.** RO^- **b.** RS^-
8-6. a. aprotic **b.** aprotic **c.** protic **d.** aprotic **8-8. a.** $CH_3CH_2Br + HO^-$
b. CH_3CHCH_2Br + HO^- **c.** $CH_3CH_2Cl + CH_3S^-$ **d.** $CH_3CH_2Br + I^-$
|
CH_3

8-11. a. $CH_3CH_2OCH_2CH_2CH_3$ **b.** $CH_3CH_2C\equiv CCH_3$ **c.** $CH_3CH_2\overset{+}{N}(CH_3)_3Br^-$
d. $CH_3CH_2SCH_2CH_3$

           CH_3
           |
8-14. CH_3CBr > CH_3CHBr > $CH_3CH_2CH_2Br$ > CH_3Br
           | |
           CH_3 CH_3

8-16.

           CH_3
           |
$CH_3CH_2CCH_2CH_3$ > $CH_3CHCH_2CH_2CH_3$ > $CH_3CHCH_2CH_2CH_3$ >
           | | |
           Br   Br   Cl

           $ClCH_2CH_2CH_2CH_2CH_3$

8-17. a, b, c, e **8-18.** 3-aceto-3-methyl-1-butene and 1-aceto-3-methyl-2-butene
8-20. trans-4-bromo-2-hexene

8-21. a. CH_3CHBr **e.** ⬡—CH_2Br
      |
      CH_3

           CH_3
           |
b. CH_3CH_2CHBr **f.** $CH_3CH=CHCHCH_3$
               CH_3 |
               | Br
c. $CH_3CH_2CHCH_2Br$ **g.** CH_3OCH_2Cl

d. ⬡—CH_2CH_2Br

8-22. a. CH_3CHBr **e.** ⬡—CH_2Br
      |
      CH_3

b. equally reactive   **f.** $CH_3CH=CHCHCH_3$
                  CH_3 |
                  | Br
c. $CH_3CH_2CH_2CHBr$ **g.** CH_3OCH_2Cl

d. ⬡—CH_2CHCH_3
                 |
                 Br

8-25. a and b **8-27. a.** decrease **b.** decrease **c.** increase

8-28. a. CH_3Br + HO^- ⟶ CH_3OH + Br^-
b. CH_3I + HO^- ⟶ CH_3OH + I^-

c. CH_3Br + NH_3 ⟶ $CH_3\overset{+}{N}H_3$ + Br^-

d. CH_3Br + HO^- $\xrightarrow{\text{DMSO}}$ CH_3OH + Br^-

e. CH_3Br + NH_3 $\xrightarrow{\text{EtOH}}$ $CH_3\overset{+}{N}H_3$ + Br^-

8-30. dimethyl sulfoxide **8-31.** HO^- in 50% water/50% ethanol

8-32. a. $HO\diagdown\diagup\diagdown\diagup\diagdown Br$
b. $HO\diagdown\diagup\diagdown\diagup\diagdown\diagup Br$
c. $HO\diagdown\diagup\diagdown\diagup\diagdown\diagup\diagdown Br$

CHAPTER 9

9-1. a. CH$_3$C=CHCH$_3$ (with CH$_3$ on the first C) **b.** CH$_3$C=CHCH$_3$ (with CH$_3$) **c.** CH$_3$CH$_2$C(CH$_3$)CH=CH$_2$

9-2. a. CH$_3$CH$_2$CHCH$_3$ with Br **c.** CH$_3$CH$_2$CH$_2$C(CH$_3$)CH$_3$ with Br

b. cyclohexane—Br **d.** CH$_3$CH(CH$_3$)CH$_2$CH$_2$Cl

9-4. a. CH$_3$CH=CHCH$_3$ **d.** CH$_2$=CHCH$_2$CH$_3$

b. CH$_3$C(CH$_3$)=CHCH$_2$CH$_3$ **e.** cyclohexene

c. CH$_3$CH=CHCH=CH$_2$ **f.** CH$_3$CH(CH$_3$)CH=CHCH$_3$

9-5. a. CH$_3$CH(Br)CHCH$_2$CH$_3$ (CH$_3$ branch) **c.** CH$_3$CH$_2$CHCH$_2$CH$_3$ with Br

b. cycloheptene with Br **d.** phenyl—CH$_2$CHCH$_2$CH$_3$ with Br

9-6.

CH$_3$C(CH$_3$)(CH$_3$)—CH$_2$CH$_3$ > [(H$_3$C)(CH$_3$CH(CH$_3$))C=C(CH$_3$)(H)] > [(H$_3$C)(CH$_3$CH(CH$_3$))C=C(H)(CH$_3$)] > CH$_3$CHCH(CH$_3$)CH$_2$CH$_3$ (=CH$_2$)

9-7. yes; elimination occurs through the carbocation **9-8. a.** B **b.** B **c.** B **d.** A

9-10. a. E2 CH$_3$CH=CHCH$_3$ **d.** E2 CH$_3$C(CH$_3$)=CH$_2$

b. E1 CH$_3$CH=CHCH$_3$ **e.** E1 CH$_3$C(CH$_3$)=CCH$_3$ (CH$_3$)

c. E1 CH$_3$C(CH$_3$)=CH$_2$ **f.** E2 CH$_3$C(CH$_3$)CH=CH$_2$ (CH$_3$)

9-11. a. 96% **b.** 1.2%

9-12. a. 1. CH$_3$CH$_2$CH=C(CH$_3$)CH$_3$ **3.** CH$_3$CH$_2$ / H C=C / H / phenyl

2. CH$_3$CH$_2$ / H C=C / H / CH=CH$_2$

b. no

9-13. b. 2-methyl-2-pentene **c.** (E)-4-methyl-3-heptene **d.** 1-methylcyclohexene
9-14. the eliminated atoms have to be in axial positions **9-15.** elimination only from the *cis* isomer occurs through the more stable conformer **9-17.** left to right 2, 5, 3, 1, 4 (1 is the most reactive, 5 the least reactive) **9-18.** ~1
9-19. it will increase **9-20.** When Br is in an axial position, both adjacent H's are in equatorial positions. **9-21. a.** 1-bromopropane **b.** iodocyclohexane
c. 2-bromo-2-methylbutane **d.** 3-bromocyclohexene **9-22. a. 1.** no reaction
2. no reaction **3.** substitution and elimination **4.** substitution and elimination
b. 1. primarily substitution **2.** substitution and elimination **3.** substitution and

elimination **4.** elimination **9-23. a.** S$_N$2 difficult because of steric hindrance; no S$_N$1 because cannot form a primary carbocation **b.** No E2 because there are no hydrogens on a β-carbon; no E1 because cannot form a primary carbocation.
9-24. a. CH$_3$CH=CH$_2$ **b.** CH$_3$CH$_2$CH=CH$_2$

9-26. CH$_3$C(CH$_3$)(CH$_3$)—OH + CH$_3$C(CH$_3$)(CH$_3$)—OCH$_2$CH$_3$ + CH$_3$C(CH$_3$)=CH$_2$

CHAPTER 10

10-1. no lone pair **10-4.** 3° > 1° > 2°

10-5. a. CH$_3$CH$_2$CHCH$_3$ with Br **c.** CH$_3$C(CH$_3$)—CHCH$_3$ with Br, CH$_3$

b. cyclopentane with CH$_3$ and Cl **d.** cyclohexane with CH$_3$, CH$_2$CH$_3$, Cl

10-8. D **10-10.** 1-bromo-1-methylcyclopentane **10-11.** #2 > #3 > #1

10-13. a. CH$_3$CH$_2$CH(CH$_3$)CH$_2$OH **c.** CH$_3$CH$_2$CHCH$_2$CH$_3$ with OH

b. cyclohexane with CH$_3$ and OH **d.** cyclohexane with CH$_2$OH

10-15.

a. CH$_3$CH$_2$C(CH$_3$)=CCH$_3$ **b.** cyclopentene with CH$_2$CH$_3$ **c.** cyclohexadiene

d. cyclohexene with CH$_3$ **e.** CH$_3$CH$_2$C(CH$_3$)=CCH$_3$ with CH$_3$ **f.** CH$_3$CH$_2$ / H C=C / H / CH$_3$

10-17. a. CH$_3$CH$_2$C(=O)CH$_2$CH$_3$ **d.** CH$_3$C(=O)CH$_2$C(=O)CH$_2$CH$_3$

b. CH$_3$CH$_2$CH$_2$CH$_2$C(=O)OH **e.** cyclohexanone

c. no reaction **f.** HOC(=O)CH$_2$CH$_2$C(=O)OH

10-21. No, F$^-$ is too poor a nucleophile.

10-23. a. HOCH$_2$C(CH$_3$)(OCH$_3$)CH$_3$ **c.** HOCH(OCH$_3$)—C(CH$_3$)(CH$_3$)

b. CH$_3$OCH$_2$C(OH)(CH$_3$)CH$_3$ **d.** CH$_3$OCH(OH)—C(CH$_3$)(CH$_3$)

10-24. noncyclic ether

10-26. phenol (OH) with CH$_3$ para

10-27. The products are the same. **10-32.** The first one is too insoluble; the second is too reactive; the third is too unreactive.

10-33. a. $CH_3CH_2CH_2CH_2CH_2OH$

b. [phenyl]—$CH_2CH_2CH_2OH$

c. [cyclohexyl]—CH_2CH_2OH

10-36. They all occur. **10-37.** $(CH_3)_4Si$

10-40.

a. [bromobenzene] **b.** [(E)-cinnamyl bromide structure] **c.** [2-(bromomethyl)allylbenzene structure]

10-41. 1-pentyne

10-42.

$CH_3\overset{O}{\overset{\|}{C}}$—[benzene]—Br   and CH_2=CH—[benzene]—OCH_3

$CH_3\overset{O}{\overset{\|}{C}}$—[benzene]—CH=$CH_2$ and   Br—[benzene]—OCH_3

CHAPTER 11

11-3. a. the one on the third carbon from the left **b.** 6 **11-5. a.** 3 **b.** 3 **c.** 5 **d.** 1 **e.** 5 **f.** 4 **g.** 2 **h.** 1 **i.** 4

11-6.

a. $CH_3CH_2CH_2CH_2CH_2Cl$ $CH_3CH_2CH_2\overset{Cl}{\underset{|}{C}H}CH_3$ $CH_3CH_2\overset{Cl}{\underset{|}{C}H}CH_2CH_3$

   21%   53%   26%

b. $ClCH_2\overset{CH_3}{\underset{|}{C}H}CH_2CH_2\overset{CH_3}{\underset{|}{C}H}CH_3$ $CH_3\overset{CH_3}{\underset{|}{\overset{|}{C}}}CH_2CH_2\overset{Cl}{\underset{|}{C}H}CH_3$ $CH_3\overset{CH_3}{\underset{|}{C}H}\overset{Cl}{\underset{|}{C}H}CH_2\overset{CH_3}{\underset{|}{C}H}CH_3$

   32%   27%   41%

c. $ClCH_2\overset{CH_3}{\underset{|}{C}H}CH_2CH_2CH_3$ $CH_3\overset{CH_3}{\underset{|}{\overset{|}{C}}}CH_2CH_2CH_3$ $CH_3\overset{CH_3}{\underset{|}{C}H}\overset{Cl}{\underset{|}{C}H}CH_2CH_3$

   21%   17%   26%

$CH_3\overset{CH_3}{\underset{|}{C}H}CH_2\overset{Cl}{\underset{|}{C}H}CH_3$ $CH_3\overset{CH_3}{\underset{|}{C}H}CH_2CH_2CH_2Cl$

   26%   10%

11-8. a. $CH_3CH_2\overset{Br}{\underset{|}{C}H}CH_3$ $CH_3CH_2CH_2CH_2Br$

   $4 \times 82 = 328$ $6 \times 1 = 6$

   $\dfrac{328}{328+6} = \dfrac{328}{334} = 0.98 = 98\%$

b. $CH_3\overset{CH_3}{\underset{\overset{|}{Br}}{\overset{|}{C}}}CH_2CH_2\overset{CH_3}{\underset{CH_3}{\overset{|}{C}}}CH_3$ other products

   $1 \times 1600 = 1600$

   $\begin{cases} 9 \times 1 = 9 \\ 2 \times 82 = 164 \\ 2 \times 82 = 164 \\ 6 \times 1 = 6 \end{cases}$ 1600

   1943

   $\dfrac{1600}{1943} = 0.82 = 82\%$

11-9. a. chlorination **b.** bromination **c.** no preference **11-11.** 99.4% (vs. 36%)

11-12.

a. $CH_3CH_2CH_2CH_2CH_2Br$ $CH_3CH_2CH_2\overset{Br}{\underset{|}{C}H}CH_3$ $CH_3CH_2\overset{Br}{\underset{|}{C}H}CH_2CH_3$

   1%   33%   66%

b. $BrCH_2\overset{CH_3}{\underset{|}{C}H}CH_2CH_2\overset{CH_3}{\underset{|}{C}H}CH_3$ $CH_3\overset{CH_3}{\underset{\overset{|}{Br}}{\overset{|}{C}}}CH_2CH_2\overset{CH_3}{\underset{|}{C}H}CH_3$ $CH_3\overset{CH_3}{\underset{|}{C}H}\overset{Br}{\underset{|}{C}H}CH_2\overset{CH_3}{\underset{|}{C}H}CH_3$

   0.3%   90.4%   9.3%

c. $BrCH_2\overset{CH_3}{\underset{|}{C}H}CH_2CH_2CH_3$ $CH_3\overset{CH_3}{\underset{\overset{|}{Br}}{\overset{|}{C}}}CH_2CH_2CH_3$ $CH_3\overset{CH_3}{\underset{|}{C}H}\overset{Br}{\underset{|}{C}H}CH_2CH_3$

   0.3%   82.6%   8.5%

$CH_3\overset{CH_3}{\underset{\overset{|}{Br}}{\overset{|}{C}}}CH_2\overset{CH_3}{\underset{|}{C}H}CH_3$ $CH_3CH_2CH_2CH_2CH_2CH_2Br$

   8.5%   0.2%

11-14. a. $CH_3\overset{CH_3}{\underset{\overset{|}{Br}}{\overset{|}{C}}}CH_2CH_3$ **c.** $CH_3\overset{CH_3}{\underset{|}{C}H}\overset{Br}{\underset{|}{C}H}CH_3$

b. $CH_3\overset{CH_3}{\underset{\overset{|}{Cl}}{\overset{|}{C}}}CH_2CH_3$ **d.** $CH_3\overset{CH_3}{\underset{\overset{|}{Cl}}{\overset{|}{C}}}CH_2CH_3$

11-15. a. 2-methylpropane **b.** butane

CHAPTER 12

12-1. #2, #3, #5 **12-3.** $m/z = 57$ **12-6.** 8 **12-7.** $1:2:1$ **12-8.** C_6H_{14}
12-10. a. 2-methoxy2-methylpropane **b.** 2-methoxybutane **c.** 1-methoxybutane
12-11. $CH_2 \overset{+}{=} \overset{..}{O}H$ **12-13. a.** 2-pentanone **b.** 3-pentanone
12-15. a. 3-pentanol **b.** 2-pentanol **12-16. a.** IR **b.** UV **12-17. a.** $2000\ cm^{-1}$
b. $8\ \mu m$ **c.** $2\ \mu m$ **12-18. a.** $2500\ cm^{-1}$ **b.** $50\ \mu m$ **12-19.** C=O
12-20. a. 1. C≡C **2.** C—H stretch **3.** C=N **4.** C=O **b. 1.** C—O **2.** C—C
12-21. a. carbon–oxygen stretch of a phenol **b.** carbon–oxygen double-bond stretch of a ketone **c.** carbon–oxygen stretch **12-22.** sp^3

12-23. a.

$H\overset{O}{\overset{\|}{C}}H$ > $H_3C\overset{O}{\overset{\|}{C}}H$ > $H_3C\overset{O}{\overset{\|}{C}}CH_3$

b. [δ-valerolactone] > [cyclohexanone] > [δ-valerolactam]

c. [2(5H)-furanone] > [γ-butyrolactone] > [2(3H)-furanone]

12-24. ethanol dissolved in carbon disulfide **12-25.** C—O bond of pentanoic acid has partial double bond character **12-26. a.** O—H has a greater change in dipole moment **b.** extent of hydrogen bonding greater in the carboxylic acid
12-27. C—N stretch would be less intense **12-28. a.** ketone **b.** tertiary amine
12-31. 2-butyne, H_2, Cl_2, ethene **12-32.** *trans*-2-hexene **12-33.** methyl vinyl ketone **12-34.** does not have a π^* antibonding molecular orbital
12-35. $4.1 \times 10^{-5}\ M$ **12-36.** $10{,}000\ M^{-1}cm^{-1}$
12-37.

a. [phenyl]—CH=CH—[phenyl] > [biphenyl] > [phenyl]—CH=CH_2 > [phenyl]

b. [phenyl]—$N(CH_3)_2$ > [phenyl]—$\overset{+}{N}(CH_3)_3$ > [cyclohexene]—$N(CH_3)_2$ > [cyclohexyl]—$N(C$

12-38. left = purple; right = blue **12-39.** monitor increase in absorption at 340 nm **12-40.** 5.0

CHAPTER 13

13-1. 43 MHz **13-2. a.** 8.46 T **b.** 11.75 T **13-3. a.** 2 **b.** 1 **c.** 3 **d.** 3 **e.** 1 **f.** 4 **g.** 3 **h.** 3 **i.** 5 **j.** 4 **k.** 2 **l.** 3 **m.** 3 **n.** 4 **o.** 3

13-5. [cyclopropane with Cl, Cl] [cyclopropane with H, Cl; Cl, H] [cyclopropane with H, H; Cl, Cl]

   1   2   3

13-6. a. 300 Hz **b.** 500 Hz **13-7. a.** 2.0 ppm **b.** 2.0 ppm **c.** 200 Hz
13-8. a. 1.5 ppm **b.** 30 Hz **13-9.** to the right of the TMS signal
13-10. a. in each structure it is the proton(s) on the carbon on the right-hand side

of the structure **b.** in each structure it is the protons on the methyl group on the left-hand side of the structure

13-11.

a. $CH_3CHCHBr$ with Br Br below **c.** $CH_3CH_2CHCH_3$ with Cl below **e.** $CH_3CH_2CH=CH_2$

b. CH_3CHOCH_3 with CH_3 below **d.** $CH_3CHCCH_2CH_3$ with CH_3 below and O above **f.** $CH_3OCH_2CH_2CH_3$

13-12. a. $CH_3CH_2CH_2Cl$ **b.** $CH_3CH_2CHCH_3$ with Cl below **c.** CH_3CHCH with O above

13-14. 9.25 ppm = hydrogens that protrude out; -2.88 ppm = hydrogens that point in **13-15.** The compounds have different integration ratios: $2:9$, $1:3$, and $2:1$ **13-17.** B **13-18.** first spectrum = 1-iodopropane **13-20. a.** 2-chloropropanoic acid **b.** 3-chloropropanoic acid **13-25. a.** propyl benzene **b.** 3-pentanone **c.** ethyl benzoate

13-27. a. $CH_3OCH_2CH_2OCH_3$

b. $CH_3CCH_2CH_2CCH_3$ (with two O above)

or

$CH_3OCH_2C\equiv CCH_2OCH_3$

or

[structure: dioxane ring with two CH$_3$ groups]

c. HC—[benzene ring]—CH (with O above each HC/CH)

13-30. CH_3O—[benzene ring]—CH_3

13-32. a. A **b.** B and D **13-35.** pure ethanol

13-37.

CH_3CH_2—C—NH_2 (with O above C)

13-41.

a. $CH_3(CH_2)_4C(CH_2)_4CH_3$ (with O above C) **b.** Br—[benzene ring]—CH_3CH_2 **c.** [cyclohexane ring]=O

d. $CH_3CH_2CH=CHCH_2CH_3$

CHAPTER 14

14-1. a. 4 **b.** It will be aromatic if it is cyclic, if it is planar, and if every carbon in the ring has a p orbital. **14-3. a.** cyclopropenyl cation **b.** cycloheptatrienyl cation **14-4.** d and e **14-5. b.** 5 **14-8.** quinoline $=sp^2$; indole $= p$; imidazole = 1 is sp^2 and 1 is p; purine = 3 are sp^2 and 1 is p; pyrimidine = both are sp^2 **14-10.** Cyclopentadiene has the lower pK_a. **14-13. a.** Cyclopropane has the lower pK_a. **b.** 3-bromocyclopropene **14-14.** c **14-15.** one bonding, two nonbonding, one antibonding; two π electrons in the bonding MO and one in each of the nonbonding MOs **14-16.** no

14-18. a. $CH_3CHCH_2CH_2CH_2CH_3$ (with phenyl below) **c.** $CH_3CH_2CHCH_2CH_3$ (with CH_2 and phenyl below)

b. [benzene ring]—CH_2OH **d.** [benzene ring]—CH_2Br

14-26. a. ethylbenzene **b.** isopropylbenzene **c.** *sec*-butylbenzene **d.** *tert*-butylbenzene **e.** *tert*-butylbenzene **f.** 3-phenylpropene

14-28. a. [benzene ring]—COOH **b.** [benzene ring with COOH and COOH]

c. [benzene ring]—CH_2OCH_3 **d.** [benzene ring]—$CH_2CH_2NH_2$

CHAPTER 15

15-1. a. *ortho*-ethylphenol **b.** *meta*-bromochlorobenzene **c.** *meta*-bromobenzaldehyde **d.** *ortho*-ethyltoluene

15-2.
a. [benzene ring with CH_3 top, NH_2 bottom] **b.** [benzene ring with CH_3 top, OH] **c.** [benzene ring with CH_3 top, CH_3 bottom] **d.** [benzene ring with SO_3H and Cl top]

15-3. a. [benzene ring with CH_3 top, Cl] **b.** [benzene ring with OH top, Br bottom] **c.** [benzene ring with NH_2, NO_2] **d.** [benzene ring with $C\equiv N$ top, Cl]

e. [benzene ring with OH top, Br, I bottom] **f.** [benzene ring with Cl, Cl] **g.** [benzene ring with $HC=O$, NO_2, O_2N] **h.** [benzene ring with NH_2 top, Cl, Br bottom]

15-4. a. 1,3,5-tribromobenzene **b.** 3-nitrophenol **c.** *para*-bromotoluene **d.** 1,2-dichlorobenzene **15-5. a.** donates electrons by resonance and withdraws electrons inductively **b.** donates electrons by hyperconjugation **c.** withdraws electrons by resonance and withdraws electrons inductively **d.** donates electrons by resonance and withdraws electrons inductively **e.** donates electrons by resonance and withdraws electrons inductively **f.** withdraws electrons inductively

15-6. a. phenol > toluene > benzene > bromobenzene > nitrobenzene **b.** toluene > chloromethylbenzene > dichloromethylbenzene > difluoromethylbenzene

15-9. a. [benzene ring with $CH_2CH_2CH_3$, NO_2 below] + [benzene ring with $CH_2CH_2CH_3$, NO_2] **d.** [benzene ring with $C\equiv N$, NO_2]

b. [benzene ring with Br, NO_2] + [benzene ring with Br, NO_2] **e.** [benzene ring with SO_3H, NO_2]

c. [benzene ring with $HC=O$, NO_2] **f.** [cyclohexyl-benzene ring with NO_2] + [cyclohexyl-benzene ring with NO_2]

15-10. They are all meta directors.

15-12. a. $ClCH_2COH$ (O above) **d.** [benzene ring]—COOH (O above) **g.** FCH_2COH (O above)

b. O_2NCH_2COH (O above) **e.** $HOCCH_2COH$ (O above each C) **h.** [benzene ring with COOH top, Cl bottom]

c. H_3NCH_2COH (+ on N, O above C) **f.** $HCOH$ (O above)

15-15. a. no reaction   **c.** no reaction

b.

2,4,6-tribromoaniline (NH$_2$ with Br at 2,4,6 positions)

d. CH$_3$-substituted benzene (1,4-dimethylbenzene) + (1,2-dimethylbenzene)

15-17. a. 4-methyl-3-bromobenzoic acid (COOH, CH$_3$, Br)

d. OCH$_3$, NO$_2$, F substituted benzene

b. benzene with COOH, COOH, Cl

e. HC=O, NO$_2$, OCH$_3$ substituted benzene

c. COOH, Br, Cl substituted benzene

f. CH$_3$, NO$_2$, C(CH$_2$)$_3$ substituted benzene

15-18. a. 2 **b.** 1 **c.** 2   **15-22.** It would complex with the amino group, converting it into a meta director.

15-27. CH$_3$CHCH$_3$ + CH$_3$CHCH$_3$ + CH$_3$CH=CH$_2$ + N$_2$
           | |
           Cl   OH

15-30. a. 1-chloro-2, 4-dinitrobenzene > p-chloronitrobenzene > chlorobenzene **b.** chlorobenzene > p-chloronitrobenzene > 1-chloro-2,4-dinitrochlorobenzene

CHAPTER 16

16-2. a. butanenitrile, propyl cyanide **b.** ethanoic propanoic anhydride, acetic propionic anhydride **c.** potassium butanoate; potassium butyrate **d.** pentanoyl chloride, valeryl chloride **e.** isobutyl butanoate, isobutyl butyrate **f.** N,N-dimethylhexanamide, N,N-dimethylcaproamide **g.** 2-azacyclopentanone, γ-butyrolactam **h.** cyclopentanecarboxylic acid **i.** 5-methyl-2-oxacyclohexanone, β-methyl-δ-valerolactone

16-5. The shortest bond has the highest frequency.

a. CH$_3$—C—O—CH$_3$ **b.** CH$_3$—C—O—CH$_3$

    1 = longest   1 = highest frequency
    3 = shortest   3 = lowest frequency

16-6. acyl chloride (1800 cm^{-1}); acid anhydride (1800 cm^{-1} and 1750 cm^{-1}); ester (1730 cm^{-1}); amide (1640 cm^{-1})   **16-7. a.** acetic acid **b.** no reaction **16-8.** true **16-9. a.** no reaction **b.** acetic acid **c.** acetic anhydride **d.** no reaction **16-10. a.** new **b.** no reaction **c.** mixture of two

16-14. a. CH$_3$CH$_2$CH$_2$OH   **c.** (CH$_3$)$_2$NH   **e.** CH$_3$CO$^-$

b. CH$_3$CH$_2$NH$_2$ **d.** H$_2$O   **f.** HO—C$_6$H$_4$—NO$_2$

16-18. a. carbonyl group of the ester is relatively unreactive, nucleophile is relatively unreactive, leaving group is a strong base **b.** aminolysis **16-20. a.** benzoic acid and ethanol **b.** butanoic acid and methanol **c.** 5-hydroxypentanoic acid **16-22. a.** protonated carboxylic acid, tetrahedral intermediate I, tetrahedral intermediate III, H$_3$O$^+$, CH$_3$OH$^+$ **b.** Cl$^-$, carboxylic acid, tetrahedral intermediate II, H$_2$O, CH$_3$OH **c.** H$_3$O$^+$ **d.** H$_3$O$^+$ if excess water was used; CH$_3$OH$_2$$^+$ if not. **16-26.** methyl butanoate and methanol **16-27. a.** the alcohol **b.** the carboxylic acid **16-30. a.** butyrate ion and iodomethane **b.** acetate ion and 1-iodooctane **16-36.** C > B > D > A **16-37. a.** pentyl bromide **b.** isohexyl bromide **c.** benzyl bromide **d.** cyclohexyl bromide **16-39. a.** 1-bromopropane **b.** 1-bromo-2-methylpropane **c.** bromocyclohexane **16-41. a.** propanoic acid + thionyl chloride followed by phenol **b.** benzoic acid + thionyl chloride followed by ethylamine **16-43.** PDS

CHAPTER 17

17-1. if elsewhere would not be a ketone   **17-2. a.** 3-methylpentanal, β-methylvaleraldehyde **b.** 4-heptanone, dipropyl ketone **c.** 2-methyl-4-heptanone, isobutyl propyl ketone **d.** 4-phenylbutanal, γ-phenyrlbutyraldehyyde **e.** 4-ethylhexanal, γ-ethylcaproaldehyde **f.** 1-hepten-3-one, butyl vinyl ketone   **17-3. a.** 6-hydroxy-3-heptanone **b.** 2-oxocyclohexylmethanenitrile **c.** 3-formylpentanamide **17-4. a.** 2-heptanone **b.** $para$-nitroacetophenone   **17-5. a.** 2-hexanol **b.** 2-methyl-2-pentanol **c.** 1-methylcyclohexanol

17-6.

CH$_3$CCH$_2$CH$_3$ + CH$_3$CH$_2$CH$_2$MgBr

CH$_3$CH$_2$CCH$_2$CH$_3$ + CH$_3$MgBr

17-7. a. two; (R)-3-methyl-3-hexanol and (S)-3-methyl-3-hexanol **b.** one; 2-methyl-2-pentanol   **17-10.** A and C

17-12.

RC≡C—C—C≡CR
        |
        R
(with OH on central carbon)

17-13. a. CH$_3$CHCH$_2$OH   **c.** C$_6$H$_5$—CH$_2$OH
              |
              CH$_3$

b. cyclohexyl—OH   **d.** C$_6$H$_5$—CHCH$_3$
                                    |
                                    OH

17-14. a. C$_6$H$_5$—CNHCH$_3$ (with C=O)   **c.** CH$_3$CNHCH$_2$CH$_3$ (with C=O)

b. CH$_3$CNH$_2$ (with C=O)   **d.** CH$_3$CN(CH$_2$CH$_3$)(CH$_2$CH$_3$) (with C=O)

17-16. no   **17-17.** The conjugate bases of the strong acids are weak and readily eliminated.   **17-19.** ~ 8.5   **17-19.** 2-propanol and 3-pentanol **17-20. a.** 1 × 10^{-12} **b.** 1 × 10^{-9} **c.** 3.1 × 10^{-3}

17-23. a. cyclopentane=NCH$_2$CH$_3$ + H$_2$O

b. cyclopentene—N(CH$_2$CH$_3$)(CH$_2$CH$_3$) + H$_2$O

c. C$_6$H$_5$—C(CH$_3$)=N(CH$_2$)$_5$CH$_3$ + H$_2$O

d. C$_6$H$_5$—C(CH$_3$)=N—cyclohexyl + H$_2$O

17-26. a secondary amine and a tertiary amine   **17-30.** the ketone with the nitro substituents   **17-31. a.** 7 **b.** 2, 3 **c.** 1,8 **d.** 5 **e.** 4, 6

17-34. a. cyclohexane with OH and CH$_2$OH   **b.** NaBH$_4$

17-37. a. 26%   **b.** 17%

17-40. a. C$_6$H$_5$—C(OH)(CH$_3$)—CH$_2$CH$_3$ S   **c.** CH$_3$CCH$_2$CH$_2$CH$_3$ (with OH and CH$_3$)
compound does not have an asymmetric center

b. CH$_3$—C(OH)(H)—C$_6$H$_5$ R   **d.** CH$_3$CH$_2$CH$_2$—C(CH$_3$)—CH$_2$CH$_3$ (with OH) S

17-43. Nonsterically hindered

17-44. a. $CH_3CHCH_2\overset{O}{\overset{\|}{C}}OCH_3$ (with Br on the CH)   **c.** $CH_3CH=CH\overset{O}{\overset{\|}{C}}NH_2$

b. $CH_3CH=CH\overset{O}{\overset{\|}{C}}OCH_3$ **d.** $CH_3CHCH_2\overset{O}{\overset{\|}{C}}NH_2$ (with NH_3 on the CH)

CHAPTER 18

18-1. The conjugate base of propene has delocalized electrons but they are delocalized onto a carbon.

18-2. a. $CH_3\overset{O}{\overset{\|}{C}}CH_2C\equiv N$ **b.** $CH_3O\overset{O}{\overset{\|}{C}}CH_2\overset{O}{\overset{\|}{C}}OCH_3$
   a β-keto nitrile   a β-diester

18-3. The proton on the nitrogen is more acidic than the proton on the α-carbon.

18-4. Competing electron delocalization by the lone pair on N or O is more important for the amide than for the ester.

18-5. a. $CH_3\overset{O}{\overset{\|}{C}}H$ > $HC\equiv CH$ > $CH_2=CH_2$ > CH_3CH_3

b. $CH_3\overset{O}{\overset{\|}{C}}CH_2\overset{O}{\overset{\|}{C}}CH_3$ > $CH_3\overset{O}{\overset{\|}{C}}CH_2\overset{O}{\overset{\|}{C}}OCH_3$ > $CH_3O\overset{O}{\overset{\|}{C}}CH_2\overset{O}{\overset{\|}{C}}OCH_3$ > $CH_3\overset{O}{\overset{\|}{C}}CH_3$

c. lactone > ester-lactone > N-methyl lactam

18-7. a. enol/enolate $= 1 \times 10^{12}$ **b.** enol/enolate $= 1 \times 10^8$
   c. enol/enolate $= 1 \times 100$

18-8. a. $CH_3CH=\overset{OH}{\overset{|}{C}}CH_2CH_3$

b. Ph$\overset{OH}{\overset{|}{C}}=CH_2$

c. cyclohexenol

d. 3-hydroxycyclohexenone (more stable) and 3-hydroxycyclohexenone

   more stable

e. $CH_3CH_2\overset{OH}{\overset{|}{C}}=CHCH_2CH_3$ and $CH_3CH=\overset{O}{\overset{\|}{C}}CH_2CH_2CH_3$ (with OH)
   more stable

f. Ph$-CH-\overset{OH}{\overset{|}{C}}CH_3$ and Ph$-CH_2\overset{OH}{\overset{|}{C}}=CH_2$
   more stable

18-10. You need to replace three H's with halogens to get a leaving group that is a weaker base than HO^-.   **18-11.** The rate-determining step must be removal of the proton from the α-carbon of the ketone.

18-15. cyclohexanone with D's (2,2,6,6-tetradeuterio)

18-19. a. cyclohexenone $CH_3CH_2\overset{O}{\overset{\|}{C}}CCH_3$ HO^-

b. $CH_3\overset{O}{\overset{\|}{C}}CH=CH_2$ $CH_3CH_2O\overset{O}{\overset{\|}{C}}CH_2\overset{O}{\overset{\|}{C}}OCH_2CH_3$ $CH_3CH_2O^-$

18-21. a. $CH_3CH_2CH_2\overset{O}{\overset{\|}{C}}H$ **c.** cyclohexyl$-CH_2\overset{O}{\overset{\|}{C}}H$

b. $CH_3\overset{O}{\overset{\|}{C}}CH_3$ **d.** $CH_3CH_2\overset{O}{\overset{\|}{C}}CH_2CH_3$

18-23. cyclohexylidene cyclohexanone

18-28. a. $CH_3CH_2CH_2\overset{O}{\overset{\|}{C}}CH\overset{O}{\overset{\|}{C}}OCH_3$ (with CH_2CH_3 on CH)   **b.** $CH_3CHCH_2\overset{O}{\overset{\|}{C}}CH\overset{O}{\overset{\|}{C}}OCH_2CH_3$ (with CH_3 and $CHCH_3$/CH_3 branches)

18-29. A, B, and D

18-33. a. cyclobutane with OH, CH₃ and COCH₃ substituents   **b.** cyclooctane with OH, CH₃ and ketone

18-36. a. decalin with OH and ketone   **c.** cyclohexane with OH and CHO

b. fused bicyclic with HO and ketone   **d.** fused bicyclic with OH and COCH₃

18-38. A and D   **18-39. a.** methyl bromide **b.** methyl bromide (twice)
c. benzyl bromide **d.** isobutyl bromide   **18-42. a.** ethyl bromide **b.** pentyl bromide **c.** benzyl bromide   **18-45.** 7   **18-46. a.** 3 **h.** 7

CHAPTER 19

19-1. a. reduction **b.** neither **c.** oxidation **d.** oxidation **e.** reduction **f.** neither

19-2. a. $CH_3CH_2CH_2CH_2CH_2OH$ **e.** no reaction
b. $CH_3CH_2CH_2CH_2NH_2$ **f.** CH_3CH_2OH

c. $\underset{H}{\overset{CH_3CH_2CH_2}{C}}=\underset{H}{\overset{CH_3}{C}}$ **g.** $CH_3\overset{O}{\overset{\|}{C}}H$

d. cyclohexyl$-OH$ **h.** cyclohexyl$-NHCH_3$

19-3. Na will cause a terminal alkyne to be converted to an acetylide ion.

19-4. a. Ph$-CH_2NH_2$ **d.** cyclohexyl$-CH_2OH$ + CH_3CH_2OH

b. Ph$-CH_2OH$ **e.** $CH_2CH_2CH_2NHCH_2CH_3$

c. $CH_3CH_2\overset{OH}{\overset{|}{C}}HCH_2CH_3$ **f.** $CH_3CH_2CH_2CH_2OH$

19-5. Yes, because the bonds are polar.

19-12. $\overset{OH}{\underset{\text{(cyclohexane)}}{}}$ with ,,OCH₃   and $\overset{OCH_3}{\underset{\text{(cyclohexane)}}{}}$ with ,,OH

19-13. It is stereoselective and stereospecific, but not enantioselective.

19-14. a. CH$_3$C—CHCH$_2$CH$_3$ (with CH$_3$ above, OH OH below)   **b.** (cyclohexane with CH$_2$OH and OH)

19-16. (bicyclohexyl structure)

19-17. a. (cyclohexane with OH, OH, C(CH$_3$)$_3$)   **b.** (cyclohexane with OH, OH, C(CH$_3$)$_3$)

19-18. 2,3-diemthyl-2-butene   **19-20. a.** 2,3-dimethyl-2-butene **b.** *cis*-4-octene and *trans*-4-octene   **19-22.** whether the alkene has the cis or the trans configuration

19-25. a. (cyclohexane)—C≡CH

b. CH$_3$CH$_2$C≡CCH$_2$CH$_2$CH$_2$C≡CCH$_2$CH$_3$

CHAPTER 20

20-1. a. 2,2-dimethylaziridine **b.** 4-ethylpiperidine **c.** 3-methylazacyclobutane **d.** 2-methylthiacyclopropane **e.** 2,3-dimethyltetrahydrofuran **f.** 2-ethyloxacyclobutane   **20-2.** The electron-withdrawing oxygen stabilizes the-conjugate base.

20-3. a. (quinuclidinone structure)   **b.** pK_a ~ 8 **c.** conjugate acid of 3-chloroquinuclidine

20-5. The only difference is the leaving group.

20-6. CH$_3$C=CH$_2$ (with CH$_3$) + NCH$_2$CH$_2$CH$_3$ (with CH$_3$, CH$_3$)

20-7. a. CH$_3$CH=CH$_2$ + NCH$_3$ (with CH$_3$, CH$_3$)   **c.** (methylenecyclohexane) + N(CH$_3$)$_3$

b. CH$_2$=CHCHCH$_2$CH$_2$NCH$_3$ (with CH$_3$, CH$_3$)   **d.** CH$_3$NCH$_2$CHCH$_2$CH=CH$_2$ (with CH$_3$, CH$_3$)

20-8. a. 2-methyl-3-pentanamine **b.** 3-methylpyrrolidine
20-10. carbanion-like

20-11. a. CH$_3$N (with CH$_3$, OH) + CH$_2$=CHCH$_3$ **c.** CH$_2$=CH$_2$ + NCH$_2$CHCH$_3$ (with CH$_3$, OH, CH$_3$)

b. CH$_3$N (with OH, phenyl) + CH$_2$=CHCH$_3$ **d.** CH$_2$=CHCH$_2$CH$_2$CH$_2$NCH$_3$ (with OH)

20-15. Both acidic and basic forms of pyrrole are aromatic; only the basic form of cyclopentadiene is aromatic.

20-17. (pyridinium, N$^+$ with CH$_2$CH$_3$)   **20-20.** (4-methylpyridinium, N$^+$CH$_2$CH$_3$ I$^-$) > (4-methylpyridine) > (3-methylpyridine)

20-21. (imidazole with Br and NCH$_3$)

20-22. (pyrrole, NH) > (imidazole, N NH) > (benzene)

20-23. Imidazole forms intramolecular hydrogen bonds that *N*-methylimidazole cannot form.   **20-24.** 24%   **20-25.** yes

CHAPTER 21

21-1. D-Ribose is an aldopentose. D-Sedoheptulose is a ketoheptose. D-Mannose is an aldohexose.   **21-3. a.** L-glyceraldehyde **b.** L-glyceraldehyde **c.** D-glyceraldehyde
21-4. a. enantiomers **b.** diastereomers   **21-5. a.** D-ribose **b.** L-talose **c.** L-allose
21-6. a. (2R,3S,4R,5R)-2,3,4,5,6-pentahydroxyhexanal **b.** (2S,3R,4S,5S)-2,3,4,5,6-pentahydroxyhexanal   **21-7.** D-psicose   **21-8. a.** A ketoheptose has four asymmetric centers (2^4 = 16 stereoisomers). **b.** An aldoheptose has five asymmetric centers (2^5 = 32 stereoisomers). **c.** A ketotriose has no asymmetric centers; therefore, it has no stereoisomers.   **21-11.** D-tagatose, D-galactose, and D-talose
21-12. a. D-iditol **b.** D-iditol and D-gulitol   **21-13. a. 1.** D-altrose **2.** L-galactose
b. 1. D-tagatose **2.** D-fructose   **21-14. a.** L-gulose **b.** L-gularic acid **c.** D-allose and L-allose, D-altrose and D-talose, L-altrose and L-talose, D-galactose and L-galactose
21-15. a. D-arabinose and D-ribulose **b.** D-allose and D-psicose **c.** L-gulose and L-sorbose **d.** D-talose and D-tagatose   **21-16.** D-gulose and D-idose
21-17. a. D-gulose and D-idose **b.** L-xylose and L-lyxose   **21-18. a.** D-glucose and D-mannose **b.** D-erythrose and D-threose **c.** L-allose and L-altrose
21-20. A = D-glucose   B = D-mannose   C = D-arabinose   D = D-erythrose
21-24. a. the OH group at C-2 **b.** the OH group at C-2, C-3, and C-4 **c.** the OH group at C-3 and C-1   **21-25.** A protonated amine is not a nucleophile.
21-26. b. α-D-talose (reducing) **c.** methyl α-D-galactoside (nonreducing) **d.** ethyl β-D-psicoside (nonreducing)
21-29. a. They cannot receive type A, B, or AB blood because these have sugar components that type O blood does not have. **b.** They cannot give blood to those with type A, B, or O blood because AB blood has sugar components that these other blood types do not have.

CHAPTER 22

22-2. a. (R)-alanine **b.** (R)-aspartate **c.** The α-carbon of all the D-amino acids except cysteine has the R-configuration.   **22-4.** Ile   **22-5.** because of the electron-withdrawing ammonium group

22-7.

a. HOCCH$_2$CH$_2$CHCOH (with $^+$NH$_3$)   **c.** $^-$OCCH$_2$CH$_2$CHCO$^-$ (with $^+$NH$_3$)

b. HOCCH$_2$CH$_2$CHCO$^-$ (with $^+$NH$_3$)   **d.** $^-$OCCH$_2$CH$_2$CHCO$^-$ (with NH$_2$)

22-10. a. 5.43 **b.** 10.76 **c.** 5.68 **d.** 2.98   **22-11. a.** Asp **b.** Arg **c.** Asp **d.** Met
22-13. 2-methylpropanal   **22-14.** Leucine and isoleucine have similar polarities and pI values, so they show up as one spot.   **22-17.** His > Val > Ser > Asp   **22-18.** One equivalent of ammonia will be protonated by the carboxylic acid.   **22-19. a.** L-Ala, L-Asp, L-Glu **b.** L-Ala and D-Ala, L-Asp and D-Asp, L-Glu and D-Glu   **22-20. a.** leucine **b.** methionine
22-21. a. 4-bromo-1-butanamine **b.** benzylbromide   **22-22. a.** alanine
b. isoleucine **c.** leucine   **22-25.** A-G-M A-M-G M-G-A M-A-G G-A-M
G-M-A   **22-27.** the bonds on either side of the α-carbon   **22-29.** Leu-Val and Val-Val   **22-32. a.** 5.8% **b.** 4.4%   **22-35.** Gly-Arg-Trp-Ala-Glu-Leu-Met-Pro-Val-Asp   **22-36. a.** His-Lys Leu-Val-Glu-Pro-Arg Ala-Gly-Ala
b. Leu-Gly-Ser-Met-Phe-Pro-Tyr Gly-Val   **22-39.** Leu-Tyr-Lys-Arg-Met-Phe-Arg-Ser   **22-40.** Edman's reagent would release two amino acids in approximately equal amounts.   **22-41.** 110 Å in an α-helix and 260 Å in a straight chain   **22-42.** nonpolar groups on the outside and polar groups on the inside   **22-43. a.** cigar-shaped protein **b.** subunit of a hexamer

CHAPTER 23

23-1. $\Delta H^\ddagger$, E_a, $\Delta S^\ddagger$, $\Delta G^\ddagger$, k_{rate} **23-3.** specific acid catalyzed   **23-7.** They do not have a negatively charged oxygen on one carbon and a carbonyl group on an adjacent carbon.   **23-9.** close to one

23-10. (cyclohexane with OH and SC$_6$H$_5$)   (cyclohexane with SC$_6$H$_5$ and OH)   (cyclohexane with OCH$_2$CH$_3$ and SC$_6$H$_5$)   (cyclohexane with SC$_6$H$_5$ and OCH$_2$CH$_3$)

23-12. The nitro groups cause the phenolate ion to be a better leaving group than the carboxylate ion. **23-15.** 2, 3, and 4 **23-16.** Ser-Ala-Phe **23-18.** Arginine forms a direct hydrogen bond; lysine forms an indirect hydrogen bond. **23-20.** NAM **23-21.** The acid denatures the enzyme. **23-23.** lysine **23-24.** 2 **23-27.** Putting a substituent on cysteine with iodoacetic acid could interfere with binding or catalysis of the substrate.

CHAPTER 24

24-1. It increases the susceptibility of the carbonyl carbon to nucleophilic attack, increases the nucleophilicity of water, and stabilizes the negative charge on the transition state.

24-2.

$$^-OCCCHCH_2CO^- + NADH + H^+$$

24-3.

$$CH_3CH-CO^- + NAD^+$$

24-4. pyrophosphate **24-5. a.** 7 **b.** 3 isolated from 2 others **24-14. a.** alanine **b.** aspartate **24-16.** the one on the right
24-20.

$$CH_3CH_2CSCoA \xrightarrow[\text{biotin}]{E} CH_3CHCSCoA \xrightarrow[\text{coenzyme B}_{12}]{E} CH_2CH_2CSCoA$$

24-21. the methylene group of N^5,N^{10}-methylene-THF

CHAPTER 25

25-1. It is a stable base because the negative charge is delocalized.
25-3. phosphocreatine **25-8.** The β-carbon has a partial positive charge.
25-9. eight **25-10.** seven **25-12. a.** conversion of glucose to glucose-6-phosphate; conversion of glucose-6-phosphate to glucose-1,6-diphosphate **b.** conversion of 1,3-diphosphoglycerate to 3-phosphoglycerate; conversion of 2-phosphoenolpyruvate to pyruvate **25-13.** two **25-14.** a ketone **25-15.** thiamine pyrophosphate **25-16.** an aldehyde **25-18.** pyridoxal phosphate **25-19.** pyruvate **25-20.** a secondary alcohol **25-21.** citrate and isocitrate **25-22.** 11

CHAPTER 26

26-2. hexanoic acid, 3 malonic acids, and glutaric acid **26-3.** glyceryl tripalmitate **26-8.** no **26-9.** Integral proteins will have a higher percentage of nonpolar amino acids. **26-10.** The bacteria could synthesize phosphoacylglycerols with more saturated fatty acids. **26-14.** the one on the left **26-16.** The two halves are synthesized in a head-to-tail fashion and then joined together in a tail-to-tail linkage. **26-24.** cis fused; trans fused **26-25.** a β-substituent **26-28.** Two are axial substituents and one is an equatorial substituent.

CHAPTER 27

27-7. a. 3′—C—C—T—G—T—T—A—G—A—C—G—5′
b. guanine **27-11.** Met-Asp-Pro-Val-Ile-Lys-His **27-12.** Met-Asp-Pro-Leu-Leu-Asn **27-14.** 5′—G-C-A-T-G-G-A-C-C-C-C-G-T-T-A-T-T-A-A-A-C-A-C—3′
27-16.

xanthine   hypoxanthine

27-18. a

CHAPTER 28

28-1. a. $CH_2{=}CHCl$ **b.** $CH_2{=}CCH_3$ **c.** $CF_2{=}CF_2$
      $COCH_3$

28-2. poly(vinyl chloride) **28-5.** beach balls
28-7. a.

$OCH_3 > CH_3 > NO_2$

b. $CH_2{=}CHOCCH_3 > CH_2{=}CHCH_3 > CH_2{=}CHCOCH_3$
c. $CH_2{=}CCH_3$ $CH_2{=}CH$

$>$

28-8. a.

$NO_2 > CH_3 > OCH_3$

b. $CH_2{=}CHC{\equiv}N > CH_2{=}CHCl > CH_2{=}CHCH_3$
28-11. a. $CH_2{=}CCH_3 + BF_3 + H_2O$ **c.** $\triangle O + CH_3O^-$
      CH_3
b. $CH_2{=}CH + BF_3 + H_2O$ **d.** $CH_2{-}CH + BuLi$
                                    $COCH_3$
28-12.

$H_3C{-}O{-}CH_2CCH_2OCH_2CCH_2OCH_2CCH_2OH$

28-15. a. $-NHCH_2CH_2CH_2CNHCH_2CH_2CH_2C-$
b. $-NH(CH_2)_4NHCCH_2CH_2CNH(CH_2)_4NHCCH_2CH_2C-$

CHAPTER 29

29-1. a. electrocyclic reaction **b.** sigmatropic rearrangement **c.** cycloaddition reaction **d.** cycloaddition reaction **29-2. a.** bonding orbitals $= \psi_1, \psi_2, \psi_3$; antibonding orbitals $= \psi_4, \psi_5, \psi_6$; **b.** ground-state HOMO $= \psi_3$; ground-state LUMO $= \psi_4$ **c.** excited-state HOMO $= \psi_4$; excited-state LUMO $= \psi_5$ **d.** symmetric orbitals $= \psi_1, \psi_3, \psi_5$; antisymmetric orbitals $= \psi_2, \psi_4, \psi_6$ **e.** The HOMO and LUMO have opposite symmetries. **29-3. a.** 8 **b.** ψ_4 **c.** 7 **29-6. a.** conrotatory **b.** trans **c.** disrotatory **d.** cis **29-7. a.** correct **b.** correct **c.** correct **29-8. 1. a.** conrotatory **b.** trans **2. a.** disrotatory **b.** cis **29-11.** yes

29-13.

29-18. a. inversion **b.** retention **29-19.** antarafacial

Glossary

absolute configuration the three-dimensional structure of a chiral compound. The configuration is designated by *R* or *S*.

absorption band a peak in a spectrum that occurs as a result of the absorption of energy.

acetal

acetamidomalonic ester synthesis a method used to synthesize an amino acid that is a variation of the *N*-phthalimidomalonic ester synthesis.

acetoacetic ester synthesis synthesis of a methyl ketone, using ethyl acetoacetate as the starting material.

achiral (optically inactive) an achiral molecule has a conformation identical to (superimposable upon) its mirror image.

acid (Brønsted) a substance that donates a proton.

acid anhydride

acid–base reaction a reaction in which an acid donates a proton to a base or accepts a share in a base's electrons.

acid catalyst a catalyst that increases the rate of a reaction by donating a proton.

acid-catalyzed reaction a reaction catalyzed by an acid.

acid dissociation constant a measure of the degree to which an acid dissociates in solution.

activating substituent a substituent that increases the reactivity of an aromatic ring. Electron-donating substituents activate aromatic rings toward electrophilic attack, and electron-withdrawing substituents activate aromatic rings toward nucleophilic attack.

active site a pocket or cleft in an enzyme where the substrate is bound.

acyclic noncyclic.

acyl adenylate a carboxylic acid derivative with AMP as the leaving group.

acyl–enzyme intermediate an intermediate formed when an amino acid residue of an enzyme is acetylated.

acyl group a carbonyl group bonded to an alkyl group or to an aryl group.

acyl halide

acyl phosphate a carboxylic acid derivative with a phosphate leaving group.

acyl pyrophosphate a carboxylic acid derivative with a pyrophosphate leaving group.

1,2-addition (direct addition) addition to the 1- and 2-positions of a conjugated system.

1,4-addition (conjugate addition) addition to the 1- and 4-positions of a conjugated system.

addition polymer (chain-growth polymer) a polymer made by adding monomers to the growing end of a chain.

addition reaction a reaction in which atoms or groups are added to the reactant.

adrenal cortical steroids glucocorticoids and mineralocorticoids.

alcohol a compound with an OH group in place of one of the hydrogens of an alkane; (ROH).

alcoholysis reaction with an alcohol.

aldaric acid a dicarboxylic acid with an OH group bonded to each carbon. Obtained by oxidizing the aldehyde and primary alcohol groups of an aldose.

aldehyde

alditol a compound with an OH group bonded to each carbon. Obtained by reducing an aldose or a ketose.

aldol addition a reaction between two molecules of an aldehyde (or two molecules of a ketone) that connects the α-carbon of one with the carbonyl carbon of the other.

aldol condensation an aldol addition followed by the elimination of water.

aldonic acid a carboxylic acid with an OH group bonded to each carbon. Obtained by oxidizing the aldehyde group of an aldose.

aldose a polyhydroxyaldehyde.

aliphatic a nonaromatic organic compound.

alkaloid a natural product, with one or more nitrogen heteroatoms, found in the leaves, bark, or seeds of plants.

alkane a hydrocarbon that contains only single bonds.

alkene a hydrocarbon that contains a double bond.

alkoxymercuration addition of alcohol to an alkene, using a mercuric salt of a carboxylic acid as a catalyst.

alkylation reaction a reaction that adds an alkyl group to a reactant.

alkyl halide a compound with a halogen in place of one of the hydrogens of an alkane.

alkyl substituent (alkyl group) formed by removing a hydrogen from an alkane.

alkyl tosylate an ester of *para*-toluenesulfonic acid.

alkyne a hydrocarbon that contains a triple bond.

allene a compound with two adjacent double bonds.

allyl group $CH_2{=}CHCH_2{-}$

allylic carbon an sp^3 carbon adjacent to a vinylic carbon.

allylic cation a species with a positive charge on an allylic carbon.

alpha olefin a monosubstituted olefin.

alternating copolymer a copolymer in which two monomers alternate.

ambident nucleophile a nucleophile with two nucleophilic sites.

amide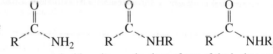

amine a compound with a nitrogen in place of one of the hydrogens of an alkane; (RNH_2, R_2NH, R_3N)

amine inversion the configuration of an sp^3 hybridized nitrogen with a nonbonding pair of electrons rapidly turns inside out.

amino acid an α-aminocarboxylic acid. Naturally occurring amino acids have the L configuration.

amino acid analyzer an instrument that automates the ion-exchange separation of amino acids.

amino acid residue a monomeric unit of a peptide or protein.

aminolysis reaction with an amine.

amino sugar a sugar in which one of the OH groups is replaced by an NH_2 group.

amphoteric compound a compound that can behave either as an acid or as a base.

anabolic steroids steroids that aid in the development of muscle.

anabolism reactions living organisms carry out in order to synthesize complex molecules from simple precursor molecules.

anchimeric assistance (intramolecular catalysis) catalysis in which the catalyst that facilitates the reaction is part of the molecule undergoing reaction.

androgens male sex hormones.

angle strain the strain introduced into a molecule as a result of its bond angles being distorted from their ideal values.

angstrom unit of length; 100 picometers $= 10^{-8}$ cm $= 1$ angstrom

angular methyl group a methyl substituent at the 10- or 13-position of a steroid ring system.

anion-exchange resin a positively charged resin used in ion-exchange chromatography.

anionic polymerization chain-growth polymerization in which the initiator is a nucleophile; the propagation site therefore is an anion.

annulation reaction a ring-forming reaction.

annulene a monocyclic hydrocarbon with alternating double and single bonds.

anomeric carbon the carbon in a cyclic sugar that is the carbonyl carbon in the open-chain form.

anomeric effect the preference for the axial position shown by certain substituents bonded to the anomeric carbon of a six-membered-ring sugar.

anomers two cyclic sugars that differ in configuration only at the carbon that is the carbonyl carbon in the open-chain form.

antarafacial bond formation formation of two σ bonds on opposites sides of the π system.

antarafacial rearrangement rearrangement in which the migrating group moves to the opposite face of the π system.

anti addition an addition reaction in which two substituents are added at opposite sides of the molecule.

antiaromatic a cyclic and planar compound with an uninterrupted ring of p orbital-bearing atoms containing an even number of pairs of π electrons.

antibiotic a compound that interferes with the growth of a microorganism.

antibodies compounds that recognize foreign particles in the body.

antibonding molecular orbital a molecular orbital that results when two atomic orbitals with opposite signs interact. Electrons in an antibonding orbital decrease bond strength.

anticodon the three bases at the bottom of the middle loop in tRNA.

anti conformer the most stable of the staggered conformers.

anti elimination an elimination reaction in which the two substituents that are eliminated are removed from opposite sides of the molecule.

antigene agent a polymer designed to bind to DNA at a particular site.

antigens compounds that can generate a response from the immune system.

anti-periplanar parallel substituents on opposite sides of a molecule.

antisense agent a polymer designed to bind to mRNA at a particular site.

antisense strand (template strand) the strand in DNA that is read during transcription.

antisymmetric molecular orbital a molecular orbital in which the left (or top) half is not a mirror of the right (or bottom) half.

antiviral drug a drug that interferes with DNA or RNA synthesis in order to prevent a virus from replicating.

apoenzyme an enzyme without its cofactor.

applied magnetic field the externally applied magnetic field.

aprotic solvent a solvent that does not have a hydrogen bonded to an oxygen or to a nitrogen.

aramide an aromatic polyamide.

arene oxide an aromatic compound that has had one of its double bonds converted to an epoxide.

aromatic a cyclic and planar compound with an uninterrupted ring of p orbital-bearing atoms containing an odd number of pairs of π electrons.

Arrhenius equation relates the rate constant of a reaction to the energy of activation and to the temperature at which the reaction is carried out ($k = Ae^{-E_a/RT}$).

aryl group a benzene or a substituted-benzene group.

asymmetric center an atom bonded to four different atoms or groups.

atactic polymer a polymer in which the substituents are randomly oriented on the extended carbon chain.

atomic number the number of protons (or electrons) that the neutral atom has.

atomic orbital an orbital associated with an atom.

atomic weight the average mass of the atoms in the naturally occurring element.

aufbau principle states that an electron will always go into that orbital with the lowest available energy.

automated solid-phase peptide synthesis an automated technique that synthesizes a peptide while its C-terminal amino acid is attached to a solid support.

autoradiograph the exposed photographic plate obtained in autoradiography.

autoradiography a technique used to determine the base sequence of DNA.

auxochrome a substituent that when attached to a chromophore, alters the λ_{max} and intensity of absorption of UV/Vis radiation.

axial bond a bond of the chair conformation of cyclohexane that is perpendicular to the plane in which the chair is drawn (an up–down bond).

aziridine a three-membered-ring compound in which one of the ring atoms is a nitrogen.

azo linkage an —N≡N— bond.

back-side attack nucleophilic attack on the side of the carbon opposite the side bonded to the leaving group.

bactericidal drug a drug that kills bacteria.

bacteriostatic drug a drug that inhibits the further growth of bacteria.

Baeyer–Villiger oxidation oxidation of aldehydes or ketones with H_2O_2 to form carboxylic acids or esters, respectively.

banana bond the σ bonds in small rings that are weaker as a result of overlapping at an angle rather than overlapping head-on.

base¹ a substance that accepts a proton.

base² a heterocyclic compound (a purine or a pyrimidine) in DNA and RNA.

base catalyst a catalyst that increases the rate of a reaction by removing a proton.

base peak the peak with the greatest abundance in a mass spectrum.

basicity the tendency of a compound to share its electrons with a proton.

Beer–Lambert law relationship among the absorbance of UV/Vis light, the concentration of the sample, the length of the light path, and the molar absorptivity ($A = cl\varepsilon$).

bending vibration a vibration that does not occur along the line of the bond. It results in changing bond angles.

benzoyl group ring bonded to a carbonyl group.

benzyl group

benzylic carbon an sp^3 hybridized carbon bonded to a benzene ring.

benzylic cation a compound with a positive charge on a benzylic carbon.

benzyne intermediate a compound with a triple bond in place of one of the double bonds of benzene.

bicyclic compound a compound containing two rings that share at least one carbon.

bifunctional molecule a molecule with two functional groups.

bile acids steroids that act as emulsifying agents so that water-insoluble compounds can be digested.

bimolecular reaction (second-order reaction) a reaction whose rate depends on the concentration of two reactants.

biochemistry (biological chemistry) the chemistry of biological systems.

biodegradable polymer a polymer that can be broken into small segments by an enzyme-catalyzed reaction.

bioorganic compound an organic compound found in biological systems.

biopolymer a polymer that is synthesized in nature.

biosynthesis synthesis in a biological system.

biotin the coenzyme required by enzymes that catalyze carboxylation of a carbon adjacent to an ester or a keto group.

Birch reduction the partial reduction of benzene to 1,4-cyclohexadiene.

blind screen (random screen) the search for a pharmacologically active compound without any information about which chemical structures might show activity.

block copolymer a copolymer in which there are regions (blocks) of each kind of monomer.

blue shift a shift to a shorter wavelength.

boat conformation the conformation of cyclohexane that roughly resembles a boat.

boiling point the temperature at which the vapor pressure from a liquid equals the atmospheric pressure.

bonding molecular orbital a molecular orbital that results when two in-phase atomic orbitals interact. Electrons in a bonding orbital increase bond strength.

bond length the internuclear distance between two atoms at minimum energy (maximum stability).

bond strength the energy required to break a bond homolytically.

brand name (proprietary name, trade name) identifies a commercial product and distinguishes it from other products. It can be used only by the owner of the registered trademark.

bridged bicyclic compound a bicyclic compound in which rings share two nonadjacent carbons.

Brønsted acid a substance that donates a proton.

Brønsted base a substance that accepts a proton.

buffer a weak acid and its conjugate base.

carbanion a compound containing a negatively charged carbon.

carbene a species with a carbon that has a nonbonded pair of electrons and an empty orbital (H_2C:).

carbocation   a species containing a positively charged carbon.

carbocation rearrangement   the rearrangement of a carbocation to a more stable carbocation.

carbohydrate   a sugar or a saccharide. Naturally occurring carbohydrates have the D configuration.

α-carbon   a carbon bonded to a leaving group or adjacent to a carbonyl carbon.

β-carbon   a carbon adjacent to an α-carbon.

carbon acid   a compound containing a carbon that is bonded to a relatively acidic hydrogen.

carbonyl addition (direct addition)   nucleophilic addition to the carbonyl carbon.

carbonyl carbon   the carbon of a carbonyl group.

carbonyl compound   a compound that contains a carbonyl group.

carbonyl group   a carbon doubly bonded to an oxygen.

carbonyl oxygen   the oxygen of a carbonyl group.

carboxyl group   COOH

carboxylic acid

$$\underset{R}{}\overset{\overset{\displaystyle O}{\|}}{C}\underset{}{}OH$$

carboxylic acid derivative   a compound that is hydrolyzed to a carboxylic acid.

carboxyl oxygen   the single-bonded oxygen of a carboxlic acid or an ester.

carotenoid   a class of compounds (a tetraterpene) responsible for the red and orange colors of fruits, vegetables, and fall leaves.

catabolism   reactions living organisms carry out in order to break down complex molecules into simple molecules and energy.

catalyst   a species that increases the rate at which a reaction occurs without being consumed in the reaction. Because it does not change the equilibrium constant of the reaction, it does not change the amount of product that is formed.

catalytic antibody   a compound that facilitates a reaction by forcing the conformation of the substrate in the direction of the transition state.

catalytic hydrogenation   the addition of hydrogen to a double or a triple bond with the aid of a metal catalyst.

cation-exchange resin   a negatively charged resin used in ion-exchange chromatography.

cationic polymerization   chain-growth polymerization in which the initiator is an electrophile; the propagation site therefore is a cation.

cephalin   a phosphoacylglycerol in which the second OH group of phosphate has formed an ester with ethanolamine.

cerebroside   a sphingolipid in which the terminal OH group of sphingosine is bonded to a sugar residue.

chain-growth polymer (addition polymer)   a polymer made by adding monomers to the growing end of a chain.

chain transfer   a growing polymer chain reacts with a molecule XY in a manner that allows X to terminate the chain, leaving behind Y to initiate a new chain.

chair conformation   the conformation of cyclohexane that roughly resembles a chair. It is the most stable conformation of cyclohexane.

chemical exchange   the transfer of a proton from one molecule to another.

chemically equivalent protons   protons with the same connectivity relationship to the rest of the molecule.

chemical shift   the location of a signal in an NMR spectrum. It is measured downfield from a reference compound (most often, TMS).

chiral (optically active)   a chiral molecule has a nonsuperimposable mirror image.

chiral auxiliary   an enantiomerically pure compound that, when attached to a reactant, causes a product with a particular configuration to be formed.

chirality center   a tetrahedral atom bonded to four different groups.

cholesterol   a steroid that is the precursor of all other animal steroids.

chromatography   a separation technique in which the mixture to be separated is dissolved in a solvent and the solvent is passed through a column packed with an absorbent stationary phase.

chromophore   the part of a molecule responsible for a UV or visible spectrum.

cine substitution   substitution at the carbon adjacent to the carbon that was bonded to the leaving group.

s-cis conformation   the conformation in which two double bonds are on the same side of a single bond.

cis fused   two cyclohexane rings fused together such that if the second ring were considered to be two substituents of the first ring, one substituent would be in an axial position and the other would be in an equatorial position.

cis isomer   the isomer with the hydrogens on the same side of the double bond or cyclic structure.

cis–trans isomers   geometric isomers.

citric acid cycle (Krebs cycle)   a series of reactions that converts the acetyl group of acetyl-CoA into two molecules of CO_2.

Claisen condensation   a reaction between two molecules of an ester that connects the α-carbon of one with the carbonyl carbon of the other and eliminates an alkoxide ion.

Claisen rearrangement   a [3,3] sigmatropic rearrangement of an allyl vinyl ether.

α-cleavage   homolytic cleavage of an alpha substituent.

Clemmensen reduction   a reaction that reduces the carbonyl group of a ketone to a methylene group using Zn(Hg)/ HCl.

codon   a sequence of three bases in mRNA that specifies the amino acid to be incorporated into a protein.

coenzyme   a cofactor that is an organic molecule.

coenzyme A   a thiol used by biological organisms to form thioesters.

coenzyme B_{12}   the coenzyme required by enzymes that catalyze certain rearrangement reactions.

cofactor   an organic molecule or a metal ion that certain enzymes need to catalyze a reaction.

coil conformation (loop conformation)   that part of a protein that is highly ordered, but not in an α-helix or a β-pleated sheet.

combination band   occurs at the sum of two fundamental absorption frequencies $(v_1 + v_2)$.

combinatorial library   a group of structurally related compounds.

combinatorial organic synthesis   the synthesis of a library of compounds by covalently connecting sets of building blocks of varying structure.

common intermediate   an intermediate that two compounds have in common.

common name   nonsystematic nomenclature.

competitive inhibitor   a compound that inhibits an enzyme by competing with the substrate for binding at the active site.

complete racemization   the formation of a pair of enantiomers in equal amounts.

complex carbohydrate   a carbohydrate containing two or more sugar molecules linked together.

concerted reaction   a reaction in which all the bond-making and bond-breaking processes occur in one step.

condensation polymer (step-growth polymer)   a polymer made by combining two molecules while removing a small molecule (usually water or an alcohol).

condensation reaction   a reaction combining two molecules while removing a small molecule (usually water or an alcohol).

conducting polymer   a polymer that can conduct electricity.

configuration   the three-dimensional structure of a particular atom in a compound. The configuration is designated by R or S.

configurational isomers   stereoisomers that cannot interconvert unless a covalent bond is broken. Cis–trans isomers and optical isomers are configurational isomers.

conformation   the three-dimensional shape of a molecule at a given instant that can change as a result of rotations about σ bonds.

conformational analysis   the investigation of the various conformations of a compound and their relative stabilities.

conformers   different conformations of a molecule.

conjugate acid   a species accepts a proton to form its conjugate acid.

conjugate addition   1,4-addition to an α,β, unsaturated carbonyl compound.

conjugate base   a species loses a proton to form its conjugate base.

conjugated double bonds   double bonds separated by one single bond.

conrotatory ring closure   achieves head-to-head overlap of p orbitals by rotating the orbitals in the same direction.

conservation of orbital symmetry theory   a theory that explains the relationship between the structure and configuration of the reactant, the conditions under which a pericyclic reaction takes place, and the configuration of the product.

constitutional isomers (structural isomers)   molecules that have the same molecular formula but differ in the way their atoms are connected.

contributing resonance structure (resonance contributor, resonance structure)   a structure with localized electrons that approximates the structure of a compound with delocalized electrons.

convergent synthesis   a synthesis in which pieces of the target compound are individually prepared and then assembled.

Cope elimination reaction   elimination of a proton and a hydroxyl amine from an amine oxide.

Cope rearrangement a [3,3] sigmatropic rearrangement of a 1,5-diene.

copolymer a polymer formed from two or more different monomers.

corrin ring system a porphyrin ring system without one of the methine bridges.

COSY spectrum a 2-D NMR spectrum that shows coupling between sets of protons.

coupled protons protons that split each other. Coupled protons have the same coupling constant.

coupling constant the distance (in hertz) between two adjacent peaks of a split NMR signal.

coupling reaction a reaction that joins two alkyl groups.

covalent bond a bond created as a result of sharing electrons.

covalent catalysis (nucleophilic catalysis) catalysis that occurs as a result of a nucleophile forming a covalent bond with one of the reactants.

Cram's rule the rule used to determine the major product of a carbonyl addition reaction in a compound with an asymmetric center adjacent to the carbonyl group.

cross-conjugation nonlinear conjugation.

crossed (mixed) aldol addition an aldol addition in which two different carbonyl compounds are used.

cross-linking connecting polymer chains by intermolecular bond formation.

crown ether a cyclic molecule that contains several ether linkages.

crown–guest complex the complex formed when a crown ether binds a substrate.

cryptand a three-dimensional polycyclic compound that binds a substrate by encompassing it.

cryptate the complex formed when a cryptand binds a substrate.

crystallites regions of a polymer in which the chains are highly ordered.

C-terminal amino acid the terminal amino acid of a peptide (or protein) that has a free carboxyl group.

cumulated double bonds double bonds that are adjacent to one other.

Curtius rearrangement conversion of an acyl chloride into a primary amine with the use of azide ion ($^-N_3$).

cyanohydrin
$$R-\underset{\underset{C\equiv N}{|}}{\overset{\overset{OH}{|}}{C}}-R(H)$$

cycloaddition reaction a reaction in which two π-bond-containing molecules react to form a cyclic compound.

[4 + 2] cycloaddition reaction a cycloaddition reaction in which four π electrons come from one reactant and two π electrons come from the other reactant.

cycloalkane an alkane with its carbon chain arranged in a closed ring.

deactivating substituent a substituent that decreases the reactivity of an aromatic ring. Electron-withdrawing substituents deactivate aromatic rings toward electrophilic attack, and electron-donating substituents deactivate aromatic rings toward nucleophilic attack.

deamination loss of ammonia.

decarboxylation loss of carbon dioxide.

degenerate orbitals orbitals that have the same energy.

dehydration loss of water.

dehydrogenase an enzyme that carries out an oxidation reaction by removing hydrogen from the substrate.

dehydrohalogenation elimination of a proton and a halide ion.

delocalization energy (resonance energy) the extra stability a compound achieves as a result of having delocalized electrons.

delocalized electrons electrons that are shared by more than two atoms.

denaturation destruction of the highly organized tertiary structure of a protein.

deoxygenation removal of an oxygen from a reactant.

deoxyribonucleic acid (DNA) a polymer of deoxyribonucleotides.

deoxyribonucleotide a nucleotide in which the sugar component is D-2'-deoxyribose.

deoxy sugar a sugar in which one of the OH groups has been replaced by an H.

DEPT ^{13}C NMR spectrum a series of four spectra that distinguishes among $-CH_3$, $-CH_2$, and $-CH$ groups.

depurination elimination of a purine ring.

detergent a salt of a sulfonic acid.

deuterium kinetic isotope effect ratio of the rate constant obtained for a compound containing hydrogen and the rate constant obtained for an identical compound in which one or more of the hydrogens have been replaced by deuterium.

dextrorotatory the enantiomer that rotates polarized light in a clockwise direction.

diastereomer a configurational stereoisomer that is not an enantiomer.

diastereotopic hydrogens two hydrogens bonded to a carbon that when replaced in turn with a deuterium, result in a pair of diastereomers.

1,3-diaxial interaction the interaction between an axial substituent and the other two axial substituents on the same side of the cyclohexane ring.

diazonium ion $Ar\overset{+}{N}\equiv N$ or $R\overset{+}{N}\equiv N$.

diazonium salt a diazonium ion and an anion ($Ar\overset{+}{N}\equiv NX^-$).

dideoxy method a method used to determine the sequence of bases in restriction fragments.

Dieckmann condensation an intramolecular Claisen condensation.

dielectric constant a measure of how well a solvent can insulate opposite charges from one another.

Diels–Alder reaction a [4 + 2] cycloaddition reaction.

diene a hydrocarbon with two double bonds.

dienophile an alkene that reacts with a diene in a Diels–Alder reaction.

β-diketone a ketone with a second carbonyl group at the β-position.

dimer a molecule formed by the joining together of two identical molecules.

dinucleotide two nucleotides linked by phosphodiester bonds.

dipeptide two amino acids linked by an amide bond.

dipole–dipole interaction an interaction between the dipole of one molecule and the dipole of another.

dipole moment (μ) a measure of the separation of charge in a bond or in a molecule.

direct addition 1,2-addition.

direct displacement mechanism a reaction in which the nucleophile displaces the leaving group in a single step.

direct substitution substitution at the carbon that was bonded to the leaving group.

disaccharide a compound containing two sugar molecules linked together.

disconnection breaking a bond to carbon to give a simpler species.

disproportionation transfer of a hydrogen atom by a radical to another radical, forming an alkane and an alkene.

disrotatory ring closure achieves head-to-head overlap of p orbitals by rotating the orbitals in opposite directions.

dissociation energy the amount of energy required to break a bond, or the amount of energy released when a bond is formed.

dissolving-metal reduction a reduction brought about by the use of sodium or lithium metal dissolved in liquid ammonia.

distribution coefficient the ratio of the amounts of a compound dissolving in each of two solvents in contact with each other.

disulfide bridge a disulfide ($-S-S-$) bond in a peptide or protein.

DNA (deoxyribonucleic acid) a polymer of deoxyribonucleotides.

double bond a σ bond and a π bond between two atoms.

doublet an NMR signal split into two peaks.

doublet of doublets an NMR signal split into four peaks of approximately equal height. Caused by splitting a signal into a doublet by one hydrogen and into another doublet by another (nonequivalent) hydrogen.

drug a compound that reacts with a biological molecule, triggering a physiological effect.

drug resistance biological resistance to a particular drug.

drug synergism when the effect of two drugs used in combination is greater than the sum of the effects obtained when the drugs are administered individually.

eclipsed conformation a conformation in which the bonds on adjacent carbons are aligned as viewed looking down the carbon–carbon bond.

E conformation the conformation of a carboxylic acid or carboxylic acid derivative in which the carbonyl oxygen and the substituent bonded to the carboxyl oxygen or nitrogen are on opposite sides of the single bond.

Edman's reagent phenyl isothiocyanate. A reagent used to determine the N-terminal amino acid of a polypeptide.

effective magnetic field the magnetic field that a proton "senses" through the surrounding cloud of electrons.

effective molarity the concentration of the reagent that would be required in an intermolecular reaction for it to have the same rate as an intramolecular reaction.

E isomer the isomer with the high-priority groups on opposite sides of the double bond.

elastomer a polymer that can stretch and then revert to its original shape.

electrocyclic reaction a reaction in which a π bond in the reactant is lost so that a cyclic compound with a new σ bond can be formed.

electromagnetic radiation   radiant energy that displays wave properties.

electron affinity   the energy given off when an atom acquires an electron.

electronegative element   an element that readily acquires an electron.

electronegativity   tendency of an atom to pull electrons toward itself.

electronic transition   promotion of an electron from its HOMO to its LUMO.

electron sink   site to which electrons can be delocalized.

electrophile   an electron-deficient atom or molecule.

electrophilic addition reaction   an addition reaction in which the first species that adds to the reactant is an electrophile.

electrophilic aromatic substitution   a reaction in which an electrophile substitutes for a hydrogen of an aromatic ring.

electrophilic catalysis   catalysis in which the species that facilitates the reaction is an electrophile.

electrophoresis   a technique that separates amino acids on the basis of their pI values.

electropositive element   an element that readily loses an electron.

electrostatic attraction   attractive force between opposite charges.

electrostatic catalysis   stabilization of a charge by an opposite charge.

elemental analysis   a determination of the relative proportions of the elements present in a compound.

α-elimination   removal of two atoms or groups from the same carbon.

β-elimination   removal of two atoms or groups from adjacent carbons.

elimination reaction   a reaction that involves the elimination of atoms (or molecules) from the reactant.

empirical formula   formula giving the relative numbers of the different kinds of atoms in a molecule.

enamine   an α,β-unsaturated tertiary amine.

enantiomerically pure   containing only one enantiomer.

enantiomeric excess (optical purity)   how much excess of one enantiomer is present in a mixture of a pair of enantiomers.

enantiomers   nonsuperimposable mirror-image molecules.

enantioselective reaction   a reaction that forms an excess of one enantiomer.

enantiotopic hydrogens   two hydrogens bonded to a carbon that is bonded to two other groups that are nonidentical.

endergonic reaction   a reaction with a positive $\Delta G°$.

endo   a substituent is endo if it is closer to the longer or more unsaturated bridge.

endopeptidase   an enzyme that hydrolyzes a peptide bond that is not at the end of a peptide chain.

endothermic reaction   a reaction with a positive $\Delta H°$.

enediol rearrangement   interconversion of an aldose and one or more ketoses.

enkephalins   pentapeptides synthesized by the body to control pain.

enolization   keto–enol interconversion.

enthalpy   the heat given off ($-\Delta H°$) or the heat absorbed ($+\Delta H°$) during the course of a reaction.

entropy   a measure of the freedom of motion in a system.

enzyme   a protein that is a catalyst.

epimerization   changing the configuration of an asymmetric center by removing a proton from it and then reprotonating the molecule at the same site.

epimers   monosaccharides that differ in configuration at only one carbon.

epoxidation   formation of an epoxide.

epoxide (oxirane)   an ether in which the oxygen is incorporated into a three-membered ring.

epoxy resin   substance formed by mixing a low-molecular-weight prepolymer with a compound that forms a cross-linked polymer.

equatorial bond   a bond of the chair conformer of cyclohexane that juts out from the ring in approximately the same plane that contains the chair.

equilibrium constant   the ratio of products to reactants at equilibrium or the ratio of the rate constants for the forward and reverse reactions.

equilibrium control   thermodynamic control.

E1 reaction   a first-order elimination reaction.

E2 reaction   a second-order elimination reaction.

erythro enantiomers   the pair of enantiomers with similar groups on the same side as drawn in a Fischer projection.

essential amino acid   an amino acid that humans must obtain from their diet because they cannot synthesize it at all or cannot synthesize it in adequate amounts.

essential oils   fragrances and flavorings isolated from plants that do not leave residues when they evaporate. Most are terpenes.

ester

$$\begin{array}{c} \text{O} \\ \| \\ \text{R}-\text{C}-\text{OR} \end{array}$$

estrogens   female sex hormones.

ether   a compound containing an oxygen bonded to two carbons (ROR).

eukaryote   a unicellular or multicellular body whose cell or cells contain a nucleus.

excited-state electronic configuration   the electronic configuration that results when an electron in the ground-state electronic configuration has been moved to a higher energy orbital.

exergonic reaction   a reaction with a negative $\Delta G°$.

exhaustive methylation   reaction of an amine with excess methyl iodide to form a quaternary ammonium iodide.

exo   a substituent is exo if it closer to the shorter or more saturated bridge.

exon   a stretch of bases in DNA that are a portion of a gene.

exopeptidase   an enzyme that hydrolyzes a peptide bond at the end of a peptide chain.

exothermic reaction   a reaction with a negative $\Delta H°$.

experimental energy of activation ($E_a = \Delta H^{\ddagger} - RT$)   a measure of the approximate energy barrier to a reaction. (It is approximate because it does not contain an entropy component.)

extrusion reaction   a reaction in which a neutral molecule (e.g., CO_2, CO, or N_2) is eliminated from a molecule.

fat   a triester of glycerol that exists as a solid at room temperature.

fatty acid   a long-chain carboxylic acid.

Favorskii reaction   reaction of an α-haloketone with hydroxide ion.

fibrous protein   a water-insoluble protein in which the polypeptide chains are arranged in bundles.

fingerprint region   the right-hand third of an IR spectrum where the absorption bands are characteristic of the compound as a whole.

first-order rate constant   the rate constant of a first-order reaction.

first-order reaction (unimolecular reaction)   a reaction whose rate depends on the concentration of one reactant.

Fischer esterification reaction   the reaction of a carboxylic acid with alcohol in the presence of an acid catalyst to form an ester.

Fischer projection   a method of representing the spatial arrangement of groups bonded to an asymmetric center. The asymmetric center is the point of intersection of two perpendicular lines; the horizontal lines represent bonds that project out of the plane of the paper toward the viewer, and the vertical lines represent bonds that point back from the plane of the paper away from the viewer.

flagpole hydrogens (transannular hydrogens)   the two hydrogens in the boat conformation of cyclohexane that are closest to each other.

flavin adenine dinucleotide (FAD)   a coenzyme required in certain oxidation reactions. It is reduced to $FADH_2$ which can act as a reducing agent in another reaction.

flavin mononucleotide (FMN)   a coenzyme required in certain oxidation reactions. It is reduced to $FMNH_2$, which can act as a reducing agent in another reaction.

formal charge   the number of valence electrons $-$ (the number of nonbonding electrons $+$ 1/2 the number of bonding electrons).

Fourier transform NMR   a technique in which all the nuclei are excited simultaneously by an rf pulse, their relaxation is monitored, and the data are mathematically converted to a spectrum.

free energy of activation ($\Delta G^{\ddagger}$)   the true energy barrier to a reaction.

free-induction decay   relaxation of excited nuclei.

frequency   the velocity of a wave divided by its wavelength (in units of cycles/s).

Friedel–Crafts acylation   an electrophilic substitution reaction that puts an acyl group on a benzene ring.

Friedel–Crafts alkylation   an electrophilic substitution reaction that puts an alkyl group on a benzene ring.

frontier orbital analysis   determining the outcome of a pericyclic reaction with the use of frontier orbitals.

frontier orbitals   the HOMO and the LUMO.

frontier orbital theory   a theory that, like the conservation of orbital symmetry theory, explains the relationships among reactant, product, and reaction conditions in a pericyclic reaction.

functional group   the center of reactivity in a molecule.

functional group interconversion   the conversion of one functional group into another functional group.

functional group region   the left-hand two-thirds of an IR spectrum where most functional groups show absorption bands.

furanose   a five-membered-ring sugar.

furanoside   a five-membered-ring glycoside.

fused bicyclic compound   a bicyclic compound in which the rings share two adjacent carbons.

Gabriel synthesis   conversion of an alkyl halide into a primary amine, using phthalimide as a starting material.

gauche   X and Y are gauche to each other in this Newman projection:

gauche conformer   a staggered conformer in which the largest substituents are gauche to each other.

gauche interaction   the interaction between two atoms or groups that are gauche to each other.

***gem*-dialkyl effect**   two alkyl groups on a carbon, the effect of which is to increase the probability that the molecule will be in the proper conformation for ring closure.

***gem*-diol (hydrate)**   a compound with two OH groups on the same carbon.

geminal coupling   the mutual splitting of two nonidentical protons bonded to the same carbon.

geminal dihalide   a compound with two halogen atoms bonded to the same carbon.

gene   a segment of DNA.

general-acid catalysis   catalysis in which a proton is transferred to the reactant during the slow step of the reaction.

general-base catalysis   catalysis in which a proton is removed from the reactant during the slow step of the reaction.

generic name   a commercially nonrestricted name for a drug.

gene therapy   a technique that inserts a synthetic gene into the DNA of an organism that is defective in that gene.

genetic code   the amino acid specified by each three-base sequence of mRNA.

genetic engineering   recombinant DNA technology.

geometric isomers   cis–trans (or *E,Z*) isomers.

Gibbs standard free-energy change (ΔG°)   the difference between the free-energy content of the products and the free-energy content of the reactants at equilibrium under standard conditions (1 M, 25 °C, 1 atm).

Gilman reagent   an organocuprate, prepared from the reaction of an organolithium reagent with cuprous iodide, used to replace a halogen with an alkyl group.

globular protein   a water-soluble protein that tends to have a roughly spherical shape.

gluconeogenesis   the synthesis of D-glucose from pyruvate.

glycol   a compound containing two or more OH groups.

glycolysis (glycolytic cycle)   the series of reactions that converts D-glucose into two molecules of pyruvate.

glycoprotein   a protein that is covalently bonded to a polysaccharide.

glycoside   the acetal of a sugar.

***N*-glycoside**   a glycoside with a nitrogen instead of an oxygen at the glycosidic linkage.

glycosidic bond   the bond between the anomeric carbon and the alcohol in a glycoside.

α-1,4′-glycosidic linkage   a linkage between the C-1 oxygen of one sugar and the C-4 of a second sugar with the oxygen atom of the glycosidic linkage in the axial position.

β-1,4′-glycosidic linkage   a linkage between the C-1 oxygen of one sugar and the C-4 of a second sugar with the oxygen atom of the glycosidic linkage in the equatorial position.

graft copolymer   a copolymer that contains branches of a polymer of one monomer grafted onto the backbone of a polymer made from another monomer.

Grignard reagent   the compound that results when magnesium is inserted between the carbon and halogen of an alkyl halide (RMgBr, RMgCl).

ground-state electronic configuration   a description of which orbitals the electrons of an atom or molecule occupy when all of the electrons of atoms are in their lowest-energy orbitals.

Hagemann's-ester   a compound prepared by treating a mixture of formaldehyde and ethylacetoacetate with base and then with acid and heat.

half-chair conformation   the least stable conformation of cyclohexane.

haloform reaction   the reaction of a halogen and HO⁻ with a methyl ketone.

halogenation   reaction with halogen (Br_2, Cl_2, I_2).

halohydrin   an organic molecule that contains a halogen and an OH group on adjacent carbons.

Hammond postulate   states that the transition state will be more similar in structure to the species (reactants or products) that it is closer to energetically.

Haworth projection   a way to show the structure of a sugar; the five- and six-membered rings are represented as being flat.

head-to-tail addition   the head of one molecule is added to the tail of another.

heat of combustion   the amount of heat given off when a carbon-containing compound reacts completely with O_2 to form CO_2 and H_2O.

heat of formation   the heat given off when a compound is formed from its elements under standard conditions.

heat of hydrogenation   the heat ($-\Delta H°$) released in a hydrogenation reaction.

Heck reaction   couples an aryl, benzyl, or vinyl halide or triflate with an alkene in a basic solution in the presence of $Pd(PPh_3)_4$.

Heisenberg uncertainty principle   states that both the precise location and the momentum of an atomic particle cannot be simultaneously determined.

α-helix   the backbone of a polypeptide coiled in a right-handed spiral with hydrogen bonding occurring within the helix.

Hell–Volhard–Zelinski (HVZ) reaction   heating a carboxylic acid with Br_2 + P in order to convert it into an α-bromocarboxvlic acid.

hemiacetal

hemiketal

Henderson–Hasselbalch equation $pK_a = pH + \log[HA]/[A^-]$

heptose   a monosaccharide with seven carbons.

HETCOR spectrum   a 2-D NMR spectrum that shows coupling between protons and the carbons to which they are attached.

heteroatom   an atom other than carbon or hydrogen.

heterocyclic compound (heterocycle)   a cyclic compound in which one or more of the atoms of the ring are heteroatoms.

heterogeneous catalyst   a catalyst that is insoluble in the reaction mixture.

heterolytic bond cleavage (heterolysis)   breaking a bond with the result that both bonding electrons stay with one of the atoms.

hexose   a monosaccharide with six carbons.

high-energy bond   a bond that releases a great deal of energy when it is broken.

highest occupied molecular orbital (HOMO)   the molecular orbital of highest energy that contains an electron.

high-resolution NMR spectroscopy   NMR spectroscopy that uses a spectrometer with a high operating frequency.

Hofmann degradation   exhaustive methylation of an amine, followed by reaction with Ag_2O, followed by heating to achieve a Hofmann elimination reaction.

Hofmann elimination (anti-Zaitsev elimination)   a hydrogen is removed from the β-carbon bonded to the most hydrogens.

Hofmann elimination reaction   elimination of a proton and a tertiary amine from a quaternary ammonium hydroxide.

Hofmann rearrangement   conversion of an amide into an amine by using Br_2/HO.

holoenzyme   an enzyme plus its cofactor.

homogeneous catalyst   a catalyst that is soluble in the reaction mixture.

homolog   a member of a homologous series.

homologous series   a family of compounds in which each member differs from the next by one methylene group.

homolytic bond cleavage (homolysis)   breaking a bond with the result that each of the atoms gets one of the bonding electrons.

homopolymer   a polymer that contains only one kind of monomer.

homotopic hydrogens   two hydrogens bonded to a carbon bonded to two other groups that are identical.

Hoogsteen base pairing   the pairing between a base in a synthetic strand of DNA and a base pair in double-stranded DNA.

Hooke's law   an equation that describes the motion of a vibrating spring.

hormone an organic compound synthesized in a gland and delivered by the bloodstream to its target tissue.

Hückel's rule states that, for a compound to be aromatic, its cloud of electrons must contain $(4n + 2)\pi$ electrons, where n is an integer. This is the same as saying that the electron cloud must contain an odd number of pairs of π electrons.

human genome the total DNA of a human cell.

Hund's rule states that when there are degenerate orbitals, an electron will occupy an empty orbital before it will pair up with another electron.

Hunsdiecker reaction conversion of a carboxylic acid into an alkyl halide by heating a heavy metal salt of the carboxylic acid with bromine or iodine.

hybrid orbital an orbital formed by mixing (hybridizing) orbitals.

hydrate (gem-diol)

$$R - \underset{\underset{\displaystyle OH}{|}}{\overset{\overset{\displaystyle OH}{|}}{C}} - R(H)$$

hydrated water has been added to a compound.

hydration addition of water to a compound.

hydrazone $R_2C{=}NNH_2$

hydride ion a negatively charged hydrogen.

1,2-hydride shift the movement of a hydride ion from one carbon to an adjacent carbon.

hydroboration–oxidation the addition of borane to an alkene or an alkyne, followed by reaction with hydrogen peroxide and hydroxide ion.

hydrocarbon a compound that contains only carbon and hydrogen.

α-hydrogen usually, a hydrogen bonded to the carbon adjacent to a carbonyl carbon.

hydrogenation addition of hydrogen.

hydrogen bond an unusually strong dipole–dipole attraction (5 kcal/mol) between a hydrogen bonded to O, N, or F and the nonbonding electrons of an O, N, or F of another molecule.

hydrogen ion (proton) a positively charged hydrogen.

hydrolysis reaction with water.

hydrophobic interactions interactions between nonpolar groups. These interactions increase stability by decreasing the amount of structured water (increasing entropy).

hyperconjugation delocalization of electrons by overlap of carbon–hydrogen or carbon–carbon σ bonds with an empty p orbital.

imine $R_2C{=}NR$

inclusion compound a compound that specifically binds a metal ion or an organic molecule.

induced-dipole–induced-dipole interaction an interaction between a temporary dipole in one molecule and the dipole the temporary dipole induces in another molecule.

induced-fit model a model that describes the specificity of an enzyme for its substrate: The shape of the active site does not become completely complementary to the shape of the substrate until after the enzyme binds the substrate.

inductive electron donation donation of electrons through σ bonds.

inductive electron withdrawal withdrawal of electrons through a σ bond.

inflection point the midpoint of the flattened-out region of a titration curve.

informational strand (sense strand) the strand in DNA that is not read during transcription; it has the same sequence of bases as the synthesized mRNA strand (with a U, T difference).

infrared radiation electromagnetic radiation familiar to us as heat.

infrared spectroscopy uses infrared energy to provide a knowledge of the functional groups in a compound.

infrared (IR) spectrum a plot of percent transmission versus wave number (or wavelength) of infrared radiation.

initiation step the step in which radicals are created, or the step in which the radical needed for the first propagation step is created.

in-line displacement mechanism nucleophilic attack on a phosphorus concerted with breaking a phosphoanhydride bond.

interchain disulfide bridge a disulfide bridge between two cysteine residues in different peptide chains.

intermediate a species formed during a reaction and that is not the final product of the reaction.

intermolecular reaction a reaction that takes place between two molecules.

internal alkyne an alkyne with the triple bond not at the end of the carbon chain.

intimate ion pair pair such that the covalent bond that joined the cation and the anion has broken, but the cation and anion are still next to each other.

intrachain disulfide bridge a disulfide bridge between two cysteine residues in the same peptide chain.

intramolecular catalysis (anchimeric assistance) catalysis in which the catalyst that facilitates the reaction is part of the molecule undergoing reaction.

intramolecular reaction a reaction that takes place within a molecule.

intron a stretch of bases in DNA that contain no genetic information.

inversion of configuration turning the configuration of a carbon inside out like an umbrella in a windstorm, so that the resulting product has a configuration opposite that of the reactant.

iodoform test addition of I_2/HO^- to a methyl ketone forms a yellow precipitate of triiodomethane.

ion–dipole interaction the interaction between an ion and the dipole of a molecule.

ion-exchange chromatography a technique that uses a column packed with an insoluble resin to separate compounds on the basis of their charges and polarities.

ionic bond a bond formed through the attraction of two ions of opposite charges.

ionization energy the energy required to remove an electron from an atom.

ionophore a compound that binds metal ions tightly.

iron protoporphyrin IX the porphyrin ring system of heme plus an iron atom.

isoelectric point (pI) the pH at which there is no net charge on an amino acid.

isolated double bonds double bonds separated by more than one single bond.

isomers nonidentical compounds with the same molecular formula.

isoprene rule rule expressing the head-to-tail linkage of isoprene units.

isopropyl split a split in the IR absorption band attributable to a methyl group. It is characteristic of an isopropyl group.

isotactic polymer a polymer in which all the substituents are on the same side of the fully extended carbon chain.

isotopes atoms with the same number of protons, but different numbers of neutrons.

iterative synthesis a synthesis in which a reaction sequence is carried out more than once.

IUPAC nomenclature systematic nomenclature of chemical compounds.

Kekulé structure a model that represents the bonds between atoms as lines.

ketal

$$R - \underset{\underset{\displaystyle OR}{|}}{\overset{\overset{\displaystyle OR}{|}}{C}} - R$$

keto–enol tautomerism (keto–enol interconversion) interconversion of keto and enol tautomers.

keto–enol tautomers a ketone and its isomeric α,β-unsaturated alcohol.

β-keto ester an ester with a second carbonyl group at the β-position.

ketone

$$\underset{R}{\overset{\overset{\displaystyle O}{\|}}{C}}{}_{R}$$

ketose a polyhydroxyketone.

Kiliani–Fischer synthesis a method used to increase the number of carbons in an aldose by one, resulting in the formation of a pair of C-2 epimers.

kinetic control when a reaction is under kinetic control, the relative amounts of the products depend on the rates at which they are formed.

kinetic isotope effect a comparison of the rate of reaction of a compound with the rate of reaction of an identical compound in which one of the atoms has been replaced by an isotope.

kinetic product the product that is formed the fastest.

kinetic resolution separation of enantiomers on the basis of the difference in their rate of reaction with an enzyme.

kinetics the field of chemistry that deals with the rates of chemical reactions.

kinetic stability chemical reactivity, indicated by $\Delta G^{\ddagger}$. If $\Delta G^{\ddagger}$ is large, the compound is kinetically stable (not very reactive). If $\Delta G^{\ddagger}$ is small, the compound is kinetically unstable (highly reactive).

Knoevenagel condensation a condensation of an aldehyde or ketone with no α hydrogens and a compound with an α-carbon flanked by two electron-withdrawing groups.

Kolbe–Schmitt carboxylation reaction a reaction that uses CO_2 to carboxylate phenol.

Krebs cycle (citric acid cycle, tricarboxylic acid cycle, TCA cycle) a series of reactions that convert the acetyl group of acetyl-CoA into two molecules of CO_2

lactam   a cyclic amide.

lactone   a cyclic ester.

λ$_{max}$ the wavelength at which there is maximum UV/Vis absorbance.

lead compound   the prototype in a search for other biologically active compounds.

leaning   when a line drawn over the outside peaks of an NMR signal points in the direction of the signal given by the protons that cause the splitting.

leaving group   the group that is displaced in a nucleophilic substitution reaction.

Le Châtelier's principle   states that if an equilibrium is disturbed, the components of the equilibrium will adjust in a way that will offset the disturbance.

lecithin   a phosphoacylglycerol in which the second OH group of phosphate has formed an ester with choline.

levorotatory   the enantiomer that rotates polarized light in a counterclockwise direction.

Lewis acid   a substance that accepts an electron pair.

Lewis base   a substance that donates an electron pair.

Lewis structure   a model that represents the bonds between atoms as lines or dots and the valence electrons as dots.

ligation   sharing of nonbonding electrons with a metal ion.

linear combination of atomic orbitals (LCAO)   the combination of atomic orbitals to produce a molecular orbital.

linear conjugation   the atoms in the conjugated system are in a linear arrangement.

linear synthesis   a synthesis that builds a molecule step by step from starting materials.

lipid   a water-insoluble compound found in a living system.

lipid bilayer   two layers of phosphoacylglycerols arranged so that their polar heads are on the outside and their nonpolar fatty acid chains are on the inside.

lipoate   a coenzyme required in certain oxidation reactions.

living polymer   a nonterminated chain-growth polymer that remains active. This means that the polymerization reaction can continue upon the addition of more monomer.

localized electrons   electrons that are restricted to a particular locality.

lock-and-key model   a model that describes the specificity of an enzyme for its substrate: The substrate fits the enzyme as a key fits a lock.

London forces   Induced-dipole–induced-dipole interactions.

lone-pair electrons (nonbonding electrons)   valence electrons not used in bonding.

long-range coupling   splitting of a proton by a proton more than three σ bonds away.

loop conformation (coil conformation)   that part of a protein that is highly ordered, but not in an α-helix or β-pleated sheet.

lowest unoccupied molecular orbital (LUMO)   the molecular orbital of lowest energy that does not contain an electron.

Lucas test   a test that determines whether an alcohol is primary, secondary, or tertiary.

magnetic anisotropy   the term used to describe the greater freedom of a π electron cloud to move in response to a magnetic field as a consequence of the greater polarizability of π electrons compared with σ electrons.

magnetic resonance imaging (MRI)   NMR used in medicine. The difference in the way water is bound in different tissues produces a variation in signal between organs as well as between healthy and diseased tissue.

magnetogyric ratio   a property (measured in rad $T^{-1}s^{-1}$) that depends on the magnetic properties of a particular kind of nucleus.

major groove   the wider and deeper of the two alternating grooves in DNA.

malonic ester synthesis   the synthesis of a carboxylic acid, using diethyl malonate as the starting material.

Mannich reaction   condensation of a secondary amine and formaldehyde with a carbon acid.

Markovnikov's rule   the actual rule is "When a hydrogen halide adds to an asymmetrical alkene, the addition occurs such that the halogen attaches itself to the sp^2 carbon of the alkene bearing the lowest number of hydrogen atoms." A more universal rule is "The electrophile adds to the sp^2 carbon that is bonded to the greater number of hydrogens."

mass number   the number of protons plus the number of neutrons in an atom.

mass spectrometry   provides a knowledge of the molecular weight, molecular formula, and certain structural features of a compound.

mass spectrum   a plot of the relative abundance of the positively charged fragments produced in a mass spectrometer versus their m/z values.

materials science   the science of creating new materials to be used in place of known materials such as metal, glass, wood, cardboard, and paper.

Maxam–Gilbert sequencing   a technique used to sequence restriction fragments.

McLafferty rearrangement   rearrangement of the molecular ion of a ketone. The bond between the α- and β-carbons breaks, and a γ-hydrogen migrates to the oxygen.

mechanism-based inhibitor (suicide inhibitor)   a compound that inactivates an enzyme by undergoing part of its normal catalytic mechanism.

mechanism of a reaction   a description of the step-by-step process by which reactants are changed into products.

melting point   the temperature at which a solid becomes a liquid.

membrane   the material that surrounds a cell in order to isolate its contents.

mercaptan (thiol)   the sulfur analog of an alcohol (RSH).

meso compound   a compound that contains asymmetric centers and a plane of symmetry.

metabolism   reactions living organisms carry out in order to obtain the energy and to synthesize the compounds they require.

meta-directing substituent   a substituent that directs an incoming substituent meta to an existing substituent.

metal-activated enzyme   an enzyme that has a loosely bound metal ion.

metal-ion catalysis   catalysis in which the species that facilitates the reaction is a metal ion.

metalloenzyme   an enzyme that has a tightly bound metal ion.

methine hydrogen   a tertiary hydrogen.

methylene group   a CH_2 group.

1,2-methyl shift   the movement of a methyl group with its bonding electrons from one carbon to an adjacent carbon.

micelle   a spherical aggregation of molecules, each with a long hydrophobic tail and a polar head, arranged so that the polar head points to the outside of the sphere.

Michael reaction   the addition of an α-carbanion to the β-carbon of an α,β-unsaturated carbonyl compound.

minor groove   the narrower and more shallow of the two alternating grooves in DNA.

mixed (crossed) aldol addition   an aldol addition in which two different carbonyl compounds are used.

mixed anhydride   an acid anhydride with two different R groups.

mixed Claisen condensation   a Claisen condensation in which two different esters are used.

mixed triacylglycerol   a triacylglycerol in which the fatty-acid components are different.

molar absorptivity   the absorbance obtained from a 1.00 M solution in a cell with a 1.00-cm light path.

molecular ion (parent ion)   peak in the mass spectrum with the greatest m/z.

molecular modeling   computer-assisted design of a compound with particular structural characteristics.

molecular modification   changing the structure of a lead compound.

molecular orbital   an orbital associated with a molecule.

molecular orbital theory   describes a model in which the electrons occupy orbitals as they do in atoms, but with the orbitals extending over the entire molecule.

molecular recognition   the recognition of one molecule by another as a result of specific interactions; for example, the specificity of an enzyme for its substrate.

molozonide   an unstable intermediate containing a five-membered ring with three oxygens in a row that is formed from the reaction of an alkene with ozone.

monomer   a repeating unit in a polymer.

monosaccharide (simple carbohydrate)   a single sugar molecule.

monoterpene   a terpene that contains 10 carbons.

MRI scanner   an NMR spectrometer used in medicine for whole-body NMR.

multiplet   an NMR signal split into more than seven peaks.

multiplicity   the number of peaks in an NMR signal.

multistep synthesis   preparation of a compound by a route that requires several steps.

mutarotation   a slow change in optical rotation to an equilibrium value.

$N + 1$ rule   an ^{1}H NMR signal for a hydrogen with N equivalent hydrogens bonded to an adjacent carbon is split into $N + 1$ peaks. A ^{13}C NMR signal for a carbon bonded to N hydrogens is split into $N + 1$ peaks.

natural-abundance atomic weight the average mass of the atoms in the naturally occurring element.

natural product a product synthesized in nature.

neurotransmitter a compound that transmits nerve impulses.

nicotinamide adenine dinucleotide (NAD$^+$) a coenzyme required in certain oxidation reactions. It is reduced to NADH, which can act as a reducing agent in another reaction.

nicotinamide adenine dinucleotide phosphate (NADP$^+$) a coenzyme required in certain oxidation reactions. It is reduced to NADPH, which can act as a reducing agent in another reaction.

NIH shift the 1,2-hydride shift of a carbocation (obtained from an arene oxide) that leads to an enone.

nitration substitution of a nitro group (NO_2) for a hydrogen of a benzene ring.

nitrile a compound that contains a carbon–nitrogen triple bond ($RC≡N$).

nitrosamine (N-nitroso compound) $R_2NN=O$

NMR spectroscopy the absorption of electromagnetic radiation to determine the structural features of an organic compound. In the case of NMR spectroscopy, it determines the carbon–hydrogen framework.

node that part of an orbital in which there is zero probability of finding an electron.

nominal mass mass rounded to the nearest whole number.

nonbonding electrons (lone-pair electrons) valence electrons not used in bonding.

nonbonding molecular orbital the p orbitals are too far apart to overlap significantly, so the molecular orbital that results neither favors nor disfavors bonding.

nonpolar covalent bond a bond formed between two atoms that share the bonding electrons equally.

nonreducing sugar a sugar that cannot be oxidized by reagents such as Ag^+ and Cu^+. Nonreducing sugars are not in equilibrium with the open-chain aldose or ketose.

normal alkane (straight-chain alkane) an alkane in which the carbons form a contiguous chain with no branches.

N-phthalimidomalonic ester synthesis a method used to synthesize an amino acid that combines the malonic ester synthesis and the Gabriel synthesis.

N-terminal amino acid the terminal amino acid of a peptide (or protein) that has a free amino group.

nucleic acid the two kinds of nucleic acid are DNA and RNA.

nucleophile an electron-rich atom or molecule.

nucleophilic acyl substitution reaction a reaction in which a group bonded to an acyl or aryl group is substituted by another group.

nucleophilic addition–elimination–nucleophilic addition reaction a nucleophilic addition reaction that is followed by an elimination reaction that is followed by a nucleophilic addition reaction. Acetal formation is an example: An alcohol adds to the carbonyl carbon, water is eliminated, and a second molecule of alcohol adds to the dehydrated product.

nucleophilic addition–elimination reaction a nucleophilic addition reaction that is followed by an elimination reaction. Imine formation is an example: An amine adds to the carbonyl carbon, and water is eliminated.

nucleophilic addition reaction a reaction that involves the addition of a nucleophile to a reagent.

nucleophilic aromatic substitution a reaction in which a nucleophile substitutes for a substituent of an aromatic ring.

nucleophilic catalysis (covalent catalysis) catalysis that occurs as a result of a nucleophile forming a covalent bond with one of the reactants.

nucleophilic catalyst a catalyst that increases the rate of a reaction by acting as a nucleophile.

nucleophilicity a measure of how readily an atom or a molecule with a pair of nonbonding electrons attacks an atom.

nucleophilic substitution reaction a reaction in which a nucleophile substitutes for an atom or a group.

nucleoside a heterocyclic base (a purine or a pyrimidine) bonded to the anomeric carbon of a sugar (D-ribose or D-2′-deoxyribose).

nucleotide a heterocycle attached in the β-position to a phosphorylated ribose or deoxyribose.

observed rotation the amount of rotation observed in a polarimeter.

octet rule states that an atom will give up, accept, or share electrons in order to achieve a filled shell. Because a filled second shell contains eight electrons, this is known as the octet rule.

off-resonance decoupling the mode in ^{13}C NMR spectroscopy in which spin–spin splitting occurs between carbons and the hydrogens attached to them.

oil a triester of glycerol that exists as a liquid at room temperature.

olefin an alkene.

oligomer a protein with more than one peptide chain.

oligonucleotide 3 to 10 nucleotides linked by phosphodiester bonds.

oligopeptide 3 to 10 amino acids linked by amide bonds.

oligosaccharide 3 to 10 sugar molecules linked by glycosidic bonds.

open-chain compound an acyclic compound.

operating frequency the frequency at which an NMR spectrometer operates.

optical isomers stereoisomers that contain chirality centers.

optically active rotates the plane of polarized light.

optically inactive does not rotate the plane of polarized light.

optical purity (enantiomeric excess) how much excess of one enantiomer is present in a mixture of a pair of enantiomers.

orbital the volume of space around the nucleus in which an electron is most likely to be found.

orbital hybridization mixing of orbitals.

organic compound a compound that contains carbon.

organic synthesis preparation of organic compounds from other organic compounds.

organometallic compound a compound containing a carbon–metal bond.

oriented polymer a polymer obtained by stretching out polymer chains and putting them back together in a parallel fashion.

orphan drugs drugs for diseases or conditions that affect fewer than 200,000 people.

ortho-para-directing substituent a substituent that directs an incoming substituent ortho and para to an existing substituent.

osazone the product obtained by treating an aldose or a ketose with excess phenylhydrazine. An osazone contains two imine bonds.

overtone band an absorption that occurs at a multiple of the fundamental absorption frequency (2ν, 3ν).

oxidation loss of electrons by an atom or a molecule.

β-oxidation a series of four reactions that removes two carbons from a fatty acyl-CoA.

oxidation reaction a reaction in which the number of C—H bonds decreases or the number of C—O, C—N, or C—X (X = a halogen) increases.

oxidative cleavage an oxidation reaction that cuts the reactant into two or more pieces.

oxidative phosphorylation a series of reactions that converts a molecule of NADH and a molecule of FADH$_2$ into 3 and 2 molecules of ATP, respectively.

oxime $R_2C=NOH$

oxirane (epoxide) an ether in which the oxygen is incorporated into a three-membered ring.

oxonium ion a compound with a positively charged oxygen.

oxyanion a compound with a negatively charged oxygen.

oxymercuration addition of water using a mercuric salt of a carboxylic acid as a catalyst.

ozonide the five-membered-ring compound formed as a result of the rearrangement of a molozonide.

ozonolysis reaction of a carbon–carbon double or triple bond with ozone.

packing the fitting of individual molecules into a frozen crystal lattice.

paraffin an alkane.

parent hydrocarbon the longest continuous carbon chain in a molecule.

parent ion (molecular ion) peak in the mass spectrum with the greatest m/z.

partial hydrolysis a technique that hydrolyzes only some of the peptide bonds in a polypeptide.

partial racemization formation of a pair of enantiomers in unequal amounts.

Pauli exclusion principle states that no more than two electrons can occupy an orbital and that the two electrons must have opposite spin.

pentose a monosaccharide with five carbons.

peptide polymer of amino acids linked together by amide bonds. A peptide contains fewer amino acid residues than a protein does.

peptide bond the amide bond that links the amino acids in a peptide or protein.

peptide nucleic acid (PNA) a polymer containing both an amino acid and a base designed to bind to specific residues on DNA or mRNA.

Perkin condensation a condensation of an aromatic aldehyde and acetic acid.

pericyclic reaction a concerted reaction that takes place as the result of a cyclic rearrangement of electrons.

peroxyacid a carboxylic acid with an OOH group instead of an OH group.

perspective formula a method of representing the spatial arrangement of groups bonded to an asymmetric center. Two bonds are drawn in the plane of the paper; a solid wedge is used to depict a bond that projects out of the plane of the paper toward the viewer, and a hatched wedge is used to represent a bond that projects back from the plane of the paper away from the viewer.

pH the pH scale is used to describe the acidity of a solution (pH = $-\log[\text{H}^+]$).

pH-activity profile a plot of the activity of an enzyme as a function of the pH of the reaction mixture.

phase-transfer catalysis catalysis of a reaction by providing a way to bring a polar reagent into a nonpolar phase so that the reaction between a polar and a nonpolar compound can occur.

phase-transfer catalyst a compound that carries a polar reagent into a nonpolar phase.

phenone

phenyl group

phenylhydrazone $R_2C{=}NNHC_6H_5$

pheromone a compound secreted by an animal that stimulates a physiological or behavioral response from a member of the same species.

phosphatidic acid a phosphoacylglycerol in which only one of the OH groups of phosphate is in an ester linkage.

phosphoacylglycerol (phosphoglyceride) a compound formed when two OH groups of glycerol form esters with fatty acids and the terminal OH group forms a phosphate ester.

phosphoanhydride bond the bond holding two phosphoric acid molecules together.

phospholipid a lipid that contains a phosphate group.

phosphoryl transfer reaction the transfer of a phosphate group from one compound to another.

photochemical reaction a reaction that takes place when a reactant absorbs light.

photosynthesis the synthesis of glucose and O_2 from CO_2 and H_2O.

pH-rate profile a plot of the observed rate of a reaction as a function of the pH of the reaction mixture.

pi (π) bond a bond formed as a result of side-to-side overlap of p orbitals.

pi-complex a complex formed between an electrophile and a triple bond.

pinacol rearrangement rearrangement of a vicinal diol.

pK_a describes the tendency of a compound to lose a proton (p$K_a = -\log K_a$, where K_a is the acid dissociation constant).

plane of symmetry an imaginary plane that bisects a molecule into mirror images.

plasticizer an organic molecule that dissolves in a polymer and allows the polymer chains to slide by each other.

β-pleated sheet the backbone of a polypeptide that is extended in a zigzag structure with hydrogen bonding between neighboring chains.

point of symmetry any line through a point of symmetry encounters identical environments at the same distance.

polar covalent bond a covalent bond between atoms of different electronegativites.

polarimeter an instrument that measures the rotation of polarized light.

polarizability an indication of the ease with which the electron cloud of an atom can be distorted.

polarized light light that oscillates only in one plane.

polar reaction the reaction between a nucleophile and an electrophile.

polyamide a polymer in which the monomers are amides.

polycarbonate a step-growth polymer in which the dicarboxylic acid is carbonic acid.

polyene a compound that has several double bonds.

polyester a polymer in which the monomers are esters.

polymer a large molecule made by linking monomers together.

polymerase chain reaction (PCR) a method that amplifies segments of DNA.

polymer chemistry the field of chemistry that deals with synthetic polymers; part of the larger discipline known as materials science.

polymerization the process of linking up monomers to form a polymer.

polynucleotide many nucleotides linked by phosphodiester bonds.

polypeptide many amino acids linked by amide bonds.

polysaccharide a compound containing more than 10 sugar molecules linked together.

polyunsaturated fatty acid a fatty acid with more than one double bond.

polyurethane a polymer in which the monomers are urethanes.

porphyrin ring system consists of four pyrrole rings joined by one-carbon bridges.

primary alcohol an alcohol in which the OH group is bonded to a primary carbon.

primary alkyl halide an alkyl halide in which the halogen is bonded to a primary carbon.

primary alkyl radical a radical with the unpaired electron on a primary carbon.

primary amine an amine with one alkyl group bonded to the nitrogen.

primary carbocation a carbocation with the positive charge on a primary carbon.

primary carbon a carbon bonded to only one other carbon.

primary hydrogen a hydrogen bonded to a primary carbon.

primary structure (of a nucleic acid) the sequence of bases in a nucleic acid.

primary structure (of a protein) the sequence of amino acids in a protein.

principle of microscopic reversibility states that the mechanism for a reaction in the forward direction has the same intermediates and the same rate-determining step as the mechanism for the reaction in the reverse direction.

prochiral carbonyl carbon a carbonyl carbon that will become an asymmetric center if it is attacked by a group unlike any of the groups already bonded to it.

prochirality center a carbon bonded to two hydrogens that will become an asymmetric center if one of the hydrogens is replaced by deuterium.

prodrug a compound that does not become an effective drug until it undergoes a reaction in the body

promoter site a short sequence of bases at the beginning of a gene.

propagating site the reactive end of a chain-growth polymer.

propagation step in the first of a pair of propagation steps, a radical (or an electrophile or a nucleophile) reacts to produce another radical (or an electrophile or a nucleophile) that reacts in the second to produce the radical (or the electrophile or the nucleophile) that was the reactant in the first propagation step.

proprietary name (trade name, brand name) identifies a commercial product and distinguishes it from other products.

pro-R hydrogen replacing this hydrogen with deuterium creates an asymmetric center with the R configuration.

pro-S hydrogen replacing this hydrogen with deuterium creates an asymmetric center with the S configuration.

prostacyclin a lipid, derived from arachidonic acid, that dilates blood vessels and inhibits platelet aggregation.

prosthetic group a tightly bound coenzyme.

protecting group a reagent that protects a functional group from a synthetic operation that it would otherwise not survive.

protein a polymer containing 40 to 4000 amino acids linked by amide bonds.

protic solvent a solvent that has a hydrogen bonded to an oxygen or a nitrogen.

proton a positively charged hydrogen (H^+); a positively charged particle in an atomic nucleus.

proton-decoupled ^{13}C NMR spectrum a ^{13}C NMR spectrum in which all the signals appear as singlets because there is no coupling between the nucleus and its bonded hydrogens.

proton transfer reaction a reaction in which a proton is transferred from an acid to a base.

protoporphyrin IX the porphyrin ring system of heme.

proximity effect an effect caused by one species being close to another.

pseudo-first-order reaction a second-order reaction in which the concentration of one of the reactants is much greater than the other, allowing the reaction to be treated as a first-order reaction.

pyranose a six-membered-ring sugar.

pyranoside a six-membered-ring glycoside.

pyridoxal phosphate the coenzyme required by enzymes that catalyze certain transformations of amino acids.

quantitative structure–activity relationship (QSAR) the relation between a particular property of a series of compounds and their biological activity.

quantum numbers numbers arising from the quantum mechanical treatment of an atom that describe the properties of the electrons in the atom.

quartet an NMR signal split into four peaks.

quaternary ammonium ion an ion containing a nitrogen bonded to four alkyl groups (R_4N^+).

quaternary ammonium salt a quaternary ammonium ion and an anion ($R_4N^+X^-$).

quaternary structure a description of the way the individual polypeptide chains of a protein are arranged with respect to each other.

racemic mixture (racemate, racemic modification) a mixture of equal amounts of a pair of enantiomers.

radical an atom or a molecule with an unpaired electron.

radical addition reaction an addition reaction in which the first species that adds is a radical.

radical anion a species with a negative charge and an unpaired electron.

radical cation a species with a positive charge and an unpaired electron.

radical chain reaction a reaction in which radicals are formed and react in repeating propagating steps.

radical inhibitor a compound that traps radicals.

radical initiator a compound that creates radicals.

radical polymerization chain-growth polymerization in which the initiator is a radical; the propagation site is therefore a radical.

radical reaction a reaction in which a new bond is formed by using one electron from one reagent and one electron from another reagent.

radical substitution reaction a substitution reaction that has a radical intermediate.

random coil the conformation of a totally denatured protein.

random copolymer a copolymer with a random distribution of monomers.

random screen (blind screen) the search for a pharmacologically active compound without any information about what chemical structures might show activity.

rate constant a measure of how easy or difficult it is to reach the transition state of a reaction (to get over the energy barrier to the reaction).

rate-determining step (rate-limiting step) the step in a reaction that has the transition state with the highest energy.

rational drug design designing drugs with a particular structure to achieve a specific purpose.

R configuration after assigning relative priorities to the four groups bonded to an asymmetric center, if the lowest priority group is on a vertical axis in a Fischer projection (or pointing away from the viewer in a perspective formula), an arrow drawn from the highest priority group to the next-highest-priority group goes in a clockwise direction.

reaction coordinate diagram describes the energy changes that take place during the course of a reaction.

reactivity–selectivity principle states that the greater the reactivity of a species, the less selective it will be.

receptor site the site at which a drug binds in order to exert its physiological effect.

recombinant DNA DNA that has been incorporated into a host cell.

reduction reaction a reaction in which the number of C—H bonds increases or the number of C—O, C—N, or C—X (X = a halogen) decreases.

red shift a shift to a longer wavelength.

reducing sugar a sugar that can be oxidized by reagents such as Ag^+ or Br_2. Reducing sugars are in equilibrium with the open-chain aldose or ketose.

reduction gain of electrons by an atom or a molecule.

reductive amination the reaction of an aldehyde or a ketone with ammonia or with a primary amine in the presence of a reducing agent (H_2/Raney Ni).

reference compound a compound added to a sample whose NMR spectrum is to be taken. The positions of the signals in the NMR spectrum are measured from the position of the signal given by the reference compound.

Reformatsky reaction reaction of an organozinc reagent with an aldehyde or a ketone.

regioselective reaction a reaction that leads to the preferential formation of one constitutional isomer over another.

relative configuration the configuration of a compound relative to the configuration of another compound.

relative rate obtained by dividing the actual rate constant by the rate constant of the slowest reaction in the group being compared.

replication the synthesis of identical copies of DNA.

replication fork the position on DNA at which replication begins.

resolution of a racemic mixture separation of a racemic mixture into the individual enantiomers.

resonance a compound with delocalized electrons is said to have resonance.

resonance contributor (resonance structure, contributing resonance structure) a structure with localized electrons that approximates the true structure of a compound with delocalized electrons.

resonance electron donation donation of electrons through p orbital overlap with neighboring π bonds.

resonance electron withdrawal withdrawal of electrons through p orbital overlap with neighboring π bonds.

resonance energy (delocalization energy) the extra stability associated with a compound as a result of its having delocalized electrons.

resonance hybrid the actual structure of a compound with delocalized electrons; it is represented by two or more structures with localized electrons.

resonances NMR absorption signals.

restriction endonuclease an enzyme that cleaves DNA at a specific base sequence.

restriction fragment a fragment that is formed when DNA is cleaved by a restriction endonuclease.

retro Diels-Alder reaction a reverse Diels-Alder reaction.

retrosynthesis (retrosynthetic analysis) working backwards (on paper) from the target molecule to available starting materials.

retrovirus a virus whose genetic information is stored in its RNA.

rf radiation radiation in the radiofrequency region of the electromagnetic spectrum.

ribonucleic acid (RNA) a polymer of ribonucleotides.

ribonucleotide a nucleotide in which the sugar component is D-ribose.

ribosome a particle composed of about 40% protein and 60% RNA on which protein biosynthesis takes place.

ribozyme an RNA molecule that acts as a catalyst.

ring current the movement of π electrons around an aromatic benzene ring.

ring-expansion rearrangement rearrangement of a carbocation in which the positively charged carbon is bonded to a cyclic compound and, as a result of rearrangement, the size of the ring increases by one carbon.

ring-flip (chair–chair interconversion) the conversion of the chair conformer of cyclohexane into the other chair conformer. Bonds that are axial in one chair conformer are equatorial in the other.

ring opening polymerization a chain-growth polymerization that involves opening the ring of the monomer.

Ritter reaction reaction of a nitrile with a secondary or tertiary alcohol to form a secondary amide.

RNA (ribonucleic acid) a polymer of ribonucleotides.

RNA splicing the step in RNA processing that cuts out nonsense bases and splices informational pieces together.

Robinson annulation a Michael reaction followed by an intramolecular aldol condensation.

Rosenmund reduction reduction of an acyl chloride to an aldehyde by using and a deactivated palladium catalyst.

Ruff degradation a method used to shorten an aldose by one carbon.

Sandmeyer reaction the reaction of an aryl diazonium salt with a cuprous salt.

saponification hydrolysis of an ester (such as a fat) under basic conditions.

saturated hydrocarbon a hydrocarbon that is completely saturated (i.e., contains no double or triple bonds) with hydrogen.

Schiemann reaction the reaction of an arenediazonium salt with HBF_4.

Schiff base $R_2C{=}NR$

S configuration after assigning relative priorities to the four groups bonded to an asymmetric center, if the lowest priority group is on a vertical axis in a Fischer projection (or pointing away from the viewer in a perspective formula), an arrow drawn from the highest priority group to the next-highest priority group goes in a counterclockwise direction.

secondary alcohol an alcohol in which the OH group is bonded to a secondary carbon.

secondary alkyl halide an alkyl halide in which the halogen is bonded to a secondary carbon.

secondary alkyl radical a radical with the unpaired electron on a secondary carbon.

secondary amine an amine with two alkyl groups bonded to the nitrogen.

secondary carbocation a carbocation with the positive charge on a secondary carbon.

secondary carbon a carbon bonded to two other carbons.

secondary hydrogen a hydrogen bonded to a secondary carbon.

secondary structure a description of the conformation of the backbone of a protein.

second-order rate constant the rate constant of a second-order reaction.

second-order reaction (bimolecular reaction) a reaction whose rate depends on the concentration of two reactants.

sedimentation constant designates where a species sediments in an ultracentrifuge.

selection rules   the rules that determine the outcome of a pericyclic reaction.

selenenylation reaction   conversion of an α-bromoketone into an α,β-unsatured ketone via the formation of a selenoxide.

semicarbazone $R_2C{=}NNHCNH_2$

semiconservative replication   the mode of replication that results in a daughter molecule of DNA having one of the original DNA strands plus a newly synthesized strand.

sense strand (informational strand)   the strand in DNA that is not read during transcription; it has the same sequence of bases as the synthesized mRNA strand (with a U, T difference).

separated charges   a positive and a negative charge that can be neutralized by the movement of electrons.

sesquiterpene   a terpene that contains 15 carbons.

shielding   phenomenon caused by electron donation to the environment of a proton. The electrons shield the proton from the full effect of the applied magnetic field. The more a proton is shielded, the farther to the right its signal appears in an NMR spectrum.

sigma (σ) bond   a bond with a cylindrically symmetrical distribution of electrons.

sigmatropic rearrangement   a reaction in which a σ bond is broken in the reactant, a new σ bond is formed in the product, and the π bonds rearrange.

Simmons–Smith reaction   formation of a cyclopropane using CH_2I_2 + Zn(Cu).

simple carbohydrate (monosaccharide)   a single sugar molecule.

simple triacylglycerol   a triacylglycerol in which the fatty acid components are the same.

single bond   a σ bond.

singlet   an unsplit NMR signal.

site-specific mutagenesis   a technique that substitutes one amino acid of a protein for another.

site-specific recognition   recognition of a particular site on DNA.

skeletal structure   shows the carbon–carbon bonds as lines and does not show the carbon–hydrogen bonds.

S_NAr reaction   a nucleophilic aromatic substitution reaction.

S_N1 reaction   a unimolecular nucleophilic substitution reaction.

S_N2 reaction   a bimolecular nucleophilic substitution reaction.

soap   a sodium or potassium salt of a fatty acid.

solid-phase synthesis   a technique in which one end of the compound being synthesized is covalently attached to a solid support.

solvation   the interaction between a solvent and another molecule (or ion).

solvent-separated ion pair   the cation and anion are separated by a solvent molecule.

solvolysis   reaction with the solvent.

specific-acid catalysis   catalysis in which the proton is fully transferred to the reactant before the slow step of the reaction takes place.

specific-base catalysis   catalysis in which the proton is completely removed from the reactant before the slow step of the reaction takes place.

specific rotation   the amount of rotation that will be caused by a compound with a concentration of 1.0 g/mL in a sample tube 1.0 dm long.

spectroscopy   study of the interaction of matter and electromagnetic radiation.

sphingolipid   a lipid that contains sphingosine.

sphingomyelin   a sphingolipid in which the terminal OH group of sphingosine is bonded to phosphocholine or phosphoethanolamine.

spin-coupled ^{13}C NMR spectrum   a ^{13}C NMR spectrum in which each signal of a carbon is split by the hydrogens bonded to that carbon.

spin coupling   the atom that gives rise to an NMR signal is coupled to the rest of the molecule.

spin decoupling   the atom that gives rise to an NMR signal is decoupled from the rest of the molecule.

spin–spin coupling   the splitting of a signal in an NMR spectrum described by the $N + 1$ rule.

α-spin state   nuclei in this spin state have their magnetic moments oriented in the same direction as the applied magnetic field.

β-spin state   nuclei in this spin state have their magnetic moments oriented opposite the direction of the applied magnetic field.

spirocyclic compound   a bicyclic compound in which the rings share one carbon.z

splitting diagram   a diagram that describes the splitting of a set of protons.

squalene   a triterpene that is a precursor of steroid molecules.

stacking interactions   van der Waals interactions between the mutually induced dipoles of adjacent pairs of bases in DNA.

staggered conformation   a conformation in which the bonds on one carbon bisect the bond angle on the adjacent carbon when viewed looking down the carbon–carbon bond.

step-growth polymer (condensation polymer)   a polymer made by combining two molecules while removing a small molecule (usually of water or an alcohol).

stereochemistry   the field of chemistry that deals with the structures of molecules in three dimensions.

stereoelectronic effects   the combination of steric effects and electronic effects.

stereogenic center (stereocenter)   an atom at which the interchange of two substituents produces a stereoisomer.

stereoisomers   isomers that differ in the way their atoms are arranged in space.

stereoselective reaction   a reaction that leads to the preferential formation of one stereoisomer over another.

stereospecific reaction   a reaction in which the reactant can exist as stereoisomers and each stereoisomeric reactant leads to a different stereoisomeric product or set of products.

steric effects   effects due to the fact that groups occupy a certain volume of space.

steric hindrance   refers to bulky groups at the site of a reaction that make it difficult for the reactants to approach each other.

steric strain (van der Waals strain, van der Waals repulsion)   the repulsion between the electron cloud of an atom or a group of atoms and the electron cloud of another atom or group of atoms.

steroid   a class of compounds that contains a steroid ring system.

Stille reaction   couples an aryl, a benzyl, or a vinyl halide or triflate with a stannane in the presence of $Pd(PPh_3)_4$.

stop codon   a codon at which protein synthesis is stopped.

Stork enamine reaction   uses an enamine as a nucleophile in a Michael reaction.

straight-chain alkane (normal alkane)   an alkane in which the carbons form a continuous chain with no branches.

Strecker synthesis   a method used to synthesize an amino acid: An aldehyde reacts with NH_3, forming an imine that is attacked by cyanide ion. Hydrolysis of the product gives an amino acid.

stretching frequency   the frequency at which a stretching vibration occurs.

stretching vibration   a vibration occurring along the line of a bond.

structural isomers (constitutional isomers)   molecules that have the same molecular formula but differ in the way their atoms are connected.

structural protein   a protein that gives strength to a biological structure.

α-substituent   a substituent on the side of a steroid ring system opposite that of the angular methyl groups.

β-substituent   a substituent on the same side of a steroid ring system as that of the angular methyl groups.

α-substitution reaction   a reaction that puts a substituent on an α-carbon in place of an α-hydrogen.

substrate   the reactant of an enzyme-catalyzed reaction.

subunit   an individual chain of an oligomer.

suicide inhibitor (mechanism-based inhibitor)   a compound that inactivates an enzyme by undergoing part of its normal catalytic mechanism.

sulfide (thioether)   the sulfur analog of an ether (RSR).

sulfonate ester   the ester of a sulfonic acid (RSO_2OR).

sulfonation   substitution of a hydrogen of a benzene ring by a sulfonic acid group (SO_3H)

suprafacial bond formation   the formation of two σ bonds on the same side of the π system.

suprafacial rearrangement   rearrangement in which the migrating group remains on the same face of the π system.

Suzuki reaction   couples an aryl, a benzyl, or a vinyl halide with an organoborane in the presence of $Pd(PPh_3)_4$.

symmetrical anhydride   an acid anhydride with identical R groups:

symmetrical ether   an ether with two identical substituents bonded to the oxygen.

symmetric molecular orbital   a molecular orbital in which the left half is a mirror image of the right half.

symmetry-allowed pathway   a pathway that leads to overlap of in-phase orbitals.

symmetry-forbidden pathway   a pathway that leads to overlap of out-of-phase orbitals.

syn addition   an addition reaction in which two substituents are added to the same side of the molecule.

syndiotactic polymer   a polymer in which the substituents regularly alternate on both sides of the fully extended carbon chain.

syn elimination   an elimination reaction in which the two substituents that are eliminated are removed from the same side of the molecule.

syn-periplanar   parallel substituents on the same side of a molecule.

synthetic equivalent   the reagent actually used as the source of a synthon.

synthetic polymer   a polymer that is not synthesized in nature.

synthetic tree   an outline of the available routes to get to a desired product from available starting materials.

synthon   a fragment of a disconnection.

systematic nomenclature   nomenclature based on structure.

target molecule   desired end product of a synthesis.

tautomerism   interconversion of tautomers.

tautomers   rapidly equilibrating isomers that differ in the location of their bonding electrons.

template strand (antisense strand)   the strand in DNA that is read during transcription.

terminal alkyne   an alkyne with the triple bond at the end of the carbon chain.

termination step   when two radicals combine to produce a molecule in which all the electrons are paired.

terpene   a lipid, isolated from a plant, that contains carbon atoms in multiples of five.

terpenoid   a terpene that contains oxygen.

tertiary alcohol   an alcohol in which the OH group is bonded to a tertiary carbon.

tertiary alkyl halide   an alkyl halide in which the halogen is bonded to a tertiary carbon.

tertiary alkyl radical   a radical with the unpaired electron on a tertiary carbon.

tertiary amine   an amine with three alkyl groups bonded to the nitrogen.

tertiary carbocation   a carbocation with the positive charge on a tertiary carbon.

tertiary carbon   a carbon bonded to three other carbons.

tertiary hydrogen   a hydrogen bonded to a tertiary carbon.

tertiary structure   a description of the three-dimensional arrangement of all the atoms in a protein.

tetraene   a hydrocarbon with four double bonds.

tetrahedral bond angle   the bond angle (109.5°) formed by adjacent bonds of an sp^3 hybridized carbon.

tetrahedral carbon   an sp^3 hybridized carbon; a carbon that forms covalent bonds by using four sp^3 hybridized orbitals.

tetrahedral intermediate   the intermediate formed in a nucleophilic acyl substitution reaction.

tetrahydrofolate (THF)   the coenzyme required by enzymes that catalyze reactions that donate a group containing a single carbon to their substrates.

tetraterpene   a terpene that contains 40 carbons.

tetrose   a monosaccharide with four carbons.

therapeutic index   the ratio of the lethal dose of a drug to the therapeutic dose.

thermal cracking   using heat to break a molecule apart.

thermal reaction   a reaction that takes place without the reactant having to absorb light.

thermodynamic control   when a reaction is under thermodynamic control, the relative amounts of the products depend on their stabilities.

thermodynamic product   the most stable product.

thermodynamics   the field of chemistry that describes the properties of a system at equilibrium.

thermodynamic stability   is indicated by $\Delta G°$. If $\Delta G°$ is negative, the products are more stable than the reactants. If $\Delta G°$ is positive, the reactants are more stable than the products.

thermoplastic polymer   a polymer that has both ordered crystalline regions and amorphous noncrystalline regions; it can be molded when heated.

thermosetting polymers   cross-linked polymers that, after they are hardened, cannot be remelted by heating.

thiamine pyrophosphate (TPP)   the coenzyme required by enzymes, which catalyze a reaction that transfers a two-carbon fragment to a substrate.

thiirane   a three-membered-ring compound in which one of the ring atoms is a sulfur.

thin-layer chromatography   a technique that separates compounds on the basis of their polarity.

thioester   the sulfur analog of an ester:

thioether (sulfide)   the sulfur analog of an ether (RSR).

thiol (mercaptan)   the sulfur analog of an alcohol (RSH).

threo enantiomers   the pair of enantiomers with similar groups on opposite sides when drawn in a Fischer projection.

titration curve   a plot of pH versus added equivalents of hydroxide ion.

Tollens test   an aldehyde can be identified by observing the formation of a silver mirror in the presence of Tollens' reagent (Ag_2O/NH_3).

trademark   a registered name, symbol, or picture.

trade name (proprietary name, brand name)   identifies a commercial product and distinguishes it from other products.

transamination   a reaction in which an amino group is transferred from one compound to another.

transannular hydrogens (flagpole hydrogens)   the two hydrogens in the boat conformation of cyclohexane that are closest to each other.

s-trans conformation   a conformation in which two double bonds are on opposite sides of a single bond.

transcription   the synthesis of mRNA from a DNA blueprint.

transesterification reaction   the reaction of an ester with an alcohol to form a different ester.

trans fused   two cyclohexane rings fused together such that if the second ring were considered to be two substituents of the first ring, both substituents would be in equatorial positions.

transimination   the reaction of a primary amine with an imine to form a new imine and a primary amine derived from the original imine.

transition metal catalyst   a catalyst containing a transition metal, such as $Pd(PPh_3)_4$, that is used in coupling reactions.

trans isomer   the isomer with the hydrogens on opposite sides of the double bond or cyclic structure. the isomer with identical substituents on opposite sides of the double bond.

transition state   the highest point on a hill in a reaction coordinate diagram. In the transition state, bonds in the reactant that will break are partially broken and bonds in the product that will form are partially formed.

transition state analog   a compound that is structurally similar to the transition state of an enzyme-catalyzed reaction.

translation   the synthesis of a protein from an mRNA blueprint.

transmetallation   metal exchange.

triacylglycerol   the compound formed when the three OH groups of glycerol are esterified with fatty acids.

triene   a hydrocarbon with three double bonds.

trigonal planar carbon   an sp^2 hybridized carbon.

triose   a monosaccharide with three carbons.

tripeptide   three amino acids linked by amide bonds.

triple bond   a σ bond plus two π bonds.

triplet   an NMR signal split into three peaks.

triterpene   a terpene that contains 30 carbons.

twist-boat conformation (skew-boat conformation)   a conformation of cyclohexane.

ultraviolet light   electromagnetic radiation with wavelengths ranging from 180 to 400 nm.

umpolung   reversing the normal polarity of a functional group.

unimolecular reaction (first-order reaction)   a reaction whose rate depends on the concentration of one reactant.

unsaturated hydrocarbon   a hydrocarbon that contains one or more double or triple bonds.

unsymmetrical ether   an ether with two different substituents bonded to the oxygen.

urethane a compound with a carbonyl group that is both an amide and an ester.

UV/Vis spectroscopy the absorption of electromagnetic radiation in the ultraviolet and visible regions of the spectrum; used to determine information about conjugated systems.

valence electron an electron in an unfilled shell.

valence shell electron-pair repulsion (VSEPR) model combines the concept of atomic orbitals with the concept of shared electron pairs and the minimization of electron pair repulsion.

van der Waals forces (London forces) induced-dipole–induced-dipole interactions.

van der Waals radius a measure of the effective size of an atom or a group. A repulsive force occurs (van der Waals repulsion) if two atoms approach each other at a distance less than the sum of their van der Waals radii.

vector sum takes into account both the magnitudes and the directions of the bond dipoles.

vicinal dihalide a compound with halogens bonded to adjacent carbons.

vicinal diol (vicinal glycol) a compound with OH groups bonded to adjacent carbons.

vinyl group $CH_2\!=\!CH\!-\!$

vinylic carbon a carbon in a carbon–carbon double bond.

vinylic cation a compound with a positive charge on a vinylic carbon.

vinylic radical a compound with an unpaired electron on a vinylic carbon.

vinylogy transmission of reactivity through double bonds.

vinyl polymer a polymer in which the monomers are ethylene or a substituted ethylene.

visible light electromagnetic radiation with wavelengths ranging from 400 to 780 nm.

vitamin a substance needed in small amounts for normal body function that the body cannot synthesize at all or cannot synthesize in adequate amounts.

vitamin KH$_2$ the coenzyme required by the enzyme that catalyzes the carboxylation of glutamate side chains.

vulcanization increasing the flexibility of rubber by heating it with sulfur.

wave equation an equation that describes the behavior of each electron in an atom or a molecule.

wave functions a series of solutions of a wave equation.

wavelength distance from any point on one wave to the corresponding point on the next wave (usually in units of μm or nm).

wavenumber the number of waves in 1 cm.

wax an ester formed from a long-chain carboxylic acid and a long-chain alcohol.

wedge-and-dash structure a method of representing the spatial arrangement of groups. Wedges are used to represent bonds that point out of the plane of the paper toward the viewer, and dashed lines are used to represent bonds that point back from the plane of the paper away from the viewer.

Williamson ether synthesis formation of an ether from the reaction of an alkoxide ion with an alkyl halide.

Wittig reaction the reaction of an aldehyde or a ketone with a phosphonium ylide, resulting in the formation of an alkene.

Wohl degradation a method used to shorten an aldose by one carbon.

Wolff–Kishner reduction a reaction that reduces the carbonyl group of a ketone to a methylene group with the use of NH_2NH_2/HO^-.

Woodward–Fieser rules allow the calculation of the λ_{max} of the $\pi \longrightarrow \pi^*$ transition for compounds with four or fewer conjugated double bonds.

Woodward–Hoffmann rules a series of selection rules for pericyclic reactions.

ylide a compound with opposite charges on adjacent covalently bonded atoms with complete octets.

Zaitsev's rule the more substituted alkene product is obtained by removing a proton from the β-carbon that is bonded to the fewest hydrogens.

Z conformation the conformation of a carboxylic acid or a carboxylic acid derivative in which the carbonyl oxygen and the substituent bonded to the carboxyl oxygen or nitrogen are on the same side of the single bond.

Ziegler–Natta catalyst an aluminum–titanium initiator that controls the stereochemistry of a polymer.

Z isomer the isomer with the high-priority groups on the same side of the double bond.

zwitterion a compound with a negative charge and a positive charge on nonadjacent atoms.

Photo Credits

Index

Page numbers in **boldface** indicate definitions; page numbers followed by f indicate illustrations; page numbers followed by t indicate tables.

A

Absolute configuration, 229–232, 251
Absorbance, 551
Absorption bands, **532,** 558
  absence of, 544
  characteristic, 533–543
  intensity of, 534
  position of, 535–543
    bond order and, 535–536
    electron donation/withdrawal and, 537–538
    hydrogen bonding and, 540
  shape of, 544
  in UV/Vis spectroscopy, 550
  wavenumber of, calculation of, 535
Acceptable daily intake (ADI), 1011
Acesulfame potassium (Sweet and Safe, Sunette, Sweet One), 1011
Acetaldehyde (ethanal), 790f
Acetal(s), **816,** 836
  hydrolysis of, 817
  mechanism of formation of, 817–818
  synthesis of, A-13
Acetamide, structure of, 758
Acetamidomalonic ester synthesis, of amino acids, 1034, 1034f
Acetic acid, 44, 47
  activation of, 767f
  dielectric constant of, 376t
  pK_a of, 312f
Acetic anhydride, 742f
Acetoacetate, 789f
  decarboxylation of, 894–895
Acetoacetic ester synthesis, 888–889, 895
Acetone, 789f, 200f
  dielectric constant of, 376t
  molar absorptivity of, 551–552
  UV spectrum of, 550, 550f
Acetonitrile (ethanenitrile)
  dielectric constant of, 376t
Acetophenone, 668f
Acetylacetone (2,4-pentanedione), naming of, 792f
Acetyl chloride (ethanoyl chloride), 739f
Acetylcholine, 771f
  receptors for, 1305
Acetylcholinesterase, in neurotransmission, 771

Acetyl-CoA, 770
  in acetylcholine synthesis, 771f
  in a biological Claisen condensation, 893
  carboxylation of, mechanism of, 1116–1117
  in citric acid cycle, 1138–1139
  conversion of pyruvate to, 1112
  fatty acyl-CoA conversion to, in fat catabolism, 1147–1148
Acetyl-CoA carboxylase, carboxylation of acetyl-CoA by, mechanism for, 1116–1117
Acetylene (*See* Ethyne)
Acetylenes, substituted, 260
Acetylide ions, 274, 276–277, 282
  formation of, 276–277
  reactions of alkyl halides with, 276–281
  reactions of carbonyl compounds with, 800
Acetylsalicylic acid (aspirin)
  physiological action of, 745, 1176
  structure of, 678f, 865f
  synthesis of, 745f, 785, 865
Acetyl thioester, in a biological Claisen condensation, 893
Achiral compound, 204
  meso compound as, 222
Achiral molecule, 203, 204, 204f, 250
Achiral object, **202,** 202f
Achiral reagents, enantiomers reacting with, 247
Acid(s), 44–65, **66**
  Brønsted–Lowry, **44**
  conjugate, 44
  dissociation constant of, 45
  Lewis, 64–65, **65**
  organic, 47–50
  pK_a of, 46
  pK_a table of, A-8
  strength of, 51
    factors determining, 58–60
    relative, 275
    substituents affecting, 55–57
  structure of, acidity and, 51–55
Acid anhydrides, 775
  conversion of, into ester, mechanism for, 743
  conversion of acyl chlorides into, mechanism for, 740

formation of, 739f
  in Friedel–Crafts acylation of benzene, 660
  nomenclature of, 726
  physical properties of, A-5
  reactions of, 742–743
  relative reactivity of, 737
  synthesis of, A-13
Acid–base reactions, 44, **66**
  predicting outcome of, 50–51
Acid catalysis, 1066–1069
  of acetal or ketal formation, mechanism of, 817–818
  of ester hydrolysis, mechanism of, 746–747, 1066–1067
  of addition of water and alcohols, mechanism of addition to alkenes, 169–172
  mechanism of addition to alkynes, 269–270
  in halogenation of α-carbon of aldehydes and ketones, 859
  of hydrate formation, mechanism of, 814–816
  in hydrolysis of amides, 760–761
  in keto-enol interconversion, 856
  in transesterification, 750
Acid-catalyzed proton exchange, mechanism of, 606–607
Acid dissociation constant, 45, **66**
Acidic amino acids, 1021
Acidity, **44, 66**
  of α-hydrogen, 851–854
  and structure, 51–55
Acidosis, 64
Acid rain, 46
Acquired immunodeficiency syndrome (AIDS), buckeyballs and, 644
Acrylic acid, structure of 725f
Acrylonitrile (propenenitrile), 728f
  uses of, 1248t
Activating substituents, **681,** 685–688
Activation
  experimental energy of, 151
  free energy of activation compared with, 151
  free energy of, **147,** 148, 155
    (*See* Free energy of activation)

Active site, **1079,** 1094
Acyclic alkenes, 125
Acyclic compounds, 125
  as meso compounds, 223
Acyclovir, 1316
Acyl adenylate(s), **768,** 775, 1159
  formation of, 770f
  in phosphoryl transfer reaction, 1143
Acylation (*See also* Friedel–Crafts acylation)
  of α-carbon, 878
  of α-carbon via enamine intermediate, 867–868
  Friedel–Crafts, **654,** 670
Acylation-reduction, alkylation of benzene by, 664
Acyl bromide(s), 726
Acyl chloride(s)
  conversion of
    into acid anhydride, mechanism for, 740
    into ester, mechanism for, 740–741
    into a primary alcohol, mechanism for, 801
  in Friedel–Crafts acylation of benzene, 660
  nomenclature of, 726
  physical properties of, 730, A-5
  reactions of, 739–742,
  reduction of, 801–802, 912, 912f
  relative reactivity of, 737
  synthesis of, A-13
Acyl-enzyme intermediate, 1085
Acyl group, **722,** 775
Acyl halides (*See also* acyl chlorides)
Acyl phosphates, **769,** 775, 1159
  formation of, 769f, 1142
Acyl pyrophosphates, **768,** 775, 1159
  formation of, 769f, 1142
Acyl transfer reaction, 734
Adams, Roger, 188
Adams catalyst, 188
Addition(s)
  aldol, **871**
    biological, 891–892
    crossed, 875
    intramolecular, 881
    mechanism of, 872
    mixed, 874–876